ELECTRONIC AND ATOMIC COLLISIONS

Electronic and Atomic Collisions

Abstracts of Contributed Papers

Thirteenth International Conference on the Physics of Electronic and Atomic Collisions
Berlin · 1983

Edited by

J. EICHLER, W. FRITSCH,
I. V. HERTEL, N. STOLTERFOHT, U. WILLE

1983

NORTH-HOLLAND
AMSTERDAM – OXFORD – NEW YORK – TOKYO

ISBN: 444 86801 1

Published by:

North-Holland Physics Publishing

a division of

Elsevier Science Publishers B.V.

P.O. Box 103
1000 AC Amsterdam
The Netherlands

Sole distributors for the U.S.A. and Canada:

Elsevier Science Publishing Company, Inc.

52 Vanderbilt Avenue
New York, N.Y. 10017
U.S.A.

This conference is indebted to the following sponsors:

Deutsche Forschungsgemeinschaft (DFG)
Senator für Wirtschaft und Verkehr, Berlin (West)
International Union for Pure and Applied Physics (IUPAP)
Balzers Hochvakuum GmbH (Wiesbaden-Nordenstadt)
Coherent GmbH (Neu-Isenburg)
Digital Equipment (Berlin)
KLM Royal Dutch Airlines (Amsterdam)
Lange & Springer (Berlin)
Leybold-Heraeus (Köln)
ORTEC GmbH (München)
Spectra-Physics GmbH (Darmstadt)
Springer-Verlag (Heidelberg)
Tektronix GmbH (Köln)
Varian GmbH (Stuttgart)
Carl Zeiss (Oberkochen)

The conference has been generously supported by:

Freie Universität Berlin
Hahn-Meitner-Institut für Kernforschung Berlin

Printed in Germany F.R. 1983

Preface

The Thirteenth International Conference on the Physics of Electronic and Atomic Collisions (XIII ICPEAC) coincides with the twenty-fifth anniversary of this international conference which started out as a small gathering of about one hundred scientists in New York (1958) and, in course of time, developed into the central conference for atomic collision physics. Subsequent ICPEAC meetings were held at Boulder (1961), London (1963), Quebec (1965), Leningrad (1967), Cambridge/Mass. (1969), Amsterdam (1971), Beograd (1973), Seattle (1975), Paris (1977), Kyoto (1979), and Gatlinburg (1981). The increasing activity in the field is reflected in the number of Contributed Papers as displayed in the figure. For the present ICPEAC, a total number of 751 Contributed Papers has been accepted.

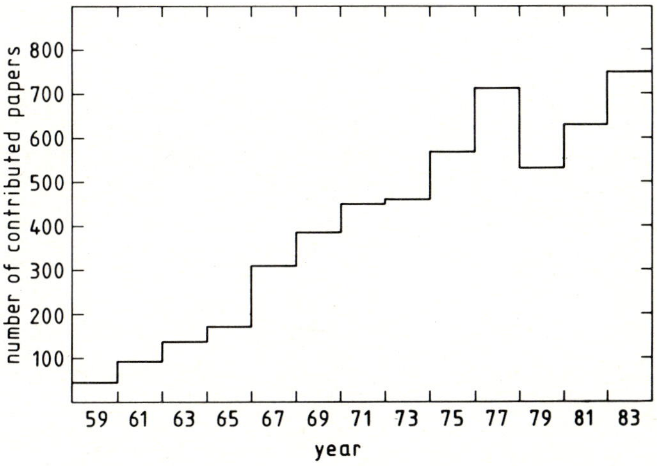

After a quarter of a century of intensive research, it is evident that the field of atomic collisions is still prospering. In fact, a closer look into this Book of Abstracts reveals a variety of new developments as well as advances in subjects of long-standing interest. The use of high-energy accelerators, molecular- and ion-beam techniques, lasers, and large-scale computers has furnished a deeper insight into the dynamics of electronic and atomic collisions. The knowledge gained in this way is of immediate importance for other fields of physics and technology. Experimental data and the results of model calculations reported at the ICPEAC conferences are prerequisite for applications in such diverse fields as gas discharge physics, gas phase chemistry, photochemistry, physics of the earth's atmosphere, astrophysics, controlled-fusion research, laser isotope separation and laser development, solid-state and surface physics, and many more.

The possibility of such applications provides much of the motivation behind a large portion of the Contributed Papers collected in this volume. All of these papers will be presented in poster sessions which during the past conferences have evolved into lively "market-places" for exchanging the latest results and ideas. Besides the poster presentations, there are 78 Invited Papers including papers reported at Symposia and, as a novelty, "Hot Topics" which have been selected from the Contributed Papers for a brief oral presentation.

The Book of Abstracts for this ICPEAC has changed in format as compared to previous conferences. By limiting each abstract to one two-column page it has been possible to accommodate all abstracts in a single volume without loss of space available for each contribution. The Book of Invited Papers will be published and mailed to all regular conference participants early in 1984.

The editors of this book are indebted to S. Datz, H.B. Gilbody, D.C. Lorents, W. Raith, F.H. Read, J.S. Risley, H. Schmidt-Böcking, T. Watanabe, and G.C. Watel for their help in arranging the Contributed Papers into a coherent program.

Berlin, May 1983

J. Eichler, W. Fritsch, I.V. Hertel, N. Stolterfoht, U. Wille

International Conference on the Physics of Electronic and Atomic Collisions

Organization 1981-1983

Executive Committee

CHAIRMAN

Frank H. Read - United Kingdom

VICE CHAIRMAN

Benjamin Bederson - USA

SECRETARY

John S. Risley - USA

TREASURER

Guy C. Watel - France

MEMBERS

Sheldon Datz - USA
Jörg Eichler - Germany F.R.
Franco A. Gianturco - Italy
H. Brian Gilbody - United Kingdom
Yuri S. Gordeev - USSR
Ingolf V. Hertel - Germany F.R.
Donald C. Lorents - USA
Walter E. Meyerhof - USA
Ronald F. Stebbings - USA
Nikolaus Stolterfoht - Germany F.R.
Tsutomu Watanabe - Japan

General Committee

AUSTRALIA

Ian E. McCarthy

AUSTRIA

Tilmann D. Märk

BELGIUM

Jacques Momigny

CANADA

Chris E. Brion
William J. McConkey

DENMARK

Knud Taulbjerg

FRANCE

Jean F. Delpech
Thomas R. Govers
Vo Ky Lan
Victor Sidis
Guy C. Watel

GERMANY

Jörg Eichler
Ingolf V. Hertel
Franz Linder
Werner Mehlhorn
Wilhelm Raith
Horst Schmidt-Böcking
Nikolaus Stolterfoht

INDIA

Satya P. Khare

ITALY

Franco A. Gianturco
Anna Giardini-Guidoni

JAPAN

Michio Matsuzawa
Nobuo Oda
Tsutomu Watanabe

NETHERLANDS

Hank G.M. Heideman
Johannes Los

SWITZERLAND

Willy Wölfli

UNITED KINGDOM

Brian H. Bransden
H. Brian Gilbody
Colin J. Latimer
M.R.Coulter McDowell
Roy W. Newell
Frank H. Read

USA

Benjamin Bederson
Roy L. Champion
Sheldon Datz
David E. Golden
Donald C. Lorents
Keith B. MacAdam
Walter E. Meyerhof
James R. Peterson
John S. Risley
Ivan A. Sellin
Ronald F. Stebbings
Sandor Trajmar

USSR

Yurii N. Demkov
Yuri S. Gordeev
E.E. Nikitin
E.A. Yukov

YUGOSLAVIA

Slobodan V. Cvejanović

Local Committee

CHAIRMEN

Jörg Eichler
Bereich Kern- und Strahlenphysik
Hahn-Meitner-Institut für Kernforschung Berlin

Ingolf V. Hertel
Institut für Molekülphysik
Freie Universität Berlin

Nikolaus Stolterfoht
Bereich Kern- und Strahlenphysik
Hahn-Meitner-Institut für Kernforschung Berlin

MEMBERS

Helmut Baumgärtel - Freie Universität Berlin
Bernhard Brutschy - Freie Universität Berlin
Adalbert Ding - Hahn-Meitner-Institut Berlin
Wolfgang Fritsch - Hahn-Meitner-Institut Berlin
Helmut Gabriel - Freie Universität Berlin
Ulrich Heinzmann - Fritz-Haber-Institut Berlin
Eugen Illenberger - Freie Universität Berlin
Georges Jamieson - Freie Universität Berlin
Wolfgang Kamke - Freie Universität Berlin
Philip J. Kuntz - Hahn-Meitner-Institut Berlin
Klaus Lacmann - Hahn-Meitner-Institut Berlin
Günther Nolte - Hahn-Meitner-Institut Berlin
Gebhard von Oppen - Technische Universität Berlin
Dirk Poppe - Technische Universität Berlin
Herbert Rinneberg - Freie Universität Berlin
Hartmut Schmidt - Freie Universität Berlin
Dieter Schneider - Hahn-Meitner-Institut Berlin
Uwe Wille - Hahn-Meitner-Institut Berlin

Program Committee

Executive Committee

and

Wolfgang Fritsch
Horst Schmidt-Böcking
Wilhelm Raith
Uwe Wille

CONTENTS: SYNOPSIS

Photon Impact

Photoionization of Atoms I 1
Photoionization of Atoms II 19
Photoionization of Molecules 34
Photodissociation of Neutral Molecules 47
Photodissociation of Molecular Ions 53
Multiphoton Ionization of Atoms 60
Multiphoton Processes in Molecules and Clusters 67

Electron-Atom Collisions

Elastic Collisions: Differential Scattering 78
Elastic Collisions: Integral Cross Sections 88
Elastic Collisions: Theoretical Methods 96
Resonances 101
Inelastic Collisions: Low Energies 112
Alignment and Orientation 126
Spin-Dependent Processes 143
Inelastic Collisions: High Energies 151
Inelastic Collisions: Bremsstrahlung 159
Electron Impact Ionization of Atoms: General, Threshold Studies 164
Electron Impact Ionization of Atoms: (e,2e) Processes, Post-Collision Interactions 174

Electron-Ion Collisions

Excitation 187
Ionization 198
Dielectronic Recombination 208

Electron-Molecule Collisions

General Aspects of Electronically Elastic Processes 216
Resonances in Electronically Elastic Processes 226
Electronically Elastic Processes in Polyatomics 240
Excitation and Ionization of Diatomics 255
Excitation and Ionization of Polyatomics 268
Dissociation 282
Dissociative Attachment 292

Positron-Atom(Molecule) Collisions 301

Atom-Atom Collisions

Theory 312
Ionization 325
Excitation and Excimer Formation 335
Fine-Structure Transitions and Polarization 341
Excitation Transfer 345

Ion-Atom Collisions

Elastic Scattering 349
Outer-Shell Excitation 357
Spectroscopy of Continuum Electrons 363
Capture into the Continuum 374
Autoionization and Auger Spectroscopy 382
Electron Emission by Negative-Ion Impact 397
Multiple Ionization 413
Direct Ionization of Inner Shells 422
Quasimolecular Excitation of Inner Shells I 435
Quasimolecular Excitation of Inner Shells II 448
Alignment and Orientation 458
X-Ray Spectroscopy 467
De-excitation of Quasimolecular States 472
Electron Exchange with Singly Charged Ions, Experiment I 478
Electron Exchange with Singly Charged Ions, Experiment II 486
Electron Exchange with Singly Charged Ions, Theory I 496
Electron Exchange with Singly Charged Ions, Theory II 508
Electron Exchange with Multiply Charged Ions I 523
Electron Exchange with Multiply Charged Ions II 539
Electron Exchange with Multiply Charged Ions III 553
Electron Exchange with Multiply Charged Ions IV 570

Ion(Atom)-Molecule Collisions

Electronically Elastic Processes 583
Theory of Rotational/Vibrational Excitation 590
Fine-Structure Transitions and Quenching at Thermal Energies . 599
Electronic Excitation at Suprathermal Energies 607
Penning Ionization 615
Electron Exchange 622
Dissociation and Recombination in Reactions 633
Reactive Scattering 642
Ion-Molecule Reactions 651

Collisions Involving Rydberg Atoms I 660
II 668

Field-Assisted Collisions

Electron-Atom Collisions in Radiation Fields 677
Laser-Assisted Collisions and Collision-Assisted Radiative Transitions 688

Experimental Techniques 703

Theoretical Methods 716

Miscellaneous 727

Post-Deadline Papers 733

Author Index 753

CONTENTS

Photoionization of Atoms I

Photoionization of the $3p^5$ 4s Excited States of Argon 1
P.C. Ojha, P.G. Burke

One and Two Electron Atomic Photoionization Processes 2
P. Scott, P.G. Burke, A.E. Kingston

Radiative Recombination Coefficients for Complex Atomic Ions 3
G. Peach

Double-Electron Photoionization of Helium 4
S.N. Tiwary

A New QDT Analysis of Two-Channel Interactions 5
Annick Guisti-Suzor

Photoionisation of Si II: an Application of Quantum Defect Theory 6
K.T. Taylor, C.J. Zeippen

Simple Dependences of Bound-Free Dipole Matrix Elements and Their Zeros on the Properties of the Initial State 7
David Salzmann, R.H. Pratt, M.S. Wang, R.Y. Yin

Zeros in Dipole Matrix Elements of Photoionization of Excited Alkali Atoms 8
Jayanti Lahiri, Steven T. Manson

Cross Sections for Photoionization of Excited d-States of Alkali Atoms: Hartree-Fock Calculations 9
Alfred Z. Msezane, Steven T. Manson

The Resonance Photoionisation of Helium to $He^+(n=2)$ 10
S.M. Burkov, S.I. Strakhova

Theoretical Investigations of Spin Polarisation in Mercury 11
F. Keller, F. Combet Farnoux

Polarization of Atomic Photoelectrons in Relativistic and Non-relativistic Theories 12
N.A. Cherepkov

Photoionization of High-Z Atoms: Relativistic Effects on Cooper Minima in the 6p Subshell 13
S.T. Manson, C.J. Lee, R.H. Pratt, I.B. Goldberg, B.R. Tambe, Akiva Ron

Z n ℓ Dependence of Cancellation of Relativistic and Retardation Effects in Photoionization 14
R.H. Pratt, Akiva Ron, S.T. Manson, Sung Dahm Oh

Relativistic Effects in the Photoionization of the 5p Subshell of High-Z Elements 15
B.R. Tambe, Steven T. Manson

Relativistic Study of Multichannel Interactions in Photoionization of High-Z Atoms: RRPA Applied to Radium and Radon 16
Pranawa C. Deshmukh, Steven T. Manson, Vojislav Radojevic

Relativistic Many-Body Approach to the Photoionization of Alkali Atoms 17
J.E. Hansen, W.R. Johnson, G. Soff

Energy Shift of Emitted X-Rays Following Inner Shell Photoionization Near Threshold 18
J. Mizuno, T. Ishihara, T. Watanabe

Photoionization of Atoms II

Near-Threshold Measurement of the Photo-Doubledetachment of $He^-(^4P^0)$ 19
Y.K. Bae, M.J. Coggiola, J.R. Peterson

Photoionization of Atomic Oxygen, Cesium, and Rubidium 20
J.A.R. Samson, P. Pareek, H. Suemitsu

Photoionization of Excited Cs 6p and Cs 7p Subshells 21
Alfred Z. Msezane

Alignment of $Cd^+(4d^{-1}\ {}^2D_{5/2})$ after Photoionization with Synchrotron Radiation 22
W. Kronast, W. Mehlhorn

Strong Variations of Angular Distribution and Polarization of Fluorescence Radiation in the Region of an Autoionizing Resonance of Photoabsorption 23
V.V. Balashov, N.M. Kabachnik, V.S. Senashenko

Observation of the 1s2p2p' $^4P^e$ Resonance in He^- Photodetachment 24
J.R. Peterson, Y.K. Bae, M.J. Coggiola

Resonance Photoabsorption by the Two Electron Systems near the n=2 and n=3 Thresholds 25
A. Wague

Electron Spectroscopy of Atomic Fe, Co and Ni Excited by VUV Photon Impact 26
E. Schmidt, H. Schröder, B. Sonntag, H. Voss, H.-E. Wetzel

Photoionization of Laser Excited Sodium and Barium Atoms 27
J.M. Bizau, B. Carré, P. Dhez, D. Ederer, P. Gérard, J.C. Keller, P. Koch, J.L. Legouet, J.L. Picqué, G. Wendin, F. Wuilleumier

High Resolution Study of the $Ne(2p^5{}_{1/2}ns', nd')$ Autoionization Resonances 28
J. Ganz, A. Siegel, W. Bussert, K. Harth, M.-W. Ruf, H. Hotop

Multistep Excitation of Autoionizing States of Ba 29
T.F. Gallagher, N.H. Tran, P. Pillet, R. Kachru

Investigation of Autoionising Levels in Ga I, In I and Tl I 30
M. Müller, M. Schmidt, P. Zimmermann

Resonant Photoemission Study of the Xenon - 4d - Excitations 31
U. Becker, E. Schmidt, B. Sonntag, H.E. Wetzel, A. Winogradow

Theoretical Study of the Shake-up in the Core Photoelectron Spectra of Atoms 32
G. De Alti, P. Decleva, A. Lisini

Photoionization of Aligned 6snd $^{1,3}D_2$ Rydberg States in Atomic Barium: The Role of the 5d7d Perturbing State 33
S.J. Smith, E. Matthias, P. Zoller, D.S. Elliott, N.D. Piltch, G. Leuchs

Photoionization of Molecules

Application of the Reaction Matrix Method to Molecular Photoionization Treatment of Final-State Correlation in $3\sigma_g$ and $2\sigma_u$ Photoionization of N_2* 34
Jeffrey Stephens, Dan Dill

Ab-Initio Approach to the Multichannel Quantum Defect Calculation of the Electronic Autoionization in the Hopfield Series of N_2 35
M. Raoult, H. Le Rouzo, G. Raseev, H. Lefebvre-Brion

A Continuum Spectrum Multichannel Finite Volume Variational Method 36
G. Raseev, H. Le Rouzo

On the Influence of Electron-Electron Correlations in H_2 Molecule on the Characteristics of the (γ,2e) Process 37
V.G. Levin, V.G. Neudatchin, A.V. Pavlitchenkov, Yu.F. Smirnov

Manifestations of the Optical Activity of Molecules in the Dipole Photoeffect 38
N.A. Cherepkov

Absolute Dipole Oscillator Strengths for the Photoabsorption, Photoionization and Fragmentation of HCl, HBr and NO 39
C.E. Brion, F. Carnovale, S. Daviel, Y. Iida

Multi-Photon Electron Spectroscopy of Nitric Oxide 40
J. Kimman, M. Lavollée, M. Spruit, M.J. van der Wiel

Ultrahigh Resolution Photodetachment Study of C_2^- 41
U. Hefter, R.D. Mead, P.A. Schulz, W.C. Lineberger

Triply Differential Photoelectron Studies of Molecular Photoionization 42
J.L. Dehmer, A.C. Parr, S.H. Southworth, D.M.P. Holland

Fluorescence Polarization as a Probe of Molecular Autoionization 43
E.D. Poliakoff, J.L. Dehmer, A.C. Parr, G.E. Leroi

Experimental and Theoretical Study of CO Photoionization in the 15-40 eV Energy Range 44
B. Leyh, G. Raseev, M.-J. Hubin-Franskin, J. Delwiche, I. Nenner, H. Lefebvre-Brion, P. Morin, M.-Y. Adam

Photoionization Cross Sections of the Valence Orbitals of CO 45
E.P. Leal, Lee Mu-Tao, F.J. da Paixao, R.R. Lucchese, V. McKoy

Theoretical Studies of Photoexcitation and Ionization in Molecules of Astrophysical Interest 46
G.H.F. Diercksen, P.W. Langhoff

Photodissociation of Neutral Molecules

Semiclassical Theory of Predissociation Induced by Rotational (Coriolis) Coupling 47
Hiroki Nakamura

Photodissociation of OH Through Non-Adiabatic Interaction 48
Ewine F. van Dishoeck, Marc C. van Hemert, A.C. Allison, A. Dalgarno

The Photodissociation of H_2O at 157 nm: Full Internal State Distributions and Alignment of Nascent $OH(X^2\Pi)$ Radicals 49
P. Andresen, G.S. Ondrey, E.W. Rothe, B. Titze

The Photodissociation of Cyanogen 50
Joshua B. Halpern, Xiao Tang, William M. Jackson

Fluorescence Excitation Spectra of Neutral Fragments from Photodissociation of OCS Molecules by VUV Incident Photons 51
J. Delwiche, M.-J. Hubin-Franskin, A. Tabché-Fouhaile, H. Frohlich, K. Ito, P.-M. Guyon, I. Nenner

Laser Multiphoton Ionization and Dissociation of Polyatomic Molecules 52
I. Dimicoli, J. Jaraudias, J. Lemaire, R. Botter

Photodissociation of Molecular Ions

Dissociation of Energy Selected States of O_2^+ and CO^+ by Threshold Photoelectron-Photoion Coincidence 53
M. Richard-Viard, O. Dutuit, T. Govers, P.M. Guyon, H. Frohlich, M. Lavollée

Photofragment Spectroscopy of Molecular Ions 54
E. Solarte, V. Hermann, R. Anselmann, F. Linder

Photofragment Spectroscopy of Cs_2^+ 55
H. Helm, R. Möller, P.C. Cosby, D.L. Huestis

Photodissociation of N_2^{++} 56
P.C. Cosby, R. Möller, H. Helm

UV-Laser Photodissociation of Molecular Ions 57
R.E. Kutina, A.K. Edwards, J. Berkowitz

Laser Photodissociation of CO_3^- and its Hydrates 58
A.W. Castleman, Jr., D.E. Hunton, M. Hofmann, T.G. Lindeman

Fast Neutral Beam Measurements of Molecular Photodissociation 59
L.D. Gardner, J.L. Kohl

Multiphoton Ionization of Atoms

A Basis Set Approach to Multiphoton Ionization Including Free-Free Transitions 60
John T. Broad

Semiclassical Many-Mode Floquet Theory 61
Shih-I Chu, Tak-San Ho, James V. Tietz

Influence of the Dynamic Stark Effect on Electron Angular Distributions in Multiphoton Ionization 62
W. Ohnesorge, F. Diedrich, D.S. Elliott, G. Leuchs, H. Walther

Angular Distribution of Photoelectrons from Multiphoton-Ionization (λ = 532 nm) 63
R. Hippler, H.-J. Humpert, H. Schwier, H.O. Lutz

Ponderomotive Force and A.C. Stark Shift in Multiphoton Ionization by Strong Fields 64
H.G. Muller, P. Kruit, A. Tip, M.J. van der Wiel

Photoelectron Energy and Angular Distributions from Multiphoton Ionization of Cesium Beams 65
J.A.D. Stockdale, R.N. Compton, C.D. Cooper

Microwave Ionization of Highly Excited Helium Atoms 66
W. van de Water, D.R. Mariani, P.M. Koch

Multiphoton Processes in Molecules and Clusters

Double Resonance Multiphoton Ionization Studies of High Rydberg States in NO 67
M. Seaver, W.A. Chupka, S.D. Colson, D. Gauyacq, Ch. Jungen

A Multiphoton Ionization Photoelectron Spectroscopic Study on Autoionization Processes of NO Molecule 68
Y. Achiba, K. Sato, K. Kimura

Photoelectron Studies of Resonant Multiphoton Ionization of CO via the A $^1\Pi$ State 69
S.T. Pratt, E.D. Poliakoff, P.M. Dehmer, J.L. Dehmer

Resonant Multiphoton Ionization of H_2 via the B $^1\Sigma_u^+$, v=7, J=2 and 4 Levels with Photoelectron Energy Analysis 70
S.T. Pratt, P.M. Dehmer, J.L. Dehmer

One- and Two-Color Multiphoton-Ionization-Experiments on Molecular Clusters 71
W. Bronner, P. Oesterlin, M. Schellhorn

Raman Analysis of SF_6 Molecular Beams Excited with a CW CO_2-Laser 72
G. Luijks, S. Stolte, J. Reuss

Spectroscopy of Fluorobenzene van der Waals Clusters by Resonant Two-Photon Ionization 73
K. Rademann, B. Brutschy, H. Baumgärtel

Multiple Photon Excitation of CF_3Br 74
A. Giardini-Guidoni, E. Borsella, R. Fantoni

Two-Photon Dissociation of Vibrationally Excited H_2^+ and HD^+ Molecules 75
Cecil Laughlin, Shih-I Chu, Krishna K. Datta

Two-Photon "Dissociation" to the Quasicontinuum 76
E. Kyrölä, J.H. Eberly

Λ-Doublet Populations of $CH(A^2\Delta)$ from CH_3 Radicals Produced by the UV Multiphoton Dissociation of CH_3- Containing Molecules 77
Takashi Nagata, Mutsumi Suzuki, Kaoru Suzuki, Tamotsu Kondow, Kozo Kuchitsu

Elastic Electron-Atom Collisions: Differential Scattering

Measurements of Differential Cross Sections for e-Ar, Kr, Xe Scattering at E = 50 meV - 2 eV 78
M. Weyhreter, B. Barzick, F. Linder

On the Elastic Electron-Neon Scattering in the 16 eV Region 79
E. Naslenas, P. Zvirblis

Elastic Scattering of Electrons from Neon at Intermediate Energies 80
B.B. Srivastava, S.S. Dhal

Elastic Small-Angle Electron Scattering by Helium 81
Ortwin Müller, J. Geiger

Pseudostate Model of Electron Shadow Scattering 82
S. Geltman, R.K. Nesbet

Small Angle (e^-,Na) Scattering 83
B. Jaduszliwer, P. Weiss, A. Tino, B. Bederson

Absolute Measurements of Small Angle Elastic Scattering of Electrons 84
R.W. Wagenaar, T. v. Tubergen, F.J. de Heer

Relative Differential Cross Sections for Elastic and Inelastic Scattering of Electrons by Xenon in the Energy Range of 15 to 80 eV 85
B. Marinkovic, V. Pejcev, D. Filipovic, L. Vuskovic

Elastic Scattering of Electrons by Sodium Atom 86
S.N. Singh, A.N. Tripathi, M.K. Srivastava

Relativistic Phase Shift Analysis for Elastic Scattering of Electrons by Xenon below the First Inelastic Threshold 87
D. Register, L. Vuskovic, L.T. Sin Fai Lam, S. Trajmar

Elastic Electron-Atom Collisions: Integral Cross Sections

Electric Polarizability of Neon Atom and Electron-Neon Scattering Length from Electron Cyclotron Resonance Absorption Spectrum 88
A.P. Kabilan

Low Energy Scattering from Krypton and Xenon 89
R.P. McEachran, A.D. Stauffer

Precision Measurements of Ramsauer Minima in Electron Total Cross Sections 90
J. Ferch, C. Masche, W. Raith

Total Cross Sections for Electron Scattering from N_2, Xe, Kr and Ar 91
K. Jost, P.G.F. Bisling, F. Eschen, M. Felsmann, L. Walther

Total Collisional Cross-Sections for $e^{\pm}$ Scattering from Li Atoms 92
S.P. Khare, Vijayshri

Total Electron Scattering Cross Sections for He, Ne, Ar, Xe, and Selected Molecules: 4-300 eV 93
J.C. Nickel, K. Imre, D.F. Register, S. Trajmar

Total Cross Sections for Electrons Scattered by Ar in the Intermediate Energy Range 94
J.C. Nogueira, I. Iga, E. Chaguri, M.T. Lee

Theoretical Cross-Sections for Electron Scattering from C, N and O-Atoms 95
H.S. Desai, K.N. Joshipura

Elastic Electron-Atom Collisions: Theoretical Methods

Scattered Wave Function Using Extended Numerov Algorithm 96
Joseph M. Paikeday

A Treatment of the Elastic Scattering of Electrons on H and He^+ Using Hyperspherical Coordinates and Incorporating the Fock Expansion 97
Steven Alston

Functional Analysis in the e-H Scattering Problem 98
P.A. Massaro

Inclusion of the Fock Expansion in Hyperspherical Coordinate Calculations 99
Jim Feagin, Joseph Macek, Anthony F. Starace

Eikonal Exchange Amplitudes for Electron-Helium Collisions 100
B. Padhy, R. Srivastava, D.K. Rai

Resonances in Electron-Atom Collisions

Hilbert Space Representation of Resonance Wavefunctions 101
A. Macias, A. Riera

Complex Virial Theorem and Complex Scaling 102
B.R. Junker

The $6s6p^2$ Resonances in e-Hg Scattering 103
N.S. Scott, P.G. Burke, K. Bartschat

High Resolution Electron Impact Studies of Mercury: Threshold Excitation and the Measurement of Metastable Excitation Functions 104
D.S. Newman, G.C. King, M. Zubek

Wannier-Ridge He^- Resonances 105
S.J. Buckman, P. Hammond, G.C. King, F.H. Read, C.D. Warner

Grandparent and Non-valence Resonances in Electron-Impact Excitation of the Noble Gases 106
S.J. Buckman, P. Hammond, G.C. King, F.H. Read

Resonances in Electron Scattering by Neon 107
K.T. Taylor, W.C. Fon, C.W. Clark

An Eleven State Electron-Helium Scattering Calculation 108
P.G. Burke, L.C.G. Freitas, A.E. Kingston, K.A. Berrington

Hyperspherical Description of Electronic Correlations in Atomic Systems I - Bound and Continuum States of Two-Electron Systems H^-, He 109
L. Pelamourgues, S. Watanabe, M. Le Dourneuf

Hyperspherical Description of Electronic Correlations in Atomic Systems II - Graphical Representation of Radial and Angular Correlations in Doubly-Excited Atoms 110
M. Le Dourneuf, S. Watanabe, L. Pelamourgues

Hyperspherical Description of Electronic Correlations in Atomic Systems III - Radial and Angular Correlations of Triply-Excited States 111
S. Watanabe, M. Le Dourneuf

Inelastic Electron-Atom Collisions: Low Energies

Variational Linear Algebraic Equations Method 112
B.L. Moiseiwitsch

A Comparative Study of Multiple Scattering Approximations to the Excitation of Atomic Hydrogen by Electron Impact 113
H.E. Fargher, M.J. Roberts

Pseudostate Expansions in Electron Hydrogen Scattering 114
D.H. Oza, J. Callaway

Applications of the Coupled-Channels Optical Calculation for Electron-Atom Scattering 115
B.H. Bransden, I.E. McCarthy, A.T. Stelbovics

Differential Cross Sections of $e^- + H(1s) \rightarrow e^- + H(2p)$ 116
S.N. Tiwary

High Resolution Total Electron Impact Excitation of Individual He $2\ ^3S$ and $2\ ^1S$ States 117
David Spence, Dorothy Stuit, M.A. Dillon, R.-G. Wang

Low Energy Electron Scattering from Interstellar Gas Molecules 118
V.M. Chhaya, H.S. Desai

Analytical Calculation of Mean Excitation Energies of Hydrogen and Other Light Atoms 119
S. Rosendorff

Inelastic Differential Electron Scattering from Metastable Helium Atoms 120
R. Müller-Fiedler, P. Schlemmer, K. Jung, H. Hotop, H. Ehrhardt

Electron-Impact Excitation from 2 ^1S and 2 ^3S States of Helium 121
David C. Cartwright, George Csanak, Fernando J. da Paixao

2 ^3S Excitation of Heliumlike Ions by Electron Impact 122
C.S. Singh, D.K. Rai

Scattering of Slow Electrons by Li Atoms 123
D.L. Moores

Stepwise Electron/Laser Excitation Studies of Atomic Collisions 124
C.J. Webb, W.R. MacGillivray, M.C. Standage

Electron Impact Photoemission Cross Sections for the VUV Light Standard Project 125
Armon McPherson, N. Rouze, W.B. Westerveld, J.S. Risley

Electron-Atom Collisions: Alignment and Orientation

Measurement of Complex Scattering Amplitudes for Electron Impact Excitation of the 3 ^1P and 3 ^1D States of Helium 126
H.B. van Linden van den Heuvell, E.M. van Gasteren, J. van Eck, H.G.M. Heideman

The Orbital Angular Momentum Transfer in the Excitation of the 2 ^1P State of Helium by Electrons 127
H.B. van Linden van den Heuvell, M.A.M. de Jong, J. van Eck, H.G.M. Heideman

Scattering Parameters for the Electron Impact Excitation of the 3 ^2P State of Sodium 128
J.E. Furst, J.L. Riley, S.J. Buckman, P.J.O. Teubner

Stokes Parameter Analysis of Electron Photon Coincidence Experiments in Sodium 129
P.J.O. Teubner, J.E. Furst, J.L. Riley

Electron-Photon Angular Correlation with Spin-Orbit Interaction in the Excitation of Kr and Xe 130
H. Nishimura, A. Danjo, T. Koike, K. Kani, H. Sugahara, A. Takahashi

Orientation and Alignment for 1s→2p Excitation of Hydrogen by Electrons and Differential Cross Sections for 1s-2p and 1s→2p(m=0) and 2p(m=±1) Excitation 131
J.F. Williams

Symmetry Properties of Electron-Photon Angular Correlations 132
J.F. Williams

The λ-Parameter of the 3 ^1P State of Helium Using the 5016 Å Photon 133
C.R. Hummer, Donal J. Burns

Orientation of the 2 ^1P State of Helium Excited by Electron Impact 134
N.C. Steph, W.H. Kloepping, D.E. Golden

Excitation and Decay of Stark Mixed n = 2 States of Hydrogen Observed in an Electron Photon Coincidence Experiment 135
S. Watkin, C.G. Back, M. Eminyan, K. Rubin, J. Slevin, J.M. Woolsey

Electron-Photon Angular Correlations for the 2 ^1P State of Helium within 1 eV of Threshold 136
P.A. Neill, A. Crowe

Alignment of the $2p_{3/2}$ Shell of Argon by Electron Impact 137
E.C. Sewell, A. Crowe

Angular Correlations for Excitation of the Lowest Excited States of Krypton by Electron Impact 138
S.J. King, A. Crowe

Superelastic Electron Scattering on Sodium Atoms in the Distorted-Wave Born Approximation 139
V.V. Balashov, A.N. Grum-Grzhimailo, O.I. Zaitseva

An Investigation of Superelastic Electron Scattering by Laser-Excited Ba Atoms 140
S. Trajmar, D.F. Register, Gy. Csanak

Attractive and Repulsive Forces in Electron Impact Excitation of Atoms 141
H.-J. Beyer, H. Kleinpoppen

Electron-Electron Coincidence Studies of the L_3-Shell Alignment Tensor in Argon 142
M. Völkel, W. Sandner

Electron-Atom Collisions: Spin-Dependent Processes

Determination of Spin Exchange Scattering Amplitudes for e-Li Elastic Scattering 143
H.S. Desai, N.S. Rao

Ionization and ns-np Excitation Asymmetry in Polarized Electron - Polarized Alkali Atom Collisions 144
G. Baum, M. Moede, W. Raith, W. Schröder

Spin Dependent Effects in Electron-Hydrogen Scattering with Polarized Beams: A Status Report 145
A. Vasilakis, A. Anan, C. Back, M. Eminyan, M.S. Lubell, K. Rubin, J. Slevin, W. Stirling, F.C. Tang, M. Turner

"Triple" Electron Scattering on Mercury and Xenon for Complete Evaluation of the Scattering Amplitudes 146
O. Berger, B. Kessler, J. Kessler, W. Wübker

The Polarization of 254 nm Mercury Line Radiation after Impact Excitation by Polarized Electrons 147
A. Wolcke, J. Goeke, W. Vollmer, G.F. Hanne, J. Kessler

Impact Excitation of the 7 $^3S_1$-State of Mercury by Polarized Electrons 148
H. Wolf, E. Reichert

Excitation of Hg-Atoms by Electron Impact 149
K. Bartschat, N.S. Scott, K. Blum, P.G. Burke

Electron Impact Excitation of Polarized Na-Atoms 150
W. Jitschin, S. Osimitsch, H. Reihl, H. Kleinpoppen, H.O. Lutz

Inelastic Electron-Atom Collisions: High Energies

How Do We Decide Whether the First Born Approximation Applies to Inelastic Collisions of Charged Particles with an Atom or Molecule? 151
Mitio Inokuti, S.T. Manson

Absolute Triple Differential Cross Sections for High Energy Electron Impact Ionization of Helium 152
A. Lahmam-Bennani, H.F. Wellenstein, A. Duguet

Absolute Triple Differential Cross Sections for High Energy Electron Impact Ionization of Argon 3p and 2p 153
A. Lahmam-Bennani, H.F. Wellenstein, A. Duguet, A. Daoud

Differential Cross Sections for the $^1S \rightarrow {}^1S$, ^3S Transitions in He by Electron Impact 154
T. Takayanagi, K. Wakiya, H. Suzuki, S. Ito, K. Hoshiba, S. Kano, H. Takuma

Measurements of Inner Shell Excitation Cross Sections in Ne by Means of Auger Electron Spectroscopy by Electron Impact 155
S. Kihara, Y. Iketaki, T. Takayanagi, K. Wakiya, H. Suzuki

Electron Impact Excitation of Lithium 3s State in the Intermediate Energy 156
Mukesh Kumar, S.S. Tayal, A.N. Tripathi

Cross Sections of Inert Gases for VUV Emissions Following Inner-Shell or Subshell Ionization by Electron Impact 157
Y. Akagi, K. Morita, T. Takayanagi, K. Wakiya, H. Suzuki

K-Shell Ionization Cross Sections of Si and Ar by Electrons with Impact Energies 4 to 10 keV 158
H. Platten, G. Schiwietz, N. Stolterfoht, G. Nolte

Inelastic Electron-Atom Collisions: Bremsstrahlung

Energy Dependence of the High Energy Tip of the Bremsstrahlung Spectrum of La, W, and Co 159
F. Riehle

Electron Bremsstrahlung - Effect of Electron Screening and Vacuum Polarisation 160
Lali Chatterjee, Sujata Bhattacharyya

Validity of Classical Approach to the Characterization of the Electron Bremsstrahlung Spectrum: Extension of its Applicability to Screened Atomic Cases 161
M. Lamoureux, R.H. Pratt

Atomic-Field Bremsstrahlung from Uranium 162
L. Estep, J. Altman, R. Ambrose, S. Salehkoutahi, G. Westbrook, C.A. Quarles

Two Photon Processes in Electron-Atom Collisions 163
J.C. Altman, C.A. Quarles

Electron Impact Ionization of Atoms: General, Threshold Studies

Eikonal Exchange Amplitude for Electron-Impact Ionization of Atomic Hydrogen 164
A.C. Roy, N.C. Sil

Ionization of Atomic Hydrogen and Helium by Fast Electron Impact 165
B. Piraux, C.J. Joachain, F.W. Byron, Jr.

Interferences Due to Competing Mechanisms for Double Excitation and Ionization of Atoms by Charged Particles at High Velocities 166
J.H. McGuire

Direct Contributions to Electron Induced Multiple Ionization of Inner Shells 167
W. Löw, H. Genz, A. Richter

Collapse of 3d Orbital in Ca Observed in Anomalies of the 2p Auger Spectrum 168
B. Breuckmann, R. Huster, W. Menzel, W. Weber, W. Mehlhorn, K.W. Dyall

Electron Impact Excitation of the nsnp(n+1)p Configurations of Autoionizing States in Indium and Thallium Vapour Atoms 169
M. Wilson, G.K. James, K.J. Ross

Branching Ratios and Cross Sections for Ion and Electron Impact Produced $2s^m2p^n$ - Configurations of Ne Using Synchrotron Radiation as a Radiometric Standard 170
H.-J. Flaig, M. Eckhardt, K.-H. Schartner, H. Kaase

The K-Shell Electron Double Ionization Cross Section of Sodium. A Power Law Z-Dependence 171
O. Keski-Rahkonen, J. Lahtinen

Threshold Behaviour of Ar-K and Xe-L_3 Inner Shell Ionisation by Electron Impact 172
R. Hippler, K. Saeed, I. McGregor, A.J. Duncan, H. Klar, H. Kleinpoppen

Threshold Law for the Triple Ionization Function 173
P. Grujic

Electron Impact Ionization of Atoms: (e,2e) Processes, Post-Collision Interactions

Coulomb Interaction in Final State of (e,2e) Reaction on Atoms in Terms of Three-Body Formalism 174
A.R. Ashurov, G.V. Avakov, V.G. Levin, A.M. Mukhamedzhanov

The Observation of an Even Parity $^1D_2$ Autoionizing Level in Cd Using (e,2e) Spectroscopy 175
N.L.S. Martin, K.J. Ross

(e,2e) Cross Sections for Helium 176
A.C. Roy

"Energy-Sharing" in Relation to Double Differential Cross-Sections for Electron Impact Ionization of Helium 177
J. Kimman, Pan Guang-Yan, C.W. McCurdy, F.J. de Heer

(e,2e) Experiments on He in Asymmetric Conditions at Intermediate Energy 178
A. Giardini-Guidoni, V. Di Martino, R. Fantoni, R. Tiribelli

Observation of the Helium Ion n=1,2,3 States in (e,2e) Collisions 179
J.P.D. Cook, E. Weigold

Triple Differential Cross Section Measurements for the Electron Impact Ionization of Atomic Hydrogen: Comparison with Theory 180
B. Lohmann, E. Weigold

Corrections to the Impulse Approximation for (e,2e) Reaction 181
Yu.V. Popov

Exchange Effects in Resonance Ionization of Atoms by Intermediate-Energy Electrons 182
V.V. Balashov, A.N. Grum-Grzhimailo, A.I. Magunov

Triple Differential Cross Section for the Electron Impact Ionization of Hydrogen 183
A.S. Ghosh, P.S. Mazumder, Madhumita Basu

Energy Exchange between Two Outgoing Electrons in the Post Collision Interaction Processes 184
J. Mizuno, T. Ishihara, T. Watanabe

Threshold Excitation of Rydberg States of Atoms and Diatomic Molecules 185
P. Hammond, G.C. King, F.H. Read, J. Jureta

Angular Momentum Exchange in Post Collision Interaction 186
A. Niehaus, C.J. Zwakhals

Electron-Ion Collisions: Excitation

Angular Distribution for Electron Excitation of the 4 $^2S \rightarrow 4\ ^2P$ Transition in Zn II: Comparison of Experiment and Theory 187
A. Chutjian, A.Z. Msezane, R.J.W. Henry

Experimental Electron Energy-Loss Spectra and Cross Sections for the 5 $^2S \rightarrow 5\ ^2P$ Transition in Cd II 188
A. Chutjian

Calculation of Electron-Ion Scattering Cross-Sections: A Perturbative Treatment of Exchange 189
M.A. Hayes

Excitation of Hydrogenlike Ions in Modified Glauber Approximations 190
T.T. Gien

Electron-Impact Excitation Collision Strengths and Excitation Rates for He-Like Ions 191
S.S. Tayal, A.E. Kingston

Electron-Impact Excitation of O V 192
S.S. Tayal, K.A. Berrington, A.E. Kingston

Angular Distributions for Electron Impact Excitation of Mg II, Zn II, and Cu I 193
Alfred Z. Msezane, Ronald J.W. Henry

Excitation of Positive Ions in Coulomb-Born Approximation 194
N.C. Sil, N.C. Deb

Measurement of the Electron Impact Excitation Cross Section for $C^+(2p\ ^2P^0 - 2p^2\ ^2D)$ 195
John L. Kohl, Gregory P. Lafyatis

Electron Impact Excitation of Fine Structure Transitions in C-Like Ions 196
K.M. Aggarwal

Resonant Excitation Double Autoionization in Xe^{6+} 197
K.J. LaGattuta, Y. Hahn

Electron-Ion Collisions: Ionization

Electron-Impact Ionisation of Rare Gas Ions 198
A. Matsumoto, S. Ohtani, A. Danjo, H. Hanashiro, T. Hino, Y. Kondo, H. Suzuki, H. Tawara, K. Wakiya, M. Yoshino

Electron-Impact Double-Ionization of Xenon Ions 199
M.S. Pindzola, D.C. Griffin, C. Bottcher

Cross Section Measurements of Single and Multiple Ionization of Xe^{i+} (i=1,2,3,4) Ions and Double Ionization of J^+ Ions by Electron Impact 200
C. Achenbach, A. Müller, E. Salzborn, R. Becker, J. Peschina, H. Klein

Cross Sections for Single and Multiple Ionization of Alkali Ions and Alkaline-Earth Ions by Electron Impact 201
T. Hirayama, K. Oda, T. Ono, Y. Morikawa, K. Wakiya, H. Suzuki

Auger Ionization Cross Section for Ions of the Li and Be Sequences 202
K.J. LaGattuta, D.J. McLaughlin, Y. Hahn

Electron Impact Ionisation of Boron-Like Ions 203
K. Butler, D.L. Moores

Absolute Cross Section Measurements for Electron-Impact Ionization of Twice Charged Ions Ti^{2+}, Fe^{2+}, Ar^{2+}, Cl^{2+} and F^{2+} 204
D.W. Mueller, T.J. Morgan, Gordon H. Dunn, D.C. Gregory, D.H. Crandall

Ionization of Singly-Charged Metallic Ions by Electron Impact 205
R.G. Montague, M.J. Diserens, M.F.A. Harrison, A.C.H. Smith

Electron Impact Excitation-Autoionization of Ni Ions 206
S.N. Tiwary, P.G. Burke, A.E. Kingston

Classical Binary-Encounter Collision Theory for Ionization in Electron-Ion Scatterings 207
Y.T. Lee

Electron-Ion Collisions: Dielectronic Recombination

Observation of KLL Dielectronic Recombination Resonance in Ar^{14+} 208
J.P. Briand, P. Charles, H. Arianer, H. Laurent, C. Goldstein, J. Dubau, M. Loulergue, F. Bely-Dubau

Dielectronic Recombination Measured with a Crossed Electron and Calcium Ion Beams Coincidence Technique 209
J.F. Williams

Dielectronic Recombination: A Crossed Beams Observation and Measurement of Cross Section 210
D.S. Belic, Gordon H. Dunn, T.J. Morgan, D.W. Mueller, C. Timmer

Dielectronic Recombination of B^{2+} and C^{3+} via $1s^22s \rightarrow 1s^22p$ Excitation 211
S. Datz, P.F. Dittner, P.D. Miller, C.D. Moak, N. Nescovic, P.H. Stelson, C. Bottcher

Effect of Extrinsic Electric Fields upon Dielectronic Recombination 212
K.J. LaGattuta, Y. Hahn

Electron Scattering with Very Highly Charged Ions: Effects of Relativistic Deviation, Autoionization and Dielectronic Recombination on Cross Section for Fe^{24+}, Se^{32+} and Mo^{40+} 213
A.K. Pradhan

The Effect of Resonances on the He-Like Electron Ion Scattering Cross Sections 214
L. Steenman-Clark, P. Faucher, J. Dubau

Dielectronic Recombination Process for Highly Ionised Argon 215
F. Bely-Dubau, J. Dubau

Electron-Molecule Collisions: General Aspects of Electronically Elastic Processes

Ab-Initio Optical Potentials Applied to Low Energy Electron-Molecule Collisions in the Linear Algebraic Approach 216
B.I. Schneider, L.A. Collins

A Linear Algebraic Approach to Electron-Molecule Scattering 217
L.A. Collins, B.I. Schneider

Variational R-Matrix Calculations of Electron-Molecule Scattering 218
C.J. Noble, R.K. Nesbet, L.A. Morgan

Vibrational Excitation of Positive Molecular Ions by Electron Impact. Exactly Solvable Model 219
A.K. Kazansky, V.V. Ponomarenko

Exactly Solvable Models in Theory of Vibrational Excitation of Molecules by Electron Impact 220
A.K. Kazansky

Electron Hydrogen Molecule Scattering at Low Energies 221
Sukanya Sur, A.S. Ghosh

Integral Cross-Sections for e-N_2 and CO Elastic Scattering at Intermediate and High Energies: A Coherent-Renormalized-Multicentre-Potential-Model (CRMPM) Approach 222
L.C.G. Freitas, Ashok Jain, Lee Mu-Tao

Electron Scattering by N_2: Calculations Using MCSCF Target Wavefunction 223
John Rumble, Jr., Donald G. Truhlar, Walter J. Stevens

Vibrational-Rotational Excitation of N_2 by Electrons at 5-50 eV 224
John Rumble, Jr., Donald G. Truhlar

Independent United Atom Model for Electron-Molecule Scattering 225
S.P. Khare, B.L. Jhanwar

Electron-Molecule Collisions: Resonances in Electronically Elastic Processes

Quantum Chemical Study of the $^2\Sigma_u^+$ Resonance of H_2^- 226
Eric A. Gislason, Nora H. Sabelli

The Adiabatic Partial Wave Method in Electron Molecule Processes I - Developments in Fixed Nuclei Electronic Continuum Processes 227
Vo Ky Lan, M. Le Dourneuf, J.M. Launay, S. Hara

The Adiabatic Partial Wave Method in Electron Molecule Processes II - Eigenphase Analysis of Electronuclear Correlations in Resonant Vibrational Excitation 228
J.M. Launay, M. Le Dourneuf, Vo Ky Lan

The Adiabatic Partial Wave Method in Electron Molecule Processes III - Eigentime Analysis of Electronuclear Correlations in Dissociative Attachment 229
M. Le Dourneuf, J.M. Launay, Vo Ky Lan

Nuclear Dynamics in Resonant Electron-Molecule Scattering Beyond the Local Approximation 230
Hernán Estrada, Michael Berman, L.S. Cederbaum, W. Domcke

Electron-Molecule Scattering Using the Optical Potential Approach: Surpassing Second Order 231
Michael Berman, O. Walter, L.S. Cederbaum

Wave Packet Formulation of the Boomerang Model for Resonant Electron-Molecule Scattering 232
C. William McCurdy, Julia L. Turner

Improved Semiclassical Approximation for Vibrational Excitation of Molecules by Slow Electrons 233
A.K. Kazansky, I.S. Yelets

Electron Scattering by Diatomic Molecules 234
S. Salvini, C.J. Noble, P.G. Burke

Absolute Cross Sections for Resonant Vibrational Excitation Processes of N_2 Molecules by Electron Impact 235
K. Onda, A. Temkin

Comparison of the 2 and the 20 eV Resonances in Vibrational Excitation of N_2 and CO by Electron Impact 236
Edward S. Chang

Vibronic Excitation of CO_2 by Electron Resonant Scattering 237
Edward S. Chang

Decay of the 22 eV Shape Resonance in N_2 238
L. Malegat, M.F. Fontaine, A. Colin, M. Tronc

Observation of Inner Shell Resonance in Elastic Scattering of Electrons from N_2 239
D. Mathur, F.A. Rajgara, A. Roy

Electron-Molecule Collisions:
Electronically Elastic Processes in Polyatomics

Threshold Behaviour in the Cross Section of Electron Scattering on CO_2 240
W. Sohn, K.-H. Kochem, N. Hebel, K. Jung, H. Ehrhardt

Rotational and Vibrational Excitation of CO_2 by Slow Electrons 241
K. Jung, T. Antoni, R. Müller, H. Ehrhardt

Absolute Total Electron Scattering Cross Sections for Triatomic Molecules in Low Energy Region 242
Czeslaw Szmytkowski

Non Adiabatic Effects in Low Energy e^--Molecule Scattering 243
E. Ficocelli Varracchio, U.T. Lamanna

Elastic Differential Scattering and Vibrational Excitation of CH_4 and C_2H_6 by Low Energy Electrons 244
P.J. Curry, W.R. Newell, A.C.H. Smith

Rotational Excitation of Methane by Electron Impact 245
H. Tanaka, N. Onodera, L. Boesten

Low Energy Electron Scattering by Hydrogen Sulphide (H_2S) 246
Ashok Jain, D.G. Thompson

Elastic Scattering of Electrons and Positrons by CH_4 at Intermediate Energies 247
Ashok Jain

(e^-,CsBr) Scattering: A Benchmark Experiment 248
B. Jaduszliwer, A. Tino, P. Weiss, B. Bederson

Electron-Polar-Molecule Collisions: Spherically Symmetric Approach 249
B. Stefanov

The Electron Gas Correlation Model: Applications to HCl and Other Molecules 250
N.T. Padial, D.W. Norcross

Effective Hamiltonian Approach to Electron-Molecule Collisions 251
N. Chandra

Small-Angle Elastic Scattering of Electrons from NH_3 Molecule at 300, 400 and 500 eV 252
Ashok Jain

The Elastic Electron Scattering Cross Section of SF_6 by Ab Initio SCF Calculation 253
Péter Pulay, Richard Mawhorter, D.A. Kohl, M. Fink

Elastic High Energy Electron Scattering as a Tool to Study Bonding and Correlation in Atoms and Molecules 254
M. Fink, R. Mawhorter, J.J. McClelland

Electron-Molecule Collisions: Excitation and Ionization of Diatomics

Calculated Cross Sections for Resonant Electron Scattering by Metastable Nitrogen 255
Iztok Cadez

Rotational Branch Structure in the Electronic Excitation of O_2 by Electron Impact 256
Edward S. Chang

Resonances in Low-Energy e-H_2^+ Collisions 257
B.I. Schneider, L.A. Collins

Threshold Electron Spectrum of I_2 258
J. Jureta, V. Bocvarski, S. Cvejanovic, M. Kurepa

Electron Impact Excitation of the B and C States of Deuterium 259
K. Becker, W. van Wijngaarden, J.W. McConkey

Electron-Photon Coincidence Experiments with Diatomic Molecules 260
K. Becker, H.W. Dassen, J.W. McConkey

Electron Energy-Loss Spectroscopy of Nitric Oxide Using a Position-Sensitive Multidetector 261
Richard J. Stubbs, Trevor A. York, John Comer

Electron Energy-Loss Studies of N_2 Using a Spectrometer with a Position Sensititve Detector 262
Tim Reddish, John Comer

Theoretical Studies of Bethe Surfaces in Small Molecules 263
K. Greenwald, P.W. Langhoff

The Electron Energy-Loss Spectrum of Hydrogen Chloride in the Region 12.748 eV to 16.254 eV 264
Trevor York, John Comer

Electron Energy Loss Spectroscopy in Molecular Chlorine 265
David Spence, R.-G. Wang, M.A. Dillon

Vibrational Excitation and Associative Detachment in the e^--HF System 266
J.P. Gauyacq

Studies of Vacuum Ultraviolet Emission from Rydberg Series of H_2 by Electron Impact 267
J.M. Ajello, D. Shemansky

Electron-Molecule Collisions: Excitation and Ionization of Polyatomics

Threshold Excitation of H_2O, D_2O and H_2S 268
J. Jureta, S. Cvejanovic, D. Cvejanovic, D. Cubric

Excitation of Optically Forbidden States in CO_2 269
S. Cvejanovic, J. Jureta, D. Cvejanovic, Dj. Srajer

Electron Scattering by Methane 270
L. Vuskovic, S. Trajmar

Photoabsorption Spectrum of CF_3I Generated by Electron Impact 271
S.K. Srivastava, S. Trajmar

Electron Impact Excitation Cross Sections for Excitation of Electronic States in UF_6 for Incident Electron Energies of 10, 20 and 40 eV 272
David C. Cartwright, S. Trajmar, A. Chutjian, S. Srivastava

The Rydberg Series of CO_2 in the Energy Range 11-14 eV 273
D. Roy, B. Leclerc, J. Delwiche, M.-J. Hubin-Franskin

DCS for the Excitation of OCS Molecules by Electron Impact 274
S. Ito, K. Hoshiba, S. Kano, H. Takuma, T. Takayanagi, K. Wakiya, H. Suzuki

Electron Attachment of CO_2 Clusters 275
K. Stephan, A. Stamatovic, H. Helm, T.D. Märk

Absolute Partial and Total Electron Impact Ionization Cross Section Functions for CF_4 and CCl_4 276
K. Stephan, K. Leiter, H. Deutsch, T.D. Märk

(e,2e) Reaction in Monoaloderivatives of C_2H_4 Molecule: Comparison between Experiments and Theory 277
A. Giardini-Guidoni, V. Di Martino, I.E. McCarthy, R. Fantoni

The Use of Electron Impact Excitation to Excite Electric-Dipole Forbidden Transitions in Atoms and Molecules 278
David A. Shaw, George C. King, F.H. Read

Probing of Orbitals by Binary (e,2e) Spectroscopy and Orbital Density Topography: A View of Chemical Bonding in Momentum Space 279
K.T. Leung, C.E. Brion

A New High Performance Electron Energy Loss Spectrometer for Valence and Inner-Shell Molecular Electronic Spectroscopy 280
C.E. Brion, S. Daviel, A.P. Hitchcock

Measurement of Partial Generalized Oscillator Strengths for Ionization of Nitrogen Molecule by 1-keV Electron Impact 281
H. Shibata, K. Kuroki, F. Nishimura, N. Oda

Electron-Molecule Collisions: Dissociation

Studies of Dissociative Excitation of N_2 and O_2 by Electron Impact with Extreme Ultraviolet Emission 282
J.M. Ajello

Angular Distribution of Balmer-α Emission Excited by Electron Impact on H_2 283
N. Kouchi, N. Takahashi, S. Arai, M. Morita, N. Oda, Y. Hatano

Translational Energy and Angular Distributions of H* and D* Produced in e-H_2, D_2 Collisions 284
Teiichiro Ogawa, Junichi Kurawaki

Angular Distributions of High-Rydberg Nitrogen Atoms Produced by Electron Impact on Nitrogen Molecule 285
T. Kondow, S. Ohshima, T. Fukuyama, K. Kuchitsu

Threshold Capture Widths Determined from Bound Rydberg States 286
Steven L. Guberman

Semiclassical Calculations of Dissociative Attachment in Collisions of Electrons with H_2, HD, D_2 287
I.S. Yelets, A.K. Kazansky

Transition State Effects in the Rotational Predissociation of H_2^+ Ions of Different Precursor Origins 288
Gareth Brenton, Paul G. Fournier, Elizabeth G. Richard, John H. Beynon

The Kinetic Energy Spectrum of Protons Produced by the Dissociative Ionization of H_2 by Electron Impact 289
M.A. Khakoo, S.K. Srivastava

The O^+ and N^+ Formation by Low-Energy Electron Impact on Nitrous Oxide 290
J.L. Olivier, R. Locht, J. Momigny

Metastable Dissociations of Small Cluster Ions 291
K. Stephan, A. Stamatovic, A.W. Castleman, Jr, J.H. Futrell, T.D. Märk

Electron-Molecule Collisions: Dissociative Attachment

Production of O^- Ions by Dissociative Electron Attachment to CO_2 292
O.J. Orient, S.K. Srivastava

Negative Ions Observed in Electron Transmission and Electron Attachment Spectroscopy 293
M. Heni, G. Kwiatkowski, E. Illenberger

Na^- Formation by Electron Impact in Na_2 294
D. Teillet-Billy, L. Bouby, J.P. Ziesel

Dissociative Electron-Attachment Processes in $HgBr_2$ 295
M. Tronc, R. Azria, R. Abouaf, L. Bouby, J.P. Ziesel

Dissociative Attachment in Highly Polar Molecules : Sodium Halides 296
J.P. Ziesel, R. Azria, D. Teillet-Billy, R. Abouaf, P. Girard

Dissociative Attachment to Polar Molecules : Angular Distribution 297
D. Teillet-Billy, J.P. Gauyacq

Formation and Dissociation of Negative Ions under Low Energy Electron Impact 298
S. Süzer, E. Illenberger, H. Baumgärtel

Thermal Electron Attachment to van der Waals Molecules as Studied by the Pulse Radiolysis Microwave Conductivity Technique 299
M. Toriumi, E. Suzuki, Y. Hatano

Dissociative Attachment of H_2 and its Isotopes by Low Energy Electron Impact 300
S. Bhattacharyya, L. Chatterjee

Positron-Atom(Molecule) Collisions

Electron and Positron Total Cross Sections on H_2 301
A. Deuring, J. Ferch, K. Floeder, D. Fromme, B. Granitza, J. Krug, C. Masche, W. Raith, A. Schwab, G. Sinapius, P.W. Zitzewitz

Measurements of Total Scattering Cross Sections for Positrons and Electrons Colliding with Potassium 302
T.S. Stein, R.D. Gomez, Y.-F. Hsieh, W.E. Kauppila, Ch.K. Kwan, S.J. Smith

Positron and Electron Total Scattering by N_2O, CH_4, and SF_6 303
W.E. Kauppila, M.S. Dababneh, Y.-F. Hsieh, Ch.K. Kwan, S.J. Smith, T.S. Stein, M.N. Uddin

Application of the Kohn Variational Method to the Calculation of Cross Sections for Low Energy Positron Hydrogen Molecule Scattering 304
E.A.G. Armour, M. Lavender

Elastic Scattering of Positron by Lithium Atom at Intermediate Energies 305
R.S. Pundir, K.C. Mathur

A Close Coupling Study of Positron Scattering by Nitrogen Molecule 306
A.K. Pande, D.N. Tripathi

Rotational Excitation of CH_4 Molecules by Low-Energy Positrons 307
Ashok Jain, D.G. Thompson

An Investigation of the Use of Inexact Wave Functions in Calculations of Positronium Formation in Positron-Atom Scattering 308
J.W. Humberston, C.J. Brown

Ideal Space Theory of the $e^+ + H \rightarrow Ps + H^+$ Process 309
E. Ficocelli Varracchio, M.D. Girardeau

One-Positron Triple Escape Threshold Behaviour 310
P. Grujic

Application to $\mu^+e^-e^+$ and $p\mu^-e^+$ of a New Method for Taking into Account Finite Nuclear Mass in the Determination of the Absence of Bound States 311
E.A.G. Armour

Atom-Atom Collisions: Theory

Use of l-Dependent Pseudopotentials in Molecular-Structure Calculations of Alkali-He Systems 312
J. Pascale

Total Cross Sections for $j_1m_1 \rightarrow j_2m_2$ Transitions within the First $n\,^2P$ States of Alkali Atoms Induced in Collisions with He 313
J. Pascale

Model Potential Calculations for the Ground and Excited (Rydberg) States of the NaAr and KAr Molecular Systems 314
A. Chebanier de Guerra, F. Masnou-Seeuws

Two Electron Model Potentials and Two Center Basis Sets 315
O. Mo, A. Riera

Model-Potential Calculations as Extended Multiproperty Analysis 316
R. Düren, E. Hasselbrink

A Two Electron Correlated Model for the Theoretical Determination of the Na_2, Li_2 and K_2 Potential Energy Curves 317
A. Henriet, M. Aubert-Frécon, C. Le Sech, F. Masnou-Seeuws

$^{1,3}\Sigma_{u,g}$ Excited States of Na_2 and K_2 (Hellmann Modified Pseudo-Potential) 318
A. Valance

Diabatic Potential Terms He(2 ^1S) - Rare Gas 319
H. Rahal, S. Runge, A. Valance

Atomic Fues Potential in He*: Oscillator Strengths and He_2* Potentials 320
S. Runge

A Simple Model for the van der Waals Potential between Two Closed Shell Atoms 321
K.T. Tang, J.P. Toennies

Wave Packet Solution of the Time-Dependent Hartree-Fock Equation for Atomic Collisions 322
Eberhard Teubner, Norbert Grün, Werner Scheid

On Ground States, at Potential Minima, of Neutral Quasi-Molecules Formed from Neutral Atoms or Ions 323
Ray Hefferlin, Ken Parker, Rosalie Parrish, Henry Kuhlman, Kevin Shaw, Ken Priddy, Mike Seaman

Quantum and Semi-Classical Calculations of the Phase-Shift in the Three Turning Points Case for Analytical Potential 324
O. Vallée, J. Picart, S. Avrillier, N. Tran Minh

Atom-Atom Collisions: Ionization

Associative Ionization of Hornbeck-Molnar Type in Thermal Energy Collisions between Laser Excited He(5 ^3P) and He Atoms 325
A. Pesnelle, S. Runge, G. Watel

Autoionization Width for He(3 ^1P)+Ne: Penning and Associative Ionization Cross Sections 326
A. Pesnelle, S. Runge

Associative Ionization and 2-Photon Laser Induced Collisional Ionization in Crossed Beam Alkali and Alkaline Earth Systems 327
J. Weiner, J. Boulmer, J. Keller, R. Bonanno

Collisional Ionisation and Energy Pooling Process in Rb Vapour 328
L. Barbier, M. Cheret

Product Angle-Velocity Distribution for Ionization of Argon Atoms by Triplet Metastable Helium 329
P.R. Jones, K.T. Gillen, M.J. Coggiola

Energy Transfer and Associative Ionization in Na(3P) + Na(3P) Collisions 330
J. Huennekens, S. Davidson, A. Gallagher

Investigation of Ionization Processes in Excited Atoms Slow Collisions by Using the Methods of Plasma Electron Spectroscopy 331
A.Z. Devdariani, V.I. Demidov, N.B. Kolokolov, V.I. Rubtsov

Calculation of Penning Ionization for He($2\ ^{1,3}S$)+Na 332
James S. Cohen, Richard L. Martin, Neal F. Lane

Collisional Ionization of H(3p) Atoms on Rare Gas Atoms and H_2 Molecules 333
A. Cornet, W. Claeys, V. Lorent, J. Jureta, D. Fussen

Ion Pair Production in H(1s) + H(2s) Collisions 334
D. Fussen, W. Claeys, A. Cornet, J. Jureta, P. Defrance

Atom-Atom Collisions: Excitation and Excimer Formation

Excitation of Hydrogen Atoms to the n = 3 and 4 Levels in Collisions with Rare-Gas Atoms 335
B. Van Zyl, M.W. Gealy, H. Neumann, R.C. Amme

Theory of Li(2s-2p) and Na(3s-3p) Excitation in Li-Na High Energy Collisions 336
Svend Erik Nielsen, Martin Larsen, John S. Dahler

Hydrogen Atom Excitation in H + He, Ne, Ar Collisions below 1 keV 337
J. Grosser, W. Krueger

Simultaneous $2\ ^1P$ Excitation of Two Colliding He Atoms to Various Substate Combinations 338
L. Moorman, K.P.J. Linnartz, J. van Eck, H.G.M. Heideman

Emission from Outer Turning Points of High Vibrational States of the Krypton Excimer after keV-Electron Impact Excitation 339
P. Wollenweber, K. Barzen, H. Schmoranzer

Emission of the Krypton Excimer at Small Interatomic Distances after keV-Electron Impact Excitation 340
K. Barzen, P. Wollenweber, H. Schmoranzer

Atom-Atom Collisions: Fine Structure Transitions and Polarization

Polarization Effects in Alkali Atoms Fine Structure Transitions (F.S.T.) Induced by He and Ar 341
J.M. Mestdagh, J. Pascale

Quantum-Mechanical Calculations of Cross Sections for Transitions within the Second and Third n ^2P Levels of Rb and Cs, Induced in Collisions with He 342
J. Pascale

Coherence Observed in Thermal K($4\ ^2P$)-Rare Gas Collisions 343
R. Düren, E. Hasselbrink, H. Tischer

Scattering of State Selected, Electronically Excited Neon Atoms 344
W. Beyer, H. Haberland, D. Hausamann

Atom-Atom Collisions: Excitation Transfer

Nonresonance Excitation Transfer in He(2 ^1S,2 ^3S) + Ne Collisions 345
A.Z. Devdariani, A.L. Zagrebin

Spectroscopic Investigation of Inelastic Scattering Channels in He^m - He Collisions 346
D.V. Elakhovsky, Yu.V. Zaitsev, A.D. Khakhaev

Atom-Atom and Atom-Molecule Collisions in Sodium Vapour Excited by Laser Resonance Radiation at 3 ^2S - 3 ^2P Transition Wavelength 347
Zh.L. Shvegzhda, S.M. Papernov, M.L. Jansons

New Approach to Multi-State Problem and Application to Laser Induced Transition: Sr(5p ^1P)+Ca(4s^2 ^1S)$\rightarrow$Sr(5s^2 ^1S)+Ca(4d ^1D) 348
H. Yagisawa

Ion-Atom Collisions: Elastic Scattering

Elastic Scattering of Slow Hydrogen Atoms on Ions with Large Impulse Transfer 349
A.K. Kazansky, I.V. Komarov

Elastic Differential Cross Sections for Proton Scattering by Atomic Hydrogen 350
J.T. Park, D.M. Blankenship, T.J. Kvale, J.L. Peacher, E. Redd, E. Rille

Measurements of Orbiting Structures in Differential H^+ - He and H^+ - Ar Scattering 351
M. Konrad, F. Linder

Semiclassical Regge-Pole Description of Low-Energy H^+-He Scattering 352
Karl-Erich Thylwe

Elastic Scattering, Excitation and Ionisation of Helium by Proton Impact 353
C. Bergnes, D. Bordenave-Montesquieu, A. Boutonnet, R. Dagnac

Elastic Scattering and Charge Exchange in He^+-He, H^+-Kr and H^+-Xe Collisions at E_{cm}= 0.5 - 30 eV 354
P. Reinig, G. Bischof, F. Linder

Potential Energy Surfaces for Di-atomic Molecules 355
H. Hartung, B. Fricke, W.-D. Sepp

Development of an Accurate Numerical Dirac-Fock-Slater Program for Di-atomic Molecules 356
W.-D. Sepp, D. Kolb, H. Hartung, W. Sengler, B. Fricke

Ion-Atom Collisions: Outer-Shell Excitation

Correction Terms for the Bethe Straggling Expression 357
Hans Bichsel

Proton Impact Excitation of Hydrogen Atom 358
S. Saxena, G.P. Gupta, K.C. Mathur

Excitation of He I Triplet States by Proton Impact 359
A.S. Aynacioglu, G. von Oppen, G. Weber

Light Ion + He Collisions in Time-Dependent Hartree-Fock Theory 360
K.R. Sandhya Devi, J.D. Garcia

A Several-Electron Atomic-Basis Calculation of Be^+ (2s-2p) Excitation in Be^+(2s)-He($1s^2$) Collisions 361
Svend Erik Nielsen, John S. Dahler

Resolution of the Glauber Inconsistency for Inelastic Scattering by Heavy Particles 362
J.H. McGuire

Ion-Atom Collisions: Spectroscopy of Continuum Electrons

δ-Electron Spectroscopy of Multiple Ionization in H^+-Ar Collisions 363
J. Bossler, R. Hippler, H.O. Lutz

Study of Inner Quasiatomic Shells (Z_U = 132-171) by Means of δ-Electron Spectroscopy in Asymmetric Collisions 364
F. Güttner, W. Koenig, N. Lutz, B. Martin, H. Skapa, J. Soltani, H. Banda, A.V. Ramayya, F. Bosch, Ch. Kozhuharov

Fast Electrons from Slow Atomic Collisions 365
Raul A. Baragiola, Eduardo V. Alonso

Calculated Double Differential Ionization Cross-Sections for Molecules Bombarded by Protons 366
B. Senger

Binary Encounter Approximations with Screening to DDCS of Electrons from Atomic Collisions 367
G. Hock

Doubly Differential Ionization Cross Sections in Fast Ion-Atom Collisions 368
A.K. Kaminsky, M.I. Popova

Electron Losses in H_2^+, H_3^+ and He^+ Collisions with Ar 369
N. Oda, F. Nishimura, K. Komatsu, H. Shibata

Collisional Electron Loss into the Continuum of Protons Using the $H^0 \rightarrow He$ System; a Comparison with Theory 370
R. Vidal, W. Meckbach, E. González Lepera

Charge State Dependence of Electron Loss Peak Measured under 0° for Fast Argon Ions 371
A. Itoh, T. Schneider, G. Schiwietz, Z. Roller, H. Platten, G. Nolte, D. Schneider, N. Stolterfoht

Origin of the Double Peak in Electron Loss in the Forward Direction 372
Victor H. Ponce, Raul A. Baragiola

Stripping Cross Sections of Multi-Charged Ions by Neutral Atoms 373
S. Karashima, T. Watanabe

Ion-Atom Collisions: Capture into the Continuum

Continuum-Electron Capture by Protons in Helium 374
Poul Dahl

Effect of a Screened Electron-Projectile Interaction on the ECC Peak Shape 375
R.O. Barrachina, C.R. Garibotti

A Comparison of Electron Capture and Electron Loss into the Continuum with H^+ and H^0 Projectiles Interacting with He 376
R. Vidal, P. Focke, E. González Lepera, I.B. Nemirovsky, W. Meckbach

δ-Electron Emission in Strong Projectile Fields 377
D.H. Jakubassa-Amundsen

The Linewidth of Electron Loss to Continuum Cusps 378
J. Burgdörfer, M. Breinig, S.B. Elston, I.A. Sellin

Electron Capture to the Continuum at Asymptotically High Velocities 379
S.D. Berry, I.A. Sellin, L.H. Andersen, M. Breinig, S.B. Elston, M.M. Schauer, K.-O. Groeneveld, D. Hofmann, N. Stolterfoht, H. Schmidt-Böcking, G. Nolte, G. Schiwietz

Description of Interaction in Final State on the Basis of the Faddeev-Merkuriev Equations in the Processes of Ionization and Charge Transfer 380
A.L. Godunov, Sh.D. Kunikeev, V.N. Mileev, V.S. Senashenko

Convoy Electron Production and Total Electron Yield in High Velocity (24 au) Heavy Ion-Solid Collisions 381
R. Latz, M. Burkhard, H.J. Frischkorn, D. Hofmann, P. Koschar, J. Schader, K.O. Groeneveld, M. Breinig, S.D. Berry, I.A. Sellin

Ion-Atom Collisions: Autoionization and Auger Spectroscopy

Auger Electron Emission Following Fast (MeV) Molecular- and Atomic-Ion Impact on Thin C-Foils 382
D. Schneider, E.P. Kanter, B.J. Zabransky

Autoionization Spectra of He Excited by Fast (MeV) H^+, He^+, and Li^{n+} (n=1,2,3) Ions 383
D. Schneider, P. Arcuni, R. Bruch, W. Stöffler

Autoionization of Fast (MeV) Li-Ions Incident on Gases and C-Foils 384
D. Schneider, P. Arcuni, R. Bruch, W. Stöffler, C.F. Moore

Position and Widths of Autoionizing States in the Helium Isoelectronic Sequence above the n=2 Continuum 385
H. Bachau

A Study of the Velocity Dependence of the Yield of the $(1s2s2p)^4P_{5/2}$ Metastable Fraction in MeV C, O and F Ions Following Excitation by C Foils 386
J.K. Swenson, D. Brandt, M. Clark, S.M. Shafroth, J.R. Huddle

Ar L-MM Auger Spectra in $Ar^{3,4+}$ + Ar Collisions 387
T. Matsuo, J. Urakawa, A. Yagishita, Y. Awaya, T. Kambara, M. Kase, H. Kumagai, J. Takahashi

Kr M-NN Auger Spectra in Ar^{4+} + Kr Collisions 388
J. Urakawa, T. Matsuo, H. Shibata, A. Yagishita, Y. Awaya, T. Kambara, M. Kase, H. Kumagai, J. Takahashi

Selective Production of Auger Electrons from Fast Ar^{q+} Ions Studied by Zero-Degree Auger Spectroscopy 389
A. Itoh, T. Schneider, G. Schiwietz, Z. Roller, H. Platten, G. Nolte, D. Schneider, N. Stolterfoht

A Study of Auger Spectra from $Ne^{3+,10+}$, $Ar^{6+,17+}$ (5.6 MeV/amu) → Ne Collisions 390
D. Berényi, G. Hock, I. Kádár, S. Ricz, V.A. Shchegolev, B. Sulik, D. Varga, J. Végh

Formation and Autoionization of Ne** at Mg Surfaces 391
G.E. Zampieri, F. Meier, R.A. Baragiola

The Barker-Berry Effect on the Ejected Electron Spectra in Rb^+-Ar and Cs^+-Kr Collisions 392
K. Wada, A. Wada, K. Wakiya, H. Suzuki

Measurement of Forward Directed Electron Spectra in Coincidence with Emergent Charge States for Collisions of 20 MeV Au^{17+} with He 393
L.H. Andersen, H. Cederquist, S. Datz, M. Frost, P. Hvelplund, H. Knudsen, L. Liljeby

Spectroscopy of Low-Energy Autoionization Electrons Emitted from the Ions in Highly Charged Ion-Atom Collisions 394
L.H. Andersen, M. Frost, P. Hvelplund, H. Knudsen, L. Liljeby

Formation and Decay of Autoionization States in Ar^{3+} - Xe Collisions 395
V.M. Mikoushkin, I.P. Flaks, G.N. Ogurtsov

Calculations of the Excitation Cross Sections of the Parity Unfavored of Autoionizing States of He Atoms by Protons 396
V.A. Sidorovich

Energy Spectra of Detached Electrons Produced in $H^-(D^-)$ Collisions on Rare Gases and Diatomic Molecules 397
Y. Itoh, U. Hege, F. Linder

Electrostatic versus Dynamical Coupling in Detachment Collisions of $H^-(D^-)$ Ions with Rare Gas Atoms 398
U. Hege, Y. Itoh, F. Linder

A Scaling Law for Electron Detachment in keV Collisions of H^-, Li^-, Na^-, K^- with He, Ne, and Ar 399
N. Andersen, T. Andersen, L. Jepsen

Electron Detachment in H^- - Na Collisions and Free Electron Scattering Approximation 400
I.T. Serenkov, V.I. Sakharov, R.N. Il'in

Electron Detachment in Negative Ion - Molecule Collisions 401
M.S. Huq, L.D. Doverspike, R.L. Champion

Born Cross Sections for H^- Collisional Electron Detachment Leading to H^0 in the 1s, 2s and 2p Final States 402
George H. Gillespie, Ralph S. Janda, David L. Moores

Energy Distribution of Electrons Detached in Negative Ion Collisions 403
Y. Sato, T. Okamoto, H. Inouye

Electron Detachment in Collisions of Cl^- and Ti^- Ions with Atoms Ar, Na and Mg 404
I.T. Serenkov, V.I. Sakharov, E.A. Solovyev, R.N. Il'in

Theory of Electron Detachment in Slow Anions Impact on Atoms 405
Fumihiro Koike

Electron Detachment in Negative-Ion Collisions: New Theoretical Methods 406
T.S. Wang, J.B. Delos

Electron Detachment in Negative-Ion Collisions: Cross Sections 407
T.S. Wang, J.B. Delos

Electron Detachment in Collisions of H^- and Several Targets in the Energy Range from 500 to 2000 eV 408
P.E. van der Leeuw, W. Koot, A.W. Kleyn, J. Los

Excitation of Autodetaching States in H^-Kr Collisions 409
V. Esaulov, F. Pichou, C. Schermann, J.P. Grouard, R.I. Hall, M. Landau, J.L. Montmagnon

F^- Collisions with Atomic and Molecular Targets 410
Vu Ngoc Tuan, V.A. Esaulov

Charge Exchange to Shape Resonances in H^--CO_2 Collisions 411
Vu Ngoc Tuan, V. Esaulov, J.P. Gauyacq

Observation of Electron Spectra Produced in F^- Collisions 412
J.P. Grouard, V. Esaulov, R.I. Hall, M. Landau, J.L. Montmagnon, F. Pichou, C. Schermann

Ion-Atom Collisions: Multiple Ionization

Ionization Cross Sections for 5-4000 keV Protons in Gases 413
M.E. Rudd, T.V. Goffe, R.D. DuBois, L.H. Toburen, C.A. Ratcliffe

Double Ionization Mechanisms in H^+ - Ne Collisions 414
R.D. DuBois, L.H. Toburen, S.T. Manson

Calculations of the Cross Sections for the Double Ionization of Helium by Light Nuclei 415
V.A. Sidorovich, V.S. Nikolaev

Measurements of the Ratio between the Double- and Single-Ionization Cross Sections of Helium in Collisions with Fast, Bare Nuclei 416
H. Knudsen, L.H. Andersen, P. Hvelplund, G. Astner, H. Cederquist, H. Danared, L. Liljeby, K.-G. Rensfelt

Coincidence Measurements of Recoil-Ion Charge-State Spectra Produced by Ionization and Charge Transfer in Collisions of 1.4 MeV/u Highly Charged Ions and Rare-Gas Atoms 417
B. Schuch, W. Groh, A. Müller, E. Salzborn, H.F. Beyer, W.A. Schönfeldt, P.H. Mokler

Influence of the Projectile Charge State on Multiple Ionization of Rare Gas Atoms 418
H.-Ch. Werner, H. Schmidt-Böcking, N. Stolterfoht, G. Nolte

Multiple Ionization of Slow Recoil Ions in Fast Heavy-Ion Atom Collisions 419
J. Ullrich, S. Kelbch, W. Schadt, H. Schmidt-Böcking, R. Schuch, H. Ingwersen

Ionisation of Atomic Hydrogen by Fast Multiply Charged Ions 420
M.B. Shah, H.B. Gilbody

Energies and Couplings for Infinitely Excited States of $He\text{-}Li^{3+}$ Quasi-molecule 421
A. Macias, R. Mendizabal, F. Pelayo, A. Riera, M. Yánez

Ion-Atom Collisions: Direct Ionization of Inner Shells

New Experimental Investigations of Density Effect in Inner-Shell Excitations 422
S.P. Møller, A.H. Sørensen, J.F. Bak, F.E. Meyer, J.B.B. Petersen, E. Uggerhøj, K. Østergaard

Total K-Shell Ionisation Cross Sections for Heavy Elements Induced by Protons 423
S. Divoux, B. Raith, B. Gonsior

Total K-Shell Ionisation Cross Sections for Elements $24 \leq Z_T \leq 60$ Induced by Li Ions 424
B. Raith, S. Divoux, B. Gonsior

Total K-Shell Vacancy Production Cross Sections for 200 to 1600 keV/amu ^3He on Ti and Cu 425
Donald G. Simons, David J. Land

K-Shell X-Rays and Multiple Vacancy Production in $_{19}K$, $_{22}Ti$, $_{12}Mg$, and $_{35}Br$ Resulting from 20 - 80 MeV $_{17}Cl$ Ion Bombardment 426
J.A. Tanis, S.M. Shafroth, T. McAbee, G. Lapicki

Measurement of the K X-Ray Production Cross Section in 50-88 MeV Si^{11+} + He Collisions 427
D. Brandt, M. Clark, T. McAbee, J. Swenson, S. Shafroth

Ionization and Electron Transfer for the K-, L-, and M-Shells for 1.86 MeV/u Ions on Selected Target Systems 428
F.D. McDaniel, J.L. Duggan, R. Mehta, M.C. Andrews, A. Toten, J.D. Gressett, D. Johnson, S.R. Wilson, P.D. Miller, G. Lapicki, G. Basbas, L.A. Rayburn, A.R. Zander, R.M. Wheeler, R.P. Chaturvedi, R.S. Peterson

L-Subshell Ionization of Au by Light Ion Impact 429
K. Finck, W. Jitschin, R. Hippler, H.O. Lutz

L-Shell X-Ray Production Cross Sections of $_{48}Cd$, $_{50}Sn$, $_{52}Te$, $_{53}I$ and $_{56}Ba$ for Protons and Alpha Particles 430
L. Avaldi, I.V. Mitchell, M. Milazzo

Survey of M-Shell X-Ray Production for 19 Elements by H^+ and $He^{+,++}$ Ions 431
R. Mehta, J.L. Duggan, F.D. McDaniel, P.M. Kocur, J.L. Price, G. Lapicki

Nonperturbative Effects in Inner-Shell Ionization 432
David J. Land

The Binding Correction for Inner Shell Ionisation in Asymmetric Ion-Atom Collisions 433
L. Kocbach

K-Shell Ionization at Large Scattering Angles with Light Projectiles 434
E. Morenzoni, R. Anholt, S. Andriamonje, W.E. Meyerhof, O.K. Baker, J.D. Molitoris

Ion-Atom Collisions: Quasimolecular Excitation of Inner Shells I

Pair Creation in "Overcritical" Coulomb Fields 435
M. Clemente, E. Berdermann, P. Kienle, H. Tsertos, W. Wagner, F. Bosch, C. Kozhuharov, W. Koenig

K-Shell Excitation by K- to L-Shell Charge Transfer in Slow Kr^{q+}-Kr and Xe^{q+}-Xe Collisions 436
R. Hoffmann, R. Schuch, E. Justiniano, W. Schadt, H. Schmidt-Böcking, P.H. Mokler, F. Bosch, W.A. Schönfeldt, Z. Stachura

K-Ionization Probability in High Energy U + U and U + Pb Collisions 437
J.D. Molitoris, R. Anholt, S. Andriamonje, E. Morenzoni, W.E. Meyerhof, O.K. Baker

X-Ray Emission Probabilities and Nuclear Time Delay in the Deep Inelastic Collision U+U at 7.5 MeV/amu 438
Ch. Stoller, M. Nessi, W. Wölfli, W.E. Meyerhof, J.D. Molitoris, E. Morenzoni, E. Grosse, Ch. Michel

Influence of the Nuclear Reaction Time onto the δ-Electron Emission Studied by the Reaction J → Au, Bi at 795 and 840 MeV 439
F. Güttner, W. Koenig, N. Lutz, B. Martin, H. Skapa, J. Soltani, H. Banda, A.V. Ramayya, F. Bosch, Ch. Kozhuharov

Charge Transfer and Simultaneous Excitation in Ion-Atom Collisions: A Comment on RTE 440
Tricia Reeves, Jim Feagin, John Briggs

Double K Excitation Cross Section of Fe^{24+} Ions at Intermediate Velocity 441
K. Wohrer, J.P. Rozet, A. Chetioui, A. Jolly, C. Stephan

Microscopic Analysis of Equilibrium Charge State Distribution for Ions in the MeV/u Range 442
D. Vernhet, J.P. Rozet, P. Legagneux-Piquemal, A. Chetioui, L. Tassen-Got, C. Stephan

Vacancy Sharing Following Resonant-Transfer-and-Excitation in S+Ar Collisions 443
J.A. Tanis, E.M. Bernstein, W.G. Graham, M. Clark, S.M. Shafroth, B.M. Johnson, K. Jones, M. Meron

Rotational Coupling in Asymmetric Ion-Atom Collisions 444
D. Maor, Z. Stachura, P.H. Mokler, B. Liu, D. Liesen

Inner-Shell Vacancy Production in Asymmetric Heavy Ion - Atom Collisions 445
A. Warczak, H.D. Dohmann, D. Liesen, B. Liu

Inner Shell Vacancy Transfer Studied with Highly Ionized, Decelerated Heavy Ions 446
P.H. Mokler, D.H.H. Hoffmann, W.A. Schönfeldt, Z. Stachura, A. Warczak

The Electron's Path Through Phase Space During Heavy Ion Impact 447
J. Krause, M. Kleber

Ion-Atom Collisions: Quasimolecular Excitation of Inner Shells II

L-Shell Excitation in Slow Ion-Atom Encounters 448
R. Shanker, R. Hippler, R. Bilau, U. Wille, H.O. Lutz

Direct Ionization of Highly Promoted Molecular Orbitals in Slow Ion-Atom Collisions 449
U. Wille

Observation of Al 2s Vacancy Production in 100 keV Al^+ on Ar Gas Collisions: Significance for MO Correlations 450
M.L. Furst, H.C. Hayden, W.W. Smith

Electron Transitions from Discrete Levels into Continuum due to Radial Coupling in Slow Atomic Collisions 451
A.N. Zinoviev, S.Yu. Ovchinnikov, Yu.S. Gordeev

Inner Shell Direct Ionization in Slow Atomic Collisions 452
S.Yu. Ovchinnikov, A.N. Zinoviev, Yu.S. Gordeev

Coupled Multichannel Calculation for Ne^+ + Ne Collisions in the Energy Range of 5-300 keV 453
A. Toepfer, B. Jacob, H.J. Lüdde, R.M. Dreizler

Calculations of Coupling Matrix Elements for Di-atomic Systems in an Atomic Basis 454
W.-D. Sepp

Semiclassical Approximation of the Time-Dependent Dirac Equation with First Order Perturbation Theory and Finite Difference Method 455
S.R. Valluri, U. Becker, N. Grün, W. Scheid

Time Dependent Screened Potentials for Atomic Scattering Problems 456
A. Henne, R.M. Dreizler

Bound Electron Energy of Superheavy Atom 457
Jiben Sidhanta, S.C. Mukherjee

Ion-Atom Collisions: Alignment and Orientation

Coherence Study of He(2 ^1P) and Li(3 ^2D) Excitation in Li^+-He Collisions 458
H.-P. Neitzke, N. Andersen, T. Andersen

Coherent Excitation of L>1 States: Multipole Conversion in External Fields 459
H.-P. Neitzke, T. Andersen

Alignment and Orientation Studies for the Ion-Impact Excitation of the 3P → 3D Transition in Na^+-Na Collisions 460
A. Bähring, E. Meyer, I.V. Hertel

A Study of the Excitation Mechanism of the $(2p^2)$ ^1D State Excited in He^+ + He Collisions 461
E. Boskamp, R. Morgenstern, A. Niehaus, P. van der Straten

Alignment of the 2p States of Helium-Like and Hydrogen-Like Ions 462
D.A. Church, R.L. Watson, R.A. Kenefick, D.-W. Wang, G. Pedrazzini

Anisotropic L X-Ray Emission in Ion-Atom Collisions 463
W. Jitschin, R. Hippler, R. Schuch, H.O. Lutz

The L_3-Vacancy Alignment in 32 MeV S + Au Collisions at Small Impact Parameters 464
A. Berinde, C. Ciortea, Al. Enulescu, Daniela Fluerasu, I. Piticu, V. Zoran

Angular Distribution of Au L X-Rays by Heavy Ion Impact 465
J. Takahashi, Y. Awaya, T. Kambara, M. Kase, H. Kumagai, J. Urakawa, T. Matsuo, H. Shibata

Effect of Screening by Projectile Electrons on the Alignment of a Vacancy in Ion-Atom Collision 466
N.M. Kabachnik, O.B. Maksimova

Ion-Atom Collisions: X-Ray Spectroscopy

Wavelength of Transitions in Few Electron Spectra of Titanium 467
H.D. Dohmann, D. Liesen, E. Pfeng

The Effect of Chemical Bonds on Fluorine K X-Ray Spectra Produced by 80 MeV Ar and 48 MeV Mg Ions 468
O. Benka, R.L. Watson

High Resolution EUV Spectra of Core-Excited $(1s2pnp)\,^2P$, $(1s2pnd)\,^2D^0$ and $(1s2pnf)\,^2F$ States of Doubly Ionized Boron 469
R. Bruch, K.T. Chung, E. Träbert, P.H. Heckmann, B. Raith

High Resolution Measurements of K X Rays from Ar Ions Impinging on Foils 470
Y. Awaya, T. Kambara, M. Kase, H. Kumagai, J. Takahashi, J. Urakawa, T. Matsuo, M. Namiki

Target Gas Pressure Dependence of Relative Yield of K X-Rays from 110 MeV Ne Ions 471
T. Kambara, Y. Awaya, M. Kase, H. Kumagai, I. Kohno, T. Tonuma, A. Hitachi

Ion-Atom Collisions: De-excitation of Quasimolecular States

Interference Effect in the Quasimolecular K-Radiation Induced by Hydrogenlike Low-Velocity Projectiles 472
R. Schuch, J. Barrette, R. Hoffmann, B.M. Johnson, K.W. Jones, M. Meron, H. Schmidt-Böcking, I. Tserruya

Impact Parameter Dependence of $1s\sigma$ and $2p\sigma$ MO Radiation in 90 MeV Ni + Ni Collisions 473
H. Richter

The Emission and Absorption of Light During Atomic Collisions 474
P.T. Greenland

Molecular Autoionization Spectra in Low-MeV Kr^+-Kr Collisions 475
P. Clapis, A. Antar, S. Kuptsov, R. Roser, R. Rubino, Q. Kessel

Molecular-Autoionization Spectra from He^+-He Collisions 476
N. Tokoro, S. Takenouchi, N. Oda

3p-Vacancy Produced due to Two-Electron Transitions in Kr-Kr and Kr-Ar Collisions 477
N.A. Guschina, G.G. Meskhi, V.K. Nikulin, A.P. Shergin

Ion-Atom Collisions: Electron Exchange with Singly Charged Ions, Experiment I

Experimental Observation of the Thomas Peak in Electron Capture 478
E. Horsdal-Pedersen, C.L. Cocke, M. Stöckli

Experimental Determination of the Density Matrix Describing Electron Transfer Collisions for H^+ on He to the n=3 State of H 479
C.C. Havener, N. Rouze, W.B. Westerveld, J.S. Risley

Differential Cross Sections in Ion-Neutral Charge Exchange Collisions 480
J.H. Newman, Y.S. Chen, P.S. Gibner, K.A. Smith, R.F. Stebbings

Angular Correlation Measurements of He(3 ^{3}P) Resulting from He^+ + Ne Collisions 481
M. Natarajan, A.L. Goldberger, O. Yenen, D.H. Jaecks

Charge Exchange between H_2^+ and H^- 482
S. Szücs, M. Karemera, M. Terao

Electron Capture, Loss and Excitation in Collisions of H^+, H(1s), H(2s) and H^- in Atomic Oxygen 483
I.D. Williams, J. Geddes, H.B. Gilbody

Excitation in Inelastic Collisions of H^+ (2 - 15 keV) with Li 484
F. Aumayr, A. Brazuk, U. Wutte, H. Winter

Polarization Studies of H(2p) Excitaton in H^+ - He, Ar Charge Changing Collisions 485
R. Hippler, M. Faust, R. Wolf, H. Kleinpoppen, H.O. Lutz

Ion-Atom Collisions:
Electron Exchange with Singly Charged Ions, Experiment II

Charge-Transfer Collisions of Ne^+ and Metastable Helium 486
R.H. Neynaber, S.Y. Tang

Multiple Ionization of Argon by Charge Transfer and Direct Ionization 487
R.D. DuBois, L.H. Toburen

Charge Transfer and Fine Structure Transitions in 2-20 keV Xe^+ Collisions 488
R.F. King, C.J. Latimer

The Charge Transfer Cross Section for Kr^+-Kr at Ion Energies between 0.08 and 2 eV 489
M.T. Elford, O.M. Williams

Experimental Cross Sections for Lyman-Alpha Emission in Charge Transfer Collisions of H^+ with Cs, Rb, K and Na Atoms 490
T. Nagata

Charge Exchange of (0.2 - 5.0) keV Protons and Hydrogen Atoms in Sodium-, Potassium- and Rubidium-Vapor Targets 491
F. Ebel, E. Salzborn

Electron Capture by C^+, O^+, In^+, Sn^+ and Pb^+ Ions in H_2 in the Energy Range 10 - 150 keV 492
F. Melchert, K. Rinn, A. Müller, E. Salzborn

Ionization and Charge Exchange Processes in Collisions of Alkali Metal Ions with Rare Gas Atoms in the Energy Region 0.5-7.0 keV 493
B.I. Kikiani, R.A. Lomsadze, S.V. Martinov, N.O. Mosulishvili, M.R. Gochitashvili, V.M. Lavrov

Differential Large Angle Scattering in Collisions of K^+ Ions with Ar Atoms 494
S. Kita, M. Izawa, H. Inouye

Collisions between Li^+ Ions 495
G.C. Angel, K.F. Dunn, M.F. Watts, H.B. Gilbody

Ion-Atom Collisions:
Electron Exchange with Singly Charged Ions, Theory I

Theory of Electron Capture in Intermediate-to-High-Velocity Collisions 496
Knud Taulbjerg, John S. Briggs

Close Coupling Calculation of Electron Capture at High Energies 497
K. Fujiwara, N. Toshima, T. Watanabe

Electron Capture by Fast Protons Scattered at Large Angles 498
L. Kocbach, J.S. Briggs

Exact Second Born Calculations for Electron Capture 499
J.H. McGuire, J. Eichler, P.R. Simony

A New Variational Principle for Charge Exchange at Arbitrary Energies 500
Dz. Belkic

The 'Classical Deflection Function' in the Continuum Distorted Wave Approximation 501
P.T. Greenland

Representation of Continuum Channels in the Description of Collisionally Induced Electronic Transitions 502
W. Fritsch, C.D. Lin

Proton-Hydrogen Scattering at High Energies 503
Roberto D. Rivarola

Electron Capture by Fast Protons in Gases 504
A.M. Popova, Ya.A. Teplova, Yu.A. Shurigina, O.S. Erkovitch

Proton-Hydrogen Electron Capture in Second Order Brinkman-Kramers Approximation 505
M.K. Srivastava, A.K. Sharma

Effect of Orthogonality of Initial and Final States on p-He Electron Capture 506
A.K. Sharma, M.K. Srivastava

K-K Shell Electron Capture in Asymmetric Proton - Atom Collisions at High and Intermediate Energies 507
Roberto D. Rivarola, Antoine Salin

Ion- Atom Collisions: Electron Exchange with Singly Charged Ions, Theory II

The Strong Potential Born Approximation for Charge Transfer at Large Scattering Angles 508
P.A. Amundsen, D.H. Jakubassa-Amundsen

Electron Capture into Highly Excited States 509
S.C. Mukherjee, Shyamal Datta, C.R. Mandal

Ion Atom Collision for Systems with Two Electrons 510
W. Stich, H.J. Lüdde, R.M. Dreizler

Electron Transfer and Excitation in p-H Collisions Using a Triple-Center Basis 511
T.G. Winter, C.D. Lin

Charge Transfer and Ionization Processes in He^+ + He^+ and He^+ + H Collisions 512
M.R.C. McDowell, G. Peach, S.L. Willis

Electron Capture and Excitation in Proton-Sodium Collisions at Energies $E \leq 10$ keV 513
R.J. Allan, A.S. Dickinson, R. McCarroll

Charge Transfer in Collisions between Protons and Lithium Atoms 514
A.M. Ermolaev

Core-Independent Parameters for Charge Transfer Reactions 515
K. Bartschat, H.J. Andrä, K. Blum

Phase Integrals and Perturbed Stationary States 516
A. Bárány, D.S.F. Crothers, J.G. Hughes

Charge Exchange between He^+-Ions and Lithium Atom 517
K. Roy, R. Shingal

Molecular State Calculation of Charge Transfer in H^+ + Li Collisions 518
M. Kimura, H. Sato, J. Pascale, R.E. Olson

Molecular Treatment of Charge Transfer in Li^+ + Ca Collisions 519
M. Kimura, H. Sato, R.E. Olson

Dynamical-State Representation and its Application to the $(Li\text{-}Na)^+$ Collision System 520
Reiko Hirokawa, Hiroki Nakamura, Eiichi Ishiguro

A Theoretical Study of Coherence Effects in Charge Transfer Collisions: Application to Na-Li^+ 521
A.E. Orel, K.C. Kulander

H^+ + H^- Neutralization 522
F. Borondo, A. Macias, A. Riera

Ion-Atom Collisions: Electron Exchange with Multiply Charged Ions I

A Quantum Electrodynamic Approach to Charge Transfer between Fully Stripped Light Ions and Hydrogen Atom 523
S. Bhattacharyya, L. Chatterjee, K. Sen Gupta

Charge Transfer for Completely Stripped Boron and Carbon Ions from Atomic Hydrogen 524
S.C. Mukherjee, C.R. Mandal, Shyamal Datta

Electron Capture for Fast Highly Charged Ions in Gas Targets 525
A.S. Schlachter, J.W. Stearns, W.G. Graham, K.H. Berkner, R.V. Pyle, J.A. Tanis

Applications of Variational Continuum Distorted Waves 526
D.S.F. Crothers, J.F. McCann

Electron Capture and Loss Cross Sections for Si^{11+} + He from 50 - 80 MeV 527
M. Clark, D. Brandt, S. Shafroth, J. Swenson

Effects of an Off-Shell Coulomb Wavefunction on Radiative Electron Capture 528
J.S. Briggs, M. Gorriz

Effect of L-Shell Non-Equilibrium on Radiative Electron Capture for 30 MeV S+C Collisions 529
J.A. Tanis, E.M. Bernstein

A Study of the Charge-Exchange of Carbon and Oxygen Ions in Various Media 530
I.S. Dmitriev, N.F. Vorobiev, G.E. Bugrov, Zh.M. Konovalova, E.A. Kral'kina, V.S. Nikolaev, Ya.A. Teplova, Yu.A. Fainberg

Experimental Study of the Formation of Metastable Lithium Ions in Collisions in Gases 531
Ya.A. Teplova, I.A. Nevostrueva, Yu.A. Fainberg, I.D. Koshevoi

Delayed Emission of X-Rays after Electron Capture into Metastable Few-Electron Oxygen and Neon Ions 532
F. Folkmann, B.J. Larsen, N.H. Eisum, K.M. Cramon

Electron Capture into Highly Charged, Metastable S, Ar, and Kr Recoil Ions Studied by Delayed Auger-Electron Measurements 533
Kurt M. Cramon, Finn Folkmann

Comparison of Various Multiple Scattering Approaches to Electron Capture 534
L.J. Dubé, J.K.M. Eichler

Collisions of Multiply Charged Projectiles with Light Targets: I. Theory 535
L.J. Dubé, R. Bruch

Collisions of Multiply-Charged Projectiles with Light Targets: II. Experiment 536
R. Bruch, L.J. Dubé, E. Träbert, P.H. Heckmann, B. Raith

Capture Cross Sections in Highly Excited p States of Ar^{18+} in High Velocity Collisions of 250 MeV Ar^{18+} on N 537
J.P. Rozet, P. Legagneux-Piquemal, A. Chetioui, P. Chevallier

Capture Cross Sections in High p Rydberg States by 400 MeV Bare Fe^{26+} and One-Electron Fe^{25+} Ions 538
A. Chetioui, D. Vernhet, J.P. Rozet, P. Legagneux-Piquemal, C. Stephan

Ion-Atom Collisions: Electron Exchange with Multiply Charged Ions II

Translational Spectroscopy of Electron Capture by Multiply Charged Ions — 539
B.A. Huber, H.J. Kahlert, K. Wiesemann

Energy Loss Spectra of Ar^{3+} - He Collisions — 540
E.Y. Kamber, J.B. Hasted

Final-State-Analysis of Electron Capture Pocesses in Collisions of Highly Stripped C, N and O Ions with He Atoms — 541
M. Kimura, T. Iwai, Y. Kaneko, N. Kobayashi, A. Matsumoto, S. Ohtani, K. Okuno, S. Takagi, H. Tawara, S. Tsurubuchi

Final-State-Analysis of Electron Capture Processes in Collisions of Highly Stripped F and Ne Ions with He Atoms — 542
H. Tawara, T. Iwai, Y. Kaneko, M. Kimura, N. Kobayashi, A. Matsumoto, S. Ohtani, K. Okuno, S. Takagi, S. Tsurubuchi

Low-Energy Electron-Capture in Ne^{2+}-He, Ar^{2+}-He and Kr^{2+}-He Collisions — 543
Kazuhiko Okuno, Yozaburo Kaneko

Ion Energy-Loss Spectroscopy of One Electron Capture Processes in the System Kr^{2+} - Ne — 544
T. Nakamura, N. Kobayashi, Y. Kaneko

Ion Energy Loss Spectroscopy for Transitions among Low-Lying States of Kr^{2+} in the Collisions with He and Ne — 545
N. Kobayashi, T. Nakamura, Y. Kaneko

Population of Electronic States of Multiply Charged Ar Ions Formed in Electron Capture from Hydrogen Atoms — 546
V.V. Afrosimov, A.A. Basalaev, K.O. Lozhkin, M.N. Panov

State-Selective Electron Capture by C^{2+}, C^{3+}, N^{2+} and Ar^{2+} in Rare Gases — 547
M. Lennon, R.W. McCullough, H.B. Gilbody

State-Selective Electron Capture by Slow Multiply Charged Ions in Atomic Hydrogen — 548
R.W. McCullough, M. Lennon, F.G. Wilkie, H.B. Gilbody

Lyman Spectra of O^{7+} and N^{6+} Produced by Low Energy Charge Exchange Collision on H_2 — 549
S. Bliman, M. Bonnefoy, J.J. Bonnet, S. Dousson, A. Fleury, D. Hitz, B. Jacquot

Electron Capture into Different (n,ℓ)-States in Slow C^{6+},N^{6+},O^{6+},Ne^{6+}-He,H_2 Collisions — 550
Yu.S. Gordeev, D. Dijkkamp, A.G. Drentje, F.J. de Heer

State-Selective Electron Capture Cross Sections for Impact of C^{q+} (q = 2,3,4) and O^{q+} (q = 2,3,6) on Li — 551
D. Dijkkamp, R.L. van der Woude, F.J. de Heer, A.G. Drentje, A. Brazuk, H. Winter

Electron Capture into Excited Projectile States in 6-100 keV Ne^{4+}-Ne Collisions — 552
D. Dijkkamp, V.K. Nikulin, Yu.S. Gordeev, A.V. Samoilov, F.J. de Heer

Ion-Atom Collisions: Electron Exchange with Multiply Charged Ions III

Extraction of Total Capture Probabilities in Atomic Many Electron Collisions 553
H.J. Lüdde, R.M. Dreizler

Dynamics of Collective Charge Flow in Diatomic Collision Systems 554
J. Eichler, T.S. Ho

Charge Capture by Multicharged Ions: Fully Quantal Calculations with Larger Basis Sets 555
C. Bottcher, T.G. Heil

Molecular Calculations of the Cross Section for Charge Transfer He^{2+} + H in the 20 eV to 10 keV E_{CM} Region 556
Marc C. van Hemert, Ewine F. van Dishoeck, Fumihiro Koike

Impact Parameter Dependence of Charge Exchange in the Scattering of Li^{3+} on H at 10.5 keV (Direct Integration of the Schrödinger Equation) 557
Norbert Grün, Werner Scheid

Theoretical Study of Li^{2+} - H Collisions at the keV Energy Range 558
J. Hanssen, C. Harel

Treatment of Charge Transfer Collisions with Translation Factors. Be^{4+} + H 559
L.F. Errea, L. Méndez, A. Riera

Stark Mixing of Sublevels in Multicharged Ion-Atom Collisions 560
R. Gayet, J. Hanssen, C. Harel, A. Salin

A Simple Account of Core Electrons in the Theory of Electron Capture in Slow Collisions of Highly Charged Ions with Atomic Hydrogen 561
O.G. Larsen, K. Taulbjerg

Coincidence Measurements of Electron Transfer in an Ion-Atom Crossed Beams Experiment 562
F.W. Meyer

Electron Capture Cross Sections for Low Velocity Ne^{q+} and Ar^{q+} Ions on Atomic and Molecular Hydrogen (50 eV/q to 3000 eV/q) 563
Tom J. Gray, C. Can, L. Tunnell, J.M. Hall, S.L. Varghese

Nonmonotonic Behaviour in the Charge Dependence of Total Electron-Capture Cross Sections for Medium-Velocity, Partly Stripped Ions on Atomic Hydrogen 564
P. Hvelplund, H. Knudsen, L.H. Andersen, S.K. Bjørnelund, L. Liljeby

Electron Capture from Hydrogen Atoms by Multiply Charged C,N,O,Ne Ions at Low keV Energies 565
V.V. Afrosimov, A.A. Basalaev, K.O. Lozhkin, M.N. Panov

Multichannel Model Study of Collisions between Many-Electron Atoms and Highly Charged Ions 566
V.K. Nikulin, A.V. Samoylov

Charge Transfer in He^{++} + Li Collisions 567
A.M. Ermolaev, B.H. Bransden

Electron-Capture in He^{2+} + Li Collision 568
H. Sato, M. Kimura

Two Electron Capture in High Energy He^{2+} + Li Collisions 569
M. Sasao, A. Matsumoto, A. Nishizawa, K.N. Sato, S. Takagi, S. Amamiya, T. Masuda, Y. Tsurita, Y. Kanamori, Y. Haruyama, F. Fukuzawa

Ion-Atom Collisions:
Electron Exchange with Multiply Charged Ions IV

Exponential Distorted Wave Approximation in Charge Transfer 570
H. Suzuki, N. Toshima, T. Watanabe

Continuous Energy State Model for Charge Transfer in Multiply Charged Ions Impact on Atoms 571
Fumihiro Koike

Semiclassical Studies of Slow Charge Transfer Reactions 572
Anders Bárány

Target Effect on the n Population for Electron Capture Processes Involving Multicharged Ions (1<E<10 keV/q) 573
M. Barat, M. Laurent, J. Pommier, S. Dousson, D. Hitz

Electron Capture by Multiply Charged Ions: Influence of Target Ionization Potential 574
M. Gargaud, P. Duthoit, R. McCarroll

Interference Structure in the Impact Parameter Dependence of K-K Charge Transfer in Slow S-Ar Collisions 575
E. Justiniano, R. Schuch, M. Schulz, H.J. Specht, H. Schmidt-Böcking, H. Ingwersen

Transfer Ionization and Coulomb Ionization in Collisions of Multiply Charged Ions with Atoms 576
W. Groh, A. Müller, B. Schuch, A.S. Schlachter, E. Salzborn

Charge Transfer Cross Sections for Multiply Charged Krypton and Xenon Ions on Various Gas Targets 577
T. Kusakabe, H. Hanaki, T. Horiuchi, N. Nagai, I. Konomi, M. Sakisaka

Electron Capture into N-, O-, and F-Like Xe and Sm Ions at 3.6 MeV/u 578
D.H.H. Hoffmann, P.H. Mokler, W.A. Schönfeldt, A. Warczak, Z. Stachura

M-Subshell Effects in Electron-Capture to Sm^{q+} Ions 579
W.A. Schönfeldt, D.H.H. Hoffmann, P.H. Mokler, A. Warczak

Charge Equilibrium of Fast Heavy Ions Penetrating through Gaseous Media 580
S. Karashima, T. Watanabe

Z_T Oscillation of Equilibrium Mean Charge of MeV He Ion 581
F. Fukuzawa, Y. Haruyama, Y. Kanamori, A. Itoh

Collisions of Multiply Charged Neon Ions with Water Vapor 582
C.-S. O, S.B. Elston, M. Breinig, B. Thomas, S. Berry, R.T. Short, I.A. Sellin, D.A. Church, R.A. Kenefick, W.S. Burns

Ion(Atom)-Molecule Collisions: Electronically Elastic Processes

Energy Dependence of Mode Resolved Vibrational Excitation in H^+ - CF_4 and SF_6 Collisions 583
U. Gierz, M. Noll, J.P. Toennies

The Anisotropic Neon-Methane Interaction from Differential Energy Loss Spectra 584
U. Buck, A. Kohlhase, T. Phillips, D. Secrest

Time of Flight Analysis of Ar_n Cluster Scattering 585
U. Buck, H. Meyer

Transitions between K 4 ^2P Zeeman Substates Induced by Collisions with N_2 and H_2 586
R. Berends, P. Skalinski, L. Krause

He^0+D_2 Collisions at Low keV Energies 587
J. Jakacky, Jr., V. Heckman, E. Pollack

Differential Cross Sections in Neutral-Neutral Collisions 588
J.H. Newman, Y.S. Chen, K.A. Smith, R.F. Stebbings

Effects of Translational and Superfine Relaxation of I Atoms in Oxigen-Iodine Laser 589
M.V. Zagidullin, V.I. Igoshin, N.L. Kuprianov

Ion(Atom)-Molecule Collisions: Theory of Rotational/Vibrational Excitation

Use of Ehrenfest's Principle for Inelastic Collisions 590
W. Elberfeld, M. Kleber

Computed Vibrational and Rotational Inelasticity in Collisions via Realistic Interactions 591
F.A. Gianturco, U.T. Lamanna, G. Petrella

Scattering of Rare Gas Atoms (Ne,Ar) by Alkali Halide Molecules (KCl,CsCl,KF,CsF) 592
D.N. Tripathi, D.K. Rai

Ion-Molecule Differential Cross Sections and Energy Transfer Distributions 593
F.E. Budenholzer, C.C. Lee

A Comparison of Close Coupled Scattering Results with Measurements on He - CO at $E_{c.m.}$ = 27.7 meV 594
W. Dilling, J. Schaefer

Excitation Functions for Rotational Excitation in Atom Rigid-Rotor Collisions 595
D. Poppe

On Rotational Excitation of Initially Rotating Molecules 596
H.J. Korsch, Z.V. Lewis, D. Poppe

Determining Intermolecular Potentials from Atom-Molecule Scattering Data 597
R. Snyder, A. Russek

Energy Loss Scaling in Ion-Molecule Scattering: 3, 10, 15, and 25 keV Ne^+ - H_2 598
B. Fastrup, P. Dahl

Ion(Atom)-Molecule Collisions: Fine-Structure Transitions and Quenching at Thermal Energies

Emission Spectrum of the Na^*-N_2 Collision Complex 599
W. Kamke, B. Kamke, I.V. Hertel, A. Gallagher

Polarization Studies in the Nonadiabatic Quenching Process $Na^*(3p) + H_2$ 600
G. Jamieson, W. Reiland, C.P. Schulz, H.-U. Tittes, I.V. Hertel

Nonresonant Electronic to Vibrational Energy Transfer in the Collision of Laser Excited $Na^*(3p)$ with Triatomic Molecules 601
G. Jamieson, C.P. Schulz, H.-U. Tittes, I.V. Hertel

A Molecular Beam Study of Electronic to Electronic, Vibrational, and Rotational Energy Transfer (E-EVRT) in the Collision of Two Step Laser Excited Sodium with N_2 602
G. Jamieson, C.P. Schulz, H.-U. Tittes, I.V. Hertel

Influence of the Rotational Energy of Molecular Perturbers on Rb(5P) Fine Structure Transition (FST) 603
J. Cuvellier, J.M. Mestdagh, M. Ferray, P. de Pujo, J. Berlande

Collisional Quenching of Selectively Excited Rovibronic States of H_2^* by H_2, Ne, Ar, and Kr 604
H. Schmoranzer, J. Imschweiler, T. Noll

Radiative Quenching of Vibronically Excited CO^+ at T = 100 ^{0}K 605
D.H. Katayama, J.A. Welsh

Temperature Dependence of De-excitation Rate Constants of He(2 ^3S), Ar($^1P_1$), and Ar($^3P_1$) by Atoms and Molecules 606
H. Koizumi, M. Ukai, Y. Tanaka, K. Shinsaka, Y. Hatano

Ion(Molecule)-Collisions: Electronic Excitation at Suprathermal Energies

Vibrational Interference Structures in Differential Cross Sections for Alkali-Atom Molecule Collisions 607
M.R. Spalburg, M.G.M. Vervaat, A.W. Kleyn, J. Los

Non Adiabatic Processes in Alkali Metal - Alkyl Halide Molecule Collisions: Neutral Excitation and Ion Pair Production Channels 608
A.M.C. Moutinho, A. Praxedes, E. Cowan, M.A.D. Fluendy

Electron Transfer in Alkali Atom-CH_3NO_2 Collisions 609
M.A.D. Fluendy, S. Lunt

Emission Cross Sections and Polarization of the Resonance Radiation in H^+,H_2^+,H_3^+ - Na Collisions at 2 - 14 keV 610
V.M. Lavrov, R.A. Lomsadze

Excitation of Hydrogen Atoms in Collisions of Slow K^+ - Ions with H_2 - Molecules 611
B.I. Kikiani, M.R. Gochitashvili, R.V. Kviszhinadze, V.A. Ankudinov

Excitation in Collisions of N_2^+ with He, Ne and Ar in the Energy Range 1 - 2000 eV (2000 - 8000 Å) 612
Ingrid Kuen, Branislav Jelenkovic, Franz Howorka

Cross Sections for Excitation of Balmer Beta Radiation in an Equilibrated Beam of H and H^+ in O_2 613
J. Roger Sheridan

Excitation in $He^0 + D_2$ Collisions at Low keV Energies 614
J. Jakacky, Jr., A. Russek

Ion(Atom)-Molecule Collisions: Penning Ionization

Penning Ionization Processes Clarified by Electron-Ion-Coincidence Measurements 615
Jürgen Baus, Arnulf Benz, Harald Morgner

A New Apparatus for the Measurement of Penning Electron Angular Distributions 616
H. Morgner, G. Zimmermann

The Reaction of Metastable Helium with Molecules Investigated by Optical and Electron Spectroscopy 617
Oskar Leisin, Harald Morgner, Hubert Seiberle, Joachim Stegmaier

Relative Orbital-Ionization Probabilities of Gaseous and Condensed Phase Molecules by Penning and Photo Processes 618
Andrew J. Yencha, Hiroyasu Kubota, Tomohiko Hirooka, Tsutomu Fukuyama, Tamotsu Kondow, Kozo Kuchitsu

Strong Variation in the Thermal Energy Ionization Cross Section of Argon Atoms Colliding with Different Laser-Excited $Ne(2p^5\,3p\ J=1,2,3)$ Atoms 619
W. Bussert, J. Ganz, M.-W. Ruf, A. Siegel, H. Hotop, H. Morgner

Penning Ionization of Ne and Ar by $He^-(1s2s2p)\ ^4P^0$ at 1-2 keV 620
M.J. Coggiola, K.T. Gillen

Ionization of CS_2 Clusters in Collision with Electrons and Long-Lived Excited Rare-Gas Atoms 621
Tamotsu Kondow, Koichiro Mitsuke, Kozo Kuchitsu

Ion(Atom)-Molecule Collisions: Electron Exchange

Charge Transfer Reaction between Ar^+ and Diatomic Molecules at the 10 eV Region 622
T. Matsuo, N. Kobayashi, Y. Kaneko

Molecular Dynamics of Electron Transfer Reactions at Low Energy: Ar^+ (N_2,Ar) N_2^+, N_2^+ (N_2,N_2) N_2^+, and Ar^+ (Ar,Ar) Ar^+ 623
J.H. Futrell, B. Friedrich

Ab Initio Calculations on the $(ArN_2)^+$ System 624
P. Archirel, B. Levy

Vibronic and FS State Selected Charge Transfer Reactions: $O_2^+ + Ar \rightleftharpoons Ar^+ + O_2$ and $NO^+ + Ar \rightleftharpoons Ar^+ + NO$ Systems 625
Inosuke Koyano, Kenichiro Tanaka, Tatsuhisa Kato

Charge Transfer Reaction between $Ar^+(^2P_{3/2}, ^2P_{1/2})$ and N_2 below 0.1 eV 626
T. Tobita, N. Kobayashi, Y. Kaneko

State Selected Ion-Molecule Reactions: $N_2^+ + Ar \rightarrow N_2 + Ar^+$ 627
T.R. Govers, P.M. Guyon, T. Baer, K. Cole, H. Frohlich, M. Lavollée

Dissociative Charge Transfer in He^+-O_2 Collisions: Energy Spectra and Angular Distributions of the O^+ Product Ions 628
G. Bischof, P. Reinig, F. Linder

Dissociative Charge Exchange of H_2^+, Studied with Translational Spectroscopy 629
D.P. de Bruijn, J. Neuteboom, J. Los, V. Sidis

Theory of Near Resonant-Dissociative Charge Exchange Collisions at keV Energies 630
V. Sidis, D. De Bruijn

Electron Transfer Processes between Electronically Excited Na and SO_2 631
B. Auschwitz, K. Lacmann

Differential Cross Sections for Electron Transfer with Selection of the Ionic Exit Channel for K, Cs + CH_3I Collisions 632
A.J.F. Praxedes, M.J.P. Maneira, A.M.C. Moutinho

Ion(Atom)-Molecule Collisions:
Dissociation and Recombination in Reactions

Alignment in Molecular Excitation in Dissociative States of H_2^+ and H_3^+ 633
O. Yenen, M. Natarajan, D.H. Jaecks

Predissociation of the $c\,^3\Pi_u$- and $d\,^3\Pi_u$-States of H_2, Studied with Translational Spectroscopy 634
D.P. de Bruijn, J. Neuteboom, J. Los, T.R. Govers

Translational Spectroscopy of D_2^+ and HD^+ from D_3^+ and HD_2^+ Dissociation 635
C. Cisneros, I. Alvarez, A. Morales, J. de Urquijo, A. Russek

Differential Positive-Fragment Cross Sections of HD_2^+ in H_2 at Energies in the Range 1.5 to 5 keV 636
A. Morales, J. de Urquijo, C. Cisneros, I. Alvarez, A. Russek

Collisional Electron-Ion and Ion-Ion Recombination in a Dense Neutral and Ion Gas 637
M.R. Flannery

Dissociative Recombination of e + H_3^+. An Analysis of Reaction Product Channels 638
H.H. Michels, R.H. Hobbs

Vibrational Predissociation of Ne...I_2(B) van der Waals Molecule: A Quasiclassical Treatment 639
G. Delgado-Barrio, P. Mareca, P. Villarreal

Vibrational Predissociation of the T-Shape He...I_2(B) van der Waals Molecule: A Close-Coupling Calculation 640
P. Villarreal, G. Delgado-Barrio, P. Mareca

Complex-Coordinate Coupled-Channel Formalism to Vibrational and Rotational Predissociation of van der Waals Molecules 641
Shih-I Chu, Krishna K. Datta

Ion(Atom)-Molecule Collisions: Reactive Scattering

Study of the Reaction of Na*(3P) + HCl by the Crossed Molecular Beam Method 642
H. Schmidt, M.F. Vernon, P.S. Weiss, M.H. Covinsky, Y.T. Lee

M(IIa)-X_2 Chemiionisation as a Function of Collision Energy 643
U. Ross, H.-J. Meyer, T. Schulze, D. Beck

Collision Energy Dependence of Vibrational/Rotational Distribution of BaBr Produced in the Crossed Beam Reaction Ba + CH_3Br 644
Toshiaki Munakata, Yutaka Matsumi, Takahiro Kasuya

Dynamics of van der Waals Bond Exchange 645
D.R. Worsnop, S.J. Buelow, D.R. Herschbach

Chemiluminescence in Reaction of Potassium Dimers with Oxygen Molecules 646
J. Berlande, J. Cuvellier, P. de Pujo, J.M. Mestdagh

A New Quantum Mechanical Theory of Chemical Reactions 647
E. Ficocelli Varracchio

Resonances and Interferences in the Theory of Chemical Reactions 648
V. Aquilanti, S. Cavalli, G. Grossi, A. Laganà

Dependence of State-to-State Reactive Scattering on Relative Angular Momentum 649
S.H. Suck, R.W. Emmons, C. Klein

An Invariant Imbedding Aproximate Solution - Type Algorithm for Solving Coupled Equations for Scattering 650
F. Mrugala

Ion-Molecule Reactions

Atom Capture and Atom Loss at Asymptotically High Speeds 651
M. Breinig, G.J. Dixon, P. Engar, S.B. Elston, I.A. Sellin

Proton-Oxygen Potential Hypersurfaces Relevant to Vibrational Energy Transfers During Collisions 652
F.A. Gianturco, V. Staemmler

A New Method for the Measurement of Vibrational State Selected Ion-Molecule Reactions at Thermal Energies 653
D. van Pijkeren, J. van Eck, A. Niehaus

B $^2\Delta$-State NH^+ (ND^+) Emission from Reactions of Metastable N^+ Ions with $H_2(D_2)$: Spectroscopy and Dynamics 654
I. Kusunoki, Ch. Ottinger

Collisonal Relaxation and Reaction of Vibrationally Excited O_2^+ 655
H. Böhringer, M. Durup-Ferguson, E.E. Ferguson

Some Production and Destruction Reactions of Protonated Formic Acid, $HC(OH_2)_2^+$ 656
H. Villinger, A. Saxer, E. Ferguson, H. Bryant, R. Richter, W. Lindinger

Ion Molecules Association at Very Low Temperatures (20 K and 70 K): $N_2^+ + 2N_2 \rightarrow N_4^+ + N_2$ and $O_2^+ + 2O_2 \rightarrow O_4^+ + O_2$ 657
B.R. Rowe, J.B. Marquette, G. Dupeyrat

Initialisation of Ion Reactions inside Methane Clusters by Electron Impact 658
A. Ding, J. Hesslich

Product State Distribution of HF in the Reaction $F+H_2$ at Enhanced Collision Energies 659
J. Barnes, J.C. Polanyi, W. Reiland, D. Thomas

Collisions Involving Rydberg Atoms I

Charge Transfer from Rydberg Atoms: M and L Dependence of Final State Distributions 660
Alan D. MacKellar, Richard L. Becker

Charge Transfer from Rydberg Atoms: Velocity Dependence of Final-State Distributions 661
Keith B. MacAdam, Richard G. Rolfes

Penning Ionization in Collisions of Rydberg Rb Atoms 662
M. Kimura, R.E. Olson

Kinematics of Collisional Ionization in $Xe(nf)-SF_6$ Interactions 663
C. Higgs, M.P. Slusher, K.A. Smith, F.B. Dunning, R.F. Stebbings

Excitation and Ionization of High Rydberg Atoms in Collision with Polar Molecules -- Classical Calculations 664
S.C. Preston, N.F. Lane

Detection of High-Rydberg Atoms and Molecules by CO_2-Laser Radiation 665
A.A. Perov, A.Yu. Zayats, A.N. Stepanov, A.P. Simonov

Model Potential Calculations for the Ground and Excited (Rydberg) States of the Li_2^+, Na_2^+, K_2^+ Ions: Core Polarization Effects 666
A. Henriet, F. Masnou-Seeuws

Electric Field-Ionization of Rydberg States of Fast Foil-Excited Sulfur Ions 667
E.P. Kanter, D. Schneider, Z. Vager

Collisions Involving Rydberg Atoms II

Collisional Angular-Momentum Mixing of Na Rydberg States by a K^+ Ion Beam 668
W.W. Smith, P. Pillet, R. Kachru, N.H. Tran, T.F. Gallagher

Collisional Transfers in Sodium Rydberg States 669
A.L. Roche, P. Valiron, F. Masnou-Seeuws

Quasi-Elastic State-Changing Collisions of High-Rydberg Atoms with Heavy Rare-Gas Atoms 670
Yutaka Sato, Michio Matsuzawa

Quenching of Rubidium Rydberg States : A Tool for Investigation of the Elastic Scattering of Ultra-Low Energy Electrons ($E \simeq 0.01$ eV) by Rubidium Atoms 671
M. Hugon, F. Gounand, P.R. Fournier, J. Berlande

Application of the Faddeev Watson Expansion to Thermal Collisions of Rydberg Atoms with Neutral Particles 672
E. de Prunelé

Sodium D State Fine-Structure in an Electric Field, and Fine-Structure Quantum Beat Echoes 673
T.H. Jeys, M.C. Copel, G.B. McMillian, F.B. Dunning, R.F. Stebbings

Collisions of Rydberg Atoms in an Electric Field: Calculations Using Hydrogenic Wave Functions in Parabolic Coordinates 674
A.P. Hickman

Analytical Expressions for Born Electron Impact Excitation Cross Sections between Rydberg States nl - nl' 675
I. Beigman, L. Bureyeva, M. Syrkin

Quantum State Distributions in Rydberg Atomic Beams Created by Electron Impact 676
R. Barbe, J.-P. Astruc, A. Lagreze, J.-P. Schermann

Electron-Atom Collisions in Radiation Fields

Electron Scattering from a Potential in a Radiation Field 677
Robin Shakeshaft

Electron-Atomic-Hydrogen Elastic Exchange Collisions in the Presence of a Laser Field 678
F. Trombetta, C.J. Joachain, G. Ferrante

Elastic Electron-Helium Collisions in Electric Field 679
C. Foglia

Effect of the Laser Polarization on Electron-Atom Collisions 680
P. Cavaliere, C. Leone, G. Ferrante

Particle Atom Scattering in the Presence of a Nonresonant Laser Field 681
S. Bivona, D. Valenza, G. Ferrante

Electron Atom Scattering in the Field of a CO_2 Laser 682
P.J. Curry, W.R. Newell, A.C.H. Smith

Radiative Electron-Atom Collision Cross Sections 683
L. Dimou, F.H.M. Faisal

Simultaneous Electron-Photon Excitation of Atoms 684
S. Jetzke, F.H.M. Faisal, R. Hippler, H.O. Lutz

Quadrupole Effect in the Excitation of He by Electron Impact in the Presence of Laser 685
R.S. Pundir, K.C. Mathur

Laser Assisted Electron Impact Ionization 686
M. Zarcone, D.L. Moores, M.R.C. McDowell

Laser Modified Electron Scattering from a Slowly Ionizing Atom 687
Emilio Fiordilino, Marvin H. Mittleman

Laser-Assisted Collisions and Collision-Assisted Radiative Transitions

Theory of Laser-Induced Chemi-ionization 688
H.P. Saha, J.S. Dahler, S.E. Nielsen

Electron Spectroscopy Study of Ionization in an Alkali Vapor Resonantly Excited with a C.W. Laser 689
J.M. Bizau, B. Carré, P. Dhez, D. Ederer, J.C. Keller, P. Koch, J.L. Le Gouet, J.L. Picqué, G. Spiess, F. Wuilleumier

Electron Spectroscopy Study of Ionization in an Alkali Vapor Excited with a High Intensity Laser 690
J.M. Bizau, B. Carré, P. Gérard, J.C. Keller, J.L. Le Gouet, J.L. Picqué, F. Roussel, G. Spiess, F. Wuilleumier

He^{2+} + H Collisions in Presence of a Laser Field 691
L. Méndez, L.F. Errea, A. Riera

Charge Transfer in the Presence of a Magnetic Field 692
S. Bivona, B. Spagnolo, G. Ferrante

Application of the Eigenvalue Method to CARE and RAIC 693
E.J. Robinson

Optical Collisions in Strong Laser Fields 694
D.S. Bakaev, Yu.A. Vdovin, V.M. Yermachenko, S.I. Yakovlenko

Nonadiabatic Theory of Final State Distributions in Laser Assisted Atomic Collisions 695
P.S. Julienne, F.H. Mies

Phase Control of Atomic Scattering States in Two-Photon Radiative Collisions 696
Munir H. Nayfeh, G.B. Hillard

Dipole Moments and Interaction Potentials in Collision-Induced Absorption 697
A. Raczynski

Multiphoton-Assisted Radiative Collisions in a Strong Field Regime 698
P. Pillet, R. Kachru, N.H. Tran, W.W. Smith, T.F. Gallagher

Temperature Dependence of the Collision Induced Absorption of CO_2 at 0.091 cm^{-1} 699
G. Kouroupetroglou, G. Boudouris

Dynamics of Positronium in a Laser Field 700
F.H.M. Faisal, P.S. Ray

Soft-Photon Interactions Generating Symmetry-Broken Pre- and Postcollision States of Molecules 701
Peter Pfeifer

Collision Induced Dipole Radiation of Normal Hydrogen Gas in the Frequency Range of the Cosmic Background 702
Joachim Schaefer, Wilfried Meyer

Experimental Techniques

A New Electron-Electron Multi-Conicidence Apparatus 703
M. Völkel, W. Sandner

Reduction of Kinematic Line Broadening in the Auger Electron Spectra of Fast Ion Beams 704
R. Bruch, A. Gutenkunst, K.H. Müller

A Compact High Current Pulsed Electron Gun with Sub-Nanosecond Electron Pulse Widths 705
M.A. Khakoo, S.K. Srivastava

ACO: A Source of Monoenergetic Electrons for Collisions with Atoms and Molecules 706
J.P. Ziesel, D. Field, P.M. Guyon, T.R. Govers, M. Lavollée, O. Dutuit

A New, Intermediate Energy, Electron Energy Loss Spectrometer 707
G. Gerson B. de Souza, A.C. de A. e Souza, R. de B. Faria

Use of a Soft X-Ray Diagnostic for Highly Charged Ion Current Measurement 708
J.J. Bonnet, M. Bonnefoy, A. Fleury, S. Bliman, S.Dousson, D. Hitz, B. Jacquot

Compact Plane-Crystal X-Ray Spectrometer 709
W. Jitschin, B. Wisotzki, U. Werner, H.O. Lutz

X-Ray Spectroscopy of Tokamak Plasmas: An Experimental and Theorectical Study 710
E. Källne, J. Källne, A.K. Pradhan

A New Technique for the Study of Charge Changing Collisions Using Recoil Ions 711
L. Liljeby, H. Cederquist, H. Danared, G. Astner, A. Bárány, A. Johnson

Construction Performance of a Position-Sensitive Particle Detector with Delay Line Readout 712
B. Liu, W. Enders, D. Liesen, R. Schulze

An Alternative Method for the Measurement of the Efficiency of an Ion Detector 713
F. Brouillard, S. Chantrenne, W. Claeys, A. Cornet, P. Defrance

Measurement of Dissociation Fraction in an Electron Impact Heated Atomic Hydrogen Oven 714
G.A. Glass, L.H. Andersen, R. Holmes, J.-P. Rozet, D.A. Taylor, R.S. Thoe, S.B. Elston, M. Breinig, I.A. Sellin

Total Electron-Atom and Electron-Molecule Scattering Cross-Sections by Photoelectron Spectroscopy 715
Vijay Kumar, E. Krishnakumar, K.P. Subramanian

Theoretical Methods

New Concepts for Treating Relativistic Collisions 716
Felix T. Smith

A Fast Asymptotic Package 717
H.E. Saraph, M.J. Seaton

On the Calculation of Quantal Matrix Elements of a Function between Unbound States 718
Staffan Yngve

The Normalized Wavefunction Close to the Centre of Force for Bound and Unbound States 719
Nanny Fröman, Per Olof Fröman, Staffan Yngve

Asymptotic Expansions of Coulomb Wavefunctions Obtained by Means of the Saddle-Point Method 720
Staffan Yngve

Recent Results on the Multiplets $He^-2p^3\ ^4S^0$, $Be^-3d^5\ ^6S$, and $C^-4f^7\ ^8S^0$ Belonging to the Half Capacity Configurations 721
Erling Holøien

Atomic and Molecular Properties from Finite Element Method Calculations 722
W. Schulze, D. Kolb

Adiabatic and Diabatic Molecular States of H_2 at Small Internuclear Distances 723
F. Borondo, A. Macias, A. Riera

Large Order Perturbation Theory for Ion H_2^+ and for Stark Effect in Hydrogen 724
R.J. Damburg, R.Kh. Propin

Threshold Breakup of Three Coulomb-Interacting Particles: Analysis with Wannier Theory and Jacobi Coordinates 725
Jim Feagin

Stieltjes Methods for Photoabsorption and Ionization - A Merging of Spectral and Collision Theories 726
M.R. Hermann, P.W. Langhoff

Miscellaneous

Classical-Trajectory Monte Carlo Calculation of Negative-Muon Capture 727
James S. Cohen

Positronium Formation from Muonium 728
Lali Chatterjee, S. Bhattacharyya

Mesomolecule Formation Studied Quantum Electrodynamically 729
Lali Chatterjee, Sujata Bhattacharyya

Mesic Molecules and Muon Catalyzed Fusion 730
A.K. Bhatia, Richard J. Drachman

The Atomic and Molecular Data Base at Belfast and Daresbury 731
K.M. Aggarwal, K.A. Berrington, J.G. Hughes, F.J. Smith, M. Elder

Ion and Electron Collision Processes of Importance in the Jovian and Saturnian Magnetospheric Plasmas 732
R.E. Johnson

Post-Deadline Papers

Resonant Photoemission Near 3d Thresholds in Lanthanum: Theoretical Predictions 733
F. Combet Farnoux

Threshold Cross Sections of Excited Atomic States 734
N.B. Avdonina, A.Z. Devdariani

Observation of ℓ-Structure in Hydrogenic States of High-Z Four-Electron Ions Produced by Ion-Solid Collisions 735
A.E. Livingston

Charge Transfer and Fine-Structure Excitation in Collisions of O^{3+} with Atomic Hydrogen at Thermal Energies 736
S. Bienstock, A. Dalgarno, E. Roueff

Radiative Decay of Li-Like Ions Following Charge Exchange of 3.3 keV/amu Ions with H_2 737
S. Bliman, M. Bonnefoy, J.J. Bonnet, S. Dousson, A. Fleury, D. Hitz, E. Knystautas, M.F. Politis, F. Zadworny

Study of the H_3 Molecules Produced by Collision of Fast H_3^+ Beams with Argon 738
M.J. Gaillard, A.G. de Pinho, J.-C. Poizat, J. Remillieux, R. Saoudi

A Method for Estimating Properties of Small Clusters: Theoretical Aspects 739
Ray Hefferlin, G.V. Zhuvikin, Ken Caviness, Penny Duerksen

The Role of the Tilt Angle in Fluorescent Measurements of the Guest-Host System in Anisotropic Molecular Liquids 740
J. Fisz, H.W. Kunert

Calculation of Collision Strengths for Fine-Structure Transitions in the Spectra of Rare Gas Atoms 741
F. Alan Bayes

Anomalous Effects in Very Small Angle Electron Potassium Differential Scattering Measurements 742
K. Rubin, I. Efremov

Excitation of the Mg II Resonance Doublet by Monoenergetic Electrons 743
I.P. Zapesochnyi, A.I. Imre, V.I. Frontov, A.N. Gomonaj, A.I. Dashchenko

Near Threshold Electron Impact Excitation of Potassium Atom Lower Levels 744
F.F. Papp, N.I. Romanjuk, O.B. Shpenik

Ultrasoft X-Ray Emission of Barium Study in Electron-Atomic Collisions 745
V.S. Vukstich, Yu.V. Zhmenyak, I.P. Zapesochnyi

Electron Impact Excitation of Thulium and Gadolinium Atoms 746
L.L. Shimon

Excitation of Metastable Levels of Noble Gas Atoms by Electron Impact in Intersecting Beams 747
O.B. Shpenik, A.N. Zavilopulo, A.V. Snegursky, I.I. Fabricant

The Calculation of Excitation of 2S and 2P Levels of He^+ Ion by Electron Impact 748
M.I. Haysak, V.I. Lengyel, V.T. Navrotsky, E.P. Sabad

Calculation of Di-electron Recombination Cross-Section on He^+-Ion 749
O.I. Zatsarinny, V.I. Lengyel, E.P. Sabad

The Influence of Electron Capture on the Elastic Cross-Section of Electron - Mg^+-Ion Scattering 750
V.I. Lengyel, V.T. Navrotsky, E.P. Sabad

The Autoionizing States of a Magnesium Atom 751
O.I. Zatsarinny, V.I. Lengyel, V.T. Navrotsky, E.P. Sabad, M. Salak

Author Index 753

Abstracts of Contributed Papers

PHOTOIONIZATION OF THE $3p^5$ 4s EXCITED STATES OF ARGON

P. C. Ojha and P. G. Burke

The Queen's University of Belfast, Belfast, Northern Ireland.

We have calculated the cross sections for photoionization of the $3p^5 4s\ ^3P^o$ and $^1P^o$ states of Argon between the thresholds for ionization to the $3s^2\ 3p^5\ ^2P^o$ and $3s\ 3p^6\ ^2S^e$ states of the positive ion. R-matrix formulation of Burke and Taylor[1] and the LS coupling approximation was used. The present work overcomes the limitations of previous calculations[2-6] by including electron correlation to a high and consistent degree-of-approximation in both initial and final states and by allowing for the existence of resonances in the close-coupling approximation for the final-state wavefunction.

Two ionic states, $3s^2\ 3p^5\ ^2P^o$ and $3s\ 3p^6\ ^2S^e$ were included in the close-coupling expansion of the initial and final-states. These were built from 1s, 2s, 3s, 4s, 2p, 3p, $\overline{4p}$ and $\overline{3d}$ orbitals allowing one and two-electron excitations.

The partial cross sections for photoionization of the $3p^5\ 4s\ ^3P^o$ state to the $3p^5$ kp $^3P^e$ and $^3D^e$ final states show Bates-Seaton-Cooper minima and the cross section for the $3p^5$ kp $^3S^e$ final state indicates a minimum in the discrete part of the spectrum. Due to the anisotropic interaction of the quadrupole moment of the residual ion with the quadrupole moment of the ejected p-electron, the photoabsorption minimum occurs at different energies in these final states. Two Rydberg series of resonances, $3s\ 3p^6$ ns $^3S^e$ and $3s\ 3p^6$ nd $^3D^e$ are found. The total cross section, which is fairly small ($\sim 10^{-19}$ cm^2) for non-resonant photoionization, is dominated by the inner-shell-excitation resonance $3s\ 3p^6\ 4s\ ^3S^e$ in a transition in which the outer 4s electron is a spectator. Similar results are obtained for the photoionization of the $3p^5\ 4s\ ^1P^o$ state.

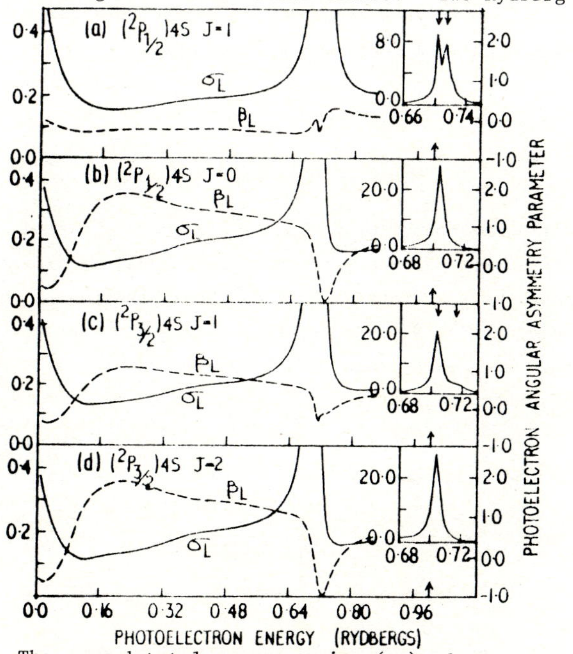

Fig. 1 The summed total cross section (σ_L) and averaged photoelectron angular asymmetry parameter for the photoionization of the fine-structure levels of the $3p^5$ 4s manifold of Argon. The zero of the energy scale is the weighted average of the $3p^5\ ^2P_{1/2}$ and $^2P_{3/2}$ thresholds. The insets show the cross sections in the resonance region.

The resonances of the $^{1,3}S^e$ and $^1D^e$ Rydberg series are not split by the spin-orbit interaction. In the relatively weak resonances of the $^3D^e$ series, the valence electron does not penetrate the core and the spin-orbit interaction is expected to be small. The predominant effect of the spin-orbit interaction is allowed by a unitary transformation from the LS coupling scheme to the jls coupling scheme appropriate to Argon. In this transformation, we have accounted for the mixing of the J = 1 states of the neutral atom but disregarded the splitting of the $3p^5\ ^2P_{3/2,1/2}$ states of the positive ion. The total cross section for photoionization of all the fine-structure levels of the $^3p^5$ 4s manifold, summed over the $3p^5\ ^2P_{3/2,1/2}$ states of the residual ion, is then readily calculated. We have also calculated the photoelectron angular asymmetry parameter averaged over the two states of the residual ion. These results are shown in figure 1.

The cross section is seen to be small except in the region of the $3s\ 3p^6$ 4s resonances. These, however, are at sufficiently high energies so as not to cause a serious depletion of the metastable states $3p^5$ 4s J=0,2 in an excimer laser cavity. The photo-electron angular asymmetry parameter shows sharp variation in the resonance region and its measurement would provide a particularly stringent test of the theory.

We are grateful to Dr. K.T. Taylor and Dr. A. Hibbert for frequent discussions and advice.

References

1. P.G. Burke and K.T. Taylor, J. Phys. B. 8 2620 (1975).
2. A.U. Hazi and T.N. Rescigno, Phys.Rev. A16 2376 (1977).
3. K.J. McCann and M.R. Flannery, App. Phys. Lett. 31 599 (1977).
4. T. W. Hartquist, J. Phys. B. 12 2101 (1978).
5. P. Ranson and J. Chappelle, J. de Phys. 40(C7) 25 (1979).
6. C. Duzy and H.A. Hyman, Phys. Rev. A 22 1878 (1980).

ONE AND TWO ELECTRON ATOMIC PHOTOIONIZATION PROCESSES

P. Scott, P.G. Burke and A.E. Kingston

Department of Applied Mathematics and Theoretical Physics,
The Queen's University of Belfast, Northern Ireland.

In this contribution we describe recent work at Queen's University, Belfast on the photoionization of atoms. The emphasis of this work is to provide cross sections of importance in astrophysical applications and to provide data which can be compared with experiments carried out using synchrotron radiation sources.

The R-matrix method as described by Burke and Taylor[1] has been applied to the photoionization of neutral calcium in the ^1S ground state leaving the Ca^+ ion in the ^2S, ^2D and ^2P states. Configuration interaction wave functions were used to describe the initial and final states. Total photoionization cross sections have been calculated between the ^2S and ^2P thresholds of Ca^+. The results obtained between the ^2S and ^2D thresholds are presented in fig. I. A Bates-Seaton-Cooper minimum is observed at 7.4 eV and its effect on the neighbouring series of Rydberg resonances is illustrated by the rapid change in the line profile indices of these resonances as this minimum is crossed.

The above calculation has been extended to include photoionization of Ca to the $..3p^54s^2\ ^2P$ state of Ca^+. This is an autoionizing state which decays to $Ca^{++}(3p^6)$. This provides a mechanism by which double photoionization can occur.

$$Ca(..3p^64s^2\ ^1S) + h\nu \rightarrow Ca^+(..3p^54s^2\ ^2P) + e_1^-$$
$$\hookrightarrow Ca^{++}(3p^6) + e_2^-$$

By considering the ratio of the photoionization cross section leaving the Ca^+ ion in the $3p^54s^2\ ^2P$ autoionizing state to the sum of the photoionization cross sections leaving the Ca^+ ion in the ^2S, ^2D and ^2P states we see that the $3p^54s^2\ ^2P$ state contributes a factor of 3.25 to the Ca^{++}/Ca^+ ratio compared with the experimental measurements of 8 by Holland and Codling[2].

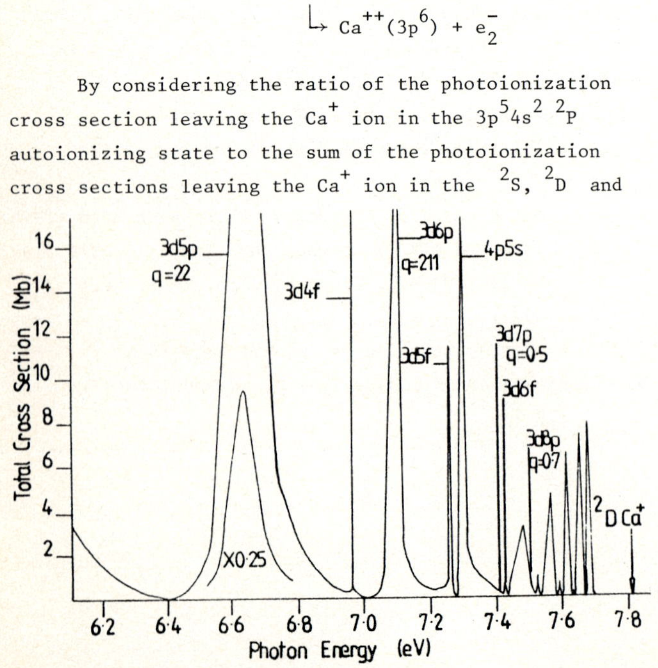

Fig. 1: Photoionization of neutral calcium.

In the previous calculation we did not include the post collision interaction between the two ejected electrons. In order to incorporate this into the calculation the existing theory requires modification. We are considering the process

$$e^- + A \rightarrow A^* + e^-$$
$$\hookrightarrow A^+ + e^-$$

where A^* is an intermediate resonance state which decays by autoionization. We propose to carry out two calculations. In the first we consider the $e^- + A^+$ system and obtain a set of R-matrix eigenstates $\Psi_k(1...N+1)$. The eigenstates Ψ_k, which form a complete basis for expanding an $N+1$ electron system with one electron in the continuum, are then used in the second calculation where we consider the $e^- + A^*$ system. This model will be applied initially to electron scattering from He.

As a further extension to the photoionization problem we are considering the process

$$h\nu + A \rightarrow A^{+*} + e^-$$
$$\hookrightarrow A^+ + h\nu$$

Following the work of Klar[3,4] we have obtained an expression for the electron photon angular correlation which we are applying to the photoionization of He leaving the He^+ ion in the ^2P state which decays to the ^2S ground state. We are also calculating the asymmetry parameter β.

The latest results in these areas will be reported at the conference.

References

1. P.G. Burke and K.T. Taylor, J. Phys. B8, 2620 (1975).
2. D.M.P. Holland and K. Codling, J. Phys. B14, 2345 (1981).
3. H. Klar, J. Phys. B13, 2037 (1980).
4. H. Klar, J. Phys. B15, 4535 (1982).

RADIATIVE RECOMBINATION COEFFICIENTS FOR COMPLEX ATOMIC IONS

G. Peach

Department of Physics and Astronomy, University College, Gower Street, London WC1E 6BT

In the radiative recombination process, an atomic ion A^{+z}, with overall charge z, combines with an electron and a photon of frequency ν is emitted, so that

$$A^{+z} + e \rightarrow A^{+z-1}(i) + h\nu \;, \qquad (1)$$

and an atomic ion $A^{+z-1}(i)$ of charge $(z-1)$ in a state i is produced. We shall assume throughout the following analysis that the ion A^{+z} is in its ground state. The reverse process of photoionization is given by

$$A^{+z-1}(i) + h\nu \rightarrow A^{+z} + e \;, \qquad (2)$$

and if $a_i(\nu)$ is the cross section for process (2), the radiative recombination coefficient for process (1) is given by

$$\alpha_i(z,T) = \frac{1}{c^2}\left(\frac{2}{\pi}\right)^{\frac{1}{2}}(mkT)^{-\frac{3}{2}}\frac{\omega_i}{\omega_+}\exp[I_i/(kT)]$$

$$\times \int_{I_i}^{\infty}(h\nu)^2 a_i(\nu)\exp[-h\nu/(kT)]\,d(h\nu), \qquad (3)$$

where I_i is the ionization potential of the ion $A^{+z-1}(i)$, T is the temperature and ω_i and ω_+ are the statistical weights for A^{+z-1} in level i and for the ion A^{+z} in its ground state respectively. The coefficients $\alpha_i(z,T)$ have the dimensions $(\text{length})^3(\text{time})^{-1}$. We have assumed in (3) that the free electrons, e, in (1) have velocities that can be described by a Maxwell distribution.

Reliable values of the coefficients $\alpha_i(z,T)$ are required for many atomic ions of interest in astrophysical problems, see for example Boksenberg *et al*[1], where the temperature T is in the range $10\,\text{K} \le T \le 10^6\,\text{K}$. Data are also required for the total recombination coefficient given by

$$\alpha_t(z,T) = \sum_i \alpha_i(z,T) \;, \qquad (4)$$

where the sum is over all states i of A^{+z-1}. Many cross sections have to be evaluated in order to obtain $\alpha_t(z,T)$, and it is not possible to carry out elaborate calculations for each one as has been done by Saraph[2], for example. The method developed in this paper involves writing expression (4) in the form

$$\alpha_t(z,T) = \alpha_t^H(z,T) + \alpha_t^C(z,T) \;, \qquad (5)$$

where $\alpha_t^H(z,T)$ is the total recombination coefficient for a hydrogenic ion of charge z and $\alpha_t^C(z,T)$ is a correction term that allows for the fact that levels i of A^{+z-1} with angular momentum quantum numbers $l \le 3$ may be appreciably non-hydrogenic. The coefficient $\alpha_t^H(z,T)$ can be evaluated essentially exactly, since analytic expressions for the corresponding cross sections are known for the hydrogenic case. The cross sections required for the correction term are generated by using either the general formula for photoionization cross sections[3], or in some cases by using model potential methods that have been developed in another connection, see Peach[4]. Several papers have been published in which tables for $\alpha_t(z,T)$ have been given for various isoelectronic sequences, see Tarter[5], Aldrovandi and Péquignot[6] and Gould[7]. However they all suffer from the defect that allowance for departures from the pure hydrogenic case is made in only a very limited way, and is often confined to states that are in the ground configuration of the ion A^{+z-1} only. Also it has not always been realised that the study of isoelectronic sequences can be simplified considerably by tabulating the scaled quantity $\alpha_t(z,T)/z$ as a function of the scaled temperature T/z^2. For hydrogenic ions such a tabulation is independent of the value of z, and this was pointed out many years ago by Seaton[8]. Although of course the effects of resonances cannot be included by these methods, it has been shown that the general formula[3] gives very good results for the background cross section in many cases[2]. The effects of resonances can be added in as a separate contribution if necessary, see Storey[9].

Some preliminary results from this work were included in the paper by Boksenberg *et al*[1] some years ago. Various improvements to the program have since been incorporated, and results for the He, Li, Na and Mg isoelectronic sequences will presented at the conference.

References

1. A. Boksenberg, B. Kirkham, E. Michelson, M. Pettini, B. Bates, P.P.D. Carson, G.R. Courts, P.L. Dufton and C.D. McKeith, *Phil. Trans. R. Soc. Lond.* A **279**, 303 (1975).
2. H.E. Saraph, *J. Phys.* B *Atom. Molec. Phys.* **13**, 3129 (1980).
3. G. Peach, *Mem. R. Astron. Soc.* **71**, 13 (1967).
4. G. Peach, *Comments Atom. Molec. Phys.* **11**, 101 (1982).
5. C.B. Tarter, *Astrophys. J.* **168**, 313 (1971); erratum, **181**, 607 (1973).
6. S.M.V. Aldrovandi and D. Péquignot, *Astron. Astrophys.* **25**, 137 (1973); erratum, **47**, 321 (1976).
7. R.J. Gould, *Astrophys. J.* **219**, 250 (1978).
8. M.J. Seaton, *Mon. Not. R. Astron. Soc.* **119**, 81 (1959).
9. P.J. Storey, *Mon. Not. R. Astron. Soc.* **195**, 27P (1981).

DOUBLE-ELECTRON PHOTOIONIZATION OF HELIUM

S.N. Tiwary

Department of Applied Mathematics and Theoretical Physics
The Queen's University of Belfast
BELFAST BT7 1NN NORTHERN IRELAND

*Permanent address: Department of Physics
Bihar University
BIHAR INDIA

Recently I have explored the possibility of using the correct asymptotic form of the double-electron continuum wave function proposed by Altick[1] to calculate the double-electron photoionization cross sections of helium atomic system[2] in the low-bombarding energy region where experimental observations are available for the meaningful comparison. My investigation clearly indicates that the Altick wave function which takes account of only the leading term of the Neumann expansion of the Coulomb interaction $1/r_{12}$ is not enough to reproduce the reliable experimental data in the close vicinity of the double electron ionization threshold where both continuum electrons have nearly the same velocity. A part of the discrepancy is also due to the ground state wave function of Byron-Joachain type which does not allow the sufficient correlation. It is well known that the electron-correlation plays an important role to obtain reliable cross sections especially in the low-energy region.

The main objective of this work is to see the impact of using the configuration interaction (CI) wave function for the ground state and the Altick wave function for the final state on the double-electron photoionization cross sections of helium. The CI wave function used in this work is described by expansions of the following form

$$\Psi(LS) = \sum_{i=1}^{M} a_i \, \Phi_i(\alpha_i LS)$$

where the single configurational functions $\{\Phi_i\}$ are constructed from one-electron orbitals, each consisting of the product of a spin function, a spherical harmonic and a radial function, whose angular momenta are coupled in a manner prescribed by α_i to give total L and S. Detailed discussion is given elsewhere[3-4].

Figure 1 shows the ratio of partial cross sections He^{2+}/He^{+} along with other available experimental and theoretical results. It is clear from the figure that the present curve tends to lie close to the reliable experimental data but still, the discrepancy exists considerably. It reflects that the higher order terms of the Neumann expansion of $1/r_{12}$ are essential to include in the Altick wave function in order to obtain reliable cross sections. This can be done by employing an R-matrix formulation for the internal (or core) region, and the correct asymptotic form of the Altick function for the external region of the entire configuration space, and to match on a boundary in the r_1 and r_2 plane.

References

1. P.L. Altick, Phys. Rev. A25, 128 (1982).
2. S.N. Tiwary, J.Phys.B:Atom.Molec.Phys. 15, L323 (1982).
3. S.N. Tiwary, Chem. Phys. Letters 93, 47 (1982).
4. S.N. Tiwary, Astrophys. J. June 15, 1983.

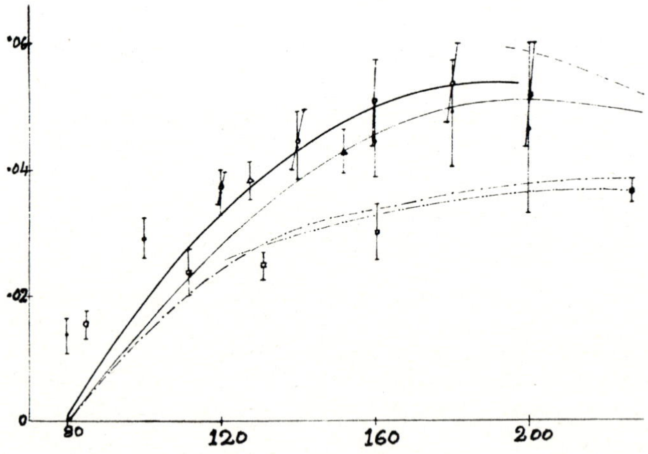

Figure 1. The ratio of photoionization cross sections He^{++}/He^{+}. ______, Present results along with other available experimental and theoretical data (see reference 2).

A NEW QDT ANALYSIS OF TWO-CHANNEL INTERACTIONS

Annick Giusti-Suzor

Laboratoire de Photophysique Moléculaire, Bât. 213, Université Paris-Sud 91405 Orsay

The behaviour of a two (open or closed) channel system is studied graphically and analytically, starting from the prototype case of two interacting Rydberg series (i.e. two closed channels in a Coulombic long range field) and from the "Lu-Fano" plots[1] largely used in atomic spectroscopy. Each level E_n of the two perturbed series is represented by a point in the (ν_1,ν_2) plane, where $\nu_i = \sqrt{2(E_i^+ - E_n)}$ is the effective principal quantum number associated with the threshold E_i^+. All points lie on a curve representing the relation[2]

$$(\tan\pi\nu_1 + R_{11})\ (\tan\pi\nu_2 + R_{22}) = R_{12}^2 \qquad (1)$$

where R is the short range reaction matrix, assumed to be energy independent. This doubly periodic relation is usually represented in the unit square [0,1] x [0,1] (zone 1 of Fig. 1), but interesting features appear more clearly on the extended plot[3] of Fig. 1 :

- the wavy lines visualize the regular (in the ν_i scale) oscillations of the system between the two channels, owing to the relation[4]

$$(Z_2/Z_1)^2 = d(-\nu_1)\ /\ d\nu_2 \qquad (2)$$

between the amplitudes Z_i of each channel and the slope of the curve.

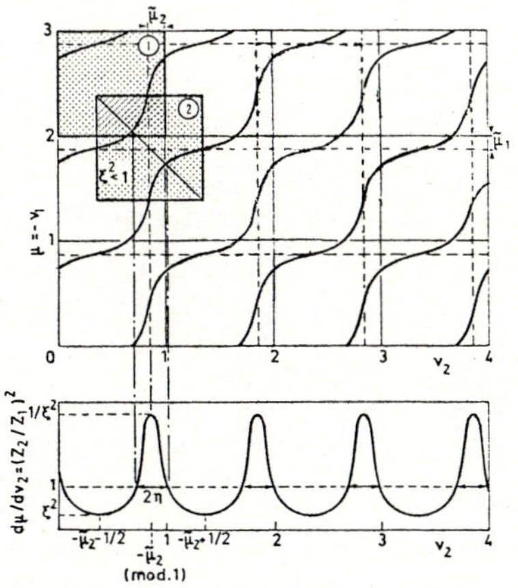

Fig. 1: (a) Extended plot of the doubly periodic equations (1) or (3) (adapted from Ref.3). The full line grid corresponds to the usual axes of the Lu-Fano plot, lying at integer values of ν_i as shown in the reduced zone 1. The dashed line grid corresponds to $\nu_i = -\tilde{\mu}_i$ (mod.1) and intersects the plot at its points of maximum or minimum slope.
(b) Slope $d\mu/d\nu_2$ of the curves in (a), equal to the squared ratio of channel amplitudes $(Z_2/Z_1)^2$, as a function of ν_2. The arrows mark the points of equal admixture of both channels at the intersection of the plot with the diagonal $\nu_1+\tilde{\mu}_1 = \nu_2+\tilde{\mu}_2$.

- by translating the unit square as shown in Fig. 1 (zone 2) one obtains a symmetric plot which can be parametrized by an equation equivalent to (1)

$$\tan\pi(\nu_1 + \tilde{\mu}_1)\ \tan\pi(\nu_2 + \tilde{\mu}_2) = \xi^2 \qquad (3)$$

where the $\tilde{\mu}_i$ correspond to the inflection points of the curve and $\xi = \tan\eta$, related to the closest approach of the two branches (see Fig. 1), measures the strength of the interaction between the two channels.

- the new parameters $\tilde{\mu}_i$ and ξ are especially significant for the continuous energy range between the two thresholds, where $\pi\mu = -\pi\nu_1$ is the short-range phase-shift in the open channel 1. There one easily identifies the <u>energies</u> $E_n = E_2^+ - \dfrac{1}{2(n-\tilde{\mu}_2)^2}$ and <u>widths</u> $\Gamma_n = \dfrac{2}{\pi}\ \dfrac{\xi^2}{(n-\tilde{\mu}_2)^3}$ of the resonances due to the discrete levels in the closed channel 2. The cross-section for elastic scattering or photoionization takes the compact form

$$\sigma = \sigma_0 \frac{(q+\varepsilon)^2}{1+\varepsilon^2} \qquad (4)$$

where $\varepsilon = \tan\pi(\nu_2+\tilde{\mu}_2)/\xi^2$ is a "reduced energy" which vanishes at each resonance ($\nu_2 = n - \tilde{\mu}_2$) and ranges from $-\infty$ to $+\infty$ between two successive resonances. This single expression (already obtained by Dubau et Seaton[5] within a different approach) depicts the whole series of resonances at once ; around each of them it may be approximated by the well-known Fano formula[6], akin to (4) with $\varepsilon = 2(E-E_n)/\Gamma_n$. The expression of the profile parameter q, common to the whole series, depends on the physical process studied.

This behaviour applies to long-range potentials other than Coulombic, the only change being the relation of the quantities ν_i to the energy. This extension as well as theoretical identification of the parameters $\tilde{\mu}_i$ and ξ will be developed at the conference.

References

1. K.T. Lu and U. Fano, Phys. Rev. A2, 81 (1970)
2. M.J. Seaton, Proc. Phys. Soc. London, 80, 804 (1966)
3. G. Herzberg and Ch. Jungen, J. Mol.Spect. 41, 425 (1972)
4. U. Fano, Phys. Rev. A2, 353 (1970)
5. J. Dubau and M.J. Seaton, J. Phys. B., to be submitted, see M.J. Seaton, Rep. on Progress in Phys. 46, 167 (1983)
6. U. Fano, Phys. Rev. 124, 1866 (1961).

PHOTOIONISATION OF SiII : AN APPLICATION OF QUANTUM DEFECT THEORY

K.T. Taylor[1] and C.J. Zeippen[2]

1 Daresbury Laboratory, SERC, Daresbury, Warrington, WA4 4AD, England
2 Observatoire de Paris, Section d'Astrophysique, 92190 Meudon, France

Quantum Defect Theory (QDT) is applicable when a scattering problem can be split into two parts, namely an inner part $r \le r_0$ in which the potential experienced by the scattering electron (radial co-ordinate r) is complicated and non-local and an outer part, $r \ge r_0$ where the potential has some simple known analytic form. Seaton[1] developed Multichannel Quantum Defect Theory (MQDT) for the fundamental case where the potential is Coulombic for $r \ge r_0$, and recently other long-range potentials have been considered[2] also. A new computer program employing ideas from both these sources is being written[3] and we report here its first application in a calculation of some astrophysical significance. This is the photoionisation of SiII, a problem in which the long-range potential is Coulombic.

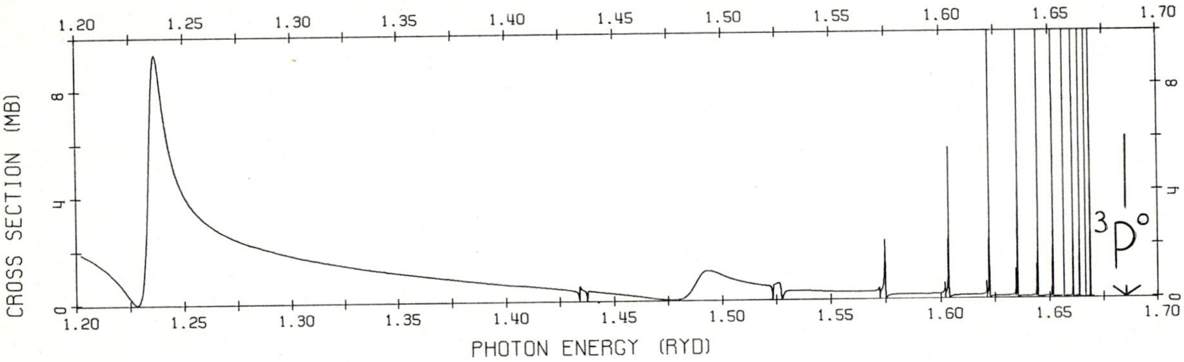

Fig. 1: SiII photoionisation cross section contribution from $^2D^e$ symmetry (length form)

The process is described by

$$h\nu + \mathrm{SiII}(3s^2 3p\,^2P^0) \rightarrow \{\mathrm{SiIII} + e^-\}\,^2S^e, ^2P^e, ^2D^e$$

where the dipole selection rules are taken to hold within an LS coupling scheme. For the inner region the calculation proceeds in the usual way for atomic photoionisation using the R-matrix method[4], thus leading to the evaluation of a wavefunction logarithmic derivative matrix at $r=r_0$ and corresponding dipole matrix elements for each photon energy over a specified range and spacing. The multi-configurational wavefunctions of Baluja and Hibbert[5] are used to describe the 12 lowest states of SiIII retained in this calculation and these fix the choice of r_0 at 15.6 Bohr radii.

The new MQDT program takes in this data. *Strongly* closed channels are dropped and, using Coulomb functions from r_0 outwards, quantities with, in general, a slow energy variation are obtained. Interpolation over energy in these quantities allows the photoionisation cross section to be calculated economically at any desired photon energy within the range spanned by the input mesh.

Figure 1 above displays the length form of the contribution to the photoionisation cross section from the residual ion plus outgoing electron in overall $^2D^e$ symmetry over a range of photon energies whose maximum is just insufficient to reach the first excited state threshold $(3s3p)\,^3P^0$ of SiIII. In this energy range and symmetry it proved adequate to retain only 3 of the closed channels in the outer region.

Two broad resonances dominate the figure. These are designated $(3s3p)\,^3P^0 4p\,^2D^e$ near 1.23 Ryd photon energy and an interloping $(3s3p)\,^1P^0 4p\,^2D^e$ near 1.49 Ryd photon energy.

Calculations are proceeding for higher incident photon energies where we find further broad interloping resonances can also dominate the cross section. These results will be presented at the conference.

The inner region calculation took about 2 hours CPU time on the GETIA CRAY-1S computer, in France.

References

1. M.J. Seaton, Proc. Phys. Soc. 88, 801 (1966).
2. C. Greene, U. Fano and G. Strinati, Phys. Rev. A 19 1485 (1979).
3. K.T. Taylor, Comput. Phys. Commun., to be submitted (1983).
4. P.G. Burke and K.T. Taylor, J. Phys. B 8, 2620 (1975).
5. K.L. Baluja and A. Hibbert, J. Phys. B 13, L327 (1980).

SIMPLE DEPENDENCES OF BOUND-FREE DIPOLE MATRIX ELEMENTS AND THEIR ZEROES ON THE PROPERTIES OF THE INITIAL STATE

David Salzmann, R. H. Pratt, M. S. Wang and R. Y. Yin
Department of Physics and Astronomy, University of Pittsburgh, Pittsburgh, Pennsylvania 15260 U.S.A.

Results are presented for the dependence of bound-free matrix elements and their zeroes on the state of excitation or ionization of the initial atom. The calculations were carried out using our FOTO code,[1] which solves the Dirac equation in a Hartree-Slater self-consistent potential and uses the resulting wave-functions to compute photoelectric cross sections, including relativistic, retardation and higher multipole effects.[2] Here we report three related studies of these matrix elements:

(A) We have examined the variation of the zeros of dipole and quadrupole bound-free matrix elements for an $nS_{\frac{1}{2}}$ ground or excited state as a function of the principal quantum number n. Neutral uranium was chosen to illustrate this problem. The quadrupole matrix elements have multiple zeros, one of which occurs even in a point Coulomb potential. The position of dipole and quadrupole zeros approach a limiting value for large n. In fact, this limit is already achieved for rather low excitation states. More generally, it can be stated that, for sufficiently high n, both the photoelectron angular distribution and the shape of the cross-section vary only slightly with principal quantum number except very close to threshold. This occurs because when the photon energy is large compared to the binding energy, the process occurs at distances close enough to the nucleus that the shape of the bound state wave-function is independent of the principal quantum number.

(B) Zeros in the dipole matrix elements and the associated Cooper minima in cross sections do not occur for the point Coulomb potential and are normally observed in neutral atoms. One generally expects that these zeros will not appear in ions, where the electrons see a more Coulomb-like potential. However, our calculations show zeros in the bound-free matrix elements of the 6p states in multiply ionized high Z atoms, where the Cooper minimun in the neutral atom occurs at a high photoelectron energy. In our results for Fermium (Z=100) the position of the zeros decreases steadily with increasing ionization from a photoelectron kinetic energy E_e=454.7 eV for the neutral atom down to threshold for Z^*=21.[3] Evidently this results from the reduction in the screening of the nuclear potential seen by the 6p electron. However, there is little change when the position of the zeros is plotted versus the photon energy, showing that the matrix element considered as a function of photon energy is almost independent of Z^*, and the main effect is the change in the threshold energy.

(C) The effect of screening on the photoionization of excited atoms was studied for low Z elements by calculating the photoelectric cross-sections for various excitation modes. The photoionization of a 3p electron of potassium (Z=19) was chosen for this study. We have compared cases for which the hole produced by the excitation is outer (4s→4p; 4s→εp), inner (1s→4s; 2p→4s) or in the same shell (3p→4s) as the ionized electron. The results are shown in Fig. 1. Evidently, holes in external shells do not significantly alter the photoionization cross-sections, while inner shell holes reduce the screening and, consequently, increase the cross-sections. Only for inner shell excitation was any significant shift of the Cooper minima observed for the excitation modes under study.

References

1. I. B. Goldberg, PITT-291, Internal Report, U. of Pittsburgh, (1982).
2. R. H. Pratt, A. Ron and H. K. Tseng, Rev. Mod. Phys. 45, 273 (1973).
3. For Z^* > 13 a 6p electron is in an excited level, as inner shell electrons are already stripped off the ion.

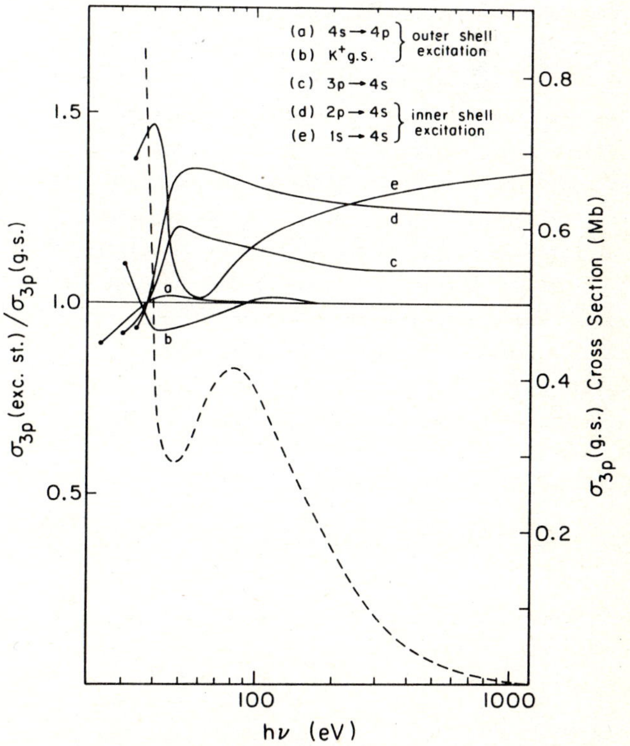

Fig. 1: Photoelectric cross section of $3p_{\frac{1}{2}}$ electron of ground state Potassium (dashed line) and the ratio of the cross section of the same electron in excited atoms to the ground state case (solid lines).

ZEROS IN DIPOLE MATRIX ELEMENTS OF PHOTOIONIZATION OF EXCITED ALKALI ATOMS*

Jayanti Lahiri

Department of Chemistry and Physics, Southern Technical Institute, Marietta, Georgia

Steven T. Manson

Department of Physics and Astronomy, Georgia State University, Atlanta, Georgia

Studies of the photoionization of excited atomic states have revealed that there are many more zeros in the dipole matrix elements for such transitions than exist for ground states.[1-3] Such zeros are found in $\ell \rightarrow \ell-1$ transitions in excited states but not for ground states[1,2]; Zeros are found for nodeless excited discrete state wave functions, but not for ground states[1,2]; Excited states exhibit as many as three zeros in a single $\ell \rightarrow \ell + 1$ transition while ground states have no more than a single zero.[1,3] The importance of these zeros is that they profoundly affect the spectral distribution of oscillator strength of the excited state.

We report here on a study of the alkali atoms, Li, Na, K, Rb, and Cs, in a Hartree-Slater atomic model, with the focus upon the zeros in the dipole matrix elements. All excited states with $\ell \leq 3$ up to n = 5 in Li, n = 6 in Na, and so on up to n = 9 in Cs have been studied from threshold to photoelectron energy of 5 Rydbergs.

Zeros are found in s → p, p → d, and d → f $\ell \rightarrow \ell + 1$ channels, as well as d → p and f → d $\ell \rightarrow \ell - 1$ transitions. The dependence of these zeros of Z and n have been explored. It is found that only the d → f transitions in Cs have more than one zero in the continuum. It is also found that zeros in the f → d transitions exist only in Rb. Further, the zeros in some of the channels are quite Z and n dependent while others are not.

The behavior of the $\ell \rightarrow \ell + 1$ zeros is explainable in terms of the quantum defects of the initial discrete states relative to the threshold phase shifts of the final continuum states. This only partially explains the details of the $\ell \rightarrow \ell - 1$ zeros and the matter is still under study.

* Work supported by the US Army Research Office

References

1. A. Z. Msezane and S. T. Manson, Phys. Rev. Letters 35, 364 (1975).
2. A. Z. Msezane and S. T. Manson, Phys. Rev. Letters 48, 473 (1982).
3. J. Lahiri and S. T. Manson, Phys. Rev. Letters 48, 614 (1982).

CROSS SECTIONS FOR PHOTOIONIZATION OF EXCITED d-STATES OF ALKALI ATOMS: HARTREE-FOCK CALCULATIONS*

Alfred Z. Msezane

Department of Physics, Morehouse College, Atlanta, Georgia 30314

Steven T. Manson

Department of Physics and Astronomy, Georgia State University, Atlanta, Georgia 30303

Calculations of the photoionization cross sections of the excited atomic Na 3d, K 3d, Rb 4d, and Cs 5d states have been carried out using Hartree-Fock (HF) wave functions, i.e., HF wave functions for the initial discrete states, for the final ionic discrete states, and for the final continuum states. The initial state HF wave functions are extremely sensitive to the numerical methods employed owing to the very long tail of the excited orbital. It has been found that various techniques which work extremely well for HF ground state wave functions, fail for excited states in the sense that the resulting single-particle wave function for the excited orbital can have inaccuracies which dramatically affect the dipole matrix elements (and, thus, the photoionization cross section) even though the energy is correct to six or seven places. In other words, the energy is insensitive to the tail of the excited state wave function but the dipole matrix element, on the other hand, is quite sensitive. It is found that several procedures which generate analytic HF wave functions[1,2] are simply too inaccurate to treat the photoionization of excited states properly. Only a completely numerical procedure[3] has been found to give the tail of the excited state wave function with sufficient accuracy to be useful for excited state photoionization.

We report here on the photoionization of excited states which have been found in simple calculations to have zeros in at least one of the $\ell \rightarrow \ell \pm 1$ channels.[4-6] These zeros are not present in ground state photoionization and they have not yet been confirmed experimentally. Thus it is of interest to investigate them employing a more accurate calculation.

Our results show that the minima are indeed in evidence, although they are moved around somewhat from their positions in the simple calculations. We also find that, using the most accurate discrete HF wave functions gives rather good agreement between the "length" and "velocity" formulations. While this is not conclusive, it is indicative of the accuracy of these results.

In addition to cross sections, photoelectron angular distribution β parameters have been calculated. These β's show the effects of the zeros in the dipole matrix elements more clearly than do the cross sections.

References

1. B. Roos, C. Salez, A. Veillard and E. Clementi, A General Program for Calculations of Atomic SCF Orbitals by Expansion, Tech. Rep. RJ518IBM Research, 1968.
2. A. Hibbert, Comput. Phys. Commun. 9, 141 (1975).
3. C. Froese Fischer, Comput. Phys. Commun. 7, 236 (1974).
4. A. Msezane and S. T. Manson, Phys. Rev. Letters 35, 364 (1975).
5. A. Msezane and S. T. Manson, Phys. Rev. Letters 48, 473 (1982).
6. J. Lahiri and S. T. Manson, Phys. Rev. Letters 48, (1982).

* Work supported by NSF and US ARO

THE RESONANCE PHOTOIONISATION OF HELIUM TO $He^{+}(N = 2)$

S.M. Burkov, S.I. Strakhova

Institute of Nuclear Physics, Moscow State University,Moscow 117234, USSR

The total and partial cross sections for resonance photoionisation of He have been calculated in the region below the N = 3 threshold. Using the method of configuration interactions as applied to autoionisation by Fano, complex representation of the energy have been used to obtain expressions for the atomic resonance photoionisation cross sections having no singularities in the general case of an arbitrary number of open and closed channels[1].

We took into account four open channels corresponding to the states 1s, 2s and 2p in He^{+}.

The open channels subspace was preliminary diagonalized in the integral formulation of the close-coupling method[2]. The integral over the open channels subspace included the energy range from the first threshold to 20 atomic units upwards. In the calculations we took into account the 3s np, n = 3-7; 3pns, n = 4-7; 3pnd, n = 3-6; 3dnp; n =4-6; 3dnf, n = 4-7 states of He. The Coulomb functions with Z = 2 were taken to be the basis functions for this states. The ground-state wavefunction were assumed to be of the Tweed form[3] with 41 parameters. The positions and widths of the resonances have been obtained as the eignvalues of the complex matrix, calculated with the preliminary diagonalized open channels subspace. In the case of resonances converging to the threshold N= 3, we numerated resonances according to their position on the scale of energy, without indicating the Rydberg characteristics. Table 1 presents the resonance parameters for the two states of He in the cross section for producing $He^{+}(N = 2)$ calculated by us and the experimental results published by Woodruff and Samson[4]. The calculated cross section for producing $He^{+}(N = 2)$ is shown in Fig. 1 together with experimental data[4].

Table 1

Calculated and experimental parameters for the resonances below the N = 3 threshold

	State 1		State 4	
	This work	Experiment[4]	This work	Experiment[4]
E, eV	69,86	69,92+0,012	71,66	71,60+0,018
Γ, eV	0,212	0,178+0,012	0,080	0,096+0,015
q	0,49	0,48+0,09	0,41	0,47+0,10
ρ^2	0,99	$0,98^{+0,02}_{-0,26}$	0,99	$0,98^{+0,02}_{-0,29}$

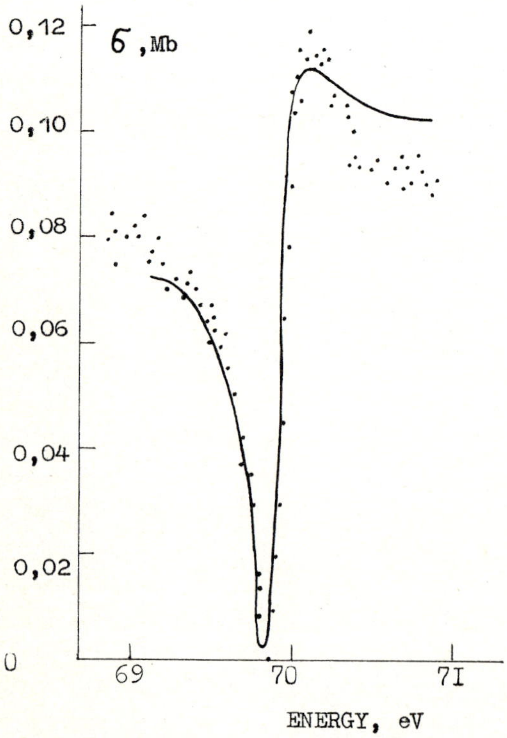

Fig. 1: The cross sections for producing $He^{+}(N=2)$, calculated (full line) and experimental[4] (dots).

References

1. S. Burkov and S. Strakhova, Proc. XII ICPEAC (Gatlinburg TN, USA),p.33 (1981).
2. S. Strakhova and V. Shakirov. J.Phys.B:Atom. Molec.Phys. 15, 2149(1982).
3. R. Tweed, J.Phys.B: Atom.Molec.Phys.5, 810 (1972).
4. P. Woodruff and J. Samson, Phys.Rev. A25,848 (1982).

THEORETICAL INVESTIGATIONS OF SPIN POLARISATION IN MERCURY

F. Keller and F. Combet Farnoux

E.R.A. Spectroscopie Atomique et Ionique ; Bâtiment 350 Université Paris-Sud, 91405 Orsay France.

Spin polarisation of photoelectrons emitted by unpolarised atoms is now a very useful tool for the study of photoemission processes. Mercury is the heaviest element for which experimental and theoretical cross sections, asymmetry parameters and spin polarisation parameters are available. As an example, we will focus here on the 5p subshell of mercury, for which we have performed calculations in relativistic Dirac-Slater model, using our own code[1]. In this case, because of the low angular momentum and the rather strong binding energy, spin polarisation parameters, in addition to cross sections and angular distributions are necessary to assess the validity of such a model.

For j = 0.5, the spin parameter[2] η describing the degree of transverse polarisation of photoelectrons ejected by unpolarised light from a np subshell can be expressed as for ns^2 subshells[3] in terms of two parameters : x the ratio of the dipolar matrix elements for the two transitions allowed by selection rules in jj coupling $np_{1/2} \rightarrow \varepsilon d_{3/2}$ and $np_{1/2} \rightarrow \varepsilon s_{1/2}$ and Δ, the total phase shift difference of the two continuum wave functions.

$$x = R_d / R_s \qquad \Delta = \delta_s - \delta_d$$

$$\eta = \frac{3x}{1 + 2x^2} \sin \Delta$$

The other spin polarisation parameters δ, ξ can be expressed as functions of the three independent dynamical parameters σ, β and η. The variation of η, δ and ξ versus photoelectron energy ε is shown in figure 1.

Near threshold, a first annulation of η is induced by the rapid variation of the phase shift Δ ; since $\sin \Delta \rightarrow 0$, $\cos \Delta \rightarrow 1$, the asymmetry parameter β is maximum and $\xi = 0$. The second zero of η is connected with the cancellation of the matrix element R_d. It is associated with a zero of β and a maximum of ξ. As R_s never vanishes and $\sin \Delta$ is an increasing function of the energy, η cannot show additional zeros, and its maximum observed at intermediate energy is smaller than the maximum which happens at higher energy.

The spin polarisation parameter δ of the photoelectrons produced by circularly polarised radiation is given by :

$$\delta = \frac{Rs^2 - Rd^2}{Rs^2 + 2Rd^2} = \frac{1 - x^2}{1 + 2x^2}$$

The greatest value $\delta = 1$ is reached when $|R_d| \ll |R_s|$ or $\sigma_d \ll \sigma_s$ where σ_s and σ_d are the partial

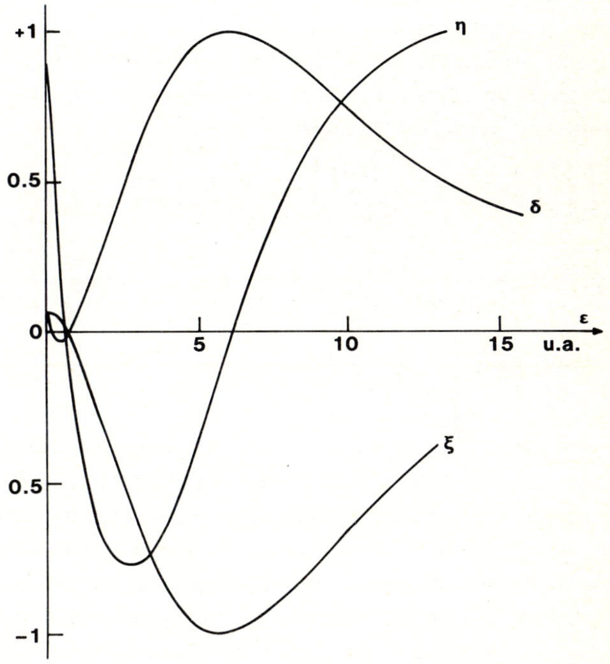

fig 1 : Hg $5p_{1/2}$ spin polarisation parameters.

photoionisation cross sections.

For j = 1.5, three final states of the photoelectron are allowed so that the theoretical expressions for η, δ, ξ cannot be reduced to a simple form. As predicted in a non relativistic theory $|\eta_{3/2} / \eta_{1/2}| < 1$, but not exactly in the statistical ratio 0.5.

Large values of spin polarisation parameters ξ and δ can be obtained for j = 0.5 when the matrix element R_d vanishes i.e. when $\beta = 0$ and $\eta = 0$. As for η, a first zero must happen near threshold for all subshells of angular momentum $l \neq 0$, as $\sin \Delta$ in our model always vanishes around 0.6 a u.

References

1 F. Keller and F. Combet Farnoux, J. Phys. B 15, 2657 (1982)

2 K. N. Huang, Phys. Rev. A 22, 223 (1980)

3 G. Schönhense, U. Heinzmann, J. Kessler and N. A. Cherepkov, Phys. Rev. Lett. 48, 603 (1982)

POLARIZATION OF ATOMIC PHOTOELECTRONS IN RELATIVISTIC AND NON-RELATIVISTIC THEORIES

N.A.Cherepkov

A.F.Ioffe Physical-Technical Institute, 194021 Leningrad, USSR

As it was shown previously[1], in the non-relativistic (dipole) approximation an angular distribution of photoelectrons with defined spin polarization, ejected from arbitrary atomic subshell by circularly polarized light, is characterized by five independent parameters and has the following form:

$$I_j^{\pm 1}(\vec{æ},\vec{s}) = \frac{\sigma_j(\omega)}{8\pi}\Big\{1-\frac{\beta^j}{2}P_2(\vec{æ}\vec{s}_\gamma) + A^j\cdot(\vec{s}\vec{s}_\gamma) - \gamma^j\Big[\frac{3}{2}(\vec{æ}\vec{s})(\vec{æ}\vec{s}_\gamma)-\frac{1}{2}(\vec{s}\vec{s}_\gamma)\Big] - \eta^j\cdot(\vec{s}[\vec{æ}\vec{s}_\gamma])(\vec{æ}\vec{s}_\gamma)\Big\} \quad (1)$$

where $\vec{æ}$, $\vec{s}$ and $\vec{s}_\gamma$ are unit vectors in the directions of the electron momentum, electron spin and photon spin respectively. For absorption of linearly polarized or unpolarized light terms linear in $\vec{s}_\gamma$ disappear, while in other terms it is necessary to substitute $\vec{e}$ (the polarization vector for linearly polarized light) or $\vec{k}$ (a unit vector in the direction of a photon momentum for unpolarized light) for $\vec{s}_\gamma$. The parameters β^j, A^j, γ^j, η^j ($j = \ell \pm 1/2$ is the total angular momentum of the ionized subshell), as well as the photoionization cross section $\sigma_j(\omega)$, are functions of a photon energy ω only, whereas the angular dependence is given explicitly by the vector products.

The angular distribution of photoelectrons with defined spin polarization in relativistic consideration (with the contribution of higher multipoles included) has been presented in the form[2]:

$$I_j(\vec{æ},\vec{\xi},\vec{\zeta}) = \frac{\sigma(\theta)}{2}\sum_{\mu,\nu=0}^{3}\xi_\mu\zeta_\nu C_{\mu\nu} \quad (2)$$

where vectors $\vec{\xi}(\xi_1,\xi_2,\xi_3)$ and $\vec{\zeta}(\zeta_1,\zeta_2,\zeta_3)$ characterize the polarization state of a photon and a photoelectron respectively, $\xi_0 = \zeta_0 = 1$, $\sigma(\theta)$ is the differential photoionization cross section, and $C_{\mu\nu}$ are some parameters, which depend on the photon energy and the photoelectron ejection angle. There are eight different from zero parameters $C_{\mu\nu}$ including $C_{00} = 1$.

Comparing (1) and (2) one can conclude that:

(i) In the cases of absorption of circularly polarized and unpolarized light the relativistic expression (2) is equivalent to the non-relativistic expression (1) with the parameters β^j, A^j, γ^j, η^j being functions of the photoelectron ejection angle, that is higher multipoles give only additional degrees of the product $(\vec{æ}\vec{s}_\gamma)$ or $(\vec{æ}\vec{k})$.

(ii) For absorption of linearly polarized light in the relativistic expression (2) there is one additional term which is absent in the non-relativistic expression (1) even if the parameters β^j, A^j, γ^j, η^j are functions of the photoelectron ejection angle. This term is proportional to the product of $(\vec{s}[\vec{e}\times\vec{k}])(\vec{æ}\vec{e})$ and some coefficient which depends on the photoelectron ejection angle. It leads to appearance of a longitudinal polarization of photoelectrons for absorption of linearly polarized light, whereas in the dipole approximation photoelectrons can be polarized only transversely.

(iii) In the non-relativistic limit the coefficient before the term discussed in (ii) (the parameter C_{23}) tends to zero as $h\nu/mc^2$ while all other parameters $C_{\mu\nu}$ in (2) as well as the parameters β^j, A^j, γ^j, η^j, remain finite. If photoelectrons corresponding to ionization of a definite fine-structure level are separated in energy, then all parameters but C_{23} are of order 1, otherwise they are of order $(\alpha Z)^2$. For ns and ns^2 subshells these parameters are also of order $(\alpha Z)^2$ [3] except for the region of the Cooper minimum where due to the Fano effect[4] they again can be of order 1. But in this region the cross section is of order $(\alpha Z)^4$.

References

1. N.A.Cherepkov. Zh. Eksp. Teor. Fiz. 65, 933 (1973).
2. R.H.Pratt, R.D.Levee, R.L.Pexton, W.Aron. Phys. Rev. 134A, 916 (1964).
3. B.Nagel, P.Olsson. Arkiv för Fysik 18, 29 (1960).
4. U.Fano. Phys. Rev. 178, 131 (1969).

PHOTOIONIZATION OF HIGH-Z ATOMS: RELATIVISTIC EFFECTS ON COOPER MINIMA IN THE 6P SUBSHELL*

S. T. Manson and C. J. Lee

Department of Physics and Astronomy, Georgia State University, Atlanta, Georgia 30303

R. H. Pratt and I. B. Goldberg†

Department of Physics and Astronomy, University of Pittsburgh, Pittsburgh, Pennsylvania 15260

B. R. Tambe

Department of Chemistry and Physics, Southern Technical Institute, Marietta, Georgia 30060

Akiva Ron

Racah Institute of Physics, Hebrew University of Jerusalem, Jerusalem, Israel 91904

Studies of the photoionization of 6p electrons from heavy atoms, $82 \leq Z \leq 100$ have been performed within the framework of a relativistic Dirac-Slater formulation.[1,2] Attention was focused particularly on the effect of relativistic interactions on the zeros (Cooper minima) in the $6p \rightarrow \varepsilon d$ dipole matrix elements. The single non-relativistic $p \rightarrow d$ matrix element splits into three matrix elements under the influence of relativistic effects, $p_{\frac{3}{2}} \rightarrow d_{\frac{5}{2}}$, $p_{\frac{3}{2}}, \rightarrow d_{\frac{3}{2}}$, and $p_{\frac{1}{2}} \rightarrow d_{\frac{3}{2}}$; each of these has its own Cooper minima. Our primary interests are the location of these minima compared to the non-relativistic value, the energy splittings between them, and how the behavior changes as a function of Z.

Our results are shown in Fig. 1 indicating the magnitude of the relativistic effects. The "trajectory" of each of the relativistic zeros is shown, as a function of Z, along with the trajectory of single non-relativistic zero obtained in Hartree-Slater (HS) calculations.[3] There is a huge splitting between the $6p_{\frac{1}{2}}$ minimum and those arising from the $6p_{\frac{3}{2}}$ state amounting to ∿ 100 eV for Z=82 and increasing to ∿ 300 eV by Z=100. Also shown in Fig. 1 is the discrete spin-orbit splitting between the $6p_{\frac{3}{2}}$ and $6p_{\frac{1}{2}}$ states, a splitting which is more than an order of magnitude smaller than the splitting of the relativistic "Cooper minima." The explanation of this phenomenon lies in the fact that the spin-orbit interaction displaces the $6p_{\frac{1}{2}}$ wave function inward with respect to the $6p_{\frac{3}{2}}$ while, owing to the centrifugal barrier for d-waves, it takes far more energy for an εd wave function to move in that amount.

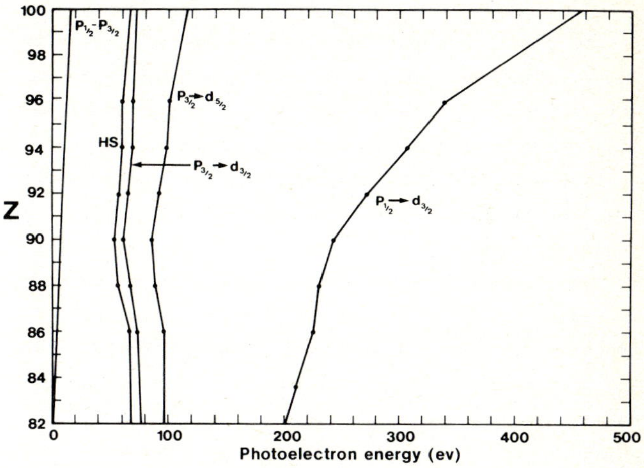

Fig. 1: Trajectory (photoelectron energy) of the "Cooper" zeros in the $6p \rightarrow d$ dipole matrix elements as a function of Z. The relativistic matrix elements are labelled and HS refers to the non-relativistic Hartree-Slater results. Also shown, for comparison, is the spin-orbit splitting of $6p_{\frac{1}{2}} - 6p_{\frac{3}{2}}$ as a function of Z.

This phenomenon has important implications for the $6p_{\frac{3}{2}}:6p_{\frac{1}{2}}$ branching ratios, making them anomolously low at low energies where the $6p_{\frac{3}{2}} \rightarrow \varepsilon d$ zeros are, and anomolously high at higher energies in the vicinity of the $6p_{\frac{1}{2}}$ minimum. Photoelectron angular distributions are also strongly affected.

* Work supported by NSF and the US Army Research Office

† Permanent Address: Racah Institute of Physics, Hebrew University of Jerusalem, Jerusalem, Israel 91904

References

1. R. H. Pratt, A. Ron, and H. K. Tseng, Rev. Mod. Phys. 45, 273 (1973).
2. W. Ong and S. T. Manson, Phys. Rev. A 19, 688 (1979).
3. S. T. Manson, Adv. Electronics Electron Phys. 41, 73 (1976).

Z n ℓ DEPENDENCE OF CANCELLATION OF RELATIVISTIC AND RETARDATION EFFECTS IN PHOTOIONIZATION

R. H. Pratt

Department of Physics, University of Pittsburgh, Pittsburgh, Pennsylvania 15260 U.S.A.

Akiva Ron

Racah Institute for Theoretical Physics, Hebrew University of Jerusalem, Jerusalem 91904 Isreal

S. T. Manson

Department of Physics, Georgia State University, Atlanta, Georgia 30303 U.S.A.

Sung Dahm Oh

Department of Physics, Sook Myung Women's University, Seoul, Korea

We wish to report a study of the (Z n ℓ) dependence of the thus far unexplained cancellation of relativistic and retardation effects observed in photoionization. For some years it has been known that there are situations in which nonrelativistic dipole approximation continues to correctly predict total cross sections for subshell photoionization well into the relativistic regime. Thus Oh, McEnnan, and Pratt[1] found that 1s and 2s nonrelativistic dipole total cross sections for uranium remained reasonably accurate even for photoelectron energies 100 keV above threshold, while the 2p cross section (summed over the two relativistic subshells to compare with the nonrelativistic prediction) showed major deviations by 30 keV above threshold. By contrast, in some outer shells, such as the 6p subshells of uranium, significant relativistic effects on the total cross section persist to threshold.[2] It should be noted that significant relativistic effects, even near threshold, have been seen in angular distributions of intermediate and high Z elements, both for inner and outer shell photoionization.

Similar cancellations of relativistic and retardation effects have been seen in other processes. In bremsstrahlung the nonrelativistic dipole Sommerfeld formula remains valid for incident electron energies as high as 100 keV in high Z elements, while it is failing by 30 keV in low Z elements.[3] Forward Rayleigh scattering from the K and L subshells show behaviors similar to those of photoionization.[4] Remarkably, in internal conversion the cancellation persists to an MeV,[5] presumably reflecting the fact that the transition involves a definite multipole and, unlike photoeffect, higher multipoles do not enter with increasing energy.

Our purpose here is to provide further information on the behavior of this cancellation in photoionization. This data may be of some assistance as one tries to identify the salient elements of the dynamics which underly the cancellation phenomena. We have studied the n-dependence, the ℓ-dependence, and the Z dependence of the phenomenon. Our data for uranium displays for various subshells the ratio of relativistic to nonrelativistic cross sections as a function of photon energy. Relativistic cross sections have been summed over the $j=\ell\pm\frac{1}{2}$ subshells to permit comparison with the corresponding nonrelativistic cases (we have also examined these subshells individually).

The n-dependence of the relativistic/nonrelativistic ration shows the standard merging toward a common curve, with lower n falling on the curve as photon energy increases, which has already been observed in other contexts. This follows from the fact that the shape of electron wave functions of given (j ℓ) is common for their interior regions, that the common shape also includes lower principal quantum numbers n, at more interior distances, and that the photoelectric matrix element is determined at more interior distances. In the s-state case the persistance of validity of nonrelativistic forms is remarkable. On the other hand, it is now quite clear that the cancellation is an s-wave feature: for higher shells the deviations occur when one would expect.

We will also discuss the Z-dependence of this phenomena, and insights regarding its origin which are available from various analytic calculations.

This work has been supported in part by NSF and in part by USARO.

References

1. Sung Dahm Oh, James McEnnan, and R. H. Pratt, Phys. Rev. A 14, 1428 (1976).
2. Young Soon Kim, Akiva Ron, R. H. Pratt, B. R. Tambe, Steven T. Manson, Phys. Rev. Lett. 46, 1326 (1981).
3. I. J. Feng and R. H. Pratt, PITT-266, July 1981 (unpublished), "Parametrization of the Bremsstrahlung Spectrum".
4. S. C. Roy and R. H. Pratt, Phys. Rev. A 26, 651 (1982).
5. Sung Dahm Oh and R. H. Pratt, Phys. Rev. A 13, 1463 (1976).

RELATIVISTIC EFFECTS IN THE PHOTOIONIZATION OF THE 5P SUBSHELL OF HIGH-Z ELEMENTS*

B. R. Tambe

Department of Chemistry and Physics, Southern Technical Institute, Marietta, Georgia

Steven T. Manson

Department of Physics and Astronomy, Georgia State University, Atlanta, Georgia

Calculations of the photoionization of 5p electrons in high-Z atoms have been performed within the framework of an explicitly relativistic Dirac-Slater (DS) formulation, a formulation which has proved quite reasonable in treating the 5d subshell of high-Z elements.[1] In this study we have considered the range of Z from Hg(Z=80) to Fm(Z=100). We have calculated cross sections, photoelectron angular distributions, and $5p_{\frac{3}{2}}$:$5p_{\frac{1}{2}}$ branching ratios.

In one case, Hg, experimental data exists. A comparison of $5p_{\frac{3}{2}}$ and $5p_{\frac{1}{2}}$ cross sections and photoelectron angular distribution asymmetry parameters show rather good agreement with our calculations. Furthermore, the branching ratio is also in reasonably good agreement as well. The agreement, surprisingly, is somewhat better than that of a recent relativistic-random-phase-approximation (RRPA) calculation.[3] Whether this is just accidental, or it is indicative of something significant, we cannot yet say.

A most interesting feature of our results involves the zero (Cooper minimum) in the $5p \rightarrow \varepsilon d$ dipole matrix elements. The single non-relativistic matrix element is split into three, $5p_{\frac{3}{2}} \rightarrow \varepsilon d_{\frac{5}{2}}$, $5p_{\frac{3}{2}} \rightarrow \varepsilon d_{\frac{3}{2}}$, and $5p_{\frac{1}{2}} \rightarrow \varepsilon d_{\frac{3}{2}}$, with each one having its own Cooper minimum. For Hg, the $5p_{\frac{3}{2}}$ minima are close together at $h\nu \sim 150$ eV while the $5p_{\frac{1}{2}}$ minimum is at $h\nu \sim 250$ eV, a splitting of 100 eV. Furthermore, with increasing Z, the $5p_{\frac{3}{2}}$ minima move into the discrete while the $5p_{\frac{1}{2}}$ minimum remains in the continuum. The behaviour of the 5p minima are similar to the 6p which have been reported earlier.[4] Except that, for the 6p subshell, all of the minima are in the continuum.

In the elements where the $5_{\frac{3}{2}}$ minima are in the discrete while the $5p_{\frac{1}{2}}$ minimum is in the continuum, the photoelectron angular distribution β parameters are differ significantly for the two different j-states. In addition the $5p_{\frac{3}{2}}$:$5p_{\frac{1}{2}}$ branching ratio shows major effects, being anomolously small near threshold and anomolously large in the vicinity of the zero in the $5p_{\frac{1}{2}} \rightarrow \varepsilon d$ transition.

References

1. B. R. Tambe and S. T. Manson, Phys. Rev. A.
2. P. H. Kobrin, P. A. Heimann, H. G. Kerkhoff, D. W. Lindle, C. M. Truesdale, T. A. Ferrett, and D. A. Shirley, to be published.
3. V. Radojevic and W. Johnson, private communication.
4. Y. S. Kim, A. Ron, R. H. Pratt, B. R. Tambe, and S. T. Manson, Phys. Rev. Letters 46, 1326 (1981).

* Work supported by NSF

RELATIVISTIC STUDY OF MULTICHANNEL INTERACTIONS IN PHOTOIONIZATION OF HIGH-Z ATOMS: RRPA APPLIED TO RADIUM AND RADON*

Pranawa C. Deshmukh and Steven T. Manson

Department of Physics and Astronomy, Georgia State University, Atlanta, Georgia 30303

and Vojislav Radojevic†

Department of Physics, University of Notre Dame, Notre Dame, Indiana 46556

The importance of relativistic effects on atomic photoionization processes has been known for quite some time.[1] Furthermore, the importance of including electron correlations has also become increasingly evident, necessitating inclusion of both relativity and many body effects in photoionization studies. To account for the relativistic and the many body effects, we have employed the relativistic random phase approximation[2] (RRPA).

As illustrative of the present investigation, we show in Fig. 1 the partial cross-section and the asymmetry parameter for Ra 7s. The curves in this figure are broken between 0.5 a.u. and 1.0 a.u. In this region, autoionization resonances occur in 7s photoionization due to coupling with the discrete photoexcitation channels originating in the 6p orbitals. These resonances can be studied within the RRPA framework[3], but have been excluded herein. Following the onset of photoionization, one can see a rapid fall in the 7s cross-section toward a "Cooper" minimum and a corresponding response of β to the relative strengths in the two relativistic channels.

Fig. 2 shows the branching ratio for photoionization from Rn 6p orbitals, and also the result[4] of a Dirac-Slater (DS) calculation. It is interesting to observe that the gross features of photoionization from Rn 6p orbitals were satisfactorily predicted by the DS calculation, although the RRPA offers a more complete investigation yielding a more detailed structure encompassing multichannel interactions. In particular, it was shown in the DS investigation[5] that the single "Cooper" minimum in the non-relativistic $p \rightarrow d$ channel breaks down into 3 separate minima relativistically, and that the minima in the $p_{\frac{3}{2}} \rightarrow d_{\frac{5}{2}}, d_{\frac{3}{2}}$ channels occur close together while the one in $p_{\frac{1}{2}} \rightarrow d_{\frac{3}{2}}$ channel occurs at a much higher energy. These observations have also been borne out in the present study, in which we have found that the first two of the above mentioned "Cooper" minima occur in the neighborhood of 7.5 a.u. and the latter at ∿ 12.5 a.u.

* Work supported by NSF

† Permanent Address: Boris Kidric Institute-Vinca, P.O. Box 522, 11001 Beograd, Yugoslavia.

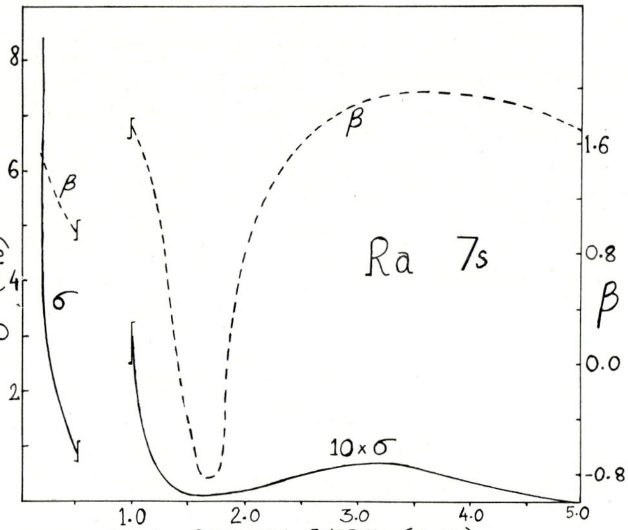

Fig. 1. PHOTON ENERGY (a.u.)

References

1. U. Fano and J. W. cooper, Rev. Mod. Phys. 40, 441 (1968).
2. W. R. Johnson, C. D. Lin, K. T. Cheng and C. M. Lee, Physica Scripta 21, 409 (1980).
3. C. M. Lee and W. R. Johnson, Phys. Rev. A22, 979 (1980).
4. S. T. Manson, C. J. Lee, R. H. Pratt, I. B. Goldberg, B. R. Tambe and A. Ron, Phys. Rev. A (to be published).
5. Y. S. Kim, A. Ron, R. H. Pratt, B. R. Tambe and S. T. Manson, Phys. Rev. Lett. 46, 1326 (1981).

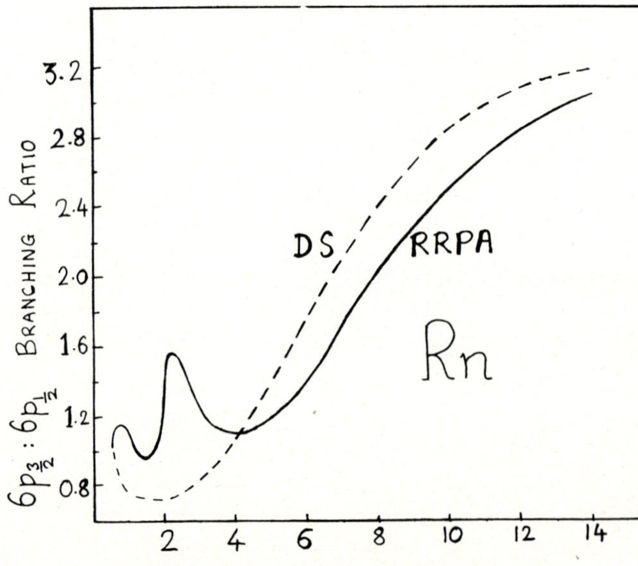

Fig. 2 Photon Energy (a.u.)

RELATIVISTIC MANY-BODY APPROACH TO THE PHOTOIONIZATION OF ALKALI ATOMS

J.E. Hansen[1] + W.R. Johnson[2] + G. Soff[3]

1 Zeeman-Laboratory, University of Amsterdam, NL-1018 TV Amsterdam, The Netherlands

2 Institut für Theoretische Physik der Johann Wolfgang Goethe-Universität, Robert-Mayer-Straße 8-10, D-6000 Frankfurt am Main, West Germany
Permanent address: Department of Physics, Notre Dame University, Notre Dame, In. 46556, U.S.A.

3 GSI , Planckstraße 1, D-6100 Darmstadt, West Germany

Low energy photoionization of alkali atoms, which is known to be sensitive to both spin-orbit and correlation effects is studied from the point of view of many-body perturbation theory. Equations are derived generalizing those of the relativistic random-phase approximation to alkali atoms, and these equations are solved to determine photoionization amplitudes. Valence shell cross sections, angular distributions and spin-polarization parameters are presented and compared with previous calculations and with available experimental data. Agreement with measured cross-sections and total spin polarization is considerably improved over previous ab-initio calculations. The present techniques are being applied to studies of inner-shell photoionization in the alkalis.

ENERGY SHIFT OF EMITTED X-RAYS FOLLOWING INNER SHELL PHOTOIONIZATION NEAR THRESHOLD

J. Mizuno*, T. Ishihara** and T. Watanabe***

*Japan Business Automation Co. Ltd., 4-2 Nihonbashi-Honcho, Chuo-ku, Tokyo 103 Japan

**Institute of Applied Physics, University of Tsukuba, Sakura-mura, Ibaraki 305 Japan

***Atomic Process Lab., The Inst. of Physical and Chemical Res.(RIKEN), Wako-shi, Saitama 351 Japan

Energy shifts of Auger electrons following inner-shell photoionization near the threshold have been observed experimentally by Schmidt et al[1)] and Hanashiro et al[2)]. This phenomenon has been interpreted as the effects of the sudden change in the effective ionic charge seen by the slow photoelectron due to the Auger decay. There seems to exist some misunderstanding in the literature that the above interpretation is equivalent to the effect of the electron-electron correlation in the final state, i.e. the screening effect of the slow electron on the ionic potential seen by the fast Auger electron. That this is not the case can be seen by considering the X-ray emission following inner-shell photoionization near the threshold, where the fast electron does not exist in the final state. Although there is no change in the net ionic charge before and after the X-ray emission, the potential seen by the slow photoelectron suffers a sudden change inside the ion. Therefore, for ions with short life-time, the slow electron is well inside the ion at the time of transition and the energy shift may be observed in the spectrum of the emitted X-ray.

Let us consider the K-shell photoionization of an atom near the threshold followed by the radiative 2p→1s transition. This process is represented by the diagram shown in Fig.1. The transition amplitude is proportional to the overlap integral

$$A = \langle \phi_k | \psi \rangle \qquad (1)$$

where ϕ_k is the wave function of the photoelectron interacting with the ion with a 2p-hole, while ψ satisfies the following inhomogeneous Schroedinger equation

$$(\Delta E + i\frac{\Gamma}{2} - h^+)\psi = h_\gamma \phi_{1s} \qquad (2)$$

Here h^+ is the single-particle Hamiltonian for the electron in the field of the ion with 1s-hole, ΔE the excess energy i.e. the incident photon energy above the threshold of the inner-shell ionization, $1/\Gamma$ the life-time of the 1s-hole, h_γ the electron-photon interaction, ϕ_{1s} the wave function of the 1s state.

In the numerical calculation, we use Hartree-Fock wave functions of Ar atom for ϕ_{1s} and in the Hamiltonian h^+. Fig.2 shows the calculated X-ray spectrum for a fixed excess energy ΔE = 0.1 a.u. with varying Γ as a parameter. For Γ = 0.02 a.u., the intensity peaks at the nominal energy. For larger values of Γ, the peaks shift as we have expected. The direction of the energy shift is towards the lower energy side, which is opposite to that in the Auger electron spectrum. As regards the magnitude of the energy shift, it is much smaller than that of the Auger electron. In fact, for the same parameters ΔE = 0.1 a.u. and Γ = 0.02 a.u., the energy shift of the Auger electron is calculated to be about 0.03 a.u., while there is no shift of the X-ray as seen in Fig.2.

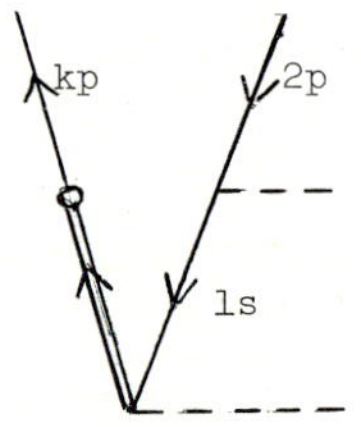

Fig.1. K-shell photoionization followed by Kα emission

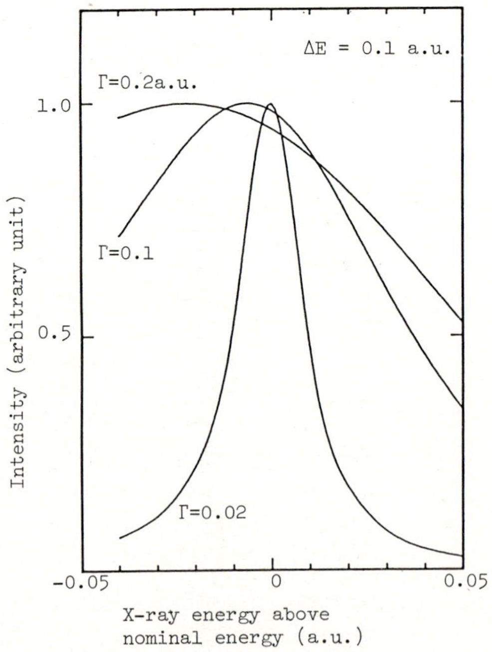

Fig.2. X-ray spectrum

References

1) V. Schmidt, S. Krummacher, F. Wuilleumier and P. Dhez, Phys. Rev. A24, 1803 (1981)

2) H. Hanashiro et al, J. Phys. B12, L775 (1979)

NEAR-THRESHOLD MEASUREMENT OF THE PHOTO-DOUBLEDETACHMENT OF He^- ($^4P^o$)

Y. K. Bae, M. J. Coggiola, and J. R. Peterson

Molecular Physics Laboratory, SRI International, Menlo Park, CA 94025

We have recently undertaken a preliminary study of the near-threshold energy dependence of the photo-doubledetachment of the metastable He^- (1s2s2p) $^4P^o$ negative ion. The threshold energy E_o for the process

$$He^- (1s2s2p) + h\nu \rightarrow He^+ (1s) + 2e^- \quad (1)$$

is expected to be 4.844 eV, corresponding to photon wavelengths shorter than 256.0 nm. Using a coaxial ion-laser beam configuration we have measured absolute cross sections for (1) from this threshold up to $h\nu = E_o + 0.41$ eV, where it reached only 2×10^{-20} cm^2. Due to these small cross sections, He^+ production tends to be dominated by the sequential, two-step process

$$He^- + h\nu \rightarrow He^* + e^-; \; He^* + h\nu \rightarrow He^+ + e^- \quad (2)$$

which can occur at photon energies below threshold, and we have not as yet obtained an accurate description of the threshold behavior of (1).

A 2 keV He^- beam was formed by electron-capture of He^+ in Cs vapor. The first of two electrostatic quadrupole deflectors, Q_1, was used to separate He^- from the other beam components, and merge it coaxially with the photon beam. Following a field-free interaction length L ~ 10 cm, the second quadrupole, Q_2, directed the product He^+ ions into a channeltron multiplier. The laser system consisted of a Quanta-Ray Nd/YAG pumped dye laser operating with DCM dye. The frequency-doubled output of the dye laser was mixed with 1.06 μ fundamental to produce the desired wavelengths. The unfocussed output pulses were directed through two 1.4 x 2.2 mm rectangular apertures that defined the spatial extent of both beams. The photon flux at each wavelength was measured by a photodiode, calibrated at 235.5 nm, whose output was integrated during each experimental run.

Because the ion transit time in L (330 ns) is long compared to the laser pulse (~ 3 ns), the complete ion beam in L was exposed to each pulse. A TAC-PHA system was used to record the arrival time spectrum of product He^+ ions using the laser Q-switch trigger as the start and the multiplier pulses as the stop signal. In this way, the desired ion signal could be distinguished from the constant background signal (A) arising primarily from collisional doubledetachment. For each wavelength studied, a number of measurements were made at varying laser powers using neutral density filters and least squares fits were made to a form $A + Bf + Cf^2$, where B yields the cross section σ_{-+} for (1) and C yields the product $\sigma_{-o}\sigma_{o+}$ for the processes in (2). An additional contribution to the linear portion can come from electric field ionization of highly excited He^* neutrals produced via single photon detachment, as occurred in the recent double detachment study of H^-, where field ionization of H^* states with $n \geq 14$ produced an anomalously low threshold.[1] Under our experimental conditions, only states with $n \geq 20$ could contribute to this background, so it should be relatively small.

The results are shown in Fig. 1 with the error bars determined from the least squares fit uncertainty. The absolute cross sections were determined by using the measured He^+ signals and photon flux densities and by assuming that the counting efficiency of each 2 keV ion was unity. Our measurements yielded also $(\sigma_{-o}\sigma_{o+})^{1/2} \cong 7 \times 10^{-18}$ cm^2 at 235.5 nm, in good agreement with the value $\sigma_{-o} \simeq 5 \times 10^{-18}$ cm^2 from an extrapolation of our σ_{-o} measurements,[2] and $\sigma_{o+} = 6 \times 10^{-18}$ cm^2 obtained by Stebbings et al.[3] for $He(2^3S)$ photoionization.

Supported by the National Science Foundation. The laser system was loaned by the NSF San Francisco Laser Center.

References

1. J. B. Donahue, et al., Phys. Rev. Lett. 48, 1538 (1982).
2. R. V. Hodges, M. J. Coggiola, and J. R. Peterson, Phys. Rev. A 23, 59 (1981).
3. R. F. Stebbings, F. B. Dunning, F. K. Tittel, and R. D. Rundall, Phys. Rev. Lett. 30, 815 (1973).

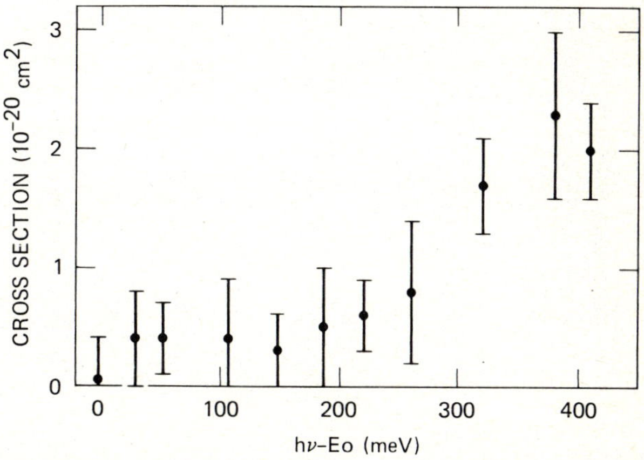

Fig. 1: Cross sections for $h\nu + He^- \rightarrow He^+ + 2e$ near the threshold E_o = 4.844 eV.

PHOTOIONIZATION OF ATOMIC OXYGEN, CESIUM, AND RUBUDIUM

J. A. R. Samson, P. Pareek, and H. Suemitsu

Behlen Laboratory of Physics, University of Nebraska, Lincoln, NE 68588

Studies of atomic species, other than the rare gases, are complicated by the uncertainty in their number densities and by the presence of dimers. In the present study photoionization-mass-spectrometry was used to identify the atomic species uniquely. With the low pressures required in a mass spectrometer measurements of ions produced per incident photon give the relative photoionization cross section for a given atom. For this method to be successful we need to know the absolute or relative intensity of the incident radiation as a function of wavelength. This can be achieved by measuring the number of ions produced from a rare gas, for example neon or argon, as a function of wavelength then dividing by the known total photoionization cross section of the gas.

We have measured the relative photoionization cross sections of O, Cs, Rb, Cs_2, and Rb_2 by the above technique. In the case of atomic oxygen we were able to measure the actual number densities within the ionization region by studying the O_2^+ signal with the atomic generator (microwave discharge) on and off. This gave a measure of the percent dissociation of O_2. The results are shown in Fig. 1 together with previously published data.[1-3] The solid circles represent our present data.

The relative photoionization cross sections of Rb are shown with the solid data points and solid line in Fig. 2. The data were normalized to the semiempirical calculations of Weisheit[4] (dashed line) at 2850Å. The other data points represent previously published data[5,6] normalized to the present data.

The relative cross sections for Cs, normalized at threshold to theory, are shown in Fig. 3 with the open circle data points. The solid and dashed lines represent the semiempirical calculations of Norcross[7] and Weisheit,[4] respectively.

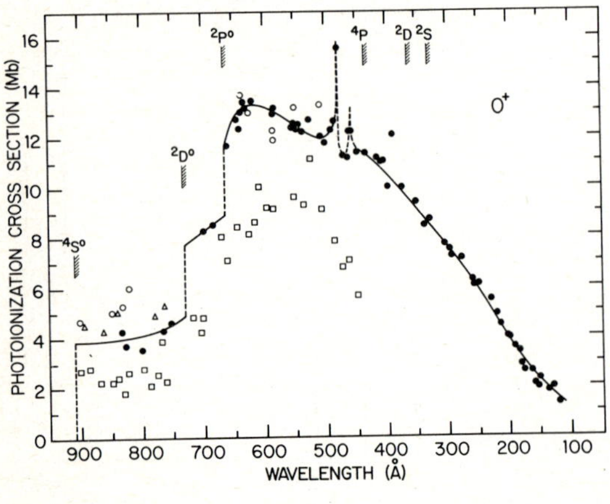

Fig. 1. Photoionization cross section of atomic oxygen. ● present results; ○ ref. 1; □ ref. 2; △ ref. 3.

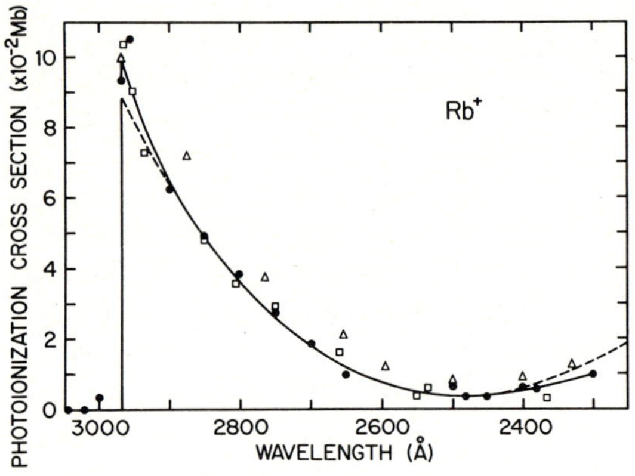

Fig. 2: Photoionization cross section of Rb. ● present data; --- ref. 4; □, ref. 5; △, ref. 6.

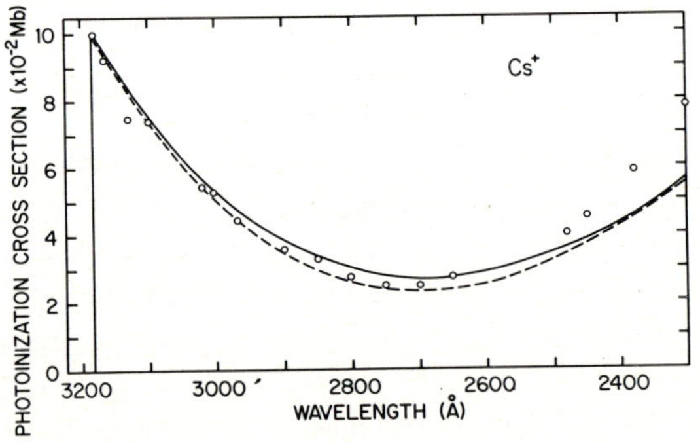

Fig. 3: Photoionization cross section of Cs. ○ present data; —— ref. 7; --- ref. 4.

1. R. B. Cairns and J. A. R. Samson, Phys. Rev. 139, A1403 (1965).
2. F. J. Comes, F. Speier, and A. Elzer, Z. Naturforsch. 23a, 125 (1968).
3. J. L. Kohl, G. P. Lafyatis, H. P. Palenius, and W. H. Parkinson, Phys. Rev. A 18, 571 (1978).
4. J. C. Weisheit, Phys. Rev. A 5, 1621 (1972).
5. F. L. Mohler and C. Boeckner, Bur. Std. J. Res. 3, 303 (1929).
6. G. V. Marr and D. M. Creek, Proc. Roy. Soc. (London) A304, 233 (1968).
7. D. W. Norcross, Phys. Rev. A 7, 606 (1973).

PHOTOIONIZATION OF EXCITED Cs 6p AND Cs 7p SUBSHELLS

Alfred Z. Msezane

Department of Physics, Morehouse College, Atlanta, Georgia 30314 U.S.A.

Photoionization cross sections and asymmetry parameters for excited Cs 6p and Cs 7p subshells have been calculated in Hartree-Fock (HF) approximation with full exchange. Excited states photoionization cross section calculations are generally very sensitive to the initial discrete excited state wave function, particularly in the asymptotic region[1,2]. Consequently, we have used very accurate single-configuration numerical HF wave functions from MCHF[3] for the initial excited discrete atomic states in both Cs 6p and Cs 7p to assess the importance of underlying physical phenomena such as correlation effects which are not incorporated in the present calculation. Continuum HF equations with full exchange were solved. The ionic orbitals were fixed and each one-electron HF orbital for the photoelectron was calculated in the field of the fully relaxed orbitals of the positive ion.[4] The Cs^+ ionic HF wave functions are Clementi-type and were obtained via the Roos et. al.[5] computer code.

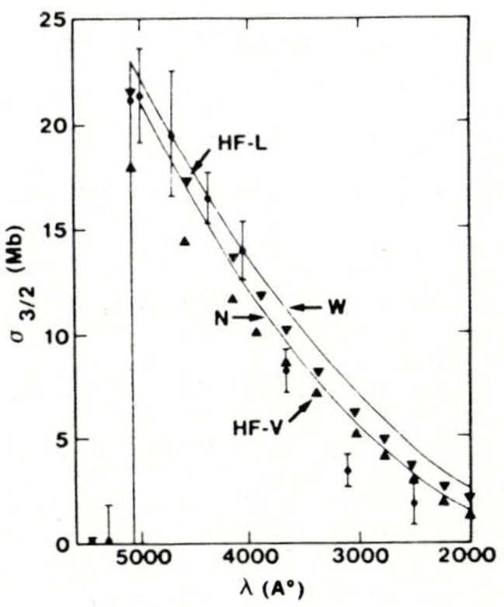

Fig. 1

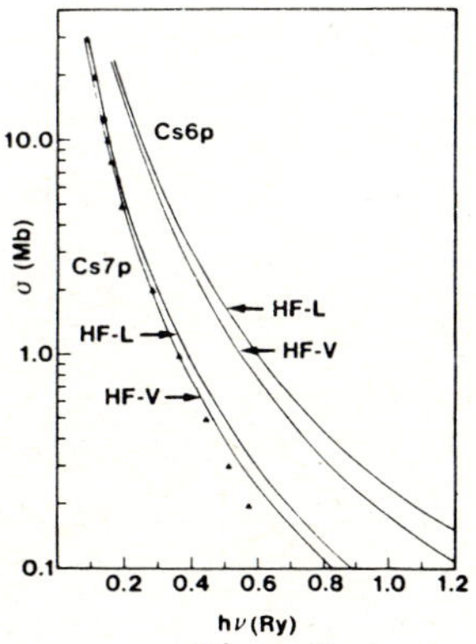

Fig. 2

Fig. 1 gives cross sections for photoionization of Cs 6p versus photon wavelength λ . HF-L and HF-V refer to the present HF length and velocity results, respectively. N and W are the Norcross[6] and Weisheit[7] parametric potential results. The experiment of Nygaard et. al.[8] is shown by ‡. The present HF-L (▼) is in excellent agreement with the Norcross calculation throughout and favor theory over experiment in the region of the lowest wavelength reported.

Fig. 2 shows the calculated photoionization cross sections for Cs 6p and Cs 7p against photon energy rather than photoelectron energy for a meaningful comparison. The triangles (▲) on the Cs 7p curve are the random phase approximation result of Cherepkov[9] and are in very good agreement with ours around threshold. The measurement of Kaminski et. al.[10] is also shown ◎: the agreement with our HF-L is excellent. At higher photon energy, however, correlation effects arising from the $4d^{10}$ and $5p^6$ terms become important and cause deviation between this calculation and Cherepkov's. The asymmetry parameter displays a similar trend.

The low-wavelength deviation between measurement and theory in Cs 6p is probably due to correlation effects arising from the $4d^{10}$ and $5p^6$ terms discussed by Cherepkov. Another experiment may throw light. We conclude that very accurate single-configuration HF wave functions are adequate around threshold for the systems Cs 6p and Cs 7p. Asymmetry parameters and cross sections will be presented and discussed.

* Work supported by National Science Foundation

References

1. M. J. Seaton, Proc. Roy. Soc. A208, 418 (1951).

2. A. Z. Msezane and S. T. Manson, To be published.

3. Froese Fischer C., Comput. Phys. Commun. 7, 236 (1974).

4. S. T. Manson et. al., Phys. Rev. A20, 1005 (1979)

5. B. Roos et. al. A. General Program for Calculation of Atomic SCF Orbitals by the Expansion Method (Tech. Rep. RJ518 IBM Research). (1968)

6. D. W. Norcross, quoted in Ref. 8.

7. J. C. Weisheit, J. Quant. Spectrosc. Radiat. Transfer 12, 1241 (1972).

8. K. J. Nygaard et. al., Phys. Rev. A12, 1440 (1975).

9. N. A. Cherepkov, ABSTRACTS XII ICPEAC 1, 26 (1981).

10. H. Kaminski et. al., Phys. Rev. Lett. 45, 1161 (1980).

ALIGNMENT OF $Cd^+(4d^{-1}\ ^2D_{5/2})$ AFTER PHOTOIONIZATION WITH SYNCHROTRON RADIATION

W. Kronast and W. Mehlhorn

Fakultät für Physik, Universität Freiburg, D-7800 Freiburg, FRG

Photoionization of atoms A

$$\gamma + A \rightarrow A^+(n\ell^{-1}, JM) + e^-(\varepsilon\ell) \quad (1a)$$

$$\hookrightarrow A^+(J'M') + \gamma'' \quad (1b)$$

$$\hookrightarrow A^{++}(J''M'') + e^- \quad (1c)$$

leads in general to an alignment $\mathcal{A}_{20}$ of ions A^+. This alignment can be measured either by the linear polarization of fluorescence radiation γ'' (1b) or by the non-isotropic emission of photons γ'' or of Auger electrons (1c). The measurement of the alignment of photoions gives an additional independent information on the photoionization process besides the subshell cross section $\sigma_{n\ell}$ and the angular parameter $\beta_{n\ell}$. Furthermore, it has been shown[1,2] that the alignment depends only on the squares of the radial dipole matrix elements, $R^2_{\ell\pm1}$.

In the present experiment we measured the alignment of $Cd^+(4d^{-1}\ ^2D_{5/2})$ after photoionization using synchrotron radiation (of DORIS at HASYLAB) via the linear polarization of radiation γ'' of the transition $Cd^+(4d^9 5s^2\ ^2D_{5/2}) \rightarrow Cd^+(4d^{10}5p\ ^2P_{3/2}) + \gamma''$ (4416 Å). The energy range of primary photons γ was from threshold (17.6 eV) to about 40 eV.

The degree of linear polarization $P_{lin,\gamma''}$ of radiation γ'' is given theoretically in the non-relativistic limit and for linearly polarized incident radiation γ by[3]

$$P_{lin,\gamma''} = -\frac{7\,R_p^2 + 3\,R_f^2}{19\,R_p^2 + 26\,R_f^2}. \quad (2)$$

Experimentally, the primary photon beam transverses the Cd-vapour oven and ionized the Cd atoms inside the oven. After passing the oven the degree of polarization p of primary photons is measured by means of a Rabinovitch analyzer[4], we obtained p = 0.92 in the energy range used. The fluorescence radiation γ'', emitted perpendicular to the beam direction of γ, passes through a interference filter and a linear polarizer and is detected by a photomultiplier.

The measured values of $P_{lin,\gamma''}$ were corrected for the depolarization effect[5] the HFS of Cd isotopes with I = 1/2 and for second order contributions in the monochromatized incident radiation (Seya monochromator with holographic grating) and are plotted in Fig. 1. Also shown in Fig. 1 are the results of various theoretical calculations, relativistic random-phase approximation (velocity form) = RRPA-V[6], many-body perturbation theory = MBPT[7] and Dirac-Fock approximation = DF[8]. All theoretical curves have been corrected for a linear polarization degree of p = 0.92 of the incident radiation. We

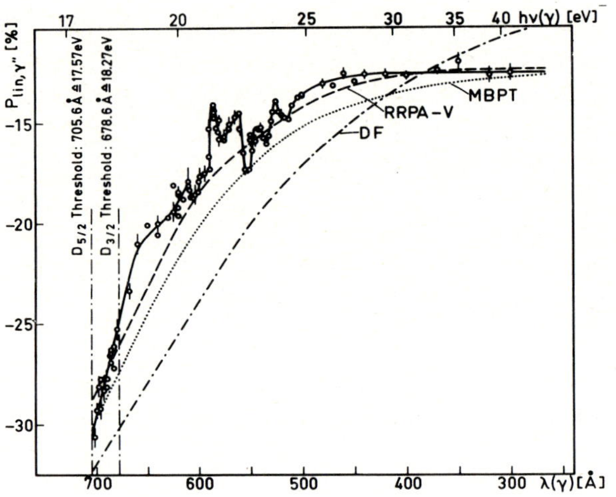

Fig. 1: Comparison of experimental values (ϕ) of $P_{lin,\gamma''}$ (the solid line through ϕ is to guide the eye) with theory (for explanation see text).

have also found resonance structures in $P_{lin,\gamma''}$ (see Fig. 1) which will be discussed more quantitatively at the conference.

The discrepancy between earlier experimental[9] and theoretical values was due to a depolarization of transition γ'' (with long decay time τ = 773 ns) caused by a spurious magnetic field of the oven.

The support by the Bundesministerium für Forschung und Technologie is gratefully acknowledged.

References

1. E.G. Berezhko, N.M. Kabachnik and V.S. Rostovsky, J. Phys. B 11, 1749 (1978).
2. K.-N. Huang, Phys.Rev. A 25, 3438 (1982).
3. H. Klar, J.Phys. B 13, 2037 (1980).
4. K. Rabinovitch, L.R. Canfield and R.P. Madden, Appl. Opt. 4, 1005 (1965).
5. Ch.H. Greene and R.N. Zare, Phys.Rev. A 25, 2031 (1982).
6. W. Johnson and K.-N. Huang, private communication, 1982.
7. S.L. Carter and H.P. Kelly, J.Phys. B 11, 2467 (1978) and private communication.
8. C.E. Theodosiou, A.F. Starace, B.R. Tambe and S.T. Manson, Phys.Rev. A 24, 301 (1981).
9. W. Mauser and W. Mehlhorn, VI. VUV-Conference, (Charlottesville, 1980), Extended Abstracts II-7.

STRONG VARIATIONS OF ANGULAR DISTRIBUTION AND POLARIZATION OF FLUORESCENCE RADIATION IN THE REGION OF AN AUTOIONIZING RESONANCE OF PHOTOABSORPTION

V.V. Balashov, N.M. Kabachnik, V.S. Senashenko

Institute of Nuclear Physics, Moscow State University, Moscow 117234, USSR

An excitation of autoionizing states in photon-atom interaction can lead to strong variations of the polarization and correlation characteristics of photoionization in the narrow spectral interval near the autoionization resonance. This effect is well known for the angular distribution and polarization of photoelectrons[1,2], and recently it was established also for the relative intensities of the photoelectrons (i.e. branching ratios) corresponding to the different states of the ion[3,4]. A similar effect is possible also for the polarization and angular distribution of fluorescence radiation produced in photoionization of inner atomic shells. It is analogous with the strong variations of the fluorescence radiation polarization observed in inelastic electron scattering by atoms when quasistationary states of the negative ion are formed[5,6].

The polarization and angular distribution of the fluorescence radiation are determined by the alignment of the excited state of the ion[7]:

$$\mathcal{A}_{20} = \sum_j \gamma_j |M_j|^2 / \sum_j |M_j|^2$$

where M_j is a matrix element which describes the atomic photoionization process leading to the channel j . The necessary condition of a strong variation of the A_{20} parameter in the region of an autoionizing resonance is the existence of several decay subchannels leading to the same excited state of the ion. The behaviour of the channel matrix elements Mj in such conditions was investigated earlier[1,2,8]. Our analysis and calculations show that the behaviour of the A_{20} parameter depends strongly on the spin-orbit interaction in continuum. In particular, the well-known approximation treating the radial parts of the Mj matrix elements as independent of the total angular momentum of photoelectrons[9], narrows considerably the limits of variation of A_{20}, which is very important when the theory is compared with the experiment.

As a particular application of the discussed theory we have considered the polarization of the fluorescence radiation following the resonance photoionization of helium in the region between n = 2 and n = 3 ionization thresholds and the resonance photoionization of cadmium in the region of 20-25 eV.

References

1. V.V. Balashov, N.M. Kabachnik ,I.P. Sazhina, VIII ICPEAC, Abstr.p.527 (1973), Vestnik MGU 6, 733(1973)
2. N.M. Kabachnik. I.P. Sazhina, J.Phys.B:At.Mol. Phys. 9, 1681(1976)
3. J.A.R. Samson, J.L. Gardner, A.F. Starace, Phys.Rev.A12, 1459 (1975)
4. P.H. Kobrin et al. Phys.Rev. A26, 842(1982)
5. A. Defrance, M. Vacher , C.R. Acad.Sc.Paris, 276B, 917(1973)
6. V.V. Balashov et al. Proc.IV Conf. on VUV (USSR), Abstr. p.68 (1975).
7. E.G. Berezhko, N.M. Kabachnik, J.Phys.B: At. Mol.Phys.10, 2467 (1977)
8. P.B. Ivanov, V.S. Senashenko, Proc. VI Conf. on VUV (USSR), Abstr. p.30 (1982)
9. J. Cooper, R.V. Zare, J.Chem.Phys. 48,942 (1968)

OBSERVATION OF THE 1s2p2p' $^4P^e$ RESONANCE IN He^- PHOTODETACHMENT

J. R. Peterson, Y. K. Bae, and M. J. Coggiola

Molecular Physics Laboratory, SRI International, Menlo Park, CA 94025

In a recently completed survey of the photodetachment spectrum of the metastable He^- $^4P^o$ negative ion, we found evidence of a significant increase in the cross section at or near the energy threshold for leaving the neutral He in the excited 2^3P state.[1] A subsequent photodetachment calculation by Hazi and Reed[2] obtained quite good agreement with our original data in both shape and absolute magnitude, and further predicted a large peak (~ 25 Å^2) located at 1.233 eV, just above the 2^3P threshold. This shape resonance is associated with the (1s2p2p') $^4P^e$ state and corresponds to the first optically allowed transition from the metastable (1s2s2p) $^4P^o$ "ground state" of He^-. We have now extended our photodetachment measurements to include the observation of this feature.

The experimental apparatus is shown schematically in Figure 1. The He^- ions are formed from a 1.3 keV He^+ beam by two-step electron capture in Na or K vapor. In a separately-pumped chamber, the He^- is directed along a 15-cm field-free drift path where it is intersected at 60° by the laser beam. The drift path is terminated by a 2.4 mm dia defining aperture, beyond which the ions are directed into a Faraday cup. Neutral atoms formed along the drift path are then counted synchronously with the mechanically chopped laser. The required photon wavelengths between 930 and 1012 nm were generated using a Coherent 590 cw dye laser pumped by the ir lines of a CR3000K Kr^+ laser. IR-140 dye was used throughout, with the longest wavelengths requiring the use of a parabolic pump mirror and short-focus folding mirror in the dye laser.

Figure 2 shows a linear plot of a portion of the photodetachment data between 1.22 and 1.27 eV. The error bars are probable errors based on statistics from a number of 10-15 minute runs. Absolute cross section values are obtained as in the earlier work[1] by normalizing the photodetachment rate to the constant autodetachment rate. Also shown is the calculated[2] cross section for the $He^- + h\nu \rightarrow He\ (2^3P) + e^-\ (kp)$ channel. There is very good agreement in both location and general shape of the peak, however, we find the maximum cross section to exceed 70 Å^2. The experimentally determined peak is located at 1.2344 ± 0.0004 eV with a FWHM of 7.0 ± 0.4 meV, in excellent agreement with the values of 1.233 eV and 7 meV found theoretically.

This work was supported by the National Science Foundation and the Air Force Office of Scientific Research.

References

1. R. V. Hodges, M. J. Coggiola, and J. R. Peterson, Phys. Rev. A23, 59 (1981).
2. A. V. Hazi and K. Reed, Phys. Rev. A24, 2269 (1981).

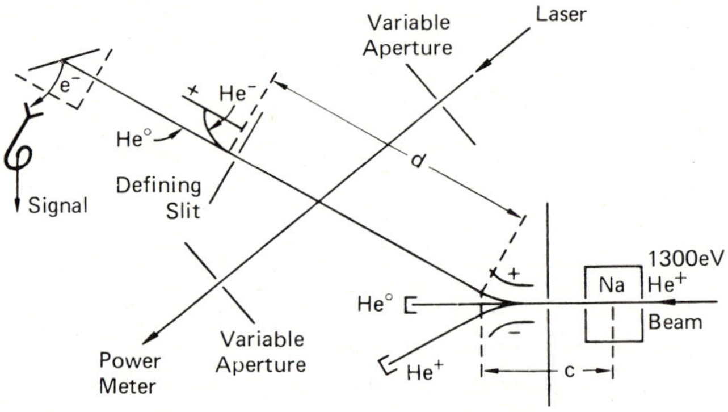

Fig. 1: Schematic diagram of the aparatus.

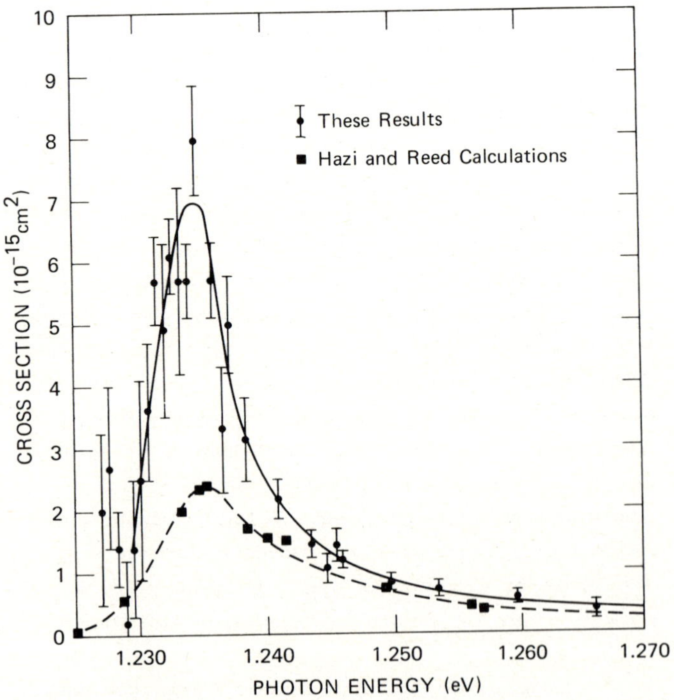

Fig. 2: Photodetachment cross section in the region of the 1s2p2p' shape resonance. Squares are the p-wave cross sections from Ref. 2. Smooth curves are visual fits to the data.

RESONANCE PHOTOABSORPTION BY THE TWO ELECTRON SYSTEMS NEAR THE n = 2 AND n = 3 THRESHOLDS

A. Wague

Université de Dakar, Faculté des Sciences, Departement de Physique, Dakar-Fann (Senegal)

The interaction of electromagnetic radiation with the two-electron systems is of great interest because these systems represent a useful subject for the creation and verification of new experimental methods and theoretical approximations in the investigation of different photo-processes. The use of synchrotron radiation and continuous radiation of high temperature plasma to obtain intensive photon-beams in the region of far U.V., and use of tunable lasers, make possible a large kind of experimental studies (photoabsorption measurement[1,2], photoionisation of excited atoms and ions,[3] photoelectron spectroscopy[4]...) which give a great impulse to the development of theoretical investigations of photoprocesses in two-electron systems.

In the present report a theoretical analysis is made of the resonance photoabsorption by the two-electron systems near the n = 2 and n = 3 thresholds in the frame work of diagonalisation approximation.[5]

1. The theoretical relativistic description of photoabsorption is made in the approximation of intermediate coupling when the photoabsorption by the two-electron systems in the region corresponding to the excitation of autoionising levels converging to the n = 2 threshold of H-like ions, is a multichannel problem: the spin-orbital interaction provides the mixing of the singlet and triplet autoionising levels and the decay of ${}^1P^{(-)}$ level is possible to the continuums $(1skp)\,{}^1P^{(-)}$ and $(1skp)\,{}^3P^{(-)}$. For the number of multicharged He-like ions with charge $Z < 30$, the eigenenergies as well as the total and partial eigenwidths are obtained near the n = 2 threshold of H-like residual ion. Our calculations show the Z dependence of the q index values for the $(23)\,{}^{1,3}P_1{}^{(-)}$ resonances converging to the n = 2 threshold while in earlier calculations[6] in the LS coupling approximation it was shown that this dependence does not take place. For Fe^{+24} ions the comparison of our results with those of the reference[7] shows that for the relativistic calculations of resonance parameters it is enough to take account only of spin own orbit interaction in the full operator of Breit.

2. Energies, total and partial eigenwidths of a few ${}^1P^{(-)}$ and ${}^3P^{(-)}$ autoionising levels converging to the n = 3 threshold of H-like ions are calculated for the He-atom and the Li^{+}, Be^{2+}, B^{3+} and C^{4+} ions. The predominant decay of these autoionising levels into the n = 2 levels of residual ions is shown and that is in good agreement with experiment[8].

3. For the metastable 2^1S and 2^3S states of He, the calculation of resonance photoabsorption is made near the n = 2 and n = 3 threshold. Our calculation of the q indexes values of the ${}^1P^{(-)}$ and ${}^3P^{(-)}$ resonances near the n = 2 threshold shows that $q \gg 1$ for all concerning resonances, which is in good agreement with those obtained in the close coupling approximation[9], and in variational[10] and perturbation[11] calculations. Near the n = 3 threshold it is shown that the q indexes values of the ${}^1P^{(-)}$ and ${}^3P^{(-)}$ resonances vary in the large interval which means a great difference in the form of concerning autoionising resonances.

References

1. R.P.Madden and K.Codling, Astrophys.J.141, 364 (1965)
2. P.Dhez and D.L.Ederer, J.Phys.B 6, L 59 (1973)
3. P. K. Carroll, E. T. Kennedy, Phys. Rev. Lett. 38, 1068 (1977)
4. F.Wuilleumier, Atomic physic with synchrotron radiation, edited by N.Oda and K.Takayanagi (North Holland, Amsterdam 1980)
5. V.V.Balashov, S.G.Grishanova, I.M.Kruglova, and V.S.Senashenko, Optica Spectrosc. 28, 859 (1970)
6. U.I.Safronova, V.S.Senashenko, Optica Spectrosc. 45, 9 (1978)
7. L.A.Vainshtein, U.I.Safranova, Inst.Spektrosk., Akad. Nauk SSSR, Preprint 6, Moscow (1975)
8. P.R.Woodruff, J.A.R.Samson, Phys.Rev. A 25, 848(1982)
9. D.W.Norcross. J.Phys. B 4, 652 (1971)
10. A.Dalgarno, H.Doyle, M.Oppenheimer, Phys.Rev.Lett. 29, 1051 (1972)
11. U.I.Safranova, V.S.Senashenko, S.V.Khristenko, Optica Spectrosc. 45, 833 (1978)

ELECTRON SPECTROSCOPY OF ATOMIC Fe, Co AND Ni EXCITED BY VUV PHOTON IMPACT

E. Schmidt, H. Schröder, B. Sonntag, H. Voss, H.-E. Wetzel

II. Inst. f. Experimentalphysik der Univ. Hamburg, Luruper Chaussee 149, 2000 Hamburg 50, F.R.G.

A high temperature oven was operated between 1450 and 1650°C in order to produce a beam containing about 10^{11} atoms/cm^3 of Fe, Co and Ni. The metal vapour was crossed by monochromatized and focussed synchrotron radiation of the DORIS II storage ring obtained by means of a TGM beam line at HASYLAB.[1] The kinetic energy of the ejected electrons was determined by a cylindrical mirror analyzer with 0.8 % relative energy resolution (FWHM). The geometry of this spectrometer, which accepts only electrons near the magic angle (54.7°) relative to the light's polarization vector renders it insensitive to the anisotropic distribution of photoelectrons, a prerequisite for direct determination of partial cross sections.[2,3]

In fig. 1 we show - on a binding energy scale - the photoelectron spectrum of Fe taken at 56.65 eV photon energy. Fe II term assignments are based on recently compiled tabulations[4], revealing two groups of photolines due to 3d- and 4s-emission, respectively. Lines 1 to 4 belong to the $3d^5 4s^2$ ionic terms ^{4}(F,D,G) and ^{6}S, whereas lines 5,6 are $3d^6 4s$ ^{4,6}D.

The photoemission lines are resonantly enhanced in the region of the asymmetric bands dominating the

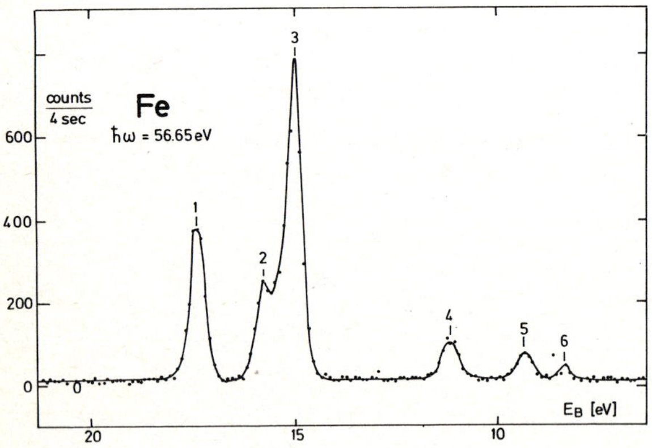

Fig. 1: Resonant photoelectron spectrum of Fe

VUV absorption spectra[5] below the 3p ionization limit. These bands are caused by autoionization due to the coupling of $3p^6 3d^6 4s^2 (^5D) \rightarrow 3p^5 3d^7 4s^2$ excitations and the 3d and 4s valence shell photoemission. In the case of Fe, Co and Ni the 3p-3d oscillator strength is distributed over several multiplet lines. The gross absorption features are in reasonable agreement with model calculations (see fig. 2).[6] Atomic 3p-3d intershell interactions are also at the roots of the resonant photoemission satellites of metallic Fe, Co and Ni.[7]

Fig. 2

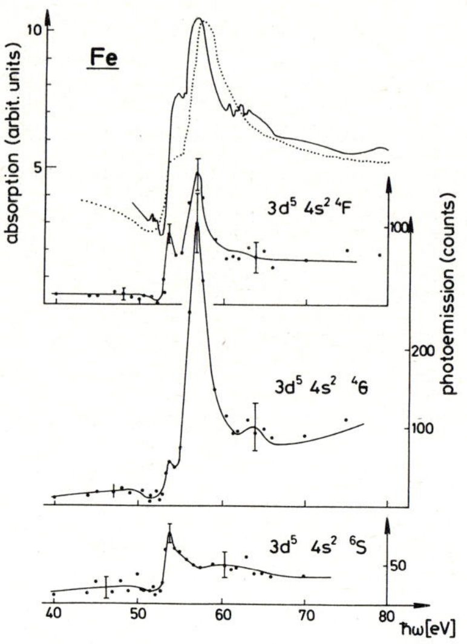

Fig. 2 shows the atomic VUV absorption of Fe (ref. 5 with dotted model curve from ref. 6) and the partial cross sections of 3d-photoelectron lines. Is is evident by the superposed asymmetric profiles that the two main 3p excited states at 53.7 and 56.65 eV are driving the 3d-photoemission via autoionization. However, it is obvious that the coupling differs for the channels in fig. 2. For instance, the $3d^5 4s^2$ ^{6}S state cannot be reached via super Coster Kronig decay of the 56.65 eV excitation because of LS selection rules.

References

1. R. Bruhn, E. Schmidt, H. Schröder, B. Sonntag, A. Thevenon, G. Passereau, J. Flamand, Nucl. Instr. and Meth. 1983 (in print)
2. J.A.R. Samson in Handbuch der Physik Vol. 31, (ed. W. Mehlhorn), Berlin 1982
3. H. Schröder, Thesis, Universität Hamburg 1982
4. C. Corliss, J. Sugar, J. Phys. Chem. Ref. Data 11, 135 (1982)
5. R. Bruhn, B. Sonntag, H.W. Wolff, J. Phys. B12, 203 (1979)
6. L.C. Davis, L.A. Feldkamp, Solid State Comm. 19, 413 (1976)
7. D. Chandesris, J. Lecante, Y. Petroff, Phys. Rev. B (to be published) and references therein

PHOTOIONIZATION OF LASER EXCITED SODIUM AND BARIUM ATOMS

J.M. Bizau,[a] B. Carré,[b] P. Dhez,[a] D. Ederer,[c] P. Gerard,[a] J.C. Keller,[d] P. Koch,[e]
J.L. Legouët,[d] J.L. Picqué,[d] G. Wendin,[f] F. Wuilleumier[a]

LURE, University Paris Sud, B.209c, 91405-Orsay, France

After the first observation of inner-shell photoionization in laser excited atoms by synchrotron radiation in 1981,[1] extensive studies have been carried out in Orsay, using a new ring dye laser and an improved experimental set up. Electrons ejected from sodium and barium vapors under the impact of laser and synchrotron radiations were analyzed, using a cylindrical mirror analyzer. In this second series of experiments, a higher proportion of sodium atoms excited in the $2p^6 3p\ ^2P$ state was routinely available (about 25%, compared to 10% during the first experiments). In Ba, up to 50% of the atoms were brought into the 6s5d ^2D metastable excited state.

In Fig.1, the electron spectrum displays the main features observed in Na: photoionization in the 2p subshell of atoms either in the excited state (photoline at 40 eV binding energy) or in the ground state (photoline at 38 eV), autoionization from doubly excited states of the type $2p^5 3s3p$ produced by absorption of two photons, one from the laser, one from the synchrotron (electron line at 3 eV BE, equivalent to resonant photoemission of the 3p excited electron). Oscillator strengths for excitation of a 2p electron in the laser excited atoms were measured for the first time and will be reported on.

Fig.2 shows a typical photoelectron spectrum ejected from barium atoms either in the ground state (upper part) or in the excited state (lower part). One observes simultaneously photoionization in the 5p, 6s and 5d subshells produced by 1st order photons and in the 4d subshell produced by 3rd order photons diffracted by the monochromator. One should note, in particular, the enhancement of the photoionization cross section for the 5d excited electron compared to the 6s cross section and the splitting of all 5p and 4d electron lines. The variation of 4d and 5d cross sections in the excited atom will be presented at the Conference

1. J.M. Bizau, F. Wuilleumier, P. Dhez, D. Ederer, J.L. Legouët, J.L. Picqué, P. Koch, in " Laser Techniques for Extreme Ultraviolet Spectroscopy", ed. K. Smith, AIP Proceedings n°90, 1982, p.331.

a) Lab. Spectroscopie Atomique et Ionique, Orsay, France
b) Service Physique Atomes et Surfaces, Saclay, France
c) National Bureau of Standards, Washington, USA
d) Lab. Aimé Cotton, Orsay, France
e) SUNY, Stony Brook, USA
f) Chalmers Institute, Göteborg, Sweden

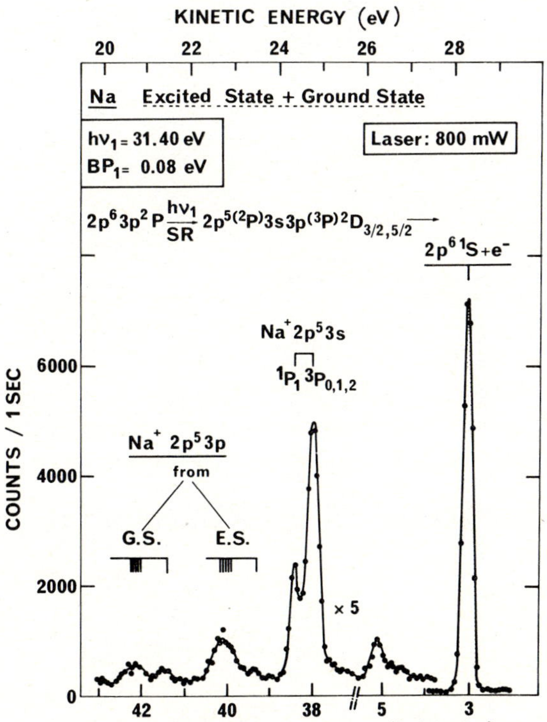

Fig.1.- Electron spectrum of sodium

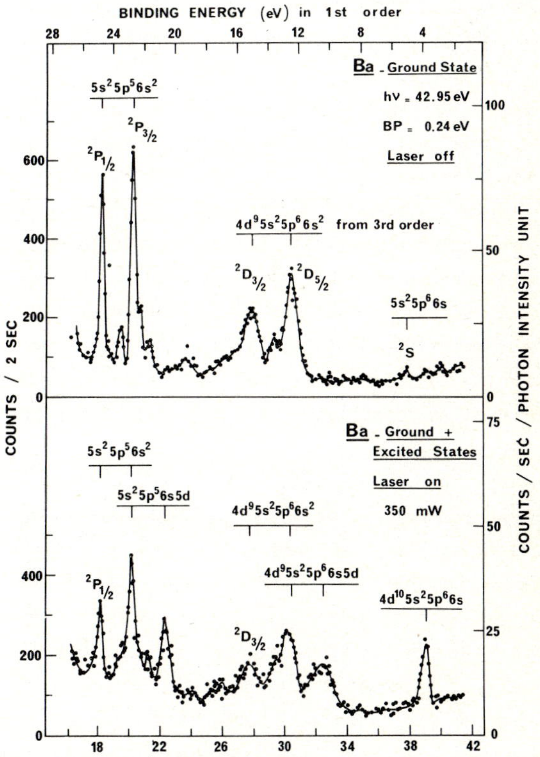

Fig.2.- Photoelectron spectrum of barium

HIGH RESOLUTION STUDY OF THE Ne($2p^5_{1/2}$ ns', nd') AUTOIONIZATION RESONANCES

J. Ganz, A. Siegel, W. Bußert, K. Harth, M.-W. Ruf and H. Hotop

Fachbereich Physik der Universität, 6750 Kaiserslautern, FRG

Using two-photon cw dye laser excitation of a collimated metastable Ne($2p^5$ 3s $^3P_{2,0}$) beam, we have studied the autoionization resonances Ne($2p^5_{1/2}$ ns', nd') with high resolution (≈4 GHz). One of our goals was to reveal the source of discrepancy between the VUV photoionization results of Radler and Berkowitz[1] and the much smaller theoretical values of Johnson and Le Dourneuf[2] for the width of these resonances.

Starting with an intense thermal energy beam of metastable Ne(3s $^3P_{2,0}$) atoms from a differentially-pumped cold cathode dc discharge source, we reach the autoionization states in two steps: first, either one of the metastable components is excited to the intermediate Ne($2p^5$ 3p)-state of interest with a stabilized cw single mode dye laser; in the second step a narrow-band blue cw dye laser (Stilben 1 or 3, power 10-50 mW, bandwidth about 4 GHz) excites atoms from the intermediate state to the ionization continuum. The product ions are detected through a quadrupole mass spectrometer with an electron multiplier and fast counting electronics. This way of detection effectively eliminates background, e.g. that due to Penning ionization of the rest gas. Although the Ne(3p J=1,2) states have short effective residence times prior to their decay to the Ne ground state via the Ne(3s J=1) states, adequate ionization signals were obtained with the rather weak ionizing laser.

As reported previously[3], no autoionization resonances were found when ionizing the Ne(3p $^3D_3$) intermediate state: core-switching processes are very improbable, as corroborated by electron spectrometry[3]. With any of the Ne(3p J=1,2) states, however, sharp autoionization structure was found with widths in qualitative agreement with the theoretical predictions of Johnson and Le Dourneuf[2]. Choosing the Ne(3p $^1D_2$) state as intermediate, we have looked at the Ne($2p^5$ 14s' J=1) and the (unresolved) Ne($2p^5$ 12d' J) resonances in detail. Fitting Beutler-Fano profiles (convoluted with the laser lineshape) to the measured resonances, we have obtained the widths Γ given in Table 1. Our data agree within a factor of two with the theoretical values of Johnson and Le Dourneuf[2]; excellent agreement is observed between our data and the widths deduced by Geiger and Fink[4] in a recent MQDT analysis of existing spectroscopic information. The VUV results of Radler and Berkowitz[1] were obtained with a quoted resolution of 0.0045 nm (400 GHz) around 57.5 nm (ionization threshold of ground state Ne); one is led to conclude that their actual photon bandwidth must have been somewhat larger than quoted.

Detailed spectra and other related results, also of the absolute resonance energies, will be presented at the conference.

This work has been supported by the Bundesministerium für Forschung und Technologie and by the Deutsche Forschungsgemeinschaft (SFB 91). We thank J. Geiger and H. Klar for useful discussions.

References

1. K. Radler and J. Berkowitz, J. Chem. Phys. 70, 216 (1979)
2. W.R. Johnson and M. Le Dourneuf, J. Phys. B13, L13 (1980)
3. J. Ganz, B. Lewandowski, A. Siegel, W. Bußert, H. Waibel, M.-W. Ruf and H. Hotop, J. Phys. B15, L485 (1982)
4. J. Geiger and M. Fink, J. Phys. B... (to be submitted)

Table 1: Resonance width Γ(GHz) of the Beutler-Fano profile for the Ne($2p^5$ 14s' J=1) and the Ne($2p^5$ 12d' J=1) autoionization resonances

Reference	Ne($2p^5$ 14s' J=1)	Ne($2p^5$ 12d' J=1)
1	-	245
2	10.6	3.8
4	5.7	2.5
this work	6.0(5)	2.0(5)‡

‡The measured resonance contains four unresolved components Ne($2p^5$ 12d' J=1,2;2,3); the dominant contribution (about 80%) is due to the J=3 state. The J-dependence of the autoionization width is expected to be small.

MULTISTEP EXCITATION OF AUTOIONIZING STATES OF Ba

T. F. Gallagher, N. H. Tran, P. Pillet,+ and R. Kachru

Molecular Physics Laboratory, SRI International, Menlo Park, CA 94025

Several years ago a multistep laser excitation method for exciting autoionizing states of alkaline earth atoms was introduced by Cooke et al.[1] in which each electron is separately excited. An example of this is shown in Figure 1 which shows the excitation of the autoionizing Ba 6p17d states. The first two lasers excite one of the two valence electrons by the transition 6s6s → 6s6p → 6s17d. The third laser then excites the 6s17d → 6p17d transition. In the third transition we are essentially driving the strong resonance transition of Ba^+ while the outer electron remains a spectator in the 17d state. The excitation to the 6sεf continuum is far weaker and may be neglected. This corresponds to the Fano q parameter → ∞.[2] Thus as the third laser is scanned across the 6s17d-6p17d transition we see a Lorentzian line, the position and width of which are easily determined. The experiment is done using an atomic beam of Ba and we detect the ions resulting from the atoms excited to autoionizing states. Thus the ion signal is proportional to the photoabsorption.

At higher powers for the third laser we observe very asymmetric satellite features corresponding to the excitation to other 6pnd, n≠17 states. In spite of the apparent interference exhibited by these structures they are the result of a single excitation amplitude. Tran et al[3] have shown, using a quantum defect theory approach, that the optical cross section is given by the product of the overlap integral between the initial 17d state and the final 6pnd channel and the spectral density of the 6pnd autoionizing channel. The effective quantum number of the 6pnd autoionizing channel is a continuous variable.

The overlap integrals produce zeroes when the effective quantum number of the autoionizing channel differs from that of the initial bound state by an integer. In practice this usually occurs near the peak in the spectral density of the autoionizing state leading to very asymmetric structures in the spectrum. The observed spectrum is slightly different from the cross section due to an easily accounted for saturation effect described by Cooke et al.[4]

These high power studies have led to several interesting observations and applications. First, just above the $6p_{1/2}$ limit, at the location of the $6p_{3/2}nd$ states we see almost entirely $6p_{1/2}$ εd excitation, corresponding to q=0,[2] yet the photoabsorption cross section increases. In addition perturbations of the autoionizing series stand out clearly because the maxima in the spectral densities move relative to the overlap integrals. Finally it is straightforward to see very clearly the mixing of the two $6p_{3/2}ns$ and $6p_{3/2}nd$ channels using this technique.

This work is supported by the National Science Foundation under grant PHY 800-70041.

+Permanent address Laboratoire Aime Cotton Orsay, France

References
1. W. E. Cooke, T. F. Gallagher, S. A. Edelstein, and R. M. Hill, Phys. Rev. Lett 40, 178 (1978).
2. U. Fano, Phys. Rev. 124, 1866 (1961).
3. N. H. Tran, R. Kachru, and T. F. Gallagher, Phys. Rev. A 26, 3016 (1982).
4. W. E. Cooke, S. A. Bhatti, and C. L. Cromer, Opt. Lett. 7, 69 (1982).

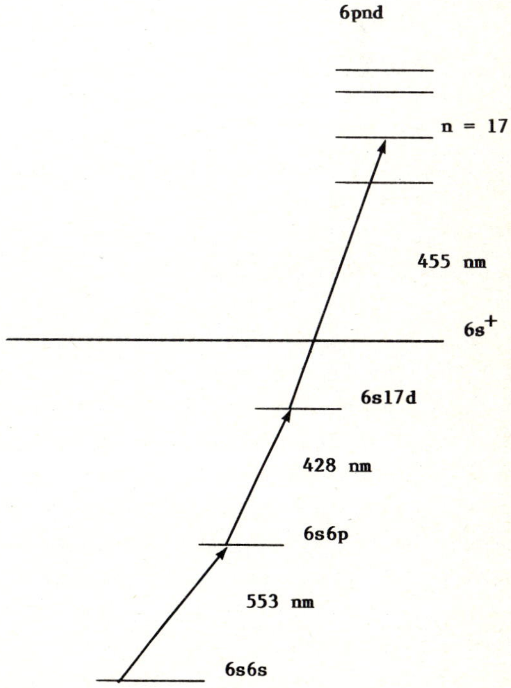

Fig. 1: Laser excitation of the Ba 6pnd states through the 6s17d state. The three lasers are shown by the arrows.

INVESTIGATION OF AUTOIONISING LEVELS IN GA I, JN I AND TL I

M. Müller, M. Schmidt, P. Zimmermann

Institut für Strahlungs- und Kernphysik, Technische Universität Berlin

The atomic energy levels of the third group elements are characterized by one-electron configurations $ns^2n'l$ which are formed by the excitation of the p-electron in the ground state configuration ns^2np. This simple structure, however, is perturbed by complex configurations like $nsnp^2$ caused by the excitation of one of the 'inner' s-electrons. This interconfiguration mixing has a strong influence on atomic parameters like fine structure intervals or hyperfine constants[1]. The identification of the perturbing levels above the ionisation limit is difficult as the perturbation contains many levels of one Rydberg series with the result that the complex states like $3s3p^2\ ^2D$ in Al I can loose their identity in some theoretical calculations[2]. The interconfiguration mixing is not restricted to bound states but also applies to the continuum states and causes autoionisation of the complex states.

We have investigated the autoionising levels ^2S, ^2P and ^2D of the $nsnp^2$ configuration in Ga I (n=4), In I (n=5) and Tl I (n=6). The synchrotron radiation of BESSY was used for the photoionisation of an atomic beam. Fig. 1 gives the example of the ion counting rate between 150 and 220 nm for In I showing the sharp signals of the ^2S and ^2D excitation and the broad resonances of the strong autoionising ^2D-levels.

Under the assumption that the main contribution of the interconfiguration interaction is due to the electrostatic interaction e^2/r_{ik}, in the limit of pure LS-coupling only ^2S and ^2P states of the complex configuration $nsnp^2$ can mix with the continuum states $ns^2\varepsilon s\,^2S$ resp. $ns^2\varepsilon d\,^2D$. Autoionisation of the complex ^2P levels can then be explained by the ^2S resp. ^2D composition of their wavefunctions which is caused by the spin-orbit interaction of the p-electrons.

If one writes the wavefunctions of the complex states in the following way, where the primed states are the real states,

$$|^2S'_{1/2}\rangle = a\ |^2S_{1/2}\rangle + b\ |^2P_{1/2}\rangle$$
$$|^2P'_{1/2}\rangle = c\ |^2S_{1/2}\rangle + d\ |^2P_{1/2}\rangle$$
$$|^2P_{3/2}\rangle = e\ |^2P_{3/2}\rangle + f\ |^2D_{3/2}\rangle$$

the mixing amplitudes should reflect the different degree of the interaction with the continuum states and can be deduced from the measured width of the photoionisation signals. The values are compared with theoretical calculations[3].

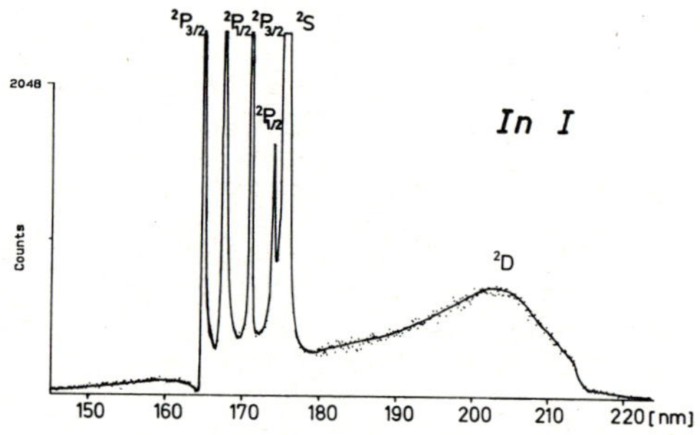

Fig. 1: Photoionisation of In I between 150 and 220 nm. The resonances are due to autoionising levels of the $5s5p^2$ configuration

References

1. B.Falkenburg and P.Zimmermann, Z.Naturforsch.34a, 1249(1979)
2. A.W.Weiss, Phys.Rev.A9, 1524 (1974)
3. J.P.Connerade and M.A.Baig,J.Phys.B.: At. Mol.Phys.14, 29 (1981)

RESONANT PHOTOEMISSION STUDY OF THE XENON - 4d - EXCITATIONS

U.Becker

Institut für Strahlungs- und Kernphysik, Technische Universität Berlin, Berlin,FRG

E.Schmidt, B. Sonntag, H.E.Wetzel and A. Winogradow*

II. Institut für Experimentalphysik, Universität Hamburg and DESY, Hamburg,FRG

Inner shell excitation studies like Xe: 4d--np by photoelectron spectroscopy provide detailed information about the subsequent decay processes. The creation of the Xe 4d-hole opens new decay channels in addition to the direct photoionization. Complex interactions among the various decay channels and their associated continua are involved. A recent study of the Xe 4d-excitations has shown that resonant Auger transitions are the dominant decay mode of the 4d-excitations.[1] Coupling of the 4d-hole state to the direct 5s and 5p photoionization continua by autoionization plays a minor role.

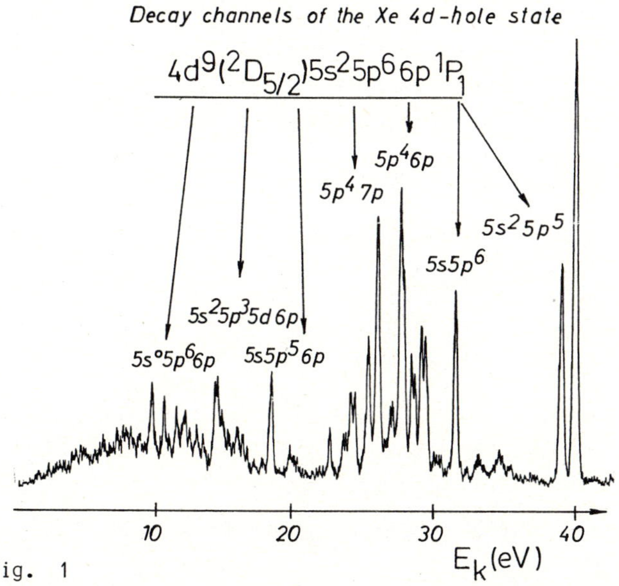

Fig. 1 shows the photoelectron spectrum at the position of the first resonance: 4d--6p at 65.1 eV. It shows numerous resonantly produced Auger lines[2]. At lower kinetic energies appears a broad maximum of electrons indicating resonant shake off processes. These resonantly dominating transitions are negligible off resonance, accounting for the nearly Lorentzian profile of the absorption lines. The autoionization in the 5s and 5p-channels is characterized by the coupling of the 4d-hole states to the corresponding continua as well as by the coupling among the various decay channels[3]. The partial photoionization cross sections and branching ratios show the strength of interchannel coupling in their resonance profiles. In case of Lorentzian absorption lines corresponding to a large Fano parameter q this effect can be studied almost independently of this profile parameter. The Xe 4d-excitations are therefore especially suited for an IC-sensitive resonance photoemission study.

Photoelectron spectra of atomic Xe have been recorded for excitation energies between 63 and 68 eV using synchrotron radiation of the storage ring DORIS. The bandpass of the toroidal monochromator was 0.16 eV, the resolution of the cylindrical mirror analyzer was about 0.74% of the kinetic energy. This resolution was sufficient to resolve all prominent features of the $5p_{3/2}:5p_{1/2}$ branching ratio.

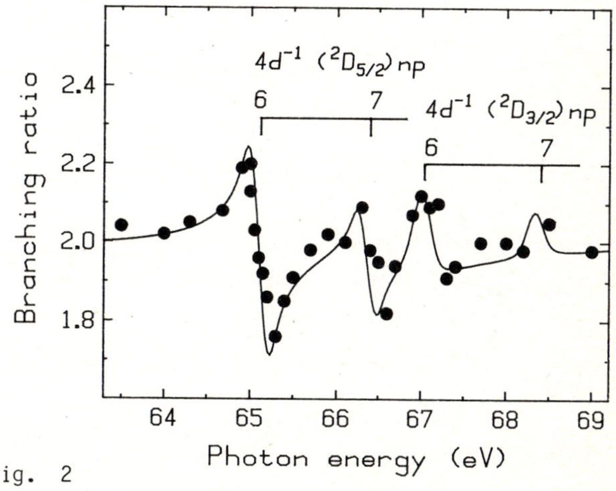

Fig. 2 shows the experimental branching ratio data together with a curve fitted to the theoretical expression for the partial photoelectron cross sections[3]. Additional off resonant data have been taken at higher energies to prove theoretical predictions of RRPA calculations considering intrashell and intershell correlations between the 5p,5s and 4d shells.

References:

1. S.Southworth, U.Becker, C.M.Truesdale, P.H.Kobrin, D.W.Lindle, S.Owaki and D.A.Shirley, submitted to Phys.Rev.A (1982)
2. W.Eberhardt, G.Kalkoffen and C.Kunz,Phys.Rev.Lett.41, 156 (1978)
3. F.Combet Farnoux, Phys.Rev.A25, 287 (1982)

*permanent address: Institute of Physics, University of Leningrad, USSR

THEORETICAL STUDY OF THE SHAKE UP IN THE CORE PHOTOELECTRON SPECTRA OF ATOMS

G.De Alti, P.Decleva and A.Lisini

Istituto di Chimica, Università di Trieste
P.le Europa 1, I-34127 Trieste, Italy.

Shake up processes in the core photoionization have been studied mostly in the noble gases. In these systems theoretical studies have shown the limits of the one particle approximation for the calculation of the intensities. Recently experimental results on alkali and alkaline earth atoms[1,2] have appeared. We have studied the adequacy of the HF approach in these atoms both for the energies and the intensities. Relativistic and non relativistic calculations have been performed using a numerical approach[3,4], while the intensities have been computed with the sudden approximation. The results obtained for the alkali atoms are reported in the Table.

The intensities show a general good agreement with the experiments, being possibly a little overestimated as is apparent in the case of Na, where the experimental value is more accurate and a previous CI calculation[5] shows a decrease of the intensity with respect to the HF result. Anyway the quality of the agreement seems fairly constant along the series. As concerns the energies, the agreement deteriorates in going to the heaviest atoms, even when relativistic effects are taken into account, showing an increased importance of correlation effects. An analogous trend has been detected also for the ns ionization potential and the first electronic transitions in the neutral systems. An anomalous behaviour is shown by Li^+, where for all ^1S states the inclusion of the lower lying $1s^2$ configuration is needed in order to obtain the same agreement.

TABLE

Experimental and theoretical energies and intensities for the ns→(n+1)s shake up[a]

atoms	E exp	I exp[b]	E_{HF}[c]	I_{HF}[c]
Li (1s)	8.35 (^1S)		9.59	32.97
	9.76 (^3S)		9.73	30.94
Na (1s)	8.4	19	8.11	21.98
		(21)	8.21	23.32
K (2p)	6.3	(19)	5.87	19.46
			5.94	20.63
Rb (3p)	6.0		5.27	17.87
			5.32	18.90
			5.41[d]	
Cs (3d)	5.2 ($3d_{3/2}$)	(18)	4.55	18.32
	5.4 ($3d_{5/2}$)		4.56	18.60
			4.75[d]	

[a] Energies in EV and intensities in percentage with respect to the relative primary peak.
[b] Our estimates from the experiment in parentheses.[c] Singlet and triplet values.
[d] DF result.

References

1.M.S.Banna, B.Wallbank, D.C.Frost, C.A. Mc Dowell and J.S.H.Q.Perera, J.Chem.Phys. 68, 5459 (1978).

2.J.S.H.Q.Perera, D.C.Frost, C.A. Mc Dowell, C.S.Ewig, R.J.Key and M.S.Banna, J.Chem. Phys. 77, 3308 (1982).

3. C.Froese Fischer, Computer Phys.Comm. 14, 145 (1978).

4.I.P.Grant, B.I. Mc Kenzie, P.H.Norrington, D.F.Mayers and N.C.Pyper, Computer Phys. Comm. 21, 207 (1980); B.J. Mc Kenzie, I.P.Grant and P.H. Norrington, Computer Phys. Comm. 21, 233 (1980).

5.N.Kosugi and H.Kuroda, Chem.Phys.Lett. 87, 365 (1982).

PHOTOIONIZATION OF ALIGNED 6snd $^{1,3}D_2$ RYDBERG STATES IN ATOMIC BARIUM: THE ROLE OF THE 5d7d PERTURBING STATE

S. J. Smith,* E. Matthias,† P. Zoller,‡ D. S. Elliott, and N. D. Piltch

Joint Institute for Laboratory Astrophysics, University of Colorado and National Bureau of Standards, Boulder, Colorado 80309 U.S.A.

and

G. Leuchs

Sektion Physik der Universität München, D8046 Garching, Federal Republic of Germany

We have measured photoionization from the 6snd $^{1,3}D_2$ states of atomic barium, resonantly excited and aligned by linearly polarized laser radiation, in the range $19 \leq n \leq 30$, using 1.06 μm radiation. Photoionization of this Rydberg series is strongly influenced by configuration mixing due to the presence of the doubly excited 5d7d state which lies between n = 26 and n = 27 of the Rydberg series. This influence is apparent in the n-dependences of the cross sections for total photoion production, in cross sections for production of photoelectrons to the Ba^+ 6s and 5d continua (distinguished by photoelectron time-of-flight), and also in the measured angular distributions of photoelectrons to each of these continua. Since the 6s and 5d continua are structureless in this regime, these effects can be analyzed in terms of configuration mixing and the associated singlet-triplet mixing in the bound Rydberg state.

We photoionized with pulsed YAG-laser radiation (duration ~7 ns) intersecting a barium atomic beam at right angles. Photoelectrons ejected into a fixed solid angle, along an axis perpendicular to the laser-atom-beam plane, were detected with a gated high-gain electron multiplier. The aligned Rydberg states were produced by two pulsed dye lasers, tuned to the $6s^2\ ^1S_0 \rightarrow 6s5p\ ^1P_1 \rightarrow 6snd$ sequence. These two lasers were pumped synchronously by second and third harmonics of the YAG laser. All three lasers were incident from the same direction and were linearly polarized along the same axis, which could be rotated synchronously with λ/2 plates in order to measure photoelectron angular distributions with respect to the fixed detector. With this type of excitation no azimuthal dependence occurs.

Figure 1 shows measured angular distributions obtained by photoionizing the 6s19d $^{1,3}D_2$ states, essentially unaffected by the presence of the 5d7d perturber, and by photoionizing the strongly perturbed 6s26d $^{1,3}D_2$ states. Such observed angular distributions, as well as the measured photoion and photoelectron spectra are interpreted on the basis of a three-channel quantum defect theory in good agreement with the results of previous multichannel quantum defect calculations for barium.[1] The angular distribution proves to be an extremely sensitive tool for probing state mixing, and to our knowledge has not been used previously for this purpose in a resonant multiphoton ionization measurement.

The work was supported in part by a National Science Foundation grant (PHY82-00805). The participation of one of us (G.L.) was made possible by a travel grant from the Deutsche Forschungsgemeinschaft.

[1]M. Aymar, P. Camus, M. Dienlin and C. Morillon, Phys. Rev. A 18, 2173 (1978); H. Rinneberg and J. Neukammer, Phys. Rev. A, in press.

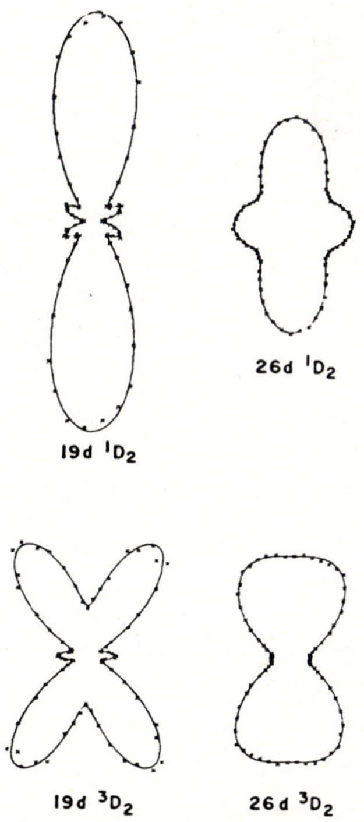

Fig. 1. Photoelectron angular distributions from perturbed (6s26d $^1D_2$ and $^3D_2$) and unperturbed (6s19d $^1D_2$ and $^3D_2$) barium Rydberg states, obtained from multiphoton ionization with parallel linearly polarized lasers.

*Staff Member, Quantum Physics Division, National Bureau of Standards.
†JILA Visiting Fellow 1982-83, permanent address: Freie Universität Berlin, 1000 Berlin 33, West Germany.
‡JILA Visiting Fellow 1982-83, permanent address: Institute for Theoretical Physics, University of Innsbruck, 6020 Innsbruck, Austria.

APPLICATION OF THE REACTION MATRIX METHOD TO MOLECULAR PHOTOIONIZATION TREATMENT OF FINAL-STATE CORRELATION IN $3\sigma_g$ and $2\sigma_u$ PHOTOIONIZATION OF N_2^*

Jeffrey Stephens and Dan Dill

Department of Chemistry, Boston University, Boston, MA 02215 U.S.A.

The reaction matrix method[1,2] has been used successfully in treating electron correlation in a number of atomic photoionization problems, in which initial calculations using an independent particle model resulted in substantial disagreement with experimental cross sections, especially in the threshold region and near continuum. In this work we describe the application of the reaction matrix method to molecular photoionization, using the complete set of one-electron wavefunctions provided by the multiple scattering model[3] (MSM) to construct the determinantal basis set of the final state.

As a prototype study, the method is applied to photoionization of the $3\sigma_g$ and $2\sigma_u$ levels of N_2, anticipating two types of correlation phenomena which are known to be important in atomic photoionization.[1,2] First, fixed-nuclei MSM calculations of $3\sigma_g$ photoionization predict an $\varepsilon\sigma_u$ shape resonance too narrow in comparison with experiment.[4] While averaging over the ground-state nuclear motion accounts in part for the disagreement,[4] the remaining discrepancy can in large part be removed by inclusion of intrachannel correlation in the final state configurations $2\sigma_u^2 1\pi_u^4 3\sigma_g \varepsilon\sigma_u (^1\Sigma_u^+)$ and $2\sigma_u^2 1\pi_u^4 3\sigma_g \varepsilon\pi_u (^1\Pi_u)$. This correlation spreads the continuum oscillator strength relative to the single-particle calculation. Second, while calculations for the $2\sigma_u$ level by both the MSM[4] and the Schwinger variational method[5] give fairly good agreement with measured cross sections, these calculations markedly disagree with angular distribution measurement from threshold (18.8 eV) to greater than 40 eV photon energy. Except within a few eV above threshold, the measured asymmetry parameter is much lower than calculated values over this whole energy range. This is probably not an intrachannel correlation effect, since such correlation is included automatically in the calculations of Ref. 5 which use a Hartree-Fock final state wavefunction.[1,2] Rather, and since the total cross section for $2\sigma_u$ photoionization is weak (< 3 Mb everywhere), it is likely that the variation of the angular distribution reflects interchannel interaction between the weak $2\sigma_u 1\pi_u^4 3\sigma_g^2 \varepsilon\pi_g (^1\Pi_u)$ and $2\sigma_u 1\pi_u^4 3\sigma_g^2 \varepsilon\sigma_g (^1\Sigma_u^+)$ channels and the strong[4,5] $2\sigma_u^2 1\pi_u^3 3\sigma_g^2 \varepsilon\delta_g (^1\Pi_u)$ and $2\sigma_u^2 1\pi_u^4 3\sigma_g \varepsilon\sigma_u (^1\Sigma_u^+)$ channels. The $\varepsilon\delta_g$ channel provides the strong, non-resonant background in $1\pi_u$ photoionization,[4,5] and its interaction with the $\varepsilon\pi_g$ channel could account for the overall depression seen in the measured asymmetry parameter. The $\varepsilon\sigma_u$ channel contains the f-wave shape resonance[4] peaked at 30 eV photon energy, and its interaction with the $\varepsilon\sigma_g$ channel could account for the modulation seen experimentally near 30 eV.

Calculations are in progress to implement these proposed intra- and interchannel interaction mechanisms.

Acknowledgement

*This work is supported by the National Science Foundation under Grant CHE-8203267.

References

1. U. Fano and J.W. Cooper, Rev. Mod. Phys. 40 , 441 (1968), Sec. 4-8 and references therein.

2. A.F. Starace, "Theory of Atomic Photoionization," in Handbuch der Physik, Vol. 31, edited by W. Mehlhorn, (Springer, Berlin, 1980): A.F. Starace, Applied Optics, 19, 4051 (1980).

3. J.L. Dehmer and D. Dill in Electron-Molecule and Photon Molecule Collisions, edited by T. Rescigno, V. McKoy, and B. Schneider (Plenum, New York, 1979), p. 225: K.H. Johnson, in Advances in Quantum Chemistry, edited by P.O. Löwdin (Academic, New York, 1973), Vol. 7, p. 143.

4. J.L. Dehmer, D. Dill, and A.C. Parr, in Photophysics and Photochemistry in the Vacuum Ultraviolet, edited by S. McGlynn, G. Findley, and R. Huebner (D. Reidel Publishing Company, Dordrecht, Holland, 1983) and references therein, in press; R.S. Wallace, Ph.D. thesis, Boston University, 1980.

5. R.R. Lucchese, G. Raseev, and B.V. McKoy, Phys. Rev. A 25, 2572 (1982).

AB-INITIO APPROACH TO THE MULTICHANNEL QUANTUM DEFECT CALCULATION OF THE ELECTRONIC AUTOIONIZATION IN THE HOPFIELD SERIES OF N_2

M. RAOULT, H. LE ROUZO, G. RASEEV and H. LEFEBVRE-BRION
Laboratoire de Photophysique Moléculaire*, Bât. 213,
Université de Paris-Sud, 91405 - ORSAY Cedex (France).

We have chosen the Hopfield (1) series in N_2 for an ab-initio study of the electronic autoionization process, because these series are among the most regular autoionized molecular Rydberg series known and much recent experimental information is available concerning them. Total and partial (electronic and vibrational) photoionization cross section measurements of the autoionization profiles have been carried out (2,3,4,5,6) and the angular distribution of the ejected photoelectron has been determined (7).

The Hopfield series correspond to Rydberg states converging to the $B^2\Sigma_u^+$ state of N_2^+. Three Rydberg series may be expected in photoabsorption : two series of Σ symmetry, namely $B^2\Sigma_u^+$ nsσg and $B^2\Sigma_u^+$ ndσg, and one series of Π symmetry $B^2\Sigma_u^+$ ndπg. More than fifty years ago, Hopfield observed two of these series in the photoabsorption spectrum and he found that they have very different behaviour. One series presents strong absorption lines, the other series presents apparent "emission" lines, which correspond for the photoionization spectrum to the peaks and the windows respectively. Several assignments have been made, but none is unequivocally established.

The aim of this work is to clarify these attributions, to explain such a large difference in the behaviour of the series. In order to do this, we must take into account the electrostatic interactions of these Rydberg series with the adjacent ionization continua. For the Σ symmetry three continua must be considered : $X^2\Sigma_g^+$ $\varepsilon p\sigma u$, $X^2\Sigma_g^+$ $\varepsilon f\sigma u$, $A^2\Pi_u$ $\varepsilon d\pi g$, and five for the Π symmetry : $X^2\Sigma_g^+$ $\varepsilon p\pi u$, $X^2\Sigma_g^+$ $\varepsilon f\pi u$, $A^2\Pi_u$ $\varepsilon s\sigma g$, $A^2\Pi_u$ $\varepsilon d\sigma g$, $A^2\Pi_u$ $\varepsilon d\delta g$. In the Multichannel Quantum Defect Theory (MQDT), the resolution of this problem needs the knowledge of 47 electronic quantities which are calculated by ab initio methods. The energy levels of the three Rydberg series involved, and their transitions moments from the ground state of N_2 have been calculated by the ion virtual orbital method using the Alchemy program. The transition moments from the ground state of N_2 to the ionization continua of the two $X^2\Sigma_g^+$ and $A^2\Pi_u$ states of N_2 have been obtained by using the one center static exchange method (8). The electrostatic interactions between these discrete states and all the continua have been also calculated (9). All these quantities have been introduced in the MQDT treatment of electronic autoionization (10).

The calculated total cross-section is in a satisfactory agreement with the photoionization data (see fig.1a)

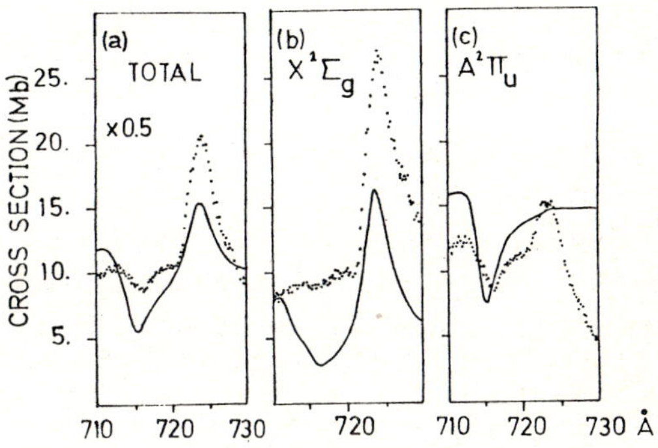

Fig. 1

• Experimental points from reference 6

▬ This work.

the partial cross section are in a qualitative agreement with the experimental results obtained by photoelectron spectroscopy (see Fig. 1b and 1c). The results clearly show that the absorption lines correspond to the ndσg series and the apparent emission lines to the ndπg series.

References

1. J.J. Hopfield, Phys. Rev. 35, 1133 (1930) and Phys. Rev. 36, 789 (1930)
2. P. Gurtler, V. Saile and E.E. Kock, Chem. Phys. Lett. 48, 245 (1977)
3. P.M. Dehmer and W.A. Chupka, Argonne National Laboratory Report ANL-77-65 (1977)
4. E.W. Plummer, T. Gustafsson, W. Gudat and D.E. Eastman, Phys. Rev. A, 15, 2339 (1977)
5. P.R. Woodruff and G.V. Marr, Proc. Roy. Soc. London A.358, 87 (1977)
6. P. Morin, I. Nenner, N.Y. Adam, J. Delwiche and M.J. Hubin-Franskin, to be published
7. J.B. West, K. Codling, A.C. Parr, D.L. Ederer, B.E. Cole, R. Stockbauer, and J.L. Dehmer, J. Phys. B 14, 1791 (1981)
8. G. Raseev, H. Le Rouzo and H. Lefebvre-Brion, J.Chem. Phys. 72, 5701 (1980)
9. H. Le Rouzo, to be published
10. A. Giusti-Suzor and H. Lefebvre-Brion, Chem. Phys. Letters, 76, 132 (1980).

* Laboratoire associé à l'Université Paris-Sud.

A CONTINUUM SPECTRUM MULTICHANNEL FINITE VOLUME VARIATIONAL METHOD

G. Raşeev[+*] and H. Le Rouzo[*]

[+] Laboratoire de Chimie Quantique, Bât. B.6, Université de Liège, B-4000 Sart-Tilman par Liège 1, Belgium

and

[*] Laboratoire de Photophysique Moléculaire du C.N.R.S., Bât. 213, 91405 Orsay, France

A Finite Volume Variational Method (FVVM) is presented for the calculation of molecular continuum wavefunctions[(1)]. It is the straighforward generalization of the Kohn's FVVM one of the two variational principles appeared in his famous paper[(2)]. A similar approach is developed by C.H. Green[(3)]. As in R matrix method the coordinate space is divided into two regions. An inner region where the N electrons are indistinguishable and an outer region where the N-1 electrons can be distinguished from the Nth excited one. The separate N-1 electron and Nth electron wavefunctions can easily be obtained by structure and collision calculations. The inner region FVVM N electron wavefunction is expended on appropriately chosen basis set. Then symmetrization of the kinetic energy operator, integration over the basis functions and application of the variational principle lead to the following generalized eigenvalue equation :

$$(\underset{\sim}{T} + \underset{\sim}{U} + \underset{\sim}{k}^2 \quad) \underset{\sim}{C} = \underset{\sim}{S} \underset{\sim}{C} \underset{\sim}{b} \qquad (1)$$

where $\underset{\sim}{C}$ is the matrix of the unknown expression coefficients

$T_{ij} = \langle \vec{\nabla}\psi_i^* | \vec{\nabla}\psi_j \rangle_v$ is an element of the gradient matrix

$\underset{\sim}{U}$ and $\underset{\sim}{A}$ are potential energy and overlap matrices

k^2 is the continuum energy of the Nth excited electron

$S_{ij} = \langle \psi_i | \psi_j \rangle_s$ is an element of the surface matrix

$\underset{\sim}{\ell}$ is a diagonal matrix representing the unknown logarithmic derivative at the surface

and all the volume integrations are performed in the internal region. Solving the equation (1) we obtain the expansion coefficients $\underset{\sim}{C}$ and logarithmic derivative $\underset{\sim}{\ell}$. Therefore the solution in the internal region is completly determined and by continuity at the surface we can extract the corresponding collision matrices which determines the solution all over the coordinate space.

Comparison between the present method and the precedingly developed Electronic Iterative Variational Eigenchannel Approach (EIVEA) in the framework of Electronic Ab-initio Quantum Defect Theory (EAQDT)[(4)] reveals that the present method keeps the stability properties of EIVEA but is much more efficient being interation free. Moreover the present method, used with energy independent basis set is as efficient as the Bloch operator R matrix allowing calculation at several energis with the same basis but free of any arbitrary parameters or inversion of nearly singular matrix.

Application of FVVM to a one electron molecular system photoionization (H_2^+, $1\sigma_g \to \varepsilon\sigma_u$)[(1)] gives very encouraging results.

Extension of FVVM to full N electron multichannel case in the framework of EAQDT is now in progress[(5)]. It will first be applied to the H_2 photoionization in the region 30-35 eV where seemingly a shape resonance has been seen in the kinetic energy of fragments spectrum performed with (e, 2e) technique[(6)]. Another application concerns the electronic autoionization near 27 eV where anomalous behaviour of the asymmetry parameter has been found experimentaly.

References

(1) H. Le Rouzo and G. Raşeev - submitted to Phys. Rev. A.

(2) W. Kohn - Phys. Rev. 74, 1763 (1948)

(3) C.H. Green - submitted to Phys. Rev. A.

(4) G. Raşeev and H. Le Rouzo - Phys. Rev. A27, 268 (1983).

(5) G. Raşeev, in preparation.

(6) B. Van Wingerden, Ph.E. Van der Leeuw, F.J. de Heer, M.J. Van der Wiel, J. Phys. B12, 1559 (1979).

ON THE INFLUENCE OF ELECTRON-ELECTRON CORRELATIONS IN H_2 MOLECULE ON THE CHARACTERISTICS OF THE (γ,2e) PROCESS

V.G. Levin, V.G. Neudatchin, A.V. Pavlitchenkov, Yu. F. Smirnov

Institute of Nuclear Physics, Moscow State University, Moscow 117234, USSR

In Ref.[1] we suggested to use the (γ,2e) and (e,3e) processes as a generalization of well-known (e,2e) method. The investigation of the double ionization processes under the conditions of measurements of full kinematics (double and triple coincidences) allows one to obtain the two-electron Fourier-amplitudes of wave functions $M(\vec{k}_1, \vec{k}_2)$ in contrast to the one-electron momentum distribution obtained by the (e,2e) method. These two-electron amplitudes directly reflect the electron correlations in the atom or molecule under consideration. Because, for example, the photon absorption operator in the (γ,2e) process is a single-particle one, it is impossible to describe the two electron emission disregarding the electron-electron interaction. Therefore, the amplitude of the process is completely associated with the ee-correlation effects. Earlier we analyzed the case of He atom. Here, we study the double photoionization process (γ,2e) for the simplest molecular system H_2 at the photon energies $E_\gamma \sim 1$ keV and the energies of final electrons detected in coincidence $E_1 \sim E_2 \sim E_\gamma/2$. In dipole approximation, the amplitude of the process is

$$M(\vec{k}_1, \vec{k}_2) = \langle \Psi_{\vec{k}_1\vec{k}_2} | \sum_{j=1,2} \hat{u}\hat{\nabla}_j | \Psi_0 \rangle$$

where the final electrons in the state $\Psi_{\vec{k}_1\vec{k}_2}$ are described by the plane waves orthogonalized to the electron wave function of the ground state of the H_2^+ ion at the equilibrium internuclear distance R_0 for the H_2 molecule. The latter condition is connected with the consequence of the adiabatic approximation, so that the cross section of the process

$$\frac{d\sigma}{dE_1 d\Omega_1 d\Omega_2} = 4\pi^2\alpha \cdot \frac{k_1 k_2}{E_\gamma} \langle |M|^2 \rangle$$

is peaking when the equality $E_\gamma - I^{++} = E_1 + E_2 + R_0^{-1}$ is valid (I^{++} is the "adiabatic" potential of double ionization of the molecule; R_0^{-1} is the energy of Coulomb repulsion of nuclei).

Fig. 1 shows the squared amplitude $\langle |M(\vec{k}_1,\vec{k}_2)|^2 \rangle$ averaged over the orientations of the molecule and polarizations of photons as a function of the angle Θ_{12} between the electron momenta $\vec{k}_1$ and $\vec{k}_2$ in the case where both electrons are detected in the plane of photon polarization at $\vec{k}_1 = \vec{k}_2 = 4\,a_0^{-1}$. The presented curves correspond to the wave functions Ψ_0 of the ground state of the molecule with different degrees of the allowance for interelectronic correlations[2]: the solid, dash-dotted, dotted, and dashed curves correspond to the wave functions of 5-configuration, MO, Heitler-London, and Weinbaum approximations, respectively. It can be seen that the results for these functions are essentially different and, therefore, the relevant experiments are much desirable. In contrast to the (e,2e) process, where the ee-correlations are mainly manifested as satellite lines[3], these correlations in the (γ,2e) process give a direct determination of the dependence of the amplitudes $M(\vec{k}_1, \vec{k}_2)$ on the energies and directions of ejected electrons.

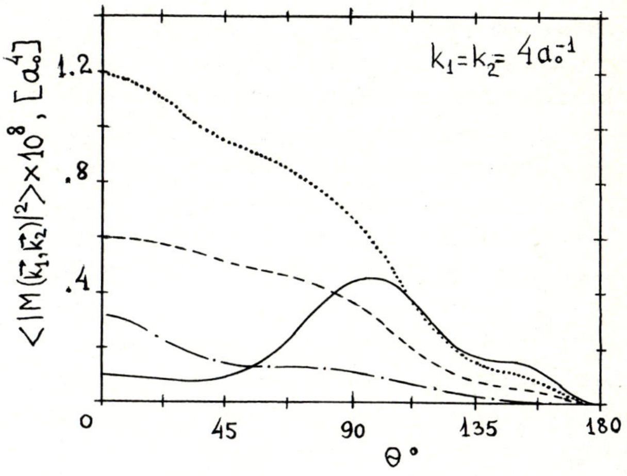

Fig. 1.

References

1. Yu.F. Smirnov et al., J.Phys.B:Atom.Molec.Phys., 11, 3587 (1978).
2. A.D. McLean et al., Rev.Mod.Phys., 32, 211 (1960).
3. A.J. Dixon et al., J.Phys.B:Atom.Molec.Phys., 9, L195 (1976);
V.G. Levin, Phys.Lett., A39, 125 (1972).

MANIFESTATIONS OF THE OPTICAL ACTIVITY OF MOLECULES IN THE DIPOLE PHOTOEFFECT

N.A.Cherepkov

A.F.Ioffe Physical-Technical Institute, 194021 Leningrad, USSR

The well known phenomena of optical rotation and circular dichroism in photoabsorption by optically active (chiral) molecules[1] appear in the electric-dipole - magnetic-dipole interference term and are the consequences of the lack of a plane of symmetry and inversion symmetry in their structure. It is shown here that the lack of a plane of symmetry in chiral molecules leads to appearance of some phenomena in the angular distribution of photoelectrons with defined spin polarization already in pure dipole approximation.

Using the method described in[2] it can be shown that the angular distribution of photoelectrons, ejected from chiral molecules in some direction $\vec{\varkappa}$ with spin oriented along another direction $\vec{s}$ by circularly polarized light with spin $\vec{s}_\gamma$ ($\vec{\varkappa}$, $\vec{s}$ and $\vec{s}_\gamma$ are unit vectors), has the following form:

$$I_\kappa^{\pm 1}(\vec{\varkappa},\vec{s}) = \frac{\sigma_\kappa(\omega)}{8\pi}\Big\{1-\frac{\beta^\kappa}{2}P_2(\vec{\varkappa}\vec{s}_\gamma) + A^\kappa\cdot(\vec{s}\vec{s}_\gamma) - \eta^\kappa(\vec{s}[\vec{\varkappa}\vec{s}_\gamma])(\vec{\varkappa}\vec{s}_\gamma) - \gamma^\kappa\Big[\frac{3}{2}(\vec{\varkappa}\vec{s})(\vec{\varkappa}\vec{s}_\gamma)-\frac{1}{2}(\vec{s}\vec{s}_\gamma)\Big] + D^\kappa\cdot(\vec{\varkappa}\vec{s}_\gamma) + C^\kappa\cdot(\vec{s}[\vec{\varkappa}\vec{s}_\gamma]) + B_1^{\kappa,1}\cdot(\vec{\varkappa}\vec{s}) + B_2^{\kappa,1}\cdot(\vec{\varkappa}\vec{s}_\gamma)(\vec{s}\vec{s}_\gamma) + B_3^{\kappa,1}\cdot(\vec{\varkappa}\vec{s}_\gamma)^2(\vec{\varkappa}\vec{s})\Big\}$$

where $\sigma_\kappa(\omega)$ is the photoionization cross section with κ being the set of quantum numbers characterizing the residual ion state. The first five terms with the parameters β^κ, A^κ, γ^κ, η^κ coincide with the analogous terms in atoms and non-chiral molecules and have been considered previously[2]. The last five terms are characteristic of chiral molecules.

The term with $(\vec{\varkappa}\vec{s}_\gamma)$ leads to a circular dichroism in the angular distribution of photoelectrons, that is to a difference between the intensity of electron fluxes ejected at a definite angle by right and left circularly polarized light. The existence of this term was first predicted by Ritchie[3].

The term with $(\vec{s}[\vec{\varkappa}\vec{s}_\gamma])$ gives the transverse polarization of photoelectrons which has a different sign for left and right circularly polarized light. The last three terms give the longitudinal polarization of photoelectrons, which also appears for absorption of linearly polarized and unpolarized light. Thus, the angular distribution of photoelectrons with given spin polarization ejected from chiral molecules is defined by ten independent parameters, whereas in atoms and non-chiral molecules it is defined by only five independent parameters.

Coefficients of all terms specific to chiral molecules are proportional to the differences between pairs of dipole matrix elements which differ from each other by the signs of all the projections of orbital momenta and spins. For non-chiral molecules these differences are identically equal to zero. Their realistic estimation is quite complicated, but they should be proportional to an asymmetry factor η discussed in[4], which depends on the degree of dissymmetry in the structure of the molecule. If the photoelectrons corresponding to different fine-structure components of the final molecular ion state can be separated, there are no other small parameters in the coefficients, and the effects discussed above should be proportional to the factor η. If the multiplet structure of the ionic state is not resolved, the polarization effects will be of order $\eta(\alpha Z)^2$. After integration over the electron ejection angles all five new terms disappear. The helicity induced in an initially unpolarized electron beam after elastic scattering from chiral molecules is of order $\eta(\alpha Z)^2$ [4], therefore angle- and spin-resolved photoelectron spectroscopy is the most promising way of searching for new effects connected with the absence of mirror symmetry in the structure of chiral molecules.

References

1. A.E.Hansen and T.D.Bouman. Adv. Chem. Phys. 44, 545 (1980).
2. N.A.Cherepkov. J. Phys. B 14, 2165 (1981).
3. B.Ritchie. Phys. Rev. A 13, 1411 (1976).
4. A.Rich, J.Van House and R.A.Hegstrom. Phys. Rev. Lett. 48, 1341 (1982).

ABSOLUTE DIPOLE OSCILLATOR STRENGTHS FOR THE PHOTOABSORPTION, PHOTOIONIZATION AND FRAGMENTATION OF HCℓ, HBr and NO

C.E. Brion, F. Carnovale†, S. Daviel and Y. Iida

Dept. of Chemistry, University of British Columbia, Vancouver, Canada

Dipole (e,2e) and dipole (e, e+ion) spectroscopies [1,2,3] have been used to measure the absolute dipole oscillator strengths (cross-sections) for the photoabsorption and partial photoionization (electronic states) as well as the molecular and dissociative photoionization of HCℓ, HBr and NO. These studies are part of a continuing programme of systematic measurements of absolute oscillator strengths for photoionization processes in total and partial channels. The relative fast electron (3 or 8 keV) impact spectra are converted to absolute dipole oscillator strengths via the Bethe-Born transformation and TRK sum rule normalization [1,2]. Some typical results are shown in the accompanying figures. The photoionization efficiency is also measured. The results of the two experiments are combined to provide quantitative assessments of the dipole breakdown pattern of these molecules in the energy ranges shown. The HCℓ partial oscillator strengths are compared with the calculations reported by Faegri and Kelly [4].

This work received finanical support from The Natural Science and Engineering Research Council of Canada and The Petroleum Research Fund administered by The American Chemical Society.

† Present address: Dept. of Chemistry, University of Melbourne.

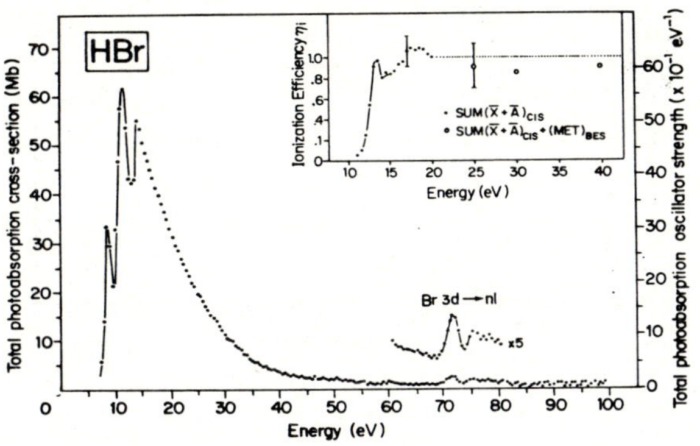

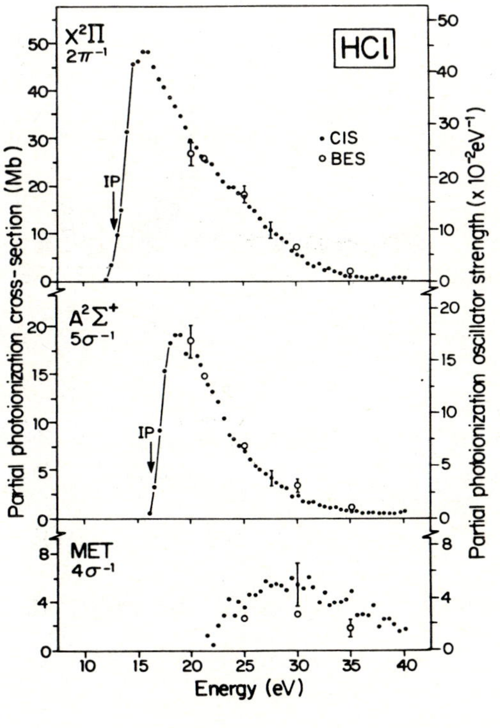

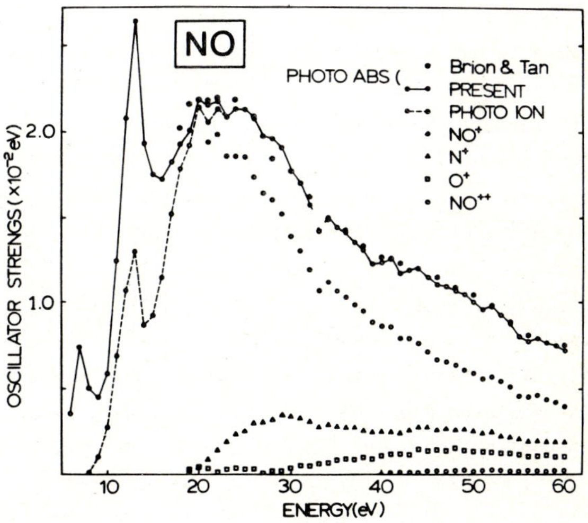

References

[1] C.E. Brion & A. Hamnett, Adv. Chem. Phys. 45, 1 (1981).

[2] C.E. Brion in "Physics of Electronic and Atomic Collisions", Ed. S. Datz (North-Holland, 1982) pages 579-593.

[3] F. Carnovale and C.E. Brion, J. Phys. B. in press, 1983.

[4] K. Faegri, Jr. and H.P. Kelly, Chem. Phys. Letters 85, 472 (1982).

MULTI-PHOTON ELECTRON SPECTROSCOPY OF NITRIC OXIDE

J. Kimman, M. Lavollee*, M. Spruit and M.J. van der Wiel

FOM-Institute for Atomic and Molecular Physics, Kruislaan 407, 1098 SJ Amsterdam, The Netherlands.
*L.C.A.M., Bât.351, Université de Paris-Sud, Orsay, France

As has been shown by Miller and Compton [1] and Kimman et al. [2] resonantly enhanced 4-photon ionization of NO via the $A^2\Sigma^+$ state ($\lambda = 452$ nm) does not follow the Franck-Condon overlap between the intermediate $A^2\Sigma^+$ state and the $X^1\Sigma^+$ state of the ion.

Because the intermediate A state is the first member of a Rydberg series converging to the ionic state, one expects in the case of direct ionization only $\Delta v = 0$ transitions. However, also the highest energetically allowed Δv transitions are observed with comparable intensities. Mechanisms like electronic auto-ionization from the repulsive part of valence states or double resonances occurring for the third absorbed photon, have been proposed to explain the observed features. We present evidence for the latter mechanism which involves the mixing between a valence state and a highly vibrationally excited Rydberg state. Electron spectra are obtained by using a TOF spectrometer with 20 meV resolution and 50% collection efficiency for all electron energies [3].

In order to investigate the third-photon region in the $\lambda = 452$ nm experiment, we now performed a 3-photon experiment using frequency doubled Rh 101 dye ($\lambda = 310$ nm). Now we only observe electron signal if two photons are resonant with states in the 60000 cm^{-1} region. In this experiment a third photon causes ionization at an energy just above $v = 12$ of NO^+. In fig. 1 we show an electron spectrum recorded at $\lambda = 302.8$ nm. We observe one strong electron peak corresponding to the production of ions with $v = 6$. The fact that only one peak is strongly enhanced indicates that $v = 6$ Rydberg states are involved in the third step. However, according to Franck-Condon factors they cannot be excited either from the A state in the 4 photon experiment or from the ground state in the 3 photon experiment. The additional small peaks in the spectrum are the subject of further studies.

In fig. 2 we show a wavelength scan for the $v = 6$ peak only. The bandheads of the two-photon transitions show the structure of the $L^2\Pi$ valence state [4]. We conclude that a mixing of a valence state, $L^2\Pi$ ($v = 4$) with the Rydberg state $C^2\Pi$ ($v = 6$) is responsible for the production of the sharp electron peak in fig. 1 and also for the unexpected vibrational distribution in the case of 4 photon ionization via the $A^2\Sigma^+$ state.

Finally we will present results on autoionization of $v = 1$ Rydberg states into the $v = 0$ state of NO^+. In this experiment NO molecules are prepared in the $v = 1$ of the $A^2\Sigma^+$ state by a two-photon transition with two 430 nm photons. Then two 590 nm photons allow the transition to high $v = 1$ Rydberg states which can autoionize to $v = 0$ of the ionic state. Thus, we obtain a well-resolved Rydberg structure without any background due to direct ionization, since this is Franck-Condon forbidden.

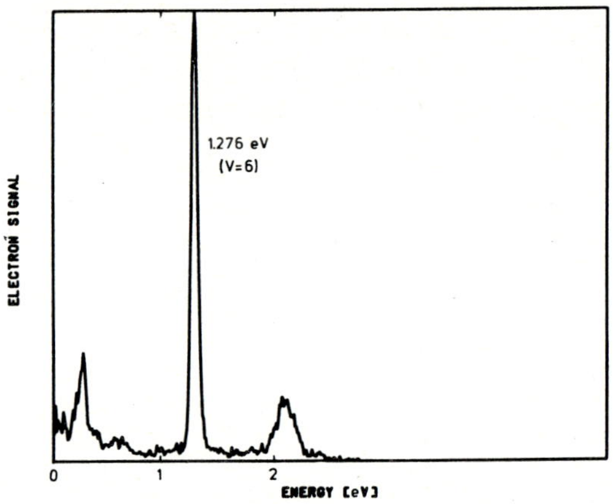

Fig. 1: Electron energy spectrum for three-photon ionization of NO recorded at $\lambda = 302.8$ nm.

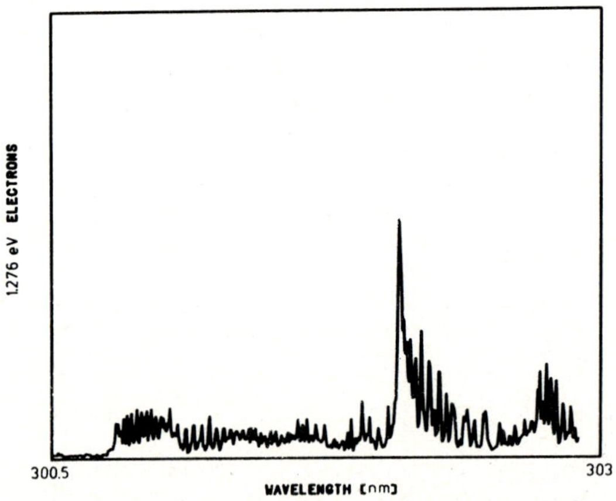

Fig. 2: Wavelength scan for the electron signal corresponding to the $v = 6$ final state of NO^+.

References

1. J. Miller and R. Compton, J.Chem.Phys. 75 (1981) 22.
2. J. Kimman, P. Kruit and M.J. van der Wiel, Chem.Phys. Lett. 88 (1982) 576.
3. P. Kruit and F.H. Read, J.Phys.E: Sci.Instrum. 16 (1983)
4. A. Laserquist and E. Miescher, Can.J.Phys. 44 (1966) 1525.

ULTRAHIGH RESOLUTION PHOTODETACHMENT STUDY OF C_2^-

U. Hefter[a], R.D. Mead, P.A. Schulz[b] and W.C. Lineberger
Department of Chemistry, University of Colorado and JILA
University of Colorado and NBS, Boulder, Colorado 80309, USA

Fundamental limitations of the crossed beam photodetachment technique preclude resolution much better than one part in 10^5 (1). Improvement in resolution up to three orders of magnitude can be achieved due to the recent advent of coaxial beam techniques, which in the ion and photon beams are merged. The velocity spread of the ions is greatly reduced along the beam direction, greatly reducing the Doppler broadening of the ion signal. In fact, this is one of the few methods where a sub-Doppler resolution is obtainable using single-photon spectroscopic techniques.

The first study of negative ions using the coaxial beam technique has been of perturbations and autodetachment in C_2^-.

In analogy to N_2^+, CN, and CO^+ the three energetically lowest states of C_2^- are expected to be a $X^2\Sigma_g^+$ groundstate, a low-lying $A^2\Pi_u$ state and a $B^2\Sigma_u^+$ excited state. In earlier experiments (2) (1) transitions between B and X states of C_2^- have been observed - but the presence of the A state could not be confirmed. Nevertheless, the A and B states interact, and give rise to the presence of highly perturbed lines in the B-X transitions. The availability of high resolution, high accuracy data from our coaxial photodetachment apparatus and upon the advanced analysis techniques developed over the last decade, the character and the molecular constants of the $A^2\Pi_u$ state could be determined without having to observe direct transitions to the state itself.

The crossed beam photodetachment experiments of Jones et al. (1) provided limits for the autodetachment rates of C_2^- levels. Arguments based upon apparative linewidth and results from a three niveau model bounded the C_2^- autodetachment rates between 10^7 sec^{-1} and 10^{11} sec^{-1}.

Direct resolution of uncertainty broadening of C_2^- levels has been accomplished (3) by using the ultrahigh resolution coaxial beam apparatus to study linewidths of C_2^- transitions. Lifetimes, i.e. autodetachment rates were obtained for many rotational and vibrational states allowing a picture of the influence of nuclear motion upon autodetachment to be formed. As shown in Fig. 1, autodetachment rates were observed to increase rapidly with vibrational excitation, and to increase at higher rotational levels. Whereas vibration to electronic energy coupling is responsible for the vibrational dependence of the rates, their rotational dependence is much too strong to be understood in terms of the available density of

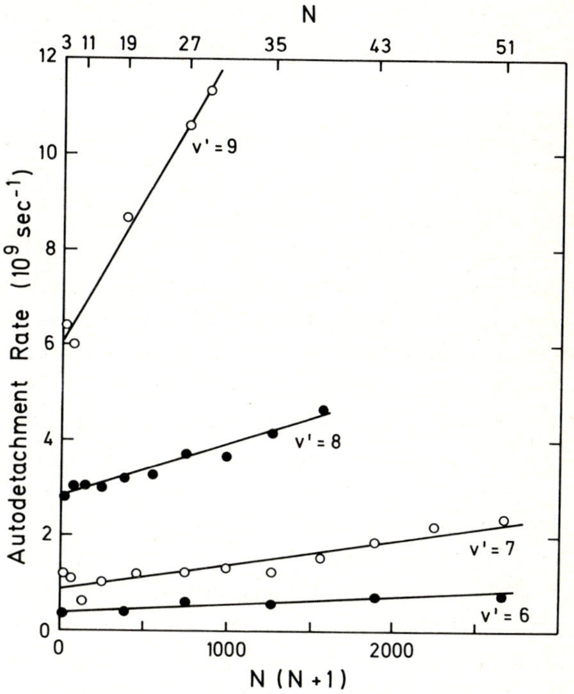

Fig. 1: Vibrational and rotational dependence of the C_2^- autodetachment rates of levels of the $B^2\Sigma_u^+$ state. Typical errors are $1 \cdot 10^8$ sec^{-1}.

final states. This and the linear increase of the rates v.s. rotational energy leads to the assumption that rotation to electronic energy coupling leads to autodetach- of C_2^- as well. In addition, autodetachment rates for states of predominantly $^2\Pi$ character were found to be much smaller than for states of $^2\Sigma$ character. These data will allow a better understanding of autoionization processes, particularly since rotationally resolved lifetime data have been so difficult to obtain for autoionization.

* Supported by NSF Grants CHE 78-18424 and PHYS 79-04928
a Present address: FB Physik der Universität Kaiserslautern, Postfach 3049, D-6750 Kaiserslautern, W.-Germany
b Present address: School of Physics, Georgia Institute of Technology, Atlanta, Georgia 30332, USA

References

1. P.L. Jones, R.D. Mead, B.E. Kohler, S.D. Rosner, and W.C. Lineberger, J. Chem. Phys. 73, 4419 (1980)
2. G. Herzberg and A. Lagerquist, Can. J. Phys. 46, 2363 (1968)
3. U. Hefter, R.D. Mead, P.A. Schulz, and W.C. Lineberger submitted to Phys. Rev. A (1983)

TRIPLY DIFFERENTIAL PHOTOELECTRON STUDIES OF MOLECULAR PHOTOIONIZATION*

J. L. Dehmer

Argonne National Laboratory, Argonne, IL 60439

A. C. Parr and S. H. Southworth

Synchrotron Ultraviolet Radiation Facility
National Bureau of Standards, Washington, D.C. 20234

D. M. P. Holland†

Institute of Physical Science and Technology
University of Maryland, College Park, MD 20742

We report studies of molecular photoionization dynamics using triply differential photoelectron spectrometry. That is, photoelectron intensity is measured as a function of three independent variables — incident wavelength, electron kinetic energy, and ejection angle. In the last few years, emphasis has been placed on shape and autoionizing resonances (see, e.g., Ref. 1 and references therein). These resonant processes are important probes of photoionization for various reasons, the most obvious one being that they are usually displayed prominently against nonresonant behavior in such observables as the total photoionization cross section, photoionization branching ratios, and photoelectron angular distributions. More importantly, the study of resonant features has repeatedly led to a deeper physical insight into the mechanisms of excitation, resonant trapping of the photoelectron, and decay of the excited complex that occur during the photoionization process. The technique employed in this work yields branching ratios and angular distributions which are very effective in studying these dynamical issues. In particular, we emphasize resolution of these quantities into alternative vibrational components as this displays the effects of the resonances much more sensitively than does the vibrationally summed results.

Recent work to be reported at this time includes triply differential photoelectron measurements on SF_6,[2] BF_3, HCN,[3] and C_2N_2,[4] and a constant-photoelectron-energy spectroscopy measurement[5] on C_2H_2. The work on SF_6 and BF_3 emphasizes the role of shape resonances in the photoionization dynamics of these highly symmetric molecules. The results contain evidence for strong channel interaction in the vicinity of shape resonances and also, taken with other evidence such as x-ray data and theoretical calculations, help clarify the ordering of certain valence levels. The work on HCN[3] and C_2N_2[4] contains vibrationally-resolved data for some channels, which indicate strongly non-Franck-Condon behavior in certain wavelength ranges. The original motivation was to study similar molecules containing the CN group, which is closely related to N_2 and CO for which so much is known about the photoionization dynamics.[1] Analysis of the data has been limited so far, owing to the lack of supporting theoretical calculations. The constant-photoelectron-energy experiments[5] attempt to examine autoionizing features just above ionic thresholds. By monitoring the photoelectron signal for a fixed, small (0.1-1.0 eV) kinetic energy as a function of incident wavelength, we attempt to bridge the gap in information obtained by zero-energy photoelectron spectroscopy and standard photoelectron spectroscopy. Finally, we will describe a new generation triply differential photoelectron spectrometer[6] designed to give orders of magnitude more sensitivity than that previously used. This instrument has just been completed and is presently being used to gather higher-resolution data on resonant molecular photoionization.

---

*Work supported by the U.S. Department of Energy, the Office of Naval Research, and NATO Grant No. 1939.

†Present address: Daresbury Laboratory, Science Research Council, Daresbury, Warrington WA4 4AD, England.

References

[1] J. L. Dehmer, D. Dill, and A. C. Parr, "Photoionization Dynamics of Small Molecules," in Photophysics and Photochemistry in the Vacuum Ultraviolet, edited by S. McGlynn, G. Findley, and R. Huebner (D. Reidel Publ. Co., Dordrecht, Holland, 1983), in press.

[2] J. L. Dehmer, A. C. Parr, S. Wallace, and D. Dill, Phys. Rev. A 26, 3283 (1982).

[3] D. M. P. Holland, A. C. Parr, and J. L. Dehmer, J. Phys. B, submitted for publication.

[4] D. M. P. Holland, A. C. Parr, D. L. Ederer, J. B. West, and J. L. Dehmer, Int. J. Mass Spect. Ion Phys., in press.

[5] D. M. P. Holland, J. B. West, A. C. Parr, D. L. Ederer, R. Stockbauer, R. D. Buff, and J. L. Dehmer, J. Chem. Phys. 78, 124 (1983).

[6] A. C. Parr, S. H. Southworth, J. L. Dehmer, and D. M. P. Holland, Nucl. Instr. Meth., in press.

FLUORESCENCE POLARIZATION AS A PROBE OF MOLECULAR AUTOIONIZATION*

E. D. Poliakoff

Boston University, Boston, Massachusetts 02215

J. L. Dehmer

Argonne National Laboratory, Argonne, Illinois 60439

A. C. Parr

National Bureau of Standards, Washington, D.C. 20234

G. E. Leroi

Michigan State University, East Lansing, Michigan 48824

Extensive effort in VUV spectroscopy has gone into developing an understanding of the spectroscopy and dynamics of autoionizing Rydberg states.[1,2] We show, using initial results on CO_2 autoionization, that the polarization of fluorescence from excited-state molecular photoions can be a significant tool in ascertaining both the symmetry signatures and dynamical properties of autoionizing resonances.[3] The process studied in the present work is given below.

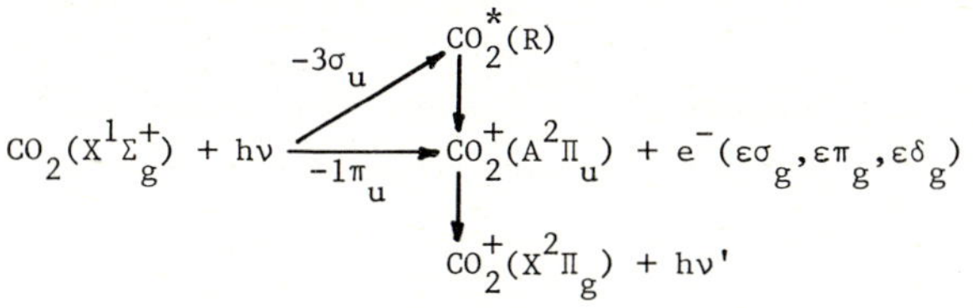

R denotes a Rydberg state; $-3\sigma_u$ and $-1\pi_u$ indicate which electron is excited; ε denotes a continuum electron. The experiment is carried out by scanning the excitation photon energy, hν, and measuring the polarization of the undispersed CO_2 $A^2\Pi_u \rightarrow X^2\Pi_g$ fluorescence, hν'.

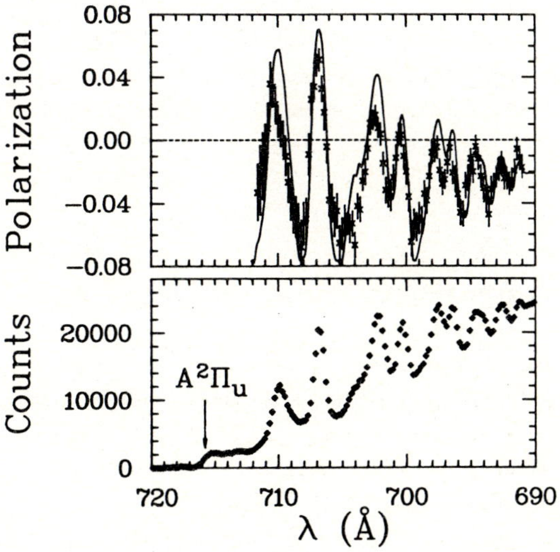

Fig. 1: Excitation spectrum (bottom) and fluorescence polarization spectrum (top). The solid line is the prediction of a simple mixing model.

The fluorescence polarization from molecular photoions reflects the degree of alignment of the molecular ion in the laboratory-fixed frame, which is, in turn determined by the relative dipole strengths for degenerate photoionization channels which have different symmetries in the molecule-fixed frame. This basic framework is used to infer, from fluorescence polarization measurements, the symmetry properties of autoionizing resonances which are superimposed on a nonresonant background of known symmetry.

The results are shown in Fig. 1. The excitation spectrum (bottom frame) shows extensive autoionization structure, and the fluorescence polarization data (top frame) exhibit features analogous to the structure in the excitation spectrum. The physical basis for the structure in this fluorescence polarization spectrum is that the absorption transition dipole moves into the plane of molecular rotation as the resonance pathway becomes enhanced, leading to a greater degree of alignment between the absorption and fluorescence transition dipoles than in the case of nonresonant ionization. The key point is that the degree of alignment of the molecular ion in the laboratory frame is dependent on the dipole strengths for the ionization channels populating the excited ionic state. These results underscore the conclusion that this method serves as a direct probe of symmetry information on autoionizing Rydberg states, thus providing experimental guideposts which test the symmetry signatures of autoionizing resonances.

References

*Work supported by the U.S. Department of Energy, the Office of Naval Research, and the National Science Foundation.

1. U. Fano, J. Opt. Soc. Am. 65, 979 (1975)
2. M. Raoult and Ch. Jungen, J. Chem. Phys. 74, 3388 (1981)
3. E. D. Poliakoff, J. L. Dehmer, A. C. Parr, and G. E. Leroi, J. Chem. Phys. 77, 5243 (1982)

EXPERIMENTAL AND THEORETICAL STUDY OF CO PHOTOIONIZATION IN THE 15-40 eV ENERGY RANGE

B. Leyh[a,*], G. Raşeev[a,c], M-J. Hubin-Franskin[a,**], J. Delwiche[a,**],
I. Nenner[b], H. Lefebvre-Brion[c], P. Morin[b] and M-Y. Adam[d]
L.U.R.E., Bât. 209c, Université de Paris-Sud, 91405 Orsay, France

a. Institut de Chimie, Université de Liège, Bât. B.6, Sart-Tilman, 4000 par Liège 1, Belgium.
b. Département de Physico-Chimie, Centre d'Etudes Nucléaires de Saclay, 91490 Gif-sur-Yvette, France.
c. Laboratoire de Photophysique Moléculaire du C.N.R.S., Bât. 213, Université de Paris-Sud, 91405 Orsay, France.
e. Equipe de Physique Atomique, Université de Paris VII, Place Jussieu, 75005 Paris, France.
*. Boursier de l'I.R.S.I.A. of Belgium.
**. Chercheur qualifié du F.N.R.S. of Belgium.

The CO photoionization has been the subject of extensive experimental work[1]. There remains, however, several uninterpreted features particularly the dip and the enhancement in the $X^2\Sigma^+$ and $B^2\Sigma^+$ photoionization cross sections at $\sim$ 22 eV near the shape resonance of the X state.

We report here a preliminary experimental and theoretical study of the photoionization spectrum between the threshold of the $X^2\Sigma^+$ state and 40 eV (above the B state shape resonance).

Angular and vibrationally resolved partial cross sections are measured using the pulsed ACO synchrotron radiation at Orsay (France)[2]. The resolution of the grazing incidence monochromator is of 0.07 nm. The ejected electrons are detected by a 127° electrostatic analyzer whose energy resolution is roughly 7 % of the pass energy, allowing the resolution of the vibrational levels of the CO^+ states. Spectra are recorded in the Constant Ionic State (CIS) mode[1b], i.e. keeping $h\nu-\varepsilon_k$ constant, to obtain directly the variation with energy of the partial cross section of a given vibronic state.

Calculations of the photoionization cross sections are performed using a method we have already developed[3] for direct processes and the Perturbational Quantum Defect Theory (PMQDT) for autoionizations[4]. The direct photoionization method is based on the calculation of the final state wavefunction in the One Configuration Frozen Core Static Exchange approximation with exact exchange. The electronic continuum wavefunction is obtained as the solution of a system of coupled second order differential equations in the one center expanded molecular field. This function is also used to calculate resonances by PMQDT. This method is based on perturbational expansion of the autoionization reactance matrix used in the framework of a MQDT formulation.

Vibrationally resolved cross section of $X^2\Sigma^+$ (v' = 0 to 2) and of $B^2\Sigma^+$ (v' = 0, 1, 2) have been measured from their respective threshold up to about 40 eV. They show many interesting features. Particularly the above mentioned structure at 22 eV appears in all the vibrational levels of the X and B states. It is slightly shifted to lower energies as v increases for the X state but appears at the same energy for the B state.

We have calculated the direct photoionization cross section at the equilibrium internuclear distance of the ground neutral state for X, A and B ionic state and the autoionization of the $3p\sigma$ v = 0 Rydberg state converging to the B ionic state. Next, we shall calculate the uninterpreted feature at 22 eV which can result from an interaction between the X and B continua or from the autoionization in the two continua of a doubly excited state as suggested by Wendin[5] for N_2 isoelectronic system.

We thank the technical staff of LURE and Laboratoire de l'Accélérateur Linéaire. The "Fonds National de la Recherche Scientifique" of Belgium and NATO (contrat n° 096.82) are gratefully acknowledged for their financial support.

References

1. (a) J.A.R. Samson and J.L. Gardner, J. Electron Spectrosc. Related Phenom. 8, 35 (1976) ; ibid., 13, 7 (1978) ; (b) E.W. Plummer, T. Gustaffson, W. Gudat and D.E. Eastman, Phys. Rev. A, 15, 2339 (1977) ; (c) G.V. Marr, J.M. Morton, R.M. Holmès and D.G. McCoy, J. Phys. B, 12, 43 (1979) ; (d) B.E. Cole, D.L. Ederer, R. Stockbauer, K. Codling, A.C. Parr, J.B. West, E.D. Poliakoff and J.L. Dehmer, J. Chem. Phys. 72, 6308 (1980).
2. P. Morin, M-Y. Adam, I. Nenner, J. Delwiche, M-J. Hubin-Franskin and P. Lablanquie, Nuclear Instruments and Methods, in press.
3. G. Raşeev, H. Le Rouzo and H. Lefebvre-Brion, J. Chem. Phys. 72, 5701 (1980).
4. A. Giusti-Suzor and H. Lefebvre-Brion, Chem. Phys. Lett. 76, 132 (1980).
5. G. Wendin, Int. J. Quantum Chem. Symposium 13, 659 (1979).

PHOTOIONIZATION CROSS SECTIONS OF THE VALENCE ORBITALS OF CO

E. P. Leal and Lee Mu-Tao

Universidade Federal de São Carlos, São Carlos, 13560, Brasil

F. J. da Paixão

Departamento de Física, UNICAMP, Campinas, 13100, Brasil

R. R. Lucchese

Department of Chemistry, Princeton University, Princeton, N. Jersey 08540

V. Mckoy

A. A. Noyes Lab. Phys. Chem. California Institute of Technology Pasadena CA. 91125, U.S.A.

The partial and total cross sections and the asymmetry parameters of the photoionization from the valence orbitals of CO (3σ, 4σ, 1π and 5σ) are calculated in the fixed-nuclear frame and frozen-core HF approximation.

In this calculation, the HF single determinantal function is used to describe the ground-state of CO (X $^1\Sigma^+$), which is obtained at the equilibrium internuclear distance of 2.132 a.u. by a SCF calculation with a standard[1] |9s5p| primitive Cartesian Gaussian functions contracted to a |4s3p| basis set.

The continuum orbital which describes the ejected photoelectron in the molecular ion (CO^+) field is obtained by iterative procedure[2] of Schwinger variational method.

Our results have been compared with the available experimental[3-4] data and show good agreement in both cross sections and asymmetry parameters.

Comparison has also been made with the other theoretical[5] results.

For $5\sigma^{-1}$ channel, we have also made the same calculation replacing the GTO molecular wavefunction by the STO expansion of Mclean and Yoshimine[6]. It has been shown that the cross section of $5\sigma^{-1}$ channel is quite sensitive with the changing of the bound orbital, however both results agree well with the experimental data.

References

1. T. H. Dunning, J. Chem. Phys. 53, 2823 (1970)
2. D. K. Watson, R. R. Lucchese, V. McKoy and T. N. Rescigno, Phys. Rev. A 4, 1482 (1980)
3. E. W. Plummer, T. Gustafsson, W. Gudat and D. E. Eastman, Phys. Rev. A15, 2339 (1977)
4. A. Hamnett, W. Stoll and C. E. Brion, J. Elec. Spec. Related Phen. 8, 367 (1976)
5. N. Padial, G. Csanak, B. V. McKoy and P. W. Langhoff, J. Chem. Phys. 69 2992 (1978)
6. A. Mclean and M. Yoshimine, "Table of Linear Molecule Wave Functions" (San Jose, California: IBM San Jose Research Laboratory) p. 27

THEORETICAL STUDIES OF PHOTOEXCITATION AND IONIZATION IN MOLECULES OF ASTROPHYSICAL INTEREST[1]

G.H.F. Diercksen and P.W. Langhoff[2]

Max-Planck-Institut für Astrophysik, 8046 Garching bei München, Federal Repulic of Germany

Cross sections for molecular excitation, ionization, and fragmentation are of considerable astrophysical importance, particularly with reference to photochemical models of the interstellar medium, the opacities of cool stars, and the spectra and composition of comets, to mention some representative examples. As part of a continuing program of ab-initio studies of molecular properties of astrophysical importance, a collaborative project has been initiated with Indiana University in computations of relevant photo cross sections in selected compounds. Studies of photoexcitation and partial-channel photoionization cross sections have been completed to date for the stable gas-phase compounds CO, CO_2, H_2O, H_2S, H_2CO, and H_2CS in their ground vibronic states, and work is presently in progress on CS_2, SO_2, and OCS. The general methodology employed in these studies and selected results of spectroscopic interest are reported in this Abstract.

The calculations performed employ Hartree-Fock ground-state electronic wave functions constructed at experimental equillibrium geometries in contracted Cartesian Gaussian basis sets. Vertical electronic dipole excitation and partial-channel ionization spectra of body-frame point group symmetry are obtained in both single and coupled channel static-exchange approximations, corresponding to single-excitation CI or TDHF calculations in the absence of significant ground-state correlation. The relevant bound excited and continuum pseudostate electronic wave functions are constructed entirely in large Cartesian Gaussian basis sets chosen to span the appropriate spatial and spectral intervals. All electronic calculations, including computations of one- and two-electron integrals, transformation to Fock symmetry orbitals, and construction and diagonalization of the Hamiltonian matrix, are accomplished with the code system MUNICH at MPI, Garching. Continuum states are constructed from the resulting discrete pseudospectra employing expilcit Hilbert space Stieltjes methods that have been described in considerable detail elsewhere and have been widely employed for this purpose.[3] Vibrational degrees of freedom are treated in the present studies in the Franck-Condon approximation, with complete neglect of rotational structure. Appropriate Hönl-London factors can be incorporated into the development, if required, and the electronic calculations performed at additional nuclear configurations in specific cases, topics that are appropriate subjects for subsequent study.

The calculated discrete excitation spectra in single and coupled-channel approximations provide a basis for clarification and assignment of experimentally determined photoabsorption and electron energy-loss spectra. Correspondingly, the partial-channel photoionization cross sections clarify quantitatively the overall natures of branching ratios obtained from synchrotron radiation, line source, and (e,2e) photoelectron spectra and from fluorescence yield measurements, and account for the presence of particularly strong features in specific instances. Graphical representation of the continuum functions obtained from the calculations are found to be helpful in this latter connection.

The single-channel results obtained from the present calculations are found to be in generally good accord with corresponding values for CO, CO_2, H_2O and H_2CO obtained previously employing Stieltjes methods. Configuration-mixing and channel-coupling effects are found to be significant in certain cases in these compounds, however, and in H_2S and H_2CS, as well. Of particular interest is the presence of strong V_π-V_σ coupling in CO, CO_2, H_2CO, and H_2CS, responsible for positioning of the strong $\pi\rightarrow\pi^*$ transition in the discrete spectral region in each case, and the appearance of $\sigma\rightarrow\sigma^*$ resonance features in the appropriate photoionization continua.

Detailed results are contained in separate reports, which describe on a common basis the natures of both discrete and continuum electronic spectra and wave functions in these compounds.[4]

References

[1]Work supported in part by the International Programs Division of the National Science Foundation, Grant #INT-8116534.

[2]Permanent address: Department of Chemistry, Indiana University, Bloomington, IN 47405, USA.

[3]P.W. Langhoff, in Proceedings of the NATO Advanced Study Institute on Methods in Computational Molecular Physics, G.H.F. Diercksen and S. Wilson, Editor (Reidel, Dordrecht, Holland, 1983).

[4]G.H.F. Diercksen, M.R. Hermann, and P.W. Langhoff (to be published).

SEMICLASSICAL THEORY OF PREDISSOCIATION INDUCED BY ROTATIONAL (CORIOLIS) COUPLING

Hiroki Nakamura

Division of Theoretical Studies, Institute for Molecular Science, Okazaki 444, Japan

The dynamical-state representation and the multichannel quantum defect theory (MQDT) of predissociation are linked to formulate a semiclassical theory of predissociation of a diatomic molecule induced by rotational (Coriolis) coupling. It is demonstrated that the simple perturbation theory becomes invalid when the rotational angular momentum quantum number is large, and that the nonperturbative formula developed here should be employed. Rotational coupling causes transitions which can not be induced by radial coupling. Typical example is a transition between the states with different Λ of a diatomic molecule. The rotational coupling problem can not be dealt with in a straightforward way by the conventional semiclassical theory because of its peculiar analytical structure. The dynamical-state representation has been proposed in order to analyze this problem analytically[1-2].

Predissociation of case I in Herzberg's notation has been formulated analytically by using the semiclassical collision theory[3]. Line widths in the strong and weak coupling limits are obtained. Recently this problem was reformulated in terms of the multichannel quantum defect theory by Colle[4]. A general analytical expression for the line width is obtained, which holds for any intermediate strength of coupling. Predissociation induced by rotational coupling has been analyzed only by the first order perturbation theory[5]. The formula thus obtained is, therefore, valid only in a weak coupling limit. We can analyze this problem in a nonperturbative way by recasting the MQDT formulation in the dynamical-state representation. All we have to do is just replace the adiabatic energies in the various expressions by the dynamical-state potential energies. A big advantage of the dynamical-state representation lies in the fact that both radial and rotational nonadiabatic transitions are made to occur locally at the avoided crossing points of the dynamical-state potential energies. Thus, once the dynamical-state potential energies are obtained, a wide variety of dynamic problems can be dealt with uniformly and analytically by using the conventional semiclassical theory; uniformly in the sense that a common way of treatment of nonadiabatic transitions is possible irrespective of the type of coupling and its strength.

Numerical application is made to the predissociation in a model potential system shown in figure 1. The rotational coupling matrix element is assumed to be unity in atomic unit, and the reduced mass is taken to be one half of proton mass. Figure 2 shows the dependence of the resonance line width Γ upon $K(K+1)$ for the n=1 vibrational level, where K is the total (electronic plus rotational) angular momentum quantum number.

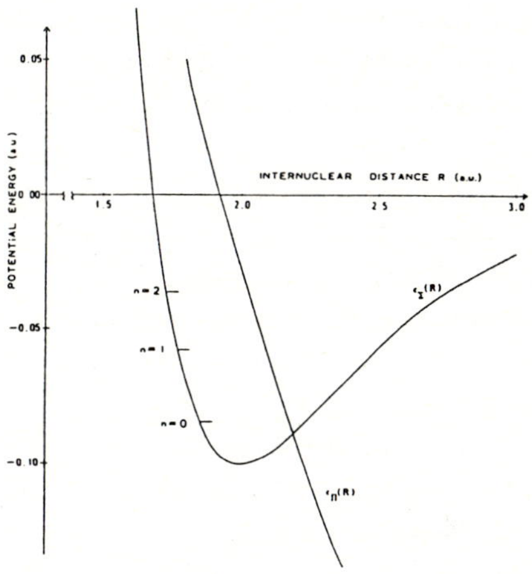

Fig. 1: Model adiabatic potentials. n represents the vibrational levels of the adiabatic Σ-state potential.

Fig 2

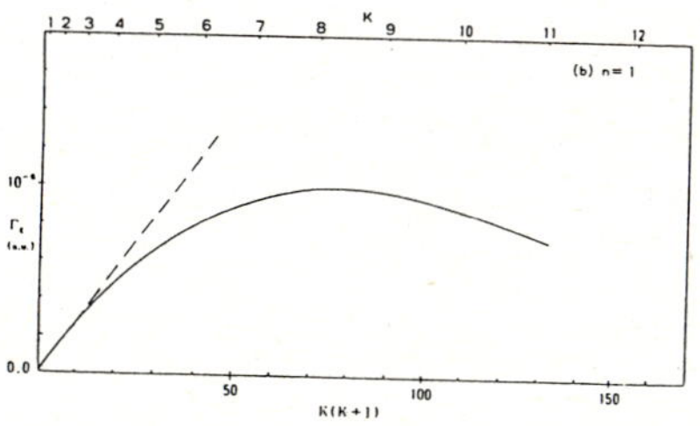

Fig. 2: Dependence of Γ on $K(K+1)$.

References

1. H. Nakamura and M. Namiki, J. Phys. Soc. Japan 49, 843 (1980); Phys. Rev. A24, 2963(1981).
2. H. Nakamura, Phys. Rev. A26, 3125 (1982).
3. See for instance, A. D. Bandrauk and M. S. Child, Mol. Phys. 19, 95 (1970).
4. R. Colle, J. Chem. Phys. 74, 2910 (1981).
5. See for instance, P. S. Julienne, Chem. Phys. Lett. 8, 27 (1971).

PHOTODISSOCIATION OF OH THROUGH NON-ADIABATIC INTERACTION

Ewine F. van Dishoeck[a], Marc C. van Hemert[b], A.C. Allison[c] and A. Dalgarno[c]

a. Sterrewacht Leiden, The Netherlands
b. Department of Physical Chemistry, University of Leiden, The Netherlands
c. Harvard-Smithsonian Center for Astrophysics, Cambridge, Mass. 02138

Potential energy curves and (electric dipole) transition moments for various electronic states of OH have recently been calculated with the aim of identifying the important photodissociation channels in this molecule[1]. The cross sections for several direct channels were obtained, but it was found that photodissociation may also occur by absorption into bound states. The oscillator strength of the $3^2\Pi$-$X^2\Pi$ transition is comparatively large[1], so the $3^2\Pi$ state will be an important photodissociation channel if absorption into it is followed by some dissociation process. Here we consider the interaction of the bound $3^2\Pi$ state with the dissociative $2^2\Pi$ state through the nuclear kinetic energy operator. No experimental information is available on these states.

The potential energy curves (see fig. 1) and transition moments were calculated by various ab initio SCF+CI methods[1,2]. The radial nuclear coupling term, $A = \langle 2^2\Pi|d/dR|3^2\Pi\rangle$, was computed by finite differences. The coupling has a sharp peak around 2.25 a_o, where the two states undergo an avoided crossing[2].

Cross sections for photodissociation were obtained by first determining the nuclear wave functions χ for the problem of the two interacting excited $^2\Pi$ states, characterized by one open and one closed channel. Both the coupled differential equations in the adiabatic formulation and the, mathematically equivalent, coupled equations in the diabatic formulation (cf. Ref. 3) were solved exactly by numerical integration. The computed photodissociation cross section for absorption into the (coupled) excited $^2\Pi$ states from the $X^2\Pi(v''=0)$ state is presented in fig.2. The two calculations gave identical results. The resonances in the cross section occur at the energies of the vibrational levels of the bound *diabatic* curve, and they have the asymmetric Fano-Beutler lineshape[4]. If the term $\underline{B}=\underline{A}^2+d\underline{A}/dR$ in the adiabatic formulation[3] is neglected, the absorption spectrum differs considerably from that shown in fig. 2.

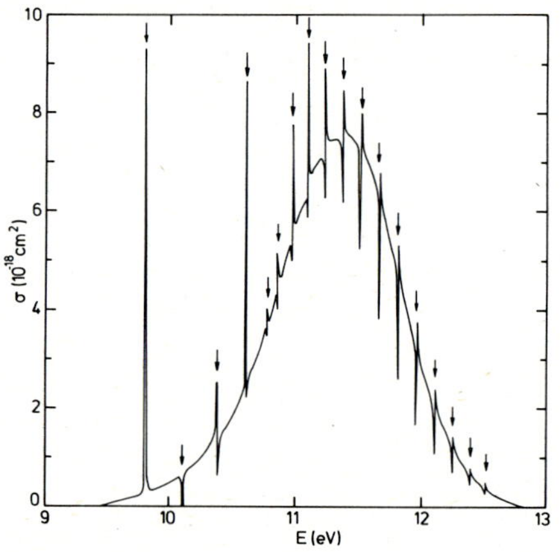

Fig. 2. Photodissociation cross section for absorption into the coupled excited $^2\Pi$ states from the $X^2\Pi(v''=0)$ state. The spectrum has been convoluted with a Gaussian lineshape of width 0.002 eV. The arrows indicate the positions of the vibrational levels of the bound diabatic curve.

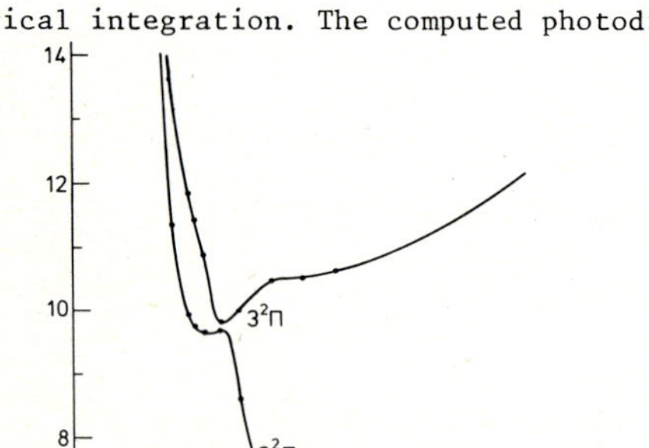

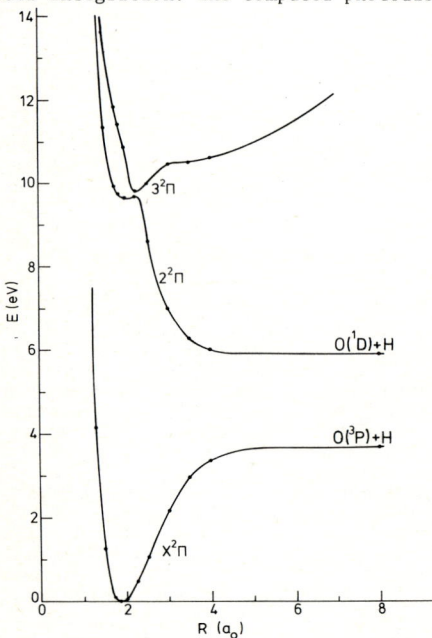

Fig. 1. Adiabatic potential curves for the $^2\Pi$ states.

The dissociation process was also investigated by the following approximate methods:

(i) The predissociation rate for a vibrational level v' of the adiabatic $3^2\Pi$ state by the continuum of the $2^2\Pi$ state can be calculated in first order by the Fermi Golden rule. Rates of the order of 10^{12}-10^{14} s^{-1} were obtained, strongly dependent on v', indicating a 100% dissociation probability upon absorption. The absorption spectrum, however, is quite different from that shown in fig. 2.

(ii) Direct photodissociation cross sections for absorption into the repulsive diabatic $^2\Pi$ potential from the $X^2\Pi(v''=0)$ state have been computed, neglecting couplings with the bound diabatic state. The result is identical to the *background* cross section in fig.2.

References

1. E.F. van Dishoeck and A. Dalgarno, J.Chem.Phys.in press.
2. E.F. van Dishoeck, M.C. van Hemert, A.C. Allison and A. Dalgarno, in preparation.
3. T.G. Heil, S.E. Butler and A. Dalgarno, Phys.Rev.A 23, 1100 (1981).
4. U. Fano, Phys.Rev. 124, 1866 (1981).

THE PHOTODISSOCIATION OF H_2O AT 157 nm: FULL INTERNAL STATE DISTRIBUTIONS AND ALIGNMENT OF NASCENT $OH(X^2\Pi)$ RADICALS

P. Andresen, G.S. Ondrey, E.W. Rothe and B. Titze

MPI für Strömungsforschung, D3400 Göttingen, Fed.Rep. of Germany

The photodissociation of H_2O in its first absorption band in the VUV-range

$$H_2O(^1A_1) + h\nu\ (157\ \text{nm}) \rightarrow H_2O(^1B_1) \rightarrow OH(^2\Pi) + H$$

is one of the rare direct dissociation processes that occur on a single, well defined potential surface. The clear physical conditions met here make a full quantum treatment possible and a comparison of experiment and theory should give a much better insight into the dynamics of this important class of dissociation processes. Most important however, is that this dissociation explains for the first time the pump mechanism of the astronomical OH-maser.

In our experiment we dissociate water with an F_2-Excimer laser at 157 nm and probe the nascent internal state distribution of OH by LIF via the $^2\Pi$-$^2\Sigma$-absorption band. To study the effect of initial rotations of H_2O prior to dissociation, water can be admitted to the machine either as a pulsed, seeded nozzle beam ($T_{rot} \sim 10$ K) or as an effusive beam ($T_{rot} \sim 300$ K). In a polarization experiment both laser were polarized to measure the alignment of OH after dissociation.

Some very interesting results are found in the OH state distributions.

1. We do find vibrational excitation in OH, the ratio of v=1/v=0-population is 0.6. This effect cannot be explained by an impulse or Franck Condon model. It is due to the symmetry of the excited a_1^* orbital at 157 nm.
2. Very little of the available energy is partitioned into OH rotation. The rotational distribution for both vibrational states are nearly identical. The rotational excitation is due to the collinearity of the $H_2O(^1B_1)$ potential surface. The rotational distribution in the Q-branch is slightly colder (~ 100 K) when rotationally cold water is dissociated, indicating a transfer of the parent rotation to the products.
3. The population of the Λ-doublets in OH is highly inverted. This is explained by an electronic correlation from the excited state $H_2O(^1B_1)$ to the product OH ("reflection symmetry" of the electronic orbitals relative to the H_2O-plane). This population inversion explains the pump mechanism for the astronomical OH-maser.
4. Large alignment of the OH product is found, demonstrating the perpendicular nature of the $H_2O(^1A_1 \rightarrow {}^1B_1)$ excitation and the directness of the dissociation.

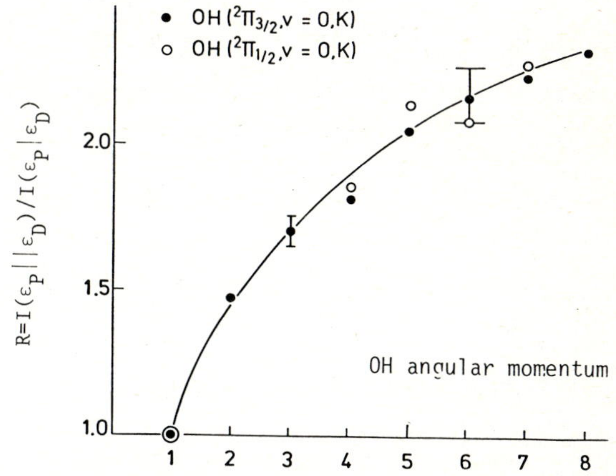

Fig. 1: Polarization ratio R as function of OH angular momentum. ϵ_D and ϵ_P are the electrical vectors of dissociation and probe laser respectively. The LIF intensity I is measured ⊥ to both laser beams for $\epsilon_P||\epsilon_D$ and $\epsilon_P\perp\epsilon_D$.

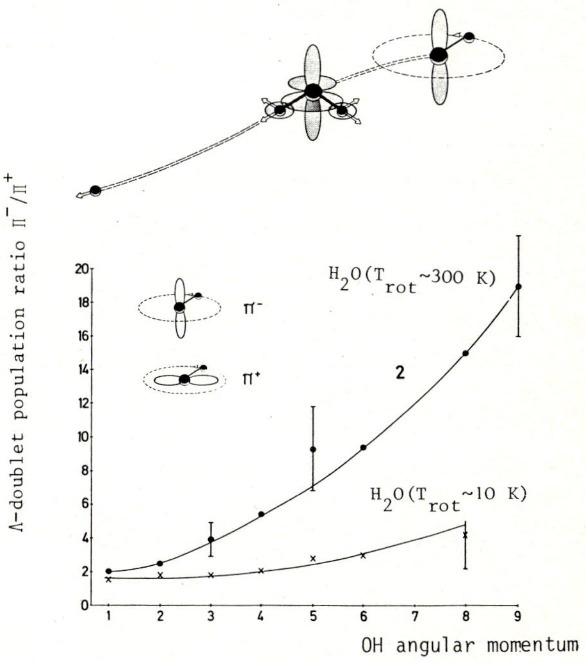

Fig. 2: Λ-doublet population inversion. The population in a Λ-doublet is given by the LIF intensity on Q-lines (Π^-) and R-lines (Π^+). The upper part explains the correlation of $H_2O(^1B_1)$ to the Π^--Λ-doublet.

THE PHOTODISSOCIATION OF CYANOGEN

Joshua B. Halpern, Xiao Tang and William M. Jackson

Department of Chemistry, Howard University, Washington, D.C., USA

We report here on new results in the photodissociation of cyanogen. It has been established conclusively that the primary process in the dissociation of the $C^1\Pi_u$ state of cyanogen is $C_2N_2 \rightarrow CN(A^2\Pi_i) + CN(X^2\Sigma^+)$. The quantum state distribution of the $A^2\Pi_i$ state fragment has been measured as a function of the photolysis energy. Clear evidence is found for the opening of a new channel, $C_2N_2 \rightarrow 2\ CN(A^2\Pi_i)$ at 154 nm. These results complement previous measurements of the $CN(X^2\Sigma^+)$ fragment state distributions.

Quantum state distributions of $A^2\Pi_i$ CN radicals can be easily measured by laser excitation of the $A^2\Pi_i \rightarrow B^2\Sigma^+$ system and measurement of the resulting fluorescence in the Violet system. The photolysis source is a capillary discharge in argon dispersed through a VUV monochromator. The capped end of the capillary forms the entrance slit of the monochromator.

The rotational distribution of the v"=0 $CN(A^2\Pi_i)$ fragments can be well described as Boltzmann. They can be parameterized as having a "temperature"ranging from 2600 to 3000 K. This increase is correlated to the increase of the photolysis energy from 60,600 to 65,000 cm^{-1}. However, the most striking feature is the appearance of a second, linear component, in the Boltzmann plot of the fragment state distribution. The second component has a higher slope, meaning that on average the fragments have less rotational energy. This second component only appears in the highest energy photolysis. State yield measurements have shown that there is measureably more $A^2\Pi_i$ CN at this higher energy.

The $A^2\Pi_i$ fragments have on average less vibrational and more rotational energy than the $X^2\Sigma^+$ fragments. The complete set of results are parsed using surprisal analysis to separate the statistical and dynamical parts of the photolysis process. Comparisons are also made with our results on the predissociation of the $B^1\Delta_u$ state of cyanogen into the ground state continuum.

Experiments are currently underway to measure the rotational distribution of the higher vibrational levels of the A state fragments.

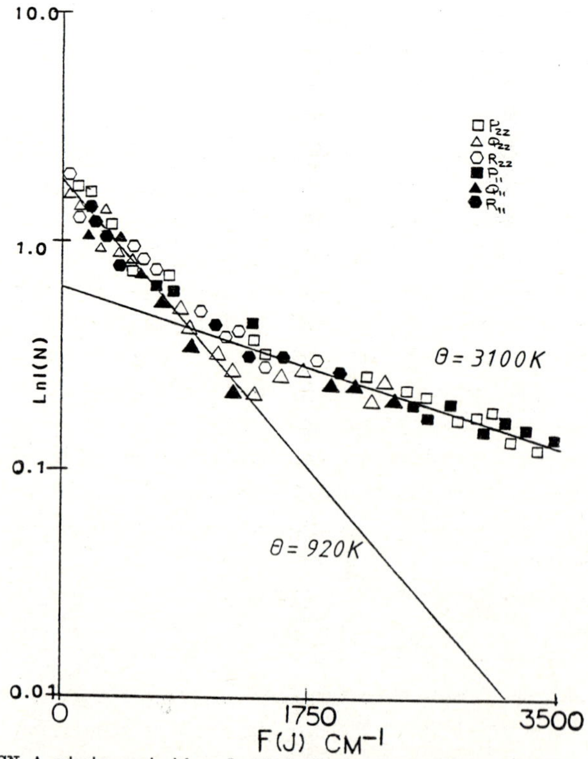

CN A state rotational distribution plotted as a function of rotational energy. The photolysis wavelength was 153.5 nm. The cyanogen pressure was 20 mtorr.

FLUORESCENCE EXCITATION SPECTRA OF NEUTRAL FRAGMENTS FROM PHOTODISSOCIATION OF OCS MOLECULES BY VUV INCIDENT PHOTONS

J. Delwiche*[a], M-J. Hubin-Franskin*[a], A. Tabché-Fouhaile[b], H. Fröhlich[c], K. Ito[d], P-M. Guyon[c], I. Nenner[e]

L.U.R.E., Bât. 209c, Université de Paris-Sud, 91405 Orsay, France

a. Institut de Chimie, Bât. B.6, Université de Liège, Sart-Tilman par 4000 Liège 1, Belgique.

b. Université Libanaise, Faculté des Sciences, -el Hadeth, Beyrouth, Liban.

c. Laboratoire des Collisions Atomiques et Moléculaires, Bât. 351, Université de Paris-Sud, 91405 Orsay, Cédex, France.

d. National Laboratory High Energy Physic, Tokyo, Japan.

e. Centre d'Etudes Nucléaires de Saclay, 91191 Gif-sur-Yvette, Cédex, France.

*. "Chercheur Qualifié" of the "Fonds National de la Recherche Scientifique" of Belgium.

The undispersed fluorescence emitted by OCS has been observed for λλ 67-112 nm incident photons issued from A.C.O., Orsay's storage ring.

The experimental setup has been described previously(1). Briefly, the incident photon beam after monochromatization by a 2400 g/nm grating enters the fluorescence cell ; the bandpass is set at 0.1 and 0.2 nm. The fluorescence observed at right angle of the incident photon beam is detected in three different ranges, the UV (160-300 nm) with an RTC56 SBUVP photomultiplier, the visible (280-630 nm) with an RTC XP20-20 photomultiplier and the near infra-red (530-630 nm) by restricting the RTC XP20-20 photomultiplier bandpass by a J5300 filter.

The fluorescence excitation spectra (FES), recorded in a more extended range than previously(2) shows that below the $\tilde{A}^2\Pi$ ionic threshold the fluorescence is partly due to the CO and CS triplets decay in addition to the well known CS $A^1\Pi \rightarrow X^1\Sigma^+$ emission(3). The fluorescence yield of these triplets is estimated not to exceed 2-3 % of the absorption cross section. The dissociation neutral states leading to these fragments are shown to predissociate all the Rydberg states, members of the series converging to the $\tilde{B}$ ionic level.

In all emission ranges studied here, the fluorescence excitation spectra are very similar to the threshold energy photoelectron spectrum(4). Thus, like in N_2O(5), there is an interconnection between resonant autoionization and the dissociation into fluorescent neutral fragments

We are grateful to the L.U.R.E. and the "Laboratoire de l'Accélérateur Linéaire" staffs for operating the ACO storage ring and general facilities and the "Fonds National de la Recherche Scientifique" of Belgium for financial support.

References

1. A. Tabché-Fouhaile, K. Ito, I. Nenner, H. Fröhlich and P-M. Guyon, J. Chem. Phys. 77, 182 (1982).
2. G.R. Cook and M. Ogawa, J. Chem. Phys. 51, 647 (1969). L.C. Lee and C.C. Chiang, Chem. Phys. Lett. 92, 425 (1982).
3. L.C. Lee and D.L. Judge, J. Chem. Phys. 63, 2782 (1975).
4. J. Delwiche, M-J. Hubin-Franskin, P-M. Guyon and I. Nenner, J. Chem. Phys. 74, 4219 (1981).
5. P-M. Guyon, T. Baer and I. Nenner, to be published in J. Chem. Phys.

LASER MULTIPHOTON IONIZATION AND DISSOCIATION OF POLYATOMIC MOLECULES.

I. Dimicoli, J. Jaraudias, J. Lemaire and R. Botter

Département de Physico-Chimie, CEN/Saclay, 91191 Gif-sur-Yvette Cedex (France)

Introduction.

One or two color laser multiphoton ionization of polyatomic molecules in the gas phase is rapidly becoming a promizing spectroscopic method (1, 2).
When ionization occurs via a real intermediate excited state the process is resonance enhanced. The scanning of the wavelength of one of the two lasers enables the spectroscopic study of this intermediate or of the ionized state. The ions produced by resonant two photon ionization (R2PI) may be analysed by mass spectrometry. This technique combined with the possibility to delay one of the laser pulse enables the determination of the various relaxation pathways and their kinetics (dissociation, conversion processes...). The photoion-photoelectron coincidence (PIPECO) technique coupled with R2PI offers to the experimentalist a high sensitivity and is a good basis for the study of the formation and photofragmentation of an isolated ionic state of the molecule.

Experimental.

The experimental set up has been described in detail elsewhere (3).

The beam of two pulsed lasers (2 UV or UV + visible) are focussed in the ionization region of a PIPECO time of flight mass spectrometer (PIPECO-TOFMS), where they cross an effusive jet of molecules. The laser pulse can be delayed one with respect to the other. The ions produced can be analysed by two methods :

- PIPECO technique (4). In the present apparatus threshold photoelectrons are selected by a steradiancy analyzer and electronically gated. The flight duration of the corresponding ions with well defined internal energy is determined with a time-to-digital converter which is connected for subsequent analysis to a mini-computer.
- Recording of the whole mass spectrum for every laser pulse. The current is processed by a transient digitizer triggered by the laser pulse.

Results and discussion.

The present experiments have been performed on some benzene derivatives.

The R2PI - technique and our experimental device can be used to several ends.

- The spectroscopy of the variety of real intermediate molecular states. This spectrum appears when the ionization current is measured as a function of the first laser wavelength λ_1 with λ_2 constant .
- The spectroscopy of the ionic states of the molecule when the second laser pulse is scanned.
- The fragmentation of both the ionic and neutral states of the molecule as a function of the wavelength and intensity of the laser.
- Kinetic studies of the dissociation or relaxation processes of excited neutrals and ions.

Taking advantage of the high wavelength selectivity of the first step the identification of molecules can be performed not only by their mass but also in addition by their typical optical molecular spectrum. Different components of a gas mixtures can be selectively ionized at different wavelengths. This yield a new dimension to mass spectrometry.

The R2PI mass spectrum of benzylchloride with trace of toluene (< 1 %) shows, for example, a different (by a factor of 5) overall ionization probability for the parent ions of these molecules at 262 and 267 nm ($\lambda_1 = \lambda_2$).

In general, two UV photons are necessary to form the parent ion. The higher the laser flux the greater the probability to absorb further photons.

By measuring the ion yield of given mass in function of the laser intensity (I) the mechanism of different processes is determined. In the case of benzylamine principally five peaks are observed at 262 nm : masses 107 (parent ion), 91 ($C_6H_5CH_2$), 79, 77 and ∿ 29.

A nonlinear intensity dependence is unambigously determined for the formation of the different ions (I^2 for the parent ion, I^3 for the mass 79, and I^{6-7} for the low mass fragments). Different states of the ion are reached. They evolve via a slow or a fast dissociation giving different ionic products. The ions produced by a slow decomposition (time range of 0.1 to 1 µs) appear as metastable (non-symmetric peak) in the TOF-MS. Aniline exibits such a metastable peak with mass 66 ($C_6H_5NH_2^+ \rightarrow C_5H_6^+ + CHN$). A rate constant of 1 µs has been determined for this process. In contrast with other work, the laser output intensity in the present experiment is low (some µJ in the UV) avoiding optical cracking or extensive fragmentation of ions.

Studies of the dissociation and relaxation of neutral excited states by time delayed tandem laser pulses of different wavelength are in progress.

References.

1. T.G. Dietz, M.A. Duncan, M.G. Liverman and R.E. Smalley Chem. Phys. Lett. 70, 246 (1980) and ref. herein.
2. W. Dietz, H.J. Heusser, U. Boesl, E.W. Schlag, Chem. Phys. 66, 105 (1982) and ref. herein.
3. I. Dimicoli, J. Jaraudias, J. Lemaire and R. Botter Intern. J. Mass Spectrom. Ion Phys. 46, 281 (1983)
4. T. Baer in Gas phase ion Chemistry, vol. 1, Academic Press, N.J. 1979, p. 153.

DISSOCIATION OF ENERGY SELECTED STATES OF O_2^+ AND CO^+ BY THRESHOLD PHOTOELECTRON-PHOTOION COINCIDENCE

M. RICHARD-VIARD[a], O. DUTUIT[b], T. GOVERS[c], P.M. GUYON[a], H. FROHLICH[a] AND M. LAVOLLEE[a]
LURE (Univ. Paris-Sud and CNRS), Bât. 209, Université Paris-Sud, 91 405 ORSAY, FRANCE

The unimolecular dissociation of O_2^+ and CO^+ is stutied through a threshold photoelectron-photoion coincidence technique (TPEPICO)[(1)]. The parent ion is prepared by photoionisation of neutral O_2 or CO, using monochromatised synchrotron radiation from the storage ring ACO as tunable photon source ($h\nu \simeq 12 - 35$ eV). Its internal energy is specified as being equal (within 30 meV, fwhm) to the incident photon energy by selective detection of threshold electrons. This selection is achieved by combined angular and time of flight discrimination. The threshold electron signal triggers the extraction of the corresponding ion and starts a multichannel time analyser which is stopped by the ion signal (TOF $\leqslant 2\ \mu s$). From the ion TOF spectrum (fig. 1) one obtains the masses, the kinetic energies and the relative abundances of the parent and fragment ions. The electronic states of the fragments and the branching ratios between the dissociation channels have been obtained through a comparison of the experimental TOF spectra with simulated ones, using Monte-Carlo trajectories.

The dissociation of the O_2^+ states has been investigated between 20 and 26 eV and the results have been compared with ab initio calculations[(2)]. The present data supersede an earlier report [(1)].

(i) The B $^2\Sigma_g^-$ state is predissociated to the first dissociation limit O (^{3}P) + O$^+$ (^{4}S°) at 18.73 eV. Spin-orbit coupling with the $d^4\Sigma_g^+$, f $^4\Pi_g$ and/or $1^2\Sigma_g^+$ states can explain this predissociation. The level v = 4 (20.815 eV) predissociates also to O (^{1}D) + O$^+$ (^{4}S°) at 20.70 eV, to an extent of $\sim$ 10 %, the level v = 5 does not. The particular behavior of v = 4 has been ascribed to interaction with the 2 $^4\Pi_g$ state [(3)]; we proposed that this perturbation may proceed through an intermediate 2 $^2\Pi_g$ or 2 $^4\Sigma_g^-$ state[(2)].

(ii)In the region of the III $^2\Pi_u$ we observe not only fragments correlated with the latter, but also dissociation into the other accessible limits. Angular distribution with negative anisotropy parameter have been observed.

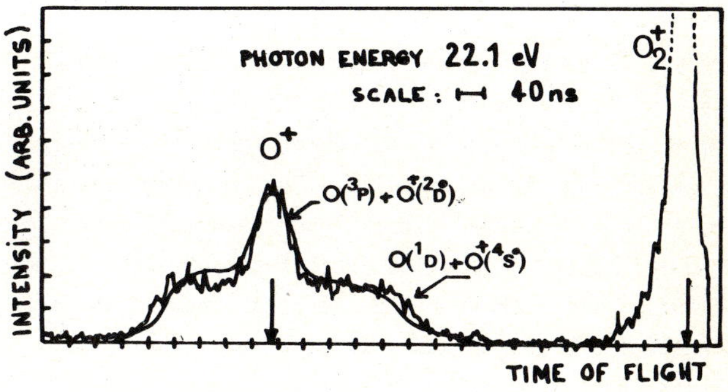

Fig. 1 : TOF spectrum in the III^2 $_u$ region. The dissociation to the limit O (^{3}P) + O$^+$ (^{2}D°) at 22.06 eV appears. The O_2^+ peak is essentially due to " hot " electrons.

(iii) The $c^4\Sigma_u^-$ (v = 0) dissociates mainly to the limit O (^{1}D) + O$^+$ (^{4}S°), possibly by tunneling through a potential barrier as predicted by theoretical study[(4)]. In addition it predissociates to an extent of about 33% to O (^{3}P) + O$^+$ (^{4}S°) at 18.73 eV.

In the case of CO^+ we have measured the branching ratios between the different dissociation channels for eight values of energy ranging from the 22.37 eV (C$^+$ + O) threshold to 32 eV. The branching ratios depend rather erratically on energy. Below 28 eV the channels C$^+$ + O dominates although many channels C + O$^+$ are accessible. At 32 eV the channel C + O$^+$ appears clearly, representing about 34% of the (pre-) dissociative decay.

REFERENCES

1. P.M. Guyon et al., J. Phys. B. 11 (1978) L 141
2. N.H.F. Beebe et al., J. Chem. Phys. 64 (1976) 2080
3. R.G.C.B. Blyth et al., Chem. Phys. Lett.84 (1981) 272
4. K. Tanaka and M. Yoshimine, J. Chem. Phys. 70 (1979 1626

(a) L.C.A.M., Bât 351, Univ. Paris-Sud, 91405 ORSAY
(b) L.R.E.I., Bât 350, Univ. Paris-Sud, 91405 ORSAY
(c) LPCR, Orsay ; present address : LPPM, Bât 213 Univ. Paris-Sud, 91405 Orsay

PHOTOFRAGMENT SPECTROSCOPY OF MOLECULAR IONS

E. Solarte, V. Hermann, R. Anselmann and F. Linder

Fachbereich Physik, Universität Kaiserslautern
D-6750 Kaiserslautern, West Germany

A new apparatus (shown schematically in Fig. 1) has been built to study the photofragment spectroscopy of molecular ions. The first section of the machine is used to produce a beam of molecular ions with a narrow energy and angular distribution. The ion beam is then accelerated and merged with a laser beam to produce the molecular dissociation. The ionic fragments are deflected out of the beam axis and after collimation and deceleration are energy analyzed by a double hemispherical condenser and counted by a multiplier. In the beginning phase of this project, complementary to other developments in this field[1-4], the emphasis will be placed on measurements of energy spectra of the photofragments. The machine has been designed with sufficient overall energy and angular resolution in order to allow us to obtain a center-of-mass energy resolution of a few meV in these spectra. These measurements will provide detailed information on the potential curves of the molecular ions, in particular on repulsive potentials, as well as on the dissociation dynamics.

The first measurements on this apparatus showed encouraging results. Sufficiently intense beams of simple molecular ions (e.g. H_2^+, Ar_2^+) were produced and guided through the ion optics. The measured energy and angular widths were found to be in agreement with expectation. First results of the photofragment spectra, which are intended to demonstrate the energy resolution of the spectrometer, will be reported at the conference.

References

1. P.C. Cosby, J.B. Ozenne, J.T. Moseley, D.L. Albritton, J. Mol. Spectrosc. 79, 203 (1980)
2. H. Helm, P.C. Cosby, M.M. Graff, J.T. Moseley, Phys. Rev. A 25, 304 (1982)
3. A. Carrington and J. Buttenshaw, Mol. Phys. 44, 267 (1981)
4. M. Carre, M. Druetta, M.L. Gaillard, H.H. Bukow, M. Horani, A.L. Roche, M. Velghe, Mol. Phys. 40, 1453 (1980)

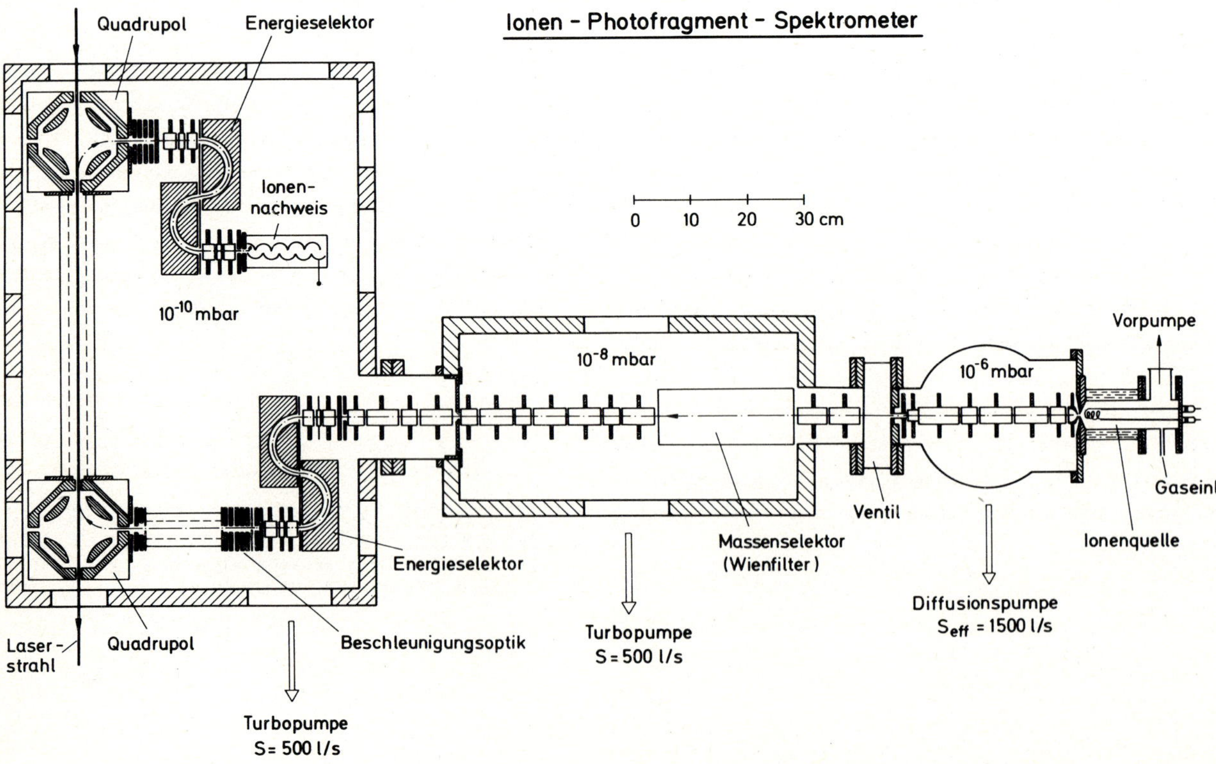

Fig. 1

PHOTOFRAGMENT SPECTROSCOPY OF Cs_2^+

H. Helm, R. Möller, P. C. Cosby, and D. L. Huestis

Molecular Physics Laboratory, SRI International, Menlo Park, CA 94025

A first experimental analysis of the electronic structure of Cs_2^+ has been made using a fast ion-beam photofragment-spectrometer. Molecular cesium ions were formed by field ionization from a liquid cesium droplet which is suspended on a tungsten needle. The beam of mass-selected Cs_2^+ ions at energies between 2 and 4 keV is photodissociated using dye lasers and the fixed laser lines of argon and krypton ion lasers covering portions of the wavelength range between 9500 and 4067 Å. The photodissociation of Cs_2^+ is observed by monitoring the intensity, the energy distribution and the angular distribution of the charged photofragment Cs^+. Over this wavelength range three dissociation channels were identified of the type

$$Cs_2^+(X^2\Sigma_g^+) + h\nu \rightarrow Cs_2^{+*} \rightarrow Cs^+(^1S_o) + Cs(n\ell) + E_{kin} \cdot$$

involving four intermediate excited states Cs_2^{+*}.

In the wavelength range around 9400 Å photodissociation of "cold" Cs_2^+ $(X^2\Sigma_g^+)$ produces ground state atoms through the bound-free transition (see Figure 1)

$$Cs_2^+ (X^2\Sigma_g^+) \xrightarrow{9400Å} Cs_2^+(1^2\Sigma_u^+) \rightarrow Cs^+ + Cs(6s) + .72 \text{ eV}.$$

Using linearly polarized laser light this transition leads to a strongly anisotropic angular distribution of photofragments which is consistent with the theoretical $(3/4\pi)\cos^2\Theta$ distribution expected for a parallel bound-free transition. (Θ is the angle of ejection of photofragments with respect to the laser polarization in the laboratory frame).

In the wavelength range between 7700 and 8100 Å predissociation of vibrational levels in the $1^2\Pi_u$ state of Cs_2^+ by interaction with the $1^2\Sigma_u^+$ state is observed

$$Cs_2^+(X^2\Sigma_g^+) \xrightarrow[8100 Å]{7700-} Cs_2^+(1^2\Pi_u) \rightarrow Cs^+ + Cs(6s) + 1 \text{ eV}.$$

The excitation transition shows two relatively narrow, weakly-structured peaks associated with transitions into the $^2\Pi_u(\Omega = 1/2)$ and $^2\Pi_u(\Omega = 3/2)$ substates. The spin-orbit splitting in the $^2\Pi_u$ state is measured to be 280 cm^{-1}. Unexpected and yet unexplained is the observation of a nearly isotropic angular distribution of photofragments in this reaction.

In the wavelength range between 5309 and 4545 Å excited state atoms Cs(6p) are produced. Beginning at 4579 Å photodissociation leads to the formation of both Cs(6p) and Cs(5d). The Cs(5d) limit dominates at wavelengths around 4100 Å. The measured angular distribution of photofragments indicates that both the $2^2\Pi_u$ and $2^2\Sigma_u^+$ state are involved in the formation of fragments Cs(6p). This may be taken as indicating the action of nonadiabtic coupling between these two electronic states as suggested by the theoretical[1] potential energy curves in Figure 1.

The analysis of the photofragment separation energies leads to a lower limit for the bond energy of $Cs_2^+(X^2\Sigma_g^+)$ of 0.59 ± 0.06 eV. This places an upper limit of 3.76 ± 0.06 eV for the ionization potential of $Cs_2(X^1\Sigma_g^+)$. A lower limit for the bond energy of Cs_2^+ $(1^2\Pi_u)$ is found to be 0.39 ± 0.1 eV.

*This research was supported by the National Science Foundation under Grant No. PHY 8112534.

References
1. A. Wetmore, F. K. Men, M. Kimura, and R. E. Olson, J. Chem. Phys., to be published.

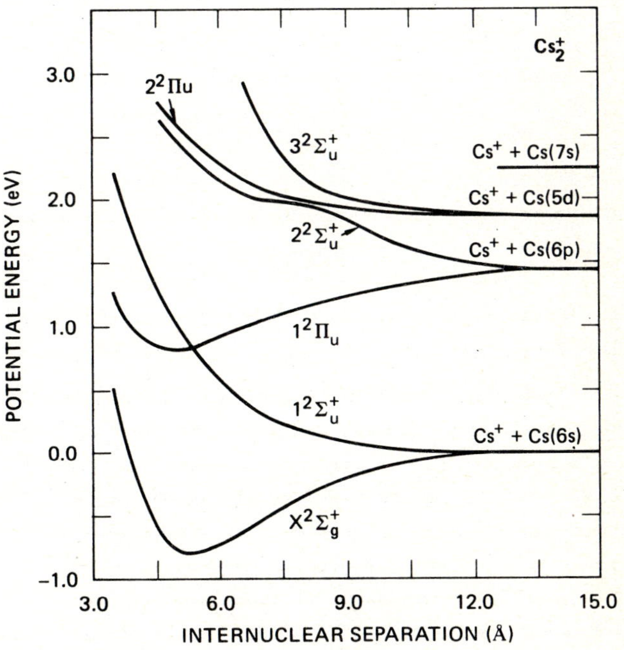

Fig. 1: Theoretical potential energy curves for the ground state and the lowest ungerade states of Cs_2^+ taken from Reference 1.

PHOTODISSOCIATION OF N_2^{++}

P. C. Cosby, R. Möller, and H. Helm

Molecular Physics Laboratory, SRI International, Menlo Park, CA 94025, USA

Electronic states of doubly-charged diatomic molecules dissociating to the single-charged atoms are mainly characterized by Coulomb repulsion of the atomic ions. However, at small internuclear distances, the chemical forces can lead to a local minimum in the potential energy curve behind the coulombic barrier which may support quasi-bound states with nearly infinite lifetime. Experimental evidence for such long-lived metastable states is well known from electron-impact, ion-bombardment, and Auger-electron experiments. However, the only detailed information on the structure of a doubly-charged ion has come from a single emission band observed by Carroll[1] in a N_2 - He discharge which has been attributed to the $\underline{D}^1\Sigma_u^+(v=0)\rightarrow\underline{X}^1\Sigma_g^+(v=0)$ transition in N_2^{++}.

We have measured the photodissociation of N_2^{++} into $N^+ + N^+$ over the photon energy range of 14900-19500 cm^{-1}. A series of rotationally-resolved bands appears in the spectrum due to predissociation of an excited electronic state of N_2^{++}. Rotational analysis of the bands and the observed alternation of their line intensities demonstrates that they arise from the $1^1\Pi_u \leftarrow \underline{X}^1\Sigma_g^+(v=0,1,2)$ absorption.

Molecular constants are obtained for the $\underline{X}^1\Sigma_g^+$ and $1^1\Pi_u$ states. In addition, kinetic energy analysis of the N^+ photofragments arising from the photodissociation fixes the energy of $\underline{X}^1\Sigma_g^+(v=0)$ at 4.8±0.2 eV above the $N^+ + N^+$ dissociation limit. The Rydberg-Klein-Rees (RKR) potential energy curve for the $\underline{X}$ state constructed from these constants is shown by the lower solid curve in Fig. 1. For comparison the theoretical curve calculated by Thulstrup and Andersen[2] is given by the lower long-dash curve.

Three vibrational levels of the $1^1\Pi_u$ state are observed to predissociate. The two highest energy levels dissociate by tunnelling through the coulombic barrier in this state's potential energy curve. The linewidths of the absorptions into the highest energy level give a predissociation lifetime of $>1.6 \times 10^{-10}$s. In contrast, the lowest of the three levels is predissociated by rotational coupling to the $1^1\Sigma_u^-$ state.

The vibrational numbering of the three observed levels in the $1^1\Pi_u$ state is not known. Consequently, an unique potential energy curve cannot be constructed for this state. The theoretical curve from Thulstrup and Andersen, shown by the upper long-dashed curve in Fig. 1, is found to support four vibrational levels. But the semiempirical potential curve of Hurley and Maslin[3] predicts a deeper well with 10 vibrational levels. Two RKR potential curves for the $1^1\Pi_u$ state are shown in Fig. 1. The upper solid curve assumed a vibrational numbering of v = 1,2,3 for the predissociated levels while the short-dashed curve assumed a numbering of v = 6,7,8. Thus, the range of possible forms of the $1^1\Pi_u$ potential energy curve consistent with the experimental data will lie somewhere between these two extremes.

References
1. P. K. Carroll, Can. J. Phys. 36, 1585 (1958).
2. E. W. Thulstrup and A. Andersen, J. Phys. B 8, 965 (1975).
3. A. C. Hurley and V. W. Maslen, J. Chem. Phys. 34, 1919 (1961).

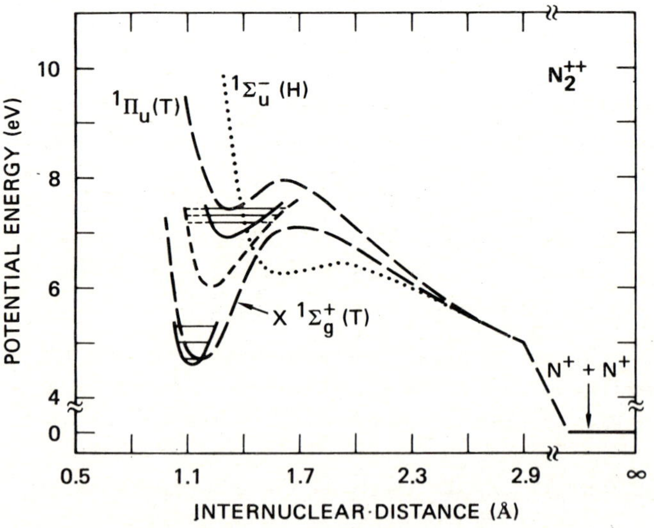

Fig. 1: Potential energy curves of N_2^{++}.

UV-Laser Photodissociation of Molecular Ions

R. E. Kutina, A. K. Edwards† and J. Berkowitz

Argonne National Laboratory, Argonne, IL 60439

We report here the first results from a new experiment, in which fragment ions produced by UV laser radiation are momentum analyzed. This experiment, performed with $h\nu = 6.42$ eV, is a significant extension of earlier studies using $h\nu \cong 2\text{-}3$eV. A much larger class of molecular ions now become subject to photodissociation consequent upon single photon absorption.

The apparatus consists of an electron impact ion source, a quadrupole mass filter for selecting target ions, an excimer laser and a magnetic mass spectrometer for momentum analysis of the fragment ions. The C.M. recoil velocity is measured in two ways—momentum analysis and time of flight.

A few of our preliminary results are summarized below.

(1) $$h\nu + D_2O^+ \rightarrow OD^+ + D$$

With $h\nu = 6.42$ eV, the exothermicity should be 0.96 eV if D_2O^+ is in its ground electronic and vibrational state. We observe a most probable recoil energy of 2.6 ± 0.3 eV. A tentative explanation is that the target ion is vibrationally hot.

(2) $$h\nu + D_2O^+ \rightarrow D^+ + OD$$

Although this reaction should be exothermic by at least 0.33 eV, it is not observed.

(3) $$h\nu + OD^+ \rightarrow O^+ + D$$

The calculated exothermicity is 1.34 eV. We observe 1.25 ± 0.2 eV.

(4) $$h\nu + OD^+ \rightarrow D^+ + O$$

Although the calculated exothermicity is almost the same as for (3), this process is not observed.

(5) $$h\nu + ND_3^+ \rightarrow ND_2^+ + D$$

The calculated exothermicity is 0.82 eV; the observed is 1.95 ± 0.4 eV. A probable interpretation is that the parent ion is formed with substantial vibrational excitation, since the ionization process involves a pyramidal-to-planar transition.

(6) $$h\nu + ND_3^+ \rightarrow ND^+ + D_2.$$

This reaction is nominally endothermic by 0.35 eV. We observe this reaction with a kinetic energy release of 0.41 ± 0.14 eV. Our tentative explanation is similar to that given in (5).

These preliminary results demonstrate that factors beyond simple energetics are involved, as might have been anticipated. Studies at other wavelengths and power levels are in progress, which should clarify the tentative interpretations presented.

---

†Permanent address: Dept. of Physics, University of Georgia, Athens, GA.

LASER PHOTODISSOCIATION OF CO_3^- AND ITS HYDRATES

A. W. Castleman, Jr.*, D. E. Hunton*, M. Hofmann+ and T. G. Lindeman†

*Department of Chemistry, The Pennsylvania State University, University Park, PA 16802

+Ciba-Geigy, Basel, Switzerland

†Chemistry Department, Middlebury College, Middelbury, VT 05753

The technique of laser-induced photodissociation combined with analysis of the kinetic energy distributions of the dissociation products provides a sensitive probe of the structure of ion clusters and of the mechanisms of energy deposition and release following laser excitation. In a recent series of experiments with CO_3^- and its hydrates, the effect of the degree of clustering on the electronic structure of the ion as well as the nature of the sequences of steps leading to laser induced dissociation were investigated.

The photodissociation of CO_3^- is found to yield O^- in accordance with previous results published in the literature, where there is considerable discrepancy concerning the magnitude of the cross sections in the energy range 1.8 to 2.3 eV. in particular, questions have arisen in interpretation due to the very low energy photon required to yield photodissociation (below the accepted bond dissociation energy of CO_2-O^-) and involve issues concerning possible effects due to multiphoton dissociation of CO_3^- or the possible influence of excited ions. Recent studies in our laboratory show that unclustered CO_3^- is initially excited by a weakly allowed transition to a bound excited state which has been identified as the source of the structure in the photodissociation spectrum. Dissociation following this excitation occurs by two mechanisms: absorption of a second photon to a purely repulsive state which correlates with O^-+CO_2, the observed dissociation products, and by CID following a non-radiative transition to a long-lived excited state (Fig. 1).

In the case of $CO_3^-\cdot(H_2O)_{1,2}$, the major photodissociation pathway is shown to be loss of all water ligands leading to CO_3^- as the product ion. Similarities between the photodissociation spectra of the $CO_3^-\cdot(H_2O)_n$, n=0,1,2 species over the energy range 1.95 to 2.2 eV indicate that the initial excitation of the clusters involves the same bound-bound electronic transition that occurs in the bare ion. The major dissociation channel for the clusters is loss of all water molecules, a process which has a significantly higher cross section than that for photodissociation of the bare ion. These observations are consistent with a model in which a radiationless transition from the initially excited electronic state, similar to that experienced by the bare ion, takes the clusters to high-lying vibrational levels of the ground state. This vibrational energy, which is redistributed to the cluster modes, is sufficient to dissociate the clusters (Fig. 2).

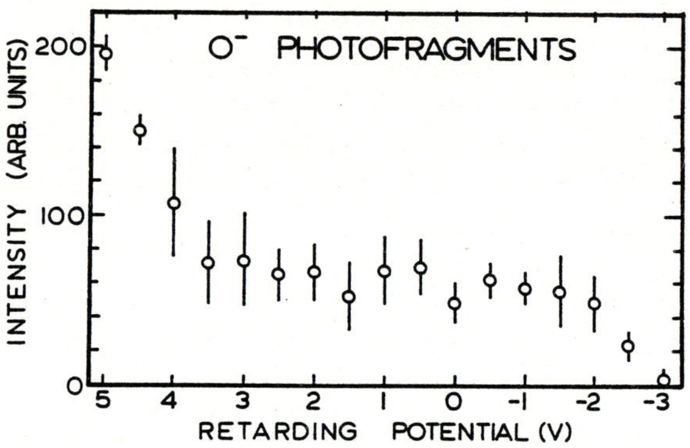

Fig. 1. Integral of the energy distribution of O^- photofragments taken with a retarding field energy analyzer. The two peaks in the distribution indicate that two distinct pathways lead to production of O^-.

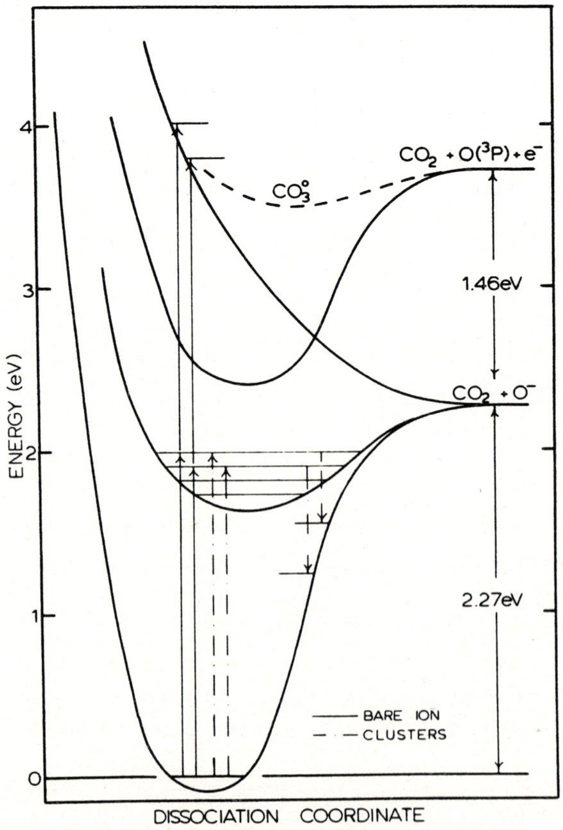

Fig. 2. Model of potential surfaces and relevant transitions of the core CO_3^- ion.

FAST NEUTRAL BEAM MEASUREMENTS OF MOLECULAR PHOTODISSOCIATION

L. D. Gardner and J. L. Kohl

Harvard-Smithsonian Center for Astrophysics, 60 Garden St., Cambridge, MA 02138 U.S.A.

We are using a new technique to measure photodissociation cross sections for neutral molecules that is especially promising for chemically unstable species. The experimental arrangement (see Fig. 1) produces negative ions of the molecule of interest which are accelerated, mass analyzed and subsequently undergo photodetachment by a coaxial pulsed laser beam. The fast neutral beam of molecules so prepared is then photodissociated by a second pulsed laser beam which is perpendicular to the molecular beam. Fragments from the photodissociation can be detected away from the beam axis with either an array of discrete secondary electron type particle multipliers or a single large area position sensitive particle detector such as a channelplate with a multiple anode array.[1,2] The detection of all fragments from a single dissociation event together with measurements of their impact positions on the detector and the time of detection, permit the mass of each fragment as well as the kinetic energy of each fragment to be determined.

To date we have constructed the apparatus and made measurements on the photodetachment of several molecular and atomic ions.[3] Additional measurements on the photodetachment of CH^- and OH^- will be reported here. Recently, to test the apparatus, we have observed the photodissociation of excited states of H_2^+ using a pulsed Nd:YAG pumped dye laser that is operated near 640 nm. Fast neutral fragments, i.e., H atoms, are detected away from the beam axis with an array of discrete particle multipliers. Charged fragments, i.e., H^+ ions, are deflected out of the beam by a low resolution parallel plate electrostatic analyzer, and are detected with a particle multiplier. Efforts to record both the H^+ and H fragments in coincidence are underway.

Experiments are also being carried out on the photodissociation of the CH radical through its predissociating $C^2\Sigma^+$ state.[4] This state can be reached from the CH ground state with photon wavelengths near 314 nm.[5] Progress on these experiments will be reported.

This work was supported by the Smithsonian Institution through its Scholarly Studies Program and by the Office of Naval Research Contract N00014-83-K-0134 to the Smithsonian Astrophysical Observatory.

References

1. J.G. Timothy, G.H. Mount, and R.L. Bybee, IEEE Trans. Nucl. Sci. NS-28, 689 (1981).
2. D.P. de Bruijn and J. Los, Rev. Sci. Instr. 53, 1020 (1982).
3. L.D. Gardner, Electronic and Atomic Collisions, Abstracts of Contributed Papers to the XII International Conference on the Physics of Electronic and Atomic Collisions, S. Datz, ed. (Gatlinburg, TN, 1981), p. 52.
4. J. Brzozowski, P. Bunker, N. Elander, and P. Erman, Astrophys. J. 207, 414 (1976).
5. G. Herzberg and J.W.C. Johns. Astrophys. J. 158, 399 (1969).

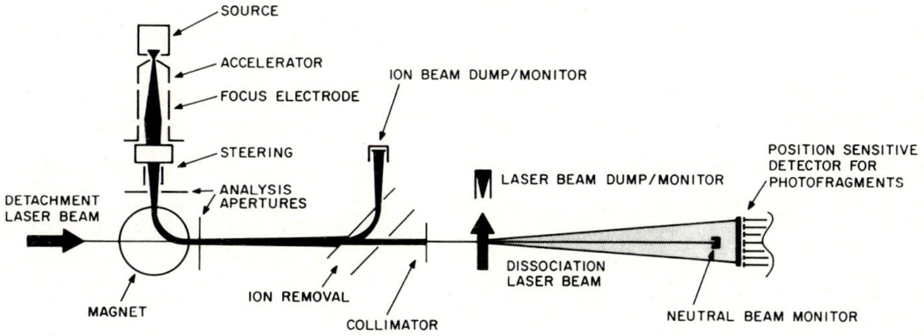

Fig. 1: Schematic of the Experiment.

A BASIS SET APPROACH TO MULTIPHOTONIONIZATION INCLUDING FREE-FREE TRANSITIONS

John T. Broad
Fakultät für Chemie, Universität Bielefeld, D4800 Bielefeld, Fed. Rep. Germany

While L^2 basis set calculations of the total multiphoton ionization probability have been successfully carried out[1] using the Floquet formalism and rotated coordinates to represent the ionization as a resonant transition in the atom-field system, finite basis set projections to compute the flux into particular continuum states do not in general converge. A solution using an expansion[2] of the Coulomb wave function and Green's function in infinitely many basis functions is presented here along with test results on the two photon ionization of the hydrogen atom above the single photon ionization threshhold.

In the basis,

$$\phi_n(r;\lambda) = (\lambda r)^{\ell+1} e^{-\lambda r/2} L_n^{2\ell+1}(\lambda r), \qquad (1)$$

the radial Coulomb Hamiltonian, $H = -1/2d^2/dr^2 + \ell(\ell+1) - 1/r$, is an infinite tridiagonal symmetric matrix. The three term recursion relation inherent herein leads to an explicit expression for the expansion coefficients of the radial Coulomb wave function,

$$\psi_n^{\ell}(E;\lambda) = \psi_0^{\ell}(E;\lambda) p_n / \begin{matrix} n+2\ell+1 \\ 2\ell+1 \end{matrix} \quad , \qquad (2)$$

and for the Green's matrix, $(E-H)^{-1}$,

$$G_{nn'}^{\ell}(E;\lambda) = \frac{-\lambda\, n!\, n!\, p_{n_<} q_{n_>}}{2(E+\lambda^2/8)(n+2\ell+1)!n'+2\ell+1)!} \qquad (3)$$

where p_n is a known polynomial in $x = (E-\lambda^2/8)(E+\lambda^2/8)$ and q_n is a known hypergeometric function[2] having poles at the Rydberg states and a branch cut from E=0 to ∞.

To explore the usefulness of the basis set ansatz in describing free-free transitions, consider the cross section for two photon ionization of the hydrogen atom by circularly polarized light, defined in perturbation theory as,

$$\sigma_2 = \pi\omega |T_{ds}|^2 / \mathcal{E}^4 \qquad (4)$$

where $\mathcal{E}$ is the field strength of the light, ω the frequency and T_{ds} is the transition matrix having the form,

$$T_{ds} = \langle E_d | \vec{\mathcal{E}}\cdot\vec{r}\; G(E_p) \vec{\mathcal{E}}\cdot\vec{r} | 1s \rangle \quad , \qquad (5)$$

in atomic units, with $E_p = E_{1s} + \omega$ and $E_d = E_{1s} + 2\omega$. Carrying out the angular integrations and introducing the basis set with $\lambda = 2$ appropriate to the 1s state of hydrogen yields the single infinite sum,

$$T_{ds} = \frac{-\mathcal{E}^2\, {}^{d}_{o}(E_d)}{8\sqrt{5}\,\omega^3} \sum_{n=o}^{\infty} \frac{n!\, q_n^p(E_p)}{(n+3)!} \{p_n^d(E_d) - p_{n-2}^d(E_d)\} \qquad (6)$$

When E_p is negative, q_n includes poles at real intermediate states and dies exponentially with increasing n to converge the sum with a few terms, while at positive E_p (free-free transition) the terms in the sum oscillate more and more rapidly with increasing n. Then summing to some n=N≈10 explicitly and using a large n asymptotic expression generated iteratively from the limiting behavior of the p_n and q_n approximates the remainer to desired accuracy, and in agreement with other methods.[3]

The basis set ansatz described above provides a procedure for calculating multiphoton ionization probabilities into particular continuum states when used in conjunction with the Floquet representation and an atomic basis set C1 code. Although choosing λ complex in the basis in Eq. 1) then rotates the branch cut of G and makes the resonance corresponding to the dressed state being ionized square integrable,[4] the asymptotic summing to infinitely many basis functions is still needed to converge transition amplitudes of the type in Eq. 5) because the free Coulomb states explode exponentially in the rotated basis.

References

1. S.-I. Chu and W.P. Reinhardt, Phys. Rev. Lett. 39, 1195 (1977).
2. H.A. Yamani and L. Fishman, J. Math. Phys. 11, 41o (1975). J.T. Broad, Phys. Rev. A26, 3o78 (1982) and "Calculations with L^2 Basis Functions" in Electron-Atom and Electron-Molecule Collisions, J. Hinze Ed. (Plenum, London, 1983).
3. S. Klarsfeld and A. Macquet, J. Phys. B12, L553 (1979); M. Aymar and M. Crance, J. Phys. B13, L287 (198o).
4. M. Reed and B. Simon, Methods of Modern Mathematical Physics (Academic Press, New York, 198o).

SEMICLASSICAL MANY-MODE FLOQUET THEORY

Shih-I Chu#, Tak-San-Ho and James V. Tietz

Department of Chemistry, University of Kansas, Lawrence, Kansas 66044 USA

The use of the (single-mode) semiclassical Floquet theory[1,2] for nonperturbative treatments of the multiphoton dynamics of finite-level systems, involving periodic time-dependent Hamiltonians, has attracted considerable attention in the last few years. A detailed review of the theoretical investigations for two-level systems has been given by Dion and Hirschfelder[2]. Recently the conventional (single-mode) finite-level Floquet theory[1,2] has been extended to infinite-level (to include both bound as well as continuum states) non-Hermitian Floquet or complex quasi-energy theories[3-6], employing the use of complex-coordinate transformation[7] and L^2-continuum discretization. This yields practical techniques for the study of intense field multiphoton ionization of atoms[3], Stark-Zeeman[4] and laser-Zeeman[5] effects as well as multiphoton dissociation (MPD) of small molecules[6], using entirely only bound-state technology.

Lately, there is also much interest in the study of atomic and molecular processes in the presence of two monochromatic radiation fields. Multiphoton double resonance experiments, collisions in two laser fields, MPD of polyatomic molecules by two infrared lasers etc. have been studied extensively.

In view of the usefulness of the semiclassical Floquet approaches for strong field problems, we have looked into the feasibility of generalizing the single-mode Floquet theory to the many-mode[8]. There have been several previous conjectures that this generalization is not possible, as the Hamiltonians for several monochromatic field problems are not explicitly periodic in time.

In this conference we shall show that the single-mode Floquet formalism of Shirley[1] can be rigorously extended to a generalized many-mode Floquet theory, yielding a practical and powerful nonperturbative technique for the semiclassical treatment of the interaction of a quantum system with several monochromatic oscillating fields[8]. Figure 1 shows the structure of the generalized Floquet Hamiltonian for the case of two monochromatic fields of frequencies ω_1 and ω_2. Details of the many-mode theory and applications of the theory to several two-color intense field multiphoton processes of current interests will be presented.

*Work supported by DOE and ACS-PRF.

#Alfred P. Sloan Foundation Fellow

References

1. J. H. Shirley, Phys. Rev. B138, 979 (1965)
2. D. R. Dion and J. O. Hirschfelder, Adv. Chem. Phys. 35, 265 (1976) and references therein.
3. S. I. Chu and W. P. Reinhardt, Phys. Rev. Lett. 39, 1195 (1977); S. I. Chu, Chem. Phys. Lett. 54, 367 (1978).
4. S. I. Chu, Chem. Phys. Lett. 58, 462 (1978).
5. S. I. Chu, Chem. Phys. Lett. 64, 178 (1979).
6. S. I. Chu, J. Chem. Phys. 75, 2215 (1981); S. I. Chu, C. Laughlin, and K. K. Datta, J. Chem. Phys. (submitted).
7. B. Simon, Ann. Math. 97, 247 (1973); E. Balslev and J. M. Combes, Comm. Math. Phys. 22, 280 (1971).
8. T.S. Ho, S. I. Chu and J. V. Tietz, Chem. Phys. Lett. (in press, 1983).

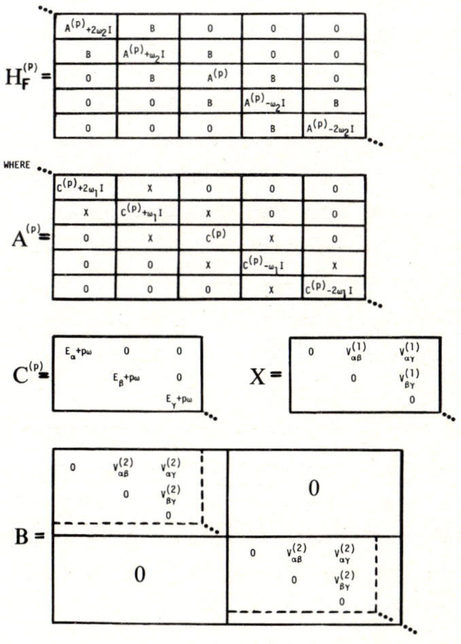

INFLUENCE OF THE DYNAMIC STARK EFFECT ON ELECTRON ANGULAR DISTRIBUTIONS IN MULTIPHOTON IONIZATION

W. Ohnesorge[1], F. Diedrich[1], D.S. Elliott[1,3], G. Leuchs[1], and H. Walther[1,2]

[1]Sektion Physik der Universität München, D-8046 Garching, [2]Max Planck Institut für Quantenoptik, D-8046 Garching, Fed. Rep. Germany, [3]Joint Institute for Laboratory Astrophysics of the University of Colorado and the National Bureau of Standards, Boulder, Colorado 80309, U.S.A.

The angular distribution of electrons ejected during the process of multiphoton ionization can be used to study various features of the atomic electronic structure, for example, the phase difference between continuum state wave functions[1], quantum interference effects[2], and perturbations of intermediate levels by configuration mixing[3]. Since measurements involving multiphoton ionization in general employ high intensity laser fields, a study of the influence of the dynamic Stark effect on these angular distributions is important. We present in this contribution measurements of the intensity dependence of the angular distribution of electrons ejected from the 4d fine structure levels of sodium atoms in a collimated thermal beam excited by a single mode narrow band pulsed dye laser system.

The angular distribution of electrons ejected in the plane perpendicular to the propagation direction of the laser beam at an angle Θ (measured from the direction of laser polarization) can be expanded in a series of Fourier functions[4] $\cos(2n\Theta)$, $n=0,\dots,N$, where N is the number of photons involved in the ionization process.

Since high-intensity fields may shift atomic energy levels in or out of resonance with the laser frequency, the photoelectron angular distribution measured may be modified[5]. Dixit and Lambropoulos[6] have calculated this effect for two photon resonant three photon ionization of sodium via the $4^2D_{5/2}$ intermediate state. They predict an intensity range where the angular distribution, which has two side lobes for low intensities, changes to a single and then back to a double side lobe structure with increasing intensity. This drastic change of the photoelectron angular distribution is less pronounced if the laser intensity is not constant in time and space.

If the laser frequency at low intensity is tuned to populate the $4^2D_{3/2}$ state the angular distribution shows one side lobe. An increase of the intensity shifts the $4^2D_{3/2}$ state out of resonance, the side lobe vanishes, and at even higher intensities two side lobes are expected to appear. This effect is less sensitive to laser intensity inhomogeneities and therefore easier to observe.

The measurements are performed using a pulsed dye amplifier chain which amplifies the output of a commercial, highly-stabilized cw ring dye laser. The amplifier chain is pumped with the second harmonic of a Nd:YAG laser. The optical pulse has a duration of 5 ns, a bandwidth of 150 MHz and a power of 5 kW. Ionization of atoms experiencing a relatively uniform dye laser intensity is produced by a second, time-delayed, nearly collinear, tightly-focussed optical pulse at 532 nm. Angular distributions are measured by rotating the polarization of the radiation and detecting single photoelectrons passing through a 10 degree-wide aperture. Since the angular distribution changes with intensity, data were accumulated only when the intensity of the laser pulse was within 10 % of a preset value.

Figure 1 shows two photoelectron angular distributions measured with the laser frequency tuned to the $3^2S_{1/2}$ (F=2) $\rightarrow$ $4^2D_{3/2}$ two photon transition in sodium. Figure 1a is measured at an intensity of about 1 MW/cm^2, and shows the expected single side-lobed structure. At a higher intensity of 3 MW/cm^2 (Fig. 1b) the cos 6Θ term is beginning to appear, and the side lobe vanishes. This result demonstrates how sensitive photoelectron angular distributions may depend on the laser intensity. Whenever atomic data are to be extracted from the angular distribution measurements such intensity effects have of course to be considered. But, on the other hand, the intensity dependence of the angular distributions opens an interesting possibility to study the change of the atomic structure when the atom is exposed to high intensity laser fields.

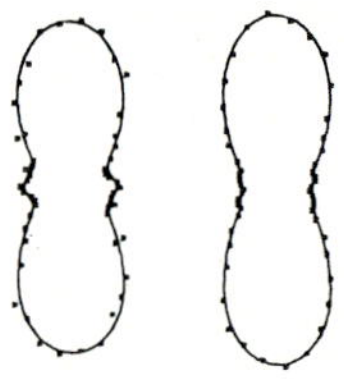
a b

Fig. 1

The financial support of the Deutsche Forschungsgemeinschaft is gratefully acknowledged. One of us (DSE) thanks the Alexander von Humboldt Stiftung for a stipend and also acknowledges assistance from NSF grant number INT 8120128 as part of a United States/Federal Republic of Germany cooperative science program.

References

1. H. Kaminski, J. Kessler, and K. J. Kollath, Phys. Rev. Lett. 45, 1161 (1980)
2. G. Leuchs, S.J. Smith, E.E. Khawaja and H. Walther, Opt. Commun. 31, 313 (1979)
3. G. Leuchs, E. Matthias, D. S. Elliott, S. J. Smith, and P. Zoller, Proc. of the Sixth Int. Conf. on Laser Spectroscopy, Interlaken, Switzerland (1983)
4. C. N. Yang, Phys. Rev. 74, 764 (1948)
5. G. Leuchs, J. Reif and H. Walther, Appl. Phys. B28, 87 (1982)
6. S. N. Dixit and P. Lambropoulos, Phys. Rev. Lett. 46, 1278 (1981)

ANGULAR DISTRIBUTION OF PHOTOELECTRONS FROM MULTIPHOTON-IONIZATION (λ = 532 nm)

R. Hippler, H.-J. Humpert, H. Schwier, H. O. Lutz

Fakultät für Physik, Universität Bielefeld, F. R. Germany

Multiphoton ionization of xenon atoms has been investigated recently in some detail[1,2]. Using a frequency-doubled Q-switched Nd:YAG laser (λ = 532 nm) it was found that xenon atoms may be ionized by absorption of six photons (photon energy $\hbar\omega$ = 2.34 eV). Moreover, by means of photoelectron spectroscopy, it was discovered that in addition to absorption of six photons required to ionize the xenon atom also 7, 8, 9, ... photon absorption during a single laser pulse occurs. Figure 1 shows a photoelectron spectrum of xenon. The four lines correspond to absorption of six and seven photons, with the doublet character of the lines being caused by the fine-structure splitting (energy difference 1.3 eV) of the $Xe^{+}(^{2}P_{1/2;3/2})$ ground state.

In the following we report on an angular distribution measurement of the ejected photoelectrons for these four transitions; similar measurements have been reported recently by Fabre et al.[3] for two transitions. The measurements were performed by measuring photoelectrons ejected at 90° with respect to the laser beam with a 45° parallel plate spectrometer. The angle between the direction of the detected electrons and the electric vector of the laser light (λ = 532 nm) was varied by rotation of a half wave plate. The laser was a Q-switched Nd:YAG laser (pulse width 8 ns) with unstable oscillator configuration and two amplifiers.

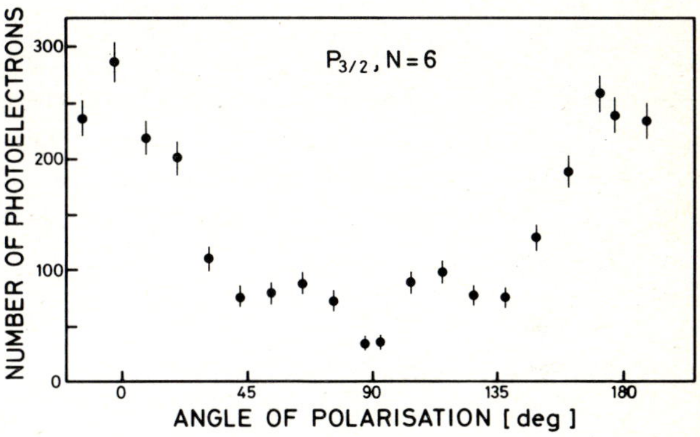

Fig. 2: Angular distribution of photoelectrons from 6-photon-ionization of xenon. The final ionic state is $Xe^{+}(^{2}P_{3/2})$.

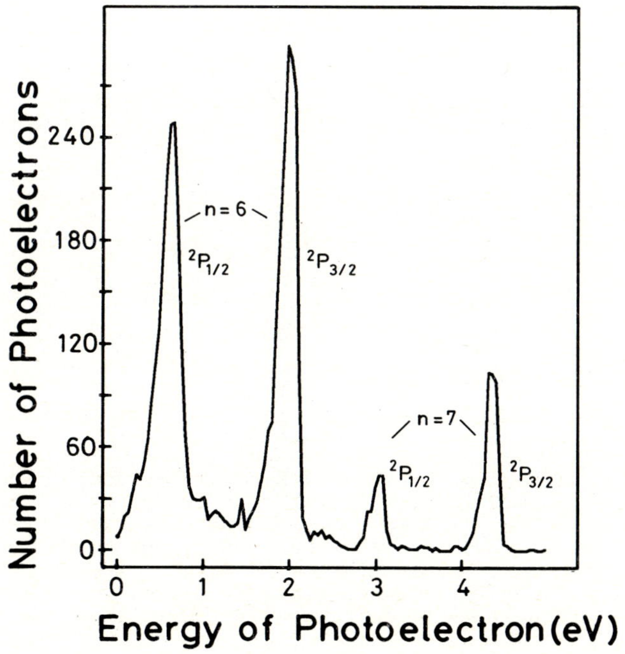

Fig. 1: Energy distribution of photoelectrons from 6- and 7-photon-ionization of xenon.

Results of such an angular distribution measurement are given in figure 2 for the $Xe^{+}(^{2}P_{3/2})$ transition. A pronounced dependence on the angle between the electric field vector and the ejected electron is observed. The angular dependence is rather complicated, giving a maximum at 0° (and 180°) and a second maximum at 65° / 115°. No quantitative theoretical description is yet available.

References

1. P. Agostini, M. Clement, F. Fabre, G. Petite, J. Phys. B 14, L 491 (1981)
2. P. Kruit, J. Kimman, M. J. Van der Wiel, J. Phys. B 14, L 597 (1981)
3. F. Fabre, P. Agostini, G. Petite, M. Clement, J. Phys. B 14, L 677 (1981)

PONDEROMOTIVE FORCE AND A.C. STARK SHIFT IN MULTIPHOTON IONIZATION BY STRONG FIELDS

H.G. Muller, P. Kruit, A. Tip and M.J. van der Wiel

FOM-Institute for Atomic and Molecular Physics, Kruislaan 407, 1098 SJ Amsterdam, The Netherlands.

A free electron in an inhomogeneous oscillating electric field (such as the field in a laser focus), experiences a force which pushes the electron out of the field. The magnitude of this so-called ponderomotive force is given by the formula

$$F = -\frac{e^2}{4m\omega^2}\nabla|E|^2 \qquad \text{(I)}$$

Suppose the intensity of the laser does not vary appreciably in the time an electron, produced by multiphoton ionization inside this focus, needs to travel to the field free region. Then the energy gain due to this force is dependent only on the intensity I at which the electron is formed. For the light of a Nd-YAG laser (1.064 μm) the energy gain is:

$$\Delta E\ (\text{eV}) = 10^{-13}\ I\ (\text{W/cm}^2) \qquad \text{(II)}$$

We have performed multiphoton ionization experiments at intensities up to $7\cdot10^{13}$ W/cm^2, and therefore expected a profound influence of this effect on our electron spectra.

We recorded the electron spectrum of xenon under multiphoton ionization by 1.064 μm radiation at $2\text{-}7\cdot10^{13}$ W/cm^2 (figure 1). At this wavelength, ionization to the $^2P_{3/2}$ continuum can be accomplished by absorption of 11 photons. At the high light intensity used, however, there is a strong tendency towards so-called above-threshold ionization: The atom absorbs (11+p) photons, where the extra energy is found as kinetic energy of the photoelectrons. In our experiment p can be as large as 10 or more.

The intensity dependence of the peaks corresponding to different p is measured, and reveals an apparent suppression of the p = 0 peak with respect to the other ones at higher intensities. Although the peaks are slightly broadened in energy (≈ 100 meV), there is no shift observed whatsoever, a fact which leads to the conclusion that the energy gain due to the ponderomotive force is cancelled exactly by some compensating effect.

A model calculation for multiphoton ionization is presented in which the essential features of the experiment are reproduced: We model the atom as a single electron in a projection operator potential, leading to a spectrum that consists of a single bound state and continuum. The laser field (which is taken circularly polarized, making it possible to remove the time dependence), couples this bound state to the continuum, causing it to become a resonance. An implicit expression for the energy of this resonance is obtained by means of the complex dilatation method, and computed numerically. The nonperturbative nature of this method makes it possible to handle very large field strengths.

The result of this calculation makes it possible to draw the following conclusions:

1) The ponderomotive energy gain in this model is cancelled by the A.C. Stark shift of the continuum limit, which turns out to shift upward by the amount given in formula (I). The only shift observed in the electron energies is therefore due to the A.C. Stark shift of the ground state, and is expected to be quite small with respect to the ponderomotive effect due to the small spatial extension of this state.

2) The increase of the ionization energy obtained in this way can cause the minimum number of photons required for ionization to rise, thereby obliterating the lowest energy peaks in the electron spectrum.

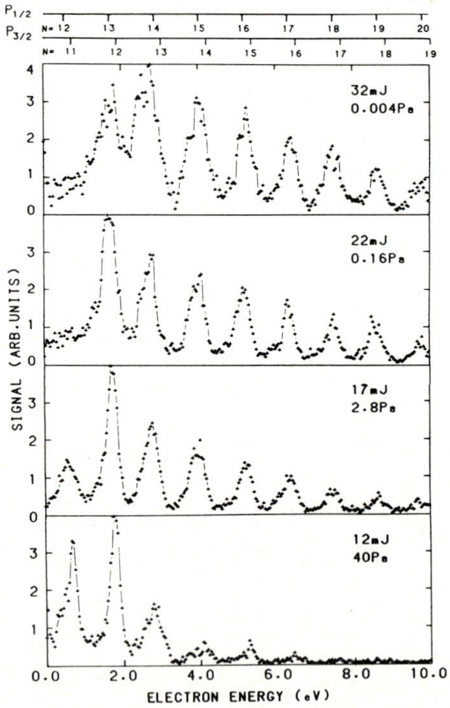

Fig. 1: Electron spectra of multiphoton ionization of Xe by 1064 nm radiation. The intensity is equal to $F\times2\cdot10^{12}$ W/cm^2, where F is the pulse energy in mJ, indicated with each of the spectra. The target pressure, indicated in Pa, has been varied with the intensity, in order to keep the total space charge produced in the laser focus, constant.

PHOTOELECTRON ENERGY AND ANGULAR DISTRIBUTIONS FROM MULTIPHOTON IONIZATION OF CESIUM BEAMS*

J.A.D. Stockdale, R. N. Compton, and C. D. Cooper**

Chemical Physics Section, Health and Safety Research Division,
Oak Ridge National Laboratory, Oak Ridge, Tennessee 37830 U.S.A.

Thermal cesium beams have been crossed by the focused light from a pulsed tunable dye laser which delivered ~10^8 W/cm^2 in the focal region. The laser beam was linearly polarized by a Glan-air prism and its polarization orientation was controlled by a double Fresnel rhomb polarization rotator. Photoelectrons produced in the electric and magnetic field-free interaction region were energy analyzed by a hemispherical sector electrostatic energy analyzer of energy resolution ~0.1 eV and angular resolution of ~3°, and were then detected by a multichannel plate detector. Figure 1 illustrates the experiment.

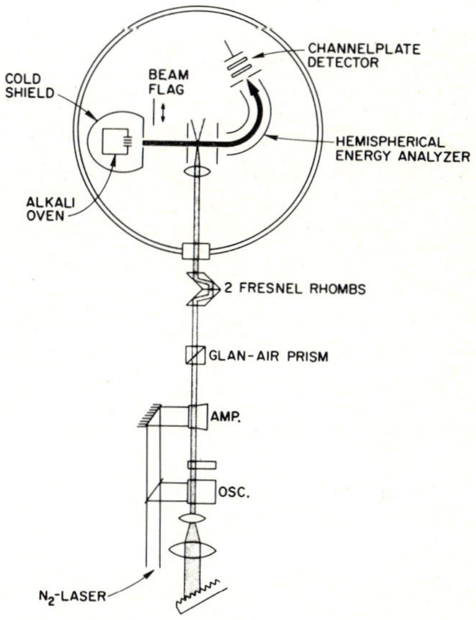

Fig. 1. Experimental arrangement for energy and angular distribution measurements for multiphoton ionization of alkali atoms.

Angular distributions of photoelectrons corresponding to two-photon ionization via the one-photon allowed $7p\,^2P_{1/2,3/2}$ and $8p\,^2P_{1/2,3/2}$ intermediate states and three-photon ionization via the two-photon allowed $8d\,^2D_{5/2,3/2}$ resonant intermediate states were recorded by tuning the analyzer transmission to the predicted electron kinetic energy and rotating the polarization vector of the laser beam. In Figs. 2 and 3 the experimental data points are compared with the theoretical predictions (solid line) of Tang and Lambropoulos[1] for the $7p\,^2P_{1/2}$ and $8d\,^2D_{5/2}$ cases. The angular distribution for the $7p\,^2P_{3/2}$ shows a subsidiary peak at 90° which was not seen in the data of Kaminski et al.[2] Allowing for the finite angular resolution of the analyzer, the agreement between experiment and theory is good.

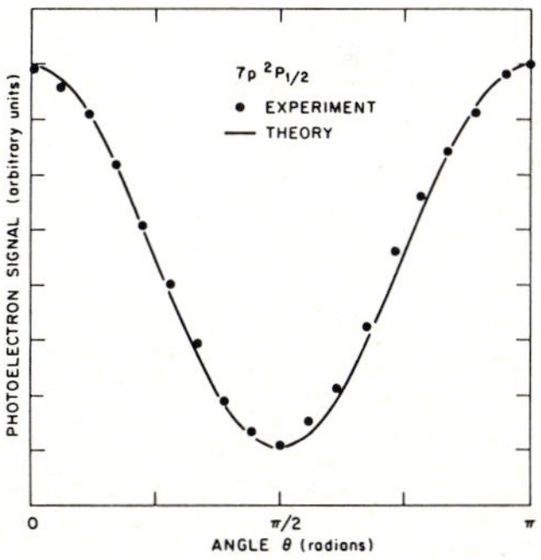

Fig. 2. Photoelectron angular distribution for two-photon ionization of Cs via the one-photon allowed $7p\,^2P_{1/2}$ state.

In addition, we have obtained energy-resolved photoelectron angular distributions for two-photon ionization via the hybrid resonances $6p\,^2P_{1/2} \rightarrow 6d\,^2D_{3/2}$ and $6p\,^2P_{1/2} \rightarrow 7p\,^2P_{3/2}$ (quadrupole allowed transition). Here the $6p\,^2P_{1/2}$ atoms are thought to be formed by dissociation of Cs_2 molecular or quasimolecular states.

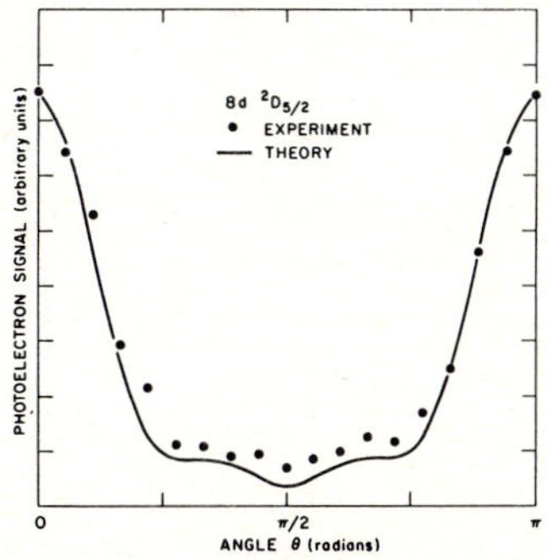

Fig. 3. Photoelectron angular distribution for three-photon ionization of Cs via the two-photon allowed $8d\,^2D_{5/2}$ state.

*Research sponsored by the Office of Health and Environmental Research, U.S. Department of Energy under contract W-7405-eng-26 with the Union Carbide Corporation.

**Consultant, Department of Physics, University of Georgia, Athens, Georgia 30601.

[1]X. Tang and P. Lambropoulos, private communication.

[2]H. Kaminski, J. Kessler, and K. J. Kollath, Phys. Rev. Lett. 45, 1161 (1980).

MICROWAVE IONIZATION OF HIGHLY EXCITED HELIUM ATOMS

W. van de Water, D.R. Mariani and P.M. Koch

Department of Physics, State University of New York, Stony Brook,NY 11794,USA

Because of the influence of the 1s (core)electron, the energy levels of an highly excited helium atom in an electric field do not cross. When the electric field changes as a function of time, the anticrossings can be traversed partly diabatically and partly adiabatically. For isolated anticrossings traversed linearly in time, the probability for transitions from one adiabatic curve to another is given by the Landau-Zener-Stückelberg formula[1]. We show that these anticrossings play a crucial role in mediating the first few excitation steps in (multiphoton) ionization of highly excited He atoms by a microwave electric field.

A fast beam of He (28^3S) atoms was prepared by a collision-laser method[2]. They were then exposed to 10^3 cycles of the axial electric field in a microwave cavity operating at 9.915 GHz in the TM_{020} mode. The electric field amplitude was constant along the axis and varied by less than 10% over the beam radius. The influence of the azimuthal magnetic field of the cavity mode was negligible. The ions emerging from the cavity were registered as function of the power incident on the cavity.

The Figure shows the ion signal as function of the peak field amplitude F_o inside the cavity. The upper n^4F_o-axis is added to aid comparison with measured static electric field ionization thresholds, which for He(1sns)^3S atoms are at the saddle point limit, approximately given by $n^{*4}F \cong 0.06$ a.u. ($n^* = n - \delta$, where $\delta = 0.297$ is the quantum defect for the 28^3S level). The microwave ionization threshold occurs at a much lower field strength than the corresponding static field ionization threshold. The Figure also shows a pronounced resonance structure at $F_o \cong 110$ V/cm. These findings contrast strikingly with results of similar experiments on highly excited hydrogen atoms. The threshold for H(n≅30) ionization is approximately equal to the lower end of the range of static ionization thresholds[2] ($n^4F \cong 0.1$ a.u-0.3 a.u. depending on the parabolic substate).

The difference between the behavior of highly excited He and H atoms in an oscillating electric field is obviously a manifestation of the finite extent of the core in the He atom. In a quasistatic picture in which the energy levels adjust adiabatically to the changing electric field, the He levels exhibit anticrossings whereas the (nonrelativistic) H levels cross exactly.

A calculation of the relevant triplet He $M_L=0$ energy levels in a static electric field reveals that the first anticrossing of the downward going adiabatic 28^3S state with the most upward going state of the n=27 manifold occurs at a field strength F=110 V/cm. This strongly suggests that the onset of the signal in the Figure is related to the occurrence of anticrossings.

Presumably, when the field reaches out to the region of anticrossings of the n=28 manifold with neighboring manifolds, transitions to states of the n=29 and higher manifolds take place. Since the field where anticrossings between states of adjacent manifolds occur scales as n^{-5}, this is indeed possible and provides a mechanism for the system to reach the ionization continuum.

To estimate the transition probabilities, we numerically integrated the time dependent Schrödinger equation. He states for constant field were approximated by projection onto a set of 45 parabolic basis states in the n=27,28,29 manifolds. The system was started with unit population of the 28^3S state at zero field and the transition probabilities were found by examining the redistribution of population after one cycle of the field.

The calculation showed that at any time during the field evolution *many* states interact. The rate of change of the field is so high that the region over which transitions take place is of the order of the field amplitude itself.[3] Consequently, the anticrossings may not be treated as isolated and use of the Landau-Zener-Stückelberg formula is not justified.

The inset of the Figure shows the probability for transition from the 28^3S state to the 29^3S state as function of F_o. Because many of the n=29 states receive an almost equal share of the transferred population, choice of the 29^3S state is arbitrary. Since all levels are excited coherently, the resonance feature in the data is most probably caused by an interference mechanism.

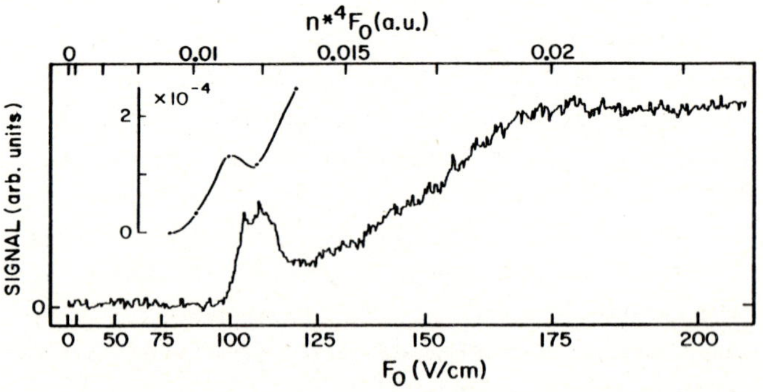

Figure: Ionization signal of He 28^3S atoms as function of the peak electric field inside the cavity. The size of the vertical scale corresponds to about 1/3 of the saturated signal. The inset shows the calculated transition probability 28^3S-29^3S after one cycle of the field.

References

1. J.R. Rubbmark et al, Phys. Rev. A23, 3107(1981)
2. P.M. Koch and D.R. Mariani, Phys. Rev. Lett. 46,1275 (1981)
3. D.R. Bates, Proc. R. Soc. London A257, 22 (1960).

DOUBLE RESONANCE MULTIPHOTON IONIZATION STUDIES OF HIGH RYDBERG STATES IN NO.

M. SEAVER, W.A. CHUPKA and S.D. COLSON
Dept. of Chemistry, Yale University, NEW HAVEN Ct 06511 (U.S.A.)

D. GAUYACQ and Ch. JUNGEN
Laboratoire de Photophysique Moléculaire*, Bât. 213,
Université de Paris-Sud, 91405 - ORSAY Cedex (France).

Optical-optical double resonance multiphoton ionization (OODR-MPI) has been used to study highly excited Rydberg states of the NO molecule in the energy region below the first ionization limit. The experimental details have been described in earlier publications [1,2].

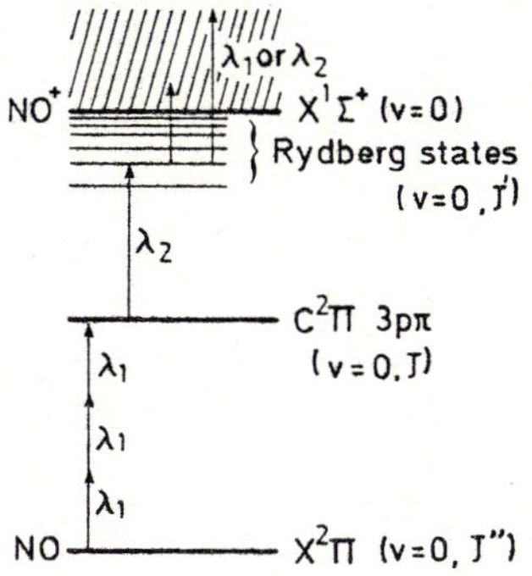

In the present study Rydberg states converging to NO^+ ($X^1\Sigma^+$, $v^+=0$, N^+) have been probed by excitation from the intermediate $C^2\Pi$ state which was in turn populated by a three-photon step and then ionized by a further photon. The OODR-MPI signal versus the probe wavelength λ_2 shows a resonance structure characteristic of Rydberg-Rydberg transitions from the $C^2\Pi$, 3pπ state. In figure 1 a typical spectrum displays part of a spectrum corresponding to the highest energy region.

Two sets of spectra have been recorded and correspond to observation of series converging to the rotational levels $N^+=2$ and $N^+=6$ of the ionic core, respectively, from n 4≃up to n≃38. The spectra can be roughly divided into 3 regions, a "low energy" region (n≃4 to n≃10), an intermediate region (n≃10 to n≃20) and a "high energy" region (n≃20 to ∞). In the low energy region, the structure of the n complexes can be described by Hund's coupling case (b) (i.e. within the Born-Oppenheimer approximation) except for the ndσ and ndπ which are in strong interaction [3]. The structure of this region has been analyzed in terms of the known Rydberg series of NO[3]. The intermediate region appears very complicated and is not a straightforward continuation of the low energy retion. Strong ℓ-uncoupling occurs and mixes several Rydberg states of different ℓ and λ within a given n manifold, so called supercomplex [4]. In contrast the high energy region has a simple structure corresponding to a strong regular series converging to a given rotational level of the ionic core. Multichannel Quantum Defect calculations (MQDT) are in progress with the aim to interpret the intermediate and high energy regions.

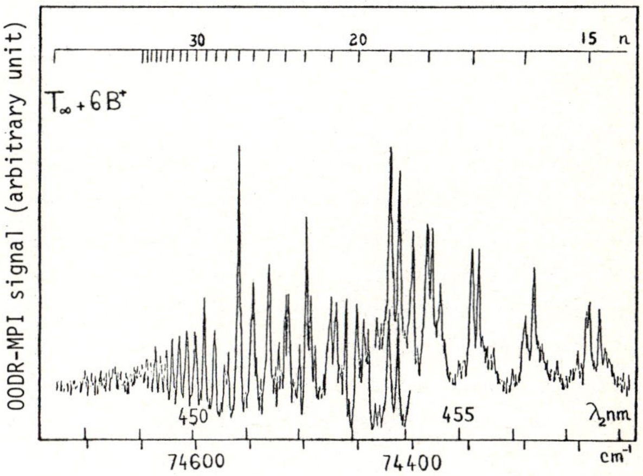

Fig. 1- OODR-MPI signal versus the probe wavelength λ_2 (intermediate state $C^2\Pi$,3pπ,v=0,J=1½).

Evidence for predissociation of some Rydberg states is also present in the spectra. The present study, while limited to the region below the IP, constitues an essential step for the understanding of the dynamical processes occuring in the Rydberg states above the ionization limit where preionization also occurs. Indeed the parameters obtained by an MQDT analysis in the discrete region contain in principal information not only on the structure of the system but also on its dynamical evolution when several decay mechanisms are in competition [5].

References

1. W.Y. Cheung, W.A. Chupka, S.D. Colson, D. Gauyacq, Ph. Avouris and J.J. Wynne, J. Chem. Phys. in press.
2. M. Seaver, W.A. Chupka, S.D. Colson and D. Gauyacq, to be published in J. Phys. Chem.
3. E. Miescher and K.P. Huber, in International Review of Sciences, Phys. Chem. Ser. 2, vol. 3 (Butterworths 1976).
4. E. Miescher, Can. J. Phys. 54, 2074 (1976).
5. A. Giusti, Invited paper, XII ICPEAC Conference, Gatlinburg, Tennessee, U.S.A., July 1981 ; Ch. Jungen Invited paper XII ICPEAC Conference, Gatlinburg, Tennessee, U.S.A., July 1981 ; A. Giusti and Ch. Jungen, to be published.

* Laboratoire associé à l'Université de Paris-Sud.

A MULTIPHOTON IONIZATION PHOTOELECTRON SPECTROSCOPIC STUDY ON AUTOIONIZATION PROCESSES OF NO MOLECULE

Y. Achiba, K. Sato and K. Kimura

Institute for Molecular Science, Okazaki 444, Japan

1. The use of multiphoton ionization (MPI) technique has undoubtedly unique advantages in molecular photoionization experiments comparing with a vacuum UV one-photon ionization method. One of the advantages is the resonant enhancement of ionization efficiency, by which it is possible to easily select a specific rotational level to be studied even for rotationally hot target molecules in the ground-state. In this sence, the possible photoionization transition also considerably depends on the electronic character of resonant states.

In the present paper we show the first experimental results on the branching ratios and angular dependences of photoelectrons produced from the autoionization levels selectively excited by resonantly enhanced multiphoton absorption. Non-Rydberg $B,{}^2\Pi$ state of NO molecules was chosen for this purpose, i.e.,

$$X,\ NO \xrightarrow{h\nu} \xrightarrow{h\nu} B,{}^2\Pi\ NO^* \xrightarrow{h\nu} NO^{**} \rightsquigarrow X,NO^+ + e^-$$

According to this scheme, the direct ionization forming X, NO^+ from a valence excited state such as B state is forbidden by a one-electron dipole operator. Thus it can be expected that absorption of an additional photon by B state forms an optically allowed superexcited state NO^{**} leading to autoionization.

2. UV lights of a Nd-YAG pumped dye laser were focused on the interaction region, crossing with an effusive jet of NO. Photoelectron kinetic energies were measured by a time-of-flight method, and angular distributions were obtained by changing the direction of the polarization vector with a 1/2 λ plate[1].

3-A. In Fig. 1, an MPI spectrum is shown on the laser wavelength of 375∼382 nm. From available spectroscopic data of NO, it was found that two-photon laser energies are well coincident with the energy required to excite NO to $B,{}^2\Pi, v'=9$ level. The MPI spectrum shows that the intensity is anomalously enhanced at several specific rotational lines, deviating from Bortzmann distributions.

3-B. Photoelectron spectra were measured by fixing the laser wavelength at each rotational line shown in Fig. 1. Two types of spectral patterns were found (Fig. 2). The first type A consists of mainly two bands whose kinetic energies are well explained in terms of X, NO^+, $v^+=2$ and 1 ions, and the branching ratios are 70 % and 30 % for $v^+=2$ and 1, respectively. The other type B accompanies an additional band attributable to $v^+=0$ ions. The latter spectrum (type B) is obtained only when the anomalous intense bands are excited. Furthermore, it should also be noted that by increasing the laser power, the type B photoelectron spectrum is followed by a further additional band in the higher kinetic energy region. This band may be interpreted in terms of two-photon ionization of B state which produces $v^+=6$, X,NO^+.

3-C. Figure 3 shows angular distributions of the photoelectrons obtained by excitation at Q_{22}, $J=6^1/_2$ and Q_{11}, $J=1^1/_2$. It is seen that the distributions for the $v^+=2$ and 1 ions are almost identical showing a large β value, while the $v^+=0$ ions show no angular dependence.

4. The ionization mechanism giving the type A spectrum may reasonably be explained by considering the $B',{}^2\Delta$ state as an autoionized state. On the other hand, the angular independence and the occurence of additional photon absorption above ionization threshold observed in the type B spectrum strongly indicate that a relatively long-lived discrete level is located at the third-photon state. The possible explanation for this state is $4d\delta$, $v'=6$ which is accidentally resonant with some of the specific rotational lines of the B state.

Reference

1. Y. Achiba, K. Sato, K. Shobatake and K. Kimura, J. Chem. Phys.,77,2709(1982);*ibid.*(1983) in press.

Fig. 1: MPI spectrum of $NO,B^2\Pi$ $v'=9$

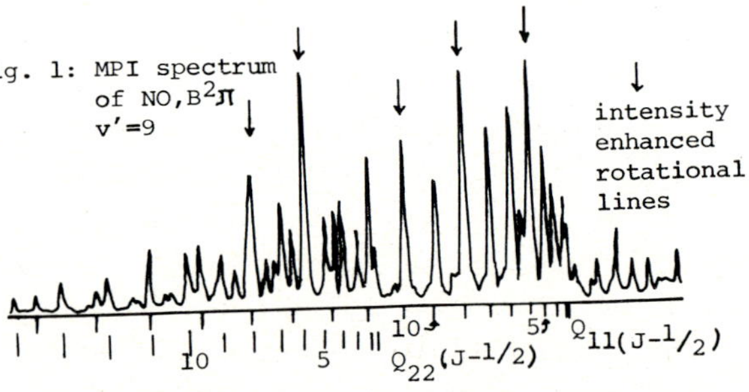

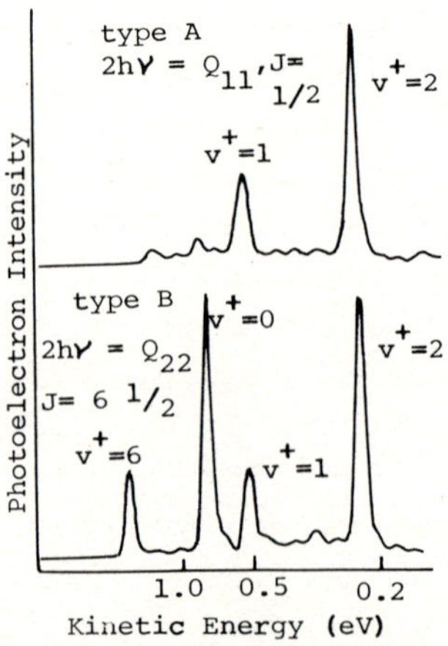

Fig. 2: Photoelectron spectra of NO.

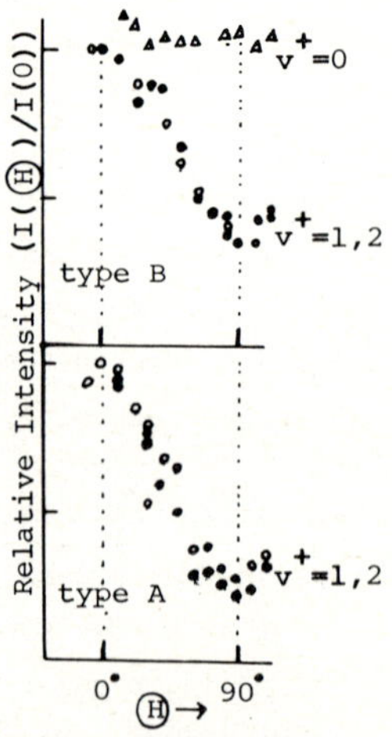

Fig.3: Angular distributions.

PHOTOELECTRON STUDIES OF RESONANT MULTIPHOTON IONIZATION OF CO VIA THE A $^1\Pi$ STATE*

S. T. Pratt, E. D. Poliakoff,† P. M. Dehmer, and J. L. Dehmer

Argonne National Laboratory, Argonne, Illinois 60439

Recently, pulsed dye lasers have been combined with mass and electron energy analyzers to create a new tool to study molecular photoionization dynamics. By such means, it is now possible to excite well-defined multiphoton ionization proceses in small molecules and to observe photoelectron branching ratios and angular distributions, i.e., the standard dynamical observables used to study single-photon processes. In the multiphoton case, however, we are probing excited states of the neutral molecule that are resonant or quasiresonant at an intermediate stage of the multiphoton process. Moreover, the high resolution of laser probes permits the excitation of particular vibrational and rotational quantum states of the excited molecule. Although the main prototype system for this type of study has been NO[1-4] (partly due to its low ionization potential), the present study[5] on CO serves to demonstrate the broad applicability of this approach.

We report measurements of multiphoton ionization of CO involving a three photon resonance to the v=1-3 levels of the CO A $^1\Pi$ state followed by the absorption of either two (v=3) or three (v=1,2) additional photons to reach the ionization continua. In one measurement, the CO^+ ion intensity is measured as a function of wavelength, yielding the rotational structure of the intermediate resonant states (see, e.g., Figure 1). In a second measurement, the laser frequency is set at the v=1-3 bandheads, and the kinetic energy spectra of the ejected electrons are measured (see, e.g., Figure 2). The observed vibrational branching ratios in these photoelectron spectra[5] do not

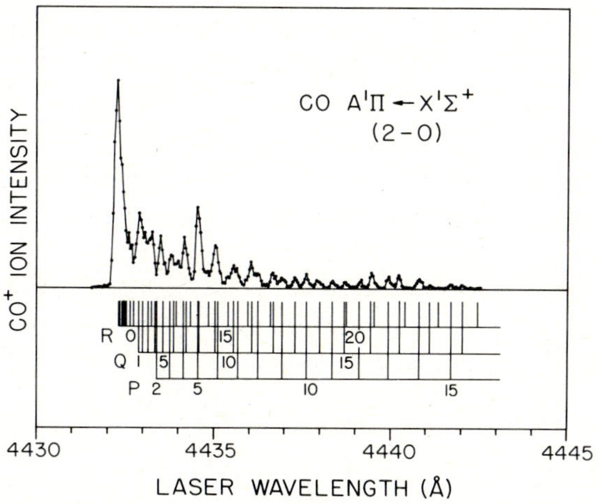

Fig. 1: Total cross section for multiphoton ionization of CO via the A $^1\Pi$, v=2 level.

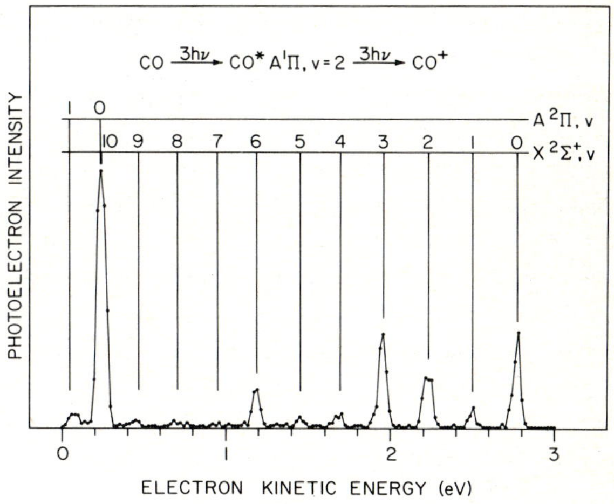

Fig. 2: Photoelectron spectrum of CO determined at the wavelength of the three photon resonance to the R bandhead of the A $^1\Pi$, v=2 level (4432.5 Å).

follow the pattern predicted by the Franck-Condon overlap between the intermediate A $^1\Pi$ state and the ionization continua. Several possible causes for this (not unexpected) deviation from Franck-Condon behavior will be presented, including: (1) perturbations of the A $^1\Pi$ state; (2) accidental resonances at the four, five, and six photon levels; (3) electronic and vibrational autoionization; and (4) the validity of the Franck-Condon approximation in a multiphoton framework.

It is interesting to note that the observation of the CO^+ A $^2\Pi$ final state in the photoelectron spectrum obtained via the A $^1\Pi$, v=2 level (Figure 2) requires a two electron transition from the intermediate state to the ionization continuum. Such two electron transitions are normally weak in single photon absorption from the ground state; however, extensive configuration interaction in the intermediate state can lead to enhanced excitation accompanying ionization.

References

*Work supported by the Office of Naval Research and the U.S. DOE.
†Present address: Department of Chemistry, Boston University, Boston, MA 02215.
1. J. C. Miller and R. N. Compton, J. Chem. Phys. 75, 22 (1981).
2. J. Kimman, P. Kruit, and M. J. van der Wiel, Chem. Phys. Lett. 88, 576 (1982).
3. M. G. White, M. Seaver, W. A. Chupka, and S. D. Colson, Phys. Rev. Lett. 49, 28 (1982).
4. J. C. Miller and R. N. Compton, Chem. Phys. Lett. (submitted).
5. S. T. Pratt, E. D. Poliakoff, P. M. Dehmer, and J. L. Dehmer, J. Chem. Phys. 78, 65 (1983).

RESONANT MULTIPHOTON IONIZATION OF H_2 VIA THE B $^1\Sigma_u^+$, v=7, J=2 AND 4 LEVELS WITH PHOTOELECTRON ENERGY ANALYSIS*

S. T. Pratt, P. M. Dehmer, and J. L. Dehmer

Argonne National Laboratory, Argonne, Illinois 60439

Resonantly enhanced multiphoton ionization of molecules traditionally has been used to obtain detailed spectroscopic information on neutral intermediate states and to study fragmentation mechanisms. However, with the addition of kinetic energy analysis of the ejected electron, it is possible to determine the branching ratios into different vibrational levels of the product ion and to focus directly on both the dynamics of multiphoton ionization and the photoionization of excited state species. We present here photoelectron spectra of H_2 obtained by three-photon absorption from the ground electronic state to specific rotational levels of the B $^1\Sigma_u^+$, v = 7 level, followed by the absorption of a single photon to reach the ionization continuum. The photoelectron spectra exhibit partially resolved rotational structure, which provides information on the partial waves of the ejected electron. This study represents the first attempt to gain the type of detailed information on a rovibrationally state-selected, electronically excited molecule that is normally associated with photoionization studies of ground state molecules.

The apparatus consists of a commercial Nd:YAG pumped dye laser system, a hemispherical electron energy analyzer, and a time-of-flight mass spectrometer.[1] The photoion spectrum was recorded by monitoring the H_2^+ ion signal as a function of the frequency doubled output of the dye laser. The laser then was tuned to the wavelength of a specific rovibronic transition and the photoelectron spectrum was recorded at a resolution of ~30 meV.

The structure in the photoelectron spectrum obtained by ionizing a specific rotational level of the B $^1\Sigma_u^+$ state is determined by the single photon transition from that intermediate level to the ionization continua. The selection rule governing this process is $N(X\ ^2\Sigma_g^+) - J(B\ ^1\Sigma_u^+) = \pm 1, \pm 3$. Because the range of allowed H_2^+ rotational values differs for s and d waves, a rotationally resolved photoelectron spectrum will give detailed information on the relative importance of the different partial waves.

The photoelectron spectra obtained by pumping the P(3) and R(3) transitions of the B $^1\Sigma_u^+$, v' = 7 ← X $^1\Sigma_g^+$, v" = 0 band are shown in Figure 1. In the photoelectron spectrum obtained by pumping the R(3) transition, the N = 3 and 5 rotational levels of H_2^+ both are observed. Although it is allowed by selection

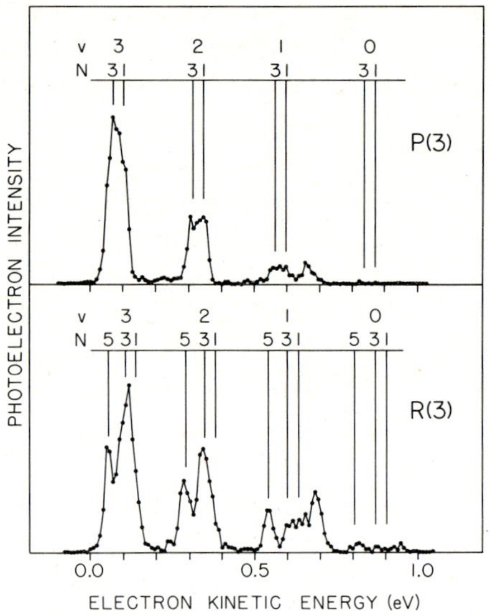

Fig. 1: Photoelectron spectra of H_2 determined at the wavelengths of the three photon resonances to the J=2 (P(3)) and the J=4 (R(3)) levels of the B $^1\Sigma_u^+$, v=7 state. The structure at approximately 0.69 eV kinetic energy in each spectrum is interpreted as resulting from photoionization of atomic hydrogen that is formed from photodissociation of H_2.

rules, the N = 7 level is not observed in any of the vibrational bands, indicating that the $\Delta J = +1$, $\Delta N = +3$, $j_t = 3$ ionizing transition, which corresponds to the ejection of a d-wave electron, is weak. Similarly, only the N = 1 and 3 levels are observed in the photoelectron spectrum obtained by pumping the P(3) transition, whereas the N = 5 level, which again requires $\Delta J = +1$, $\Delta N = +3$, $j_t = 3$ and the ejection of a d-wave electron, is not observed.

The N = 1 rotational level of H_2^+ appears to be missing in the photoelectron spectrum obtained by pumping the R(3) transition. The N = 1 level is allowed by selection rules, but requires $\Delta J = -1$, $\Delta N = -3$, $j_t = 3$, and the ejection of a d-wave electron. Together with the absence of the N = 7 level in the R(3) spectrum and the N = 5 level in the P(3) spectrum, this indicates that either j_t values of 3 or the ejection of a d-wave electron are very weak in the photoionization processes studied here.

References

*Work supported by the U.S. Department of Energy and the Office of Naval Research.

[1] S. T. Pratt, E. D. Poliakoff, P. M. Dehmer, and J. L. Dehmer, J. Chem. Phys. 78, 65 (1983).

ONE- AND TWO- COLOR MULTIPHOTON - IONIZATION - EXPERIMENTS ON MOLECULAR CLUSTERS

W. Bronner, P.Oesterlin, M.Schellhorn

Fakultät für Physik der Universität Freiburg, Freiburg, W.-Germany

During the last years the interest in the investigation of atomic and molecular clusters has strongly increased. These particles have properties in between those of atoms or molecules and those of bulk materia. It is this transition from single particles to solids which causes the large attention from both atomic (and molecular) and solid state physics.

The exploration of small clusters, consisting of less than 10 particles, has mostly been done spectroscopically, for example with microwave spectroscopy, laser induced fluorescence, IR excitation, UV absorption, and VUV ionization. Two main problems are encountered with these methods. Free small clusters can only be generated with densities smaller than 10^{12} cm^{-3}, except for very few special cases. Thus a high sensitivity is necessary. The second problem lies in the production of the desired clusters. Normally they are produced in an expansion of gas into vacuum, where the adiabatic cooling leads to condensation. But in this way clusters of different sizes are created, and with the methods cited above it is not easy to attribute a specific spectroscopic feature to the appropriate cluster species.

One method to overcome these drawbacks is resonant multiphoton ionization. It has an extremely high sensitivity, making it a suitable tool for the investigation of clusters in molecular beams. On the other hand it creates cluster ions which can easily be mass selected. So it is possible to correlate spectroscopic features and cluster size, assuming that fragmentation does not play a significant role.

We constructed an apparatus to measure one- and two-color multiphoton ionization spectra of free clusters in a molecular beam. It consists mainly of a vacuum system with a pulsed molecular beam, which enters the main chamber via a skimmer. Ions created by the interaction with photons are accelerated and mass selected in a 1 m time-of-flight tube.

A commercial excimer laser pumps two home build pulsed dye lasers. The repetition rate is limited to 40 Hz by the pump laser. Both dye laser beams are focused into the molecular beam, appr. 35 mm behind the tip of the skimmer.

The figure shows a one-color 4-photon ionization spectrum of benzene with 2-photon resonances of the electronically excited $\tilde{A}(B_{2u})$ state. This spectrum was taken very recently without mass selection of the ions. It shows the high resolution and the high sensitivity of the apparatus. The partial pressure of benzene in the interaction region was about $4 \cdot 10^{-5}$ Torr.

We plan to carry out one-color multiphoton ionization of clusters, which shows shifts and alterations of vibrations in electronically excited states of the molecules in the clusters, as well as two-color experiments where one laser excites an electronic state and a second one ionizes the clusters. In this way it is possible to measure the dependence of the ionization limit on cluster size. On the other hand it is likely that a cluster, ionized just above its threshold, fragments less compared to one-color ionization where the cluster ion is vibronically excited due to excess photon energy. Such a behavior was observed in benzene clusters resonantly ionized by two UV photons[1]. This can perhaps lead to a better understanding of fragmentation processes of clusters, caused by the ionization process, which is the most important problem in the detection of clusters.

A third type of experiment is a combination of resonant multiphoton ionization and stimulated Raman effect, known as "Ion-Dip-Spectroscopy"[2]. This method allows to take Raman spectra of mass selected clusters with the high resolution of stimulated Raman scattering and the high sensitivity of resonant multiphoton ionization.

Combinations of all these methods are likely to reveal several properties of clusters, as geometrical structures, binding energies, interaction energies of the constituents, and so on. We will present examples of all three types of measurements at the meeting.

1) J.B.Hopkins, D.E.Powers, R.E.Smalley, J.Phys. Chem.85, 3739 (1981)

2) D.E.Cooper, C.M.Klimcak, J.E.Wessel, Phys.Rev. Lett.46, 324 (1981)

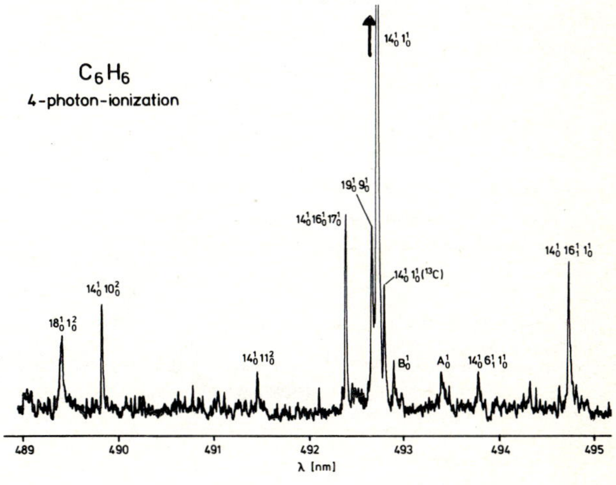

RAMAN ANALYSIS OF SF_6 MOLECULAR BEAMS EXCITED WITH A CW CO_2-LASER

G. Luijks, S. Stolte and J. Reuss

Fysisch Laboratorium, K.U. Nijmegen, Toernooiveld, Nijmegen, The Netherlands

A beam of SF_6 molecules, expanding freely from a nozzle (D = 0.25 mm) is crossed by a focused Ar^+-laser, operating in a folding intracavity configuration (fig. 1). Intracavity laser powers up to 500 Watts can easily be achieved. The scattered Raman light is analysed by a double monochromator (spectral resolution ≥ 1 cm^{-1}) and a photomultiplier system. The direct relationship existing between the intensity of a Raman transition and the population of the initial state provides us with a monitor of the state distribution in the beam as a function of x_R/D, where x_R is the distance between the Raman probing Ar^+-laser and the nozzle. A line-tunable CW CO_2-laser can also cross the molecular beam at x_E/D and is focused below the nozzle on the beam-axis by a lens (L_2).

The excitation of the Raman-inactive ν_3-mode of SF_6 (948 cm^{-1}) by the CO_2-laser is analysed through the Raman active ν_1-hotbands (774 cm^{-1}), i.e. the ν_1-resonances redshifted due to the anharmonic shift of the higher vibrational levels of SF_6. Two vibrational relaxation processes play an important role in the ν_3-deëxcitation:

$$h\nu_3 + h\nu_3 \rightarrow 2h\nu_3 + 3.5\ cm^{-1} \qquad (1)$$

$$h\nu_3[948\ cm^{-1}] + \Delta E_{rot} \rightarrow h(\nu_4 + \nu_6)[962\ cm^{-1}] \qquad (2)$$

where (1) represents the near-resonant V-V rate and (2) the rate determining step for the non-resonant V-V rate, since $\nu_4 + \nu_6$ is the level nearest to ν_3.

The observed spectra give rise to a distinction between three cases, based upon the local rotational temperature (T_{rot}) and the number of collisions at x_E/D.

A. Pure ν_3-excitations (10% SF_6 in He, $x_E/D \geq 4$)
Due to seeding, T_{rot} is low, especially at some distance from the nozzle ($x_E/D \geq 4$, $T_{rot} \leq 20$ K). Therefore, independently of the number of collisions, the excitation of the ν_3-mode can not relax to other modes as insufficient rotational energy is available to overcome the gap to the nearest vibrational level $\nu_4 + \nu_6$ via (2). The only channel open to the excited molecules is the near-resonant ν_3-ladder via (1), leading to a ν_1-Raman spectrum consisting of equidistant hot bands, separated by the anharmonic constant $X_{13} = -2.9$ cm^{-1} (fig. 2).

B. Thermalization (100% SF_6, $x_E/D \leq 1$)
In a pure SF_6-beam, close to the nozzle ($x_E/D \leq 1$), T_{rot} is so high ($\geq$ 150 K) that all vibrational levels are easily accessible via non-resonant ν_3-relaxation (e.g. via (2)). Moreover, the collisions in this region are so numerous that the ν_3-vibrational energy gets redistributed to an equilibrium in which all vibrational levels are thermally populated. At most we found T_{vib}= 950 K, equivalent to an average absorption of 6.4 photons per molecule (see fig. 2).

C. Intermediate case (100 % SF_6, $x_E/D > 1$)
(10% SF_6 in He, $1 \leq x_E/D < 4$)
Neither pure ν_3-excitations (T_{rot} too high) nor thermalization (number of collisions too low) are observed under these intermediate conditions.

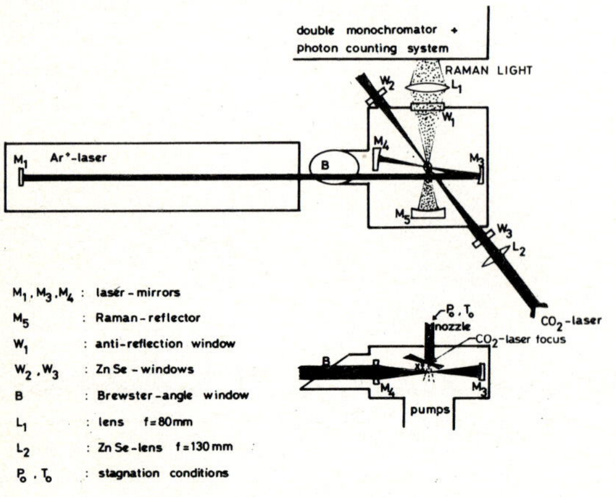

Fig. 1. The intracavity experimental set-up.

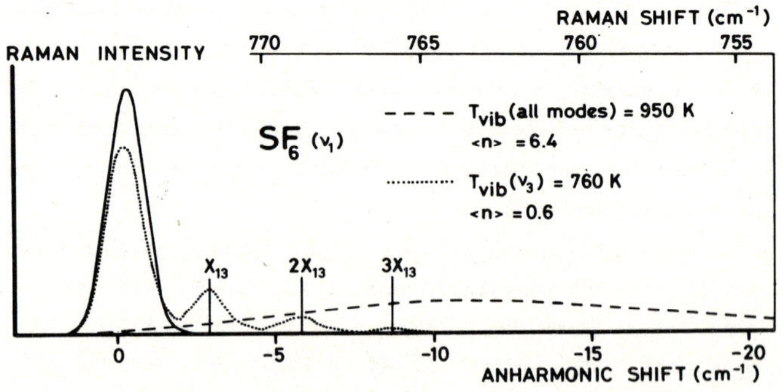

Fig. 2. The ν_1 Raman spectrum
a) (————) without CO_2-laser
b) (·······) with CO_2-laser (20 Watt at P(16)) at $x_E/D = 4$, $x_R/D = 4$, 10% SF_6 in He, P_o= 6 atm, T_o= 300 K
c) (- - - - -) with CO_2-laser (22 Watt at P(22)) at $x_E/D = 0.7$, $x_R/D = 2$, pure SF_6, P_o= 2 atm, T_o= 300 K.
<n> is the average number of ν_3 photons absorbed per SF_6-molecule.

SPECTROSCOPY OF FLUOROBENZENE VAN DER WAALS CLUSTERS BY RESONANT TWO-PHOTON IONIZATION

K. Rademann, B. Brutschy and H. Baumgärtel

Institut für Physikalische Chemie der Freien Universität Berlin, Takustraße 3, 1000 Berlin 33

In a seeded supersonic expansion with a rare gas as carrier gas we produced van der Waals (vdW) clusters of fluorobenzene (FB) and ionized them by laser-induced, resonant two-photon ionization[1] (R2PI) combined with TOF-mass spectrometry. As the excitation transition is the only resonant step of this two-step process, the wavelength dependence of the ion yield curve reflects the electronic state energetics. We excited the FB molecules near the vibronic origin of the S_1 ($\pi\pi^*$) $\leftarrow S_0$ transition of the monomer (λ_{00} = 2644 Å), the energy of which is 89 meV more than half of the ionization limit. For low laser intensity ($I < 3\cdot 10^6$ W$\cdot$cm^{-2}) no chemical fragments could be observed in the TOF-mass spectrum, providing evidence for the "soft" character of this ionization method. For even medium stagnation pressure in the continuous nozzle source, broad clustersize distributions could be observed, containing homogeneous clusters $(FB)_n^+$ with n < 30 and with Ar as seeding gas also heterogeneous clusters $FB\cdot Ar_n^+$ with n < 6. In a one-color-R2PI experiment we studied[2] after careful choice of the expansion parameters the electronic spectra of the clusters $FB\cdot Ar_n$ (n < 3) near the origin-band at ν_{00}. We observed cluster-specific shifts of the 0_0^0-band due to vdW interaction (FB·Ar: -23 cm^{-1}, $FB\cdot Ar_2$: -46 cm^{-1}, $FB\cdot Ar_3$: +4.8 cm^{-1} relative to ν_{00}). In addition satellite bands appeared due to intermolecular photofragmentation. The spectra of the FB-dimer and trimer were more complex. For the dimer progressions were observed, which were slightly blue-shifted relative to ν_{00} and which were tentatively assigned to transitions between vibrationally excited vdW states with a frequency of 20 cm^{-1} in the ground and 15 cm^{-1} in the excited state. The trimer showed prominent structure which is so far not assigned.

In a two-color-R2PI experiment with two separately tunable lasers, we measured the ionization potential of the cold FB and FB·Ar molecule. The latter was selectively excited. Its IP was red-shifted by 218 cm^{-1} relative to that of the FB-monomer (IP = 9.1989 ± 0.0005 eV).

Further results of ionization and fragmentation potentials of different FB-clusters will be presented. The experiments, we report on, were undertaken to investigate the feasibility of clusterspecific spectroscopy with two-color R2PI.

References

1. U. Boesl, H.J. Neusser and E.W. Schlag Z. Naturforschung, 33a, 1546 (1978)
2. K. Rademann, B. Brutschy and H. Baumgärtel submitted for publication in Chem. Phys.

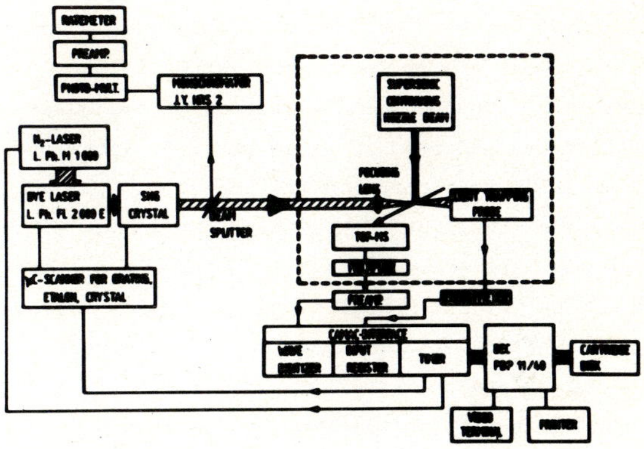

Fig. 1: Blockdiagram of the experimental set-up for one-color R2PI.

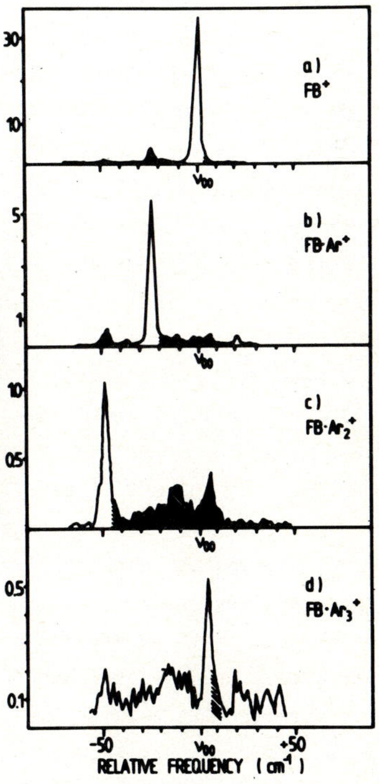

Fig. 2a-d): one-color R2PI spectra near the monomers vibronic origin of the S_1^* ($\pi\pi^*$) $\leftarrow S_0$ transition for $FB\cdot Ar_n$ (n ≤ 3) clusters. The shaded spectral features are due to intermolecular photofragmentation.

MULTIPLE-PHOTON EXCITATION OF CF_3Br

A. Giardini-Guidoni, E. Borsella and R. Fantoni

ENEA, Dip. TIB, Divisione Fisica Applicata, C.P. 65, 00044 Frascati (Rome), Italy

In recent years the process of multiple photon excitation (MPE) and dissociation (MPD) of polyatomic molecules has been the subject of extensive investigation. Usually the process of MPE is divided into three steps. The first stage is the coherent pumping of the discrete energy levels of the I.R. active vibrational mode. As the vibrational energy increases, the oscillator strength of the pumped mode is spread over a quasi-continuum of states and resonant absorption of laser photons is possible up to the region of the true continuum of states above the first dissociation threshold.

In our laboratory we have undertaken a systematic study of the MPE and MPD of simple molecules (i.e., CF_3Br and C_2F_5Cl) both in a cell and in a supersonic beam by using one or two laser frequencies. In the last case, separate information can be obtained on the discrete and quasi continuum regions [(1)].

The most characteristic feature of the first region (coherent pumping) is the occurrence of resonances whenever the laser frequency is equal to 1/N of the energy difference between the ground state and the level with N quanta. However, most of the molecules studied up-to-date have a nearly continuous excitation spectrum, while one-frequency MPA and MPD spectra of CF_3Br show several prominent structures which can be interpreted as multiphoton resonances [(2)].

The absorption measurements reported in Fig. 1 have been performed in an opto-acoustic cell at p = 0.5 Torr by using as a radiation source a high pressure continuously tunable laser (Lumonics mod. TE 281) with a resolution of about 0.12 cm^{-1} and a fluence of 0.3 j/cm^2. All the peaks have been assigned as multiphoton resonances of different order and evidence has been found of the influence of combination bands (quasi-degenerate with the harmonics of the pumped mode) in determining the excitation spectrum. The results of MPD experiments performed on expansion cooled molecular beam of CF_3Br are reported in Fig. 2. One and two-frequency measurements show a series of peaks which confirm the MPA results.

Two-frequency MPD measurements are now in progress at wavelengths of the second laser much lower than the frequency of the first one (ω_1 = 1084.76 cm^{-1} and ω_2 = 1030 ÷ 1050 cm^{-1}). Aim of this experiment is to investigate if the q.c. of CF_3Br shows strong spectral structures as in the case of C_2F_5Cl [(1)]. In fact it is still matter of controversy whether the occurrence of peaks in the q.c. of C_2F_5Cl is peculiar of this molecule (being associated with its low intensity spectral features) or it is a general phenomenon (i.e. Fano-like resonances of the q.c. with the states of the true continuum or statistical fluctuations).

(1) E. Borsella, R. Fantoni, A. Giardini-Guidoni and C.D. Cantrell: Chem.PHys.Lett. 87, 284 (1982)

(2) E. Borsella, R. Fantoni, A. Giardini-Guidoni, D. Masci, A. Palucci and J. Reuss: Chem.Phys.Lett. 93, 523 (1982)

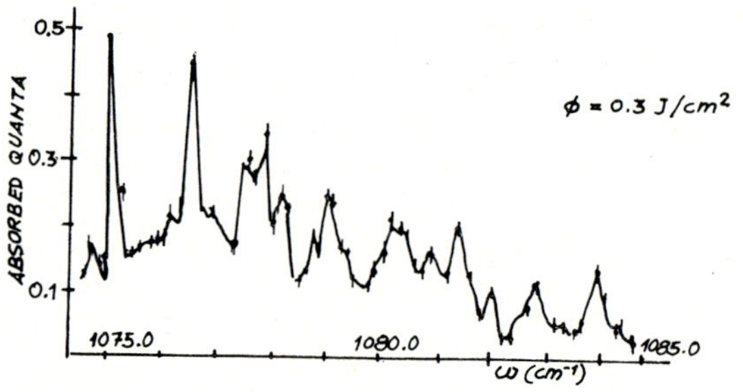

Fig. 1 - MPA spectrum measured at room temperature

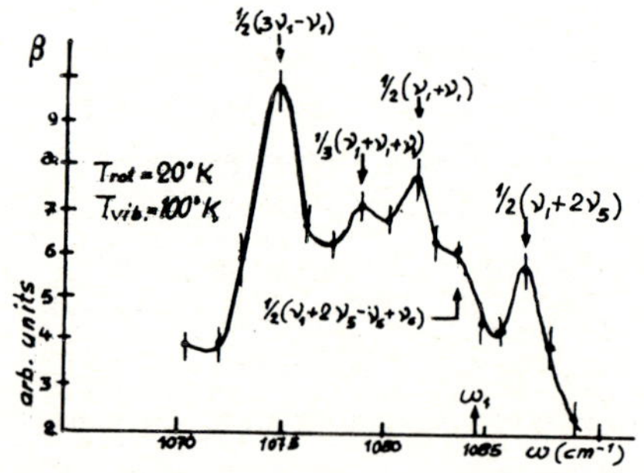

Fig. 2 - Two frequency MPD measured in supersonic molecular beam ω_1 = 1084.76 cm^{-1}, ϕ_1 = 1.5 J/cm^2, ϕ_2 = = 3.5 J/cm^2

TWO-PHOTON DISSOCIATION OF VIBRATIONALLY EXCITED H_2^+ AND HD^+ MOLECULES*

Cecil Laughlin#, Shih-I Chu## and Krishna K. Datta

Department of Chemistry, University of Kansas, Lawrence, Kansas 66045 USA

There is currently much interest both experimentally and theoretically in the problem of multiphoton dissociation (MPD) of molecules[1]. From the theoretical point of view, perturbative technique[2] has been developed for the case of weak fields. However, there remains the difficulty of carrying out the summation over the complete vibrational intermediate states in a converged way even for the simplest molecule like H_2^+.[3] At higher fields, perturbative theory may not be applicable, and development of practical nonperturbative approach[4] is a subject of current interest.

For diatomic molecules, the MPD rates from ground vibrational states are expected to be extremely small and in fact have never been observed. However, MPD from excited vibrationally states are substantially large and may become appreciable. Recently, two-photon infrared dissociation of HD^+ from highly excited vibrational states have been observed experimentally[5].

In this conference we shall present the theories and results of our recent study of the MPD of H_2^+ (ref. 6) and HD^+ (ref. 7) from vibrationally excited states. Two different theoretical techniques are employed. The first method is the nonperturbative complex quasi-vibrational energy (QVE) approach recently developed by one of us[4], based on the generalization of the conventional finite-level Floquet theory[8] to include the complete set of continuum as well as bound vibronic states. This has the effect of giving each of the dressed vibronic levels an intensity dependent part (width). Examination of the frequency and intensity dependence of these complex QVE's gives rise to rates for MPD processes and is equivalent to infinite-order perturbation theory, self-consistent in that shifts and widths of all levels are simultaneously determined.

The second approach is the inhomogeneous-differential-equation (IDE) method of Dalgarno and Lewis[9]. We have developed an efficient code which is capable of obtaining converged MPD results rapidly. This method, however, is a perturbative technique and is therefore applicable to weak field problems only.

Table I shows some results for the weak-field two-photon dissociation cross sections of H_2^+ ($1s\sigma_g$) from several excited vibrational states. The agreement of the two sets of data is in general within a few percents, providing a mutual check of the numerical convergency. The cross sections are pretty small for the low-lying states but increase rapidly with vibrational quantum number v. The H_2^+ cross sections are largest nearby two-photon dissociation thresholds and decrease monotonically with increasing photon energy. The pattern is different for HD^+ where the dominant features are multiphoton resonant structures. Details will be presented in the conference.

Table I. Two-photon Dissociation Cross Sections σ_2 (in cm^4 sec) of H_2^+

λ(Å)	v	σ_2(QVE)	σ_2(IDE)
20,000	6	1.55(-60)**	1.68(-60)
19,600		1.20(-60)	1.25(-60)
18,800		6.81(-61)	7.04(-61)
18,000		3.67(-61)	3.78(-61)
16,800		1.30(-61)	1.35(-61)
16,000		6.05(-62)	6.35(-62)
74,044	12	2.57(-55)	2.78(-55)
71,670		1.92(-55)	1.82(-55)
65,383		1.15(-55)	1.18(-55)
58,535		5.26(-56)	5.24(-56)
50,589		1.60(-56)	1.73(-56)

**1.55(-60) = 1.55×10^{-60}

*Work supported by DOE and ACS-PRF.
##Alfred P. Sloan Foundation Fellow.
#Permanent address: Mathematics Department, Univ. of Nottingham, Nottingham, England.

References

1. See, for example, N. Bloembergen and E. Yablonovitch, Phys. Today 31, 23 (1978), and references therein.
2. F. V. Bunkin and I. I. Tugov, Phys. Rev. A8, 601 (1973).
3. See, for example, A. Lami and N. K. Rahman, Chem. Phys. Lett. 68, 324 (1979).
4. S. I. Chu, J. Chem. Phys. 75, 2215 (1981).
5. A. Carrington and J. Buttenshaw, Mol. Phys. 44, 267 (1981).
6. S. I. Chu, C. Laughlin and K. K. Datta, J. Chem. Phys. (submitted).
7. C. Laughlin and S. I. Chu (in preparation).
8. J. H. Shirley, Phys. Rev. B138, 979 (1965).
9. A. Dalgarno and J. T. Lewis, Proc. R. Soc. Lond., A233, 70 (1955).

TWO-PHOTON "DISSOCIATION" TO THE QUASICONTINUUM+

E. Kyrölä* and J. H. Eberly

Department of Physics & Astronomy, University of Rochester, Rochester, N.Y., U.S.A.

Present understanding of laser-induced dissociation of polyatomic molecules rests strongly on the existence of a so-called quasicontinuum (QC) in the energy spectrum. It is common to approximate the QC as a true continuum. This approximation may be inappropriate[1] and we avoid it.

We consider transitions into the QC by a two-photon process.[2] The level diagram we have in mind is shown in Fig. 1. [If we replace the QC by an ordinary continuum we obtain a system similar to that studied in the theory of two-photon ionization.[3]] The transition from the ground state $|g\rangle$ to the QC can go via two different channels: One nearly resonant and the second completely nonresonant, as shown. We assume[3] that we can describe the second simply by introducing an effective two-photon coupling constant β.

If we assume that the QC consists of evenly spaced levels (spacing = δ) and that the coupling constants α and β are independent of the QC index m, we are able to solve the Schrödinger equation exactly within the RWA. Both amplitudes $c_g(t)$ and $c_a(t)$ can be written in the form

$$c(t) = \sum_{\pm} A_{\pm}(\Delta_a,\gamma_\alpha) e^{s_\pm t} + \sum_{\pm} \sum_{n=1}^{M} e^{-in\Delta_p T_{QC}} \times e^{s_\pm (t-nT_{QC})} \sum_{p=0}^{n} L_{n-p}^{p-1}(t_\pm^n)\, B_\pm^{n,p}(\Delta_a,\beta_a,\gamma_\alpha,\gamma_\beta) \tag{1}$$

Here $\gamma_\alpha = \pi\alpha^2/\delta$, $\gamma_\beta = \pi\beta^2/\delta$ are the conventional Weisskopf-Wigner linewidths and $s_\pm$ are the eigenpoles of the system. The time variables in the associated Laquerre-polynomials are defined by $t_\pm^n = -(s_\pm + s_\pm^*) \times (t - nT_{QC})$ and the general time scale of the problem is determined by the QC coherence time $T_{QC} = 2\pi/\delta$. Also, M is an integer determined by $M < t/T_{QC} \leq M + 1$.

The "A" term in Eq. (1) is the same as we obtain for transitions to a continuum ($\delta \to 0$), whereas the "B" terms represent the effect of the discreteness of the QC. They give rise to a sequence of coherence intervals. We will show differences among these intervals in two limiting cases.

1) The transition from the level $|a\rangle$ to the QC dominates the other two transitions. If $\gamma_\alpha T_{QC} >> 1$, we obtain rate-like behaviour for the level populations and an absorption cross section can be defined as usual. We find

$$\sigma = \sigma_0 \frac{(q-\varepsilon)^2}{1+\varepsilon^2} \frac{\sin[(M+\frac{1}{2})\tilde{\Delta} T_{QC}]}{\sin[\frac{1}{2}\tilde{\Delta} T_{QC}]} \tag{2}$$

where $q = \beta_a/\sqrt{\gamma_\alpha\gamma_\beta}$, $\varepsilon = \Delta_a/\gamma_\alpha$ and $\tilde{\Delta} = \Delta_p - \delta/2 + (\delta/\pi)\tan^{-1}\varepsilon$. We show in Fig. 1 how the cross section behaves in the first (M = 0) and second (M = 1) coherence intervals. If M = 0 we obtain the familiar asymmetric Fano lineshape, but if M = 1 the cross section oscillates wildly and even changes its sign.

2) In this case levels $|g\rangle$ and $|a\rangle$ are strongly coupled to each other ($\beta_a >> \alpha, \beta$). The dominant feature is strong Rabi oscillations between $|g\rangle$ and $|a\rangle$, and much slower change in the sum of QC level populations, denoted Σ in Fig. 2. If the damping is slow enough the "A" and "B" terms can interfere with each other, leading to an interesting oscillation of the population between the two-level part and the QC part.

+Research partially supported by the US AFOSR, and the Finnish Cultural Foundation.

*Permanent address: Department of Physics, University of Helsinki, 00170 Helsinki 17, Finland.

1) H.W. Galbraith and J.R. Ackerhalt, Chem. Phys. Letters 84, 458 (1981).

2) One-photon effects have been already discussed J.J. Yeh, C.M. Bowden and J.H. Eberly, J. Chem. Phys. 76, 5936 (1982).

3) B.L. Beers and L. Armstrong, Jr., Phys. Rev. A. 12, 2447 (1975).

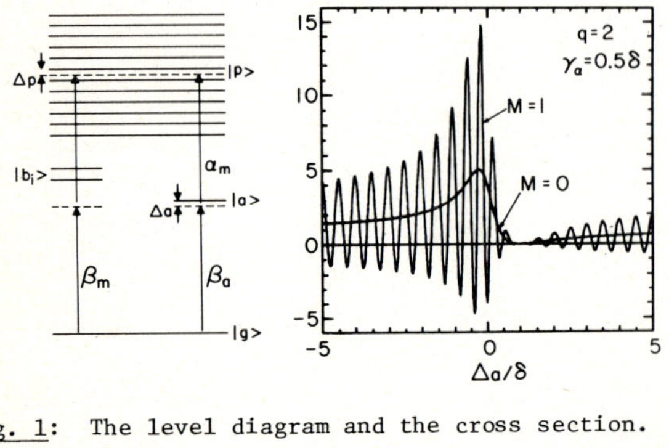

Fig. 1: The level diagram and the cross section.

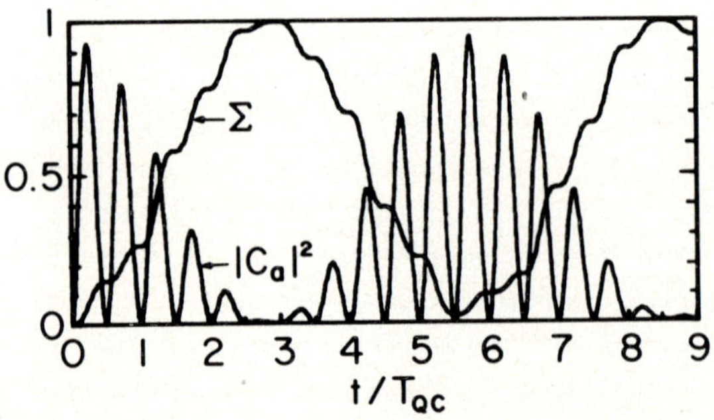

Fig. 2: Here $\Delta_a = \Delta_p = \beta = 0$, $\beta_a = 8\alpha = \delta$.

Λ-DOUBLET POPULATIONS OF $CH(A^2\Delta)$ FROM CH_3 RADICALS PRODUCED BY THE UV MULTIPHOTON DISSOCIATION OF CH_3-CONTAINING MOLECULES

Takashi NAGATA, Mutsumi SUZUKI, Kaoru SUZUKI, Tamotsu KONDOW and Kozo KUCHITSU

Department of Chemistry, Faculty of Science, The University of Tokyo, Bunkyo-ku, Tokyo 113, Japan

Λ-components were observed in the $CH(A^2\Delta - X^2\Pi)$ emission following multiphoton dissociation of $(CH_3)_2CO$, CH_3NO_2 and $(CH_3)_2S$ by 193 nm radiation. These Λ-components appeared at unequal intensities, but the intensity ratios were found to be consistent among the three individual cases. An analysis of the Λ-doublet populations has revealed that the degree of disparity in the relative population of the Λ-components depends on the rotational quantum number of $CH(A^2\Delta)$. These findings indicate that the fragmentation occurs along a preferential trajectory (or trajectories) on the relevant repulsive potential surface. The observed population inversion in the Λ-components and laser power dependence of the emission intensity can be explained by a stepwise multiphoton process via CH_3 as an intermediate:

$$CH_3R + h\nu \rightarrow CH_3^{\dagger} + R, \quad \text{or } 2CH_3^{\dagger} + R' \qquad \left\{ \begin{matrix} R = CH_3CO,\ NO_2 \text{ or } CH_3S \\ R' = CO,\ S \end{matrix} \right\} \qquad (1a)$$

$$CH_3^{\dagger} + 2h\nu \rightarrow CH(A^2\Delta) + H_2 \text{ or } 2H. \qquad (1b)$$

Experimental

The excitation source was a 193 nm excimer laser operated at a repetition rate of 4 - 6 Hz. The laser beam (~40 mJ / pulse) was focused by a suprasil lens ($f \simeq 500$ mm) into a vacuum chamber evacuated to 5×10^{-8} Torr. An effusive beam of the sample gas was introduced into the chamber with an effective pressure of $\lesssim 1 \times 10^{-3}$ Torr. The CH(A-X) emission was viewed at right angles to the laser beam with a 0.5-m grating monochromator (spectral resolution ≃0.065 nm fwhm). The signal was processed by a gated integrator and recorded on a strip chart.

Power dependence

The intensity of the CH(A-X) emission was examined as a function of the laser power, the power dependence was nearly cubic when relatively low power was applied, however, saturation occurred as the power was increased. This cubic dependence indicates that at least three photons are required to produce $CH(A^2\Delta)$ from a parent molecule. The feasibility of the three photon process (1) is also supported by consideration of energetics.

Λ-Doublet populations

The emissions from the Λ-components for $A^2\Delta$, $v' = 0$, $N' \geq 11$ were resolved, and their relative populations, P_c and P_d in the upper c and the lower d levels, respectively, were estimated. Figure 1 shows the normalized population difference, $\delta = (P_c - P_d)/(P_c + P_d)$, obtained in the case of $(CH_3)_2CO$. The δ value depends on the rotational quantum number; the population inversion increases with N' up to 16 and then decrease giving a preferred population in the d level. Nearly identical results were obtained for CH_3NO_2 and $(CH_3)_2S$; this observation provides further evidence for process (1), which is shown to proceed via the formation of CH_3.

It seems possible to account for the observed population inversion in terms of process (1). The nascent CH_3 produced in process (1a) retaining a pyramidal geometry of the parent molecule relaxes toward a planar equilibrium geometry of the ground electronic state. When such a transient CH_3 radical further absorbs 193 nm photons and dissociates, (process 1b), the torque exerted by the recoiling H atoms tends to rotate the CH fragment about an axis perpendicular to the lobe of the $1a_2''(3a_1)$ nonbonding orbital of the dissociating CH_3. Under these circumstances, the promotion of an electron from the singly occupied $1a_2''(3a_1)$ orbital into a high-lying valence orbital, or a Rydberg orbital, results both in a vacant lobe of the CH 1π orbital perpendicular to the axis of rotation, and in a doubly occupied 1π lobe parallel to it. In this situation, the CH radical is populated in the upper Λ level, which corresponds to a smaller moment of inertia of the electrons occupying the Δ orbital. This preference for one of the Λ-doublet levels is analogous to that of $NH(c^1\Pi)$ produced in the photodissociation of NH_3[1].

Process (1) can thus explain the inverted Λ-doublet populations observed in the $11 \leq N' \leq 19$ levels, but the Λ-doublet populations in the higher levels cannot be accounted for simply by the mechanism mentioned above. Other processes may also contribute to the product distributions.

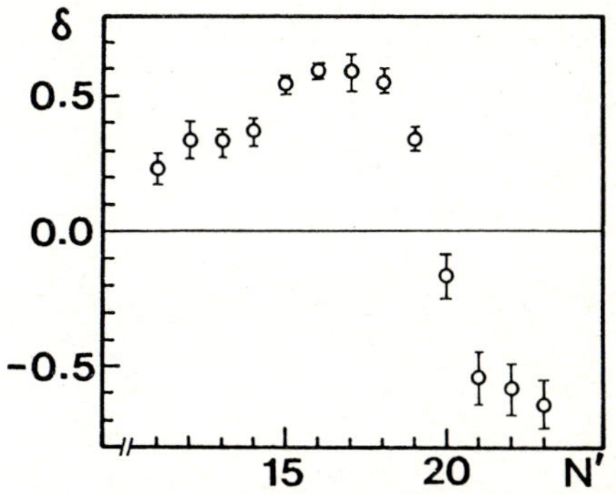

Fig. 1: Normalized population difference in the Λ-doublet components of $CH(A^2\Delta)$ from $(CH_3)_2CO$, plotted against the rotational quantum number.

Reference

1. A. M. Quinton and J. P. Simons, Chem. Phys. Letters 81 (1981) 214.

MEASUREMENTS OF DIFFERENTIAL CROSS SECTIONS FOR e-Ar, Kr, Xe SCATTERING AT E = 50 meV - 2eV

M. Weyhreter, B. Barzick and F. Linder

Fachbereich Physik, Universität Kaiserslautern
D-6750 Kaiserslautern, West Germany

We report on differential scattering experiments for the system e - Ar, Kr, Xe in the energy range E = 50 meV - 2 eV using a crossed-beam arrangement. The measurements cover the region of the Ramsauer-Townsend minimum (for the first time in differential scattering using modern techniques) and show the drastic increase of the cross section towards zero energy which is normally treated by extrapolation techniques using MERT methods. For these experiments, a new electron spectrometer has been built which is especially designed to allow measurements at very low collision energies under well controlled conditions. In the present work, the measurements could be extended down to 50 meV. The energy resolution was typically 20 meV (FWHM). The transmission properties of the systems were controlled using elastic e-He scattering as a reference system[1]. The energy scale was calibrated by measuring narrow resonances in electron-rare gas scattering.

The present measurements for the systems e - Ar, Kr, Xe have been performed in the angular range $\vartheta = 20^0 - 100^0$. Two examples of the measurements are shown in Fig. 1 and 2. Absolute cross sections are determined using the method of Srivastava et al.[2]. The evaluation of the data is in progress and a full set of the results will be presented at the conference. Comparison with recent theoretical work[3] will be made.

References

1. R.K. Nesbet, Phys. Rev. A 20, 58 (1979)
2. S.K. Srivastava, A. Chutjian, S. Trajmar, J. Chem. Phys. 64, 2659 (1975)
3. A.W. Yau, R.P. McEachran, A.D. Stauffer, J. Phys. B 13, 377 (1980) and private communication

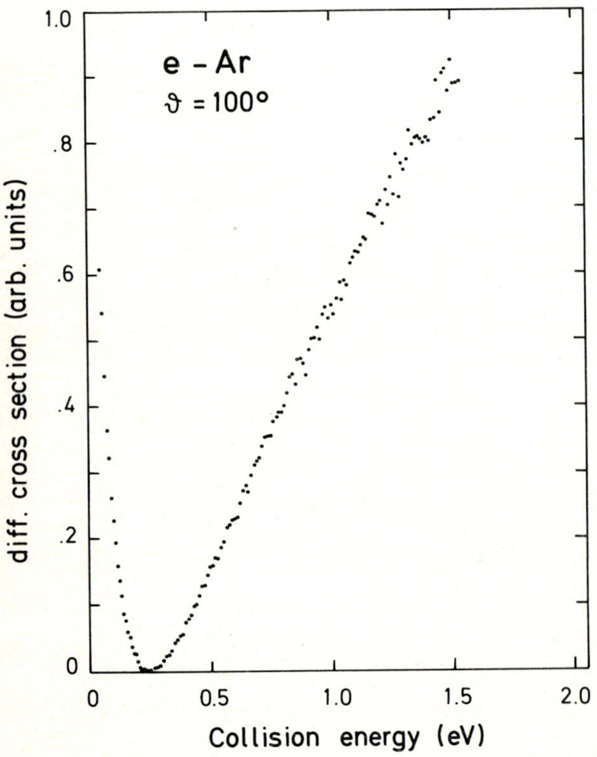

Fig. 1

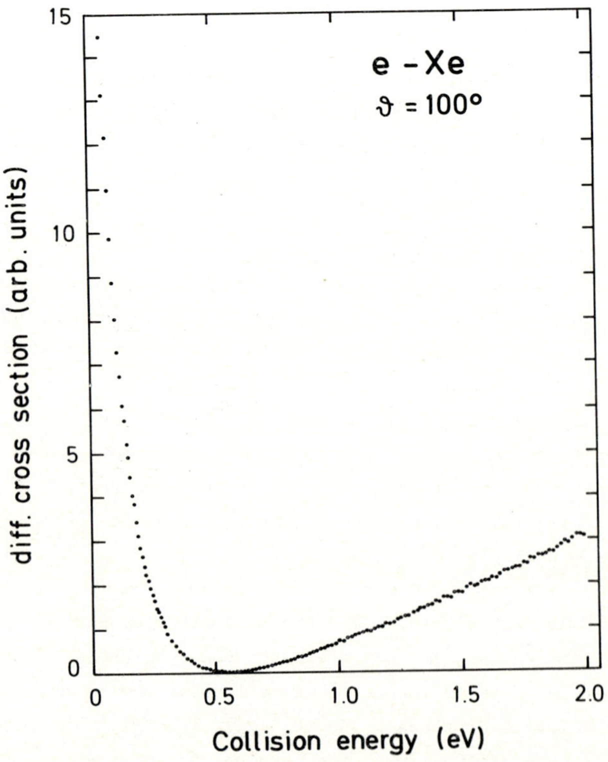

Fig. 2

ON THE ELASTIC ELECTRON-NEON SCATTERING IN THE I6 eV REGION

E.Našlenas

Institute of Physics,Academy of Sciences of the Lithuanian SSR, Vilnius,232600,USSR

P.Žvirblis

Kaunas A.Sniečkus Polytechnic Institute, Kaunas,233006,USSR

In electron-neon scattering with the energy about I6 eV the formation of the compound state

$$1s^2 2s^2 2p^5 3s^2 (^2P^o) \quad (I)$$

is observed. By analogy with the target of argon and electrons of the energy around II eV [I], we can hope for the appearance of a feature (minimum) in the energy dependence of differential cross section of elastic scattering of electrons at large angles ($\theta \approx I60^o$). For the modelling of this process we shall apply the close coupling approximation.

Since the compound state decays by radiating an electron, one electron in (I) must have a radial wave function oscillating in asymptotics. But there is no shell in (I) with one electron. Therefore two different radial functions ought to be ascribed to equivalent electrons. The situation reminds us the one considered by the extended method of calculation in the theory of atomic spectra [2]. The shells of equivalent electrons are like the configurations $\{ns\ k's\}$ and $\{2p^4 k''p\}$ in this approximation. For simplification let's consider only a 3s-shell out of three s-shells (the rest two are connected with the appearance of deeply closed channels). In that way we obtain the following superposition:

$$(2p^6 kp + 2p^5\{3s\ k's\} + 3s^2\{2p^4 k''p\})(^2P^o). \quad (2)$$

The elastically scattered s-wave enters into the configurations of opposite parity and therefore the related subsystem of equations of close coupling is solved independently on (2). The differential cross section of angular distribution of the elastically scattered electrons can be presented as

$$\frac{d\sigma}{d\Omega} = a + b\cos\theta + c\cos^2\theta, \ (a_0^2\ sr^{-1}). \quad (3)$$

The values of the coefficients are given below in the table

E (eV)	a	b	c
I2.24	0.925	4.083	5.II4
I4.96	0.82I	3.000	3.653
I6.32	0.775	2.669	3.285
I7.68	0.73I	2.4I3	3.026

For the ground state of atom a self-consisted Hartree-Fock field and frozen-core approximation for the excited states were used.

As easily seen, (3) has a minimum, the angular position θ' of which depends on the energy. θ' reaches the maximum value at $E \approx$ I5 eV ($\theta' \approx$ II4^o). This feature, however, does not show itself noticeably in the energy dependence of the differential cross section of elastically scattered electrons at large angles. Only a steep fall in the curve, which takes place at smaller energies, in the region I6 eV becomes more flat the sign of derivative remaining the same.

The given model is of greater accuracy for small angles of scattering. However, the results in the region I6 eV are in disagreement with those in [3] for small angles. On the other hand, the agreement is satisfactory for other energies at small angles. For illustration we can compare the value of (3) at E=2.2 eV (a=I.425, b=-I.047, c=0.272) with experimental data, refered to in [4]. Let us discuss in brief an example for higher energy E=50 eV. The coefficients a=0.225, b=0.622, c=I.642 and (3) agree with [5] within the limits of I0% for angles less than θ' ($\theta' \approx$ I0I^o). For large angles difference rises till 50%. The disagreement in the total elastic cross section is about I5%.

The errors in the last example mainly result from insufficient estimation of higher partial waves. They become smaller in the I6 eV region. However, for such energies a greater correction can be found after specifying the polarizability of the target.

References

I. D.Andrick,L.Langhans,X ICPEAC(Paris,I977).
2. A.Jucys,E.Našlėnas,P.Žvirblis,Int.J.Quantum Chem.6,465(I972).
3. D.F.C.Brewer,W.R.Newell,S.F.W.Harper, A.C.H.Smith,J.Phys.B,I4, L749(I98I).
4. D.G.Thompson,J.Phys.B,4,468(I97I).
5. J.F.Williams,A.Crowe,J.Phys.B,8,2233(I975).

ELASTIC SCATTERING OF ELECTRONS FROM NEON AT INTERMEDIATE ENERGIES

B.B. Srivastava and S.S. Dhal

Physics Department, Meerut University, Meerut-250005, INDIA

Differential cross sections (DCS) for the elastic scattering of electrons from neon atoms have been calculated at incident energies of 200,300,400,500,750 and 1000 eV using an optical potential which governs the direct scattering to second order in a multiple scattering expansion. The theoretical procedure[1] is based on a two-potential treatment of the first-order static and the second-order polarization-absorption parts of this optical potential. According to this procedure, under reasonable assumptions, the direct scattering amplitude can be written as $f_{ST}+f_E-f_{EST}$, where f_{ST} is the scattering amplitude due to the static potential, V_{oo}, computed from the scattering equation corresponding to V_{oo} using a partial wave procedure, and f_E and f_{EST} are the eikonal scattering amplitudes due to the optical and the static potential, respectively. In the present calculations f_E and f_{EST} have been calculated using instead of the usual eikonal wave function the somewhat better Blankenbecler and Goldberger[2] (BG) wave function. The exchange contribution is included by means of an exchange phase factor in f_E calculated from the exchange potential obtained following a procedure similar to that described by Vanderpoortan[3]. For the ground state of neon atom the orthogonalized product wave function given by Sheorey[4] has been used. The calculation of f_E including the polarization and absorption parts of the optical potential has been carried out using a procedure described earlier[1].

Many of the calculated DCS are shown by solid curves in Fig.1 along with the experimental data of Gupta and Rees[5] (O),William and Crowe[6] (Δ), Jansen et al[7] (□), and Bromberg[8] (◇). It is seen that the present procedure yields values of DCS in good agreement with experiment throughout the full angular range at these energies.

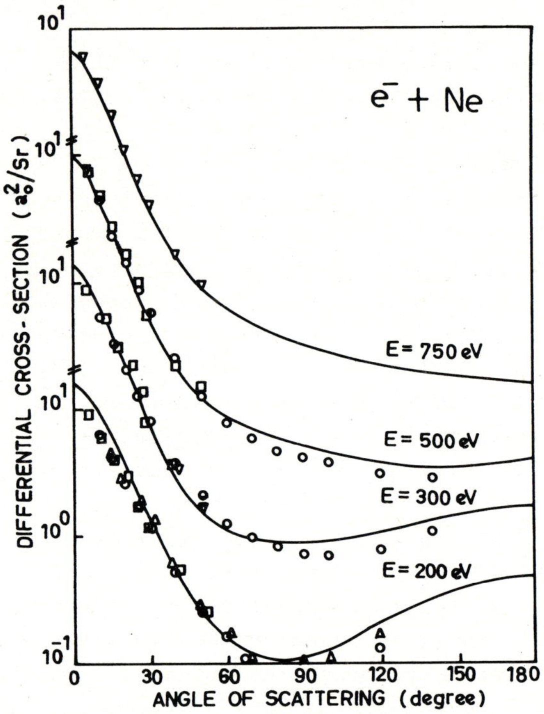

Fig.1: DCS for e-Ne elastic scattering at 200,300,500 and 750 eV are shown by solid lines. The experimental data are shown by O, Δ , □ , and ◇ .

References

1. B.B.Srivastava,S.S.Dhal and R.Shingal, XII ICPEAC Abstracts of contributed papers (ed.S.Datz)p.137 and the references therein.
2. R.Blankenbecler and M.L.Goldberger,Phys. Rev.126, 766 (1962).
3. R.Vanderpoortan,J.Phys.B19,L535 (1976).
4. V.B.Sheorey,J.Phys.B2,442 (1969).
5. S.C.Gupta and J.A.Rees,J.Phys.B8,417 (1975); 8, 1267 (1975).
6. J.F.Williams and A.Crowe,J.Phys.B8,2238 (1975).
7. R.H.J.Jansen,F.J.de Heer,H.J.Luyken, B. van Wingerden,and H.J.Blaauw,J.Phys.B9 185 (1976).
8. J.P.Bromberg,J.Chem.Phys.61,963 (1974).

ELASTIC SMALL-ANGLE ELECTRON SCATTERING BY HELIUM

Ortwin Müller and J. Geiger

Fachbereich Physik, Universität Kaiserslautern, PO Box 3049, D-6750 Kaiserslautern

Since the classical work by Hughes, McMillen and Webb[1] in 1932 it is well known that the elastic scattering of low-energy electrons into small angles can not be described by the first Born approximation taking only the static atomic potential into account. Further important and clarifying contributions to this problem in particular to the scattering of electrons by the rare gas atoms were published by Bromberg[2] and Jansen et al.[3]. According to these papers the first Born approximation seemed to describe the elastic differential cross section for vanishing wave number transfer $k \to 0$ correctly if only the electron energy is sufficiently high.

Since helium should be most easily accessible to a theoretical treatment, scattering measurements have been carried out for He with 10 - 25 keV electrons. The experimental set-up was essentially the same as in ref.[4]. The geometry of the five-electrodes intermediate-image filter lens[5] was maintained, but an improved ray path was chosen. This way the maximum angle accepted by the filter lens could be almost doubled to θ_{max} = 15 mrad without changing the cut-off potential very much.

The experimental results are shown in Fig. 1 as a function of the wavenumber transfer k in a semilogarithmic representation. The experimental scattering distribution for 25 keV electrons was normalized at $a_o k$ = 0.52 to the first Born approximation which is constant within the angular range considered. According to our experience this procedure is justified to obtain reliable cross sections for 25 keV electrons. The curves for the lower electron energies were fitted to the 25 keV distribution at large k values. The angular distribution for the 25 keV electrons agrees within the experimental accuracy with the previous results[4].

From Fig. 1 the following conclusions have to be drawn:

(a) At all primary energies used the measured elastic differential cross section of helium cannot be described in terms of the first Born approximation.

(b) The angular distributions show a linear behaviour towards small wavenumber transfers. The linear extrapolation is shown in Fig. 1 for 15 - 25 keV. It is omitted for 10 keV to avoid confusion. The differential cross section may, therefore, in a certain range of k have the form

$$\frac{d\sigma}{d\Omega} = \sigma_o e^{-\beta k} \quad ,$$

and it could be speculated whether the scattering is caused by a polarization potential (Bromberg[6]). The value of the polarizability α = 15 a.u. for 25 keV, however, was unrealistic high compared to the static polarizability of helium α = 1.49 a.u. . On the other hand the optical model successfully introduced in ref.[4] might suggest the slope of this linear portions of the distributions in Fig. 1 to depend on the primary electron energy E_i.

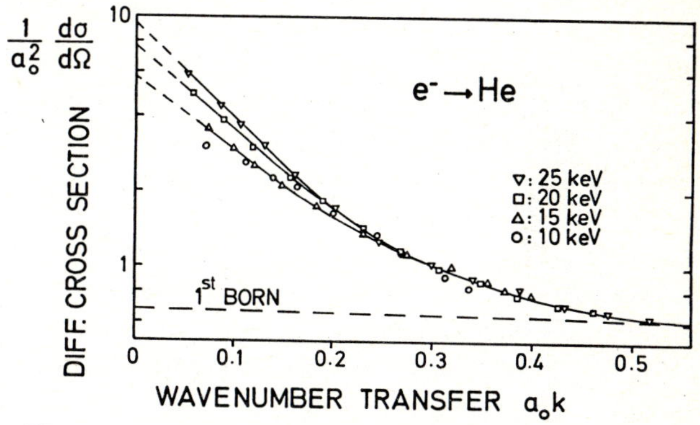

Fig. 1: Elastic small-angle electron scattering by helium. a_o: Bohr radius.

Empirically we find $\beta \propto \sqrt{E_i}$ in a good approximation:

E_i	=	25	20	15	10	keV
β_{exp}	=	8.6	7.7	6.8	5.1	a.u.
β_{calc}	=	(8.6)	7.7	6.7	5.5	a.u. ,

where the value β_{calc} is related to its value at 25 keV.

This work was supported in part by the Deutsche Forschungsgemeinschaft

References

1. A.L. Hughes, J.H. McMillen, G.M. Webbs, Phys. Rev. 41, 154 (1932)
2. J.P. Bromberg, J. Chem. Phys. 61, 963 (1974)
3. R.H.J. Jansen, F.J. de Heer, H.J. Luyken, B. van Wingerden, H.J. Blaauw, J. Phys. B 9, 185 (1976); R.H.J. Jansen, F.J. de Heer, J. Phys. B 9, 213 (1976)
4. J. Geiger, D. Morôn-Leôn, Phys. Rev. Letters 42, 1336 (1979)
5. D. Morôn-Leôn, J. Geiger, Electron Microscopy 1980, Vol. 1, p. 42
6. J.P. Bromberg, J. Chem. Phys. 60, 1717 (1974)

PSEUDOSTATE MODEL OF ELECTRON SHADOW SCATTERING

S. Geltman[+] and R. K. Nesbet[++]

[+]JILA, University of Colorado, Boulder, Colorado 80309, USA

[++]IBM Research Laboratory, San Jose, California 95193, USA

A strong forward peak and rapid angular variation, essentially a Fraunhofer diffraction pattern, have been observed in extreme forward scattering of 15-25 keV electrons by rare gas atoms.[1] Geiger and Morón-León interpret their data as a diffraction pattern due to shadow scattering. The removal of forward electron flux, due to inelastic scattering, which dominates elastic scattering by a factor $k^2a_0^2$ in the forward direction, is expected to affect the elastic cross section analogously to the scattering effect of a black disk, whose shadow would be described by a Fraunhofer pattern. Despite this simple physical picture, no published theoretical work leads to such a diffraction pattern for electron-atom scattering. The first Born elastic cross section is nearly independent of angle in the angular range of interest (below 10 mrad).

As shown by Joachain[2], a strongly absorptive potential (complex square well) with a finite radius produces a Fraunhofer diffraction pattern. The purpose of the present work is to see if such a result can emerge from a detailed study of electron scattering by an atom. To simplify the theory here, an atom is characterized by a pseudostate model, in which electric dipole excitation from the ground state is described by transitions to a single pseudostate, constructed to give the exact static dipole polarizability. Using the $\ell=1$ pseudostate for hydrogen[3], the 2nd Born scattering amplitude has been computed explicitly for the coupled elastic and inelastic channels. Results of these calculations are discussed below. A real polarization potential function, valid for all values of r in the low-energy limit, is extracted from these results. With a further simplification of the model problem (replacing the short-range potential by a hard-core boundary condition), this polarization potential at large r is shown[4] to be the leading term of a complex expansion in powers of r^{-1}.

The 2nd Born elastic cross section is found to have no strong forward peak, in contrast to coupled partial wave calculations of Mohr[5], and no oscillations at small angles. The imaginary part of the forward scattering amplitude is determined in 2nd order by the relatively large inelastic cross section, in agreement with the optical theorem.

These results can be reconciled by considering the two leading terms of the complex optical potential, deduced from the coupled differential equations by an analytic technique[4]. The first two terms are

$$\tfrac{1}{2}\alpha(-r^{-4} + (2ik/\Delta E)r^{-5}), \qquad (1)$$

where α is the static dipole polarizability, k is the incident wave vector, and ΔE is the pseudostate excitation energy. The WKB phase shifts due to this potential are complex. For the ground state of atomic hydrogen α is $4.5\ a_0^3$ and ΔE is $(18/43)e^2a_0^{-1}$. Taking the core radius r_0 to be $1.0\ a_0$, the imaginary term in the optical potential introduces a damping factor exp (-5.375) into all partial wave S-matrix elements with $\ell+\tfrac{1}{2}$ less than kr_0. The resulting effective shadow in the forward direction must produce a diffraction pattern similar to that derived for a complex square well model potential[2].

Both terms in Eq. (1) are quadratic in the transition amplitude, compatible with the 2nd Born approximation. However, the damping effect is so large in the shadow region that a low-order Born approximation to the extreme forward elastic scattering amplitude cannot be valid, essentially a power series expansion of an exponential function of large negative argument. Hence specific effects of shadow scattering, such as the forward diffraction pattern, cannot usefully be treated by a low-order Born approximation.

[+]Work at University of Kaiserslautern, W. Germany, sponsored by the Alexander von Humboldt Foundation.

[++]Research supported in part by the U.S. Office of Naval Research.

References

1. J. Geiger and D. Morón-León, Phys. Rev. Letters 42, 1336 (1979).
2. C.J. Joachain, "Quantum Collision Theory" (North-Holland, Amsterdam, 1975), v. 1, pp. 197-199.
3. R. Damburg and E. Karule, Proc. Phys. Soc. (London) 90, 637 (1967).
4. R.K. Nesbet, to be published.
5. C.B.O. Mohr, J. Phys. B 2, 166 (1969).

SMALL ANGLE (e^-,NA) SCATTERING

B. Jaduszliwer, P. Weiss, A. Tino and B. Bederson

New York University, New York, N. Y. 10003, U.S.A.

We report here on measurements of small angle elastic and forward inelastic electron scattering by sodium atoms, at energies between 10 and 25 eV. These are the first measurements taken using a high resolution atomic beams apparatus recently built at the New York University Atomic Beams Laboratory. Such measurements are of particular interest for probing the long-range part of the electron-atom interaction.

The atomic beam, formed using a two-chamber oven to minimize dimer population in the beam, travels along the y-axis of the apparatus. It is focussed and velocity-selected by an hexapole magnet, and then enters the collision region. The electron beam, travelling along the z-axis, is ribbon shaped; its energy width is about 0.4 eV (FWHM), and typical currents are of the order of 500 μA. The electron energy is corrected for contact potential differences and space charge effects.

Observation of the scattering is made on the recoiled atom, rather than on the scattered electron[1]. After crossing the collision region, the atoms travel down a 335 cm tube, at the other end of which they are detected by surface ionization. To perform scattering-in experiments, in which one collects atoms which have been recoiled away from the atomic beam (y-) axis, the detector can be moved, and precisely positioned, in the x-z plane.

The atomic recoil angles are $\psi=\Delta z/L$ and $\chi=\Delta x/L$, where Δz and Δx are the detector displacements from the beam axis, and L is the distance between interaction region and detector. For elastic scattering, ψ and χ are kinematically related to the electron scattering angles θ,ϕ by $\psi=\alpha(1-\cos\theta)$, $\chi=\alpha\sin\theta\sin\phi$, where α is the electron-to-atom momentum ratio. For inelastic scattering, $\psi=\alpha-\beta\cos\theta$, and $\chi=\beta\sin\theta\sin\phi$. β is the electron-to-atom momentum ratio after the collision. Only atoms which undergo elastic collisions will be recoiled into the range $0<\psi<\alpha-\beta$, and we have exploited that fact to measure the differential cross sections for small angle (e^-,Na) scattering at 10 eV.

The cross section for recoiling an atom into the detector set at $z=z_D$, $x=0$, is given by

$$\Delta Q=g(z_D)\frac{hv}{2I_e}\frac{S'-S}{S}=\int\frac{d\sigma}{d\Omega}\,g(z_D-\alpha L+\alpha L\cos\theta)\,f(-\alpha L\sin\theta\sin\phi)\,d\Omega$$

where S' and S are the atomic detector signals with the electron beam on and off respectively; I_e is the electron current (e^-/sec); h is the height of the collision region, V the atomic beam velocity (about 850 m/sec), and g(z) and f(x) are the horizontal and vertical beam-shape functions, normalized to unity at the centroid.

If an "effective collection solid angle" is defined by $\Delta\Omega=\int g(z_D-\alpha L+\alpha L\cos\theta)f(-\alpha L\sin\theta\sin\phi)d\Omega$, extensive consistency checks have proven that in the conditions of our experiment, the "average differential cross section" $\langle d\sigma/d\Omega\rangle=\Delta Q/\Delta\Omega$ is essentially identical with the actual differential cross section $d\sigma/d\Omega$ for $\theta>8^{o}$. Differential cross sections determined in this way are absolute, in the sense that they do not require any normalization.

Our results are presented in Fig. 1, together with previous measurements by Srivastava and Vuskovic[2], and calculations by Issa[3].

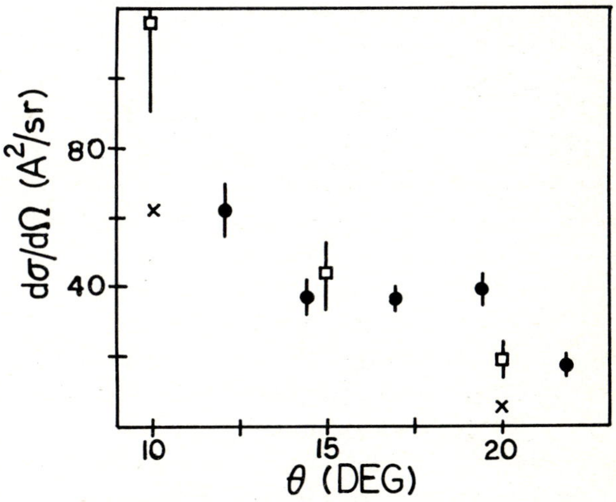

Fig. 1: (e^-,Na) differential cross section at 10 eV vs. electron polar angle. Black dots: this work. Squares: Srivastava and Vuskovic. Crosses: Issa's calculation.

We have also studied the $3^2S\rightarrow3^2P$ electron impact excitation of sodium in the 10 to 25 eV range, by setting the detector to collect atoms recoiled by $\psi=\alpha-\beta,\chi=0$. The quantity determined in this case is the excitation differential cross section in the forward ($\theta\simeq0$) direction, averaged over the apparatus azimuthal form factor $\gamma(\theta,\theta_0)$ where θ_0 is a cut-off angle. The interval $\Delta\theta$ over which the averaging is performed is small, varying from about 4.5^{o} at 10 eV to about 2.5^{o} at 25 eV. Results will be presented at the meeting and compared with Born approximation calculations[4].

Research supported by U.S.A. National Science Foundation

References

1. K. Rubin, B. Bederson, M. Goldstein and R.E. Collins, Phys.Rev.182,201(1969).

2. S.K. Srivastava and L. Vuscovic, J.Phys.B13,2633(1980)

3. M. Issa, Ph.D. Thesis, University of Durham(1977).

4. M. Inokuti, Rev.Mod.Phys.43,297(1971).

ABSOLUTE MEASUREMENTS OF SMALL ANGLE ELASTIC SCATTERING OF ELECTRONS

R.W. Wagenaar, T. v. Tubergen, F.J. de Heer

FOM-Institute for Atomic and Molecular Physics, Kruislaan 407, 1098 SJ Amsterdam, The Netherlands.

This experiment is part of an extensive study on the validity of forward dispersion relations for electron scattering at atoms. On a semi-empirical basis we could already show [1,2] that the Guerjoy-Krall formulation was incorrect, due to non-analytic properties of the exchange amplitude. For more quantitative insight in this exchange, we had to measure accurately total cross sections (t.c.s.) [3,4] as well as elastic differential cross sections (d.c.s.) at as small as possible angles.

For the elastic d.c.s. measurements we developed a new type parallel plate electrostatic analyser with a curved entrance and exit slit. In this particular geometry all electrons enter the analyser under 45^0, independent of their scattering angle, see fig. 1. In this way the parabolic path length of the electrons in the analyser depends only on their energy. With channelplates over the exit slit as position sensitive detector, the angular distribution between -10^0 and $+10^0$ is obtained simultaneously with a resolution of 0.1^0. The energy resolution can be enhanced to about 1/300 by pre-retarding the electrons by a set of spherically shaped electrodes, without severely deforming the scattering angles.

A small box-like Faraday cage just in front of the analyser, used to intercept and monitor the primary beam, limited the smallest angle to 1.5^0 below ~100 eV, whereas 0.5^0 is reached at higher energies. By sweeping a low intensity beam over the entrance slit we could calibrate the channelplate efficiency absolutely.

Fig. 1: Apparatus for small angle electron scattering.

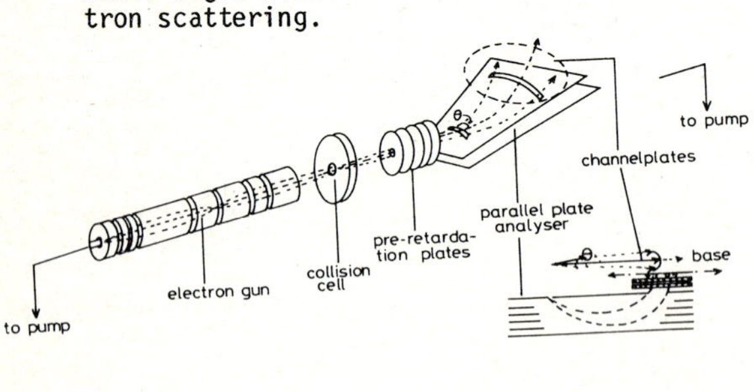

Measurements have been carried out for noble gases and molecules (N_2 and H_2) between 17.5-750 eV. Fig. 2 shows agreement with the accurate matrix variational calculation of Nesbeth [5] within 7% at 17.5 eV and with the extrapolated experimental values of Srivastava et al. [6] within 10%. Scott and Taylor [7], in their many body calculation, clearly underestimate elastic forward scattering.

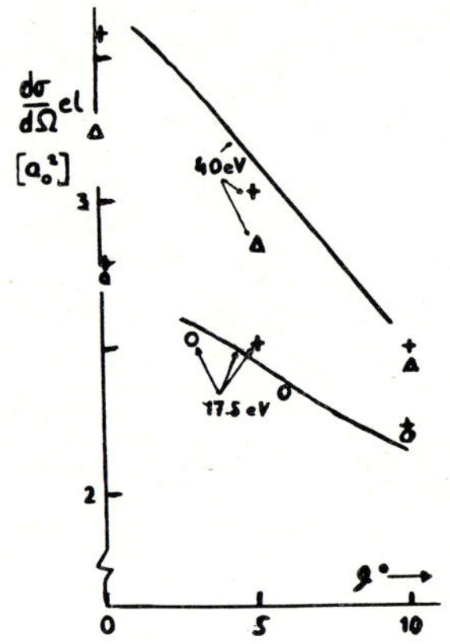

Fig. 2: Our experimental results for electrons on helium, compared with other data.
o : Nesbeth [5]
Δ : Scott & Taylor [6]
+ : Srivastava et al. [7]
—— : this experiment

Scanning the impact energy over the ionization threshold, we observed a dip, which increases towards smaller angles, see fig. 3. No explanation is available yet, but strong channel coupling near the ionization threshold might be important. For neon the same structure appears, although less pronounced.

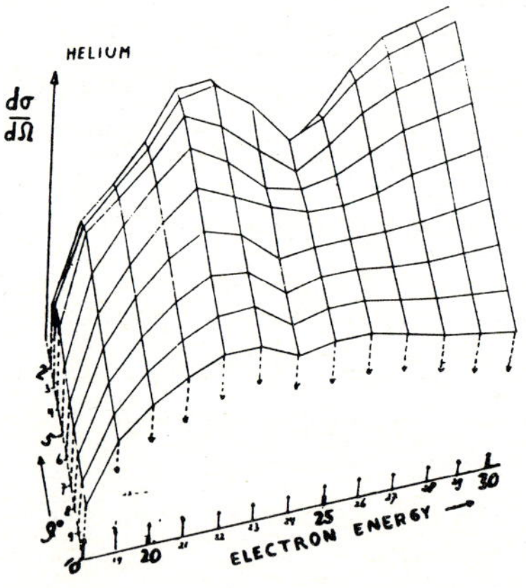

Fig. 3: Experimental results for elastic scattering on helium.

References

1. F.J. de Heer, R.W. Wagenaar, H.J. Blaauw and A. Tip, J.Phys.B 9, L 269-74 (1976).
2. F.J. de Heer, M.R.C. McDowell and R.W. Wagenaar, J.Phys.B 10, 1945-1953 (1977).
3. H.J. Blaauw, R.W. Wagenaar, D.H. Barends and F.J. de Heer, J.Phys.B 13, 359-76 (1980).
4. R.W. Wagenaar and F.J. de Heer, J.Phys.B 13, 3855-66 (1980).
5. R.K. Nesbeth, Phys.Rev. A 20, 58-70 (1979).
6. T. Scott and H.S. Taylor, J.Phys.B 12, 3385-97 (1979).
7. D.F. Register, S. Trajmar and S.K. Srivastava, Phys. Rev. A 21, 1134-51 (1980).

RELATIVE DIFFERENTIAL CROSS SECTIONS FOR ELASTIC AND INELASTIC SCATTERING OF ELECTRONS BY XENON IN THE ENERGY RANGE OF 15 TO 80 eV

B.Marinković, V.Pejčev, D.Filipović and L.Vušković

Institute of Physics, P.O.Box 57, 11001 Beograd, Yugoslavia

A crossed electron - atom beam collision technique has been used to study the electron scattering by Xenon atoms. Relative differential cross sections (RDCS) for elastic and inelastic scattering have been measured at electron impact energies of 15,20,30 and 80 eV. In addition,only elastic scattering have been measured at 50 and 63 eV.

Tipical energy-loss spectrum with inelastic features is shown in Fig. 1 with impact energy and scattering angle indicated. We studied inelastic region between 8.0 and 10.8 eV. Since the energy resolution of the spectrometer was approximatly 40 mV, all transitions to individual states could not be resolved. Peaks indicated on Fig. 1 are explained in the Table. Similar spectra were obtained at other scattering angles ranging from 0° to 150°. From each spectrum ratio of the intensity of a spectral feature to the intensity of the transition $6s\,[3/2]_1^o$ was obtained. Special attention was paid in obtaining the angular distribution of the intensity of transitions $6s\,[3/2]_1^o$ and results are shown on Fig. 2. Ratios between this transition and elastic scatting for same impact energy has been determined in the separate experiment. So, all data of RDCS for elastic and inelastic scattering have the same relative units at each impact energy of electrons.

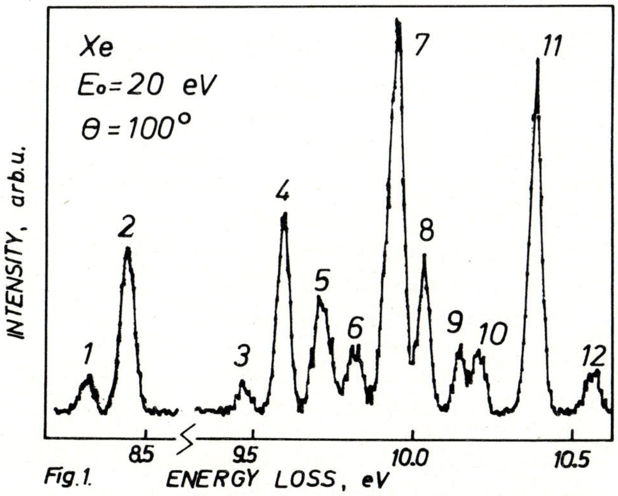

Fig.1.

Number	Desig.	J	eV	Number	Desig.	J	eV
1	6s[3/2]	2	8.315	7	6p[1/2]	0	9.934
2	6s[3/2]	1	8.437		5d[1/2]	0	9.891
3	6s'[1/2]	0	9.447		5d[1/2]	1	9.917
4	6s'[1/2]	1	9.570		5d[7/2]	4	9.943
	6p[1/2]	1	9.580		5d[3/2]	2	9.959
5	6p[5/2]	2	9.686	8	5d[7/2]	3	10.039
	6p[5/2]	3	9.721	9	5d[5/2]	2	10.158
6	6p[3/2]	1	9.789	10	5d[5/2]	3	10.220
	6p[3/2]	2	9.821	11	5d[3/2]	1	10.401
				12	7s[3/2]	2	10.562
					7s[3/2]	1	10.593

In order to put results on absolute scale calibration of elastic scattering is in progress in our laboratory. We compare our data with previous measurements of elastic scattering[1,2] as well as inelastic measurements of Williams et al.[3]

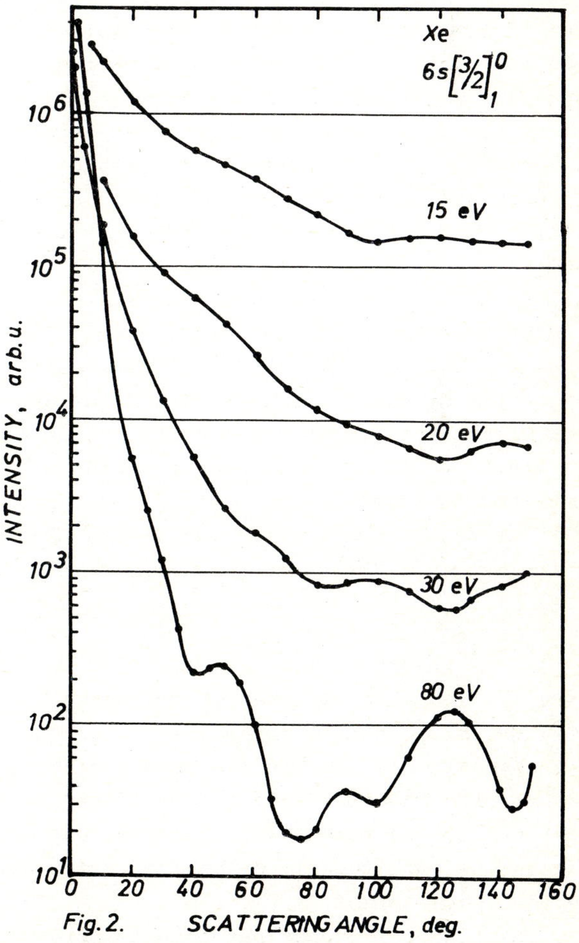

Fig.2.

References

1. D.F.Register, L.Vušković and Trajmar, 7th International Conference on Atomic Physics, Massachusetts Institute of Tech., August 4-8, 1980, Book of Abstracts; private communication.
2. M.Klewer, M.J.M.Beerlage and M.J.van der Wiel, J.Phys.B:Atom.Molec.Phys.13,(1980)571
3. W.Williams,S.Trajmar and A.Kuppermann, J.Chem.Phys. 62, (1975) 3031.

ELASTIC SCATTERING OF ELECTRONS BY SODIUM ATOM

S.N.Singh, A.N.Tripathi and M.K.Srivastava

Department of Physics, University of Roorkee, Roorkee-247667, India

Recent absolute measurements of angular distribution by Srivastava and Vuskovic[1], Teubner et al[2] and Shuttelworth et al[3] for elastic and inelastic scattering of electrons by sodium atom in the intermediate energy range have necessitated theoretical studies for these processes. The sodium atom with its core and a single valence electron resembles in electronic configuration to a hydrogen atom, although its nature enormously differs from the later due to its size and threshold excitation. There exists a strong coupling between the ground and the first excited state in the sodium atom. In such a situation the loss of flux from the elastic channel should play an important role in any description of collision processes. In recent years there have been many attempts to explore the feasibility of modifications of high energy methods to incorporate the important features of the dynamics of collisions. In most of these methods, a common feature is that they pay special attention to second order terms of the multiple scattering series and treat them more accurately than the higher order terms. In the present study, we have examined the elastic scattering of electrons by sodium in an approach which involves an exact contribution of the static part of the interaction and on shell contribution of the remaining part of the multiple scattering series (i.e. in the Glauber approximation). The scattering amplitude is given by

$$f = f^{st}(V_{st}) + f_G(V) - f_G$$

Here, V is the total interaction and $V_{st} = \langle 0|V|0\rangle$ is the static potential. The exchange has been taken following the prescription of Furness and McCarthy[4] and is considered along with the static potential. The resulting scattering amplitude is calculated using a partial wave format.

The figure shows our results for differential scattering cross-section (DCS) at 54.4 eV. We have also shown the results of other calculations[2,5]. The present results agree well upto the scattering angle $\theta \simeq 40^\circ$ with the experimental data of Teubner et al[2] and Srivastava and Vuskovic[1]. In the region

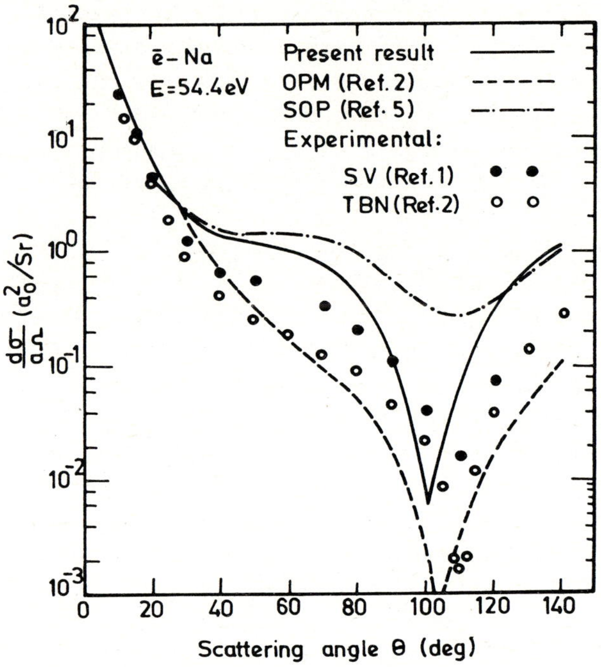

$\theta > 40^\circ$, the two experimental results differ appreciably in magnitude. Our results in this region compare favourably with the results of Srivastava and Vukovic, Both measurements show a similar variation in DCS with a deep minima around 110°. The present results exhibit qualitatively the same shape as predicted by the data. The results for other energy will be presented in the Conference.

References :

1. S.K.Srivastava and L.Vuskovic, J.Phys.B, 13, 2633 (1980).
2. P.J.O.Teubner, S.J.Buckman and C.J.Noble J.Phy. B 11, 2305 (1978).
3. T.Shuttleworth, W.R.Newell and A. Smith, J.Phys. B, 10, 1601 (1977).
4. J.B. Furness and I.E.McCarthy, J.Phys.B6, 2280 (1973).
5. M.R.Issa, Ph.D.Thesis, Durham University (1977)

RELATIVISTIC PHASE SHIFT ANALYSIS FOR ELASTIC SCATTERING OF ELECTRONS BY XENON BELOW THE FIRST INELASTIC THRESHOLD

D. Register[+], L. Vuskovic[*], L. T. Sin Fai Lam[++] and S. Trajmar

Jet Propulsion Laboratory, California Institute of Technology, Pasadena, California 91109

Differential cross sections (DCS) for elastic scattering of electrons by Xe have been reported previously for the impact energy range 1.75 eV to 100 eV.[1] The phase shift analysis used in the previous case was non-relativistic and ignored any spin-orbit coupling effects. In the present work, relativistic effects are included in a formalism developed by Sin Fai Lam[2] for elastic scattering below threshold.

For the case of Xe, relativistic effects will be most pronounced near the Feshbach resonances just below the first excitation threshold. At these energies, the η_ℓ^+ and η_ℓ^- phase-shifts will exhibit different energy dependencies. For the 7.75 eV Xe resonance, the η_ℓ^+ ($\ell = 1$, $j = 3/2$) character has been identified and will be analyzed using the relativistic formalism. In this case:

$$\frac{d\sigma_{el}}{d\Omega} = |f(\theta)|^2 + |g(\theta)|^2$$

where:

$$f(\theta) = \frac{1}{2ik} \sum_{\ell=0}^{\infty} ((\ell + 1)[\exp(i2\eta_\ell^+) - 1] + \ell[\exp(i2\eta_\ell^-)-1])\, P_\ell(\cos\theta)$$

$$g(\theta) = \frac{1}{2ik} \sum_{\ell=0}^{\infty} (\exp[i2\eta_\ell^-] - \exp[i2\eta_\ell^+])\, P_\ell^1(\cos\theta)$$

and for $\ell = 0$, $\eta_o^+ = \eta_o^-$ whereas for $\ell >> 0$, $\eta_\ell^+ \simeq \eta_\ell^-$ and $g_\ell(\theta) \simeq 0$. In the non-relativistic limit, $\eta_\ell^+ \equiv \eta_\ell^-$ for all ℓ and the usual phase-shift formalism is recovered.

Since the coupling for Xe below threshold is dominated by the $^2P_{1/2,3/2}$ resonances, relativistic effects have been included only in the p-wave. The solution is based on an iterative method in which the non-relativistic least-square fitted[3] phase shifts are used as input to the relativistic calculation. Results for Xe both with and without these corrections will be reported for 10 energies below threshold.

This work is supported by NASA.

[+]Permanent address: Phillips Petroleum Company, Bartlesville, Oklahoma 54003

[*]Permanent address: Institute of Physics, Belgrade, Yugoslavia

[++]Permanent address: Australian National University, Canberra, Australia

References

1. D. F. Register, L. Vuskovic and S. Trajmar, 7th ICAP, Boston, MA (1980).
2. L. T. Sin Fai Lam, Aust. J. Phys. 33, 261 (1980), and private communication.
3. D. F. Register, S. Trajmar and S. K. Srivastava, Phys. Rev. A 21, 1134 (1980).

ELECTRIC POLARIZABILITY OF NEON ATOM AND ELECTRON-NEON SCATTERING LENGTH FROM ELECTRON CYCLOTRON RESONANCE ABSORPTION SPECTRUM

A.P. Kabilan

Department of Physics, College of Engineering, Guindy, Madras 600 025, India.

The electric polarizability α of neon atom and the scattering length A for electron-neon collision have been determined by fitting the experimental electron cyclotron resonance (ECR) absorption curve, registered in a neon after glow, with a modified effective range theory (MERT) representation of the partial-wave phase shifts choosing α and A as the adjustable parameters. The ECR spectroscopy technique has been described in previous papers[1,2] by the author together with Golovanivsky. The analysis of ECR spectrum was carried out on the basis of the well known integrals of the kinetic theory for ECR absorption intensity and electron energy distribution function. The energy dependence of momentum-transfer cross section $\sigma_M(w)$ in these integrals was expressed by the low-energy partial-wave expansion,

$$\sigma_M(w) = 4\pi \sum_{L=0}^{3} (\eta_L/q - \eta_{L+1}/q)^2(L+1) \qquad (1)$$

where $q = (2\,mw)^{\frac{1}{2}}/\hbar$ is the de Broglie wave number of an electron in units of a_0^{-1} ; a_0 is the Bohr radius; m is the electronic mass;

The phase shifts η_L for sufficiently low energies were represented by MERT expressions,

$$\eta_0/q = -A-0.2839\alpha w^{\frac{1}{2}} - 0.0490A\alpha w \ln w + Bw \qquad (2)$$

$$\eta_1/q = 0.05679\alpha w^{\frac{1}{2}} - 0.07353Cw \qquad (3)$$

$$\eta_L/q = 0.8517\alpha w^{\frac{1}{2}} / [(2L+3)(2L+1)(2L-1)] \qquad (4)$$

for L = 2,3,4.

Here, the scattering length A is expressed in units of a_0 and the polarizability α and the variable parameters B and C, in units of a_0^3. In the previous work[2] the value of α was fixed (the value α =2.65 was used) and the parameters A,B and C were adjusted so that the cross-section σ_M calculated by (1)-(4) together with the computed distribution function (Eq.5 of Ref.2) yielded the best simulating ECR spectrum (Eq.6 of Ref.2) compatible with the experimental spectrum within the error bars. A similar procedure was adopted by O'Malley and Crompton[3] in their MERT analysis of drift-velocity measurements with the difference that they included a $w^{3/2}$ term in the S-wave expansion (2), thereby introducing another variable parameter, meanwhile fixing the values of the much smaller P-and D-wave phase shifts (3) and (4). In the present work, on the contrary, the S-wave was represented by a two-parameter expansion, α being treated as an adjustable parameter at that. This was done not by eliminating the w term, but by making use of the following expression for B in terms of α,A, and r_0 (the effective range)[4],

$$B = A\,\alpha(0.1076 - 0.4900\ln\alpha) - 0.0368\,rA \qquad (5)$$

$$\text{where,}\quad r = r_0 + 2\pi/3\,(\alpha^{\frac{1}{2}} - \alpha^{3/2}/A^2) \qquad (6)$$

A comparative analysis of the different terms in (2) after substitution of B by (5) in the energy region concerned (10^{-3} - 1 eV) showed that the term containing r_0 contributed less than 0.5% to S-wave phase shift. Hence one could safely take a fixed value for r_0 thereby expressing B in terms of α and A. As for the P-wave, it turned out that the w term was less than a few percent of η_0, which meant, an approximate value for the parameter C would suffice to calculate η_1. Thus, ultimately only α and A were retained in the expansions (2)-(4) and both were treated as adjustable parameters in fitting procedure. (The inclusion of a $w^{3/2}$ term in the S-wave expansion was considered and rejected, as it did not result in any noticeable improvement in the simulating ECR spectrum). Having taken the values for r_0 (corresponding to the given values of α, A and B) and C from the previous work[2], the optimum values of the adjustable parameters α and A were found to be 2.73 ± 0.05 and 0.205 ± 0.005 respectively.

The main result of the present work is the determination of α (instead of assuming its value from other experiments) along with A. In this sense, the partial-wave phase shifts (including higher order phase shifts) and momentum transfer cross section determined by these two quantities in the energy region 10^{-3} -1 eV are entirely independent measurements. Besides the small number of parameters (50 points in the ECR curve were fitted by just two parameters) highly restricts the manoevravbility in the fitting procedure thus enhancing the uniqueness of the obtained values for α and A.

References

1. K.S. Golovanivsky and A.P. Kabilan, Phys. Lett. 80 A, 249 (1980).
2. K.S. Golovanivsky and A.P. Kabilan, Sov. Phys. JETP 53 (6), 1153 (1981).
3. T.F. O'Malley and R.W. Crompton J.Phys. B: Atom. Molec. Phys. 13, 3451 (1980).
4. T.F. O'Malley, L. Rosenberg and L. Spruch Phys. Rev. 125, 1300 (1962).

LOW ENERGY SCATTERING FROM KRYPTON AND XENON

R.P. McEachran and A.D. Stauffer

Physics Department, York University, Downsview, Ontario, Canada M3J 1P3

We have previously calculated elastic scattering of electrons from helium[1], neon and argon[2]. Here we shall present results for scattering of electrons from krypton and xenon over the energy range of 0-50 eV. These calculations are carried out in a polarized orbital approximation used previously for both electron and positron scattering.

Detailed results will be presented for phase shifts as well as differential, total and momentum transfer cross sections. These will be compared to existing theoretical and experimental results

References

1. R.P. McEachran and A.D. Stauffer, J. Phys. B 16, 255 (1983)
2. R.P. McEachran and A.D. Stauffer, J. Phys. B (submitted)

PRECISION MEASUREMENTS OF RAMSAUER MINIMA IN ELECTRON TOTAL CROSS SECTIONS

J. Ferch, C. Masche, and W. Raith

Fakultät für Physik, Universität Bielefeld, D-4800 Bielefeld 1

The location and size of pronounced minima in the energy dependence of cross sections are important parameters for comparison with theory. Precise measurements must deal with specific sources of systematic errors.

For total cross section measurements at very low electron energies with high energy resolution and reliable energy scale calibration we developed the time-of-flight (TOF) spectrometer described previously.[1] In recent years this spectrometer was further improved: We eliminated the influence of the target gas on the emission characteristics of the cathode by introducing additional differential pumping in the cathode region. Furthermore, for absolute measurements of the target gas pressure over a wider range we added a spinning-rotor gas-friction gauge[2] to the Baratron used thus far.

The design of the gas target was modified in such a way that now the gas emerging from the target orifices is differentially pumped in an adjacent region in which the electrons have the same energy as in the target. Before, electrons were decelerated and re-accelerated just outside of the target orifices. In the gas effusing from those orifices some of the electrons were scattered outside the target at higher energies. This effect led to measured minimum cross sections which were too high. The design shown in Fig.1 now eliminates this source of error.

A systematic error which lowers the cross section values is the incomplete account of forward scattering. The solid angle within which forward-scattered electrons can leave the target is extremely small for our spectrometer: $\Omega/4\pi \approx 2\cdot10^{-5}$. Therefore, we can neglect this error even in case of enhanced forward scattering.

The first measurements were performed with Ar which has a Ramsauer minimum of about $3x10^{-17}$ cm^2 near 0.3 eV and a maximum of $2.3x10^{-15}$ cm^2 near 14 eV. We measured the absolute total cross section between 0.07 and 20 eV. In the region of the Ramsauer minimum our results are about a factor of two higher than those of Golden and Bandel[3] and Gus'kov et al.[4] Our data agree satisfactorily with calculations of Stauffer et al.[5] up to 5 eV; between 5 eV and 14 eV our cross sections are considerably lower.

Measurements are in progress with methane which also has a pronounced Ramsauer minimum.[6]

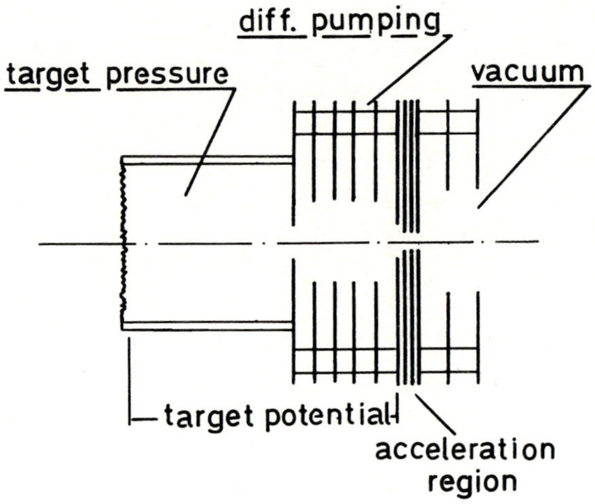

Fig. 1: Design of target at the beam-exit region. The gas emerging from the orifice is differentially pumped before the electrons are re-accelerated. The beam-entrance region of the target was modified accordingly.

References

1. J. Ferch et al., J. Phys. B: Atom. Molec. Phys. 13, 1481 (1980)
2. G. Comsa et al., J. Vac. Sci. Technol. 17, 642 (1980)
3. D.E. Golden, H.W. Bandel, Phys. Rev. 149, 58 (1966)
4. Yu.K. Gus'kov et al., Sov. Phys. Tech. Phys. 23, 167 (1978)
5. A. Stauffer, York University, Toronto, private communication (1982)
6. F.A. Gianturco and D.G. Thompson, J. Phys. B 13, 613 (1980)

TOTAL CROSS SECTIONS FOR ELECTRON SCATTERING FROM N_2, Xe, Kr AND Ar

K. Jost, P.G.F. Bisling, F. Eschen, M. Felsmann and L. Walther

Physikalisches Institut der Universität Münster, D-4400 Münster, Germany

Absolute total cross sections (Q) for electron scattering have been measured by means of a linearized Ramsauer technique. The apparatus is shown schematically in Fig. 1. Electrons from a tungsten hairpin cathode are focussed to the entrance slit of a monochromator. After passing the monochromator at a constant energy of about 2 eV, the electrons are either decelerated or accelerated to the desired energy by a seven-element afocal system. Except for the improved energy resolution ($\Delta E \approx$ 25 meV) due to a new monochromator[1], the apparatus is essentially the same as used for measurements of Q(E) for electron-mercury scattering[2].

By measuring the currents I_s and I_c to scattering cell and collector at two different target densities n_1 and n_2, one obtains

$$Q = [\ell(n_2-n_1)]^{-1} \cdot \ln\{[1+(I_{s2}/I_{c2})]/[1+(I_{s1}/I_{c1})]\},$$

where ℓ denotes the length of the scattering cell. As emphasized in Ref. 2, the data have to be corrected for finite angular resolution of the apparatus. In the case of the (older) measurements at molecular nitrogen[3] and xenon[4], the correction method described in Ref. 2 has been employed. The (recent) measurements at krypton[5] and argon[6] have been corrected experimentally; by employing a set of different exit apertures in the scattering cell, one can extrapolate to an exit aperture of zero diameter or "ideal" angular resolution.

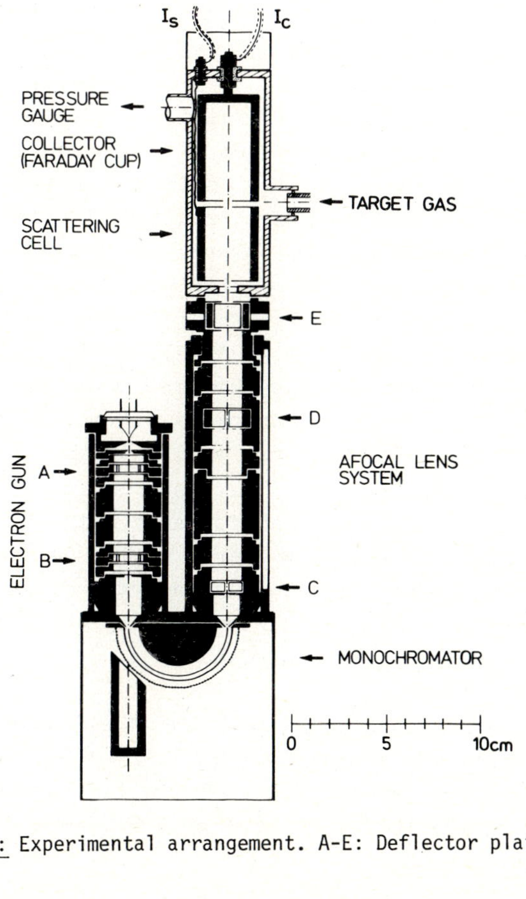

Fig. 1: Experimental arrangement. A-E: Deflector plates.

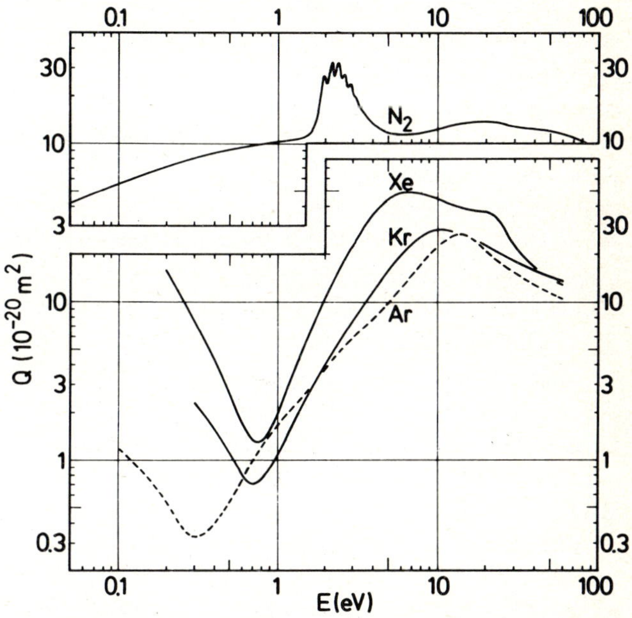

Fig. 2: Results of absolute cross section measurements.

The corrected data are shown in Fig. 2. In the range of overlap at higher energies, our data show very good agreement with measurements performed at the FOM-institute[7,8]. The "Ramsauer-minima" of our measurements are higher by about a factor of 2 in comparison to measurements of a russian group[9], whereas our low-energy N_2-data are again in excellent overall agreement with data of the same group[9] and with data of Kennerly[10] and Golden[11]. Resonance structures in N_2 below 1.8 eV, as observed by Golden[11], could not be verified, however.

References

1. K. Jost, J. Phys. E 12, 1001 and 1006 (1979)
2. K. Jost and B. Ohnemus, Phys. Rev. A 19, 641 (1979)
3. P.G.F. Bisling, Diplomarbeit, Univ. Münster (1978)
4. L. Walther, Diplomarbeit, Universität Münster (1980)
5. M. Felsmann, Diplomarbeit, Univ. Münster (1982)
6. F. Eschen, Diplomarbeit, Universität Münster (1983)
7. H.J. Blaauw, R.W. Wagenaar, D.H. Barends and F.J. de Heer, J. Phys. B 13, 359 (1980)
8. R.W. Wagenaar and F.J. de Heer, J. Phys. B 13, 3855 (1980)
9. Yu.K. Gus'kov, R.V. Savvov and V.A. Slobodyanyuk, High Temp. (USA) 16, 351 (1979) and Sov. Phys.-Tech. Phys. 23, 167 (1978)
10. R.E. Kennerly, Phys. Rev. A 21, 1876 (1980)
11. D.E. Golden, Phys. Rev. Letters 17, 847 (1966) and Adv. At. Mol. Phys. 14, 1 (1978)

TOTAL COLLISIONAL CROSS-SECTIONS FOR $e^{\pm}$ SCATTERING FROM Li ATOMS

S.P. Khare and Vijayshri

Department of Physics, Institute of Advanced Studies, Meerut University, Meerut-250005, India

It is well known that the Glauber approximation suffers from the drawback that its imaginary part diverges logarithmically in the forward direction. Thus, in itself the GA cannot be used to obtain total collisional cross-sections via the optical theorem. Recently, Jhanwar et al[1] developed an analytical expression for $f_G - f_{G2}$ in the forward direction for a hydrogenic atom and employed the Modified Glauber Approximation to evaluate the total collisional cross-sections for $e^{\pm}$-H, He and H_2[1] scattering. Their results are in excellent agreement with the experimental data. Hence the extension of this technique to complex atoms was thought to be of interest and the next atom in the periodic table i.e. Li was selected for a similar investigation.

So far only Inokuti and McDowell[2] and Walters[2] have reported total collisional cross-sections for e^--Li scattering in this energy range. They, however, did not employ any ab-initio theoretical method but simply summed the integral elastic cross-sections and certain reasonable estimates of the total inelastic cross-sections using different methods The experimental data as quoted by Walters[2] are available for only e^--scattering at 20 and 50 eV. The paucity of theoretical and experimental data makes the present investigation interesting especially in the light of earlier successes of MGA[1].

The direct scattering amplitude for the scattering of a charged structureless particle $Z_i e$ of mass m_i by the ground state Li atoms (represented by the wave function of Veselov et al[3]) in the MGA is given as

$$f_{MG} = f_G - f_{G2} + f_{B2} \quad \text{where}$$

$$f_G = \frac{ik_i}{2\pi}\int e^{i\underline{q}.\underline{b}}\, d^2b\, \Phi_i(\underline{r}_1,\underline{r}_2,\underline{r}_3)\left[\Gamma_1+\Gamma_2+\Gamma_3 - \Gamma_1\Gamma_2-\Gamma_2\Gamma_3-\Gamma_3\Gamma_1+\Gamma_1\Gamma_2\Gamma_3\right]\Phi_i(\underline{r}_1,\underline{r}_2,\underline{r}_3)\, d\underline{r}_1 d\underline{r}_2 d\underline{r}_3$$

where $\Gamma_i = 1-\left(\frac{|\underline{b}-\underline{s}_i|}{b}\right)^{2i\eta}$, $\eta = -\frac{Z_i\mu}{k_i}$, μ is the reduced mass of the system and k_i^2 is the incident energy in a.u.. $f_G - f_{G2}$ and f_{B2} are evaluated following the methods described by Jhanwar et al[1]. The results for the total collisional cross-sections for $e^{\pm}$-Li in the 20 eV to 1 KeV energy range are presented in Table 1 alongwith the same obtained by using the frozen (inert) core approximation where scattering by only the valence electron is considered i.e. only Γ_3 is retained in f_G and the corresponding terms in f_{G2} and f_{B2}. Calculations have also been carried out in EBS and EBS frozen (inert) core.

Table 1: Total cross-sections for $e^{\pm}$-Li scattering (πa_0^2)

E(eV)	EBS	MGA	EBS(FC)	MGA(FC)
20	7.51(1)	5.19(1)	6.76(1)	5.20(1)
30	5.49(1)	4.24(1)	4.94(1)	4.16(1)
50	3.65(1)	3.09(1)	3.28(1)	2.97(1)
60	3.15(1)	2.73(1)	2.83(1)	2.61(1)
100	2.06(1)	1.88(1)	1.85(1)	1.77(1)
200	1.14(1)	1.09(1)	1.03(1)	1.01(1)
400	6.26(0)	6.11(0)	5.67(0)	5.61(0)
700	3.83(0)	3.78(0)	3.48(0)	3.46(0)
1000	2.79(0)	2.77(0)	2.54(0)	2.53(0)

$a(b) \equiv a\times10^b$.

The cross-sections for both e^+ and e^- scattering are the same as we have neither included Ps formation nor exchange effects in our calculations. The effect of including higher order terms via $f_G - f_{G2}$ is to reduce the EBS results significantly throughout the intermediate energy region; towards the tail end of the spectrum, both results get very close to each other. Secondly, the difference between the frozen core (FC) and full calculations increases with the increase in the energy but is only 9% at 1 KeV which is not much. The theoretical results agree well with the experimental data[2] at 20 eV but underestimate them at 50 eV. Hence the need for such data in the intermediate and high energy region is acutely felt. Similar investigation for H^+ impact is also in progress and the results shall be reported at the Conference.

References

1. B.L.Jhanwar, S.P.Khare and M.K.Sharma, Phys. Rev. A25, 1993(1982); A26, 1392 (1982)
2. M.Inokuti and M.R.C. McDowell, J.Phys.B7, 2382 (1974); H.R.J.Walters, B9, 227(1976)
3. F.T.Chan and C.H.Chang, Phys.Rev.A14, 189(1976)

TOTAL ELECTRON SCATTERING CROSS SECTIONS FOR He, Ne, Ar, Xe, and SELECTED MOLECULES: 4-300 eV

J. C. Nickel and K. Imre

Department of Physics, University of California, Riverside, California 92521

D. F. Register[a] and S. Trajmar

Jet Propulsion Laboratory, California Institute of Technology, Pasadena, California 91109

Total electron scattering cross sections are reported for the rare gases and selected molecules (H_2, CO_2, CH_4, SO_2) in the energy range of 4 eV to 300 eV. The results are obtained in a linear transmission apparatus with a relatively long (14.43 cm) scattering cell. The entrance and exit apertures of the scattering cell are 1 mm in diameter and both sides of the scattering cell are differentially pumped to minimize the target gas density outside the cell. The pressure in the scattering cell is measured with a thermally stabilized MKS model 310 capacitance manometer and the results are corrected for thermal transpiration. The electron gun, enclosed in a separate differentially pumped chamber, is a 6 element cylindrical tube-lens gun capable of providing adequate current in the 4-300 eV range with a resolution of .35 eV FWHM. The detector is a gridded Faraday cup whose entrance aperture is 2.2 mm in diameter. The entrance aperture of the Faraday cup is located 7.1 cm from the exit aperture of the scattering cell. The effective solid angle subtended by the detector is $\langle \Delta\Omega \rangle = 2.3 \times 10^{-4}$ steradians thus providing good discrimination against electrons scattered at small angles. The biased grid in the Faraday cup serves to further discriminate against electrons inelastically scattered in the forward direction.

As examples, results for helium and argon are presented in Figures 1 and 2. The results are compared with those obtained in References 1-6. A conservative estimate of the accuracy of the present results is ±5%.

Work supported by University of California Intramural Research Fund, NASA and DOE (Los Alamos Scientific Laboratory).

[a]Present address: Phillips Petroleum Research, Bartlesville, Oklahoma 54003

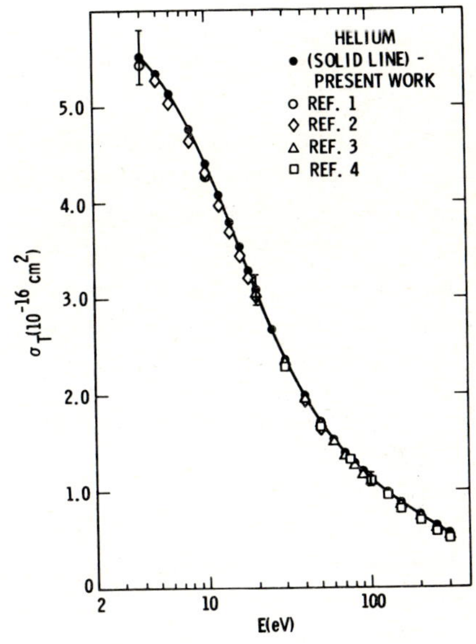

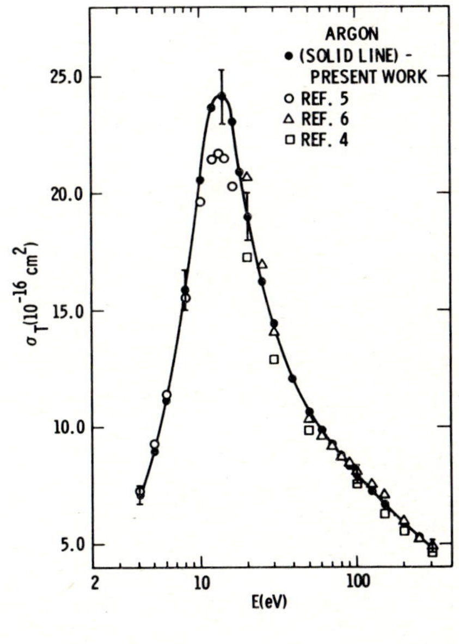

References

1. T. W. Stein, W. E. Kauppila, V. Pol, J. H. Smart and G. Jesion, Phys. Rev. A 17, 1600 (1978).
2. R. E. Kennerly and R. A. Bonham, Phys. Rev. A 17, 1844 (1978).
3. H. J. Blaauw, R. W. Wagenaar, D. H. Barends and F. J. de Heer, J. Phys. B: Atom. Molec. Phys. 13, 359 (1980).
4. W. E. Kauppila, T. W. Stein, J. H. Smart, M. S. Dababneh, Y. K. Ho, J. D. Downing and V. Pol, Phys. Rev. A 24, 725 (1981).
5. W. E. Kauppila, T. S. Stein, G. Jesion, Phys. Rev. Lett. 36, 580 (1976).
6. R. W. Wagenaar and F. J. de Heer, J. Phys. B: Atom. Molec. Phys. 13, 3855 (1980).

TOTAL CROSS SECTIONS FOR ELECTRONS SCATTERED BY AR IN THE INTERMEDIATE ENERGY RANGE

J.C. Nogueira, I. Iga, E. Chaguri and M.T. Lee

Universidade Federal de São Carlos, Departamento de Química, 13560 - São Carlos, Brasil

In our former paper[1], we have shown that the forward - peaked inelastic scattering contributes significantely to reduce the measured Total Cross Section of electron - gas scattering in the intermediate and high impact energy range. Our experimental apparatus has been modified in the attemption to eliminate this systematic error. The modified apparatus is shown in Fig. I; basically it consist on introduction of filter system of energy retarding type that repels the inelastic scattered electrons from the detector (Faraday Cup).

A set of preliminar measurement has been done in the impact energy range between 500 - 700 eV. The comparison with our former measurement and the experimental results of Kauppila et al[2] and Wagenaar and de Heer[3] show good agreement. It can be shown that the error introduced by the inelastic scattering contribution to the measured Total Cross Section is small in this energy range

References

1. J.C. Nogueira, Ione Iga and Lee Mu-Tao, J. Phys. B: At. Mol. Phys. 15, 2539 (1982).

2. W.E. Kauppila, T.S. Stein, J.H. Smart, M.S. Dababneh, Y.K. Ho, J.P. Downing and V. Pol, Phys. Rev. A 24, 725 (1981).

3. R.W. Wagenaar and F.J. de Heer, J. Phys. B: At. Mol. Phys. 13, 3855 (1980).

(FAPESP, CNPq)

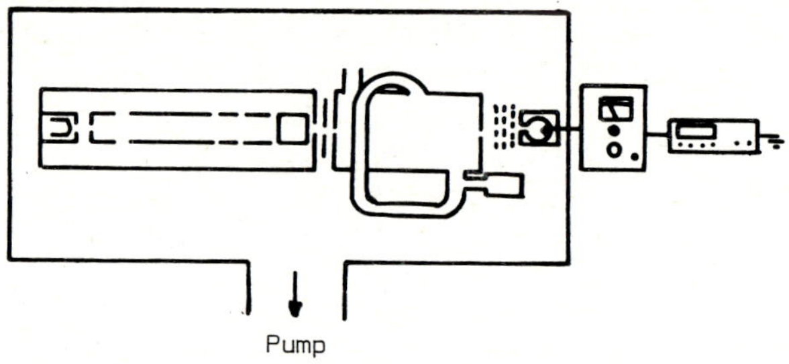

Fig. 1 - Schematic picture of the apparatus

THEORETICAL CROSS - SECTIONS FOR ELECTRON SCATTERING FROM C, N and O - ATOMS

H. S. Desai

Physics Department Faculty of Science
M.S. University of Baroda-India

K. N. Joshipura
M. B. Patel Science College, ANAND, INDIA

Most of the investigations on fast electron Scattering by Carbon, Nitrogen and Oxygen atoms do not include absorption effects[1]. A simple way of doing this is through the imaginary part of the second Born amplitude which is straightforward with the use of static potentials of Cox and Bonham,[2] for these atoms. Optical theorem is used to obtain total cross-sections therefrom. These quantities, calculated using the Strand and Bonham[3] potentials, give higher results, not shown here. No other data, experimental or theoretical, are available for the cases considered. Independent atom model may be used to estimate these quantities for relevent molecules.

Table : Total Cross - Sections ( )

Incident energy (ev)	N - Atom	O - Atom
100	2.2	2.5
300	7.8	8.9
500	4.8	5.4
700	3.4	3.9

REFERENCES :

1. Y.D.Kaushik, S.P.Khare and DeoRaj, Ind.J. Pure Appl. Phys., 20, 466 (1982).
2. H.L.Cox and R.A.Bonham, J. Chem. Phys., 47, 2599 (1967).
3. T.G.Strand and R.A.Bonham, J.Chem.Phys.40, 1686 (1964).

SCATTERED WAVE FUNCTION USING EXTENDED NUMEROV ALGORITHM

Joseph M. Paikeday

Southeast Missouri State University, Cape Girardeau, MO., U.S.A.

An extended Numerov algorithm is derived and tested for the computation of scattered wave functions of electrons scattered by an atom in its ground state. For Helium and Neon atoms, the effective interaction potential is approximated in the form:

$$V(r)=\sum_{\mu\nu}(A_{\mu\nu}r^{\mu-1})e^{-C_{\mu\nu}r}+(1-e^{-Br^2})\alpha(r^4+f(E,A))^{-1}$$

where α is the dipole polarizability of the target atom and the parameters A and B are determined from recently available experimentl differential cross section in the angular range $5 < \theta < 10^{o}$, using Glauber approximation for the scattering amplitude. Results are compared with those obtained by the ordinary Numerov method for the incident electron energy in the range $100<E<700$ eV and angular momentum in the range $0< \ell<15$. It is found that the extended Numerov Algorithm (ENA) constructed in the present method yield approximately an order of magnitude improvement in the accuracy of the computed phase shifts for the electron-atom potential in the intermediate energy range. The stability of the algorithm is comparable to that of the ordinary Numerov algorithm (ONA). This is further analysed by calculating the complex roots of the first and second characteristic polynomials of the difference operator representing the ENA used in the present study. The algorithm is shown to satisfy the condition of 'zero stability' by determining the modulus and multiplicity of the roots of the associated first characteristic polynomial. The ENA is in a form suitable for generalization for the multi-channel matrix difference algorithm for electron-atom scattering in which long range polarization contribute significantly to the S-matrix elements.

For the single channel case studied in the present paper, the algorithm is represented by

$$\sum_{i=1}^{N+1}\frac{A_i(\Phi_{n+i-1}+\Phi_{n-i+1})}{(1+\delta_{i1})}=(\Delta r)^2\sum_{i=1}^{N+1}\frac{A_{N+1+i}(\Phi''_{n+i-1}+\Phi''_{n-i+1})}{(1+\delta_{i1})}$$

in which Φ is the radial part of the scatterd wave function corresponding to a given angular momentum ℓ, δ is the Kronecker symbol and the 2N+2 values of A_i are determined by inverting a matrix B satisfying certain optimization conditions. For the electron-atom interaction potential, the wavefunctions $\Phi_\ell(k,r)$ satisfy the condtions

$\Phi_\ell(k,r)\sim Ar^{\ell+1}$ near r=0 and, for $r\to\infty$,

$$\Phi_\ell(k,r)\sim k_i r j_\ell(k_i r)+k_i r S_\ell(k_i)n_\ell(k_i r)\ .$$

where j_ℓ and n_ℓ are the spherical Bessel and Neumann functions and k_i is the incident wave number of the electron. The value of A_i is set =1 for i=2N+2 and the other 2N+1 values are determined from the equation

$$A_i^{(N)}=\sum_{j=1}^{2N+1}(B^{-1})_{ij}B_{j,2N+2}$$

with the conditions;

$$\sum_{i}^{N}(2-\delta_{1i})A_i=0 \text{ and } |\Phi_\ell^E(r_\mu)-\Phi_\ell^A(r_\mu)|=\text{minimum}$$

for $0<r<20a_o$ and $0<\ell<12$ where a_o is the Bohr radius and subscripts E and A correspond to the 'exact' and approximate functions for Δr in the range $0.05a_o< \Delta r <0.5a_o$ for a fixed incident energy of 100eV. and V(r) replaced by the electron-Helium interaction potential.

For N=2, one set of A_i are given by

A_1=-16.0273972602739
A_2= -6.5753424657534
A_3= 14.5890410958904
A_4=-19.9726027397260
A_5=-14.9041095890410

A comparison of the accuracy of computation of phase shifts for electron-Neon interaction is shown in Fig.(1). Curve A is the percent diff. for the present method and curve B shows the same for the Numerov algorithm as a function of Δr (in atomic units).

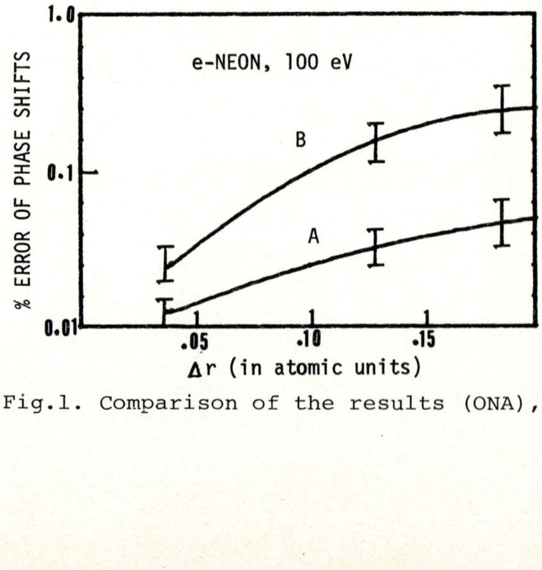

Fig.1. Comparison of the results (ONA), (ENA).

A TREATMENT OF THE ELASTIC SCATTERING OF ELECTRONS ON H AND He^+ USING HYPERSPHERICAL COORDINATES AND INCORPORATING THE FOCK EXPANSION

Steven Alston

Behlen Laboratory of Physics, University of Nebraska-Lincoln
Lincoln, Nebraska 68588-0111, USA

The hyperspherical coordinate approach has significantly advanced our understanding of two-electron correlations in atoms;[1] however, accurate calculations have proved difficult to obtain owing either to the necessity of coupling several channels or possibly to the presence of an unrealistically large angular momentum barrier in the hyperradius R. The present work studies the importance of this second problem by incorporating the exact solution into the adiabatic one at small R. The elastic scattering of electrons on H and He^+ at energies near threshold is chosen as a simple test case.

In the adiabatic approximation the full wavefunction is written as $F(R)\phi(R,\Omega)$ with R treated parametrically in ϕ. The hyperspherical coordinates are defined as $R=(r_1{}^2+r_2{}^2)^{1/2}$, $\alpha=\tan^{-1}(r_2/r_1)$ and $\Omega=(\alpha, \hat{r}_1, \hat{r}_2)$, with $\vec{r}_1$, $\vec{r}_2$ the usual electron coordinates. F is the solution of a one-dimensional Schrödinger equation with potential U(R). When one of the electrons is free, F is a continuum wave. Consequently, it samples more of the large potential barrier contained in U(R) for small R as the free electron's energy increases above threshold. Since U(R) is so steep, the wave is progressively expelled and its phase reduced. This observation is supported by the calculations of Miller and Starace.[2] They find, using an adiabatic treatment, that the cross section for the photoionization of the helium ground state falls increasingly below experiment as the photoelectron's energy increases. This situation is not significantly improved by including coupling to other channels.[3]

Lin has also performed adiabatic calculations of the elastic ^1S phase shift for the electron-hydrogen system.[4] His values fall below the (essentially) exact ones of Schwartz,[5] the more so as the energy increases, but these results are improved by including 1s - 2s coupling.

In the Fock expansion[6] the exact wavefunction is expanded in powers of R and ln(R) with the coefficients determined so as to satisfy the Schrödinger equation. We include a truncated form of it by writing the total wavefunction as $\psi(R) = F(R)\phi(R,\Omega) + \chi(R,\Omega)$ with χ the Fock term. An inhomogenous Schrödinger equation for F is then derived in the manner of Burke and Taylor,[7] i.e. we include χ as a correlation function. This wavefunction is used to calculate ^1S elastic phase shifts for energies up to the first excitation threshold. A comparison of the hydrogen and helium ion cases gives an indication of the relative importance of channel coupling and barrier repulsion.

This work was supported by the National Science Foundation under Grant No. PHY 82-03400.

References

1. J. Macek, J. Phys. B 1, 831 (1968); C. D. Lin, Phys. Rev. A 10, 1986 (1974); C. Greene, Phys. Rev. A 23, 661 (1981); S. Watanabe, Phys. Rev. A 25, 2074 (1982)

2. D. L. Miller and A. F. Starace, J. Phys. B 13, L525 (1980)

3. D. L. Miller and A. F. Starace, (private communication)

4. C. D. Lin, Phys. Rev. A 12, 493 (1975)

5. C. Schwartz, Phys. Rev. 124, 1468 (1961)

6. J. H. Macek, Phys. Rev. 160, 170 (1967); V. Fock, Kgl. Norske Videnskab. Selskabs, Fohr. 31, 145 (1958)

7. P. G. Burke and A. J. Taylor, Proc. Phys. Soc. 88, 549 (1966)

FUNCTIONAL ANALYSIS IN THE e-H SCATTERING PROBLEM

P.A. Massaro

Dipartimento di Fisica dell'Università,Sezione INFN,Bari,Italy

Various integral equations are used to treat the elastic scattering of electrons by hydrogen atoms at high and intermediate energies,but the existence and uniqueness properties of their solution are usually ignored. Recently[1] these properties have been considered for the Faddeev equations with the on-shell and one-channel approximations, and a more careful treatment is done in this communication.

For the elastic e-H channel the Faddeev equations for the singlet and triplet amplitude can be written as[1,2]

$$f^{\pm}(\vec{k}'\cdot\vec{k}) = \lambda\, f_1^{\pm}(\vec{k}'\cdot\vec{k}) + \lambda\frac{ik}{4\pi}\int d\hat{k}''\, f_1(\vec{k}'\cdot\vec{k}'')\, f^{\pm}(\vec{k}''\cdot\vec{k}) \qquad (1)$$

where $f_1^{\pm} = f^B \pm g^{Ock}$, being f^B the Born amplitude and g^{Ock} the Ockur exchange amplitude[3]; λ is the coupling constant for the interacting particles (λ =1 is the physical value). By choosing the vector $\vec{k}$ in the z-axis direction and the (x,z)-plane as scattering plane, equations (1) take the one-dimensional form

$$f^{\pm}(k;u) = \lambda f_1^{\pm}(k;u) + \lambda\int_{-1}^{1} dv\, \tilde{K}^{\pm}(k;u,v) f^{\pm}(k;v) \qquad (2)$$

where

$$\tilde{K}^{\pm}(k;u,v) = \frac{ik}{4\pi}\int_0^{2\pi} dw\, f_1^{\pm}(k;u,v,w) \qquad (3)$$

$$f_1^{\pm}(k;u,v,w) = f_1^{\pm}(\vec{k}'\cdot\vec{k}'')$$

$$u = \cos\theta, v=\cos\theta'', w=\phi''$$

Evaluating the integral in the formula (3) we see that the kernels $\tilde{K}^{\pm}(k;u,v)$ are continuous functions in the domain $[-1,1]\times[-1,1]$ and,therefore,the integral operators $\tilde{U}^{\pm}$ with kernels $\tilde{K}^{\pm}(k;u,v)$ are completely continuous operators from the space of continuous functions $C[-1,1]$into the same space[4]. We can conclude that eqs.(2) have a unique continuous solution for $|\lambda|<|\lambda_1^{\pm}|$, being $\lambda_1^{\pm}$ the minimum modulus characteristic values of the integral operators $\tilde{U}^{\pm}$.

Since we have

$$|\lambda_1^{\pm}| \geq 1/||\tilde{U}^{\pm}||$$

it is important to evaluate the norms $||\tilde{U}^{\pm}||$ .After a few calculations we find the formula

$$||\tilde{U}^{\pm}|| = \frac{1}{2k}\,[\ln(1+k^2) + \frac{k^2\mp 4}{k^2+1}],\; k\geq 2 \qquad (4)$$

We have $||\tilde{U}^{\pm}|| < 1$ in the energy range $k\geq 2$, for every scattering angle, and therefore the physical value $\lambda = 1$ lies in the disk $|\lambda|<|\lambda_1^{\pm}|$.

The elastic singlet and triplet amplitudes can be evaluated exactly from eq.(2) without partial wave decomposition.

The method can be generalized by including the 2s and 2p virtual states.

Numerical results will be presented at the conference.

References

1. P.A.Massaro,Nuovo Cimento D 1, 9 (1982)
2. I.H.Sloan and E.J.Moore, J.Phys. B 1, 414 (1968)
3. V.I.Ockur, Sov.Phys. JETP 18, 503 (1964)
4. L.V.Kantorovich and G.P.Achilov: Functional Analysis in Normed Spaces(Pergamon Press,Oxford 1964),ChaptXIII

INCLUSION OF THE FOCK EXPANSION IN HYPERSPHERICAL COORDINATE CALCULATIONS*

Jim Feagin, Joseph Macek, and Anthony F. Starace

Behlen Laboratory of Physics, The University of Nebraska, Lincoln, Nebraska 68588-0111 U.S.A.

The hyperspherical framework has much conceptual appeal and has been introduced frequently in connection with the analysis of the states of the helium atom and other two electron systems.[1] Fock[2] made a key contribution by introducing log terms in a power series expansion appropriate for ^1S states of He, i.e., he showed that the two electron wavefunction may be written in the form

$$\psi = \sum_n \sum_j C_{nj} R^n (\ell n R)^j , \tag{1}$$

where $R \equiv (r_1^2 + r_2^2)^{\frac{1}{2}}$ and where the coefficients C_{nj} depend on five hyperspherical angular variables. Demkov and Ermolaev[3] generalized the Fock expansion in the hyperspherical coordinates R to an N-electron system having any symmetry. In addition, Macek[4] proved that, for sufficiently small values of the mean square radius R of the two electrons, the Fock expansion converges and thus could indeed represent a physical solution. No practical means to impose boundary conditions at large values of the mean radius was found, however, so that despite the basic importance of the hyperspherical coordinates and the associated expansion of the wavefunction, their use in atomic physics was limited to analytical studies[5,6] and to variational calculations.[7]

Only in 1968, when the concept of approximate separability was introduced,[8] was a means provided to treat two-electron eigenstates quantitatively. The quantitative experience obtained since then indicates that while the separability approximation allows one to describe a large part of the dynamics of two electron systems, it is only a first order approximation. The hyperspherical coordinate method, if it is to be a truly comprehensive framework for treating two electron atomic states, must provide a means for systematically improving numerically calculated results.

To this end we have obtained numerically the coefficients C_{nj} of the Fock expansion (1) as combinations of hyperspherical harmonics. We are thus able to represent a two electron wavefunction in the inner region $R \leqslant R_0$ as

$$\Psi_\mu{}^i(R \leqslant R_0) = \sum_\nu a_{\mu\nu}\, \psi_\nu{}^F(R,\alpha,\hat{r}_1,\hat{r}_2) \tag{2}$$

where the basis functions $\psi_\nu{}^F$ are Fock series expansions having the specified behavior:

$$\psi_\nu{}^F \xrightarrow[R \to 0]{} R^\nu \tag{3}$$

For $R \geqslant R_0$ we represent the outer wavefunction $\Psi_\mu{}^o$ as a linear combination of separable or adiabatic channel wavefunctions:

$$\Psi_\mu{}^o(R \geqslant R_0) = \sum_{\mu'} F_{\mu\mu'}(R)\phi_{\mu'}(R;\alpha,\hat{r}_1,\hat{r}_2) \tag{4}$$

Note that $\Psi_\mu{}^o(R \to \infty)$ must satisfy asymptotic boundary conditions appropriate to the collision process under study.

We match the inner and outer wavefunctions at $R = R_0$ by a variational procedure, viz., $\delta(N/D) = 0$, where

$$N = \int d\Omega\, |\Psi_\mu{}^o(R_0) - \Psi_\mu{}^i(R_0)|^2 \tag{5a}$$

and

$$D = \int d\Omega\, |\Psi_\mu{}^i(R_0)|^2 \tag{5b}$$

This procedure is being applied first to the calculation of e^- - H elastic scattering phase shifts at energies where methods coupling only a few adiabatic hyperspherical channels are inadequate. Our method would also test the small R representation of the wavefunction obtained by these previous hyperspherical calculations.

References

*This work was supported in part by U.S. Department of Energy Contract No. DE-AC02-82ER12081.

1. See A. F. Starace in *Physics of Electronic and Atomic Collisions*, edited by S. Datz (North-Holland, Amsterdam, 1982) pp. 431-446, and references therein.
2. V. Fock, Izvest. Acad. Nauk USSR Ser Fiz. 18, 161 (1954) [Eng. Transl.: Kong. Norske Videnskabers Selskabs Forh. 31, 138, 145 (1958).]
3. Yu. N. Demkov and A. M. Ermolaev, Zh. Eksp. Teor. Fiz. 36, 896 (1959) [Sov. Phys. - JETP 36, 633 (1959)].
4. J. H. Macek, Phys. Rev. 160, 170 (1967).
5. F. T. Smith, Phys. Rev. 120, 1058 (1960).
6. W. Zickendraht, Annals of Physics 35, 18 (1965).
7. K. Frankowski and C. L. Pekeris, Phys. Rev. 146, 46 (1966); K. Frankowski, Phys. Rev. 160, 1 (1967).
8. J. H. Macek, J. Phys. B 2, 831 (1968).

EIKONAL EXCHANGE AMPLITUDES FOR ELECTRON-HELIUM COLLISIONS

B.Padhy, R.Srivastava and D.K. Rai

Department of Physics, Banaras Hindu University, Varanasi-221005, India

Recently considerable attention[1-4] has been given to the calculation of Glauber and/ or eikonal exchange amplitudes for scattering of electrons by hydrogen atom. As regards scattering of electrons by helium target, the only eikonal exchange amplitude calculation is that by Byron and Joachain[5] who studied the $1^1S \rightarrow 2^3S$ excitation for incident electron energy of 225 eV. The calculation involved the evaluation of a six-dimensional integral which was carried out by the Monte-Carlo method. The extensive numerical computation involved precludes any straight forward extension of such techniques to more widespread applications. We, therefore, thought it worthwhile to examine if the recent developments in e-H scattering can be invoked to improve this situation. Following the work of Halpern and Franco[3] and of Gau and Macek[6], we have succeeded in reducing the nine-dimensional integral appearing in the case of e-He exchange amplitude to a two-dimensional one which is likely to be more amenable to numerical evaluation. In this note we present our analysis for the elastic scattering of electrons by helium in its ground state.

Let $\vec{r}_0$ denote the position vector of the incident electron and $\vec{r}_1$ and $\vec{r}_2$ that of the two bound electrons (in the initial channel). The target nucleus is assumed to be stationary and at the origin of the coordinate system. If, as a result of collision, the incident electron 0 having a propagation vector $\vec{k}_+$ exchanges its role, say, with the bound electron 1 which becomes the outgoing electron having a propagation vector $\vec{k}_-$ and the target helium undergoes a transition from the initial state $\psi_i(\vec{r}_1,\vec{r}_2)$ to the final state $\psi_f(\vec{r}_0,\vec{r}_2)$, the exact eikonal exchange T matrices in the post and prior interactions are respectively given by (atomic units will be used throughout)

$$T^+_{fi}=\int d\vec{r}_0 d\vec{r}_1 d\vec{r}_2 \exp(-i\vec{k}_-.\vec{r}_1)\psi^*_f(\vec{r}_0,\vec{r}_2)\psi_i(\vec{r}_1,\vec{r}_2) \times \exp(i\vec{k}_+.\vec{r}_0)\left(\frac{r_{01}-z_{01}}{r_0-z_0}\right)^{i\eta_+}\left(\frac{r_{02}-z_{02}}{r_0-z_0}\right)^{i\eta_+}\left(\frac{1}{r_{01}}+\frac{1}{r_{12}}-\frac{2}{r_1}\right) \quad (1)$$

and

$$T^-_{fi}=\int d\vec{r}_0 d\vec{r}_1 d\vec{r}_2 \exp(-i\vec{k}_-.\vec{r}_1)\psi^*_f(\vec{r}_0,\vec{r}_2)\psi_i(\vec{r}_1,\vec{r}_2) \times \exp(i\vec{k}_+.\vec{r}_0)\left(\frac{r_{01}-z_{01}}{r_1+z_1}\right)^{i\eta_-}\left(\frac{r_{12}+z_{12}}{r_1+z_1}\right)^{i\eta_-}\left(\frac{1}{r_{01}}+\frac{1}{r_{02}}-\frac{2}{r_0}\right) \quad (2)$$

where

$$r_{ij}=|\vec{r}_i-\vec{r}_j|,\quad z_{ij}=z_i-z_j \text{ and } \eta_\pm=1/k_\pm .$$

For elastic scattering in the 1^1S ground state we express the target wave function[7] as

$$\psi_i(\vec{r}_1,\vec{r}_2) = \phi_{1s}(\vec{r}_1)\phi_{1s}(\vec{r}_2) \quad (3)$$

with

$$\phi_{1s}(\vec{r}) = (4\pi)^{-1/2}\sum_{i=1}^{2} C_i \exp(-\alpha_i r). \quad (4)$$

We have chosen the direction of $\vec{k}_+(\vec{k}_-)$ as the z-axis for space quantisation of the target wave functions in the post (prior) interaction. We obtain the following expressions for $T^\pm_{ii}$:

$$T^\pm_{ii}=(4\pi)^{-1}\, 2^{i\eta_\pm}\Gamma(1+i\eta_\pm)\exp(\pi\eta_\pm/2)k_\pm^{-2-i\eta_\pm} \times\Bigg[\sum_{i,j,k,l=1}^{2} C_iC_jC_kC_l\left\{I^\pm_1+\eta^2_\pm(3+i\eta_\pm)(I^\pm_3-2I^\pm_2)\right\} + \eta_\pm(2+i\eta_\pm)\sum_{k,l=1}^{2} C_kC_l(\partial/\partial q_{z\pm})\left\{\alpha_1C_1^2 I^\pm_2\Big|_{\alpha_{ij}=\alpha_{11}} +\alpha_2C_2^2I^\pm_2\Big|_{\alpha_{ij}=\alpha_{22}} + \alpha_{12}C_1C_2I^\pm_2\Big|_{\alpha_{ij}=\alpha_{12}}\right\}\Bigg] \quad (5)$$

where

$$\vec{q}=\vec{k}_+-\vec{k}_-,\quad q_{z\pm}=\vec{q}.\hat{z}_\pm,\quad \hat{z}_\pm=\hat{k}_\pm,\quad \alpha_{ij}=\alpha_i+\alpha_j ,$$

$$I^\pm_1=-2D^\pm\int_0^\infty d\lambda\, \lambda^{-i\eta_\pm-1}\int_0^1 du\, u^{-2}\alpha_{kl}(d/d\alpha_{kl})^3F^\pm(1,0,0,1),$$

$$I^\pm_2= D^\pm\int_0^\infty d\lambda\, \lambda^{-i\eta_\pm-1}\int_0^1 du\, u^{-1}\alpha_{kl}(d/d\alpha_{kl})^2 F^\pm(1,0,0,1),$$

$$I^\pm_3= D^\pm\int_0^\infty d\lambda\, \lambda^{-i\eta_\pm-1}\int_0^1 du\, u^{-1}\alpha_{kl}(d/d\alpha_{kl})^2 F^\pm(1,0,0,0),$$

$$D^\pm=\pi^2\, 2^{5-2i\eta_\pm}\Gamma(1-2i\eta_\pm)\left[\Gamma(-i\eta_\pm)\right]^{-1},$$

$$F^\pm(m,p,r,s)=\lambda^s(1-u)^s\Lambda^{-p}(\beta^2+Q^2_\pm)^{2i\eta_\pm-m} \times(\beta\mp i\vec{Q}_\pm.\hat{z}_\pm)^{-2i\eta_\pm-r},$$

$$\Lambda^2= u\alpha^2_{kl}+ \lambda^2(1-u)^2,$$

$$\beta=\alpha_{ij}+\Lambda ,$$

and

$$\vec{Q}_\pm= \vec{q} \mp i\lambda(1-u)\,\hat{z}_\pm .$$

Since for elastic scattering $|\vec{k}_+|=|\vec{k}_-|$ and $\vec{q}.\hat{z}_+ = -\vec{q}.\hat{z}_-$, it can be shown that $T^+_{ii}=T^-_{ii}$. Thus there is no post-prior discrepancy in our analysis.

References

1. G.Foster and W.Williamson Jr.,Phys.Rev.A 13, 2023 (1976).
2. V.Franco and A.M.Halpern,Phys.Rev.A 21,11 (1980) and references 1 to 8 therein.
3. A.M.Halpern and V.Franco,Phys.Rev.Lett. 46, 714 (1981).
4. T.T.Gien, Phys.Rev.A 26, 658 (1982).
5. F.W.Byron Jr. and C.J.Joachain,Phys.Lett. 38A, 185 (1972).
6. J.N.Gau and J.Macek, Phys.Rev.A 10, 522 (1974).
7. F.W.Byron and C.J.Joachain, Phys.Rev. 146, 1 (1966).

HILBERT SPACE REPRESENTATION OF RESONANCE WAVEFUNCTIONS

A. Macías and A. Riera

Departamento de Química Física y Química Cuántica. Centro Coordinado. CSIC-UAM. Universidad Autónoma de Madrid. CANTOBLANCO (Madrid, 34) (Spain).

Under many different headings most calculations of resonance wavefunctions for molecular systems rely on stability properties of (real or complex) eigenvalues. An eigenvalue is called stable if it changes little when the basis set is varied; for example, non linear parameters in real or complex STO's or GTO's can be varied quasi continuously.

A point never mentioned, the basis of these stabilization properties, has been studied in our Laboratory for the last few years and we have reached some quite general conclusions.

The esential physics behind the method is that an eigenvalue corresponding to a resonance stabilizes for two reasons (notice that we make no distintion between shape and Feshbach resonances):

a) the resonance has a maximum amplitude in a certain region of configuration space and it is much smaller elsewhere, in contrast with neighbouring continuum states,

b) the basis set of L^2 integrable functions is chosen so as to span that particular region of configuration space.

An illuminating proof is presented for a very simple case (Dirac δ shell potential) where the exact resonance wavefunctions are known.

We plot the norm of the exact continuum wavefunctions projected onto a basis of 1s STO's in Fig. 1 a (for $H = -1/2\,\nabla^2$) and Fig. 1 b (for $H = -1/2\,\nabla^2 + \delta\,(r-1)$) When the exponents of the STO's (α_o) are varied the eigen values show a pattern determined by the position of the resonances in Fig. 1 b.

It is clearly seen in

$$\langle\Psi_v|H|\Psi_v\rangle = \int |\langle\Psi_v|\Psi_E\rangle|^2 E\, dE$$

that a peak in $|\langle\Psi_v|\Psi_E\rangle|^2$ for $E = E_r$ (which means that because of reasons a) and b) Ψ_v represents well Ψ_E) results in

$$\langle\Psi_v|H|\Psi_v\rangle = E_r$$

and small variations in the basis set will yield the same eigenvalue. On the other hand when $|\langle\Psi_v|\Psi_E\rangle|^2$ is a flat function the value $\langle\Psi_v|H|\Psi_v\rangle$ will vary quickly with the basis set. See Fig. 2 a, b, c.

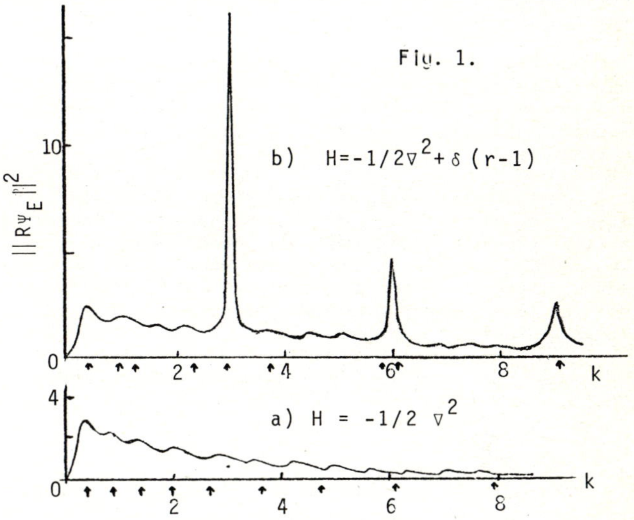

Fig. 1. R is the projector onto the manifold spanned by the basis set and Ψ_E the eigenfunctions of H. (↑) Positions of eigenvalues of H

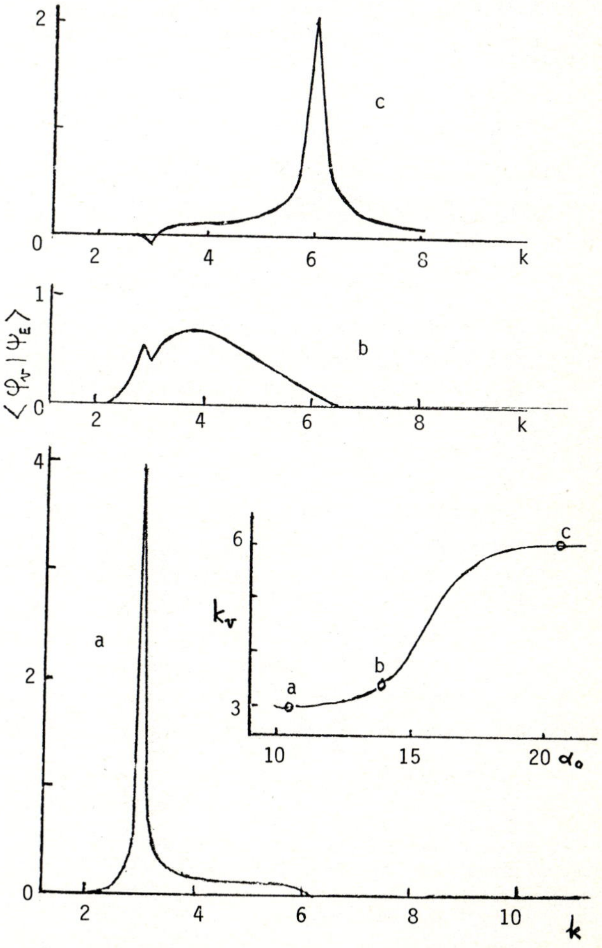

Fig. 2. Overlap between variational wavefunctions Ψ_v and exact eigenfunctions Ψ_E of $H = -1/2\,\nabla^2 + \delta\,(r-1)$

Complex Virial Theorem and Complex Scaling

B. R. Junker

Office of Naval Research-Code 412, 800 N. Quincy Street, Arlington, VA 22217, USA

The complex scaling theorems[1] stimulated considerable research to develop computational techniques using square-integrable basis functions for calculating resonance parameters for many different resonant phenomena. Froelich et al[2], Brandas and Froelich[3], and Winkler[4] showed that requiring the wavefunction to satisfy the appropriate complex virial theorem forced the first order error in the variational expression for the energy to be zero. They suggested using this as the criterion for selecting the complex scale factor.

On the other hand, in bound state calculations for atoms, the virial constraint has been used to determine a final global scaling which for bound states necessarily provides a lower upper bound. We use this latter approach to define a final global complex scaling for resonant wavefunction determined using the complex stabilization method.

The complex virial theorem for resonant states for Coulomb potentials takes the form

$$(V_R^V/T_R^V) = -2 \tag{1a}$$

$$(V_I^V/T_I^V) = -2 \tag{1b}$$

where $V^V_{I(I)}$ and $T^V_{R(I)}$ are the real (imaginary) parts of the potential and kinetic energies, respectively. If $V_{R(I)}$ and $T_{R(I)}$ are the real (imaginary) parts of the potential and kinetic energies from a complex stablization calculation, the global scale factor

$$\eta = \eta_R + \eta_I i = \gamma^{-1} \exp(-i\chi) \tag{2}$$

is given by

$$\eta_R = -(T_R V_R + T_I V_I)/[2(T_R^2 + T_I^2)] \tag{3a}$$

and

$$\eta_I = (T_I V_R - T_R V_I)/[2(T_R^2 + T_I^2)] \tag{3b}$$

The energy satisfying the virial theorem is given by

$$E_r^V = -[(T_R V_R^2 + 2 T_I V_I V_R - T_R V_I^2) + i(T_R V_R V_I + T_I V_I^2 - T_I V_R^2)]/[4(T_R^2 + T_I^2)] \tag{4}$$

Note that if T_R, V_R, T_I, and V_I satisfy the virial theorem

$$\eta_R = \gamma = 1 \tag{5a}$$

$$\eta_I = \chi = 0 \tag{5b}$$

Almost all resonances fall into one of two special cases. If $|T_R| \gg |T_I|$, $|V_R| \gg |V_I|$, and $V_R/T_R \approx -2$,

$$\gamma \approx 2|T_R/V_R| \tag{6a}$$

$$\chi \approx \tan^{-1}[(T_I/T_R)(V_R/T_R)^{-1}((V_R/T_R)-(V_I/T_I))] \tag{6b}$$

$$E_r^V \approx \tfrac{1}{2}V_R + i(T_I + V_I) \approx E_r \tag{6c}$$

If $|T_I| \gg |T_R|$, $|V_I| \gg |V_R|$, and $V_I/T_I \approx -2$,

$$\gamma \approx 2|T_I/V_I| \tag{7a}$$

$$\chi \approx \tan^{-1}[(T_R/T_I)(V_I/T_I)^{-1}((V_R/T_R)-(V_I/T_I))] \tag{7b}$$

$$E_r^V \approx (V_R + T_R) + \tfrac{1}{2}V_I \approx E_r \tag{7c}$$

In the first case γ, χ, and E_R^V are virtually independent of the imaginary part of the virial theorem, while in the second case, they are virtually independent of the real part.

Results of the use of Equations (3)-(7) for computing the resonance parameters for $(1s2s^2)\ ^2S$ He^-, $(1s2s2p)\ ^2P°$ He^-, and $(1s^2 2s^2 kp)$ Be^- with a variety of wavefunctions will be reported. Table I gives the results for three wavefunctions for the $(1s2s2p)\ ^2P°$ He^- resonances. The number in parenthesis corresponds to the number of terms in the configuration interaction wavefunction[5].

Table I

	$-V_R/T_R$	$-V_I/T_I$	γ	χ*	γ #	χ#*
Ψ (72)	2.003	1.637	0.9985	0.1016	0.9985	0.1016
Ψ (94)	2.008	7.445	0.9961	-0.1331	0.9961	-0.1331
Ψ (126)	2.006	3.879	0.9962	-0.0987	0.9962	-0.0987

* Values for χ are to be multiplied by 0.01
\# γ and χ here are computed using Equation (6).

Table I (cont'd)

	E_R	E_I*	$-E_R^V$	$-E_I^{V*}$
Ψ (72)	2.1574	0.7615	2.1574	0.7609
Ψ (94)	2.1565	0.6778	2.1565	0.6801
Ψ (126)	2.1565	0.6526	2.1565	0.6542

* Values for E_I are to be multiplied by 0.01.

As can be seen from Table I, this resonance falls into the first special case noted above. While the complex stabilized wavefunction yields very poor results for the imaginary part of the virial theorem, forcing the satisfaction of the virial theorem has a negligible effect on the resonant energy.

In conclusion, several comments concerning the use of the complex virial theorem to define a final global complex scaling summarize the results above. Unlike the case for variational bound state calculations, a final global scaling does not necessarily yield an improved energy nor does it imply that the complex energy is stabilized with respect to variations in other nonlinear parameters in the wavefunction. Neither does constructing an energy satisfying the virial theorem provide information concerning convergence with respect to configuration interaction. Finally, as shown above, all resonances essentially fall into one of two cases in which just the real or imaginary part of the virial theorem determines the virially constrained energy and global scaling parameters. Since this is the dominant part of the complex energy, it generally satisfies its part of the complex virial theorem quite well in the variational calculation.

References

1. J. Aguilar and J. M. Combes Commun. Math. Phys., 22, 269 (1971); E. Balslev and J. M. Combes Commun. Math. Phys. 22, 280 (1971); B. Simon, Commun. Math. Phys. 27, 1 (1972); and B. Simon Ann. Math. 97, 247 (1973).
2. P. Froelich, M. Hehenberger, and E. Brandas, Int. J. of Quantum Chem.: Quan. Chem. Symp. 11, 295 (1977).
3. E. Brandas and P. Froelich Phys. Rev. A16, 2207 (1977).
4. R. Yaris and P. Winkler J. Phys. B Atom. Molec. Phys. 11, 1475 (1978).
5. B. R. Junker J. Phys. B: At. Mol. Phys. 15, 4495 (1982).

THE $6s6p^2$ RESONANCES IN e-Hg SCATTERING

N.S. Scott[1], P.G. Burke[1], K. Bartschat[*]
[1] The Queen's University Belfast, [*] Universitat Münster

In this contribution we present results on the low energy elastic and inelastic scattering of electrons by Hg atoms. This system is of importance since it provides an example of a collision involving a heavy atom where strong channel coupling, electron exchange, relativistic effects and resonances all play an important role in the collision. Experimentally this system has been extensively studied with recent work by Albert et al[1], Bartschat et al[2] and Ottley and Kleinpoppen[3]. However theoretical work by Sin Fai Lam[4] and Walker[5] has been restricted to elastic scattering.

Using the relativistic R-matrix method, introduced by Scott and Burke[6] and programmed by Scott and Taylor[7], the elastic scattering and excitation cross sections for the (6s6p) $^3P^o_{0,1,2}$ states in e-Hg scattering have been calculated from 4-7 eV. Strong coupling between the $(6s^2)$ $^1S^e_0$, (6s6p) $^3P^o_{0,1,2}$ and (6s6p) $^1P^o_1$ target states, electron exchange and relativistic effects are all included for the first time.

The five target states included in our R-matrix expansion are represented by an intermediate coupling wavefunction constructed from 6s and 6p orbitals. The influence of the remaining core electrons is approximated by a model potential calculated from Thomas-Fermi target orbitals. Both core and valence orbitals were calculated using the SUPERSTRUCTURE program of Eissner et al[8]. Partial cross sections were determined for the total angular momenta J=1/2, 3/2, 5/2, 7/2, and 9/2 for both parities which gives converged elastic scattering and excitation cross sections up to about 7eV.

In the figure 1 we present the total cross sections for the $^1S^e_0 \rightarrow {}^1S^e_0$, $^3P^o_0$, $^3P^o_1$, $^3P^o_2$ transitions. The cross sections are seen to be dominated by resonant peaks which we attribute to the eight $(6s6p^2)$ $^4P^e_{1/2,3/2,5/2}$, $(6s6p^2)$ $^2P^e_{1/2,3/2}$, $(6s6p^2)$ $^2D^e_{3/2,5/2}$ and $(6s6p^2)$ $^2S^e_{1/2}$ states. However the assignment of one of these states to each resonance is not always unique owing to the strong mixing between the resonant states and the continuum corresponding to the same J and Π. In the table 1 we give the position of the peaks and, where possible, an assignment based on the dominant contribution to the nearest R-matrix pole.

At present the resonance structure is being analysed in detail and the theory is being compared with other experimental observables including angular distributions and spin polarisation measurements. Current progress will be presented at the conference.

References

1. Albert K, Christian C, Heindorff T, Reichert E and Schön S, J.Phys.B. 10 3733 (1977)
2. Bartschat K, Hanne GF, Wolcke A and Kessler J, Phys. Rev. Lett. 47 997 (1981)
3. Ottley TW and Kleinpoppen H, J.Phys.B.8 621 (1975)
4. Sin Fai Lam LT, Aust.J.Phys.33 261 (1980)
5. Walker DW, J.Phys.B. 8 L161 (1975)
6. Scott NS and Burke PG, J.Phys.B. 13 4299 (1980)
7. Scott NS and Taylor KT, Comp.Phys.Com.25 347 (1982)
8. Eissner W, Jones M and Nussbaumer H, Comp. Phys. Com. 8 270 (1974)

J = 1/2+	J = 3/2+	J = 5/2+
4.7 eV (^{4}P)	4.7 eV (^{4}P)	
		4.9 eV (^{4}P)
	5.0 eV (^{4}P)	
	5.5 eV (^{2}D)	5.5 eV (^{2}D)
5.8 eV (?)		
6.7 eV (?)	6.7 eV (?)	

Table 1

Figure 1

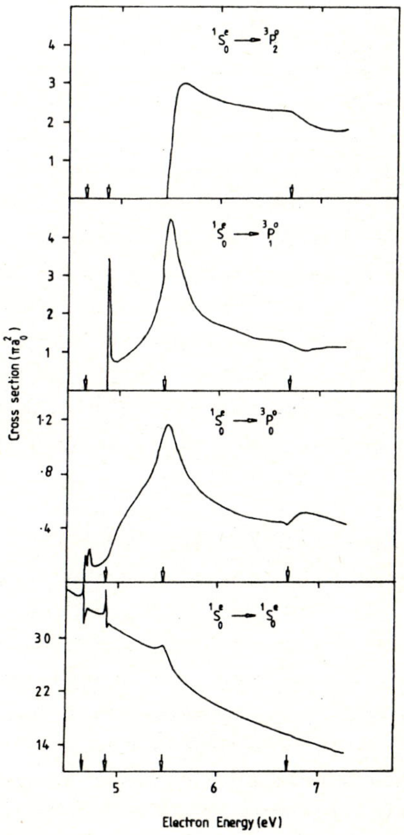

HIGH RESOLUTION ELECTRON IMPACT STUDIES OF MERCURY: THRESHOLD EXCITATION AND THE MEASUREMENT OF METASTABLE EXCITATION FUNCTIONS

D S Newman, G C King and M Zubek†

Physics Department, Manchester University, UK.

A powerful technique for the study of temporary negative ions (resonances) is the measurement of metastable excitation functions by electron impact, in which structure due to resonances occurs. The advantages of the technique are its high sensitivity, the direct information about the shapes and widths of the resonances it provides, and the ability to calibrate the positions of the observed structure against the onsets for metastable production which occur at the spectroscopically known neutral state energies. The technique has been used previously to obtain valuable information about resonances in atoms, for example Brunt et al[1] and molecules, for example Brunt et al[2]. These earlier high resolution measurements were limited to the study of metastable states of excitation energy greater than about 9eV because of the relatively high work function of the detector for the metastable species, a channel electron multiplier. In the present work we have extended the range of the channel electron multiplier by using it in conjunction with a heated tantalum ribbon of low work function[3], and are able to detect metastable species of energy down to approximately 4eV.

The metastable states of mercury are the 6p ^3P_o (4.667eV), 6p $^3P_2$ (5.461eV) and 6p' $^3D_3$ (8.795eV) states and these can all be detected in our measurements. We report the metastable excitation function of mercury over the incident electron energy range from 4.0 to 17.0eV, obtained with a resolution of approximately 25meV. This high resolution and the high sensitivity of the detector have enabled a wealth of new resonance structure to be observed. This may be seen in the figure which shows the measured metastable excitation function over the range from 8.5eV to 11.0eV. The measurements also give information about the shapes and relative sizes of the cross sections of the metastable states in mercury and in particular of the 6p ^3P_o and 6p $^3P_2$ states.

In conjunction with these metastable measurements we have also obtained threshold excitation spectra in mercury using the penetrating field technique, for example Cvejanović and Read[4]. These measurements will also be presented. They provide a useful comparison of the relative cross sections near threshold and provide valuable information to aid in the classification of the observed resonance structure in the metastable excitation functions.

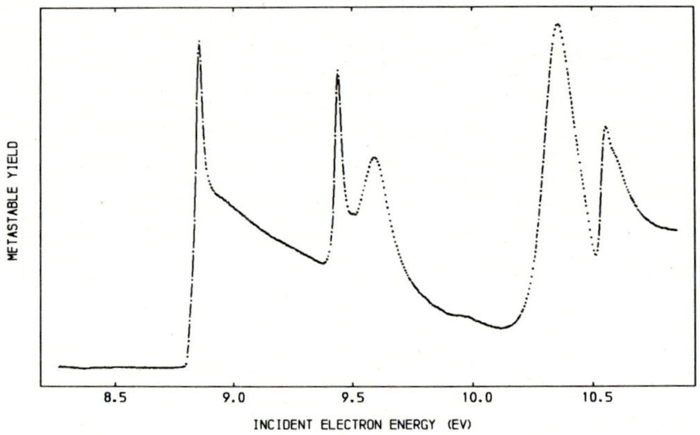

Metastable excitation function for mercury over the incident energy range from 8.5 to 11eV.

REFERENCES

1. J.N.H. Brunt, G.C. King and F.H. Read, J.Phys.B 9, 2195 (1976).
2. J.N.H. Brunt, G.C. King and F.H. Read, J.Phys.B 11, 173 (1978).
3. M. Zubek and G.C. King, J.Phys.E 15, 511 (1982).
4. S.Cvejanović and F.H. Read, J.Phys.B 7, 1180 (1974).

† Present address: Institute of Physics, Technical University 80-952, Gdansk, Poland.

WANNIER-RIDGE He^- RESONANCES

S J Buckman*, P Hammond, G C King, F H Read and C D Warner

Department of Physics, University of Manchester, UK.

Fano[1] has recently pointed out a connection between the behaviour of an atomic Rydberg electron in a strong magnetic field, giving rise to Landau standing waves, and the behaviour of two slow electrons escaping from the field of an atomic ion in the process of near-threshold electron-impact ionization of an atom. In both examples there is unstable motion along a potential ridge. In the case of two electrons escaping from an atomic field the ridge exists at the electron configuration $r_1 = -r_2$, and the potential energy in the vicinity of the ridge is given by

$$V=-\frac{4R}{r_{1,2}}\{(Z-\tfrac{1}{4})+\frac{12Z-1}{8}(\alpha-\tfrac{1}{4}\pi)^2-\frac{1}{32}(\theta_{12}-\pi)^2+\ ...\} \qquad (1)$$

where $\alpha=\tan^{-1}(r_2/r_1)$ and θ_{12} is the angle between $\underline{r}_1$ and $\underline{r}_2$. On the ridge, $\alpha = \pi/4$ and $\theta_{12} = \pi$. The motion near the ridge is unstable because V decreases as α diverges from $\pi/4$ (i.e. as r_1 and r_2 become different in magnitude). Fano[1] has suggested that in analogy with the formation of Landau standing waves the form of the two-electron potential ridge might give rise to series of quasi-standing wave patterns formed when the wave-packet representing the two-electron system propagates along the ridge (i.e. keeps $\alpha = \pi/4$ and $\theta_{12} = \pi$ as r changes) and becomes reflected at the radius at which V is equal to the total energy of the system. These standing waves could manifest themselves as resonances in electron-atom scattering at energies near to, and below, the ionization energy. We have looked for such resonances in electron-helium scattering, using an apparatus similar to that described by Brunt et al[2]. The yield of metastable helium atoms, in either the 2^1S or 2^3S states, is measured. The result is shown in the figure. The positions of the more prominent resonance features are indicated in the figure, together with the energies of the 1sns ^3S and 1snd ^1D states of neutral helium. The resonances are seen to exist and to be narrow, which is surprising in view of the fact that there are many channels into which they can decay.

The energies of the resonances can be parametrized with a formula that has proved useful[3] when applied to two-electron correlated states having the configuration $[\text{core}](ns^2\ ^1S)$,

$$E(ns^2) = I - \frac{2R(Z_{core} - \sigma)^2}{(n - \delta_{ns})^2}, \qquad (2)$$

where σ is a screening parameter and δ_{ns} is the energy-averaged quantum defect for the configuration [core]ns. For the lowest resonance in each multiplet the required value of σ ranges from 0.233 at n=3 to 0.212 at n=7, and so is near to the value 0.25 that appears in equation (1). We therefore identify these lowest resonances with those predicted by Fano. As well as forming Rydberg series that are characterized by a non-integral charge, the resonance states are also unconventional in that they form multiplets that begin to overlap at high values of n, as can be seen from the figure. Since Wannier[4] (1953) was the first to realise the importance of the potential ridge region for determining the behaviour of low-energy two-electron systems, we refer to the lower resonances in each multiplet as "Wannier-ridge" resonances.

REFERENCES

1. U. Fano, J.Phys.B 13, L519 (1980).
2. J.N.H. Brunt, F.H. Read and G.C. King, J.Phys.E 10, 134 (1977).
3. F.H. Read, J.Phys.B 12, 449 (1977), Australian J.Phys. (1982).
4. G. Wannier, Phys.Rev. 90, 817 (1953).

* Present address: Research School of Physical Sciences, The Australian National University, Canberra, Australia.

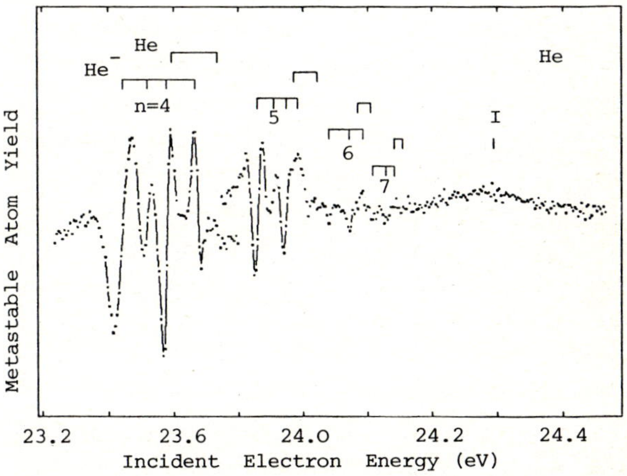

Yield of metastable helium atoms resulting from electron impact, as a function of the incident electron energy. In each of the regions a sloping background has been subtracted and the vertical scale has been expanded by an arbitrary factor. The energies of the more prominent He^- resonances are indicated, as well as those of the 1sns ^3S and 1snd ^1D states of He.

GRANDPARENT AND NON-VALENCE RESONANCES IN ELECTRON-IMPACT EXCITATION OF THE NOBLE GASES

S J Buckman*, P Hammond, G C King and F H Read

Department of Physics, University of Manchester, UK.

Several recent high-resolution electron-impact experiments have firmly established the existence of a large number of negative-ion resonances in the noble gases neon, argon, krypton and xenon below the first ionization potential (see for example Sanche and Schulz[1], Swanson et al[2], Brunt et al[3], and references therein). Although the energies of many of these resonances are well catalogued little is yet known about their spectroscopic classifications or decay modes. The purpose of the present study has been to obtain further experimental information that will help in the elucidation of the nature of these resonances. We have used an energy resolution higher than that presented by Brunt et al[3], and have also achieved a much improved sensitivity, enabling many new features to be seen. The apparatus used is similar to that described previously[4], and the yields of metastable atoms resulting from electron-impact have been measured. The result for neon is shown in the figure.

A dominant feature of the metastable excitation functions of neon, argon, krypton and xenon is the appearance of many pairs of strong resonances having an energy separation close to that of the $np^5\ {}^2P_{\frac{1}{2},3/2}$ ion core. Six such pairs are indicated in the figure, and a total of 21 have been found for the four noble gases. These are almost certainly resonances of the grandparent type, having the dominant configurations $[A^+](n'\ell^2\ {}^1S)$, and their energy separations indicate that they represent nearly pure external coupling.

A second interesting observation is the appearance of nine isolated peaks that have the same energy, within experimental error, as neutral excited states having the configuration $[A^+]n'p$ with J=0. Three such peaks in neon are indicated in the figure. Since these neutral states have the same spin and parity as the target state, excitation can take place with s-waves in the entrance and exit channels. There are no other excited states for which this is true (except for auto-ionizing states). Similar near-coincidences in energy have been observed in the metastable excitation spectrum of helium[5] at the thresholds of the 1s2s ^3S and 1s3s ^1S states, and have been shown by Nesbet[6], Berrington et al[7] and Fon et al[8] to be due to s-wave virtual states or to non-valence resonances consisting of a Rydberg electron weakly bound to the excited state in question. We assume that the structures seen in the present work can be similarly explained.

The present metastable measurements, together with recent high-resolution measurements of elastic-scattering spectra for the four noble gases, have also enabled other resonance features to be tentatively classified.

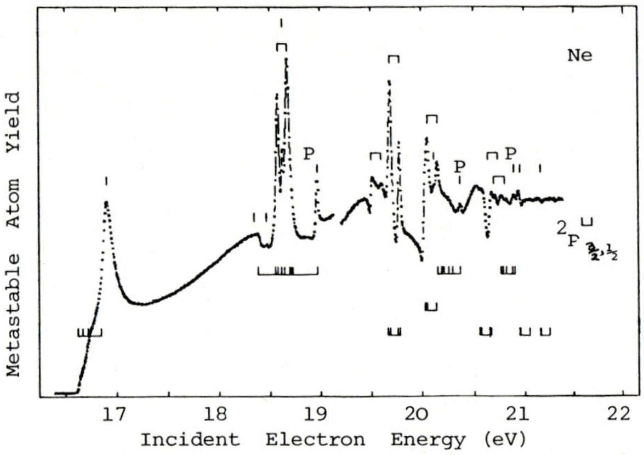

Yield of metastable atoms resulting from electron impact on neon. Assignment bars above the spectrum indicate the positions of the resonance pairs, the non-valence resonances (labelled p) and other structures; those below mark the energies of neutral excited states.

REFERENCES

1. S.L. Sanche and G.J. Schulz, Phys.Rev.A5, 1672 (1972).
2. N. Swanson, J.W. Cooper and C.E. Kuyatt, Phys.Rev. A8, 1823 (1973).
3. J.N.H. Brunt, G.C. King and F.H. Read, J.Phys.B9, 2195 (1976).
4. J.N.H. Brunt, F.H. Read and G.C. King, J.Phys.E10, 134 (1977).
5. J.N.H. Brunt, G.C. King and F.H. Read, J.Phys.B10, 433 (1977).
6. R.K. Nesbet, Phys.Rev.A12, 444 (1975), J.Phys.B8, 1459 (1975).
7. K.A. Berrington, P.G. Burke and A.L. Sinfailam, J.Phys.B8, 1459 (1975).
8. W.C. Fon, K.A. Berrington, P.G. Burke and A.E. Kingston, J.Phys.B11, 325 (1978).

* Present address: Research School of Physical Sciences, The Australian National University, Canberra, Australia.

RESONANCES IN ELECTRON SCATTERING BY NEON

K. T. Taylor and W. C. Fon*
Daresbury Laboratory, Daresbury, Warrington WA4 4AD, ENGLAND

C. W. Clark†
Atomic and Plasma Radiation Division, National Bureau of Standards, Washington, D.C. 20234, U.S.A.

We have carried out calculations of electron scattering by neon using the latest version of the programs described by Berrington et al.[1] The energy range extends to 2 Rydbergs above the ground state, and all states of the two lowest excited manifolds $2p^5 3s$ and $2p^5 3p$ are explicitly included. A previous report[2] on this work dealt with the establishment of a classification scheme for several resonances observed in this range; here we present converged cross sections for a number of inelastic and superelastic processes. Our collision calculations have been carried out entirely in LS coupling; this restriction limits the degree of detail to which we can make comparison with experiment, and practical means of lifting it are under investigation.[3]

Figure 1 shows the total cross section for electron impact excitation of the $2p^5 3s$ ^3P state as a function of the electron energy above its threshold. The "b" and "c" resonances, which were treated in ref. 2, are seen to appear prominently. The experimental curve is the total yield of metastable atoms

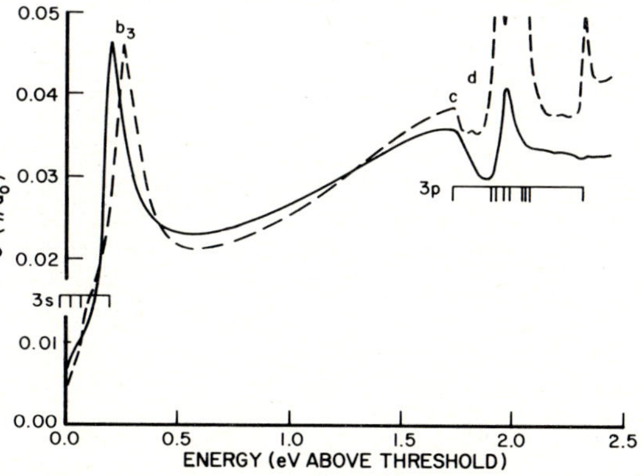

Fig. 1. Cross section for the $^1S \rightarrow {}^3P$ transition as a function of excess electron energy. Solid line : calculated ; dashed line : experimental yield of metastables, from ref. 4 courtesy Stephen Buckman. Insets show the locations of Ne excited states; b, c, d label resonance features.

produced by electron impact, from ref. 4. It has been normalized to match the theoretical curve at the peak of the b feature and the energies have been aligned so that the calculated ^3P threshold coincides with the experimental center of gravity of the three ^3P states. Up to about 1.76 eV above threshold, the experimental data is proportional to the sum of the cross sections for excitation of the $^3P_2$ and $^3P_0$ levels; the comparison therefore suggests that the excitation mechanism is not sensitive to the value of j, and thus the absolute metastable production cross section should be 2/3 of the value indicated on the ordinate of this figure. This idea is substantiated by comparison of experimental data on ultraviolet photon yields, with the appropriately weighted sum of calculated ^3P and ^1P excitation cross sections. At energies above 1.76 eV there are cascade contributions to the total metastable yield; these are not reflected in the calculated cross section shown here, but our calculations show, e.g., the d feature appearing prominently in the excitation of the $2p^5 3p$ ^3S state.

As an example of the importance of resonance features in transitions among excited states we show in Fig. 2 the cross section for the process $2p^5 3s$ $^3P \rightarrow 2p^5 3s$ ^1P as a function of incident electron energy. The rather large cross section near threshold may make this a significant mechanism for quenching of the metastable population in a neon plasma with an electron temperature of $\sim$ 1/4 eV.

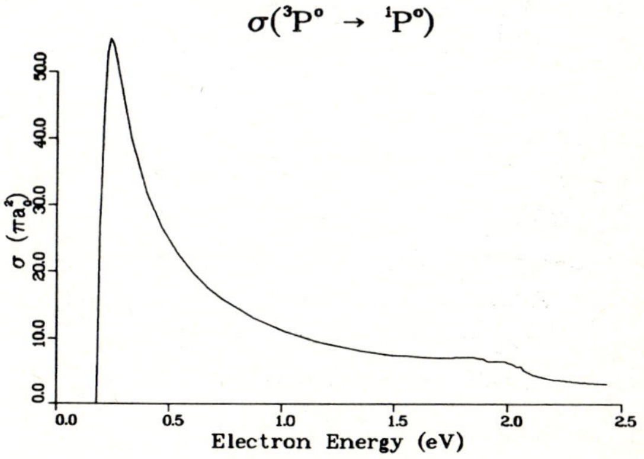

Fig. 2. Calculated cross section for the $^3P \rightarrow {}^1P$ transition as a function of incident electron energy.

*Permanent address: Department of Mathematics, Univ. of Malaya, Kuala Lumpur, MALAYSIA
†National Research Council postdoctoral fellow

1. K. A. Berrington, P. G. Burke, M. LeDorneuf, W. D. Robb, K. T. Taylor, and Vo Ky Lan, Comp. Phys. Comm. 14, 367 (1978).
2. C. W. Clark and K. T. Taylor, J. Phys. B 15, L213 (1982).
3. N. S. Scott, K. L. Bell, P. G. Burke, and K. T. Taylor, J. Phys. B 15, L627 (1982).
4. S. J. Buckman, P. Hammond, G. C. King, and F. H. Read, to be published; J. N. H. Brunt, G. C. King, and F. H. Read, J. Phys. B 9, 2195 (1976).

AN ELEVEN STATE ELECTRON-HELIUM SCATTERING CALCULATION

P.G. Burke, L.C.G. Freitas[1], A.E. Kingston and K.A. Berrington

Department of Applied Mathematics and Theoretical Physics
The Queen's University of Belfast
BELFAST BT7 1NN NORTHERN IRELAND

1 Permanent address: Departmento de Quimica
Universidade Federal de São Carlos
via Washington Linz kM 234
13560 São Carlos, SP BRASIL

In this abstract we report some preliminary electron-Helium scattering results obtained in an eleven-state R-matrix[1] calculation. The eleven target states included (1^1S, 2^3S, 2^1S, 2^3P, 2^1P, 3^3S, 3^1S, 3^3P, 3^3D, 3^1D, 3^1P) were represented by configuration interaction wave functions, calculated using the CIV3 package of Hibbert[2]. We show in Fig. 1 the ^2S wave contribution to the ground state elastic cross section. Two large resonance effects below the 2^3S and 3^3S threshold are clear. In Table 1 we compare the paramters Γ (width) and E_R (position) associated to the ^2S resonance below the 2^3S threshold obtained in the present calculation with other theoretical and experimental ones.

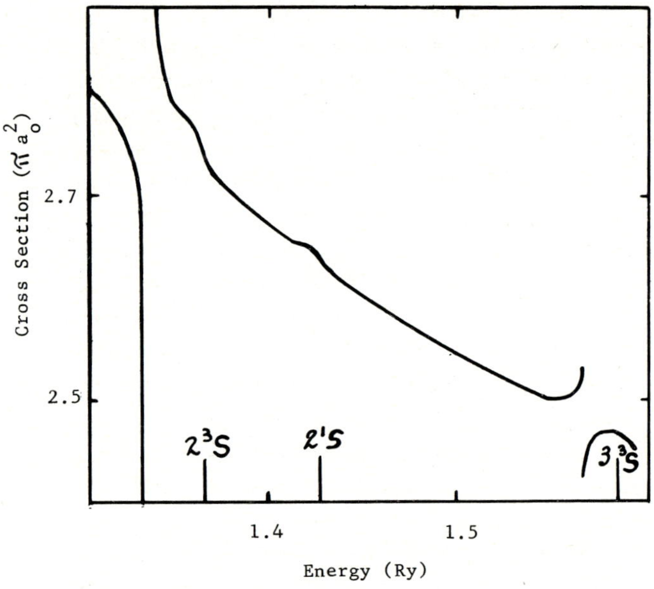

TABLE 1

	Theoretical			Experimental		
	Ref.3	Ref.4	Present	Ref.5	Ref.6	Ref.7
Γ(meV)	15.1	14.4	11.72	11.$\pm$0.5	13.0	9.0
E_R(eV)	19.38	13.9	19.37	19.37	19.35	19.37

As we can see in table 1 the parameter estimated by the present calculation is somewhat smaller in agreement with most of the recent experimental results. Calculation of the elastic and inelastic cross sections involving the above eleven states as well as other resonance analysis are under way and will be reported later.

References

1. K.A. Berrington, P.G. Burke, M. Le Dourneuf, W.D. Robb, Comp.Phys. Commun. 14, 367 (1978)
2. A. Hibbert, Comp. Phys. Commun. 9, 141 (1975)
3. K.A. Berrington, P.G. Burke and A.L. Sinfailam, J.Phys.B:At.Mol.Phys. 8, 1459 (1975)
4. A. Temkin, A.K. Bhatia and J.N. Bardsley, Phys. Rev. A 5, 1663 (1972)
5. R.E. Kennerly, R.J. Van Brunt and A.C. Gallagher, Phys. Rev. A 23, 2430 (1981)
6. D.E. Golden, F.D. Schowengerdt and J. Macek, J.Phys.B:At.Mol.Phys. 7, 478 (1974)
7. S. Cvejanovic, J. Comer and F.H. Read, J.Phys.B:At.Mol.Phys. 7, 468 (1974)

HYPERSPHERICAL DESCRIPTION OF ELECTRONIC CORRELATIONS IN ATOMIC SYSTEMS
I- BOUND AND CONTINUUM STATES OF TWO-ELECTRON SYSTEMS H^-, He

L. Pelamourgues, S. Watanabe and M. Le Dourneuf

Observatoire de Paris, 92190 Meudon, France

In recent years, the hyperspherical method proved effective in accounting for characteristics and trends of electronic correlations in doubly-excited atomic states[1]. The success of pilot calculations made an efficient implementation of the method desirable for quantitative studies. Developments along this direction and tests of the method with the fundamental H^-,He systems will be presented at the conference.

The key feature of the hyperspherical description of two-electron atoms is the definition of a single size parameter $R = (r_1^2 + r_2^2)^{1/2}$ and a symmetrical description of radial and angular correlations by angle $\hat{\Omega}$. The most convenient choice of angles to describe numerically the low energy region of the two-electron spectrum[2] retains the individual angles and spin — and therefore the standard concept of angular configurations $\ell_1 \ell_2\ ^{2S+1}L$ — but replaces the two independent radial variables r_1 and r_2 by their polar representation. Besides the global size parameter R, the radial correlation angle $\alpha = \tan^{-1}(r_2/r_1)$ characterises the evolution from uncorrelated particles ($\alpha \rightarrow 0,\pi/2$) to the strong correlation region ($\alpha \rightarrow \pi/4$).

The solution of the fixed-R Hamiltonian ,

$$\left[\frac{\Lambda^2 (\hat{r}_1,\hat{r}_2,\alpha)}{2 R^2} + \frac{C (\hat{r}_1,\hat{r}_2,\alpha)}{R} - E_\mu \right] \Phi_\mu = 0. \quad (1)$$

provides a locally optimal description of correlation eigenmodes determined by the competition between kinetic $(1/R^2)$ and potential (1/R) effects. In the standard adiabatic approach, this R-dependent basis is used to expand the exact wavefunction,

$$\Psi(R,\Omega) = \sum_\mu F_\mu (R) \Phi_\mu (R ,\Omega) \quad (2)$$

and the radial functions $F_\mu(R)$ satisfy second order differential equations coupled exclusively by velocity and base-dependent non-adiabatic couplings. At low energy these couplings are usually negligible, except near localised avoided crossings where they vary rapidly and are numerically difficult to handle. This drawback is easily overcome by the diabatic-by-sector method[3]. By dividing the radial domain into sectors and using a constant angular basis within each sector,

$$\Psi(R,\Omega) = \sum_\mu F_\mu (R,R_i) \Phi_\mu (R_i,\Omega) \quad (3)$$

The couplings induced by the basis evolution are represented by a limited number of weak couplings within each sector and by a redistribution of channel amplitudes embodied by a frame transformation between adjacent sectors.

The results will be presented at the conference along three lines:

(a) Optimal representation of the hyperspherical potential adapted basis $\Phi_\mu(R;\Omega)$ using various choices of basis functions of the radial correlation cordinate α.

(b) Convergence rate of elastic and inelastic phaseshifts with the increase in the number of quasi-adiabatic channels included.

(c) Range of applicability of the present approach to energy ranges near double ionisation threshold.

References

1. U. Fano, Rep.Prog.Phys.46, 97(1983).
2. J.H. Macek,J.Phys.B1,831(1968);C.D.Lin,Phys.Rev.A10, 1986(1974).
3. Vo Ky Lan, M. Le Dourneuf and J.M. Launay,in Electron-Atom and Electron-Molecule Collisions (ed. by J. Hinze) p.161 (Plenum,1983).
4. C.Schwartz, Phys.Rev.124,1468(1961).

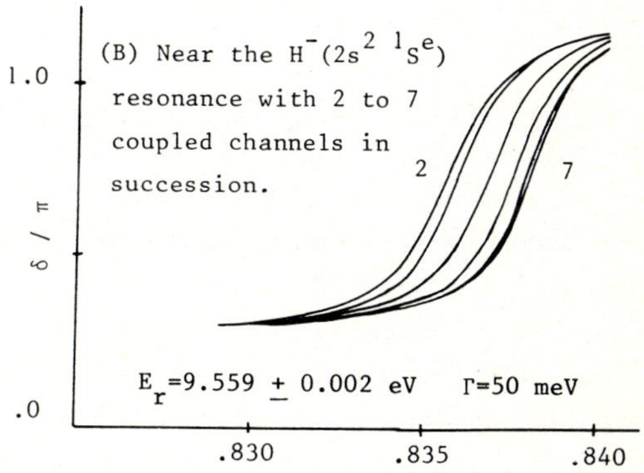

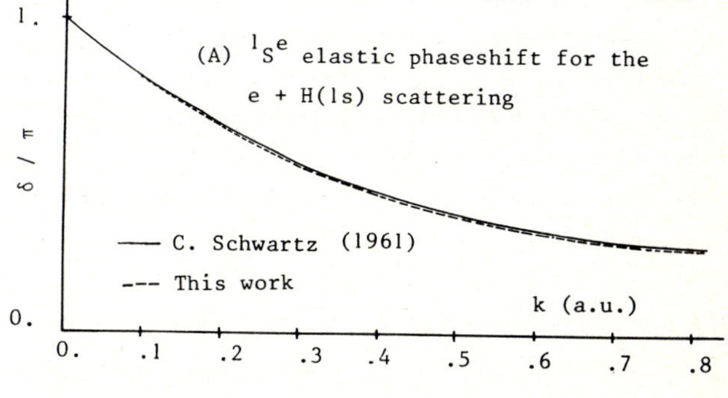

HYPERSPHERICAL DESCRIPTION OF ELECTRONIC CORRELATIONS IN ATOMIC SYSTEMS
II- GRAPHICAL REPRESENTATION OF RADIAL AND ANGULAR CORRELATIONS IN DOUBLY-EXCITED ATOMS

M. Le Dourneuf, S. Watanabe and L. Pelamourgues

Observatoire de Paris, 92190 Meudon, France

Graphical displays of two-electron correlations introduced recently[1] have been restricted to states with zero orbital momentum. In this case the electron wavefunction depends only on variables internal to the pair's body-frame, namely, the radial distances from the nucleus (r_1, r_2) and the angular variable $\hat{r}_1 \cdot \hat{r}_2 = \cos\theta_{12}$. For non-zero orbital momenta the wavefunction depends also on the body-frame's rotation about one of its inertial axes, being represented by

$$\Psi = \Sigma_K D^{(L)}_{M\,K}(\phi,\theta,\psi)\; f_K(r_1,r_2,\theta_{12}). \qquad (1)$$

The summation over K values of the same parity is generally required because the electron pair constitutes an asymmetric top.

Displaying different components f_K separately reduces their usefulness, but this difficulty may be eased by seeking representations (1) in which one component may have a dominant role. Searching for a suitable display, one may utilize alternative choices of the body-frame axis of the azimuthal angle ψ. We report here results obtained by taking this axis, $\hat{\zeta}$, parallel to the inter-electron distance $\vec{r}_1 - \vec{r}_2$. This selection implies that

$$|\hat{\zeta} \times \vec{r}_1| = |\hat{\zeta} \times \vec{r}_2| \qquad (2)$$

whereby both electrons contribute equally to the amount of inertia I_ζ. The Hamiltonian term representing the kinetic energy of rotation about ζ,

$$I_\zeta\, \dot{\psi}^2 / 2 \qquad (3)$$

remains thus symmetric with respect to the two electrons.

We have decomposed hyperspherical adiabatic channel functions in the form of Eq(1). A simple illustration of this development is provided by lowest two doubly-excited channels of $H^-(^1P°)$ in Fig's 1 and 2, Fig.1 corresponds to the well-known shape-resonant channel[2], in which the radial motion of the pair is in-phase (maximum amplitude at $\alpha=\pi/4$) and the angular correlation is predominated by a state with one unit of bending (v=1) and one unit of orbital momentum about ζ. Fig.2 corresponds to the Feshbach resonant channel with out-of-phase type radial correlation (node near $\alpha=\pi/4$) and a predominance of the linear bending mode (v=0,K=0).

Results on the N=2 and N=3 manifold will be presented at the conference.

References

1. C.D. Lin, Phys.Rev.A25,76,1535 (1982), Phys.Rev.A26, 2305(1982); G.S. Ezra and R.S. Berry, Phys.Rev. A25,1513 (1982); also U.Fano, Rep.Prog.Phys.46,97(1983)
2. C.D.Lin, Phys.Rev.Lett.35,1150(1975).

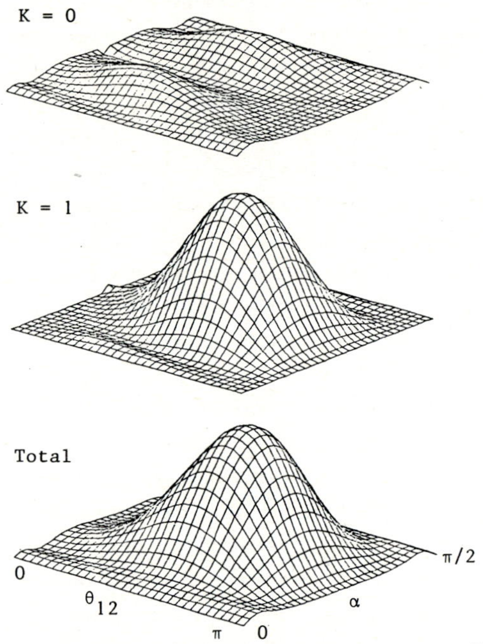

Fig.1: Three dimensional representation of the $H^-(2\,^1P°)$ shape resonance eigenmode: K=0,1 partial densities and total density. '+' radial correlations and dominant K=1,v=1 bending angular correlation.

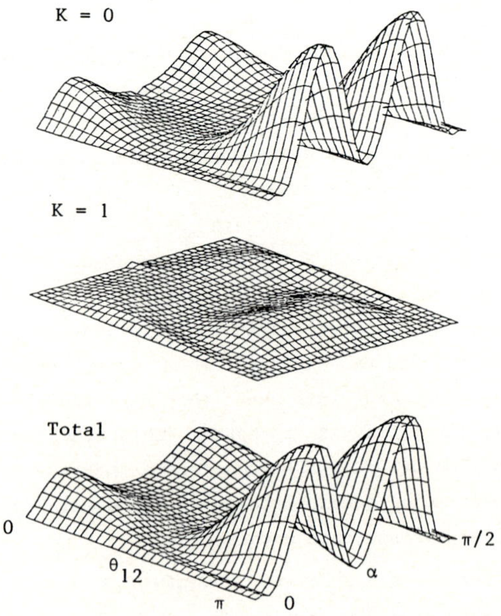

Fig.2: Three dimensional representation of the $H^-(2\,^1P°)$ Feshbach eigenmode: K=0,1 partial denstities and total density. '-' radial correlations and dominant K=0,v=0 angular correlation.

HYPERSPHERICAL DESCRIPTION OF ELECTRONIC CORRELATIONS IN ATOMIC SYSTEMS

III- RADIAL AND ANGULAR CORRELATIONS OF TRIPLY-EXCITED STATES

S. Watanabe and M. Le Dourneuf
Observatoire de Paris, 92190 Meudon, France

Advances in the hyperspherical analysis of doubly-excited states have prompted its extension to triply-excited states. Applying the formulation put forward by Clark and Greene[1], we have carried out preliminary studies on He^-.

As in the two-electron case, the hyperspherical method analises radial and angular correlations at various values of fixed hyperradius $R = (r_1^2 + r_2^2 + r_3^2)^{1/2}$. The relative radial motion of the three electrons is parametrised by 2 pseudo-angles (α_1, α_2). In actual computations we use the independent particle angular coordinates (θ,ϕ) while the analysis favors the 3 relative angles $\theta_{12}, \theta_{23}, \theta_{31}$, and a set of Euler angles Ω which relate the lab-frame to the body-frame of the three electrons.

Fig.1 displays the lowest hyperspherical potential of the $He^-(^4S^o)$ symmetry in three approximations of increasing accuracy. The curves converge asymptotically to the energy of $He^{**}(2p^2\ ^3P^e)$. A quasibound state is confined by the well at moderate R, $5 \leqslant R \leqslant 10a_o$ with large electron affinity of $0.28 \leqslant E_b \leqslant 0.54$ eV. As illustrated by the figure, this large affinity is due to the quadrupole interactions amongst p-electrons pairs. The angular wavefunction of the dominant $p^3\ ^4S^o$ configuration is proportional to $(\hat{r}_1 x \hat{r}_2).\hat{r}_3$ which is automatically antisymmetric. The corresponding radial function is necessarily symmetric at moderate values of R; this represents the inphase radial oscillations of all three electrons just like the '+' motion found in doubly-excited states. The function $(\vec{r}_1 x \vec{r}_2).\vec{r}_3$ equals the volume of a pyramid formed by the electrons and the nucleus. It attains maximum when the pyramid becomes cubic and vanishes when it collapses to a plane, giving a nodal plane.

Fig. 2 displays the lower adiabatic potentials for the $^4P^e$ symmetry. There are two distinct groups of curves, the lower group represents doubly excited states formed outside the $He^+(1s)$ core. The upper group asymptotically converges to various doubly excited states and forms triply excited eigenmodes at moderate R. One finds a triply excited state bound to $He(2sp^+\ ^3P^o)$ with a large affinity of $0.58 \leqslant E_a \leqslant 1.00$eV. The angular wavefunction of the dominant configuration $2sp\ ^3P^o$ is proportional to the vector $\vec{P} = \hat{r}_1 x \hat{r}_2 + \hat{r}_2 x \hat{r}_3 + \hat{r}_3 x \hat{r}_1$ projected onto a suitable axis. $\vec{P}$ is fully antisymmetric and the distribution maximises for the equilateral triangular configuration ($\theta_{12} = \theta_{23} = \theta_{31} = 120°$) with the nucleus at the center. The corresponding radial function, which minimises the radial correlation kinetic energy, is symmetric and represents inphase radial motions of the 3 electrons. In this configuration, the vector $\vec{P}$ coincides withe the vector $\vec{N}$ perpendicular to the plane of the 3 electrons, meaning no net rotation about $\vec{N}$ (K = 0).

Three electron correlations are known to influence the process[2]

$$A\ell(^2P^o) + h\nu \longrightarrow A\ell(^2S^e, ^2P^e, ^2D^e)$$

one example amenable to the hyperspherical analysis.

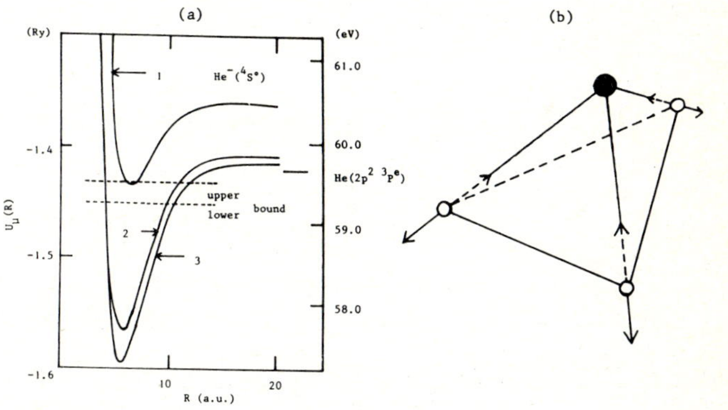

Fig. 1 (a) Hyperspherical adiabatic potential for the lowest channel of $He^-(^4S^o)$
1 p^3 alone and monopole interactions only
2 same as 1, adding quadrupole interactions amongst p-electron pairs
3 full calculation, including angular couplings p^3 and pd^2
Dashed lines are upper and lower bounds for the bound state
(b) Schematic representation of three-electron eigenmodes.

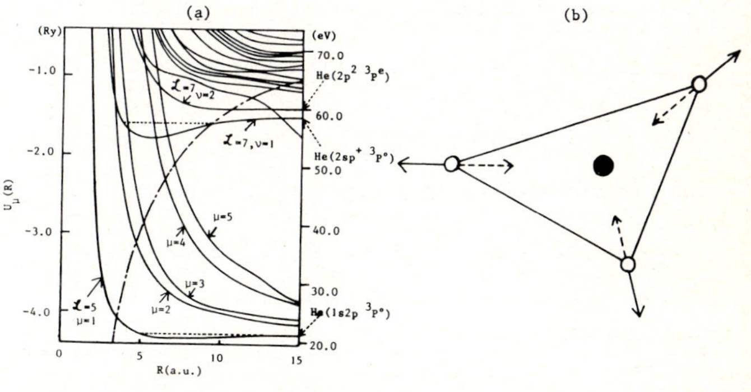

Fig. 2 (a) Hyperspherical adiabatic potentials for $He^-(^4P^e)$. Dashed lines are resonances. The chain curve is the locus of the ridge.

References

1. C.W. Clark and C.H.Greene,Phys.Rev.A21,1786(1980)
2. C.D. Lin,Astrophys.J.187,385(1974);also M.Le Dourneuf, VoKy Lan,P.G. Burke, and Ken Taylor,J.Phys.B8,2640(1975)

VARIATIONAL LINEAR ALGEBRAIC EQUATIONS METHOD

B.L. Moiseiwitsch

Department of Applied Mathematics and Theoretical Physics
The Queen's University of Belfast, Belfast BT7 1NN,
Northern Ireland

It has been shown previously[1] that the linear algebraic (LA) equations method can be expressed in a Schwinger variational form for the tangent of the phase shift in the case of the elastic scattering of particles by a local potential U(r). The variational linear algebraic (VLA) equations method has now been generalised to include exchange by the use of non-local potentials, and to treat excitation collisions involving coupled channels.

In the latter case, if we represent the matrix of the interaction potentials by $\underline{U}$, the $\underline{K}$ matrix can be written in the Schwinger variational form

$$\underline{K} = -\underline{k}^{-\frac{1}{2}} \int_0^\infty \underline{f}\,\underline{U}\,\underline{v}^t\,dr\,\underline{M}^{-1} \int_0^\infty \tilde{\underline{v}}^t\underline{U}\,\underline{f}\,dr\,\underline{k}^{-\frac{1}{2}} \qquad (1)$$

where

$$\underline{M} = \int_0^\infty \tilde{\underline{v}}^t\,\underline{U}\,\underline{v}^t\,dr + \int_0^\infty \tilde{\underline{v}}^t\underline{U}\,dr \int_0^\infty \underline{G}(r,r')\underline{U}\,\underline{v}^t\,dr' \qquad (2)$$

Here $\underline{v}^t$ is a square matrix composed of trial functions v^t_{nm} for the channels n with wave numbers k_n, $\underline{f}$ is a diagonal matrix composed of the terms $k_n r\, j_\ell(k_n r)$ involving spherical Bessel functions j_ℓ, $\underline{k}^{-\frac{1}{2}}$ is a diagonal matrix with elements $k_n^{-\frac{1}{2}}$ and $\underline{G}(r,r')$ is a diagonal matrix composed of the appropriate Green's functions.

Now we choose the ansatz

$$\underline{v}^t = \sum_i \underline{v}_i^t\, I(h_i, r-r_i) \qquad (3)$$

where the r_i form a suitable net of points over the range of integration with $h_i = r_{i+1} - r_i$,

$$\begin{aligned} I(h_i, r-r_i) &= 1 \qquad (r_i < r < r_i + h_i) \\ &= 0 \qquad \text{otherwise} \end{aligned} \qquad (4)$$

and $\underline{v}_i^t$ is the square matrix composed of the values of the trial functions in the interval $r_i < r < r_i + h_i$.

Substitution of (3) into (1) and optimisation with respect to the $\underline{v}_i^t$ produces the VLA formula

$$\underline{K} = -\underline{k}^{-\frac{1}{2}}\,\tilde{\underline{F}}\,(\underline{V}\underline{w}^{-1} + \underline{L})^{-1}\,\underline{F}\,\underline{k}^{-\frac{1}{2}} \qquad (5)$$

where $\underline{F} = (\underline{F}_i)$, $\underline{V} = (\underline{V}_i)$, $\underline{L} = (\underline{L}_{ij})$ and $\underline{w} = (\underline{w}_i)$ are composed of the matrices

$$\underline{F}_i = h_i^{-1} \int_{r_i}^{r_i+h_i} \underline{U}(r)\,\underline{f}(r)\,dr \qquad (6)$$

$$\underline{V}_i = h_i^{-1} \int_{r_i}^{r_i+h_i} \underline{U}(r)\,dr \qquad (7)$$

$$\underline{L}_{ij} = (h_i h_j)^{-1} \int_{r_i}^{r_i+h_i} \underline{U}(r)dr \int_{r_j}^{r_j+h_j} \underline{G}(r,r')\underline{U}(r')dr' \qquad (8)$$

and $\underline{w}_i$ with diagonal elements h_i respectively.

Calculations have been performed using the VLA method for the static exchange scattering of $\ell = 0$ wave electrons by H atoms, for the scattering of $\ell = 0$ wave electrons by H atoms in the 1s-2s coupled channels case, and also for a coupled channels schematic model involving square well potentials. Encouraging results have been obtained for these simple illustrative cases.

References

1. B.L. Moiseiwitsch, J.Fhys. B15, L863 (1982).

A COMPARATIVE STUDY OF MULTIPLE SCATTERING APPROXIMATIONS TO THE EXCITATION OF ATOMIC HYDROGEN BY ELECTRON IMPACT

H. E. Fargher and M. J. Roberts

Department of Physics, University of Stirling,
Stirling FK9 4LA, Scotland, United Kingdom.

There are two fundamental perturbation expansions of the transition operator for electron-atom collisions. The most widely used is the Born series, which can be obtained by iterating the Lippmann-Schwinger equations, and the other is the Faddeev-Watson series, which is an iteration of the Faddeev equations. These series are multiple scattering approximations[1] in the sense that, for electron-hydrogen collisions for example, they approximate the full three-body interaction by a succession of two-body interactions. Formally, each series is a rearrangement of the other. Since both series, when truncated, could be expected to be valid only at incident electron energies considerably larger than the binding energy of the target, it is instructive to investigate their relative merits over the entire angular range of the scattered electron at high energies.

For low angle scattering the Born series should be superior since it can account for polarization of the target, induced by the scattered electron, in second order whereas such an effect cannot be treated by the Faddeev-Watson until third order. For high angle scattering, however, the interaction between the incident electron and the proton should be dominant and there is no reason why the Born series should be superior in this angular region.

In this presentation we compare three second order multiple scattering approximations - one based on the Faddeev-Watson series, one based on the Born series and the third obtained by different approximate simplifications of the first two:

1. THE ON-SHELL FADDEEV-WATSON APPROXIMATION

This has been derived previously by the authors[2]. The basic assumption is that the major contribution to the second order matrix elements in the Faddeev-Watson series comes from near the three-body energy shell.

2. THE FREE PARTICLE SECOND BORN APPROXIMATION

For high angle scattering it is assumed that the interaction between the incident electron and the atom dominates the interaction which binds the atom. Consequently the Green's operator describing the intermediate virtual states of the target may be replaced by the free particle Green's operator.

3. THE ZERO PHASE FADDEEV-WATSON APPROXIMATION

This is a further approximation to 1. obtained by neglecting the phases of the half-shell two-body Coulomb T-matrices and gives an idea of the importance of these phases. This approximation can also be obtained from 2. by neglecting two of the four second order terms. The importance of these neglected terms can thus be evaluated.

The three approximations are applied to the excitation of the n=2 levels of atomic hydrogen by 100 eV and 680 eV incident electrons and their relative merits, in comparison with experiment, are discussed.

Acknowledgement

Mr H E Fargher would like to thank the University of Stirling for a studentship.

References

1. Joachain C. J., 1975, Quantum Collision Theory, Chapter 19 (North-Holland, Amsterdam)
2. Fargher H.E. and Roberts M. J. to be published in J.Phys.B: Atom. Molec. Phys.

PSEUDOSTATE EXPANSIONS IN ELECTRON HYDROGEN SCATTERING

D. H. Oza and J. Callaway
Department of Physics, Louisiana State University, Baton Rouge, LA 70803 U.S.A.

A close-coupling expansion in which only bound states are included does not yield accurate cross-sections for the scattering of electrons by hydrogen atoms. Inclusion of pseudostates improves the theoretical results but it has the serious drawback of introducing unphysical pseudoresonances near the thresholds for the excitation of the pseudostates[1]. Since the effects of the spurious structure can extend over a relatively large range of energies (see figure 1), one needs to remove them in order to extract meaningful results.

We consider here a physical model in which the hydrogen atom has only s states and where the total angular momentum of the collision is zero. We have applied four different pseudostate basis sets of rather different characters to this model. The scattering calculations were performed using the algebraic variational method[2]. Complicated pseudo-resonance structure was obtained.

We removed the spurious structure in cross-section by using the T matrix averaging technique used by Burke et. al.[3] In this procedure, a linear least squares fit is made to the transition amplitudes for the physical channels using a low order polynomial in energy. Let $T_{ij}(E)$ be an element of the T matrix for a transition between channels i and j. We represent this as

$$T_{ij}(E) = \sum_{m=0}^{N} a_m E^m \qquad (1)$$

where the complex coefficients a_m are determined by the linear least squares fit procedure. Quadratic, cubic and quartic powers of energy were used in the expansion.

We have considered both singlet and triplet states. Only the singlet state calculation showed appreciable pseudothreshold structure so the T matrix fitting was not employed for triplet states. An example of a singlet transition depicting pronounced structure is the 1s->2s excitation. The cross section computed from one of the four basis sets is presented in figure 1 in short dashed curve. The relevant pseudo-threshold for this basis is at 1.68 Ry. The

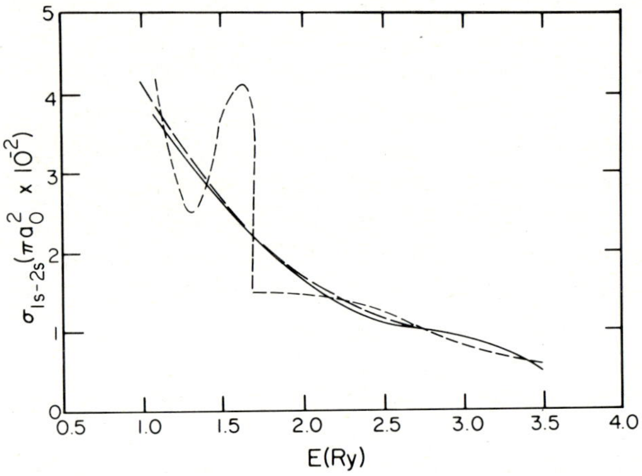

Figure 1. Cross sections for the 1s-2s transition in the ^1S state.

directly calculated results agree reasonably well with the exact values[4] (long dashed curves) above about 2.1 Ry. A least squares fit was made to T_{1s-2s} using a quartic polynomial in energy. The resulting cross sections are shown by the solid curve. The errors with respect to the exact results are of the order of 3%.

Reasonably accurate cross sections were obtained from all four basis sets after the spurious threshold structure was removed (errors less than 3% for elastic scattering and less than 8% for 2s excitation for three out of four cases). If this standard of accuracy can be maintained in the full electron-hydrogen problem, it should be possible to improve the comparison between theory and experiment for processes such as 2s excitation, where uncertainties of the order of 20% still exist.

References

1. P. G. Burke and J. F. Mitchell, J. Phys. B 6, 320 (1973)
2. J. Callaway, Phys Repts. 45, 89 (1978)
3. P. G. Burke, K. A. Berrington, and C. V. Sukumar, J. Phys B 14, 289(1981).
4. R. Poet, J. Phys. B 11, 3081(1978); J. Phys. B 13, 1995 (1980).

APPLICATIONS OF THE COUPLED-CHANNELS OPTICAL CALCULATION FOR ELECTRON-ATOM SCATTERING

B.H. Bransden, I.E. McCarthy and A.T. Stelbovics

Institute for Atomic Studies, The Flinders University of South Australia, Bedford Park, S.A. 5042, Australia.

The coupled-channels optical (CCO) method has been very successful in describing electron scattering from hydrogen[1]. The momentum-space coupled (Lippmann-Schwinger) equations are solved numerically for a finite set P of electron-target channels. The remaining channels Q are explicitly taken into account by ab initio calculations of the corresponding optical potential $V^{(Q)}$. The optical potential for the excitation of a particular channel depends on an integral over the corresponding kinematic space of a hermitian product of excitation amplitudes for that channel. Continuum channels and relevant discrete channels are included.

The numerical methods may be summarized as follows. The coupled equations are solved for the T-matrix elements by matrix methods[2]. Potential matrix elements consist of first-order (Born) terms and, in the diagonal cases, polarization terms. Polarization potential matrix elements are usually small when they couple different channels, so they are neglected in those cases. The excitation amplitudes for discrete polarization potentials[3] are computed in the Born approximation, using the Bonham-Ochkur peaking approximation for the exchange amplitudes. For continuum excitations[4] the amplitudes are calculated in the extreme screening approximation (with anti-symmetry), in which the slower of the continuum electrons is represented by a Coulomb wave orthogonalized to all m-components of the ground-state Hartree-Fock orbital, and the faster electron is represented by a plane wave. Orbitals are represented by a linear combination of Slater functions.

Computation time is saved by using the fact that the polarization potential matrix element for channel i $\langle \underline{q}' | V_{ii}^{(Q)} - V_{ii} | \underline{q} \rangle$ depends to a good approximation only on the total energy E and $P = |\underline{q}-\underline{q}'|$. This is the momentum-space expression of the equivalent local central approximation for coordinate space. The value of Im $\langle \underline{q} | V_{ii}^{(Q)} | \underline{q} \rangle$ is proportional to the total cross section for the excitation of Q-space at incident momentum q. This provides an essential check on the approximations used, particularly for the continuum where an excellent description of the total ionization cross section is given for a wide range of incident energies and ions[5].

The application of the method will be described in the following cases:

1. Positron excitation of hydrogen at energies where positronium formation may be neglected,
2. Electron excitation of sodium,
3. Electron excitation of hydrogen from excited states,
4. Electron excitation of the He^+ ion.

References

1. I.E. McCarthy and A.T. Stelbovics, J. Phys. B. (in press)
2. I.E. McCarthy and A.T. Stelbovics, Aust. J. Phys. 35, 543 (1982)
3. I.E. McCarthy, B.C. Saha and A.T. Stelbovics, J. Phys. B. 14, 2871 (1981)
4. I.E. McCarthy and A.T. Stelbovics, Phys. Rev. A 22, 502 (1980)
5. I.E. McCarthy and A.T. Stelbovics, Phys. Rev. A (in press)

DIFFERENTIAL CROSS SECTIONS OF $e^- + H(1s) \rightarrow e^- + H(2p)$

S.N. Tiwary
Department of Physics, L.S. College, Muz., Bihar University, Bihar, India.

A number of different theoretical schemes to electron-atom, electron-molecule and electron-ion scattering problems have been introduced in the literature which are broadly grouped into three categories depending on the energy range (low, intermediate and high) over which they are applicable. Several experimental techniques have also been developed for the measurements of sensitive scattering quantities such as the target parameters, i.e., the alignment parameter λ, the orientation parameter x, and the corresponding differential cross sections $\frac{d\sigma}{d\Omega}$. From a review of the literature it is clear that none of the theoretical models are capable of producing the sensitive experimental measurements especially λ and x parameters even in the simplest case of $1s \rightarrow 2p_m$ transitions in the hydrogen atom[1]. The close-coupling method has been successfully applied to electron-atom collision problems but it is very complicated from the computational point of view because in many cases, the convergence is very slow and many equations have to be solved. In the light of the convergence as well as the cumbersome computational problem we have recently developed a new linear-algebraic approach to electron-atom scattering which allows the incorporation of any number of target states in order to obtain reliable theoretical results. This method is easily extendable to electron-molecule collision problems.

The approximation consists in replacing the exact bilinear Green's function of the unperturbed "electron + atom" system.

$$G^{(0)}(\vec{r}_1,\vec{r}_1';\vec{r},\vec{r}')=\sum_j \phi_j(\vec{r}_1)\phi_j^*(\vec{r}_1')\frac{e^{iK_j|\vec{r}-\vec{r}'|}}{|\vec{r}-\vec{r}'|}\left[\frac{-1}{2\pi}\right] \quad (1)$$

by its asymptotic form for $r \gg r'$ (Faisal and Tiwary)[2]

$$G^{(0)}(\vec{r}_1,\vec{r}_1';\vec{r},\vec{r}')\simeq\sum_j \phi_j(\vec{r}_1)\phi_j^*(\vec{r}_1')\frac{e^{iK_j r}}{r}\left[\frac{-1}{2\pi}\right] \quad (2)$$

where $\{|\phi_j>\}$ is a complete set of target states and $K_j = \sqrt{2(E_o - E_j)}$ where E_o is the total energy and E_j is the binding energy of the state $|\phi_j>$. With the use of equation (2) the multichannel T-matrix equations can be written as follows:

$$<j\vec{K}_j|T|0> \simeq <j\vec{K}_j|V|0> + \sum_{j'} S_{jj'}(\vec{K}_{jj'}) <j'\vec{K}_{j'}|T|0> \quad (3)$$

Equation (3) has been discussed in details elsewhere[1,3]. We have calculated the differential cross sections of $e^- + H(1s) \rightarrow e^- + H(2p)$ using equation (3).

Figure 1 displays the present asymptotic Green's function approximation (AGFA) results for the differential cross sections of $e^- + H(1s) \rightarrow e^- + H(2p)$ at $K_i^2 = 2.25$ and 4.00 Rydberg with the other available theoretical data of Kingston et al[4], and Brandt and Truhlar[5]. The theoretical predictions of Kingston et al, and Brandt and Truhlar differ qualitatively with each other. The results of Brandt and Truhlar exhibit oscillations in the differential cross section curve whereas Kingston et al results show smooth behaviour. In order to resolve the existing discrepancy between these two sets of theoretical calculations we have performed calculations for $\frac{d\sigma}{d\Omega}$ at exactly the same energies using our multi-channel T-matrix equation (3). It is seen from the figure that our present results favour the results of Kingston et al but differ quantitatively from both theoretical predictions. The quantitative disagreement may be, probably, due to the neglect of exchange as well as the unphysical situation for small values of r.

References

1. S.N. Tiwary, Phys. Rev. A. 1983 (in press)
2. F.H.M. Faisal and S.N. Tiwary, German Phys. Soc. (1980)
3. S.N. Tiwary, J. Phys. B.: 14 2951 (1981)
4. A.E. Kingston, W.C. Fon and P.G. Burke, J. Phys. B: 9, 605 (1976)
5. M.A. Brandt and D.G. Truhlar, Phys. Rev. 11, 1340 (1975)

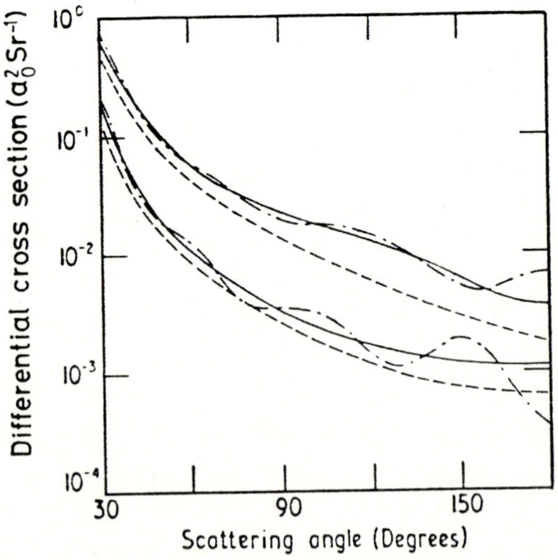

Fig. 1: $\frac{d\sigma}{d\Omega}$ for the process $e^- + H(1s) \rightarrow e^- + H(2p)$ at 2.25 and 4.00 Rydberg. ---, present; ——, Kingston et al; -.-, Brandt and Truhlar.

HIGH RESOLUTION TOTAL ELECTRON IMPACT EXCITATION OF INDIVIDUAL He 2^3S and 2^1S STATES*

David Spence, Dorothy Stuit, M. A. Dillon, and R.-G. Wang†

Radiological and Environmental Research Division, Argonne National Laboratory, Argonne, IL 60439

Because of their importance as a testing grounds for new theoretical models,[1,2] we have reinvestigated the total electron impact excitation of the He2^3S and ^1S states to 3 eV above threshold. The only previous such experiments performed with sufficient resolution to be meaningfully compared with modern theories in this energy region were those of Brongersma et al.[3]

The technique used by Brongersma et al.[3] was the double retarding potential difference (DRPD)[4] trapped electron technique, which has the ability to energy select the scattered electrons. We have used a similar apparatus but have replaced the RPD electron gun with a trochoidal monochromator, so that a single modulation technique can be used.[5] As with all electron analyzers, there are two possible modes of operation with this technique for obtaining cross sections.

1. The first mode is to set the analyzer to accept electrons of a fixed energy loss at some threshold and sweep the incident and analyzer energy to plot out an excitation function. Using this mode of operation with the DRPD technique has three potentially serious defects i.e., (a) "slipping out of tune" due to potential well end-effects at high energies, which produces an underestimate of the cross section.[4] (b) Degradiation of resolution at higher analysis energy, resulting in an additional underestimate of the cross section. (c) Because of (b) above, contributions from neighboring states may be included in the signal. To some degree, effects (a) and (b) are offset by effect (c), though by an unknown amount. There appear to be no criteria for producing unique excitation functions in this mode.

2. Alternatively, the analyzer is set to accept a fixed final energy of the scattered electrons for variable incident energy. This method produces energy loss spectra whose peak areas are proportional to the cross sections at a fixed energy above threshold. From a series of such spectra one can then measure peak areas and plot a cross section. Though more tedious than mode 1,mode 2 avoids all three defects, and we use this mode as opposed to earlier works.[3]

Our data for the He 2^3S and 2^1S excitation are shown in Figures 1 and 2 respectively, together with a comparison with two theories.[1,2] Our data, at present relative, are normalized to the first peak of the

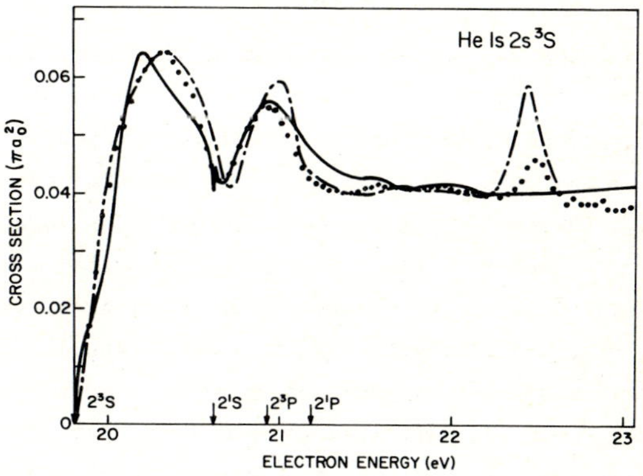

Fig. 1: Comparison of theoretical and experimental values of the He $1s2s\,^3S$ excitation cross section. Present experimental data (••••); theory (Ref. 2) (——); theory (Ref. 1) (—— • —— •).

He 2^3S theoretical cross section. Note, however, that our $^3S/^1S$ ratios are not independent, and as such very good agreement is achieved between experiment and both theories. Though the magnitudes of the theoretical cross[1,2] sections are somewhat larger than most earlier experimental cross sections, they are in good agreement with the recent experiments of Johnston and Burrow which measure, however, the sum of the ^3S and ^1S cross sections.

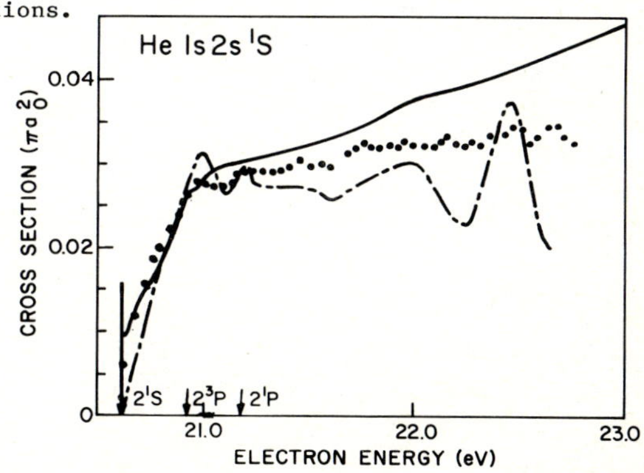

Fig. 2: Comparison of theoretical and experimental values of He $1s2s\ ^1S$ excitation cross section. Present experimental data (••••); theoretical curve of Berrington et al. (Ref. 2) (——); theoretical curve of Oberoi et al. (Ref. 1) (—— • —— •).

References

[1] R. S. Oberoi and R. K. Nesbet, Phys. Rev. A 8, 2969 (1973).

[2] K. A. Berrington, P. G. Burke, and A. L. Sinfailam, J. Phys. B: Atom. Molec. Phys. 8, 1459 (1975).

[3] H. H. Brongersma, F. W. E. Knoop, and C. Backx, Chem. Phys. Lett. 13, 16 (1972).

[4] F. W. E. Knoop, H. H. Brongersma, and A. J. H. Boerboom, Chem. Phys. Lett. 5, 450 (1970).

[5] D. Spence, Phys. Rev. A 12, 2353 (1975).

[6] A. R. Johnston and P. D. Burrow, submitted to J. Phys. B.

* Work performed under the auspices of USDOE.

† Dept. Phys., Chengdu Univ. of Sci. and Tech., China

LOW ENERGY ELECTRON SCATTERING FROM INTERSTELLAR GAS MOLECULES

V.M.Chhaya
Dept.of Physics,MNV Science College,Saurashtra University,Rajkot,India,360 005
and
H.S.Desai
Dept.of Physics,Faculty of Science,M.S.University, Vadodara,India,390 002

One of the most exciting aspect of the interstellar physics is the continued discovery of a wide range of interstellar molecular species.The study of these molecules may throw more light on the formation of stars and the cooling of interstellar gas. The initial formation of stars of low mass might lead to the production of large amounts of molecular species,which then allowed the formation of more massive stars. The knowledge of density of such molecules in a particular region of space may thow light on the formation of stars. It is known that when the density of the gas is low, rotationally and/or vibrationally exited molecule decays via emission of radiation. Thus if there exist free electpons, then their collisions with interstellar molecules might be an important mechanism to cool the gas. Thus the study of rotational,rotational -vibrational and vibrational excitations of such molecules may lead to better understanding of the cooling cooling of the interstellar gas,its densities and finally the formation of stars.

The polar molecules such as CN, CH, OH, CO, H_2O, NH_3 and H_2CO are found in interstellar gas. In order to calculate rotational excitation cross sections and mean fractional energy loss per collision due to low energy electron impact, a method simpler than the close coupling method;but more accurate than the First Born Approximation is used. In the present investigations Modified Born Eikonal Series method is employed. The interaction potential is electron-dipole potential having cut-off parameter super--imposed by incident electron energy dependent term. The results for CN molecules are first compared with the close coupling results of Itikawa[1]. Having found quite a good agreement in that comparision, the method is extended to remaining molecules.

References

1. Itikawa, Y. J.Phys. Soc. Japan, 27, 44 (1969).

ANALYTICAL CALCULATION OF MEAN EXCITATION ENERGIES OF HYDROGEN AND OTHER LIGHT ATOMS

S. Rosendorff

Department of Physics, Technion, Haifa, Israel

There are two sum rules which play a central role in the theory of stopping power and total cross section. They are determined solely by the properties of the atoms. In case of stopping power, the sum rule in question is the logarithmic mean over the excitation energies weighted according to the corresponding oscillator strengths f_n, and in case of total cross section it is the logarithmic mean of the excitation energies divided by the energies weighted as above. They are defined by ($\nu = 0$ for stopping power, and $\nu = -1$ for total cross section)

$$L(\nu) = \sum_n f_n(E_n-E_o)^\nu \ \ell n(E_n-E_o), \tag{1}$$

where

$$f_n = \frac{2m}{\hbar^2}(E_n-E_o)\,|z_{on}|^2 \ . \tag{1'}$$

The sum is over the complete set of states of the atom. The numerical values of these sum rules may not be well known and are often obtained by fitting the theory to experimental results, (except for hydrogen, where analytic expressions for f_n and E_n are available and the sum is performed numerically).

We have developed a method by which the above sums $L(\nu)$ can be calculated analytically, the only input being the ground-state wave function of the atom. The method is based on a generalization of an idea by Dalgarno and Lewis.[1] We find

$$L(-1) = \int_o^\infty \langle 0|zF(\vec{r},\lambda)|0\rangle \, d\lambda, \tag{2}$$

$$L(0) = \int_o^\infty \langle 0|\vec{\nabla}z\cdot\vec{\nabla}F(\vec{r},\lambda)|0\rangle \, d\lambda. \tag{2'}$$

Here the function $F(\vec{r},\lambda)$ is a solution of a certain linear second-order, inhomogeneous differential equation. The equation has been solved exactly, and F is known for every value of $\vec{r}$ and λ. The integral over λ is performed numerically. As F is defined by an infinite series, it is advantageous to express it by Padé approximants in order to obtain higher accuracy.

References

1. A. Dalgarno and J.T. Lewis, Proc. R. Soc. London A233, 70 (1956).

INELASTIC DIFFERENTIAL ELECTRON SCATTERING FROM METASTABLE HELIUM ATOMS

R. Müller-Fiedler, P. Schlemmer, K. Jung, H. Hotop, H. Ehrhardt

Fachbereich Physik der Universität Kaiserslautern, 6750 Kaiserslautern, W.-Germany

While electron scattering from helium atoms in the ground state is very well known, relatively little work has been done to study the scattering of electrons from excited helium atoms. Until now only transmission measurements of the total cross sections of excited helium atoms with electrons have been performed[1,2]. On the theoretical side Fon et al [3] and Oberoi and Nesbet[4] have determined the total cross sections for electron transitions between the n=2 states of helium. Flannery and McCann[5] have calculated in an Eikonal approximation the differential cross sections for the electron impact excitation of helium from 2^1S to 2^1P, to the various n=3 singlet states and for the analogous transitions in the triplet system, but the triplet results have not been published.

In a crossed beam experiment we have now measured for the first time angular dependences of the differential cross sections for the process $e^- + He(2^3S) \rightarrow e^- + He(2^3P)$ as well as for the transitions from the 2^3S to the 3^3S, 3^3P and 3^3D states at detection energies of 15 eV, 20 eV and 30 eV. The transitions to the n=4 triplet states have also been observed in the energy loss spectra, but due to the small energy separation the individual final states could not be resolved.

The apparatus used is a conventional electron spectrometer with a double energy selector in the detector which enables a good suppression of background scattering. The overal resolution of the spectrometer is about 200 meV, the primary electron current is about 5×10^{-8} A. The metastable atoms are produced in a differentially-pumped cold cathode dc discharge source, the density of helium atoms in the 2^3S state is about 6×10^7 cm^{-3} in the scattering region. By determining the target densities of helium in the 1^1S and 2^3S states and by comparing the elastic and inelastic count rates we could obtain the absolute values of the cross sections.

Figure 1 shows the measured differential cross sections for the transitions from 2^3S to 2^3P and to the n=3 triplet states as a function of the scattering angle at a detection energy of E'=20 eV. Similar to other optically allowed transitions the cross section for the $2^3S \rightarrow 2^3P$ excitation falls off very steeply with increasing scattering angle. Between 15 eV and 30 eV the cross sections decrease with increasing energy for all angles studied. For the transitions to the n=3 states it is striking that the optically forbidden 3^3D excitation dominates the optically allowed 3^3P excitation for all three energies. This effect is caused by the abnormally small line strength for the $2^3S \rightarrow 3^3P$ transition in helium. The structure in the angular dependence for the 3^3P excitation is due to the strong 3^3D-3^3P coupling. At the conference we shall present a comparison between our measurements and the calculations of Flannery and McCann.

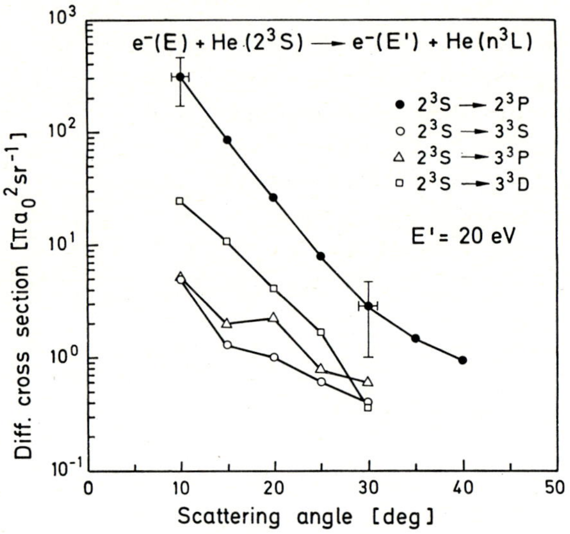

Fig. 1: Differential cross sections for the transitions from 2^3S to 2^3P, 3^3S, 3^3P, 3^3D at the detection energy E'=20 eV.

References

1. W.G. Wilson and W.I. Williams, J. Phys. B9, 423 (1976)

2. R.H. Neynaber, S.M. Trujillo, L.L. Marino, E.W. Rothe, Proc. 3rd ICPEAC, 1089 (1964)

3. W.C. Fon, K.A. Berrington, P.G. Burke, A.E. Kingston, J. Phys. B 14, 2921 (1981)

4. R.S. Oberoi and R.K. Nesbet, Phys. Rev. A8, 2969 (1973)

5. M.R. Flannery and K.J. McCann, Phys. Rev. A12, 846 (1975)

ELECTRON-IMPACT EXCITATION FROM 2 ^{1}S AND 2 ^{3}S STATES OF HELIUM

David C. Cartwright
University of California, Los Alamos National Laboratory
Los Alamos, New Mexico 87545 (USA)

George Csanak
154 Cold Spring Road, #67, Stamford, Connecticut 06905

Fernando J. da Paixão
Instituto de Fisica, Universidade Estadual de Campinas
Campinas, S. P. Brazil

A Many-Body formalism for electron impact excitation from metastable targets (MBT-MT) was recently presented by Paixão, Csanak and Cartwright.[1] In its first order approximation (FOMBT-MT), the following formula is obtained for the T-matrix

$$T_{nq,mp}=\int dr_2 dr_3 \left[f_q^{m(-)*}(r_2)f_p^{(+)}(r_2)+f_q^{(-)*}(r_2)\ f_p^{m(+)}(r_2)\right]$$

$$*V(\bar{r}_2-\bar{r}_3)\ \tilde{X}_n(r_3,r_3)\ -$$

$$-\int dr_2 dr_3 \left[f_q^{m(-)*}(r_3)\ f_p^{(+)}(r_2)+f_q^{(-)*}(r_3)\ f_p^{m(+)}(r_2)\right]*$$

$$*V(\bar{r}_2-\bar{r}_3)\ \tilde{X}_n(r_2,r_3)\ + \qquad (1)$$

$$+\int dr_2 dr_3 f_q^{(-)*}(r_2)f_p^{(+)}(r_2)V(\bar{r}_2-\bar{r}_3)X_n^m(r_3,r_3)-$$

$$-\int dr_2 dr_3 f_q^{(-)*}(r_3)f_p^{(+)}(r_2)V(\bar{r}_2-\bar{r}_3)X_n^m(r_2,r_3)\ ,$$

where the notation used here is that discussed in detail in Ref. 2, and also used in Ref. 1. The quantities $f_p^{(+)}(r)$ and $f_q^{(-)}(r)$ are the Hartree-Fock (static-exchange) continuum orbitals with outgoing-wave and incoming-wave boundary conditions, respectively. In order to calculate $T_{nq,mp}$ as given by Eq. (1), the $f_p^{m(+)}(r)$ and $f_q^{(m(-)}(r)$ functions need to be determined. For the case of helium initially in a metastable S-state (i.e. 2 ^{3}S or 2 ^{1}S) these functions can be expanded in the following form

$$f_p^{m(+)}(r) = \eta_{m_s}(\sigma)\sqrt{\frac{8}{p}}\ \frac{\Pi^{3/2}}{r}\sum_{\ell,m}(-i)^\ell \frac{e^{i\delta_\ell(p)}}{(2\ell+1)^{1/2}}F_{p\ell}(r)*$$

$$*Y_{\ell m}(\hat{r})Y^*_{\ell m}(\hat{p})\ , \qquad (2)$$

$$f_q^{m(-)*}(r) = \eta_{m_s'}(\sigma)\sqrt{\frac{8}{q}}\ \frac{\Pi^{3/2}}{r}\sum_{\ell,m}(-i)^\ell \frac{e^{i\delta_\ell(q)}}{(2\ell+1)^{1/2}}G_{q\ell}(r)*$$

$$*Y_{\ell m}(\hat{r})Y^*_{\ell m}(\hat{q})\ . \qquad (3)$$

The radial functions $F_{k\ell}$ and $G_{k\ell}$ are "channel functions" that can be shown to satisfy the following inhomogeneous differential equations

$$\left(\frac{1}{2}\frac{d^2}{dr^2}+\frac{2}{r}-\frac{\ell(\ell+1)}{2r^2}+\varepsilon_k+\omega_m\right)F_{k\ell}(r)\ -$$

$$-\frac{2}{r}Y_o(1s,1s/r)\ F_{k\ell}(r)+\frac{1}{r}Y_o(F_{k\ell},1s/r)P_{1s}(r) =$$

$$= 2\ \frac{\delta_{S^m,0}}{r}\ P_{k\ell}(r)Y_0(ms,1s/r)\ - \qquad (4)$$

$$-\frac{1}{r}\ Y_\ell(k\ell,1s/r)\ \frac{P_{m(r)}}{2\ell+1}\ ,$$

$$\left(\frac{1}{2}\frac{d}{dr^2}+\frac{2}{r}-\frac{\ell(\ell+1)}{2r^2}+\varepsilon_k-\omega_m\right)G_{k\ell}(r)\ -$$

$$-\frac{2}{r}Y_0(1s,1s/r)\ G_{k\ell}(r)+\frac{1}{r}\ \frac{Y_\ell(G_{k\ell},1s/r)}{2\ell+1}\ P_{1s}(r) =$$

$$= \frac{2\delta_{S^m,0}}{r}\ P_{k\ell}(r)\ Y_0(ms,1s/r)\ - \qquad (5)$$

$$-\frac{1}{r}\ Y_\ell(k\ell,1s/r)\ \frac{P_{m(r)}}{2\ell+1}$$

where

$$Y_k(m\ell,m'\ell'/r) \equiv \frac{1}{r^k}\int_0^r dr'\ P_{m\ell}(r')\ P_{m'\ell'}(r')r'^k\ +$$

$$+\ r^{k+1}\int_r^\infty dr'\ \frac{1}{r'^{k+1}}\ P_{m\ell}(r')\ P_{m'\ell'}(r')\ . \qquad (6)$$

Differential and integral cross sections will be presented for excitation from both the 2 ^{1}S and 2 ^{3}S states to higher-lying n ^{3,1}P and n ^{3,1}S states.

References

1. F. J. Da Paixão, G. Csanak and D. C. Cartwright, abstracts XII ICPEAC, Gatlinburg, TN, edited by S. Datz (1981) p. 178.

2. N. T. Padial, G. D. Meneses, F. J. da Paixão, G. Csanak and D. C. Cartwright, Phys. Rev. 23, 2194 (1981).

2^3S EXCITATION OF HELIUMLIKE IONS BY ELECTRON IMPACT

C.S. Singh and D.K. Rai

Department of Physics, Banaras Hindu University, Varanasi-221005, INDIA

Electron impact excitation of 1^1S-2^3S transition in heliumlike ions Li II, C V, O VII and Si XIII is studied in an energy range from near threshold to five times threshold energy using a distorted wave (DW) model in which the prior form of T-matrix

$$T = \langle \Psi_f^- | V_i - \overline{V} | \Psi_i \rangle \qquad (1)$$

is used. Ψ_f^- is the properly symmetrized wavefunction in the final channel, V_i is the initial interaction potential and $\overline{V}$ is taken as the static potential in the final state of the target ion. Mathematical details of the model are given elsewhere[1]. The radial solution to the final channel wavefunction is distorted by the potential $\overline{V}$ plus the antisymmetrization potential (also called Hartree-Fock exchange potential) and in the initial channel it is distorted by $\overline{V}$ only.

The application of the above method to the 1^1S-2^3S excitation of helium atom[1,2] by electron impact has already given very good results which prompted us to extend these studies to other systems of helium-isoelectronic sequence whose collision strengths are of great importance[3,4] to its various applications in fusion research, ion lasers, astrophysics and atmospheric physics. Experimental difficulties[5] in the determination of electron impact excitation cross sections for positive ions (especially multiply ionized species) also makes a correct theoretical description of importance.

We have calculated the total as well as the differential cross sections and compared it with other theoretical results[6,7,8].

The wavefunctions used in these studies are identical to those in Bhatia and Temkin[6].

References

1. C.S.Singh and D.K.Rai (To be communicated)
2. C.S.Singh, R.Srivastava and D.K.Rai (Communicated to J.Phys.B)
3. R.J.W.Henry, Phys.Rept. 68, 1 (1981)
4. W.D.Robb, "Atomic and Molecular Processes in Controlled Thermonuclear Fusion", ed. M.R.C.McDowell and A.M.Ferendeci (Plenum Press, New York 1980) p. 245
5. K.T.Dolder and B.Peart, Rept.Prog.Phys. 39, 693 (1976)
6. A.K.Bhatia and A.Temkin, J.Phys.B 10, 2893 (1977)
7. J.A.Tully, J.Phys.B 11, 2923 (1978)
8. W.L. Van Wyngaarden, K.Bhadra and R.J.W. Henry, Phys.Rev.A 20, 1409 (1979)

SCATTERING OF SLOW ELECTRONS BY LI ATOMS

D.L. Moores

Dept of Physics & Astronomy, University College London, Gower St, London WC1E 6BT, UK

Cross sections for elastic and inelastic scattering of electrons in the energy range 0-10 ev by neutral Li atoms have been obtained from a 5-state (2s - 2p - 3s - 3p - 3d) close coupling calculation with correlation, using the published program IMPACT[1]. Wave functions for the target Li atom were calculated by the method described by Mendoza[2]. Provisional results obtained for elastic scattering from the 2s state and excitation of 2p, 3s, 3p and 3d states are given in Table 1, together with the asymmetry factor

$$A = (Q(S=0) - Q(S=1)) / Q \qquad (1)$$

for 2s - 2p excitation. At the meeting results of calculations of cross sections for transitions between excited states : polarisation of line radiation : differential cross sections and elastic and inelastic scattering amplitudes will be presented and compared with experimental data available.

(E(ryd)	2s-2s	2s-2p	2s-3s	2s-3p	2s-3d	A(2s-2p)
0.15	133.8	13.7				0.873
0.2	102.4	31.5				0.231
0.265	75.7	39.0				0.136
0.3	65.7	41.0	2.68			0.088
0.3675	53.6	41.3	2.49	2.04	4.47	0.049
0.45	43.6	42.4	2.74	2.59	5.77	0.020
0.5	38.3	43.5	2.67	2.55	5.91	0.019
0.55	35.2	44.6	2.39	2.52	6.07	0.023
0.6	31.9	45.3	2.41	2.29	5.57	0.028
0.65	28.7	45.8	2.30	2.17	5.30	0.032
0.7	26.5	45.8	2.43	2.04	5.17	0.031
0.75	24.9	45.9	2.11	2.01	4.76	0.035

Table 1 : Cross sections (in πa_o^2) as a function of incident electron energy in rydbergs. In the final column we give the asymmetry factor for 2s-2p excitation defined by equation (1).

References

1. M.A. Crees, P.M.H. Wilson and M.J. Seaton, CPC 15, 23-83 (1978).
2. C. Mendoza, J.Phys. B 14, 397-409 (1981).

STEPWISE ELECTRON/LASER EXCITATION STUDIES OF ATOMIC COLLISIONS

C.J. Webb, W.R. MacGillivray & M.C. Standage

School of Science, Griffith University, Nathan, Qld, 4111, Australia

The application of lasers to the field of electron-atom collision studies has provided a range of new techniques for investigating collision processes. In this paper we present results from an experimental and theoretical study of one of these techniques, the stepwise excitation of atoms which are first excited by electron impact followed by a second excitation step using C.W. single-mode laser radiation. Measurements of the intensity and polarisation of fluorescence emitted following stepwise excitation allow the excitation cross-section and line polarization of the electron impact excited transition to be determined.

An interesting feature of this technique is that the high spectral resolution of the laser permits the fine and hyperfine structure of the target atoms to be resolved in the laser excitation step. This has been used by us to study the electron-impact excitation of the 6^1S_0 - 6^1P_1 transition of mercury for both even and odd isotopes of mercury [1,2]. Besides providing excitation cross-section and line polarisation data for specific isotopes, such measurements also provide a means of testing the validity of the Percival-Seaton hypothesis concerning the role nuclear spin plays in collision processes. In the course of this work, the first evidence of breakdown in this hypothesis was obtained.

We have also used stepwise excitation techniques to investigate the electron-impact excitation of metastable levels. Following excitation of the metastable level by electron impact, laser radiation is used to excite an optically allowed transition, and fluorescence emitted from the stepwise excited atoms is analysed for intensity and polarisation. We have investigated the electron excitation of the 6^1P_2 metastable state of mercury using this technique. Because of the high resolution of the laser excitation step it was possible to obtain data for selected isotopes. The excitation cross-section as a function of incident electron energy was obtained from fluorescent intensity measurements normalized to the electron beam current. The relative total partial cross-sections of the metastable sublevels were obtained by making polarisation measurements on the fluorescence as a function of the polarisation of the laser radiation.

In summary, the stepwise excitation techniques presented in this paper provide new methods for:

1) the study of electron-impact excited VUV lines
2) the investigation of isotope effects in electron-atom collisions
3) the measurement of excitation cross-section and line polarisation data for selected isotopic species
4) the measurement of metastable partial and total cross-sections.

The talk will also include a discussion of the extension of these techniques to coincidence studies between inelastically scattered electrons and photons emitted following stepwise excitation.

References

1) C.W. McLucas, W.R. MacGillivray and M.C. Standage 1982 Phys. Rev. Lett. 48, 88-92.
2) C.W. McLucas, W.R. MacGillivray and M.C. Standage 1982 J.Phys.B. 15, 1883-98.

ELECTRON IMPACT PHOTOEMISSION CROSS SECTIONS FOR THE VUV LIGHT STANDARD PROJECT

Armon McPherson, N. Rouze, W. B. Westerveld† and J. S. Risley

Department of Physics, North Carolina State University, Raleigh, NC 27650 USA

A primary absolute light standard for the vacuum ultraviolet region based upon electron impact photoemission cross sections is under development. By measuring the emission cross sections in the region of 30 to 150 nm a standard intensity source can be built for which the photon fluxes can be accurately determined. Such an intensity standard has practical applications in the calibration of vuv spectrometers for space research, fusion diagnostics and laboratory measurements.

Figure 1 shows the observed spectrum between 50 and 110 nm for 100 eV electrons on argon using a 0.2m Seya-Namioka spectrometer at 0.25 nm resolution. The axis of the electron beam was perpendicular to the grating grooves. Polarization effects in the line intensity can be checked by tilting the electron beam axis. The rich spectrum of calibration lines shown in Fig. 1 can be augmented and expanded in the 30 to 150 nm region by using emission lines from other atomic or molecular targets, e.g. He, Ne, Kr, N_2 and O_2.

Accurate measurement of the photoemission cross sections requires the absolute calibration of the sensitivity of our spectrometer-detector system. The Seya-Namioka spectrometer has a Jobin-Yvon type IV holographic grating coated with MgF_2. The detector is an EMI#9642/4B venetian blind photomultiplier with a BeCu cathode. Synchrotron radiation from the U.S. National Bureau of Standards SURF II is used to determine the detection efficiency as a function of wavelength. The spectrometer and detector system are mounted on a large manipulator that allows rotation about a horizontal and a vertical axis to determine the effective solid angle of the system as viewed from the axis of the electron beam.

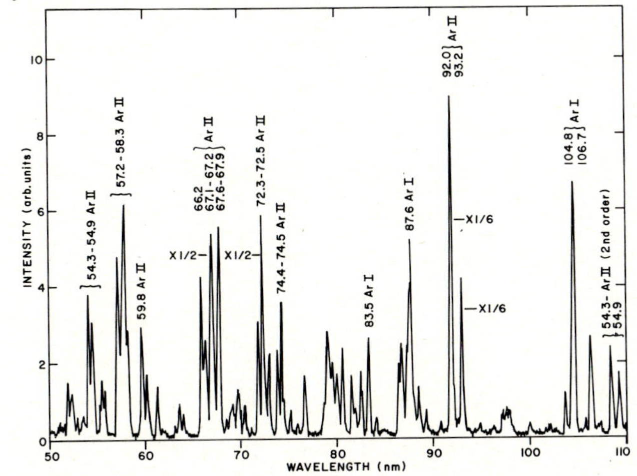

Fig. 1. Argon spectrum for 100 eV electrons.

Figure 2 shows the efficiency of the spectrometer-detector system as a function of wavelength using 280 MeV electrons. The notation ‖ refers to the grooves of the grating oriented parallel to the orbital plane of the electrons in SURF II, ⊥ refers to the grooves oriented perpendicular to the orbital plane. These two efficiency curves were taken in the middle of the grating. Other scans at different locations exhibit similar structure but the absolute efficiency is position dependent. The efficiency curves in Fig. 2 have not been corrected for 2nd order reflections. Accuracies to within a few percent are expected. The significant structure in the efficiency curves appears to be caused by a combination of effects due to the reflectance of the MgF_2 overcoating, the photoelectric yield of the BeCu cathode, and the effective blaze angle of the holographic grating.

This work was supported in part by the Aeronomy program of the U.S. National Science Foundation.

1. W. B. Westerveld et al. At. Data Nucl. Data Tables 28, 21-105 (1983).

†Present address: Department of Physics, University of Windsor, Windsor, ON, Canada

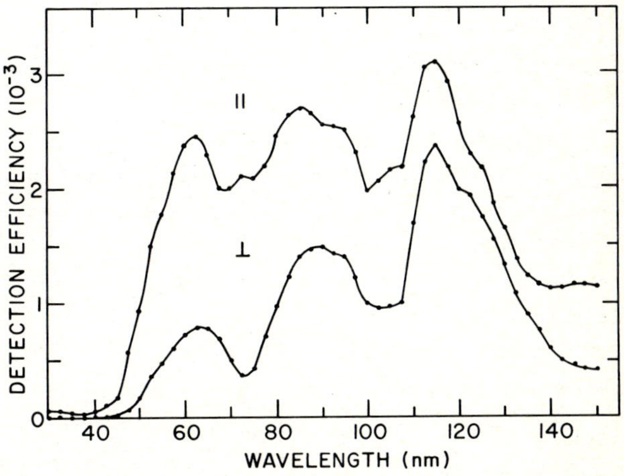

Fig. 2. Detection efficiency.

MEASUREMENT OF COMPLEX SCATTERING AMPLITUDES FOR ELECTRON IMPACT EXCITATION OF THE 3^1P AND 3^1D STATES OF HELIUM

H.B. van Linden van den Heuvell, E.M. van Gasteren, J. van Eck and H.G.M. Heideman

Fysisch Laboratorium, Rijksuniversiteit Utrecht, the Netherlands

In a scattered-electron cascaded-photon coincidence experiment[1] we studied the electron impact excitation of the 3^1P and 3^1D states of helium for incident electron energies between 28 and 46 eV and for a scattering angle of 35°. This was done by measuring the angular distribution of the $3^1P \rightarrow 1^1S$ and $2^1P \rightarrow 1^1S$ photons (the latter resulting from the $3^1D(3^1S) \rightarrow 2^1P \rightarrow 1^1S$ cascade) in coincidence with the n = 3 scattered electrons, both in the scattering plane and in the plane perpendicular to the scattering plane. Due to the differences in lifetime of the excited states the various contributions to the coincidence peak can be separated. Because the angular distribution of the $3^1D \rightarrow 2^1P$ photons is identical to that of the $2^1P \rightarrow 1^1S$ cascade photons direct information on the 3^1D excitation can be obtained in addition to that on the 3^1P excitation.

Taking the quantisation axis perpendicular to the scattering plane[2] we define the following parameters for the 3^1P excitation:

$\mu = (|a_1|^2 - |a_{-1}|^2)/\sigma$, $\eta = \arg(a_{-1}/a_1)$ and

$\sigma = |a_1|^2 + |a_{-1}|^2$

Similarly for the 3^1D excitation:

$\xi = |a_0|^2/\sigma$, $\zeta = \arg[(a_0a_2^* + a_{-2}a_0^*)/a_0]$,

$\nu = |a_0a_2^* + a_{-2}a_0^*|/\sigma$, $\omega = (|a_2|^2 - |a_{-2}|^2)/\sigma$ and

$\sigma = |a_2|^2 + |a_0|^2 + |a_{-2}|^2$,

Here the a_m are the complex excitation amplitudes for the various magnetic substates and σ is the differential cross section. For symmetry reasons $a_0 = 0$ for the 3^1P excitation and $a_1 = a_{-1} = 0$ for the 3^1D excitation. The reason why we take the quantisation axis perpendicular to the scattering plane, rather than along the incident beam is the fact that the expressions for the angular distributions of the emitted photons take on a much simpler form and display a much more transparant relation between the observed radiation pattern and the anisotropy parameters of the excited atom[3].

By measuring the angular distributions of the emitted photons the parameters μ and η for 3^1P excitation and ξ, ζ and ν for 3^1D excitation can directly be determined. The parameter ω for 3^1D excitation can only be obtained when a circular polarisation measurement is performed.

Our results are summarized in the tables 1 and 2. The contribution of the 3^1S excitation to the coincidence signal appeared to be negligible. A discussion of the results and a comparison with other data will be presented at the conference.

Table 1. Measured orientation and alignment parameters for the 3^1P excitation in helium at various energies of the incident electrons and for a fixed scattering angle of 35^0

E (eV)	λ	$\lvert\chi\rvert$ (rad)	$\lvert\mu\rvert$ ($\hbar$ Js)	η (rad)
28.5	0.798 ± 0.019	$0.110 \pm^{0.202}_{0.110}$	0.161 ± 0.104	2.211 ± 0.014
31.5	0.651 ± 0.021	0.552 ± 0.077	0.454 ± 0.073	1.920 ± 0.037
34.6	0.572 ± 0.031	0.489 ± 0.109	0.425 ± 0.091	1.741 ± 0.061
34.6†	(0.568 ± 0.017)	(0.745 ± 0.037)	(0.670 ± 0.025)	(1.759 ± 0.041)
45.6	0.455 ± 0.029	0.998 ± 0.047	0.863 ± 0.023	1.406 ± 0.112
45.6†	(0.460 ± 0.019)	(1.06 ± 0.030)	(0.871 ± 0.013)	(1.412 ± 0.013)

† Values between parathesis are derived under the assumption that only the 3^1P excitation gives a contribution to the observed coincidence peaks.

Table 2. Measured anisotropy parameters for the 3^1D excitation in helium at various energies of the incident electrons and for a fixed scattering angle of 35^0. The last column gives the experimentally derived cross section ratios $\sigma_{3^1D}/\sigma_{3^1P}$.

E (eV)	ξ	ν	ζ (rad)	$\sigma_{3^1D}/\sigma_{3^1P}$
28.5	0.57 ± 0.20	1.00 ± 0.23	2.18 ± 0.20	0.46 ± 0.09
31.5	0.35 ± 0.20	0.42 ± 0.13	2.49 ± 0.15	0.48 ± 0.07
34.6	$0.00 \pm^{0.53}_{0.00}$	0.81 ± 0.45	2.43 ± 0.37	0.32 ± 0.06
45.6				$0.17 \pm^{0.04}_{0.08}$

References

1. H.B. van Linden van den Heuvell, G. Nienhuis, J. van Eck and H.G.M. Heideman, J. Phys. B: At. Mol. Phys. 14, 2667 (1981)
2. H.W. Hermann and I.V. Hertel, J. Phys. B: At. Mol. Phys. 13, 4285 (1980)
3. H.B. van Linden van den Heuvell, Ph.D. Thesis, Univ. of Utrecht, 1982

THE ORBITAL ANGULAR MOMENTUM TRANSFER IN THE EXCITATION OF THE 2^1P STATE OF HELIUM BY ELECTRONS

H.B. van Linden van den Heuvell, M.A.M. de Jong, J. van Eck and H.G.M. Heideman

Fysisch Laboratorium, Rijksuniversiteit Utrecht, the Netherlands

In an electron-photon coincidence experiment one may determine the absolute value $|\langle\vec{L}\rangle|$ of the expectation value of the orbital angular momentum transfer to the atom. In the case of 2^1P excitation in helium at 80 eV it appears[1,2] that $|\langle\vec{L}\rangle|$ tends to zero at a certain electron scattering angle ($\approx 60^o$). This may imply that $\langle\vec{L}\rangle$ changes sign near that scattering angle. From the analysis of Kohmoto and Fano[3] it follows that $\langle\vec{L}\rangle$ changes sign if the interaction between the electron and the atom changes sign. Thus the experimental results at 80 eV may point to a sign reversal of the interaction at a scattering angle of 60°. This may be explained in a semi-classical picture[4]. At small scattering angles the long-range attractive potential due to the atomic polarisability is dominant and hence $\langle\vec{L}\rangle$ should be positive at the smaller angles. (The latter conclusion is based on a classical argument and does not follow from the quantum mechanical analysis of Kohmoto and Fano[3].) As the scattering angle increases the repulsive potential of the atomic electrons becomes more and more important, balances the attractive polarisation potential at some intermediate angle, and finally dominates at the larger scattering angles. Thus at some angle the interaction will change sign and, according to Kohmoto and Fano, so will the angular momentum transfer.

We have now made a systematic study (for 2^1P excitation) of the behaviour of $|\langle\vec{L}\rangle|$ as a function of the scattering angle, more specifically directed towards possible nodes in these curves in order to reveal sign reversals of $\langle\vec{L}\rangle$. We have done this for various energies below 80 eV. Unless very rapid changes occur in $|\langle\vec{L}\rangle|$ as a function of the electron scattering angle our results, together with those of Slevin et al[1] and McAdams et al[5], indicate that there is an energy range where $\langle\vec{L}\rangle$ keeps the same (probably positive) sign over the whole angular range. The upper limit of this energy range is likely to be around 60 eV. We conclude therefore that in this energy range the interaction is dominated by the attractive potential over the full angular range from 0 to π. At very low energies near the excitation threshold there are again strong indications for nodes (or at least minima) in $\langle\vec{L}\rangle$ as a function of the scattering angle. In view of the results between 40 and 50 eV it is very unlikely that these nodes result from a ballance between attractive and repulsive forces. They are most probably caused by typical quantum mechanical interference effects due to the fact that only very few partial waves contribute at these low energies.

References

1. J. Slevin, H.Q. Porter, M. Eminyan, A. Defrance and G. Vasilev, J. Phys. B: At. Mol. Phys. 13 3009 (1980)
2. M.T. Hollywood, A. Crowe and J.F. Williams, J. Phys. B: At. Mol. Phys. 12, 819 (1979)
3. M. Kohmoto and U. Fano, J. Phys. B: At. Mol. Phys. 14, L447 (1981)
4. M.C. Steph and D.E. Golden, Phys. Rev. A 21, 759 (1980)
5. R. McAdams, M.T. Hollywood, A. Crowe and J.F. Williams, J. Phys. B: At. Mol. Phys. 13, 3691 (1980)

SCATTERING PARAMETERS FOR THE ELECTRON IMPACT EXCITATION OF THE 3^2P STATE OF SODIUM

J.E. Furst, J.L. Riley, S.J. Buckman and P.J.O. Teubner

Institute for Atomic Studies, Flinders University of South Australia, Bedford Park, South Australia 5042.

The relationship between the scattering parameters λ and χ and the components of the Stokes vector of the radiation emitted in the decay of an excited atomic state can be deduced from the work of Fano and Macek[1]. At incident energies where exchange may be neglected these relationships are for the 3^2P state of sodium,

$$P_1 = 0.1411\ (2\lambda-1)$$

$$P_2 = -0.2822\ [(\lambda(1-\lambda)]^{\frac{1}{2}} \cos\chi$$

$$P_3 = -1.117\ [(\lambda(1-\lambda)]^{\frac{1}{2}} \sin\chi$$

Thus the measurement of each component P_n determines the value of λ and the absolute value of χ.

(e,e'γ) experiments have been performed in which the polarisation state of the emitted photon has been analysed before the photon was detected in coincidence with the inelastically scattered electron; the radiation was viewed normal to the scattering plane. Values of λ and χ have been deduced for scattering angles of 3°, 5° and 10° at incident energies of 100, 54.4 and 30eV. These data will be compared with theoretical predictions from a distorted wave polarised orbital calculation[2], a four state close coupling calculation[3], a coupled channels optical model calculation[4] and with the Born approximation.

Data will also be presented at an incident of 22.1eV and a scattering angle of 5°. These data will be compared with those from the time reversed experiment of Hermann et al[5].

References

1. U. Fano and J. Macek, Rev. Mod. Phys. 45, 553 (1973)
2. J.V. Kennedy, V.P. Myerscough and M.R.C. McDowell, J. Phys. B. 10 3759 (1978)
3. D.L. Moores and D.W. Norcross, J. Phys. B. 5, 1482 (1972)
4. I.E. McCarthy and A. Stelbovics to be published
5. H.W. Hermann, I.V. Hertel, W. Reiland, A. Stamatovic and W. Stoll, J. Phys. B. 10, 251, (1977)

STOKES PARAMETER ANALYSIS OF ELECTRON PHOTON COINCIDENCE EXPERIMENTS IN SODIUM

P.J.O. Teubner, J.E. Furst and J.L. Riley

Institute for Atomic Studies, The Flinders University of South Australia, Bedford Park, South Australia 5042.

Electron photon coincidence experiments have been performed on the decay of the 3^2P state of sodium which was excited by 100eV electrons. In these experiments the linear and circular components of the emitted radiation were measured. The three components P_1, P_2 and P_3 of the Stokes vector have been obtained at electron scattering angles of 3°, 5° and 10°. The total polarisation $|P|$ of the radiation is given by

$$|P| = \sqrt{P_1^2 + P_2^2 + P_3^2}$$

In an earlier experiment on the 3^1P state in helium, Standage and Kleinpoppen[1] have demonstrated that $|P| = 1$ independent of the electron scattering angle. This feature is consistent with the fact that the state is coherently excited. The 3^2P state in sodium on the other hand possesses significant fine and hyperfine structure which depolarises the decay radiation such that $|P| < 1$. Kempter and coworkers[2,3] have observed this feature in the collision of lithium and potassium with inert gases and have introduced the concept of reduced degree of polarisation[3] which assists in the description of the excitation process.

We find that, for the 3^2P state of sodium

$$0.14 \leqslant |P| \leqslant 0.56 \qquad (1)$$

However the presence of zero field quantum beats in the decay radiation from the 3^2P state excited by 100eV electrons clearly demonstrates the coherent preparation of the excited state. Thus any departure of $|P|$ from $|P| = 1$ in this case can be attributed to the influence of the hyperfine structure on the decay radiation.

Values of P_1, P_2 and P_3 will be presented for three electron scattering angles.

The measured values of $|P|$ are

Scattering Angle	$\|P\|$
3°	0.157 ± 0.015
5°	0.167 ± 0.017
10°	0.133 ± 0.026

References

1. M.C. Standage and H. Kleinpoppen, Phys. Rev. Letts. 36, 577 (1976)
2. L. Zehnle, E. Clemens, P.J. Martin, W. Schäuble and V. Kempter, J. Phys. B. 11, 2865 (1978)
3. B. Menner, Th. Hall, L. Zehnle and V. Kempter, J. Phys. B. 14, 3693 (1981)
4. P.J.O. Teubner, J.E. Furst, M.C. Tonkin and S.J. Buckman, Phys. Rev. Letts, 46, 1569 (1981)

ELECTRON-PHOTON ANGULAR CORRELATION WITH SPIN-ORBIT INTERACTION IN THE EXCITATION OF Kr AND Xe

H. Nishimura, A. Danjo, T. Koike, K. Kani, H. Sugahara and A. Takahashi

Department of Physics, Niigata University, Ikarashi, Niigata 950-21 Japan

In recent years, electron-photon angular correlation technique has been applied for the study of the radiative decay of an inelastically excited atom by electron impact. The target parameters like orientation vector and alignment tensor of the excited atom are derived from these experiments.[1] However, most of these experiments have been carried out for atoms that are well described by an LS-coupling scheme but a few works were reported for heavy atoms where a spin-orbit coupling effect can not be neglected.[2-7]

Recently, Blum et al.[8] and da Paixão et al.[9] introduced new parameters to describe the electron-photon coincidence rate for atoms where spin-orbit coupling effect exists. The new parameters for J=1 state are

$$\lambda = \sigma_0/(\sigma_0 + 2\sigma_1)$$

$$\cos\Delta = |\langle a(0)a(1)\rangle|/(\sigma_0\sigma_1)^{1/2}$$

$$\cos\bar{\chi} = \mathrm{Re}\langle a(0)a(1)\rangle/|\langle a(0)a(1)\rangle|$$

$$\cos\varepsilon = -\langle a(-1)a(1)\rangle/\sigma_1$$

where $a(M_J)$ and σ_{M_J} are the scattering amplitude and the excitation cross section for the M_J magnetic sublevel respectively. Following the theory of Blum et al. and da Paixão et al, the angular correlation function is expressed for the J=1 excitation from J=0 ground state with new four parameters;

$$I = \frac{CS}{4}(1+\lambda+(1-3\lambda)\cos^2\theta+2[\lambda(1-\lambda)]^{1/2}\cos\bar{\chi}\cos\Delta\sin 2\theta\cos\phi + (\lambda-1)\cos\varepsilon\sin 2\theta\cos 2\phi) \quad (1)$$

where θ and ϕ are the polar and azimuthal angle of an emitted photon. As pointed out by da Paixão et al, we can determine three parameters, λ, ε and $\cos\bar{\chi}\cos\Delta \equiv \cos\chi$ combining independent angular correlation experiments for different photon azimuth angles $\phi = 3\pi/4$ and $\phi = \pi$.

In this experiment, the angular correlation parameters were determined for Krypton $5s(1\frac{1}{2})$ and $5s'(0\frac{1}{2})$ and Xenon $6s(1\frac{1}{2})$ excited states at the incident electron energy from 40 to 80 eV and the electron scattering angle from 10° to 30°. The experimental angular correlations which give common λ, χ and ε for $\phi = 3\pi/4$ and $\phi = \pi$ were obtained by least square fitting of the experimental data to the equation (1). Correlation parameters λ, χ and ε for Kr at the incident electron energy of 60 eV are shown in Fig. 1. For the excitation of Xe $6s(1\frac{1}{2})$, $\lambda = 0.18 \pm 0.03$, $|\chi| = 1.24 \pm 0.03^*$ and $|\varepsilon| = 1.48 \pm 0.1^*$ were obtained at the incident electron energy of 50 eV and the scattering angle of 15°. The detailed experimental results will be presented.

* rad.

References

1. K. Blum and H. Kleinpoppen, Phys. Rept. 52, 203 (1977).
2. H. Arriola, P.T.O. Teubner, A.Ugbabe and E. Weigold, J. Phys. B: Atom. Molec. Phys. 8, 1275 (1975).
3. I.C. Malcolm and J.W. McConkey, J. Phys. B: Atom. Molec. Phys. 12, 511 (1979).
4. A. Pochat, F. Gelebart and J. Peresse, J. Phys. B: Atom. Molec. Phys. 13, L79 (1980).
5. A.A. Zaidi, S.M. Khalid, J. McGregor and H. Kleinpoppen, J. Phys. B: Atom. Molec. Phys. 14 L503 (1981).
6. J. McGregor, D. Hills, R. Hippler, N.A. Malid, J.F. Williams, A.A. Zaidi and H. Kleinpoppen, J. Phys. B: Atom. Molec. Phys. 15, L411 (1982).
7. H. Nishimura, A. Danjo, T. Koike, K. Kani, H. Sugahara, A. Takahashi, US. Japan Seminar on Electron-Molecule Collisions and Photoionization Process. 163 (1982).
8. K. Blum, F.J. da Paixão and Gy. Csanak, J. Phys. B: Atom. Molec. Phys. 13 L257 (1980).
9. F.J. da Paixão, N.T. Padial, Gy. Csanak, Phys. Rev. Letters. 45, 1164 (1980).

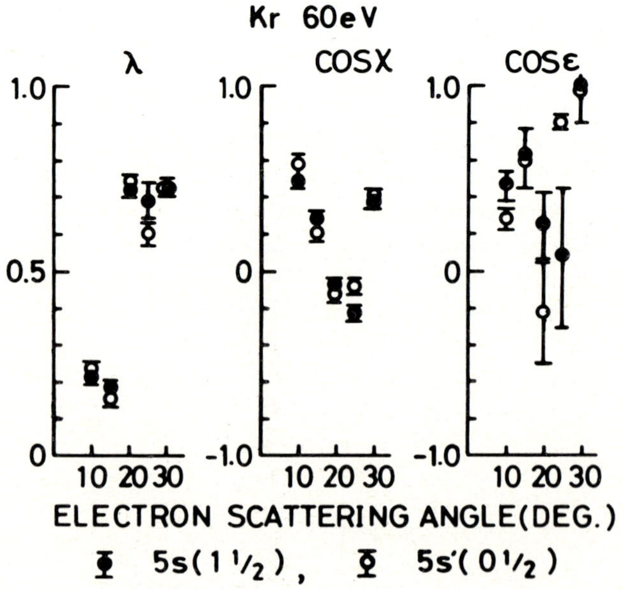

Fig. 1 Correlation parameters for the $5s(1\frac{1}{2})$ and $5s'(0\frac{1}{2})$ states of Kr at 60 eV electron impact energy.

ORIENTATION AND ALIGNMENT FOR 1s→2p EXCITATION OF HYDROGEN BY ELECTRONS AND DIFFERENTIAL CROSS SECTIONS FOR 1s-2p AND 1s→2p(m=0) AND 2p(m=±1) EXCITATION

J.F. Williams

Physics Department, University of Western Australia, Perth, Australia

The experimental method[1] involves measuring the angular correlations between the radiated 121.6 nm photons and the scattered 10.2 eV energy loss electrons for a given incident electron energy. The techniques include crossed electron-atom beams, electron energy loss spectroscopy and time coincidence detection. Specific details include incident electron energies from 13.0 to 30.6 eV with resolutions of 0.220 eV, beam current of usually 2μA, a 127° electrostatic electron energy analyzer with an acceptance angle of 10^{-2} ster., two channeltron photon detectors with an acceptance angle of about 10^{-2} ster. which cover the coplanar angular range +15° to 135° and -20° to -150° depending upon the location of the electron detector. The coincidence resolving time was within 4.0 nsec.

The alignment parameter λ has an energy and angular dependence similar to that well established for helium. There are two distinct minima; the one at forward scattering angles becomes shallower and moves towards 60° as the energy decreases to 13.0 eV while the minima at large scattering angles becomes deeper and also moves towards longer scattering angles as the energy decreases to 13.0 eV.

The orientation parameter R shows no such regular behaviour.

Measurements of the electron-photon coincidence rate for the photon detector located normal to the scattering plane permit the 2p magnetic substates and the 2s differential cross sections to be determined. The 1s→2p (m = ±1) cross section displays two maxima, the first appears to decrease with energy while the second rises with decreasing energy. At large angles and low energy the 2p (m = ±1) cross section is smaller than the 2p (m = 0), but at 54 eV (Williams, 1981)[1] the 2p (m = ±1) cross section becomes larger at angles less than about 40° although the very small angle, less than about 10°, behaviour has not been studied.

In the backward direction the 1s-2s cross section is larger than the 1s→2p (m = 0) and 1s→2p (m = ±1) cross sections.

The λ parameter is in good agreement with the 3-state, close coupling calculated values by Kingston, Liew and Burke (1982)[2] for angles less than about 50°. At higher angles theory does not predict the observed deep minimum. All the differential cross sections are in good agreement with the close coupling predictions.

Graphed and tabulated results will be presented at the conference.

References

1. J.F. Williams, J.Phys.B: At.Mol.Phys. 14, 1197-1217 (1981)
2. A.E. Kingston, Y.C. Liew and P.G. Burke, J.Phys.B: At.Mol.Phys. 16, 2755-67 (1982)

SYMMETRY PROPERTIES OF ELECTRON-PHOTON ANGULAR CORRELATIONS

J.F. Williams

Physics Department, University of Western Australia, Perth, Australia

Symmetry properties of electron impact excitation collisions with atoms have been discussed frequently since the initial papers of Macek and Jaecks[1] and Fano and Macek[2] concerning electron-photon angular correlations. Recently Bartschat and Blum[3], Blum[4], Beyer et al[5], Gien[6] and Kohmoto and Fano[7] have discussed the attractive or repulsive nature of the electron-atom interaction and the sign of the atomic orientation. For the case of L = 0 to L = 1 excitation for a process well described by LS coupling, as for 1s→1p in helium and 1s→2p in hydrogen, the angular correlations between the scattered, energy-loss electrons and the decay photons have been measured[8,9] and analyzed to yield the atomic orientation, O_{1-}^{col}, or angular momentum, $\langle L_y \rangle$, transferred in the excitation. These quantities for He are related through the angular correlation alignment, λ and relative phase χ, parameters by

$$-2\, O_{1-}^{col} = +2\{\lambda(1-\lambda)\}^{\frac{1}{2}}\sin\chi = -\langle L_y\rangle = +P_{circ} = \frac{I_{RH} - I_{LH}}{I_{RH} + I_{LH}}$$

where I_{RH} and I_{LH} are the intensities of right and left handed circularly polarized light. While the sign of the orientation is established by the sign of the measured circular polarization, P_{circ}, such measurements have not previously been made for U.V. 21.2 eV or 10.2 eV photons because of the difficulty of constructing an efficient circular polarizer. This has been overcome using a double reflection technique.

Two surfaces are needed for a circular polarizer. A single surface produces linear polarization on reflecting unpolarized light by virtue of having different reflectivities for the components in, and normal to, the plane of incidence and has been used by McConkey in the analysis of linear polarization for He 2^1P radiation. The second surface reflects both incident components but with a 90° difference in the phase shifts for the two components. The materials for the surfaces were chosen so that, for the second surface the reflectivities of both polarization components be as close as possible to being equal, while for both surfaces the reflectivity be high. $A\ell_2O_3$ and Aℓ about 600°A thick were selected after many tests. The incident light was reflected at 60° off the first surface and reflected from the second with the incident linear polarization vector at 45° with the normal to the plane of incidence for the Aℓ reflection. The angle of incidence for the Aℓ reflection was 55°.

Standage and Kleinpoppen[10] showed, for e - He(3^1P) excitation, that the 501.6 nm radiation, detected normal to the scattering plane, in the positive y direction, and in coincidence with the energy loss electrons, is circularly polarized and that the circular polarization changes sign as the electron scattering angle, θ, changes sign i.e. $P_{circ}(+\theta) = -P_{circ}(-\theta)$. This result has been confirmed in the present work also for e - He(2^1P) excitation at an incident electron energy of 81.2 eV and at θ = ±10°, where $P_{circ}(+10°) = -0.30 \pm 0.06$.

Theory[11] predicts that the sign of the orientation parameter, O_1^{col}, changes as the electron scattering angles increases beyond about 60°. Measurement shows that the circular polarization $P_{circ} = + 0.89 \pm 0.2$ at θ = 90° and confirms that prediction.

Measurements have also been made of P_{circ} for e + H(2p) excitation at 54 eV and have shown $P_{circ}(+10°) = -0.26$; $P_{circ}(-10°) = -0.30$ and $P_{circ}(+90°) = +0.1$ again verifying the theoretical predictions for the changes of sign of P_{circ} to be similar to those e + He(2 P) excitation. These measurements are believed to be the first to show these features for these excitation processes.

References

1. J.H. Macek and D.H. Jaecks. Phys.Rev.A4 2288 (1971)
2. U. Fano and J.H. Macek. Rev.Mod.Phys. 45,553 (1973)
3. K. Bartschat and K. Blum. J.Phys.B:At.Mol.Phys. 15, 2747 (1982)
4. K. Blum (1981) Density Matrix Theory and Applications (New York, Plenum Press)
5. H.J. Beyer, H. Kleinpoppen, I.McGregor and L.C. McIntyre. J.Phys.B:At.Mol.Phys. 15 L545 (1982)
6. T.T.Gien. J.Phys.B:At.Mol.Phys. 15 2481 (1982)
7. M.Kohmoto and U.Fano. J.Phys.B.At.Mol.Phys. 14 L477 (1981)
8. M.T.Hollywood, A.Crowe and J.F.Williams. J.Phys.B: At.Mol.Phys. 12 819 (1979)
9. J.F.Williams. J.Phys.B:At.Mol.Phys. 14 1197 (1981)
10. M.C.Standage and H.Kleinpoppen. Phys.Rev.Lett. 36 577 (1976)
11. W.C. Fon, K.A.Berrington and A.E.Kingston. J.Phys.B:At.Mol.Phys. 13 2309 (1980)

THE λ-PARAMETER OF THE 3^1P STATE OF HELIUM USING THE 5016Å PHOTON*

C. R. Hummer and Donal J. Burns

Department of Physics and Astronomy, University of Nebraska, Lincoln, Nebraska 68588 U.S.A.

Much interesting work has been done on the measurement of the λ-parameter for the 2^1P state of helium. The usual technique involves the detection of a coincidence between the scattered electron and the 584Å photon emitted subsequent to the exciting collision. For each scattering angle, analysis of the isotropy of the radiation pattern in the scattering plane allows the determination of λ. This method has also been used to study the 3^1P state,[1,2] which can decay directly to the ground state with the emission of a 537Å photon. However, without the benefit of wavelength selection or high beam energy resolution, such measurements may be affected by additional correlations from the other n = 3 singlet states, notably the 3^1D state. This state can decay via the 2^1P state to the ground state producing a 587Å photon that is also correlated to the inelastically scattered electron. Van Linden van den Heuvell et al.[3] have shown that such an effect is clearly present.

Such problems may be eliminated by performing an experiment in which the visible $3^1P \rightarrow 2^1S$ 5016Å photon is detected, rather than the resonance radiation. The difficulty is that the branching ratio reduces the photon flux by a factor of about 40.

In the present experiment, the scattering geometry was chosen so that photons are detected in the scattering plane and at 90° to the electron beam axis. For such a configuration, Macek and Jaecks[4] have shown that the coincidence rate for a ^1P to ^1S transition in helium is proportional to the differential cross section of $\sigma_0(\theta)$, where the subscript represents the $m_\ell = 0$ substate.

Such an experiment has been performed over a range of scattering angles from 20° to 50° at an energy of 70 eV. The relative values of $\sigma_0(\theta)$ were then converted to λ by using the total differential cross section data of Chutjian[5] and are shown in Fig. 1 when normalized to the theory of Scott and McDowell.[6]

Further data will be presented at the conference, over an extended angular range, including comparison of the shape of $\sigma_0(\theta)$ with a variety of theories, since this may provide a more stringent test than a comparison with λ.

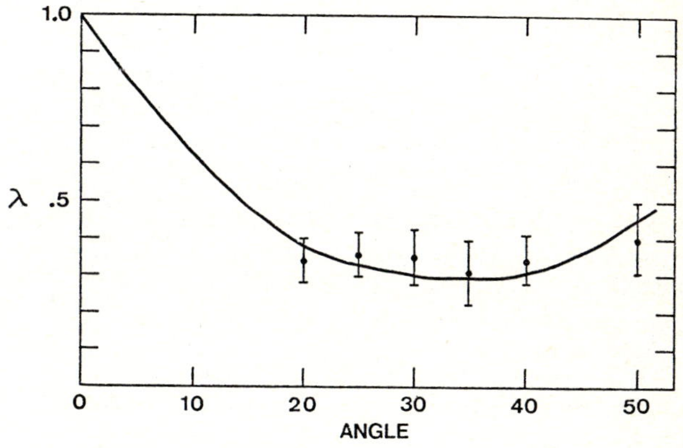

Fig. 1: The present results are shown with the λ-parameter of the DWPO 1 theory by Scott and McDowell.[6]

REFERENCES

*Work supported by the National Science Foundation.

1. M. Eminyan, K. B. MacAdam, J. Slevin, M. C. Standage, and H. Kleinpoppen, J. Phys. B 8, 2058 (1975).
2. A. Crowe, T. C. F. King, and J. F. Williams, J. Phys. B 14, 1219 (1981).
3. H. B. van Linden van den Heuvell, G. Nienhuis, J. van Eck, and H. G. M. Heideman, J. Phys. B 14, 2667 (1981).
4. Joseph Macek and D. H. Jaecks, Phys. Rev. A 4, 2288 (1971).
5. Ara Chutjian, J. Phys. B 9, 1749 (1976).
6. T. Scott and M. R. C. McDowell, J. Phys. B 9, 2235 (1976).

ORIENTATION OF THE 2^1P STATE OF HELIUM EXCITED BY ELECTRON IMPACT

N. C. Steph, W. H. Kloepping and D. E. Golden

Department of Physics and Astronomy, University of Oklahoma, Norman OK

Electron-photon angular correlations between electrons which have excited the 2^1P state of helium and photons from the $2^1P \rightarrow 1^1S$ transition have been studied for 27-, 30-, 35-, and 40-eV incident electrons for a range of electron scattering angles from 15° to 110°. The values of λ and $|\chi|$ have been combined to examine the behavior of the orientation of the 2^1P state. The results are compared to the values obtained in distorted wave calculations by Stewart and Madison, and to values obtained in R-matrix calculations by Fon et al. The distorted wave results are in fair agreement with the data at angles $< 90°$ for energies > 30 eV. The R-matrix calculation is in good qualitative agreement with the data at both 27 and 30 eV. In particular, the calculation predicts an increase in $|\chi|$ with decreasing scattering angle at 26.5 eV. We have observed such an increase at 27 eV. This results in an increase in the orientation at small angles. The behavior of the orientation as the energy decreases from 40 to 30 eV is in agreement with the prediction of the semiclassical model discussed by Steph and Golden. As the energy decreases, the position of the small angle maximum moves from 50° to 70°. At 27 eV, we observe two small angle maxima, one at 80° and the other at $\sim 30°$.

EXCITATION AND DECAY OF STARK MIXED n = 2 STATES OF HYDROGEN OBSERVED IN AN ELECTRON PHOTON COINCIDENCE EXPERIMENT

S. Watkin, C.G. Back, M. Eminyan, K. Rubin, J. Slevin and J.M. Woolsey

Department of Physics, University of Stirling, Stirling, Scotland.

s-p coherence is an important phenomenon in atomic excitation[1]: recent measurements of charge-transfer collisions[2,3] and atom-atom collisions[4] in hydrogen have demonstrated the existence of a high degree of excitation coherence between s and p levels, and measurements of the s-p coherence parameter $Re<f_{oo}f_{1o}^{*}>$ have been reported. We report here the first measurements of the decay of Stark-mixed n=2 states in atomic hydrogen using the electron-photon coincidence technique, where the experimental observables are related to the various cross-sections for magnetic sublevel excitation and include a term proportional to $Re<f_{oo}f_{1o}^{*}>$.

A beam of hydrogen atoms, dissociated by an R.F. discharge, is excited by a 350 eV electron beam in a region in which an electric field of 250 V cm^{-1} parallel to the electron beam is present. The Lyman-α radiation emitted perpendicular to the electron beam is observed in coincidence with electrons scattered with an energy loss of 10.2 eV into a small range of angles near the forward direction.

In the presence of an electric field, the n=2 s and p states are mixed and transitions from these mixed states contribute to the Lyman-α coincidence signal. The transition probability for the decay of the Stark mixed states excited by electrons scattered through a particular scattering angle can be written in the form

$$P(t) = K\left[F_{3/2}e^{-t/t_p}+F_-e^{-t/t_-}+F_+e^{-t/t_+}\right]$$

where $F_{3/2} = (1 + {}^{3\lambda}/_2)\,\sigma_p$

$$F_{\pm} = \left[3(\gamma\pm)^2\sigma_s+\sigma_p\mp 2\sqrt{3}(\gamma\pm)|f_{00}||f_{10}|\cos\delta\right] / \left[1+(\gamma\pm)^2\right]^2$$

and $\gamma\pm = 2r/\left[-1 \pm (1+4r^2)^{\frac{1}{2}}\right]$, where r is the ratio of the Stark shift to the Lamb shift.

Interference terms give rise to quantum beats which are too fast for the present time resolution of our apparatus.

The experimental data allow the determination of $(F_{3/2} + F_-)$ and F_+ by least squares fitting, and these results are compared with various theories. The parameters λ and R have also been measured at this energy over a similar range of angles.

References

1. T.G. Eck, Phys.Rev.Lett. 31, 270 (1973)
2. C.C. Havener, W.B. Westerveld, J.S. Risley, N.H. Tolk and J.C. Tulley, Phys.Rev.Lett, 48, 926 (1982)
3. I.A. Sellin, L. Liljeby, S. Mannervik and S. Hultberg, Phys.Rev.Lett. 42, 570 (1979)
4. R. Krotkov and J. Stone, Phys.Rev.A, 22 473 (1980).

ELECTRON-PHOTON ANGULAR CORRELATIONS FOR THE 2'P STATE OF HELIUM WITHIN 1eV OF THRESHOLD

P.A. Neill and A. Crowe

Department of Pure and Applied Physics, The Queen's University of Belfast, Belfast BT7 1NN, Northern Ireland.

The measurement of electron-photon angular correlations is now well established as a most valuable technique in the study of fundamental aspects of electron excitation processes. Measurements, particularly those over a wide range of electron scattering angles, provide a sensitive test of theoretical models.

For excitation of the 2'P state of helium, most previous data of this type have been reported over the 30-80 eV incident electron energy range[1-5]. Very recent measurements in this laboratory have concentrated on the incident electron energy region closer to the excitation threshold. By contrast with the situation at higher energies there is good agreement between theory and experiment in this energy region. At 26.5 eV, the λ parameters measured by Crowe and Nogueira[6] are in excellent agreement with the 5-state R-matrix predictions of Fon et al[7] at all scattering angles, while there are only small discrepancies in the measured and calculated $|\chi|$ parameters. This has been confirmed by the measurements of van Linden van den Heuvell et al[5]. The experimental data and 5-state R-matrix calculations of Crowe et al[8] below the ionisation threshold at 24.0 eV are in complete agreement.

The latest measurements in this series have been carried out at an incident electron energy of 22.0 eV i.e. 0.8 eV above the 2'P excitation threshold, and for scattering angles in the range 40°-120°. The angular correlations are measured by observing the coincidence signal between the 0.8 eV scattered electron and the 58.4 nm photon from the decay of the 2'P state as a function of the photon detector angle, for a particular electron scattering angle. From the measured correlations the parameters $\lambda = |a_0|^2/(|a_0|^2 + 2|a_1|^2)$, where a_0 and a_1 are the excitation amplitudes for sub states with M = 0, M = ± 1 respectively, and $|\chi|$, the relative phase between a_0 and a_1, are extracted.

Figure 1 shows the λ and $|\chi|$ parameters obtained at 22.0 eV as a function of electron scattering angle. Theoretical calculations are not yet available at this energy. The 5-state R-matrix calculations of Crowe et al[8] at 24.0 eV, which are in complete agreement with experiment at that energy, are shown only to highlight the significant change which has taken place, particularly in the λ parameters, in this 2 eV energy interval.

References

1. M.T. Hollywood, A. Crowe and J.F. Williams, J. Phys. B. 12, 819 (1979).
2. R. McAdam, M.T. Hollywood, A. Crowe, J.F. Williams, J. Phys. B. 13, 3691 (1980).

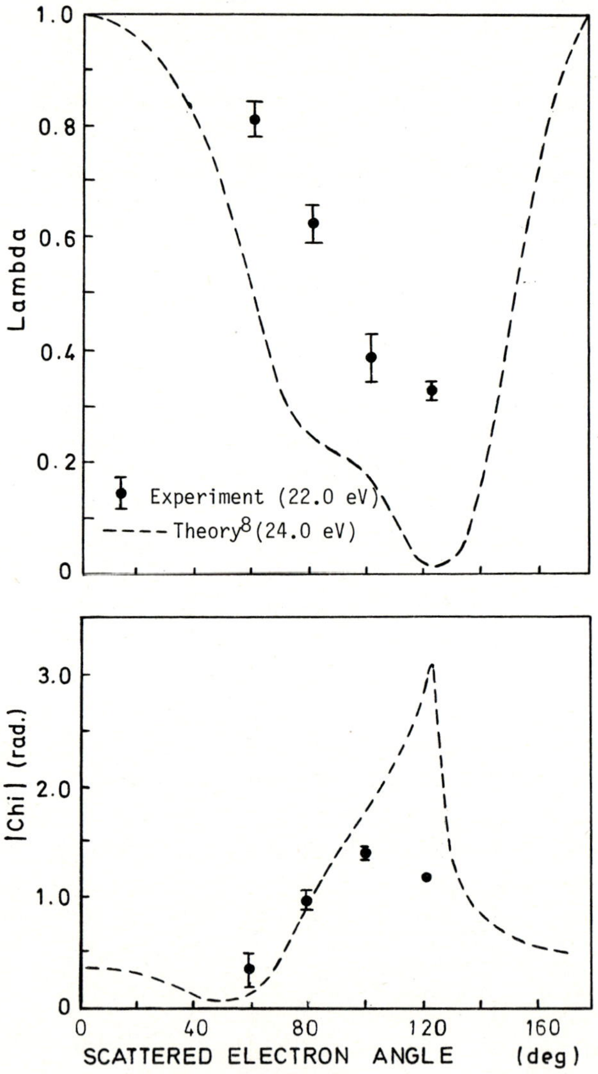

Figure 1

3. J. Slevin, H.Q. Porter, M. Eminyan, A. Defrance and G.Vassilev, J. Phys. B. 13, 3009 (1980).
4. N.C. Steph and D.E. Golden, Phys. Rev. A. 21, 1848 (1980).
5. H.B. van Linden van den Heuvell, J. van Eck and H.G.M. Heideman, J. Phys. B. 15, 3517 (1982).
6. A. Crowe, J.C. Nogueira, J. Phys. B. 15, L501 (1982).
7. W. C. Fon, K.A..Berrington and A.E. Kingston, J. Phys. B. 13, 2309 (1980).
8. A. Crowe, J.C. Nogueira and Y.C. Liew, J. Phys. B. 16, 481 (1983).

ALIGNMENT OF THE $2p_{3/2}$ SHELL OF ARGON BY ELECTRON IMPACT

E.C. Sewell and A. Crowe

Department of Pure and Applied Physics, The Queen's University of Belfast, Belfast BT7 1NN, Northern Ireland.

Recent measurements in this laboratory[1] have demonstrated the feasability and value of coincidence techniques in the study of inner shell ionization by electron impact. In that work the angular correlations between scattered electrons which had created a $2p_{3/2}$ vacancy state and the L_3-$M_{23}M_{23}(^1S_0)$ Auger electrons were measured for an incident electron energy of 1000 eV.

In those measurements it was assumed that the small background electron signal, due to other ionization processes, in the Auger electron spectrum could be ignored in the angular correlation measurements. This assumption is readily tested by measuring the angular correlation between scattered electrons which have created a $2p_{\frac{1}{2}}$ vacancy and the L_2-$M_{23}M_{23}(^3P_{012})$ Auger electrons which are emitted isotropically. The results of such a measurement are shown in Figure 1(a) for an incident electron energy of 1000 eV, an electron scattering angle of 15^0 and a mean ejected electron energy of 5 eV. Clearly this is not an isotropic angular correlation. The increase in the coincidence signal for Auger emission angles less than approximately 80^0 is due to an (e, 2e) type process from the outer shells of argon, i.e. 3s and 3p orbitals, and not to any experimental artefact. By consideration of the kinetics of the measurement, the (e, 2e) contribution is associated with ionization of argon states with binding energies of approximately 50 eV. Such high lying states of argon have not previously been studied in detail[2].

We have measured the (e, 2e) contribution to the present data which is shown in Figure 1(b), and by subtraction have obtained the angular correlation between the scattered electrons and the L_2-$M_{23}M_{23}$ $(^3P_{012})$ Auger electrons, shown in Figure 1(c). Use of a least-squares fitting procedure shows that the data of Figure 1(c) is isotropic within experimental error.

The (e, 2e) contribution to the angular correlations resulting from a $2p_{3/2}$ vacancy has also been determined at 1000 eV for small electron scattering angles, and will be presented.

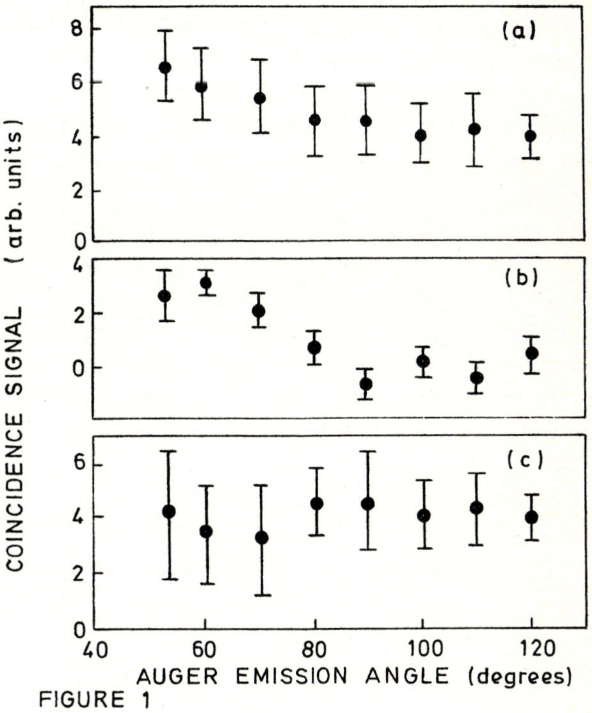

FIGURE 1

References

1. E.C. Sewell and A. Crowe. J. Phys. B. 15, L357 (1982).
2. J.E. Hansen. Comments At. Mol. Phys. 12, 197 (1982).

ANGULAR CORRELATIONS FOR EXCITATION OF THE LOWEST EXCITED STATES OF KRYPTON BY ELECTRON IMPACT

S.J. King and A. Crowe

Department of Pure and Applied Physics, The Queen's University of Belfast, Belfast BT7 1NN, Northern Ireland.

Excitation of krypton by low energy electrons has received little attention both experimentally and theoretically. The extent of previous studies has been discussed by Trajmar et al[1] who measured differential cross sections for a number of excited states in the energy range 15-100 eV.

First attempts at a more complete study of the lowest lying excited states viz $4p^5(^2P_{3/2})5s\,^3P_1$ and $4p^5(^2P_{\frac{1}{2}})5s\,^1P_1$ using the electron-photon angular correlation technique (Kleinpoppen and McGregor[2], Nishimura et al[3]) ignored the effects of spin-orbit coupling in the analysis of the data. Blum et al[4] and da Paixao et al[5] have shown that the angular correlations for excitation of a J = 1 from a J = 0 state for heavy atoms can be characterised by four parameters (λ, $\overline{\chi}$, ε, Δ) compared with only two (λ, χ) for an LS-coupled system. These parameters are defined as

$$\lambda = \sigma_0/(\sigma_0 + 2\sigma_1) \qquad \cos\overline{\chi} = \mathrm{Re} < a_0\ a_1 >/< a_0\ a_1 >$$

$$\cos\varepsilon = - < a_{-1}\ a_1 >/\sigma_1 \qquad \cos\Delta = |< a_0\ a_1 >|(\sigma_0\ \sigma_1)^{\frac{1}{2}}$$

where σ_0, σ_1, a_0 and $a_{\pm1}$ are the differential cross sections and excitation amplitudes for the magnetic sublevels, respectively.

Experimentally λ, $\overline{\chi}$ and ε can be determined by measuring electron-photon angular correlations for two different azimuthal angles of the photon detector. Such measurements have been reported very recently by McGregor et al[6] and Nishimura et al[7] for electron scattering angles $\leqslant 30^o$ and in the energy range 36-60 eV. At 60 eV, where measurements have been reported by both groups, the data are not in agreement. Indeed correlations measured in the scattering plane have opposite phases at a scattering angle of 20°. Both groups indicate no significant difference between the correlations measured for the two states. This is in contrast to earlier work on the corresponding states of argon using an electron-polarised photon technique by Malcolm and McConkey[8].

Results will be presented at an incident electron energy of 60 eV to resolve and clarify these discrepancies.

References

1. S. Trajmar, S.K. Srivastava, H. Tanaka and H. Nishimura, Phys. Rev. A., 23, 2167 (1981).
2. H. Kleinpoppen and I. McGregor in Coherence and Correlation in Atomic Collisions, edited by H. Kleinpoppen and J.F. Williams (Plenum, New York, 1980).
3. H. Nishimura, A. Danjo and Y. Koike in Proc 12th ICPEAC edited by S. Datz, (Gatlinburg, Tennessee, 1981).
4. K. Blum, F.J. da Paixao, Gy. Csanak, J. Phys. B., 13, L257 (1980).
5. F. J. da Paixao, N.T. Padial and Gy. Csanak, Phys. Rev. Lett. 45, 1164 (1980).
6. I. McGregor, D. Hils, R. Hippler, N.A. Malik, J.F. Williams, A.A. Zaidi and H. Kleinpoppen, J. Phys. B. 15, L411 (1982).
7. H. Nishimura, A. Danjo, Y. Koike, K. Kani, H. Sugahara and A. Takahashi in Proc 1st US-Japan Seminar on Electron Molecule Collisions and Photoionisation Processes, (Pasadena, California 1982).
8. I.C. Malcolm and J.W. McConkey, J. Phys. B. 12 511 (1979).

SUPERELASTIC ELECTRON SCATTERING ON SODIUM ATOMS IN THE DISTORTED-WAVE BORN APPROXIMATION

V.V. Balashov, A.N. Grum-Grzhimailo, O.I. Zaitseva

Institute of Nuclear Physics, Moscow State University, Moscow 117234, USSR

The distorted-wave Born approximation with a semiphenomenological optical potential[1] allows one to get a complete enough description for the $^1S \rightarrow {}^1P$ excitation of rare gas atoms during the inelastic scattering of intermediate-energy electrons[2] namely, the integral and differential cross sections, the correlation parameters λ and χ. There are statements[3], however, that in the case of alkali metals the optical potential[1] is not adequate even for elastic scattering. In order to examine further the range of applicability of the DWBA method presented in[2] we consider here the correlation parameters of the 3P ↔ 3S transition in sodium. They have been measured by Hertel et al[4] in the superelastic electron scattering experiment. Earlier theoretical analysis of these data was performed in the close-coupling and in the distorted-wave polarised-orbital (DWPO) methods.

Bearing in mind general limitations for the DWBA-method we restrict ourselves only by the small angle scattering ($\theta_{sc} \leqslant 25°$). On the other hand the superelastic data are available also only for this angle range. The following characteristics have been calculated: the multipole moments T^1_1, T^2_o, T^2_1, T^2_2 ; the correlation parameters λ, χ as well as the differential and integral cross sections for the 3S↔3P transition in Na at incident electron energies 20-55 eV. An example of our results is shown in Fig. 1. As in[2] we are specially interested in the problem of the imaginary part of the optical potential as well as in the exchange scattering problem. The discussion of this points well be presented at the conference.

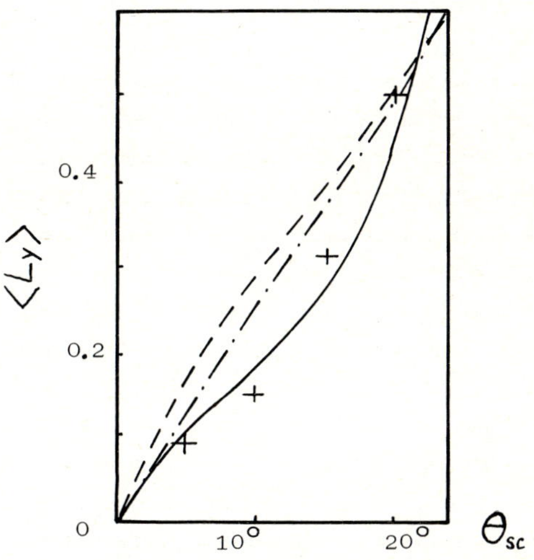

Fig. 1: Orientation value $\langle Ly \rangle$ for the $^2P \rightarrow {}^2S$ transition in Na induced by 20 eV electons; + experiment; – – two-state close-coupling calculations by Moores, Norcross; — · — · — distorted-wave polarised-orbital calculations by Kennedy (all these results from [4b]); ——— our DWBA calculations.

References

1. J.B. Furness, I.E. McCarthy, J.Phys.B6,2280(1973)
2. V.V. Balashov et al. J.Phys.B14, 357 (1981)
3. P.J.O. Teubner, S.J. Buckman, C.J. Noble, J.Phys. B11, 2345 (1978)
4. I.V. Hertel et al. a) J.Phys.B10, 251(1977)
   b) J.Phys.B13, 3465(1980);
   c) Z.Phys. A307, 89 (1982).

AN INVESTIGATION OF SUPERELASTIC ELECTRON SCATTERING BY LASER-EXCITED Ba ATOMS

S. Trajmar

Jet Propulsion Laboratory, California Institute of Technology, Pasadena, California 91109

D. F. Register

Phillips Petroleum Research, Bartlesville, Oklahoma 54003

Gy. Csanak

154 Cold Spring Road, Stamford, Conneticut 06905 (Consultant to Jet Propulsion Laboratory, Space Sciences Division)

Electron-photon coincidence experiments[1] and the corresponding experiments on superelastic electron scattering by laser excited atoms[2] yield information on coherence and correlation in electron-atom collisions. The analysis and interpretation of these experiments in terms of orientation and alignment properties of the excited atom have been based on the theoretical treatment of Fano and Macek[3] and Macek and Hertel.[4] Symmetry for the superelastic scattering intensity with respect to reflection through the scattering plane has always been assumed and predicted by theory unless some presently unknown symmetry violating term exists in the interaction Hamiltonion or the electron beam posses longitudinal polarization.

We report here observations made during the past four years concerning superelastic electron scattering by laser excited ^{138}Ba (^1P) atoms which indicate an asymmetry in the superelastic scattering signal with respect to the scattering plane at small scattering angles.

The experimental arrangement is shown in Fig. 1 A collimated Ba beam was crossed at 90° with a linearly polarized laser beam which excited the ^{138}Ba atoms from the $...6s^2\ ^1S_0$ state to the $6s6p\ ^1P_1$ level. The laser

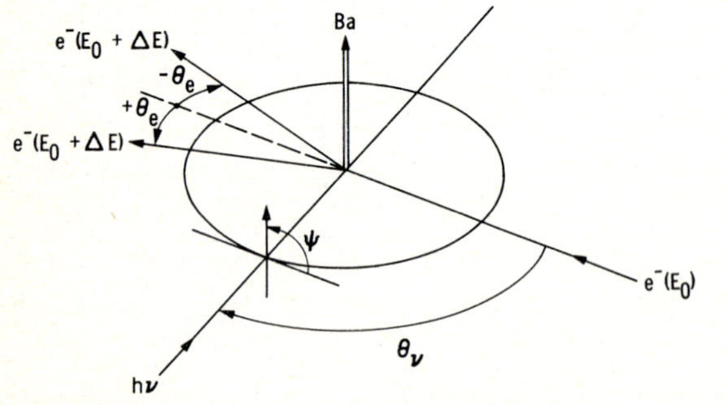

Fig. 1: Schematic diagram of the experimental arrangement.

beam was generated by a single-mode, tunable, cw, dye laser and was located in the scattering plane. An energy selected electron beam also crossed the Ba beam at 90°. The superelastic electron scattering intensity (I^S) was measured as a function of laser beam polarization angle (ψ) with respect to the scattering plane at fixed impact energies (E_o), electron scattering angles (θ_e) and laser beam directions (θ_ν) with respect to the electron beam.

A number of checks were made to eliminate possible experimental conditions which could cause the observed asymmetry. We investigated the effect of magnetic field, laser intensity, target density and collimation, small out-of-scattering-plane laser position and elliptical polarization of the pumping light but found that the asymmetry persisted during the variation of these conditions.

The polarization and angular distribution of the fluorescence light was found to be consistent with the predictions for pumping with linearly polarized light. This indicates that the asymmetry is not associated with the pumping but with the electron scattering process. The hyperfine spectrum was produced both in the fluorescence and in the superelastic channels by sweeping the laser frequency. In these spectra no serious overlap from other hyperfine level occurs.

At the present time we have no explanation for the observed asymmetry. Clearly further work is required to clarify the situation.

References

1. K. Blum and H. Kleinpoppen, Phys. Rep. 52, 204 (1979).
2. I. V. Hertel and W. Stoll, Advances in Atomic and Molecular Phys. 13, 113 (1977).
3. U. Fano and J. H. Macek, Rev. Mod. Phys. 45, 553 (1973).
4. J. Macek and I. V. Hertel, J. Phys. B: At. Mol. Phys. 7, 2173 (1974).

Work is supported by NASA and NSF.

ATTRACTIVE AND REPULSIVE FORCES IN ELECTRON IMPACT EXCITATION OF ATOMS

H.-J. Beyer and H. Kleinpoppen

Atomic Physics Laboratory, University of Stirling
Stirling, Scotland

Kohmoto and Fano[1] have recently shown that a classical grazing model can be applied to the problem of atomic orientation produced by electron scattering. Beyer et al.[2] extended this model by connecting the amplitudes f_A for attractive and f_R for repulsive forces to the amplitudes for the excitation of the magnetic substates in a $^1S_o \rightarrow {}^1P_1$ excitation process:

$$f_R = \frac{1}{\sqrt{2}}(f_o + i\sqrt{2}\, f_1) \qquad f_A = \frac{1}{\sqrt{2}}(f_o - i\sqrt{2}\, f_1) \quad ,$$

f_o and f_1 are the amplitudes for the excitation of the $^1P_1$ substates with $m_L=0$ and $m_L = \pm 1$, respectively; f_o real, $f_1 = |f_1|e^{i\chi}$.

Using the relations $\lambda = f_o^2/\sigma$ and $\cos\chi = Re(f_1)/|f_1|$, $|f_A|^2$ and $|f_R|^2$ can be expressed as:

$$|f_A|^2 = \tfrac{1}{2}\sigma(1 + \sqrt{\lambda(1-\lambda)}\, \sin\chi)$$

$$|f_R|^2 = \tfrac{1}{2}\sigma(1 - \sqrt{\lambda(1-\lambda)}\, \sin\chi)$$

Furthermore

$$\tan(\delta_R - \delta_A) = \frac{2\sqrt{\lambda(1-\lambda)}}{2\lambda - 1} \cos\chi$$

where $\delta_R - \delta_A$ is the phase difference between f_A and f_R. Fig.1 shows the values $|f_A|^2/\sigma$ and $|f_R|^2/\sigma$ and Fig.2 the values $\delta_R - \delta_A$ computed from previously reported measurements[3,4]. The structure will be discussed in more detail.

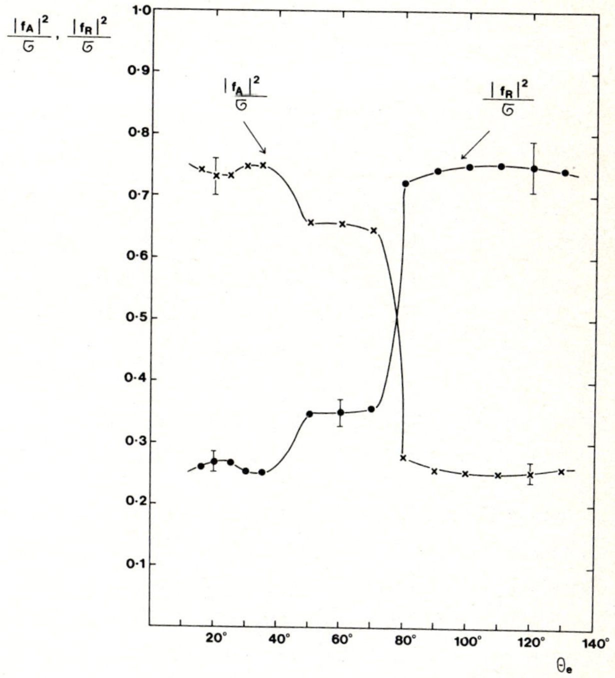

Fig.1: $|f_A|^2/\sigma$ and $|f_R|^2/\sigma$ for 80 eV - electron impact excitation of He - 2^1P_1.

References

1 M Kohmoto, U Fano, J.Phys.B14, L477 (1981)

2 H J Beyer, H Kleinpoppen, I McGregor, L C McIntyre, J.Phys.B15, L545 (1982)

3 M Eminyan, K MacAdam, J Slevin, H Kleinpoppen, J.Phys.B7, 1519 (1974)

4 M T Hollywood, A Crowe, J F Williams, J Phys. B12, 819 (1979).

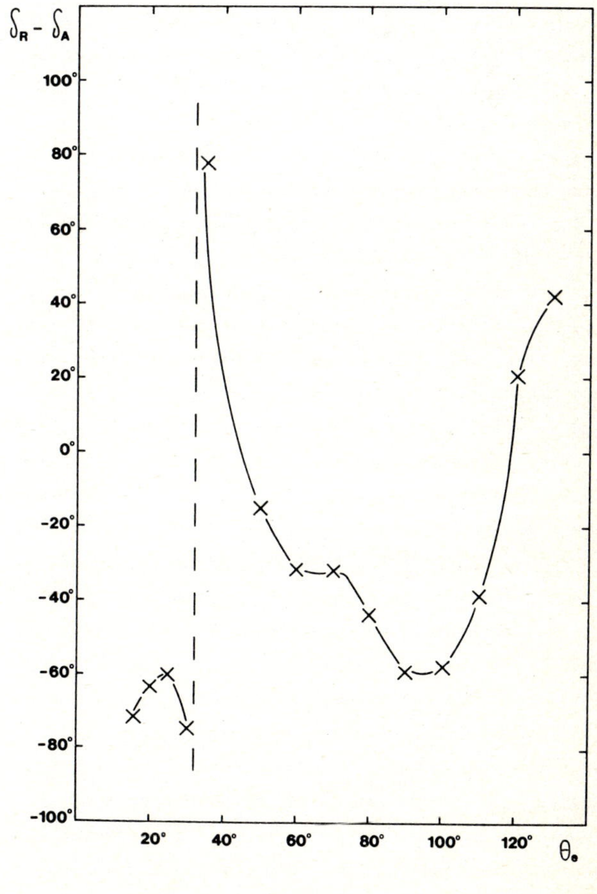

Fig.2: $\delta_R - \delta_A$ for 80 eV - electron impact excitation of He - 2^1P_1. →

ELECTRON-ELECTRON COINCIDENCE STUDIES OF THE L_3-SHELL ALIGNMENT TENSOR IN ARGON

M. Völkel and W. Sandner

Fakultät für Physik, Hermann-Herder-Str. 3, D-7800 Freiburg, FRG

Measurement of the angular correlation between scattered electrons and Auger electrons determines the alignment of ions after inner shell ionization with fixed scattering geometry[1]. The angular distribution of the Auger electrons is given by

$$W(\theta,\phi) = 1 + \sum_k \alpha_k \sum_\kappa \mathcal{A}_{k\kappa} \left(\frac{4\pi}{2k+1}\right)^{1/2} Y_{k\kappa}(\theta,\phi)$$

where $\mathcal{A}_{k\kappa}$ are the alignment tensor components and the α_k are analytically known coefficients. We define $\theta = 0$ to coincide with the incident beam direction and the plane $\phi = 0$ to be the scattering plane.

In a new electron-electron coincidence apparatus[2] we measured the angular distribution of argon $L_3M_{2,3}M_{2,3}(^1S_0)$ Auger electrons following electron impact ionization. Incident energy was 1 keV, energy transfer was 253 eV and scattering angle θ_s was 15°. Auger electrons were detected under 12 directions (θ,ϕ) simultaneously in coincidence with scattered electrons, with the 12 directions lying on a common conical surface $(\theta, \phi = \arccos\{\cos(39.2°)/\sin\theta\})$. The observed angular correlation is shown on Fig. 1a as a function of the independent detection angle θ.

Our geometry allows only the determination of two combinations of the three nonzero tensor components $\mathcal{A}_{k\kappa}$. A complete determination of the alignment tensor has been obtained by including the previously measured[3] angular correlation within the scattering plane (Fig. 1b), measured under otherwise identical conditions. The result of one least squares fit (solid line) to both experiments yields $\mathcal{A}_{20} = -0.90 \pm 0.16$; $\mathcal{A}_{21} = 0.19 \pm 0.08$; $\mathcal{A}_{22} = 0.01 \pm 0.1$. The resulting angular distribution $W(\theta,\phi)$ is rotationally symmetric around $\theta = 9.6° \pm 4°$. This value indicated the total breakdown of the first Born approximation (BA), which strictly requires any angular distribution to be symmetric around the momentum transfer direction $\hat{K}$ ($\theta = 53°$ in our case). Not surprisingly, a calculation of $\mathcal{A}_{k\kappa}$ in BA with Hartree Slater wave functions[1] yields $\mathcal{A}_{20} \approx \mathcal{A}_{21} \approx \mathcal{A}_{22} < 0.01$, (no alignment), in contrast to the experimental result of an almost totally aligned ion ($\mathcal{A}_{20} \approx -1$).

Our two values measured below $\theta = 60°$ were omitted from the fit, since their inclusion consistently increased the χ^2-value by at least a factor of 2. Their direction θ coincides with the momentum transfer direction, around which one might expect a certain background from (e, 2e) processes in outer shells[4]. We note that part of this background leads to the same final ionic state as the Auger process and thus may interfere, which would explain the observed sharp decrease in the angular distribution. Similar interferences are well known in autoionization, but have never before been observed in Auger transitions. Further studies of this effect are under way.

Support from the Deutsche Forschungsgesellschaft is gratefully acknowledged.

References

1. E.G. Berezhko, N.M. Kabachnik and V.V. Sizov, J.Phys. B 11, 1819 (1978)
2. M. Völkel and W. Sandner, this book
3. E.C. Sewell and A. Crowe, J.Phys. B 15, L357 (1982)
4. E.C. Sewell, private communication

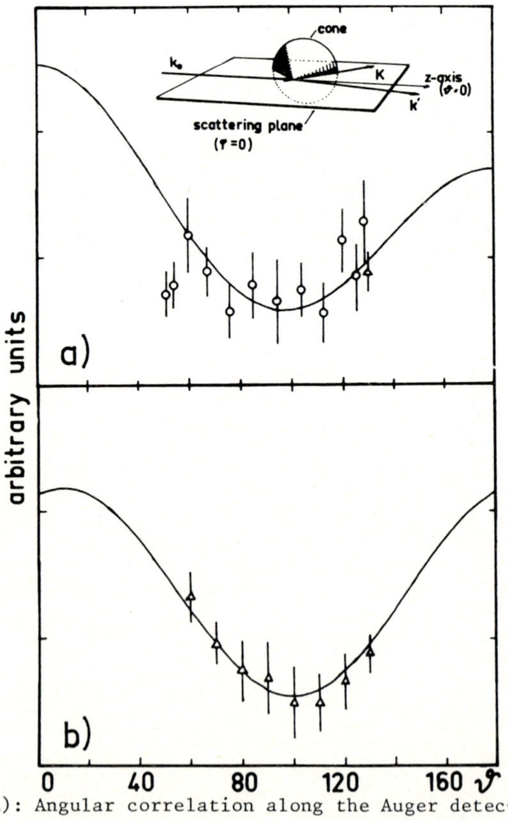

Fig.1) a): Angular correlation along the Auger detection cone (the geometrical arrangement is shown on the inset), b): Angular correlation in the scattering plane. **O** : this work, **Δ**: ref. 3, solid line: common fit to both experiments.

DETERMINATION OF SPIN EXCHANGE SCATTERING AMPLITUDES FOR e-Li ELASTIC SCATTERING

H. S. Desai and N. S. Rao

Department of Physics, Faculty of Science, M. S. University of Baroda, Baroda - 390 002

INDIA.

Three electron system of the lithium atom is reduced in to one electron hydrogen atom like system by using the core, potential approximation. The exchange scattering amplitudes are derived for elastic scattering of the electrons by the lithium atoms. Two Techniques are used for these derivations. One the Ochkur[5] type approximation and secondly the Lewis[4] integral technique.

An anlytical study is made between these two types of exchange scattering amplitudes. Walter's[1] type of wave function is used for these exchange derivations. These exchange scattering amplitudes are included in the direct scattering amplitudes of Rao and Desai[2]. Considerable improvements are obtained by the inclusion of these exchange amplitudes. Finally using the analytical expression of Rao and Desai[3] for e-Li elastic scattering, differential scattering cross sections at incident energies 100 to 700 eV are calculated. These results are found to be in good agreement with the compared data.

REFERENCES :

1. H.R.J. Walters, J.Phys. B.6. 1003 (1973).
2. N.S.Rao and H.S.Desai, In. Jour.Pure. and App. Phys. (In Press).
3. N.S.Rao and H.S.Desai, Current Science. (In Press).
4. R.R.Lewis, Phys. Rev. 102, 537 (1956).
5. V.I.Ochkur. Soviet Physics, JETP 18, 503 (1963).

IONIZATION AND ns-np EXCITATION ASYMMETRY IN POLARIZED ELECTRON-POLARIZED ALKALI ATOM COLLISIONS

G. Baum, M. Moede, W. Raith, and W. Schröder

Universität Bielefeld, Fakultät für Physik, D-4800 Bielefeld, FRG

Using a crossed beam arrangement we studied the spin dependence of the total ionization cross section and the total resonance-line excitation cross section for lithium, sodium, and potassium from threshold to about ten times the threshold value. For alternating antiparallel ($\uparrow\downarrow$) and parallel ($\uparrow\uparrow$) spin configurations of the two beams we measured signal rates in rapid succession at fixed energy and obtained in this way an asymmetry $\Delta = (N_{\uparrow\downarrow}-N_{\uparrow\uparrow})/(N_{\uparrow\downarrow}+N_{\uparrow\uparrow}) = P_e P_a A$, where N denotes the average counting rate, $P_e(P_a)$ the electron (atom) beam polarization, and A the physical asymmetry of interest. The asymmetry A can be related to angle (and energy) integrated cross sections:
$A = (\sigma_S-\sigma_T)/\sigma_S+3\sigma_T) = (\sigma_S-\sigma_T)/4\sigma = \sigma_{INT}/\sigma$, with $\sigma = 1/4\ \sigma_S + 3/4\ \sigma_T$. The asymmetry is thus seen to be a relative measure of singlet (S) versus triplet (T) cross sections or of the importance of interference (INT) effects between direct and exchange amplitudes.

For the ionization studies the ions were extracted by a weak electric field from the collision region and counted by an electron channel multiplier. They served as signal for the total ionization cross section. For the excitation studies the fluorescent light intensity emitted under 90^0 to the two beams was detected by a photomultiplier. Due to fine structure and hyperfine structure coupling after excitation this intensity can be shown[1] to be almost proportional to the total excitation cross section $\sigma_{S,T}=(\sigma_0+2\sigma_1)_{S,T}$.

Compared to our earlier studies on the ionization of lithium[2] the apparatus was changed with regard to the methods for producing the two spin polarized beams. The improvement mainly came through long term stability and ease of operation, but also through a considerable reduction of systematic error sources, allowing more precise data. The polarized electrons were obtained from a GaAs-photoemission source (P_e=0.35). The atomic beam was state-selected in a magnetic sextupole field (P_a(Li) = 0.3, P_a(Na) = 0.19, P_a(K) = 0.22) and the spin direction could be reversed with a spin flipper.[3] The polarization of each beam was measured directly, by Mott scattering for the electrons and state-selection for the atoms.

Our results for the ionization asymmetry, A_{ION}, of Li, Na, and K are shown in Fig. 1. The data will be compared with the experimental results of other researchers for sodium[4] and potassium[5] and with theoretical approximations in the case of lithium (see also reference 2). The GaAs-source allowed asymmetry measurements at the ionization threshold with a higher

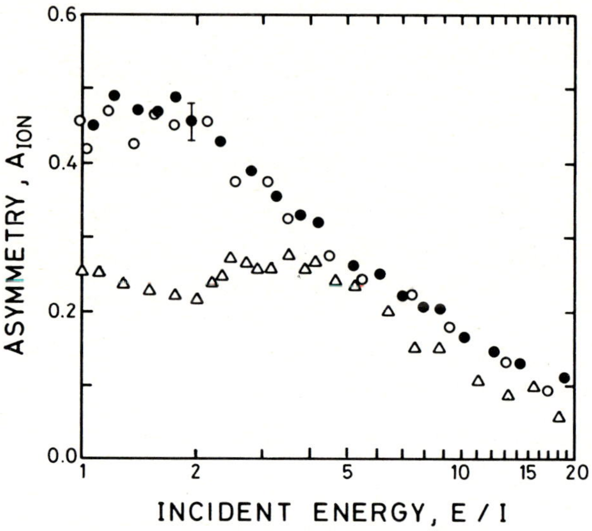

Fig. 1: Ionization asymmetry for Li(•), Na(o), and K(Δ) as function of incident energy E in units of threshold energy I. A typical systematic error is shown.

resolution (ΔE=150 meV) than before. The new data still show a flat behavior of the asymmetry with energy in the threshold region, confirming the conclusion [2] that the singlet and triplet ionization cross section have the same energy dependence close to threshold.

The measured asymmetry for the excitation of the resonance transitions, A_{ns-np} shows dominant singlet excitation at threshold (A_{ns-np} = +1), but drops rather rapidly to A_{ns-np} = 0.2 (within 0.9 eV for Li, 0.7 eV for Na, and 0.5 eV for K), and from there gradually approaches zero at about seven times the threshold energy. These measurements will be compared with close-coupling calculations.

References

1. W. Schröder, Thesis, Universität Bielefeld 1982
2. G. Baum, E. Kisker, W. Raith, W. Schröder, U. Sillmen and D. Zenses, J. Phys. B 14, 4377 (1981)
3. W. Schröder and G. Baum, J. Phys. E 16, 52 (1983)
4. D. Hils, W. Jitschin, and H. Kleinpoppen, J. Phys. B 15, 3347 (1982)
5. D. Hils and H. Kleinpoppen, J. Phys. B 11, L283 (1978)

SPIN DEPENDENT EFFECTS IN ELECTRON-HYDROGEN SCATTERING WITH POLARIZED BEAMS: A STATUS REPORT

A. Vasilakis, A. Anan, C. Back, M. Eminyan, M. S. Lubell, K. Rubin, J. Slevin, W. Stirling, F. C. Tang, M. Turner

The City College of CUNY, New York, USA; University of Stirling, Scotland; University of Paris VII, France

As has been demonstrated by exploratory experiments,[1,2] the use of polarized beams to study electron-hydrogen scattering provides a sensitive method for examining the three-body problem involving long-range forces. In order to probe more fully the nature of electron-hydrogen collisions we have prepared an experiment to make precise measurement of cross section asymmetries over a wide range of energies and scattering angles and for a number of scattering channels.

In the absence of relativistic considerations, the scattering of an electron by a hydrogen atom is characterized completely by the direct and exchange amplitudes, f and g respectively, both of which are complex quantities. Thus the scattering process is described by three parameters, $|f|$, $|g|$, and the relative phase θ. Experiments which do not include polarization determine a spin-averaged cross section

$$\frac{d\bar{\sigma}}{d\Omega} = \frac{1}{4}|f+g|^2 + \frac{3}{4}|f-g|^2. \qquad (1)$$

By contrast, our experiment will measure the spin-antiparallel (↑↓) spin-parallel cross-section asymmetry $A=[\sigma(\uparrow\downarrow) - \sigma(\uparrow\uparrow)]/[\sigma(\uparrow\downarrow) + \sigma(\uparrow\uparrow)]$, where σ denotes either a differential or a total cross section depending upon the process studied. For differential cross-section measurements, A is related to f and g by the expression

$$A = \frac{|f|\,|g|\,\cos\theta}{d\bar{\sigma}/d\Omega}\,, \qquad (2)$$

while for total cross-section measurements, the numerator and denominator must be averaged angles, and, if necessary, final-state energies.

We will measure A for elastic scattering over an angular range of $20^\circ - 150^\circ$ and over an energy range of 1 eV - 250 eV, with particular emphasis on the n = 2 resonance region and the intermediate energy region (30 eV - 100 eV). In addition we will explore impact ionization in the threshold region with high resolution to elucidate the dynamics of low-energy two-electron escape, a problem of considerable current interest.[3,4] Finally, with the use of electron-photon coincidence techniques we will be able to determine asymmetries of the form given by Eq. 2 where the $M_L = 0$ and $M_L = \pm 1$ Zeeman amplitudes for 2P excitation are completely separated.

The experimental method is illustrated in Fig. 1. The polarized hydrogen beam is produced by rf dissociation[5] and high-field state selection in a hexapole magnet, while the polarized electrons are generated in

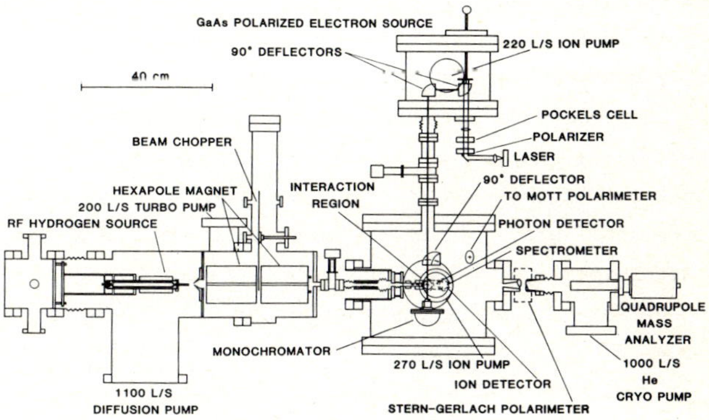

FIG. 1. Scale layout of experiment.

a conventional GaAs photoemission source.[6] A high resolution electron monochromator-spectrometer pair is employed together with a fast Lyman-α photon detector, an H^+ ion detector and appropriate beam monitors and polarimeters, all under the control of an LSI 11/23 computer. Experimental operating parameters are given in Table I.

TABLE I. Experimental design parameters.

Electron Beam		
GaAs Source		
•Intensity	>10 μA	
•Polarization	0.4	
•Emittance (at 1 eV)	<7 mrad cm	
Monochromator		
•Resolution	70 meV	30 meV
•Output current at stated resolution (space charge limited)	0.2 μA	0.03 μA
•Acceptance (at 1 eV) at stated resolution	33 mrad cm	9 mrad cm
Spectrometer		
•Resolution	70 meV	30 meV
•Solid angle subtended at stated resolution	2 msr	2 msr
Hydrogen Beam		
Characteristics 1 cm from nozzle		
•Density	3×10^{12} atoms/cm^3	
•Molecular fraction	<0.05	
Characteristics at interaction region		
•Density	3 x 10 atoms/cm	
•Molecular fraction	<0.01	
•Polarization	0.50	

*Work supported by NSF, CUNY, SERC, CNRS, and NATO.

[1]G.D. Fletcher et al., Phys. Rev. Lett. 48, 1671(1982).

[2]T.J. Gay et al., Phys. Rev. A 26, 3664(1982).

[3]C. Green and A.R.P. Rau, Phys. Rev. Lett. 48, 533(1982)

[4]A. Temkin, Phys. Rev. Lett. 48, 533(1982).

[5]J. Slevin and W. Stirling, Rev. Sci. Instrum. 52, 1780 (1981).

[6]D. T. Pierce et al., Rev. Sci. Instrum. 51, 478(1980).

"TRIPLE" ELECTRON SCATTERING ON MERCURY AND XENON FOR COMPLETE EVALUATION OF THE SCATTERING AMPLITUDES

O. Berger, B. Kessler, J. Keßler and W. Wübker

Physikalisches Institut der Universität Münster, D 4400 Münster, Germany

Elastic electron scattering from spinless atoms is theoretically described by two scattering amplitudes if Coulomb interaction and spin-orbit interaction in the continuum are taken into account[1]. Contrary to more complicated processes like inelastic scattering where a larger number of amplitudes is required, "complete" elastic scattering experiments in the sense that the moduli and the relative phase of the two scattering amplitudes can be determined[2] are feasible today.

Besides the differential cross section σ and the Sherman function S which can be obtained by scattering of initially unpolarized electrons, two additional observables, T and U, must be measured in such an experiment. The latter parameters can only be determined by scattering polarized electrons and observing the change of the polarization vector as caused by the scattering process.

In our experiments the polarization parameters S, T and U have been measured at various fixed scattering angles in an energy range between 20 eV and 360 eV for mercury and xenon[2,3]. The polarized electrons have been produced by a Fano effect source and a GaAs source, respectively. With the Fano effect source a current of typically 5 nA and polarization values of 84% could be obtained[4] whereas the GaAs source yielded a beam of about 400 nA with 30% polarization.

The polarized electrons were scattered by the target consisting of either mercury vapour or xenon gas. In order to measure the change of the electron polarization in the elastic scattering process the electrons pass through a filter lens and a Wien filter and, after acceleration, are scattered by a gold foil in the Mott detector for polarization analysis (see Fig. 1). With two pairs of counters the transversal polarization components are measured simultaneously while the Wien filter is switched off, whereas the longitudinal component must be rotated by the Wien filter through 90° for detection in the Mott analyzer.

From the measured polarization values the parameters S, T and U can be derived. In the special case where the primary polarization $\vec{P}$ lies in the scattering plane the polarization $\vec{P}'$ of the scattered electrons can be described by

$$\vec{P}' = S\hat{n} + T\vec{P} + U\hat{n}\times\vec{P} \qquad (1)$$

where $\hat{n}$ is the unit vector normal to the scattering plane.

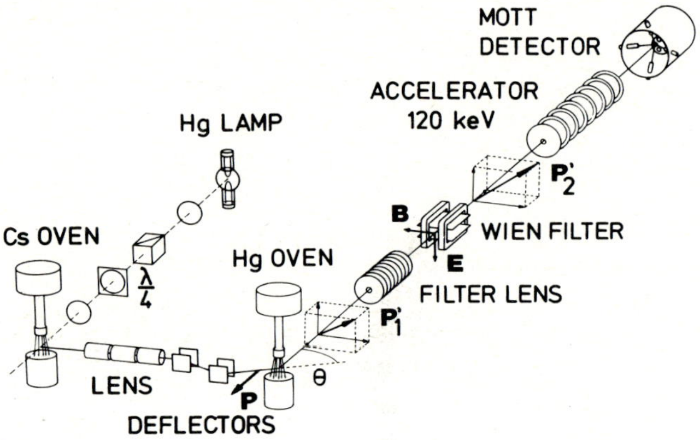

Fig. 1: Schematic diagram of the apparatus.

The results obtained in the experiments are compared with theoretical calculations[5] and show considerable deviations from theory in some cases. It must be considered, however, that the measurement of the polarization parameters is a very sensitive tool for the test of theoretical models.

The complete evaluation of the scattering amplitudes can be presented for xenon since the absolute differential cross section has already been investigated by different groups in the energy range of interest. For mercury only one absolute cross section measurement at 300 eV has so far been published, in that energy range but measurements of the absolute differential cross section at low and medium energies are in progress in Münster so that the determination of the scattering amplitudes will be possible in the future.

The apparatus described in our recent papers has now been modified for measurement of angular distributions of the polarization parameters. In a preliminary stage the scattering angle can be varied continuously in a limited angular range with the help of deflector plates as shown in Fig. 1, and first results obtained with this arrangement will be presented. The construction of a rotatable source of polarized electrons is in progress by which angular distributions of the polarization parameters can be measured in a wide angular range.

References

1. J. Keßler, Polarized Electrons (Springer, Berlin, (1976)
2. W. Wübker, R. Möllenkamp, and J. Keßler, Phys. Rev. Letters 49, 272 (1982)
3. O. Berger, J. Keßler, K.J. Kollath, R. Möllenkamp, and W. Wübker, Phys. Rev. Letters 46, 768 (1981)
4. R. Möllenkamp and U. Heinzmann, J. Phys. E 15, 692 (1982)
5. D.W. Walker, Adv. Phys. 20, 257 (1971), and private communication

THE POLARIZATION OF 254 nm MERCURY LINE RADIATION AFTER IMPACT EXCITATION BY POLARIZED ELECTRONS

A. Wolcke, J. Goeke, W. Vollmer, G.F. Hanne and J. Kessler

Physikalisches Institut der Universität Münster, D 4400 Münster, W. Germany

We report on an electron-photon coincidence experiment with initially polarized electrons. Such experiments yield direct information about magnitudes and relative phases for scattering amplitudes describing spin-dependent inelastic collisions. The spin-dependence of electron-atom collisions may be caused, for instance, by spin-orbit interaction or exchange. In previous experiments with unpolarized electrons[1,2] only spin averaged parameters could be measured. Spin effects were observed in these experiments indirectly by obtaining a reduction of the degree of coherence[1], i.e., the emitted light was only partially polarized.

In our new experiment, parameters can be measured that are nonzero only for initially polarized electrons; if, e.g., the spin polarization vector $\underset{\sim}{P}$ has a component in the scattering plane reflection invariance does not hold[3]. As a result, circular light polarization may occur for photons emitted in the scattering plane.

We have investigated the polarization of 254-nm-photons emitted from mercury atoms (6^3P_1 - 6^1S_0 transition) after collisional excitation by polarized electrons (6^1S_0 - 6^3P_1). The photons are detected in coincidence with the scattered electrons without observing the final electron polarization. A schematic diagram of the apparatus is shown in Fig. 1. Longitudinally polarized electrons are emitted from a GaAs photocathode which is irradiated with circularly polarized laser light and placed into ultrahigh vacuum. After deflection of the electrons by 90° their polarization may be rotated by two magnetic coils through 90° to be perpendicular to the xz plane. The electrons then pass through a differential pumping stage where they are again deflected by 90°. A lens system focuses the polarized electron beam onto the mercury target. At the target the electrons are either longitudinally polarized (magnetic spin rotator off) or transversely polarized (spin rotator on). Some of the mercury atoms are excited by electron impact. A photon analyser system measures the polarization of the 6^3P_1 - 6^1S_0 line radiation (254 nm) as a function of collision energy between 4.5 and 15 eV.

In the present experiment the photon analyser detects photons which are emitted in the direction of the electron spin polarization (y axis) in coincidence with electrons scattered in the forward direction (z axis). The collision is then described by only two independent scattering amplitudes. According to a discussion by Bonham[4] the relative phase of these two amplitudes may tend to zero for zero scattering angle. This would

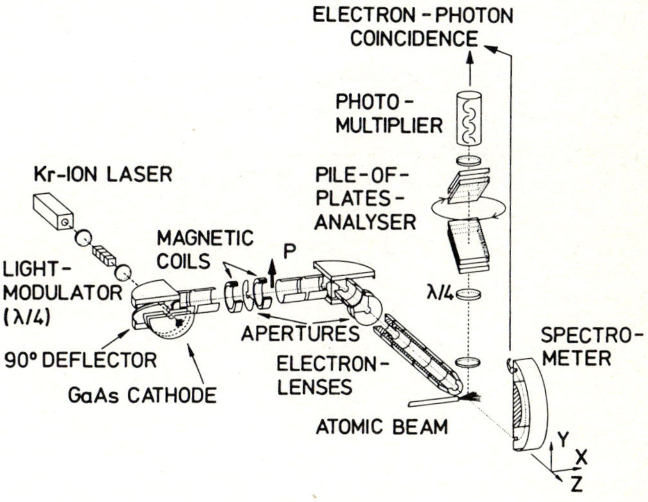

Fig. 1: Schematic diagram of the apparatus

mean that from the three Stokes parameters that may in principle be obtained in such experiments[3], one, the Stokes parameter $\eta_1 = [I(45°)-I(135°)]/[I(45°)+I(135°)]$ is zero, where $I(\alpha)$ is the fraction of intensity transmitted by a linear polarization filter which is inclined with respect to the z axis by an angle α. Our experiment will test this argument, giving also a complete determination of the two scattering amplitudes involved. In such experiments, the influence of hyperfine interaction needs consideration, too[5]. First results will be given.

References

1. A.A. Zaidi, I. McGregor and H. Kleinpoppen, Phys. Rev. Letters 45, 1168 (1980)
2. G.F. Hanne, K. Wemhoff, A. Wolcke and J. Kessler, J. Phys. B 14, L 507 (1981)
3. K. Bartschat, K. Blum, G.F. Hanne and J. Kessler, J. Phys. B 14, 3761 (1981)
4. R.A. Bonham, J. Phys. B 15, L 361 (1982)
5. C.W. McLucas, H.J.E. Wehr, W.R. MacGillivray and M.C. Standage, J. Phys. B 15, 1883 (1982)

IMPACT EXCITATION OF THE 7 $^3S_1$-STATE OF MERCURY BY POLARIZED ELECTRONS

H. Wolf, E. Reichert

Institut für Physik, Johannes Gutenberg-Universität, Mainz, W-Germany

Following a proposal of J. Wykes[1] the circular polarization P_c of light emitted by mercury atoms excited to the 7 $^3S_1$-state by impact with polarized electrons is studied. The primary electron beam transversely polarized to a degree of P_e = 0.4 is produced using a negative affinity GaAsP-photocathode irradiated with circurlar polarized 647 nm light from a Kr^+-ion laser- The mercury 7 $^3S_1 \rightarrow 6\ ^3P_o$ = 404.7 nm line is analysed emitted in direction parallel to the polarization vektor of the primary electrons. At a collision energy of 20 eV P_c/P_e = 0.1 is observed. Measurements will be extended down to collision energies near 7 $^3S_1$-threshold.

References

1. J. Wykes, J.Phys. B. Atom.Molec.Phys. 4, L91 (1971)

EXCITATION OF Hg-ATOMS BY ELECTRON IMPACT

K.Bartschat[+], N.S.Scott[++], K.Blum[+] and P.G. Burke [++]

[+]Inst. f. Theor. Physik I, Universität Münster,Domagkstr., W.-Germany

[++]Dept. of Appl. Math. and Theor. Physics, Queen´s University, Belfast BT7 1NN, N. Ireland

Recently the R-matrix method describing the scattering of low energy electrons by complex atoms has been extended to include terms of the Breit-Hamiltonian (Pauli approximation) [1] . This theory has been applied to the excitation of the 6^3P_1 state of Hg [2]. First numerical results for observable quantities (spin polarisation, asymmetry function, integrated state multipoles) will be presented and compared to recent experimental results [3,4]. Of particular interest are those parameters which can directly be related to specific spin-dependent interactions [5] . An analysis of the numerical results will be performed in order to obtain detailed information on specific aspects of the spin-coupling mechanism.

References

1. N.S. Scott and P.G. Burke, J.Phys. B 13, 4299 (1980)
2. N.S. Scott, P.G. Burke and K.Bartschat J.Phys.B (1983), to be published; see also Book of Abstracts, IPEAC 1983
3. K. Bartschat, G.F.Hanne, A.Wolcke and J. Kessler, Phys.Rev.Lett.47, 997 (1981)
4. A. Wolcke, K.Bartschat, K.Blum, H.Borgmann, G.F. Hanne and J. Kessler, J.Phys.B 16, 639 (1983)
5. K. Blum and H. Kleinpoppen in: "Advances in Atomic and Molecular Physics" (1983), to be published.

ELECTRON IMPACT EXCITATION OF POLARIZED NA-ATOMS

W. Jitschin, S. Osimitsch, H. Reihl, H. Kleinpoppen[+], H. O. Lutz

Fakultät für Physik, Universität Bielefeld, F. R. Germany
[+]Atomic Physics Laboratory, University of Stirling, Stirling, Scotland

Collisional excitation can occur by direct excitation of the atomic electron and by electron exchange; the corresponding amplitudes in general also interfere. Decisive information on the different interaction mechanisms can be obtained experimentally by employing appropriate combinations of the use of spin-polarized collision partners and polarization analysis after the collision[1].

In the present experiment, unpolarized electrons were impinging on polarized Na-atoms:

$$e + Na(\uparrow) \rightarrow e + Na^*(3p)$$
$$\downarrow$$
$$Na(3s) + h\omega$$

In this case four incoherent channels can yield excitation (nomenclature as in ref. 2):

- exchange excitation to Na(3p m= 0), cross section E_0
- exchange excitation Na(3p m=±1), E_1
- direct/interference Na(3p m= 0), D_0+I_0
- direct/interference Na(3p m=±1), D_1+I_1

The cross section difference of the magnetic substates m=0, ±1 reflects the alignment of the excited 3p state and can experimentally be observed by the linear polarization of the induced fluorescence light. The spin polarization of the 3p electron yields information about the relative contributions of exchange and direct/interference interaction to the total interaction since the exchange results in an unpolarized 3p electron in contrast to the other types of interaction. A spin-polarization of the 3p electron is converted to an unequal population of the $-m_j$ and $+m_j$ excited states by the spin-orbit coupling. This *orientation* of the 3p state can be measured via the circular polarization of the induced fluorescence light.

The apparatus is shown in Fig. 1. Na-atoms are polarized by optical pumping of the 3s(F=2) to 3p(F=3) transition. A commercial frequency stabilized dye laser

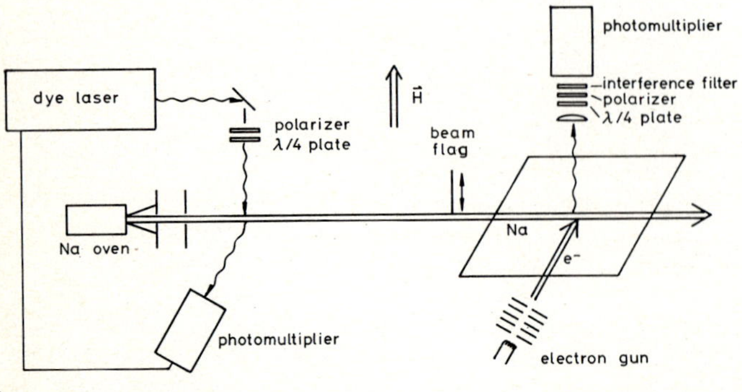

Fig. 1: Experimental setup.

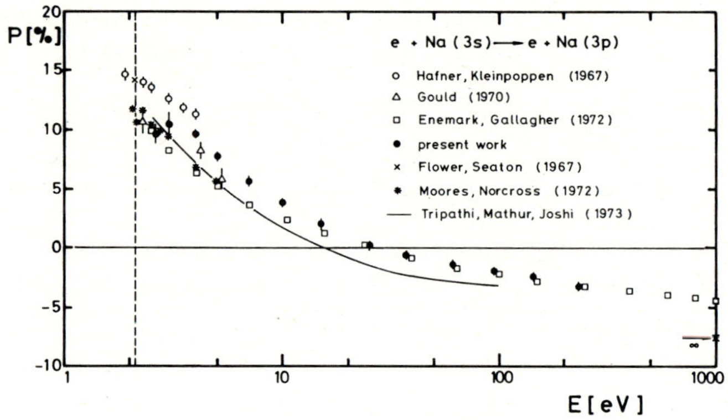

Fig. 2: Linear polarization P of the induced light.

is additionally locked to the atomic transition by a servo loop; this ensures an exact tuning of the laser frequency for a period of several hours. The Na polarization amounts to ca. 60%. In a test run the linear polarization of the fluorescence light was measured for unpolarized atoms (Fig. 2). The agreement with other experimental work and theoretical predictions seems satisfactory.

Measurements with polarized Na-atoms yielded a circular polarization of the induced light of 30±5% for electron energies ranging from 10 to 250 eV. There is no simple relation between measured circular polarization and the different excitation cross sections due to the complex hyperfine coupling. A preliminary theoretical analysis shows that the circular polarization originates partly from the nuclear spin polarization (the nuclear spin is also polarized by the optical pumping process) and thus bears no information on the electronic excitation process. Nevertheless, one expects a drop of the circular polarization close to threshold where the exchange contribution becomes significant[2]. Corresponding experiments are in preparation.

The work has been supported by the Deutsche Forschungsgemeinschaft. Technical support by Dr. Ġ. Meisel is gratefully acknowledged.

References

1. H. Kleinpoppen, Phys. Rev. A 3, 2015 (1971)
2. D. L. Moores and D. W. Norcross, J. Phys. B 5, 1482 (1972)

HOW DO WE DECIDE WHETHER THE FIRST BORN APPROXIMATION APPLIES TO INELASTIC COLLISIONS OF CHARGED PARTICLES WITH AN ATOM OR MOLECULE?*

Mitio Inokuti

Argonne National Laboratory, Argonne, Illinois 60439, U.S.A.

and

S. T. Manson

Department of Physics & Astronomy, Georgia State University, Atlanta, Georgia 30303, U.S.A.

Since the 1930 work by Bethe,[1] we all know that, if the first Born approximation (FBA) applies, then the scattering amplitude is a function of the single scalar variable, i.e., the magnitude of the momentum transfer $\hbar K$, apart from a kinematic factor. In other words, the angular distribution of inelastic scattering is completely governed by $\hbar K$, and does not depend explicitly upon the incident-particle energy. This is true so long as a target atom or molecule is either spherical or is randomly oriented so that one may assume the rotational symmetry of the target. The above statement is a straightforward consequence of the result that the FBA amplitude is essentially determined by the matrix element of the operator $\Sigma_j \exp(i\vec{K}\cdot\vec{r}_j)$ taken between the initial eigenstate and the final eigenstate of the target.[2] Lassettre and co-workers[3-5] used the statement skillfully and systematically in their analysis of electron-impact data.

A motivation of our study is to help resolve a general issue in atomic-collision physics. There are two major sources of uncertainties in the evaluation of cross sections. First, one uses an approximation for treating the collision process, e.g., the FBA, the distorted-wave approximation, or the close-coupling approximation. Second, explicit evaluation of cross sections within any of these approximations must use as input eigenfunctions for the target in the initial state and in the final state at least, and possibly in the intermediate states. It is important to distinguish these two sources of uncertainties as clearly as possible. For instance, once we are sure that the FBA holds, then uncertainties in the cross-section evaluation are fully attributable to the uncertainties in the target eigenfunctions.

The well-known result stated in the first paragraph is a necessary condition for the applicability of the FBA, and has not been regarded as a sufficient condition.[2-5] In other words, if the FBA applies, then the magnitude $\hbar K$ of the momentum transfer determines all the dynamics of the inelastic collision.

We have studied for some time the logical converse of the well-known result. In other words, if the magnitude of the momentum transfer completely determines the scattering amplitude or the angular distribution of inelastic scattering, then are we sure that the FBA holds? To paraphrase it further, we have seriously asked whether a sufficient condition for the applicability of the FBA is the dependence of the scattering amplitude solely on the magnitude of the momentum transfer. There are many other ways to pose virtually the same question. For instance, if an observed angular distribution of secondary electrons of fixed kinetic energy is axially symmetric around the vector momentum transfer $\hbar\vec{K}$, then can we say that the scattering amplitude should be fully described within the FBA? Another example concerns the polarization of excited states resulting from electron or other charged-particle impact.[6] If the FBA holds, then the target polarization must be completely determined by the momentum-transfer direction; in other words, the final state of the target must have the magnetic quantum number $M = 0$, with respect to the quantization axis along vector $\hbar\vec{K}$. Is the logical converse of this statement true?

We now have a tentative answer to the question in the affirmative. That is to say, if the angular distribution of inelastic scattering is fully determined by the magnitude of the momentum transfer $\hbar K$ (but is independent of its direction), then the same angular distribution will be given by the FBA. We have no completely rigorous proof for this, but have strong plausibility arguments.

*Work performed under the auspices of the U.S. Department of Energy, and of the National Science Foundation.

References

[1] H. Bethe, Ann. Phys. (Leipzig) 5, 325 (1930).
[2] M. Inokuti, Rev. Mod. Phys. 43, 297 (1971).
[3] E. N. Lassettre and S. A. Francis, J. Chem. Phys. 40, 1208 (1964).
[4] E. N. Lassettre, Can. J. Chem. 47, 1733 (1969).
[5] E. N. Lassettre, A. Skerkele, and M. A. Dillon, J. Chem. Phys. 50, 1829 (1969).
[6] U. Fano and J. H. Macek, Rev. Mod. Phys. 45, 553 (1973).

ABSOLUTE TRIPLE DIFFERENTIAL CROSS SECTIONS FOR HIGH ENERGY ELECTRON IMPACT IONIZATION OF HELIUM

A. Lahmam-Bennani, H.F. Wellenstein*, A. Duguet
L.C.A.M., Bât. 351, Université Paris-Sud, 91405 ORSAY Cedex, FRANCE
*Phys. Dept., Brandeis University, Waltham, Mass. 02154, USA

An 8 keV electron beam ionizes a helium gas jet, and the two outgoing electrons (denoted a for the scattered and b for the ejected) are detected in coincidence. All the kinematical variables of the collision are determined : energies E_o, E_a, E_b, momenta $\vec{k}_o$, $\vec{k}_a$, $\vec{k}_b$, and the scattering angles θ_a, θ_b. The *triple* differential cross sections (TDCS) for the single ionization of He, $d^3\sigma/dE_a d\Omega_a d\Omega_b$, are measured by fixing all variables but θ_b and recording an angular distribution of the ejected electron.

The TDCS are made absolute by setting the integral of the coincidence intensity over all ejection directions equal to the *double* differential cross section (DDCS), $d^2\sigma/dE_a d\Omega_a$, which we have separately measured onto an absolute scale for the E_a and θ_a values of the experiment. The overall uncertainty in this normalization procedure is estimated to be ≤ 8 %.

Each angular distribution is characterized by 2 lobes, the binary and the recoil lobes, respectively peaking in the $\vec{K}$ and $-\vec{K}$ direction (k= momentum transfer)

Several such angular distribution have been obtained with different values of the couple of parameters (θ_a, E_b). These couples have been chosen to be representative of different interaction mechanisms.

In the impulse approximation, (large θ_a and E_b values), the TDCS are proportional to the electron momentum density in the target, $\rho(p)$. As θ_a or K increases, the quantity $\rho(p)$ so deduced from our measurements progressively approaches the theoretical $\rho(p)$ obtained by Fourier inversion of the He wave function tabulated by Froese-Fischer. The agreement becomes excellent at K=5.1 a.u. Under these conditions the high energy (e, 2e) reactionis a probe for the target structure.

At smaller θ_a or E_b values, the (e, 2e) reaction becomes a sensitive test for the description of the collision mechanism.

Several theoretical models have been used to calculate the TDCS in our experimental conditions and will be discussed at the Conference time.

ABSOLUTE TRIPLE DIFFERENTIAL CROSS SECTIONS FOR HIGH ENERGY ELECTRON IMPACT IONIZATION OF ARGON 3p AND 2p

A. Lahmam-Bennani, H.F. Wellenstein*, A. Duguet, A. Daoud

L.C.A.M. Bât. 351, Université Paris-Sud, 91405 ORSAY Cedex, FRANCE
*Phys. Dpt., Brandeis University, Waltham, Mass. 02154, USA

The TDCS for the 3p- and 2p- ionization of Argon have been measured (for the first time on an inner orbital) by recording the in-plane angular distribution of the ejected electrons at several fixed values of the scattering angle θ_a (For notation, see preceding paper). The energy balance allows to choose the orbital where the ionization takes places.

As for He, two lobes are observed, on the binary and recoil sides. The relative intensity of the recoil lobe increases as K (or θ_a) decreases, becoming more important than the binary one for the smallest K values observed here.

The data are made absolute by setting the integral over all ejection directions equal to the corresponding double differential cross section (DDCS), assuming a cylindrical symmetry about the $\vec{K}$ direction as suggested from the in-plane symmetry of the data. The DDCS are either theoretically determined at large K values using an impulse Hartree-Fock model (Ar-3p case), or separately measured onto an absolute scale for the E_a and θ_a values of the experiment (Ar-2p case). The overall uncertainty is this normalization procedure is $\sim$10% and $\sim$20%, respectively.

For the Ar-3p, the TDCS are converted into momentum distribution, $\rho(p)$, and found to be in very good agreement for $K \gtrsim 3$ a.u. with the Ar-3p Clementi wavefunction. The agreement rapidly vanishes at smaller K Values.

For the Ar-2p, the K and E_b values used in this work are too small for the impulse approximation to be satisfied, hence no attempt was made to determine $\rho(p)$.

Several theoretical models have been used to calculate the TDCS in our experimental conditions and will be discussed at the Conference time.

DIFFERENTIAL CROSS SECTIONS FOR THE $^1S \to ^1S, ^3S$ TRANSITIONS IN He BY ELECTRON IMPACT

T.Takayanagi, K.Wakiya and H.Suzuki

Department of Physics, Sophia University, Chiyoda-ku, Tokyo 102, JAPAN

S.Ito, K.Hoshiba, S.Kano and H.Takuma

Institute for Laser Science, University of Electro-Communication, Chofu-shi, Tokyo 182 JAPAN

In the case of excitations to the ^1S and the ^3S states in helium by electron impact, it is known that the first order plane-wave approximations (Born or Born-Oppenheimer approximation) fail to give accurate differential cross sections at small scattering angles for rather high electron impact energies such as 400 and 500 eV[1)-6)].

Huo found that the second terms in Born series contain effective transition potentials like $1/r^4$, which contribute largely to small angle scattering in the case of S→S transitions[7)]. Many kinds of theoretical methods were proposed to calculate accurate differential cross sections, and they were compared with one another and with experimental results in the reviews of Bransden and McDowell[8)]. But more accurate experiments are needed especially for the scattering angles smaller than 10° to compare theoretical methods more precisely. Because, in this range of scattering angle, there exist very small numbers of experiments and theories give results which differ largely with one another.

We measured the differential cross sections of 2^3S, 2^1S, 3^3S and 3^1S excitations for the impact energy E=50-500eV and scattering angle θ=2° to 25° using the electron energy-loss method. The cylindrical electro-static energy selector[9)] with the mean radius of 50mm is used as the monochromator. The same type selector with the mean radius of 80mm is used as the energy analyzer for scattered electrons. The electron gun and the monochromator can be rotated around the gas nozzle. The overall energy resolution is 40-60meV (FWHM), and typical electron current is 10-50nA. The He gas beam is effused through a 10mm long 1mm diameter tube.

Typical energy loss spectrum is given in Fig.1. (E=50eV, θ=21°). The differential cross sections of 2^3S, 2^1S, 3^3S and 3^1S were determined relative to that of 2^1P, whose experimental value at scattering angle larger than 5° can be extrapolated to 0° using optical oscillator strength.

References.

1) L.Vriens, J.A.Simpson and S.R.Mielczarek, Phys. Rev. 165 7 (1968)
2) A.Skerbele and E.N.Lassettre, J. Chem. Phys. 45 1077 (1966)
3) H.Suzuki and T.Takayanagi, Abstr. VIII ICPEAC (Beograd, Inst. of Physics, 1973) I, 286
4) A.Skerbele, W.R.Harshbarger and E.N.Lassettre, J. Chem. Phys. 58 4285 (1973)
5) M.A.Dillon, J. Chem. Phys. 63 2035 (1975)
6) A.Yagishita, T.Takayanagi and H.Suzuki, J. Phys. B9 L53 (1976)
7) W.M.Huo, J. Chem. Phys. 56 3468 (1972)
W.M.Huo, J. Chem. Phys. 60 3544 (1974)
8) B.H.Bransden and M.R.C.McDowell, Phys. Report 30 207 (1977)
B.H.Bransden and M.R.C.McDowell, Phys. Report 46 249 (1978)
9) K.Jost, J. Phys. E : Sci. Instrum. 12 1006 (1979)

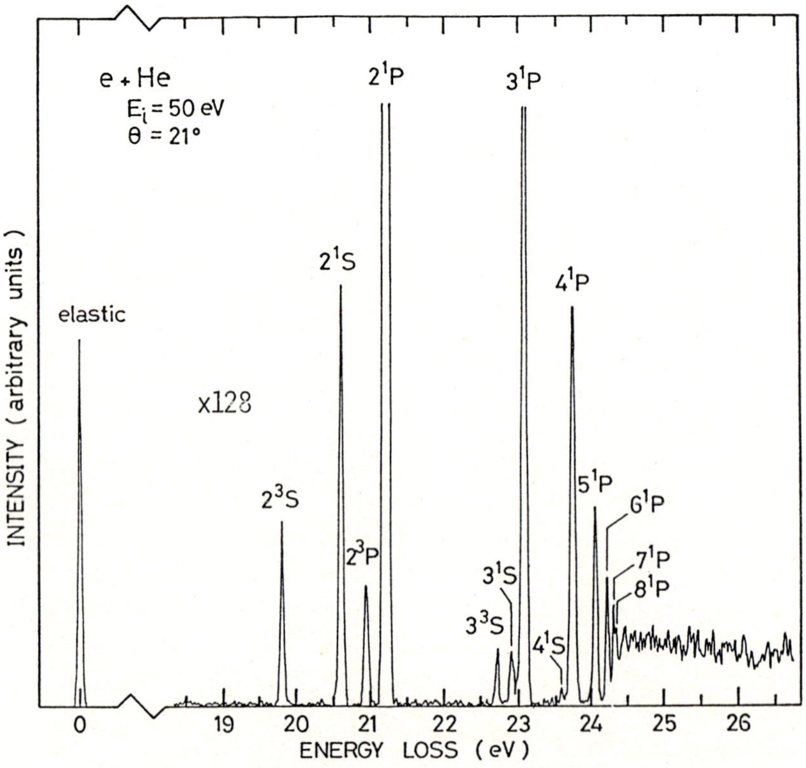

Fig.1: Energy loss spectrum of He

MEASUREMENTS OF INNER SHELL EXCITATION CROSS SECTIONS IN Ne BY MEANS OF AUGER ELECTRON SPECTROSCOPY BY ELECTRON IMPACT

S.Kihara, Y.Iketaki, T.Takayanagi, K.Wakiya and H.Suzuki

Department of Physics, Sophia University, 7-1, Kioi-cho, Chiyoda-ku, Tokyo, 102, Japan

We report a new experimental procedure and results of measurements of the Ne K-shell excitation (1s→3s,1s→3p) cross sections based on the Auger electron spectroscopy.

We noticed the excitation-Auger-satellite lines $(1s^2 2s^2 2p^4)^1D(3s)^2D$ and the $(1s^2 2s^2 2p^4)^1D(3p)^2D$, which appear at about 8eV higher energy side of the K $L_{23}L_{23}(^1D)$ normal Auger line. We measured intensity ratios of these satellite peaks relative to the K $L_{23}L_{23}(^1D)$ normal Auger line. Absolute cross sections for the K-shell excitations were obtained by normalizing the intensity ratios to the absolute cross section of the normal Auger process which was determined by H.Klar[1)], H.Tawara[2)] and G.Glupe[3)].

Schematic view of apparatus is shown in Fig.1. An accelerated and collimated electron beam collides in a scattering centre with Ne atoms. Impact energy was from about 900eV to 2keV. The Auger electron spectra were measured by a 127° cylindrical electrostatic analyzer against the direction of the electron beam. The analyzer was operated by a constant resolution mode and resolution of about 300meV was employed through this experiment.

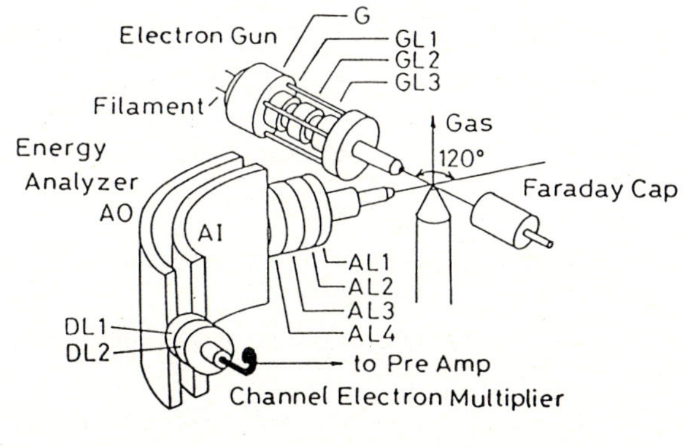

Fig.1: Schematic view of the electron spectrometer

A typical spectrum is shown in Fig.2. In order to reduce the statistical error in the determination of the intensity ratios, we performed a data smoothing procedure using Moving Average Method in the spectral intensities of the excitation-Auger-satellite lines.

Experimental results are shown in Fig.3. The 1s→3s excitation cross sections which characterlize the optically forbidden transition rise rapidly near the threshold. In contrast to this, the 1s→3p excitation cross sections show a slower rise that is peculiar to the optically allowed transition, and show a broad maximum at about 1300eV.

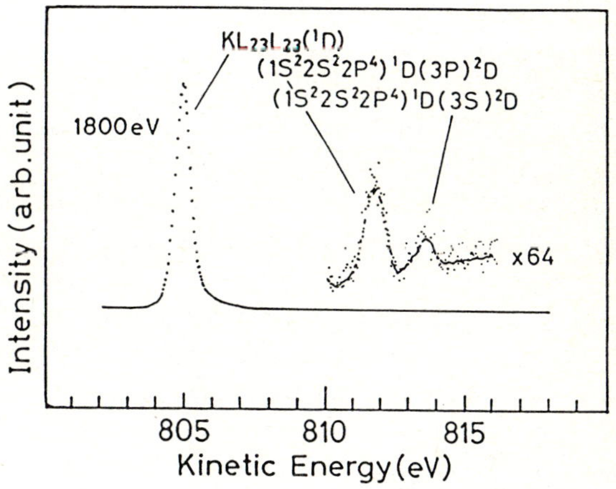

Fig.2: A typical spectrum of the KLL(^{1}D) normal Auger line and the excitation-Auger-satellite lines.

The 1s→3p excitation cross sections are about three times larger than the 1s→3s excitation cross sections at the higher impact energy, although the 1s→3s excitation cross sections are larger than the 1s→3p ones at very close region to the threshold.

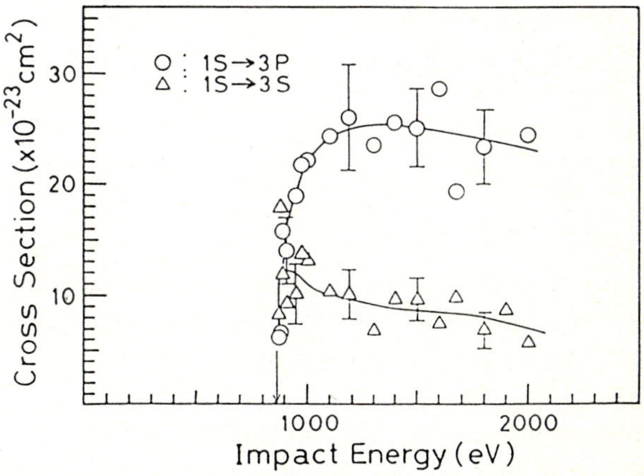

Fig.3: Cross sections for the 1s→3p and the 1s→3s excitation in Ne as functions of the impact energy.

References

1) H.Klar, J.Phys. B 14 (1981) 3265

2) H.Tawara, G.Harrison and F.J.de Heer, Physica 63 (1973) 351

3) G.Glupe and W.Mehlhorn, J.Physique 32 (1971) C4-40

ELECTRON IMPACT EXCITATION OF LITHIUM 3S STATE IN THE INTERMEDIATE ENERGY

Mukesh Kumar, S.S. Tayal and A.N. Tripathi

Department of Physics, University of Roorkee,Roorkee 247667,INDIA

Considerable attention has been devoted in past to study the discrete excitation of sodium,potassium and caesium alkali atom but relatively less effort was made to study discrete higher excitation of lithium atom. There is only one measurement due to Williams et al[1] exist to date for differential scattering cross section(DCS)of electrom impact excitation of higher states of lithium. The study of DCS of optically forbidden transition at intermediate energy needs careful examination from the theoretical point of view. There are various methods used in this energy region. Among them, the eikonal Born series(EBS) and its variants,distorted wave Born approximation (DWBA) and the second order potential method are a few worth mentioning.

We have in the present investigation used the modified Glauber (MG), EBS, simplified second Born (SSB), Glauber and first Born (FBA) to study the electron impact excitation of 2s→3s transition in lithium. The DCS in MG is given by:

$$f_{MG} = f_G - f_{G2} + f_{B2}$$
$$= f_{G1} + f_{B2} + f_{G3} + \sum_{n=4}^{\infty} f_{Gn}$$
$$= f_{EBS} + \sum_{n=4}^{\infty} f_{Gn}$$

The SSB amplitude is evaluated by considering the lithium atom as one electron system with inert core. We have used a simple wave function for 2s, 3s state provided by Hibbert (private communication).The mean excitation energy was chosen following the prescription of Byron and Latour[2] and Winter and Vanderpoorten[3].

We present our results at 20 eV for which the experimental data is available over wide angles. We show on our curve the other theoretical results obtained in DWBA[3] and two potential modified Born(TPMB)[4]. It is seen that in the region of small scattering angle, prior to the dip the MG results show a good agreement with the data where DWBA underestimate. At larger angles the present

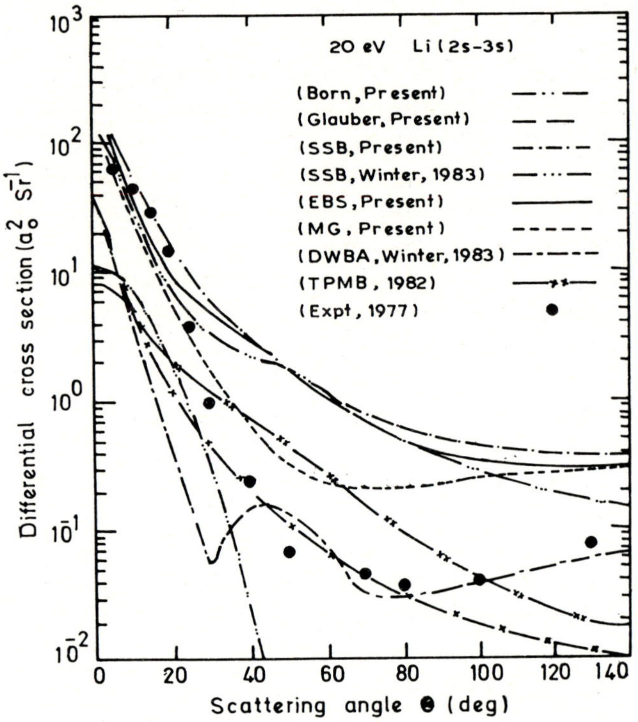

calculation in MG, EBS and SSB stand higher, showing the poor convergence of the Born series. A detailed discussion and results will be presented at the conference.

References:

1. W.Williams, S.Trajmar and D. Bozinis, J.Phys. B9, 1529 (1976).

2. F.W.Byron Jr., and Latour, Phys.Rev. A13, 649 (1976).

3. K.H. Winters and R.Vanderpoorten,J.Phys. B15, 3945 (1982).

4. R.K.Sharma and K.C.Mathur, Phys. Rev. A26, 1122 (1982).

CROSS SECTIONS OF INERT GASES FOR VUV EMISSIONS FOLLOWING INNER-SHELL OR SUBSHELL IONIZATION BY ELECTRON IMPACT

Y.Akagi, K.Morita, T.Takayanagi, K.Wakiya and H.Suzuki

Department of Physics, Sophia University, 7-1 Kioi-cho, Chiyoda-ku, Tokyo 102, JAPAN

In order to investigate radiative transitions whose initial states are the inner-shell or subshell excited states, we have measured light emissions in the VUV wavelength region.

The apparatus which we used for the present measurments is as follows. An electron-collision chamber is attached to the entrance slit of a 50cm VUV monochromator of the Seya-Namioka mounting. The incident angle of the electron beam was fixed to 55° with respect to the entrance axis of the monochromator, in order to avoid the anisotropy effect on the light emission from atomic beam by the electron impact. The photons were detected by channel electron multiplier using pulse counting techniques. Wavelength is scanned by rotating the grating on the axis through the grating plane. The counting rate of the phton detector is recorded as a functions of the wavelength to obtain a whole spectrum.

A typical emission spectrum for Xe is shown in Fig.1. The lines of interest, which are caused by the inner-shell transitions $(5s^0 5p^6)\,^1S \rightarrow (5s^1 5p^5)\,^1P$ and $(5s^1 5p^5)\,^1P \rightarrow (5s^2 5p^4)\,^3P$ [1] are indicated in Fig.1, in addition to the resonance lines which are caused by the outer shell transitions $(5p^6)\,^1S_0 \rightarrow 5p^5(^2P_{3/2})6s$, $5p^5(^2P_{3/2})6p$ and $5p^5(^2P_{3/2})5d$. The $(5s^0 5p^6)\,^1S_0 \rightarrow (5s^1 5p^5)\,^1P_1$ and $(5s^1 5p^5)\,^1P_1 \rightarrow (5s^2 5p^4)\,^3P$ transitions follow the 4d-vacancy state through the $N_{4,5}OO$ Auger transitions. We measured the excitation functions of these lines for impact energies from the 4d ionization threshold to 1.5keV. These excitation functions are shown in Fig.2. In this figure the $N_{4,5}O_1O_1$ Auger electron emission cross sections which were measured in our laboratory previously are also plotted.

The excitation function of the $(4s^0 4p^6)\,^1S_0 \rightarrow (4s^1 4p^5)\,^1P_1$ transition in Kr, which arises from the 3d-vacancy state through the $M_{4,5}N_1N_1$ Auger transition, is shown in Fig.3. The $M_{4,5}N_1N_1$ Auger electron emission cross sections of Kr are also plotted. In both cases the VUV emission cross sections show good agreements with Auger electron emission cross sections within the experimental errors.

We gave the absolute scale of the excitation cross sections by the following procedure: 1) We determined a relative efficiency of the system of the monochromator and the detector as a function of wavelength, using the relative intensities of Lyman series of H atom dissociated from NH_3 by electron bombardment.[2] 2) Absolute scale of the cross section was deduced from the intensity of the transitions of interest rerative to that of the resonance radiations from the neutral atoms, using absolute cross sections for resonance lines measured by Williams and Trajmar [3] by the methode of electron energy loss.

We are doing same type of measurments with respect to the 2s ionization of Ne and the 3s ionization of Ar.

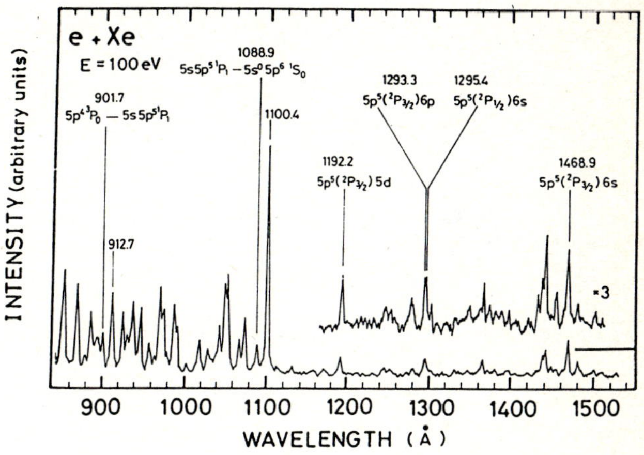

Fig.1: A typical VUV spectrum of Xe.

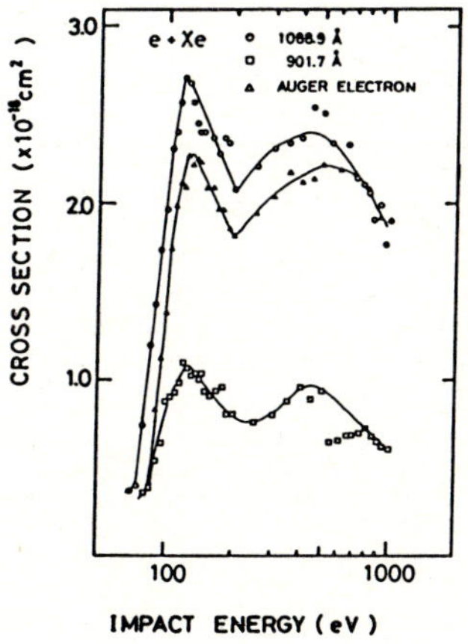

Fig.2: Excitation function in Xe.

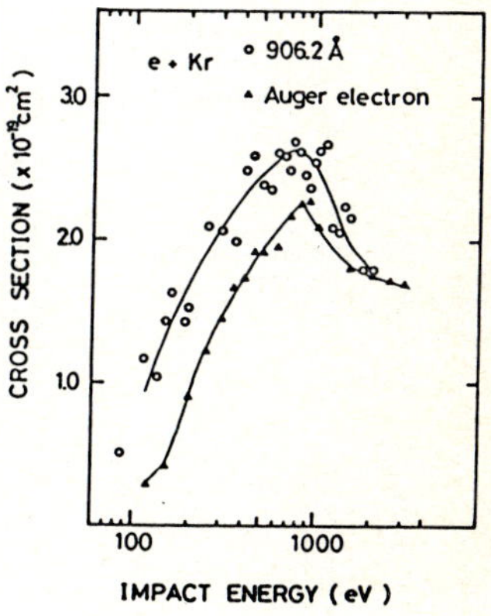

Fig.3: Excitation function in Kr.

References

1) H.Hertz, Z.Physik, A272, 289 (1975)
2) N.Bose and W.Sroka, Z,Naturforsch, 26a, 1491 (1971)
3) W.Williams and S.Trajmar, J.Chem.Phys, 62, 3031(1975)

K-SHELL IONISATION CROSS SECTIONS OF SI AND AR BY ELECTRONS WITH IMPACT ENERGIES 4 TO 10 KEV

H. Platten, G. Schiwietz, N. Stolterfoht and G. Nolte

Hahn-Meitner-Institut für Kernforschung Berlin GmbH, Glienickerstr. 100, D-1000 Berlin 39

The K-Auger electron emission of Si (SiH_4) and AR induced by 4 to 10 keV electrons was measured using an electrostatic spherical condensor (McPherson ESCA 36) with an energy resolution of 0.15%. The large effective energy range of the spectrometer (5eV to 50keV) allows for normalisation of the Auger intensities to the intensity of the elastic scattered primary electrons. Using the elastic scattering cross sections from Riley et al.[1] absolute K-shell ionisation cross sections were deduced. This method was tested by measuring cross sections for K-shell ionisation of O and Ne and by comparing them with the results of Glupe and Mehlhorn[2]. Good agreement was obtained.

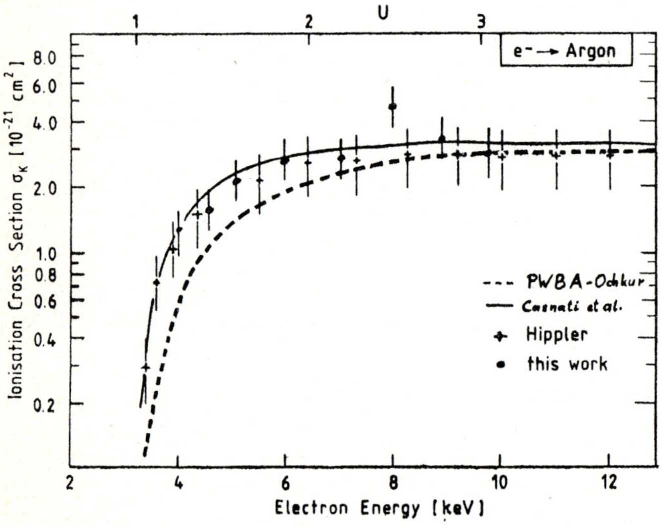

Fig. 1: K-shell excitation function of Ar induced by electrons. U=E/I: reduced energy, E: projectile energy, I: K-shell binding energy.

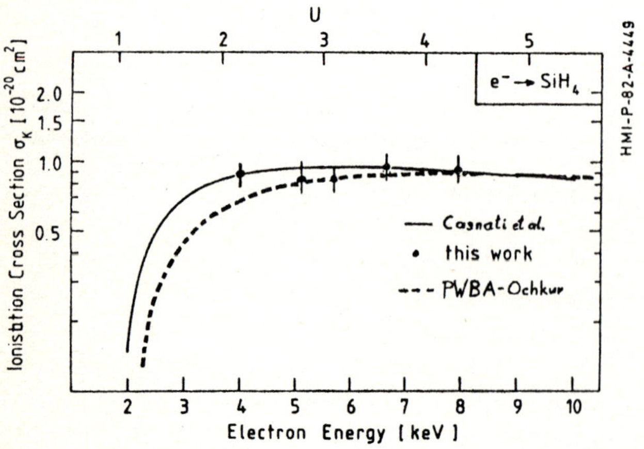

Fig. 2: K-shell excitation function of Si induced by electrons.

Fig. 1 shows the results for the Ar-K shell. The data of Hippler et al.[3] are also shown. They are obtained from x ray measurements normalizing the measured Bremsstrahlung intensity to theoretical results. Within the limits of error both data sets agree well. Also Fig. 1 shows absolute K-shell ionisation cross sections calculated from the analytical fit-formula of Casnati et al.[4] based on the Bethe theory[5]. The experimental data are well described by this formula even in the range of very low impact energies. The broken line in Fig. 1 gives the K-shell ionisation cross sections calculated within the PWBA taking into account the electron exchange in the Ochkur approximation[6]. The cross sections based on the PWBA-Ochkur describe the experimental data for higher energies quite well, while at lower energies the experimental data are systematically underestimated.

Fig. 2 shows the measured absolute cross sections for K-shell ionisation of Si. The analytic fit-formula again agrees well with the experimental data. Also the PWBA-Ochkur calculations are shown in Fig. 2.

References

1. M.E. Riley, J. Crawford, Mac Callum, F. Biggs, Atomic Data and Nuclear Tables 15, 443 (1975)
2. G. Glupe, W. Mehlhorn, J. Physique C4-40 (1971)
3. R. Hippler, K. Saeed, I. McGregor, H. Kleinpoppen, Z. Phys. 307, 83 (1982)
4. E. Casnati, A. Tartari, C. Baraldi, J. Phys. B 15, 155 (1982) and J. Phys. B 16, 505 (1983)
5. H. Bethe, Ann. Phys. 5, 325 (1930)
6. R. Hippler, W. Jitschin, Z. Phys. 307, 287 (1982)

ENERGY DEPENDENCE OF THE HIGH ENERGY TIP OF THE BREMSSTRAHLUNG SPECTRUM OF La, W, AND Co.

F. Riehle

Physikalisch Technische Bundesanstalt, Abbestr. 2 -12, 1ooo Berlin, Germany

Bremsstrahlung is one of many possibilities of an atom to respond the excitation by electrons. The bremsstrahlung spectrum of a material shows features characteristic of this material. These structures result from scattering of the impinging electrons into states of different symmetries and densities. The inset of Fig. 1 shows such a structure of metallic lanthanum in the vicinity of the high energy limit $h\nu_0$. The linelike stucture above an unstructured continuum results from transitions into empty 4f-states and the continuum from those into mainly 5d-states.

The experiments were performed in such a way that bremsstrahlung spectra of metallic La, W, and Co were measured in the vicinity of the high energy cut-off by means of the isochromat method [1] for electron energies between ~2oo and ~3ooo eV: The relative intensities of these structures varied with the kinetic energy of the incident electrons. This can be explained by different dependences on the energy of the cross sections for radiative capture of electrons into states of different angular momenta [2]

For each isochromat a ratio of the hatched areas a and b has been constructed and plotted in Fig. 1 together with calculations of cross sections for the radiative capture of electrons by atomic or ionic targets. Similar plots have been obtained for W and Co.

Hahn and Rule [3] (curve 3 of Fig. 1) estimate the cross sections using a scaling property and a Coulomb potential with an effective charge. Similarly as did Keßler and Ulmer [4] for tungsten the ratio of the transition probabilities were calculated for the transitions from the continuum into the 4f- and 5d-states of the isolated atom resulting in slightly different ratios compared with [3] as can be seen from Fig. 1. Moreover the calculations of Lee and Pratt [5] using a Hartree - Slater potential have been used to compute the ratio of the transition probabilities. The three calculations converge for higher energies but they show marked differences at lower energies.

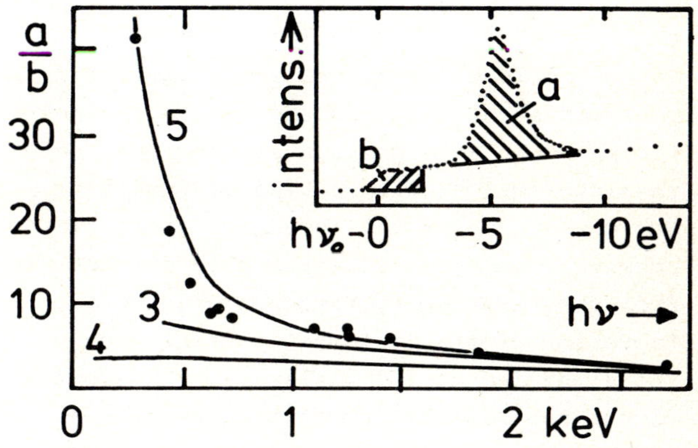

Fig. 1: Experimental ratio of transition probabilities into 4f- and 5d- states (dots) of lanthanum and calculated ones (full curves) after ref. 3, 4, and 5 vs photon energy.

Even though an absolute comparison between theories and experiment is not possible (curve 5 is normalized to the experiment), the relative dependence on energy of the experiment is well reproduced by the calculation of Lee and Pratt.

For tungsten the experimental data show a minimum of the transition ratio into 5d - and other states around 2oo eV and a maximum above 1 keV. Again the energy dependence can be reproduced by atomic calculations.

The experimental data of cobalt show a monotonous increase of the transitions into 3d - states compared with other (mostly 4p) transitions towards lower energies and can be explained by the calculations of ref.[3].

References

1. F.Riehle,phys.stat.sol.(b) 98, 245 (1980)
2. F.Riehle,Jap.J.Appl.Phys. 17, 314 (1978)
3. Yukap Hahn and D.W.Rule,J.phys.B 10, 2689 (1977)
4. J.Keßler and K.Ulmer,Z.Phys. 159, 443 (1960)
5. C.M.Lee and R.H.Pratt,Phys.Rev.A 12, 707 (1975)

ELECTRON BREMSTRAHLUNG - EFFECT OF ELECTRON SCREENING AND VACUUM POLARISATION

Lali Chatterjee *

Department of Physics, Jadavpur University, Calcutta - 700 032, India

Sujata Bhattacharyya

Gokhale College, Calcutta - 700 020, India

Bremstrahlung radiation of electrons in the field of atoms and atomic nuclei is of importance in fusion and plasma applications, and in some biological studies. The characteristics of this type of radiation are determined by the 'braking' field. Thus bremstrahlung from free atoms is distinct from that induced by free nuclei, due to the reduction of the Coulomb potential due to the screening of the nucleus by the atomic electrons. Several investigations of this interesting problem have been recently performed theoretically[1] and experimentally[2]. However to our knowledge, the effect of Vacuum Polarisation has not yet been studied, and it may have a not insignificant role for medium and high z elements, specially for 'close encounters' involving large momentum transfer. Previously theoretical investigations have generally invoked first order quantum electrodynamics.

We report on investigation of electron bremstrahlung in material media in the framework of second order field theory, taking electron screening and vacuum polarisation into account. Both the effects are introduced through modifications of the 'braking' nuclear potential. In particular, we treat electron screening by using a cut off type potential with a cut-off parameters 'a' such as was used by Allis and Morse to explain successfully electron scattering data[3].

Thus the coulomb potential $V_c(r)$ is replaced by

$$V_s(r) = \begin{cases} -ze^2\left(1/r - 1/a\right) & \text{for } r \le a \\ 0 & \text{for } r \ge a \end{cases} \qquad (1)$$

where a = the cut-off parameter.

The Uehling term to be added to the Coulomb term to account for Vacuum Polarisation is written as

$$V_{v.p.} = e^2 \int \frac{\rho(r)\,\rho(r')}{|\underline{r} - \underline{r}'|} \cdot \left\{\frac{\alpha}{\pi} Z_0(r - r')\right\} d^3\underline{r}\, d^3\underline{r}' \qquad (2)$$

with

$$Z_0(R) = \frac{2}{3}\left\{\ln(mR) + \gamma\right\} - \frac{\pi}{2}(mR) + O(R^2) \qquad (3)$$

γ = Euler's constant, $\rho(r)$ = charge density of the electronic field.

Numerical values will be presented at the conference and compared with experiment. The constant 'a' serves as an useful parameter for fitting with experimental data.

L. Chatterjee thanks the C.S.I.R. financial support.

References

1. R.H. Pratt et al. At Data Nuc. Data Tables 20, 175 (1977).
2. R. Hippler et al. Phys. Rev. Lett. 46, 1622 (1981).
3. Allis and Morse - Z. Fum. Phys. 70, 567 (1931).
4. C.S. Brown et al. P.R. Lett. 33 No. 26 (1974).

* Mailing Address

Dr. Mrs. Lali Chatterjee
84/SB, Block E, New Alipur
Calcutta- 700 053, India.

VALIDITY OF CLASSICAL APPROACH TO THE CHARACTERIZATION OF THE ELECTRON BREMSSTRAHLUNG SPECTRUM: EXTENSION OF ITS APPLICABILITY TO SCREENED ATOMIC CASES

M. Lamoureux* and R. H. Pratt

Department of Physics, University of Pittsburgh, Pittsburgh, Pennsylvania 15260 U.S.A.

The classical approach[1,2] is quite successful in accounting for the electron Bremsstrahlung spectrum at low energies in the Coulomb case. We wish to show that it can become valid for the general screened cases as well, when one defines proper limit impact parameters. For this purpose the Gaunt factors G are evaluated in terms of their departure from the Coulomb case assuming that (as in the Coulomb case) they are proportional to the square of some limit impact parameter b. This parameter is defined by $[(Ze^2)/(b^{cb})] \times (2/v^2) = p$ for the Coulomb case, whereas we take it as determined by $\left.\frac{dV(r)}{dr}\right|_{r=b} \frac{2b}{v^2} = p$ for the screened central potential V(r), with Z and v being the atomic number and the velocity of the incident electron. The parameter p may be adjusted along the spectrum so as to obtain an exact fit of the calculated Coulomb values to the actual rigorous Coulomb[1] results.

Predictions for Gaunt factors are given in the table for incident electron energies of 1 keV, both for the soft and the tip ends of the bremsstrahlung spectrum (i.e. near zero energy radiation, or all of the energy emitted as radiation). We show sample isolated neutral atom results and also results for cesium atoms in a high temperature and high density plasma environment, using parametric central potentials.[3,4] The Gaunt factors we obtain from these simple classical ideas differ by 0 - 15% from those obtained from similar central potentials with the elaborate complete relativistic partial wave expansion method.[5,6] Traditional simple evaluations are far less satisfactorily than the present method. For example, the Born Elwert approximation overestimates the G's by a factor of 2 or 3 in this energy range if Z > 20.

The present method appears thus to be very efficient, especially in view of its simplicity. It should be valid also for lower energies, as around a few 100 eV. Its use in dense plasmas would contribute to the understanding of the Bremsstrahlung emission in such an environment.

References

*On leave from: "Spectroscopie Atomique et Ionique", Bât 350, Université Paris-Sud, 91405 Orsay, France.

1. I. J. Feng and R. H. Pratt, "Parametrization of the Bremsstrahlung Spectrum", Internal report prepared for Lawrence Livermore Laboratory, PITT-266, University of Pittsburgh (1981).
2. W. H. Tücker, Radiation Processes in Astrophysics, the MIT Press (1975), Chapter 5.
3. J. D. Jackson, Classical Electrodynamics, Wiley and Sons (1975), Chapter 13.
4. B. F. Rozsnyai, J. Quant. Spectrosc. Radiat. Transfer 22, 337 (1979).
5. R. H. Pratt, H. K. Tseng, C. M. Lee, L. Kissel, C. MacCallum and M. Riley, Atomic and Nucl. Data Tables 20, 175 (1977); and erratum 26, 477 (1981).
6. M. Lamoureux, I. J. Feng, R. H. Pratt and H. K. Tseng, J. Quant. Spectrosc. Radiat. Transfer 27, 227 (1982); and I. J. Feng, M. Lamoureux, R. H. Pratt and H. K. Tseng, Phys. Rev. A, to be published.

G		Soft End C1	Soft End PWE	Tip End C1	Tip End PWE
Neutrals					
Z=					
10		.85	.76	.95	.81
30		.30	.30	.53	.50
55		.18	.18	.36	.31
70		.10	.10	.21	.24
90		.08	.08	.19	.19
Cesium Plasma*					
T_e=	ρ/ρ_o=				
1	100	.21	.22	.36	.35
1	1	.59	.68	.65	.75
0.1	100	.15	.15	.30	.27
0.1	1	.23	.24	.41	.40

Table: Comparison of electron Bremsstrahlung Gaunt factors G obtained with the present classical method (Cl) and with the partial wave expansion method (PWE) for isolated neutral atoms and for cesium atoms in a plasma at the temperature T_e and at the density ρ/ρ_o (expressed in terms of the solid density ρ_o). Incident electrons of 1 keV.

*In this case, we did not use an adjusted parameter since its variation had only a small effect on the Gaunt factors.

ATOMIC-FIELD BREMSSTRAHLUNG FROM URANIUM

L. Estep, J. Altman, R. Ambrose, S. Salehkoutahi
G. Westbrook and C. A. Quarles

Texas Christian University, Fort Worth, TX 76129

In a recent experiment on the atomic-field bremsstrahlung spectrum produced by 10 keV electrons on gaseous uranium hexafluoride, Hippler et al[1] have reported an enhancement of low energy photons over that expected from the theory. In contrast to these results for Z=92, good agreement between theory and experiment has been reported for Z up to 54[1,2] and for Z=80 using gaseous dimethyl-mercury as a target.[3] The work reported here represents an effort to extend our measurement to Z=92.

The apparatus has been described[2] and consists of an electron gun mounted in an aluminum scattering chamber at 90° to a capillary array gas inlet. The electron beam is collected in a carbon Faraday cup. In the present experiments, solid UF_6 in a metal tube is heated to about 70° C causing sublimation. The gaseous UF_6 is then pumped through the microvalve into the chamber. The ambient pressure monitored by an ion gauge is kept below $2x10^{-5}$; however, the pressure in the region crossed by the electron beam is expected to be as much as an order of magnitude higher than the ambient pressure. A Si(Li) x-ray detector views the interaction region at 90° to the electron beam. The detector is collimated to reduce background from bremsstrahlung produced by electron scattering into the chamber walls. In previous experiments on the rare gases and dimethyl-mercury, the background has been less than a few percent. The data has been corrected for background by making a background run with the gas flow turned off and subtracting a background photon spectrum normalized to the same total electron charge collected as the data run when the target gas was flowing.

We have made several runs with UF_6 and to date have not been able to either reproduce the results of reference 1 or to observe a bremsstrahlung spectrum which we could unambiguously attribute to a thin target. Our initial runs were made at an electron energy of 5 keV. At this bombarding energy, the observed photon spectrum is almost entirely attributable to M-shell x-rays from uranium. While the dominant M_α, M_β and M_γ x-rays range fom 3.15 to 3.55 keV, we also see M-shell x-rays at 2.4-2.5 keV, and at 4.2-4.4 keV. In addition we see silicon escape peak effects in the region below 2 keV.

At an electron energy of 7 keV, we see evidence for the onset of ionization of the M_I edge of 5.5 keV. This is not seen in the 10 keV data of reference 1. Furthermore, the region from 5.5 keV to 7 keV does not exhibit the shape characteristic of a thin target. Instead of the usual sharp end point, there is an approximate linear rise in the photon energy spectrum from 7 keV to 5.5 keV that is more typical of a "thick target" bremsstrahlung spectrum. Such a spectrum might result if we were observing a large background from scattering in the chamber walls, the Faraday cup, or the photon detector collimator. We do find that background with UF_6 is significantly different from that observed with previous gases studied. The UF_6 is very reactive and after only a few minutes of gas flow the aluminum interior of the chamber is blackened by reaction with the gas. The background observed after once filling the chamber with gas and then evacuating it is different from that observed before the gas has been introduced and includes M-shell x-rays from uranium due to electron scattering from the chamber walls which have been coated with uranium. Hence, the background is potentially a much more serious problem than we had observed with the rare gases or dimethyl-mercury. Because of this and because the presence of any background which does not completely subtract out in a target empty run can easily produce the effect of an enhancement in the number of low energy photons as seen by reference 1, it appears that additional work is needed to produce a bremsstrahlung spectrum from UF_6 which can confidently be compared with theory.

In addition, experiments on thin uranium targets (less that 1 $\mu m/cm^2$) deposited on thin carbon foils are underway as an alternative approach to studying bremsstrahlung from high Z atoms. We expect to have preliminary results for presentation at the meeting.

This research was supported by the TCU Research Foundation and the Robert A. Welch Foundation.

References

1. R. Hippler, K. Saeed, I. McGregor and H. Kleinpoppen, Phys. Rev. Letters 46, 1622 (1981).
2. M. Semaan and C. A. Quarles, Phys. Rev. A 24, 2280 (1981).
3. M. Semaan and C. A. Quarles, Phys. Rev. A 26, 3152 (1982).

TWO PHOTON PROCESSES IN ELECTRON-ATOM COLLISIONS

J. C. Altman and C. A. Quarles

Texas Christian University, Fort Worth, TX 76129

Experiments are underway to study a variety of processes in electron-atom collisions which lead to the production of two photons in coincidence. The incident electrons have energy variable from 40 to 300 keV generated by an electrostatic accelerator. Targets are thin foils of silver and gold mounted in a scattering chamber. Photons produced at angles from 0° to 90° in 15° increments, and at 270° to the incident beam direction are observable using Si(Li) and HpGe solid state detectors. The lithium drifted silicon detector has a planar geometry with active area of 30 mm^2, a thickness of 3 mm, and a 1 mm Be entrace window. The high purity germanium detector also has a planar geometry with 100 mm^2 active area, 10 mm thickness, and 1 mm Be window. Output signals are processed in a fast-slow coincidence arrangement. In the "fast" leg of the coincidence circuit signals are conditioned by timing amplifiers and constant fraction discriminators which act as start and stop signals for a time-to-amplitude converter. Converter output is a positive square wave whose amplitude is proportional to the temporal separation of the two imput signals. A resolving time of 100 to 200 nsec is typical with some variation dependent upon discriminator threshold settings, increasing with lower detected photon energy. Energy discrimination of detector outputs is also performed to allow selection of specific portions of the full energy spectrum in each detector, using timing single channel analyzers. The "slow" leg of the coincidence circuit compares these energy sensitive outputs and acts as a gate to allow passage of the converter output to a multichannel analyzer.

With this arrangement the timing spectrum can be studied as a function of selected regions of interest. The energy gating also aids in reducing the accidental coincidence rate (background) which can be particularly helpful at the lower discriminator threshold settings required to detect low energy photons.

When electrons strike a thin target (defined so that only single interactions between an electron and the target are probable and typically 10-50 $\mu gm/cm^2$), x-ray photons are produced by atomic field bremsstrahlung and by decay of atomic excited states produced by innershell ionization. An important two-photon process which can be observed and studied is the Kα-L x-ray cascade which occurs when the K shell of a target atom is ionized. Such studies yield information on the average L-shell fluorescence yield corresponding to an L vacancy distribution produced by K x-ray emission. The previous work in this area is summarized by Bambynek, et al.[1] The Kα-L cascade from electron bombardment of gold and silver has been observed and is being used to check out the experimental apparatus and to establish its sensitivity to the detection of other low cross section two-photon coincidence processes. One such process is the production of two bremsstrahlung photons in a single interaction of an electron with a target atom. Direct observation of this double bremsstrahlung process has not been reported as yet; however, it was suggested by Hippler et al[2] as a possible explanation for the observed enhancement of low energy photons over that expected from theory in the bremsstrahlung spectrum produced by 10 keV electrons in gaseous UF_6.[2]

We are currently attempting to observe two photon bremsstrahlung emission from electron bombardment of a gold target by 50 keV electrons. Preliminary results are expected to be presented at the meeting.

References

1. W. Bambynek, B. Crasemann, R. Fink, H. Freund, H. Mark, C. Swift, R. Price and P. Venugopala Rao, Rev. Mod. Phys. 44, 716 (1972).
2. R. Hippler, K. Saeed, I. McGregor and H. Kleinpoppen, Phys. Rev. Letters 46, 1622 (1981).

EIKONAL EXCHANGE AMPLITUDE FOR ELECTRON-IMPACT IONIZATION OF ATOMIC HYDROGEN

A. C. Roy

Department of Physics, University of Kalyani, Kalyani 741235, West Bengal, India.

N. C. Sil

Department of Theoretical Physics, Indian Association for the Cultivation of Science, Jadavpur, Calcutta 700032, India.

Recently, Franco and Halpern[1] have developed an approximate eikonal exchange amplitude for the elastic scattering of electrons from atomic hydrogen. It has been pointed out that the method is expected to be useful in the energy region above ~ 100 eV for e^--H scattering. We have extended the method to calculate the exchange amplitude for the $H(e,2e)H^+$ process. As in the case of elastic scattering we also obtain a closed form for the eikonal exchange amplitude for electron-impact ionization of atomic hydrogen.

In the Franco-Halpern approximation, the eikonal exchange T matrix for e^--H scattering in the post form is given by

$$T_{fi} = \frac{4\pi}{k^2} \left(\tfrac{1}{2}k\right)^{-i\eta} \Gamma(1+i\eta) e^{\pi\eta/2} \times \int d\vec{r}\, e^{i\vec{q}\cdot\vec{r}} \Phi_i(\vec{r}) \Phi_f^*(\vec{r}) (r-z)^{-i\eta} \quad (1)$$

where $\vec{q} = \vec{k} - \vec{k}_1$ and $\eta = 1/k$. In Eq. (1), $\vec{k}$ and $\vec{k}_1$ denote respectively, the initial and final momenta of the free electron. $\vec{q}$ represents the momentum transfer and z is chosen to be along the direction of $\hat{k}$. $\Phi_i(\vec{r})$ and $\Phi_f(\vec{r})$ denote, respectively, the initial and the final states of the target, and are given by

$$\Phi_i(\vec{r}) = \lambda^{3/2} \pi^{-1/2} e^{-\lambda r} \quad (2)$$

and

$$\Phi_f(\vec{r}) = (2\pi)^{-3/2} e^{\pi\alpha/2} \Gamma(1+i\alpha) \times e^{i\vec{k}_2\cdot\vec{r}} {}_1F_1(-i\alpha, 1, -i(k_2 r + \vec{k}_2\cdot\vec{r})) \quad (3)$$

with $\alpha = 1/k_2$ and $\lambda = 1$. In Eq. (3), $\vec{k}_2$ represents the momentum of the ejected electron.

Using Eqs. (2) and (3) in Eq. (1), we have been able to reduce Eq. (1) to the following closed form :

$$\begin{aligned} T_{fi} = {} & -C_1 C_2 \pi\, 2^{2-i\eta} \Gamma(1-i\eta) D^{-i\eta} \\ & \times \Big[B^{i\eta-1} \mu^{-i\eta} \nu^{i\eta-1+i\alpha} (\nu+1)^{-i\alpha} \\ & \times (i\alpha)(i\eta) \frac{-\mu\nu' + \mu'\nu - \mu^2\nu' + \mu'\nu^2}{\mu^2(\nu+1)^2} \\ & \times {}_2F_1(1+i\alpha; 1+i\eta; 2; (\mu-\nu)/(\mu(\nu+1))) \\ & + {}_2F_1(i\alpha; i\eta; 1; (\mu-\nu)/(\mu(\nu+1))) \Big\{ B^{i\eta-1} \mu^{-i\nu} \\ & \times (\nu^{i\eta-1+i\alpha} (-i\alpha)(\nu+1)^{-i\alpha-1} \nu' \\ & + (i\eta-1+i\alpha) \nu^{i\eta-2+i\alpha} \nu' (\nu+1)^{-i\alpha}) \\ & - i\eta\, B^{i\eta-1} \mu^{-i\eta-1} \mu' \nu^{i\eta-1+i\alpha} (\nu+1)^{-i\alpha} \\ & + (i\eta-1) B^{i\eta-2} B' \mu^{-i\eta} \nu^{i\eta-1+i\alpha} (\nu+1)^{-i\alpha} \Big\} \Big], \end{aligned} \quad (4)$$

where

$A = \lambda^2 + q^2 + k_2^2 - 2\vec{q}\cdot\vec{k}_2$, $B = -2k_2^2 + 2\vec{q}\cdot\vec{k}_2 - 2ik_2\lambda$,

$C = \lambda - iq_z + ik_{2z}$, $D = -(ik_2 + ik_{2z})$,

$\mu = C/D$, $\nu = A/B$

$C_1 = (4\pi/k^2)(k/2)^{-i\eta} \Gamma(1+i\eta) e^{\pi\eta/2}$

$C_2 = \lambda^{3/2} \pi^{-1/2} (2\pi)^{-3/2} e^{\pi\alpha/2} (1-i\alpha)$

and the prime indicates differentiation with respect to λ .

Eq. (4) can now be combined with the direct eikonal amplitude to obtain (e,2e) cross sections for atomic hydrogen. We intend to combine this exchange amplitude with the corresponding direct amplitude for the Glauber case ($\vec{q}\cdot\hat{z}=0$) and to make detailed comparison with the measured triply-differential cross sections of Weigold et al.[3]

References

1. V. Franco and A. M. Halpern, Phys. Rev. A21, 1118 (1980).
2. A. C. Roy, A. K. Das and N. C. Sil, Phys. Rev. A 23, 1662 (1981).
3. E. Weigold, C. J. Noble, S. T. Hood and I. Fuss, J. Phys. B 12, 291 (1979).

IONIZATION OF ATOMIC HYDROGEN AND HELIUM BY FAST ELECTRON IMPACT

B. Piraux and C.J. Joachain

Physique Théorique, Université Libre de Bruxelles, Belgium
Institut de Physique Corpusculaire, Université Catholique de Louvain, Belgium

and

F.W. Byron, Jr.

Department of Physics and Astronomy, University of Massachusetts, Amherst, Mass. 01003, U.S.A.

In previous work[1-3], we have analyzed the triple differential cross section for the ionization of both atomic hydrogen and helium using the Eikonal-Born series (EBS) method[4]. Following this method, the direct scattering amplitude was obtained by adding the first Born term f_{B1}, the second Born term f_{B2}, calculated by closure, and the third order Glauber term f_{G3}. Exchange effects were included by using the Peterkop theorem[5].

In the present work, we have studied the exact contribution of a few dominant intermediate states to f_{B2}. Calculations have been performed in the case of coplanar geometries for both "asymmetric" and "symmetric" kinematical situations.

In the asymmetric geometry, which corresponds to Ehrhardt-type experiments[3,6], we have included the contribution of the 1s, 2s and 2p intermediate states of atomic hydrogen. The relative importance of these states will be discussed for several values of the scattering angle θ_A of the scattered (fast) electron and the energy E_B and angle θ_B of the (slow) ejected electron.

For the case of the energy-sharing ($E_A = E_B$), symmetric geometry, with $\theta_A = \theta_B = \theta$, we have found, for both atomic hydrogen and helium, that the contribution to f_{B2} due to the ground state acting as an intermediate state is much more important than the first Born term f_{B1}. This is confirmed by analytical calculations performed in the limit of large incident energies. As an example, we display in Fig. 1 the results we have obtained for the case of atomic hydrogen, for an incident electron energy E_0 = 250 eV, and with $E_A = E_B$ = 118.2 eV. The dashed line corresponds to the first Born approximation and the solid line is obtained by adding to f_{B1} the contribution to f_{B2} arising from the 1s state acting as an intermediate state. The difference between the two curves is striking, particularly in the region $\theta = 130°$, where the second order cross section exhibits a maximum.

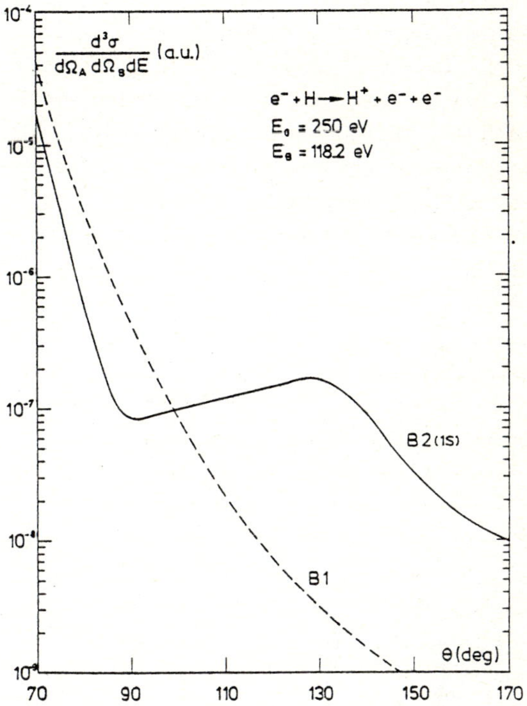

Fig. 1. The triple differential cross section (in a.u.) for the ionization of atomic hydrogen by electron impact, for an energy-sharing, symmetric geometry, with $\theta_A = \theta_B = \theta$.

References

1. F.W. Byron, Jr., C.J. Joachain and B. Piraux, J. Phys. B 13, L 673 (1980).
2. F.W. Byron, Jr., C.J. Joachain and B. Piraux, J. Phys. B 15, L 293 (1982).
3. H. Ehrhardt, M. Fischer, K. Jung, F.W. Byron, Jr., C.J. Joachain, B. Piraux, Phys. Rev. Letters, 48, 1810, (1982).
4. F.W. Byron, Jr. and C.J. Joachain, Phys. Rev. A 8, 1267, (1973).
5. R.K. Peterkop, Proc. Phys. Soc. (London) 77, 1220 (1961).
6. H. Ehrhardt, H. Hesselbacher, K. Jung and K. Willmann in Case Studies in Atomic Physics, 2, 159 (1971).

INTERFERENCES DUE TO COMPETING MECHANISMS FOR DOUBLE EXCITATION AND IONIZATION OF ATOMS BY CHARGED PARTICLES AT HIGH VELOCITIES

J. H. McGuire

Department of Physics, Kansas State University, Manhattan, Kansas, USA 66506

It has long been recognized that at sufficiently high velocities the dominant mechanism for multiple excitation and ionization of atoms by charged particles may be described in terms of rearrangement of the target electrons remaining following single excitation or ionization. In photoionization this is called shakeoff. On the other hand at intermediate velocities in ion-atom collisions it is well established that multiple ionization occurs predominantly due to independent interaction of each electron with the projectile. The independent electron mechanism gives rise to a binomial distribution of the final charge states of the target.

Recently, a factor of two difference observed in the double ionization of helium by electrons and protons at high velocity has been explained[1] as due to an interference between these two mechanisms. This is illustrated in Figure 1 where the shakeoff (SO) mechanism predicts a ratio, R, of double to single ionization cross sections which is independent of velocity, v. The two step (TS) mechanism, where both electrons are ejected due to direct interaction with the projectile, gives a $(v^2 \ell n v)^{-1}$ dependence for R. Since the SO amplitude, a_{SO}, is linear in the projectile charge, Z_p, and the TS amplitude, a_{TS}, is quadratic in Z_p, the square of the combined amplitude

$$|a|^2 = |a_{SO} + a_{TS}|^2 = |-Z_p C_1 + Z_p^2 C_2|^2$$

can depend on the sign of Z_p. Hence the cross sections for double ionization of helium by electrons can differ from the proton impact case when the constants C_1 and C_2 are comparable, as shown in Fig. 1.

Here it is pointed out that this effect can be viewed as an interference between a first Born type mechanism, where the amplitude is linear in Z_p, and a second Born type mechanism, quadratic in Z_p. Furthermore for double excitation of helium it is relatively straightforward to evaluate the coefficients C_1 and C_2. For a transition from $1s^2$ to $2p^2$, the coefficients C_1 and C_2 have both real and imaginary components. However, for a $1s^2$ to $2s^2$ transition the coefficient C_1 is purely imaginary and C_2 is purely real. Consequently for the $1s^2$ to $2s^2$ double excitation $|a|^2 = |a_{SO}|^2 + |a_{TS}|^2$ and no interference is expected. Hence the ratio R is expected to be the same for both protons and electrons for the $1s^2$ to $2s^2$ transition. However, a difference, similar to that observed for double ionization, is expected for $1s^2$ to $2p^2$ excitation by electrons and by protons.

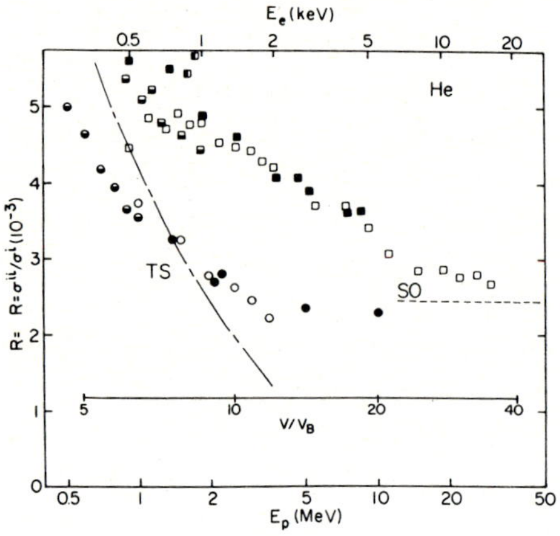

Fig. 1: Ratio, R, of double to single ionization cross sections in helium versus projectile velocity (in units of $v_B = 2.2 \times 10^8$ cm/sec). Various proton data are represented by circles and electron data by squares. The curves TS represents a $(v^2 \ell n v)^{-1}$ velocity dependence of the two-step mechanism, and SO a constant velocity dependence of shakeoff. Amplitudes for TS and SO interfere near $v/v_B = 10$.

Finally it is noted that the differences between $1s^2$ to $2p^2$ excitation by protons and electrons will be larger in the differential cross sections than in the total cross sections. In particular, since SO is expected to dominate at large impact parameters (or small momentum transfer), and TS is expected to dominate at small impact parameters (large momentum transfer), the interference effect is expected to be largest at intermediate momentum transfers, $q \sim Z_T$, where Z_T is the nuclear charge of the target.

This work was supported by the U.S. Department of Energy Division of Chemical Sciences.

Reference

1. J. H. McGuire, Phys. Rev. Letters 49, 1153 (1982).

DIRECT CONTRIBUTIONS TO ELECTRON INDUCED MULTIPLE IONIZATION OF INNER SHELLS*

W. Löw, H. Genz and A. Richter

Institut für Kernphysik, Technische Hochschule Darmstadt, 6100 Darmstadt, Germany

The energy and Z dependence for electron induced multiple ionization has systematically been studied for a variety of elements ($9 \leq Z \leq 29$) and electron impact energies of $10 \leq E \leq 200$ keV. The measurements were performed at the electron gun of the superconducting pilot accelerator by bombarding thin selfsupporting targets with electrons. The produced X rays were detected under $\Theta=90^o$ with respect to the beam axis by means of a high resolution crystal spectrometer. A typical spectrum (Fig.1) shows the diagram line KL^o and the satellites KL^1 and KL^2 that have been obtained for the bombardment of Mg with 150-keV electrons. The identification of the satellite lines and their substructure is achieved by comparison of their energy with Dirac-Fock calculations by Desclaux [1]. The groups KL^1 resp. KL^2 thus stem from transitions with one resp. two additional vacancies in the L shell.

For a better understanding of the origin for these multiple vacancies the ratio KL^1/KL^o has been measured as a function of electron impact energy. From Fig.2 it becomes apparent that the present results are in agreement with former data [2-4] based on electron and photon

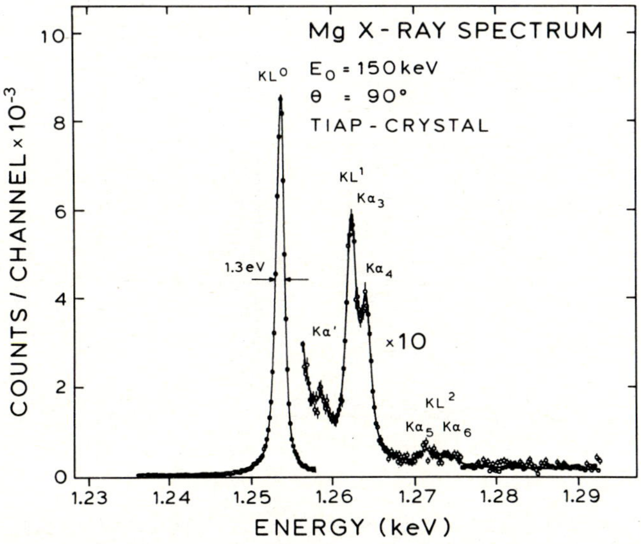

Fig. 1: Diagram and satellite lines

impact at low energies but also that the ratio is decreasing with increasing energy. This behaviour is observed for all investigated elements.

In order to separate the energy dependent contribution from the constant part we have fitted the function [5]

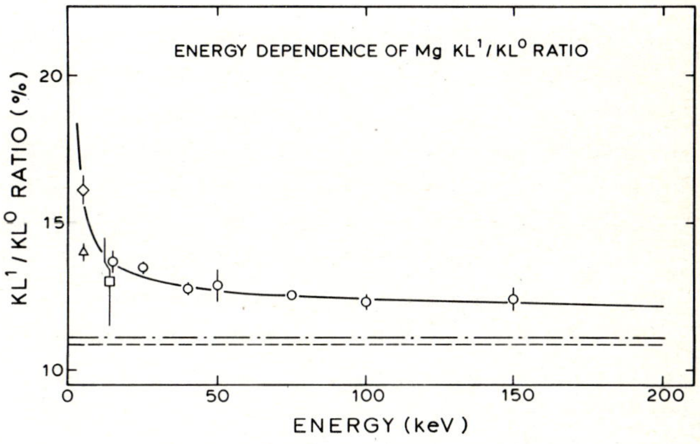

Fig. 2: Satellite intensity vs energy and comparison with electron and photon induced data [2-4]

$$KL^1/KL^o = a + b(\ln E)^{-1} \quad (1)$$

to the data points (Fig.2, full line), where a (dash-dotted line) and b are constants ($a \sim 2.5$ b) and E denotes the electron impact energy in units of the K-shell ionization energy. Assuming that the constant contribution to the multiple ionization is caused by the shake-off process we have calculated this probability in the sudden approximation utilizing the radial part of the Dirac-Fock wave functions[1] (broken line). The excellent agreement between the shake-off probability and the experimentally deduced energy independent part of the multiple vacancy production indicates that - contrary to former understanding - there are in addition energy dependent contributions which may be caused by direct double ionization. It is currently investigated if parts of this behaviour originate from an energy dependent alignment.

References

1. J.P. Desclaux, Comput. Phys. Commun. 9, 31 (1975)
2. J. Utriainen, Z. Naturforschung 23a, 1178 (1968)
3. M.O. Krause, T.A. Carlson and R.D. Dismukes, Phys. Rev. 170, 37 (1968)
4. D.W. Fischer and W.L. Baun, see T. Åberg, Phys. Lett. 26A, 515 (1968)
5. T. Åberg, Ann. Acad. Sci. Fennicae, A.VI.308, 1 (1969)

*Work supported by Deutsche Forschungsgemeinschaft

COLLAPSE OF 3d ORBITAL IN Ca OBSERVED IN ANOMALIES OF THE 2p AUGER SPECTRUM

B. Breuckmann, R. Huster, W. Menzel, W. Weber, W. Mehlhorn

Fakultät für Physik, Universität Freiburg, D-7800 Freiburg, FRG

K.G. Dyall

Theoretical Chemistry Department, University of Oxford, Oxford WX1 3TG, England

In Fig. 1 the 2p-MM Auger spectra of Ar, K and Ca, excited by 2 keV electrons, are shown[1]. The energy axis of the K and Ca spectra have been multiplied by factors in order to have the Auger groups $3s^2$, 3s3p and $3p^2$ falling on top of each other. Also the ratio of the diagram intensities of fine-structure levels $r = I(2p_{3/2}\text{-MM})/I(2p_{1/2}\text{-MM})$ is given. In going from Ar and K to Ca we note the following dramatic changes:

1) The ratio r varies from the near statistical value of 1.75(5) for Ar and 1.7(1) for K to the low value of 0.9(1) for Ca.

2) The intensity at the high-energy side of the $2p\text{-}3p^2$ Auger group increases strongly when going from K to Ca.

In the experimental $2p\text{-}4s^2$ spectrum of Ca (not shown) the final state is a closed shell ion $3s^23p^6\ ^1S$ and one doublet according to the $2p_{1/2,3/2}$ splitting is expected. At least 2 doublets with comparable intensity have been found.

These facts have been interpreted as due to a very strong CI in the initial ionic state 2p of Ca[1]. A MCHF calculation of the configurations $4s^2$ and $3d^2\ ^1S$ mixed into the 2p-vacancy state confirmed that ISCI is indeed very strong. But in order to explain the small value of r it has been assumed[1] that strong ISCI occurs for the $2p_{3/2}$ vacancy state but not for the $2p_{1/2}$ state.

In order to get quantitative results also for the fine-structure vacancy states $2p_j$, we have now performed MCDF calculations using the program by Grant et al.[2]. For the J = 0 neutral atomic states the CI of $4s^2$ and $3d^2$ was considered and admixtures of $c^2(4s^2) = 0.9914$ and $c^2(3d^2) = 0.0085$ into the ground state were obtained. For the $2p_j$ ionic states the configurations $4s^2$, $3d^2$ and 3d4s were included, yielding a total of 12 and 19 states with J = 1/2 and 3/2. Taking the $2p_j$ states with largest $4s^2$ admixture as initial diagram states, the ratio of diagram intensities is given by

$$r = 2c^2(4s^2)_{3/2}/c^2(4s^2)_{1/2} = 2\cdot 0.2810/0.5574 = 1.01,$$

which is in essential agreement with the experiment. The shaded area below the high-energy part of the Ca spectrum in Fig. 1 indicates the expected energy range of ISCI satellite transitions due to HF calculations[1]. More results of the MCDF calculations and a comparison with experimental spectra, measured with better statistics than earlier[1], will be presented at the conference.

This very strong CI between $4s^2$ and $3d^2$ configurations is due to the crossing of state energies which occurs for the neutral atom for Z around 21 but for the 2p vacancy atoms already for Z = 20. This cross-over is known as collapse of d orbitals. A similar collapse and strong ISCI has been found for the 5d orbital in the 5p ionization of Ba[3].

The support by the Deutsche Forschungsgemeinschaft is gratefully acknowlegded.

References

1. B. Breuckmann, Ph.D. Thesis, Universität Freiburg, 1978, unpublished.
2. I.P. Grant, B.J. McKenzie, P.H. Norrington, D.F. Mayers and N.C. Pyper, Comp.Phys.Comm. 21, 207 (1980).
3. S.J. Rose, I.P. Grant and J.P. Connerade, Phil.Trans. Roy.Soc. London 296, 527 (1980).

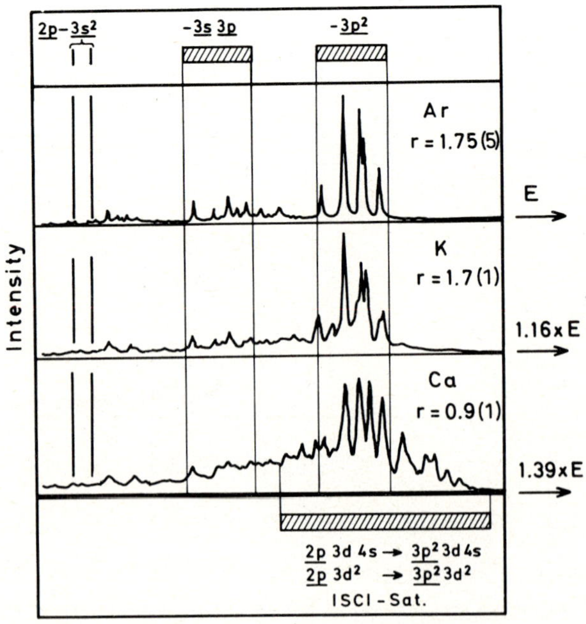

Fig. 1: 2p-MM Auger spectra of Ar, K and Ca excited by 2 keV electrons. For explanation see text.

ELECTRON IMPACT EXCITATION OF THE nsnp(n+1)p CONFIGURATIONS OF AUTOIONIZING STATES IN INDIUM AND THALLIUM VAPOUR ATOMS

M. Wilson
Physics Department, Chelsea College, London, England.
and
G. K. James and K. J. Ross
Physics Department, Southampton University, England

We have studied the ejected electron spectra of indium and thallium metal vapour atoms resulting from inner-shell excitation to discrete autoionizing levels produced by electron impact excitation.

The measurements were performed using a crossed beam electron spectrometer[1] over the range of incident electron kinetic energies 15 to 500 eV.

The position of the nsnp(n+1)p configurations of autoionizing states in indium and thallium has been established by ultraviolet absorption studies[2]; the corresponding ejected electron spectra of these configurations obtain-in our work are shown in Fig. 1 and Fig. 2.

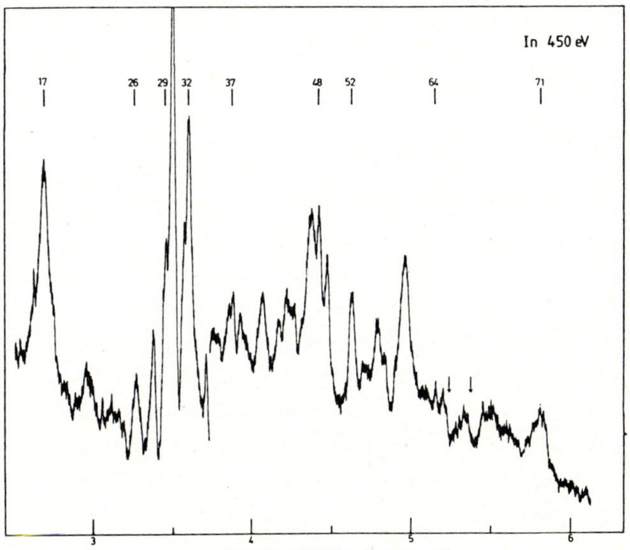

Fig. 1: Ejected electron spectrum of In between 2.5 and 6 eV ejected electron energy.

In order to be able to assign the levels of the above configurations which are not observable by photoabsorption we have used a least squares fitting of the eigenvalues of the matrices of electrostatic and spin-orbit interaction to the known energy levels to obtain estimates of the positions of the remaining levels. Reasonable estimates of the required constraints on the fitting parameters were obtained from comparisons with other members of the homologous sequence and their isoelectronic configurations.

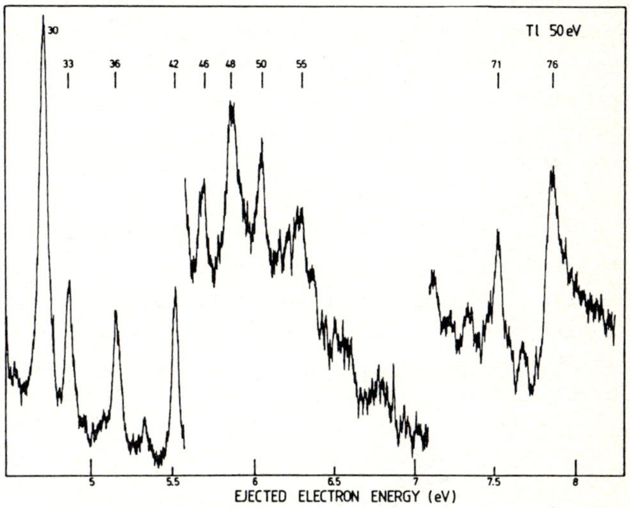

Fig. 2: Ejected electron spectrum of Tl between 4.5 and 8 eV ejected electron energy.

Using this method some of the assignments made by Connerade and Baig[2] for the nsnp(n+1)p configurations of indium and thallium have been revised. In addition, revised assignments have been made to the optically-forbidden levels observed in our earlier work[1].

This work is supported by a grant from the Science and Engineering Research Council.

References

1. G. K. James, D. Rassi, K. J. Ross and M. Wilson, J. Phys. B. 15, 275 (1982)
2. J. P. Connerade and M. A. Baig, J. Phys. B. 15, L587 (1982)

BRANCHING RATIOS AND CROSS SECTIONS FOR ION AND ELECTRON IMPACT PRODUCED $2s^m2p^n$-CONFIGURATIONS OF Ne USING SYNCHROTRON RADIATION AS A RADIOMETRIC STANDARD*

H.-J. Flaig, M. Eckhardt, and K.-H. Schartner

I. Physikalisches Institut der Justus-Liebig-Universität, 6300 Giessen, W.-Germany

H. Kaase

Physikalisch-Technische Bundesanstalt, 3300 Braunschweig, W.-Germany

We are studying the proton and electron impact production of $2s^m2p^n$-configurations of Ne (m = 0,1; n = 5,6) in the 50 keV to 1 MeV and the 200 eV to 10 keV energy range, respectively. The optical method in the spectral range of the VUV is used for the determination of total ionization cross sections. In former measurements we have applied a number of theoretical branching ratios[1] for evaluation of the relative quantum efficiency of the monochromator and detection system. These branchings need experimental verification, testing at the same time atomic structure calculations. We are reporting first results of branching ratio measurements of

$$\text{Ne } 2s^m2p^n \rightarrow 2s^{m+1}2p^{n-1}$$

transitions wherein we use synchrotron radiation as radiometric standard for the determination of the quantum efficiency of monochromator and detector[2]. We will discuss the implication of the measured branching ratios on our earlier cross section data based on theoretical branchings.

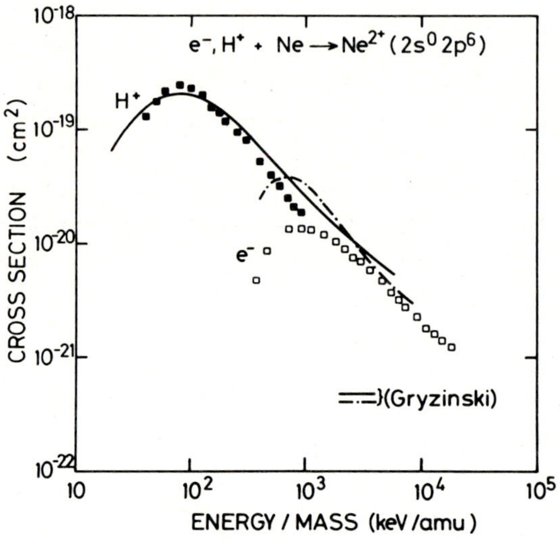

Fig. 1: Cross sections for double ionization of Ne 2s-electrons as function of projectile energy/projectile mass.

I. Branching Ratios

In table 1 we compare theoretical values for the branching according to the decay of the Ne IV $2s2p^4$ configuration with our experimental value. The wavelengths of this branching are 52.2 nm and 47.0 nm. Our experimental value is smaller than the theoretical values, though the difference is still within the combined error limits.

Measurements of the respective branching for O II, F III, and Na V are intended in order to study in a more systematic manner the relation between experimental and theoretical values.

Table 1: Calculated and measured branching ratio for the $(2s2p^4)^2D \rightarrow (2s^22p^3)\ ^2D/^2P$ transition of Ne IV. + theoretical.

Cohen and Dalgarno[3]	6.42⁺
Sinanoğlu[4]	7.1 ± 1.4⁺
Fawcett[5]	6.3 ± 1.3⁺
Cheng et al.[6]	6.7⁺
this experiment	4.8 ± 1

II. Cross Sections

We have measured proton- and electron impact cross sections for single and double ionization of Ne-L shell electrons. Using the experimental branching ratio given in table 1, we find now a good agreement with the electron impact cross sections for production of the $2s2p^6$ configuration measured by Dijkkamp and de Heer[7].

Cross sections for double ionization of two 2s-electrons are shown in fig. 1. In the absence of quantum mechanical calculations only the binary encounter model suggested by Gryziński[8] can be used for comparison. The qualitative agreement is rather satisfactory supporting the model of double binary encounters as the dominating mechnism for double ionization in the investigated energy range.

References

1. H.F. Beyer, R. Hippler, K.-H. Schartner, R. Albat, Z. Physik A 289, 239 (1979)
2. H.-J. Flaig, K.-H. Schartner, H. Kaase, Nucl. Instr. Meth., to be published
3. N. Cohen, A. Dalgarno, Proc. Roy. Soc. A 280, 258 (1964)
4. O. Sinanoğlu, Nucl. Instr. Meth. 110, 193 (1973)
5. B.C. Fawcett, At. Data Nucl. Data Tables 22, 473 (1978)
6. K.T. Cheng, Y.-K. Kim, J.P. Desclaux, At. Data Nucl. Data Tables 24, 111 (1979)
7. D. Dijkkamp, F.J. de Heer, J. Phys. B: At. Mol. Phys. 14, 1327 (1981)
8. M. Gryziński, Phys. Rev. 138, 336 (1965)

*Work is supported by the Deutsche Forschungsgemeinschaft.

THE K-SHELL ELECTRON DOUBLE IONIZATION CROSS SECTION OF SODIUM. A POWER LAW Z-DEPENDENCE.

O. Keski-Rahkonen[+] and J. Lahtinen

Laboratory of Physics, Helsinki University of Technology, 02150 Espoo 15, Finland

We have measured the $K\alpha^h$ hypersatellite spectrum from thin and thick metallic sodium targets varying the energy of exciting electrons. Using experimental data for slowing down of electrons in matter we have been able to extract from the relative intensities the electron double ionization cross section (EDC) of the K-shell for electron energies from 4 to 25 keV[1].

From the $K\alpha^h$ hypersatellite and other K-satellite energies we obtain also the threshold energy for the double ionization $E(K^2)$ relative to the Fermi-level according to the formula

$$E(K^2)=E(K\alpha_2^h)+E(K\alpha_4)+\Delta E(K\text{-}KLL,{}^1D) \quad (1)$$

where the energies on the right hand are: $K\alpha_2^h$ hypersatellite energy and $K\alpha_4$ satellite energy; ΔE is obtained as the difference of Fermi edge energies of the $K\beta_x$ band and the KLL,^1D radiative Auger edge all of which were recorded simultaneously[2]. The binding energy obtained is 2271.7(6) eV.

Thin targets were made by evaporating a thin layer of sodium, as compared to the range of exciting electrons, on an aluminum substrate in a demountable x-ray tube. Cross sections from a thick target were obtained as in Cr, Fe[3] and Mg[4]. Comparing sodium cross sections from thick and thin targets showed the results to be equivalent within the estimated systematic errors (10...30%). The thick target data show a higher maximum which probably is a slight artifact due to inaccurate slowing down data.

The shape of the cross section curve, when the energy scale is normalized by the double ionization energy $E(K^2)$, becomes similar to Mg, Cr and Fe which all have a broad maximum at two to four times of the threshold energy. Our measurement were made at so low energies that we could not yet observe any linear parts in the Fano-Bethe plot of the cross section.

We calculated theoretically the cross section[5] applying Gryzinski's method. That yields a result which agrees in magnitude with observation but fails to reproduce the energy behaviour. The theoretical cross section grows too fast at the threshold, has too high a maximum and decreases faster toward high energies than the observed cross section. The last phenomenon is probably caused by the wrong asymptotic behaviour of the binary encounter cross section utilized in Gryzinski's formulae.

At constant relative energy the theory yields a simple power law dependence of the target[4,5]

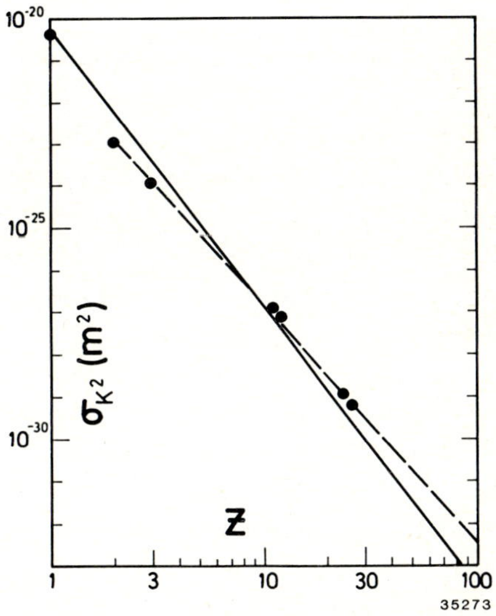

Fig. 1: Z-dependence of the EDC. Exponents: th. -6.6 (full line), exp. - 5.6 (dashed line)

$$\sigma_{K^2} \sim Z^{-6.6} \quad (2)$$

In Fig. 1 this is drawn in full line. The experimental data at an energy $4 \times E(K^2)$ are denoted by dots. The power law seems to give an order of magnitude description for elements $1 \leq Z \leq 26$ when the absolute cross section varies by ten orders of magnitude ! Omitting the anomalous H^- data allows much better fitting with an exponent of -5.6 drawn dashed in Fig. 1, on which the experimental points fall within their experimental errors. This power law behaviour was found to be approximately valid in the energy region $(2...5) \times E(K^2)$ for elements indicated in Fig. 1. New data are needed to test the validity of this law for more elements and within a broader energy range.

+Presently an Alexander von Humboldt Fellow at HASYLAB, DESY, Notkestr. 85, 2000 Hamburg 52, FRG.

References

1. J.Lahtinen and O.Keski-Rahkonen, Phys.Scr. 26, (1983) in press
2. O.Keski-Rahkonen, K.Reinikainen, and E.Mikkola, Phys. Scr. (to be published)
3. J.Saijonmaa and O.Keski-Rahkonen, Phys.Scr. 17, 451 (1978)
4. E.Mikkola, O.Keski-Rahkonen, and R.Kuoppala, Phys.Scr. 19, 29 (1979); 5. J.Saijonmaa, Phys.Scr. 17, 457 (1978)

THRESHOLD BEHAVIOUR OF Ar - K AND Xe - L_3 INNER SHELL IONISATION BY ELECTRON IMPACT

R. Hippler[+], K. Saeed, I. McGregor, A.J. Duncan, H. Klar[*] and H. Kleinpoppen

Atomic Physics Laboratory, University of Stirling
Stirling, Scotland

We report a measurement of the Ar-K and Xe-L_3 ionisation cross sections by electron impact 10 eV to 1 keV above threshold. The measurements have been performed by directing an electron beam on to a thin thermal gas target. Characteristic x-rays resulting from the decay of collision-induced inner shell vacancies were detected by a HP(Ge) x-ray detector. The energy dependence of the characteristic line radiation was used to determine the threshold behaviour of the ionisation of the above inner shells[1]. The energy dependence of the bremsstrahlung associated with the spectra was used for calibration.

In Figs. 1 and 2, the experimental data for Ar-K and Xe-L_3 ionisation are plotted versus the excess energy. The dashed line in Fig.1 represents the Wannier[2] law $Q(E) \propto (E-I)^{1.127}$ and the other curves various theoretical approximations. Note the satisfactory agreement between the experimental data and the Wannier law near threshold. In contrast, the Xe-L_3 data are inconsistent with the Wannier law (Fig.2). The experimental data for this ionisation are close to the linear threshold law $Q(E) \propto (E-I)$ (dashed line in Fig.2). A detailed theoretical argument for this behaviour will be presented.

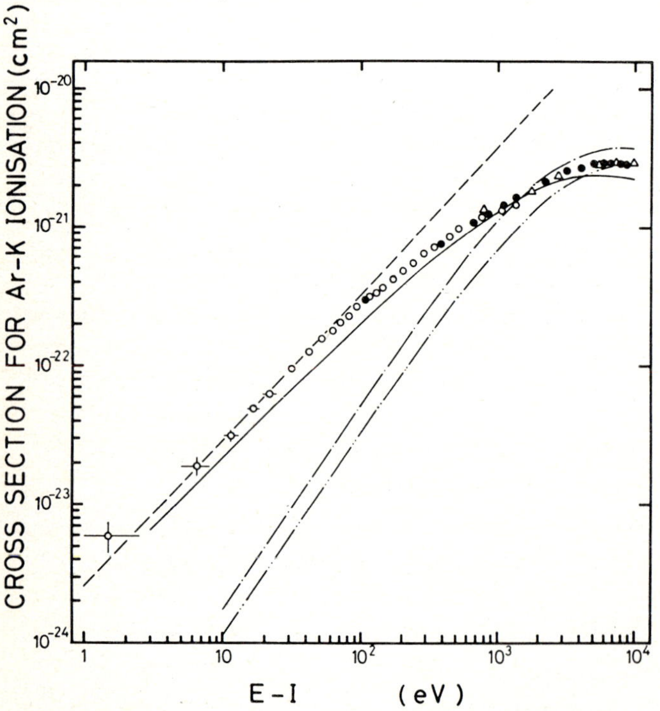

Fig.1: Cross section for Ar-K ionisation versus excess energy E-I; I ionisation energy of Ar-K shell, E electron impact energy. o present result, ---- Wannier law[2], · Hippler et al.[1] Δ Tawara et al.[3], ——— CBE, Moores et al.[4], -·-·-· PWBA, Hippler, Jitschin[5], -·····-·· PWBA Ochkur approximation.

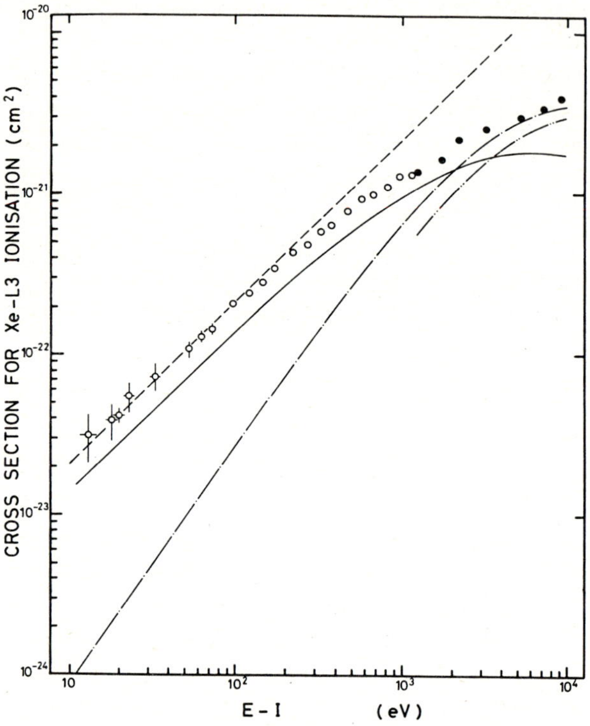

Fig.2: Cross section for Xe-L_3 ionisation versus excess energy; I ionisation energy of Xe-L_3 shell, E electron impact energy.
o present result, ----- linear threshold law,
· Hippler et al.[1], ——— CBE Moores et al.[4]
-·-·-· PWBA, Hippler, Jitschin[5], -·-·-·-·· modified Born approximation, McGuire[6]

References

1 R Hippler, K Saeed, I McGregor, H Kleinpoppen Phys.Rev.A23, 1730 (1981)

2 G H Wannier, Phys.Rev.90, 817 (1953)

3 H Tawara, K G Harrison, F J de Heer, Physica 63, 351 (1973)

4 D L Moores, L B Golden, D H Sampson, J.Phys.B13, 385 (1980)

5 R Hippler , W Jitschin, Z.Physik.A307, 287 (1982)

6 E J McGuire, Phys.Rev.A16, 62 (1977)

+ Permanent address: Fakultät für Physik, Universität Bielefeld, West Germany

* Permanent address: Fakultät für Physik, Universität Freiburg, West Germany.

THRESHOLD LAW FOR THE TRIPLE IONIZATION FUNCTION

P. Grujić

Institute of Physics, P.O. Box 57, 11001 Belgrade, Yugoslavia

The method which has been successfully applied in deriving threshold behaviour of a number of processes,[1),2),3)] is used to obtain the threshold law for the triple ionization cross section by electron impact.

The skeleton configuration at the zero energy possesses tetrahedral symmetry. If the centre of symmetry is taken as the origin of the coordinate system, where also the residual ion with charge Z (in atomic units) is at rest, one can write for the electrons radius vectors

$$\begin{aligned} \vec{r}_1^{(o)} &= r\vec{k} \\ \vec{r}_2^{(o)} &= (2/3)^{\frac{1}{2}}(\vec{i} + \vec{j}/3^{\frac{1}{2}} - \vec{k}/6^{\frac{1}{2}})r \\ \vec{r}_3^{(o)} &= (2/3)^{\frac{1}{2}}(-\vec{i} + \vec{j}/3^{\frac{1}{2}} - \vec{k}/6^{\frac{1}{2}})r \\ \vec{r}_4^{(o)} &= -1/3\ (8^{\frac{1}{2}}\vec{j} + \vec{k})r \end{aligned} \tag{1}$$

At $E = 0$ one has an approximate solution for the escaping electrons (cf. Ref. 4)

$$r_i^{(o)} = r(t) = (9[Z - (27/32)^{\frac{1}{2}}]t^2/2)^{1/3} \tag{2}$$

At a small, positive E, we write

$$\vec{r}_i = (r_i^{(o)} + \Delta_i)\hat{\Delta}_i + \vec{\delta}_i + \vec{\nabla}_i \tag{3}$$

where $\vec{\Delta}_i$ are small deviations collinear with $\vec{r}_i^{(o)}$ and $\vec{\delta}_i, \vec{\nabla}_i$ are mutually orthogonal and orthogonal to $\vec{r}_i^{\,o}$, too. Within the linear approximation one obrains a set of twelve differential equations for deviations $\Delta_i, \delta_i, \nabla_i$, in the Coulomb zone, where the classical dynamics holds. Accounting for the constraints, imposed by the conservation of the angular momentum of the system, at constant radius of the hypersphere,[1)] one can eliminate deviations of, say, fourth electron and thus gets a matrix equation[2)]

$$t^2 \frac{d^2}{dt^2} \mathbf{F} = \mathbf{B}\,\mathbf{F} \tag{4}$$

where $\mathbf{F}$ is a column vector, with components : $\Delta_1, \Delta_2, \Delta_3, \delta_1$, etc., and $\mathbf{B}$ is a real matrix, whose diagonal elements depend on Z. In solving Eq. (4) we confine ourselves to zero - approximation, in which all off-diagonal elements of $\mathbf{B}$ are neglected. Further, as only Δ_i are important for the fourfold escape probability,[2)] we retain only first three equations and get a solution in the form

$$\Delta_i^{(o)} = c_1^{(i)} r^{\gamma_1^{(i)}} + c_2^{(i)} r^{\gamma_2^{(i)}} \tag{5}$$

$$\gamma_{1,2}^{(i)} = \tfrac{3}{4}[1 \pm (1+4b_{ii})^{\frac{1}{2}}],\quad i = 1,2,3 \tag{6}$$

Exponents $\gamma_{1,2}^{(i)}$ have been evaluated for a number of Z-values. Numerical results for $\gamma_1^{(i)} = \gamma_i$ are shown in Table 1.

The threshold law for the fourfold escape then reads

$$\sigma_{ion} \sim E^{\gamma},\quad \gamma = \gamma_1 + \gamma_2 + \gamma_3 - 3 \tag{7}$$

Numerical values for γ are presented in Table 1, too. For the triple ionization of a neutral target one has : $\gamma = 3.525$, what is to be compared with $\gamma = 2.270$ for double-ionization threshold exponent.[2)]

As will be argued in a more detailed presentation of the calculations expounded here (to be published elsewhere), one can go beyond the zero-order approximation, with negligible consequences on final results.

Table 1. Exponents γ_i for the deviations Δ_i

Z	γ_1	γ_2	γ_3	γ
1	3.462	3.504	3.111	10.278
2	1.326	1.333	1.299	3.958
3	1.178	1.183	1.163	3.525
4	1.123	1.126	1.112	3.362
5	1.094	1.096	1.086	3.276
10	1.043	1.044	1.039	3.126

References

1. P. Grujić, J. Phys. B 15, 1913 (1982)
2. P. Grujić, J. Phys. B, to be published.
3. P. Grujić, these Proceedings, and to be published.
4. I. Vinkalns and M. Gailitis, Latvian Academy of Science Report No 4 (Riga : Zinatne) (1967), p. 17 (In Russian).

COULOMB INTERACTION IN FINAL STATE OF (e,2e) REACTION ON ATOMS IN TERMS OF THREE-BODY FORMALISM

A.R. Ashurov, G.V. Avakov, V.G. Levin, A.M. Mukhamedzhanov

Institute of Nuclear Physics, Moscow State University,Moscow 117234, USSR

Impulse approximation is usually used to calculate the cross sections of (e,2e) reactions on atoms in a symmetric kinematics at ~1 keV. The wave distortion of the incident and outgoing electrons is described in this case by the optical potentials taking into account for polarizational interactions. At the same time, the Coulomb interaction is of major importance in the final state and requires special techniques for itself to be allowed for. To the approximation of pure Coulomb electron-ion interaction, a consistent treatment of the Coulomb effect necessitates that the formalism of three-body theory should be used. Such approach is realized in the present report which deals with the effect of the Coulomb scattering of electrons by a residual ion on the form and absolute value of the (e,2e) reaction differential cross section.

The formalism developed has been based on the application of so-called Coulomb asymptotic states (CAS) which were first introduced by van Haeringen[1] in the theory of scattering of two charged particles and are analogues of the conventional asymptotic states in the case of scattering by short-range potentials. We have generalized the concept of CAS for the case of three charged particles and obtained the following expression of a three-body CAS in impulse representation (in the C-system)

$$\langle \vec{P}_1, \vec{P}_2 | \vec{K}_1, \vec{K}_2 \infty \rangle = $$
$$= (2\pi)^{-3} \int d\vec{p}\, \langle \vec{P}_1 - \vec{K}_1 + \vec{K}_{12} - \vec{p} + \vec{K}_{13} | \vec{K}_{13} \infty \rangle \times$$
$$\times \langle \vec{P}_2 - \vec{K}_2 + \vec{p} - \vec{K}_{12} + \vec{K}_{23} | \vec{K}_{23} \infty \rangle \langle \vec{p} | \vec{K}_{23} \infty \rangle$$

where $\vec{K}_\alpha$ is the momentum of particle α ; $\vec{K}_{\alpha\beta}$ is relative momentum ; $\langle \vec{P}_{\alpha\beta} | \vec{K}_{\alpha\beta} \infty \rangle$ is the two-body CAS of the pair α and β . The expression was used to discriminate the main member of the amplitude of the reaction $1 + (23) \rightarrow 1 + 2 + 3$ which makes the major contribution to the cross section near the quasi-elastic peak and is a generalization of the impulse approximation for the case allowing for scattering of electrons 1 and 2 by residual ion 3.

In Figs 1 and 2, the solid lines show the results of our calculations for He and Ne, and the dashed lines show the PWIA. It is seen that the inclusion of the Coulomb rescattering effects shits the quasielastic peak to the right, thereby giving a better agreement with experimental results[2,3] and reducing the absolute value of the differential cross section at low energies.

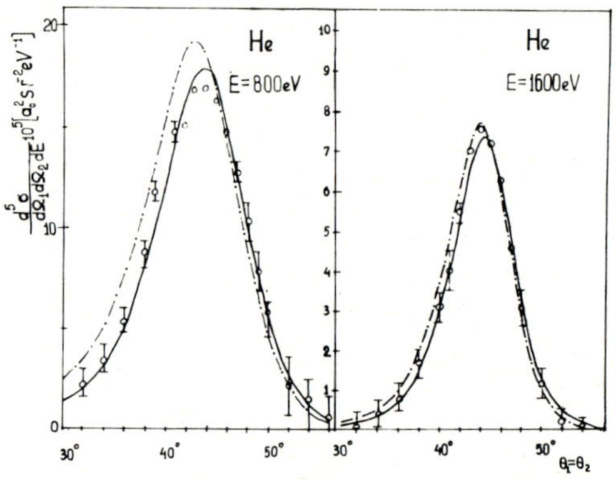

Fig. 1

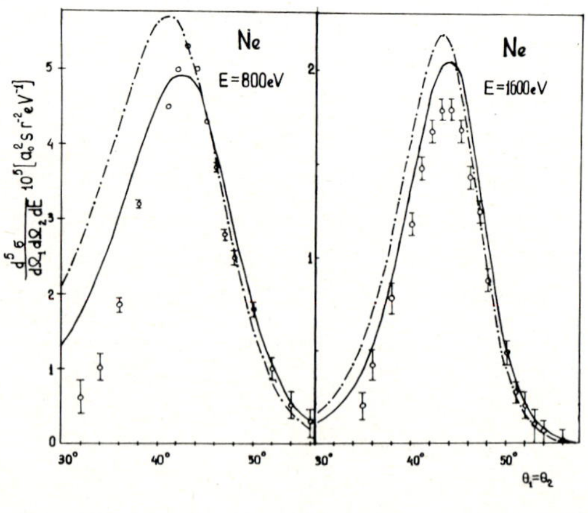

Fig.2

References

1. Van Haeringen H. J.Math.Phys. 17, 995(1976).
2. G. Stefani and R. Camilloni. Phys.Lett.,64A,364, (1978).
3. G. Stefani et al. J.Phys.B: Atom.Molec. 12, 2583(1979).

THE OBSERVATION OF AN EVEN PARITY $^1D_2$ AUTOIONIZING LEVEL IN Cd USING (e,2e) SPECTROSCOPY

N. L. S. Martin and K. J. Ross

Physics Department, Southampton University, England.

We wish to report the detection, by means of electron-electron coincidence spectroscopy, of a previously unobserved J=2 even parity autoionizing level in atomic cadmium.

During an investigation of the Cd ($4d^9 5s^2 5p$) 12.06 eV autoionizing level, excited by a 150 eV electron beam the coincidence angular distribution shown in Fig. 1 was obtained. This level is of mainly $^3P_1$ character[1] and from an analysis[2] of the photoabsorption data of Marr and Austin[3], the angular distribution is expected to be of the form:-

$$I_1(\Theta_{ej})\Big|_{\substack{\Theta_{sc}=+3^o \\ E_o=150\ eV}} \propto P_1^{\,2}(\cos(\Theta_{ej}+110^o)) = A \qquad (1)$$

where P_n is the Legendre polynomial of order n.

In order to explain the observed angular distribution of Fig. 1 it is necessary to assume the existence of an overlapping level of strong $^1D_2$ character, which has an angular distribution[4]:-

$$I_2(\Theta_{ej})\Big|_{\substack{\Theta_{sc}=+3^o \\ E_o=150\ eV}} \propto P_2^{\,2}(\cos(\Theta_{ej}+47^o)) = B \qquad (2)$$

Because the two levels have different J and parity, they cannot interfere; the expected angular distribution is therefore simply the sum of (1) and (2). A good fit to the experimental data was obtained with :-

$$I(\Theta)\Big|_{\substack{\Theta_{sc}=+3^o \\ E_o=150\ eV}} \propto 0.44A + 0.56B \qquad (3)$$

It is difficult to assign the new level to a single configuration; the $^1D_2$ level of Cd ($4d^{10}5p^2$) is expected to lie well below 11 eV whilst that of Cd ($4d^{10}5p6p$) is expected to lie at about 13 eV[5]. Assignment must therefore await theoretical calculations of the even parity autoionizing levels of cadmium.

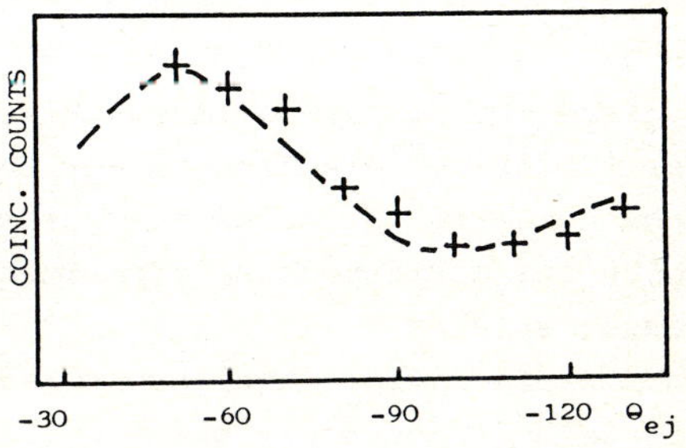

Fig. 1: Ejected electron angular distribution in coincidence with electrons scattered through $+3^o$. +, experimental points with their associated errors; ----, fit given by (3). Incident energy = 150 eV, scattered energy = 137.94 eV, ejected electron energy = 3.07 eV.

References

1. M. Wilson, J. Phys. B. 1, 734 (1968)
2. N. L. S. Martin, to be published
3. G. V. Marr and J. M. Austin, Proc. Roy. Soc. A 310, 137 (1969)
4. N. L. S. Martin, T. W. Ottley and K. J. Ross J. Phys. B. 13, 1867 (1980)
5. M. Wilson, private communication

(e,2e) CROSS SECTIONS FOR HELIUM

A. C. Roy

Department of Physics, University of Kalyani, Kalyani 741235, West Bengal, India

We have applied the method of Roy et al[1] to calculate the triply-differential cross sections (TDCS) for electron-impact ionization of He and compared the calculated cross sections with the corresponding experimental data of Ehrhardt et al.[2]

The triply-differential cross section for the He(e,2e)He$^+$ process is given by (atomic units will be used throughout)

$$\frac{d^3\sigma}{d\hat{k}_1 d\hat{k}_2 dE_2} = \frac{k_1 k_2}{k}\left|F(\vec{q},\vec{k}_2)\right|^2, \qquad (1)$$

where $d\hat{k}_1$ and $d\hat{k}_2$ denote, respectively, the elements of solid angle for the scattered and ejected electrons and dE_2 represents the energy interval of the ejected electron. In Eq. (1), $\vec{k}$, $\vec{k}_1$ and $\vec{k}_2$ are the momenta of the incoming, scattered and ejected electrons, respectively, $F(\vec{q},\vec{k}_2)$ is the scattering amplitude in the Glauber approximation[3] (GA) and is given by

$$F(\vec{q},\vec{k}_2) = \frac{ik}{2\pi}\int d\vec{b}\, d\vec{r}_1 d\vec{r}_2 \Phi_f^*(\vec{r}_1,\vec{r}_2) \times \Gamma(\vec{b};\vec{r}_1,\vec{r}_2)\Phi_i(\vec{r}_1,\vec{r}_2)e^{i\vec{q}\cdot\vec{b}}, \qquad (2)$$

where

$$\Gamma(\vec{b};\vec{r}_1,\vec{r}_2) = 1-\left(\frac{|\vec{b}-\vec{s}_1|}{b}\right)^{2i\eta}\left(\frac{|\vec{b}-\vec{s}_2|}{b}\right)^{2i\eta},$$

and $\eta = 1/k$. Here $\vec{q}$ represents the momentum transfer, $\vec{b}$, $\vec{s}_1$ and $\vec{s}_2$ are the respective projections of the position vectors of the incident particle and the two bound electrons onto the plane perpendicular to the direction of the Glauber path integration. In Eq. (2), $\vec{q}$, $\vec{b}$, $\vec{s}_1$ and $\vec{s}_2$ are all coplanar. $\Phi_i(\vec{r}_1,\vec{r}_2)$ and $\Phi_f(\vec{r}_1,\vec{r}_2)$ represent, respectively, the wave functions of the initial and the final states of the target and are taken to be of the form :

$$\Phi_i(\vec{r}_1,\vec{r}_2) = \lambda^3\pi^{-1}e^{-\lambda(r_1+r_2)} \qquad (3)$$

and

$$\Phi_f(\vec{r}_1,\vec{r}_2) = 2^{-1/2}\left[v(\vec{r}_1)\chi^-_{\vec{k}_2}(\vec{r}_2)+v(\vec{r}_2)\chi^-_{\vec{k}_2}(\vec{r}_1)\right], \qquad (4)$$

where

$$\chi^-_{\vec{k}_2}(\vec{r}) = (2\pi)^{-3/2}e^{\gamma\pi/2}\Gamma(1+i\gamma)e^{i\vec{k}_2\cdot\vec{r}}\, {}_1F_1(-i\gamma,1,-i(k_2 r+\vec{k}_2\cdot\vec{r}))$$

$$v(\vec{r}) = \lambda'^{3/2}\pi^{-1/2}e^{-\lambda' r}$$

and $\gamma=Z/k_2$. In the present formalism, we have chosen $Z=\lambda=1.618$ so that $\Phi_i(\vec{r}_1,\vec{r}_2)$ and $\Phi_f(\vec{r}_1,\vec{r}_2)$ are orthogonal to each other.

Using the technique of Roy et al.[1] the eight-dimensional integral in Eq. (2) has been reduced to a two-dimensional integral. The integrand of this integral, however, contains a sum of two one-dimensional integral functions which are computed numerically.

Table I presents our results for the coplanar TDCS in the GA for the ionization of He by electron impact at the incident energy of 250 eV with E_1=220.42 eV, E_2=5 eV, $\theta_1=3.5^\circ$, $\Phi_1=0^\circ$ and $\Phi_2 = \pi$.

TABLE I. Magnitudes of TDCS, $d^3\sigma/d\hat{k}_1 d\hat{k}_2 dE_2$, (in atomic units) versus the angle of ejection θ_2. The format AB stands for Ax10^B .

θ_2 (deg)	GA	θ_2 (deg)	GA
0.0	9.27 -1	120.0	3.79 -2
30.0	2.18	140.6	2.12 -1
60.0	2.02	160.0	6.06 -1
88.9	7.50 -1	180.0	9.99 -1

We find that the angular distributions of the ejected electron obtained in the GA are in close agreement with the experimental results of Ehrhardt et al.[2]

References

1. A.C.Roy, A.K.Das and N.C.Sil, Phys. Rev.A23, 1662 (1981).
2. H.Ehrhardt (private communication).
3. See, for example, E.Gerjuoy and B.K.Thomas, Rep. Prog. Phys. 37, 1345 (1974).

"ENERGY-SHARING" IN RELATION TO DOUBLE DIFFERENTIAL CROSS-SECTIONS FOR ELECTRON IMPACT IONIZATION OF HELIUM

J. Kimman, Pan Guang-Yan*, C.W. McCurdy** and F.J. de Heer

FOM-Institute for Atomic and Molecular Physics, Kruislaan 407, 1098 SJ Amsterdam, The Netherlands.

* Institute of Physics, Chinese Academy of Sciences, Beijing, China

** Permanent address: Department of Chemistry, The Ohio State University, Columbus, Ohio 43210

We have measured triple differential cross-sections for electron impact ionization of helium at impact energies of 200 eV - 1500 eV for "energy sharing" kinematics. In an "energy sharing" measurement one varies the energies of the two outgoing electrons while keeping their sum constant. We have also measured absolute double differential cross-sections by measuring the electron signal at only one analyzer. For both types of cross-sections we restrict the scattering angles to values close 45^0.

A comparison of the measurements is made with theoretical calculations in the simplified eikonal impulse approximation (EIA) as given by McCarthy et al. [1] and Van Wingerden et al. [2]. Good agreement is found for situations in the triple differential cross-sections in which the outgoing electrons share the available energy nearly equally. Energies further away from symmetrical energy sharing than 25% are difficult to measure.

For the double differential cross-sections the agreement between theory and experiment is good in the symmetrical energy region where the detected electron has energy near half the primary energy, as can be seen in figure 1 at impact energy 1500 eV and 45^0 scattering angle.

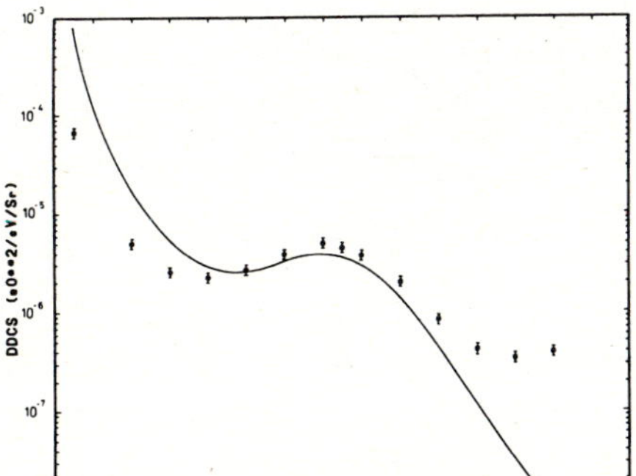

Fig. 1: The experimental results for double differential cross-sections of electron impact ionization of helium at scattering angle 45^0 as a function of secondary energy (o) together with the theoretical curve for the integrated simplified eikonal impulse approximation (——) at impact energy of 1500 eV.

At this high impact energy one can clearly distinguish the so-called impulse peak in the symmetrical region due to the minimization of the recoil momentum in that configuration. At low secondary energies the double differential cross-section is large because exchange plays a large role. From the figure we conclude that the magnitude is over estimated in EIA.

At large secondary energies the experimental cross-section rises again while the EIA decreases rapidly because the recoil momentum is large in this case. We propose a theoretical model to explain this increase in the cross-section which involves scattering from the nucleus with large momentum transfer in addition to scattering from the target electrons. Scattering from the nucleus is not present in the EIA or the first Born approximation. An approximation which will take nuclear scattering into account is the second order Born approximation. We present preliminary results in an "on the energy shell" approximation to the second Born approximation which is greatly simplified in this form and allows us to make a preliminary evaluation of our model. The rise of the cross-section at high secondary energies is seen clearly in these calculations.

A similar argument has been given by Inokuti [3] and by Geltman and Hidalgo [4] in a different context. Full second Born calculations are in progress in our laboratory.

References

1. I.E. McCarthy, E. Weigold, 1976, Phys.Rep. 27C.
2. B. van Wingerden, J.T.N. Kimman, M. van Tilburg and F.J. de Heer, 1981, J.Phys.B: At.Mol.Phys. 14, 2475.
3. M. Innokuti, 1971, Rev.Mod.Phys. 43, 297.
4. S. Geltman and M.B. Hidalgo, 1972, J.Phys.B, 617.

(e,2e) EXPERIMENTS ON He IN ASYMMETRIC CONDITIONS AT INTERMEDIATE ENERGY

A. Giardini-Guidoni, V. Di Martino [(+)], R. Fantoni and R. Tiribelli

ENEA, Dip. TIB, Divisione Fisica Applicata, C.P. 65, 00044 Frascati (Rome), Italy

It is now well established that (e,2e) experiments performed in suitably chosen kinematic conditions allow to obtain information either on the structure of the target or on the dynamics of the interaction [(1)]. First experiments [(2)] on the dynamics of the interaction performed on He at low incident energy E_o (up to 250 eV) were not fully reproduced by theoretical predictions. Recent measurements [(3)] taken at higher incident energy (up to 500 eV) are only in qualitative agreement with calculations in the second Born approximation. Aim of this work is to extend to higher incident energies the investigation on the interaction mechanism in asymmetric conditions. Present measurements are taken in the range $500 \leqslant E_o \leqslant 2000$ eV on He. The energy of the slow electron, whose angular distribution is measured, has been fixed at 20 eV and 50 eV in data reported in Figs. 1 and 2. Data are compared with first order distorted wave Born approximation and eikonal averaged distorted wave impulse approximation [(4)] (DWIA). It can be seen that at 500 eV data, as expected, are not reproduced by calculations performed in the first order Born approximation. Data taken at higher energies are instead quite adequately described by the DWIA also in asymmetric conditions, provided that the incident energy is at least a factor 50 higher than the binding energy of the struck electron (∿ 24 eV). This agreement found in the shape of the angular distributions must be checked also in the absolute value. Work is in progress in order to measure the absolute value of the (e,2e) cross section in these conditions. Since the residual ion field originates distortions on the slow outcoming electron which become more relevant as its energy decreases, work is in progress also to detect electrons ejected at lower energies.

(1) I.E. Mc Carthy and E. Weigold: Phys.Reports 27C, 275 (1976); A. Giardini-Guidoni, R. Fantoni, R. Camilloni and G. Stefani: Comm.Atom.Mol.Phys. 10, 107 (1981)

(2) H. Ehrhardt, K.H. Hasselbacher, K. Jung and K. Willmann: Case Studies in Atom.Phys. 2, 159 (1971)

(3) H. Ehrhardt, M. Fisher and K. Jung: Z.Phys. A304, 119 (1982)

(4) R. Camilloni, A. Giardini-Guidoni, I.E. Mc Carthy, and G. Stefani: Phys.Rev. A17, 1634 (1978)

(+) Guest.

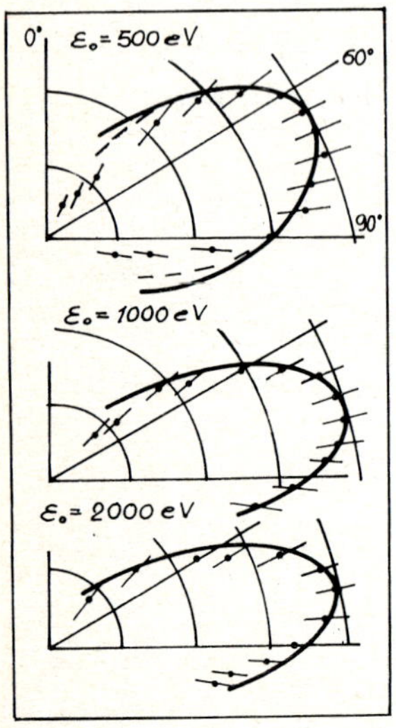

Fig. 1 - (e,2e) cross section (arbitrary units) measured at different E_o, E_B = 20 eV
$\theta_A = 9°$ (——) PWIA and (- - -) DWIA V = 10 eV calculations

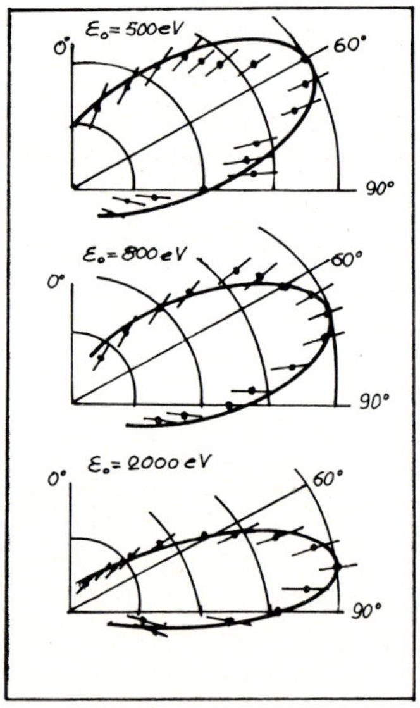

Fig. 2 - (e,2e) cross section (arbitrary units) measured at different E_o, E_B = 50 eV
$\theta_A = 9°$ (——) PWIA calculations

OBSERVATION OF THE HELIUM ION n = 1,2,3 STATES IN (e,2e) COLLISIONS.

J.P.D. Cook and E. Weigold

School of Physical Sciences, The Flinders University of South Australia, Bedford Park, 5042, Australia.

Helium provides an excellent target for testing binary (e,2e) coincidence spectroscopy[1,2]. In the plane wave impulse approximation the binary (e,2e) cross section should depend directly on the square of the momentum space overlap function of the initial manybody target and final ion wave functions. In the case of helium the final wave functions of the states are known exactly, and hence the overlap provides a sensitive measure of the helium ground state wave function. The cross section leading to the n = 2 and 3 ion states is particularly sensitive to ground state correlations. In the HF model excitation of the ns ion states can occur, since the atom 1s and ion ns orbitals are not orthonormal. However, the shape of the cross section to all ns ion states is given simply by the square of the 1s atom wave function and is independent of n. Correlation severely affects both the magnitudes and shapes for transitions to ion states with $n \geq 2$.

In order to carry out the measurements an existing out of plane symmetric (e,2e) coincidence spectrometer was modified. The single particle channel electron multiplier detectors have been replaced by commercial (Surface Science Labs) microchannel plate position-sensitive detectors which intersect the energy dispersing planes of the hemispherical analysers. Positions are decoded from the relative amplitudes of pulses taken from resistive film anodes which have RC transmission line characteristics. The two resultant DC voltages are summed and digitised, giving directly the separation energy spectrum between target initial and final states. A wide energy range is sampled concurrently and resolution is limited by only two analyser apertures instead of four. The efficiency is improved by a factor of approximately 400.

Initial trials of the method were done on a helium target with about 3.5eV FWHM energy resolution at a total energy of 1200eV. Separation energies of the He^+ n = 1,2,3 states (24.5, 65.3, 72.9eV), and a 45eV background point were sampled over a wide angular (i.e. momentum) range. Results for n = 1 and n = 2 ion states are in agreement with and an improvement on previous work.[1,2]. The cross-section for n = 3 is three orders of magnitude below n = 1, yet is distinctly visible above the background with this method. The results are compared with several accurate correlated helium ground state wave functions.

References

1. I.R. McCarthy, A. Ugbabe, E. Weigold, and P.J.O. Teubner, Phys. Rev. Lett. 33, 459, (1974).
2. A.J. Dixon, I.E. McCarthy and E. Weigold, J. Phys. B: 9, L195 (1976).

TRIPLE DIFFERENTIAL CROSS SECTION MEASUREMENTS FOR THE ELECTRON IMPACT IONIZATION OF ATOMIC HYDROGEN : COMPARISON WITH THEORY.

B. Lohmann and E. Weigold

School of Physical Sciences, The Flinders University of South Australia, Bedford Park, 5042, Australia.

Since the first experimental measurements were reported by Ehrhardt et al[1], the (e,2e) differential cross section has provided the most detailed information about the ionization process. The simplest ionization problem is that of atomic hydrogen, involving only two electrons and a proton. Whereas other ionization problems are often complicated by the use of different bound-state wave functions and potentials, these are known exactly for atomic hydrogen, therefore permitting direct comparison of the different scattering approximations.

In the present work we have extended our earlier work on atomic hydrogen[2] to more asymmetric kinematics. The measurements were carried out with a RF discharge tube atomic hydrogen source with dissociation $\gtrsim$ 80%.

Triple differential cross sections for the ionization of atomic hydrogen by electrons of 250eV impact energy were measured in coplanar geometry for ejected electron energies of 5, 10, 14, and 20eV. The angles of emission of the high energy electrons were fixed at 5°, 8° and 10°. A comparison is made between the experimental data and the second Born calculations of Byron, Joachain and Piraux[3] and various distorted wave approximations.

References

1. H. Ehrhardt, M. Schultz, T. Tekaat and K. Willman, Phys. Rev. Lett. 22, 89 (1969).
2. E. Weigold, C.J. Noble, S.T. Hood and I. Fuss, J. Phys. B: 12, 291 (1979).
3. F.W. Byron, Jr., C.J. Joachain, and B. Piraux, J. Phys. B. 13, 2673 (1980).

CORRECTIONS TO THE IMPULS APPROXIMATION FOR (e,2e) REACTION

Yu.V. Popov

Institute of Nuclear Physics, Moscow State University, Moscow 117234, USSR

A great number of triple differential cross section measurements for the electron impact ionization of atoms used to carry out under the following kinematics:

1. All momenta of the electrons are in the same plane;

2. The energy Eo of the incident electron and the energies E_1 and E_2 of the two emitted electrons are much more than the ionization potential of the atom.

3. The angles of the outgoing electrons are not too small.

In this case the form of the binary (correlative) peak can be described well in terms of the symplest plane wave impuls approximation (PWIA), but its angle position and absolute value differ from the experimental data[1]. So we need in corrections physically stipulated to reproduce all features of the peak.

If (e,2e) processes are analysed on the basis of the perturbation theory which is fitted only for short-range potentials, then it is necessary to take into account the whole totality of diagrams. Higher diagrams are known to be divergent for the Coulomb forces. The suitable renormalization procedure[2] factorises the scattering amplitude into the product of two terms. One term is an infinite sum of convergent diagrams and the first addend is the usual PWIA. The other one is a smooth function of angles of the electrons and should be taken into account if we are interested in the absolute value of the differential cross section. The equations for the definition of this factor have been formulated in papers[2,3]. The factor suppresses PWIA for small relative angle between electrons.

A sum of convergent diagrams mentioned above is important for the explanation of the correct angle position of the peak. It was shown earlier[4] that the Second Born approximation moves the maximum of the peak to larger angles but not too much. Some idea about the influence of the high order Born approximations on the position of the peak one can get with the help of semiclassical considerations[5]. If non-central post-collision Coulomb interaction of outgoing electrons is absent they would reached the counters without any dis-

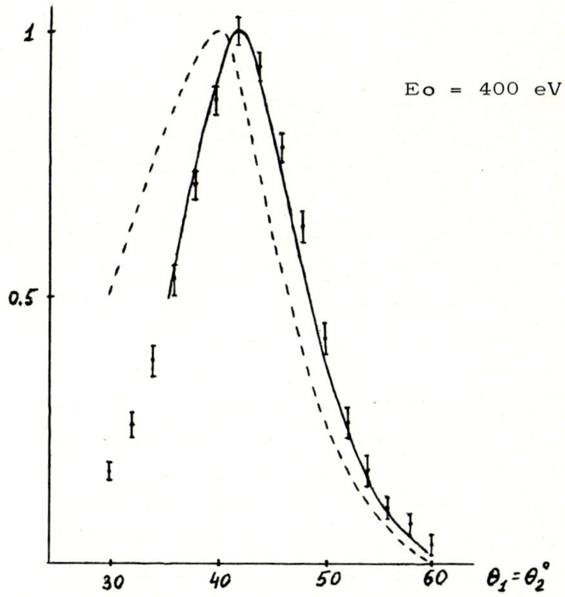

Fig.1: Relative cross section for the ejection of 1s electrons from He, calculated in PWIA (broken line) and corrected PWIA (full line).

tortion. But long-range post-interaction makes them to deflect their paths, i.e. the triple differential cross section calculated in terms of PWIA has to be moved by the definite angle in any point. For example PWIA and corrected PWIA calculations of the differential cross section for He[6] are presented on Fig. 1. Earlier the analogous coincidence has been obtained for H[5].

The investigation presented permits us to keep at least the metodological importance of impulsive (e,2e) experiments becouse we can still describe such experiments in terms of PWIA. In turn it gives us the information on the wavefunction of an atom directly from the observable data.

References

1. The rather full servey of the modern experimental and theoretical situation one can find in:J.Stefani. NATO A.S.I. Progress report, Maratea, Italy,1982.
2. Yu.V. Popov,J.Phys.B:At.Mol.Phys.14,2449(1981).
3. Yu.V.Popov,J.Bang, J.J.Benayoun. J.Phys.B:At.Mol. Phys.14, 4637(1981).
4. A.M. Popova, Yu.V.Popov, V.F. Erokhin. Izv.Akad. Nauk SSSR, ser.fiz. 44, 2351(1980).
5. Yu.V. Popov, J.J. Benayoun. J.Ph.:At.Mol.Phys. 14, 3513 (1981).
6. The experiment and PWIA are from R. Camilloni et al. Phys.Rev. A17, 1634,(1978).

EXCHANGE EFFECTS IN RESONANCE IONIZATION OF ATOMS BY INTERMEDIATE-ENERGY ELECTRONS

V.V. Balashov, A.N. Grum-Grzhimailo, A.I. Magunov

Institute of Nuclear Physics, Moscow State University, Moscow 117234, USSR

Application of the intermediate-energy electrons to the studies of resonance ionization of atoms in the (e,2e) experiments[1,2] requires the theory of such processes should be generalized to include the exchange scattering.

We proceed from a certain expression of angular correlation function for the autoionization (e,2e) process obtained by extending our earlier expressions[3] in two lines. First, allowance is directly made for the exchange scattering amplitudes simultaneously with the distortions by the corresponding optical potential. Second, the autoionization state wave function can be constructed in the intermediate coupling approximation, i.e. the singlet-triplet mixing can be included. The two aspects are of great importance when describing the polarization properties (statistical tensors) of autoionization states and hence the shape of the angular correlation function between the scattered and ejected electrons.

Tentative quantitative estimates of the exchange effects have been obtained in the Ochkur approximation for the optically allowed and optically forbidden states of the $4d^{-1}5p$ configuration in Cd (Fig.1). The shapes of the differential cross section and angular correlation function may prove to be much different in these two cases .

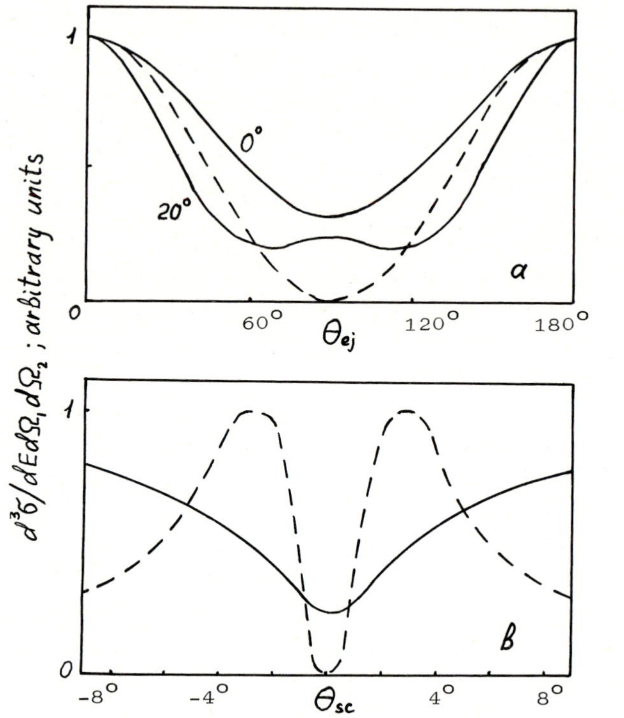

Fig. 1.: Angular correlation functions for $4d^{-1}5p$ J= 2 state of Cd (E_{ex}=12.3 eV) in the (e,2e) process calculated at 150 eV (solid curves). a) At the scattering angles 0° and 20° depending on θ_{ej} b) At the ejection angle 90° depending on θ_{sc} . The dashed lines show the angular correlation function for $4d^{-1}5p\,^1P$ state[3].

References

1. N.L.S. Martin, T.W. Ottley, K.J. Ross, J.Phys.B13, 1867 (1980)
2. N.L.S. Martin, K.J. Ross, J.Phys.B15,3959(1982)
3. V.V. Balashov et al. J.Phys.B13, L269(1980).

TRIPLE DIFFERENTIAL CROSS SECTION FOR THE ELECTRON IMPACT IONIZATION OF HYDROGEN

A. S. Ghosh, P. S. Mazumder and Madhumita Basu

Department of Theoretical Physics, Indian Association for the Cultivation of Science, Calcutta 700032, India.

The triple differential cross section (TDCS) for the ionization of hydrogen atom by electron impact has been calculated using three distorted wave models. In our calculations, the wave functions of the scattered and the ejected electrons are represented by the Coulomb function of unit charge, following the work of Smith et al.[1].

The total wave function $\Psi(r_1,r_2)$ in the incident channel is written as

Model - I

$$\Psi^+(r_1,r_2) = \Phi_{1s}(r_2)F(r_1) \quad (1)$$

Model - II

$$\Psi^+(r_1,r_2) = (\Phi_{1s}(r_2)+\Phi^{pol}(r_1,r_2))F(r_1) \quad (2)$$

and Model - III

$$\Psi^+(r_1,r_2) = (1\pm P_{12})(\Phi_{1s}(r_2)+\Phi^{pol}(r_1,r_2))F(r_1) \quad (3)$$

Here $\Phi_{1s}(r_2)$ is the ground state wave function of the hydrogen atom and Φ^{pol} is the perturbed target state.

In the case of Model-I and II, the scattering function $F(r_1)$ satisfies the differential equation

$$(\nabla_r^2+k_i^2-v_s(r)-v_p(r))F(r) = 0 \quad (4)$$

In other words, the effect of Pauli principle has been neglected in Model - I and II. The scattering function F(r) in Model - III satisfies the full polarized orbital equation as obtained by Sloan[2]. The effect of exchange and exchange polarization are included in Model - III.

The TDCS at the incident energy E_i=100 eV, ejected electron energy E_o = 25 eV and the scattering angle θ_1 = 30° has been plotted in Fig. 1 . The present three sets of results are normalised to give the experimental peak height as the measurements[3] are relative. The qualitative agreement between the present theoretical results and those of experiment is fair. The present results are also in fair agreement with those of Smith et al. (not shown in the figure) above the ejected angle θ_2 = 20°. The effect of exchange is found to be about 8% when comparisons are made between Model - II and III at the incident energy

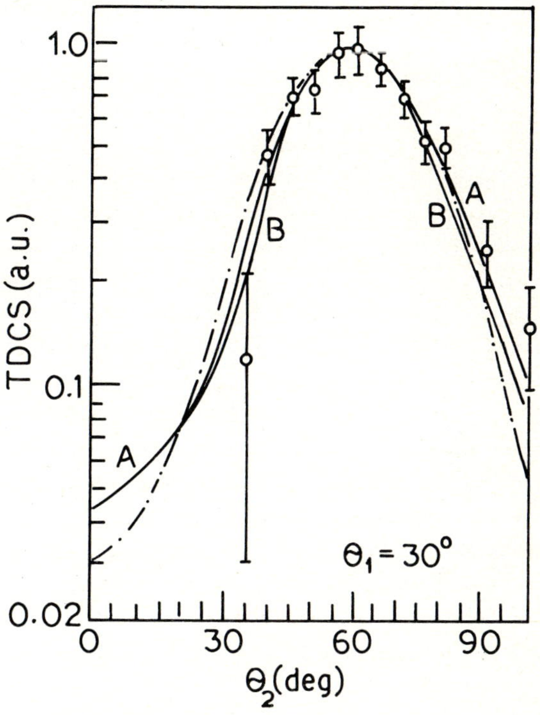

Fig. 1: Triple Differential cross sections at E_1=100 eV, E_2=25 eV and θ_1=30°. A = Model II, B= Model III and Chained curve = Model I.

E_i= 100 eV. It has been found from the numerical values of TDCS that the effect of the matrix elements involving the target distorted wave functions is appreciable.

It is very difficult to prefer one theoretical method over others in predicting the TDCS by comparing the unnormalised data. The present results using Model-III is supposed to be most elaborate calculations for this process. Some absolute measurement is essential to find the validity of the theoretical methods.

References

1. J. J. Smith, K. H. Winters and B. H. Bransden, J. Phys. B 13 1723 (1979).
2. I. H. Sloan, Proc. R. Soc. London Ser. A 281, 151 (1964).
3. E. Weigold, C. J. Noble, S. T. Hood and I. Fuss, J. Phys. B 12, 291 (1979).

ENERGY EXCHANGE BETWEEN TWO OUTGOING ELECTRONS IN THE POST COLLISION INTERACTION PROCESSES

J. Mizuno*, T. Ishihara** and T. Watanabe***

*Japan Business Automation Co. Ltd., 4-2 Nihonbashi-Honcho, Chuo-ku, Tokyo 103 Japan

**Inst. of Applied Physics, University of Tsukuba, Sakura-mura, Ibaraki 305 Japan

***Atomic Process Lab., The Inst. of Physical and Chemical Res.(RIKEN), Wako-shi, Saitama 351 Japan

Energy shifts, the so-called post collision interaction (PCI) effects, have been observed in the spectrum of ejected electrons from autoionizing states excited by electron impact[1)] and in the spectrum of Auger electrons emitted following inner-shell ionization by electron or photon impact[2,3)]. So far in all the theories[4-7)] of PCI, which deal only with the energy region close to the threshold, the "fast" ejected electron is assumed to disappear at the moment of its creation, thus changing suddenly the effective ionic charge seen by the "slow" electrons. The electron-electron interaction in the final state is entirely neglected. The energy shifts of the ejected electrons predicted by these theories show the Barker-Berry classical behaviour when the excess energy $\Delta\varepsilon$ becomes large. On the other hand, the experimental results at higher energies show that the shift should vanish much faster than the classical $(\Delta\varepsilon)^{-1/2}$ law. In order to explain this fact, the electron-electron interaction must be fully taken into account. We are investigating this problem by solving the classical Coulomb three-body problem numerically.

Let us consider the autoionizing reaction

$$e + A \rightarrow A^{**} + e_1, \quad A^{**} \rightarrow A^+ + e_2 \qquad (1)$$

At time t=0, the atom A is excited to the autoionizing state A^{**} with the life time $\tau=1/\Gamma$ and the inelastically scattered electron e_1 with the excess energy $\Delta\varepsilon$ begins to recede from the atom with the constant velocity $v_{10}=\sqrt{2m\,\Delta\varepsilon}$. At some later time $t=t_0$, the autoionizing state decays and the electron e_2 is ejected with the initial velocity v_{20} given by

$$v_{20} = \sqrt{2m\,(\,\varepsilon_0 + 1/r_{20}\,)} \qquad (2)$$

where $\varepsilon_0 = E(A^{**}) - E(A^+)$ is the energy difference of the autoionizing state and the ionic state, r_{20} the radial position of the ejection. We assume that the electron e_2 is ejected only in the radial direction and isotropically. The initial situation is illustrated in Fig.1, where $r_{10} = V_{10}t_0$. Now that the initial conditions are set up, the Newton's equations given below for the two electrons can be solved numerically.

$$m\frac{d^2\vec{r}_1}{dt^2} = -\frac{1}{r_1^2}\hat{r}_1 + \frac{1}{r^2}\hat{r} \qquad (3)$$

$$m\frac{d^2\vec{r}_2}{dt^2} = -\frac{1}{r_1^2}\hat{r}_2 - \frac{1}{r^2}\hat{r} \qquad (4)$$

where $\vec{r} = \vec{r}_1 - \vec{r}_2$ and the vectors with hats are unit vectors. The energy $\varepsilon_2(\hat{r}_{20}, t_0)$ of the ejected electron at $t=\infty$ depends on the initial direction $\hat{r}_{20}$ of the ejection. Averaging over the direction, we obtain

$$\varepsilon(t_0) = \frac{1}{4\pi}\int \varepsilon_2(\hat{r}_{20}, t_0)\, d\hat{r}_{20} \qquad (5)$$

Finally taking into account of the decay probability $\Gamma\exp(-\Gamma t_0)$ at time $t=t_0$, the energy spectrum $P(\varepsilon)$ of the ejected electron is given by

$$P(\varepsilon) = \Gamma\exp(-\Gamma t_0)\,\frac{dt_0}{d\varepsilon} \qquad (6)$$

The computation is in progress along this line and the results will be presented at the conference.

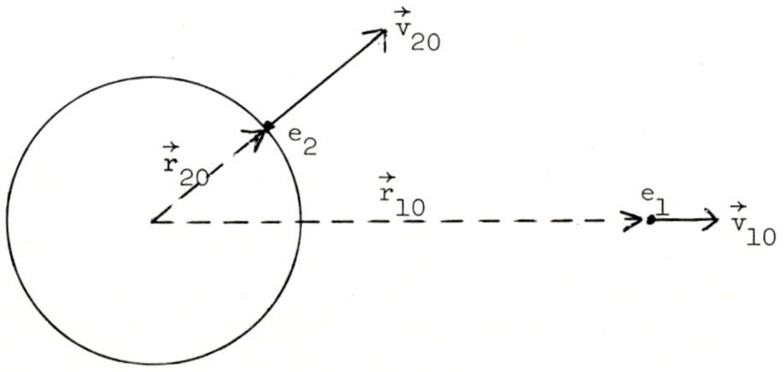

Fig.1. Initial conditions.

References

1) D. Spence, J. Phys. B11, 1243 (1978)
2) S. Ohtani, H. Nishimura, H. Suzuki and K. Wakiya, Phys. Rev. Letters 36, 863 (1976)
3) V. Schmidt and S. Krummacher, Phys. Rev. A24, 1803 (1981)
4) G. C. King, F. H. Read and R. C. Bradford, J. Phys. B8, 2210 (1975)
5) A. Niehaus, J. Phys. B10, 1845 (1977)
6) F. H. Read, J. Phys. B10, L207 (1977)
7) M. Ya. Amusia, M. Yu. Kuchiev and S. A. Sheinerman, Sov. Phys. JETP 49, 238 (1979)

THRESHOLD EXCITATION OF RYDBERG STATES OF ATOMS AND DIATOMIC MOLECULES

P Hammond, G C King, F H Read and J Jureta*

Department of Physics, University of Manchester, UK.

We have carried out two different sets of experiments to study near-threshold electron impact excitation of Rydberg states. In the first set the yield of electrons of very low energy ($\lesssim$15meV) in the process

$$e + A \rightarrow A^* + e_{thresh}$$

is measured for the Rydberg states of Kr and Xe that lie between the $^2P_{3/2}$ and $^2P_{\frac{1}{2}}$ ionization energies. The results are shown in figure 1. Autoionization of the Rydberg states causes the scattered electrons to lose energy by a post-collision interaction, so that the incident energy required to produce scattered electrons that have a nearly-zero final energy is greater than the spectroscopic energy. The values of the shifted energies, where these can be identified, together with the spectroscopic energies, are indicated on the figure. The shifts are greatest for the shortest-lived states, which are those having the lowest values of n and ℓ. Using the shake-down model of PCI[1] we have been able to deduce approximate values of the lifetimes. We have also estimated the dependence of the relative yield of threshold electrons on n and ℓ, for the non-overlapping peaks that can be unambiguously identified, and have corrected these yields for the reduction caused by the post-collision interaction, again using the shake-down model. Excitation of the higher ℓ-values becomes more favourable as the binding energy decreases, as predicted by Fano[2].

In the second set of experiments the yield of metastable high-n Rydberg states formed by electron impact is measured. The Rydberg states are detected by field ionization, using two regions of electric field, the range of n-values being determined by the two field strengths. Figure 2 shows the yield of Xe atoms having n from approximately 18 to 40. The form of the yield is essentially that of a step-function at threshold (as found previously by Tarr et al[3] in a low-resolution experiment), followed by structure due to autoionizing states, as in figure 1. The threshold step-function is found also for high-n states of Ar, Kr, H_2 and N_2. No detailed theoretical model is yet available to explain this phenomenon.

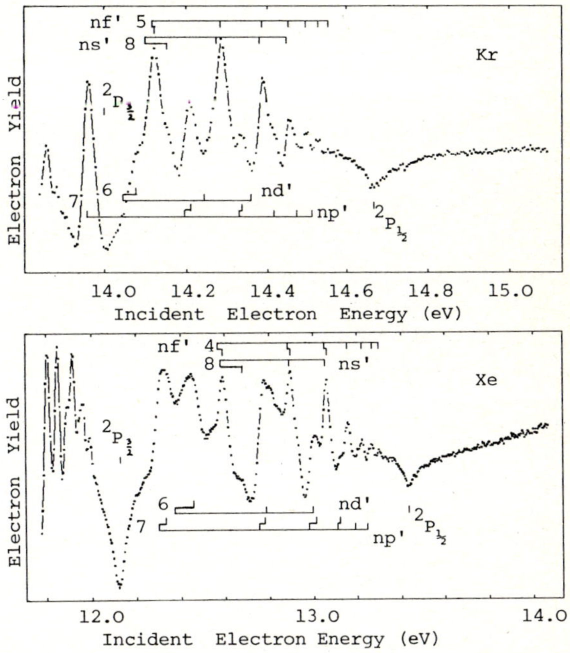

Figure 1. Yield of threshold electrons resulting from electron impact on krypton and xenon. The spectroscopic energies of the autoionizing states are indicated, together, in some cases, with the PCI-shifted energies of the observed peaks.

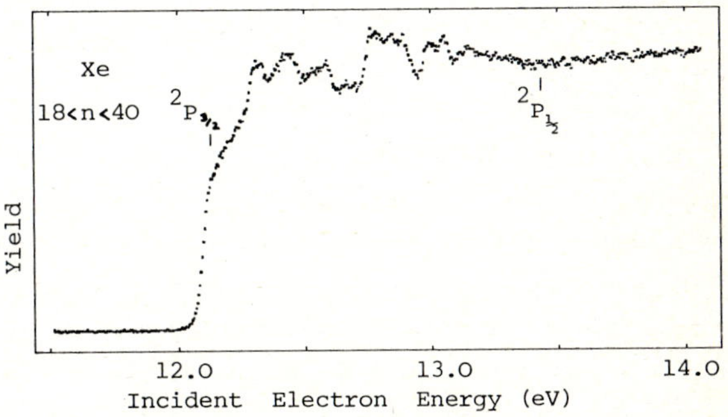

Figure 2. Yield of Rydberg atoms having n from approximately 18 to 40 resulting from electron impact on xenon.

REFERENCES

1. F.H.Read, J.Phys.B 10, L207 (1977)
2. U.Fano, J.Phys.B 7, L401 (1974)
3. S.M.Tarr, J.A.Schiavone and R.S.Freund, J.Chem.Phys. 74, 2869 (1981)

* Present address; Institute of Physics, Belgrade, Yugoslavia.

ANGULAR MOMENTUM EXCHANGE IN POST COLLISION INTERACTION

A. Niehaus and C.J. Zwakhals

Fysisch Laboratorium, Rijksuniversiteit Utrecht, the Netherlands

PCI-influenced Auger electron line shapes and line shifts observed for the processes of inner shell photoionization[1] or inner shell electron impact ionization[2] have in the past been very satisfactorily explained on the basis of a simple semiclassical theory[3], in which only energy exchange between the slow electron and the fast Auger electron is taken into account, but in which possible angular momentum exchange is neglected. ArL-line shapes recently observed for electron impact ionization[4] show a significant and characteristic deviation from the normal PCI-line shape: most of the Ar-$L_{23}M_{23}M_{23}$-lines develop a high energy shoulder as the electron impact energy is decreased towards the L-ionization threshold. We interpret these shoulders as a kind of a rainbow, caused by a maximum in the difference of the Born-Oppenheimer-type potentials describing the motion of the slow electron before and after emission of the Auger electron. Such a maximum arises for the angular momentum dependent effective potentials if the slow electron increases its angular momentum by $\ell_f - \ell_i = \Delta\ell$. For given ℓ_f, ℓ_i the rainbow occurs at positions ε_* given by

$$\varepsilon_* = \varepsilon_0 + [2\ell_f(\ell_f + 1) - 2\ell_i(\ell_i + 1)]^{-1}$$

with ε_0 the nominal Auger energy.

We have developed an improved semiclassical theory which accounts for the effect of angular momentum exchange in the case of inner shell photoionization. For given ℓ_f and ℓ_i this theory yields an analytical expression for the line shape. In fig. 1 we show as an example the line shape for $\ell_f = 4$, $\ell_i = 3$, and for various excess energies of the photoelectron ε_i. The line position ε_0, and the lifetime - which enters the calculation - are the ones for the Ar-$L_2M_{23}M_{23}(^3P)$ transition[4]. The position of the rainbow, and its behaviour as a function of ε_i agrees very well with the position and the behaviour of the unexplained shoulders observed recently[4].

Our new formulation of the theory is not only improved regarding the inclusion of angular momentum exchange, but also regarding the behaviour of the resulting line shape at large excess energy ε_i. While in the earlier formulation[3] the line became asymptotically a δ-function at ε_0, the present formulation yields a Lorentzian of width $\Gamma = \tau^{-1}$, which is the correct behaviour.

The earlier formulation has been very successful in predicting the observed shifts of the main Auger peaks. It is important to note that regarding these shifts the predictions of the present formulation are virtually identical.

References

1. V. Schmidt, N. Sandner, W. Mehlhorn, M.Y. Adam and F. Wuilleumier, Phys. Rev. Lett. 38, 63 (1977)
2. R. Huster and W. Mehlhorn, Z. Phys. A 307, 67 (1982)
3. A. Niehaus, J. Phys. B 10, 1845 (1977)
4. S. Hedman, K. Helenelund, L. Asplund, K. Gelius and K. Siegbahn, J. Phys. B 15, L799 (1982).

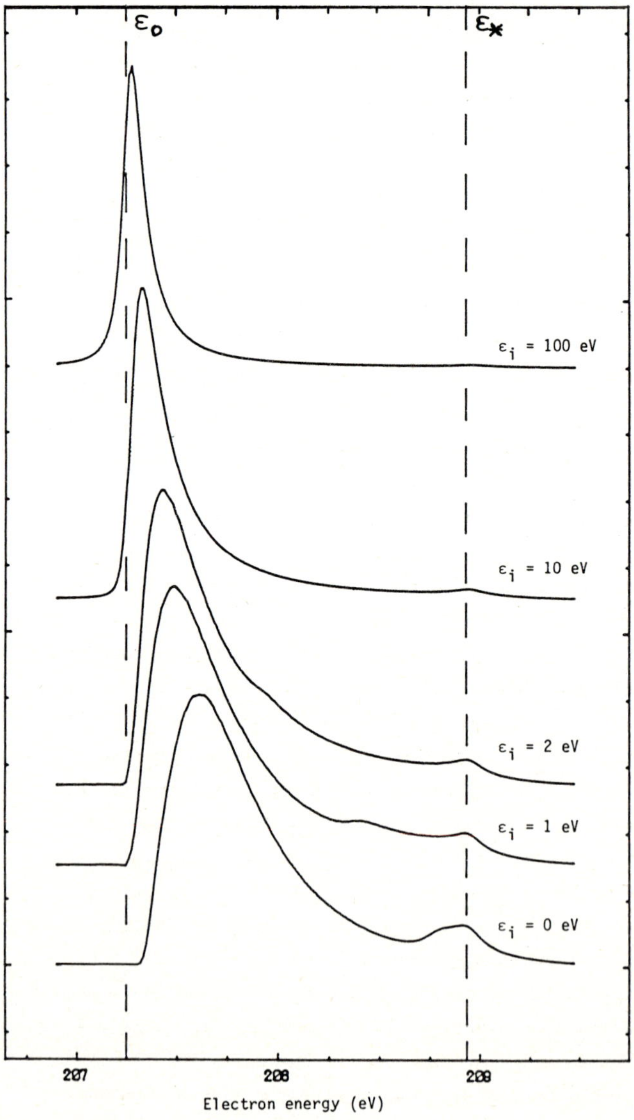

Fig. 1: Calculated PCI-influenced Auger line shapes for $\ell_i = 3$, $\ell_f = 4$, and for various energies of the photoelectron. Line position and lifetime correspond to Ar-$L_2M_{23}M_{23}(^3P_{0,1,2})$[4]. Also indicated the rainbow position ε_*.

ANGULAR DISTRIBUTION FOR ELECTRON EXCITATION OF THE $4^2S \rightarrow 4^2P$ TRANSITION IN ZnII: COMPARISON OF EXPERIMENT AND THEORY

A. Chutjian

Jet Propulsion Laboratory, California Institute of Technology, Pasadena, California 91109

A. Z. Msezane

Morehouse College, Atlanta, Georgia 30314

R. J. W. Henry

Lousiana State University, Baton Rouge, Lousiana 70803

Differential electron scattering cross sections for inelastic excitation of an ion have been measured for the first time. Experiments were carried out in a crossed electron-ion beam geometry for the $4^2S \rightarrow 4^2P$ transition in ZnII at 75 eV. In addition, differential cross sections were calculated at energies between 15 eV and 100 eV in a five-state close-coupling approximation in which 4s, 4p, $3d^94s^2$, 5s and 4d states were included.

A beam of ZnII was generated from a DC discharge in Zn vapor produced in a heated boron-nitride oven. The ions were velocity selected, electrostatically focused, and magnetically deflected prior to the collision region. After the collision region the ions were deflected and collected in a deep Faraday cup. A beam of electrons of low angular divergence was focused onto the ion beam. Inelastic electrons were detected over a range of angles in the forward scattering direction using a hemispherical electrostatic analyzer and lens system which were rotated relative to the electron gun. Electron and ion currents, overall resolution in the energy-loss spectra, and typical gun and analyzer focusing procedures were the same as given earlier.[1]

Spectra corresponding to the unresolved $4^2S_{1/2} \rightarrow 4^2P^o_{1/2,3/2}$ fine-structure transitions were measured at one-degree intervals over the range of scattering angles $4° \le \theta \le 16°$. Peak areas were measured and normalized to unity at $\theta = 14°$ at which a cross section of 1.30×10^{-16} cm^2/sr was reported.[1] Results of these differential cross section measurements are shown in Fig. 1.

From the theoretical standpoint, Msezane and Henry[2] had previously calculated cross sections for excitation of Zn^+ from the 4s ground state to the resonance 4p state in a close-coupling approximation in which the 4s, 4p, $3d^94s^2$, 5s, and 4d states were retained. Very good agreement with measurements of the absolute emission cross sections of Rogers et al.[3] was obtained for the energy range $15 \le E_o \le 100$ eV when cascade contributions were included. We used the reactance matrices obtained by Msezane and Henry[2] in a program developed by Brandt et al.[4] to calculate angular differential cross sections.

In Fig. 1 we show the calculated differential cross section for excitation from 4s to 4p in ZnII at an incident electron energy of 75 eV. There is excellent agreement in magnitude and shape between present measurements and calculations. One notes that at small angles the scattering intensity falls very rapidly with angle. At large angles, diffraction maxima and minima are noted. The same qualitative behavior was observed in a five-state close-coupling calculation for e-MgII scattering[5] at an electron energy (50 eV) corresponding to a fractional energy loss comparable to ZnII.

Work supported by NASA and the Department of Energy.

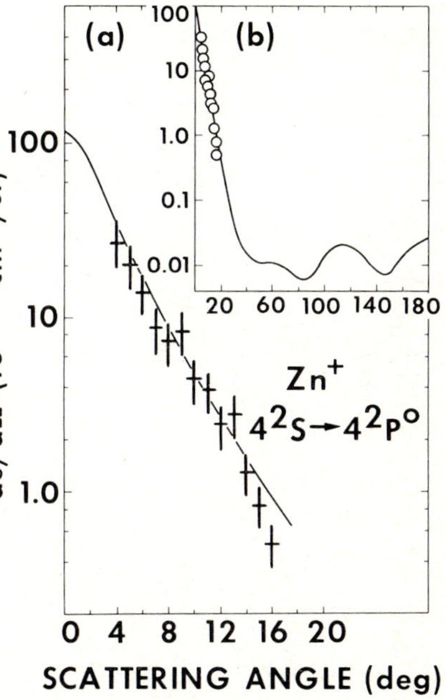

Fig. 1: Differential cross sections for the $4^2S \rightarrow 4^2P$ transition in ZnII at 75 eV. Shown are comparisons of experimental results (crosses) and theoretical calculations (solid lines).

References

1. A. Chutjian and W. R. Newell, Phys. Rev. A 26, 2271 (1982).
2. A. Z. Msezane and R. J. W. Henry, Phys. Rev. A 25, 692 (1982).
3. W. T. Rogers, G. H. Dunn, J. Østgaard-Olsen, M. Reading, and G. Stefani, Phys. Rev. A 25, 681 (1982).
4. M. A. Brandt, D. G. Truhlar and R. L. Smith, Comput. Phys. Commun. 5, 456 (1973).
5. A. Z. Msezane and R. J. W. Henry (unpublished results). The 3s, 3p, 3d, 4s and 4p states in MgII were included in this calculation.

EXPERIMENTAL ELECTRON ENERGY-LOSS SPECTRA AND CROSS SECTIONS FOR THE $5^2S \rightarrow 5^2P$ TRANSITION IN CdII

A. Chutjian

Jet Propulsion Laboratory, California Institute of Technology, Pasadena, California 91109

Experimental energy-loss spectra for excitation of the $5^2S \rightarrow 5^2P^o$ resonance transition in CdII are reported for the first time. Measurements were carried out in a crossed (90°) electron-beam - ion-beam measurement in CdII.[1,2] Spectra as a function of electron scattering angle were obtained at 75 eV electron energy, and in the angular range $4° \leq \theta \leq 16°$. From these spectra, shown in Fig. 1, differential scattering cross sections were obtained using appropriate values of electron and ion currents, interaction velocity, signal rates, beam and analyzer geometries, and detector efficiency. These differential cross sections are shown in Fig. 2. Also shown for comparison is the calculated cross section for the analogous $4^2S \rightarrow 4^2P$ resonance transition in ZnII at a corresponding energy (80 eV) above threshold.[2] One notes a steeper differential cross section in the CdII case, giving also a larger integral cross section than in ZnII.

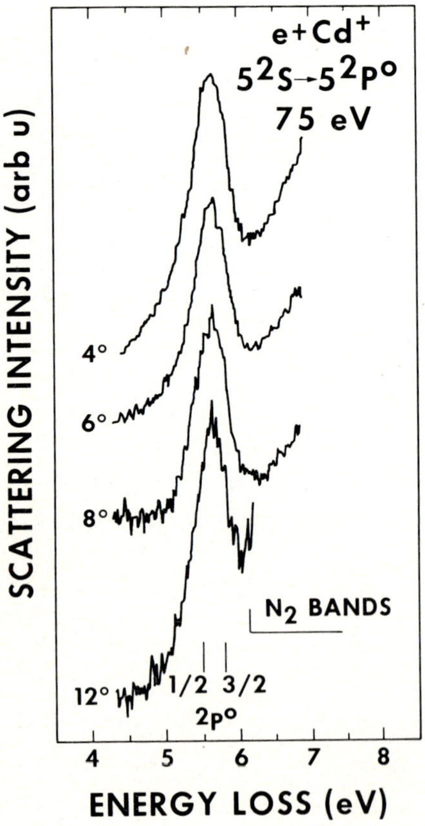

Fig. 1: Energy-loss spectra at the indicated scattering angles for the $^2S \rightarrow {}^2P$ resonance transition in CdII.

Comparisons will be made to recent electron excitation cross section obtained from photon emission data,[3] and with results of semiempirical formulas of Mewe.[4]

This work is supported by NASA.

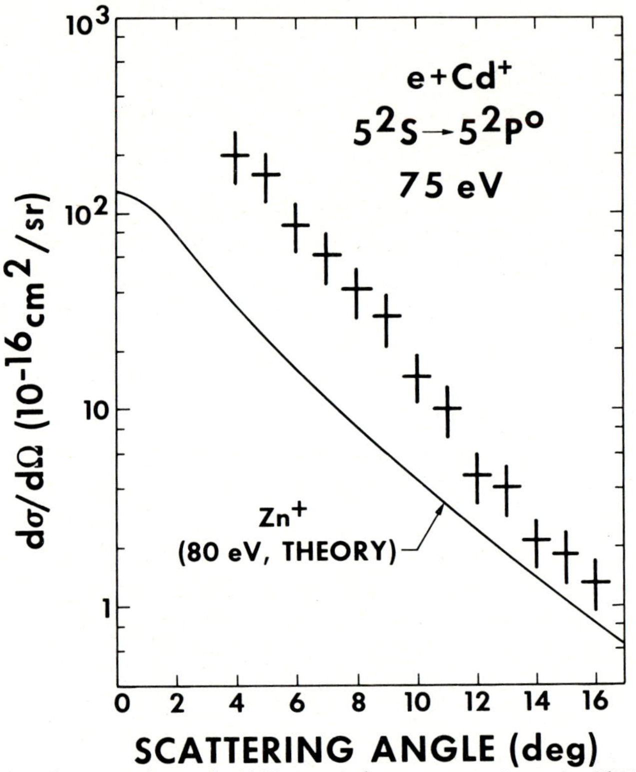

Fig. 2: Experimental differential cross sections in CdII (crosses) for the $5^2S \rightarrow 5^2P$ transition, with comparison to the calculated cross sections for the $4^2S \rightarrow 4^2P$ transition in ZnII at an electron energy (80 eV) corresponding to a comparable fractional energy loss.

References:

1. A. Chutjian and W. R. Newell, Phys. Rev. A 26, 2271 (1982).
2. A. Chutjian, A. Z. Msezane and R. J. W. Henry, Phys. Rev. A, in press.
3. K. Hane, T. Goto and S. Hattori, Phys. Rev. A 27, 124 (1983).
4. R. Mewe, Astron. Astrophys. 20, 215 (1972).

CALCULATION OF ELECTRON-ION SCATTERING CROSS-SECTIONS: A PERTURBATIVE TREATMENT OF EXCHANGE

M.A. Hayes

Institut für Astronomie, ETH-Zentrum, CH-8092 Zürich, Switzerland

A method of calculating electron-ion scattering cross-sections, designed to handle cases where there is strong potential coupling between the target states and hence a distorted wave approximation is not valid but where the problem is too large for the full close-coupling method, is presented.

In the close-coupling method, the set of coupled integro-differential equations has to be solved[1]

$$\left(-\frac{d^2}{dr^2} + \frac{l_i(l_i+1)}{r^2} - \frac{2Z}{r} - k_i^2\right)F_{ii'} + \sum_j (V_{ij}+W_{ij})F_{ji'} = 0 \qquad (1)$$

where V_{ij} and W_{ij} are the direct and exchange potential operators respectively; i and j denote scattering channels and i' denotes a particular solution. W_{ij} is an integral operator. The required boundary conditions are

$$F_{ii'}(0) = 0$$

$$F_{ii'}(r) \rightarrow 0 \quad \text{as } r\rightarrow\infty \quad ; \; k_i^2 < 0$$

We define two regions of space; $r < r_a$ and $r > r_a$, where for $r > r_a$ we assume the target functions and hence W_{ij} are zero.

Equation (1) is solved for $r < r_a$ neglecting W_{ij}. This gives a set of coupled differential equations which are easier to solve numerically than the full equation.

Let $\underline{F}_t$ be the solution matrix of the non-exchange equations and let the exact solution and its first derivative be written as

$$\underline{F}(r) = \underline{F}_t(r) + \delta\underline{F}(r)$$

$$\underline{F}'(r) = \underline{F}_t'(r) + \delta\underline{F}'(r)$$

From Green's theorem we have

$$\underline{F}_t(r_a)\delta\underline{F}'(r_a) - \underline{F}_t'(r_a)\delta\underline{F}(r_a) = \langle \underline{F}_t|\underline{H}-E|\underline{F}_t\rangle + O(\delta\underline{F})^2 \qquad (2)$$

where $\underline{H}$ is the full Hamiltonian including the exchange operator. Using (2) we evaluate $\delta\underline{F}(r_a)$ and $\delta\underline{F}'(r_a)$ and hence estimate $\underline{F}(r_a)$ and $\underline{F}'(r_a)$.

Equation (1) has solutions which behave asymptotically as

$$s_{ii'}(r) \sim \delta_{ii'} k_i^{-1/2} \sin \eta_i$$

$$c_{ii'}(r) \sim \delta_{ii'} k_i^{-1/2} \cos \eta_i \qquad \text{as } r\rightarrow\infty$$

$$e_{ii'}(r) \sim 0$$

These solutions are obtained by the program of Crees[2] and integrated in to $r = r_a$ where they are matched with the solutions $\underline{F}$ to obtain the coefficient matrices $\underline{a}$, $\underline{b}$ and $\underline{d}$ such that

$$\underline{F} = \underline{s}\,\underline{a} + \underline{c}\,\underline{b} + \underline{e}\,\underline{d}$$

The reactance matrix $\underline{R} = \underline{b}\,\underline{a}^{-1}$ and the cross-sections can then be evaluated.

References

1. Seaton M.J., J.Phys.B: Atom.Molec.Phys.7, 103.(1974)
2. Crees M.A., Comput.Phys.Commun.19, 103.(1980)

EXCITATION OF HYDROGENLIKE IONS IN MODIFIED GLAUBER APPROXIMATIONS

T.T. Gien

Physics Department, Memorial University of Newfoundland, St. John's, NFLD, Canada A1B 3X7

The excitation of hydrogenlike ions by electron impact is investigated, using the modified Glauber approximations [1]. At present, the study is focused on the 1s - 2s excitation process of He^+ by electron impact, where experimental data of integrated cross section were available [2] for comparison. As usual, the essential idea of the modified Glauber approximation[1] is to repair some of the very serious deficiencies which experience in the second-order scattering term of the conventional Glauber amplitude, due to the consideration of the eikonal approximation for this term. In the modified Glauber approximation, the eikonal approximation is considered only for the third-order scattering terms onward, as it is believed that at the intermediate energy range, the energies are not high enough to justify an eikonal approximation for its second-order scattering term. The second-order scattering term is calculated by using instead its counter-part prior to eikonalization, i.e. the second-order Born term. By this way, some serious deficiencies in the second-order eikonal term can be avoided. In the modified Glauber approximation, contributions from higher order scattering are not totally neglected but are included via their approximate eikonal forms. In some collision processes[3], the contributions from these higher order terms were shown quite significant at intermediate energies and there-by non-negligible. Thus, the modified Glauber amplitude, as usual[1], is in the form

$$f_{MG} = f_G - f_{G2} + f_{B2} \qquad (1)$$

In my study of $e^- - He^+$ excitation to be presented here, relevant formulae of the second order eikonal and Born terms, which are required for the calculation of the modified Glauber amplitude, are derived and reduced to appropriate forms which can be accessed with ease to numerical computations. The second-order scattering terms in this case are classified into two types, one is due to scattering by "core atom" only, while the other, to both residual Coulomb potential and "core atom",

$$f_{G2} = f_{G2}^{(I)} + f_{G2}^{(II)} \qquad (2)$$

Modified Glauber approximations in which either $f_{G2}^{(I)}$ or $f_{G2}^{(II)}$ or both are corrected with the use of the appropriate corresponding second-Born expressions are all considered for the excitation of He^+ by electron impact. The second Born terms $f_{B2}^{(II)}$ and $f_{B2}^{(I)}$ are either calculated exactly or approximately by using the well-known closure approximation. The Glauber amplitude of e^- - ion excitation is calculated with the use of its closed form.

At present, the method is applied to the 2s excitation of He^+ from its ground state. The differential and integrated cross sections are calculated at several collision energies. Results of the modified Glauber approximations are compared, with discussion, to experimental data, as well as to those calculated in other methods of approximation such as conventional Glauber, first Born approximations etc. More complete results of my investigation and details of the derivation of the relevant formulae are to be presented at the conference.

This research work is supported by the Natural Science and Engineering Research Council of Canada.

1. T.T. Gien, J. Phys. B: Atom. Molec. Phys. 9, 3203-3211 (1976).
2. K.T. Dolder and B. Peart, J. Phys. B: At. Mol. Phys. 6, 2415 (1973).
3. T.T. Gien, J. Phys. B: Atom. Molec. Phys. 12, 3987-3992 (1979).

ELECTRON-IMPACT EXCITATION COLLISION STRENGTHS AND EXCITATION RATES FOR He-like IONS

S.S. Tayal and A.E. Kingston

Department of Applied Mathematics and Theoretical Physics,
The Queen's University of Belfast, Belfast BT7 1NN, Northern Ireland.

Relative emission line strengths for He-like ions can provide important information in the study of solar flares conditions[1-3]. Very accurate collision excitation rates are required in order to obtain reliable results from such observations.

We have carried out accurate eleven state calculations for collision strengths for a number of He-like ions for all transitions involving ground state, n=2 and n=3 states using R-matrix method. Eleven target eigenstates consisting of the five n=1 and n=2 states with configurations $1s^2$, 1s2s and 1s2p and six n=3 states with configurations 1s3s, 1s3p and 1s3d are included in the expansion of the total wavefunction. These eigenstates are represented by CI wavefunction constructed from eight orthonormal basis orbitals: 1s, 2s, 2p, 3s, 3p, 3d, 4s and 4p. The 1s orbital is taken to be hydrogenic and the other orbitals are optimised on the n=2 and n=3 states using the CIV3 program of Hibbert[4]. Good agreement between length and velocity formulations for the oscillator strengths was obtained. The collision calculations are carried out at a very fine energy mesh in the threshold regions using R-matrix package described by Berrington et al[5]. The contribution from all significant partial waves is included.

The collision strengths for most transitions are found to have a complicated resonance structure. The quantity which is of interest in astronomy is the excitation rate. These are obtained by averaging the cross sections over a Maxwellian distribution. The results are compared with solar observations.

In figure 1 we have plotted the collision strengths for the electron collisional excitation of O VII for the transition $1s^2\,{}^1S^e - 1s2s\,{}^3S^e$ as a function of energy. There is a complicated structure of resonances in the collision strengths.

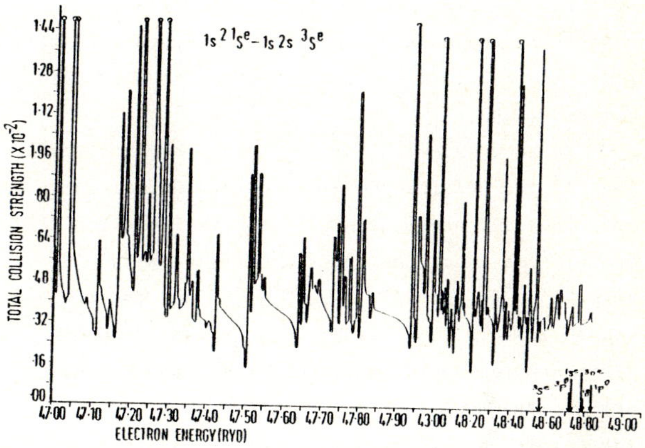

Fig. 1: The Collision strength for the $1s^2\,{}^1S^e - 1s2s\,{}^3S^e$ transition in O VII

References

1. A.H. Gabriel and C. Jordan, Mon. Not. R. Astron. Soc. 145, 241, (1969).
2. G.R. Blumenthal, G.W.F. Drake and W.H. Tucker Astrophys. J. 172, 205 (1972).
3. D.L. McKenzie, R.M. Broussard, P.B. Landecker, H.R. Rugge, R.M. Young, G.A. Doschek and U. Feldman, Astrophys. J. Lett. 238, L43 (1980).
4. A. Hibbert, Comput. Phys. Commun. 9, 14 (1975).
5. K.A. Berrington, P.G. Burke, M. LeDourneuf, W.D. Robb, K.T. Taylor and Vo Ky Lan, Comput. Phys. Commun. 14, 367 (1978).

ELECTRON-IMPACT EXCITATION OF OV

S.S. Tayal, K.A. Berrington and A.E. Kingston
Department of Applied Mathematics and Theoretical Physics
The Queen's University of Belfast
BELFAST BT7 1NN NORTHERN IRELAND

Emission lines of Be-like ions are particularly important as electron-density diagnostics in the solar transition region and corona[1]. Accurate rate coefficients are required for the interpretation of solar observations[2].

We have extended the earlier R-matrix calculations[3,4] for Be-like OV to calculate the collision strengths and excitation rate coefficients for all transitions involving the six n=2 states $2s^2\ ^1S^e$, $2s2p\ ^3P^o$, $^1P^o$ and $2p^2\ ^3P^e$, $^1D^e$, $^1S^e$ and the six n=3 states $2s3s\ ^3S^e$, $^1S^e$, $2s3p\ ^3P^o$, $^1P^o$ and $2s3d\ ^3D^e$, $^1D^e$. Elaborate configuration interaction wave functions are used to describe the target state wave functions. The collision strengths for most transitions are found to have a very complicated resonance structure. Cross sections are averaged over a Maxwellian distribution to give excitation rate coefficients. The resonances result in a significant enhancement of excitation rates.

In table 1 we show few of our results for effective collision strength for $2s^2\ ^1S^e$ - $2s3s\ ^3S^e$, $2s^2\ ^1S^e$ - $2s3p\ ^3P^o$ and $2s^2\ ^1S^e$ - $2s3d\ ^3D^e$ transitions as a function of electron temperature.

Table 1. Effective collision strength as a function of electron temperature

Electron temperature (10^5k)	2^1S^e-3^3S^e	2^1S^e-3^3P^o	2^1S^e-3^3D^e
0.2	1.18(-1)	9.40(-2)	1.31(-1)
0.6	8.06(-2)	7.20(-2)	1.19(-1)
1.0	6.19(-2)	6.29(-2)	1.15(-1)
4.0	2.81(-2)	4.17(-2)	1.03(-1)
6.0	2.22(-2)	3.57(-2)	9.83(-2)
10.0	1.64(-2)	2.84(-2)	8.95(-2)
20.0	1.03(-2)	1.91(-2)	7.06(-2)
40.0	5.66(-3)	1.11(-2)	4.48(-2)

References

1. A. Gabriel and C. Jordan, in Case Studies in Atomic Collision Physics, edited by E.W. McDaniel and M.R.C. McDowell (North-Holland, Amsterday 1972) vol. II p.209
2. P.L. Dufton, K.A. Berrington, P.G. Burke and A.E. Kingston, Astron.Astrophys. 62, 111 (1978)
3. K.A. Berrington, P.G. Burke, P.L. Dufton and A.E. Kingston, J.Phys.B 10, 1465 (1977)
4. K.A. Berrington, P.G. Burke, P.L. Dufton, A.E. Kingston and A.L. Sinfailam, J.Phys.B 12, L275 (1979)

ANGULAR DISTRIBUTIONS FOR ELECTRON IMPACT EXCITATION OF MGII, ZNII, AND CUI

Alfred Z. Msezane
Morehouse College, Atlanta, GA 30314, USA

Ronald J.W. Henry
Louisiana State University, Baton Rouge, LA 70803, USA

Electron impact excitation of atoms and ions is important in a wide variety of fields such as the design and feasibility of new lasers, Tokamak-type discharges, and observations of solar and stellar spectra. Measurements for either integral or differential cross sections are difficult to obtain due to the low target ion densities involved. Thus, calculations play a useful role in both interpretation of experiments and in the prediction of cross sections for targets that are experimentally inaccessible at present. Further, experimental determinations of angular distributions provide a more sensitive test of theory than integral cross sections.

Measurements of emission cross sections have been reported by Zapesochnyi et al.[1] and Dunn et al.[2] for MgII, and by Rogers et al.[3] for ZnII. Trajmar et al.[4] have deduced elastic and inelastic cross sections for CuI from energy loss spectra at fixed impact energies and scattering angles. Differential electron scattering cross sections for inelastic excitation of an ion were measured for the first time by Chutjian et al.[5] for ZnII, as a follow-up to measurements of the excitation function taken at 14° by Chutjian and Newell.[6]

Msezane and Henry[7] calculated cross sections for excitation of ZnII in a close coupling approximation in which the 4s, 4p, $3d^9 4s^2$, 5s, and 4p states were retained. Very good agreement with measurements of the absolute emission cross sections of Rogers et al.[3] was obtained for the energy range $15 < E < 100$eV, when cascade contributions were included.

For MgII, excitation cross sections were calculated by Msezane and Henry[8] in a close coupling approximation in which the 3s, 3p, 4s, 3d and 4p states were retained. Very good agreement with measurements of the absolute emission cross sections of Dunn et al.[2] was obtained for the energy range $15 < E < 100$eV, again when cascade contributions were included.

For CuI, information available on electron impact excitation cross sections shows conflicts between various experiments. Msezane and Henry[9] have performed a four state close coupling calculation in which the 4s, $3d^9 4s^2$, 4p, and 4d states were retained. This calculation used an accurate description of the target wave functions and reported cross sections in reasonable agreement with those deduced from models[10] for copper lasers.

We use the reactance matrices obtained by Msezane and Henry for MgII, ZnII, and CuI in a program developed by Brandt et al.[11] to calculate angular differential cross sections in the energy range $15 < E < 100$eV.

We will compare and contrast differential cross sections for elastic scattering and for excitation of s-p and s-d transitions in MgII, ZnII, and CuI at energies where the incident electron has suffered approximately the same energy loss in the inelastic collision.

References

1. I.P. Zapesochnyi, V.A. Kel'man, A.I. Daschenko, and F.F. Danch, Sov. Phys. JETP 42, 989 (1976).
2. G.H. Dunn (private communication, 1982).
3. W.T. Rogers, G.H. Dunn, J.O. Olsen, M. Reading, and G. Stefani, Phys. Rev. A25, 681 (1982).
4. S. Trajmar, W. Williams, and S.K. Srivastava, J. Phys. B10, 3323 (1977).
5. A. Chutjian, A.Z. Msezane, and R.J.W. Henry, Phys. Rev. Letters (submitted 1983).
6. A. Chutjian and W.R. Newell, Phys. Rev. A26, 2271 (1982).
7. A.Z. Msezane and R.J.W. Henry, Phys. Rev. A25, 629 (1982).
8. A.Z. Msezane and R.J.W. Henry, Phys. Rev. A (submitted, 1983).
9. A.Z. Msezane and R.J.W. Henry, Phys. Rev. A (submitted, 1983).
10. M.J. Kushner and B.E. Warner, private communication (1980).
11. M.A. Brandt, D.G. Truhlar, and R.L. Smith, Comput. Phys. Commun. 5, 456 (1973).

EXCITATION OF POSITIVE IONS IN COULOMB-BORN APPROXIMATION

N. C. Sil and N. C. Deb

Indian Association for the Cultivation of Science, Calcutta 700032, India.

We present a method for the evaluation of the Coulomb-Born matrix element between arbitrary initial and final Slater orbitals for application to electron impact excitation of positive ions. The matrix element can be expressed in terms of the type integral

$$J = \int I(\vec{r}_2)\Psi_{\vec{k}_f}^{(-)*}(\vec{r}_2)\Psi_{\vec{k}_i}^{(+)}(\vec{r}_2)d\vec{r}_2 \quad (1)$$

where $\Psi_{\vec{k}_i}^{(+)}$ and $\Psi_{\vec{k}_f}^{(-)}$ represent the incident and outgoing Coulomb waves and

$$I(r_2) = (-\frac{\partial}{\partial\lambda})^{n+1}\int e^{-\lambda r_1} r_1^{\ell-1} Y_{\ell m}(\hat{r}_1) r_{12}^{-1} d\vec{r}_1 \quad (2)$$

Making use of the Fourier transforms of $e^{-\lambda r_1} r_1^{\ell-1} Y_{\ell m}(\hat{r}_1)$ and r_{12}^{-1} and Feynman identity we perform the r_1 integration and applying Leibniz theorem for the differentiation of a product we obtain

$$I(\vec{r}_2) = 2\pi \int_0^1 \left\{ \sum_{s=0}^{n+1} \frac{(n+1)!}{s!\lambda^{n+2-s}} e^{-\mu_1 r_2} r_2^{\ell+s} Y_{\ell m}(\hat{r}_2) d\vec{r}_2 \right\} x^{\ell+s/2-1/2}\, dx \quad (3)$$

with $\mu_1 = \lambda\sqrt{x}$.

Substituting Eqn. (3) in Eqn. (1) and using the integral representation[1]

$${}_1F_1(i\alpha_j, 1; Z) = \frac{1}{2\pi i}\oint_{\Gamma_j}^{(0^+,1^+)} dt_j\, p(\alpha_j, t_j) e^{zt_j} \quad (4)$$

with $p(\alpha_j, t_j) = t_j^{i\alpha_j - 1}(t_j - 1)^{-i\alpha_j}$, $j = 1, 2$

for the confluent hypergeometric function appearing in the Coulomb waves, we can carry out the r_2 integration. The results may be expressed in terms of the type integrals

$$H = -(2\pi i)^{-2}\oint_{\Gamma_1}\oint_{\Gamma_2} \frac{q^\ell Y_{\ell m}(\hat{q})\, \mu^{s+1-2r} p(\alpha_1, t_1) p(\alpha_2, t_2)}{(\mu^2+q^2)^{s+\ell+2-r}} x^{\ell+s/2-1/2}\, dt_1 dt_2 \quad (5)$$

with $\mu = \mu_1 - ik_i t_1 - ik_f t_2$, $\vec{q} = (1-t_1)\vec{k}_i - (1-t_2)\vec{k}_f$.

We now take the binomial expansion of $(\mu^2+q^2)^{-(s+\ell+2-r)}$ in the form $(E-Ft_1)^{-B}$, trinomial expansion of μ^A and use the addition theorem for regular spherical harmonics[2]. The t_1-integration gives us a terminating Gauss hypergeometric series and the results are obtained in terms of the integrals of the type

$$G = (2\pi i)^{-1}\oint_{\Gamma_2} t_2^{s_1 - 1 + i\alpha_2}(t_2 - 1)^{\ell'' - i\alpha_2} E^{-(B-r-v-i\alpha_1)} (E-F)^{-(r+v+i\alpha_1)} dt_2 \quad (6)$$

where $\ell'' = \ell - \ell'$, $0 \leqslant \ell' \leqslant \ell$.

E and F are linear functions of t_2. Taking the binomial expansion of $E^{-(B-r-v-i\alpha_1)}$ and $(E-F)^{-(r+v+i\alpha_1)}$ in the forms $(X-Yt_2)^{-Q}$ and $(U-Vt_2)^{-R}$ respectively we obtain after the t_2-integration in Eqn. (6) an Appell hypergeometric function F_1 of two variables. This F_1 function can be recast as a terminating series each term of which is a ${}_2F_1$ function. Thus we finally obtain

$$G = (-1)^{\ell''} X^{-Q} U^{-R} (i\alpha_2)_{s_1} (1-i\alpha_2)_{\ell''} (1-V/U)^{-\alpha} \sum_{k=0}^{P} \frac{(\alpha)_k (-P)_k (\varepsilon_2)^k}{(\gamma)_k k!} {}_2F_1(\alpha+k, \beta, \gamma+k, \varepsilon_1) \quad (7)$$

where

$\varepsilon_1 = \frac{V/U - Y/X}{V/U - 1}$, $\varepsilon_2 = \frac{V/U}{V/U-1}$, $\alpha = i\alpha_2 + s_1$, $\beta = Q$, $\gamma = s_1 + \ell'' + 1$, $P = B - (s_1 + \ell'' + 1)$ and $(x)_N$ is the Pochhammer symbol.

As an example we have calculated the cross section for the electron impact excitation of He^+ ion from ground state to 4f-states. The results will be presented at the time of the conference.

References

[1] N. Nordsieck, Phys. Rev. 93, 785 (1954).

[2] M. J. Caola, J. Phys. A 11, L23 (1978).

MEASUREMENT OF THE ELECTRON IMPACT EXCITATION CROSS SECTION FOR $C^+(2p\ ^2P^o - 2p^2\ ^2D)$

John L. Kohl and Gregory P. Lafyatis

Harvard-Smithsonian Center for Astrophysics, 60 Garden St., Cambridge, MA 02138 U.S.A.

We are using an inclined beams apparatus to study processes involving electron collisions with singly and multiply charged ions. Several experiments to study electron impact excitation and dielectronic recombination processes are in progress. Results have been obtained for electron impact excitation of $C^+(2p\ ^2P^o - 2p^2\ ^2D)$.[1] There have been no previous measurements of this cross section, no excitation measurements for a boron-like ion and no measurements on any non-lithium-like isoelectronic species that is in boron's row of the periodic table.

For the C^+ measurement, a 4.2 keV ion beam intersected an electrostatically focused electron beam at an angle of 45°. The cross section was determined from measurements of the beam fluxes and energies, the spatial distributions of ions and electrons in the beams and the rate of photon production from the radiative decay of the excited $2p^2\ ^2D$ state. It was also necessary to determine the number of metastable $C^+(2p^2\ ^4P)$ ions in the incident beam. This was accomplished by measuring the absolute photon flux at 233 nm which results from the radiative decay of the $2p^2\ ^4P$ state. A calculated radiative lifetime[2] for the $2p^2\ ^4P$ metastable term average was used along with the measured flux at 233 nm to determine a fractional population of 10% in the $2p^2\ ^4P$ state.

The results of this work together with several calculated values are shown in Fig. 1. There is an uncertainty in the absolute energy scale of about ±0.3 eV due to the unknown offset potential of the electron gun cathode. The vertical lines at the bottom of the graph indicate the thresholds for ground state excitation of the following terms (from left to right): $2s2p^2\ ^2D$, $2s2p^2\ ^2S$, $2s2p^2\ ^2P$, $3s\ ^2S$, and $3p\ ^2P^o$. The dark error bars are relative uncertainties and the light error bars are absolute uncertainties that are evaluated at a 68% confidence level. The dashed line is the result obtained by using the Gaunt factor estimator formula[3] with g equal to 0.2 and the oscillator strength of the optical transition taken equal to 0.12.[1] Two Coloumb-Born calculations of Mann[4] are shown. They differ only in that exchange is included in the cross section plotted as dots and is not included in the one with x's. The open squares show the results of a close-coupling calculation of Robb.[4]

It is premature to draw firm conclusions from the comparison of theoretical values to this work because no detailed calculations have been made on resonance structure. Henry[5] has used the (continuum) cross section data of Robb to estimate that the net energy-averaged effect of resonances for collision energies just below 12.0 eV is a 24% enhancement in the cross section. He estimates that near the threshold for excitation of the $2s2p^2\ ^2P$ state, the cross section should be enhanced an average of about 15%.

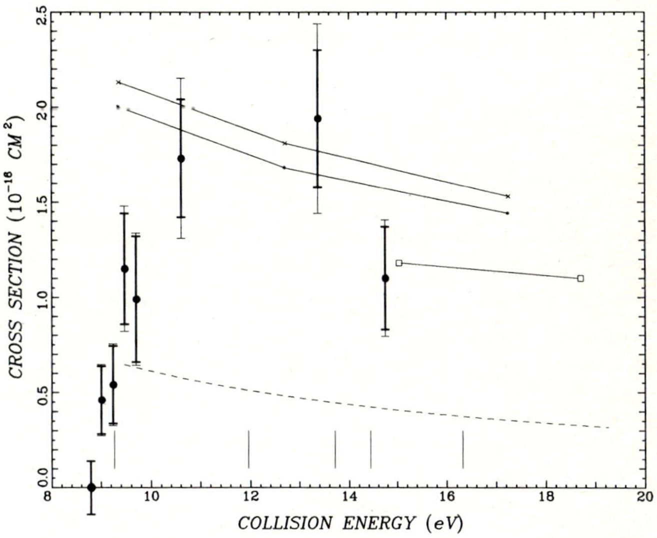

Fig. 1: Electron Impact Excitation Cross Section for $C^+(2s^22p\ ^2P^o - 2s2p^2\ ^2D)$.

Clearly the present measurements have not been made at enough energies to justify strong statements about effects of resonances in the experimental data. However, two items suggest that the measurements at 10.61 eV and 13.37 eV may reflect a cross section enhanced by resonances, specifically, the rapid rise in the measured cross section between the 9.70 eV and 10.61 eV data points and the rapid decrease in the measured cross section between the 13.37 eV and 14.75 eV data points. The first may be otherwise explained by an unusually large energy spread in the electron beam.

In the time frame of this meeting, we expect to report additional data for C^+ and also discuss progress toward measurements of dielectronic recombination.

References

1. G.P. Lafyatis, Ph.D. Thesis, Harvard (1982).
2. H. Nussbaumer and P.J. Storey, Astron. Astrophys. 96, 91 (1981).
3. H. van Regemorter, Astrophys. J. 136, 906 (1962).
4. N.H. Magee, J.B. Mann, A.L. Merts, and W.D. Robb, Los Alamos informal report LA-6691-MS (1977).
5. R.J.W. Henry, personal communication.

ELECTRON IMPACT EXCITATION OF FINE STRUCTURE TRANSITIONS IN C-LIKE IONS

K. M. Aggarwal

Department of Applied Mathematics and Theoretical Physics
The Queen's University of Belfast, Belfast BT7 1NN, Northern Ireland

A series of calculations is in progress at the Queen's University of Belfast to generate electron collision strength data using the R-matrix method for the excitation of ions in the C-isoelectronic sequence, viz. O III, Ne V, Si IX, Ca XV and Fe XXI. In this series the results for the first three ions have already been reported[1-4] and here we present our preliminary results for Ca XV and Fe XXI.

Both Ca XV and Fe XXI ions are of astrophysical interest as their abundance in the sun and other stars can be relatively high, and collision data for these ions can be used in density and temperature diagnostics of the astrophysical plasma. Existing data available[5,6] for these ions is only at a few energies, well above the excitation thresholds, from which rate coefficients (an important parameter for plasma diagnostics) cannot be calculated accurately because the collision strengths are resonance dominated, especially in the threshold energy region. In the present calculations, the data is being computed at several hundred energy points in a very fine energy mesh in order to account for the contribution of resonances.

For both Ca XV and Fe XXI, which have a relatively high nuclear charge, we have observed that relativistic effects are more important than configuration interaction in calculating excitation thresholds as well as oscillator strengths. Therefore, only HF + $2p^4$ wave functions calculated from the CIV3 program of Hibbert[7] are used to compute the electron collision strengths. All channels arising from the $(1s^2)2s^22p^2$, $2s2p^3$ and $2p^4$ configurations, and all partial waves with $L \leqslant 9$ of doublet, quartet and sextet spin multiplicities, and of even and odd parities are used in the calculations. The R-matrix code of Berrington et al[8] is used to calculate the reactance matrices and hence the collision strengths. The calculations have been done in the L-S coupling scheme and the reactance matrices obtained are stored in the Atomic and Molecular Databank developed at the Daresbury Laboratory and at the Queen's University of Belfast. We shall transform these reactance matrices to pair coupling through the JAJOM program of Saraph[9] to calculate the collision strengths for the fine structure transitions. Some relativistic effects will be included through the term coupling coefficients, as defined by Saraph[9]. These calculations are, however, in progress and here we report the collision strengths in Table 1 for transitions in the $1s^22s^22p^2$ configuration at a few energies. These results are, however, preliminary as no relativistic effects have so far been included. The final results including these effects for the excitation rate coefficients will be presented at the conference, prior to publication. We intend to make a comparative assessment of the importance of configuration interaction and relativistic effects in determining the electron collision strengths in case of Ca XV.

Table 1. Collision strengths for the transitions in the $1s^22s^22p^2$ configuration of Ca XV and Fe XXI.

	Transition					
Energy (Ryd)	Ca XV			Fe XXI		
	$^3P^e \to {}^1D^e$	$^3P^e \to {}^1S^e$	$^1D^e \to {}^1S^e$	$^3P^e \to {}^1D^e$	$^3P^e \to {}^1S^e$	$^1D^e \to {}^1S^e$
2.0	0.186	0.028	0.032	0.088	0.014	0.018
4.0	0.179	0.025	0.026	0.097	0.014	0.019
6.0	0.178	0.025	0.031	0.085	0.012	0.024
8.0	0.163	0.023	0.035	0.149	0.012	0.015
10.0	0.154	0.022	0.035	0.088	0.012	0.019
15.0	0.137	0.019	0.035	0.082	0.011	0.019
20.0	0.125	0.017	0.036	0.077	0.011	0.019
30.0	0.102	0.013	0.037	0.068	0.009	0.020
50.0	0.070	0.008	0.040	0.054	0.007	0.021
75.0	0.045	0.005	0.041	0.042	0.005	0.022
100.0	0.037	0.004	0.043	0.033	0.004	0.022

References

1. K.M. Aggarwal, Astrophys. J. Suppl., in press (1983).
2. K.L. Baluja, P.G. Burke and A.E. Kingston, J. Phys. B13, 4675 (1980).
3. K.M. Aggarwal and K.L. Baluja, J. Phys. B16, 107 (1983).
4. K.M. Aggarwl, J. Phys. B16, L59 (1983).
5. K.P. Dere, H.E. Mason, K.G. Widing and A.K. Bhatia, Astrophys. J. Suppl. 40, 341 (1979).
6. H.E. Mason, G.A. Doschek, U. Feldman and A.K. Bhatia, Astron. Astrophys. 73, 74 (1979).
7. A. Hibbert, Comput. Phys. Commun. 9, 141 (1975).
8. K.A. Berrington, P.G. Burke, M. Le Dourneuf, W.D. Robb, K.T. Taylor and Vo Ky Lan, Comput. Phys. Commun. 14, 367 (1978).
9. H.E. Saraph, Comput. Phys. Commun. 15, 247 (1978).

RESONANT EXCITATION DOUBLE AUTOIONIZATION IN Xe^{6+}

K. J. LaGattuta and Y. Hahn

Physics Department, University of Connecticut, Storrs, CT 06268, U. S. A.

Excitation autoionization of positive ions, also called Auger ionization (AI) is a subject of widespread current research interest[1-3]. In this process, a scattering electron strikes an ion of charge Z, promoting an inner shell target electron to a higher lying bound state. Subsequently, an autoionization event occurs as the vacancy is filled and a target electron is ejected; i.e.,

$$e^- + A^Z \rightarrow (A^Z)^{**} + e^{-\prime} \hookrightarrow A^{(Z+1)} + e^{-\prime\prime}.$$

Recently, it was suggested[4] that a similar process, involving initial projectile capture to a metastable state of the ion may occur at energies just below the threshold for ion vacancy production; viz.,

$$e^- + A^Z \rightarrow (A^{(Z-1)})^{**} \hookrightarrow (A^Z)^{**} + e^{-\prime} \hookrightarrow A^{(Z+1)} + e^{-\prime\prime}.$$

This is the resonance excitation double autoionization (REDA) process. So far, evidence for the existence of this effect has been confined to work done on sodium sequence targets of charge Z = 1, 2 and 3, using two and three state close-coupling[5]. Also, new measurements of cross sections for ionization of Xe^{6+} ions[6] show indications of a possible contribution from REDA.

In this report we will describe a calculation of the REDA cross section for Xe^{6+} ions, using a distorted wave, isolated resonance approximation[7], and in LS coupling. Non-relativistic Hartree-Fock single configuration wave functions will be employed. Principally, the process to be considered is,

$$4d^{10}5s^2 + k_c\ell_c \rightarrow 4d^9 5s^2 4f(^1P)n\ell,\ n \gtrsim 10$$
$$\downarrow$$
$$4d^9 5s^2 5d(^1S) + k_c'\ell_c'$$
$$\downarrow$$
$$4d^{10}5s + k_c''\ell_c'',$$

for continuum electron energies, $k_c^2/2$, such that 95 eV $\lesssim k_c^2/2 \lesssim$ 101 eV. Comparison will be made between REDA, AI, direct ionization and experimental data.

References

1. Y. Hahn, Phys. Lett. 78A, 57 (1980)
2. Y. Hhan, Phys. Rev. Lett. 39, 82 (1977)
3. R. D. Cowan and J. B. Mann, Astrophys. J. 232, 940 (1979)
4. K. J. LaGattuta and Y. Hahn, Phys. Rev. 24A, 2273 (1981)
5. R. J. W. Henry and A. Z. Msezane, Phys. Rev. 26A, 2545 (1982)
6. M. S. Pindzola, D. C. Griffin and C. Bottcher, ORNL preprint (1982)
7. J. N. Gau and Y. Hahn, J. Quant. Spectrosc. Radiat. Transfer 23, 121 (1980)

Acknowledgement: this work was supported in part by a grant from the United States DOE. Also, we thank the experimental group at ORNL for supplying us with their data.

ELECTRON-IMPACT IONISATION OF RARE GAS IONS

A. Matsumoto, S. Ohtani, A. Danjo*, H. Hanashiro, T. Hino, Y. Kondo+, H. Suzuki#, H. Tawara, K. Wakiya# and M. Yoshino"

Institute of Plasma Physics, Nagoya Univ., Nagoya, 464 Japan

Absolute cross sections for single ionisation of Ar^{2+} and Kr^{2+} by electron impact have been measured at electron energies from below threshold to 1000 eV using a crossed beams technique.

An ion beam extracted from an ECR ion source is accelerated to 6 keV, is mass-selected, and crossed with an electron beam at right angles. After colliding with electrons, the ions are charge-analysed with a 45° electrostatic analyser. Parent ions are measured with a Faraday cup and product ions are detected with a micro-channel plate(MCP). The electron gun and the analyser are mounted in an ultra-high vacuum chamber($< 1\times10^{-7}$Pa). In order to separate signals from backgrounds, both the electron and the ion beam (typical currents are 1 mA and 100 nA, respectively) are modulated with 500 Hz. Signal to background ratio is typically 2. The counting rate of product ions is increased with their impinging energy onto the MCP and when the impinging energy is higher than 10 keV, the counting rate becomes constant. Thus, the detection efficiency of the MCP is assumed to be unity at the impinging energy of 13 keV used in the present study. The form factors are determined at every two or three data points of electron impact energies.

Figure 1(a) shows the measured cross section for the electron impact ionisation of Ar^{2+}. Non-zero cross sections below the ground state ionisation threshold at 40.9 eV indicate that there is a possible metastable fraction in the ion beam. The present cross sections are in good agreement with those measured by Müller et al[1], except for low energies near the threshold. The cross sections calculated by Lotz formula[2] are also shown in Fig. 1, for the direct ionisation process of the outermost 3p sublevel electrons. Figure 1(b) shows the ionisation cross sections of Kr^{2+}. In comparison with the Ar^{2+} case, the measured cross sections are clearly different from the Lotz values, which are expected to give a reasonable estimate of the direct ionisation of the outermost 4p sublevel electrons. For energies up to 70 eV the cross sections rise more rapidly than expected and at around 100 eV they show clearly a shoulder. This implies that, excitation and ionisation of inner-sublevel electrons are expected to make a significant contribution to the total single ionisation cross section.

The measurements for Ne^{2+} and Xe^{2+} are underway, and double ionisation processes for various ions are being investigated.

References

1. A. Müller, E. Salzborn, R. Frodl, R. Becker, H. Klein and H. Winter, J. Phys. B 13, 1877 (1980)
2. W. Lotz, Z. Physik 216, 241 (1968)
Y. Itikawa and T. Kato, Empirical formulas for electron collisions, IPPJ-AM-17, Institute of Plasma Physics, Nagoya Univ., Japan

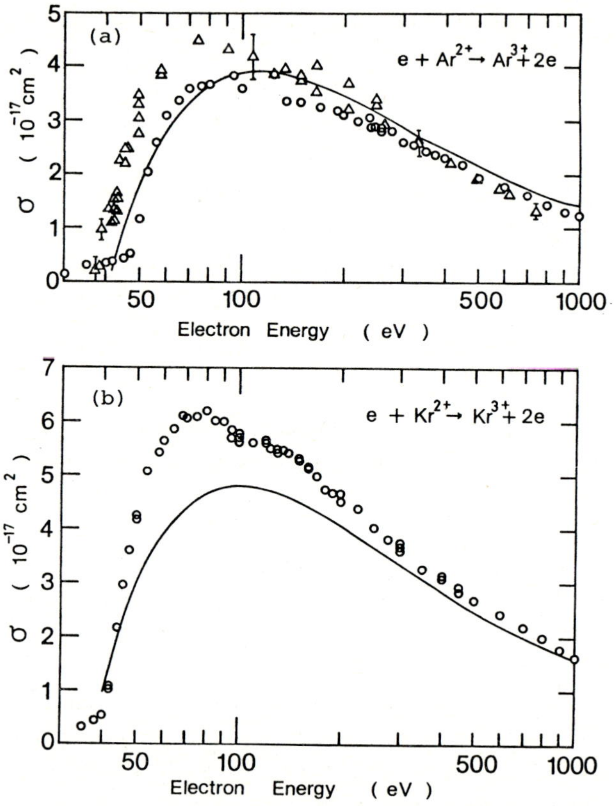

Fig. 1: Absolute ionisation cross sections as a function of electron impact energy for (a) $Ar^{2+} \rightarrow Ar^{3+}$ and (b) $Kr^{2+} \rightarrow Kr^{3+}$. Circles are the present data; triangles are data of Müller et al[1]. The solid line is the Lotz formula calculation for direct ionisation from the outermost subshell.

Permanent addresses
* Dept. of Phys.,Niigata Univ.,Niigata 950-21
+ Dept. of Elect. Eng., Toyota Technical College, Toyota 471
# Dept. of Phys., Sophia Univ., Tokyo 102
" Fac. of Gen. Educ., Shibaura Inst. of Tech., Omiya 330

ELECTRON-IMPACT DOUBLE-IONIZATION OF XENON IONS

M. S. Pindzola

Auburn University, Auburn, Alabama 36849 USA

D. C. Griffin

Rollins College, Winter Park, Florida 32789 USA

and

C. Bottcher

Oak Ridge National Laboratory, Oak Ridge, Tennessee 37830 USA*

Electron-impact ionization of inner-shell electrons followed by autoionization can make significant contributions to the total double-ionization cross section of certain ions. For Xe ions the indirect contribution from the $4d^{10}5s^{2}5p^{x} \rightarrow 4d^{9}5s^{2}5p^{x}$ ionization followed by Auger decay is calculated to be approximately 2-3 times the direct "double-knockouts" contribution. The distorted-wave inner-shell ionization cross section is found to be strongly influenced by both ground state correlations and term-dependence in the ejected-electrons continuum. Our Xe results are compared with the recent experimental crossed-beams measurements by groups at the University of Geissen[1] and Oak Ridge National Laboratory.[2]

---

*Operated by Union Carbide Corporation under contract W-7405-eng-26 with the U.S. Department of Energy.

[1] C. Achenback, A. Müller, and E. Salzborn, unpublished.

[2] D. C. Gregory and D. H. Crandall, unpublished.

CROSS SECTION MEASUREMENTS OF SINGLE AND MULTIPLE IONIZATION OF Xe^{i+} (i=1,2,3,4) IONS AND DOUBLE IONIZATION OF J^{+} IONS BY ELECTRON IMPACT

C. Achenbach, A. Müller and E. Salzborn
Institut für Kernphysik, Strahlenzentrum, Universität Giessen, 6300 Giessen, West Germany

R. Becker, J. Peschina, H. Klein
Institut für Angewandte Physik, Universität Frankfurt, 6000 Frankfurt, West Germany

We have used a crossed-beams technique to measure absolute ionization cross sections of Xe^{i+} (i=1,2,3,4) ions and J^{+} ions by electron impact from threshold up to electron energies of 700 eV. The single ionization cross sections $\sigma_{i,i+1}$ of Xe^{i+} ions are compared with distorted wave born approximation (DWBA) calculations of Younger[1] and the semiempirical Lotz formula[2]. The experimental results exceed the theoretical predictions up to a factor of 4 for the higher charge states (i=2,3,4). This shows the increasing importance of indirect ionization mechanisms due to excitation-autoionization.

Measurements of the double ionization cross sections $\sigma_{i,i+2}$ of Xe^{i+} ions are compared with the Gryzinski[3] cross section for direct ionization of two outer electrons added to the Lotz[2] cross section for the ionization of an inner 4d electron. The latter process is followed by the emission of an outer electron via Auger decay and thus contributes to the double ionization cross section. The experimental cross sections for the double ionization of Xe^{+} and Xe^{2+} show a broad resonance like structure centered at the ionization energy of an inner 4d electron. This enhancement of the cross sections is also seen for the double ionization of J^{+} which is isoelectronic to Xe^{2+} and less pronounced for Xe^{3+}. In order to extract the resonance structure, we have fitted the cross sections $\sigma_{i,i+2}$ with a Lotz-type formula (Fig. 1).

Additionally we have measured the triple ionization cross sections $\sigma_{i,i+3}$ of Xe^{i+} (i=1,2,3) and the quadruple ionization cross sections $\sigma_{i,i+4}$ of Xe^{i+} (i=1,2) ions. The data clearly demonstrate the importance of multiple ionization processes for this heavy ion (Z=54). As a consequence multiple ionization processes should be considered in plasma diagnostics and modeling.

References

1. S.M. Younger 1982 (private communication)
2. W. Lotz, Z. Phys. 232 (1970) 101
3. M. Gryzinski, Phys. Rev. A138 (1965) 136

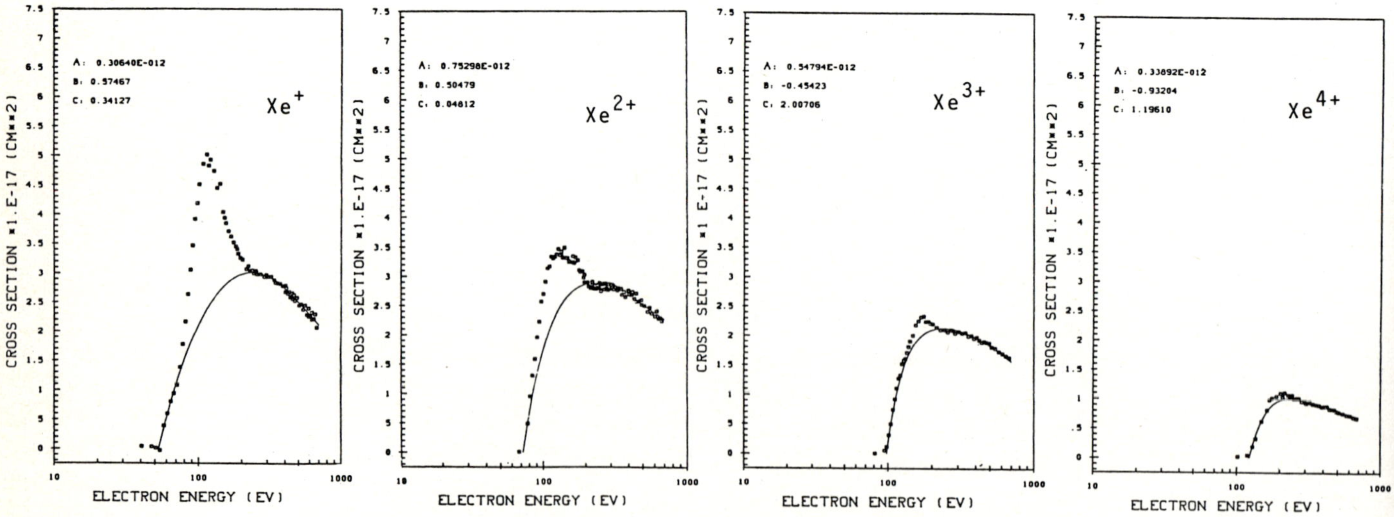

Fig. 1: Double ionization cross sections $\sigma_{i,i+2}$ of Xe^{i+} (i=1,2,3,4) ions by electron impact.

CROSS SECTIONS FOR SINGLE AND MULTIPLE IONIZATION OF ALKALI IONS AND ALKALINE-EARTH IONS BY ELECTRON IMPACT

T.Hirayama, K.Oda, T.Ono, Y.Morikawa, K.Wakiya and H.Suzuki

Department of Physics, Sophia University, 7-1 Kioi-cho, Chiyoda-ku, Tokyo 102, JAPAN

A new crossed-beam apparatus for the measurements of ionization cross sections of atomic ions by electron impact has been constructed. We considered all the essential difficulties pointed out in the literatures[1,2,3,4], in the process of designing of our apparatus. For example, these difficulties are (i) pumping systems of ultra-high vacuum, (ii) the beam modulation techniques, (iii) the measurements of a form factor, (iv) a determination of the overall efficiency of the charge state analyzer and the detector, (v) the exact measurements of the parent ion and electron currents, (vi) consideration of metastable contents in the target ion beam.

A schematic view of the apparatus is shown in Fig.1. The apparatus consists of an ion source chamber, an ion chopping chamber and a collision chamber, which are evacuated differentially with each other. The ion source chamber is evacuated with a 10-inch oil diffusion pump (Edwards, Diffstak 250/1700p) and is maintained within 10^{-7} to 10^{-8} Torr during operations. An ion source of an alkali ion or an alkaline earth ion of the thermionic-emission type is used in order to minimise the possibility of metastable content in the beam.

The ion chopping chamber is evacuated with a turbomolecular pump (Balzers TPU 270), and is kept within 5×10^{-9} Torr. In order to separate the true signal from the variety of background noise, we need the double chopping technique of both the ion and the electron beams. We designed a gating circuit suitable to this purpose, which enables us to treat very faint signals against the signal to background ratio less than 10^{-3}.

The collision chamber of about 40cm diameter is evacuated with a cryogenic pump (Sargent-Welch, 7370CW) and is kept within 4×10^{-10} Torr during the normal operation. A slit system that is movable very precisely operated from the outside of the vacuum chamber is especially designed in order to measure the form factor in the beams at the crossing region.

Systematic measurements of absolute cross sections for single and multiple ionization of the singly ionized ions of alkali atoms (rare-gas atom-like ions) and the singly ionized ions of alkaline earth atoms (alkali atom -like ions) are now in progress. Excitation of an inner shell electron, followed by the excitation-auto-ionization processes can in some instances be the dominant mechanism for ionization in alkali atom-like ions[2]. These cross sections will be reported at the conference.

1) M.F.A.Harrison, Methods of Experimental Physics Vol. 7B (revised volume), eds.B.Bederson and W.L.Fite (Academic Press,New York, 1968) pp95-115

2) K.T.Dolder and B.Peart, Rep. Prog. Phys. 39 697 (1976)

3) G.H.Dunn, Electron-Ion Collisions in the Physics of Ionized Gases eds. M.Matic and Boris Kidric, Institute of Nuclear Sciences, Beograd, pp49-95(1981)

4) D.H.Crandall, Electron-Ion Collisions (Prepared for the NATO Advanced Study Summer Institute on the "Atomic Physics of Highly Ionized Atoms" at Cargese, Corsica, France, June 7-18 1982) ORNL/TM-8453 (1982)

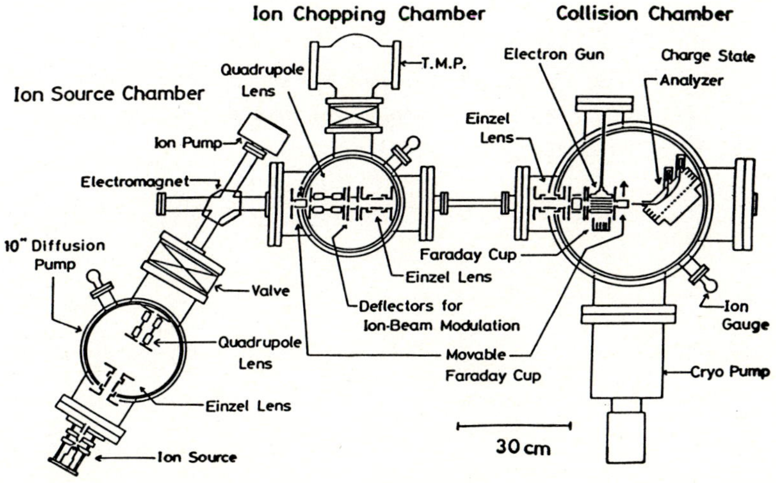

Fig.1. Schematic view of the crossed-beam apparatus used in our experiments.

AUGER IONIZATION CROSS SECTION FOR IONS OF THE Li AND Be SEQUENCES

K. J. LaGattuta, D. J. McLaughlin and Y. Hahn

Physics Department, University of Connecticut, Storrs, CT 06268, U. S. A.

Together with the radiative capture of continuum electrons by ionic targets, collisional ionization is important in understanding the behavior of high temperature plasmas. It has become increasingly clear in recent years that the higher-order effects in these processes often play dominant roles. In this report, we present a theoretical investigation of the ionization cross section for Li and Be-like ionic targets, with emphasis on the contributions of the higher-order effects. The charge dependence of the cross section, as well as the variation in going to different isoelectronic sequences will be investigated. The result then should be useful in constructing a phenomenological formula for the total ionization cross section and also for the reaction rate coefficient.

Electron impact ionization of positive ions can proceed either (1) by direct excitation of an outer-shell electron to the continuum (DI), (2) by collisional excitation of an inner-shell electron to one of the autoionization states which subsequently decays by Auger electron emission[1-3] (AI), or (3) by a resonant process in which the incoming electron is captured while the target ion is excited to form an intermediate resonance state, which then decays by the double Auger electron emission[4] (REDA). (This latter process should also be effective in collisional double and triple electron ionization.)

Thus we have

$$e^- + A^{z+} \longrightarrow e^{-\prime} + e^{-\prime\prime} + A^{(z+1)+} \qquad \text{(DI)}$$

$$\longrightarrow e^{-\prime} + (A^{z+})^{**}$$
$$\quad \hookrightarrow e^{-\prime\prime} + (A^{(z+1)})^{*} \qquad \text{(AI)}$$

$$\longrightarrow (A^{(z-1)+})^{***}$$
$$\quad \hookrightarrow (A^{z+})^{**} + e^{-\prime}$$
$$\qquad \hookrightarrow (A^{(z+1)+})^{*} + e^{-\prime\prime} \qquad \text{(REDA)}$$

Extensive theoretical[5,6] and experimental[6,7] works on Li-like ionic targets are already available and some work on the Be-like ions has also been done. For lighter ions, the Auger yield is nearly unity and thus the AI contribution to the ionization is often replaced by the inner-shell excitation cross section. This is not valid for heavier ions[5]. A coherent theoretical investigation of the systems will be presented. We employ the calculational procedure developed earlier for treating dielectronic recombination[8] (DR); the nonrelativistic single configuration Hartree-Fock approximation is used throughout, with LS coupling. The intermediate states relevant to AI for the Li and Be-like targets correspond respectively to the He and Li-like ion dielectronic recombination. Thus the Auger yields are already available from the DR work. The collisional excitation and Auger probabilities are evaluated in the distorted wave Born approximation. Our result will be compared in detail with the available experimental and theoretical data.

Acknowledgment: This work was supported in part by US DOE contract DE-AC02-76ET3035.

References

1. B.Peart and K.Dolder, J.Phys.B1,872(1968).
2. L.Goldberg et al, Ann.Astr. 28,589 (1965).
3. Y. Hahn, Phys.Rev. Lett. 39,82 (1977).
4. K.LaGattuta and Y.Hahn, Phys. Rev. A24, 2273 (1981).
5. Y. Hahn, Phys.Lett. 78A, 57 (1980).
6. D.H. Crandall, ORNL/TM-8453 (1982).
7. R.A.Falk et al, Phys.Rev.Lett.47,494 (1981) and Phys. Rev. A27, 762 (1983).
8. K.LaGattuta and Y.Hahn, J.Phys. B15,2101, (1982); D. McLaughlin and Y. Hahn, JQSRT 28, 343 (1982).

ELECTRON IMPACT IONISATION OF BORON-LIKE IONS

K. Butler and D.L. Moores

Dept of Physics & Astronomy, University College London, Gower St, London WC1E 6BT, UK.

Cross sections for electron impact ionisation of the boron-like positive ions C^+ and N^{++} (ground configuration $2s^2$ 2p) have been calculated in the Coulomb-Born-Exchange approximation, using an extended version of the computer program COBION developed by Jakubowicz and Moores[1] and used previously by those authors to obtain data for Li- and Be-like ions. A 3-state ($2s^2$ ^1S, 2s2p $^{1,3}P$) close coupling approximation with added correlation was used to calculate the wave functions for the ground state of the ion and for the final state of ionised ion plus ejected electron. By this means, the complex resonance structure caused by excitation of autoionising states, which occurs at energies down to the ionisation threshold, is automatically included in the cross section in the form of both the closed and the bound (correlation) channels. Techniques for handling the highly complex structure created by the multiplicity of resonances of the form 2s2p ($^{1,3}P$) nl ^2L will be described. Comparison will be made with previous calculations of these cross sections[2,3].

References

1. H. Jakubowicz and D.L. Moores, J.Phys.B 14, 3733-60 (1981).
2. D.L. Moores, J.Phys.B 12, 4171-8 (1979).
3. M. Chidichimo, J.Phys.B 15, 3333-46 (1982).

ABSOLUTE CROSS SECTION MEASUREMENTS FOR ELECTRON-IMPACT IONIZATION OF TWICE CHARGED IONS Ti^{2+}, Fe^{2+}, Ar^{2+}, $C\ell^{2+}$ and F^{2+} *

D. W. Mueller, T. J. Morgan,† and Gordon H. Dunn‡

Joint Institute for Laboratory Astrophysics, University of Colorado and
National Bureau of Standards, Boulder, Colorado 80309

and

D. C. Gregory and D. H. Crandall

Oak Ridge National Laboratory, Oak Ridge, Tennessee 37830

Electron-impact ionization cross sections have been measured recently for the doubly charged ions Ti^{2+}, Fe^{2+}, Ar^{2+}, $C\ell^{2+}$, and F^{2+}. The measurements were made using a crossed beams technique[1] with ions from the ORNL-PIG ion source. The measurements cover a region from below threshold to 1500 eV with an energy resolution of about 2 eV at the lowest energies. Only for Ar^{2+} are data available[2] in the literature.

Cross section estimates using the Lotz formula[3] are in excellent agreement with the measurements for F^{2+}, but not for $C\ell^{2+}$. In the $C\ell^{2+}$ case the measured cross section rises more quickly from threshold than estimated to a maximum value of 46×10^{-15} cm^2 near 100 eV. Above 200 eV the measured value is somewhat smaller than the Lotz estimate. For F^{2+} the cross section maximizes near 200 eV with a value of 18×10^{-18} cm^2.

A number of features appear in the Ti^{2+} ionization cross section. Sharp rises in the ionization cross section for Ti^{2+} near 33 eV and 44 eV suggest excitation autoionization as a major mechanism for ionization of this ion, reminiscent of the dramatic case of Ti^{3+}.[4] By comparison, the Fe^{2+} cross section appears to be relatively plain. Comparison with distorted wave calculations by Younger[5] shows reasonable agreement in this case, while the Lotz formula overestimates the cross section. Onset of ionization below the ground state 35 eV threshold suggests a significant metastable component in the beam.

Present measurements of the Ar^{2+} ionization cross section agree with those of Müller et al.[2] below 100 eV, with an offset in magnitude at higher energies. In these newer data there are contributions to the cross section below threshold, probably arising from a small metastable component in the beam.

*This work supported in part by Office of Fusion Energy, U.S. Department of Energy.

†JILA Visiting Fellow, 1982-83, on leave from Wesleyan University, Middletown, Connecticut 06457.

‡Staff Member, Quantum Physics Division, National Bureau of Standards.

References

1. D. H. Crandall, R. A. Phaneuf and P. O. Taylor, Phys. Rev. A 18, 1911 (1978).
2. A. Müller, E. Salzborn, R. Frodl, R. Becker, H. Klein and H. Winter, J. Phys. B 13, 1877 (1980).
3. W. Lotz, Z. Phys. 216 (1968).
4. R. A. Falk, G. H. Dunn, D. C. Gregory and D. H. Crandall, Phys. Rev. A 27, 762 (1983).
5. S. M. Younger, J. Quant. Spectrosc. Radiat. Transfer 27, 541 (1982).

IONISATION OF SINGLY-CHARGED METALLIC IONS BY ELECTRON IMPACT

R G Montague, M J Diserens*, M F A Harrison and A C H Smith*

Culham Laboratory, Abingdon, Oxon OX14 3DB, UK
(Euratom/UKAEA Fusion Association)

Metallic elements likely to be used in the fabrication of containment vessels for fusion reactors include aluminium, iron and tungsten. It is necessary (see Harrison, 1983)[1] to know the electron ionisation cross sections of atoms and lowly-charged ions in order to evaluate the interactive processes which give rise to radiative power losses in the plasma boundary and to sputter erosion of the vessel walls. There is only a limited amount of data available for relevant species and so a programme of measurements at Culham using the crossed ion and electron beams technique is in progress. The ionisation of Al^+ has already been reported (Montague and Harrison, 1983).[2] In this paper measurements are reported of absolute cross sections for the process

$$e + M^+ \longrightarrow 2e + M^{2+}$$

where M includes Fe and W. The incident electron energy range is from threshold to 750eV.

A target ion beam of 2 to 4KeV energy is extracted from a sputter ion source. The beam is mass and charge selected in a sector magnetic field and is collimated before passing through the interaction region. Here it is crossed at 90° by a variable-energy planar electron beam. Doubly-charged products are separated from the remaining singly-charged target ions in a second sector magnetic field and are detected with a calibrated electron multiplier used in a pulse-counting mode. The target ion beam and the electron beam enter Faraday cup collectors and the currents are measured. The vertical intensity profiles of the beams are determined with the aid of a scanning slit. The absolute ionisation cross section is calculated from the product ion count rate, the currents of the two crossed beams, the effective height of the larger (ion) beam, which is calculated from the intensity profiles, and the beam velocities. The apparatus and data collection are controlled by an on-line computer, which also performs all calculations. The absolute accuracy of the measured cross sections at impact energies greater than 20eV is $\pm 5\%$. In the cases of Fe^+ and W^+, ionisation contributions from metastable ions have been identified and the accuracy of the derived ground state cross sections depends upon estimates of these contributions. The additional error due to the presence of the metastable ions is not more than $\pm 3\%$ at the peak of the cross section. At the conference we expect also to present data for some other elements which are currently under study at the time this abstract is written. One of the authors, M J Diserens, is in receipt of a S.E.R.C. CASE Studentship.

References

1. M F A Harrison, Applied Atomic Collision Physics, Volume II ed. H S W Massey, B Bederson and E W McDaniel (New York: Academic Press) 1983.

2. R G Montague and M F A Harrison, J.Phys. B; At. Mol. Phys. 16 (submitted for publication 1983).

*Department of Physics and Astronomy, University College London, London WC1E 6BT, UK

ELECTRON IMPACT EXCITATION-AUTOIONIZATION OF Ni IONS

S.N. Tiwary*, P.G. Burke and A.E. Kingston

The Queen's University of Belfast, Northern Ireland.

* now at Bihar University, Bihar, India.

It was shown by Peart and Dolder[1] that the electron impact ionization cross section of Ba^+ has an abrupt increase at low energies due to the onset of autoionization. Since this early work, this excitation -autoionization process has been shown to be important for many ions (e.g. see review by Crandall[2]). At Queen's University, calculations have been carried out, or are underway, for a number of ions of interest in applications in fusion plasmas and in astrophysics including Sc^{2+}, Ti^{3+} and V^{4+} by Hibbert et al[3] and Tiwary et al[4] and Ca^+ by Burke et al[5]. In this contribution we describe work which we are carrying out on ions of Ni.

We are interested in the process where the Ni ion is first excited to an autoionizing state by the incident electron

$$e^- + Ni^{n+}(3p^6 3d^q) \rightarrow Ni^{n+\,*}(3p^5 3d^{q+1}) + e^-$$

and then the autoionizing state decays with the emission of an electron

$$Ni^{n+\,*}(3p^5 3d^{q+1}) \rightarrow Ni^{(n+1)+}(3p^6 3d^{q-1}) + e^-$$

where n = 1, 2, 3, 4 or 5 and q = 10-n. In order to determine the relative importance of this contribution to ionization compared with the ionization process where the 3d electron is ejected directly into the continuum it is first necessary to know the location of the terms of the autoionizing configuration $3p^5 3d^{q+1}$ relative to the ionization threshold and to determine the number of terms which occur. We have therefore carried out calculations for these terms using the C.I. atomic structure program of Hibbert[6] and the orbitals of Clementi and Roetti[7].

In figure 1 we show the energy level structure of the $3p^6 3d^q$ and $3p^5 3d^{q+1}$ terms of Ni^{n+} for n=3,4 and 5. Owing to the large number of terms which arise, which we give in table 1, we have indicated the energy regions where they occur by appropriate shading. For Ni^+ there is only one autoionizing term $3p^5 3d^{10}\ ^2P^o$ and this lies far in the continuum and is thus not likely to contribute appreciably to the ionization cross section. For Ni^{2+} there are 6 autoionizing terms corresponding to the configuration $3p^5 3d^9$ which also lie quite far from the ionization threshold. However, in the case of Ni^{3+} and Ni^{4+} there are 19 and 42 autoionizing terms respectively which lie close to the ionization threshold and these are expected to play a dominant role in ionization. Finally in the case of Ni^{5+} the 68 terms corresponding to the configuration $3p^5 3d^6$ all lie in the discrete spectrum and thus do not contribute to ionization.

The next step in our work is to calculate the cross sections for excitation of these autoionizing states of Ni^{3+} and Ni^{4+} and this is underway at present using our R-matrix program package.

References

1. B. Peart and K.T. Dolder, J. Phys.B 1, 872 (1968).
2. D.H. Crandall, Physica Scripta 23, 153 (1981).
3. A. Hibbert, A.E. Kingston and S.N. Tiwary, J. Phys. B. 15, L643 (1982).
4. S.N. Tiwary, A.E. Kingston and A. Hibbert, to be published in J.Phys.B. (1983).
5. P.G. Burke, A.E. Kingston and A. Thompson, submitted to J.Phys.B. (1983).
6. A. Hibbert, Comp. Phys. Commun. 9, 141 (1975).
7. E. Clementi and C. Roetti, Atom. and Nuc. Data 14, 177 (1974).

Table 1. Number of terms corresponding to the $3p^6 3d^q$ and $3p^5 3d^{q+1}$ configurations of Ni ions.

Configuration	Ni^+	Ni^{2+}	Ni^{3+}	Ni^{4+}	Ni^{5+}
$3p^6\ 3d^q$	1	5	8	16	16
$3p^5\ 3d^{q+1}$	1	6	19	42	68

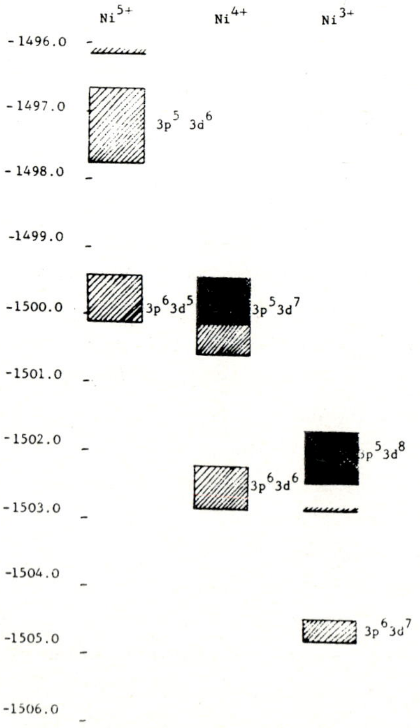

Figure 1. Energy level structure of the terms of the $3p^6\ 3d^q$ and $3p^5\ 3d^{q+1}$ configurations of Ni^{3+}, Ni^{4+} and Ni^{5+}. Autoionizing state regions are represented by cross hatching and bound state regions are represented by diagonal lines.

CLASSICAL BINARY-ENCOUNTER COLLISION THEORY FOR IONIZATION IN ELECTRON-ION SCATTERINGS*

Y. T. Lee

University of California, Lawrence Livermore National Laboratory, Livermore, California 94550, U.S.A.

Electron ionization cross sections are used to determine nonequilibrium ionization state of the plasmas. For these studies cross sections for highly ionized ions are needed over a wide range of incident electron energies. Since quantum approximations for many ionization cross sections are not available, cross sections employed in many applications are usually derived from semi-empirical formulas.

In this paper we apply classical binary-encounter collision theory to calculate ionization in electron-ion scattering. The basic approximations used are: (1) the incident electron interacts with only one target electron at a time, (2) the target electron is assumed to be ionized whenever the energy transfer is greater than its binding energy.

For classical binary-encounter collision we derive a differential cross section per unit energy and momentum transfer. To calculate ionization cross section, we integrate this differential cross section over the appropriate energy and momentum transfer and bound electron velocity distributions.

In the calculation the bound electrons are described in the Thomas-Fermi approximation. We obtain both kinetic energy and binding energy of the bound electron from the Thomas-Fermi potential, $e\varphi(r)$. Using this approach we derive the following scaled total ionization cross section for electron-ion scatterings,

$$\sigma(E,Q) = \frac{4\sqrt{2}\ (b/a_o)^{5/2}\ a_o^{\ 2}}{Z^{1/3}(E/2I_H)}\quad F(r_m/r_o,Q)$$

where

E = incident electron energy
Q = Z^*/Z (Z^* = charge state)
Z = nuclear charge
r_o = ion radius
a_o = Bohr radius
I_H = ionization potential of hydrogen atom
b = 0.468479×10^{-8} cm

At the position r_m, the potential energy of the incident electron equals its initial kinetic energy.

The universal function $F(r_m/r_o, Q)$ is given in terms of an integral involving the Thomas-Fermi potential and the differential cross section per unit energy transfer for binary-encounter collision.

Comparison of our results to available quantum approximation and experimental measurement will be presented. We will also discuss applications of the theory to study plasma screening effects on the ionization cross section.

---

*Work performed under the auspices of the U.S. Department of Energy by Lawrence Livermore National Laboratory under Contract No. W-7405-Eng-48.

OBSERVATION OF KLL DIELECTRONIC RECOMBINATION RESONANCE IN Ar14+

J.P. Briand, P. Charles
Université P & M Curie and Institut Curie, 11, rue P & M Curie 75231 Paris Cedex 05

J. Arianer, H. Laurent, C. Goldstein
Institut de Physique Nucléaire, 91405 Orsay

J. Dubau, M. Loulergue
Observatoire de Meudon, 92190 Meudon

and F. Bely-Dubau
Observatoire de Nice, 06700 Nice Cedex, France

The dielectronic recombination (DR) process has been observed very recently in cross electron-ion beam experiments on singly charged C and Mg ions[1,2]. We present a new technique which allowed us to observe a DR resonance effect in the n=1 and n=2 shells in the collisions of an electron beam and a trapped Ar14+ target. The principle of the experiment is to use an EBIS ion source in which various argon ions have been trapped and to bombard them by a tunable energy electron beam after the equilibrium charge state has been reached. The DR process involving a K hole has been observed with a SiLi detector when looking at the Ar characteristic Kα x rays through a small hole drilled into the cathode of the electron gun.

The experiment was performed with a classical EBIS source named SILFEC III built at the Institute of Nuclear Physics in Orsay. In this source, the equilibrium charge state distribution is broad and extends up to Ar15+.

Three different kinds of processes in the interaction of an electron beam and an ion target have to be considered: ionization, excitation, and DR. The cross sections of the first two processes vary smoothly with the energy of the electron beam above a 2.95keV threshold (absolute minimum threshold for the excitation of the least charged ions in the source). The DR process has a resonant behavior and occurs for KLL or KLX transitions at energies below the 2.95keV threshold (we use in this paper the terminology of the Auger effect : the KLL DR process is then the simultaneous excitation of a K electron in a L shell and capture of an electron in the L shell). We have observed, above 2.7keV, the overall contributions of the KLM and KLX DR processes for all the trapped ions up to all the Rydberg states. We have also observed a very sharp resonance at 2.3keV which can be attributed without any ambiguity to the KLL peak of Ar14+. This assignment has been made on the basis of the energy of the KLL Auger lines of Ar14+[3] which peaks at 2.3keV (and is separated from the corresponding energies for the other ions by a value larger than the experimental resolution), and of the value of the cross sections for KLL DR process[3] in all the argon ions trapped in the source which leads to a maximum for Ar14+. The experimental value for this cross section is in reasonable agreement with the calculations.

References

1. J.B.A. Mitchell, C.T. Ng, J.L. Forand, D.P. Levac, R.E. Mitchell, A. Sen, D.B. Miko, and J. Wm Mc Gowan, Phys. Rev. Lett. 50 335 (1983).
2. D.S. Belic, Gordon, H. Dunn, T.J. Morgan, D.W. Mueller, and C. Timmer, Phys. Rev. Lett. 50 339 (1983).
3. F. Bely-Dubau et al, this conference.

DIELECTRONIC RECOMBINATION MEASURED WITH A CROSSED ELECTRON AND CALCIUM ION BEAMS COINCIDENCE TECHNIQUE

J.F. Williams

Physics Department, University of Western Australia, Perth, Australia

Direct observations of dielectronic recombination using merged[1] and crossed[2] electron-ion beams have recently been made for singly charged carbon[1] and magnesium[2] ions. This collision process involves the resonant capture of the incident electron to form a doubly excited state of the neutral atom which is "radiatively stabilized" with the emission of a photon and the atom is left in a high $n\ell$ Rydberg level. The alternative decay mode of autoionization usually dominates for low n values.

The present study measures the dielectronic recombination collision cross section

$$e + Ca^{+}(4s^{2}S_{1/2}) \rightarrow Ca^{**}(4p\ n\ell) \rightarrow Ca(4s\ n\ell) + h\nu$$

as well as the ion excitation collision cross sections

$$e + Ca^{+}(4s^{2}S_{1/2}) \rightarrow Ca^{+}(4p\ ^{2}P_{3/2,1/2}) \rightarrow Ca(4s^{2}S_{1/2}) + h\nu$$

(393.4 and 396.8 nm)

The measurement technique consists of crossed, modulated electron and calcium ion beams with a photomultiplier to detect de-excitation photons, an open particle multiplier to detect fast neutral atoms and normal pulse counting and coincidence detection systems. Details are as follows. The electron beam of up to 100μA in 1 cm^2 has an energy resolution of about 300 meV FWHM and variable energy. The $^{40}Ca^{+}$ ion beam of 1μA had a cross section of 4 x 9 mm at an energy of 500 eV. Photons were collected by a pair of 52 mm diam lenses with a 27 mm aperture stop between them at a distance of 54 mm from the ion-electron collision volume. For the ion excitation cross section measurements separate filters with 2.0 nm FWHM were used. The calibration and measurement procedures of Taylor and Dunn[3] were followed.

For the D.R. coincidence measurements, the photons passed through an interference filter centred at 410 nm with 34 nm FWHM at a nominal transmission of 23% which allowed both the ion de-excitation photons as well as the D.R. stabilizing radiation to pass. The fast ^{40}Ca atoms were detected by an open particle multiplier then via amplification and discrimination to the stop input of a TAC while the photons similarly reached the start of the TAC.

Background photons (~45 sec^{-1}) and neutrals (~10^5 sec^{-1}) gave rise to the background coincidence rate of about 1 per 30 sec. Numerous tests were made to ensure the validity of the true coincidence signal, especially that it did not arise from either charge transfer to excited states or excitation of neutral atoms.

The absolute emission cross section for 393.4 nm radiation from threshold to 14 eV has a general step function shape and a value of 1.8 (±0.2) x 10^{-16} cm^2 at threshold which is about 10% higher than the measurement of Taylor and Dunn but within overlapping experimental accuracies.

The dielectronic recombination cross section is peaked about the threshold energy of 3.1 eV with a maximum value of 1.2 ± 0.3 x 10^{-17} cm^2. There are no theoretical values available. Graphed data will be presented at the conference.

References

1. J.B.A. Mitchell, C.T. Ng, J.L. Forand, D.P. Levac, R.E. Mitchell, A. Sen, D.B. Miko and J.Wm. McGowan. Phys.Rev.Lett. 50, 335 (1983)
2. D.S. Belic, Gordon H. Dunn, T.J. Morgan, D.W. Mueller and C. Timmer. Phys.Rev.Lett. 50, 339 (1983)
3. Paul O. Taylor and Gordon H. Dunn. Phys. Rev. A8, 2304 (1973)

DIELECTRONIC RECOMBINATION: A CROSSED BEAMS OBSERVATION AND MEASUREMENT OF CROSS SECTION*

D. S. Belić,[†] Gordon H. Dunn,[‡] T. J. Morgan,[§] D. W. Mueller and C. Timmer

Joint Institute for Laboratory Astrophysics, University of Colorado and National Bureau of Standards, Boulder, Colorado 80309 U.S.A.

Dielectronic recombination can be visualized with the help of Fig. 1. In Fig. 1a an electron is incident on a target Mg^+ ion, a sodium-like structure with a single 3s electron in the outer shell. At infinity the incoming electron has ε less energy than $\Delta E = 4.43$ eV, the energy needed to excite the bound 3s electron to the 3p level. However, in the Coulomb field of the ion, the electron gains kinetic energy, so that at small distances it has more than enough energy to excite the 3p level. Having excited the 3p level, the incident electron no longer has enough energy to leave the ion, and is trapped in a Rydberg level $n\ell$, with an energy ε below the continuum, resulting in an intermediate doubly excited neutral atom Mg^{**} (Fig. 1b). The doubly excited atom can autoionize, leaving a ground state ion and a continuum electron; or it may radiatively stabilize [$\tau_{3p} \sim 3.7$ ns] as shown in Fig. 1c, leaving an atom in a Rydberg level and a photon -- completing the DR process.

The present experiment[1] is outlined schematically in Fig. 2. A beam of 2 keV mass-selected $^{24}Mg^+$ ($\sim$1 μA) is crossed by a magnetically confined (0.02T) variable energy beam of electrons ($\sim$10 μA; $\sim$300 meV FWHM). Photons from the 3p-3s stabilizing transition are collected in a lens system, passed through an interference filter F (12 nm FWHM, 278 nm peak), and focused onto the cathode of a photomultiplier PM. Pulses from the PM are delayed (typically 3.9 μs) and they then initiate the START gate of a time-to-amplitude converter (TAC). The stabilized neutral atom travels 54 cm to a particle multiplier in about 4.2 μs, generating a pulse which stops the TAC. The output of the TAC is fed into a pulse height analyzer (PHA) yielding a coincidence spectrum. Background photons ($\sim$30 s^{-1}) primarily from electron-ion excitation and neutrals ($\sim 10^5$ s^{-1}) primarily from charge transfer give rise to the accidental coincidence background. A least-squares fit to this background is subtracted from the spectrum.

*This work supported through the Office of Fusion Energy, U. S. Department of Energy.

†Permanent address: Department of Physics, Faculty of Natural and Mathematical Science, P.O. Box 57, Belgrade, Yugoslavia.

‡Staff Member, Quantum Physics Division, National Bureau of Standards.

§JILA Visiting Fellow, 1982-83. On leave from Wesleyan University, Middletown, Connecticut 06457.

Resulting measured cross sections are shown in Fig. 3. The error bars here represent relative uncertainty only. The dashed curve shows calculated cross sections from Ref. 2 modified to account only for contributions from n < 64, which are not field ionized and convoluted with our electron energy distribution. There is more than a factor of 5 difference between the measured peak value and that calculated. Even considering the lower uncertainty limit of the measurements, there is still a factor of 3.5 discrepancy.

1. D.S. Belić, G.H. Dunn, T.J. Morgan, D.W. Mueller & C. Timmer, Phys. Rev. Lett. 50, 339 (1983).
2. K. LaGattuta & Y. Hahn, J. Phys. B 15, 2101 (1982).

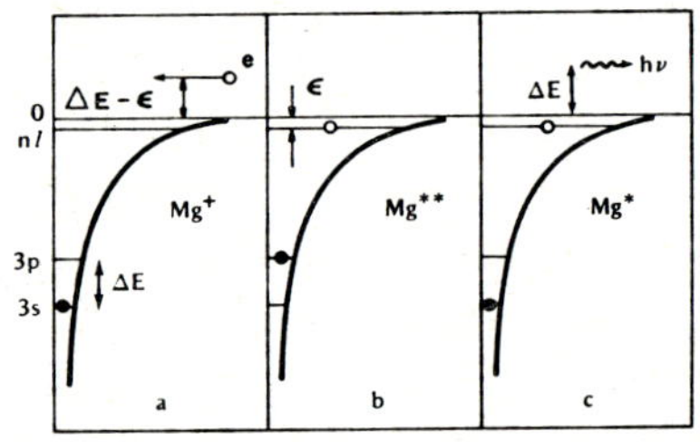

Fig. 1. Sequence of events in e + Mg⁻ DR.

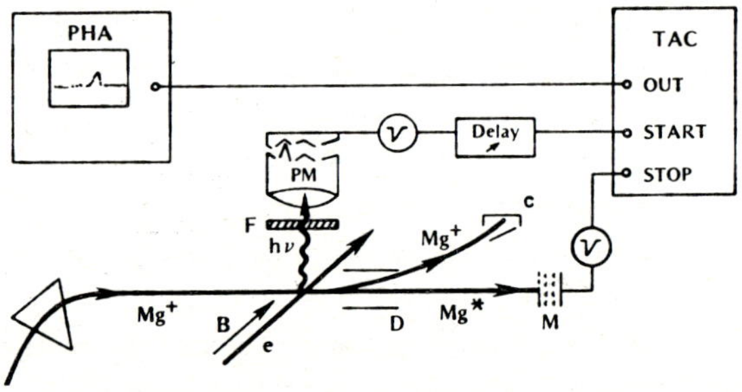

Fig. 2. Schematic representation of the experimental apparatus.

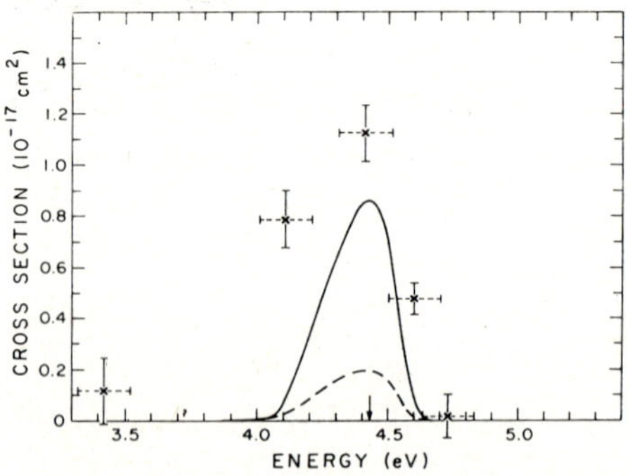

Fig. 3. DR cross section vs. energy for e + Mg^+; crosses, experiment; dashed curve, convolution of theory (Ref. 2) for n ≲ 64 with experimental electron energy distriution; solid curve, same as dashed curve, including all n. Arrow indicates excitation threshold energy. Bars are relative uncertainty only. The absolute uncertainty in cross section is ±58%.

DIELECTRONIC RECOMBINATION OF B^{2+} and C^{3+} via $1s^22s \rightarrow 1s^22p$ EXCITATION

S. Datz, P. F. Dittner, P. D. Miller, C. D. Moak, N. Nescovic, P. H. Stelson and C. Bottcher
Oak Ridge National Laboratory, Oak Ridge, Tennessee 37830 USA

Dielectronic recombination (DER) is the process by which electron capture from the continuum to a bound state is facilitated via excitation of a previously bound electron followed by radiative stabilization. Using a merged electron-ion beam technique we have measured the DER cross section for Li like systems A^{q+} $(1s^22s) + e^- \rightarrow [A^{(q-1)+} (1s^22p\ n\ell)] \rightarrow A^{(q-1)+} (1s^22s\ n\ell)$ $+ h\nu$, where A^{q+} is either B^{2+} or C^{3+}.

In this experimient, B^{2+} beams (14-18 MeV) and C^{3+} beams (17-24 MeV) from the ORNL EN tandem are merged with a magnetically-confined, space-charge-limited electron beam over an ~ 80 cm path. Following the merged region the ion beam is subjected to a charge analysis through magnetic deflection. The A^{q+} beam enters a Faraday cup and the $A^{(q-1)+}$ ions are counted in a position sensitive detector. DER is seen as an increase of the B^{1+} signal above the level caused by charge capture from residual gas. The cross sections for DER obtained are shown in Figs. 1 and 2, and compared with Burgess-Merts[1] (B-M) theoretical predictions. To make this comparison one must take into account the voltage drop across the space charge limited electron beam and the Stark stripping of electrons in high $n\ell$ states on the $A^{(q-1)+}$ ions by the field in the charge state analyzer. The observed peak is a convolution of the remaining set of $n\ell$ resonances with the instrumental resolution. The data are fit very well (in shape) with theory in both cases if one folds in a 2 eV energy width for the electron beam.

In Table I we list for B^{2+} and C^{3+}; the maximum n value retained; the peak value of the effective cross section obtained for n<n max; from B-M theory, from the

Table I. Theoretical and experimental DER effective cross sections for B^{2+} and C^{3+}.

ion	n_{max}	$\sigma(10^{-18}\ cm^2)$ B-M[1]	DW	Expt.
B^{2+}	22	0.21	0.18[2]	0.15±0.03
C^{3+}	26	0.14	0.25[3]	0.35±0.05

more detailed distorted (DW) wave calculation's of Hahn et al. and our experimental values. The stated errors are purely statistical and the values are the average of two runs for each ion. Systematic errors, not yet fully assessed, could be as much as 30% but the error in the ratio of the C^{3+} to B^{2+} cross sections should be small. Thus the results are clearly in better agreement with the DW theory.

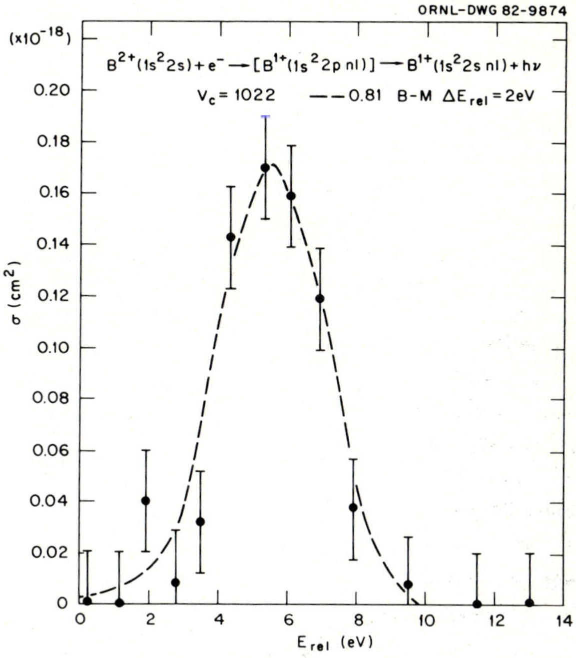

Fig. 1: Effective cross section for DER in B^{2+}.

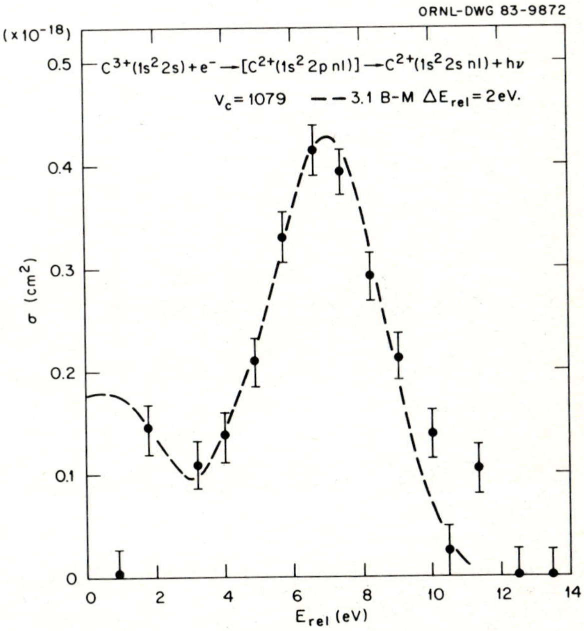

Fig. 2: Effective cross section for DER in C^{3+}.

References

1. R. D. Cowan, The Theory of Atomic Structure and Spectra, Univ. Calif. Press, Berkeley, 1981.
2. D. McLaughlin and Y. Hahn, Phys. Rev. A, in press.
3. Y. Hahn, private communication.

EFFECT OF EXTRINSIC ELECTRIC FIELDS UPON DIELECTRONIC RECOMBINATION

K. J. LaGattuta and Y. Hahn

Physics Department, University of Connecticut, Storrs, CT 06268, U. S. A.

Dielectronic recombination (DR) cross sections (σ^{DR}) have been calculated[1,2] for the singly charged target ions Mg^{1+} and C^{1+} in a distorted wave and isolated resonance approximation[3], using single configuration non-relativistic LS coupled Hartree-Fock wave functions. For the case of Mg^{1+}, the process considered was $3s + k_c\ell_c \rightarrow 3pn\ell \rightarrow 3sn\ell + \gamma$, where capture of the continuum electron, $k_c\ell_c$, to a high Rydberg state (HRS) was found to be important; i.e., $n \gg 1$ was dominant. For C^{1+}, the process was $2s^2 2p + k_c\ell_c \rightarrow 2s2p^2(^2D)n\ell \rightarrow 2s^2 2pn\ell + \gamma$, where again captures to HRS were important. Recent crossed beam[4] (for Mg^{1+}) and merged beam[5] (for C^{1+}) experiments have suggested values of σ^{DR} several times larger than the theoretically determined cross sections. However, the experimental procedures are such that measurements may not have been free of the influence of intrabeam electric microfields; i.e., neighboring beam ions may have influenced the recombination process. The effect of stray electric fields upon DR has been discussed[6,7] and may lead, generally, to an increase in σ^{DR}.

We report here the results of a calculation of σ^{DR} for Mg^{1+} in a Stark representation. Auger probabilities defined previously in LS coupling[3],

$$A_a = \frac{2\pi}{\tau_0} \left| \langle \overbrace{3s + k_c\ell_c}^{LS} \right| \frac{1}{r_{12}} \left| \overbrace{(3p)(n\ell)}^{LS} \rangle \right|^2$$

are rewritten using Stark coordinates for the outermost electron as,

$$A_a = \frac{2\pi}{\tau_0} \sum_{\ell_c} (2\ell_c+1) \left| \langle 3s + k_c\ell_c m_c \right| \frac{1}{r_{12}} \left| (3p)(n\lambda m_n) \rangle \right|^2$$

where[8],

$$|n\lambda m_n\rangle \equiv \sum_{\ell} \sqrt{2\ell+1}\,(-1)^{-2\lambda - m_n} \begin{pmatrix} \frac{1}{2}(n-1), & \frac{1}{2}(n-1), & \ell \\ \frac{1}{2}(m_n-\lambda), & \frac{1}{2}(m_n+\lambda), & -m_n \end{pmatrix} |n\ell m_n\rangle$$

and $\lambda = n_1 - n_2$ is the electric field quantum number. Radiative probabilities, A_r, for the $3p \rightarrow 3s + \gamma$ stabilizing transition are nearly independent of the state of the outermost electron and are taken to be the same in both representations. Values of σ^{DR} vs. $k_c^2/2$ will be presented in Stark representation and in LS coupling. Comparison with experiment will be made.

References

1. K. J. LaGattuta and Y. Hahn, J. Phys. B 15, 2101 (1982)
2. K. J. LaGattuta and Y. Hahn, Phys. Rev. Lett. 50, 700 (1983)
3. J. N. Gau and Y. Hahn, J. Quant. Spectrosc. Radiat. Transfer 23, 121 (1980)
4. D. S. Belic, G. H. Dunn, T. J. Morgan, D. W. Mueller and C. Timmer, Phys. Rev. Lett. 50, 339 (1983)
5. J. B. A. Mitchell, C. T. Ng, J. L. Forand, D. P. Levac, R. E. Mitchell, A. Sen, D. B. miko and J. W. McGowan, Phys. Rev. Lett. 50, 335 (1983)
6. A. Burgess and H. P. Summers, Astrophys. J. 157, 1007 (1969)
7. V. L. Jacobs, J. Davis and P. C. Kepple, Phys. Rev. Lett. 37, 1390 (1976)
8. B. W. Shore and D. H. Menzel, Principles of Atomic Spectra, J. Wiley & Sons, Inc., NY (1968) p. 502

Acknowledgement: this work was supported in part by a grant from the United States DOE.

ELECTRON SCATTERING WITH VERY HIGHLY CHARGED IONS: EFFECTS OF RELATIVISTIC DEVIATION, AUTOIONIZATION AND DIELECTRONIC RECOMBINATION ON CROSS SECTION FOR Fe^{24+}, Se^{32+} AND Mo^{40+}.

A.K. Pradhan

Department of Physics, University of Windsor, Windsor, Ontario, Canada N9B 3P4.

In view of the need for large-scale and accurate atomic data for highly charged ions present in fusion and astrophysical plasmas, detailed and systematic calculations are reported taking into account the effects of intermediate coupling, autoionization and dielectronic recombination (DER). The latter two atomic effects have recently been shown to be quite important[1,2] and some new and general techniques are described in order to determine precisely their role in electron scattering.

Calculations[3,4] are carried out in a 9-state distorted wave and 9-state close-coupling approximations ℓ, $\ell' \leq 15$ and ℓ, $\ell' \leq 4$ respectively, and cross sections are computed for all 78 transitions, with fine structure, between the states $1s^2(^1S_0)$, $1s2s(^3S_1, {}^1S_0)$, $1s2p(^3P_{0,1,2}, {}^1P_1)$, $1s3s(^3S_1, {}^1S_0)$ and $1s3p(^3P_{0,1,2}, {}^1P_1)$ included in the target representation of the He-like ions under consideration. Higher partial waves ℓ, $\ell' > 15$ are summed over with the Coulomb-Bethe approximation employing relativistic target energies and relativistic oscillator strengths for the permitted transitions. Detailed analysis of autoionization resonance structures lying in the large energy region between the n = 2 and n = 3 states is carried out, in intermediate coupling, using multi-channel quantum defect theory and numerical techniques. The effect of DER through these resonances is investigated including all allowed radiative decay channels to the n = 2 states and the ground state, and all autoionization channels. It is found that for some transitions DER takes place not directly to the ground state but through a radiative cascade mechanism via autoionizing levels. For Fe^{24+} the effect of DER in reducing autoionization enhancement (up to a factor of two), is found to be small (<10%); although for the heavier ions it is larger. Rate coefficients and line ratios are also computed and compared with tokamak measurements.

This work is supported by the Natural Sciences and Engineering Research Council of Canada.

References

1. A.K. Pradhan, D.W. Norcross and D.G. Hummer, Phys. Rev. A 23, 619 (1981).
2. A.K. Pradhan, Phys. Rev. Letts. 47, 79 (1981).
3. - , Phys. Rev. A (1983) submitted.
4. Ibid, submitted.

THE EFFECT OF RESONANCES ON THE He-LIKE ELECTRON ION SCATTERING-CROSS SECTIONS

L. Steenman-Clark, P. Faucher
Observatoire, B.P. 252, 06007 Nice Cedex, France

J. Dubau
Observatoire de Paris, DAPHE, 92190 Meudon, France.

The helium-like resonance, intercombination and forbidden lines observed from hot, low density plasmas such as coronal or Tokamak plasmas are important for the measurement of the plasma parameters e.g. electron and ion temperatures, the electron density, ionisation equilibrium [1,2]. These diagnoses are sensitive to the accuracy of the calculations of the collision cross-sections and excitation rates for these lines.

Resonances due to doubly excited lithium-like states in the collision cross-section are numerous enough to enhance the effective rate coefficients. These doubly excited states can also decay radiatively which decreases the enhancement.

New calculations are presented for the collision cross-sections and excitation rates for O VII, Mg XI, Ca XIX and Fe XXV in which the effect of the resonances and the radiative decay are taken into account. A comparison is made with previous calculations [3,4] for O VII and Fe XXV.

References.

1. A.H. Gabriel, Mon. Not. R. Astron. Soc. 160, 99 (1972)
2. A.H. Gabriel and C. Jordan, Mon. Not. R. Astron. Soc. 145, 241 (1969)
3. A.E. Kingston, S.S. Tayal, J. Phys. B 16, L 53 (1983)
4. A.K. Pradhan, Phys. Rev. Lett. 47, 79 (1981)

DIELECTRONIC RECOMBINATION PROCESS FOR HIGHLY IONISED ARGON

F. Bely-Dubau
Observatoire, B.P. 252, 06007 Nice Cedex, France

J. Dubau
Observatoire de Paris, DAPHE, 92190 Meudon Cedex, France

The inner-shell radiation transitions of the type

$$1s\ 2s^q\ 2p^m - 1s^2\ 2s^q\ 2p^{m-1} \quad (1)$$

are usually called satellite lines [1]. These lines are generally formed by the process of di-electronic recombination from the helium-like ion, lithium-like ion, etc., ... and by inner-shell collisionnal excitation. The atomic parameters, energies, autoionisation rates, radiative decay rates and excitation rates have been calculated for the transitions represented by (1). The method used involves the program SUPER STRUCTURE [2] and the associated DISTORTED WAVE program and follows the procedure described in [3,4]. The object of the present work is to extend our recent work on chromium [5] to argon including six stages of ionisation (Ar^{16+} - Ar^{12+}). This element is going to be injected for diagnostic purposes in Tokamak plasma discharges in 1983 at Princeton - USA (Plasma Physics Laboratory) as well as at Fontenay-aux-Roses - France (Centre d'Etudes Nucléaires). Electron temperatures and ion abundances will be derived from relative line intensities - Tokamak plasmas are the best diagnosed low density high temperature plasmas and are therefore suitable sources of comparison with results from advanced theoretical calculations.

References

1. A.H. Gabriel, Mon. Not. R. Astron. Soc. 160, 99 (1972)
2. W. Eissner, M. Jones, H. Nussbaumer, Comput. Phys. Commun. 8, 270 (1974)
3. F. Bely-Dubau, A.H. Gabriel, S. Volonté, Mon. Not. R. Astron. Soc. 186, 405 (1979)
4. F. Bely-Dubau, J. Dubau, P. Faucher, A.H. Gabriel, Mon. Not. R. Astron. Soc. 198, 239 (1982)
5. TFR Group, J. Dubau, M. Loulergue, J. Phys. B 15, 1007 (1982)

AB-INITIO OPTICAL POTENTIALS APPLIED TO LOW ENERGY ELECTRON-MOLECULE COLLISIONS IN THE LINEAR ALGEBRAIC APPROACH

B. I. Schneider and L. A. Collins

Theoretical Division, Los Alamos National Laboratory, Los Alamos, NM 87545

The need to accurately treat polarization and correlation effects in low energy electron collisions is well established. These effects control the shapes of both differential and integral cross sections and the positions and widths of scattering resonances. We have develped an optical potential technique[1] based on the Feshbach projection operator method which permits us to make use of powerful multiconfiguration self consistent field and configuration interaction methods used by quantum chemists. The essential feature of our approach is the introduction of a finite, square integrable basis set to span the open channel space of the optical potential. Practical considerations require that this basis not be too large. We have reduced the number of basis functions required by removing the strong static interaction from the Hamiltonian before defining the optical potential. The result of these manipulations is to define the optical potential as a finite sum of separable terms,

$$U_{cc'}(r,r',E) = \sum_{\lambda} \phi^*_{c\lambda}(r,E)\,\lambda(E)\,\phi_{c'\lambda}(r',E) \qquad (1)$$

where $\lambda(E)$ are the eigenfunctions and $\phi_{c\lambda}(r,E)$ the projections of the eigenfunctions onto the channel states which diagonalize the optical potential. These eigenfunctions and eigenvalues may be calculated using matrix methods and the separable form inserted into the linear algebraic method developed by the authors. The separable form of the optical potential reduces the time necessary to solve the scattering equations by 3 to 10 times. The method has been applied to low energy elastic collisions with H_2 and N_2. For the former system our results are in substantial agreement with those of other workers as well as experiment. In the case of N_2, our calculations are the first to treat polarization in a fundamental, ab-initio manner and are in good agreement with the experiments of Kennerly at low energy (Table I). Applications to other systems such as Li_2 and HCl should be available by the time of the conference.

TABLE I. A COMPARISON OF LOW ENERGY, STATIC EXCHANGE (SE), EFFECTIVE OPTICAL POTENTIAL (POL) AND EXPERIMENTAL (KENNERLY) $e+N_2$ CROSS SECTIONS

k^2 (Ry)	δ^a(S.E.)	δ(Pol)	σ^b(S.E.)	σ(Pol)	σ(Exp)
.02	- .3392	-.2330	72.28	33.45	31.10
.1	- .7429	-.5749	60.57	40.50	38.09
.2	- .9984	-.7691	48.72	36.73	
.3	-1.1732	-.9036	41.95	35.24	

[a] δ is phase shift in radians

[b] All cross sections, σ, are in $bohr^2$

References

1. B. I. Schneider and L. A. Collins, J. Phys. B 15, 1335 (1982); B. I. Schneider and L. A. Collins, Phys. Rev., in press.

A LINEAR ALGEBRAIC APPROACH TO ELECTRON-MOLECULE SCATTERING

L. A. Collins and B. I. Schneider

Theoretical Division, Los Alamos National Laboratory, Los Alamos, NM 87545

The accurate calculation of electron-molecule scattering cross sections has been hampered by the difficulty of representing the multi-center electrostatic potential and the short-range exchange and correlation interaction in a representation which is optimal for both types of interaction. We have developed a new approach to the electron-scattering problem which removes many of the difficulties of the older methods.[1] The problem is formulated using the R-matrix philosophy which allows us to separate configuration space into two regions. In the internal or strong interaction region, a set of coupled integral equations for the R-matrix Green's function is reduced to a set of linear algebraic equations by introducing suitable quadratures. The solution to the linear algebraic problem gives us the R-matrix on the surface of a sphere bounding the two regions. The R-matrix is then propagated from the surface of the sphere to the asymptotic region using the R-matrix propagation techniques developed by Light and Walker. By proceeding in this fashion we may make optimal use of numerical techniques for each region of space. Thus, in the internal region where exchange and correlation play a central role, the linear algebraic method allows us to treat non-local interactions with no more difficulty than local potentials. The need to include large numbers of angular terms to represent the multi-center static-potential is made practical by adapting the quadratures to the channels under consideration. Thus for the low ℓ-partial waves it is necessary to use a reasonably large mesh (40-60 points) distributed rather uniformly in space. The high partial waves on the other hand are concentrated near the nuclear singularity and may be represented with few points (10-15). Exchange[2] and correlation effects[3] are accounted for using bound state techniques which are uniquely suited to these short-range phenomena. Our approach makes use of the projection operator formalism developed by Feshbach to obtain an optical potential which is then approximated as a finite sum of separable terms. This technique permits us to employ the multiconfiguration self consistent field and configuration interaction techniques developed by quantum chemists in the scattering problem. Once we have solved the linear algebraic problem in the internal region, the R-matrix is propagated to the asymptotic region using purely local and long range interaction potentials. Proceeding in this fashion we have been able to treat a host of diatomic and triatomic molecules at the static-exchange level (H_2,N_2, LiH,LiF,HCL,Li_2,CO_2). In addition we have incorporated polarization and correlation in low energy $e+H_2^+$, $e+H_2$ and $e+N_2$ scattering with considerable success (see Fig. 1 and 2). Progress in the treatment of electronically and vibrationally inelastic processes is underway and numerical results will be presented at the meeting.

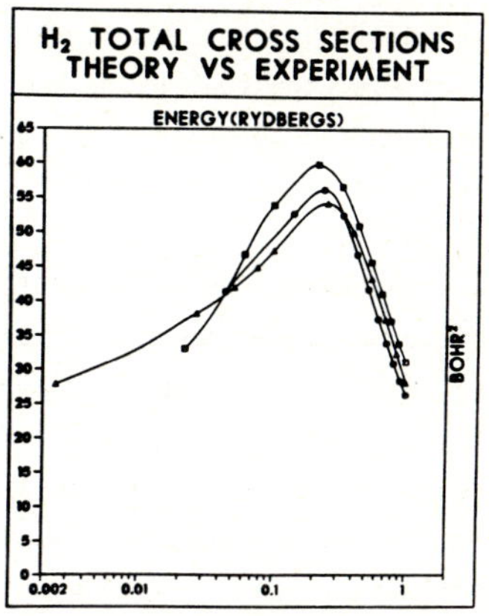

Fig. 1: 1) Triangle (Δ)-effective optical potential, 2) Circle (O)-experiment Golden et al., 3) Square (O)-experiment Dalba et al.

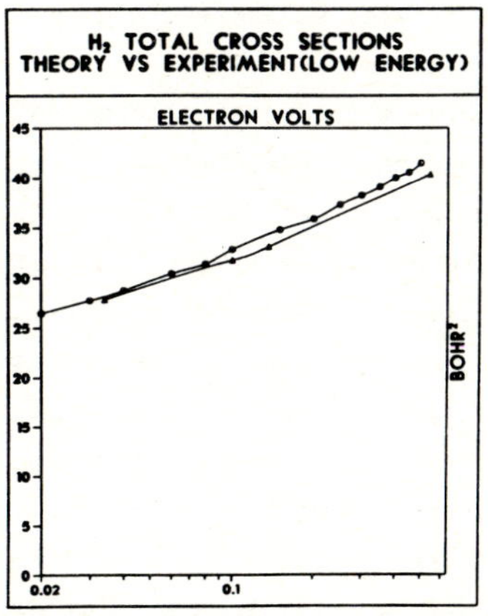

Fig. 2: 1) Triangle (Δ)-effective optical potential, 2) Circle (O)-experiment Firch.

References

1. B. I. Schneider and L. A. Collins, J. Phys. B 14, L101(1981); L. A. Collins and B. I. Schneider, Phys. Rev. 27, 101(1983).
2. B. I. Schneider and L. A. Collins, Phys. Rev. 24, 1264(1961).
3. B. I. Schneider and L. A. Collins, J. Phys. B 15, L335(1982); B. I. Schneider and L. A. Collins, Phys. Rev., in press.

VARIATIONAL R-MATRIX CALCULATIONS OF ELECTRON-MOLECULE SCATTERING

C.J. Noble[+], R.K. Nesbet[++] and L.A. Morgan[+++]

[+]SERC Daresbury Laboratory, Warrington WA4 4AD, England

[++]IBM Research Laboratory, San Jose, California 95193, USA

[+++]Dept. of Statistics and Computer Science, Royal Holloway College, Egham, Surrey TW20 0EX, England

Computer programs designed for efficient quantitative calculations of electron-molecule scattering cross sections have been constructed in a collaboration involving several research groups. The purpose of the present subproject is to carry e^-+H_2 calculations to effective convergence with respect to both the orbital basis set and the representation of target molecule and scattering wave functions. New results for e^-+H_2 obtained with these programs for a range of fixed internuclear distances are reported here. Converged fixed-nuclei electron scattering results of this kind are required for subsequent calculations of nonadiabatic effects in vibrational excitation and dissociative attachment.

The Belfast/Daresbury group[1] has modified the ALCHEMY system of bound state electronic wave function programs[2] for molecular R-matrix calculations. Results obtained with Slater orbitals on an artificial center within the molecule[3] indicate that this approach is feasible only for low impact energies, because of slow convergence. Oscillatory continuum basis orbital functions are needed to improve convergence. In calculations of e^-+N_2 scattering[4], numerical solutions of a diagonal spherical model potential scattering problem were used as basis orbitals, with satisfactory results up to intermediate energies.

The present work uses two forms of continuum basis orbitals. Numerical asymptotic functions (NAFs) are generated by solving the coupled multichannel potential problem defined by the static and transition multipole moments of a chosen set of target molecular states and pseudostates. Variational R-matrix calculations include two independent NAFs for each open scattering channel, as energy-dependent basis functions. Effective completeness within the R-matrix boundary r_1 is achieved by including spherical Bessel functions that vanish at r_1, which by themselves define a complete set for each orbital ℓ-value. NAFs are integrated inwards to some inner point r_0, inside the target charge distribution, where they are matched to regular spherical Bessel functions. With two NAFs per open channel, the exact multichannel solution of the asymptotic scattering problem is brought inside the R-matrix boundary. Hence the energy-independent continuum basis functions (Bessel functions) are needed only to represent short-range and nonlocal effects.

New programs have been written to compute static and transition multipole moments and the corresponding multipole pseudostates. Pseudostate orbitals are used to describe the dynamical polarization response of the target molecular states. New techniques are used to integrate the asymptotic equations with strongly closed channels due to pseudostates. Closed-channel components of NAFs are included in the orbital basis.

References

1. P.G. Burke, I. Mackey and I. Shimamura, J. Phys. B 10, 2497 (1977);
J. Kendrick and B.D. Buckley, Daresbury Lab. Rept. DL/SCI/TM22T, May 1980;
C.J. Noble, Daresbury Lab. Rept. DL/SCI/TM33T, November 1982.
2. A.D. McLean, M. Yoshimine, P. Bagus and B. Liu, ALCHEMY programs, IBM San Jose Research Laboratory.
3. C.J. Noble, P.G. Burke and S. Salvini, J. Phys. B 15, 3779 (1982).
4. P.G. Burke, C.J. Noble and S. Salvini, J. Phys. B, in the press.

[++]Research supported in part by U.S. Office of Naval Research

VIBRATIONAL EXCITATION OF POSITIVE MOLECULAR IONS BY ELECTRON IMPACT. EXACTLY SOLVABLE MODEL

A.K.Kazansky, V.V.Ponomarenko
Dept. of Theoret. Phys., Institute of Phys., Leningrad State University, Leningrad, 198904;
Leningrad Technological Institute, Leningrad; USSR

We consider the system with Hamiltonian

$$\hat{H} = -\frac{1}{2M}\frac{\partial^2}{\partial R^2} + \frac{M\omega^2 R^2}{2} - \frac{1}{2}\frac{\partial^2}{\partial r^2} - \frac{1}{r} \qquad (1)$$

(R - internuclear distance, R=0 corresponds to the equilibrium of the ion; r - electron coordinate; ω - the ion frequency). The electronic and nuclear motions are connected via the boundary condition

$$\frac{1}{\Psi}\frac{\partial \Psi}{\partial r}\bigg|_{r=a} = U - \beta R . \qquad (2)$$

The model might be defined as a particular case of the multichannel quantum defect theory. Paper [1] should also be mentioned in connection with the problem.

In the fixed nuclei approximation system (1),(2) has an electronic continuum with the lower boundary $U_0 = M\omega^2 R^2/2$ and a set of discrete states. This set could be described in terms of a diabatic state, intersecting the Rydberg diabatic series. The complexity of the adiabatic states structure shiould be stressed.

The model describes the following mechanism of the vibrational excitation of the molecular ion. Firstly, the incident electron is attached to the ion and the "double-excited molecular configuration" is formed. Then the nuclei move in conformity with the appropriate energy curve and pass a series of pseudocrossings with the Rydberg states. At each pseudocrossing the Landau-Zinner transition takes place, and the system with some probability is detained in the appropriate diabatic well.

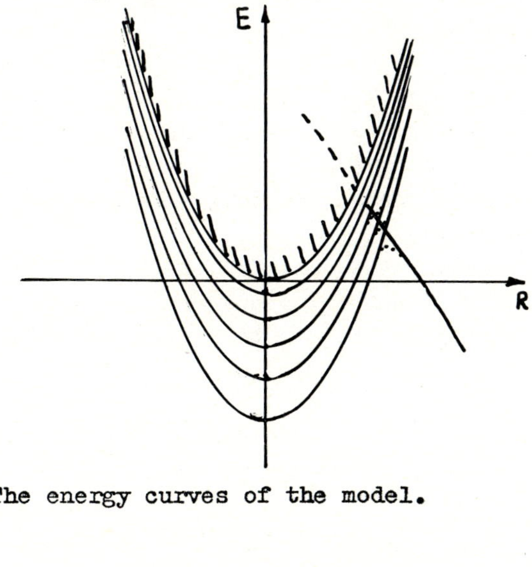

The energy curves of the model.

There are two final channels: electron may detach from the molecule, and the final vibrational state of the ion will be formed; or the atoms go away and dissociative recombination occurs. It should be noted that the model does not involve any final Rydberg state and the Rydberg series appears only as resonant structure of all cross sections. The considered mechanism is usually marked as "direct process".

In any real system some nonadiabatic effects, which are not considered in the present study, occur. Firstly, the long-range interaction between the Rydberg electron and the ion core (i.e. indirect mechanism) should be mentioned. Secondly, only one state of the Rydberg multiplet is involved in the Landau - Zinner transition. During the nuclearmotion (in conformity with the diabatic Rydberg energy curve) the state gets mixed with other states of the multiplet and thus the width of the diabatic state decreases to a marked degree.

The problem (1),(2) is solved by the method of [2] and some characteristics are computed. The Rydberg states show themselves as the spikes in the cross sections. The widths of the spikes are small and the states could be revealed only in photoionization experiments.

References

1. Dube L., Herzenberg A. Phys.Rev.Lett., 38, 820 (1977)
2. Kazansky A.K. Teoret. i Matem. Fyz., 36, 414, (1978)

EXACTLY SOLVABLE MODELS IN THEORY OF VIBRATIONAL EXCITATION OF MOLECULES BY ELECTRON IMPACT

A K Kazansky

Dept.of Theor.Phys.,Institute of Phys., Leningrad State University, Leningrad,198904,USSR

It is generally accepted now that vibrational excitation of some molecules by electron impact proceeds via virtual intermediate state. The study of the role of such a state in the processes, using the Bardsley-Herzenberg resonant theory, is hardly possible It is almost obvious that the threshold bumps could be described only with the long-lived resonance and, consequently, the N_2-type oscillatory structure in the cross sections would be revealed without fail.

To describe the processes, some models have been proposed. One of the models has been suggested by Domcke and Cederbaum [1]. It is useful to write this model in terms of usual resonant theory. The Hamiltonian of the system is

$$\hat{H}^{DC} = -\frac{1}{2M}\frac{\partial^2}{\partial R^2} + U_0(R) + v(R)|d\rangle\langle d| + \int dk\, w(k)|k\rangle\langle k| + \iint dk\,dk'\, W_1(k,k')|k\rangle\langle k'| + \int dk \left(V_d(k)|d\rangle\langle k| + V_d^*(k)|k\rangle\langle d|\right),$$

$$\langle d|d\rangle = 1,\ \langle d|k\rangle = 0,\ \langle k|k'\rangle = \delta(k-k')$$

The eigenfunction problem is formally a solvable one and thus the basic equation of the Bardsley theory will be reproduced. Therefore any unformal results could be obtained in framework of the model only in some particular cases.

The alternative model has been introduced in [2]. Here we give an account of generalized version of the model, considering the systems with the Hamiltonians

$$\hat{H}^a = -\frac{1}{2M}\frac{\partial^2}{\partial R^2} + \frac{M\omega^2 R^2}{2} - \frac{1}{2}\frac{\partial^2}{\partial r^2} + V(r) + (\beta R + U)|\varphi\rangle\langle\varphi|, \quad (1)$$

$$\hat{H}^b = -\frac{1}{2M}\frac{\partial^2}{\partial R^2} + \frac{M\omega^2 R^2}{2} + \frac{J(J+1)}{2MR^2} + \hat{H}_0(r) + \beta(R^2 - R_0^2)|\varphi\rangle\langle\varphi|. \quad (2)$$

The electronic Hamiltonian

$$\hat{H}_0(r) = -\frac{1}{2}\frac{\partial^2}{\partial r^2} + V(r)$$

takes into account long-range electron-molecules interactions. The models are exactly solvable, and it will be possible to calculate all essential characteristics of the system, including dissociative attachmeht cross sections, if the Franck-Condon function $R(\varepsilon)$ of the discrete state:

$$|ad(R)\rangle = \hat{G}^+(\varepsilon(R))|\varphi\rangle$$

where

$$\hat{G}^+(\varepsilon) = (\hat{H}_0 - \varepsilon + i0)^{-1},$$

is known. The function $R(\varepsilon)$ could be determined for any energy curve $T(R)$ of the discrete state by solution of the equation

$$\varepsilon = T(R) - M\omega^2 R^2/2$$

in the case (1), or

$$\varepsilon = T(R) - M\omega^2 R^2/2 - J(J+1)/2MR^2$$

in the case (2). Thus the models provide the base for the consideration of any resonant state.

The method of determination of eigenfunctions of the Hamiltonians (1) and (2) is described in [2,3]. The model (2) has been applied to HCl + e^- problem in [3]. The results obtained for vibrational excitation and dissociative attachment cross sections are qualitatively satisfactory. To improve them further, the long-range electron molecule interaction, omitted in [3], should be taken into account.

References

1. Domcke W., Cederbaum L.S. Phys.Rev. A16, 1465 (1977)
   Domcke W., Cederbaum L.S. Invited report at 12th ICPEAC (1980)
2. Kazansky A.K. Teoret. i Matem. Fyz., 36, 414 (1978)
3. Kazansky A.K. Zh. Eks. i Teoret. Fyz., 82, 1422 (1982)
   Kazansky A.K., to be published in J.Phys.B

ELECTRON HYDROGEN MOLECULE SCATTERING AT LOW ENERGIES

Sukanya Sur and A. S. Ghosh

Department of Theoretical Physics, Indian Association for the Cultivation of Science, Calcutta 700032, India.

The elastic scattering and rotational excitation of hydrogen molecule by electron impact have been studied in the low energy region. A local model exchange potential is used following Hara[1] and is given by

$$V_{ex}(\bar{r}) = -(2/\pi)k_F(\bar{r})F(\eta) \qquad (1)$$

Here

$$F(\eta) = 1/2 + (1-\eta^2/4\eta)\ln|(1+\eta)/(1-\eta)| \qquad (2)$$

and $\eta = k(\bar{r})/k_F(\bar{r})$. In the present case we represent $k(\bar{r})$, the wave vector of the scattering electron, as follows

$$k^2(\bar{r}) = k_o^2 + k_F^2(\bar{r}) + 2I/(1+\lambda k_o) \qquad (3)$$

Here λ is the parameter used by Sur and Ghosh[2]. The charge density $\rho(\bar{r})$, involved in the Fermi momentum $k_F(\bar{r})$, can be expanded in terms of the Legendre polynomials as follows,

$$\rho(\bar{r}) = \sum_\lambda a_\lambda(r)P_\lambda(\cos\theta) \qquad (4)$$

This model exchange potential predicts reliable elastic cross sections for all the symmetries in the energy range 0.04 to 0.25 Ryd. This is evident from Table I in which the present results are compared with the exact results (as quoted by Gibson and Morrison[3]) and also with the other model exchange results.

The rotational excitation results in the close coupling approximation are obtained by solving the following set of coupled equations (Land and Geltman[4])

$$\left[\frac{d^2}{dr^2} - \frac{\ell'(\ell'+1)}{r^2} + k_{j'}^2\right] u_{j'\ell'}^{Jj\ell}(r) = \sum_{j''}\sum_{\ell''}\sum_{\lambda} v_\lambda(r) f_\lambda(j'\ell';j''\ell'';J)u_{j''\ell''}^{Jj\ell}(r) \qquad (5)$$

The static and the exchange potentials are evaluated by using Wang's wave function[5] of hydrogen molecule. The polarization potentials are taken from Henry and Lane[6]. We have retained the rotational states with j = 0, 2, 4 and j = 1, 3, 5 in our calculations. The results will be presented at the conference.

References

1 S. Hara, J. Phys. Soc. Jpn. 22, 710 (1967).

2 Sukanya Sur and A. S. Ghosh, Abstract, IV National Workshop on Atomic and Molecular Physics, India.

3 T. L. Gibson and M. A. Morrison, J. Phys. B 14, 727 (1981).

4 N. F. Lane and S. Geltman, Phys. Rev. 160, 53 (1967).

5 S. C. Wang, Phys. Rev. 31, 579 (1928).

6 R. J. W. Henry and N. F. Lane, Phys. Rev. 183, 221 (1969).

Acknowledgement

One of us (SS) is thankful to the Council of Scientific and Industrial Research, India for providing financial support.

Table I. Electron-hydrogen molecule SEP cross sections (a_o^2) for the three dominant symmetries.

Energy (Ryd)		AAFEGEP[3]	TFEGE[3]	ESEP[3]	Present
0.04	Σ_g	18.91	31.53	33.35	28.82
	Σ_u	38.73	7.47	4.404	3.98
	π_u	3.82	1.43	1.026	1.05
0.09	Σ_g	27.18	33.94	34.509	36.22
	Σ_u	65.62	22.26	13.038	17.27
	π_u	7.91	3.92	2.965	3.22
0.25	Σ_g	28.70	30.64	29.978	31.25
	Σ_u	43.81	35.27	30.303	32.66
	π_u	12.21	8.74	8.23	8.48

INTEGRAL CROSS-SECTIONS FOR e-N_2 and CO ELASTIC SCATTERING AT INTERMEDIATE AND HIGH ENERGIES: A COHERENT-RENORMALIZED-MULTICENTRE-POTENTIAL-MODEL (CRMPM) APPROACH

L.C.G. Freitas[1*], Ashok Jain[1] and Lee Mu-Tao[*]

1 Department of Applied Mathematics and Theoretical Physics, The Queen's University of Belfast, Belfast BT7 1NN, Northern Ireland.

* Departmento de Quimico
Universidade Federal de São Carlos
via Washington Luiz kM 234, 13560, São Carlos, SP Brasil

In order to study the scattering of electrons (or positrons) by molecules at intermediate and high-energies, the two-potential coherent approach of Hayashi and Kuchitsu[1] has recently been investigated for diatomic[2,3] (N_2, CO and O_2) and linear polyatomic[4] (CO_2) molecules in the range 40-800 eV. The approach of Hayashi and Kuchitsu represents the total interaction between the incoming electron and the target molecule by the sum of a short-range potential V_s ($V_s = \sum_{i=1}^{N} V_i$, where V_i is the spherical potential located at the ith atom and N is the number of atoms in the molecule) and a long-range potential V_L (centred at the centre-of-mass of the molecule). The contributions to the cross section come from the coherent sum of these two potentials and, in additon, the effects of multiple scattering within the molecule are considered. If one uses simple atomic scattering functions, the above approach should be a modified version of the independent-atom-model (IAM). However, the approach may be made more realistic if one derives the short-range potential V_s from molecular wavefunctions: this has recently been done by Jain et al[5] by incorporating the approach of Lee and Freitas[6,7], who derive the multicentre e-molecular interaction potential from molecular wavefunctions. They found a considerable difference between the two scattering amplitudes, f^i_{YTP} (due to Yukawa -type potential) and f^i_{RMPM} (due to renormalised-mutlicentre-potential-model (RMPM)) for the ith atom.

Thus, in the new model, the individual atomic scattering amplitude is a function of internuclear separation. The new approach of Jain et al[5] (which was tested for a linear polyatomic molecule C_2H_2) is now extended to calculate vibrationally elastic (and inelastic) cross sections for diatomic species. In this abstract, we report our preliminary data on the integral (σ_i) vibrationally elastic cross sections for e-N_2 and CO systems in the range 50-800eV. In table 1, we have shown this data along with the old calculations of Jain[2] and Lee and Freitas[7] and the experimental data. The new results are now in better agreement with experiment than the old calculations[2,7]; the improvement is about 10% as compared to Jain results. Our results on the differential cross sections will be reported elsewhere.

Table 1. Elastic σ_i cross sections (in units of a_o^2) for e-N_2 and e-CO scattering.

N_2 Energy (eV)	Theory Ref.7	Theory Ref.2	Theory Present	Experiment Ref.8
50	35.90	28.47	26.14	25.8
75	25.12	23.13	20.85	20.7
100	20.20	20.38	19.14	17.7
200	12.90	14.40	12.66	12.6
400	7.94	9.51	8.15	7.54
500	-	8.17	6.97	-
800	4.33	5.81	4.87	3.8
CO				
50	35.10	30.81	28.08	19.6[a]
75	26.20	24.43	23.02	14.2[a]
100	21.10	21.43	20.29	10.7[a]
200	13.70	14.51	13.62	11.3[b]
500	6.59	8.14	7.54	6.15[b]
800	4.94	5.78	5.30	3.74[b]

a. Ref. 9

b. obtained through the numerical integration of the experimental data of Ref. 8.

References:

1. S. Hayashi and K. Kuchitsu, J.Phys.Soc. Japan 41, 1724 (1976)
2. A. Jain, J.Phys.B15, 1533 (1982)
3. A. Jain (in press, Phys. Rev. A (1983))
4. A. Jain and S.S. Tayal, J.Phys.B15, L867 (1982)
5. A. Jain, S.S. Tayal, L.C.G. Freitas and Lee Mu-Tao, J.Phys.B16 (in press, 1983).
6. Lee Mu-Tao and L.C.G. Freitas, J.Phys.B14, 4691 (1981)
7. Lee Mu-Tao and L.C.G. Freitas, J.Phys.B16, 233 (1983)
8. R.D. Dubois and M.E. Rudd, J.Phys.B9, 2657 (1976)
9. H. Tanaka, S.K. Srivastava and A. Chutjian, J. Chem. Phys., 69, 5329 (1978)

ELECTRON SCATTERING BY N_2: CALCULATIONS USING MCSCF TARGET WAVEFUNCTION

John Rumble, Jr.*, Donald C. Truhlar+, and Walter J. Stevens*

*National Bureau of Standards, Washington, DC 20234 USA
+Department of Chemistry, University of Minnesota, Minneapolis, MN 55455 USA

We report on the first calculations done for the scattering of electrons by N_2 in which the target electron density is described by a MCSCF wavefunction.

The interaction potential was divided into two parts: a static and an exchange term. Both were calculated from extended-basis-set MCSCF level wavefunctions which gave about 30 percent of the correlation energy. No polarization potential was used. The internuclear distance was 2.068 a_0. The scattering calculations were done by using standard computer codes previously described[1] and were converged to about 5 percent.

Two sets of calculations have been done, one using a static potential, the other a static-exchange-potential where the exchange was obtained from the Hara free-electron gas model.

Results

Results for the static level calculations are given in Table 1 for impact energies of 5 to 30 eV, where they are compared to identical calculations which differed only in the use of an extended-basis-set Hartree-Fock level target electron density. Calculations on the static-exchange level have also been done, and results for 13.6 and 30 eV elastic scattering are given in Table II. The exchange was calculated using the Hara free-electron gas model. Though the electron density does differ significantly in various regions of space, the elastic scattering cross sections change only on the order of 5 percent when some correlation is included in the target wavefunction. The Π_g channel does show greater changes than for any other channel, which probably is a reflection of a better description of the N_2 target wavefunction by including correlation with the unoccupied $1\pi_g$ orbital of N_2 at the MCSCF level.

These preliminary results thus seem to indicate that a good Hartree-Fock target wavefunction for closed shell molecules is probably sufficient for accurate elastic electron scattering results. For inelastic scattering, a similar conclusion cannot yet be drawn. At the meeting, more complete results for static-exchange level of calculations will also be presented.

Table I

e - N_2

Cross Sections - (a_0^2) - Elastic Scattering

Static Potential Only

Channel Energy (eV)	Σ_g	π_g	Σ_u	π_u	All[c]
5	11.9[a]	.197	34.2	62.1	108.5
	11.0[b]	.278	33.7	59.4	104.5
10	10.7	4.80	16.9	30.1	62.8
	11.4	6.47	17.1	29.4	64.7
13.6	9.43	15.5	12.0	22.9	60.4
	9.70	18.5	12.3	22.5	63.5
20	7.27	17.1	7.17	16.4	48.9
	7.30	16.8	7.43	16.3	48.9
30	5.47	10.4	4.15	11.7	33.4
	5.44	10.1	4.48	11.7	33.5

a top entry - MCSCF target wavefunctions
b bottom entry - Hartree-Fock target wavefunctions
c sum for $m \leq 4$

Table II

e - N_2

Cross Sections - (a_0^2) - Elastic Scattering

Static Exchange Potential

Channel Energy (eV)	Σ_g	π_g	Σ_u	π_u	All[c]
13.6	11.63[a]	9.75	11.71	12.74	47.4
	11.90[b]	8.80	11.44	12.47	46.2
30.0	5.53	6.19	10.45	10.81	36.6
	5.63	5.91	10.66	10.85	36.8

a top entry - MCSCF target wavefunctions
b bottom entry - Hartree-Fock target wavefunctions
c sum for $m \leq 4$

1. J. R. Rumble and D. C. Truhlar, J. Chem. Phys. 70, 4101 (1979), 72, 3441 (E) (1980).

VIBRATIONAL-ROTATIONAL EXCITATION OF N_2 BY ELECTRONS AT 5-50 eV

John Rumble, Jr.* and Donald G. Truhlar+

*National Bureau of Standards, Washington, DC 20234 USA

+Department of Chemistry, University of Minnesota, Minneapolis, MN 55455 USA

We have calculated differential, integral, and momentum-transfer cross sections for vibrational-rotational excitation as well as pure rotational excitation, pure vibrational excitation, and elastic scattering for electron collisions with N_2 at 5-50 eV impact energy.[1] The interaction potential has three terms: static and local exchange potentials calculated from extended-basis-set Hartree-Fock wavefunctions as functions of internuclear distance and a semiempirical polarization potential. The results are compared to previous calculations and to experiment when available.

State-to-state differential cross sections are presented for

$$e^- + N_2(X\ ^1\Sigma_g^+, v, j) \rightarrow e^- + N_2(X\ ^1\Sigma_g^+, v', j')$$

where v and j are the initial vibrational and rotational quantum numbers, respectively, and v' and j' are the final values. Cross sections are reported here for v=0, j=0, and for v'=0, 1, 2 and j'=0, 2, 4, 6, 8.

The static potential was calculated from extended-basis-set Hartree-Fock calculations by converged procedures with no approximations. Internuclear distances used were R=1.85, 1.95, 2.068, 2.15, and 2.45 a_0. Contributions up to λ_{max}=14 for electron-electron terms and up to λ_{max}=28-48 for electron-nucleus terms where λ is the order of a Legendre expansion of the potential. The exchange potential is evaluated by the Hara free-electron-gas exchange (HFEGE) approximation. The polarization potential used included the spherical and asymmetric component of the static dipole polarizability tensor and involved a cutoff function.

The scattering calculations were performed in the fixed nuclei approximation in which the T-matrix is parameterized by the internuclear distance and is block diagonal in M, the projection of the electron angular momentum along the internuclear axis. Close-coupling matrix elements were used up for M=0-3 and the unitarized polarized Born approximation for M=4-9. The close-coupling calculations were well converged.

The T-matrix elements were converted to the vibrational-quantum-number representation by the vibrational sudden approximation. Finally, state-to-state differential, integral, and momentum-transfer cross sections were calculated using standard codes.

Results

The rotationally summed DCS cross section for 10 eV electrons (figure 1) for excitation to $v'=1_\lambda$ are given. These results agree fairly well with those of Trajmar's laboratory[2].

State-to-state DCS for v'=0 and various j' for 10 eV shows fair agreement in shape to other calculations. The same DCS for 5 eV (figure 2) is compared to the only available experimental data, from Tanaka[3], which resolve individual j'. His j'=0 results are in substantial disagreement with the calculations.

In addition, the present calculations confirm the intermediate-energy (20-30 eV) resonance in the vibrational excitation cross section.

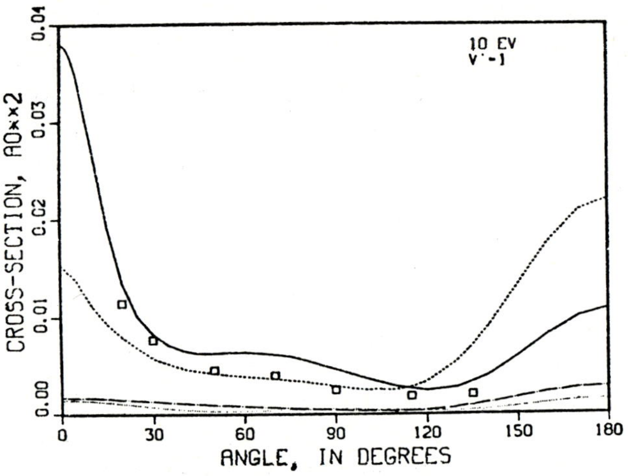

Figure 1. Rotationally summed inelastic DCS. Solid-Present, □ - Trajmar et al., other curves - other calculations.

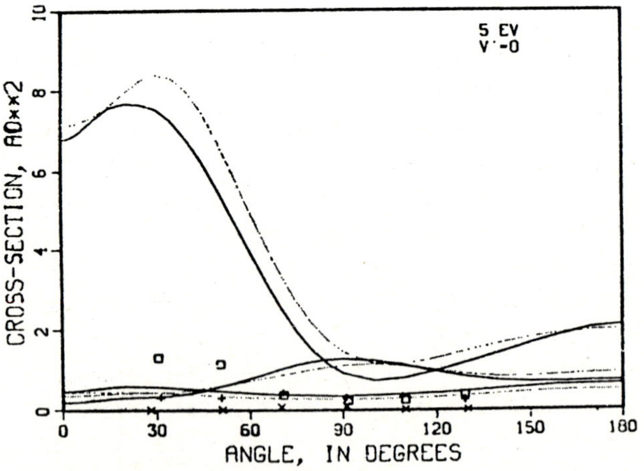

Figure 2. State-to-state DCS. Solid-Present. At 0 degrees, results are (from top) j'=0,2,4. □,+,X - Tanaka, j'=0,2,2, respectively.

References

1. J. R. Rumble, D. G. Truhlar, and M. A. Morrison, J. Chem. Phys. (to be published).
2. D. G. Truhlar et al., J. Chem. Phys. 66, 655 (1977) and references therein.
3. H. Tanaka et al., Seventh Int. Conf. on At. Phys., p. 43 (1980).

INDEPENDENT UNITED ATOM MODEL FOR ELECTRON-MOLECULE SCATTERING

S.P. Khare and B.L. Jhanwar

Department of Physics, Institute of Advanced Studies, Meerut University, Meerut-250005, India

In recent years a number of investigations[1] dealing with the elastic scattering of electrons by molecules have been carried out in independent atom model (IAM) in which the differential cross section averaged over the orientations of the molecule is given by

$$\bar{I}_M(K) = \sum_{i,j=1}^{N} f_i(K)\, f_j^*(K)\, j_o(KR_{ij}) \qquad (1)$$

where N represents the number of atoms in the molecule, $f_i(K)$ is the scattering amplitude due to i^{th} atom, $\underline{K}$ is the change in the momentum vector of the free electron due to scattering and R_{ij} is the equilibrium internuclear distance between the ith and j^{th} atoms. The derivation of (1) in the first Born approximation is well known. Recently, Narasimham et al[2] have given an approximate derivation of (1) for e^--H_2 elastic scattering in the static field approximation.

Massey[3] has given a derivation to show that (1) is valid in the first Born approximation even for excitation of H_2 by electron impact. However, in this investigation we have shown that the derivation of Massey is incorrect and that eq. (1) has limited validity.

In the present paper, instead of IAM, we propose a more general method, to be called as independent united atom model (IUAM). In the proposed model (IUAM) the scattering is considered in terms of the scattering by independent atoms again centred at the various nuclei of the molecule but the atom is obtained by reducing R_{ij} to zero. For example in the present model e^--H_2 scattering is considered in terms of two helium like atoms, centred at the two nuclei, each scattering independently. Half of the scattering amplitude from each atom is added coherently to obtain scattering amplitude due to molecule. Such a model yields

$$\bar{I}(H_2) = 0.5\ I(He)\ (1+j_o(2KR)); \qquad (2)$$

2R being the equilibrium internuclear distance.

To derive (2), let us consider the elastic scattering of the incident electron of the momentum $\underline{k}_o$ by H_2. Then the exact T-matrix is given by $T(H_2) = \mathop{S}_n T_n(H_2)$; with (3)

$$T_n(H_2) = \langle F_{\underline{k}_f}(1)\psi_o(2,3)|V(1;2,3)|F_n(1)\psi_n(2,3)\rangle \quad (4)$$

where the symbols have their usual significance.

Now we represent the molecular states by

$$\psi_n(2,3) = N_n[\,X_n(2)\phi_n(3) + X_n(3)\phi_n(2)\,] \qquad (5)$$

where X_n represents the core orbital and is orthogonal to valence orbital ϕ_n. However, for the ground state $X_o = \phi_o$. Further, each molecular orbital is given by LCAO, like

$$\phi_n(\underline{r}) = P_n\ [\ u_n(A) + u_n(B)\ ] \qquad (6)$$

where A and B are two centres of the H_2 molecule. Use of (6) and (5) in (4) yields, without introducing any appreciable error,

$$T_n(H_2) = \mathop{S}_n D_n[\ T_n(A) + T_n(B)\ ]\ \langle\phi_o|X_n\rangle, \qquad (7)$$

where $D_n = 1/\sqrt{2}$ for $n \neq 0$ and is unity for n=0. $T_n(A)$ is the T-matrix for the inelastic scattering of the electron by a hydrogenic atom centered at A. Shifting the origin to the centre of the molecule we obtain

$$T_n(H_2) = \mathrm{Cos}\{(\underline{k}_n - \underline{k}_f).\underline{R}\}\ [2D_n.T_n(H)\langle\phi|X_n\rangle] \qquad (8)$$

Since $D_n \neq D_o$, it is not possible to obtain (1) from (8). However, it is easy to see that the term in the square bracket is identical to $T_n(He)$ with He states represented by (5). Thus $T_n(H_2) = \mathrm{Cos}\{(\underline{k}_n - \underline{k}_f).\underline{R}\}\ T_n(He)$. Approximating $\underline{k}_n$ by $\underline{k}_o$ for all n, we then obtain (2) from (8). Thus (2) is valid for inelastic scattering as well as for elastic scattering. Another advantage of the present model is that multiple scattering is automatically included in (2). It may be shown that (1) and (2) both are valid in polarized orbital method. The validity of (1) in POM arises due to the fact that in POM, $F_o(\underline{r}_1)$ is obtained correct to all orders of interaction and $F_n(\underline{r}_1)$ are obtained correct to first order only, whereas in the close coupling all $F_n(\underline{r}_1)$ are treated on equal footing.

References

1. B.L.Jhanwar, S.P.Khare and M.K.Sharma, Phys. Rev.A22, 2451(1980); S.P.Khare and Deo Raj, Ind.J.Pure and App.Phys.20, 538(1982) and the references given in them.
2. V.L.Narasimham, A.S.Ramchandran and C.S. Warke, Phys.Rev.A23, 641 (1981).
3. H.S.W.Massey, Electronic and Ionic Impact Phenomena, Vol.2 (Clarendon, Oxford), 1969.

QUANTUM CHEMICAL STUDY OF THE $^2\Sigma_u^+$ Resonance of H_2^-

Eric A. Gislason and Nora H. Sabelli

Department of Chemistry and Computer Center, University of Illinois at Chicago, Chicago, Illinois, 60680

We have developed a novel technique for carrying out quantum chemical calculations on low-lying molecular negative ion resonances. The method allows us to obtain the real part V(R) of the complex energy function which is commonly used to describe such resonances. In this technique we temporarily convert the resonance to the ground state of the system; this allows us to optimize the negative ion basis set and to calculate the wave function using standard programs developed for bound states. In the first part of the calculation we embed the system of electrons and nuclei inside a positively charged spherical cage of total charge +1. The cage is transparent; that is, the wave function extends outside the cage and is continuous across the surface. If the cage radius is S, the potential inside the cage has a constant value 1/S (all units here are a.u.), whereas outside the cage it falls off as 1/r. The cage converts all states of the system, both resonances and neutral molecule-free electron (NMFE) states, into bound states. If S is small enough, the resonance becomes the ground electronic state of the system, because its wave function is more compact than any NMFE state. This allows us to compute an accurate SCF (or MCSCF) wave function for the resonance. The energy of the state is then computed by freezing the resonance orbital and recalculating the energy in the absence of the cage. This gives the desired SCF energy of the resonance.

The method has certain problems which we believe we have overcome. The first is that it is not desirable to use an actual sphere in the calculation. Rather we used 20 charges of +0.05 located at the vertices of a dodecahedron. These charges can be added to existing quantum chemistry programs as additional "nuclei." The second problem is that the resonance will not be the ground state of the system unless S is fairly small. For the $^2\Sigma_u^+$ state of H_2^- we were able to do calculations for S=2.0, 2.5, and 3.0, but larger values of S were not practical. Consequently, the σ_u^* orbital of H_2^- is somewhat perturbed by even the largest cage. We can partially correct the SCF energy for this distortion by carrying out two calculations (with and without the cage) of the wave function and energy at an internuclear distance R=3.0, where H_2^- is stable relative to H_2. Energies at other values of R are then adjusted by the difference in energy between these two calculations.

For our SCF calculations on H_2^- we used a large basis set of Gaussion orbitals optimized for H_2^-. The final SCF basis was a (9s, 5p) set contracted to (6s, 5p). The most diffuse S and two most diffuse P exponents were optimized at R=1.8; their values are 0.0013 (s) and 0.001697 (p). Results were obtained for R values between 1.2 and 3.0. The $^2\Sigma_u^+$ state appears as the ground state of the system (with the cage present) and is easily identified by its electronic expectation values such as $\langle z^2 \rangle \sim 10$. By comparison the NMFE states have $\langle z^2 \rangle$ values an order of magnitude larger and H_2 an order of magnitude smaller. Our calculated dissociation energy $D_e \approx 0.05$ agrees well with both the complex SCF results of McCurdy[1] and the stabilization computations of Taylor.[2] The minimum occurs at Re ≈ 1.4, but this result is very sensitive to any mixing of NMFE orbitals into the resonance wave function by the cage. An accurate determination of R_e must await our CI calculations.

We are presently carrying out full CI calculations on H_2^- which include the d basis set space. We expect to present these results at the conference.

We gratefully acknowledge a gift of computer time from the UIC Computer Center.

References

1. C. W. McCurdy and R. C. Mowrey, Phys. Rev. A 25, 2529(1982).
2. (a) H. S. Taylor and F. E. Harris, J. Chem. Phys. 39, 1012 (1963); (b) I. Eliezer, H. S. Taylor, and J. K. Williams, J. Chem. Phys. 47, 2165 (1967).

THE ADIABATIC PARTIAL WAVE METHOD IN ELECTRON MOLECULE PROCESSES
I- DEVELOPMENTS IN FIXED NUCLEI ELECTRONIC CONTINUUM PROCESSES

Vo Ky Lan, M. Le Dourneuf, J.M. Launay and S. Hara
Observatoire de Paris, 92190 Meudon, France

In fixed nuclei electron molecule scattering, the standard single center partial wave expansion remains slowly convergent at low energies, because of the strong anisotropy of the interaction potential which culminates near the nuclear singularities. An alternative adiabatic partial wave expansion has recently been proposed to circumvent this difficulty[1]. The combined effect of the anisotropic interaction and the centrifugal potential is indeed compactly described by the eigenstates at fixed r of the effective potential $V_{eff} = V(r,\hat{r}) + \frac{\vec{\ell}^2}{2r^2}$. These potential-adapted partial waves, which depend parametrically on r, depart strongly from spherical harmonics and vary rapidly mainly near the nuclear singularities. The qualitative and practical advantage of this approach is that only a few locally-open adiabatic partial waves need to be included in the scattering calculations. Typically, a single dominant asymptotic partial wave suffices to reproduce qualitatively shape resonances in diatomic systems (Fig. 1) and a few asymptotically significant partial waves in heteronuclear systems (Fig. 2). Converged results are obtained with a small number of potential-adapted partial waves (typically 2 and 5 in the previous examples), the coupling of which is easily handled numerically, using a diabatic-by-sector expansion[2], in which the basis is assumed to be piecewise constant and a frame transformation from one sector to the other is performed to account for the variation of the basis.

The first applications of these methods[1-3] were conducted using an approximate local potential to model the effect of exchange forces on closed shell diatomic molecules in their ground electronic state. Several extensions have been investigated since and will be presented at the conference:

1) The inclusion of the non local effect of exchange, either by the exact conventional methods[4] or in the separable expansion approach[5] are computationnally lightened by the adiabatic partial wave expansion.

2) The inclusion of correlation effects and electronic excitation is straightforward, either in the standard close coupling approach[6] or in its R-matrix variant.

3) The evolution of properties with increasing anisotropy of the nuclear potential, typically along isoelectronic sequences such as Ar, HCl, H_2O, is most conveniently studied with this approach.

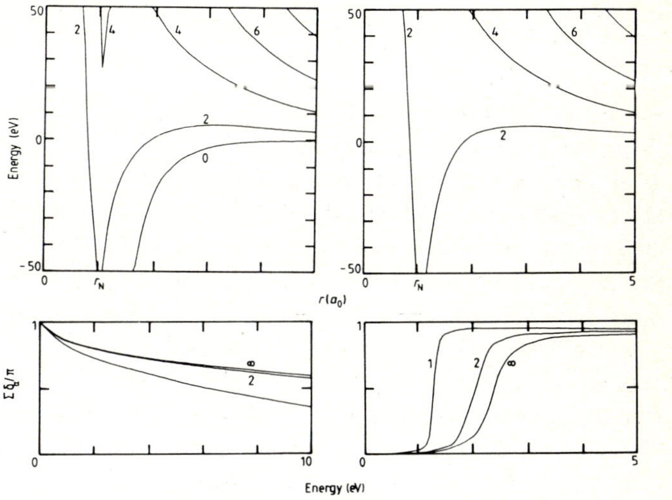

Fig. 1 : e + N_2 fixed nuclei elastic scattering in the $^2\Sigma_g$ (left) and $^2\Pi_g$ (right) symmetries. Adiabatic potentials (top) and eigenphase sum convergence (bottom).

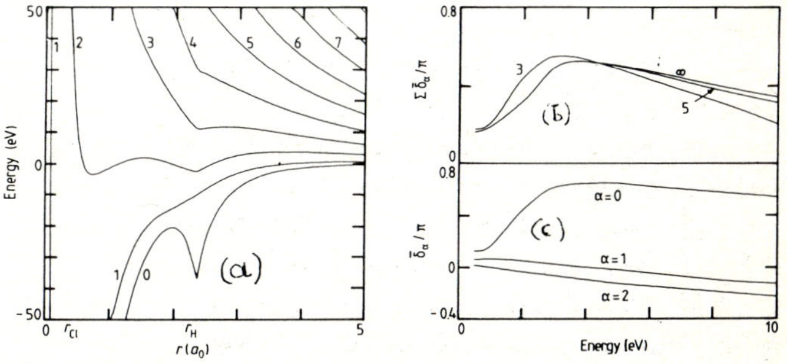

Fig. 2: e + HCl fixed nuclei elastic scattering in the $^2\Sigma$ symmetry. Adiabatic potentials (a), eigenphase-sum convergence (b) and eigenphaseshifts in the dipole field representation (c).

References

1. M. Le Dourneuf, Vo Ky Lan and J.M. Launay, XIIth ICPEAC Abstracts, p. 375 (1981)
2. M. Le Dourneuf, Vo Ky Lan and J.M. Launay, J.Phys.B 15, L685-90 (1982)
3. Vo Ky Lan, M. Le Dourneuf and J.M. Launay, in Electron atom and electron molecule collisions (Plenum Press, 1983) p. 161-200
4. W.D. Robb and L.A. Collins, Phys. Rev. A 22, 2474 (1980)
5. T.N. Rescigno and A.E. Orel, Phys. Rev. A 24, 1267 (1981)
6. P.G. Burke and M.J. Seaton, Methods in Computational Physics, 10, 1-80 (1971)

THE ADIABATIC PARTIAL WAVE METHOD IN ELECTRON-MOLECULE PROCESSES
II- EIGENPHASE ANALYSIS OF ELECTRONUCLEAR CORRELATIONS IN RESONANT VIBRATIONAL EXCITATION

J.M. Launay, M. Le Dourneuf and Vo. Ky Lan

Observatoire de Paris, 92190 Meudon, France

Vibrational excitation of molecules by electrons $e + AB_v \longrightarrow e + AB_{v'}$, usually proceeds through the formation of an intermediate complex AB^-, which can decay by electronic detachment. A detailed understanding of this mechanism requires the analysis of the wavefunction $\Psi(\vec{r},R)$ which describes the correlated motion of the electrons and nuclei ($\vec{r}$ for the electron-center of mass distance and R for the internuclear distance).

The basic characteristics of these processes are modelled by an approximate local potential V(r,R) and a factorisation of the wavefunction $\Psi(\vec{r},R)$. The angular motion of the electron is represented through an adiabatic partial wave[1,2] $\Phi(\hat{r};r,R)$ which is an eigenfunction of the effective potential operator $\vec{\ell}^2/2mr^2 + V(r,R)$ with eigenenergy $\varepsilon(r,R)$. They are adapted to the potential anisotropy and improve the usual spherical harmonics expansion. The wavefunction is thus written as $\Psi(\vec{r},R) = F(r,R)\ \Phi(\hat{r};r,R)$ where F(r,R) solves the two-dimensional Schrödinger equation:

$$\left[-\frac{1}{2m}\frac{\partial^2}{\partial r^2}-\frac{1}{2M}\frac{\partial^2}{\partial R^2}+\varepsilon(r,R)-E\right]F(r,R)=0$$

This equation is then solved using an adiabatic R-matrix formalism[3]. In resonant vibrational excitation of N_2 ($^2\Pi_g$ shape resonance), the electronuclear potential energy surface $\varepsilon(r,R)$ exhibits two important features

1) The line of nuclear singularities r = R/2
2) The saddle point under which the tunnelling effect is responsible for formation of the ionic complex.

The oscillatory structure of experimental vibrational cross-sections[4] is reproduced semi-quantitatively. When eigenphases boundary conditions are imposed, the electronuclear wavefunctions $F_\alpha(r,R)$ show a remarkably simple structure (Fig. 2). The short range behaviour of $F_\alpha(r,R)$ is clearly characteristic of vibrational levels of the complex (nodes parallel to the r axis), while the moving oblique node is an earmark of the adiabatic factorisation of the electronic wavefunction at short distances.

The essence of this formalism will be displayed at the conference. In addition, a detailed analysis of the electronuclear interaction in terms of non-adiabatic couplings will be presented at the conference in order to clarify the connection with various other approaches.

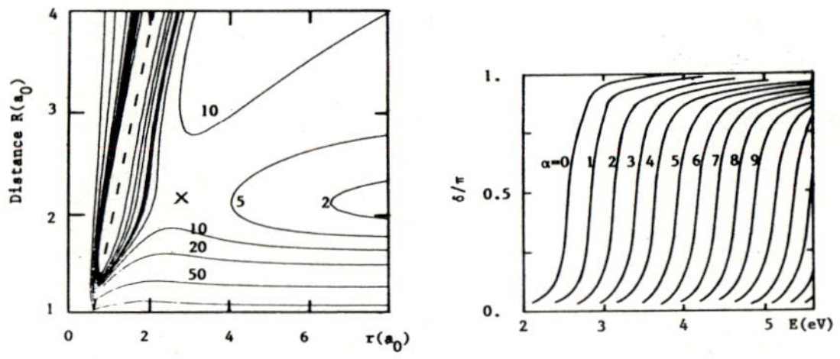

Fig. 1 : Electronuclear potential energy surface $\varepsilon(r,R)$ for N_2^- ($^2\Pi_g$) (left) and vibrational eigenphase-shifts $\delta_\alpha(E)$. Energies are in eV.

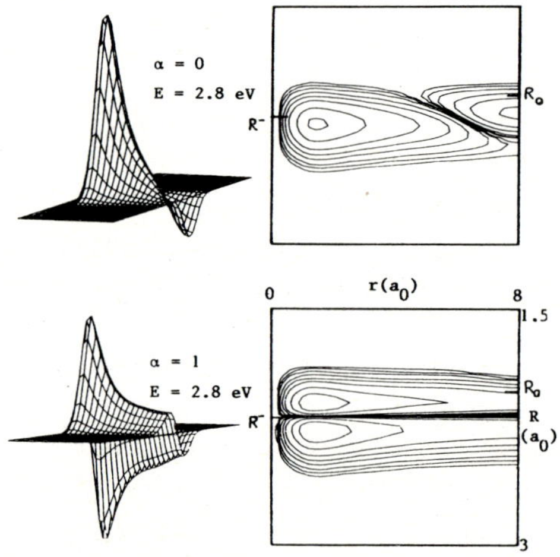

Fig. 2 : Electronuclear eigenmodes $F_\alpha(r,R)$ for $N_2^-(^2\Pi_g)$ The functions are not normalised. R_o (R_-) are the equilibrium positions of $N_2(N_2^-)$.

References

1. Vo Ky Lan, M. Le Dourneuf and J.M. Launay in Electron Atom and Electron Molecule Collisions (Plenum, 1983) p. 161
2. M. Le Dourneuf, Vo Ky Lan and J.M. Launay, J. Phys. B 15, L685 (1982)
3. B.I. Schneider, M. Le Dourneuf and P.G. Burke, J. Phys. B 12, L365 (1979)
4. H. Ehrardt and K. Willmann, Z. Phys., 204, 462 (1967)

THE ADIABATIC PARTIAL WAVE METHOD IN ELECTRON MOLECULE PROCESSES
III- EIGENTIME ANALYSIS OF ELECTRONUCLEAR CORRELATIONS IN DISSOCIATIVE ATTACHMENT

M. Le Dourneuf, J.M. Launay and Vo Ky Lan

Observatoire de Paris, 92190 Meudon, France

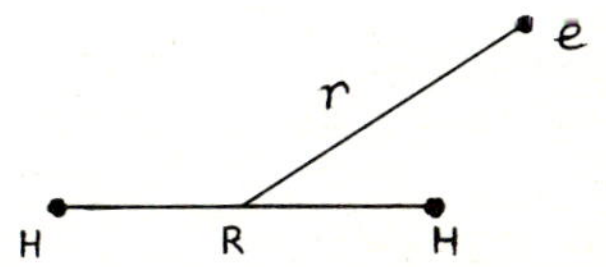

Dissociative attachment in low energy electron-molecule collisions provides a simple prototype of rearrangement collisions. The fundamental role of the intermediate electronuclear complex, which fragments into alternative electronic or nuclear continua, has already been treated using resonant formalisms. The Feshbach formalism has been successfully been applied to an ab-initio study of F_2^- [1]. In this abstract, we illustrate the first application of the R-matrix formalism[2], to the first dissociative shape resonance of H_2^- [3].

$$e + H_2\ (v) \longrightarrow H_2^-\ (^2\Sigma_u) \begin{array}{l} \nearrow e + H_2\ (v') \\ \searrow H + H^- \end{array}$$

As in the case of resonant vibrational excitation described in the preceding abstract, the compact description of low energy homonuclear diatomic shape resonances in terms of a single adiabatic partial wave, reduces the description of the $H_2^-(^2\Sigma_u)$ dissociative attachment to a two-dimensional problem[4].

By imposing separable boundary conditions at the limit of the electronic and nuclear fragmentation zones, the standard R-matrix formalism[5] allows a splitting of the exact inner region solution in two subproblems: the determination of the fixed-nuclei R-matrix electronic states followed by the determination of nuclear R-matrix states, eventually coupled by non-adiabatic effects.

The present study provides three types of results:

1) A strictly adiabatic calculation within the R-matrix reaction zone properly matched to the electronic and nuclear fragmentation waves in the outer regions, provides qualitative agreement with experiment.
2) The residual influence of non-adiabatic couplings, specially near the stabilisation point $R_s = 3a_o$ has been investigated.
3) The characterisation of the rearrangement process in terms of eigenfunctions of the S-matrix seems awkward. Instead, the eigenfunctions of the lifetime matrix $Q = \frac{\hbar}{i} S^+ \frac{dS}{dE}$ [6] provide insights to the complex formation and decay mechanisms. Figure 1 shows two eigenmodes which are responsible for the dissociative process. The enhancement of dissociative attachment cross-sections with increasing initial vibrational quantum number the diatomic is clearly visualised.

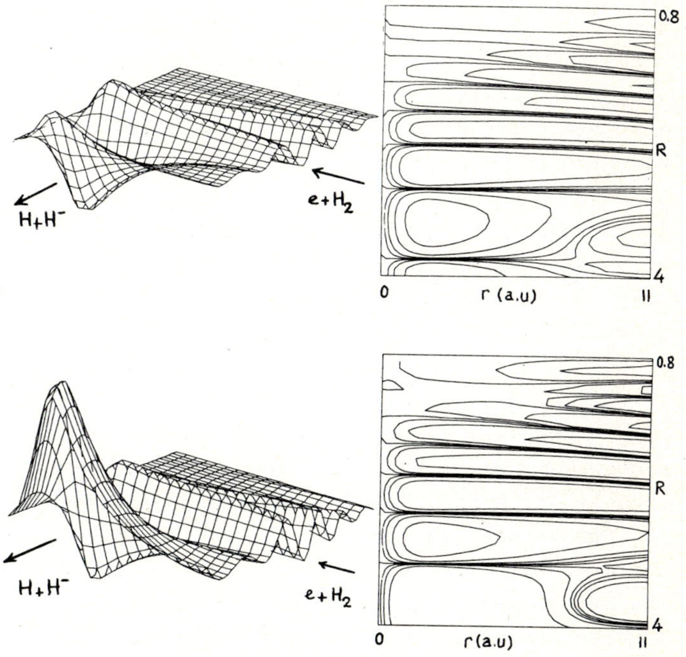

Fig. 1 : The two eigenmodes of the time-delay matrix which lead to dissociative attachment. The total energy is 0.5 eV above the threshold for $H + H^-$.

References

1. A.U. Hazi, A.E. Orel and T.N. Rescigno, Phys. Rev. Lett. 46, 918 (1981)
2. B.I. Schneider, M. Le Dourneuf and P.G. Burke, J. Phys. B12, L365 (1979)
3. G.J. Schulz and R.K. Asundi, Phys. Rev. 158, 25 (1967)
   M. Allan and S.F. Wong, Phys. Rev. Lett. 41, 1971 (1978)
4. J.M. Launay, M. Le Dourneuf and Vo Ky Lan, preceeding abstract.
5. C. Bloch, Nuclear Physics 4, 503 (1957)
6. F.T. Smith, Phys. Rev. 118, 349 (1960)
   F.T. Smith, Phys. Rev. 130, 394 (1963)

NUCLEAR DYNAMICS IN RESONANT ELECTRON-MOLECULE SCATTERING BEYOND THE LOCAL APPROXIMATION

Hernán Estrada, Michael Berman, L.S. Cederbaum and W. Domcke

Theoretische Chemie, Universität Heidelberg, D-6900 Heidelberg, W.-Germany

The object of the present work[1] is to present a full non-local calculation of the vibrational excitation functions for the 2.3 eV resonance in N_2. Using Feshbach's projection operator formalism the resonance is described as a discrete state embedded in and interacting with a continuum.

We solve for the nuclear dynamics in the resonance state by using a basis set representation of the complex, energy-dependent and non-local optical potential. The fixed-nuclei input data required in the calculation are the potential energy curve of the projected discrete state as well as the corresponding width function. The width function of the 2.3 eV shape resonance is taken from previous *ab initio* calculations.[2] The potential energy curve of the discrete electronic state giving rise to the resonance is adjusted to obtain quantitative agreement with experiment[3] for the v = 0 → 1 vibrational excitation channel (fig. 1a).

A detailed comparison of calculated cross sections in a non-local treatment with available experimental data[3,4] has been performed for vibrational excitation of N_2. Calculated differential cross sections for the inelastic channels compare very well both in shape and in absolute magnitude with measured data. The total cross section (summed over all channels) $\sigma_{tot} = \sum_v \sigma_{0 \to v}$ has also been calculated. The theoretical σ_{tot} compares well with the experimental measurements of Kennerly[4] (fig. 1b).

In addition a detailed comparison between the cross sections obtained within a non-local theory and those calculated in the local approximation is provided. We have found that in the case of the 2.3 eV shape resonance in N_2 the local approximation is very accurate for vibrationally inelastic channels. For elastic channels v = 0→0, 1→1, 2→2, 3-3 the local approximation is found to be less satisfactory.

We conclude that the method used in this work can lead to cross sections in quantitative agreement with experiment.

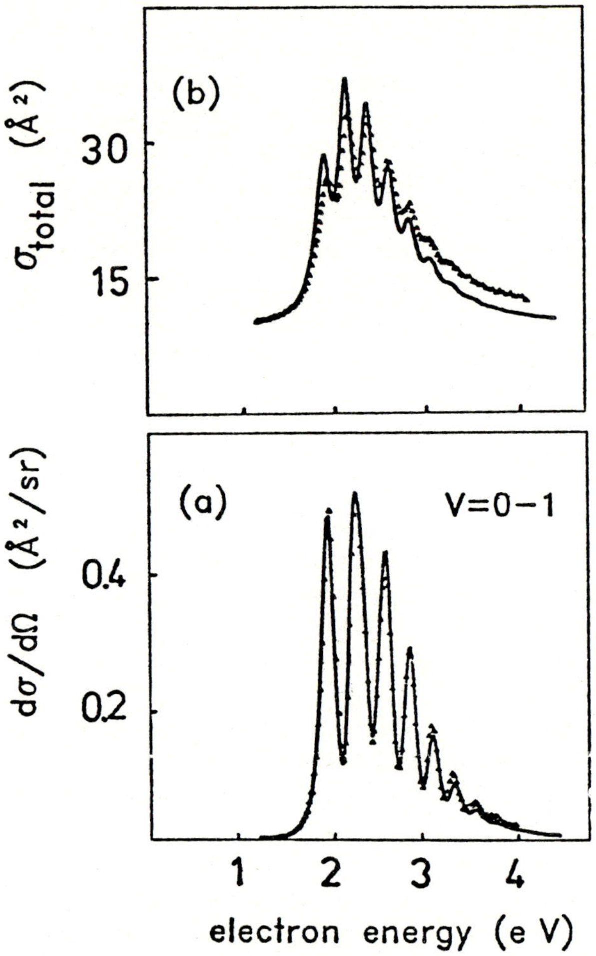

Fig. 1: Calculated and experimental vibrational excitation cross sections for electron -N_2 scattering (— theory, ▵▵▵ experiment). a) Calculated differential cross section for vibrational excitation at 90° for the channel v=0→1 together with the measurement of Wong et al.[3] b) Total cross section compared to the experiment of Kennerly.[4]

References:

1. M. Berman, H. Estrada, L.S. Cederbaum and W. Domcke, in press (Phys. Rev. A)
2. A.U. Hazi, in Electron-Atom and Electron-Molecule Collisions, edited by J. Hinze, Plenum 1983, pg 103
3. S.F. Wong, J.A. Michejda, A. Stamatovic (unpublished)
4. R.E. Kennerly, Phys. Rev. A 21, 1879 (1980)

ELECTRON-MOLECULE SCATTERING USING THE OPTICAL POTENTIAL APPROACH: SURPASSING SECOND ORDER

Michael Berman, O. Walter and L.S. Cederbaum

Theoretische Chemie, Universität Heidelberg, D-6900 Heidelberg, W.-Germany

This work reports the first application of an optical potential beyond second order for electron-molecule scattering. The optical potential is calculated in the two-particle-hole Tamm-Dancoff approximation[1] (2ph-TDA). The scattering equations are solved in the T-matrix expansion method using the 'truncated' approximation as well as the Schwinger variational principle.[2]

The only calculation which include polarization via a perturbative expansion of the optical potential in electron-molecule scattering was performed by Klonover and Kaldor.[3] These authors have included polarization up to <u>second order</u> in the optical potential. A CI approach to the optical potential, which neglects ground state correlations was recently reported by Schneider and Collins.[4]

Fig. 1 shows the fixed-nuclei eigenphase sum of the $^2\Pi_g$ resonance of N_2. The static exchange results are compared to the R-matrix calculations of Morrison and Schneider[5] and Noble et al.[6] This comparison demonstrates the importance of the use of the Schwinger rather than the 'truncated' expression for the T-matrix.

To have an indication for the quality of the eigenphase sum in the various approximations to the optical potential we include in figure 1 the eigenphase sum of ref.[7] In that study a purely resonant process has been assumed and a non-local theory has been applied to the vibrational motion of the nuclei. Experimental vibrational excitation cross sections together with the energy dependent width function calculated by Hazi have been used to fit the resonance position. In this fitting procedure, the cross sections have been found to be very sensitive to changes in the resonance position. Therefore, we consider the resonance position extracted from the eigenphase sum of ref.[7] as an 'experimental' reference value.

The effect of including the second order contribution in the optical potential is to shift (to lower energy) the resonance position by about 1 eV compared to the static-exchange

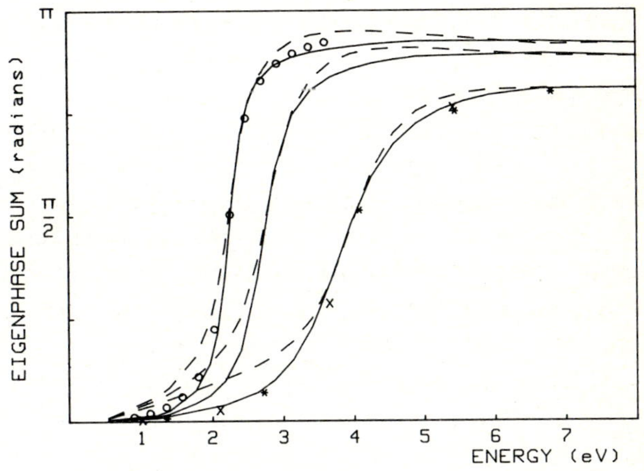

Fig. 1: Eigenphase sums of the $^2\Pi_g$ resonance of e-N_2 scattering. Dashed and full lines correspond to calculations done with a truncated potential and the Schwinger expression, respectively. From high energy to low energy the pairs of dashed and full lines correspond to static-exchange, second order and 2ph-TDA approximations, respectively. The crosses correspond to the R-matrix results of Morrison and Schneider[5] and the stars to those of Noble et al.[6] The circles are the purely resonant 'experimental' reference values[7] (see text).

position. The contribution of the 2ph-TDA leads to an additional shift of ~0.5 eV, bringing it to excellent agreement with the 'experimental' reference value. In addition, the overall shape of the 2ph-TDA eigenphase sum agrees well with the purely resonant eigenphase sum of ref.[7]

References

1. J. Schirmer and L.S. Cederbaum, J. Phys. B <u>11</u>, 1889 (1978)
2. D.W. Watson and V. McKoy, Phys. Rev. A <u>20</u>, 1974 (1979)
3. A. Klonover and U. Kaldor, J. Phys. B <u>11</u>, 1623 (1978)
4. B.I. Schneider and L.A. Collins, J. Phys. B <u>15</u>, L335 (1982)
5. M.A. Morrison and B.I. Schneider, Phys. Rev. A <u>16</u>, 1003 (1977)
6. C.J. Noble, P.G. Burke and S. Salvini, J. Phys. B <u>15</u>, 3779 (1982)
7. M. Berman, H. Estrada, L.S. Cederbaum and W. Domcke, in press (Phys. Rev. A)

WAVE PACKET FORMULATION OF THE BOOMERANG MODEL FOR RESONANT ELECTRON-MOLECULE SCATTERING

C. William McCurdy and Julia L. Turner

Department of Chemistry, Ohio State University, Columbus, Ohio 43210

Resonances in electron-molecule scattering are characterized in the simplest theoretical description by a resonance energy, E_r, and width, Γ, which are functions of the nuclear positions, R, and provide a complex potential surface, $W(R) = E_r(R) - i\Gamma(R)/2$, for motion in the metastable anion (resonance) state. The "boomerang" model, or local complex potential approach, makes use of this information to compute vibrational excitation cross sections [1]. The usual formulation of the boomerang model involves an inhomogeneous differential equation and is difficult to apply to multidimensional systems.

We present a time-dependent formulation which features a wave packet propagating on the complex potential surface, and which can be applied more easily to the polyatomic case. In this approach the T-matrix for excitation from vibrational state ν_i to state ν_f is given by the Fourier transform of the overlap of the moving wave packet with a stationary packet determined by the final state.

$$T_{\nu_f\nu_i} = -\frac{i}{\hbar}\int_0^\infty e^{iEt/\hbar}\langle\phi_{\nu_f}|\Psi_{\nu_i}(t)\rangle\, dt$$

where

$$\phi_{\nu_f} = (\Gamma(R)/2\pi)^{\frac{1}{2}}\,\chi_{\nu_f}(R)$$

$$\Psi_{\nu_i}(t)\Big|_{t=o} = (\Gamma(R)/2\pi)^{\frac{1}{2}}\,\chi_{\nu_i}(R)$$

and χ_{ν_i} and χ_{ν_f} are the initial and final state vibrational wave functions.

In this form the boomerang model is mathematically almost identical to the wave packet approaches to photodissociation and Raman scattering developed by Heller and coworkers [2]. Thus we can apply their semiclassical techniques for propagating the wave packet which have been used succesfully for polyatomic molecules. We have tested this method on the well studied $^2\Pi_g$ resonance in N_2^- and found that it performs very well as is shown in figure 1.

We also discuss observed structure, and the lack thereof, in polyatomic vibrational excitation cross sections from the time-dependent point of view.

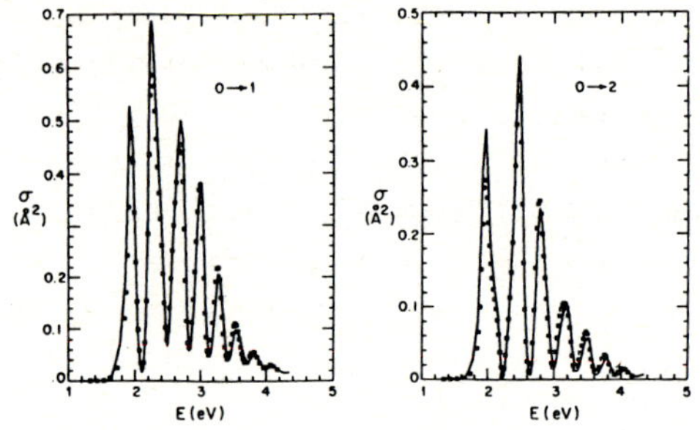

Fig. 1: Vibrational excitation cross sections for N_2. Solid line is semiclassical time-dependent result, and points are full quantum results from the usual approach.

References

1. L. Dubé and A. Herzenberg, Phys.Rev. A 20, 194 (1979).
2. See for example D.J. Tannor and E.J. Heller, J.Chem. Phys. 77, 202 (1982); and S. Lee and E.J. Heller, J.Chem.Phys. 76, 3035 (1982).

IMPROVED SEMICLASSICAL APPROXIMATION FOR VIBRATIONAL EXCITATION OF MOLECULES BY SLOW ELECTRONS

A.K.Kazansky and I.S.Yelets

Dept.of Theor.Phys., Institute of Theor.Phys., Leningrad State University, Leningrad, 198904, USSR

D.V.Efremov Scientific Research Institute of Electrophysical Apparatus, Leningrad, 188631,USSR

The following semiclassical representation for the vibrational excitation cross sections was offered by the authors[1,2] at the 12-th ICPEAC within the resonant theory[3]:

$$\sigma_{n_0\to n}=\sqrt{E_n/E_{n_0}}\,(2\pi)^{-3}|I_{n_0}|^2|I_n|^2/|Q|^2 \qquad (1)$$

where

$$Q=-\sin\left\{\int_{a^-}^{b^-}\sqrt{2M[E-W(z)]}\,dz-\pi/2\right\}$$

$$I_i=\frac{1}{2}\left(\frac{\omega}{\pi}\right)^{\frac{1}{2}}\frac{V(z_i)}{[\mathcal{E}_i-U_0(z_i)]^{\frac{1}{2}}}\left(\frac{2\pi}{\Psi_i''(z_i)}\right)^{\frac{1}{2}}\cos\left\{\Psi_i(z_i)-\frac{\pi}{4}\right\}, \quad i=n_0,n \qquad (2)$$

$$\Psi_{n_0}(z)=\int_{a_{n_0}}^{z}\sqrt{2M[\mathcal{E}_{n_0}-U_0(z)]}\,dz-\int_{a^-}^{z}\sqrt{2M[E-W(z)]}\,dz$$

$$\Psi_n(z)=\int_{z}^{b^-}\sqrt{2M[E-W(z)]}\,dz-\int_{z}^{b_n}\sqrt{2M[\mathcal{E}_n-U_0(z)]}\,dz$$

Here $a_{n_0}, b_{n_0}; a_n, b_n$ are turning points of a molecule in n_0-th and n-th vibrational states with the energies $\mathcal{E}_{n_0}, \mathcal{E}_n$; a^-, b^- are those of a temporary ion with the energy E. $U_0(R)$ and $W(R)=U_I(R)-i\Gamma(R)/2$ are the potential curves of the molecule and the ion. z_{n_0}, z_n are the Franck-Condon transition points:

$$E-W(z_i)=\mathcal{E}_i-U_0(z_i)\ ;\quad i=n_0,n \qquad (3)$$

The representations (2) fail when the transition points are close to either the turning points, or to each other. To modify the representation (2) we used[4] the linear approximation for the potential curves:

$$U_0(R)=U_0(a_i)-F_0(R-a_i);\ W(R)=W(a^-)-F_I(R-a^-) \qquad (4)$$

For instance, in the vicinity of the left turning points the modified representation gets the form:

$$I_{n_0}=K_{n_0}\,Ai(X_{n_0})$$
$$I_n=\tfrac{1}{2}K_n\{w^+(X_n)\exp(i\lambda)+w^-(X_n)\exp(-i\lambda)\} \qquad (5)$$

Here $Ai(z), w^{\pm}(z)$ are the Airy functions with the asymptotic form at $z\to-\infty$:

$$Ai(z)\sim(-z)^{-1/4}\cos\left\{\tfrac{2}{3}(-z)^{3/2}-\pi/4\right\}$$
$$w^{\pm}(z)\sim(-z)^{-1/4}\exp\left\{\pm i\left[\tfrac{2}{3}(-z)^{3/2}+\pi/4\right]\right\} \qquad (6)$$

and

$$\lambda=\int_{a^-}^{b^-}\sqrt{2M[E-W(z)]}\,dz\ ;$$

$$X_i=(a_i-a^-)(2M\tilde{F})^{\frac{1}{3}}\ ,\quad \tilde{F}=\frac{F\cdot F_I}{F_I-F} \qquad (7)$$

$$K_i=\left(\frac{\omega}{F-F_I}\right)^{\frac{1}{2}}V(z_i)(2M\tilde{F})^{-\frac{1}{6}}$$

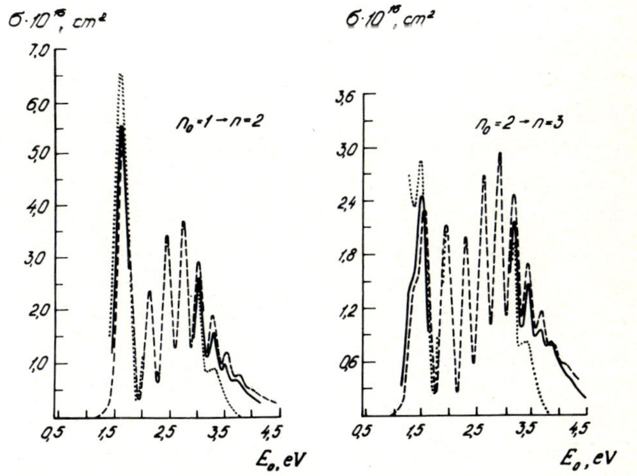

The vibrational excitation cross sections $\sigma_{n_0\to n}$ for the nitrogen molecule; the dashed curves represent the results of the numerical calculations[3]; the dotted curves are those calculated from the semiclassical formulas (1), (2); the solid lines are the curves computed from the formulas (1), (5),(8).

As the linear approximation for $U_0(R)$ and $W(R)$ is valid only in the small R-region, the further improvement of (2),(5),(7) is of interest. Here, the following representations are proposed:

$$K_i=\frac{1}{2}\left(\frac{\omega}{\pi}\right)^{\frac{1}{2}}\frac{V(z_i)}{[\mathcal{E}_i-U_0(z_i)]^{\frac{1}{2}}}\left\{\frac{2\pi}{\Psi_i''(z_i)}\right\}^{\frac{1}{2}}\left\{-X_i(z_i)\right\}^{\frac{1}{4}},\ i=n_0,n$$

$$X_i(z)=-\left\{\frac{3}{2}\left(\int_{a_i}^{z}\sqrt{2M[\mathcal{E}_i-U_0(z)]}\,dz-\int_{a^-}^{z}\sqrt{2M[E-W(z)]}\,dz\right)\right\}^{\frac{2}{3}} \qquad (8)$$

Some precautions should be taken while calculating the Airy functions arguments, as the identity $(z^3)^{\frac{1}{3}}=z$ is not valid for any complex z.

1. Kazansky A.K., Yelets I.S. Abstracts of the 12-th ICPEAC, 1176 (1981).
2. Yelets I.S., Kazansky A.K. Abstracts of the 12-th ICPEAC, 1178 (1981).
3. Dubé L. and Herzenberg A. Phys.Rev.A, 20, 194 (1979).
4. Yelets I.S., Kazansky A.K. Sov.Phys.JETP, 55, 258 (1982).

ELECTRON SCATTERING BY DIATOMIC MOLECULES

S. Salvini*, C. J. Noble† and P. G. Burke*

* Queen's University Belfast and † SERC Daresbury Laboratory England.

In this paper we describe the current status of our work on electron scattering by diatomic molecules using the R-matrix method. This work was initiated by Burke et al[1] who developed a fixed-nuclei theory of rotational and electronic excitation and was extended by Schneider et al[2] to enable vibrational excitation and dissociative attachment cross sections to be calculated.

In the initial computational developments, Slater basis functions (STO's) were used to represent the scattered electron. Preliminary calculations in the static exchange approximation for H_2 and N_2 were reported by Buckley et al[3] and more complete results, based on a modification of the ALCHEMY molecular structure package[4] by Noble[5], were published by Noble et al[6]. This work showed that while STO's were capable of giving accurate results for electron impact energies below about 1 Ryd. they could not accurately represent the oscillations in the continuum wave functions at higher energies.

In recent work by Burke et al[7] it has been shown that this difficulty can be overcome using numerical basis functions to describe the scattered electron. The radial basis functions $u_{\ell i}(r)$ are obtained by solving numerically the zero-order equation

$$\left(\frac{d^2}{dr} - \frac{\ell(\ell+1)}{r^2} + V_o(r) + k_i^2\right) u_{\ell i}(r) = \sum_j \lambda_{ij} P_{\ell j}(r)$$

subject to fixed R-matrix boundary conditions at r = a. The Lagrange undetermined multipliers λ_{ij} are determined so that the $u_{\ell i}(r)$ are orthogonal to the single-centre expansion components $P_{\ell j}(r)$ of the occupied molecular orbitals of the target over the range $0 \leq r \leq a$. Finally the zero-order potential $V_o(r)$ is chosen to be the spherically symmetric component of the static potential of the target.

We present in figure 1 results calculated at the equilibrium nuclear separation for the $e^- - N_2$ total cross section over an energy range from 1 to 31.6 eV compared with measurements of Kennerly[8]. The target was represented by the H.F. wave function calculated by Nesbet[9] and electron exchange was included exactly. Also non-adiabatic charge polarization effects were included by allowing one-electron excitations out of the H.F. sea into a virtual orbital basis. The low energy $^2\Pi_g$ resonance position and width is in good agreement with other recent calculations[10] over a wide range of internuclear separations giving us confidence that our theory will correctly describe the vibrational structure in this peak. We also found a pronounced peak at intermediate energies due to a $^2\Sigma_u$ resonance

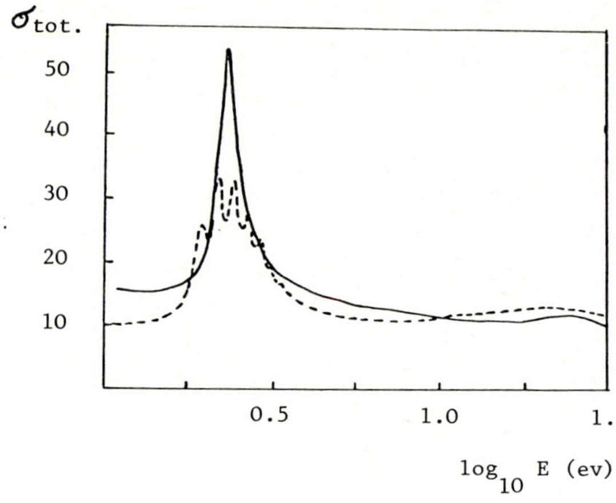

Fig. 1: The total cross section in $Å^2$ for $e^- - N_2$ scattering between 1 and 31.6 eV. Solid line; R-matrix calculations including exchange polarization. Dashed line: measurements of Kennerly[8].

in accord with experiment.

Work is now underway on the calculation of the angular distribution of the vibrational excitation cross sections in the neighbourhood of the $^2\Sigma_u$ resonance. We are also investigating the importance of electron correlation effects in the target. Finally, work is underway to extend the calculations to other diatomic molecules and to calculate electronic excitation cross sections at low energies. It is intended to present our latest results in these areas at the Conference.

References

1. P. G. Burke, I. Mackey and I. Shimamura, J.Phys. B. (Atom.Molec.Phys.) 10 2497 (1977).
2. B. I. Schneider, M. LeDourneuf and P. G. Burke, J. Phys. B. (Atom.Molec.Phys.) 12, L365 (1979).
3. B. D. Buckley, P. G. Burke and Vo Ky Lan, Comp. Phys. Commun. 17 175 (1979).
4. A. D. McLean, Conf. on Potential Energy Surfaces in Chemistry (Ed. W. A. Lester Jr. IBM San Jose, California) 87, (1971).
5. C. J. Noble, Daresbury Laboratory Technical Memorandum DL/SCI/TM33T (1982).
6. C. J. Noble, P. G. Burke, S. Salvini, J. Phys. B. (Atom.Molec.Phys.) 15 3779 (1982).
7. P. G. Burke, C. J. Noble and S. Salvini, J.Phys. B. (Atom.Molec.Phys.) to be published (1983).
8. R. E. Kennerly, Phys. Rev. A21 1876 (1980).
9. R. K. Nesbet, J. Chem. Phys. 40 3619 (1964).
10. A. V. Hazi, T. N. Rescigno and M. Kurilla, Phys. Rev. A23 1089 (1981).

ABSOLUTE CROSS SECTIONS FOR RESONANT VIBRATIONAL EXCITATION PROCESSES OF N_2 MOLECULES BY ELECTRON IMPACT

K. Onda* and A. Temkin

NASA-Goddard Space Flight Center, Greenbelt, Maryland 20771, U. S. A.

We report here absolute integrated cross sections for resonant vibrational excitation processes of N_2 molecules by electrons with incident energy in the range of 1.8 - 3.0 eV.

PDE approach[1-3] is applied to the vibrational close coupling equations of the hybrid theory[4] as is appropriate for the resonant Π_g partial wave, and to equations derived in the fixed nuclei approximation being good for nonresonant partial waves. An effective interaction potential employed here consists of a sum of static, Hara free electron gas exchange[5] (V_{FEG}^{ex}) multiplied by an adjustable parameter, and polarization potentials. A static potential is obtained from the wave function of Cade, Sales, and Wahl[6]. A polarization potential is calculated by generalizing the method of polarized orbitals to molecular targets. An r, θ, and R dependence of polarization potential constructed from polarized molecular orbitals is found[7] to be significantly different from the well-known phenomenological one. Here r and θ are, respectively, radial and angular coordinates of scattered electrons, and R the internuclear distance of the N_2 molecule. An adjustable parameter multiplied by V_{FEG}^{ex} for resonant Π_g partial wave is chosen so that the first peak of the vibrational excitation cross section occurs at the observed position 1.98 eV.

Results obtained by 7-state, 10-state, and 13-state vibrational close coupling calculations are presented in Table I. As can be seen our results indicate reasonably good convergence in the vibrational close coupling expansions in the energy range investigated here. In Fig. 1 a sum of resonant and nonresonant contributions to integrated cross section obtained here is compared with an experimental total cross section of Kennerly[8], and with a theoretical one obtained by Schneider, Le Dourneuf, and Vo Ky Lan[9]. It is seen that our results are in satisfactory accord with experimental ones. Our results[10] for vibrational excitation processes from v = 0 to v' = 1 and 2 also agree reasonably well with experimental ones normalized by Wong[11]. Reasonably good accord of our results with experimental ones is lost if a polarization potential calculated by the method of polarized orbitals is replaced by a phenomenological one[7].

Further details will be presented at the meeting.

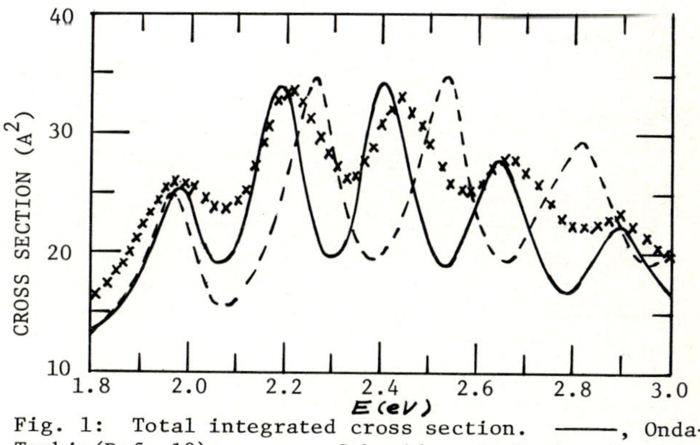

Fig. 1: Total integrated cross section. ——, Onda-Temkin(Ref. 10); - - -, Schneider, Le Dourneuf, and Vo Ky Lan(Ref. 9); x x x , Kennerly(Ref. 8).

Table I: Π_g partial wave contribution to the sum $\sum_{v=0}^{v_{max}} \sigma_{0\to v}$ (A^2). The rows correspond to the results obtained by solving 7-state, 10-state, and 13-state vibrational close coupling equations.

E(eV)	7-state	10-state	13-state
1.8	3.58	3.58	
1.9	8.63	8.67	
2.0	15.64	15.03	
2.1	9.90	10.91	
2.2	24.47	24.54	24.44
2.3	15.34	10.37	10.38
2.4	10.54	23.88	24.64
2.5	19.39	12.96	11.44
2.6	17.85	10.75	14.40
2.7	9.96	16.99	13.38
2.8	7.57	9.74	7.83
2.9	9.14	7.18	13.03
3.0	11.31	9.69	7.26

* NAS-NRC/NASA Senior Resident Research Associate, 1980 - 1982

References

1. A. Temkin in Symposium on Electron-Molecule Collisions(I. Shimamura and M. Matsuzawa eds., University of Tokyo, 1979) p. 55.
2. K. Onda and A. Temkin, Abstracts XII ICPEAC (S. Datz editor, Gatlinburg, 1981) p. 293.
3. E. C. Sullivan and A. Temkin, Comp. Phys. Comm. 25, 97 (1982)
4. N. Chandra and A. Temkin, Phys. Rev. A13, 188(1976).
5. S. Hara, J. Phys. Soc. Japan 22, 710 (1967).
6. P. E. Cade, K. D. Sales, and A. C. Wahl, J. Chem. Phys. 44, 1973 (1966).
7. K. Onda and A. Temkin, (to be published).
8. R. E. Kennerly,, Phys. Rev. A 21, 1876 (1980).
9. B. Schneider, M. Le Dourneuf, and Vo Ky Lan, Phys. Rev. Lett. 43, 1926 (1979).
10. K. Onda and A. Temkin, (in preparation).
11. Quoted by G. J. Schulz in Principle of Laser Plasmas (G. Bekefi ed., Interscience, New York, 1976) Chap. 2.

COMPARISON OF THE 2 AND THE 20eV RESONANCES IN VIBRATIONAL EXCITATION OF N_2 AND CO BY ELECTRON IMPACT

Edward S. Chang

Department of Physics & Astronomy, University of Massachusetts, Amherst, MA 01003 U.S.A.

It is generally agreed that low-lying antibonding molecular orbitals with a vacancy will capture scattering electrons into shape resonances which enhance the vibrational cross sections. These resonances with their associated partial waves in N_2 and CO are identified in tabular form below:

e + molecule

Incident energy		N_2	CO
2eV	resonance	$^2\Pi g$	$^2\Pi$
	partial waves	$d\pi$	p and $d\pi$
20eV	resonance	$^2\Sigma u^+$	$^2\Sigma^+$
	partial waves	f and $p\sigma$	f and $d\sigma$

At the 2eV resonances, detailed measurements of the vibrational cross section with resolution of the rotational branches and with angular distribution were recently performed.[1] Good agreement with theory[2] was found in both N_2 and CO.

For the higher energy resonance in N_2, the angular distribution data of Tronc *et al*[3] and Tanaka *et al*[4] are shown in Fig. 1 at the specified energies.

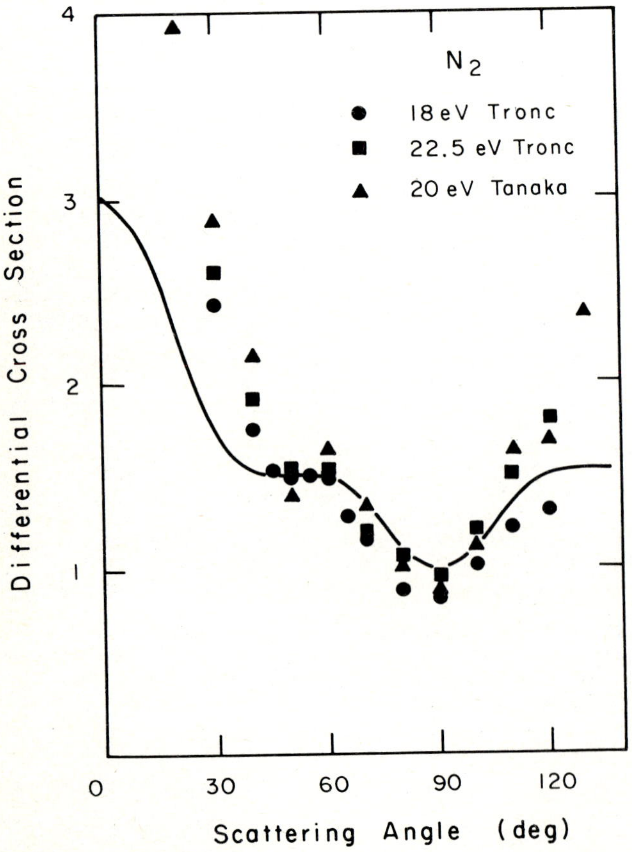

Similar data for CO can be seen in Fig. 2.

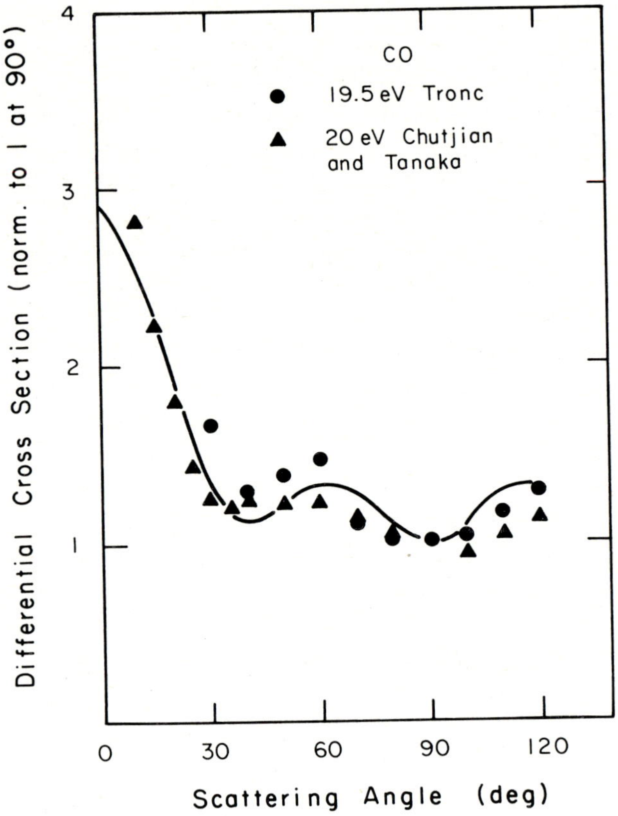

Theory[5] represented by the solid curves in both Figs. 1 and 2 is seen to agree well with experiment. Further, theory can now predict the cross sections with resolution of the rotational branches. These results are verifiable with the high resolution technique[1] developed by Jung *et al*.

References

1. K. Jung, Th. Antoni, R. Muller, K.H. Kochem, and H. Ehrhardt, J. Phys. B 15, 3535 (1982).
2. E.S. Chang, J. Phys. B 15, L873 (1982).
3. H. Tanaka, T. Yamamoto, and T. Okada, J. Phys. B 14, 2081 (1981), and A. Chutjian and H. Tanaka, J. Phys. B 13, 1901 (1980).
4. M. Tronc, R. Azria, and Y. LeCoat, J. Phys. B 13, 2327 (1980).
5. E.S. Chang, Phys. Rev. A 27, 709 (1983).

VIBRONIC EXCITATION OF CO_2 BY ELECTRON RESONANT SCATTERING

Edward S. Chang

Department of Physics & Astronomy, University of Massachusetts, Amherst, MA 01003 U.S.A.

Existence of the $^2\Pi_u$ resonance at 3.5 eV in the vibronic excitation of CO_2 seems well established[1]. Unfortunately, the experimental data on the angular distributions remain unpublished[2], and the theory[1] contains enough parameters to render a good fit unconvincing.

For the 83 meV energy loss process, Andrick and Read[1] correctly pointed out that the final molecular state is $(\nu_1, \nu_2^{\ell}, \nu_3) = (0, 1^{\pm 1}, 0)$ i.e., $^1\Pi_u$, and that the dominant incident partial waves are $(\ell, m) = (0, 0)$, $(1, \pm 1)$, and $(3, \pm 1)$ and the dominant outgoing partial waves are $(0, 0)$, $(2, 0)$ and $(2, \pm 2)$. In fact, a simple assumption of $(1, \pm 1)$ in the incident and of $(2, 0)$ in the outgoing partial in the present theory leads to a parameter-free expression for the angular distributions

$$\frac{d\sigma}{d\Omega}\,(0\ 0^{\circ}\ 0 \rightarrow 0\ 1^{\pm 1}\ 0) \propto (1 + 0.6\sin^2\theta).$$

Fitted to the data at 90°, this curve appears to agree well with experiment as shown in the figure below.

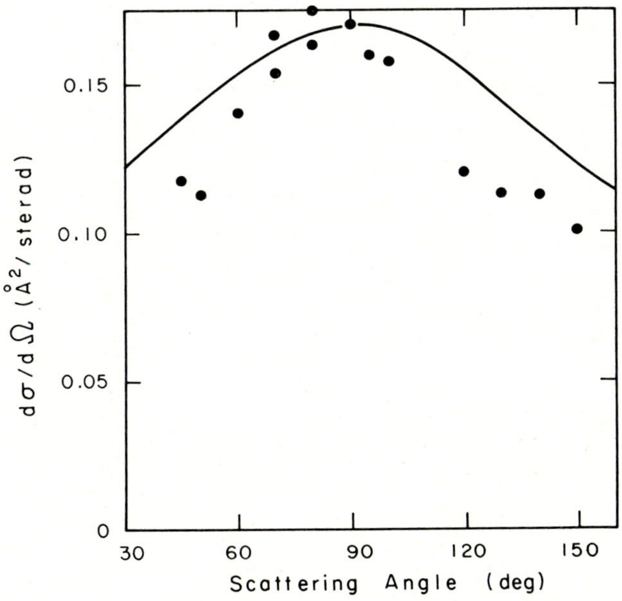

The situation at 172 meV energy loss is more complicated. The (1 0° 0) state is in Fermi resonance with the (0 2° 0) state at 159 meV, both of symmetry $^1\Sigma g^+$. Further, the $(0\ 2^{\pm 2}\ 0)$ states lie only 7 meV away at 165 meV and are of $^1\Delta g$ symmetry. Since these complications were not discussed in the previous work[1], the theoretical fit to the data there may not be valid.

Hopefully the high resolution experiment being performed at Kaiserslautern[3] will produce refined and unambiguous data. Further theoretical predictions can then be made on the new angular distribution data.

References

1. D. Andrick and F.H. Read, J. Phys. B 4, 389 (1971).
2. D. Andrick and D. Danner, unpublished.
3. H. Ehrhardt (private communication).

DECAY OF THE 22 eV SHAPE RESONANCE IN N_2

L.Malegat, M.F.Fontaine, A.Colin and M.Tronc

Laboratoire de Chimie Physique, Université Pierre et Marie Curie
11, rue Pierre et Marie Curie, 75231 Paris, Cedex 05, France

Accurate angular distributions have been obtained for electrons having excited the v = 1, v = 2 and v = 3 states of N_2 through the $^2\Sigma_u^+$ shape resonance in the 18 - 30 eV energy range and from 0 to 130° scattering angle.

A cross electron-beam molecular-beam spectrometer with hemispherical filters has been used and special attention was given to get angular distributions normalised at 90° with smaller error bars than in previous measurements [1,2] :

i) the resolution was 60-80 meV , good enough for a good separation of the low intensity v = 1 peak and the high-energy-loss-tail of the elastic peak even at small angle where the elastic increases faster than the v = 1, but not too high ($<$ 30 meV) with exclusion of some rotational levels which will result in distorsion of the measured vibrational DCS[3].

ii) on elastic the signal with molecular beam on was 150 times the signal with beam of at the same background pressure of 1 x 10^{-5} Torr for angles higher than 30°, and 50 times at 10° so that no correction has to be done for collisional volume variation with angle. Moreover good signal to noise condition especially at small scattering angle was obtained with tandem hemispherical analysers.

The fitting of the experimental data following Read's theory [4] shows clearly the contribution of the f (l = 3) and p (l = 1) partial waves and the slow evolution of the angular distribution with energy is related to the variation of the relative contribution of these two partial waves. The counterpart of this shape resonance in the photoionization of the K-shell of N_2 is also dominated by a resonant f - like channel, and a non-resonant p-like channel nearly constant throughout the resonant region [5].

Moreover results will be presented on the decay of the resonance in the valence excited states of N_2 in the 8 - 18 eV energy loss range with the possibility of underlying resonances derived from multiply excited states of N_2 and in connection with the role of multielectron transitions in the near continuum part of the K-shell electron energy loss spectrum[6]

References

1. M.Tronc, R.Azria and Y.Le Coat, J.Phys.B 13, 2327 (1980)
2. H.Tanaka, T.Yamamoto, T.Okada.J .Phys.B 14,2081(1981)
3. E.S.Chang, Phys.Rev. A 27, 709 (1983).
4. F.H.Read, J.Phys. B 1, 893, (1968).
5. J.L.Dehmer, J.Siegle, J.Welch and D.Dill, Phys. Rev. A 21, 101 (1980).
6. R.Arneberg, H.Agren, J.Muller and R.Manne, Chem.Phys. Let. 91, 362,(1982).

OBSERVATION OF INNER SHELL RESONANCE IN ELASTIC SCATTERING OF ELECTRONS FROM N_2

D. Mathur, F.A. Rajgara and A. Roy

Tata Institute of Fundamental Research, Bombay 400005, India

The existence of resonances associated with electron impact excitation of K-shell electrons in diatomic molecules was established a few years ago[1] by observing structure in the total ionisation cross section at electron energies close to the K-shell binding energy. These structures were attributed to the autoionisation of the temporary negative ion state formed by the capture of the incoming electrons. In these measurements the resonances appeared as an extremely small (~0.1%) effect superimposed on a large continuum.

We report the first observation of the resonance associated with excitation of a ($1s\sigma$) electron in N_2 to a ($2p\pi$) orbital in the **differential elastic scattering cross section** at 401.9 eV in the angular range from 40° to 140°. The experiment was performed using a crossed electron and molecular beam apparatus. The incident electrons had a typical energy spread of 0.25 eV and the scattered electrons were analysed by a cylindrical mirror analyser which is an apt device for high energy electrons. By choosing large scattering angles we reduce the direct scattering contribution and this enables the resonances to appear as prominent structures in the elastic scattering excitation function.

The excitation function as obtained at 51° scattering angle is presented in figure 1. The effect (~ 15%) is large enough and the signal to noise ratio is good enough to enable us to carry out a polynomial line fit and deduce the following parameters:

Resonance energy	401.9 eV
FWHM	1.1 eV
Total lifetime	6×10^{-16} sec.

The relatively large resonance effect observed in this channel opens up the possibility of determining the angular momentum quantum number of the partial wave involved in the formation of the shape resonance.

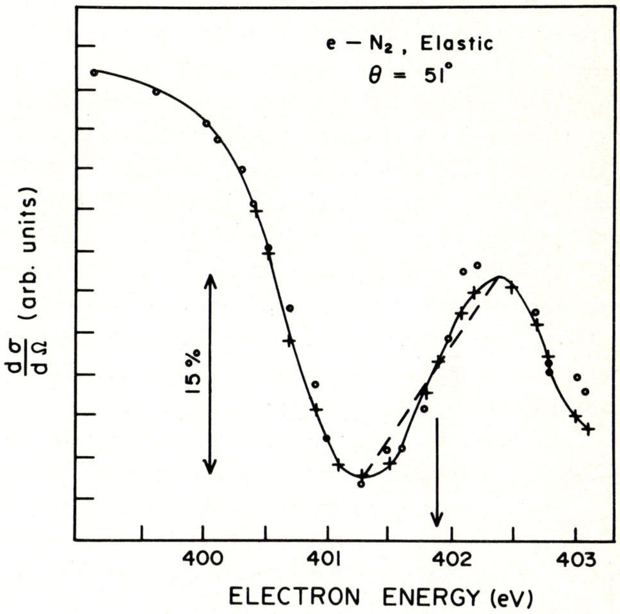

Fig.1: e-N_2 elastic scattering at 51° scattering angle. The solid line is a polynomial fit to the experimental data. The arrow points to the resonance energy.

References

1. G.C. King, J.W. McConkey and F.H. Read, J. Phys. B10, L541 (1977).

THRESHOLD BEHAVIOUR IN THE CROSS SECTION OF ELECTRON SCATTERING ON CO_2

W. Sohn, K. - H. Kochem, N. Hebel, K. Jung, H. Ehrhardt

Fachbereich Physik der Universität Kaiserslautern, D-6750 Kaiserslautern, W.-Germany

The purpose of the present paper is to contribute experimentally to the research of the influence of a virtual state pole to different reaction channels in low energy electron - molecule scattering. CO_2 is an interesting example since i) the virtual state influence in the elastic channel is well established[1], ii) CO_2 has three inelastic channels with low threshold energies (83 meV bending mode 010, 172 meV symmetric stretch mode 100, and 291 meV asymmetric stretch mode 001), iii) below 2 eV collision energy all three states are unperturbed, since overtones and mixed modes are not excited and iv) the three vibrational channels are excited by different interaction potentials as dipole, quadrupole and polarizability. The molecular constants are known.

In a crossed beam experiment energy- and angular dependencies have been determined for the elastic e^- - CO_2 scattering and the excitation of the three fundamental vibrations (0, 0, 1), (0, 1, 0) and (1, 0, 0) in the angular range between 15° and 100° and the range of collision energies from 0.1 eV to about 2 eV. The transmissions of the electron optical systems are controlled by comparison with the well known e^- - helium cross sections. Absolute cross sections are determined using the flux - rate-method[2].

All four channels (elastic and three inelastic) exhibit a peak of the cross section close to the corresponding threshold, i.e. the influence of the virtual state pole is existent in all four channels and qualitatively similar. In figure 1 there are shown two typical excitation functions for the asymmetric stretch mode between 1.1 eV and the corresponding threshold energy. Obviously the width, the energetic position of the maximum of the threshold structure and the value of the differential cross section strongly depend on the scattering angle. This anisotropic angular behaviour of those modes which are coupled to an appreciable high dipole moment in the exit channel also becomes obvious in the angular dependencies shown in figure 2.

The strong forward peaking of the (0, 0, 1), (0, 1, 0) and elastic mode is similar over the whole energy range, wheras the cross section for the excitation of the symmetric stretch vibration is nearly isotropic.

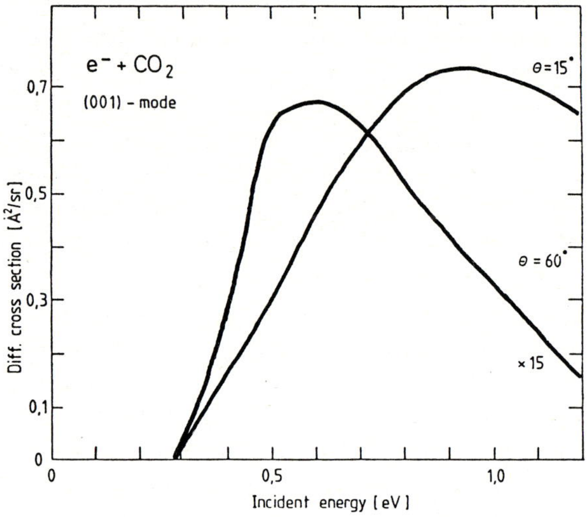

Fig. 1:
Energy dependencies of the differential cross section of the (0, 0, 1)-mode at scattering angles of 15° and 60°.

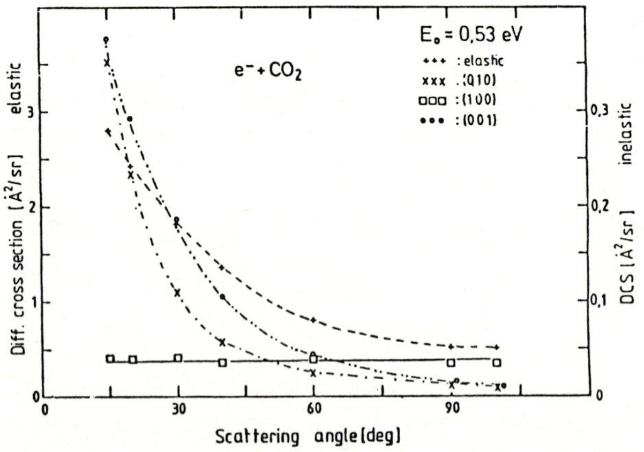

Fig. 2:
Angular dependencies at 0.55 eV collision energy of the elastic and the three inelastic channels.

References

1. M.A. Morrison, Phys. Rev. A3, 1445 (1982)
2. W. Sohn, K. Jung, H. Ehrhardt, J. Phys. B (1983), in press

ROTATIONAL AND VIBRATIONAL EXCITATION OF CO_2 BY SLOW ELECTRONS

K. Jung, T. Antoni, R. Müller, H. Ehrhardt

Fachbereich Physik der Universität Kaiserslautern, D-6750 Kaiserslautern, W.-Germany

The scattering of low-energy electrons by CO_2 has been studied frequently[1,2], as several types of interaction (dipole, strong quadrupole, resonant) are possible and different vibrational channels show different symmetry properties. Experimental studies, however, have been limited by an energy resolution of about 23 meV. The present measurements were performed for the primary energies 1.2, 2, and 3.8 eV.

At the lower energies, rotational excitation by dipole and quadrupole interaction can be expected. Especially in the 1 - 3 eV region the coupled channel calculation in fixed nuclei-approximation of Morrison and Lane[3] predicts that $\sigma(J=0 \rightarrow 2)$ exceeds $\sigma(0 \rightarrow 0)$ by more than an order of magnitude.

In spite of that no rotational broadening has been measured in the spectrum of the symmetric stretch mode where transitions with even ΔJ are allowed. Furthermore, no excitation of the 0 2^0 0-vibrational mode which is coupled to the 1 0 0-mode by a Fermi resonance, and of the 0 $2^{\pm 2}$ 0 mode was found.

The excitation of the asymmetric stretch mode is always coupled to rotational transitions by symmetry properties of the rovibrational levels. In the energy loss spectrum, a rotational broadening of 2 meV independent of the scattering angle could be measured. For a rather good fit of the energy loss peak only $\Delta J = \pm 1$ transitions had been taken into account.

Within the region of the $^2\pi_u$ resonance, rovibrational excitation has been studied for all fundamental modes.

Small rotational broadening was found in the energy loss spectra of the 0 1 0 mode and the elastic peak.

With special interest we looked for the energy loss region around $\Delta E = 172$ meV. In CO_2 the ground mode of the symmetric stretch coincides within 1 meV with the second harmonic of the bending mode. The sublevel 0 2^0 0 belongs to the same symmetry as the 0 0 1 vibrational level and therefore the correlation interaction shifts the energy levels by about 8 meV.

The substantial mixing of these states suggests that angular dependence and rotational broadening of their spectra should be quite similar.

Figure 1 shows an energy loss spectrum at E_0 = 3.8 eV for the scattering angle of 5°. It can be seen that the excitation of higher harmonics of the bending mode is possible by resonant scattering. For small angle scattering, however, only the three vibrational lines 0 2^0 0, 0 $2^{\pm 2}$ 0 and 1 0 0 are necessary to build up the measured intensity profile. At large angles, the 0 $2^{\pm 2}$ 0 contribution disappears, which can be understood, if this transition is mainly caused by dipole interaction. Both the other channels show considerable $\Delta J = \pm 2$ excitation, in agreement with theoretical calculations.

Higher rotational transitions seem to be of less importance and have been neglected in a first analysis.

References

1. A. Andrick, D. Danner, H. Ehrhardt, Phys. Letters 29A (1969) 346
2. D.F. Register, H. Nishimura, S. Trajmar, J. Phys. B 13 (1980) 1651
3. M.A. Morrison, N.F. Lane, Phys. Rev. A16 (1977) 975

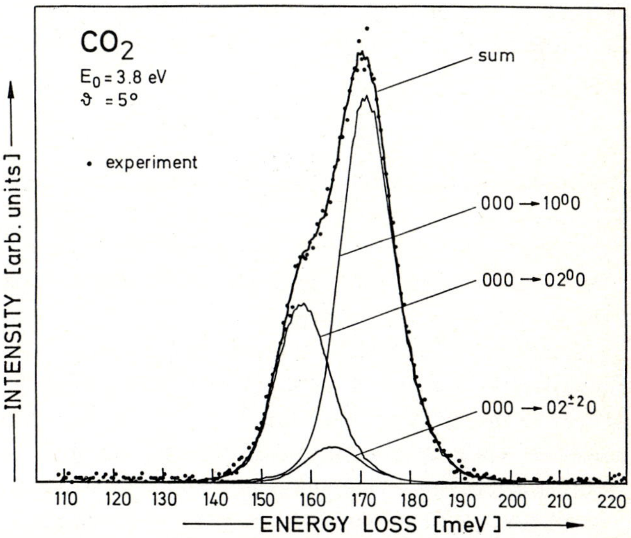

Fig. 1:

Energy loss spectrum of CO_2 at E_0 = 3.8 eV and θ = 5° around ΔE = 170 meV, showing intensities of the three vibrational modes 0 2^0 0, 0 $2^{\pm 2}$ 0 and 1 0^0 0.

ABSOLUTE TOTAL ELECTRON SCATTERING CROSS SECTIONS FOR TRIATOMIC MOLECULES IN LOW ENERGY REGION

Czesław Szmytkowski

Institute of Physics, Technical University of Gdańsk, Majakowskiego 11/12, 80-952 Gdańsk, Poland

Absolute total electron scattering cross sections for OCS, N_2O, CO_2 and CS_2 have been determined from transmission experiments at collision energies between 0.4 and 30 eV. The reported measurements were performed with the 127° electrostatic monochromator coupled with a collision chamber and Faraday cup detector. The monochromator was operated at a typical beam intensity of 10^{-11} A and 0.07 eV FWHM energy resolution. The Faraday cup accepted electrons from an angle of $6\cdot10^{-3}$ sr.

The absolute total cross sections at given energy were obtained from the attenuation of electron current intensity as function of the target gas pressure, and the cell length (30 mm) by using the Beer-Lambert relation. The target gas pressure (range of 10^{-3} Torr) was initially measured with the McLeod gauge and then using a capacitance manometer head from MKS.

The resonance structure in transmission current in N_2 was used for determination of the energy scale. The Earth's magnetic field was compensated through the use of Helmholtz coils.

The cross section values obtained in the same energy range with different runs were averaged. Taking into account random (<3%) as well as expected systematic errors, the total cross section uncertainty is estimated to be ±20% at the lowest energies decreasing to ±10% with an increase of energy.

Figure 1 shows the results of present measurements compared with theoretical CMSM calculations of Lynch et al.[1] for $e^- + OCS$ process. Besides the short-lived $^2\Pi$ shape resonance giving structureless intense peak[2] at 1.15 eV, a weak enhancement of the experimental curve around 4 eV is observed. At 3.8eV the d-wave dominated resonance was already observed[3] in excitation of the |100| vibrational mode. The broad bump centered at 11 eV can be attributed to a short-lived shape resonance or to the existence of a number of overlapping resonances. Large visible nonresonant contribution to the cross section is attributable to the permanent dipole moment (0.71 D) of the OCS molecule.

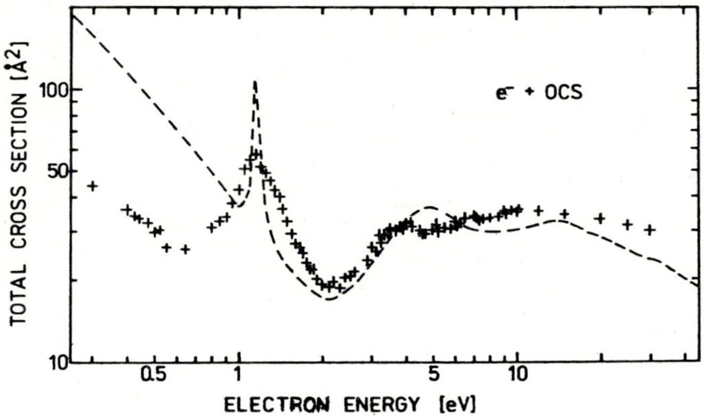

Fig. 1: Total electron scattering cross section for OCS. Crosses, +++, depict the present results and the dashed curve, ---, is the result of the theoretical calculations by Lynch et al.[1]

The present experimental and theoretical[1] results are in quite good agreement although several substantial differences are evident. The discrepancy between shapes of the first resonant peak can be diminished by including nuclear motion in calculations. The shift between the experimental and theoretical positions of the resonant features at higher energies is probably due to the neglection of polarisation potential in the calculations.

The present absolute values of total cross section for OCS deviate from the normalized[2] ones measured previously due to the incorect calibration of ionisation gauge used earlier for determining target gas pressure.

Work supported in part by the Institute of Experimental Physics of Warsaw University.

References

1. M. G. Lynch, D. Dill, J. Siegel and J. L. Dehmer, J. Chem. Phys. 71 4249-54 (1979)
2. Cz. Szmytkowski and M. Zubek, Chem. Phys. Letters 57 105-8 (1978)
3. M. Tronc and R. Azria, Symposium on Electron-Molecule Collisions (University of Tokyo) Invited Papers, eds. I. Shimamura and M. Matsuzawa, (Sept. 1979) 105-9.

NON ADIABATIC EFFECTS IN LOW ENERGY e^-- MOLECULE SCATTERING.

E. Ficocelli Varracchio and U. T. Lamanna

Department of Chemistry, Centro Chimica Plasmi, C.N.R., University of Bari, 70100 Bari, Italy

It is well known that the adiabatic nuclei (A.N.) approximation of e^-- molecule scattering fails to reproduce experimental cross sections, for vibrational and rotational processes, either in the presence of resonant mechanisms or close to thresholds[1]. It has recently been shown that the A.N. approximation can be generalized to include, in an ab initio fashion, the efect of nuclear relaxation[2]. The final form of the generalized A.N. approximation can be written as

$$T_{f;i} = \langle \xi_f | \exp(\hat{C}) T^{AN}(\varepsilon_f, \varepsilon_i; \vec{R}) | \xi_i \rangle \qquad (1)$$

where $|\xi\rangle$ represents the vibrorotational structure of the target, T^{AN} is an "off-shell", fixed nuclei, transition amplitude, involving the energies of the projectile both in the entrance (ε_i) and exit (ε_f) channels, and the $\hat{C}$ operator, correcting for the nuclear relaxation, essentially involves spatial and energy differentiations. Series expansion of the exponential, in (1), leads to

$$T_{f;i} = \langle \xi_f | T^{AN}(\varepsilon_f, \varepsilon_i; \vec{R}) | \xi_i \rangle + \langle \xi_f | [\hat{C} + \frac{1}{2} \hat{C}^2 + \ldots] T^{AN}(\varepsilon_f; \varepsilon_i; \vec{R}) | \xi_i \rangle \qquad (2)$$

showing that, in particular, the lowest order form of the theory (first term on the r.h.s. of (2)), has the general structure of the conventional A.N. approximation, once the substitution of off-shell $T^{AN}(\varepsilon_f, \varepsilon_i; \vec{R})$ amplitudes, in place of the (currently used) simpler $T^{AN}(\varepsilon_i, \varepsilon_i; \vec{R})$, on-shell, counterparts, has been effected.

We are presently evaluating this term (first on the r.h.s. of (2)), for rotational excitation processes, in order to investigate wether the off-shell constraint will be able to remedy the breakdown of the A.N. approximation (particularly at threshold), or the higher order terms of (2) will have to be introduced. Computations are being performed for the e^-- H_2 system, for which non adiabatic effects have recently been put in evidence[3]. The essential quantity, for a numerical implementation of the theory, is the off-shell T^{AN} amplitude. In order to evaluate this quantity, we directly solve the integral equation for the T-matrix, in a fixed nuclei approximation, according to

$$T^{AN}(\varepsilon_f, \varepsilon_i; \vec{R}) = \Sigma(\varepsilon_f, \varepsilon_i) + \int d\varepsilon \, \frac{\Sigma(\varepsilon_f, \varepsilon) T^{AN}(\varepsilon, \varepsilon_i; \vec{R})}{\varepsilon_i^2 - \varepsilon^2 + i\eta} \qquad (3)$$

In (3), $\Sigma(\varepsilon', \varepsilon)$ are matrix elements of the e^-- H_2 interaction, evaluated between plane wave states. The following explicit form for the interaction has been assumed

$$\Sigma = V_{st} + V_{ex} + V_{pol} \qquad (4)$$

that completely represents the static, exchange and polarization effects. Besides, the electronic ground state of H_2 has been described in terms of the Hartree-Fock SCF wave function of Fraga and Ransil[4], at the equilibrium separation. Our computations then essentially involve: a) determining the $\Sigma(\varepsilon_f, \varepsilon)$ matrix elements, numerically, using a single center expansion for the molecular orbital of H_2 ($^1\Sigma_g$); b) solving the integral equation (3) by a Gaussian quadrature scheme.

Our computer codes have been checked by reproducing the exact results of Collins et al.[5], based on $T^{AN}(\varepsilon_i, \varepsilon_i; \vec{R})$ on-shell T-matrices, that are also automatically obtained from the numerical solution of (3). We are presently analyzing differential and total scattering cross sections, in a wide energy range, for general multi-quanta rotational processes. Detailed numerical results will be reported at the Conference.

References

1. N.F. Lane, Rev. Mod. Phys. 52, 29 (1980)
2. E. Ficocelli Varracchio, J. Phys. B 14, L511 (1981)
3. A.N. Feldt and M.A. Morrison, J. Phys. B 15, 301 (1982)
4. S. Fraga and B. Ransil, J. Chem. Phys. 35, 1967(1961)
5. L.A. Collins, W.D. Robb and M.A. Morrison, Phys. Rev. A21, 488 (1980)

ELASTIC DIFFERENTIAL SCATTERING AND VIBRATIONAL EXCITATION OF CH_4 AND C_2H_6 BY LOW ENERGY ELECTRONS

P.J. Curry, W.R. Newell and A.C.H. Smith

Department of Physics and Astronomy, University College London, Gower Street, London WC1E 6BT, England.

In this paper we present new measurements for the elastic scattering of monochromatic electrons from the polyatomic molecules methane and ethane for incident electron energies of 7.5, 10, 12.5, 15, 17.5 and 20eV and for scattering angles of $\theta = 30^o$ to 130°. In addition to the measurement of elastic differential cross sections vibrational excitation cross sections for the lowest normal modes in both CH_4 and C_2H_6 are obtained.

A double hemispherical electron spectrometer with an energy resolution of 40 meV and an electron beam current of 5 nA is used. A second analyser is employed as a reference detector at a fixed scattering angle to correct for any fluctuations in the product of the primary electron beam current and the gas beam number density.

Figure (1) shows the elastic differential cross section in CH_4 compared to the work of Tanaka et al [1]. The earlier work[2] done using electron beams of low monochromaticity is not included. Good agreement is obtained with the calculations of Gianturco and Thompson[3] which include the effects of exchange and polarization.

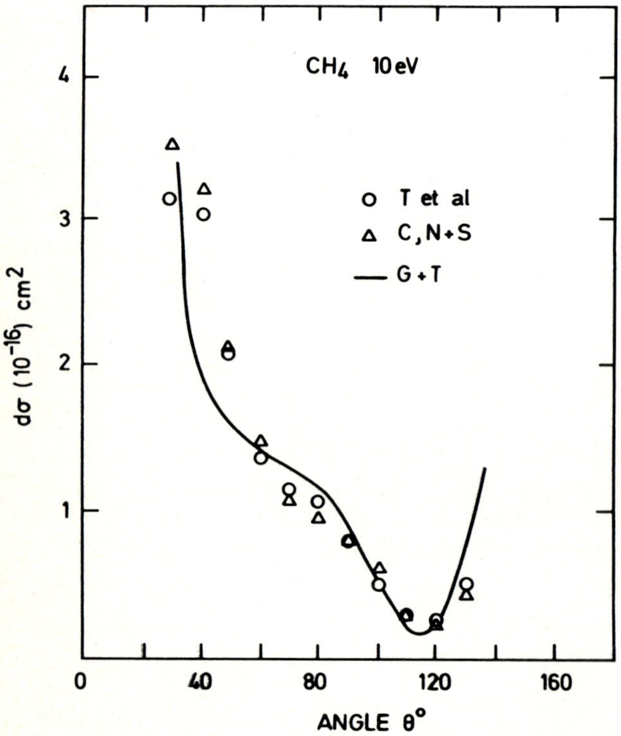

Figure 1

Figure (2) shows the low lying vibrational excitation spectra obtained at an incident electron energy of 7.5 eV for a scattering angle of 110°.

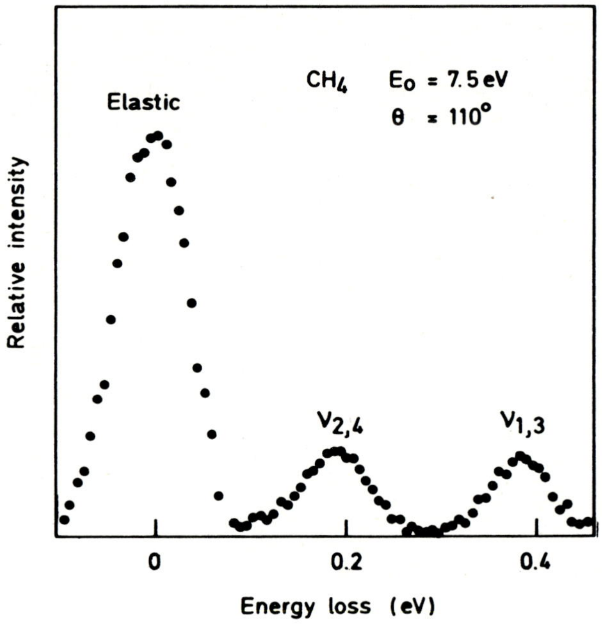

Figure 2

Diffenerential cross sections for the vibrational excitations and elastic scattering for both CH_4 and C_2H_6 will be presented and compared with current calculations.

References

1. H. Tanaka, T. Okadu, L. Bresten, T. Suzuki, T. Yamamato and M. Kubo, J. Phys. B15, 3305 (1982).

2. A.L. Hughes and H.H. McMillen, J. Phys. Rev. 44, 876 (1933).

3. F.A. Gianturco and D.G. Thompson, J. Phys. B13, 613 (1980).

ROTATIONAL EXCITATION OF METHANE BY ELECTRON IMPACT

H. Tanaka, N. Onodera, and L. Boesten

Deparetments of General Sciences and Physics, Sophia University, Chiyoda-ku, Tokyo 102, Japan

Rotational excitation of CH_4 have been investigated with a crossed-beam electron-impact apparatus. Although extremely high resolution is required to separate out rotational branches in CH_4 (rotational constant B= 0.65 meV), broadening of the vibrationally elastic energy loss spectra for incident energies of 5 to 20 eV has been observed even with an apparatus of energy-resolution ΔE=20 ∿ 24 meV. This broadening is due to the superposition of unresolved lines corresponding to many rotational transitions with different initial and final rotational states (selection rules ΔJ=0, ±1, ±2, ±3, ±4, J=0 1, 2). As a preliminary investigation, we have experimentally obtained the rotational branches with ΔJ =0, ±1, ±2, ±3, and ±4 of CH_4 by using the high J approximation proposed by Read[1].

In the high J approximation, the intensity profile for rotational transition from J_1 to J_2 at scattering angle θ is given by

$$I(E_0,\theta,T) \propto \int dE\ f(E,E_0,\theta) \sum_{\Delta J} P(E,\Delta J,T) \times \sigma(|\Delta J|,\theta)\,[(2J_2+1)/(2J_1+1)] \qquad (1)$$

where $\sigma(|\Delta J|,\theta)$ is a cross section depending only on $|\Delta J|$ and the fixed angle θ, $f(E,E_0,\theta)$ is the apparatus profile, and $P(E,\Delta J,T)$ is the population density distribution. In order to determine the branching ratios of the pure rotational transition, line-shape-analysis was used. The sum of the cross sections was normalized to the rotationally unresolved cross sections measuered by Tanaka et al[2].

Fig.1 shows a typical result of the line-shape-analysis at an incident energy of 10 eV for a scattering angle of 50°. Comparison of the results for 5 to 20 eV in the scattering angles of 30 to 130° show that the main components are branches of ΔJ=±1, ±4 at 5 eV; ΔJ=±1 ±2,±3, ±4, at 10 eV; and ΔJ=±2, ±3, ±4 at 20 eV. The ΔJ =±4 appear at all energies investigated while ΔJ=±1 show up only at energies 5 and 10 eV but not at 20 eV. The ΔJ= ±2 and ±3 do not appear in the spectrum at 5 eV. This could mean that the transition ΔJ=±1 at 5 eV, ΔJ=±1, ±2 at 10 eV, and ΔJ=±2 at 20 eV are the effect of a long range interaction. The appearence of the ΔJ=±4 througout the whole energy range is due to (a) the effect of a resonance[2] and (b) a direct influence of the weak octopole field (a short range interaction). Fig.2 gives the angular distribution of the DCS at 10 eV. We have estimated the order of the cross sections as 10^{-18} cm^2.

References

1.F. H. Read, J. Phys. B: At. Mol. Phys. 1 893 (1968)
—— Ibid.5 255 (1972)
2.H.Tanaka et al., Ibid. 15 3305 (1982)

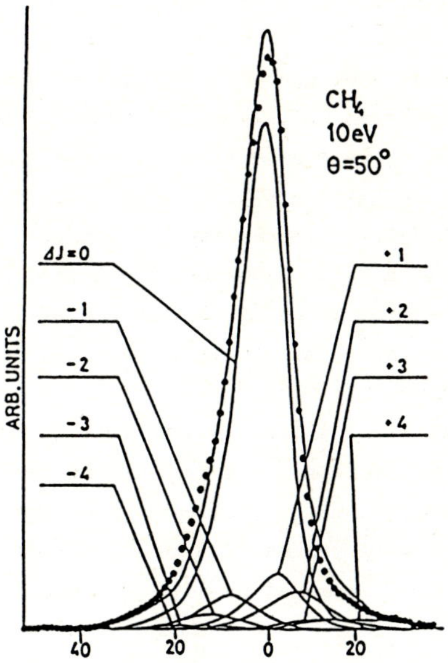

Fig. 1

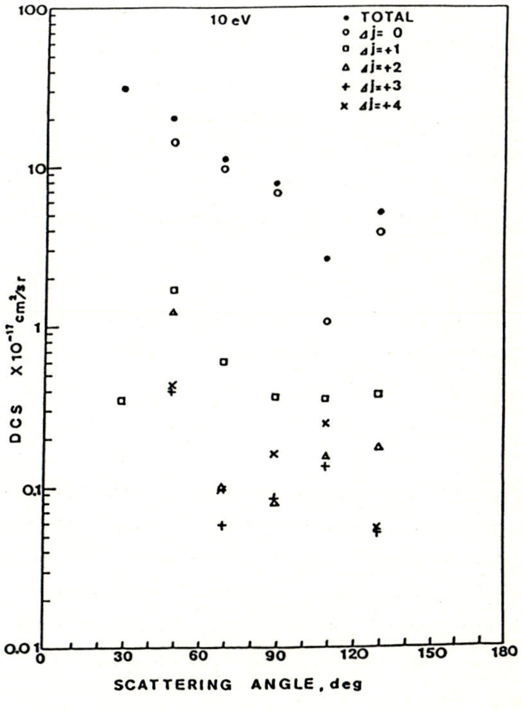

Fig. 2

LOW ENERGY ELECTRON SCATTERING BY HYDROGEN SULPHIDE (H_2S)

Ashok Jain and D. G. Thompson

The Department of Applied Mathematics and Theoretical Physics

The Queen's University of Belfast, Belfast BT7 1NN, Northern Ireland.

The fixed-nuclei (FN) approximation has been used to investigate low energy (0 - 10 eV) electron scattering by H_2S molecules. A model potential approach has been employed (cf. Jain and Thompson[1]; Gianturco and Thompson[2]). Gianturco and Thompson[2] have reported some preliminary calculations on this system and enumerate two A_1 resonance states around 3 and 6 eV respectively and also a B_1 resonance state around 5 eV. Note that there is an error in this paper: B_1 should be read as B_2. However, the theoretical designation of the B_1 and B_2 states has been redefined in this paper, and will be discussed further below.

There are several improvements over the work of Gianturco and Thompson[2]. A much larger basis set has been used for the H_2S molecular wave function. Its quality can be guaged from the new dipole moment of .43 (in a.u.) compared to an experimental value of .38 and the Gianturco and Thompson value of .679. Exchange has been represented by a Hara local exchange term rather than the orthogonalisation technique only in the previous work and polarisation has been treated in a much less ad hoc way, using the Pople-Schofield and Temkin treatments (cf. Jain and Thompson[1] for full details); the polarisability was found to be 33.5 a.u. compared to the experimental value 25.55 a.u.

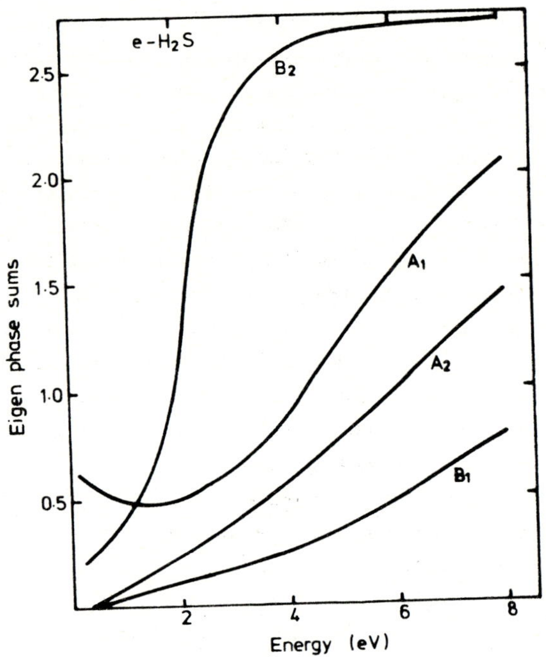

The new results for eigenphase sums are shown in the figure. There is a clear B_2 resonance at 2.2 eV with width 1.28 eV. Our notation is the opposite of that of Gianturco and Thompson[2] - the symmetry of a B_2 function is such that a rotation of π about the symmetry axis results in change of sign; a reflection in the molecular plane leaves the function unchanged but the reflection in a plane perpendicular to the molecular plane does change the sign of the function. From this figure, there is also evidence of a broad A_1 resonance feature at 6 - 10 eV. The large difference between these results and those of Gianturco and Thompson[2] is due to the strength of the potential. We have been able to reproduce the eigenphase structure of this figure with the Gianturco and Thompson[2] model but we found that we had to use a very much weaker polarisation potential than used in the previous work[2].

Experiments[3] also observed a strong shape resonance at about 2.3 eV but assign it to $^2A_1$ symmetry. Our results for the elastic, rotational and vibrational excitation processes will be published elsewhere.

References

1. A. Jain and D. G. Thompson J. Phys. B15 L631 (1982).
2. F. A. Gianturco and D. G. Thompson J. Phys. B13 613 (1980).
3. K. Rohr J. Phys. B. B11, 4109 (1978).

ELASTIC SCATTERING OF ELECTRONS AND POSITRONS BY CH_4 AT INTERMEDIATE ENERGIES

Ashok Jain

Department of Applied Mathematics and Theoretical Physics,

The Queen's University of Belfast, Belfast BT7 1NN, Northern Ireland.

The CH_4 molecule is a highly symmetrical molecule and it has no dipole or quadrupole moments. In the $e^{\pm}$ - CH_4 scattering problem, if we assume methane molecules to be spherical and retain only the spherical (ℓ = 0) term in the expansion of the total interaction potential,

$$V(\underline{r}) = V_{st}(\underline{r}) + V_p(\underline{r}) + V_{ex}(\underline{r}) \qquad (1)$$

then the usual close-coupling (CC) techniques[1,2] reduce simply to a partial wave analysis problem. At intermediate and higher energies, the conventional CC methods fail. Recently, we[3] have used such a spherical approximation (SA) to investigate the $e^{\pm}$ - CH_4 elastic scattering in the range 25 - 800 eV. In order to evaluate the static potential V_{st} (Eq.1), we employed an analytic form derived from the simple spherical wavefunctions for CH_4 molecule[4]. A semi-empirical form of the polarisation potential V_p was invoked and for the third interaction, the exchange potential V_{ex}, we used free-electron-gas-exchange (FEGE) approximation of Hara[5].

In the present new calculations, we obtain V_{st} from more accurate single-centre wavefunctions[6] and, in addition, an ab initio parameter-free polarisation potential[2] is employed for V_p. The total optical potential (Eq.1) was then treated in a partial-wave-analysis method to yield differential, integral and momentum transfer cross sections in the range 15 - 800 eV. The results are compared with the close-coupling calculations of Jain and Thompson[2] and experimental measurements.

In Fig. 1, we have shown the present differential cross sections at 30 eV along with the calculations of Jain and Thompson[2] and the experimental data of Tanaka et al[7]. Also shown in this figure are the relative cross sections of Arnot[8] (divided by a factor of 32.5). The present SA calculations are equally good with the CC calculations (these numbers are not converged, since at this energy, a large number of channels are required). The integral cross sections for electron case are, for example (in units of 10^{-16} cm^2)

	σ_i(SA)	σ_i(CC)	σ_i(Expt[7])
20 eV	17.644	18.467	14.3
30 eV	13.725	14.278	

For positron scattering: σ_i(SA) at 20 and 30 eV are respectively 2.975 and 2.642 in above units.

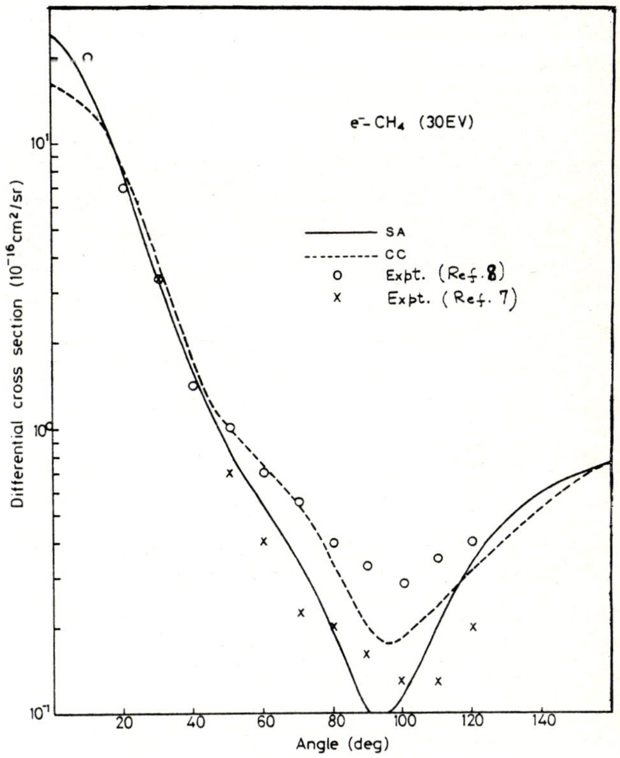

References:

1. F. A. Gianturco and D. G. Thompson J. Phys. B13, 613 (1980).
2. A. Jain and D. G. Thompson J. Phys. B15, L631 (1982).
3. A. Jain (in press, J. Chem. Phys. 1983).
4. A. F. Saturno and R. G. Parr, J.Chem. Phys. 33 22 (1960).
5. S. Hara J. Phys. Soc. Japan 22 710 (1967).
6. See, for example, F. A. Gianturco and D. G. Thompson, Chem. Phys. 14, 111 (1976).
7. H. Tanaka, T. Okada, L. Beosten, T. Suzuki, T. Yamamoto and M. Kubo, J. Phys. B15 3305 (1982).
8. F. L. Arnot, Proc. R. Soc. London Ser. A133 615 (1931).

(e^-,CsBr) SCATTERING: A BENCHMARK EXPERIMENT

B. Jaduszliwer, A. Tino, P. Weiss and B. Bederson

New York University, New York, N. Y. 10003 U.S.A.

The electron-dipole force is the longest range interaction between an electron and a neutral particle, giving rise to huge small-angle differential cross sections for highly polar molecules. Thus, scattering of electrons by such molecules has received substantial theoretical attention. On the other hand, very few single-collision cross section measurements have been reported. Slater and coworkers[1] used the molecular beam recoil technique to study electron scattering by CsF, CsCl and KI. Their analysis involved folding a parametrized model cross section with the molecular and electron momentum distributions, and adjusting the parameters to provide the best fit to the measured recoiled-molecule distribution; these "best fit" cross sections, are subject to questions relating to the choice of model and the uniqueness of the fitting procedure. Trajmar and coworkers[2] performed conventional differential measurements for electron scattering by LIF and KI. Those were relative measurements, requiring normalization to theory at some angle. In the KI case, their results were much higher than those of Slater's group.

We have studied electron scattering on CsBr in the 1 to 22.5 eV energy range. Our aim was to produce bench mark measurements, independent of any normalization or modelling of the differential cross section, against which theory could be tested.

The CsBr molecular beam is produced in a two-chamber stainless steel oven, keeping the dimer concentration to about 1%. In the interaction region the molecular beam is crossed at right angles by a ribbon-shaped electron beam with an energy resolution of about 0.4 eV (FWHM). The electron energy is corrected for contact potential differences and space charge effects.

After leaving the interaction region, the molecules travel 335 cm down an evacuated tube, and are then detected by surface ionization. In the present work, performed using the molecular recoil technique in the scattering-out mode , the molecular detector is positioned at the flat top of the beam profile, and the changes in the detector signal as the electron gun is turned on and off are measured.

The effective cross section for recoiling molecules out of the detector is given by

$$Q' = \frac{h}{I_e}\frac{S-S'}{S} = 2\pi \int \frac{dV}{V} f(V) \int \Gamma(\theta,E,V)\sigma(\theta,E)\sin\theta d\theta \quad (1)$$

where h is the height of the interaction region, I_e the electron current (e^-/sec), S' and S are the detector signals with the electron gun on and off respectively, f(V) the velocity distribution in the molecular beam and $\sigma(\theta,E)$ the electron scattering differential cross section at energy E. $\Gamma(\theta,E,V)$ is the apparatus form-factor for scattering-out experiments, known accurately in closed form. (If $\Gamma(\theta,E,V)\equiv 1$, then $Q'=Q_T$, the total scattering cross section.) Fig. 1 shows two examples; it can be seen that the effect of $\Gamma(\theta,E,V)$ is to cut-off the small-angle contribution to the total cross section, thus making Q' a more sensitive test of theory than Q_T , since suppressing the very large small-angle contribution, determined mainly by the gross features of the long range interaction, enhances shorter range contributions.

Fig. 2 shows our results for Q', as well as the corresponding values calculated solving the integral in eq. (1) with an analytical expression for the first Born approximation.

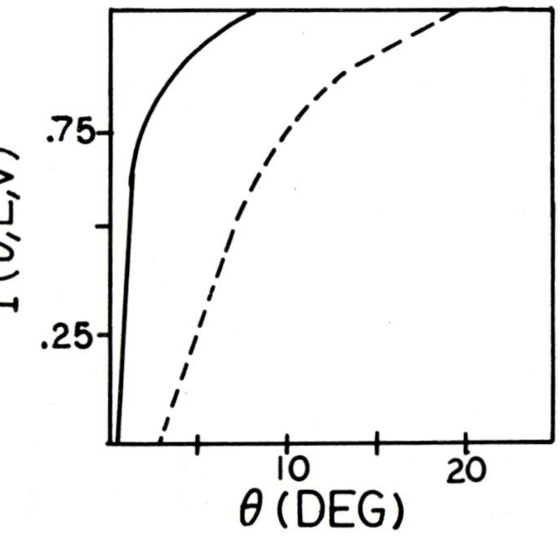

Fig. 1: $\Gamma(\theta,E,V)$ vs. θ. Full line: E=22.5 eV, V=200m/sec
Dash line: E=5 eV, V=525 m/sec.

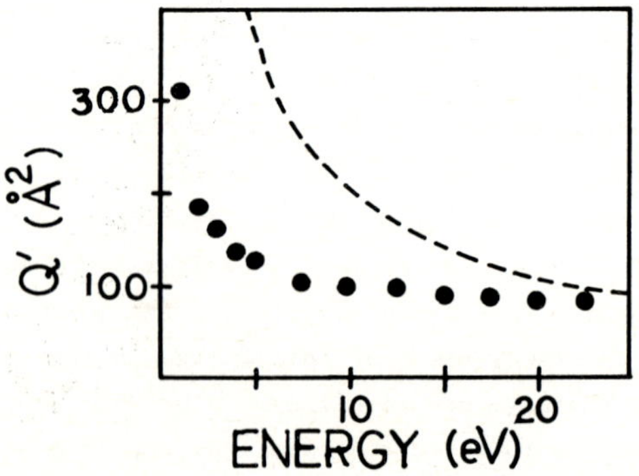

Fig. 2: Effective cross section vs. electron energy. Dots: Our data, Dash line: Born approximation.

Research supported by U.S.A. Department of Energy

References

1. R.C. Slater, M.G. Fickes, W.G. Becker and R.C. Stein, J.Chem.Phys. 60, 4697 (1974); ibid, 61, 2283 (1974); ibid, 61 2290
2. L. Vuskovic, S.K. Srivastava and S. Trajmar, J.Phys. B11, 1643 (1978).

ELECTRON - POLAR-MOLECULE COLLISIONS: SPHERICALLY SYMMETRIC APPROACH

B. Stefanov

Institute of Electronics, Bd. Lenin 72, Sofia 1184, Bulgaria

A method of reducing electron - polar molecule scattering to problems of spherical symmetry is proposed referred hereafter as RSSM.

Model potentials $U=-e\mu\cos\vartheta/r^2$ $(r>\sigma)$, $U\to\infty$ $(r\leq\sigma)$, hard sphere cut-off (HSCOP), and $U=-(e\mu\cos\vartheta/r^2)\{1-\exp[-(r/\ )^6]\}$, exponential cut-off (ECOP), are considered. Here μ is the dipole moment, ϑ is the angle between $\vec{r}$ and $\vec{\mu}$ and σ is the cut-off radius.

The electronically elastic differential cross section summed over all final rotational states is assumed arbitrarily to be $I=I_a/4+I_i/2+I_r/4$, where the indices a, i, r denote the solutions of spherically symmetric problems corresponding to constant values of $\cos\vartheta=1, 0, -1$. Similar equation holds for the momentum-transfer cross section Q^m.

The scaling of the problem in general without the restriction of using RSSM shows that for HSCOP and ECOP $Q^m\mathcal{E}/\mu$ is function only of $\sigma^2\mathcal{E}/\mu$ in a classic approach (CA) and of both $\sigma^2\mathcal{E}$ and μ in quasiclassic (QCA) and quantum (QMA) approaches ($\mathcal{E}$ is the energy of the incident electron).

In a classic approach RSSM has been checked by comparison with Monte Carlo (MC) calculations. The results shown in Table 1 are in a fairly good agreement.

Table 1. RSSM and MC calculations of Q^m in the classic limit

$\sigma^2\mathcal{E}/\mu$, Å^2.eV/D		0.02	0.04	0.13	0.50	4.50
$Q^m\mathcal{E}/\mu$, Å^2.eV/D	RSSM	11.2	11.2	11.2	11.5	17.3
	MC	12.4	12.4	11.8	11.0	16.2

Fig. 1 shows the dependence of $Q^m\mathcal{E}$ on $\lg(\sigma^2\mathcal{E})$. Our RSSM results are compared with close-coupling (CC) calculations of Collins and Norcross[1] who used ECO potential.

In the case of large μ or small $\mathcal{E}$ the scattering is dominated by high-order phases δ_l, $l\gg1$. For $\delta_l\ll1$ both QCA and QMA within the frame of RSSM lead to a formula valid for a structureless dipole:

$$Q^m\mathcal{E}=(\pi^2e\mu/4)\left[\mathrm{si}(2d)-\sin^2 d/d\right],\ d=2\pi me\mu/3\hbar^2 \quad (1)$$

shown together with other approximations on Fig. 2.

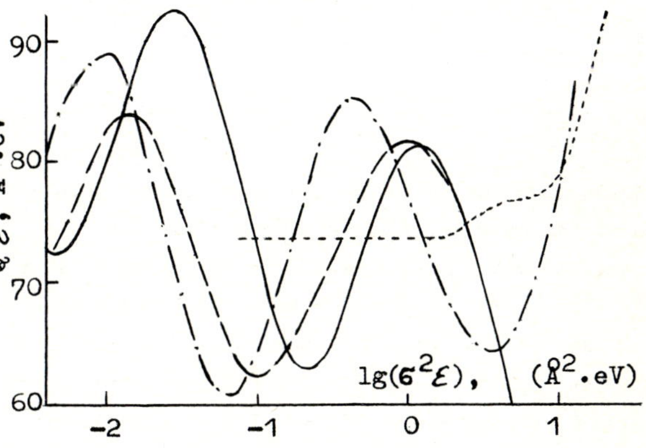

Fig. 1. Momentum-transfer cross section in different approaches. ECOP,QMA: - - - CC results[1], —— RSSM; HSCOP, RSSM: -.-.- QMA, ... CA

The results of Table 1 and Fig. 1 and 2 allow to make a conclusion that for large μ the reducing-to-spherical-symmetry method (RSSM) adequately describes electron - polar molecule momentum transfer cross section within an error less than 20 percent.

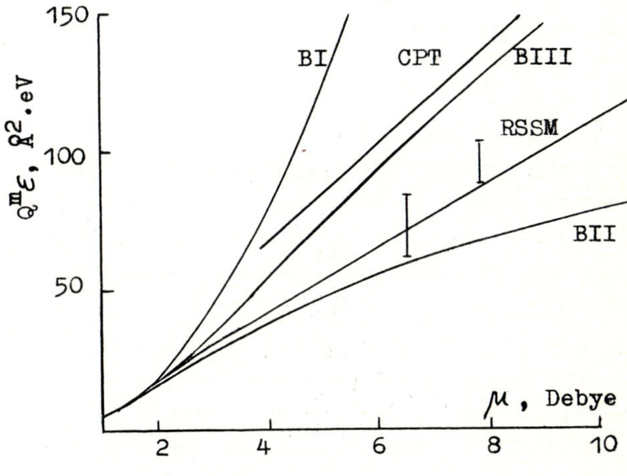

Fig. 2. Momentum-transfer cross section for a structureless dipole. BI, BII, BIII - first second and third Born approximations[1], CPT - classical perturbation theory[2], RSSM - equation (1). Bars - CC calculations[1] for LiF, CsF and KI.

References

1. L. A. Collins and D. W. Norcross, Phys. Rev. A 18, 467 (1978)
2. A. S. Dickinson, J. Phys. B 10, 967 (1977)

THE ELECTRON GAS CORRELATION MODEL: APPLICATIONS TO HCℓ AND OTHER MOLECULES*

N. T. Padial and D. W. Norcross†

Joint Institute for Laboratory Astrophysics, University of Colorado and National Bureau of Standards, Boulder, Colorado 80309 U.S.A.

The importance of polarization in electron scattering from molecules, even polar molecules, is illustrated by the results, e.g. Fig. 1, of recent close-coupling calculations for HCℓ.[1] These employed the Hara electron gas model for exchange and, when polarization was included, the familiar adjustible functional multiplier of the asymptotic form. The scattering calculations were carried out at R_e in the body-fixed molecular frame with the fixed-nuclei approximation, and cross sections for the ten lowest rotational states were obtained using the MEAN approximation.[2] The agreement of calculated cross sections with measured values[3,4] over the range 0.01-10.0 eV is good. It is also clear that polarization cannot be neglected in any interpretation of the observed[4] near-threshold vibrational excitation peak.

Extension of these calculations to vibrational excitation is in progress. Zero-point (over internuclear distance) averaging has been found to have no significant effect on cross sections when polarization is neglected. Inclusion of polarization presents a difficult problem, even were semi-empiricism to be retained. Powerful optical-potential techniques might be possible and successful, but we prefer to view HCℓ as an ideal test case for simpler models with perhaps wider applicability.

The short-range counterpart of the asymptotic polarization potential is correlation. It has been shown[5] that a nonadjustable model based on an energy-independent electron gas correlation potential $2E_c(\vec{r})$, joined onto the long-range polarization potential, yields remarkably accurate results for electron scattering by the noble gases. $E_c(\vec{r})$ is a function of the molecular charge density $\rho(r)$ alone, and has simple forms in the high- and low-density limits.[6]

$E_c(\vec{r})$ is, strictly speaking, the correlation energy density, but was interpreted,[5] by analogy with the Coulomb (and Slater exchange) potentials, as a potential. We are exploring, for several molecules, the usefulness of an alternate interpretation, more consistent with the variational principle. The potential is defined as $V_c(\vec{r}) = \delta[\rho\, E_c(\vec{r})]/\delta\rho$, leading to, for $r_s \leq 0.7$,

$$V_c(\vec{r}) = 0.0311\, \ell n r_s - 0.0584 + 0.006\, r_s \ell n r_s - 0.015\, r_s ,$$

and, for $r_s \geq 10.0$,

$$V_c(\vec{r}) = -0.584 r_s^{-1} + 1.988 r_s^{-3/2} - 2.450 r_s^{-2} - 0.733 r_s^{-5/2} ,$$

where $r_s = [3/4\pi\, \rho(\vec{r})]^{1/3}$. The high- and low-density limits are bridged by the interpolation formula

$$V_c(\vec{r}) = -0.07356 + 0.02224\, \ell n r_s , \quad 0.7 \leq r_s \leq 10.0 .$$

We have tested this approximation, in combination with the Hara exchange potential, at R_e against essentially exact[7] fixed-nuclei treatments of H_2 and N_2. For H_2 we find that results for the total scattering cross section in the range 0.1 - 10.0 eV are everywhere within 20% of the exact results. For N_2 the π_g resonance falls within a few tenths of an eV of the correct position, and has the correct shape. It is intriguing to note that for these, and for HCℓ, the correlation-polarization joining point occurs consistently at about 0.9 eV. Similar results were noted[5] for the noble gases. Tests for other polar and non-polar molecules are under way, including internuclear dependence.

We are sufficiently encouraged to have in progress a complete recalculation of vibrationally elastic and inelastic cross sections for HCℓ using this model.

1. N. T. Padial and D. W. Norcross, Phys. Rev. A 27, 141 (1983).
2. D. W. Norcross and N. T. Padial, Phys. Rev. A 25, 227 (1982).
3. E. Brüche, Ann. Phys. 82, 25 (1927); D. K. Davies, personal communication.
4. K. Rohr and F. Linder, J. Phys. B 9, 2521 (1976); and personal communication.
5. J. K. O'Connell and N. F. Lane, Phys. Rev. A (in press).
6. W. J. Carr, Jr., R. A. Coldwell-Horsfall, and A. E. Fein, Phys. Rev. 124, 747 (1961); W. J. Carr, Jr., and A. A. Maradudin, Phys. Rev. 133, A371 (1964).
7. L. A. Collins and B. I. Schneider, personal communication.

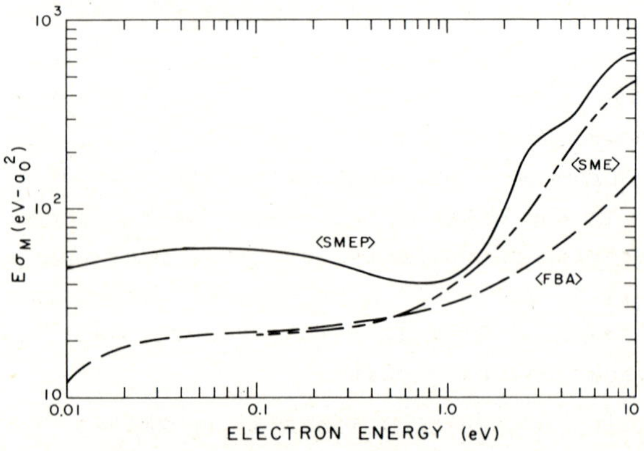

Fig. 1. Total momentum transfer cross section, without (SME) and with (SMEP) polarization, and the first Born approximation (FBA).

*Supported by Dept. of Energy (Basic Energy Sciences).
†Staff Member, Quantum Physics Division, NBS.

EFFECTIVE HAMILTONIAN APPROACH TO ELECTRON-MOLECULE COLLISIONS

N. Chandra

Department of Physics, Institute of Advanced Studies, Meerut University, Meerut-250005, India

We have been trying to perform ab-initio studies of electron scattering from polar molecules[1-3] using the frame-transformation theory[4] (FT). While in the inner-region one has to solve a many-particle problem, the outer-region potential is in a single asymptotic form of the multipole expansion. But the r^{-2} tail of the dipole interaction and narrow spacing of the rotational energy levels give rise to serious convergence problems in the solution of the scattering problem[3] in the outer-region even for as weakly a polar system as CO. The numberical integration of a large set of coupled radial equations over hundreds or thousands of Bohr radii imposes heavy demands on computational efforts and obscures the physics.[3]

In order to over come these difficulties a non-perturbative approximation, which enables one to write down simple analytic solutions of the scattering equations in the outer-region, was proposed.[5-7] All the results — except those of thermal-energy cross-sections for $\Delta j=0$ and ± 2 transitions — obtained from this method were in excellent agreement[6] with those calculated earlier accurately.[3]

To rectify those short-comings[6] of this approximation which give rise to above-mentioned discrepancies in the cross-sections and with a view to be able to study electron impact vibrational excitation of polar targets, the effective Hamiltonian method — introduced by Rabitz[8] in molecule-molecule collisions — has been adapted to electron scattering from molecules. This adaplation, which, unlike the previous approximation[6], neither introduces any change in the molecular energy levels nor requires neglect of any multipole terms in the electron-molecule interaction potential, is found to be easiest in the frame-work of FT. Although, one now needs to perform numerical integration of the coupled equations is considerably smaller than that required in the usual close-coupling theories[9,10] of electron -molecule collisions. Results would be presented at the Conference.

References

1. N. Chandra and F.A. Gianturco, Chem. Phys. Lett. 24, 326 (1974)
2. N. Chandra, Phys. Rev. A12, 2342 (1975)
3. N. Chandra, Phys. Rev. A16, 80 (1977)
4. E.S. Chang and U.Fano, Phys. Rev. A6, 173 (1972)
5. N. Chandra, Proc.VI Int.Conf. on Atomic Physics, Riga (Riga: Phys. Int. Latvian Acad. Sci; 1978) Abstracts pp 230-1
6. N. Chandra, J. Phys. B15, 4465 (1982)
7. N. Chandra, J. Phys. B(submitted)
8. H. Rabitz, J. Chem. Phys. 57, 1718 (1972)
9. A.M. Arthurs and A. Dalgarno, Proc. Roy. Soc. A256, 540 (1960)
10. N. Chandra and A. Temkin, Phys. Rev. A13, 188 (1976).

SMALL-ANGLE ELASTIC SCATTERING OF ELECTRONS FROM NH_3 MOLECULE AT 300, 400 and 500 eV

Ashok Jain

Department of Applied Mathematics and Theoretical Physics,
The Queen's University of Belfast, Belfast BT7 1NN, Northern Ireland.

The scattering of electrons by ammonia molecules has recently received renewed interest,both experimental[1] and theoretical[2,3]. Very recently, we[4] have investigated low-energy e - NH_3 scattering for the elastic and the rotational excitation processes. Excluding the low-energy measurements, the only earlier experimental study on e - NH_3 is due to Harshbarger et al[5], who measured absolute differential cross sections (dcs) at 300, 400 and 500 eV in a very small angular range ($\theta \leqslant 10°$). At very high energies, Bennani et al[1] have measured e - NH_3 dcs as a function of momentum transfer; in the large momentum transfer region, the first Born calculations compare well with these measurements.

We demonstrate here that if one includes long-range effects in the Born calculations, a direct comparison with the measurements of Harshbarger et al is possible even in this small angular region. For this purpose, we first assume that the ammonia molecules are spherical, so that the interaction between the e - NH_3 system can be written as[7]

$$V_i = -\frac{7}{r} - \frac{3}{r_>} + \sum_{i=1}^{10} \frac{1}{|\underline{r} - \underline{r}_i|} \qquad (1)$$

Here, $r_>$ is the greater of r and R(the internuclear separation) and $\underline{r}_i$ are the position vectors of bound electrons. Employing a single determinant wavefunction of the one-centre form[6], the static potential can be derived analytically (see, for example, Jain[7]).

$$V_{st} = -\frac{3}{r_>} + \frac{3}{r} - \sum_{i=1}^{4} \sum_{j=1}^{4} A_{ij}\, r^{j-2} \exp(-\alpha_i r) \qquad (2)$$

where A_{ij} and α_i are some constants[7]. Similarly, we use the following polarization potential,

$$V_p = \frac{-\alpha_o r^2}{2(r^2 + r_c^2)^3} \qquad (3)$$

where α_o is the dipole polarisability (15.0 a.u.) and r_c is an energy-dependent cut-off parameter determined from 0.375 k/ Δ. The Δ is the mean excitation energy obtained in the closure approximation[7](for NH_3, we obtain Δ= .482 a.u.). Since the NH_3 molecule has a permanent dipole moment D (.5744 a.u.), the corresponding asymptotic form of the dipole potential,

$$V_D = -\frac{D}{r^2} P_1(\cos\theta) \qquad (4)$$

is used to calculate the scattering amplitude f_D due to V_D. Thus obtaining all the three Born scattering amplitudes, f_{st}(due to V_{st}), f_p(due to V_p) and f_D, we calculate the dcs from the relation,

$$\frac{d\sigma}{d\Omega} = |f_{st} + f_p + f_D|^2 \qquad (5)$$

In table 1, we have shown our Born dcs at 300 eV only in the range 2° - 10° with and without long-range effects. It is clear from this figure that the inclusion of the dipole and the polarisation terms makes the results in good agreement with measurement. However, the dipole contributions are more important, which is quite obvious for a polar molecule. Results at 400 and 500 eV (not shown) are also in very good agreement.

Table 1. The dcs for e - NH_3 elastic scattering at 300 eV.

Angle (deg)	First Born Calculations			Expt.[5]
	ST + P	ST + D	ST + P + D	
2	20.99	57.25	74.45	66.42 ± 2.52
4	14.88	29.62	34.58	41.62 ± .30
6	11.71	21.25	22.76	26.82 ± 1.12
8	9.63	16.57	16.81	18.46 ± .51
10	8.00	13.24	13.24	12.49 ± .37

References

1. A.L. Bennani, A. Duguet and M.R. Wellenstein, J. Phys. B12, 461 (1979).
2. C. Tavard, Cah. Phys. 17, 165 (1963).
3. A. Szabo and N.S. Ostlund, J. Chem. Phys. 60, 946 (1974).
4. A. Jain and D.G. Thompson (in press, J. Phys. B (1983)).
5. W.R. Harshbarger, A. Skerbele and E.N. Lassettre J. Chem. Phys. 54, 3784 (1971).
6. D.M. Bishop, J.R. Hoyland, R.G. Parr, Mol. Phys. 6, 467 (1963).
7. A. Jain (in press, J. Chem. Phys. (1983)).

THE ELASTIC ELECTRON SCATTERING CROSS SECTION OF SF_6 BY AB INITIO SCF CALCULATION

Péter Pulay, Richard Mawhorter, D.A. Kohl, and M. Fink

Dept. of Physics & Dept. of Chemistry, The University of Texas at Austin, Austin, Texas 78712 USA

The ability to calculate electron-molecule scattering cross sections in the first Born approximation is important for several reasons, not the least of which are inferences that can be made from these about the charge density distribution of the molecule. Improvements in high energy (30-60 keV) gas phase electron diffraction have made this information accessible in addition to the vibrationally-averaged structural parameters. While the structure is extracted by a straightforward Fourier analysis of the measured cross section, the analagous Fourier transform for the charge density is not unique. Some reasonable physical assumptions allow the construction of experimental charge density maps for diatomic molecules, but this is already beyond feasibility for triatomics, not to mention larger polyatomics such as SF_6. Thus the comparison of theory and experiment at least for polyatomics must be made through the cross sections themselves, requiring the computation of these from various ab initio wave functions. In this work, basis set sensitivity was first investigated on N_2, and a triple-zeta plus polarization basis set was found to be sufficiently close to the Hartree-Fock limit. Correlation effects were not considered, as for the elastic channel they should be quite small.

Another use of these theoretical cross sections is to check the efficacy of the independent atom model or IAM commonly used in analyzing the structural component of the electron diffraction data. The IAM superimposes spherical, atomic electron densities to describe the molecular electron density. This of course neglects the reorganization of the charge due to bonding, but gives excellent agreement with experiment except at small angles. Heuristically structural parameters are derived from the experimental cross sections by adjusting the geometry and mean amplitudes of vibration of the corresponding IAM molecule until the easily calculated IAM cross section matches the experimental one. The resulting difference is due to the bonding charge density, and is customarily discussed in terms of a quantity related to the experimental intensity, namely

$$\Delta I(s) = I(s) - I_{IAM}(s), \text{ where } I = s^4 \cdot (d\sigma/d\Omega) \qquad (1)$$

and the angular variable s is the momentum transfer. As the precision of electron diffraction structure determination approaches the .0001Å niveau, questions arise concerning the effect of this difference on the structure determination itself. In fact, Bartell and coworkers[1] suggest this as a probable cause for the systematic residuals they observed for SF_6 even after correcting for multiple scattering.

This coupled with a high level of symmetry and the general interest in SF_6 made it a natural choice. Symmetry is important because it is necessary to do a rotational average over the random orientations of the gas molecules. Exploiting this octahedral symmetry, the equivalent of 4560 orientations over the scattering sphere was obtained by calculating the SF_6 molecular form factors for 95 orientations over a sector of the solid angle. The corresponding IAM form factors were then subtracted from each of these to smooth out the highly oscillatory behavior. The spherical average itself was carried out numerically, first fitting this surface with a high order spherical harmonic expansion, followed by analytic integration. The result is the spherically-averaged elastic electron cross section. Significant differences were found between this and the IAM cross section, showing an especially strong sensitivity to the bonding charge density. As the inelastic contribution is expected to be a smooth function of s, the results also qualitatively confirm Bartell's hypothesis (see Fig. 1), and spur us on to measure these effects directly using the new elastic electron diffraction apparatus in Austin.

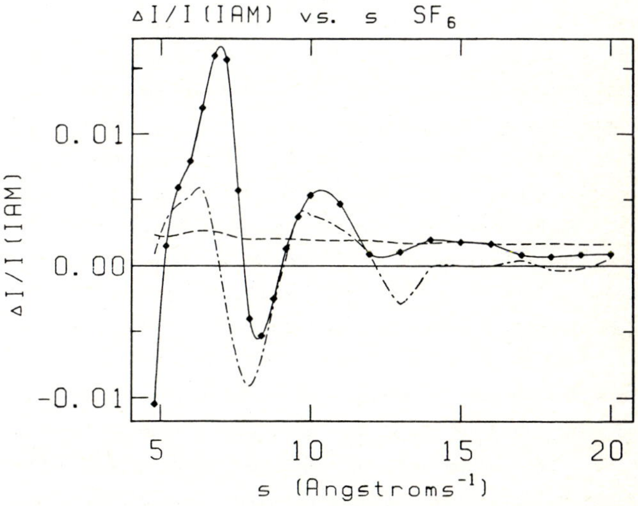

Fig. 1: Relative deviation from the IAM model, showing the present calculation (solid line), Bartell's residuals (dot-dashed line), and the effect of a -0.01 change in the atomic form factors for the fluorine atom (dashed line).

References

1. L.S. Bartell, M.A. Kacner, and S.R. Goates, J. Chem. Phys. 75, 2730 (1981)

Elastic High Energy Electron Scattering As a Tool to Study Bonding and Correlation in Atoms and Molecules

M. Fink, R. Mawhorter and J. J. McClelland

The University of Texas, Physics Department, Austin, Texas, 78712

Precise differential cross sections for electron scattering from atoms or molecules have proven to be sensitive to atomic correlation and bonding.[1] The interpretation of the data is possible in the first BORN approximation when the atomic number is low and the incident electron energy is high. The validity can be confirmed by the momentum scaling of data recorded at various energies. The basic concepts of the theory (wavefunctions, sudden approximation and scattering) and the limits of the apparatus have been investigated[2] with 40 KeV electrons incident on He and H_2. The scaling properties have been confirmed on N_2 for 30, 40 and 50 KeV electron energy. Extensive data have been collected on several small molecules and the cross sections have been compared first with the predictions of the independent atom model and recently with calculated cross sections derived from near Hartee Fock ab initio wavefunctions. Additional experimental and theoretical studies have been carried out to gain more insight into the importance of vibrational averaging, a feature unavoidable in our experiment. Theory and experiment show that a small averaging effect is observable only for H_2. The more rigid molecules such as N_2, O_2 or CF_4 show no effect at room temperature, but the structure parameters do shift by a fraction of a percent of the equilibrium values as the temperature is increased.[3] High precision measurements of these values give additional information on the molecular potential function, in particular on the anharmonicity.

Most of the data recorded until now has been total (elastic plus inelastic), but recently a new scattering unit has been built, equipped with a Moellenstedt analyzer to record the elastic scattering only. New data for SF_6 will be presented to demonstrate the importance of the effect of bonding on the elastic cross sections. According to our Hartee Fock theory and previous measurements, the influence of the chemical bond is so strong that the basic structure parameters are affected and comparisons between different methods become impossible. In addition, extensive calculations for several molecules show that the basis set selection for the wavefunction calculations is of importance when elastic cross sections are used as an evaluation tool. These results will be highlighted with results on C_2H_4.

As has been pointed out by Bonham,[4] the elastic and inelastic cross sections give direct access to various potential energies, such as the electron-nuclear potential V_{ne} or the electron-electron value V_{ee}. With the availability of these potentials it is possible to study the contribution of each term to the binding energy. Since this energy is the difference between the potential energies, great sensitivity to the leading contributor to the bond results. The cross section data presented, when compared with the best available theory, give a better insight into the detailed components of the calculations and a better understanding of the chemical bonding in the potential energy representations.

Finally the elastic cross sections have been analyzed with respect to the charge density difference between the molecule and its constituent atoms. These charge density difference maps lead to a set of moments which are compared with existing values. The same data lead directly to electrostatic field distributions, which link our results with studies in x-ray diffraction and the semi-empirical theory by Politzer.[5]

References

1. R.A. Bonham and M. Fink, "High Energy Electron Diffraction", Van Nostrand-Reinhold, New York (1974)
2. S.N. Ketkar, M. Fink and R.A. Bonham, Phys. Rev. A 27, 806 (1983)
3. M.H. Kelley and M. Fink, J. Chem. Phys. 77, 1813 (1982)
4. R.R. Goruganthu and R.A. Bonham, Phys. Rev. A 26, 1 (1982)
5. P. Politzer, J. Chem. Phys. 72, 3027 (1980).

CALCULATED CROSS SECTIONS FOR RESONANT ELECTRON SCATTERING BY METASTABLE NITROGEN

Iztok Čadež

Institute of Physics, P. O. BOX 57, 11001 Belgrade, Yugoslavia

In this contribution the calculated cross sections for the electron scattering by nitrogen in the metastable $A^3\Sigma_u^+$ state through the $A^2\Pi_u$ resonance are presented. The cross sections for the following processes were calculated:

$$e + N_2(A_{v_i}) \rightarrow N_2^-(A) \begin{cases} e + N_2(A_{v_f}) \\ e + N_2(X_{v_f'}) \end{cases}$$

v_i being 0,3 and 6, v_f from 0 to 15, and v_f' from 0 to 30. In the parenthesis the lebel of the corresponding electronic state (A or X) is indicated together with the vibrational number (v_i, v_f or v_f') for neutral molecule. The $A^2\Pi_u$ resonant state is formed by trapping of the incident electron to the excited nitrogen molecule in the $A^3\Sigma_u^+$ state. This resonance is thus coupled to the ground and excited parent state through two - and one - electron transition, respectively.

In present calculation the boomerang model[1] was applied assuming similar approximations as in Huetz et al.[2]. In this modified boomerang model one assumes that the resonant state is well described by tha local complex potential and that the nuclear wave function of this state is determined by electronic transition to the parent state only. The electronic transition amplitude which couples resonant and electronic ground state is taken to be constant as a function of the internuclear separation although experimental evidence and some theoretical considerations indicate that it should increase with internuclear separation[3,4]. This and some other shortcomings of the applied model will be descussed at the conference.

The obtained cross sections are large for both electronic channels. The mean value of the autodetachment width is smaller in the case of the electron scattering by the metastable N_2 as compared to the scattering through the same resonance but starting from the ground state because initial internuclear separation is larger in the former case. Due to this the relative probability for quenching decay channel is increased. In Fig. 1 the total quenching CS for A_3 level is shown. The CS for individual transitions from A_3 at 0.86 eV are presented in Fig. 2.

These calculations were performed in order to estimate possible atmospheric implications[4].

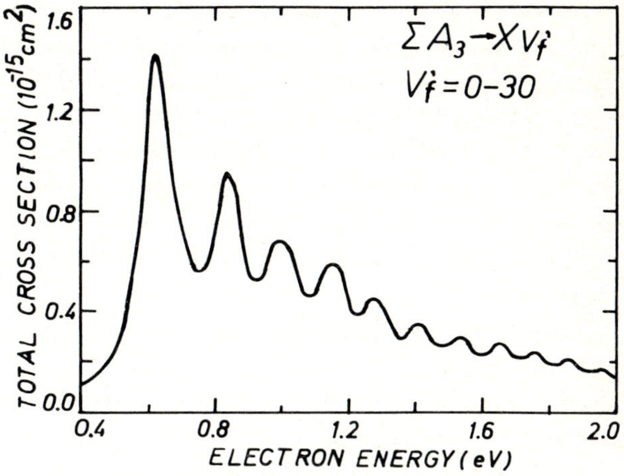

Fig. 1

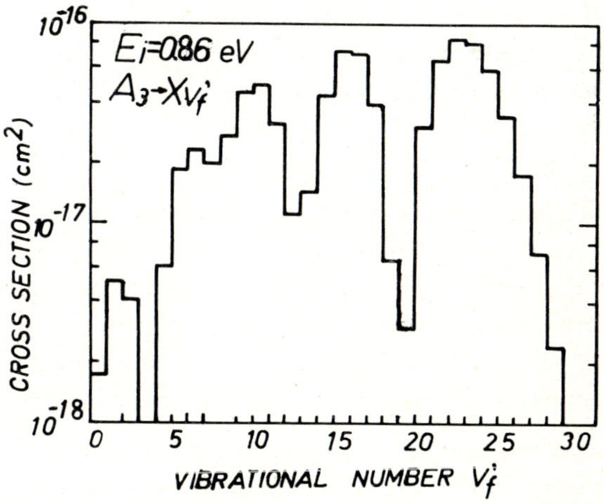

Fig. 2

Acknowledgment

The collaboration with Prof.A.W.Castleman Jr. and his laboratory and Visiting Fellows Program of CIRES (University of Colorado,Boulder) are gratefully acknowledged.

References

1. Dubé L. and Herzenberg A., Phys.Rev.A,20, 194 (1979)
2. Huetz A., Čadež I., Gresteau F., Hall R.L., Vichon D. and Mazeau J., Phys.Rev.A., 21, 622 (1980)
3. Huetz A., These de Doctorat d'Etat, Univ. P. et M. Curie, Paris (1980)
4. Čadež I., Planet.Space Sci., in press

ROTATIONAL BRANCH STRUCTURE IN THE ELECTRONIC EXCITATION OF O_2 BY ELECTRON IMPACT

Edward S. Chang

Department of Physics & Astronomy, University of Massachusetts, Amherst, MA 01003 U.S.A.

In a previous study of the angular distributions,[1] the processes

(a) $e + X\ ^3\Sigma g^-(\nu=0) \rightarrow e + a\ ^1\Delta g(\nu=0)$ and

(b) $e + X\ ^3\Sigma g^-(\nu=0) \rightarrow e + b\ ^1\Sigma g^+(\nu=0)$

at about 5eV had been shown to proceed via the $^2\Pi g\ O_2^-$ complex. Without any parameter, the theoretical expressions with only the dπ partial wave fit the experimental data[2] to an accuracy of 10%. However, the $^2\Pi g$ complex assumed appeared to contradict calculations showing a $^2\Pi u$ complex which was responsible for dissociative attachment. With the capability of resolving the rotational branch structure by the new technique of Jung *et al*,[3] it may be possible to settle this issue unambiguously. Accordingly, theoretical expressions for the rotational branch resolved angular distributions are derived.

The following features are noteworthy:

<u>for the process (a)</u>

(1) The partial sum over ΔK = odd and the partial sum over ΔK = even are equal.

(2) Undulation is the most marked for $\Delta K = 0$ and gradually disappears as $|\Delta K|$ increases.

(3) The cross section for $\Delta K = -3$ exceeds the one for $\Delta K = 3$, unlike the situation[3] in N_2 and CO.

(4) The angular distributions for $\Delta K = -2$ and $\Delta K = 2$ are not even proportional, again in contrast to N_2 and CO.

<u>for the process (b)</u>

(1) All cross sections with ΔK = even vanish.

(2) All cross sections with ΔK = odd are parity unfavored, i.e., vanishing at 0 and 180°.

(3) Undulation is larger for $|\Delta K| = 1$ than $|\Delta K| = 3$.

(4) $\Delta K = +3$ has the largest cross section around 90°, exceeding the one for $\Delta K = 1$ by a factor of two.

Absolute cross sections may be obtained by normalizing theory to an experimental cross section, e.g., Hall and Trajmar.[2] All differential cross sections described above in the accessible angular range of 30 to 130° are of the order of $10^{-17}\ cm^2/sr$.

<u>References</u>

1. E.S. Chang, J. Phys. B <u>10</u>, L677 (1977).
2. R.I. Hall and S. Trajmar, J. Phys. B <u>8</u>, L293 (1975).
3. K. Jung, Th. Antoni, R. Muller, K.H. Kochem, and H. Ehrhardt, J. Phys. B <u>15</u>, 3535 (1982).

RESONANCES IN LOW-ENERGY $e-H_2^+$ COLLISIONS

B. I. Schneider and L. A. Collins

Los Alamos National Laboratory, Los Alamos, NM 87545 USA

We have applied two techniques,[1,2] based on a linear algebraic (LA) approach,[3] to the study of low-energy collisions of electrons with H_2^+. The first technique[1] is developed along traditional close-coupling (CC) lines in which the total system wavefunction is expanded in terms of a complete set of target electronic states. Removing the dependence on target and angular coordinates, we derive a set of radial, coupled integral equations, whose solution is the scattering orbital. These radial equations are converted to a set of LA equations by imposing a discrete quadrature on the integrals. This system of equations can then be solved by standard linear systems packages. Coupling and correlation effects are introduced by including closed electronic states in the expansion.

The second technique[2] involves the introduction of an optical potential to the elastic static-exchange scattering equations in order to treat correlation and electronic channel-coupling effects. We partition the function space into two parts,[4] P and Q, with P-space containing a two-electron function representing the electron incident on the ground state of H_2^+ and with Q-space containing all functions that describe virtual excitations of the system. The effect of the Q-space part on the scattering electron is contained in an optical potential term. We approximate the optical potential by placing it on a square integrable basis and evaluate it with standard configuration interaction bound state codes.

In Table 1, we present our results for the widths and positions of the two lowest $^1\Sigma_g$ resonances at equilibrium ($2.a_o$) in 2-state ($1\sigma_g, 1\sigma_u$) and 4-state ($1\pi_u^{\pm}$) CC and in the optical potential (OP) approach and compare them with the results of other methods such as projection operator (PO)[5,6] and Kohn variational (KV).[7] The small differences between the CC and the OP and PO methods probably arise from a poorer representation of correlation effects in the CC case due to the slow convergence of the target-state expansion.

References

1 L. A. Collins and B. I. Schneider, Phys. Rev. A 27, 101 (1983).

2 B. I. Schneider and L. A. Collins, J. Phys. B 15, L334 (1982); ibid Phys. Rev. A (in press).

3 L. A. Collins and B. I. Schneider, Phys. Rev. A 24, 2387 (1981); ibid 24, 1264 (1981).

4 H. Feshbach, Ann. Phys. (NY) 5, 357 (1958).

5 A. Hazi, C. Derkits, and J. N. Bardsley, Phys. Rev. A (in press).

6 H. Sato and S. Hara (private communication).

7 H. Takagi and H. Nakamura, J. Phys. B 13, 2619 (1980); ibid Phys. Rev. A 27, (in press).

Table 1. Widths and positions for lowest two $^1\Sigma_g$ resonances for $e-H_2^+$ scattering ($R_e = 2.0a_o$).

Method	E_r (eV)	Γ (eV)
OP	5.37	1.40
4CC	5.87	1.55
2CC	6.04	1.38
PO[5]	5.57	1.32
PO[6]	5.53	1.35
KV[7]	5.47	1.60
OP	9.92	0.151
4CC	9.95	0.160
KV	9.96	

THRESHOLD ELECTRON SPECTRUM OF I_2

J. Jureta, V. Bočvarski, S. Cvejanovic and M. Kurepa

Institute of Physics, P.O.Box 57, 11001 Beograd, Yugoslavia

By use of an electron impact threshold spectrometer[1] the threshold spectrum of I_2 has been investigated. The lowest part of the spectrum is shown in Fig. 1.

Of ten valence states dissociating to two $I(^2P_{3/2})$ atoms the following six: $^3\Pi_{2u}$, $^3\Pi_{1u}$, $^3\Pi_{0^-u}$, $^1\Pi_{1u}$, $^3\Pi_{2g}$ and $^3\Pi_{1g}$ are located in the investigated energy range, as well as the state $^3\Pi_{0^+u}$ dissociating to $I(^2P_{3/2})$ and $I(^2P_{1/2})$ atoms[2].

There are many optical absorption measurements in the visible-infrared spectrum[3-7]. For the present analysis of threshold electron spectrum the results of Tellinghuisen[8] have been used, since the absorption spectrum was successfully interpreted by three well separated structures corresponding to $^3\Pi_{1u}$, $^3\Pi_{0^+u}$ and $^1\Pi_{1u}$ states, respectively. We used the $^3\Pi_{0^+u}$ and $^1\Pi_{1u}$ absorption curves to approximate the maximum in the threshold spectrum at 2.5 eV. The same was done with the $^3\Pi_{1u}$ absorption curves at 1.850 eV. The remaining of our spectrum could be interpreted by gaussian-like curves corresponding to excitation of other valence states.

The vertical excitation energy of the $^3\Pi_{2u}$ state should be 1.706 eV according to Mulliken[2], while the corrected prediction of Tellinghuisen[8] is 1.692. This state can be excited to bound vibrational levels at energies lower than 1.543 eV, the dissociation limit. Above it the excitation should lead to dissociation with a cuntinuum-like spectrum. Although the peak of this continuum could not be determined with great precission, it seems to us that it is at an energy of 1.62 eV.

Mulliken[2] predicted that the vertical excitation energy of the $^3\Pi_{0^-u}$ state is 2.141 eV, the value of Tellinghuisen[8] being 2.128 eV. The interpretation of our experimental curve gives a continuum-like structure with the maximum at 2.10 eV, in better agreement with the predicted value of Tellinghuisen[8].

To our knowlege, these are the first experimental evidences for excitation of $^3\Pi_{2u}$ and $^3\Pi_{0^-u}$ states in I_2.

The threshold signal above incident energies of 2.5 eV could be interpreted by at least one continuum-like spectrum. In this energy range states $^3\Pi_{2g}$ and $^3\Pi_{1g}$ should be located, their predicted vertical excitation energies being 3.2 eV and 3.4 eV, respectively[2]. Our signal indicates that the maximum of the first of them should be located much lower than expected, around 2.85 eV, while for the second nothing definite could be said.

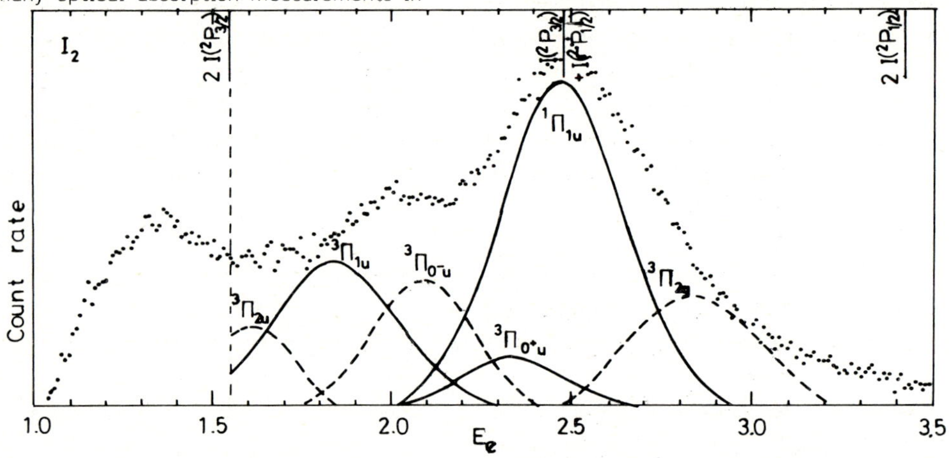

Proposed interpretation of the rest of the threshold electron spectrum in I_2 will be presented at the Conference.

REFERENCES

1. J.Jureta, S. Cvejanović, J.N.H. Brunt and F.H. Read; J. Phys. B., **11** L 347 (1978)
2. R.S. Mulliken; J. Chem. Phys., **55** 288 (1971)
3. A. Chutjian and T.C. James; J. Chem. Phys. **51** 1242 (1969)
4. A. Chutjian; J. Chem. Phys., **51** 5414 (1969)
5. M. Kroll; J. Mol. Spectrosc., **36** 44 (1970)
6. R.J. Oldman, R.K. Sander and K.R. Wilson; J. Chem. Pjys., **54** 4127 (1971)
7. R.J. Le Roy; J. Mol. Spectrosc., **39** 175 (1971)
8. J. Tellinghuisen; J. Chem. Phys., **58** 2821 (1973)

This work was supported by RZN Srbije, Yugoslavia.

ELECTRON IMPACT EXCITATION OF THE B AND C STATES OF DEUTERIUM

K. Becker, W. van Wijngaarden and J.W. McConkey

Department of Physics, University of Windsor, Windsor, Ontario, Canada N9B 3P4.

The Lyman [B $^1\Sigma_u^+ \rightarrow$ X $^1\Sigma_g^+$] and Werner [C $^1\Pi_u \rightarrow$ $^1\Sigma_g^+$] band emissions of D_2 in the 1100-1800Å region following electron impact excitation have been studied using a crossed electron-gas beam set-up in conjunction with a 0.5 m Seya-Namioka monochromator. The VUV photons were detected by a caesium iodide photocathode deposited on the inner surface of a channeltron cone. A MgF_2 input window limits the detectable impact radiation to wavelengths above 1140Å. The emission cross-sections of the CO fourth positive band system and the NI and Ly α line radiation produced by electron impact on N_2 and H_2 were used to calibrate the entire optical system with an accuracy of 15%.

Fig. 1 shows a wavelength scan at 100 eV incident electron energy. No correction for variation in detector sensitivity has been made. The instrumental resolution of 3.5Å (fwhm) largely resolves the vibrational but not the rotational structure of the bands. The emissions above 1350Å are due to the B → X transition, whereas the Werner bands (C → X) dominate the spectrum below 1300Å. The maximum in the Ly α emission cross-section (corrected for molecular contributions) was found to be 1.12 x 10^{-17} cm^2 at 75 eV. At 100 eV an intensity ratio of 1.20 ± 0.04 was measured for Ly α from H_2 to Ly α from D_2. Assuming that cascading into the D_2 C-state is negligible and that unmeasured bands can be taken into account using published transition probabilities[1] we calculate a total C-state excitation cross section of (3.55 ± 0.75) x 10^{-17} cm^2 at 100 eV. For the Lyman (B-X) bands of H_2 cascading was demonstrated to play an important role.[2] In the case of the D_2 B-state we estimate a total emission cross-section including cascade of (4.0 ± 1.2) x 10^{-17} cm^2 at 100 eV. The larger error margin (±30%) reflects additional uncertainties such as perturbations in the vibrational population distribution due to cascading and the emission continuum associated with B → X transition.[3]

Relative excitation functions from 20-500 eV for the strongest vibrational bands indicate a B-state cross-section that peaks at 45 eV and gradually falls off with increasing impact energy to half its maximum value around 250 eV. The energy dependence of the C-state cross-section is very similar with a maximum around 50 eV and a half value slightly above 300 eV.

We have also measured the Lyman and Werner band emissions produced by electron impact on H_2, and at 100 eV we obtain emission cross-sections which agree reasonably well with cross-section data reported by other groups (see e.g. Ajello et al.[2] and references given therein).

Financial assistance from the Natural Sciences and Engineering Research Council of Canada is gratefully acknowledged. One of us (K.B.) wishes to acknowledge financial aid from the Deutsche Forschungsgemeinschaft (DFG).

References

1. A.C. Allison and A. Dalgarno, At. Data 1, 289 (1970).
2. J.M. Ajello, S.K. Srivastava and Y.L. Yung, Phys. Phys. Rev. A 25, 2485 (1982).
3. A. Dalgarno, G. Herzberg and T.L. Stephens, Astrophys. J. 162, L49 (1970).

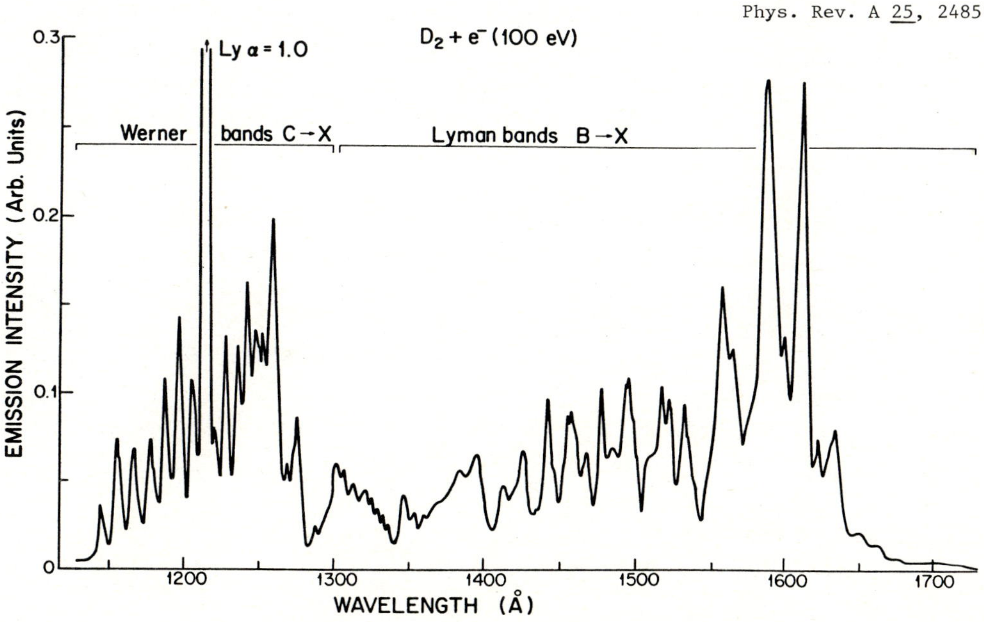

Fig. 1: Emission spectrum in the 1100-1700Å range produced by 100 eV electrons on D_2. No correction for variation in detector sensitivity has been made. The instrumental resolution was 3.5A (fwhm). The three important electronic transitions - Lyman (B → X) and Werner (C → X) bands and the atomic 2p → 1s Ly α - are indicated.

ELECTRON-PHOTON COINCIDENCE EXPERIMENTS WITH DIATOMIC MOLECULES

K. Becker, H.W. Dassen and J.W. McConkey

Department of Physics, University of Windsor, Windsor, Ontario, Canada N9B 3P4.

The electron-polarized-photon coincidence technique has been used to study the excitation of N_2, H_2 and D_2 in the VUV spectral range. A monoenergetic electron beam is crossed with a gas beam at right angles and electrons scattered inelastically in a direction θ_e with respect to the axis of the incident electron beam are detected in coincidence with the VUV photons emitted perpendicular to the scattering plane. Polarization analysis is made either by a double-reflection polarizer or by a recently developed single-reflection device which has an 80% polarization efficiency at Lyman-α.

Let $N(\beta)$ denote the number of coincidences for a polarizer orientation β with respect to the electron beam quantization axis, then the polarization correlation $P(\beta)$ is defined by

$$P(\beta) = (N(\beta) - N(\beta+\tfrac{\pi}{2}))/(N(\beta) + N(\beta+\tfrac{\pi}{2})) \qquad (1)$$

We have measured the polarization correlation parameters $P(0)$ and $P(\pi/4)$ for impact energies in the 40 - 90 eV range and for electron scattering angles θ_e from 0° to 15°. $P(0)$ and $P(\pi/4)$ are equivalent to two Stokes parameter of the radiation; the third Stokes parameter, the circular polarization, was not accessible because the measurements were performed in the VUV.

In N_2, the $C_4' \, ^1\Sigma_u^+(v'=0)$ state was studied and the results may be summarized as follows:[1]

1) A pseudo-threshold polarization (i.e. $P(0)$ for $\theta_e = 0°$) of 0.18 has been measured in good agreement with theoretical predications. Discrepancies between theory and previous straightforward polarization measurements of the $C_4' \rightarrow X$ emission[2] may be attributed to non-dipole routes in the near threshold excitation.

2) At small electron scattering angles, the correlation parameters are close to zero and thus the level of coherence observed is small. At larger scattering angles, $P(0)$ stays small and fairly constant, but $P(\pi/4)$ displays significant variations, see Fig. 1. For example at $\theta_e = 15°$, $P(\pi/4)$ changed sign when the incident electron energy was changed from 40 to 75 eV.

3) $P(\pi/4)$ changes sign at about 7.5° for both impact energies. $P(0)$ is always positive, but has a shallow minimum at about the same angle and possibly also goes to zero.

A detailed explanation of these findings is very difficult, particularly because it was not possible to isolate individual rotational transitions in either the electron or photon channel.

In H_2, the preliminary data on the excitation of the $C\,^1\Pi_u(v'=0)$ state reported by Malcolm and McConkey[3] have been extended to various impact energies and larger scattering angles. In general, the level of coherence in H_2 is found to be higher than in N_2, probably because the rotational development in the $C_o \rightarrow X_{v''}$ Werner band excitation is less pronounced and so less interference is occurring. The pseudo-threshold polarization (0.22) is in very good agreement with theory and previous polarization studies.[4] These measurements are currently being extended to D_2 and full details of the H_2 and D_2 data will be presented at the Conference.

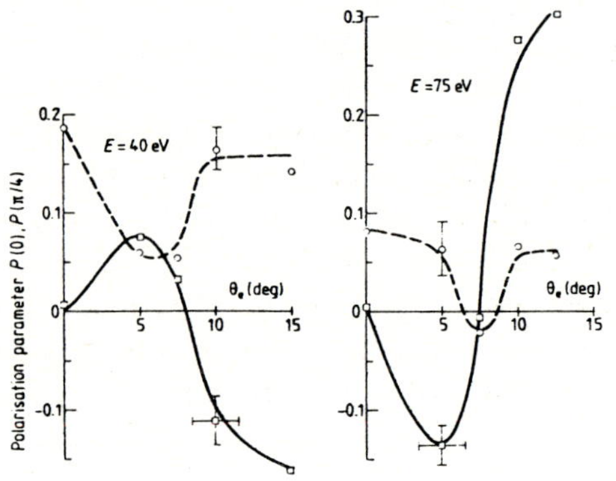

Fig. 1: Polarization-correlation parameter P(0),Φ; and P(π/4), ϕ; as a function of electron scattering angle θ_e for incident energies of 40 and 75 eV, respectively. Vertical error bars represent statistical uncertainties, the horizontal bars reflect the angular resolution of the electron spectrometer.

Financial assistance from the Natural Sciences and Engineering Research Council of Canada is gratefully acknowledged. One of us (K.B.) wishes to acknowledge financial aid from the Deutsche Forschungsgemeinschaft (DFG).

References

1. See also K. Becker, H.W. Dassen and J.W. McConkey, J. Phys. B 16 (1983), in press.
2. J.C. Huschilt, H.W. Dassen and J.W. McConkey, Can. J. Phys. 59, 1893 (1981).
3. I.C. Malcolm and J.W. McConkey, J. Phys. B 12, L67 (1979).
4. H.W. Dassen and J.W. McConkey, J. Phys. B 14, 3777 (1981).

ELECTRON ENERGY-LOSS SPECTROSCOPY OF NITRIC OXIDE USING A POSITION-SENSITIVE MULTIDETECTOR

Richard J Stubbs, Trevor A York and John Comer

Physics Department, The University, Manchester M13 9PL, England, UK.

The electron energy-loss spectrum of nitric oxide has been studied between 4eV and 20eV, using a spectrometer incorporating a position-sensitive multidetector. The detector employs a charge-coupled imaging device, which allows the accumulation of signals from a wide range of electron energies simultaneously[1].

High resolution spectra have been obtained at electron impact energies between 1eV and 100eV above the excitation threshold and at scattering angles between 0° and 90°. The spectra confirm the results of Frueholz et al[2] in the 4eV to 10eV range and show clearly their $X^{2}\pi \rightarrow b^{4}\Sigma^{-}$ assigned transition. Many previously unobserved transitions have been found above the first ionization potential, in particular a number of spin-forbidden transitions between 10eV and 14eV. Above 14eV the dipole-allowed Rydberg series of Metzger et al[3] and Huber[4] have been seen by electron impact spectroscopy for the first time.

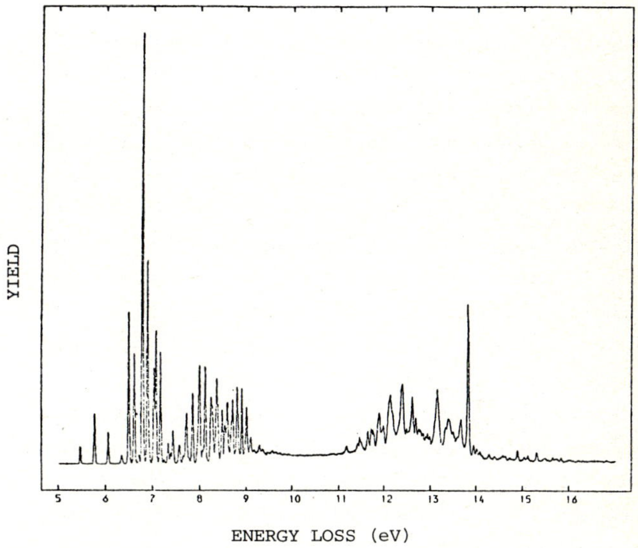

Sample spectrum at 10eV impact energy and 5° scattering angle.

REFERENCES

1. Hicks P.J., Daviel S., Wallbank B., Comer J., J.Phys.E., 13, 712 (1980).
2. Frueholz R.P., Randa R., Kuppermann A., J.Chem. Phys. 31, 315 (1978).
3. Metzger P.H., Cook G.R., Ogawa M., Can.J.Phys. 45, 203 (1967).
4. Huber K.P., Helv.Phys.Acta. 34, 929 (1961).

ELECTRON ENERGY-LOSS STUDIES OF N_2 USING A SPECTROMETER WITH A POSITION SENSITIVE DETECTOR

Tim Reddish and John Comer

Physics Department, The University, Manchester M13 9PL, UK.

Electron energy-loss spectra have been measured using an electron spectrometer fitted with a position sensitive multidetector. This detector has been described by Hicks et al[1] and incorporates microchannel plates, a phosphor, and a charge coupled imaging device that enables electrons with a range of energies to be detected simultaneously.

In the present work studies are made of molecular nitrogen and the use of low electron energy and non-zero electron scattering angles has revealed a large number of spin and symmetry forbidden transitions. Figure 1 shows the energy region of the lowest lying electronic states of N_2. It has been studied previously by electron impact for example by Wilden et al[2] but the present work shows a substantial improvement in quality. The narrowest features have a measured width of 14meV and most of the peaks display significant rotational effects. The improvement in resolution over previous measurements has enabled the relative intensities of the low lying vibrational series to be measured much more accurately.

A second sample spectrum with a measured resolution of 21meV is shown in figure 2 and includes a number of Rydberg series converging on the $B^2\Sigma_u^+$ state of N_2^+. In photoabsorption measurements (Gürtler et al[3]), the spectra in this energy region (17-19eV) is dominated by Hopfield's emission series which is indicated in figure 2. The electron impact measurements have also been made by Wilden et al[2] who observed a number of isolated features in the energy range 14-16eV. These have been clarified in the present work in which a large number of new Rydberg levels have been observed. Their identification will be discussed.

REFERENCES

1. P.J. Hicks, S. Daviel, B. Wallbank, J. Comer, J.Phys.E 13, 713 (1980).
2. D.G. Wilden, P.J. Hicks, J. Comer, J.Phys.B 12, no. 9, 1579 (1979).
3. P. Gürtler, V. Saile and E. Koch, Chem.Phys.Letts. 48, no. 2, 245 (1977).

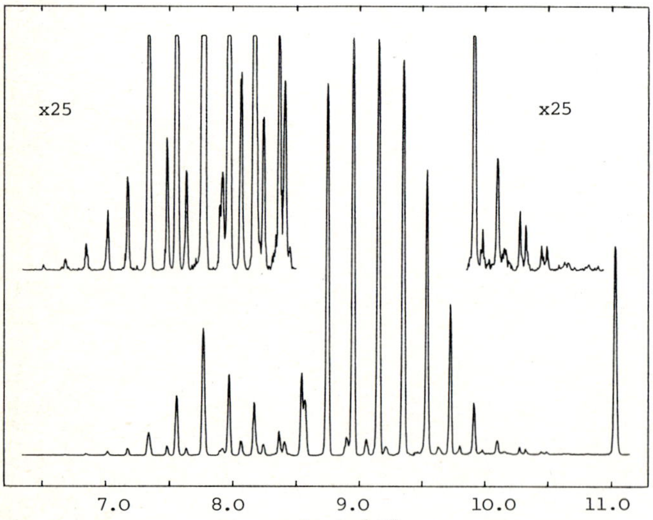

Figure 1: Electron energy loss spectrum of N_2 displaying low-lying **triplet states.** The scattered electron energy and the scattering angles were 10eV and 45°.

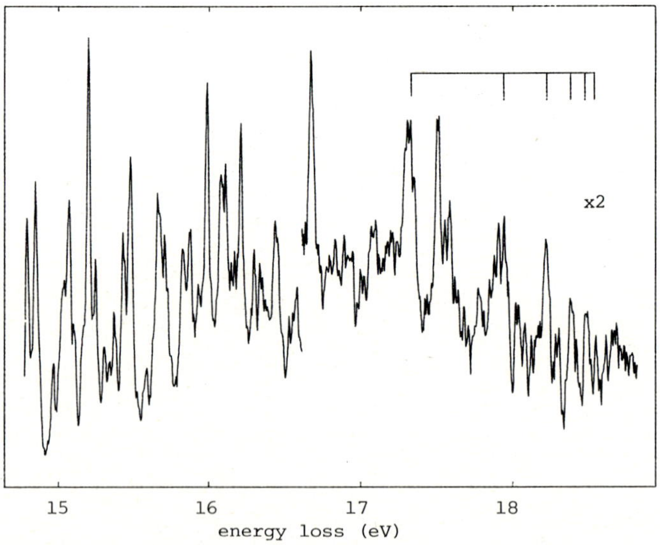

Figure 2: Electron energy loss spectrum of N_2 displaying the Rydberg states. The scattered electron energy and the scattering angle were 5eV and 45°.

THEORETICAL STUDIES OF BETHE SURFACES IN SMALL MOLECULES[1]

K. Greenwald and P. W. Langhoff

Department of Chemistry, Indiana University, Bloomington, IN 47405, USA

Computational studies have been performed of Bethe surfaces in a variety of small diatomic and polyatomic molecules employing static-exchange and TDHF approximations, and explicit Hilbert space methods in construction of discrete and continuum electronic wave functions. The calculated generalized oscillator strengths and continuum densities are useful in the interpretation and assignment of electron energy-loss spectra, and may help motivate corresponding experimental studies of partial-channel cross sections. A general account of the calculations performed and some illustrative results are reported in the present Abstract.

Hartree Fock target functions are constructed in Cartesian Gaussian basis sets at the appropriate experimental ground-state nuclear configurations employing standard computational procedures. Vertical electronic static-exchange and TDHF excitation spectra are obtained from conventional basis-set methodology and matrix diagonalization. These provide the discrete electronic excited states required in forming generalized oscillator strengths, and furnish the pseudostates used in conjunction with Stieltjes methods in construction of continuum wave functions and oscillator strength densities.[2] Three-dimensional graphical representations are constructed of the partial-channel Bethe surfaces, and of selected final-state continuum wave functions associated with resonance features. Detailed comparisons are made of the calculated spectra with available electron energy-loss measurements and assignments are provided.

The diatomic molecules N_2, O_2, and HF provide representative illustrations of aspects of energy-loss spectra in small molecules. In Fig. 1 are shown measured[3] and calculated values of the Lyman-Birge-Hopfield ($3\sigma_g \rightarrow 1\pi_g$) transition strength at 9.35 eV in N_2 as a function of momentum transfer. Evidently, the measured values and static-exchange calculations are in satisfactory accord, results that are typical of dipole forbidden transitions in other light diatomic and polyatomic molecules. At large momentum transfer values the measured and calculated energy-loss spectra in N_2 are dominated by intravalence transitions of both dipole allowed and forbidden character. These results suggest the use of large-scattering-angle electron energy-loss measurements as a convenient form of Rydberg quenching spectroscopy.

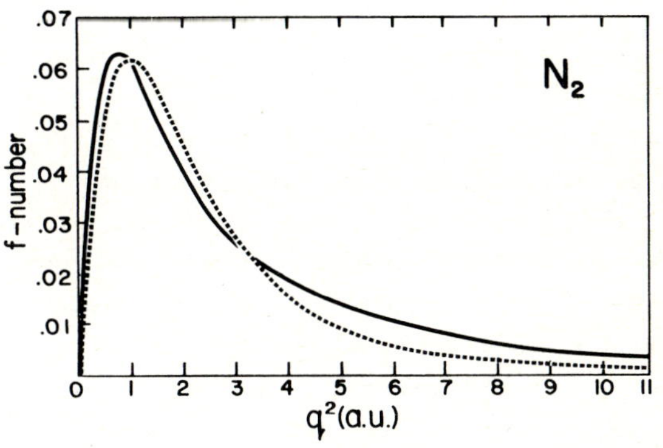

Fig. 1: Measured (——) and calculated (---) Lyman-Birge-Hopfield generalized oscillator strengths in N_2

The spectrum of O_2 is found to include strong $2\sigma_u \rightarrow 1\pi_g$ autoionizing features at 15 to 17 eV that have previously gone unassigned.[4] In addition, recently identified $2\sigma_u \rightarrow n\sigma_u$ Rydberg excitations are verified,[5] and the $2\sigma_u \rightarrow \sigma^*$ resonance transition is found above threshold in the ionization continuum, clarifying earlier measured values.[6] Detailed assignments are provided on basis of TDHF calculations of recent high-resolution forward scattering energy-loss measurements in HF.[7] Strong configuration mixing is found between $3\sigma \rightarrow 4\sigma$ and $1\pi \rightarrow np\pi$ excitations, accounting for the irregular intensity pattern in the measured spectrum. Use of sufficiently large basis sets insures proper description of Rydberg-valence mixing in this case. Further details of these studies and the results of additional calculations are reported separately.

References

[1]Work supported in part by ACS-PRF Grant #12342-AC6.

[2]M.R. Hermann and P.W. Langhoff, Abstract in this Proceedings.

[3]T.C. Wong, J.S. Lee, H.F. Wellenstein, and R.A.Bonham, J. Chem. Phys. 63, 1538 (1975).

[4]E.N. Lassettre, S.M. Silvermann, and M.E. Krasnow, J. Chem Phys. 40, 1261 (1964).

[5]M. Dillion and D. Spence, J. Chem. Phys. 74, 6070 (1981).

[6]J.S. Lee, J. Chem. Phys. 67, 3998 (1977).

[7]A.P. Hitchcock and C.E. Brion, Chem. Phys. 61, 281 (1981).

THE ELECTRON ENERGY-LOSS SPECTRUM OF HYDROGEN CHLORIDE IN THE REGION 12.748eV TO 16.254eV.

Trevor York and John Comer

Physics Department, The University, Manchester M13 9PL, UK.

The electron energy-loss spectrum of HCl has been studied at incident energies between 10eV and 110eV, and scattering angles ranging from 2° to 90°. Corrosion of the spectrometer has been reduced by the use of a position-sensitive multidetector like the one described by Hicks et al[1]. This enables us to obtain high quality spectra in relatively short accumulation times. In addition, the spectrometer is differentially pumped, allowing us to operate with high target gas pressures in a molecular beam, whilst subjecting the spectrometer components to a reduced background gas pressure.

A sample energy-loss spectrum of HCl obtained under conditions of high incident electron energy and 2° scattering angle is shown in figure 1. The spectrum can be split into two regions, terminating at the first ionization potential $X^2\pi$, and the second ionization potential $A^2\Sigma^+$, respectively. The present work concentrates on the observed structure in the upper region from 12.748eV to 16.254eV. Previously the technique of photoabsorption has been used to observe many transitions in this region, including Rydberg progressions converging on the second ionization potential[2]. However, this is the first report concerning HCl, using the technique of low energy electron-impact. The spectrum obtained under conditions of high incident electron energy and low scattering angle closely resembles that from photoabsorption. However, when low energies and large scattering angles are employed, the spectrum changes in appearance, and previously unobserved transitions which are forbidden by electric-dipole selection rules become enhanced, in agreement with the predictions of Trajmar et al[3]. Many of these transitions have been grouped into spin forbidden Rydberg progressions, whose members overlap each other. By measuring the corresponding spectra of DCl and using the technique of isotope shift analysis the lowest lying members of each vibrational series are determined. This technique employs the fact that corresponding electronic transitions in the two isotopes display different vibrational characteristics. This enables us to calculate accurate values for the associated vibrational constants and quantum defects of the new series.

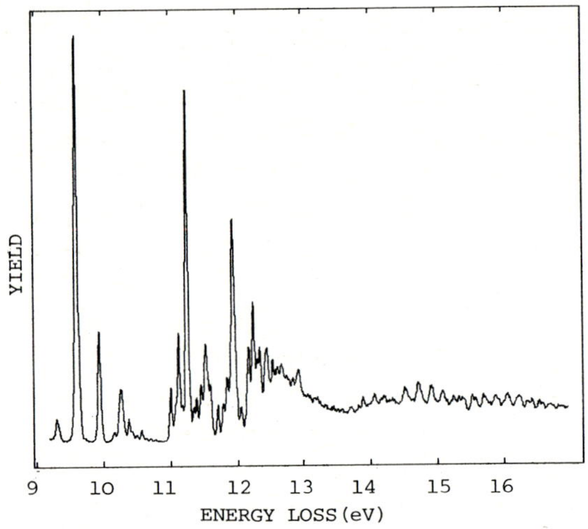

Figure 1: Electron energy-loss spectrum of hydrogen chloride at a constant scattered electron energy of 100eV, and 2° scattering angle.

REFERENCES

1. P.J. Hicks, S. Daviel, B. Wallbank, J. Comer, J.Phys.E: Sci.Instrum. 13, 713 (1980).
2. D.T. Terwilliger, A.L. Smith, J.Chem.Phys. 45, 366 (1973).
3. S. Trajmar, J.K. Rice, A. Kuppermann, "Advances in chemical physics", New York: Wiley (1970).

ELECTRON ENERGY LOSS SPECTROSCOPY IN MOLECULAR CHLORINE*

David Spence, R.-G. Wang,† and M. A. Dillon

Radiological and Environmental Research Division, Argonne National Laboratory, Argonne, IL 60439

We have obtained electron energy loss spectra in molecular chlorine in the energy-loss range of 5 to 14 eV for incident electron energy of 200 eV and scattering angles betwen 3° and 15°. The spectra obtained for low scattering angles (e.g., Fig. 1) correspond closely to optical absorption spectra, the most prominent features arising from excitation of optically allowed Rydberg states.[1,2] The bars shown in Figure 1, indicate the approximate expected location of Rydberg states with configuration 4s 4p and 4d, based on the calculated atomic quantum defects of Dehmer and Saxon[3](a) and the approximate molecular quantum defects of Rabalais et al.[4] (b) The most surprising feature of Figure 1 is band A, which arises from an excited configuration 4sσ and is parity forbidden from the ground state.[2] The energy of this state had previously been predicted by Spence[5] to be 8.017 eV, and was subsequently observed in the threshold electron spectrum of Jureta et al.,[6] as two unresolved vibrational progressions with term symbols $^{3,1}\Pi_g$. Here we are almost certainly observing only the $^1\Pi_g$ state as our incident energy is too high to expect spin forbidden transitions. Though our experimental energies of band A are in excellent agreement with those of Jureta et al.,[6] we believe (based on relative intensity plots of the Cl_2^+ $^2\Pi$, Cl_2 4sσ $^{3,1}\Pi$ and Cl^- $(4s\sigma)^2$ $^2\Pi$ systems) their vibrational levels to be mislabeled by 1 quantum, their v=0 being a hot band.

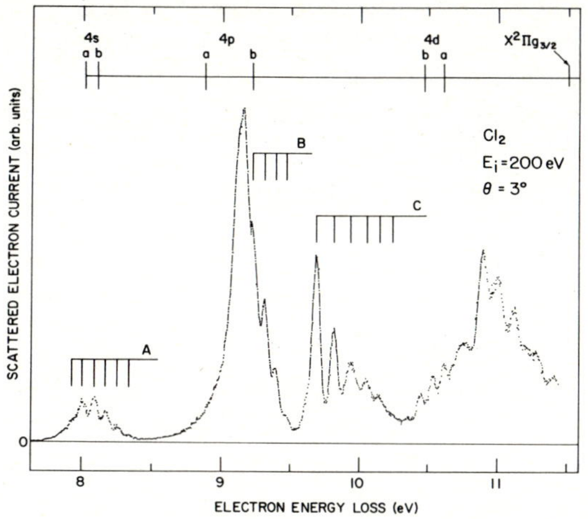

Fig. 1: Electron energy loss spectrum in molecular chlorine at incident electron energy 200 eV, scattering angle 3°, and energy loss between 7.5 and 11.5 eV.

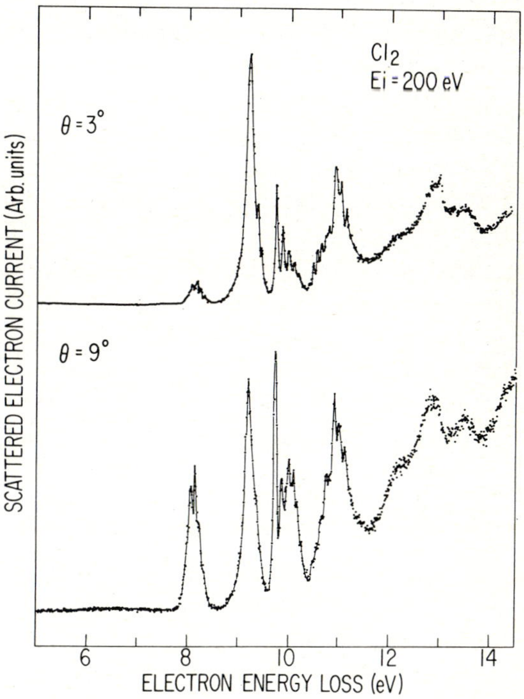

Fig. 2: Electron energy loss spectra in Cl_2 for incident electron energy 200 eV, scattering angles 3° and 9° for electron energy loss between 5 and 14 eV.

At higher scattering angles optically forbidden transitions become relatively more prominent, as demonstrated in Figure 2, which shows a relatively large increase in intensity of the 4sσ $^1\Pi_g$ state. Higher-resolution, large-angle spectra (not shown here) indicate band C to be a super-position of at least three states, one being optically allowed and two forbidden.

References

[1] T. Moeller, B. Jordan, P. Gurtler, G. Zimmerer, D. Haaks, J. LeCalve and M. C. Castex, Institut für Experimentalphysik der Universität Hamburg, D-2000, Hamburg, July 1982, unpublished.

[2] G. Herzberg, Spectra of Diatomic Molecules, Van Nostrand, NY, 1950.

[3] J. L. Dehmer and R. P. Saxon, Argonne National Laboratory, Radiological and Environmental Research Division Annual Report ANL-8060, Part I, July 1972-June 1973, p. 102.

[4] J. W. Rabalais, J. M. McDonald, V. Scheer, and S. P. McGlynn, Chem. Rev. 71, 73 (1971).

[5] D. Spence, Phys. Rev. 10, 1045 (1974).

[6] J. Jureta, S. Cvejanovic, M. Kurepa, and D. Cvejanovic, Z. Phys. A 304, 143 (1982).

* Work performed under the auspices of the USDOE.

† Dept. Phys., Chengdu Univ. of Sci. and Tech., China.

VIBRATIONAL EXCITATION AND ASSOCIATIVE DETACHMENT IN THE e^--HF SYSTEM

J.P. Gauyacq

L.C.A.M., Bât. 351, Université Paris-Sud, 91405 ORSAY, FRANCE

The experimental finding by Rohr and Linder[1] of threshold peaks in vibrational excitation in electron HCl, HF collisions, was followed by numerous theoretical investigations of this problem[2]. Various interpretations were proposed for these threshold peaks which either invoked the effect of short range forces via a virtual state of the e^--fixed R molecule scattering or the effect of the long range dipolar field.

This problem is investigated in an effective range theory which represents the low energy e^--HF interaction by a dipolar field at large r (only one angular mode is considered) and a boundary condition at $r=r_c$ in the electron wavefunction, independant of the energy, which represents the short range forces. This formalism is analogous to the zero range potential (ZRP) approximation which only considers short range forces[3]. The boundary condition at r_c is extracted from the *ab initio* calculations of Segal and Wolf[4]. Within the two formalisms (effective range and ZRP), the vibrational motion can be treated exactly during the collision ; and both can treat vibrational excitation (VE) in e^--HF collisions as well as associative detachment AD in F^--H collisions[3].

Table 1 presents the relative population of the final vibrational states of HF, in the AD reaction ($F^-+H \rightarrow HF+e^-$). The two sets of results are in good agreement with the experimental data[5] ; they are almost identical and one has to examine individual collisions close to a threshold to find differences. The main characteristic of these results is the dominance of high v levels.

TABLE 1. Relative population of final vibrational states in F^--H collisions at 300°K.

	V=5	V=4	V=3	V=2	V=1	V=0
ZRP	.25	.48	.21	.05	.01	<.01
Present	.23	.48	.22	.06	.01	<.01
Experiments[5]	.30	.41	.21	.09	.0	

Figure 1 presents the total cross section for VE in e^--HF collisions ; the two sets of results (with dipole DP and without : ZRP) are very similar, both presenting threshold peaks. Both formalisms treat the nuclear motion exactly and have been used to test the adiabatic nuclei approximation for the vibration. The corresponding results (AN) do not display any threshold peak ; in contrast the simple version of the energy-modified adiabatic approximation (EMA) ($k=\sqrt{k_i\, k_f}$) provides quite satisfying results, at least for $\sigma_{0\rightarrow 1}$. These comparisons stress the importance of formalisms such as the present ones which extract their parameters from *ab initio* calculations and which can treat exactly the nuclear motion.

In conclusion, one can say that the pecularities of these results (high v populated in ED and threshold peaks in VE) are not due to the singularities of the dipolar field ; they are associated with a HF^- potential energy curve coming close to the HF curve near the HF equilibrium distance.

Reference

1. K. Rohr and F. Linder 1976 J.Phys.B 9 2521
2. Taylor et al. 1977 J.Phys.B 10 2253 ; Dubé and Herzenberg 1977 PRL 38 820 ; Rudge 1980 J.Phys.B 13 1269 ; Domcke and Cederbaum 1980 J.Phys.B 14 149
3. Gauyacq JP 1982 J.Phys.B 15 2721
4. Segal GA and Wolk K. 1981 J.Phys.B 14 2291
5. Zwier et al. 1982 J.Chem.Phys. 75 4885 ; Smith and Leone 1983 J.Chem.Phys. 78 1325

Figure 1 : Vibrational excitation in e^--HF collisions : —— : present effective range results ; -- : ZRP ; Δ : adiabatic nuclei ; * : EMA and -·- : experimental results[1]

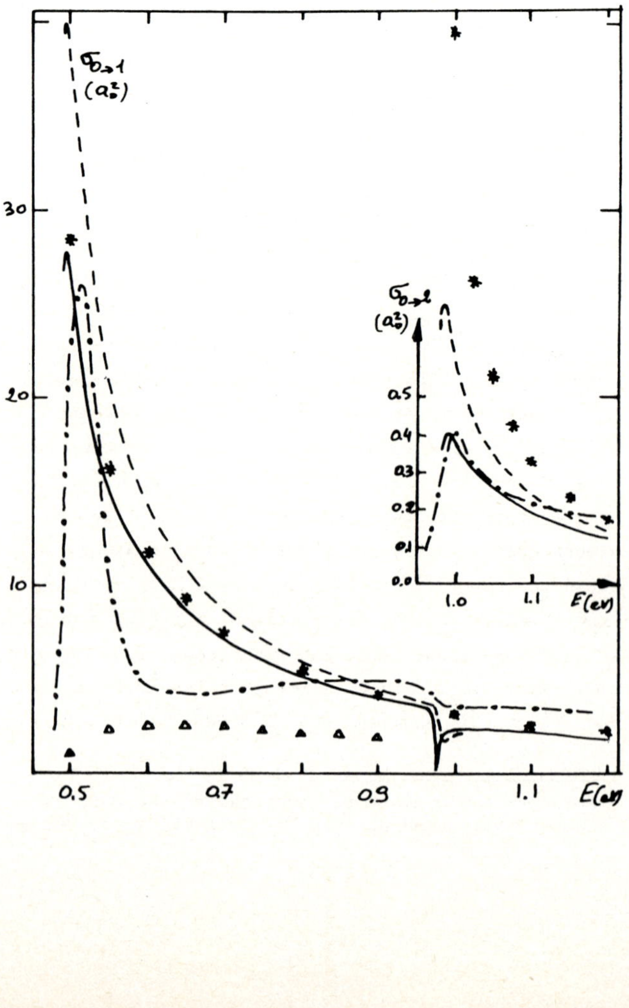

STUDIES OF VACUUM ULTRAVIOLET EMISSION FROM RYDBERG SERIES OF H_2 BY ELECTRON IMPACT*

J. M. Ajello and D. Shemansky+

Jet Propulsion Laboratory, California Institute of Technology, Pasadena, California 91109

We have completed a comprehensive study of electron impact emissions of H_2 in the vacuum ultraviolet (VUV) extending from 70 to 170 nm. H_2 is the simplest molecule to study from a theoretical point of view and yet until the appearance of our work there existed limited experimental cross sections in H_2 for comparison. In particular we have measured in the laboratory the excitation cross sections of the following two Rydberg series of H_2: ${}^1\Sigma_u^+$ 1sσ npπ(B, B', B" states with principal quantum numbers n = 2, 3, 4 respectively) and ${}^1\Pi_u$ 1sσ npπ(C, D, D' states with principal quantum numbers n = 2, 3, 4 respectively) by electron impact over the energy range of 0 to 350 eV. We also estimate the predissociation (autoionization is small) and emission yields of the vibrational levels of the D, D' and B" states whose band systems exhibit strong "breaking off in emission" for wavelengths below 85 nm. Furthermore, we report the first dissociative excitation cross section for production of Lyman-β.

The instrumentation used in this experiment has been described in a previous paper.[1] In brief the instrument consists of an electron impact emissions chamber in tandem with a VUV spectrometer. We show in Fig. 1 a calibrated experimental spectrum at 100 eV for the wavelength range 70 to 130 nm together with a best fit to the laboratory spectrum. The references for the modeling parameters for the molecular band systems are given in Shemansky and Ajello.[2] The preliminary cross section data used in fitting the laboratory spectrum are plotted in Fig. 2 over the complete energy range 0 to 300 eV for the Rydberg series ${}^1\Sigma_u^+$ npσ, n = 2, 3, 4 and ${}^1\Pi_u$ npπ, n = 2, 3, 4 and for the dissociative excitation transitions of Lyman-α and Lyman-β. The excitation cross sections for the higher members ($n \geq 3$) of the Rydberg series are not equal to the emission cross section because of predissociation.[3] For this reason the branching ratio yields for emission and predissociation for these excitation cross sections are also given in Fig. 2 for the B", D and D' states. A review of the experimental and theoretical cross section data for the B- and C-states is given in Ajello et al.[4]

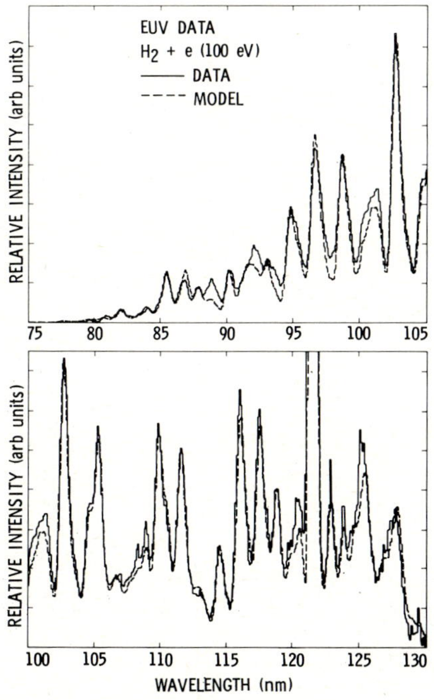

Fig. 1: Calibrated laboratory spectrum of H_2 at 100 eV electron impact energy and 0.5 nm resolution together with best-fit model.

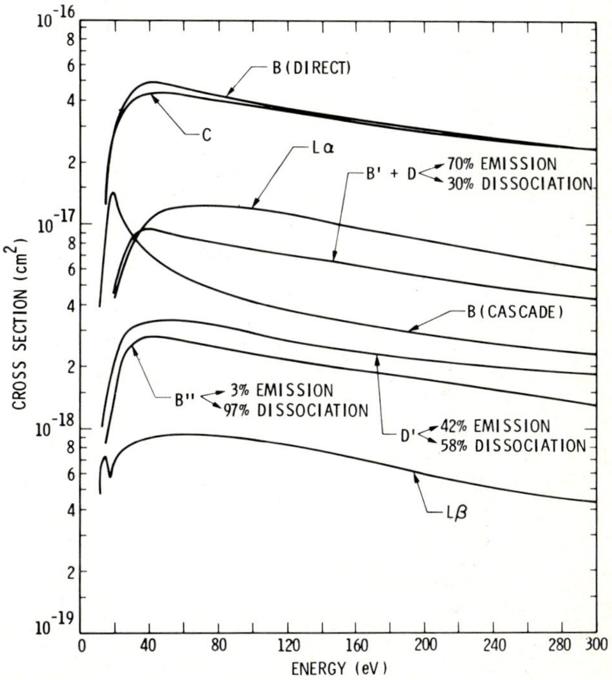

Fig. 2: A complete set of electronic excitation cross sections for H_2 in the VUV.

References

1. J. M. Ajello and S. K. Srivastava, J. Chem. Phys. 75, 4544 (1981).
2. D. Shemansky and J. Ajello, J. Geophys. Res., In press (1983).
3. P. M. Guyon, J. Breton and M. Glass-Maujean, Chem. Phys. Lett. 68, 314 (1979).
4. J. M. Ajello, S. K. Srivastava and Y. L. Yung, Phys. Rev. A. 25, 2485 (1982).

*This work was supported by Air Force Office of Scientific Research (AFOSR), NASA Planetary Atmospheres, and NASA Astronomy/Astrophysics Program Offices.

+Permanent address: Earth & Space Sciences Institute, Tucson, Arizona 85713

THRESHOLD EXCITATION OF H_2O, D_2O and H_2S

J. Jureta, S. Cvejanović, D. Cvejanović*, D. Čubrić
Institute of Physics, Beograd, POB 57
*and Faculty of Natural and Mat. Sci. POB 550, Beograd, Yugoslavia

Threshold electron impact spectrometer described previously[1] has been used to study optically forbidden transitions in H_2O,D_2O and H_2S. Both H_2O (D_2O) and H_2S belong to the C_{2v} symmetry group and have similar electronic structure, resulting in the similarity of their excitation spectra as well. The spectrum of H_2O is shown in Fig. 1. The first unoccupied orbital for these molecules is nsa_1, with n=3 for H_2O and n=4 for H_2S. It supports $^{3,1}B_1$ states when populated from the last occupied b_1 orbital, or $^{3,1}A_1$ states, when populated from the more strongly bound a_1 orbital. These states together with the lowest $^{3,1}A_2$ state, appear as strong continua in our threshold spectra, followed by-or interfering with-very strong first members of numerous Rydberg series converging onto the H_2O^+ , $\tilde{X}^2B_1$. The highest Rydberg peaks coincide with the energies of the resonances observed by Sanche and Schulz[2].Another example of strong resonant contribution characteristic for the threshold spectra appears bellow the first excited state in H_2O, the $^3B_1$ state, through isoenergetic decay of the $^2B_1$ resonance at 6.5 eV, which has been already observed in dissociative attachment experiments[3]. Another threshold feature common to both molecules is an unusually strong onset of the partial ionisation cross section.

References:

1. J. Jureta, S. Cvejanović, J.N.H. Brunt, F.H. Read, J. Phys. B, 11, L347 (1978)
2. L. Sanche, and G.J. Schulz, J. Chem. Phys. 58, 479 (1973)
3. D. Belić, M. Landau and R.I. Hall, J Phys B, 14, 175 (1981)

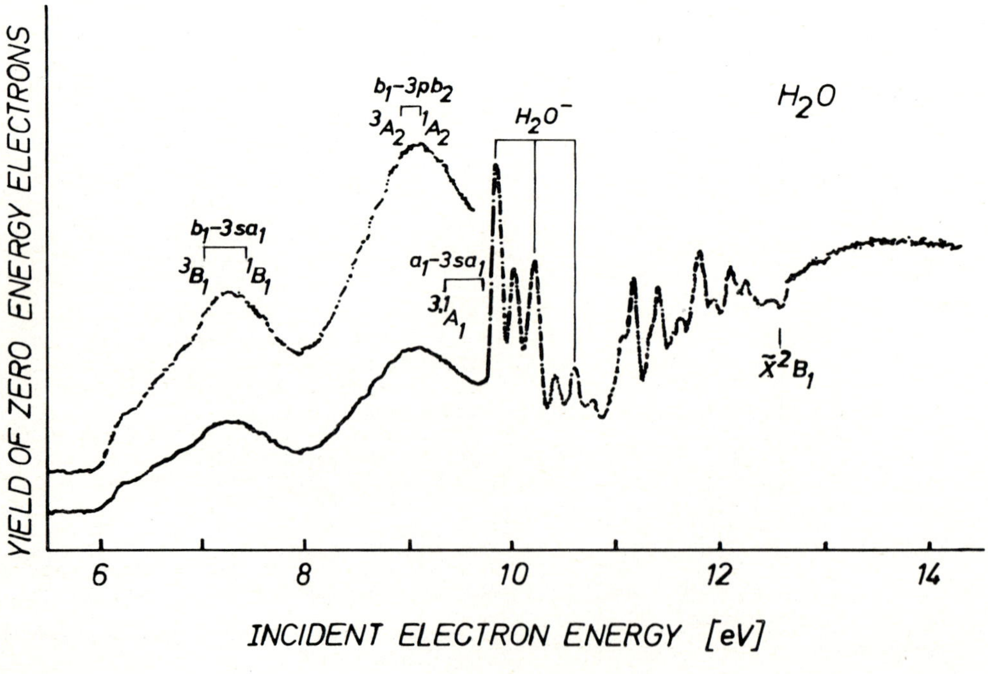

Fig. 1. Threshold electron impact spectrum of water.

EXCITATION OF OPTICALLY FORBIDDEN STATES IN CO_2

S. Cvejanović, J. Jureta, D. Cvejanović[+] and Dj. Šrajer
Institute of Physics, Beograd, POB 57
[+]Faculty of Natural and Math. Sciences. Beograd POB 550

An electron impact threshold spectrometer described by Jureta et al[1] was used to study the optically forbidden states and resonant threshold phenomena in CO_2. The part of the CO_2 threshold spectrum is shown in Fig. 1. All structures observed in the threshold spectrum can be divided into two classes. First class consists of strong continua mainly with undetectable or very weak vibrational structures. Second class consists of discrete peaks most of which can be correlated with excitation into the Rydberg orbitals. Inspection of Fig. 1. indicates that continuum - like structures dominate in threshold spectrum. The continuum like states above 10 eV can be, at least up to the fist ionization threshold, explained in terms of excitation into a series of valence states. This valence states arise when one of the electrons from the four outer molecular orbitals is subsequently promoted into the first unoccupied $2\pi_u$ orbital. The strong threshold excitation of valence states in CO_2 relative to the Rydberg states we atribute to the specific properties of the $2\pi_u$ orbital, which have been already manifested in the $^2\Pi_u$ shape resonance and related effects observed in the excitation of an C_K inner shell electron of CO_2[2].

All the Rydberg states observed in the threshold spectrum are associated with the excitation of a $1\pi_g$ electron into the np and nd Rydberg orbitals. Energy difference between states observed in the present experiment and optical absorption indicate that different multiplet components are excited at threshold.

References:

1. J. Jureta, S. Cvejanović, JNH Brant, F.H. Read, J. Phys.B, 11, L347 (1978)
2. M. Tronc, G.C. King and F.H. Read, J. Phys. B: Atom.Molec.Phys. 12, 137 (1979)

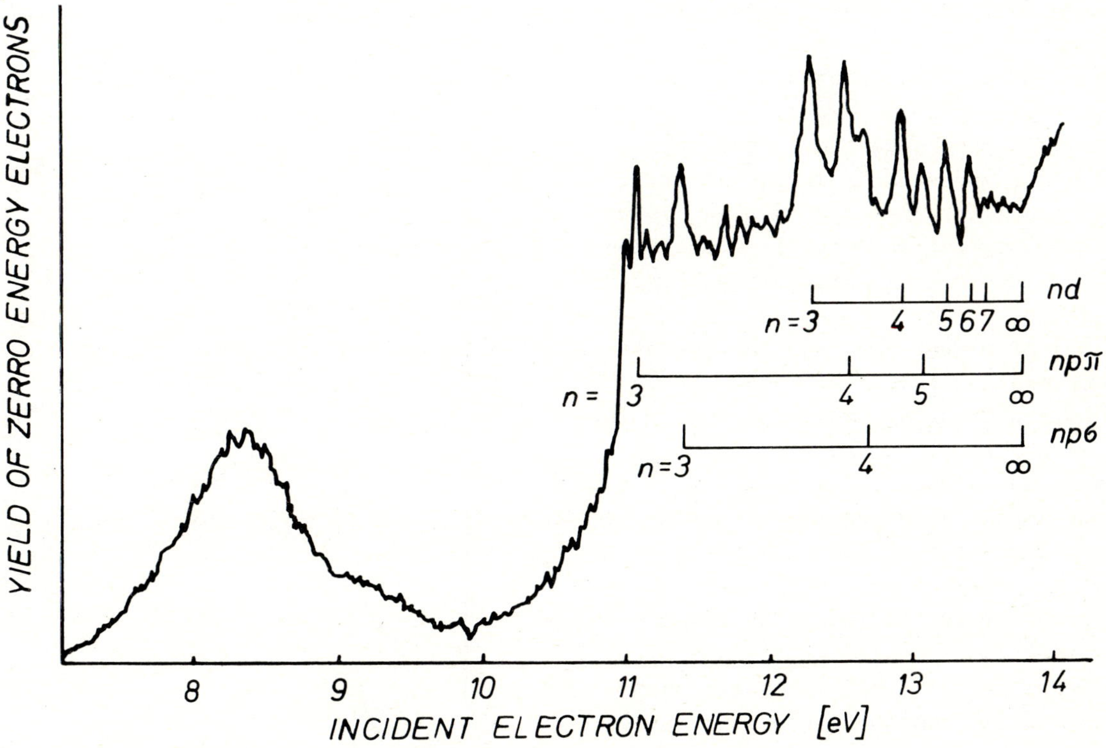

Fig. 1: Threshold spectrum of CO_2

ELECTRON SCATTERING BY METHANE

L. Vušković

Institute of Physics, P.O.Box 57, 11001 Beograd, Yugoslavia

S. Trajmar

Jet Propulsion Laboratory, Pasadena, CA 91109 USA

Electron energy-loss spectra covering the elastic and inelastic region up to 15.0 eV were obtained at 20, 30 and 200 eV impact energies at scattering angles ranging from 8° to 130°. Relative differential elastic scattering cross sections were determined at each impact energy.

The measurements were carried out in the crossed molecular beam - electron beam arrangement. The electron scattering intensities were measured as a function of energy loss (ΔE) at fixed impact energies (E_o) and scattering angles (θ) by multichannel scaling technique.

Relative angular distributions for elastic scattering was corrected for effective path length variation with scattering angle to obtain the relative DCS. These DCS were normalized to the absolute measurements of Tanaka et al.[1] at 100° for 20 and 30 eV impact energies. For 200 eV impact energy normalization has been made using data[2] calculated with two--potential approximation.

Typical energy-loss spectra with inelastic features are shown in Fig. 1. with impact energy and scattering angle indicated. The spectrum changes appreciably with scattering angle. At low angles some structure can be recognized but assigment to specific transitions even with the theoretical guide of reference 3 is not possible.

The inelastic region of the energy-loss spectrum was divided into five ranges. From the integrated scattering intensities over these ranges with respect to the elastic scattering and from the normalized elastic cross section, the inelastic cross sections were calculated. The cross sections were extrapolated to 0° and 180° in order to obtain integral and momentum transfer cross sections. The total electron scattering cross sections obtained by summing the present elastic and inelastic cross sections and the ionization cross sections of Rapp and Englander-Golden[4] are in excelent agreement with the recent total electron scattering cross sections measured by Nickel et al[5].

References

1. H.Tanaka, T.Okada, L.Boesten, T.Suzuki, T. Yamamoto and M.Kubo, J.Phys.B: Atom.Molec. Phys.15,3305(1982) and private communication.
2. S.S.Dahl, B.B.Srivastava and R.Shingal, J.Phys.B:Atom.Molec.Phys.12,2727 (1979).
3. B.Schurmann and R.J.Buenker, private communication (1981); B.Schurmann, Ph.D.Thesis, 1981, Lehrstuhl fur Theoretische Chemie der Gesamthochschule Wuppertal, Wuppertal, W. Germany.
4. D.Rapp and P.Englander-Golden, J.Chem.Phys. 43,1464 (1954); B.L.Schram, M.J.Van der Wiel, F.J. De Heer and H.R. Moustafa, J. Chem.Phys. 44, 49 (1966).
5. J.C.Nickel, D.F.Register and S.Trajmar (private communication; to be published).

Experiment has been performed at Jet Propulsion Laboratory, California Institute of Technology, Pasadena, California, USA.

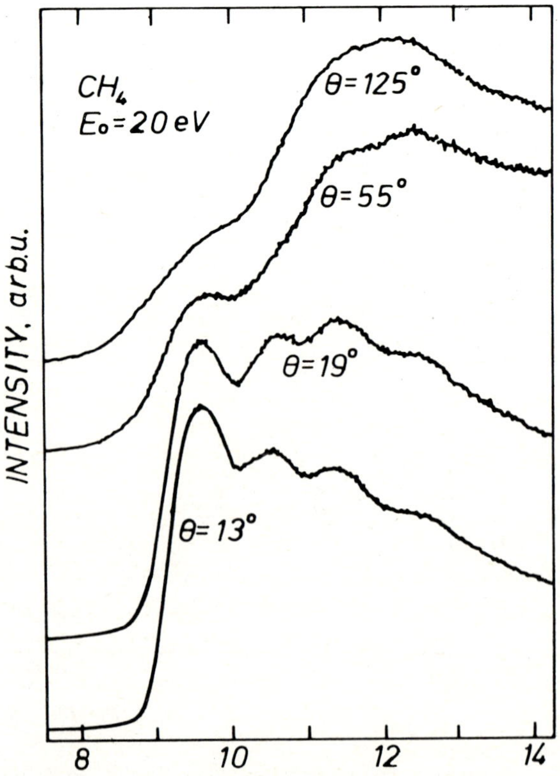

Fig. 1.

PHOTOABSORPTION SPECTRUM OF CF_3I GENERATED BY ELECTRON IMPACT*

S. K. Srivastava and S. Trajmar

Jet Propulsion Laboratory, California Institute of Technology, Pasadena, California 91109

In the case of electron-atom/molecule collisions it has been shown by Lassettre et al.[1] that the generalized oscillator strength, f^G, approaches the optical oscillator strength, f^O, in the limit of zero momentum transfer regardless whether the Born scattering approximation holds or not. This applies not only to the discrete transitions in the target but also to the continuum region as shown by Inokuti.[2] We have utilized these properties of the generalized oscillator strength and generated a photoabsorption spectrum of CF_3I. It is shown in Fig. 1. This spectrum was recorded for 0 degree scattering angle and 200 eV electron impact energy. We could only measure the relative values of the photoabsorption cross sections. Therefore, the absolute values were obtained by normalizing the spectrum to a measurement at 686Å by Judge and Shyn.[3]

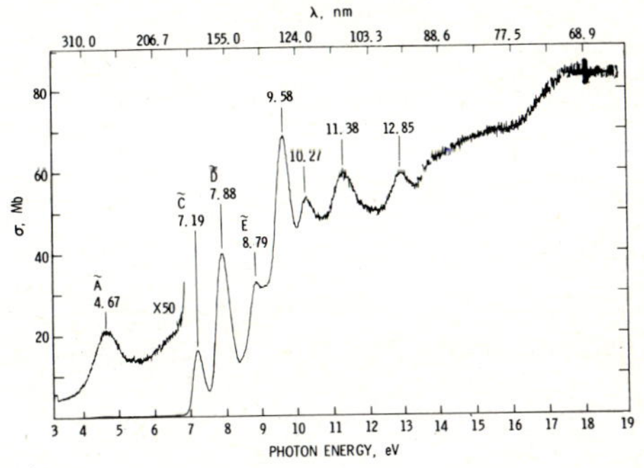

Fig. 1: Photoabsorption spectrum of CF_3I. The normalization point is indicated by +.

CF_3I has an electronic state which dissociates with an almost hundred percent yield of excited iodine atoms and gives rise to the laser emission at 1.315μ. In order to understand the kinetics and to improve the efficiency of a CF_3I laser, one would like to know the excited states of CF_3I, their symmetries, their transition moments, electron impact excitation cross sections, and photoabsorption cross sections. Only fragmentary data are available at the present time.

The photoabsorption spectrum (Fig. 1) shows several spectral features. Among them $\tilde{A}$, $\tilde{C}$, $\tilde{D}$ and $\tilde{E}$ have been observed spectroscopically in the past by Shutcliffe and Walsh.[4] Other strong features at 9.58 eV, 10.27 eV, 11.38 eV and 12.85 eV are new. In general it has been concluded that the upper state for the laser transition in iodine is formed by the dissociation of the $\tilde{A}$ state. But it is clear from Fig. 1 that there are other much stronger transitions in CF_3I which may be able to generate this excited state in iodine. This can be answered only if we have a clear understanding of the symmetries and transition moments. It is hoped that this work will stimulate interest in this direction.

*This work is supported by the OAST office of NASA.

References

1. E. N. Lassettre, A. Skerbele and M. A. Dillon, J. Chem. Phys. 50, 1829 (1969).
2. M. Inokuti, Rev. Mod. Phys. 43, 297 (1971).
3. D. L. Judge and N. Shyn, private communication.
4. L. H. Shutcliffe and A. D. Walsh, Trans. Far. Soc. 57, 873 (1961).

ELECTRON IMPACT EXCITATION CROSS SECTIONS FOR EXCITATION OF ELECTRONIC STATES IN UF_6 FOR INCIDENT ELECTRON ENERGIES OF 10, 20 and 40 eV

David C. Cartwright
University of California, Los Alamos National Laboratory
Los Alamos, New Mexico 87545 (USA)

S. Trajmar, A. Chutjian, and S. Srivastava
Jet Propulsion Laboratory, California Institute of Technology
Pasadena, California 91103

Electron energy-loss spectroscopy has been utilized to identify[1] optically forbidden features in the electron energy-loss spectra at 10, 20 and 40 eV incident electron energy.

The apparatus used consisted of an electron gun which produces a collimated, energy-selected beam of electrons of impact energy E_o which crosses the target UF_6 beam, generated by flowing the UF_6 through a capillary array, at 90°. The incident energy scale was calibrated against the He 19.35 eV resonance and the true zero scattering angle was determined from the symmetry of the scattering intensity around the nominal zero scattering angle.

The elastic scattering intensity (which includes rotational and vibrational contributions within the instrumental resolution) was measured as a function of scattering angle at fixed impact energies ranging from 5 to 75 eV. The same procedure was applied to He and the ratio of the UF_6 scattering intensity to He scattering intensity was used to determine the absolute elastic scattering cross sections for UF_6 using the He elastic scattering cross sections as secondary standards.

Figure 1 shows electron energy-loss data (the points), for 40 eV incident electron energy, and scattering angles of 20, 60, 90 and 135 degrees. Also shown are model fits to the spectra, and the individual components of the total spectra, used to extract the absolute excitation cross sections for the various features in the energy-loss spectra. The model assumes that the line-shape for an inelastic transition is Gaussian and that the width of a single feature is inversely proportional to the shape of the excited state potential energy curve in the Franck-Condon region. Since the density of the excited states per eV is fairly high, states were combined together into effective states for the purposes of fitting the electron energy-loss spectra.

In the energy-loss spectra, no transitions were found below the onset of the first feature near 3.0 eV. The lowest feature (~3.3 eV) region must correspond to the lowest triplet and singlet charge-transfer excitations from a $t_{1u}\sigma$ orbital centered on the fluorine atoms to the a_{2u} uranium 5f orbital. In general, spin-and dipole-allowed transitions are strongly forward peaked in θ, while spin-forbidden transitions tend to be nearly isotropic. In the optical spectra, the ratio of the (optically allowed) 5.8 eV feature to the 3.3 eV (lowest-energy) feature is $\sim 10^3$, while in the 90° spectrum (Fig. 1) it is ~50. This fact suggests that the 3.3 eV feature is optically forbidden.

Other differences between the optical and electron scattering spectra exist in the intensities of features in the 3.6- to 5.0-eV region relative to the 5.8-eV excitation. Firstly, the 4.2-eV excitation is greatly enhanced relative to the 5.8-eV feature in the electron-scattering spectra, and is clearly separated from the optically-allowed transition for θ > 20°. This fact indicates that the 4.2-eV excitation is optically forbidden. Secondly, the fairly strong optical absorption at 4.8 eV is observed to "fill in" at θ = 20° but is practically absent at higher angles. This is indicative of a weak dipole-allowed transition and has been interpreted as a vibrationically allowed transition. We also note that the features observed above ΔE = 6.2 eV have not previously been reported. Abolute differential and integral cross sections will be reported at the conference.

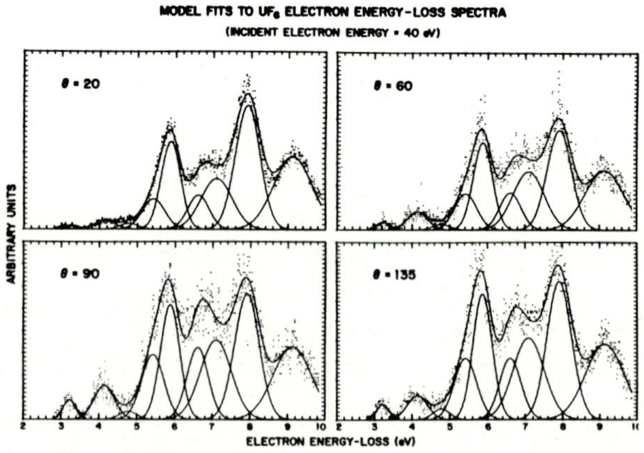

Fig. 1: Electron energy-loss spectra in UF_6, for an incident energy of 40 eV, at scattering angles of 20, 60, 90, 135°. Dots are experimental data, solid lines are model fits.

Reference

1. Preliminary results for 20 eV reported by S. K. Srivastava, D. C. Cartwright, S. Trajmar, A. Chutjian, and W. Williams, J. Chem. Phys. 65, 208 (1976).

THE RYDBERG SERIES OF CO_2 IN THE ENERGY RANGE 11-14 eV

D. Roy, B. Leclerc, J. Delwiche*, and M.-J. Hubin-Franskin*

Département de Physique et CRAM, Université Laval, Québec, Canada G1K 7P4

We have studied the CO_2 molecule by inelastic electron scattering spectroscopy. The spectrometer used is composed of an electrostatic electron monochromator and an energy analyzer of the 127° cylindrical type (energy resolution of 30-40 meV).

For about a decade, many works were made on this molecule in order to find the excited states and their characteristics. In Fig. 1, are shown a few electron energy loss spectra in the range to 10.5 to 14.5 eV. These spectra were selected among a series of measurements involving a great variety of incident electron energies and scattering angles. Thus one may distinguish the "allowed" and "forbidden" transitions.

Foo et al.[1] published a spectrum in the 8-18 eV energy range, but with a medium resolution. The energy range below 14 eV was studied with a good energy resolution by Krauss et al.[2] (inelastic scattering), and by Greening and King[3] (photoabsorption) among others. These date were afterwards analyzed by Fridh et al.[4] In our spectrum we observe many progressions and Rydberg series already identified in the optical spectra, v.g. the "Rathenau strong and broad" bands, marked α and β, and located about between 11.5 and 13 eV. In addition

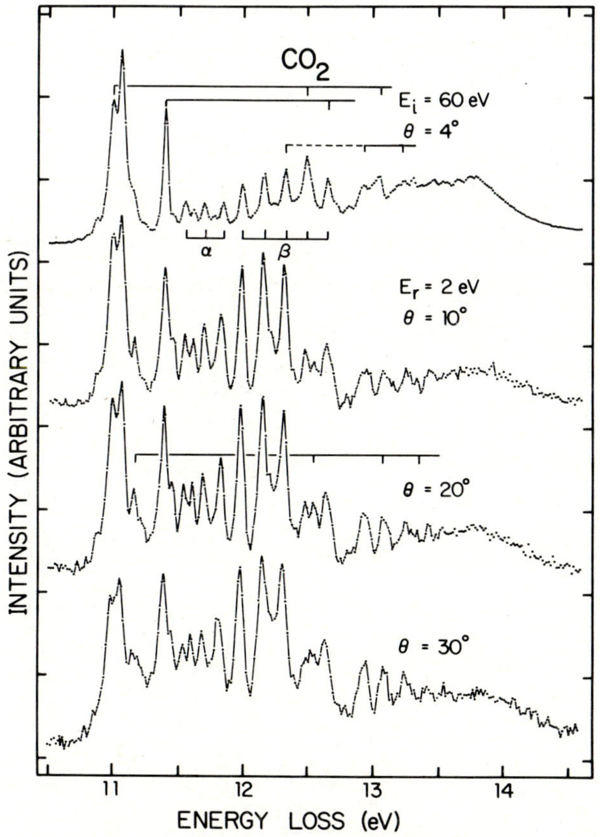

Fig. 1. Electron energy loss spectra in CO_2.

to these bands, we observe the Rydberg series "main 1" which converges to the first ionization potential ($\tilde{X}\ ^2\Pi_g$), with the members n = 3,4,5, as indicated in Fig. 1 with E_i = 60 eV. Another series, "minor 1", is observed with a good intensity. A third series is also present; it was previously indentified[4] as a nf series and could have a first number n = 3 at 12.307 eV. The quantum defect is in a good agreement with this assignment and the angular behavior for the different members of this series in consistent.

All these series are observable as well with a low residual electron energy as with a high incident electron energy (60 eV). However the assignments are supported by the angular behavior of the differential cross sections. A different case is that of series "a" of which only the first member is apparent as a small feature at 11.142 eV, in the curve with 60 eV incident electron energy; the second member seems too weak to be identifiable at this energy. Now in the spectrum obtained with 2 eV residual energy (or less) and especially at 20° scattering angle, we observe an increase in the relative intensity of the features at 11.142, 12.535 and 13.061 eV which could be the members of the same Rydberg series. In this case, their effective quantum numbers are 2.27, 3.31 and 4.37, respectively, which gives a mean quantum defect of 0.72. These transitions are not allowed by all the selections rules. In reason of the magnitude of its quantum defect, the series could be of type npσ or npπ. The fact that these features are weak but present at high incident energy, indicates that they involve symmetry forbidden transitions rather than spin forbidden transitions.

* Permanent address: Lab. de Spectroscopie d'Electrons, Université de Liège, Sart Tilman, Liège, Belgique.

References

1. V.Y. Foo, C.E. Brion, and J.B. Hasted, Proc. R. Soc. London Ser. A 322, 535 (1971)
2. M. Krauss, S.R. Mielczarek, D. Neuman, and C.E. Kuyatt, J. Geoph. Res. 76, 3733 (1971)
3. F.R. Greening and G.W. King, J. Mol. Spectrosc. 59 312 (1976)
4. C. Fridh, L. Åsbrink, and E. Lindholm, Chem. Phys. 27, 169 (1978)

DCS FOR THE EXCITATION OF OCS MOLECULES BY ELECTRON IMPACT

S.Ito, K.Hoshiba, S.Kano and H.Takuma
Institute for Laser Science, University of Electro-Communication, Chofu-shi, Tokyo 102 Japan

T.Takayanagi, K.Wakiya and H.Suzuki
Department of Physics, Sophia University, Chiyoda-ku, Tokyo 102 Japan

About the $^1S \rightarrow {}^1D$ transitions in the ground state configuration of group VI elements, a possibility of high energy laser transition is pointed out, because of generally favorable energy storage characteristics of the ^1S state.[1] And the quantum yield of producing the ^1S state of sulphur atom from OCS molecule by photodissociation is known as 0.8 - 1.0 over the wavelength from 1400 to 1600 Å (7.8 -8.9 eV),[2] where the strong absorption band of the $^1\Sigma^+$ state exists.[3] To find the possibility to realize the population inversion in the ^1S state of sulphur atom dissociated from OCS by electron bombardment or discharge, many kinds of cross sections of OCS for electron impact excitations are needed, including the excitation to the $^1\Sigma^+$ state.

Some studies[4)-6)] were performed about OCS involving electron scattering, but the cross sections for the laser related excitations were not reported. We will report the differential cross sections (DCS) for the states whose excitation energies are from 4.5 to 8.9 eV. The DCS were determined form the electron energy-loss spectra. The whole apparatus is schematically shown in Fig.1. The cylindrical electrostatic energy selector[7] with the mean radius of 50 mm is used as the monochromator. The same type selector with the mean radius of 80 mm is used as the energy analyzer for scattered electrons. The electron gun and the monochromator can be rotated from -30 to 120° around the gas nozzel. The whole system is pumped with a 10" diffusion pump. The electron source and the analyzer are differentially pumped separately with two 4" diffusion pumps to maintain a stable operation.

Typical energy loss spectra at electron impact energy E=10 eV are shown in Fig.2. Overall energy resolution in the spectra is about 30 meV(FWHM). In the high impact energy region from 300 to 500 eV, we determined the DCS of the $^1\Sigma^+$, $^1\Pi$ and the $^1\Delta$ excitations relative to that of the He $2\,^1P$ excitation. To determine the absolute values of DCS, we used the main chamber as a static gas-target chamber by closing the 10" main valve and filling a gas in the main chamber. From measured DCS, we deduced effective generalized oscillator strength (GOS). We also measured DCS of the $^1\Sigma^+$, $^1\Pi$, $^1\Delta$ and the $^3\Sigma^+$ excitations at impact energies from 10 to 60 eV. In these case, we determined them relative to elastic scattering cross sections of He.

References.
1) J.R.Murray and C.K.Rhodes, J.Appl.Phys. 11 5041 (1976)
2) G.Black, R.L.Sharpless, T.G.Slanger and D.C.Lorents, J.Chem.Phys. 62 4274 (1975)
3) J.W.Rabalais, J.M.McDonald, V.Scherr and S.P.Mcglynn, Chem.Rev. 71 73 (1971)
4) V.Y.Foo, C.E.Brion and J.B.Hasted, Proc.R.Soc. London Ser. A322 535 (1971)
5) W.M.Fliker , O.A.Mosher and A.Kuppermann, J.Chem. Phys. 69 3910 (1978)
6) B.Leclerc, A.Poulin and D.Roy, J.Chem.Phys. 75 5329 (1981)
7) K.Jost, J.Phys.E: Sci.Instrum. 12 1006 (1979)

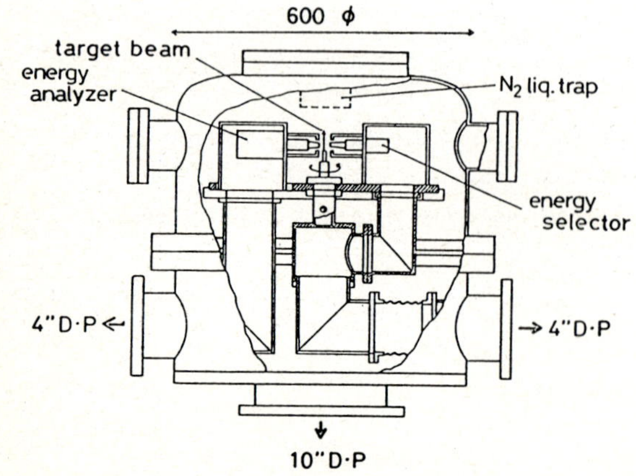

Fig.1. Schematic view of the apparatus

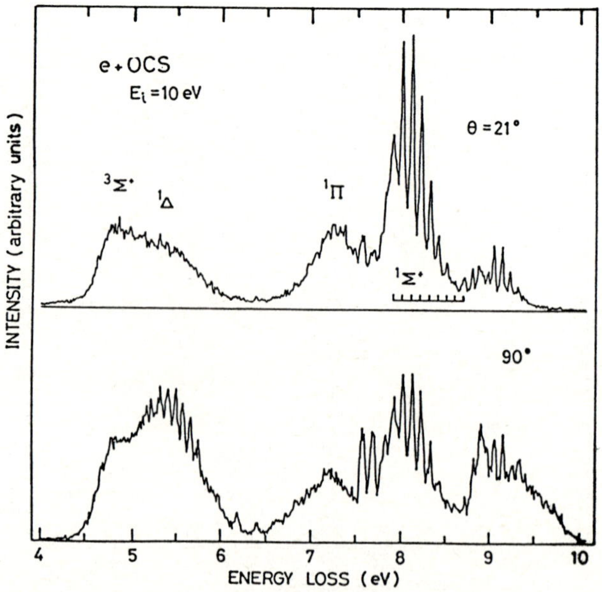

Fig.2. Energy loss spectrum of OCS at E=10 eV

ELECTRON ATTACHMENT OF CO_2 CLUSTERS

K. Stephan, A. Stamatovic[a)], H. Helm[b)] and T. D. Märk

A.f. Kernphysik und Gaselektronik, Inst. f. Experimentalphysik, Leopold Franzens Universität, A 6020 Innsbruck Austria

Recently a variety of methods has been used to investigate molecular cluster beams. Most commonly the formation of positive cluster ions by either electron or photon impact with subsequent mass spectrometric analysis is used. In addition, negative cluster ion formation has been reported by Klots and coworkers (e. g. Ref. 1). The formation of negative ions for CO_2 monomers has been investigated by several authors and appears to be well characterized.[2] The present study is intended to provide related information on the electron attachment processes for CO_2 clusters.

The molecular beam electron impact mass spectrometer system has been described previously[3]. In brief, the molecular beam containing the neutral clusters is formed by expanding up to 5 bar of CO_2 through a 10 μm nozzle. The stagnation gas temperature can be varied and a typical value used is - 30°C. The neutral cluster beam is crossed at right angles by an electron beam of variable energy. Negative ions formed are analyzed in a double focussing sector field mass spectrometer (reversed geometry).

Under these conditions negative cluster ions of the type $(CO_2)_n^-$ and $(CO_2)_nO^-$ for $1 \leq n \leq 10$ could be detected, and the electron energy dependence has been measured for CO_3^-, $(CO_2)_2O^-$ and $(CO_2)_2^-$, and also of O^- from CO_2 in the energy range 2 - 25 eV. The peak positions of the first maxima of CO_3^-, $(CO_2)_2O^-$ and $(CO_2)_2^-$ are shifted towards energies lower than that of O^-. A similar observation was reported by Klots and Compton[1]. The overall energy dependence of negative ions of the $(CO_2)_nO^-$ type is similar to that of O^- from CO_2. By contrast the attachment cross section for $(CO_2)_2^-$ ions shows only a single peak at low electron energy.

The dependence of negative and positive cluster ion intensity on the stagnation pressure (and temperature) has been measured to obtain information about possible fragmentation and is illustrated in Fig. 1. From the different onsets it can be concluded that the $(CO_2)_2^-$ clusters are formed predominantly by dissociative attachment of higher neutral clusters. It can also be seen that the pressure dependence of positive ions is similar to that for negative ions. However it has to be taken into account that the different onsets may be obscured by the different fragmentation patterns.

Work supported by Österr. Forschungsfonds, S-18/05 and 08.

1. C.E. Klots and R.N. Compton, J. Chem. Phys. 69, 1636 (1978)
2. A. Stamatovic and G.J. Schulz, Phys. Rev. A7, 589 (1973) and references therein
3. K. Stephan et al., J. Chem.Phys. 77, 2408 (1982)

a) PMF Kragujevac, Jugoslavia
b) SRI International, Menlo Park, Ca 94025, USA

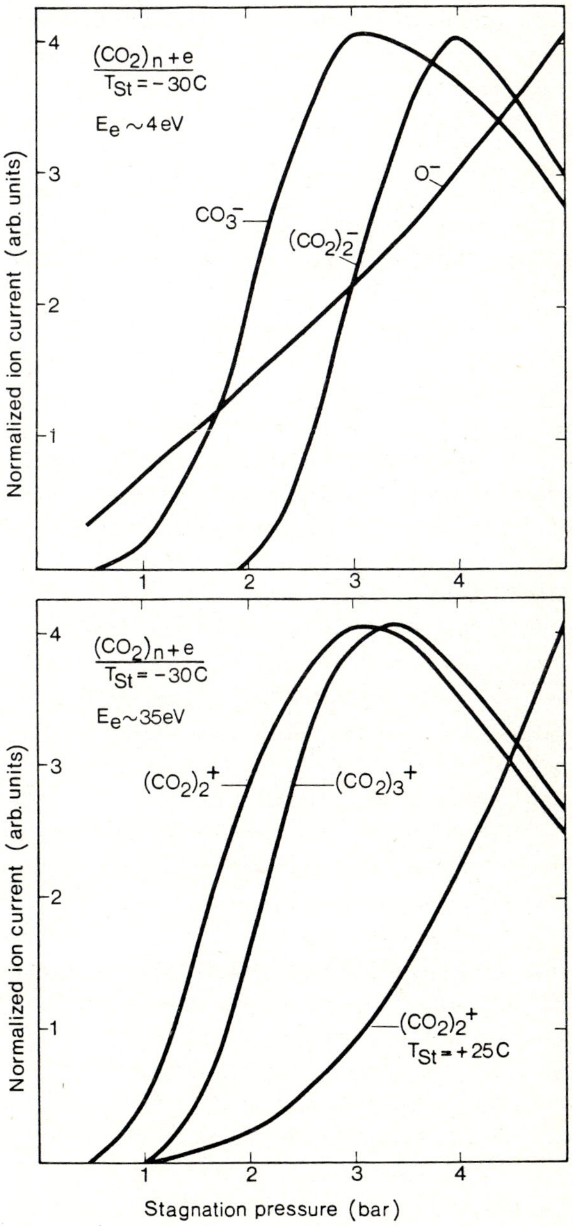

FIG.1

ABSOLUTE PARTIAL AND TOTAL ELECTRON IMPACT IONIZATION CROSS SECTION FUNCTIONS FOR CF_4 and CCl_4

K. Stephan, K. Leiter, H. Deutsch[a] and T. D. Märk

Institut für Experimentalphysik, Leopold Franzens Universität, A 6020 Innsbruck, Austria

A number of electron and photon impact ionization studies have been made on halo-methanes; and although various positive and negative ions have been observed, there remains considerable interest in the quantitative determination of partial and total electron impact ionization cross section functions. The ionization of these halocompounds is not restricted to the interests of fundamental physics, but has become recently of applied relevance in the frame of plasma etching[1]. The emphasis of this paper will be on the quantitative study of the electron impact ionization process of CF_4 and CCl_4 including a preliminary study of the production of fragment ions with excess kinetic energy.

All experiments were performed on a double focussing, reverse geometry, sector field mass spectrometer in combination with an improved electron impact ion source. The experimental details are given elsewhere[2-4]. In order to alleviate problems in the extraction and detection efficiency of fragment ions with excess kinetic energy some modifications have been made, improving and extending the previous extraction and detection modes[2-4]. Absolute calibration of measured relative partial ionization cross section functions was made with help of measured cross section ratios and by normalization against the Ar^+/Ar cross section with help of the method of effusive flow[5,6].

Using this method we have determined from threshold up to 180 eV partial ionization cross section functions for the production of CF_3^+, CF_2^+, CF^+, C^+, F^+, CF_3^{2+}, CF_2^{2+} and CF^{2+} in CF_4, and CCl_3^+, CCl_2^+, CCl^+, C^+, Cl_2^+, Cl^+, CCl_3^{2+}, CCl_2^{2+} in CCl_4. Fig. 1 shows a comparison between experiment and theory (full dot : total ionization cross section by Ref.7; dashed line: present counting ionization cross section; full line 1: semiclassical binary encounter approximation; full line 2: classical binary encounter approximation).

In agreement with earlier investigations we have not detected any stable parent ion. However, the unimolecular dissociation process $CF_4^+ \rightarrow CF_3^+$ and $CCl_4^+ \rightarrow CCl_3^+$ was investigated using the technique[4] of decoupling the acceleration and analyzer fields. Moreover, using n-th root extrapolation the appearance energy of the various singly and doubly charged ions was determined, i.e. $AP(CF_3^{2+})$ = 41.8 eV, $AP(CF_2^{2+})$ = 42.9 eV, $AP(CF^{2+})$ = 52.1 eV, $AP(CCl_3^{2+})$ = 30.4 eV, $AP(CCl_2^{2+})$ = 31.8 eV.

Work supported by Österreichischer Forschungsfonds, Projekt S-18/05 and 08.

References

1. E.g. see: R. d'Agostino, F. Cramarossa and S. De Benedictis, Plasma Chem. Plasma Processing, 2, 213 (1982); H.U. Poll and H. Schlemm, Beitr. Plasmaphysik, 22, 195 (1982); H.J. Tiller, R. Mohr, D. Bery, F.W. Breitbarth, K. Dumke and R. Göbel, Beitr. Plasmaphysik, 21, 329 (1981)
2. K. Stephan, H. Helm and T.D. Märk, J. Chem. Phys., 73, 3763 (1980)
3. Y.B. Kim, K. Stephan, E. Märk and T.D. Märk, J. Chem. Phys., 74, 6771 (1981)
4. E. Märk, T.D. Märk, Y.B. Kim and K. Stephan, J. Chem. Phys., 75, 4446 (1981)
5. T.D. Märk, J. Chem. Phys., 63, 3731 (1975)
6. D. Rapp and P. Englander-Golden, J. Chem. Phys., 43, 1464 (1965)
7. J.A. Beran and L. Kevan, J. Phys. Chem. 73, 3866 (1969)

a Inst. f. Physik/Elektronik, Univ. Greifswald, DDR

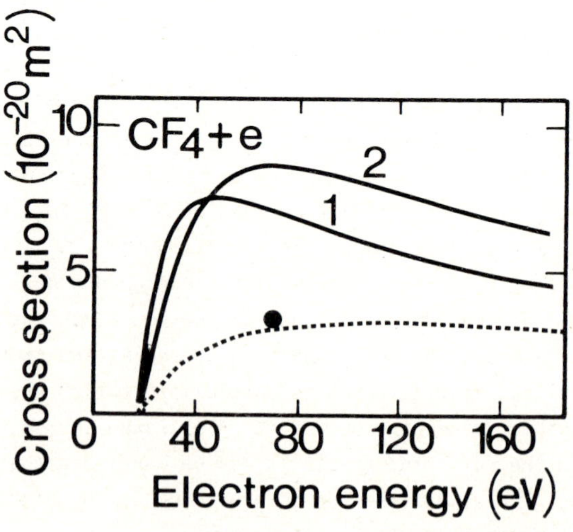

Fig. 1

(e,2e) REACTION IN MONOALODERIVATIVES OF C_2H_4 MOLECULE: COMPARISON BETWEEN EXPERIMENTS AND THEORY

A. Giardini-Guidoni, V. Di Martino [(+)], I.E. Mc Carthy [(=)] and R. Fantoni

ENEA, Dip. TIB, Divisione Fisica Applicata, C.P. 65, 00044 Frascati (Rome), Italy

The binary (e,2e) experiments, which can give information on the reaction mechanism [(1)], allow to fully investigate the electronic structure of the target molecule by measuring the ionization potentials, the relevant satellite structure and the momentum distribution of the one-electron orbitals. The experimental results can be theoretically interpreted on the basis of the Green's function formalism by extending to (e,2e) reactions the treatment developed by Cederbaum and Domcke [(2)] for photoionization. Provided that impulse approximation holds [(1)] the (e,2e) cross section is factorized in:

$$\frac{d^5\sigma}{dK_A\, dK_B dE_B} = (2\pi)^4 \frac{K_A K_B}{K_o} t_{ee} S(\varepsilon,q)$$

where the labelling of electron energies and momenta uses the obvious convention [(1)]. t_{ee} is the half-off--shell Mott electron-electron scattering cross section while the spectral function S (ε , q) is a property of the molecular target. Considering both the initial neutral state ψ_o^N and the final ionic state ψ_s^{N-1} as products of single particle wave functions, S (ε ,q) can be written as:

$$S(\varepsilon, q) = F_s(q)\ \delta(\varepsilon - \varepsilon_s)\ P_1(s)$$

where ε_s is the energy of the s ionic state and $F_S(q)$ is the momentum distribution of the electron ejected to originate the s state. P_1 (s) is the pole strength of each transition $P_1(s) = |\langle \psi_s^{N-1} | a_1 | \psi_o^N \rangle|^2$

This term appears in the advanced Green's function

$$G_{e,n}(\omega - i\eta) = \Sigma \frac{\langle \psi^N | a_n^\dagger | \psi_s^{N-1} \rangle \langle \psi_s^{N-1} | a_1 | \psi_o^N \rangle}{\omega + E_s^{N-1} - E_o^N - i\eta}$$

$$\omega - \omega_o = - E_o + E_A + E_B = \varepsilon_s$$

whose poles gives the ionization potentials. Analysis of (e,2e) experiments by Green's function method has been essentially successful in several applications to small molecules, including the monofluoroethylene [(3)]. Here we report results on two larger molecules of the same series: C_2H_3Cl and C_2H_3Br . Ionization potentials measured in the energy spectra reported in Fig. 1 are in good agreement with theoretical predictions, even for the innermost valence orbital (4a') which gives origin to very many satellites. Angular distributions measured for C_2H_3Br (Fig. 2) show, however, the poor single particle description of the 2a" and 10a' molecular orbitals involving the large Bromine atom (a, b).

(1) R. Camilloni, A. Giardini-Guidoni, I.E. Mc Carthy, G. Stefani: Phys.Rev. 17A, 1634 (1978)

(2) L.S. Cederbaum, W. Domcke: Adv.Chem.Phys. 36, 205 (1977)

(3) R. Fantoni, A. Giardini-Guidoni, R. Tiribelli, R. Cambi, G. Ciullo, A. Sgamellotti, F. Tarantelli: Mol.Phys. 45, 839 (1982)

(+) Guest

(=) Permanent address: Flinders University, Adelaide, South Australia.

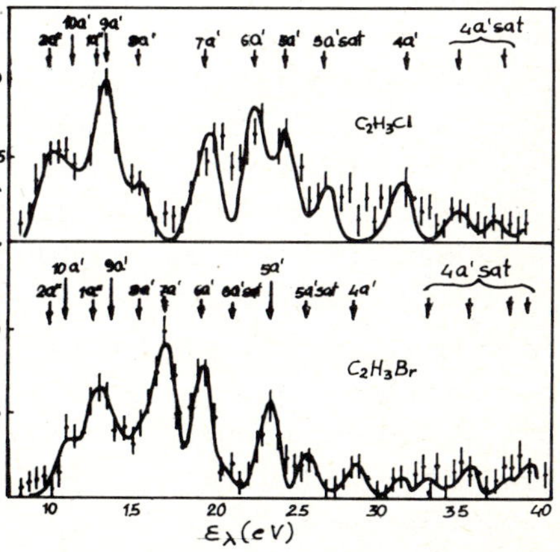

Fig. 1 - (e,2e) energy spectra measured at Ø = 0° (q = 0.1 a_o^{-1}). Full curve is best fit of data. Arrows mark IP's obtained through Gaussian deconvolution of data.

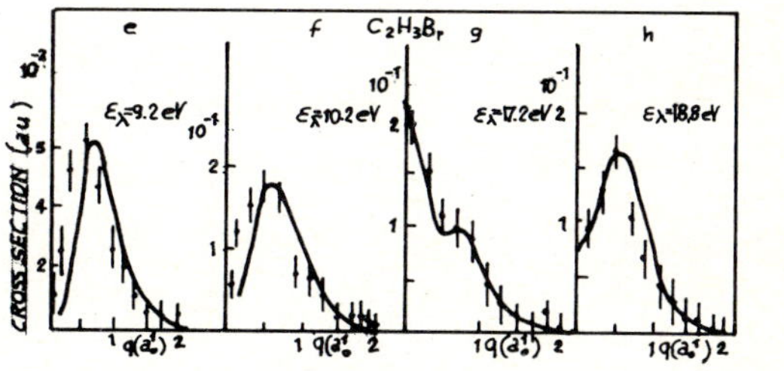

Fig. 2 - (e,2e) angular distributions measured at different ε_λ. Full curves are calculated from SCF wave functions of C_2H_3Br

THE USE OF ELECTRON IMPACT EXCITATION TO EXCITE ELECTRIC-DIPOLE FORBIDDEN TRANSITIONS IN ATOMS AND MOLECULES

David A Shaw, George C King and F H Read

Physics Department, Manchester University, UK.

In the study of inner-shell transitions in atoms and molecules the electron energy-loss technique has some advantages over photon absorption measurements using synchroton radiation sources. Thus for states with high values of excitation energy (≳200 eV) electron impact can provide superior energy resolution. In the past we have exploited this advantage and have used the technique to obtain a wealth of new information about inner-shell excitation in atoms and molecules, for example King et al[1], Tronc et al[2], Shaw et al[3]. Another and perhaps more important advantage of electron-impact excitation is its ability to induce electric-dipole forbidden transitions at low values of electron incident energy. We have recently reported[4] the first use of electron impact to study such dipole forbidden inner-shell transitions, and here we present the application of the technique to a number of atoms and molecules.

We present energy loss spectra obtained near the $M_{4,5}$ edges of Kr and the $N_{4,5}$ edges of Xe. In these measurements the incident electron energy is varied from 1500 eV to typically 150 eV. We resolve individual J values of the energy-loss structure and observe dramatic changes in the relative intensities of these J levels as the incident electron energy is varied.

We also present energy-loss spectra obtained in CO, CO_2 and N_2O again as a function of incident electron energy. At the lower values of incident electron energy, typically twice the excitation energy of the molecular state, forbidden transitions to inner-shell triplet states can be observed and we report the first direct observation of these in CO, CO_2 and N_2O. The figure shows energy loss spectra obtained in CO where the presence of the $(1s)^{-1}\ \pi 2p\ ^3\Pi$ state may be seen at the lower value of electron incident energy.

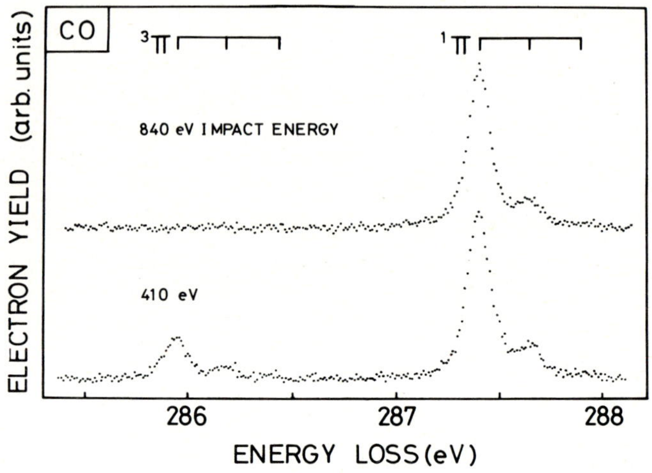

Fig. 1: Electron energy loss spectra of CO at incident electron energies of 840 and 410 eV

References

1. G C King, M Tronc, F H Read and R C Bradford, J Phys B 10 2479 (1977)

2. M Tronc, G C King and F H Read J Phys B 12 137 (1979)

3. D A Shaw, G C King, F H Read and D Cvejanovic J Phys B 15 1785 (1982)

PROBING OF ORBITALS BY BINARY (e,2e) SPECTROSCOPY AND ORBITAL DENSITY TOPOGRAPHY: A VIEW OF CHEMICAL BONDING IN MOMENTUM SPACE

K.T. Leung and C.E. Brion

Dept. of Chemistry, University of British Columbia, Vancouver, Canada

Recent developments in binary (e,2e) spectroscopy [1] have clearly demonstrated the importance of the (e,2e) reaction in the study of molecular electronic structure. Using an improved, high momentum resolution noncoplanar symmetric binary (e,2e) spectrometer [2], we have measured the binding energy spectra and valence orbital momentum distributions of the noble gases He, Ne, Ar, Kr and Xe [2], as well as a number of small molecules including H_2 [3], F_2, CO_2, OCS, CS_2 and CF_4 [4]. Figure 1 shows the binding energy spectra (ϕ=0°) for the noble gases. Significant population splittings among several ion states have been found for ns ionization in the noble gases. Similar breakdown of the independent particle ionization picture in the valence regime has also been found for CO_2, OCS and CS_2 [4]. Figure 2 compares the spherically averaged orbital momentum distribution of the H_2 $1\sigma_g$ orbital measured by binary (e,2e) spectroscopy with theoretical momentum distributions calculated from a number of literature SCF electronic wave-

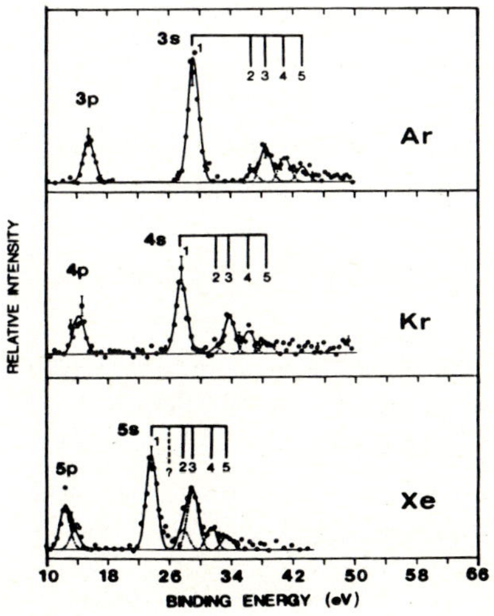

Figure 1

functions of different qualities. It can be seen that the measured momentum distribution is in good agreement with that obtained by Weigold et al [5] and also with results of Compton scattering [7]. The present result does not agree with the work of Migdall et al [6].

Computer generated contour maps of orbital density and density difference functions in both position and momentum space have been used to interpret experimental (e,2e) momentum distributions. We have

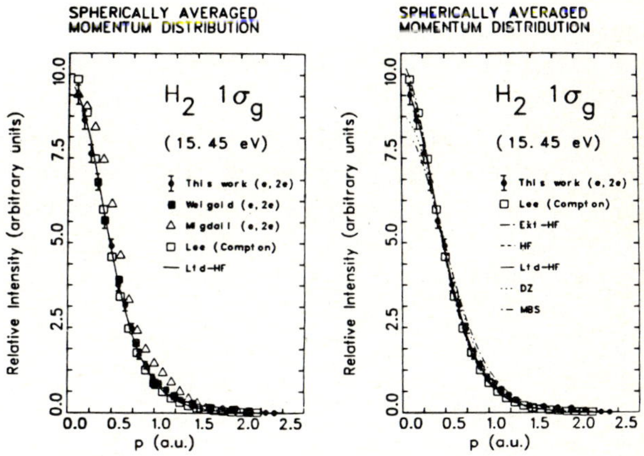

Figure 2

recently developed a 3-D orbital density visualization procedure [3] to further facilitate the understanding of chemical bonding and molecular structural properties in the momentum space representation. Such a topographical study of momentum space chemical

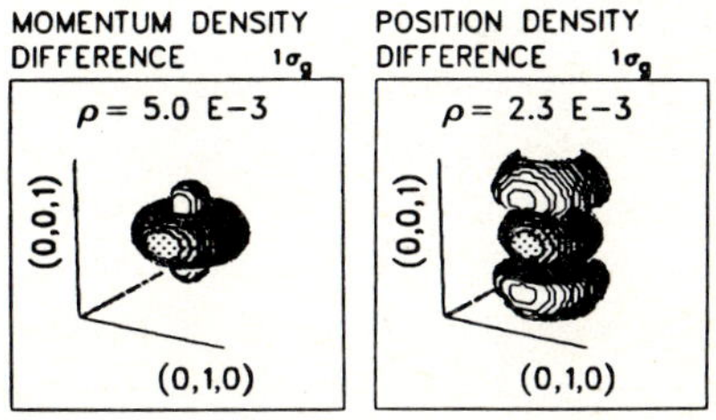

Figure 3

properties provides a new perspective for chemical bonding phenomena [3]. The procedure has been used to investigate the single covalent sigma bond in H_2 (see Figure 3) and the dynamics of chemical binding for the process: H + H → H_2 in momentum space as a function of the internuclear separation.

This work was supported by NSERC (Canada).

[1] E. Weigold (ed.); "Momentum Wave Functions", AIP Conference Proceedings No. 86, New York, 1982.

[2] K.T. Leung and C.E. Brion; submitted to Chem. Phys.

[3] K.T. Leung and C.E. Brion; submitted to Chem. Phys.

[4] K.T. Leung and C.E. Brion; to be published.

[5] E. Weigold, I.E. McCarthy, A.J. Dixon & S. Dey, Chem. Phys. Letters 47 , 209 (1977).

[6] J.N. Migdall, M.A. Coplan, D.S. Hench, J.H. Moore, J.A. Tossell, V.H. Smith Jr. and J.W. Liu, Chem. Phys. 57, 141 (1981).

[7] J.S. Lee, J. Chem. Phys. 66, 4906 (1977).

A NEW HIGH PERFORMANCE ELECTRON ENERGY LOSS SPECTROMETER FOR VALENCE AND INNER-SHELL MOLECULAR ELECTRONIC SPECTROSCOPY

C.E. Brion, S. Daviel and A.P. Hitchcock†

Dept. of Chemistry, University of British Columbia, Vancouver, Canada

A large radius electron energy loss spectrometer [1] has been designed, constructed and tested. Special features include high sensitivity, separate differential pumping of the gun, monochromator, collision chamber and analyser/detector regions as well as sophisticated electron optics for focussing and beam transport. The differential pumping permits introduction of a wide variety of gas samples including corrosive and reactive species without requiring re-tuning

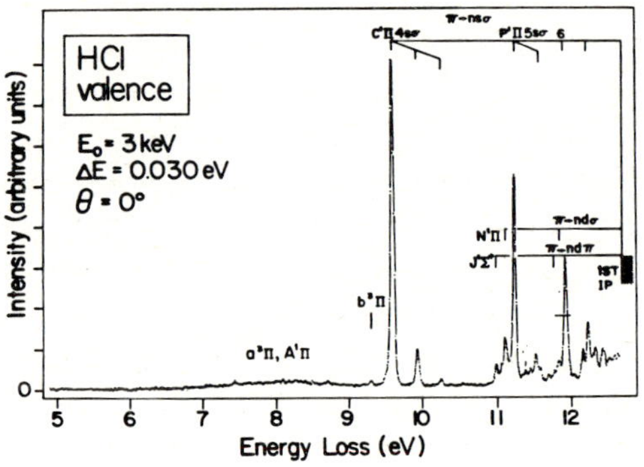

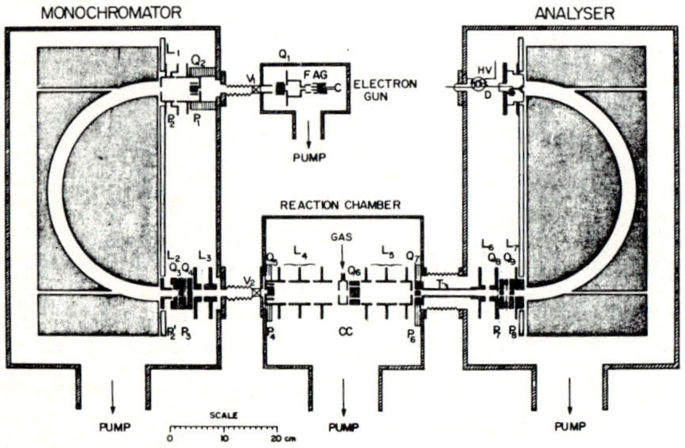

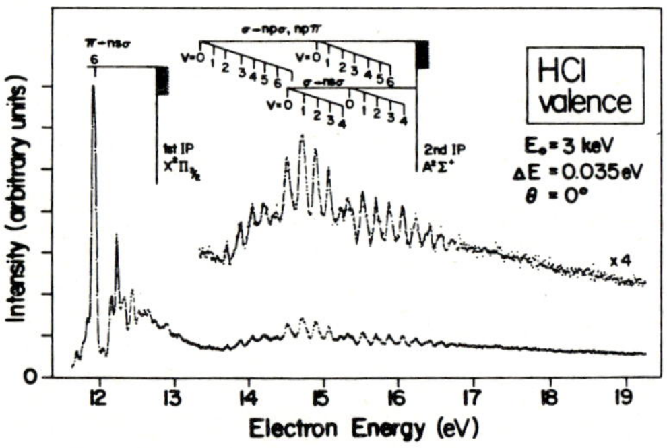

of the gun, monochromator or analyser. The large radius (20 cm) of the hemispheres permits use of high impact energies (1-10keV) with high energy resolution since the lens acceleration/deceleration ratios are kept within reasonable limits. At the present interim stage of testing resolution has been as good as 0.030 ev (FWHM) but this does not represent the expected limit of performance. The instrument functions well at zero degrees scattering angle and it is planned for use in the measurement of absolute dipole oscillator strengths. Typical high resolution spectra for core and valence shell excitation of N_2 [1] and HCℓ [1,2] are shown in the figures. These and other results will be presented. A multidetector [3] incorporating microchannel plates, phosphor screen, fibre optics and a photodiode array is being installed. This should result in further gains in sensitivity. This work was supported by NSERC (Canada).

References

[1] S. Daviel, C. Brion and A. Hitchcock, to published.

[2] S. Daviel and C.E. Brion, to be published.

[3] P.J. Hicks, S. Daviel, B. Wallbank & J. Comer, J. Phys. E 13, 713 (1980).

† Permanent address: Dept. of Chemistry, McMaster University, Hamilton, Canada.

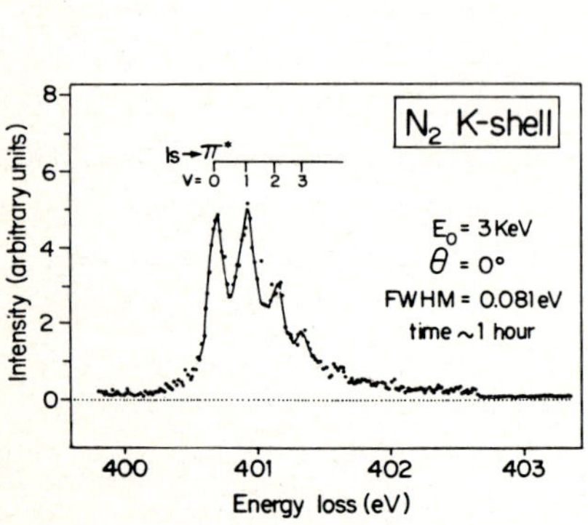

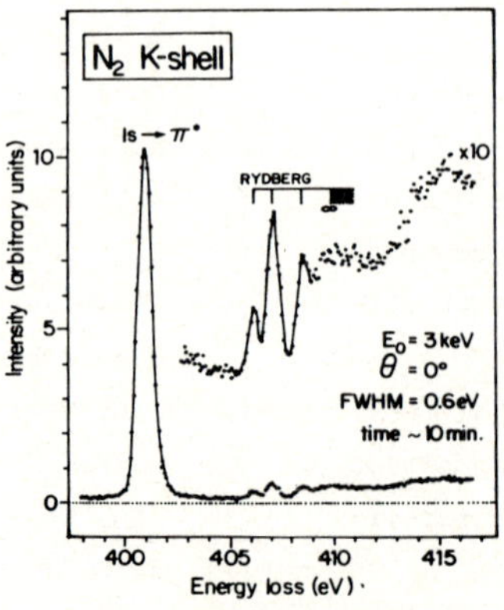

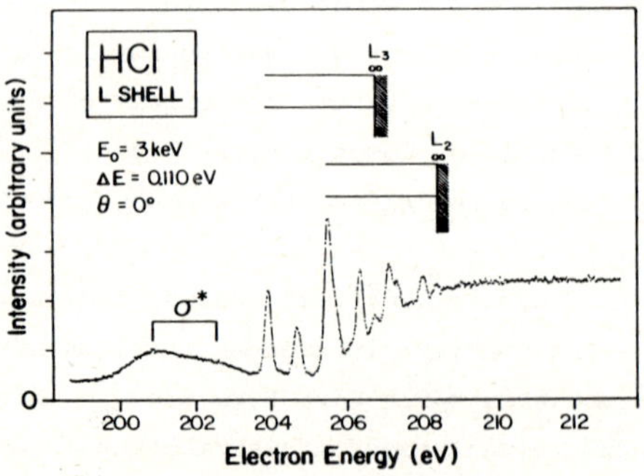

MEASUREMENT OF PARTIAL GENERALIZED OSCILLATOR STRENGTHS FOR IONIZATION OF NITROGEN MOLECULE BY 1-KEV ELECTRON IMPACT

H.Shibata, K.Kuroki#, F.Nishimura and N.Oda

Research Laboratory for Nuclear Reactors, Tokyo Institute of Technology, Meguro-ku, Tokyo, Japan

We have measured the double-differential cross sections (DDCS) for ionizing collisions by electron impacts on many kinds of atoms and molecules in the last decade, and derived the generalized oscillator strengths (GOS) from these double-differential cross sections. However, the initial as well as the final states of the transitions of intrest can not be uniquely determined from the double-differential electron spectra only if many orbitals take part in these transitions.

The aim of this study is to identify the initial and final states for ionization processes of atoms and molecules by using the electron-electron coincidence method and thereby to obtain more detailed information on the structure of the GOS, that is, the partial generalized oscillator strengths (PGOS).

Under coplanar and asymmetric experimental conditions, only a few measurements have been so far reported in the energy range above 500 eV and for molecules. Lahmam-Bennani et al.[1)] have recently measured the triple-differential cross sections (TDCS) for Ar at 8 keV incident energy, and Jung et al.[2)] measured those for H_2 and N_2 at 100 and 250 eV. In the present work, the TDCS have been measured for nitrogen molecule bombarded by

Fig.1

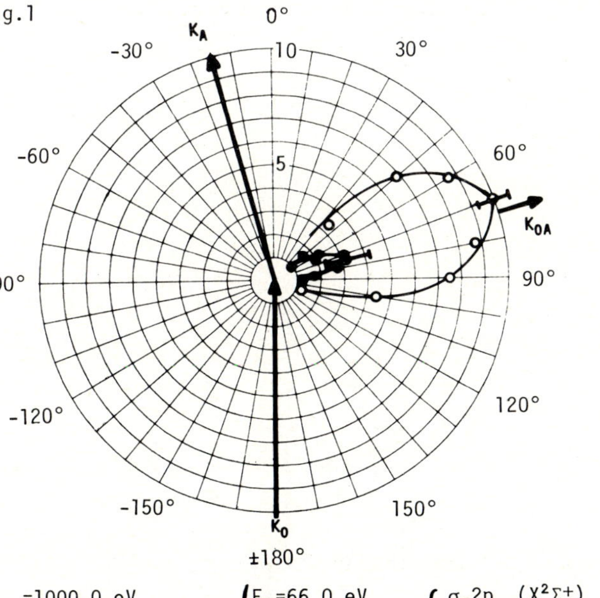

E_0 =1000.0 eV E_{loss}= 83.0 eV	○	E_B=66.0 eV ε_i=17.0 eV *	$\sigma_g 2p$ ($X^2\Sigma_g^+$) $\pi_u 2p$ ($A^2\Pi_u$) $\sigma_u 2s$ ($B^2\Sigma_u^+$)
E_A = 917.0 eV θ_A = -15 deg	●	E_B=45.0 eV ε_i=38.0 eV *	$\sigma_g 2s$ ($^2\Sigma_g^+$)

* The values of ε_i are the binding energies corresponding to the maxima in the measured energy distributions of ejected electrons.

1 keV electrons using the crossed-beam method. In the measurements of the PGOS, when the energy loss of the incident electrons is fixed, the energy of ejected electrons varies, corresponding to the individual molecular orbitals taking part in the ionizing collision.

An electron beam from an electron gun with momentum $\mathbf{k}_0$ and energy E_0 collides with a target gas beam from a single-tube nozzle at right angle. The energy analysis is performed using two 45° parallel-plate mirror energy analyzers which have the same dimensions, and the energy and angular resolutions are 0.5% and 2° for the scattered electron analyzer, and 1.4% and 2° for the ejected electron analyzer, respectively. The present work was performed for the coplanar arrengement where the azimuthal angle ϕ_B equals 0 or π.

Fig.1 shows an example of angular distribution of ejected electrons which are in coincidence with scattered electrons at the scattered angle θ_A=-15°. The scattered electron energy E_A is 917 eV corresponding to the energy loss of 83.0 eV, and the ejected electrons consist of two groups with energies of 66.0 eV (open circle ○) and 45.0 eV (closed circle ●). The former group corresponds to the electrons ejected from the orbitals of $\sigma_g 2p$ (15.59 eV, $X^2\Sigma_g^+$), $\pi_u 2p$ (16.96 eV, $A^2\pi_u$), and $\sigma_u 2s$ (18.78 eV, $B^2\Sigma_u^+$),and the latter from $\sigma_g 2s$ (37.3 eV, $^2\Sigma_g^+$). In the parentheses are indicated the ionization potentials ε_i and the final ionic states. Three states of residual ions in the former group could not be resolved because of the poor resolution of the analyzers. Three arrows in the figure indicate the directions of the incident electrons ($\mathbf{k}_0$), scattered electron ($\mathbf{K}_A$) and the momentum transfer ($\mathbf{K}_{OA}$), respectively. The recoil peak was not found in our experimental condition, because the collisions may be regarded to be due to the binary electron-electron collision process for such high momentum transfers, K_{OA}= 2.2 a.u. The DDCS's were calculated by integrating the TDCS over the ejected angle θ_B at ϕ_B=0, resulting in that the DDCS for the final ionic states of the $X^2\Sigma_g^+$, $A^2\pi_u$, and $B^2\Sigma_u^+$ group is about five times larger than that for the $^2\Sigma_g^+$ state.

Further measurements are in progress.

1) A.Lahmam-Bennani, H.F.Wellenstein, A.Duguet and M.Rouault J.Phys.B:At.Mol.Phys. 16 121 (1983)

2) K.Jung, E.Schubert, D.A.L.Paul and H.Ehrhardt J.Phys.B:At.Mol.Phys. 8 1330 (1975)

On leave from National Research Institute of Police Science

STUDIES OF DISSOCIATIVE EXCITATION OF N_2 AND O_2 BY ELECTRON IMPACT WITH EXTREME ULTRAVIOLET EMISSION*

J. M. Ajello

Jet Propulsion Laboratory, California Institute of Technology, Pasadena, California 91109

We have measured in the laboratory the extreme ultraviolet spectrum of O_2, N_2 and other molecules (CO, NO, H_2, CO_2) from 40 to 130 nm at electron impact energies from 0 to 350 eV. It is found that the extreme ultraviolet (EUV) portion of the spectrum is rich in atomic emission features from atoms in various stages of ionization, e.g., NI, NII and NIII for N_2 and OI, OII and OIII for O_2. For these features we have measured absolute emission cross sections. These cross sections are important in defining the total amount of energy deposition by electrons into excited dissociative states. The weakness of emission band systems provides an indication of preionization and/or predissociation yields of Rydberg states. The observed atomic lines arise partially from predissociation of Rydberg states. For energies above the ionization potential (within 0.5 eV) the Rydberg states of a molecule are 100% preionized and/or predissociated.

The experimental apparatus has been described in detail elsewhere.[1] The instrument consists of an electron impact collision chamber in tandem with a UV spectrometer. The relative wavelength sensitivity was accomplished by a double monochromator optical calibration. We show in Fig. 1 a calibrated spectrum of N_2 at 100 eV from 50 to 190 nm. Additionally we show in Fig. 2 a calibrated spectrum of N_2 at 200 eV. The molecular features are identified in Fig. 1 and the atomic features are identified generally according to their Rydberg series in Fig. 2. For the atomic emissions of NI the strongest features fit into Rydberg series which decrease in intensity with increasing principal quantum number. The total dissociative vs. molecular emission cross sections can be compared as a function of energy.

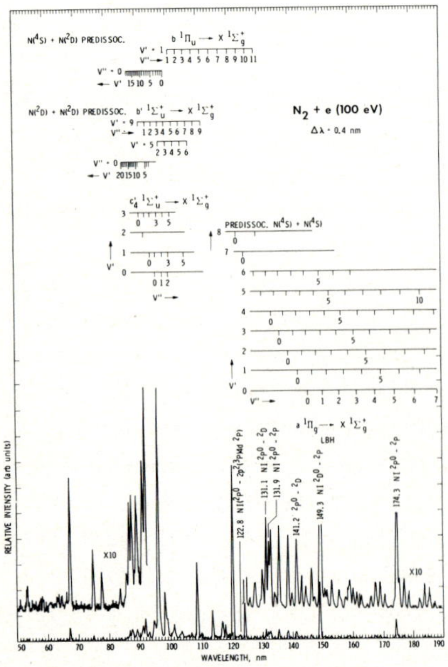

Fig. 1: Calibrated optically thin spectrum of N_2 at 100 eV, 0.5 nm resolution.

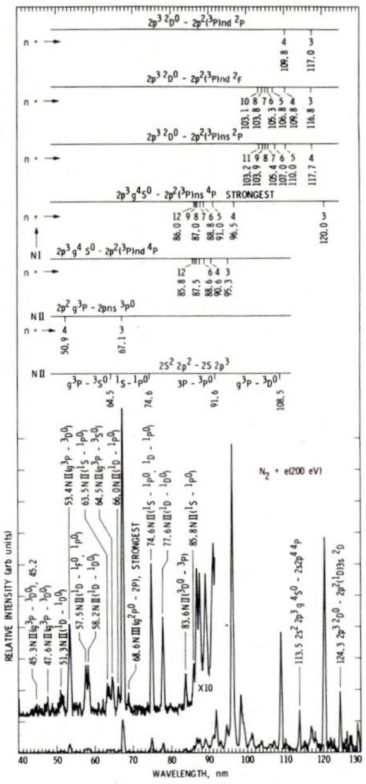

Fig. 2: Calibrated spectrum of N_2 at 200 eV.

We have measured the EUV spectrum of O_2 + e in Fig. 3. In distinction to N_2 the O_2 spectrum consists entirely (>98%) of atomic emission features. A few unidentified features with question marks may be weak molecular features at energies below the ionization potential. The strongest feature is the 83.3 nm resonance line of OII ($^4P-^4S^0$). The strongest OI features fit into Rydberg series as shown in Fig. 3.

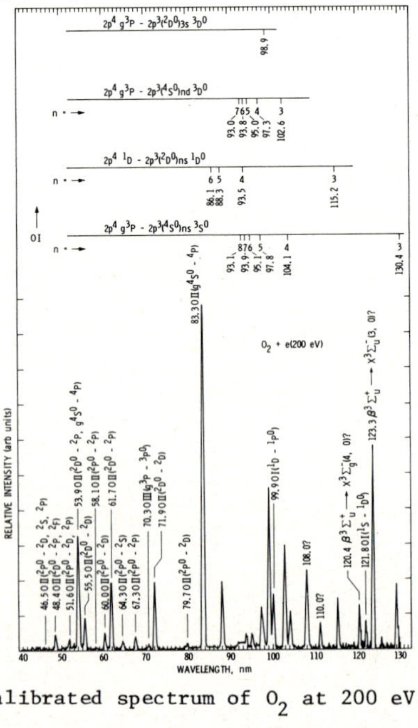

Fig. 3: Calibrated spectrum of O_2 at 200 eV.

*This work has been supported by AFOSR, NASA Planetary Atmospheres and Astronomy/Astrophysics Program Offices.

Reference

1. J. M. Ajello and S. K. Srivastava, J. Chem. Phys. 75, 4544 (1981); J. M. Ajello, S. K. Srivastava, Y. L. Yung, Phys. Rev. A. 25, 2485 (1982).

ANGULAR DISTRIBUTION OF BALMER-α EMISSION EXCITED BY ELECTRON IMPACT ON H_2.

N. Kouchi[1)], N. Takahashi, S. Arai, M. Morita, N. Oda* and Y. Hatano

Department of Chemistry and Research Laboratory of Nuclear Reactors*,
Tokyo Institute of Technology, Meguro-ku, Tokyo 152, Japan

Observation of optical emission from fragment atoms produced by dissociative excitation of molecules is of great importance for understanding highly excited moleoular states and their dissociation processes. In recent several years, we investigated the electron impact dissociation of hydrogen and other simple molecules by Doppler profile measurements of Balmer-α emission,[2,3)] and indicated that Doppler profile depends not only on kinetic energy distribution of fragment atoms but also on their population to magnetic sublevels and that more direct information about highly excited states and their dissociation processes may be obtained by measuring the anisotropy of the emission[4)]. This paper presents the angular distribution of Balmer-α emission excited by electron impact on H_2.

The anisotropy of optical emission has been mainly described using the polarization degree Π:

$$\Pi = \frac{I_{\parallel} - I_{\perp}}{I_{\parallel} + I_{\perp}}, \qquad (1)$$

where $I_{\parallel}$ and $I_{\perp}$ are emission intensities whose electric vector are parallel and perpendicular to incident electron beam, respectively. The value of Π is also obtained from eq. (2) by measuring the angular distribution of optical emission I(θ),

$$I(\theta) = I(90°)(1-\Pi\cos^2\theta), \qquad (2)$$

where θ is the detecting angle with regard to the electron beam. Since the measurement of $I_{\parallel}$ and $I_{\perp}$ has been generally affected by some difficulties in determining the detection efficiency of $I_{\parallel}$ and $I_{\perp}$, the latter method based on eq. (2) might have an advantage for obtaining the value of Π. Thus we have measured the angular distribution I(θ)/I(90°) instead of $I_{\parallel}$ and $I_{\perp}$.

Fig. 1 shows I(θ)/I(90°) vs $\cos^2\theta$. The value of Π is obtained from the slope of this straight line. Fig. 2 shows the value of Π obtained in this way as a function of incident electron energy. Our result is compared with the results[5,6)] obtained from $I_{\parallel}$ and $I_{\perp}$ measurements. There is a large discrepancy among the values in Fig. 2. A large discrepancy between the result of Karolis & Harting and that of Glass-Maujean may reflect a difficulty in determining the detection efficiency of $I_{\parallel}$ and $I_{\perp}$.

The anisotropy of emission as shown in Fig. 2 is ascribed to the anisotropy of H*(3) formation and the anisotropy of the population to the magnetic sublevels of H*(3). The fact that the emission becomes near isotropic with increasing electron energy is also ascribed to the following two reasons. The first is that angular distribution of H*(3) becomes near isotropic. Secondly, the emission from H*(3s) may become predominant with increasing electron energy.

References

1. Present adress: Res. Center for Nucl. Sci. and Eng., Univ.Tokyo, Tokai-mura, Ibaraki 319-11, Japan
2. K.Ito, N.Oda, Y.Hatano and T.Tsuboi, Chem.Phys.17, 35 (1976); ibid.21, 203 (1977); N.Kouchi, K.Ito, Y.Hatano, N.Oda and T.Tsuboi, ibid.36, 239 (1979); M.Ohno, N.Kouchi, K.Ito, N.Oda and Y.Hatano, ibid.58, 45 (1981); N.Kouchi, M.Ohno, K.Ito, N.Oda and Y.Hatano, ibid.67, 287 (1982).
3. Y.Hatano, Invited Papers, Symposium on Electron Molecule Collisions (Tokyo, 1977) p.135 and references therein; Y.Hatano, Comments on At.Mol.Phys., to be published.
4. N.Kouchi, K.Ito, N.Oda and Y.Hatano, Chem.Phys.70, 105 (1982).
5. C.Karolis and E.Harting, J.Phys. B 11, 357 (1978).
6. M.Glass-Maujean, J.Phys. B 11, 431 (1978).

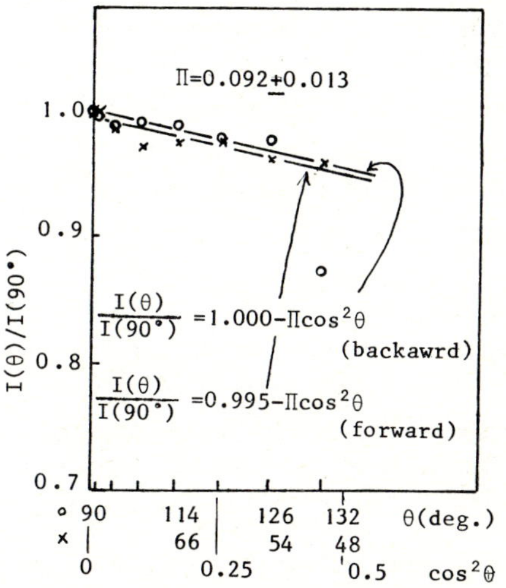

Fig. 1: Angular distribution of Balmer-α emission at the incident electron energy of 80eV.

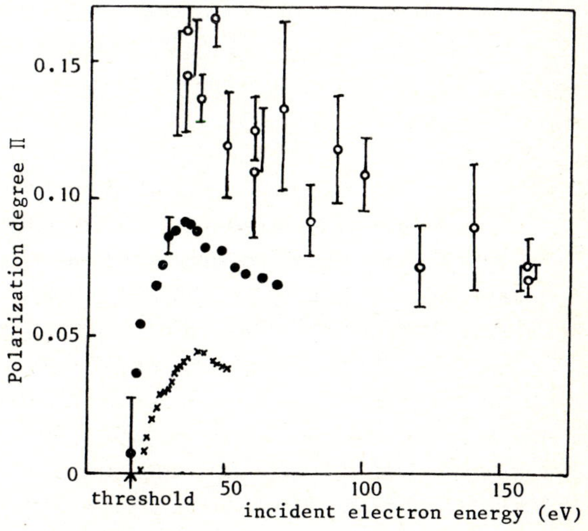

Fig. 2: Polarization degree Π as a function of incident electron energy. ○: this work, ●: Karolis & Harting (Ref. 5), ×: Glass-Maujean (Ref. 6), ● and × are obtained from $I_{\parallel}$ and $I_{\perp}$ measurements.

TRANSLATIONAL ENERGY AND ANGULAR DISTRIBUTIONS OF H* AND D* PRODUCED IN e-H_2,D_2 COLLISIONS

Teiichiro Ogawa and Junichi Kurawaki

Department of Molecular Science and Technology, Kyushu University, Kasuga-shi, Fukuoka 816, Japan

Electron impact excitation and high resolution spectroscopy are useful techniques for the investigation of the dissociation dynamics of highly-excited molecules. The spectral line shape of the Balmer lines of H* and D* produced in collisions of hydride molecules with electrons is determined by the Doppler effect and, thus, represents the motion of H* and D*. If the line shape is isotropic, its differentiation gives the translational energy distribution of the emitting species.[1)]

The line shapes produced in e-H_2,D_2 collisions have been measured at 55° and 90° with respect to the electron beam and anisotropy is concluded to be smaller than experimental uncertainties. The translational energy (TE) distributions of H* and D* shows there are three components of the excited hydrogen atoms and their assignments are as follows.[2)]

Component 1: TE peak 0 eV, threshold energy 17.1 eV. Direct dissociation and predissociation through the Rydberg states converging to the $^2\Sigma_g^+(1s\sigma_g)$ state of H_2^+.

Component 2: TE peak 4-6 eV, threshold energy 24-26 eV. Excitation to doubly excited states such as $H_2(2p\sigma_u)^2$ and dissociation through the above Rydberg states after curve crossing.

Component 3: TE peak 7-8 eV, threshold energy 26-27 eV. Direct dissociation through the Rydberg states converging to the $^2\Sigma_u^+(2p\sigma_u)$ state of H_2^+.

However, if the dissociation proceeds through a repulsive potential curve with a well-defined symmetry, the dissociation process should be anisotropic. We have recently constructed a collision apparatus to measure both the translational energy distribution and the angular distribution. The schematic diagram of this apparatus is shown in Fig. 1. The electron gun can be rotated from 35° to 145°; an aperture limits acceptance angles to 1 - 3°. The base pressure is of the order of 10^{-7} Torr and the operating pressure is of the order of 10^{-4} - 10^{-5} Torr. The optical resolution in the line shape measurements is about 0.3 Å with the use of a Fabry-Perot interferometer.

The angular distributions of two groups of H* produced in e-H_2 collisions at 100 eV are shown in Fig. 2. The slow group, component 1, is isotropic. This finding is consistent with its assignment; the dissociation through bonding potential curves takes enough time for rotation. The fast group, components 2 and 3, is anisotropic with a peak at 90°. This is consistent with the assignment of component 3, since the $2p\sigma_u$ curve is highly repulsive. The separation of the angular distribution of the fast group into components 2 and 3, and the analysis is now in progress.

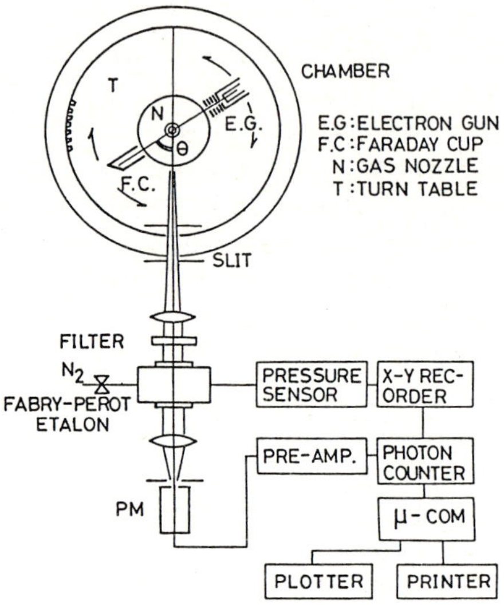

Fig. 1. The schematic diagram of the apparatus.

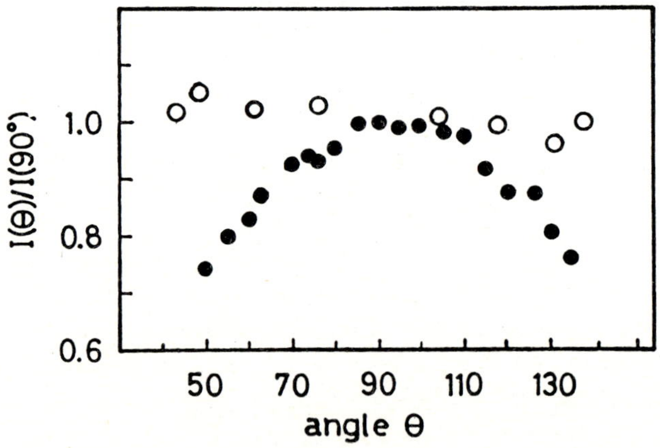

Fig. 2. Angular distributions of the excited hydrogen atom produced in e-H_2 collisions at 100 eV.
○ slow group (component 1)
● fast group (components 2 and 3)
Resolution: optical 0.26 Å, angle 1°.

References

1)T.Ogawa and M.Higo, Chem. Phys. Lett. 65,610 (1979).

2)M. Higo, S. Kamata and T. Ogawa, Chem. Phys. 73, 99 (1982) and references cited therein.

ANGULAR DISTRIBUTIONS OF HIGH-RYDBERG NITROGEN ATOMS PRODUCED BY ELECTRON IMPACT ON NITROGEN MOLECULE

T. Kondow,* S. Ohshima, T. Fukuyama and K. Kuchitsu

Department of Chemistry, Faculty of Science, The University of Tokyo, Bunkyo-ku, Tokyo 113, Japan
*and Institute for Molecular Science, Okazaki 444, Japan

When N_2 molecule is excited by electrons of more than 50 eV to a molecular high-Rydberg state, it dissociates promptly and produces a high-Rydberg nitrogen atom, N^{**}. The energies and symmetries of the precursor states are estimated in the present study by measurement of the kinetic-energy and angular distributions of the N^{**} atoms.

A supersonic molecular beam of N_2 was excited by pulsed electrons of 50 - 100 eV, and fragments, N^{**}, were observed by a rotatable high-Rydberg detector at 60-120° with respect to the electron beam axis, as shown in Fig. 1. The voltage applied to a pre-ionizer was so set that all the N^{**} atoms ($15 \leq n \leq 45$) were admitted into the detector. The time-of-flight distribution of the N^{**} atoms was obtained and was transformed to the distribution of released kinetic energy, E_k, estimated as twice the kinetic energy of the N^{**} atom.

Figure 2 shows typical kinetic-energy distributions measured at an impact energy of 80 eV at three scattering angles, θ. At 90°, there is one intense peak at $E_k = 8.5$ eV, one shoulder at $E_k = 7.4$ eV, and a broad tail

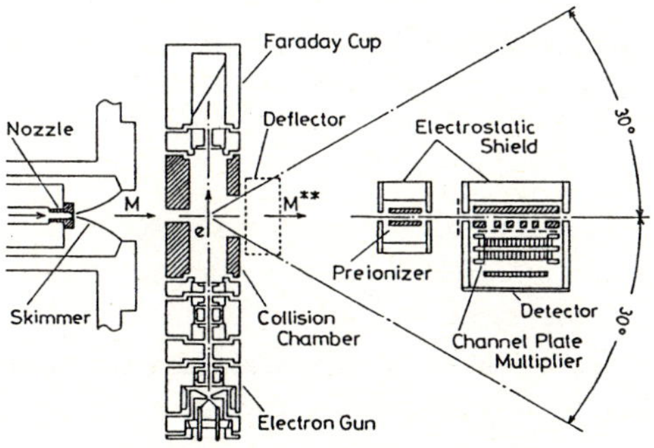

Fig. 1: A shematic diagram of the apparatus.

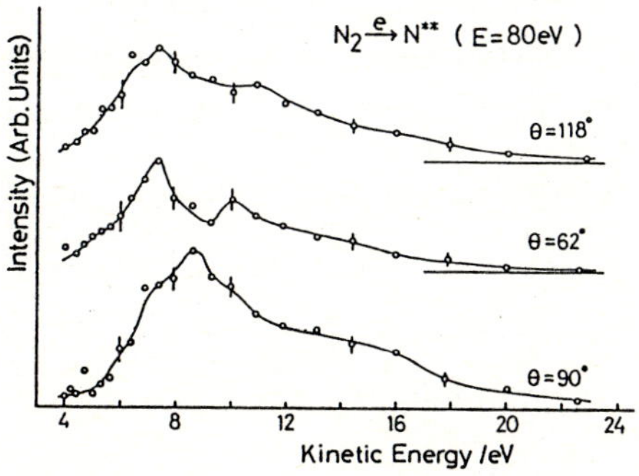

Fig. 2: Kinetic-energy distributions of the N^{**} atoms fragmented from N_2 by electron impact (80 eV) at three scattering angles.

at $E_k = 10 - 16$ eV. The present kinetic-energy distributions agree with those measured by Smyth et al.[1]

The angular distributions of the N^{**} atoms were derived from the θ-dependences of the kinetic-energy distributions, as shown in Fig. 3 for an impact energy of 80 eV. The distributions are angle-independent for $E_k < 7.4$ eV, whereas the distributions for $E_k \geq 8.5$ eV have a distinct peak at $\theta = 90°$, and the peak intensity decreases as E_k increases.

The observed kinetic-energy and angular distributions are interpreted on the basis of Dunn's selection rule[2], a core-ion model[1], and theoretical potential curves[3] as follows. The N^{**} atoms are fragmented mainly from molecular high-Rydberg states, $[N_2]^{**}$, which converge to N_2^{2+}; they are located at about 43 - 60 eV above the ground state of N_2 ($X^1\Sigma_g^+$) state in the Franck-Condon region. Since the N^{**} atoms having $E_k \lesssim 8$ eV show essentially angle-independent distributions as described above, the atoms, N^{**}, are expected to originate from the $X^1\Sigma_g^+$ state of $[N_2]^{**}$. On the other hand, the N^{**} atoms having $E_k \gtrsim 8$ eV (peaked at 90°, see Fig. 3) are contributed significantly by the $b^1\Pi_u$ state of $[N_2]^{**}$. The measured distributions at different impact energies indicate that the threshold for excitation of $[N_2]^{**}$ converging to N_2^{2+} is slightly lower than 50 eV, which is consistent with the result from double-charge transfer spectroscopy which gave 43.1 ± 0.5 and 45.2 ± 0.5 eV for the energies of $X^1\Sigma_g^+$ and $b^1\Pi_u$ states, respectively.[4]

Fig. 3: Angular distributions of the N^{**} atoms shown as a function of the released kinetic energy, E_k.

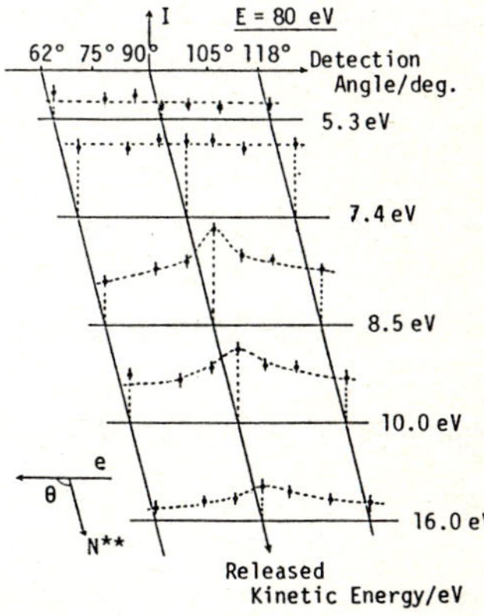

References

1. K.C. Smyth, S.A. Schiavone and R.S. Freund, J. Chem. Phys. 59, 5225 (1973)
2. G.H. Dunn, Phys. Rev. Lett. 8, 62 (1962)
3. E.W. Thulstrup and A. Anderson, J. Phys. B8, 965 (1975)
4. J. Appell, J. Durup, F.C. Fehsenfeld and P. Fournier, J. Phys. B6, 197 (1973)

THRESHOLD CAPTURE WIDTHS DETERMINED FROM BOUND RYDBERG STATES

Steven L. Guberman

Boston College, 885 Centre St., Newton, MA 02159

An L^2 approach is described for the calculation of threshold molecular capture/autoionization widths needed for the determination of dissociative recombination cross sections. The widths are calculated using Fermi's golden rule:

$$\Gamma = 2\pi\rho|\langle\Psi_f|H|\Psi^R_{n^*}\rangle|^2 \qquad (1)$$

where ρ is the density of states, Ψ_f is the wave function for the dissociative state and $\Psi^R_{n^*}$ is an appropriately antisymmetrized product of an ion and a high Rydberg orbital with an effective principal quantum number, n^*. In this approach, a bound Rydberg orbital replaces the coulomb orbital often used for the free electron in (1). Taking the high Rydberg orbitals to be hydrogenic, the orbital energy can be expressed as

$$E=-1/2(n^*)^2, \qquad (2)$$

where $n^*=n-\delta$. n is the principal quantum number of the Rydberg orbital and δ is the quantum defect. The density of Rydberg states can be written as[1]

$$\rho = 1/(E(n^*-.5)-E(n^*+.5)). \qquad (3)$$

Inserting (2) into (3) we have

$$\rho = 2[(1/(n^*-.5)^2)-1/(n^*+.5)^2]^{-1} \qquad (4)$$

which can be closely approximated by

$$\rho \approx (n^*)^3\exp(-.5/(n^*)^2). \qquad (5)$$

The exponential factor is a small correction for large n^*.

For $n^* \gg l$, where l is the angular momentum quantum number, the radial portion of the Rydberg orbital takes the form[2]:

$$R_{n^*l}(r)=2(Z/n^*)^{1.5}[(2Zr)^l/(2l+1)!]\exp(-Zr/n^*)F \qquad (6)$$

where F is the confluent hypergeometric function which is dependent on r but independent of n^* for $n^* \gg l$. Taking Z=1, the n^* dependence of the Rydberg orbital is given by $(n^*)^{-1.5}\exp(-r/n^*)$. Since the matrix element in (1) involves a region of coordinate space close to the molecule the exponential factor can be expected to be small. Calculations on O_2 indicate that the square of the matrix element in (1) is well represented by $(n^*)^{-3}K\exp(-C/n^*)$ where K and C are constants. For O_2 C is found to be about 0.2 and the exponential factor is small for large n^*. The total width in (1) is then given by:

$$\Gamma = 2\pi K\exp-((C/n^*)+.5/(n^*))$$

The limit $n^*\rightarrow\infty$ gives the threshold capture width, $\Gamma=2\pi K$. Applications of this approach to the calculation of widths needed for the description of the dissociative recombination of O_2^+ [3] will be described. Studies of the dependence of the width on configuration interaction in the entrance and exit channels will be presented.

References

1. J. N. Bardsley, J. Phys. B 1, 365(1968); A. P. Hickman, A. D. Isaacson amd W. H. Miller, Chem. Phys. Letters 37, 63(1976)
2. H. A. Bethe and E. E. Salpeter, Quantum Mechanics of One and Two Electron Atoms(Plenum, New York, 1977) pp 15-18.
3. S. L. Guberman, Potential Energy Curves for Dissociative Recombination, in Physics of Ion-Ion and Electron-Ion Collisions ed. by F. Brouillard(Plenum, New York,1983)pp 167-200.

SEMICLASSICAL CALCULATIONS OF DISSOCIATIVE ATTACHMENT IN COLLISIONS OF ELECTRONS WITH H_2,HD,D_2

I S Yelets and A K Kazansky

D V Efremov Scientific Research Institute of Electrophysical Apparatus, Leningrad, 188631,USSR

Dpt of Theor.Phys., Institute of Phys., Leningrad State University, Leningrad, 198904, USSR

Semiclassical approach to dissociative attachment (DA) was formulated by Herzenberg[1], but there were no systematic calculations based on this method. Recently, a numerical calculation of DA process was made [2,3]. Here we present the results of semiclassical consideration of the process.

The fundamental equation for the nuclear wave function $\xi(R)$ has the form (we use the local Herzenberg theory):

$$\left\{-\frac{1}{2M}\frac{d^2}{dR^2}+W(R)-E\right\}\xi(R)=-V(R)\zeta_n(R), \quad (1)$$

$$V(R)=[\Gamma(R)]^{\frac{1}{2}}\{2[U_I(R)-U(R)]\}^{-\frac{1}{4}}$$

with the boundary conditions

$$\xi(0)=0,\ d\ln\xi/dR|_{R\to\infty}=i\sqrt{2M[E-W(\infty)]}$$

For the DA cross section the following expression is used:

$$\sigma_{DA}=-\frac{i}{2M}\left(\xi^*(R)\frac{d\xi(R)}{dR}-\xi(R)\frac{d\xi^*(R)}{dR}\right)\Big|_{R\to\infty} \quad (2)$$

The semiclassical approach [4,5] gives:

$$\sigma^n_{DA}=8\pi^2(2E_n)^{-1/2}MA^2|I_n|^2, \quad (3)$$

where

$$A=\exp\left\{-I_m\int_{a_I}^{R_{st}}\sqrt{2M[E-W(R)]}\,dR\right\},$$

$$I_n=V(z_n)\,\omega^{\frac{1}{2}}\{2[\varepsilon_n-U(z_n)]\}^{-1/2}\{\Psi''(z_n)\}^{\frac{1}{2}}, \quad (4)$$

$$\cos\{\Psi(z_n)-\pi/4\},$$

$$\Psi(z)=\int_{a_I}^{z}\sqrt{2M[E-W(z)]}\,dz-\int_{a}^{z}\sqrt{2M[\varepsilon_n-U(z)]}\,dz$$

Here R_{st} is the intersection point of the potential curves of the molecule ($U(R)$) and the ion ($W(R)=U_I(R)-i\Gamma(R)/2$, $\Gamma=C\{2[U_I(R)-U(R)]\}^{3/2}$); z_n is the transition point: $E-W(z_n)=\varepsilon_n-U(z_n)$; a_I and a are the turning points for the ion and the molecule: $W(a_I)=E$; $U(a)=\varepsilon_n$.

The factor $|I_n|^2$ may be interpreted [4,5] as the probability for the compound system formation. In the case of hydrogen-isotope molecules the width Γ is large enough and only the rising exponent in trigonometric function (4) is essential:

$$\sigma^n_{DA}=\frac{4\pi^2}{\sqrt{2E_n}}\frac{M\omega V(z_n)}{\sqrt{2M[\varepsilon_n-U(z_n)]}}\left|\frac{dU}{dR}-\frac{dW}{dR}\right|^{-1}_{R=z_n} S\cdot\exp\left\{-2I_m\int_a^{z_n}\sqrt{2M[\varepsilon_n-U(R)]}\,dR\right\} \quad (5)$$

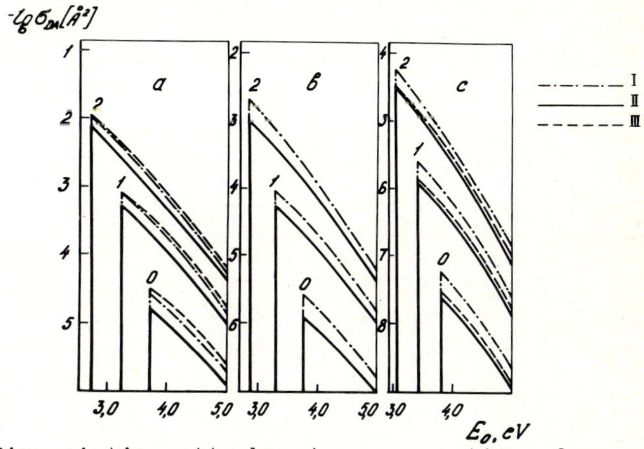

Dissociative attachment cross sections for various rotationless vibrational states of H_2 (a), HD (b), D_2 (c):

I, II - the semiclassical results computed with the formula (5) with C=2.65 a.u. and C=2.966 a.u.; III, a) - the results of numerical calculations[2], C=2.65 a.u.; III,c) - the results of calculations[3], C=2.966 a.u.

$$S=\exp\left\{-2I_m\int_{z_n}^{R_{st}}\sqrt{2M[E-W(R)]}\,dR\right\}$$

The difference (especially in Γ-large case) between S and the classical survival factor

$$S_1=\exp\left\{-\int_{R_0}^{R_{st}}\Gamma(R)\frac{dR}{v(R)}\right\},\ v=\sqrt{2[E-U_I(R)]/M} \quad (6)$$

should be noted.

Discussing the isotope effect one can conclude from (5) that the quantity

$$q=\ln(\sigma_{H_2}/\sigma_{HD})/\ln(\sigma_{H_2}/\sigma_{D_2})$$

is close to the value $(\sqrt{3}-\sqrt{4})/(\sqrt{6}-\sqrt{2})=0.373$

The semiclassical calculations give $q=0.371-0.373$, whereas the experimental value $q=0.38$ [6] .

References.

1. Herzenberg A, Phys.Rev.,160, 80 (1967)
2. Wadehra J M,Bardsley J N, Phys.Rev.Lett., 41, 1795 (1978)
3. Bardsley J N, Wadehra J M, Phys.Rev.A,20, 1398 (1979)
4. Yelets I S, Kazansky A K, Sov.Phys.JETP, 53, 499 (1981)
5. Yelets I S, Kazansky A K, Sov.Phys.JETP, 55, 258 (1982)
6. Schulz G J, Asundi R K, Phys.Rev.Lett., 15, 946 (1965)

TRANSITION STATE EFFECTS IN THE ROTATIONAL PREDISSOCIATION OF H_2^+ IONS OF DIFFERENT PRECURSOR ORIGINS

Gareth Brenton*, Paul G. Fournier+, Elizabeth G. Richard+ and John H. Beynon.*

*Royal Society Research Unit, University College Swansea, Swansea SA2 8PP, U.K.

+L.C.A.M., (Associated with C.N.R.S.), Groupe de Spectroscopie de Translation, Université de Paris-Sud, 91405 Orsay, France.

In this investigation we are concerned with highly excited ro-vibronic states of ground state $(1s\sigma_g)H_2^{+\cdot}$ that can fragment by the mechanism of rotational predissociation. These states have been predicted theoretically and we have recalculated them together with their lifetimes and widths with a new programme.

A mass spectrometric method has been employed in which the states are characterised by measuring the translational energy release (ΔE) during the unimolecular (i.e. spontaneous) fragmentation $H_2^{+\cdot} \rightarrow H^+$, where $\Delta E = (16\ m_2 m_3 WV/m_1{}^2)^{\frac{1}{2}}$ and m_1, m_2 and m_3 are the masses of $H_2^{+\cdot}$, the proton and neutral respectively, V the kinetic energy of m_1 and W the excess internal energy release which is related to the energy level of the quasi-bound state. In a previous study[1] six states were identified where $H_2^{+\cdot}$ was formed from hydrogen gas in a monoplasmatron source. We have used a novel method of changing the ro-vibronic population of $H_2^{+\cdot}$ by forming it as a fragmentation product, in an electron impact source, using a variety of hydrogen containing precursor molecules.

Translational energy spectra of $H_2^{+\cdot} \rightarrow H^+$ were obtained on a modified double-focusing mass spectrometer which has three sectors (magnet/electric sector/ electric sector). Ions extracted from the ion source pass into a large magnetic sector which is set to transmit only $H_2^{+\cdot}$. Unimolecular fragmentation of $H_2^{+\cdot}$ occurs continuously along the flight path of the ion beam. Fragment ions formed after the magnetic sector but before the first electric sector were energy analysed by scanning the voltages on the electric sector plates. An initial survey of the thirteen different samples was carried out at a moderate setting of energy resolution and fine structure was observed corresponding to predissociation from different quasi-bound states. The relative pattern of intensities of this structure changed upon the choice of sample. Four main states were identified clearly. The energy resolution was then increased to resolve states of smaller energy separation. Methane gas was found to give the most intense spectra from which at least nine states were assigned. There are a total of 56 ro-vibronic energy levels above the dissociation limit. However, because the time scale for observation is very limited (ca. 1 to 10 μ.sec.) only states with lifetimes in the range 10^{-7} to 10^{-3} sec. are potentially observable and these total 14. Table 1 lists these states together with their lifetimes (τ) calculated energy release (W_C) and measured energy release (W_E). The relative population of states ($I_{v,J}$), see Table 1, as formed in the ion source was calculated using the lifetime data and from a knowledge of the apparatus collection efficiency.

Confirmation of our assignment of states was made by varying the time scale of observation (i.e. by changing the kinetic energy of $H_2^{+\cdot}$ and by selecting the next adjacent field free region in our 3-sector instrument). Here we noted the relative changes in the pattern of peak intensities and found it to agree in the manner predicted from their lifetimes.

An interesting result of this work has been that highly rotationally excited states were not observed when hydrogen gas was used and also that we were able to vary the ro-vibrational population of states by choosing various precursors. Certain groups of samples were found to produce similar spectra. We have made qualitative arguments to account for the above behaviour in terms of i) the formation of a highly excited complex of $H_2^{+\cdot}$; and ii) as a transition state phenomena of the precursor.

TABLE 1

v,J	τ(μ.sec.)	W_c(meV)	W_E(meV)	$I_{v,J}$
0,39	3.4	233.5	220	1
1,37	46.5	177.7	171	.3
2,35	3960	127.8	123	<.05
3,34	.48	143.4	139	.25
4,32	36.1	102.2	104	.15
5,30	18400	66.2	-	<.05
6,29	.21	83.8	-	<.05
7,27	21.8	55.6	51	.9
8,25	17900	32.4	-	<.05
10,22	.14	33	33	.4
11,20	2.6	20.6	18	2.9
12,18	34.4	12.3	12	.5
13,16	83.6	7.5	~7	<.2
14,14	3.4	5.5	5	2.7

References

1. J.G. Maas, N.P.F.B. Van Asselt and J.Los, Chem. Phys. 8, 37 (1975).

THE KINETIC ENERGY SPECTRUM OF PROTONS PRODUCED BY THE DISSOCIATIVE IONIZATION OF H_2 BY ELECTRON IMPACT

M. A. Khakoo[+] and S. K. Srivastava

Jet Propulsion Laboratory, California Institute of Technology, Pasadena, California 91109

The kinetic energy spectra of protons generated by the electron impact on H_2 have been observed at the scattering angles of 90° and 30° to the incident electron beam for electron energies ranging from 18 eV to 120 eV. Two independent experimental methods have been used to observe the proton spectra.

In the first method a time-of-flight (TOF) arrangement was used in conjunction with a pulsed electron gun. This electron gun produced electron pulses of widths 50 ns and peak intensities ∿4 μA. The TOF system was 20 cms long. The energy resolution of this system was estimated to be 0.4 meV for 10 eV protons and better than 0.15 meV for protons of 3 eV or less.

In the second method, an electrostatic hemispherical energy analyzer was used in place of the TOF system with the electron gun operated in D.C. mode. The energy resolution of this analyzer was approximately 100 meV.

In most cases, the TOF proton spectra and the proton spectra obtained by the hemispherical energy analyzer were in good agreement, especially around proton kinetic energies of ∿3 eV where the energy resolutions of the two systems were comparable (see Fig. 1 for an example of this). However, despite adequate energy resolution, we could not find any of the deep structures, in the proton distributions, comparable to that reported by Crowe and McConkey[1] and Landau et al.[2] In general, the data were found to be in good agreement with the recent experimental observations of Kollmann[3] and Burrows et al.[4] and are shown in Fig. 1. The difference in Kollmann's data at 55° and ours at 90° for protons with kinetic energies at approximately 4 eV and higher can be explained in terms of the angular distributions of these energetic protons as observed by Johnson & Franklin,[5] i.e. their angular distributions show a deeper minimum at 90° for higher kinetic energy protrons than for lower kinetic energy protons.

In this experiment a consistent peak in the proton spectra was observed at the kinetic energy value of between 1.3 eV to 1.8 eV for electron impact energies of 19 eV and above. This feature has been previously observed by Landau et al.[2] although they report its threshold to be at 22 eV (see Figure 2). At the present time it is not clear what process is responsible for this feature. It seems possible that at higher energies this feature is predominantly produced by a low-lying autoionizing state of H_2 i.e. possibly the $H_2^{**}\ {}^1\Sigma_g^+$ state.[6] Details will be presented at the conference.

[+]NASA-NRC Resident Research Associate.

Work is supported by NASA.

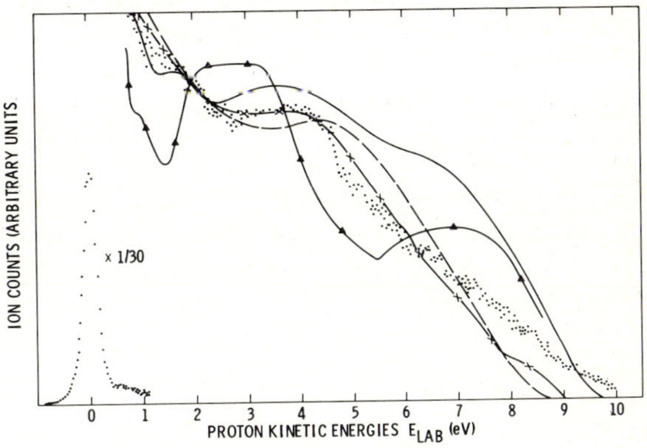

Fig. 1: Proton energy distributions for the incident electron energy of 40 eV. Present work (90°): .... hemispherical energy analyzer, --- TOF; ━━ Kollmann (55°)[3]; x-x- Burrows et al. (90°)[4]; -Δ- Landau et al. (120°)[5]. Comparison made by normalizing distributions at E_{LAB} = 2.0 eV.

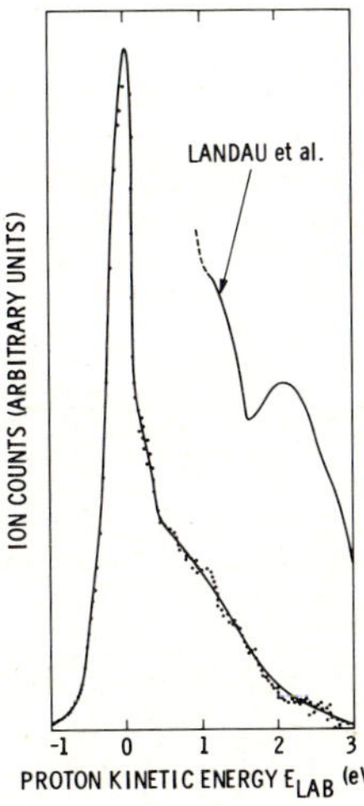

Fig. 2: Proton energy distributions at low impact energies showing the feature at 1.3 eV to 1.8 eV kinetic energy. ⋯ This work 20 eV, 30° (hemi-spherical energy analyzer); ━━ Landau et al.[2] 20 eV, 120°.

References

1. A. Crowe and W. McConkey, Phys. Rev. Lett. 31 192 (1973).
2. M. Landau, R. I. Hall and F. Pichou, J. Phys. B. 14 1509 (1981).
3. K. Kollmann, J. Phys. B. 11 339 (1978).
4. M. D. Burrows, L. C. McIntyre, Jr., S. R. Ryan and W. E. Lamb, Jr., Phys. Rev. A 21 1841 (1980).
5. J. P. Johnson and J. L. Franklin, Int. J. Mass. Spec. Ion. Phys. 33 393 (1980).
6. A. U. Hazi, Chem. Phys. Lett. 25 259 (1974).

THE O^+ AND N^+ FORMATION BY LOW-ENERGY ELECTRON IMPACT ON NITROUS OXIDE

J.L. Olivier, R. Locht, J. Momigny

Département de Chimie Générale et de Chimie Physique, Université de Liège,
Institut de Chimie, Bât. B.6, Sart-Tilman par B-4000 Liège 1, Belgium

In an earlier paper we extensively discussed the NO^+ and N_2^+ production by dissociative electroionization of N_2O[1]. We recently completed the dissociative electroionization study of this system by examining the appearance of O^+ and N^+.

The experimental set-up used in this experiment has been described earlier[2]. Kinetic energy (K.E.) distributions and onset energies (A.P.) are measured for the two mass selected ions. K.E. - versus - A.P. diagrams are obtained and shown in fig. 1 and fig. 2 for O^+ and N^+ respectively.

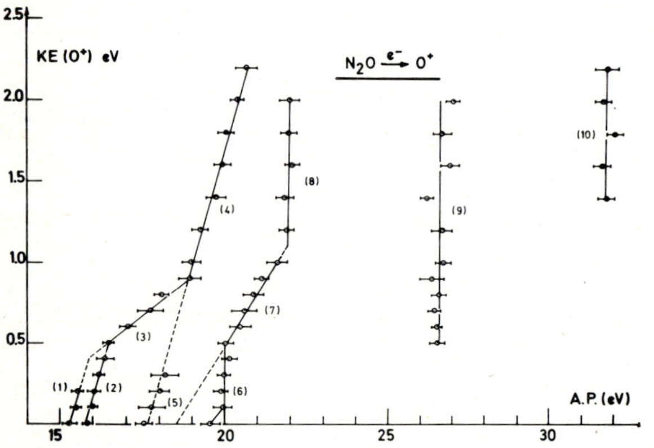

Fig. 1 : K.E. - versus - A.P. diagram for O^+/N_2O.

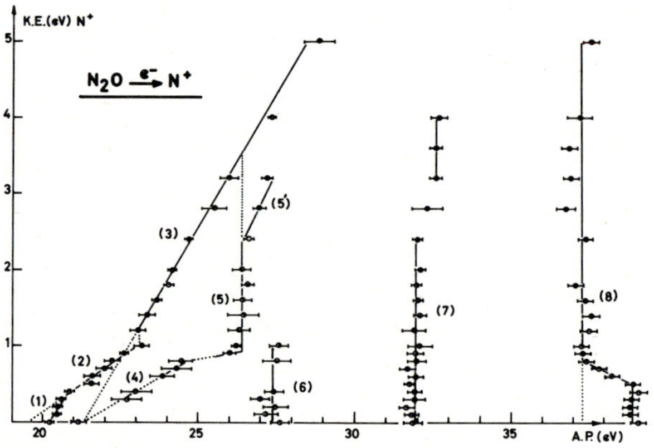

Fig. 2 : K.E. - versus - A.P. diagram for N^+/N_2O.

For the O^+ ions, thresholds are measured at 15.27 eV, 15.83 eV and 19.6 eV. Autoionization maxima are observed at 17.5 eV and around 23.4 eV. From the diagram in fig. 1, two salient features are observed below 21 eV :

(i) two dissociation limits are concerned, i.e.

$$N_2O + e^- \begin{cases} \rightarrow O^+(^4S) + N_2(X^1\Sigma_g^+, v = 0) \\ \rightarrow O^+(^2D) + N_2(X^1\Sigma_g^+, v = 0) \end{cases}$$

calculated at 15.29 eV and 18.61 eV. Mostly the $N_2(X^1\Sigma_g^+)$ appears vibrationally excited.

(ii) the main O^+ producing mechanism is the predissociation of the successive ionized states of N_2O by the repulsive $N_2O^+(^4\Sigma^-)$ and $(^2\Pi)$ states.

Behind 21 eV, doubly excited states of N_2O are shown to dissociate by the same mechanism.

For the N^+ ions, onset energies are measured at 20.26 eV, 21.36 eV, 27.4 eV, 31.9 eV and 38.9 eV. The N^+ ions production mainly occurs in the energy range of doubly excited states of N_2O^+ known by He(II) photoelectron[3] and dipole (e, 2e) spectroscopy[4]. Only the onset at 20.36 eV is ascribed to the predissociation of the $N_2O^+(\tilde{C}^2\Sigma^+)$ state producing $N^+(^3P)$ + $NO(X^2\Pi, v = 4)$. At higher energies the $NO(X^2\Pi)$ fragment is vibrationally excited upto v = 8.

A second dissociation limit is populated, i.e., $N^+(^1D) + NO(X^2\Pi, v = 0)$, calculated at 21.36 eV, through the predissociation of two doubly states of N_2O^+.

These mechanisms will be discussed in detail at the meeting and a comparison will be made with previous experiments[5].

We wish to thank the F.R.F.C. for financial support and the A.R.C. for a research grant.

References

(1) J.L. Olivier, R. Locht, J. Momigny, Chem. Phys. 68, 201 (1982).
(2) R. Locht, J. Schopman, Int. Mass Spectrom. Ion Phys. 15, 361 (1974).
(3) A.W. Potts, T.A. Williams, J. Electr. Spectry. Rel. Phenom. 3, 3 (1974).
(4) C.E. Brion, K.H. Tan, Chem. Phys. 34, 141 (1978).
(5) J. Berkowitz, J.H.D. Eland, J. Chem. Phys. 67, 2740 (1977).

METASTABLE DISSOCIATIONS OF SMALL CLUSTER IONS

K. Stephan, A. Stamatovic[a], A.W. Castleman, Jr[b], J.H. Futrell[c] and T.D. Märk

Institut f. Experimentalphysik, Leopold Franzens Universität, A 6020 Innsbruck, Austria

Molecular ions which have dissociation lifetimes in the range $\gtrsim 10^{-7}$s are usually referred to as metastable ions. There are, in principle, three different dissociation mechanisms possible in these metastable ions: electronic predissociation (and intersystem crossing), tunneling through a barrier, and vibrational (statistical) dissociation. Recently, the existence of metastable cluster ions has been confirmed for a number of systems and statistical dissociation has been invoked as the likely dissociation mechanism[1]. The emphasis of this paper will be on metastable transitions of small cluster ions, extending previous studies on He_2^+, ArN_2^+ and Ar_3^+ cluster ions[2-4] to Ar_2^+ and N_4^+ cluster ions. Metastable transitions observed in the present study for these ions can be interpreted in terms of electronic predissociation and/or tunneling through rotational barriers with help of recently calculated potential energy curves using the methods of ab initio molecular quantum mechanics[5,6].

All experiments were performed on a molecular beam-electron impact ion source- high resolution, reverse geometry, double focussing mass spectrometer. The details are given elsewhere[3,4,7]. Neutral clusters are produced by expanding neat or seeded gas (at room or liquid nitrogen temperature) through a 10 µm nozzle. The molecular beam is crossed at right angles by an electron beam and ions are extracted at right angles. Using the technique of decoupling the acceleration and analyzer fields, it is possible to investigate dissociations occurring in a specific time window (i.e. in the first field free region).

An important goal was the separation of a possible metastable dissociation process from collision induced dissociations occurring in the field free region. For this purpose we measured the precursor and product ion intensity as a function of gas pressure in the ion source and field free region. Fig. 1 shows as an example the ion intensity of N_4^+, N_2^+ and the intensity ratio N_2^+/N_4^+. The approximate straight line obtained for this ratio indicates that thin target conditions pertain. This allows the calculation of the collision induced cross section from the slope of this line, i.e yielding ~ 4 . 10^{-17} cm^2 for N_4^+ (mixed target) and ~ 2 . 10^{-16} cm^2 for Ar_2^+ (at 3 keV collision energy).

Extrapolating the intensity ratio to zero pressure yields finite intercepts in both cases implying that the dissociative product ions are produced not only by collision induced dissociation, but also by metastable decay (e.g. see Fig. 1). The number of metastable transitions compared to the stable precursor ions in the present time window is 6 . 10^{-4} for $Ar_2^{+*} \rightarrow Ar^+$ between 3 and 8 µs and 5 . 10^{-4} for $N_4^{+*} \rightarrow N_2^+$ between 2.6 and 6.6 µs.

Work supported by Österreichischer Forschungsfonds, Projekt S-18/05 and 08.

References

1. See e.g.: A.J. Stace and A.K. Shukla, Int. J. Mass Spectrom. Ion Phys., 36, 119 (1980); J. Sunner and P. Kebarle, J. Phys. Chem., 85, 327 (1981); J.H. Futrell, K. Stephan and T.D. Märk, J. Chem. Phys., 76, 5893 (1982); A. J. Illies, M.F. Jarrold and M.T. Bowers, Int. J. Mass Spectrom. Ion Phys., 47, 93 (1983)
2. J.P. Flamme, T.D. Märk and J. Los, Chem. Phys. Lett., 75, 419 (1980)
3. K. Stephan and T.D. Märk, Chem. Phys. Lett., 87, 226 (1982)
4. K. Stephan and T.D. Märk, Chem. Phys. Lett., 90, 51 (1982)
5. W.R. Wadt, J. Chem. Phys., 68, 402 (1978)
6. S.C. de Castro, H.F. Schaeffer III and R.M. Pitzer, J. Chem. Phys., 74, 550 (1982)
7. H. Helm, K. Stephan, T.D. Märk and D.L. Huestis, J. Chem. Phys., 74, 3844 (1981)

a. PMF Kragujevac, Jugoslavia
b. Dept. Chemistry, Penn State University, University Park, PA 16802, USA
c. Dept. Chemistry, Univ. Utah, Salt Lake City, UT 84112 USA

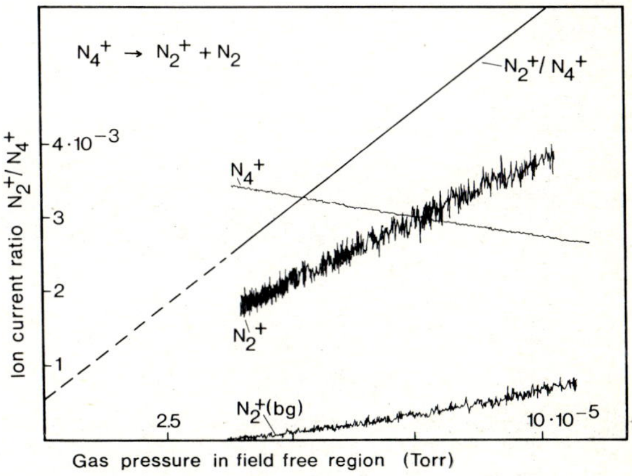

FIG.1

PRODUCTION OF O^- IONS BY DISSOCIATIVE ELECTRON ATTACHMENT TO CO_2

O. J. Orient* and S. K. Srivastava

Jet Propulsion Laboratory, California Institute of Technology, Pasadena, California 91109 USA

Dissociative electron attachment cross section measurements for the production of O^- from CO_2 have been performed utilizing a crossed target beam-electron beam collision geometry and a quadrupole mass spectrometer. The relative flow technique is employed to determine the absolute values of cross sections.[1]

The beam of molecules is produced by flowing the gas through a capillary array. The energy selected electron beam is generated by a trochoidal gun which has an energy resolution of approximately 50 meV (FWHM). The electron beam is collimated by an axial magnetic field of about 120 G. The energy of the electron beam is calibrated by utilizing maximum electron attachment energies to H^-/H_2 and O^-/O_2. The negative ions produced by attachment of electrons to the target molecules are extracted out of the collision region by a homogeneous electric field. The direction of this extracting electric field is normal to both the electron beam and the molecular beam. The extracted negative ions are accelerated and focused at the entrance aperture of a quadrupole mass spectrometer. The mass analyzed ions are detected by a spiraltron multiplier.

The accurately known H^-/H_2 and O^-/O_2 cross section data are employed for normalization. The transmission efficiency of the detecting system (extracting field, accelerating field, ion optics, quadrupole mass spectrometer and particle detector) is also determined. The dissociative electron attachment cross section values of O^-/CO_2 as a function of the electron beam energy are shown in Fig. 1. Five peaks are discernible in the cross section curve. The attachment energies corresponding to the five cross section peaks are 4.4, 8.2, 13.0, 16.9 and 19.4 eV. The cross sections at these maxima are: 1.43×10^{-19} cm^2, 4.48×10^{-19} cm^2, 8.1×10^{-21} cm^2, 1.2×10^{-20} cm^2 and 1.2×10^{-20} cm^2 respectively. In the case of peak at 4.4 eV, our cross section value agrees very well with the published data[2,3,4] when the measurement error is taken into consideration. For the peak at 8.2 eV we have similarly good agreement with the other data,[2,3,4] except for that of Craggs and Toser,[5] whose peak cross section value is higher than ours by about 10%. Our ratio of peak cross section values at 8.2 eV to 4.4 eV (3.13) agrees very well with most of the data found in the literature.[2,3,4] In the case of peaks at 13.0 eV and 16.9 eV the maximum energy and the cross section data found in the literature[4,6] are believed to be rather inaccurate ($\sim 2 \times 10^{-20}$ cm^2 at 12 eV,[4] $\sim 6 \times 10^{-21}$ cm^2 at 13 eV,[6] $\sim 2 \times 10^{-20}$ cm^2 at 17 eV[4]). We have an additional fifth cross section peak at 19.4 eV energy which can not be found in the literature.

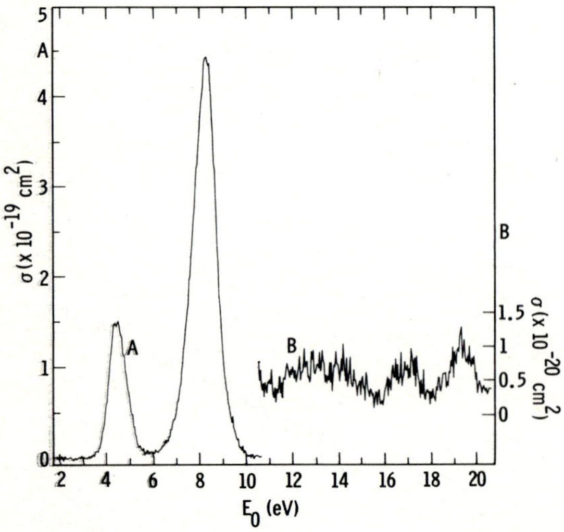

Fig. 1: Dissociative electron attachment cross section for O^-/CO_2 as a function of the electron beam energy. The y axis for A is on the left and for B is on the right side of the figure.

In a previous paper we reported[7] a sixth dissociative attachment peak at 15.0 eV for the production of O^- from CO_2. The experiment described in that paper was performed for the purpose of observing the dissociative attachment from vibrationally excited CO_2. There, an enhancement in the intensity of the dissociative attachment occured which allowed us to identify a new peak at 15.0 eV. This peak is not clearly seen in the present experiment.

This work was supported by NASA.

*NRC-NASA Senior Research Associate.

References

1. O. J. Orient and S. K. Srivastava, J. Chem. Phys. March (1983).
2. G. J. Schulz, Phys. Rev. 128, 178 (1962).
3. R. K. Asundi, J. B. Craggs and M. V. Kurepa, Proc. Phys. Soc. 82, 967 (1963).
4. D. Rapp and D. D. Briglia, J. Chem. Phys. 43, 1480 (1965).
5. J. D. Craggs and B. A. Tozer, Proc. Roy. Soc. (London) A254. 229 (1960).
6. P. J. Chantry, J. Chem. Phys. 57, 3180 (1972).
7. S. K. Srivastava and O. J. Orient, Phys. Rev. A., Rapid Communications, February 1983.

NEGATIVE IONS OBSERVED IN ELECTRON TRANSMISSION AND ELECTRON ATTACHMENT SPECTROSCOPY

M. Heni, G. Kwiatkowski and E. Illenberger

Institut für Physikalische Chemie der Freien Universität, Takustraße 3, D-1000 Berlin 33

We studied negative ions in ethylene and the fluoroethylenes applying electron transmission and dissociative electron attachment spectroscopy. In these molecule the negative ion states lie energetically above the ground states of the corresponding neutrals. By convention the electron affinity is then considered to be negative and the negative ion state is directly accessible in electron impact.
In polyatomic molecules the lifetimes of those anions (resonances) with respect to autodetachment vary in a wide range.
Resonances generally can be observed in electron transmission[1] where the transmitted current of electrons through a gas cell versus their energy is measured.
This is shown in fig.1 for nitrogen and ethylene. In both molecules the resonance is formed by the temporary accommodation of the incident electron into the lowest normally unoccupied molecular orbital. In nitrogen the structure in the derivative of the transmitted current reflects the vibrational energy levels of the nitrogen anion $N_2^-(^2\Pi_g)$[2].

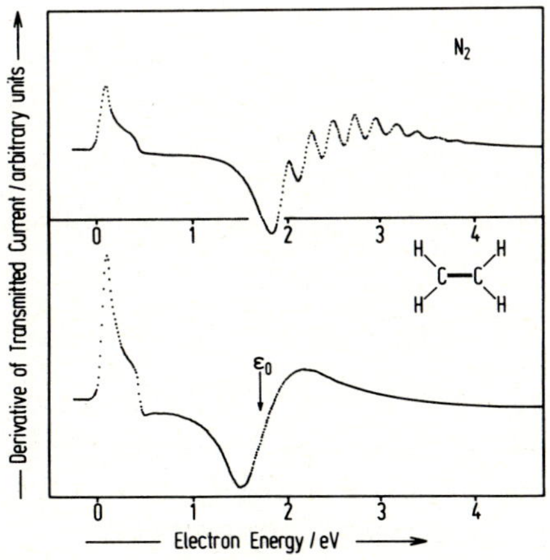

Fig.1: Electron transmission spectra of nitrogen and ethylene

In the fluoroethylenes dissociation channels into stable negative and neutral fragments below the resonance energy exist so that the molecular negative ion may dissociate along repulsive potential energy surfaces prior to rejection of the additional electron. In such cases the resonance and the dissociation channels can be observed mass spectrometrically. As an example fig.2 shows various ion fragments generated within the first resonance of difluoroethylene.

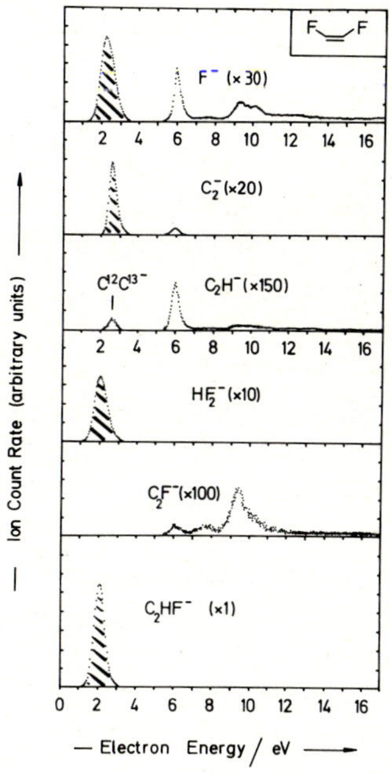

Fig.2: Ion yield for the formation of negative fragments from cis-1,2- difluoroethylene

We studied the formation and dissociation of negative ions in ethylene and the fluoroethylenes. Results obtained in electron transmission and dissociative electron attachment are presented.

References

1. L.Sanche and G.J.Schulz, Phys.Rev. A6, 69 (1972)
2. M.Berman, H.Estrada, L.S.Cederbaum and W. Domcke, Phys. Rev. A (in press)

Na^- FORMATION BY ELECTRON IMPACT IN Na_2

D. Teillet-Billy, L. Bouby and J.P. Ziesel

Laboratoire des Collisions Atomiques et Moléculaires, Université Paris-Sud, 91405 ORSAY Cedex, FRANCE

Most experimental and theoretical works on the sodium dimer have dealt with the neutral molecule and the positive ion. The negative ion Na_2^- has been observed as secondary ion in a sputtering experiment[1]. The bound character of the $Na_2^-(^2\Sigma_u^+)$ ground state has been further demonstrated by Bardsley et al.[2] and Shepard et al.[3], who calculated a positive adiabatic electron affinity for Na_2, of 0.15 and 0.42 eV respectively. In the calculation of Bardsley et al., the Na_2^- $^2\Sigma_u^+$ potential energy curve crosses the Na_2 curve at R=2.9 Å, close to the equilibrium distance, while Shepard et al. computed it to be bound at every R. The Na_2^- $(^2\Sigma_g^+)$ excited state[2,4], correlated to the Na^- ^1S + Na ^2S dissociation limit at 0.18 eV, crosses the Na_2 curve near R=4 Å. The existence of two negative-ion states at very low energy should show up in electron scattering.

We have studied Na^- formation by dissociative electron attachment in Na_2 from 0 to 10 eV. The experiment is made using a crossed-beam magnetic mass spectrometer. The electron beam is energy selected by a trochoïdal monochromator and crosses the atomic and molecular sodium beam issuing from an effusive oven. This oven was operated at temperatures between 350 and 425°C, where the dimer concentration[5] is estimated to be between 3.3 and 5.3%.

The Na^- cross-section is shown in Fig.1. Only one peak, near zero energy, has been observed from 0 to 10eV. The energy scale is calibrated either from the O^-/CO onset, at 9.62 eV, or with the SF_6^-/SF_6 peak at zero energy. The O^-/CO measurements yielded values slightly higher than those with SF_6 ; we then set the maximum at 0.05 $\pm$ 0.05 eV. This peak energy is smaller than the minimum energy to dissociate Na_2^- in Na^- ^1S + Na ^2S from the Na_2 ground state level (v=0, J=0). Dissociation then proceeds thru electron attachment to rovibrationnally excited $Na_2(X^1\Sigma_g^+)$. The FWHM of the Na^- cross-section is comparable to the width of the electron beam and of the SF_6^- peak. This indicates a width for Na^- not larger than a few tenth of eV.

Because of the small width, the absolute cross section is better given by the energy integrated measurement. The Na^- cross section was determined relatively to the Na ionisation cross section at 50 eV, measured by Mac Farland[6], at three different temperatures. The energy integrated cross section for Na^- formation is estimated to be 4 $\pm$ 2 10^{-18} cm^2.eV.

References

1. M. Leleyter and P. Joyes, J.Phys. (Paris) 35, L85 (1974)
2. J.N. Bardsley, B.R. Junker and D.W. Norcross, Chem. Phys. Lett. 37 502 (1976)
3. R. Shepard, K.D. Jordan and J. Simons, J. Chem. Phys. 69, 1788 (1978)
4. K.D. Jordan, private communication
5. JANAF Thermodynamical Tables, Nat. Stand. Ref. Data Ser. NBS 37 (1971)
6. L.J. Kieffer and G.H. Dunn, Rev. Mod. Phys. 38, 1 (1966)

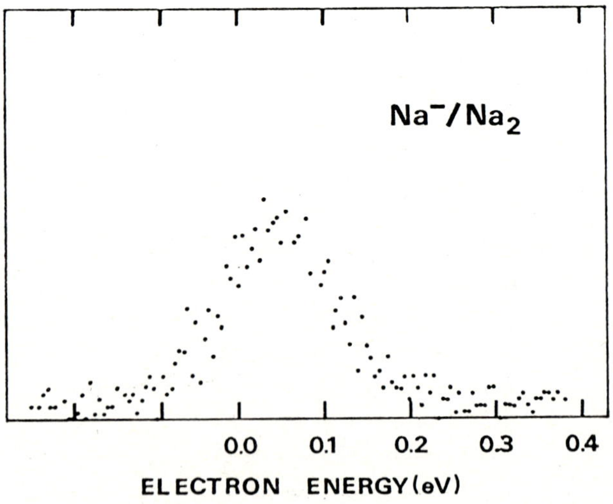

Fig. 1. Cross section for Na^- formation from Na_2. This data was taken with an electron beam FWHM of about 150 meV.

DISSOCIATIVE ELECTRON-ATTACHMENT PROCESSES IN $HgBr_2$

M.Tronc*, R.Azria**, R.Abouaf**, L.Bouby** and J.P.Ziesel**

*Laboratoire de Chimie Physique, 11, rue Pierre et Marie Curie
75 231 Paris Cedex 05 (France)

**Laboratoire des Collisions Atomiques et Moléculaires
Bât.351, Université Paris-Sud, 91405 Orsay (France)

Br formation in $HgBr_2$ has been observed in the 0 - 6eV energy range both with a magnetic mass spectrometer and a quadrupole mass filter with facility for kinetic energy analysis and angular distribution of the ions.[1] The cross section shows two peaks at 0.0 + 0.05 eV and 4.15 ± 0.10 eV. The zero energy peak has a width of 0.15 eV and corresponds to the $HgBr_2^-$ ground state correlated to HgBr and Br^- in their ground state. The 4.15 eV peak is 1 eV broad and has an onset at 2.4 ± 0.15 eV.

Kinetic energy distributions of Br^- ions in the energy of the second peak exibit two broad maxima, the positions of which are shifted toward higher energy by half the increase of incident electron energy[2]. The formation of Br^- ions proceed through Π compound states of $HgBr_2^{-*}$ (as shown by the angular distribution of fig.1) correlated to Br^- (^1S) plus HgBr (A $^2\Pi$) and leading to the atomisation limits Hg (^1S) + Br ($^2P_{3/2}$) + Br^- (^1S) and Hg (^1S) + Br ($^2P_{1/2}$) + Br^- (^1S) at 0.42 and 0.88 eV respectively, owing to the strictly repulsive character of the HgBr ($A^2\Pi$) state. This atomisation of $HgBr^-$ on an avalanche corridor potential energy surface may explain the kinetic energy distribution and its evolution with energy[3] with the Br ions and the Br atoms carrying out the same amount of translational energy and the Hg atoms being left with zero energy.

Similar results for dissociative attachment in $HgCl_2$ will be presented at the conference.

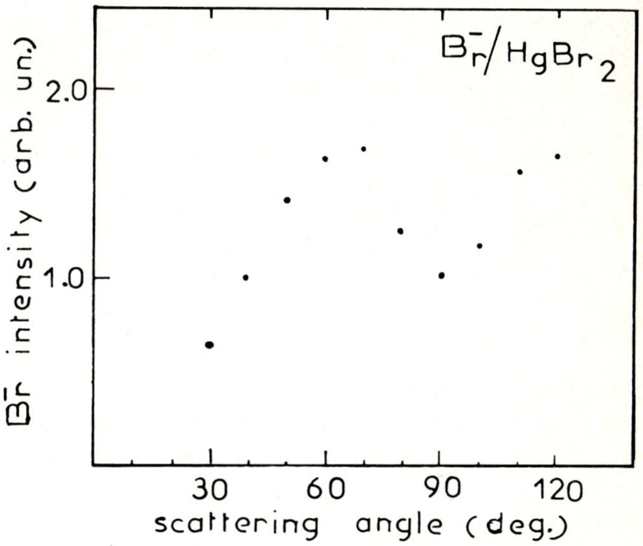

Fig.1 : angular distribution of Br^- ions at incident electron energy E_i 4.1 eV and Br^- kinetic energy E_k 1.8 eV.

References

1.Y.Le Coat, R.Azria and M.Tronc, J.Phys.B 15,1569(1982).

2.R.Azria,J.P.Ziesel, R.Abouaf, L.Bouby and M.Tronc, J.Phys.B 16 L 7 (1982).

3.M.Tronc et al, to be published.

DISSOCIATIVE ATTACHMENT IN HIGHLY POLAR MOLECULES : SODIUM HALIDES

J.P. Ziesel, R. Azria, D. Teillet-Billy, R. Abouaf and P. Girard
Laboratoire des Collisions Atomiques et Moléculaires, Université Paris-Sud,
91405 ORSAY Cedex, FRANCE

There is a renewed interest in the study of electron polar molecule interaction[1], particularly in highly polar alkali halides MX[2].

Dissociative electron attachment (D.A.) gives information on the dissociative resonant NaX^- states. At the last ICPEAC[3], the shape of the Na^- and X^- cross sections, on an absolute electron energy scale, were determined in a crossed beam magnetic mass-spectrometer.

We have extended this work by measuring the kinetic energy and angular distributions of the fragment negative ions. The apparatus is a crossed beam electrostatic spectrometer[4], with quadrupole mass analysis ; the molecular beam is generated by resistively heating a tantalum two-stage oven fitted with a 1mm diameter effusing channel[5].

In NaBr, the yields of Na^- ions associated with the dissociation to Na^-+Br ^2P are shown in Fig.1. One peak is observed in each dissociation channel, at 3.75eV to the Br $^2P_{3/2}$ limit and at 4.4eV to the Br $^2P_{1/2}$ limit. A third Na^- peak, at 5.1 eV, was observed in the previous experiment[3], but the present work shows that it is not correlated to dissociation in NaBr. We then assign this peak to D.A. in the dimer as well as the 5.6eV peak in sodium chloride. In NaCl, the spin-orbit splitting of Cl ^2P is too small to get separate ion yields. However, two different processes, like in NaBr, are also observed[5] at observation angles around 70°.

The angular behaviors of the Na^- intensity for the two peaks are shown in Fig. 2. The two distributions are quite different but are characterized by a strong asymmetry with respect to θ=90°. A similar behavior is observed in Na^-/NaCl. Teillet-Billy and Gauyacq[6] have shown that in highly polar molecules, the dipole field must be taken in account and consequently spherical harmonics in the angular distribution have to be replaced by dipolar angular modes. The lower Na^- process in NaCl and NaBr can then be attributed to dissociation of the NaX^- $^2\Pi_{3/2}$ state ; the higher energy state is still in question.

The kinetic energy distribution of the Cl^- ion in sodium chloride, observed[3] at incident electron energy around 5.2eV, was measured. The kinetic energy has a maximum near zero and a tail on the high energy side ; the distribution maximum does not change for distributions taken at different electron incident energies. This Cl^- ion then arises from D.A. in $(NaCl)_2$ with the excess energy being distributed in the neutral fragment. In NaBr, no Br^- ion was even detected in the present work.

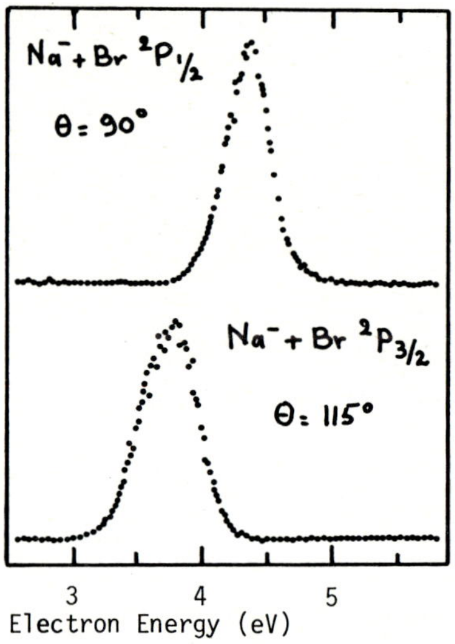

Fig. 1. Na^-/NaBr ion yields

References

1. D.W. Norcross and L.A. Collins, in "Advances in Atomic and Molecular Physics" (D.R. Bates and B. Bederson eds) vol. 18, 341-397 Academic Press (1982)
2. Alkali Halide Vapors (ed. P. Davidovits and DL Mc Fadden) Academic Press (1979)
3. D. Teillet-Billy, L. Bouby and JP Ziesel, 12th ICPEAC Abstracts, p. 407, Gatlinburg (1981)
4. Y. Le Coat, R. Azria and M. Tronc, J.Phys.B 15, 1569 (1982)
5. JP Ziesel, R. Azria and D. Teillet-Billy, to be published
6. D. Teillet-Billy and JP Gauyacq, to be published; also at this conference.

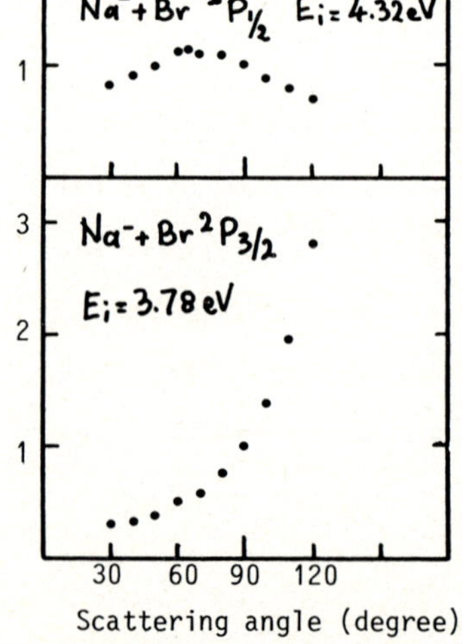

Fig. 2. Angular distributions I(θ)/I(90°) at incident electron energies E_i

DISSOCIATIVE ATTACHMENT TO POLAR MOLECULES : ANGULAR DISTRIBUTION

D. Teillet-Billy and J.P. Gauyacq

Laboratoire des Collisions Atomiques et Moléculaires
Université Paris-Sud, 91405 ORSAY Cedex (France)

The angular distribution of negative ions formed by dissociative attachment (D.A.) of electrons on a molecular target AB provides some information about the symmetry of the dissociating negative ion state. This problem was examined by O'Malley and Taylor[1] with the following assumptions :

(i) the molecular axis does not rotate during the collision ;

(ii) the conservation of the projection of the electronic angular momentum on the molecular axis (Λ_r, Λ_t and m for the dissociating state, the initial target state, and incident electron) selects a m value equal to $\Lambda_r-\Lambda_t$;

(iii) the electron fixed-R molecule scattering can be separated in angular modes, and only one mode is active ;

(iv) if the electron molecule interaction is not too far from spherical symmetry (H_2,CO) the active mode can be approximated by one spherical harmonics Y_ℓ^m, or a superposition of a few Y_ℓ^m (same m). With these assumptions, the angular distribution of D.A is a spherical harmonics (or a superposition of a few of them)[1 2].

In the case of asymmetric molecules, such as polar molecules, assumption (iv) is not expected to hold any longer. However, in the case of highly polar molecule, the e^--molecule interaction at large distance is dominated by the dipolar field and one can expect that the role of spherical harmonics in almost spherically symmetric molecules will be played by the dipolar angular modes ϕ_n^m [3] in highly polar molecules. For a vanishing dipole, the ϕ_n^m dipolar modes reduce to spherical harmonics. Thus, assuming that the active angular mode is a dipolar mode, one finds angular distributions for D.A. expressed in terms of $|\phi_n^m|^2$. The angular distribution for $e^- + HF \rightarrow H^- + F$ at 9.5 eV [4] appears to be well

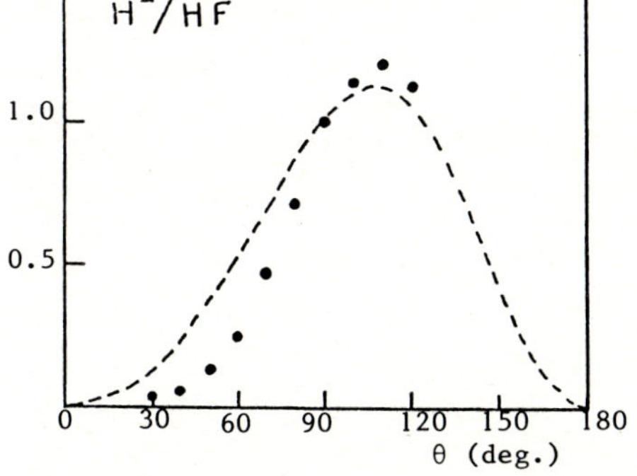

Figure 1 : • experimental[4] ; _______ theoretical $|\phi_1^1(\theta)|^2$ for μ = 1.82D.

described by ϕ_1^1, the first angular mode corresponding to a Π resonant HF^- state as shown in figure 1. It is noteworthy that for a small dipole as HF, the ϕ_1^1 mode is not very much different from Y_1^1, it is only slightly shifted backwards. In the case of a larger dipole moment, the asymmetry around 90° is much more pronounced. The ϕ_1^1 eigen mode corresponding to the NaCl dipole moment (μ = 9D) acounts very well for the observed asymmetry[5] as shown in figure 2. On figure 2, we present both $|\phi_1^1|^2$ and the result of the averaging by the thermal motion of the target (1000°K).

References

1. T.F. O'Malley and H.S. Taylor, Phys. Rev. 176 207 (1968)
2. M. Tronc, F. Fiquet-Fayard, C. Schermann and R.I. Hall, J. Phys. B 10 305 (1977)
3. M. Mittleman and R.E. Von Holdt, Phys. Rev. 140 726 (1965)
4. R. Azria, Invited paper of the XII Intern. Conf. on the Phys. of Electronic and Atommic Collisions. Gatlinburg (1981)
   R. Abouaf, R. Azria, D. Teillet-Billy to be published
5. J.P. Ziesel, R. Azria, D. Teillet-Billy to be published ; see also at this conference.

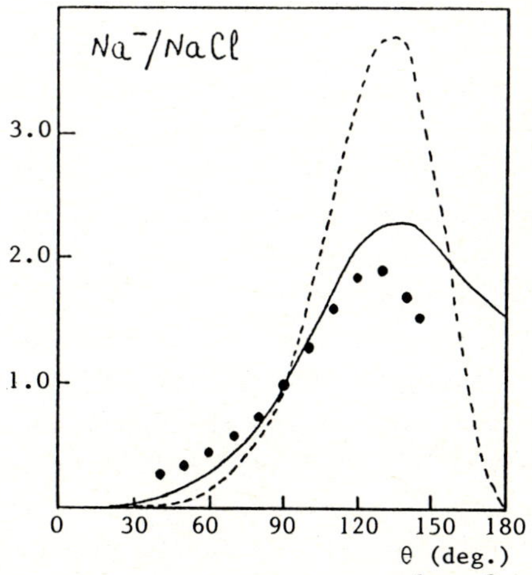

Figure 2 : ● experimental[5] ; ______$|\phi_1^1(\theta)|^2$; ________ calculated with thermal spreading.

In the calculated results, we take into account the smoothing effect[2] of thermal kinetic energy of the target at a 1000°K temperature on a theoretical $|\phi_1^1(\theta)|^2$ angular distribution.

FORMATION AND DISSOCIATION OF NEGATIVE IONS UNDER LOW ENERGY ELECTRON IMPACT

S. Süzer
Middle East Technical University, Department of Chemistry
Ankara, Turkey

E. Illenberger and H. Baumgärtel
Institut für Physikalische Chemie der Freien Universität Berlin
Takustraße 3, 1000 Berlin 33, FRG

Electron attachment spectroscopy is applied to study the formation and dissociation of small perfluoro-hydrocarbons in the energy range between 0.0eV and 15eV.
If an electron interacts with a molecule AB a molecular negative ion may be formed via the resonant electron attachment process

$$e^- + AB \rightleftharpoons AB^{-(*)}.$$

The molecular ion can directly be observed with a mass spectrometer if its lifetime with respect to autodetachment of the additional electron is larger than its flight time from the source to the detector.
In fluorinated hydrocarbons the common reaction for electron energies larger than thermal energies is the dissociation of the molecular negative ion

$$AB^{-(*)} \longrightarrow A + B^-$$

or autodetachment of the extra electron.
At very low electron energies the rejection of the attached electron may be drastically delayed so that the molecular negative ion, if to be observed at all in a mass spectrometer, can preferably be formed at thermal energies. Furthermore, whether a long lived molecular ion can be generated depends on the type and structure of the molecule.
For a series of unsaturated perfluoro hydrocarbons, CF_2=CF_2, CF_3-CF=CF_2, CF_2=CF-CF=CF_2 and cyclo perfluoro-butene the usual channel to be observed is the dissociative attachment channel. In cyclo perfluoro-butene, however, the molecular negative ion can be observed as well as different fragments (fig. 1).

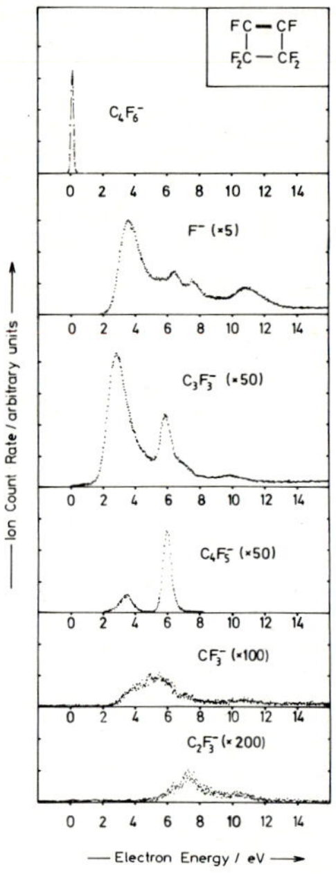

Fig. 1: Negative ion formation in c-C_4F_6

THERMAL ELECTRON ATTACHMENT TO VAN DER WAALS MOLECULES AS STUDIED BY THE PULSE RADIOLYSIS MICROWAVE CONDUCTIVITY TECHNIQUE

M. Toriumi, E. Suzuki, and Y. Hatano

Department of Chemistry, Tokyo Institute of Technology, Meguro-ku, Tokyo, Japan

Thermal electron attachment to oxygen molecule is explained by the Bloch-Bradbury(B-B) mechanism or a two-stage mechanism involving vibrationally excited molecular negative ion $O_2^{-*}(X^2\Pi_g, v'=4)$ as follows;

$$O_2(X^3\Sigma_g^-, v=0) \ + \ e^- \rightleftarrows O_2^{-*}(X^2\Pi_g, v'=4)$$

$$O_2^{-*}(X^2\Pi_g, v'=4) \ + \ M \longrightarrow O_2^-(X^2\Pi_g, v'\leqq 3) \ + \ M$$

where M indicates another oxygen molecule or a third-body molecule.[1,2)]

We showed that the pulse radiolysis microwave conductivity technique is very useful for studying such a low energy electron attachment process,[2)] and that the electron attachment to oxygen seemed to follow the B-B mechanism in many binary gas mixtures.[1)] We obtained the initial electron attachment cross section and the autoionization lifetime of $O_2^{-*}(X^2\Pi_g, v'=4)$, and compared the results with theories.[2,3)]

But these values were quite different from those obtained by the electron swarm experiments at higher densities of M, in which a much larger attachment cross section and a much shorter lifetime were obtained.[4)] We extended the experimental conditions to higher densities using the pulse radiolysis microwave conductivity technique, and showed that the electron attachment could be explained by the B-B mechanism at relatively low densities but that a new electron attachment to van der Waals (vdW) molecule $(O_2{\cdot}M)$ dominates at higher densities.[5)] The experiments of third-body density dependence[5,6)] and temperature dependence[7,8)] showed the existence of vdW molecule mechanism in many O_2-M mixtures.

In this experiment thermal electron attachment rates in O_2-C_2H_6 mixtures have been measured at various temperatures and pressures in the wide range using the pulse radiolysis microwave conductivity technique; in which time-resolved measurements of the number density of thermal electrons have been made after a nsec pulsed X-ray or e-beam irradiation on O_2-C_2H_6 mixtures. A general vdW-molecule mechanism involving autoionization, dissociation, and collisional stabilization of $(O_2{\cdot}M)^{-*}$, where $M=C_2H_6$, is assumed.

$$(O_2{\cdot}M) \ + \ e^- \rightleftarrows (O_2{\cdot}M)^{-*}$$

$$(O_2{\cdot}M)^{-*} \longrightarrow O_2 \ + \ M \ + \ e^-$$

$$(O_2{\cdot}M)^{-*} \longrightarrow O_2^- \ + \ M$$

$$(O_2{\cdot}M)^{-*} \ + \ M \longrightarrow O_2^- \ + \ 2M$$

The best-fitting of this mechanism to the experimental data (Fig. 1 and 2) shows an evident contribution of electron attachment to vdW molecules $(O_2{\cdot}M)$, and

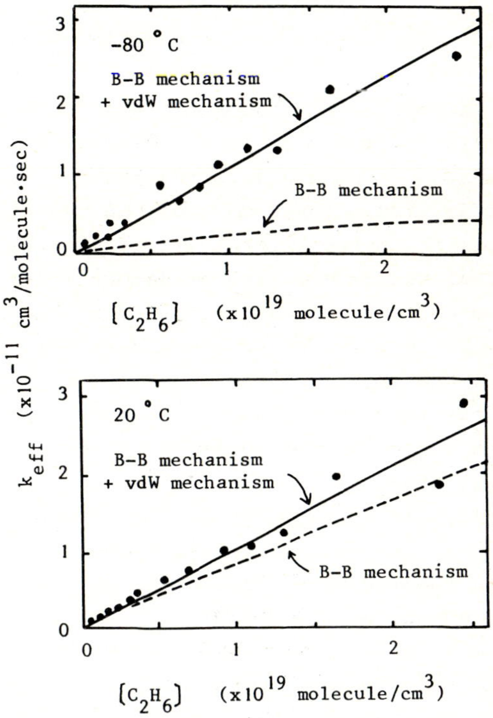

Fig. 1 and 2: The observed electron attachment rate as a function of C_2H_6 density

presents interesting results of electron attachment to van der Waals molecules. The rate constant for $e^- + (O_2{\cdot}M) \longrightarrow (O_2{\cdot}M)^{-*}$ is larger than that for $e^- + O_2 \longrightarrow O_2^{-*}$ by a factor of 10^2-10^3, which is ascribed to a decrease in the resonance energy and an increase in the resonance width due to the formation of van der Waals molecules.

References

1. H.Shimamori and Y.Hatano, Chem.Phys.Lett.38,242(1976); Chem.Phys.21,187(1977).
2. Y.Hatano and H.Shimamori, "Electron Attachment in Dense Gases", in "Electron and Ion Swarms", ed. by L.G.Christophorou, Pergamon (1981), pp.103.
3. A.Herzenberg, J.Chem.Phys.51,2942(1969); F.Koike, J. Phys.Soc.Jpn.35,1166(1973); G.Parlant and F.Fiquet-Fayard, J.Phys.B9,1617(1976).
4. R.E.Goans and L.G.Christophorou, J.Chem.Phys.60,1036 (1974).
5. Y.Kokaku, Y.Hatano, H.Shimamori and R.W.Fessenden, J.Chem.Phys.71,4883(1979).
6. Y.Kokaku, M.Toriumi and Y.Hatano, J.Chem.Phys.73,6167 (1980).
7. M.Toriumi, A.Nishikawa and Y.Hatano, 12th ICPEAC, Gatlinburg, 1981, pp.415.
8. H.Shimamori and R.W.Fessenden, J.Chem.Phys.74, 453 (1981).

DISSOCIATIVE ATTACHMENT OF H_2 AND ITS ISOTOPES BY LOW ENERGY ELECTRON IMPACT

S. Bhattacharyya

Gokhale College, Calcutta - 700020, India

L. Chatterjee*

Department of Physics, Jadavpur University, Calcutta - 700032, India

Dissociative attachment (DA) cross-sections for Hydrogen and its isotopes are computed using quantum electrodynamics. The S-matrix itself takes care of the reaction channel of interest between initial and final state. It consists of second order static Coulomb interactions between electron-electron field and electron-proton field. The states of the interacting systems are written in the manner of our previous papers[1]. The DA time is known to be much shorter than rotational time of the target and also it is assumed that the internuclear axis of the molecule remains fixed during the entire collision process. In this state of lowest rotational and vibrational level, hydrogen molecule consists almost exclusively of para form.

DA cross-section are computed for electron energies from 4 eV to 16 eV for the formation of H^- and D^- in the ground state (Figs. 1 and 2). This work shows a fair agreement with the experimental results of Rapp et al.[2] and Schulz et al.[3] The peaks may be due to formation of resonance. However, the life time of negative molecular ion states are very short, between 10^{-13} to 10^{-15} seconds. In the case of HD the dominant compound state near 3.8 to 3.9 eV corresponds to $(1s\sigma_g)^2\,(2p\sigma_u)^2\ {}^2\Sigma_u^+$ shape resonance. For H_2 and D_2 the maxima near to 12 eV energetically corresponds to valency excited resonance $(1s\sigma_g)\,(2p\sigma_u)^2\ {}^2\Sigma_g^+$.

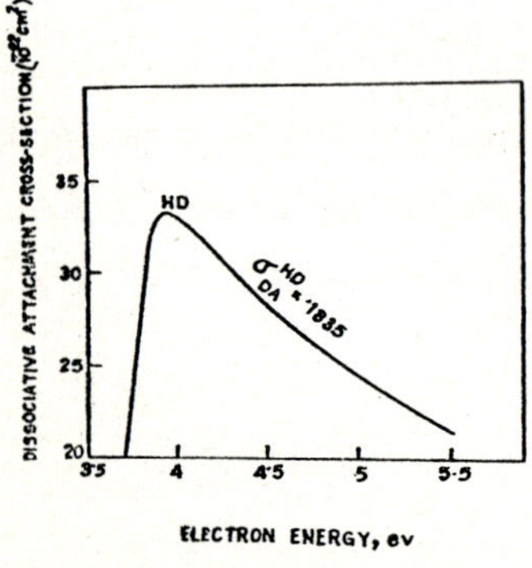

Fig. 1. H^- from HD

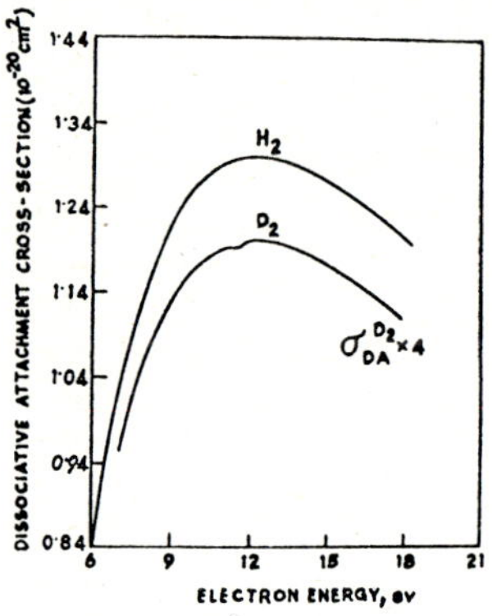

Fig. 2. H^- from H_2
D^- from D_2

Authors would like to thank Professor T. Roy for his interest in the work. L. Chatterjee thanks CSIR for award of Research Associateship.

References

1. S. Bhattacharyya, L. Chatterjee and T. Roy
   Physica 106C (1981) 135.
   L. Chatterjee and S. Bhattacharyya
   Phys. Lett. 93A7 (1983) 360.
2. D. Rapp, T.E. Sharp and D.D. Briglia
   Phys. Rev. Lett. 14 (1965) 533
3. G.J. Schulz and R.K. Asundi
   Phys. Rev. 158 (1967) 25.

* Mailing Address
Dr. Mrs. Lali Chatterjee
84/SB, Block 'E', New Alipur
Calcutta - 700 053, India.

ELECTRON AND POSITRON TOTAL CROSS SECTIONS ON H_2

A. Deuring, J. Ferch, K. Floeder, D. Fromme, B. Granitza, J. Krug[+], C. Masche, W. Raith, A. Schwab, G. Sinapius, and P.W. Zitzewitz[++]

Fakultät für Physik, Universität Bielefeld, D-4800 Bielefeld
Federal Republic of Germany

A renewed interest in electron total cross sections has developed in connection with positron total cross section measurements. Experimental arrangements for low-energy positron total scattering studies can als be used with electrons. Such experiments yield absolute cross sections for electrons and positrons as well as ratios $\sigma^-(E)/\sigma^+(E)$ for which some of the systematic errors (e.g. of gas-manometer calibration) cancel. The $\sigma^-(E)$ data obtained in e^-/e^+ experiments and those from "electron-only" experiments are compared and the existing deviations stimulate further investigations. The recent measurements on H_2 are a good example for the feedback from positron to electron studies.

Our e^+H_2 absolute total cross section measurements[1] cover the range of 8 to 400 eV and can be compared with measurements of the groups at London[2,3] and Detroit.[4] The agreement is satisfactory. The most noticeable discrepancies are in the region of the steep rise of the cross section where inelastic reaction channels come in. Possibly, these discrepancies result from errors in absolute positron energy scale; they would disappear almost entirely if energy shifts on the order of 1 eV were applied. But the source of these discrepancy is not known yet.

It is significant that positron measurements with three technically quite different apparatuses agree so well over a wide energy range. This consistency demonstrated for total e^+H_2 scattering is essential for accessing the accuracy of future measurements on a great variety of target molecules, some of which might only be investigated in one laboratory. In our laboratory measurements with SF_6 and some hydrocarbons are in progress.

The three groups at London, Detroit and Bielefeld measured absolute e^+H_2 and e^-H_2 cross sections of comparable accuracy. The latter can also be compared with e^-H_2 data of four other groups. At energies above 15 eV the electron data agree within the errors, but at lower energies systematic differences appear. Particularly large deviations of almost 20% exist at energies around 3 eV (not including pre-1966 data in this comparison).

Our e^-H_2 measurements accompanying the positron studies extended down to 6 eV. Previously, we measured e^-H_2 total cross sections at energies of 0.02 to 1.5 eV in a time-of-flight (TOF) spectrometer.[4] In order to bridge the gap between 1.5 eV and 6 eV not yet covered

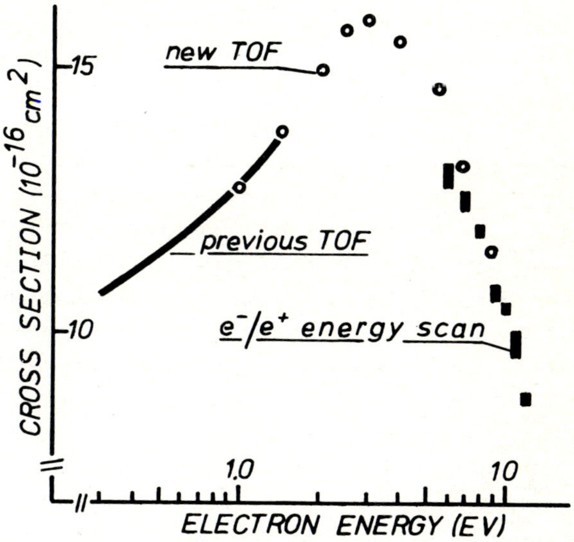

Fig. 1: Absolute e^-H_2 total cross sections measurements from our laboratory obtained in three different studies

by our own measurements we recently performed additional TOF measurements. All our e^-H_2 measurements between 0.5 and 10 eV are shown in Fig. 1. The new TOF data are in excellent agreement with our other data on either end of the energy range. The error of our new TOF data was estimated as 3%.

The reliability of our previous TOF measurements in the very-low energy range was confirmed by a comparison with results derived from swarm experiments using the formalism of the "modified effective range theory" for extrapolating to zero energy.[5] The agreement with the swarm results is excellent. The now established connection of our e^-/e^+ energy-scan data with these TOF results gives additional confidence in the accuracy of our electron and positron cross section measurements at higher energies.

+ Institut für Experimentalphysik, Universität Bochum, D-4630 Bochum

++ Department of Natural Science, University of Michigan, Dearborn, Michigan 48128, USA

References

1. A. Deuring et al., J. Phys. B, to be published
2. M. Charlton et al., J. Phys. B 13, L353 (1980)
3. M. Charlton et al., J. Phys. B 16, 323 (1983)
4. J. Ferch et al., J. Phys. B 13, 1481 (1980)
5. E.S. Chang et al., J. Phys. B 14, 898 (1981)

MEASUREMENTS OF TOTAL SCATTERING CROSS SECTIONS FOR POSITRONS AND ELECTRONS COLLIDING WITH POTASSIUM*

T.S. Stein, R.D. Gomez, Y.-F. Hsieh, W.E. Kauppila, Ch.K. Kwan, and S.J. Smith

Department of Physics and Astronomy, Wayne State University, Detroit, Michigan 48202, U.S.A.

Up to the present time, total scattering cross section (Q_T) measurements have been reported in the literature for positrons colliding with only room-temperature gases (the inert gases and several molecules).[1] We have set up an experimental system to measure absolute Q_T values for positrons and electrons colliding with alkali atoms using a beam transmission technique.

Our positrons are produced by an ^{11}C source created by the $^{11}B(p,n)^{11}C$ reaction using a 4.75 MeV proton beam from a Van de Graaff accelerator. Low energy positrons emitted by the ^{11}C source are guided by a curved axial magnetic field to our alkali scattering apparatus shown schematically in Fig. 1.

We have measured e^--K Q_T values in preparation for the corresponding measurements with positrons in the same apparatus and using the same technique. Our preliminary Q_T measurements are shown in Fig. 2 along with other experimental[2,3] and theoretical[4] results. Although we have many different experimental tests to perform before we report our final results and assign experimental uncertainties to our measurements, the pre-

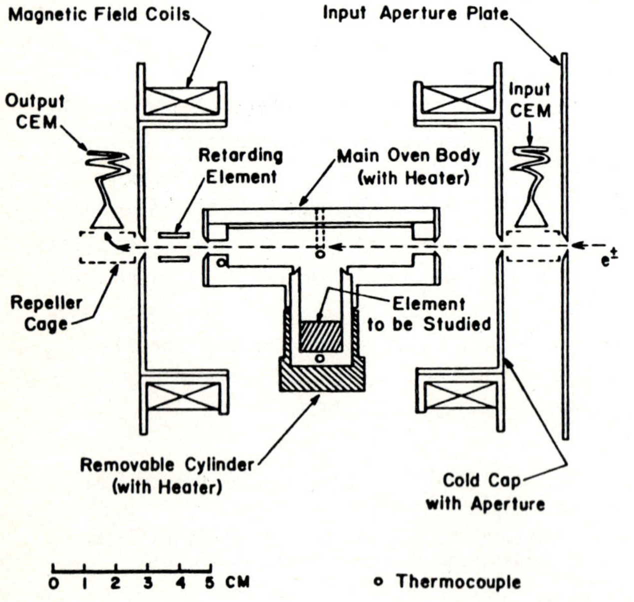

Fig. 1: Schematic diagram of alkali scattering system.

liminary indications are that our e^--K Q_T values are in good agreement (within 12% at the energies of overlap) with the total cross sections determined by Vuskovic and Srivastava[2] and are also in good agreement (within 17%

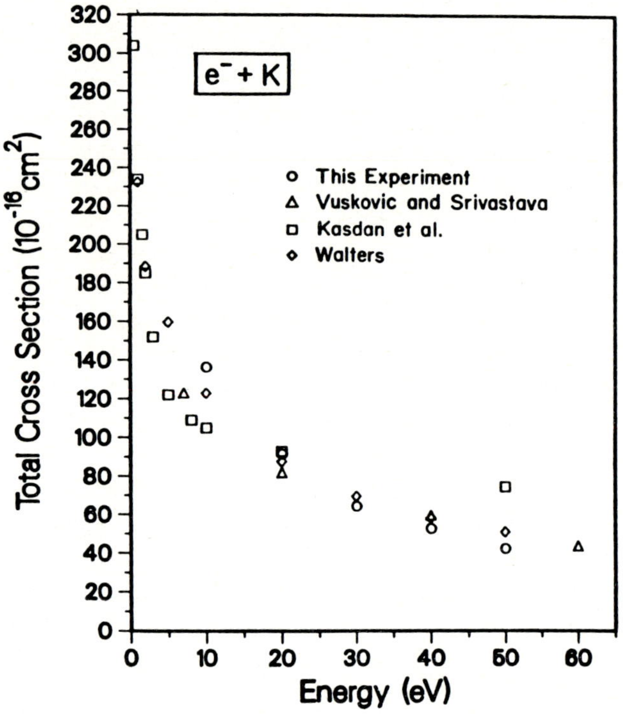

Fig. 2: e^--K total scattering cross sections.

at all energies of overlap) with the estimates made by Walters[4]. Our Q_T values (like Walters' estimates and the values obtained by Vuskovic and Srivastava) tend to have a somewhat different shape than the measured values of Kasdan et al.[3] from 10 to 50 eV.

We are now in the process of measuring e^+-K Q_T values and we hope to have preliminary values to report at the conference. We feel that the direct comparisons of e^+- and e^--K Q_T values may be of considerable interest based upon our earlier comparisons between the scattering of positrons and electrons by the inert gas atoms and several molecules, which have revealed many interesting differences and similarities.[1]

*This work supported by the National Science Foundation.

References

1. T.S. Stein and W.E. Kauppila, Adv. At. Mol. Phys. 18, 53 (1982).
2. L. Vuskovic and S.K. Srivastava, J. Phys. B 13, 4849 (1980).
3. A. Kasdan, T.M. Miller, and B. Bederson, Phys. Rev. A 8, 1562 (1973).
4. H.R.J. Walters, J. Phys. B 9, 227 (1976).

POSITRON AND ELECTRON TOTAL SCATTERING BY N_2O, CH_4, AND SF_6*

W.E. Kauppila, M.S. Dababneh**, Y.-F. Hsieh, Ch.K. Kwan, S.J. Smith, T.S. Stein, and M.N. Uddin

Department of Physics and Astronomy, Wayne State University, Detroit, Michigan 48202 USA

In this paper we report total cross section (Q_T) measurements for 1-500 eV positrons and electrons scattered by N_2O, CH_4, and SF_6. Comparisons, such as these, between the scattering of positrons and electrons by the inert gas atoms[1,2] and several molecules (H_2, N_2, CO, and CO_2)[3,4] have already revealed many interesting differences and similarities.

Our experimental approach is to use a beam transmission technique where the projectile beam is passed through a gas scattering region. Our positron source is ^{11}C, which is produced by the $^{11}B(p,n)^{11}C$ reaction using a 4.75 MeV proton beam from a Van de Graaff accelerator.

Our positron and electron Q_T results for N_2O are shown in Fig. 1. Some notable features of the Q_T curves are the narrow shape resonance in the vicinity of 2.3 eV for electrons and the increase in the positron Q_T curve at the positronium formation threshold (6.1 eV). The electron curve is everywhere higher than the positron curve, except possibly at the lowest energies, with both curves having a maximum between 10-50 eV. At higher energies both curves are decreasing and appear to be gradually approaching each other. It is very interesting that the shapes and absolute magnitudes of these comparison Q_T curves for N_2O are remarkably similar to corresponding comparison Q_T curves our group has made[4] for CO_2. It is even more interesting that we earlier had found[3,4] remarkably similar sets of positron and electron comparison curves for N_2 and CO. In both cases, the target molecules are isoelectronic.

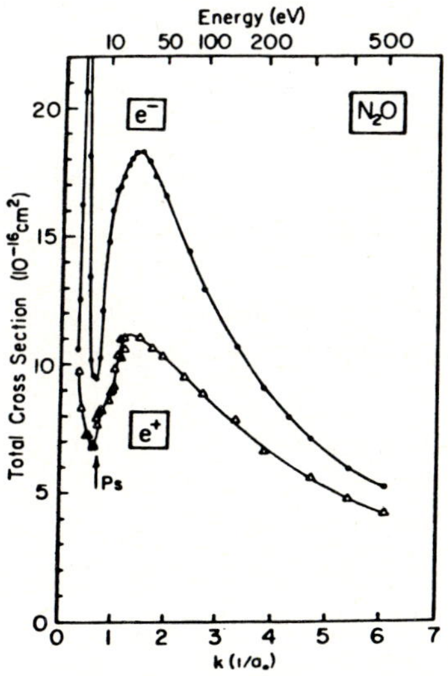

Fig. 1. Q_T measurements for N_2O.

The present Q_T measurements for CH_4 and SF_6 are displayed in Fig. 2. It is intriguing that in many respects the comparison curves for CH_4 are quite similar to the shapes of our earlier comparison measurements[1] for Ar. The e^--SF_6 Q_T curve is dominated by a series of narrow, low energy shape resonances, which have previously been observed in the Q_T measurements of Kennerly and Bonham[5] at 2.56 eV, 7.05 eV, and 11.87 eV. Dehmer et al.[6] have made orbital assignments of a_{1g}, t_{1u}, and t_{2g}, respectively, for these resonances. Our e^--SF_6 Q_T measurements are in very good agreement in both magnitude (within a few percent) and shape with the measurements of Kennerly and Bonham.

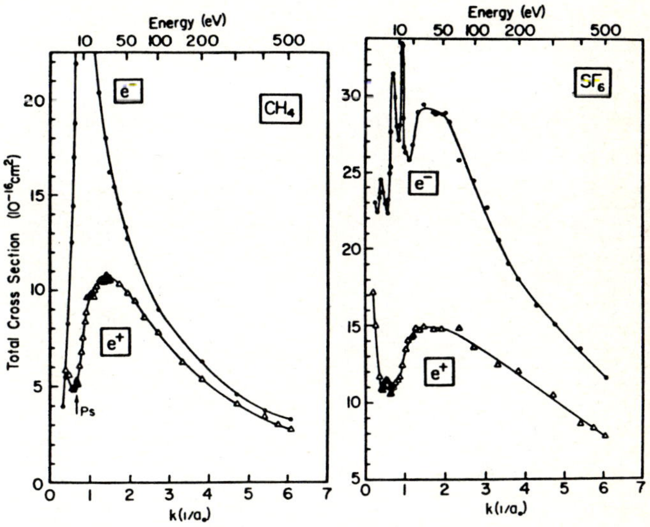

Fig. 2. Q_T measurements for CH_4 and SF_6.

*This work supported by the National Science Foundation.

**Physics Dept., Yarmouk University, Irbid, Jordan.

References

1. W.E. Kauppila, T.S. Stein, J.H. Smart, M.S. Dababneh, Y.K. Ho, J.P. Downing, and V. Pol, Phys. Rev. A 24, 725 (1981).
2. M.S. Dababneh, Y.-F. Hsieh, W.E. Kauppila, V. Pol, and T.S. Stein, Phys. Rev. A 26, 1252 (1982).
3. K.R. Hoffman, M.S. Dababneh, Y.-F. Hsieh, W.E. Kauppila, V. Pol, J.H. Smart, and T.S. Stein, Phys. Rev. a 25, 1393 (1982).
4. Ch.K. Kwan, Y.-F. Hsieh, W.E. Kauppila, S.J. Smith, T.S. Stein, M.N. Uddin, and M.S. Dababneh, Phys. Rev. A 27, XXXX (1983). (to be published)
5. R.E. Kennerly and R.A. Bonham, J. Chem. Phys. 70, 2039 (1979).
6. J.L. Dehmer, J. Siegel, and D. Dill, J. Chem. Phys. 69, 5205 (1978).

APPLICATION OF THE KOHN VARIATIONAL METHOD TO THE CALCULATION OF CROSS SECTIONS FOR LOW ENERGY POSITRON HYDROGEN MOLECULE SCATTERING

E.A.G. Armour and M. Lavender

Mathematics Department, University of Nottingham, Nottingham NG7 2RD, England

The cross section for the elastic scattering of low energy positrons by hydrogen molecules is calculated using the Kohn variational method. The nuclei in the target hydrogen molecule are taken to be fixed in their equilibrium position with internuclear separation, R = 1.4 bohr and in their lowest rotational state. The electronic wavefunction, Ψ_G, for the hydrogen molecule is taken to be of the form

$$\Psi_G = \sum_{i=1}^{N} c_i \phi_i \qquad (1)$$

where

$$\phi_i = \frac{1}{2\pi}\left(\lambda_1^{m_i}\lambda_2^{n_i}\mu_1^{j_i}\mu_2^{k_i} + \lambda_1^{n_i}\lambda_2^{m_i}\mu_1^{k_i}\mu_2^{j_i}\right)e^{-\delta(\lambda_1+\lambda_2)} \qquad (2)$$

m_i, n_i, j_i and k_i are non-negative integers. λ_1 and μ_1, for example, are the confocal elliptical coordinates

$$\lambda_1 = \frac{r_{1A}+r_{1B}}{R} \qquad (3)$$

$$\mu_1 = \frac{r_{1A}-r_{1B}}{R} \qquad (4)$$

and r_{1A} and r_{1B} are the distances of electron 1 from nuclei A and B, respectively. The coefficients $\{c_i\}$ are determined by the variational method.

The energy expectation value associated with Ψ_G is lower than the SCF value but is considerably above the highly accurate values obtained using the wavefunctions of James and Coolidge [1] and Kołos and Roothaan[2]. This is due to the omission from the ϕ_i of powers of the interelectronic distance, r_{12}.

This omission is not considered likely to affect the accuracy of the results very much, as positron-electron correlation is expected to be much more important than electron-electron correlation, and it greatly simplifies the calculation.

The trial wavefunction, Ψ_T, in the Kohn method is taken to be of the form

$$\Psi_T = \frac{N}{(\lambda_3-1)}[\sin\{c(\lambda_3-1)\} + a\{1-e^{-\gamma(\lambda_3-1)}\} \times \cos\{c(\lambda_3-1)\}][1+g\mu_3^2]\Psi_G + \sum_{i=1}^{M} d_i\chi_i \Psi_G \qquad (5)$$

N is a normalization constant,

$$c = \tfrac{1}{2}kR \qquad (6)$$

particle 3 is the positron with wave number k in atomic units and the functions $\{\chi_i\}$ in the closed channels are taken to be suitably chosen functions of the electronic and positron coordinates. They have the important role of taking into account positron-electron correlation. This trial wavefunction is a generalization of the form of the wavefunction used by Massey and Ridley[3] in their variational treatment of the low energy elastic scattering of electrons by hydrogen molecules.

The parameters a, g and the $\{d_i\}$ are determined by the Kohn method. The use of an approximate wave-function for the target molecule is allowed for using the method of models[4]. At low energies the scattering cross section is given, to a good approximation, by[3]

$$\frac{4\pi}{k^2}\sin^2\zeta \qquad (7)$$

where

$$\zeta = (\arctan a) - c. \qquad (8)$$

Results obtained by this method will be presented at the conference and comparison made with the results of other theoretical calculations and with experiment[5,6].

References

1. H.M. James and A.S. Coolidge, J. Chem. Phys. 1, 825 (1933).
2. W. Kołos and C.C.J. Roothaan, Rev. Mod. Phys. 32, 219 (1960).
3. H.S.W. Massey and R.O. Ridley, Proc. Phys. Soc. A69 (1956).
4. R.J. Drachman, J. Phys. B. 5, L30 (1972).
5. W.E. Kauppila, T.S. Stein, J.H. Smart and V. Pol, Abstracts of the Proceedings of the Xth ICPEAC (Paris 1977) p826.
6. T.C. Griffith and G.R. Heyland, Phys. Rep. 39, 169 (1978).

ELASTIC SCATTERING OF POSITRON BY LITHIUM ATOM AT INTERMEDIATE ENERGIES

R.S. Pundir and K.C. Mathur

Department of Physics, University of Roorkee, Roorkee (U.P.) 247 667, India

The elastic scattering of positron by lithium atom is studied at intermediate energies in the framework of a two potential approach. The T-matrix is given by[1]

$$T = \langle \Phi_f | U | \chi_i^{(+)} \rangle + \langle \chi_f^{(-)} | W | \Psi_i^{(+)} \rangle \qquad (1)$$

where† $U = \frac{\zeta(r_2)}{r_2}$, $W = -\frac{1}{r_{12}} + \frac{(1-\zeta(r_2))}{r_2} + V_c(r_2)$.

$\vec{r}_1$ and $\vec{r}_2$ are the position vectors of the valence electron and incident positron respectively. $\zeta(r_2)$ is a position dependent screening parameter[2] and $\Phi_f = \exp(\vec{k}_f \cdot \vec{r}_2) u_f(\vec{r}_1)$ where $u_f(\vec{r}_1)$ is the final state atomic wave function and $\vec{k}_f$ is the momentum of scattered particle. $V_c(r_2)$ is the core potential[3].

Considering the polarisation effect we write

$$\Psi_i^{(+)} = F_i^{(+)}(\vec{r}_2)(u_i(\vec{r}_1) + X(\vec{r}_1, \vec{r}_2)) \qquad (2)$$

$X(\vec{r}_1, \vec{r}_2)$ is the target distortion term[4]. The distorted waves for the projectile $F_n(\vec{r}_2)$ in the initial and final channels are obtained from the following equation.

$$\left[\frac{1}{2}\nabla_2^2 - \frac{\zeta(r_2)}{r_2} + \frac{1}{2}k_n^2\right] F_n(\vec{r}_2) = 0 \qquad (3)$$

subject to the usual boundary conditions.

The differential cross section is given by

$$\frac{d\sigma}{d\Omega} = \frac{k_f}{k_i} \frac{1}{4\pi^2} |T|^2 \qquad (4)$$

Figure shows our result for $e^+ + Li(2s)$ elastic differential cross section at 100 eV energy and we compare them with the only other available calculations of Sarkar et al[5] based on the eikonal approximation (EA). Both the calculations show a dip at about 10°. The dip is more pronounced in Sarkar et al calculation. Beyond 60° angles EA results tend to agree with our results. Detailed results will be presented at the time of the conference.

One of us (RSP) thanks CSIR, India for the award of Senior Research Fellowship.

References

1. G.P. Gupta and K.C. Mathur, Phys.Rev.A 22 1455 (1980)
2. J.A. Schaub - Shaver and A.D. Stauffer, J. Phys. B13, 1457 (1980)
3. H.R.J. Walters, J. Phys. B6, 1003 (1973)
4. V.D. Ob'edkov, Sov. Phys. JETP16, 463 (1963)
5. K. Sarkar, B.C. Saha and A.S. Gosh Phys.Rev. A8, 236 (1973)

† Here we treat the lithium atom as a one electron system and include the effect of core in interaction potential V(=U+W).

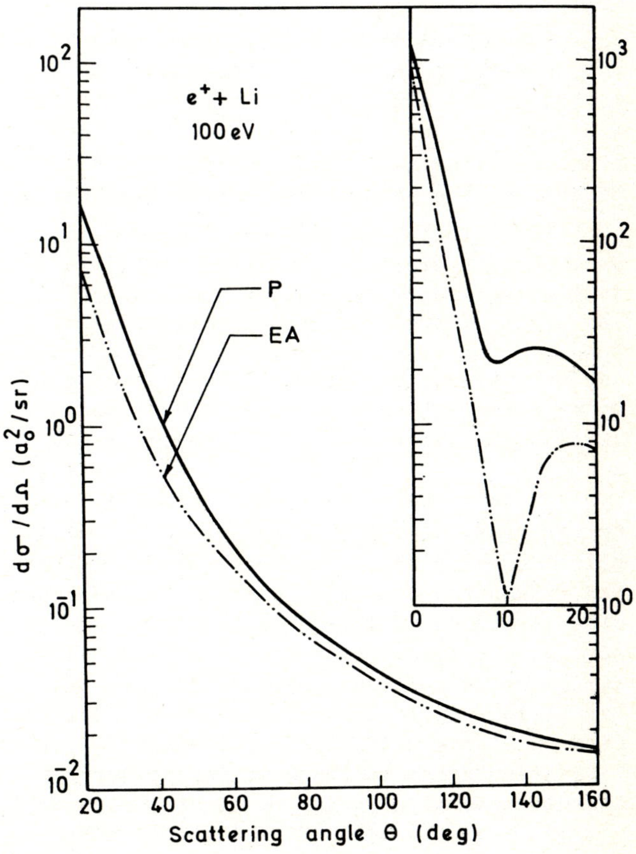

A CLOSE COUPLING STUDY OF POSITRON SCATTERING BY NITROGEN MOLECULE

A.K. Pande and D.N. Tripathi

Department of Physics, Banaras Hindu University, Varanasi-221005, India

The study of low energy positron scattering by the nitrogen molecule is far from satisfactory. The experimental results of Coleman et al.[1] and Kauppila et al.[2] differ from each other both in magnitude as well as in shape. Theoretical results of previous workers[3,4] differ appreciably from each other as well as from the experimental results. Therefore a systematic and thorough investigations of the problem appears desirable.

In the present study we have used quantum mechanical close-coupling formulation of Arthurs and Dalgarno[5] the details of which are given elsewhere[6]. The coupled radial equations are solved using the integral equation algorithm of Sams and Kouri[7] to yield the scattering T-matrix elements. These T-matrix elements, inturn, have been used to obtain the various cross section of the interest[8,9]. The j=10 rotor-basis-set has been found sufficient to yield the converged cross sections. The convergence with respect to the number of total angular momentum (j+1) as well as the range of integeration has been fully ensured.

The interaction potential is taken to be the long-range one mainly constituted of the spherical and non-spherical polarizabilities and the quadrupole moment. An attempt has also been made to include indirectly, the effects of the short range correlation forces.

The integral and differential cross sections for elastic as well as inelastic scattering are presented and are analysed in the light of the available results[1,2,3,4]. The momentum transfer cross sections and the relative transition probability corresponding to each partial wave considered have been given. The agreement of the present close coupling results with that of the experimentally determined cross sections in the low energy range ($\lesssim$5 eV) is quite good. The annihilation being not taken into account may be the reason for a slight deviation of computed cross sections in the higher positron energy region. Calculation of the rotational excitation cross sections corresponding to the transition j=10 have been made in order to find out the scaling law for predicting the excitation cross sections of the higher rotational transition on the basis of the lower transitions. Elastic cross sections are found to depend on the rotator state of the molecule. The present study provides useful information for further work in this regard.

References

1. P.G.Coleman, T.C.Griffith and G.R.Heyland Appl.Phys. 4, 89 (1974).

2. W.E.Kauppila, T.S.Stein, J.H.Swart and V.Pol X ICPEAC, 826 (1977).

3(a) J.W.Darewych and P.Baille, J.Phys.B 7, L1 (1974).

3(b) J.W.Darewych, P.Baille and S.Hara, J.Phys.B 7, 2407 (1974).

4. E.S.Gilleespie and D.G. Thompson, J.Phys.B 8, 2858 (1975).

5. A.M.Arthurs and A.Dalgarno, Proc.Roy.Soc. (London) A256, 540 (1960).

6. A.K.Pande, Ph.D. Thesis, Banaras Hindu University, Varanasi, India.

7(a) W.N.Sams and D.J.Kouri,J.Chem.Phys. 51, 4809 (1969).

7(b) W.N.Sams and D.J.Kouri,J.Chem.Phys. 51, 4815 (1969).

8. M.A.Brandt, D.G.Truhlar and R.L.Smith, Comp.Phys.Comm. 7, 172 (1974).

9. J.M.Blatt and L.C.Beidenharn, Rev.Mod.Phys. 24, 258 (1952).

ROTATIONAL EXCITATION OF CH_4 MOLECULES BY LOW-ENERGY POSITRONS

Ashok Jain and D. G. Thompson

Department of Applied Mathematics and Theoretical Physics,

Queen's University, Belfast BT7 1NN, Northern Ireland.

The subject of positron-molecule scattering has recently received a renewed interest from theoretical as well as experimental points of view (see recent reviews by Griffith and Heyland[1] and Ghosh et al[2]). The rotational excitation of molecules by positron impact is likely to be quite different from that of electron one and much bigger differences are to be expected for vibrational excitation[3]. There are not many calculations on the rotational and/or vibrational excitation of molecules (diatomic or polyatomic) by positron impact: Hara[4] has reported cross sections for the rotational excitation of H_2, N_2 and O_2 in the distorted wave approximation; in the fixed-nuclei approximation with a one-centre formalism, Baille and Darewych[5] and Baille, Darewych and Lodge[6] have studied vibrational and rotational excitations of H_2 molecule, respectively.

Recently, Jain and Thompson[7] have carried out extensive calculations on the rotational excitation of CH_4 molecule by electron impact in the fixed-nuclei and the adiabatic-nuclei-rotation approximations. In their calculation, Jain and Thompson[7] found that for the $e^- - CH_4$ case,

$$\sigma_i\,(0 \to 4) >> \sigma_i(0 \to 3) \qquad (1)$$

where, $\sigma_i(n \to n')$ is the integral cross section from an initial rotational state n to the final state n'. We have carried out similar calculations (neglecting all possibilities of the positronium formation) for $e^+ - CH_4$ scattering. In the approach of Jain and Thompson[7] the total interaction potential is a sum of the exact static and an ab initio parameter-free polarisation[8] potentials. The method of solving the close-coupling (one-centre) equations has been described elsewhere[7].

Here, we report integral (σ_i) and momentum transfer (σ_m) cross sections for the $0 \to 0$, $0 \to 3$ and $0 \to 4$ transitions in an energy range from threshold to about the positronium threshold. In the figure, we have shown these σ_i and σ_m for all three processes. We see a remarkable difference between the electron and positron cross sections. Here (see equation 1 for the electron scattering),

$$\sigma_i(0 \to 3) >> \sigma_i(0 \to 4) \qquad (2)$$

(the same is true for σ_m). There is a deep minimum around 0.7 eV in the σ_m (0 → 0) cross section. In the differential cross section curves (not displayed), there is a deep minimum around 50° at 3 eV in the elastic (0 → 0) process; this minima shifts towards lower angles as the energy increases (around 40° at 5 eV and around 30° at 7.5 eV).

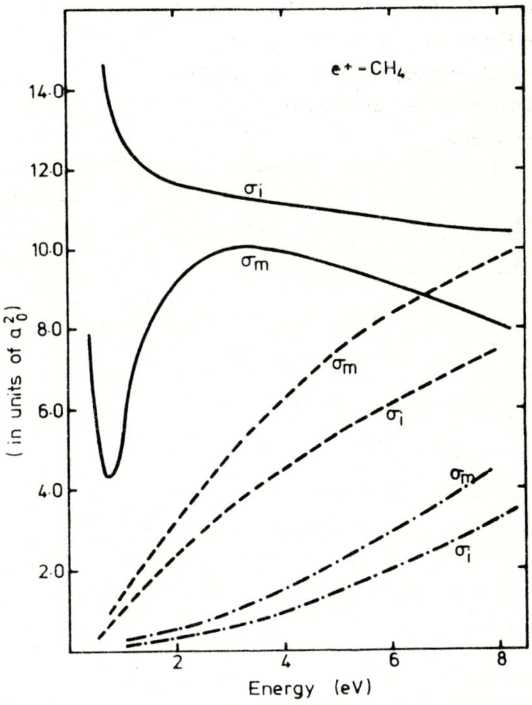

Fig 1 $e^+ - CH_4$, σ_i and σ_m cross section for various transitions: ____, 0 → 0; - - - - , (0 → 3) (multiplied by a factor of 5); -•--•-•, (0 → 4) (multiplied by a factor of 20).

References

1. T. C. Griffith and G. R. Heyland, Phys. Rep. 39C 169 (1978).
2. A. S. Ghosh, N. C. Sil and P. Mandal, Phys. Rep. 87 313 (1982).
3. H.S.W. Massey, Cand. J. Phys. 60 461 (1982).
4. S. Hara, J. Phys. B5 589 (1972).
5. P. Baille and J. W. Darewych, J. Phys. Lett. (France) 35 L243 (1974).
6. P. Baille, J. W. Darewych and J. G. Lodge, Cand. J. Phys. 52 667 (1974).
7. A. Jain and D. G. Thompson (submitted in J. Phys. B).
8. A. Jain and D. G. Thompson J. Phys. B15 L631 (1982).

AN INVESTIGATION OF THE USE OF INEXACT TARGET WAVE FUNCTIONS IN CALCULATIONS OF POSITRONIUM FORMATION IN POSITRON - ATOM SCATTERING

J.W. Humberston and C. J. Brown

Department of Physics and Astronomy, University College London, Gower Street, London WC1E 6BT. England.

It is well known that the use of inexact target wave functions in calculations of the parameters describing the elastic scattering of positrons by atoms can produce very inaccurate results. [1,2,3] Several prescriptions have been recommended for dealing with the problem. The most satisfactory of these is probably the 'method of models' [4] in which the inexact target wave function is assumed to be an exact energy eigenfunction of a model target hamiltonian. The model hamiltonian differs from the exact hamiltonian in having a model potential function in place of the exact potential. Provided the total wave function for the positron-atom system is written as

$$\Psi(\underline{r}_1, \underline{R}) = \phi(\underline{R})\, \psi(\underline{r}_1, R),$$

where $\underline{r}_1$ is the position vector of the positron, $\underline{R}$ represents the position vectors of all the atomic electrons and $\phi(\underline{R})$ is the inexact target wave function, it is not even necessary to know the form of the model potential or the energy eigenvalue.

In calculations of positronium formation the method of models cannot easily be used consistently. The model potential will not in general be a sum of two body potentials and its form might be a very inappropriate representation of the exact potential function of the residual ion plus the electron-ion interaction potential. A possible alternative, which is still within the spirit of the method of models, is to use different model potentials in the two channels so that the wave function of the residual ion is regarded as an energy eigenfunction for an ion model potential. The interaction potential between the positronium and the ion is, however, taken to have its exact form. There is an inconsistency between the two model potentials, but some such inconsistency is unavoidable except in positron-hydrogen scattering. In positron-helium scattering the inconsistency might be expected to have a relatively slight effect because the appropriate model potential for the residual hydrogenic He^+ ion is the exact coulomb potential.

Before attempting to calculate the parameters for positronium formation is positron-helium scattering, we are testing the technique on the positron-hydrogen system using an inexact hydrogen atom wave function

$$\phi(r) = (\alpha^3/\pi)^{\frac{1}{2}}\, e^{-\alpha r} \qquad \text{with } \alpha \neq 1.$$

The formulation is very similar to that used by Humberston [5] in his detailed investigation of positronium formation in s-wave positron-hydrogen scattering, namely a two channel form of the Kohn variational method with trial functions containing many linear variational parameters. By setting $\alpha = 1$ we are able to reproduce his results, which are believed to be very accurate. We are investigating the sensitivity of the scattering parameters to small variations in α about the value 1. This will need to be slight if the technique is to be used successfully in positron-helium scattering. Results will be presented at the Conference.

References

1. Peterkop R. and Rabik L. J. Phys. B 4, 1440 (1971)
2. Page B.A.P. J. Phys. B 8, 2486 (1975); J. Phys. B 9, 2221 (1976)
3. Ho Y.K., Fraser P.A. and Kraidy M. J. Phys. B8, 1289 (1975)
4. Drachman R.J. J. Phys. B 5, L30 (1972)
5. Humberston J.W. Can. J. Phys. 60, 591 (1982)

IDEAL SPACE THEORY OF THE $e^+ + H \dashrightarrow Ps + H^+$ PROCESS

E. Ficocelli Varracchio and M.D. Girardeau*

Department of Chemistry, Centro Chimica Plasmi, C.N.R., University of Bari, 70100 Bari, Italy
*Institutes of Theoretical Science and Chemical Physics, Department of Physics, University of Oregon, Eugene, OR 97403, U.S.A.

Positronium (Ps) formation in e^+- H scattering is a fundamental process that deserves a deep investigation on different accounts. First of all, it represents one of the few three-body systems for which a rearrangement channel is available; secondly, the study of this process, unencumbered by the approximate knowledge of the target states, is a useful testing ground for the development of new ideas, and numerical techniques, to be extended later to more complicated targets.

For such reasons we have recently investigated, in depth, this system, showing that, in particular, it may be very convenient to perform a change of "representation" for the system Hamiltonian, before considering dynamical computations[1-3]. This change of representation can be effected by means of a unitary transformation of the system Hamiltonian and it characteristically leads to a transformed Hamiltonian, having the structure

$$\hat{H} = \hat{H}_o + \hat{V}_{diss} + \hat{V}^{\dagger}_{diss} + \hat{V}_{in} \tag{1}$$

The $\hat{H}_o$ term, in (1), contains the kinetic energies of the e^+, e^- and Ps species, respectively, while the three $\hat{V}$ interactions refer to the channels available to the system. In particular, $\hat{V}_{diss}$ and $\hat{V}^{\dagger}_{diss}$ correspond to the processes of Ps break-up and formation, respectively, while $\hat{V}_{in}$ presides over elastic and inelastic collisions.

While we refer to our recent papers for details of (1), here we wish to consider some numerical applications of the new Hamiltonian, to the study of the rearrangement process

$$e^+(\vec{k}_i) + H(\alpha_i) \dashrightarrow Ps(\vec{k}_f\beta_f) + H^+ \tag{2}$$

with α and β the quantum numbers characterizing the H and Ps species, in the entrance and exit channels, respectively. We have shown[4] that, in particular, enforcement of the Hamiltonian (1), for the process (2), leads, to first order, to the following expression for the rearrangement T-matrix

$$T^{(1)}_{\vec{k}_f\beta_f;\vec{k}_i\alpha_i} = T^B + T^S \tag{3}$$

In eq. (3), T^B is the conventional first Born approximation, while the T^S term, characteristic of the transformation leading to (1), is explicitly given by

$$T^S = -(2\pi)^{-3/2} \int d\vec{x}_1 d\vec{x}_2 d\vec{x}_3 d\vec{x}_4 \; \Phi^*_{\vec{k}_f\beta_f}(\vec{x}_1\vec{x}_2)\left[\frac{1}{x_2} - \frac{1}{x_1}\right] \times$$
$$\times \; \Delta^{Ps}(\vec{x}_1\vec{x}_2,\vec{x}_3\vec{x}_4)\chi_{\alpha_i}(x_3)\exp(i\vec{k}_i\cdot\vec{x}_4) \tag{4}$$

In (4), Φ describes the translational and internal motions of Ps, while χ is the bound state wave function of H. Besides, the Δ^{Ps} kernel can conveniently be represented as

$$\Delta^{Ps} = \Sigma^b_n \, |\Phi_{\vec{k}_n\beta_n}><\Phi_{\vec{k}_n\beta_n}| \tag{5}$$

with the "b" upper index denoting that only "bound" Ps states will contribute to the sum, in (5). Characteristically, the T^S term decreases the contribution of the simple first Born approximation, T^B, and it can be readily interpreted as a "screening" effect for the bare Coulomb interactions, present in the system.

We have evaluated the T^S term, in (4), numerically, and we have calculated differential and total scattering cross sections, in the $T^{(1)}$ approximation, according to (3). Our results are in extremely good agreement with distorted wave computations of the literature[5], and will be presented and discussed at the Conference.

References

1. M.D. Girardeau, Phys. Rev. A26, 217 (1982)
2. E. Ficocelli Varracchio, Lett. Nuovo Cimento 34, 137 (1982)
3. E. Ficocelli Varracchio, Ann. Phys. (N.Y.)(in press)
4. E. Ficocelli Varracchio and M.D. Girardeau, J. Phys. B (in press)
5. P. Mandal, S. Guha and N.C. Sil, J. Phys. B 12, 2913 (1979).

ONE-POSITRON TRIPLE ESCAPE THRESHOLD BEHAVIOUR

P. Grujić

Institute of Physics, P.O. Box 57, 11001 Belgrade, Yugoslavia

Up to now, both experimentally and theoretically, almost exclusively elastic positron-atom scattering has been studied.[1] Here we present results of analytical calculations of threshold law for the break-up processes, when two electrons and one positron are in the continuum in the final channel, just above the ionization threshold. These final states can be achieved via several types of processes, as examplified by

$$e^+ + H^- \rightarrow H^+ + 2e^- + e^+ \qquad (1.a)$$

$$h\nu + [H^-; e^+] \rightarrow H^+ + 2e^- + e^+ \qquad (1.b)$$

$$e^- + [H\ ; e^+] \rightarrow H^+ + 2e^- + e^+ \qquad (1.c)$$

where $[A;e^+]$ denotes one-positron atomic system.[2] We have investigated the ionization function behaviour just above the break-up energy limit, making use of the classical theory. The method has been based on Wannier's model and is developed within the formalism due to Vinkalns and Gailitis, as described elsewhere.[3),4)]

In the near-zero total energy region the phase space available for the triple escape is small and the particles move along the potential ridge. In the limiting case E = 0 the escaping trajectories cluster around the leading configuration, with zero phase space volume. Numerical examinations reveal that the only possible arrangement is the linear configuration, with electrons moving in opposite directions and positron advancing in front of one of them. We write the position vectors as

$$\vec{r}_i = (\beta_i r + \Delta_i)\vec{k} + \delta_i \vec{i} + \nabla_i \vec{j}, \quad i = 1,2,3 \qquad (2)$$

where Δ_i, δ_i and ∇_i are small deviations along the leading paths (Δ_i) and perpendicular to them (δ_i, ∇_i). Correlation parameters β_i are determined from the condition of stationarity of the potential at : $R = [r_1^2 + r_2^2 + r_3^2]^{\frac{1}{2}}$ hypersphere (see Fig. 1), with $\beta_1 = -1$, $\beta_{2,3} > 0$.

Substituting Eqs. (2) into the corresponding Newton's equations and retaining terms linear in Δ_i/r, etc., one arrives at differential equations for $\Delta_i, \delta_i, \nabla_i$. Because of the linear geometry of the leading configuration, equations for the longitudinal and transfersal deviations decouple and one has for the electrons

$$r^3 \frac{d^2}{dt^2}\begin{Bmatrix}\Delta_1\\ \Delta_2\end{Bmatrix} = \underset{\sim}{A}\begin{Bmatrix}\Delta_1\\ \Delta_2\end{Bmatrix} \qquad (3)$$

the third deviation Δ_3 being eliminated by the condition : $R = 1\ a_o$. Elements a_{ij} of matrix $\underset{\sim}{A}$ are functions of β_i and Z (charge of the ion, in atomic units). Eqs. (3) are solved to give

$$\Delta_1 = T_{11}(C_1 r^{\varkappa_1} + C_2 r^{\varkappa_2}) + T_{12}(C_3 r^{\varkappa_3} + C_4 r^{\varkappa_4}) \qquad (4)$$

$$\Delta_2 = T_{21}(C_1 r^{\varkappa_1} + C_2 r^{\varkappa_2}) + T_{22}(C_3 r^{\varkappa_3} + C_4 r^{\varkappa_4}) \qquad (5)$$

where C_i are arbitrary constants and T_{ij} are the elements of matrix $\underset{\sim}{T}$, which quasidiagonalizes $\underset{\sim}{A}$.

By a procedure appropriate for Wannier's theory,[3),4)] one obtains the threshold law

$$\sigma_{ion} \sim E^{\varkappa}, \quad \varkappa = \varkappa_1 + \varkappa_3 - 2 \qquad (6)$$

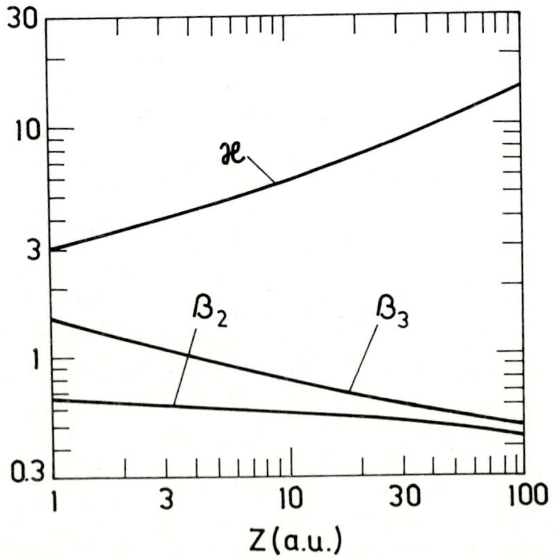

Fig. 1 : Correlation parameters β_i and exponent $\varkappa$ for the one-positron triple escape near the threshold.

References

1. A. S. Ghosh, N. C. Sil and P. Mandal, Physics Reports 87, 313 (1982)
2. D. C. Clary, J. Phys. B 9, 3115 (1976)
3. P. Grujić, J. Phys. B 15, 1913 (1982)
4. P. Grujić, J. Phys. B, to be published.

APPLICATION TO $\mu^+e^-e^+$ AND $p\mu^-e^+$ OF A NEW METHOD FOR TAKING INTO ACCOUNT FINITE NUCLEAR MASS IN THE DETERMINATION OF THE ABSENCE OF BOUND STATES

E.A.G. Armour

Mathematics Department, University of Nottingham, Nottingham NG7 2RD, England

Two distinct hydrogen-like atoms can be formed by replacing one or other of the particles in the hydrogen atom by a muon. In the first the proton is replaced by a positive muon (μ^+) to form muonium (μ^+e^-). In the second the electron is replaced by a negatively charged muon (μ^-) to form a mesic atom.

The presence or absence of bound states involving an atom and a positron is very important in theoretical considerations of low energy scattering of positrons by these atoms. In particular, proof of the absence of a bound state below the continuum makes possible the calculation of a variational upper bound on the scattering length[1].

It has been proved by Aronson, Kleinman and Spruch[2] and Armour[3] that no bound state of a system made up of a hydrogen atom and a positron exists below the continuum if the proton is taken to be infinitely massive. The proof has recently been extended by Armour[4] to take into account the finite mass of the proton. The essential idea of the proof was to reduce the case of finite proton mass to a form which could be dealt with by the method used in the case of infinite proton mass. This was done by proving that no bound state exists for finite mass if a system made up of a proton of infinite mass, an 'electron' of mass $0.9995\,m_e$ and a 'positron' of charge $1.0011\,e$ (where m_e is the mass of an electron and e is the charge on the proton) do not form a bound state. This system was then shown not to form a bound state by the already established method[2,3] for the case of infinite proton mass.

The effect of including the finite mass of the proton is very small as the mass of the proton $M = 1836\,m_e$. However, in the case of muonium (μ^+e^-), where the proton is replaced by the muon (μ^+), whose mass $M_u = 207\,m_e$ is considerably smaller, the effect of including the finite mass of the muon can be expected to be more important.

The question of the effect of the inclusion of the finite nuclear mass also arises in the case of the mesic atom ($p\mu^-$) in which the electron is replaced by a muon (μ^-) also having mass M_u. This very much reduces the ratio of the mass of the proton to the mass of the negatively charged particle. It is to be expected on account of the very much reduced polarizability of the mesic atom as compared with the hydrogen atom that no bound state will exist in this case (Gertler, Snodgrass and Spruch[5], Pauling[6]).

It is shown[7], using the new method described above, that no bound state below the continuum exists in the case of either the muonium-positron system ($\mu^+e^-e^+$) or the muonic-atom-positron system ($p\mu^-e^+$). This is the first time that these results have been proved.

References

1. L. Spruch and L. Rosenberg, Phys. Rev. 116, 1034 (1959).
2. I. Aronson, C.J. Kleinman and L. Spruch, Phys. Rev. A 4, 841 (1971).
3. E.A.G. Armour, J. Phys. B 11, 2803 (1978).
4. E.A.G. Armour, Phys. Rev. Letters 48, 1578 (1982).
5. F.H. Gertler, H.B. Snodgrass and L. Spruch, Abstracts of the Proceedings of the Vth ICPEAC (Leningrad 1967) and Phys. Rev. 172, 110 (1968).
6. L. Pauling, Private Communication (1982).
7. E.A.G. Armour, J. Phys. B 16, in press.

USE OF 1-DEPENDENT PSEUDOPOTENTIALS IN MOLECULAR-STRUCTURE CALCULATIONS OF ALKALI-He SYSTEMS

J. PASCALE

Service de Physique des Atomes et des Surfaces, CEN/Saclay, 91191 Gif-sur-Yvette Cedex, France.

Molecular-structure calculations using l-dependent pseudopotentials have been performed to obtain the adiabatic potentials for ground state and numerous excited states of all alkali-He systems. Ab initio calculations for these systems become rapidly difficult and costly as the number of electrons in the atomic core increases. The model potential and pseudopotential techniques therefore offer a very interesting alternative to treat the problem. In both techniques effective potentials are defined to represent the e^{-}-alkali^{+} and e^{-}-rare gas atom interactions. The pseudopotential technique appears to be the most convenient for molecular structure calculations because the core orbitals do not have to be included in the atomic basis set expansion of the molecular wave function. In this technique, the antisymmetry effects due to the Pauli principle are simulated by a repulsive potential which must be l-dependent.

The l-dependent pseudopotentials were defined from spectroscopy or electron scattering data. Standard variational calculations were made to obtain the adiabatic potentials and a large basis set of slater-type orbitals was used in order to ensure accuracy and stability of the results. The method of calculations and the results will be discussed at the Conference. Our calculations are shown to agree much more closely with all available experimental data than previously published calculations, as illustrated for example in Figs 1-2, indicating that a large improvement in the calculation of the adiabatic potentials for all the alkali-He systems has been achieved by using the l-dependent pseudopotential technique[6].

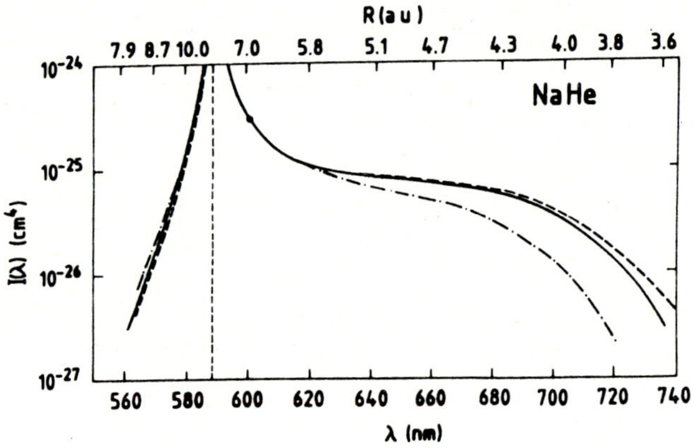

Fig. 1 - Normalized emission spectra of the Na3P state in the presence of He at T = 403 K. Dashed line : experimental data[1]; full curve : present results; dot-long-dashed line : model potential calculations[2]. The calculated spectra are normalized to the experimental ones at λ= 600 nm. Our calculated spectra are also reported versus the internuclear distance R at which the light emission occurs.

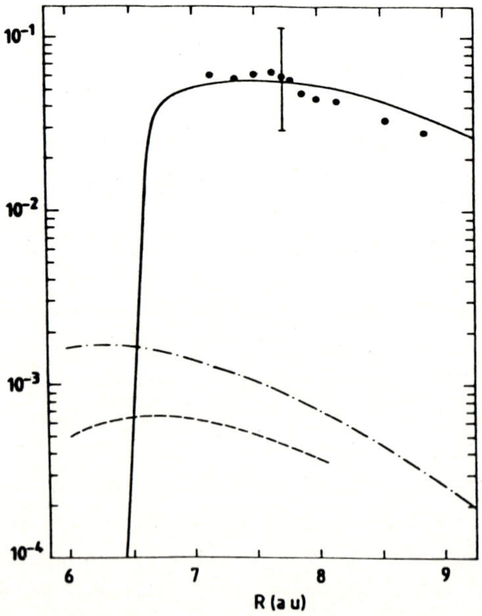

Fig. 2- Dipole-induced oscillator strength versus the internuclear distance R for the $^{2}\Sigma^{+}$6S- $^{2}\Sigma^{+}$ 5D transition in CsHe. Full line : present results; dashed and dot-long-dashed lines are the l-independent pseudopotential calculations of Refs.[4] and 3,respectively; full circles : experimental data[5].

References.

1. G. York, R. Scheps and A. Gallagher, J.Chem. Phys. 63, 1052 (1975).
2. J. Hanssen, R.Mc Carroll and P. Valiron, J.Phys. B 12, 899 (1979).
3. J. Pascale, J. Chem. Phys. 67, 204 (1977)
4. E. Czuchaj, Z. Phys.A 292, 109 (1979)
5. M. Ferray, J.P. Visticot, J.Lozingot and B. Sayer, J. Phys.B 13, 2571 (1980)
6. J. Pascale, Phys. Rev. A 26, 3709 (1982).

TOTAL CROSS SECTIONS FOR $j_1m_1 \rightarrow j_2m_2$ TRANSITIONS WITHIN THE FIRST n^2P STATES OF ALKALI ATOMS INDUCED IN COLLISIONS WITH He

J. PASCALE

Service de Physique des Atomes et des Surfaces,CEN/Saclay,91191 Gif-sur-Yvette Cedex,France.

We have recently determined adiabatic potentials for all alkali-He systems from l-dependent pseudopotential calculations which have been shown reliable[1]. They are used to calculate quantum-mechanically the energy dependence of cross sections for the Zeeman transitions within the first n^2P levels of alkali atoms induced in collisions with He, in the energy range from thresholds of the transitions up to 0.5-1 eV. The cross sections $\sigma_o(j_1m_1 \rightarrow j_2m_2)$ and $\overline{\sigma}(j_1m_1 \rightarrow j_2m_2)$ which are relevant to crossed-atomic beam and cell type experiments,respectively, have been obtained. These calculations, first provide further test of our adiabatic potentials by comparisons with all available experimental data; second they allow the interpretation of numerous crossed-atomic beam or cell-type experimental data. In particular, the $\sigma_o(j_1m_1 \rightarrow j_2m_2)$ cross sections are needed to interpret polarization effects which can be observed in crossed-atomic beam experiments[2]. The results of our calculations for all alkali-He systems will be presented and discussed at the Conference. Two examples are presented in Figs. 1-2, showing excellent agreement between our results and experimental data. In both cases our calculations significantly improve the agreement between previous calculations and experimental data.

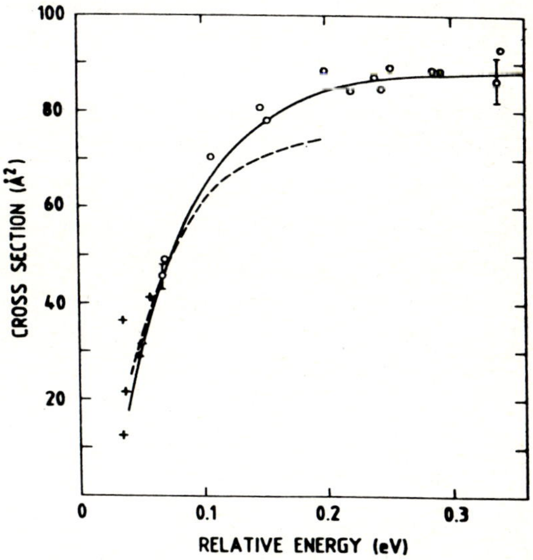

Fig. 1 - Energy dependence of the cross section for the $4^2P_{1/2} \rightarrow 4^2P_{3/2}$ transition in K induced in collisions with He. Open circles : crossed-atomic beam experimental data[3] normalized on cell-experimental data[4]; crosses : crossed-atomic beam experimental data of Ref. 5; full line : present results; dashed line : model potential calculations of Ref.6.

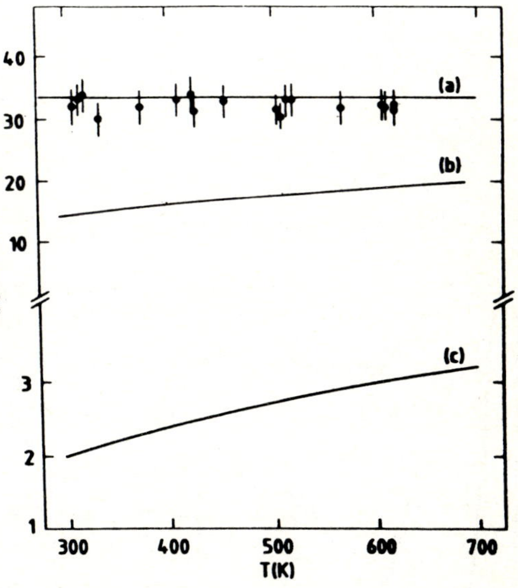

Ref. 2 - Disorientation cross section $\sigma^{(1)}_{1/2} = 2\overline{\sigma}(\frac{1}{2}\frac{1}{2} \rightarrow \frac{1}{2} -\frac{1}{2})$ for the Rb (5^2P) + He collision. Full circles with error bars : experimental data of Ref.7. Full lines are the calculated cross sections reported versus $T=E\,(\frac{3}{2}\,k)^{-1}$,where E is the relative energy; (a) present results; (b) calculations of Ref.8 using potentials of Ref.9; (c) calculations of Ref. 7 using potentials of Ref. 10.

References

1. J. Pascale, Phys.Rev.A 26 3709 (1982)
2. J.M. Mestdagh, J.Berlande,P. de Pujo,J.Cuvellier and A.Binet, Z. Physik A 304 3 (1982)
3. J.M.Mestdagh, Thesis,Université de Paris-Nord (1982) Unpublished.
4. R. Boggy and F.A. Franz,Phys. Rev.A 25 1887 (1982)
5. R.W. Anderson, J.P.Goddard,C. Parravano and J. Warner, J. Chem.Phys. 64 4037 (1976)
6. F. Masnou-Seeuws, J.Phys.B 15, 883 (1982)
7. H. Doebler and B.Kamke, Z.Physik A 280 111 (1977)
8. B. Brouillaud and R. Gayet,J.Phys.B 10,2143 (1977)
9. W.E. Baylis, J.Chem. Phys. 51, 2665 (1969)
10. J. Pascale and J.Vandeplanque,J. Chem.Phys. 60, 2278 (1974).

MODEL POTENTIAL CALCULATIONS FOR THE GROUND AND EXCITED (RYDBERG) STATES OF THE NaAr and KAr MOLECULAR SYSTEMS

A. Chebanier de Guerra+, F. Masnou-Seeuws
Laboratoire des Collisions Atomiques et Moléculaires, Université Paris-Sud, Bât. 351
91405 ORSAY Cedex, FRANCE

Accurate results for systems such as NaHe, NaNe, KHe etc. have been obtained owing to the model potentiel method of Valiron et al[1]. In the present work we discuss the application of the method to heavier systems.
We have solved a one electron Schrödinger equation :

$$\{T + V_A(r_A) + V_B(r_B) + V_3(\underset{\sim}{R}, \underset{\sim}{r_A}, \underset{\sim}{r_B})\}\, \phi_i^\lambda = E_i^\lambda \phi_i^\lambda , \lambda = 0,1,2 \quad (1)$$

The electron-alkali ion potentiel $V_A(r_A)$ was taken from ref. 2. We *have fitted an attractive model potential* $V_B(r_B)$ to reproduce *experimental electron-argon phaseshifts* for collision energies from 0 up to 15 eV. $V_B(r_B)$ contains bound states. V_3 is a three body tensorial term which depends upon the dipole α_d and quadrupole α_q polarizabilities of the rare gas. From the solution of eq(1) we obtain the ionization energy. The molecular energy is then

$$\mathcal{E}_i^\lambda(R) = E_i^\lambda(R) + \mathcal{E}_o(R) \sim E_i^\lambda(R) + \mathcal{S}(R) - \alpha_d/(2R^4) - \alpha_q/(2R^6) \quad (2)$$

$\mathcal{E}_o(R)$ is the energy of the $NaAr^+$ or KAr^+ ground states. $\mathcal{S}(R)$ is the short range repulsive part of the core-core interaction. Accurate results should be obtained for quantities which are not sensitive to the choice of $\mathcal{S}(R)$.

i) Our results for the depth D_e and position R_e of the ground state can be considered as satisfactory considering that *nothing in our model is fitted to molecular data*

$NaAr\ X^1\Sigma_g$	D_e cm^{-1}	R_e (Å)
Present results	38	5.08
Spectroscopy experiment[3,4]	40.4(1.0)	4.991(2)
$KAr\ X^1\Sigma_g$		
Present results	33.14	5.56
Scattering experiment[5]	(39.07) 42.35	5.55

However, the accuracy is not so good as in the case of lighter rare gases.

ii) The well in the $A^2\Pi$ excited state strongly depends upon the choice of the core-core term (Fig.1)

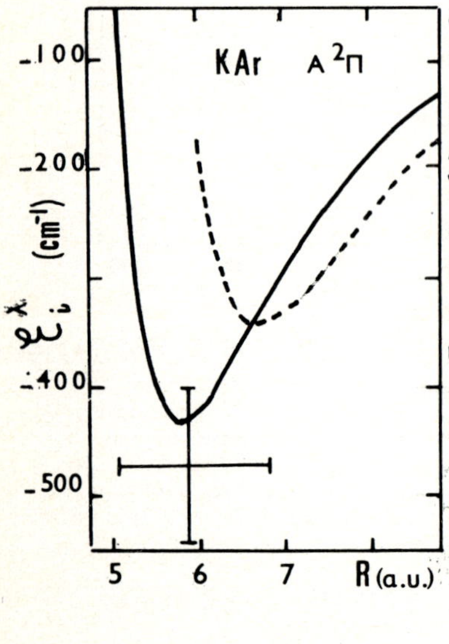

Fig.1. Well in the KAr $A^2\Pi$ state choosing for $\mathcal{E}_o(R)$: Full line : the estimation of Ref. 6., broken line : the potential deduced from mobility measurements[7] . Cross : experimental error bar [5]

We suggest that alkali rare gas experiments could be used to clarify the problem of the core-core term.

iii) For Rydberg states, the comparison of E_i^λ for the NaHe, NaNe and NaAr systems show that the predictions of the Fermi model are *qualitatively* but not *quantitatively* reproduced. However it is important to note that the asymptotic methods give a good estimation of the results[8]

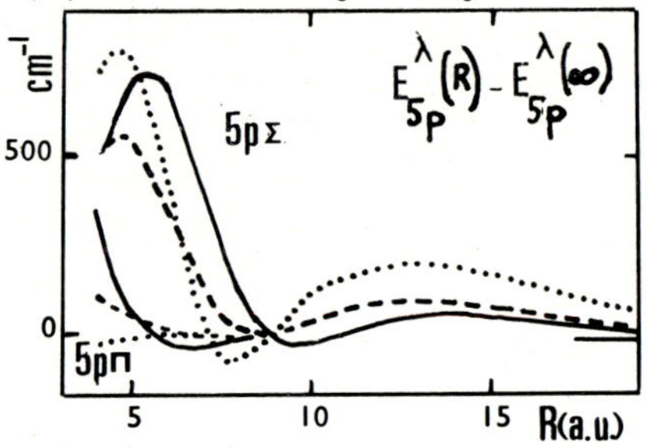

Fig. 2. Variation of the ionisation energies of the $5p^2\Sigma$ and $5p^2\Pi$ states: KHe (dotted line) KNe (broken line) KAr (full line).

iv) In the NaAr case, we have checked that the molecular quantum defects are fairly n-independant and stay within the shadowed area in Fig. 3.

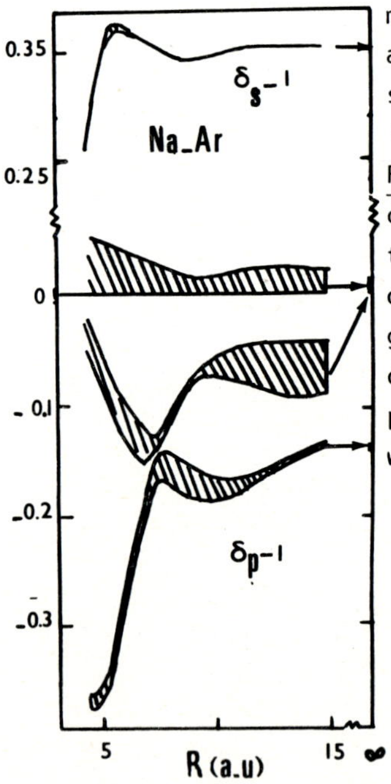

Fig. 3. Molecular quantum defects for the Σ states of the NaAr system. The states correlated to Na $n\ell, \ell \geq 2$ levels give rise to two quantum defect curves, one of which presents an avoided crossing with the p curve.

References

1. Valiron, McCarroll et al. J.Phys.B 12 53 (1979)
2. M. Klapisch Comput.Phys.Commun. 2 239 (1971)
3. R.E. Smalley et al. J.Chem.Phys. 70 2051 (1979)
4. J. Tellinghuisen et al. J.Chem.Phys. 71 1283 (1979)
5. R. Düren, E. Hasselbrink, G. Moritz Z.Phys. 307 1 (1982) and private communication
6. N.A. Sondergaard, E.A. Mason J.Chem.Phys.62 1299 (1975)
7. D.R. Lamm et al. J.Chem.Phys. 74 5 (1981)
8. F. Masnou-Seeuws J.Phys.B 15 883 (1982)

+Present address: University of Caracas, Venezuela.

TWO ELECTRON MODEL POTENTIALS AND TWO CENTER BASIS SETS

O. Mó and A. Riera

Departamento de Química Física y Química Cuántica. Centro Coordinado CSIC-UAM. Universidad Autónoma de Madrid. CANTOBLANCO (Madrid,34) (Spain)

Model potential calculations of molecular energies are usually carried out employing effective potentials evaluated for separated atoms.

In a molecular calculation, however, one should take into account that in a two center basis sets the orbitals are not orthogonal and this has important consequences in the description of the "phantom" states of the model.

We shall illustrate these consequences with a calculation carried out for the NaH molecule. A model potential of the form:

$$V = -\frac{10}{r}(1+\alpha r)\, e^{-2\alpha r}$$

has been ajusted to describe the valence 3s orbital of the sodium atom. In this procedure one solves in a finite representation the effective eigenvalue equation:

$$\left[-\frac{1}{2}\nabla^2 + V - \frac{1}{r}\right]\Psi = \varepsilon\,\Psi$$

one then obtains a series of eigenvalues ε_{1s}, ε_{2s}, ε_{2p}, ε_{3s}... and one optimizes α such that ε_{3s} = the first ionization potential of sodium atom.

The next step is to set up a two-electron effective molecular hamiltonian:

$$H_{eff} = \left[-\frac{1}{2}\nabla^2(1)-\frac{1}{2}\nabla^2(2)-\frac{1}{r_{1A}}-\frac{1}{r_{1B}}-\frac{1}{r_{2A}}-\frac{1}{r_{2B}}+V(1) + V(2) + \frac{1}{r_{12}}\right]$$

where now

$$V(1) = \frac{10}{r_{1A}}(1+\alpha r_{1A})\, e^{-2\alpha r_{1A}}$$

$$V(2) = \frac{10}{r_{2A}}(1+\alpha r_{2A})\, e^{-2\alpha r_{2A}}$$

(Polarization and other long-range terms may be added, if necessary, without affecting our conclusions).

One then solves the eigenvalue equation of this hamiltonian in a finite representation. At very large internuclear distances one obtains a series of "phantom" states

$$1S_{Na}(1)\ \phi_H(2) + \phi_H(1)\ 1S_{Na}(2)$$

$$2S_{Na}(1)\ \phi_H(2) + \phi_H(1)\ 2S_{Na}(2)$$

$$2P_{Na}(1)\ \phi_H(2) + \phi_H(1)\ 2P_{Na}(2)$$

and the state of interest (we shall call it "real")

$$3S_{Na}(1)\ \ 1S_H(2) + 1S_H(1)\ 3S_{Na}(2)$$

The usual procedure is that one calculates the energies for different values of R, forgetting about "phantom" states and concentrating on the "real" one. Our point is that this may be feasible in weakly bonding states. However in strongly or intermediate bonding cases such as the one considered here, the potential energy curve has a minimum at relatively short distances; at these distances it may happen that a two center basis set is able to describe many more "phantom" states (these is a infinity of them for the two electron case in the limit of a complete basis set) than it is able to do at large distances. As a consequence the real state collapses and for $R \to 0$ it represents a "phantom" state.

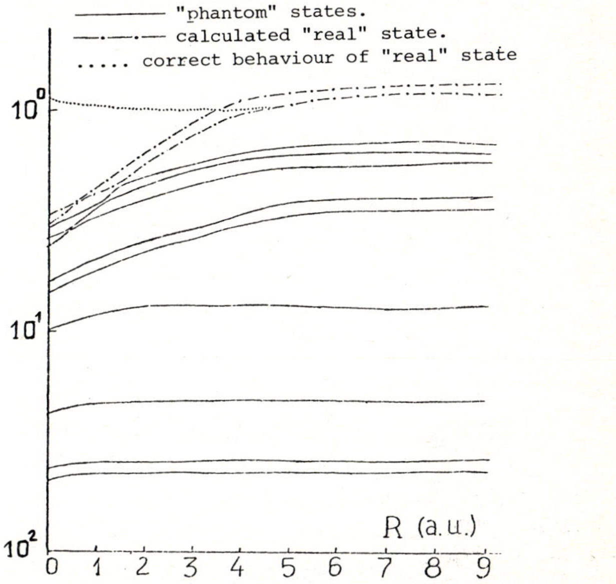

Using a non local pseudo-potential to eliminate "phantom" states as is commonly used for atoms does not solve the problem precisely because it stems from the properties of two center basis set.

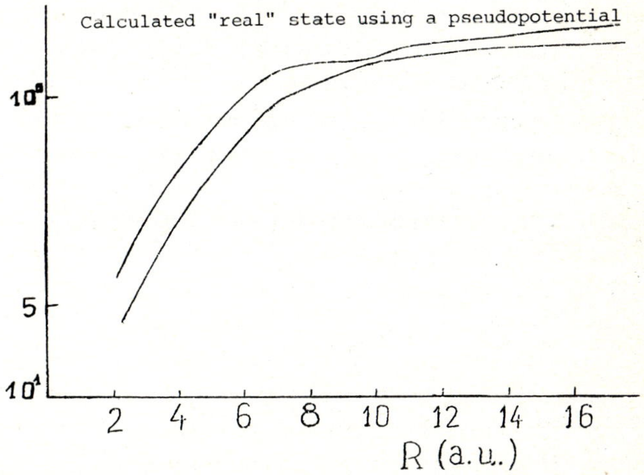

More details of the problem and posible solutions will be discussed at the conference.

MODEL-POTENTIAL CALCULATIONS AS EXTENDED MULTIPROPERTY ANALYSIS

R. Düren and E. Hasselbrink

Max-Planck-Institut für Strömungsforschung, D3400 Göttingen, West Germany

In a earlier paper[1] we have pointed out that model potential calculations can and should be used by experimentalist to evaluate there results in the spirit of a multiproperty analysis i.e. as a uniform representation of many different experimental data. This approach differs from the usual multiproperty analysis (see f.i. ref. 2) by its indirect determination of the potential matrix with the potential calculation.

As main advantages of this approach one may mention the extension of the experimental basis of the multiproperty analysis, the reduction of the number of free parameters, the possibility of an interpolation in states (i.e. to determine the potential for the A^{**}-B interaction from A^*-B experiments) and the interpolation in the interatomic distance R (i.e. to determine the potential for some R_2 from the value given at R_1).

We have realized such an analysis for alkali-rare gas interactions restricting our attention to low energy collisions, the lowest 42 alkali m_j states and the Na, K - Ar,Kr systems. These boundary conditions allow the use of a simple model potential, which is described in detail in ref. 3. It is based on results from P. Gombas, W.E. Baylis, C. Bottcher and A. Dalgarno and G. Peach (see ref. 3 for the original references).

The results that we have obtained for Na-Ar are discussed here, a system for which a large number of different experiments are available. Let us point out that after the adjustment of 2 parameters($r_o' = 0.304$, $r_o'' = 0.1588$) the model potential and hence the matrix of interatomic potentials is completely defined. All the results described below are obtained with this same set of parameters.

The results, which are considered in the comparison of experimental and calculated data are 2-body interactions: (1) the spectra of Na-atom[4]; (2) the e^--Ar phase shifts[5]; (3) the Na^+-Ar interaction[6][7]. As a second group we compare the interactions, which take reference to the total model, i.e. 2- and 3-body interactions; (4) the van der Waals constants[8][9]; (5) differential cross sections for the groundstate and the first excited state[10][11]; (6) the satellite in the line profile[12]; (7) the rovibronic spectra of the van der Waals molecule[13][14][15].

Since these latter data are the most critical ones we present in the table below samples of worst and best cases for these data to give an illustration of the agreement achieved. Some other data are discussed in ref. 3, which together with more recent results will be discussed in this contribution.

Equilibrium points

State/Ref.		13	14	15	this work
$X^2\Sigma$	R_e(a.u.)	9.432	9.464	9.464	9.59
	$D_e(cm^{-1})$	40.4	41.7	44.61	41.5
$A^2\Pi_{1/2}$	R_e(a.u.)	5.493	-	5.503	5.27
	$D_e(cm^{-1})$	568.2	-	5.776	599.1

Band origins (cm^{-1})

	v' - v''			Δ
$A^2\Pi_{1/2}$ - X		ref. 13	this work	Δ
	8 - 0	16879.9	16885.8	5.9
	8 - 1	16868.6	16874.5	5.9
	8 - 2	16859.7	16865.5	5.8
	8 - 3	16853.3	16858.7	5.4
	8 - 4	16849.1	16854.1	5.0
$A^2\Pi_{1/2}$ - X		ref. 15	this work	Δ
	11 - 0	16958.9	16959.5	0.6
	12 - 0	16973.6	16973.6	0.0
	12 - 1	16962.3	16962.3	0.0
	13 - 2	16964.9	16964.6	0.3
$B^2\Sigma_{1/2}$ - X		ref. 15	this work	Δ
	1 - 0	16988.5	16993.8	5.3
	1 - 1	16977.3	16982.6	5.3
	1 - 2	16968.2	16973.5	5.3
	6 - 0	17007.0	17008.4	1.4
	6 - 1	16995.7	16997.2	1.5

Spacing of vibrational levels in the $X^2\Sigma_{1/2}$-potential (cm^{-1})

	ref. 14	this work	Δ
ΔE_{1-0}	11.240	11.230	0.010
ΔE_{2-1}	9.061	9.040	0.021
ΔE_{3-2}	6.824	6.810	0.014
ΔE_{4-3}	4.585	4.570	0.015

References

1. R. Düren, Adv.Atom.Molec.Phys. 16, 55 (1980)
2. F. Battaglia, F.A. Gianturco, P. Casavecchia, F. Pirani and F. Vecchiocattivi, Disc.Faraday Soc. 73, 257 (1982)
3. R. Düren, E. Hasselbrink and G. Moritz, Z.Phys.A 307, 1 (1982)
4. C.E. Moore, NBS(US)circ. 467 (1949)
5. J.F. Williams, J.Phys.B 12, 265 (1979)
6. A. Chebanier de Guerra and F. Masnou-Seeuws, J.Chem.Phys. (1983)
7. M. Waldmann and R.G. Gordon, J.Chem.Phys. 71, 1325 (1979)
8. A. Dalgarno and A.E. Kingston, Proc.Roy.Soc.London A 259, 424 (1961)
9. K.T. Tang, J.M. Norbeck and P.R. Certain, J.Chem.Phys. 64, 3063 (1976)
10. R. Düren and W. Gröger, Chem.Phys.L. 56, 67 (1978)
11. R. Düren, W. Gröger, E. Hasselbrink and R. Liedtke, J.Chem.Phys. 74, 6806 (1981)
12. J.P. Woerdmann and J.J. de Groot, J.Chem.Phys. 76, 5653 (1982)
13. J. Tellinghuisen, A. Ragone, Soo Kim Mying, D.A. Auerbach, R.E. Smalley, L. Wharton and D.H. Levy, J.Chem.Phys. 71, 1283 (1979)
14. G. Aepfelbach, A. Nunnemann and D. Zimmermann, Chem. Phys.L. (in press) (1983)
15. W.P. Lapatovich, A. Marjatta Lyyra, P.E. Moskowitz, M.D. Havey and D.E. Pritchard (private communication) (1983)

A TWO ELECTRON CORRELATED MODEL FOR THE THEORETICAL DETERMINATION OF THE Na_2, Li_2 AND K_2 POTENTIAL ENERGY CURVES

A. Henriet*, M. Aubert-Frécon**, C. Le Sech*** and F. Masnou-Seeuws*

*Laboratoire des collisions Atomiques et Moléculaires, Bât.351
Université Paris-Sud, 91405 ORSAY Cedex (FRANCE)

**Laboratoire de Spectrométrie ionique et moléculaire, Bât. 205
Université de Lyon I, Campus de la Doua, 69622 VILLEURBANNE Cedex (FRANCE)

***Laboratoire d'Astrophysique fondamentale, Observatoire de Meudon, 92190 MEUDON (FRANCE)

Accurate potential energy curves are now obtainable owing to pseudo-potential and model potential techniques for molecular systems with one active electron such as alkali dimer ions[1,2] or alkali rare gas systems[4]. The extension of such techniques to systems with two active electrons is highly desirable. Up to now, it has been performed in the framework of configuration interaction [1,2,3]. Recently two of us have proposed[5], as an alternative way for the description of H_2, a monoconfiguration approach using correlated wavefunctions easily derivable from the known solutions of H_2^+. We report here a generalization for diatomic systems with two active electrons such as alkali dimers.

We write a two electron Schrödinger equation :

$$\{h(1)+h(2)-E_{core}+V_{int}(1,2)\}\Upsilon(1,2)=\mathcal{E}\,\Upsilon(1,2) \qquad (1)$$

In (1) the coordinates of the electrons 1 and 2 are denoted by (1) and (2) h(i) is a *one electron model hamiltonian* which has been chosen in the form

$$h(i)=-\frac{1}{2}\nabla_i^2+V(r_{Ai})+V(r_{Bi})+V_3(\mathbf{R},\mathbf{r}_{Ai},\mathbf{r}_{Bi})+E_{core}(R) \qquad (2)$$

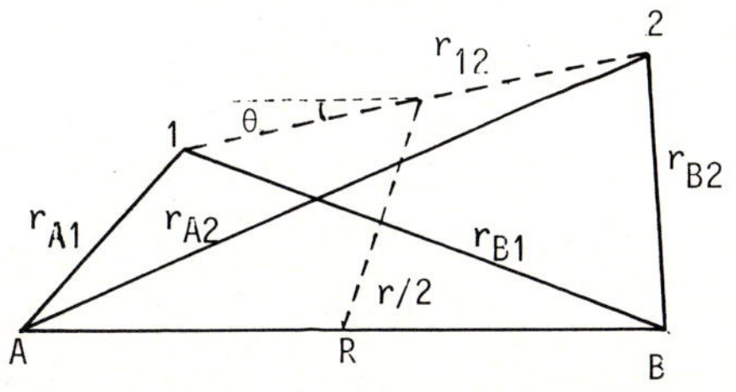

The internuclear distance R and the distances r_{Ai} and r_{Bi} of electron i to the cores are indicated on the figure V(r) is a parametric model potential fitted[6] to the energy spectrum of the atom (Li,Na,K). V_3 is a cross polarization tensorial term.

In eq(1) the bielectronic interaction term is given by

$$V_{int}(1,2)=1/r_{12}+V_{diel}(1,2) \qquad (3)$$

The dielectric term V_{diel}[7] is also a cross polarization term.

We solve eq(1) by a variational method using as trial function the product :

$$\Upsilon(1,2)=\phi(1,2)\,\Omega(1,2) \qquad (4)$$

$\phi(1,2)$ is a solution of

$$(h(1)+h(2)-E_{core})\,\phi(1,2)=E_\phi\,\phi(1,2) \qquad (5)$$

and can be obtained as a product of two orbitals of the molecular ion Li_2^+, Na_2^+, K_2^+ solutions of the equation :

$$h(i)\,\chi(i)=\varepsilon\,\chi(i) \qquad (6)$$

Eq(6) was solved using the numerical procedure developped by Valiron[4]. $\phi(1,2)$ describes the motion of two *independant* electrons in the field of the two cores.

The correlated function $\Omega(1,2)$ is a solution of :

$$\left\{-\frac{1}{2}(\nabla_1^2+\nabla_2^2)+1/r_{12}\right\}\Omega(1,2)=E_\Omega\,\Omega(1,2) \qquad (7)$$

and describes the motion of two electrons interacting *in the lack of an external field*. It can be expanded as :

$$\Omega(1,2)=g(tr)\sum_{\ell=0}^{\infty} d_\ell\, P_\ell(\cos\theta)\, F_\ell(kr_{12})/r_{12} \qquad (8)$$

The coordinates r_{12} and θ of the relative distances $\vec{r}_{12}$ of the two electrons appear on the figure. r is twice the distance between the centre of mass of the two nuclei and the centre of mass of the two electrons. g(tr) is a modified Bessel function of the first kind, $P_\ell(\cos\theta)$ a Legendre polynomial and $F_\ell(kr_{12})$ a regular spherical coulombic function. In the present work, expansion (8) was limited to $\ell=0,1$.

It appears that eq(1) can be separated in (5) and (7) provided we may neglect in the total hamiltonian both the cross kinetic term $\sum_{i=1}^{2}\nabla_i\,\phi(1,2).\nabla_i\Omega(1,2)$ and the dielectric term V_{diel}. We can therefore express the total energy as :

$$\mathcal{E}=E_\phi+E_\Omega+\Delta E \qquad (9)$$

$$\Delta E=\left\{\langle\Upsilon|\sum_{i=1}^{2}\nabla_i\phi.\nabla_i\Omega\rangle+\langle\Upsilon|V_{diel}|\Upsilon\rangle\right\}/\langle\Upsilon|\Upsilon\rangle \qquad (10)$$

A variational procedure is then developped, the parameters t, k and d_1/d_0 being varied, for each internuclear distance, in order to mimize the total energy $\mathcal{E}$ for the ground state.

Preliminary results show that a monoconfiguration calculation limited to the $\ell=0$ partial wave and neglecting the cross polarization terms V_3 and V_{diel} is capable of predicting *the position* R_e of the well of the Na_2 ground state : we obtain a well of 5223cm^{-1} at R=5.8 a.u. the experimental determination being 5988 ± 20 cm^{-1} at R=5.83a.u. Full calculations will be presented at the conference.

References

1. A. Dalgarno in Atomic Physics 4 ed. by Zu Putlitz et al. (Plenum Press New York and London 1975) p. 325
2. J.N. Bardsley in Case Studies in Atomic Physics 4 (North Holland, Amsterdam, 1974) p.299
3. A. Valance and Nguyen Tuan J.Phys.B 15 17 (1982)
4. P. Valiron, R. McCarroll et al. J.Phys.B 12 53 (1979)
5. M. Aubert-Frécon and C. Le Sech J.Chem.Phys.74 2931(1981)
6. M. Klapisch Comput.Phys.Commun.2 239 (1971)
7. G.A. Victor, C. Laughlin Chem.Phys.Let. 14 74 (1972)
8. P. Kusch and M.M. Hessel J.Chem.Phys. 68, 2591 (1978)

$^{1,3}\Sigma_{u,g}$ EXCITED STATES OF Na_2 AND K_2 (HELLMANN MODIFIED PSEUDO-POTENTIAL)

A. Valance

Service de Physique des Atomes et des Surfaces, CEN/Saclay, 91191 Gif-sur-Yvette Cedex, France

We consider a new pseudo-potential to describe the valence electron interacting with atomic core of Na.

This new pseudo potential is well adapted to give the high excited states of Na_2. The previous results on Na_2 given in ref. /1/ were obtained with a simple Hellmann pseudo-potential (A'=0 in the following eq.).

The pseudo-potential presented here has the : following form :

$$V_H = -\frac{1}{r} + (\frac{A}{r} + A')\, e^{-Kr} + V_p$$

This form was previously used /2/ with the basic difference that it was treated in a model potential formalism for NaHe case.

We recall that in the pseudo-potential formalism the 3s, 3p and 4s atomic state of Na are represented by a 1s, 2p and 2s "hydrogenlike" wave function respectively. To determine the pseudo-potential parameters (A, A' and K) and the pseudo-wave functions characterized by the exponential parameters (ε_{3s}, ε_{3p} and ε_{3d}), we use a variational procedure, so we get an analytical basis set functions.

The varionational procedure used leads us to solve 3 non-linear equations for the energies and 3 other for the derivative of energy versus the parameter governing the corresponding wave function.

We notice that for $l = 0$ the pseudo wave functions (P.W.F) are (n-2) mode-hydrogenlike wave function (H.L.W.F.), for $l=1$ the (P.W.F.) are (n-1) H.L.W.F. and for $l \geqslant 2$ the P.W.F. are exact H.L.W.F.

In the procedure presented here we get the exact energies for the 3s, 3p and 4s states of Na. The error percentages for the 5s, 6s, 3p $\rightarrow$ 5p, 3d $\rightarrow$ 5d and 4f are 0.6 %. The polarization term V_p in (1) can be considered as a perturbation for the atomic problem, but il will be really considered in the molecular Hamiltonian;because for the excited state it does not remain a perturbation term via the cross-term /3/.

The results for this pseudo-potential are :
$A = -97.569$, $A' = 349.226$, $K = 2.607$
$\varepsilon_{3s} = 0.38090$, $\varepsilon_{3p} = 0.40626$, $\varepsilon_{4s} = 0.28023$.

The formalism for obtaining Na_2 curve are the same as reported in /1/, except for the pseudo-potential, and for the inclusion of the polarisation terms.

We find that the $^1\Sigma_u^+$ (3s + 4s) state presents- a double minima at $R \simeq 6$ and 13 a_o of 0.38 eV and 0.32 eV. A the present time no experimental data confirm these predictions.

For K_2 we predict the same double minima for the $^1\Sigma_u^+$ (4s + 5s) state at $R \simeq 7$ and 15 a_o of 0.36 eV and 0.26 eV. In this case experimental data /4/, /5/ seem to confirm the predicted minima.

The configration interaction calculations of the adiabatic $^{1,3}\Sigma_u$ potentials for K_2 may possibly serve to explain the origin of the 518 nm experimental band. By inspection of the experimental data /4/ and forming all available theroretical difference potential curves we obtained that the difference $^1\Sigma_u$ (4S + 5S) - $X^1\Sigma_g$(4S+4S) has a minimum at 522° nm and a maximum at 457.0 nm. The latter was observed because of strong C-X band. The minimum at 522 nm is very close to the band 518 nm (experimental) indicating the possible mutual connection.

References

1. A. Valance and Q. Nguyen, J. Phys. B 15, 1733 (1982).
2. F. Masnou et al. Phys. Rev. Lett. 41, 6, 395 (1978).
3. C. Bottcher and A. Dalgarno, Proc. Roy. Soc. London A340, 187 (1974).
4. S. Milosevic, G. Pichler, Fizika 13, 4, 377 (1981).
5. B. Stefanov, to be published in High Temperature.

DIABATIC POTENTIAL TERMS He (2^1S) - RARE GAS

H. Rahal, S. Runge, A. Valance

Service de Physique des Atomes et des Surfaces, 91191 Gif-sur-Yvette Cedex, France

The differential cross sections for elastic and excitation transfer scattering have been extensively measured for He (2^1S) + R.G (Rare Gas) collisions at very low energies (25-225 meV) /1, 2/, and potential interaction were completely determined by fitting procedure on the experimental data. Theoretical calculations on this system have been done /2, 3/. We present in this abstract a new method to determine the diabatic potential interaction near the well-region.

We use both a model potential (M.P) and a pseudopotential (P.P) formulation for the active electron of He(2^1S) atom.

The main features of the method can be recalled. As in the alkali-rare gas systems, an effective one-electron molecular hamiltonian for the active electron is constructed in two steps :

(i) Firstly atomic M.P or P.P. hamiltonians are obtained which describe the active electron in the field of each core (He^+ and R.G.). A semi-empirical procedure is adopted : a reasonable parametrized form of a P.P. is optimized so as to reproduce electron scattering data in the case of the neutral core Ne.

$$V_{e-\,RG} = \sum_{\ell} A_{\ell} \exp(-\alpha_{\ell} r^2) \cdot P_{\ell} + V_p(r)$$

where P_{ℓ} are angular projection operators, V_p the polarization term.

For the interaction V_{e-He^+} our method is flexible so as to study the difference between the P.P and the M.P. formalism, in each case the method is "ℓ" dependent and the spectroscopic date are exactly reproduced for the first state in each different ℓ subspace /4/.

(ii) Secondly the model hamiltonian is constructed from the atomic P.P or the M.P. according to the procedure of Bottcher and Dalgarno /5/.

For V_{e-He^+} we propose the following form $-\frac{Z}{r} + \frac{B_{\ell} P_{\ell}}{r^2}$ where B_{ℓ} is an adjustable constant for a given ℓ valence state. This potential has simple eigenvalues and eigenfunctions. The eigenvalue equation has the solutions

$$E_{n,\ell} = -2Z^2 / [n - \ell - a_{\ell}]^2 \qquad (1)$$

where a_{ℓ} is a function of B_{ℓ}, and the radial wavefunction is

$$\Psi_{n,\ell} = W\left(Z, \sqrt{-2E_{n,\ell}}, \ell + \tfrac{1}{2}, 2\sqrt{-2E_{n,\ell}}\, r\right)$$

where W is the second Whittaker function. When n = 2 is used in eq. (1) to fit the n = 2 energy state of He (2^1S) we adopt the M.P. formalism, when n = 1 in eq.(1) P.P. formalism is adopted.

We have observed that the crucial points in the calculations are the cross-term and the cutt-off function in the region of small r values.

At the present time some improvements on Ar, Kr and Xe well depths are obtained.

References

1. H. Haberland, W. Konz, P. Oesterlin, J. Phys. B 15, 2969 (1982).
2. P.E. Siska, J. Chem. Phys. 73, 2372 (1980).
3. G. Peach, J. Phys. B 11, 2107 (1978).
4. S. Runge, A. Valance, to be published in Chem. Phys. Lett.
5. C. Bottcher, A. Dalgarno, Proc. Roy. Soc. A340, 187 (1974).

ATOMIC FUES POTENTIAL IN He^* : OSCILLATOR STRENGTHS AND He_2^* POTENTIALS

S. Runge

Service de Physique des Atomes et des Surfaces, CEN/Saclay, 91191 Gif-sur-Yvette Cedex, France

The interpretation of processes involving He^*-He collisions, such as associative ionization[1], requires He_2^* potential curves with an accuracy of a few 10^{-3} a.u. We present here a calculation using a model potential (MP) description of He^* and a pseudo-potential (PP) description of the interaction e-He.

The excited states He^* are described as a single electron interacting with the core He^+ by the Atomic Fues Potential[2] :

$$V_l(r) = \frac{-Z}{r} + \sum_{l,m} (B_l/r^2) \left| \mathbf{Y}_l^m(\hat{r}) \right\rangle \left\langle \mathbf{Y}_l^m(\hat{r}) \right| \qquad (1)$$

where $\mathbf{Y}_l^m(\hat{r})$ is a spherical harmonic and Z=1.

The stationary states in this potential are given by the energies :

$$E_{n,l} = -2\, Z^2/\left[\, n-1 + a_l \right]^2 \qquad (2)$$

and the eigenfunctions :

$$\Phi_{n,l} = W\left(Z, \sqrt{-2E_{n,l}},\ l+\frac{1}{2},\ 2\sqrt{-2E_{n,l}}\ r\right) \qquad (3)$$

where W is the second Whittaker function. a_l is a parameter related to B_l, adjusted so that the first level of each l series is exactly fitted by (2). The use of n=1 or n=2 to fit the energy of He (2S) determines the PP or MP formulation of the atomic problem. For $l \geq 1$, n is the principal quantum number. In any case the agreement between the calculated energies and the spectroscopic data is better than $5\cdot10^{-4}$ a.u.

As a further check of the accuracy of this model we have calculated the oscillator strengths for some transitions in He. The use of the PP formalism lead to completely wrong results, but the MP formalism gives an agreement generally better than 12% with precise calculations[3] (Table I and Ref.2)

This MP description of the He^* states is then applied to the calculation of the $^3\Pi$ states of He_2. The interaction between the excited electron, in the field of He^+, and the He(1S) atom is described by the l-dependent pseudo-potential and the crossed polarization term, proposed by Pascale[4]. The wavefunctions with the correct u or g symmetry are constructed from (3) and the ion core wavefunctions:

$$\left| 1s'_a\, 1s_b\, \overline{1s_b} \right| \pm \left| 1s'_b\, 1s_a\, \overline{1s_a} \right| \qquad (4)$$

where 1s and 1s' are Slater-type orbitals with exponents 1.6875 and 2. The adiabatic potentials are obtained following the formalism given by Peach[5]. The secular equation is solved in a basis set including all the n=2, 3, 4 states of He^* . The analytical expressions[5] for the first $^2\Sigma_u^+$ and $^2\Sigma_g^+$ potentials of He_2^+ are used to calculate the difference between these two potentials.

First calculations around the wells of the $f^3\Pi_u$ and $b^3\Pi_g$ states, correlated to $He(2^3P)+He(1S)$, are presented in Table II. They are compared, for the same choice of $^2\Sigma_u^+$ (He_2^+) potential values whenever possible, with experimental results[6], previous PP[5] and ab initio [7,8] calculations. It should be noted that the ab initio calculation[8] explicity used the value -0.085 for $^2\Sigma_u^+$. Comparison with the PP calculation[5], where the e-He potential was not l-dependent, shows the determinant effect of the l-dependence.

References

1. A.Pesnelle, S.Runge, M. Perdrix, D.Sevin, and G.Watel to be published in J. Phys. B. Lett.
2. S. Runge, A. Valance to be published in Chem.Phys.Lett
3. J.C. Green, N.C. Johnson, E.K. Kolchin, Astrophys.J 144, 369 (1966).
4. J. Pascale, Phys. Rev. A26, 3709 (1982).
5. G. Peach, J. Phys. B11, 2107 (1978).
6. M.L. Ginter, R. Battino, J. Chem. Phys.52,4469 (1970).
7. B.K. Gupta, F.A. Matsen, J.Chem.Phys.50, 3797 (1969).
8. L. Lenamon, J.C. Browne, R.E. Olson, Phys.Rev.A8,8380 (1973).

Table I - Oscillator strengths in He.

Transition	Singlet spectrum This work	Singlet spectrum Ref.3	Triplet spectrum This work	Triplet spectrum Ref.3
2S-2P	0.3748	0.3773	0.5636	0.5398
-3P	0.1666	0.1513	0.0576	0.0644
-4P	0.0538	0.0493	0.0233	0.0259
2P-3D	0.7027	0.7106	0.6501	0.6105
-4D	0.1200	0.1203	0.1288	0.1232

R	$^2\Sigma_u^+$	$b^3\Pi_g$ this calc.	$b^3\Pi_g$ previous work	$f^3\Pi_u$ this calc.	$f^3\Pi_u$ previous work
2.06	-0.086	-0.082	-0.086[6] (exp)	-0.0076	-0.008[6] (exp)
2.05	-0.085				-0.0083[7] (calc.)
2.0	-0.091	-0.087	-0.058[5] (calc.)	-0.0130	-0.0166[5] (calc.)
2.0			-0.0808 (calc.)		

Table II. Adiabatic potentials $b^3\Pi_g$ and $^3\Pi_u$ of He^*_2 (atomic units).

A SIMPLE MODEL FOR THE VAN DER WAALS POTENTIAL BETWEEN TWO CLOSED SHELL ATOMS

K.T. Tang[+] and J.P. Toennies
Max-Planck-Institut für Strömungsforschung, 3400 Göttingen, Federal Republic of Germany

The ab initio calculation of van der Waals potentials is still essentially restricted to closed shell systems with about 4 electrons. The recent availability of very accurate experimental data on potentials from molecular beam scattering experiments, spectroscopy and bulk properties has stimulated the development of simple models. The long range attractive potential is well known from perturbation theory to be given by $-\sum_{2n=6} \frac{C_{2n}}{R^{2n}}$. At small distances this expression breaks down since it does not take into account the overlap of the atomic charge distributions. This can be corrected for by introducing a damping function $f_{2n}(R)$ so that the attractive potential for all distances becomes $-\sum_{2n=6}^{\infty} \frac{f_{2n}(R) C_{2n}}{R^{2n}}$. Previously[1] we derived from the Drude model a semiclassical expression for an effective damping function, which in contrast to other models[2], has no adjustable parameters.

A more general damping function for individual terms in the dispersion series can be obtained by taking account of the following boundary conditions:

$$f_{2n} = 1 - P_{2n}(R) \exp(-bR), \quad (1)$$

$$f_{2n} \to 1 \quad , R \to \infty ; \quad (2)$$

$$f_{2n} \to 0 + O(R^{2n+1}), \; R \to 0. \quad (3)$$

Eq. (1) is the form imposed by the previous Drude model calculation. Eq. (2) and (3) are the realistic boundary conditions with (3) suggested by ab initio calculations for $^3\Sigma$ H_2[3]. The final result is

$$f_{2n} = 1 - \left(\sum_{k=0}^{2n} \frac{(bR)^k}{k!} \right) \exp(-bR) \; , \quad (4)$$

which depends only on b, the range parameter in the repulsive potential, which can be accurately fitted to A exp(-bR). Fig. 1 compares this simple expression with a recent very accurate[4] ab initio calculation for $^3\Sigma$ H_2; the b in our $f_{2n}(R)$ was taken from an SCF calculation. This excellent agreement and other comparisons with less accurate approximations for He_2 suggest that Eq. (4) may be generally valid.

The new potential model is then given by

$$V(R) = A \exp(-bR) - \sum_{2n}^{\infty} \left[1 - \sum_{k=0}^{2n} \frac{(bR)^k}{k!} \exp(-bR)\right] \frac{C_{2n}}{R^{2n}} \quad . \quad (5)$$

The effect of exchange dispersion and other neglected interaction terms for $^3\Sigma$ H_2 has been determined by fitting Eq. (5) to the accurately known potential[5]. The best fit A value was found to be 17% greater and b was the same as the SCF values. The size of this correction agrees with an ab initio study[6]. To test Eq. (5) further we have compared the model potential with the best available potentials for the following partners[7]: $^1\Sigma$ systems: 5 rare gas homogeneous and 10 heterogeneous dimers, and Mg_2 and Ca_2; $^2\Sigma$ systems: 5 H-rare gas systems. LiHg and NaAr; $^3\Sigma$ systems: H_2 and NaK. In each case effective values of A and b were determined from the experimental well depth, ε and well location R_m. In all cases the agreement is within experimental error. In the case of Ar_2, for example, the agreement with the best empirical potential is better than $2 \cdot 10^{-3}$ ε in the attractive region. For a number of systems the A values so determined could be compared directly with accurate SCF values. With the exception of Ne_2 they were found to be on the average of 19% larger. Eq. (5) thus provides a simple universal model for all van der Waals potentials, with a minimum of only 5 parameters A, b, and C_6, C_8, C_{10}, since C_{12} etc. can be accurately estimated.

[+]Permanent address: Department of Physics, Pacific Lutheran University, Tacoma, WA 98447, USA.

References

1) K.T. Tang and J.P. Toennies, J.Chem.Phys. 66, 1496 (1977)
2) Ahlrichs, R. Penco, and G. Scoles, Chem. Phys. 19, 119 (1977); C. Douketis, G. Scoles, S. Mordetti, M. Zen, and A.J. Thakkar, J. Chem. Phys. 76, 3057 (1982); K.C. Ng, W.J. Meath and A.R. Allnatt, Chem. Phys. 32, 175 (1978), A.J.C. Varandas and J. Brandâo, Mol. Phys. 45, 857 (1982).
3) A. Koide, J. Phys. B9, 3173 (1976)
4) A. Koide, W.J. Meath and A.R. Allnatt, Chem. Phys. 58, 105 (1981)
5) W. Kolos and L. Wolniewicz, Chem. Phys. Lett. 24, 457 (1970)
6) W. Kolos, Int. J. Quantum Chem. Symp. 8, 241 (1974)
7) K.T. Tang and J.P. Toennies, to be published

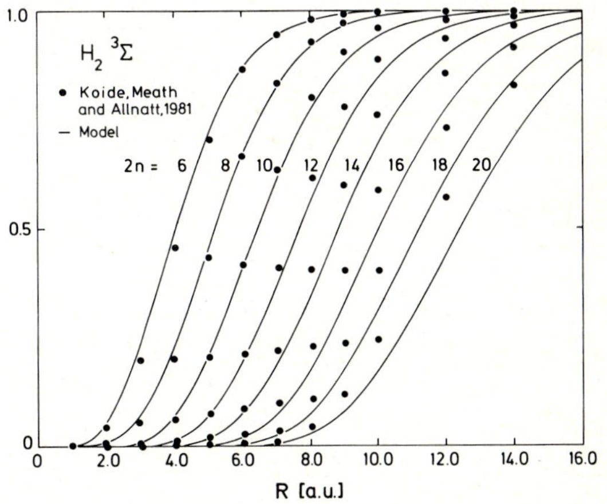

Fig. 1. The present approximation (-) is compared with an accurate[4] ab initio calculation (points).

WAVE PACKET SOLUTION OF THE TIME-DEPENDENT HARTREE-FOCK EQUATION FOR ATOMIC COLLISIONS

Eberhard Teubner, Norbert Grün and Werner Scheid

Institut für Theoretische Physik der Justus-Liebig-Universität Giessen, West Germany

In order to treat the problem of a collision of atomic shells with many electrons we assume that each electron moves in the Coulomb fields of the two nuclei and in the self-consistent field generated by the other electrons. The motion of the electrons is approximately described by the one-body density matrix $\rho(\vec{r},\vec{r}\,',t)$ solving the time-dependent Hartree-Fock equation:

$$\left(-\frac{\hbar^2}{2m}(\vec{\nabla}^2-\vec{\nabla}'^2)+V(\vec{r},t)-V(\vec{r}\,',t)-i\hbar\frac{\partial}{\partial t}\right)\rho(\vec{r},\vec{r}\,',t)$$
$$-e^2\int\left(\frac{1}{|\vec{r}-\vec{x}|}-\frac{1}{|\vec{r}\,'-\vec{x}|}\right)\rho(\vec{r},\vec{x},t)\rho(\vec{x},\vec{r}\,',t)d^3\vec{x}=0 \qquad (1)$$

where $V(\vec{r},t)=-\frac{Z_1e^2}{|\vec{r}-\vec{R}_1(t)|}-\frac{Z_2e^2}{|\vec{r}-\vec{R}_2(t)|}-e\varphi(\vec{r},t) \qquad (2)$

$$\Delta\varphi = 4\pi e\rho(\vec{r},\vec{r},t) \qquad (3)$$

In this report we outline a new method for solving Eq.(1), which is based on the procedure of Heller[1,2], who expanded wave functions into semi-classically moving Gaussian wave packets with time-dependent widths and applied this formalism to problems of molecular physics.

First we introduce the Wigner transformed function of the density matrix defined by

$$F(\vec{r},\vec{p},t)=\frac{1}{(2\pi\hbar)^3}\int\rho(\vec{r}+\frac{\vec{x}}{2},\vec{r}-\frac{\vec{x}}{2},t)\exp(-\frac{i}{\hbar}\vec{p}\vec{x})d^3\vec{x} \qquad (4)$$

This function is expanded into Gaussian wave packets with time-dependent parameters determined later:

$$F(\vec{r},\vec{p},t) = \sum_{n=1}^{N} A_n(t)f_n(\vec{r},\vec{p},t), \qquad (5)$$
$$f_n(\vec{r},\vec{p},t)=\exp[-\{(\vec{r}-\vec{r}_n(t))^T\alpha_n(t)(\vec{r}-\vec{r}_n(t)) + 2(\vec{r}-\vec{r}_n(t))^T\beta_n(t)(\vec{p}-\vec{p}_n(t)) + (\vec{p}-\vec{p}_n(t))^T\gamma_n(t)(\vec{p}-\vec{p}_n(t))\}] \qquad (6)$$

where $\alpha_n(t),\beta_n(t)$ and $\gamma_n(t)$ are real 3x3 matrices with $\alpha_n^T=\alpha_n$ and $\gamma_n^T=\gamma_n$.

The functions $f_n(\vec{r},\vec{p},t)$ are chosen as solutions of the Liouville equation for a potential $V_n(\vec{r},t)$, which is the Taylor expansion of the Hartree potential $V(\vec{r},t)$ defined in Eq.(2) about $\vec{r}=\vec{r}_n(t)$ up to quadratic terms:

$$V_n(\vec{r},t)=V(\vec{r}_n(t),t)+(\vec{r}-\vec{r}_n(t))(\vec{\nabla}V(\vec{r},t))_{\vec{r}=\vec{r}_n(t)} + \frac{1}{2}(\vec{r}-\vec{r}_n(t))^TV_n''(\vec{r}_n(t),t)(\vec{r}-\vec{r}_n(t)) \qquad (7)$$

Here, V_n'' denotes the symmetric 3x3 matrix formed by the second partial derivatives of V with respect to the coordinates. Since the potential (7) is of oscillator type, Liouville's equation is identical with the quantum mechanical equation of motion for the Wigner function f_n:

$$\left(\frac{\partial}{\partial t}+\frac{\vec{p}}{m}\vec{\nabla}_r-\vec{\nabla}V_n\vec{\nabla}_p\right)f_n=0 \qquad (8)$$

With the condition that f_n fulfills Eq.(8), we obtain equations for the parameters:

$$\dot{\vec{r}}_n=\frac{1}{m}\vec{p}_n,\quad \dot{\vec{p}}_n=-\vec{\nabla}V|_{\vec{r}=\vec{r}_n(t)} \qquad (9)$$

$$\dot{\alpha}_n=\beta_nV_n''+V_n''\beta_n^T,\quad \dot{\beta}_n=V_n''\gamma_n-\frac{\alpha_n}{m},\quad \dot{\gamma}_n=-\frac{1}{m}(\beta_n+\beta_n^T) \qquad (10)$$

Hence the evolution of the packets f_n is equivalent to the motion of a pseudo-particle with coordinate $\vec{r}_n(t)$ and momentum $\vec{p}_n(t)$ in the potential $V(\vec{r},t)$. Next we transform the Wigner function back into the density matrix:

$$\rho(\vec{r},\vec{r}\,',t) = \sum_{n=1}^{N} A_n(t)g_n(\frac{\vec{r}+\vec{r}\,'}{2},\vec{r}-\vec{r}\,',t) \qquad (11)$$

$$g_n(\vec{s},\vec{q},t) = \int f_n(\vec{s},\vec{p},t)\exp(\frac{i}{\hbar}\vec{p}\vec{q})d^3\vec{p} \qquad (12)$$

The functions g_n fulfill Eq.(1) with the potential V_n defined in Eq.(7), but without the exchange terms. The coefficients $A_n(t)$ are obtained by solving the following set of equations derived from a variational principle analogous to that given by McLachlan[3]:

$$\sum_{k=1}^{N}[S_{nk}\dot{A}_k(t)-M_{nk}A_k(t)]=0,\quad n=1,\dots,N, \qquad (13)$$

where $S_{nk}=i\hbar\int g_n^*(\vec{s},\vec{q},t)g_k(\vec{s},\vec{q},t)d^3\vec{s}d^3\vec{q}, \qquad (14)$

$$M_{nk}=\int g_n^*(\vec{s},\vec{q},t)(V(\vec{r},t)-V(\vec{r}\,',t)-V_k(\vec{r},t)+V_k(\vec{r}\,',t))\cdot g_k(\vec{s},\vec{q},t)d^3\vec{s}d^3\vec{q} + \text{exchange terms} \qquad (15)$$

with $\vec{r}=\vec{s}+\vec{q}/2$, $\vec{r}\,'=\vec{s}-\vec{q}/2$.

The time variation of the coefficients $A_n(t)$ is small if the potential $V(\vec{r},t)$ is well approximated by an oscillator potential inside the width of a Gaussian wave packet. The number N of coefficients depends on the initial density matrix and on the type of the considered reaction.

In conclusion, the suggested method allows to solve the TDHF equation (1) on the basis of a finite number of Gaussian wave packets, which move on self-consistent classical trajectories. When we have developed analytical approximations for the matrix elements (15), we will obtain a fast computational procedure for solving the TDHF-equation for problems in atomic or nuclear physics.

References

1 E.J.Heller, J.Chem.Phys. 62 (1975) 1544
2 S.Y.Lee and E.J.Heller, J.Chem.Phys. 76 (1982) 3035
3 A.D.McLachlan, Mol.Phys. 8 (1964) 39

ON GROUND STATES, AT POTENTIAL MINIMA, OF NEUTRAL QUASI-MOLECULES FORMED FROM NEUTRAL ATOMS OR IONS

Ray Hefferlin, Ken Parker, Rosalie Parrish, Henry Kuhlman, Kevin Shaw, Ken Priddy and Mike Seaman

Southern College, Collegedale, TN 37315 USA

The results presented are an extension of those shown at ICPEAC XII.[1,2] They relate to atom-atom collisions, including those where the quasi-molecule is superheavy[3,4] and where one or both atoms is superheavy.[5]

We employ the periodic system of diatomic (quasi-) molecules, wherein each species has coordinates R_1, R_2, C_1, C_2. (R is the row number, and C is the column number, of either atom in the chart of the atoms; e.g. Li is in row 2 and F is in column 7.)[6-8] This system is based on trends observed in critically analyzed data bases for 18 collision-related properties and five other properties (to date); it also has group-theoretical support.[8,9]

We next plot graphs for (1) atoms with fixed R, (2) molecules with fixed R_1 and R_2, (3) atoms with fixed C, and (4) molecules with fixed C_1 and C_2.

We then seek functions of (2) C_1 and C_2 and (4) R_1 and R_2, and determine those portions of the relevant planes have statistically good distributions of points (molecular data for a property). The search for functions is guided by semi-emperical formulae[10-12] and by the graphs. The graphs also show where the data are distributed, as well as being of theoretical interest *per se*.[1,2]

On the chart we state which results are now available in published or publishable form ("a"), which are complete but in rough-draft form ("b"), and which are conservatively planned for completion by Conference time ("c"). Results are available from R.H.

	GRAPHS				PREDICTED DATA	
	FIXED C_1,C_2		FIXED R_1,R_2		SPECIES	EST. ERROR
r_e	□a[1,2]	○c	□b	○b	□a 176	median 1.6%
D_o	□b	○b	□b	○b		
I.P.	□a[8]	○c	□b	○b		
ω_e	□c	○c	□b	○b	□c	
S_{298K}	□c	○c	□b	○b	□a 419	median .36%

The squares indicate molecules formed from s and p atoms; the circles indicate molecule groups containing at least one d or f atom. For r_e the functions are

(4) $r_e = K_o + K_1\log(R_1) + K_2\log(R_2)$ and

(2) $r_e = K_3 + K_4C_1 + K_5C_1^2 + K_6C_2 + K_7C_2^2 + K_8C_1C_2$.

For entropy at 298K, the functions are (4) precisely the same as for r_e and (2) r_e = constant. Estimated errors are based on cases where the same data are predicted twice (i.e., from (2) and (4)).

Since our purpose is also to improve understanding of the periodic system of diatomic (quasi-) molecules, we note that the graphs uniformly make clear the order in which species with given C_1 and C_2 should be stacked to make a block.[6-9] The order which guarantees that any property will vary monotonically (with few, and somewhat understood, exceptions) is that of increasing ($R_1 x R_2$).

References

1. Ray Hefferlin and Mickey Kutzner, ICPEAC XII TH5.
2. Ray Hefferlin and Mickey Kutzner, J. Chem. Phys. 75, 1035 (1981).
3. B. Müller, G. Soff, W. Greiner and V. Ceausescu, Z. Phys. A 285, 27 (1978).
4. G. Soff, B. Müller and W. Greiner, Phys. Rev. Let. 40, 540 (1978).
5. A. Rosen, B. Fricke and T. Morović, Phys. Rev. Let. 40, 856 (1978) (F and non-main-group atom 110).
6. Ray Hefferlin *et al*, J. Quant. Spectrosc. Rad. Transfer (JQSRT) 21, 315 and 337 (1979).
7. Ray Hefferlin and Henry Kuhlman, JQSRT 24, 379 (1980).
8. Ray Hefferlin and Wendy Innis, JQSRT 29, 97 (1983).
9. G. V. Zhuvikin and Ray Hefferlin, Vestnik Leningradskovo Universitata, scheduled for August, 1983.
10. G. Herzberg, "Spectra of Diatomic Molecules," Van Nostrand, New York, 1950, page 456.
11. A. V. Rajulu, S. R. Ratman and R. R. Reddy, JQSRT 29, 85 (1983).
12. J. L. Gazquez and R. G. Parr, Chem. Phys. Let. 66, 419 (1979).

QUANTUM AND SEMI-CLASSICAL CALCULATIONS OF THE PHASE-SHIFT IN THE THREE TURNING POINTS CASE FOR ANALYTICAL POTENTIAL

O. Vallée[+], J. Picart[++], S. Avrillier[+++], N. Tran Minh[++]

[+]C.R.P.H.T. - 1D Ae de la Recherche Scientifique - 45045 Orléans Cédex, France
[++]Département d'Astrophysique Fondamentale - Observatoire de Meudon - 92190 Meudon, France
[+++]Laboratoire de Physique des Lasers - Université Paris Nord - Ae J.B. Clément - 93430 Villetaneuse, France

The Numerov method is used for solving the Schrödinger equations in reduced coordinates[1]. By scaling the potential with the capacity parameter ($12<B_z<1500$), it is shown that the resonances appear in a cyclic way.

On the other hand, improving the Miller result[2], using the uniform Airy approximation, in the case of three turning points, we propose a semi-classical formula for the phase shift, which simplify in many cases the quantum calculation.

References

1. A.S. Dickinson and R.B. Bernstein, Mol. Phys. 18, 305 (1970)
2. W.H. Miller, J. Chem. Phys. 48, 464 (1968)

ASSOCIATIVE IONIZATION OF HORNBECK-MOLNAR TYPE IN THERMAL ENERGY COLLISIONS BETWEEN LASER EXCITED He(5^3P) AND He ATOMS

A. Pesnelle, S. Runge and G. Watel

Service de Physique des Atomes et des Surfaces, CEN/Saclay, 91191 Gif-sur-Yvette Cedex, France

In contrast with Penning ionization arising in systems $A^* + B$ where the internal energy of $A(E_{A^*})$ is higher than the ionization potential of $B(E_{B^+})$ producing either a B^+ ion or a molecular AB^+ ion, Hornbeck-Molnar (H-M) ionization can occur in systems where $E_{A^*} < E_{B^+}$ provided a well exists in the $V^+(R)$ potential at small R's leading to the formation of a stable molecular AB^+ ion by A^*+B $AB^+ + e$.

Although Penning and H-M processes both generate stable AB^+ ions, the basic mechanisms which govern the ionization are completely different. In H-M processes, the initial discrete electronic state A^*B is a stationary state, eigen state of the electronic hamiltonian in contrast with the situation encountered with Penning processes. Ionization can occur only at the expense of the instantaneous kinetic energy; the dynamic coupling of electronic and nuclear motion is believed to be responsible for the reaction.

We report here the first measurement of the associative ionization cross section of the H-M system He(5^3P_2) +He, as a function of relative velocity. This measurement is performed with the three-crossed-beam apparatus described in our study on Penning ionization of the He(3^1P)+ Ne system[1]: one He(2^3S) metastable atoms beam at thermal energies, one He target beam cooled at 80°K and one c.w. U.V. laser beam tuned on the transition $2^3S_1 \rightarrow 5^3P_2$ at 294.5 nm generated by intracavity frequency doubling[2].

In the present experiment, velocity dependence is investigated using the time-of-flight (TOF) method previously used in our Penning ionization studies on metastable species (Fort et al in Ref 1).

In the present case of a radiative level, the velocity distribution (and then the TOF spectrum) of the 5^3P atoms is not expected to be exactly similar to the metastable one for the following reasons : (i) the average U. V. laser intensity available for several hours is 3mW; the saturation parameter is therefore such that the saturation plateau in the 5^3P population is not instantaneously reached. (ii) the life time of the 5^3P(2.3×10^{-7}s) is long enough for 5^3P atoms to exist outside the laser spot. (iii) the excitation takes place in an elementary He^* beam the divergence of which is a function of v: this modifies the thickness of the interaction volume colinear with the laser beam. These effects are introduced in the derivation of the 5^3P velocity distribution from the 2^3S one, and the H.M. ionization cross section $\sigma(v_r)$ is derived using the He_2^+ TOF spectrum. The most striking features of this cross section is a pronounced maximum and an absolute value of the order of some $10^{-16}cm^2$ (Fig.1).

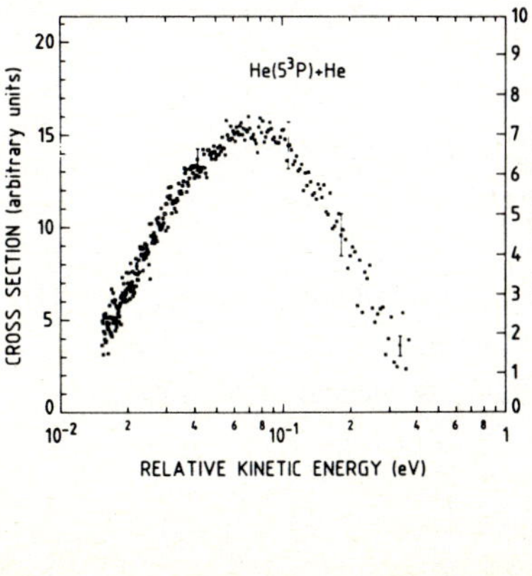

Fig 1: Associative ionization cross-section (normalized using the rate constant of Collins et al (in Ref 3)).

In the frame of diabatic curves representation, qualitatively we can understand that every time the kinetic energy E_c is sufficient to overcome the repulsive-attractive diabatic potential curve crossing of height E_o, a new trajectory through all the states towards He_2^+ becomes possible and a maximum is therefore expected at $E_c = E_o$. A theoretical approach was made by Cohen[4] for the He(3,4 ^{3}S,P,D)+He systems in a multistate diabatic curve-crossing model. The cross sections are derived from the electronic coupling between the diabatic states corelated with He(nl)+He. According to Mulliken, the basic feature responsible for the ionization is the crossing of the repulsive states into the continuum delimited by the attractive potential curve $He_2^+(^2\Sigma_u^+)$ (Fig.2). Intuitively and in the light of Cohen's discussion (the n=5 state is not included in his study), the maximum observed in the present experiment around 0.08 eV could be assigned to the opening of a new pathway from the He (5^3P)+He molecular state in the $^3\Pi_u$ symetry towards the continuum He_2^+ ($^2\Sigma_u^+$) via 5p→6 or 7d→np (n=4,3,2)→ He_2^+.

Previously, Nielsen et al[5] and Koike et al[6] had calculated σ for H + H(n=3) as a result of the vibronic coupling between adiabatic states and continuum: they obtained very small absolute values ($3 \times 10^{-18}cm^2$ at 300°K) and a monotonously decreasing energy dependence which is not observed here.

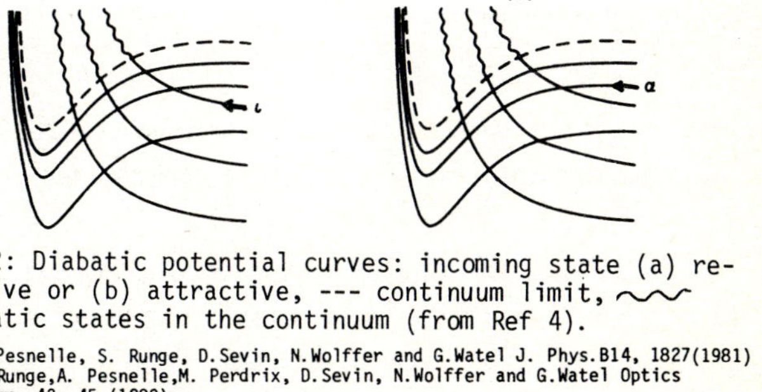

Fig 2: Diabatic potential curves: incoming state (a) repulsive or (b) attractive, --- continuum limit, ∿∿ diabatic states in the continuum (from Ref 4).

1 - A.Pesnelle, S. Runge, D.Sevin, N.Wolffer and G.Watel J. Phys.B14, 1827(1981)
2 - S.Runge,A. Pesnelle,M. Perdrix, D.Sevin, N.Wolffer and G.Watel Optics Comm. 42, 45 (1982).
3 - A. Pesnelle,S.Runge, M.Perdrix, D.Sevin and G.Watel to be published.
4 - J.S. Cohen Phys. Rev A13, 99 (1976).
5 - S.E. Nielsen and R.S. Berry Phys. Rev A4, 865 (1971).
6 - F. Koike and H. Nakamura J. Phys. Soc. Japan 33, 1426 (1972).

AUTOIONIZATION WIDTH FOR He(3^1P)+Ne : PENNING AND ASSOCIATIVE IONIZATION CROSS SECTIONS

A. Pesnelle and S. Runge

Service de Physique des Atomes et des Surfaces, CEN/Saclay, 91191 Gif-sur-Yvette Cedex, France

Penning ionization arises in systems $A^* + B$ where the internal energy of A^* is higher than the ionization potential of B, producing either a B^+ ion or a molecular AB^+ ion :

$$A^* + B \begin{cases} \nearrow AB^+ + e \\ \searrow A + B^+ + e \end{cases} \qquad (1)$$

In these systems, the decay of the autoionizing state $(AB)^*$ formed during the collision can proceed via direct or exchange mechanism. In the case of triplet excited species, the exchange channel is the only open channel : this feature is due to the conservation of the total spin before and after the collision. In the case of singlet excited species, both channels are open. However the direct process contribution is expected to be different whether the excited atom A^* is in a metastable state (the electric dipole transition $A^* \rightarrow A$ is forbidden; the magnetic dipole or the electric quadrupole transition is allowed) or a radiative state (the electric dipole transition $A^* \rightarrow A$ is allowed).

In our interpretation of associative ionization for the He(3^1P)+Ne system [1], measured in our laboratory as a function of relative velocity, these channels have been considered separately. Ionization cross sections have been calculated independently using the autoionization widths Γ_{di} (R) (direct) published by Katsuura[2] or Γ_{ex}(R) (exchange) published by Zhdanov et al[3]. A comparison of the experimental associative ionization cross section $\sigma_{AI}(v_r)$ with these calculations provided a useful test for the incoming channel V^* (R) and outgoing channel V^+(R) potential curves.

In the present and more realistic approach, both mechanisms-direct and exchange-are considered to take place simultaneously. In case A^* and B are close enough for the charge distributions to overlap one another the exchange channel will be preponderant ; otherwise, at large interatomic distances, the direct process will predominates. The autoionization width Γ is related to the interaction matrix element V_{if} by

$$\Gamma(R) = 2\pi\rho\,|Vif|^2 \qquad (2)$$

where ρ is the density of final continuum states. The total autoionization width can be taken equal to

$$\Gamma_T(R) = (\Gamma_{di}^{1/2} + \Gamma_{ex}^{1/2})^2 \qquad (3)$$

Until now, the interpretation of experimental data in Penning ionization with metastable states He^*+Ar/H[4,5] has been successful using a saturated Γ(R) in the small R region; ab initio calculations for the simplest H^*+H system[5] corroborated this saturation. Only the widths Γ_{di} for metastable and radiative states are basically different. Since Γ_{ex} is several order of magnitude larger than Γ_{di} at small R, it seemed to us reasonable to saturate Γ_T(R) (which is close to Γ_{ex} in the short R range) for He(3^1P)+Ne. Therefore, we used the following functions :

$$\Gamma_T(R) = \begin{cases} \Gamma_T(R) & \text{defined in formula (3) for } R > R_m \\ \Gamma_T(R_m) & \text{for } R \leqslant R_m \end{cases}$$

Our experimental results[1] are relative; however we obtain from the $HeNe^+$ counting rate an order of magnitude for σ_{AI} close to 10^{-16} cm^2. The critical distance R_m can be determined by fitting the calculated $\sigma_{AI}(v_r)$ cross section to the slope of the experimental one in the expected absolue value range : R_m = 5.5 a.u. provides a good agreement (Fig.2). We observe that the total ionization cross section $\sigma_{TI}(v_r)$ is not sensitive to this saturation; this is due to the fact that the whole $[R_0, \infty]$ range (R_0 is the classical turning point) contributes to σ_{TI} in contrast with σ_{AI}, and that R_m is close to R_0.

References

1. A. Pesnelle, S. Runge, D. Sevin, N. Wolffer and G. Watel, J. Phys.B 14, 1827 (1981).
2. K. Katsuura, J. Chem. Phys. 42, 3771 (1965).
3. V.P. Zhdanov and M.I. Chibisov, Sov. Phys. JETP 43, 1089 (1976).
4. H. Nakamura, J. Phys. B9, L59 (1976).
5. A.P. Hickman, A.D. Isaacson and W.H. Miller, J. Chem. Phys. 66, 1483 (1977). A.P. Hickman, H. Morgner, J. Chem. Phys. 67, 5484 (1977).

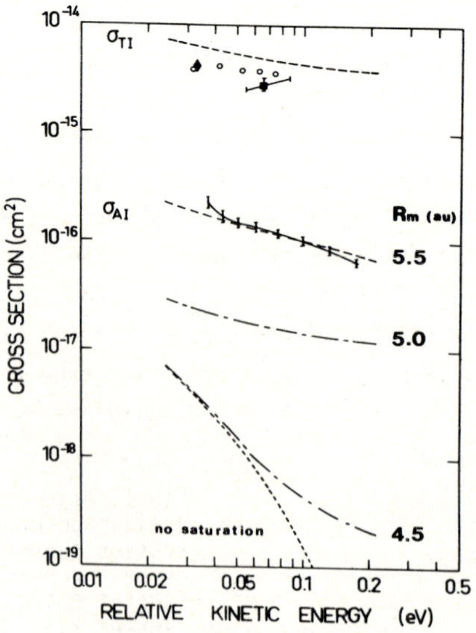

Figure 1 - Ionization cross sections σ_{AI} and σ_{TI}. Full curve with error bars, experiment[1] . Broken curves, present calculations. ◆ , ○ , ■ energy averaged quenching cross sections obtained in discharges.

ASSOCIATIVE IONIZATION AND 2-PHOTON LASER INDUCED COLLISIONAL IONIZATION IN CROSSED BEAM ALKALI AND ALKALINE EARTH SYSTEMS

J. Weiner, J. Boulmer, J. Keller and R. Bonanno

Department of Chemistry, University of Maryland, College Park, Maryland, 20742

We have studied associated ionization in thermal collisions between Na*(np) levels and Na(3s) ground state in a crossed-beams experiment. Excited Na*(np) levels are prepared by laser photoexcitation, and ionic species are analyzed by time-of-flight mass separation. Absolute rate coefficients exhibit a pronounced dependence on n with a maximum of approximately $3 \times 10^{-9} cm^3 sec^{-1}$ near n=11. We compare our results with predictions from a theoretical model developed by Duman and Shmatov, and Mihajlov and Janev (DSMJ). The global collisional process may be expressed as

$$Na^*(np) + Na(3s) \rightarrow Na_2^+ + e^-$$

in which Na*(np) levels ($5 \leq n \leq 15$) are directly photoexcited from the ground state by 12 nsec laser pulses. Taking advantage of the time resolution of the excitation, we are able to eliminate contributions to the ion production rate from excited states populated by radiative caseade from Na*(np). The results show a rapid rise in rate coefficient with principal quantum number n until a broad maximum is reached around n=11. We find that the DSMJ model predicts the correct magnitude but overestimates the sensitivity to principal quantum number. Our results are in better agreement with the simple assumption that the probability for associative ionization is nearly unity over the region of phase space where the process is permitted.

In a second set of experiments we have measured the rate constant for associative ionization in collisions between two sodium atoms excited to the $Na^*(3p\,^2P_{3/2})$ level,

$$Na^*(^2P_{3/2}) + Na^*(^2P_{3/2}) \rightarrow Na_2^+ + e$$

New measurement techniques greatly improve the accuracy of rate constant determinations in crossed -beam experiments. Previously reported cross sections have varied over several orders magnitude from 10^{-15} to $10^{-18} cm^2$. Two principal sources of error account for such wide disparity: (1) Apparatus constants inherent in the extraction and detection of ions are notoriously difficult to characterize accurately. (2) Determination of excited $Na^*(^2P_{3/2})$ number density requires great care due to unexpected effects arising principally from radiation trapping. We present a new approach to these old problems and report an absolute rate constant $k = 1.8 \pm 0.7 \times 10^{-11} cm^3\ sec^{-1}$.

Finally we present new results on two-photon laser-induced collisional ionization in crossed-beam Na + Ba experiments,

$$Na^*(^2P_{3/2}) + Ba(^1S_0) \rightarrow [Na(^2S_{1/2}) + Ba(^1P_1)] + 2\hbar\omega$$

$$Na(^2S_{1/2}) + Ba^+ + e$$

These results show that long-range dipole collisions can trigger a subsequent resonant two-photon ionization in one of the collision partners and that the process is closely allied to radiative and optical collisions discussed earlier by Yakavlenko and Harris. Finally we report the first direct observation of $NaBa^+$ formed associative ionization between $Na(^2S_{1/2}) + Ba^*(7d\,^3D_2)$ collisions.

COLLISIONAL IONISATION AND ENERGY POOLING PROCESS IN Rb VAPOUR

L. Barbier and M. Cheret

Service de Physique des Atomes et des Surfaces, CEN/Saclay, 91191 Gif-sur-Yvette Cedex, France

Various collisional mechanisms involving excited rubidium atoms have been investigated in a cell /1/, /2/ /3/ :

1. Hornbeck-Molanr associative ionisation Rb(nl) + Rb(5s) $\longrightarrow Rb_2^+ + e^-$.
2. Penning atomic ionisation Rb(nl)+ Rb(5p)$\longrightarrow Rb^+$+Rb(5s) + e^-.
3. Penning molecular ionisation Rb(nl)+Rb(5p)$\longrightarrow Rb_2^+ + e^-$
4. Energy pooling reaction Rb(5p $P_{3/2}$) + Rb(5p $P_{3/2}$)$\longrightarrow$ Rb(5d) + Rb(5s) - 68 cm^- .

The first mechanism is caracterized by an internal energy of the reactants lower than the ionisation potential of Rb^+ and the second and the third by a higher one. In reaction (4) internal energies of two excited atoms are put together to give a high lying atom and a ground state atom.

The excited states are created in the cell at low Rb density (10^{13} cm^{-3}) by two C.W. multimode lasers, first Rb(5s) + $h\nu_1 \longrightarrow$ Rb(5p), then Rb(5p)+ $h\nu_2 \longrightarrow$ Rb(nl) with l=0 or 2.

Photoionisation, Rb(nl) $+h\nu \longrightarrow Rb^+ + e^-$, also plays an important role and an additional photoionization reaction at a frequency different from ν_1 and ν_2 is used to determine the concentration of excited species, the photoionisation cross sections being known with a rather good precision /4/. Total ion current are measured with a nanoameter between 2 plates and $Rb^+$and Rb_2^+ ratio by mass analysis through a quadrupole mass spectrometer. To separate the contribution of photoionisation and collisional (reaction 2) mechanism to the atomic ion current the laser beams are chopped and the Rb^+ current is synchronously detected / 1 /.

The interpretation of the experimental results in terms of individual rate coefficients need the use of a model which takes into account the radiative cascading deexcitation of the pumped level.

For obvious energy considerations Hornbeck-Molnar associative ionisation (reaction 1) is only possible for levels higher than 7p and Penning atomic ionisation (reaction 2) for levels higher than 6s. Penning associative ionisation (reaction 3) is hidden by Hornbeck-Molnar associative ionisation as soon as this reaction is possible. Consequently results are avalaible for reaction 3 only for two levels (5d and 7s).

Hornbeck-Molnar associative rate coefficients (fig. 1) do not depend on l value and vary versus the effective quantum number n^* approximatively as predicted by Mihajlov /5/.

Penning atomic ionisation rates are smaller for the lowest s levels than for the neighbouring d levels and converge to the same value when n is increased (fig. 2). For the lowest levels the difference between s and d levels can be related to the difference of behaviour of photoionisation cross section of these levels /4/(dipole dipole interaction) whereas for high n value l-mixing occurs before ionisation and consequently increases the transition probability for s levels.

Experimental study /3/ of the energy pooling process give a reaction rate coefficient of (1.5 0.3) $10^{-9} cm^3$ s^{-1}. This value agree with a theoretical calculation based on the Borodin model /6/ giving a cross section of 9.6 10^{-14} cm^2.

References

1. M. Cheret, L. Barbier, W. Lindinger and R. Deloche J. Phys. B 15, 3463 (1982).
2. L. Barbier and M. Cheret, submitted to J. Phys. B.
3. L. Barbier and M. Cheret, submitted to J. Phys. B
4. M. Aymar, J. Phys. B11, 1413 (1978).
5. A.A. Mihajlov and R.K. Janev, J. Phys. B 14, 1639 (1981).
6. V.M. Borodin and I.V. Komarov, Opt. Spectr. 36, 145 (1974).

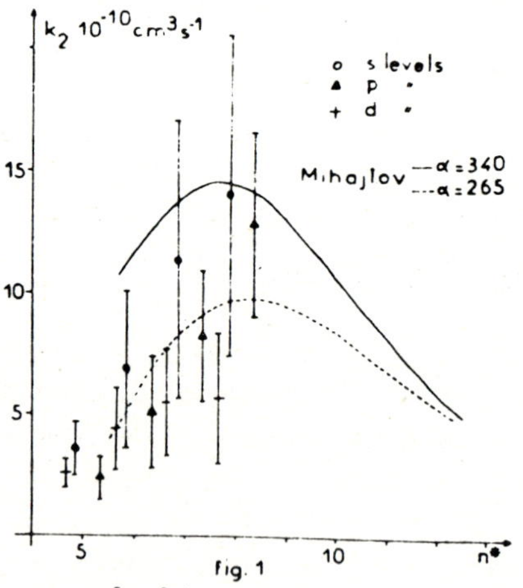

fig. 1

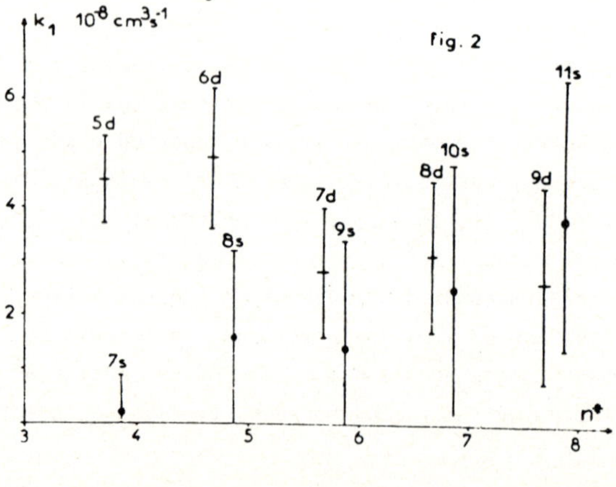

fig. 2

PRODUCT ANGLE-VELOCITY DISTRIBUTION FOR IONIZATION OF ARGON ATOMS BY TRIPLET METASTABLE HELIUM

P. R. Jones*, K. T. Gillen+, and M. J. Coggiola+

*Physics Department, University of Massachusetts, Amherst, MA 01003
+Molecular Physics Laboratory, SRI International, Menlo Park, CA 94025

We have measured directly the double differential cross section for the Penning ionization process $He^* (2^3S) + Ar \rightarrow He + Ar^+ + e^-$. The He product was isolated from the background of elastically and inelastically scattered He^* by detection in delayed coincidence with slow Ar^+ ion products.

A fast He^* (2^3S) beam intersects a crossed beam of Ar effusing from a capillary array. A Channeltron rotates around the intersection zone and detects scattered He and He^* atoms. Extraction plates perpendicular to both beams, pull product Ar^+ ions to a second Channeltron. Time delay spectra between Ar^+ and He pulses yield the distribution of He product velocities (and hence the collision exothermicity values, Q). Scattered He^* is not time-correlated to the Ar^+ and only contributes a uniform background to a coincidence spectrum.

Figure 1 shows a coincidence spectrum at 12° for a He^* beam energy of 296 eV. Excitation transfer to doubly-excited Ar^{**} $3s^2 3p^4 n\ell n'\ell'$ states with the Ar^{**} autoionizing after the products separate, produces the dominant coincidence peak at Q ~ -11 eV. Penning ionization is responsible for the other coincidence feature near Q=0. Also shown in Fig. 1 is a conventional time-of-flight (TOF) spectrum, produced by pulsing the He^* beam and detecting the arrival time distribution of all fast neutral products.[1] Comparing the two distributions demonstrates the existence of non-ionizing channels at Q ≈ 0 (elastic scattering) and at positive Q values from the excitation transfer process to singly excited Ar^* states. A small fraction of the Q=-11 eV feature may also be due to collisional excitation of Ar^* states without ionization.

The angular distributions of the two major features in the coincidence spectra are shown in Figure 2. Clearly the Penning channel becomes the dominant ionization feature at smaller angles and its contribution remains important at angles lower than our current apparatus resolution permits measurement.

Our results are the first reported determinations of heavy particle product differential cross sections for Penning ionization at energies above the thermal range. At thermal energies, Siska's group[2] has recently measured double differential cross sections for the product ions produced in a Penning reaction.

This investigation received financial support from the National Science Foundation.

References

1. K. T. Gillen and M. J. Coggiola, Abstracts XII ICPEAC, Gatlinburg (1981), p. 970.
2. A. Khan, H. R. Siddiqui, D. W. Martin, and P. E. Siska, Chem. Phys. Lett. 84, 280 (1981).

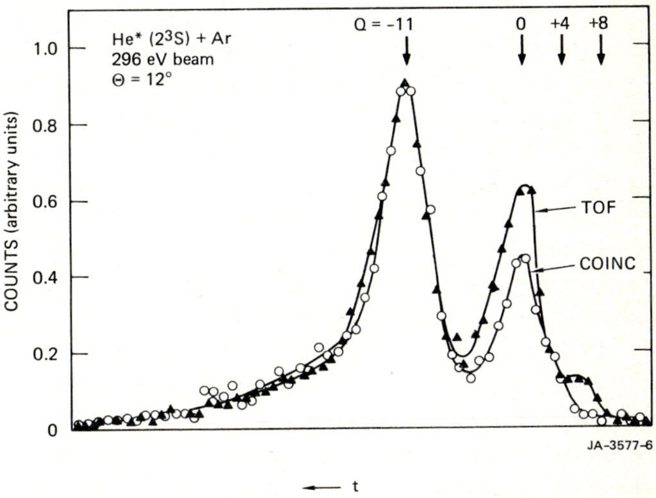

Fig. 1: Coincidence and TOF distributions normalized at peak.

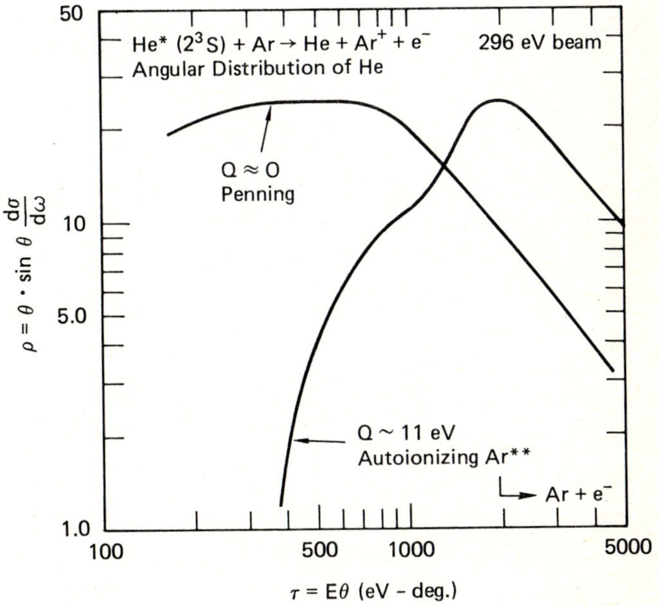

Fig. 2: ρ-τ angular distributions of two coincidence features.

ENERGY TRANSFER AND ASSOCIATIVE IONIZATION IN Na(3P) + Na(3P) COLLISIONS

J. Huennekens, S. Davidson and A. Gallagher*

Joint Institute for Laboratory Astrophysics, University of Colorado and
National Bureau of Standards, Boulder, Colorado 80309 U.S.A.

Rate coefficients have been measured for several outgoing channels that result from collisions between two Na atoms excited to the 3P state. Specifically, we have studied the processes of energy transfer

$$Na(3P_J) + Na(3P_{J'}) \xrightarrow{k_{nL}} Na(nL_{J''}) + Na(3S_{1/2}) \qquad (1)$$

and associative ionization

$$Na(3P_J) + Na(3P_{J'}) \xrightarrow{k_{AI}} Na_2^+ + e \quad . \qquad (2)$$

The largest rate coefficients for process (1) are to the nL = 5S, 4D and 4F states, for which the internal energy change in process (1) is smallest ($\sim$600 cm^{-1}; see Fig. 1). The rate coefficients for a statistical mixture of $3P_{3/2}$ to $3P_{1/2}$ states are $k_{5S} = 1.6 \times 10^{-10}$ cm^3 s^{-1} and $k_{4D} = 2.5 \times 10^{-10}$ cm^3 s^{-1}. The associative ionization process requires no internal energy transfer, but has a smaller rate coefficient of 5.8×10^{-12} cm^3 s^{-1} for the same mixture of $3P_J$ states. In order to obtain these results from optically thick (10^{13}–10^{15} cm^{-3}) Na vapor, it was necessary to carefully analyze radiation diffusion and the spatial distribution of excited atoms. Furthermore, the absolute value of the excited-atom density had to be determined. This has been done, redundantly for increased reliability, by a series of measurements that included measuring the resonance-line absorption coefficient before and after pulsed excitation of the vapor, the radiation decay after pulsed excitation, and the cw-power absorption.

These rate coefficients for process (1) are also observed to be strongly dependent on the values of J and J', with the J = J' = 3/2 case transferring preferentially to the 4D state and J = J' = 1/2 to the 5S state. Temperature and full J-J' dependences of these rate coefficients are under investigation, to more definitively characterize the behavior of the long-range excited Na_2 states responsible for these energy-transfer processes.

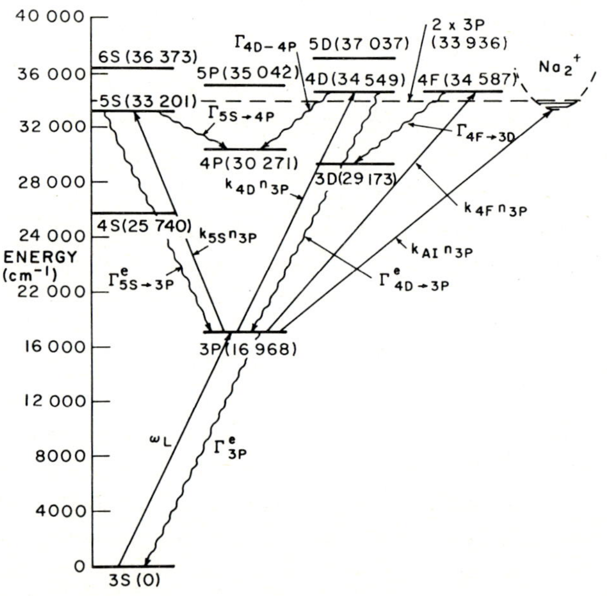

Fig. 1. Energy-levels and collision processes in Na vapor.

*Staff Member, Quantum Physics Division, National Bureau of Standards.

INVESTIGATION OF IONIZATION PROCESSES IN EXCITED ATOMS SLOW COLLISIONS BY USING THE METHODS OF PLASMA ELECTRON SPECTROSCOPY

A.Z.Devdariani, V.I.Demidov, N.B.Kolokolov, V.I.Rubtsov

Physical Institute, Leningrad State University, SU-Leningrad

The purpose of the present work is to investigate the relative yield Q of molecular ions in the reactions such as:

$$A^* + A^{**} \begin{cases} \to A^+ + A + e \\ \to A_2^+ + e \end{cases} \qquad (1)$$

in which the atoms of inert gases take part. To measure Q it is suggested to use the method based on the determination of the electrons energy distribution in afterglow plasma for the energy interval corresponding to the "fast" electrons. Under certain conditions[1] the distribution function is in good agreement with the spectrum of the electrons produced in the reaction such as (1). Therefore, this method has the same possibilities as Penning electron spectroscopy.

The electron spectra obtained can be used to determine Q if there is a method of separating the spectral components corresponding to the two channels of the reaction (1) from the total electron spectrum. It can be shown that the maximum width of the associative and Penning ionization spectra overlap is equal to

$$\Delta\varepsilon = 0.9\left(\frac{R_{ie}\, D_i^{1/6}}{R_{fe}\, D_f^{1/4}}\right)^4 E_i^{4/3}$$

where R_{ie}, D_i, R_{fl}, D_f are the equilibrium distances and well depths of the potential curves for the initial and final states, E_i is the energy of an excited atoms collision. The value of $\Delta\varepsilon$ obtained for inert gases is small as compared to the total spectrum width. Thus the Q values can be obtained by means of spectrum separation at the energy $\varepsilon^* = 2\varepsilon_m - \varepsilon_i$, where ε_m, ε_i are the excitation and ionization energies of A atom. The calculation of the electron spectrum (1) based on the potential curves Ref.2 for He(2^3S) atoms at the temperature 300K gives Q=14% for the "classical approach and Q=10% for the "quantum" approach[2]. This indicates that Q value obtained in the theoretical work[2] is incorrect.

The above conclusion concerning the possibility of spectrum separation at the energy ε^* has been used to determine Q experimentally. The figure shows the electron spectrum obtained for He(2^3S)-He(2^3S) reaction. Associative ionization is indicated by the dashed part of the figure. The Q value is equal to (7±4)% and agrees well with our results and those of Ref.3 done in merging beams. The molecular ions yield Q=(16±6)% for the He(2^3S)-He(2^3S) reaction has been determined for the first time. It has been found that Q < 25% for the Ne($^3P_2$)-Ne($^3P_2$) interaction.

References

1. N.B.Kolokolov, P.M.Pramatarov. Sov. Phys. Tech. Phys. 23, 176 (1978).
2. B.J.Garrison, W.H.Miller, H.F.Schaefer. J. Chem. Phys. 59, 3193 (1973).
3. R.H.Neynaber. Proc. Xl Int. Conf. on the Phys. of Electr. and At. Col. Inv. Pap. and Pr. Rep., Kyoto, p. 287 (1980).

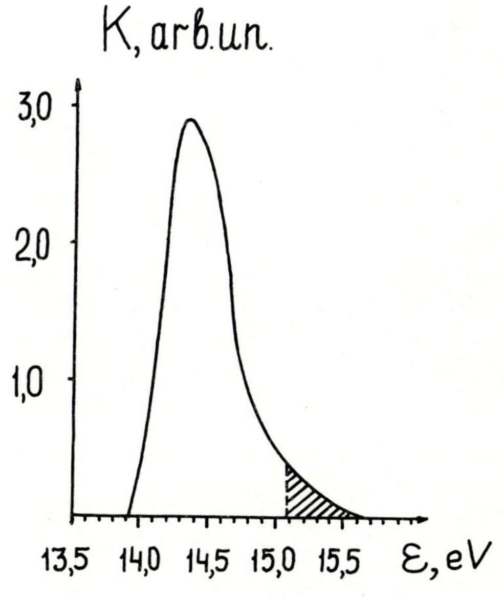

Fig. 1.

CALCULATION OF PENNING IONIZATION FOR $He(2^1,{}^3S)+Na$

James S. Cohen and Richard L. Martin*

Theoretical Division, Los Alamos National Laboratory, Los Alamos, NM 87545 USA

Neal F. Lane**

Physics Department, Rice University, Houston, TX 77251 USA

Cross sections have been calculated for the Penning ionization reactions

$$He(2^1S \text{ or } 2^3S)+Na \rightarrow He(1^1S)+Na^{+}+e, \quad (1)$$

which are usually of interest in thermal collisions. The interaction is described by a complex potential

$$W(R) = V(R) - \frac{i}{2}\Gamma(R), \quad (2)$$

where V(R) is the usual potential energy curve and Γ(R) is the autoionization width. The triplet atom forms both $^2\Sigma$ and $^4\Sigma$ states with Na, but only the doublet state can undergo Penning ionization.

The electronic structure calculations utilized a large contracted gaussian basis set consisting of (10s,8p) orbitals on He and (9s,8p,3d) orbitals on Na including diffuse functions needed to describe the "near" continuum character of the interaction. All configurations generated by single excitations with respect to the Hartree-Fock ground state were included in the calculation (113 spacial configurations/187 spin eigenfunctions). Following the prescription of Hazi,[1] square-integrable representations of the resonance and the continuum states were obtained in configuration-interaction (CI) calculations, and the autoionization widths were extracted using Stieltjes moment theory.[2]

The CI calculation is believed to represent the resonant part of the He*-Na interaction fairly accurately, but is not adequate for the van der Waals interaction. The latter contribution to V(R) is quite large [C_6=3660 a.u. for $He(2^1S)+Na$ and C_6=2220 a.u. for $He(2^3S)+Na$].[3] To improve on the ab initio potentials we introduced a single parameter R_o, which was determined such that the adjusted potential curves,

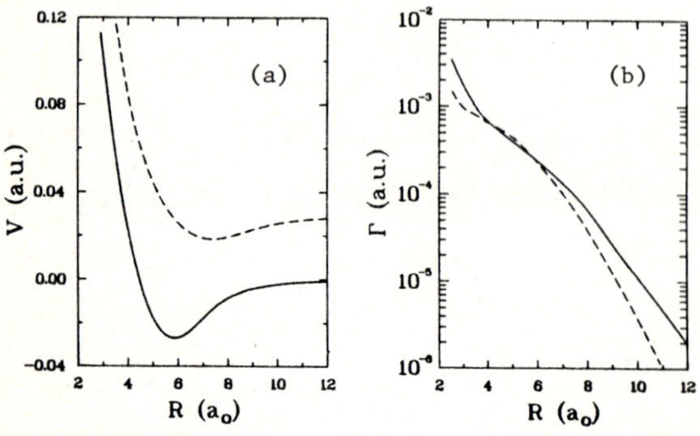

Fig. 1: Potential energy (a) and width (b) for $He(2^3S)+Na$ (solid lines) and $He(2^1S)+Na$ (dashed lines).

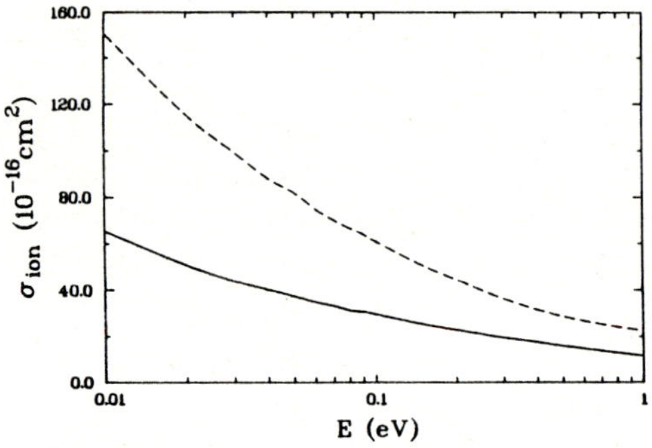

Fig. 2: Penning ionization cross sections for $He(2^3S)+Na$ (solid line) and $He(2^1S)+Na$ (dashed line).

shown in Fig. 1a, have the experimentally observed well depths.[4] The functional form was

$$V(R) = V_{ab\ initio}(R)\ f(R)-(C_6/R^6)[1-f(R)] \quad (3a)$$

$$f(R) = \exp[-(R/R_o)^8] \ . \quad (3b)$$

The widths Γ(R), shown in Fig. 1b, were determined by a cubic-spline least-squares fit to the energy-dependent Stieltjes derivatives through 10th order. The behavior of Γ(R) at $R \gtrsim 8$ is clearly exponential decay as expected. Phase shifts for scattering by the complex potential (2) were calculated in the JWKB approximation.

The cross sections are shown in Fig. 2. The singlet-to-triplet ratio is 2.2 at 0.04 eV (the ratio would be 3 if purely statistical). A value of 1.2 has been determined experimentally (by comparison with He*+Xe).[4] The calculated absolute cross section for the triplet at ~0.04 eV is 40×10^{-16} cm^2 and may be compared with 33×10^{-16} cm^2 measured in a stationary afterglow[5] and 14×10^{-16} cm^2 measured in an earlier beam experiment.[4]

*Work performed under the auspices of the U.S. DOE.

**Supported in part by the U.S. DOE Office of Basic Energy Sciences and by the Robert A. Welch Foundation.

1. A. U. Hazi, J. Phys. B 11, L259(1978).

2. P. W. Langhoff, Int. J. Quantum Chem. Symp. 8, 347(1974).

3. A. Dalgarno and G. A. Victor, J. Chem. Phys. 49, 1982(1968).

4. H. Hotop and A. Niehaus, Z. Phys. 238, 452(1970).

5. C. E. Johnson, C. A. Tipton, and H. G. Robinson, J. Phys. B. 11, 927(1978).

COLLISIONAL IONIZATION OF H(3p) ATOMS ON RARE GAS ATOMS AND H_2 MOLECULES

A. Cornet, W. Claeys, V. Lorent, J. Jureta and D. Fussen

Institut de Physique, Université Catholique de Louvain
B-1348 Louvain-la-Neuve - BELGIUM

A beam of H(3p) atoms has been formed by laser excitation of metastable hydrogen atoms. The H(2s) beam is formed by charge exchange of protons in a cesium cell. Three components are present in the beam : H(1s), H(2s) and H(3p). The absolute cross sections for collisional ionization :

$$H + X \rightarrow H^+$$

of these three components will be written, σ_1,σ_2,and σ_3 respectively. In the present work we measure the ratios σ_2/σ_1 and σ_3/σ_2 for He,Ne,Ar,Kr,Xe and H_2 as target gas, in the energy range 1.2-3.6 keV.

1. Production of a H(3p) beam : A proton beam is partially neutralized in a Cs cell (1) (see fig. 1), the remaining ions are removed by a weak electric field deflector (2). A known fraction (F. Brouillard et al (1977)) of the atoms is in the metastable 2s state. The metastable atoms can be quenched by Stark effect in the electric field of a deflector (4). The atom beam interacts with a laser beam in a biased (voltage V_0) gas cell (6), they intersect at right angles. A mirror (5) allows to double the laser intensity. The evolution of the 2s component is observed in a Lyman α detector formed of an axial quenching field (7), a LiF window (8) and a channeltron (9). This device allows to determine absolutely the 3p fraction in the neutral beam through the depopulation of the 2s component when the laser is on. Neutrals are monitored on a secondary emission probe (11).

2. Production of an intense light beam inducing the (2s-3p) transition at 6563 Å : the green (5145 Å) light of an Ar laser pumps the dye (DCM) of a monomode ring laser. A discharge absorption cell, filled with H_2, is used as a reference for the (2s-3p) transition. The optics were set-up to accomodate the beam divergence and the laser divergence for maximal excitation efficiency. Transitions towards $3p_{1/2}$ and $3p_{3/2}$ are perfectly resolved and 50% efficiency was obtained for the $(2s-3p_{3/2})$ transition.

3. Measurement of the ratio σ_2/σ_1. By measuring the count rate of protons formed in the bias voltage V_0 on the detector (12) (see fig. 1) with the quench field (4) on and off, one can easily determine the ratio σ_2/σ_1, knowing the metastable fraction in the beam. Acceptable agreement is found with the measurements of Roussel et al (1977) and Dose and Gunz (1972).

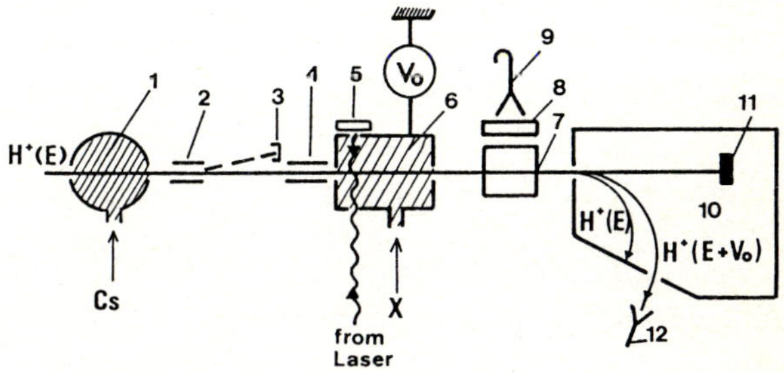

Fig. 1

4. Measurement of the ratio σ_3/σ_2. This ratio is obtained from the proton count rate with laser frequency at and away from the transition (2s-3p) excitation frequency. In fact the laser frequency is continuously swept, the proton and Lyman α(characteristic of excitation efficiency) count rates are recorded in phase. The ratio σ_3/σ_2 is derived from : σ_2/σ_1, the excitation efficiency, the initial metastable fraction, the total interaction length in the target gas and the fraction of it before the laser excitation and the 3p radiative transition probabilities. Figure 2 shows the results for the different targets. It is usefull to note that σ_3/σ_2 is independent of the absolute determination of laser power, neutral intensity, detection efficiency, laser and atom beam interaction geometry. The angular acceptance of the system is 10 mrad.

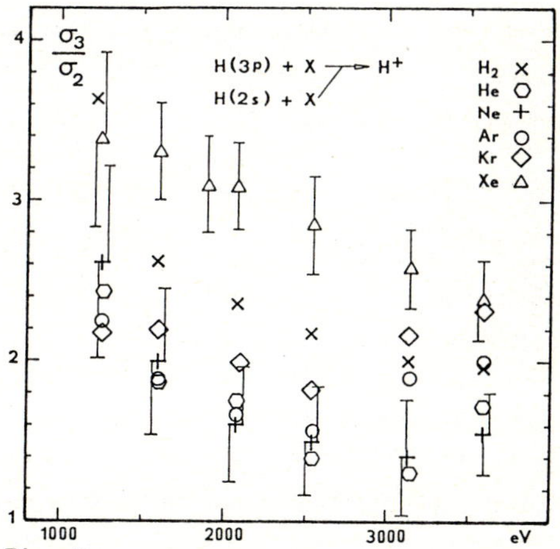

Fig. 2

References

F. Brouillard, W. Claeys and G. Van Wassenhove, J. Phys. B, 10, 687, (1977).

F. Roussel, P. Pradel and G. Spiess, P.R.A., 16, 1854, (1977).

V. Dose and R. Gunz, J. Phys. B, 5, 636, (1972).

ION PAIR PRODUCTION IN H(1S)+H(2S) COLLISIONS

D. Fussen, W. Claeys, A. Cornet, J. Jureta and P. Defrance

Institut de Physique Corpusculaire, Université Catholique de Louvain
B1348 Louvain-la-Neuve - Belgium

The total cross section of the charge transfer reaction

$$H_A(2s) + H_B(1s) \rightarrow H_A^- + H_B^+ \qquad (1)$$

has been experimentally investigated by Claeys et al (1977) in the energy range 80 to 400 eV. No evidence of ion pair production was found and only an upper limit of the cross section was assigned

$$\sigma < 2.8\ 10^{-17}\ cm^2.$$

This result was surprising as it was in apparent disagreement with the detailed balancing for the neutralisation reaction :

$$H_A^+ + H_B^- \rightarrow H(2s) + H(1s). \qquad (2)$$

The question is now better understood in view of recent theoretical approaches (Sidis et al, 1981; Borondo et al, 1981). Reaction (1) is in fact a "two-electron process" (core exchange of the 1s orbital followed by the excitation of the 2s electron to the diffuse orbital of H^-) while reaction (2) is a direct charge transfer.

In a one-active electron model, Sidis et al (1983) have predicted a cross section of 5 10^{-16} cm^2 for the "direct" ion pair production :

$$H_A(2s) + H_B(1s) \rightarrow H_A^+ + H_B^- . \qquad (3)$$

Measurements of the total cross section of (3) have been conducted with the same apparatus as Claeys et al (1977) except that ways of ion detection were interchanged (Fussen et al 1982).
Results are presented below for four energies. A good agreement is obtained between theory and experiment, confirming the important distinction to make between "direct" and "indirect" reactions in both neutralisation and ion pair production.

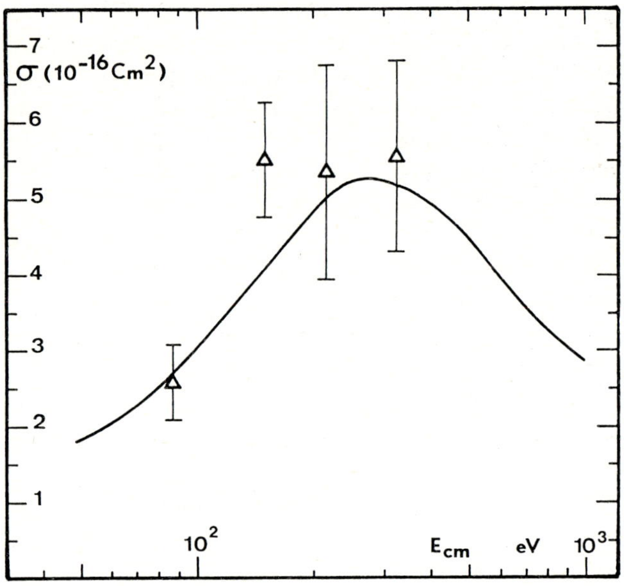

Fig. 1 : Total cross section for ion pair production $H_A(2s) + H_B(1s) \rightarrow H_A^+ + H_B^-$.
Δ : present experimental result.
— : theoretical calculation of Sidis et al (1983).

References

1. Claeys, Brouillard and Van Wassenhove, Abstract of papers, ICPEAC X, Paris (1977).
2. Sidis, Kubach and Fussen, Phys. Rev. Letters, vol. 47 n° 8, (1981).
3. Sidis, Kubach and Fussen, Phys. Rev. A (1983), in press.
4. Borondo, Macias and Riera, Phys. Rev. Letters, vol. 46, n° 4,(1981)
5. Fussen, Claeys, Cornet, Jureta and Defrance, J. Phys. B, n° 19, (1982).

EXCITATION OF HYDROGEN ATOMS TO THE n = 3 AND 4 LEVELS IN COLLISIONS WITH RARE-GAS ATOMS

B. Van Zyl, M. W. Gealy, H. Neumann, and R. C. Amme

Department of Physics, University of Denver, Denver, Colorado 80208, USA

Excitation of ground-electronic-state hydrogen atoms (H) to the n = 3 and 4 levels in collisions with rare-gas atoms was examined by observation of the Balmer-alpha (H_α) and Balmer-beta (H_β) emissions resulting from the interactions. The H-atom energy range covered was from 2.5 keV down to below 50 eV.

The H-atom beam was produced by photodetachment of electrons from H^- ions, which were collimated into a beam of the desired trajectory and ion energy prior to the photodetachment region. The photodetaching light source employed was a yttrium-aluminum-garnet laser (1064 nm) operated in a 50 % duty-cycle mode, with the H^- ion beam passing through the laser cavity.[1]

After deflecting the remaining H^- ions into a suitable collector, the fast H atoms entered a target cell containing low-density rare-gas atoms. The H_α and H_β photons resulting from collisionally excited H atoms were observed at 90° relative to the H-beam axis with a detector calibrated absolutely by observation of these same emissions produced in electron collisions with H_2 molecules using the known cross sections for these reactions.[2]

To separate those components of the H_α and H_β emissions resulting from decay of the relatively long-lived 3s and 4s states, the entire photon detector was moved along the H-atom-beam axis through the target cell. The exponential growth of these components of the photon signal as a function of distance into the target cell allowed the excitation cross sections for production of H atoms in these ns states to be separated from the total measured emission cross sections.

Because excited H atoms in the 3p and 4p states decay primarily to the 1s state via Lyman-line emissions, the remaining components of the H_α and H_β emissions can largely be attributed to decay of the 3d and 4d states. Thus the cross sections for excitation of these nd states were also approximately determined.

For all the rare-gas target atoms, the cross sections for excitation to the 3s and 4s states were found to reach their maximum values near 1 keV H-atom energy. The 3s-state excitation cross sections have maximum values ranging from 0.56×10^{-17} cm^2 for He targets to 1.7×10^{-17} cm^2 for Ar targets. The 4s-state excitation cross sections were found to have maximum values of about 0.3 times those for 3s-state excitation, a value not far below that of 0.42 predicted by the n^{-3} scaling law for ns-state excitations.

The considerable similarity of these ns-state excitation cross sections was not expected in this low H-atom energy range. Here, where the collision velocity is well below the electronic Bohr velocity, one might have guessed that these excitation cross sections would have been highly dependent on the transient molecular complexes existing during the interactions.

In sharp contrast, the cross sections for excitation of the 3d and 4d states of H in these collisions varied markedly from one rare-gas atom to another. For Kr targets, these nd-state excitation cross sections reached large maximum values for H-atom energies in the 100 eV range, similar to the cross sections for H + Ar collisions reported earlier.[2] For He and Ne targets, the nd-state excitation cross sections were found to be almost 2 orders of magnitude smaller in this H-atom energy range, with the cross sections for Xe targets lieing about midway between these extremes. The nd-state excitation cross sections thus do appear to depend strongly on the details of the interactions leading to the populations of these states.

The polarizations of the H_α and H_β radiations resulting from the collisions were also measured during these studies. Because of the spherical symmetry of the ns states, the observed radiation polarizations must result largely from decay of the nd states. The large cross sections for excitation of the nd states in low-energy H + Ar and H + Kr collisions result in highly positively polarized emissions, suggesting that the m_ℓ = 0 and ± 1 sublevels of the nd states are predominantly populated in these interactions. For Ne targets, the radiation polarizations were quite small, consistent with the fact that ns-state excitations are highly dominant for this target.

For the case of H + Ar collisions, it was suggested earlier[2] that an intermediate collision complex of the form $H^- + Ar^+$ provided the mechanism for nd-state excitations in this reaction. A similar explanation now appears plausible for H + Kr collisions. Because of the high ionization potentials of He and Ne, and the lower ionization potential of Xe, such intermediate Coulomb states do not appear to play as important a role in these interactions.

References

1. B. Van Zyl, N. G. Utterback, and R. C. Amme, Rev. Sci. Instrum. 47, 814 (1976).
2. B. Van Zyl, H. Neumann, H. L. Rothwell, Jr., and R. C. Amme, Phys. Rev. A 21, 716 (1980).

THEORY OF Li(2s-2p) AND Na(3s-3p) EXCITATION IN Li-Na HIGH ENERGY COLLISIONS

Svend Erik Nielsen[+], Martin Larsen[+] and John S. Dahler[++]

[+]Chemistry Laboratory III, H.C.Ørsted Institute, University of Copenhagen, 2100 Copenhagen Ø, Denmark
[++]Department of Chemistry, University of Minnesota, Minneapolis, Minnesota 55455, USA

Quasi-one-electron collision systems, e.g. Li-He and Na-He, have been investigated in detail in recent years, experimentally and theoretically.[1] Observed impact excitations of the alkali projectile have been successfully predicted and interpreted in terms of a direct excitation mechanism in the important impact energy range about and above the Massey maxima of the excitation cross sections, i.e. typically at energies 1-100 keV. Simple one-electron models based upon local potentials as well as more refined several-electron models, generating important non-local interactions,[2] have been invoked in atomic-basis impact-parameter studies of these systems.

In the present study we have extended this approach to the quasi-two-electron collision system Li(2s)-Na(3s) in order to obtain theoretical estimates of the excitation cross sections, polarization of impact radiation and coherence parameters associated with Li(2s-2p) and/or Na(3s-3p) excitation for comparison with experimental results.[3]

The Li-Na system is modelled as a two-electron system with the two valence electrons interacting with each other ($1/r_{12}$), and with the two closed-core ions Li^+ and Na^+ by local pseudopotentials $V^A(r_A)$ and $V^B(r_B)$, respectively. The complete hamiltonian operator has the form,

$$H = H^A(1) + H^B(2) + V^A(2) + V^B(1) + 1/r_{12} + V^{core}(R)$$

with the quasi-one-electron hamiltonian H^A for Li (and analogous expressions for H^B of Na),

$$H^A(1) = -\tfrac{1}{2}\nabla_1^2 + V^A(r_{1A}) \qquad (H^A(1)-\varepsilon_j^A)\phi_j^A(1) = 0$$

$$V^A(r) = -\frac{Z_N-Z_A}{r}\left(1 + \frac{r}{2\alpha_A}\right)\exp(-r/\alpha_A) - Z_A/r$$

where Z_A is the charge of A (Li^+) and Z_N the charge of the A nucleus (Li^{3+}), and the constant α_A is determined such that the eigenvalue ε_{2s}^A reproduces the spectroscopic ionization potential of Li(2s).[1]

We obtain the two-electron scattering states, Ψ^{SM}, of total spin S(M) solving the time-dependent Schrödinger equation $(H - i\frac{\partial}{\partial t})\,\Psi^{SM}(1,2;t) = 0$ by the impact-parameter method, assuming a rectilinear trajectory and constant relative velocity v of the heavy particles, and expanding Ψ^{SM} in terms of an antisymmetrized, translational phase factor (ETF) modified atomic basis.

$$\Psi^{SM} = \sum_{jk} a_{jk}^{SM}(t)\,\Psi_{jk}^{SM}(1,2;t)\,\exp\{-i(\varepsilon_j^A + \varepsilon_k^B)t\}$$

$$\Psi_{jk}^{SM} = 2^{-\frac{1}{2}}\{\phi_j^{Av}(1)\phi_k^{Bv}(2) \pm \phi_j^{Av}(2)\phi_k^{Bv}(1)\}\,\Lambda_{SM}(1,2)$$

where Λ_{SM} are the singlet/triplet spin states (S=0/S=1) and standard atomic ETF's have been included in ϕ_j^{Av} and ϕ_k^{Bv}.

The resulting close-coupled equations for the state amplitudes $a_{jk}^{SM}(t)$ assume the form

$$i\sum_{j'k'} \dot{a}_{j'k'}^{SM}\{\delta_{jj'}\delta_{kk'} \pm S_{jk'}^{v}S_{kj'}^{v}\}\exp(i\Delta\varepsilon t)$$

$$= \sum_{j'k'} a_{j'k'}^{SM}\{D(jk|j'k') \pm E^v(jk|j'k')\}\exp(i\Delta\varepsilon t)$$

with $\Delta\varepsilon = (\varepsilon_j^A + \varepsilon_k^B) - (\varepsilon_{j'}^A + \varepsilon_{k'}^B)$ and matrix elements in standard notation

$$D = \langle k_B|V^A|k_B'\rangle\delta_{jj'} + \langle j_A|V^B|j_A'\rangle\delta_{kk'} + \langle j_Ak_B|\frac{1}{r_{12}}|j_A'k_B'\rangle$$

$$E^v = \langle k_B|V^B|j_{Av}'\rangle S_{jk'}^v + \langle j_{Av}|V^A|k_B'\rangle S_{kj'}^v + \langle k_Bj_{Av}|\frac{1}{r_{12}}|j_{Av}'k_B'\rangle$$

$$S_{jk}^v = \langle j_{Av}|k_B\rangle \qquad \text{and} \qquad \langle \underline{r}|j_{Av}\rangle = \phi_j^A(r_A)\exp(-ivz)$$

to be solved with the initial conditions $a_{jk}^{SM}(-\infty) = \delta_{j,2s}\delta_{k,3s}$.

The direct couplings lead (by the three terms of D) to Na excitation, Li excitation and to Na,Li single/double excitation, respectively. The exchange couplings of E^v on the other hand connect to singly and doubly excited channels by all three interactions V^A, V^B and $1/r_{12}$. We may note that the high energy limit is particularly simple since all exchange terms and overlap terms vanish due to the rapidly oscillating electron translation factor exp(-ivz), and the singlet/triplet set of solutions degenerate to one set of amplitudes.

Preliminary calculations involve the high and low energy limits of the theory, and close-coupling solutions are obtained for the alkali atomic states 3s, $3p_0$, $3p_{\pm1}$ of Na and 2s, $2p_0$, $2p_{\pm1}$ of Li, corresponding to 10 independent state amplitudes $a_{jk}^{SM}(t)$ for each spin state. Numerical solutions are in progress and results for single and double excitation cross sections will be reported at the conference.

References

1. N.Andersen and S.E.Nielsen, Adv.Atom.Mol.Phys. 18, 265 (1982).
2. S.E.Nielsen and J.S.Dahler, this conference.
3. B.Bisgård, T.Andersen, B.V.Sørensen, S.E.Nielsen and J.S.Dahler, J.Phys.B 13, 4441 (1980).

HYDROGEN ATOM EXCITATION IN H + He, Ne, Ar COLLISIONS BELOW 1keV

J. Grosser and W. Krueger

Institut für Atom- und Molekülphysik, Universität Hannover, Federal Republic of Germany

The radiative lifetimes of excited hydrogen atoms show a considerable variation in a weak electric field. We make use of this property for an investigation of the inelastic collision processes

$$H + M \rightarrow H(nl) + M$$

Neutral hydrogen atoms of 20-1200 eV are formed by H^- laser photodetachement in our apparatus. They pass through a scattering cell, in which an electric field of variable intensity can be applied. Excited hydrogen atoms are detected in our experiment by the emission of Lyα(2p→1s) radiation, which can follow their formation. The Lyα intensity without an electric field serves in the first instance as a measure for the cross section σ_{2p}. Atoms, which are formed in the state 2s are metastable (o.6s lifetime) in the absence of an electric field. In a field of some hundred V/cm, however, they emit Lyα radiation as well. Thus the Lyα intensity in such a field contains information about the cross section σ_{2s}.

Hydrogen atoms in states with n>2 can decay by cascades like 3d→2p→1s and thereby contribute to the Lyα signal. The cascade intensities vary in an electric field as well. For example, the cascade 3d→2p→1s is expected to show an intensity reduction by about 40%, when the 3d atoms are formed in an electric field of 2V/cm instead of a zero field. The field dependence of the Lyα̃ signal can be used to identify cascade contributions and even to measure n>2 cross sections.

We apply this technique to collisions of hydrogen and deuterium atoms with noble gas atoms. With He and Ne as the second collision partners, no appreciable cascade contributions are found. We report the energy dependence of the integral cross sections σ_{2s} and σ_{2p} in these cases (fig.1), and we obtain upper limits to some of the n=3 cross sections. For H + Ar, we find a considerable contribution from atoms produced in the state 3d. We report a preliminary evaluation yielding σ_{2s}, σ_{2p}, σ_{3s}, and $\sigma_{3d}-0.61\sigma_{3p}$. The results are compared to directly obtained n=3 cross sections[2].

A detailed measurement of the field dependence of the Lyα intensity can further yield[3] the value of an offdiagonal density matrix element $\sigma_{2s/2p0}$, which characterizes the correlation between the 2s and 2p wavefunctions immidiatly after the collision. This quantity cannot exceed the value $\sqrt{\sigma_{2s}\cdot\sigma_{2p}}$. When it has this order of magnitude, however, our experimental arrangement is sufficiently sensitive for the measurement of this quantity. We did not yet find a nonzero value of $\sigma_{2s/2p0}$. We discuss the implications of this result for the mechanism of the scattering process.

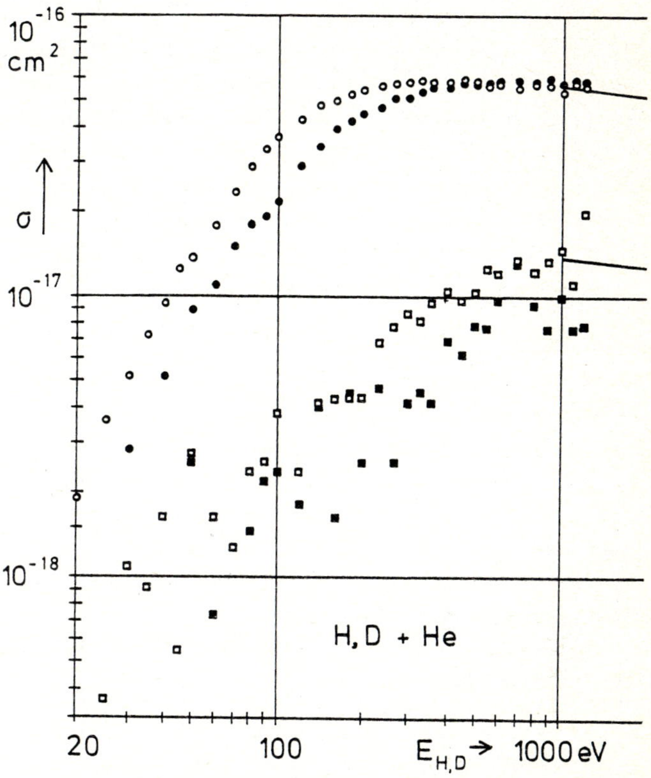

Fig. 1: Energy dependence of the integral cross sections for the process

H + He → H(2p)

H + He → H(2s)

D + He → D(2p)

D + He → D(2s)

The solid lines are the experimental results of Birely and Mc Neal[4].

References

1. W. Aberle, J. Grosser, W. Krueger, Chem.Phys.41, 245 (1979); J.Phys.B.13, 2083 (1980)
2. B. VanZyl, H. Neumann, H.L. Rothwell, R.C. Amme, Phys.Rev.A 21, 716 (1980)
3. R.Krotkov, J. Stone, Phys.Rev.A 22. 473 (1979)
4. J.H. Birely, R.J. Mc Neal, Phys.Rev.A 5, 257 (1972)

SIMULTANEOUS 2^1P EXCITATION OF TWO COLLIDING He ATOMS TO VARIOUS SUBSTATE COMBINATIONS

L. Moorman, K.P.J. Linnartz, J. van Eck and H.G.M. Heideman

Fysisch Laboratorium, Rijksuniversiteit Utrecht, the Netherlands

In a previous publication[1] we presented for the first time photon-photon angular correlation measurements on the simultaneous excitation of two colliding helium atoms to the 2^1P state. Schematically:

$$\vec{He}+He \to \vec{He}^*(2^1P)+He^*(2^1P) \to \vec{He}(1^1S)+He(1^1S)+p_1+p_2$$

where the (fast) projectile is indicated by an arrow. The angular correlation between the emitted photons p_1 and p_2, which are measured in coincidence, provides information on the excitation of the different magnetic substate combinations. Denoting by $|M_1M_2\rangle$ the final state where, after the collision, one atom is in the $|LM\rangle = |1M_1\rangle$ state and the other in the $|1M_2\rangle$ state, the density matrix for the two excited atoms can be written as

$$\langle M_1M_2|\rho|M_1'M_2'\rangle = \int f(M_1M_2;\hat{v})f^*(M_1'M_2';\hat{v})\, d\hat{v}$$

where $f(M_1M_2;\hat{v})$ is the amplitude for the collision process leading to excitation of atom 1 to the state $|1M_1\rangle$ and atom 2 to the state $|1M_2\rangle$, and where the integration is to be performed over all directions of the relative velocity $\hat{v}$ between the two atoms after the collision. The quantisation axis is taken along the incident beam. Due to the various symmetries in our experiment[1] and to the Hermiticity of the density matrix there are in our case of 2^1P excitation only five independent parameters, namely: $\sigma(00)$, $\sigma(01) = \sigma(10) = \sigma(-10) = \sigma(0-1)$, $\sigma(11) = \sigma(-1-1)$, $\sigma(1-1) = \sigma(-11)$ and $\rho(1-1;00)$, where

$$\sigma(M_1M_2) = \langle M_1M_2|\rho|M_1M_2\rangle = \int |f(M_1M_2;\hat{v})|^2\, d\hat{v}$$

$$\rho(1-1;00) = \langle 1-1|\rho|00\rangle = \int f(1-1;\hat{v})f^*(00;\hat{v})\, d\hat{v}$$

The first four are the cross sections for excitation of the various possible substate combinations integrated over the relative velocity $\hat{v}$ after the collision; the last one involves an integral (over $\hat{v}$) of the phase between the amplitudes $f(1-1;\hat{v})$ and $f(00;\hat{v})$. It is convenient to make the above parameters dimensionless by dividing them by σ, the total cross section for simultaneous excitation of both atoms to the 2^1P state.

$$\alpha = \sigma(00)/\sigma,\quad \beta = \sigma(01)/\sigma,\quad \lambda = \sigma(1-1)/\sigma,\quad \mu = \sigma(11)/\sigma$$

$$\chi = \mathrm{Re}(\langle 1-1|\rho|00\rangle)/\sigma = \sigma^{-1}\int |f(1-1;\hat{v})|\,|f(00;\hat{v})|\cos\psi(\hat{v})\, d\hat{v}$$

where $\psi(\hat{v})$ is the phase between the amplitudes $f(1-1;\hat{v})$ and $f(00;\hat{v})$.

We have derived a theoretical expression[1] which relates the above five parameters to the angular distribution of the coincident photons. The experimental procedure is the same as in ref. 1. We measure the angular distribution of photon 1 at a few fixed detection angles of photon 2. By fitting the theoretical distribution to the experimental distributions the parameters α, β, λ, μ and χ can be determined. We have performed measurements at 500, 600, 750, 1250, 2500 and 3500 eV. The results are summarized in table 1.

Table 1: Measured excitation parameters for simultaneous excitation of two colliding He-atoms to various substate combinations. Errors represent statistical ones only.

E (eV)	α	β	$\lambda + \mu$	χ
500	0.32±0.05	0.09±0.02	0.17±0.06	0.05±0,05
600	0.18±0,03	0.09±0.02	0.22±0.05	0.09±0.02
750	0.11±0.03	0.12±0.01	0.21±0.02	0.11±0.02
1250	0.01±0.02	0.15±0.01	0.19±0.01	-
2500	-0.02±0.05	0.19±0.015	0.13±0.09	-
3500	0.00±0.03	0.18±0.01	0.13±0.03	-

In the present experiment we are unable to determine λ and μ separately, because for the photon-detector configurations we can realise at present the angular distribution functions are rather insensitive for the value of λ. In particular at the higher energies the results are not in agreement with what one expects on the basis of a simple correlation diagram.

Reference

1. G.J.N.E. de Vlieger, H.G.M. Heideman, J. van Eck and G. Nienhuis, J. Phys. B: At. Mol. Phys. 15, L345 (1982).

EMISSION FROM OUTER TURNING POINTS OF HIGH VIBRATIONAL STATES OF THE KRYPTON EXCIMER AFTER KEV-ELECTRON IMPACT EXCITATION

P. Wollenweber, K. Barzen and H. Schmoranzer

Fachbereich Physik, Universität Kaiserslautern, D-6750 Kaiserslautern, Germany

Vacuum-uv emission from high vibrational levels of the $0_u^+(^3P_1)$ state of the Kr excimer has been investigated at 0.01 nm spectral resolution after continuous bombardment by 250 μA of 20 keV electrons[1]. Radiation observed in the vicinity of the first resonance line $^3P_1$-$^1S_0$ at λ = 123.584 nm can be related to excimer emission from the region of the outer vibrational turning points near the dissociation limit.

Besides the red satellite of the first resonance line observed at 123.83 nm and discussed previously[1,2] emission from the region of the outer turning points appears to be involved in the structures observed on the long wavelength side of the red satellite[3]. The spectral portion extending from about 124 nm to 125 nm, i.e. between the red satellite and the onset of the $1_u(^3P_2)$ - $0_g^+(^1S_0)$ emission, exhibits rather weak and diffuse structures shown in Fig. 1 after subtracting a fitted background decreasing exponentially with increasing wavelength. The structures observed here could not be interpreted as being primarily the envelopes of bound-bound band progressions (system II 0_u^+ - 0_g^+ of Ref. 4), as was also the case for argon[5].

Therefore bound-free Franck-Condon factors were calculated for transitions from all upper vibrational levels v' of $0_u^+(^3P_1)$ into the 0_g^+ continuum. The simulated spectrum displayed in Fig. 2 was generated by equally weighted superposition of all bound-free Franck-Condon factors and shows an oscillatory structure within a narrow spectral range. The calculated spectrum is similar to the experimental one as far as width, spacing and position of the modulations are concerned and reminds of structures calculated for free-free transitions[6]. The simulation predicts the structures at the long wavelength side to shift to shorter wavelengths if the vibrational population is changed in favour of the lower v'. This was experimentally confirmed by increasing the pressure and thus the vibrational relaxation by collisions (see Fig. 1). A more detailed comparison between simulated and measured spectral structures is under way and can be expected to provide a refined empirical potential curve of 0_u^+ in the range of large interatomic distances corresponding to the Franck-Condon region of the outer turning points of high vibrational states.

Financial support from the Deutsche Forschungsgemeinschaft under Sonderforschungsbereich 91 is gratefully acknowledged.

References

1. K. Barzen, H. Schmoranzer and P. Wollenweber, Abstracts 8th ICAP (Eds. I. Lindgren, A. Rosén, S. Svanberg, Göteborg 1982), A 97
2. F.X. Gadea, F. Spiegelmann, M.C. Castex and M. Morlais, preprint, submitted 1982
3. H. Schmoranzer, P. Wollenweber and K. Barzen, Abstracts 5th Gen. Conf. of the EPS (Istanbul 1981),103
4. Y. Tanaka, K. Yoshino and D.E. Freeman, J. Chem. Phys. 59, 5160 (1973)
5. Y. Tanaka, W.C. Walker and K. Yoshino, J. Chem. Phys. 70, 380 (1979)
6. F.H. Mies and A.L. Smith, J. Chem. Phys. 45, 994 (1966)

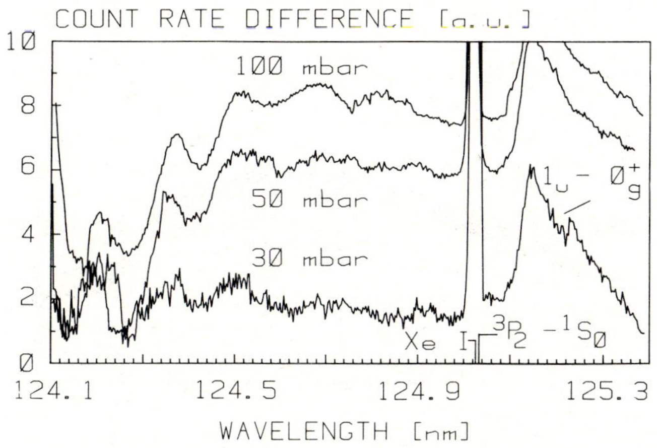

Fig. 1: Pressure dependence of diffuse structures in Kr excimer emission spectra at 300 K

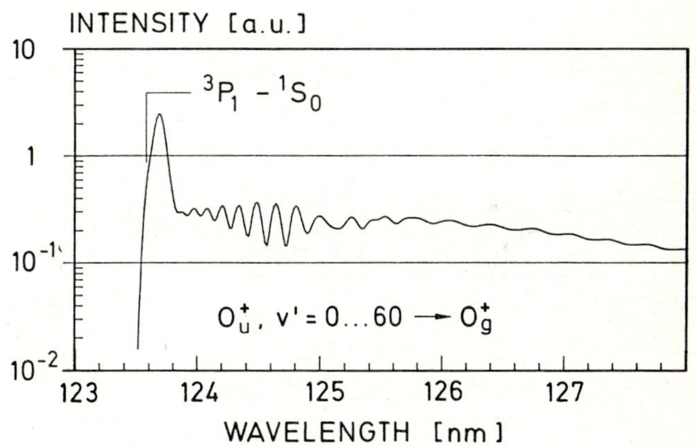

Fig. 2: Computed bound-free emission spectrum of Kr_2 $0_u^+(^3P_1)$ - $0_g^+(^1S_0)$

EMISSION OF THE KRYPTON EXCIMER AT SMALL INTERATOMIC DISTANCES AFTER KEV-ELECTRON IMPACT EXCITATION

K. Barzen, P. Wollenweber and H. Schmoranzer

Fachbereich Physik, Universität Kaiserslautern, D-6750 Kaiserslautern, Germany

The vacuum-uv emission from the $^1\Sigma_u^+$, $0_u^+(^3P_1)$ state of the krypton excimer is expected to extend from the first continuum until a marked cut-off at long wavelengths. The spectrum near the cut-off wavelength can be related to emission from the region of the inner turning points of the highest vibrational levels to the repulsive ground state $^1\Sigma_g^+$, $0_g^+(^1S_0)$.

Within the pertinent Franck-Condon region, i.e. at small interatomic distances corresponding to a collision energy of $Kr(^1S_0)$ atoms of several eV, the potential curves of the electronic states involved are rather uncertain, so that evaluating the spectral cut-off observed is considered a promising way of testing or empirically refining the potential curves.

In search of experimental evidence of the spectral cut-off the apparatus described previously[1,2] was improved by extending the spectral sensitivity up to 320 nm. Krypton (30 ppm impurities) at room temperature was bombarded by 250 μA of 20 keV electrons. The emission spectrum was recorded at a resolution of 0.01 nm in the wavelength range from the second continuum[1] to 300 nm at pressures from 10 to 300 mbar.

Besides lines and bands arising from impurities the spectrum exhibits four continuous structures, two narrower ones observed previously[1] centered around 174 nm and about 180 nm, a broad one starting at about 200 nm and sharply cut off between 225 nm and 227 nm and a less pronounced one peaked around 248 nm.

Superimposed radiation from XeKr[3] has been hampering a quantitative evaluation of the first two structures. Therefore the present study was concentrated on the third one described previously[4]. Its maximum was found to shift from 225 nm at 10 mbar (Fig. 1) to about 220 nm at 75 mbar (Fig. 2), where the structure is also broadened. The pressure dependent shift indicates that the ground state potential curve must have a steeper slope than the upper one. The position of the cut-off was found independent of pressure as is expected for the emission from the uppermost inner vibrational turning point. Although the qualitative behaviour of the observed third continuum can be anticipated, spectral cut-offs calculated from the potential curves[5,6] lie at significantly shorter wavelengths.

Financial support from the Deutsche Forschungsgemeinschaft under Sonderforschungsbereich 91 is gratefully acknowledged.

References

1. H. Schmoranzer and R. Wanik, Abstracts XIth ICPEAC (Eds. K. Takayanagi and N. Oda, Kyoto, 1979), 418
2. K. Barzen, H. Schmoranzer and P. Wollenweber, Abstracts 8th ICAP (Eds. I. Lindgren, A. Rosén, S. Svanberg, Göteborg, 1982) A 97
3. Y. Salamero et al., J. Chem. Phys. 74, 288 (1981)
4. E.T. Verkhovtseva, E.A. Katrunova, A.E. Ovechkin, and Ya.M. Fogel, Chem. Phys. Letters 50, 463 (1977)
5. H. Schmoranzer, R. Wanik and H. Krüger, Abstracts XIth ICPEAC, (Kyoto 1979), 416
6. F.X. Gadea, F. Spiegelmann, M.C. Castex and M. Morlais, preprint, submitted 1982

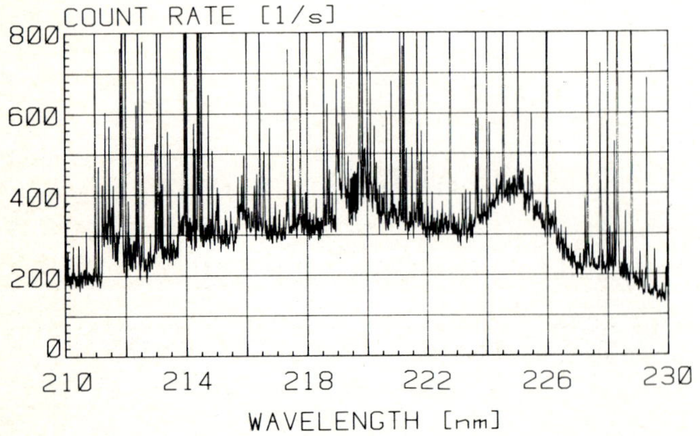

Fig. 1: Kr emission at 10 mbar

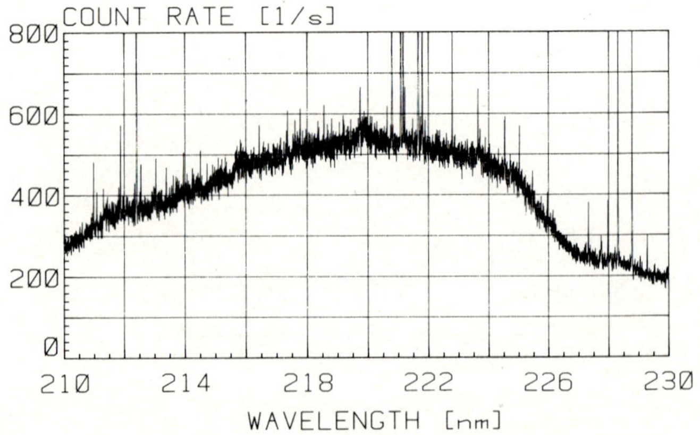

Fig. 2: Kr emission at 75 mbar

POLARIZATION EFFECTS IN ALKALI ATOMS FINE STRUCTURE TRANSITIONS (F.S.T) INDUCED BY He AND Ar

J.M. Mestdagh, J. Pascale

Service de Physique des Atomes et des Surfaces, CEN/Saclay, 91191 Gif-sur-Yvette Cedex, France

It has been predicted theoretically[1] and recently confirmed experimentally[2] that in a crossed beam experiment both the absolute value and the energy dependence of $P_{3/2} \rightarrow P_{1/2}$ F.S.T. cross-sections for alkali-rare gas pairs are strongly affected by the polarization of the laser light used to excite the alkali atoms. This will be referred to as polarization effect. Such effects can arise because i) two entrance channels, corresponding to the $\Sigma_{1/2}$ or $\Pi_{3/2}$ potential curves are possible when the alkali atom is excited in a $P_{3/2}$ level ii) the behaviour of the colliding system is not the same whether the entrance channel is the $\Sigma_{1/2}$ or $\Pi_{3/2}$ potential curve. By studying an F.S.T. process with different polarization schema of the excited $P_{3/2}$ level, it is therefore possible to have insight into the dynamics of the system. This has been done explicitely for the process

$$K(4P_{3/2}) + Ar \longrightarrow K(4P_{1/2}) + Ar \qquad (1)$$

When He is substituted to Ar as perturber in process (1), we have looked at polarization effects to get further information on the reliability of the relevant adiabatic potentials which have been calculated recently[3]. In view of the excellent agreement between experimental and theoretical data, for all the excitation schema we have investigated, we have concluded to a good accuracy of these potentials, as already pointed out under other circumstances [3].

Turning to process (1), the experimental data can be interpreted in terms of two parameters $A_{1/2}$ (full circles in fig.1) and $A_{3/2}$ (open circles) which are related respectively to the partial cross sections with $\Sigma_{1/2}$ and $\Pi_{3/2}$ potential curves as entrance channels[4] (For example, with unpolarized $K(P_{3/2})$ atoms $\sigma_{3/2 \rightarrow 1/2} = (A_{1/2} + A_{3/2})/2$). These data are in good agreement with close-coupling calculations using K-Ar potentials given in ref[5] (full curves in fig.1). These potentials are known to be relatively accurate at medium and large internuclear distances ($R \sim 10$ u.a). Therefore the agreement observed in fig.1 between the experimental and theoretical results indicate that the $A_{1/2}$ and $A_{3/2}$ parameters test mainly this domain of R-values. In other respects these results indicate that the radial coupling $\Sigma_{1/2}-\Pi_{1/2}$ is predominant over rotational coupling $\Pi_{3/2}-\Pi_{1/2}$. Comparison of $A_{1/2}$ and $A_{3/2}$ parameters in the case of K-He and K-A collisions will allow us to discuss further the coupling mechanisms responsible for F.S.T. This discussion will be presented at the conference and will include also results for the Rb(5P)-He collision.

Our work illustrate how crossed beam experiments involving laser polarized atoms can be used to detail the dynamics of a collision. The method is not restricted to F.S.T processes induced by rare gas perturbers. It will be particularly interesting to apply it to molecular perturbers since coupling mechanisms in that case are yet poorly known[6].

References

1. J. Pascale, M.Y. Perrin, J.Phys. B 13, 1839 (1980).
2. J.M. Mestdagh, J. Berlande, P. de Pujo, J. Cuvellier, A. Binet, Z. Phys. A 304, 3 (1982).
3. J. Pascale, Phys. Rev. A26, 3709 (1982) See other abstracts.
4. J.M. Mestdagh, J. Pascale (to be published).
5. J. Pascale, J. Vandeplanque, J. Chem. Phys. 60, 2278 (1974).
6. J.M. Mestdagh, Thèse d'Etat Paris (1982).

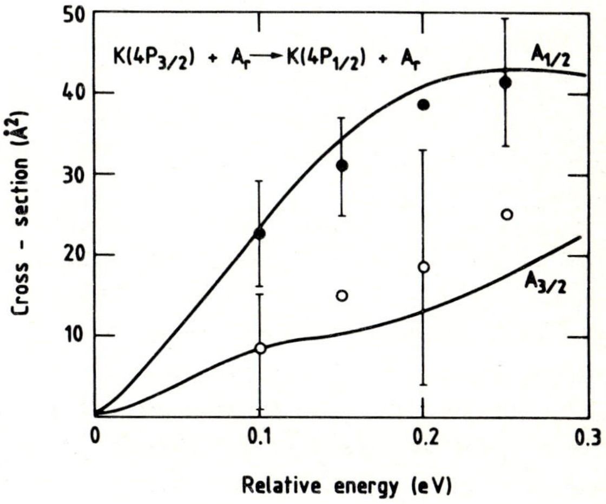

Fig.1: Energy dependence of the parameters $A_{1/2}$ and $A_{3/2}$. Experimental and theoretical data.

QUANTUM-MECHANICAL CALCULATIONS OF CROSS SECTIONS FOR TRANSITIONS WITHIN THE SECOND AND THIRD n^2P LEVELS OF Rb AND Cs, INDUCED IN COLLISIONS WITH He.

J. Pascale

Service de Physique des Atomes et des Surfaces, CEN/SACLAY, 91191 Gif-sur-Yvette Cedex France

Several cell-type experiments have been made in the past years dealing with fine-structure transitions of the second and third n^2P levels of Rb and Cs, induced in collisions with rare gas atoms [1-6]. For these transitions there are a few theoretical data [7-8]. In Ref.8 molecular semiclassical calculations which coupled only molecular states relevant to given n^2P levels were reported for collisions between Rb(6^2P) or Cs(8^2P) and He, Ar or Xe; except for Cs(8^2P)-He collisions, the results agreed with the experimental data[1,2] within a factor 1.5-3.0 which might be due either to the neglect of coupling with the nearest neighboring states (n^2D) or inaccuracy in the adiabatic potentials used in these calculations. In particular, these calculations greatly reduced large discrepancies (one to two orders of magnitude) between experiment[1] and previous theory [7] for the Cs (8^2P) rare gas collisions, except for collisions with He.

In the present work we have made, for the first time, quantum-mechanical calculations to determine the cross sections for transitions within the second and third n^2P levels of Rb and Cs, induced in collision with He. They use recent adiabatic potentials obtained from l-dependent pseudopotential calculations which have been shown to be reliable[9]. As for the first n^2P levels, we have assumed that coupling with neighboring levels can be neglected, and therefore the same method[10] has been used to determine effective interaction matrix elements for the close-coupling calculations. As long as this approximation is justified (it will be discussed at the Conference) the quantum-mechanical approach is very useful to obtain energy dependences of the cross sections near threshold energies and information on the various Zeeman cross sections.

The results we have obtained for the fine-structure transition cross sections are reported in Tables I-III. The present calculated cross sections are within the experimental uncertainties, except for Cs(8^2P)-He collisions for which the calculated cross sections are a factor of about 2.0 larger than the experimental data. Because it is for the case of Cs(8^2P) that the energy gap between the ^2P and ^2D levels is the smallest, we conclude that for the Cs(8^2P)-He collisions coupling with the 7^2D levels cannot be ignored. Calculations and results will be discussed at the Conference.

References

1. M. Pimbert, J. Phys. B(Paris) 33, 331 (1972).
2. I. Siara, E.S. Hrycyshyn and L. Krause, Can. J. Phys. 50, 1826 (1972).
3. I.N. Siara, H.S. Kwong, and L. Krause, Can. J. Phys. 52, 945 (1974).
4. J. Cuvellier, P.R. Fournier, F. Gounand, J. Pascale and J. Berlande, Phys. Rev. A 11, 846 (1975).
5. J. Kielkopf, J. Chem. Phys. 62, 4809 (1975).
6. P. Munster and J. Marek, J. Phys. B 14, 1009 (1981).
7. E.E. Nikitin and A.J. Reznikov, Chem. Phys. Lett. 8 161 (1971).
8. J. Pascale and P.M. Stone, J. Chem. Phys. 65, 5122 (1976).
9. J. Pascale, Phys. Rev. A 26, 3709 (1982).
10. R.H.G. Reid, J. Phys. B 6, 2018 (1973).

		Theory		Experiment	
n	T(K)	Present	Ref.8	Ref.2	Ref.6
6	337	15.6	7.8		17.1 ± 3.4
6	430	17.0	8.7	19.0±1.7	
7	353	63.9			70.8±14.2

Table I. Maxwellian-averaged cross sections (in Å^2) for the $n^2P_{3/2} \rightarrow n^2P_{1/2}$ transition in Rb.

T(K)	Present results	Ref.3	Experiment Ref.4	Ref.6
320	11.0			12.8±2.6
405	11.7	15.2± 4.6		
450	12.2		11.0± 2.0	
520	12.5	14.9± 4.5		
615	13.0		11.0± 2.0	
630	13.1	15.6± 4.7		

Table II. Maxwellian-averaged cross sections (in Å^2) for the $7^2P_{3/2} \rightarrow 7^2P_{1/2}$ transition in Cs.

	Theory		Experiment
T(K)	Present	Ref.8	Ref.1
420	59	3.2	34 ± 10
620	70	3.2	34 ± 10

Table III. Maxwellian averaged cross sections (in Å^2) for the $8^2P_{1/2} \rightarrow 8^2P_{3/2}$ transition in Cs.

COHERENCE OBSERVED IN THERMAL $K(4^2P)$ - RARE GAS COLLISIONS

R. Düren, E. Hasselbrink and H. Tischer

Max-Planck-Institut für Strömungsforschung, D3400 Göttingen, West Germany

At an earlier conference we have reported scattering experiments with laser excited $K(4^2P)$ interacting with rare gas targets[1]. In these the m_j distribution in the entrance channel has been modified by changing the orientation of the E vector of linearly polarized light - i.e. the quantization axis for the m_j - to be parallel and perpendicular to the beam. At the same time calculations of the scattering cross sections with σ^+ and σ^- light directed perpendicular to the scattering plane have been presented[2]. These calculations were based on the interatomic potentials calculated with the model potential method fitting ground- and excited state experiments[3)4)5]. The most interesting results predicted then were (1) a left/right symmetry of the differential cross sections complementary for σ^+ and σ^- excitation and (2) a difference between the K-Ar and the K-Kr interaction. For both systems one cross section is larger with increasing difference for smaller angles and for K-Ar the smaller one shows additional oscillations around the other. This phenomena, which are a direct consequence of the coherent superposition in the entrance channel have now been observed experimentally and are displayed in fig. 1 for K-Ar and figure 2 for K-Kr. The measurements clearly exhibit both the above mentioned predicted features: the left/right asymmetry and the difference between K-Ar and K-Kr. Quantitatively the comparison between calculated and measured cross sections is satisfactory but will be considered further.

References

1. R. Düren and H. Tischer, Verhandl.DPG(VI) 17, 379 (1982)
2. R. Düren and H. Tischer, Verhandl.DPG(VI) 17, 374 (1982)
3. R. Düren, E. Hasselbrink and H. Tischer, J.Chem.Phys. 77, 3286 (1982)
4. R. Düren, E. Hasselbrink and G. Moritz, Z.Phys.A 307, 1 (1982)
5. S. Milosevič and G. Pichler (private communication)

Fig. 1: Differential cross section for $K(4^2P_{3/2})$-Ar. K is excited with circularly polarized light σ^+ and σ^- directed perpendicular into the scattering plane E_{cm} = 125 meV.

Fig. 2: Same as figure 1 for $K(4^2P_{3/2})$-Kr with E_{cm} = 147 meV.

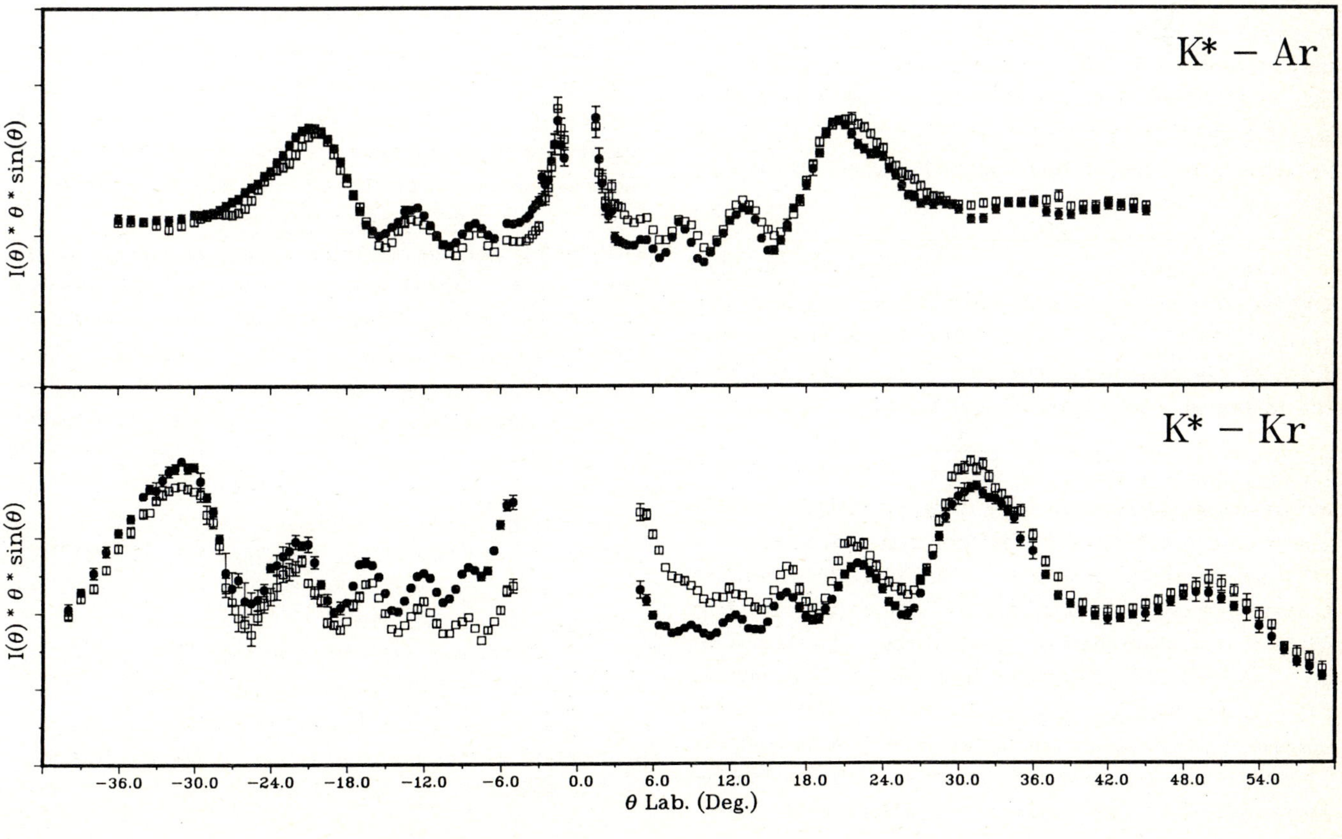

SCATTERING OF STATE SELECTED, ELECTRONICALLY EXCITED NEON ATOMS

W. Beyer, H. Haberland, and D. Hausamann

Fakultät für Physik, D-7800 Freiburg, Germany

The first excited configuration of the neon atom is $2p^5 3s$, which leads in LS-coupling to four states: $^1P_1$, $^3P_0$, $^3P_1$, and $^3P_2$. Differential cross sections in the energy range from 20 to 150 meV have been measured separately for the two metastable states (J=0 and J=2) using the apparatus shown in Fig. 1.

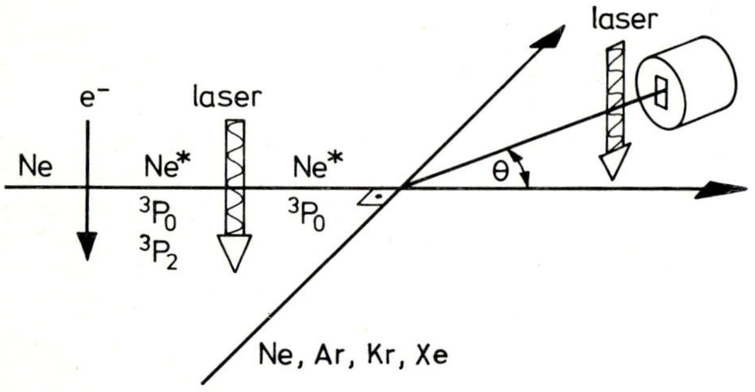

Fig. 1: Schematic of the apparatus. A supersonic Ne beam is excited by electron impact. The population of the two metastable Ne states is probed before the scattering center and before the detector by two separately chopped single mode laser beams. Inducing the optical transitions in a magnetic field, a specified M_J level of the $^3P_2$ state can be labeled. Save for the laser system the apparatus has been described in detail earlier[1].

Natural neon contains 90.5% ^{20}Ne and 9.2% ^{22}Ne. The isotope effect of the optical pumping transition allows an isotope selective measurement. The hyperfine structure of the 0.3% ^{21}Ne could be measured with a line width of 12 MHz. The following results have been obtained:

RESULTS FOR Ne^* + Ne

1) The sign and magnitude of the g-u splitting of the J=0 excimer potentials is obtained from an analysis of the interference pattern of the differential cross sections. It is standardly assumed in the literature, that the sign cannot be obtained in this way. The sign of the splitting differs from that obtained experimentally by Baudon et al.[2], and from that calculated by Cohen and Schneider[3].

2) An upper limit for the cross section for finestructure changing collisions has been obtained a) from an analysis of the differential cross sections and b) from a Doppler shift experiment. The value is compatibel with earlier optical pumping experiments[4], but at least a factor of four smaller than the value of Baudon et al.[5].

3) The differential cross sections of the J=2 state are very difficult to analyze, because of the many (at least 12) scattering amplitudes, which contribute to the intensity at one scattering angle. The symmetry oscillations can be suppressed for the isotope specific excitation transfer process. This allows an unambiguous identification of the rainbow structure, which gives a value of 25 meV for the height of the intermediate maxima of the 1_u and 0_u^- potentials[2].

4) Neon atoms in the $^3P_2$ state can easily be polarized in their M_J = 0, 1, and 2 states, using the Zeeman splitting of the optical pumping transition. By suitably orienting the quantisation axis with respect to the asymptotic internuclear axis, a specific Ω-value of the interaction potentials could preferentially be selected. The angular distributions show large differences for the different M_J states.

RESULTS FOR Ne^* + Ar, Kr, Xe

5) The angular resolution of the apparatus had to be increased to 0.3^0, in order to resolve all the interference patterns of the differential cross sections. The rainbow maxima have been resolved even for Ne^*+Ar. Even at the lowest kinetic energy there are unsuspected differences in the cross sections for the two fine structure states.

RESULTS FOR AR^* + Ar

6) With the improved resolution we were able to resolve the symmetry and g-u oscillations also for non state resolved Ar^* - Ar collisions. An analysis will have to wait for a state selective measurement. As the J=1 states are known from UV absorbtion experiments, a combination of the two data will allow a first check on the much used assumption[3], that the spin orbit matrix element is independent of the internuclear distance for the rare gas excimer states.

References

1. B.Brutschy, H.Haberland, Phys.Rev. A19, 2232 (1979)
2. I.Colomb de Daunant, G.Vassilev, J.Baudon and B.Stern J.Physique 43, 591 (1982)
3. J.S.Cohen, B.Schneider, J.Chem.Phys. 61, 3230 (1974)
4. R.A.Sierra, J.D.Clark, and A.J.Cunningham J.Phys.B21, 4113 (1979), and references therein
5. I.Colomb de Daunant, G.Vassilev, M.Dumont, J.Baudon Phys.Rev.Lett. 46, 1322 (1981)

NONRESONANCE EXCITATION TRANSFER IN He(2^1S,2^3S)+Ne COLLISIONS

A.Z.Devdariani, A.L.Zagrebin

Physical Institute, Leningrad State University, SU-Leningrad

The main results of the theoretical investigation of excitation transfer in the collisions

$$He(2^1S)+Ne \rightarrow He+Ne(2p^5 5s) \qquad (1)$$

$$He(2^3S)+Ne \rightarrow He+Ne(2p^5 4s) \qquad (2)$$

are presented.

For the distances which are essential for the collisions the type of quasi-molecular He Ne($2p^5ns$) n = 4, 5 terms is basically dependent on the ion interaction of Ne^+-He[1].This interaction has been determined on the basis of $X^2\Sigma^+$ and $A^2_2\Pi_{1/2}$ terms from [2]. The influence of a weakly bound electron was taken into account using the method described in [3]. Excitation transfer mechanism in (1) which mainly causes the population of Ne($5s^1P_1$) term is dependent on nonadiabatic interaction of terms 0^+Ne($5s^1P_1$) and 0^+He(2^1S) at R = $5{\cdot}8a_o$. Experimental date on the temperature dependence of the rate constant K(T)[4] were analysed by means of the formula taking into account the orbiting. This made it possible to determine the value of the interaction matrix element and to calculate the cross section (Fig. 2).

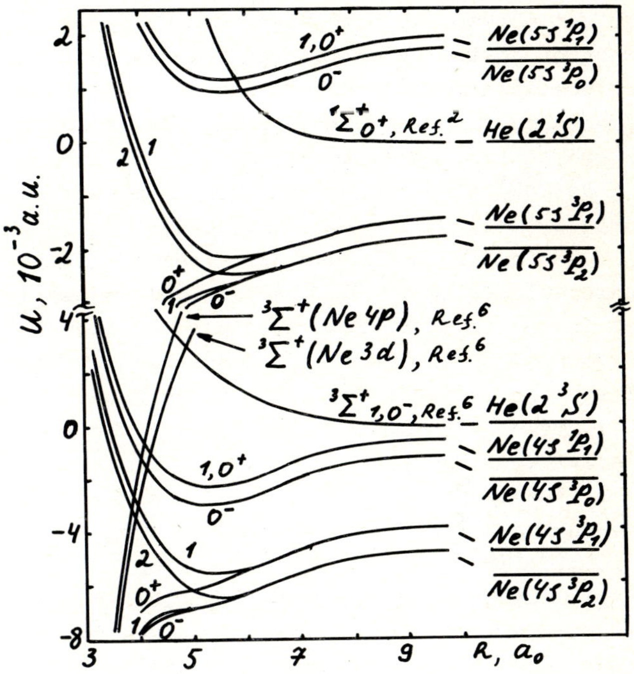

Fig. 1

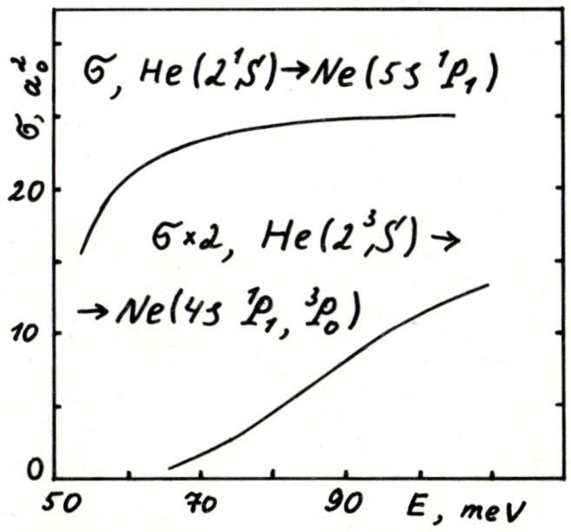

Fig. 2

As it seen from Fig. 1 the main feature of the collision (2) is determined by the potential barrier caused by the nonadiabatic interaction of terms[5,6]. Taking into account the vibrations in the terms $^3\Sigma^+$Ne(3d,4p) the transition probability in the collision (2) is

$$W = \frac{TP}{T + P - TP} \qquad (3)$$

where P is the probability of the transition between the term $^3\Sigma^+$Ne(3d,4) and the group of the terms Ne(4s). P is nearly independent of the collision energy. T is the probability of the nonadiabatic tunelling [7]. The temperature dependence of reaction (2) K(T) has been calculated by means of Eq. (3) and compared with experimental data [8], which made it possible to determine P = 0.25. The cross section is shown in Fig. 2.

The potential curves (Fig.1) and their symmetry offer an explanation of the discrepances in the population of the levels Ne(ns, $^{1,3}P_{o,1,2}$) in the collisions (1) and (2).

References

1.A.K.Belyaev et al.Opt.Spektrosk.49,633(1980).
2.C.H.Chen et al. J.Chem.Phys.61, 3095 (1974).
3. G.K.Ivanov, Teor.I Eksp.Chim.15,642 (1979).
4. V.A.Kostenko and Tolmachev, Opt. Spektrosk. 47, 1050 (1979).
5. A.K.Belyaev and A.Z.Devdariani, Opt.Spektrosk. 53, 610 (1982).
6. H.Haberland and P.Oesterlin, Z. Phys. 304, 11 (1982).
7. E.E.Nikitin, Opt. Spektrosk. 11, 452 (1961).
8. R.A.Zhitnikov et al., Zh. Eksp. Teor. Fiz. 80, 992 (1981).

SPECTROSCOPIC INVESTIGATION OF INELASTIC SCATTERING CHANNELS IN He^m - He COLLISIONS

D.V.Elakhovsky, Yu.V.Zaitsev, A.D.Khakhaev

Petrozavodsk State University, Petrozavodsk, USSR.

The excitation of helium by 0,2-1,1 Kev He^m atom beam has been investigated by the method of optical spectroscopy for the analysis of possible realisation channels öf inelastic processes with the participation of metastable atoms in reactions of the type:

$He(2^3S) + He(1^1S) \text{--} He(n^sX) + He(m^sX')$

n=3-12, m=1, 2, n; X=S, P, D; s=1,3.

Using the technique described in[1] we were able to carry out absolute measurements of the excitation cross sections for the 20 states of He in the energy range 0,2-1,0 KeV. The cross sections for some states investigated are plotted as the functions of the relative motion energy in Fig.1.

The technique proposed in[2] was used for measuring differential excitation cross sections. It was based on the connection of spectral-line profile intensity distribution and the differential scattering cross section. It enabled us to obtain Abel's integral equation which was solved with the help of an algorythm proposed in[3] to find the sought cross sections

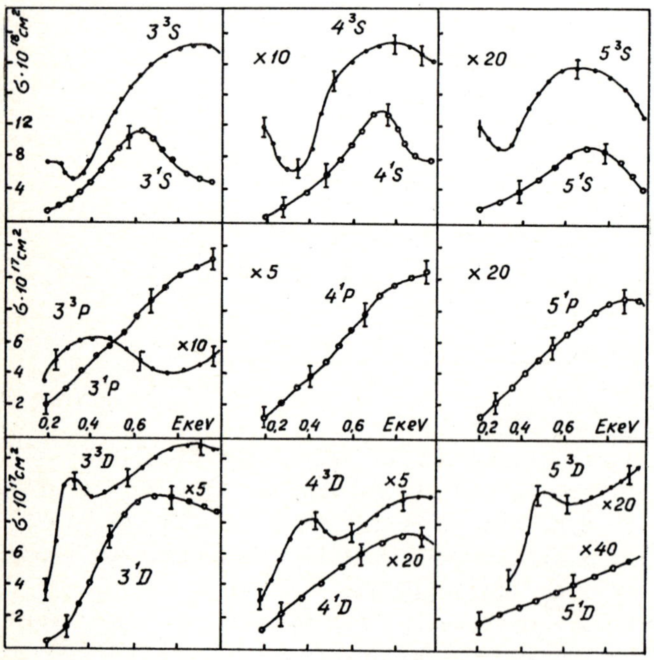

Fig.1: Excitation function for S, P and D states with n=3-5.

Apparatus distortions and spectral-line profile broadening induced by angular discrepancy of the projectiles didn't exceed 17%.

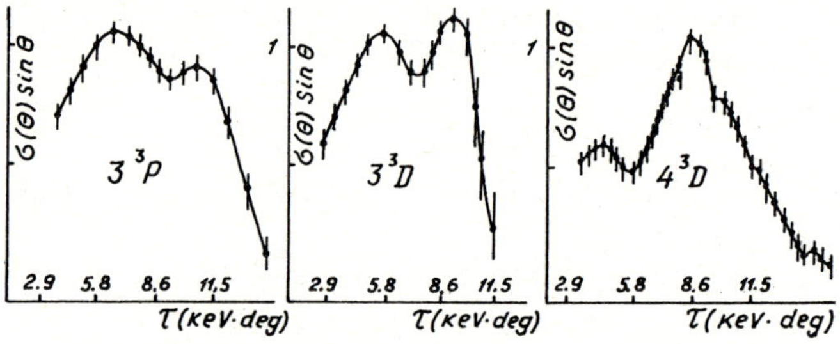

Fig.2: Differential excitation cross sections with a 1,1 KeV energy of metastable atoms

There is an essential difference in the dependence course of singlet and triplet excitation cross section on energy. In additional, the role of the final state orbit momentum is obvius both in the cross section magnitude and in their relative change in connection with growth of the principal quantum number. The analysis of the course of cross sections magnitudes dependens on their adherence to a definite series and on the principal quantum number inside the series showed that in the energy range 0,4-1,0 KeV $\sigma = A \cdot n^{-\alpha}$, while the values α for the states n^3D, n^1D, n^3S, n^1S are relatively equal to: 5,7±0,13; 4,58±0,12; 5,95±0,18; 4,80±0,22.

The presence of singlet lines in the He--radiation spectrum excited by metastable 2^3S atom indicates the possibility of the excitation channels correspoding to the excited states of both co-partners. The obtained values of differential cross sections and their comparison with another results[4] confirm this conclusion.

In accordace with the correlation diagram the formation of two excited states for the investigation reaction may be caused by the formation of quasimolecular states $g\Pi_g$ and $e\Pi_u$ on small internuclear distances owing to rotational coupling of terms with their subsequent interactional with series resulting in the sought excited states of atoms after their scattering

References

1. В.А.Гостев и др. Оптика и спектр., т.48, I980
2. В.А.Гостев и др. ЖТФ, т.50, в.I0, I980
3. Э.С.Парилис. ЖТФ, т.44, I974
4. K.T.Gillen. Inv.Paper X ICPEAC, 1978, p.473.

ATOM-ATOM AND ATOM-MOLECULE COLLISIONS IN SODIUM VAPOUR EXCITED BY LASER RESONANCE RADIATION AT $3^2S - 3^2P$ TRANSITION WAVELENGTH

Zh. L. Shvegzhda, S. M. Papernov, M. L. Jansons

P.Stuchka Latvian State University, Riga, Latvian SSR, USSR

Collision processes with participation of one or two 3^2P - sodium atoms have been studied. $Na(3^2P)$ atom concentration was determined by direct measurement of relative absorption, accounting for contour functions fine and hyperfine structure.

I. Excitation of high atomic states:

$$Na(3^2P) + Na(3^2P) \xrightarrow{Q_1} Na(n^2L) + Na(3^2S) \quad (1)$$

where $n^2L = 4^2S, 3^2D, 4^2P, 5^2S, 4^2D, 5^2P, 6^2S, 5^2D, 7^2S, 6^2D$. Cross sections Q_1 for n^2L-levels lie within the range : $2\cdot 10^{-16}$ cm^2 (4^2D) - 10^{-20} cm^2 (7^2S)

II. Population of molecular $A^1\Sigma_u^+$ and $B^1\Pi_u$ states:

$$Na(3^2P) + Na_2(X^1\Sigma_g^+) \xrightarrow{Q_2} Na(3^2S) + Na_2(A^1\Sigma_u^+, B^1\Pi_u) \quad (2)$$

Using correlational diagrams and measurements of molecular fluorescence dependence on concentration of selectively excited $3^2P_{3/2}$ and $3^2P_{1/2}$ atoms, the dominating role of process (2) with respect to collisional photorecombination in $Na(3^2P) + Na(3^2S)$ collisions is demonstrated. Effective cross sections Q_2, averaged over the ensemble of non-excited molecules $Na_2(X^1\Sigma_g^+)$ $Q_2(A^1\Sigma_u^+) = (1.6\pm0.6)\ 10^{-14}$ cm^2 and $Q_2(B^1\Pi_u) = (1.4\pm0.4)\ 10^{-17}$ cm^2. The low value of the latter cross section is due to the energy threshold for process (2).

III. The stepwise mechanisms of excitation of the diffuse molecular band in the 420-450 nm region has been studied.

$$Na(3^2P) + Na_2(X^1\Sigma_g^+) \rightarrow Na(3^2S) + Na_2(b^3\Pi_u) \quad (3)$$

$$Na_2(b^3\Pi_u) + \hbar\omega_{las.} \rightarrow Na_2(^3\Sigma_g^+ \text{ or } ^3\Pi_g) \quad (4)$$

$$Na_2(^3\Sigma_g^+, ^3\Pi_g) \rightarrow Na_2(x^3\Sigma_u^+) + \hbar\omega_{dif.} \quad (5)$$

Mechanisms (3 - 5) are confirmed by quadratic dependence of band intensity on excitation power, and by strictly linear dependence on concentration of $Na(3^2P)$ and of $Na_2(X^1\Sigma_g^+)$.

The role of $Na_2(b^3\Pi_u)$ in collisional and radiative processes is discussed.

Results obtained are compared with results of [1-6].

References

1. M. Allegrini, G. Alzetta, A. Kopystynska, L. Moi, G. Orriols, Opt. Commun. 22, 329 (1977)
2. L. K. Lam, T. Fujimoto, A. C. Gallagher, M. M. Hessel, J.Chem.Phys. 68, 3553 (1978)
3. D. J. Krebs, L. D. Schearer, J.Chem.Phys. 75, 3340 (1981)
4. V. S. Kushawaha, J. J. Leventhal, Phys. Rev. 25, 570 (1982)
5. M. Allegrini, P. Bicchi, L. Moi, in: Abstracts of the 8 th International Conference on Atomic Physics, Göteborg A 88 (1982)
6. J. Huennekens, A. Gallagher, in: Abstracts of the 8 th International Conference on Atomic Physics, Göteborg A89 (1982)

NEW APPROACH TO MULTI-STATE PROBLEM AND APPLICATION TO LASER INDUCED TRANSITION: $Sr(5p\,^{1}P)+Ca(4s^{2}\,^{1}S)\longrightarrow Sr(5s^{2}\,^{1}S)+Ca(4d\,^{1}D)$

H. Yagisawa

Div. of General Education, Takachiho Univ., Ohmiya 2-19-1, Suginamiku, Tokyo

Theory

Making a theoretical study of atomic collision is reduced to solve n-channel coupled equations,

$$\dot{a}_k = i\sum_{m}^{n}{}' u_{km}\, a_m \quad , \quad a_k(t=-\infty)=\delta_{k1} \quad . \tag{1}$$

With unitary matrix T which satisfies

$$\dot{T} = i\begin{pmatrix} 0 & \cdots & u_{1n-1} \\ \vdots & & \vdots \\ u_{n-11} & \cdots & u_{n-1n-1}\end{pmatrix} T \quad , \tag{2}$$

the equation (1) is reduced to "2-channel" as

$$\dot{a}_n = iVx, \qquad \dot{x} = iV^{+}a_n \tag{3}$$

where $V=(u_{n1},\ldots\ldots,u_{nn-1})$ and $x = T^{+}\begin{pmatrix}a_1\\ \vdots \\ a_{n-1}\end{pmatrix}$.

From (3) we have

$$\dot{a}_n = -V\int_{-\infty}^{t} V^{+}a_n\, dt' , \tag{4}$$

which is just the result of 2-channel problem[1] except that V is a vector($V\int V^{+}$ is a scalar). Introducing such a complex phase as to satisfy

$$D = -iV\exp\left(i\int_{-\infty}^{t} D\right)\int_{-\infty}^{t} V^{+}\exp\left(-i\int_{-\infty}^{t'} D dt''\right)dt' \quad , \tag{5}$$

the solution becomes

$$a_n = \exp\left(-i\int_{-\infty}^{t} D dt'\right), \qquad \begin{pmatrix}a_1\\ \vdots \\ a_{n-1}\end{pmatrix} = -T^{+}\int_{-\infty}^{t} V^{+}a_n dt' \quad . \tag{6}$$

Finally ,since T is just the solution of the n-1 states we can say " With the knowledge of n-1 states n-channel is solved".

Application

The title process is an example of 3-state problem[2] and the Hamiltonian, states and the coupled equations are as followings:

$$H=-eE\cos(wt)(y_A+y_B)+\frac{e^2}{R^3}(x'_A x'_B+y'_A y'_B-2z'_A z'_B) \quad , \tag{7}$$

$$|1\rangle = |a_2\rangle\,|b_1\rangle \ , \ |2\rangle = |a_1\rangle\,|b_2\rangle \ , \ |3\rangle = |a_1\rangle\,|b_3\rangle \ ,$$

$$\dot{c}_1=iu_{12}c_2,\ \dot{c}_2=iu_{12}^{*}\,c_1+iu_{23}\,c_3,\ \dot{c}_3=iu_{23}^{*}\,c_2$$

where

$$u_{12}=\frac{2\sqrt{3}\,\mu^1\mu^2\exp(-idwt)}{3}\,R^{-3},\quad u_{23}=\frac{\mu^3 E\exp(idwt+idst)}{2}$$

$$\mu^1=\langle b_1|\, z_B\,|b_2\rangle\ ,\quad \mu^2=\langle b_2|\, z_B\,|b_3\rangle\ ,\quad \mu^3=\langle a_2|\, z_A\,|a_1\rangle\ ,$$

w is laser frequency, dw is the separation between $|1\rangle$ and $|2\rangle$ and ds is a detuning from $|3\rangle$.

After some calculations and neglect of very small terms we have

$$|a_3|^2 = 4|u_{12}\,u_{23}|^2\, G^2K_1^2(G)w^{-2}v^{-2}r^{-4} \quad , \tag{8}$$

with $G=rds/v+ru_{12}u_{12}R^3r^{-5}/v$ and K_1 is the second order modified Bessel function. Except the second term of G the formula (8) is same with that from the usual perturbataion theory which diverges at r=0. In the following figures a few examples of calculation are shown. The parameters used are $w=1954cm^{-1}$, $v=9.6\times10^4 cm/sec$, $\frac{4}{3}\left(\frac{\mu^1\mu^2}{dw}\right)^2=2.1\times10^4$ Å^6 and $u_{23}^2/dw=2.75\times10^{-4}cm^{-1}$.

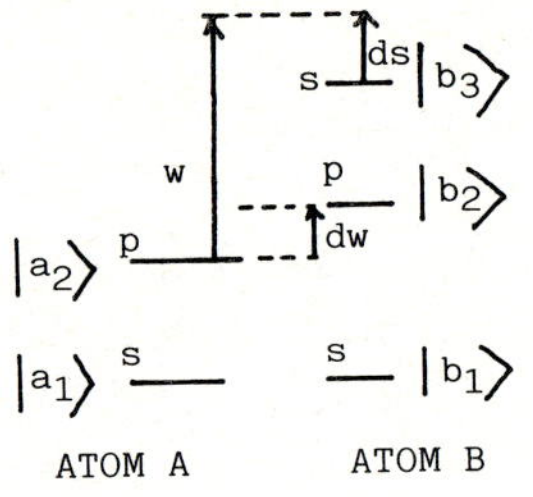

Fig. 1.

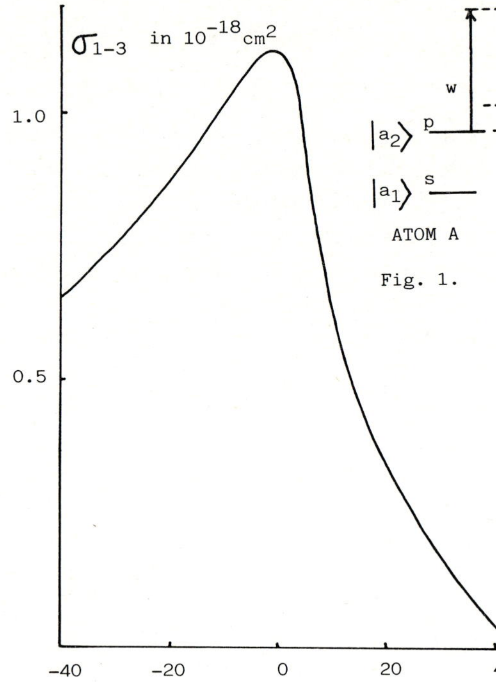

Fig. 3. Transition cross section vs detuning.

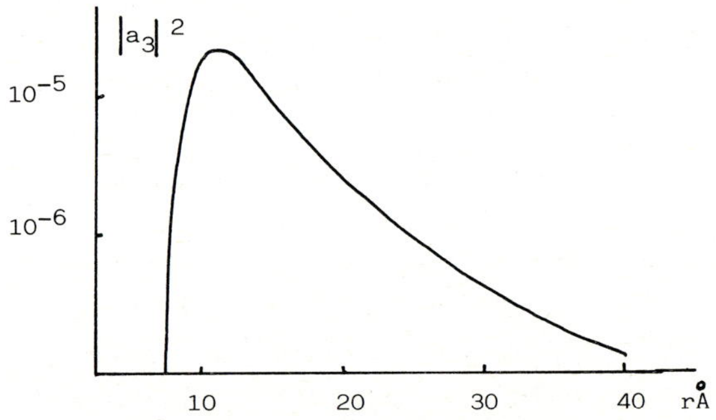

Fig. 4. Probability vs impact parameter r, velocity(v) is 9.6×10^4cm/sec.

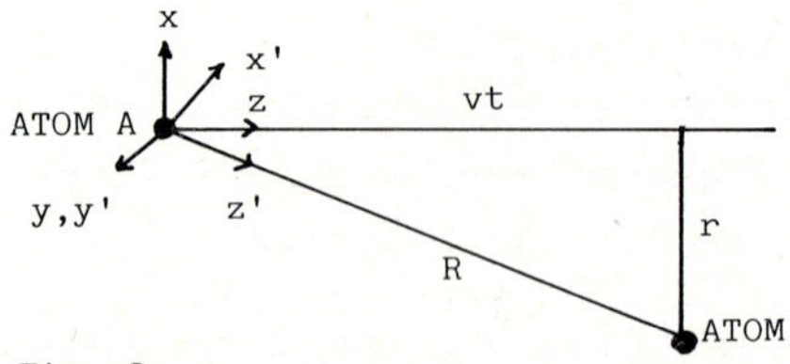

Fig. 2.

References

1. H. Yagisawa, J.Phys. B9, 2757(1976)
2. S. Harris et al, Phys. Rev. 33, 674(1974) and the related their papers.

ELASTIC SCATTERING OF SLOW HYDROGEN ATOMS ON IONS WITH LARGE IMPULSE TRANSFER

A K Kazansky, I V Komarov

Dept.of Theor.Phys., Institute of Physics, Leningrad State University, Leningrad,198904,USSR

Elastic scattering of neutral atoms at a fixed angle in laboratory frame is of importance for beam diagnostics of thermonuclear plasma. We studied theoretically the scattering of hydrogen atoms with the energy $E = 5 \div 15$ keV at the angles $\theta_{lab} = 5 \div 15^{\circ}$ on protons as well as multicharge ions C^{+6} and O^{+8}. In this case the relative velocity v is still in adiabatic region, while the impulse transfer $q = 2\mu v \sin(\theta_{lab}/2)$ is sufficiently high (here μ is reduced mass of the nuclei).

The process occurs due to the very close collisions between the nuclei of a projectile and a target because only in this way it is possible to change considerably the impulse of the incident particle. The process can be naturally described in the three - stage model:

i) the nuclei move on the straight line coinciding with the axis of collision,

ii) the suden flip of the internuclear axis at the scattering angle θ occurs near united atom,

iii) the nuclei move away on the straight line coinciding with the rotated axis of collision.

Obviously on the i) and iii) stages only radial nonadiabatic coupling should be taken into account.

As a result for elastic differential scattering $H + A^+$ one gets

$$\frac{d\sigma}{d\Omega} = \left[\frac{d\sigma}{d\Omega}\right]_{Ruth} \cdot W(v,\theta),$$

where $[d\sigma/d\Omega]_{Ruth}$ is the well-known Rutherford cross section for nuclei and $W(v,\theta)$ is electronic probability factor.

The simplest case is realized in H + p collisions, where evolution of the electronic system in adiabatic region can be treated in the two state ($1s\sigma_g$, $2p\sigma_u$) approximation. The populations of these states up to the moment of closest approach of the nuclei can be calculated with the help of usual formulae of resonant charge transfer. In the center-of-mass frame we arrive at

$$W^{Hp}(v,\theta) = (1+\cos^2\theta)/4 + \cos\theta \cos\left\{\frac{2}{v}\int_{-\infty}^{0}(E_g - E_u)\,dt\right\}/2.$$

In considered interval of the scattering angles $\cos\theta = 1 - \theta^2/2$, thus

$$W^{Hp} = (1+\cos(4.90/v))/2 - \theta^2\cos(4.90/v)/4.$$

When evolution of electronic shell can be described by the two states Landau-Zinner model, the electronic probability factor is

$$W^{LZ} = |P_{\ell_1}(\cos\theta)p + P_{\ell_2}(\cos\theta)(1-p)e^{iS}|^2.$$

Here p is the probability of transition, S is the Stueckelberg phase for zero impact collision, ℓ_1, ℓ_2 are angular momentum in the electronic states near united atom.

For scattering H on C^{+6} and O^{+8} it is necessary to take into account few molecular states. The corresponding calculations were performed by the method described earlier [1]. It was shown that in a wide range of scattering angles the electronic probability factor can be approximately factorized [2]

$$W(v,\theta) = W(v,0)\cdot y(\theta)$$

The velocity dependent function $W(v,0)$ was fitted as

$$W(v,0) = (1+\cos(a/v+b))/2$$

with the parameters a =8.05 a.u., b =0.4 a.u. for H + C^{+6} and a =8.49 a.u., b =0.4 a.u. for H+ O^{+8} collisions. The angle - dependent function $y(\theta)$ is practically the same for both reactions. It is shown in the table.

θ rad	0.125	0.195	0.265	0.335
C^{+6}	0.75÷0.82	0.46÷0.59	0.19÷0.37	0.03÷0.15
O^{+8}	0.75÷0.82	0.44÷0.51	0.20÷0.33	0.04÷0.14

References

1. Kazansky A K, Komarov I V Zh.Tehn.Fyz., 52, 1734 (1982)
2. Kazansky A K, Komarov I V ibid., 52, 2456 (1982)

ELASTIC DIFFERENTIAL CROSS SECTIONS FOR PROTON SCATTERING BY ATOMIC HYDROGEN*

J. T. Park, D. M. Blankenship, T. J. Kvale, J. L. Peacher, E. Redd, and E. Rille

Physics Department, University of Missouri-Rolla, Rolla, MO 65401

The differential ion-energy-loss spectrometer at the University of Missouri-Rolla provides both the high resolution in energy loss which permits positive identification of elastically scattered ions, and the high angular resolution, which makes measurements of differential cross sections at small scattering angles possible. The present measurements are of elastic differential cross sections for proton scattering from an atomic hydrogen target. The experimental method employed is discussed in Reference 1. The angular distributions of incident, elastically, and inelastically scattered protons are measured by recording the transmitted proton current as a function of the proton ion energy loss while pivoting the apparatus about the scattering center. The elastically scattered proton has undergone a small kinetic energy loss due to the recoil of the target atom even if no inelastic energy loss is involved. The recoil energy loss is calculated and then set by the controlling minicomputer prior to each measurement. The data is analyzed using techniques discussed previously.[1,2,3]

The differential cross sections for elastic scattering of 25 keV protons by atomic hydrogen is shown in Fig. 1. The curve shape of the measured cross section at 25 keV is typical for the experimentally obtained elastic scattering data. The error bars represent rms statistical errors. The differential cross sections are normalized using the Born approximation cross section for excitation of the n=2 state of atomic hydrogen by 200 keV protons[4] to establish the target hydrogen atom density. Systematic errors which arise from the normalization and data analysis process are not shown. The figure also includes the results of our Born approximation, and Eikonal approximation, and the multistate calculation (M.S.) of Wadehra and Shakeshaft.[5] Only the multistate calculation is in good agreement with our experimental results. An estimation of the elastic differential cross section based on the Glauber approximation,[6] however, points to a better agreement with the experimental data than with the Born and Eikonal results shown in the graph. This is probably the result of the failure of the Born and Eikonal calculations to adequately account for the effect of the various inelastic channels; particularly ionization and electron capture.

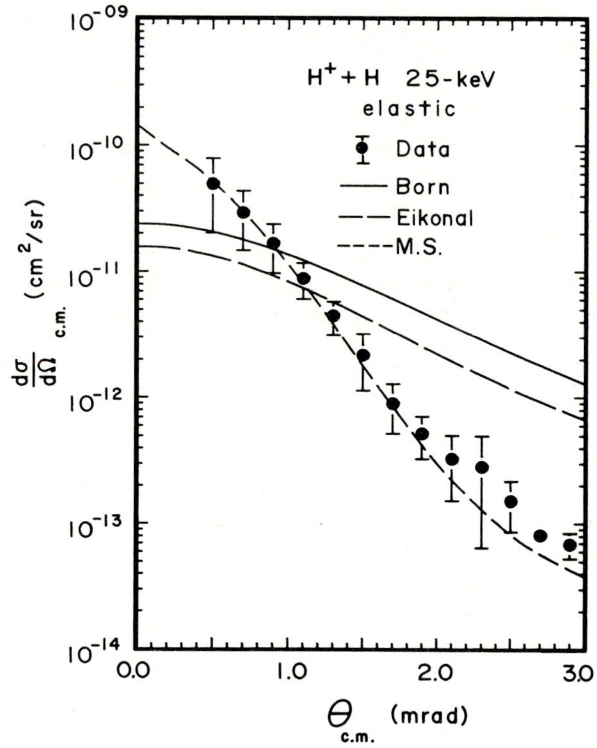

Fig. 1: Elastic differential cross section for proton-atomic hydrogen scattering for a proton with laboratory energy of 25 keV. The error bars indicate random errors only. The curve labeled M.S. is taken from Ref. 5.

References

1. J. L. Peacher, T. J. Kvale, E. Redd, P. J. Martin, D. M. Blankenship, V. C. Sutcliffe and J. T. Park, Phys. Rev. A 26, 2476 (1982).
2. J. T. Park, IEEE Trans. Nucl. Sci. NS-26, 1012 (1979).
3. J. T. Park, J. M. George, J. L. Peacher and J. E. Aldag, Phys. Rev. A 18, 48 (1978).
4. J. T. Park, J. E. Aldag, J. L. Peacher and J. M. George, Phys. Rev. 21, 751 (1980).
5. J. M. Wadehra and R. Shakeshaft, Phys. Rev. A 26, 1771 (1982).
6. V. Franco and B. K. Thomas, Phys. Rev. A 4, 945 (1971).

Acknowledgement

*This work was supported in part by the National Science Foundation.

MEASUREMENTS OF ORBITING STRUCTURES IN DIFFERENTIAL H^+ - He AND H^+ - Ar SCATTERING

M. Konrad and F. Linder

Fachbereich Physik, Universität Kaiserslautern
D-6750 Kaiserslautern, West Germany

The transition region from rainbow to orbiting scattering represents an important and interesting energy regime in low-energy atomic and molecular collision physics. In the present work, we have investigated this transition region by high-resolution differential scattering experiments using crossed beams. As first examples we have chosen the system H^+-He and H^+-Ar. The apparatus used has been described earlier[1], but considerable improvements have been made. A first report on these measurements has been published recently[2].

Fig. 1 (ref. 3) shows the position of the classical rainbow angle (minimum of the deflection function) on an E_{cm} - ϑ_{cm} plot for the system H^+-He. When we measure the energy dependence of the differential cross section for a fixed scattering angle (corresponding to a vertical cut in the diagram), we see that there is an infinite number of classical rainbows at E_1, E_2, E_3, ... converging to the critical energy E_{crit} where the minimum of the classical deflection function goes to infinity. In the vicinity of E_{crit} and below, we expect orbiting or shape resonances in the cross sections and a general change of the interference pattern known from rainbow scattering.

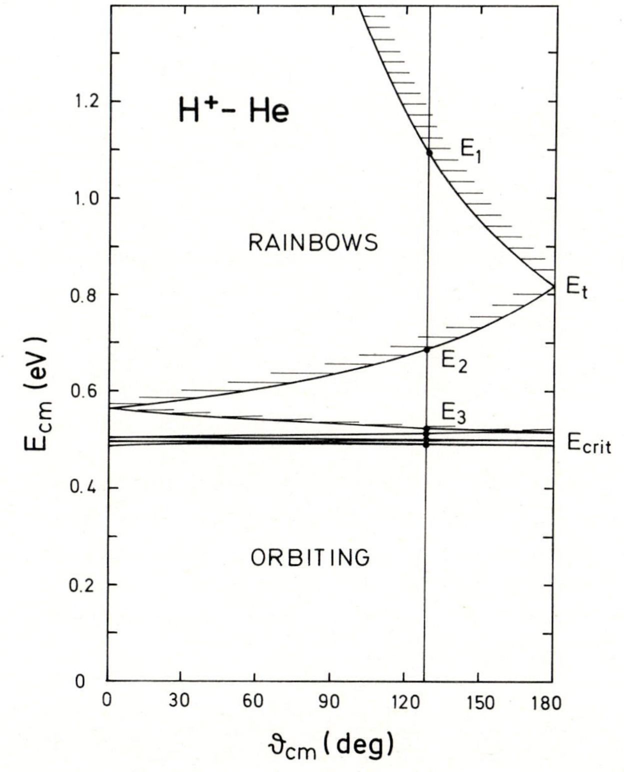

Fig. 1 (from reference 3)

Fig. 2 and 3 give two examples of the measurements. In the H^+-He measurement, we can identify two orbiting resonances with quantum numbers (n, l) = (0.28) and

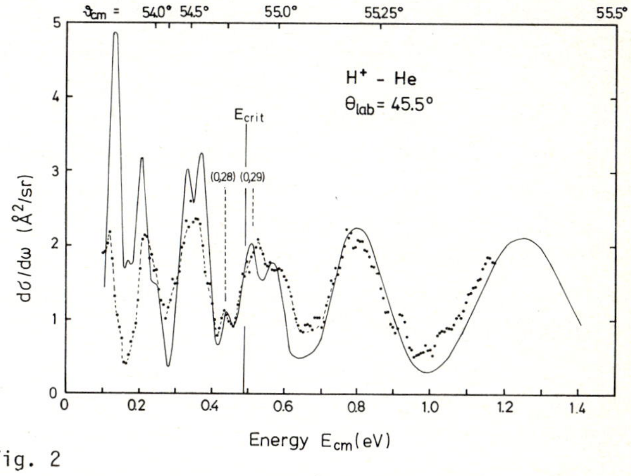

Fig. 2

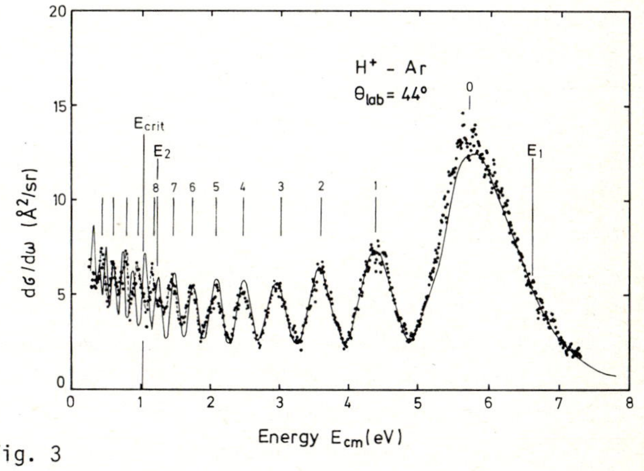

Fig. 3

(0.29) which have calculated widths of Γ = 13 and 36 meV, respectively. Further results and a detailed analysis will be presented at the conference. In addition we present refined measurements for rainbow scattering in the systems H^+-He, H^+-Ne and H^+-Ar which are used to test recently calculated potentials for these systems.

References

1. V. Hermann, H. Schmidt, F. Linder, J. Phys. B 11, 493 (1978)
2. M. Konrad and F. Linder, J. Phys. B 15, L 405 (1982)
3. H.J. Korsch and K.E.Thylwe, J. Phys. B 16, in press (1983)

SEMICLASSICAL REGGE-POLE DESCRIPTION OF LOW-ENERGY H^+-He SCATTERING

Karl-Erich Thylwe

Fachbereich Physik, Universität Kaiserslautern, D-6750 Kaiserslautern, W.-Germany

A series of successful H^+-He collision experiments by Konrad and Linder[1] provides the first detailed information of differential cross sections in the orbiting regime. Below the critical energy (E_{crit}) a very complex pattern in the energy dependence of the differential cross sections is observed. Above E_{crit} the structure gradually simplifies and beyond a certain transition energy (E_t) well known rainbow oscillations are visible.

Although H^+-He is a light system the semiclassical (SC) trajectory picture was found to be adequate[2]. The singularity ($E \leq E_{crit}$) or deep minimum ($E_{crit} < E < E_t$) of the classical deflection function gives rise to orbiting trajectories which produce a complicated interference pattern in the differential cross section. There are, however, three main drawbacks in that description: (i) the resonances are not rigorously accounted for; instead, pronounced multi-trajectory structures appear at approximately the same energies as broad quantum resonances

(ii) the contributions from higher-order orbiting trajectories are difficult to compute numerically

(iii) the infinite number of folded rainbows in the transition region gives rise to singularities in the classical cross sections and a complicated uniformication procedure is required.

In a recent paper[3] the author outlines a semiclassical (or semiquantal) method which in the low-energy regime avoids problems of uniformization and which further is capable of describing resonance phenomena in an unquestionable manner. The method utilizes the complex angular momentum (CAM) (or Regge-) poles of the elastic scattering matrix S, and the calculations involved in the method are based on semiclassical phaseintegral techniques. Some calculated pole positions for the H^+-He system, using a model potential fit to <u>ab initio</u> data[4], are shown in Fig. 1.

The scattering amplitude is conveniently divided into two parts:

$$f(E,\theta) = f_d(E,\theta) + f_0(E,\theta) \quad ,$$

the direct part f_d is described by the three well-known non-orbiting SC trajectories, while the orbiting part is described by a small number (2 - 5) of CAM poles. No uniformization procedures were used but could easily be applied to increase the accuracy. In Fig. 2 the differential cross section at fixed angle $\theta = 128°$ is analysed and compared with exact computations.

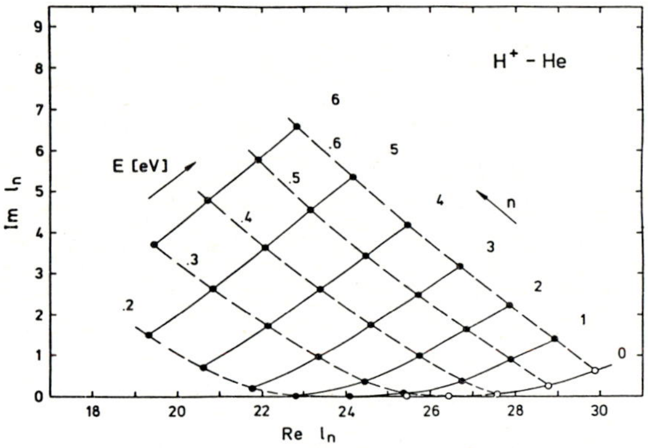

<u>Fig. 1</u>: Numerically calculated (SC) pole strings (---) and pole trajectories (——). The positions for the leading pole are given as small circles for clarity.

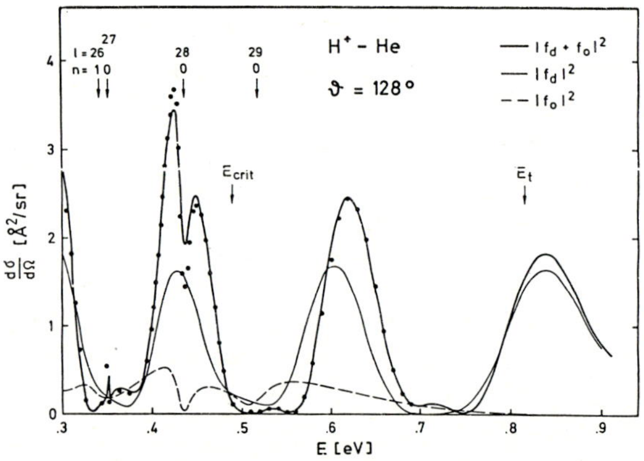

<u>Fig. 2</u>: Energy dependence of the SC CAM cross section together with its direct and orbiting components at fixed scattering angle $\theta = 128°$. Exact partial wave calculations are represented by dots. In the figure is indicated the positions of E_{crit} and E_t together with some resonances derived from the pole trajectories in Fig. 1.

It is important to notice that differential cross sections for "heavy" particles are most sensitive to the broad resonances (e.g. (l,n) = (28,0)) near the top of the centrifugal barrier. Sharp resonances due to tunnelling through the barrier and back again give negligible contribution (e.g. (l,n) = (27,0)).

<u>References</u>

1. Konrad M. and Linder F., J.Phys. <u>B15</u>, L405 (1982)
2. Korsch H.J. and Thylwe K-E. J.Phys.<u>B16</u>(to appear 1983)
3. Thylwe K-E., J. Phys. <u>B17</u>, (to appear 1983)
4. Kotos W. and Peek J.M., Chem. Phys. <u>12</u>, 381 (1976)

ELASTIC SCATTERING, EXCITATION AND IONISATION OF HELIUM BY PROTON IMPACT

C. Bergnes, D. Bordenave-Montesquieu, A. Boutonnet, R. Dagnac

ERA 598, Centre de Physique Atomique, Université Paul Sabatier, 31062 Toulouse (France)

The differential cross sections for elastic scattering, excitation and ionization for proton colliding with an helium target have been measured at several incident energies from 7 to 40 keV and over the angular range 0.25° to 1.5°.

A typical energy loss spectra is shown in figure 1.

The peak labelled A represents elastically scattered ions through an angle of 10'. Several processes contribute to peak B because the energy resolution is not suffisent : excitation of He n = 2 and n   3 and the beginning of ionization.

So we have been led to use a numerical treatment for resolving our spectra.

We have chosen a n^{-3} scaling law for the excitation of He on n   3 and a parabolic form for ionization (figure 2).

The experimental spectra obtained is a convolution of the energy loss distribution with the energy effect for the source distribution and dispersive effects introduced by the apparatus. These effects can be accurately approximated by a truncated Lorentzian function   , fitted on the elastic peak.

For the convoluted spectra :

$$F(y) = \int_{-\infty}^{+\infty} \left(k_1 \Delta E + \sum_{i=2}^{\infty} k_i \quad \delta(\Delta E - c_i) + g(\Delta E)\right) \mathcal{L}(y - \Delta E)\, d(\Delta E) \qquad (1)$$

We made a least square fitting of eq. (1) to the experimental data for from the ionization threshold where the influence of the excitation of He has not to be taken into account ; we thus obtained the aforementioned values of α β, β and γ . With these values we calculated ionization in the all spectra and deduced the excitation peak contribution k_2.

A very good agreement was found between the fitted curves and the experimental data in the spectra (figure 3).

Relative cross sections were obtained from the peak heights in the deconvoluted spectra.

The results were normalized using the probability of charge exchange, we have also measured.

A typical energy is presented on figure 4. The cross sections vary rather smoothly with energy.

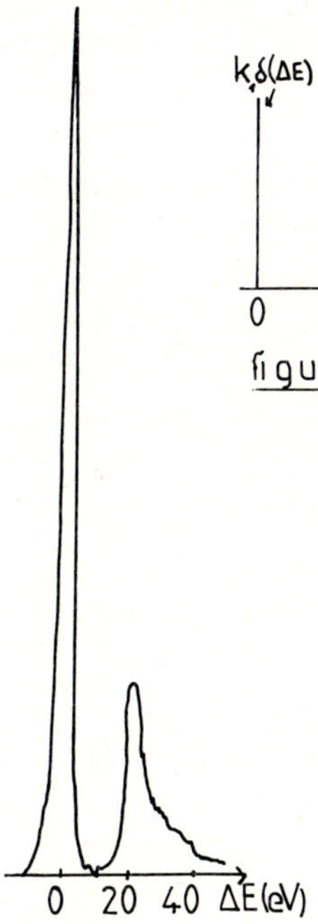

figure 1 : energy loss spectra
E = 25 keV
θ = 10'

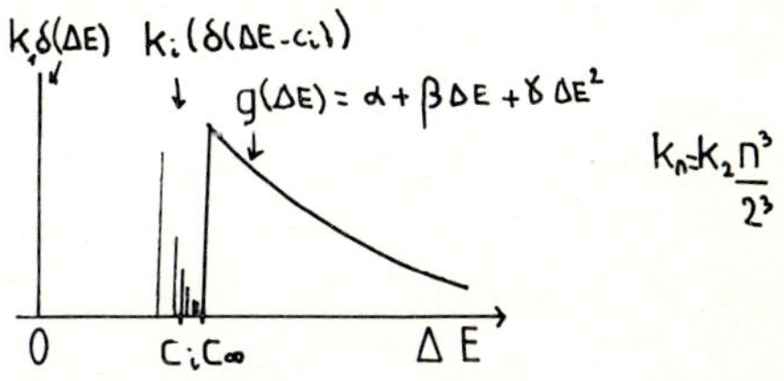

figure 2 : theorical spectra

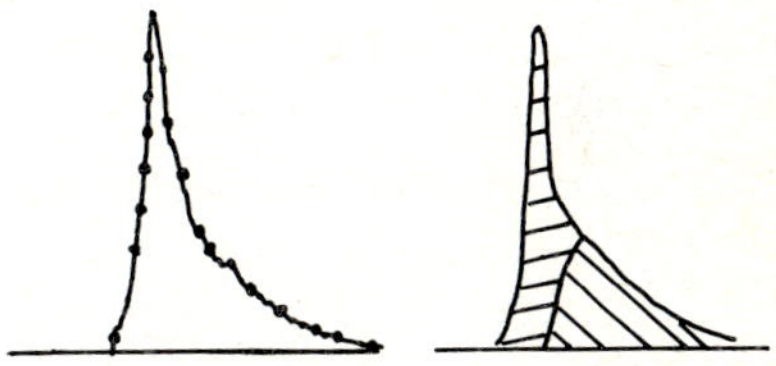

figure 3 :
— experiment
. calcul
▨ excitation
▧ ionization

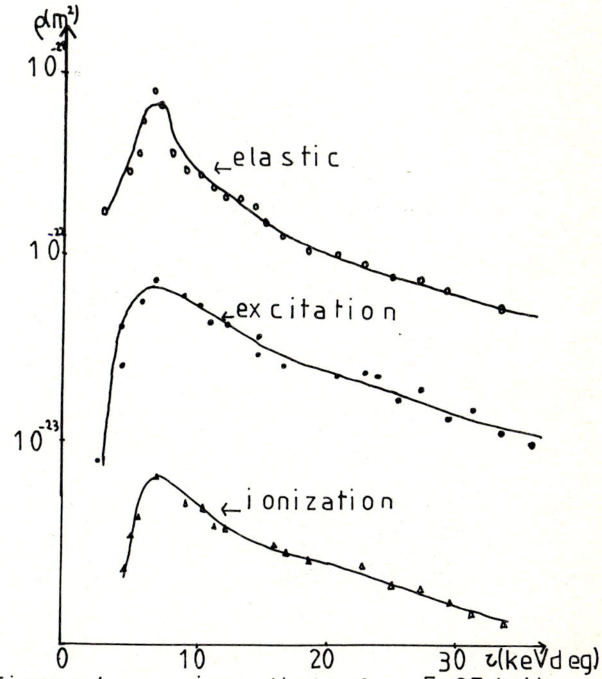

figure 4 : cross sections for E = 25 keV

ELASTIC SCATTERING AND CHARGE EXCHANGE IN He^+-He, H^+-Kr AND H^+-Xe COLLISIONS AT E_{cm} = 0.5 - 30 eV

P. Reinig, G. Bischof and F. Linder

Fachbereich Physik, Universität Kaiserslautern
D-6750 Kaiserslautern, West Germany

The present study is part of our program to investigate charge transfer reactions of ion-neutral systems in the eV range by high-resolution differential scattering experiments using crossed beams. As a prestep to the study of ion-molecule systems, we have selected three simple ion-atom systems representing different situations: symmetric energy-resonant (He^+-He), near-resonant endothermic (H^+-Kr) and near-resonant exothermic (H^+-Xe) charge transfer. Search of the literature shows that the understanding of the processes is quite unsatisfactory at low energies[1-3].

Fig. 1 gives a sample of our measurements for the He^+-He system. The measured cross section has been transformed into the CM system. Calculations based on available potential curves for this system are not in satisfactory agreement with the measurements.

The systems H^+-Kr and H^+-Xe are known to have large cross sections for charge transfer in the eV range[4]. In Fig. 2 and 3, we present some measurements of differen-

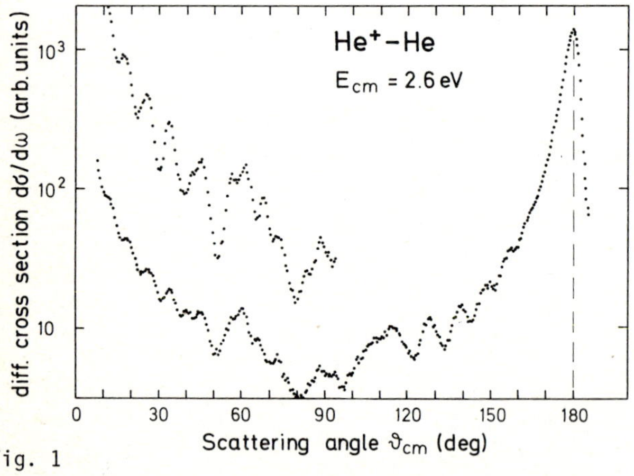

Fig. 1

tial cross sections for elastic scattering in these systems. In H^+-Kr, we observe rainbow scattering very similar to the H^+-Ar case, whereas the differential elastic cross section for the H^+-Xe system shows a quite different behaviour (a monotonic decrease with superimposed oscillations and a sudden fall-off for ≳ 500 eV · deg at the higher collision energies). A possible explanation has been proposed in reference 3, however, a more detailed analysis of elastic scattering and charge exchange at low energies seems necessary for these systems.

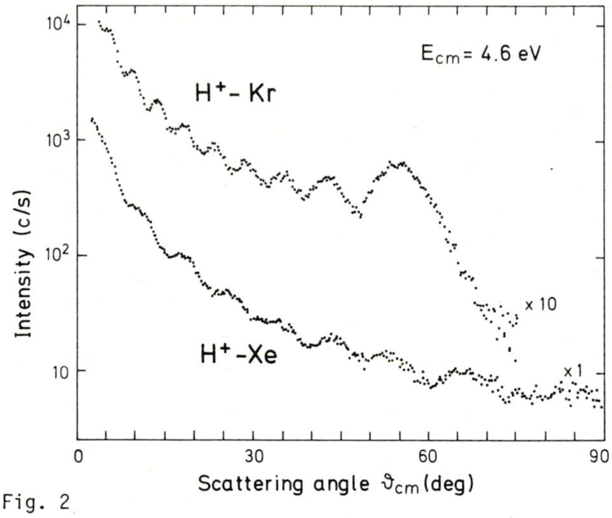

Fig. 2

References

1. C.R. Hsieh and M.L. Vestal, Abstracts XII. ICPEAC Gatlinburg/USA 1981 (ed. S. Datz), p. 619
2. H.P. Weise, H.P. Mittmann, A. Ding, A. Henglein, Z. Naturf. 26 a, 1112 (1971)
3. C. Kubach, V. Sidis, J. Durup, J. Phys. B 8 1129 (1975)
4. W.B. Maier II, Phys. Rev. A 5, 1256 (1972)

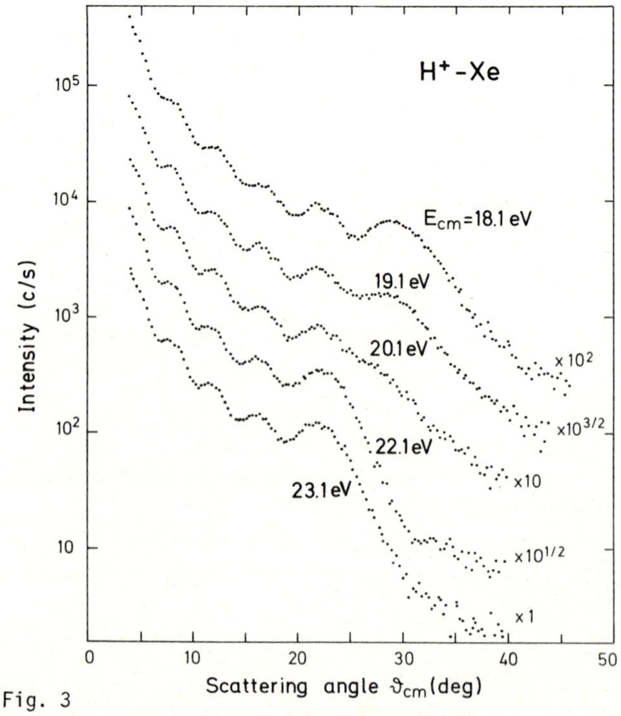

Fig. 3

POTENTIAL ENERGY SURFACES FOR DI-ATOMIC MOLECULES

H. Hartung, B. Fricke, and W.-D. Sepp

Physics Department, University of Kassel, D-35 Kassel, W.-Germany

For heavy ion scattering experiments it is interesting to have detailed information on the elastic scattering process, e.g. trajectories, scattering angles, and scattering cross-sections. These quantities can be obtained from the interatomic potential of the scattering system. For our potential calculations we use two different methods.

The first method is a density functional approach[1]. For a given internuclear distance $R = |\underset{\sim}{r}_1 - \underset{\sim}{r}_2|$ the total system energy is

$$E = V_{NN} + V_{Ne} + V_{ee} + T_e$$

where $V_{NN} = Z_1 \cdot Z_2 / R$ is the potential energy of the bare nuclei,

$$V_{Ne} = - \int \left(\frac{Z_1}{|\underset{\sim}{r}-\underset{\sim}{r}_1|} + \frac{Z_2}{|\underset{\sim}{r}-\underset{\sim}{r}_2|} \right) \rho(\underset{\sim}{r}) d^3r$$

the electron-nucleus interaction energy,

$$V_{ee} = \frac{1}{2} \int \frac{\rho(\underset{\sim}{r}) \cdot \rho(\underset{\sim}{r}')}{|\underset{\sim}{r}-\underset{\sim}{r}'|} d^3r d^3r' - \kappa_a \int \rho^{4/3} d^3r$$

the electron-electron energy. The first term is the direct part, the second is the exchange term in Slater-approximation with

$$\kappa_a = \alpha_x \cdot \frac{3}{2} \left(\frac{3}{\pi}\right)^{1/3}, \quad \alpha_x = 0.7.$$

$$T_e = \kappa_k \cdot \int \rho^{5/3} d^3r$$

is the kinetic energy with $\kappa_k = \frac{3}{10}(3\pi^2)^{2/3}$. For the electron density ρ of the system we take the sum of the densities of the separated atomic systems

$$\rho = \rho_1^{DFS} + \rho_2^{DFS},$$

where the atomic densities ρ_i^{DFS} are calculated from SCF Dirac-Fock-Slater wave-functions. The interatomic potential as function of R is given by

$$V(R) = E(R) - E_1 - E_2$$

where E_i are the total energies of the separated atoms.

The second method is based on self-consistent Dirac-Fock-Slater calculations of the two-centre many-electron system. Here the total system energy is

$$E(R) = \sum_i \varepsilon_i(R) - \frac{1}{2} \int \rho(\underset{\sim}{r},R) V_c(\underset{\sim}{r},R) d^3r - \frac{1}{4} \int \rho(\underset{\sim}{r},R) V_{ex}(\underset{\sim}{r},R) d^3r + V_{NN}(R).$$

ε_i are the electron eigenvalues of all occupied states, ρ is the electron density, V_c is the direct part of the electronic potential and V_{ex} is the local exchange potential in the Slater approximation.

For the numerical calculation of the Fock-, overlap-, and potential matrix elements in the Dirac-Fock-Slater code we use a spherical one-centre Gauß-Laguerre product integration with a two-centre coordinate transformation. Due to this transformation the integration points are concentrated around the nuclei in such a way that the point density $D(\underset{\sim}{r}) = D_1(r) \cdot D_2(\theta) \cdot D_3(\varphi)$, which is the inverse of the volume element of the integral, has the same structure and magnitude as the integral. The transformed coordinates are obtained by indefinite integration

$$U_1(r) = \int_0^r D_1(r') dr'$$

$$U_2(\theta) = \int_0^\theta D_2(\theta') d\theta'$$

$$U_3(\varphi) = \int_0^\varphi D_3(\varphi') d\varphi'$$

For the point density function in each coordinate direction we use linear combinations of analytically integrable functions with Gaussian or Wood-Saxon type peak structure. The hights and widths of these peaks correspond to the dimensions of inner shell atomic orbital electron densities. This procedure enables us to get the accuracy needed for the calclulation of structures in the interatomic potential which have magnitudes of about 1o eV.

Interatomic potential calculations and differential scattering cross-sections like I-I, Ni-Ni, and asymmetric systems like I-Cu are presented.

References

1. H. Hartung, B. Fricke, Physica Scripta T3, 244 (1983)

DEVELOPMENT OF AN ACCURATE NUMERICAL DIRAC-FOCK-SLATER PROGRAM FOR DI-ATOMIC MOLECULES

W.-D. Sepp, D. Kolb, H. Hartung, W. Sengler and B. Fricke

Physics Department, University of Kassel, D-35 Kassel, W.-Germany

Since several years a numerical Dirac-Fock Slater program for molecules is used which is able to calculate approximate energy eigenvalues[1]. This information is taken as an input in the physical interpretation of spectra, binding energies, bond lengths, etc. Due to two reasons the accuracy is poor. First the integrations performed so far are good only within a few percent and second the calculation of the potential in the program is performed using a Mullikan population analysis which also is good only to the order of percent.

In heavy ion physics correlation diagrams calculated with this method give a first interpretation of inner shell couplings in ion-atom collisions. As long as only inner shells of heavy colliding systems are discussed it is probably sufficient to use even simpler approximate calculations like the one discussed by Soff et al.[2] or the calculation of Eichler and Wille[3] with their variable screening model.

The smaller the impact energy becomes and the higher (less bound) the levels of interest are, the more is it necessary to do accurate selfconsistent field Dirac-Fock-calculations. Since Dirac-Fock-calculations are not yet possible we discuss here the development of an accurate numerical Dirac-Fock-Slater calculation for di-atomic collision systems which is able to give good wave-functions, energy eigenvalues, and total energies which are necessary for potential energy curves.

The calculations can be described briefly as follows. The relativistic one-electron Hamiltonian is given by

$$h = c\,\alpha\cdot p + \beta mc^2 + V(r)$$

where $c\,\alpha\cdot p + \beta mc^2$ are the kinetic and rest energy operators. V(r) is the potential energy operator which can be divided as usual into the Coulomb and exchange terms. This last part is obtained from the molecular charge density using Slater Xα-method with the exchange parameter 0.70. A variational method is used to find the molecular wave-function of symmetry orbitals constructed from atomic numerical basis functions. To avoid linear dependencies especially at small internuclear distances the atomic basis states are prediagonalized so that the linear dependent part can be taken out. This avoids spurious states. The Dirac-Fock-Slater and overlap matrix elements uo to now were evaluated by the discrete variational method[4]. Since this is only accurate to within a few percent we have changed the whole integration procedures in the case of di-atomic molecules. Since the azimuth variable along the internuclear axis is separable a 2-dimensional integration remains. After appropriate transformations we are able to integrate with a Gauß-Laguerre scheme with an accuracy better than 10^{-5} using only about 1000 grid points. (More details about this integration is given in our abstract on potential energy curves.)

The second essential improvement is the calculation of the inter-atomic potential from the molecular density. We avoid the Mullikan population analysis by using a numerical solution of the 2-dimensional Poisson-equation. In elliptic hyperbolic coordinates (η,Θ,Ψ) the 2-dimensional Poisson-equation takes the form:

$$\frac{\partial^2\phi}{\partial\eta^2} + \cosh\eta\,\frac{\partial\phi}{\partial\eta} + \frac{\partial^2\phi}{\partial\Theta^2} + \cos\Theta\,\frac{\partial\phi}{\partial\Theta} =$$

$$-a^2\,(\sinh^2\eta + \sin^2\Theta)\cdot 4\pi\rho .$$

This equation can be solved by discretisation where the resulting difference equation reduces the solution of the differential equation to an algebraic problem[5].

The boundary conditions are of Neumann type with $\frac{\partial\phi}{\partial\eta} = 0$ at $\eta = 0$ and $\Theta = 0, \infty$. For large η we have boundary conditions of Dirichlet type which can be determined by multipole expansion. Using 100 x 100 mesh points, an accuracy of $5\cdot 10^{-5}$ in the potential and $2\cdot 10^{-5}$ in the total energy has already been reached.

References

1. A. Rosén, D.E. Ellis, Chem.Phys.Lett. 27, 595 (1974); J.Chem.Phys. 62, 3039 (1975)
2. G. Soff, W. Greiner, W. Betz and B. Müller, Phys.Rev. A20, 169 (1979), and refs. therein
3. J. Eichler, U. Wille, B. Fastrup and K. Taulbjerg, Phys.Rev. A14, 707 (1976)
4. D.E. Ellis, G.S. Painter, Phys.Rev. B2, 2887 (1970)
5. P.N. Swarztrauber and R.A. Sweet, ACM Trans. Math. Software 5, 352 (1979)

CORRECTION TERMS FOR THE BETHE STRAGGLING EXPRESSION

Hans Bichsel

1211 - 22nd Ave E   Seattle WA 98112

Bethe[1] gave the stopping power M_1 (the first moment of the spectrum for the energy loss E in the collisions of particles with charge ze and velocity v with atoms Z) in a first, nonrelativistic approximation as $M_1 = (k/mv^2) \ln (2mv^2/I)$, where $k = 4\pi z^2 e^4 N Z$, I is the mean excitation energy and N the number of atoms per unit volume. He noted that correction terms of order $1/v^2$ would be needed. In the same approximation, he also gave the second moment describing the straggling:

$$M_2 = k\,[1 + (4/3)\,(\langle K\rangle/mv^2)\,\ln (2mv^2/I_1)] \qquad (1)$$

where $\langle K\rangle$ is approximately the kinetic energy of the electrons in the atom, per electron, and I_1 is weighted differently from I.

The correction terms to this expression for K-shell electrons described by screened hydrogenic wavefunctions have been calculated. Two further terms proportional to v^{-2} and three terms proportional to v^{-4} have been obtained. Using Walske's notation[2], $\eta = mv^2/2\epsilon$, $\theta = I_K/\epsilon$, where $\epsilon = R_y\,(Z-0.3)^2$ and I_K is the observed ionization potential for the K-shell, we obtain for $\eta \gg 1$ the expression:

$$M_2(\theta,\eta)/k = 1 + \{[4/3 + S(\theta)]/4\eta\} \ln 4\eta - 2\,L(\theta)/4\eta - [D(\theta) + H]/4\eta - [b + h + P(\theta)]/(4\eta)^2 \qquad (2)$$

with $S(\theta) = \int_\theta^\infty E\, f(E,0)\, dE$,

$L(\theta) = \int_\theta^\infty E\, f(E,0) \ln E\, dE$ and

$P(\theta) = \int_\theta^\infty E^3\, [df(E,Q)/dQ]_{Q=0}\, dE$

where f(E,Q) is the generalized, f(E,0) the dipole scillator strength. Note that[3] $\ln I_1 = L(\theta)/S(\theta)$. Bethe found that for a hydrogen atom $S(\theta) = 4/3$. The expressions for $D(\theta)$, H, h and b are more complex. For $\theta=0.9$ ($Z>50$), I calculated S=0.97, L=0.857, D=1.57, H=2.0, P=37.3, h=21.3 and b=6.0.

In the use for comparison with experiments[4], only the terms with $S(\theta)$ and $L(\theta)$ have been included in the calculations of M_2 so far.

Various approximations for Eq.(2) are compared in the figure:

a) Bohr approximation: $M_2/k = 1$

b) Bethe approximation: using $S(\theta)$ and $L(\theta)$

c) setting $b=h=P(\theta)=0$

d) including all correction terms given.

Finally, the calculation without approximations[5] is given by e).

It is easy to understand why the Bethe approximation b) was not successful: the correction terms $D(\theta)$ and H are of the same order of magnitude as $L(\theta)$.

While for the stopping power the sum $D(\theta)$ for all shells is equal to zero, this is not the case for M_2. The determination of $L(\theta)$ (or I_1) therefore is not sufficient for a determination of M_2 for large particle velocities.

References

1. H. A. Bethe, Rev. Mod. Phys. 9, 245 (1937)

2. M. C. Walske, Phys. Rev. 88, 1283 (1952)

3. U. Fano, Ann. Rev. Nucl. Sci 13, 1 (1967)

4. H. Bichsel, Phys. Rev. A 9, 571 (1974)

5. H. Bichsel, Phys. Rev. B 1, 2854 (1970)

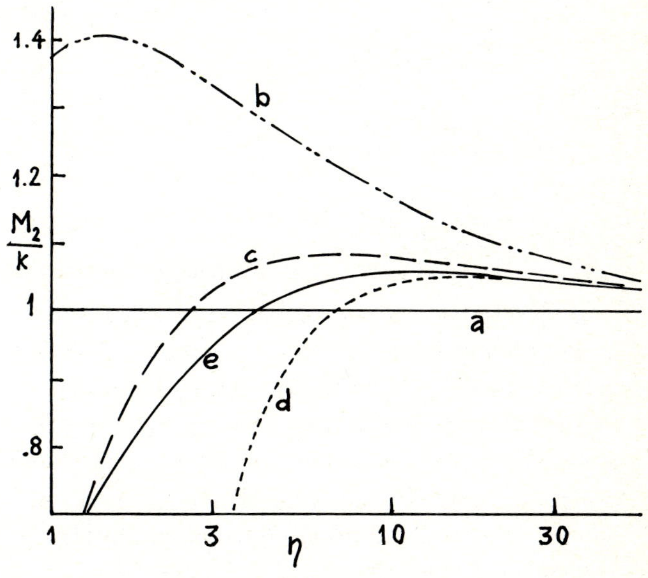

PROTON IMPACT EXCITATION OF HYDROGEN ATOM

S. Saxena, G.P. Gupta* and K. C. Mathur

Department of Physics, University of Roorkee, Roorkee (U.P.) 247667 India.
* Department of Physics, S.D. (P.G.)College, Muzaffarnagar (U.P.) 251001 India.

The differential cross sectionsfor the $1s\to 2s$ excitation of hydrogen atom by proton impact are computed at intermediate and high energies, in the distorted wave approximation. A comparative study is made by taking distorted waves (Coulomb waves) either in initial channel[1] or in final channel[2] or in both the channels[3].

The differential cross section for a collision in which the target atom is excited from an initial state i to a final state f is given by[1-3]

$$\left(\frac{d\sigma}{d\Omega}\right)_{i\to f} = \frac{\mu^2 k_f}{4\pi^2 k_i}\left|T_{i\to f}\right|^2 \qquad (1)$$

where for the distortion in the initial channel,

$$T_{i\to f} = \Gamma(1+ia_i)\exp(-\pi a_i/2)\int d\vec{r}_2 \exp(i\vec{q}\cdot\vec{r}_2) \times {}_1F_1(-ia_i,1,ik_ir_2-i\vec{k}_i\cdot\vec{r}_2)\langle U_f(\vec{r}_1)|W|U_i(\vec{r}_1)\rangle \qquad (2)$$

in the final channel,

$$T_{i\to f} = \Gamma(1+ia_f)\exp(-\pi a_f/2)\int d\vec{r}_2 \exp(i\vec{q}\cdot\vec{r}_2) \times {}_1F_1(-ia_f,1,ik_fr_2+i\vec{k}_f\cdot\vec{r}_2)\langle U_f(\vec{r}_1)|W|U_i(\vec{r}_1)\rangle \qquad (3)$$

and in both the channels,

$$T_{i\to f} = \Gamma(1+ia_i)\Gamma(1+ia_f)\exp\left[-\pi(a_i+a_f)/2\right]\int d\vec{r}_2 \exp(i\vec{q}\cdot\vec{r}_2) \times {}_1F_1(-ia_i,1,ik_ir_2-i\vec{k}_i\cdot\vec{r}_2)\,{}_1F_1(-ia_f,1,ik_fr_2+i\vec{k}_f\cdot\vec{r}_2) \times \langle U_f(\vec{r}_1)|W|U_i(\vec{r}_1)\rangle \qquad (4)$$

With

$$W = \frac{Z-\delta}{r_2} - \frac{1}{r_{12}}.$$

$a_i = \mu\delta/k_i$ and $a_f = \mu\delta/k_f$. δ is a screening parameter[1] and μ is the reduced mass of the system. $\vec{k}_f$ and $\vec{k}_i$ are the momenta of the scattered and the incident proton respectively. $\vec{q} = (\vec{k}_i - \vec{k}_f)$ is the momentum transfer vector. ${}_1F_1$ is the confluent hypergeometric function. U_i and U_f are the atomic wave functions for the initial and final states respectively. $\vec{r}_1$ and $\vec{r}_2$ are the position coordinates of atomic electron and incident proton respectively. Z is the nuclear charge.

The integrals in the T-matrix elements are calculated analytically following the technique of Nordsieck[4].

Figure shows the differential cross section(DCS) for the $1s\to 2s$ excitation of H atom at an incident proton energy of 100 keV. From the figure, it is seen that all the calculations (Coulomb wave in the initial channel (IC) with $\delta = 0.346$ and 1.0; Coulomb wave in the final channel (FC)[5] with $\delta = 1.0$ and Coulomb waves in both channels (BC) with $\delta = 1.0$) give higher cross sections compared to first Born (FB, $\delta = 0.0$) results at large scattering angles. The Born calculation highly underestimates the cross section in the large angle region.

Detailed results will be presented at the Conference.

One of us (SS) thanks UGC ,India for the award of SRF.

References

1. B.R.Junker, Phys. Rev. A 11, 1552(1975).
2. S.Geltman and M.B.Hidalgo;J.Phys.B4,1299(1971).
3. G.P.Gupta and K.C.Mathur,J. PhysB12,1733(1979).
4. A.Nordsieck,Phys.Rev.93, 785(1954).
5. Shyamal Dutta and S.C.Mukherjee,Phys.Rev. A23,1780(1981).

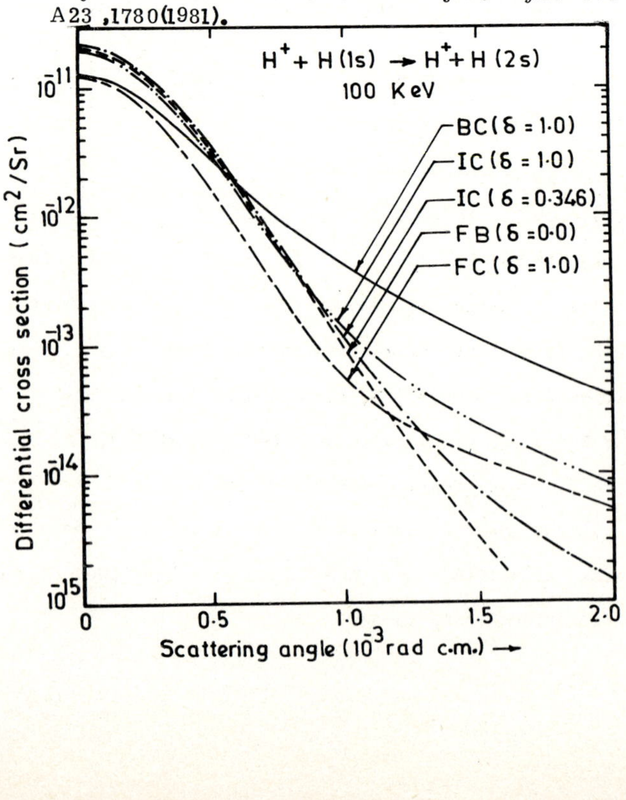

EXCITATION OF HE I TRIPLET STATES BY PROTON IMPACT

A.S.Aynacioglu, G.von Oppen, G.Weber

Institut für Strahlungs- und Kernphysik, Technische Universität Berlin

We measured magnetic and electric depolarization of impact radiation emitted by He atoms after excitation by 2-20 keV hydrogen ion-He atom collisions. In particular, we wanted to investigate the excitation of He I triplet states by proton impact. According to the spin conservation rule of Wigner, these excitation processes are forbidden. Nevertheless, several authors[1] observed significant intensities of He I-triplet lines after impact excitation with protons and partly questioned the validity of Wigner's spin rule.

This discrepancy between theory and experiment may be caused in various ways. We mention the following:

a.) Due to charge exchange, a small amount of H-atoms may be present in the ion beam. However, by studying the pressure dependence of excitation cross sections, the conribution of H-atom excitation has usually been determined and could be subtracted.

b.) Spin-orbit coupling causes a mixing of singlet and triplet states. This S-T-mixing is less than 1 % for the S, P and D levels of He I, but is extremely strong for the levels with orbital angular momentum L = 3. The decay lines of the latter ones are in the infrared region and are usually not observed. However, due to cascade processes the S-T-mixing of these levels may well be responsible also for the excitation of levels with L = 2.

Recently we have shown that magnetic and electric depolarization techniques are well suited to the determination of cascade conributions to the polarization of impact radiation[2,3]. In particular, the width (FWHM) $\Delta H \approx h/(g_J \cdot \mu_B \cdot \tau)$ of magnetic depolarization signals depends sensitively on the radiative lifetimes τ of the cascade levels. By investigating the depolarization of impact radiation induced by proton impact it could be shown that indeed cascade processes are the predominant mechanism for the excitation of triplet levels by proton impact. In detail, we investigated proton excitation of the $1s3d^3D$ level. A signal is shown in fig. 1. According to the width of the magnetic depolarization signal, the strong excitation of the $1s3d^3D$ level is mainly due to the mixing of the 1s4f $^1F_3$ and $^3F_3$ states.

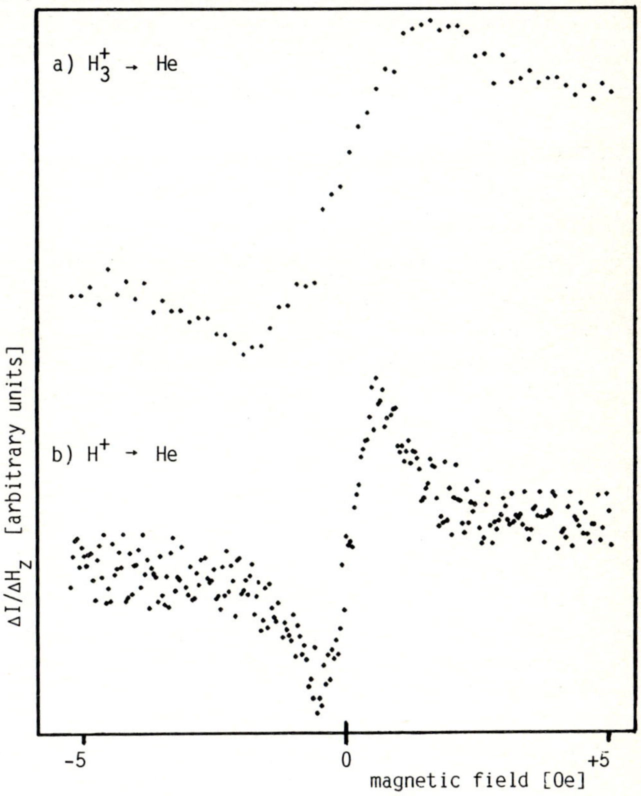

Fig. 1: Magnetic depolarization signals of the 588 nm (3 D-2) line of He I (impact energy: 15 keV)

References

1. E.W.Thomas: Exciation in Heavy Particle Collisions, Wiley Interscience (1972)
2. A..S.Aynacioglu, G.v.Oppen, W.-D.Perschmann, D.Szostak: J.Phys.B.14, 2611 (1981)
3. A.S.Aynacioglu, G.v.Oppen, W.-D.Perschmann, D.Szostak: Z.Physik A303, 97 (1981)

LIGHT ION + He COLLISIONS IN TIME-DEPENDENT HARTREE-FOCK THEORY*

K. R. Sandhya Devi and J. D. Garcia
Physics Department, University of Arizona, Tucson, Arizona 85721, USA

Collision phenomena resulting from the impact of light ions on He atoms are studied in the framework of Time-Dependent Hartree-Fock Theory.[1,2] The equations governing the evolution of single-electron orbitals are solved self-consistently on a discretized space-time mesh in the rotating coordinate frame. Excitation and charge exchange cross-sections are obtained by projecting the asymptotic, time-evolved state onto final channel states which are atomic states with appropriate translation factors.

Transformation to a rotating coordinate system gives rise to a Coriolis coupling term which breaks the axial symmetry with respect to the internuclear axis. Imposition of this symmetry not only results in underestimating transitions to certain classes of excited states, but also renders the results of calculations ambiguous at incident speeds greater than the characteristic speeds of the electrons in the target atom. In this work, the effect of the Coriolis term on the transition rates is carefully investigated.

Excitation and charge transfer cross-sections are compared with basis expansion results and with experimental data.

*Work supported by Grant No. 815301 at the National Science Foundation.

References

1. K. R. S. Devi and S. E. Koonin, Phys. Rev. Lett. 47, 27 (1981); K. C. Kulander, K. R. S. Devi and S. E. Koonin, Phys. Rev. A25, 2968 (1982).
2. K. R. S. Devi and J. D. Garcia, J. Phys. B. (to be published).

A SEVERAL-ELECTRON ATOMIC-BASIS CALCULATION OF Be^{+}(2s-2p) EXCITATION IN Be^{+}(2s)-He($1s^2$) COLLISIONS

Svend Erik Nielsen[+] and John S. Dahler[++]

[+]Chemistry Laboratory III, H.C.Ørsted Institute, University of Copenhagen, 2100 Copenhagen Ø, Denmark
[++]Department of Chemistry, University of Minnesota, Minneapolis, Minnesota 55455, USA

The present investigation includes a numerical test of our recently proposed several-electron theory of collisional excitation in quasi-one-electron systems.[1]

In previous studies we have used local model potentials to represent the interaction of the single valence electron and the closed-shell cores of the alkalilike projectile and a rare gas target. The electronic state has been expanded in one-electron single-center eigenstates, and scattering amplitudes have been obtained solving the close-coupled impact-parameter equations assuming a constant velocity rectilinear trajectory for the relative motion of the heavy particles.[2] This approach has offered a simple description of the direct excitation mechanism in which transitions are induced during large impact parameter collisions by the interaction of the valence electron and the rare gas atom. Indeed, the results obtained have been able to account for observed excitation cross sections and polarization of emission of low lying projectile states at impact energies about and above the Massey energy,[2] typically for energies from a few keV to 100 keV.

The conventional way of handling excitation at low energy (10 eV - 1 keV) is to represent the electronic state in terms of an adiabatic (diabatic) molecular basis, including core as well as valence orbitals in order to describe the highly distorted charge distributions produced by close encounters important at these energies. It is the purpose of the present contribution to provide support for our conviction that the atomic expansion approach used in the high energy theory of quasi-one-electron systems can be usefully extended to lower energies provided that the semi-empirical model potentials are replaced by the more realistic interactions generated by a several-electron theory.[1]

We adopt as a three-electron atomic basis antisymmetrized products of ETF-modified two-electron target (He) HF-states and one-electron model projectile (Be^{+}) states. We regain a quasi-one-electron theory assuming a frozen He($1s^2$) ground state, and we have obtained close-coupled solutions within a $Be^{+}(2s,2p_0,2p_{\pm1})$ basis. The low energy limit results in a potential coupling matrix

$$\underline{V}_{tot}= \underline{(1-Q)}^{-1}\ \{\underline{(1-Q)(-2/r_B+2J_o^B-K_o^B)(1-Q)} - \underline{(1-Q)H^A Q}\}$$

where Q, J_o^B and K_o^B are the projection, Coulomb and exchange operators, respectively, of the He ground state orbital, $H^A=-\frac{1}{2}\nabla^2+V^A$ and V^A is the electron-Be^{++} model potential. The high energy limit of the theory results in the simple electrostatic potential $V_{HF}=-2/r_B+J_o^B$ of the He ground state. The figures show the Be^{+}(2s-2p) excitation cross section and the polarization of the 2^2S-2^2P emission as functions of impact energy E (E_{CM}) together with the experimental results.[3] V_{tot}^{sym} stands

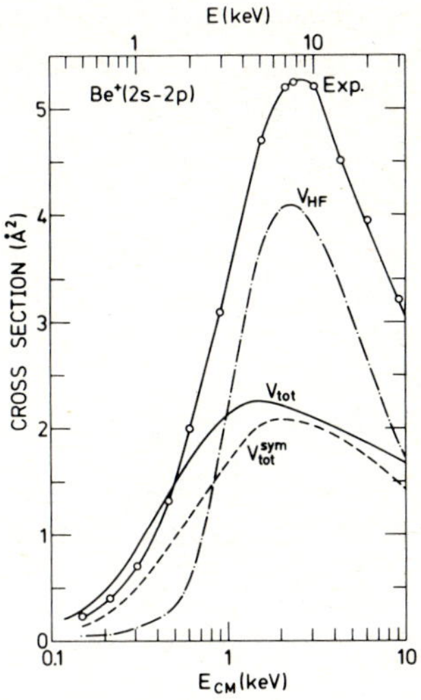

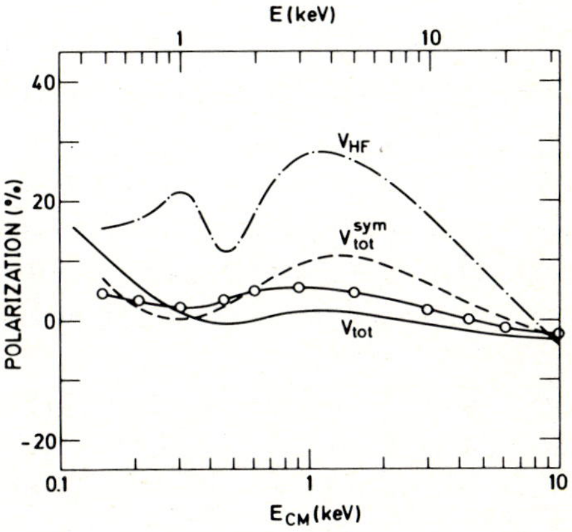

for a symmetrized approximation to V_{tot}. The well studied high energy limit of the theory, V_{HF}, accounts for the results at and above the cross section maximum but fails at lower energies. The low energy limit of the theory, V_{tot}, succeeds very well in the range E≃0.4-2 keV. The excitation probability and coherence parameter predictions obtained also compare well with experiments.

References

1. S.E.Nielsen and J.S.Dahler, Phys.Rev.A23, 1193 (1981)
2. N.Andersen and S.E.Nielsen, Adv.Atom.Molec.Phys. 18, 265 (1982).
3. N.Andersen, T.Andersen, J.Ø.Olsen and E.H.Pedersen, J.Phys.B 13, 2421 (1980).

RESOLUTION OF THE GLAUBER INCONSISTENCY FOR INELASTIC SCATTERING BY HEAVY PARTICLES

J. H. McGuire

Physics Department, Kansas State University, Manhattan, Kansas, USA 66506

Although the widely used Glauber approximation has been, on the whole, successful in comparison with observed data, some question has remained concerning the consistency of the Glauber approximation itself. In particular reduction from eikonal wavefunctions to the Glauber approximation has not been justified for inelastic collisions, where the choice of the Z axis necessary for the Glauber approximation itself is inconsistent with the general arguments supporting the more exact eikonal approximation.

The inconsistency is as follows. The eikonal approximation consists of replacing some exact, but unknown, wavefunction, ψ, by an unperturbed plane wave times an eikonal phase, $\exp\{-\frac{i}{v}\int^{Z} V dZ'\}$, so that the scattering amplitude is given by

$$f(q) = -\frac{M}{2\pi}\int d\vec{R}\, e^{i\vec{q}\cdot\vec{R}}\, V(R)\, e^{-\frac{i}{v}\int^{Z} V dZ'} \tag{1}$$

The Glauber approximation follows when $\vec{q}\cdot\vec{R} = \vec{q}\cdot\vec{B} + q_{\parallel} Z$ is replaced by $\vec{q}\cdot\vec{B}$, so that the Z integration can be done and

$$f(q) = -\frac{M}{2\pi}\int d^2B\, e^{i\vec{q}\cdot\vec{B}}\,[1 - \exp(-\frac{i}{v}\int_{-\infty}^{\infty} V dZ] \tag{2}$$

If the Z axis is chosen parallel to the initial momentum, $\vec{k}_i$, then $q_{\parallel}$ must be ignored in the Glauber approximation. This parallel momentum transfer, however, is never zero in inelastic collisions since $q_{\parallel} \geq q_0 = \Delta E/v$ where ΔE is the energy loss of the projectile. Furthermore, since the total cross section usually depends critically on q_0, taking $q_{\parallel}(\geq q_0) \to 0$ cannot be done consistently for inelastic collisions where $\Delta E > 0$.

Alternatively one may choose $\hat{Z}$ perpendicular to $\vec{q}$. Then, however, $\vec{B}$ does not correspond to a classical impact parameter unless $\hat{Z} = \hat{k}_i$ (as for elastic scattering in the forward direction). Indeed in inelastic collisions for $\theta = 0$, $\vec{q} = \vec{q}_{\parallel}$ so that with $\hat{Z}\cdot\vec{q} = 0$, $\vec{B}$ is perpendicular to the classical impact parameter to which it should correspond in the classical limit.

This apparent inconsistency may be resolved for heavy projectiles in the following way. An exact expression for the scattering amplitude is given by

$$f = -\frac{M}{2\pi}\int d\vec{R}'\, e^{i(\vec{k}_i - k_f\hat{R})\cdot\vec{R}'}\int d\vec{r}'\, u_f^{*}(\vec{r}')V(\vec{r}',\vec{R}') \times u_i(\vec{r}')\phi(\vec{r}',\vec{R}') \tag{3}$$

where $\psi_i = e^{i\vec{k}_i\cdot\vec{R}'}u_i(\vec{r})\,\phi\,(\vec{r}'\,\vec{R}')$. Here $\vec{q} = \vec{k}_i - k_f\hat{R}$ is the usual momentum transfer. For heavy incident projectiles, only contributions of O(1/M) come from scattering angles greater than O(1/M) except at very low projectile velocities. It is now reasonably assumed that changing $\vec{k}_f(=k_f\hat{R})$ to $\vec{k}_i$, a relative change of O(1/M), changes $\phi(\vec{r}', \vec{R}')$ by O(1/M) at most. Now further assume that f depends only on the magnitude of $\vec{q}$, i.e. $f = f(q)$, as is the case if there is azimuthal symmetry in the collision. There exists a $\tilde{q} = |\vec{k}_i - k_i\hat{\tilde{R}}|$ equal to $q = |\vec{k}_i - k_f\hat{R}|$. Changing $(k_f,\hat{R})$ to $(k_i\hat{\tilde{R}})$ in ψ_i^{+} corresponds in Eqn. (4) to changing $\vec{q}$ to $\tilde{q}$ and a change of O(1/M) in $\phi(\vec{r}', \vec{R}')$ (with no change in the state u_f onto which ψ_i^{+} was projected). Defining

$$f_{in} \equiv f(\vec{q}) = f(q) \text{ (with } k_f \neq k_i)$$

and

$$f_{e\ell} \equiv f(\vec{\tilde{q}}) = f(\tilde{q}) \text{ (with } k_f \to k_i),$$

then for $q = \tilde{q}$,

$$f_{in}(q) = f_{e\ell}(q) + O(1/M) \tag{4}$$

The scattering amplitude for the Glauber approximation may now be easily derived for inelastic collisions. Using the eikonal approximation, namely $\phi(\vec{r}',\vec{R}') = \exp\{-i/v\int^{Z} V(\vec{r}';\vec{B}',Z'')dZ''\}$ in Eqn. (3),

$$f(q) = -\frac{M}{2\pi}\langle u_f|\int d\vec{R}'\, e^{i(\vec{k}_i - k_f\hat{R})\cdot\vec{R}'}\, V(\vec{r}',\vec{R}') \times e^{-i/v\int^{Z'} V(\vec{r}';\vec{B}',Z'')dZ''}|u_i\rangle \tag{5}$$

where $\langle u_i|\ldots|u_f\rangle$ implies an integration over $\vec{r}'$. With $\vec{q} = \vec{k}_i - k_f\hat{R}$, one has $\vec{q}\cdot\vec{R}' = \vec{q}_{\parallel}\cdot\vec{B}' + \vec{q}\cdot\vec{Z}'$ where $q \geq \Delta E/v \neq 0$, and the Z' integration is not easily done. However, $\vec{q} = \vec{k}_i - k_f\hat{R}$ may be replaced by $\vec{\tilde{q}} = \vec{k}_i - k_i\hat{\tilde{R}}$ where $\hat{\tilde{R}}$ is chosen so that $\tilde{q} = q$. Then $\vec{\tilde{q}}\cdot\vec{R}' = \vec{\tilde{q}}\cdot\vec{B}' + O(1/M)$ at forward angles that are O(1/M) or smaller, since the magnitude of $k_i\hat{\tilde{R}}$ is now precisely the same as the magnitude of $\vec{k}_i$ and the directions differ by O(1/M) at most. Consequently Eqn. (5) becomes

$$f(q) = f(\tilde{q}) = -\frac{M}{2\pi}\langle u_f|\int d^2B'\, dZ'\, e^{i\vec{\tilde{q}}\cdot\vec{B}'}\, V(\vec{r}';\vec{B}',Z') \times e^{-i/v\int^{Z'} V(\vec{r}';\vec{B}',Z'')}|u_i\rangle$$

$$= -\frac{M}{2\pi}\langle u_f|\int d^2B'\, e^{i\vec{\tilde{q}}\cdot\vec{B}'}\,\{1 - e^{i/v\int_{-\infty}^{\infty} V(\vec{r}',\vec{R}')dZ'}\}|u_i\rangle \tag{6}$$

This is the standard Glauber expression.

δ-ELECTRON SPECTROSCOPY OF MULTIPLE IONIZATION IN H^+-AR COLLISIONS

J. Bossler, R. Hippler, H. O. Lutz

Fakultät für Physik, Universität Bielefeld, F. R. Germany

In binary collisions of protons with neutral target atoms multiple ionization events may occur. As an example we consider the following collision process

$$H^+ + Ar \rightarrow [H^+ + (n-1)e] + Ar^{n+} + e_\delta$$

where we can distinguish between ionization processes resulting in differently charged target ions by detecting emitted δ-electrons in coincidence with slow Ar^{n+} ions. (Information about the system within the square bracket can only be obtained in a triple coincidence experiment.)

In the following we report on first results of an Ar^{n+} ion - δ-electron coincidence experiment for impact of 350 keV protons on an argon gas target. The Ar^{n+} ions have been extracted from the collision region by a small (20V/cm) electric field. They have been detected in coincidence with emitted δ-electrons; their charge state was determined by their time-of-flight. The δ-electrons are detected perpendicular to the extracting electric field by a 45° parallel plate analyser (energy resolution ± 15%); their kinetic energy scale was corrected for the increase caused by the applied electric field. No significant change of the δ-electron angular distribution with and without electric field was observed.

Experimental results for the δ-electron ejection at 53° (with respect to the incident proton beam) are given in Fig. 1 versus the kinetic energy E_δ of the electrons. The three curves in Fig. 1 correspond to final ionic charge states 1+, 2+, and 3+, respectively. The peak around $E_\delta \cong 200$ eV which is clearly visible in the Ar^{3+} spectrum (and to a lesser extent in the other two spectra) is caused by Ar-L Auger transitions. The data have been made absolute by normalizing to previously published (total, i.e. without specification of the ionic charge state) doubly differential cross sections (DDCS) of Gabler et al.[1]. As can be seen from Fig. 1, production of Ar^{3+} is at low δ-electron energies about one order of magnitude smaller than for Ar^{2+}, which again is about one order of magnitude smaller than Ar^{1+}. This changes at large δ-electron energies. For instance, at about $E_\delta \cong 600$ eV, about 40% and 15% of all produced ions are Ar^{2+} and Ar^{3+}, respectively.

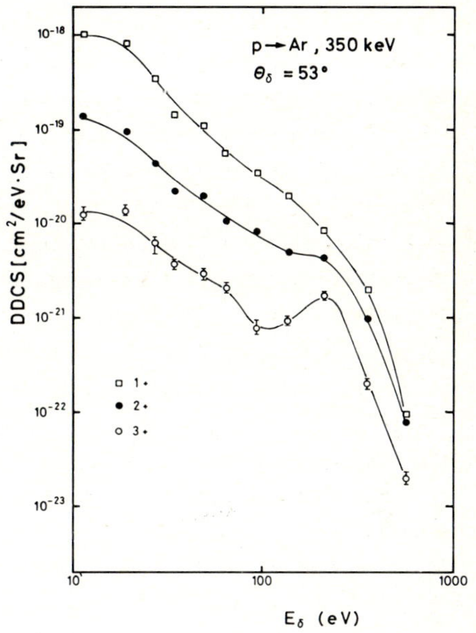

Fig. 1: Energy distribution of δ-electrons ejected at $\theta_\delta = 53^o$, in coincidence with □ Ar^+, ● Ar^{2+}, and o Ar^{3+} ions.

Multiply ionized target atoms may result from single ionization in inner shells and the subsequent decay of inner shell vacancies by Auger and/or Coster-Kronig transitions. However, the DDCS for L- and K-shell ionization in H^+-Ar collision is known to be one order of magnitude smaller than required to explain the experimental data[2,3]. More likely is therefore that multiple M-shell ionization is dominant. While ejection of δ-electrons with large kinetic energy becomes more likely for collisions with small impact parameters, the same holds for multiple ionization events. This gives some explanation why multiple ionization is more likely for large energies of ejected δ-electrons.

This work was supported by the Deutsche Forschungsgemeinschaft (DFG).

References

1. H. Gabler, N. Stolterfoht, U. Leithäuser, unpublished data given in M. E. Rudd, L. H. Toburen, N. Stolterfoht, At. Data Nucl. Data Tabl. 23, 405 (1979)
2. D. H. Madison, S. T. Manson, Phys. Rev. A 20, 825 (1979)
3. L. Sarkadi, J. Bossler, R. Hippler, H. O. Lutz, J. Phys. B 16, 71 (1983)

STUDY OF INNER QUASIATOMIC SHELLS (Z_U = 132-171) BY MEANS OF δ-ELECTRON SPECTROSCOPY IN ASYMMETRIC COLLISIONS

F. Güttner, W. Koenig, N. Lutz, B. Martin, H. Skapa, J. Soltani, H. Banda (MPI für Kernphysik, Heidelberg)
A.V. Ramayya (Vanderbilt University, USA)
F. Bosch, Ch. Kozhuharov (GSI, Darmstadt)

The emission of high energetic electrons (50-1000 keV) was measured for the systems Br→Pb, J→Au, Pb, Bi, U and Au→U. From electron-K x-ray coincidences K-L binding energy differences of the quasiatom were determined. Whereas in the range Z_u < 145 the results show good agreement with calculations in the adiabatic approximation, the 1s-2s binding energy difference obtained for the heaviest system (Z_u = 171) is already at 3.2 MeV/u incident energy smaller. The impact parameter dependence of the emission from the quasiatomic K-shell of the Au-U system indicates in addition a distinctly smaller increase of the electron density near the nuclei than expected. Fig. 1 and 2 show electron spectra measured for the systems Br→Pb (Z_u = 117) and Au→U (Z_u = 171). In Fig. 3 the binding energy difference between the 1s and 2s state is shown as a function of the united atom charge.

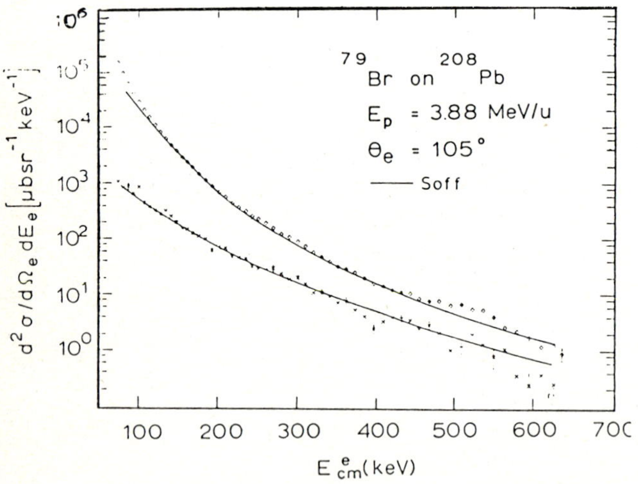

Fig. 1 electron spectra measured for the system Br→Pb (Z_u = 117)

◇ : total spectrum

x : electrons in coincidence with Pb-K x-ray

—: theory Soff et al.. Coupled channel calculations including screening

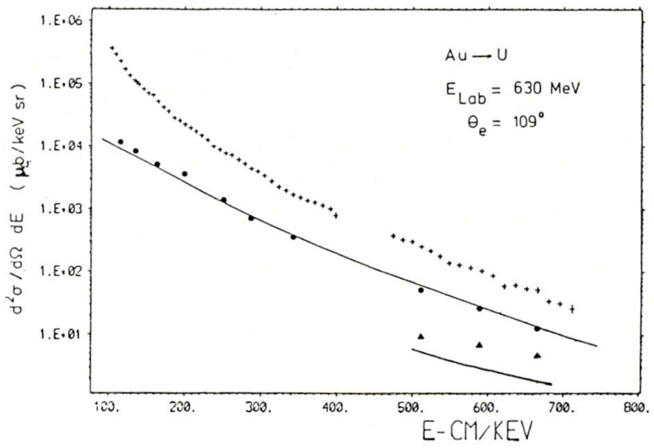

Fig. 2: Electron spectra measured for the system Au→U (Z_u = 171)

x : total spectrum

•, ▲ : contribution from the $2p_{1/2}$ (•) and 1s shell (▲) of the united atom measured via e^--K x-ray coincidences. Data are corrected for uncorrelated coincidences and vacancy sharing.

—: Theory Soff et al.. The calculation includes partially uncorrelated coincidences and represents therefore only an upper limit.

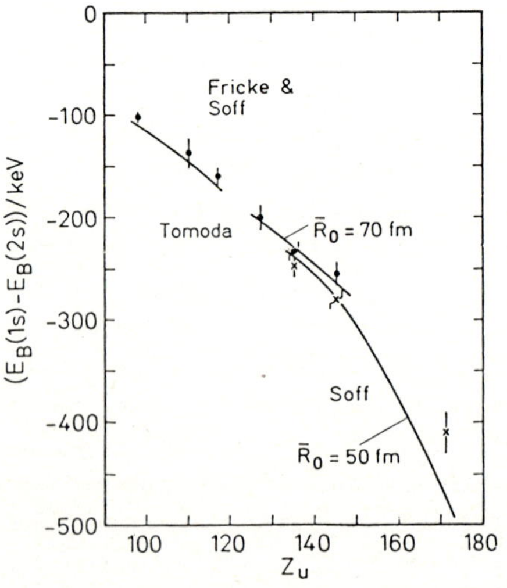

Fig. 3: Binding energy difference between the 2s and 1s state of the united atom.

$\bar{R}_0$: mean distance, where the electron emission takes place

—: binding energy differences calculated in monopol approximation including screening.

FAST ELECTRONS FROM SLOW ATOMIC COLLISIONS

Raúl A. Baragiola and Eduardo V. Alonso

Centro Atómico Bariloche* and Instituto Balseiro#
8400 - Bariloche, Argentina

We have studied very violent collisions between atoms at keV energies in which a large fraction of the center-of-mass energy E_{cm} is spent in the emission of one electron.

Previous observations on the continuum background of electrons ejected in atomic collisions have been restricted to relatively small electron energies. At high impact velocities, energetic electrons originate in a binary collision between the projectile and a target electron. At low velocities, the continuum has been interpreted in some cases to result from the decay of quasimolecular autoionization states during the collision or from direct coupling of a bound state with the continuum.

In this work we have used solid targets to increase the density of target atoms and so be able to study very unlikely processes where continuum electrons are ejected with very high energies. We do not expect in our case any effect specific of the solid state but rather that the electrons originate from single collisions. The electrons emitted with high energies can only travel a few nm inside the solid; it is expected that the energy degradation of the projectile in a region of this thickness is not very important. The experiments were made under computer control using a hemispherical electrostatic analyzer and momentum analyzed primary ions in the energy range 0.5-6 keV. The count time per channel was set proportional to electron energy to obtain better statistics in the high energy region. The spectra were then corrected for this, and also for the energy dependent analyzer transmission.

Fig.1 shows the high-energy electron tail for Ar on Au, with the ion energy as a parameter. We find electrons with extremely high energies, up to 60% of E_{cm}. The tail decays roughly exponentially but a high-energy cutoff is apparent in most cases. The large fraction of E_{cm} which goes into electronic excitation can only occur in nearly head-on collisions between the projectile and a target atom. The results for He projectiles show that fast electrons also occur for light-ion impact, albeit with a smaller cross section. The high energy cutoff occurs at somewhat smaller energies than for Ar ions at the same impact energy. In both cases, the energy cutoff grows approximately linear with ion energy.

Experiments with other targets (Be, Cu, Ce) show that the continuum tail is more intense and decreases less rapidly with energy as the atomic number of the target increases.

The origin of these fast electrons has not at yet been clarified. Any model for the process must take into account or explain the following crucial facts:

1) The transfer of a large fraction of the center-of mass energy to a single electron.
2) The occurrence of very high energy electrons both for light and heavy projectiles.
3) The increase of the number of high energy electrons and the cutoff energy with target atomic number.
4) The lack of evidence for Auger electrons signalling the ionization of deep inner shells.

Notes

* Comisión Nacional de Energía Atómica
# Comisión Nacional de Energía Atómica and Universidad Nacional de Cuyo.

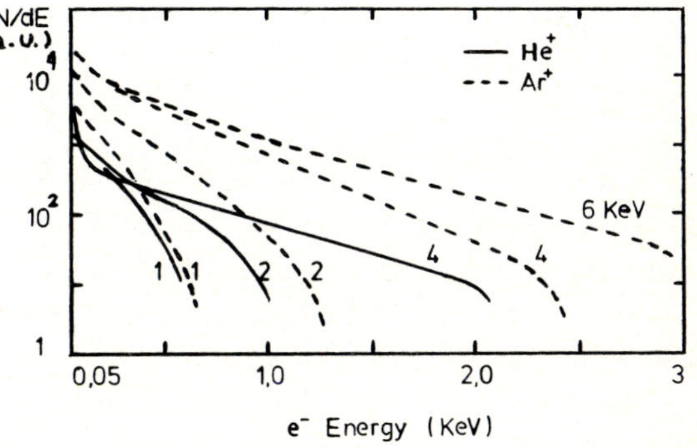

Fig.1. Electron spectra from Au bombarded with He^+ and Ar^+ ions at different incident energies.

CALCULATED DOUBLE DIFFERENTIAL IONIZATION CROSS-SECTIONS FOR MOLECULES BOMBARDED BY PROTONS

B. Senger

Laboratoire de Biophysique des Rayonnements et de Méthodologie, INSERM U.220, Université Louis Pasteur, 11, rue Humann, 67085 Strasbourg Cedex, France

Linear density distributions of secondary electrons ejected by α particles have been determined in nuclear emulsions of various compositions, which were submitted to the so-called activation treatment[1]. Over the whole incident particle's energy range (4-12 MeV), good agreement with these data could not be obtained by using single differential cross-sections (SDCS) of the Rutherford or the Binary Encounter Approximation (BEA) type, and this whatever the ejection angle θ chosen. A better agreement could be reached by means of a so-called mixed treatment, where quantum mechanical expressions were used for the inner, and a classical formalism for the outer shells[2]. In order to describe more realistically the heavy charged particle's track pattern, we have tried BEA type double differential cross-sections (DDCS)[3], as well as quantum mechanical DDCS[4], with the result that the calculations could not be made to agree with our experimental δ-ray distributions in the energy range considered.

We have therefore developed a quantum mechanical DDCS formula for the K shell and the L- and M-subshells on the basis of existing expressions[4,5]. For the N- and higher shells, this formula can be completed by the DDCS-BEA[3], thus constituting a DDCS mixed treatment (DDCS-MT)[6]. This treatment led to a satisfactory reproduction of our experimental results as well as of experimental DDCS in helium, argon and methane[6].

In this paper, it will be shown that the application of the treatment can be extended to the case of molecules by taking the water molecule as an example. The same approach leads to satisfactory results in the case of other molecules, notably methane and benzene. The binding energies of the five molecular orbitals of the water molecule have been taken into account as well as the breaking up of these orbitals in terms of H1s, O1s, O2s and O2p atomic orbitals[7], and the increase of the binding energies owing to the supplementary positive charge introduced by the passing ion[8]. On fig. 1 and 2, the calculated DDCS are compared with the corresponding measured DDCS[9] for 0.3 and 1.5 MeV incident protons, respectively. It appears that the accord between calculated and experimental data stays acceptable for ejection energies T as low as ≃ 10 eV, considering the experimental uncertainties[9]. The results are still improved at the small ejection angles θ if we take into account the final state interaction between the ejected electron and the proton[10].

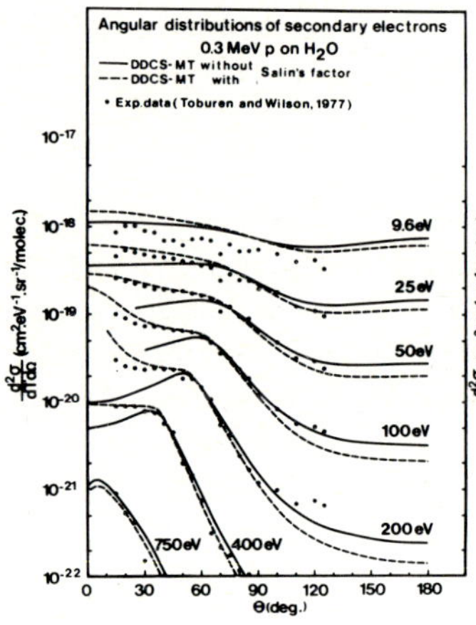

Fig. 1

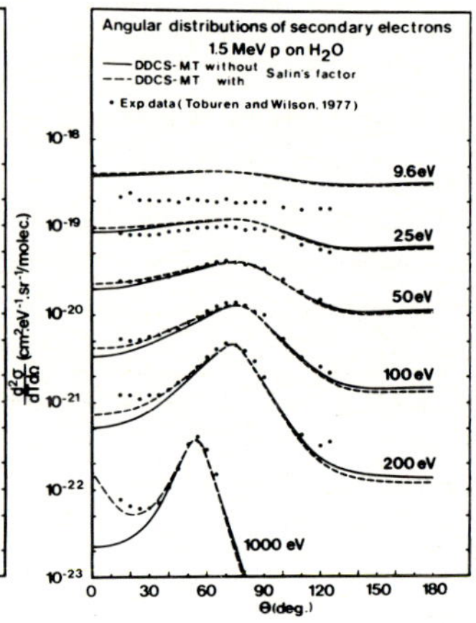

Fig. 2

This example demonstrated that molecular DDCS can be quite accurately reproduced by the DDCS-MT, provided that a detailled description of the target is introduced in the calculations. These determinations are especially useful in the evaluation of the spatial distribution of the secondary electrons and of the associated energy deposition, e.g. in biological tissues.

The author is particularly indebted to Dr. R.V. Rechenmann for many stimulating discussions. This work was partly supported by a contract (BIO 294 F(G)) with the European Communities.

References:

[1] R.V. Rechenmann, E. Wittendorp and B. Senger, Rad. Effects 34,87(1977).

[2] B. Senger, E. Wittendorp and R.V. Rechenmann, 6th Symp. Microdos., p.361; ed. Harwood Acad. Publ. (1978).

[3] T.F.M. Bonsen and L. Vriens, Physica 47,307(1970).

[4] C.E. Kuyatt and T. Jorgensen, Phys. Rev. 130,1444(1963).

[5] E. Merzbacher and L.H. Lewis, Handb. der Physik 34,166 (1958); G.S. Khandelwal and E. Merzbacher, Phys. Rev. 151,12(1966); B.-H. Choi, E. Merzbacher and G.S. Khandelwal, At. Data 5,291(1973).

[6] B. Senger, E. Wittendorp-Rechenmann and R.V. Rechenmann, Nucl. Instr. Meth. 194,437(1982).

[7] K. Siegbahn et al., in *ESCA applied to free molecules*, North Holland Publ. Comp., Amsterdam (1969).

[8] G. Basbas, W. Brandt and R. Laubert, Phys. Rev. A7,983 (1973); W. Brandt and G. Lapicki, Phys. Rev. A10,474 (1974).

[9] L.H. Toburen and W.E. Wilson, J. Chem. Phys. 66,5202(1977).

[10] A. Salin, J. Phys. B2,631(1969).

BINARY ENCOUNTER APPROXIMATIONS WITH SCREENING TO DDCS OF ELECTRONS FROM ATOMIC COLLISIONS

G. Hock

Institute of Nuclear Research of the Hungarian Academy of Sciences (ATOMKI), H-4001 Debrecen Pf. 51, Hungary

For comparison or due to their relative simplicity compared to the Born approximation (PWBA), the binary encounter approximations (BEA) to the DDCS for electrons ejected in atomic collisions, $\partial^2\sigma/\partial\varepsilon\partial\Omega$, are often used for the ionization of either the target atom or the projectile[1-3]. In such cases, it is necessary to introduce screening of the nucleus by its own electrons.

Every kind of approximations called as BEA to ionization can be regarded[4,5] as a classical impulse approximation (IA) and obtained on the basis of the quantal IA within PWBA if the wave function of the ejected electron is replaced by a plane wave[6], in accordance with the physical picture. In this way one arrives immediately at the cross sections in BEA for bare projectiles

$$d\sigma^{BEA} = \frac{(2\pi)^4}{V_i}\tilde{V}_c^2(Q)|\phi_i(\vec{k}_e-\vec{Q})|^2\delta(\text{en.balance})\,d\vec{K}_f\,d\vec{k}_e \qquad (1)$$

in general, where

$$\tilde{V}_c(Q) = -Z_1/2\pi^2Q^2 \quad \text{and} \quad \phi_i(\vec{k}_e-\vec{Q})$$

are the Fourier transforms of the projectile - electron Coulomb potential and the atomic bound state, respectively, $\vec{Q}=\vec{K}_i-\vec{K}_f$ is the momentum transferred to the electron ejected out with momentum $\vec{k}_e = m_e\vec{v}_e$ by the incident projectile of momentum $\vec{K}_i = m_p\vec{V}_i$ and charge Z_1. Atomic units and δ-function normalization of plane waves are used.

The different binary encounter approximations in literature, apart from the particular type of the cross sections considered, are distinguished from each other in that to what extent they adhere to the elastic or inelastic kinematics. Bonsen and Vriens[5,7] maintained the binding energy defect with simultaneous fully elastic kinematics allowing thereby only the same range of the elastic energy transfer to contribute as in the inelastic process. In this case (1) should be written as

$$d\sigma^{BV} = \frac{(2\pi)^4}{V_i}\tilde{V}_c^2(Q)|\phi_i(\vec{k}_i)|^2\delta(E_{inel}-\varepsilon_f+\varepsilon_i)\,d\vec{K}_f\,d\vec{k}_f \qquad (2)$$

where $\vec{K}_i, \vec{K}_f$ and $\vec{k}_i, \vec{k}_f$ are subject to energy and momentum conservation but $\hat{k}_f = \hat{k}_e$ $(\hat{k} = \vec{k}/k)$. Transforming the variables one has

$$d\vec{k}_f = m_e k_f\, d\Omega_{\hat{k}_f} = m_e k_f\, d\varepsilon_f\, d\Omega_{\hat{k}_e} \;;\; \cos\vartheta_e = \hat{K}_i\hat{k}_e$$

$$d\vec{K}_f = m_p K_f\, d\Omega_{\hat{K}_f} = m_p K_f\, d\varepsilon_e (k_i dk_i/K_f S)\, d\chi \qquad (3)$$

where the auxiliary vector $\vec{S} = \vec{K}_i - \vec{k}_f = \vec{K}_f - \vec{k}_i$ has been introduced and χ is the azimuthal angle of $\vec{K}_f$ and $\vec{k}_i$ around $\vec{S}$, measured from the $(\vec{K}_i, \vec{k}_f)$ plane. The integration limits follow from the fact that certain angles have to be regular ones. The allowed ranges[7] of k_i apply, for given K_f/K_i, m_p/m_e, ϑ_e $(0 \le \vartheta_e \le \pi - \cos^{-1}(K_f/K_i))$, setting out at the same time the possible values of other variables such as k_f, S and $\cos\alpha = (\hat{k}_f\hat{S}) = (K_i\cos\vartheta_e - k_f)/S$; $\cos\alpha' = (\hat{k}_i\hat{S}) = [\vec{k}_f\vec{S} - (m_p - m_e)E]/k_iS$. Thus,

$$Q^2 = [k_f^2 + k_i^2 - 2k_fk_i(\cos\alpha\cos\alpha' + \sin\alpha\sin\alpha'\cos\chi)] = [a - b\cos\chi]$$

and for later use one still has

$$d\chi = 2[(Q_2^2 - Q^2)(Q^2 - Q_1^2)]^{-1/2}\, Q\, dQ \qquad (4)$$

where $Q_{1,2}^2 = \min, \max[k_f^2 + k_i^2 - 2k_fk_i\cos(\alpha \mp \alpha')]$; $Q_1 \le Q \le Q_2$.

The DDCS is obtained from (2) by averaging over the magnetic states of the bound electron, $\Sigma_m|\Phi_{nlm}(\vec{k}_i)|^2 = (4\pi)^{-1}|\Phi_{nl}(k_i)|^2$, and with (3) and (4) for bare projectiles reads

$$\frac{\partial^2\sigma^{BV}}{\partial\varepsilon_e\partial\Omega_{\hat{k}_e}} = \frac{Z_1^2 m_p m_e}{\pi V_i}\int_{k_{i1}}^{k_{i2}} \frac{k_f}{Sk_i}[k_i^2|\Phi_{nl}(k_i)|^2]\,dk_i \;\; *$$

$$* \begin{cases} \int_0^{2\pi} d\chi\,[a - b\cos\chi]^{-2} & (5a) \\ 4\int_{Q_1}^{Q_2} dQ\, Q^{-3}[(Q_2^2-Q^2)(Q^2-Q_1^2)]^{-1/2} & (5b) \end{cases}$$

$$= \frac{2Z_1^2 m_p m_e}{V_i}\int_{k_{i1}}^{k_{i2}} \frac{k_f}{Sk_i}\frac{a}{[a^2-b^2]^{3/2}}[k_i^2|\Phi_{nl}(k_i)|^2]\,dk_i\,. \qquad (5c)$$

(5b) and (5c) are now the suitable forms of DDCS to introduce screening in this construction.

The conception involved in IA directly suggests how to introduce the screening of the projectile nucleus by its electrons. The pure Coulomb $T^B_{\vec{k}_i\vec{k}_f} = \tilde{V}_c(Q)$ has to be replaced by the appropriate screened potential $\tilde{V}_s(Q)$. In case of using[1,3] the Thomas-Fermi potential (if it applies) it simply means the substitution of $(Q^2+\mu^2)$ for Q^2, $\mu \sim 1.13\, Z^{1/3}$ i.e. $(a+\mu^2) \to a$ in (5c).

For a given arbitrary static charge distribution $\varrho(\vec{r})$ of the projectile, the field $\varphi(\vec{r})$ experienced by the target electron is known from electrostatics. Its Fourier transform is[8]

$$\varphi(\vec{q}) = (4\pi/q^2)\varrho(\vec{q}) = (1/2\pi^2q^2)\int d\vec{r}\, e^{-i\vec{q}\vec{r}}\varrho(\vec{r})$$

with $\varrho(\vec{r}) = Z_1\delta(\vec{r}) - \Sigma_i|\Psi_i(\vec{r})|^2$, the $\Psi_i(\vec{r})$ are the projectile electron wave functions. The interaction potential between the target electron and the projectile corresponding to elastic electron-projectile scattering

$$\tilde{V}_s^{el}(Q) = -[Z_1 - F_{el}(Q)]/2\pi^2Q^2 \equiv -Z_1(Q)/2\pi^2Q^2 \qquad (6)$$

defines an effective projectile nucleus charge, $F_{el}(Q)$ is the elastic atomic formfactor summed up for the electrons and their states, while if allowance is made for inelastic processes the corresponding inelastic formfactors contribute, as surveyed[9] and demonstrated[10] in earlier works. (5b) makes it possible to use screening in this BEA by moving back a $Z_1(Q)$ like in (6) into integral over Q.

References

1. F. Drepper and J.S. Briggs, J. Phys. B: Atom. Mol. Phys. 9, 2063 (1976)
2. M. Prost et al., Proc. Int. Seminar on High-Energy Ion-Atom Coll., Debrecen, 1981, eds. D. Berényi and G. Hock, Budapest, Akadémiai Kiadó, p. 99
3. Á. Kövér et al., J. Phys. B: Atom. Mol. Phys., in print
4. M.R.C. McDowell and J.P. Coleman, Introd. to the Theory of Ion-Atom Collisions, North-Holl. Amsterdam, 1970, p. 292
5. L. Vriens, Case Studies in Atomic Collision Physics, eds. E.W. McDaniel and M.R.C. McDowell, North-Holl. Amsterdam, 1969, Ch. 6
6. D.R. Bates and W.R. McDonough, J. Phys. B: Atom. Mol. Phys. 5, L107 (1972)
7. T.F.M. Bonsen and L. Vriens, Physica 47, 307 (1970)
8. L.D. Landau and M.E. Lifshitz, Quantum Mechanics, Pergamon P. 1965, 2nd ed. Ch. XVII § 137
9. J.S. Briggs and K. Taulbjerg, in Topics in Current Physics, ed. I.A. Sellin, Springer Vlg. Berlin, 1978, Vol. 5 p. 105
10. J.H. McGuire, N. Stolterfoht and P.R. Simony, Phys. Rev. A24, 97 (1981)

1KL6

DOUBLY DIFFERENTIAL IONIZATION CROSS SECTIONS IN FAST ION-ATOM COLLISIONS

A.K. Kaminsky , M.I. Popova

Institute of Nuclear Physics, Moscow State University, Moscow 117234, USSR

The results are reported of calculations on the doubly differential cross sections $\frac{d^2\sigma}{dE_L d\Omega_L}$ for ionization in rapid ion-atom collisions when both colliding partners have their own electrons. Here E_L and χ_L are the energy and angle of electron ejection in the lab.system, $d\Omega_L = 2\pi \sin\chi_L\, d\chi_L$. The method of calculation is based on the first Born approximation and the sum rule for summation over the final states of the ionizing particle[1-3].

The previous analysis [1-4] of the applicability of these approximations served to stimulate the calculations of $\frac{d^2\sigma}{dE_L d\Omega_L}$ for systems with more electrons, for which matrix element and form factor expressions are known.

Figs. 1,2 illustrate the results of calculating the doubly differential cross sections $\frac{d^2\sigma}{dE_L d\Omega_L}$ for ionization in collision of 10 MeV hydrogen atoms with those of a carbon target. The contributions of electrons of the H atom and those from various subshells of the carbon-atom L-shell are shown separately.

The ejected electron energy distributions (figs. 1 and 2) have maxima at low electron velocity v_L and also at $v_L \simeq V$, $v_L \simeq V\cos\chi_L$,and $v_L \simeq 2\,V\cos\chi_L$, V being the collision velocity. The origin of these maxima has been discussed in[1-4]. The low energy region E_L is outlined in the upper right-hand corner of fig. 1.

In binary peaks of electrons ejected at small angles from the $2p_0$-state of a target atom there are pits when the quantization axis is taken along the direction of electron ejection. At $E_L \to 0$, the contribution of electrons from this state is dominant.

The peak at $v_L = V$ in the projectile electron distributions results from the electron ejection without any change of the target atom state, while the projectile electron peak at $v_L = V\cos\chi_L$ is due to the ionization of the projectile with a simultaneous change of the target atom state. Thus the structure of projectile electron spectrum exhibits itself when χ_L increases.

The relatively large value of the peak due to the H-atom electrons in Fig.1 is explained by the fact that most of the H-atom electrons and a small fraction of C-atom electrons are found in the low χ_L range. As χ_L increases, the contribution of the projectile electrons to the total spectrum decreases.

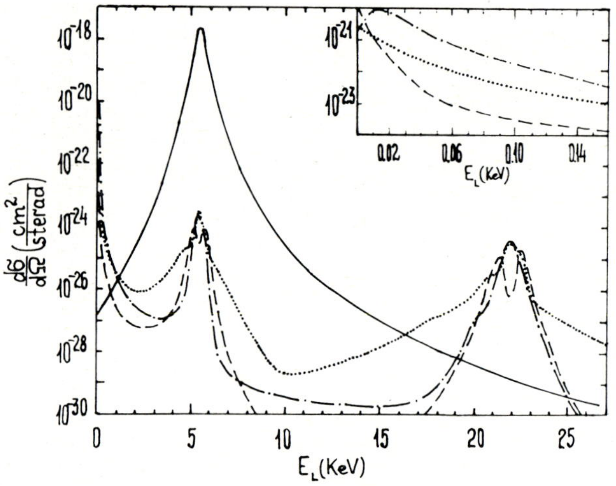

Fig. 1: Energy spectra $\frac{d^2\sigma}{dE_L d\Omega_L}$ of electrons ejected at an angle $\chi_L = 1°$ in collisions of 10-MeV hydrogen with carbon. The solid curve corresponds to the contribution of electrons ejected from H. Dotted curve describes the contribution of electrons ejected from the 2s =state of C, dashed-from $2p_0$-state, dash-dot- from $2p_1$--state.

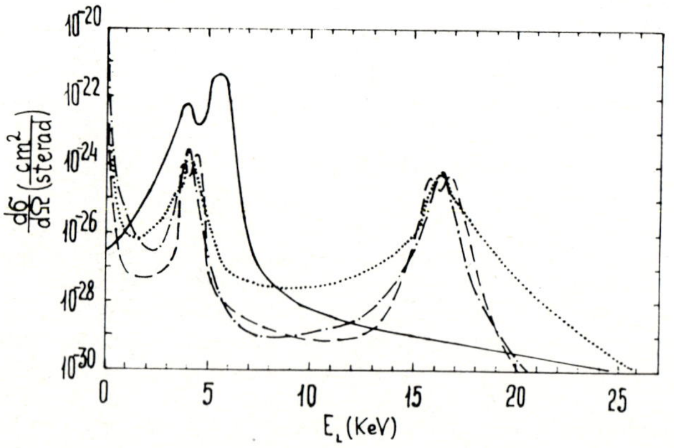

Fig. 2: The same as in fig. 1 except for χ_L being $30°$.

References.

1. A.K. Kaminsky et al. J.Phys.B.13, 1160 (1980)
2. M.H. Day , J.Phys.B14, 231(1980)
3. A.K. Kaminsky, M.I. Popova, J.Phys.B15, 403 (1982)
4. S.T. Manson, IEEE Trans. Nucl. Sci.NS-28, 1084 (1981).

ELECTRON LOSSES IN H_2^+, H_3^+ AND He^+ COLLISIONS WITH Ar

N. Oda, F. Nishimura, K. Komatsu and H. Shibata

Research Laboratory for Nuclear Reactor, Tokyo Institute of Technology, O-okayama, Meguro-ku, Tokyo 152, Japan

To investigate the electron loss process, angular dependencies of the peak positions, the peak widths, and the single differential cross sections (SDCS) of the electron loss peaks have been determined for collisions of 0.5 MeV/amu H_2^+, H_3^+ and He^+ and 0.25 - 1.0 MeV/amu H_2^+ with Ar.

An electron analyzer, which can analyze ejected electrons at angles from 0° to 180°, consists of two sets of electrostatic cylindrical analyzers of 32 mm radius. Electron deflector angles for the first and second analyzers are 30° and 56°, respectively, and deflection planes of two analyzers are mutually perpendicular. The energy and angular resolutions were confirmed to be $\Delta E/E \simeq 0.01$ and $\Delta\theta \simeq 0.7^o$, with a nearly parallel electron beam.

Doubly differential cross sections (DDCS) differential in energy and electron ejection angle were measured over energy range from 10 to 2000 eV at angles from 0° to 120° for the above collision systems, as well as for 0.5 MeV/amu H^+ - Ar collision system. The measured DDCS's were put on the absolute scale using the absolute DDCS for 0.5 MeV/amu H^+ - Ar by Toburen et al.[1]

Fig.1 gives the peak shifts for H_2^+, H_3^+ and He^+, $E_i - E_1$, as functions of electron ejection angle, where E_i and E_1 are electron energy corresponding to projectile velocity and the peak position, respectively. In the figure, it is seen that the angular dependencies for H_2^+ and H_3^+, as well as for H^0 by Duncan et al.[2], are similar to each other within experimental error.

For 0.25 - 1.0 MeV/amu H_2^+ - Ar collision, the peak shift was independent on projectile energy in angular range less than 30°. And, one may compare the present peak shift with that for 0.8 MeV/amu H_2^+ - Ar by Kövér et al.[3], which was calculated according to the frame transformation of the DDCS for projectile ionization described by scaled-plane-wave Born approximation. Our peak shifts for H_2^+ and H_3^+ are well reproduced by that calculated by Kövér et al., but are about twice that measured by them. It seems that our peak shifts have a bump at about 45° similar to that calculated by Jakubassa with the electron impact approximation[4]. The peak shift for He^+ is about twice that for H_2^+, reflecting tight binding of the projectile electron.

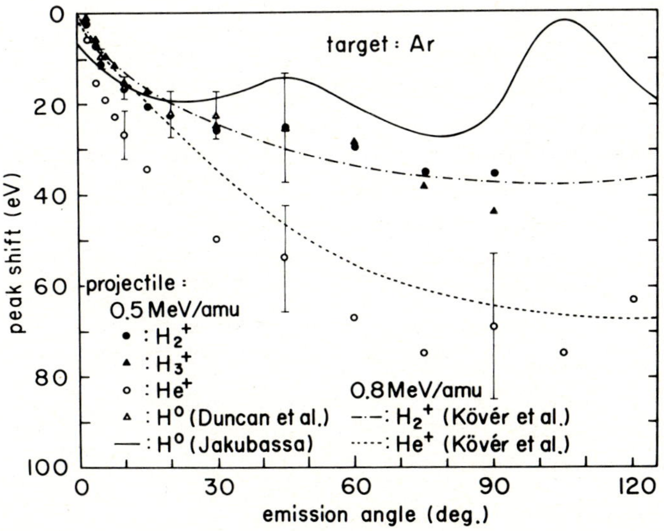

Fig. 1

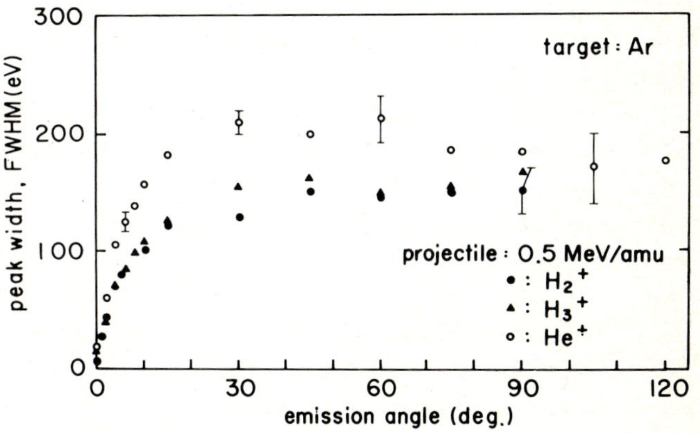

Fig. 2

Fig.2 shows the peak widths as functions of electron ejection angle for 0.5 MeV/amu H_2^+, H_3^+ and He^+ impacts on Ar. The peak widths increase rapidly with ejection angle and approach nearly constant values, which are related to the initial momentum distributions of ejected electrons, at angles larger than 30°. The peak width for H_2^+ is same with that for H_3^+.

References

1. L. H. Toburen, N. Stolterfort, Z. Zien and D. Schneider, Phys. Rev. A 24, 1741 (1981)
2. M. M. Duncan and M. G. Menendez, Phys. Rev. A 19, 49 (1979)
3. A. Kövér, D. Varga, Gy. Szabó, D. Berényi, I. Kádár, S. Ricz, J. Végh and G. Hock, to be appeared in J. Phys. B: Atom. Molec. Phys. (1983)
4. D. H. Jakubassa, J. Phys. B: Atom. Molec. Phys. 13, 2099 (1980)

COLLISIONAL ELECTRON LOSS INTO THE CONTINUUM OF PROTONS USING THE $H^0 \rightarrow He$ SYSTEM; A COMPARISON WITH THEORY.

R. Vidal, W. Meckbach and E. González Lepera

Centro Atómico Bariloche* and Instituto Balseiro#
8400 - Bariloche, Argentina

Electron loss into the continuum is a simple ionization process of projectiles which bring bound electrons into the collision. It leads to a final state that contains a Coulomb wave centered at the projectile.[1] As in the case of electron capture into the continuum this results in a cross section $d\sigma/d\vec{v}$ that diverges when the electron velocity $\vec{v}$ matches the velocity $\vec{v}_i$ of the emerging ion. The double differential distribution in speed and angle, measured experimentally, is a cusp shaped peak, centered at v_i, that results as a convolution of $d\sigma/d\vec{v}$ with the instrumental acceptance in solid angle and speed.[2]

A first theoretical discussion of the process by Drepper and Briggs[1] led to a spherically symmetrical cross section in velocity space. Subsequent theories by Day[3] and Briggs and Day[4] led to anisotropies that can be described by introducing an additional term into the cross section which contains a second spherical harmonic $(P_2(\cos\alpha))$ multiplied by a weighing factor ß. Here α is the polar angle defined in the moving system. In order to facilitate a comparison of this theory with experiment we measured the two dimensional cusps as a function of electron energy (speed) and angle at three different ion energies (E_i= 105, 187 and 286 keV; v_i= 2.05, 2.74 and 3.38 a.u.). The incident projectile was H^0, the target He. The design of our coaxial cylindrical electron spectrometer[5] was such that the direction or angle θ with respect to the ion beam axis as well as the angular acceptance θ_0 were determined at the exit side of the deflected electrons. This permitted the measurement of electron spectra at small θ without intercepting the ion beam. Such spectra, taken at intervals of 0.5 deg between θ = 0 and 8 deg allowed the construction of three dimensional cusps. The absence of a negative cusp skewness, observed at $\theta=0^0$ is in agreement with the theory and permits a comparison of measured and calculated anisotropies. For this purpose contour lines taken at different levels of the experimental cusps and those resulting from a convolution of theoretical cross sections[3,4] with the instrumental acceptances in solid angle and speed were determined and compared.

It results that the experimentally observed contours as a function of α contain a contribution that can be attributed to a second spherical harmonic $P_2(\cos\alpha)$ with a positive weight ß $\simeq$ +0.15, but that a more complicated structure is superposed which needs further theoretical interpretation.

Partially supported by the Multinational Program of Physics of the Organization of American States.

* Comisión Nacional de Energía Atómica.
# Comisión Nacional de Energía Atómica and Universidad Nacional de Cuyo.

References

1 F. Drepper and J. Briggs, J.Phys.B 9, 2063 (1976).
2 W.Meckbach, K.C.R.Chiu, H.H. Brongersma and J.Wm.McGowen, J.Phys.B 10m 3255 (1977).
3 M. Day, J.Phys.B 13, L 65 (1980).
4 J. Briggs and M. Day, J.Phys. B 13, 4797 (1980).
5 W. Meckbach, I.B. Nemirovsky and C.R. Garibotti, Phys.Rev.A 24, 1793 (1981).

CHARGE STATE DEPENDENCE OF ELECTRON LOSS PEAK MEASURED UNDER 0° FOR FAST ARGON IONS

A.Itoh, T.Schneider, G.Schiwietz, Z.Roller, H.Platten, G.Nolte, D.Schneider and N.Stolterfoht

Hahn-Meitner-Institut für Kernforschung Berlin GmbH, Glienicker Str. 100, D-1000 Berlin 39

In ion-atom collisions it is well known that electron spectra observed near zero degree with respect to the beam direction show a sharp peak centered at electron velocity equal to that of the projectile.[1] This so-called electron loss to the continuum (ELC) peak of the projectile originates from electrons with near zero energies in the projectile rest frame. Recent theoretical calculations[2] made for simple collision systems predict an anisotropic angular distribution of the double differential cross section in the laboratory rest frame. For the precise study of ELC it is favourable to use high velocity ion beams, since the low energy electrons emitted from projectiles are more "stretched" kinematically and can easily be separated from continuous background electrons.

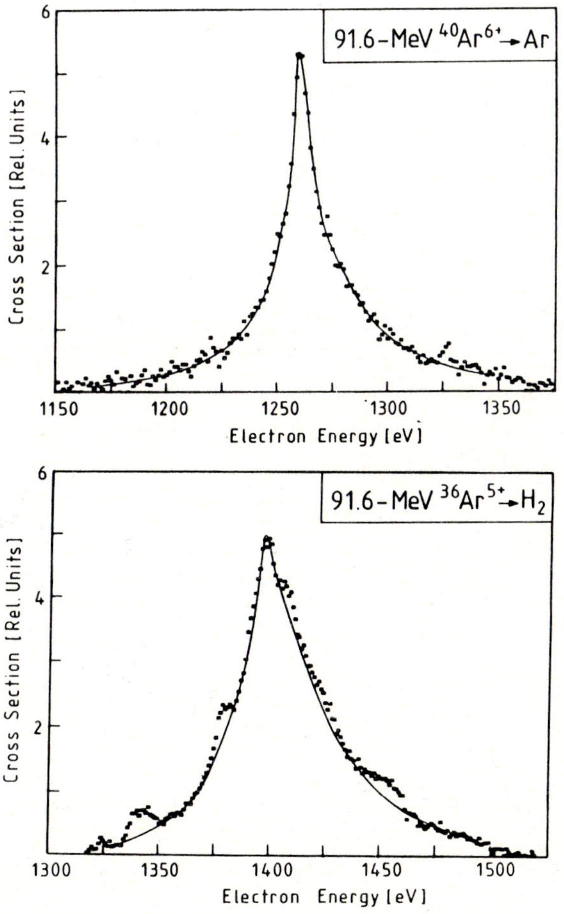

Fig.1: ELC spectra for 91.6 MeV $^{40}Ar^{6+}$ (upper figure) and $^{36}Ar^{5+}$ (lower figure) on Ar and H_2, respectively.

In our laboratory, 0° electron spectroscopy measurements have been performed for 91.6 MeV $^{36}Ar^{5+}$ and $^{40}Ar^{6+}$ ions on H_2, He and Ne. A 90° parallel-plate spectrometer[3] was slightly modified to allow for the 0° measurements. The electron acceptance angle was chosen to be relatively small (±0.3°), in order to minimize broadening effects. The electron deceleration method[3] was used to achieve improved energy resolution (1%). Examples of ELC spectra are shown in Fig.1.

Contrary to the ion charge state q = 6 (electron configuration is $1s^2 2s^2 2p^6 3s^2$) where the cusp shape is almost symmetric, a considerable asymmetry can be seen for q = 5 ($1s^2 2s^2 2p^6 3s^2 3p$) which is far beyond predictions given for light collision systems[2]. Furthermore, for q = 5 the full width at half maxima of the peak is about twice of that for q = 6. These results differ from conclusions in previous work using C, O and Si ions[1]. The origin of the observed asymmetry in our experiment is not known at present. It should be noted that Jakubaßa[4] pointed out the enhancement of the asymmetry when higher order terms are included in SCIA calculations. Further measurements with various charge states are required to study the asymmetry of the ELC peak in more detail.

In addition, autoionization electrons are observed in wings of the ELC peak. The rest frame energies of these lines are smaller than 1 eV. A comparison between results of q = 5 and 6 indicates the strong influence of the 3p electron on the production of autoionization electrons. From simple calculations based on energy considerations, the two strong peaks seen for q = 5 may be attributed to the excitation process $1s^2 2s^2 2p^6 3s^2 3p \longrightarrow 1s^2 2s^2 2p^6 3s 3p 6\ell$ and the decay to $1s^2 2s^2 2p^6 3s^2$.

References

1. M.Breinig, S.B.Elston, S.Huldt, L.Liljeby, C.R.Vane, S.D.Berry, G.A.Glass, M.Schauer, and I.A.Sellin; Phys.Rev.A 25, 3015 (1982)

2. J.S.Briggs and M.H.Day; J.Phys.B 13, 4797 (1980)

3. N.Stolterfoht, D.Schneider, D.Burch, B.Aagaard, E.Bøving, and B.Fastrup; Phys.Rev.A 12, 1313 (1975)

4. D.H.Jakubaßa-Amundsen; J.Phys.B 14, 3139 (1981)

ORIGIN OF THE DOUBLE PEAK IN ELECTRON LOSS IN THE FORWARD DIRECTION

Victor H. Ponce and Raúl A. Baragiola

Centro Atómico Bariloche* and Instituto Balseiro#
8400 - Bariloche, Argentina

The purpose of this work was to give a physical explanation to features which appear[1] in the doubly-differential cross sections (DDCS) for electron loss from fast H^-, in the forward direction.

Experiments show two peaks near zero degrees. One (P_0) is centered ar ~ V_i, the projectile velocity, and is observed only close to the forward direction. The second (P_1), occurs at a lower velocity. The difference ΔE in electron energy between these peaks depends on the target and observation angle.

Recently, there have been attempts to explain this double-peak structure; all of them assume they come from single-electron loss (SEL). Franz et al[2] have obtained a fair agreement between their Born approximation calculations and experiment and suggested that both peaks result from single electron loss collisions in which momentum is transferred mainly to the resultant atom (Peak P_0) or to the ejected electron (P_1). Day[3] has performed similar calculations using different final state wave functions for the ejected electron. He does not improve the agreement with experiment but shows that most electron loss collisions lead to simultaneous excitation of the target, in agreement with previous work by Bell et al.[4] Similar results were obtained by Maleki and Macek[5] who pointed out the need of correct wave function shape for $v_e \sim V_i$. None of the work mentioned above gave a physical interpretation of the experimental results, which could provide direct guidelines for further experiments.

We have reproduced first Born calculations for SEL in fast H^- + He collisions, and separated processes in which the target scatters elastically and inelastically.

Furthermore, we have added calculations for double-electron loss (DEL); where we have assumed no correlation between the electrons. The resulting diverging cross section was then integrated over the experimental detector resolution.

Our results for SEL and DEL are shown in the figure. We obtain good agreement with experiment except that the peaks are narrower, possibly as a result of the closure approximation. We find that peak P_0 comes mainly from DEL; this process is strongly peaked in the forward direction due to the Coulomb potential of the resulting proton. A small contribution comes from SEL in which the target is excited by the H core while the ejected "diffuse" electron acts as an spectator; this process also decays fast with the ejection angle, since it involves only small energy transfers. Peak P_1 results from SEL where the target is excited by the diffuse electron of H^- acting as a nearly free electron. The difference ΔE in energy with peak P_0 is the mean inelastic energy transfer to the target. This explains the experimental dependence of ΔE with target species and observation angles.

From these results one should expect that ΔE decreases with decreasing impact velocity, particularly close to the ionization threshold for target ionization by a free electron: $V_i = (2Ei/m)^{1/2}$. We also propose that the relative role of DEL be determined from measurements at different angular and energy resolutions. Due to the cusp-shaped behaviour of the cross section, the measured DEL distribution should depend critically on the resolution volume of the detector.

References

* Comisión Nacional de Energía Atómica

Comisión Nacional de Energía Atómica and Universidad Nacional de Cuyo.

1 M.M.Duncan and M.G.Menendez, Phys.Rev.A 16, 1799 (1977); Phys.Rev.A 19,49 (1979); Phys.Rev.A 23, 1085 (1981); M.G. Menendez and M.M.Duncan, Phys.Rev. Letters 40, 1642 (1978); Phys.Rev.A 20, 2327 (1979).

2 M.R. Franz, L.A.Wright and T.C. Genoni, Phys.Rev.A 24,1135 (1981).

3 K.L. Bell, A.E. Kingston and P.J. Madden, J.Phys.B: 11, 3357 (1978).

4 M.Day, Phys.Rev.A 26, 1260 (1980).

5 N.Maleki and J.Macek, Phys.Rev. A 26, 3198 (1982).

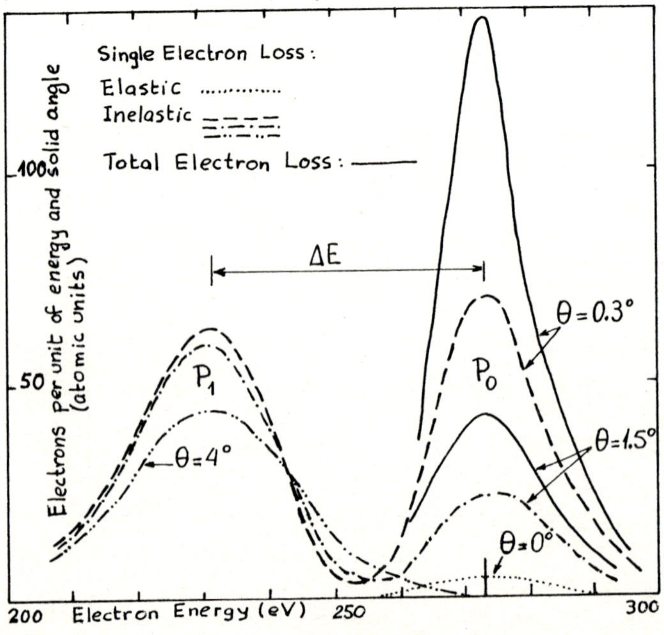

DDCS for electron loss from 0.5 MeV H on He.

STRIPPING CROSS SECTIONS OF MULTI-CHARGED IONS BY NEUTRAL ATOMS

S.Karashima

Department of Electrical Engineering,Faculty of Engineering,Science University of Tokyo, Sinjuku-ku,Tokyo 162

T.Watanabe

Department of Applied Physics,Faculty of Engineering,The University of Tokyo, Bunkyo-ku,Tokyo 113

The mechanism of charge-stripping processes of energetic multi-charged ions passing through matter has played an important role for the following problems of the charge stripper in the accelerator science, of the average equilibrium charges of heavy ions in matter and of atomic processes in controlled thermonuclear fussion reactors. However only a few theoretical investigations[1)] are available at present time for this charge-stripping process.

The cross sections for the processes:

$$X^{(Z-1)+}(1s) + H(1s) \rightarrow X^{Z+} + e + H(\Sigma), \quad (1)$$

where Σ denotes all the accessible electronic states of H, was calculated by Shirai et al[1)] based on the classical binary-encounter approximation (BEA) and the plane-wave Born approximation (PWBA).

The characteristic feature of these processes lies in the large momentum transfer given to the electrons in the interacting ion on a occasion of collisions with a target particle (electron or nucleus). Such a collision with the large momentum transfer can well approximated by the semi-classical description In this sence, it is considered that BEA is one of the proper method to calculate the charge-stripping cross section.

In the present paper, we calculate these cross sections using BEA: the incident ion interacts with only electron or nucleus of the target atom and the mutual interaction between the atomic electrons and the nucleus can be disregarded during the collision. The evaluation of the cross section requires information about the electron momentum distribution. As the first step, the momentum distribution of electrons in the incident ions which interact with the hydrogen atoms is obtained using the Thomas-Fermi (TF) model. In next step, we will extend the target atom from hydrogen to a many-electron atom in general by the use of TF model again.

Let $\rho(\vec{p}, \varepsilon_i)$ be the momentum distribution with momentum $\vec{p}$ and binding energy ε_i in the atomic field of nucleus charge $Z_i e$ derived from the statistical (TF) method:

$$\rho(\vec{p}, \varepsilon_i) = \left(\frac{a_0}{\hbar}\right)^3 8\pi \lambda^4 Z_i^{-7/3} x^2 \frac{dx}{dv} \quad (2)$$

where a_0 is Bohr velocity, x satisfies the TF equation; that is $d^2\chi(x)/dx^2 = x^{-1/2}\chi(x)^{3/2}$ with the inner atomic potential $V(r) = -Z_i e^2 \chi(x)/r$ λ = 0 8834 and denotes dimensionless values $v = \chi(x)/x$.

The charge-stripping cross section for a many-electron ion by a proton as follows:

$$\sigma_p(u) = \int_0^\infty \sigma_p^{(i)}(u, w) f(w)\, dw \quad (3)$$

where $\sigma_p^{(i)}(u,w)$ is the ionization cross section in the laboratory system for the collision between a proton with velocity u and electrons (velocity w, distribution f(w)). We can obtain the analytical formula $\sigma_p^{(i)}$ by Vriens[2)]

For the charge-stripping cross section between an ion and the neutral hydrogen atom with a velocity vector $\vec{u}_e$ is defined as

$$\sigma_e(u) = \int_0^\infty \int_0^\infty \int \sigma_e^{(i)}(u', w, \mu) f(\vec{u}_e) f(w) \times du_e\, dw \frac{d\mu}{2} \quad (4)$$

where

$$\vec{u}' = |\vec{u} + \vec{u}| = \left\{ (u^2 - u_e^2 + \mu^2 u_e^2)^{\frac{1}{2}} + \mu u_e \right\}$$

and

$$\mu = (\vec{u}_e \cdot \vec{u}') / |\vec{u}_e| \cdot |\vec{u}'|$$

for the distribution of 1s wave function. Extending eq.(4) for hydrogen atom to a charge Z_t atom, we can write

$$f(u_e) = \left(\frac{a_0}{\hbar}\right)^3 8\pi \lambda^4 Z_t^{-7/3} F(v_e) P^3, \quad (5)$$

$$F(v_e) = \frac{1}{v_e^4} \frac{f^3(v_e)}{1 - \chi'(v_e)/v_e}, \quad v_e = \chi(x)/x$$

The preliminary results of the cross section $\sigma_p(u)$ are shown in the figure.

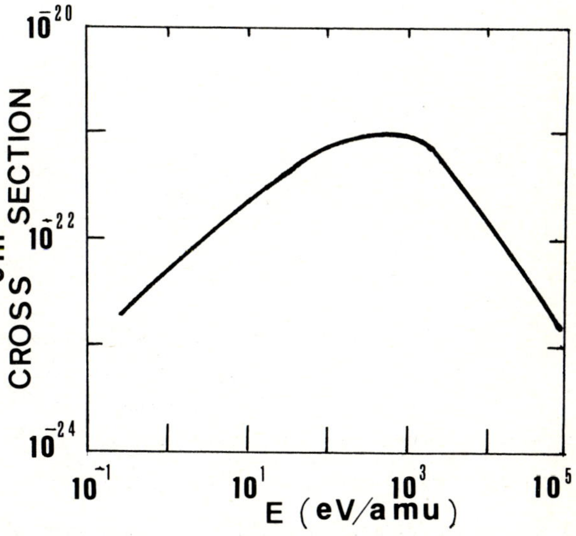

References

1. T.Shirai,K.Iguchi and T.Watanabe, J. Phys. Soc. Jpn. 42, 238 (1977).
2. Vriens, Case Studies in Atomic Collisions (E.W.McDaniel and M.C.McDowell eds. North-Holland Publ. Amsterdam) ch.6.

CONTINUUM-ELECTRON CAPTURE BY PROTONS IN HELIUM

Poul Dahl

Institute of Physics, University of Aarhus, DK-8000 Aarhus C, Denmark

Charge transfer to the continuum in the proton-helium collision is studied by measuring (i) energy spectra $\sigma(E)_{\theta_0}$ for electrons ejected into a narrow forward cone $\theta < \theta_0 = 0.38^{\circ}$, and by measuring (ii) the double-differential cross sections $\sigma(E,\theta)$ at small observation angles θ, 0.4°-4°. The asymmetric peak structure reflects anisotropy of $\sigma(E',\theta')$ in the projectile frame, and this anisotropy may yield detailed information on the transfer process[1].

In the first experiment, the cone angle θ_0 is sufficiently small to ensure that the angle θ' in the projectile frame becomes well defined in the wing regions of the $\sigma(E)_{\theta_0}$ spectrum, $\theta' \simeq 180^{\circ}$ for $E < \varepsilon$ and $\theta' \simeq 0^{\circ}$ for $E > \varepsilon$, where ε is the mass-reduced projectile energy. Thus, the ratio between $\sigma(E',\theta')$ at $\theta' = 180^{\circ}$ and 0° may be determined. The results in Fig. 1 show that the highest ratios are obtained at an impact energy of about 50 keV, i.e., at the energy where also the maximum yield of the transfer process is obtained.

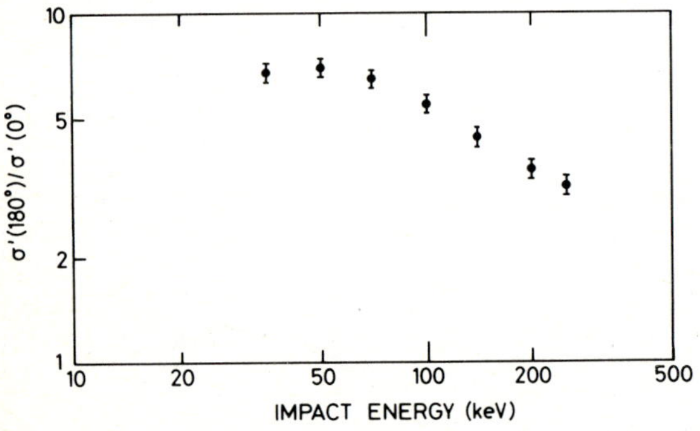

Fig.1

In the second experiment, where $\theta \neq 0^{\circ}$, the observed volume element in electron-velocity space is differential both in the laboratory frame and the projectile frame. The cross-section transformation is

$$\sigma(E',\theta') = \frac{v'}{v}\sigma(E,\theta) = \frac{\sin\theta}{\sin\theta'}\sigma(E,\theta) ,$$

where v is the electron velocity in the laboratory and v' in the projectile frame. Here, it may be assumed that

$$\sigma(E',\theta') = \frac{\sigma(E')}{4\pi}W(\theta') ,$$

where $W(\theta')$ is the angular distribution, and $\sigma(E')$ is a nearly constant singly differential cross section, which

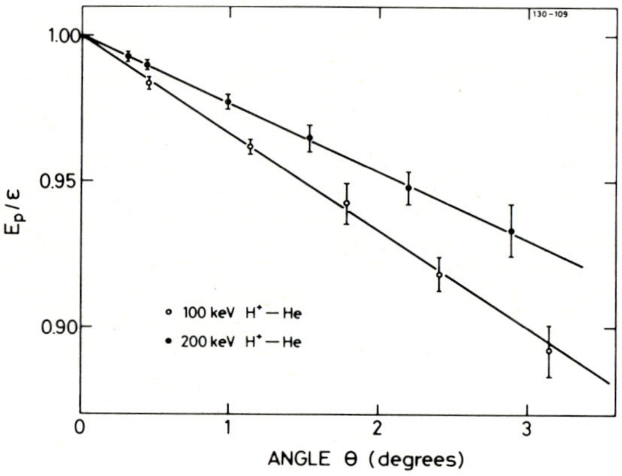

Fig.2

in terms of the reduced cross section σ_c is given by $\sigma(E') = \sigma_c/27.2$ eV. For any observation angle θ, the peak position E_p of $\sigma(E,\theta)$ then corresponds to the same angle θ'_p in the projectile frame, namely the angle where $\sin\theta' W(\theta')$ has its maximum. Since $v \simeq u(1+\theta\cot\theta')$, where u is the proton velocity, it follows that

$$\varepsilon - E_p = -2\varepsilon\cot\theta'_p\cdot\theta ,$$

so that proportionality between $\varepsilon - E_p$ and θ should be expected when $\varepsilon - E_p \ll \varepsilon$. This is confirmed by the data shown in Fig. 2, which give $\theta'_p = 133.6^{\circ} \pm 1.5^{\circ}$ at an impact energy of 100 keV, and $\theta'_p = 124.0^{\circ} \pm 1.5^{\circ}$ at 200 keV. Furthermore, it is seen that when $\sigma(E',\theta')$ is plotted against θ', coinciding curves should be obtained for different observation angles θ, and data confirm this in a broad θ' range, which makes a rather complete analysis of the anisotropy possible.

REFERENCE

(1) J. Macek, J.E. Potter, M.M. Duncan, M.G. Menendez, M.W. Lucas, and W. Steckelmacher, Phys.Rev.Lett. 46 (1981) 1571

EFFECT OF A SCREENED ELECTRON - PROJECTILE INTERACTION ON THE ECC PEAK SHAPE

R. O. Barrachina and C. R. Garibotti *

Instituto Balseiro,8400 S.C. de Bariloche,RN,Argentina

The usual description of the effect of electron capture to continuum states (ECC) in the atomic ionization by collision with stripped ions considers the ejected electron as lying in a Coulomb continuum state of the ion[1]. In the electron double differential cross section (DDCS), the normalization of that state introduces a factor $|f_c(k)|^2$ which diverges as k^{-1},when the electron - ion relative moment k goes to zero. This explain qualitatively the ECC forward peak[1].

In practice the scattering of charged particles occurs via potentials which are somehow screened. Under experimental conditions,the finite range R of the electron -ion interaction may be interpreted as arising from the finite separation between the incident ions or target atoms, or could be considered as a distorsion effect produced on the final ion-centered Coulomb wave by the target nucleus.

Considering a screened electron - ion interaction (with a range R), and relying on the final state interaction theory[2], the Coulomb factor $f_c(k)$ is replaced by the inverse of the s - wave Jost function $f_0(k,R)$. When $k^2/2\mu \gg Z/R$ (atomic units), where z is the ion charge and μ is the reduced mass of the electron - ion system, $|f_0(k,R)|^{-1}$ tends to the modulus of the Coulomb factor[3]

$$|f_0(k,R)|^{-1} \sim |f_c(k)|$$

However the previous condition is not verified necessarily into an interval around k = 0. There the DDCS diverges as k^{-2} or reaches a bounded maximun depending on the value of R. This means a notable deviation from the Coulomb factor like peak. For $k < k_0$ where $k_0 = (2\mu Z/R)^{1/2}$ the k dependence of the cross section is like that resulting from a short range potential. A characteristic value for the distance between the ions in the experimental conditions[4] is $R = 10^5$ Å, then $k \sim 0.0032$ a.u..However, if the detector velocity resolution Δv is greater than k_0, effects due to the finite distance between ions should be completely hidden. For particular experimental situations it is possible that $\Delta v < k_0$ and an anomaly near the ECC peak could be observed.

For gas targets R has not a well defined value,e.g. it must have a dynamical variation during the collision. Then, it is necessary to replace $|f_0(k,R)|$ by an average over the range, in the DDCS. When $R \gg 1$ we obtain:

$$\frac{1}{\Delta R}\int_R^{R+\Delta R} |f_0(k,R)|^{-2}dR \approx |f_c(k)|^2 \qquad (R \gg 1)$$

This result shows that when the range of the screened electron - ion interaction is large and no well defined the consequences of the usual ECC description are recovered as an average effect over the potential cut off.

For solid targets a defined value of R could be suppose. A natural range for the potential is of the order of half the distance between the atoms in the solid. For carbon foils this distance should be $R \sim 3$ a.u.,and some anomalies might appear on the peak.

References

1. K. Dettmann,K. G. Harrison and N. Lucas,J. Phys.B 7, 269 (1974).
2. K. M. Watson,Phys. Rev. 88,1163 (1952); J. Gillespie: Final State Interaction,Holden Day Inc.,San Francisco (1964).
3. C.R. Garibotti and R.O. Barrachina,Phys.Rev. submitted for publication (1983).
4. W. Meckbach,I. Nemirovsky and C. R. Garibotti,Phys. Rev. A 24,1793 (1981).
   C. Vane,I. Sellin,M. Suter,J.Aton,S.Elston,P. Griffin and R. Thoe,Phys.Rev. Lett. 40,1020 (1978).

* Consejo Nacional de Investigaciones Científicas y Técnicas,Argentina.

A COMPARISON OF ELECTRON CAPTURE AND ELECTRON LOSS INTO THE CONTINUUM WITH H^+ AND H^0 PROJECTILES INTERACTING WITH He.

R. Vidal, P. Focke, E. González Lepera, I.B. Nemirovsky and W. Meckbach

Centro Atómico Bariloche* and Instituto Balseiro#
8400 - Bariloche, Argentina

There are two collisional processes of electron transfer which lead to projectile centered continuum states of electrons traveling with velocities $\vec{v}$ close to the velocity $\vec{v}_i$ of the emerging ions: electron capture (ECC) and electron loss (ELC) into the continuum. The essential difference between these processes is that in the case of ECC the origin of the electron is the target atom; a momentum transfer of the order of v_i is needed. Contrarily ELC is a simple ionization process of the incoming projectile. ECC is the only possible process if the incoming projectile is a bare ion; ELC appears as a competing process if the the projectile carries bound electrons into the collision. Earlier theories[1] describing the two processes led, through a series of approximations to a double differential cross section that diverges like $1/|\vec{v}-\vec{v}_i|$ and whose shape in the projectile frame of reference is only determined by this simple spherically symmetrical dependence.

The first experimental comparison of the two processes was made by Menéndez et al.[2] using 2 MeV He^{++} and He^+-projectiles. They observed no difference in the shape of longitudinal ECC and ELC spectra obtained by measuring electrons emitted at $\theta=0$, that is along the axis determined by the ion beam. Subsequent measurements of longitudinal spectra[3] performed with heavy ion beams of different charge in gas targets showed a transition from strongly negatively skewed electron peaks towards narrower and more symmetrical peaks. When the charge of the impinging ions was changed progressively from $i=z$ to $i<z$. In this laboratory such a negative skewness was also observed and studied in detail for the case of the $H^+\rightarrow He$ system;[4] a spectrum obtained with $He^+\rightarrow He$ resulted in an almost symmetrical cusp.[5]

It was the purpose of the present work to perform a comparison of the double differential electron distribution resulting from ECC and ELC by using the simple systems of H^+ and H^0 interacting with He. Measurements were performed at different angles θ as a function of electron energy resp. speed. The spectrometer[4] used permitted measurements at small angles ($0<\theta<8$ deg) without intercepting the ion beam; in this manner we were able to obtain complete pictures of the resulting three dimensional electron distributions. An evaluation of these distributions led to the folowing conclusions: a) The negative skewness of the ECC cross section ($H^+\rightarrow He$) and the almost summetrical distributions for ECC ($H^0\rightarrow He$) are confirmed. b) As θ increases the position of ECC peaks of spectra taken at different angles is displaced toward lower electron velocities. For ELC this displacement is lower, but does not agree with that predicted by a convolution of instrumental resolutions with a simple $1/|\vec{v}-\vec{v}_i|$ shape of the cross section. c) A comparison of relative peak heights as a function of θ leads to a good agreement between ECC and ELC. d) Peak widths as a function of θ increase more rapidly for ECC than for ELC, but in no case agree with those predicted by a $1/|\vec{v}-\vec{v}_i|$ shaped cross section. e) The yield of the ECC process decreases quickly with increasing ion speed v_i. For ELC a larger yield is obtained which tends towards a constant value, as predicted by theory.[6] The yield as a function of instrumental angular acceptance follows exactly the theoretical dependence by Briggs and Drepper.[6]

* Comisión Nacional de Energía Atómica.

\# Comisión Nacional de Energía Atómica and Universidad Nacional de Cuyo.

References

1 J.Macek, Phys.Rev. A 1, 235 (1970); A.Salin, J.Phys.B 5, 979 (1972); K.Dettmann, K.G. Harrison and M.W.Lucas, J.Phys.B 7, 269 (1974); F.Drepper and J.Briggs, J.Phys.B 9, 2063 (1976).

2 M.G.Menéndez, M.M.Duncan, F.L.Eisele and B.R.Junker, Phys.Rev.A 15, 80 (1977).

3 C.R.Vane, I.A.Sellin, M.Suter, G.D.Alton, S.B.Elston, P.M.Griffin and R.S.Thoe, Phys. Rev.Lett. 40, 1020 (1978).

4 W.Meckbach, I.B.Nemirovsky and C.R. Garibotti, Phys.Rev.A 24, 1793 (1981).

5 I.A.Sellin, Physics of Electronic and Atomic Collisions, Invited Papers, S. Datz editor, p.213, North Holland (1982).

6 J.Briggs and F.Drepper, J.Phys.B 11, 4033 (1978).

δ-ELECTRON EMISSION IN STRONG PROJECTILE FIELDS

D.H.Jakubaßa-Amundsen

Physics Department, Technical University of München, 8046 Garching, Germany

Electrons which are emitted from light target atoms (T) in fast collisions with heavy, highly stripped projectiles (P) may be used to investigate the influence of strong perturbing fields. Thereby one may profit by the fact that the spectrum of the ejected electrons at fixed emission angle ϑ_f shows two distinct structures: the classical binary encounter peak at an energy $E_f = 2v^2 \times \cos^2\vartheta_f$, where v is the projectile velocity, and the forward peak at $v^2/2$ if ϑ_f is near zero.

For heavy projectiles ($Z_P \gg Z_T$) we suggest a description of the δ-electrons by means of charge transfer to the projectile continuum states, a theory which is well suited to explain the forward peak[1]. Thereby, in contrast to the first-order Born approximation, multiple interactions with the strong projectile field are taken into account. In the framework of the semiclassical impulse approximation[2], the transition amplitude results from the ejection of the target electron into a continuum projectile state $\psi^P_{\vec{q}-\vec{v}}$ with momentum $\vec{q}-\vec{v}$ and a subsequent inelastic scattering on the target field V_T:

$$a_{fi} = -i\int_{-\infty}^{\infty} dt \int d\vec{q}\; \varphi_i^T(\vec{q}) \langle \psi^P_{\vec{k}_f-\vec{v}} | V_T | \psi^P_{\vec{q}-\vec{v}} \rangle \times e^{i\vec{q}\vec{b}}\, e^{i(\varepsilon_f - E_i - \frac{v^2}{2} + \vec{q}\vec{v})t}$$

Here, a straight-line path with impact parameter b is used, $\varphi_i^T(\vec{q})$ is the initial electronic state in momentum space, E_i the initial and $\varepsilon_f = (\vec{k}_f-\vec{v})^2/2$ the final electronic energy in the projectile frame.

Fig.1 shows the electron spectrum at zero emission angle from collisions of bare Kr with H. While the first order Born theory describes the binary encounter peak quite well, it fails for both higher and lower E_f. Also a simple inclusion[3] of the normalisation constant of the projectile Coulomb wave is incorrect for asymmetric systems.

Fig.2 shows the forward peak for electrons from Ar^{18+}-He collisions. Its strong asymmetry results from a discontinuity at $\vec{k}_f = \vec{v}$ which reflects the influence of the target field on the projectile scattering states, dependent on the direction of their momentum.

The validity of the above theory extends down to $k_f \gtrsim vZ_T/Z_P$, independent of the emission angle ϑ_f. It fails for $k_f=0$ as the influence of the target field on the ejected electron can then no longer be treated in first order.

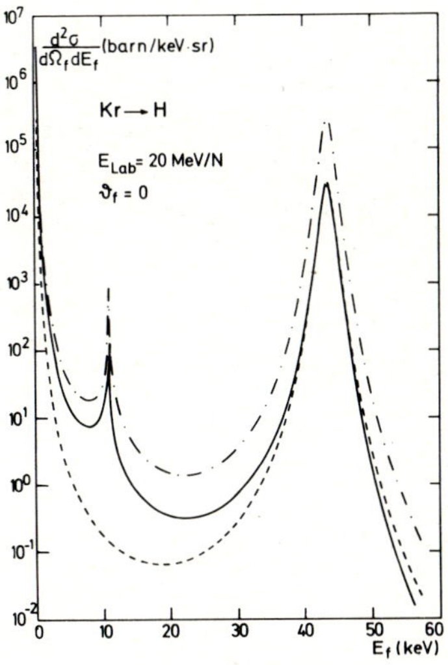

Fig.1: Differential cross section versus electron energy; full line, present theory; broken line, Born approximation; chain curve, theory from Ref.3.

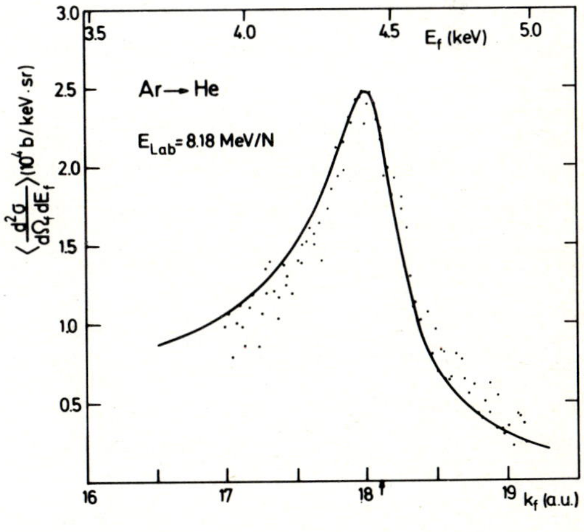

Fig.2: Differential cross section in the forward peak region; full line, present theory; data points from Ref.4

References

1. J.Macek, Phys.Rev.A1, 235 (1970)
2. J.S.Briggs, J.Phys.B10, 3075 (1977)
3. J.E.Miraglia and V.H.Ponce, J.Phys.B13, 1195 (1980)
4. M.Breinig et al, Phys.Rev.A25, 3015 (1982)

THE LINEWIDTH OF ELECTRON LOSS TO CONTINUUM CUSPS

J. Burgdörfer,* M. Breinig, S. B. Elston and I. A. Sellin

University of Tennessee, Knoxville TN 37996, USA, and
Oak Ridge National Laboratory, Oak Ridge TN 37830 USA

The velocity spectra of projectile electrons emitted in ionizing fast ion-atom collisions show a cusp at electron velocities $\vec{v}_e$ near the projectile velocity $\vec{v}_p$. This corresponds to an enhancement of the cross section for ionization to low-lying continuum states in the Coulomb field as seen in the rest frame of the projectile.

Recent experiments[1,2] focused on the shape of the electron loss to continuum (ELC) cusp as described by the linewidth (FWHM) of the singly differential cross section $\frac{d\sigma}{dv_e}(v_e)$ for electrons emitted within a narrow cone of half angle Θ_o around the forward direction. It is found that the ELC cusps are symmetric and narrow in comparison with electron capture to continuum (ECC) cusps.

We have initiated a systematic theoretical study of the ELC linewidth as a function of the initial bound state of the electron, its binding energy, the projectile velocity and the target structure. Previous calculations for the ground state[3] are extended by evaluating the bound-free transition form factor $\langle \vec{k}^- | \exp(i\vec{q}\vec{r}) | n\ell m\rangle$ from arbitrary hydrogenic initial states $|n\ell m\rangle$ to low-lying continuum states $|\vec{k}^-\rangle$ with $k \ll (na)^{-1}$. This form factor determines the ELC cusp in Born approximation. The calculation can be done most conveniently by using an algebraic 0(4, 2) approach taking explicitly into account the dynamical symmetry of the Coulomb field. The result permits a systematic investigation of the angular distribution of the ejected electron.

In the rest frame of the projectile, the doubly differential cross section can be written as:

$$\frac{d\sigma}{d\vec{v}_e'}(\vec{v}_e') = \sum_{\substack{k=0 \\ (k\ \text{even})}}^{2n} a_k(n\ell m) P_k^0(\cos\Theta), \qquad (1)$$

where $\cos\Theta = \hat{v}_e' \cdot \hat{v}_p$, and the coefficients a_k contain information about the initial state and the scattering dynamics. The highest-order anisotropy is determined by $k_{max} = 2n$ (not 2ℓ!). Therefore, the angular distribution becomes highly anisotropic with increasing principal quantum number n. Transformed into the laboratory frame, Eq. (1) leads to drastic changes of the ELC linewidth compared with Γ for an isotropic cusp (Figure 1). In the calculation for $\Gamma(v_p)$, we have used hydrogenic wave functions with effective charges for the O^{5+} projectile. The argon target is treated in closure approximation. The resulting cusp shape is primarily determined by the loosely bound 2s electron yielding a narrow linewidth in agreement with the experimental data.

This research was supported in part by the National Science Foundation and by the Fundamental Interactions Branch, Division of Chemical Sciences, Office of Basic Energy Sciences, U. S. Department of Energy, under contract W-7405-eng-26.

References

1. M. Breinig, S. B. Elston, S. Huldt, L. Liljeby, C. R. Vane, S. D. Berry, G. A. Glass, M. Schauer, I. A. Sellin, G. D. Alton, S. Datz, S. Overbury, R. Laubert, and M. Suter, Phys. Rev. A 25, 3015 (1982).
2. R. Cranage, W. Steckelmacher, and M. Lucas, Nucl. Instr. Meth. 194, 419 (1982).
3. J. Briggs and M. Day, J. Phys. B 13, 4797 (1980); and M. Day, J. Phys. B 13, L65 (1980).

* Permanent address: Freie Universität Berlin, Institut für Atom- u. Festkörperphysik, Arnimallee 14, 1 Berlin 33, Germany.

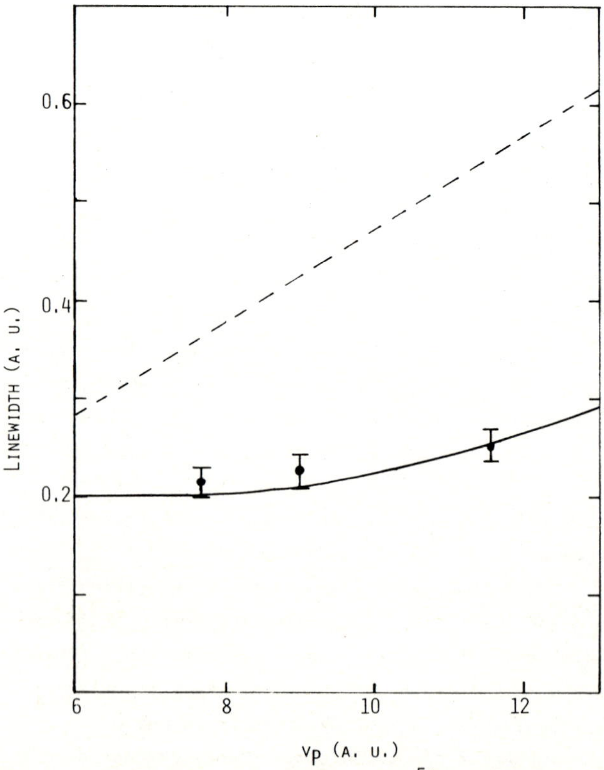

Figure 1: ELC cusp linewidth Γ for O^{5+} on argon (Θ_o = 1.8°). ——: present calculation; ----: $\Gamma(v_p)$ for isotropic electron emission; ⫿: experimental data.

ELECTRON CAPTURE TO THE CONTINUUM AT ASYMPTOTICALLY HIGH VELOCITIES

S. D. Berry, I. A. Sellin, L. H. Andersen, M. Breinig, S. B. Elston and M. M. Schauer

University of Tennessee, Knoxville TN 37996-1200, USA, and
Oak Ridge National Laboratory, Oak Ridge, TN 37830, USA

K.-O. Groeneveld and D. Hofmann

Institut für Kernphysik, Universität Frankfurt, D 6 Frankfurt/Main 90, F. R. Germany

N. Stolterfoht, H. Schmidt-Böcking, G. Nolte and G. Schiwietz

Hahn-Meitner-Institut für Kernforschung, D 1 Berlin 39, Germany

A sharp cusp is observed in the ejected electron velocity spectrum in ion-atom collisions when the electron velocity $\vec{v}_e$ matches the emergent ion velocity $\vec{v}$ in both direction and speed. If the electron originates from excitation of a projectile electron to a low-lying continuum state, the process is termed electron loss to the continuum (ELC), and if a target electron is transferred to a low-lying (projectile) continuum state, the process is called electron capture to the continuum (ECC). If the projectile is bare only ECC can occur, and the resulting cusp is asymmetric with higher intensity on the low-velocity side of the peak with the width observed to be roughly proportional to the ion velocity[1]. When several loosely bound electrons are available, ELC is the dominant process, producing a narrower, symmetric cusp peak.

To account for the ECC cusp shape asymmetry, Shakeshaft and Spruch[2] (SS) proposed inclusion of second Born expansion terms in calculation of the ECC cusp shape. In the second Born term all partial waves are thought to be important, and an asymmetry results from odd-ℓ (p, f, ...) partial waves for which $d\sigma/dv$ is asymmetric under the transformation $(\vec{v}_e - \vec{v}) \to -(\vec{v}_e - \vec{v})$. Their ECC cusp shape calculation for C^{6+} on H is characterized by a sheer drop in intensity on the high-velocity side of the peak. A modified first Born approach retaining terms linear in this difference has been proposed by Chan and Eichler[3] (CE). Because of a $\sim v^{-2\ell}$ partial wave dependence at asymptotically high velocities for first Born theories, the CE shape is dominated by s partial wave contributions and lacks the above-mentioned drop of the SS shape.

We present detailed results of recent measurements[4] of ECC for O, Ne and Ar projectiles colliding with He, with velocities 15-18 au. Using a partial wave expansion-fitting method[5], we find that for ECC at these velocities the p wave term gives a significant contribution to the cusp shape (Figure 1), producing a drop in the ECC cross section for $v_e \gtrsim v$ qualitatively more similar to the SS shape than to the CE shape. However, the predicted strong projectile Z dependence of the asymmetry within the SS theory is not found for the projectiles studied here, indicating a need for extension of their second Born calculations to helium targets.

We would like to thank the staffs of the HMI VICKSI cyclotron in Berlin and the Holifield Heavy Ion Research Facility at Oak Ridge National Laboratory for their generous assistance. This research was supported by the National Science Foundation; the Fundamental Interactions Branch, Division of Chemical Sciences, Office of Basic Energy Sciences, U.S. Department of Energy, under contract W-7405-eng-26; by DFG and BMFT in West Germany.

References

1. M. Breinig, S. B. Elston, S. Huldt, L. Liljeby, C. R. Vane, S. D. Berry, G. A. Glass, M. Schauer, I. A. Sellin, G. D. Alton, S. Datz, S. Overbury, R. Laubert, and M. Suter, Phys. Rev. A 25, 3015 (1982).
2. R. Shakeshaft and L. Spruch, Rev. Mod. Phys. 51, 369 (1979); and Phys. Rev. Lett. 41, 1037 (1979).
3. F.T. Chan and J. Eichler, Phys. Rev. A 20, 367 (1979).
4. S. D. Berry, I. A. Sellin, K.-O. Groeneveld, D. Hofmann, L. H. Andersen, M. Breinig, S. B. Elston, M. M. Schauer, N. Stolterfoht, H. Schmidt-Böcking, G. Nolte, and G. Schiwietz, IEEE Transactions on Nuclear Science, Vol. NS 30, No. 2 (1983).
5. W. Meckbach, I. B. Nemirovsky, and C. R. Garibotti, Phys. Rev. A 24, 1793 (1981).

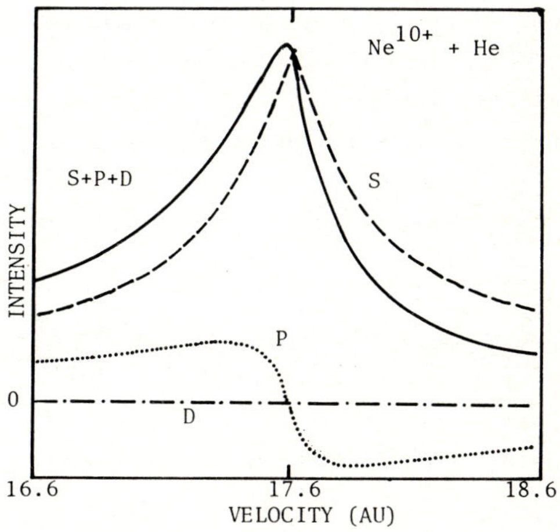

Figure 1: Fitted ECC cusp shape (solid line) for 17.6 au bare neon-He gas collision. Other curves are component partial waves of the fit.

DESCRIPTION OF INTERACTION IN FINAL STATE ON THE BASIS OF THE FADDEEV-MERKURIEV EQUATIONS IN THE PROCESSES OF IONIZATION AND CHARGE TRANSFER

A.L. Godunov, Sh.D. Kunikeev, V.N. Mileev, V.S. Senashenko

Institute of Nuclear Physics, Moscow State University, Moscow 117234, USSR

The experimental studies of the processes of ionization and charge transfer in the collisions of atoms with intermediate and high energy protons have revealed a number of interesting features in doubly-differential cross sections of ionization[1,2] and differential cross sections of charge transfer[3]. The nature of these features is determined by the structure of the final state formed as a result of collision. Therefore, an adequate description of the interaction in the final state between various atomic particles is of great importance when examining the above mentioned processes.

Here, we discuss the method for consistent description of the final states produced as a result of ionization and charge transfer. The method is based on the Faddeev equations modified by Merkuriev for the Coulomb potentials[4].

The amplitude determing the cross sections of ionization and charge transfer for proton-atom collisions is of the form

$$t_{fi} = \frac{\mu}{2\pi} \langle \Psi_f | V_i | X_i \rangle \qquad (1)$$

where μ is the reduced mass; V_i is the proton-target atom interaction. If the one-configuration wave functions of the initial and final states of atoms are used, then the initial state wave function of the three-particle system X_i, is described by the asymptotic wave function, and the final state wave function Ψ_f is the solution for the Faddeev-Merkuriev equations.

The approximate solution for the Faddeev-Merkuriev equations for three asymptotically free charged particles, which corresponds to the final state of the ionization process, is of the form[5]

$$\Psi_f = \frac{1}{(2\pi)^{9/2}} \exp i(\vec{k}_1\vec{r}_1+\vec{k}_2\vec{r}_2+\vec{k}_3\vec{r}_3) \prod_{k=1}^{3} \Phi_k \qquad (2)$$

In the case when two of the particles form a bound state (which corresponds to charge transfer) one has

$$\Psi_f = \frac{1}{(2\pi)^3} \exp i(\vec{k}_3\vec{r}_3+\vec{K}_{12}\vec{R}_{12}) \varphi_f(\vec{r}_{12}) \prod_{k=1}^{2} \Phi_k \qquad (3)$$

Here

$$\Phi_k = \exp\left(-\frac{n_{ij}\pi}{2k_{ij}}\right) \Gamma\left(1-\frac{n_{ij}}{k_{ij}}\right) F\left(\frac{i n_{ij}}{k_{ij}}, 1, -i(k_{ij}r_{ij}+\vec{k}_{ij}\vec{r}_{ij})\right)$$

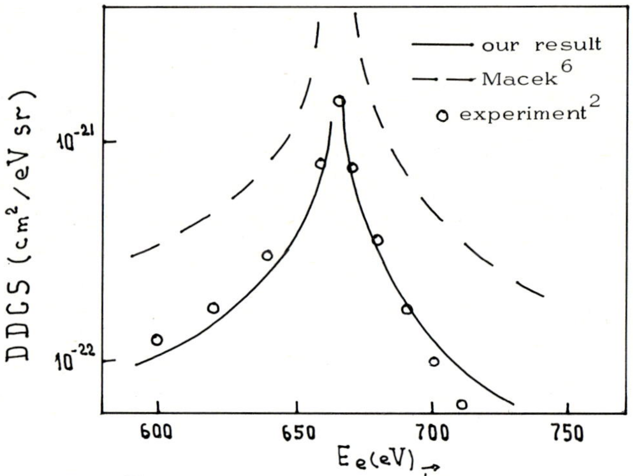

$n_{ij} = Z_i Z_j \mu_{ij}$, $\vec{k}_{ij}$ is the relative momentum; $\vec{r}_{ij}$ is relative coordinate of particles i and j; φ_f is the wave function of the pair (1,2) in bound state. The formulas (2)-(3) give the simple relationship between the cross sections of charge transfer and ionization which was observed experimentally in[2]. The correct asymptotic behaviour of the functions (2) and (3) is their important property, in contrast to the solution for the Faddeev equations of short-range potentials which was used in[6] to describe the ionization.

Fig. 1 illustrates the results of the calculations of the energy spectrum of the electrons emitted by He atoms at zero ejection angle in the case of ionization by 1225 keV protons. A marked improvement of the agreement of the presented calculations with experiment can be seen as compared with the calculations in[6]. Thus, the approach based on the Faddeev-Merkuriev equations with interaction in final state can well give an adequate description of the differential cross sections of the studied processes.

References .

1. M.E. Rudd, J. Macek, Case Studies in Atomic Phys. 3, 48 (1972).
2. M. Rødbro, F.D. Andersen , J.Phys.12B,2883 (1979).
3. P.J. Martin et al., Phys.Rev.,A23,3357(1981).
4. S.P. Merkuriev, Ann.Phys.(N.Y.)130, 395(1980).
5. A.L. Godunov et al., Zh.Tekh.Fiz.,53,436(1983).
6. J. Macek, Phys.Rev.,A1, 235(1970).

CONVOY ELECTRON PRODUCTION AND TOTAL ELECTRON YIELD IN HIGH VELOCITY (24 au) HEAVY ION-SOLID COLLISIONS*

R. Latz, M. Burkhard, H.J. Frischkorn, D. Hofmann, P. Koschar, J. Schader, K.O. Groeneveld

J.W. Goethe Universität 6000 Frankfurt/M, Germany

M. Breinig, S.D. Berry, I.A. Sellin

University of Tennessee and Oak Ridge National Laboratory, Oak Ridge, Tennessee 37830, USA

A sharp cusp-shaped peak in the velocity spectrum of electrons, ejected in ion-atom and ion-solid collisions, is observed when the ejected electron velocity $\vec{v}$ matches that of the emergent ion $\vec{v}_p$ in both speed and direction. Recent studies[1] of these cusp electrons produced by bare heavy ions on gaseous targets at high velocities ($v_p \simeq 17$ au) revealed a strong asymmetry which was explained in terms of a double collision process requiring second or higher order Born expansion contributions[2] to electron capture to the continuum (ECC). It is predicted that these higher order terms become more important with increasing v_p. The measured asymmetry also depends strongly[1] on the initial projectile charge q_i when $q_i < Z_i$ because of additional contributions to forward electron production from electron loss processes.

To compare ECC and convoy electron production in solid targets further we measured the cusp peak at higher velocities $v_p = 24.8$ a.u. We chose the collision system Ni (882 MeV) → C, Al using target thicknesses ρx from 3 to 500 μg/cm^2, and $q_i = 24$ and 26. The cusp electrons were recorded with a magnetic analyzer.[3] The cusp electron yield Y_c was obtained from the spectra by subtracting electron and nuclear background contributions Y_i from the measured data, fitted in the wings by a third order polynomial. The following results were obtained:

1. The cusp shape can be well represented by the fit approach of Meckbach et al.[4]
2. Under the present experimental conditions no significant asymmetry of the cusp peak has been observed.
3. The shape is independent of a. the target thickness, b. the target atomic number, and c. the initial ion charge state q_i for the values studied in this experiment.
4. The cusp electron yield Y_c(Al) rises faster than Y_c(C) with increasing ρx.

In addition to the measurement of the cusp electrons we recorded simultaneously the total electron yield γ per projectile by a method described in Ref. 5. It was found that the ρx-dependence of Y_c, Y_i and γ is similar:

1. At ρx < 10 μg/cm^2 all three quantities increase slowly.
2. At 10 < ρx < 200 μg/cm^2 all three quantities increase more rapidly.
3. At ρx > 400 μg/cm^2, Y_i saturates independent of q_i.
4. At ρx > 200 μg/cm^2 both Y_c and γ saturate, with $\gamma_\infty(C) = 95 \pm 10$ and $\gamma_\infty(Al) = 105 \pm 10$ independent of q_i. The observed γ_∞-values are in quantitative agreement with the linear dependence of the electronic stopping power.[6]

* Partially supported by DFG/Bonn, BMFT/Bonn, GSI/Darmstadt, and NSF/Washington.

References

1 M. Breinig, S. Elston, I.A. Sellin, L. Liljeby, R. Thoe, C. Vane, H. Gould, R. Marrus and R. Laubert, Phys.Rev.Lett. 45, 1689 (1980).

2 R. Shakeshaft, L. Spruch, Rev.Mod.Phys. 51, 369 (1979).

3 R. Latz, G. Astner, H.J. Frischkorn, P. Koschar, J. Pfennig, J. Schader, K.O. Groeneveld, Nucl.Instr.Meth. 194,315 (1982).

4 W. Meckbach, I.B. Nemirovsky, C.R. Garibotti, Phys.Rev. A 24, 1795 (1981).

5 J. Schader, B. Kolb, K.D. Sevier, K.O. Groeneveld, Nucl.Instr.Meth. 151, 563 (1978).

6 H.J. Frischkorn, P. Koschar, R. Latz, J. Schader, M. Burkhard, D. Hofmann, K.O. Groeneveld, IEEE Trans. NS ___, (1983) to be published.

Auger Electron Emission Following Fast (MeV) Molecular- and Atomic-Ion Impact on Thin C-foils

D. Schneider, E. P. Kanter, B. J. Zabransky
Argonne National Laboratory, Argonne, IL 60439

Auger electron emission following foil-excitation of fast molecular and atomic ion-beams (CO_2^+, N_2^+ N^+, C^+) has been measured. Relative Auger electron yields from molecular-ion impact have been compared to that from atomic ion impact at equal ion velocities. Fig. 1 shows a comparison of continous electron spectra from 1 MeV C^+ and 2.33 MeV CO^+ ion impact on 5 μg/cm^2 C-Foils. Within the electron energy range the electron yield decreases by about 5 orders of magnitude. The structures are due to Auger electrons caused by target and projectile excitation. The spectrum produced with a 2.35-MeV CO^+ beam shows two distinct peaks due to Auger-electron emission from the two molecular partners.

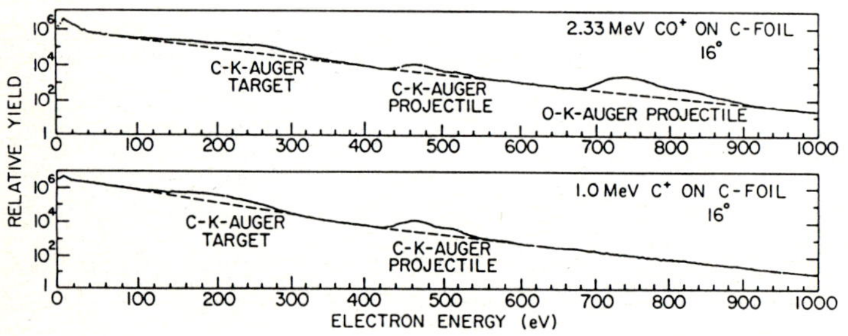

Fig. 1. Relative yield of secondary electrons versus electron energies up to 1000 eV. The electron emission is observed at 16° observation angle (w.r.t. beam axis).

The yields per incident ion are determined after background subtraction and integration of the remaining measured intensities from measurements at forward angle. It is found that the Auger yields from molecular ion impact are up to 40% lower than from the corresponding atomic beam impact.[1] For the cases studied the inner-shell vacancy production and decay in the solid can be interpreted within the framework of the Fano-Lichten model.[2] Previous studies[3] have shown that in order to produce a K vacancy via rotational coupling, the distance of closest approach in the collision must be smaller than approximately 0.1 Å. The molecular bond length of e.g. N_2 is approximately 1 Å, this means that the chance of making two violent K-vacancy-producing collisions with the nitrogen atoms in a nitrogen molecule in a single collision event is negligible. This assumption, does not imply that the probability of producing a K vacancy in such collisions is insensitive to the molecular nature of the target.[3] The quasimolecule formed in the collision of a molecular ion with a target atom is rather complex. The decrease in the Auger-yield and therefore the inner shell vacancy fraction can be assumed to be due to the vicinity of the molecular partner atoms. The influence of the fragment atoms in the molecule is felt through the mechanism determining the number of 2pπ MO vacancies in the diatomic ion-atom system at small internuclear distances. Considering the nitrogen case, the internuclear distance of an N_2^+ molecule is about 2.2 a.u. This distance varies due to the Coulomb explosion in the foil target. A 1.5-MeV N_2^+ molecule extends to an internuclear distance of about 4.4 a.u. when the cluster molecule emerges from a 5 μgr/cm^2 carbon foil. These distances are comparable to internuclear distances where the radial coupling mechanism between MO's becomes effective in the individual quasimolecule (e.g.N + C) formed in the collision. An increase of the effective nuclear charge in the individual nitrogen atom due to the proximity of the partner atom can explain a decrease in the vacancy production via radial coupling. Therefore a decrease of vacancy transfer from the 2pπ into the 2pσ and the increased effective nuclear charge in nitrogen could cause a reduction of the vacancy transfer probability into the K-shell of nitrogen.

This work was supported by the U.S. Department of Energy, Office of Basic Energy Sciences, under Contract No. W-31-109-Eng-38.

References

1 Y. D. Garcia, R. J. Fortner, H. C. Werner, D. Schneider, N. Stolterfoht, D. Ridder, Phys. Rev. A 22, 5 1884 (1980).

2 D. Schneider, N. Stolterfoht, Phy. Rev. A 18, 55 (1979).

3 B. Fastrup, A. Crone, Phy. Rev. Lett. 25, 825 (1972).

AUTOIONIZATION SPECTRA OF He EXCITED BY FAST (MeV) H^+, He^+, AND Li^{n+} (n = 1,2,3) IONS

D. Schneider, P. Arcuni, R. Bruch*), W. Stöffler

Argonne National Laboratory, Argonne, Illinois 60439

The study of high-resolution electron spectroscopy and angular distributions of He autoionization structures following energetic light ion bombardment yield fundamental information regarding ionization, electron transfer and double excitation mechanism. In particular, asymmetries of the He autoionization lines reflect the interference of the transition amplitudes for autoionization and direct ionization processes. So far such studies have only been performed at rather low projectile velocities[1-4]

In this contribution, we present autoionization spectra of He following excitation by 1 to 3 MeV H^+ and He^+. In addition, we have analyzed for the first time the Li^{n+} (n = 1,2,3) → He collision system. A detailed study of the $(2p^2)^1D$ and $(2s2p)^1P^0$ resonances have been made and a strong dependence on projectile velocity, charge state and observation angle is found.

The measurements were performed in a crossed-beam scattering chamber. The target-gas pressure was in the order of 10^{-4} Torr. The secondary electrons have been energy analyzed by a 45° parallel-plate analyzer. The resolution was typically 0.12 eV FWHM (obtained by applying deceleration mode). The observation angle of the spectrometer could be varied from 20° to 160° (w.r.t. beam-axis).

Fig. 1 shows spectra measured at various observation angles for different projectiles. The line shapes show strong asymmetries at forward angles which disappear towards backward angles. The interference is particularly strong if the transition amplitudes are comparable; the interference pattern disappears at backward angles where the direct ionization cross section is small. A preliminary analysis shows that the phase between the transition amplitudes depends sensitively on the projectile parameter (velocity, Z, charge state) and the observation angle.

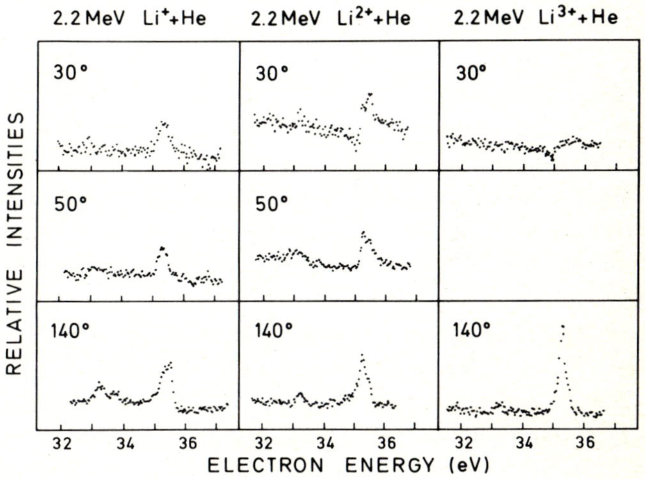

Fig. 1: He-autoionization spectra produced by fast Li^{n+}(n = 1,2,3) beams and measured as a function of observation angles.

References

*) Permanent address, Fakultät für Physik der Universität Freiburg, Freiburg, West Germany.

1. N. Stolterfoht, D. Ridder, and P. Ziem, Phys. Lett. 42A, 240 (1972).
2. A. Bordenave-Montesquieu, A. Gleizes, M. Rodiere, and P. Benoit-Cattin, J. Phys. B: 6, 1997 (1973).
3. N. Stolterfoht, Z. Phys. 248, 81 (1971).
4. M. Prost, Ph.D. Thesis (Hahn-Meitner-Institut, Berlin) unpublished.

AUTOIONIZATION OF FAST (MeV) Li-IONS INCIDENT ON GASES AND C-FOILS

D. Schneider, P. Arcuni, R. Bruch*, W. Stöffler, C.F. Moore†

Argonne National Laboratory, Argonne, Illinois 60439

Measurement of excitation cross sections of energetic light ions in collisions with gas targets and thin foils is of considerable importance for future developments in fusion research[1,2] and in extreme UV lasers. In this study we report for the first time relative cross sections for multiple excitation of LiI**, LiI*** and LiII** following beam-gas and beam-foil excitation of .75 and 2.2 MeV Li^+ ions. The cross sections are deduced from measurements of electron yields following the decay of the core-excited states via autoionization. The measurements have been performed at the Dynamitron accelerator at Argonne National Laboratory. Typical beam-currents applied in the experiments were about 30 nA. The well-collimated and defined ion-beam was directed onto a gas jet or a 5 μgr/cm^2 carbon foil. The electrons emitted at the target were analyzed with a 45° parallel plate analyzer. Electrons resulting from the prompt decay of autoionizing states in Li were measured at 15° observation angle with an energy resolution of 2.6 %. Due to the kinematic effect the Li-autoionization lines are strongly shifted in energy (Fig. 1). The continuous electron background was subtracted and the spectra were transformed into the projectile rest frame[3]. The intensities were integrated and relative cross sections and yields were deduced. Some general trends can be observed. For example it was found that the cross section increases drastically with increasing target Z's and decreasing beam velocity. For .75 MeV Li an increase of about a factor 20 is observed when going from He to Ar as a target.

The measurement seems to indicate that electron capture plays a dominant role in the excitation of Li as higher target Z are used. At the higher velocity (2.2 MeV Li^+) no increase in the cross section is observed when going from Ne to Ar, the cross section increase going from He to Ar is about the same as for 750 keV Li^+ impact. The insert in Fig. 1 shows the Li autoionization spectra at rest-frame energies. The intensities in the first peak labelled I is due to singly core-excited states in LiI, whereas peaks labelled II and III originate from doubly core-excited states in LiII and LiI[4]. In the 750 keV Li^+ case the integrated intensities in the three peaks I, II, III are not very different for the Ne and Ar targets, while with the He target the intensity in peak I is higher by a factor 2 indicating that neutral LiI states are formed more efficiently in this system. For 2.2 MeV impact energy peak I disappears, i.e. only LiII** and LiI*** states with double K-vacancies are formed in the collision process. Finally in the case of foil excitation the relative intensities in the three observable peaks are again quite different. It is found that the total relative yield decreases considerably when going from ~ 50 keV to 2.2 MeV impact energy. In the 750 keV case almost all intensity is due to doubly core-excited states in LiII and LiI.

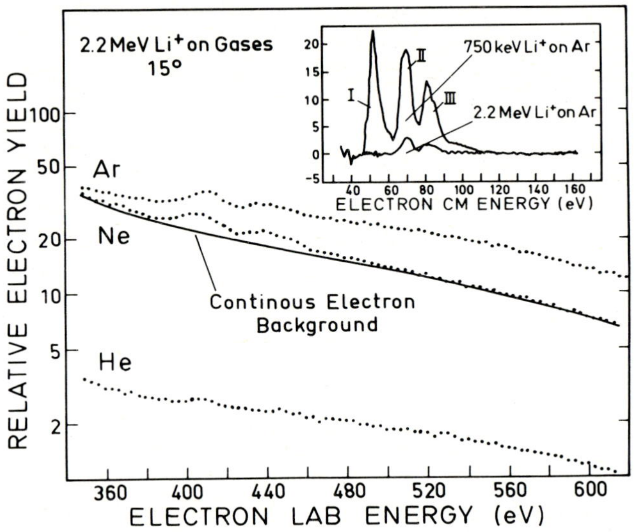

Fig. 1: Gas excited Li-autoionization spectra following 2.2 MeV Li^+ impact on He, Ne and Ar. The insert shows background subtracted spectra in the center of mass system for .75 MeV Li^+ and 2.2 MeV Li^+ incident on Ar (the 2.2 MeV lines are magnified by factor of three).

References

*) Permanent address, Fakultät für Physik, Universität Freiburg, 7800 Freiburg, West Germany.

†) Permanent address, University of Texas, Austin, Texas.

1. H. Bluhm, H.U. Karow, D. Rusch, K.W. Zieher KfK Nachrichten 2, 89 (1982)
2. H.U. Karow, Materieforschung mit Leichtionen-Strahlen, Primärbericht, Kernforschungszentrum Karlsruhe (1982) and private communication.
3. N. Stolterfoht, D. Schneider, D. Burch, B. Aagaard, E. Bøving, B. Fastrup, Phys.Rev., A12, 1313 (1975).
4. M. Rødbro, R. Bruch, P. Bisgaard, J.Phys. B12, 2413 (1979).

POSITION AND WIDTHS OF AUTOIONIZING STATES IN THE HELIUM ISOELECTRONIC SEQUENCE ABOVE THE n = 2 CONTINUUM

H. Bachau

Laboratoire des Collisions Atomiques, Université de Bordeaux I, 33405 Talence, France

Collisions between multicharged ions and helium-like atoms can produce atoms where the two electrons are excited (autoionizing states). For example, collision between O^{8+} and He produces autoionizing states of O^{6+} below the levels n = 3 and n = 4 of O^{7+}. There are many studies of autoionization in the helium isoelectronic sequence below the n = 2 hydrogenic thresholds, but few papers treat the higher series. In this work, we study these series. Position and widths of levels are calculated using the Feshbach projection operator theory[1].

We use the truncated diagonalization method[2] to calculate eigenvectors χ and associated eigenenergies of the Q H Q operator. We define $P = P_1 + P_2 - P_1 P_2$, where P_K is a projection operator onto the space of energetically accessible eigenstates of the k th electron.

Example : consider the autoionizing states below the n = 3 hydrogenic thresholds, we use :

$$P_i = |U_{1s}(i)><U_{1s}(i)| + |U_{2s}(i)><U_{2s}(i)| + |U_{2p}(i)><U_{2p}(i)|$$

where $U_{n\ell}(i)$ is the hydrogenic wavefunction which represents the n, ℓ orbital of electron i.

$$Q = 1 - P$$

The non resonant part of the total wavefunction is solution of equation :

$$(P\, H^{LS}\, P - E)\, \psi = 0$$

H^{LS} : non relativistic hamiltonian of the system where L and S are respectively the total orbital momentum and spin.

Application of the static exchange approximation gives :

$$\Psi \simeq \Psi^{EX}_{n_2\ell_2,\ell_1} = U_{n_2\ell_2}(1)\, F_{\ell_1}(2) + (-1)^S\, U_{n_2\ell_2}(2)\, F_{\ell_1}(1)$$

In fact, we have to solve the problem of elastic scattering of an electron by a hydrogenic target (in a state $n_2\ \ell_2$), where ℓ_1 is the orbital momentum of the scattered electron. This approximation is good for large target charges.

We utilize the Kohn-Hulthen variationnal principle and obtain the well-known integro-differential equations[3] of the form :

$$\left(\frac{d^2}{dr^2} - \frac{\ell_1(\ell_1+1)}{r^2} + k^2 + 2\frac{Z}{r}\right) F_{\ell_1}(r) = \sum_K V^K_{n_2\ell_2,n_2\ell_2}(r) + (-1)^S R_{n_2\ell_2}(r) \sum_{K'} Y_{K'}(F_{\ell_1}, r) + \delta_{\ell_1\ell_2} (-1)^{S+L} R_{n_2\ell_2}(r)\, X\,(F_{\ell_1}, r)\, E_1$$

$n_2\ell_2$	ℓ_1	Width (eV)
2p	2	0.09
	0	0.22
2s	1	0.13
1s	1	0.004
Total		0.44 (0.41)
energy (eV)	- 180.24 (- 180.21)	

Tableau I - Resonance parameter of the first $^1P^o$ state under n = 3 hydrogenic threshold. The number inside parentheses are the HO values.

We use and generalize the procedure applied in[4] (see also[5]) by MRC McDowell et al. to solve equation E1. In our case, we can obtain more than two coupled equations (see equat. C9 and C10 in[5]), the number $N^{K'}$ of values K' determines the number of coupled equations :

- We have $N^{K'} + 1$ coupled second order equations if $\ell_1 \neq \ell_2$
- We have $N^{K'} + 2$ coupled second order equations if $\ell_1 = \ell_2$

Application of the golden rule formula gives the width $\Gamma_{n_2\ell_2,\ell_1}$:

$$\Gamma_{n_2\ell_2,\ell_1} = 2\pi\, |<\chi\, |\, Q\, H^{LS}\, P\, |\, \psi^{EX}_{n_2\ell_2,\ell_1}>|^2$$

Finally, we determine the spontaneous emission probability using the dipole approximation.

Tableau I gives an example of calculation for O^{6+}. We consider the first state of the ^1P sequence below the n = 3 hydrogenic threshold. All possible final orbital momentum ℓ_1 and ℓ_2 are considered. Results are in good agreement with those of YK Ho [6]. However, this is not the case in general, because the angular orbital expansion given by Ho (see formula[6] p. 388) does not cover all possible open channels. The lifetime against radiative decay to the 1s 3s and 1s 3d states is of the order of 10^{-12}s, the total autoionization lifetime is about 10^{-15}s.

Reference

1. H. Feshbach, Annals of Physics, 19 287-383 (1962)
2. L. Lipsky, M.J. Connely, Phys. Rev. A 14 2193-05 (1976)
3. P.G. Burke, H.M. Schey, Phys. Rev. 126 147-62 (1962)
4. MRC. McDowell, L. Morgan and V.P. Myerscough, J. Phys. B 6, 1435-51 (1973)
5. MRC. McDowell, L. Morgan and V.P. Myerscough, Comp. Phys. Com. 7, 38-49 (1974)
6. YK. Ho, J. Phys. B 12, 387-99 (1979

A Study of the Velocity Dependence of the Yield of the $(1s2s2p)^4P_{5/2}$ Metastable Fraction in MeV C, O and F Ions Following Excitation by C Foils*

J. K. Swenson, D. Brandt, M. Clark, and S. M. Shafroth

University of North Carolina at Chapel Hill, Chapel Hill NC and
Triangle Universities Nuclear Laboratory (TUNL) Durham, NC

and

J. R. Huddle

North Carolina State University, Raleigh, NC

We studied the velocity dependence of Li-like $(1s2s2p)^4P_{5/2}$ metastable fraction in $C^{1,2,3+}$, $O^{1,2,3,4+}$ and $F^{2,3,4+}$ ions incident on 2-100 $\mu g/cm^2$ C foils over the velocity range $.02 \leqslant v/c \leqslant .06$. By measuring the decay in flight spectra of the various beams after foil excitation, the metastable fraction could be determined from the observed yield of Auger electrons resulting from the decay of the $^4P_{5/2}$ state[1].

The electrons were analysed with the new high efficiency Position Sensitive 30° Parallel Plate Analyser (PSPPA)[2] shown in figure 1. This analyser can detect and energy analyse electrons that differ in energy by 20% at one plate voltage setting. This gives the PSPPA an electron detection efficiency of over 100 times that of conventional parallel plate analysers. The parallel plate geometry allows angular distributions to be done in contrast to other high efficiency analysers.

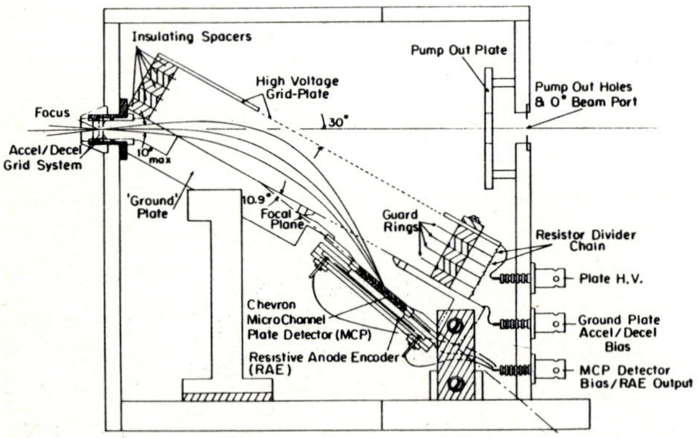

Fig. 1. Schematic drawing of the 30° PSPPA.

Smith, et. al.[1], made similar measurements for O and F and found a maximum of nearly 1% of the beam in the $^4P_{5/2}$ state upon emerging from their foils. We cannot make a direct comparison between our data and theirs as we have yet to determine the absolute efficiency of the PSPPA. The general trend of the data and relative maximum yields between O and F do agree quite well.

The results shown in figure 2 show the metastable fraction for C, O and F in relative units. The metastable fraction is found to exhibit a maximum just below the L-shell velocity, given by $Z\alpha/2$, for the ion species studied and falls off rapidly in either direction. The behavior below the L shell velocity is not as well observed in the systems studied thus far. However, we will be making measurement of the Si $^4P_{5/2}$ metastable fraction which should span a larger velocity range than studied thus far.

Angular distribution measurements and foil thickness studies have also been done and show interesting results. We have also done gas target feasibility studies for measurement of the $^4P_{5/2}$ metastable fraction as a function of incident charge state under single collison conditions and preliminary results are encouraging.

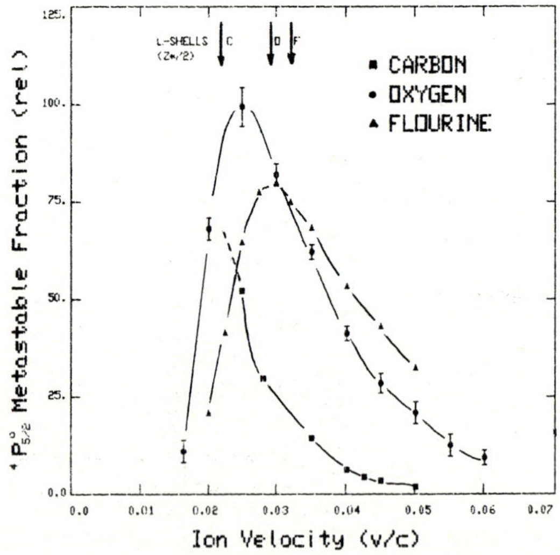

Fig. 2. $^4P_{5/2}$ metastable fraction for beams emerging from carbon foils.

[1]W. W. Smith, et. al., Phys. Rev. A 7, 487, (1973).
[2]J. K. Swenson and D. Br andt, IEEE Trans. on Nucl. Sci. NS 30 No. 2 (1983) (to be published).

*Work supported by U.S.D.O.E. Chem. Science Division.

AR L-MM AUGER SPECTRA IN $AR^{3,4+}$ + AR COLLISIONS

T. Matsuo,* J. Urakawa,** A. Yagishita,*** Y. Awaya, T. Kambara, M. Kase, H. Kumagai, J. Takahashi

The Institute of Physica and Chemical Research, Hirosawa, Wako-shi, Saitama, 351 JAPAN

The energy distributions of Ar L-MM Auger electrons produced in $Ar^{3,4+}$ + Ar collisions have been measured for several projectile energies between 800 keV and 27 MeV. The incident $Ar^{3,4+}$ beams were accelerated by the heavy-ion lenear accelerator of the Institute of Physical and Chemical Research. The electrons ejected at angle of 132 degree with respect to the direction of the primary beam were retarded and focussed with an electrostatic lens, and energy= analyzed in a hemishherical electrostatic analyzer. The analyzer was operated in a constant resolution mode. The present energy resolution was 0.7 eV FWHM. The target gas was effused from multichannel capillary positioned a few milli-meters below the collision region.

An example of the obtained spectra is shown in Fig.1. It was measured with incident Ar^{4+} energy of 14.3 MeV. The electrons were energy-analyzed in the energy range from 85 to 345 eV. The L-MM Auger lines from projectile Ar ions shift to the energy region below 85 eV according to the Doppler energy shift in the present experimental condition. The peaks which appeared are, therefore, attributed to the Auger electrons originated from the targets. The spectrum is characterized by sharp peaks positioned at about 101 and 112 eV, and by a broad-peaked structure extended through 120 eV to 260 eV. This structure is different from that obtained by electron impact, and is considered to be explained in terms of the L-MM Auger transitions in highly ionized Ar.[1] Tow remarkable features were found from this result.

The peaks around 101 eV, which were observed for the first time in the present measurement, are considered to be due to the L_{23}-M_1M_1 Auger electrons ejected from Ar ions with six vacancies in the M-shell; $Ar^{7+}(2p^5 3s^2; {}^2P_{3/2}, {}^2P_{1/2})$. Larkins[2] has predicted an energy value of 104 eV for the L_{23}-M_1M_1(N=6) transitions, where the number(N) indicates the number of vacancies in the M-shell. We found that they consist of two peaks and their intensity ratio is about 2:1. This value is almost the same as the ratio of the degeneracy of the electric total angular momentum (2J+1) between the initial states (${}^2P_{3/2}$ and ${}^2P_{1/2}$). The peaks around 112 eV may be due to the L_{23}-M_1M_1(N=5) transitions, and the structure between 120 and 200 eV is ascribed to the superposition of various transitions, such as L_{23}-M_1M_1(N=4,3,2,1,0), L_{23}-M_1M_{23}(N=5,4,3,2,1,0) and L_{23}-$M_{23}M_{23}$(N=4,3,2,1,0).

A broad structure appeared in the energy range between 200 and 260 eV. In the spectrum by electron impact, no significant peak has been observed in the energy range between 220 and 300 eV, corresponding to the transition energies for the Ar L_1-MM Auger because of the high Coster-Kronig transition probability for the filling a 2s-vacancy by a 2p-electron. Here, it should be noted that as the number of vacancy (N) in the M-shell increases, the L_1-L_{23}M Coster-Kronig transitions become energetically forbidden.[2] The spectrum between 200 and 260 eV can be explained to be due to the L_1-MM Auger electrons ejected from multiply-ionized atoms.

Reference

1) N. Storterfoht, D. Schneider and H. Galber: Phys. Lett., 47A, 271 (1974).

2) F.P. Larkins:J. Phys. B, 4, 1 (1971).

* Tokyo University of Agriculture and Technology.

** Tokyo Institute of Technology.

*** National Laboratory for High Energy Physics.

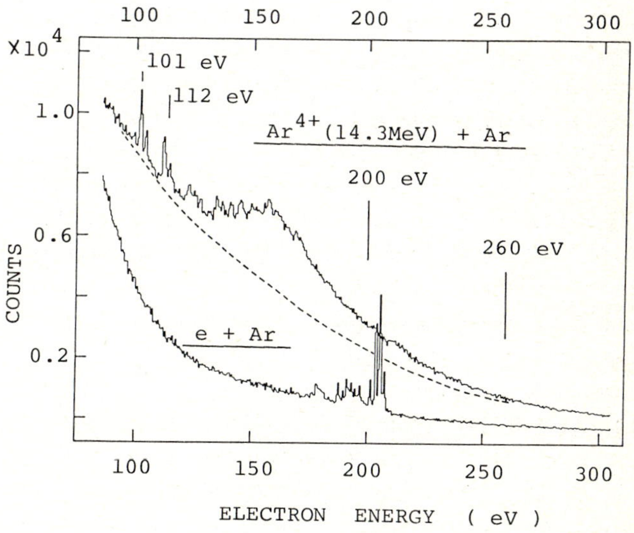

Fig.1 Electron spectrum obtained for (14.3 MeV-Ar^{4+} + Ar) system compared with for (1.2 keV-e + Ar) system, measured at 132° with respect to the direction of the incident beam. The dotted line indicates the contineous background.

KR M-NN AUGER SPECTRA IN AR^{4+} + KR COLLISIONS

J. Urakawa*, T. Matsuo**, H. Shibata, A. Yagishita***, Y. Awaya, T. Kambara, M. Kase, H. Kumagai, J. Takahashi

The Institute of Physical and Chemical Research, Hirosawa, Wako-shi, Saitama, 351 JAPAN

In order to obtain detailed information on both energy levels of highly ionised atoms and multiple-ionisation processes, some measurements have been done for the L-Auger electrons in $Ar^{3,4+}$ + Ar collisions with incident energies from 800 keV to 27 MeV, using a high-resolution electron spectrometer and the beams from heavy-ion linear accelerator(RILAC) of IPCR.[1] It has been shown that many satellite lines resulting from multi-vacancy states appear dominantly. Similar measurements have been made for Kr targets. In this report, experimental results for M-shell Auger electron from multiply ionised Krypton are presented.

The strong Krypton M-vacancy production is expected to be induced by the projectile with the same velocity as the orbital velocity of the M bound electrons; corresponding energy of Argon ions is about 7 MeV for the Kr M_{45}-shell and 16 MeV for the Kr M_{23}-shell. Furthermore, the strong electron stripping of the outer shell is expected to be produced by sufficiently heavy projectile such as Argon ion. Therefore, electron spectra in Ar^{4+} + Kr collisions with incident energies from about 8 MeV to 18 MeV were measured in the energy range from 36 to 260 eV at an emission angle of 132° with respect to the direction of the beam. This experimental condition is chosen such that the peaks appeared are attributed to the electrons from the targets by considering the Doppler energy shift of Auger electrons emitted from Ar ions and that kinematic line-broadening effects and line blending are negligible. The experimental apparatus and procedure have been described elsewhere.[1] The energy scale of the electron was calibrated by Kr M-NN and Ar L-MM Auger peaks induced by 1.2 keV electron impact. The energy values of each M-NN and L-MM Auger peak were referred to the results by Werme et al.[2]

The spectrum obtained is shown in Fig.1. It was measured with an incident energy of 8.3 MeV. The measurement was made in the energy range from 50 to 260 eV, scanning with 0.36 eV step intervals. This structure is very different from that obtained by electron impact and is considered to be explained in terms of the M-NN Auger, M-MN Coster-Kronig and M-MM super-Coster-Kronig transitions in highly ionised Kr. For an investigation of the structure, we have compared the spectrum with the theoretical result by Larkins,[3] who showed that the Auger lines shift to lower energies as the degree of ionisation increases. The lines above spectrum in Fig.1 indicate the energies calculated by Larkins with the adiabatic model. The spectrum is characterised by the complicated structure from 50 to 58 eV, the broad one with a sharp peak from 58 to 65 eV and other one with a long high-energy tail from 65 to 90 eV. These structures are superimposed upon a nearly exponential continuous background resulting from the direct ionisation process. The structure from 50 to 65 eV is mainly ascribed to the superposition of many M_{23}-$M_{45}N$ Coster-Kronig transition lines stemming from various multiplets of the initial and final states. On the other hand, the structure from 65 to 95 eV is mainly ascribed to the superposition of many M_1-$M_{45}N$ Coster-Kronig transition lines. The measurements for various incident energies of the projectile ions are in progress.

Reference

1) T.Matsuo et al: J.Phys.B,16(1983) in press.
2) L.O.Werme et al: Phys.Scr.,6,161(1972) and 8,146 (1973).
3) F.P.Larkins: J.Phys.B,6,2450(1973).

* Tokyo Institute of Technology.
** Tokyo University of Agriculture and Technology.
*** National Laboratory for High Energy Physics.

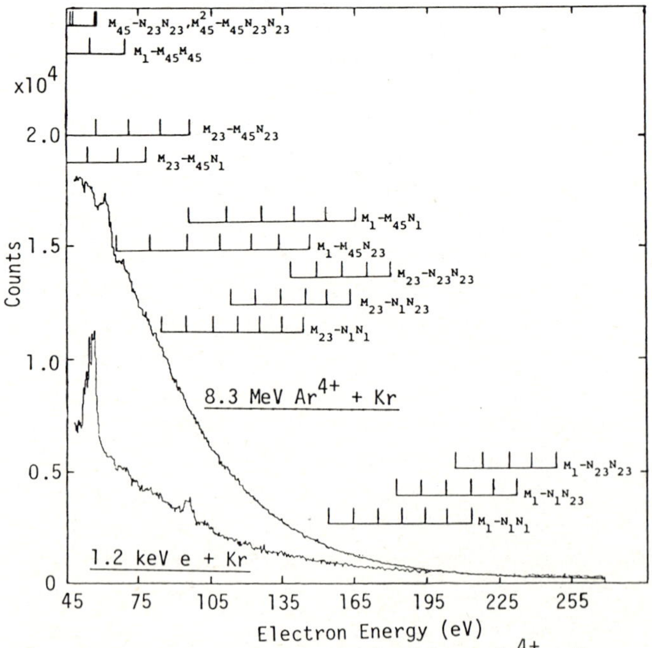

Fig.1 An electron spectrum obtained for a Ar^{4+} + Kr system (at an incident energy of 8.3 MeV) compared with that for a e + Kr system (at an incident energy of 1.2 keV), measured at 132° with respect to the direction of the incident beam.

SELECTIVE PRODUCTION OF AUGER ELECTRONS FROM FAST Ar^{q+} IONS STUDIED BY ZERO-DEGREE AUGER SPECTROSCOPY

A.Itoh, T.Schneider, G.Schiwietz, Z.Roller, H.Platten, G.Nolte, D.Schneider, and N.Stolterfoht

Hahn-Meitner-Institut für Kernforschung Berlin GmbH, Glienicker Str.100, D-1000 Berlin 39

High-resolution spectroscopy of Auger electrons in ion-atom collisions is often hampered by kinematic line broadening effects.[1] The Auger electrons from fast projectiles are primarily affected by kinematic broadening due to the finite acceptance angle of the spectrometer. Hence, most spectroscopic studies[2] of projectile Auger electrons have been limited to incident energies smaller than a few MeV.

In this work Ar-L Auger electrons were measured for 91.6-MeV Ar^{5+} and Ar^{6+} ions in collisions with H_2, He, and Ne. To reduce the kinematic line broadening effects, the electrons were observed under zero degree with respect to the incident beam. The experiments were performed using the heavy-ion accelerator facility VICKSI at Hahn-Meitner-Institut Berlin. The apparatus[3] was equipped with a gas cell and the ion beam was guided through the electron spectrometer to allow for the zero-degree measurements. The Auger spectra measured in the energy range from 1800 eV to 2800 eV were transformed into the projectile rest frame. Examples are given in Fig. 1 for Ar^{5+}.

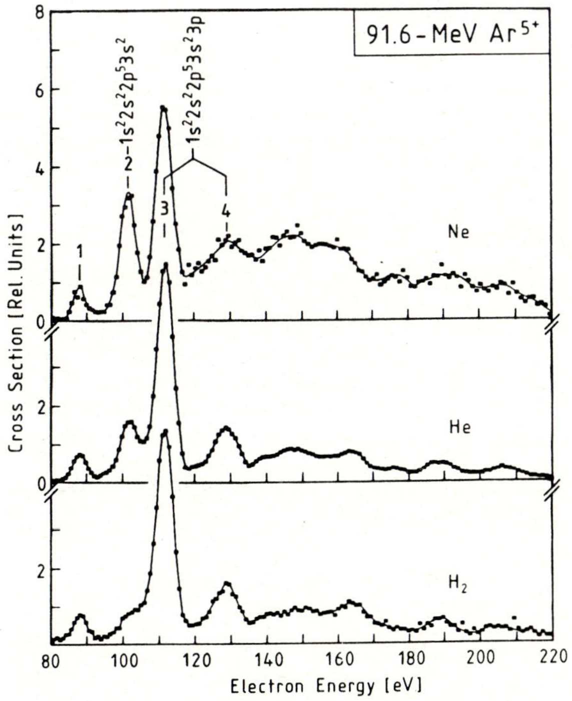

Fig.1: Ar-L Auger Spectra from 91.6-MeV Ar^{5+} incident on H_2, He, and Ne. The electron energy refers to the projectile rest frame.

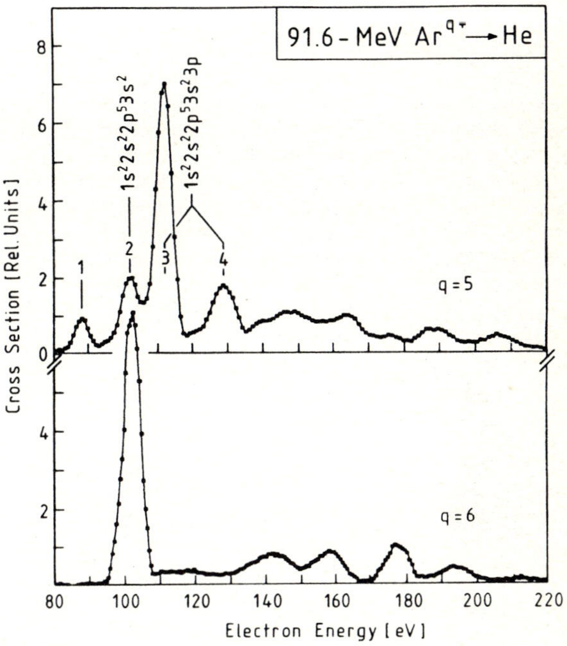

Fig.2: Ar-L Auger spectra from 91.6-MeV Ar^{5+} and Ar^{6+} incident on He. Energy scale as in Fig.1.

The Auger line identification is based on extrapolated theoretical data for transition energies.[4] Apart from the removal of the 2p electron line 3 and 4 are attributed to the ground state configuration of Ar^{5+}. Line 2 refers to the additional ionization of the 3p electron. The additional ionization is seen to loose significance for He and H_2. Hence, by using light target atoms, it is possible to excite predominantly lines which are characteristic for the charge state of the projectile. This is clearly demonstrated in Fig.2 where Auger spectra for q=5 and 6 are compared. It is seen that selective excitations of the configurations $1s^2 2s^2 2p^5 3s^2 3p$ and $1s^2 2s^2 2p^5 3s^2$ are achieved for incident Ar^{5+} and Ar^{6+}, respectively.

References

1. N.Stolterfoht, D.Schneider, D.Burch, B.Aagaard, E.Bøving, and B.Fastrup, Phys.Rev.A 12, 1313 (1975)
2. D.Schneider, R.Bruch, W.Butcher, and W.H.Schwarz, Phys.Rev.A 24, 1223 (1981)
3. D.Schneider, M.Prost, B.DuBois, and N.Stolterfoht, Phys.Rev. A 25, 3102 (1982)
4. P.Dahl, M.Rødbro, G.Hermann, B.Fastrup, and M.E.Rudd, J.Phys.B 9, 1581 (1976).

A STUDY OF AUGER SPECTRA FROM $Ne^{3+, 10+}$, $Ar^{6+, 17+}$, (5.6 MeV/amu) → Ne COLLISIONS

D. Berényi and G. Hock

Institute of Nuclear Research (ATOMKI) of the Hungarian Academy of Sciences, Debrecen, Hungary

I. Kádár[+], S. Ricz[+], V.A. Shchegolev, B. Sulik[+], D. Varga[+], J. Végh[+]

Joint Institute of Nuclear Research, Dubna, USSR

The K Auger spectra of Ne from high energy heavy ion collisions on Ne have been studied at various conditions in the last decade (recently in ref. 1; further references see there). However several aspects of the issues in connection with the processes involved have not been clarified.

In this study[2] energetic Ne and Ar ions (5.6 MeV/amu) were used as projectiles carrying either relatively many (Ne^{3+}, Ar^{6+}) or no and only one electron (Ne^{10+}, Ar^{17+}). The ion beams were obtained from the heavy ion cyclotron U-300 of the Nuclear Reaction Laboratory of JINR, Dubna.

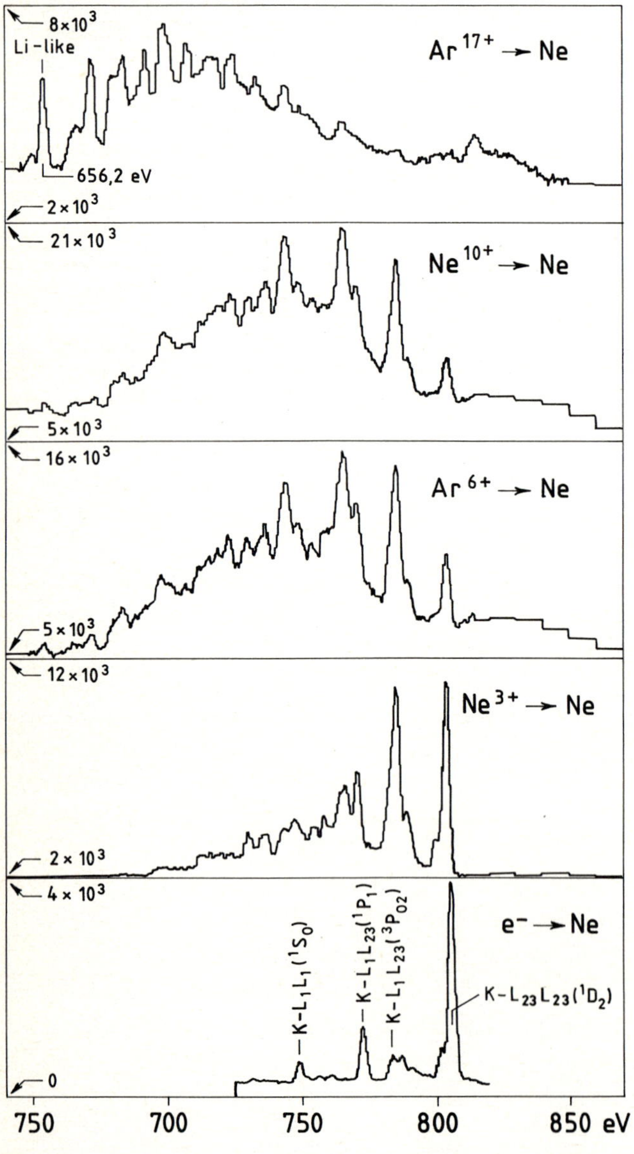

Fig. 1: Ne K Auger electron spectra produced by Ne^{3+}, Ne^{10+}, Ar^{6+}, Ar^{17+} of the same velocity (5.6 MeV/amu) and by 1.5 keV electrons

The spectra concerned were measured with an electrostatic electron spectrometer which is a combination of a spherical and a double pass cylindrical mirror type spectrometer (ESA - 21). The simultaneous measurement of the energy and the angular distributions (at thirteen angles from 0 to 180°) of the electrons ejected in ion-atom collisions is possible by means of this equipment (see a preliminary report on the spectrometer in ref. 3; a detailed publication is in preparation by Varga et al.[4]). The instrumental energy resolution (FWHM) was about 0.3 % (at 800 eV) in the present measurements.

The diameter of the heavy ion beam was 2.5 mm at target region. The ions of high charge state were produced by passage of Ne^{3+} and Ar^{6+} ions through a carbon foil (60-80 $\mu g/cm^2$). Finally, the intensity of the beam of the ions of different charge states is as follows : Ne^{3+} - 2×10^{11}/sec, Ne^{10+} - 2.5×10^{10}/sec, Ar^{6+} - 2×10^{11}/sec, Ar^{17+} - 1×10^{10}/sec. As target a Ne atom beam was used. The pressure in the atomic beam was about 10^{-3} mbar and outside the collision region $3 \div 5\times10^{-5}$ mbar. Without the target-beam a value lower than 10^{-7} mbar was measured in the collision chamber.

In. Fig. 1 the K Auger electron spectra of Ne are shown which were obtained by 5.6 MeV/amu $Ne^{3+, 10+}$ and $Ar^{6+, 17+}$ projectiles. For comparison the same spectrum was also taken by 1.5 keV electron impact. Here the spectra from the thirteen angular channels are summed up. The 804 keV diagram line can be observed in all the spectra except that one which was induced by Ar^{17+} projectile. Here however the line from Li-like configuration is visible. A tendency of the decreasing intensity of diagram lines (804 eV) can be observed as the charge state of the projectile increases.

Having the ratio of the areas under the spectra concerned by using Ne^{10+} - where there is no screening by electrons - as a reference case, one can get information on the deviation from Z^2 scaling rule or on the effective Z of the projectile ions.

References

1. D. Schneider, M. Prost, B. DuBois and N. Stolterfoht, Phys. Rev. A25 (1982) 3102
2. D.Varga, J. Végh, I. Kádár, S. Ricz, B. Sulik, V.A. Shchegolev, Contribution at Int. School-Seminar on Heavy Ion Physics, Alushta, 14-21 April, 1983
3. D. Berényi, " High-Energy Ion-Atom Collisions " (ed. D. Berényi and G. Hock). Akadémiai K. and Elsevier Sci. Publ. Co. Budapest-Amsterdam, 1982. p. 131
4. D. Varga, I. Kádár, S. Ricz, J. Végh, Á. Kövér, I. Cserny, A. Domonyi, to be published

[+] Permanent address : ATOMKI, Debrecen, Hungary

FORMATION AND AUTOIONIZATION OF NE** AT MG SURFACES

G. E. Zampieri, F. Meier* and R. A. Baragiola

Centro Atómico Bariloche, 8400 Bariloche, Argentina

We have studied the decay of autoionizing states of Ne formed in two-body collisions of 0.4-5 keV Ne^+ and Ne with Mg atoms at clean surfaces. The binary collision produces reflection of the Ne and the formation of two 2p-vacancies. In the outgoing part of the collision the Ne^{++} thus formed captures two 3s-electrons; the decay occurs outside the solid as shown by the energies of the autoionization lines and their small widths.

Our purpose in this work was to determine to what extent the excitation and decay of Ne^{**} is affected by the surface. The experiments were performed under UHV and the electron spectra were taken with a hemispherical analyzer with an energy resolution of 0.5 %.

In Fig. 1 we compare electron spectra from the decay of Ne^{**} produced in collisions with free atoms[1] and with a clean Mg surface. The assignment of the lines in solids was published[2]. The experiments with surfaces differ from those with gases in that:

a) Only two strong lines appear

b) The lines are broader

c) The lines peak at slightly larger energies

d) The triplet line is stronger than the singlet.

The excitation proceeds similarly to gas phase collisions except that in the incoming path all projectiles are neutralized by long-range resonant capture of electrons from the valence band of the solid. We confirmed this by observing the same spectra with incident neutrals or ions. The two 2p-vacancies are formed by the promotion of the 4fσ MO and coupling to empty states. This excitation requires distances of closest approach smaller than ∿0.6 Å, as follows from the threshold energy for excitation[3]. In our energy range, these distances result in most cases in reflection of the projectile into vacuum; this was confirmed by the energy-sependent Doppler broadening observed at different angles of observation with respect to the target normal.

The Ne^{++} may then capture two electrons to shells with n≥3. In contrast with the case of free atoms, the surface presents a continuum of empty states with the same energy as most of these Rydberg levels. The process of resonant ionization to empty conduction-band states thus competes with autoionization, and being faster (one-electron vs. two-electron transition) it depopulates all states of Ne^{**} lying above the Fermi level. Thus only final states $Ne^{**}(1s^2 2s^2 2p^4 3s^2)$ can survive; this explains the smaller number of peaks in experiments on surfaces.

Autoionization occurs in vacuum at a distance from the surface given by the velocity of the Ne^{**} perpendicular to the surface and the moment of decay. The interaction with the surface causes broadening of the peaks and their shift to high energies.

The larger intensity of the triplet line with respect to the singlet can be explained with the aid of a model by Olsen et al.[4] Triplet cores can be formed by an interchange of one core electron with a Rydberg electron at intermediate distances. This two-electron process, which is inefficient with free atoms, is more probable in solids due to the large density of available states and possibly by long range couplings with several target atoms

References

*Permanent address: Lab. Festkörperphysik, ETH Zürich, CH-8093 Zürich, Switzerland

1. N. Andersen and J. Olsen, J.Phys. B10, L719 (1977)
2. G.Zampieri and R.Baragiola, Surf.Sci.114, L15 (1982)
3. J.Ferrante and S.Pepper, Surf.Sci.57, 420 (1976)
4. J. Olsen et al, Phys. Rev. A19, 1457 (1979)

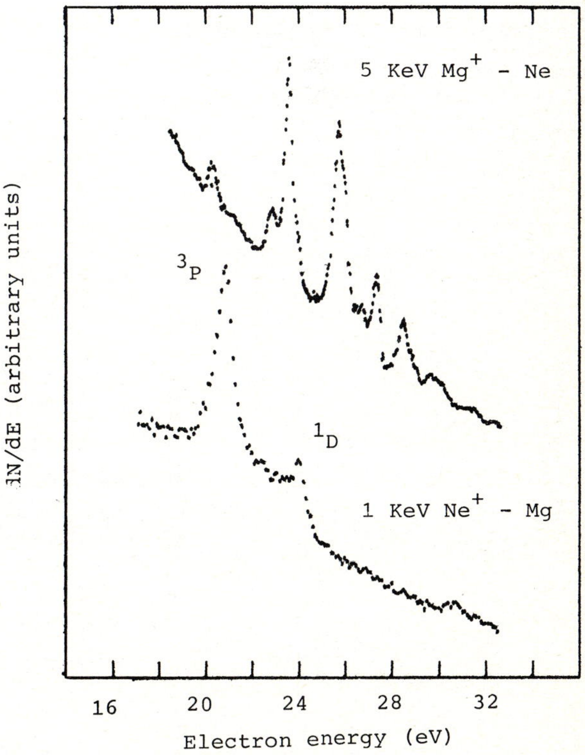

Fig. 1 - Electron spectra from gas-phase[1] Mg^+ - Ne collisions and Ne^+ - Mg surface collisions.

THE BARKER-BERRY EFFECT ON THE EJECTED ELECTRON SPECTRA IN Rb^+- Ar AND Cs^+- Kr COLLISIONS

K. Wada, A. Wada, K. Wakiya and H. Suzuki

Department of Physics, Sophia University, 7-1 Kioicho, Chiyoda-ku, Tokyo, 102 Japan

In the heavy particle collision, a detail analysis of the Barker-Berry effect[1)] on the ejected-electron spectra has not been done so far, except in a few papers,[2,3)] because the spectra are strongly affected by the Doppler effect. In this work the typical spectra in Rb^+- Ar and Cs^+- Kr systems can be reproduced by taking account for the classical model of the Barker-Berry effect and the Doppler effect.

Ejected-electron spectra have been measured in the range of the impact energies from 500 eV to 8 keV and at the observation angles from 25 deg to 150 deg with respect to the incident ion beam. Electron energies are analyzed by a concentric hemispherical analyzer with the resolution of 120 meV in FWHM. Detail of apparatus used here were given in ref.4.

At first, in Rb^+- Ar collisions many peaks were observed corresponding to the autoionizing levels of Rb formed by the charge-transfer from the target Ar. Figure 1 (a) and (b) show the measured spectra for different impact energies at the observation angle 25 deg and 90 deg in Rb^+- Ar collisions. It is remarkable that the peak $Rb(4p^5 5s^2)\,^2P_{3/2}$ shows an asymmetric shape and shifts toward the lower energy side as the impact energy decreases. These phenomena could be caused by the so-called Barker-Berry effect. When the Barker-Berry effect is discussed on the heavy particle collision, it is very important to estimate the Doppler effect correctly. By improving the treatment of the scattering angle on the incident particle, we could reduce the number of arbitrary parameters involved in the Doppler function.[5)] The distribution function used here is constructed in the following way. The classical Barker-Berry function is convoluted with a new Doppler function and furthermore with the apparatus function. Using this function, we tried to reproduce the whole spectra measured at any observation angles and impact energies with only the two parameters; τ and Rx. A symbol τ is the lifetime of the Rb $(4p^5 5s^2)\,^2P_{3/2}$ state and Rx is the internuclear distance where the avoided crossing exists. The best pair of the parameters has been determined to reproduce most closely the eight spectra measured for the different conditions, as shown Fig.1 (a) and (b). The most suitable parameters are determined; the lifetime τ of the $Rb(4p^5 5s^2)\,^2P_{3/2}$ state is 3500 ± 100 (a.u.) and the avoided crossing point Rx is 2.3 ± 0.1 (a.u.).

The same phenomenon caused by the Barker-Berry effect was observed in Cs^+- Kr collision. The spectra ejected from the $Cs(5p^5 6s^2)\,^2P_{3/2}$ state have been analyzed by the same method and the best pair of the parameters, τ and Rx, has been determined; the lifetime of the Cs $(5p^5 6s^2)\,^2P_{3/2}$ state is 3300 ± 100 (a.u.) and Rx is 2.3 ± 0.1 (a.u.).

These results support that such a phenomenon is well explained quantitatively by the classical Barker-Berry model. This method proves itself to be useful to determine a long lifetime of the autoionizing state.

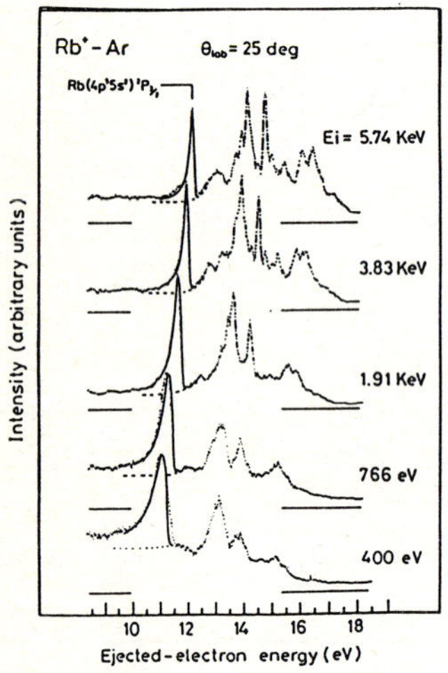

(a)

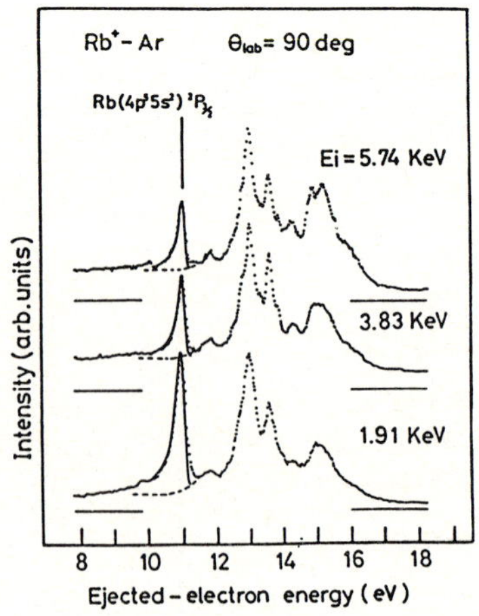

(b)

Fig.1: Ejected-electron spectra of Rb^+- Ar collision.
The solid curves represent the results of the calculation.
Observation angles (a) 25 deg; (b) 90 deg

References

1) R.B. Barker and H.W. Berry, Phys.Rev. 151, 14 (1966).
2) R. Morgenstern et al., J.Phys.B 13, 4811 (1980).
3) A. Bordenave-Montesquieu et al., Phys.Rev.A 25, 245 (1982).
4) "Atomic Collision Handbook" (1981), P.60, (in Japanese).
5) A. Gleizes et al., J.Phys.B 9, 473 (1976).

MEASUREMENT OF FORWARD DIRECTED ELECTRON SPECTRA IN COINCIDENCE WITH EMERGENT CHARGE STATES FOR COLLISIONS OF 20 MEV Au^{17+} WITH He

L.H.Andersen, H.Cederquist[§], S.Datz[$], M.Frost, P.Hvelplund, H.Knudsen, and L.Liljeby[§]
Institute of Physics, University of Aarhus, DK-8ooo Denmark

We have previously reported on the measurement of partial cross sections $\sigma^{o,n}_{q,q'}$ for the processes [1)]

$$Au^{q+}\ (2o\ MeV) + He^{o} \rightarrow Au^{q'+} + He^{n+} \qquad (q=5\ to\ 24)$$

where we observed large cross sections for transfer ionization (i.e. $\sigma^{o,2}_{q,q-1}$) which increased with q to values approaching half the simple charge transfer cross section ($\sigma^{o,1}_{q,q-1}$). We have also reported on the spectra of forward directed electrons accompanying these collisions [2)]. For large values of q where electron loss cross sections are very small the spectra consisted of an electron capture to continuum (ECC) peak and projectile Auger lines.

In this paper we present measurements of electron spectra in coincidence with specific exit charge states for 2o MeV Au^{17+} ions on He. Figure 1. shows a singles spectrum and Figure 2. a spectrum in coincidence with Au^{16+}.

Since He contains only two electrons , coincidences with Au^{16+} must correspond to transfer ionization. Contrary to initial expectations, we find that the Auger electrons are not associated with Au^{16+} but with Au^{17+} (spectrum not shown). Moreover, we find that the number of ECC-electrons associated with Au^{16+} is 1.4 times that obtained with Au^{17+}.

Using the same procedure employed by Vane et.al. [3)] to estimate the total ECC cross section we find that much of the transfer ionization cross section can be accounted for by

$$Au^{17+} + He \rightarrow Au^{16+} + He^{2+} + ECC$$

Since the Auger electrons are associated with Au^{17+}, they do not come from double capture to autoionizing states of Au^{15+}. They could arise from single capture onto metastable Au^{17+} ions which might be present in the beam. This possibility was checked by preparing the Au^{17+} ions in two ways: By foil-stripping Au^{5+}, and by charge capture from Au^{18+}. Both beams gave the same result, thus indicating that metastable contributions are not significant. An alternative explanation is exitation of a core electron and electron transfer occuring in a single collision. This possibility remains to be assesed.

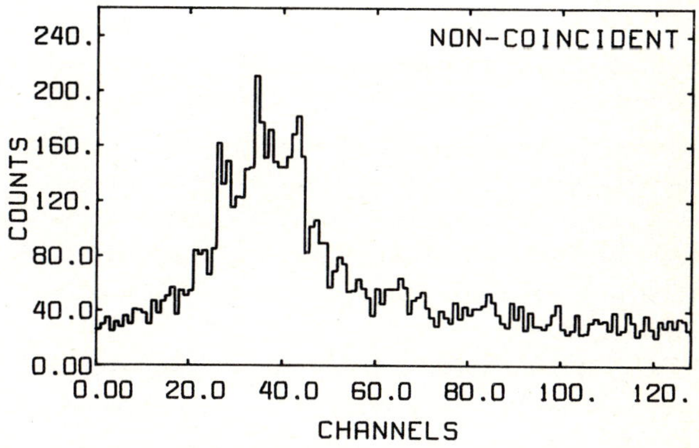

Figure 1. Singles spectrum.

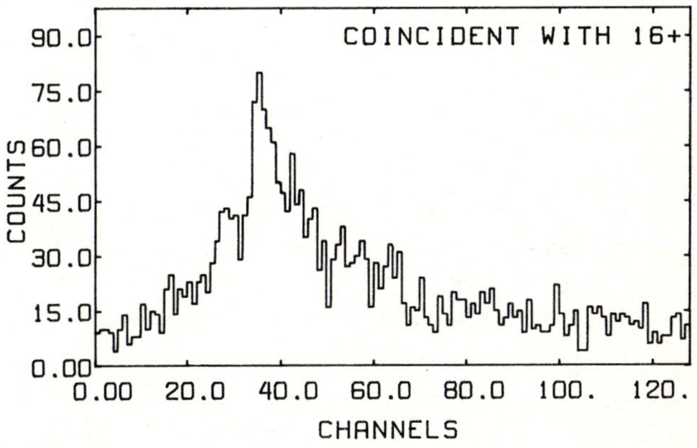

Figure 2. Spectrum in coincidence with Au^{16+}.

$ Oak Ridge National Laboratory , Oak Ridge Tenn. 3783o U.S.A

§ Research Institute for Atomic Physics, University of Stockholm S-1o4o5 Stockholm, Sweden.

References:

1) H.Damsgaard, H.K.Haugen, P.Hvelplund, and H.Knudsen Phys.Rev. A 27 ,112, (1983).

2) H.Knudsen, P.Hvelplund, L.H.Andersen, S.Bjørnelund, M. Frost, H.K.Haugen, and E.Samsøe, Physica Scripta, T3, 1o1, (1983).

3) C.R.Vane, I.A.Sellin, S.B.Elston, M.Suter, R.S.Thoe, G.D. Alton, S.D.Berry, and G.A.Glass, Phys.Rev.Lett, 43, 1388 (1979)

SPECTROSCOPY OF LOW-ENERGY AUTOIONIZATION ELECTRONS EMITTED FROM THE IONS IN HIGHLY CHARGED ION-ATOM COLLISIONS

L.H. Andersen, M. Frost, P. Hvelplund, and H. Knudsen
Institute of Physics, University of Aarhus, DK-8000 Aarhus C, Denmark

L. Liljeby
Research Institute for Atomic Physics, University of Stockholm, S-10405 Stockholm, Sweden

Recently, coincidence measurements of electron capture and target ionization in multiply charged Au^{q+} + He collisions ($5 \le q \le 21$) have been reported[1]. Determined was the so-called partial cross sections for charge change, where charges of both projectile and target are known before and after the collision. The cross sections are $\sigma^{on}_{q,p}$, where 0 refers to the initial charge of the target and n to the charge state of the target after the collision. q and p are the corresponding charges of the projectile.

A striking feature of these partial cross sections is the relatively large transfer-ionization cross section $\sigma^{02}_{q,q-1}$ (Ref. 2). This is of special interest because it leads to single-electron capture while involving two target electrons. The understanding of this process is therefore essential for the understanding of charge-exchange processes.

A number of possible mechanisms may lead to $\sigma^{02}_{q,q-1}$. At low ion velocities, it has been established that autoionization of the quasimolecule is important[3]. At medium velocities, autoionization after capture[4,5] as well as single capture with capture of another electron to the continuum[6] contributes to $\sigma^{02}_{q,q-1}$. Finally, direct ionization and single capture contribute at higher velocities[7].

Since the different mechanisms leading to transfer ionization result in different energy spectra of the emitted electrons, it is useful to study these electrons. We have used a 30° parallel-plate spectrometer and detected electrons emitted within a solid angle of 4.5 mstr around the beam direction. The spectrometer transmission/detection efficiency was found through normalization to measurements of K-Auger emission from CH_4[8]. With the 0° spectrometer, we detect the 'cusp' due to continuum-captured electrons. Further, we detect electrons emitted from the projectile via an autoionization process. Figure 1 shows as an example a spectrum from 20-MeV Au^{15+} on helium. The spectrum is presented as the double-differential cross section in both laboratory and projectile frames. The cusp is clearly seen in the laboratory representation together with some lines, which are due to autoionization. In the projectile frame, these lines are easily integrated to yield total cross sections.

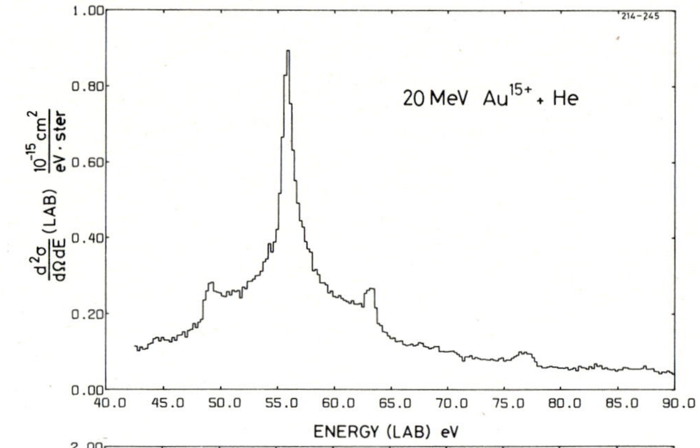

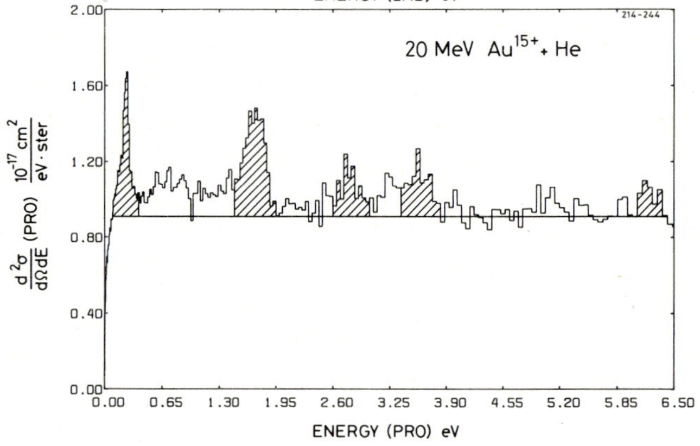

Further, we report on measurements of autoionization electrons from collisions between oxygen ions and argon ions.

REFERENCES

(1) H. Damsgaard, H.K. Haugen, P. Hvelplund, and H. Knudsen, Phys.Rev.A 27, 112 (1983)

(2) L.M. Kishinevskii and É.S. Parilis, Sov.Phys.JETP 28, 1020 (1969)

(3) A. Niehaus, Comm.Atom.Molec.Phys. 9, 153 (1980)

(4) C.L. Cocke, R. DuBois, T.J. Gray, E. Justiniano, and C. Can, Phys.Rev.Lett. 46, 1671 (1981)

(5) P.H. Woerlee, T.M. El-Sherbini, F.J. de Heer, and F.W. Saris, J.Phys.B:Atom.Molec.Phys. 12, L235 (1979)

(6) C.R. Vane, I.A. Sellin, S.B. Elston, M. Suter, R.S. Thoe, G.D. Alton, S.D. Berry, and G.A. Glass, Phys. Rev.Lett. 43, 1388 (1979)

(7) T.J. Gray, C.L. Cocke, and E. Justiniano, Phys.Rev. A 22, 849 (19870)

(8) M. Rødbro, E. Horsdal-Pedersen, and C.L. Cocke, Phys. Rev.A 19, 1936 (1979)

FORMATION AND DECAY OF AUTOIONIZATION STATES IN Ar^{3+}-Xe COLLISIONS

V.M.Mikoushkin, I.P.Flaks, G.N.Ogurtsov

A.F.Ioffe Physical-technical Institute of the Academy of Sciences of the USSR, Leningrad, USSR

Energy spectra of electrons ejected in Ar^{q+}-Xe collisions (q=I,2,3) have been studied experimentally at an incident ion energy E_o=5÷60 kev and at an electron ejection angle 54,5°.

In Ar^{3+}-Xe collisions, intensive peaks appeared in the energy spectra which were absent in the case of Ar^{+} and Ar^{2+} as projectiles (Fig.I). A shift of structure with varying E_o is caused dy Doppler effect, and this indicates that autoionization states (AIS) of the incident ion are excited. Positions of the peaks observed coincides whith energies of autoionization transitions from the states $3s3p^5(^3P)4p(\ D)$ (the lines I and 2) and $3s3p^5(^3P)4p(\ P)$ (the lines 3 and 2) to the states $3p^4(^3P_2)$ and $3p^4(^3P_{I,0})$ of Ar.*) Transitions to the different final states lead to the doublet structure of the peaks.

Appearance of the intensive AI-lines in Ar^{3+}-Xe case is naturally caused by large potential energy of the multiply charged ion and Coulomb interaction of particles that gives rise to the crossing of the relevant curves at large internuclear distances R and to transitions between these curves. It follows from the diagram of potential curves for the Ar^{3+}-Xe system (Fig.2) that multicharged nature of the projectile leads mainly to the possibility of exothermic charge exchange to the excited level 4p of the Ar^{2+} (Curve 2). In the case of projectiles of lesser charge the exothermic charge exchange to excited states is improbable. The Ar^{3+}-Xe system has also another, somewhat accidental, feature. The curve 2 corresponding to charge exchange almost coincides with the curve corresponding to excitation of AIS of Ar^{+} in a broad range of R. So one can arrive at a conclusion that the most probable mechanism for formation of AIS is follows:

$$Ar^{3+}+Xe \xrightarrow{I} Ar^{*2+}(3p^3(^4S)4p(\ P)+Xe^{+} \xrightarrow{2}$$

$$Ar^{**+}(3s3p^5(^3P)4p(\ D,P)+Xe^{2+}$$

$$\llcorner\!\!\rightarrow Ar^{2+}(3p^4(^3P_2,{}^3P_{I,0}) + e$$

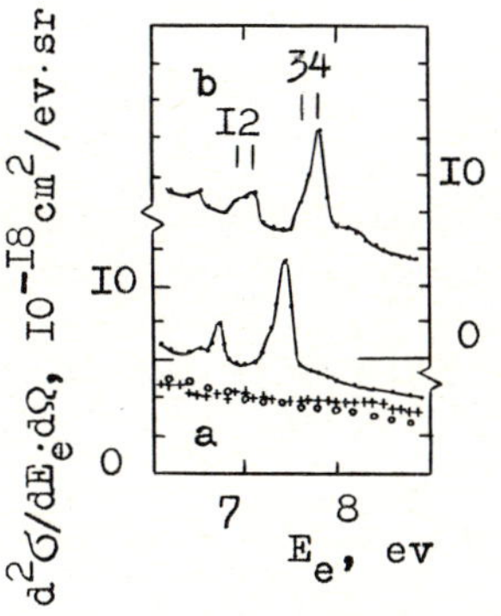

Fig.I Energy spectra of electrons ejected in collisions between :
+++ Ar^{+}-Xe,
ooo Ar^{2+}-Xe,
••• Ar^{3+}-Xe;
a) E_o=IO kev,
b) E_o=20 kev.

On the first step exothermic charge exchange to the excited state $Ar^{2+}(4p)$ occurs, and on the second step an electron of the Xe^{+} ion fills a vacancy in 3p-shell of Ar^{2+} and 3s-electron is transferred to the 3p-state (Fig.3). Thus, in Ar^{3+}-Xe collisions, the AIS $3s3p^54p$ Ar^{+} is formed as a result of two electron transition in the colliding system accompanied by demotion of vacancy.

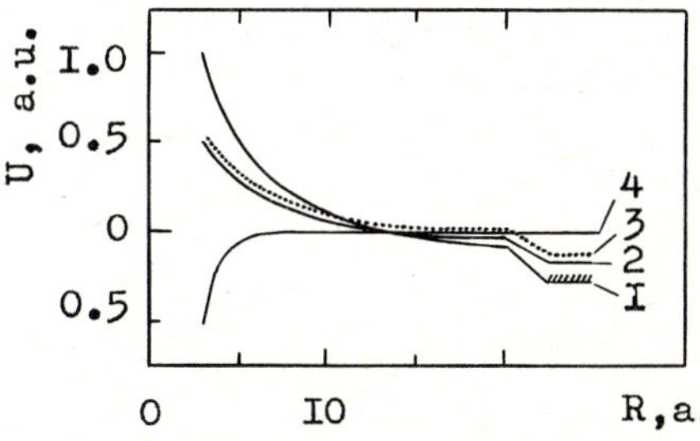

Fig.2 Potential energy curves for the systems
I. $Ar^{2+}+Xe^{2+}$, the boundary of continuum,
2. $Ar^{2+}(4p)+Xe^{+}$,
3. $Ar^{+}+Xe^{2+}$,
4. Ar^{3+}+Xe, the initial state.

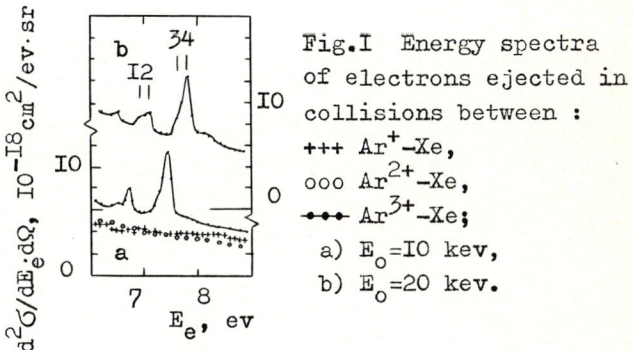

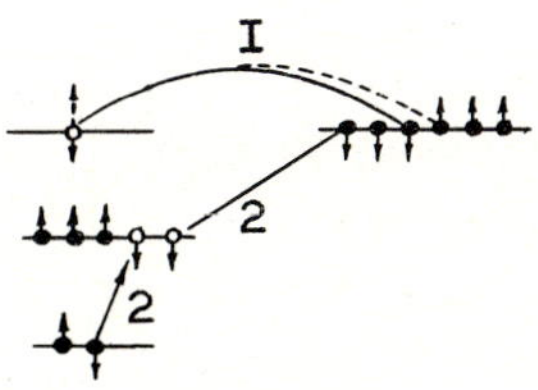

Fig.3 Schematic diagram of electron transition in the Ar^{3+}+Xe system.

*) Energies of autoionization states have been estimated using method of extrapolation of quantum defect.

CALCULATIONS OF THE EXCITATION CROSS SECTIONS OF THE PARITY UNFAVORED OF AUTOIONIZING STATES OF He ATOMS BY PROTONS

3KL30

V. A. Sidorovich

Institute of Nuclear Physics, Moscow State University, Moscow 117234, USSR

The double electron transition from the $1s^2\,{}^1S$ to the $n_1l_1n_2l_2\,{}^1L$ state in helium induced by protons is considered in the independent electron approximation. Using the results of[1] for the excitation probability of the $n_1l_1n_2l_2\,{}^1L$ state as a function of impact parameter ρ, we obtain

$$W(1s^2\,{}^1S \to n_1l_1n_2l_2\,{}^1L;\rho) = \sum_{M_L} \Big| \sum_{m_1,m_2} (l_1m_1\,l_2m_2|LM_L) \prod_{j=1}^{2} \frac{2\,i^{m_j+1}}{Z^*V} \sqrt{\frac{(l_j-|m_j|)!}{(l_j+|m_j|)!}} \int \frac{q_\perp dq_\perp}{q^2} J_{|m_j|}(q_\perp\rho)\, P_{l_j}^{|m_j|}(\cos\beta)\, M^{1s}_{n_jl_j0}(q) \Big|^2 \quad (1)$$

where n_j, l_j, m_j are the principal, orbital and magnetic quantum numbers for the final state of j-th electron; L and M_L are the total orbital angular momentum of final state electrons and its projection upon the direction of the relative motion velocity $\vec{V}$; $(l_1m_1\,l_2m_2|LM_L)$ the Clebsch-Gordan coefficients; $P_l^m(\cos\beta)$ are the associated Legendre polynomials; $J_\mu(x)$ is the Bessel function of order μ; $\cos\beta = q_\parallel/\sqrt{q_\parallel^2 + q_\perp^2}$; $\vec{q}$ is the momentum-transfer vector; $q_\parallel = \Delta E/V$ and $q_\perp$ are the parallel and orthogonal components of the vector $\vec{q}$ relative to the direction of $\vec{V}$; ΔE is the excitation energy per one electron; $M^{1s}_{nl0}(q)$ is the Born matrix element for the $1s$–$nl0$ transition calculated with the Coulomb wave functions $\varphi_{nlm}(Z^*|\vec{r})$ for nlm states in the field of the nucleus with effective charge Z^* given in the coordinate system with the Z-axis directed along the vector $\vec{q}$. In formula (1) the orbital angular momenta projections are determined on the direction of $\vec{V}$. The Coulomb units are used here.

From formula (1) it follows that the substates with $M_L = 0$ for the parity unfavored autoionizing states[2] ($l_1 + l_2 - L$ is odd) are unexcited and the cross sections for excitation of the substates with $M_L = \pm k$ are the same.

Fig. 1 presents the results of the calculations on the excitation cross sections for the $2p3d\,{}^1D$ autoionizing state of helium by protons and its substates with $M_L = 1$ and 2 as functions of proton energy. The excitation energy ΔE has been found from the effective charge Z^*, which has been put equal to 1.69. From Fig. 1 it follows that the main contribution to the cross section for excitation of the $2p3d\,{}^1D$ autoionizing state of helium by protons comes from the transitions into the substates with $|M_L| = 1$. This is due to the predominant role of single-electron transitions to the states with $m=-l, -(l-2), \ldots, +(l-2), +l$.

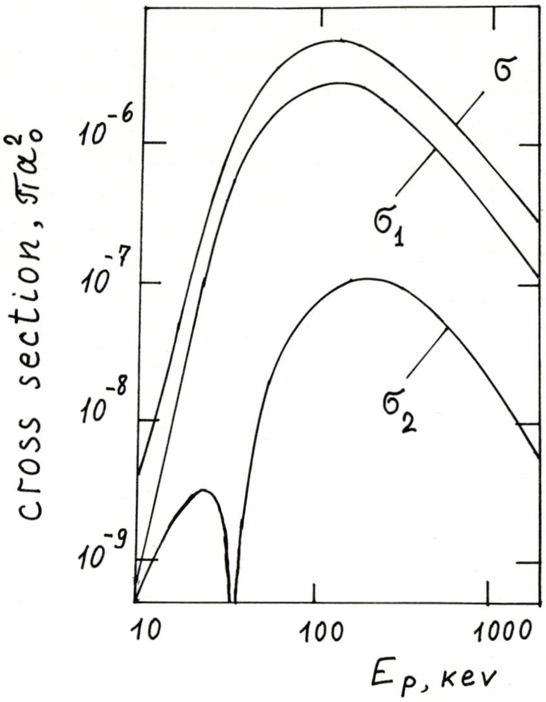

Fig.1: Cross sections for excitation of the $2p3d\,{}^1D$ autoionizing state – σ and its substates with $M_L = 1$ and 2 – σ_1 and σ_2 induced in helium by protons.

References.

1. V.A. Sidorovich, J. Phys. B14, 4805(1981).
2. U. Fano, Phys.Rev. B135, 863 (1964).

ENERGY SPECTRA OF DETACHED ELECTRONS PRODUCED IN $H^-(D^-)$ COLLISIONS ON RARE GASES AND DIATOMIC MOLECULES

Y. Itoh, U. Hege and F. Linder

Fachbereich Physik, Universität Kaiserslautern
D-6750 Kaiserslautern, West Germany

Measurements have been performed of the energy distribution of detached electrons resulting from $H^-(D^-)$ collisions on rare gases (He to Xe) and diatomic molecules (H_2, N_2, O_2) for collision energies E_{lab}=10-250 eV. In this collision energy range, most of the detached electrons are distributed below 1 eV, so a careful check of the transmission properties of the electron spectrometer at these low energies was required. For this purpose, we have studied (i) the continuum electrons produced in e - He ionization, (ii) the excitation function of He(2^3S), and (iii) the elastic scattering in e-He. As a result of these measurements, the transmission function of the detector system is known from 50 meV to 6 eV. The energy resolution of the electron detector was typically 40 meV (FWHM).

Two examples of the measured spectra are shown in Fig. 1 and 2. In Fig. 1, a structure is observed around 1.3 eV superimposed on a continuous spectrum. This structure may be attributed to the interference of the electrons produced in the incoming and outgoing part of the trajectory[1-2]. In Fig. 2, we see discrete peaks which correspond to the process $H^- + O_2$ (v=0) $\rightarrow H + O_2^-(v') \rightarrow H + O_2(v'') + e$. The peaks are assigned to v' - v" transitions as indicated in the figure. It is obvious from this spectrum that charge transfer to the autodetaching O_2^- states is dominant over simple direct detachment at this collision energy.

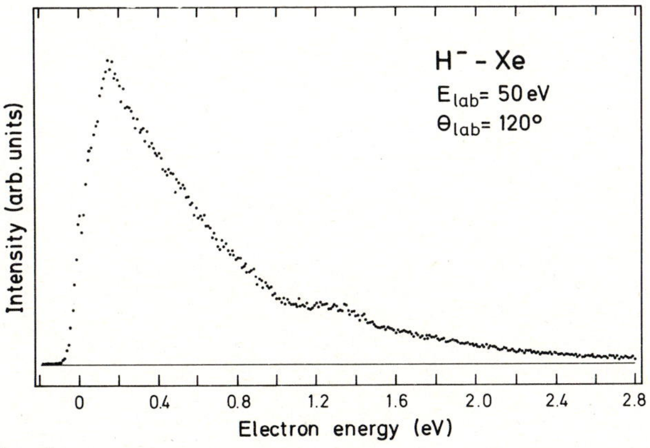

Fig. 1

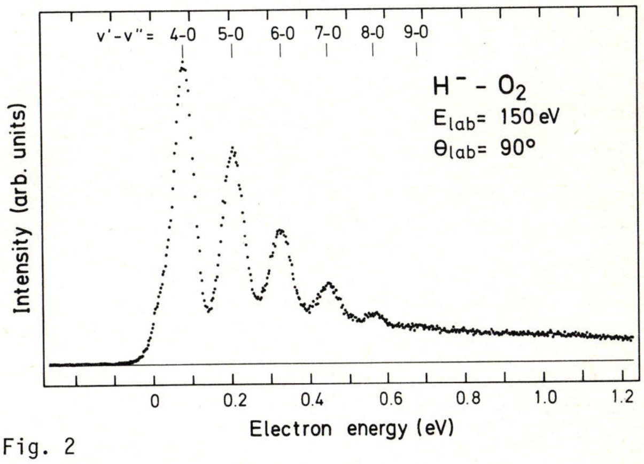

Fig. 2

References

1. R.D. Taylor and J.B. Delos, Proc. Roy. Soc. A 379, 209 (1982)
2. A. Herzenberg and P. Ojha, Phys. Rev. A 20, 1905 (1979)

ELECTROSTATIC VERSUS DYNAMICAL COUPLING IN DETACHMENT COLLISIONS OF $H^-(D^-)$ IONS WITH RARE GAS ATOMS

U. Hege, Y. Itoh and F. Linder

Fachbereich Physik, Universität Kaiserslautern
D-6750 Kaiserslautern, West Germany

In a crossed beam apparatus[1] we have performed measurements of the angular distribution of neutral H(D) atoms produced in detachment collisions of $H^-(D^-)$ ions with rare gas atoms (He, Ne, Ar, Kr, Xe). Collision energies ranged from 40 eV to 250 eV, the angular range studied was from 0° to 17°. The neutrals are detected by secondary electron emission. A first report on these measurements for the $H^-(D^-)$-Ar system is in press[2].

Some examples of the measurements are shown in Fig. 1. A pronounced structure is observed indicating two components in the angular distribution of the neutral atoms. One component is located around 0° and decreases rather sharply with increasing scattering angle. The second component is characterized by a distinct maximum at a finite angle with constant $E \cdot \theta$. The relative magnitude of the two components varies with the velocity of the projectile. Within one system, the relative importance of the zero-angle peak increases with the collision energy. When it is compared for two isotopic systems at the same energy, it is always larger for the H^- than for the D^- case.

The observed structure is attributed to the competing influence of electrostatic and dynamical coupling in the detachment process. The electrostatic mechanism involves a curve-crossing of the negative-ion system AB^- with the corresponding neutral system AB and is usually described in the local-complex-potential model[3]. The dynamical mechanism[4-5] is induced by the nuclear motion and is represented by velocity-dependent non-adiabatic coupling terms between the bound state and the continuum. Evidently, the small-angle component is characteristic of dynamical coupling at large internuclear distances outside the curve-crossing, whereas the second component located at constant $E \cdot \theta$ shows the characteristics of a curve-crossing process.

References

1. J. Krutein and F. Linder, J. Phys. B 10, 1363 (1977)
2. Y. Itoh, U. Hege, F. Linder, J. Phys. B 16, in press (1983)
3. S.K. Lam, J.B. Delos, R.L. Champion, L.D. Doverspike, Phys. Rev. A 9, 1828 (1974)
4. A. Herzenberg and P. Ojha, Phys. Rev. A 20 1905 (1979)
5. J.P. Gauyacq, J. Phys. B 13, 4417 (1980)

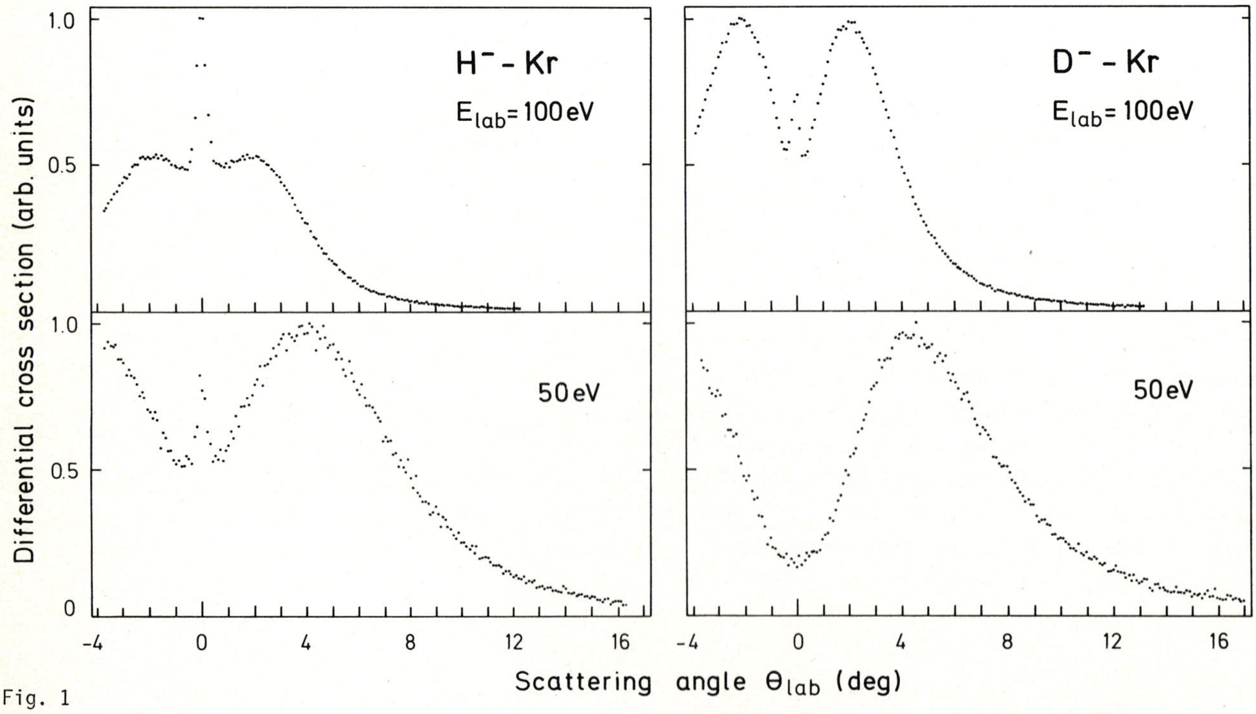

Fig. 1

A SCALING LAW FOR ELECTRON DETACHMENT IN keV COLLISIONS OF H^-, Li^-, Na^-, K^- WITH He, Ne, AND Ar

N. Andersen, T. Andersen, and L. Jepsen

Institute of Physics, University of Aarhus, DK-8000 Aarhus C, Denmark

We have studied electron-detachment processes for negative (quasi-) two-electron H^-, Li^-, Na^-, K^- ions in collisions with He, Ne, and Ar in the energy range 2-200 keV to investigate whether it is possible to establish scaling laws for electron-detachment cross sections at medium and high energies.

Absolute total cross sections have been measured for the processes

(i) Electron detachment, σ_{-10}.

(ii) One-electron loss with simultaneous excitation of the other electron to the resonant p state, $\sigma_{-10}(p)$.

(iii) Two-electron loss, σ_{-11}.

The cross sections were determined by means of the beam-attenuation method and optical spectrometry. The negative-ion beam was passed through a gas cell, electrostatically separating the exciting ion beam X and measuring the negative X^- and positive X^+ currents as a function of gas pressure. The attenuation of the X^- as the gas pressure is increased gives rise to the single -electron-detachment cross section σ_{-10}, and the growth of the X^+ current as the pressure is increased gives rise to the double-electron-detachment cross section σ_{-11}, The $\sigma_{-10}(p)$ cross section is determined by measuring the optical emission from the excited-resonance p state.

The resonant p state is the dominantly populated excited state in the electron-detached ions. The ratio $\sigma_{-10}(p)/\sigma_{-10}$ is nearly constant ($\sim$0.3) for all projectile and target combinations and independent of the ion velocities.

The absolute σ_{-10} cross sections for H^- colliding with He, Ne, or Ar are found to be in good agreement with the values reported by Risley[1]. The σ_{-10} cross sections for H^-, Li^-, Na^-, and K^- all exhibit the same curve form when plotted as a function of the beam velocity. Thus, the σ_{-10} cross sections for Li^-, Na^-, or $K^- = X^-$ can be related to σ_{-10} for H^- at constant velocity through the empirical scaling law,

$$\sigma_{-10}(X^-) = K_{X^-} \cdot \sigma_{-10}(H^-),$$

where K_{X^-} is a constant, being 1.4 for Li^-, 1.8 for Na^-, and 2.6 for K^-. A plot of K_{X^-} versus the binding energy E_B of the weakly bound electron shows that K_{X^-} is proportional to $E_B^{-\alpha}$, α being 2.2±0.3.

REFERENCES

(1) J.S. Risley, Proc.XI ICPEAC, Kyoto, 1979, Invited Papers and Progress Reports (North-Holland) p. 619

ELECTRON DETACHMENT IN H^- - Na COLLISIONS AND FREE ELECTRON SCATTERING APPROXIMATION

I.T.Serenkov, V.I.Sakharov, R.N.Il'in

A.F.Ioffe Physical-Technical Institute. Leningrad, USSR

The electron detachment cross section σ_{I0} can be calculated by means of the free electron scattering approximation in case of the collision velocty exceeding the velocity of the outer electron. This approximation had been developed by Bates and Walker and applied to the collisions of negative ions in gases[1,2]. Results obtained in these works were in agreement with the experimental ones. But attempts to use the Bates and Walker formulas for a negative ion - alkaline atom collisions gave cross section values exceeding those obtained in our experiment on H^- - Na collision and measured by Anderson et al[3] (fig 1). We believe that this disagreement is caused by the assumtion[1] that the electron scattering is near-isotropic. In contrary to weak angular electron scattering dependence for gas atoms and molecules the electron scattering differential cross section $d\sigma/d\Omega$ falls very rapidly with the scattering angle increasing for alcaline metal atoms. Thus it is necessary to exclude small angle scattering resulting in momentum transfer being deficient for electron detachment. Instead of the Bates and Walker assumption of $d\sigma/d\Omega$ being constant[1] we substituted in their intermediate expression the $d\sigma/d\Omega$ dependence and made numerical integration. We used the dependence for e-Na scattering obtained by Srivastava and Vuskovic[3] for the velocity equal to the projectile velocity. This velocity in the energy range under investigation exceeds considerably the orbital electron velocity. The results of our calculations given in fig 1 agree with experiment much better then the results obtained by using Bates and Walker formula[1]. The comparison of both methods of cross section calculation for H^- - H_2 collision with our experiments shows that for targets with weak scattering anisotropy both methods give similar results.

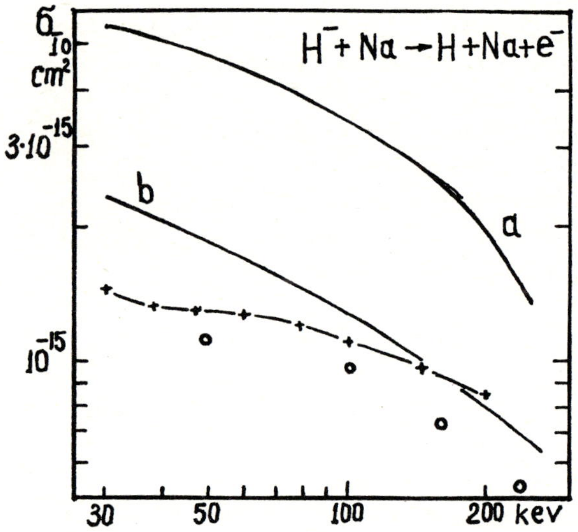

fig 1. The electron detachment σ_{I0} cross section for H^- - Na collisions. Experiment: circles - this work, crosses - data of Anderson et al[3], a - calculations made by Bates and Walker formula[1], b - our calculations taking into account electron scattering anisotropy.

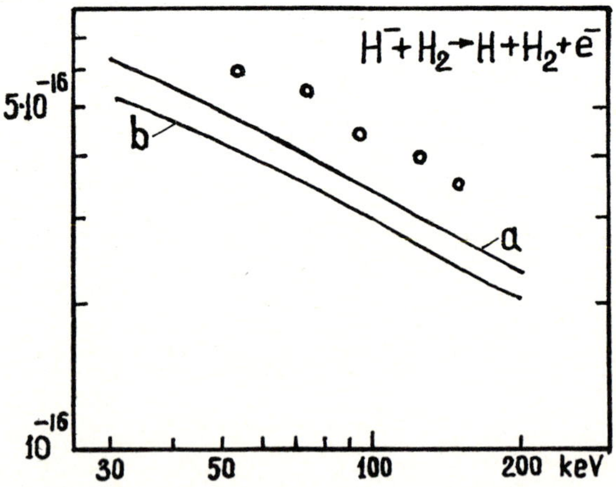

fig 2. The same for H^- - H_2 collisions.

References

1. D.R.Bates, J.C.O.Walker. Planet. Space Sci. 14. 1367. (1966).
2. R.Snyder. J. Phys. B6 L8. (1973).
3. C.J.Anderson, R.J.Givnius, A.M.Howald, L.W.Anderson. Phys. Rev. A22, 822 (1980).
4. S.K.Srivastava, L.Vuskovic. J. Phys. B13, 2633 (1980).

ELECTRON DETACHMENT IN NEGATIVE ION-MOLECULE COLLISIONS

M. S. Huq, L. D. Doverspike, and R. L. Champion

Department of Physics, College of William and Mary, Williamsburg, VA 23185 USA

The collisional dynamics for a number of negative ion-molecule systems have been studied for collision energies ranging from threshold up to 500 eV. Measurements of total electron detachment cross sections and those for charge transfer and dissociative charge transfer will be presented. Differential and doubly-differential cross sections and the kinetic energy spectra of the detached electrons have also been measured for selected reactants.

An example of the total electron detachment cross section, $\sigma_{-0}(E)$, is given for $D^- + H_2$, D_2 and HD in Fig. 1, for collision energies in the near threshold region. The solid lines in the figure are convolutions for apparatus broadening of an assumed form, $\sigma_{-0}(E) = Q(E - E_{th})$, which have been fitted to the experimental points. The threshold values which result from such a fitting procedure are 1.23 ± 0.10 eV for the D^- projectile. Interestingly the same procedure gives a higher value, 1.45 ± 0.10 eV, for electron detachment of H^- by H_2, D_2 and HD. This difference in the observed thresholds may be related to the large isotope effects which have been observed in the rearrangement

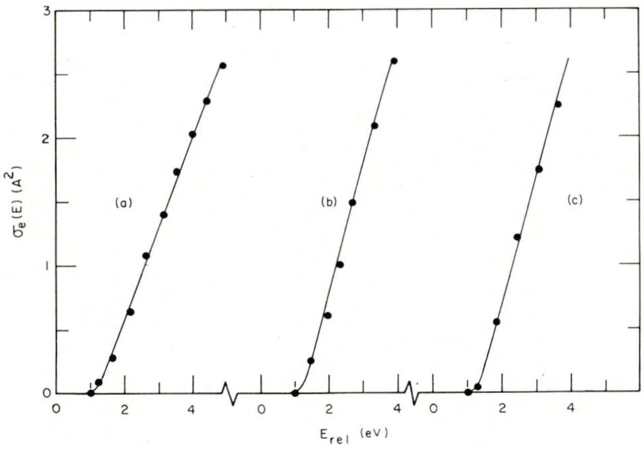

Fig. 1. $\sigma_{-0}(E)$ for D^- + (a) H_2, (b) D_2, and (c) HD. The solid line is a convolution fitted to the experimental points.

channel[1] and competes with electron detachment at low collision energies. Moreover, diffraction effects (which would be dependent upon whether the projectile is H^- or D^-) may be important at the lowest collision energies for these systems.

For $H^-(D^-) + O_2$, it is found that charge transfer is the dominant process at low energies. A striking feature of those cross sections is the large oscillations which are observed to scale remarkably well with relative collision velocity, as is seen in Fig. 2. A

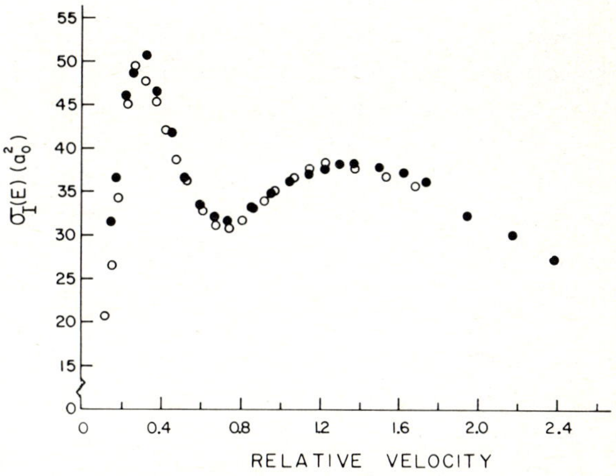

Fig. 2. Charge transfer cross section for H^- (solid points) and $D^- + O_2$. The relative velocity is in units of 10^7 cm/sec.

detailed calculation is in progress to explain the basic dynamics for these reactants. The detachment cross sections are also found to display structure, which is probably due to competition between associative detachment and charge transfer.

[1] H. H. Michels and J. F. Paulson, "Potential Energy Surfaces and Dynamics Calculations", edited by D. G. Truhlar, P 535 ff, Plenum Press, New York.

BORN CROSS SECTIONS FOR H^- COLLISIONAL ELECTRON DETACHMENT LEADING TO H^0 IN THE 1S, 2S, AND 2P FINAL STATES

George H. Gillespie and Ralph S. Janda

Physical Dynamics, Inc., P. O. Box 1883, La Jolla, CA 92038

David L. Moores

University College London, Gower St., London WCIE 6BT, England

Collisional electron detachment from H^- primarily results in a hydrogen atom in the ground (1S) state. However, there is considerable experimental[1-4] and theoretical[5-8] evidence that an appreciable fraction of single electron-detachment collisions also lead to H^0 in the 2S and 2P excited states. For high-speed collisions of H^- with H and He, Bell et al[7] have estimated that the sum of the cross sections leading to these excited states (denoted here by $\sigma_{-1,2S}$ and $\sigma_{-1,2P}$) may be as large as 30-50% of that which leads to the ground state ($\sigma_{-1,1S}$). We have carried out computations of the three cross-sections $\sigma_{-1,1S}$, $\sigma_{-1,2S}$, and $\sigma_{-1,2P}$ employing a high-speed form of the Born approximation[9] which utilizes H^- matrix elements obtained from 3-state close-coupling calculations.[10] We report here some of the principal results of these computations which we believe are the first to utilize H^- wave functions of comparable accuracy for all three detachment processes.

Our calculations have been carried out for H^- colliding with low-Z ($\leq$ 11) atomic targets as well as for a molecular H_2 target. In each case we have computed the cross sections to two orders in an expansion in v^{-2} where v is the initial H^- velocity: Here we will discuss results for the leading-order collision strengths $I_{-1,1S}$, $I_{-1,2S}$ and $I_{-1,2P}$ which, to within an overall constant, are v^2 times the corresponding cross sections at high speeds.

The collision strengths for excited state production are each comparable in magnitude for all the targets considered. For Z $\geq$ 5, $I_{-1,2S}$ is greater than $I_{-1,2P}$, but is never more than 40% larger (for Ne). For Z < 5, the reverse is generally the case with $I_{-1,2P}$ exceeding $I_{-1,2S}$ a maximum of 20% (for Li). An exception to this is for a He target, where $I_{-1,2S}$ is about 10% greater than $I_{-1,2P}$. $I_{-1,2S}$ shows a monotontic increase with increasing Z although shell-structure effects are apparent. $I_{-1,2P}$ shows a more pronounced dependence on the target shell structure and is not a monotonic function of Z. The values range from about 0.17 to 2.2.

The collision strength $I_{-1,1S}$, corresponding to ground-state production, is roughly an order-of-magnitude larger than $I_{-1,2S}$ or $I_{-1,2P}$. Similar to $I_{-1,2S}$, it is a monotonically increasing function of Z and a dependence on the shell structure of the target is also clear. The ratio $I_{-1,2S}/I_{-1,1S}$, which determines the relative production of H(2S) and H(1S) at high speeds, varies between 0.061 and 0.11 for the targets considered. The ratio $I_{-1,2P}/I_{-1,1S}$ shows a weaker dependence on Z, varying between 0.073 and 0.090.

1. A. L. Orbeli, E. P. Andreev, V. A. Ankudinov and V. M. Dukel'ski, Zh. Eksp. Teor. Fiz, 58, 1938 (1970) [Sov. Phys. - JETP 31, 1044 (1970)].
2. M. Harnois, R. A. Falk, R. Geballe and J. S. Risley, Phys. Rev. A 16, 2256 (1977).
3. J. S. Risley, F. J. deHeer and C. B. Kerkdijk, J. Phys. B11, 1783 (1978).
4. J. Geddes, J. Hill and H. B. Gilbody, J. Phys. B14, 4837 (1981).
5. G. F. Drukarev, Zh, Eksp. Teor, Fiz. 58, 2210 (1970) [Sov. Phys. - JETP 31, 1193 (1970)].
6. M. R. C. McDowell and G. Peach, Proc. Phys. Soc. (London) 74, 463 (1959).
7. K. L. Bell, A. E. Kingston and J. P. Madden, J. Phys. B11, 3357 (1978).
8. T. C. Genoni and L. A. Wright, J. Phys. B13, L61 (1980).
9. G. H. Gillespie, Phys. Rev. A15, 563 (1977).
10. D. L. Moores, "Continuum Generalized Oscillator Strengths for H^-", Los Alamos Scientific Laboratory Report (1980).

ENERGY DISTRIBUTION OF ELECTRONS DETACHED IN NEGATIVE ION COLLISIONS

Y. Sato, T. Okamoto, and H. Inouye

Research Institute for Scientific Measurements
Tohoku University, Katahira 2-1-1, Sendai 980, Japan

Energy distribution has been measured of the electrons detached in the collisions of H^- and O^- ions with rare gas atoms (He, Ne and Ar) and diatomic molcculcs (N_2, CO, H_2 and O_2) at collision energies between 0.4 and 3 keV.

Measurements are made by means of an electrostatic analyzer at the right angle to both the directions of an incident ion beam and an effusive beam of neutral targets. Examples of the results are shown in Fig.1 and Fig.2 for atomic and diatomic targets respectively.

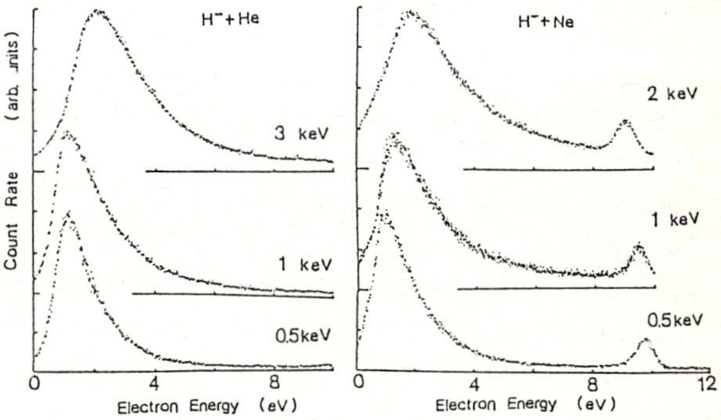

Fig. 1

The distribution of low-energy electrons produced by "simple detachment" (SD) is similar in shape and position for the H^- + He and H^- + Ne collisions even though the relative yield of electrons detached is found to be much larger for the He case than for the Ne case.

Detachment via "excitation" to the n=2 Feshbach resonance of H^- (DFR)[1] appears as a peak at 9-10 eV in the Ne case. The DFR is quite small compared to the SD process in the He case.

For the molecular targets, detachment via "charge transfer" to shape resonance (DSR) is more or less superimposed on the SD process. In the H^- + N_2 system, the DSR $N_2^-(^2\Pi_g)$[2,3] appears as a shoulder on the high-energy side of the SD peak more intensely with increasing collision energy. The DSR feature does not appear in the O^- + N_2 system. In the H^- + CO, the DSR $CO^-(^2\Pi)$ gives a broad feature on the high-energy side of the SD peak. In the H^- + O_2, detachment via vibrationally excited $O_2^-(^2\Pi_g)$ seems to be superimposed on the low-energy side of the SD electrons.

The integrated relative role of the SD and DSR processes observed here is different from the results of differential time-of-flight study[3], suggesting that the SD process is more important in small-angle scatterings.

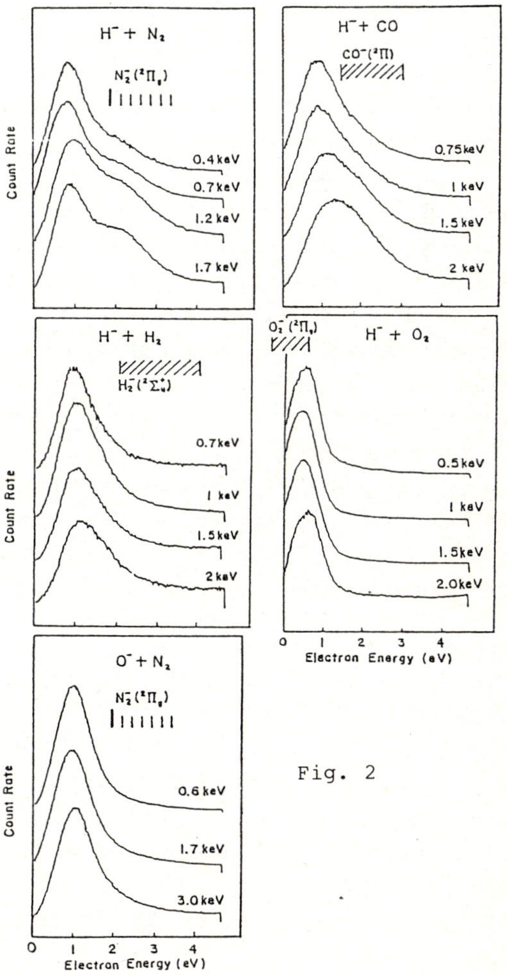

Fig. 2

References

1) J.S. Risley and R. Geballe, Phys. Rev. A10, 2206 (1974).
2) J.S. Risley, Phys. Rev. A16, 2346 (1977).
3) Vu Ngoc Tuan and V.A. Esaulov, XII ICPEAC, Abstracts of contributed papers, Gatlinburg, p.531 (1981).

ELECTRON DETACHMENT IN COLLISIONS OF Cl^- AND Ti^- IONS WITH ATOMS Ar, Na AND Mg

I.T.Serenkov, V.I.Sakharov, E.A.Solovyev, R.N.Il'in

A.F.Ioffe Physical-Technical Institute. Leningrad, USSR

The electron detachment cross section $\sigma_{\bar{1}0}$ was measured for Ti^- and Cl^- collision with Ar, Na and Mg atoms at velocities $0.3-1.1 \cdot 10^8$ cm/s (20-200keV). These projectiles were chosen because of extreme values of EA (Ti^- - 0.08eV and Cl^- - 3.62eV). The comparison of the measured $\sigma_{\bar{1}0}$ values with calculated ones is given in fig1. The free electron isotropic scattering approximation (EIS) proposed by Bates and Walker[1] and free electron anisotropic scattering approximation (EAS) discussed in our next abstract were used for calculations. The outer electron velocities were obtained from Hartree-Fock wavefunctions[2,3]. For Cl^- - Ar case potential energy curve and position of its crossing with continuum Cl - Ar is known[4] and local complex potential approximation (LCP) was used for $\sigma_{\bar{1}0}$ calculation. For this case the sum of EAS and LCP calculations fits the experiment better then any separate approximation. For Cl^- - Na having no potential energy curve crossing with continuum EAS approximation is rather satisfactory, but EIS approximation gives overestimated cross-section value.

As to Ti^- projectiles the large values of $\sigma_{\bar{1}0}$ that exceed ones obtained by EAS approximation and their weak dependence on velocity suggest far potential curve crossing with continuum existence. Estimations of this curve crossing position was made by means of zero-radius potential approximation[5]. Obtained $\sigma_{\bar{1}0}$ value for Ti^- - Na ($6 \cdot 10^{-15} cm^2$) agreed with experimental one.

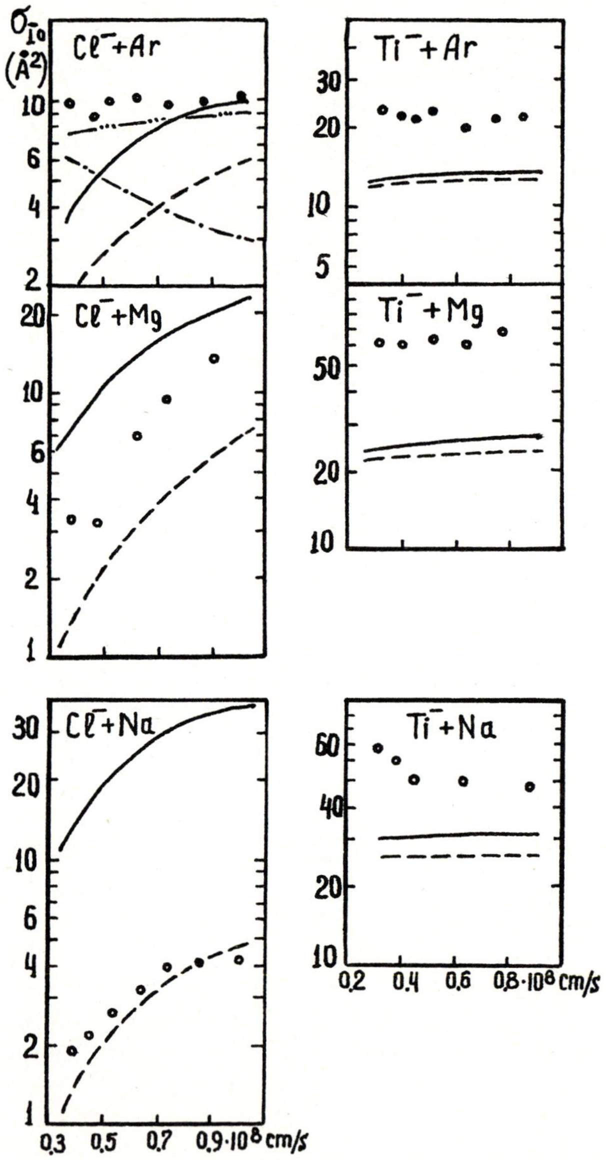

Fig 1. Electron detachment cross section
experiment - o
calculations:
IES ————
AES --------
LCP ——·——
LCP+AES ——···——

References

1. D.R.Bates, J.C.G.Walker. Proc. Phys. Soc. 90. 333. (1967).
2. E.Clementi. Phys. Rev, 135, A981 (1964).
3. E.Clementi, A.D.McLean, D.L.Raimondi, M.Yoshimine. Phys.Rev., 133, A 1274 (1964)
4. R.L.Champion, L.D.Doverspike. Phys. Rev. 13, A 607 (1976).
5. Yu.N.Demkov. Sov.Phys.JETP 39, 410 (1964).

THEORY OF ELECTRON DETACHMENT IN SLOW ANIONS IMPACT ON ATOMS

Fumihiro Koike*

FOM-Institute for Atomic and Molecular Physics, Kruislaan 407, 1098SJ Amsterdam

The electron detachment from negative ionic atoms A^- by collisions with neutral atoms B,

$$A^- + B \to A + B + e, \qquad (1)$$

is investigated in the semiclassical framework. We restrict ourselves to slow collisions in which the quasi-molecular description works and the intermediate excitation channels can be neglected. The two major difficulties of the local complex potential model[1] are removed. First, the decay rate of the initial ionic state, Γ, is defined without requiring its energy curve to be embedded in the detachment continuum. Second, the effect of dynamical couplings is included in Γ. Furthermore, the time dependent coupled channel equations are solved by means of the perturbation expansion technique; the effect of recapture of the detached electrons is formulated.

The total electronic wavefunction Ψ is expanded in terms of the eigenfunctions Φ_{Ek} of the following stationary state equation,

$$[H_{e\ell} - i\underset{\sim}{v}\cdot\nabla_{\underset{\sim}{R}} - E]\Phi_{Ek} = 0, \qquad (2)$$

as

$$\Psi = \int_E \sum_k C_{E,k}(t)\Phi_{E,k}(\underset{\sim}{r},t,\underset{\sim}{R})\, e^{-i\int^t E\,dt'}, \qquad (3)$$

where $H_{e\ell}$ is the total electronic Hamiltonian, $\underset{\sim}{v}$ is the collision velocity, ∇_R is the gradient operator with respect to the internuclear vector $\underset{\sim}{R}$, E is the energy eigen value, k is a set of quantum numbers, and $C_{Ek}(t)$ is the expansion coefficient. Equation (2) is solved using a basis set which includes one discrete state wavefunction, u_d, and numbers of continuum wavefunctions, u_{Ek}. As the lowest order solution of the time dependent coupled-channel equations that is given by substituting eq.(3) into the time dependent total electronic Schroedinger equation: $H_{e\ell}\Psi = i\partial/\partial t\Psi$, we obtained the following exponentially decaying formula for the survival probability of the initial state. That is

$$P_s = \exp[-\int_{-\infty}^{+\infty}\Gamma dt]. \qquad (4)$$

The decay rate Γ is given by the twice of the imaginary part of the root $E_* \equiv E_r - i\Gamma/2$ of the following equation, that is

$$E_* - E_a - \int_{E_0' \, C^+}^{\infty} \frac{W(E')}{E' - E_*}\, dE' = 0, \qquad (5)$$

where E_a is the unperturbed energy of the initial ionic state, $W(E') \equiv \sum_k \langle u_d|H_{e\ell} - i\underset{\sim}{v}\nabla_{\underset{\sim}{R}}|u_{E'k}\rangle\langle u_{E'k}|H_{e\ell} - i\underset{\sim}{v}\nabla_{\underset{\sim}{R}}|u_d\rangle$ is the strength of the coupling with the detachment continuum. The quantity $E_r - E_a$ (which is always < 0) gives the shift of the initial ionic state energy due to the coupling with the detachment continuum, and the quantity Γ/2 is its imaginary part, which can be finite even when $E_r < E_0'$.

For the demonstration, we calculated the root of eq.(5), E_*, for variable E_a, i.e., $E_* = E_*(E_a)$, by assuming the model case that

$$E_0' = 0, \qquad W(E') = 0.01\cdot E'^{1/2}\exp(-4E'), \qquad (6)$$

in atomic units. The result is shown below. Each pair of points on the real axis and on the curve connected by a straight line shows the correspondence of E_a and E_*. This result suggests the possibility of obtaining the electron detachment from outside the potential crossing point.

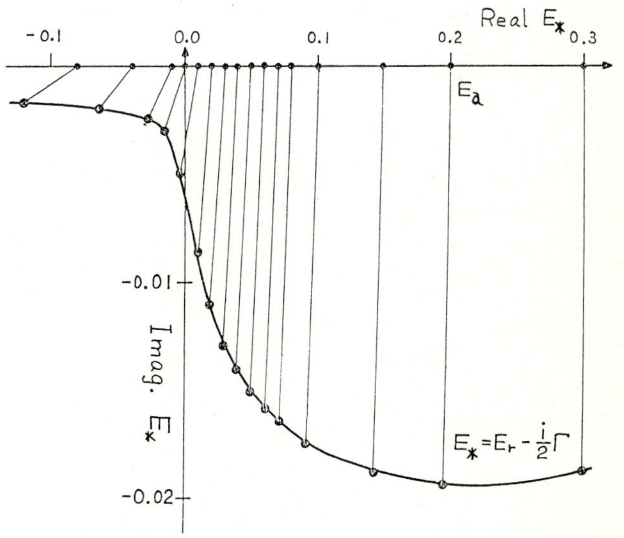

*Present address: School of Medicine, Kitasato University, Sagamihara, Kanagawa, 228 Japan.

References

1. S. K. Lam, J. B. Delos, R. Champion, and L. D. Doverspike: Phys. Rev. A9, 1828 (1974)

ELECTRON DETACHMENT IN NEGATIVE-ION COLLISIONS: NEW THEORETICAL METHODS

T. S. Wang and J. B. Delos

Physics Department, College of William and Mary, Williamsburg, VA 23185 USA

We develop the theory of electron detachment in slow collisions of negative ions with atoms using a semiclassical close-coupling framework. The electronic wave-function is expanded in an orthogonal diabatic basis, with ϕ_{-1} representing the bound state and ϕ_ε a free state with asymptotic kinetic energy ε. We consider only coupling between the bound state and the continuum, and we neglect intra-continuum couplings. Then the expansion coefficients in this basis satisfy a non-denumerably infinite set of coupled equations:

$$i\hbar\, \dot{c}_{-1} = \Delta(t)\, c_{-1} + \int V_{-1,\varepsilon}(t)\, c_\varepsilon(t)\, \rho_\varepsilon\, d\varepsilon \tag{1a}$$

$$i\hbar\, \dot{c}_\varepsilon = \varepsilon\, c_\varepsilon + V_{\varepsilon,-1}(t)\, c_{-1}(t) \tag{1b}$$

$\Delta(t)$ is the energy gap between negative-ion and neutral states and $V_{-1,\varepsilon}(t)$ is the coupling between them.

With the boundary conditions $c_{-1}(t_0) = 1$, $c_\varepsilon(t_0) = 0$, eqs. (1b) can be solved for $c_\varepsilon(t)$:

$$c_\varepsilon(t)=(i\hbar)^{-1}\int_{t_0}^{t} V_{\varepsilon,-1}(t)\exp[\, i\varepsilon(t-t')/\hbar]c_{-1}(t')dt' \tag{2}$$

and when this is put into (1a) a single integrodifferential equation is obtained for $c_{-1}(t)$:

$$[\, i\hbar \frac{d}{dt} - \Delta(t)]c_{-1}(t) - \int_{t_0}^{t} \mathcal{G}\,(t,t')\, c_{-1}(t')dt'=0 \tag{3}$$

Electron detachment cross-sections are therefore found by three steps: (A) Calculate the propagator $\mathcal{G}\,(t,t')$ exactly or approximately; (B) Solve the integrodifferential equation (3) exactly or approximately; (C) Use the resulting $c_{-1}(t)$ in eq. (2) to find $c_\varepsilon(t)$.

General properties of the propagator have been determined, and it has been calculated for a simple model.

The coefficient $c_{-1}(t)$ can be written as

$$c_{-1}(t) = \exp[-i\int_0^t \mathcal{E}\,(t')dt'/\hbar] \tag{4}$$

$\mathcal{E}\,(t)$ plays the role of a "dynamical complex potential". From eq. (3) it is easy to show that $\mathcal{E}\,(t)$ satisfies an integral equation

$$\mathcal{E}\,(t)=\Delta(t)+\int_{t_0}^{t} \mathcal{G}\,(t,t')\exp[\, i\int_{t'}^{t} \mathcal{E}\,(t'')dt''/\hbar]dt' \tag{5}$$

for which a first-order solution is

$$\mathcal{E}\,(t)=\Delta(t)+\int_{t_0}^{t} \mathcal{G}\,(t,t')\exp[\, i\int_{t'}^{t} \Delta(t'')dt''/\hbar]dt' \tag{6}$$

It can be shown that for slow collisions the dynamical complex potential generally approaches the static complex potential,

$$\mathcal{I}m\, \mathcal{E}\,(t) \simeq -\pi|V_{-1,\varepsilon}(t)|^2\rho_\varepsilon\,, \qquad v\to 0$$

but in general it oscillates about this value, and it goes continuously through regions in which the static complex potential has sudden changes.

The figure shows a comparison between the "dynamical" and the "static" complex potentials.

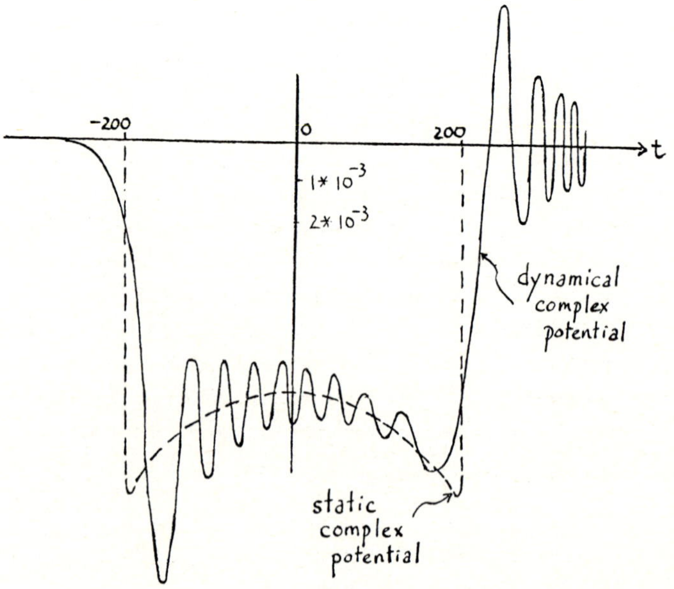

ELECTRON DETACHMENT IN NEGATIVE ION COLLISIONS: CROSS SECTIONS

T. S. Wang and J. B. Delos

Department of Physics
College of William and Mary
Williamsburg, VA 23185

Using methods related to those presented in the accompanying contribution, we have calculated cross sections for some H^--rare gas systems.

1. H^--Ne

Potential curves were taken from a calculation by Gauyacq, and the coupling $V_{-1\varepsilon}$ was assumed to be $A\sqrt{\varepsilon}\exp(-\beta R)$. Using first-order perturbation theory, the total cross section for detachment is easily calculated. Results are shown in Figure 1. At low energies it is found that the total detachment cross section for H^- is larger than that for D^-, while at high energies, the opposite is found. Experiments by Huq et al. show the same effect.

2. H^--He

Potential curves were taken from a calculation by Olson and Liu, and the coupling matrix element was inferred from earlier fits to experimental data using the local complex potential model. The dynamical complex potential was calculated, and used to obtain the electron energy spectrum. Reasonable agreement is found with experiments by Taun, Esaulov and Gauyacq.

3. A Model

Modifying potential curves to increase the penetration of the negative ion into the continuum can lead to oscillations in the electron energy spectrum. (Fig. 2)

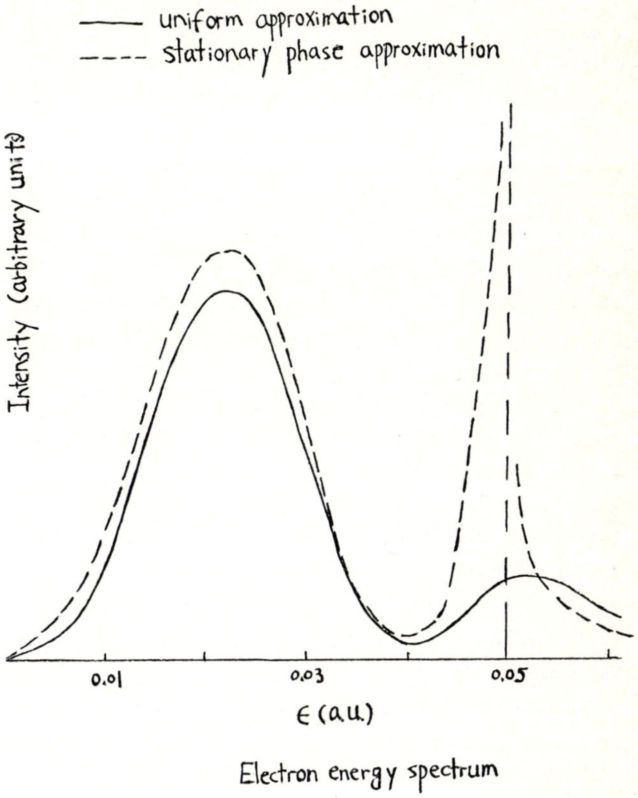

Figure 2

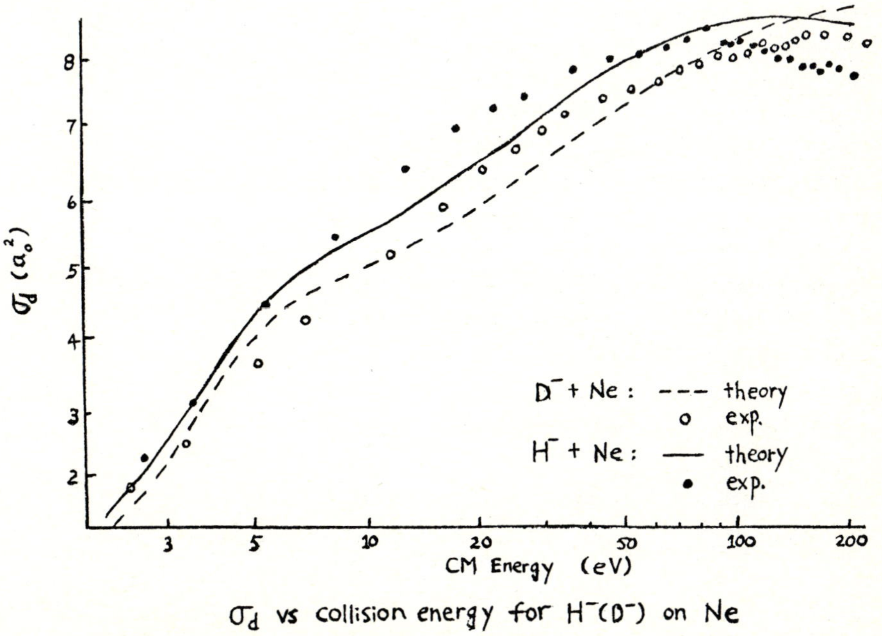

Figure 1

ELECTRON DETACHMENT IN COLLISIONS OF H^- AND SEVERAL TARGETS IN THE ENERGY RANGE FROM 500 TO 2000 eV

P.E. van der Leeuw, W. Koot, A.W. Kleyn and J. Los

FOM-Institute for Atomic and Molecular Physics, Kruislaan 407, 1098 SJ Amsterdam, The Netherlands.

The differential cross section for elastic scattering and H atom formation has been determined for collisions of H^- and the rare gas atoms He, Ar and Xe and some simple molecules like N_2, O_2 and SF_6.

The H^- beam is produced in an electron impact source and is focused through a collision chamber on a position sensitive channelplate detector, which employs a computer controlled two dimensional (x-y) readout system. The angular resolution obtained is about 0.02 degrees.

Typical differential cross sections for H atom formation in collisions of H^- and He are presented in Fig. 1. The data is given on a $\rho = \sigma(\theta) \times \theta^2$ versus $\tau = E \times \theta$ scale. E is the primary energy and θ the scattering angle. From the figure it appears that there are two processes leading to detachment, one around $\tau = 200$ eV * degree and the other around $\tau = 800$ eV * degree. The contribution at small τ increases with increasing energy and dominates $\sigma(\theta)$, since the other contribution becomes of equal intensity after multiplication of $\sigma(\theta)$ by θ^2. The spectrum at $E = 500$ eV is in good agreement with earlier measurements by Esaulov et al. [1]. In this work only the large τ feature of the differential cross section is observed, because the small τ feature was inaccessible in the experimental set-up used.

Itoh et al. [2] have investigated $H^- + Ar$ collisions in the energy range of 50-250 eV and angular range of 0-14 degrees. In their differential cross section for H atom production a double structure can be observed, but only if the data is plotted on a $\sigma(\theta)$ versus θ scale. At 50 eV one feature is found around $\tau = 0$ eV * degree, while the other is located around $\tau = 200$ eV * degree. The second one is like the large τ feature of the present study, and it completely dominates $\sigma(\theta)$. The same feature has been observed and explained by Esaulov et al. [1], using a model assuming electrostatic coupling.

From the present investigation and the work of Itoh et al. [2] it is clear that the process observed and explained by Esaulov et al. is mostly important at low energies and large τ values. At higher energies a new detachment process becomes apparent, and this may be attributed to dynamic coupling.

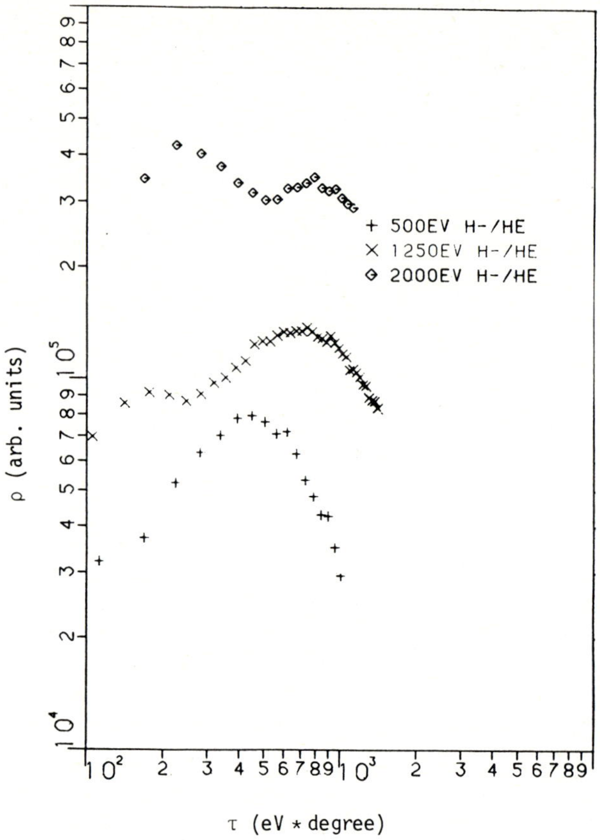

Fig. 1: Differential cross section multiplied by for H atom production in $H^- + He$ collisions at several collision energies versus the reduced scattering angle τ.

References

1. V. Esaulov, D. Dhuicq and J.P. Gauyacq, J.Phys.B 11 (1978) 1049.
2. Y. Itoh, U. Hege and F. Linder, submitted for publication in J.Phys.B.

EXCITATION OF AUTODETACHING STATES IN H⁻Kr COLLISIONS

V. Esaulov+, F. Pichou*, C. Schermann*, J.P. Grouard*, R.I. Hall*
M. Landau* and J.L. Montmagnon*

+L.C.A.M. Bât. 351, Université Paris-Sud, 91405 ORSAY Cedex, FRANCE
*L.P.O.C. T.12-E5, Université Pierre et Marie Curie, 4 Pl. Jussieu, 75005 PARIS, FRANCE

We have observed by electron spectroscopy excitation processes in H^-Kr collisions in the laboratory energy range from 20 eV to 500 eV. The electron spectra were obtained in an angular range from 15° to 135° using an apparatus described by Montmagnon et al.[1]. A typical electron energy spectrum showing all the observed lines in the energy range studied is shown in fig.1. Similar spectra have been reported by Risley[2]. The lines corresponding to autodetaching states of H^- are shifted and broadened due to kinematic effects[3] and for small energies and scattering angles are asymmetric in shape. Typical angular distributions corrected for kinematic effects and transmission function of the analyser[1] and normalised to unity at 80° are shown in fig. 2.

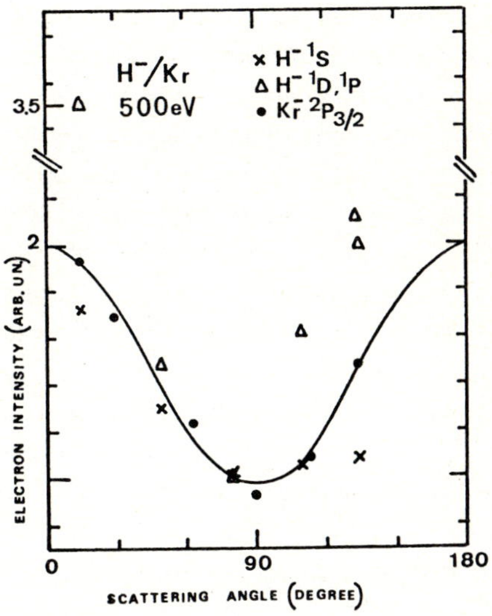

The $Kr^-(^2P_{3/2})$ angular distribution follows a $\cos^2\theta$ dependence which (neglecting H atom scattering into a finite angular range) may be expressed in terms of the magnetic sublevel population $P(J,M_J)$[4]

$$\sigma(\theta) = \frac{3}{4\pi}\left\{\left(P(\tfrac{3}{2},\tfrac{3}{2}) + \tfrac{1}{3}P(\tfrac{3}{2},\tfrac{1}{2})\right) + \left(P(\tfrac{3}{2},\tfrac{1}{2}) - P(\tfrac{3}{2},\tfrac{3}{2})\right)\cos^2\theta\right\}$$

The solid line in the figure corresponds to a least squares fit to the data giving at 500 eV $\sigma(\theta)=0{,}966\left(1+1.15\cos^2\theta\right)$. From this distribution we find at 500 eV $\frac{P(3/2,3/2)}{P(3/2,1/2)} = 0{,}29$. Measurements were performed at 125.3° in order to obtain the population of the $^2P_{3/2}$ state as a function of energy $\left(\sigma(\theta=125{,}3°)=(2\pi)^{-1}\left(P(\tfrac{3}{2},\tfrac{3}{2})+P(\tfrac{3}{2},\tfrac{1}{2})\right)\right)$. The resulting cross section is shown in fig.3. The cross section for the $^2P_{1/2}$ state is obtained *assuming* an isotropic distribution[4]. The ratio of these two cross sections is roughly *independent* of energy.

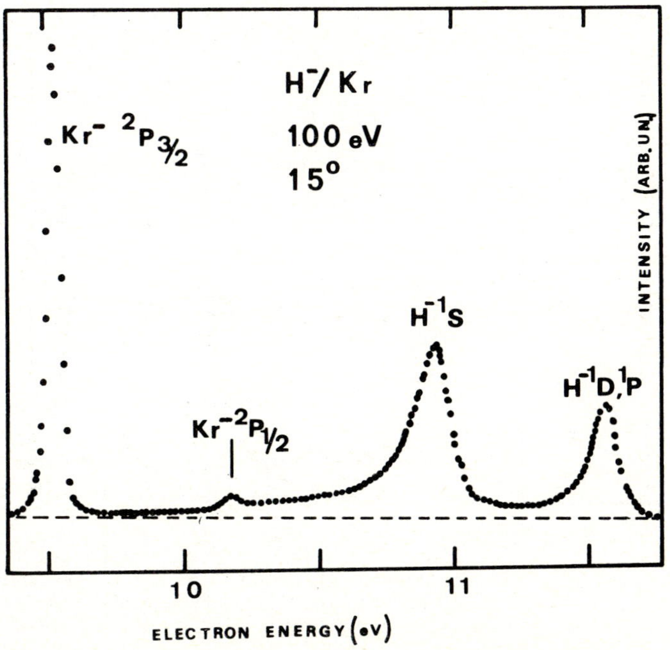

The angular distribution of the $H^-(2s^2)\,^1S$ state is found to be anisotropic at our collision energies as in the case of H^-Ar and H^-H_2 collisions[3]. Integration over angle of the angular distribution of the H^- states yields after normalisation to the $Kr^-\,^2P_{3/2}$ state, the cross section shown in fig. 3. The Kr^- and H^- peaks were observed down to H^- energies of 20 eV and will be presented later.

References

1. Montmagnon et al. J.Phys.B 16, L143 (1983)
2. Risley J.S. 6th ICAP, Riga (1978)
3. Risley J.S. and Geballe R. Phys.Rev. A10, 2206 (1974)
4. Edwards A.K. Phys. Rev. A12, 1830 (1975) ; Macek J. and Edwards A.K. Phys.Rev. A25, 881 (1982)

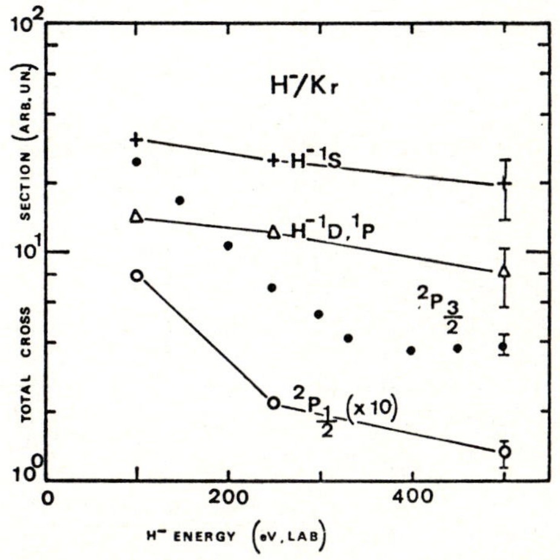

F^- COLLISIONS WITH ATOMIC AND MOLECULAR TARGETS

Vu Ngoc Tuan and V.A. Esaulov
Laboratoire des Collisions Atomiques et Moléculaires, Bât. 351
Université Paris-Sud, 91405 ORSAY Cedex, FRANCE

We present results of a differential time of flight energy loss[1] study of F^- collisions with He, N_2 and CO_2 for energies below 2 keV. This work is motivated by two facts: (a) the electron detachment mechanism in halogen negative ion atom collisions is not well understood[2] and (b) insofar as collisions with molecules are concerned some controversy exists[3,4] about the importance of charge exchange to shape resonances (CESR) (eg. $X^-+N_2 \rightarrow X+N_2^-(^2\Pi_g)$; $N_2^- \rightarrow N_2+e^-$). This process is important in energetic H^- collisions[4,5], but our preliminary work on F^- and Cl^- collisions[4] suggested that this process was not important at energies below 1 keV.

The scattered F atom energy loss spectra consisted of two peaks : a high energy loss peak corresponding to detachment accompanied by excitation (D.E.) and a low energy loss peak which for F^-He collisions corresponds to direct detachment (D.D.). The D.E. peak is attributable to target excitation for molecular targets. In case of F^-He it appears to correspond mainly to F atom excitation in agreement with what may be expected from a molecular description of the collision. Fig. 1 shows reduced differential cross sections (DCS) for the observed processes. Integration of these over angles allows an estimate of the contribution of the D.E. process to the total detachment cross section. For F^-He collisions this contribution is of about 16% at 2 keV and for F^-CO_2 collisions of about 40% (lower limit) at 1 keV.

Fig. 2 shows the low energy loss peak for F^-CO_2 and N_2 collisions. In case of F^-CO_2 collisions no sign of CESR is observed for 1000 eV and 500 eV. At 2000 eV the energy resolution is insufficient to make an unambiguous statement about its existence, but in any event it appears to be small. In case of F^-N_2 collisions a small hump in the 1000 eV spectrum (which was not resolved in our previous measurements[4]) suggests that this process does exist but is quite small. This result seems in qualitative agreement with our observation for H^- collisions[4,5] and the existing theoretical treatments[5] which suggest that the cross sections for this process attains its maximum at energies such that the translational energy of the outer electron is equal to the resonance energy (e.q. 4 keV for H^-N_2). For F^-N_2 collisions this would correspond to energies of about 80 keV.

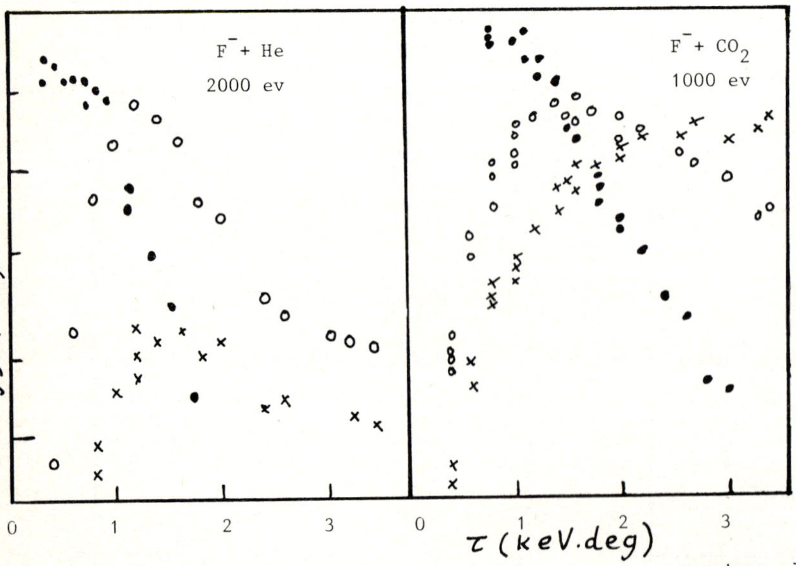

Fig.1. Reduced DCS :• elastic ; o D.D. ; x D.E.

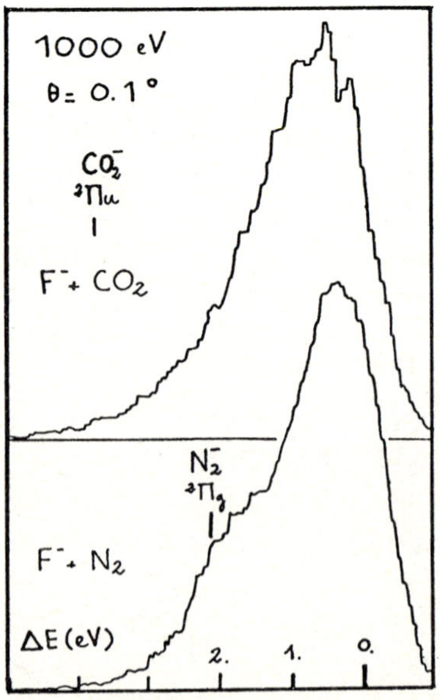

Fig. 2. F atom energy loss spectrum. The position of the resonance is taken as the energy difference between the N_2 and N_2^- and CO_2 and CO_2^- states at the equilibrium separation the parent molecule.

References

1. Esaulov et al. 1978 J.Phys.B 11, 1049
2. De Vreugd et al. 1982 Chem.Phys. ; Esaulov V.A. 1981 J.Phys.B 14, 1303
3. Annis B. et al. 1980 Phys. Rev. Lett. 45, 1554
4. Vu Ngoc Tuan and Esaulov V.A. 1982 J.Phys.B 15, L95
5. Vu Ngoc Tuan et al. 1983 in this conference

CHARGE EXCHANGE TO SHAPE RESONANCES IN H^--CO_2 COLLISIONS

Vu Ngoc Tuan, V. Esaulov and J.P. Gauyacq

L.C.A.M., Bât. 351, Université Paris-Sud, ORSAY, FRANCE

In negative ion-molecule collisions, besides the direct detachment process (DD) (e.e. $H^-+N_2 \to H+N_2+e^-$), charge exchange to shape resonances CESF (e.g. $H^-+N_2 \to H+N_2$ $(^2\Pi_g) \to H+N_2+e^-$) constitutes an important detachment channel[1]. Electron spectroscopy measurements[2] showed that this process becomes important in the few hundred eV energy range, onsetting above 50eV for H^-N_2.

We performed a differential time of flight energy loss study of H^-collisions with CO_2. The relative narrow $(\Gamma \sim 0.2\ eV)^2\Pi_u$ shape resonance lies around 3.8 eV leading to an energy defect (4.55eV) for CESR larger than in the case of the N_2 target (3eV). H atom energy loss spectra (fig.1) display three peaks attributed to DD, CESR and excitation processes. Note that the peak we attribute to CESR lies at 3.4eV and not at 3.8eV which corresponds to the energy difference between CO_2 and CO_2^- at CO_2 equilibrium. Fig.2 shows the energy dependence of the spectrum. The CESR process appears to onset for energies above 150 eV in agreement with preliminary results of electron spectroscopy studies[2].

A theoretical description of CESR has recently been proposed[3]. The method consists in developping the wave-packet of the bound electron of H^- over the energy spectrum of the ejected electron and the detachment probability in a first order perturbation theory. For fast collisions, the target molecule is assumed not to vibrate nor rotate during the collision and the spectra are averaged over the vibrational and rotational coordinates as classical variables. Fig.3 shows the theoretical spectrum corresponding to CESR at 1 keV, compared with the experimental spectrum. Both spectra peak at 3.6eV. This shift from 3.8eV can be related to the large energy defect for CESR, which results in a fast increase of the CESR probability with a decrease of the resonance energy. CESR is thus favored for nuclear positions where the CO_2^- and CO_2 energies are closer than at equilibrium. The CESR factor is not determined only by a nuclear Franck-Condon factor, but by the superposition of this factor and the CESR probability, that shifts the spectrum to small energy losses. As the energy is increased, the CESR probabilities dependence on the energy defect decreases and the shift of the peak's maximum disappears. The maximum position in the theoretical spectrum is found at 3.4 eV, 3.6eV and 3.8eV for collision energies of 500eV,1keV and 5keV respectively. As for the CESR total cross section, it reaches its maximum around the energy where the electron translational energy is equal to the resonance energy.

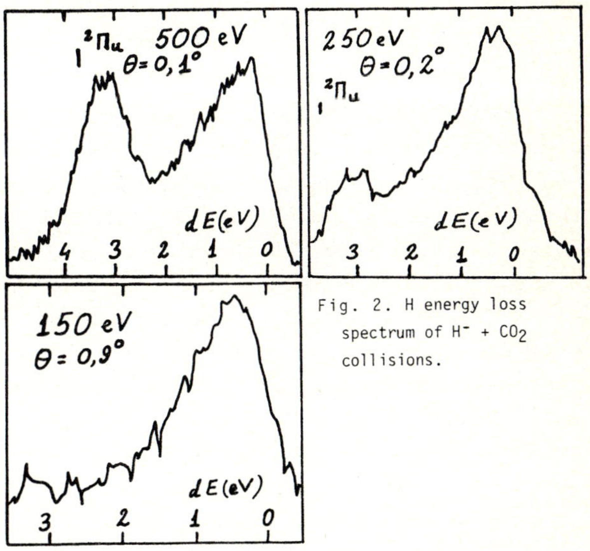

Fig. 2. H energy loss spectrum of H^- + CO_2 collisions.

References

1. J. Risley 1977 Phys.Rev.A 16 2346 ; Vu Ngoc Tuan and V. Esaulov 1982 J.Phys.B 15 L95
2. Montmagnon et al. 1983 J.Phys.B L16 and to be published
3. J.P. Gauyacq and A. Herzenberg to be published ; Vu Ngoc Tuan et al SPIG 1982 p.101

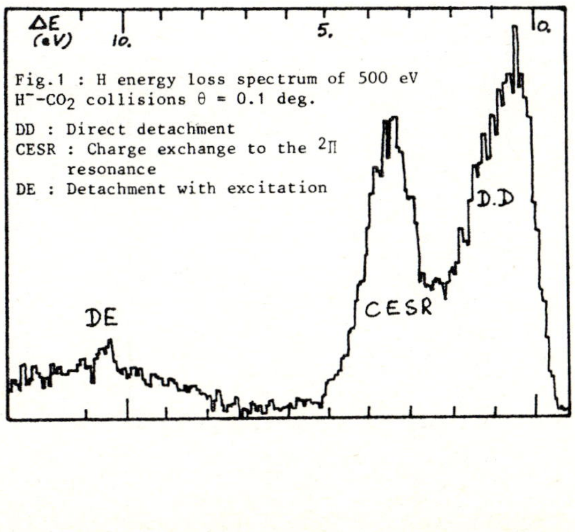

Fig.1 : H energy loss spectrum of 500 eV H^--CO_2 collisions θ = 0.1 deg.
DD : Direct detachment
CESR : Charge exchange to the $^2\Pi$ resonance
DE : Detachment with excitation

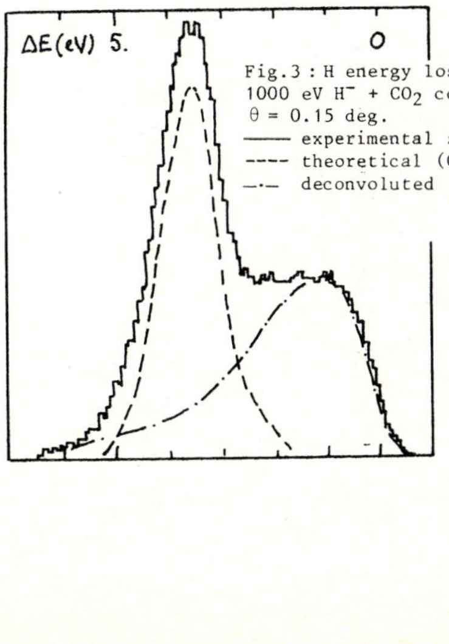

Fig.3 : H energy loss spectrum of 1000 eV H^- + CO_2 collisions θ = 0.15 deg.
—— experimental spectrum
---- theoretical (CESR)
—·— deconvoluted DD spectrum

OBSERVATION OF ELECTRON SPECTRA PRODUCED IN F^- COLLISIONS

J.P. Grouard, V. Esaulov, R.I. Hall, M. Landau, J.L. Montmagnan, F. Pichou and C. Schermann

L.P.O.C. T.12-E5, Université Pierre et Marie Curie, 4 Pl. Jussieu, 75005 PARIS, FRANCE

L.C.A.M. Bât. 351, Université Paris-Sud, 91405 ORSAY Cedex, FRANCE

Firstly, the formation of autodetaching states yielding electrons with energies between 6 and 17eV was studied in F^- collisions with rare gases, N_2, H_2 and CO_2 at 500eV (lab). For both He and Ne projectile excitation is observed to the $2p^4(^1D)\ 3s^2$ state of F^- at 14.85eV[1]. In the case of Ar and Kr, formation of the lowest Feshbach resonance of the target ($^2P_{3/2,1/2}$) is detected but with no F^- excitation. The Ar^- $^2P_{3/2}$: $^2P_{1/2}$ intensity ratio was about statistical (2:1) whereas the $^2P_{1/2}$ level of Kr^- was only weakly excited relative to $^2P_{3/2}$. F^- collisions with H_2 and CO_2 revealed no autodetaching structure in this electron energy range, however the N_2 target produced prominent structure due to formation of the lowest Feshbach resonance $(3s\sigma_g)^2\ ^2\Sigma_g^+$ at 11.50eV (see figure).

Secondly, the intensity of detached electrons with energies below 7eV was studied at an observation angle of 90° for F^- impinging on various gases with impact energies up to 500eV (lab). The spectra obtained for He and N_2 are displayed in the adjacent figures. These spectra have been corrected for instrumental effects by a calibration procedure described recently by Montmagnon et al [2]. The helium spectra show that the distribution peaks below 0.2eV spreading out in energy as the F^- energy increases. Of interest here is the low structure around 2eV seen on the upper two curves.

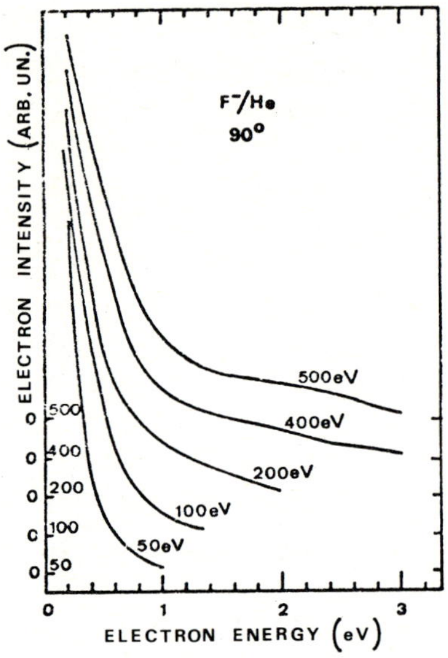

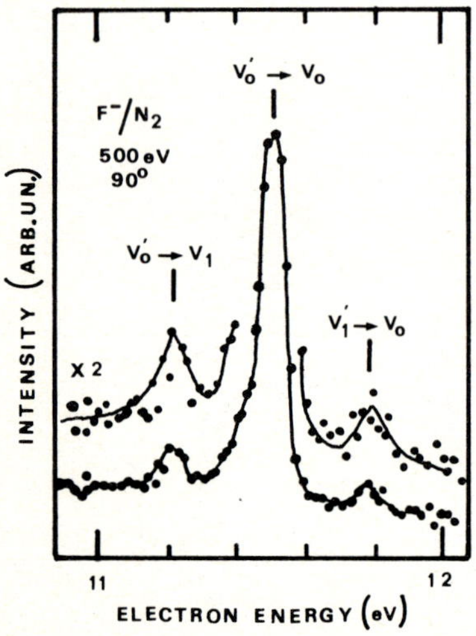

Preliminary TOF spectra[3] did not reveal a similar behaviour. Note that our spectra represent a sum of the energy distribution from all detachment channels, which then would lead us to attribute this structure to detachment accompanied by excitation. Structure in the electron spectra near 2eV might be expected for F^- collisions with N_2 corresponding to formation of the $N_2^-\ ^2\Pi_g$ state by charge exchange. In fact this process was not observed at 500eV or below as the figure shows. Population of this resonance would only appear to occur at higher energies[3] in contrast to the H^- projectile where the $^2\Pi_g$ state was detected down to 50eV[2].

1. A.K.Edwards and D.L.Cunningham, Phys.Rev.A9,1011,1974
2. J.L.Montmagnon et al, J. Phys.B.16, L143, 1983
3. V.N.Tuan and V.Esaulov, 1983 this conference

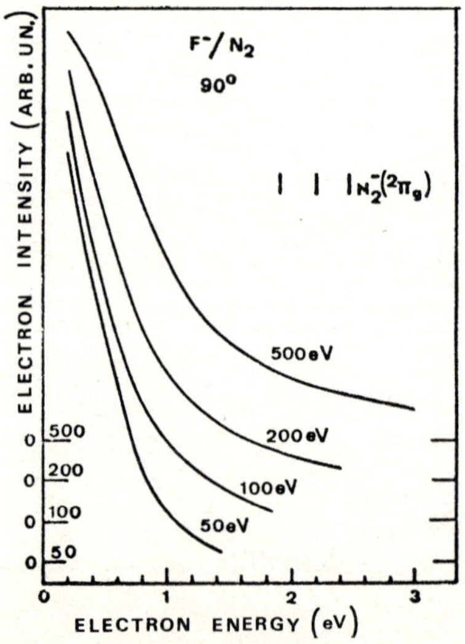

IONIZATION CROSS SECTIONS FOR 5-4000 keV PROTONS IN GASES*

M. E. Rudd and T. V. Goffe, University of Nebraska, Lincoln, NE 68588-0111, R. D. DuBois, L. H. Toburen and C. A. Ratcliffe, Battelle Pacific Northwest Laboratory, Richland, WA 99352

Measurements of electron production for 5-400 Kev protons in He, Ne, Ar, Kr, H_2, N_2, CO, O_2, CH_4, and CO_2 have been made with a parallel plate electrode apparatus utilizing four different accelerators at two laboratories. Target gas pressures were measured with capacitance manometers. Ionization cross sections were corrected for beam neutralization, beam scattering, thermal transpiration, change of grid transparence with electrode potentials, secondary electrons from the grid, and background currents.

Overlapping data runs have been averaged by making a least squares fit to the equation

$$\sigma_- = A\sigma_0 \sum_i \frac{N_i}{I_i^2} \frac{(T/I_i)^B}{C+(T/I_i)^{B+1}} \ln(1+DT/I_i). \quad (1)$$

The sum is over the subshells of the target which contain N_i electrons having binding energies I_i, T is the kinetic energy an electron would have at the velocity of the incident proton and $\sigma_0=4\pi a_0^2 I_H^2$. Parameters for the fit are given in Table I with the standard deviations of the data points from the fitting curve. Uncertainties are estimated to be 25% at 5 keV, decreasing to 15% at 25 keV and to 8% above 500 keV.

Fig. 1 shows the present data for three gases, represented by Eq. 1, compared to high energies for all gases except hydrogen where the present data are 18% above the earlier data. The greatest disagreement at the lower energies is for helium where our data below 30 keV are much lower than any of the earlier values although agreement is good with data of deHeer[3] at other energies. For all of the gases for which data is available, the high energy cross sections agree with electron impact ionization data at the same projectile velocities.

Table I. Parameters for Fitting Eq. (1) to σ_- Data.

	A	B	C	D	Ave Dev
He	0.413	1.16	2.18	10.6	8.7%
Ne	0.503	0.412	1.19	0.992	8.3
Ar	0.474	0.291	2.72	14.0	10.8
Kr	0.765	0.158	3.24	5.58	9.4
H_2	0.314	1.01	3.24	94.4	9.2
N_2	0.461	0.165	2.18	7.62	9.2
CO	0.443	0.241	2.30	9.05	9.9
O_2	0.517	0.180	1.66	2.68	8.7
CH_4	0.515	0.298	2.68	5.32	9.7
CO_2	0.515	0.160	1.94	3.00	7.4

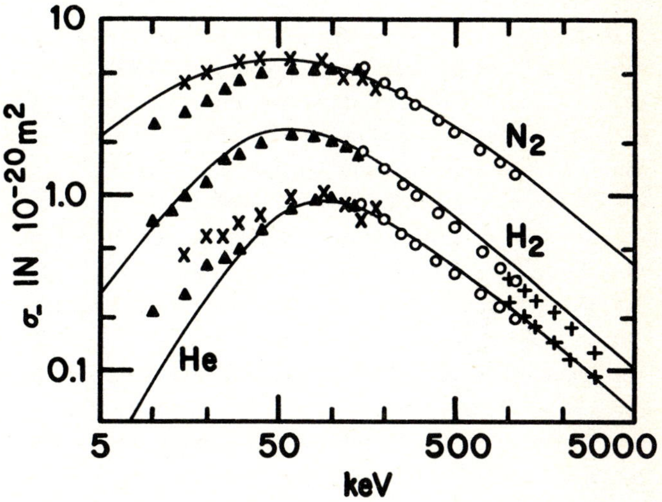

Fig. 1: Ionization cross sections vs. proton energy. Solid lines are the present data represented by Eq. (1); data of McDaniel, et al, ⊙ ; data of Pivovar and Levchenko, +; data of deHeer, et al, ▲; data of Solov'ev, et al, x.

* This paper is based on work performed under National Science Foundation Grant No. PHY 80-25599 and the United State Department of Energy Contract DE-AC06-76RLO-1830.

1. E. W. McDaniel, J. W. Hooper, D. W. Martin, and D. S. Harmer, Proc. Fifth Int. Conf. on Ionization Phenomena in Gases, Munich, 1961, North-Holland, Amsterdam, p. 60.
2. L. I. Pivovar and Yu. Z. Levchenko, Sov. Phys. JETP 25, 27 (1967).
3. F. J. deHeer, J. Schutten and H. Moustafa, Physica 32, 1766 (1966).
4. E. S. Solov'ev, R. N. Il'in, V. A. Oparin and N. V. Fedorenko, Sov. Phys. JETP 15, 459 (1962).

DOUBLE IONIZATION MECHANISMS IN H^+ - Ne COLLISIONS*

R. D. DuBois, L. H. Toburen and S. T. Manson[†]

Pacific Northwest Laboratory, Richland, WA 99352 U.S.A.
[†]Georgia State University, Atlanta, GA 30303 U.S.A.

Proton impact on atomic targets results primarily in single target ionization; this process is fairly well understood theoretically for high impact energies. The production of multiply charged target species is less well understood. Multiple ionization can arise from several mechanisms, the simplist being inner shell ionization followed by Auger decay. Competing mechanisms include direct double ionization and processes involving electron correlation and shake-off. Electron capture can also result in multiple ionization, particularly if the electron is captured from an inner shell.

We have measured absolute single and double ionization cross sections for neon resulting from 10 to 4000 keV proton impact. The cross sections illustrated in Fig. 1 shows that double ionization peaks at about 100 keV and is about 8% of the single ionization value. As the proton energy increases the contribution of double ionization decreases to about 2% at 4 MeV. Double ionization of neon by fast protons can result from K-shell ionization ($1s2s^22p^6$) followed by K-Auger emission, direct double outer shell ionization ($1s^22s^22p^4$, $1s^22s2p^5$, $1s^22s^02p^6$), and production of certain autoionizing states (i.e. $1s^22s2p^5nl$). The relative strengths of these ionization channels can be determined from experimental and theoretical results. Our measurements of the K-shell ionization cross sections for H^+ - Ne collisions agree well with our Born calculations and with previous work.[1] This contribution to double ionization via Auger cascades is at most 25% of the total as shown in Fig. 1. Coincidence measurements were also performed which indicate that double ionization resulting from electron capture is negligible for proton energies greater that a few hundred keV. Thus the dominant Ne^{2+} production mode for energies above a few hundred keV is direct double outer shell ionization and autoionization.

One would expect the total strength of autoionization to be a small contribution to the Ne^{2+} production. Furthermore, Beyer et al.[2] have measured cross sections for the $1s^22s2p^5$ and $1s^22s^02p^6$ vacancy production. These vacancy configurations lead to Ne^{2+} following deexcitation via photon emission. We have extrapolated their data to higher energies and combined them with the K-shell contribution. These processes then represent 17 to 45% of the Ne^{2+} production cross sections for proton energies from 100 to 4000 keV, respectively. The remainder is attributed to direct $1s^22s^22p^4$ double ionization. The ratios of the direct double ionization cross sections ($2p^4$: $2s2p^5$: $2s^02p^6$) are approximately constant with proton energy and are found to be 1 : 0.24 : 0.02 at 1 MeV. This is consistent with results based on statistical weighting of theoretical cross sections which yield ratios of 1 : 0.27 : 0.007. We plan to use similar analysis on understanding multiple ionization in other systems.

*Research supported in part by the U.S. Department of Energy Contract DE-AC06-76RLO-1830 and the U.S. Army Research Office Contract DAA6-29-80-C-0027.

References

1. M. Rødbro et al. Phys. Rev. A 19, 1936 (1979).
2. H. F. Beyer, R. Hippler, and K-H Shartner, Z. Phys. 292, 353 (1979).

Figure 1. Total, single, and double ionization cross sections for protons on neon. Contributions to double ionization from initial $1s^22s^02p^6$, $1s^22s2p^5$, $1s^22s^22p^4$, and $1s2s^22p^6$ were obtained as described in the text.

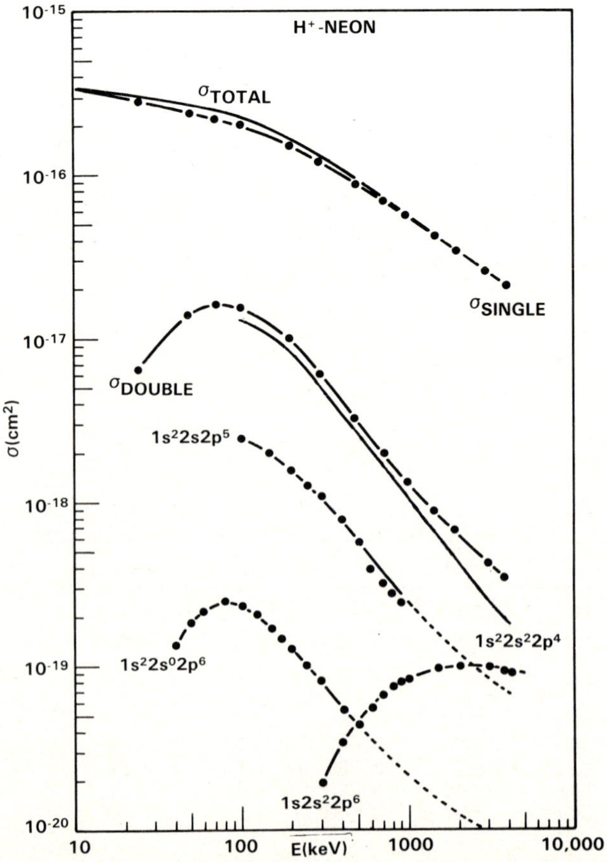

4KL8 CALCULATIONS OF THE CROSS SECTIONS FOR THE DOUBLE IONIZATION OF HELIUM BY LIGHT NUCLEI

V.A. Sidorovich , V.S. Nikolaev

Institute of Nuclear Physics, Moscow State University,Moscow 117234, USSR

The cross sections σ for the double ionization of He atoms by the nuclei of H^+, He^{2+} and Li^{3+} at energies E/A from 0.01 to 5 MeV/amu have been calculated in the independent electron approximation, in which we have

$$\sigma = 2\pi \int w^2(\rho)\,\rho\, d\rho \qquad (1)$$

where w is the probability of a transition of one of the electrons into the continuum , ρ the impact parameter. As the values of w we have used the probabilities w_B for the ionization of a hydrogen-like system with nuclear charge Z_t=1.69,which have been calculated in the first Born approximation. Since it is not infrequent the probabilities exceed unity, the cross sections σ have been calculated also using the probabilities $w_p = 1 - \exp(-w_B)$ and $w_\alpha = [1 - \exp(-\sqrt{w_B})]^2$. The former corresponds to the decay model, which is similar to the absorption model used in[1] to calculate the charge transfer cross sections, while the latter corresponds to the decay model for the effective probability amplitude $\sqrt{w_\alpha}$.

A comparison of the results of the calculations with the experimental data[2-5] shows (Fig. 1) that the independent electron approximation leads to a qualitatively correct energy dependence of the cross sections in the range of E/A values under consideration. At E/A > 0.1 to 0.2 MeV/amu when the single-electron ionization is well described by the first Born approximation, the calculated double ionization cross sections are close to experiment. At E/A =0.15 to 1 MeV/amu the experimental cross sections for the double ionization of helium by protons coincide to within 30% with the calculated values of σ_α and that for double ionization of helium by the nuclei of He^{2+} and Li^{3+} coincide to within 15-20% with σ_p values. At E $\gtrsim$ 2 MeV the experimental cross sections for proton-induced helium double ionization are close to the cross sections calculated in[6] in the first Born approximation, which describes double ionization resulting from a change of the atom field after an electron has been removed from it (shake up).

The results of the calculations show that the mechanism of the independent electron removal is dominant in the double ionization of He by light nuclei at E/A < 2 Z^2 MeV/amu.

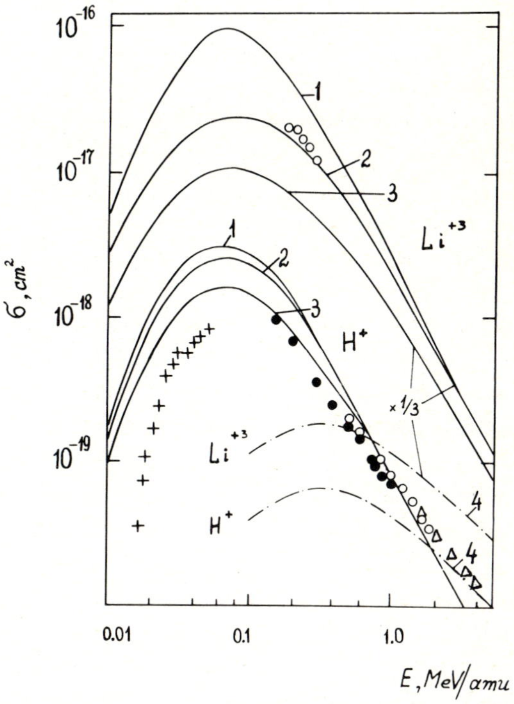

Fig. 1: Cross sections for double ionization of He by nuclei of H^+ and Li^{3+}, 1,2,3- present calculation for w_B, w_p, and w_α respectively. 4-from the results of[6]. Experiment: + - from[2] , ● -from[3] , o-from[4], Δ -from[5].

References

1. H. Ryufuku, T. Watanabe, Phys.Rev., A18, 2005 (1979).
2. V.V. Afrosimov, Yu.A. Mamaev, M.N. Panov , N.V. Fedorenko, ZhTF, 39, 159 (1969).
3. L.J. Puckett, D.W. Martin, Phys.Rev., A1, 1432 (1970).
4. Yu.Z. Levchenko, Theses, Phys.Tech.Inst. Ukr. Acad.Sci. Kharkov, 1974.
5. S. Wexler, J. Chem.Phys. 44, 2221 (1966).
6. W.J.B. Oldham, Phys.Rev. 166, 34 (1968).

MEASUREMENTS OF THE RATIO BETWEEN THE DOUBLE- AND SINGLE-IONIZATION CROSS SECTIONS OF HELIUM IN COLLISIONS WITH FAST, BARE NUCLEI

H. Knudsen, L.H. Andersen, and P. Hvelplund
Institute of Physics, University of Aarhus, DK-8000 Aarhus C, Denmark

G. Astner, H. Cederquist, H. Danared, L. Liljeby, and K.-G. Rensfelt
Research Institute for Atomic Physics, University of Stockholm, S-10405 Stockholm, Sweden

For collisions between fast, fully stripped ions and He, the cross section for single ionization is fairly well known, both experimentally and theoretically[1]. For small values of the parameter $\kappa = 2qv_0/V$ (where q is the ion charge and V the ion velocity), the cross section can be calculated in the first Born approximation, and one obtains $\sigma^+ \propto q^2V^{-2}\ell nV$.

For double ionization, the situation is less well known. For very small values of κ, the first Born approximation predicts that the cross section σ^{++} is proportional to σ^+. The mechanism behind double ionization in this region is the so-called 'shake-off' mechanism, where one electron is removed during particle impact, while the other electron relaxes onto the electronic states (including continuum states) of He^+. The calculated ratio $R \equiv \sigma^{++}/\sigma^+$ depends critically on the amount of electronic correlation incorporated in the calculations, and it is not known to better than a factor of ∿3 (Ref. 2).

At larger κ values, it was suggested recently by McGuire[3] that another mechanism becomes dominant. Here double ionization occurs due to two consequetive, close particle-electron encounters, and McGuire found that this 'two-step' model leads to $\sigma^{++} \propto (q/V)^4$. However, he was not able to calculate the absolute magnitude of the cross section.

Using a time-of-flight secondary-ion spectrometer[4], we have measured R for bare ions of atomic number between 1 and 8 and (0.64-15) MeV/amu. The data are shown in the figure. We find that all the measured ratios agree well with the form

$$R = 2.2\times10^{-3} + 4.6\times10^{-3}q^2[E\ell n(13.123\sqrt{E})]^{-1},$$

where the ion energy E is measured in MeV/amu. This empirical result is shown as the straight line in the figure. The first term gives the contribution from the 'shake-off' mechanism. The magnitude is close to that expected theoretically[3]. In the second term, we have used the logarithmic dependence of the single-ionization cross section, as given by Kim and Inokuti[5]. The absolute magnitude of this 'two-step' contribution is determined within ±10%.

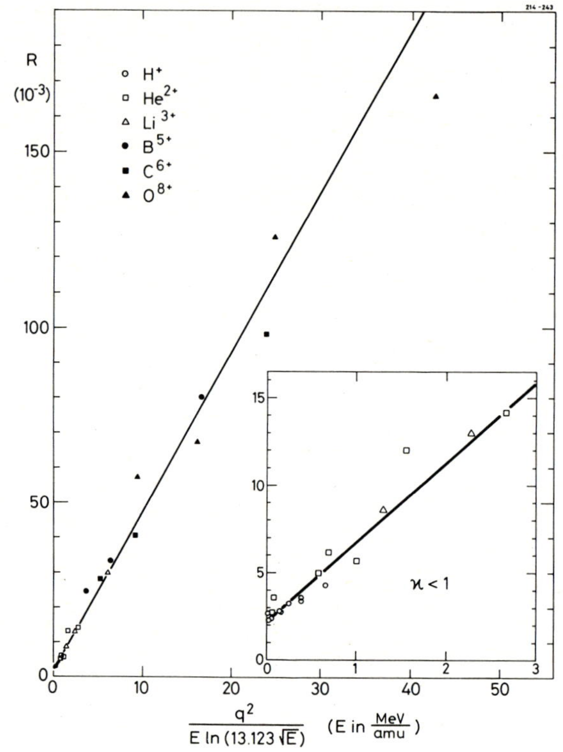

REFERENCES

(1) H. Knudsen, Book of Invited Papers XII ICPEAC (1981) p. 657
(2) F.W. Byron and C.J. Jochain, Phys.Rev.Lett. 16, 1139 (1966)
(3) J.H. McGuire, Phys.Rev.Lett. 49, 1153 (1982)
(4) H.K. Haugen, L.H. Andersen, P. Hvelplund, and H. Knudsen, Phys.Rev.A 26, 1962 (1982)
(5) Y.-K. Kim and M. Inokuti, Phys.Rev.A 3, 665 (1971)

COINCIDENCE MEASUREMENTS OF RECOIL-ION CHARGE-STATE SPECTRA PRODUCED BY IONIZATION AND CHARGE TRANSFER IN COLLISIONS OF 1.4 MeV/u HIGHLY CHARGED IONS AND RARE-GAS ATOMS

B. Schuch, W. Groh, A. Müller, E. Salzborn
Institut für Kernphysik, Strahlenzentrum, Universität Giessen, 6300 Giessen, West Germany

and

H.F. Beyer, W.A. Schönfeldt, P.H. Mokler
Gesellschaft für Schwerionenforschung (GSI), D-6100 Darmstadt

Previous investigations have demonstrated the outstanding capability of fast highly stripped ions to produce slow highly charged recoil ions when colliding with atoms[1,2]. Further experiments were devoted to investigate correlations between the charge state of the outgoing fast projectile and the charge state spectrum of the slow recoil ions produced in high-energy collisions[3,4]. At low energies (about 10 to 100 keV) transfer ionization has been extensively studied in collisions of highly charged rare gas ions and atoms[5]. The present investigation extends this work to a wide spectrum of collision systems at a (fixed) specific energy of 1.4 MeV/u ($\hat{=}$ 7.5 $\mathbf{v}_0$).

We have measured the charge-state spectra of recoil ions produced in collisions of Fe^{q+}(q=12,...,21), Kr^{18+}, Zr^{20+}, U^{q+}(q=30,...,48) with He, Ne, Ar, Kr and Xe. By using a time-of-flight coincidence technique[5] we could measure correlations of the charge spectra with the number k of electrons captured or lost by the projectile where k was 0, ∓1 or ∓2 in most of the cases, however, cases up to +4 were also observed.

Since our previous measurements[2] showed already that high recoil-ion charge states are produced in fast ion-atom collisions (e.g. up to about Xe^{20+} by U^{44+} impact on Xe) we had to take care for sufficient resolution to obtain complete separation also of high recoil-ion charge states. For this purpose a time-of-flight spectrometer was set up with a time resolution of better than 5ns which is good enough to separate all isotopes of natural xenon.

As an example of the measured data Fig. 1 shows a comparison of Ar recoil-ion spectra produced by ionization and capture of k=1, 2, or 3 electrons in Fe^{15+} + Ar collisions. Similar to the results of Gray et al.[3] the charge states of the recoil ions are strongly shifted to higher values when charge transfer to the projectile occurs instead of a pure ionization process. The shift, however, does not increase significantly when the number of captured electrons is further increased.

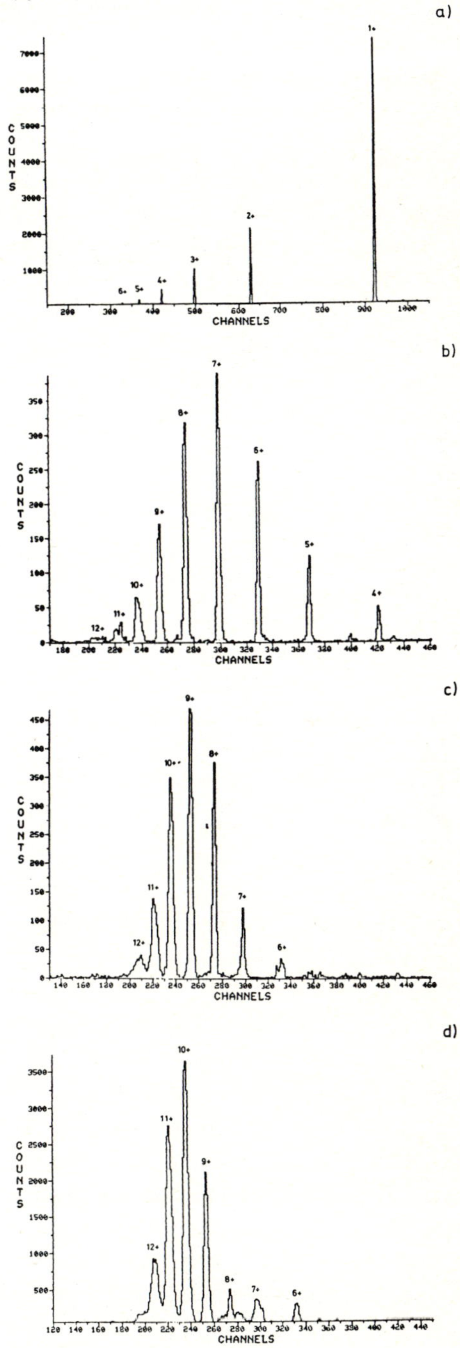

Fig. 1: Charge-state spectra of Ar recoil ions, created by 1.4 MeV/u Fe^{15+} projectiles in a thin Ar gas target
a) $Fe^{15+} \rightarrow Fe^{15+}$ (direct ionization)
b) $Fe^{15} \rightarrow Fe^{14+}$ (one-electron capture)
c) $Fe^{15+} \rightarrow Fe^{13+}$ (two-electron capture)
d) $Fe^{15+} \rightarrow Fe^{12+}$ (three-electron capture)

References

1. C.L. Cocke, Phys.Rev. A 20, 749 (1979)
2. A.S. Schlachter et al., Phys.Rev. A 26, 1373 (1979)
3. T.J. Gray et al., Phys.Rev. A 22, 849
4. H.-C. Werner et al., XII ICPEAC (Gatlinburg 1981), Abstracts of Contributed Papers, 876
5. W.Groh et al., J.Phys. B, in print

INFLUENCE OF THE PROJECTILE CHARGE STATE ON MULTIPLE IONIZATION OF RARE GAS ATOMS

H.-Ch. Werner, H. Schmidt-Böcking[+], N. Stolterfoht, G. Nolte

Hahn-Meitner-Institut, 1000 Berlin 39, FR Germany
[+)] IKF, Universität Frankfurt, FR Germany

We have measured the charge state distributions of recoil ions produced in collisions of 221-MeV Kr^{q+} with Ne, Ar, and Kr. The measurements were performed using the time-of-flight technique[1] improved by geometrical considerations[2]. Charge states up to Ne^{9+}, Ar^{16+}, and Kr^{19+} could be resolved in the spectra. A typical spectrum is displayed in fig. 1.

From these spectra relative intensities were determined and corrected due to the channeltron detection efficiency for ions of different energy. The efficiency was determined by means of an electron-ion coincidence method[3]. The resulting fractions f_n of recoil ions in charge state n (the sum is normalized to unity) are displayed in fig. 2.

The projectile charge states were chosen to be q = 13, 25, and 27. By comparison of the q=13 and the q=25 results we can see the effect of a big difference in q. Kr^{27+} differ from Kr^{25+}, that the latter has a filled L shell while there is a L vacancy in Kr^{27+}. Thus we could study the influence of inner shell vacancies present prior to the collision.

For the target Ne one can observe a drastic effect when q is increased from 13 to 25. The fraction of highly charged recoil ions ($n \geq 7$) is increased by more than a factor of 3. This increase may be explained by the higher projectile charge. In the transition from q=25 to q=27 minor effects appear. Only the fraction of highly charged recoil ions is enlarged. To give some quantitative conclusions we put our data on an absolute scale due to the scaling calculations of Schlachter et al.[4]. It is found that the increase of partial cross sections for highly charged recoil ions is in the same order of magnitude as we expect vacancy capture from the Kr-L shell to the Ne-K shell. Thus we conclude, that at least partially this inner shell charge transfer is responsible for the enhancement of the fractions of highly charged recoil ions.

The opposite result is obtained when heavier target gases (Ar, Kr) are used (see fig. 2). There is only a small increase of the fractions of highly charged recoil ions and even no enhancement for the highest charge states. However, a strong enlargement of these fractions is observed when Kr^{27+} is used as a projectile. For $n \geq 9$ they increase by a factor of 4. Estimates show that the corresponding partial cross sections are more than one order of magnitude larger than the values expected for inner shell charge transfer processes. Here, further work is needed to explain the production of high charge states for Kr^{27+} ions.

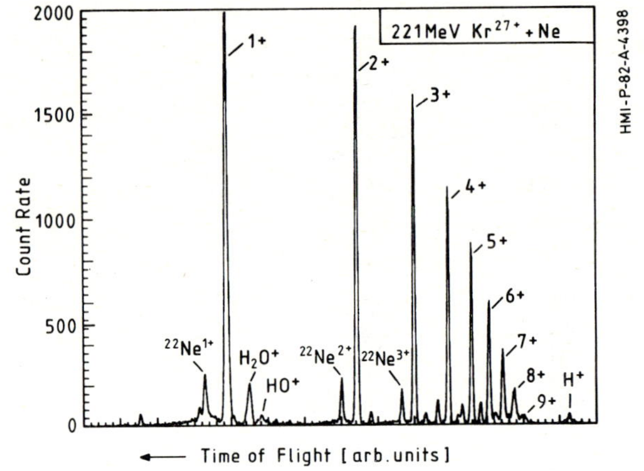

Fig. 1: Time-of-flight spectrum of recoil ions produced in collisions of 221-MeV Kr^{27+} with Ne.

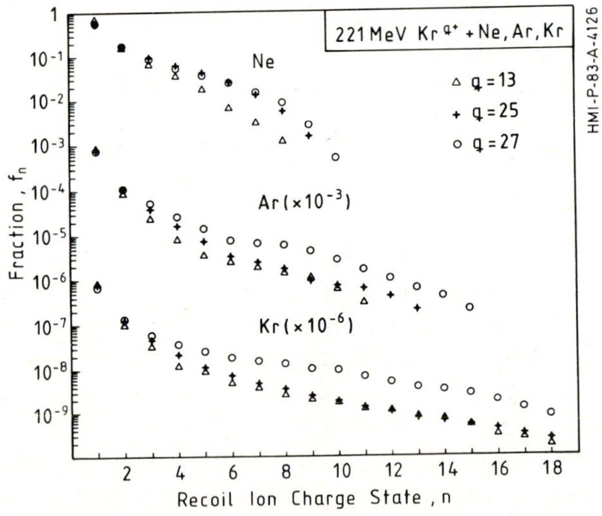

Fig. 2: Fractions f_n of recoil ions in charge state n produced in collisions of 221-MeV Kr^{q+} (q=13, 25, and 27) with Ne, Ar, and Kr.

References

1. C. L. Cocke; Phys. Rev. A 20, 749 (1979)
2. H.-Ch. Werner, D. Schneider, H. Schmidt-Böcking, G. Nolte, N. Stolterfoht; XIIth ICPEAC, Abstracts of Contributed Papers, 876 (Gatlinburg 1981)
3. H.-Ch. Werner; Thesis (Freie Universität Berlin 1982) and to be published
4. A. S. Schlachter, K. H. Berkner, W. G. Graham, R. V. Pyle, P. J. Schneider, K. R. Stalder, J. W. Stearns, J. A. Tanis, R. E. Olson; Phys. Rev. A 23, 2331 (1981)

Multiple ionization of slow recoil ions in fast heavy-ion atom collisions[+]

J. Ullrich, S. Kelbch, W. Schadt, H. Schmidt-Böcking
Institut für Kernphysik der Universität Frankfurt

R. Schuch
Physikalisches Institut der Universität Heidelberg

H. Ingwersen
MPI für Kernphysik, Heidelberg

For the collision systems 156 MeV $Br^{n+} + Ne^{0+} \rightarrow Br^{n'+} + Ne^{q+}$ and (1.5 - 7 MeV) $Ne^{n+} + Ne^{0+} \rightarrow Ne^{n'+} + Ne^{q+}$ the differential cross sections $\sigma(n,n',q)$ were measured. Using a coincidence technique, both outgoing charge states n' and q, respectively, were determined for the same collision. In both collision systems the direct ionization process of target L-shell electrons to the continuum yields the predominating contribution to the ionization of the slow recoil ions and extends to very large impact parameters.

For 156 Br^{n+} + Ne the capture of target electrons by the projectile is very important for the production of highly charged recoil ions (q > 6) e.g. for $\sigma(27, 26, 8) = 4,4 \cdot 10^{-18} [cm^2]$ and for $\sigma(27, 25, 8) = 5.9 \cdot 10^{-18} [cm^2]$ were measured. From the recoil charge state distributions one can derive that the capture of Ne-K-electrons with subsequent Ne-K-Auger emission is the dominating capture channel, whereas the contribution of Ne-L-electron capture seems to be negligible. The Ne-K-electron capture probability increases strongly with the projectile charge state.

For the $Ne^{n+} + Ne^{0+}$ collision system, however, the K-shell excitation (by capture or quasimolecular promotion) yields only a small contribution to multiple ionization of the recoil ion, whereas the capture process of L-electrons of the recoil ion into the projectile shell is important for the multiple ionization of the recoil ions. This capture process decreases strongly with increasing projectile velocity. The production of highly charged recoil ions is mainly due to ionization processes at small impact parameters and its cross section increase dramatically with increasing projectile energy (see figure 1).

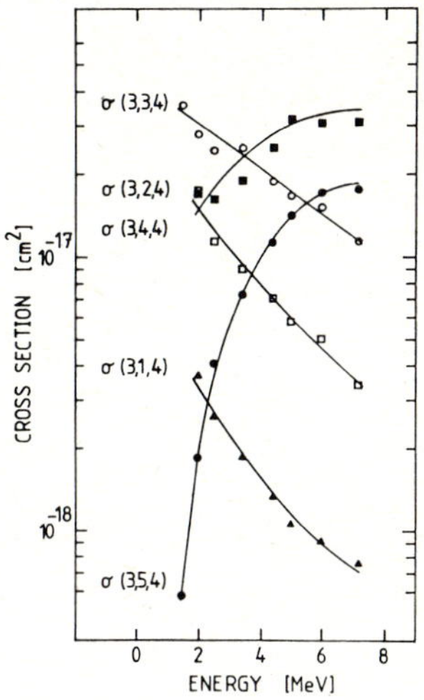

Fig. 1 Energy dependence of cross sections $\sigma(n, n', 4)$ for the production of recoil ions with charge state q = 4 in the collision system $Ne^{n+} + Ne^{0+} \rightarrow Ne^{n'+} + Ne^{4+}$

+ Supported by BMFT

IONISATION OF ATOMIC HYDROGEN BY FAST MULTIPLY CHARGED IONS

M.B. Shah and H.B. Gilbody

Department of Pure and Applied Physics
The Queen's University of Belfast
Belfast, United Kingdom

In recent work[1,2,3] we have developed and used a crossed beam technique to determine ionisation cross sections for the process

$$X^{q+} + H(1s) \rightarrow X^{q+} + H^{+} + e$$

with high precision over a wide energy range. The H^+ products formed in collisions with a thermal energy target beam of highly dissociated hydrogen are selectively distinguished from background ions by time-of-flight analysis and counted in coincidence with electrons from the same events.

We have used the same technique to obtain data for $C^{(3-6)+}$, $O^{(3-6)+}$ and $Ar^{(4-9)+}$ impact up to energies of 400, 300 and 250 keV amu^{-1} respectively. The data for carbon and oxygen ions extend our previous measurements[2] which were carried out for $C^{(2-4)+}$ in the range 16-195 keV amu^{-1} and for $O^{(2-5)+}$ in the range 12-145 keV amu^{-1}.

The measured ionisation cross sections are also considered in relation to those for the charge transfer process

$$X^{q+} + H(1s) \rightarrow X^{(q-1)+} + H^{+}$$

which have also been determined using a coincidence counting technique[4].

The dependence of the measured ionisation cross sections on charge state q is considered. The results are also compared with scaled cross sections obtained by Gillespie[5] from the Bethe asymptotic cross section at high velocities and with recent predictions by Hardie and Olson[6] developed from the classical-trajectory-Monte-Carlo treatment used by Olson and Salop[7].

Data for fully stripped C^{6+} ions are compared with predictions based on the first Born and Bethe approximations[8], the Glauber approximation[9], the unitarized-distorted-wave approximation[10] and the continuum-distorted-wave approximation[11].

References

1. M B Shah and H B Gilbody, J Phys B 14, 2361 (1981).
2. M B Shah and H B Gilbody, J Phys B 14, 2831 (1981).
3. M B Shah and H B Gilbody, J Phys B 15, 413 (1982).
4. M B Shah and H B Gilbody, J Phys B 15, 3441 (1982).
5. G H Gillespie, J Phys B 15, L729 (1982).
6. D J Hardie and R E Olson - private communication.
7. R E Olson and A Salop, Phys Rev A 16, 531 (1977).
8. G H Gillespie, J Phys B, Phys Rev A 24, 608 (1981).
9. J H McGuire, Phys Rev A 26, 143 (1982).
10. H Ryufuku, Phys Rev A 25, 720 (1982).
11. J F McCann and D S Crothers - private communication.

ENERGIES AND COUPLINGS FOR INFINITELY EXCITED STATES OF He-Li^{3+} QUASI MOLECULE

A. Macías, R. Mendizabal, F. Pelayo, A. Riera and M. Yáñez.

Departamento de Química Física y Química Cuántica. Centro Coordinado.
CSIC-UAM. Universidad Autónoma de Madrid.
CANTOBLANCO (Madrid,34) (Spain).

To treat the charge-exchange processes in Li^{3+}-He Collisions[1] we are interested in the representation of states of the quasi-molecule $HeLi^{3+}$ which are embedded, for an infinite range of internuclear distances, in the continium corresponding to $Li^{++}(1) + He^{++} + 1e^-$.

These states are not only autoionizing but also they are infinitely excited since there is a finity of lower lying levels corresponding to the two Rydberg series $Li^+(1snl) + He^{++}$ and $Li^{++}(1s) + He^+(n)$. This means that the usual techniques of quantum chemistry are not directly applicable to represent these states and a new approach is needed.

In this work we report an application of the so-called stabilization techniques to a molecular case which allow us to calculate those autoionizing states. In short, this stabilization technique consists in using a straightforward CI method and then, by selection of the basis set and by variation of parameters in the bases, decide which eigenvalues of the electronic hamiltonian are stable and represent resonances.

We present in Fig. 1 the qualitative correlation diagram corresponding to the states under study and in Fig. 2 a and b the result of varying the basis set in the range of intermediate internuclear distances. When the basis set is indefinitely enlarged collapse of the eigenvalues embedded in the continium occurs but this takes place through avoided crossings with other unstable eigenvalues in a very similar manner to that of using the stabilization method to calculate Stark resonance parameters[2].

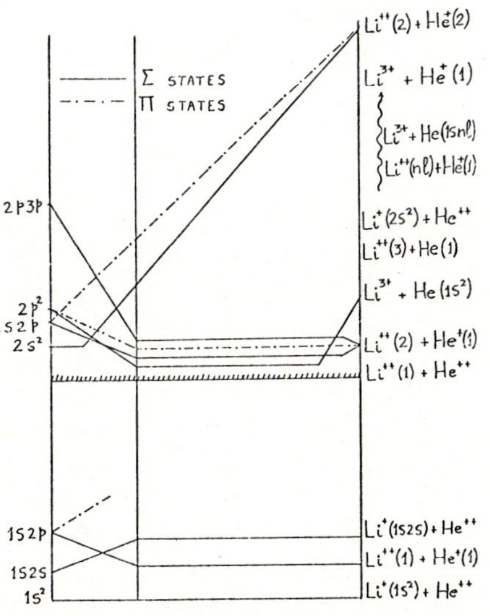

Fig. 1.

In Fig. 3 we show the importance of the choice of the configuration basis set; when the same basis set which is selected at large internuclear distances to avoid the appearence of unstable eigenvalues, is used at small internuclear distances, one does no obtain a correct representation of the resonances and it is necessary to include in the basis configurations which were unimportant for large distances.

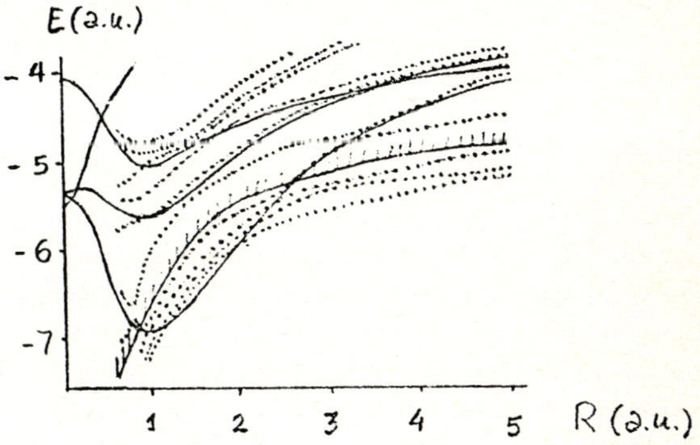

Fig. 2 a.

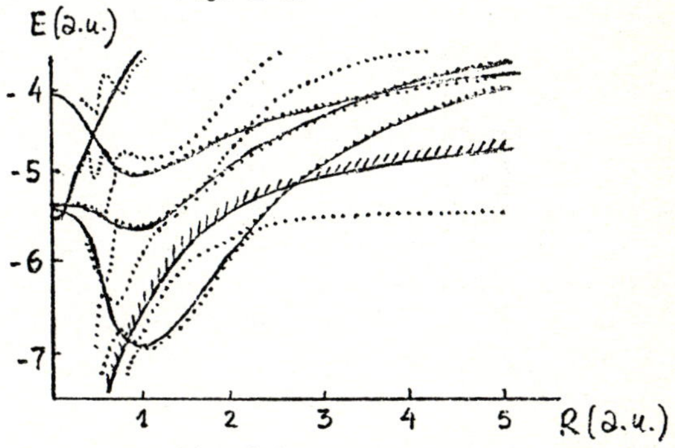

Fig. 2 b.

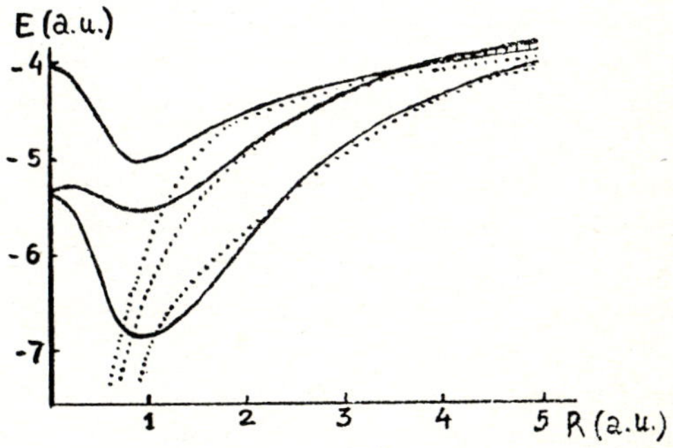

Fig. 3.

A further advantage of the stabilization method is that all radial couplings can be analytically evaluated as in the CI method[3]. The couplings between stable states can then be obtained by eliminating those due to the sharp avoided crossing with unstable eigenvalues.

References

1. I. Wirkner-Bott, W. Sein, A. Müller, P.Kester and E. Salzborn. J. Phys. B., 14, 3987 (1981).
2. A. Macías and A. Riera, J. Phys. B., 11, 3827 (1978); 13, L449(1980).
3. A. Macías and A. Riera, J. Phys. B., 10, 861 (1977); 11, 1077 (1978).

NEW EXPERIMENTAL INVESTIGATIONS OF DENSITY EFFECT IN INNER-SHELL EXCITATIONS

S.P.Møller and A.H.Sørensen
CERN, Geneva, Switzerland

J.F.Bak, F.E.Meyer, J.B.B.Petersen, E.Uggerhøj and K.Østergaard
Institute of Physics, University of Aarhus, Aarhus, Denmark

In recent years a large amount of work has been devoted to measure inner-shell excitation cross sections for relativistic charged particles[1]. With increasing values of the relativistic γ-factor, excitations may occur at still larger distances. In condensed materials the interaction extends over many interatomic distances, and, consequently, the projectile field will be screened, leading to a reduced cross section. This so-called density effect is well-known in energy loss measurements, but has only been claimed observed in one inner-shell excitation cross section experiment[2]. The present work describes a new, accurate way to measure relative inner-shell excitation cross sections over a large span of γ-values, and the first results are presented, indicating a reduced density effect.

The experiment was performed at the CERN 28 GeV proton synchrotron in a non-separated secondary beam, which at 5 GeV/c in the positive polarity mainly consists of positrons, pions and protons. The principle of the experiment was to simultaneously measure the K X-ray yield from 5 GeV/c e^+, π^+ and p traversing a Cu or an Al target. Relative inner-shell cross sections for $\gamma \approx 1\text{-}10^4$ can in this way be measured very accurately, since the large uncertainties in absolute cross section measurements from target thickness, detector solid angle, detector dead-time etc. are eliminated. On the other hand, the present experiment is dependent on particle identification, which, however, can be done to the percent level, using threshold Cerenkov counters and lead glass detectors.

In fig. 1 is shown measured K-shell excitation cross sections for 5 GeV/c e^+, π^+ and p incident on a Cu and Al target, all normalized to the proton value. Also shown are results of previous experiments together with theoretical estimates primarily drawn to indicate the expected magnitude of the density effect. The theoretical curves are obtained in a simple Kolbenstvedt-type approach. The saturating density effect curves are calculated using the simple dielectric function corresponding to a medium where all but the K-electrons are free. In fig. 1 the theoretical curves as well as the cross sections measured in other laboratories are normalized to the value calculated for γ=5.422 corresponding to 5 GeV/c protons.

For Cu the results of previous experiments show a clear tendency to lie above the theoretical curve for no density effect. For Al the points fall around the full-drawn curve. In contrast, our relative measurements show clearly a density effect in both Al and Cu K-shell excitation, since the measured cross sections for high γ increase considerably slower than proportional to logγ. However, the measured positron points are significantly above the theoretical 'full density effect' curve. At the present stage it is not possible to judge whether a saturation has occured, or if the cross section keeps growing beyond the positron point.

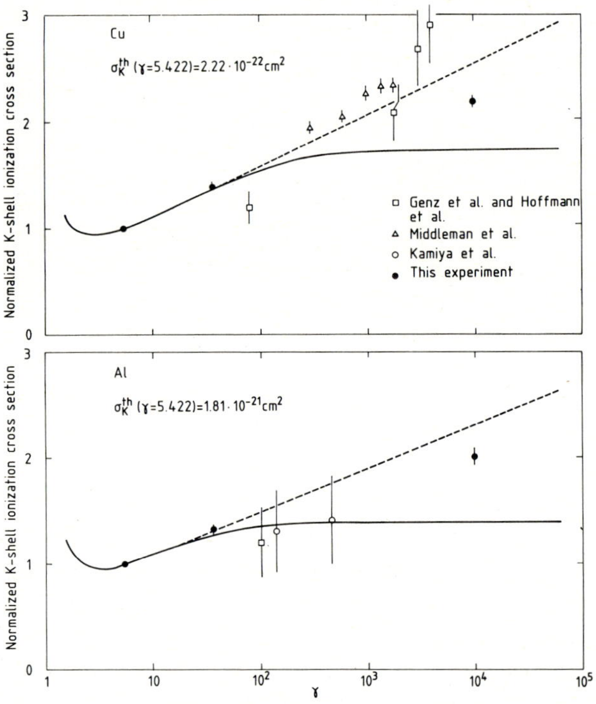

Fig. 1: Normalized K-shell excitation cross section in Al and Cu as a function of γ. The full-drawn curves are calculated by inclusion of the dielectric response of the medium, whereas the dashed curves correspond to neglect of density effects.

The inner-shell excitation results have been compared to new energy loss measurements in Si and Ge. These measurements are in agreement with the Bethe-Bloch formula including full density effect on all shells. A reduced density effect on the K-shell excitations can not be excluded, due to the small contribution from these electrons to the stopping power. However, a strong influence of density effect on excitations of L- and outer-shell electrons is required.

References

1. H.Genz et al., Z.Phys. A305, 9 (1982) and ref. therein.
2. M.Kamiya et al., Phys.Rev. A22, 413 (1980).

TOTAL K-SHELL IONISATION CROSS SECTIONS FOR HEAVY ELEMENTS INDUCED BY PROTONS

S. Divoux, B. Raith and B. Gonsior

Institut für Experimentalphysik III, Ruhr-Universität Bochum, D4630 Bochum 1, FR Germany

We measured total K-shell ionisation cross sections for the irradiation of thin solid targets ($\leq$ 5 μg/cm^2) of ^{92}Mo, ^{104}Pd, ^{120}Sn, ^{138}Ba, ^{140}Ce, ^{142}Nd and ^{144}Sm with protons of 0.5-6.0 MeV/u. Only isotopes were chosen as targets whose first excited states do not decay by internal conversion. The characteristic x rays of the target elements were detected by a 500 mm^2-Ge(I) detector. Its efficiency times solid angle were determined immediately before and after the cross-section measurements by placing twelve radioactive standards at the target position. The fluorescence yields were taken from the work of Krause[1]. The target thickness times projectile number were determined by measuring the elastically scattered projectiles with two detectors at different angles small enough to ensure pure Rutherford scattering. The data for the lower proton energies were corrected for screening effects following the work of Andersen et al[2].

The measured ionisation cross sections are compared with PSS calculations by Brandt et al[3] and with SCA calculations by Lägsgaard et al[4]. Fig. 1 shows the comparison for the lower proton energies of the collision system p → ^{138}Ba. Included are the data by Khelil and Gray[5]. Within the experimental errors (8-14 %) our results agree with both theories mentioned.

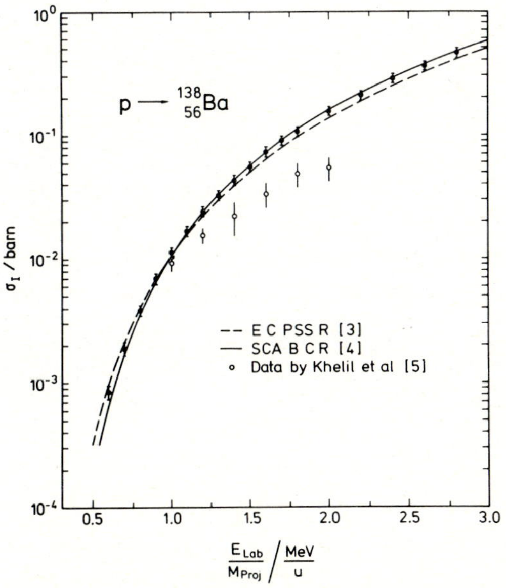

Fig. 1: Ionisation cross sections for the collision system p → ^{138}Ba as a function of the projectile energy

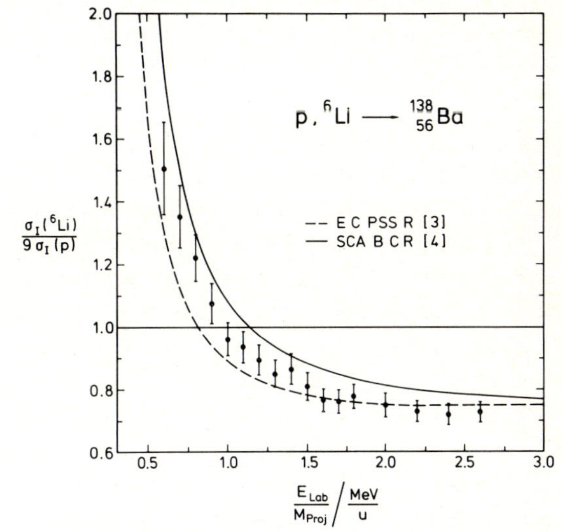

Fig. 2: Ionisation cross-section ratios for protons and ^{6}Li ions impinging on ^{138}Ba targets as a function of E_{Lab}/M_{Proj}

Furthermore, we measured for ^{138}Ba and ^{142}Nd in one beam time the ionisation cross-section ratios for protons and ^{6}Li having the same velocity. Fig. 2 shows the comparison of measured ratios with those calculated in the framework of the theories mentioned above.

References

1. M.O. Krause, J. Phys. Chem. Ref. Data 8, 307 (1979)
2. H.H. Andersen, F. Besenbacher, P. Loftager and W. Möller, Phys. Rev. A21, 1891 (1980)
3. W. Brandt and G. Lapicki, Phys. Rev. A23, 1717 (1981)
4. E. Lägsgaard, J.U. Andersen and M. Lund, in: Proceedings of the 10th International Conference on the Physics of Electronic and Atomic Collisions, Paris, July 1977, ed. G. Watel, p. 353
5. N.A. Khelil and T.J. Gray, Phys. Rev. A11, 893 (1975)

TOTAL K-SHELL IONISATION CROSS SECTIONS FOR ELEMENTS $24 \leq Z_T \leq 60$ INDUCED BY LI IONS

B. Raith, S. Divoux and B. Gonsior

Institut für Experimentalphysik III, Ruhr-Universität Bochum, D4630 Bochum 1, FR Germany

Total K-shell ionisation cross sections were measured for the irradiation of very thin ($\leq$ 5 μg/cm^2) solid targets of ^{52}Cr, ^{58}Ni, ^{89}Y, ^{104}Pd, ^{120}Sn, ^{138}Ba and ^{142}Nd with ^{6}Li ions in the energy range 1-16 MeV. The x rays were detected by a Si(Li) detector for ^{52}Cr, ^{58}Ni and ^{89}Y and by a Ge(I) detector for the other elements. All isotopes chosen as targets fulfil the prerequisite that internal conversion of nuclear excited states does not contribute to the K-shell ionisation. The efficiency times solid angle of the x-ray detectors were experimentally determined immediately before and after the cross-section measurements by using 16 radioactive standard sources placed at the target position. The fluorescence yields which are needed to convert the measured x-ray-production cross sections into ionisation cross sections were taken from the work of Krause[1] without modification in regard to multiple ionisation. The product of target thickness times projectile number was determined by measuring the elastically scattered projectiles with two detectors at different angles. These were chosen small enough to ensure pure Rutherford scattering. For the lower projectile energies the data were corrected with respect to screening effects following the work of Andersen et al[2].

The measured ionisation cross sections are compared with PSS calculations by Brandt et al[3] and with SCA calculations by Lägsgaard et al[4]. Fig. 1 shows the comparison for the collision system ^{6}Li → ^{52}Cr. Included are the data for ^{7}Li → Cr by McDaniel et al[5]. According to the calculations by Lapicki and McDaniel[6] capture of a target K electron by a bare Li projectile was neglected; for Cr this effect is smaller than 3 %, for all other elements smaller than 1 %. The measured cross sections agree within the experimental errors (9-14 %) with both theories. Only for the higher energies for ^{52}Cr and ^{58}Ni ($\xi \gtrsim 0{,}6$) the SCA calculations according to Lägsgaard et al[4] yield too low results because the (Z_P+Z_T)-approximation for the binding effect used in this work does no longer hold.

In order to investigate further the effects of Coulomb deflection and repulsion we measured in one time for ^{89}Y and ^{138}Ba the ionisation cross-section ratios for ^{6}Li and ^{7}Li ions having the same velocity. The data agree with both theories within the experimental error.

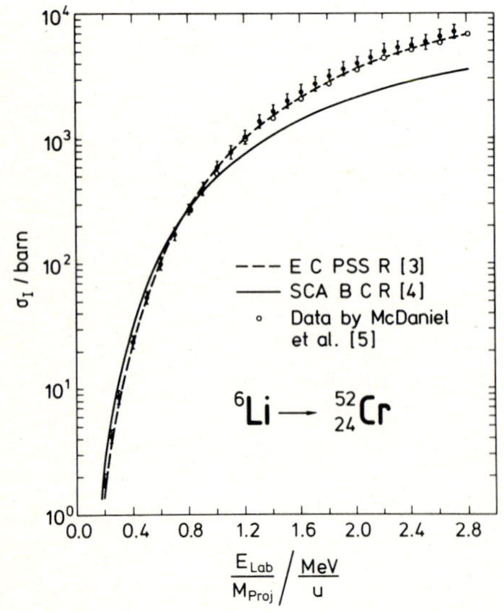

Fig. 1: Ionisation cross sections for the collision system ^{6}Li → ^{52}Cr as a function of the projectile energy

References

1. M.O. Krause, J. Phys. Chem. Ref. Data 8, 307 (1979)
2. H.H. Andersen, F. Besenbacher, P. Loftager and W. Möller, Phys. Rev. A21, 1891 (1980)
3. W. Brandt and G. Lapicki, Phys. Rev. A23, 1717 (1981)
4. E. Lägsgaard, J.U. Andersen and M. Lund, in: Proceedings of the 10th International Conference on the Physics of Electronic and Atomic Collisions, Paris, July 1977, ed. G. Watel, p. 353
5. F.D. McDaniel, T.J. Gray, R.K. Gardner, G.M. Light, J.L. Duggan, H.A. Van Rinsvelt, R.D. Lear, G.H. Pepper, J.W. Nelson and A.R. Zander, Phys. Rev. A12, 1271 (1975)
6. G. Lapicki and F.D. McDaniel, Phys. Rev. A22, 1896 (1980)

TOTAL K-SHELL VACANCY PRODUCTION CROSS SECTIONS FOR 200 TO 1600 keV/amu ^{3}He ON Ti AND Cu

Donald G. Simons and David J. Land

Naval Surface Weapons Center, White Oak, Silver Spring, Maryland, USA 20910

Total K-shell vacancy production cross sections have been measured for 200 to 1600 keV/amu ^{3}He ions incident on thin-foil targets of Ti and Cu. The ionization cross sections are inferred from measurements of the x-ray production cross section and tabulated values of fluorescence yields.[1] The x-ray production cross section is obtained by normalizing the x-ray yield to the scattered particle yield. These yields are determined by simultaneously measuring the x-rays and the scattered particles for incident monoenergetic ^{3}He ions on the foil targets. This procedure eliminates the need for accurate determination of target thicknecesses and beam currents. In this work it is assumed that the particle scattering is purely Rutherford. Details of our experimental procedure have been previously reported for our measurements of proton-induced K-shell vacancy cross sections.[2] By paying close attention to experimental detail, the values of the total K-shell vacancy production cross sections reported in the work are accurate to about 5% over most of the energy range studied.

The experimental values of the total cross sections are compared with the values obtained from ECPSSR theory of Brandt and Lapicki.[3] The results of this comparison are shown in Fig. 1 for the Ti target. In the figure, comparisons are made by plotting the ratio of the experimental values of the cross sections to the theoretical values as a function of the ion energy in keV/amu. In order to cover the entire energy range presented here, both singly and doubly charged ^{3}He ions were used. The values obtained from $^3He^+$ ions are presented by the closed circles in Fig. 1., while those obtained from $^3He^{++}$ ions are shown by the open circles.

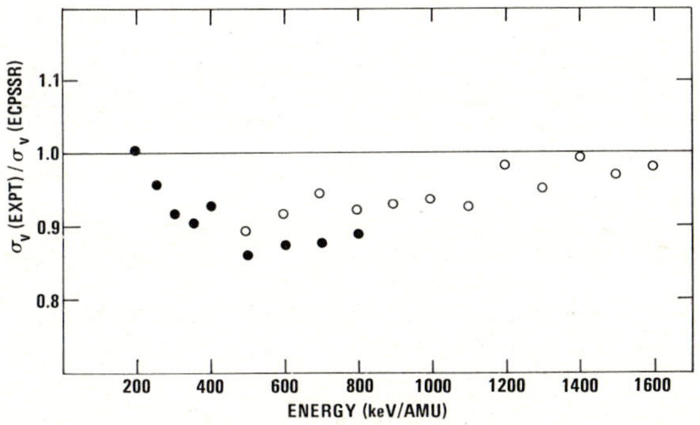

Fig. 1: Ratio of the experimental values to the ECPSSR theoretical results as a function of energy per amu for ^{3}He-induced K-shell ionization cross sections in Ti.

Measurements were made at four energies (500 through 800 keV/amu) for both charge states. In this overlap region, values of cross sections obtained from the $^3He^{++}$ ions are on the average 5% higher than those obtained from the $^3He^+$ ions. The energy dependence of these differences is not known and, hence, it is not possible to say at the present time how the 5% difference extends to the higher and lower energy regions. Nevertheless, we see that the ECPSSR values are generally larger than the experimental values for almost the entire energy range studied. The greatest discrepancies appear at the intermediate projectile energies with a tendency for ECPSSR values to approach the experimental values at both ends of the energy range. Similar discrepancies are also observed in comparison between ECPSSR and experimental vacancy production cross sections induced by protons in the same energy range.[2] However, the discrepancies are not as pronounced for protons. The results of these comparisons will also be presented. The total K-shell vacancy production cross sections were measured on thin-foil Cu targets for ^{3}He ions in the same energy range. The results of these measurements will also be presented and compared with ECPSSR theory. Differences are also observed between the values of the total cross sections induced by $^3He^+$ and by $^3He^{++}$ and are similar to those for Ti.

For the Cu measurements targets thicknesses of 63 $\mu g/cm^2$ and 32 $\mu g/cm^2$ were used. The values of the cross sections obtained from the thicker target are consistently lower than those for the thinner target. This difference appears to hold for both singly and doubly charged ^{3}He ions. Further studies concerning the effect of target thickness on the K-shell vacancy production cross sections are currently being made for both Ti and Cu and will be reported.

References

1. W. Bambynek, B. Craseman, R. W. Fink, H. U. Freund, H. Mark, C. D. Swift, R. E. Price and P. V. Rao, Rev. Mod. Phys. 44, 716 (1972).
2. M. D. Brown, D. G. Simons, D. J. Land and J. G. Brennan, Phys. Rev. A 25, 2935 (1982).
3. W. Brandt and G. Lapicki, Phys. Rev. A 23, 1717 (1981).

K-SHELL X-RAYS AND MULTIPLE VACANCY PRODUCTION IN $_{19}K$, $_{22}Ti$, $_{12}Mg$, and $_{35}Br$ RESULTING FROM 20-80 MeV $_{17}Cl$ ION BOMBARDMENT*

J. A. Tanis
Western Michigan University, Kalamazoo, MI 49008 USA

S. M. Shafroth and T. McAbee, University of North Carolina, Chapel Hill, N. C. 27514 USA
and Triangle Universities Nuclear Lab, Durham, N. C. USA

G. Lapicki, East Carolina University, Greenville, N. C. 27834 USA

Target K x-ray production cross sections have been determined for 20-80 MeV Cl^{q+} (q = 3-10) ions incident on thin targets of KBr, Ti, Mn, evaporated onto carbon backings. This work was done at Triangle Universities Nuclear Laboratory using the FN tandem Van de Graff. X-rays were detected in air at 90° to the beam with a Si(Li) detector. Elastically scattered Cl ions were detected at 60° to the beam direction and used for normalization.

Cross sections deduced from the present measurements are listed in Table I. Also shown in the table are results previously obtained[1] for Cu targets. All data were obtained for ions with incident charge states $q < Z_1 - 2$, where Z_1 and Z_2 are projectile and target atomic numbers respectively. In this way the possibility of target K vacancy production via electron transfer to the projectile K-shell which is known to be large[1,2], was minimized.

In addition to cross sections, target K_β/K_α ratios and the average K_α x-ray energy shifts were measured. Combining Hartree-Fock-Slater calculations and statistical scaling methods for multiple ionization, with the K_α/K_β ratios and K_α shifts, the numbers of 2p and 3p vacancies at x-ray emission were determined. The results show that the number of 2p vacancies increases with beam energy while the number of 3p vacancies decreases.

The number of 2p and 3p vacancies gives a prediction of the fluorescence yields ω_K in the presence of this multiple ionization. The modified values of ω_K, which are as much as 50% larger than the single vacancy fluorescence yields, were used to compare the measured x-ray cross sections with theoretical predictions of ionization from the ECPSSR theory of Brandt and Lapicki[3]. Reasonable agreement with theory is obtained for Cl + Br ($Z_1/Z_2 \sim 1/2$). As expected, the agreement becomes progressively worse as $Z_1/Z_2 \rightarrow 1$. The present work provides evidence for the validity of the ECPSSR theory for systems like Cl on Br where the ratio of projectile to tarket K electron velocity, $V_1/V_{2K} > 0.3$. (See Fig. 1.)

*Supported by U. S. D.O.E. Chemical Sciences Div.

Table I. Target K x-ray production cross sections, resulting from $_{17}Cl^{q+}$ ions bombardment. Relative errors are ±10%; absolute errors are estimated to be ±30%. The copper data are from Ref. 1. Effective target thicknesses are 7.6, 3.9, 7.3 and 15.5 μg/cm² for $_{19}K$, $_{22}Ti$, $_{25}Mn$ and $_{35}Br$, respectively.

	Cross Section (kb)				
E(MeV)/q	$_{19}K$	$_{22}Ti$	$_{25}Mn$	$_{29}Cu$	$_{35}Br$
20/3	10.8	1.6	0.20	0.007	--
30/6	26.1	4.8	0.9	--	--
40/7	44.6	11.7	2.9	0.2	0.020
50/7	57.3	14.7	5.3	--	0.056
60/8	103.00	32.5	10.1	1.1	0.145
80/10	164.00	--	--	3.5	0.420

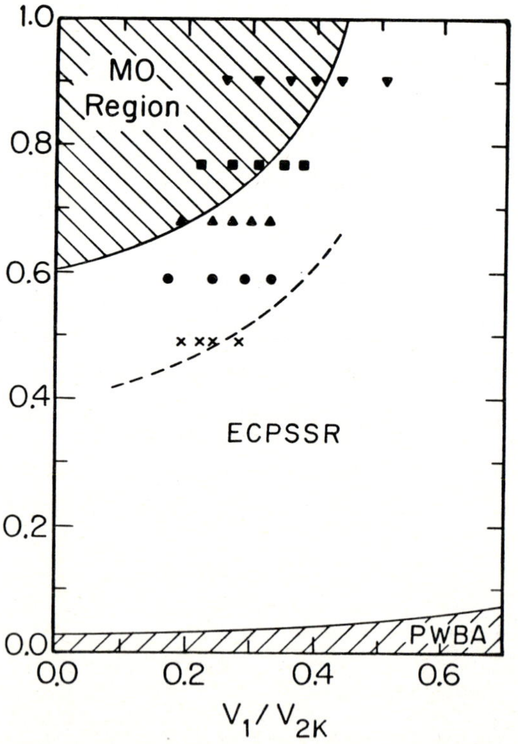

Fig. 1. Plot showing the region where data was taken in the present experiment. The region below the dashed curve is where the ECPSSR theory of Brandt and Lapicki[3] is considered to be applicable.

References

[1] A. Tanis, W. W. Jacobs, and S. M. Shafroth, Phys. Rev. A 22 483 (1980).

[2] J. Gray, P. Richard, K. A. Jamison, and J. M. Hall, Phys. Rev. A 14, 1333 (1976).

[3] W. Brandt and G. Lapicki, Phys. Rev. A 23, 1717 (1981).

MEASUREMENT OF THE K X-RAY PRODUCTION CROSS SECTION IN 50-88 MeV Si^{11+} + He COLLISIONS*

D. Brandt, M. Clark, T. McAbee, J. Swenson and S. Shafroth
University of North Carolina at Chapel Hill, N. C. 27514 USA
and Triangle Universities Nuclear Lab, Durham, N. C. USA

We measured the projectile K x-ray production in collisions of lithium-like Si^{11+} and He under single collision conditions. In its $1s^2\ 2s\ ^2S$ ground state, the incoming projectile has a filled K-shell and no 2p electrons. Thus, the predominant mechanism for K α x-ray production is 1s-2p excitation. There are, however, two step processes, which will also result in K α x-rays, for example K ionization and simultaneous 2s-2p excitation, K ionization and capture to the 2p orbital, and 1s-2s excitation simultaneous with 2s-2p excitation.[1] Since all these processes depend on the projectile energy in different ways, our measurement of the total K α and K β x-ray production may help to identify their contributions.

The experiment was carried out at the Triangle Universities Nuclear Laboratory at Durham, N. C. Silicon ions from a sputter ion source were accelerated with a model FN tandem Van de Graaff to energies between 50 MeV and 88 MeV with charge states 7-10 and momentum analyzed. The beam was then passed through a thin carbon foil to produce higher charge states and the desired lithium-like fraction of the beam was magnetically deflected in our beamline. There, it was focussed and collimated by a defining aperture (Ø 1.5 mm) upon entering a differentially pumped gas cell with the He target gas. After pasing through the collision region, the beam was collected in a suppressed Faraday cup. The beam currents varied from 0.2 nA to 2 nA for the different beam energies. Projectile x-rays from the collision region were observed with a Kevex Si(Li) detector mounted at 90° with respect to the beam direction. The detector resolution of about 160 eV allowed to distinguish between Si K α -and K β x-rays. A low energy tail apparent in the spectra was attributed to insufficient charge collection in the detector crystal. At each beam energy the x-ray yield was measured for at least three different He gas pressures between 0 and 30 mTorr, where single collision conditions should prevail. To avoid contaminations of condensible vapors, the He was passed through a liquid nitrogen cooled trap before entering the gas cell through an electronically regulated leak valve. The pressure in the gas cell as measured with a capacitive manometer (MKS Baratron) was stable within 2%.

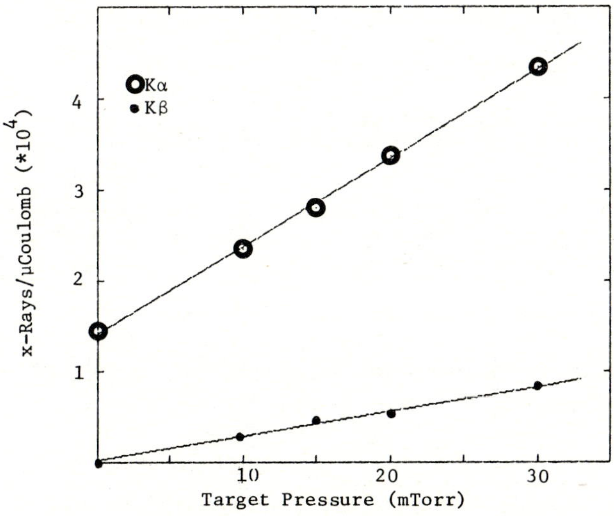

Fig. 1: Normalized x-ray yield as a function of target gas pressure for 75 MeV Si^{11+} + He collisions.

The measured x-ray spectra were fitted to Gaussian distributions for the K α -and K β lines and the low energy tail was added to the K α intensity. Fig. 1 shows the normalized x-ray yield as a function of target gas pressure for 75 MeV projectile energy. The K α intensity at 0 gas pressure is attributed to the decay in flight of metastable projectile ions, which have been produced by slit scattering at the defining aperture. The straight lines are the result of a linear least square fit of the K α and K β data points.

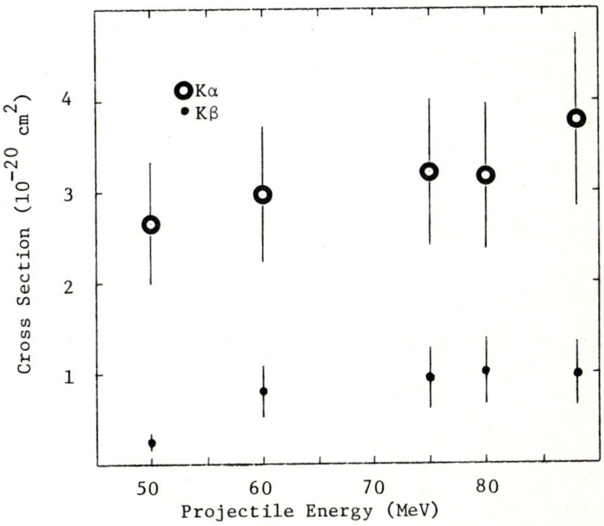

Fig. 2: K x-ray production cross section in Si^{11+} + He collisions.

The uncertainties in the slope of these lines and thus of the relative cross section are less than 5% for the K α data and less than 15% for the K β data for all beam energies investigated. The absolute cross section (Fig. 2) has been normalized at 60 MeV to the total K x-ray data of Doyle et al.,[2] since the solid angle and the efficiency of our Si(Li) detector has not yet been accurately determined.

[1]H. Tawara, M. Terasawa, P. Richard, T. J. Gray, P. Pepmiller, J. Hall, and J. Newcomb, Phys. Rev. A 20, 2340 (1979).
[2]B. L. Doyle, U. Schiebel, J. R. Macdonald and L. D. Ellsworth, Phys. Rev. A 17, 523 (1978).

*Work supported by U.S.D.O.E. Chem. Science Division.

IONIZATION AND ELECTRON TRANSFER FOR THE K-, L-, AND M-SHELLS FOR 1.86 MeV/u IONS ON SELECTED TARGET SYSTEMS

F. D. McDaniel, J. L. Duggan, R. Mehta,[+] M. C. Andrews, A. Toten, J. D. Gressett, D. Johnson,[++] and S. R. Wilson[+++]
Department of Physics,[*#] North Texas State University, Denton, TX 76203, USA

P. D. Miller
Physics Division,[**] Oak Ridge National Laboratory, Oak Ridge, TN 37830, USA

G. Lapicki
Department of Physics, East Carolina University, Greenville, NC 27834, USA

G. Basbas
Editor, Physical Review Letters, P.O. Box 1000, Ridge, NY 11961, USA

L. A. Rayburn
Department of Physics,[#] University of Texas at Arlington, Arlington, TX 76010, USA

A. R. Zander
Department of Physics,[*#] East Texas State University, Commerce, TX 75428, USA

R. M. Wheeler and R. P. Chaturvedi
Department of Physics,[#] State University of New York College, Cortland, NY 13045, USA

R. S. Peterson
Department of Physics,[#] University of Tennessee at Chattanooga, Chattanooga, TN 37402, USA

The direct ionization (DI) and electron capture (EC) contributions to inner-shell ionization have been inferred from x-ray production cross section measurements for selected heavy ion-atom collision systems. The x-ray production cross sections were measured for vanishingly thin solid targets approximating single-collision conditions.

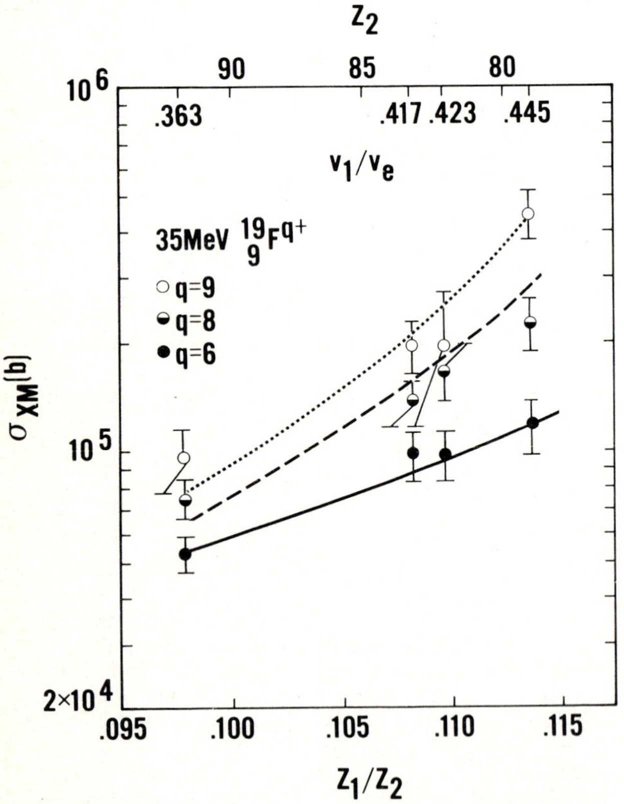

Figure 1. The sum of direct ionization (DI) and electron capture (EC) obtained from x-ray production measurements. Cross sections for 1.86 MeV/u $^{19}_{9}F^{q+}$ in charge states q = 4,5,or 6 consists of the sum of DI and EC (M→L, M, etc.) and are compared to perturbed stationary state (PSS) theory shown as the solid line. Cross sections for hydrogen-like $^{19}_{9}F^{q+}$ (q=8) and fully stripped $^{19}_{9}F^{q+}$ (q=9) are compared to PSS theory with the dashed and dotted curves, respectively.

DI and EC results will be presented for the K-shell for 1.86 MeV/u $^{28}_{14}Si$ ions incident upon Sc, Ti, Cu, and Ge; for the L-shell for 1.86 MeV/u $^{19}_{9}F$ and $^{28}_{14}Si$ ions and for 1.80 MeV/u $^{35}_{17}Cl$ ions incident upon Nd, Ho, and Au; and for 1.84 MeV/u $^{19}_{9}F$ ions incident upon Au, Pb, Bi, and U.

The inferred DI and EC cross sections have been compared to Coulomb ionization theories consisting of the first Born[1,2] and the perturbed stationary state (PSS) approaches.[3,4,5] In Fig. 1 the DI and EC (M→L, M, etc.) data obtained from x-ray cross section measurements are compared to the perturbed stationary state calculations with energy loss, Coulomb deflection and relativistic effects (solid curve). The targets were ≤ 1 μg/cm and the $^{19}_{9}F$ ion charge state was q=4,5, or 6. Figure 1 also shows target K-shell x-ray cross sections due to DI + EC to hydrogen-like ions (q=8, dashed curve) and fully stripped ions (q=9, dotted curve) versus Z_1/Z_2 and v_1/v_e. The comparison provides evident of the validity of the PSS theory for the range of ion-target systems and ion velocity to electron velocity ratios studied.

[+]Present address, Department of Physics, Univ. of Texas at Arlington, Arlington, TX 76010, USA

[++]Present address, Fermi Lab, Batavia, IL 60510, USA

[+++]Present address, Semiconductor Research & Development Laboratories, Motorola Inc., 5005 E. McDowell Rd., Pheonix, AZ 85008, USA

[*]Supported in part by the Robert A. Welch Foundation and the State of Texas Organized Research Fund.

[#]Travel to Oak Ridge National Laboratory provided in part by Oak Ridge Associated Universities.

[**]Supported by DOE, Division of Basic Energy Science, Contract No. W-7405-ENG-26 with Union Carbide Corp.

[1]R. Rice, G. Basbas, and F.D. McDaniel, At. Data Nucl. Data Tables 20, 503 (1977).
[2]D.E. Johnson, G. Basbas, and F.D. McDaniel, At. Data Nucl. Data Tables 24, 1 (1979).
[3]G. Basbas, W. Brandt, R. Laubert, Phys. Rev. A7, 983 (1973); A17, 1655 (1978).
[4]G. Lapicki and W. Losonsky, Phys. Rev. A15, 896(1977).
[5]G. Lapicki and F.D. McDaniel, Phys. Rev. A22, 1896 (1980); A23, 475 (1981).

L-SUBSHELL IONIZATION OF AU BY LIGHT ION IMPACT

K. Finck, W. Jitschin, R. Hippler, H. O. Lutz

Fakultät für Physik, Universität Bielefeld, F. R. Germany

The three L-subshells L_1, L_2 and L_3 have different binding energies and electronic wavefunctions. This feature can be used to study separately various aspects of the collisional interaction, as, e.g., the slowing down and deflection of the projectile in the target field (*Coulomb deflection*) or the perturbation of the atomic wavefunctions by the projectile.

In case of H impact one expects a good description of the interaction by first-order theories (Born approximation PWBA, semiclassical approach SCA). For the absolute L_3 ionization cross section σ_{L3} reasonable agreement is obtained between experiment and those theories which take the important effect of Coulomb deflection into account (Fig. 1). Ratios of subshell cross sections can be determined with high accuracy. For the σ_{L1}/σ_{L3} ratio a maximum is found experimentally at ca. 250 keV impact energy. This is due to the different influence of the projectile deflection on the ionization of the corresponding subshells. A reasonable theoretical description has only recently been achieved by SCA calculations which employ a hyperbolic projectile trajectory[1,2]. The wavefunctions of the L_2- and L_3-subshells

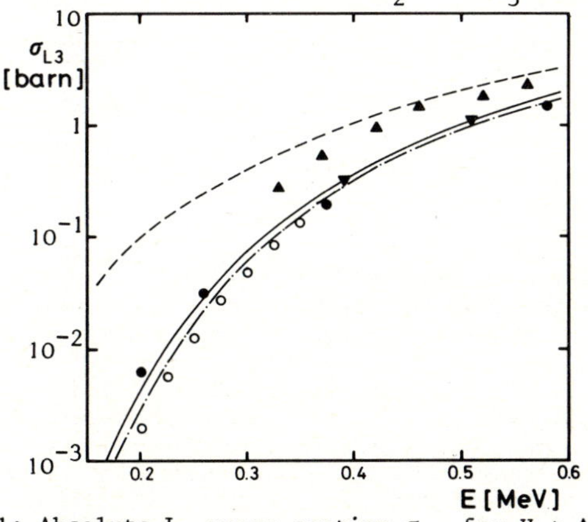

Fig. 1: Absolute L_3 cross section σ_{L3} for H → Au. Experimental points are from several groups, o present work. Curves: --- PWBA, —— CPSSR[4], -·- SCA[2].

are very similar and thus the cross section ratio σ_{L2}/σ_{L3} shows a simple dependence on the impact energy (Fig. 2) which is well described by different theories. The experimental data for σ_{L2}/σ_{L3} are in general somewhat larger than the predictions; this might be due to systematic deviations of the atomic parameters used in the data evaluation.

In case of He impact the experimental σ_{L2}/σ_{L3} data appear not to decrease at small impact energies; this behavior becomes pronounced for C, N and O impact where the data even increase steeply (Fig. 3). Note that in the small impact energy regime the theoretical L_2 cross

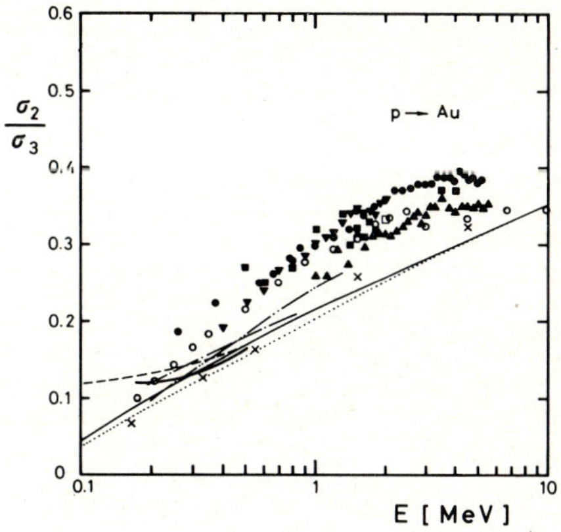

Fig. 2: Cross section ratio σ_{L2}/σ_{L3} for H → Au. Experimental points are from several groups, o present work. Curves: —— CPSSR[4], ━━ SCA[2].

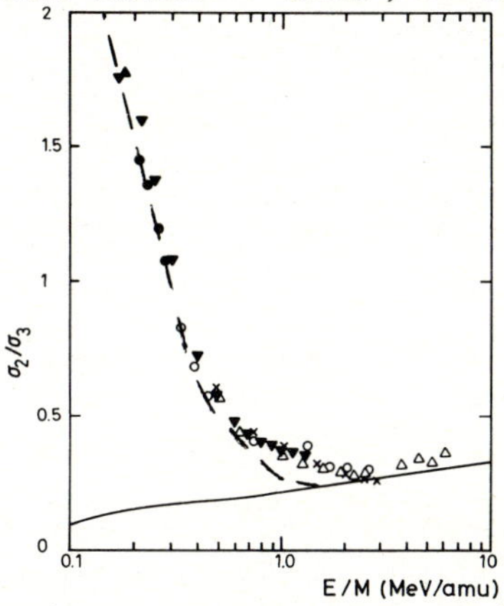

Fig. 3: σ_{L2}/σ_{L3} for C, N, O → Au. Experimental points are from several groups, o Δ present work. Curves: —— CPSSR[4], --- two step model.

section is about one order of magnitude smaller than the L_1 and L_3 cross sections. It has, therefore, been suggested that the unexpected behavior is due to a rearrangement of the created vacancies during the collision[3]. Ionization occurs dominantly at very close nuclear distances whereas the vacancy transfer may occur at larger distances. Thus in first approximation it seems reasonable to separate the collisional interaction into two steps: firstly, vacancies are created in one subshell and secondly, they are transferred to other subshells on the outgoing part of the trajectory. Corresponding calculations which include all interaction multipoles are in progress; they will clarify the unknown contribution of the quadrupole term. Preliminary values (Fig. 3) show reasonable agreement with the experimental data.

References

1. D. Trautmann, private communication (1981)
2. K. Aashamar and P. A. Amundsen, preprint (1982)
3. L. Sarkadi and T. Mukoyama, J. Phys. B 14, L255 (1981)
4. W. Brandt and G. Lapicki, Phys. Rev. A 20, 465 (1979)

L-SHELL X-RAY PRODUCTION CROSS SECTIONS OF $_{48}Cd$, $_{50}Sn$, $_{52}Te$, $_{53}I$ and $_{56}Ba$ FOR PROTONS AND ALPHA PARTICLES

L. Avaldi+, I.V. Mitchell+ and M. Milazzo++

+ J.R.C., Central Bureau for Nuclear Measurements, B-2440 Geel, Belgium
++ Istituto di Fisica Generale Applicata, Università degli Studi di Milano, Italy

The measurements of L-shell cross sections are important for two main reasons. Firstly, there is a need for a reliable set of L-shell cross section data for analytical work involving medium and high atomic number elements, and secondly such measurements provide a means of testing the validity of current theories on inner shell ionization by charged particles. In this work we report on the L-shell X-ray production cross sections of $_{48}Cd$, $_{50}Sn$, $_{52}Te$, $_{53}I$ and $_{56}Ba$ for alpha particles of energy between 2.2 and 2.8 MeV and of $_{52}Te$ for protons of energy between 1.6 and 3.0 MeV.

The 3.7 MeV KN vertical Van de Graaff of JRC-Geel has been used for the experiments. Thin targets were employed for all the measurements. The cross sections are obtained by comparing X-ray and Rutherford backscattering (RBS) yields simultaneously from the thin targets. By so doing, the need to know absolutely the target thicknesses and the projectile flux is avoided. The RBS particles are detected at 165° with respect to the beam direction. The surface barrier detector subtends a solid angle of 1.5 10^{-3} steradian to the target. The X-ray Si(Li) detector is set at 157.5° to the beam direction, its solid angle is 2.06 10^{-3} steradian. Its efficiency, in the energy range of these element X-ray lines ($2.9 \leqslant E \leqslant 5$ keV), has been extracted from the curve given by Barfoot [1]. This curve has been experimentally derived using calibrated X-ray sources and taking into account a polyethylene absorber in front of the detector.

Following the current trend in the L-shell studies [2-3], we have derived from our experimental data the X-ray production cross section for the $L_{\alpha_{1-2}}$, L_β, $L_{\beta_{2-15}}$ and L_{γ_1} lines. Therefore we present the $\sigma_{L\alpha_{1-2}}$, $\sigma_{L\beta}$, $\sigma_{L\beta_{2-15}}$ and $\sigma_{L\gamma_1}$ production cross sections and their ratios rather than the individual subshell ionization cross sections (L_I, L_{II}, L_{III}). Moreover, with the available energy resolution of our Si(Li) detector and under our experimental conditions, the spectral peaks, resulting from the individual subshell ionization, are not completely resolvable.

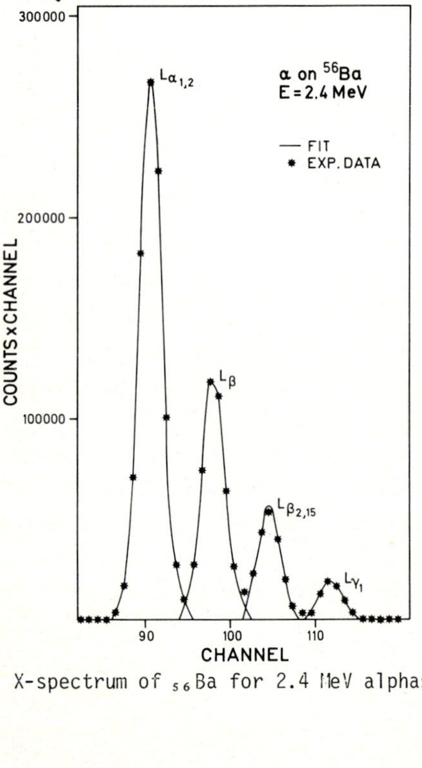

Fig. 1: X-spectrum of $_{56}Ba$ for 2.4 MeV alphas

The Teoh [4] code, in the version which is running on IBM computer of the JRC-Geel, has been used for data analysis, including background subtraction, peak fitting and deconvolution. In Fig. 1 we show, as an example, the spectrum of $_{56}Ba$ for alpha particles of 2.4 MeV. The different spectral L-lines are well separated.

The total uncertainty of the cross sections, which is obtained by quadratic addition of partial contributions, due to the peak statistics and Si(Li) efficiency, is always less than 12 %. In general a large part of this is due to the uncertainty in Si(Li) efficiency determination.

The results of this work are compared with the theoretical predictions of the plane wave Born approximation including binding and Coulomb deflection corrections (PWBA + BC) [5] and of the perturbed stationary state theory including Coulomb deflection and relativistic effects (CPSSR) [6]. The theoretical values for the individual production cross sections were determined using the following relationships, relating the subshell ionization cross sections σ_{L_I}, $\sigma_{L_{II}}$ and $\sigma_{L_{III}}$ to the production cross sections:

$$\sigma_{L\alpha_{1-2}} = \omega_3 \left[\sigma_{L_{III}} + f_{23}\,\sigma_{L_{II}} + (f_{12}f_{23} + f_{13} + f'_{13})\,\sigma_{L_I}\right] \times \frac{\Gamma_{3\alpha_1} + \Gamma_{3\alpha_2}}{\Gamma_3}$$

$$\sigma_{L\beta_1} = \omega_2 \left[\sigma_{L_{II}} + f_{12}\,\sigma_{L_I}\right] \frac{\Gamma_{2\beta_1}}{\Gamma_2}$$

$$\sigma_{L\beta_{3-4}} = \omega_1\,\sigma_{L_I}\,\frac{\Gamma_{1\beta_3} + \Gamma_{1\beta_4}}{\Gamma_1}$$

$$\sigma_{L\beta_{2-15}} = \omega_3 \left[\sigma_{L_{III}} + f_{23}\,\sigma_{L_{II}} + (f_{13} + f'_{13} + f_{12}f_{23})\,\sigma_{L_I}\right] \times \frac{\Gamma_{3\beta_2} + \Gamma_{3\beta_{15}}}{\Gamma_3}$$

$$\sigma_{L\gamma_1} = \omega_2 \left[\sigma_{L_{II}} + f_{12}\,\sigma_{L_I}\right] \frac{\Gamma_{2\gamma_1}}{\Gamma_2}$$

The values of subshell fluorescent yields ω_1, ω_2 and ω_3 and the Coster-Kronig yields f_{12}, f_{13}, f_{23} and f'_{13} were taken from the compilation by Krause [7]. The relative radiative widths were extracted from the compilation of Scofield [8]. Due to the large uncertainties in the fluorescence yields [7] (between 5 and 30 %) and in the Coster-Kronig yields [7] (between 5 and 20 %) care must be taken in making comparisons with the experimental results and theory. The agreement between experimental results and theoretical predictions are discussed in some details.

References

1. K.M. Barfoot, Ph.D. Thesis (1980), University of Surrey, U.K.
2. R.S. Sokhi and D. Crumpton, Nucl. Instr. Meth. 192, 121 (1982)
3. S. Fast, J.L. Flinner, A. Glick, F.W. Inman, L. Oolman, P. Pearson and D. Wickelgren, Phys. Rev. A26, 2417 (1982)
4. W. Teoh, Nucl. Instr. Meth. 109, 509 (1973)
5. W. Brandt and G. Lapicki, Phys. Rev. A10, 474 (1974)
6. W. Brandt and G. Lapicki, Phys. Rev. A20, 465 (1979)
7. M.O. Krause, J. Phys. Chem. Ref. Data 8, 307 (1979)
8. J.M. Scofield, Phys. Rev. A10, 1507 (1974)

SURVEY OF M-SHELL X-RAY PRODUCTION FOR 19 ELEMENTS BY H^+ and $HE^{+,++}$ IONS†

R. Mehta*, J. L. Duggan, F. D. McDaniel, P. M. Kocur, J. L. Price

Department of Physics, North Texas State University, Denton, Texas 76203, U.S.A.

and

G. Lapicki

Department of Physics, East Carolina University, Greenville, N.C. 27834, U.S.A.

Most of the measurements of innershell ionization phenomena that have been made to date pertain to the K-shell or L-shell. There have been only a limited number of experiments for the more complex M-shell. The M-shell x-ray production cross section data is limited in terms of ion-atom-energy combinations, and the uncertainties have been larger than in K- and L-shell measurements. Theoretical investigations also follow this trend. With the recent extension of the ECPSSR theory to M-shell ionization, a survey of the present theories will add to the understanding of the systematics involved in the M-shell ionization.

This survey covers the elements Xe, Pr, Nd, Sm, Eu, Gd, Tb, Dy, Ho, Yb, Hf, Ta, W, Pt, Au, Hg, Pb, Bi and U. The projectiles, H^+ and $He^{+,++}$, are used for an energy range from 300 keV to 40 MeV for the H^+ ions and 250 keV to 2.5 MeV for $H^{+,++}$ ions. Recent measurements done at North Texas State University of the M-shell x-ray production cross sections of some rare earth elements for incident H^+ and He^+ ions are also reported. The range of energies of these ions were 0.25 to 2.5 MeV.

The two processes responsible for M-shell ionization are the direct ionization to the continuum (DI) and target electron capture (EC) by the projectile ion. The first Born calculations includes the DI in the plane wave Born approximation (PWBA) and the EC in the Oppenheimer-Brinkman-Kramers (OBK) approximation of Nikolaev. The ECPSSR approach that goes beyond the first Born is based on the perturbed stationary state (PSS) formalism and accounts for the Energy loss, Coulomb deflection and Relativistic effects. The M-shell ionization cross sections calculated in the first Born and the ECPSSR theories are converted to x-ray production cross sections using single-hole fluorescence yields for each individual subshell. This allows for comparison between the prediction of the theories and the experimental data.

Figure 1 displays proton induced M-shell x-ray production cross section measurements performed over the last 20 years versus target atomic number (Z_2). The predictions of the ECPSSR theory are represented by a solid curve. The energy of the H^+ ions depicted here range from 0.25 to 2.5 MeV. A majority of the experimental data points fall below the solid curve.

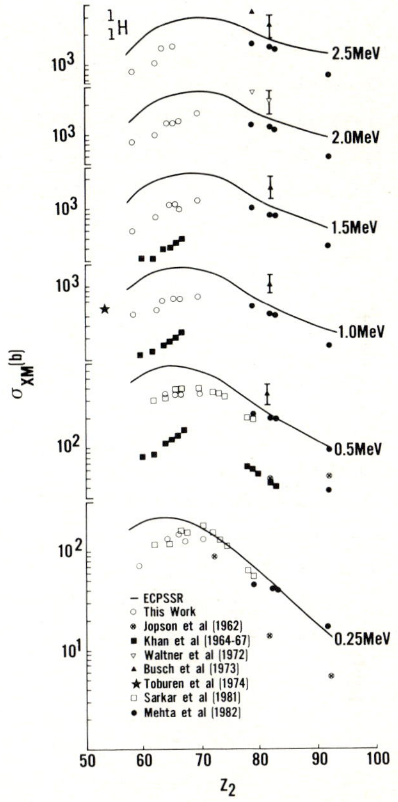

Fig. 1. M-shell x-ray production cross sections for protons at various energies from 0.25 to 2.5 MeV incident on various targets, of atomic numbers $54 \leq Z_2 \leq 92$. At $60 \leq Z_2 \leq 70$, the data lie below the theory while at $70 \leq Z_2 \leq 92$, the theory has better agreement with recent measurements.

The lack of available experimental M-shell x-ray production cross sections data is more evident for $He^{+,++}$ induced data. We have calculated the ECPSSR theory for targets of $60 < Z_2 < 92$ for incident He ions from 0.25 to 2.5 MeV. These calculations are compared to the available experimental data in a manner similar to Figure 1 for protons. As in the case for protons, the general trend for the ^4He theory compares favorably with the available experiments.

*Present Address: Department of Physics, Univ. of Texas, Arlington, Texas 76019, U.S.A.

†Supported in part by the Robert A. Welch Foundation and North Texas State University Faculty Research Organization.

NONPERTURBATIVE EFFECTS IN INNER-SHELL IONIZATION

David J. Land

Naval Surface Weapons Center, White Oak, Silver Spring, Maryland, USA 20910

This investigation is concerned with the application of perturbation-theoretic techniques to calculations of the ion-induced (p and He) inner-shell ionization probability for asymmetric collisions. In particular, we wish to identify aspects of the problem that are nonperturbative in nature and to model these suitably within the formalism of perturbation theory. The goal is to develop a method of calculation that is fast and accurate and that leads to physical insight into the process. This work is intended as complementary to treatments that are fully nonperturbative. Our work to date has been restricted to K-shell ionization but the ideas should be applicable to other inner shells as well.

The Hamiltonian which governs inner-shell ionization in the semiclassical approximation is given by

$$H = H_a + V(t) = H_a - Z_1/|\underline{r} - \underline{R}(t)| ,$$

where H_a is the Hamiltonian for the target atom and $V(t)$ describes the Coulomb interaction between the projectile of atomic number Z_1 with the prescribed trajectory $\underline{R}(t)$ and the inner-shell electron with coordinate $\underline{r}$. The Schrödinger equation, $i\ \partial\psi/\partial t = H(t)\ \psi$, is solved by setting

$$\psi(t) = \sum_n a_n(t)\ u_n(t) ,$$

where $u_n(t)$ form a complete set of time-dependent basis states but are otherwise unspecified except that, at $t = \pm\infty$, they must be eigenstates of H_a with eigenvalues, $E_n(\pm\infty)$, and hence are clearly orthogonal in this limit. At intermediate times they need not be orthogonal. In addition, a phase transformation is performed by writing the coefficients $a_n(t)$ as

$$a_n(t) = e^{i\phi_n(t)}\ b_n(t) .$$

The phases $\phi_n(t)$ are usually chosen by $\phi_n(t) = -E_n(\pm\infty)\ t$. However, another choice is given by

$$\phi_n(t) = - \int_0^t dt'\ H_{nn}\ (t') , \qquad (1)$$

which has the effect of (nearly) removing the secular terms from the coupled set of equations for $a_n(t)$.

It is clear that departures from conventional first-order perturbation theory are contained within the choice of the time-dependent, perturbed basis states, $u_n(t)$, and the energy-distorted matrix elements, $H_{nn}(t)$; it is here where nonperturbative effects are brought into the calculation.

The choice that the $u_n(t)$ be eigenfunctions of the atomic Hamiltonian, H_a, and the phases $\phi_n = E_n t$ leads to the usual Born approximation and is valid at high projectile velocities. The choice that u_n be eigenfunctions of the total Hamiltonian $H(t)$ and that the phase ϕ_n be given by Eq. (1) leads to the adiabatic approximation, sometimes called the perturbed-stationary-state theory in the present context,[1] and is valid at low projectile velocities. The adiabatic approximation has been applied with some success in calculations of the K-shell ionization probability.[2] However, it is important to bridge the gap between these two velocity limits and, in particular, to determine how high in velocity the adiabatic approximation is valid. Basis states appropriate at intermediate velocities have been constructed[2] through the use of a dynamical variational principle introduced by Theis, Reinhardt, and Müller.[3]

In addition to the choice of the u_n, there is also a choice involved in the determination of the matrix elements, $H_{nn}(t)$, in Eq. (1). In our previous calculations[2] we have demanded that the ground-state energy be correct in the united-atom limit. However, at intermediate projectile velocities for which the ground state may display adiabatic effects, the outer atomic shells can be nearly completely nonadiabatic. In this case the united-atom ground-state energy is not correct, since the outer electrons have not had the time to relax to the correct electronic configuration appropriate to the united atom. Hence we realize the necessity for taking into account the role played by the outer electrons in the inner-shell ionization, and this influences the construction of $H_{nn}(t)$.

Results of calculations reflecting the nonperturbative effects of different choices of both the basis states, $u_n(t)$, and matrix elements, $H_{nn}(t)$, will be shown for incident protons and ^{3}He ions within the range of 0.2 to 2.5 MeV/amu in targets of Sc (Z = 21) to Zn (Z = 30). Comparisons with data taken at the Naval Surface Weapons Center[2] and with other data will also be presented.

References

1. G. Basbas, W. Brandt, and R. H. Ritchie, Phys. Rev. A 7, 1971 (1973).

2. D. J. Land, D. G. Simons, and M. D. Brown, Nucl. Instrum. Methods (to be published); M. D. Brown, D. G. Simons, D. J. Land, and J. G. Brennan, Phys. Rev. A 25, 2935 (1982).

3. J. Theis, J. Reinhardt, and B. Müller, J. Phys. B 12, L479 (1979).

THE BINDING CORRECTION FOR INNER SHELL IONISATION IN ASYMMETRIC ION-ATOM COLLISIONS

L. Kocbach*

Det fysiske Institut, Aarhus Universitet, DK-8000 Århus-C, Denmark

Inner shell ionization in highly asymmetric ion-atom collisions has been studied for many years, using as a useful reference the theoretical predictions based on first order perturbative approaches (PWBA, SCA, c.f. ref.[1]).

One of the bridges between the simple theoretical framework and the experimental observations has been the concept of increased electronic binding ('binding correction') introduced[2] and extensively studied by W. Brandt and coworkers (reviews e.g. in ref.[1]).

Shortly, this correction procedure in many of its variants is a method of simulating higher order effects in the framework of first order perturbation theoretical calculations (c.f. also ref.[3]).

Recently, the region of very low projectile velocities has become an active area of investigations. There are some indications[4] that the usual binding correction procedure might not be sufficient under such extreme conditions. Our unpublished SCA-like calculations[5] with the distortion approximation[6] gave lower ionization probabilities than corresponding binding-corrected results.

The validity and justification of such procedures can in principle be studied with the help of nonperturbative numerical calculations (e.g. J.F. Reading in ref.[1]). In the region of very small velocities and extremely low transition probabilities such calculations become very time-consuming, if at all feasible.

An alternative approach to this task might be through understanding of possible general features of time dependent Schrödinger equation, based on studies of similar, but very simple model systems. As a first step we try to demonstrate, that also for very weakly time-dependent problems one can approach an adiabatic region where the perturbative methods, even with binding correction, overestimate strongly the transition probabilities.

As an example, fig. 1 shows results for a simple two state problem. The transition probabilities reach values as low as 10^{-10}, so that it is convenient to divide all the results by the first order perturbation transition probabilities. Numerical results for full solution (EX), distortion approximation (DA) and first order perturbation theory with the binding correction (BC) are compared. We also show results for 'excitation by antiparticles' by switching the sign of the interaction (curves above 1.0).

We intend to study a wide class of two state problems (and more complicated ones) and relate their properties to the properties of coupling matrix elements in the socalled adiabatic approximation[7]. In this way one can obtain a new prescription for simulation of higher order effects in regions where the large numerical calculations might be too difficult.

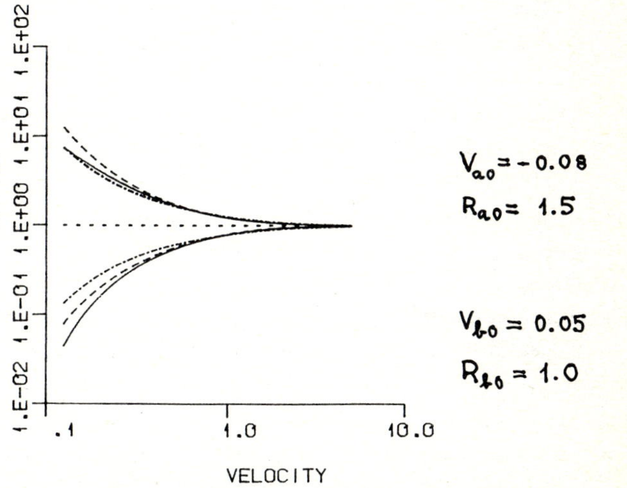

Fig. 1 Transition probabilities divided by 1. order result. —(EX),---(DA),-·-·-·-(BC)

Details of the presented model

$$H(t) = \begin{pmatrix} -1.+V_a(t) & V_b(t) \\ V_b(t) & 0 \end{pmatrix}$$

$$V_i(t) = V_i(R(t)) = V_{i0} / \cosh(R(t)/R_{i0}) \qquad i = a, b$$

$$R(t) = v \cdot t$$

References

* NORDITA Guest

1. Proc. of 2. Linz Workshop, Nucl.Instr. and Meth. 192, 1-127 (1982)
2. W. Brandt et al. Phys.Rev. 151, 56 (1966)
3. J.U. Andersen et al. Nucl.Instr. and Meth. 132, 505 (1976)
4. H. Paul, Linz, private communication
5. L. Kocbach, unpublished
6. e.g. eq.(26) of L. Kocbach et al., Nucl. Instr. and Meth. 169, 281 (1980)
7. L.I. Schiff, Quantum Mechanics (McGraw&Hill (1955) Section 35.

K-SHELL IONIZATION AT LARGE SCATTERING ANGLES WITH LIGHT PROJECTILES[+]

E. Morenzoni[*], R. Anholt, S. Andriamonje[‡], W. E. Meyerhof, O. K. Baker, J. D. Molitoris

Department of Physics, Stanford University, Stanford, California 94305, USA

The ionization probability at large scattering angles, i.e. small impact parameters, by light projectiles is very sensitive to the details of the ionization mechanism, in particular to the contribution of two effects: the rapid change of the direction of the Coulomb force on the electron as the projectile passes by the target nucleus, and the recoil of the target nucleus.[1,2] Both effects give rise coherently to dipole electronic transitions from the initial 1s state to the continuum. The K-shell ionization probability as a function of the projectile scattering angle θ is expected to have the simple form $P_K(\theta) = P(90°)(1 + B\cos\theta)$ for $\theta \gtrsim 10°$. The anisotropy coefficient B, which depends essentially on v_p/v_K (v_p = projectile velocity, v_K = Bohr orbital velocity), gives information about the dipole transitions terms.[1]

Since ionization probabilities and Rutherford scattering at large scattering angles are very small, only few measurements of $P_K(\theta)$ have been made.[1,3,4] In order to reduce the measuring time, and facilitate a systematic study of the ionization probability at large scattering angles, an 8-angle detector was used to determine large angle K-ionization probabilities $P_K(\theta)$ for p + Cu, p + Mo and p + Ag collisions at energies of 0.6-, 1- and 2-MeV. The detector consists of 8 plastic scintillator foils curved around the beam axis sustaining 2π in the azimuthal direction and with different widths, such that the counting rate in each particle detector is approximately equal. The light emitted by each scintillator is brought through a light pipe to a photomultiplier. The eight particle detectors cover scattering angles ranging from 10° to 135°. The x-rays are detected in a large solid angle NaI detector.

In all cases the K-ionization probability was found to have a cosθ scattering angle dependence. The corresponding anisotropy coefficient B is plotted against the reduced projectile velocity $\xi = \theta_K(v_p/v_K)$ in Fig. 1. ($\theta_K = Z_t^2 I_H/E_K$, Z_t = target Z, I_H = hydrogen ionization energy, E_K = target K binding energy). The experimental points are compared with the prediction for B(ξ) from the atomic theory of Andersen et al. (with and without recoil contribution)[1] and from the molecular theory of Anholt[5] which implicitly takes into account the recoil term. Our results clearly demonstrate the presence of a recoil effect.

Data obtained with Oxygen projectiles will also be presented.

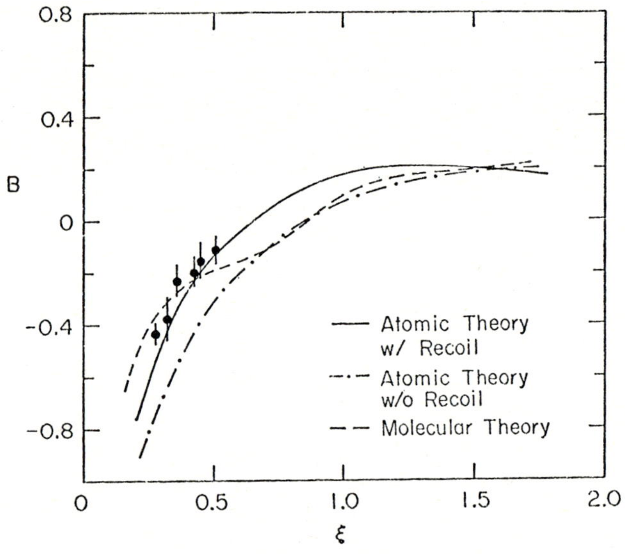

Fig. 1: Anisotropy coefficient B in the K-ionization probability as a function of the reduced projectile velocity ξ. Comparisons with theoretical estimates by Andersen et al.[1] (with or without recoil) and by Anholt[2] (molecular theory).

[+]Supported in part by NSF (Grant PHY 80-15348)
[*]Holder of Swiss National Fonds Fellowship
[‡]Permanent address: C.E.N.B.G., Bordeaux, France

References

1. J. U. Andersen, L. Kocbach, E. Laegsgaard, M. Lund, C. D. Moak, J. Phys. B9, 3247 (1976).
2. G. Ciochetti, A. Molinari, Nuovo Cimento 40B, 69 (1965).
3. J. F. Chemin, S. Andriamonje, J. Roturier, B. Saboya, R. Gayet, A. Salin, Phys. Lett. 67A, 116 (1978).
4. G. Gruber, K. Bethge, S. Kelbach, N. Löchter, W. Schadt, K. E. Stiebing, H. Schmidt-Böcking, R. Schuch, F. Rösel, D. Trautmann, to be published.
5. R. Anholt, Z. Physik A295, (1980).

PAIR CREATION IN "OVERCRITICAL" COULOMB FIELDS

M. Clemente, E. Berdermann, P. Kienle, H. Tsertos, W. Wagner
Physik-Department, Technische Universität München, 8046 Garching, FRG
F. Bosch, C. Kozhuharov
Gesellschaft für Schwerionenforschung (GSI), 6100 Darmstadt, FRG
W. Koenig
Max-Planck-Institut für Kernphysik, 6900 Heidelberg, FRG

During recent years we investigated the positron production in the strong field of heavy nuclear molecules with the orange-type ß-spectrometer of the GSI. We measured positron spectra in ^{238}U-^{238}U and ^{238}U-^{232}Th-collisions at bombarding energies 5.7, 5.9 and 6.2 MeV/u, which is close to the Coulomb barrier, and found unexpected results: The positron spectra in coincidence with projectile or target-like fragments revealed rather narrow lines superimposed on a continuum, which is well understood as being produced by the time dependent variation of the strong Coulomb field. The energies of the lines (226 keV to 311 keV) and their widths (61 keV to 220 keV) are strongly dependent on the kinematical conditions of the collision. For both systems and all bombarding energies the line energy and width seems to decrease somewhat for smaller scattering angles. The width increases with bombarding energy from 61 keV at 5.7 MeV/u to 220 keV at 6.2 MeV/u.

Fig. 1 and fig. 2 show positron spectra obtained at 5.9 MeV/u in coincidence with particles detected in the angular ranges indicated. Curves a) are calculated spectra for dynamically induced positron production, curves b) are fits to the experimental data as described below.

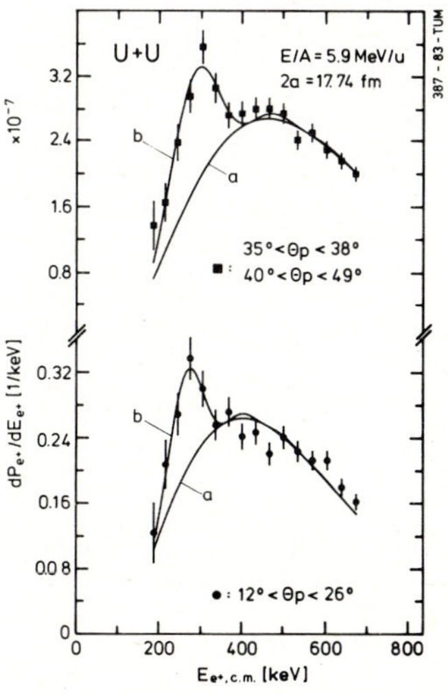

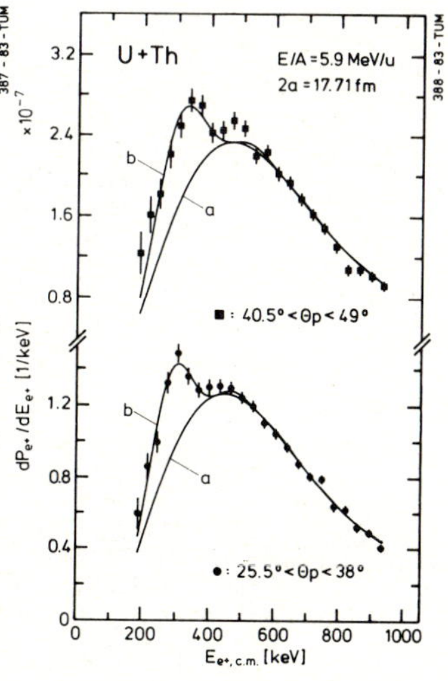

Fig. 1: Fig. 2

Spectra of positrons emitted in 5.9 MeV/u U-U- (Fig. 1) and U-Th-collisions (Fig. 2) in coincidence with particles detected in the angular ranges indicated
a) theoretical spectrum assuming Rutherford scattering only; b) fit to the experimental data assuming additionaly emitted positrons with a spectral shape given by (1)

The origin of the unexpected positron line is not clear. One explanation could be that the line originates from the decay of a nuclear state excited strongly at about 1.3 MeV. However a search for corresponding γ- and conversion electron lines dit not reveal any indication of the expected strength for E0, E1, M1 and E2-multipolarity.

Reinhardt et al.[1] suggest that the line originates from the decay of K-vacancies, formed in the initial phase of the collision. The binding of the electrons and the spontaneous emission of the positrons result in a line with an energy corresponding to the diving energy E(1s) of the K-hole in the antiparticle continuum of a rather long living nuclear molecule formed in central collisions. In this case the exact energy is expected to depend sensitively on the mean distance of the two centers of the nuclear molecule, and on the united charge of the system. Due to the finite lifetime T of the nuclear molecule the positron line with natural width $\Gamma \approx 2$ keV is broadenend with a spectral shape given by

$$\frac{dP}{dE} \sim \frac{\Gamma}{2\pi} \cdot T^2 \cdot (\sin x/x)^2 \qquad (1)$$

$$\text{with } x = (E - E(1s)) \cdot T/2\hbar$$

We extracted the K-hole diving energy and the nuclear lifetime T from a fit to the experimental data using eq.(1) shown in fig. 1 and fig. 2 (curves b). For both systems the line energy is found to agree with a value calculated for the U-U-system; the calculated 40 keV difference between the U-Th (lower energy) and U-U-system is not found. This results might indicate a larger two center distance in the U-U nuclear molecule.

The nuclear lifetime T is found to be between $1.6 - 6.0 \times 10^{-20}$s. This extremely large contact time is uncommon in view of the fact that more than 97 % of the observed positrons are found in kinematic coincidence with two U-like products from a binary reaction. This means that very little or no energy and/or angular momentum was exchanged between the two colliding nuclei during such an extremely long time.

References

1. J. Reinhardt et al., Z. Phys. A 303, 173 (1981)

K-SHELL EXCITATION BY K- TO L-SHELL CHARGE TRANSFER IN SLOW Kr^{q+}-Kr AND Xe^{q+}-Xe COLLISIONS

R. Hoffmann, R. Schuch, E. Justiniano, W. Schadt*, H. Schmidt-Böcking*, P.H. Mokler§, F. Bosch§, W.A. Schönfeldt§, Z. Stachura§

Physikalisches Institut, Universität Heidelberg, 6900 Heidelberg, W.-Germany
*Institut für Kernphysik, Universität Frankfurt, 6000 Frankfurt, W.-Germany
§G S I Darmstadt, 6100 Darmstadt, W.-Germany

In collisions of slow highly charged ions with atoms vacancies in the inner atomic shells are predominantly produced by charge transfer instead of ionization. We investigated in this work the mechanism for capture of K-electrons into projectile L-vacancy states for heavy collision systems ($Z_1+Z_2 \simeq 70$-110). From measurements of impact parameter dependent K-vacancy probabilities ($P_K(b)$) for light symmetric collision systems it is well known that this transfer process occurs with a high probability (≃ 10% per projectile L-vacancy) by rotational coupling between 2pπ, 2pσ quasimolecular states.

Because of the increasing fine structure splitting with increasing Z, it is not clear whether 2pπ-2pσ rotational coupling is still the predominant L- to K-shell vacancy transfer mechanism in heavy collision systems. Calculations[1] have shown that in very heavy collision systems radial coupling between states of equal parity is predominant. A sensitive tool to determine the charge transfer mechanism is the measurement of $P_K(b)$ with slow highly charged projectiles incident upon a gas target. Such a measurement did not exist to date in the region of heavy collision systems.

The projectiles of 2.8 MeV/u Kr^{q+} (q=16,26,29-33) and 3.6 MeV/u Xe^{q+} (q=43,45,47), respectively were obtained by the Unilac accelerator by poststripping at 5.9 MeV/u and 8.6 MeV/u, respectively and deceleration to low velocities. A thin Kr and Xe gas target, respectively was used in order to have single collision conditions. The $P_K(b)$ for Kr on Kr (fig.1) shows for q=16 and 26 (no L-vacancy) a flat behaviour at small b and a strong decrease at b > 2000 fm. This can be understood as ionization of the 2pσ state as shown by the SCA prediction (dashed line). The shape of $P_K(b)$ changes drastically as soon as L-vacancies are brought into the collision. In this case one observes a clear peak at about 1300 fm which increases with increasing q and reaches a saturation at an empty 2p-shell. The full line represents the calculation[2] of the 2pπ-2pσ rotational coupling for q=32. The agreement is good especially when considering that in this calculation a nonrelativistic description of the states is used.

A better agreement of the $P_K(b)$ data with calculations[3] done with relativistic quasimolecular 2p wavefunctions is obtained for the Xe-Xe collision system.

This work is supported by the BMFT.

References

1. G. Soff, W. Greiner, W. Betz, B. Müller, GSI Report VT-5-78
2. K. Taulbjerg, J.S. Briggs, J. Vaaben, J. Phys. B9, 1451 (1976)
3. D.H. Jakubassa, K. Taulbjerg, J. Phys. B13, 757 (1980) and D.H. Jakubassa, private communication.

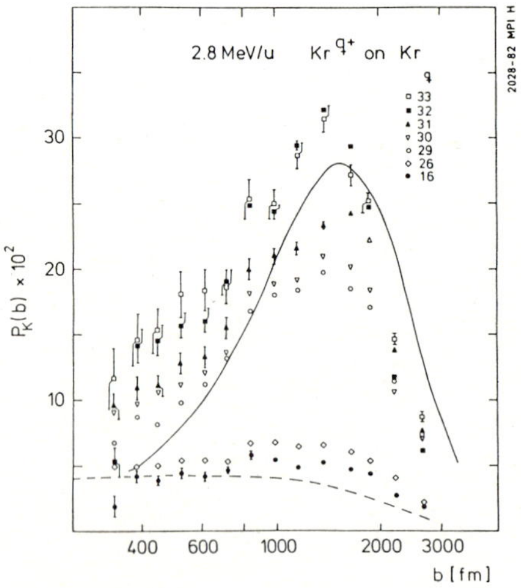

Fig. 1: K-vacancy production probability as function of impact parameter.

K-IONIZATION PROBABILITY IN HIGH ENERGY U + U AND U + Pb COLLISIONS*

J.D. Molitoris, R. Anholt, S. Andriamonje,+ E. Morenzoni,‡ W.E. Meyerhof, and O.K. Baker

Department of Physics, Stanford University, Stanford, California 94305

In this work we have determined K-ionization probabilities P_K for U + U and U + Pb collisions in the impact parameter range from 10 to 60 fm. P_K was measured below and above the nuclear Coulomb barrier at energies of 4.6-, 5.8-, and 7.3-MeV/a.m.u. The experiment was motivated not only by the intrinsic interest in these K-ionization probabilities, but also by their use in nuclear time-delay studies.

In symmetric or near symmetric collisions such as U + Pb or U + U, the K-ionization probability is determined predominantly by electron excitation from the 2pσ molecular orbital.[1,2] If an atomic collision is accompanied by a nuclear reaction with a time delay τ, a phase change is introduced between the incident and outgoing ionization amplitudes which affects P_K. Furthermore, if a deep inelastic nuclear reaction occurs, the nuclear reaction time τ is related to the total kinetic energy loss (-Q) of the reaction products. A rough criterion to assure a strong influence of the nuclear sticking time τ on the inner-shell ionization probability is given by $\omega_{UA}\tau \simeq 1$, where $\hbar\omega_{UA}$ represents the united-atom binding energy of the relevant MO. For U + U, theoretical calculations yield $\hbar\omega_{UA} \simeq 850$ keV (ref. 3) and $\tau \simeq 1 \times 10^{-21}$ sec (ref. 4), so that $\omega_{UA}\tau \simeq 1.3$.

An experiment determining the nuclear time delay in 7.5-MeV/a.m.u. U + U collisions has been undertaken and analyzed by Stoller et al.[5] One of the important backgrounds arises from internal conversion of the highly excited reaction products, which produce x-rays indistinguishable from the x-rays generated in the atomic excitation. Therefore, we made an independent measurement of the K ionization probability P_K in U + U collision at Q = 0, in which the internal conversion contribution was reduced by an anticoincidence requirement with gamma-rays. Since most internal conversion transitions occur in gamma cascades, this method is quite effective in reducing the internal conversion contribution to P_K.[6]

In our measurement, gamma rays emitted in the collision were put into anticoincidence with the detected x-ray-particle coincidences and with the detected particles.[6] The target was bombarded by ^{238}U ions from the Lawrence Berkeley Laboratory SuperHILAC. Scattered U particles were detected by Si surface barrier detectors placed at seven different scattering angles θ_{LAB} between 7.5° and 35° to the incident beam.

Preliminary analysis has shown that in the present case the gamma-ray suppression had only a very small effect on P_K in the elastic-particle region. This is to be expected, since even at the highest bombarding energy, P_K consists mainly of the atomic contribution.

The ionization probabilities have the rough form $P_K(b) \simeq P_0\exp(-b/a)$ as a function of impact parameter b. One preliminary result is shown in Fig. 1, where an exponential fit has been made. For 7.3-MeV/a.m.u. U + U collisions, we found $P_0 \simeq 1.6 \pm 0.1$ and $\underline{a} \simeq 36 \pm 5$ fm^{-1}. Data for all three energies in U + U and U + Pb collisions will be presented.

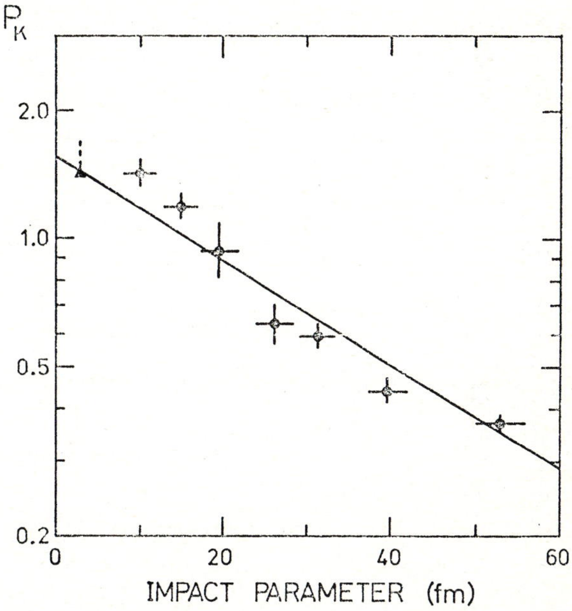

Fig. 1. K-ionization probability in 7.3-MeV/a.m.u. U + U collisions as a function of the impact parameter b. The triangle shows a calculation by U. Müller (ref. 1), without screening which would increase P_K by ~20%.

References

*Supported in part by NSF (Grant PHY 80-15348).

+Permanent Address: C.E.N.B.G., Bordeaux, France.

‡Holder of a Swiss National Fonds Fellowship.

1. U. Müller, et al, Z. Phys. A 297, 357 (1980).
2. T. de Reus, et al, in Quantum Electrodynamics of Strong Fields, W. Greiner, ed., (Plenum, New York, 1983) to be published and GSI preprint 81-83 (October, 1981).
3. G. Soff, et al, Phys. Scripta 17, 417 (1978).
4. R. Schmidt, et al, Nucl. Phys. A 311, 247 (1978).
5. Ch. Stoller, et al, (this meeting).
6. R. Anholt, et al, Z. Phys. A 308, 189 (1982).

X-RAY EMISSION PROBABILITIES AND NUCLEAR TIME DELAY IN THE DEEP INELASTIC COLLISION U+U AT 7.5 MeV/amu

Ch. Stoller, M. Nessi, W. Wölfli
Laboratorium für Kernphysik, Eidg. Technische Hochschule, CH-8093 Zürich
W.E. Meyerhof, J.D. Molitoris, E. Morenzoni
Stanford University, Stanford, Ca 94305, USA
E. Grosse, Ch. Michel
GSI Darmstadt, D-6100 Darmstadt

In 1962 Gugelot[1] suggested that the ionization probability of the atomic inner shells in an atomic collision may be changed if a nuclear reaction takes place which delays the scattered particle. Recent measurements of resonant elastic scattering of protons from various nuclei have confirmed this prediction[2,3]. Anholt[4] and Müller[5] have suggested that a similar effect should exist in deep inelastic collisions of heavy nuclei. If the ionization probability P_K is measured as a function of the total kinetic energy loss in the reaction -Q, the average time delay associated with a certain Q-value can be deduced.

In a first experiment[6] the reactions Xe+Pb and Xe+Th at 8.6 MeV/amu were investigated without significant results, because in this case the K-shell vacancy formation process is dominated by the internal conversion of γ-rays from the highly excited nuclei. We therefore chose the U+U system which should exhibit an ionization probability which is at least ten times higher than for Xe+Pb, whereas the internal conversion contribution should be similar to the one of the Xe+Pb and Xe+Th cases. The experimental set up is shown in fig.1. The U beam was directed on a 500 μg/cm² U target. The particles emerging from the reaction were measured in two parallel-plate-avalanche-detectors (PPAC) positioned symmetrically to the beam axis. The particle detectors were slightly modified: A second plane was added which allowed a better determination of the energy loss ΔE, and the position sensing delay line was equipped with amplifiers on both sides. The latter modification allowed us to recognize fission events in which both fission fragments reach the counter. The major part of the events in which one

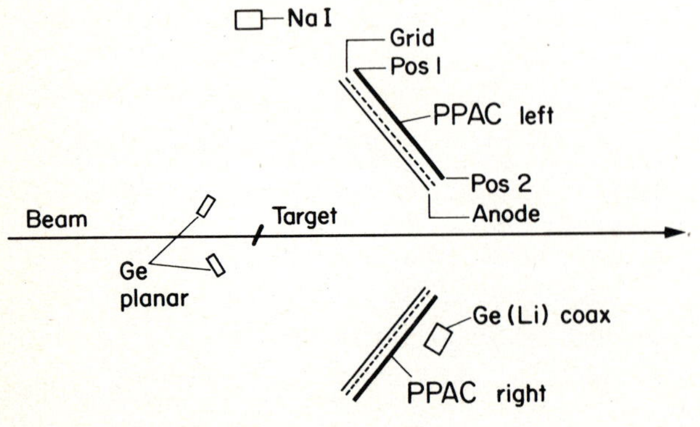

Fig. 1. Experimental set up.

nucleus had undergone fission was suppressed by requiring that the position deduced from the time signals on both sides of the delay line be the same. The ΔE-counter cannot distinguish between the two types of event since the energy loss of two fission fragments equals the energy loss of one U nucleus at the particle energies of interest. Different Q-values were chosen by requiring the corresponding kinematic conditions. The results are shown in fig. 2. The contribution from internal conversion has already been subtracted. The vertical error bars indicate the statistical error and the uncertainty of the background and of the IC contribution. One sees that the

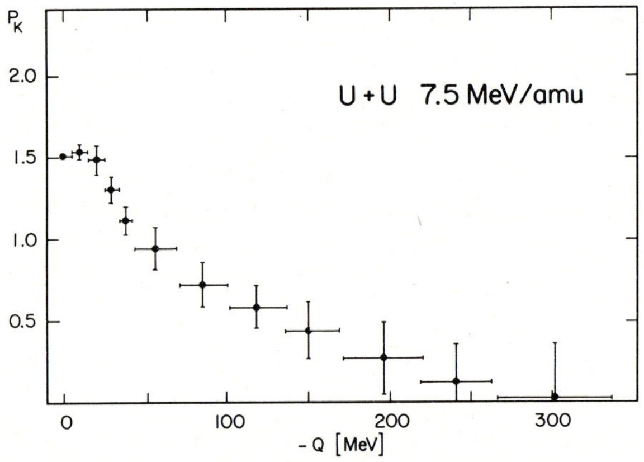

Fig. 2. K X-ray emission probability $P_K(-Q)$ for the lab scattering angles between 32 and 58° after subtraction of IC.

ionization probability declines as a function of Q. From the calculations of Müller et al.[5] one can infer a delay time of ca 10^{-21}s at Q = -180 MeV. Further evaluation of the data is necessary to confirm these results and to rule out background effects, such as unrecognized fission events, which might influence the observed probability.

References

1. P.C. Gugelot, in "Direct Interactions and Nuclear Reaction Mechanisms", p. 382 (Gordon and Breach, New York, 1962)
2. J.S. Blair et al., Phys. Rev. Lett. 41, 1712 (1978)
3. J.F. Chemin et al., Phys. Rev. A24, 1218 (1981)
4. R. Anholt, Z. Phys. A292, 123 (1979)
5. A. Müller et al., Z. Phys. A297, 357 (1980)
6. Ch. Stoller et al., Z. Phys. A, in press

INFLUENCE OF THE NUCLEAR REACTION TIME ONTO THE δ-ELECTRON EMISSION STUDIED BY THE REACTION J→Au, Bi AT 795 AND 840 MeV

F. Güttner, W. Koenig, N. Lutz, B. Martin, H. Skapa, J. Soltani, H. Banda (MPI für Kernphysik, Heidelberg)
A.V. Ramayya (Vanderbilt University, USA)
F. Bosch, Ch. Kozhuharov (GSI, Darmstadt)

Spectra of high energetic atomic electrons (δ-electrons) were recorded in case of quasielastic and deep-inelastic scattering of J→Au and Bi at incident energies of 6.25 and 6.6 MeV/u, respecitvely. Total kinetic energy, mass difference and total mass of the binary reaction products were measured via kinematical coincidences by a parallel plate avalanche counter with annular geometry (Θ_{ion} = 10°-60°).

In case of a nuclear reaction a change in height and slope of the atomic electron spectrum is expected due to interference effects between the electron emission amplitudes of the entrance and the time delayed exit channel. This change contains a direct information of the nuclear reaction time[1,2].

Electron spectra were taken for pure elastic scattering at large and small impact parameters (lowest two curves in fig. 1) and for deep inelastic reactions (Q-value≅100 MeV, upper two curves). The deep inelastic nuclear reactions were divided in addition into to classes, one nearly without a mass drift (crosses) and the other with a large mass drift to symmetry (upper full line).

The basic difficulty of the experiment results from the contribution of conversion electrons stemming from the highly excited nuclei. Whereas single Au conversion lines were directly subtracted in case of elastic heavy ion scattering, the conversion continuum - which dominates the upper two spectra - was estimated from measured γ-ray spectra. First results indicate that - in case of deep inelastic collisions - the remaining atomic electron spectra show a steeper fall-off compared to the emission without time delay by the nuclear reactions.

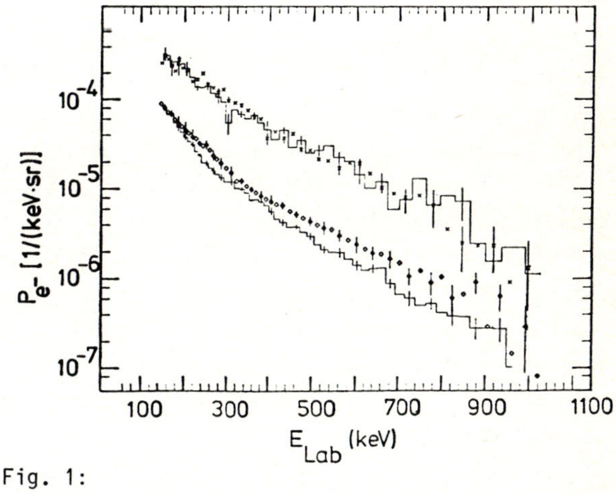

Fig. 1:

Electron spectra measured in coincidence with scattered heavy ions for the system J→Au at 795 MeV incident energy.

lower two curves: elastically scattered projectiles at impact parameters b=17 fm (full line) and b=9 fm (squares) The spectra were corrected for conversion lines of the Coulomb excited Au ions.

upper two curves: deep inelastic reactions (Q-value≅100 MeV) with (full curve) and without (crosses) large mass drift in the heavy ion exit channel. Conversion electrons are not subtracted.

1) J. Reinhardt, B. Müller, W. Greiner and G. Soff, Z. Physik A292 (1979) 211.
2) W. Koenig, K.W. McVoy, H.A. Weidenmüller and P. Kienle to be published

CHARGE TRANSFER AND SIMULTANEOUS EXCITATION IN ION-ATOM COLLISIONS: A COMMENT ON RTE.

Tricia Reeves[a)], Jim Feagin[b)], John Briggs[c)]

a)Univ. North Carolina-Chapel Hill U.S.A.; b)Univ. Nebraska-Lincoln, U.S.A.; c)Universität Freiburg, W. Germany

In a recent letter[1], Tanis and coworkers presented experimental evidence linking an inverse-Auger transition with the production of inner-shell vacancies in light ion-atom collisions. They bombarded argon gas targets with sulfer 13+ and observed the sulfer Kα x rays in coincidence with the projectiles, sulfer 12+, which had captured a target electron. Thus they identified those events in which excitation of and capture to the projectile occurred in the same collision. They observed an anomalous peak in the cross section as a function of the incident energy.

Based on a calculation by Brandt[2], they interpeted the peak in their cross section as a resonance due to a "correlated", or inverse Auger, excitation and capture. By analogy with dielectronic recombination[3], in which a target ion captures a free electron via the inverse-Auger interaction, Brandt inferred the cross section for a correlated excitation and capture event in a heavy-ion collision. He termed the process resonant excitation and transfer and labeled it RTE.

Pepmiller and coworkers at Kansas State University made independent observations with fluorine projectiles which tended to confirm those of Tanis et al.[4] Rather than performing a coincidence experiment, however, the KSU group resolved the various Kα transitions to distinguish those projectiles which had captured an electron. The success of their technique has been hampered somewhat by the ambiguity of theoretical calculations of x ray energies. Nevertheless, it appears that the KSU high-resolution work could better discern correlated and uncorrelated vacancy production mechanisms.

These experiments remain only partially analyzed since Brandt's model excludes the amplitude for excitation and electron transfer directly as an uncorrelated event. The probability, however, for this uncorrelated event was discussed by Brandt and estimated by McGuire (see Ref. 4) to check that such events would not contribute to the observed resonance anomaly. The KSU group labeled this mechanism NTE for nonresonant transfer and excitation.

From the superposition principle, we expect that the amplitudes for the correlated and uncorrelated events should first be added together and then squared to determine the cross section. Furthermore, it is not a priori clear that just the correlated events would lead to a resonant transfer and excitation. Generally, both uncorrelated and correlated events can be thought of as exciting projectile resonances: in the former, discrete resonance states are formed which may then decay by photon emission; in the later, continuum resonance states are formed which may first decay by an inverse-Auger transition followed by photon emission. Thus the notion of RTE is appropriate only to the extent that the correlated and uncorrelated events resonantly couple.

To investigate this possibility in a systematic way, we have invoked the semiclassical impact-parameter formalism to show that the transition amplitude, to lowest order in the electron-electron interaction r_{12}^{-1}, is indeed a sum of correlated and uncorrelated amplitudes. We use an independent electron model in order to retain single particle transfer and excitation amplitudes and thereby to provide a practical starting point for calculating cross sections in these inherently two-electron processes.

Our uncorrelated amplitude alone, when squared and integrated over all impact parameters, is just the NTE cross section estimated in Ref. 4. Within an impulse approximation (IA)[5], our correlated amplitude alone, when squared and integrated over all impact parameters, gives the starting point (eq. 8, Ref. 2) on which Brandt based his calculations.

The IA clearly shows that, since the inverse-Auger transition (dielectronic recombination) occurs in the ionic field of the projectile, a Coulomb wavefunction is required for the captured electron. Furthermore, since the IA treats the target electron as free and therefore implicitly assumes conservation of energy between the projectile and target electron alone during the impulsive interaction, the Coulomb wavefunction of the captured target electron in the field of the projectile ion is on-energy-shell. Recent studies[6] of off-shell corrections to simple capture cross sections indicate that their incorporation could alter the IA cross section by a factor between 1 and 5. Such corrections could account for the factor of two discrepancy between Brandt's calculation of the strength of the RTE peak and the data of Tanis et al.

1. J.A. Tanis et al., Phys. Rev. Lett. 49, 1325 (1981).
2. D. Brandt, to appear Phys. Rev. A.
3. D.J. McLaughlin and Y. Hahn, Phys. Lett. 88A, 394 (1982).
4. P.L. Pepmiller et al. 7th Conf. Appl. of Accel. in Research and Industry, Denton, Texas, 8 - 10 Nov., 1982.
5. J.S. Briggs, J. Phys. B10, 3075 (1977).
6. J. Macek and S. Alston, Phys. Rev. A26, 250 (1982).

DOUBLE K EXCITATION CROSS SECTION OF Fe24+ IONS AT INTERMEDIATE VELOCITY

K. Wohrer, J.P. Rozet, A. Chetioui, and A. Jolly

Institut Curie and Université P & M Curie, 11, rue P & M Curie 75231 Paris Cedex 05

and

C. Stephan

Institut de Physique Nucléaire, 91405 Orsay, France

We have measured $1s^2 \rightarrow 2p^2$ double excitation cross-section for 400MeV heliumlike Fe24+ ions colliding with argon atoms in the intermediate velocity range ($v/v_K \sim 0.6$ where v is the projectile velocity and v_K the mean orbital velocity of the Fe K electron). The method (see figure 1) consists in observing in coincidence the two "Lyman" x-ray transitions deexciting the Fe24+ after a double excitation. The first transition is a satellite of the hydrogenlike transition, the second one is a pure heliumlike transition.

Figure 1

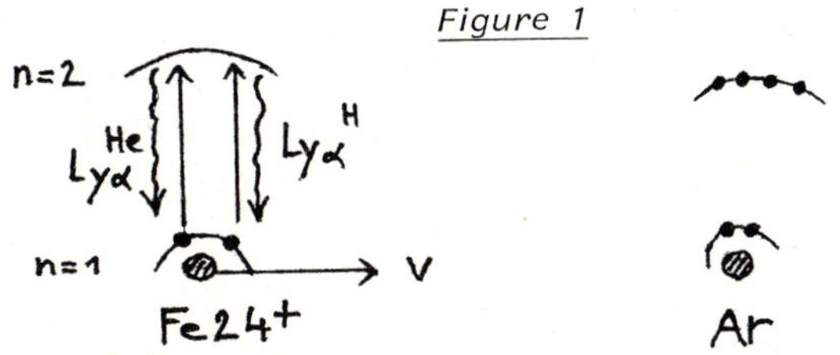

The experiment was performed at ALICE in Orsay. A gaseous jet was used operating under single collision condition. X rays emitted simultaneously were detected by a Si(Li) detector (resolution of 200eV at 6 keV and a photomultiplier (resolution 1keV at 6keV). The SiLi spectrum is shown on figure 2. Both Lyman transitions are superimposed on a large background coming from the fringe of projectiles striking the walls of the collision chamber. Correction due to the autoionization of the $2p^2$ state has been made and is about 20%[1]. We found $\sigma_{II}(1s^2 \rightarrow 2p^2) = (3,5 \pm 1,6) 10^{-20}$ cm² which is (50±20)% of the single $\sigma_I(1s \rightarrow 2p)$ excitation cross sections measured by Rozet et al[2]. An estimation of σ_{II}/σ_I can be obtained in the independent electron model which should be valid for such a high nuclear charge[3] : assuming furthermore a constant excitation probability versus the impact parameter, $\sigma_{II}/\sigma_I = P^2/2P(1-P)$. On the other hand, a crude estimation of P can be given from the relation $\sigma_I/\pi r_K{}^2 = 2P(1-P)$ and gives the value P=0.5. Then $\sigma_{II}/\sigma_I = 0.5$, in good agreement with the experimental value.

Figure 2

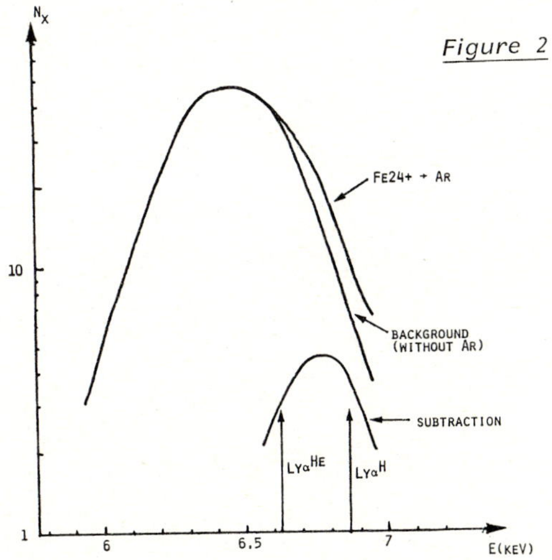

References

1. Dielectronic satellite spectra for hydrogenlike iron in low density plasmas. J. Dubau, A.H. Gabriel, M. Loulergue, L. Steenman-Clark, and S. Volonté. Mon. Not. R. Astr. Soc. (1981) 195, 000.
2. Excitation of 400MeV heliumlike Fe ions in the intermediate energy range. J.P. Rozet et al, to be published in J. Phys. B.
3. Double electron capture by fast nuclei. R. Gayet, R.D. Rivarola, and A. Salin, J.Phys. B: Atom. Molec. Phys. 14, 2421 (1981).

MICROSCOPIC ANALYSIS OF EQUILIBRIUM CHARGE STATE DISTRIBUTION FOR IONS IN THE MeV/u RANGE

D. Vernhet, J.P. Rozet, P. Legagneux-Piquemal, A. Chetioui

Institut Curie and Université P & M Curie, 11, rue P & M Curie 75231 Paris Cedex 05

and

L. Tassen-Got, C. Stephan

Institut de Physique Nucléaire, 91405 Orsay, France

Charge state distribution for 400MeV Fe ions passing through 150μg/cm² and 200μg/cm² beryllium, carbon, aluminium and copper foils have been measured by magnetic analysis*. We have developed a statistical model taking into account competition between capture, ionization+excitation, and Auger+radiative deexcitation within each shell ; precise cross sections have been introduced in the calculation. For each shell, the mean hole number is calculated and then a binomial law is applied.

The mean hole number at equilibrium inside the foil is obtained by solving the system of coupled differential equations between the various charge state fractions of each shell. Radiative and Auger couplings were taken into account between L and K shells but not between M, N... and lower shells since the mean occupation number in these outer shells is very low. The system of equations is :

$$\frac{dY_L^n}{dx} = (9-n)(\sigma_{iL}+\sigma_{RL})Y_L^{n-1} - (8-n)(\sigma_{iL}+\sigma_{RL})Y_L^n + (n+1)\sigma_{CL} Y_L^{n+1} - n\sigma_{CL}Y_L^n$$

$$\frac{dY_K^n}{dx} = (3-n)\sigma_{iK}Y_K^{n-1} - (2-n)\sigma_{iK}Y_K^n + (n+1)(\sigma_{CK}+\sigma_{RK})Y_K^{n+1} - n(\sigma_{CK}+\sigma_{RK})Y_K^n$$

etc...

where σ_i, σ_C and σ_R stand respectively for ionization+excitation, capture and Auger+radiative cross sections. Mean hole numbers inside the foil are then :

$$\overline{n}_K = \frac{2\sigma_{iK}}{\sigma_{iK}+\sigma_{CK}+\sigma_{RK}} \qquad \overline{n}_L = \frac{8\sigma_{iL}+\sigma_{RL}}{\sigma_{iL}+\sigma_{RL}+\sigma_{CL}} \qquad .....$$

These formulae are valid when only single processes occur and they can only be applied for stripping by light target atoms (beryllium or carbon) for which double ionization processes are negligible.

Charge state percentages outside the target, corrected for autoionization, are presented in Table I for beryllium strippers and compared with experimental results.

In the calculation experimental cross sections have been used for iron ionization and excitation[1] and for capture in K or outer shells[2,3]. Radiative and Auger cross sections have been calculated by a statistical law.

Table I. Charge state percentages for 400MeV iron ions stripped by 200μg/cm² Be foils.

charge state	calculated %	experiment. %
26+	2	2
25+	22	20
24+	53	46
23+	19	25
22+	3	6
21+	0.3	0.8
20+	0	0

References

1. K. Wohrer, F. Fernandez, J.P. Rozet, A. Jolly, A. Chetioui, and C. Stephan. International Conference on X-Ray and Atomic Inner-Shell Physics, Eugene, Oregon, August 1982 and to be published in J. Phys. B.
2. K. Wohrer, A. Chetioui, J.P. Rozet, A. Jolly, C. Stephan, to be published in J. Phys. B.
3. A. Chetioui et al, to be published in J. Phys. B.

* *work performed at the CEVIL Orsay.*

VACANCY SHARING FOLLOWING RESONANT-TRANSFER-AND-EXCITATION IN S + Ar COLLISIONS

J. A. Tanis* and E. M. Bernstein*
Physics Dept., Western Michigan University, Kalamazoo, MI 49008 USA

W. G. Graham+
Physics Dept., The New University of Ulster, Coleraine, N. Ireland

M. Clark‡ and S. M. Shafroth‡
Dept. of Physics, University of North Carolina, Chapel Hill, NC 27514 USA

B. M. Johnson, K. Jones, and M. Meron
Brookhaven National Laboratory, Upton, Long Island, NY 11973 USA

Target K x rays associated with single electron capture and loss events have been measured for 70-160 MeV S^{13+} ions incident on Ar under single collision conditions. This work was done at Brookhaven using the MP tandem. Coincidences were detected between K x rays and ions which undergo single electron capture or loss.

Projectile K x-ray coincidences with capture and loss events, which were also measured, provided the first strong experimental evidence[1] for the existence of resonant-transfer-and-excitation (RTE) in ion-atom collisions, a process which is analogous to dielectronic recombination. As expected for RTE, a peak was observed in the cross section $\sigma_x^{q-1}(S)$ (q = incident charge state) for sulfur projectile K x ray emission associated with capture. The position (~120 MeV) and width (~60 MeV) of the peak are in good agreement with theoretical calculations[2] of the RTE process; the absolute magnitude of the cross section is about a factor of two larger than the predicted value.

For argon target K x rays coincident with single capture, the results of which are reported here for the first time, the cross section $\sigma_x^{q-1}(Ar)$ also exhibits a peak as shown in Fig. 1. The maximum occurs at an energy of ~125 MeV with a width of ~60 MeV. The maximum in $\sigma_x^{q-1}(Ar)$ is attributed to K vacancy sharing[3] following RTE by the projectile. Due to the fact that the collision system is nearly symmetric there is a high probability (~35-40%) for transfer of a K vacancy from the projectile to the target during such a collision.

Vacancy transfer from the projectile to the target relates the projectile σ_x^{q} (S) to the target $\sigma_x^{q-1}(Ar)$. Although the projectile σ_x^{q} (S) was not measured directly, it can be estimated from $\sigma_x^{q-1}(S)$ and $\sigma_{K\alpha}(S)$ (see Ref. 1). By applying the Meyerhof vacancy transfer probability[3] to the "resonant" part of the projectile σ_x^{q} (S), the "resonant" part of the target $\sigma_x^{q-1}(Ar)$ can be calculated. Using the assumed linear background shown in Fig. 1, the predicted target $\sigma_x^{q-1}(Ar)$ can then be obtained. The curve labeled ω_K = 0.078 in Fig. 1 shows the predicted $\sigma_x^{q-1}(Ar)$ for the target using single K-vacancy fluorescence yields and K-shell binding energies. The curve labeled ω_K = 0.117 was calculated assuming that the fluorescence yield of the projectile relative to the target is increased by 50%. (Due to the high incident projectile charge state, it is expected that the projectile fluorescence yield is increased relative to the target fluorescence yield.) The shape of the calculated curve in each case is in reasonable agreement with the experimental results; for ω_K = 0.117 the absolute magntiude also agrees well with the data.

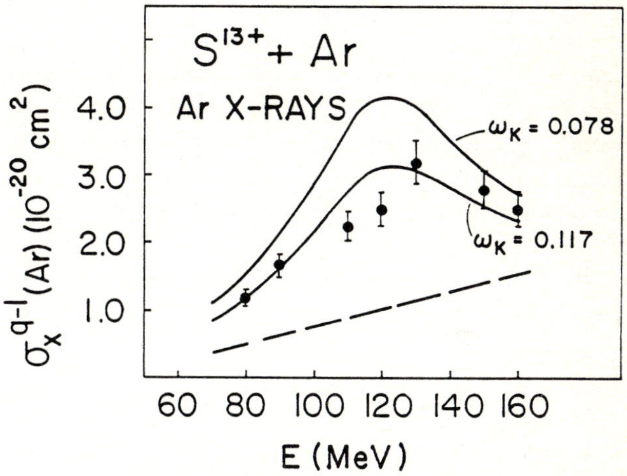

Fig. 1: Cross section $\sigma_x^{q-1}(Ar)$ for target K x rays coincident with single electron capture.

The results presented here emphasize the necessity of measuring "partial" x-ray production cross sections in order to probe the individual processes of ionization, excitation, and charge transfer in ion-atom collisions.

The BNL portion of this research was supported by the U.S. Department of Energy, Division of Basic Energy Sciences under Contract No. DE-AC02-76CH00016.

*Supported in part by grants from the Precision Kawasaki Co., Sinotest Shoji K.K., and Nakano Transportation Co.

+Supported in part by NATO Research Grant No. 1910.

‡Supported in part by the U.S. Department of Energy, Division of Chemical Sciences.

1. J. A. Tanis, E. M. Bernstein, W. G. Graham, M. Clark, S. M. Shafroth, B. M. Johnson, K. Jones, and M. Meron, Phys. Rev. Lett. 49, 1325 (1982).

2. D. Brandt, Phys. Rev. A27, (to be published).

3. W. E. Meyerhof, Phys. Rev. Lett. 31, 1341 (1973).

ROTATIONAL COUPLING IN ASYMMETRIC ION-ATOM COLLISIONS

D. Maor[+], Z. Stachura*, P.H. Mokler, B. Liu**, D. Liesen

Department of Physics, Technion, Haifa, Israel[+] and GSI Darmstadt

In almost symmetric ion-atom collisions, vacancy production in shells for which the orbital velocity u is higher than the collision velocity v is considered to be due to electron promotion, in particular via the 2pπ-2pσ rotational coupling. In more asymmetric collisions using the lighter collision partner as projectile, the adiabatic peak in the vacancy production probability as a function of impact parameter P(b) disappears, and the main mechanism there is direct ionization from the molecular orbitals, near the united atom limit[1].

In the present experiment, P(b) measurements for asymmetric systems ($Z_2/Z_1 \gtrsim 1.8$) were extended to very small impact parameters, where a second, the kinematic peak, is expected due to rotational coupling. This peak has been measured for near-symmetric I - Ag collisions by Anholt et al.[2]. In the sudden approximation this kinematic peak is a geometrical effect which requires only that the electrons follow the molecular orbitals. Here, P(b) was determined by coincidences between K x-rays of the lighter collision partner and scattered particles. The heavier collision partner was chosen as projectile. Thus, more 2pπ vacancies are introduced into the collision by a multiple-collision mechanism and the rotational coupling should be enhanced. Additionally, large center-of-mass angles, (Θ_{cm}) can be measured at moderate lab. angles (Θ_{lab}), and, due to the dual kinematics, up to three impact parameters can be measured at the same Θ_{lab}. ^{136}Xe and ^{120}Sn ion beams, at 1.4 MeV/u impinged upon thin Ni and Cu targets, respectively. The x rays were recorded by two Si(Li) detectors both at 90° to the beam. The particles were detected by one annular, or by an array of several normal surface barrier detectors.

Fig. 1 displays P(b) for the Cu K-shell. In spite of the fact that more 2pπ vacancies are brought into the collision compared with the $Z_P < Z_T$ data, the adiabatic peak in the large impact parameter region is barely discernible above a constant background. If it exists, it peaks at ~ 500 fm, in general agreement with the systematics found by Schuch et al.[3]. The constant background is again due to direct excitation and its decrease at ~ 1000 fm is in agreement with the calculations of Ref. 1.

At the lowest b values, a prominent peak appears, its maximum being at $\Theta_{cm} = 90°$, exactly where it should be for the kinematic peak. No other mechanism is supposed to cause such a shape in a region corresponding to distances of closest approach from 65 fm to 36 fm,(cf.) the K-shell radius of the united atom is ~ 600 fm. The solid curve in the figure is $A\sin^2 \Theta_{cm}+B$ with the parameters, A and B, fitted to the data. B is the assumed constant contribution of the direct ionization. An asymmetry of the data compared to the calculated shape can be seen. This cannot stem from the change in the shape of the wavefunctions due to relativistic effects[4].

It seems that the kinematic peak appears even in asymmetric systems, provided the MO model is still applicable. On the other hand, the adiabatic part of the rotational coupling is strongly diminished in these asymmetric collision systems. One sees that an upper limit of ~ 10 can be set to the ratio of the amplitudes of the kinematic to the adiabatic peak, after the subtraction of the direct ionization contribution. This should be compared to a value of ~ 2 for the same ratio in lighter and more symmetric systems[5].

---

*Permanent address: Inst. Nucl. Phys., Cracow, Poland
**Permanent address: Inst. Mod. Phys., Academia Sinica, Lanzhou, China

References

1. D. Maor et al., Phys. Rev. A, in press
2. R. Anholt et al., J. Phys. B13, 3807 (1981)
3. R. Schuch et al., Phys. Rev. A22, 1447 (1980)
4. B. Fricke, private communication
5. K. Taulbjerg et al., J. Phys. B9 1351 (1976)

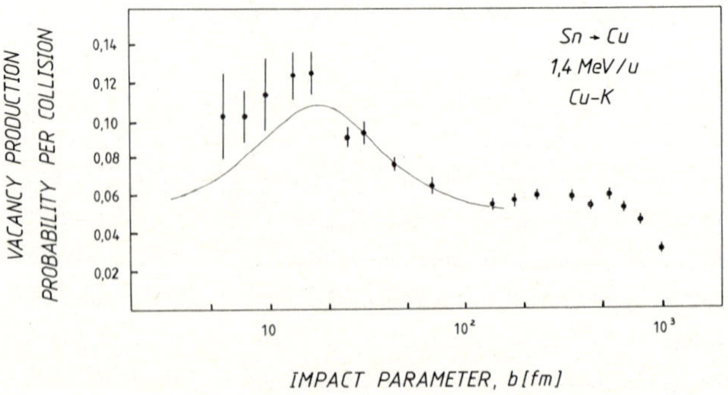

Fig. 1: Vacancy production probability for Cu-K excitation in 1.4 MeV/u Sn-Cu collisions. The solid line gives the expected shape for the kinematic peak of the 2pπ-2pσ coupling.

INNER-SHELL VACANCY PRODUCTION IN ASYMMETRIC HEAVY ION-ATOM COLLISIONS

A. Warczak*, H.D. Dohmann, D. Liesen, B. Liu**

Gesellschaft für Schwerionenforschung, D-6100 Darmstadt
Permanent Addresses: *Jagellonian Univ., Cracow, Poland; **Inst. of Modern Physics, Academia Sinica, Lanzhou, China

In order to study inner-shell vacancy production in collision systems near K-L level matching the impact parameter dependence of the corresponding excitation probabilities has been measured in U+Sn, Xe+Ni, and Kr+U collisions. Targets of several thicknesses were bombarded with ions from the UNILAC at the GSI, Darmstadt at 1.4 MeV/N projectile energy.

The scattered ions, registered by a position sensitive avalanche detector[1] were recorded in coincidence with characteristic X-rays. In Fig. 1 Sn-K excitation curves are shown at two target thicknesses. The Sn-K excitation increases by almost a factor of 2 for 40 μg/cm² target compared to 5 μg/cm². This is probably due to the increasing role of multistep processes which are probably more effective in a higher excited collision system which is produced in a thicker target. Nevertheless, in both cases the Sn-K vacancy production is well described by an exponential function after an appropriate correction of the contributions coming from the K-L vacancy sharing (open points in Fig. 1). The slope of solid curves (Fig. 1) suggests that the direct excitation of the $2p_{3/2}\sigma$ MO dominates the Sn-K vacancy production. The existence of a strong radial coupling between $2p_{3/2}\sigma$ and $2p_{1/2}\sigma$ MO is of importance to explain this finding.

At impact parameters between 70 and 150 fm both curves (Fig. 1) exhibit a steep increase with oscillating tendences.

Similar oscillatory structures of the excitation function are also observed in other collision systems e.g. Xe+Ni (Fig. 2), and Kr+U (Fig. 3).

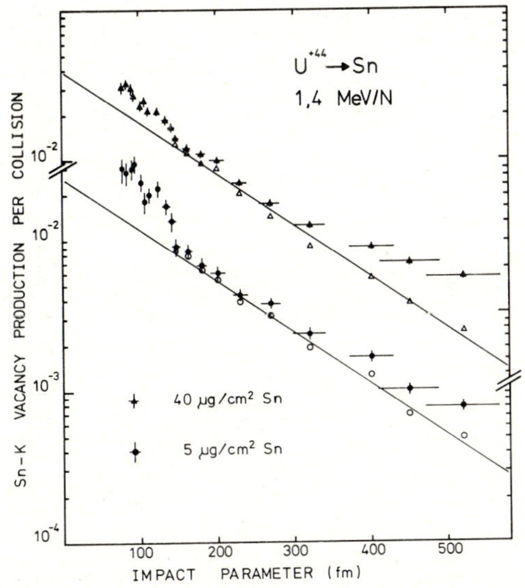

Fig. 1: Sn-K vacancy production in U+Sn collisions.

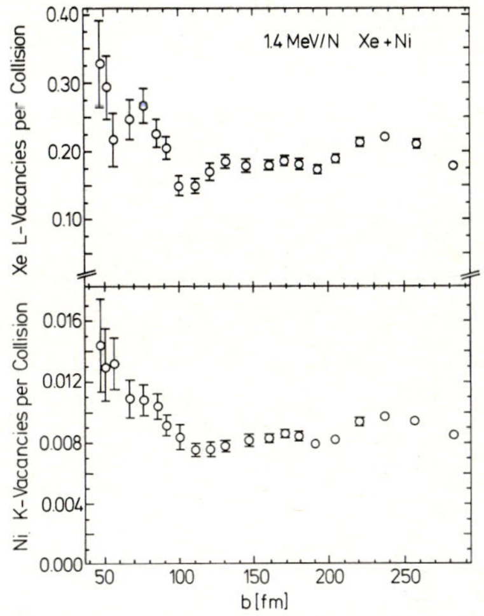

Fig. 2: Xe-L and Ni-K vacancy production in Xe+Ni collisions.

Also in these systems the vacancy production probability was determined via the observation of characteristic x-rays which are emitted with the decay of the excited states. Thereby, only the final states are fixed, while the many initially excited states can not be separated. Thus, the coherent superposition of all these excitation channels which lead to the same final state, may result in structures observed in our experiments. Particularly, in U+Sn collisions the $2p_{3/2}\sigma$, $2p_{1/2}\sigma$, and $2s_{1/2}\sigma$ excitation channels are favourable candidates for such a coherent excitation of the Sn-K shell. In how far our interpretation is relevant to the experimental data is an open question. It requires strong theoretical verification.

Nevertheless, further investigations, also with gas targets, are planned in order to find a possible systematics.

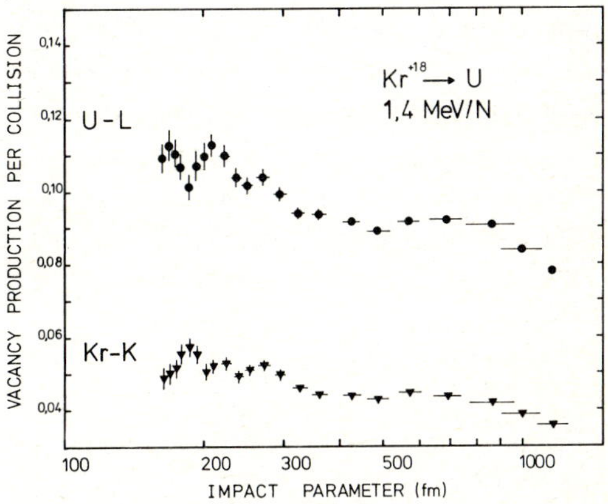

Fig. 3 U-L and Kr-K vacancy production in Kr+U collisions.

1. B. Liu, W. Enders, D. Liesen, R. Schulze, contribution to this conference.

INNER SHELL VACANCY TRANSFER STUDIED WITH HIGHLY IONIZED, DECELERATED HEAVY IONS

P.H. Mokler, D.H.H. Hoffmann, W.A. Schönfeldt, Z. Stachura*, and A. Warczak*
Gesellschaft für Schwerionenforschung, D-6100 Darmstadt
*Permanent Addresses: Inst. of Nucl. Phys. and Jagellonian University, Cracow, Poland

Measuring the dependence of inner shell excitation on the charge state of the incoming ion gives a unique possibility to extract information on vacancy transfer between inner shells in a collision molecule [1,2]. This is especially true if already an inner shell of the projectile can be opened in the experiment and on the other hand the collision is reasonably slow. In order to provide relatively slow, highly ionized heavy ions the acceleration/stripping/deceleration method [3] was used to provide 2.2 to 3.6 MeV/u Xe^{q+} ions and 3.6 to 4.7 MeV/u Sm^{q+} ions with charge states ranging up q = 47 and q = 53, respectively. That means up to 3 projectiles shell vacancies can be provided in collisions with the used Xe gas target atoms. The charge transfer to the (Xe-) K shell was studied for the collision molecules with a united nuclear charge Z_1+Z_2 of 108 and 116, respectively. In such heavy systems the spin orbit splitting in the u.a. L-shell exceeds already 1/3 of the corresponding L binding energies and the question arises whether the normal 2pπ-2pσ rotational coupling still works, cf. also Ref. 4.

From the yield increase in Xe-K x-ray emission for 1, 2, and 3 projectile L shell vacancies - see Figs. 1 and 2 - the L-K vacancy transfer cross sections given in Table 1 are extracted. For comparison the 2pπ-2pσ cross sections according to Taulbjerg et al. [5] are given assuming a statistical factor 1/6. These values are roughly a factor of 2 larger than the experimental ones. For relativistic calculations [6] only a small reduction in the rotational cross section results.

Before opening the projectile L shell a slight increase in Xe-K x-ray emission is observed already. This is attributed to couplings to higher lying vacant MO. Considering the $Sm^{q+} \rightarrow Xe$ data we can confine these couplings mainly to MO correlating to the Sm 3s and 3p levels but obviously not to the Sm 3d levels. Probably both radial sharing and rotational coupling contribute to this M-K vacancy transfer.

1. A. Warczak et al., J. Phys. B14 (1981) 1315
2. W.N. Lennard et al., Phys. Rev. A23 (1981) 2260
3. P.H. Mokler et al., annual report 1981 - GSI-82-1
4. R. Schuch et al., to be publ. J. Phys. B. (1983)
5. K. Taulbjerg et al., J. Phys. B9 (1976) 1351
6. D. Jakubassa et al., J. Phys. B13 (1980) 757

VACANCY TRANSFER CROSS SECTION, Δσ/Δq (kbarn)

Collision System $Z_1 \rightarrow Z_2$,	MeV/u	L → K exp	L → K TBV	M → K exp
$Xe^{q+} \rightarrow Xe$	2.2	2.5±0.3	5.6	0.08±0.05
	3.6	2.9±0.1	7.1	0.09±0.06
Sm^{q+} Xe	3.6	2.3±0.3	5.7	0.12±0.03
	4.7	2.3±0.7	6.5	-

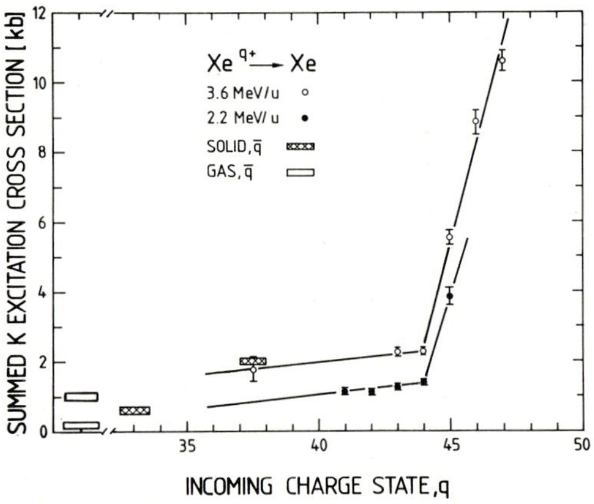

Fig. 1

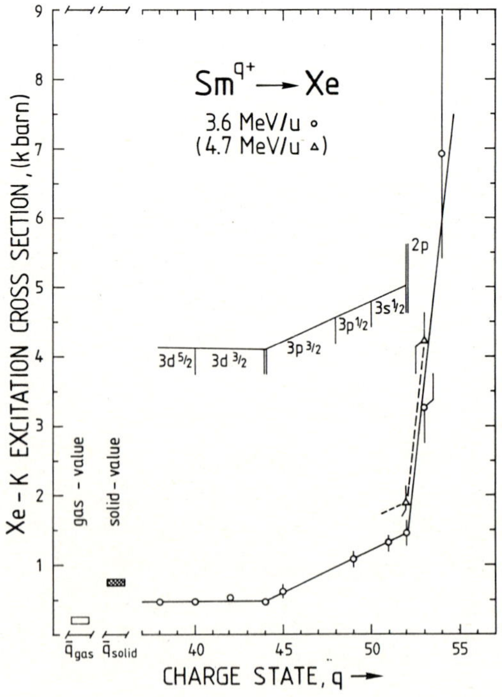

Fig. 2

THE ELECTRON'S PATH THROUGH PHASE SPACE DURING HEAVY ION IMPACT

J. Krause and M. Kleber
Physik-Department, T.U. München, West Germany

The concept of trajectories (or paths) may be introduced into quantum mechanics through Ehrenfest's principle which states the correspondence between classical trajectories and quantal expectation values of the corresponding operators. In order to obtain closed equations of motion for average position and momentum of a particle, one has to use wave functions which are optimized with respect to their time-dependence. This is achieved[1] by parametrizing the wave function in terms of the various multipole excitation modes (l=0,1,2,...).

Monopole excitation is depicted in Fig. 1 for a strongly bound Dirac electron. The plotted trajectory is the one obtained by saddle-point evaluation of the path integral. Fluctuations around the optimum path are obtained in the usual way[2] by evaluation of

$$|\phi(t)\rangle = \int d^2\lambda \; F(\lambda-\lambda_o(t))\,|\lambda\rangle \qquad (1)$$

in harmonic approximation. Here $\lambda_o(t)$ is the (complex) generator coordinate for monopole excitation which makes the action stationary and F is a weighting function of gaussian type. Fig. 2 shows the importance of fluctuations for heavy collision systems.

It is clear that intrinsic generator coordinates will generate a mean field. One is therefore left with the problem of extracting transition probabilities from this mean field. This is in fact possible by modifying Demkov's approach[3] to transition amplitudes. A careful analysis together with a comparison of experiments will be reported on elsewhere.

[1] J. Krause and M. Kleber in: Quantum electrodynamics of strong fields (Nato Advanced Study Inst. series B80, Plenum Press, 1982)

[2] B. Jancovici and D.H. Schiff, Nucl.Phys. 58 (1964) 678

[3] Yu. N. Demkov, Sov.Phys. JETP 11 (1960) 1351

Fig. 1: Phase space trajectory of a K-shell electron in Cm during a head-on collision between Pb and Cm. Starting point (t=-∞) and point of closest internuclear distance are marked by small circles. Sense of circulation is clockwise.

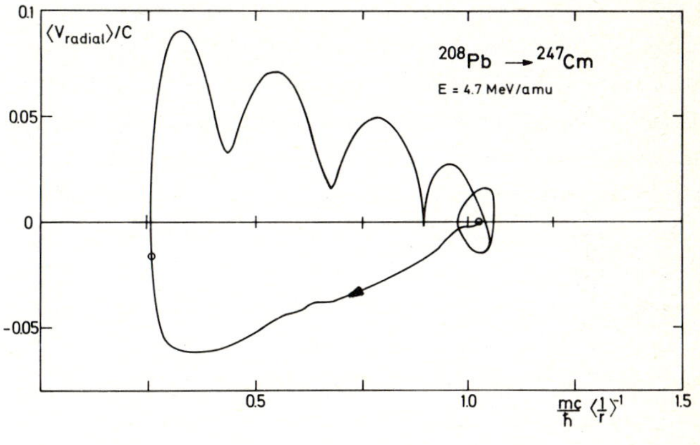

Fig. 1

Fig. 2: Relative increase of K-shell binding due to fluctuations of the monopole generator coordinate. All values for slow head-on collisions at the point of closest approach.

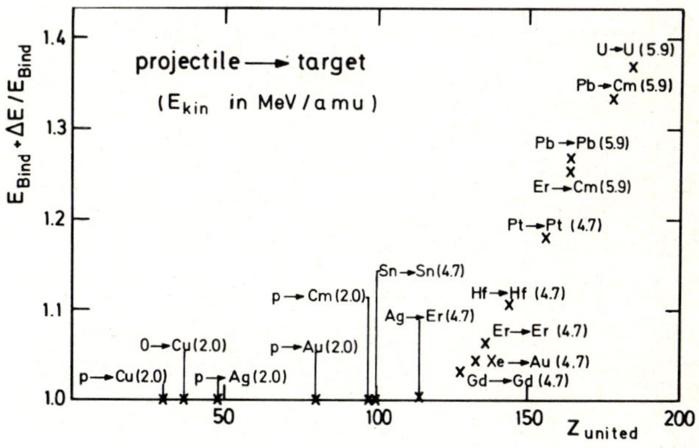

Fig. 2

L-SHELL EXCITATION IN SLOW ION-ATOM ENCOUNTERS

R. Shanker*, R. Hippler*, R. Bilau*, U. Wille[+], H. O. Lutz*

* Fakultät für Physik, Universität Bielefeld, F. R. Germany
+ Hahn-Meitner-Institut, Berlin, F. R. Germany

In slow ion-atom collisions inner-shell vacancies may be produced by promotion of molecular orbitals (MO) in the transiently formed quasi-molecule and subsequent couplings to empty orbitals. Prototype for such an excitation mechanism in light symmetric collision systems is K-shell excitation via 2pσ-2pπ rotational coupling[1,2].

Similarly, L-shell excitation may occur via promotion of 3dσ and 4fσ MOs. For example, in the Ar-Ar collision system, the 4fσ MO is strongly promoted at an internuclear distance R = 0.5 a.u.; the excitation probability as a function of impact parameter shows a step function behaviour at that distance[3]. The strong promotion of the 4fσ MO is also responsible for L-shell excitation in much heavier collision systems, e.g., Kr-L excitation in slow Kr-Kr and Kr-Xe collisions. However, in contrast to Ar-Ar (where the 4fσ MO couples to a large number of weakly bound or unbound empty orbitals), in much heavier systems the united-atom 4f level is strongly bound. As a result, the 4fσ-4fπ-4fδ-4fφ rotational coupling with its characteristic kinematic-peak structure in the impact-parameter dependence is found to be responsible for Kr-L excitation[4].

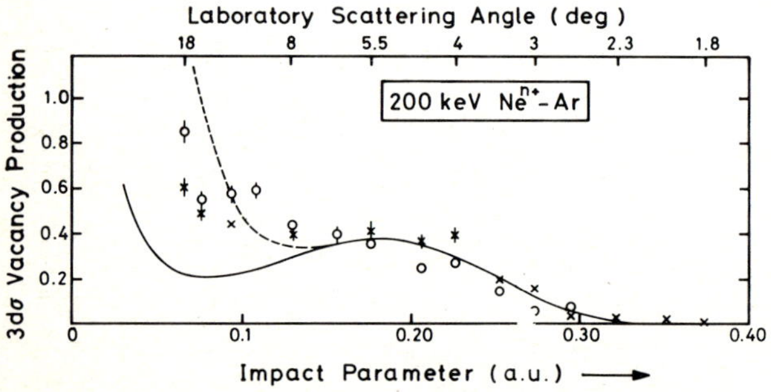

Fig. 1: Impact-parameter dependence of Ar-L vacancy production in 200 keV o Ne^+ and x Ne^{++} collisions with Ar. —— 3dσ-3dπ-3dδ rotational coupling calculation, --- sum of direct and rotational coupling calculations.

At considerably smaller internuclear distances, the 3dσ-3dπ-3dδ rotational coupling provides an additional excitation channel. This has been shown for Ar-L excitation in Ar-Ar collisions at small impact parameters[3]. In slightly asymmetric collision systems excitation via the 3dσ MO may even be the dominant mechanism for L-shell excitation in the heavier collision partner. In sufficiently light collision systems where the united atom 3d level is only weakly bound, the highly promoted 3dσ MO may also be ionized directly (as, e.g., the 4fσ MO in Ar-Ar). In order to study the competition between rotationally induced excitation and direct ionization in more detail, we have measured the impact parameter dependence of Ar-L vacancies in Ne-Ar collisions. Figure 1 shows the result of such an investigation for 200 keV Ne^+, Ne^{++} → Ar collisions. The experimental values for the number of Ar-L vacancies versus impact parameter are compared with theoretical calculations based on (a) the 3dσ-3dπ-3dδ rotational coupling model as outlined by Shanker et al.[3] and (b) the 3dσ direct-ionization model calculated in first Magnus approximation (for details see Wille[5]). While the rotational-coupling calculation gives a satisfactory agreement only at large impact parameters (at the so-called adiabatic maximum), it deviates considerably from experiment at smaller impact parameters. These discrepancies become larger when the 3dσ-3dπ-3dδ rotational coupling calculation is compared to experiment at larger incident energies of 400 keV and 700 keV. In contrast, the direct-ionization model fails to explain satisfactorily the larger-impact-parameter regime. However, an (incoherent) sum of direct and rotational-coupling probabilities describes the experiment reasonably well. This indicates that both excitation mechanisms cause comparable contributions to Ar-L excitation in Ne-Ar collisions.

In summary, the molecular model of inner-shell excitation is found to carry surprisingly far as it provides in an increasing number of cases a quantitative description of the experimental data in terms of the interaction of individual quasimolecular orbitals.

This work was supported by the Deutsche Forschungsgemeinschaft (DFG).

References

1. S. Sackmann, H. O. Lutz, J. S. Briggs, Phys. Rev. Letters 32, 805 (1974)
2. N. Luz, S. Sackmann, H. O. Lutz, J. Phys. B 12, 1973 (1979)
3. R. Shanker, R. Bilau, R. Hippler, U. Wille, H. O. Lutz, J. Phys. B 14, 997 (1981)
4. R. Shanker, R. Hippler, U. Wille, R. Bilau, H. O. Lutz, J. Phys. B 15, L 495 (1982)
5. U. Wille, J. Phys. B 16, L xxx (1983)

DIRECT IONIZATION OF HIGHLY PROMOTED MOLECULAR ORBITALS IN SLOW ION-ATOM COLLISIONS

U. Wille

Bereich Kern- und Strahlenphysik, Hahn-Meitner-Institut für Kernforschung
Berlin GmbH, and Fachbereich Physik, Freie Universität Berlin,
D-1000 Berlin 39, West Germany

Among the primary processes leading to inner-shell vacancy production in slow ion-atom collisions, the excitation of promoted molecular orbitals (MO) by direct coupling to the continuum[1,2] and by rotational coupling among near-degenerate orbitals[3] are likely to play a dominant role. In this work, we study the direct ionization of the highly promoted 4fσ MO within the united atom ionization model of Briggs[1]. Quantitative calculations of direct 4fσ ionization are required in the analysis of experimental data on L-shell excitation in near symmetric, heavy collision systems[3,4].

According to the model of Briggs, the amplitude for ionization of a united-atom orbital is given by the superposition of the (first-order) amplitudes corresponding to the perturbation due to projectile and target nucleus, respectively. In the complete evaluation of the amplitude for 4fσ ionization, we use the partial-wave formalism outlined by Meyerhof[2]. The key quantities to be calculated are the radial formfactors corresponding to transitions of definite multipolarity from (hydrogen like) initial bound orbitals to final continuum orbitals. We have bypassed possible difficulties in the numerical computation of these formfactors by using the analytic integrals given by Tweed[5], which allow the formfactors to be expressed entirely in terms of elementary (algebraic and transcendental) functions.

We have explicitly calculated the impact-parameter dependence of direct 4fσ ionization for Ag+Ag collisions[3] at 17, 26 and 40 MeV collision energy and for I+Ag collisions[4] at 63 MeV collision energy. In these calculations, the binding energy of the initial orbital has been set equal to the binding energy[3,4] of the 4fσ MO at the distance of closest approach corresponding to an (unscreened) Coulomb trajectory. The results for Ag+Ag collisions are displayed in Fig. 1. The (first-order) ionization probabilities calculated from the Briggs model are shown as dashed curves. It is seen that the results depend strongly on collision energy. At 40 MeV (collision velocity v = 3.854 a.u.) and small impact parameters, the first-order probability exceeds unity.

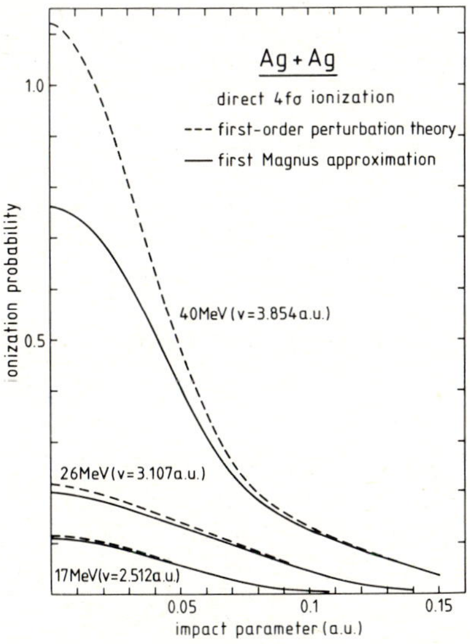

Fig. 1: Impact-parameter dependence of the probability for direct 4fσ ionization in Ag+Ag collisions.

In order to remove the latter deficiency, we have devised a simple non-perturbative generalization of the Briggs model. We use the Magnus expansion[6] of the time evolution matrix and make two simplifying assumptions: (i) time ordering is neglected, i.e., we employ the first Magnus approximation; (ii) direct couplings among different continuum orbitals are neglected. The exponential series defining the time-evolution matrix can then be summed up in closed form, and the (total) ionization probability P in the Magnus approach can be expressed in terms of the first-order probability $P^{(1)}$ simply as $P = \sin^2\sqrt{P^{(1)}}$. The numerical results corresponding to the first Magnus approximation are shown in Fig. 1 as full curves.

With regard to the analysis of experimental data on L-shell excitation[3,4], our calculations show that a substantial fraction of the enhancement observed in the vacancy production at small impact parameters may be caused by direct 4fσ ionization. Particularly for the larger collision energies, the inclusion of the 4fσ contribution into the analysis would require the conclusions reached[3,4] on the relative contributions of 3p and 3d rotational coupling to be modified to some extent. For a full quantitative analysis of the data, a reduction of the systematic experimental errors is desirable.

References

1. J.S. Briggs, J. Phys. B: At. Mol. Phys. 8, L485(1975).
2. W.E. Meyerhof, Phys. Rev. A18, 414(1978).
3. G. Presser, J. Stähler, R. Werner, and U. Wille, J. Phys. B: At. Mol. Phys. 16, 197(1983).
4. E. Morenzoni, M. Nessi, P. Bürgy, Ch. Stoller, and W. Wölfli, Preprint ETH Zürich (1983).
5. R.J. Tweed, J. Phys. B: At. Mol. Phys. 5, 820(1972).
6. U. Wille, Z. Phys. A - Atoms and Nuclei 308, 3(1982).

OBSERVATION OF Al 2s VACANCY PRODUCTION IN 100keV Al^+ ON Ar GAS COLLISIONS: SIGNIFICANCE FOR MO CORRELATIONS

M.L. FURST, H.C. HAYDEN and W.W. SMITH

Physics Department, University of Connecticut, Storrs, CT 06268

Previous high resolution grazing-incidence measurements of soft x-ray spectra from near-symmetric ion-atom collisions in the 100keV range[1] have indicated that L-x-rays arise predominately from single or double 2p vacancy production in the lower-Z collision partner. A series of measurements of L-x-ray single-collision spectra from more and more asymmetric collisions (Cl^+, S^+, P^+ on Ar) revealed little or no intensity due to initial 2s vacancy production. The Barat-Lichten molecular-orbital correlation rule[2] relates the 2s orbital in the lower-Z collision partner to the 3p orbital in the higher-Z target. Since this orbital is normally occupied, 2s vacancy creation would be inhibited. Eichler and co-workers[3] have proposed an alternative diabatic correlation scheme in which the 2s correlates with different MO's of the quasimolecule, leading to possible mechanisms for 2s vacancy production at moderate energies.

Figure 1 shows an L-x-ray spectrum from 100keV Al^+ collisions with Ar, in which we have identified several prominent features due to 2s vacancy production in the 250-350Å wavelength range. These arise from transitions of the type $2s2p^n$--$2s^2 2p^{n-1}$. In particular we see strong transitions from such states in Al V to Al VIII ions arising from configurations with unit fluorescence yield. In addition, we see the usual neon-like doublet from $2p^5 3s$ as well as radiation from projectile ion configurations such as $2p^5 3s^2 3p$, corresponding to previously-observed transitions in P^+ and S^+ collisions with Ar. No strong Ar L (target) satellite lines between 48 and 56Å were observed in any of the collisions studied, though weak target lines could be present. We are presently studying the energy dependence of the transitions arising from 2s vacancies to attempt to shed light on the specific vacancy-production mechanism.

In near-symmetric collisions with Ar, 2s vacancy-production is difficult to observe in the x-ray spectrum because of the high probability of a Coster-Kronig branch (2s vacancy conversion to a 2p vacancy). The collisional production of 2s vacancies might be inferred from the spectrum of emitted Auger electrons, but the complexity of the Auger spectrum makes state identification ambiguous.[4] The kind of high resolution soft x-ray emission spectra observed in this work removes much of this ambiguity.

By reducing the number of outer shell electrons in the projectile from 7 in the case of Ar^+ on Ar to 2 in

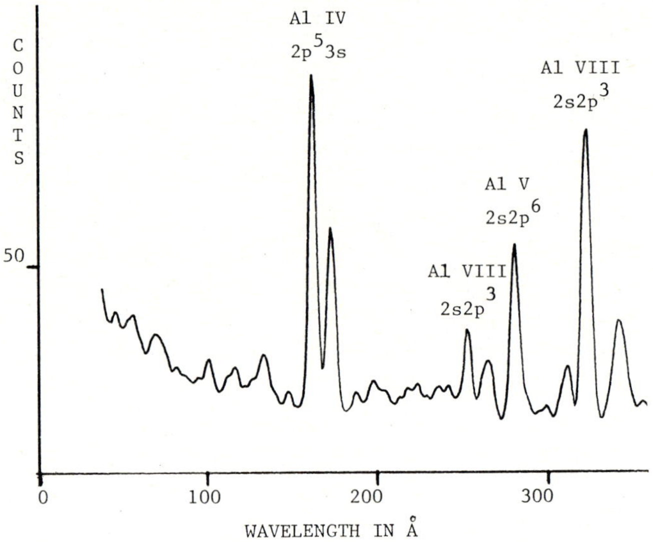

Figure 1. L x-ray satellite spectrum from 100keV Al+ on Ar gas collisions.

the Al^+ on Ar collision, the number of possible satellites in the x-ray spectrum is greatly reduced, while the mean fluorescence yield is enhanced.[5] We are examining these new data in the light of the types of MO previously suggested by Larkins and by Eichler and Wille.[3]

* Research supported by the University of Connecticut Research Foundation. We gratefully acknowledge the technical support of Jim Koch.

References

1. See for example R.S. Peterson, W.W. Smith, H.C. Hayden, and M. Furst, IEEE Transactions on Nuclear Science, NS-28, 1114 (1981).
2. M. Barat and W. Lichten, Phys. Rev. A6, 211 (1972).
3. J. Eichler, U. Wille, B. Fastrup and K. Taulbjerg, Phys. Rev. A14, 707 (1976).
4. We note that the Aarhus group has previously reported an unexpected Auger peak from Mg^+ on Ar collisions at 70keV which they suggest might be due to an L_1-$L_{2,3}M$ Coster-Kronig transition: P. Dahl et al, J. Phys. B9, 1581 (1976). Auger electrons from asymmetric collisions with L-shell vacancies have also been studied by Schneider et al, XII ICPEAC, Abstracts, p.829.
5. F. Larkins, J. Phys. B4, L29 (1971).

ELECTRON TRANSITIONS FROM DISCRETE LEVELS INTO CONTINUUM DUE TO RADIAL COUPLING IN SLOW ATOMIC COLLISIONS

A.N.Zinoviev, S.Yu.Ovchinnikov, Yu.S.Gordeev

A.F.Ioffe Physical-Technical Institute of the Academy of Sciences of the USSR,Leningrad,USSR

Electron transitions from discrete state into continuum in slow atomic collisions can sometimes give the major contribution to a continuous part of electron energy spectra[1,2]. In this paper an attempt has been made to obtain an analytical expression for the probability of such transitions into continuum when the electron energy ω is large.

For adiabatic collisions the expansion coefficients $a_K(t)$ and $b_\omega(t)$ of nonstationary wave function of quasimolecule on molecular basis are found from

$$\sum_K \frac{\partial}{\partial t}\left[a_K(t)\varphi_K(\vec{r},t)\right]\exp\left[-i\int^t E_K(t')dt'\right] + \int d\omega \frac{\partial}{\partial t}\left[b_\omega(t)\varphi_\omega(\vec{r},t)\right]\exp(-i\omega t)=0 \quad (1)$$

where $\vec{r}$ are electron coordinates. Here $\varphi_K(\vec{r},t)$ corresponds to the bound state with the energy $E_K(t)$ (molecular term) and $\varphi_\omega(\vec{r},t)$ is the state in continuum with energy ω. If prior to the collision the only populated state is $\varphi_p(\vec{r},t)$, the first order of pertubation theory gives the probability for an electron to occur in a continuum state $\varphi_\omega(\vec{r},t)$ as

$$|b(\omega)|^2=|b_\omega(+\infty)|^2=\left|\int dt \langle\varphi_\omega(\vec{r},+\infty)|\varphi_p(\vec{r},t)\rangle \times \left[E_p(t)-\omega\right]\exp\left\{-i\int dt'\left[E_p(t')-\omega\right]\right\}\right|^2 \quad (2)$$

Using the general properties of Schrödinger equation solutions [3] the matrix element in (2) is equal to :

$$\langle\varphi_\omega(\vec{r},+\infty)|\varphi_p(\vec{r},t)\rangle = -\left[E_p(t)-\omega\right]^{-1}\times \lim_{r\to\infty}\frac{r^2}{2}\left[\varphi_\omega(\vec{r},+\infty)\nabla\varphi_p(\vec{r},t)-\nabla\varphi_\omega(\vec{r},+\infty)\varphi_p(\vec{r},t)\right] \quad (3)$$

Then, the integral in (2) is calculated using the method of a stationary phase

$$|b(\omega)|^2=\left|\lim_{r\to\infty}\frac{r^2}{2}\left[\varphi_\omega(\vec{r},+\infty)\nabla\varphi_p(\vec{r},t_p)-\nabla\varphi_\omega(\vec{r},+\infty)\varphi_p(\vec{r},t_p)\right]\times\left(\frac{\pi}{2}\frac{\partial t_p}{\partial\omega}\right)^{1/2}\exp\left\{-i\int^{t_p(\omega)}\left[E_p(t')-\omega\right]dt'\right\}\right|^2 \quad (4)$$

where the point of the stationary phase $t_p(\omega)$ is the solution of the equation $E_p(t_p)=\omega$ and has a complex value. In our case an electron at large r is in the Coulomb field , therefore we can use the well known asympthotic form of the Coulomb wave functions at $r\to\infty$ [3] and obtain

$$|b(\omega)|^2= 2^{1/2}\pi^{-1}\omega^{3/2}\left[\exp\left(\pi\left(\frac{2}{\omega}\right)^{1/2}\right)-1\right]\left|\frac{dt_p}{d\omega}\right| \times \exp\left\{-2\,\mathrm{Im}\int_{-\infty}^{t_p(\omega)}\left[E_p(t')-\omega\right]dt'\right\} \quad (5)$$

For short lengh potential our expression (4) agrees well with the results obtained in[4] for electron detachment in collisions of negative ion with atom.

It is well known that in slow atomic collisions the adiabatical approximation is invalid at some local points, for exsample at crossing points of terms. The results (4,5) can be also used for this case if we replace $\varphi_p(\vec{r},t)$ and $E_p(t)$ by a diabatical state and diabatical term, respectively.

In collisions of heavy atoms we must summarize the ionization cotributions of all populated molecular states and at large ω dominant cotributions are made from the terms , which strongly change their energy during the collision.

References

1. Yu.S.Gordeev et el J.Phys.B 14,527 (1981).
2. A.N.Zinoviev,S.Yu.Ovchinnikov,Yu.S.Gordeev Abstracts XII ICPEAC,Gatlinburg, 1981.
3. L.D.Landau,E.M.Lifshiz.Quantum mechanics, Moscow, 1974, p.79, 152.
4. E.A.Solov'ev.Z.E.T.P. 70, 872 (1976).

INNER SHELL DIRECT IONIZATION IN SLOW ATOMIC COLLISIONS

S.Yu.Ovchinnikov, A.N.Zinoviev, Yu.S.Gordeev

A.F.Ioffe Physical-Technical Institute of the Academy of Sciences of the USSR,Leningrad,USSR

The problem of direct ionization of K-shell in slow collisions of heavy atoms has been considered in [I].In this work the difficulties in the describtion of L or M-shell ionization are shown to be related to the model of the parabollic term used in [I] and they can be removed.

Actually, for the fixed collision trajectory the probability of the electron transition into continuum with small energy ω is given by Demkov-Osherov model [2]

$$P(\omega,b) = \frac{2}{v} Jm\, R(\omega) \times$$
$$\times \exp\{-\frac{2}{v} Jm \int^{\omega} R(\omega)[1+\frac{b^2}{|R(\omega)|^2}]\} \qquad (I)$$

where $R(\mathcal{E})$ is the inverse function for the internuclear distance dependence of the term $\mathcal{E}(R)$, $R(\omega)$ is the analytical continuation of $R(\mathcal{E})$ into the region $\mathcal{E}=\omega>0$, V is the collision velocity and b is the impact parameter.In the parabollic term model [I] $|Jm R(\omega)| \approx |Re R(\omega)|$ .For L or M-shell ionization this condition is not satisfied, therefore the model must be used where $Jm R(\omega)$ and $Re\, R(\omega)$ are two independent parameters.Let us assume:

$$\mathcal{E}(R) = \mathcal{E}_K\, \Phi[(\frac{R}{R_o})^{\frac{\pi R_o}{R_1}}] + \mathcal{E}_o,$$
$$R(\omega) = R_o[\Phi^{-1}(\frac{\omega-\mathcal{E}_o}{\mathcal{E}_K})]^{\frac{R_1}{\pi R_o}} \qquad (2)$$

where $\Phi(x)$ is an analytical function without singularities and $\Phi(0)=0$, $\Phi(1)=1$, $\Phi'(1)=1$. $\mathcal{E}_K$, $\mathcal{E}_o$, R_1 , R_o are parameters of the term.For sharply rising term $|Jm R(\omega)| \ll |Re R(\omega)|$ and $Re\, R(\omega) \approx R_o$, $Jm R(\omega) \approx R_1$. In our case $Jm R(\omega)$ is inversely proportional to the slope of the term at the point $R=R_o$. Rewriting the formula (I) gives

$$P(\omega,b) \approx \frac{2R_1}{V} \exp\{-\frac{2}{v}(\omega-\mathcal{E}_o)(1+\frac{b^2}{2R_o^2})\} \qquad (3)$$

The total ionization cross section is obtained by integrating (3) over all b 's and ω 's :

$$\sigma(v) \approx \pi R_o^2 \frac{V}{R_1|\mathcal{E}_o|} \exp(-\frac{2}{v}R_1|\mathcal{E}_o|)(1-\exp(-\frac{R_1|\mathcal{E}_o|}{V})) \qquad (4)$$

The expression (4) describes well experimental data (**fig.I**).

In [3] the calculations of the terms for H_2^+ system for a complex plane of R were made and the region of strong state coupling was found at internuclear distances

$$Re\, R_c \approx \frac{L(L+1)}{Z} , \; Jm R_c \approx \frac{2L}{Z} \qquad (5)$$

Using (5) we see that for the ionization of 2pσ - molecular orbital (L = I) $Jm R_c \approx Re\, R_c$ and this allows the model described in [I] to be applied for the K-shell ionization.

References

I. A.N.Zinoviev, S.Yu.Ovchinnikov,Yu.S.Gordeev Zh.Tech.Phys. Letters 7, I39 (I98I).

2. E.A.Solov'ev Zh.Exp.Theor.Phys.70,872 (I976).

3. E.A.Solov'ev Zh.Exp.Theor.Phys.8I, I68I (I98I).

4. P.H.Woerlee,R.J.Fortner,F.W.Saris J.Phys. B I4 (I98I),3173.

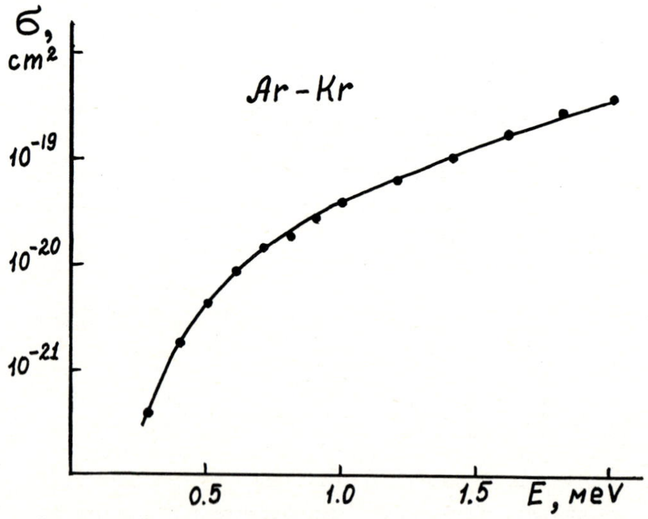

Fig.I. Cross sections for L-shell ionization of Kr in Ar - Kr collisions (line - our calculations, points - experimental data [4]).

COUPLED MULTICHANNEL CALCULATION FOR Ne^{+}+Ne COLLISIONS IN THE ENERGY RANGE OF 5-300 keV

A. Toepfer, B. Jacob, H.J. Lüdde and R.M. Dreizler

Institut für Theoretische Physik der Universität Frankfurt/Main, BRD

We investigate the total cross section for the vacancy transitions

$$(2p)_{\text{projectile}} \rightarrow (2s)_{\text{proj., target}}$$

$$(2p)_p \rightarrow (2p)_{p,t}, \quad (2p)_p \rightarrow (1s)_{p,t}$$

in the energy range of 5-300 keV. The basis of the investigation is the solution of an effective time-dependent Schrödinger equation in the impact parameter approximation

$$i(\partial_t + \dot{\underline{R}} \cdot \underline{\nabla}_R)\, \Psi_j\, (\underline{r},R(t),t) =$$

$$H_{\text{eff}}\, \Psi_j\, (\underline{r},R(t),t) \qquad (1)$$

by expanding the one-particle wavefunctions in molecular orbitals including electron translation factors (according to the prescription of Taulbjerg et al.[1]).

$$\Psi_j(\underline{r},R(t),t) = \qquad (2)$$

$$\sum_n c_n^j(t)\phi_n(\underline{r},R)\; e^{if\underline{v}\cdot\underline{r}}\; e^{-i\int^{t'} \varepsilon_n dt}$$

Many particle aspects are incorporated by the use of effective potentials with the aid of extended Thomas-Fermi methods and by working in a determinantal representation.[2]
In particular we investigate the dependence of the results on the size of the basis used. Being principly interested in the 2s vacancy production we enlarge an initial basis (A) consisting of four MO's, $2\sigma_g, 2\sigma_u, 3\sigma_g, 1\pi_u$, systematically to a set of 14 basis states, including the 1s-shell as well as the unoccupied 3s- and 3p-shells.

For lower impact energies one finds a marked dependence of the predicted cross section on the choice of the classical trajectory (standard straight line trajectory versus trajectory of a screened Bohr potential).

Results obtained with the initial configuration

$$(\underbrace{1s^2\ 2s^2\ 2p^6\ 3s^0}_{\text{target}},\ \underbrace{1s^2\ 2s^2\ 2p^5\ 3s^0}_{\text{projectile}})$$

are shown in the figure. The final configurations on which we projected in this case are

$$\Phi_{dir} \rightarrow (1s^2\ 2s^2\ 2p^6\ 3s^0,\ 1s^2\ 2s^1\ 2p^6\ 3s^0)$$

$$\Phi_{tr} \rightarrow (1s^2\ 2s^1\ 2p^6\ 3s^0,\ 1s^2\ 2s^2\ 2p^6\ 3s^0)$$

for direct excitation and electron exchange respectively.

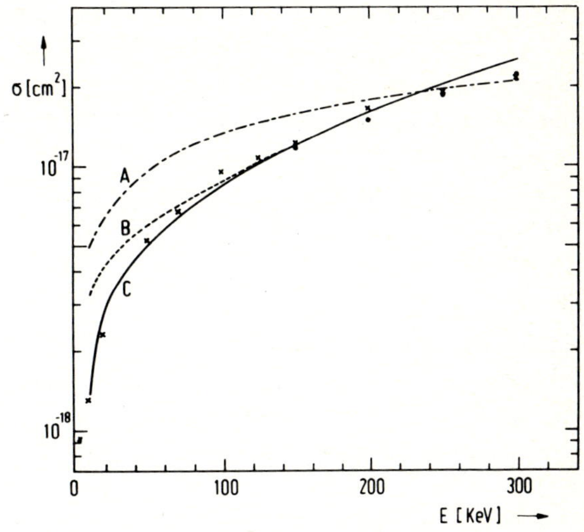

Curve A shows the total cross section for L-shell excitation within the basis A, curve B is calculated for the straight line trajectory, in a basis consisting of 10 MO's,

$$1\sigma_g, 1\sigma_u, 2\sigma_g, 2\sigma_u, 3\sigma_g, 3\sigma_u, 4\sigma_g, 4\sigma_u, 1\pi_g, 1\pi_u,$$

and is compared with the screened trajectory calculation (curve C). The experimental points are taken from Andersen et al.[3] In the higher energy range we also show the data of Hippler et al.[4]

References

1. K. Taulbjerg, J. Vaaben, B. Fastrup, Phys. Rev. A 12 (1975) 2325
2. A. Toepfer, B. Jacob, H.J. Lüdde, R.M. Dreizler, Phys. Lett. 93A (1982) 18
3. T. Andersen, E. Bøving, P. Hedegard, J. Østgaard, J. Phys. B 11 (1978) 1449, and private communication
4. R. Hippler, K.-H. Schartner, J. Phys. B8 (1975) 2528
   H.F. Beyer, R. Hippler, K.-H. Schartner, Z. Phys. A 289 (1979) 239, and private communication.

CALCULATIONS OF COUPLING MATRIX ELEMENTS FOR DI-ATOMIC SYSTEMS IN AN ATOMIC BASIS

W.-D. Sepp

Department of Physics, University of Kassel, D-35 Kassel, W.-Germany

Good rotational and radial coupling matrix elements are essential to calculate via a coupled channel calculation the time evolution for a heavy ion-atom colliding system. The matrix elements are calculated within a numerical Dirac-Fock-Slater molecular program which uses numerical DFS-atomic wavefunctions as the basis.

The dynamic coupling matrix element between two adiabatic molecular orbitals $|\Psi_i\rangle$ and $|\Psi_j\rangle$ is given by

$$\langle\Psi_i|\frac{\partial}{\partial t}|\Psi_j\rangle = \dot{R}\langle\Psi_i|\frac{\partial}{\partial R}|\Psi_j\rangle + \frac{i}{h}\dot{\varphi}\langle\Psi_i|j_y^{CM}|\Psi_j\rangle \tag{1}$$

$$= M_{ij}^{Rad} + M_{ij}^{Rot}$$

where $\dot{R}$ is the radial velocity between the two nuclei, $\dot{\varphi}$ the angular velocity of the internuclear axis, and j_y^{CM} the y-component of the electronic angular momentum operator in the CM system.

The MO-wavefunctions $|\Psi_i\rangle$ are expanded in atomic wavefunctions $|\varphi_\mu\rangle$, centered at the nuclei

$$|\Psi_j\rangle = \sum_\mu c_{j\mu}|\varphi_\mu\rangle \tag{2}$$

The expansion coefficients $c_{j\mu}$ depend only on the internuclear distance R, whereas the atomic states $|\varphi_\mu\rangle$ depend on the coordinates of the nuclei as well as on the orientation of the internuclear axis which is used as quantization axis for the atomic states. Inserting (2) in (1), the radial coupling matrix may be written as

$$M_{ij}^{Rad} = M_{ij}^{MO} + M_{ij}^{kin} \tag{3}$$

where the MO-part of the coupling is given by

$$M_{ij}^{MO} = \dot{R}\sum_{\mu\nu} c_{i\nu}^{*} \left(\frac{\partial}{\partial R} c_{j\mu}\right) \cdot S_{\nu\mu}^{AO} \tag{4}$$

with the atomic overlap matrix

$$S_{\nu\mu}^{AO} = \langle\varphi_\nu|\varphi_\mu\rangle \tag{5}$$

The kinematic part can be written as

$$M_{ij}^{kin} = \dot{R}\sum_{\mu\nu} c_{i\nu}^{*} c_{j\mu} \cdot \frac{M_{I(\mu)}}{M_A+M_B} \cdot \frac{i}{\hbar} \langle\varphi_\nu|p_z|\varphi_\mu\rangle \tag{6}$$

where $M_{I(\mu)}$ is M_A or $-M_B$ if the atomic state $|\Psi_\mu\rangle$ is centered at the nucleus A or B, respectively. The resulting p_z-matric element can be calculated without any numerical differentiation using the DF-equation for the atomic basis states. Since the MO-coefficients $c_{j\mu}$ are independent of the orientation of the colliding system, the rotational coupling matric element is given by

$$M_{ij}^{Rot} = \frac{i}{\hbar}\dot{\varphi}\sum_{\mu\nu} c_{i\nu}^{*} c_{j\mu} \langle\varphi_\nu|j_y^{CM}|\varphi_\mu\rangle.$$

Shifting the origin of the angular momentum operator from the centre of mass to the centre of the atom $I(\mu)$

$$j_y^{CM} = j_y^{I(\mu)} + [p \times X^{I(\mu)}]_y \tag{7}$$

we get two contributions to the rotational coupling matrix element

$$M_{ij}^{Rot} = M_{ij}^{Orient.} + M_{ij}^{Displ.} \tag{8}$$

where the orientation part is given by

$$M_{ij}^{Orient.} = \frac{i}{\hbar}\dot{\varphi}\sum_{\mu\nu} c_{i\nu}^{*} c_{j\mu} \langle\varphi_\nu|j_y^{I(\mu)}|\varphi_\mu\rangle \tag{9}$$

and the displacement part by

$$M_{ij}^{Displ.} = -\frac{i}{\hbar}\dot{\varphi}\sum_{\mu\nu} c_{i\nu}^{*} c_{j\mu} X_z^{I(\mu)} \cdot \langle\varphi_\nu|p_x|\varphi_\mu\rangle. \tag{10}$$

Inserting a complete set of states $|\varphi_\lambda^{I(\mu)}\rangle$ of the atom $I(\mu)$ in the orientation matrix element

$$\langle\varphi_\nu|j_y^{I(\mu)}|\varphi_\mu\rangle = \sum_\lambda \langle\varphi_\nu|\varphi_\lambda^{I(\mu)}\rangle\langle\varphi_\lambda^{I(\mu)}|j_y^{I(\mu)}|\varphi_\mu\rangle \tag{11}$$

the calculation of (9) is reduced to a normal one-center angular momentum matrix element and a two-center overlap matrix element $\langle\varphi_\nu|\varphi_\lambda\rangle$. The advantage of the method is that all calculations except (4) can be done at a fixed internuclear distance R avoiding numerical differentiation. Up to now no translational factors are included in the coupling matrix elements. These are expected to modify mainly the kinematic part of the radial coupling.

As examples we present results of a low Z system like Ne-Ne where the matrix elements are known, and a high Z system like Pb-Cm to study the various unusual couplings which probably do occur between the lowest MO-levels at small internuclear distances.

SEMICLASSICAL APPROXIMATION OF THE TIME-DEPENDENT DIRAC EQUATION WITH FIRST ORDER PERTURBATION THEORY AND FINITE DIFFERENCE METHOD*

S.R. Valluri**, U. Becker, N. Grün and W. Scheid
Institut für Theoretische Physik der Justus-Liebig-Universität Giessen, West Germany

We have studied the impact parameter dependence of K-shell excitation and ionization of heavy atoms by relativistic ions by an extension of the semiclassical approximation (SCA)[1]. We derived analytical expressions for the probabilities of these processes in the framework of the first order time-dependent perturbation theory. In the independent-electron approximation this theory leads to the following probability of ionization of a $1s_{1/2}$-electron with quantum numbers $n_i=1$, $\kappa_i=-1$, $j_i=1/2$ which is induced by a projectile with charge Z_1e moving on a classical straight-line trajectory $R_b(t)$ with velocity v_1 and impact parameter b(see Ref.1):

$$\frac{dI_b}{dE_f} = \left(\frac{Z_1e^2}{\hbar v_1}\right)^2 \frac{1}{2j_i+1} \sum_{\kappa_f m_f m_i} |M_{fi}|^2 \qquad (1)$$

with the abbreviations

$$M_{fi}=i\Gamma \int_{-\infty}^{+\infty} dT e^{iqT} \langle E_f,\kappa_f,m_f|\frac{1-\beta\alpha_3}{r'}|n_i,\kappa_i,m_i\rangle \qquad (2)$$

$$r' = [(x-b)^2+y^2+\Gamma^2(z-T)^2]^{1/2} \qquad (3)$$

$$q = (E_f-E_i)/\hbar v_1 \qquad (4)$$

$$\beta = v_1/c, \; \Gamma = (1-\beta^2)^{-1/2} \qquad (5)$$

$$\alpha_3 = \begin{pmatrix} 0 & \sigma_3 \\ \sigma_3 & 0 \end{pmatrix} \qquad (6)$$

Here, E_f is the energy of the ejected electron and α_3 the Dirac matrix. For the evaluation of the transition matrix elements (2) with the bound state i and the continuum state f it is advantageous to introduce a multipole expansion of the interaction[1]:

$$\Gamma \int_{-\infty}^{+\infty} dT\, e^{iqT} \frac{1}{r'} =$$

$$8 \sum_{\ell m} i^{\ell+m} \int_q^{\infty} \frac{s^2 ds}{s^2-(\beta q)^2} j_\ell(sr) Y_{\ell m}(\hat{r}) A_{\ell m}(b,q;s) \qquad (7)$$

$$A_{\ell m}(b,q;s) = \frac{\pi}{s}\Theta(s-q) Y_{\ell m}(\cos^{-1}\frac{q}{s},0) J_m(b\sqrt{s^2-q^2}) \qquad (8)$$

Here, Θ is the step function and J_m are Bessel functions. The matrix elements

$$\langle E_f,\kappa_f,m_f|(1-\beta\alpha_3) j_\ell(sr) Y_{\ell m}(\hat{r})|n_i,\kappa_i,m_i\rangle \qquad (9)$$

can be analytically calculated for relativistic Coulomb wave functions. Finally the integral over s can be performed by using the series expansion of $Y_{\ell m}$ and hypergeometric functions.

Excitation probabilities for ionization are calculated for a $Ca^{20+}-U^{91+}$ collision at a velocity of the Ca projectile of 0.8c and 0.2c. The results will be compared with those obtained by a numerical solution of the time-dependent Dirac equation[2] (finite difference method) for an impact parameter b=0. Fig.1 shows the time evolution of the density of the electron during the collision. In Ref.2 we find an ionization probability of 0.063 for the collision with β=0.8 after projecting with continuum wave functions on the numerical ones.

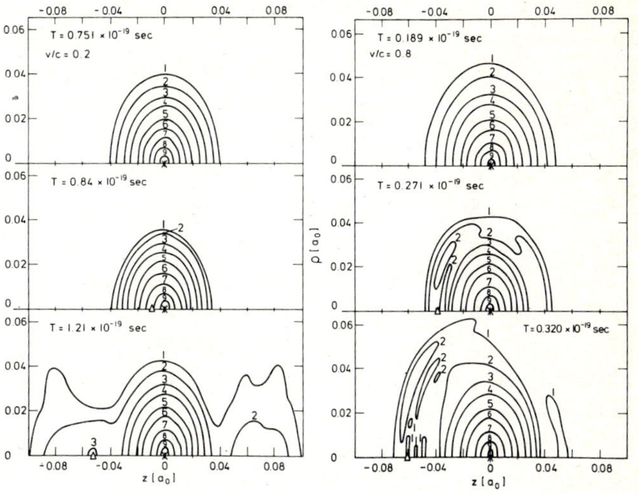

Fig.1 Contour maps of the probability density of the $1s_{1/2}$-electron. The position of the nuclei are denoted by crosses (U) and triangles (Ca). The factor between two contour lines is $\sqrt{10}$.

* Work supported by GSI
** Alexander von Humboldt fellow

References

1. P.A. Amundsen, K. Aashamar, J.Phys.B 14 (1981) 4047
2. U. Becker, N. Grün and W. Scheid, Solution of the time-dependent Dirac equation by the finite difference method and application for $Ca^{20+}+U^{91+}$, Preprint Giessen (1983), to be published in J.Phys.B

TIME DEPENDENT SCREENED POTENTIALS FOR ATOMIC SCATTERING PROBLEMS

A. Henne, R.M. Dreizler

Institut für Theoretische Physik der Universität Frankfurt/Main, BRD

The standard treatment of atomic collision systems in the impact parameter approximation relies heavily on the reduction of many body aspects to an effective one particle problem. As a first step for an acceptable inclusion of screening effects we suggest the calculation of time dependent electron - densities from which effective single particle potentials can be constructed.

As a vehicle for the construction of such densities (and the corresponding velocity fields) we rely on the time dependent Thomas-Fermi model in a two component version. A derivation of the model on the basis of variation of the Lagrange functional in the time dependent Thomas-Fermi approximation with respect to the densities and velocity potentials will be outlined. The set of differential equations that emerges consists of a set of continuity equations and a set of hydrodynamic Eulerian-equations of motion for the two electron fluids.

We attempted the solution of this set of differential equations for atom-atom and ion-atom collisions with the aid of difference procedures and the assumption of rotational symmetry with respect to the internuclear axis.

The time development of the densities and velocity fields indicates a strong dependence on both the impact energy and impact parameter.

BOUND ELECTRON ENERGY OF SUPERHEAVY ATOM

Jiben Sidhanta* and S. C. Mukherjee**

* Physics Department, Charuchandra College, Calcutta - 29, India

** Indian Association for the Cultivation of Science, Calcutta-32, India.

The present work employs a currently - growing group-theoretic technique of the field theory to see if, theoretically superheavy atom (SHA) can exist - a question of growing importance in heavy ion collisions. This paper has two distinct objectives. First, we indicate some serious theoretical difficulties of the charged vacuum conjecture recently developed by Reinhardt, Greiner et al[1] (hereafter called RG theory) for heavy ion collisions. Second and more important, we point out that using a noncompact group theory viz. O(4, 2) in the manner of Barut et al.[2] even K-level energy of an atom with atomic number $Z > 137$ (fine structure constant α being $\frac{1}{137}$) turns out to be quite acceptable.

We have noticed the following major difficulties in RG theory. (1) It overlooks the fact that Dirac eq. (which is useful as a single charge field eq.) is not a valid relativistic single particle eq. because no relativistic single particle eq. can be constructed within the Poincare group. Yet RG theory resorts to Dirac eq. for making the K-level of SHA dive into the Dirac sea and hence building the charged vacuum hypothesis. (2) RG theory at places uses conventional field theory but subverts a cardinal notion of the same field theory viz. the vacuum notion. Indeed both Lagrangian and axiomatic versions of the field theory require a unique nondegenerate vacuum $|0\rangle$ which must provide a one dimensional representation of all symmetry groups. RG theory's charged vacuum being nonunique contradicts the cluster decomposition theorem of Wightman. (3) RG theory does not establish the maximality of its overcritical operator and yet projects it on to the undercritical basis, making identity resolution questionable for that reason. (4) Case[3] had long proved that (even when one conceded to a use of the Dirac eq.) K-level diving was not inevitable.

Regarding our second objective we point out that proceeding from the minimal conserved current of O(4,2) group i.e.

$$j^{\mu} = a_1\Gamma_{\mu} + a_2P_{\mu} + a_3P_{\mu}L_{46} \quad \ldots (1),$$

the notation whereof may be seen in reference 2, one finally arrives at the following result for the bound electron energy :

$$E_n = -\frac{Z^2\alpha^2 m}{2n^2} - \frac{3Z^4\alpha^4 m}{8n^4} \quad \ldots (2)$$

where m = electron mass, Z = atomic number α = fine structure cst, n = Bohr quantum number. Eq. (2) shows that whatever be Z, E_n is well-behaved and hence there is no a priori reason against the existence of SHA. Just as undetected quarks explain a lot of particle physics, so also this SHA (whose observability even as an ultratransient resonance has not been considered here) may give insight concerning UNILAC heavy ion collisions (and, hopefully, even the mysterious primary spectra of cosmic rays).

References

1. J. Reinhardt and W. Greiner, Rep. Prog. Phys. 40, 221 (1977).

2. A. O. Barut, D. Corrigan and H. Kleinert, Phys. Rev. 167, 1527 (1968).

3. Case, Phys. Rev. 80, 797 (1950).

COHERENCE STUDY OF He(2^1P) AND Li(3^2D) EXCITATION IN Li^+ - He COLLISIONS

H.-P. Neitzke, N. Andersen, and T. Andersen

Institute of Physics, University of Aarhus, DK-8000 Aarhus C, Denmark

The Li^+ - He collision system, representing the interaction between two atoms both having closed K shells with no outer electrons, has been the subject of extensive theoretical and experimental studies[1-3] before since the 2pσ-2pπ rotational-coupling mechanism is of central interest in both inner- and outer-shell excitation studies.

We have focussed the attention on the orbital shapes and orientation by the excitation mechanisms vs impact parameter and velocity for the following reaction channels:

$$Li^+ + He \rightarrow \begin{cases} Li(2^2P) + He^+ & \text{(i)} \\ Li^+ + He(2^1P) & \text{(ii)} \\ Li(3^2D) + He^+ & \text{(iii)} \end{cases}$$

Results obtained for process (i) with the polarized photon-scattered particle-coincidence technique have already been reported[4]. For the investigation of process (ii), a new equipment for the measurement of the photon -scattered ion angular correlation was constructed. This investigation aims at the clarification of the role, which secondary transitions at large internuclear distances play in the process of sharing the excitation between Li(2^2P) and He(2^1P). Now, five months before the conference, the first data points have been obtained with promise of detailed data in the near future.

Process (iii) has been studied over a large range of impact energies and impact parameters. The full set of Stokes parameters P_1, P_2, P_3, P_4 has been determined in the energy range 2.5-15 keV for selected scattering angles 0.3-4 degrees. A polarization analysis of the D → P transition by the photon-particle coincidence method is not sufficient to completely determine the excitation amplitudes for the D state since two solutions are obtained[5]. The electron distributions associated with these two are, however, closely related since they are mirror images of one another in the plane by the normal to the scattering plane and the direction of maximum intensity of linearly polarized light, as measured from the direction of the normal. The ambiguity can be resolved by performing the experiments in the presence of a symmetry-breaking electromagnetic field[6].

The figure illustrates the variation of the D-electron charge distribution at a constant impact parameter (0.55 a.u.) as a function of velocity. Note: The experiment cannot distinguish between these shapes and their mirror images with respect to the dashed line. Figure (a) (2.5 keV): M_L = 2 and 0 dominate; fig. (b) (7.5 keV): Large M_L = 1 contribution; fig. (c) (15 keV): Nearly pure M_L = 0 contribution.

REFERENCES

(1) V. Sidis, N. Stolterfoht, and M. Barat, J.Phys.B: Atom.Molec.Phys. 10 (1977) 2815

(2) Q.C. Kessel, R. Morgenstern, B. Müller, and A. Niehaus, Phys.Rev.A 20 (1979) 804

(3) J.A. Fayeton, J.C. Houver, J.C. Brenot, and M. Barat, J.Phys.B:Atom.Molec.Phys. 14 (1981) 2599

(4) N. Andersen, T. Andersen, and E. Horsdal-Pedersen, 8th ICPEAC Conf. (1982) Book of Abstracts, B68

(5) N. Andersen, T. Andersen, J.S. Dahler, S.E. Nielsen, G. Nienhuis. and K. Refsgaard, J.Phys.C:Atom.Molec. Phys (1983) (in press)

(6) H.-P. Neitzke and T. Andersen, this conference

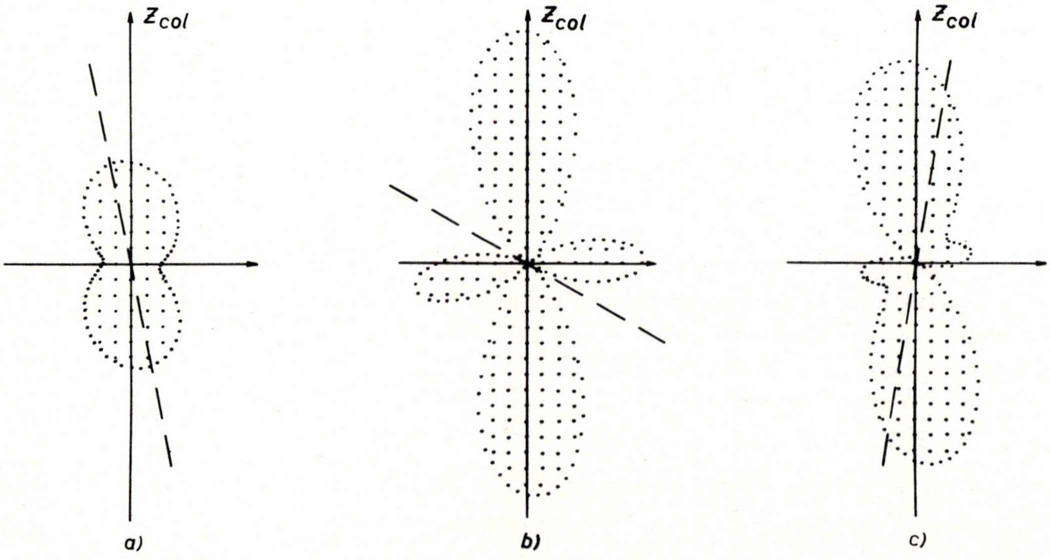

COHERENT EXCITATION OF L > 1 STATES: MULTIPOLE CONVERSION IN EXTERNAL FIELDS

H.-P. Neitzke and T. Andersen

Institute of Physics, University of Aarhus, DK-8000 Aarhus C, Denmark

For an atom in an excited state with angular momentum L, the information about the contribution of all multipoles to the order k = 2L is necessary for a complete description of the state. By measuring the dipole characteristics of the radiation emitted after the excitation, it is possible to determine the state multipoles with k = 0 (total intensity), k = 1 (orientation), and k = 2 (alignment). With no additional effort, an unambiguous determination of the excited-state wave function is thus possible only for P states.

These facts explain why it is not possible by the photon-particle coincidence method alone to determine the relative-scattering amplitudes for a collisional-induced coherent S→D excitation from a polarization analysis of the subsequent D→P transition. Andersen et al.[1] have shown that two distinct solutions are obtained. The electron distributions associated with these two are, however, closely related since they are mirror images of one another in the plane determined by the normal to the scattering plane and the direction of maximum intensity of linearly polarized light, as measured from the direction of the normal.

The necessary additional constraints contained in the distributions from the higher multipoles (k > 2) can be obtained by the investigation of the emission of photons with higher multipolarity, or through the 'mixing' of tensors of rank k > 2 with tensors of rank k ≤ 2. The conversion of higher-to-lower multipole contributions can be studied by the observation of the emitted radiation in an appropriate combination of external fields or by combining the effects of internal couplings with those caused by external fields[2].

We have investigated the time evolution of the state multipoles for the Li(3d) state under the combined action of internal coupling and an external magnetic field in the case, where the fine structure was not resolved. Figure 1 shows the different behaviour of the two solutions (+,-) for the D wave function. Measurement of the linear-polarization vector P_1 in the region of the sublevel crossings with $\Delta M_L = 2$ gives information about the octupole contribution. Hereby the missing constraint for an unambiguous fixation of the charge distribution can be obtained.

An experimental verification of the Li(3d) state in the Li+He collisions is in progress.

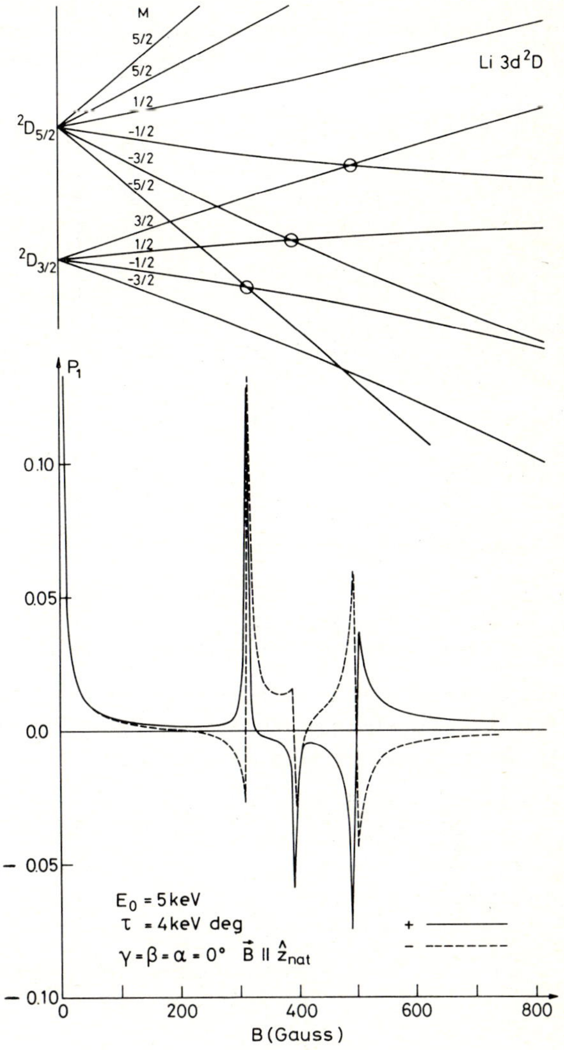

Figure 1

REFERENCES

(1) N. Andersen, T. Andersen, J.S. Dahler, S.E. Nielsen, G. Nienhuis, and K. Refsgaard, J.Phys.B:Atom.Molec. Phys. (1983) (in press)

(2) U. Fano and J. Macek, Rev.Mod.Phys. 45 (1973) 553

ALIGNMENT AND ORIENTATION STUDIES FOR THE ION-IMPACT EXCITATION OF THE 3P→3D TRANSITION IN Na^+-Na COLLISIONS

A. Bähring, E. Meyer, and I.V. Hertel

Freie Universität Berlin, Fachbereich Physik, Institut für Molekülphysik, Arnimallee 14, D-1000 Berlin 33, FRG

In a crossed beam experiment alignment and orientation effects of Na^+ ions scattered from laser optically pumped $Na(3^2P)$ atoms are studied for the inelastic 3P→3D transition shown in Fig. 1.

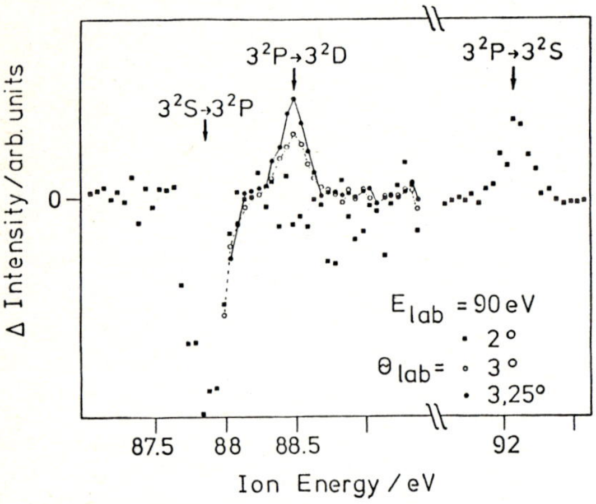

Fig. 1: Difference of energy loss spectra with and without laser excitation for 3 scattering angles.

From the sharply peaked angular distribution of the scattering cross section shown in Fig. 2 we conclude that the 3P→3D transition is mainly induced by rotational coupling. By varying the linear polarization of the exciting light, we prepare asyptotically $|p\pi^{\pm}\rangle$ and $|p\sigma\rangle$ atomic states, which develop into the corresponding molecular states.

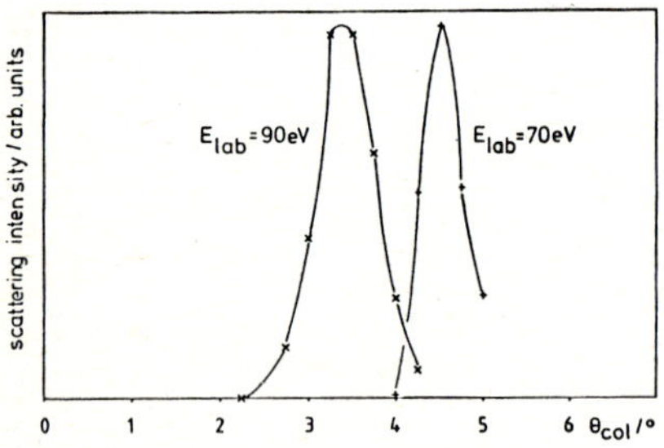

Fig. 2: Angular dependence of the scattering signal for 2 primary kinetic ion energies.

We observe the highest possible anisotropy of the scattering signal, displaying its maximum, when the electric vector of the laser light is approximately parallel to the incoming ion beam. We thus conclude that the process proceeds via the $2\Sigma_u$ potential correlating to the $|3p\sigma\rangle$ atomic state in contrast to the $3^2P \to 3^2S$ transition, previously studied[1], which proceeds via a crossing of the excited $1\Pi_u$ potential with the $1\Sigma_u$ ground state. Since rotational coupling is responsible for the $3^2P \to 3^2D$ excitation as discussed above we further conclude that a crossing of the $2\Sigma_u$ potential with

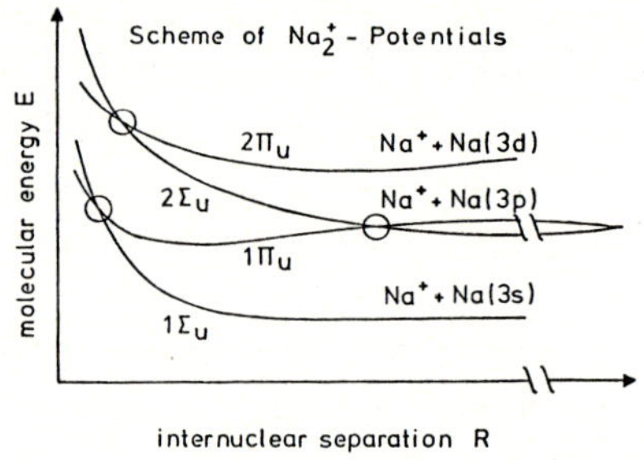

Fig. 3: Potential curves for Na_2^+ (schematic)

the $2\Pi_u$ is accessible as schematically illustrated in Fig. 3.

Upon closer inspection we find that the maximum scattering intensity is found for a polarization angle $\gamma \sim -15°$ differing slightly from exact $|p\sigma\rangle$ preparation, as illustrated in Fig. 4. Assuming that the atomic orbital is locked to the molecular body fixed frame at an internuclear distance R_L one may deduce that an effectively repulsive force is acting on the particles, as seen in Fig. 4.

Additional information is obtained from studies with left (LHC) and right (RHC) hand circulary polarized light, giving a circular asymmetry of the scattering signal $[I(LHC) - I(RHC)]/[I(LHC) + I(RHC)] \sim -0.3$ to -0.4.

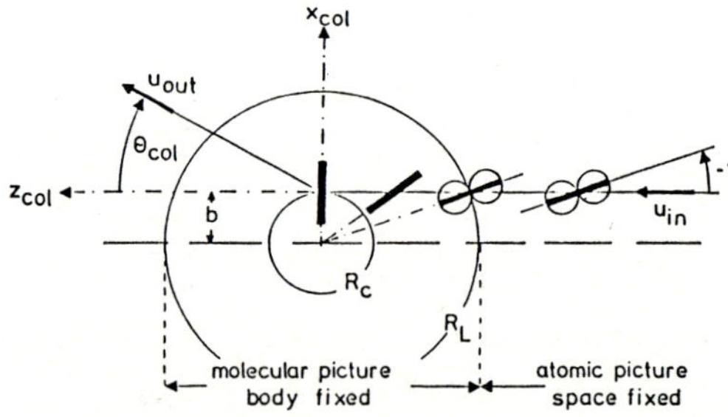

Fig. 4: Transition from the atomic to the molecular frame

Detailed considerations of the phase-evolution on the long-range parts of the Π and Σ-potentials allow to conclude that in the present case this orientation is negative or positive depending on whether the forces being effectively repulsive or attractive. Thus the observed sign of the orientation confirms the repulsive nature of the potentials, as deduced above from γ. A complete density matrix[2] for the transition P→D in the energy range E_{LAB} = 90 - 40 eV will be reported at the conference.

References

1. A. Bähring, I.V. Hertel, E. Meyer, H. Schmidt, submitted to Zeitschrift für Physik
2. H. Hermann, I.V. Hertel, Comments At. Mol. Phys. 12, 61-84 (1982)

A STUDY OF THE EXCITATION MECHANISM OF THE $(2p^2)^1D$ STATE EXCITED IN He^+ + He COLLISIONS

E. Boskamp, R. Morgenstern, A. Niehaus and P. van der Straten

Fysisch Laboratorium, Rijksuniversiteit Utrecht, the Netherlands

We have studied the process

$$He^+ + He \rightarrow He^+ + He^{**} \rightarrow He^+ + He^+ + e^-$$

at various collision energies between 1.2 and 4 keV, and at laboratory scattering angles between $\theta_p = 6^o$ and 10°. In this energy range the $(2p^2)^1D$ state is excited for more than 95%. The emitted electrons were measured in coincidence with the scattered ions. In this way electron angular distributions for well defined excitation events are obtained.

Some measured angular distributions together with the result of the fitcalculations are shown in figs. 1 and 2, whereas fig. 3 shows a calculated 3 dimensional view of the total distribution at 2 keV, $\theta_p = 6^o$ as obtained from the fitparameters.

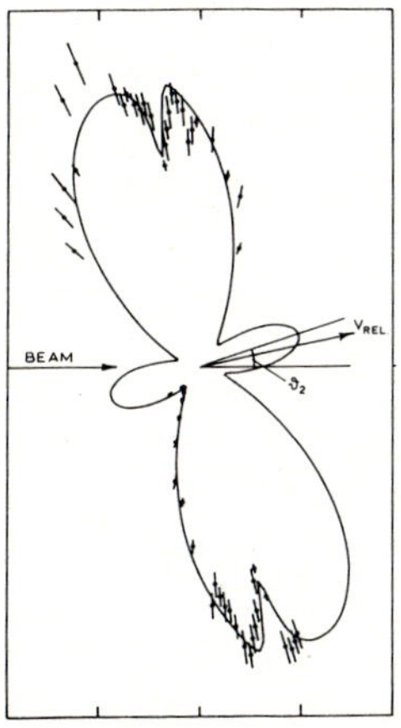

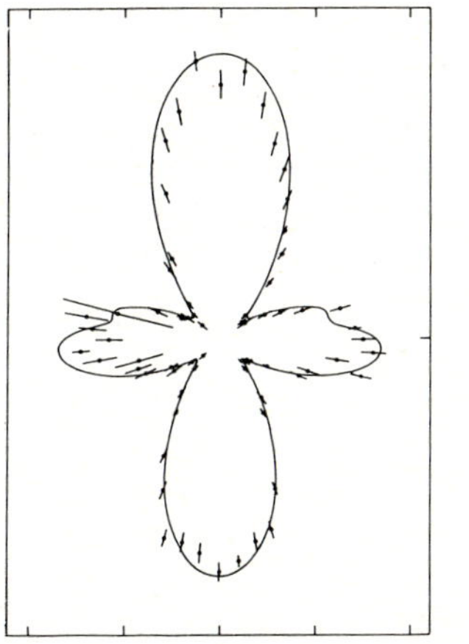

Fig. 1: Angular distribution in the scattering plane, 2 keV, $\theta_p = 6^o$. The angle between beam direction and the axis where d_2 is in its maximum is ϑ_2.
Fig. 2: Angular distribution perpendicular to the beam direction, 2 keV, $\theta_p = 8^o$

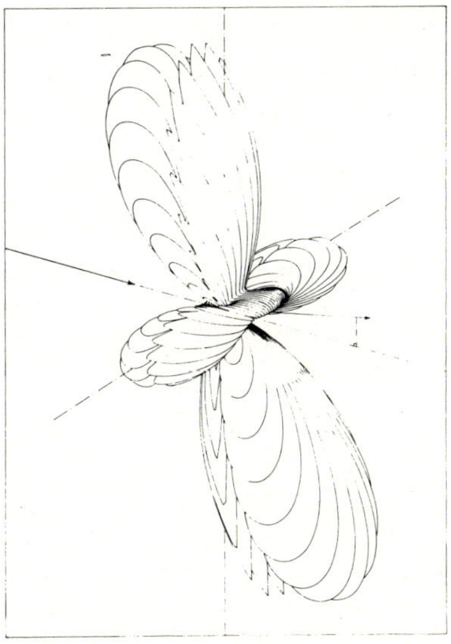

Fig. 3: Calculated distribution at 2 keV, $\theta_p = 6^o$.

In fig. 3 the sharp structure near angles perpendicular to the relative velocity of the collision partners is due to interference of target and projectile contributions and can be understood quantitatively[1,2]. These interferences are taken into account in the fitcalculations. The fitting results are shown in fig. 4 where the moduli of the population amplitudes d and their phases ρ are plotted versus v/b (v: the relative velocity, b: the impact parameter). The quantisation axis is an axis in the scattering plane at which d_2 is in its maximum.

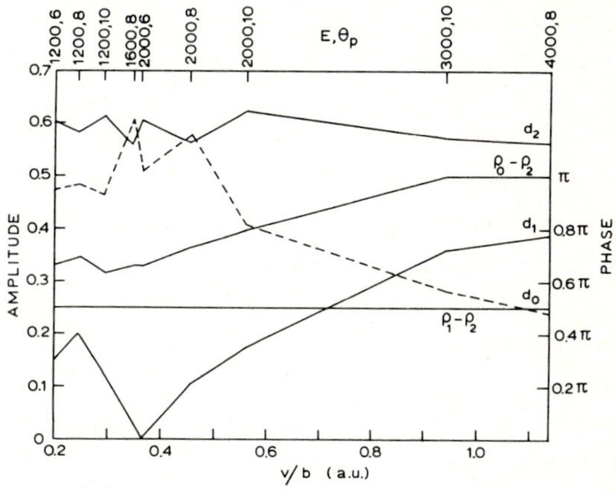

Fig. 4: Fitting results

From these results we can obtain information about the collision dynamics. If we compare the axis where d_0 or d_2 is in its maximum with 3 axes defined by the kinematics, it appears that the axis where d_2 is in its maximum nearly always coincides with the asymptotic molecular axis (except for 4 keV). We can explain this in the following way: during the collision 2 electrons are promoted via the $2p\sigma_u$ molecular orbital. Since, at small distances, the rotation of the internuclear line is fast, the excited electron "cloud" can not follow and remains fixed in space. The excited atom is in an M = 0 state with respect to the distance of closest approach. Later, however, when the rotational energy has decreased the "cloud" is again coupled to the internuclear axis. With respect to this axis the M = 2 population is in its maximum and the electron configuration stays constant with respect to the slowly rotating internuclear axis.

References

1. E. Boskamp, R. Morgenstern, A. Niehaus, J. Phys. B: At. Mol. Phys. 15, 4577 (1982)
2. A. Niehaus, "Trends in Physics 1981", Inv. Papers 5th Gen. Conf. Europ. Phys. Soc., Inst. Phys. (I.A. Dorobantu, ed.), Bucharest (1982) p.556.

ALIGNMENT OF THE 2P STATES OF HELIUM-LIKE AND HYDROGEN-LIKE IONS

D. A. Church, R. L. Watson, R. A. Kenefick, D.-W. Wang+, and G. Pedrazzini

Physics Department and Cyclotron Institute, Texas A&M University, College Station, Texas 77843, U.S.A.

Recent high-resolution angular distribution measurements of K x-rays emitted in the decay of the 2^1P and 2^3P states of 2 MeV/amu foil-excited helium-like sulfur[1] indicated a high degree of alignment. This alignment is thought to result primarily from capture of K shell electrons from carbon target atoms. The cross section ratio[2] $\sigma(m=0)/\sigma(m=\pm1) = 1.8(0.3)$ corresponded to a polarization fraction for the 2^1P state $P(^1P) = 0.28(0.05)$ and an alignment tensor component[3] $A_0^{col} = 4P/(3-P)h^{(2)}(1,0) = -0.2(0.04)$, where the ratio of 6j symbols $h^{(2)}(1,0) = -2$.

Similar measurements on P states of H-like and He-like Mg ions are now in progress. Any Z or velocity dependence of the alignment may provide a significant test of the charge capture theory, due to the near velocity-matching of the projectile with the carbon K-shell electron[4]. Further, a search is being conducted for coherence effects produced in the collision.

K x-rays are detected at angles $30° \leq \theta \leq 161°$ relative to the Mg beam direction, and energy analyzed using a rotatable plane-crystal Bragg spectrometer. The data are normalized to the total Mg K x-ray production from an upstream foil, as detected with a flowing gas proportional counter of the type used with the Bragg spectrometer. Due to the Doppler effect on the projectile x-rays, an angular distribution measurement samples the reflection and absorption properties of the Bragg crystal over a range of Bragg angles determined by the ion velocity and the observation angle. Calibration through use of the measured x-ray intensities is possible[1], but other means are being tested. For the current measurements, an ADP crystal is used. The 2 MeV/amu spectra show comparable intensities of H-like and He-like Mg transitions, with better resolution than was obtained for the sulfur spectra using a NaCl crystal. Initial results for the intensity ratio of the helium-like Mg transitions clearly show polarization of the 1^1S-2^1P and 1^3S-2^3P transitions, with an apparent magnitude of $P(^1P)$ slightly smaller than that observed for the helium-like sulfur transition. However, these initial data have not yet been corrected for the crystal reflectivity functional dependence. The reflectivity calibration is essential to determine the H-like transition polarization.

An initial search for coherent alignment using foil inclination angles up to 60 degrees indicates that the magnitudes of the alignment parameters A_{1+}^{col} and A_{2+}^{col} are not large.

This research is supported by the National Science Foundation and the Texas A&M Center for Energy Resources (DAC) and the Department of Energy, Office of Basic Energy Sciences and the Robert A. Welch Foundation (RLW).

+now at Peking Normal University, Peking, People's Republic of China.

References

1. D.-W. Wang, R. L. Watson, R. A. Kenefick, and D. A. Church, Nuc. Instrum. Meth 202, 355 (1982).
2. D. A. Church, R. A. Kenefick, D.-W. Wang, and R. L. Watson, Phys. Rev. A26, 3093 (1982).
3. J. Macek and D. Burns in "Beam-Foil Spectroscopy", S. Bashkin, Ed. (Springer, Berlin (1976)), p. 237.
4. J. Burgdörfer, (private communication).

ANISOTROPIC L X-RAY EMISSION IN ION-ATOM COLLISIONS

W. Jitschin, R. Hippler, R. Schuch*, H. O. Lutz

Fakultät für Physik, Universität Bielefeld, F. R. Germany
*Physikalisches Institut, Universität Heidelberg, F. R. Germany

For proton impact on heavy atoms a considerable anisotropy (ca. - 15%) of the $L\ell$ x-ray line has been observed at small impact energies (ca. 300 keV). The anisotropy is caused by the anisotropy (alignment) of the collisionally excited vacancy state. Good agreement between the numerous experimental data and advanced calculations of the direct ionization is obtained[1].

For Si and S impact, as for protons, the dominant ionization mechanism is direct ionization; other processes as, e.g., electron capture by the projectile, are only of minor importance. Thus, considerable anisotropy may still be expected. In contrast, the experimental data show only small anisotropy (Fig. 1).

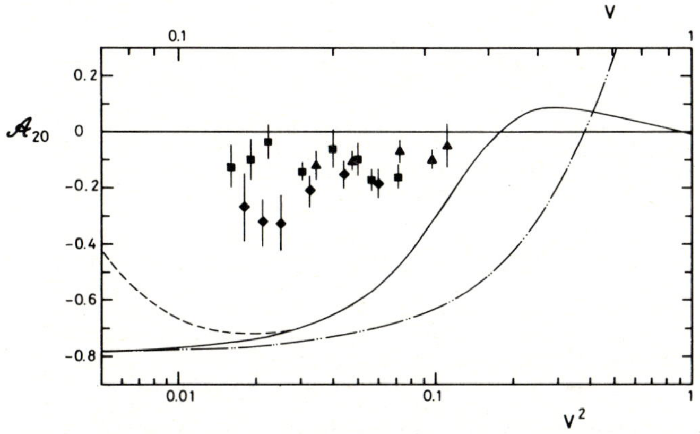

Fig. 1: L_3-subshell alignment $\mathcal{A}_{20}$ for Si impact on Dy (▲) and Au (■) and S impact on Au (◆). Theoretical curves: —— direct ionization, -··- electron capture.

For Br and I impact ionization of Au the primary ionization process is vacancy transfer from higher shells to the inner shells via the $3d\sigma$ molecular orbital[2]. Since σ-states correspond to m=0 atomic states and radial coupling at the outgoing part of the trajectory does not change the magnetic quantum number one expects the creation of only m=0 vacancies, i.e. again large anisotropy effects. This is not found experimentally (Fig. 2,3).

We suppose that the discrepancy is mainly due to multiple ionization of the target atom. Calculations of the electrostatic interaction energies for atoms with one L vacancy and an additional M or N vacancy show that the corresponding multiplet splitting is partly larger than the Au L_3 level width of 6 eV. Thus the additional vacancy does not merely act as a spectator but effectively couples to the L_3 vacancy. This coupling causes considerable dealignment of the unresolved multiplet x-ray lines.

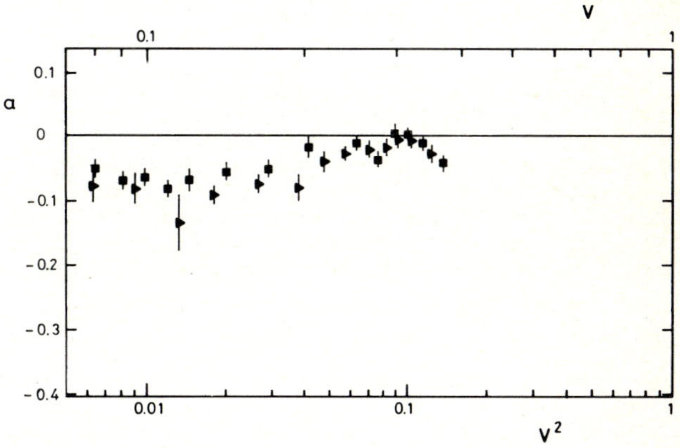

Fig. 2: Anisotropy of the $L\ell$ line for Br impact on Au (■) and Pb (▶).

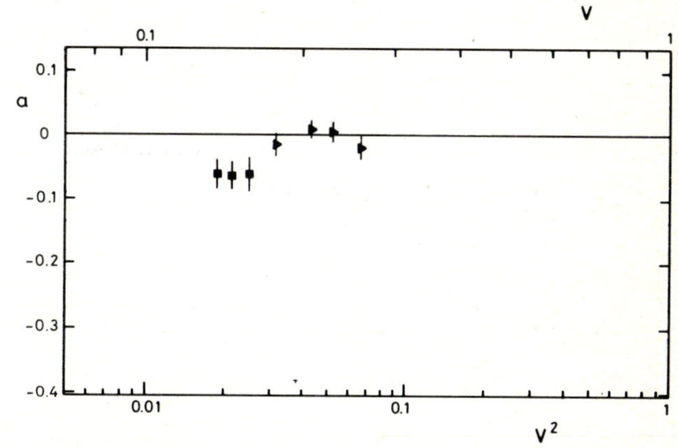

Fig. 3: Anisotropy of the $L\ell$ line for I impact on Au (■) and Pb (▶).

Experimental evidence for multiple ionization can be obtained from the observed energy shifts of the x-ray lines. One expects a shift of the $L\alpha$ line of ca. 170 and 32 eV per additional L_3 and M_5 vacancy, respectively. Experimentally we observed a shift of, e.g., 150 eV for I+Au at 70 MeV. It may be added that the known strong anisotropy of molecular-orbital x-radiation is less influenced by multiple ionization than the anisotropy of characteristic lines: Molecular-orbital radiation is emitted during the collision within a time too short for efficient angular momentum coupling between vacancies.

References

1. W. Jitschin, A. Kaschuba, H. Kleinpoppen, and H. O. Lutz, Z. Phys. A 304, 69 (1982)
2. W. E. Meyerhof, R. Anholt, J. Eichler, and A. Salop, Phys. Rev. A 17, 108 (1978)

THE L_3-VACANCY ALIGNMENT IN 32 MeV S+Au COLLISIONS AT SMALL IMPACT PARAMETERS

A. Berinde, C. Ciortea, Al. Enulescu, Daniela Flueraşu, I. Piticu and V. Zoran

Central Institute of Physics, P.O. Box MG-6, Bucharest, Romania

The study of the differential (i.e. impact parameter dependent) alignment of the vacancies produced in ion-atom collisions is expected to reveal details of the inner shell ionization mechanism. In the present work, the alignment of the L_3 subshell in Au induced by S^{6+} ions at 32 MeV bombarding energy has been investigated as a function of the impact parameter in the range b = 70-250 fm. The X-rays in coincidence with the scattered particles were detected in a plane perpendicular to the beam axis. For this case, the angular distribution of the L_ℓ line relative to the reaction plane has the form:

$$(I_\ell)_c = (\bar{I}_\ell)_c[1+\beta_\perp(b)\cdot\cos 2\varphi] \quad (1)$$

where $(\bar{I}_\ell)_c$ is the coincident emission rate integrated over φ. The coefficient $\beta_\perp$ depends in a simple way[1] on the $A_{20}(b)$ and $A_{22}(b)$ components of the statistical tensor describing the alignment. For experimental convenience we measured the coincident $(I_\ell/I_\alpha)_c$ intensity ratio at $\varphi = \pi/2$, $3\pi/4$ and π. The results are plotted in Fig. 1 for three impact parameters representing weighted averages over the particle detector acceptance angle. Also shown in this figure is a least squares fit of equation (1) to the experimental data.

The fitted values of $(\overline{I_\ell/I_\alpha})_c$ and $\beta_\perp(b)$ were used to obtain the components of the alignment tensor multiplied by an attenuation factor $a^{CK}(b)$ due to the Coster-Kronig transitions and the finite solid angle of the detectors:

$$a^{CK}(b)A_{20}(b)=a^{CK}A_{20} + \frac{2}{\alpha_2-\alpha_2'}\left[1 - \frac{(\overline{I_\ell/I_\alpha})_c}{(I_\ell/I_\alpha)_s}\right] \quad (2)$$

$$a^{CK}(b)A_{22}(b)=6^{-1/2}\beta_\perp(b)\left[\frac{2}{\alpha_2-\alpha_2'} - a^{CK}(b)A_{20}(b)\right].$$

Here $a^{CK}A_{20}$ describes the integral alignment which was measured in a separate experiment to be $-(6\pm1)\cdot10^{-2}$, $\alpha_2 = 0.5$ for L_ℓ and $\alpha_2' = 0.0489$ for L_α (as calculated from unperturbed atomic parameters), and $(I_\ell/I_\alpha)_s$ is the L_ℓ/L_α intensity ratio in singles spectra. The results are presented in Fig. 2. The vertical bars reflect the statistical uncertainty; the experimental arrangement being more appropriate for measurements of $A_{22}(b)$, this component resulted with better accuracy.

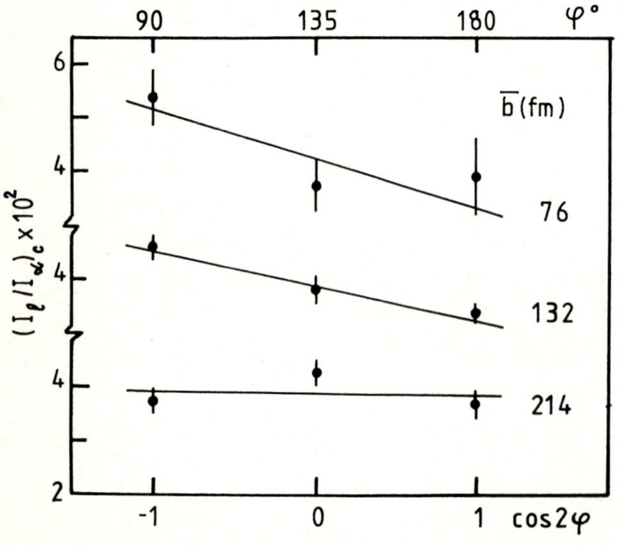

Fig. 1: The angular distribution of the L_ℓ/L_α intensity ratio in coincidence with the scattered particles

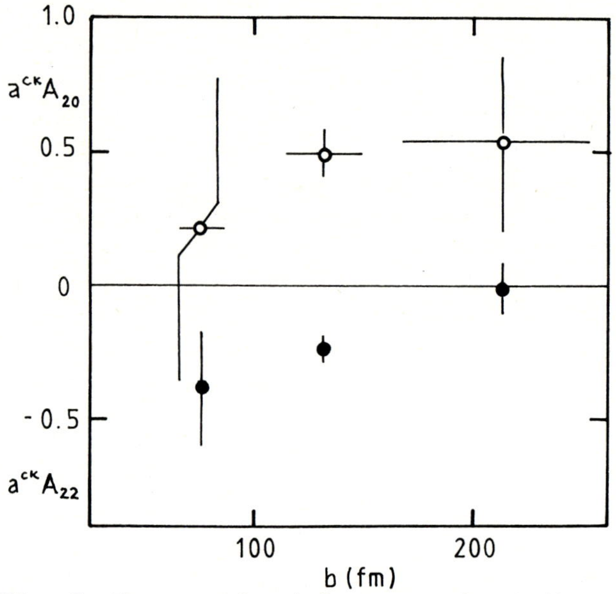

Fig. 2: The experimental components of the alignment tensor versus the impact parameter

In order to obtain the values of $A_{20}(b)$ and $A_{22}(b)$, the data of Fig. 2 should be corrected for a geometrical attenuation of 0.92 and for the Coster-Kronig transitions. The correction factor for the latter was evaluated to 1.3 from the coincident emission rates I_α, I_β and I_γ and to 1.8 from I_α, I_{γ_1} and $I_{\gamma_{2,3}}$; these values remain practically constant over the whole range of b studied.

A direct comparison of our results with the few other existing differential data (see e.g. the X82 Conference, Eugene, Oregon) is not possible. On the other hand, the present values of $A_{20}(b)$ seem to confirm the trend of the alignment in integral measurements[2,3] at average ionizing distances not far from our range of b, which in turn is at variance with the SCA calculations.

References

1. E.G. Berezhko et al., J. Phys. B 11, 1819 (1978)
2. J. Pálinkás et al., J. Phys. B 15, L451 (1982)
3. W. Jitschin et al., Proc. XII. ICPEAC (1981, Gatlinburg) p. 843

ANGULAR DISTRIBUTION OF Au L X-RAYS BY HEAVY ION IMPACT

J. Takahashi, Y. Awaya, T. Kambara, M. Kase, H. Kumagai, J. Urakawa*, T. Matsuo** and H. Shibata*

The Institute of Physical and Chemical Research, Hirosawa, Wako-shi, Saitama, 351 JAPAN

Recently, the alignment of the L3 subshell for several targets (Xe, Dy, Au and U) has been studied by measuring the anisotropy of the proton-induced L_ℓ X-rays[1]. In this report is presented the study of the angular distribution of L_ℓ line for Au target by heavy ion impact(N, Ar).

The incident ions N^{2+} and Ar^{4+} were accelerated by the linear accelerator of IPCR. The Au target was evapolated onto a 4 μm Mylar foil and the thickness was 200 μg/cm^2. The angular distribution of Au L X-rays were measured by a Si(Li) detector as a function of the photon emission angle (θ). For the forward and backward angles, the target was tilted to be 32.5° and 147.5°, respectively, with repect to the beam direction. Fig.1 shows the photon spectrum at θ = 90° obtained by 0.87 MeV/amu N^{2+} impact. The L_ℓ (M_1-M_3 transition), L_α(M_4,M_5-L_3), L_{β_1}(M_4-L_2), L_{γ_1}(N_4-L_2) lines were observed. The L_ℓ lines is expected to have the highest anisotropy of all resolved lines and the assumption of a not-aligned upper state is automatically fulfilled since it originates from 3s level. In Fig.2 is shown the angular distribution of L_ℓ line relative to L_α line assuming that the anisotropy of L_α line is small.

The angular dependence of the emitted dipole radiation I(θ) is expressed as

$$I(\theta) = (I_0/4\pi)(1 + \alpha A_2 P_2(\cos\theta))$$

where I_0 is the total intensity, P_2 the second Legendre Polynomial and αA_2 the anisotropy. A_2 is the alignment and the coefficient α depending on the angular momenta j of the initial and final states, is calculated to be α = 0.5 for the L_ℓ line[2]. The A_2 deduced from the 0.87 MeV/amu N^{2+} bombard was -0.18 ± 0.05 (not corrected for Coster-Kroning transitions). Further analysis and experiments are in progress.

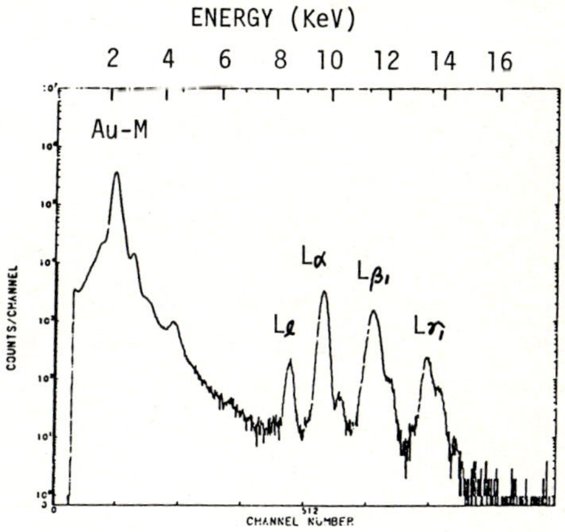

Fig.1 The photon spectrum for Au target at θ = 90° bombarded by 0.87 MeV/amu N^{2+}.

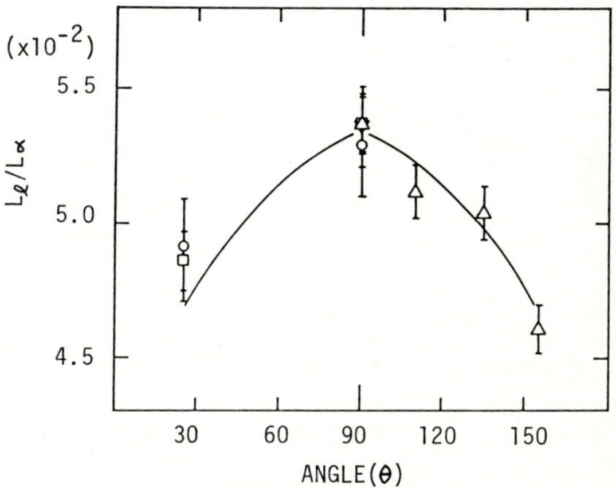

Fig.2 The angular distribution of the L_ℓ/L_α intensity ratio. 0.87 MeV/amu N^{2+} bombard.

References

1) W. Jitschin et al: Z. Phys. A304, 69(1982).

2) E. G. Berezhko and N. M. Kabachnik, J. Phys. B. : Atom. Molec. Phys. 10, 2467 (1977).

* Tokyo Institute of Technology.

** Tokyo University of Agriculture and Technology.

EFFECT OF SCREENING BY PROJECTILE ELECTRONS ON THE ALIGNMENT OF A VACANCY IN ION-ATOM COLLISION

N.M. Kabachnik, O.B. Maksimova

Institute of Nuclear Physics, Moscow State University, Moscow 117234, USSR

Effect of screening by projectile electrons on the alignment of an inner-shell vacancy formed in fast ion-atom collision is theoretically studied. The alignment of the $2p_{3/2}$ and $3d_{3/2,5/2}$ vacancies in a number of atoms is calculated for the cases of collisions with the He^+, O^+ and Ne^+ ions. The calculations have been made within the Born approximation using the Hartree-Slater atomic wave functions. The projectile electrons are described by hydrogen-like wave functions. Two approximations are used to account for screening by projectile electrons[1].

(1) Static screening model, in which the screening of the nuclear charge by the static charge distribution of the electronic shell has been taken into account

(2) Dynamic screening model in which the possibility of exciting the projectile electronic shell has been taken into account using the closure approximation.

The results, of our calculations may be summerized as follows:

1. The effect of screening on the alignment is considerable for not very tightly bound subshells in the region of reduced velocities $V > 0.3$.
2. The influence of screening relatively grows as the ion velocities increase.
3. The screening effects are more important for the alignment of the subshells with small bound energies and become negligible for the deep inner shells of heavy atoms.
4. The difference between the alignments calculated using static and dynamic screening models is not significant although the cross sections differ considerably.
5. The effect of screening can not account for the descrepancy between the theoretical and experimental alignment of the $2p_{3/2}$ subshell of heavy atoms in ion-atom collisions[2].

References

1. J.S. Briggs, K. Taulbjerg. In "Structure and Collisions of Ions and Atoms", ed. by I.A. Sellin, Springer, New York (1978) p.105.
2. W. Jitschin et al. XII ICPEAC, Gatlinburg, USA (1981) p.834.

WAVELENGTH OF TRANSITIONS IN FEW ELECTRON SPECTRA OF TITANIUM

H.D. Dohmann, D. Liesen, E. Pfeng
Gesellschaft für Schwerionenforschung, D-6100 Darmstadt

Using the beam foil technique, Ti-ions with specific energies of 3.6 and 4.8 MeV/N were excited by C-foils (110 μg/cm²). Prompt and delayed emitted K x-rays have been studied with a flat-crystal spectrometer ($\lambda/\Delta\lambda \approx$ 1200 FWHM). Fig. 1 shows the prompt emitted spectrum taken at 4.8 MeV/N projectile energy in the wavelength region between 2.45 and 2.70 Å. At this specific beam energy, the emission from He-like Ti ions with one K-vacancy is dominant. Therefore, at λ = 2.6105 Å we find the He-like resonance line which is separated from the decay of Li-like configurations 1s2s2p, $1s2s^2$, and $1s2p^2$ (marked by bars in Fig. 1 between 2.62 and 2.7 Å).

In the lower wavelength region between 2.48 and 2.50 Å Fig. 1 shows the clearly resolved Lyman-α splitting. A detailed analysis of the spectrum (Fig. 2) which takes the resolution function of the spectrometer into account, reveals that the wavelength of the transitions can only be calibrated properly if the influence of spectator electrons with principal quantum number n = 3 is considered. By measurements of every line at two known beam velocities, the influence of the linear and quadratic Doppler effect on the accuracy of the absolute wavelength was drastically reduced.

A preliminary comparison of experimental and theoretical wavelengths is shown in Fig. 3. Within the absolute error bar of the experiment $\Delta\lambda$ = ± 0.0003 Å ≅ ± 100 ppm most theoretical data agree with the experimental results [1].

The aim of further experiments is the precise measurement of the fine structure splitting and of the Quantum Electrodynamics (QED) effects.

1. H.D. Dohmann et al., to be published

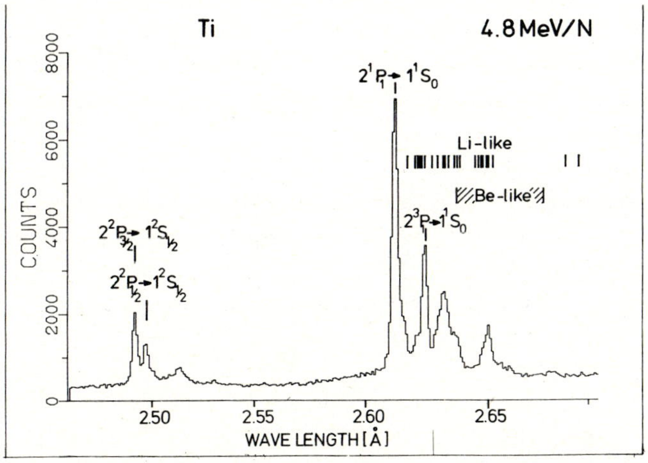

Fig. 1: Few-electron x-ray spectrum of a Ti beam with 4.8 MeV/N specific energy.

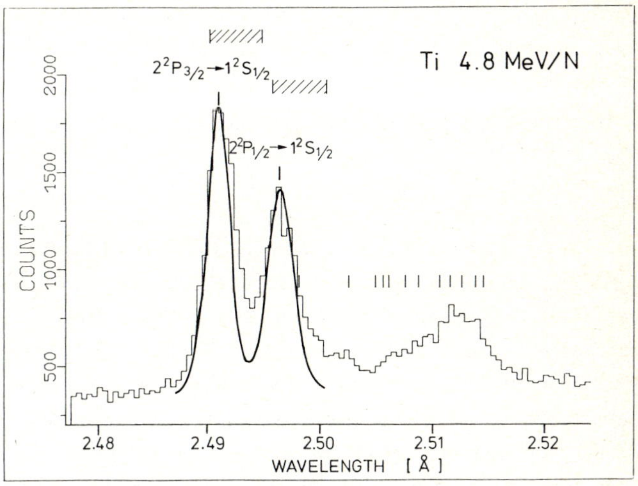

Fig. 2: Lyman-α lines of Ti. The position of the 2p2s,p → 1s2s,p transition is marked by bars and the region of the 2p3s,p,d → 1s3s,p,d wavelength is hatched. These transitions with an spectator electron in n=3 result in the observed asymmetry of the Lyman-α lines.

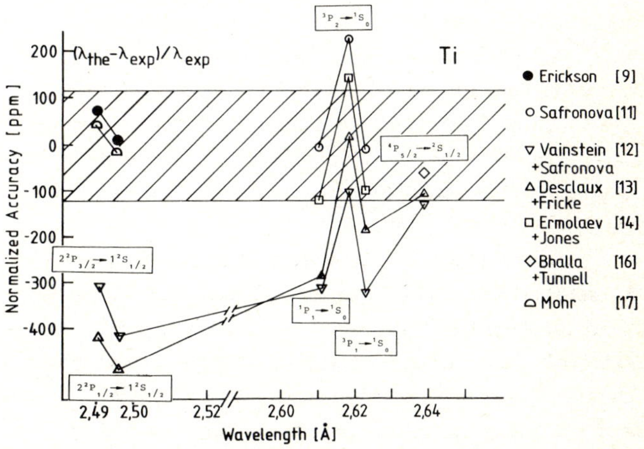

Fig. 3: Comparison of experimental wavelength and theoretical wavelength in H-, He-, and Li-like Ti ions.

THE EFFECT OF CHEMICAL BONDS ON FLUORINE K X-RAY SPECTRA PRODUCED BY 80 MeV Ar AND 48 MeV Mg IONS

O. Benka

Institut für Experimentalphysik, Universität Linz, A-4040 Linz, Austria

R. L. Watson

Cyclotron Institute and Department of Chemistry, Texas A & M University, College Station, TX 77843

In a recent work we investigated the effect of chemical bonds on fluorine K_α satellite spectra of alkali and alkaline earth fluorides, exciting the X-rays by 22 MeV C-projectiles[1]. To explain qualitatively the line intensities three effects had to be considered: a) A primary L-vacancy production in the fluorine ion like that in neon atoms, and an electron transfer mechanism from neighboring atoms to the L-holes during the following lifetime of the K-hole. This process was described by a L-hole decay constant λ_t. b) A resonant electron transfer process effecting mainly the KL^1 satellite intensity. c) An enhancement process of the KL^0 line by secondary electrons. The intensity distribution of the satellite lines KL^n could then be explained yielding a L-hole decay constant λ_t which is not significantly depending on the chemical bonding of the fluorine compounds. In another work the K_α hypersatellite spectra were studied[2], also excited by 22 MeV C-ions, giving ionic λ_t values close to the satellite values.

Here we report an extension of these fluorine K_α satellite and hypersatellite measurements to higher Z-projectiles using 80 MeV Ar ion projectiles to excite extremely high primary states of ionization. In addition the satellite spectra of gaseous fluorine compounds were studied using 48 MeV Mg ions for X-ray excitation.

Measurements of the spectra were performed using a 12.7 cm curved crystal spectrometer; details of the measurements on solid targets are given in ref. 1, for the gas targets the same set up as for previous neon target measurements[3] was used. The satellite spectra of solids showed two different regions, the groups KL^0, KL^1 and KL^2, which had a relatively high intensity compared to the higher satellite groups indicating an enhancement process, and the groups KL^3 to KL^7 which showed a broad intensity distribution centered around KL^4. We propose that these latter lines originate only from the primary Ar-ionization and not from secondary enhancement processes. An evaluation similar to the previous one applied to the C-projectile measurements, yielded L-hole decay constants λ_t which were close to the values obtained using C-projectiles.

The K_α hypersatellite spectra of solids showed a smooth distribution centered about K^2L^5, indicating relatively smaller L-refilling and no enhancement. In an analysis taking the lifetimes of double K-hole states into account λ_t is again about the same as for the satellite spectra, so in a rough approximation the L-hole decay constant (per L-hole) seems to be independent of additional L- and K-vacancies for solid ionic fluorine compounds. The resulting decay constants yield high refilling probabilities for KL^n states with high n: for KL^6 state, e.g., the chance for a decay via L-refilling to a KL^5 state is 96% and for decay via K-X-ray or Auger electron emission only 4%.

For the gaseous compounds SiF_4, SF_6, C_2F_6, CHF_2 and CHF_2CH_3, only the satellites were measured. The satellite lines KL^5 and KL^6 had the highest intensities. This indicates a less effective L-hole refilling process compared to the solid fluorine measurements. An analysis similar to the solid state analysis gave λ_t-values which were a factor 5 to 10 lower than the solid ones, depending significantly on the compounds. Comparing the time for molecular break up due to recoil of the fluorine ion or Coulomb explosion, which is about 10^{-15} sec, with the lifetime of a KL^n state for high n, which is about 10^{-14} sec, reveals that molecular break-up will take place with high probability before K-hole decay and the resulting L-vacancy distribution at the time of K-hole decay will then depend on the mechanism of this molecular break up. For understanding this process further measurements using different gaseous fluorine targets and different projectiles would be necessary.

References

1. O. Benka, R. L. Watson, K. Parthasaradhi, J. M. Sanders and R. J. Maurer, Phys. Rev. A 27, 149 (1983)
2. O. Benka and R. L. Watson, Phys. Lett., to be published
3. R. L. Watson, O. Benka, K. Parthasaradhi, R. J. Maurer, J. M. Sanders, J. Phys. B, to be published

HIGH RESOLUTION EUV SPECTRA OF CORE-EXCITED $(1s2pnp)^2P$, $(1s2pnd)^2D^0$ AND $(1s2pnf)^2F$ STATES OF DOUBLY IONIZED BORON

R. Bruch, K.T. Chung*), E. Träbert**), P.H. Heckmann**) and B. Raith**)

Fakultät für Physik, Universität Freiburg, 7800 Freiburg, West Germany

Optical transitions which originated from core-excited $(1s2pnp)^2P$ ($n \geq 2$), $(1s2pnd)^2D^0$ ($n \geq 3$) and $(1s2pnf)^2F$ ($n \geq 4$) states of the Li-isoelectronic sequence are good candidates for lasers in the extreme UV region[1]. On the other hand precision measurements of such highly excited states yield fundamental information concerning electron correlation and relativistic interactions and allow sensitive tests of atomic structure theories . Accurate theoretical calculations in turn are valuable to other fields in physics like plasma- and astrophysics.

In this contribution we present the most accurate measurement of the $(1s2p^2)^2P \rightarrow (1s^22p)^2P^0$ transition so far. For boron we find a wavelength of $\lambda = 6.2255 \pm 0.0008$ nm. This experimental accuracy even allows to test relativistic contributions. Our calculated relativistic result 6.2251 nm agrees excellently with experiment.

Fig. 1 shows a typical high-resolution beam-foil spectrum of boron, displaying the BIV $(1s2p)^1P^0 \rightarrow (1s^2)^1S$ transition as well as nearby satellite structures which lie on the long wavelength side of the parent resonance line. Due to theoretical predictions these structures can be uniquely identified. Thus we have found evidence for the first time of odd parity $(1s2pnd)^2D^0$ states. This experimental finding differs significantly from the Li-work of Willison et al.[1] who suggested that only even parity ^2P states have contributed to the observed spectrum. Detailed results of very accurate new calculations and comparision with experimental data will be presented. Moreover novel atomic systems suitable for the construction of high-frequency lasers will be discussed.

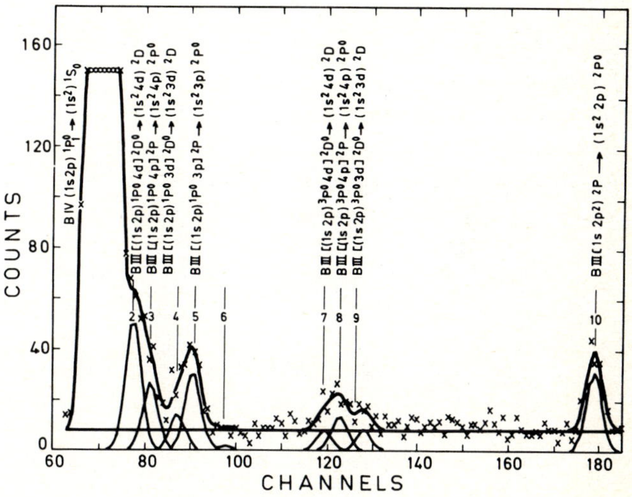

Fig. 1: High resolution beam-foil EUV-spectrum of B at 3 MeV from 5.9 to 6.3 nm.

References

*) Permanent address, North Carolina State University, Raleigh, North Carolina 27650 - supported in part by The National Science Foundation and by the Deutsche Forschungsgemeinschaft.

**) Permanent address, Ruhr-Universität Bochum, 4360 Bochum, West Germany

1. J.R. Willis, R.W. Falcone, J.C. Wang, J.F. Young and S.E. Harris, Phys.Rev.Lett. 17, 1125 (1980)

HIGH RESULUTION MEASUREMENTS OF K X RAYS FROM Ar IONS IMPINGING ON FOILS

Y. Awaya, T. Kambara, M. Kase, H. Kumagai, J. Takahashi, J. Urakawa*, T. Matsuo** and M. Namiki***

The Institute of Physical and Chemical Research, Hirosawa, Wako-shi, Saitama 351, Japan

The K X-ray spectra have been studied for Ar ions passing through various target foils in order to investigate the dependence of multiple inner-shell ionization of projectiles on the atomic number of target atom.

The Ar^{4+} ions were accelerated by the linear accelerator of IPCR. The incident energy of Ar ions was 33.6 MeV and target elements were C, Mg, Al, Ca, Ti, V, Cr, Fe, Co, Ni, Zn, Y, Mo, Sn, Ta and Au. Targets were self-supporting foils but for Ca which was evaporated onto a 10-$\mu g/cm^2$ C foil. The thickness of targets was in the range of 0.5 to 1 mg/cm^2 except for C, V, Zn, Ta and Au. The thickness of C foil was 80 $\mu g/cm^2$ and those of V, Zn, Ta and Au foils were between 2 and 2.6 mg/cm^2. The Ar K X rays emitted from the beam incident side of the target foil at right angle to the beam direction were measured by a broad range crystal spectrometer[1)] with a position sensitive proportional counter and a plane crystal of Ge(111). Examples of K X-ray spectra from Ar ions are shown in Fig. 1. The targets are Al, Ti, Fe and Ta. KL^n denotes single K-shell and n L-shell vacancy configuration in the initial state.

Assuming that the cross section of simultaneous single K-shell and n L-shell ionization, $\sigma_{K,nL}$, is expressed by a binomial distribution as

$$\sigma_{K,nL} = \binom{8}{n} P_L{}^n (1-P_L)^{8-n} \cdot \sigma_K \quad ,$$

we obtained the value of P_L for each target element.

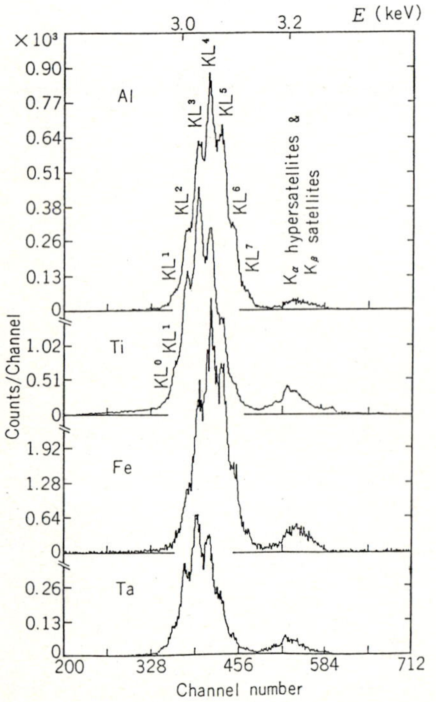

← Fig. 1.

Here, P_L is the ionization probability of an L-shell electron at the impact parameter of around zero and σ_K is the K-shell ionization cross section. A constant value of fluorescence yield is assumed for KL^n lines. The obtained P_L values are shown against Z_2 in Fig. 2.

The effect of target thickness to the P_L was examined for C, Al and Ni targets by using two foils with different thickness: 20 and 80 $\mu g/cm^2$ for C, 0.20 and 1 mg/cm^2 for Al and 0.45 and 0.89 mg/cm^2 for Ni. Open circles in Fig. 2 show the P_L values for thinner foils. It is found when the thickness of the target increases, the P_L value decreases. The change of P_L value according to the incident energy was also studied for the 80-$\mu g/cm^2$ C, 0.2-mg/cm^2 Al and 0.9-mg/cm^2 Ti targets. The P_L values for the incident energy of 21.1 MeV are shown by crosses in Fig. 2. Curves are only to guide the eye.

As is seen in Fig. 2, the P_L value displays an oscillational behaviour against Z_2. It reaches the first minimum around Z_2=22 and probably the second one around Z_2=45, where the L-shell binding energy of highly ionized Ar ions matches to the L-shell and M-shell binding energy of the target atoms, respectively. This may suggest that the electron capture cross section of the L shell of Ar ions from target atoms varies according to Z_2 and becomes large when the level-matching occurs.

Further experiments and analysis are in progress.

1) A. Hitachi, H. Kumagai and Y. Awaya, Nucl. Instr. and Meth., 195, 631 (1982).

* Tokyo Institute of Technology.
** Tokyo University of Agriculture and Technology.
***Takachiho College.

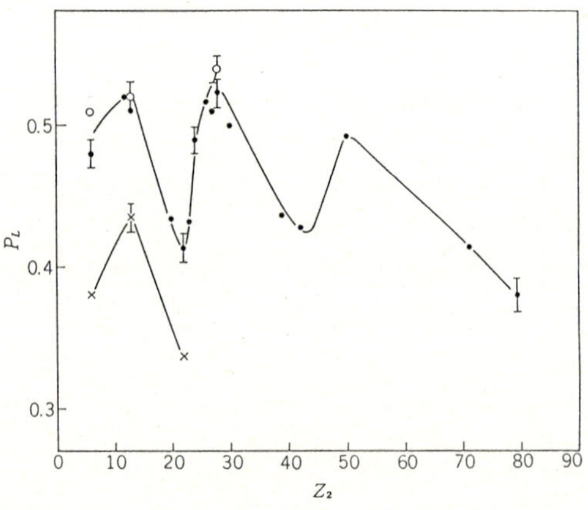

Fig. 2.

TARGET GAS PRESSURE DEPENDENCE OF RELATIVE YIELD OF K X-RAYS FROM 110 MEV NE IONS

T. Kambara, Y. Awaya, M. Kase, H. Kumagai, I. Kohno, T. Tonuma and A. Hitachi*

The Institute of Physical and Chemical Research, Hirosawa, Wako-shi, Saitama 351, Japan
*Radiation Laboratory, University of Notre Dame, Notre Dame, Indiana 46556, USA

High resolution energy spectra of X-rays are obtained for the Lyman-series 3p-1s, 4p-1s and 5p-1s transitions in 5.5 MeV/amu Ne^{9+} ions (hydrogen like) passing through gaseous targets of N_2 and Ne. It was found that the relative intensities between the transitions were dependent on the target gas pressure. This effect is attributed to collisional quenching of the 3p, 4p and 5p excited states of the ions. The quenching cross section values are estimated.

The measurements were done with a broad range crystal spectrometer with a plane crystal and a position sensitive proportional counter[1]. We used a RAP crystal with a length of 5cm and width of 2.5cm for the analyser placed at 90° to the beam direction. The range of the energy of X-rays analysed by the spectrometer was between 1.1 and 1.44 keV, where the np-1s transitions with n > 2 could be observed. The resolution was about 10eV (FWHM). The target gas of N_2 and Ne was contained in a gas cell and the X-ray spectra were obtained at several target gas pressures between 20 and 300 Torr.

An example of the spectra is shown in Fig. 1. The peaks from the transitions np-1s with n=3, 4 and 5 are well resolved. The intensity ratio between different transitions depends on the gas pressure as shown in Fig. 2(a) for the N_2 and Fig. 2(b) for the Ne target, where the intensities of the 4p-1s and 5p-1s transitions relative to that of the 3p-1s transition are plotted against the gas pressure.

This e-fect is explained as a result of collisional quenching in which a Ne^{9+} ion in an excited state changes to another state by a collision with a target atom before it decays with a radiative emission. The probability of the collisional quenching depends on the lifetime of the state τ, the velocity of the ion v, the density of the target atoms N and the cross section of the quenching σ. The intensity of the X-ray transition from a state i relative to that from a state j is approximately proportional to $(1+Nv\sigma_j\tau_j)/(1+Nv\sigma_i\tau_i)$ which causes the target gas pressure dependence of the relative radiation intensities. The experimental data shown in Fig. 2 are fitted to the expression described above and the most probable values of the quenching cross section σ_i are searched. The values of the lifetime τ_i of the excited states are obtained by multiplying the lifetime of the corresponding states of a hydrogen atom[2] by a factor of 10^{-4}.

The result is that teh values of the quenching cross section for the 4p and 5p states are in the order of 10^{-17}cm for the both targets. The values are not very different from that obtained by Matthews and Fortner[3,4] for the quenching cross section of the 2^3P_1 metastable state of 30 MeV F^{7+} ions.

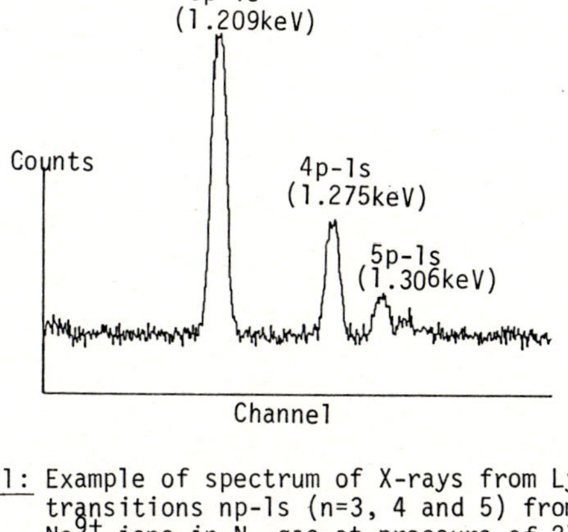

Fig. 1: Example of spectrum of X-rays from Lyman-series transitions np-1s (n=3, 4 and 5) from 110 MeV Ne^{9+} ions in N_2 gas at pressure of 210 Torr.

References

1. A. Hitachi, H. Kumagai and Y. Awaya, Nucl.Instr. and Meth., 195, 631 (1982).
2. H. A. Bethe and E. E. Salpeter, "Quantum Mechanics of One- and Two-Electron Atoms", (Plenum, New York, 1979)
3. D. L. Matthews, R. J. Fortner and G. Bissinger, Phys. Rev. Lett., 36, 664 (1976).
4. R. J. Fortner and D. L. Matthews, Phys. Rev. A 16, 1441 (1977).

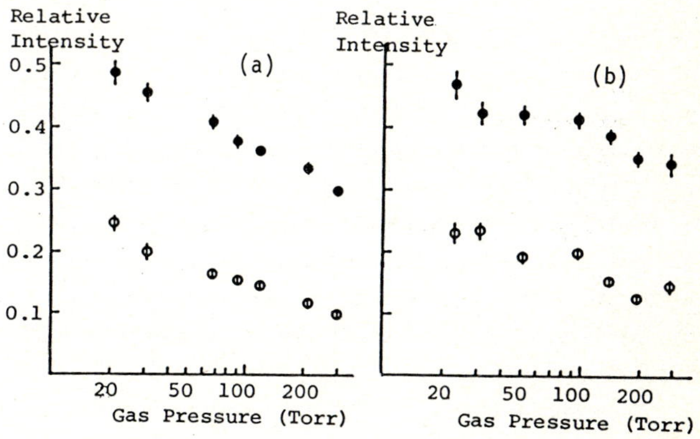

Fig. 2: Target gas pressure dependence of the X-ray intensity of the 4p-1s (closed circle) and 5p-1s (open circle) transitions relative to that of the 3p-1s transition for (a) N_2 and (b) Ne gas target.

INTERFERENCE EFFECT IN THE QUASIMOLECULAR K-RADIATION INDUCED BY HYDROGENLIKE LOW-VELOCITY PROJECTILES

R. Schuch, J. Barrette[+], R. Hoffmann, B.M. Johnson[+], K.W. Jones[+], M. Meron[+], H. Schmidt-Böcking[*], I. Tserruya[§]

Physikalisches Institut, University Heidelberg, 69 Heidelberg, FRG
[+]Brookhaven National Laboratory, Upton L.I. N.Y., USA
[*]Hahn Meitner Institut, 1000 Berlin, FRG
[§]Weizmann Institut, Rehovot, Israel

Quasimolecular K-radiation should provide a direct information about molecular orbital binding energies especially of the strongest bound state (1sσ), because the radiative transitions between these states are detected. However, at the relative high velocities where 1sσ vacancies are produced, the dynamics of the collision introduces a broadening which makes direct spectroscopy impossible[1]. This was circumvented in the present experiment by bringing the K-vacancy at low velocity into the collision. Experimental results for 2.5, 5, 10 and 20 MeV Cl^{16+} beams obtained at the accel-decel tandem facility of Brookhaven National Laboratory are reported where the quasimolecular K-radiation from collisions with an Ar gas target was measured in coincidence with the scattered particle in an impact parameter (b) region of b ≃ 400 fm (inside the united atom K-shell radius r_{ua}) up to b ≃ 3000 fm (outside r_{ua}).

The x-ray production probability as function of photon energy for the beam energies of 5, 10 and 20 MeV is shown in Fig.1 for an impact parameter b ≃ 690 fm. The coincidence spectra show a pronounced and broad maximum which moves toward higher x-ray energy as the beam energy decreases.

The solid and dashed line represents the result of a dynamical calculation which was done according to ref.3. The structure in the calculated curves arises from the interference between the 1sσ decay amplitude in the incoming part and the outgoing part of the trajectory. In the 10 and 20 MeV collision system the calculated quasimolecular K x-ray spectrum goes through only one oscillation between 6 and 12 keV. It can be shown that the position of the maximum is very sensitive to the quasimolecular energy difference used in the calculation. Furtheron two interference maxima are expected from this calculation to be seen at the low projectile energies 2.5 and 5 MeV.

Work supported by the Deutsche Forschungsgemeinschaft, BMFT, and US Dept. of Energy.

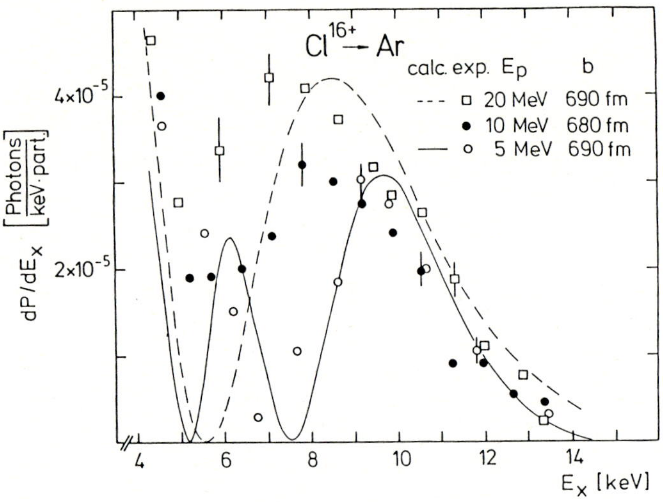

Fig. 1: Quasimolecular x ray emission probabilities for three different projectile velocities at a fixed impact parameter.

References

1. J.H. Macek, J.S. Briggs, J. Phys. B7, 1312 (1974)
2. C.K. Davis, J.S. Greenberg, P.R.L. 32, 1215 (1974)
3. B. Müller in ICPEAC, J.S. Risley and R. Geballe, Eds. Seattle 1975, p. 481.

IMPACT PARAMETER DEPENDENCE OF 1sσ AND 2pσ MO RADIATION IN 90 MEV NI + NI COLLISIONS

H. Richter

Zentralinstitut für Kernforschung, Rossendorf, PO Box 19, DDR-8051 Dresden, German Democratic Republic

For the Ni+Ni system in the x-ray energy range of about 8-40 keV three different types of molecular orbital (MO) radiation dominate: (i) 2pσ MO radiation: 2pπ vacancy production is followed by both, 2pπ-2pσ rotational coupling and radiative 2pσ decay; (ii) 1sσ-I MO radiation: production of 1sσ vacancies by ionization and its radiative decay on the outgoing part of the trajectory; (iii) 1sσ-II MO radiation: in the solid target a mean differential K vacancy fraction $\sigma^{-1}\sigma_R(\theta_1)P_K(\theta_1)d\Omega_1$ of the projectiles scattered into the solid angle $d\Omega_1$ is produced, which subsequently decays by 1sσ radiative transitions in a second close collision, with probability $(1-P_{sh})dP_{1s\sigma}(\theta_2)/dE_x$. ($P_{sh}$ is the 2pσ-1sσ sharing probability, $\sigma_R(\theta)$ the Rutherford cross section and $\sigma=\sigma_K+\sigma_q+\Gamma_K/nv^1$). The impact parameter dependence of the MO radiation for this system was investigated experimentally[1] as well as theoretically[1-3]. The theoretical results of the 2pσ and 1sσ emission probabilities (denoted by $dP^{rot}_{2p\sigma}/dE_x$ and $dP_{1s\sigma}/dE_x$) obtained from dynamical model calculations[3] has been taken from Jäger[1].

As shown in fig.1, $P^{rot}_{2p\sigma}$ and $P^{exp}_{2p\sigma}$ differ substantially, therefore the calculated 2pσ emission probability has to be improved. We used following approximation:

$$\frac{dP_{2p\sigma}}{dE_x} = K\,\frac{dP^{rot}_{2p\sigma}}{dE_x}, \quad \text{where } K=\begin{cases} P^{exp}_{2p\sigma}/P^{rot}_{2p\sigma} & \text{for } E_x \lesssim 20 \text{ keV} \\ \bar{N}_{2p\pi} & \text{for } E_x \gtrsim 20 \text{ keV} \end{cases}$$

$\bar{N}_{2p\pi}\simeq 0.45$ is the mean number of 2pπ_x vacancies.

With increasing E_x (>20 keV) $dP^{rot}_{2p\sigma}/dE_x$ shows an evident maximum in the valley of $P^{rot}_{2p\sigma}$(fig.1). In this b-range a 2pπ vacancy has to follow approximately the fast rotation of the internuclear axis (in the framework of the 2pπ-2pσ model) and thus it emits bremsstrahlung.

For the 1sσ-I emission probability we used the approximation[1] $dP_{1s\sigma-I}/dE_x = P_{1s\sigma}(dP_{1s\sigma}/dE_x)/2$, with $P_{1s\sigma}(b)=P(0)\exp(-\alpha R_o E_{1s\sigma}(R_o)/\hbar v)$. From the comparison with the experimental data we obtained $P(0)\simeq.03$ and $\alpha\simeq 1$.

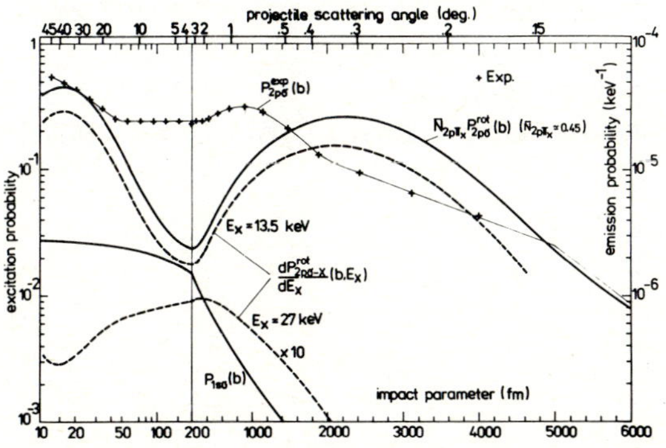

Fig.1: 1sσ and 2pσ excitation and 2pσ x-ray emission probabilities versus impact parameter $b(=\varrho\cdot ct\theta)$

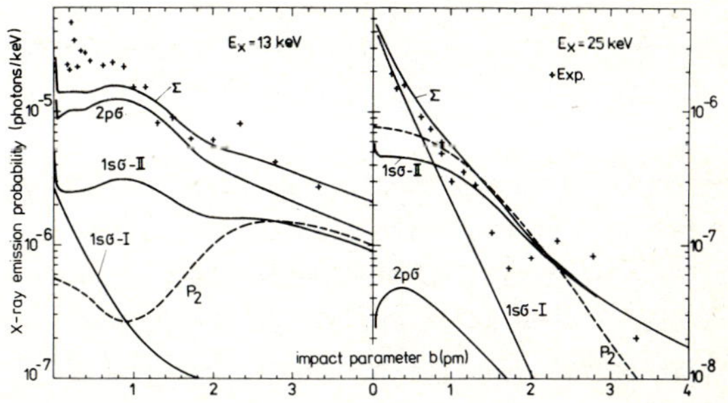

Fig.2: MO x-ray emission probabilities for two E_x values

The 1sσ-II emission probability $dP_{1s\sigma-II}/dE_x=P_o$ is determined by the integral

$$\sigma_o(\theta_o)=P_o(\theta_o)\sigma_e(\theta_o)=\sigma_1^{-1}\int d\Omega_1\,\sigma_1(\theta_1)\sigma_2(\theta_2(\theta_o,\theta_1,\varphi_1-\varphi_o))$$

where $\sigma_i(\theta)=P_i\sigma_R(\theta)$, $P_1=P_K=P_{sh}P^{exp}_{2p\sigma}$ and $P_2=\sigma^{-1}\sigma_K(1-P_{sh})\times dP_{1s\sigma}/dE_x$ is the unfolded 1sσ -II emission probability. Note, that P_o depends also on the elastic cross section $\sigma_e(\theta)$ of the double collision process, defined analogously to $\sigma_o(\theta)$. For small θ_i and $\theta_o>\theta_{max}$ (maxima of $\sigma_i(\theta)$) in 0th approximation one obtains

$$P_o(\theta_o)\simeq(P_2(\theta_o) + P_1(\theta_o)\sigma_2/\sigma_1)/2 \text{ and } \sigma_e(\theta_o)\simeq 2\sigma_R(\theta_o).$$

With decreasing $\theta_o(\lesssim\theta_{max})$ P_o becomes increasingly overestimated, but the experimental data in this range are still more inaccurate. In contrary to the numerical folding results from Anholt[2] and Jäger[1] our result show (fig.2) that in 0th folding approximation P_o is the mean value of P_2 and $P_1\cdot\sigma_2/\sigma_1$.

In fig. 3 our calculated x-ray energy differential cross sections are compared with the experimental data.

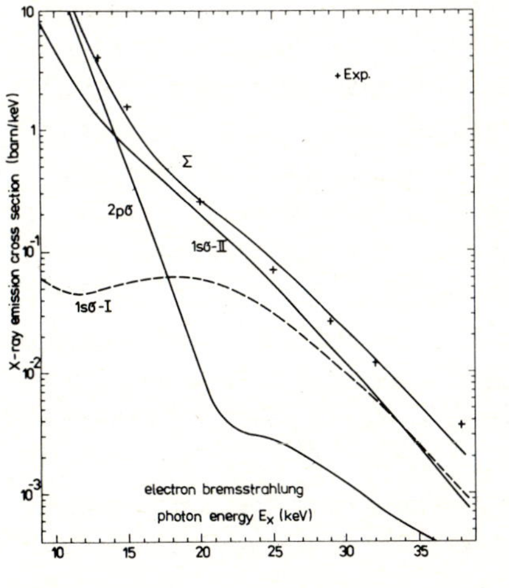

Fig. 3
MO x-ray emission cross sections. The experimental data shown here has been obtained by integration of the experimental emission probabilities[1].

References

1 H.Schmidt-Böcking, R.Anholt, R. Schuch, P.Vincent, K.Stiebing and H.U.Jäger, J.Phys. B 15, 3057 (1982)

2 R.Anholt, Nucl.Instrum.Methods 198, 567 (1982)

3 H.U.Jäger et al., J.Phys. B 14, 701 (1981)

THE EMISSION AND ABSORPTION OF LIGHT DURING ATOMIC COLLISIONS

P. T. Greenland

Theoretical Physics Division, AERE Harwell, Didcot, Oxon OX11 ORA

The emission of photons during slow atom-atom and atom-ion collisions is important both for the information it gives on the collision itself, and because levels not otherwise strongly coupled may be most efficiently populated through radiative transitions. The non-characteristic X-rays emitted during heavy ion collisions(1) provide an example of the former; the radiative charge transfer to the ground state of He in He^{++} H collisions is an important electron capture channel in very low energy collisions(2). Furthermore the availability of high power lasers has lead to interest in the reverse process of laser assisted collisions(3).

The stationary phase approximation has been applied to both continuum emission(4) and photon absorption during slow atom ion collisions(3). Physically this corresponds to assuming that light is emitted resonantly between the virtual molecular levels formed in the collision. In general two internuclear separations correspond to regions where a photon of given energy can be emitted (or absorbed). Structure results from interference between photons from these two sources. The stationary phase approximation breaks down if the two regions coalesce. In this case the uniform approximation in terms of Airy functions is known to give good results(5).

However if more than 2 points of stationary phase coalesce, the Airy function expansion breaks down. Figure 1 shows how this happens. Figure 1a compares the exact result of a model problem (solid line) with the Airy function approximation (dashed line), where only 2 points of stationary phase occur. Figure 1b shows the same comparison where there are 4 points of stationary phase, which coalesce in pairs. Figure 1c shows the breakdown of the Airy function approximation when all 4 points coalesce.

This work describes how the use of higher uniform asymptotic expansions(6) similar to those used in the semiclassical theory of rainbow scattering(7) can be used to systematically improve on the Airy function approximation. The evaluation of the appropriate canonical integrals and results derived from them will be described.

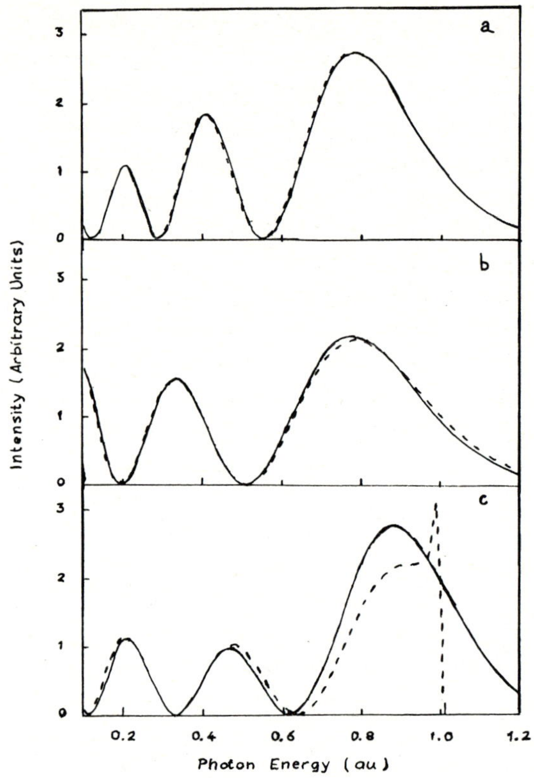

Fig. 1: Photon energy dependence in a model problem, showing the breakdown of the Airy function approximation.

References

1. D. Liesen, Comments At. Mol. Phys. 12 39 (1982).
2. B.W. West, N.F. Lane, J.S. Cohen; Phys. Rev. A 26 3164 (1982).
3. M.H. Mittleman; 'Theory of Laser Atom Interactions' Plenum New York (1982).
4. J.S. Briggs, K. Dettman; J. Phys. B Atom. Molec. Phys. 10, 1113, (1977).
5. W. Fritsch and U. Wille, Japanese Journal of Applied Physics 17 Supplement 17-2, 387 (1978).
6. N. Bleistein, J. Math. Mech. 17 533 (1967).
7. J.N.L. Connor; Molec. Phys. 31 33 (1976).

MOLECULAR AUTOIONIZATION SPECTRA IN LOW-MEV KR^+-KR COLLISIONS

P. Clapis, A. Antar**, S. Kuptsov***, R. Roser, R. Rubino and Q. Kessel

Department of Physics, The University of Connecticut, Storrs, CT 06268, USA

Molecular autoionization in Kr^+-Kr collisions was first reported by Afrosimov and coworkers[1] and in more detail by Gordeev and coworkers[2] for energies up to 0.8 MeV. This report presents spectra for collisions from 0.25 MeV to 3.0 MeV. The 1.0 MeV spectrum is shown in Figure 1 and it displays a broad peak between 150 eV and 600 eV. The electrons were detected at an angle of 107 degrees with respect to the incident beam direction and the peak is more evident than in the spectra of reference 2. The energy range of this peak falls between the expected Auger electron energies for L- or M-shell vacancy production. This unexpected broad peak has been attributed[1,2] to the filling of a vacancy in the 4pπ molecular orbital of the Kr - Kr system during the collision. Coincidence measurements on this system at 0.45 MeV were reported by Liesen and coworkers[3] and we are presently extending these measurements to the higher energies available in this laboratory. The data will be compared with impact parameter dependent measurements of ionization and inelastic energy losses for these same collisions.[4]

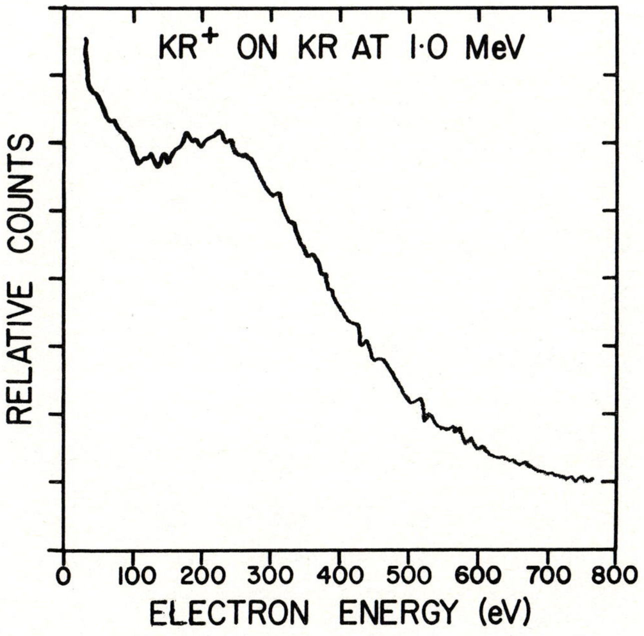

Fig. 1: Electron counts plotted versus electron energy for 1.0 MeV Kr^+-Kr collisions.

* Supported by NSF Grant PHY 81-06915.

** Central Connecticut State University.

*** IREX Scientist from Leningrad Polytechnical Institute

References

1. V. V. Afrosimov, Yu. S. Gordeev, A. N. Zinoviev, D. H. Rasulov and A. P. Shergin, Pisma Zh. E.T.F. 24, 33 (1976).
2. Yu. S. Gordeev, P. H. Woerlee, H. de Waard and F. W. Saris, J. Phys. B: At. Mol. Phys. 14, 513 (1981); 527 (1981).
3. D. Liesen, A. N. Zinoviev and F. W. Saris, Phys. Rev. Lett., 47, 1392 (1981).
4. A. Antar and Q. Kessel (to be published).

MOLECULAR-AUTOIONIZATION SPECTRA FROM He^+- He COLLISIONS

N. Tokoro, S. Takenouchi and N. Oda

Research Laboratory for Nuclear Reactors, Tokyo Institute of Technology, Meguro-ku, Tokyo 152, Japan

The study of the continuum part of ejected electron spectra produced in slow ion-atom and ion-molecule collision has been becoming one of the most interesting subject in atomic collision processes.

Recently we reported the angular distributions of continuous electrons with energies of 5-70 eV ejected from 20 keV He^+ on He[1], where we suggested that the mechanisms for production of continuum part of electron spectra are attributed to two kinds of processes; one of which is mainly responsible for low-energy part (≲30 eV) whose angular distribution is symmetrical around 90° in the laboratory system and the other dominates for high-energy part (~70 eV). The latter process was interpreted as the molecular autoionization of a quasimolecule of He_2^+ ($1s\sigma_g\cdot 2p\sigma_u^2\ {}^2\Sigma_g \rightarrow 1s\sigma_g^2\ {}^1\Sigma_g$ of He_2^{2+} + e) because the angular distribution of high energy electrons (~70 eV) is approximately isotropical in the centre-of-mass system. In the present work, measurements of ejected electrons have been done for the extended electron energy region (3-200 eV) to investigate the molecular autoionization processes more precisely. Figure 1 shows the measured electron spectra at ejection angles of 30°, 90° and 150° for 20 keV He^+ on He. As can be seen from Fig. 1, it seems that the continuum parts of spectra in common consist of two components; one of which is well described by an exponential function and the other is a broad band overlapped on the former in higher energy region. The contribution of the latter is obviously dependent on the ejection angles. This feature is very similar to that for Kr^+ impact on Kr[2]. The spectral-shape of the molrcular autoionization may be best described by the profile function P(Ee), introduced by Gerber and Niehaus[3]. The case of two points stationary phase in the Gerber-Niehaus theory is applied for He^+-He collision at near the turning point. Then we obtain

$$P(Ee)=\frac{2}{\hbar^2}\left(\frac{2\hbar}{\ddot{R}_* \varepsilon'_*}\right)^{\frac{2}{3}}\Gamma_A\cdot Ai^2(z),\quad Z=\left(\frac{\ddot{R}_*\varepsilon_*}{2\hbar}\right)^{-\frac{1}{3}},$$

where Γ_A is the autoionization width, Ai denotes the Airy function, $\ddot{R}$ is the radial acceleration, and ε and ε' are the potential difference between $1s\sigma_g\cdot 2p\sigma_u^2\ {}^2\Sigma_g$ and $1s\sigma_g^2\ {}^1\Sigma_g$ of He_2^{2+} and its derivative. The asterisk means values at the closest distance. Since the potential curves of $1s\sigma_g\cdot 2p\sigma_u^2\ {}^2\Sigma_g$ and $1s\sigma_g^2\ {}^1\Sigma_g$ for the internuclear distance from 0.5 au to 1.4 au have been calculated by Sidis[4], values for the distance less than 0.5 au were estimated by the linear extrapolation. The Coulomb trajectry is assumed for the incident ion. Hence, only Γ_A is left as an adjustable parameter to be determined by comparision with experimental spectra. The value of Γ_A has been determined according to the following equation, assuming the isotropic distribution of ejected electrons,

$$\sigma(Ee,\ 90°)-Ae^{-B\cdot Ee}=\frac{1}{4\pi}\int_0^{Rc}P(Ee)\cdot 2\pi b\cdot db,$$

where b is the impact parameter, Rc is the critical internuclear distance the molecular autoionization begins to occur, and σ(Ee,90°) is the measured electron spectrum at 90°, which is least influenced by the kinematical effect. The "back ground" electron spectrum is represented by a function $Ae^{-B\cdot Ee}$. The calculated result is compared with the experimental one in Fig. 2, resulting in a fairly good agreement, where the value of Γ_A is ~1.1×10^{15} sec^{-1}. More detailed discussions will be given at the conference.

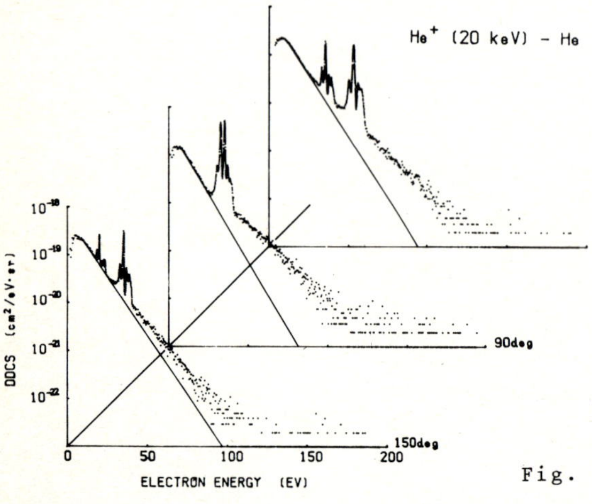

Fig. 1

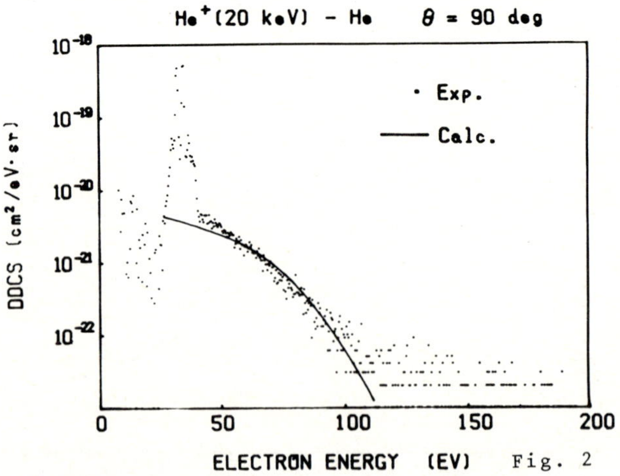

Fig. 2

1) N.Tokoro, S.Takenouchi, J.Urakawa and N.Oda J.Phys.B 15,3737(1982)
2) V.V.Afrosimov, Yu.S.Gordeev, A.N.Zinov'ev, D.Kh.Rasulov and A.P.Shergin, JETP Lett. 24,5(1976)
3) G.Gerber and A.Niehaus, J.Phys.B 9,123(1976)
4) V.Sidis, J.Phys.B 6,1188(1973)

3p-VACANCY PRODUCED DUE TO TWO-ELECTRON TRANSITIONS IN Kr-Kr AND Kr-Ar COLLISIONS

N.A.Guschina, G.G.Meskhi, V.K.Nikulin, A.P.Shergin

A.F.Ioffe Physical Technical Institute, Leningrad, USSR

In[I] the certain group of electrons produced in Kr^+-Kr collisions in the keV energy range was supposed to be connected with the creation of 3p-vacancies due to two-electron transitions $3d^2$-3pnl in quasimolecule or atom. In such transitions two 3d-vacancies are transformed to 3p-vacancy and vacancy in outer shell.

In the present work two-electron states of the Kr-Kr and Kr-Ar systems were calculated as a sum of the corresponding one-electron levels. Two-electron levels $(5g\sigma)^2$ and 5fσ5dσ, formed from $3d^2$ and 3p4p states of the separate Kr atoms, are presented in Fig.I. As seen, these levels approach at the internuclear distance $R \gtrsim 5$ a.u. and cross at $R \simeq 2$ a.u. for Kr_2^+ and Kr_2^{3+} system, respectively. Therefore outer shell ionization or excitation occuring during collision can result in significant probability of the transitions between two-electron states.

Here direct evidence of such a transition in collisions of Kr ions with Kr and Ar atoms was received. For this aim the 3p-x-ray emission was studied in dependence on the incident ion energy E_0 with proportional counter.

The experimental cross sections $\sigma_x(3p)$ are presented in Fig.2. The increase of $\sigma_x(3p)$ near E_0=35 and 75 keV in Kr-Kr and Kr-Ar collisions is connected with the known 3p-vacancy formation due to promotion of the one-electron 5fσ and 4dσ orbitals, respectively. The criti-

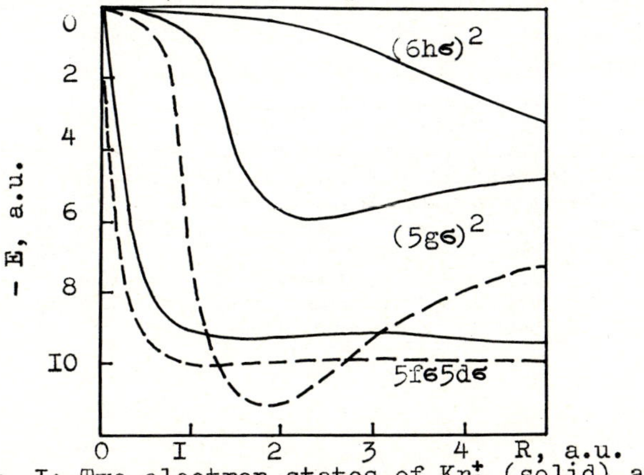

Fig. I: Two-electron states of Kr_2^+ (solid) and Kr_2^{3+} (dashed) quasimolecules.

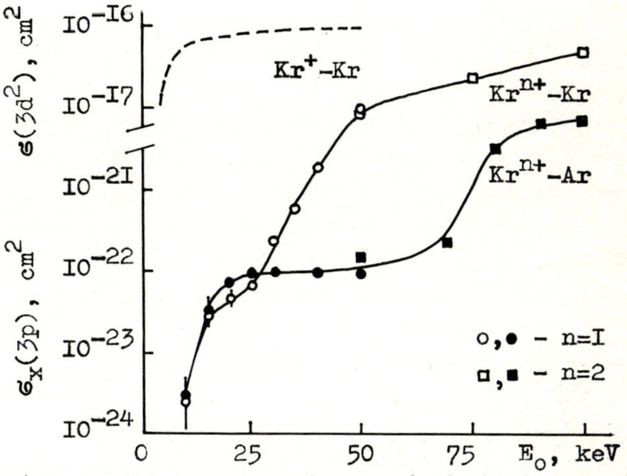

Fig.2: The cross sections $\sigma_x(3p)$ (points) and $\sigma(3d^2)$ (dashed line) versus the incident ion energy E_0.

cal internuclear distance for 5fσ promotion $R_c^{exp} \simeq 0.45$ a.u. obtained from the threshold energy E_0 agrees well with the calculated value R_c^{calc} (Fig.I). For 4dσ orbital in the case of Kr-Ar system $R_c^{exp} \simeq 0.25$ a.u. agrees with R_c^{calc} =0.2 a.u.

The behaviour of $\sigma_x(3p)$ at E_0<25 and 75 keV in the Kr^+-Kr and Kr^+-Ar cases, respectively, is connected with the 3p-vacancy formation due to $3d^2$-3pnl transitions. This conclusion follows from the similarity in the behaviours of $\sigma_x(3p)$ and cross section for two 3d-vacancy production $\sigma(3d^2)$ [2] (also shown in Fig.2). In Kr^+-Ar case R_c^{exp} for 3d vacancy production is equal to 0.8 a.u., agreeing with R_c^{calc}=0.9 a.u.

The two-electron transition probability was estimated using the relation $W=\sigma_x(3p)/\sigma(3d^2)\cdot\omega_x(3p)$ and the single vacancy fluorescence yield $\omega_x=6\cdot10^{-5}$ for 3p-vacancy decay[3]. The estimation gives $W \simeq 1\cdot10^{-2}$.

References

I. V.V.Afrosimov, Yu.S.Gordeev, A.P.Shergin, A.N.Zinoviev, Abstr. XI ICPEAC, Kyoto, I979, p.784.
2. V.V.Afrosimov, Yu.S.Gordeev, A.M.Polyansky, A.P.Shergin, Zh.Thechn.Fiz. 4I, I34 (I972).
3. E.J.McGuire, Phys.Rev. A45, I043 (I972).

EXPERIMENTAL OBSERVATION OF THE THOMAS PEAK IN ELECTRON CAPTURE

E. Horsdal-Pedersen[†], C.L. Cocke, and M. Stockli

J.R. Macdonald Laboratory, Physics Department, Kansas State University, Manhattan, KS 66506, U.S.A.

The elementary understanding of electron capture by fast, light ions has proven to present a very difficult problem even for simple one-electron target atoms or ions. It is now widely understood, on theoretical grounds, that any quantum treatment of the process based on perturbation theory must include at least terms up to second order in the two nuclear potentials[1,2]. For capture near the forward direction, the internuclear potential can be disregarded, and the only second-order term present involves the scattering of the electron off each nuclear potential[2]. Asymptotically, this term becomes the dominant single term of the expansion near the forward direction, and it is believed that the expansion converges to it in this region[1].

The present abstract reports experimental data, which support the above expectations. To understand this, it is useful to examine the classical analogue of the second-order term, which was published already in 1927 by Thomas[3] long before the first discussion[4] of the second-order term itself.

According to the classical theory, an electron may be captured by a fast, light ion as a result of two binary collisions, the first with the projectile and the second with the target nucleus. After the first collision, the electron must move with nearly the same speed as the projectile (it recoils near $\varphi_{LAB} = 60^o$), and in the second collision, it must scatter off the target nucleus to finally move with the same velocity as the projectile.

A very distinct feature of the classical theory is the small but definite deflections θ_T of the projectile in the first collisions, given by the ratio between the transverse momentum transfer, $m_e v\cos 30^o$, and the longitudinal momentum of the projectile, $m_p v$, where v is the projectile velocity and m_e and m_p are the electron and projectile masses, respectively. We find the velocity-independent expression $\theta_T = \sqrt{3}/2 \cdot m_e/m_p$, which for protons give $\theta_T = 0.47$ mrad. The same angle also emerges from an analysis of the (quantum) second-order term as a critical angle[2], at which the term has a maximum. This maximum should be observable experimentally as a peak in the differential-capture cross section[5-7].

The experimental observation of this signature of the second-order quantum theory is made difficult by the very tiny scattering angles, which must be resolved experimentally, and the extremely small reaction cross sections in the asymptotic velocity region.

We have succeeded in conquering these difficulties in a recent experiment on H^+ + He at Kansas State University. The resulting differential-capture cross sections are shown in Fig. 1. At the largest projectile velocity (7.4 MeV), the predicted peak (the Thomas peak) at $\theta = \theta_T = 0.47$ mrad is clearly seen. The trend of the data (the peak disappearing at low velocities) is also as expected. Although the experimental data thus leave little doubt that the widely held belief in the importance of the double-scattering mechanism at high velocity is correct, a quantitative comparison (Fig. 1) with the best theoretical calculations available[5,7,8] shows that the angular distribution and the size of the high-velocity differential-capture cross section is not yet fully understood.

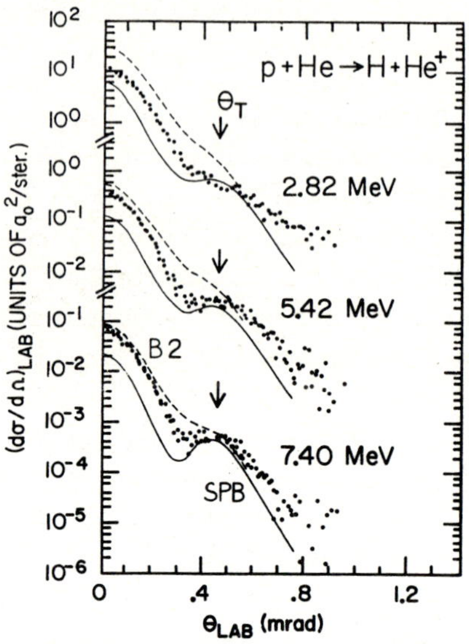

Fig. 1: Plots of $d\sigma/d\Omega$ vs θ. The experimetal data (dots) are compared with theoretical curves representing the exact second Born theory[5], B2, and the strong-potential Born theory[8], SPB.

This work was supported by the NATO-SFP (EH-P) and the Chemical Sciences Division, US DOE.

REFERENCES

[†]Permanent address: Institute of Physics, University Aarhus, DK-8000 Aarhus C, Denmark

1. J.S. Briggs et al., Comm.At.Mol.Phys. 12, 1 (1983)
2. K. Dettman and G. Liebfried, Z.Phys. 218, 1 (1969)
3. L.H. Thomas, Proc.Roy.Soc. A114, 561 (1927)
4. R.M. Drisko, Thesis, unpublished (1955)
5. P.R. Simony et al., Phys.Rev.A 26, 1337 (1982)
6. J.S. Briggs et al., J.Phys.B 15, 3085 (1982)
7. R.D. Rivarola, priv.comm. (1983); the CDW approx.
8. J.H. McGuire, priv.comm. (1982)

EXPERIMENTAL DETERMINATION OF THE DENSITY MATRIX DESCRIBING ELECTRON TRANSFER COLLISIONS FOR H^+ ON He TO THE n=3 STATE OF H

C. C. Havener, N. Rouze, W. B. Westerveld† and J. S. Risley

Department of Physics, North Carolina State University, Raleigh, NC 27650

Excited hydrogen atoms in the n=3 state were formed in electron transfer collisions of 40-80 keV protons on He inside a gas cell. Well-defined, reversible electric fields were applied in axial (parallel) and transverse (perpendicular) directions to the ion beam. A 7 mm beam length centered 27 mm downstream from the entrance aperture to the gas cell was viewed using an optical system consisting of a Balmer-alpha filter, a linear polarizer, a $\lambda/4$ retardation plate and a photomultiplier. The intensity and polarization fraction of the Balmer-alpha radiation emitted at an angle of 90° to the beam axis were measured as a function of the applied electric field.

Measurements of the total intensity and the intensity through the linear polarizer set at 45° (I_{45}) were found to increase symmetrically with respect to the direction of the applied transverse electric field except for a small asymmetry in the I_{45} measurements at 40 keV. The linear polarization fraction decreased symmetrically with the transverse field.

The circular polarization fraction for 60 keV H^+ on He shown in Fig. 1 exhibits highly oscillatory structure. The solid line is calculated from our best estimate of the 14 non-zero elements of the density matrix determined by using a linear least-squares fit to the four intensity and two polarization fraction measurements using transverse fields and to the two previous measurements of the total intensity and linear polarization fraction using axial electric fields.[1] The circular polarization fraction is influenced by the imaginary parts of the off-diagonal terms in the t=0 density matrix, which were not observable using only the axial fields.

From the previous axial electric field measurements for He we found a large dipole with the electron cloud polarized toward the He^+ ion.[1] The transverse electric field measurements are sensitive to the projection along the beam axis of the perihelion velocity of an electron in a classical orbit about the proton.[2] This velocity component can be related to the tilt or orientation of the dipole immediately after the electron transfer collision.

The electron probability distribution for the n=3 state of H is shown in Fig. 2 using our best values for the density matrix for 40 keV H^+ on He. The peaked intensity distribution reflects the large 3s cross section and the slight ridge at the left indicates the electric dipole.

This work was supported in part by the Atomic, Molecular and Plasma Physics Program of the U. S. National Science Foundation.

†Present address: Department of Physics, University of Windsor, Windsor, ON, Canada

1. C. C. Havener, Phys. Rev. Lett. 48, 926-929 (1982).
2. J. Burgdörfer, Z. Phys. A (in press).

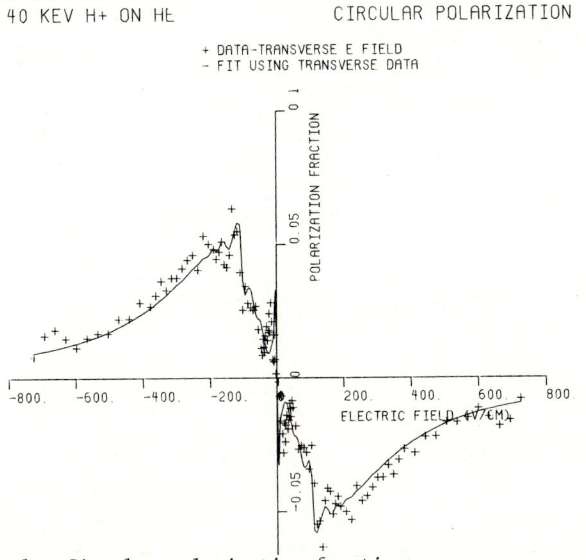

Fig. 1. Circular polarization fraction.

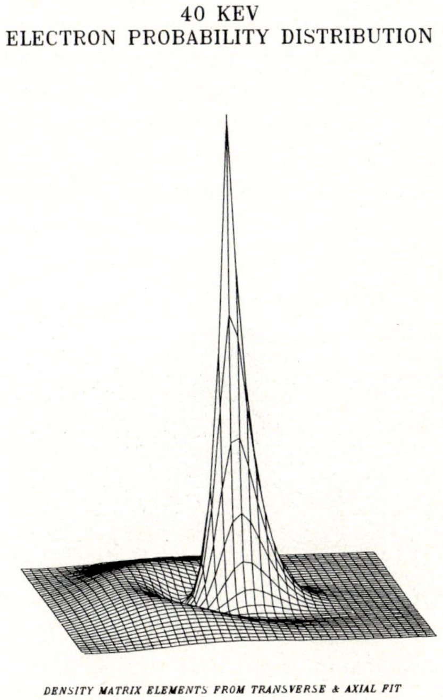

Fig. 2. Electron probability distribution.

DIFFERENTIAL CROSS SECTIONS IN ION-NEUTRAL CHARGE EXCHANGE COLLISIONS

J. H. Newman, Y. S. Chen, P. S. Gibner, K. A. Smith, and R. F. Stebbings

Department of Space Physics and Astronomy, Rice University, Houston, Texas USA 77251

We present differential cross sections for charge transfer of 0.4 to 5.0 keV H^+, He^+, and O^+ atoms with various gases. Cross sections have been measured within the angular range 0.1^o to 20°.

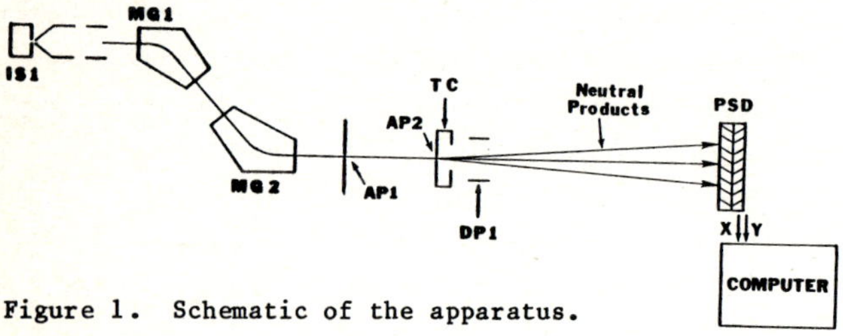

Figure 1. Schematic of the apparatus.

Ions are extracted from an electron-impact ion source and momentum analyzed by a pair of 60° sector magnets. These ions are collimated to about 0.5 mrad by two apertures (AP1 and AP2) < 25 microns in diameter and 10 cm apart. AP2 is also the entrance to the target cell (TC). Gas is admitted to the TC and the collimated ions may undergo scattering and charge transfer collisions. Deflection plates DP1 remove ions from the beam, allowing only neutrals to impact the detector. Target cells with different length to exit aperture diameter ratios are used to detect particles scattered up to 20° in the laboratory.[1]

Neutrals resulting from charge transfer are detected by a position sensitive detector (PSD) comprised of a pair of microchannel plates in a chevron configuration. The PSD has a resolution of about ± 60 microns and an active area diameter of 25 mm. Position information is stored by a microprocessor during the experiment and analyzed later. The PSD allows collection of data from all scattering angles of interest simultaneously, permitting data acquisition "in parallel", thereby greatly enhancing data rates.

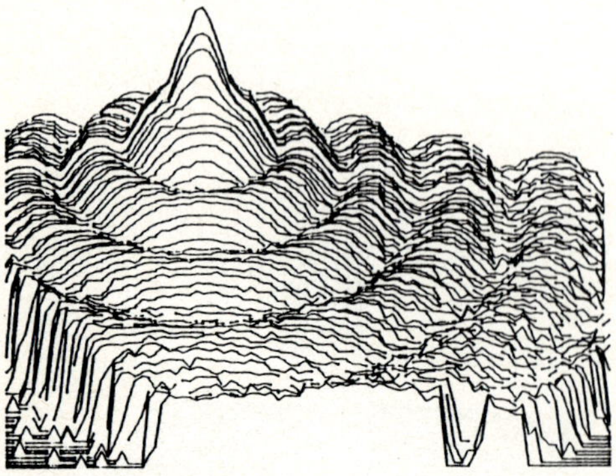

Figure 2. Logarthmic intensity as a function of position on the detector for product neutrals of reaction 1. Primary He^+ with 750eV kinetic energy.

Figure 3. Relative differential cross sections for reaction 1.

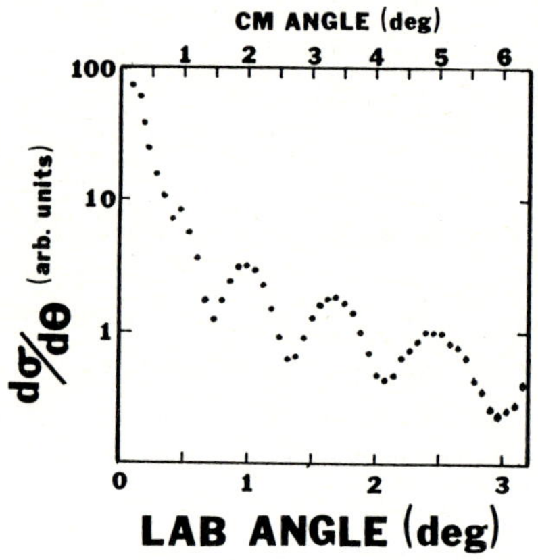

Raw data for the reaction:

$$He^+ + He \rightarrow He + He^+ \qquad (1)$$

is shown in Figure 2. Relative differential cross sections determined from this data are in agreement with previous data.[2] Figures 3 and 4 show differential cross sections for reaction (1) with a primary He+ energy of 750eV.

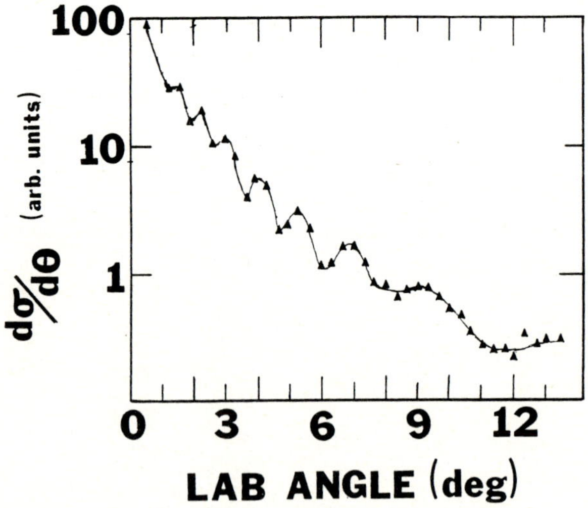

Figure 4. Differential cross sections for charge transfer of He^+ with He. Primary He^+ energy 750eV.

REFERENCES

1. J.H. Newman, et al., "Differential Cross Sections in Neutral-Neutral Collisions", ICPEAC XIII (this conference), 1983
2. S.W. Nagy, et al., Phys. Rev. A, 3,280 (1971)

Research supported by NSF Grant No. ATM 80-23219 and NASA Grant No. NSG-7386.
One of us, J.H.N., gratefully acknowledges the support of Texaco in the form of a Texaco Fellowship.

ANGULAR CORRELATION MEASUREMENTS OF He (3^3P) RESULTING FROM He^+ + Ne COLLISIONS.†

M. Natarajan, A.L. Goldberger, O. Yenen and D.H. Jaecks

University of Nebraska, Lincoln, Nebraska, USA 68588

Using photon-scattered particle coincidence techniques, measurements of linear and circular polarization of the radiation resulting from the collisions.

$$He^+ + Ne \rightarrow He\,(3^3P) + \sum_f Ne^+$$
$$\hookrightarrow He\,(2^3s) + h\nu\,(3889\ A°) + \sum_f Ne^+$$

have been made for impact parameters (b) from .65 to 1.7 a.u. The summation on Ne^+ indicates the different possible final states of the target after the collision. Projectile energies of 3, 4.5, 6 and 7 keV were used. The apparatus used is fully described elsewhere.[1]

The intensity of the emitted radiation is given by,

$$I(\beta) = \sum_f I_f(\beta) = \sum_f c\,[28\sigma_o^f + 26\sigma_1^f + 15\,(\sigma_o^f\sigma_1^f)^{1/2}\cdot$$
$$\cos\Delta\phi^f \sin 2\beta + (30\sigma_1^f - 15\sigma_o^f)\sin^2\beta]$$

where β is the angle of the polariser with respect to the incident He^+ beam, σ_o^f, σ_1^f and $\Delta\phi^f$ are the excitation cross-sections and their relative phases for the magnetic sublevels $m_\ell = 0, \pm 1$ respectively, for a particular final state f of Ne^+. Measurements were made for β = o°, 45°, 90° and 135° with and without a quarter wave plate to determine the Stoke's parameters[2] P_1, P_2, P_3 and $|P|$.

From the diabatic correlation diagram[3] for $(He - Ne)^+$, we can see that He^* + Ne^+ and He^* + Ne^{++*} levels are populated through crossings occurring at 1.5 < b < 2 a.u. and b < 1.3 a.u. respectively. There is an additional possible rotational coupling mechanism, at the united atom limit, leading to the He^* states. Figure (1) shows the observed threshold at about 1.6 a.u. for the formation of He (3^3P), indicating the $\Sigma - \Sigma$ crossing resulting in He^* + Ne^+ states. This is in good agreement with previous total cross-section measurements of Tolk[4] et al. The onset of each coupling can be seen from the behaviour of $\sigma_o'/2\sigma_1'$ and $|P|$, the degree of polarisation, as shown in Figure (2). (Primes on σ_o and σ_1 indicate that they are summed over all Ne^+ final states.)

†This research is supported by the National Science Foundation.

1. D.H. Jaecks, F.J. Eriksen, W. de Rijk and J. Macek, Phys. Rev. Let. 35, 723 (1975).
2. Born and Wolf, *Principles of Optics*, 4th Edition (1970), Pergamon, NY.
3. M. Barat, J.C. Brenot, D. Dhuicq, J. Pommier, V. Sidis, R.E. Olson, E.J. Shipsey and J. Browne, J. Phys. B9, 269 (1976).
4. N.H. Tolk, C.W. White, S.H. Dworetsky and L.A. Farrow, Phys. Rev. Lett. 25, 1251 (1970).

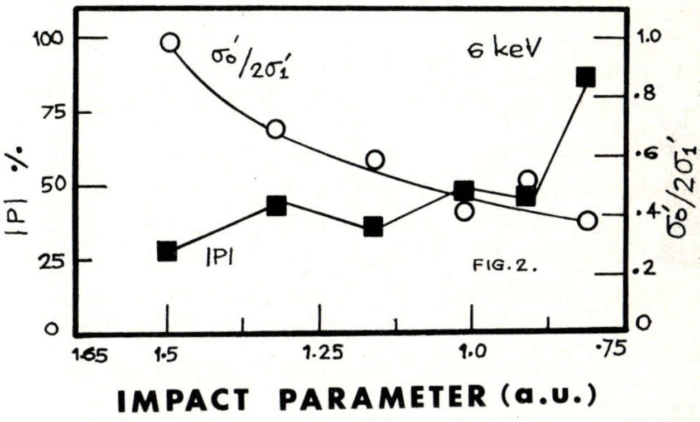

FIG. 2.

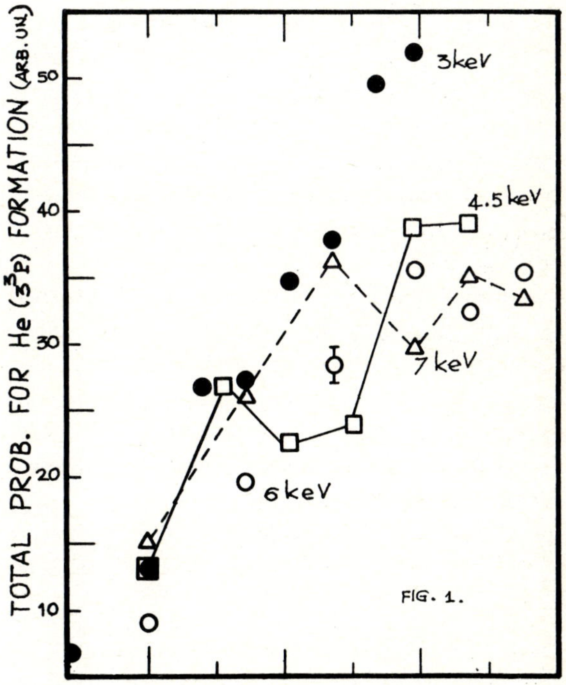

FIG. 1.

CHARGE EXCHANGE BETWEEN H_2^+ AND H^-

S. Szücs, M. Karemera and M. Terao

Institut de Physique Corpusculaire, Université Catholique de Louvain
B-1348 Louvain-la-Neuve - BELGIUM

The cross section for the charge exchange reaction :

$$H_2^+ + H^- \rightarrow H_2 + H \quad (1)$$

has been measured over the energy range 20-2000 eV.

The experimental method is essentially the same as the one used previously for the investigation of the processes :

$$H^+ + H^- \rightarrow H + H \quad (2)$$

and

$$H^+ + H^- \rightarrow H + H^+ + e \quad (3)$$

that have been reported earlier (Szücs et al, 1981 and 1982). The main features are the use of merged ion beams and the coincident detection of the products. The beams are merged over a distance of some 10cm. The precise interaction length is defined by means of an equipotential cell that surrounds the beam path and on which a voltage is applied, thus producing a characterising shift of the kinetic energies of the reactants and of the products formed there.

At the same time, this voltage controls the energy of the collision. The detection of the neutral products takes place some 120cm ahead, thus with a significant difference in the times of flight, yielding a pair of pulses with a well defined and characteristic separation. A multichannel analysis of the time intervals between successive pulses unambiguously reveals the reaction rate.

The results are shown in figure 1 (solid circles) together with those obtained previously for reaction (2) (open circles). It must be pointed out that the measurements, in the present state, do not distinguish the formation of a molecule H_2 from that of two atoms so that the dissociative electron capture is included in the cross section shown in figure 1.
It is even possible that a process as :

$$H_2^+ + H^- \rightarrow H + H^+ + H + e$$

contribute to the observed cross section.

More detailed experimental work is on the way to investigate the contributions of the dissociative channels and additional results will hopefully be presented at the conference.

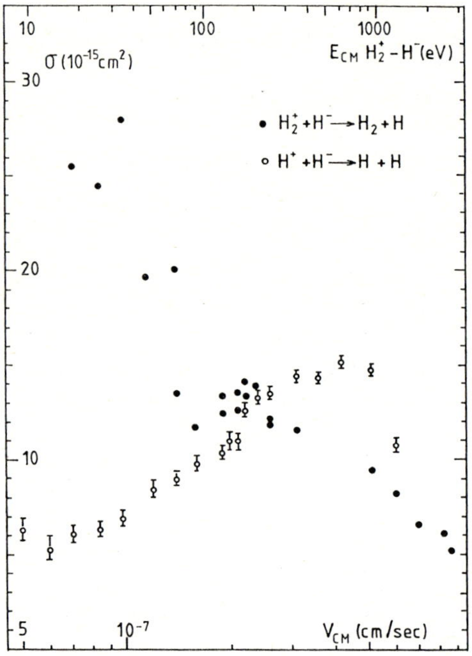

Fig. 1

References

1. S. Szücs, M. Karemera and F. Brouillard, 12th ICPEAC post deadline paper, Gatlinburg, (1981).
2. S. Szücs, M. Karemera and M. Terao, 9ème colloque Phys. Coll. Atom. et Elec., Abstract, p. 35, Nice, (1982).

ELECTRON CAPTURE, LOSS AND EXCITATION IN COLLISIONS OF H^+, H(1s), H(2s) AND H^- IN ATOMIC OXYGEN

I D Williams, J Geddes and H B Gilbody

Department of Pure and Applied Physics
The Queen's University of Belfast
Belfast, United Kingdom

The electron capture and electron loss cross sections σ_{10}, $\sigma_{0\bar{1}}$, σ_{01}, $\sigma_{\bar{1}0}$ and $\sigma_{\bar{1}1}$ have been determined for H^+, H(1s) and H^- in atomic oxygen and the results are presented in Figure 1. We have also studied the formation and collisional destruction of H(2s) atoms in atomic oxygen. Cross sections $\sigma_{1.2s}$ and $\sigma_{0.2s}$ for H(2s) formation and $\sigma_{2s.1}$ and $\sigma_{2s.\bar{1}}$ have been measured, together with the total collisional destruction cross section σ_D for H(2s) atoms. These results will be presented at the conference. The measurements are relevant to the physics of the aurora.

Measurements have been carried out within the energy range 2.5 to 25 keV. A momentum analysed primary beam was directed through a thin oxygen target, and the fractional yield of the collision products in the emergent beam were separated electrostatically and recorded. H(2s) metastable atoms were detected by observing Lyman-α radiation, induced by a quenching electric field, with an electron multiplier fitted with a LiF window. Considerable care was taken to ensure single collision conditions in the target, and to accommodate the scattered collision products within the angular acceptance of the detectors.

An iridium tube furnace was used to provide the highly dissociated oxygen target. The amount of dissociation occurring in the furnace was assessed by means of a quadrupole mass spectrometer, which directly viewed gas effusing from a 1 mm hole in the side of the iridium tube. With the furnace heated to a temperature of 2200 K, over 60% of the molecular oxygen was found to be dissociated in the well defined target region.

The measured cross sections were normalised to the σ_{10} cross section obtained by Stier and Barnett[1] for protons in O_2 at 15 keV, and the σ_{12s} cross section measured by Shah et al[2] for protons in Ar at 16 keV. The cross sections in O in general follow the energy dependence of the corresponding cross sections in O_2, but are smaller in absolute magnitude. The present measurements are compared where possible with results of previous experimental investigations[3,4] and comparison is made with the theoretical prediction of σ_{10} by Mapleton[5]. The cross section for the $H^+ + O \rightarrow H + O^+$ interaction is also shown to display the expected characteristics of accidental resonance behaviour.

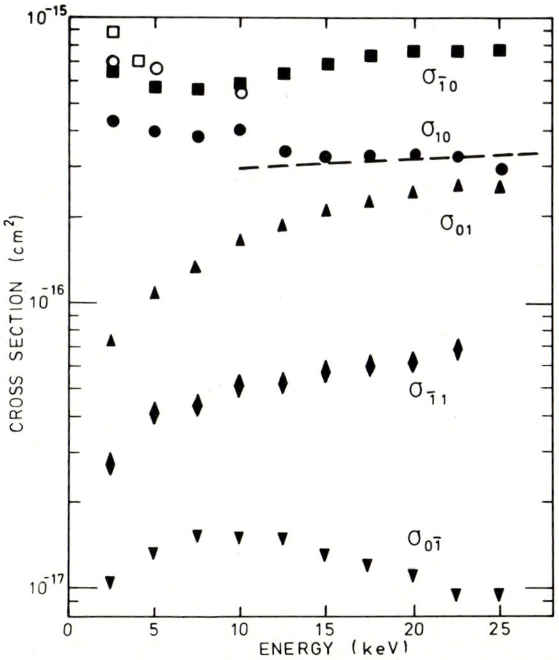

Fig. 1 Electron capture and electron loss cross sections for H^+, H(1s) and H^- in atomic oxygen
σ_{10} ● Present data; ○ Stebbings et al (1964)
---- Theory, Mapleton (1967)
$\sigma_{\bar{1}0}$ ■ Present data; □ Snow et al (1969)

References

1. P M Stier and C F Barnett. Phys Rev 103, 896, 1956.
2. M B Shah, J Geddes and H B Gilbody. J Phys B 13, 4049, 1980.
3. R F Stebbings, A C H Smith and H Ehrhardt. J Geophys Res 69, 2349, 1964.
4. W R Snow, R D Rundel and R Geballe. Phys Rev 178, 228, 1969.
5. R A Mapleton. Phys Rev 164, 51, 1967.

EXCITATION IN INELASTIC COLLISIONS OF H^+ (2 - 15 keV) WITH Li

F. Aumayr, A. Brazuk, U. Wutte and H. Winter

Institut für Allgemeine Physik, Technische Universität Wien, Austria

For collisions of protons with alkali atoms the probability of electron capture increases strongly toward lower impact velocity. Cross section maxima of typically 10^{-14} cm^2 are reached at impact energies of some keV, cf. refs. 1 and 2. This behaviour is caused by similar binding energies if the outer s-electron is transferred from the alkali atom into the n = 2 states of hydrogen. For the H^+- Li collision system, calculations show that at impact velocities above 1 a.u. electron capture proceeds increasingly by removal of Li K-shell electrons with accordingly decreasing cross section[3, 4]. In the low impact energy region below 25 keV calculations[5, 6] can be compared with only one measurement of total electron capture cross sections[1]. Similar data for heavier alkali atoms from ref. 1 differ considerably from results published more recently[2]. Because of these reasons we have undertaken detailed investigations on electron capture for collisions of protons with Li atoms in the impact energy range between 2 and 15 keV.

Applying photon spectroscopy with intensity-calibrated spectrometers for both the vuv- and visible spectral region, the following reaction channels are studied:

H^+ + Li → H(2p) + Li^+ (L_α, 121.6 nm)
-"- → H(3p) + Li^+ (L_β, 102.6 nm)
-"- → H(n=3) + Li^+ (H_α, 656.3 nm)
-"- → H^+ + Li(2p) (670.8 nm)

The observations are carried out for crossed beams of protons and Li atoms under "double magic" angles to avoid errors from possibly polarised radiation. Absolute emission cross sections are determined by direct comparison with Ne^+- Li data for emission at 74 nm in the same experimental configuration (cf. ref. 7 which also contains further experimental details).

In fig. 1 the relative dependence of the L_α emission cross section on proton impact energy is shown. From the simultaneously observed L_β emission it was found that the population of H(3p) states reached about 7 % of that for H(2p) at 5 keV, 11 % at 10 keV and 30 % at 15 keV. If the population ratio of H(3p) and H(3d) is assumed as predicted by theory[6] and H(3d) lifetime corrections appropriate to the experimental geometry are applied, observable cascade contributions from H(3d) to the L_α emission turned out to be less than 3 % at 15 keV and to reach a maximum of 6 % at 4 keV. Cascade contributions from H(3s) can be safely neglected because of the respective long lifetime. In summary, the experimental data shown in fig. 1 correspond to the H(2p) excitation cross section with minor cascade contributions from H(3d) in the region of the cross section maximum.

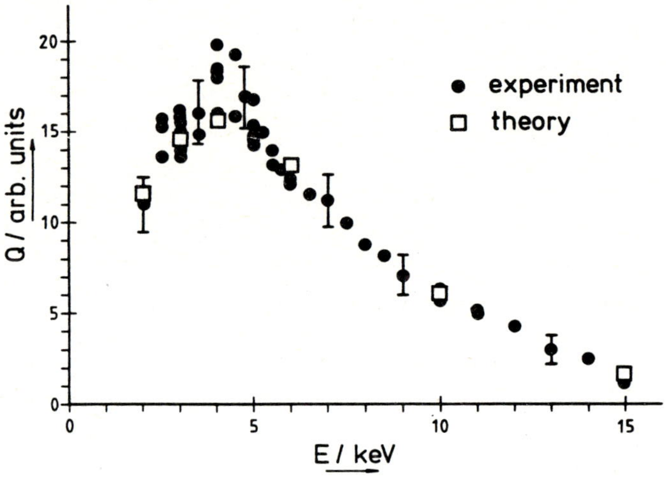

Fig. 1: Dependence of L_α emission cross section on proton impact energy (full points, this work), as compared with calculated H(2p) excitation cross sections (ref. 6, open squares). The data have been normalized at 10 keV. The errors bars show statistical errors only.

The agreement between experiment and calculation for the impact energy dependence is very good. At 7.5 keV the absolute L_α emission cross section amounted to 4 x 10^{-15} cm^2 which is almost a factor of two larger than the corresponding theoretical value from ref. 6. Better agreement is expected from measurements with improved spectrometer intensity calibration and a more accurately determined target thickness for the Li atom beam. Moreover, cross sections for H_α- and Li I resonance line emission as well as for total electron capture are being determined.

Acknowledgment: This work is supported by Austrian Fonds zur Förderung der wissenschaftlichen Forschung (Projekt Nr. 4376) and by International Atomic Energy Agency, Res. Contract No. 2497 RB

References

1. W. Grüebler, P.A. Schmelzbach, V. König and P. Marmier, Helv.Phys. Acta 43,254(1970)
2. T. Nagata, J.Phys.Soc. Japan 48,2068(1980)
3. K.E. Banyard and G.W. Shirtcliffe, J.Phys.B:At.Mol. Phys. 12,3247(1979)
4. R.E. Olson, J.Phys.B:At.Mol.Phys. 15,L163(1982)
5. R.J. Allan, A.S. Dickinson and R. McCaroll, J.Phys.B: At.Mol.Phys. 16,467(1983)
6. W. Fritsch and C.D. Lin, J.Phys.B:At.Mol.Phys. 16, (1983, in print)
7. E. Rille and H. Winter, J.Phys.B:At.Mol.Phys. 15,3489 (1982)

POLARIZATION STUDIES OF H(2P) EXCITATION IN H^+-HE, AR CHARGE CHANGING COLLISIONS

R. Hippler, M. Faust, R. Wolf, H. Kleinpoppen*, H. O. Lutz

Fakultät für Physik, Universität Bielefeld, F. R. Germany

Detailed information about atomic collision processes may be obtained from polarization or angular correlation studies of photons emitted in coincidence with scattered projectiles. For example, in the charge changing collision

$H^+ + He \rightarrow H(2p) + He^+$

the H(2p) atom (lifetime 1.6 ns) decays by emission of Lyman-α photons to H(1s). As a result of the collision process the spatial distribution of the angular momenta of the excited H(2p) state is in general non-isotropic. This anisotropy is conveniently described in terms of an orientation (dipole) vector $\langle T_{11}\rangle$ and an alignment (quadrupole) tensor $\langle T_{2Q}\rangle$ (Q = 0,1,2)[1,2].

The experimental arrangement for such an investigation consists of an ion gun producing protons with kinetic energies of a few keV. The protons are crossfired onto a thermal gas target. Scattered (fast) hydrogen atoms are detected by a position-sensitive channel plate array. Lyman-α photons are detected at 90° with respect to the proton beam by a polarization-sensitive device, consisting of a suprasil plate at Brewster's angle and a photomultiplier. Circular polarization analysis is accomplished by inserting a quarter-wave plate in front of the suprasil plate. By measuring simultaneously at up to eight different azimuthal angles ϕ with respect to the scattering plane the orientation vector $\langle T_{11}\rangle$ and the three alignment tensor components $\langle T_{2Q}\rangle$ (Q = 0,1,2) have been determined.

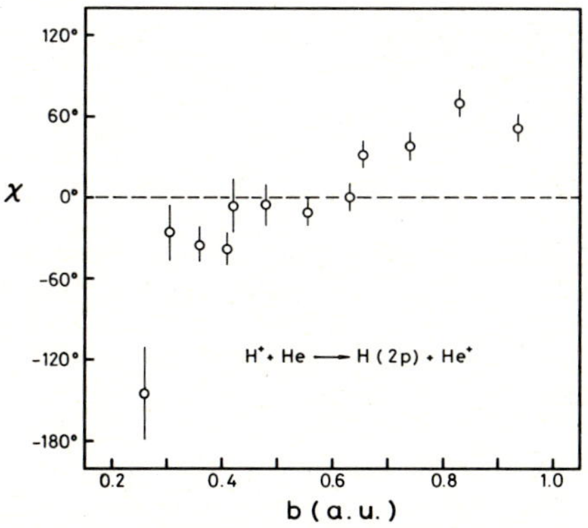

Fig. 1: Phase difference χ versus impact parameter b for 3 keV H^+-He collisions

Results of our measurements for 3 keV H^+-He collisions are given in Fig. 1. In this collision system, H(2p) excitation proceeds via a radial coupling at large internuclear distances between the 1sσ-2pσ molecular states in the transiently formed quasi-molecule. At small internuclear distances a rotational coupling between 2pσ-2pπ results in population of π molecular states which correlate to the $H(2p_{\pm1})$ atomic states. The experimental finding differs from that, as in addition strong population of $H(2p_0)$ states occurs. A similar observation was made by Mueller and Jaecks[3] for 4-8 keV H^+-He collisions. They explained $H(2p_0)$ excitation by a 1sσ-2sσ radial coupling which is crossed on both the ingoing and outgoing part of the trajectory. Another finding of Mueller and Jaecks[3] was that in 4 keV H^+-He collisions the magnitude of the phase difference between the excitation amplitudes f(0) and f(+1) for excitation of $H(2p_0)$ and $H(2p_1)$ states, respectively, is constant over a wide range of impact parameters and amounts to about $|\chi| = 45^\circ$. As is shown in Fig. 1 this differs from the present results as the phase difference χ obtained from[2]

$$\tan\chi = \langle T_{11}\rangle/\langle T_{21}\rangle$$

varies strongly in 3 keV H^+-He collisions.

In proton-argon collisions H(2p) excitation again may proceed via a radial (3pσ-3dσ) coupling at large and a rotational (3dσ-3dπ) coupling at small internuclear distances. As before, both $H(2p_0)$ and $H(2p_{\pm1})$ excitation are observed, as is illustrated in Fig. 2. This behaviour may indicate that a one-electron molecular orbital picture is insufficient to describe the observation[4]. Alternatively, decoupling mechanisms at large internuclear distances may be responsible for $H(2p_0)$ excitation[5].

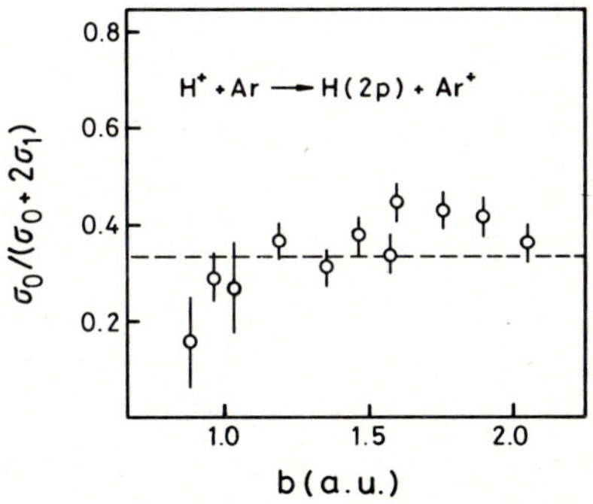

Fig. 2: Relative population of $H(2p_0)$ states versus impact parameter b for 1.5 keV H^+-Ar collisions

This work was supported by the Deutsche Forschungsgemeinschaft (DFG).

*Permanent address: Atomic Physics Laboratory, University of Stirling, Stirling FK9 4LA, Scotland

References

1. U. Fano, J. Macek, Rev. Mod. Phys. 45, 553 (1973)
2. K. Blum, H. Kleinpoppen, Phys. Reports 52, 203 (1979)
3. D. W. Mueller, D. H. Jaecks, Proc. 8th Int. Conf. Atomic Physics, Göteborg (1982), p. B55
4. R. Hippler, G. Malunat, M. Faust, H. Kleinpoppen, H. O. Lutz, Z. Physik A 304, 63 (1982)
5. J. Grosser, J. Phys. B 14, 1449 (1981)

CHARGE-TRANSFER COLLISIONS OF Ne^+ AND METASTABLE HELIUM

R. H. Neynaber and S. Y. Tang

University of California, San Diego, La Jolla, CA 92093, USA and
La Jolla Institute, La Jolla, CA 92038, USA

Absolute and relative cross sections Q for the asymmetric non-resonant charge-transfer reaction

$$Ne^+ + He^* \rightarrow Ne^{\ddagger} + He^+(1\,^2S) \qquad (1)$$

are measured by a merging-beams (mb) technique for a relative kinetic energy of the reactants W from 0.01 to 1000 eV corresponding to $7.6\times10^4 \le v_r \le 2.4\times10^7$ cm/s, where v_r is the relative velocity of the reactants. The Ne^+ is a composite of Ne^+ $(2p^5\,^2P_{3/2})$ and Ne^+ $(2p^5\,^2P_{1/2})$, the He^* is a composite of He (2^3S) and He (2^1S), and $Ne^{\ddagger}$ represents excited Ne. No state selection is made. The studies are conducted by measuring the total current of the charge-transfer product He^+.

The mb apparatus, the general technique for obtaining data, and the procedure for extracting Q from these data have been discussed previously.[1] Electron bombardment sources are used for generating beams of Ne^+ and He^+. The desired W is obtained generally by fixing the energy of the Ne^+ beam at 5000 eV and varying the energy of the He^+ beam in the range 800 to 4400 eV. The He^+ beam is converted in a charge-transfer cell containing Na vapor to a beam consisting of a mixture of less than 6% ground-state He and He^*.[2] The fraction of He (2^1S) in the He^* beam is between a few percent and 13% depending on the energy of the beam.[2,3]

The data show that Q has a $W^{-1/2}$, or v_r^{-1}, dependence for $W \leqslant 0.05$ eV and that Q increases monotonically above W = 0.15 eV to a maximum near W = 200 eV. As W increases above 200 eV, Q gradually falls. The absolute Q (0.01) ≈ (11.6 ± 33%) $Å^2$ and Q (200) ≈ (105 ± 30%) $Å^2$.

The v_r^{-1} dependence at low W agrees with the Langevin model,[4] whereas the behavior of Q for $W \geqslant 0.3$ eV is in fair agreement with the predictions of a modified Demkov approach.[5] The data in the transition region from 0.05 to 0.3 eV is quite well represented by a combination of these two theories. The modified Demkov approach is applied only to the following reactions, each of whose energy defect ΔE (+ for exothermicity) is shown:

$$Ne^+(2p^5\,^2P_{3/2}) + He(2^3S_1) \rightarrow Ne(3s\,^3P_2) + He^+(1\,^2S_{1/2}) + 0.18\ eV \qquad (1a)$$

$$Ne^+(2p^5\,^2P_{3/2}) + He(2^3S_1) \rightarrow Ne(3s\,^3P_1) + He^+(1\,^2S_{1/2}) + 0.13\ eV \qquad (1b)$$

$$Ne^+(2p^5\,^2P_{1/2}) + He(2^3S_1) \rightarrow Ne(3s\,^3P_0) + He^+(1\,^2S_{1/2}) + 0.18\ eV \qquad (1c)$$

$$Ne^+(2p^5\,^2P_{1/2}) + He(2^3S_1) \rightarrow Ne(3s\,^1P_1) + He(1\,^2S_{1/2}) + 0.05\ eV\,. \qquad (1d)$$

In general, other reactions are eliminated either because a) the reactants and products cannot form molecular electronic states of the same species, b) the angular momentum of the $2p^5$ configuration of $Ne^+(^2P_{3/2})$ and $Ne^+(^2P_{1/2})$ is not conserved,[6] or c) $|\Delta E|$ is too large. Reactions with He (2^1S) as a reactant are negligible not only because $|\Delta E|$ is excessive but also because the concentration of He (2^1S) is small compared to that of He (2^3S).

References

1. R. H. Neynaber, G. D. Magnuson, S. M. Trujillo, and B. F. Myers, Phys. Rev. A5, 285 (1972).
2. R. H. Neynaber, G. D. Magnuson, and S. Y. Tang, J. Chem. Phys. 68, 5112 (1978).
3. C. Reynaud, J. Pommier, Vu Ngoc Twan, and M. Barat, Phys. Rev. Lett. 43, 579 (1979).
4. P. Langevin, Ann. Chem. Phys. 5, 245 (1905).
5. R. E. Olson, Phys. Rev. A6, 1822 (1972).
6. For a similar application of such conservation see, for example, G. E. Ice and R. E. Olson, Phys. Rev. A11, 111 (1975).

MULTIPLE IONIZATION OF ARGON BY CHARGE TRANSFER AND DIRECT IONIZATION*

R. D. DuBois and L. H. Toburen

Pacific Northwest Laboratory, Richland, WA 99352 U. S. A.

Charge transfer and direct ionization are competing processes for producing target ionization. Previous investigations of total ionization[1] and electron capture[2] in light ion-atom collisions have shown that at low energies electron capture can dominate the total ionization cross sections. At these lower energies primarily outer shells are involved and since second order processes are generally considered to be negligible, single target ionization is expected. In previous measurements total electron capture cross sections were obtained from growth curves of the charge exchanged beam as a function of target thickness, while the total ionization cross sections were determined from the total electron charge produced. In neither case was any information about the resulting target ionization charge state measured.

In the present measurement we have a target cell of known target thickness through which H^+ or He^+ beams pass. The emerging beam is charge state analyzed and counted by particle detectors. Single electron capture cross sections are obtained from the neutral beam growth as a function of target thickness. In addition, target ions are extracted and charge state analyzed by time-of-flight techniques. When counted in coincidence with the neutral and ion beams cross sections for multiple target ionization resulting from direct target ionization and single electron capture by the projectile are obtained.

Argon charge states resulting from single electron capture and direct ionization by protons and He^+ ions are shown in Figs. 1a and 1b, respectively. The partial ionization cross sections were summed to give total ionization and charge transfer cross sections and are compared to values obtained from references 1 and 2 (dashed lines). Clearly the contribution from multiple ionization is significant. Total electron production cross sections are obtained from $\Sigma_q\, q(\sigma_{1,1,q} + \sigma_{1,0,q+1})$, where q is the target charge state. The first and second subscripts give the projectile charge state before and after the collision. Note that σ_{102} (single capture plus ionization) is a major contributor to the total electron production cross section in this energy range. In fact, for He^+ impact this transfer ionization process is the dominant electron production mechanism. These measurements are being extended to other collision systems.

Figure 1. Argon target charge states produced by electron capture (on left) and direct ionization (on right) by H^+ and He^+, respectively. Total cross sections (dashed lines) obtained from ref. 1 & 2.

(1a)

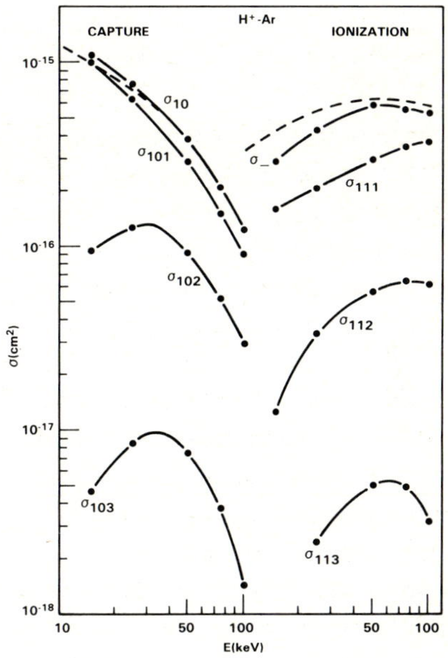

(1b)

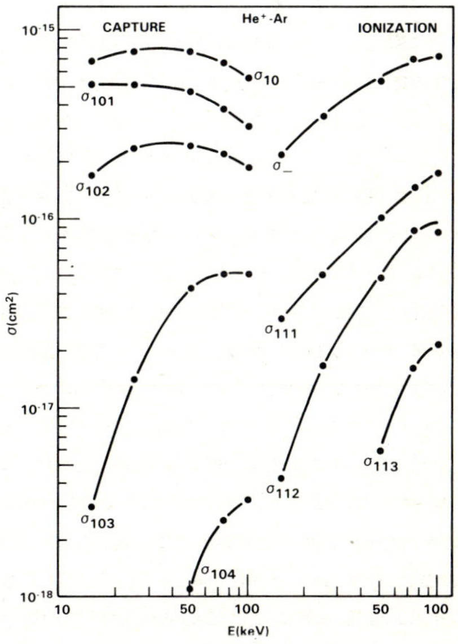

References

1. Atomic Data for Controlled Fusion Research, ORNL-5206, Vol. 1, 1977.
2. M. E. Rudd, R. D. DuBois, L. H. Toburen and C. A. Ratcliffe, in Abstracts of Papers, XII ICPEAC, ed. by S. Datz, 1981, pp. 804-805.

*This work supported jointly by the U.S. Department of Energy, Division of Magnetic Fusion Energy and the Office of Health & Environmental Research. Contract DE-AC06-76RLO-1830.

CHARGE TRANSFER AND FINE STRUCTURE TRANSITIONS IN 2-20 keV Xe^+ COLLISIONS

R.F. King and C.J. Latimer

Department of Pure and Applied Physics, The Queen's University of Belfast, Belfast BT7 1NN, Northern Ireland.

Electron capture collisions between Xe^+ ions and the inert gases, hydrogen and methane have been investigated within the energy range 2-20 keV. Although total electron capture cross sections for fast xenon atoms are in many cases well known, lack of state selection has severely hampered comparison between experiment and theory. In recent experiments[1] we have however used state selected projectiles and now in the present work cross sections for charge transfer into the $^1S_0$ ground state

$$Xe^+ (^2P_{3/2}) + X \rightarrow Xe (^1S_0) + X^+$$

and into the $^2P_{\frac{1}{2}}$ 6s metastable state

$$Xe^+ (^2P_{3/2}) + X \rightarrow Xe (^2P_{\frac{1}{2}} 6s) + X^+$$

of the xenon atom have separately determined for the first time. The $^2P_{\frac{1}{2}}$ 6s xenon atoms were unambiguously detected by a new technique involving selective laser photoionization via the autoionizing $7p^1_{11}$ state which has an autoionizing lifetime[2,3] of $\sim 10^{-13}$s which is fast enough relative to radiative decay to make the ionizing efficiency essentially 100%.

Cross sections for charge transfer into the $^1S_0$ ground state of xenon are slowly increasing with energy for all targets, except for krypton which maximizes just within the present energy range, and the special case of xenon (shown in Fig. 1) which is a symmetric charge transfer process.

Cross sections for charge transfer into the $^2P_{\frac{1}{2}}$ 6s state which involves a fine structure transition plus electron capture into an excited state are rapidly increasing functions of energy and over an order of magnitude lower than the corresponding ground state cross section. Here again the krypton target exhibits a maximum which is not understood, alghough similar maxima have been observed previously[4] in H^+ collisions. For Xe^+ - Xe collisions (Fig. 1) Johnson[5] has used the impact parameter method with a simple molecular model which includes spin orbit effects to calculate charge transfer cross sections. These calculations are in satisfactory accord with the present data.

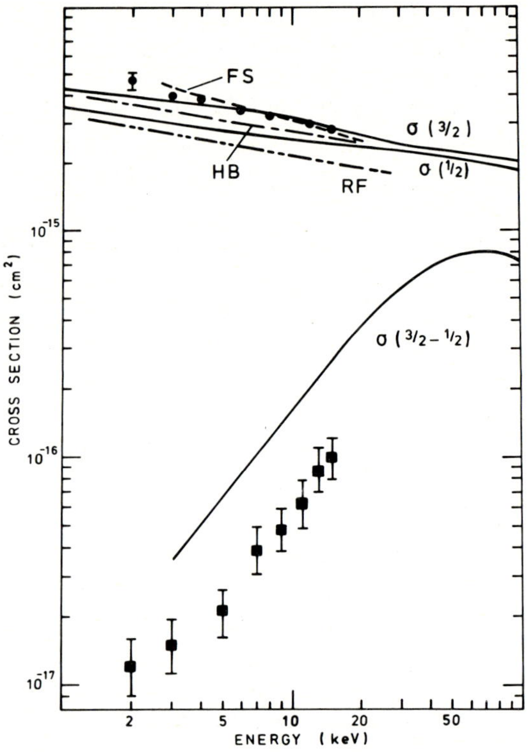

Figure 1. Measured cross sections for the formation of $^1S_0$ ground state, 0, and $^2P_{\frac{1}{2}}$ 6s excited state, , xenon atoms in $Xe^+(^2P_{3/2})$ - Xe collisions. Experiment:- FS, Flaks and Solov'ev[6], total electron capture. Theory:- HB, Hodgkinson and Briggs[7], RF, Rapp and Francis[8], total electron capture: $\sigma(3/2)$, $\sigma(\frac{1}{2})$ Johnson[5], resonant charge transfer: $\sigma(3/2 - \frac{1}{2})$ Johnson[5], charge transfer involving a fine structure transition.

References

1. F.M. Campbell, R. Browning and C.J. Latimer. J. Phys. B. 14, 1183, (1981).
2. R.D. Rundel, F.B. Dunning, H.C. Goldwire and R.F. Stebbings. J. Opt. Soc. Am. 65, 628, (1975).
3. R.F. King and C.J. Latimer. J. Opt. Soc. Am. 72, 306, (1982).
4. J.E. Bayfield. Phys. Rev. 182, 115, (1969).
5. R.E. Johnson. J Phys. Soc. Japan 32, 1612, (1972).
6. I.P. Flaks and E.S. Solov'ev. Sov. Phys. - Tech. Phys. 3, 564, (1958).
7. D.P. Hodgkinson and J.S. Briggs. J. Phys. B 9, 255, (1976).
8. D. Rapp and W.E. Francis. J. Chem. Phys. 37, 2631, (1962).

THE CHARGE TRANSFER CROSS SECTION FOR Kr^+-Kr AT ION ENERGIES BETWEEN 0.08 AND 2 eV

M.T. Elford and O.M. Williams

Australian National University, Canberra, Australia

The cross section for resonant charge transfer σ_c has been derived for Kr^+ ions in Kr by the analysis of drift velocity data obtained using the drift tube described by Hegerberg et al.[1] with a triode type ion source.[2] This modification has enabled the upper energy limit to the derived cross section to be increased by a factor of approximately 3.

The analysis[1] is based on the theory of Skullerud[3] and Forsth and Skullerud.[4] It is assumed that the momentum transfer cross section $\sigma_m = 2\sigma_c$ and that σ_c can be written as $\sigma_c = \sigma_0 v^{-\gamma}$ where v is the ionic velocity. The parameters σ_0 and γ are obtained by fitting to the drift velocity data. Although the theory assumes entirely backward scattering the drift velocity is relatively insensitive to the particular scattering distribution assumed and the error incurred in σ_c from this cause is less than 2½%.

The lower energy limit of the derived cross section (Figure 1) of 0.08 eV is set by the breakdown of the assumption that $\sigma_m = 2\sigma_c$ while the upper limit 2 eV is determined by the largest value of the electric field strength which can be used in the drift velocity measurements.

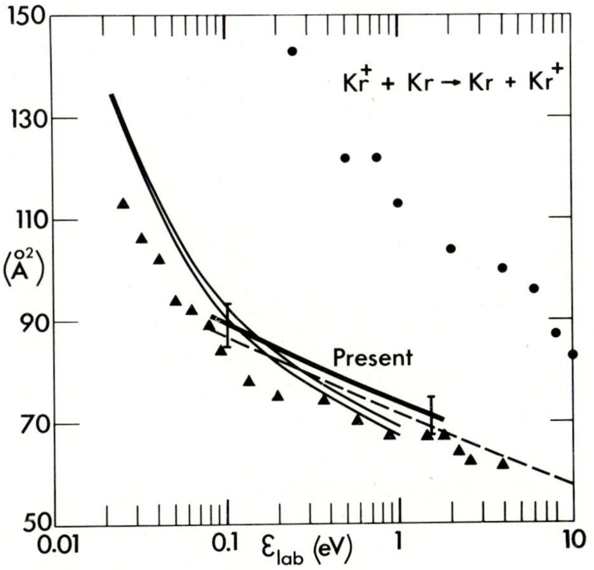

Fig. 1: The charge transfer cross section for Kr^+-Kr as a function of ion energy
▲ Koizumi, et al.[8], ● Smith and Futrell,[9] — Sinha and Bardsley[7] (theoretical, upper curve for Kr^+ ($^2P_{3/2}$), lower curve for Kr^+ ($^2P_{1/2}$)), -- Rapp and Francis[5] (as modified by Dewangan[6]).

The measurements indicated the presence of only one Kr^+ ion species although it was not possible to resolve ions with drift velocities which differed by less than about 3%. The cross section shown should therefore be regarded as being that for a mixture of Kr^+ ($^2P_{1/2}$) and Kr^+ ($^2P_{3/2}$) ions with the proportions being unknown.

The present cross section is in good agreement with the theoretical cross section of Rapp and Francis[5] as modified by Dewangan[6] and with the ab initio calculated cross sections of Sinha and Bardsley.[7] The present cross section and that of Koizumi et al.[8] also derived by the analysis of ion swarm data, agree to within the combined error limits.

References

1. R. Hegerberg, M.T. Elford and H.R. Skullerud, J. Phys. B. 15, 797, (1982)
2. G. Sejkora, M. Hilchenbach, M.T. Elford and T.D. Märk, 6th Eur. Conf. on Atom. and Molec. Phys. of Ion. Gases, 6D, T27, (1982)
3. H.R. Skullerud, J. Phys. B 2, 86, (1969)
4. L.R. Forsth and H.R. Skullerud, J. Phys. B12, 1881, (1979).
5. D. Rapp and W.A. Francis, J. Chem. Phys. 37, 2631, (1962)
6. D.P. Dewangan, J. Phys. B. 6, L20, (1973)
7. S. Sinha and J.N. Bardsley, Phys. Rev. A, 14, 104, (1976)
8. T. Koizumi, K. Okuno, N. Kobayashi and Y. Kaneko, J. Phys. Soc. Japan, 51, 2650, 1982
9. D.L. Smith and J.H. Futrell, J. Chem. Phys. 59, 463, (1973)

EXPERIMENTAL CROSS SECTIONS FOR LYMAN-ALPHA EMISSION IN CHARGE TRANSFER COLLISIONS OF H^+ WITH Cs, Rb, K AND Na ATOMS

T. Nagata

Department of Science and Technology, Meisei University, Hodokubo, Hino, Tokyo 191, Japan

Near-resonant charge transfer processes

$$H^+ + A \rightarrow H(2s \text{ or } 2p) + A^+ \quad (\text{A:alkali atom}) \qquad (1)$$

have received much practical and theoretical interests. Experimental studies of these quasi-one-electron systems are very useful to test theoretical models of atomic collisions. In this laboratory, a series of experimental studies of H^+-alkali atom charge transfer collisions including the processes (1) has been made in the last five years. Very recently, the cross section σ_L for Lyman-alpha emission in H^+-Cs charge transfer collision has been measured in the H^+ energy range 0.18-5.0 keV, and the result has been reported.[1] After that, the lowest limit of the energy range was extended to 0.06 keV, and measurement was made on Rb, K and Na targets as well. This contributed paper reports the result of the extended measurement.

The conventional crossed-beam method was used. A H^+ ion beam was obtained by the use of an electron bombardment ion source and an analyzing magnet. The beam then crossed an alkali-atom beam from an oven at right angles in a collision chamber. A part of Lyman-alpha photons resulting from spontaneous decay of formed H(2p) atoms were counted in the direction normal to the two beams. The photodetector was an electron multiplier (EMI 9643/2B) with a LiF window in front of it. The measurement at energies lower than 0.5 keV was made by using a D^+ ion beam decelerated to desired energies after mass analysis. It was assumed that H^+ and D^+ have the same cross section at the same velocity. The polarization effect of emitted radiation is neglected in this study.

Figure 1 shows the result of the present measurement. The estimated errors in the absolute scale are between 38 % and 65 % depending on the collision energy and target. The Cs cross section is in good agreement in relative variation with the measurement on Cs(0.55-2.4 keV) by Pradel et al.[2] Agreement with the calculation on Cs by Sidis and Kubach[3] and with those on Cs, Rb, and K by Kimura et al.[4] is also satisfactory in both over-all energy dependence and absolute magnitude.

One noticeable feature seen in the present result is the oscillatory structure. The structure in the Cs cross section is most apparent. The similar structure can be seen in the Rb and Na results. In Cs case, this oscillatory structure has been explained as arising from

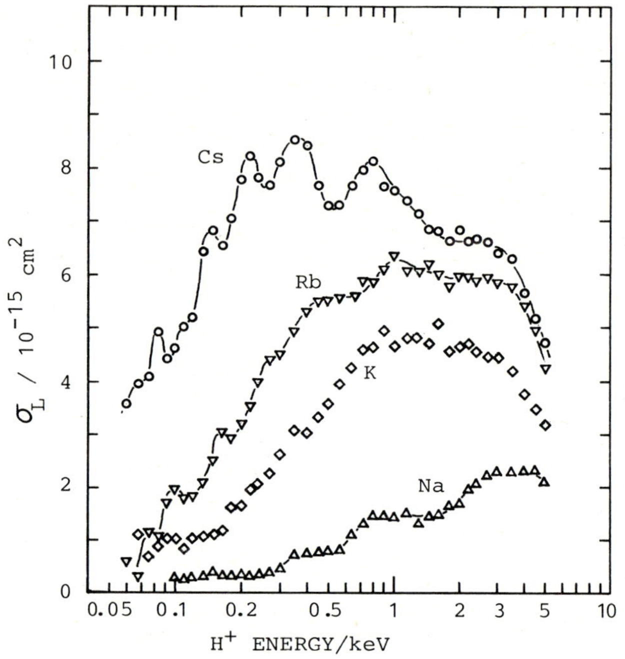

Fig.1: Lyman-alpha emission cross sections for H^+-alkali atom charge transfer collisions. Lines are drawn to clarify the oscillatory structure.

long-range interaction between the charge-exchange channels relating to $H(2s) + Cs^+$ and $H(2p) + Cs^+$ in the dissociation limit.[1] The reasons are (1) σ_L and σ_m, the cross section for H(2s) formation, oscillate with antiphase to each other and (2) the maxima in the plot of σ_L vs. v^{-1} are equidistant. However, it is not certain whether the same explanation is possible for the cases of other targets or not because of lack of experimental evidences. Kimura et al.[4] have offered a different explanation. They obtained theoretical cross-section curves having, although very weak, the similar oscillatory structure and have explained this as arising from the existence of maximum in the difference of the potential curves inside the transition point. Measurements of σ_m in wider energy range would be very useful to establish the origin of the oscillatory structure.

References

1. T. Nagata, J. Phys. Soc. Jpn. 52, 4 (1983).
2. P. Pradel, F. Roussel, A. Schlachter, G. Spiess and A. Valance, Phys. Rev. A 10, 797 (1974).
3. V. Sidis and C. Kubach, J. Phys. B 11, 2687 (1978).
4. M. Kimura, R. E. Olson and J. Pascale, Phys. Rev. A 26, 3113 (1982).

CHARGE EXCHANGE OF (0.2-5.0) keV PROTONS AND HYDROGEN ATOMS IN SODIUM-, POTASSIUM- AND RUBIDIUM-VAPOR TARGETS+

F. Ebel and E. Salzborn

Institut für Kernphysik, Strahlenzentrum, Universität Giessen, 6300 Giessen, West Germany

An important diagnostic tool for fusion plasmas is energy analysis of the neutral hydrogen atoms leaving the plasma. Commonly, the atoms are stripped in a nitrogen gas target to allow the measurement of the energy spectrum by electromagnetic fields. At energies below 1.0 keV, however, the yield of ions can be considerably enhanced by using an electron-capture reaction rather than a stripping reaction[1].

The present study was also motivated by the current interest in negative-ion formation of atomic or ionic hydrogen projectiles in collisions with alkali atoms. A detailed knowledge of charge transfer cross sections in such collisions is relevant to the production of intense $H^-(D^-)$ beams for heating magnetically confined plasmas[2].

We have measured the cross sections σ_{+0}, σ_{+-}, σ_{0-} and σ_{0+} for the following reactions: $H^+ \xrightarrow[Na,K,Rb]{} H^{0,-}$ and $H^0 \xrightarrow[Na,K,Rb]{} H^{+,-}$ in the energy range 0.2-5.0 keV.

The protons were formed in a radio-frequency gas discharge and accelerated to the energy desired. After magnetic deflection of 90° the beam enters a neutralizer cell. If an incident H^0 beam was required the proton beam was converted in this cell with methane. Also in this case a transverse electric field eliminated remaining ions and metastable atoms. After leaving the neutralizer cells the projectile beam was collimated to an angular divergence of $\pm$0.4°.

The intensity of the proton beam was measured in a Faraday cup. The intensity of the neutral beam was measured in a secondary-electron-emission detector which was calibrated using a H^+ beam, by use of the secondary-electron-coefficient ratio $\bar{\gamma}_0/\bar{\gamma}_+=1.17$[3,4].

The ions produced in the charge exchange cell were focused with an Einzel lens and measured in a Faraday cup. Up to a scattering angle of $\pm$3.9° all produced ions were detected. Neutral atoms produced in the charge exchange cell were measured in a secondary-electron-emission detector. The alkali target thickness was varied and determined by the temperature of the target cell.

The electron capture cross sections are showing a maximum behavior depending on the energy defect of the respective reaction. The stripping cross sections increase monotonically with increasing energy. The total uncertainty of the measured cross sections has been determined from the sum in quadratures of all contributing errors. It is about $\pm$20% for σ_{+0} and σ_{0-} and about $\pm$25% for σ_{0+} and σ_{+-}.

The present results for σ_{+0} are in agreement with measurements[5,6] and calculations[7,8]. Only for the Na target we find disagreement with the results of Nagata[5] and Kubach et al[7].

The results for σ_{0-} agree with measurements by Nagata[5] for Rb, however, the data for K and Na show an increasing discrepancy.

The results for σ_{+-} for Na are a factor of 20-30 lower than earlier data[6,9]. For the K target the discrepancy is about a factor of 3. These discrepancies could be explained by a too large target thickness used in earlier experiments. Hence two-step processes $H^+ \to H^0 \to H^-$ increase the apparent cross section.

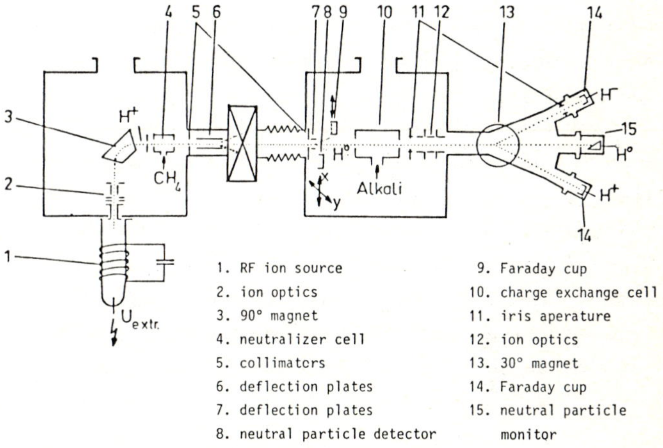

Fig. 1: Schematic diagram of the apparatus

References

1. D. Brisson, F.W. Baity, B.H. Quon, J.A. Ray and C.F. Barnett, Rev. Sci. Instr. 51 (1980) 511
2. K.H. Berkner, R.V. Pyle and J.W. Stearns, Nucl. Fusion 15 (1975) 249
3. J.A. Ray, C.F. Barnett and B. van Zyl, J. Appl. Phys. 50 (1979) 6516
4. K. Miethe, T. Dreiseidler and E. Salzborn, J. Phys. B 15 (1982) 3069
5. T. Nagata, J. Phys. Soc. Japan 48 (1980) 2068
6. C.J. Anderson, A.M. Howald and L.W. Anderson, Nucl. Instr. Meth. 165 (1979) 583
7. C. Kubach and V. Sidis, Phys. Rev. A23 (1981) 110
8. M. Kimura, R.E. Olson and J. Pascale, Phys. Rev. A26 (1982) 3113
9. W. Grübler, P.A. Schmelzbach, V. König and P. Marmier, Helv. Phys. Acta 43 (1970) 254

+Work supported by DFG

ELECTRON CAPTURE BY C^+, O^+, In^+, Sn^+ and Pb^+ IONS IN H_2 IN THE ENERGY RANGE 10-150 keV+

F. Melchert, K. Rinn, A. Müller and E. Salzborn
Institut für Kernphysik, Strahlenzentrum, Universität Giessen, 6300 Giessen, West Germany

The present study was motivated by the current interest in neutralization cross sections of singly charged metallic ions in collisions with hydrogen molecules. Such cross sections are relevant for the neutral beam injection into fusion plasmas since impurity ions from the ion source which after neutralization enter the plasma cause severe energy losses due to radiation processes.

We have measured the cross sections σ_{10} for electron capture by C^+, O^+, In^+, Sn^+ and Pb^+ ions in H_2 in the energy range 10-150 keV. Additionally, the two-electron capture cross sections σ_{20} for Sn^{2+} and Pb^{2+} ions in H_2 were measured for energies up to 300 keV.

For the production of the ions a cartridge metal ion source of Oxford Applied Research Ltd. was used. After acceleration and momentum analysis the ion beam was collimated to an angular divergence of $\pm$ 0.17° and passed through the target gas cell. A 90° electrostatic analyzer downstream separated the generated neutrals from the primary beam. The intensities of both components were measured using single particle detectors[1], however, in the case of metallic ions the primary beam could be measured with a Faraday cup due to the small cross sections. The cross sections were deduced from the "growth rate" method. The target gas pressure was measured with a Baratron capacitance manometer.

The cross sections σ_{10} measured for O^+ and C^+ ions are in good agreement with earlier data[2,3,4] (Fig. 1). For In^+, Sn^+ and Pb^+ ions the cross sections σ_{10} increase monotonically in the investigated energy range and reach values of 4.2×10^{-17} cm², 6.7×10^{-17} cm² and 6.8×10^{-17} cm², respectively, at 150 keV impact energy. The data for Sn^{2+} and Pb^{2+} ions show a broad maximum in the cross sections σ_{20} of 4.2×10^{-18} cm² and 3.3×10^{-18} cm² at 50 keV and 140 keV, respectively.

Furthermore, we have measured the apparent cross section for the production of neutrals in collisions of H_2O^+ and HO^+ ions with H_2 molecules between 10 and 150 keV.

References

1. K. Rinn, A. Müller, H. Eichenauer and E. Salzborn, Rev. Sci. Instr. 53 (1982) 829
2. R.A. Phaneuf, F.W. Meyer and R.H. McKnight, Phys. Rev. A17 (1978) 534
3. G.I. Lockwood, G.H. Miller and I.M. Hoffmann, Phys. Rev. A18 (1978) 935
4. W.L. Nutt, R.W. McCullough and H.B. Gilbody, J. Phys. B 12 (1979) L 157

+Work supported by DFG

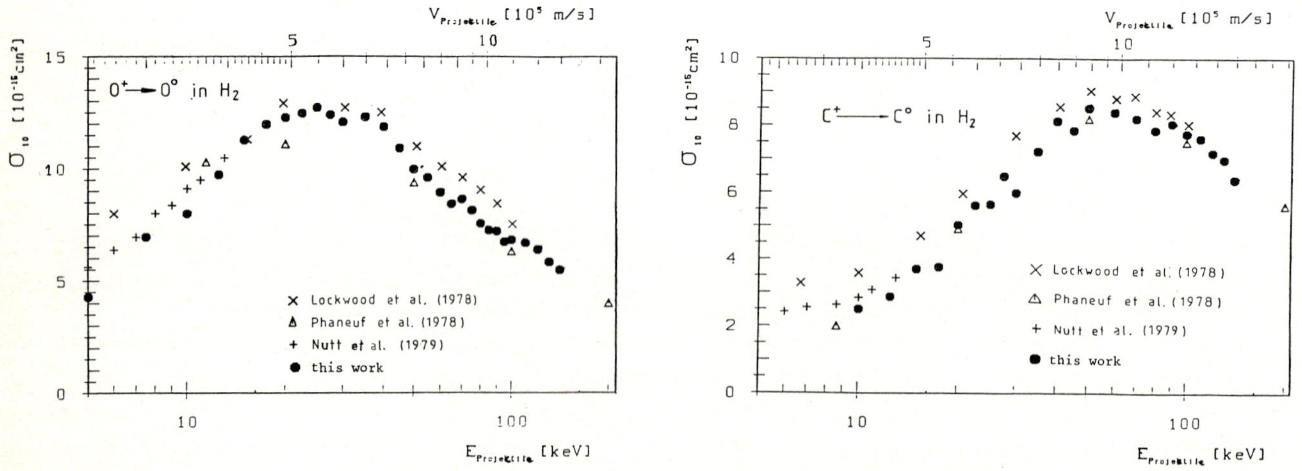

Fig. 1: Electron capture cross section σ_{10} of O^+ and C^+ ions in H_2 versus impact energy.

IONIZATION AND CHARGE EXCHANGE PROCESSES IN COLLISIONS OF ALKALI METAL IONS WITH RARE GAS ATOMS IN THE ENERGY REGION 0,5-7,0 KEV.

B.I.Kikiani, R.A.Lomsadze, S.V.Martinov, N.O.Mosulishvili, M.R.Gochitashvili, V.M.Lavrov

A.F. Ioffe Physico-Technical Institute, Leningrad, USSR.
Tbilissi State University, Tbilissi, USSR.

Ionization and charge exchange processes have been studied in collisions of Na^+ ions with Ar and Kr atoms and K^+ ions with Ne, Ar, Kr and Xe atoms in the ion energy region from 0,5 to 7,0 kev. Previously total cross sections were known only at energies above 3 kev[1], and for some pairs there are data for the differential cross sections of elastic and inelastic scattering also[2,3].

Measurements have been made by the potential method[4]. Secondary ions and electrons produced in a collision have been detected. Since inelastic processes in the collision of an alkali metal ion with a rare gas atom are accompanied by transfer of a considerable kinetic energy to target particles[2,4], the experimental procedure was improved.

The measured ionization and charge exchange cross sections are shown in Fig.1,2. The smollest charge exchange cross section was observed in K^+-Ne collision and the ionization cross section was observed in Na^+-Ar, Kr and K^+-Ne collisions.

When discussing the mechanisms of inelastic processes let us consider the lovest quasimolecular terms, presented qualitatively in Fig.3, which are characteristic for all pairs of particles but K^+-Ne one. As a result of the exchange repulsion of terms corresponding in the dissocation limit to the M^+-A^o and M^o-A^+ states, a term coresponding to the charge exchange process approaches to the ground state term. This fact leads to relatively high probability of the charge exchange process due to the Demkov-Zener transition. In the case of K^+-Ne collisions terms correspoding to the states K^o-$Ne^+(2p^5)$ and K^+-$Ne(2p^53s)$ have an inverse order as compared to that given in Fig.3. Therefore the term corresponding to the charge exchange process moves off from the ground state term with decreasing of the internuclear distance instead of aproaching to it as it is in all other cases. It provides an anomalously small charge exchange cross section this case.

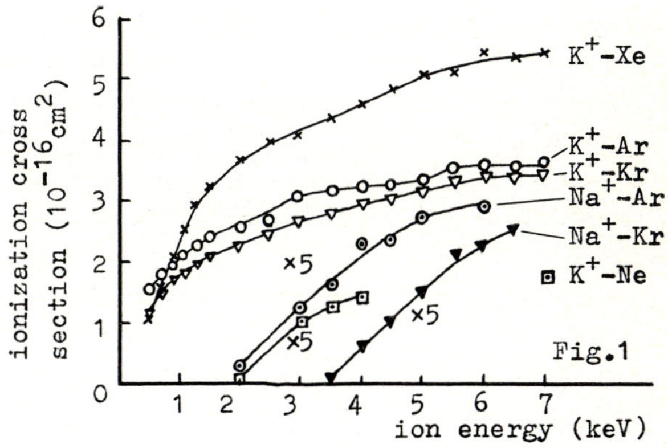

Fig.1

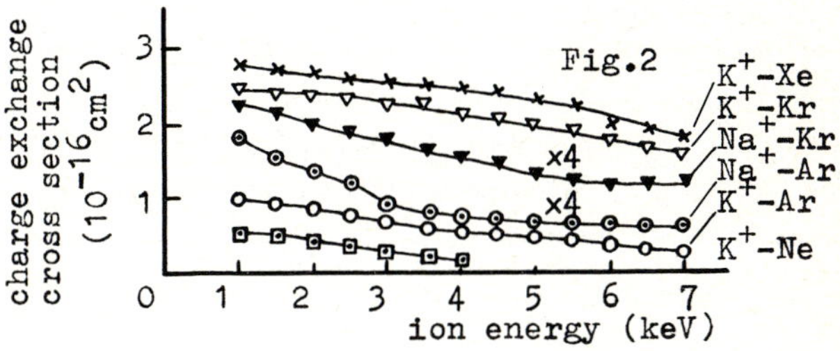

Fig.2

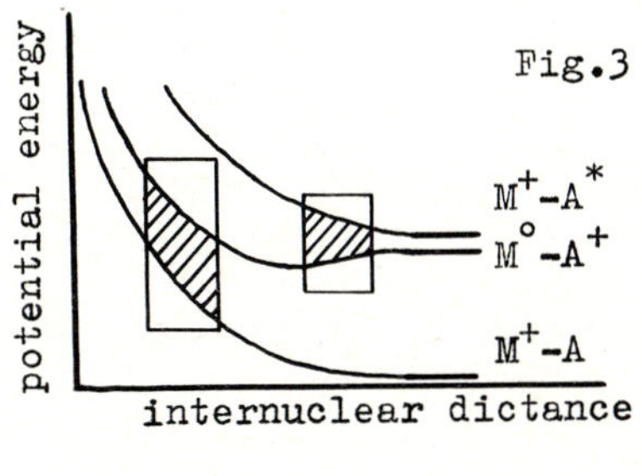

Fig.3

References

1. G.N.Ogurtsov, B.I.Kikiani, I.P.Flaks, Zh. Tekh. Fiz. 36,491(1966)
   I.P.Flaks, B.I.Kikiani, G.N.Ogurtsov, Zh. Tekh. Fiz. 35,2076-2082(1965)
2. V.V.Afrosimov, Yu.S.Gordeev, V.M.Lavrov, Zh. Eksp. Teor. Fiz. 68,1715(1975)
3. V.V.Afrosimov, Yu.S.Gordeev, V.M.Lavrov, V.K.Nikulin, Abstracts of Papers of the VII ICPEAC, Amsterdam, 143(1971)
4. N.V.Fedorenko, I.P.Flaks, L.P.Filipenko, Zh. Eksp. Teor. Fiz. 38,719(1960)

DIFFERENTIAL LARGE ANGLE SCATTERING IN COLLISIONS OF K^+ IONS WITH Ar ATOMS

S. Kita, M. Izawa, and H. Inouye

Research Institute for Scientific Measurements, Tohoku University
1-1 Katahira-Nichome, Sendai 980, Japan

Among the alkali ion - rare gas collision systems, isoelectronic colliding pairs are most interesting, since the inelasticity is best effectuated.[1,2] This report presents the results on differential scattering in collisions of K^+ ions with Ar atoms over a wide range of scattering angles in the laboratory frame ($5° \lesssim \theta \lesssim 80°$) and at incident ion energies of 200 - 350 eV. A crossed beam technique has been applied using a supersonic atomic beams.

The open circles in Fig. 1 are the total differential cross sections measured at an ion energy E = 350 eV. The broken curve expresses the elastic scattering cross sections calculated with a ground-state potential.[3] The difference between the experimental cross sections and those by the calculation in the region $16° < \theta < 40°$ is caused by the production of neutral particles in a charge exchange reaction. The neutral particles have been observed by sweeping off the scattered ions, the result for E = 350 eV being shown by the solid triangles in Fig. 1. The positions of the two peaks marked with θ_i and θ_n are dependent on the collision energy, but the sum $\theta_i + \theta_n$ is 87.3°, independently of the collision energy.

Time-of-flight (TOF) measurements on the scattered ions and neutrals were made in order to elucidate the inelastic processes involved in the differential scattering. The results revealed that a few one-electron processes,

$$K^+ + Ar \longrightarrow K, K^* + Ar^+ + Q_1, \qquad (1)$$

have occured, in addition to the two-electron processes,

$$K^+ + Ar \longrightarrow K^+ + Ar^{**} + Q_2$$
$$\qquad \qquad \qquad \qquad \hookrightarrow Ar^+ + e, \qquad (2)$$

$$K^+ + Ar \longrightarrow K, K^* + Ar^{+*} + Q_3, \qquad (3)$$

where Q_2 and Q_3 are in the range 26 - 30 eV. Figure 2 shows a TOF ion spectrum at E = 350 eV and $\theta = 65°$. The dominant signal in the spectrum is due to the recoiled Ar^+ ions produced through Reactions (1) - (3). The differential cross sections for the recoiled Ar^+ at the large angles are shown by the solid circles and open triangles in Fig. 1. Large cross sections beyond $\theta \simeq 45°$ in the differential cross section curve are due to the recoiled Ar^+ ions. The dominant signal at θ_i is Ar^+ produced in Reaction (1) with K(4s), while the dominant neutral signal at θ_n is the K(4s) atoms produced in Reaction (1). The peaks at θ_i and θ_n are a twin peaks made of the neutrals and the ions produced by Reaction (1) at a CM scattering angle, the sum $\theta_i + \theta_n$ being approximately 87° independently of collision energy.

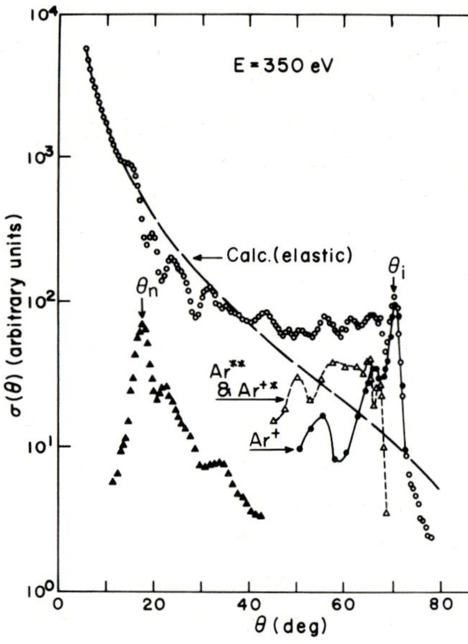

Fig. 1. Differential cross sections in collisions of K^+ ions with Ar atoms. The open circles are the total differential cross sections, while the solid triangles are the differential cross sections of the neutral atoms. The solid circles and the open triangles are the differential cross sections for the recoiled product Ar^+ ions, the formers being for Reaction (1) and the latters for Reaction (2) and (3).

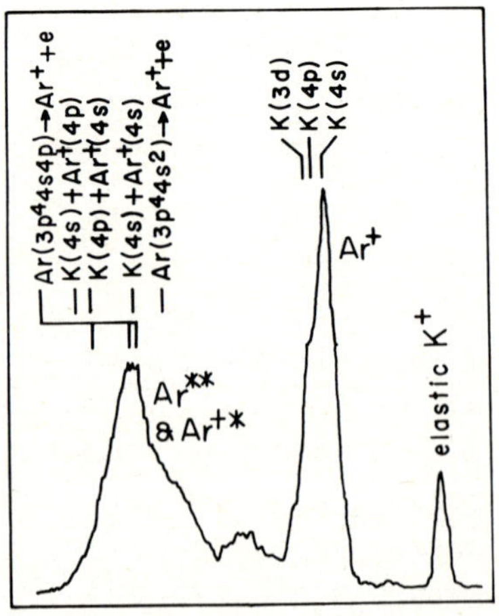

Fig. 2. A TOF ion spectrum at E = 350 eV and $\theta = 65°$. Ar^+ is the product ions of Reaction (1), and Ar^{**} and Ar^{+*} are those of Reactions (2) and (3).

References

1. J. C. Mouzon, Phys. Rev. 41, 605 (1932).
2. V. V. Afrosimov et al., Abstracts of Papers, 7th ICPEAC, p. 143 (1971).
3. H. Inouye and S. Kita, J. Chem. Phys. 56, 4877 (1972).

COLLISIONS BETWEEN Li^+ IONS

G.C. Angel, K.F. Dunn, M.F. Watts and H.B. Gilbody

Department of Pure and Applied Physics
The Queen's University of Belfast
Belfast, United Kingdom

A fast intersecting beam technique previously developed in this laboratory and used to study collisions between positive ions (see review by Gilbody[1]) has now been applied to the Li^+-Li^+ collision system. We have obtained cross sections $\sigma(Li^{2+})$ for Li^{2+} production from the combined processes of charge transfer.

$$Li^+ + Li^+ \rightarrow Li + Li^{2+}$$

and ionisation

$$Li^+ + Li^+ \rightarrow Li^+ + Li^{2+} + e$$

within the c.m. energy range 50-250 keV. The results provide an opportunity to test the range of validity of theoretical models which might then be extended to heavier ions of mass A > 100 relevant to schemes for heavy ion fusion.

The apparatus and measuring procedure was similar to that used in our recent studies of Cs^+ - Cs^+ collisions[2]. A beam of 100-500 keV Li^+ ions is arranged to intersect at 90 deg a target beam of Li^+ from a second accelerator. The energy of the Li^+ target beam is set within the range 5-12 keV. The Li^{2+} collision products from the combined processes of charge transfer and ionisation are selected by a two stage electrostatic analyser and counted by a particle multiplier. Both beams are modulated, the modulation pattern being carefully synchronised with scaler gating pulses to permit separation of the required Li^{2+} signals from those arising from interactions with the background gas. Absolute cross sections $\sigma(Li^{2+})$ for Li^{2+} formation can then be determined from a knowledge of the beam intensities and profiles in the region of intersection.

Our cross sections $\sigma(Li^{2+})$ in the c.m. range 50-250 keV are in reasonable agreement with the cross sections previously measured by Peart et al[3] in the c.m. range 19-88 keV. Calculations by Ermolaev et al[4,5] indicate that at energies in the present range, Li^{2+} ions are formed mainly by ionisation. Cross sections for ionisation calculated[5] using the first Born approximation exhibit a different energy dependence and are larger than our measured values of $\sigma(Li^{2+})$ over most of the energy range.

References

1. H B Gilbody. Physics of Electronic and Atomic Collisions - S Datz editor. North Holland 1982, pp 223-34.
2. P A Neill, G C Angel, K F Dunn and H B Gilbody. J Phys B, 15, 4219 (1982).
3. B Peart, R A Forrest and K Dolder. J Phys B 14, 3457 (1981).
4. A M Ermolaev, C J Noble and B H Bransden. J Phys B 15 457 (1982).
5. A M Ermolaev, J E Miraglia and B H Bransden. J Phys B 15, L677 (1982).

THEORY OF ELECTRON CAPTURE IN INTERMEDIATE-TO-HIGH-VELOCITY COLLISIONS

Knud Taulbjerg† and John S. Briggs‡

†Institute of Physics, University of Aarhus, DK-8000 Aarhus C, Denmark

‡Fakultät für Physik, Universität Freiburg, D-7800 Freiburg, B.R.D.

The recently developed Strong-Potential Born (SPB) approximation[1)] for electron capture in asymmetric ion-atom collisions proceeds from the equivalent post and prior forms of the exact T matrix by approximating the exact scattering waves by distorted waves,

$$|\psi_f^-\rangle \simeq |\chi_f^-\rangle = (1 + G_T^- V_T)|\phi_f\rangle \quad (1a)$$

$$|\psi_i^+\rangle \simeq |\chi_i^+\rangle = (1 + G_P^+ V_P)|\phi_i\rangle . \quad (1b)$$

Here, standard notation[1)] is used. For example, $G_P^+ = [E - H_0 - V_P + i\eta]^{-1}$ is the Greens operator, including the electron-projectile interaction $V_P(\propto Z_P)$ but not the electron-target interaction $V_T(\propto Z_T)$. In case of a strong target potential ($Z_T \gg Z_P$), the relevant SPB is obtained from the prior form of the exact T matrix,

$$T_{fi}^{SPB} = \langle\chi_f^-|V_P|\phi_i\rangle , \quad (Z_P \ll Z_T) , \quad (2a)$$

while for a strong projectile potential, the relevant form is obtained from the post form of the exact T matrix,

$$T_{fi}^{SPB} = \langle\phi_f|V_T|\chi_i^+\rangle , \quad (Z_T \ll Z_P) . \quad (2b)$$

Since the strong field is included to all orders via the Greens operator, it is clear that the SPB is a consistent first-order theory in the weak field. Consequently, the SPB is valid not only in the high-velocity region $Z_> \ll v$ but also in the intermediate-velocity region $Z_< \ll v \lesssim Z_>$. Accordingly, it forms a natural extension of the ordinary Second Born (SB) approximation, which is of combined second order in the interaction potentials.

In this work, we study a corresponding natural extension of the SB for near-symmetric collisions. Our method is based on a distorted-wave approach and is close in spirit to the Continuum Distorted-Wave (CDW) method of Cheshine[2)]. Distorted-wave theory usually proceeds by chosing a useful partitioning of the perturbations in initial and final channels in terms of distortions and residual interactions. The distortion potentials define distorted waves in the usual way via the Lippmann-Schwinger equation. Finally, the T matrix is recast into a new exact form, which basically involves distorted-wave states in place of initial and final states and residual perturbations in place of full channel perturbations. Alternatively, the distorted-wave theory may proceed by chosing appropriate distorted waves and determining the corresponding residual interactions.

Since the natural choice of distorted waves is dictated by the SPB forms (Eqs. 1), we follow this latter approach and derive the following exact prior and post forms of the T matrix,

$$T_{fi} = \begin{cases} \langle\chi_f^-|V_P|\psi_i^+\rangle - \langle\phi_f|V_P|(\psi_i^+ - \phi_i)\rangle , & (3a) \\ \langle\psi_f^-|V_T|\chi_i^+\rangle - \langle(\psi_f^- - \phi_f)|V_T|\phi_i\rangle . & (3b) \end{cases}$$

First, we note that the SPB forms (2a) and (2b) are recovered from (3a) and (3b), respectively, if the exact states $\psi_{i,f}^{\pm}$ are replaced by unperturbed states $\phi_{i,f}$. However, for symmetric collisions, it is logical to account equally for distortion of initial and final states. This is achieved by approximating the exact states in Eqs. (3) by the corresponding distorted-wave states (Eqs. 1) to give

$$T_{fi}^{DWB} = \begin{cases} \langle\chi_f^-|V_P|\chi_i^+\rangle - \langle\phi_f|V_P|(\chi_i^+ - \phi_i)\rangle , & (4a) \\ \langle\chi_f^-|V_T|\chi_i^+\rangle - \langle(\chi_f^- - \phi_f)|V_T|\phi_i\rangle . & (4b) \end{cases}$$

Although it does not appear so at first sight, the two forms (4a and 4b) of the Distorted-Wave Born (DWB) approximation may readily be shown to be identical and symmetric in the operation of the two potentials, as should be the case for a proper theory of near-symmetric collisions. Further, the DWB theory relates favourably to standard Born-series expansions and to other symmetric theories for electron capture.

REFERENCES

(1) J.H. Macek and K. Taulbjerg, Phys.Rev.Lett. 46, 170 (1981); J.H. Macek and S. Alston, Phys.Rev.A 26, 250 (1982)

(2) I.M. Chechire, Proc.Phys.Soc. 84, 89 (1964)

CLOSE COUPLING CALCULATION OF ELECTRON CAPTURE AT HIGH ENERGIES

K. Fujiwara

Micro Communications, 4-1-9 Shinjuku, Shinjuku-ku, Tokyo 160 Japan

N. Toshima

Dep. of Appl. Phys., Fac. of Engineering, University of Tokyo, Bunkyo-ku, Tokyo 113 Japan

T. Watanabe

The Institute of Physical and Chemical Research, Wako-shi, Saitama 351 Japan

A number of papers have been published in which charge transfer cross sections are calculated by the close-coupling method. However, most authors are concerned with relatively low energy regions less than a few hundred keV/amu. One reason of this scarcity of the calculation at high energies may be that the numerical calculation of the matrix elements involving two-center atomic orbitals is considerably time-consuming there because of the oscillatory behavior of the integrand caused by the electron translation factor.

On the other hand, the validity of the first-order perturbation theory is not clear in charge transfer processes even at high energy region. It has been shown by many workers that the second Born term corresponding to the Thomas double scattering process overwhelms the first Born term at extremely high energies. In the energy region less than a few MeV/amu the contribution of the double scattering process can be expected to be insignificant to the total cross sections. Nevertheless the exact calculations of the second Born approximation[1,2] have shown that the second Born terms other than the double scattering term contribute significantly. Therefore it is a physically interesting problem to investigate the contribution of intermediate states by the close-coupling method at such energy regions.

In this report we present the cross sections for the following processes,

$$He^{++} + H(1s) \rightarrow He^{+} + H^{+} \qquad (1)$$

and

$$H^{+} + H(1s) \rightarrow H + H^{+} \qquad (2)$$

in the energy region higher than 100 keV/amu and less than 5.626 MeV/amu. The numerical procedure is basically the same as Fujiwara[3] but the number of points of the Gauss-Laguerre quadruture for the integration of the two-center matrix elements is increased up to 1500 corresponding to the incident energy. The number of states included in the basis functions are

1s,2s,2p for H

1s,2s,2p,3s,3p,3d for He^{+}

Therefore nineteen states are included for He^{++} + H system and ten states are included for H^{+} + H system. In addition fifteen-state calculation in which only the gound state is included for the hydrogen atom has been performed for He^{++} + H system for the investigation of the contribution of the excited states of H.

The cross sections are given in Table 1 and 2. The cc cross section is always smaller than the BK cross section and the ratio of them approaches to unity monotonically as the energy increases. The cc cross sections by Cheshire et. al.[4] are also presented in Table 2. The basis functions employed by them are the same as ours but their cross sections do not agree with ours. This discrepancy may be come from the numerical inaccuracy of their calculation. (The BK cross sections listed by them are wrong so that their cross section ratios are revised by us using the correct BK cross sections.)

Table 1. The total capture cross sections in cm^2 for $He^{++} + H(1s) \rightarrow He^{+} + H^{+}$.

E(keV/amu)	19states	15states	BK
225	4.05(-18)	4.30(-18)	1.62(-17)
400	4.67(-19)	4.96(-19)	1.36(-18)
900	1.39(-20)	1.45(-20)	2.78(-20)
2500	8.39(-23)	8.53(-23)	1.16(-22)
5625	9.37(-25)	9.47(-25)	1.12(-24)

Table 2. The 1s capture cross sections in cm^2 for $H^{+} + H(1s) \rightarrow H + H^{+}$.

E(keV/amu)	$\sigma_{present}$	$\sigma_{present}/\sigma_{BK}$	$\sigma_{Cheshire}/\sigma_{BK}$
100	7.58(-18)	0.216	0.262
300	1.28(-19)	0.347	0.420
1000	4.46(-22)	0.638	0.695
2000	1.08(-23)	0.745	0.799
4000	2.14(-25)	0.881	1.090
5625	2.96(-26)	0.911	

References

1. P. R. Simony and J. H. McGuire, J. Phys. B14 L737 (1981).
2. J. M. Wadehra, R. Shakeshaft and J. H. Macek, J. Phys. B14 L767 (1981).
3. K. Fujiwara, J. Phys. B14 3977 (1981).
4. I. M. Cheshire, D. F. Gallaher and A. J. Taylor, J. Phys. B3 813(1970).

ELECTRON CAPTURE BY FAST PROTONS SCATTERED AT LARGE ANGLES

L. Kocbach* and J.S. Briggs**

*Det fysiske Institut, Aarhus Universitet, DK-8000 Århus-C, Denmark
**Fakultät für Physik, Universität Freiburg, D-7800 Freiburg i.Br.,Germany

The understanding of the electron capture processes might be furthered by studies of differential cross sections. A special aspect are the differential cross sections for very large scattering angles. In this case the cause of scattering is the internuclear interaction, usually neglected when total cross sections are considered.

In the cases we discuss it is justified to study the capture probability, defined as a ratio of two differential cross sections

$$I(\vartheta) = \frac{d\sigma_{CAPTURE}}{d\Omega} \Big/ \frac{d\sigma_{ELASTIC}}{d\Omega} \qquad (1)$$

Experimental measurements of capture probabilities have been performed recently[1)] for wide range of scattering angles and projectile energies. Only total capture probabilities have been measured, i.e. including all occupied target states.

We study the capture process in the framework the classical trajectory version of the impulse approximation[2)]. The main assumption of this model is that the potential of the projectile is much weaker than the target potential. The projectile potential enters thus only by defining the final bound state and is neglected when the time development of the wavefunction is described. Thus it appears only in first order in the transition amplitudes. The trajectories can be represented as two straight-line segments on which the internuclear velocity is constant. Both have impact parameter zero, the scattering happens instantaneously at zero internuclear distance.

A procedure which is a straight forward generalisation of these described in ref.[2)] and [3)] leads (using the peaking approximation (i) of ref.[3)] to an expression for the capture probability from an arbitrary target state (n,l) to the 1s state hydrogen

$$I(\vartheta) = \left|\varphi_{1s}^{P}(\vec{r}_P = 0)\right|^2 \times \left|\int_{-\infty}^{\infty} dt\, e^{i\alpha t} \int d^3\vec{r}_T\, \psi^{(-)*}_{\vec{V}(\vartheta)}(\vec{r}_T)\, V_p(\vec{r}_T - \vec{R}_\vartheta(t))\, \psi^{T}_{n\ell}(\vec{r}_T)\right|^2 \qquad (2)$$

where p and T indices denote projectile and target quantities, respectively. This expression is related to the differential probability for ionization when the electron is ejected with the same velocity vector as the outgoing projectile. We evaluate these latter quantities using multipole expansion of the electronic states in continuum.

In fig. 1 results of model calculations (with modified hydrogenlike wavefunctions and experimental binding energies, c.f.[3)] are compared with some of the experimental results of ref.[1)] for protons of indicated energies on neon target. These calculations include also target recoil amplitudes as used in ref.[4)].

These results indicate that the picture of the capture process as one related to ionization is relevant also in the cases when nuclear scattering is important. An even better agreement in the shapes of the curves should be expected when more realistic wavefunction will be used. This can be seen when the parameters of the hydrogen-like functions are varied.

A model related to the present one has been studied in ref.[5)]. These authors do not arrive at equivalent results. This is probably related to the fact that their Hamiltonian does not account for the acceleration of the projectile reference frame.

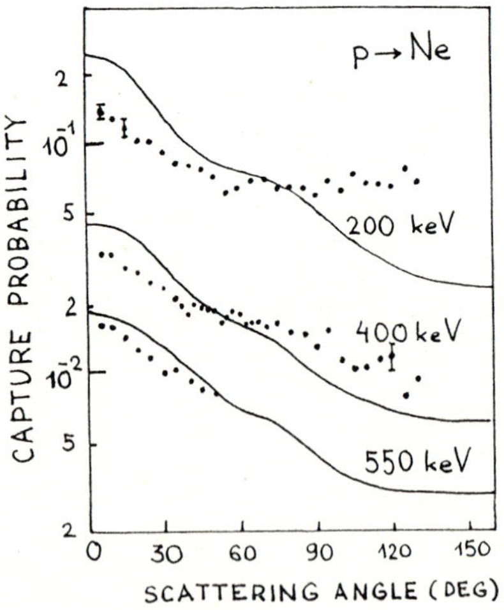

Fig. 1

References

* NORDITA Guest

1. Horsdal-Pedersen E. et al.,J.Phys.B 15, 2461 (1982)
2. Briggs J.S., J.Phys.B 10, 3075 (1977)
3. Kocbach L. J.Phys.B 13, L665 (1980)
4. Andersen J.U. et al.,J.Phys.B 9, 3247 (1976)
5. Amundsen P.A., and Jakubassa-Amundsen D.H. Physica Scripta 26, 155 (1982)

EXACT SECOND BORN CALCULATIONS FOR ELECTRON CAPTURE

J. H. McGuire and J. Eichler

Bereich Physik, Hahn Meitner Institut, Berlin, FRG

and

P. R. Simony

Department of Physics, Kansas State University, Manhattan, Kansas, USA 66506

Electron capture has the unusual feature that the second term in the Born series dominates over all other orders at high velocities. The total cross section decreases asymptotically as v^{-11} in second order, as compared to v^{-12} in the first order. The v^{-11} dependence at high velocities arises due to a simple two step mechanism suggested[1] long ago by Thomas. This Thomas mechanism, where the electron is first scattered into 60° by the projectile and then rescattered by the target nucleus, is associated with a peak in the differential scattering cross section at a center of mass angle of $\sin(60^{o})/Mp$ where Mp is the mass of the projectile (in atomic units). It is this Thomas peak that gives rise to the v^{-11} second Born behavior at high velocity. Recently this Thomas peak was observed[2] to emerge in H^{+} + He above about 5 MeV/amu as predicted.

We have completed exact numerical calculations of differential and total cross sections evaluated in the second Born approximation for various projectile and target charges, Z_p and Z_T, between 1 and 10, at velocities between 10 and 200 MeV/amu. The Thomas peak in the differential cross section, characteristic of a free wave second Born process, appears at velocities above $Z^2 \times 7$ MeV/amu for symmetric systems ($Z_p = Z_T = Z$). The shape of this Thomas peak contains information[3] about real and virtual intermediate states of the system.

In order to delineate the region in which the non-relativistic second Born approximation is applicable we have chosen the somewhat arbitrary criterion of selecting that energy at which the second Born peak appears. Specifically we choose the lowest energy where the Thomas peak lies above the dip, i.e. the lowest energy at which a relative minimum occurs. We have plotted the velocity of this point as a function of Z_T in Figure 1. For symmetric systems, this locus of these points corresponds to a constant value of v/Z, = 15, or $Z^2 \times 7$ MeV/amu. For energies below this locus we expect relativistic effects to be important. In this way we delineate a region of validity for the non-relativistic second Born approximation. It is noted that we do not expect non-relativistic second Born calculation to apply to systems with Z>10.

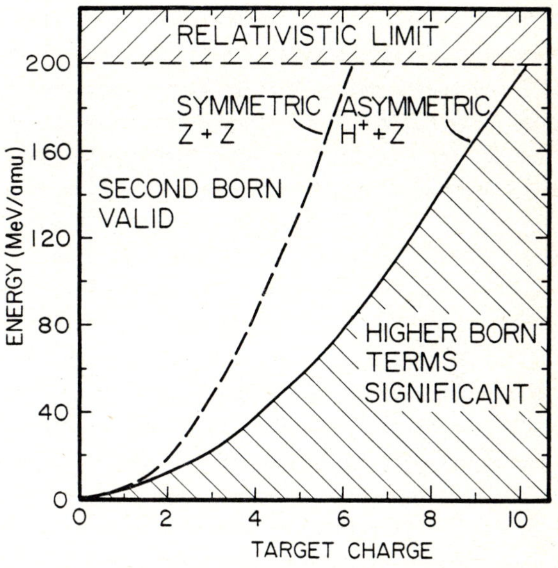

Fig. 1: Locus of minimum velocity at which Thomas peak occurs as a function of target charge, Z_T for symmetric and asymmetric systems. Below this locus the second Born cross sections tend to be too large. Above 200 MeV/amu relativistic effects become significant.

This work was supported in part by the U.S. Department of Energy, Division of Chemical Sciences.

References

1. L. H. Thomas, Proc. R. Soc. London, Ser. A 114, 561 (1927).
2. C. L. Cocke, E. Horsdal-Pedersen and M. Stoeckli, private communications.
3. J. H. McGuire, P. R. Simony, O. L. Weaver and J. Macek, Phys. Rev. A 26, 1109 (1982).

A NEW VARIATIONAL PRINCIPLE FOR CHARGE EXCHANGE AT ARBITRARY ENERGIES

Dž. Belkić

Institute of Physics, Studentski trg 12/V, 11001 Belgrade, Yugoslavia

Charge exchange has recently[1] received remarkable attention, primarily because of its fundamental importance for scattering theory. Unlike excitation, electron capture from hydrogenlike atoms by bare nuclei is a purely three body problem. Therefore, no simple theory including only the direct scattering path would be adequate. Thus, for example, the first Born approximation, besides yielding incorrect high energy behaviour of the cross section, is known to fail completely in comparison with experimental data[1]. Inclusion of intermediate states of the captured electron, such as excitation and ionization, would be of crucial importance, because these two channels dominate charge exchange at high energies. Various theories[2] have been proposed to deal with these virtual channels, and they might be considered as satisfactory when the incident velocity is sufficiently large. However, none of them is adequate at intermediate energies, mainly because of an overaccount of the continuum. Furthermore, these high energy methods are incapable of systematic improvement, since they neglect excitation of the electron, which might be bound to either of the two Coulomb centres, projectile or target nucleus. At low energies, the close coupling method[3] predicts fairly acceptable results. However, its usefulness becomes quite restricted, as the incident velocity increases. This is due to formidable numerical work required to solve a large number of coupled differential equations. These computational difficulties are persistent even at lower energies, when one attempts to account for multiple curve crossings, which occur in the collision between multiple charged ions. The above discussion can be summarized as follows:

(i) At high energies, the transition T operator is accounted for exactly, but only up to a maximum of <u>second order</u>. Subsequently, an <u>infinite</u> number of intermediate transitions is included, by means of the restricted T propagator.

(ii) At low energies, the transition T operator is included through <u>all orders</u>, whereas only a quite <u>limited</u> number of discrete intermediate states are retained.

Neither of the above two distinct groups of theories can appropriately describe the physics of the most critical intermediate energy region and, thus far, no connection between them has been established. Clearly, it would be advantageous to devise a theoretical scheme which would incorporate effects from most of the continuum intermediate states, and which may systematically be improved by successive inclusion of the more significant discrete states. To this end, we propose a new variational principle for the T matrix in the form:

$$T_{if}=\langle\Phi_f|V_f|\Phi_i\rangle+\sum_{\alpha,\beta}\langle\Phi_f|V_fG_o|\alpha\rangle B_{\alpha\beta}\langle\beta|G_oV_i|\Phi_i\rangle \qquad (1)$$

where $(B^{-1})_{\beta\alpha}=\langle\beta|G_o-G_oVG_o|\alpha\rangle$. Here, Φ_i and Φ_f are the usual unperturbed asymptotic channel states, V is the total potential, V_i and V_f are perturbation interactions and G_o is the three body free-particle Green function. Expansion trial functions $|\alpha\rangle$ and $|\beta\rangle$ need not be identical to each other; we choose the complete sets of Sturmians centered on both the incident and target nucleus. A remarkable feature of the present method is its generality and simplicity; it variationally unifies the physics of the entire scattering energy region. Required optimal characteristics of L^2 expansion methods are encompassed in Eq (1), as well, which obviates the problem of solving a system of coupled differential equations. In addition, T_{if} can be calculated analytically from (1), in the exact closed form for arbitrary Sturmian functions. This makes the method particularly tractable, since the complete evaluation of T_{if} would require only one single numerical operation, i.e., inversion of the square matrix $\underset{\approx}{B}$. Direct collision path $\langle\Phi_f|V_f|\Phi_i\rangle$ is singled out in (1) as the first Born approximation, while the remaining stationary part is responsible for higher order effects due to multiple scattering. Extensive computation for H^+-H collision are in progress and results will be reported shortly.

References

1. Dž.Belkić, R.Gayet and A.Salin,Phys.Rep.<u>56</u>,279(1979), Bransden, Adv.Atom.Molec.Phys.<u>15</u>,263(1979), R.Shakeshaft and L.Spruch, Rev.Mod.Phys.<u>51</u>,369 (1979)
2. I.M.Cheshire,Proc.Phys.Soc.<u>82</u>,113(1963),<u>84</u>,89(1964), Dž.Belkić,J.Phys.B:Atom.Molec.Phys.<u>10</u>,3491(1977)
3. D.Basu,S.C.Mukherjee and D.P.Sural,Phys.Rep.<u>42</u>,145 (1978), C.D.Lin and P.Richard,Adv.Atom.Molec.Phys. <u>17</u>,275 (1981)

THE 'CLASSICAL DEFLECTION FUNCTION' IN THE CONTINUUM DISTORTED WAVE APPROXIMATION

P. T. Greenland

Theoretical Physics Division, AERE Harwell, Didcot, Oxon OX11 ORA

The Thomas peak, produced by a well known double scattering mechanism, is the most prominent feature of the high energy charge-transfer differential cross section. The 1s - 1s transfer cross section has been examined in many approximations[1] including the continuum distorted wave approximation[2] (CDW). The most obvious feature of the differential cross section in the CDW is a pronounced dip in the centre of the Thomas peak. This has been analysed in some detail[3] and it has been shown that the differential cross section can be represented as the coherent sum of two terms, each of which separately would produce a Thomas peak, one with position and width characteristic of scattering first from the projectile and then from the target (assumed the heavier) and one with width, but not position characteristic of the time reversed process.

It has been shown[1] that it is possible to derive an analogue of the classical deflection function for high energy charge transfer which correctly describes how the contribution to the charge transfer amplitude from a range of impact parameters is focussed into a peak at the Thomas angle. The same analysis is applied to the CDW charge transfer amplitude. It is shown that for asymptotically high velocities the CDW amplitude implies that there are three distinct regions in impact parameter space. The inner and outer regions (1 and 2 in fig. 1) are responsible for the two contributions described above. The region between produces the dip. Since this treatment follows classical mechanics as closely as the physics allows, greater insight into the processes included in the CDW may be obtained.

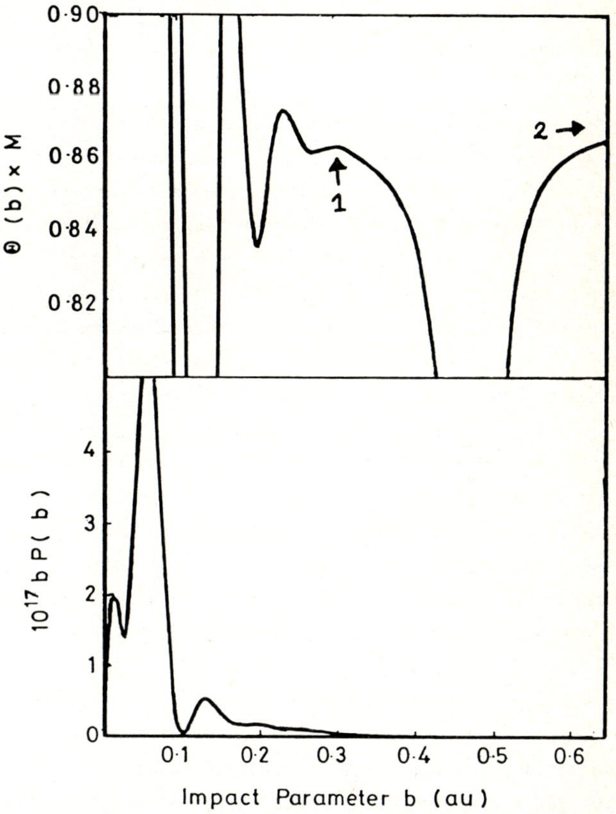

Fig. 1: The charge transfer probability P multiplied by impact parameter b (lower figure) and classical deflection function θ multiplied by the projectile mass (upper figure) as a function of impact parameter b. The two horizontal parts 1 and 2 contribute to the Thomas peak, the region between contributes to the dip. The impact velocity is 100 atomic units. This unrealisticly high value was chosen in order to illustrate the three contributing regions clearly. The target has charge 3, the projectile charge 1.

References

1. J.S. Briggs, P.T. Greenland, L. Kocbach; J. Phys. B Atom. Molec. Phys. 15, 3085, (1982).

2. I.M. Cheshire; Proc. Phys. Soc. 84, 89, (1964).

3. R.D. Rivarola, J.E. Miraglia; J. Phys. B, Atom. Molec. Phys. 15, 2221, (1982).

REPRESENTATION OF CONTINUUM CHANNELS IN THE DESCRIPTION OF COLLISIONALLY INDUCED ELECTRONIC TRANSITIONS

W. Fritsch* and C.D. Lin+

* Bereich Kern- und Strahlenphysik, Hahn-Meitner-Institut für Kernforschung, D-1000 Berlin 39, West Germany
+ Department of Physics, Kansas State University, Manhattan, Kansas 66506, USA

In this contribution we discuss the representation of continuum channels in close-coupling calculations for slow ion-atom collisions. We concentrate on collision encounters in which both excitation and transfer processes are important and which thus have been desribed within the semiclassical two-center expansion method in the past. We demonstrate that, in such calculations, continuum channels can be included with small modifications and a moderate enlargement of the expansion basis.

We start from the observation that couplings between the bound and the continuum states of the colliding system are sentitive only to that part of the continuum which lies within the range of the relevant bound orbitals. As far as continuum-continuum couplings are not all-important, therefore, a representation of the continuum with square-integrable functions is sufficient. With such a discretization of the continuum in mind as it has been used in photoionization studies, it is important to ensure that the final results do not depend critically on the exact form or size of the basis.

In this investigation, the continua of target and projectile are represented separately in basis sets of hydrogenic orbitals at, respectively, the target and projectile center. Atomic-orbital (AO) expansions of this kind, with AOs of the separated atoms and the united atom, have previously[1] been used in close-coupling calculations of excitation and transfer in a broad range of energies, in particular at low energies where 'AO+' expansions are equivalent to expansions in molecular orbitals. By adding a few more basis orbitals with suitably adjusted charge numbers, one easily arrives at separated-atom spectra such that some energies are close to the continuum threshold. As a test of the quality of the continuum pseudostates, we have calculated Coulomb ionization probabilities with various l=0,1 basis sets in perturbation theory (BA) and in close-coupling calculations (CC) and compared with SCA results, the latter involving exact continuum functions. Table 1 shows the results for l=0 H(1s) ionization by an impinging proton at 25 keV. BA and CC results are calculated with a basis set of only 8 l=0 states (1s,2s,3s bound states included). The close agreement between SCA and BA results

Table 1: p+H Coulomb ionization probabilities for 25 keV at impact parameters b

b(au)	0.2	0.6	1.0	2.2
SCA	0.480	0.283	0.122	0.029
BA	0.434	0.254	0.106	0.020
CC	0.347	0.161	0.078	0.002

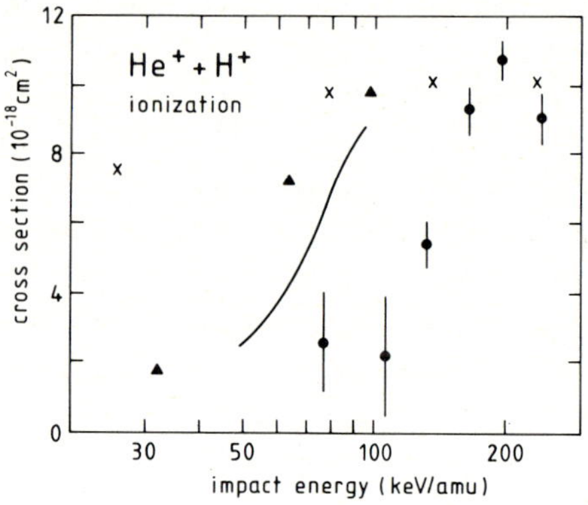

Fig. 1: Ionization cross sections in $He^+ + H^+$ collisions. For explanations, see text.

demonstrates the good representation of l=0 continua by this basis set. The deviation between BA and CC results illustrates the need of a full close-coupling calculation for arriving at realistic cross sections. Similar results are derived with other basis sets.

Excitation, transfer and ionization cross sections have been calculated for $H^+ + H$ and $He^+ + H^+$ collisions with basis sets consisting of 20-23 pseudostates (mostly l=0,1) at each center. For $H^+ + H$ collisions above 15 keV, continuum channels show some effect on transitions into bound states. For ionization, the presently calculated cross sections lie some 15% below the most recent experimental data[2] (available above 38 keV).

For $He^+ + H^+$, continuum channels have little effect on transitions into bound states below 100 keV/amu. Figure 1 shows the calculated ionization cross sections (triangles) which are reasonably close to experimental data[3] by the Newcastle group (full line) but quite different from the data[4] of the Belfast group (circles). For low energies, the deviation between the present results and those (crosses) from a 'one and a half centered' expansion[5] illustrates the importance of a proper account of transfer channels in a two-center expansion.

This work is supported in part by the US Department of Energy. C D L is also supported by the Alfred P Sloan Foundation.

References

1. W. Fritsch and C.D. Lin, J. Phys. B 15, 1255 (1982)
2. M.B. Shah and H.B. Gilbody, J. Phys. B 14, 2361(1981)
3. B. Peart, K. Rinn and K. Dolder, preprint
4. G.C. Angel et al., J. Phys. B 11, L297 (1978)
5. J.F. Reading et al., J. Phys. B 15, 625 (1982)

PROTON - HYDROGEN SCATTERING AT HIGH ENERGIES

Roberto D. Rivarola
Instituto de Física Rosario (CONICET-UNR)
Avenida Pellegrini 250, 2000 Rosario, Argentina

Resonant charge exchange $H^+ + H(1s)$ collisions at high energies has been a subject of active research in the last five years [1-3]. Probabilities for this process at 10 MeV laboratory energy as a function of the impact parameter ρ and calculated using the Continuum-Distorted-Wave (CDW) model, are compared with recent results of Briggs et al[4] obtained using a peaking Second-Born (PB2) approximation. The same qualitative behaviour is found up to $\rho^* \simeq 0.9$ au ($\rho^* v \simeq 18$ au, where v is the relative velocity between the nuclei) but differences arise for larger values. About 96.4 per cent of the total cross section σ^{CDW} is contained in the range of impact parameters between zero and ρ^*. When eikonal-differential cross sections[5] $d\sigma/d\Omega$ are calculated, the dip of CDW which overlaps the Thomas'peak[6] comes from the contribution of $\rho>\rho^*$. It must be pointed out that in this region of large encounters the CDW model has been developed in order to have a correct asymptotic behaviour[7]. The $(d\sigma/d\Omega)^{CDW}$ (integrating the eikonal expression up to $\rho>>\rho^*$) shows a close agreement with the exact-Impulse (IA) results of Briggs et al[4] except in the narrow region corresponding to the dip of CDW. These CDW and IA models give comparative values for the valley (with a minimum at $\Theta \simeq 1/2M$; where Θ is the center of mass scattering angle and M the mass of the projectile) which arises from the interference of single and double scattering contributions[4]. On the contrary the exact-Second-Born (B2) is higher by an order of magnitude in this region. Contribution of intermediate states off- and on-energy shell are also studied throught the analysis of the T-matrix element. The real and imaginary parts of T include only off-energy shell and on-energy shell states respectively. T is calculated using the Hankel transform of the acattering amplitude b (ρ). When this transform is obtained integrating on the impact parameter up to ρ^* and using the CDW model, the T-matrix element gives results in qualitative agreement with the B2 ones obtained by McGuire et al [8,9]. When the integration is extended to $\rho>>\rho^*$ new structures appear into the CDW model.

Detailed results will be given at the Conference.

References

1 R. Shakeshaft, and L. Spruch, J. Phys, B 11, L457 (1978)

2 J. E. Miraglia, R. D. Piacentini, R. D. Rivarola, and A. Salin, J. Phys. B 14, L197 (1981).

3 P. R. Simony, and J. H. McGuire, J. Phys. B 14, L737 (1981).

4 J. S. Briggs, P. T. Greenland, and L. Kocbach, J. Phys. B 15, 3085 (1982)

5 R. McCarroll, and A. Salin, J. Phys. B 1, 163 (1968).

6 R. D. Rivarola, and J. E. Miraglia, J. Phys. B 15 2221 (1982).

7 I. M. Cheshire, Proc. Phys. Soc. 84, 89 (1964)

8 J. H. McGuire, P. R. Simony, O. L. Weaver, J. Eichler, and J. Macek, private communication (1982)

9 See also, P. R. Simony, Ph. D., Kansas State University, unpublished (1981)

ELECTRON CAPTURE BY FAST PROTONS IN GASES

A.M. Popova, Ya.A. Teplova, Yu.A. Shurigina and O.S. Erkovitch

Institute of Nuclear Physics, Moscow State University, Moscow 117234, USSR

Electron capture by projectile ions versus impact parameter to targets atoms has been investigated in the present work. The geometry of the interacting particles is given in Fig. 1.

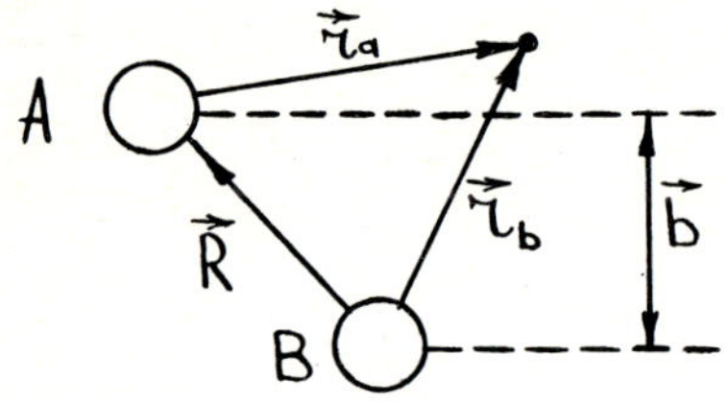

Fig. 1: Geometry of the process.

The initial ion A by velocity $\vec v$ has an impact parameter b to the target atom B. If $\vec r_a$ and $\vec r_b$ are electron vectors to the ion A^+ and the atom B, then we have the Schrödinger equation

$$i\hbar\frac{\partial\Psi(\vec r,t)}{\partial t}=\left\{-\frac{\hbar^2}{2m}\nabla^2+V(\vec r_a)+V(\vec r_b)\right\}\Psi(\vec r,t) \quad (1)$$

where $V(\vec r_a)$ and $V(\vec r_b)$ are electron-ion and electron-atom interactions. The eq. (1) can be solved by the wave function $\Psi(\vec r,t)$ expantion over a total set of initial ion wave functions [1].

$$\Psi(\vec r,t)=\sum_m a_m(t)\varphi_m^{(A)}(\vec r_a)\exp[-i\varepsilon_m^{(A)}t] \quad (2)$$

Here $\varepsilon_m^{(A)}$ and $\varphi_m^{(A)}$ are eigenvalue and eigenfunction of Hamiltonian $H^{(A)}=H_0+V(\vec r_a)$. After substituting of (2) into (1) and by using methods of the nonstationary perturbation theory we have for the capture amplitude of a first order the following expression:

$$a_{capture}=\frac{1}{i\hbar}\int_{-\infty}^{\infty}dt\int d\vec r_a\exp\left[\frac{i}{\hbar}\left(\frac{1}{2}mv^2\frac{M_A-M_B}{M_A+M_B}+\varepsilon_f-\varepsilon_i\right)t\right]* \quad (3)$$

$$*\Phi_f(\vec r_a)V(\vec r_b)\exp\left(-\frac{i}{\hbar}m\vec v\vec r\right)\Phi_i(\vec r_b)$$

where $\Phi_i(\vec r_b)$ and $\Phi_f(\vec r_a)$ are initial state wave function of the electron in the atom B and bound state wave function of the electron and the ion A^+. Bound state energies are noted by ε_f and ε_i.

We have for the hydrogen 1S wave function

$$\Phi_i(\vec r)=(\pi d_{0B}^3)^{-\frac{1}{2}}\exp\left(-\frac{r}{d_{0B}}\right);\Phi_f(\vec r)=(\pi d_{0A}^3)^{-\frac{1}{2}}\exp\left(-\frac{r}{d_{0A}}\right)$$

and the Coulomb forces $V(\vec r_b)=-Ze^2/r_b$ the following expression for the capture amplitude

$$|a_c|=\frac{8Ze^2}{\hbar v d_{0B}A\sqrt{d_{0A}^3d_{0B}^3}}\int_0^\infty x^2K_0([x+b]B)K_1(Ax)dx,$$

$$A=\sqrt{1/d_{0B}^2+\varepsilon_+^2};\ B=\sqrt{1/d_{0A}^2+\varepsilon_-^2}$$

$$\varepsilon_\pm=\frac{1}{\hbar v}\left(\varepsilon_f-\varepsilon_i\pm\frac{1}{2}mv^2\right)$$

After using Neumann's theorem for modificated Bessel functions:

$$|a_c|=\frac{8Ze^2}{\hbar v d_{0B}A\sqrt{d_{0A}^3d_{0B}^3}}\left\{L_0(b;A,B)+\sum_{n=1}^{\infty}L_n(b;A,B)\right\}$$

$$L_n(b;A,B)=\int_0^b x^2K_1(Ax)K_n(bB)I_n(Bx)dx+ \quad (5)$$

$$+\int_b^\infty x^2K_1(Ax)K_n(Bx)I_n(Bb)dx$$

The above expression is different from the one obtained in Ref. 2, where the sum is limited by two first terms of the series.

A probability of the electron capture by helium $P_c=|a_c|^2$ for initial protons by 0.5-2.0 MeV have been calculated. It was shown by that that the series in (5) is converged slowly for impact parameters b = 0 - 5Å by which there were obtained a maximum contribution into the process cross section and two terms limitation is not well justified. The results of the calculation are given in Fig. 2.

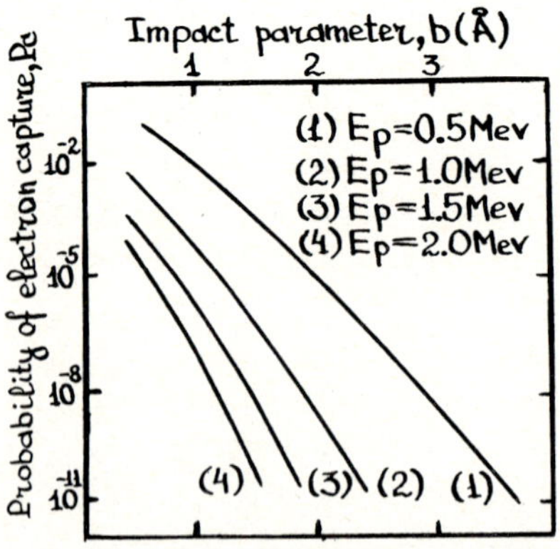

Fig. 2: Probability of the electron capture into initial ion 1S state by passing through the He target.

It is possible to conclude that the correct calculation of the series in (5) leads to essentional increasing of the process cross sections.

References

1. R.K. Janev and L.P. Presnyakov, Physics Reports, 70, No 1, 1981
2. T. Kaneko et.al., Rad. Eff. 54, 183 (1981)

PROTON-HYDROGEN ELECTRON CAPTURE IN SECOND ORDER BRINKMAN-KRAMERS APPROXIMATION

M.K.Srivastava and A.K.Sharma

Department of Physics, University of Roorkee, Roorkee - 247667, India

At asymptotically high energies ($v >> v_e$, where v is the projectile target relative velocity and v_e is the initial or final electron orbital velocity) it is well known that the second Born amplitudes are important in non-radiative electron capture cross sections. The understanding of electron capture is thus directly related to an understanding of the second Born approximation. Over the past several years a number of second Born calculations using peaking approximation and approximations related to the second Born approximation have been done. Exact numerical second Born calculations have also been reported. The differential cross section in the second Born approximation is given by

$$\frac{d\sigma}{d\Omega} = \mu_1 \mu_f k_f |T_1 + T_2|^2 / 4\pi^3 k_i , \qquad (1)$$

where

$$T_1 = \langle \phi_f | V_f | \phi_i \rangle \quad .. \qquad (2)$$

and

$$T_2 = \langle \phi_f | V_f G^+ V_i | \phi_i \rangle \quad .. \qquad (3)$$

Here ϕ_i, V_i, μ_i and $\vec{k}_i$ (ϕ_f, V_f μ_f and $\vec{k}_f$) are respectively the unperturbed wave function, interaction, reduced mass and momentum in the initial (final) state and G^+ is the Green's function.

At high energies and not too highly charged projectile/target, G^+ can be approximated by the free Green's function G_o^+. Further in the spirit of Brinkman-Kramers (BK) approach one may drop the internuclear potential V_{PT} in V_f. Simony and McGuire[1] and Simony et al[2] drop the term V_{PT} from V_i as well on the argument that it would lead to physically unsound contribution to the cross section. We however, feel that in a second order calculation of the differential cross section, V_{PT} should be retained in V_i leading to a natural second order extension of the standard BK calculation

$$T_{BK2} = \langle \phi_f | V_{Te} | \phi_i \rangle + \langle \phi_f | V_{Te} G_o^+ (V_{Pe} + V_{PT}) \quad .. \qquad (4)$$

First term (usual BK amplitude) on the R.H.S. of Eq. (4) can be reduced to the well known analytical form while the second has been evaluated by following the procedure given by Simony[1] (Lewis integral[3] followed by a peaking approximation).

A comparison of our results with the

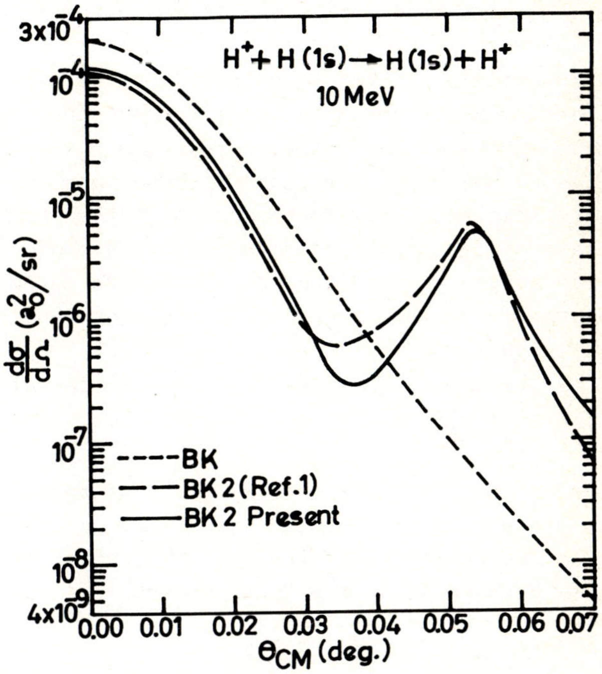

Figure 1

coresponding ones of Simony and McGuire[1] at some typical laboratory energy, say 10 MeV, shows the Thomas peak of identical heights and at the same angle ($\theta_{CM} = 0.054^\circ$) and almost no change at small angles (Fig.1). However, our results exhibit a deeper dip in the cross section at $\theta_{CM} \approx 0.035^\circ$ and a slower fall at large angles. Same general behaviour is obtained at other energies.

The results at other energies will be presented at the Conference.

References

1. P.R.Simony and J.H.McGuire, J.Phys.B14, L737 (1981); P.R.Simony, Ph.D.Thesis, Kansas State Univ., 1981 (unpublished).
2. P.R.Simony, J.H.McGuire and J.Eichler, Phys.Rev.A26,1337(1982).
3. P.R.Lewis, Phys.Rev. 102,537(1956).

EFFECT OF ORTHOGONALITY OF INITIAL AND FINAL STATES ON p-He ELECTRON CAPTURE

A.K.Shrma and M.K.Srivastava

Department of Physics, University of Roorkee, Roorkee - 247667, India

The effect of making the initial and final state wavefunctions orthogonal to each other is studied within the framework of the first Born (BK version) and the Coulomb-projected Born (CPB) approximations. We apply these to study cross sections for proton-helium non-resonant capture reaction $H^+ + He \rightarrow H(1s) + He^+(1s)$. We employ Hylleraas one parameter wavefunction to represent the ground state of the helium atom and prior form of interaction to evaluate the transition matrix element. Momentum transfer of passive electron is ignored.

The usual CPB transition matrix element

$$\langle \phi_f | - \frac{Z_P e^2}{r_{Pe}} | \chi_{iC} \rangle$$

for electron capture from state i of target to state f of projectile (charge Z_P) can be modified by replacing χ_{iC} by χ'_{iC} which is made orthogonal to the final unperturbed state ϕ_f:

$$\chi'_{iC} = \frac{\chi_{iC} - \langle \phi_f | \chi_{iC} \rangle \phi_f}{1 - |\langle \phi_f | \chi_{iC} \rangle|^2} .$$

Likewise BK transition matrix element is calculated from

$$\langle \phi_f | - \frac{Z_P e^2}{r_{Pe}} | \phi'_i \rangle ,$$

where ϕ'_i is the initial state which is made orthogonal to final state ϕ_f:

$$\phi'_i = \frac{\phi_i - \langle \phi_f | \phi_i \rangle \phi_f}{1 - |\langle \phi_f | \phi_i \rangle|^2}$$

Here ϕ_i and ϕ_f are the solutions of the unperturbed Hamiltonian satisfying the initial and the final state boundary conditions respectively and χ_{iC} is the initial state Coulomb-projected wavefunction.

We display in Fig.1 our results for 1s→1s capture in laboratory energy range 300 KeV → 1 MeV along with BK, first Born results of Mapleton[1], CPB, CDW results of Banyard and Szuster[2] and experimental results of Barnett and Reynolds[3] and Welsch et al[4]. All the theoretical results are for 1s → 1s capture while experimental results contain contributions from capture into excited states of the projectile. Our results with orthogonalised initial-final states [BK(0) and CPB(0)] show an improvement over the usual BK results over the entire energy range. In the MeV range where cross sections are small as compared to those in the KeV range, the (0) results do not differ much from usual results (not shown).

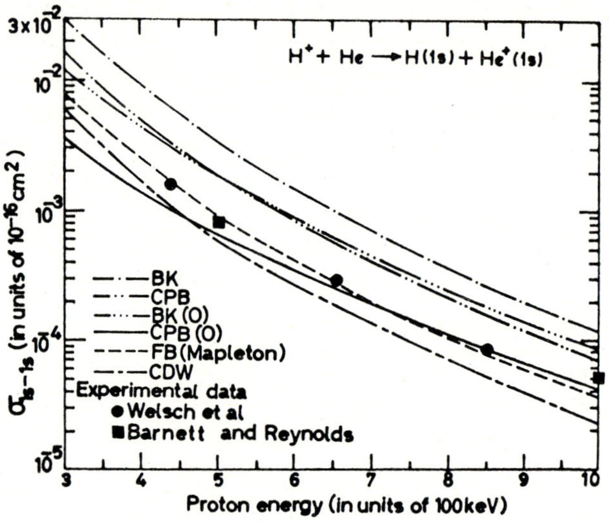

Fig. 1

To conclude : Within a first order theory, just the orthogonalisation of initial and final states leads to a substantial improvement both in magnitude and the general trend of variation with energy.

References :

1. R.A.Mapleton, Phys. Rev. 122, 528 (1961)
2. K.E.Banyard and B.J.Szuster, Phys.Rev.A 16, 129 (1977)
3. C.F.Barnett and H.K.Reynolds, Phys.Rev. 109, 355 (1958)
4. L.M.Welsch, K.H.Berkner, S.N.Kaplan and R.V.Pyle, Phys. Rev. 158, 85 (1967)

K-K SHELL ELECTRON CAPTURE IN ASYMMETRIC PROTON - ATOM COLLISIONS AT HIGH AND INTERMEDIATE ENERGIES

Roberto D. Rivarola[†], Antoine Salin[††]

[†] Instituto de Física Rosario (CONICET-UNR) Avenida Pellegrini 250, 2000 Rosario, Argentina

[††] Laboratoire des Collisions Atomiques, Université de Bordeaux I, 33405 Talence, France

The K-K shell one electron capture by protons from He, C, Ne and Ar ions (atoms) at intermediate ($v \sim v_e$, where v and v_e are the relative velocity of the nuclei and the initial orbital electron velocity respectively) and high ($v \gg v_e$) energies is theoretically studied using the Continuum-Distorted-Wave model (Cheshire, 1964). We find a close agreement with recent experimental differential cross sections $d\sigma/d\Omega$ (Horsdal-Pedersen et al, 1982) for the p + C and p + Ne systems at 0.6 MeV and 1.5 MeV laboratory energies respectively (see figures 1a,b). The sensitivity of $d\sigma/d\Omega$ to the choice of the initial wave function and its corresponding orbital energy ε_i is also analysed, showing a strong dependence on ε_i. Charge exchange probabilities against impact parameters are compared with experimental data and other available theoretical results. The use of a theoretical models which considers a classical one to one correspondence between scattering angle and impact parameter will be discussed.

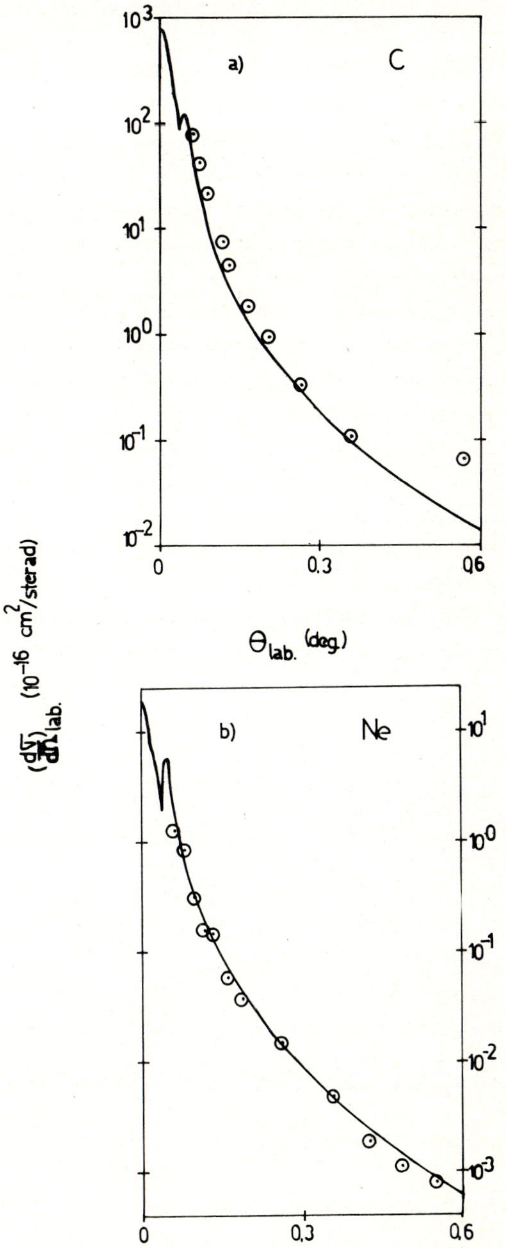

Fig. 1 : Differential cross sections against scattering angles for :
Present theoretical results calculated using initial wave functions and orbital energies of Table 1 of Clementi and Roetti (1974) : (——) ; experimental data from Horsdal-Pedersen et al : (⊙). All quantities are given in the laboratory system.
a) p + C at 0.6 MeV collision energy and
b) p + Ne at 1.5 MeV collision energy

References

1. I.M. Cheshire, Proc. Soc. 84, 89-98 (1964)
2. E. Clementi and C. Roetti, Atomic Data and Nuclear Data Tables, 14, 177-478 (1974)
3. E. Horsdal-Pedersen, F. Folkman and N.H. Pedersen, J. Phys. B 15, 739-62 (1982)

THE STRONG POTENTIAL BORN APPROXIMATION FOR CHARGE TRANSFER AT LARGE SCATTERING ANGLES

P.A. Amundsen* and D.H. Jakubassa-Amundsen**

*NORDITA, Blegdamsvej 17, DK-2100 Copenhagen Ø, Denmark

**Technische Universität München, D-8046 Garching, BRD

Recently, experimental results on charge transfer at large scattering angles in proton-atom collisions at intermediate energy have become available [1], providing a new testing ground of theories of charge transfer. In the experiments a very strong angular dependence of the scattering probabilities was found. Analyzed in terms of the impulse approximation [2] (IA), theory and experiment could only be reconciled by assuming a surprisingly large contribution from capture into excited states. Since more recent experiments strongly indicate that the angular dependence is even present for 1s-1s capture [3], we have calculated the transfer probability for this case, using the strong potential Born (SPB) approximation instead of the IA, which predicts a scattering angle (ϑ) independent transfer probability in this case.

Using a classical description of the internuclear motion, and assuming transfer from a strong (target) potential, V_T, to a weak projectile potential, V_P, the SPB amplitude can be written [2]

$$a_{fi} = \frac{-i}{2\pi} \int_{-\infty}^{\infty} dt \int d^3\underset{\sim}{k} \int_{-\infty}^{\infty} dt' \int_{-\infty}^{\infty} d\omega \, \varphi_f^P(\underset{\sim}{k} - \dot{\underset{\sim}{R}}(t)) \, e^{i(\varepsilon_f(t') - \omega t')} \times e^{i(\omega - E_i^T)t} \, {}^T\langle \underset{\sim}{k}, \omega | V_P(\underset{\sim}{r} - \underset{\sim}{R}(t)) | i \rangle^T$$

where $\underset{\sim}{R}(t)$ is the classical path, φ_f^P is the momentum-space final state wavefunction (energy E_f^P), $|i\rangle^T$ is the initial state and $|\underset{\sim}{k}, \omega\rangle^T$ is an <u>off-shell</u> target continuum state of momentum $\underset{\sim}{k}$ and energy ω $(= \omega - i\varepsilon)$. Furthermore

$$\varepsilon_f(t) \quad E_f^T + \underset{\sim}{k} \cdot \underset{\sim}{R}(t) + \frac{1}{2}\int_0^t \dot{\underset{\sim}{R}}^2(t)\,dt - \dot{\underset{\sim}{R}}(t) \cdot \underset{\sim}{R}(t)$$

To the above amplitude also a recoil amplitude should be added [2].

Thus, in the SPB-picture, the transfer is described as an excitation of the target, at time t, followed by a <u>subsequent</u> transfer to the projectile, at time t'. The $i\varepsilon$-prescription on ω ensures that t'>t. This time-delay between excitation and capture is allowed by the energy-momentum uncertainty-relation, since we have a distribution in energy of the off-shell state $|k, \omega\rangle^T$. In the IA, this off-shell state is replaced by an on-shell one $|\underset{\sim}{k}\rangle^T$ $(\omega = \frac{1}{2}k^2)$, and the ω integration gives t'=t. From this it follows that the IA excludes a potentially important process which the SPB picture includes [4]: If the projectile ionizes the target on the incoming leg of the collision, there is a large probability that an electron will be ejected at an emission angle of about 60° (the binary encounter peak). If the projectile itself is then scattered through the same angle at the nucleus, it may pick up the electron on the way out of the collision. This amplitude can then interfere with the amplitudes for both capture and excitation on the ingoing and on the outgoing part of the collision.

We have evaluated the SPB-amplitude for 1s-1s capture with hydrogenic wavefunctions and using Briggs' peaking approximation for he momentum integrals. We do indeed find a pronounced scattering angle dependence, which in general centers around $\vartheta = 60^\circ$. In the figure we compare with data from ref.3 for the system $p+CH_4$. It is seen that the agreement between theory and experiment is much improved from the IA (no ϑ-dependence). Also the energy dependence at a fixed scattering angle is qualitatively well reproduced. As for quantitative agreement, it is actually not much worse than for the SCA-calculations of K-shell ionization of the same system, using hydrogenic wavefunctions.

<u>References</u>

1. E. Horsdal-Pedersen, P. Loftager and J.L. Rasmussen, J. Phys. <u>B15</u>, 2461 (1982).
2. P.A. Amundsen and D.H. Jakubassa-Amundsen, Phys. Scr. <u>26</u>, 155 (1982).
3. E. Horsdal-Pedersen, P. Loftager and J.L. Rasmussen, J. Phys. <u>B15</u>, 4423 (1982).
4. E. Horsdal-Pedersen and J.L. Rasmussen, Invited talk at: "7th Conf. on the Application of Accelerators in Research and Industry", Denton, Texas (1982).

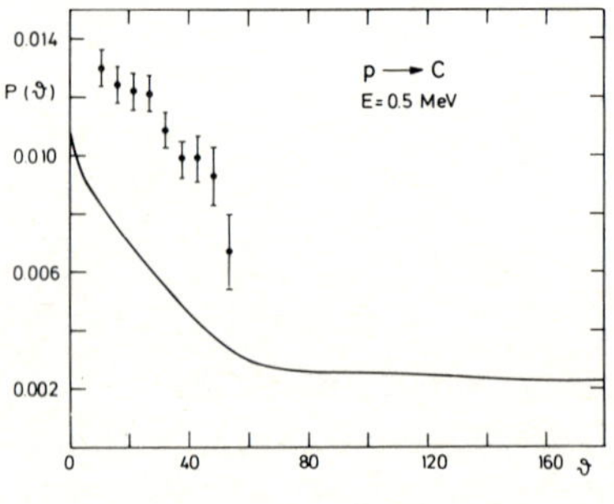

<u>Fig. 1:</u> SBP 1s-1s transfer probability for p+C. Experimental points for total capture from CH_4 from ref.3.

ELECTRON CAPTURE INTO HIGHLY EXCITED STATES

S. C. Mukherjee, Shyamal Dutta and C. R. Mandal

Indian Association for the Cultivation of Science, Calcutta 700 032, India.

The study of highly excited atoms has received attention in connection with their practical applications in plasma diagnostics as well as in problems connected with astrophysics. The viability of Rydberg atoms in Stellar atmospheres and interstellar space has encouraged the experimental workers to produce Rydberg atoms to study its various properties in the laboratory. The electron capture in ion-atom collisions is one of the few techniques[1] used for production of Rydberg atoms in the laboratory. The calculation of cross sections for electron capture into highly excited states presents formidable difficulties to a quantum mechanical treatment. This is due to the presence of a large number of oscillations in the final bound state wave functions. Recently, Belkic has introduced continuum intermediate state approximation[2] (CISA) for electron capture at large impact parameters. The capture probability at large impact parameter obtained by the use of CISA are found to be in excellant agreement with the measurements in the intermediate and high energy region. The present paper is aimed at developing a method for the calculation of cross sections for electron capture into arbitary n, ℓ and m states for fast protons in collision with the ground state hydrogen atoms in the frame work of the CISA. The scattering amplitude has been reduced to a closed analytical form which enables one to make a comparative study of the charge transfer cross sections on the quantum numbers n, ℓ and m. The asymptotic behaviour of the capture cross sections with respect to n has also been investigated. It is shown conclusively that the charge exchange cross sections asymptotically obey the inverse n-cube law throughout the entire energy range of the projectile. It is interesting to find that the n^3-times cross section for different value of ℓ tends to a constant value in a consistent manner with the increase of n at each incident energy depending only on the energy of the projectile.

In Figure 1 we compare our present results for capture in the 4S excited state with the CDWA approximation 3, as well as with the experimental data of Hughes et al[4].

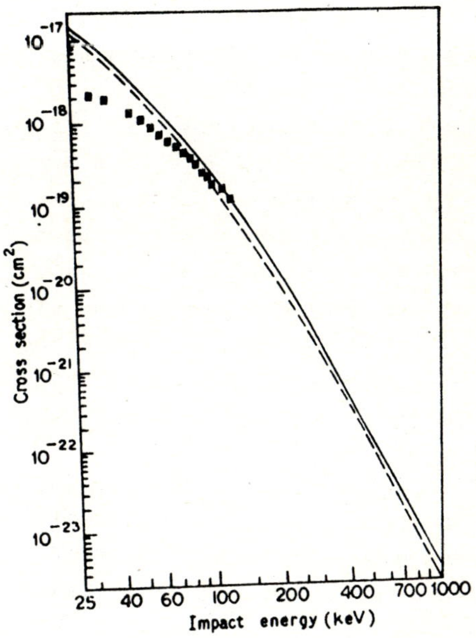

Fig. 1. Cross sections for electron capture by protons in the 4S state from the ground state of atomic hydrogen. Theory (———) present work; - - - Belkic' and Gayet[1], ■ Experiment Hughes et al[2].

References

1. R. F. Stebbings, Science 193, 537 (1976).
2. Dz' Belkic', J. Phys. B 10, 3491 (1977).
3. Dz' Belkic' and R. Gayet, J. Phys. B 10, 1911 (1977).
   ----------------, Ibid 10, 1923 (1977).
4. R. H. Hughes, H. R. Dawson and B. M. Doughty, Phys. Rev. 164, 166 (1967).

ION ATOM COLLISION FOR SYSTEMS WITH TWO ELECTRONS

W. Stich, H.J. Lüdde and R.M. Dreizler

Institut für Theoretische Physik der Universität Frankfurt/Main, BRD

In order to investigate many electron processes in ion-atom collisions (involving a small number of electrons) we attempt a solution of the time dependent Hartree Fock equations (TDHF) within the framework of the semiclassical approximation.

As in the calculation for one electron systems we expand the time dependent orbitals in a two centre pseudobasis first introduced by Hylleraas. The evaluation of one particle matrix elements can be carried out analytically. For the determination of the electron-electron interaction we require the steps

(i) Expansion of V_{EE} in terms of the Hylleraas basis set and solution of Poisson's equation in each time step

(ii) Assembly of the matrix elements from the resulting expansion coefficients.

In this fashion the two particle matrix elements can be calculated exactly within the chosen representation.

The initial state of a particular collision system is represented by projection of travelling Hartree Fock Roothaan orbitals onto the basis. Global capture probabilities (and cross sections) are determined by integrating the many electron density over projectile and target half spaces, respectively

First results for the systems

$H^+ + He\ (^1S)$

$He^{2+} + He\ (^1S)$

will be presented and compared with experimental data and results of competing theoretical approaches.

ELECTRON TRANSFER AND EXCITATION IN p-H COLLISIONS USING A TRIPLE-CENTER BASIS

T. G. Winter

Department of Physics, Pennsylvania State University, Wilkes-Barre Campus, Wilkes-Barre, PA 18708, U.S.A.

C. D. Lin

Department of Physics, Kansas State University, Manhattan, KS 66506, U. S. A.

The triple center method for electron transfer in ion-atom collisions was developed by Anderson, Antal, and McElroy[1] and applied by them and us[2] to p-H collisions using a limited basis. This coupled-state approach is one extension of the two-center, traveling-atomic-state method developed by Bates[3] and applied by others to proton-hydrogen-atom collisions[4-6] (as well as by others to other systems). In the triple-center approach, a third set of atomic basis states is added; these states are ordinarily centered on the center of charge and have effective charge equal to the sum of the nuclear charges.

The purpose of the third set of basis states is of course to add variational freedom. In the united-atoms limit these states exactly describe the bound molecular states (apart from any ambiguity in the choice of translational factors). Thus the method is an atomic-state alternative to low-energy molecular-state methods originally developed by Bates and McCarroll[7] and refined and extended by others.[8]

Another related atomic-state method, the AO+ method, has recently been developed by Fritsch and Lin[9] and applied to several systems. In this two-center method, united-atoms states are placed on each nucleus, augmenting the normal atomic-state basis. These additional orbitals thus have the same shape but different locations and, even in the united-atoms limit, different plane-wave factors than do the third-center basis states in the triple-center approach. Although the AO+ method was originally developed as a simplification of the triple-center method, it has proven to be a very effective method in itself.

A stumbling block in the triple-center method has been the velocity-dependent, triple-center, charge-exchange integrals. One way to evaluate them is to expand each atomic orbital in terms of Gaussian orbitals. We have found this to be unnecessary: One simply evaluates the matrix elements using spheroidal coordinates with one focus on one orbital's center and the other on the singularity in the potential rather than on the other orbital's center; this avoids a "numerical singularity" at the singularity in the potential.

Cross sections have been calculated for electron transfer into the 1s, 2s, and 2p states and excitation to the 2s and 2p states from about 1.5 keV to at least 11 keV. Although adiabatic energies of the lowest molecular states are well represented by a smaller basis, it has been found that the minimum basis to represent the dynamic collision processes to 10-20% may be fairly large at the higher energies. At about 5 keV, for example, a large basis of 36 states has been used: 10 states 1s, ..., $3d_{0,1,2}$ centered on each nucleus and on the center of charge C, plus the states $4p_{0,1c}$, $4f_{0,1,2,3c}$. Of these states, 16 are gerade and 20 are ungerade. The ungerade states tend to be more important, particularly at lower energies, owing to the large energy gap between the initial gerade state $1s\sigma_g$ and the other gerade states. (See also Ref. 8.)

These triple-center results will be compared with atomic-state,[4-6] AO+,[9] pseudo-state,[10-12] molecular-state,[8,13] and experimental[14] results.

References

1. D. G. M. Anderson, M. J. Antal, and M. B. McElroy, J. Phys. B 7, L118 (1974); 14, 1707(E) (1981).
2. C. D. Lin, T. G. Winter, and W. Fritsch, Phys. Rev. A 25, 2395 (1982).
3. D. R. Bates, Proc. R. Soc. London Ser. A 247, 194 (1958).
4. R. McCarroll, Proc. R. Soc. London Ser. A 264, 547 (1961).
5. D. F. Gallaher and L. Wilets, Phys. Rev. 169, 139 (1968).
6. D. Rapp, D. Dinwiddie, D. Storm, and T. E. Sharp, Phys. Rev. A 5, 1290 (1972).
7. D. R. Bates and R. McCarroll, Proc. R. Soc. London Ser. A 247, 175 (1958).
8. See M. Kimura and W. R. Thorson, Phys. Rev. A 24, 1780 (1981) and references therein.
9. W. Fritsch and C. D. Lin, Phys. Rev. A 26, 762 (1982).
10. I. M. Cheshire, D. F. Gallaher, and A. J. Taylor, J. Phys. B 3, 813 (1970).
11. W. Fritsch and C. D. Lin (private communication)
12. H. J. Lüdde and R. M. Dreizler, J. Phys. B 15, 2703 (1982).
13. D.S.F. Crothers and J.G. Hughes, Phil. Trans. R. Soc. London 292, 56 (1979).
14. T. J. Morgan, J. Stone, and R. Mayo, Phys. Rev. A 22, 1460 (1980), and references therein.

CHARGE TRANSFER AND IONIZATION PROCESSES IN $He^+ + He^+$ and $He^+ + H$ COLLISIONS

M.R.C. McDowell

Department of Mathematics, Royal Holloway College, Egham Hill, Egham, Surrey TW20 0EX,

G. Peach and S.L. Willis

Department of Physics and Astronomy, University College, Gower Street, London WC1E 6BT

Calculations have been carried out to obtain total cross sections for the processes

$$He^+ + He^+ \rightleftarrows He^{+2} + He \quad (1)$$
$$He^+ + He^+ \rightarrow He^{+2} + He^+ + e \quad (2)$$
$$He^+ + H \rightleftarrows H^+ + He \quad (3)$$
$$He^+ + H \rightarrow H^+ + He^+ + e \quad (4)$$

In each case, the colliding particles have been represented by a model in which only the active electron is treated explicitly; the residual atomic ion He^+ is considered to be simply a polarisable core. The interactions between the He^+ core and the other particles in the collision are represented by model potentials, see Peach[1], and are of modified Coulomb form. Additional terms are included in order to allow for polarisation. The collision problem is treated as a purely classical one, and Hamilton's equations of motion are integrated numerically for thousands of different trajectories that are selected randomly. By examining the final energies of the particles after the collision, the number of ionizing and charge transfer events can be determined, and hence the total cross sections for these processes can be obtained. These classical calculations using Monte Carlo methods, are the first to have been carried out in which the non-Coulombic nature of the interactions has been allowed for in a realistic way. In all previous calculations in which complex residual ions have been involved, the ion has simply been assumed to have a suitable effective charge, see for example, Olson[2]. Processes (1) and (3) have been studied in both the forward and reverse directions, partly because all these processes are of interest in their own right, and partly because such a study provides a practical test of our assumptions concerning the classical equivalent of the quantum mechanical problem. We have adopted a description for the atom in its initial state which is based on the use of the microcanonical distribution, see Abrines and Percival[3]. A criterion has been developed in which a given band of energies corresponds to a particular quantum mechanical state of the atom. Thus when the collision has been completed, the energy of the captured electron can be used to determine whether the atom is in the ground state or is in an excited state.

Cross sections for the processes (1)-(4) have been obtained for kinetic energies of the incident He^+ ion in the range 20 keV to 800 keV. These results will be presented at the conference and compared with the existing theoretical and experimental data on charge exchange, see Afrosimov *et al*[4], Harel and Salin[5] and Peart *et al*[6] for process (1) and Olson *et al*[7] and Harel and Salin[8] for process (3). Preliminary results from this present study show an encouraging level of agreement with the results of other workers. Our calculations for the ionization process (2) can be compared with the experimental results deduced from Peart *et al*[6] by subtracting the charge-exchange cross section from the total cross section for the production of He^{+2}; good agreement is obtained.

References

1. G. Peach, *Comments Atom. Molec. Phys.* **11**, 101 (1982).
2. R.E. Olson, *Phys. Rev.* A **18**, 2464 (1978).
3. R. Abrines and I.C. Percival, *Proc. Phys. Soc.* **88**, 873 (1966).
4. V.V. Afrosimov, A.A. Basalaev, G.A. Leǐko and M.N. Panov, *Soviet Phys. JETP* **47**, 837 (1978).
5. C. Harel and A. Salin, *J. Phys.* B *Atom. Molec. Phys.* **13**, 785 (1980).
6. B. Peart, K. Rinn and K.T. Dolder, *J. Phys* B *Atom. Molec. Phys.* (1983), submitted for publication.
7. R.E. Olson, A. Salop, R.A. Phaneuf and F.W. Meyer, *Phys. Rev.* A **16**, 1867 (1977).
8. C. Harel and A. Salin, *J. Phys.* B *Atom. Molec. Phys.* **16**, 55 (1983).

Electron capture and excitation in proton-sodium collisions at energies E≤10keV.

R.J.Allan,† A.S.Dickinson,Dept. of Atomic Physics, The University,Newcastle upon Tyne, UK.
and R.McCarroll, Laboratoire d'astrophysique, Université de Bordeaux I, 33270 Floirac, France.

Model-potential methods for molecular structure calculations in proton-alkali systems allow straightforward evaluation of the non-adiabatic coupling terms needed in the perturbed-stationary-state (pss) formulation of inelastic collision processes[1]. We use the potential of Klapisch[2] to model the e-Na^+ interaction, with a basis of Slater-type orbitals in prolate spheroidal coordinates. Our calculated molecular energy levels as functions of internuclear distance (R), and atomic oscillator strengths are in good agreement with more sophisticated ab initio and model potential results. This, together with internal consistency tests, suggests that our matrix elements are of sufficient accuracy.

In the collision calculation difficulties arise due to the neccessity to limit the number of states included, and from neglect of electron translation effects. The problem of coupling between states of separate atoms, and resulting non-convergance of the cross section has been overcome using an adiabatic switching factor (ASF) in the asymptotic region.[3,4] Our collisional calculation was carried out by solving the coupled straight-line impact-parameter equations with all neccessary radial and rotational coupling terms in an interaction region of radius $R_o=59a_o$. Six states were employed which correlate at large internuclear distance to Na(3s), Na(3p) and H(n=2).

The cross sections are linearly dependent on the fractional displacement of the origin along $\hat{R}$ from one of the nuclei. We will present our results for an origin on either atomic nucleus for transfer of an electron from the initial Na(3s) state to H(n=2). The largest cross section($\sigma \approx 50 Å^2$) occurs for E≈ 4.5keV. Results agree with the experiments of Grüebler et al.[5] and Anderson et al.[6]. There is however some disagreement in both magnitude and shape with the data of Nagata[7] and recent pseudopotential calculations[8]. We will also present cross sections for excitation of states correlated to Na(3p) and for charge-transfer and de-excitation from the initial Na(3p) level.

Our charge-transfer cross section with initial Na(3p) states peaks at low energy and is larger than that from Na(3s), consistent with the smaller asymptotic energy defect. We again disagree with Kimura et al.[8] who found the two cross sections to be essentially identical above 100eV.

We investigated the effect of variation in the number of molecular basis states included in the calculation. Addition of the state correlated to Na(4s) reduces slightly the origin dependence of the results with no significant change in their overall magnitude. If we reduce the basis to four states, Na(3s) and H(n=2), only the principal radial coupling mechanism remains and we obtain somewhat larger cross sections with increased origin dependence.

At low and intermediate relative speeds the molecular eigenstate expansion with ASF is a useful method to calculate total cross sections and allows the relevant excitation mechanism to be understood in a particularly simple way. Our method can also be extended to include a larger number of basis states in order to reduce further the origin dependence.

RJA acknowledges support from the SERC.

References:

1 MGargaud,JHanssen,RMcCarroll and PValiron J.Phys.B. 14 (1981) 2259-76
2 MKlapisch 1969 (Thèse de Doctorat-ès-Sciences Université de Paris, Orsay.)
3 ASDickinson and RMcCarroll J.Phys.B. 16 (1983) 469-476.
4 RJAllan,ASDickinson and RMcCarroll J.Phys.B. 16 (1983) 476-80
5 WGrüebler,PASchmelzbach,VKönig and PMarmier Helv.Phys.Acta 3 (1970) 254-71
6 CJAnderson,AMHowald and LWAnderson Nucl. Instruments and Methods 165 (1979) 583-7
7 TNagata J.Phys.Soc.Japan 46 (1979) 1622-7
8 MKimura,REOlson and JPascale Phys.Rev.A 26 (1982) 1138-41

† present address: Fachbereich Chemie der Universität Kaiserslautern,D-6750 Kaiserslautern BRD.

CHARGE TRANSFER IN COLLISIONS BETWEEN PROTONS AND LITHIUM ATOMS

A.M. Ermolaev

Department of Physics, University of Durham, South Road, Durham DH1 3LE, England.

In preceding studies[1] of charge transfer, it was established that effective one-particle central potentials derived from a SCF model were adequate in a wide range of energies within the impact parameter approximation, for describing interactions between structured atomic systems.

In the present work, a similar approach was applied to the charge transfer reaction $H^+ + Li(2s) \rightarrow H(nlm) + Li^+$ in the intermediate energy range where the experimental data is scarce and where there was no theoretical data until recently.

A corresponding one-active-electron model was obtained by replacing the full atomic potential at the lithium centre B by $U_{eff,B}$. The latter was chosen to be the SCF potential supporting the 2s electron of the lithium ground state. The AO basis was constructed from hydrogenic functions ϕ_{iA} and eigenstates χ_{jB} of a one-particle Hamiltonian with the central field $U_{eff,B}$. The basis included plane wave translational factors.

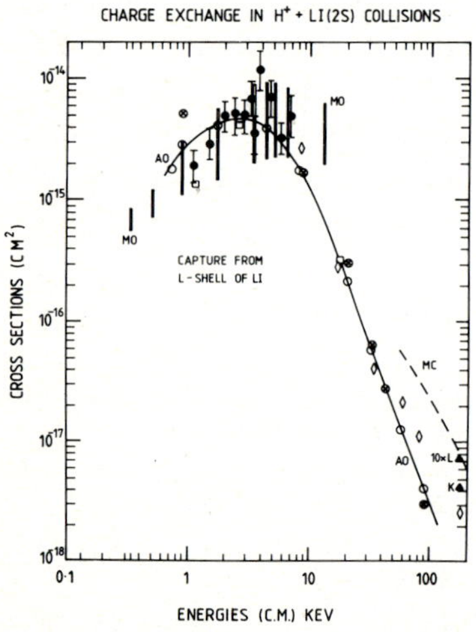

Fig. Total cross sections for production of H in H^+ + Li collisions (L-shell capture). Theoret. data: MO-molecular expansion of ref.2. Atomic expansion: ⊗ 2-state, present work; O 13-state, present work; □ 19-state, present work; MC - classical trajectory Monte Carlo, ref.3; L,K ▲ CDW, ref.4 for L and K-shell capture respectively. Expt. data: ● - ref.5; ◊ ref.6.

Calculations reported here used a 13-state AO set (10 states with $n \leqslant 3$ centred on A and 3 states with $n = 2$ centred on B) and they covered an energy range between 0.70 and 109.3 keV c.m. An enlarged set of 19 states (9 states with n = 2,3 centred on B) was used for selected energies to assess the convergence trends for this basis. In order to establish the energy range where capture from the K-shell of Li becomes dominant, a set of two-state calculations at E up to 350 keV c.m. was carried out.

The total cross sections are displayed in the figure. At low energies (0.70-5.0 keV c.m.) the present results are in good accord with recent MO calculations of Allan et al.[2] with adiabatic switching factors. However for $E \geq 5$ keV c.m. cross sections[2] diverge rapidly from the data reported here. The present calculations confirm that preferential production of H(n=2) collision takes place within a energy range up to 10-15 keV c.m.

The two-state cross sections are equal at 200 keV for both the L- and K-shell captures (6×10^{-19} cm^2). For E > 200 keV, the K-shell capture is dominant. This result does not support a previous conclusion of Olson[3] and Banyard and Shirtcliffe[4] that L-shell capture is of importance only at E < 25 keV c.m.

E keV c.m.	σ(n) n = 1	2	3	Total σ	AO
1.14	0.08	9.18	3.6	12.9	19
2.62	0.70	35.4	5.0	41.1	19
4.37	0.62	32.9	6.1	39.6	13
8.74	0.58	11.5	5.0	17.1	13
21.86	0.35	0.72	0.58	1.74	10
	0.38	0.75	0.61	1.65	13

Table. Calculated cross section σ_n for production of H(n) in H^+ + Li collisions (in units of 10^{-16} cm^2).

In conclusion, this work compares certain aspects of the present model calculations with 'exact' calculations of electron transfer between two Coulomb centres[7].

References

1. A.M. Ermolaev, C.J. Noble and B.H. Bransden, J. Phys. B: At. Mol. Phys. 15, 457 (1982); A.M. Ermolaev and B.H. Bransden, to be published (1983).
2. R.J. Allan, A.S. Dickinson and R. McCarroll, J. Phys. B: At. Mol. Phys. 16, 467 (1983).
3. R.E. Olson, J. Phys. B: At. Mol. Phys. 15, L163 (1982).
4. K.E. Banyard and G.W. Shirtcliffe, J. Phys. B: At. Mol. Phys. 12, 3247 (1979).
5. W. Gruebler, P.A. Schmelzbach, V. Konig and P. Marmier, Helv. Phys. Acta 3, 254 (1970).
6. R.N. Ilin, V.A. Oparin, E.S. Solovev and N.V. Federenko, JETP Lett. 2, 197 (1967), Sov. Phys. - Tech. Phys. 11, 921 (1967); B.A. D'yachkov, V.I. Zinenko, Sov. At. Energy, 24, 16 (1968).
7. B.H. Bransden, C.J. Noble and J. Chandler (1983), to be published and references therein.

CORE-INDEPENDENT PARAMETERS FOR CHARGE TRANSFER REACTIONS

K.Bartschat, H.J.Andrä[+], K.Blum

Institut für Theoretische Physik I, Universität Münster, Domagkstr., D-4400 Münster

[+] Institut für Kernphysik, Universität Münster, Domagkstr., D - 4400 Münster

Numerical calculations for charge transfer reactions have been performed for the system $H^+ + H \rightarrow H + H^*$. An experimental test of the theory is difficult, particularly because the nearly degenerate H-levels in the final excited states cannot be spectroscopically resolved and give rise to many coherence effects overlapping the experimental results.

These latter experimental difficulties can be overcome by using homologous alkali systems where the transfer of the valence electron can be considered as H-like while the degeneracies are removed. In this case, however, the process can be influenced by the core electrons, particularly in that core electrons may be tranfered from the atom to the ion in some events. Although it is usually assumed that this effect can be neglected it is still an open question up to which projectile energies this assumption is justified. A direct comparison between the theory of quasi - one-electron systems and the suggested alkali experiment requires therefore a clear experimental separation between the contributions of the valence and the core electrons to the total charge exchange reaction.

We show that this separation can be achieved in the following way. We consider the reaction

$$Li^+ + Na(3s\uparrow) \rightarrow Li(2p) + Na(\alpha)^+$$
$$\quad\quad \hookrightarrow Li(2s) + \gamma$$

where the Na-valence electron is spin polarised. The light emitted in the decay of the Li-atoms is analysed, the final Na-ions are not observed. α denotes the different final states of the Na-ions. We are interested in the channel which corresponds to the exchange of the valence electron (denoted by α=0). Transfer of core electrons leads to final states denoted by $\alpha \neq 0$.

We have defined observables which depend only on the transfer of the valence electron and which are exactly independent of any exchange of core electrons. These particular observables can therefore be calculated by treating the alkali atoms and ions as quasi-one-electron systems. We show that the use of polarised initial Na-atoms allows the experimental determination of the observables of interest. By comparing experimental and theoretical results an unambigous test of the theory is then possible.

PHASE INTEGRALS AND PERTURBED STATIONARY STATES

A. Bárány*, D.S.F. Crothers† and J. G. Hughes†

* Institute of Theoretical Physics, University of Uppsala, S-752 38, Uppsala, Sweden.

† School of Physics and Mathematical Sciences, The Queen's University of Belfast, Belfast BT7 1NN.

Two aspects of low-energy heavy-particle collisions are considered with reference to charge transfer and excitation.

Firstly it is shown[1] that the S matrix for non-adiabatic collisions, including pseudo-crossings, perturbed symmetric resonance, rotational coupling etc., may be parametrized in terms of complex Coulomb phases associated with the τ-plane. The approach adopted is the JWKB Green-Liouville phase integral method, in which the effective potential is a complex one given by

$$1 + T^2 + idT/d\tau \qquad (1)$$

where T is the Stueckelberg variable and τ an effective reduced time variable. An interesting pattern of logarithmic-spiral Stokes' lines occurs due to an initial pole in the complex τ-plane on the real axis. Phase integral abstraction is promoted by comparison with the exponential model. The application to problems with a Coulomb potential pole in the R-plane is considered, as in for example

$$Mg^{++} + H(1s) \rightarrow Mg^{+}(\Sigma) + H^{+}. \qquad (2)$$

This R-plane pole is a weaker singularity than the τ-plane pole.

It is also shown that when applying the method of variational perturbed stationary states[3,4] at low energies to the calculation of cross sections for processes such as

$$H^{+} + H(1s) \rightarrow \begin{cases} H(2s,\ 2p) + H^{+} & (3a) \\ H^{+} + H(2s,\ 2p) & (3b) \end{cases}$$

not only is state selection of importance in determining switching functions and convergence, but also there are two inter-related issues worthy of attention. Firstly, non-Hermitean coupled equations may be sufficient, provided all orders in v the velocity are retained (or at least one order higher than that required), but they are not necessary. By the same token, Hermitean equations are sufficient though not necessary, again provided all orders in v are retained. Secondly, and also of importance to economy of effort and computer time, Hermitean equations are necessary if we wish to perform a perturbation calculation with respect to v, that is, if we wish to avoid the inefficient and time consuming strategy of going one order higher in the perturbation series. Both these assertions follow from the well known relation between the off-diagonal matrix elements and the time derivative of the overlap matrix element, and from the symmetric orthogonalisation of the travelling molecular-orbital basis sets via the Löwdin formula[5] for a row matrix $\underline{\chi}$ of kets, namely

$$\underline{\chi}\,(\underline{\chi}^{+}\underline{\chi})^{-\frac{1}{2}} \qquad (4)$$

a perturbation form of which, with suitable choice of matrix square root, we have employed previously[3,6]. We present 8 state calculations in the 1-10 keV impact energy range for processes (3a,b), which show that terms of second order in v can not be neglected when calculating total cross sections.

References

1. A. Bárány and D.S.F. Crothers, Proc. R. Soc. Lond. A. 385, 129 (1983).
2. D.S.F. Crothers and N. R. Todd, J. Phys. B. 13, 547 (1980).
3. D.S.F. Crothers and J. G. Hughes, Proc. Roy. Soc. Lond. A. 359, 345 (1978).
4. D.S.F. Crothers and J. G. Hughes, Phil. Trans. Roy. Soc. Lond. 292, 539 (1979).
5. Y. J. I'Haya and T. Morikawa, Adv. Quant. Chem. 12, 43 (1980).
6. D.S.F. Crothers and N. R. Todd, J. Phys. B. 14, 2233 (1981).

CHARGE EXCHANGE BETWEEN He^{+}-IONS AND LITHIUM ATOM

K. Roy and R. Shingal

Department of Physics, The University of Durham, Durham DH1 3LE, England

The recent measurements of the charge transfer cross-sections for electron capture by He^{+}-ions in Lithium vapour by McCullough et al.[1] covering the energy range 6.7-400 keV and the absence of a satisfactory theoretical treatment have led us to make a theoretical investigation into this problem. The only other available experimental findings are due to Auciello et al.[2] in the range 7.5-44 keV.

It has been well established that the collisions of He^{+}-ions with the alkali metal vapours are dominated by the near resonant capture into n=2 states of Helium. We have, thus, considered the following reaction as a starting point:

$$\begin{aligned} He^{+} + Li &\rightarrow He^{+} + Li \\ &\rightarrow He(1s^{2}) + Li^{+} \\ &\rightarrow He(1s,2s) + Li^{+} \\ &\rightarrow He(1s,2p_{0}) + Li^{+} \\ &\rightarrow He(1s,2p_{\pm 1}) + Li^{+} \end{aligned} \qquad (1)$$

We have carried out the calculations in the multi-state atomic impact parameter model employing a straight line trajectory for the projectile. The electronic wave functions are the solutions of the time dependent Schrodinger equation (in atomic unit)

$$\left(H - i\frac{\partial}{\partial t}\right)\psi\left(\vec{r},t\right) = 0 \qquad (2)$$

Here, the Hamiltonian H is given by

$$H = -\tfrac{1}{2}\nabla_1^2 - \tfrac{1}{2}\nabla_2^2 - z_A'\left(\frac{1}{r_{A1}} + \frac{1}{r_{A2}}\right) - z_B\left(\frac{1}{r_{B1}} + \frac{1}{r_{B2}}\right) + \frac{1}{r_{12}} \qquad (3)$$

In equation (3), r_{Ai}, r_{Bi} is the distance of the electron e_i (i=1,2) from the target nucleus A and the projectile nucleus B, respectively, r_{12} being the distance between the electrons e_1 and e_2.

We have assumed a one electron model for the Li-atom in which the outer electron moves in the effective potential $V_{eff}(= - z_A'/r_A)$ exerted by the Li^{+}-core, z_A' is taken to be the variational value (z-5/16). The electronic wave function $\psi(\vec{r},t)$ is expanded in a basis set consisting of five states describing the reaction (1);

$$\psi(\vec{r},t) = \sum_{i=1}^{5} A_i(b,t)\,\psi_i(\vec{r},t) \qquad (4)$$

Here, $\psi_i(\vec{r},t)$ is the properly symmetrized combination of the product of the wavefunction of He^{+}-ion and the Li-atom. The other terms are the appropriate wave functions of the He-atom. The ground state of He-atom is represented by the Byron and Joachain[6] wavefunction. For 2s and 2p excited states of Helium we have considered the wavefunction given by Cohen and McEarchran[7] and Goldberg and Clogston[8] respectively. The amplitudes $A_i(b,t)$ are the solutions of a variationally determined coupled differential equations[9]

$$g\dot{A} = -iFA \qquad (5)$$

where g and F are the overlap and the interaction matrix elements. The momentum transfer of the electrons is neglected in these calculations.

The capture cross-section into the specific states $n\ell m$ is determined using the relation

$$Q_{n\ell m} = 2\pi \int_0^\infty |A_i(b,t=\infty)|^2\, b\,db \;.. \qquad (6)$$

and the total cross-section is

$$Q = \sum_{n\ell m} Q_{n\ell m} \qquad (7)$$

The results are under investigation and will be presented at the conference.

References

1. McCullough, R.W. Goffe, T.V., Shah, M.B., Lennon, M. and Gilbody, H.B., J. Phys.B:A & Mol.Phys.15, 111 (1982)

2. Auciello, O., Alonso, E.V. and Baragiola, R.A., Phys. Rev. A.13, 985 (1976)

3. Peterson, J.R. and Lorents, D.C., Phys. Rev. A. 182, 152 (1969)

4. Olson, R.E. and Smith, F.T., Phys. Rev. A. 7,1529(1973)

5. McCullough, R.W., Goffe, T.V. and Gilbody, J. Phys. B: A & Mol. Phys. 11, 2333 (1978)

6. Byron, F.E. and Joachain, C.J., Phys. Rev. 146,1(1966)

7. Cohen, M. and McEarchran, R.P. Proc. Phys. Soc., 92, 37 (1967)

8. Goldberg, L. and Clogston, A.M., Phys. Rev., 56, 696 (1939)

9. Roy Kanika and Mukherjee, S.C., Phys. Rev. A., 7, (130(1973).

MOLECULAR STATE CALCULATION OF CHARGE TRANSFER IN H^+ + Li COLLISIONS

M. Kimura, H. Sato, J. Pascale,* and R. E. Olson

Physics Department, University of Missouri-Rolla, Rolla, MO 65401, U.S.A.
*C.E.N./Saclay, Service de Physique des Atoms et des Surfaces, 91191
Gif-Sur-Yvette, France

Previously, we have calculated the charge transfer cross sections for H^+-Alkali atom (Na, K, Rb, Cs) collisions. The last species of this series of work is to compute the cross section for the system

$$H^+ + Li(2s) \rightarrow H(n) + Li^+.$$

A close coupling method with six molecular states was used in the calculation. The states included are 1Σ (corresponding to Li(2s) + H^+); 2Σ, 1π (corresponding to Li^* (2p) + H^+); and 3Σ, 4Σ, 2π (corresponding to H(n=2) + Li^+). The potential curves are shown in Fig. 1. Electron translation factors (ETF's) were taken into account to both first and second order in velocity.

The calculated cross section is shown in Fig. 2 along with experimental results[1] and other theoretical calculations[2] which neglected ETF effects. Our close coupling calculations which include ETF effects up to the second order term in velocity appear to reproduce the experimental values[1] measured using deuteron projectiles. The experimental observations employed an attenuation method which can be

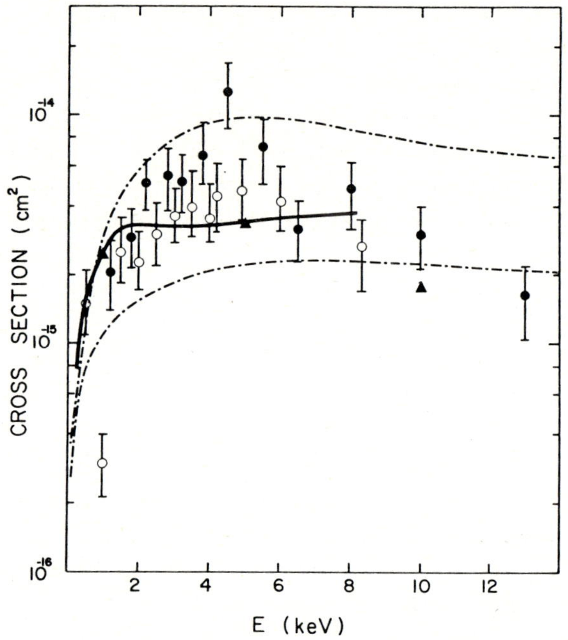

Fig. 2. Cross sections. ▲: our second order of V, solid line: our first order of V, dot-dash line: results of Ref. 2 (upper line: origin on H, lower line: origin on Li). Experiment: ● ○ Ref. 1

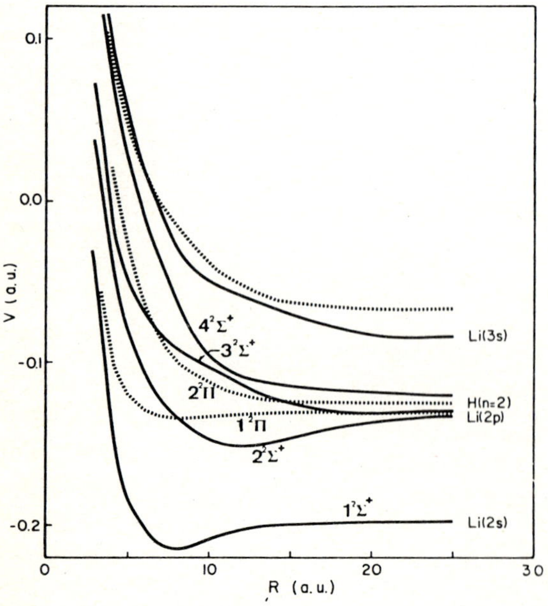

Fig. 1. Adiabatic energies of HLi^+.

complicated by elastic scattering effects. Thus, it is expected the deuteron data is more reliable than the proton data.

Although the inclusion of ETF effects avoids the difficulty of the origin-dependence of the electron coordinate on the calculated coupling matrix, the results obtained by the close coupling calculation with only a first order velocity correction to the ETF's does not behave correctly at high collision energies, E $\geq$ 5 keV. Such a result implies that in some systems, higher order ETF corrections are essential to study charge transfer collisions.

Work supported by the Office of Fusion Energy of the U.S. Department of Energy.

References

1. W. Gruebler, P. Schmelzbach, V. Konig and P. Marmier, Helv. Phys. Acta. 3, 254 (1970).
2. R. J. Allan, A. S. Dickinson and R. McCarroll, J. Phys. B (in press).

MOLECULAR TREATMENT OF CHARGE TRANSFER IN Li^+ + Ca COLLISIONS

M. Kimura, H. Sato, and R. E. Olson

Physics Department, University of Missouri-Rolla, Rolla, MO 65401, U.S.A.

The perturbed-stationary-state (PSS) method appended with the electron translation factors (ETF's)[1] has been applied to the processes:

$$Li^+ + Ca(4s^2) \rightarrow Li(2s) + Ca^+(4s) \quad (1)$$
$$\rightarrow Li^*(2p) + Ca^+(4s) \quad (2)$$
$$\rightarrow Li^+ + Ca^*(4s3d) \quad (3)$$

at collision energies of 0.1 to 20 keV/amu. Adiabatic wavefunctions and eigenvalues are determined by the pseudo-potential technique, which reduces the many electron system to a simpler two-electron system.[2]

Figure 1 shows the calculated adiabatic potential curves for the $(Li\text{-}Ca)^+$ system, including 1Σ (corresp. to Li^+ + Ca); 2Σ (corresp. Li + Ca^+); 3Σ, 1π (corresp. Li^* + Ca^+); 4Σ, 2π (corresp. Li^+ + Ca^*); 5Σ, 3π (corresp. Li^+ + Ca^*). Due to the inclusion of the ETF's, the wavefunctions satisfy the correct scattering boundary conditions and hence, the calculated coupling matrix elements are not dependent on the origin of the electronic coordinates. The calculated cross sections for the processes 1 to 3 are shown in Fig. 2. The dominant process is charge transfer to the Li(2s) state in the energy range studied. As the collision energy increases, process 2, charge transfer to the $Li^*(2p)$ state, becomes increasingly important due to strong rotational coupling at R ≅ 6a.

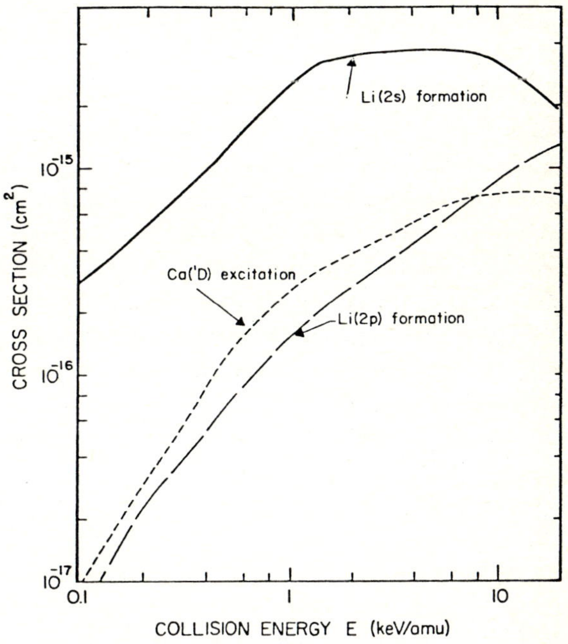

Fig. 2. The charge transfer cross section vs. ion energy.

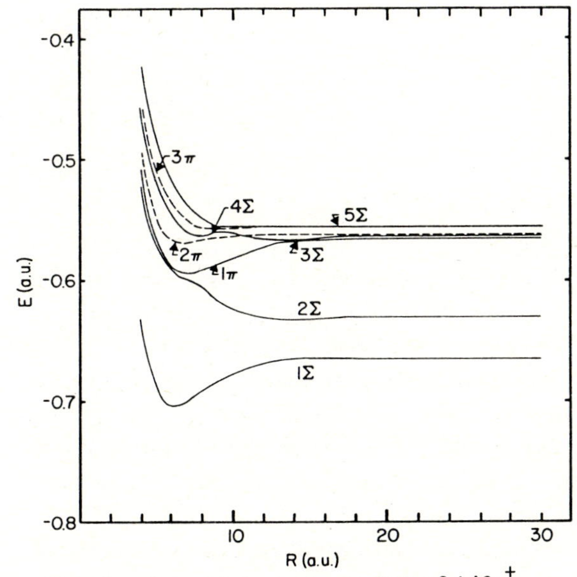

Fig. 1. Potential energy curves of $LiCa^+$.

The details of the theory along with the calculated results will be presented at the meeting.

Work supported by the Air Force Office of Scientific Research.

1. See for example, M. Kimura and W. R. Thorson, Phys. Rev. A 24, 1780 (1981).
2. J. N. Bardsley, Case Stud. At. Phys. 4, 299 (1974).

DYNAMICAL-STATE REPRESENTATION AND ITS APPLICATION TO THE $(Li - Na)^+$ COLLISION SYSTEM

Reiko Hirokawa[a,b], Hiroki Nakamura[b], and Eiichi Ishiguro[a]

a) Department of Physics, Ochanomizu University, Bunkyoku, Tokyo 112, Japan
b) Division of Theoretical Studies, Institute for Molecular Science, Okazaki 444, Japan

It is well known that rotational coupling plays an important, sometimes dominant, role in various collision processes. There are mainly three characteristic features about this coupling. First, this coupling opens a new channel whcih can not be reached by radial coupling; this causes a transition between the electronic states of different symmetry. Second, the mathematical analytical structure of this coupling problem is different from that of the radial coupling problem; this fact causes the inapplicability of the conventional semiclassical theory. Third, the coupling term itself depends on the collision velocity and impact parameter; this fact causes a peculiar velocity dependence of cross sections.

It is desirable to develop a theory which can deal with both radially and rotationally induced transitions analytically in a unified way based on the same footing. A theory which meets this requirement was proposed recently by using a new (dynamical-state) representation[1-2]. A framework of this theory is equal to the conventional semiclassical one based on the path-integral formalism except that the dynamical-state representation is employed instead of the ordinary adiabatic-state representation. The theory has been applied to some model collision systems.[1-2]

In this paper the theory is applied to the following collision processes:

$$Li^+ + Na(3s) \rightarrow Li(2s,2p) + Na^+,$$
$$Li(2s) + Na^+ \rightarrow Li^+ + Na(3s,3p).$$

These processes have been studied theoretically as well as experimentally by several authors. Differential cross sections are measured at several collision energies.[3-5] Fully numerical calculations were made by Melius and Goddard by using a linear trajectory approximation.[6] Some simplified versions of semiclassical calculations were also made, but none of them deals with the rotational coupling unambiguously.[3,5,7] Our formalism can take into account properly the multi-trajectory effect and the rotational coupling effect without any ambiguity. Three lowest electronic states (two Σ states and one Π state) are involved in the processes (see figure 1). Adiabatic potential energies are taken from Wijnaendts van Resandt et al[3]. The rotational coupling matrix element is taken from Melius and Goddard[6]. Calculations of the total as well as differential cross sections will be reported and be compared with other calculations and experiments. Not only the rainbow peak due to the attractive potential well of the lowest Σ state, but also a peak due to the rotational coupling appear clearly in the differential cross sections. Figure 2 shows an example of the results on the total cross sections as a function of collision velocity.

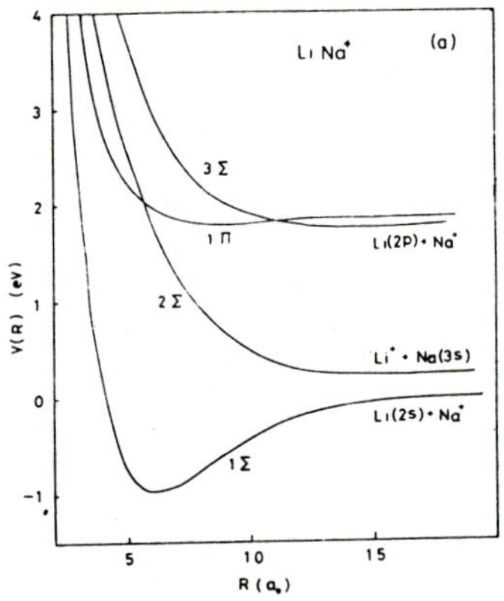

Fig. 1: Adiabatic potential energy curves of the lowest three states of $(LiNa)^+$.

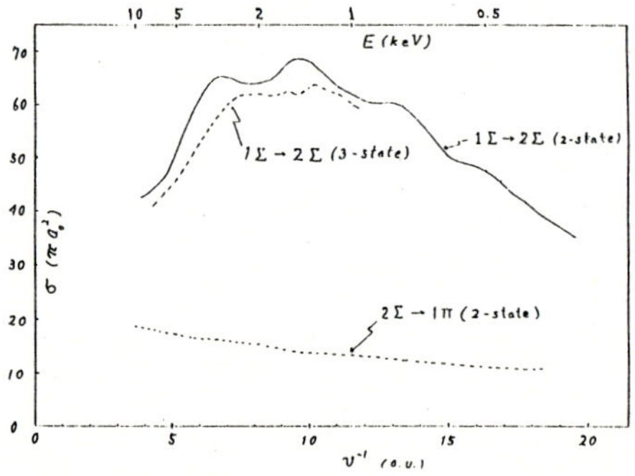

Fig. 2: Total cross sections as a function of inverse collision velocity.

References

1. H. Nakamura and M. Namiki, J. Phys. Soc. Japan 49, 843 (1980); Phys. Rev. A24, 2963 (1981).
2. H. Nakamura, Phys. Rev. A26, 3125 (1982).
3. R. W. Wijnaendts van Resandt, C. de Vreugd, R. L. Champion, and J. Los, Chem. Phys. 29, 151 (1978).
4. F. von Busch, J. Hormes, and H. D. Liesen, Chem. Phys. Lett. 34, 244 (1975); 35, 372 (1975); J. Phys. B15, 3695 (1982).
5. T. Okamoto, Y. Sato, N. Shimakura, and H. Inouye, J. Phys. B14, 2379 (1981).
6. C. F. Melius and W. A. Goddard III, Phys. Rev. A10, 1541 (1974).
7. F. von Busch, J. Phys. B15, 3707 (1982).

A THEORETICAL STUDY OF COHERENCE EFFECTS IN CHARGE TRANSFER COLLISIONS: APPLICATION TO Na-Li$^+$

A. E. Orel and K. C. Kulander

Lawrence Livermore National Laboratory, University of California, Livermore, CA 94550

There has been much interest recently in collision processes involving electronically excited atoms[1,2]. The excitation is provided by a laser tuned to a resonance transition of the target atom. If, for example, the atomic transition is s→p, and the incident ion is in an S-state, the products of the collision will depend on the polarization of the laser light relative to the collision axis due to different initial fluxes in the excited Σ and Π electronic states. Such behavior has been observed recently by Schmidt, et al.[1] in $Na^+ + Na^*$ and by Reiland, et al.[2] in $Na^* + N_2$.

In these experiments it is assumed that an incoherent superposition of substates with a known[1-3] ratio of populations is generated by pumping the target atoms for times very long compared to the spontaneous lifetimes of the excited states. In this case effects related to Rabi oscillations in the state populations are expected to be damped out. However, if shorter pumping times could be utilized it would be possible to create a coherent superposition of initial sublevel populations and additional alternations of the scattering intensities to the various final product states could be observed. These will be due to the coherent properties of the exciting radiation. We believe it should be possible to observe oscillations in cross sections for charge exchange if the excited atoms collide with incident ions before they can decay radiatively.

The system we have chosen to study is $(NaLi)^+$.with the target atom being either the Li or the Na atom. Accurate potential energy curves and non-adiabatic coupling matrix elements were obtained from ab initio, all-electron configuration-interaction calculations. These were used in determining the cross sections over a range of collision energies. Significant differences between coherent and incoherent target preparation were observed in both total and partial cross sections.

In Fig. 1 we show the charge transfer cross sections from the three lowest states. Our results agree reasonably well with the previous calculations of Melius and Goddard[4] and with the experiments of Daley and Perel[5]. In Fig. 2, the excited state cross sections are shown, both including and excluding the interference term, as functions of collision energy.

References

1. H. Schmidt, A. Bahring, E. Meyer and B. Miller, Phys. Rev. Letts. 48, 1008 (1982).
2. a) W. Reiland, H.-U. Tittes, C. P. Schulz and I. V. Hertel, Chem. Phys. Letts. 91, 329 (1982); b) W. Reiland, G. Jamieson, U. Tittes and I. V. Hertel, Zeit. Phys. A 307, 51 (1982).
3. a) A. Fischer and I. V. Hertel, Zeit. Phys. A 304, 103 (1982); b) I. V. Hertel and W. Stoll, Adv. At. Mol. Phys. 13, 113 (1978); c) J. Macek and I. V. Hertel, J. Phys. B 7, 2173 (1974).
4. C. F. Melius and W. A. Goddard III, Phys. Rev. A 10, 1541 (1974).
5. H. L. Daley and J. Perel, VI ICPEAC Abstracts (MIT, Cambridge, Mass. 1969), p.1051.

This work was supported, in part, by the Air Force Weapons Laboratory with ILIR funds and performed under the auspices of the U. S. Department of Energy at the Lawrence Livermore National Laboratory under contract No. W-7405-ENG-48.

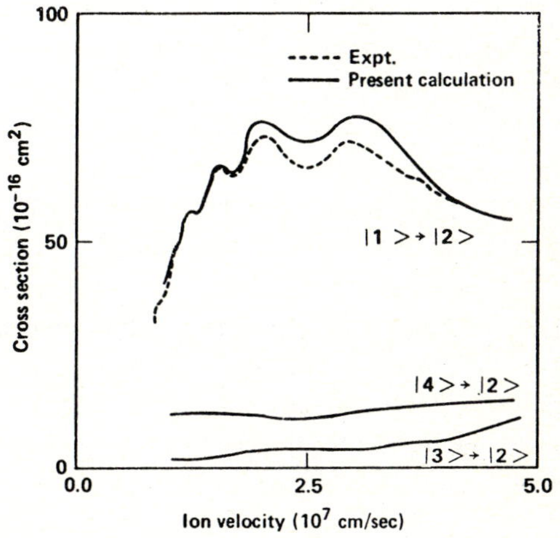

Figure 1

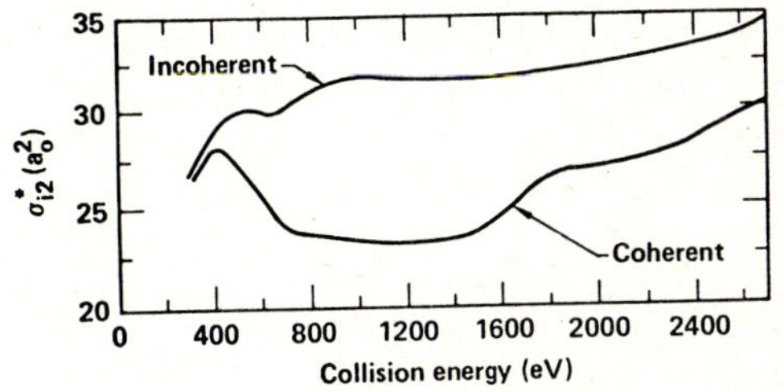

Figure 2

$H^+ + H^-$ NEUTRALIZATION

F. Borondo, A. Macías and A. Riera

Departamento de Química Física y Química Cuántica. Centro Coordinado CSIC-UAM. Universidad Autónoma de Madrid.
CANTOBLANCO (Madrid, 34) (Spain).

We have continued[1] the study of the neutralization process $H^+ + H^-(1s^2)$ including exit channels with n > 2. We have performed full CI calculations using a basis set six 1s, twelve 2p and two $3d_{z^2}$ GTO's centered on each nuclei.

From the quantitative correlation diagram (Fig.1) and the radial couplings (Fig. 2) one can conclude that the crossing of the entrance channel with the n= 3 manifold can be treated with the non-overlapping Nikitin exponential linear model.

We have found that including angular correlation in the wavefunction that describes the H^- ion has a very great influence on the calculated cross sections. As the slope of the energy curve of the ionic state relative to those of the n = 3 manifold is very small, any improvement on the energy ($R \to \infty$) of the former, causes a considerably displacement of the crossing towards larger R values (Fig. 3). This changes the value of the preexponential parameter in the equation which gives the ionic-covalent interaction so that different models would describe this crossing depending on the value of R_c.

We have estimated the interactions between the ionic state with the n = 4 manifold, and we conclude that the exponential-linear model applied to this crossing coincides with the limit of the Landau-Zener model for small coupling between diabatic states.

Neutralization cross section have been calculated using the molecular model with a semiclassical approach and straight line trajectories. The results are presented in Fig. 4. Our results do not coincide with that of Peart et al.[2] in the region below 0.7 keV, just as these of Sidis et al.[3]. Including channels with n > 3 in the calculation cannot account for the remaining structure reported by Peart et al. leaving this still an open question.

Our results, using the standard molecular model indicate that including angular correlation on the ionic state is very important and translation factors are not needed to explain the experimental results, except at very high energies.

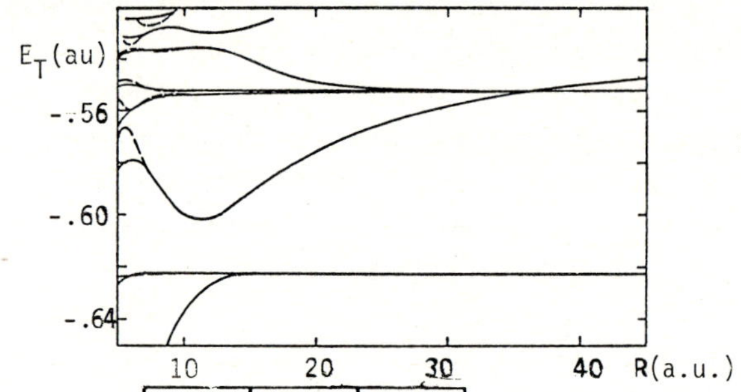

Fig. 1

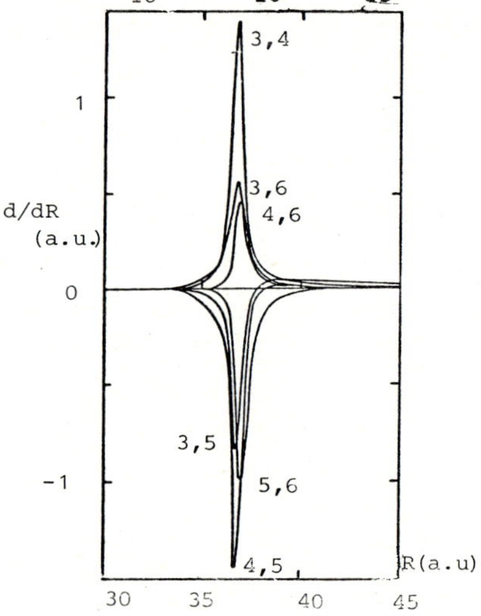

Fig. 2.

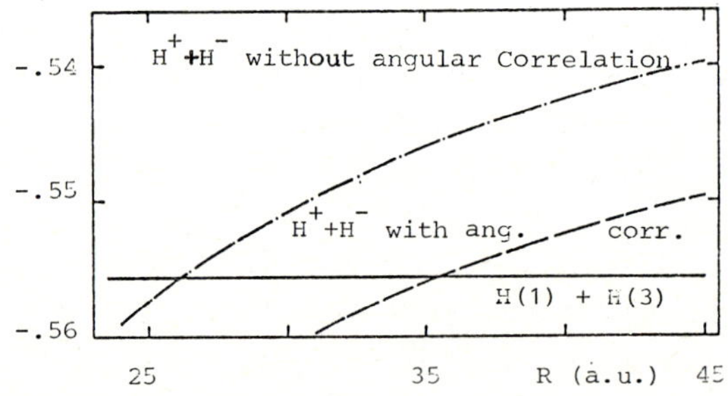

Fig. 3.

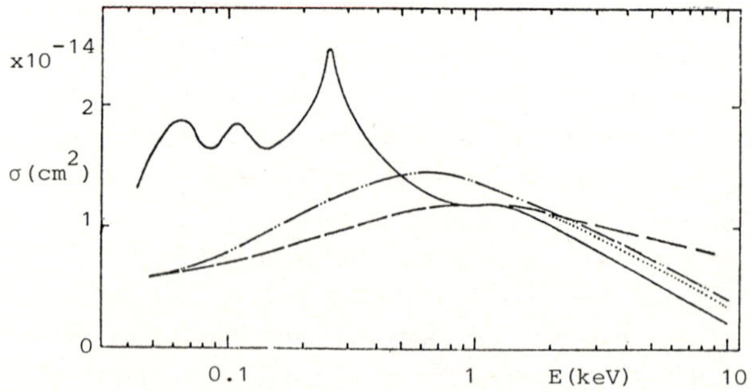

Fig. 4. —— Peart et al. (Ref. 2)
-··- Sidis et al (Ref. 3)
-- this work
.... this work taking ionization into account.

References

1. F. Borondo, A. Macías and A. Riera, Phys. Rev. Lett., 46, 420 (1981); J. Chem. Phys., 74, 6126 (1981).
2. B. Peart, R. Grey and K.T. Dolder, J. Phys. B., 9, L369 (1976).
3. V. Sidis, C. Kubach and D. Fussen, Phys. Rev. Lett., 47, 1280 (1981).

A QUANTUM ELECTRODYNAMIC APPROACH TO CHARGE TRANSFER BETWEEN FULLY STRIPPED LIGHT IONS AND HYDROGEN ATOM

S. Bhattacharyya

Gokhale College, Calcutta - 700020, India

L. Chatterjee and K. Sen Gupta

Jadavpur University, Calcutta - 700032, India

Study of charge transfer in fully stripped light ions from hydrogen atoms has assumed a tremendous proportion in Atomic collision Physics because of its implication in Fusion problem. Atomic researchers throughout the world are vigorously involved in the experimental[1,2] and theoretical[3,4,5] investigations of the processes like

$$({}^{A}M_{x})^{z+} + H\,(1S) \rightarrow ({}^{A}M_{x})^{(z-1)+}(n,l) + H^{+} \quad (1)$$

To facilitate fusion studies we shall invoke quantum electrodynamics to study these phenomena[6,7,8].

The nucleus of the stripped ion $({}^{A}M_{x})^{z+}$ is assumed to be a particle obeying a statistics according to its spin state. In Schrödinger picture and in Coulomb gauge, the equation of state for the interacting system is

$$(H_0 - V_{ep} - V_{eM^{z+}} + V_{pM^{z+}})|\Psi\rangle = i\hbar\frac{d}{dt}|\Psi\rangle \quad (2)$$

H_0 = free Hamiltonian of the interacting particle i.e. electron (e), proton (p), and the fully stipped ions (M^{z+})

$$V_{ep} = \int \frac{\rho_e(x)\,\rho_p(x')}{|x-x'|}\, d^3x\, d^3x'$$

$$V_{eM^{z+}} = \int \frac{\rho_e(x)\,\rho_{M^{z+}}(x')}{|x-x'|}\, d^3x\, d^3x'$$

$$V_{pM^{z+}} = \int \frac{\rho_p(x)\,\rho_{M^{z+}}(x')}{|x-x'|}\, d^3x\, d^3x'$$

ρ_e, ρ_p, $\rho_{M^{z+}}$ are the charge densities of the subscripted particles. Interaction Hamiltonian for the process is

$$H_e = V_{pM^{z+}} - V_{eM^{z+}} \quad (3)$$

Hence the probability matrix is

$$M_{fi} = \langle \Psi_f | H_c | \Psi_i \rangle \quad (4)$$

To get $|\Psi_i\rangle$ and $|\Psi_f\rangle$ the initial state and final state respectively, eqn. (2) is solved by method of perturbation[6].

$$|\Psi_f\rangle = \int \sum_{def} g'^{\rho\delta\lambda}_{def}(l_e', l_p', l_M')\, e_\rho^{+}(l_e')\, p_\delta^{+}(l_p')\, M_\lambda^{z+}(l_M')|0\rangle \quad (5)$$

$$|\Psi_i\rangle = \int \sum_{abc} g^{\alpha\beta\gamma}_{abc}(l_e, l_p, l_{M^{z+}})\, e_\alpha^{+}(l_e)\, p_\beta^{+}(l_p)\, M_\gamma^{z+}(l_M)\, |0\rangle \quad (6)$$

l and l' are the momentum variables of the subscripted particles. In the intermediate and high energy region of the incident ion (50 KeV to 1.5 MeV) the orthogonality relation between $|\Psi_i\rangle$ and $|\Psi_f\rangle$ is found to be justified,

$$g^{\alpha\beta\gamma}_{abc}(l_e, l_p, l_{M^{z+}}) = (2\pi)^3 \delta^3(P_c - l_e - l_p - l_M) \times \phi_H(l_e)\, \phi_{M^{z+}}(l_M - A(l_e + l_p)) \quad (7)$$

and

$$g'^{\rho\delta\lambda}_{def}(l_e', l_p', l_{M^{z+}}') = (2\pi)^3 \delta^3(Q_c - l_e' - l_p' - l_M')\, \phi^{nlm}_{M^{(z-1)+}}(l_e) \times \phi_p(l_p' - (l_M' + l_e')/A) \quad (8)$$

$\phi_H(1S)$ and $\phi_{M^{(z+1)}}$ are the orbital wave functions in momentum space for hydrogen atom in IS-state and $({}^{A}M_x)^{(z-1)+}$ ion in (nl) state respectively.

For projectile ion with momentum P

$$\phi_{M^{z+}}(l_M - A(l_e + l_p)) = \delta^3(P - \{l_M - A(l_e + l_p)\}) \quad (9)$$

and the ejected proton with momentum Q'

$$\phi_p((l_p') - (l_M' + l_e')/A) = \delta^3\left(Q' - \left\{l_p' - \frac{l_M' + l_e'}{A}\right\}\right)$$

The cross-section for charge transfer

$$\sigma = \int M\, d^3Q\, d^3Q'\, |M_{fi}|^2 / (|P|(2\pi)^6)$$

ia computed for Li^{3+}, Be^{4+} and B^{5+} at intermediate and high energies. The variation of with different n and l states of the ${}^{A}M_x^{(z-1)+}$ ion will be presented at the Conference.

References

1. M.B. Shah, T.V. Goffe and H.B. Gilbodey, J.Phys. B11 (1978) L 223.
2. W. Seim, A. Muller, E.Salzborn, Phys. Letts. 80A, 20 (1980).
3. H. Ryufuku and T. Watanabe, Phys. Rev. A19, 1538 (1979).
4. D.S.F. Crothers and N.R. Todd, J.Phys. B13, 2277 (1980).
5. Bransden, C.W. Newby and C.J. Noble, J. Phys. B13 (1980) 4245.
6. T. Roy, N. Cim. Lett 5 (1972) 1048.
7. S. Bhattacharyya et al, Physica 106C (1980) 135.
8. L. Chatterjee, S. Bhattacharyya, Phys Lett 93A (1983) 360.

CHARGE TRANSFER FOR COMPLETELY STRIPPED BORON AND CARBON IONS FROM ATOMIC HYDROGEN

S. C. Mukherjee, C. R. Mandal and Shyamal Datta

Indian Association for the Cultivation of Science, Calcutta 700 032, India.

Charge transfer cross sections in collisions of completely stripped ions of boron and carbon with ground state of atomic hydrogen are calculated in the frame work of the continuum-distorted-wave approximation for the incident energy varying between 750 to 2750 keV. The total cross sections are also evaluated by applying n^{-3} law from n = 6 levels. The present calculated results are compared with the existing theoretical results[1-3] and the experimental results of Goffe et al[4]. We have calculated the results for the capture cross sections $Q(n) = \sum_{\ell m} Q_{nlm}$ into each complete shell as well as the individual cross sections in each sublevel $Q_{nl} = \sum_{m} Q_{nlm}$ for the collisions of fully stripped boron and carbon with atomic hydrogen. In order to compare our calculated results with the existing observed results for the total cross sections $Q(tot) = \sum_{n} Q(n)$, we calculate them for each individual energies by assuming the cross sections Q(n) to be proportional to n^{-3} for $n \geqslant 6$.

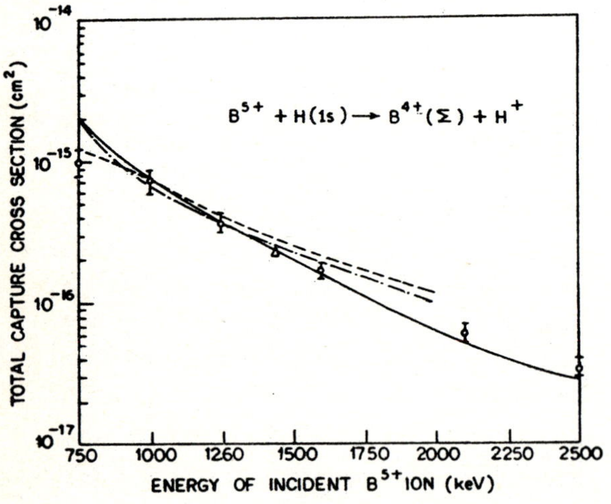

Fig. 1. Total cross sections for electron capture by B^{5+} ion from ground state atomic hydrogen. Theory (———) present CDW calculation; (- - - - -) UDWA calculation of Ryufuku and Watanabe[1]; (-.-.-) Two-state atomic expansion method, Bransden et al[2] and Δ CDW calculation of Crothers[3]; Experiment o Goffe et al[4].

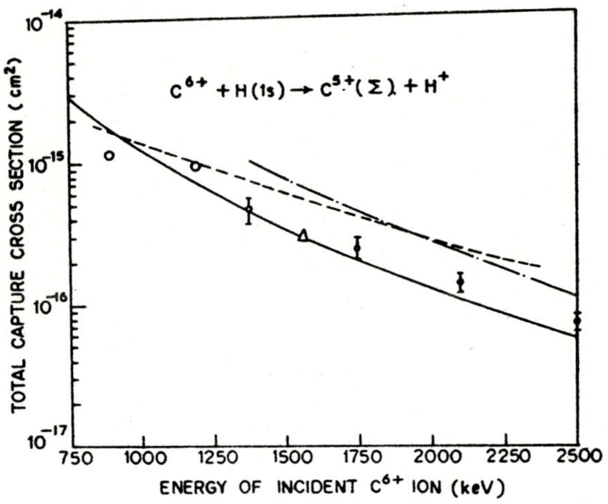

Fig. 2. Total cross sections for electron capture by C^{6+} ion from ground-state atomic hydrogen. Theory : (———) present CDW calculation; (-.-.-.-) Eikonal calculation of Ho et al[6] (---) UDWA calculation of Ryufuku and Watanabe[2]; Δ CDW calculation Crothers[3]; o CMTC calculation of Olson[5]; Experiment : 0 Goffe et al[4].

In Figs. 1 and 2 we present respectively the calculated values for the total capture cross sections for B^{5+} — H and C^{6+} — H collisions and have compared them with the existing theoretical calculations as well as with the observed findings of Goffe et al[4]. The CDW calculation of Crothers[3], obtained by the use of parabolic coordinates have been reported for a particular projectile velocity and upto the excited charge transfer level n = 15.

References

1. H. Ryufuku and T. Watanabe, Phys. Rev. A20, 1828 (1979).
2. B. H. Bransden, C.W. Newby and C.J. Noble, J. Phys. B13, 4245 (1980).
3. D. S. F. Crothers, J.Phys. B14, 1035 (1981).
4. T. V. Goffe, M. B. Shah and H. B. Gilbody, J. Phys. B12, 3763 (1979).
5. R. E. Olson, Phys. Rev. A24, 1726 (1981).
6. T. S. Ho, M. Lieber and F. T. Chan, Phys. Rev. A24, 2925 (1981).

ELECTRON CAPTURE FOR FAST HIGHLY CHARGED IONS IN GAS TARGETS*

A. S. Schlachter, J. W. Stearns, W. G. Graham, †
K. H. Berkner, R. V. Pyle, and J. A. Tanis ‡

Lawrence Berkeley Laboratory
University of California
Berkeley, CA 94721

We have measured electron-capture cross sections for fast, highly charged C, Ar, Fe, Nb, and Pb projectiles in H_2, He, N_2, Ar, and Xe targets. Projectiles range in energy from 1.3-8.5 MeV/amu, with charge states as high as 59+. We have found empirically a scaling procedure which describes our measurements plus a large number of available electron-capture data for fast highly charged ions in many gas targets. This scaling rule permits prediction of electron-capture cross sections for a wide variety of fast highly charged projectiles in gas targets.

This work is a generalization of our previous scaling rule[1] for iron ions in H_2 and the scaling of Knudsen et al[2] based on the Lenz-Jensen atomic model. We found previously[1] that electron-capture cross sections scaled as projectile charge q to the 3.15 power and energy/nucleon E to the minus 4.5 power for fast projectiles.

We find empirically the following reduced parameters

$$\tilde{\sigma} = \sigma Z^{1.8}/q^{0.5}$$
$$\tilde{E} = E/(Z^{1.25}\, q^{0.7}) \qquad (1)$$

where Z is target atomic number, E is in keV/amu, and σ is in cm^2. Our electron-capture cross sections[3,4] are shown plotted in these reduced parameters in Fig. 1a; other cross-section data are shown in Fig. 1b. Also shown in Fig. 1 is a line fit to the data: the analytic expression is

$$\tilde{\sigma} = \frac{1.1 \times 10^{-8}}{\tilde{E}^{4.8}} \left[1-e^{-(\tilde{E}^{2.2}/27)}\right] \left[1-e^{-(\tilde{E}^{2.6}/4.1\times 10^{4})}\right]. \qquad (2)$$

The limit of expression (3) for large values of E is

$$\sigma = 1.1 \times 10^{-8}\, q^{3.9}\, Z^{4.2}/E^{4.8} \qquad (3)$$

which has similar energy dependence but a higher q dependence than our previous result for iron ions in H_2. Equations 2 and 3 include the target atomic number Z, and can be used to predict cross sections for a wide variety of projectile-target systems.

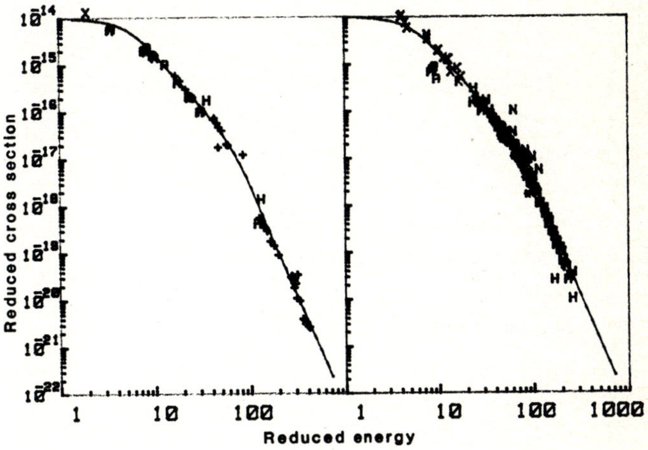

Fig. 1: Reduced plot of single-electron-capture cross sections for a fast highly charged ion in charge state q in a gas target with atomic number Z. Cross sections in molecules are divided by 2 and plotted with the atomic Z. The line is an empirical fit (Eq. 2) to the cross sections. Figure 1a shows our measurements; Fig. 1b shows other cross-section data.[4]

References

*This work was supported by the Director, Office of Energy Research, Office of Fusion Energy, Applied Plasma Physics Division of the U. S. Department of Energy under Contract No. DE-AC03-76SF11198.

†The New University of Ulster, Coleraine BT 52 1SA, Northern Ireland.

‡Western Michigan University, Kalamazoo, MI 49118

1. K. H. Berkner, W. G. Graham, R. V. Pyle, A. S. Schlachter, and J. W. Stearns, Phys. Rev. A 23, 2891 (1981).

2. H. Knudsen, H. K. Haugen, and P. Hvelplund, Phys. Rev. A 23, 597 (1981).

3. We have divided cross sections for molecular targets by 2 and used the atomic Z. This use of a factor of 2 does not imply that cross sections differ by a factor of 2 for molecular and atomic targets.

4. A. S. Schlachter, J. W. Stearns, W. G. Graham, K. H. Berkner, R. V. Pyle, and J. A. Tanis, submitted to Phys. Rev. A; and references therein.

APPLICATIONS OF VARIATIONAL CONTINUUM DISTORTED WAVES

D.S.F. Crothers and J. F. McCann

Department of Applied Mathematics and Theoretical Physics

The Queen's University of Belfast, Belfast BT7 1NN, Northern Ireland.

Cross sections for charge transfer and ionization are calculated within the continuum distorted wave (CDW) method.

In particular, a recently developed refined orthogonal variational CDW treatment[1] is applied in perturbation form[2] to the charge transfer process

$$B^{Z+} + H(1s) \rightarrow B^{(Z-1)+}(n) + H^{+} \qquad (1)$$

at the intermediate velocity of 2 a.u. and where the projectile B^{Z+} is a fully stripped ion. Values are presented for $Z \varepsilon [12,18]$ and $n \varepsilon [9,16]$ and are compared with experiment[3,4]. The question of asymptotic behaviour is considered using the method of symmetrised orthogonalisation and the question of ℓ, m distributions[5] is recalled. Calculations are also presented for the ionization process

$$B^{Z+} + H(1s) \rightarrow B^{Z+} + H^{+} + e^{-} \qquad (2)$$

in the energy range 20 - 1000 kev/amu and for $Z \varepsilon [1,6]$. The model includes distortion in both channels. The final channel is described by the CDW wave function[6]; and therefore automatically includes both direct ionization and capture to the continuum, while the initial channel is described by the eikonal wave function thus avoiding the renormalization problem. Comparison is made with recent experimental and theoretical work[7,8,9,10]. The results obtained show a significant improvement on previous high energy approximations by describing the total cross sections accurately at intermediate energies.

We also present calculated cross sections for elastic scattering and symmetric resonance charge transfer

$$H^{+} + H(1s) \rightarrow \begin{cases} H(1s) + H^{+} & (3a) \\ H^{+} + H(1s) & (3b) \end{cases}$$

down to low energies within a fully variational and unitarized non-perturbation calculation. In our adopted notation[1,2] the capture and elastic probabilities for processes (3a,b) are given by

$$P_{\substack{01 \\ 00}} = \begin{matrix} \sin \\ \cos \end{matrix}^{2} \left[\int_{-\infty}^{+\infty} dt \; \frac{\mathrm{Re}\left\{ \hat{K}^{++}_{01} - \hat{h}^{++}_{00} \hat{S}^{++}_{01} \right\}}{1 - (\hat{S}^{++}_{01})^{2}} \right] \qquad (4)$$

where $\hat{M}_{ij}$ denotes the normalized matrix element M_{ij}/S^{++}_{00} and where the superscripts indicate that both the initial and final (CDW) states are outgoing waves. The proof follows from a consideration of gerade and ungerade CDW wave functions.

References

1. D.S.F. Crothers, J. Phys. B. 15, 2061 (1982).
2. D.S.F. Crothers, Physica Scripta, T3, 236 (1983).
3. P. Hvelplund, E. Samsø, L. H. Andersen, H. K. Haugen and H. Knudsen, Physica Scripta, in press (1983).
4. H. Knudsen, P. Hvelplund, L. H. Andersen, S. Bjørnelund, M. Frost, H. K. Haugen and E. Samsø, Physica Scripta, in press (1983).
5. D.S.F. Crothers and J. F. McCann, Phys. Lett. A. 92, 189 (1982).
6. Dz. Belkić, J. Phys. B. 11, 3529 (1978).
7. M. B. Shah and H. B. Gilbody, J. Phys. B. 14, 2361 (1981).
8. M. B. Shah and H. B. Gilbody, J. Phys. B. 15, 413 (1982).
9. J. H. McGuire, Phys. Rev. A 26, 143 (1982).
10. R. E. Olson and A. Salop, Phys. Rev. A 16, 531 (1977).

ELECTRON CAPTURE AND LOSS CROSS SECTIONS FOR Si^{11+} + He from 50-80 MeV.*

M. Clark, D. Brandt, S. Shafroth, J. Swenson

University of North Carolina, Chapel Hill, NC and
Triangle Universities Nuclear Laboratory, Durham, NC

Recently a process known as RTE has been given increasing attention. The background non-resonant process is important in determining the absolute RTE cross section and a possible method for determining this background is to use

$$\sigma_{background} = 2\pi \int P_{exc}(b)\ P_{cap}(b)\ bdb$$

Now if it can be shown that $P_{exc}(b)$ is relatively constant (say $P_{exc}(0)$) over the range of $P_{cap}(b)$ then $P_{exc}(0)$ can be removed from the integral and then:

$$= 2\pi\ P_{exc}(0) \int P_{exp}(b)\ bdb$$

and

$$= P_{exc}(0)\ \sigma_{cap}$$

σ_{cap} has been calculated classically by D. Brandt using the Bohr-Lindhard model and is plotted as the solid line with the data.[1]

Total cross sections have been measured for single electron capture and single electron loss for the system Si^{11+} + He. 50-80 MeV Si^{11+} ions were produced by the TUNL FN Tandem and then collimated by two apertures, one 1.0 mm in diameter and the other 1.6 mm in diameter, spaced 2 meters apart. The beam then passed through a differentially pumped gas cell that was operated in the range of 0-40 mTorr to assure single collision conditions and to allow for correction of the residual beam line pressure. It was charge-state analyzed by electrostatic plates and then after a drift-space of 1.5 meters the Si^{10+} and Si^{13+} beams were counted by Si surface barrier detectors while the Si^{11+} beam was collected in a Faraday cup. Counting rates were maintained at about 20,000 counts/sec with beam currents on the order of 2 pA.

The yields were corrected for the gas that leaked into the beam line between the collimating apertures and in the region between the gas cell and the analyzing plates. The errors in the fits for the gas pressure dependence were on the order of 5% while the error in measuring the beam current in the Faraday cup is estimated to be 10%.

*Supported by U.S. D. O. E. Chemical Sciences Division.

[1] D. Brandt, Proc. of the Int. Workshop on Fusion (1982) Texas A & M Univ., College Sta., TX (To be published in Nucl. Inst. and Meth. (1983)).

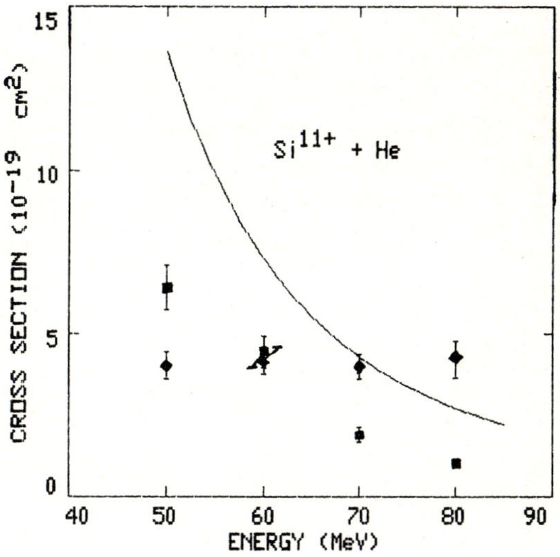

Fig. 1. Cross section for electron capture (■) and electron loss (◆) for Si^{11+} + He and the solid line is the classical calculation by Brandt.

EFFECTS OF AN OFF-SHELL COULOMB WAVEFUNCTION ON RADIATIVE ELECTRON CAPTURE

J.S. Briggs and M. Gorriz

Fakultät für Physik, Universität Freiburg, D-7800 Freiburg

The phenomenon of radiative electron capture (REC) accompanying ion-atom collisions was recognised and studied more than ten years ago[1,2]. The main features are the appearance of a continuous photon emission spectrum, centered roughly at the translational kinetic energy of the captured target electron measured in the projectile frame, and with a shape which reflects the Compton profile of the target electron in momentum space. Simplified theories of the REC process based upon an impulse approximation (IA) argument can explain these main features and the accuracy of these approximations can in principle be assessed by a more accurate theoretical treatment.

In order to facilitate the observation of the REC spectrum it is necessary to employ fast highly-charged ions, with an effective nuclear charge Z_p which is much greater than the effective nuclear charge Z_T seen by the target electrons. Then in the IA the REC T-matrix element appears as the radiative recombination matrix element for a free electron in a projectile continuum state of given momentum, weighted by the target state momentum distribution. A similar IA has been made in the case of Coulomb capture. However it has been pointed out that the IA use of a free-electron (i.e. on-shell) Coulomb wavefunction is not valid except at collision velocities v so large that $Z_p/v \ll 1$. In particular in the velocity region where experiments exist this criterion is not satisfied and the more correct off-shell continuum wavefunction[4] should be used. The theory of REC has been re-derived in the Strong Potential Born (SPB) approximation in a time-independent picture. Approximating the internuclear motion by plane waves the REC cross-section can be put into the form

$$\frac{d\sigma}{d\omega d\Omega} = \frac{(2\pi)^2\omega}{vc^3} \sin^2\theta \int d\underline{k}\; \delta(\omega-\varepsilon_p-\tfrac{1}{2}v^2+\underline{k}\cdot\underline{v})\, |\tilde{\phi}_i(\underline{k})|^2 \; |\langle\phi_f|\underline{p}\cdot\hat{\underline{v}}|\psi^+_{\underline{k}-\underline{v},\varepsilon_T}\rangle|^2$$

where $\psi^+_{\underline{k}-\underline{v},\varepsilon_T}$ is an off-shell Coulomb wavefunction, and $\underline{p}$ is the momentum operator, ε_T and ε_p being the respective target and projectile binding energies. In a peaking approximation, valid for $Z_T/v \ll 1$, the cross-section becomes

$$\frac{d\sigma}{d\omega d\Omega} = 2\pi\nu\,(1-e^{-2\pi\nu})^{-1} \left(\frac{d\sigma}{d\omega d\Omega}\right)_{IA}$$

where $\nu = Z_p/v$ and the IA cross-section is given by Briggs and Dettmann[3]. The off-shell correction is sizeable (Fig. 1) although the available experimental data does not serve to conclusively demonstrate the necessity to employ an off-shell Coulomb wavefunction in the calculation of the REC cross-section.

REC-Cross-Section with off-shell correction

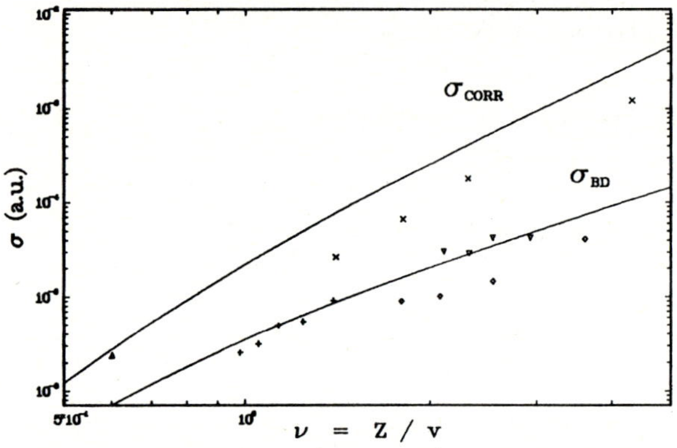

Fig. 1: REC-cross section per quasi-free target elektron, σ_{BD} = Briggs, Dettmann[3], σ_{corr} (see text) experimental values of ▲P. Kienle[2], ✖R. Schulé[5], ✚H. Tawara[6], ◆J.A. Tanis[7], ▼J. Lindskog[8].

References

1. G. Raisbeck, F. Yiou, Phys. Rev. A 4, 1858 (1971)
2. P. Kienle et al., Phys. Rev. Lett. 31, 1099 (1973)
3. J.S. Briggs, K. Dettmann, J. Phys. B 10 1113 (1977)
4. J. Macek, S. Alston, Phys. Rev. A 26, 250 (1982)
5. R. Schulé et al., J. Phys. B 10, 889 (1977)
6. H. Tawara et al., Phys. Rev. A 26, 154 (1982)
7. J.A. Tanis, S.M. Shafroth, Phys. Rev. Lett. 40, 1174 (1978)
8. J. Lindskog et al., Phys. Scr. 14, 100 (1970)

EFFECT OF L-SHELL NON-EQUILIBRIUM ON RADIATIVE ELECTRON CAPTURE FOR 30 MeV S+C COLLISIONS

J. A. Tanis* and E. M. Bernstein

Physics Dept., Western Michigan University, Kalamazoo, MI 49008 USA

Yields for radiative electron capture (REC) to the K-shell have been measured as a function of target thickness for 30 MeV S^{q+} ions incident on thin (2.4 - 150 $\mu g/cm^2$) carbon foils using the WMU EN tandem Van de Graaff. Measurements were made using projectiles with incident charge states 6^+ and 10^+. The measured yields rise steeply for thicknesses $\lesssim 5\mu g/cm^2$ and approach equilibrium for thickness $\gtrsim 80\mu g/cm^2$. Since the incident projectile has no initial K vacancies, the measured yields must extrapolate to zero for zero target thickness.

Previously it has been found[1] that non-equilibrium of the L-shell has a considerable effect on the measured characteristic K x-ray yields for small ($\lesssim 10\mu g/cm^2$) target thicknesses. This L-shell effect is due to the fact that 2p vacancies in the projectile can be transferred through $2p\pi$- $2p\sigma$ rotational coupling and vacancy sharing[2] to the K-shell of the projectile, thereby enhancing the K-vacancy production in the ion as it traverses the foil. Characteristic K x-ray yields were parametrized by generalizing the two-component model[3,4] for K x-ray production to include L-shell non-equilibrium.

The effect of non-equilibrium of the L-shell should also manifest itself in the measured REC yields. In this case, since the REC yield increases over the entire range of target thicknesses measured, the L-shell non-equilibrium should produce an initial steep rise in the REC yield for small ($\lesssim 10\mu g/cm^2$) target thicknesses, followed by a slower rise due to K-shell non-equilibrium and finally a leveling off at equilibrium. Preliminary analysis indicates that the REC yields measured in this experiment support this interpretation since a single exponential function based on K-shell non-equilibrium does not adequately describe the data.

The rate equation for the case in which L-shell vacancies contribute to K-vacancy production is:

$$\frac{dY_1}{dx} = N_p(x)\sigma_{rot}wY_0 + \sigma_{01}Y_0 - \sigma_{10}Y_1 \qquad (1)$$

where Y_1 = fraction of ions with one K vacancy

Y_0 = fraction of ions with no K vacancies

$N_p(x)$ = number of 2p vacancies per ion at x

σ_{rot} = $2p\pi$-$2p\sigma$ rotational coupling cross section

w = K vacancy transfer probability

σ_{01} = K vacancy production cross section

σ_{10} = K vacancy loss cross sections.

Furthermore, if it is assumed that the 2p vacancy production probability is the same for each electron, then

$$N_p(x) = 6\,\frac{\sigma_{vp}}{\Sigma}\,(1 - e^{-\Sigma x}) + N_p^{inc}\,\sigma^{-\Sigma x} \qquad (2)$$

where $\Sigma = \sigma_{vp} + \sigma_{dp}$

σ_{vp} = 2p vacancy production cross section

σ_{dp} = 2p vacancy loss cross section

N_p^{inc} = number of incident 2p vacancies per ion

Since REC events can only come from inside the foil, the number of such events due to an infinitesimal element of target thickness dx is:

$$dN_{REC} = I_0Y_1(x)\,\frac{d\sigma^0_{REC}}{d\Omega}\,\varepsilon dx \qquad (3)$$

where I_0 = number of incident ions

$\frac{d\sigma^0_{REC}}{d\Omega}$ = REC differential cross section

ε = detection efficiency

Then eqs. (1-3) can be used to obtain an expression for the REC yield, $\frac{d\bar{\sigma}_{REC}}{d\Omega}$, for a target foil of thickness T:

$$\frac{d\bar{\sigma}_{REC}}{d\Omega}(T) = \frac{N_{REC}}{I_0T\varepsilon}$$

$$= \frac{d\sigma^0_{REC}}{d\Omega}\left[\frac{B_3}{\sigma} - \left(\frac{B_3}{\sigma} + \frac{CB_2}{\sigma-\Sigma}\right)\left(\frac{1 - e^{-\sigma T}}{\sigma T}\right) + \frac{CB_2}{\sigma-\Sigma}\left(\frac{1 - e^{-\Sigma T}}{\Sigma T}\right)\right] \qquad (4)$$

where $B_2 = -B_1 + \frac{N_p^{inc}}{6}$; $B_1 = \frac{\sigma_{vp}}{\Sigma}$

$B_3 = CB_1 + \sigma_{01}$; $C \equiv 6\,\sigma_{rot}w$

$\sigma = \sigma_{01} + \sigma_{10}$

This formulation predicts that REC yields should rise more steeply as N_p^{inc} is increased, e.g., in going from $q = 6^+$ to 10^+. Examination of the 30 MeV data for these charge states indicates this interpretation to be consistent with the data.

*Supported in part by a Fellowship from the WMU Faculty Research Fund.

1. E. M. Bernstein, D. S. Shippy, and J. A. Tanis, in Proceedings of the International Conference on X-Ray and Atomic Inner-Shell Physics, Eugene, Oregon, 1982, Abstracts (unpublished), pp. 198-199.

2. W. E. Meyerhof, Phys. Rev. Lett. 31, 1341 (1973).

3. T. J. Gray, P. Richard, K. A. Jamison, and J. M. Hall, Phys. Rev. A14, 1333 (1976).

4. J. A. Tanis, W. W. Jacobs, and S. M. Shafroth, Phys. Rev. A22, 483 (1980).

A STUDY OF THE CHARGE-EXCHANGE OF CARBON AND OXYGEN IONS IN VARIOUS MEDIA

I. S. Dmitriev, N.F. Vorobiev, G.E. Bugrov, Zh. M. Konovalova, E.A. Kral'kina, V.S. Nikolaev, Ya.A. Teplova, Yu.A. Fainberg

Institute of Nuclear Physics, Moscow State University, Moscow 117234, USSR

The cross sections $\sigma_{i,i-1}$ for single electron capture by carbon and oxygen ions with charges i= 3 to 6 and velocities V= 4.10^8, 6.10^8 and 8.10^8 cm/s have been measured in H_2, He, N_2, and Ar.

In has been found (fig.1) that at V= 8.10^8 cm/s with the increasing nuclear charge of the atoms of a medium, Z_t, the cross sections $\sigma_{i,i-1}$ calculated per atom of medium change nonmonotonically, the cross sections in Ne exceeding those in N_2 and Ar. As the ionic charge i decreases the relative value of this difference becomes higher. At V= 4.10^8 cm/s the cross sections $\sigma_{i,i-1}$ for Z_t = 7, 10 and 18 increase with Z_t. Similar relations between the values of $\sigma_{i,i-1}$ in N_2, Ne and Ar have been observed previously for the ions of nitrogen and neon[1] and helium[2].

The relations between $\sigma_{i,i-1}$ values in nitrogen, neon and argon at V= 8.10^8 cm/s are reproduced in the calculations of these cross sections in Oppenheimer-Brinkman-Kramers approximation[2]. These calculations indicate that the Z_t-dependence of $\sigma_{i,i-1}$ is an oscillating one and that the qualitative changes in the relations between $\sigma_{i,i-1}$ values in nitrogen, neon, and argon with changing Z_t are due to these oscillations which shift toward the higher Z_t with increasing V.

At V≥ 6.10^8 cm/s, maxima in the cross sections correspond to the atoms of a medium whose binding energy I_n of electrons of one of the shells with the principal quantum number n turns out to be close to 1/6 μV^2 (μ being the electron mass) and each of the maxima in the dependence of $\sigma_{i,i-1}$ on Z_t is due to a maximum of the partial cross sections $\sigma_{i,i-1}(n)$ for electron capture from the K, L, M ... shells of the atoms of a medium. In particular, an increased value $\sigma_{i,i-1}$ in neon at V= 8.10^8 cm/s is due to the maximum of the cross section for electron capture from the L-shell.

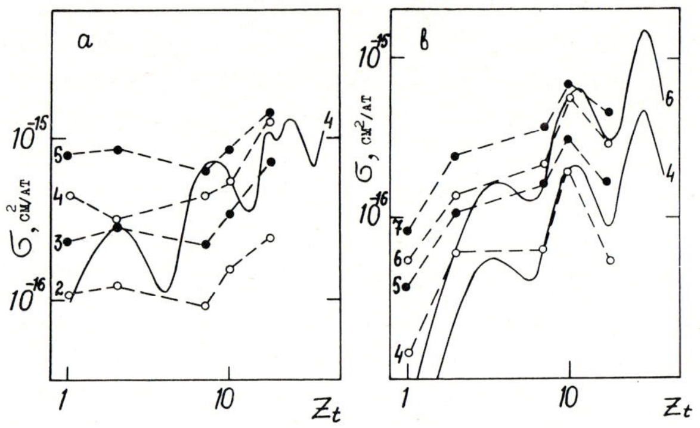

Fig. 1. Electron capture cross sections $\sigma_{i,i-1}$ by oxygen ions at V= 4.10^8 cm/s (a) and 8.10^8 cm/s (b). The broken lines connect the experimental points (○, ●) to guide the eye. The solid lines represent the values $\sigma_{i,i-1}$ in the OBK approximation (K= 0.02 and 0.10 for V= 4.10^8 and 8.10^8 cm/s, respectively). The values of ionic charge i are indicated near the lines.

References

1. I.S. Dmitriev, Yu.A. Tashaev, V.S. Nikolaev, Ya.A. Teplova, B.M. Popov, ZhETF 73, 1684 (1977).
2. I.S. Dmitriev, N.F. Vorobiev, V.P. Zaikov, Zh.M. Konovalova, V.S. Nikolaev, Ya.A. Teplova, Yu.A. Fainberg. J.Phys.B 15, L351 (1982).

EXPERIMENTAL STUDY OF THE FORMATION OF METASTABLE LITHIUM IONS IN COLLISIONS IN GASES

Ya. A. Teplova, I.A. Nevostrueva , Yu. A. Fainberg, I.D. Koshevoi

Institute of Nuclear Physics, Moscow State University, Moscow 117234, USSR

We have measured the relative number α of long-lived excited particles in beams of fast lithium ions (Li^+), the cross sections for the formation of excited ions in electron capture $\sigma^*_{i,i-1}$ and by direct excitation $\sigma^*_{i,i}$, and also the cross section for destruction through electron loss $\sigma_{i^*,i+1}$ and without a change in charge $\sigma_{i^*,i}$. The ion velocity ranged from $V \gtrsim 4.10^8$ cm/s to $V \gtrsim 8.10^8$ cm/s ($\sim$ 0.08-0.3) MeV/amu , where "i" is the initial charge of ions , i= Z-2, (Z is the nuclear charge of ions). The gaseous medium included helium, hydrogen, nitrogen, and argon. The measurements were made using the installation described elsewhere[1].

The value of α for Li^+ ions formed as a result of electron capture by Li^{++} in a collision in the gas target ($Tg \sim 10^{15}$ at/cm^2 thick) reaches on average 0.5 for various gases, which is twice as high as the similar value for helium atoms[2] and close to α for N^{+5} ions. As the target thickness increases, α decreases down to $\alpha \lesssim 0.2$ (Fig. 1).

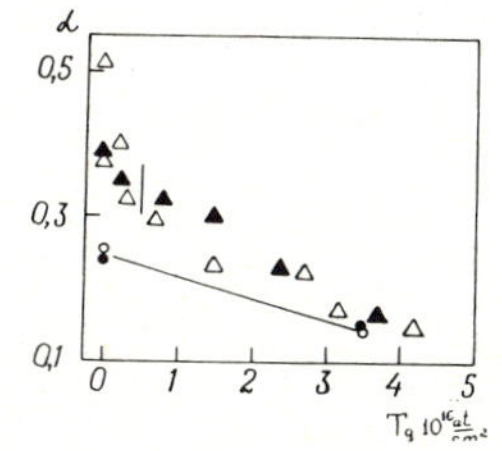

Fig.1.Relative number α (averaged for 4 gases in the target) of metastable ions for the electron capture as a function of gas pressure.Tg at/cm^2 in the target; (Δ) $-Li^+$ ions ($V \gtrsim 4.10^8$ cm/s), ($\blacktriangle$)- $-N^{+5}$ions ($V \gtrsim 8.10^8$ cm/s), $\Delta\alpha = \pm$ 20-30%, (o, $\bullet$)-He atoms in helium and nitrogen respectively[2].

Using the experimental values of α , we have determined the cross sections for electron capture by the one-electron lithium into the excited states $(1s2s)^{1,3}S$ in collisions with the atoms of helium, argon and the molecules of nitrogen. At the ion velocity $V \gtrsim 4.10^8$ cm/s, the cross sections for electron capture into the excited states $\sigma^*_{i\,i-1}$ differ from those for electron capture into the ground state $\sigma_{i\,i-1}$ by about a factor of $\sim$ 1.5. The cross sections for direct excitation of helium-like lithium are an order-of-magnitude lower.

As has been pointed out[3], the average cross section for electron capture by the metastable Li^{+m} ions are near the cross section for electron capture by an unexcited ion. This is consistent with the estimates made in ref.[4] for Li^{+m} ions with $V \sim 1.5 .10^8$ cm/s in helium. For the He atoms, Li^+ N^{+5} ions the ratio $\sigma^*_{i,i+1} / \sigma^*_{i,i}$ increases with the relative ion velocity. As an example Fig.2 presents the cross sections for the various processes, leading to both the formation of the metastable Li^{+m} and N^{+5m} component and its destruction.

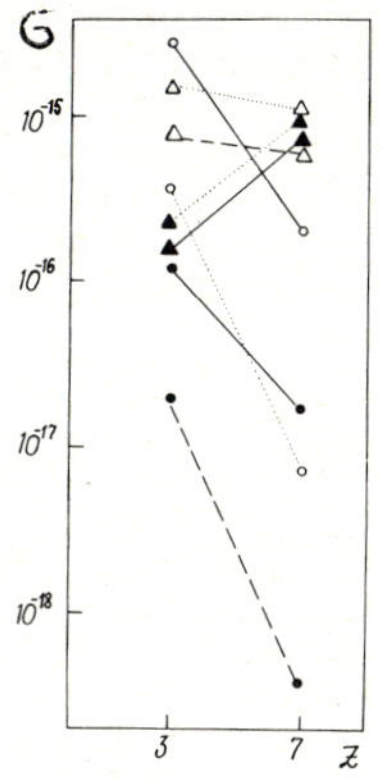

Fig. 2. Cross section for the different processes for Li^+, Li^{+m}($V \gtrsim 4.10^8$cm/s) and N^{+5}, N^{+5m}($V \gtrsim 4.10^8$cm/s) in nitrogen collisions; excitation (full line), destruction (broken like), processes for ions in unexcited states (dotted line), (Δ)- $\sigma^*_{i+1,i}$, ($\blacktriangle$)- $\sigma_{i,i-1}$, $\sigma_{i^*,i-1}$, (o)- $\sigma_{i,i+1}$, $\sigma_{i^*,i+1}$, ($\bullet$)- σ_{i,i^*}, $\sigma_{i^*,i}$.

The cross section $\sigma_{i^*,i}$ for the de-excitation an excited state without a change in charge is $>$ 10 times lower than that for electron loss by the metastable Li^{+m} and is close to that for electron capture by the metastable helium-like ion. The relation between the various de-excitation channels is close to that obtained for the He atoms and N^{+5} ions. The relation between the various processes of metastable lithium formation is rather weakly affected by the kind of gas in the target, but it may change with ion velocity.

References

1. I.S. Dmitriev , Ya.A. Teplova, V.S. Nikolaev, Lectures, MSU Publishers, M.,(1977).
2. E. Horsdal-Pedersen, J. Heinemeier, L.Larsen, J.V. Mikkelson, J.Phys.B13, 1167-1183(1980).
3. Ya.A. Teplova, I.S. Dmitriev, V.S. Nikolaev XI ICPEAC, 538-539 (1979).
4. P. Hvelplund, J.Phys.B,9No.9, 1555-1565 (1976).

DELAYED EMISSION OF X RAYS AFTER ELECTRON CAPTURE INTO METASTABLE FEW-ELECTRON OXYGEN AND NEON IONS

F. Folkmann, B.J. Larsen, N.H. Eisum, and K.M. Cramon

Institute of Physics, University of Aarhus, DK-8000 Aarhus, Denmark

To study electron capture into excited states of few-electron recoil ions, we have measured x rays emitted delayed relative to the exciting particles in the same way as has been applied to Auger electrons[1-3]. By this method, we probe states, which live long enough to capture electrons in a later collision with a cross section around 10^{-14} cm^2, and which emit radiation immediately afterwards.

X rays from oxygen and neon have been measured with a curved TlAP-crystal spectrometer after bombardment of 40 to 80 μbar H_2O or Ne targets with a pulsed 0.7-MeV/amu Cl^{11+} beam, stripped in a carbon foil. In Fig. 1 are shown the time spectra relative to the beam pulses measured for two neon x-ray lines, corresponding to initial states KM and KLN. These x-ray lines are seen in the upper 'total' spectrum in dependence of wavelength - or energy - in Fig. 2. With the delayed window indicated in Fig. 1, the corresponding sorted spectra are shown in the bottom part of Fig. 2, both for pure neon and for a mixture of neon and methane. The prompt $1s3p^1P$ peak (KM) of Fig. 2 is in Fig. 1 seen to have an experimental confinement to 68 ns, whereas the nearby three-electron line $(1s2p)^3P4p$ in addition has a delayed time distribution.

From the spectra, such as shown in Fig. 2, it is seen that mainly the KLN peak survives in the delayed spectrum for pure neon, with smaller contributions from KLM and $KL^{2\,4}P$, the latter two possibly after cascade transitions. For admixture to neon of methane, other peaks appear in the delayed spectra, indicating capture to states KLO and KLM. Also other mixtures of neon and water vapour have been studied.

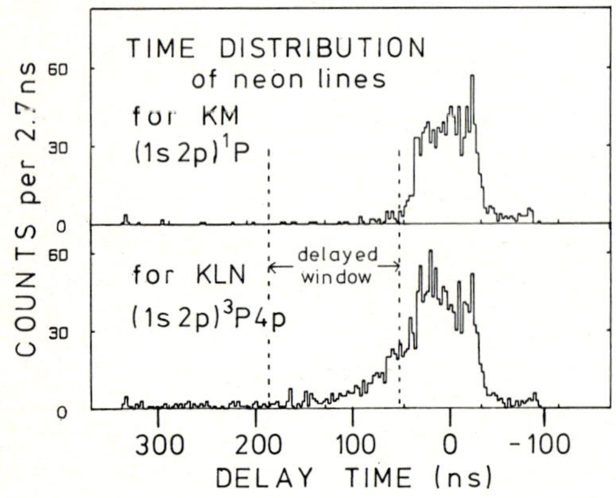

Fig. 1: Time spectra relative to a pulsed beam measured for 0.7-MeV/amu Cl^{11+} on a 80-μbar Ne target and sorted for the two-electron line KM and the three -electron line KLN, seen in Fig. 2.

The metastable state responsible for the capture lines is mainly $1s2s^3S$ because this is the only state with sufficiently long lifetime to produce the observed decay curves[1,4]. The $1s2p^3P$ has too short a lifetime, and yet the observed capture proceeds for Ne^{8+} + Ne to a state $(1s2p)^3P4p$, observed both in the x-ray and the Auger-electron channel. This is an example of a capture process, where the core structure is changed.

The excited states, to which the electron capture proceeds, is compared with a classical description, which is presented in Ref. 4 together with the first experimental indications of delayed x-ray emission.

REFERENCES

1. F. Folkmann, R. Mann, and H.F. Beyer: *Inner-Shell and x-Ray Physics of Atoms and Solids*, eds. J. Fabian, H. Kleinpoppen, and L.M. Watson (Plenum, 1981) 145
2. R. Mann, H.F. Beyer, and F. Folkmann, Phys.Rev.Lett. 46, 646 (1981)
3. F. Folkmann, H.F. Beyer, R. Mann, and K.-H. Schartner, Nucl.Instrum.Methods 181, 99 (1981)
4. F. Folkmann, R. Mann, and H.F. Beyer, Phys.Scripta T3, 88 (1983)

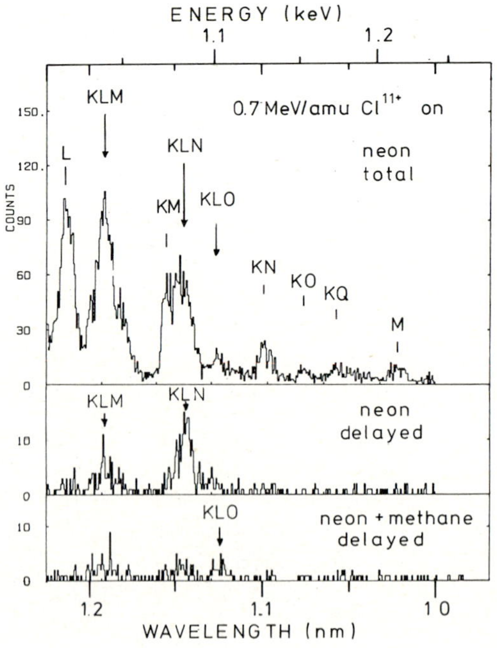

Fig. 2: Ne x-ray lines from highly excited states. Delayed spectra, sorted with the window of Fig. 1, show capture lines. Electron occupation of initial states is indicated. Top shows total spectrum for 40-μbar Ne, middle the corresponding delayed one, and bottom the delayed spectrum for 39-μbar Ne + 39- bar methane, all for 4 μC/channel.

ELECTRON CAPTURE INTO HIGHLY CHARGED, METASTABLE S, Ar, AND Kr RECOIL IONS STUDIED BY DELAYED AUGER-ELECTRON MEASUREMENTS

Kurt M. Cramon and Finn Folkmann

Institute of Physics, University of Aarhus, DK-8000 Aarhus C, Denmark

The emission of electrons from a gaseous target after bombardment with 0.4-0.9-MeV/amu Cl^{8-12+} beams has been measured. In these measurements, we have used a delayed-coincidence technique with a pulsed ion beam[1], and it was found that certain delayed Auger-electron lines appeared in the spectra when $S(H_2S)$, Ar, and Kr were used as target gas. The detailed population of various states is very sensitive to gas admixtures, as seen in Fig. 1. This highly selective population of special states, depending on the admixed gas, indicates the presence of a metastable state, to which an electron is captured.

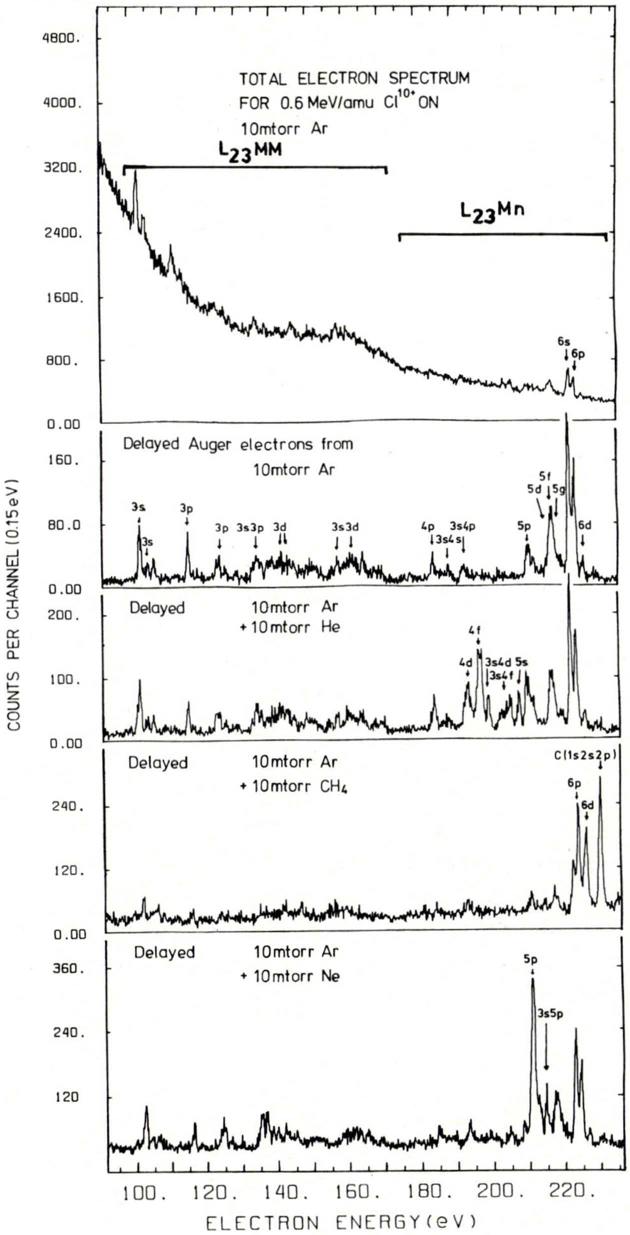

Fig. 1: Ar Auger-electron spectra after impact of 0.6-MeV/amu Cl^{10+} on Ar and Ar mixtures. The electrons added to the metastable $1s^22s^22p^53s\,^3P$ core are indicated. The delayed spectra are accumulated 40-400 ns after the beam pulses.

Calculations of many-electron states of argon and sulphur with the multiconfigurational Dirac-Fock program of Desclaux[2] show that the observed electron lines come mainly from 11-electron states consisting of $1s^22s^22p^5$ + two electrons with $n \geq 3$. The metastable 10-electron states responsible for the delayed observed lines are $1s^22s^22p^53s\,^3P_{0,2}$ with lifetimes 0.5 ms and 2.0 ms for argon and sulphur, respectively. After the metastable 10-electron states are produced in a primary collision with the ion beam, possibly after cascading from related states, they recoil and capture an electron in a second collision with a neutral atom or molecule. This electron is captured into an orbit (n,ℓ), and if for simplicity we assume that the 10-electron core remains unchanged, we expect resulting states of the form $(1s^22s^22p^53s)\,^3P(n,\ell)$. In an L-S-coupling scheme, it can be argued that the main contribution to the Auger decay of such states comes from doublet states with $L = \ell\pm1$. Using this selection rule and the transition energies calculated with the Dirac-Fock program, we have interpreted the capture orbits as indicated in Fig. 1.

The krypton lines have not yet been fully interpreted. However, Dirac-Fock calculations have shown that the main contribution to the observed lines comes from states consisting of $1s^22s^22p^63s^23p^63d^9$ + two electrons with $n \geq 4$.

The experimentally found capture orbits indicated in Fig. 1 for argon cover a couple of different n,ℓ states, which are centered around a value, in good agreement with the predictions of the classical model described in Ref. 3.

REFERENCES

1. F. Folkmann, K.M. Cramon, R. Mann, and H.F. Beyer, Physica Scripta T3, 166 (1983)
2. J.P. Desclaux, Comput.Phys.Comm. 9, 31 (1975)
3. F. Folkmann, R. Mann, and H.F. Beyer, Physica Scripta T3. 88 (1983)

COMPARISON OF VARIOUS MULTIPLE SCATTERING APPROACHES TO ELECTRON CAPTURE

L.J. Dubé* and J.K.M. Eichler+

* Fakultät für Physik, Universität Freiburg, 7800 Freiburg, W. Germany
\+ Hahn-Meitner Institut für Kernforschung, D-1000 Berlin 39, W. Germany

The lack of multiple scattering theories implemented over a large range of quantum numbers has long delayed a systematic study of the predictions of various approaches.

However such a comparison is now possible due to the recent availability of several approximations which have been generalized to include capture from and to arbitrary initial and final states: namely, the Eikonal Approximation (EA)[1], an approximate (Peaking) form of the Impulse Approximation (PIA)[2] and the Continuum Distorted Wave (CDW)[2,3] approximation.

All of these approaches are so-called multiple scattering theories. They contain (at different level of sophistication) some multiple scattering information of the interaction e - Target (V_T) and/or e - Projectile (V_P) in distinction to the first (B1) and second (B2) Born approximations which include only single and double scattering respectively.

Furthermore, in contrast to the CDW which is a symmetrical theory (and which will not be considered further here), the PIA and EA exist under two unequivalent forms, i.e. the post (+) and prior (-) forms. In the post (prior) form, V_T (V_P) is treated to first order while V_P(V_T) is included to all orders (albeit approximately) through the eikonal phase in EA or through an approximate continuum intermediate state in PIA. It is therefore clear that the post/prior forms will be appropriate to describe quite different physical situations.

The aims of the present contribution are:

i. to clarify the relationship between PIA and EA;

ii. to extract and compare the physical content of PIA and EA; and

iii. to provide a systematic comparison of PIA and EA for different velocities and quantum numbers.

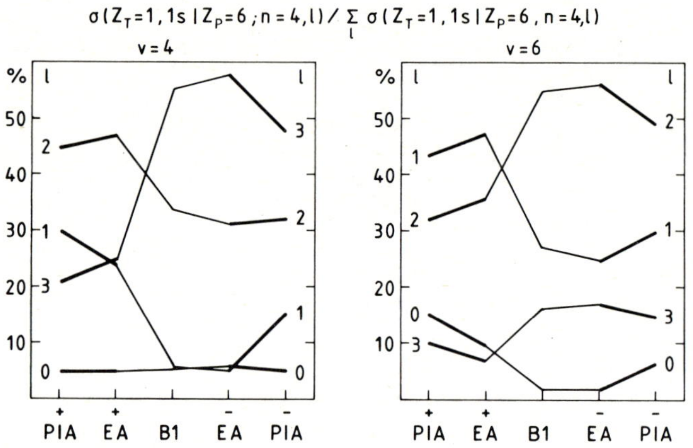

Fig. 1: Percentual distribution of $n\ell$ states for a given shell n = 4. Collision partners are Z_P=6 and Z_T=1 at v=4,6.

For example, we show in Fig.1 a representative plot of such a comparison. At 2 velocities, v= 4 and 6 , and for asymmetric collision partners (projectile with charge Z_P=6 incident on a target with charge Z_T=1), we draw the predictions of 5 different calculations: $\sigma^{\pm}_{PIA}$, $\sigma^{\pm}_{EA}$ and σ_{B1} for capture from the ground state to the n = 4 shell of the projectile. The vertical axis indicates the relative contribution in % of each ℓ quantum number within the given shell. We remark that (within 10%) the distributions from σ^{+}_{PIA} (σ^{-}_{PIA}) and σ^{+}_{EA} (σ^{-}_{EA}) agree with each other but that the post forms differ drastically from the prior forms.

The discussion of such similarities and differences will be the object of our presentation.

References

1. F.T. Chan and J. Eichler, Phys.Rev.Lett. 42, 58 (1979)
   J.K.M. Eichler, Phys.Rev. A23, 498 (1981)
2. L.J. Dubé, J.Phys.Bxx (1983) to be published
3. D.S.F. Crothers, J.Phys.B14, 1035 (1981)

COLLISIONS OF MULTIPLY CHARGED PROJECTILES WITH LIGHT TARGETS: I. THEORY

L.J. Dubé and R. Bruch

Fakultät für Physik, Universität Freiburg, 7800 Freiburg, W. Germany

Among the possible reaction channels available in the collisions of multiply charged projectiles with light targets, we study in detail the one electron capture process. Typically, we consider the transfer of an electron from an initial state $\underline{i}$ ($\equiv n'\ell'm'$) bound to a Target (T) with (effective) charge Z_T to a final state $\underline{f}$ ($\equiv n\ell m$) bound to a Projectile (P) with (effective) charge Z_P, viz.

$$P^{Z_P+} + T^{(Z_T-1)+}(\underline{i}) \longrightarrow P^{(Z_P-1)+}(\underline{f}) + T^{Z_T+}. \quad (1)$$

To be specific, the present contribution examines cases for which $\underline{i}$ is the ground state and for collision partners having $Z_P >> Z_T$.

In order to calculate cross sections corresponding to reaction (1), it has been necessary to generalize existing approaches to include capture to arbitrary final states. One of us (LJD)[1] has recently implemented the Continuum Distorted Wave (CDW) approximation, and 2 approximate (Peaking) forms of the Impulse Approximation (PIA), namely the post and prior form of the PIA, to fulfill this purpose.

We will present specific calculations based on these multiple scattering theories and will attempt to gauge the importance of multiple scattering effects by comparison with first Born (B1, single scattering) and second Born (B2, double scattering) cross sections[2]. Preliminary theoretical and experimental results[3] indicate that these effects are important not only for the absolute magnitudes but also for the relative population of final states.

Moreover, since experimental detection of capture to specific states involves the observation of emission lines from the projectile, we will discuss and quantify an inherent problem to the method, i.e. the unavoidable cascade contribution (from high lying states) to the observed line intensities. With the help of the present theoretical cross sections as input, we will show simulations of this cascade effect on the intensities of a few selected lines.

References

1. L.J. Dubé, J.Phys.Bxx (1983) to be published
2. L.J. Dubé and J.S. Briggs, J.Phys.B14,4595 (1981)
3. R. Bruch, L.J. Dubé, E. Träbert et.al., J.Phys.B15,L857 (1982)

COLLISIONS OF MULTIPLY - CHARGED PROJECTILES WITH LIGHT TARGETS: II EXPERIMENT

R. Bruch[+)], L.J. Dubé[+)], E. Träbert[++)], P.H. Heckmann[++)] and B. Raith[++)]

[+)] Fakultät für Physik Universität Freiburg, West Germany
[++)] Institut für Experimentalphysik III, Ruhr Universität Bochum, West Germany

To test the specific predictions of various multiple scattering theories (see contribution of Dubé and Bruch at this meeting), we examine experimentally the following prototype reaction:

$$C^{4+}(1s^2) + T \dashrightarrow C^{3+}(1s^2 n\ell) + T(?), \qquad (1)$$

where the target T is either H_2 or He. These systems are chosen for their quasihydrogenic nature, allowing on the one hand a meaningful comparison with purely hydrogenic theories, and avoiding on the other hand multi-electron effects such that deexcitation of Rydberg states proceeds only via known radiative transitions. We select the impact energies (E = 2,3,4 and 5 MeV) such that the relative velocities are large enough for the charge transfer theories to be applicable but small enough for the cross sections to remain sizeable.

In order to make selective studies of the capture process, a combination of high energy beam and spectroscopic technology is necessary. We have used the facilities available at the Bochum Dynamitron tandem accelerator laboratory. Intense fast $^{12}C^{2+}$ ion beams are passed through a gas stripper (gas pressure ≅ 1-2 X 10^{-2} Torr), giving rise to ions in high charge states. A C^{4+} beam (5-8 μA at 5 MeV) is then magnetically selected and directed to a differentially pumped gas cell (≅ 10^{-2} Torr) with an effective excitation length of about 25mm. The 2.2 m grazing incidence spectrometer (McPherson 247 equipped with a gold coated 600 lines/mm ruled grating) views < 1 mm length of the interaction region at right angles with respect to the beam direction. A channeltron serves as detector and all measurements are performed under single collision conditions. Beam normalization is achieved by charge integration on a large Faraday cup.

Figure 1 shows a spectrum taken in the range 20-44 nm for 5 MeV C^{4+} ions excited by collisions with H_2. From this figure it is evident that several prominent lines can be identified as transitions from C^{3+} $1s^2ns$, C^{3+} $1s^2np$ and C^{3+} $1s^2nd$ states. In particular the most intense line is asssigned to the C^{3+} $1s^22s$-$1s^23p$ transition. The contribution due to C^{3+} $1s^2n\ell$ levels results from single-electron capture. A more detailed spectroscopic line identification reveals also contributions from competing processes, e.g. lines originating from C^{2+} $1s^2n\ell n'\ell'$ and C^{4+} $1sn\ell$ configurations.

Fig. 2 shows a spectrum for 2 MeV C^{4+} + He. The mainlines originate from C^{3+} $1s^2n\ell$ and He^+np target excitation. From the observed spectra relative cross sections are extracted and compared with first Born (B1) and second Born (B2) cross sections and various multiple scattering approaches . The measured cross sections are well reproduced by the Impulse Approximation (IA) and the continuum Distorted Wave (CDW) method, whereas B1 and B2 are found to be inadequate to represent the experimental data. These results represent the first quantitative test of available charge exchange theories and provide a measure of the importance of multiple scattering effects in capture processes.

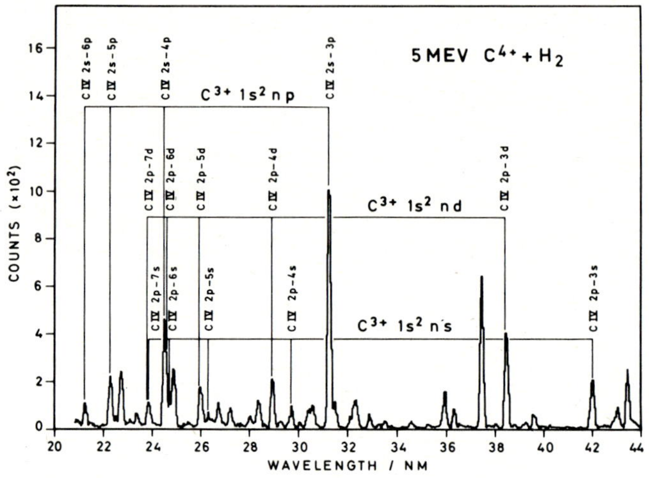

Fig. 1: Representative emission spectrum for 5 MeV C^{4+} + H_2. The C^{3+} $1s^2ns$, $1s^2np$ and $1s^2nd$ Rydberg series are indicated.

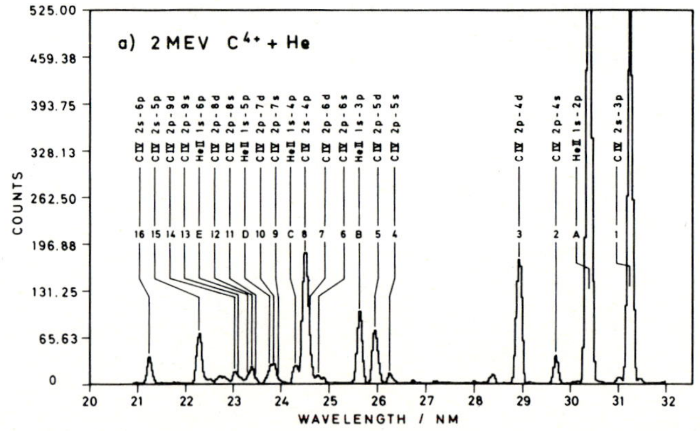

Fig. 2: Emission spectrum obtained at 2 MeV impact energy. The spectroscopic assignments of the most important C^{3+} projectile and He^+ target lines are given.

CAPTURE CROSS SECTIONS IN HIGHLY EXCITED p STATES OF Ar18+ IN HIGH VELOCITY COLLISIONS OF 250MeV Ar18+ ON N

J.P. Rozet, P. Legagneux-Piquemal, A. Chetioui, P. Chevallier

Institut Curie and Université P & M Curie, 11, rue P & M Curie 75231 Paris Cedex 05, France

Capture cross sections to excited np levels of Ar18+ ($4 \leqslant n \leqslant 10$) have been measured in collisions of 250MeV Ar18+ fully stripped ions on a gaseous nitrogen target. Beams of 250MeV Ar12+ are obtained at the CEVIL facility in Orsay. Poststripping by a thin carbon foil and magnetic analysis of the produced charge states gives us typical ion currents of about 40mA for Ar18+ ($\sim 10^{10}$pps). A gas jet has been used to produce a mean pressure in the interaction region of $\sim 3\ 10^{-3}$ Torr[1]. Lyman x rays emitted by excited hydrogenlike Ar17+, following capture to excited states of Ar18+ have been observed at an angle of 60° with respect to the beam direction. A high transmission crystal spectrometer (total transmission-efficiency $T > 10^{-6}$) has been used, allowing reasonable counting rate despite low beam intensities and target density. Typical resolution of 18eV or better at 4.5keV can be achieved. Part of the spectrum corresponding to np→1s transitions for $4 \leqslant n < 10$ is shown in figure 1 (total counting time of $\sim$70mn). Lyman x rays where also simultaneously observed with a SiLi detector at an angle of 90° with respect to the beam axis for monitoring purposes.

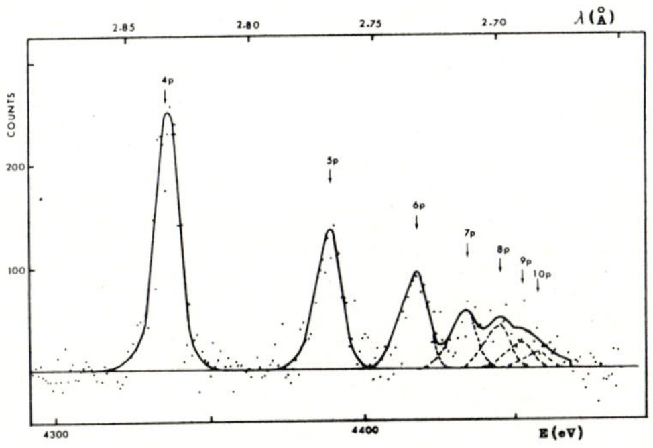

Figure 1.

Preliminary results are given in Table I for absolute Lyman x-ray emission cross sections and compared with CDW calculation of np capture cross sections[2] (for details on absolute cross section extraction, see Ref.1).

This comparison is expected to be valid for high n np levels only, since cascade effects increase population of low lying np levels and contribute significantly to Lyman x-ray emission rate especially in the case of n=2 and n=3. It has been observed for example that, in the case of 400MeV Fe26+ ions on nitrogen target, cascade contribution to the feeding of 2p, 3p and 4p levels is about 67%, 50% and 20% respectively[3]. On the other hand, cascade effects play a minor role in the case of high n np levels.

Table I. np capture cross sections

np states n=	Experimental Lyman np→1s cross sect. (cm^2)	Theoretical np capture cross sect. (cm^2)
2	$5.7\ 10^{-19}$	$4.12\ 10^{-19}$
3	$1.7\ 10^{-19}$	$1.98\ 10^{-19}$
4	$1.02\ 10^{-19}$	$1.02\ 10^{-19}$
5	$5.5\ 10^{-20}$	$5.73\ 10^{-20}$
6	$4\ 10^{-20}$	$3.50\ 10^{-20}$
7	$2.2\ 10^{-20}$	$2.27\ 10^{-20}$
8	$1.6\ 10^{-20}$	$1.56\ 10^{-20}$
9	$1.0\ 10^{-20}$	$1.11\ 10^{-20}$
10	$6.1\ 10^{-21}$	$8.17\ 10^{-21}$

Agreement between theory and experiment is found to be extremely good for $n \geqslant 4$. Only in the case of 3p levels it appears clearly that the theory overestimates the capture cross section. This had already been observed in the case of 400MeV 26+ ions[3].

References

1. A. Jolly, K. Wohrer, A. Chetioui, J.P. Rozet, C. Stephan, L.J. Dubé, J. Phys. B, to be published (1983).
2. Dz Belkic, R. Gayet, A. Salin (1982) submitted to C.P.C.
3. A. Chetioui et al, J. Phys. B, to be published (1983).

CAPTURE CROSS SECTIONS IN HIGH p RYDBERG STATES BY 400MeV BARE Fe26+ AND ONE-ELECTRON Fe25+ IONS

A. Chetioui, D. Vernhet, J.P. Rozet, P. Legagneux-Piquemal

Institut Curie, and Université P & M Curie, 11, rue P & M Curie 75231 Paris Cedex 05

and

C. Stephan

Institut de Physique Nucléaire, 91405 Orsay, France

We have measured 1s→np (n⩾5) capture cross sections for 400MeV Fe26+ and Fe25+ ions colliding with nitrogen atoms*. Hydrogenlike and heliumlike x rays were detected by a high resolution crystal spectrometer. Because of low values of cross sections and spectrometer transmission, a closed gaseous cell was used and pressure was fixed at 1 Torr. A 200μg/cm² beryllium stripper was used, producing a charge state distribution which was previously measured[1]. A part of the spectrum is shown on figure 1.

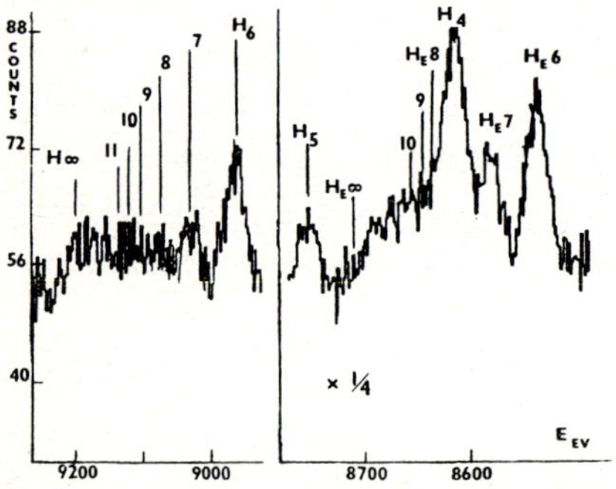

Figure 1. X-ray spectrum for 400MeV Fe→N.

Hydrogenlike Lyman x rays follow either capture by the bare ion fraction or excitation by the one-electron ion fraction : the last contribution was evaluated from previous measurements[2] and found to be less than 5%. In the same manner heliumlike Lyman x rays are mainly due to capture by one-electron ions, excitation of two electron ions being negligible.

From the hydrogenlike spectrum, one can draw the following conclusions :

- for 5⩽n⩽8 1s→np capture cross sections are in good agreement with CDW predictions ; this extends our conclusions for n<4[3];

- for 8⩽n⩽15, the observed flat spectrum could indicate a dependence in $1/n^3$ for σ(1s→np) since the level density varies as n^3. Usually, only the summed over ℓ cross sections have this dependence;

- for very high n quantum numbers, a rise in σ(1s→np) seems to appear.

References

1. D. Vernhet et al, this conference.
2. K. Wohrer, F. Fernandez, J.P. Rozet, A. Jolly, A. Chetioui, and C. Stephan. International Conference on X-ray and Atomic Inner-shell Physics, Eugene, Oregon, August 1982, and to be published in J. Phys.B.
3. A. chetioui et al, to be published in J. Phys. B.

work performed at the CEVIL Orsay.

TRANSLATIONAL SPECTROSCOPY OF ELECTRON CAPTURE BY MULTIPLY CHARGED IONS

B.A.Huber, H.J.Kahlert, K.Wiesemann

Inst.f.Exp.Physik AG II, Ruhr-Universität Bochum, West Germany

By aid of the translational spectroscopy method electron capture by multiply charged ions is studied at collision energies E between 100 eV and 1 keV at a detection angle of $(0 \pm 0{,}7)^\circ$. Particularly the following aspects have been investigated:

a) Influence of projectile ions in metastable, excited states. Fig. 1 shows the measured energy spectra of Ar^+ ions formed in Ar^{2+} + He and Ar^{2+} + Ne collisions. The exoergic ground state reactions are described by the following equations:

Ia: $Ar^{2+}(^3P_2) + He(^1S_0) \rightarrow Ar^+(^2P_{3/2}) + He^+(1s^2S_{1/2}) + 3{,}04$ eV

IIa: $Ar^{2+}(^3P_2) + Ne(^1S_0) \rightarrow Ar^+(^2P_{3/2}) + Ne^+(^2P_{3/2})$ $+6{,}07$ eV

The corresponding reactions (Ib,IIb) with Ar^{2+} projectiles in the metastable $^1D_2$-state are correlated with energy defects ΔE of +4,78 eV and +7,81 eV, respectively. As to be seen in Fig. 1, the contribution of metastable Ar^{2+} ions dominates the spectrum in the collision system $(ArHe)^{2+}$, whereas in $(ArNe)^{2+}$ these ions give a minor contribution only. This drastic difference is surprising as the primary beam composition is kept constant. It is caused by the favourable energy defect for reaction Ib and IIa, i.e. the favourable position of the corresponding potential curve crossings[1]. It clearly demonstrates the sensitivity of the reaction cross section to the energy defect. Thus relative contributions to the translational spectrum, originating from projectiles in different excited states, are no measure for the primary ion beam composition, even if the correlated energy defects are nearly the same.

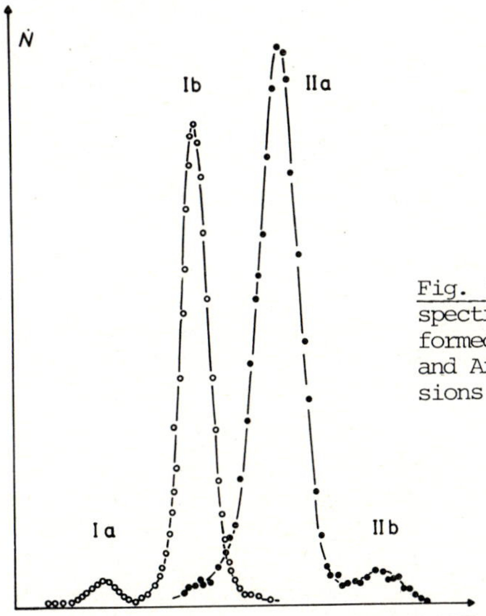

Fig. 1: Translational spectrum of Ar^+ ions formed in Ar^{2+}-He (o) and Ar^{2+}-Ne (•) collisions at E=1200 eV.

b) Transfer ionisation processes at low collision energies. The translational spectroscopy allows a determination of the reaction mechanism of the transfer ionisation process. In collision systems (e.g. Ne^{2+} + Xe or Kr) where this process is weakly exoergic ($\Delta E \lessapprox + 10$ eV), our measurements[2] show the transfer ionisation to occur as a two-step process[3]. Within one collision an electron is captured into an excited state of the projectile ion, then at a nuclear distance of about 5 atomic units a second target electron is released via a Penning-type process. Compared to the pure single electron capture the contribution of the transfer ionisation tends to increase towards lower energies. The direct transfer ionisation mechanism is not effective in weakly exoergic systems studied so far.

c) Translational spectra in triply charged collision systems. As an example Fig. 2 shows the energy spectrum of Ar^{2+} ions produced in Ar^{3+}/Ar collisions. A large number of different reaction channels can be identified, corresponding in most cases to single electron capture by ground state Ar^{3+} ions. Possible energetic positions for the formation of projectile and target ions in different states are given in the upper part of Fig. 2. The Ar^{2+} ions are produced dominantly in the ^3D state as well as in the metastable $^1D_2$ and $^1S_0$ states.

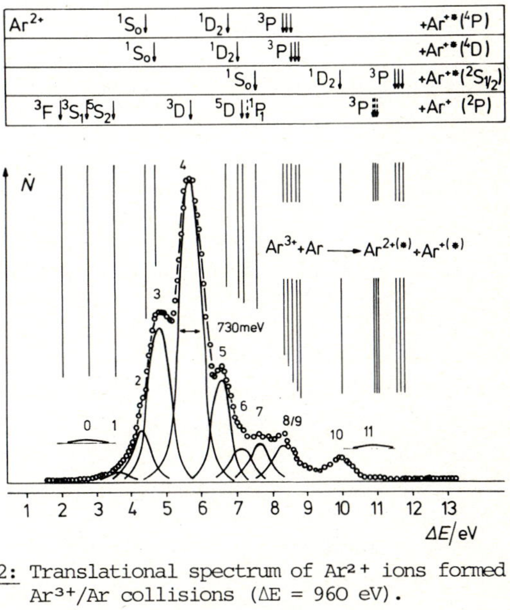

Fig. 2: Translational spectrum of Ar^{2+} ions formed in Ar^{3+}/Ar collisions (ΔE = 960 eV).

References:

1. B.A.Huber, Physica Scripta T3, 96 (1983)
2. H.J.Kahlert, B.A.Huber, K.Wiesemann, J.Phys.B16, 449 (1983)
3. A.Niehaus, M.W.Ruf, J.Phys. B9, 1401 (1976)

ENERGY LOSS SPECTRA OF Ar^{3+}-He COLLISIONS

E.Y. Kamber

Department of Physics & Astronomy, University College London, Gower Street, London WC1E 6BT, U.K.

J.B. Hasted

Department of Physics, Birkbeck College (University of London), Malet Street, London WC1E 7HX, U.K.

Differential single electron capture energy loss spectra of 1250 eV Ar^{3+} ions in collision with He are presented at 0° and 0.4°. The energy defects are measured by parallel plate electrostatic momentum analyzers before and after the collision chamber, with resolution 0.85 eV; the angular resolution is 0.07°[1,2].

Although the predominant exit channel is into the ground state $Ar^{2+}(^3P_2)$, there are contributions from the first excited state $Ar^{2+}(^1D)$ and other states. Reaction products arising from long-lived metastable Ar^{3+} ions are detected, as is shown in Fig. 1. Differential scattering data in the angular range 0° - 1.0° are presented and interpreted in terms of the avoided crossings of potential energy curves, those of the products being dominated by the Coulomb repulsion between the product ions. The probability of capture into the ground state of the product ion is deduced as a function of impact parameter, calculated from the scattering angle.

Fig. 2 illustrates the elastic differential scattering function, which shows contributions from three avoided crossings, associated with channels IαX, IβX, IγX, the arrows indicating the angles corresponding to impact parameters equal to values of the calculated nuclear separations.

References

1. E.Y. Kamber, D. Mathur and J.B. Hasted, J. Phys. B: At. Mol. Phys. 15, 263 (1982)
2. E.Y. Kamber, D. Mathur and J.B. Hasted, J. Phys. B: At. Mol. Phys. 15, 2051 (1982).

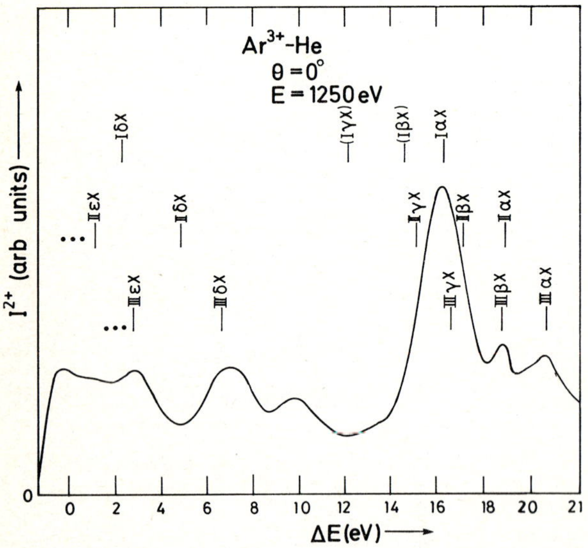

Fig. 1: Ar^{3+}-He, 1250 eV, θ = 0°, energy loss spectrum with single electron capture.

Key to assignments:

I	Ar^{3+}	$(3p^3\ ^4S^o_{3/2})$	processes
II	Ar^{3+}	$(3p^3\ ^2D^o)$	processes
III	Ar^{3+}	$(3p^3\ ^2P^o)$	processes
X	He^+	$(1s^2\ ^2S_{1/2})$	product
α	Ar^{2+}	$(3p^4\ ^3P_2)$	product
β	Ar^{2+}	$(3p^4\ ^1D)$	product

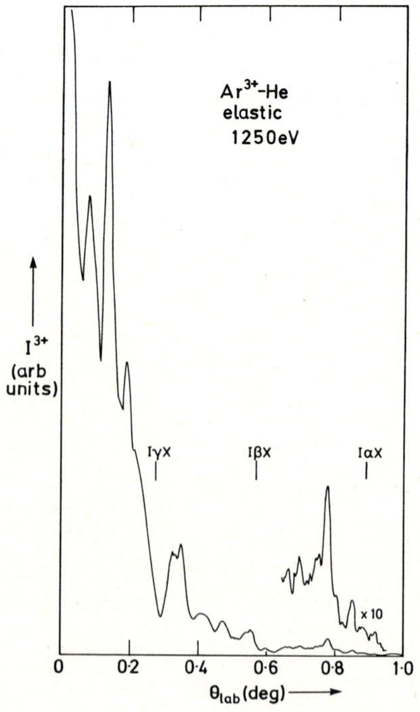

Fig. 2: The elastic differential scattering function for 1250 eV Ar^{3+}-He collisions.

FINAL-STATE-ANALYSIS OF ELECTRON CAPTURE PROCESSES IN COLLISIONS OF HIGHLY STRIPPED C, N AND O IONS WITH He ATOMS

M.Kimura, T.Iwai, Y.Kaneko, N.Kobayashi, A.Matsumoto, S.Ohtani, K.Okuno, S.Takagi, H.Tawara and S.Tsurubuchi

Institute of Plasma Physics, Nagoya University, Nagoya 464, Japan

Translation-energy-spectra of the charge-changed projectile ions scattered into the forward direction in collisions of C^{q+}(q=4,5), N^{q+}(q=5,6) and O^{q+}(q=6) ions with He atoms have been measured at collision energies of 1 x q and 2 x q keV.

Total cross sections for one-electron transfer from He atoms into multi-charged B, C, N, O, F, Ne and S ions were measured recently at low collision energies[1]. When the cross sections are measured as a function of the ionic charged q of projectile ions, strong oscillations in the cross sections are observed. This oscillatory behavior is interpreted by a classical one-electron model, whose essential assumption is that at low energies an electron is captured selectively into a level with a particular quantum number n.

We have made a series of experiments with the intention to find its evidence and to determine which shell the transferred electron does get into. The experimental method and procedure are described in ref.2 : the ion beam extracted from an ion source of EBIS type called NICE-1 is mass-analyzed and led into a cell containing He target gas. Charge-changed projectiles passing through the cell are decelerated and energy-analysed by a 127° electronstatic analyzer. Results for the projectiles C^{q+}(q=3,6), N^{q+}(q=4,7) and O^{q+}(q=5,7,8) have been published[2,3,4].

From the measured energy-gains of the product (q-1)-charged ions, it has been found that an electron is captured selectively into a single shell with a particular principal quantum number ; n=2 for the incident C^{4+} ions and n=3 for incident C^{5+}, N^{5+}, N^{6+} and O^{6+} ions. These results, together with our previous observations in other systems, are in accord with the prediction of the classical one-electron model.

It has also been revealed that there is good similarity among the energy spectral patterns obtained for the ions with the same ionic charge q, irrespectively of the ionic species. As an example, the energy-gain spectra of the product A^{5+} ions in the A^{6+} + He collisions (A= O, N and C) are shown in Fig. 1 together with energy profile of the primary ions. This figure demonstrates not only that the electron is captured selectively into the n=3 shell of the product ions but also that the energy spectral patterns of the product C^{5+}, N^{5+} and O^{5+} ions are quite similar to each other. Such similarity is also observed in other (q-1)-charged products and can be considered to result from the similarity among the diabatic potential curves for the collision systems of the q-charged ions and He.

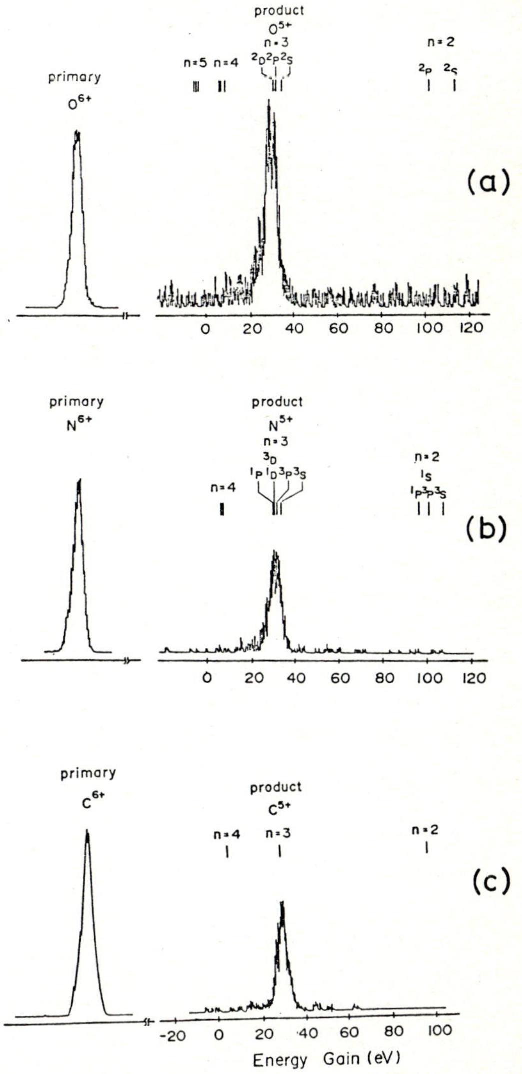

Fig. 1. Energy spectra of the charged-changed ions at forward directions from a) O^{6+} + He collision at 0.33 keV/amu, b) N^{6+} + He collision at 0.40 keV/amu and c) C^{6+} + He collision at 0.46 keV/amu.

1) T.Iwai, Y.Kaneko, M.Kimura, N.Kobayashi, S.Ohtani, K. Okuno, S.Takagi, H.Tawara and S.Tsurubuchi, Phys. Rev. A26, 105 (1982).
2) S.Ohtani, Y.Kaneko, M.Kimura, N.Kobayashi, T.Iwai, A. Matsumoto, K.Okuno, S.Takagi, H.Tawara and S.Tsurubuchi, J.Phys. B 15, L533 (1982).
3) S.Tsurubuchi, T.Iwai, Y.Kaneko, M.Kimura, N.Kobayashi, A.Matsumoto, S.Ohtani, K.Okumo, S.Takagi and H.Tawara, J.Phys. B 15, L733 (1982)
4) M.Kimura, T.Iwai, Y.Kaneko, N.Kobayashi, A.Matsumoto, S.Ohtani, K.Okuno, S.Takagi, H.Tawara and S.Tsurubuchi, J.Phys. B 15, L851 (1982).

FINAL-STATE-ANALYSIS OF ELECTRON CAPTURE PROCESSES IN COLLISIONS OF HIGHLY STRIPPED F AND Ne IONS WITH He ATOMS

H.Tawara, T.Iwai, Y.Kaneko, M.Kimura, N.Kobayashi, A.Matsumoto, S.Ohtani, K.Okuno, S.Takagi and S.Tsurubuchi

Institute of Plasma Physics , Nagoya University, Nagoya 464, Japan

To investigate the electron capture processes of highly stripped heavy ions at low energies, we made a series of measurements of the energy-gain of various projectile ions including C, N and O ions in collisions with He atoms using the translational energy spectroscopy technique. Most of the energy-gain spectra observed show only a single peak indicating that the electron is captured selectively into a particular single state of projectile ions. However, in some cases such as C^{3+} + He collisions, four peaks were observed. It was found that they correspond to the electron capture into the levels with the same principal quantum number n but different orbital angular quantum number ℓ.

Following our previous work[1], we have made measurement of the energy-gain in the electron capture processes of highly stripped F^{q+}(q=6,7,8) and Ne^{q+}(q=7,8,9) ions on He atoms. Some typical examples of the energy-gain spectra are shown in Fig.1. Some features are as follows :

1) Ne^{7+} + He (Fig.1a)

Three peaks are clearly seen. The strongest peak at the energy-gain ΔE≃20 eV is found to be due to the following one-electron capture process into the n=4 state:

$$Ne^{7+}(1s^2 2s) + He \rightarrow Ne^{6+}(1s^2 2s4\ell) + He^+ + \Delta E. \qquad (1)$$

The second peak at ΔE≃38 eV is due to the different one-electron capture process into the n=3 level :

$$Ne^{7+}(1s^2 2s) + He \rightarrow Ne^{6+}(1s^2 2p3\ell) + He^+ + \Delta E. \qquad (2)$$

It should be noted that this process (2) involves two electrons, that is, one electron in the projectile ion is excited and the other electron is captured into the excited state of the projectile ion from the target atom. This indicates that even the two electron process becomes significant if the proper crossing in the diabatic energy curves is available.

This spectrum is a clear evidence that the electron is captured into the levels with different principal quantum number n, indicating that the classical one-electron model breaks down.

2) F^{7+} + He (Fig.1b)

The stronger peak at ΔE≃18 eV is due to the following one-electron capture process into the n=4 state :

$$F^{7+}(1s^2) + He \rightarrow F^{6+}(1s^2 4\ell) + He^+ + \Delta E. \qquad (3)$$

The peak at ΔE≃66 eV is probably due to the transfer ionization :

$$F^{7+}(1s^2) + He \rightarrow F^{5+**}(1s^2 3\ell 3\ell') + He^{2+} + \Delta E \qquad (4)$$
$$\rightarrow F^{6+} + He^{2+} + e.$$

We notice that the transfer ionization peak is much stronger for F^{7}+ ions, compared with other ions such as N^{7+} and O^{7+} ions, and its intensity amounts to about 20 % of that for the main one-electron capture process(3).

The electron-capturing levels of all the one-electron capture processes invevestigated in our study are summarized in Table 1.

1) M.Kimura, T.Iwai, Y.Kaneko, N.Kobayashi, A.Matsumoto, S.Ohtani, K.Okuno, S.Takagi, H.Tawara and S.Tsurubuchi, preceeding paper in this book.

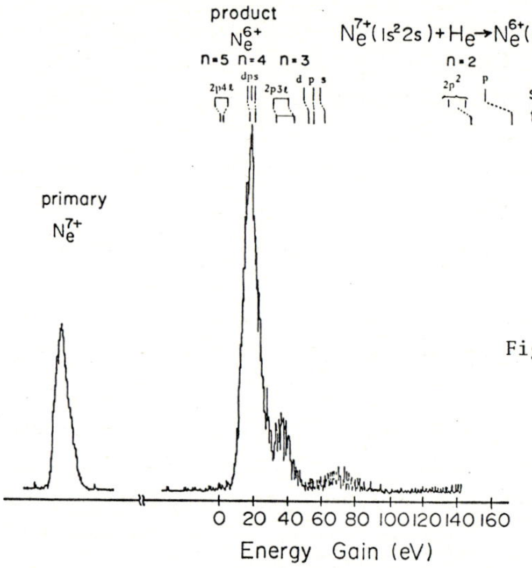

Fig. 1a

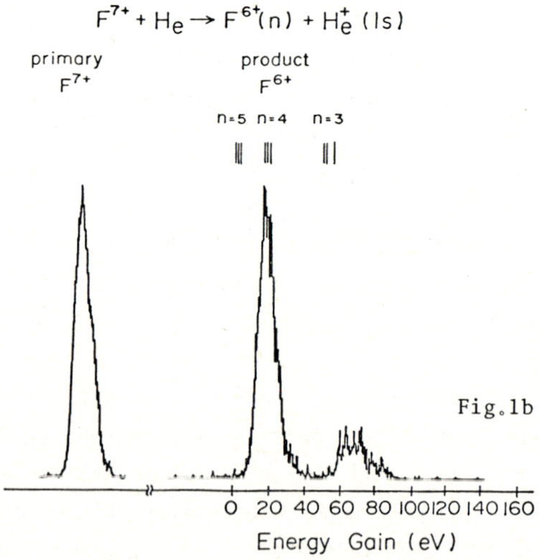

Fig.1b

Table 1. the principal quantum number n of the electron-capturing levels in A^{q+} + He collisions

q	11	10	9	8	7	6	5	4	3
S									
Ne			5(4)	4	4(3)				
F					4	3			
O				4	4	3	3		
N					4	3	3	2	
C						3	3	2	2

LOW-ENERGY ELECTRON-CAPTURE IN Ne^{2+}-He, Ar^{2+}-He AND Kr^{2+}-He COLLISIONS

Kazuhiko OKUNO and Yozaburo KANEKO

Department of Physics, Tokyo Metropolitan University, Tokyo 158, Japan

In order to study low energy charge transfer reactions, we have developed a new technique using an octopole ion-beam guide (OPIG).[1,2] The OPIG is coupled with a collision cell placed in the middle of tandem mass spectrometer as shown in Fig.1. An R-F voltage supplied to OPIG is utilized for storage and guidance of ions, and the intensity of the incident ion beam passed through the OPIG can be kept almost constant in the collision energy region studied.

We have investigated the electron capture processes in the Ne^{2+}-He, Ar^{2+}-He and Kr^{2+}-He collisions.

$$X^{2+} + He \longrightarrow X^{+} + He + \Delta E \qquad (1)$$

The electron capture cross sections in three collision systems are measured by an initial growth method. Their dependences on the collision energy are quite different and have some respective structures. The most drastic case is that for Ne^{2+}-He as shown in Fig.2, where the cross section rises up very steeply at about Ecm=16 eV (in the center-of-mass systems) and makes a saddle linked to other data at higher energies.[3,4,5] The Ne^{+} product blow Ecm=16 eV is in proportional to the second power of the target pressure and it is due to the multiple process.

The incident double charged ion beam from a conventional electron impact type of ion source used is usually composed of the low lying metastable ions in the $^1D_2$ and $^1S_0$ states together with the ground state ^3P ion. The measured cross sections are composed of the partial cross sections multiplied by the respective state fractions in incident beam, and they are represented as $\sigma_{21} = \Sigma_i f_i \cdot \sigma_i$. It is a matter of cource that the cross section data have not always the universality barring the informations about the condition of ion formation at ion source and/or about the state fraction in the incident beam. Although the data for Ne^{2+}-He are not so dependent on the electron impact energy, those for Kr^{2+}-He increase entirely with the electron energy as shown in Fig.3.

We have succeeded to determine separatly the fractions and the partial cross sections in Kr^{2+}-He collision in an attempt to analyze the dependence of the Kr^{+}formation upon the target pressure. It makes quantitatively clear that the electron capture taking place in Kr^{2+}-He results mainly from the least component of $Kr^{2+}(^1S_0)$ in the incident beam. We will present additional details and discuss the mechanisms of electron capture processes.

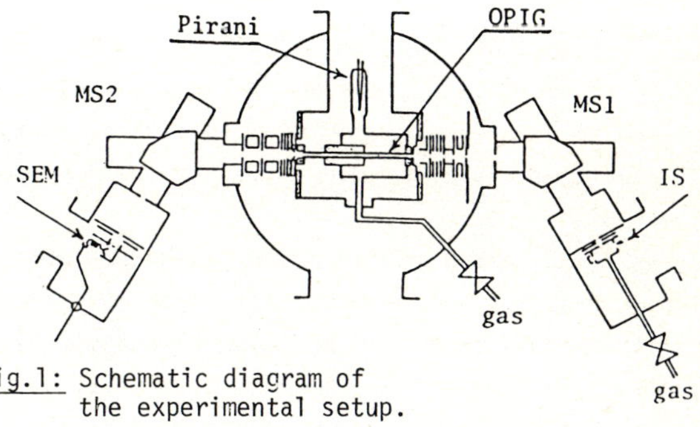

Fig.1: Schematic diagram of the experimental setup.

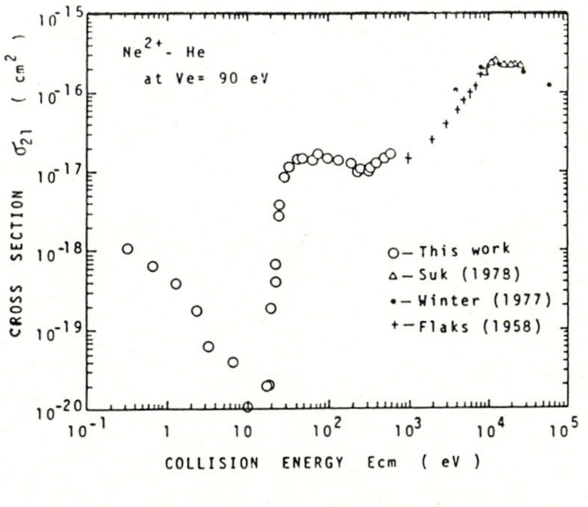

Fig.2: Electron capture cross sections in Ne^{2+}-He as a function of collision energy.

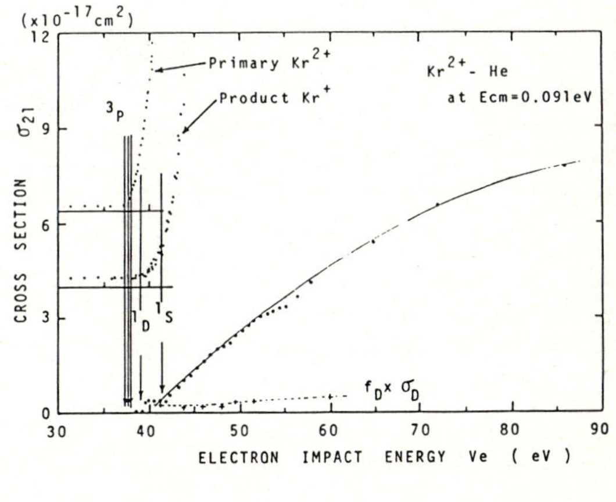

Fig.3: Electron energy dependence of electron capture cross sections in Kr^{2+}- He collision.

Reference

1. E.Teloy and D.Gerlich, Chem. Phys. 4, 417 (1974)
2. K.Okuno and Y.Kaneko, abs. XII ICPEAC, p.686 (1981)
3. H.C.Suk, A.Guilbaud and B.Hird, J. Phys. B11, 1453 (1978)
4. H.Winter, E.Bloeman and F.J.DeHeer, J. Phys. B10, L453 (1977); ibid., J. Phys. B10, 1599 (1977)
5. I.P.Flaks and E.S.Solov'ev, Sov. Phys. Tech. Phys. 3, 577 (1958)

ION ENERGY-LOSS SPECTROSCOPY OF ONE ELECTRON CAPTURE PROCESSES IN THE SYSTEM Kr^{2+} - Ne

T. Nakamura, N. Kobayashi and Y. Kaneko

Department of Physics, Tokyo Metropolitan University, Tokyo, Japan

One electron capture process

$$Kr^{2+}(^3P_{2,1,0}, {}^1D_2, {}^1S_0) + Ne \rightarrow Kr^+(^2P_{3/2,1/2}) + Ne^+(^2P_{3/2,1/2}) + \Delta E$$

was studied with an ion energy-loss spectrometer in the energy range from 500 eV to 1500 eV and at a scattering angle near 0°.

As Kr^{2+} ions are produced with an electron-impact type ion source, the ion beam extracted from the source contains the ground state ^3P_J(J=2,1 and 0) and low-lying metastable states $^1D_2$ and $^1S_0$. Translational energy of the product Kr^+ ion was measured with a high resolution energy analyzer.

A typical spectrum obtained for 1334 eV Kr^{2+} incident is shown in Fig.1 as a function of exothermicity ΔE. The spectrum was obtained by setting the energy analyzer, the acceptance angle of which is ±0.45°, at 0°. In this spectrum, peaks due to different reaction channels are separated well. The reaction channels identified are given in Table 1.

The spectrum may be devided into three regions. The weak structure observed between the exothermicity of 6 eV and 7.5 eV are due to the reactions starting from the initial states $Kr^{2+}(^1S_0)$. The most prominent peaks located between 4 eV and 5 eV are attributed to the processes from $Kr^{2+}(^1D_2)$. Three peaks observed below 4 eV are caused by the processes starting from $Kr^{2+}(^3P_j)$.

In the separate experiment, we have succeeded in determining the populations of all of the low-lying states in the Kr^{2+} ion beam by observing excitation and de-excitation among these states, and the result is presented at this conference.

The relative cross sections of the different reaction channels are, then, possible to be evaluated from the intensities of the peaks observed in the spectrum. The zero order potentials, taking into account of the polarization force and Coulomb repulsion for the initial and final states only, suggest that potential crossings exist at intermediate nuclear distances. Therefore, these one-electron capture processes are considered to arise from the potential crossings. The relative cross sections of different reaction channels are presented in Fig.2 as a function of the crossing radius Rx evaluated from the zero order potentials.

This result indicates that a cross section is large when the potential crossing exists in the nuclear distance range between 2.5 Å and 6 Å. The result of Landau-Zener calculation normalized to unity at the cross section maximum is shown by dashed curve in Fig.2. The shape of this curve reproduces well the general trend of the experimental results.

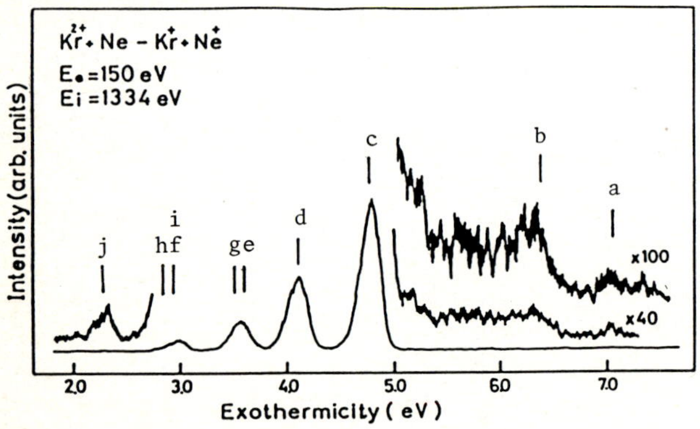

Fig.1 A typical spectrum obtained.

Kr^{2+}	$Kr^+(^2P_j)$	$Ne^+(^2P_j)$	Peak
$^1S_0$	3/2	3/2,1/2	a
	1/2	3/2,1/2	b
$^1D_2$	3/2	3/2,1/2	c
	1/2	3/2,1/2	d
$^3P_0$	3/2	3/2,1/2	e
	1/2	3/2,1/2	f
$^3P_1$	3/2	3/2,1/2	g
	1/2	3/2,1/2	h
$^3P_2$	3/2	3/2,1/2	i
	1/2	3/2,1/2	j

Table 1 Reaction channels in the Kr^{2+}- Ne system.

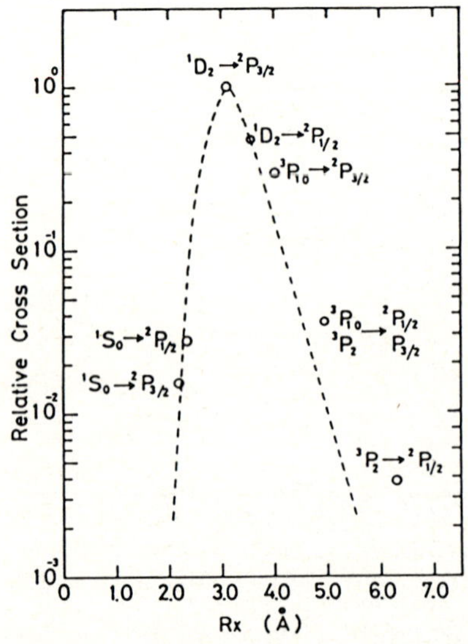

Fig.2 The relative cross sections as a function of the crossing radius Rx.

ION ENERGY-LOSS SPECTROSCOPY FOR TRANSITIONS AMONG LOW -LYING STATES OF Kr^{2+} IN THE COLLISIONS WITH He AND Ne

N. Kobayashi, T. Nakamura and Y. Kaneko

Department of Physics, Tokyo Metropolitan University, Tokyo, Japan

Excitation and de-excitation among low-lying states of Kr^{2+}, $^3P_2$, $^3P_1$, $^3P_0$, $^1D_2$ and $^1S_0$, in the collisions with He and Ne were studied, using a high resolution ion energy-loss spectrometer,[1)] in the energy range from 0.5 keV to 2.7 keV.

The ground state of Kr^{2+} is the triplet state, ^3P_J (J=2, 1 and 0), and the ions have also low-lying meta-stable states, $^1D_2$ and $^1S_0$. The ions were produced with an electron-impact type ion source. All of these states may be, then, contained in the ion beam extracted from the ion source. Ion energy-loss spectra of scattered Kr^{2+} ions from He and Ne were measured and the transitions among these states were directly observed.

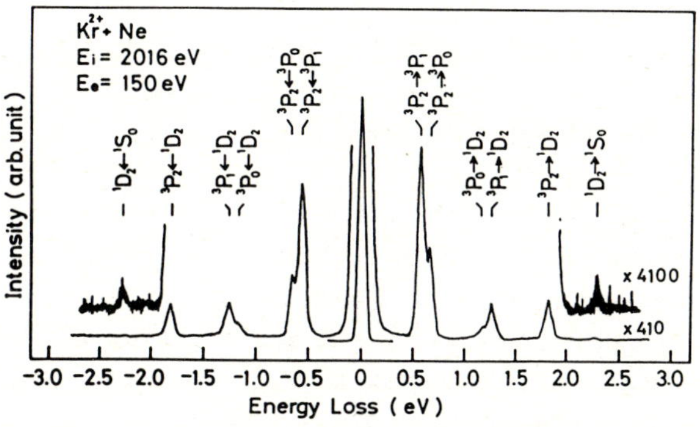

Fig.1 A typical energy-loss spectrum

A typical energy-loss spectrum for 2016 eV Kr^{2+} incident on Ne is shown in Fig.1. The spectrum was obtained by setting the energy analyzer, acceptance angle of which is $\pm$ 0.45°, at 0°. In the case of He target, almost the same spectrum was obtained except that no peak due to the transitions $^1D_2 \leftrightarrow {}^1S_0$ could be found.

According to the principle of the detailed balance between excitation and de-excitation, fractional ratio of the populations of state i and k contained in the primary beam is given by

$$F(i)/F(k) = (2j_i + 1)/(2J_k + 1) \cdot I(i \rightarrow k)/I(i \leftarrow k),$$

where F(i) is the fractional population of the state i, J_i is the total angular momentum of the state i and $I(i \rightarrow k)$ and $I(i \leftarrow k)$ are the intensities of the peaks corresponding to the excitation and reverse transition from the state i to k, respectively.

Assuming that the population of highly excited long -lived metastable states is negligibly small and only five low-lying states are contained in the ion beam, we were possible to determine the fractional populations of all of these states. The result is shown in Fig.2.

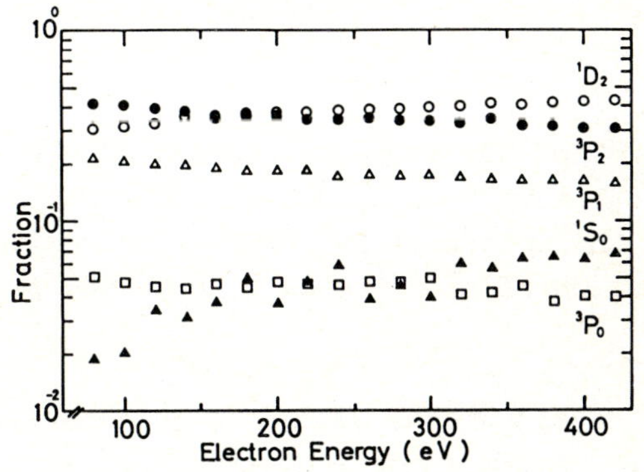

Fig.2 Fractional populations obtained.

If the production of these states occur in accordance with a simple statistical model, the fractional populations must be 0.333 for $^3P_2$ and $^1D_2$, 0.200 for $^3P_1$ and 0.066 for $^3P_0$ and $^1S_0$ and independent on the electron energy in the source. The measured fractional populations are different from the statistical values and they depend on the electron energy.

Since we succeeded in determinig the populations of the states, cross sections for the transitions are possible to be evaluated from the intensities of the peaks observed in the spectrum. Measured cross sections of the excitation in the system Kr^{2+} — He are shown in Fig. 3. The cross sections obtained are considered to

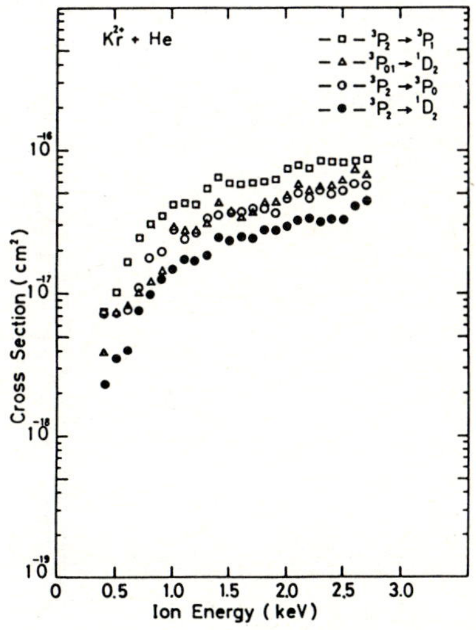

Fig.3 Cross sections for the excitation. The de-excitation cross section from state i to k can be obtained by multiplying the factor $(2J_i + 1)/(2J_k + 1)$ to the cross sections shown in the figure.

be partial ones scattered within the acceptance angle of the energy analyzer. However, in the case of He target, the measured cross sections are expected to be very close to the integral ones, because the ions may be scattered into strongly forward direction.

1) N. Kobayashi, Y. Itoh and Y. Kaneko: J. Phys. Soc. Jpn. 45 617 (1978).

POPULATION OF ELECTRONIC STATES OF MULTIPLY CHARGED Ar IONS FORMED IN ELECTRON CAPTURE FROM HYDROGEN ATOMS

V.V.Afrosimov,A.A.Basalaev,K.O.Lozhkin,M.N.Panov

A.F.Ioffe Physical-technical Institute of the Academy of Sciences of the USSR,Leningrad,USSR

Interaction of multiply charged ions with hydrogen atoms has been the subject of many theoretical investigations[I-3].However the results of the calculations[I-3] of cross sections for formation of $A^{+(Z-I)}(nl)$ ions in the process

$$A^{+Z} + H \rightarrow A^{+(Z-I)}(nl) + H^{+} \quad (I)$$

strongly depend on the method of calculation. Therefore for future development of physical models of charge transfer it is important to have experimental data concerning total capture cross sections and population of various electronic states of multiply charged ions formed in the collisions with H atoms.The energies of populated levels are tens or hundreds eV,so it is difficult to use optical methods for the experimental research.Only the case of formation of metastable $He^{+}(2s)$ ions in He^{+2}+H collisions[4-5] has been experimentally studied.

In this work we used the method of collisional spectroscopy[6] for the measurement of population of electronic states of $Ar^{+(Z-I)}$ ions formed in the electron capture.The scheme of the experimental device is shown in Fig.I. The ion source I with electron impact ionization produces Ar^{+Z} (Z=3-8) ions at E=0,5·Z - 50·Z keV energies.The primary ion beam is monokinetic within 0,5·Z eV.Electrostatic analyzer A has total energy resolution E/ΔE= 4000.For formation of atomic hydrogen target we used the process of thermal dissociation of molecular hydrogen in the tungsten chamber (Fig.2).The central part of the chamber was heated by a diffused electron beam up to the temperature T=2500°K.The degree of dissociation of H_2 was measured using the process of double electron capture by Ar ions from H_2 molecules.The population of various electronic states of $Ar^{+(Z-I)}$ ions were defined from the measured spectra of kinetic energies of fast $Ar^{+(Z-I)}$ ions.

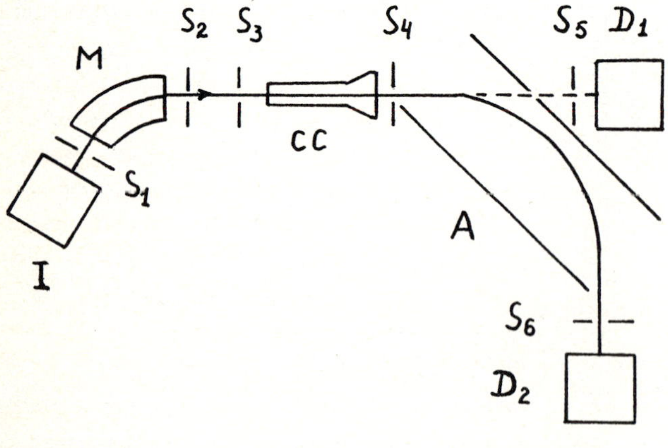

Fig.I:The scheme of the experimental device. I-ion source,M-magnetic monochromator,S_I-S_6-slits,CC-collision chamber,A-electrostatic analyzer,D_I-D_2-detectors.

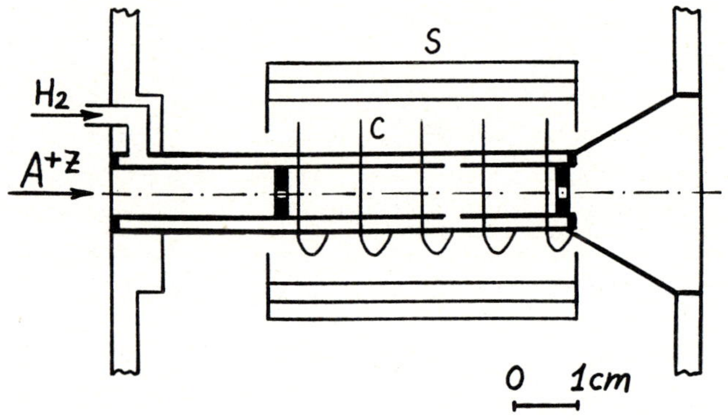

Fig.2:The scheme of the collision chamber for formation of atomic hydrogen target.C-cathodes,S-radiation screens.

References

I.L.P.Presnyakov,A.D.Ulantsev,Kvant.Electronika I,2377 (I974).
2.C.Harel,A.Salin,J.Phys.BIO,35II (I977).
3.V.A.Abramov,F.F.Baryshnikov,V.S.Lisitsa, JETP Lett.27,464 (I978).
4.J.E.Bayfield,G.A.Khayrallah,Phys.Rev.AI2, 869 (I975).
5.M.B.Shah,H.B.Gilbody,J.Phys.BII,I2I (I978).
6.V.V.Afrosimov,G.A.Leiko,M.N.Panov,IX ICPEAC Abstracts of Papers,Seatle (I975),p.I83.

STATE-SELECTIVE ELECTRON CAPTURE BY C^{2+}, C^{3+}, N^{2+} AND Ar^{2+} IN RARE GASES

M. Lennon, R.W. McCullough and H.B. Gilbody

Department of Pure and Applied Physics
The Queen's University of Belfast
Belfast, United Kingdom

At low velocities V < 1 a.u. it is well known that electron capture in collisions of the type

$$X^{q+} + Y \rightarrow X^{(q-1)+} + Y^{+}$$

may take place very effectively in reactions of moderate exothermicity through pseudo-crossings of the adiabatic potential energy curves. These occur at internuclear separations (neglecting polarisation) $R_c \simeq (q-1)/\Delta E$ a.u. where ΔE is the energy defect for the reaction channel.

We have used energy gain/loss spectroscopy to study electron capture by 1.6-6 keV C^{2+} in He, Ne and Ar, by 0.8-8 keV N^{2+} in He and Ne, by 0.14-5 keV Ar^{2+} in He and Ne and by 3-18 keV C^{3+} in He. The relative yields of collision product $X^{(q-1)+}$ corresponding to reaction channels characterised by particular values of ΔE are determined.

For convenient identification of the observed collision channels we adopt the notation of Kamber et al[1] in which ground and metastable states of the incident ion are designated by I, II, III etc; ground or higher states of the $X^{(q-1)+}$ product are designated as α, β, γ, etc.; ground or higher states of the Y^{+} target product ion are designated by X, A, B etc.

In all cases, moderately exothermic processes involving curve crossings at internuclear separations R_c in the range 4-6 a.u. are found to be dominant and all occur in accord with the Wigner total electron spin conservation rule. In the reactions involving doubly charged ions, metastable ions in the primary beam often provide the dominant collision channels.

For C^{2+} in He, Ne and Ar the dominant collision channels are IIαX, IIβX and IIγX respectively. For N^{2+} in He and Ne, the dominant channels are IαX and IβX respectively and for Ar^{2+} in He and Ne the dominant channels are IIαX and IαX respectively. In cases where comparisons are possible, good agreement is obtained with some previous workers[2], but unexplained discrepancies occur with others[3,4].

The energy change spectrum for 15 keV C^{3+} in He which exhibits the five exothermic channels IαX, IβX, IγX, IδX and IϵX corresponding to curve crossings at R_c = 2.3, 3.2, 5.1, 8.7 and 10.5 a.u. and leading to C^{2+} ions in the ^1S, $^3P^0$, $^1P^0$, ^3P and ^1D states respectively. Cross sections for each channel in the range 3-18 keV were obtained (Fig 1) by reference to known[5,6] total electron capture cross sections σ_{tot}.

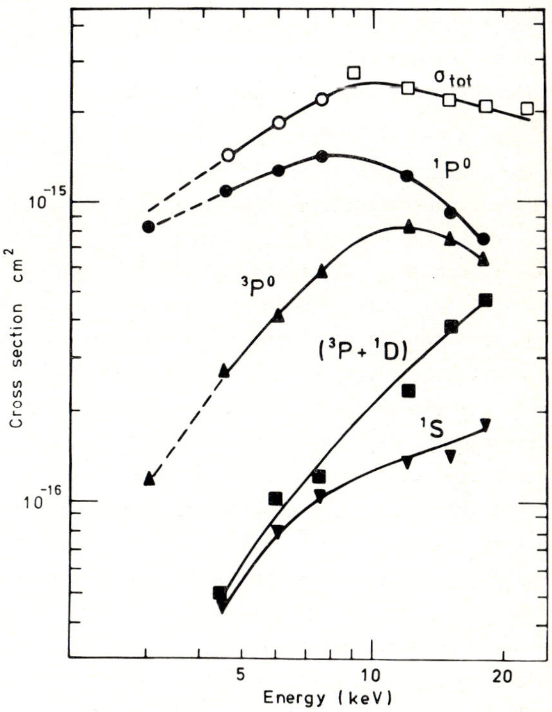

Fig 1 : Cross sections for capture into particular states of C^{2+} in C^{3+}-He collisions. Also shown are total electron capture cross sections σ_{tot}; (o)[5], (□)[6].

References

1. E Y Kamber, D Mathur and J B Hasted, J Phys B 15, 263 (1982).
2. Y Sato and J H Moore. Phys Rev A19, 495 (1979).
3. Y Y Makhdis, K Birkinshaw and J B Hasted. J Phys B 9, 111 (1976).
4. S Sharma, G L Awad, J B Hasted and D Mathur. J Phys B 12, L163 (1979).
5. T Iwai, Y Kaneko, M Kimura, N Kobayashi, S Ohtani, K Okuno, S Takagi, H Tawara and S Tsurubuchi. Phys Rev A 26, 105 (1982).
6. J Geddes and F B Yousif - private communication.

STATE-SELECTIVE ELECTRON CAPTURE BY SLOW MULTIPLY CHARGED IONS IN ATOMIC HYDROGEN

R.W. McCullough, M. Lennon, F.G. Wilkie and H.B. Gilbody

Department of Pure and Applied Physics
The Queen's University of Belfast
Belfast, United Kingdom

In the first measurements of their type we demonstrate the feasibility of the use of energy gain/loss spectroscopy in conjunction with a tungsten tube furnace for studies of the electron capture process

$$X^{q+} + H \rightarrow X^{(q-1)+} + H^+$$

leading to product ions in specified excited states. Most previous experimental studies of this process have been limited to the measurement of total cross sections for capture into all final states of $X^{(q-1)+}$. At velocities V < 1 a.u. where a molecular description of the collision is appropriate, it is well known that such collisions may occur very effectively in moderately exothermic reactions through pseudo-crossings of the adiabatic potential energy cures.

In our apparatus, ions are decelerated to $\sim$ 50 eV and then passed through two hemispherical analysers to provide a beam with an energy half-width $\sim$ 0.8 eV. After acceleration to the required energy and passage through a tungsten tube furnace containing highly dissociated hydrogen, the forward scattered $X^{(q-1)+}$ product ions formed in single collisions then undergo energy analysis in another hemispherical analyser and are counted with a particle multiplier. The voltage applied to this analyser is scanned to provide an energy spectrum (displayed on an MCA) which allows determination of the relative yields of $X^{(q-1)+}$ product ions in particular states corresponding to collision channels characterised by the different energy defects ΔE.

Fig 1 shows the energy change spectrum observed for the formation of N^+ ions from 8 keV N^{2+} ions in passage through (a) molecular hydrogen with the furnace at 300 deg K, and (b) through highly dissociated hydrogen with the furnace at 2470 deg K. The bands AB, CD and EF in the spectrum obtained in H_2 can be correlated with the formation of N^+ ions in the $^3D^0$, ^1S and ^1D states respectively, while H_2^+ ions are formed in vibrational states of $^2\Sigma_g^+$ ranging from v = 0 to the dissociation limit.

The spectrum in Fig 1b indicates the positions of the collision channels A, B, C, D and E corresponding to the formation of N^+ in $(2s^22p3s)^3P^0$, $(2s^22p3p)^1P^0$, $(2s2p^3)^1D^0$, $(2s2p^3)^3P^0$ and $(2s2p^3)^3D^0$ states. Channels A, B and C are endothermic while channels D and E are exothermic with curve crossings at internuclear separations of 11.1 a.u. and 6.0 a.u. respectively. After making allowance for the small undissociated hydrogen background in (b) we estimate that channels E and D account for (47 ± 5)% and (24 ± 5)% respectively of the total capture cross section. Using the known value[1] of the latter at 8 keV, we estimate that the cross sections for capture into the $^3D^0$ and $^3P^0$ states of N^+ are about 2.3×10^{-16} cm^2 and 1.2×10^{-16} cm^2 respectively.

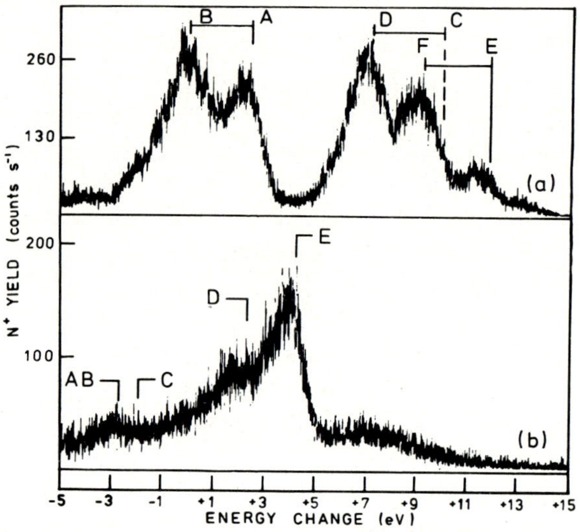

Fig 1 : Energy change spectrum for the formation of N^+ ions in one-electron capture by 8 keV N^{2+} ions in passage through hydrogen in a tungsten tube furnace at a temperature of (a) 300 deg K and (b) 2470 deg K.

For electron capture by C^{3+} in H, the energy change spectrum exhibits channels corresponding to the formation of C^{2+} in the ^1D, ^1S, ^3S and $^3P^0$ states with energy defects of 16.2, 11.6, 4.7 and 2.1 eV respectively. The results are considered in terms of known total cross sections and recent theoretical predictions for selective capture[2].

For electron capture by C^{2+} in H, the dominant collision channel is found to involve metastable $^3P^0$ ions in the primary beam, a result which indicates that our previously measured total cross sections[3] require a new interpretation.

References

1. W Seim, A Müller, I Wirkner-Bott and E Salzborn. J Phys B, 14, 3475 (1981).
2. S Bienstock, T G Heil, C Bottcher and A Dalgarno - private communication.
3. W L Nutt, R W McCullough and H B Gilbody, J Phys B 11, L181 (1978).

LYMAN SPECTRA OF O^{7+} AND N^{6+} PRODUCED BY LOW ENERGY CHARGE EXCHANGE COLLISION ON H_2

S. Bliman*, M. Bonnefoy+, J.J. Bonnet+, S. Dousson*, A. Fleury+, D. Hitz* and B. Jacquot*

* DRF-CENG 85 X 38041 Grenoble Cedex - France

+ C.N.A.M. Laboratoire de Physique des Collisions Atomiques 292 rue Saint-Martin 75141 Paris Cedex 03

Observation of electron capture into selective state by fully stripped ions have been made recently by ion beam spectroscopy (1). Such measurements indicate that, for $O^{8+} + H_2$, at low collision and velocities, an electron is captured selectively into a level with a particular quantum number n = 4.

In this work we describe a X-ray decay study of hydrogen like ions following charge exchange collisions of 80 KeV O^{8+} ions with H_2. Such processes are very interesting for hot plasma physics. The high intensity (10^{11} s^{-1}) ion beam used here is extracted from an ECR source called MINIMAFIOS (2). A molecular hydrogen gas pressure of 8.10^{-5} mbar in the target chamber ensures single collision condition. A high efficiency Bragg spectrometer supplied with lead stearate crystal and gas flow proportional counter is used to detect the X-ray photon emitted following capture. The spectrometer viewing axis is perpendicular to the incident beam direction

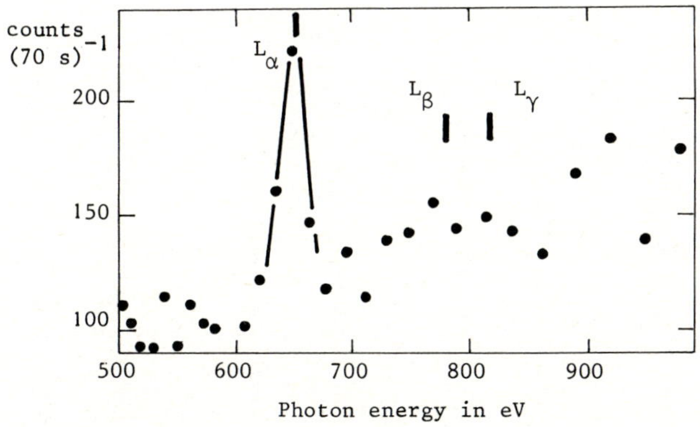

Fig. 1: Spectrum in the single electron capture collision $O^{8+}(80\ KeV) + H_2 \rightarrow O^{7+}(n,l) + H_2$.
$\hookrightarrow h\nu$

A typical spectrum is shown in Figure 1. Although these preliminary experimental data are fraught with rather large background due to the long counting time. It can be seen on Figure 1 that electron is selectively cascading by emitting Lyman α lines.

X-ray spectra with lower background are currently in progress in our laboratory. Thus the relative intensity measurement in such spectrum should give many valuable informations on the population of the states and an interpretation of this observation in term of cascading theories (3,4) is very interesting.

The measured emission rate of a spectral line can be related to the cross section for excitation of the line at a collision energy. The total cross section for charge exchange excitation of a given spectral line can be derived from the individual charge-transfer cross sections for populating specific n and l states of the hydrogen like oxygen after taking account of all the branching and cascading in the decay process (5).

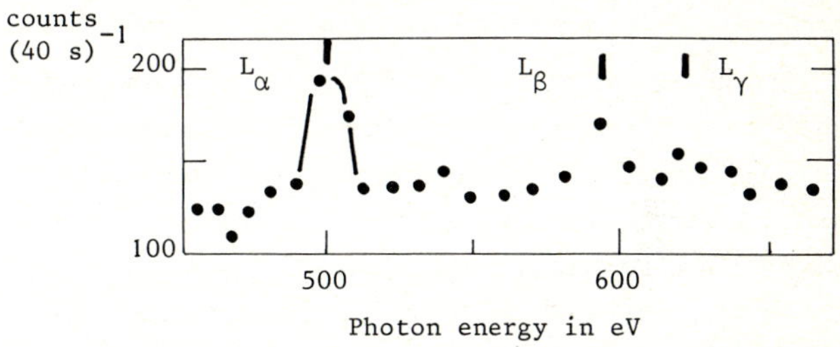

Fig. 2: Lyman spectrum in the single electron capture collision $N^{7+}(70\ KeV) + H_2 \rightarrow N^{6+}(n,l) + H_2^+$
$\hookrightarrow h\nu$

In contrast with $O^{8+} + H_2$ collision, preliminary results concerning Lyman emission in $N^{7+}(70\ KeV) + H_2 \rightarrow N^{6+} + H_2^+$ collision (see Figure 2) indicate that many different l-states are populated in this case. Much more precise experimental findings should be ready for presentation at the time of the conference.

References

1. S. Ohtami and al, J. Phys. B 15 L 533 (1982)
2. R. Geller and B. Jacquot, Nucl. Instr. Meth. 194,293 (1981)
3. R.K. Janev and D.S. Belic, preprint and private communication, unpublished (1982)
4. A.J. Salop, J. Phys. B12, 919 (1979)
5. R.C. Isler and al Phys. Rev. A, 24,5 2701 (1981)

ELECTRON CAPTURE INTO DIFFERENT (n,ℓ)-STATES IN SLOW C^{6+},N^{6+},O^{6+},Ne^{6+}-He,H_2 COLLISIONS

Yu.S. Gordeev†, D. Dijkkamp, A.G. Drentje‡, F.J. de Heer

FOM-Institute for Atomic and Molecular Physics, Kruislaan 407, 1098 SJ Amsterdam, The Netherlands.

† On leave from: Ioffe Physical Technical Institute, Leningrad, USSR
‡ KVI, University of Groningen, The Netherlands

We have measured absolute cross sections for electron capture into different excited (n,ℓ)-states in the following processes:

$$C^{6+},N^{6+},O^{6+},Ne^{6+} + He \to A^{5+*}(n,\ell) + He^+$$

$$N^{6+}, O^{6+} + H_2 \to A^{5+*}(n,\ell) + H_2^+$$

The relative collision velocity was varied between 0.15 and 0.5 a.u.

The choice of different projectiles with the same charge and of targets with different ionization potentials enabled us to study the influence of the ionic core on electron capture into different states and to look for some systematics in the (n,ℓ)-distribution.

The ion beams were produced by an Electron Cyclotron Resonance source of the MINIMAFIOS-type [1], at the KVI, University of Groningen.

The ions traversed a static gas target, under single collision conditions ($p < 0.7$ mbar). VUV-photon emission (3 nm $< \lambda <$ 60 nm), resulting from the decay of the $A^{5+*}(n,\ell)$-states was observed perpendicular to the ion beam by a grazing incidence spectrometer, absolutely calibrated on sensitivity and equipped with a position sensitive channelplate detector. Experimental set-up and calibartion-procedure have been described elsewhere [2]. From the measurements we deduced absolute emission cross sections for transitions between 2ℓ, 3ℓ, 4ℓ and 5ℓ levels of the A^{5+*}-ions, with an absolute error of about 30% and a relative error of 10-25%. From the emissions we calculated absolute $\sigma_{n,\ell}$ cross sections, taking into account cascading and using known branching ratios. Further we obtained $\sigma_n = \sum_\ell \sigma_{n\ell}$ and σ_t, the total capture cross section. As an example we show the results for the O^{6+}-He,H_2 systems in figs. 1,2. Also indicated are published total capture cross sections, measured by charge state analysis [3,4,5]. The agreement with our results is quite good. Baptist et al.[6] measured the relative intensities of the 4ℓ-3ℓ' transitions in O^{6+}-H_2 collisions at 0.4 a.u. velocity. They agree well with our measurements at this energy.

The capture process is dominated by capture into n=3 and n=4 states for He and H_2 respectively, which is in accordance with a simple classical model. σ_n and σ_t do not vary much in the studied velocity range contrary to $\sigma_{n\ell}$, resulting in strong redistribution of ℓ-sublevel population. This effect cannot be understood from presently available calculations for bare nuclei-atomic hydrogen collisions. Our measurements show large similarity in the (n,ℓ)-distribution for the C^{6+},N^{6+},O^{6+}-He systems. Only the Ne^{6+}-He system behaves rather differently. This can be understood qualitatively, since the energy levels of Ne VI are deviating much more strongly from the hydrogen-like C VI energies than those of N VI and O VI.

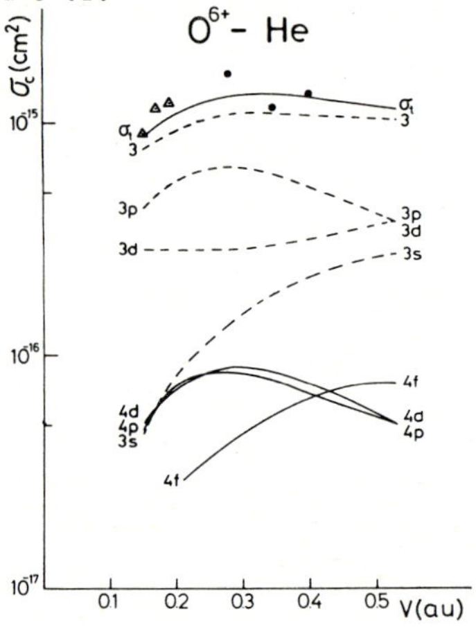

Fig. 1: $\sigma_{n\ell}$, $\sigma_n (= \sum_\ell \sigma_{n\ell})$, $\sigma_t (= \sum_{n,\ell} \sigma_{n\ell})$ as a function of velocity for O^{6+}-He collisions. o - σ_t from ref. 3, Δ - σ_t from ref. 4.

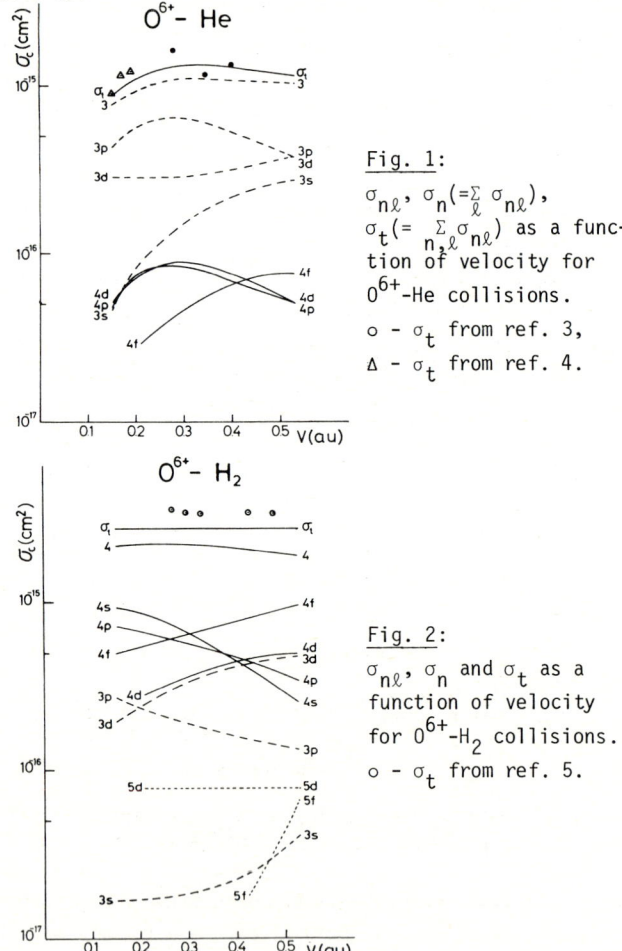

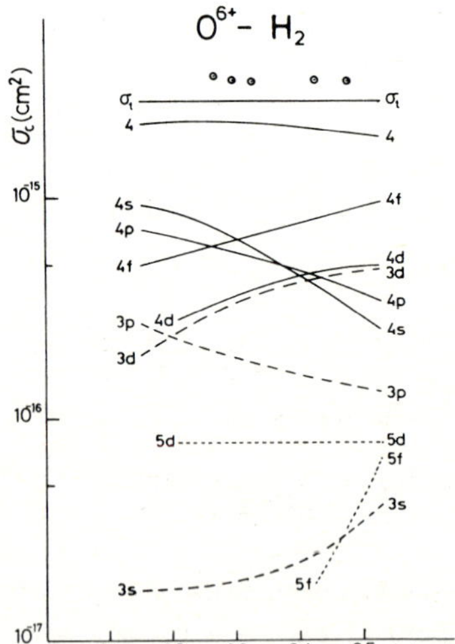

Fig. 2: $\sigma_{n\ell}$, σ_n and σ_t as a function of velocity for O^{6+}-H_2 collisions. o - σ_t from ref. 5.

References

1. R. Geller and B. Jacquot, Nucl.Instr.Meth.Phys.Res. 202, 399 (1982).
2. K. Kadota et al., J.Phys.B: At.Mol.Phys. 15, 3275 (1982).
3. D.H. Crandall, Phys.Rev. A 16, 958 (1977).
4. I. Iwai et al., Phys.Rev. A 26, 105 (1982).
5. D.H. Crandall, R.A. Phaneuf and F.W. Meyer, Phys.Rev. A 19, 504 (1979).
6. R. Baptist et al., Physics Letters 93A, 185 (1983).

STATE-SELECTIVE ELECTRON CAPTURE CROSS SECTIONS FOR IMPACT OF C^{q+} (q = 2, 3, 4) AND O^{q+} (q = 2, 3, 6) ON Li

D. Dijkkamp, R.L. van der Woude and F.J. de Heer
FOM Institute for Atomic and Molecular Physics, Amsterdam, The Netherlands

A.G. Drentje
KVI, University of Groningen, The Netherlands

A. Brazuk and H. Winter
Institut für Allgemeine Physik, Technische Universität Wien, Austria

Single electron capture into selected states has been studied by means of photon emission spectroscopy for collisions of multiply charged C and O ions with Li atoms. The doubly and triply charged ions have been produced at the AMI accelerator of the FOM Institute with acceleration voltages between 15 and 150 kV. C^{4+} and O^{6+} ions have been obtained from a MINIMAFIOS-type ion source[1,2] at KVI, University of Groningen, by applying an acceleration voltage between 16 and 20 kV. For collisions with a Li atom beam target vuv photon emission was detected at wavelengths between 10 and 160 nm by means of two intensity-calibrated grating spectrometers with channel-plate detectors. As an example, in fig. 1 all emission lines observed for the C^{4+}-Li collision system are shown. Intensity calibration, determination of Li target thickness and emission cross section measurements are described in ref. 3. Primary ion beams of C^{2+}, O^{2+} and O^{3+} contained important metastable fractions. Both for C^{4+} and O^{6+} possibly present metastable fractions could be neglected.

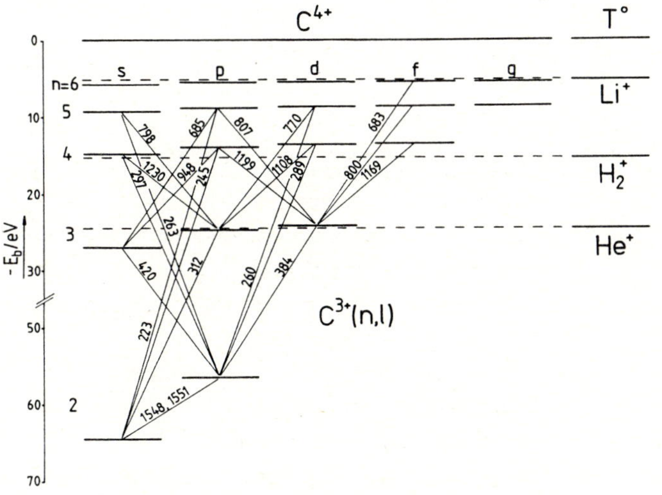

Fig. 1: Grotrian diagram of C IV states excited via electron capture in C^{4+}-Li collisions. Also shown are reaction energy defects and resulting emission lines.

The largest emission cross sections exceeded 10^{-15} cm^2 for all collision systems studied. The obtained systematics agree with the classical model for electron capture into highly charged ions[4] which predicts inverse proportionality between the square of the target ionisation potential and the total electron capture cross section. With higher primary ion charge state q the principal quantum number of dominantly populated shells increases. The electron capture takes place at selected potential energy curve crossings with internuclear distances R_c which from our measurements appear to be most important between 10 and 15 a_o.
For impact of C^{2+} and O^{2+} capture into n = 3 states was found to be of importance as already shown for Ne^{2+}- Li collisions[5]. For C^{3+} and O^{3+} capture into n = 4 states becomes probable, too. For C^{4+} the n = 4 and n = 5 states are most important and the total electron capture cross section was determined with 2 x 10^{-14} cm^2. For O^{6+} this cross section was 2.6 x 10^{-14} cm^2 and capture takes place mainly into n = 5, n = 6 and n = 7 states.

No line emission corresponding to double electron capture could be detected. Therefore, electron capture from Li by slow multiply charged ions can be well described as transfer of a single "active" electron.
In cases where different primary ion states have been present (i.e. C^{2+}, O^{2+}, O^{3+}), experimental evidence strongly suggests the conservation of the initial ion core state while the electron capture takes place. Therefore, the total spin of the projectile particle can only change by 1/2. If capture into a specific primary ion results in secondary ions with different multiplicity, the obtained emission cross sections follow a statistical law, e.g. for C^{3+}- Li collisions emission from C III triplet states is approximately three times stronger than from singlet states which are populated via potential curve crossings at similar values of R_c. Further studies for impact of C^{4+} and O^{6+} on the substate distribution among dominantly populated shells are in progress.

References

1. R. Geller and B. Jacquot, Nucl.Instr.Meth.Phys.Res. 184,293(1981) and 202,399(1982)
2. A.G. Drentje, Physica Scripta T 3(1983, in press)
3. K. Kadota, D. Dijkkamp, R.L. van der Woude, A. de Boer, Pan Guang Yan and F.J. de Heer, J.Phys.B: At.Mol.Phys. 15,3275(1982)
4. R. Mann, F. Folkmann and H.F. Beyer, J.Phys.B:At.Mol. Phys. 14,1161(1981)
5. E. Rille and H. Winter, J.Phys.B:At.Mol.Phys. 15, 3489(1982)

ELECTRON CAPTURE INTO EXCITED PROJECTILE STATES IN 6-100 keV Ne^{4+}-Ne COLLISIONS

D. Dijkkamp†, V.K. Nikulin‡, Yu.S. Gordeev‡, A.V. Samoilov‡, F.J. de Heer†

† FOM-Institute for Atomic and Molecular Physics, Kruislaan 407, 1098 SJ Amsterdam, The Netherlands.
‡ Joffe Physical Technical Institute, Leningrad, USSR.

We have measured and calculated theoretically cross sections for electron capture into excited states of the projectiles in Ne^{4+}- Ne collisions for the collision velocities $v \leq 1$ a.u.

The experimental set-up was the same as used in ref. 1. Absolute emission cross sections for $20.8 \leq \lambda \leq 29.3$ nm have been measured. This emission results from the decay of the following states of Ne IV ions: $3s\,^4P$, $3s\,^2P$, $3s'\,^2D$, $3p\,^4S^0$ and $3p\,^4P^0$. The states $3p\,^4D^0$ and $3p'\,^2F^0$ cascade to the states $3s\,^4P$ and $3s'\,^2D$ respectively and accordingly also contribute to the measured emission cross sections. Therefore the total emission cross section is equal to the cross sections for electron capture into 3s and 3p states of the projectile. Only capture into the $3p'\,^6P$ state was not measured in the present experiment but previous measurements for the Ne^{4+}- Ne, 100 keV case [2] have shown that the cross section for this process is less than 3% of the total capture cross section.

Our theoretical calculation includes both one-electron and two-electron transitions:

$$Ne^{4+}(2p^2) + Ne(2p^6) \rightarrow Ne^{3+}(2p^2 n\ell) + Ne^+(2p^5)$$

$$\rightarrow Ne^{2+}(2p^2 n\ell^2) + Ne^{2+}(2p^4)$$

with $n\ell$ equal to 3s or 3p.

The quasimolecular two-electron $^1\Sigma g$ states were calculated in a wide internuclear distance range (see fig. 1). The entrance channel was $(4f\sigma)^2$. The core state in each of the cases was the same:

$$^1\Sigma_g \; 1s\sigma_g^2 \; 2s\sigma_g^2 \; 2p\sigma_u^2 \; 2p\pi_u^2 \; 3p\sigma_u^2 \; 3d\sigma_g^2 \; .$$

These quasimolecular states were calculated using the diabatic MO basis [3] in the single configuration approximation [4]. The interaction between diabatic states of the same symmetry was obtained by calculating adiabatic energy splittings in the two configuration approximation [4]. The cross sections for one- and two-electron capture were calculated using the multichannel Landau-Zener model. All crossings were considered as independent and all possible trajectories in the diabatic state diagram (fig. 1) were taken into account. The calculated cross sections for 0.01 - 1 a.u. collision velocities together with our experimental data are shown in fig. 2. The experiment and the calculation agree very well with each other and with previous measurements [2]. The calculated two-electron capture is about three orders of magnitude less probable than one-electron capture in the whole velocity region.

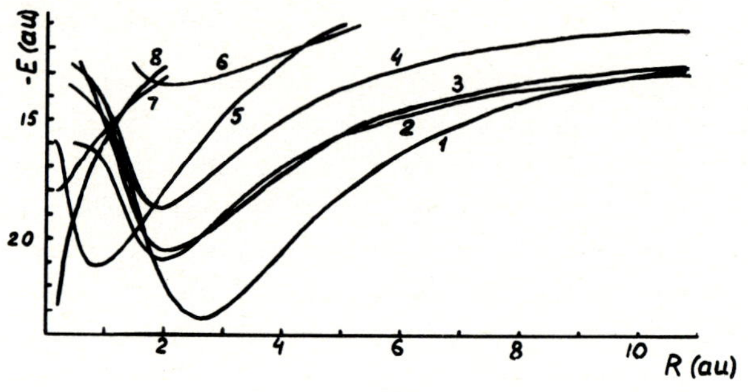

Fig. 1: Energies of $^1\Sigma g$ diabatic states of Ne_2^{4+} quasi-molecule as functions of internuclear distance: 1 - $(4f\sigma_u)^2$, 2 - $(4p\sigma_u\,4f\sigma_u)$, 3 - $(4f\sigma_u\,5f\sigma_u)$, 4 - $(4f\sigma_u\,5p\sigma_u)$, 5 - $(4d\sigma_g)^2$, 6 - $(5f\sigma_u)^2$, 7 - $(4p\sigma_u)^2$, 8 - $(3s\sigma_g)^2$.

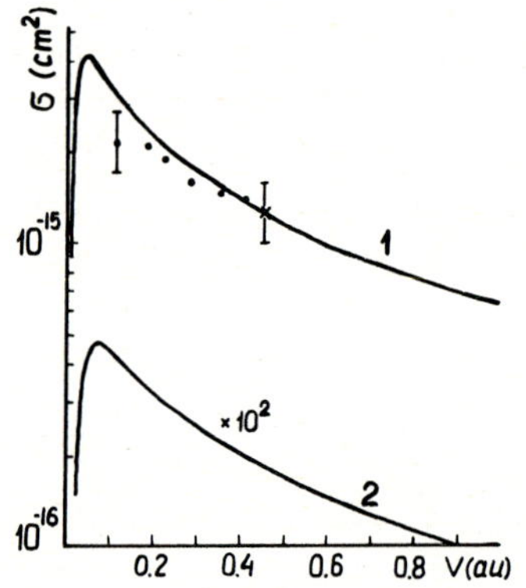

Fig. 2: Cross sections for electron capture as functions of the collision velocity.

1 - One-electron capture into (3s,3p) states. The line: our calculation, the points: our measurements, the cross: measurement [2].

2 - Two-electron capture into $(3s^2, 3p^2)$ states. Our calculation.

References

1. Yu.S. Gordeev, D. Dijkkamp, A.G. Drentje, F.J. de Heer, this book.
2. E.W.P. Bloemen, D. Dijkkamp and F.J. de Heer. J.Phys. B 15 (1982) 1391.
3. V.K. Nikulin and N.A. Guschina. J.Phys.B 11 (1978) 3553.
4. V.K. Nikulin and A.V. Samoilov. Phys.Lett. 98A (1982) 225.

EXTRACTION OF TOTAL CAPTURE PROBABILITIES IN ATOMIC MANY ELECTRON COLLISIONS

H.J. Lüdde and R.M. Dreizler

Institut für Theoretische Physik der Universität Frankfurt/Main, BRD

By solution of the time dependent Hartree Fock (TDHF) problem or approximations thereto one obtains the final wavefunction of an atomic collision system in the form of a Slater determinant $\Psi(x_1 \ldots x_N, \tau\to\infty)$. The probability of finding the electrons in a particular final configuration - described by a Slater determinant Φ - is obtained by projecting Ψ onto Φ. Effects of the exclusion principle are incorporated by the mutual projection of Slater determinants.

If one is, however, interested in global statements for the capture of q electrons an alternative to the projection on a large number of final channels and subsequent summation offers itself.

We first consider a system of two electrons and divide the space $V(x,y,z)$ into the projectile space $P(x,y,z \leq 0)$ and target space $T(x,y,z > 0)$. The normalisation integral of the two particle Slater determinant has the form

$$1 = \int_V d^4x_1 \int_V d^4x_2 \Psi^*\Psi =$$

$$\int_P d^4x_1 \int_P d^4x_2 \Psi^*\Psi + \iint_{PT} \ldots + \iint_{TP} \ldots + \iint_{TT} \ldots \quad .$$

The physical interpretation of the four individual integrals is

$P_{20} = \int_P d^4x_1 \int_P d^4x_2 \Psi^*\Psi$, probability for finding both electrons bound to the projectile

$P_{11} = 2 \int_P d^4x_1 \int_T d^4x_2 \Psi^*\Psi$, probability for finding one electron bound to the projectile and the other to the target

$P_{02} = \int_T d^4x_1 \int_T d^4x_2 \Psi^*\Psi$, probability for finding both electrons bound to the target nucleus.

According to the initial condition- as for instance the system $Q^{\Omega+}+He(^1S)$ one can interpret the quantities P_{20}, P_{11} and P_{02} as total probabilities for two electron capture, one electron capture and target excitation plus elastic scattering. For the particular system one finds with

$$P_{20} = \langle \psi | \psi \rangle_P^2$$

$$P_{11} = 2\,(\sqrt{P_{20}} - P_{20})$$

$$P_{02} = (\,1 - \sqrt{P_{20}}\,)^2 \,,$$

a connection between one and two electron capture, that has been indicated before[1].

Details and extension to N electron systems, as well as the calculation of total ionisation probabilities in the global formulation will be presented.

References

1. W. Lichten, Phys. Rev. 131,229 (1963)

DYNAMICS OF COLLECTIVE CHARGE FLOW IN DIATOMIC COLLISION SYSTEMS

J. Eichler and T. S. Ho+

Hahn-Meitner-Institut Berlin, D-1000 Berlin 39, Germany F.R.

If a large number of electrons participates in an atomic transition or reaction the corresponding many-body theory becomes complicated. It may then be advantageous to introduce collective variables for he most essential degrees of freedom and to treat the motion in these variables macroscopically. Such an approach can be based on the time-dependent Hartree-Fock theory provided that the adiabatic approximation is valid.

In a charge-changing atomic collision, the charge asymmetry (dipole deformation) ζ with respect to the midplane between the nuclei is certainly the simplest and most relevant degree of freedom in addition to the internuclear separation R. The potential energy surface $V(R,\zeta)$ can be obtained[1,2] ab initio from constrained Hartree-Fock calculations. In a similar way, one is also able to derive[3] an approximate expression for the inertia of the charge flow. Finally, if there are many possible ways to intrinsically excite the quasimolecule during the collision via rotational, radial, or charge-flow coupling or to dissipate energy by electron evaporation, the resulting leakage of energy from the collective motion into other degrees of freedom may be described by a friction force.

In order to provide an illustration, the associated equations of motion[3] have been solved for simple model systems disregarding the fact that the many-channel assumption underlying the dissipation picture are not well fulfilled in the cases considered. A variety of trajectories have been calculated for the collision system LiF and HF. Fig.1 shows the potential energy map[2] of the system HF and Fig.2 various trajectories[3] within the same R,ζ frame. The trajectories exhibit elastic and charge-changing collisions as well as trapping into a molecular state.

It is suggested that the mechanism studied here might play a role in slow collisions of multicharged ions with neutral atoms.

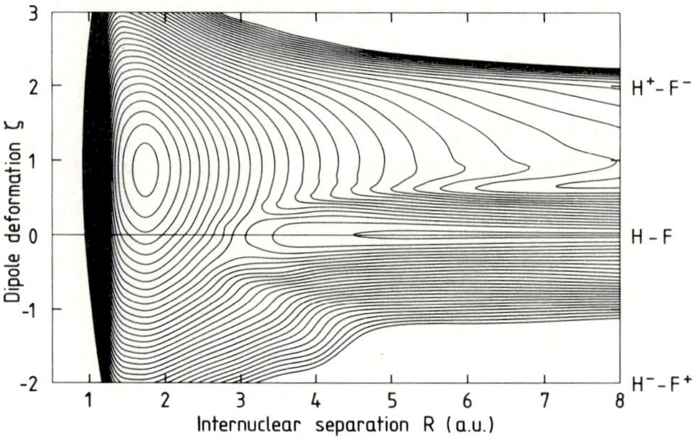

Fig. 1: Contour map of the potential energy surface for HF. Successive lines are separated by $\Delta E=0.02$ a.u.

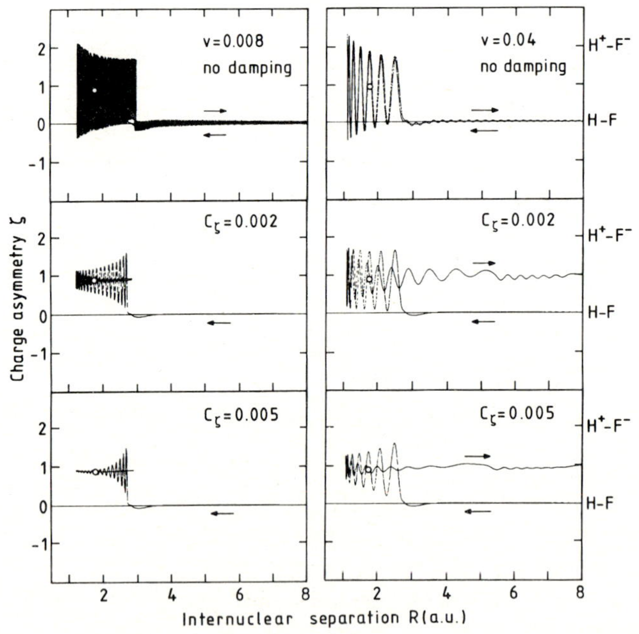

Fig. 2: Trajectories in the R,ζ plane for a H-F collision with impact parameter b=1.0 a.u. In the upper row no damping, otherwise a weak charge flow damping is assumed. The equilibrium point is indicated by an open circle.

References

+ Present address: Department of Chemistry, University of Kansas, Lawrence, Kansas 66045

1. J.K.M. Eichler, Phys. Rev. Lett. 46, 1619 (1981).
2. J. Eichler and T.S. Ho, Z. Physik A, 311, 19 (1983)
3. J. Eichler and T.S. Ho, Z. Physik A, 311, 29 (1983)

CHARGE CAPTURE BY MULTICHARGED IONS: FULLY QUANTAL CALCULATIONS WITH LARGER BASIS SETS

C. Bottcher

Oak Ridge National Laboratory, Oak Ridge, Tennessee 37830 USA*

and

T. G. Heil

Department of Physics, University of Georgia, Athens, Georgia 30602 USA

We have extended our earlier calculations[1] on one-electron and few-electron projectiles incident on H(1s) by including π as well as σ states. As before we expand in exact bare-nucleus molecular orbitals, describing the nuclear motion in a fully quantal manner. Table 1 quotes results on C^{6+} with $3\sigma + 3\pi$ states, which are about 2X the old 3σ calculation, and now agree well with measurements.[2]

Our 3σ calculations on B^{5+}, C^{5+} have now been extended to O^{5+}, which shows remarkable behavior (Table 1). The cross section has an orbiting form below 10 eV, and between 50-150 eV goes through two strong-coupling oscillations, thereafter falling rapidly. This is in qualitative record with experiment[2] which falls from 70 $Å^2$ at 100 eV to 30 $Å^2$ at 300 eV. However π-states are probably needed for quantitative agreement and this work is in progress.

We are also developing a frame transformation approach to include translation factors in the full quantal formalism. Preliminary results for B^{5+} + H will be presented at the conference.

Table 1

	C^{6+}			O^{5+}						
E eV	32	136	245	1	10	30	50	100	180	500
σ $Å^2$	1.6	7.7	23.7	118	91	62	89	61	85	45

*Operated by Union Carbide Corporation under contract W-7405-eng-26 for the U.S. Department of Energy.

References

1. C. Bottcher and T. G. Heil, Chem. Phys. Letts 86, 506-9 (1981).
2. R. A. Phaneuf, I. Alvarez, F. W. Meyer, and D. H. Crandall, Phys. Rev. A 26, 1892-1905 (1982).

MOLECULAR CALCULATIONS OF THE CROSS SECTION FOR CHARGE TRANSFER. $He^{2+} + H$ IN THE 20 eV TO 10 keV E_{CM} REGION.

Marc C. van Hemert, Ewine F. van Dishoeck and Fumihiro Koike

Department of Physical Chemistry and Sterrewacht, University of Leiden, The Netherlands
FOM-Institute for Atomic and Molecular Physics, Amsterdam, The Netherlands

The $He^{2+} + H(1s) \rightarrow He^{+}(n=2) + H^{+}$ charge transfer process has been studied theoretically by treating HeH^{2+} as a (quasi) molecule. In the collision energy range considered, only the $2s_{\sigma}$, $2p_{\sigma}$, $3d_{\sigma}$ and $2p_{\pi}$ molecular states have to be taken into account. Adiabatic potential energy curves and radial and rotational couplings, calculated[1] by expressing the one electron wavefunctions as variationally determined linear combinations of Slater type orbitals (LCAO's), agree with exact values within the required accuracy. This LCAO approach has the advantage of being directly extendable to many electron systems by application of configuration interaction methods.

Both quantummechanical close coupling (20-500 eV) and semiclassical curvilinear trajectory calculations (0.1-10 keV) were performed to obtain the cross sections. The method of ref. 2 was extended to include the rotational coupling in the quantummechanical treatment.

Electron translation factors (ETF's) were introduced to remedy the problems associated with the origin dependence of the couplings. Here this results only in a correction to the coupling matrix elements, both for the quantummechanical and for the semiclassical treatment.[3] Two simple types of ETF[4,5] were considered and proved to be well suited for LCAO-type calculations.

Fig. 1 shows that the ETF corrected radial coupling between the $2p_{\sigma}$ entrance and the $3d_{\sigma}$ exit channel is similar to the H-atom origin coupling. For the rotational $2p_{\sigma}$-$2p_{\pi}$ coupling no such privileged origin is found.

The calculated total cross section, fig. 2, agrees very well with the experimental values.[6] Furthermore there is a near perfect match between the results from the quantummechanical and semiclassical calculations. In the 1-10 keV E_{CM} region the semiclassical values are only slightly below results obtained with more elaborate ETF's.[7] At lower energies the use of the more appropriate Coulomb trajectory in this work leads to a larger discrepancy. It is also observed that ETF's have an important effect in this system above 0.2 keV. Quantummechanical and semiclassical partial cross sections agree less well, as a consequence of the use of a single potential in trajectory methods. Radial ($\rightarrow 3d_{\sigma}$ and $\rightarrow 2s_{\sigma}$) and rotational ($\rightarrow 2p_{\pi}$) couplings are about equally important. The $2s_{\sigma}$ channel -the only channel measured separately[6]- contributes least.

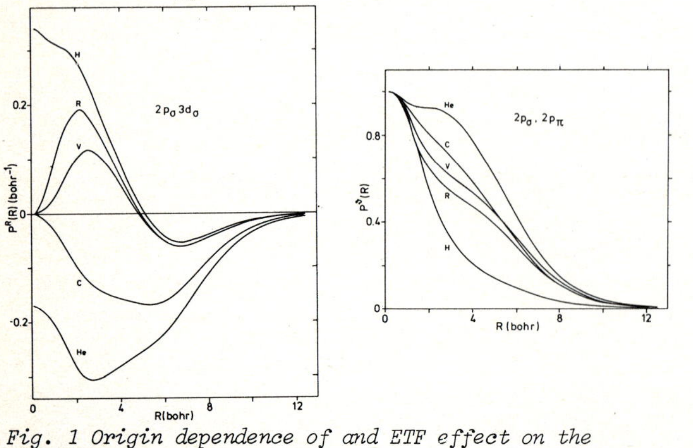

Fig. 1 Origin dependence of and ETF effect on the $2p_{\sigma}$-$3d_{\sigma}$ and $2p_{\sigma}$-$2p_{\pi}$ couplings. H, C and He correspond to H-atom, center of charge and He- atom origin. R and V indicate ETF modified couplings with switching functions according to refs. 4 and 5.

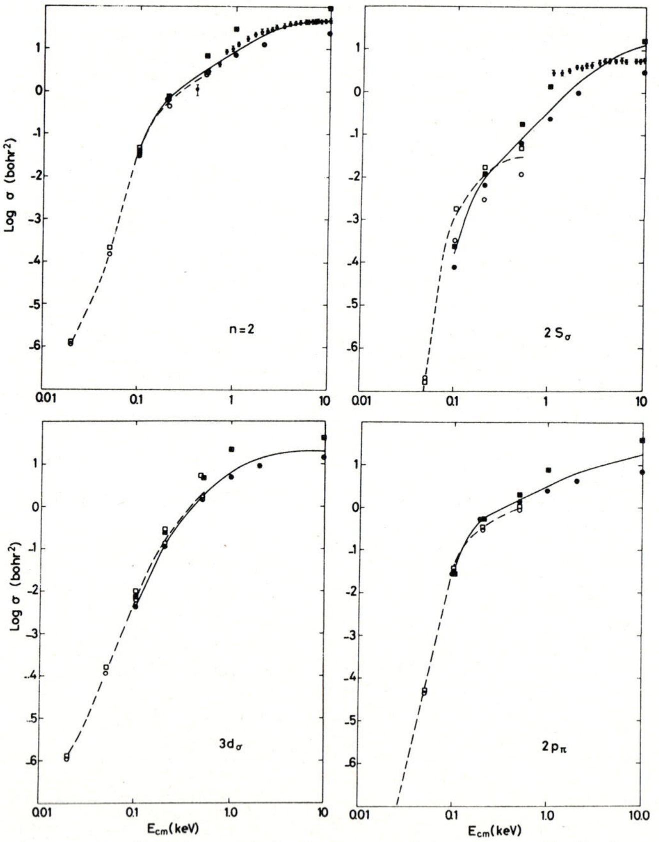

Fig. 2 Total and partial cross sections. Drawn/dashed curve semiclassical/quantummechanical calculation with ETF[5] modified coupling. ■/● semiclassical, □/○ quantummechanical results for the He/H-atom origin couplings. experimental results.[6]

References

1. M.C. van Hemert, E.F. van Dishoeck, J. van der Hart-van der Hoek and F. Koike to be published.
2. T.G. Heil, S.E. Butler and A. Dalgarno, Phys. Rev. A 23, 1100 (1981).
3. J.B. Delos, Rev. Mod. Phys. 53, 287 (1981).
4. L.F. Errea, L. Méndez and A. Riera, J. Phys. B 12, 69 (1979).
5. J. Vaaben and K. Taulbjerg, J. Phys. B 14, 1815 (1981).
6. W.L. Nutt, R.W. McCullough, K. Brady, M.B. Shah and H.B. Gilbody, J. Phys. B 11, 1457 (1978); M.B. Shah and H.B. Gilbody, J. Phys. B 11, 121 (1978).
7. M. Kimura and W.R. Thorson, Phys. Rev. A24, 3019 (1981).

IMPACT PARAMETER DEPENDENCE OF CHARGE EXCHANGE IN THE SCATTERING OF Li^{3+} ON H AT 10.5 KEV (DIRECT INTEGRATION OF THE SCHRÖDINGER EQUATION)

Norbert Grün and Werner Scheid

Institut für Theoretische Physik der Justus-Liebig-Universität Giessen, West Germany

For atomic collisions with impact energies larger than 100 eV/amu the nuclei can be assumed to move on classical trajectories. In that case, the electrons can be treated by the Schrödinger equation with Coulomb potentials depending on time. The well known and established methods for solving this problem are the analytical basis expansions. In the last years numerical methods have been developed to integrate the Schrödinger equation directly in time. The basic ideas of the latter methods are the following: We solve the Schrödinger equation at discrete space points approximating spatial differentials by differences (finite difference-method, for details see Ref.1). The resulting matrix equations are integrated in time in small time steps. Previous restrictions to collisions with a rotational symmetry of the wave function about the internuclear axis have been removed by solving the Schrödinger equation in a coordinate frame rotating with this axis and by expanding the wave function in magnetic substates[1]. This method has been first applied to H^+-H collisions.

The system Li^{3+}+H may serve as a further test for the applicability of our numerical method for the following reasons. (1) Measurements for the total charge transfer[2] have been done at relative velocities (energies ∿10keV) where our method is presently working. (2) Calculations with large basis sets are published, but differ in the 10 keV region by about 10%[3,4]. (3) The charge exchange in the asymmetric system Li^{3+}+H involves higher molecular states and, therefore, the calculations may be more informative than for the charge exchange in H^++H dominated by the resonant exchange channel[1]. We have carried out calculations for an impact energy of 10.5 keV. Fig.1 shows the density distributions of the magnetic substates for the time when the nuclei are separating again. The impact parameter dependence of the total charge exchange is given in Fig.2. From this curve we get a cross section for total exchange of $3.8\times10^{-16}cm^2$ which is equal to the one of Fritsch and Lin[3] obtained with an extended basis.

References

1. N.Grün et al., J.Phys.B 15, 4043 (1982)
2. W.Seim et al., J.Phys.B 14, 3475 (1981)
3. W.Fritsch, C.D.Lin, J.Phys.B 15, L281 (1982)
4. H.J.Lüdde, R.Dreizler, J.Phys.B 15, 2713 (1982)

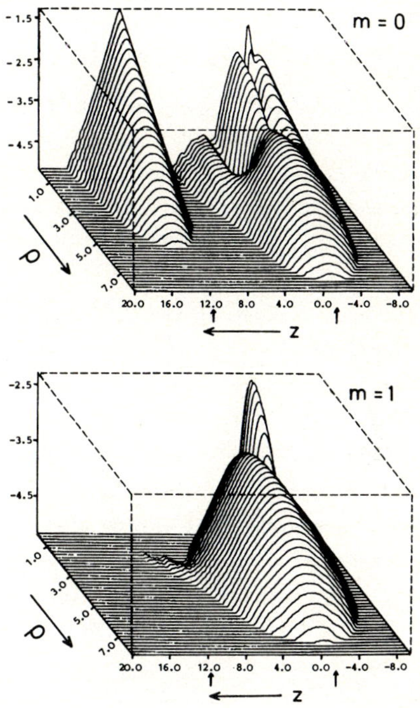

Fig.1 Density distributions for the magnetic substates m=0 and 1 are shown in 3D-plots.

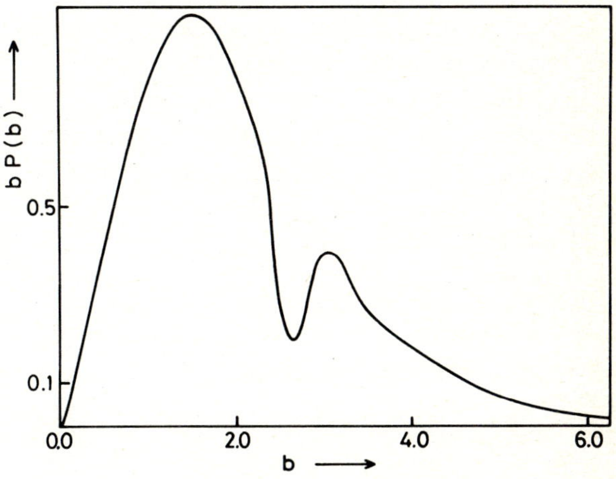

Fig.2 The impact parameter dependence of the charge exchange probability for Li^{3+}+H at 10.5 keV.

THEORETICAL STUDY OF Li^{2+} - H COLLISIONS AT THE KeV ENERGY RANGE

J. Hanssen and C. Harel

Laboratoire des Collisions Atomiques, Université de Bordeaux I, 33405 Talence, France

In a recent work Seim et al[1] have measured the total charge exchange cross sections for the reactions :

Li^{2+} (1s) + H (1s) → Li^{+} + H^{+} (a)

Li^{3+} + H (1s) → Li^{2+} + H^{+} (b)

For the reaction (a) the measured total charge exchange cross-section is nearly constant in the energy range considered (2 keV/amu - 4 keV/amu).

This behaviour is not typical of light systems (as Li^{3+} - H or He^{2+} - H). It is similar to the one encountered for highly charged systems.

We present here a study of the Li^{2+}H collisions in a molecular approach.

The method used is described in previous work (C. Harel and A. Salin[2,3]).

From the theoretical correlation diagram given in the fig. 1 one can expect :

(i) The charge exchange in the ground state of Li^{+} (singlet subsystem) must be small.

(ii) The excitation channels of H are completely negligible at the energies of the incident Li^{2+} considered here.

(iii) The charge exchange arises mostly in the Li^{+} (1s, 2ℓ) + H^{+} channels.

Our calculations confirm these hypothesis. Furthermore, the agreement of our results with the experimental ones is satisfactory (see fig. 2).

1. W. Seim, A. Müller, I. Wirkner-Bott, E. Salzborn, J. Phys. B 14, 3475-91 (1981)
2. C. Harel, A. Salin, J. Phys. B 13, 785-98 (1981)
3. C. Harel, A. Salin, J. Phys. B 16, 55-70 (1983)
4. M.B. Shah, T.V. Goffe, H.B. Gilbody, J. Phys. B 11, L233-6 (1978)

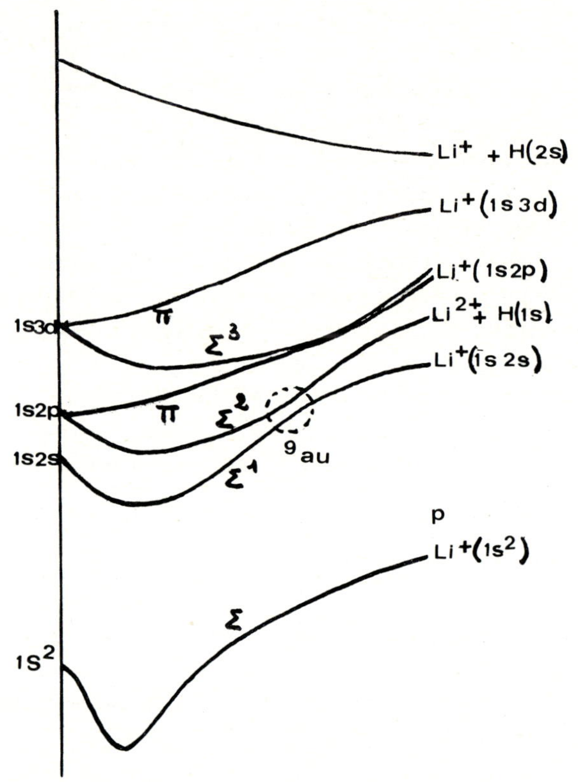

Fig. 1 : Adiabatic correlation diagram of $(LiH)^{2+}$ (singlet subsystem).

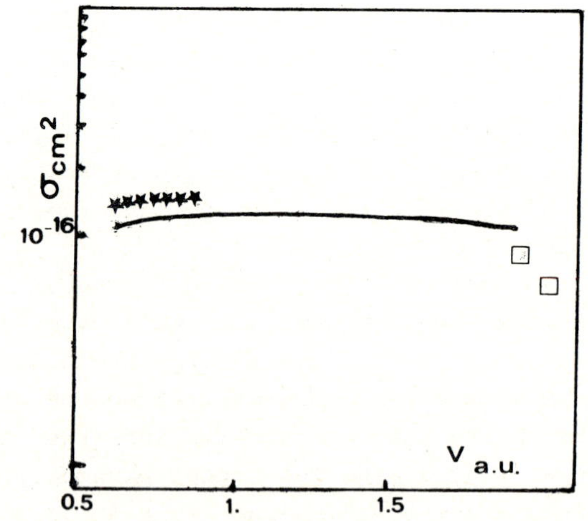

Fig. 2 : Total charge exchange cross-section

★ refer. 1

□ refer. 4

—— this work

TREATMENT OF CHARGE TRANSFER COLLISIONS WITH TRANSLATION FACTORS. $Be^{4+}+H$

L.F. Errea, L. Méndez and A. Riera.

Departamento de Química Física y Química Cuántica. Centro Coordinado.
CSIC-UAM. Universidad Autónoma de Madrid.
CANTOBLANCO.(Madrid,34) (Spain)

As is well known[1], the main difficulties of the standard molecular approach withouth translation factors (TF), are: the dependence of the calculated cross section upon the origin of electronic coordinates and the existence of residual couplings between the molecular channels at infinite internuclear separation.

However, as pointed out in ref. 2 a point that is not sufficiently emphasized in the literature is that the introduction of TF it is not free from ambiguity. There is a wide choice of these factors for small and intermediate internuclear distances and results can vary significantly from one choice to another. To study the effect of including TF we have treated the reaction

$$Be^{4+} + H(1s) \rightarrow Be^{3+}(n=3) + H^{+} \qquad (1)$$

as an specific case of a charge process between a fully stripped ion and hydrogen atoms. This kind of processes have received much attemption lately because of their relevance to nuclear fusion research.

To introduce TF we have employed the common translation factor method (CTF)[3] which is the only one that preserves the formal convergence of the molecular expansion[2]. The calculation has been performed using a three-term ($4f\sigma$-$3d\sigma$-$3d\pi$) molecular expansion

$$\Psi(\underline{r},t) = \exp\left[iU(\underline{r},t) \right] \sum_{n=1}^{3} a_n(t)\, \phi_n(\underline{r},R) \times$$

$$\times \exp\left[-i \int_{-\infty}^{t} E_n\, dt' \right] \qquad (2)$$

where

$$U(r,t) = f\, \underline{v}.\underline{r} - \frac{1}{2} f^2 v^2 t \qquad (3)$$

$$f(\underline{r},R) = \left[R/(R^2+\beta^2) \right] z$$

$\phi_n(\underline{r},R)$ are the molecular wavefunctions of energies E_n and $a_n(t)$ are the expansion coefficients, R is the internuclear distance, $\underline{v}$ the nuclear velocity and β is a paramter which determines the radius of the neighbourhood of R = 0 where U (R,t) is made to vanish.

The use of a switching function like $f(\underline{r},R)$ introduces a "privileged origin", which is the origin of electronic coordinates defining z in eq.(3). In principle, changing this origin can yield, in general, different results.

The charge transfer process (1) takes place through a (strongly origin dependent) radial coupling between the entrance channel ($4f\sigma$) and the exit channel ($3d\sigma$). In Fig. 1 we present the origin dependence of this coupling together with the parameter dependence of the modified couplings when TF are incorporated. Also, for β=3, we have represented the privileged origin dependence.

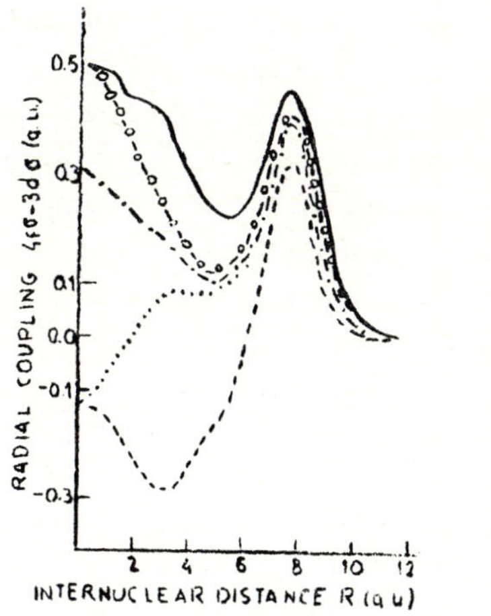

Fig. 1. $4f\sigma$-$3d\sigma$ radial coupling ——without TF, origin H^+; --- without TF origin on Be nucleus; —.—.— with TF, β=0.1, any privileged origin; —0—0— with TF, β=3, privileged origin on H^+; ..... with TF β=3, privileged origin on Be nucleus.

We present in Fig. 2 our results for the charge exchange cross section for reaction (1). It is significant that our calculated cross sections turn out to be independent of the choice of parameters in (3). To study processes like (1), this conclusion is very encouraging in order to sistematically incorporate TF in cross-section calculations.

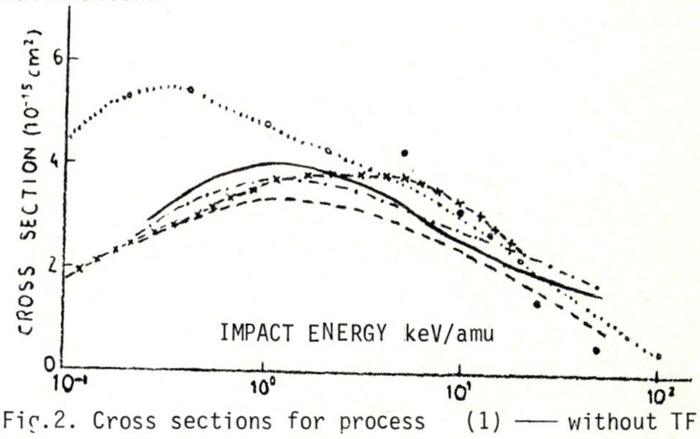

Fig.2. Cross sections for process (1) —— without TF, origin on proton;---- without TF, origin on Be nucleus; —.—.— with CTF, any β, any privileged origin;.... results of ref. 4.; -x-x-x results of ref. 5.

References

1. A. Riera and A. Salin, J.Phys.B., 9, 2877 (1976).
2. L.F. Errea, L. Méndez and A. Riera, J.Phys.B. 15, 101 (1982).
3. S.B. Schneidermann and A. Russek, Phys.Rev., 181, 311 (1969)
4. H.Ryufuku and T.Watanabe, Phys.Rev.A, 14, 1538 (1979).
5. T.Ohyama and Y.Itikawa, Nagoya University Research Report IPPJ -532 (1981)

STARK MIXING OF SUBLEVELS IN MULTICHARGED ION-ATOM COLLISIONS

R. Gayet, J. Hanssen, C. Harel and A. Salin

Laboratoire des Collisions Atomiques, Université de Bordeaux I, 33405 Talence, France

The Stark effect has been shown by Burgdöfer to play a significant role for the determination of the relative population of ℓ levels following electron capture. The case of low-energy electron capture by multicharged ions is an ideal case to study this phenomenon : electron capture occurs through a coupling between three adiabatic states at a well defined internuclear distance. The population is distributed among the neighbouring states leading at infinite nuclear distances to atomic states with the same value of the principal quantum number by subsequent non adiabatic coupling. The Stark effect plays a central role in the characteristics of this redistribution.

We have studied first this phenomenon in the case of the one electron systems : C^{6+} - H, O^{8+} - H and Ne^{10+} - H. In the adiabatic picture, the linear Stark effect at large internuclear distances enters through a diagonal term in the collision equation so that it can be easily accounted for. We study how the ℓ distribution is affected by non-adiabatic coupling between substates following the primary process at the pseudo-crossing. We show that a simple model can be built to explain the population of ℓ levels (which are degenerated in the present case).

The extension to the case of incoming ions having a closed shell core (e.g. O^{6+}, Ne^{6+}, Ne^{8+}) introduces a number of modifications into the treatment :

- the quasi selection rules that explain the primary process of charge exchange are less restrictive.
- the Stark effect competes with the interaction between the core and active electrons.

The present work gives an interpretation of the differences observed experimentally between true one-electron systems and other systems. A detailed discussion will be given at the conference.

Table I

ℓ distribution for electron capture in C^{6+} - H (a), O^{8+} - H (b), Ne^{10+} - H (c) collisions for impact energy 1 keV/amu.

	ℓ = 0	ℓ = 1	ℓ = 2	ℓ = 3	ℓ = 4	ℓ = 5
(a) C^{5+} (n = 4)	0.1	0.26	0.34	0.31		
(b) O^{8+} (n = 5)	0.08	0.19	0.24	0.26	0.23	
(c) Ne^{9+} (n = 6)	0.05	0.14	0.19	0.21	0.21	0.20

Reference

1. J. Burgdöfer, Phys. Rev. A, 24, 1756-67 (1981)

A SIMPLE ACCOUNT OF CORE ELECTRONS IN THE THEORY OF ELECTRON CAPTURE IN SLOW COLLISIONS OF HIGHLY CHARGED IONS WITH ATOMIC HYDROGEN

O.G. Larsen and K. Taulbjerg

Institute of Physics, University of Aarhus, DK-8000 Aarhus C, Denmark

We consider the electron-capture process,

$$X^{+q} - H(1s) \rightarrow X^{+(q-1)} - H^{+} \quad (1)$$

at low velocities and for high charge states $q \gg 1$. The ions may be fully stripped or carry a residual configuration of core electrons. It is known that electron capture predominantly goes into a single or a few of the high-lying levels of the ion. This selectivity plainly reflects that transitions in slow collisions are likely to occur only if they may take place under near-adiabatic conditions. The Coulomb interaction between the ionic products on the r.h.s. of Eq. (1) ensures in general that exothermic capture transitions are allowed because of level crossings during the course of the collision. However, the couplings are weak at very large separations, while curve crossings become strongly avoided at small separations. In the C^{+6}-H case, a dominant part of the capture cross section goes into the n = 4 level via a crossing at R_x = 7 a.u., while crossings at R_x = = 21 a.u. and 4 a.u., corresponding to the n = 5 and 3 levels, respectively, are only weakly populated.

Although the core of electrons is located well within 1 a.u., it has a significant influence on the electron-capture process via its effect on the wave functions and binding energies of the relevant final states of the captured electron. A change in binding energy affects the crossing position, while the wave function determines the coupling strength of the crossing. As another consequence of the core, the degeneracy in principal shells is lifted.

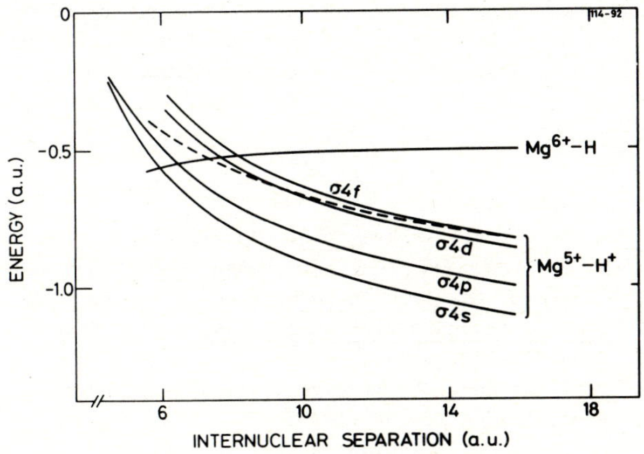

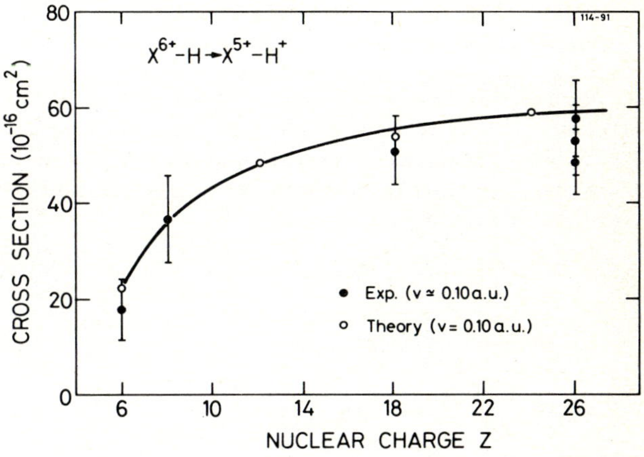

Figure 1 shows the perturbed energies of the initial state and the final σ states with n = 4 for the $(Mg-H)^{+6}$ system. Generally, we represent final states by quantum -defect functions. The perturbed energy of the relevant σ state for the $C^{+5}(n=4) - H^{+}$ case is also shown (dashed curve).

The quantum defects for Mg^{+} are smaller than $\delta_s \simeq 0.4$. For heavier ions such as Ar^{+5}, δ_s may be of order unity. Then the principal-shell structure disappears entirely, and crossings with the initial state appears rather uniformly over the relevant range of internuclear separations. In this limit, it may be expected that the capture cross section for fixed q becomes independent of the ionic species. This is confirmed by detailed close-coupling calculations. Results are presented in Fig. 2 in comparison with the experimental data of Crandall et al.[1]. The perfect agreement suggests that our simple atomic model should be preferred over molecular models of Hartree-Fock type. These are more complicated, for example with respect to translation factors, and do include fine details and often abrupt structure in the basis set, which are immaterial or perhaps even destructive in a dynamical situation.

REFERENCES

(1) D.H. Crandall, R.F. Phaneuf, and F.W. Meyer, Phys. Rev.A 22, 379 (1980)

COINCIDENCE MEASUREMENTS OF ELECTRON TRANSFER IN AN ION-ATOM CROSSED BEAMS EXPERIMENT

F. W. Meyer

Oak Ridge National Laboratory, Oak Ridge, Tennessee 37830 USA*

A crossed beams apparatus that has been previously used to measure total cross sections for electron loss from Rydberg states of atomic hydrogen during fast collisions with multicharged ions[1] has been modified, by addition of a low noise charge analyzer and a coincidence leg, to permit measurements of total electron capture cross sections.

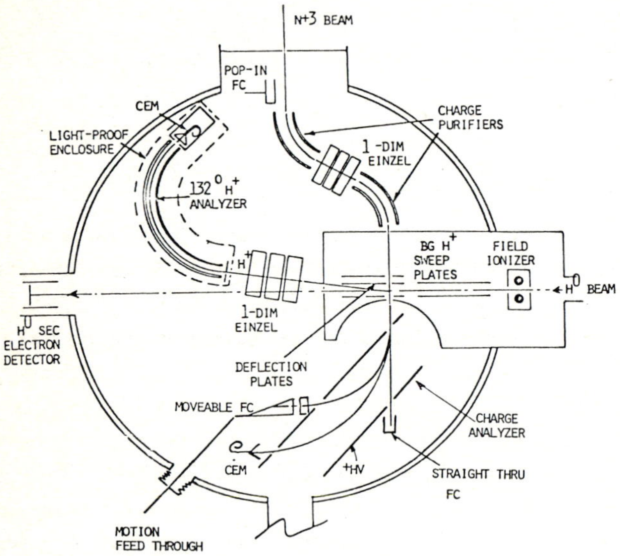

Fig. 1. Schematic of apparatus.

Referring to the schematic of the apparatus shown in Fig. 1, a neutral beam of fast hydrogen atoms, produced by charge transfer neutralization of an H^+ beam having energies in the range 20-50 keV, was crossed in mid 10^{-10} torr vacuum with a slow beam of multicharged ions of charge 'q' obtained from the ORNL PIG ion source. Charge transfer events were isolated by detection in coincidence of both collision products: the H^+ ion and the multicharged ion having charge 'q-1'.

The Faraday cup measuring the primary multicharged ion beam intensity could be moved externally relative to the CEM product ion detector over a range sufficient to allow electron transfer studies of ions having initial charges in the range 2-5. In addition, product ions were deflected 32° out of the plane of the parallel plate analyzer prior to detection on the CEM. With this deflection, there was no direct line-of-sight from the CEM to the interior of the analyzer, resulting in a significant reduction in background counts due to photons.

---

*Operated by Union Carbide Corporation under contract W-7405-eng-26 with the U.S. Department of Energy.

Due to the method of production, the neutral atomic hydrogen beam contained a small fraction (0.01%) of Rydberg states[1] having principal quantum numbers in the range 9-27. The contribution of these Rydberg atoms to the total electron capture cross section was determined to be negligible in a preliminary experiment in which coincidence spectra were acquired for two different settings of the field ionizer in the H^0 beam leg. This finding is theoretically expected,[2] but is in sharp contrast to our earlier measurements of total electron loss cross sections,[1,3] in which the contribution of the small Rydberg component in the H^0 beam dominated the electron loss signal, due to the ease with which Rydberg atoms can be collisionally ionized.

The systems studied in the present experiment are N^{+q} + H (q=2,3) and He^{++} + H in the velocity range 2.0 - 3.0 × 10^8 cm/s. The results are summarized in the Table.

System	V 10^8 cm/s	10^{-15} cm^2 σ X-Beam	σ Oven
N^{+3} + H	2.7	1.06+0.4-0.09	0.97±0.06
N^{+3} + H	2.1	1.35+0.13-0.09	1.40±0.13
N^{+2} + H	2.0	0.57+0.09-0.08	0.58±0.06
He^{++} + H	2.2	1.06+0.14-0.12	1.02±0.14
He^{++} + H	2.9	0.50+0.08-0.07	0.53±0.10

These collision systems have been previously measured by us using the ORNL atomic hydrogen oven.[4,5] As can be seen, the agreement between the present coincidence measurements and the dissociation furnace results is excellent. Since the present measurements are independently absolute, this agreement provides verification of the hydrogen oven calibration, used in all our previous atomic hydrogen electron capture data.

References

1. H. J. Kim and F. W. Meyer, Phys. Rev. A26, 1310 (1982).
2. R. E. Olson, J. Phys. B13, 483 (1980).
3. F. W. Meyer and H. J. Kim, Abstracts of Contributed Papers of the 11th Conference on the Physics of Electronic and Atomic Collisions, p. 552, Kyoto, Japan (1979).
4. R. A. Phaneuf, F. W. Meyer, and R. H. McKnight, Phys. Rev. A17, 534 (1978).
5. R. E. Olson, A. Salop, R. A. Phaneuf, and F. W. Meyer, Phys. Rev. A16, 1867 (1977).

ELECTRON CAPTURE CROSS SECTIONS FOR LOW VELOCITY Ne^{q+} AND Ar^{q+} IONS ON ATOMIC AND MOLECULAR HYDROGEN (50 eV/q to 3000 eV/q)

Tom J. Gray, C. Can, L. Tunnell and J. M. Hall

James R. Macdonald Laboratory, Kansas State University
Manhattan, Kansas USA 66506

and

S. L. Varghese

Physics Department, University of South Alabama
Mobile, Alabama USA 36688

We have used recoil ion sources to produce low velocity highly charged recoil (LEHQ) ions of Ne^{q+} (q=2-7) and Ar (q=2-10) with kinetic energies ranging from 50 eV/q to 3000 eV/q. The techniques of the production and use of the LEHQ ions in electron capture cross section measurements have been previously discussed.[1,2] The LEHQ Ne and Ar ion beams are incident on either a standard gas cell containing molecular hydrogen or a thermal hydrogen oven target. The characteristics of the thermal oven have been reported previously.[3]

The present results for the measured energy dependences of the charge exchange cross sections for Ar + H_2 are given in Fig. 1. Also shown in the figure are data from the literature.[4,5] The measured cross sections σ_{if}, where i and f represent the initial and final charge states of the Ne ion, are shown for σ_{21}, σ_{32}, σ_{43} and σ_{76}. The agreement between the various sets of data and our present results is quite good with a smooth joining of the lower energy data and the higher energy data of Crandall et al.[5] The error in the present results is estimated to be ∿13% with the major contribution associated with the end effects of the gas target cell. There is notable disagreement between the present results and those reported by Huber et al.[4] for $Ar^{2+} + H_2 \longrightarrow Ar^{1+} + H_2^{+}$. Our data suggests that the earlier results of Huber et al.[4] may include detector efficiency effects that are both charge state and energy dependent.

Measured electron capture cross sections for 400 eV/q Ne ions incident upon atomic hydrogen are given in Fig. 2. These are compared to the previous results of Huber et al.[6] The agreement between the present results and those of Huber et al. is within the errors which are estimated to be ∿15%. Once again the data points for σ_{21} do not agree. This is to be expected as the molecular hydrogen capture cross sections are used to set the absolute scale of the electron capture cross sections measured for the atomic hydrogen targets. Theoretical comparisons are given for the classical barrier model[7] and absorbing sphere model.[8] The absorbing sphere model has been renormalized to bring its predictions into agreement with the measured cross sections.

Supported by D.O.E., Division of Chemical Sciences.

References:
1) C. L. Cocke, Phys. Rev. A 20, 749 (1979).
2) T. J. Gray & C. L. Cocke, Proc. Conf. on Application of Accelerators in Research and Industry (to be published in IEEE, 1983).
3) C. Can, T. J. Gray, J. M. Hall, L. Tunnell & S. L. Varghese, Proc. Conf. on Application of Accelerators in Research and Industry (to be published in IEEE,'83).
4) B. A. Huber & H. J. Kahlert, J. Phys. B13, L159 (1980)
5) D. H. Crandall, R. A. Phaneuf & F. W. Meyer, Phys. Rev. A 22, 379 (1980).
6) B. A. Huber, Z. Phys. A 299, 307 (1981).
7) H. Ryufuku, K. Sasaki and T. Watanabe, Phys. Rev. A 21, 745 (1980).
8) R. E. Olson and A. Salop, Phys. Rev. A 14, 579 (1976).

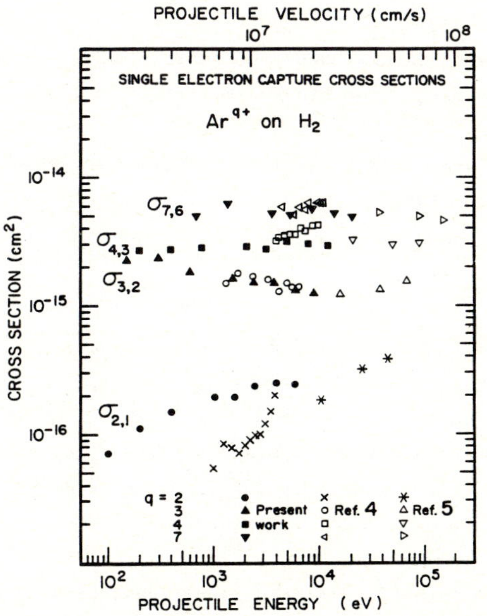

Fig. 1: Electron capture cross sections for $Ar^{q+} + H_2$.

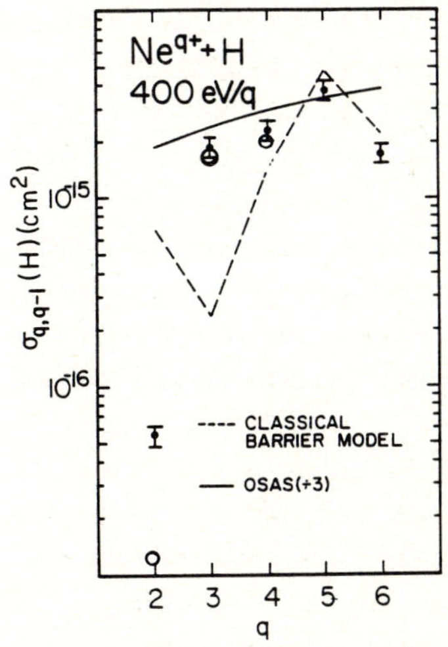

Fig. 2: Electron capture cross sections for Ne^{q+} + H. The closed symbols ● .. present work and the open symbols ○ .. Ref. 6.

NONMONOTONIC BEHAVIOUR IN THE CHARGE DEPENDENCE OF TOTAL ELECTRON-CAPTURE CROSS SECTIONS FOR MEDIUM-VELOCITY, PARTLY STRIPPED IONS ON ATOMIC HYDROGEN

P. Hvelplund, H. Knudsen, L.H. Andersen, S.K. Bjørnelund, and L. Liljeby†

Institute of Physics, University of Aarhus, DK-8000 Aarhus C, Denmark

Scaling laws have proved helpful in the understanding and description of the general behaviour of total cross sections for electron capture in collisions between highly charged ions and neutral atoms,

$$A^{q+} + B \rightarrow A^{(q-1)+} + B^{+}.$$

It is found that for a given target atom, the capture cross section divided by the ion charge depends essentially on the energy divided by the square root of the charge state. This scaling rule is found to work over a broad energy and charge-state range for totally and partly stripped ions[1,2]. However, deviations from such simple scaling rules do exist, and we shall discuss here what has normally been termed charge-state or q oscillations. These 'oscillations' manifest themselves, for example, as an oscillating behaviour of the capture cross section vs charge state at a fixed projectile energy.

At low collision velocities and not too high charge states, one electronic level is predominantly populated by the compound electron, and the cross section is very sensitive to whether or not quasiresonance is fulfilled for this level. This leads to q oscillations that are relatively well understood[3] and found experimentally also on an atomic-hydrogen target[4].

Oscillations in the charge dependence of total electron-capture cross sections have also been observed[5] for collisions between medium-velocity ($v \gtrsim v_0$), partly stripped heavy ions and atomic hydrogen. These oscillations were attributed[6] to the interference between the amplitudes arising from the scattering of the transferred electron on two potentials: The long-range Coulomb potential and the short-range screened Coulomb potential. It is, however, fair to say that the nature of these oscillations is still more or less an open question, and more theoretical as well as experimental work is needed.

In order to shed some light on this problem, we have performed measurements of total electron-capture cross sections for 100-keV/amu ($v = 2v_0$) Dy^{q+}, Ta^{q+}, Re^{q+}, Au^{q+}, and U^{q+} ions on atomic hydrogen. The charge state q varied from 4 to 25.

The experimental techniques have been discussed in detail in previous reports[2,7]. The results are shown in the figure. Here, the capture cross section $\sigma_{q,q-1}$ divided by q^2 is plotted vs projectile charge state q. Also shown as solid curves are theoretical estimates based on the Bohr-Lindhard theory[2]. It should be noted that the observed 'oscillations' do not depend on the charge state of the projectile only but also depend strongly on the projectile atomic number or, in other words, on the structure of the core of the projectile in question.

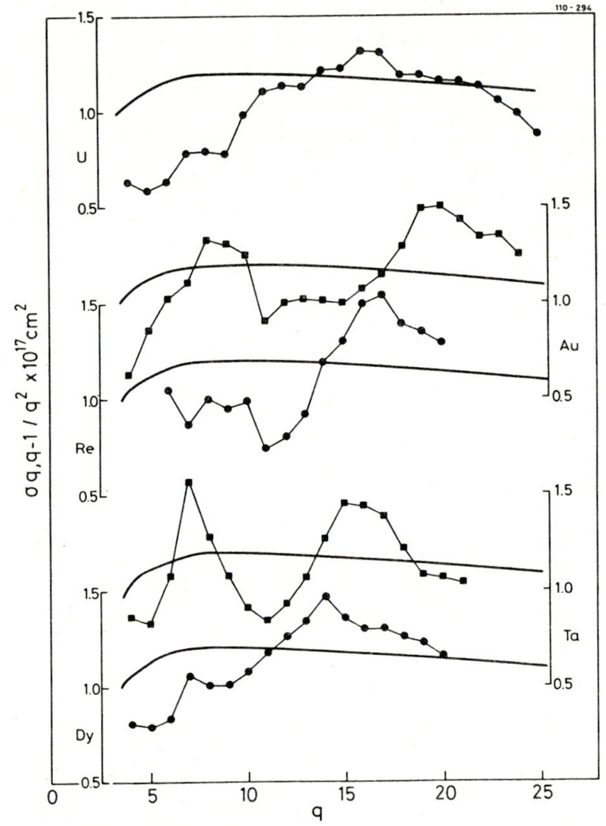

REFERENCES

†Permanent address: Research Institute for Atomic Physics, University of Stockholm, S-10405 Stockholm, Sweden

(1) R.K. Janev and P. Hvelplund, Comm.Atom.Molec.Phys. XI, 75 (1981)

(2) H. Knudsen, H.K. Haugen, and P. Hvelplund, Phys.Rev. A 23, 597 (1981)

(3) R.K. Janev and D.S. Belić, Phys.Scripta (1983) in press

(4) T. Iwai et al., Phys.Rev.A 26, 105 (1982)

(5) F.W. Meyer, R.A. Phaneuf, H.J. Kim, P. Hvelplund, and P.H. Stelson, Phys.Rev.A 19, 515 (1979)

(6) H.J. Kim, P. Hvelplund, F.W. Meyer, R.A. Phaneuf, P.H. Stelson, and C. Bottcher, Phys.Rev.Lett. 40, 1635 (1978)

(7) P. Hvelplund and A. Andersen, Phys.Scripta 26, 375 (1982)

ELECTRON CAPTURE FROM HYDROGEN ATOMS BY MULTIPLY CHARGED C,N,O,Ne IONS AT LOW KEV ENERGIES

V.V.Afrosimov,A.A.Basalaev,K.O.Lozhkin,M.N.Panov

A.F.Ioffe Physical-technical Institute of the Academy of Sciences of the USSR,Leningrad,USSR

Total cross sections for electron capture from atomic hydrogen by purely stripped,hydrogenlike and heliumlike C,N,O,Ne ions have been measured at relative velocities of colliding particles $v=3-12\cdot 10^7$cm/s.The results of available theoretical calculations[I-7] differ considerably (Fig.I) and need an experimental check-up.

In this work the ion source KRION-2 [9] was used.The atomic hydrogen target was formed using thermal dissociation of H_2.The degree of dissociation was determined by means of registration of C^{+2} ions after the process $C^{+4}+H_2 \rightarrow C^{+2}+2H^+$ [IO].For absolute calibration of cross sections we used our data on electron capture from hydrogen molecules[IO].

Our experimental data for C^{+6}+H (Fig.I) agrees best of all with the close-coupling calculations using molecular basis[I-2,6-7].The results of calculations using atomic basis[4-5] exceed the experimental values of cross sections at low velocities v while the two-level Landau-Zener calculation[3] gives much lower values.The difference between our experimental data and the close-coupling calculations[I-2,6-7] for $v> 9\cdot 10^7$cm/s is possibly connected with the violation of quasimolecular character of interaction of the colliding particles at these velocities. The observed nonmonotonic dependence of the capture cross sections on the ion charge Z and velocity (Fig.2) for Z=6-8 is caused by a small number of pseudocrossings between the initial and final state terms of the system.For higher Z the cross section becomes nearly independent of the velocity v because several final state levels are populated.

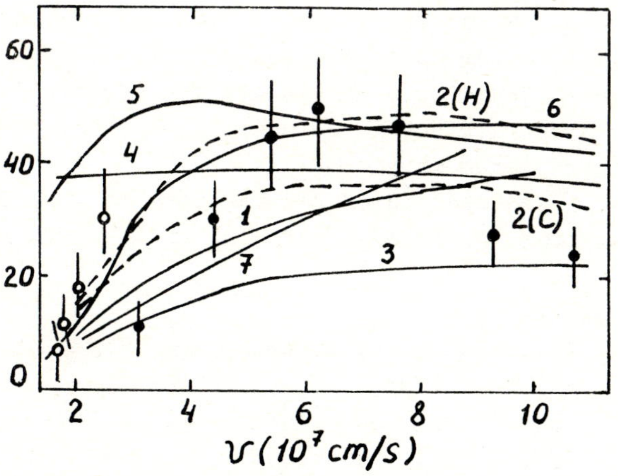

Fig.I:Electron capture in C^{+6}+H collisions. ⧫ -our data, ⧫ - [8].The numbers of theoretical papers in the list of references are indicated near corresponding curves.(H) and (C)-coordinates centered on H^+ and C^{+6}respectively.

References

I.J.Vaaben,J.C.Briggs,J.Phys.BIO,L52I (I977).
2.A.Salop,R.E.Olson,Phys.Rev.AI6,I8II (I977).
3.P.T.Greenland,J.Phys.BII,LI9I (I978).
4.H.Ryufuku,T.Watanabe,Phys.RevAI9,I538 (I979)
5.H.Ryufuku,Phys.Rev.A25,720 (I982).
6.T.A.Green,E.J.Shipsey,J.C.Browne,Phys.Rev. A25,I364 (I982).
7.A.K.Kazanski,I.V.Komarov,Zh.Tehn.Fiz.52, I734 (I982).
8.R.A.Phaneuf,I.Alvarez,F.W.Meyer,D.H.Crandall Phys.Rev.A26,I892 (I982).
9.E.D.Donets,IEEE Trans.NS 23,897 (I976).
IO.M.N.Panov,A.A.Basalaev,K.O.Lozhkin,Physica Scripta 27,№3 (I983).

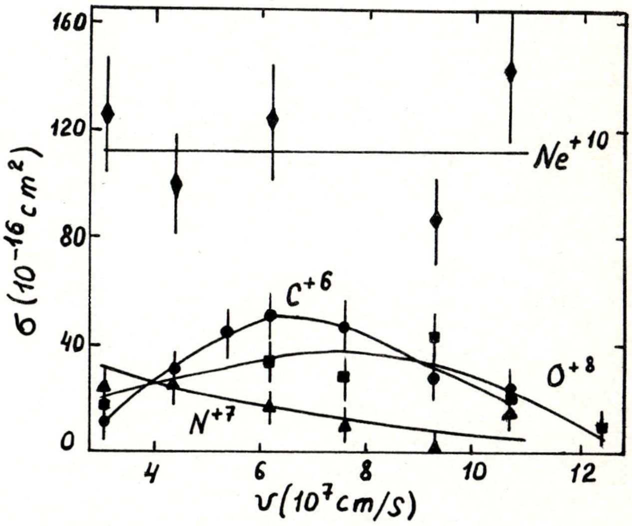

Fig.2:Electron capture by purely stripped ions from hydrogen atoms.

MULTICHANNEL MODEL STUDY OF COLLISIONS BETWEEN MANY-ELECTRON ATOMS AND HIGHLY CHARGED IONS

V.K.Nikulin and A.V.Samoylov

A.F.Ioffe Physical-Technical Institute, Academy of Sciences of the USSR, Leningrad, USSR

For many-electron systems reliable information on the excitation processes in collisions can be obtained only by the analysis of the many-electron states. In the present report the results of theoretical study of charge transfer reactions in the medium- and low-energy regions for collisions between C^{6+}, N^{7+}, O^{8+} ions and He atoms are discussed. This study is based on the consideration of two-electron quasimolecular states and on the multichannel Landau-Zener and Nikitin models with radial and rotational couplings taken into account. The method of calculation of two-electron states using the basis of screened diatomic molecular orbitals (SDMO) was given[1].

The diabatic SDMO [2] were obtained by a solution of two-center problem with an effective potential including electron screening of the nuclei and allowing the separation of variables in prolate spheroidal coordinates. The diabatic states were calculated in a single configuration approximation. The parameters characterizing the potential interaction of diabatic states in the vicinities of their crossings were obtained by constructing adiabatic states in the frame of the configuration interaction method. The matrix elements of dynamic coupling were calculated using the SDMO.

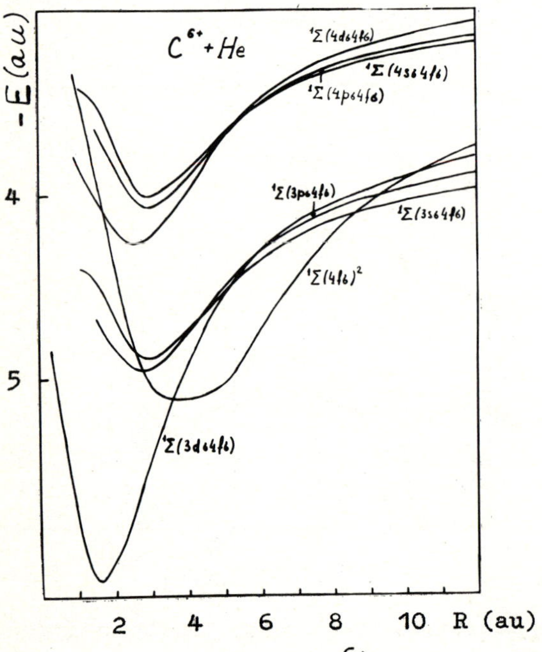

Fig. 1: Diabatic states for C^{6+} + He system.

In fig.1 the results of diabatic states calculation are given for C^{6+} + He system. For this system the capture processes are mostly induced by the one-electron transitions from $(4f\sigma)^2$ entrance state to $(3d\sigma 4f\sigma)$ and $(3p\sigma 4f\sigma)$ states in the range of 2 - 4 au.

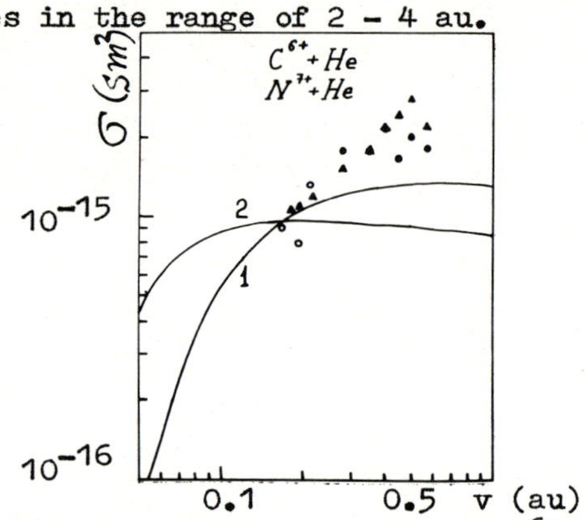

Fig. 2: Total cross sections for C^{6+}+He (1) and N^{7+}+He (2). Experiment: ● - C^{6+}, ▲ - N^{7+} - Afrosimov et al [3]; o - C^{6+}, △ - N^{7+} - Kaneko et al [4].

The calculations of the total cross sections of one-electron capture were performed within multichannel both Landau-Zener and Nikitin models. All crossings were considered as independent and all possible paths in the diabatic state picture were taken into account.

Fig.2 shows the results of such calculations within Nikitin model with potential coupling only. The results for Landau-Zener model at the cross section maximum are lower by a factor about one and a half. It should be noted that for both systems the capture into states with n=4 is as small as 1% of capture into states with n=3.

The results of calculations taking into account the dynamic coupling will be given.

References

1. V.K.Nikulin and A.V.Samoylov, Phys.Lett. 89A, 225 (1982)
2. V.K.Nikulin and N.A.Guschina, J.Phys. B11, 3553 (1978)
3. V.V.Afrosimov et al, XII ICPEAC, Abstracts of papers, Gatlinburg, 1981, p.690
4. Y.Kaneko et al, XII ICPEAC, Abstracts of papers, Gatlinburg, 1981, p.696

CHARGE TRANSFER IN He^{++} + Li COLLISIONS

A.M. Ermolaev and B.H. Bransden

Department of Physics, University of Durham,
South Road, Durham DH1 3LE, England

Possible applications of the preferential production of He^{+} (n = 3), in He^{++} + Li collisions, in soft X-ray lasers and in plasma diagnostic in the controlled thermonuclear fusion have led to intensive studies of this reaction. A theoretical prediction of this process at energies lower than 7 keV c.m. was made by Shipley et al.[1] on the basis of MO calculations and was experimentally confirmed by Barrett and Leventhal[2]. The present authors developed a high-energy AO two-state impact parameter model[3] to show that this preferential type of reaction persisted even at much higher energies (20-30 keV c.m.) than those considered by Shipley et al. For an important intermediate energy region 5-20 keV c.m. where there was no experimental data available at the time, the method employed by Bransden and Ermolaev needed extension to a multi-channel version in order to get reliable computed cross-sections.

In the present work, we report a further study of the total cross-sections as well as the distribution in ℓ in final states for single-electron transfer in He^{++} + Li collisions in the region of 5-20 keV c.m. using multi-state AO expansions with translational factors. We have also extended the energy range of our two-state model up to 1000 keV c.m. covering the energy region (E $\gtrsim$ 150 keV c.m.) where charge exchange is dominated by capture of electrons from the K-shell of Li atoms.

E c.m. keV	$\sigma_3(\ell)$	10^{-16} cm^2	
	ℓ = 0	1	2
3.18	6.7	15.9	11.6
5.08	9.8	19.3	28.0
6.99	8.5	16.4	38.8
13.97	3.7	14.6	43.3
25.40	2.5	9.5	33.7

Table. The n = 3 cross-sections for production of He^{+} (n,ℓ) in He^{++} + Li collisions. The distribution in ℓ

The multi-channel calculations reported here used 18-state AO expansion with n $\leqslant$ 4. The He^{++} + Li system was treated as a one-electron system with the active electron moving in a combined field of the Coulomb He^{++} centre and an effective potential of the Li^{+} core derived from a SCF approximation. In a preceding study[4] such potentials were found to be adequate for describing interactions between heavy ions (e.g. Cs^{+}) and one of the aims of the present work was to assess the usefulness of the effective potentials within a multi-state approach.

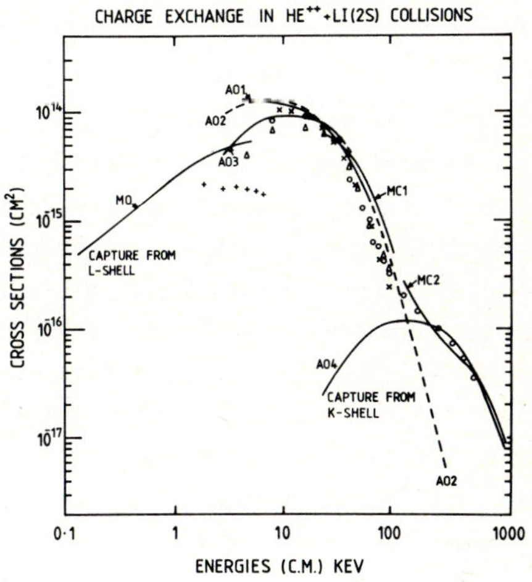

Fig. Total cross-sections for production of He^{+} in He^{++} + Li collisions. Theoret. data: MO-molecular expansion of ref. 1. Atomic expansions: AO1 - 2-state, ref. 3; AO2 - 2-state, present work; AO3 - 18-state, present work; AO4 - 2-state, present work; classical trajectory Monte-Carlo calculations: MC1 - ref. 1; MC2 - ref. 8. Expt. data: + ref. 2 for production of He^{+} (n = 3) only; X ref.5; O ref. 6; Δ ref. 7.

Computed total cross-sections presented in the figure show very good agreement with recent experimental data[5,6,7] at 5-25 keV c.m. For higher energies, our two-state results are in reasonable accord with experiment as well as with classical trajectory calculations of Olson[8], though both tend to over-estimate experimental cross-sections, especially in the region of 40-80 keV c.m. which precedes the maximum of capture from the K-shell of Li.

References

1. E.J. Shipsey, L.T. Redmon and J.C. Browne Phys. Rev. A18, 1961 (1978).
2. J.L. Barrett and J.J. Leventhal Phys. Rev. A23, 485 (1981).
3. B.H. Bransden and A.M. Ermolaev Phys. Lett. 84A, 316 (1981).
4. A.M. Ermolaev, C.J. Noble and B.H. Bransden J. Phys. B: At. Mol. Phys., 15, 457 (1982).
5. K. Kadota, D. Dijkkamp, R.L. van der Woude, P.G. Yan and F.J. de Heer Phys. Lett. 88A, 135 (1982).
6. R.W. McCullough, T.V. Goffe, M.B. Shah, M. Lennon and H.B. Gilbody J. Phys. B: At. Mol. Phys. 15, 111 (1982).
7. G.A. Murray, J. Stone, M. Mayo and T.J. Morgan Phys. Rev. A25, 1805 (1982).
8. R.E. Olson J. Phys. B: At. Mol. Phys. 15, L163 (1982).

ELECTRON-CAPTURE IN He^{2+} + Li COLLISION

H. Sato and M. Kimura

Physics Department, University of Missouri-Rolla, Rolla, MO 65401, U.S.A.

The electron-capture process has

$$He^{2+} + Li \rightarrow He^{+} + Li$$

attracted attention because it is applicable to lasers in the vuv or soft x-ray region of the spectrum and to fusion plasma diagnostics. Recently, theoretical[1,2] and experimental[3,4] work on this electron-capture process has been reported. There is a considerable discrepancy between the two experimental measurements. Shipsey et al.[1] calculated the cross section at low energies using the perturbed-stationary-state (PSS) method with two molecular states but neglected the electron translation factors (ETF's). At high energies they used the classical-trajectory Monte-Carlo method. Their result and the recent experimental data[4] are in reasonably good accord. Bransden and Ermolaev[2] performed calculations based on a two-state atomic expansion method at collision energies 5 ∿ 60 keV. Their results agree with the measurements[4] down to 15 keV. At lower energies, however, the behavior of their cross section is not verified by the experiments.

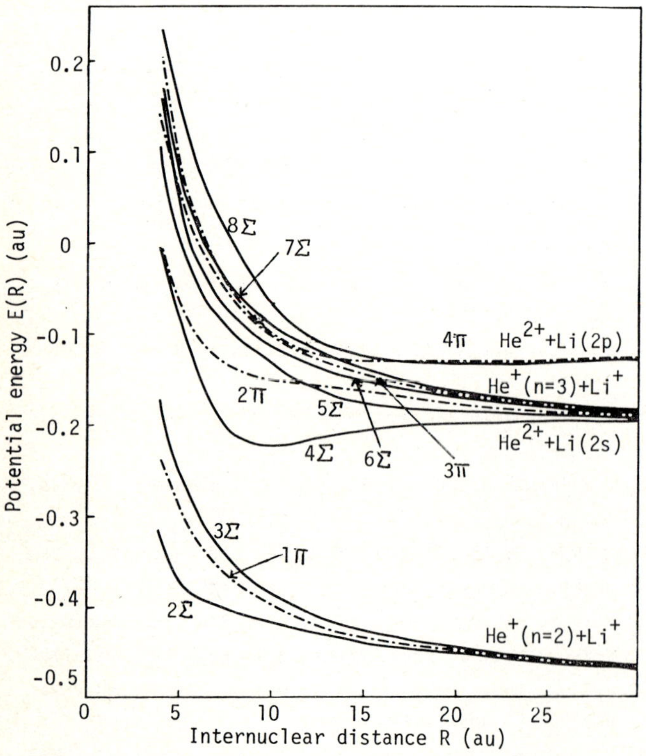

Fig. 1 Potential energy curves of $(He\text{-}Li)^{2+}$ system.

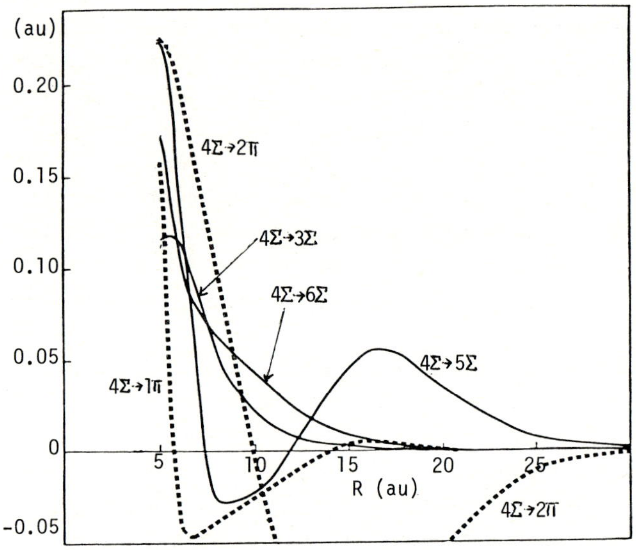

Fig. 2 Nonadiabatic coupling matrix elements of $(He\text{-}Li)^{2+}$ system. Rotational coupling matrix elements are scaled down by the factor 1/2.

We have calculated the electron-capture cross section at low to intermediate energy using the multi-state PSS method. The pseudo-potential method was used to obtain the potential energies and wavefunctions of the system. Seven σ and four π states were taken into account in the PSS calculation. Also, the first order correction due to the ETF effect[5] was included.

Potential energy curves of the $(Li+He)^{2+}$ system and several coupling matrix elements among the states are shown in Figs. 1 and 2, respectively. The calculated cross section will be given at the conference.

References

1. E. J. Shipsey, L. T. Redmon, J. C. Browne, and R. E. Olson, Phys. Rev. A 18, 1961 (1978).
2. B. H. Bransden and A. M. Ermolaev, Phys. Lett. 84A, 316 (1981).
3. J. L. Barrett and J. J. Leventhal, Phys. Rev. A 23, 485 (1981).
4. G. A. Murray, J. Stone, M. Mayo, and T. J. Morgan, Phys. Rev. A 25, 1805 (1982).
5. J. B. Delos and W. R. Thorson, J. Chem. Phys. 70, 1774 (1979).

TWO ELECTRON CAPTURE IN HIGH ENERGY He^{2+} + Li COLLISIONS

M. Sasao, A. Matsumoto, A. Nishizawa, K.N. Sato, S. Takagi,

Institute of Plasma Physics, Nagoya Univ., JAPAN

S. Amamiya, T. Masuda, Y. Tsurita

Faculty of Engineering, Nagoya Univ., JAPAN

Y. Kanamori, Y. Haruyama and F. Fukuzawa

Faculty of Engineering, Kyoto Univ., JAPAN

It has been proposed to use two electron capture processes by He^{2+} to measure α-particle distribution in magnetically confined fusion plasma[1]. Neutral He and Li beams are two of the most promising candidates as doping beams. A lot of new data in atomic physics, however, are necessary to investigate requirements for an ion source and an accelerator; for example metastable fraction of neutral He beam which are produced from negative ions, He^- life times, and so on[2]. Two electron capture cross section in He^{2+} + Li collision is one of them. Two experimental results were recently reported in the energy range of 4.8 keV - 800 keV[3), 4)]. In these experiments, relations between vapour pressure and temperature were used to determine the absolute target densities.

Experiments have been carried out at Nagoya Univ. and Kyoto Univ. to get absolute values of σ_{20} in Li vapour whose pressure is determined by Rutherford scattering, and to extend the energy range up to 2 MeV. High energy He^+ beam from the Van de Graaff is fully ionized and charge selected, and enters into the lithium vapour oven in a high vacuum region ($\sim 7 \times 10^{-5}$ Pa). The Li target density along the beam path is simultaneously monitored by SSD at 5°. The effective values of $(\frac{d\sigma}{d\Omega} d\Omega)$Ruth, have been computed by simulation considering the finite beam size and other geometrical conditions. Measurements have been performed at several vapour pressure, typically 0.1 - 2.5 x $10^{13}/cm^3$, at each He^{2+} energy. The growth curve shows that the single collision approximation is valid and errors due to background correction are not substantial in these pressure region.

Results are shown in Fig. 1. Cross sections at 0.79 and 0.8 MeV agree with that of McCullough et al[3], within the statistical errors. In high energy region of 0.8 - 2.0 MeV, cross sections exhibit a energy dependence of $E^{-5.6}$, indicating that neutral He° detection in α-particle diagnostics with Li° beam at an angle less than 30° may be as possible as in the case of He° beam.

References

1. D.E. Post, et al., J. Fusion energy 1 (1981) 129
2. M. Sasao, IPPJ-599 (1982), 355
3. R.W. McCullough et al., J. Phys. B: 15 (1982) 111
4. G.A. Murray et al., Phys. Rev. A25 (1982) 1805

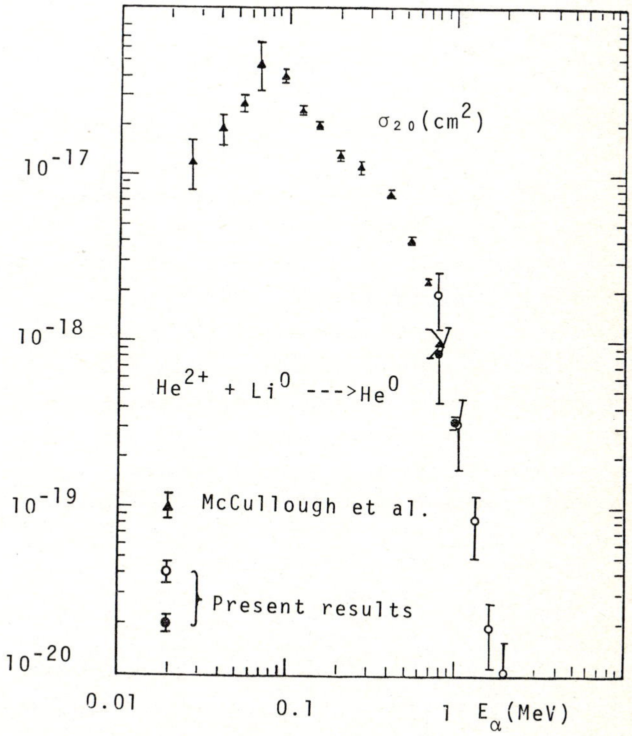

Fig. 1. Two electron capture cross sections for He^{2+} in Li

EXPONENTIAL DISTORTED WAVE APPROXIMATION IN CHARGE TRANSFER

H. Suzuki and N. Toshima

Dep. of Appl. Phys., Fac. of Engineering, University of Tokyo, Bunkyo-ku, Tokyo 113 Japan

T. Watanabe

The Institute of Physical and Chemical Research, Wako-shi, Saitama 351 Japan

Ryufuku and Watanabe[1] have developed a theoretical method named the unitarized ditorted wave approximation (UDWA) and have applied it to the electron capture processes between multicharged ions and a hydrogen atom. In the above process so many electronic states are involved that consideration of them is indispensable for the adequate treatment of the collision processes. The most noticeable feature of UDWA is that unlimited number of electronic states of multicharged ions can be incorporated in the calculation in contrast to the close-coupling method.

Although UDWA has given cross sections which agree well with experimental data over a surprisingly wide energy region, there are some theorists who are doubtful of the physical foundations of the assumptions employed in UDWA. One of the purposes of the present study lies in the quantitative investigation of the validity of the assumptions.

We make use of the S matrix representation in the framework of the impact-parameter distorted wave method proposed by Ryufuku and Watanabe[1].

$$S^{int} = T\exp(-i\int_{-\infty}^{\infty}\hat{H}^{int}\,dt), \qquad (1)$$

where

$$\hat{H}^{int} = \exp(i\int_{-\infty}^{\infty}H^{o}\,dt)\; H^{int}\exp(-i\int_{-\infty}^{\infty}H^{o}\,dt). \qquad (2)$$

H^{o} and H^{int} are the diagonal and the off-diagonal parts of the interaction matrix $H = s^{-1}h$, respectively. UDWA makes two approximations in this stage; (i) to drop the chronological operator T in the equation (1) and (ii) to neglect all the matrix elements except those involving the initial state in the series expansion of (1). The second approximation considerably simplifies the S matrix elements for charge transfer because the even-order terms disappear in the series expansion. In EDWA we abandon the second approximation and retain all the matrix elements.

$$S^{int}_{EDWA} = \exp(-i\int_{-\infty}^{\infty}\hat{H}^{int}\,dt). \qquad (3)$$

As in UDWA we simplify $\hat{H}^{int}$ neglecting the higher-order ($\geq$ the second) elements with respect to the overlap of the different atomic centers. $A(=\int_{-\infty}^{\infty}\hat{H}^{int}\,dt)$ then reduces to a Hermitian matrix and can be diagonalized by a unitary transformation.

$$U^{-1}AU = \begin{bmatrix}\lambda_1 & & 0\\ & \lambda_2 & \\ 0 & & \ddots\end{bmatrix} \qquad (4)$$

where λ_k is the eigenvalue of A. S^{int} is calculated as

$$S^{int} = U\begin{bmatrix}\exp(-i\lambda_1) & & 0\\ & \exp(-i\lambda_2) & \\ 0 & & \ddots\end{bmatrix}U^{-1} \qquad (5)$$

We have calculated the cross sections of the process $H(1s) + A^{q+} \rightarrow H^{+} + A^{(q-1)+}$ for $2\leq q\leq 6$ (from He^{++} to C^{6+}). One of the results is illustrated in Fig. 1.

The fact that the cross sections of EDWA and UDWA are close together above 10 keV/amu indicates that the couplings among the final states are not important in that energy region for the determination of the total capture cross sections. This point has been also confirmed in the cases of Li^{3+} + H through C^{6+} + H. The agreement with the close coupling result and with the experimental data guarantees that the omission of T and the simplification of $\hat{H}^{int}$ do not bring about a serious error for $v\geq 0.3$ a.u..

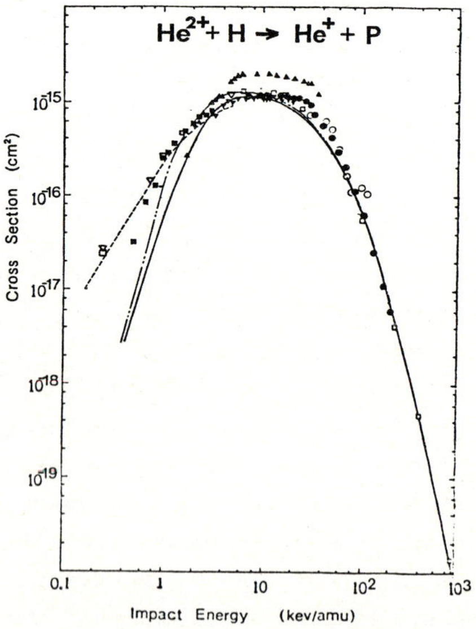

Fig. 1. Electron capture cross sections as functions of impact energy per nucleon in the case of He^{++} + H → He^{+} + H^{+}. Theories:—— EDWA, —·— UDWA, ----MO close-coupling (Winter and Lane), □ AO close-coupling (Fujiwara and Toshima) ○CTMC (Olson). Experiments: ■ Nutte et al, ▲Bayfield et al, ● Olson et al.

References

1. H. Ryufuku and T. Watanabe, Phys. Rev. A18 2005 (1979), ibid A19 1538 (1979), ibid A20 1828 (1979).

CONTINUOUS ENERGY STATE MODEL FOR CHARGE TRANSFER IN MULTIPLY CHARGED IONS IMPACT ON ATOMS

Fumihiro Koike *

FOM-Institute for Atomic and Molecular Physics, Kruislaan 407, 1098sj Amsterdam

The continuous energy state model has been introduced on the charge transferred final states in order to decouple the initial state from the others. The coupled equations are reduced to numbers of smaller set of inhomogeneous coupled equations that can be solved independently of each other. An approach that can be classified to the same category, in fact, has been proposed recently by Presnyakov et al.[1] We have removed a number of their restrictions, i.e., we have allowed the presence of the rotational coupling between the final states and also of the differences of energies between the final states with different ℓ.

We consider the charge transfer process:

$$A + B^{Z+} \rightarrow A^{+} + B^{(Z-1)+}(n\ell m). \qquad (1)$$

We take ϕ_a, ε_a, and C_a as the initial $A+B^{Z+}$ diabatic-state wavefunction, energy, and probability amplitude, respectively. We take $\phi_{bn\ell m}$, $\varepsilon_{bn\ell m}$, and $C_{bn\ell m}$ as the final $A^{+}+B^{(Z-1)+}(n\ell m)$ diabatic-state wavefunctions, energies, and probability amplitudes, respectively, where m is the magnetic quantum number. We represent the total electronic Hamiltonian by $H_{e\ell}$, the rotational coupling operator by L_y, the collision velocity by v, the impact parameter by b, and the internuclear distance by R, which is the function of time t. We employ the impact parameter formalism; we may write $R^2=b^2+v^2t^2$. Then we obtain

$$\dot{C}_a = -i\sum_{n\ell}\langle\phi_a|H_{e\ell}|\phi_{bn\ell o}\rangle C_{bn\ell o}\exp\left[-i\int^{t}(\varepsilon_{bn\ell o}-\varepsilon_a-\tfrac{1}{2}v^2)dt'\right] \qquad (2a)$$

$$\dot{C}_{b,n,\ell,m} = -i\langle\phi_{bn\ell o}|H_{e\ell}|\phi_a\rangle\delta_{m,o}C_a\exp\left[-i\int^{t}(\varepsilon_a+\tfrac{1}{2}v^2-\varepsilon_{bn\ell o})dt'\right]$$
$$-\sum_{m'}\frac{vb}{R^2}\langle\phi_{bn\ell m}|L_y|\phi_{bn\ell m'}\rangle$$
$$\times C_{bn\ell m'}\exp\left[-i\int^{t}(\varepsilon_{bn\ell m'}-\varepsilon_{bn\ell m})dt'\right] \qquad (2b)$$

We have retained only the rotational couplings between the states with the same n and ℓ, since the others are expected to be small. As the first step, we solve eqs.(2) without the terms including the $\phi_{bn\ell m}$ with $m\neq0$, because they are indirectly coupled with ϕ_a. We employ the method proposed by the present author[2] regarding that the state $\phi_{bn\ell 0}$ is continuum state. We obtain

$$C_a = \exp[-i\int_{-\infty}^{t}(\varepsilon_r-i\Gamma/2)dt']. \qquad (3)$$

The complex energy $\varepsilon_* \equiv \varepsilon_r - i\Gamma/2$ is given by the root of the following algebraic equation. That is

$$\varepsilon_* - \varepsilon_a - v^2/2 - \int_{C^+}\frac{\sum_\ell\left[\frac{d\varepsilon_{bn\ell 0}}{dn}\right]^{-1}V_{\varepsilon'\ell}V^{\dagger}_{\varepsilon'\ell}}{\varepsilon'-\varepsilon_*}d\varepsilon', \qquad (4)$$

where $V_{\varepsilon'\ell}\equiv\langle\phi_a|H_{e\ell}|\phi_{bn\ell 0}\rangle$ is the coupling between the initial and final diabatic states. The integration contour C^+ turns anti-clockwise around the pole of the integrand. Substituting eq.(3) into eq.(2b), we obtain a number of sets of coupled equations for $\phi_{bn\ell m}$ which are decoupled from ϕ_a. We solve them numerically as the second step.

We show below the n-partial cross sections, σ_n, for the process: $Ne^{10+}+H \rightarrow Ne^{9+}+H^{+}$ at 1, 10, 100, and 1000keV/amu of Ne^{10+} impact energies. Present results are given by solid lines. Dashed lines represent UDWA. Dotted dashed lines represent M-LZ-RC. And dotted lines represent CTMC. The general trend is quite nicely reproduced over the wide energy range.

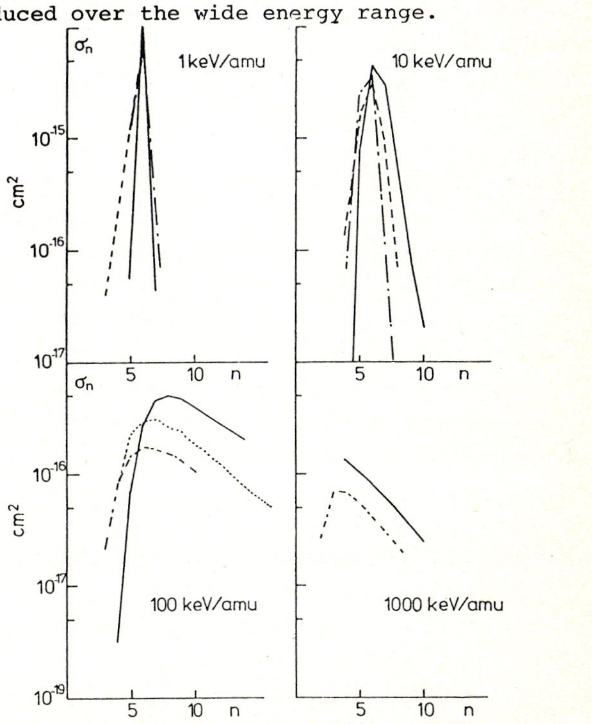

*Present address: School of Medicine, Kitasato University, Sagamihara, 228 Japan

References

1. L.P.Presnyakov, D.B.Uskov, and R.K.Janev: Phys.Lett. 84A 243 (1981)
2. F.Koike: Submitted to J.Phys.B.

SEMICLASSICAL STUDIES OF SLOW CHARGE TRANSFER REACTIONS

Anders Bárány

Institute of Theoretical Physics, University of Uppsala
Thunbergsvägen 3, S-752 36 Uppsala, Sweden

In close connection with the newly started multiply charged ion collision experiments[1] at the Research Institute of Physics (AFI), Stockholm, theoretical studies of charge transfer reactions at low velocities ($v << v_B$) are performed. Among other topics, the following two questions are being addressed:

Are there any systematic but simple ways of deriving the ion-atom and ion-ion interactions to be used for determining average classical trajectories and in particular the energy and impact parameter dependences of the average classical scattering angles? Even though most slow charge transfer reactions involving multiply charged ions proceed by capture at large impact parameters (where long-range asymptotic representations of the interactions are valid), some systems need closer contact to react. This means that also the short-range forces have to be taken into account. Experiments in Stockholm so far use argon targets and projectiles (with charges ranging from 1 to about 6), implying that multielectron systems have to be considered. Here use is made of the electron gas model expressions for the total interaction energy[2]. The electron densities have so far been taken ad hoc, but plans call for the use of modified Thomas-Fermi methods in the future. Such semiclassical methods have long been used to find the behaviour of the interaction for small internuclear separations, but here special importance will be given to the outer regions of the atomic systems[3].

What are the energy and impact parameter dependences of the partial wave scattering matrix for charge transfer proceeding along a specified reaction path? It is well known that most experimentally observed total cross sections for electron pick-up by multiply charged ions are quasivelocity independent in the region between low and moderate velocities. This is attributed to the large number of reaction paths leading to the same final charge on the projectile. By using more sophisticated experimental techniques it is possible to select projectiles that are products of a smaller class of reactions. The corresponding cross sections then show more structure. To interpret these structures one needs a tool to handle the localised transition mechanisms and interference effects that characterize slow heavy-particle collisions in general[4]. To a certain degree numerical solutions of classical trajectory equations constitute such a tool, and the development of suitable computer codes has been initiated. Earlier results show, though, that a deeper understanding can be obtained by proper analytical methods, such as the semiclassical Zwaan-Stueckelberg method[5,6]. The resulting expressions generalize and to a certain extent uniformize existing approximations. They give simple explanations for most structural cross section phenomena and are also basic for semiclassical inversion procedures. Here work is going on applying the formulas to specific systems exhibiting different transition mechanisms. According to semiclassical results the scattering matrix is to a large extent determined by the analytic structure of the adiabatic energies and not by the non-adiabatic coupling elements. The problem of a practical complex analytic continuation of numerically defined real adiabatic potentials is being given special attention.

The poster presented at the conference will give a summary of results obtained in the above-mentioned problem areas.

References

1. G. Astner, A. Bárány, H. Cederquist, H. Danared, P. Hvelplund, A. Johnson, H. Knudsen, L. Liljeby and L. Lundin, Phys. Scr. T 3, 163 (1983).
2. W.E. Baylis, in Progress in Atomic Spectroscopy, Part A, edited by W. Hanle and H. Kleinpoppen (Plenum Press, New York, 1978), p. 207.
3. B.-G. Englert and J. Schwinger, Phys. Rev. A 26, 2322 (1982).
4. M.S. Child, Adv. At. Mol. Phys. 14, 225 (1978).
5. D.S.F. Crothers, Adv. At. Mol. Phys. 17, 55 (1981).
6. A. Bárány and D.S.F. Crothers, Proc. R. Soc. Lond. A 385, 129 (1983).

TARGET EFFECT ON THE n POPULATION FOR ELECTRON CAPTURE PROCESSES INVOLVING MULTICHARGED IONS (1<E<10 keV/q)

M. Barat*, M. Laurent** and J. Pommier*
S. Dousson+ and D. Hitz+
AGRIPPA[x] CENG - Bât. 1005 85X39041 GRENOBLE Cedex FRANCE

During a collision between a multicharged ion and a target atom, the electron is initially captured into an highly excited state. First predicted theoretically, the selectivity of the process has been demonstrated only recently in a "energy gain experiment"[1,2]. In such an experiment, measurement of the kinetic energy of the fast multicharged ions after the collision allows the determination of the principal quantum number (n) in which the electron is captured.

We have undertaken such type of experiment using the "Micromafios" multicharged ion source, developped in Grenoble[3]. The incident beam is energy and q/m selected using a 180° magnetic analyzer (R=50 cm). An identical magnet is used to analyse the energy of the ions. The energy spectra are obtained by sweeping the voltage applied to the collision chamber.

In this experiment, we have focused our attention on the role of the target in the selectivity of the capture process. The energy of the A^{q+} ion beam was varied between 1 keV/q and 10 keV/q. Fig. 1 shows the energy gain spectra for $N^{6+} \rightarrow N^{5+*}(n)$ at 5 keV/q. This result demonstrates that n=3 is predominantly populated with He target, in agreement with the previous results obtained at lower energy. In contrast dominant population of n=4 is obtained with H_2 target. With N^{7+} as projectiles, the electron is captured in n=4 with He target and n=5 and 4 with H_2 target. No drastic change in the population is observed in the explored energy range.

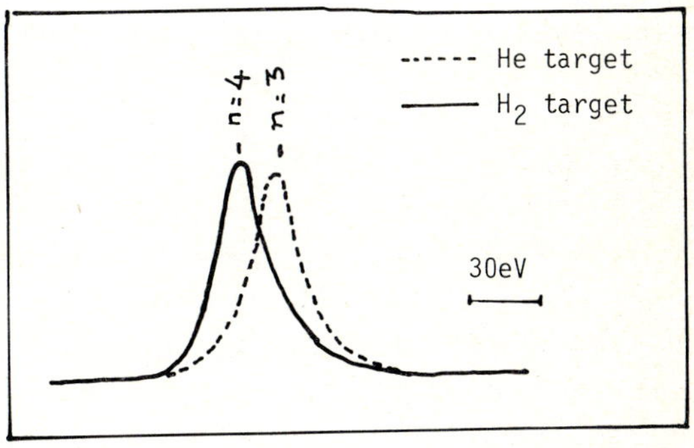

Fig. 1 Energy gain spectra for $N^{6+} \rightarrow N^{5+*}(n)$ at E=5keV/q

References

1. Ohtani et al. J. Phys. B (1982) 15 L 533 ibid L.733
2. R. Mann et al. P.R.L. 49, 1329 (1982)
3. R. Geller et al. Rev. Phys. Appl. 15 995 (1980)

*L.C.A.M., Bât. 351, Université Paris-Sud, 91405 Orsay
**I.P.N., Bât. 100, Université Paris-Sud, 91405 Orsay
+CENG GRENOBLE
[x]Association Grenobloise de Recherches Interdisciplinaires Plasmas et Physique Atomique.

ELECTRON CAPTURE BY MULTIPLY CHARGED IONS : INFLUENCE OF TARGET IONIZATION POTENTIAL

M. Gargaud, P. Duthoit and R. McCarroll

Observatoire de l'Université de Bordeaux I, B.P. 21, 33270 Floirac, France

Recent experiments[1] on N^{5+}/H_2 collisions at keV energies reveal that electron capture proceeds mainly via the n=3 states of N^{4+}. This is in contrast with the N^{5+}/H system, where capture takes place primarily via the n=4 states[2].

In the molecular model of ion-atom collisions, the cross sections are largely determined by two types of parameters, R_{xi} the position of the avoided crossing i and ΔE_{xi} the corresponding potential energy difference. Both R_{xi} and ΔE_{xi} are directly related to the target ionization potential. Of course, ΔE_{xi} also depends on the detailed structure of the target, but the strong exponential decrease of ΔE_{xi} with R_{xi} means that, at least in low energy ion atom collisions, the cross section is controlled primarily by the ionization potential of the target. The aim of this work is to investigate whether the experimental results on charge exchange with molecular H_2 targets can be interpreted in a similar way or whether a new physical mechanism must be invoked to take account of the molecular nature of the target.

The method adopted is a generalization of our previous work on C^{4+}, N^{3+}/H collisions[3], using a model potential method. In the case of systems such as C^{4+}, N^{3+}, N^{5+}, O^{6+}/H, which are effectively described by an active electron in the field of two spherically symmetric ionic cores, the model potential method is capable of high precision. Its application to open shell ionic cores is more delicate but results for He targets[4,5] has proved satisfactory. In this work, we approximate the H_2 molecule by an atomic type model potential which reproduces correctly the binding energy of the H_2 ground state.

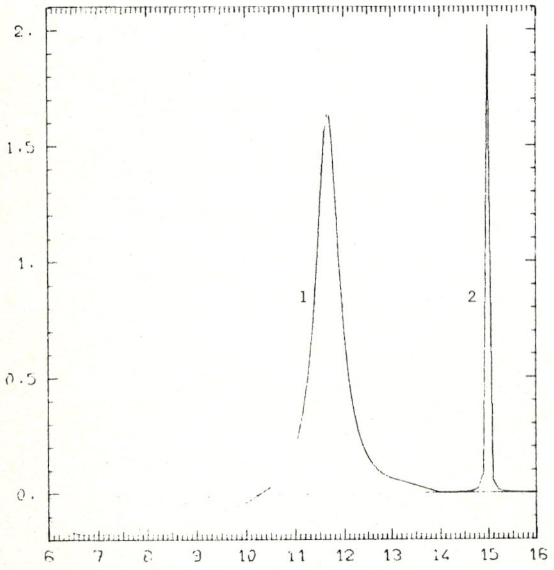

Fig. 1 : *Radial coupling matrix elements between the entry channel and the (σ4s) exit channel. Curve 1 refers to N^{5+}/H, curve 2 to N^{5+}/H_2.*

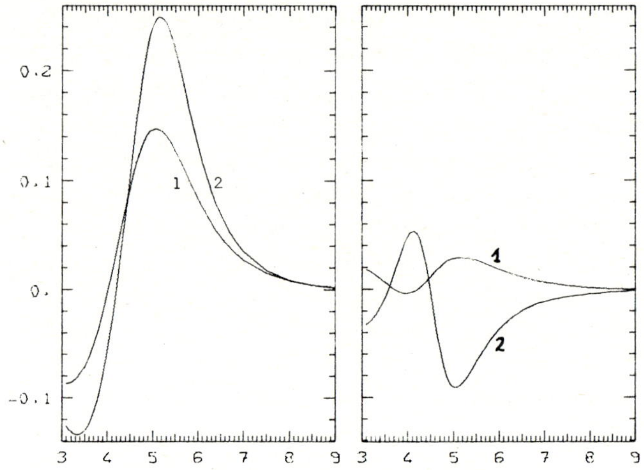

Fig. 2-3 : *Radial coupling matrix elements between the entry channel and the (σ3p) channel (Fig. 2) and the (σ3d) channel (Fig. 3). Curve 1 refers to N^{5+}/H and curve 2 to N^{5+}/H_2.*

Although the calculated energy curves for N^{5+}/H and N^{5+}/H_2 are similar, there are striking differences in the radial coupling matrix elements. Whereas for the N^{5+}/H system, the radial coupling of the entry channel with the (σ4s) state has a wide distribution around the peak at 11 a_0, for the N^{5+}/H_2 system the corresponding radial coupling is very narrowly peaked around 15 a_0. As a consequence the (σ4s) crossing is purely diabatic in N^{5+}/H_2 collisions in agreement with experiment. However, detailed calculations of the scattering equations are necessary to investigate capture into the n=3 levels. The avoided crossing involving the (σ3d) state is much more marked for N^{5+}/H_2 than for N^{5+}/H. But although the radial coupling is greater by more than a factor of four, the question is whether this coupling is sufficiently large to explain the measured total charge exchange cross sestions[6,7]. Calculations in progress will be reported at the conference.

References

1. S. Bliman, J.J. Bonnet, G. Chauvet, S. Dousson, D. Hitz, B. Jacquot, E.J. Knystautas (Private communication, 1983)
2. E.J. Shipsey, J.C. Browne, R.E. Olson, J. Phys. B 14, 869 (1981)
3. M. Gargaud, J. Hanssen, R. McCarroll, P. Valiron, J. Phys. B 14, 2259 (1981)
4. C. Bottcher, J. Phys. B 6, 2368 (1973)
5. L. Opradolce, P. Valiron, R. McCarroll, J. Phys. B 16, (to appear)
6. D.H. Crandall, R.A. Phaneuf, F.W. Meyer, Phys. Rev. A19, 504 (1979)
7. R.A. Phaneuf, F.W. Meyer, R.H. Mc Knight, R.E. Olson, A. Salop, J. Phys. B 10, L425 (1977)

INTERFERENCE STRUCTURE IN THE IMPACT PARAMETER DEPENDENCE OF K-K CHARGE TRANSFER IN SLOW S-Ar COLLISIONS

E. Justiniano, R. Schuch, M. Schulz, H.J. Specht, H. Schmidt-Böcking* and H. Ingwersen§

Physikalisches Institut, Universität Heidelberg, 6900 Heidelberg, W.-Germany
*Institut für Kernphysik, Universität Frankfurt, 6000 Frankfurt, W.-Germany
§MPI für Kernphysik, 6900 Heidelberg, W.-Germany

The importance of charge transfer processes for the K-shell excitation in highly charged ion-atom collisions at low velocities is well established[1]. In cases of near symmetric collisions K-electrons may be transferred between the K-shells of the collision partners at large internuclear distances (≈ 3 times the K-shell radius). Furthermore the coupling region is traversed twice during the collision. If one uses H-like projectiles K-K charge transfer may take place at either the incoming part of trajectory or the outgoing one. Depending on their relative phases the transition amplitudes from both coupling regions may interfere constructively or destructively. This leads to an interference structure in the transfer probability either as a function of projectile velocity for fixed trajectories or as a function of impact parameter for fixed velocities. We have investigated the impact parameter dependence of the K-K charge transfer with slow H-like ions colliding with a thin Ar gas target. The ions were produced at the MPI für Kernphysik with the MP tandem + NB combination by post-stripping a 108 MeV S beam to charge state 15 and decelerating to 7.9, 16, and 32 MeV. The K excitation probabilities were measured by x-ray scattered particle coincidence for 16 impact parameters simultaneously.

In figure 1 we present as an example the transfer probability of an Ar K-electron to S as a function of the impact parameter b for 16 MeV S^{+15}-Ar collisions. The interference structure is clearly visible. The dashed line shows the results of a calculation of Stolterfoht[2] where 2pσ and 1sσ wave functions were employed together with a model coupling matrix element. The dotted-dashed line are the results of Lin[3] where the coupling matrix elements were calculated ab initio by using two center wave functions in an atomic basis set together with a realistic (HFS) atomic potential. A comparison of both calculations indicates the sensitivity of the interference structure (phase and wavelength of oscillations) on the chosen approach. However, the height of the probabilities shows almost no dependence on the type of approach. Comparison of these calculations to the experimental results presented in figure 1 shows an agreement restricted to a rather small b range since the predicted oscillation wavelengths differ from the measured ones.

This work is supported by the BMFT.

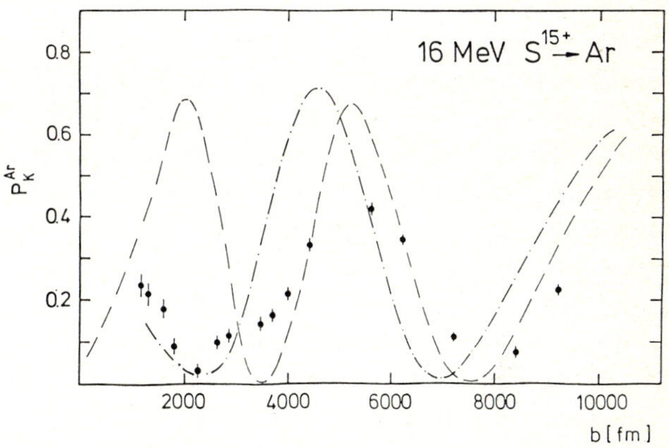

Fig. 1: Ar K-excitation probability by K-K charge transfer as a function of impact parameter. For explanation of lines see text.

References

1. see e.g. C.D. Lin and P. Richard, Adv. At. Mol. Phys. 17, 275 (1981)
2. N. Stolterfoht, J. Phys. B 13, L651 (1980)
3. C.D. Lin, private communication.

TRANSFER IONIZATION AND COULOMB IONIZATION IN COLLISIONS OF MULTIPLY CHARGED IONS WITH ATOMS+

W. Groh, A. Müller, B. Schuch, A.S. Schlachter* and E. Salzborn
Institut für Kernphysik, Strahlenzentrum, Universität Giessen,
6300 Giessen, West Germany

A time-of-flight coincidence technique was used to perform systematic experiments on transfer-ionization processes

$$A^{q+} + B \rightarrow A^{(q-k)+} + B^{i+} + (i-k)e + \Delta E$$

A,B=rare gases, q=charge state of the incident ions (q up to 15 for Xe), k=number of captured electrons found with the projectile long after the collision (k was 1,2 or 3), i=charge state of the slow recoil ion. The projectile energies ranged from 3·q keV to 15·q keV. However, with very few exceptions the charge state fractions F_i of recoil ions B^{i+} were found to be independent of the energy in the energy range studied.[1,2]

Large influences on recoil-ion spectra were observed by changing the projectile charge state q or the number of electrons k captured in a collision. They could be attributed to the maximum potential energy defect of the investigated electron capture process. The energy defect can be easily calculated for transitions between the ground states of the collision partners involved and was found to determine the recoil-ion charge-state fractions, i.e., the fractions do not depend explicitly on the projectile-ion species and charge state, but rather upon the maximum available energy defect of the collision, as long as the target is fixed. On the basis of this finding, a statistical theory was developed, which assumes that the potential energy is distributed among the outer target electrons, thus leading to a statistical degree of ionization. The recoil-ion charge-state distributions predicted by this model were found to be in good agreement with experimental charge-state distributions for two- and three-electron capture reactions.

We have performed similar measurements at higher collision energies ((30-150)q keV) with H^+-, He^+-, Ne^+-, Ne^{2+}- and Ar^+-projectiles capturing one electron (or two electrons in the case of Ne^{2+}) in various rare gas targets. Different from the low-energy measurements, the recoil-ion charge-state fractions are found to be no longer independent of the projectile energy. Furthermore, the time-of-flight spectra contain much higher recoil-ion charge states than could be expected from the maximum potential energy available in such collisions (see Fig. 1), which we attribute to Coulomb ionization by the projectile. While at low velocities the charge-state fractions are only determined by exothermal processes, the experiments at higher velocities show that more and more endothermal reaction channels become important as the collision energy increases.

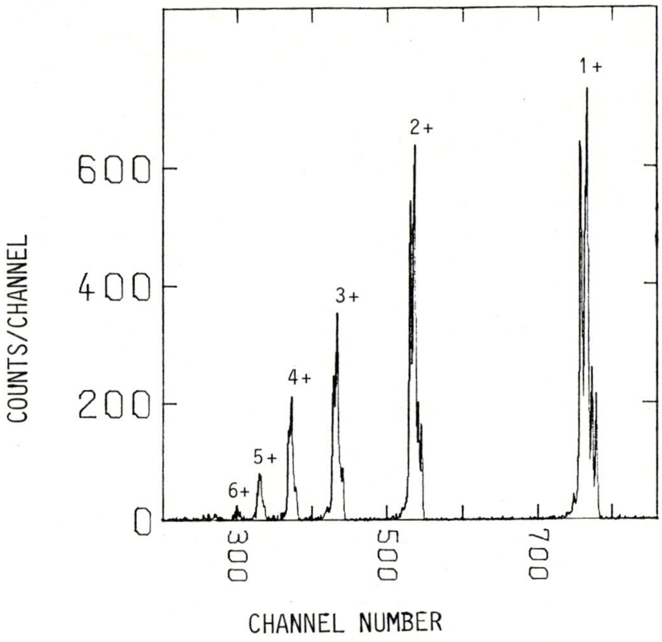

Fig. 1: Charge-state spectrum of Xe recoil ions, created by 100 keV He^+-projectiles going to He^0 in a thin Xe gas target. The structures in the charge-state peaks are due to the isotopes of natural Xe.

References

1. W. Groh, A. Müller, C. Achenbach, A.S. Schlachter and E. Salzborn, Phys. Lett. 85A (1981) 77
2. W. Groh, A. Müller, A.S. Schlachter and E. Salzborn J. Phys. B, in press

+Work supported by GSI, Darmstadt
*On leave from Lawrence Berkeley Laboratory, Berkeley/USA

CHARGE TRANSFER CROSS SECTIONS FOR MULTIPLY CHARGED KRYPTON AND XENON IONS ON VARIOUS GAS TARGETS

T. Kusakabe*, H. Hanaki, T. Horiuchi, N. Nagai, I. Konomi and M. Sakisaka

Department of Nuclear Engineering, Kyoto University, Sakyo-ku, Kyoto, Japan
*Department of Nuclear Reactor Engineering, Kinki University, Kowakae, Higashi-osaka, Japan

We have developed an "ion-impact ion source"(IIIS)[1] to produce multiply charged recoil ions, which are applicable to various charge transfer collisions in a low energy region.[2] The feature of this ion source is to use "multiple ionization" of target atoms bombarded with energetic heavy ions. In this report, we present the charge transfer cross sections ($\sigma_{q,q-k}$;k=1∿5) for 0.286 keV/amu Kr^{q+}(q=1∿9) and Xe^{q+}(q=1∿11) ions on He, Ne, Ar, Kr, Xe, H_2, N_2, CO_2, CH_4, C_2H_6 and C_3H_8.

The empirical scaling laws for single electron capture cross sections have been proposed by Salzborn et al.[3] and Bliman et al.[4] The former is

$$\sigma_{q,q-1} = 1.43\times10^{-12}\cdot q^{1.17}\cdot I_1^{-2.76}, \qquad (1)$$

where I_1 is the first ionization potential of target gas. The latter is represented as

$$\sigma_{q,q-1} = A(I_1)\cdot q\cdot I_1^{-3}. \qquad (2)$$

But applying our data, we have derived a similar rule as follows:

$$\sigma_{q,q-1} = 1.4\times10^{-13}\cdot q\cdot I_1^{-2}. \qquad (3)$$

In Fig. 1, the Salzborn's formula as well as classical one electron model (COEM)[5] are compared with one of the present results. The COEM rather well reproduces our observations. The Salzborn's scaling law deviates from our rule for smaller ionization potentials.

We have found that some of the observed $\sigma_{q,q-k}$ values against q show very similar trend irrespective of target gas. Since those for k-electron transfer decrease as the k-th ionization potential I_k becomes deeper —— roughly $\sigma_{q,q-k}\propto I_k^{-2}$, classical absorbing sphere model (CASM)[6] seems applicable. But at present, it is only a measure to know the cross section magnitudes and behaviors. Therefore, we have proposed "statistical electron transfer model"(SETM),[7] which implies that projectile ion with ionicity q captures k electrons from an electron cloud of target particle having i identical electrons and the transfer proceeds in a statistical manner via a quasimolecule formation. The transfer probability P per every electron is assumed constant against impact parameter b within an interaction range a_x. The cross section for k-electron transfer, $\sigma_{q,q-k}$, is expressed by

$$\sigma_{q,q-k} = \int_0^\infty 2\pi b\cdot {}_iC_k\cdot P^k(1-P)^{i-k}\cdot db$$

$$= {}_iC_k\cdot P^k(1-P)^{i-k}\cdot \pi a_x^2, \qquad (4)$$

where ${}_iC_k$ is binomial coefficient and ${}_iC_k\cdot P^k(1-P)^{i-k}$ is called "transfer factor". The P and a_x values were determined by a least squares method from the smoothed experimental data. A comparison is made in Fig. 2 of the present cross sections for Kr^{q+} on Kr with this SETM, and an overall agreement is seen. The interaction range a_x would be closely related with the transfer radius R_x of the COEM.

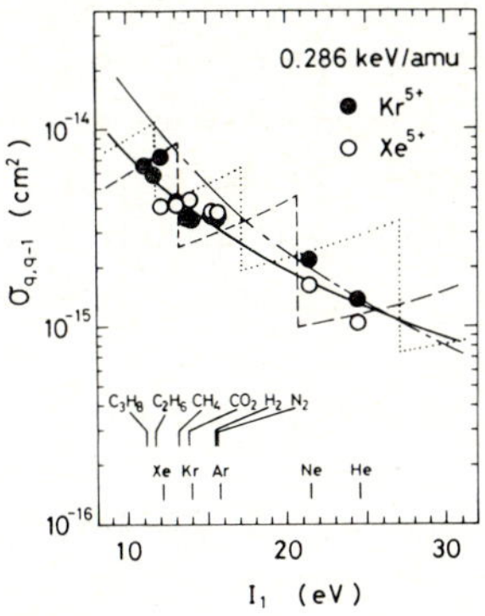

Fig. 1: Ionization potential dependence of the single electron capture cross sections for 0.286 keV/amu Kr^{5+} and Xe^{5+} ions.
———— Our law,
— - — Salzborn's law,
---------- COEM for Kr^{5+},
.... COEM for Xe^{5+}.

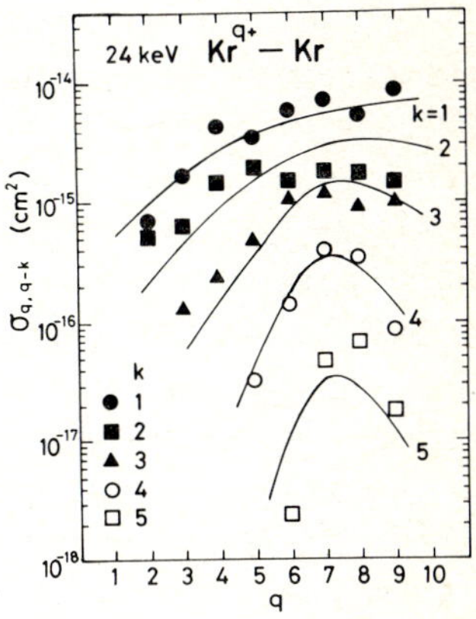

Fig. 2: q-dependence of multiple electron capture cross sections for 24 keV Kr^{q+} ions on Kr.
———— SETM.

References

1. T. Kusakabe, H. Hanaki, N. Nagai, K. Kuroda, N. Maeda, and M. Sakisaka, Nucl. Instr. and Meth. 198, 577 (1982)
2. T. Kusakabe, H. Hanaki, N. Nagai, T. Horiuchi, and M. Sakisaka, Physica Scripta (Topical Issue) 3, (1983)
3. A. Müller, and E. Salzborn, Phys. Lett. 62 A, 391 (1977)
4. S. Bliman, S. Dousson, B. Jacquot, and D. Van. Houtte, J. de Physique 42, 1387 (1981)
5. H. Ryufuku, K. Sasaki, and T. Watanabe, Phys. Rev. A 21, 745 (1980)
6. R. K. Janev, and L. P. Presnyakov, Physics Reports 70, 82 (1981)
7. M. Sakisaka, H. Hanaki, N. Nagai, T. Horiuchi, I. Konomi, and T. Kusakabe, J. Phys. Soc. Japan 52, (1983), in press

ELECTRON CAPTURE INTO N-, O-, AND F-LIKE Xe AND Sm IONS AT 3.6 MeV/u

D.H.H. Hoffmann, P.H. Mokler, W.A. Schönfeldt, A. Warczak*, and Z. Stachura*
Gesellschaft für Schwerionenforschung, D-6100 Darmstadt
*Permanent Addresses: Jagellonian University and Inst. of Nucl. Phys., Cracow, Poland

The total charge transfer into Xe^{q+} and Sm^{q+} ions at 3.6 MeV/u from Xe^{q+}-Xe and Sm^{q+}-Xe collisions was measured. Charge states q = 44-47 and q = 52-54 were selected for Xe and Sm, respectively, to study the influence of a rising number of L-shell vacancies in the projectile.

The GSI UNILAC provides the unique opportunity to produce highly ionized heavy ions at relatively low collision velocities, by first accelerating the ions up to 8.6 MeV/u, stripping in a foil to high charge states at this energy, and finally decelerating down to 3.6 MeV/u [1]. The charge state selected projectiles hit a 12 cm long gas cell which is schematically shown in Fig. 1. Target thicknesses used were of the order of 100 - 1000 μbar·cm. The target region is viewed by a Si(Li) detector and a position sensitive proportional counter. Soller slits minimized Doppler shift effects for the projectile emitted x-radiation. Particle normalization was achieved through monitoring the elastically scattered particles from a thin gold target.

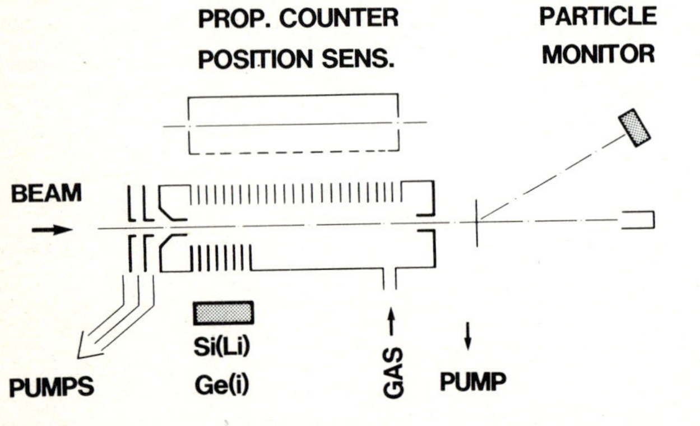

Fig. 1

For single collision conditions at low gas pressure we observe an almost charge independent L x-ray emission of low intensity when the projectile L-shell is without vacancies ($q \leq 44$ and $q \leq 52$ for Xe and Sm, respectively). A dramatic yield increase is observed as soon as L-shell vacancies in the projectile are carried into the collision. The average charge state of the ion species considered is appreciably lower ($\bar{q} \approx 30$) than the projectile charge states

$$44 \leq q(Xe) \leq 47 \text{ and } 52 \leq q(Sm) \leq 54$$

in this experiment. Thus, a yield decrease with target thickness due to multiple charge exchange collisions is expected for projectiles with L-shell vacancies, from which capture cross sections can be deduced [2]. Fig. 2 shows the normalized x-ray yield observed with the position sensitive proportional counter. The yield is normalized to the x-ray yield from projectiles with filled L-shells. This yield is practically constant over the observed range of target thickness. The full lines in Fig. 2 represent decay functions of the form

$$Y_i(x) \sum_{i=1}^{i} a_{ij}\exp(-\lambda_j x) \qquad (1)$$

which were fitted to the background corrected data, where $Y_i(x)$ (i = 1,2,3) is the x-ray yield corresponding to the number i of the L-shell vacancies, x is the target thickness, a_{ij} are weight factors, and λ_j (j=1,...,i) are the decay constants which are proportional to the charge exchange cross sections.

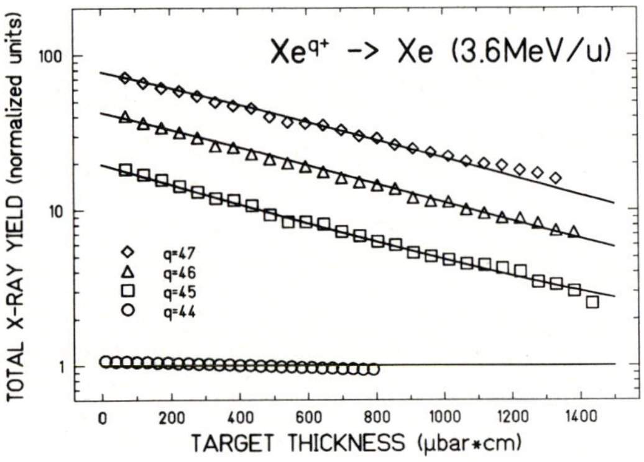

Fig. 2

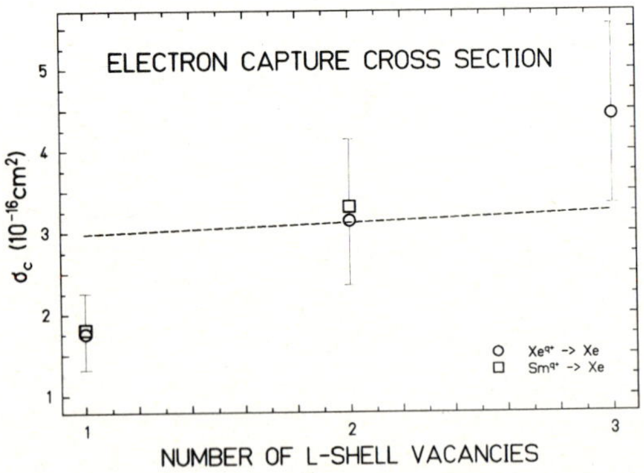

Fig. 3

This representation was chosen under the assumption that one-electron capture is the dominating charge exchange process. From this fitting procedure total capture cross sections have been deduced. The results are shown in Fig. 3 as a function of the number of projectile L-shell vacancies, and are compared to a modified capture cross section formula of Bohr and Lindhard [3,4].

1. P.H. Mokler et al., Annual Report GSI 82-1, p. 143
2. A. Schönfeldt, GSI Report 81-7
3. W. Bohr, J. Lindhard, Dan. Mat. Phys. Medd. 28, 7 (1954)
4. W. Erb, GSI-report P-7-28, p. 51

M-SUBSHELL EFFECTS IN ELECTRON-CAPTURE TO Sm^{q+} IONS

W.A. Schönfeldt, D.H.H. Hoffmann, P.H. Mokler, A. Warczak*

Gesellschaft für Schwerionenforschung, D-6100 Darmstadt
*Permanent Address: Jagellonian University, Cracow, Poland

Sm^{q+}-projectiles with only 10, 9, or 8 electrons (0, 1, or 2 incoming L-vacancies) produce L x-ray spectra (see Fig. 1) in collisions with gases. Depending on the incoming charge-state, the different L x-ray transitions could be identified by relativistic Hartree-Fock-caluclations (Ref. 1). The cross sections of these transitions are shown in Fig. 2. They exhibit for charge states $q \leq 52$ (no incoming L-vacancy) roughly a constant value, which are comparable to cross-sections for direct L-shell ionization of the projectile by the gas atoms. For charge states $q \geq 53$ (1,2 incoming L_3-vacancies) the cross sections for x-ray transitions into the L_3-shell (L_3-M_1; L_3-$M_{4,5}$) are about a factor of 50 greater than transitions to the L_2-shell (L_2-M_4) (the L_2-shell is still closed). The reason is (Ref. 2), that capture of electrons into excited states of the projectile can be recorded by these transitions. We point out that for one incoming L_3-vacancy the L_3-M_1 transition is about a factor of two greater than the L_3-$M_{4,5}$ transitions. A similar increase in the L_3-M_1 x-ray yield was observed for $Xe^{45+} \to Xe$ collisions [3]. The situation changes with more than one incoming L-vacancies (Fig. 2) and is still not fully understood. In order to study the electron capture from multi-electron systems the x-ray emission from collisions with Ar and Xe was studied, too. We find:

1. The structure of the Sm-L x-ray transitions does not depend on the target gas.
2. The cross-section for Sm-L x-ray transitions increase with the number of target-electrons and show a saturation at higher atomic numbers of the target atoms (Fig. 3).
3. The x-ray spectra of Sm → Xe collisions exhibit only a vanishing Xe-L x-ray emission for $q \geq 53$.

From this we conclude, that mainly outer target electrons are captured into the projectile.

1. J.P. Declaux, Comp. Phys. Comm. 9 (1975) 31
2. W.A. Schönfeldt, GSI Report 81-7
3. P.H. Mokler et al., 8. Int. Conf. Atomic Physics (Göteborg 1982) p. B66

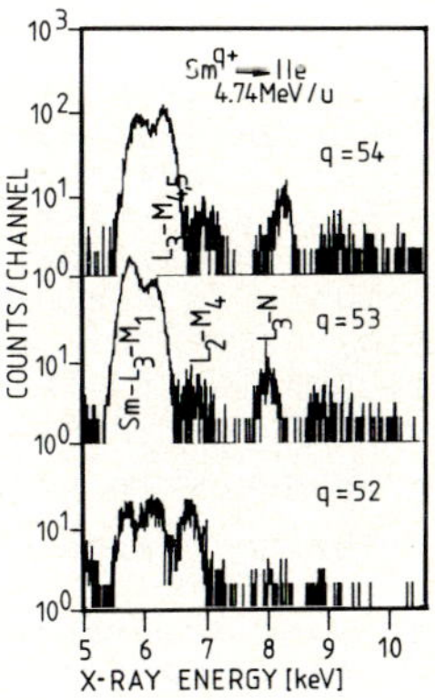

Fig. 1

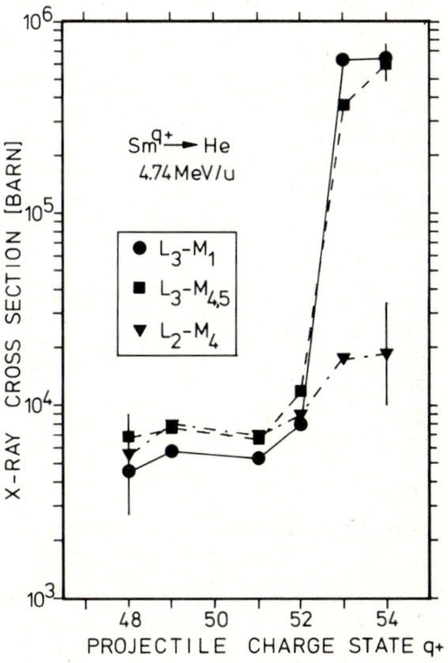

Fig. 2

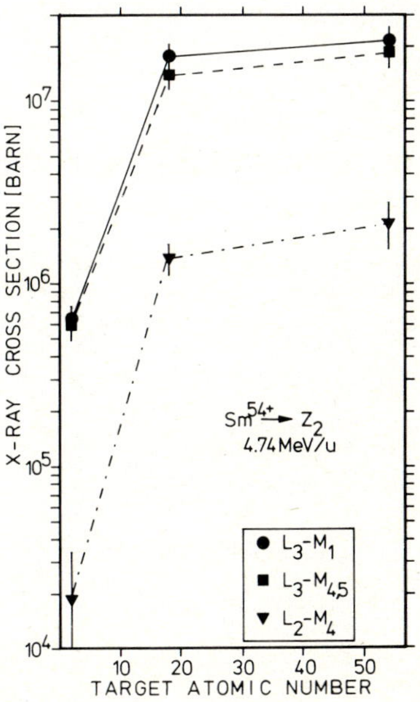

Fig. 3

CHARGE EQUILIBRIUM OF FAST HEAVY IONS PENETRATING THROUGH GASEOUS MEDIA

S.Karashima

Department of Electrical Engineering, Faculty of Engineering, Science University of Tokyo, Sinjuku-ku, Tokyo 162

T.Watanabe

Department of Applied Physics, Faculty of Engineering, The University of Tokyo, Bunkyo-ku, Tokyo 113

The average equilibrium charges, $\bar{q}$ of heavy ions passing though gaseous targets are of great theoretical interest and understanding of the processes requires the knowlege about dynamical atomic phenomena. These quantites are of decisive importance in evaluation of stopping power for many practical purposes. Stopping power is considered to the square of effective charge of the projectile as $\propto q^2_{eff}$. This q_{eff} is considered to be nearly equal to the average equilibrium charge of the projectile.

The present work theoretically develops a simple model to describe charge-changing processes of energetic heavy ions after passing through gaseous hydrogen atomic (ideal) target. The charge balance of the ions and the average equilibrium charges can dominantly determined by electron loss and capture cross sections for collisions with neutral hydrogen atoms.

When an ion of charge q traverses through matter with a velocity v_i comparable with the orbital velocity of its electron, its charge states fluctuate as a result of the electron loss and capture.

Let $\phi_q(v_i)$ be charge fraction of an incident ion with charge q at a certain velocity v_i, then $\phi_q(v_i)$ is normalized by $\sum_q \phi_q = 1$ and the average equilibrium charge is given by

$$\bar{q} = \sum_{q}^{Z} q \cdot \phi_q(v_i) \tag{1}$$

When the ion moves by a lengh along the trajectory dx in a material containing N atoms per unit volume, the change in probability $d\phi_q$ is represented by the charge-changing cross section $\sigma_{q,p}$.

Ignoring the contribution of excited states of ion and assuming the local balance in the distribution, we can obtain the following equation:

$$\frac{d\phi_q(v_i)}{N\,dx} = \sum_p \left(\phi_p(v_i)\sigma_{p,q} - \phi_q(v_i)\sigma_{q,p}\right) = 0. \tag{2}$$

Low charge-state impact with high velocity and high charge-state impact with low velocity should be taken into account unbalance effects. If multiple charge change is negligible, we can obtain under equilibrium conditions

$$\frac{\phi_q}{\phi_{q-1}} = \frac{\sigma_{q-1,q}}{\sigma_{q,q-1}} = P_q \tag{3}$$

The eq.(2) yields a formula of ϕ_q for each charge state q

$$\phi_q(v_i) = \frac{\prod_{j=1}^{q}}{1 + \sum_{q=1}^{Z}\prod_{j=1}^{q} P_j} \tag{4}$$

In order to obtain ϕ_q in eq.(4), we need $\sigma_{p,q}$ for electron loss and capture. We shall estimate the cross section for single electron loss and capture process such as

$$X^{(q-1)+} + H \rightarrow X^{q+} + e + H(\Sigma), \tag{5}$$

and

$$X^{q+} + H \rightarrow X^{(q-1)+} + H^+, \tag{6}$$

where $H(\Sigma)$ denotes all the electronic states (including ionized states) of H. For the electron loss process (5), we extend the calculation for the process of hydrogen-like ion with atomic hydrogen

$$X^{(Z-1)+}(1s) + H(1s) \rightarrow X^{Z+} + e + H(\Sigma) \tag{7}$$

by Shirai et al, where Z is the atomic number.

For the loss cross sections under the assumption of 1s-like electron for valence electrons, we adopted the scaled cross section calculated by Shirai et al based on the binary-encounter approximation and plane-wave Born approximation. The cross section for the electron capture processes (6) were obtained from the scaled results by means of unitarized-distorted wave approximation .

When we have cross sections for charge-changing processes as a function of v_i and q, we can easily calculate ϕ_q by using eqs.(3) and (4). One of the calculated results are shown in the figure and are compared with experimental data. If the momentum distribution of the outer most elecrons in the incident ion is similar to that of 1s-wave function, the ϕ_q can be scaled in terms of v_i.

We are now extending the calculation by using the Thomas-Fermi model for the momentum distribution of electrons in incident ions.

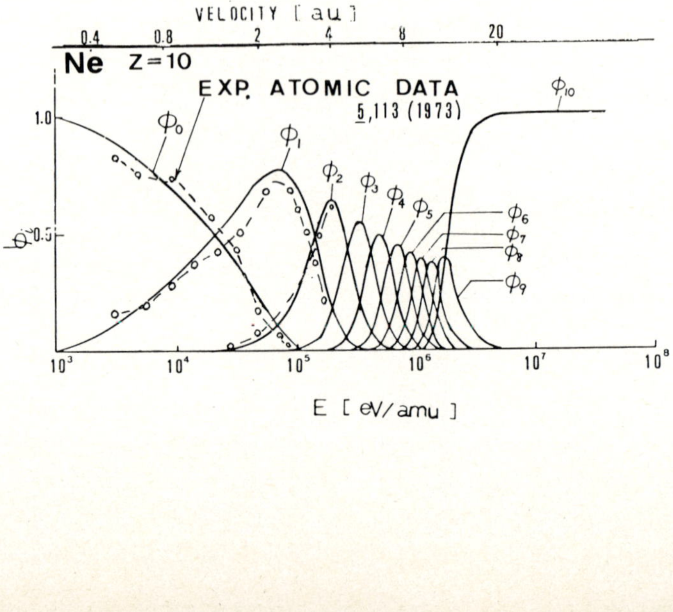

Z_T OSCILLATION OF EQUILIBRIUM MEAN CHARGE OF MeV He ION

F. Fukuzawa, Y. Haruyama, Y. Kanamori and A. Itoh

Department of Nuclear Engineering, Kyoto University, Kyoto 606, Japan

Charge state fractions of He beam emerging from various solid targets have been measured. The combination of a deflecting magnet and a solid state position sensitive detector makes it possible to measure charge fractions of backscattered beam which has continuous energy distribution[1]. With this method we can obtain charge fractions over a wide range of energy with only one incident energy. In addition to this, thick targets can be treated very easily to remove surface contamination with which the measured charge fractions of emerging ions become the same irrespective of the target material. Two methods of cleaning surfaces were used; raising target temperature and continuous evaporation during measurement. It was found that above about 1000°C of target temperature charge fractions became characteristic ones of target bulk material[2]. For the targets of low melting points which cannot be held at such high temperature, the continuous evaporation method was used, In this method, freshly evaporated surface can be produced continuously burying the sticked contamination atoms[3]. In the present energy region, 0.6-2.5 MeV, neutral fraction F_0 is very small to be neglected, and mean charge $\bar{q}$ is given in good approximation as,

$$\bar{q} = (F_1+2F_2)/(F_1+F_2) = 1+ [1+(\sigma_{21}/\sigma_{12})]^{-1}, \qquad (1)$$

where F_1 and F_2 are charge fractions of 1+ and 2+ ions, and σ_{21} and σ_{12} are electron capture and loss cross sections, respectively.

In Fig. 1, as a typical example, are plotted the $\bar{q}$ of 1 MeV He beam as a function of target atomic number Z_t. The $\bar{q}$ estimated by Eq. (1) with experimental values of σ_{21} and σ_{12} for some gas targets and metal vapor targets are also plotted. It is clearly seen that there is Z_t oscillation of $\bar{q}$, and further that the period of this oscillation is almost the same as that of the stopping cross section. From Eq. (1), it is evident that the oscillation of $\bar{q}$ is attributed to that of σ_{21}/σ_{12}. We have calculated σ_{21} on the basis of Bohr-Lindhard model[5] with taking into account of the atomic shell structure of the target[4,6]. For σ_{12}, experimental monotonic Z_t dependence has been assumed. The calculated $\bar{q}$ with these cross sections reproduces very well the experimental Z_t oscillation. Different calculations have been reported to show similar Z_t oscillation[7,8].

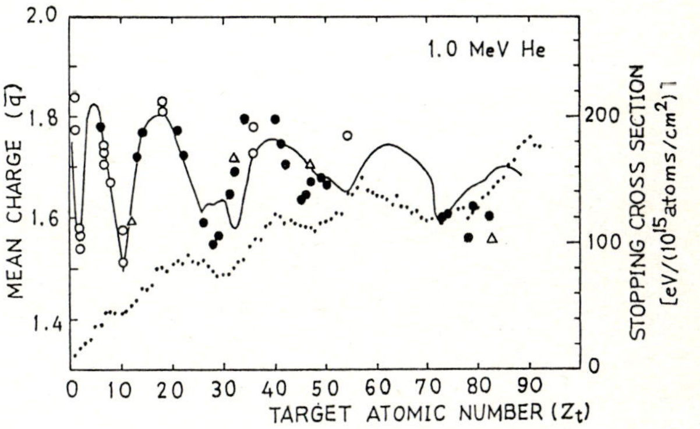

Fig. 1: Oscillation of equilibrium mean charge of 1 MeV He ions as a function of target atomic number Z_t. : solid targets[2,3,4], : metal vapor targets[9], : gas targets[9,10,11,12]. A solid curve is the calculated mean charge. The stopping cross sections of 1 MeV He ion are also plotted[13].

References

1. F. Fukuzawa, Phys. Lett. 43A, 147 (1973)
2. Y. Kido, Y. Kanamori and F. Fukuzawa, Nucl. Instr. Meth. 164, 565 (1979)
3. Y. Haruyama, Y. Kanamori, T. Kido and F. Fukuzawa, J. Phys. B: At. Mol. Phys. 15, 779 (1982)
4. Y. Haruyama, Y. Kanamori, T. Kido, A. Itoh and F. Fukuzawa, J. Phys. B: At. Mol. Phys. (in press)
5. N. Bohr and J. Lindhard, K. Dansk. Vidensk. Selsk. Mat.-Fys. Meddr. 28, 7 (1954)
6. A. Itoh, Y. Haruyama, Y. Kanamori, T. Kido and F. Fukuzawa, Bull. Inst. Chem. Res., Kyoto Univ. 60, 289 (1982)
7. I. S. Dmitriev, N. F. Vorobiev, V. P. Zaikov, V. S. Nikolaev and Ya. A. Teplova, J. Phys. B: At. Mol. Phys. L351 (1982)
8. T. Kaneko and Y. Ohtsuki, Phys. Stat. Sol. (b) 111, 491 (1982)
9. Y. Kanamori, Y. Haruyama and F. Fukuzawa, to be published
10. V. S. Nikolaev, I. S. Dmitriev, L. N. Fateeva and Ya. A. Teplova, Sov. Phys.-JETP 12, 627 (1961)
11. L. I. Pivovar, V. M. Tubaev and M. T. Novikov, Sov. Phys.-JETP 14, 20 (1962)
12. A. Itoh and F. Fukuzawa, J. Phys. Soc. Japan 50, 632 (1981)
13. J. F. Ziegler, The Stopping Powers and RAnges in All Elemental Matter, vol. 4 (Pergamon Press 1977)

COLLISIONS OF MULTIPLY CHARGED NEON IONS WITH WATER VAPOR

C.-S. O, S. B. Elston, M. Breinig, B. Thomas*, S. Berry, R. T. Short, and I. A. Sellin

University of Tennessee, Knoxville TN 37996, USA, and
Oak Ridge National Laboratory, Oak Ridge TN 37830 USA

and

D. A. Church, R. A. Kenefick and W. S. Burns

Texas A & M University, College Station TX 77843 USA

Reaction rate constants of low-energy, multiply charged neon ions with water vapor are measured using a Penning ion trap on-line with the Oak Ridge National Laboratory EN Tandem van de Graaff accelerator. Recoil neon ions are produced inside the ion trap by impact ionization of neon atoms using a pulsed 35 MeV Cl ion beam, foil-stripped to a mean charge state near 10 e. The ionization state of stored neon ions is determined by cyclotron resonance in the confining electromagnetic fields. The number of stored ions is detected by resonant axial excitation.[1] Ion loss is mainly due to electron capture from the background gas. Storage times of the order of 100 ms are observed, for neon ions of different charge states, at neon gas pressure up to 10^{-7} Torr. Addition of water vapor into the target chamber is controlled by a separate gas handling system. About one cubic centimeter of water is kept in a stainless steel tube and outgassed to remove the dissolved air. Water vapor is admitted into the target chamber via a micrometer adjusted leak valve. Partial pressures of neon gas and water vapor inside the target chamber are determined by using a quadrupole residual gas analyzer and a calibrated ionization gauge. Values of reaction rate constants of multiply charged neon ions with water vapor are obtained from variation of the storage time constants at fixed neon pressure as a function of the water vapor pressure.

Typically, a constant neon partial pressure of 10^{-7} Torr is used while the water vapor pressure is varied from 10^{-8} to 10^{-7} Torr. Previously determined rate constants[2] for Ne^{q+} on neon are used to separate the contributions to the overall rate constant from respective charge exchange collisions of Ne^{q+} with neon and water vapor. Preliminary values of the rate constants indicate they are larger by about one order of magnitude than the corresponding rate constants for Ne^{q+} on Ne collisions.

This research was supported in part by the National Science Foundation; by the Fundamental Interactions Branch, Division of Chemical Sciences, Office of Basic Energy Sciences, U. S. Department of Energy, under contract W-7405-eng-26; by the Texas A & M Center for Energy and Mineral Resources; by an NBS/NSF Precision Measurement Grant; and by the Oak Ridge Associated Universities.

References

1. R. Holmes, M. Breinig, S. Elston, S. Huldt, J.-P. Rozet, I. Sellin, W. Burns, D. Church and R. Kenefick, Bull. Am. Phys. Soc. 27, 513 (1982).
2. D. A. Church, R. A. Kenefick, W. S. Burns, I. A. Sellin, M. Breinig, S. B. Elston, S. Huldt, J.-P. Rozet, R. Holmes, D. Taylor, G. Glass and S. Berry, Physica Scripta T3, 71 (1983).

* Permanent address: Department of Physics, Carleton College, Northfield MN 55057.

ENERGY DEPENDENCE OF MODERESOLVED VIBRATIONAL EXCITATION IN H^+-CF_4 AND SF_6 COLLISIONS

U. Gierz, M. Noll and J.P. Toennies
Max-Planck-Institut für Strömungsforschung, Böttingerstraße 4-8, D-3400 Göttingen, Federal Republic of Germany

Inelastic collisions of H^+-CF_4 and SF_6 have been investigated for various collision energies E_{cm} and scattering angles θ ranging from 10 to 30 eV and 0 to 25 degrees respectively, using the time-of-flight method. In agreement with earlier experiments at 9.7 eV[1,2] we observed highly selective vibrational excitation of only the infra-red most active mode in both cases.

Due to an improved overall enery resolution, which is now better than 50 meV, it was possible to resolve pure overtone transitions of the υ_3-mode in CF_4 molecules as high as n=9. This is well within the "quasicontinuum" and provides the first direct evidence for discrete vibrational states at these high energies. Additionally at low angles the measured time-of-flight spectra show clear evidence for contributions from the second infrared less-active υ_4-mode (see Fig.1). The behaviour of SF_6 molecules in collisions with protons is found to be very similar. So far all experimental observations agreed nicely with classical trajectory calculations based on the forced oscillator model. Only the long-range attractive electrostatic interaction for vibrational coupling of ion and molecule derived from the known spectroscopic line strengths was needed[3]. According to this model the vibrational transition probabilities for overtone excitation follow a Poisson distribution, given by

$$P(0 \to n) = e^{-\varepsilon}\,\varepsilon^n/n!, \quad \varepsilon = \Delta E/h\upsilon .$$

This has been experimentally verified within 5% and better at E_{cm} = 9.7 eV and $\theta \leq 20^0$.

With increasing values of E_{cm} the measured energy loss distributions rapidly change character in a way that additional unresolved energy transfer is observed in the high energy loss part of the spectrum, whereas modeselectivity in the low energy loss region remains nearly unaffected (see Fig.2). This new behaviour which has been observed for CF_4 as well as SF_6, is not yet fully understood. It can't be ascribed do dissociation and ionization since the energy transfers are too small. A rotational coupling to such high excitation is not possible in the entrance channel because of the short range of the potential. After excitation a long range electrostatic interaction with the oscillating dipole is conceivable[3,4], but then the additional energy transfer should depend on the amount of energy transferred into the Poisson distributed overtones, which is not the case. An enhanced multimode vibrational excitation due to the short range repulsion with individual F-atoms, as observed previously in Na^+-CF_4 collisions[5], is just being investigated using classical trajectory calculations. This mechanism is not expected to be efficient enough however, since scattering angles are less than the rainbow and the mass ratio (1/19) is rather unfavourable. On the other hand, multimode vibrational excitation through a temporary charge transfer mechanism similar to the effect observed in H^+-O_2[6] can't be ruled out at the moment.

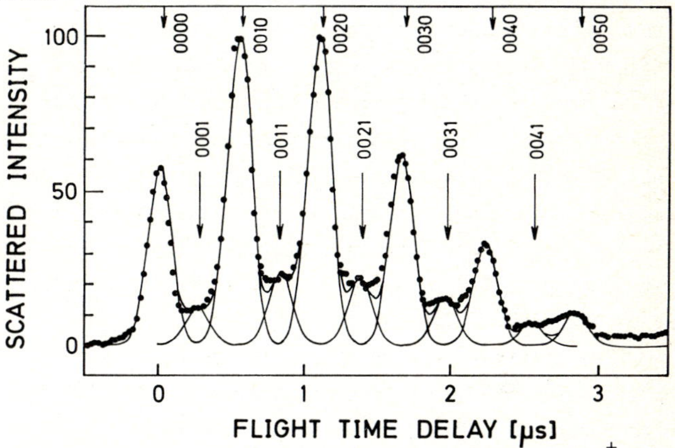

Fig.1: Time-of-flight spectrum for scattering of H^+-CF_4, taken at E = 9.8 eV and $\theta = 10^0$. Vibrational transition probabilities of υ_3-mode (00n0) are compared to a calculated Poisson-distribution for an energy transfer of $\Delta E(\upsilon_3)$ = .30 eV. The partial fill-up between adjacent υ_3-peaks can be fully explained by taking account of υ_4-mode combination band excitation (00n1).

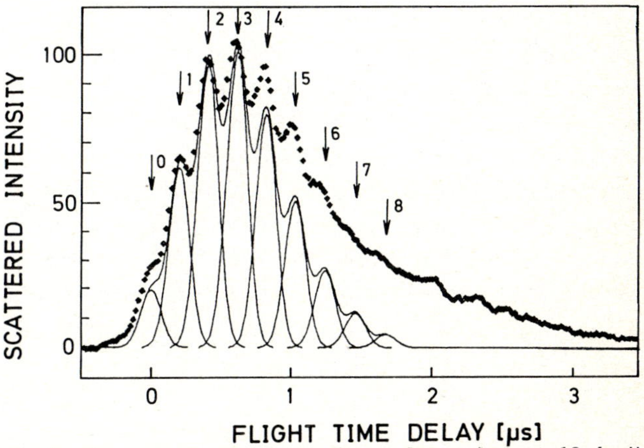

Fig.2: The same spectrum as above, but taken at 18.4 eV. The solid curve represents a calculated Poisson-distribution for $\Delta E(\upsilon_3)$ = .51 eV, which is able to reproduce the low energy loss part of the spectrum up to n=3 very accurately. However, at higher energy transfers the Poisson-distribution is by far no longer able to explain the experimentally observed distribution.

1) T. Ellenbroek, U. Gierz and J.P. Toennies, Chem. Phys. Lett. 70 (1980) 459
2) T. Ellenbroek, U. Gierz, M. Noll and J.P. Toennies, J. Phys. Chem. 86 (1982) 1153
3) T.Ellenbroek and J.P. Toennies, Chem.Phys.71 (1982)309
4) R.Tice and D.Kivelson, J.Chem.Phys.46 (1967) 4748
5) U. Gierz, J.P. Toennies and M. Wilde, accepted by Chem. Phys. Lett.
6) F.A. Gianturco, U. Gierz, and J.P. Toennies, J. Phys. B14 (1981) 667

THE ANISOTROPIC NEON-METHANE INTERACTION FROM DIFFERENTIAL ENERGY LOSS SPECTRA

U. Buck and A. Kohlhase (MPI für Strömungsforschung, D3400 Göttingen, Fed.Rep. of Germany)
T. Phillips and D. Secrest (University of Illinois at Urbana-Champaign, Urbana, Illinois, USA)

Methane is a molecule of highly symmetric structure. As a result it has no permanent dipole or quadrupole moment and its dipole polarizability is isotropic. At a first glance it appears very similar to a noble gas atom and it is interesting to study its anisotropic interactions by measuring differential cross sections for rotationally inelastic collision. Experiments on $He+CH_4$ at low energies show only weak rotational transition probabilities[1]. For $Ar+CH_4$, however, surprisingly large inelastic transition probabilities are observed in the backward direction[2]. In this work we present differential cross sections for the system $Ne+CH_4$ at the collision energy of E = 84.7 meV. A centre-of-mass angular range from 10^0 to 170^0 is covered.

The experiments have been carried out in a crossed nozzle-beam arrangement (source pressures 40 to 60 bar, nozzle diameter 20 μm). The inelastic events are detected by time-of-flight (TOF) analysis of the scattered particles. Under our experimental conditions each of three nuclear spin modifications of CH_4 relaxes to its specific ground state j = 0(A), j = 1(T), and j = 2(E) with the fractional populations 0.3125, 0.5625, 0.1250, respectively. The measured TOF spectra are transformed into energy loss spectra in the centre-of-mass system using the procedure of Ref. 3, which guarantees also the correct transformation of the amplitudes. The resulting spectra are shown in Fig. 1. The position of the elastic transition is marked by arrows. The results show remarkable inelasticities similar to those found for $Ar+CH_4$. The maximum of the transferred energy corresponds to final angular momenta between j' = 2 at $\vartheta = 45^0$ and $\vartheta = 145^0$ and j' = 5 at 85^0. Transitions up to j' = 10 are observed. Generally the transition probability increases with increasing scattering angle an exception being the transition to j' = 7, which has a maximum near 85^0.

Test calculations show the transition probabilities to depend mainly on the strength of the anisotropic repulsion corresponding to the degree of the neon atom's penetration into methane. For the analysis of the spectra a potential is constructed based on SCF-calculations for the repulsion and damped dispersion coefficients for the attraction. The scattering calculations are performed in the coupled states approximation. The final results for the potential are compared with similar results for $He+CH_4$ and $Ar+CH_4$.

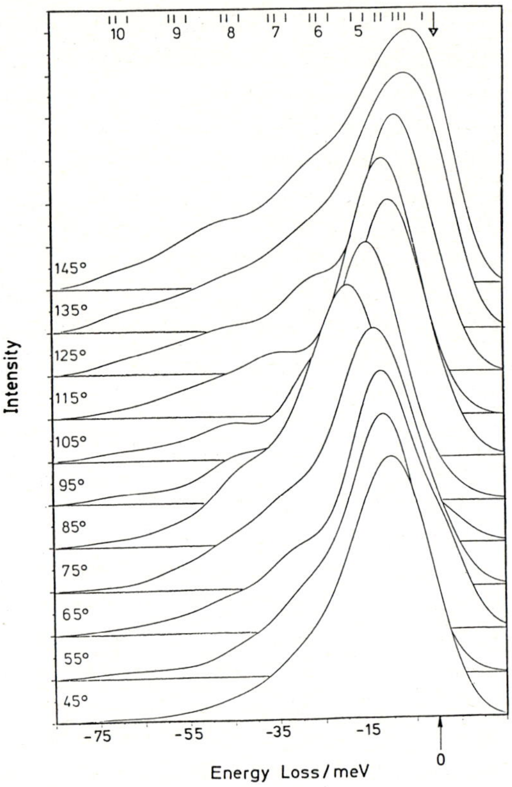

Fig. 1: Measured energy loss spectra in the centre-of-mass system for $Ne+CH_4$ collisions. The numbers in the upper part denote final rotational states of CH_4.

References

1. M. Faubel, K.H. Kohl, and J.P. Toennies, J.Chem.Phys. 73, 2506 (1980
2. U. Buck, A. Kohlhase, T. Phillips, D. Secrest, Chem.Phys.Lett. (1983)
3. J. Andres, U. Buck, H. Meyer and J.M. Launay, J.Chem.Phys. 76, 1417 (1982)

TIME OF FLIGHT ANALYSIS OF Ar_n CLUSTER SCATTERING

U. Buck and H. Meyer
MPI für Strömungsforschung, D3400 Göttingen, Fed.Rep. of Germany

The cluster population of supersonic nozzle beams is usually estimated from electron impact ionization mass spectrometry. An important problem of such a measurement is the unknown fragmentation in the ion source[1,2]. It is well known from optical methods that this fragmentation can be appreciable[3,4]. In the present work we have used an alternative method to distinguish between different cluster sizes: the kinematically different behaviour in a differential scattering experiment. Fig. 1 shows a Newton diagram for the scattering of Ar_n clusters from He at room temperature. The different circles mark the final centre of mass velocities of the different cluster sizes Ar_n. Each crossing point with the laboratory velocity arrow at a certain angle gives rise to a peak in a measured time-of-flight (TOF) distribution. Using a crossed molecular beam apparatus we have performed such a scattering experiment for Ar_n (source pressure p_0 = 1.4 bar, nozzle diameter d = 100 μm) scattered from He (p_0 = 30 bar, d = 30 μm) at room temperature. The results for the laboratory scattering angle of $\theta = 11.5^0$ are shown in Fig. 2. The upper part exhibits the TOF spectrum detected on the dimer mass (k = 2, m = 80 a.m.u). The lower part displays the result on the monomer mass (k = 1, m = 40 a.m.u.). The dimer spectrum shows the clearly resolved fast dimer peak (n = 2), the trimer peak (n = 3) and the slow dimer peak (n = 2). The same peaks are recovered in the monomer spectrum of the lower part of the figure. In addition the much larger monomer signal is now present. The measured area of the TOF spectra for the cluster size n measured on the mass k is given by

$$N_{n,k} \sim n(He)n(Ar)(p_0,T_0,d)\bar{\sigma}(Ar_n+He)(g,\theta)f_{n,k}(E_e)C_n(E_e),$$

where n is the beam density, $\bar{\sigma}$ the laboratory differential cross section, $f_{n,k}$ the fragmentation of Ar_n to Ar_k, C_n the ionization probability of Ar_n, and E_e the electron energy. For fixed source conditions the ratio of the measured areas yields directly the fragmentation $f_{n,k}$. At an electron energy of E_e = 100 eV a preliminary evaluation of the spectra gives $f_{2,2}$ = 0.3 and $f_{2,2}$ = 0.7 for the dimers and $f_{3,3}$ = 0.0, $f_{3,2}$ = 0.38, and $f_{3,1}$ = 0.62 for the trimers, where the second index gives the mass of the daughter fragment.

The experimental method has also been used to study at fixed angle θ and electron energy E_e the formation of clusters as a function of the source conditions independent of the detection mechanism. In addition, the spectra yield information on the dynamics of elastic, inelastic and dissociative scattering processes of clusters.

References

1. N. Lee and J.B. Fenn, Rev.Sci.Instrum. 49, 1269 (1978)
2. W.R. Gentry, Rev.Sci.Instrum. 53, 1292 (1982)
3. J. Geraedts, S. Setiadi, S. Stolte and J. Reuss, Chem.Phys.Lett. 78, 277 (1981)
4. T.E. Gough and R.E. Miller, Chem.Phys.Lett. 87, 280 (1982)

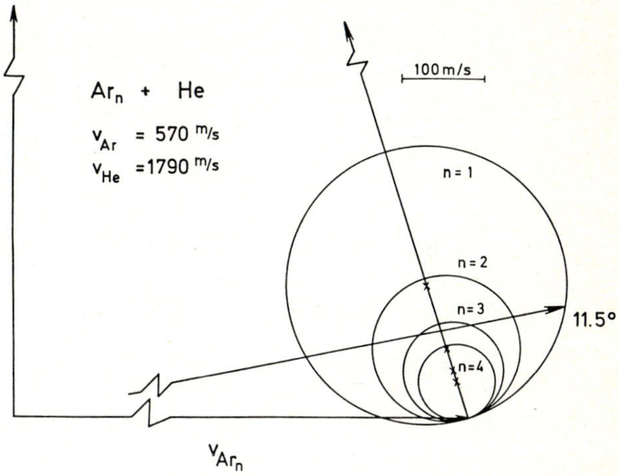

Fig. 1: Newton diagram for the scattering of Ar_n clusters from He.

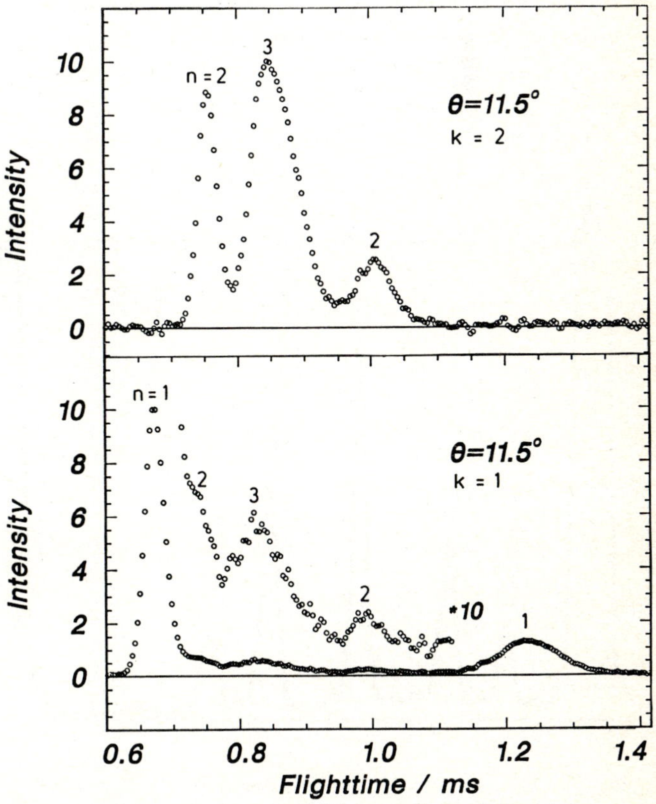

Fig. 2: Measured time of flight spectra of Ar_n+He collisions detected on the dimer (k = 2) upper part) and the monomer mass (k = 1, lower part).

TRANSITIONS BETWEEN K 4^2P ZEEMAN SUBSTATES INDUCED BY COLLISIONS WITH N_2 AND H_2

R. Berends, P. Skaliński & L. Krause

Department of Physics, University of Windsor, Windsor, Canada.

When alkali metal vapor contained in a fluorescence cell and placed in a kG magnetic field is irradiated with the corresponding monochromatic circularly polarized alkali resonance radiation, the atoms are preferentially excited to certain Zeeman sublevels of the resonance states, establishing a bulk multipole moment in the vapor. Collisions between the excited alkali metal atoms and other ground-state atoms or molecules tend to equalize the Zeeman populations, causing multipole relaxation. Cross sections for transfer between the various $^2P_{1/2}$ and $^2P_{3/2}$ Zeeman substates and for relaxation of the dipole, quadrupole and octupole moments may be determined from the relative intensities of the Zeeman components in the fluorescent spectrum.[1-3] In this investigation we studied m_J mixing and multipole relaxation induced by K-N_2 and K-H_2 collisions.

The apparatus has been described elsewhere.[3] Light from an electrodeless discharge located in a magnetic field was passed through a narrow-band interference filter and circular polarizer, and was focused in a fluorescence cell containing K-vapor + buffer gas, located in the field of a second electromagnet. The two magnetic fields were adjusted so as to permit the selective excitation of the $^2P_{1/2,-1/2}$ or $^2P_{3/2-3/2}$ Zeeman substate.[3] With N_2 or H_2 in the cell, the other Zeeman substates become collisionally populated, causing the appearance of the corresponding components in the fluoresence spectrum. The latter was resolved with a piezoelectrically scanned Fabry-Perot interferometer and was plotted out as shown in Fig. 1. Measurements of relative intensities of the Zeeman components in the fluorescent spectrum, in relation to gas pressure, yielded cross sections for m_J mixing, multipole relaxation and $^2P_{1/2}\leftrightarrow{}^2P_{3/2}$ excitation transfer as listed in Table 1. The Ar cross sections agree well with previous measurements[3] as do the $^2P_{1/2}$ - $^2P_{3/2}$ mixing cross sections.[4] The ordering of the N_2 and H_2 and Ar cross sections indicates their dependence on the polarizability of the collision partner and lack of participation in the interaction of molecular rotational and vibrational degrees of freedom.

Table 1: Cross sections $\Lambda^{(1)}$, $\Lambda^{(2)}$ and $\Lambda^{(3)}$ for dipole, quadrupole and octupole relaxation in K $4^2P_{1/2}$ and $^2P_{3/2}$ atoms and for $1/2\leftrightarrow 3/2$ mixing. All values in $Å^2$, ±15%.

Coll. Partner	$\Lambda^{(1)}_{1/2}$	$\Lambda^{(2)}_{3/2}$	$\Lambda^{(2)}_{3/2}$	$\Lambda^{(3)}_{3/2}$	$\sigma_{1/2\to 3/2}$	$\sigma_{3/2\to 1/2}$
Ar	52.0	170.1	250.0	203.2		
N_2	71.3	165.1	240.1	184.0	74.0	53.4
H_2	58.0	95.4	144.2	80.8	61.6	38.9

References

1. J. Gay and W. Schneider, Z.Phys. A278, 211 (1976).
2. R. Boggy and F. Franz, Phys.Rev. A 25, 1887 (1982).
3. P. Skaliński and L. Krause, Phys.Rev.A 26, 3338(1982).
4. J. Ciuryło and L. Krause, J.Q.S.R.T. 29, 57 (1983).

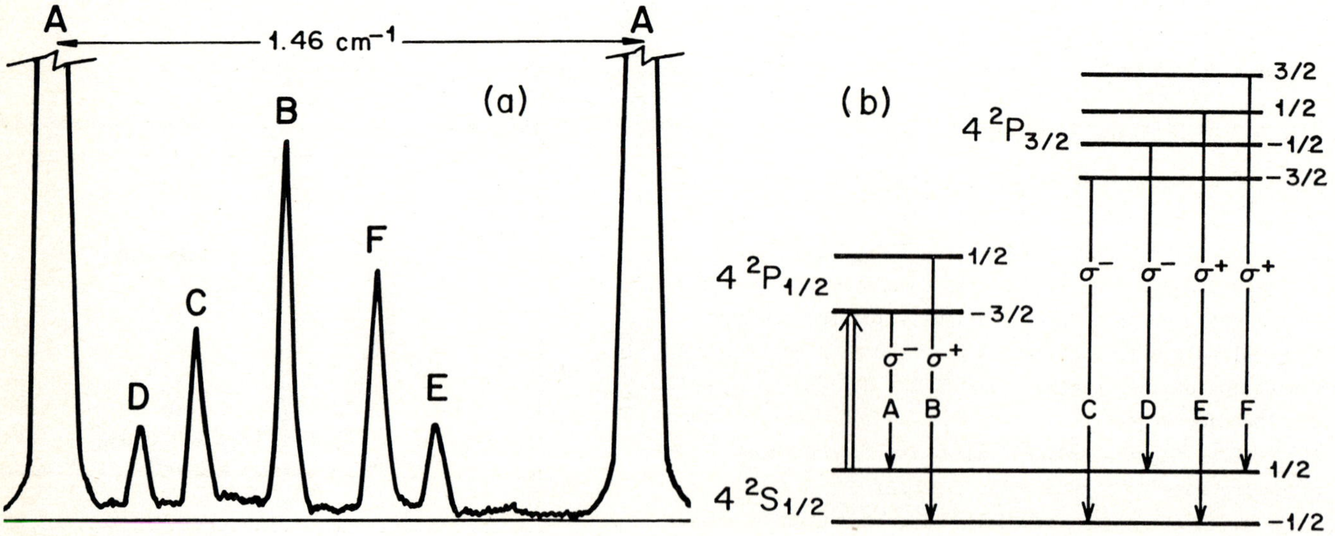

Fig. 1. Fluorescent Zeeman spectrum with 0.31 torr N_2 (a). The $4^2P_{1/2,-1/2}$ Zeeman state was optically populated. Measured intensity ratio A/B ≅ 10. The peaks correspond to transitions labelled on the energy level diagram (b).

He^o+D_2 COLLISIONS AT LOW keV ENERGIES*

J. Jakacky, Jr., V. Heckman, and E. Pollack

Dept of Physics, University of Connecticut, Storrs, CT 06268

We are studying low keV energy He^o+D_2 collisions with particular emphasis on the "quasi-elastic" scattering. This is an extension of our earlier work[1] on He^++H_2, and[2] Ne^o and Ne^+ on H_2 and D_2. The quasi-elastic channel is of particular interest since it yields information on the ground state potential energy surface.[3] In addition, in He^o+D_2, it provides a basis for the extension of a scaling law,[4,5] for the energy loss distributions in the electronically elastic ion-molecule channel, when electronically inelastic channels are present. The scaling law was derived under the assumption that the collision is electronically elastic and it was verified[2] when this condition was met. In He^o+D_2 we find (in agreement with the results of a recent study[6]) that the inelastic channels are strong in the Eθ (beam energy x lab scattering angle) range investigated. The scaling is confirmed even at large Eθ values where the inelastic channels are dominant.

Time-of-flight techniques are used for the measurements. The apparatus has not previously been described and is shown in Fig. 1. Figure 2 is a plot of the measured kinetic energy loss of He^o in electronically

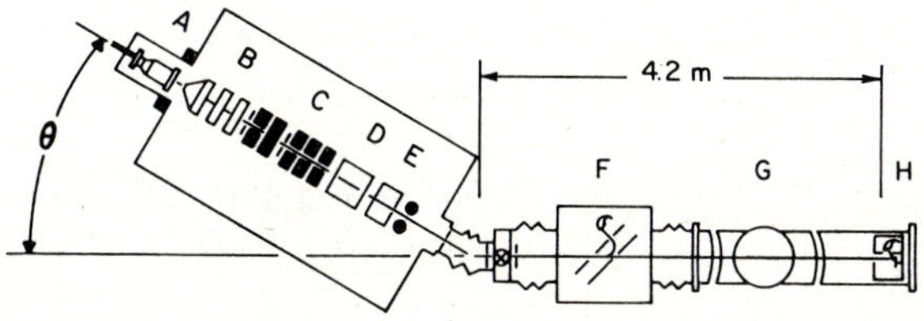

Fig. 1: The experimental arrangement.

elastic 1.5 keV He^o+D_2 collisions as a function of $E\theta^2$. The collision is seen to be elastic for $E\theta^2<2.25$ keV deg^2. Beyond this point the data "break-away" from the

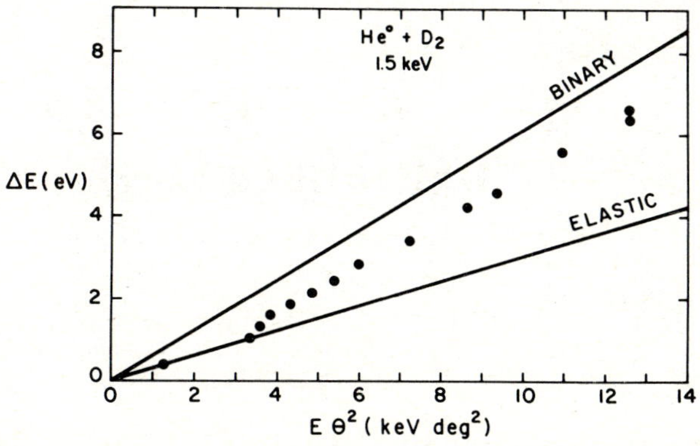

Fig. 2: Measured energy loss for the quasi-elastic channel.

elastic limit. The difference between the measured energy loss and the elastic curve represents Q, the vibro-rotational excitation energy of the D_2. Because the energy differences are small, He^o+He collisions are used to determine the position of T, the elastic channel at each scattering angle. The scaling predicts that $f = \frac{1}{2}(1+Q/T)$ is a function of Eθ only. Figure 3 shows a plot of f as a function of Eθ for 1.0, 1.5, and 2.0

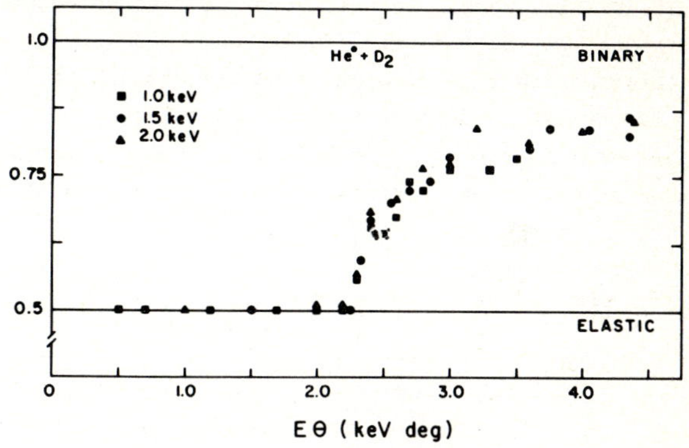

Fig. 3: A plot of f vs Eθ for He^o+D_2.

keV He^o+D_2. The data are seen to scale even though there are strong electronically inelastic channels contributing to the scattering.

We have also studied the width of the quasi-elastic peaks and our results show that there is only a limited range of vibro-rotational excitation about the maximum. Cross sections will be presented. Related work[7] is now in progress.

References

*Research supported by the University of Connecticut Research Foundation.

1. A.V. Bray, D. S. Newman, and E. Pollack, Phys. Rev. A 15, 2261 (1977).
2. N. Andersen, M. Vedder, A. Russek, and E. Pollack, Phys. Rev. A 21, 782 (1980).
3. R. Snyder and A. Russek, Phys. Rev. A 26, 1931 (1982) and R. Snyder and A. Russek, XIII ICPEAC, Book of Abstracts.
4. P. Sigmund, J. Phys. B 11, L145 (1978).
5. P. Sigmund, J. Phys. B 14, L321 (1981).
6. D. Dowek, D. Dhuicq, V. Sidis, and M. Barat, Phys. Rev. A 26, 746 (1982).
7. J. Jakacky and A. Russek, XIII ICPEAC, Book of Abstracts.

DIFFERENTIAL CROSS SECTIONS IN NEUTRAL-NEUTRAL COLLISIONS

J. H. Newman, Y. S. Chen, K. A. Smith, and R. F. Stebbings

Department of Space Physics and Astronomy, Rice University, Houston, Texas USA 77251

Energetic fluxes of oxygen, hydrogen, and helium ions precipitating into the atmosphere play an important role in magnetospheric and ionospheric Physics. Neutralized by charge transfer at high altitudes, these particles undergo collisions with atmospheric neutrals. There is a need for measured differential cross sections to accurately model the effects of neutral-neutral collisions in atmospheric heating and to determine the resulting escape flux of atoms from the atmosphere.[1]

We present differential cross sections for collisions of 0.4 to 5.0 keV H, He, and O atoms with various gases. Cross sections have been measured within the angular range 0.1° to 20°.

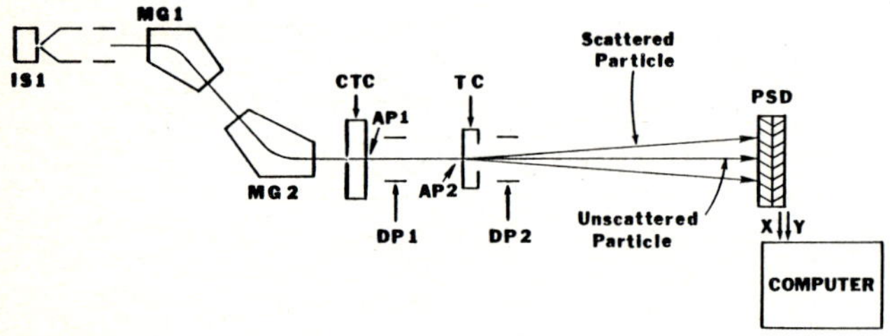

Figure 1. Schematic of the apparatus.

Ions extracted from an electron-impact ion source are momentum analyzed by a pair of 60° sector magnets and subsequently pass through a charge transfer cell (CTC), into which an appropriate gas is admitted. Neutrals formed in the CTC are collimated to about 0.5 mrad by two apertures (AP1 and AP2) < 25 microns in diameter and 10 cm apart. Residual ions are removed from the beam by deflection plates DP1.

AP2 is also the entrance to the target cell (TC). Gas is admitted to the TC and the collimated neutrals undergo scattering collisions. Three target cells are used. A "long" cell of length 1 cm and exit aperture diameter 200 microns (L/D ratio of 50:1) is used to measure differential cross sections up to 1°. An "intermediate" cell about 2 mm long with L/D ratio 7:1 permits measurement of cross sections up to about 4°. A "short" cell of length 1 mm and L/D ration 2:1 is used to measure differential cross sections to 20°. The exit aperture of the "short" cell is 500 microns and is offset by 150 microns from the beam axis, allowing detection of signal up to 20°. The effective collision path length of the "long" cell may be accurately determined because, under present operating conditions, end effects for a cell with L/D ratio 50:1 are neglible. The effective length of the other two cells may then be determined relative to the "long" cell.

Figure 2. Scattering of 3000eV He atoms by various targets.

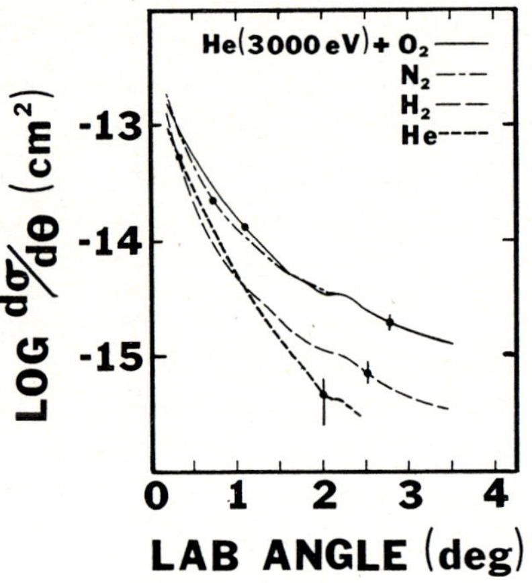

Primary and scattered particles are detected by a position sensitive detector (PSD) comprised of a pair of microchannel plates in a chevron configuration. The PSD has a resolution of about ± 60 microns and an active area diameter of 25 mm. Position information is stored by a microprocessor during the experiment and analyzed later. The PSD allows collection of data from all scattering angles of interest simultaneously, permitting data acquisition "in parallel", thereby greatly enhancing data rates.

Differential cross sections as a function of laboratory scattering angle for Helium scattered from various targets are shown in Figures 2 and 3.

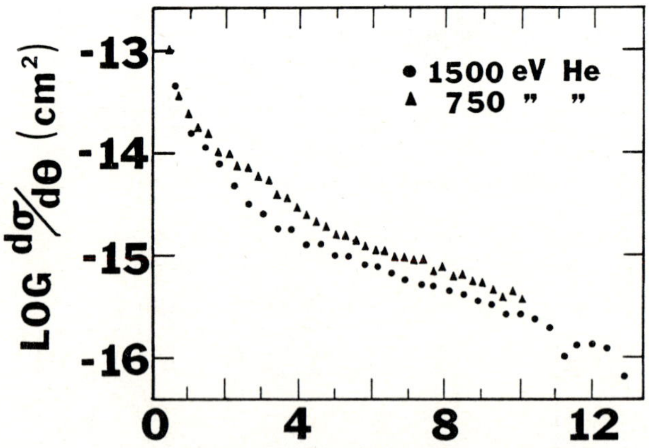

Figure 3. He - O_2 scattering as a function of energy.

REFERENCES

1. M. R. Torr and D. G. Torr, Geophys. Res. Lett., 6, 700 (1979); J. U. Kozyra, T. E. Cravens, and A. F. Nagy, J. Geophys. Res., 87, 2481, (1980)

Research supported by NSF Grant No. ATM 80-23219 and NASA Grant No. NSG-7386.
One of us, J.H.N., gratefully acknowledges the support of Texaco in the form of a Texaco Fellowship.

EFFECTS OF TRANSLATIONAL AND SUPERFINE RELAXATION OF I ATOMS IN OXIGEN-IODINE LASER

M. V. Zagidullin, V. I. Igoshin, N. L. Kuprianov

Kuibyshev Filial of Phisical Lebedev Institute of Academy of Sciences of USSR

One of the most important advances in the field of electronic transitions lasers is the creation of the chemical oxygen-iodine laser based on the resonance energy transfer from singlet oxygen to atomic iodine:

$$I(^2P_{3/2})+O_2(^1\Delta)=I(^2P_{1/2})+O_2(^3\Sigma),$$
$$k=7,6\ 10^{-11}cm^3/s.$$

Usually the stimulated emission in this laser is described within the framework of two-level exitation scheme. Due to superfine splitting the spectrum of $^2P_{1/2}-^2P_{3/2}$ transition consists of 6 separate lines. In addition to chemical reactions, energy exchange and deactivation processes the important role in kinetics of lasing play the effects of translational and superfine relaxation (cross-relaxation) of atomic iodine.

The rate of cross-relaxation processes and their influence on spectral and power performances of the oxygen-iodine laser are analysed in this work. The possibility of formation of atomic iodine due to dissotiation in threemolecular collisions

$$2O_2(^1\Delta)+I_2=O_2(^3\Sigma)+2I$$

is dis cussed.

The expression for unsaturated gain wich takes into account the finite rate of cross-relaxation is derived. The parameters of saturation are determined by electron energy transfer, translational relaxation of iodine atoms and mixing the superfine sublevels of $^2P_{1/2}$ and $^2P_{3/2}$ states. The energy transfer from singlet oxygen to atomic iodine is the Maxwellian pumping source for the upper laser level $^2P_{1/2}$. The kinetic analysis showes that mixing sublevels of $^2P_{1/2}$ state results from collisions with oxygen molecules having electron moment with the rate constant equalled to $10^{-10}cm^3/s$, while mixing the sublevels of $^2P_{3/2}$ state results from collisions with any particle with the rate constant equalled to $4\ 10^{-10}cm^3/s$. Evaluation of the rate constant translational relaxation gives the value $7\ 10^{-11}cm^3/s$. It is followed from these evaluations that character of saturation depends on active medium composition. For a large dilution of the mixture by a buffer gas the saturation of separate line of $^2P_{1/2}-^2P_{3/2}$ transition is homogeneous, that is the region of phase space, where particle velocity results in resonance with resonator mode, is furnaced by translational relaxation well inuf. In the case of pure oxygen-iodine mixture the line remains inhomogeneous broadened at high pressures and Bennet dips in frequency dependence of radiation power must be occured. It is shown that the three-four fold exeeding of the gain above the threshold on the F=3-F=4 transition results in lasing on the line F=2-F=2. The evaluating of saturation intensities are made. The regime of multimode lasing when the saturation of line may be regarded as homogeneous is considered too.

The results obtained must be taken into account in modelling oxygen-iodine chemical laser.

USE OF EHRENFEST'S PRINCIPLE FOR INELASTIC COLLISIONS

W. Elberfeld and M. Kleber
Physik-Department, T.U. München, West Germany

In a recent study[1] we showed

1) how to put Ehrenfest's principle into action for heavy particle impact;
2) how to determine transition probabilities in a unique way from trajectories.

We investigated the Landau-Teller model[2] for the collinear atom-diatomic molecule collision. In this model a harmonic oscillator CB is exposed to an exponential potential acting between A and B. We found excellent agreement between the results of the optimized Ehrenfest method and those of an exact semiclassical coupled channel calculation. In the following example the Ehrenfest results cannot be distinguished from exact coupled channel results.

The reaction dynamics is depicted in Figs.1-2. These show the deformation of the oscillator in configuration and momentum space as a function of time. Also shown is the time dependence of the uncertainty Δx^2. We have chosen a heavy scaled reduced mass (m=3.737) and a deep molecular potential width (α=0.5584) which approximately represents the collision HBr+He with Br being atom C. $\tilde{E}=E/\hbar\omega + 0.5$ is the sum of impact energy and ground state energy of the oscillator ($\tilde{E}$=6 in Fig.1). Fig. 3 shows energy transfer, excitation probability and the Cauchy-Schwarz lower bounds for the squared overlap between exact wave function $|\psi(t)\rangle$ (coupled channels) and the approximate wave function $|\phi(t)\rangle$(Ehrenfest method)

$$|\langle\psi(\infty)|\phi(\infty)\rangle|^2 \geqq 1 - \sin^2 \min\left\{ \frac{\pi}{2},\; 1/\hbar \int_{-\infty}^{\infty} dt' \langle\varepsilon(t')|\varepsilon(t')\rangle^{1/2} \right\} \qquad (1)$$

with $|\varepsilon(t)\rangle = (H - i\hbar\partial_t)|\phi(t)\rangle$. This lower bound is an extremely pessimistic estimate in view of $|\langle\psi(\infty)|\phi(\infty)\rangle|^2 \geqq 0.999$ in the whole range of impact energies. In part b of Fig.3 we compared for illustrative purposes various approximations to a complete quantal calculation in which the relative atom-molecule motion has been treated quantally. Fig.3 warns us not to believe too faithfully in semiclassical methods for head-on collisions.

Fig. 1: Average displacement (in units of the oscillator length d) and momentum (in units of $\hbar$/d) of the harmonic oscillator in a collinear collision. Starting point t=-∞, turning point t=0.

Fig. 2: Change of the collision induced fluctuation Δx^2. Length and time are in units of d and ω^{-1} respectively. Trajectory starts with Δx^2=0.5.

Fig. 3: a) Energy transfer ΔE in units of $\hbar\omega$. A is the classical result, B the result of the Ehrenfest method. C are the r.h.s. values of Eq.(1).
b) Vibrational transition probabilities for the 0→1 transition. Compared are various methods: complete quantal calculation by Clark and Dickinson (D), Ehrenfest method (E), classical (F), first order distorted wave (G).

[1]W. Elberfeld and M. Kleber, J.Phys.B16 (in print)
[2]A.P. Clark and A.S. Dickinson, J.Phys. B6, 164 (1973)

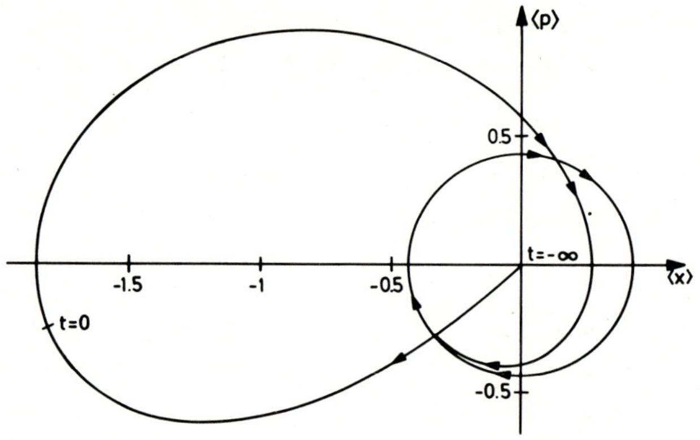

Fig. 1

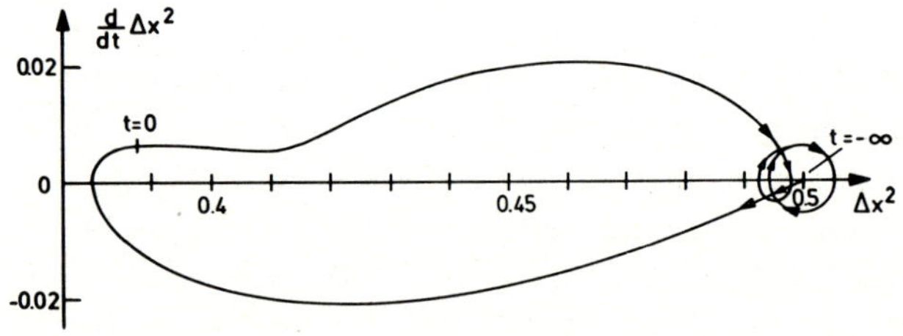

Fig. 2

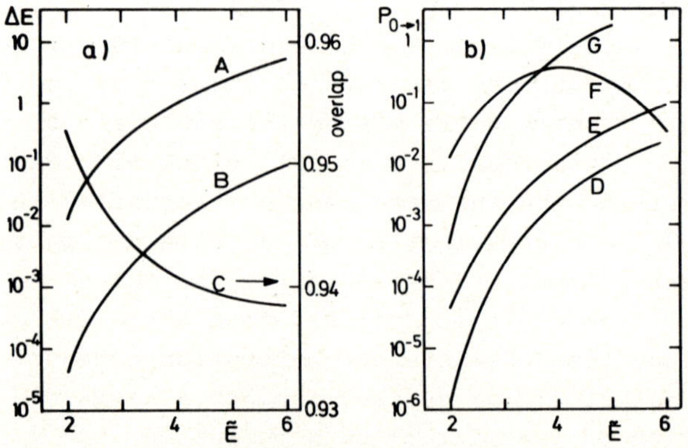

Fig. 3

COMPUTED VIBRATIONAL AND ROTATIONAL INELASTICITY IN COLLISIONS VIA REALISTIC INTERACTIONS

F.A. Gianturco[*] U.T. Lamanna[+] G. Petrella[+]

[*]Department of Chemistry , University of Rome, Rome , Italy .

[+]Department of Chemistry, University of Bari, Bari, Italy.

Accurate benchmark calculations of vibrational excitation processes taking place under collisional regimes have been restricted so far to very simple systems and have mostly used model potentials to treat either the anisotropic features of the interactions or the dependence of the potential energy surfaces on the internal coordinate.

This restriction in computational treatments mostly comes from two main bottlenecks that immediately appear as soon as one proceedes to formulate the numerical problems. One is the rapid proliferation of rotational states that need to be considered as the collisional energy increases and as more vibrational channels become energetically open,while the other is the largely scant knowledge that one still has about the detailed interplay of the structural features of the potential energy surface and the outcomes of the dynamics driven by it to yield vibrational excitation of the target and ultimately its relaxation behaviour as a function of gas temperature.

We are reporting here exact quantum mechanical calculations of the cross sections for vibrational excitation of HF in collision with He and for the vibrational excitation of CO + H_2 mixtures. In both cases the computed interactions explicitely involved their dependance on the vibrating coordinates and therefore allowed us to study in greater detail the effect of thelocal shapes of the potential energy surfaces on the vibrational coupling matrix elements, hence on the final computed total and partial inelastic cross sections.

For the case of the HF target, previous calculations had been performed via a model potential and a model form of the coupling between vibrational modes as induced by the He projectile[1,2]. They were concerned with the spherical interaction only and therefore were carried out within the Breathing sphere Approximation(BSA) , although the coupled channel problem was correctly solved till convergence only in the more recent attempt by us[2]. The present study employs instead a full potential energy surface which was recently computed, for the CO+H_2 case, by judiciusly combining the Electron Gas Model approach and the asymptotic form of dispersion forces smoothly joined with the short-range region via damping functions[3]. For the case of the HF+He system a fully ab initio surface was obtained by combining the short-range behaviour from Hartree-Fock calculations and the long-range region from dispersion coefficients that were smoothly joined in via damping functions(the HFD model)[4] . In both cases the dependance on the internal coordinate was fully considered over the whole range of interaction and the corresponding matrix elements were explicitely computed as function of the collision coordinate.

The quantum, close-coupling formalism was employed for the (He+HF) system for the vibrational degree of freedom, while the IOSA approximation was used to treat rotational inelasticity. In the case of the (CO+H_2) mixtures only the (V,T) transfer process was considered, while in both cases a detailed analysis of the collisional inelasticity as dictated by the shape of the coupling[5] was carried outsuccessfully and will be presented at the Meeting.

References

1.J.Detrich and R.W.Conn,Chem.Phys.36,407(1979).

2.F.Battaglia and F.A.Gianturco,Chem.Phys.,55,283(1980)

3.M.C.VanHemert,Jerusalem Symposia in Quant.Chem.,14,161('81)

4.W.R.Rodwell,L.T.Sin Failam,R.O.Watts,Mol.Phys.,44,225('81)

5.G.Drolshagen and F.A.Gianturco,Mol.Phys.,in press.

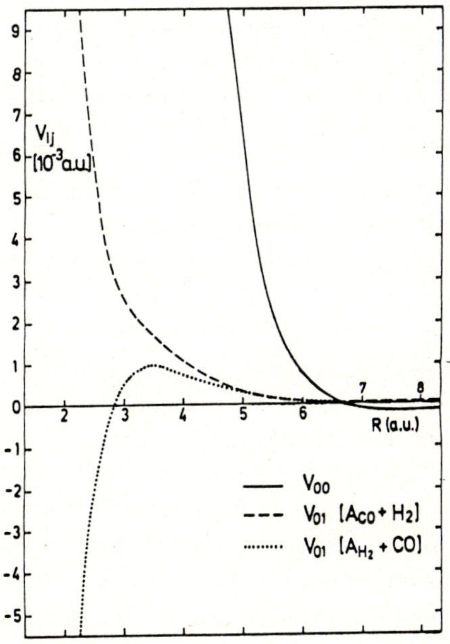

Fig.1 Coupling matrix elements as function of collision coordinates:(—)spherical interaction for CO+H_2 as rigid rotors;(---)with only CO as an averaged rotor;(...) with H_2 as an averaged rotor.

SCATTERING OF RARE GAS ATOMS (Ne, Ar) BY ALKALI HALIDE MOLECULES (KCl,CsCl,KF,CsF)

D.N. Tripathi and D.K. Rai
Department of Physics
Banaras Hindu University
Varanasi-221005
INDIA

Collisions of rare gas atoms with alkali halide molecules have been studied extensively[1]. Several model interaction potentials[2] have been used for determining the scattering cross-sections. A close scrutiny of these potentials including the latest one given by Toennies et al.[1], which was extracted from the experimental scattering data for rare gas- CsF, CsCl and KCl collisions, clearly indicates the limitations of these models in representing correctly the long-range anisotropic part of the potential. The potential energy hyper surface as obtained from ab-initio CI calculations may also be used in scattering studies and their success or otherwise in reproducing the observed cross section would indicate their suitability. A systematic evaluation of the various model potentials as compared to the potential energy surfaces on the one hand and the observed scattering cross section data on the other are however lacking.

The present paper, therefore, aims to present a critical study of the various model potentials in the light of the experimentally determined cross sections as well as the results obtained when the calculated potential energy surface is taken to represent the interaction potential. The relative importance of the various Δj transitions as a function of the colliding energy has also been investigated.

Quantum mechanical close coupling formulation in the total angular momentum $J(=j+l)$ representation is used on the lines suggested by Arthurs and Dalgarno[3]. The diatomic molecule is taken as a rigid rotor with no vibrational degrees of freedom in its ground electronic state. The coupled radial Schrödinger equations are solved by the integral equation method of Sams and Kouri[4]. The scattering T-matrix so obtained has been used to calculate the integral and differential cross sections using the expressions due to Blatt and Biedenharn[5]. Inclusion of terms upto $j=16$ in the rotator-basis-set has been found sufficient to give good convergence for the energy range of 0.05 to 5.0 eV. The convergence has also been tested with respect to the maximum total angular momentum quantum number (J_{max}) as well as the range of the integration. Alongwith, various semiempirical interaction potentials given earlier the model potential given by Toennies et al.[1] has also been used.

Ne-KCl, Ne-KF, Ar-KF, Ar-KCl, and Ne-CsF, Ne-CsCl, Ar-CsF, Ar-CsCl systems have been investigated. The calculated cross sections for eleastic and rotational excitation (integral + differential) upto $\Delta j=16$ and the relative transition probabilities with respect to the partial waves would be reported. A comparison of the present results with the available observed and other theoretically calculated results as well as our conclusions regarding the potentials will be presented.

References

1. G.Meyer and J.P.Toennies, J.Chem.Phys. 77, 798 (1982) and the references given therein.
2. R.B.Bernstein, Atom-Molecule Collision Theory - Planum Press, New York (1979).
3. A.M.Arthurs and A.Dalgarno, Proc. Phys. Soc. (London) A256, 540 (1960).
4. W.N.Sams and D.J.Kouri,(a) J.Chem.Phys. 51, 4809 (1969),(b) J.Chem.Phys. 51, 4815 (1969).
5. J.M.Blatt and L.C.Biedenharn, Rev. Mod. Phys. 24, 258 (1952).

ION-MOLECULE DIFFERENTIAL CROSS SECTIONS AND ENERGY TRANSFER DISTRIBUTIONS

F. E. Budenholzer and C. C. Lee

Department of Chemistry, Fu Jen Catholic University, Taipei, Taiwan, ROC

In a recent paper[1] we described a new method to calculate total (that is summed over all final states) differential cross sections and rotational energy transfer distributions at specified scattering angles for low energy atomic ion - linear molecule collisions. Briefly, for a potential of the form

$$V(R,\psi) = \sum_{l=0}^{n} V_l(R)\, P_l(\cos\psi)$$

where R is the distance between the linear molecule center of mass and the atomic ion, ψ the angle between R and the molecular axis, $V_l(R)$ potential terms generally in inverse powers of R and P_l the Legendre polynomials, classical perturbation scattering theory (CPST) can be used to calculate deflection angles and changes in angular momentum as a function of impact parameter b and space-fixed angles defining the orientation of the molecule.[2] Given these functional forms, differential cross sections and energy transfer distributions can be calculated using standard Monte-Carlo techniques. Within the limits of the perturbation approximation, cross sections can be calculated which agree well with the results of full trajectory calculations, but at a fraction of the computer time.

We here present the application of the method to several systems which have been previously studied experimentally: $Li^+ - H_2$, D_2 and N_2. In figure 1 are displayed the results for Li^+ scattered by N_2. The circles show the results of our calculations, the triangles are taken from the experimental report of Böttner, Ross and Toennies.[3] The agreement between our results, using potential parameters from the work of Polak-Dingels[4], and the experimental results is quite satisfactory.

Figure 2 shows the rotational energy transfer distribution at ϑ_{cm}=36.5 deg. and E_{cm}=4.23 eV. Experimental results[5] show a Gaussian like curve with a peak at a final angular momentum J_f = 17 and dropping to zero at about J_f = 36, in good agreement with our results.

The peak in the energy transfer distribution (proportional to the double differential cross section, $d^2\sigma/d\vartheta dE$) corresponds to the so-called rotational rainbow. Comments will be made on the nature of the dynamics of anisotropic ion-molecule scattering.

This work was supported by the National Science Council of the Republic of China.

References

1. F. E. Budenholzer and C. C. Lee, Chem. Phys. 73, 323 (1982).
2. E. A. Gislason and J. G. Sachs, Chem. Phys. 25, 155 (1977).
3. P. M. Polak-Dingels, Ph.D. thesis, (University of Illinois at Chicago Circle, 1979).
4. R. Böttner, U. Ross and J. P. Toennies, J. Chem. Phys. 65, 733 (1976).
5. D. Poppe and R. Böttner, Chem. Phys. 30, 375 (1978).

Figure 1. Differential Cross Section

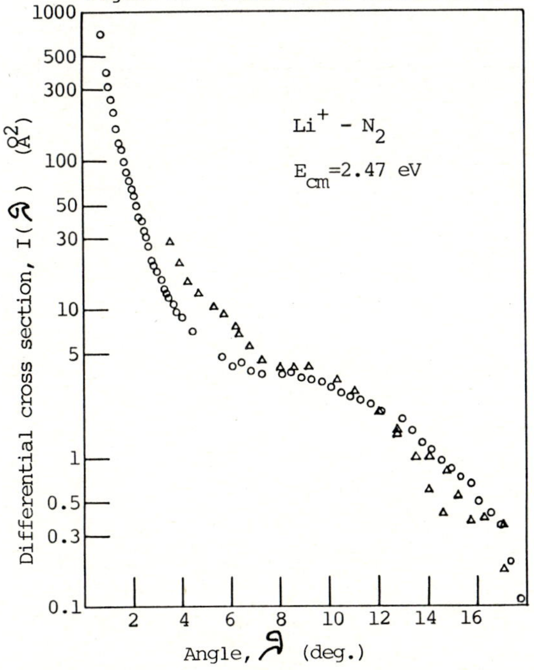

Figure 2. Final Rotational Energy Distribution

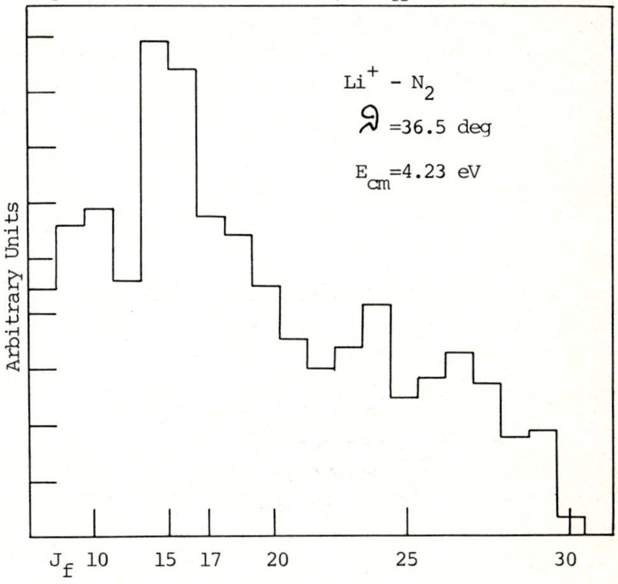

A COMPARISON OF CLOSE COUPLED SCATTERING RESULTS WITH MEASUREMENTS ON He-CO AT $E_{c.m.}$ = 27.7 meV

W. Dilling and J. Schaefer

Max-Planck-Institut für Astrophysik, 8046 Garching, Fed. Rep. Germany

Experimental results of He-CO beam scattering[1] have been used to test an ab initio interaction potential[2] in close coupled scattering calculations, and to adjust the leading three terms of the potential Legendre expansion. The experimental results available for this comparison were the total differential cross section and time-of-flight (TOF) spectra at two laboratory scattering angles (Θ_{LAB}=32.73,40.33°) measured at E_{cm}=27.7 meV. The initial rotational state distribution has been estimated as 66.1%,27.1%, and 6.9%, for j=0,1,2, respectively.

The theoretical center of mass differential cross sections have been calculated by using a sufficiently complete basis (j_{max}=9) and have been transformed to the laboratory frame for the comparison. Empirical improvements of the potential Lengendre terms have been obtained by scaling the center of mass distance, as $V = \Sigma\, V_\lambda^{emp.}(R)\, P_\lambda\,(\cos\gamma) = \Sigma\, V_\lambda^{ab\ initio}(b_\lambda R) P_\lambda\,(\cos\gamma)$, which gives smaller corrections at small distances and very sensitive corrections close to R_o determining the phase of the undulatory structure of the total differential cross section.

The best agreement with measurements has been obtained with a scaling parameter set $b_o=b_1=0.99$, $b_2=0.985$. Fig.1 shows the leading three ab initio potential terms and the corrections (points) established by the latter parameter set. Fig.2 presents the total differential cross section measured (and scaled:+) and calculated including all inelastic and the elastic contributions (x:j=0→0;•:j=1→1;□:j=2→2). Fig.3 shows the measured profile of the TOF at Θ_{LAB}=40.33°, and the profile of the sum of the weighted theoretically determined cross sections. Gaussian profiles derived from the apparatus data have been used for each transition.

Some of the features of the He-CO system important for the comparison are noted briefly: V_o, V_1, V_2 represent a strongly coupled interaction which mainly determines all the rotational transitions observed (successive excitations to j=3 and 4). The results are insensitive to comparable changes of V_3 and V_4. At Θ_{LAB}=40.33° all the absolute differential cross sections contributing to the main peak have almost the same magnitude and undulatory structure with different phases. The results are also insensitive to the form of the attractive part of the potential.

References:

1) M. Fanbel, K.H. Kohl, and J.P. Toennies, J. Chem. Phys. 73, 2506 (1980)

2) L.D. Thomas, W.P. Krämer, and G.H.F. Diercksen, Chem. Phys. 51, 131 (1980)

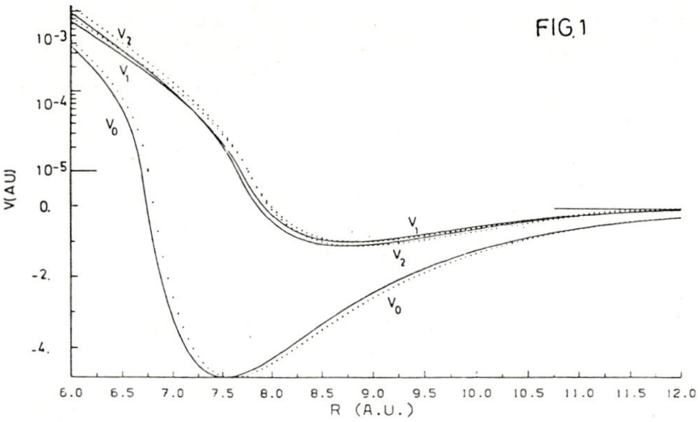

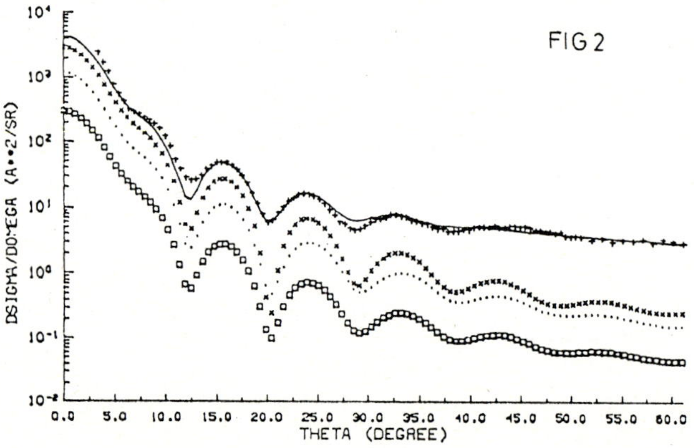

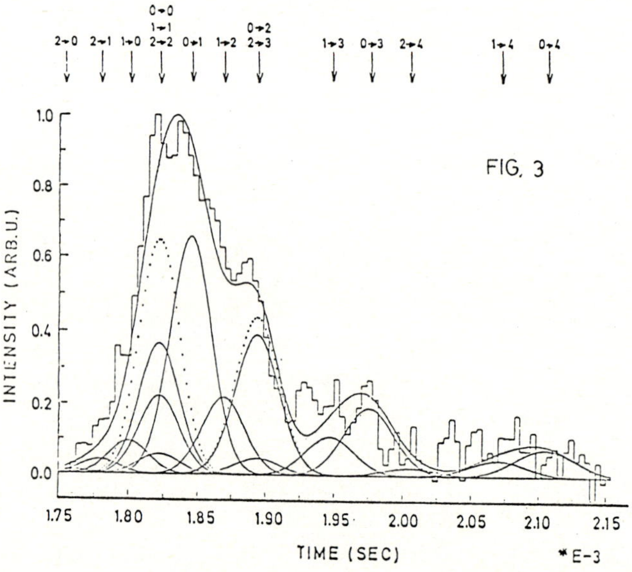

EXCITATION FUNCTIONS FOR ROTATIONAL EXCITATION IN ATOM RIGID-ROTOR COLLISIONS

D. Poppe

I.-N.-Stranski-Institut der Technischen Universität Berlin, D-1000 Berlin, FRG.

Rotational excitation of a molecule arising from collision with an atom has been the subject of many recent investigations in collision theory[1]. It is the purpose of the present paper to give some insight into the exact classical dynamics of rotational excitation. This can be most appropriately done by investigating the so-called excitation functions. They link uniquely the initial conditions of classical trajectories to their final properties. All the information necessary to construct cross sections is contained in the excitation functions for the molecular angular momentum, the orbital angular momentum, and the scattering angle.

Numerical results have been obtained using a model potential for the He-Na_2-system. This has been chosen in order to compare with IOS(infinite-order-sudden)-cross sections[2,3] which agree very well with more exact treatments. Typical contour plots for various axcitation functions are given in Fig 1. Many of their properties can be understood from a comparison with the corresponding IOS-function. Differences indicate a partial break down of the IOS approximation.

References

1. R. Schinke and J.M. Bowman in Molecular Collision Dynamics, edited by J.M. Bowman, (Springer, Heidelberg, 1982) and references herein.
2. H.J. Korsch and D. Poppe, Chem. Phys. 69,99(1982).
3. Reinhard Schinke, H.J. Korsch and D. Poppe J. Chem. Phys. 77, 6005(1982).
4. W.H. Miller, J. Chem. Phys. 54, 5386(1971).

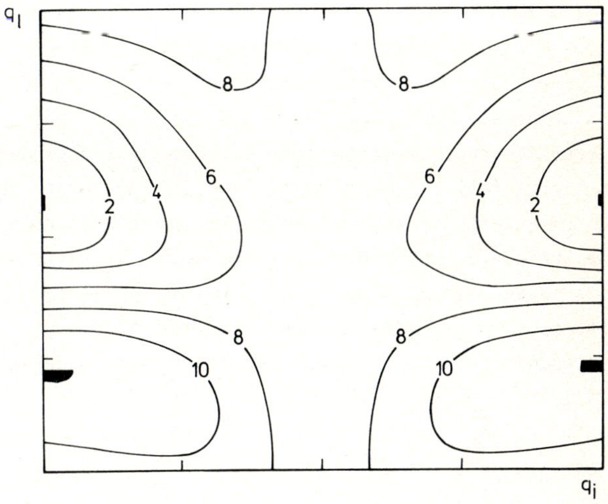

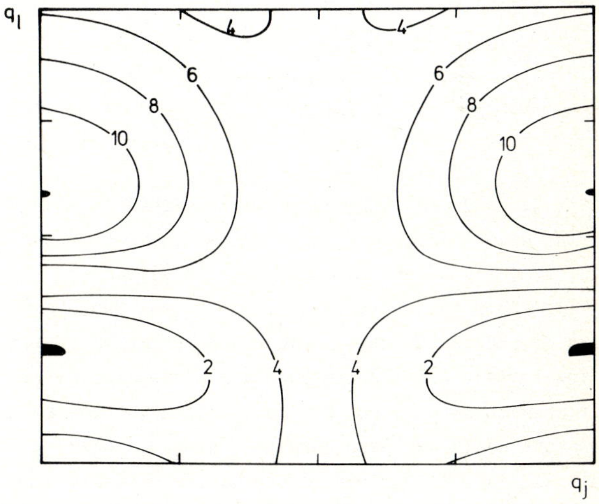

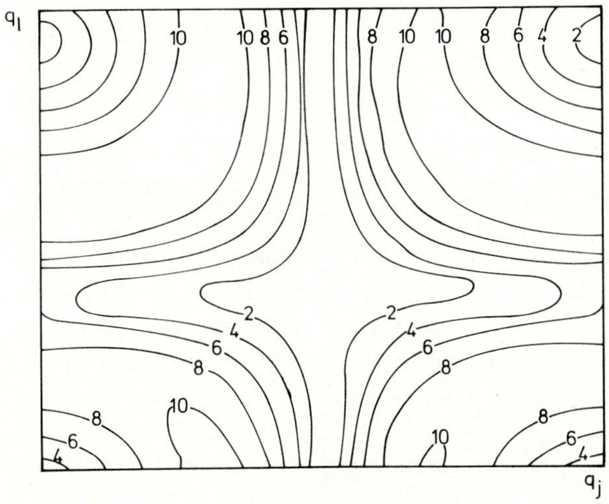

Fig. 1: Contour plots of various final properties in the (q_j, q_l)-plane[4] for total angular momentum J = 21.5.

(a) angular momentum (O, 17.3)

(b) orbital angular momentum (4.8, 34.8)

(c) scattering angle (95°, 143°).

Contour lines are given for 12 equidistantly chosen values between maximum and minimum. Their values are in dicated in brackets.

ON ROTATIONAL EXCITATION OF INITIALLY ROTATING MOLECULES

H.J. Korsch[+], Z.V. Lewis[+], D. Poppe*

[+]Fachbereich Physik der Universität, 6750 Kaiserslautern, W.-Germany

*Institut für Physikalische und Theoretische Chemie, Technische Universität, Berlin, W.-Germany

The typical features of the $j \to j'$ cross sections in atom-diatom collisions under sudden conditions are well understood for initially nonrotating (j=o) molecules with purely repulsive interactions: The classical differential cross sections show a characteristic square root singularity at the maximum allowed final rotational momentum at each scattering angle (the celebrated rotational rainbow), which is quantum mechanically softened into a maximum accompanied by a series of oscillations on the classically allowed side. The integral $o \to j'$ cross sections show a structureless decay with increasing j'.

The purpose of this paper is to understand the modifications of this behaviour for $j \neq 0$. The IOS cross sections for such transitions are directly obtained from the $o \to j'$ ones by means of the factorization formula

$$\frac{d\sigma}{d\Omega}(j' \leftarrow j;\vartheta) = (2j'+1) \sum_{j''} \begin{pmatrix} j & j' & j'' \\ 0 & 0 & 0 \end{pmatrix}^2 \frac{d\sigma}{d\Omega}(j'' \leftarrow 0;\vartheta) \qquad (1)$$

or its classical equivalent

$$\frac{d\sigma}{d\Omega}(J' \leftarrow J;\vartheta) = \frac{2J'}{\pi} \int_{|J'-J|}^{J'+J} \frac{dJ'' \, \frac{d\sigma}{d\Omega}(J'' \leftarrow 0;\vartheta)}{\sqrt{[(J'+J)^2 - J''^2][J''^2 - (J'-J)^2]}} \qquad (2)$$

(J,J', ... are classical rotational momenta in units of $\hbar$). The weight factor in (2) is a consequence of the axial symmetry of the molecule. The relations (1) and (2) also hold for integral cross sections. Numerical tests of the factorization formula have been restricted to low transitions and low energies because of difficulties with exact numerical calculations of the quantum cross sections. Such computations are, however, possible in classical mechanics and can be used as a test of the validity of the IOS approximation.

Figure 1 shows integral cross sections for the $He-Na_2$ system[1] at a collision energy of 0.1 eV and a high initial rotational state j=50 (rotational energy 0.05 eV). A comparison between classical IOS[1] results computed from the factorization formula (2), quantum IOS results obtained from (1) and exact classical trajectory calculations shows excellent agreement.

The structure of the differential cross sections is determined by the IOS selection rule $|\Delta J| \leq J_R(\vartheta)$, where $J_R(\vartheta)$ is the rotational rainbow curve for J = 0,

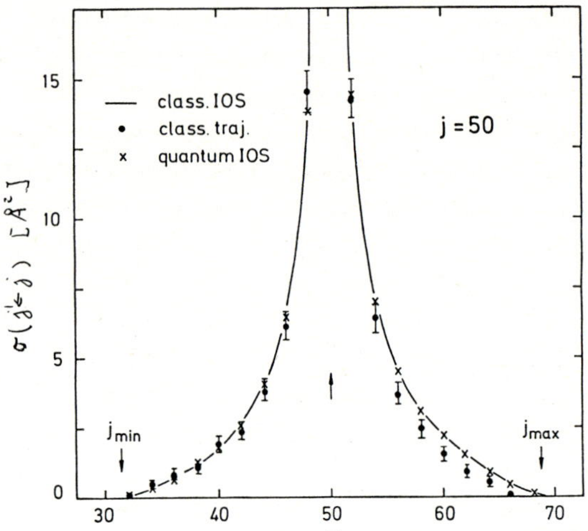

Fig. 1

i.e. the boundaries of the classically allowed region are $\max\{o, J-J_R(\vartheta)\} \leq J' \leq J+J_R(\vartheta)$. An analysis of equation (2) shows that the square root singularities of the classical cross sections at the boundaries of the classically allowed region change into steps, and in addition logarithmic singularities appear for elastic transitions and for $J' = J_R(\vartheta) - J > 0$. Similar structures have been recently observed for hard shell scattering[2].

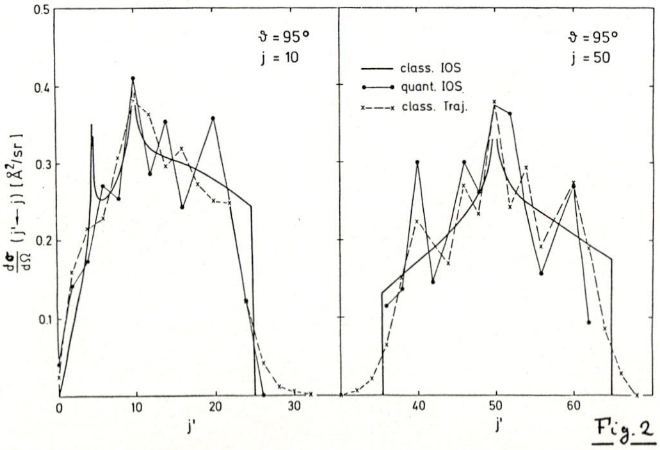

Fig. 2

Fig. 2 shows the rotational rainbow structures for two typical cases and gives a comparison of classical IOS and quantum IOS differential cross sections for $\vartheta = 95°$. Again good agreement is observed and the prediced structural changes are confirmed.

References

1. H.J. Korsch and D. Poppe, Chem. Phys. 69, 99 (1982)
2. D. Beck, U. Ross and W. Schepper, Z. Phys. A299, 97 (1981)

DETERMINING INTERMOLECULAR POTENTIALS FROM ATOM-MOLECULE SCATTERING DATA

R. Snyder and A. Russek

Department of Physics; The University of Connecticut; Storrs, CT 06268

Previous work on energy loss in medium energy rare gas + H_2 collisions[1-5] has studied the nature of the changeover from purely elastic scattering at small reduced scattering angles $\tau = E\theta$ to scattering with significant vibro-rotational excitation at large τ. Experimentally, the observed behavior, first seen by Bray *et al*[1] and confirmed by Andersen *et al*[2,3] and Vedder et al[4], is completely elastic scattering for all values of τ less than some critical value, τ_c, with a sudden onset of vibro-rotational excitation beyond τ_c. In all cases except that of H^+ on H_2,[4] the scaling law derived by Sigmund[6] is observed. *Ab initio* Hartree-Fock studies[7] of the ground state energy surface of the triatomic HeH_2 system has indicated that a good approximation to the interaction potential can be obtained by parametrizing it as a sum of relatively short range two-body screened Coulomb repulsions between the He and each H plus a longer range three-body Born-Mayer repulsive polarization potential acting between the He and the *center* of the H_2. This latter term, which dominates at larger He-H_2 separations, cannot cause vibro-rotational excitation. Only the short range two-body forces can lead to scattering beyond τ_c, and these forces bring with them vibro-rotational excitation. Calculations demonstrate that the onset of vibro-rotational excitation occurs at a value of τ_c which is very nearly the maximum τ allowed by the three-body force alone[5]. Thus, an important connection was established between the scattering data and the intermolecular potentials.

Here we extend and tighten this connection by developing a formula which expresses τ_c as an explicit function of the three-body potential parameters. Following Ref. 7, we parametrize the three-body polarization component of the interaction potential as

$$V_3(R) = A_3 \exp(-\lambda_p R) - B \exp(-bR), \qquad (1)$$

where R is the distance between the He and the center of the H_2. To preserve continuity of slope at the origin, it is necessary that $A_3 \lambda_p = Bb$. For a single Born-Mayer potential, $A \exp(-\lambda R)$, it can be shown that τ_{max} is given by $A \sqrt{(\pi/4e)}$ in the small angle, large impact parameter approximation, where e is the base of natural logarithms. Although the latter approximation is not strictly valid at the maximum itself, the expression for τ_{max} is an excellent approximation for a single exponential potential. The broadness of the maxima, along with the loose condition $b/\lambda_p \simeq 1$, which is satisfied in all cases studied, suggests that the maximum τ resulting from the three-body potential described by Eq. (1) satisfies a relation of the form:

$$\tau_{max} \simeq (A_3 - B)\sqrt{(\pi/4e)} \qquad (2)$$

Explicit numerical calculations show that τ_{max} for a wide range of potentials of the form (1) are indeed proportional to $A_3 - B$. The best fit coefficient of proportionality is found to be $1.15\sqrt{(\pi/4e)}$ rather than the value provided by Eq. (2). Converting from hartree-radians to keV-degrees, the final result is

$$\tau_{max} = 36(A_3 - B) \text{ keV-deg}, \qquad (3)$$

where the strengths A_3 and B are expressed in keV. The table shows how well Eq(3) predicts the critical τ_c as calculated for several rare gas + H_2 potentials we have previously discussed:

System	A_3(keV)	B(keV)	Eq.3	τ_c(KeV-deg)
Ne + H_2 (Ref. 5)	.163	.129	1.2	1.2
Ne^+ + H_2 (Ref. 5)	.544	.430	4.1	4.5
He + H_2 (Ref. 8)	.092	.040	1.9	1.8
He + H_2 (Ref. 5)	.075	.060	0.5	0.5

It is seen that Eq. (3) is remarkably successful in providing a direct and simple link between τ_c and the strengths A_3 and B for the three-body polarization potential. It quantifies the physical observation that the stronger the three-body interaction is, the further out in τ elastic scattering will dominate over scattering with vibro-rotational excitation, and relates experimental scattering information to the triatomic molecular energy surface.

References

1. A. V. Bray, D.S. Newman and E. Pollack, Phys. Rev. A15, 2261 (1977).
2. N. Andersen, M. Vedder, A. Russek and E. Pollack, J. Phys. B 11, 1493 (1978).
3. N. Andersen, M. Vedder, A. Russek and E. Pollack, Phys. Rev. A21, 782 (1980).
4. M. Vedder, H. Hayden and E. Pollack, Phys. Rev. A23, 2933 (1981).
5. R. Snyder and A. Russek, Phys. Rev. A26,1931(1982).
6. P. Sigmund, J. Phys.B11, L145 (1978);14,L321(1981).
7. A. Russek and R. Garcia G.,Phys.RevA26,1924(1982).
8. R. Snyder and A. Russek, Abstracts of Contributed Papers, ICPEAC XII, 947 (1981).

ENERGY LOSS SCALING IN ION-MOLECULE SCATTERING: 3,10,15, and 25 keV Ne^+-H_2.

B.Fastrup and P.Dahl

Institute of Physics, Aarhus University, DK 8000 Aarhus C

As noted by Andersen et al.[1], Ne^+-H_2 is an almost ideal collision system for testing experimentally energy-loss scaling[2]. Not only is $\gamma = M_1/M_2$ large, ensuring a substantial energy loss, even at small laboratory angles, but electronic excitations are also known to be rare events[1].

To eliminate as far as possible errors due to uncertainties in assessment of scattering angle Θ and incident energy E_0, we have carried out concurrent energy-loss studies of the Ne^+-He collision system without altering neither E_0 nor Θ, see Fig.1. Making use of a well known relation from binary encounters: $E_0 - E_{el} = \gamma\Theta^2 E_0$, where E_{el} is the kinetic energy of the elastically scattered particles, we find the following formula

$$\Delta E_{el}/\Theta = (\gamma_{eff}^{H_2} - \gamma^{He})(E_0\Theta),$$

where

$$\Delta E_{el} = E_{el}^{He} - E_{el}^{H_2}.$$

Two extreme cases occur when

i) $\gamma_{eff}^{H_2} = 10$ (the molecular target acts as one massive particle with mass 2 amu).

ii) $\gamma_{eff}^{H_2} = 20$ (scattering is effected by a binary encounter with one of the atoms forming the molecule).

Intuitively, case (i) is expected to prevail at small values of E_0 and Θ, while case (ii) predominates at larger values of E_0 and Θ. To test scaling, we have plotted $\Delta E_{el}/\Theta$ versus the reduced scattering angle $E_0\theta$ and found that within experimental uncertainty, the data points fall on a single curve, Fig.2.

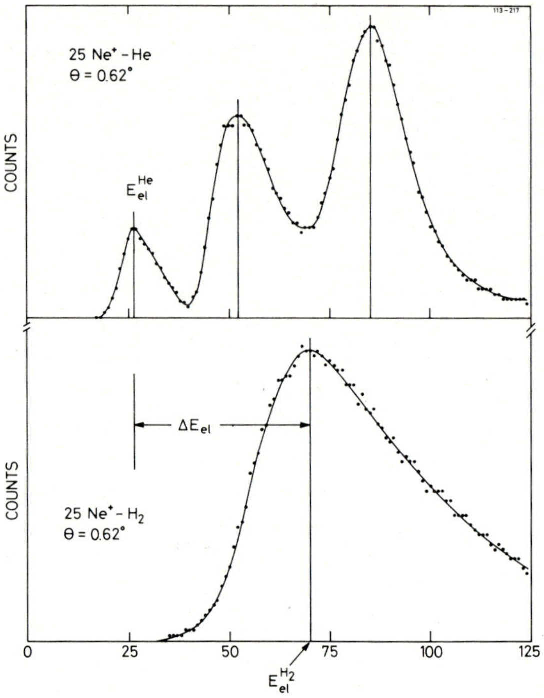

Fig.1

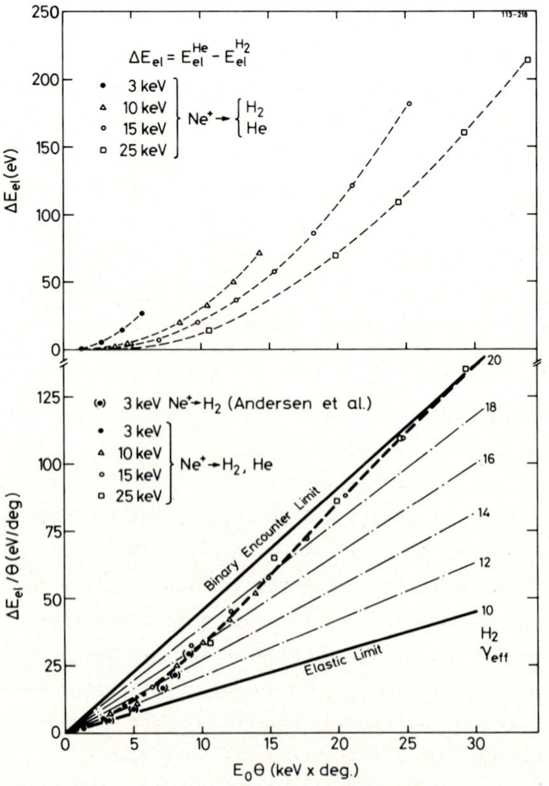

Fig.2

References

1) N.Andersen, M.Vedder, A.Russek, and E. Pollack, Phys.Rev.A 21, 782 (1980)

2) P.Sigmund, J.Phys.B 11, L145 (1978)

EMISSION SPECTRUM OF THE Na^*-N_2 COLLISION COMPLEX

W. Kamke*, B. Kamke*, I.V. Hertel*, and A. Gallagher

JILA, University of Colorado and NBS, Boulder CO 80309

Experiment. The fluorescence spectrum of Na(3p) atoms in N_2 gas has been measured in the far wings of the Na resonance lines, from 530 - 790 mm. A stainless steel cross with a sidearm was used as a resonance cell. Cell-temperature has been varied from ~ 450 K to ~800 K. Na-pressure was regulated by the sidearm temperature and was typically 10^{-5} to 10^{-6} Torr. The N_2-pressure has been varied from 2 - 800 Torr. The fluorescence spectrum was measured by setting a cw-laser to or near the Na-resonance line and observing the fluorescence while scanning a monochromator over the spectrum. The main experimental results are: We observe a broad continuum from 630 - 790 nm that is independent of N_2-pressure and gas temperature. The intensity on the blue wing decreases rapidly with decreasing wavelength and shows temperature dependence corresponding to a repulsive $Na(3p)-N_2$ interaction.

Discussion. The relation of the far wing spectrum to the atomic potentials is described in[1] for a diatomic collision process. This method can be extended to the $Na-N_2$ case, but now the atomic potentials depend on three coordinates: R for the $Na-N_2$ separation, r for the N-N separation and θ for the angle between both. The Franck-Condon principle, yielding $\nu(R,r,\theta)$, still holds for each set of coordinates (R,r,θ). Similar to the diatomic case the emission at frequency ν is proportional to the radiative rate $\Gamma(R,r,\theta)$, the volume-element $dVol(R,r,\theta)/d\nu$ and the probability $P(R,r,\theta)$ of the atoms being at (R,r,θ).

The potentials given by Habitz[2] have been used to calculate the emission spectrum. Since Habitz does not give full (R,r,θ)-dependences, some interpolating was necessary. Furthermore we assumed a constant transition moment, and for the probability $P(R,r,\theta)$ we used the following approximations: We separate $P(R,r,\theta)$ into $P_{N_2}(r,r_m)\cdot P_{Na}(R,r_m,\theta)$ and either use a normalized ground-vibrational-state distribution P_{N_2} for the N_2-molecule, centered at the local minimum r_m or collapse P_{N_2} to a deltafunction at r_m when the r-dependence is unknown. For P_{Na} we used two different assumptions: 1. $P_{Na}^{Ea} = \exp\{-(V^*(R,r_m,\theta) - V_\infty^*)/kT\}$ representing a thermal population of bound and free Na^*-N_2 states. 2. $P_{Na}^{F} = P_{Na}^{Ea}$ in repulsive regions and $P_{Na}^{F} = 1$ in attractive regions[3]. This represents only free collision states of Na^*-N_2.

The (B-X), (A'-X) and (A-X) band intensities have been calculated for $\theta = 0°$, 45°, 70° and 90° (full r-dependence only for 90°). X is the molecular ground state and A,A' and B are the molecular states corresponding to $Na(3p)+N_2$ at $R\to\infty$. The weighted sum of the results for the different θ gives the average spectrum of each band. Then each band was weighted by $\frac{1}{3}$ and summed over to yield the full spectrum that can be compared with the experimental data. Fig. 1 gives preliminary results. There is fairly poor agreement between the theoretical and experimental data, but the following conclusions can be drawn. The blue wing temperature dependence is fairly similar to that observed but the calculated intensity is too high and there is no satellite to be seen at ~ 510 nm. In the red wing the data calculated with P_{Na}^{F} are much closer to the experimental data than those obtained with P_{Na}^{Ea} concerning both the skope of the spectrum and the independence of temperature. This together with the independence of N_2-pressure indicates that only free collision states are occupied[3] and that there is no emission by bound or quasibound states. That leads to the conclusion that all bound states of Na^*-N_2 are rapidly predissociated. Compared with the measurements the theoretical spectrum is too high, which indicates that quenching also occurs at larger distances away from the actual curve crossing region. These problems will be discussed in more detail at the conference.

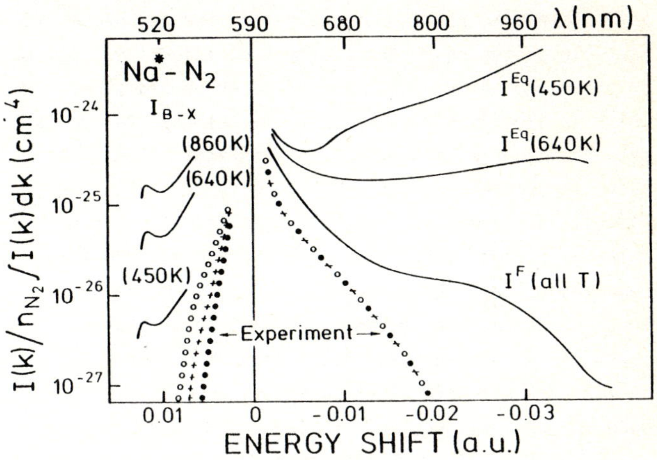

Fig. 1: Experimental and calculated (——) Na^*-N_2 spectrum. (Preliminary results).

* present address: Institut für Molekülphysik, Freie Universität Berlin

References

1. R. Hedges, D. Drummond, and A. Gallagher, Phys. Rev. A6, 1519 (1972)
2. P. Habitz, Chem. Phys. 54, 13 (1980)
3. G. York, R. Schops, and A. Gallagher, J. Chem. Phys. 63, 1052 (1975)

POLARIZATION STUDIES IN THE NONADIABATIC QUENCHING PROCESS Na*(3p) + H_2

G. Jamieson, W. Reiland, C.P. Schulz, H.-U. Tittes and I.V. Hertel

Freie Universität Berlin, Fachbereich Physik, Institut für Molekülphysik, Arnimallee 14, D-1000 Berlin 33, FRG

Continuing our alignment and orientation studies of the quenching of Na* by diatomic molecules[1] at thermal energies we report here new data for Na*(3p) + H_2. Such investigations allow one to obtain a detailed insight into the molecular dynamics at large internuclear distances. In this kind of crossed molecular beam polarization experiments one may use either the technique of particle photon coincidence after collisional excitation or the method of deexcitation of laser pumped atoms, which is used in our laboratory. In the past few years the physics of such processes in ion(atom)-atom collisions at superthermal and high energies has become fairly well understood. In molecular collisions, in particular at thermal energies, the situation is much more complicated and the physics is less well understood. Here one may neither assume conservation of atomic reflection symmetry nor even coherence among the atomic states of equal reflection symmetry, as has often been done in past. Two central problems have to be understood: i) How and at which internuclear configuration are the anisotropic states of the excited atom coupled into the triatomic molecular system. ii) Can one use such polarization studies, in particular the left right asymmetry for circular polarization, to obtain a detailed information on whether individual trajectories probe mainly the repulsive or attractive part of the potential. Here we concentrate on the second aspect.

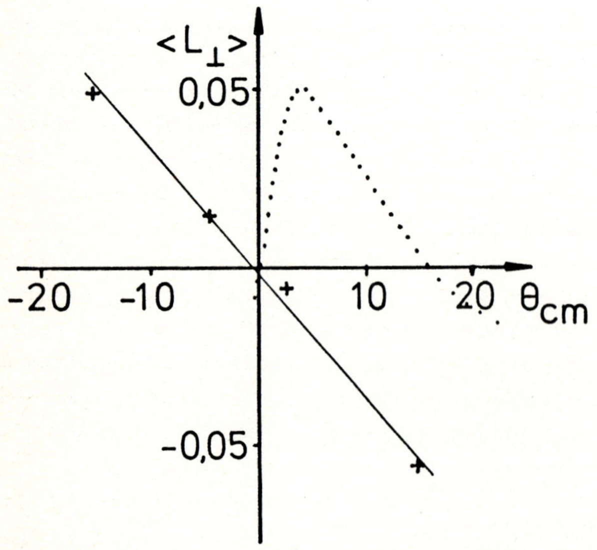

Fig. 1: Angular momentum transfer $\langle L_\perp \rangle$ for quenching of Na*(3p) by H_2(+) as a function of c.m. scattering angle. For comparison the data for N_2(....) are reproduced.[1]

Fig. 1 shows the angular momentum transfer $\langle L_\perp \rangle$ as a function of scattering angle. We compare our new data for Na*(3p) + H_2(v=0) → Na(3s) + H_2(v'=2) with previously published results for N_2.[1] It is interesting to note that for small scattering angles $\langle L_\perp \rangle$ is negative while in contrast it is positive for N_2 and changes sign only for larger scattering angles. In a simple minded picture one would explain these findings as being due to an essentially repulsive ($\langle L_\perp \rangle < 0$) or attractive ($\langle L_\perp \rangle > 0$) influence on the trajectories. However, one must be aware, that such a model ignores the development of the phases of the excited atomic states on the long ranging Π and Σ parts of the interaction potential.

We also will report studies carried out with linearly polarized light. These show significant but small anisotropies and may be evaluated in terms of the Na(3p) collision induced density matrix[2]. Apparently, as for the N_2 case, reflection symmetry changing collisions play an important role in the quenching of Na(3p) by H_2. The alignment parameters, however, depend less on scattering angle and energy transfer.

References

1. W. Reiland, G. Jamieson, H.-U. Tittes, and I.V. Hertel, Z. Phys. A - Atoms and Nuclei 307, 51-66 (1982)
2. H.W. Hermann, and I.V. Hertel, Comments At. Mol. Phys. 12, 61-84 (1982), and 12, 127-148 (1982)

NONRESONANT ELECTRONIC TO VIBRATIONAL ENERGY TRANSFER IN THE COLLISION OF LASER EXCITED Na*(3p) WITH TRIATOMIC MOLECULES

G. Jamieson, C.P. Schulz, H.U. Tittes, and I.V. Hertel

Freie Universität Berlin, Fachbereich Physik, Institut für Molekülphysik, Arnimallee 14, D-1000 Berlin 33, FRG

During the past few years we have extensively studied the quenching of excited Na*(3p) atoms by diatomic molecules. Experiments have been carried out in a new 90° crossed supersonic beam machine using a frequency stabilized ring dye laser to excite the sodium. A velocity selector is used to energy analyze the scattered sodium so that two quantities can be accurately measured; firstly the double differential quenching cross section at a fixed energy transfer and secondly the energy transfer spectrum at a fixed center of mass scattering angle.

The quenching of Na*(3p) by diatomic molecules can now be considered to be well understood while the quenching by tri and polyatomic has hardly been touched upon[1]. We have therefore expanded our measurements and report here the first double differential cross sections (angular distributions) for the scattering of Na*(3p) by N_2O and CO_2. Both processes were predicted to be resonant[2]. Our new measurements show a peak at high energy transfers and do not follow a prior translational distribution for triatomic molecules[3]. Typical energy transfer spectra are shown in fig. 1 and 3.

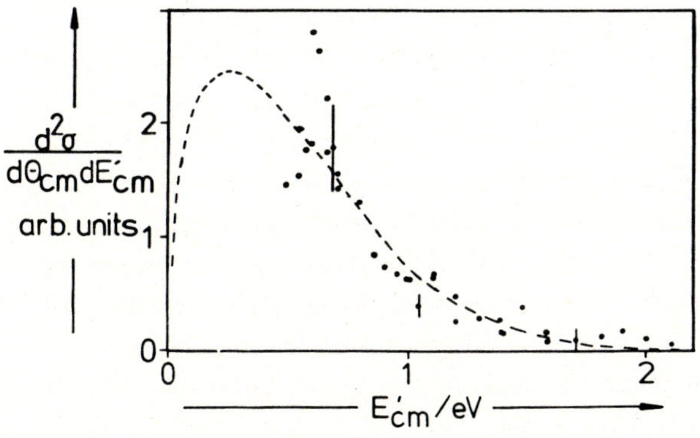

Fig. 1: Energy transfer spectrum for Na*(3p) + CO_2. The dashed line shows a prior translational distribution.

Fig. 2 and 4 show angular distributions for both N_2O and CO_2. The intensity scale is multiplied by $\sin\theta$. The N_2O data resembles somewhat N_2 angular distributions showing a rather short lived complex[4] while the CO_2 results may indicate some tendency towards complex formation.

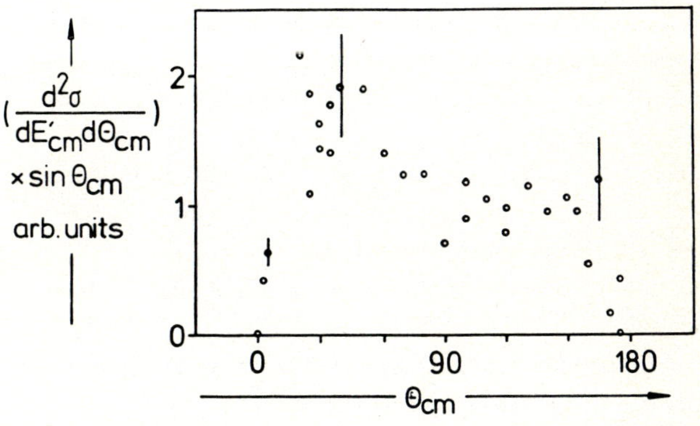

Fig. 2: Angular distribution for Na*(3p) + CO_2. The vertical scale is multiplied by $\sin\theta_{cm}$.

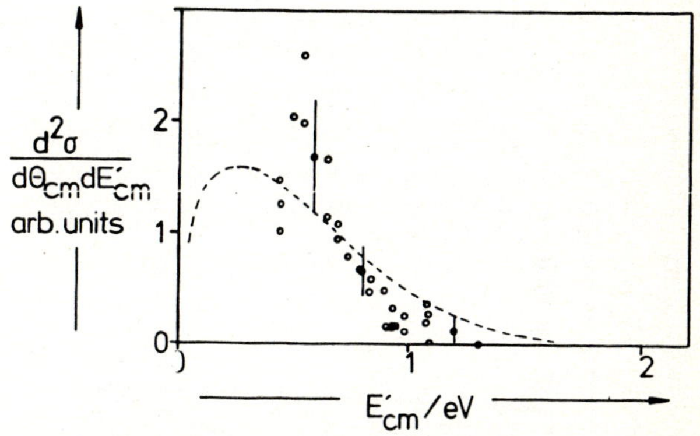

Fig. 3: Same as fig. 1 but for Na*(3p) + N_2O

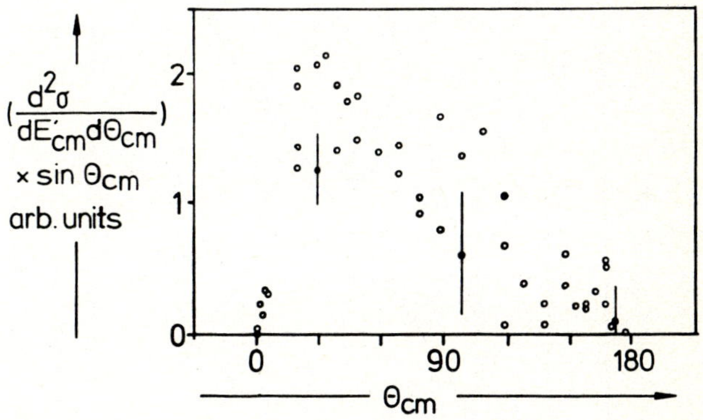

Fig. 4: Same as fig. 2 but for Na*(3p) + N_2O

References

1. I.V. Hertel: In: Advances in Chemical Physics. The dynamics of the excited state. Lawley, K. (ed.), p. 475. New York: John Wiley 1982
2. H.S. Taylor, Chem. Phys. Letters 64, 817 (1979)
3. A.P. Wilson, and R.D. Levine, Molec. Phys. 27, 1197 (1974)
4. W. Reiland, C.P. Schulz, H.U. Tittes, and I.V. Hertel, Chem. Phys. Lett. 91, 1982

A MOLECULAR BEAM STUDY OF ELECTRONIC TO ELECTRONIC, VIBRATIONAL, AND ROTATIONAL ENERGY TRANSFER (E-EVRT) IN THE COLLISION OF TWO STEP LASER EXCITED SODIUM WITH N_2

G. Jamieson, C.P. Schulz, H.U. Tittes, and I.V. Hertel

Freie Universität Berlin, Fachbereich Physik, Institut für Molekülphysik, Arnimallee 14, D-1000 Berlin 33, FRG

The quenching of laser excited Na* in the 5 $^2S_{1/2}$, 4 $^2D_{5/2}$, 4^2P, and 4^2S excited states by diatomic gases has been studied in a crossed molecular beam experiment at a collision energy of 0.16 eV. Both energy transfer spectra at several center of mass angles and double differential scattering cross sections have been measured. All of the energy transfer spectra show a strong increase at high energy transfers. The differential cross sections exhibits large forward and smaller backward peaks.

The various features are explained as being due to a normal EVR process as seen for $Na(3^2P_{3/2}) + N_2$ at lower energy transfers superimposed on an electronic to electronic energy transfer mechanism at high energy transfers. All of the possible quenching paths are shown in fig. 1.

Fig. 1: Quenching path for the collisional de-excitation of sodium in higher excited states. Solid lines show possible transitions for atoms in the two step laser excited 4d level and dashed lines show the paths from the 5s level. Due to spontaneous emission the 4p and 4s levels are strongly populated both for 5s and 4d optical pumping. Quenching from the 4s and 4p states is shown as dotted lines.

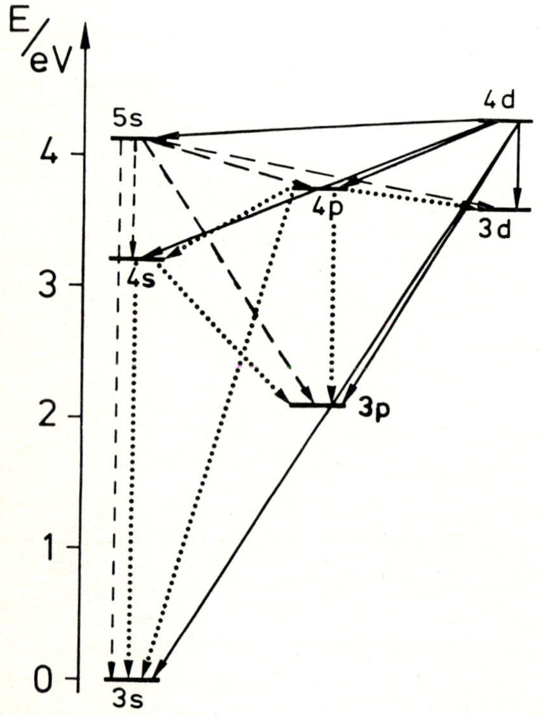

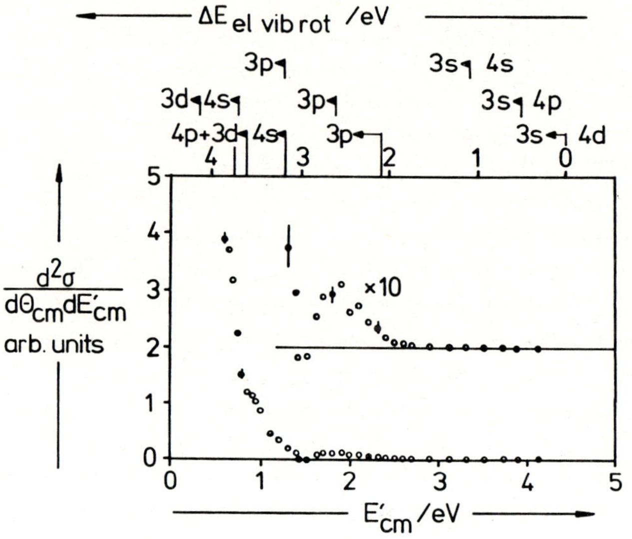

Fig. 2: Energy transfer spectrum for the case of optical pumping to the 4d level (both lasers on) at a center of mass scattering angle of 0°.

Typical experimental results are shown in fig. 2 and 3. Note that no quenching to the ground state is observed (centers possibly the N_2 is excited to high vibrational states). Energy transfer spectra for optical pumping to the 5s level are similar in shape. Both the E-VR and E-EVR processes proceed by way of dopping at crossings between the various potential hypersurface. Based on the $Na^* + N_2$ potential curves, only a few of which are known[1], we conclude that most likely in the case of higher excited sodium atoms a surface crossing with the $Na^*(3P) + N_2$ surface is reached before the crossing with the $Na(3s) + N_2$ ground state potential. This would explain the lack of scattering signal for the direct quenching to the ground state (for example 5s→3s).

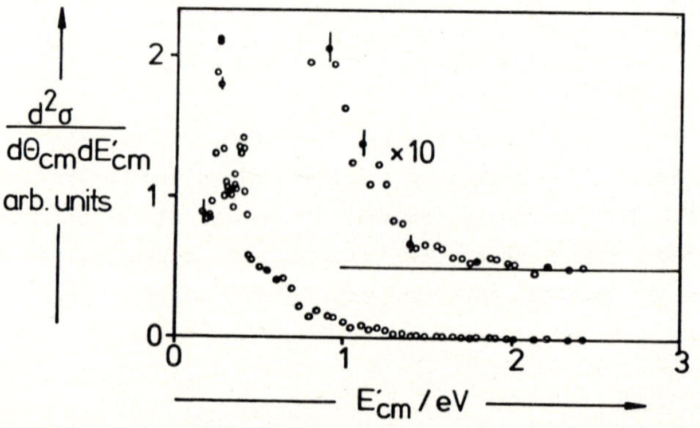

Fig. 3: Same as fig. 2 but for center of mass scattering angle Θ_{cm} = 180°. The region below 0,5 eV must be interpreted with care.

References

1. P. Habitz, Chem. Phys. 54, 131 (1980)

INFLUENCE OF THE ROTATIONAL ENERGY OF MOLECULAR PERTURBERS ON Rb(5P) FINE STRUCTURE TRANSITION (FST)

J. Cuvellier, J.M. Mestdagh, M. Ferray, P. de Pujo, J. Berlande

Service de Physique des Atomes et des Surfaces, CEN/Saclay, 91191 Gif-sur-Yvette Cedex, France

Inelastic collisions involving molecules are known to depend on both the translation energy of the reactants and the internal energy (rotation and vibration) of the molecular perturbers. An advantage of nozzle beam experiments over cell type measurements is to allow to decouple these two parameters in order to detail their relative influence on inelastic processes.

We have used a crossed nozzle-beam apparatus to study the following process

$Rb(5P_{1/2}) + X \rightarrow Rb(5P_{3/2}) + X$ where $X \equiv H_2, D_2$ (1)

as a function of both the relative energy E_R in the range 0.1 - 0.3 eV and rotational temperature T_R between 100 and 500K (Here the rotational energies of H_2 and D_2 molecules are expressed in terms of T_R[1]). Description of the apparatus is given elsewhere[2]. The results are given in figure 1 for H_2 and figure 2 for D_2 (note the calibration of the apparatus on a reference process to get the absolute values of the cross sections). For comparison purpose the dashed line in figures 1 and 2 give the theoretical results of Hickman[3], who has considered molecules as structureless particles.

It is observed that :

- For H_2, there is a slight increase of $\sigma_{1/2\rightarrow3/2}$ with T_R; the E_R-dependence of $\sigma_{1/2\rightarrow3/2}$ is that predicted by the theoretical approach.
- For D_2, the increase of $\sigma_{1/2\rightarrow3/2}$ with T_R is more important than for H_2; unlike what is observed for H_2, the E_R-dependence of $\sigma_{1/2\rightarrow3/2}$ is markedly different from that theoretically predicted.

A clear evidence of the influence of rotational energy on the inelastic process (1) has been given here for the first time. This will be discussed at the conference.

References

1. J.M. Mestdagh, P. de Pujo, J. Cuvellier, A. Binet, P.R. Fournier, J. Berlande, J. Phys. B 15, 663 (1982)
2. J. Cuvellier, J.M. Mestdagh, J. Berlande, P. de Pujo, A. Binet, Rev. Phys. Appl. 16, 679 (1981).
3. A.P. Hickman, J. Phys. B15, 3005 (1982); Phys. Rev. Lett. 47, 1587 (1981).

Fig.1

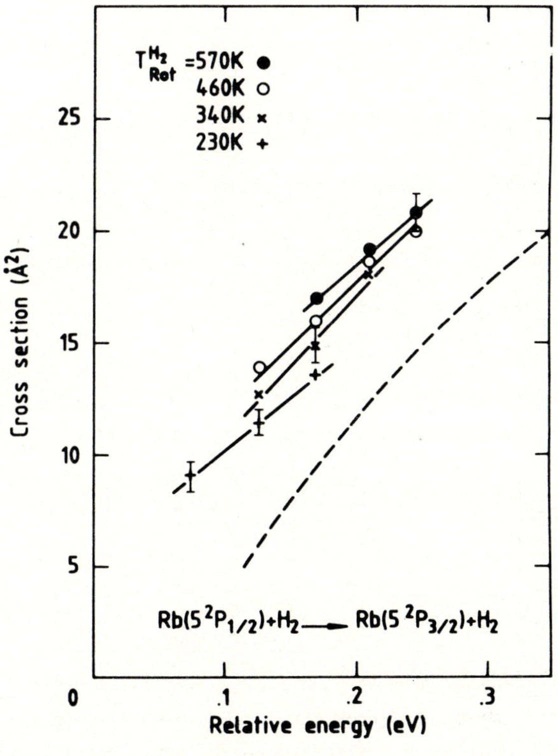

Fig.2

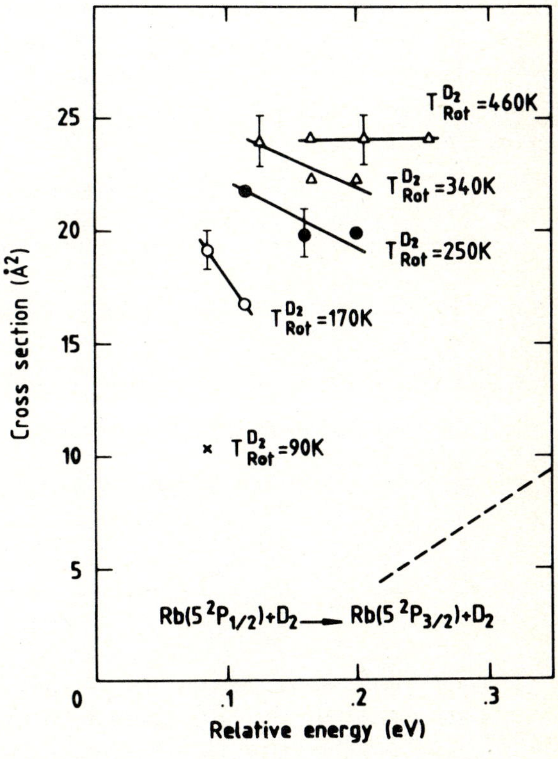

COLLISIONAL QUENCHING OF SELECTIVELY EXCITED ROVIBRONIC STATES OF H_2^* BY H_2, NE, AR, AND KR

H. Schmoranzer, J. Imschweiler and T. Noll

Fachbereich Physik, Universität Kaiserslautern, D-6750 Kaiserslautern, Germany

Radiative and collisional deactivation rates have been measured for rotationally resolved rovibronic states J' = 1...6, v' = 0...4 of the B 2p $^1\Sigma_u^+$ state of H_2. The subnanosecond lifetimes and the absolute cross sections for quenching by thermal collisions with H_2, Ne, Ar, and Kr in the ground state have been determined.

Previous measurements[1,2] were repeated and extended with an improved arrangement at reduced bandwidth (0.05 nm)[3] of the synchrotron radiation from the storage ring DORIS II at DESY, Hamburg, used for selective excitation. Rotational selectivity was enhanced further by preparation of para-H_2. The gas cell was sealed by windows made of LiF on the entrance side and of CaF_2 on the exit side of radiation.

Decay curves (Fig. 1) were recorded for individual rovibronic states at pressures from 3 mbar to 140 mbar at room temperature. The effective decay rate was evaluated by determining the best fit of an exponential decay curve convoluted with the response function also measured. For pure H_2, selected results for J' = 1 evaluated from Stern-Volmer plots (Fig. 2) have been compiled in Tab. 1, together with theoretical lifetimes[4] calculated by disregarding rotational distortion. All uncertainties quoted correspond to one standard deviation as extracted from the least-squares fit procedures.

For quenching of H_2^* by rare gases the following cross sections (10^{-15} cm^2) have been determined for B, v' = 0, J' = 1 and 2: 8.5 ± 0.2 for Kr, 6.7 ± 0.1 for Ar, 0.08 ± 0.06 for Ne.

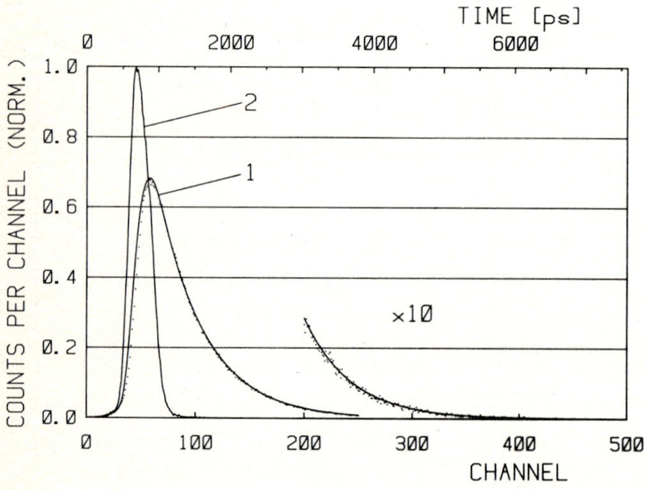

Fig. 1: Typical decay curve (dotted) of H_2 B,v'=4, J'=5 state at 3.54 mbar H_2, with (1) best fit and (2) reponse function (FWHW 334 ps). Computed decay rate is $1.54 \cdot 10^{-9}$ s^{-1}.

Table 1: Radiative lifetimes (ps) of rovibronic B states and cross sections ($10^{-15}cm^2$) for quenching by thermal collisions with $H_2(X, v''=0)$

v'	J'	τ	τ(Ref.4)	σ
0	1	535 ± 16	535	5.6 ± 0.2
1	1	591 ± 13	574	5.8 ± 0.2
2	1	627 ± 17	613	5.8 ± 0.2
3	1	664 ± 10	651	5.7 ± 0.1
4	1	731 ± 8	690	6.0 ± 0.1

Ab initio calculations show that quenching is possible on a single electronic surface predominantly resulting in products H_2 + 2H. Long range attraction on this surface appears to be sufficiently strong to account for the large quenching cross section[5].

Financial support from the Bundesministerium für Forschung und Technologie is gratefully acknowledged.

References

1. H. Schmoranzer and J. Imschweiler, Abstracts XIIth ICPEAC (Ed. S. Datz, Gatlinburg, Tenn. 1981) Vol. 2, 964
2. H. Schmoranzer and J. Imschweiler, Abstracts 8th ICAP (Eds. I. Lindgren, A. Rosén, S. Svanberg, Göteborg, 1982) A 97
3. H. Wilcke, W. Böhmer and N. Schwentner, N.I.M. 204, 533 (1983)
4. T.L. Stephens and A. Dalgarno, J.Q.S.R.T. 12, 569 (1972)
5. W. Meyer, private communication

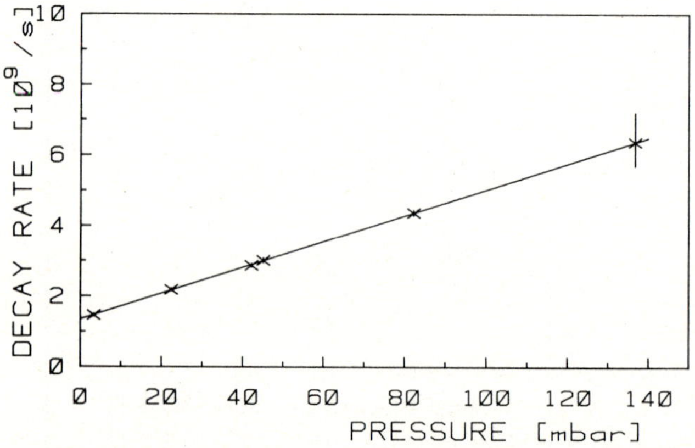

Fig. 2: Decay rate of H_2 B,v'=4, J'=2 state versus H_2 pressure (error bars drawn in correspond to three standard deviations). τ_0=(742± 6)ps, $\sigma(H_2^* - H_2) = (6.05 \pm 0.07)\ 10^{-15} cm^2$.

RADIATIVE QUENCHING OF VIBRONICALLY EXCITED CO^+ AT T = 100°K

D.H. Katayama and J.A. Welsh

Aeronomy Division, Air Force Geophysics Laboratory, Hanscom AFB, MA. 01731

There have been numerous experiments on collisional quenching of electronically excited molecules. In many of these measurements, however, the state to which the molecule is quenched is unknown. It has been determined previously[1,2] that CO^+ and the isoelectronic CN radical in their A $^2\Pi_i$ state are quenched to high lying vibrational levels of the ground state. In this investigation, vibronic state to state radiative quenching rates are obtained for v' = 0, 1, and 2 of the A state of CO^+ at T = 100°K and compared with room temperature data.

Rovibronic levels of the A state of cooled and room temperature CO^+ are selectively excited by a pulsed tunable dye laser and their fluorescence recorded as a function of time and helium pressure. The carbon monoxide cations are formed by reaction of neutral carbon monoxide with helium metastable atoms created in a dc discharge. The cooled data of CO^+ are obtained by passing the helium and CO gas through a specially constructed dewared pyrex cell filled with liquid nitrogen. An analysis of the relative intensities for the rotational excitation spectrum obtained at room and liquid nitrogen temperatures in Fig. 1(a) and (b) yields temperatures of 300 and 100°K, respectively. The spectrum is from a portion of the CO^+ A-X (0,0) band.

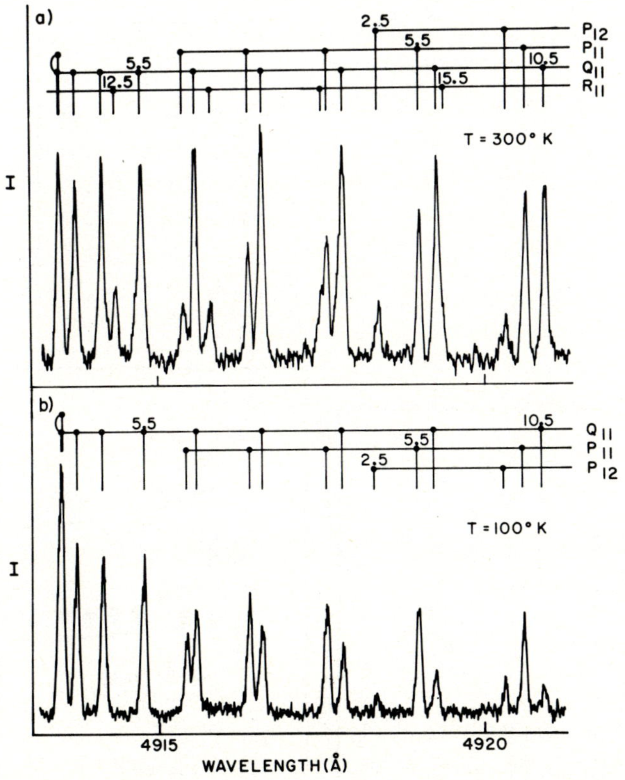

Fig. 1: Laser Excitation Spectrum of the CO^+ A-X (0,0) band at (a) T = 300°K and (b) T = 100°K.

For v' = 1 and 2, the effective radiative lifetime τ obtained from the kinetic equations is

$$1/\tau = 1/\tau_r + k_{AX}\ (\mathrm{He})$$

where τ_r is the radiative lifetime and k_{AX} is the collisional quenching rate from the A state to its adjacent vibrational ground state level. The v' = 0 level is perturbed and requires a treatment for the radiative decay which takes this mixing into consideration[3]. Fig. 2 shows plots of the reciprocal decay times for v' = 1 of the A state as a function of He number density at T = 300 and 100°K. The quenching rate at T = 100°K is approximately twice as fast as that for T = 300°K. Similar results are obtained for the other vibrational levels studied which indicate that long range dipole-induced dipole and dispersion forces are important in the radiative quenching process.

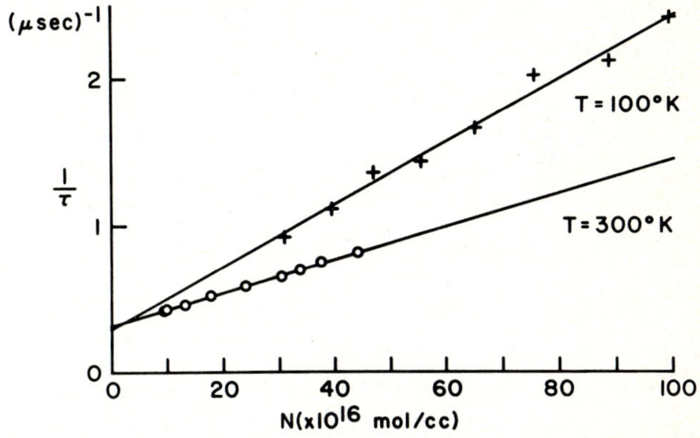

Fig. 2: Stern-Volmer plots of measured decay rates for v' = 1 versus He number density at T = 300°K and 100°K.

References

1. V.E. Bondybey and T.A. Miller, J. Chem. Phys. 69, 3597 (1978).
2. D.H. Katayama, T.A. Miller and V.E. Bondybey, J. Chem. Phys. 71, 1662 (1979).
3. D.H. Katayama and J.A. Welsh, J. Chem. Phys. 75, 4224 (1981); J. Chem. Phys. 76, 3848 (1982).

TEMPERATURE DEPENDENCE OF DE-EXCITATION RATE CONSTANTS OF $He(2^3S)$, $Ar(^1P_1)$, AND $Ar(^3P_1)$ BY ATOMS AND MOLECULES

H. Koizumi, M. Ukai, Y. Tanaka, K. Shinsaka, and Y. Hatano

Department of Chemistry, Tokyo Institute of Technology, Meguro-ku, Tokyo, Japan

The rate constants or the cross sections of de-excitation of excited rare gas atoms have been measured using a flowing afterglow method, a static afterglow method, a beam method, and a pulse radiolysis method. We have shown some advantages of a pulse radiolysis method[1-4] to obtain the absolute value of the rate constant or the cross section of this process.

This paper reports the temperature dependence of the rate constants of $He(2^3S)$, $Ar(^1P_1)$, and $Ar(^3P_1)$ by atoms and molecules, from which the collision energy dependence of the cross sections is obtained.

The experimental method is almost the same as our previous one.[1-4] A nsec 600keV electron beam pulse impinges upon a sample gas mixture in a cell. The decay of the number density of excited atoms after the pulse is observed with a time-resolved emission or absorption system, which is composed of a xenon flash lamp, a 1m grating monochromator, and a photomultiplier. The cell is kept within the error limit of ±2°C at temperatures ranging from room temperature to -140°C.

The temperature dependence of $He(2^3S)$ de-excitation rate constants by atoms and molecules M ($M=N_2$, CO, Ar, Kr, NO, O_2, C_2H_4, and CO_2) has been measured. The rate constants k_M at each temperature are converted into the cross sections σ_M at the corresponding collision energy using the equations $\sigma_M = k_M(\pi\mu/8k_BT)^{1/2}$ and $E=\frac{3}{2}k_BT$, where μ, k_B, T, and E are reduced mass, Boltzmann constant, absolute temperature, and collision energy, respectively. One of the results is shown in Fig. 1.

Since the major de-excitation process in this experiment is thought to be Penning ionization, it is assumed[5] that the transition rate from $He(2^3S)$-M state to $He-(M^++e^-)$ state to be $W(R)=A\exp(-\alpha R)$, and that the interaction potential between $He(2^3S)$ and M, $V^*(R)$, is equal to $B\exp(-\beta R)$, where A, B, α, and β are constants and R is the intermolecular distance. According to the above mentioned theory,[5] the de-excitation cross section has the collision energy dependence such as $\sigma_M \propto E^{\alpha/\beta-1/2}$ and the values of α/β are experimentally determined for various atoms and molecules M. Previously,[2] we obtained the "de-excitation probability", $P=\sigma_M/\sigma_C$, where σ_C is a gas kinetic collision cross section. They are listed in Table 1, which shows that the value of α/β decreases with increasing of the value of P.

The collision energy dependence of the de-excitation cross sections of radiative states has also been obtained in a way similar to that described above. Because of the short lifetimes of these states it has been very difficult to apply a beam method or a flowing afterglow method for the experimental determination of the cross sections or the rate constants. The de-excitation of $Ar(^1P_1)$ and $Ar(^3P_1)$ by SF_6 and N_2 has been examined. The result of SF_6 is shown in Fig. 2. Since SF_6 has an intense absorption band almost at the excitation energy of $Ar(^1P_1)$ and $Ar(^3P_1)$, we have tried to compare the data with the Watanabe-Katsuura theory.[6] The agreement with the data is fairly good. The result of N_2 is largely different from that of SF_6, and is rather similar to the result of $Ar(^3P_{0,2})-N_2$ collision. Thus, it has been concluded that a long-range dipole-dipole interaction is important for $Ar(^1P_1)$- or $Ar(^3P_1)-SF_6$ collision, while a short-range interaction with curve crossing is important for $Ar(^1P_1)$- or $Ar(^3P_1)-N_2$ collision.

References

1. T.Ueno and Y.Hatano, Chem.Phys.Lett.40, 283 (1976).
2. T.Ueno, A.Yokoyama, S.Takao, and Y.Hatano, Chem.Phys. 45, 439 (1980).
3. A.Yokoyama, S.Takao, T.Ueno and Y.Hatano, Chem.Phys. 45, 439 (1980).
4. A.Yokoyama and Y.Hatano, Chem.Phys.63, 59 (1981).
5. A.Niehaus, Proc. 5th Int. Congress Radiat.Res., Seatle, 1974, Academic Press (1975), p.227; A.Niehaus, "Physics of Electronic and Atomic Collisions", Invited Papers of 12th ICPEAC, Gatlinburg, 1981, ed.by S.Datz, North-Holland (1982), p.237.
6. T.Watanabe and K.Katsuura, J.Chem.Phys.47, 800 (1967).

Table 1

M	α/β	P
N_2	1.4	0.05
CO	1.1	0.07
Ar	1.0	0.07
Kr	1.0	0.07
NO	0.6	0.16
O_2	0.6	0.18
C_2H_4	0.3	0.31
CO_2	0.1	0.39

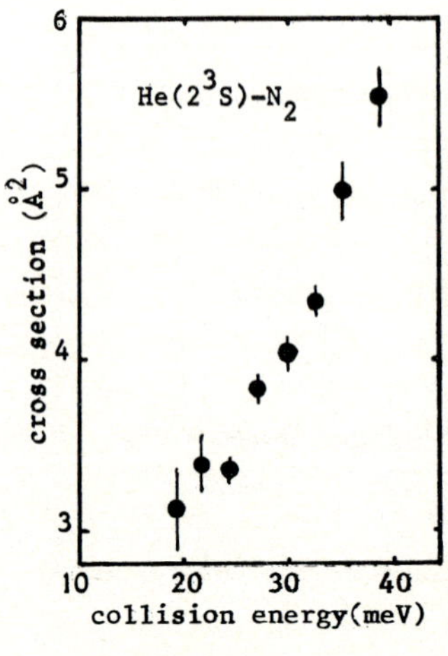

Figure 1

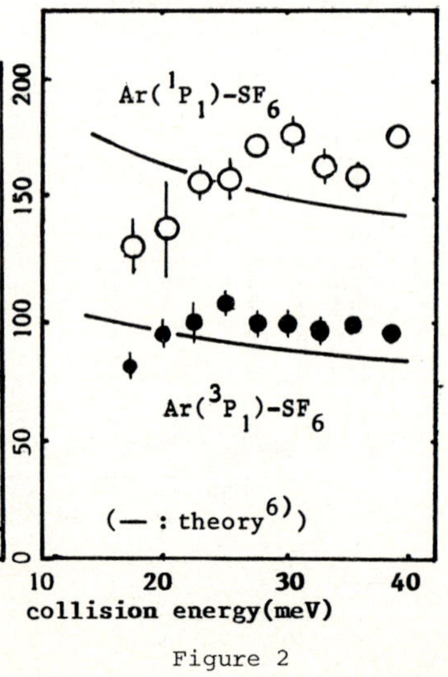

Figure 2

VIBRATIONAL INTERFERENCE STRUCTURES IN DIFFERENTIAL CROSS SECTIONS FOR ALKALI-ATOM MOLECULE COLLISIONS

M.R. Spalburg, M.G.M. Vervaat, A.W. Kleyn and J. Los

FOM-Institute for Atomic and Molecular Physics, Kruislaan 407, 1098 SJ Amsterdam, The Netherlands.

Differential cross sections for positive ion formation and (in)elastic scattering in alkali-atom molecule scattering have been determined. The projectile energies range from 80 to 500 eV. The use of channelplates significantly enhanced the experimental resolution with respect to earlier work [1]. Molecules studied include O_2, SF_6 and NO_2. For all of these systems an ionic state $M^+ + XY^-$ exists.

In the case of $M + O_2$ the differential cross sections for ion-pair formation clearly reflect the vibration of the O_2^- during the collision. In this regard the results agree completely with the differential cross sections for inelastic scattering in $M + O_2$ collisions measured by Kleyn et al. [1]. Due to the better experimental resolution additional structure is observed. Both classical and semi-classical theory explain these phenomena very well [2,3]. Classically one envisions a classical vibrating O_2^- molecule. The O-O motion, changing the vertical electron affinity, strongly influences the transition probability at the crossing between the covalent and ionic states. This results in vibrational structures in the cross sections. From a quantum mechanical point of view it is the interference between vibronic waves which results in these structures. The interference arises due to the phase differences which these waves accumulate while evolving over the ionic intermediate.

It should be mentioned that although the classical model is capable of (qualitatively) describing the vibrational effects, the model fails in the description of the coupling at large impact parameters, due to a breakdown of the linear Landau-Zener approximation. This breakdown leads in the case of $M + O_2$ to additional structure in the differential cross sections for ion-pair formation.

The differential cross sections for $M + SF_6$ do not show structures similar to those for $M + O_2$. The results point to the formation of a dissociative state of SF_6^-.

References

1. A.W. Kleyn, V.M. Khromov and J. Los, Chem.Phys. 52 (1980).
2. U.C. Klomp, M.R. Spalburg and J. Los, submitted to Chem.Phys.
3. M.R. Spalburg, V. Sidis and J. Los, accepted for publication in Chem.Phys.Letters.

NON ADIABATIC PROCESSES IN ALKALI METAL - ALKYL HALIDE MOLECULE COLLISIONS: NEUTRAL EXCITATION AND ION PAIR PRODUCTION CHANNELS

A.M.C. Moutinho and A. Praxedes

Centro de Fisica Molecular des Universidades de Lisboa, Av. Rovisco Pais, Complexo 1-IST, 100 Lisboa, Portugal.

E. Cowan and M.A.D. Fluendy

Department of Chemistry, University of Edinburgh, West Mains Road, Edinburgh, EH9 3JJ, Scotland.

Alkali metal/alkyl halide molecular collisions provide convenient systems in which to study non adiabatic effects in molecular collisions.[1,2] In contrast to halogen molecules[3] alkyl halides have a negative vertical electron affinity and curve crossing effects are expected to occur at shorter ranges in these systems. At collision energies in the eV range a large number of exit channels, including electronic, vibrational and ion production processes, are open and can yield information on the non adiabatic effects.

In this paper measured differential cross sections for vibronic excitation[4] and ion production[5] are compared with the predictions of a simple model. At the energies of these experiments the trajectories are almost straight so that the deflection of the exit atom or ion is a good probe for the atom/ion - molecule potential surface. The vibrational excitation observed in the neutral exit channels at small deflections is similarly sensitive to the intramolecular potential operating during the collisions.

Comparison of these results with the predictions of a model based upon diabatic surfaces showed electron transfer yielding a strongly attractive potential to be as essential ingredient, Fig. 1. In the close encounter region, however, the energy losses observed, Fig. 2, show that the C-I potential is considerably modified from that seen in the free molecular ion.

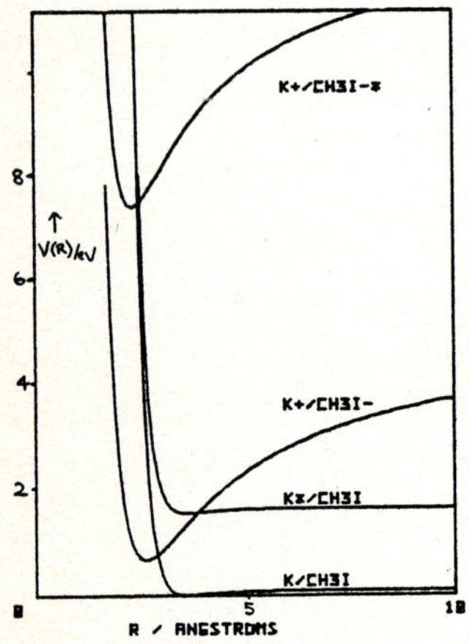

Fig. 1 K-CH_3I Potentials

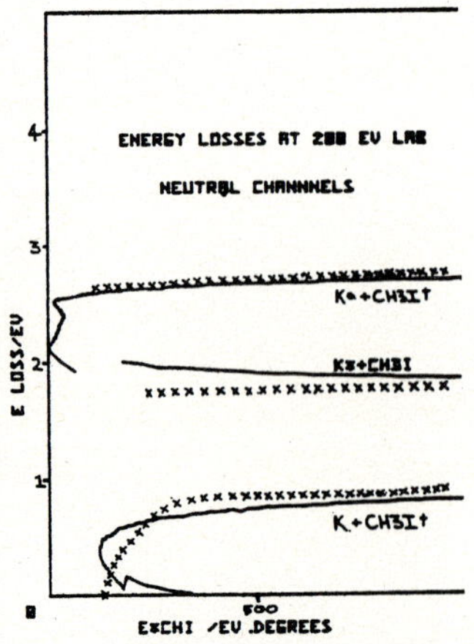

Fig. 2 Comparison of model (solid line) and experiment (XX)

The ion production cross sections, Fig. 3, are consistent with this picture but reveal the involvement of both the ground and an excited ion state, each contributing a rainbow to the observed ion scattering. The nature and possible role of this excited ion state in feeding the observed neutral excited states of the molecule - in an analogous way to the ground ion state - is under investigation.

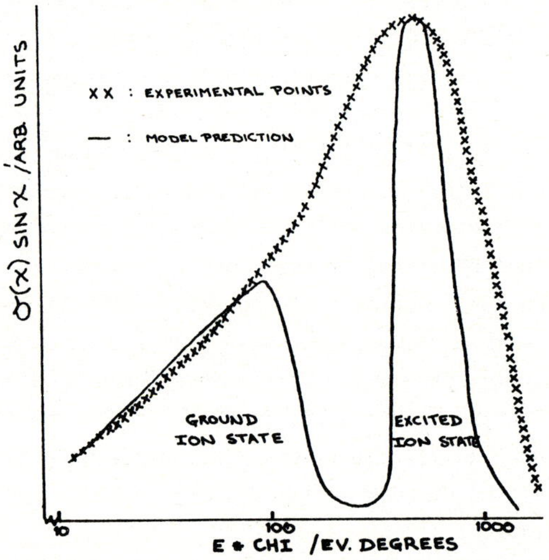

Fig. 3 Differential cross-section at 200 eV LAB Ion pair production channels

References

1. K. Lacmann, Adv. Chem. Phys. 42, 513 (1980).
2. V. Kempter, Adv. Chem. Phys. 30, 417 (1975).
3. P. Davidovits and D.L. McFadden, Alkali Halide Vapours, Academic Press.
4. M.A.D. Fluendy, K.P. Lawley, C. Sholeen and D. Sutton, Mol. Phys. 42, 1 (1981).
5. A.M.C. Moutinho and A. Praxedes, private communication.

ELECTRON TRANSFER IN ALKALI ATOM-CH_3NO_2 COLLISIONS

M.A.D. Fluendy and S. Lunt

Department of Chemistry, University of Edinburgh, Edinburgh, Scotland.

Alkali metal/alkyl halide molecular collisions provide convenient systems in which to study non adiabatic effects.[1,2] In particular, the electron harpoon model yields valuable insight into the dynamics of the possible processes. These can be substantially described as evolving via a repulsive energy release following electron capture into a C-X anti-bonding orbital.[3] This intermediate in the collision being similar to the ground state of the molecular ion. In contrast, the ground state of CH_3NO_2' is not repulsive in the C-N bond. It is thus surprising that thermal energy reactive scattering experiments suggest broadly similar dynamics for CH_3NO_2 and CH_3I.[4] In the framework of the harpoon model questions then arise as to:

(i) The initial state or states populated by the harpoon

(ii) The exit state leading to reaction.

We have investigated this system in the small angle/fast collision regime by measuring inelastic differential scattering cross sections using a time of flight technique. In this regime the reaction channel is closed by momentum constraints and the amount of vibrational excitation associated with different exit electronic states is sensitive to the changes in geometry occurring in the molecular ion. The measured contour maps for the differential inelastic cross sections showed a number of discrete features on-setting at small angles and having an energy loss almost independent of scattering angle. The energy losses associated with these features is plotted for a range of collision velocities in the figure.

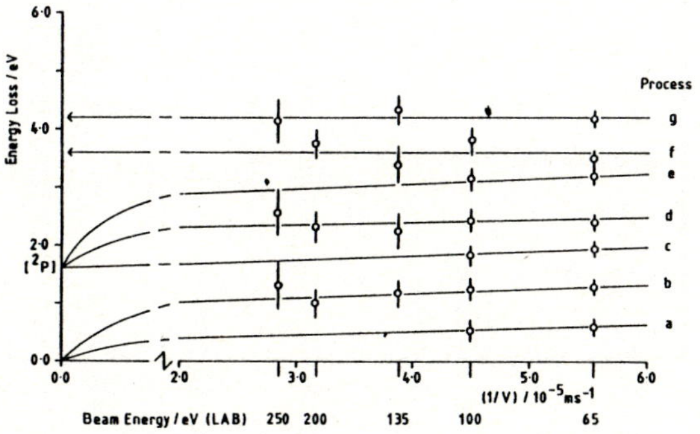

In the sudden limit the harpoon model predicts zero vibrational excitation. This pattern is very similar to that seen for the CH_3I/K system[5] and is compared in more detail in the Table at a common collision velocity.

Energy Loss/eV		
CH_3I	CH_3NO_2	Atom Exit State & Assignment
0	0	^2S Elastic
0.6	0.5	^2S Ionic bow
1.3	1.2	^2S Ionic, early crossing
-	1.8	^2P *
2.0	2.4	^2P Ionic, late crossing
3.1	3.1	^2P Ionic, early "
-	3.6	^2S *
4.1	4.3	^2S $CH_3I(A)$

The states marked *, not seen in the CH_3X measurements, account for only ∿15% of the total neutral scattering in the small angle region. Since the collision lifetime is $\sim 10^{-14}$ s in these measurements we conclude that the initial electron donation is dominantly to form an excited state of the CH_3NO_2' ion with a similar antibonding character to that in CH_3X. The initial electron transfer thus determines the subsequent dynamics.

References

1. K. Lacmann, Adv. Chem. Phys. 42, 513 (1980).
2. V. Kempter, Adv. Chem. Phys. 30, 417 (1975).
3. D.R. Herschbach, Disc. Farad. Soc. 55, 233 (1973).
4. C.M. Sholeen and R.R. Herm, J. Chem. Phys. 65, 5398 (1976).
5. M.A.D. Fluendy, K.P. Lawley, C. Sholeen and D. Sutton, Mol. Phys. 42, 1 (1981).

EMISSION CROSS SECTIONS AND POLARIZATION OF THE RESONANCE RADIATION IN H^+, H_2^+, H_3^+ - Na COLLISIONS AT 2 - 14 KEV.

V.M.Lavrov, R.A.Lomsadze

A.F.Ioffe Physical-Technical Institute, Leningrad, USSR

Cross sections for emission and polarization of the sodium resonance D_1 line ($3s^2S_{1/2}-3p^2P_{1/2}$ transition) and D_2 line ($3s^2S_{1/2}-3p^2P_{3/2}$ transition) have been measured in collisions of H^+, H_2^+, H_3^+ ions with sodium vapour over the projectile energy range from 2 to 14 keV. Experimental data on the cross sections and polarization are not available in the literature on this subject[1-3].

For the measurements the apparatus with ion (electron) and atom crossed beams was used. The sodium beam density used in the experiment was a few times 10^9 atom/cm^3. No self absorption effects were found around this beam density. The absolute cross sections were obtained by normalization our relative measurements with electron beam at 1 keV on the electron impact excitation cross sections taken from[4,5]. The accuracy of the present normalization prosedure was estimated to be within ±15%. Work is continuing in order to improve the accuracy.

The emission cross sections of the sodium resonance D_1- and D_2-lines as a function of the ions velocity are given in Fig.1. Comparison of our results with theoretical predictions for the excitation by H^+ impact[1,2] shows that calculations give cross sections much larger than experiment.

As seen from Fig.1, the experimental cross sections depend weakly on the structure of the ions. This indicates apparently that quasimolecular states arising in collision process are quasi-one-electron ones, and that the main contribution to theirs formation is provided by an electron from sodium atom. It is also possible that the coulomb excitation model is more appropriate than the quasimolecular one when discussing the excitation of resonance states of sodium. It is proposed to discuss this question at the conference.

The polarizations of emission from $3S_{1/2}-3P_{1/2}$ and $3S_{1/2}-3P_{3/2}$ transitions in H^+-Na collisions show pronounced variations with energy. In both cases the polarization is in order of 1-3% at 6 keV and higher ion energies and increases to 5-10% with the decreasing energy to 3 keV.

References

1. M.J. Seaton, Proc. Phys. Soc., A68,457(1955)
2. R.J. Bell, B.G. Skinner, Proc. Phys.Soc(L), 80, 404 (1962)
3. G.H. Bearman, S.D. Alspach, J.J. Leventhal Phys.Rev., A18, 68 (1978)
4. E.A. Enemark, A.Gallagher, Phys.Rev., A6, 192 (1972)
5. J.O. Phelps, C.C.Lin, Phys. Rev., A24, 1299 (1981)

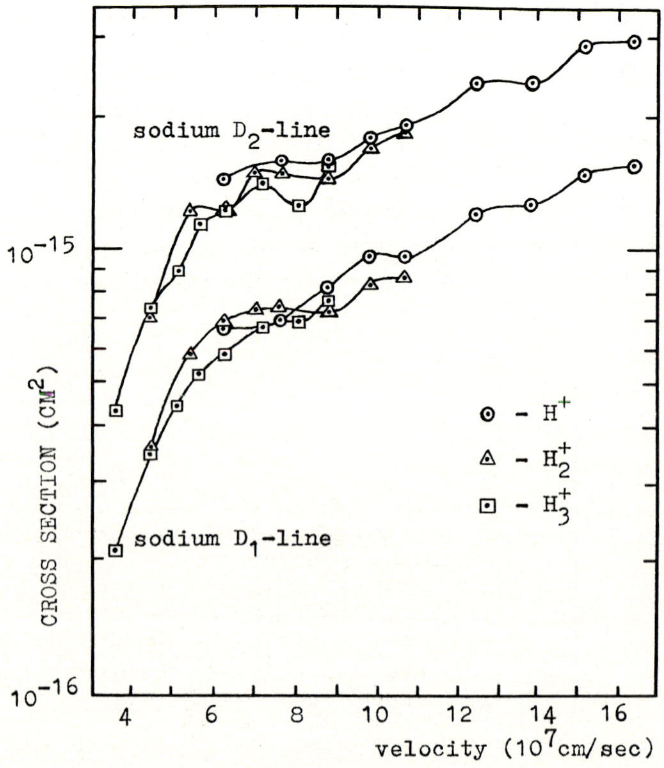

FIG.I

EXCITATION OF HYDROGEN ATOMS IN COLLISIONS OF SLOW K^+ - IONS WITH H_2 - MOLECULES.

B.I. Kikiani, M.R. Gochitashvili, R.V. Kvizhinadze, V.A.Ankudinov.

Tbilissi State University, Tbilissi, USSR.

In this work, the cross sections of H_α and H_β in $K^+ - H_2$ collisions have been measured. The results are presented in Fig. I. The slope of H - emission cross section vs. energy plot undergoes a change. This can be attributed to the fact that this line can be excited in various inelastic dissociation channels. There are three processes which can contribute to the formation of excited hydrogen atoms: direct dissociation

$K^+ + H_2 \text{ -- } K^+ + H(n = 3) + H(Is) - I6.6$ eV,

dissociation with charge exchange

$K^+ + H_2 \text{ -- } K + H(n = 3) + H^+ - 25.9$ eV

and dissociation with ionization

$K^+ + H_2 \text{ -- } K^+ + H(n = 3) + H^+ + e - 30.2$ eV.

Although these processes cannot be discriminated experimentally, we can suggest that because of the smallest energy defect involved the first process provides a major contribution to the formation of excited hydrogen atoms.

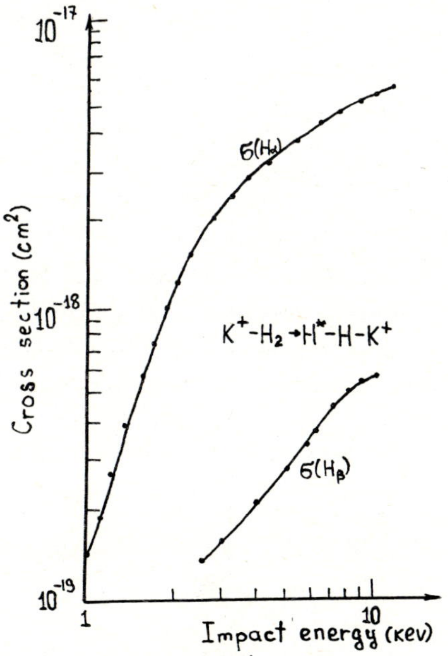

Fig. 1

The direct dissociation of H_2 involving the formation of excited hydrogen atoms can occur in two different ways. One of them is the excitation of the H_2 - molecule from the ground state to excited vibrational levels above the dissociation limit of a bound state. Another is the excitation of bound Rydberg states followed by their predissociation.

When discussing the mechanism of the formation of excited hydrogen atoms we have used the available data [I] on the excitation of hydrogen atoms by slow electrons in collisions with the H_2 - molecule. It was shown [I] that a large contribution to the Balmer α - emission comes from the predissociation of high $np\pi$- Rydberg states of the H_2 - molecule in the D $(3p\pi)\,^1\Pi_u$ state which in dissociation limit results in an excited atom of hydrogen with n = 3. It can be suggested that the mechanism of the excited hydrogen atom formation in the direct dissociation of H_2 by K^+ - ions is the same. However, an excitation of vibrational levels of the D $(3p\pi)\,^1\Pi_u$ state above the dissociation limit is also possible.

References

I. O.P.Bochkova, Opt. i Spektr. 50,668 (I98I).

EXCITATION IN COLLISIONS OF N_2^+ WITH He, Ne and Ar IN THE ENERGY RANGE 1 - 2000 eV (2000 - 8000 Å)

Ingrid Kuen, Branislav Jelenković* and Franz Howorka
Institut für Experimentalphysik, Abt. Spektroskopie und Laserphysik der Universität Innsbruck, Karl Schönherrstraße 3, A 6020 Innsbruck, Austria

The excitation of N_2^+ in collisions of He^+, Ne^+ and Ar^+ with N_2 has been studied by Simonis /1/, Brandt et al. /2/, Lipeles /3/ and Schuster /4/ in the energy range 1 - 2000 eV. There are two remarkable observations: 1) The excitation of vibrational states of the N_2^+ B state does not correspond to favorable Franck-Condon factors (FCF), 2) at low energies in He^+ - N_2 and Ne^+-N_2 encounters an undefined system comes up, having the appearance of a quasicontinuum in the wavelength region 3600 to 3900 Å. This is partly ascribed to the tailbands of the B system, partly to a new unidentified electronic excitation. The fact that higher vibrational states are preferred as compared to the ones expected by favorable FCF is explained by Lipeles /3/ by a long-distance electron jump mechanism where the ground state of N_2 is slightly shifted due to the ion-dipole interaction and so gives FCF favorable to higher vibrational levels of the B state.

The apparatus we have used to study the excitations has been described in detail /5/ and therefore a very short description is given here. It consists of a hollow cathode ion source, magnetic mass separator, collision chamber, optical spectrometer, cooled photomultiplier and single-photon counting equipment. The mass selected ion beam (in our case N_2^+) enters the collision chamber which contains the target gas under a pressure of 3 mtorr. This ensures single-collisions conditions. From the light pulses a "normalized light intensity" is derived which can be converted to absolute cross sections after a calibration process. The energy dependence of the cross section is determined by taking individual measurements at the energies adjusted by a deceleration lens equipment.

In order to clarify the excitation process we have studied the excitation of N_2^+ (B) in collisions of N_2^+ (X) with He, Ne and Ar. Here, no ground state molecule is present which could be distorted by the influence of an approaching ion and we should see the excitation corresponding to favorable FCF from the ground state of N_2^+ (X) which is practically congruent with N_2(X). The long-distance electron jump mechanism for He^+-N_2 has already been questioned by ab initio calculations of M. Gerard-Ain /6/ which showed that a nonadiabatic surface crossing at around 2.5 Å is responsible for the vibrational population of N_2^+(C). From our results of N_2^+ on He, Ne and Ar it follows that the population of N_2^+(B) vibrational levels corresponds to the excitation of the reverse case, i.e. He^+, Ne^+ and Ar^+ on N_2. The "unidentified" system also appears at corresponding c.m. energies.

The energy dependence of the excitation cross section for the 0-0 band of the N_2^+(B) system shows a threshold around the endoergicity (3 eV) and then rises very similar to the case of atomic excitations (e.g. Ar^+ on He or O_2). Landau-Zener calculations /7/ have shown that atomic ion excitation can be understood very well by considering a quasimolecular collision complex whose energy is found in the final products after separation.

The conclusion drawn is that in the case of N_2^+ on He, Ne and Ar the same surface-crossing mechanisms are utilized to populate the vibrational B levels as in the reverse case He^+, Ne^+ and Ar^+ on N_2. It is the property of the quasimolecular collision complex that determines the molecular final states rather than general collision dynamics.

Acknowledgements. The present study is supported by the Fonds zur Förderung der wissenschaftlichen Forschung in Österreich under Contract no. S-18/04.

References.

1. J. Simonis, Dissertation Bonn (1977)
2. D. Brandt, Ch. Ottinger and J. Simonis, Ber. Bunsenges. 77, 648 (1973)
3. M. Lipeles, J. Chem. Phys. 51, 1252 (1969)
4. T. Schuster, Diplomarbeit Freiburg (1979)
5. F. Howorka, I. Kuen, H. Villinger, W. Lindinger and J.H. Futrell, Phys. Rev. A 26, 93 (1982)
6. M. Gerard-Ain, T.R. Govers, B. Levy and Ph. Millie, Int. J. Mass Spectrom. Ion Phys. 47, 167 (1983)
7. F. Howorka, I. Kuen and W. Federer, Int. J. Mass Spectrom. Ion Phys. 47, 151 (1983)

*Permanent address: P.O.B. 57, Yu- 11001 Beograd

CROSS SECTIONS FOR EXCITATION OF BALMER BETA RADIATION IN AN EQUILIBRATED BEAM OF H AND H^+ IN O_2

J. Roger Sheridan

University of Alaska, Fairbanks, Alaska 99701, U.S.A.

Excitation of radiation by projectiles in an equilibrated beam can be defined in terms of an "effective emission cross section" which lumps the effects of cascade, quenching, and excitation.

The effective emission cross section can be related to emissions observed in the laboratory and to the evolution of excited states in a beam via a theoretical beam growth equation. The effective emission cross section allows prediction of projectile emissions in an equilibrated beam without a knowledge of the fundamental excitation and quenching cross sections.

Effective emission cross sections have been measured in our laboratory for Balmer beta emission in equilibrated beams of H and H^+ having energies from 1.7 to 32 keV and travelling in O_2. The results are consistent with fundamental cross section measurements under single collision conditions and are consistent with the results of Van Zyl and Neumann where valid comparisons can be made.

*This work has been supported in part by NSF Grant ATM79-16502.

1. B. Van Zyl and H. Neumann, J. Geophys. Res. 85, 6006 (1980).

EXCITATION IN $He^{o}+D_2$ COLLISIONS AT LOW KEV ENERGIES

J. Jakacky, Jr. and A. Russek
Dept. of Physics, University of Connecticut Storrs, CT 06268

Recent scattering calculations for He on H_2 collisions using an energy surface which is a parametric fit to ab initio calculations of the HeH_2 energy surface[1] has predicted a τ_c substantially less than that observed experimentally.[2] The new experimental data implies that the three-body polarization potential, which is a function of the He-H_2 separation, R, only, is substantially stronger than previously thought, prompting a re-examination of the SCF energy surfaces used to calculate τ_c. The electronic contribution to the interaction energy (Fig. 1) for $\gamma = 0^o$ and 30° exhibits pronounced minima in the vicinity of R = 0.7 bohr due to the fact that HeH_2 here approaches LiH. Excited states of the HeH_2 system are being studied with particular attention to intersections with the ground state energy surface. Figure 2 shows crossings of two excited states with the ground state. The energies are plotted as functions of R, with H-H separation fixed at 1.4 bohr and $\gamma = 90^o$. The excited states considered are those for which the H_2 is in the ($1\sigma g, 1\sigma u$) configuration and in the ($1\sigma u^2$) configuration. Both cross and ground state in the vicinity of 0.7 bohr. The threshhold values of τ for the two excitations have been classically calculated to be 1.7 and 3.0 keV-deg respectively. Experimentally, the excitation to the He($1s^2$) + H_2($1\sigma u^2$) state is found to be approximately 2.3 keV-deg. The transition to the He($1s^2$) + H_2($1\sigma g, 1\sigma u$) state is not seen experimentally, most likely due to the strong selection rule against singlet to triplet transitions.[3] Figure 3 shows an approximate cross section of the intersection surface between the ground and the He($1s^2$) + H_2($1\sigma u^2$) energy surfaces for H_2 separation 1.4 bohr.

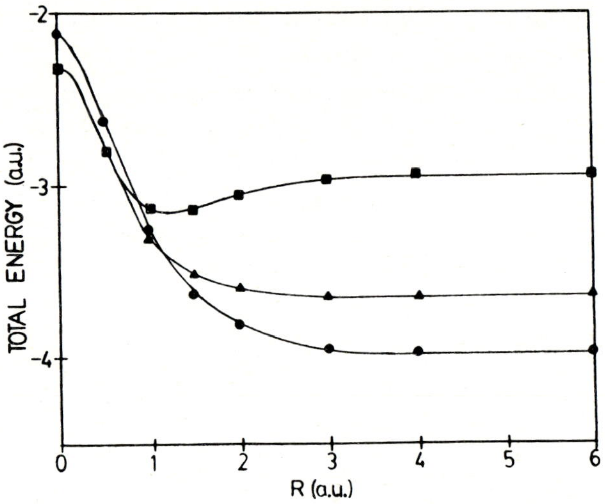

Fig. 2. Three lowest states for γ=90° and r = 1.4.

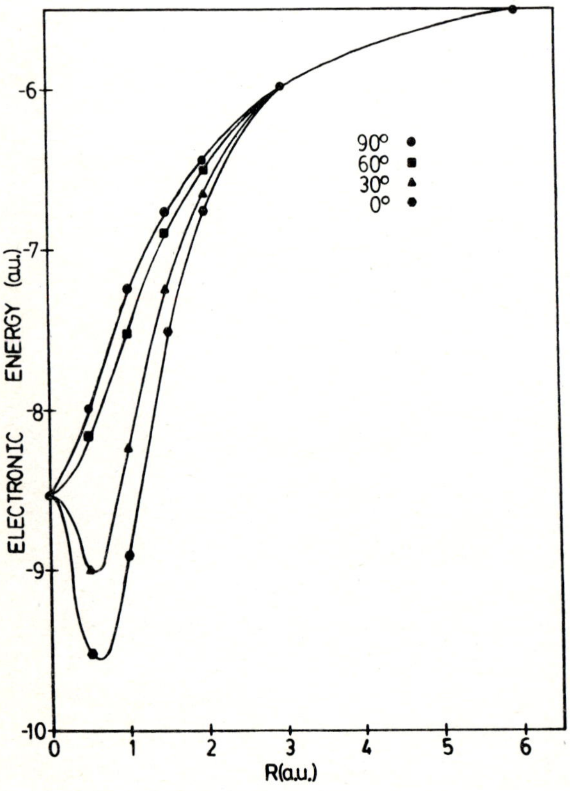

Fig. 1. Ground state electronic energy at r = 1.4 a.u.

References

1. A. Russek and R. Garcia G., Phys. Rev. A 26, 1924 (1982).
2. R. Snyder and A. Russek, Phys. Rev. A 26, 1931 (1982).
3. D. Dowek, D. Dhuicq, V. Sidis, and M. Barat, Phys. Rev. A 26, 746 (1982).

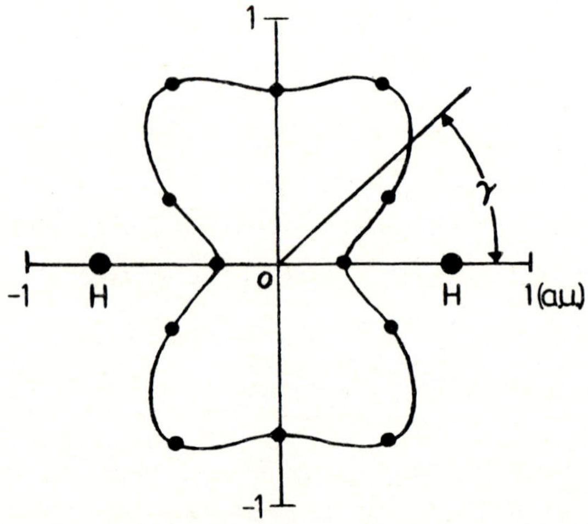

Fig. 3. Intersection surface of ground and $1\sigma u^2$ states.

PENNING IONIZATION PROCESSES CLARIFIED BY ELECTRON-ION-COINCIDENCE MEASUREMENTS

Jürgen Baus, Arnulf Benz and Harald Morgner

Fak.f. Physik, Hermann-Herder-Str. 3, 7800 Freiburg

The collision of a metastable rare gas atom with a molecule of positive electron affinity can lead to a charge exchange process prior to ionization. This has been verified for several molecules NO_2[1], SO_2[2], Cl_2[3] by electron-ion coincidence measurements. With O_2 the same process was found.[4]

Unexpected results were obtained for Penning ionization of the chlorofluoromethanes CF_nCl_{4-n} (n=0,..,4) by He (2^3S).[5] The adiabatic electron affinity of these molecules varies monotonically when changing n=0,..,4: $EA_{adiab.}$ = 2.0, 1.5, 0.4, -0,4, -0.7 eV. We found no indication that the ionic channel He^+ $CF_nCl^-_{4-n}$ plays a role in Penning ionization of the first three members of this group of molecules (n=0,1,2). However, the Penning process for CF_3Cl shows the characteristic features which signal that charge exchange occurs prior to ionization, even though this molecule has a negative electron affinity.

In the present study we have repeated these measurements with a newly designed spectrometer equipped with a position sensitive electron detector. In fig. 1 we show spectra of electrons formed in coincidence with different fragment ions. Those parts of the spectra which should be absent in case of pure PI out of the covalent entrance channel are indicated. A possible explanation and results for other molecules will be presented at the conference.

This work is supported by the Deutsche Forschungsgemeinschaft.

1) W. Goy, V. Kohls and H. Morgner, J.Electr.Spec.Rel. Phen. 23 (1981) 383
2) W. Goy, H. Morgner and A.J. Yencha, J.Electr.Spec.Rel. Phen. 24 (1981) 77
3) W. Kischlat and H. Morgner, Z.Physik A, submitted 1983
4) O. Leisin, H. Morgner and W. Müller, Z.Physik A 304 (1982) 23
5)a) W. Kischlat, Diplom thesis, Freiburg 1981.
b) W. Kischlat and H. Morgner, to be published.

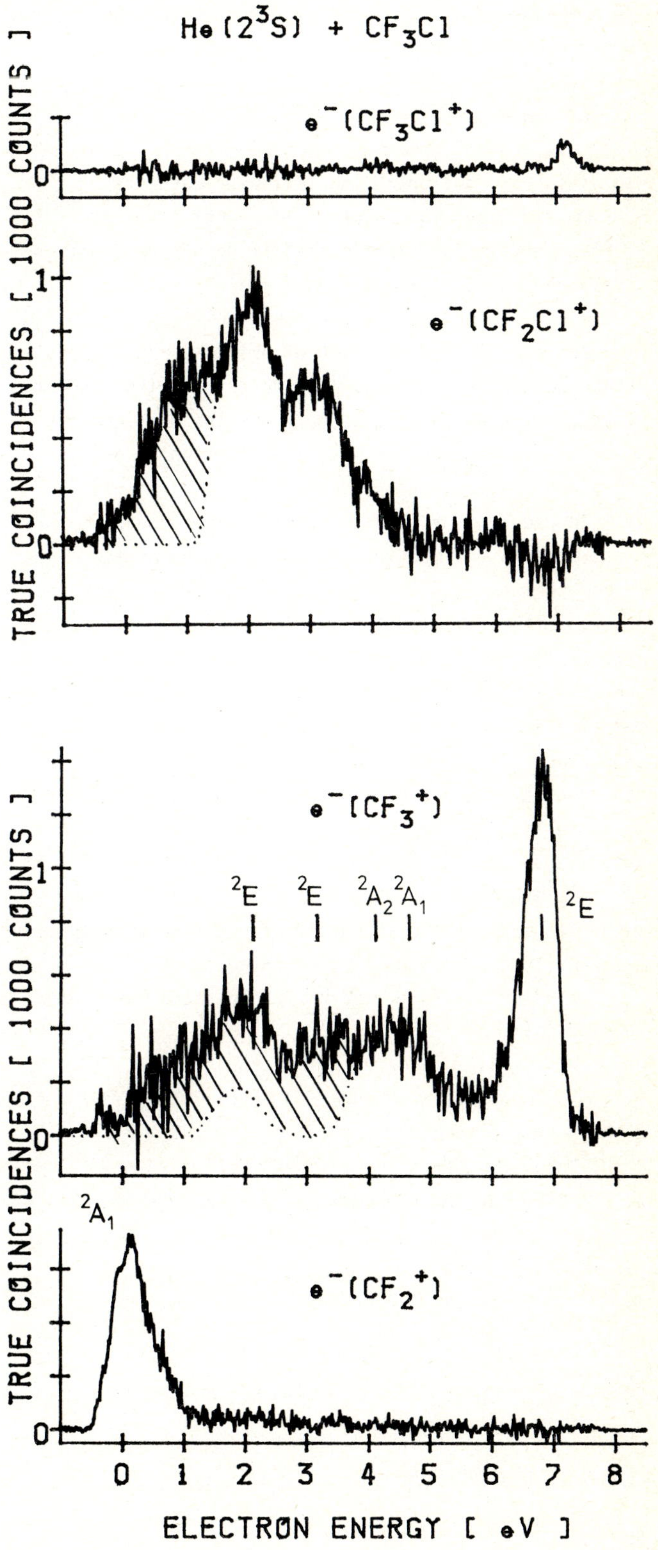

A NEW APPARATUS FOR THE MEASUREMENT OF PENNING ELECTRON ANGULAR DISTRIBUTIONS

H. Morgner and G. Zimmermann

Fakultät für Physik der Universität, Herrmann-Herder-Str. 3, D-7800 Freiburg

Measured angular distributions of electrons emitted in the Penning process

$$A^{*} + B \rightarrow A + B^{+} + e^{-}$$

can yield information in the following respects:

1. symmetry selection rules in the Penning process[1)2)]
2. phase shifts for elastic scattering of the electron off the quasi molecular ion $A + B^{+}$ [3)]
3. the excited orbital in the quasi molecule $A^{*} + B$

In ref. 1 the evaluation of existing experimental data by Ebding and Niehaus[4)] is described with respect to symmetry considerations. More precise measurements are required in order to yield information to points 2. and 3.

In order to carry out precise measurements we have constructed a new machine. A cross section of the main vacuum chamber is shown in fig. 1. An electron spectrometer, differentially pumped by a turbomolecular pump can be rotated around the reaction zone. The dispersive element is a plate condensor. It has the advantage that the focus points of all electron energies fall into the same plane. This allows to employ channelplates followed by a resistive anode as a position sensitive detector in order to improve the luminosity of the instrument. A second spectrometer of the same construction mounted at an angle of 90 with respect to the metastable beam serves to monitor the intensity of electron emission. Results will be presented at the conference.

This work is supported by the Deutsche Forschungsgemeinschaft.

1. V. Hoffmann and H. Morgner, J.Phys. B 12 (1979) 2857
2. H. Morgner, Comm.Atom.Mol.Phys. 11 (1981) 271
3. H. Morgner, J.Phys. B 11 (1978) 269
4. T. Ebding and A. Niehaus, Z.Physik 270 (1974) 43

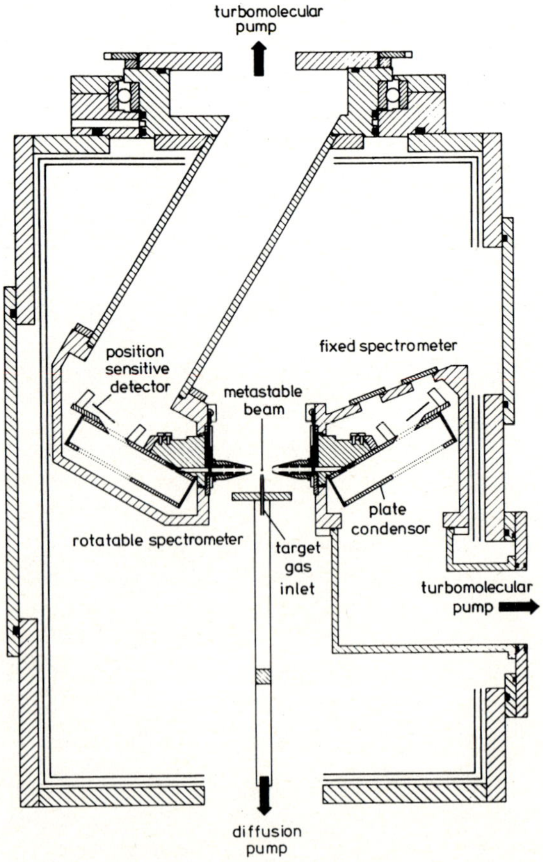

THE REACTION OF METASTABLE HELIUM WITH MOLECULES INVESTIGATED BY OPTICAL AND ELECTRON SPECTROSCOPY

Oskar Leisin, Harald Morgner, Hubert Seiberle and Joachim Stegmaier

Fakultät für Physik der Universität, Herrmann-Herder-Str. 3, D-7800 Freiburg

The reaction of metastable helium with molecules has been investigated since several years by electron spectroscopy (PIES)[1][2] and by optical spectroscopy in a flowing afterglow[3][4]. Only recently Penning ionization optical spectroscopy (PIOS) has been performed in a beam experiment[5].

We have built up a machine which is equipped with both, an electron spectrometer and an optical detection system. Thus electron spectra and fluorescence spectra can be taken simultaneously. This ensures that both spectra are taken under the same conditions and thus can be compared more directly than could be done with spectra measured in different machines. Furthermore, we have the advantage that coincidence measurements between electrons and photons can be carried out as well. The combination of these experimental facilities allows the clarification of the sometimes complex reaction mechanisms.

As an example we discuss here the reaction of He^* with SO_2. PIES displays a structure near 2.3 eV electron energy for both metastable species[6][7]. This feature can be understood with the aid of PIOS. The figures show OI emission lines for both, $He(2^3S)$ and $He(2^1S)$. The emitting O^* levels are Rydbergstates with an excitation energy between 10.740 eV and 13.405 eV. These lines indicate that energy transfer into SO_2^* with subsequent dissociation into SO and O^* occurs. Since the excitation energy of the observed O^* levels exceeds the ionization potential of SO, autoionization can occur during dissociation and thus lead to the peaks at 2.3 eV in PIES. More details and results for other molecules will be presented at the conference.

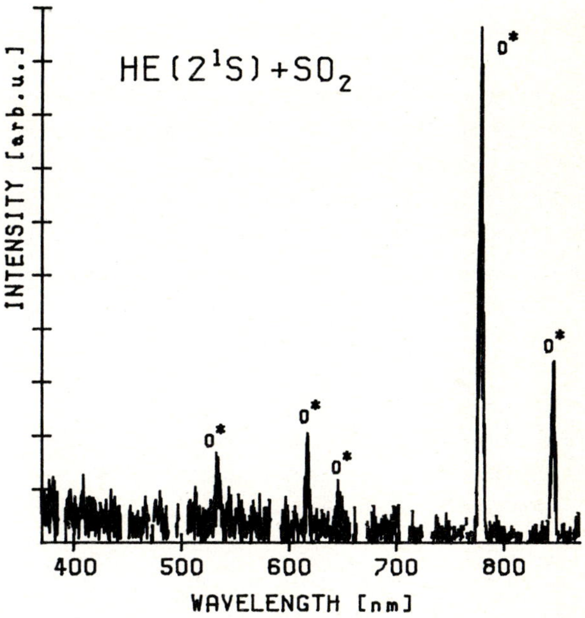

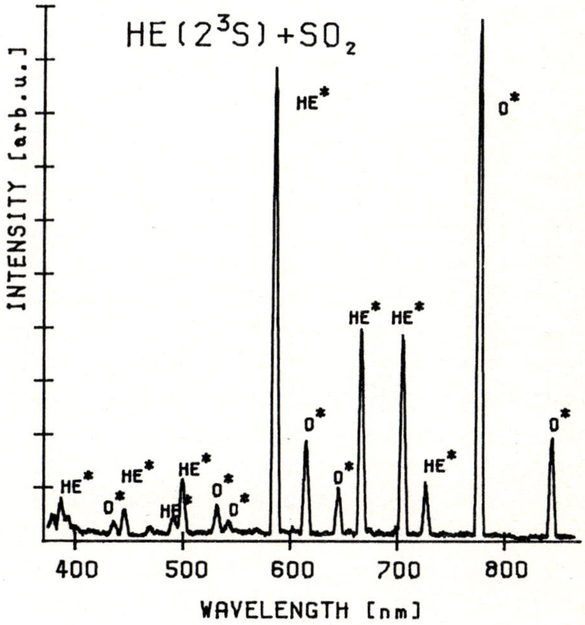

This work is supported by the Deutsche Forschungsgemeinschaft.

References

1. H. Hotop, E. Kolb and J. Lorenzen J.Electr.Spec.Rel. Phen. 16 (1979) 213
2. O. Leisin, H. Morgner and W. Müller Z.Physik A 304 (1982) 23
3. W.C. Richardson and D.W. Setzer J.Chem.Phys. 58 (1973) 1809
4. A.J. Yencha and K.T. Wu Chem.Phys. 49 (1980) 127
5. H.L. Snyder, B.T. Smith and R.M. Martin, Chem.Phys. Lett. 94 (1983) 90
6. W. Goy, H. Morgner and A.J. Yencha, J.Electr.Spec.Rel. Phen. 24 (1981) 77
7. W. Kischlat, O. Leisin, H. Morgner and W. Müller XII. ICPEAC (1981), Book of Abstracts, p. 1097

RELATIVE ORBITAL-IONIZATION PROBABILITIES OF GASEOUS AND CONDENSED PHASE MOLECULES BY PENNING AND PHOTO PROCESSES

Andrew J. Yencha,† Hiroyasu Kubota,* Tomohiko Hirooka,* Tsutomu Fukuyama,* Tamotsu Kondow,* and Kozo Kuchitsu*

†Depatments of Physics and Chemistry, State University of New York at Albany, Albany, New York 12222 (USA)
*Department of Chemistry, Faculty of Science, University of Tokyo, Bunkyo-ku, Tokyo 113 (Japan)

In recent years several electron spectroscopic studies have been conducted on the ionization behavior of gaseous and condensed molecular systems using metastable rare-gas atoms (Penning ionization - PI) and resonance photons (photoionization - PhI). In molecular systems as diverse as ammonia, water, and numerous varied hydrocarbon molecules, it has been found that these condensed materials behave as molecular crystals, thus retaining the essential ionization-potential characteristics of the gas phase systems.[1-3] We have undertaken a detailed quantitative study of a series of molecular systems in condensed phase in comparison with those in the gas phase to determine their relative orbital-ionization probabilities. We have subtracted out the low-energy electron scattering peak in condensed-phase electron spectra obtained by both PI and PhI, so as to obtain realistic values for the intensities for valence-electron ionizations.

As an example of a representative molecular-crystal system, in Fig. 1 is shown the electron energy distribution curves (EEDC) for molecular benzene using He* and Ne* metastables and HeI resonance photons in the gaseous and condensed phases. In Fig. 2 is shown a detailed comparison of the EEDC for electrons emitted from benzene condensed on a copper substrate at 150 K using He*($2^{1,3}S$) impact (PI) and HeI (58.4 nm) radiation (PhI). These spectra show a strong, broad peak at low electron energies for solid benzene (Fig. 2), which does not appear in gaseous benzene spectra, that is thought to arise from scattered secondary electrons.[2] We have estimated the intensity ratios amoung each band of benzene on a copper substrate after subtraction of a smooth background curve. The peak intensity ratios normalized to peak II are given in Table 1 for all spectra shown in Figs. 1 and 2. They indicate a strong dependence on the type of molecular-orbital ionization (e.g. π vs. σ) and on the mode of ionization (e.g. PI vs. PhI). In addition, we have measured these same peak intensity ratios for benzene condensed on a graphite substrate (spectra not shown) in several He* and HeI electron spectra. These data, some of which were obtained on the same day while others were obtained on different days, are presented in Table 2. The reproducibility of the data is rather good. Comparisons of similarly produced data (He* or HeI) in Tables 1 and 2 indicate considerable differences in the relative peak intensity ratios in the two substrate systems. Similar spectral data have also been obtained for condensed H_2O and NH_3 on both copper and graphite substrates and attempts are being made to include several organic molecular-crystal systems in this study as well. All of these data and their implications to Penning ionization processes in gaseous and condensed phases will be presented.

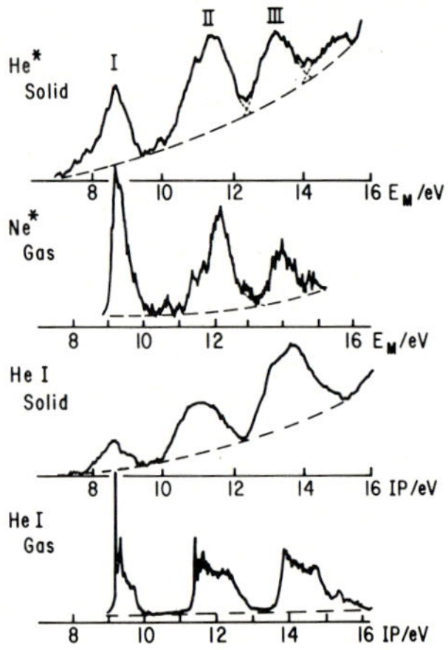

Fig. 1

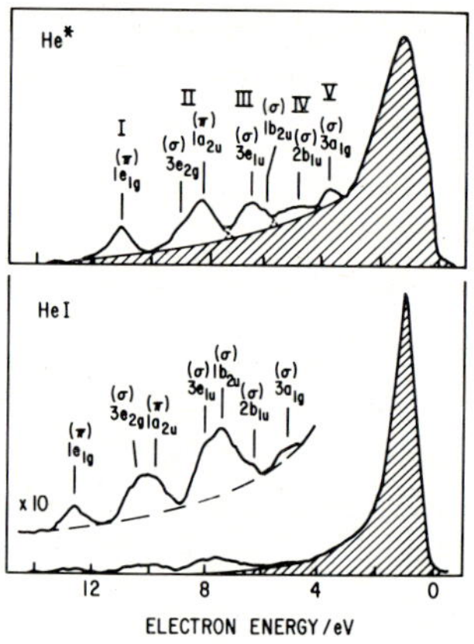

Fig. 2

Fig. 1. EEDC for PI and PhI of benzene in the condensed and gas phases.[3] The first peaks are aligned.

Fig. 2. EEDC for PI and PhI of benzene condensed on Cu.[2]

Table 1. Electron peak intensity ratios for condensed benzene on copper and benzene gas.

	Phase	Fig.	I	II	III	IV	V
He*	solid	1	0.58	1	0.55		
Ne*	gas	1	0.83	1	0.52		
HeI	solid	1	0.40	1	1.61		
HeI	gas	1	0.50	1	1.16		
He*	solid	2	0.41	1	0.60	0.29	0.18
HeI	solid	2	0.32	1	1.67		0.15

Table 2. Electron peak intensity ratios for condensed benzene on graphite.

	Run	I	II	III	IV	V
He*	1	0.86	1	0.54	0.22	0.26
	2	0.86	1	0.60	0.33	0.21
	3	0.89	1	0.64	0.29	0.26
	4	0.94	1	0.58	0.29	0.26
	5	0.96	1	0.57	0.27	0.23
HeI	1	0.33	1	1.28		
	2	0.32	1	1.35		
	3	0.36	1	1.31		
	4	0.30	1	1.23		

References

1. A. J. Yencha et al., J. Electron Spectrosc. Relat. Phenom., 23, 431 (1981).
2. H. Kubota et al., J. Electron Spectrosc. Relat. Phenom., 23, 417 (1981).
3. T. Munakata et al., Chem. Phys. Lett., 64, 409 (1979).

STRONG VARIATION IN THE THERMAL ENERGY IONIZATION CROSS SECTION OF ARGON ATOMS COLLIDING WITH DIFFERENT LASER-EXCITED Ne($2p^5$ 3p J=1,2,3) ATOMS

W. Bußert, J. Ganz, M.-W. Ruf, A. Siegel and H. Hotop

Fachbereich Physik der Universität, 6750 Kaiserslautern, FRG

H. Morgner

Fakultät für Physik der Universität, 7800 Freiburg, FRG

We have studied the state-dependence of the Penning ionization cross section of Ar in thermal energy collisions with Ne($2p^5$ 3p J=1,2,3) multiplet states under single collision conditions. The Ne(3p) states were selectively prepared in the reaction center by transverse excitation of a collimated metastable Ne(3s $^3P_{2,0}$) beam with a single mode dye laser. Background discriminating high resolution (20 meV) electron spectrometry allowed us to identify the Ne(3p)+Ar ionization processes in spite of the short effective residence time of the Ne (3p J=1,2) states prior to their decay to the Ne ground state via the Ne(3s J=1) states.

From the measured energy-integrated electron intensities and with appropriate state-dependent corrections for differences in the effective residence times and for polarization effects we have deduced the following ionization cross sections for Ne(3p J=1,2,3) + Ar collisions at an average relative collision energy of about 45 meV (Table 1):

Table 1: Ionization cross sections for Ne(3p)+Ar

Neon excited state	Ionization cross section ($10^{-16}cm^2$)
Ne(3p $^3D_3$, $2p_9$)	19.7(7)
Ne(3p $^3D_2$, $2p_8$)	15.3(8)
Ne(3p $^3D_1$, $2p_7$)	6.9(9)
Ne(3p $^1D_2$, $2p_6$)	11.6(11)
Ne(3p $^1P_1$, $2p_5$)	1.5(10)
Ne(3p $^3P_2$, $2p_4$)	3.0(9)
Ne(3p $^3P_1$, $2p_2$)	0.4(4)

The uncertainties in Table 1 only reflect the estimated error of the relative size of these cross sections. The absolute scale of the cross section is based on the Ne(3s $^3P_2$)+Ar ionization cross section (13 ± 3)·$10^{-16}cm^2$ (average of values in refs. 1-3), relative to which the value for Ne(3p $^3D_3$)+Ar was determined with an uncertainty of ±40% in an ion measurement of the type described elsewhere[4]. The listed cross sections are the average values of two different measurements with the electric vector of the linearly-polarized laser parallel and perpendicular to the neon beam. In some cases significant polarization effects have been observed.

The strong state-dependence of the measured cross sections reflects the different properties of the respective potential energy curves. Using semiempirical adiabatic potential curves for Ne(3p)+Ar, based on the ab initio CI results[5] for the Σ- and Π-states of Na(3s, 3p)+Ar, we have calculated the state dependence of the ionization cross section, ignoring transitions between the various entrance channels and using a simple autoionization width function $\Gamma(R) = A \exp(-\alpha R)$, common to all entrance channels. With an optimized choice for the critical parameter α (the factor A mainly adjusts the absolute scale of the cross sections) satisfactory agreement between the calculated cross sections and the measured values was obtained. We interprete this agreement as an indication that the used potential curves are (at least) qualitatively correct.

At the conference, we shall present the electron energy spectra and the observed polarization dependence in detail together with a description of the cross section calculation and the available potential energy curves for Ne(3p)+Ar. In the future, we plan to study the collision energy dependence of these ionization cross sections in order to elucidate more details of this many-channel energy transfer system.

This work has been supported by the Deutsche Forschungsgemeinschaft (SFB 91).

References

1. W.P. West, T.B. Cook, F.B. Dunning, R.D. Rundel and R.F. Stebbings, J. Chem. Phys. 63, 1237 (1975)
2. J.M. Brom Jr., J.H. Kolts and D.W. Setser, Chem. Phys. Lett. 55, 44 (1978)
3. A.Yokoyama and Y. Hatano, Chem. Phys. 63, 59 (1981)
4. W. Bußert, J. Ganz, H. Hotop, M.-W. Ruf, A. Siegel, H. Waibel, J. Lorenzen and P. Botschwina, Chem. Phys. Lett. ...,... (1983)
5. R.P. Saxon, R.E. Olson and B. Liu, J. Chem. Phys. 67, 2692 (1977)

PENNING IONIZATION OF Ne AND Ar BY $He^-(1s2s2p)\ ^4P^o$ at 1-2 keV

M. J. Coggiola and K. T. Gillen

Molecular Physics Laboratory, SRI International, Menlo Park, CA 94025

We have recently completed measurements of the energy dependence of the total collisional destruction of He^- (1s2s2p) $^4P^o$ in a variety of atomic and molecular targets,[1] and here report additional cross section measurements for target ionization in the targets He, Ne, and Ar. For Ne and Ar, the cross sections for target ionization are a significant fraction of the total destruction values. This observation and the existence of a large excitation energy (19.7 eV) and an inner shell vacancy in $He^-\ ^4P^o$ suggest the possible importance of Penning ionization channels in the collisional destruction of this metastable negative ion.

The fast He^- beam was produced from He^+ by double electron capture in an alkali cell, deflected by 90° in an electrostatic quadrupole, then entered a 1.4 cm long gas target cell through a 1 mm dia aperture. An exit aperture (3 mm dia) was followed by a 2 cm dia gridded Faraday cup ~8 cm beyond the target cell. A collector plate within the gas cell was biased at -20 v through an electrometer for measurements of slow positive ion signals. The effective cell length was determined by measuring the charge transfer of a He^+ beam in a He target, and normalizing to known cross sections.[2] For He^+ + He the measured attenuation of He^+ on the Faraday cup always agreed with the collected slow ion current to within 5-10%.

For beam energies between 1 and 2 keV, target ionization cross sections were determined by measuring the slow ion current relative to the incident He^- current as a function of target gas pressure. Product currents in all cases increased linearly with target gas density. The total collisional destruction cross section of He^- in He at these energies is 14-16 $Å^2$. No significant slow ion production is seen (upper limit of 0.2 $Å^2$). In contrast, for Ne, the target ionization increases from 1.6 to 2.8 $Å^2$ as the energy decreases from 2 to 1 keV, while for Ar the cross section increases from 2.9 to 4.0 $Å^2$. Total destruction cross sections for both targets are 12-13 $Å^2$ in this energy range.

These slow ion production cross sections are much larger than expected for impact ionization at these collision energies. A simple mechanism to explain these results is diagrammed in Fig. 1 for the He^- + Ar system. The first step involves a transfer from the incoming He^- + Ar curve to the He^- + Ar^+ + e ion pair curve at (a) via a Penning electron capture process. The transient He^- $(1s^22\ell)$ state is then stabilized against decay by the Ar^+ ion. As the ion pair state evolves to large R, it can either produce He + Ar^* products via crossings with a large number of excited neutral states (b), or continue on the ion pair curve to large R where the isolated He^- $(1s^22\ell)$ will rapidly autodetach (c), yielding He $(1s^2)$ + Ar^+ + $2e^-$. For Ne target, the He^- $(1s^22\ell)$ + Ne^+ + e^- curve lies asymptotically above the incoming He^- (1s2s2p) + Ne curve, but the strongly attractive ion pair curve drops below the incoming state at large R to yield a schematic diabatic picture essentially analogous to that shown in Fig. 1.

This work was supported by the National Science Foundation.

References

1. R. V. Hodges, M. J. Coggiola, and K. T. Gillen (in preparation).

2. F. L. Eisele and S. W. Nagy, J. Chem. Phys. 65, 752 (1976).

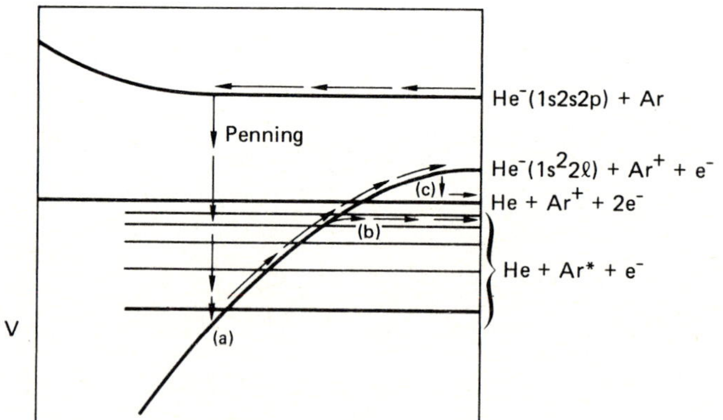

Fig. 1: Schematic potential energy curves for He^- + Ar collisions and postulated important Penning mechanism.

IONIZATION OF CS_2 CLUSTERS IN COLLISION WITH ELECTRONS AND LONG-LIVED EXCITED RARE-GAS ATOMS

Tamotsu Kondow, Koichiro Mitsuke and Kozo Kuchitsu

Department of Chemistry, Faculty of Science, The University of Tokyo, Bunkyo-ku, Tokyo 113, Japan

Ionization processes of neutral clusters of CS_2 in collision with electrons and long-lived excited states of rare gas atoms have been studied, the emphasis being laid on the effect of clustering on the following processes:

$$(CS_2)_n + e \rightarrow (CS_2)_n^+ \text{ and fragment ions,} \quad (1)$$

$$(CS_2)_n + A^M \rightarrow (CS_2)_n^+ \text{ and fragment ions,} \quad (2)$$

and $$(CS_2)_n + Ar^{**} \rightarrow (CS_2)_n^- \quad (3)$$

where A^M and Ar^{**} represent $Ar(^3P_{0,2})$ or $He(2^3S, 2'S)$ and $Ar(n,\ell)$ atoms with $n \simeq 15-40$, respectively.

Apparatus

A schematic diagram of the apparatus is shown in Fig. 1. A pulsed cluster beam of CS_2 seeded with H_2 produced in the beam chamber (N) was admitted into the collision chamber (C). An electron gun of a quadrupole mass filter (Extranuclear) was employed for electron-impact ionization (process 1). The ions thus formed were mass-analyzed and detected by an electron multiplier. A CAMAC system was used for data acquisition.

The cluster beam source consists of a solenoid valve driven typicaly at 20 Hz, a converging-diverging supersonic nozzle[1]. The dimer intensity was estimated to be about 2×10^{18} at a stagnation pressure of 300 Torr. A^M and Ar^{**} atoms were produced in chamber C by electron impact on the rare gas introduced from the gas inlet. Only neutral excited species were admitted to the collision region by applying appropriate voltages to the three concentric grids.

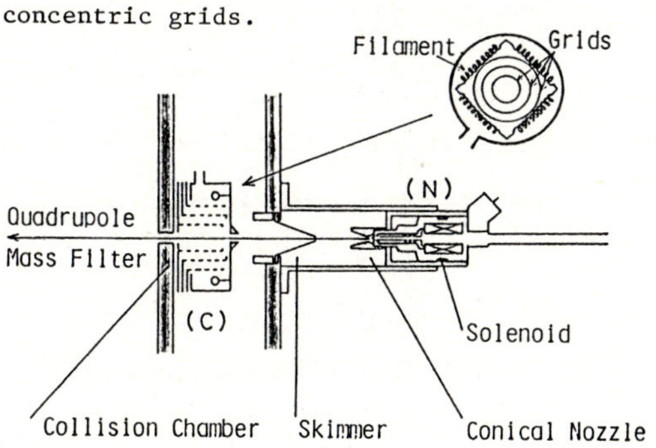

Fig. 1: A schematic diagram of the apparatus: a cluster beam source and a skimmer (N); a collision chamber (C) equipped with an electron gun (1 - 5 mA), three concentric grids (stainless steel 50 mesh).

Electron impact

Positive ions of $(CS_2)_n^+$ $(n \leq 6)$ were detected by electron impact ionization. The appearance potential for $(CS_2)_2^+$ (9.5 eV) agreed with the reported value, 9.37 eV, determined by photoionization.[2]

In the fragmentation pattern of $(CS_2)_n$ $(n = 1, 2)$, in particular, the following trends in the ion intensities observed:

$$\frac{(CS_2)S_2^+}{(CS_2)_2^+} > \frac{S_2^+}{CS_2^+} \quad \text{and} \quad \frac{(CS_2)CS^+}{(CS_2)_2^+} < \frac{CS^+}{CS_2^+}$$

These trends seem to be accounted for by a staggerd parallel structure of $(CS_2)_2$ determined by a photoionization study.[2]

Excited-atom impact

The neutral clusters were also ionized in chamber C by collision with A^M and Ar^{**} (processes 2 and 3). In the case of Ar^M impact, $(CS_2)_n^+$ $(n \leq 2)$ due to Penning ionization were observed (Fig. 2). On the other hand, only CS_2^+ and its fragment ions, CS^+ and S^+, were observed when He^* was used; this is probably because the excess energy transmitted from He^* destabilizes $(CS_2)_2^+$.

In the collisional ionization with Ar^{**}, $(CS_2)_2^-$ but no monomer ion, CS_2^-, was detected (Fig. 2). This is probably because the electron affinity of $(CS_2)_2$ is higher than that of CS_2. This process is analogous to attachment of slow electrons, and has a potential application to non-destructive ionization of clusters.

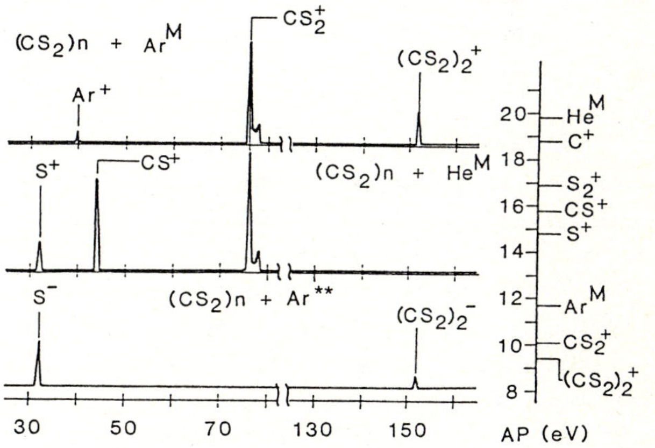

Fig. 2: Mass spectra of the ions produced by impact of A^M and Ar^{**} on $(CS_2)_n$. The appearance potentials of the ions produced from $(CS_2)_n$ are compared with the excitation energies for Ar^M and He^M.

References

1. O. F. Hagena, "Molecular Beams and Low Density Gas Dynamics", ed. P. P. Wegener, Marcel Dekker, New York, 1974, p. 93.

2. W. M. Trott, N. C. Blais and E. A. Walters, J. Chem. Phys. 71, 1692 (1979).

CHARGE TRANSFER REACTION BETWEEN Ar^+ AND DIATOMIC MOLECULES AT THE 10 EV REGION

T. Matsuo, N. Kobayashi and Y. Kaneko

Department of Physics, Tokyo Metropolitan University, Setagaya-ku, Tokyo 158

On the ion-molecule charge transfer reaction, $A^+ + M \longrightarrow A + M^+ + \Delta E$, many questions have been left unsolved in the low energy region. Though a number of cross section measurements have been reported so far, the interpretation of the results is not necessarily clear. There are always ambiguities in how the excess energy is divided into the translational motion and the internal freedom of the products, and how the Franck-Condon factors affect the magnitude of the cross section. It is necessary for the study of charge transfer mechanism to know the internal states of the product molecular ions. A precise measurements of the actual energy defect ΔE is an useful way to know the internal states of the products. The ΔE in the charge transfer reaction can be determined by the measurement of the translational energies of secondary ions and/or those of product neutrals. We adopted, here, secondary neutrals as the particles whose translational energies to be measured, because the energetic measurement can be made with better resolution.

The detail of the time-of-flight spectrometer used has been described elsewhere.[1] The apparatus consists of a cylindrical vacuum chamber and a flight tube. On an rotatable table in the chamber, an ion beam assembly which consists of an ion source of electron impact type, an ion lens system and a mass-selector is mounted. The detector of the neutrals is a secondary electron multiplier of 20-stage Cu-Be dynodes. The energy of the electrons in the ion source was kept at 30 eV throughout the measurement, at which no metastable Ar ion can be contained in the primary beam. The targets were H_2, N_2 and O_2.

Time-of-flight spectra obtained for O_2 target are shown in figure 1. They were measured for several scattering angles from 0° to 2° at a collision energy of 80 eV. The energy scale of ΔE is indicated; the sign of minus indicates endothermic. The energy reference, ΔE=0, was taken to be a peak position of the Ar^+ + Ar resonance charge transfer. The broadening of this peak gives an estimation of the energy resolution; 0.8 eV in the present measurement. The spectra for O_2 target are distributed predominantly in the endothermic region. At Θ=0°, a peak is located at about 0.25 eV endothermic, and is broadened considerably; about 1.5 times broader at FWHM in comparison with that of Ar^++ Ar. When the angle Θ is increased, the peak position shifts gradually to the larger endothermicity, and the magnitude of these decreases rapidly. This result indicates that the product O_2^+ ions are mainly in the electronic excited $a^4\Pi_u$ states with vibrational levels of v'=0-3, and some ions are excited in higher vibrational levels at relatively intimate collisions. It is known that the Franck-Condon factors between $O_2(X^3\Sigma_g, v''=0)$ and $O_2^+(a^4\Pi_u, v')$ are distributed over wide range of v' having a maximum at v'=5.[2] In the present energy region, the Franck-Condon factor alone is not sufficient to predict the vibrational distributions of the product O_2^+ in Ar^++ O_2 charge transfer reaction. Similar results have been obtained also for H_2 and N_2 targets.

REFERENCES

1) T.Matsuo et al., J. Phys. Soc. Japan 50, 3482 (1981)
2) M.Halmann et al., J. Chem. Phys. 43, 1503 (1965)

present address : Tokyo University of Agriculture and Technology.

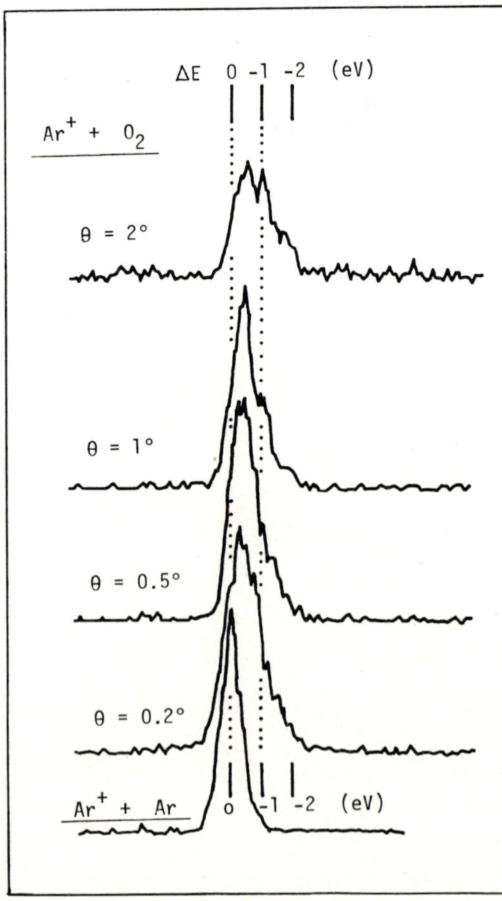

Figure 1. Time-of-flight spectra for Ar^++ O_2 collision, Note that the intensities are not scaled.

MOLECULAR DYNAMICS OF ELECTRON TRANSFER REACTIONS AT LOW ENERGY: $Ar^+ (N_2, Ar)N_2^+$, $N_2^+(N_2,N_2)N_2^+$, AND $Ar^+(Ar, Ar) Ar^+$

J.H. Futrell and B. Friedrich[1]

Department of Chemistry, University of Utah, Salt Lake City, Utah U.S.A.

Only a few molecular dynamics studies of electron transfer have been carried out using crossed beam apparatuses, largely because of experimental difficulties in detecting thermal energy product ions. The only low energy (1-10 eV) studies known to the authors have been carried out by Herman and collaborators[2,4] at the Heyrovsky Institute, capitalizing on the capabilities of that instrument to detect low energy product ions. One study has investigated the resonance charge transfer $Ar^+(Ar,Ar)Ar^+$ using supersonic beams and high angular resolution of the detector to distinguish the reactions of the $^2P_{3/2}$ and $^2P_{\frac{1}{2}}$ sub-states of Ar^+ [5]. A previous study of elastic scattering of rare gas ions from this laboratory included a brief report on resonant charge transfer[6,7].

The Utah low energy crossed beam apparatus, whose general features have been described[7,8], was used to investigate the title reactions. The ion source and electron gun was modified to operate at high ionizing currents and moderate pressures. For the present study this mode of operation results in a nearly pure ground-state population of reactant ions ($Ar^+(^2P_{3/2})$ and N_2^+ ($X^2\Sigma_g^+$,v=o) with a narrow distribution of kinetic energies. The supersonic neutral beam has a angular divergence of = 2°, comparable to that of the ion beam at several eV, resulting in excellent resolution for the scattering diagrams.

The resonant charge transfer reaction of $Ar^+(^2P_{3/2})$ with Ar clearly dominates over the endothermic process Ar^+ $(^2P_{3/2})$ (Ar,Ar) $Ar^+(^2P_{\frac{1}{2}})$ in the range of laboratory energies from 1.89 eV - 61.53 eV, contrasting with the results of McAfee, Falconer, Hozack, and McClure at 123 eV[5]. This series of experiments also illustrates the excellent resolution of the apparatus for charge-transfer processes occuring mainly by large impact parameter electron jump mechanisms.

The reaction $Ar^+(^2P_{3/2})$ (N_2, Ar) $N_2^+(X^2\Sigma_g^+)$ was investigated as an example of a quasi-resonant reaction. Production of ground-state N_2^+ in the v=o level is exothermic by 0.09 eV. However, the experimental results show that the endothermic process generating vibrationally excited N_2^+ ($X^2\Sigma_g^+$,v=1) is the dominant channel at low collision energy. This finding is consistent with results obtained using a SIFT-Drift tube apparatus[9] and an unpublished laser-induced fluorescence study[10].

The reaction $N_2^+(N_2,N_2)N_2^+$ proceeds mainly as a resonant charge-transfer reaction over the energy range investigated. However, at the lowest energy (0.7 eV CM) evidence is obtained for a competing reaction channel proceeding via an N_4^+ intermediate with a lifetime exceeding the rotational period of the complex. Both the forward-backward symmetry of the scattering pattern for the complex mechanism about the plane perpendicular to the relative velocity vector and containing the velocity of the center-of-mass and the extensive energy transfer implied by the velocity distribution argue for the existence of a chemically-bound N_4^+ intermediate. The relatively high binding energy of N_4^+, approximately 0.9 eV[11], provides further support for this interpretation.

References

1. Permanent Address: The J. Heyrovsky Institute of Physical Chemistry and Electrochemistry, Czechoslovak Academy of Sciences, Prague 2, Machova 7, Czechoslovakia.
2. B. Friedrich and Z. Herman, Chem. Phys., 60, 369, (1981).
3. Z. Herman, V. Pacak, A.J. Yencha, and J. Futrell, Chem. Phys. Lett., 37, 329 (1976).
4. P.M. Hierl, V. Pacak, and Z. Herman, J. Chem. Phys., 67, 2678, (1977).
5. K.B. McAfee, Jr., W.E. Falconer, R.S. Hozack, and D.J. McClure, Phys. Rev., A 21, 827 (1980).
6. M.L. Vestal, C.R. Blakely, and J.H. Futrell, Phys. Rev. A 17, 1321 (1978).
7. M.L. Vestal, C.R. Blakley, and J.H. Futrell, Phys. Rev. A 17, 1337 (1978).
8. C. Blakley, P.W. Ryan, M.L. Vestal, and J.H. Futrell, Rev. Sci. Instr., 47, 15 (1976).
9. W. Lindinger, F. Howorka, P. Lukac, I. Kuhn, H. Villinger, E. Alge, and H. Ramler, Phys. Rev. A 23, 2319 (1981).
10. D.R. Guyer, Ph.D. Thesis, Univ. of Colorado (1982). D.R. Guyer and S. Leone, to be published.
11. K. Stephan, M.H. Futrell, H. Helm, and T.D. Märk, to be published.

AB INITIO CALCULATIONS ON THE $(ArN_2)^+$ SYSTEM

P. ARCHIREL , B. LÉVY

Laboratoire de Chimie Quantique (ERA 470) E.N.S.J.F. 92120 MONTROUGE FRANCE

We study theoretically the charge exchange process :

$$Ar + N_2^+ \rightarrow Ar^+ + N_2$$

The crossings between the $X(^2\Sigma_g^+)$ and $A(^2\Pi u)$ states of N_2^+ (perturbed by Ar) and between these two states and the ground state of $Ar^+ + N_2$ are determined by means of a valence bond calculation using PAO's (polarized atomic orbitals).

We compare our results with the experimental work of T. GOVERS and P.M. GUYON.

VIBRONIC AND FS STATE SELECTED CHARGE TRANSFER REACTIONS: $O_2^+ + Ar \rightleftharpoons Ar^+ + O_2$ AND $NO^+ + Ar \rightleftharpoons Ar^+ + NO$ SYSTEMS

Inosuke Koyano, Kenichiro Tanaka, and Tatsuhisa Kato

Institute for Molecular Science, Myodaiji, Okazaki 444, Japan

We have recently demonstrated that the TESICO technique is a powerful means for studying state selected ion-molecule reactions.[1] In the present paper, we report the application of this technique to investigate the vibronic-state selected charge transfer reactions

$$O_2^+(X^2\Pi_g,\ v;\ a^4\Pi_u,\ v) + Ar \rightarrow Ar^+ + O_2 \quad (1)$$

$$NO^+(a^3\Sigma^+,\ v;\ b^3\Pi,\ v) + Ar \rightarrow Ar^+ + NO \quad (2)$$

and the spin-orbit state selected charge transfer reactions

$$Ar^+(^2P_{3/2},\ ^2P_{1/2}) + O_2 \rightarrow O_2^+ + Ar \quad (3)$$

$$Ar^+(^2P_{3/2},\ ^2P_{1/2}) + NO \rightarrow NO^+ + Ar. \quad (4)$$

Experiments were performed with the TEPSICO apparatus,[1] using the double chamber mode of operation. All of the above vibronic states have a sufficiently long radiative lifetime compared with the flight time between the ionization and reaction chambers, allowing the use of this mode of operation.

Experimental results for Reaction (1) are summarized in Fig. 1. The two curves correspond to two different collision energies (1.4 and 5.8 eV) and are normalized at v=5 of the $a^4\Pi_u$ state. Several interesting features are clearly seen from the figure. The cross section for Reaction (1) is extremely small for the v=19 and v=20 states of $X^2\Pi_g$ although Reaction (1) is exoergic for v=19 and above, while it becomes significant at v=0 of the $^4\Pi_u$ state which lies only 0.17 eV above the $X^2\Pi_g$, v=20 state. When vibrational quantum number is increased in the $a^4\Pi_u$ state, a sharp rise in the cross section is observed at v=2, followed by subsequent gradual decrease except at v=5, where the cross section shows a resonance-like enhancement.

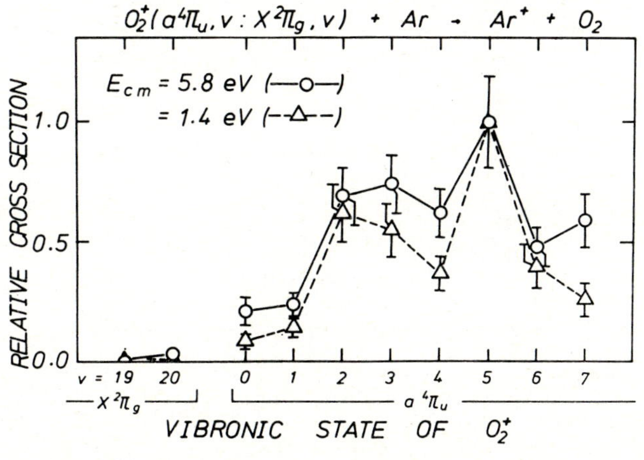

Fig. 1: State selected cross sections for Reaction (1).

Figure 2 shows similar results for Reaction (2) at a collision energy of 1.4 eV. Reaction (2) is slightly (by 0.09 eV) endoergic for v=0 of the $a^3\Sigma^+$ state and exoergic for v=1 and above. Here again, a pronounced resonance-like enhancement of the cross section is observed at v=2 of the $a^3\Sigma^+$ state. Except this point, the cross section shows a tendency to increase slightly but

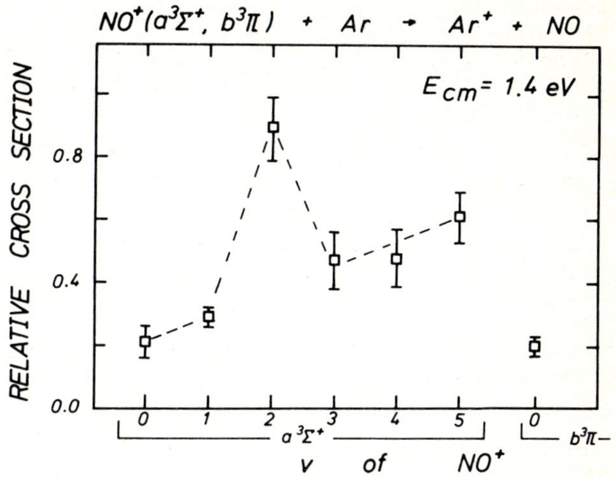

Fig. 2: State selected cross sections for Reaction (2).

steadily with increasing v in the $a^3\Sigma^+$ state, whereas it suddenly decreases when the electronic state changes from $a^3\Sigma^+$ to $b^3\Pi$.

The pronounced enhancement of the cross sections at a particular vibrational level, as observed above, can be attributed to the energy resonance between the reactant state in question and one of the possible product states. In the case of Reaction (2), for example, a close energy resonance exists between the $NO^+(a^3\Sigma^+$, v=2) + Ar reactant state and the $Ar^+(^2P_{1/2}) + NO(X^2\Pi$, v=0) product state. In this connection, it is interesting to compare the results for Reaction (2) with those for Reaction (4). If the microscopic reaction mechanism is the same for the two reactions, the enhancement of the v=2 cross section in the former reaction is expected to be reflected in the latter reaction as an enhanced J=1/2 cross section. The experimental results for the latter reaction show that this is not the case. The ratios of the J=3/2 and J=1/2 cross sections are essentially unity at all collision energies studied. This contrasts with the case of the $H_2^+ + Ar \rightleftharpoons Ar^+ + H_2$ system and would suggest that different potential energy surfaces are involved in Reactions (2) and (4).

References

1. I. Koyano and K. Tanaka, J. Chem. Phys. 72, 4858 (1980)

CHARGE TRANSFER REACTION BETWEEN $Ar^+(^2P_{3/2}, ^2P_{1/2})$ AND N_2 BELOW 0.1 eV

T. Tobita, N. Kobayashi and Y. Kaneko

Department of Physics, Tokyo Metropolitan University, Setagaya-ku, Tokyo 158

Since argon and nitrogen are the most popular gases in laboratory experiments, charge transfer reaction between Ar^+ ion and N_2 molecule;

$$Ar^+ + N_2 \rightarrow Ar + N_2 + \Delta E, \qquad (1)$$

has been studied by many investigators. In addition to that reason, the reaction (1) is especially interesting because it is near resonant and the energy defect ΔE can be the variety of small values, either positive or negative, depending on the combination of the states Ar^+ $(^2P_{3/2}, ^2P_{1/2})$ and $N_2^+(v')$. Though this reaction is considered to be a good example of near resonant asymmetric charge transfer involving a diatomic molecule, the matter is so complicated and there still be many questions unsolved. It is well known that the cross section of reaction (1) decreases monotonously with the decrease of the collision energy below several electron volts[1]. It shows a decreasing trend even at room temperature($\sim$0.04 eV) in contrast to the fact that ion=molecule reactions have generally a cross section rising up again after passing a minimum when the collision energy is decreased below 1 eV. Recently, Smith and Adams[2] showed that the cross section of reaction (1) is larger at liquid nitrogen temperature($\sim$0.01 eV) than at room temperature. It suggests that a cross section minimum should exist between 0.01 and 0.04 eV. On the other hand, Kato et al.[3] reported that $Ar^+(^2P_{1/2})$ has a little smaller cross section for reaction (1) than Ar^+ $(^2P_{3/2})$ and the cross section ratio $\sigma(1/2)/\sigma(3/2) = 0.7$ above 0.2 eV, while Lindinger et al.[4] and Smith and Adams could not see any evidence of the difference.

In order to get futher information on these unsolved questions, we have measured the cross sections of reaction (1) using a liquid-nitrogen-cooled drift tube. The apparatus used is the one used in the previous experiments[5]. The Ar^+ ions are produced in an ion source of electron impact type and injected into a drift tube. Helium gas is used as a buffer gas and a small amount of nitrogen gas is admixed with it. Cross sections are evaluated from the current ratio of Ar^+ and N_2^+, which is measured with a sampling mass spectrometer attached on the end of the drift tube. The collision energies are calculated by using Wannier's formula. Since the drift tube is cooled with liquid nitrogen, the lowest limit of collision energies is 0.01 eV.

When the percentage of N_2 gas in He gas was changed from 0-0.4%, a semi-logarithmic plot of the current ratio $I(Ar^+)/I(Ar^+ + N_2^+)$ showed a definite break indicating that at least two components of Ar^+ contribute to the production of N_2^+ at different rates. From the slopes and the extrapolated values of the linear portions, two cross sections σ_1 and σ_2 were deduced and the component 1 was found to have a smaller fractional ratio f_1 than f_2 of component 2. In a separate experiment using ion energy-loss spectroscopy[6], the fractional ratio $f(^2P_{1/2})/f(^2P_{3/2})$ was found to be the same as the statistical weight ratio 1/2. The present ratio f_1/f_2 is around 1/4 which is smaller than the statistical weight ratio but too large for upper-lying metastable states of Ar^+. Therfore, the components 1 and 2 can be refered as $Ar^+(^2P_{1/2})$ and $Ar^+(^2P_{3/2})$, respectively. In fact, f_1 depends weakly on the collision energy suggesting a possible quenching of $Ar^+(^2P_{1/2})$ to occur by collisions with He atoms.

The cross sections $\sigma(1/2)$ and $\sigma(3/2)$ deduced are shown in Fig. 1. The cross section $\sigma(1/2)$ is larger than $\sigma(3/2)$ in the whole energy range studied except at the high energy end. Above 0.1 eV, the break on the semi-logarithmic plot disappears indicating that both the cross sections come close to each other. The cross section $\sigma(3/2)$ has a sharp minimum around 0.5 eV, and increases rapidly when the collision energy is increased beyond 0.05 eV. It suggests that $\sigma(3/2)$ could exceed $\sigma(1/2)$ above 0.2 eV as reported by Kato et al. Both the cross sections show a sharp decrease at the lowest energy end. Whether it is essential or instrumental is not clear at present.

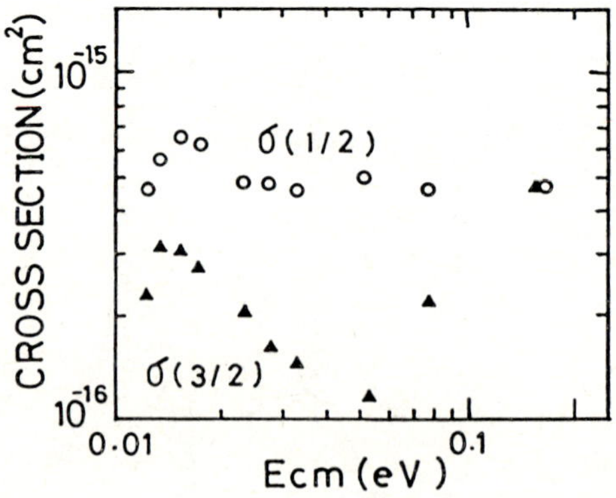

Fig. 1. Cross sections of reaction (1)

References
1) Y. Kaneko et al., J. Phys. Soc. Japn. 27, 992 (1969)
2) D. Smith and N. G. Adams, Phys. Rev. A 23, 2327 (1981)
3) T. Kato et al., J. Chem. Phys. 77, 337 (1982)
4) W. Lindinger et al., Phys. Rev. A 23, 2319 (1981)
5) T. Koizumi et al., J. Phys. Soc. Japn. 51, 2650 (1982)
6) Y. Itoh et al., J. Phys. Soc. Japn. 50, 3541 (1981)

STATE SELECTED ION-MOLECULE REACTIONS : $N_2^+ + Ar \longrightarrow N_2 + Ar^+$

T.R. Govers [a], P.M. Guyon [b], T. Baer [c], K. Cole [c], H. Frohlich[b], and M. Lavollée[b]
LURE (Univ.Paris-Sud and C.N.R.S.), Bât.209, Université Paris-Sud,91405 ORSAY,France

The threshold-photoelectron/photoion coincidence technique (TPEPICO)[1] is used to study state-selected ion-molecule collisions at laboratory energies ≈3-100eV. The parent ion is prepared by photoionisation, using monochromatised synchrotron radiation from the storage ring ACO as tunable photon source($h\nu$≈10-25eV). Its internal energy, E_i, is specified as being equal (within≈30 meV,fwhm) to the incident photon energy by selective dectection of threshold electrons. This selection is achieved by combined angular-and TOF discrimination and allows an excellent compromise between resolution and electron detection efficiency. The threshold-electron signal triggers the extraction of the corresponding ion($E_i \approx h\nu$) from the photoionisation chamber and this ion is made to collide with a effusive jet of target gas. The masses and relative abundances of reactant-and product ions are analysed by TOF measurements in delayed coincidence with the threshold electron signal. The variation of the total reaction cross-section and product branching ratios are obtained directly from a comparison of TPEPICO spectra obtained at photon energies chosen to coincide with specific vibronic levels of the reactant ion. Scanning $h\nu$ to vary E_i allows us to examine ions formed not only through "direct" ionisation(at threshold) but also through "resonant autoionisation"[2]. This mechanism widens the range accessible vibronic levels compared to fixed-wavelength experiments employing 21.2 eV photons, e.g..

As an example, table 1 lists total cross-sections obtained at a center-of-mass collisionenergy E_{cm}=8eV for the process $N_2^+ + Ar \rightarrow N_2 + Ar^+$, where the parent N_2^+ has been selected in each of the vibrational levels v=0 to 4 of the $X^2\Sigma_g^+$ ground state, and v=0 to 6 of the metastable $A^2\Pi_u$ state. The A-state data have been corrected for radiative $A \rightarrow X$ decay($\tau \approx 10\mu s$) which, during their ≈5μs flight time to the collision region, transforms a known fraction of N_2^+ ions initially formed in the (A,v)level, into ground-state ions. The necessary lifetimes and emission branching ratios were obtained from ref.[3]. Our relative measurements were put on an absolute basis by comparison with data obtained for $Ar^+ + Ar \rightarrow Ar + Ar^+$. Since we found the cross sections for this process to be equal (within 10%) for $Ar^+(^2P_{3/2})$or $(^2P_{1/2})$reactants, the state-unresoved data of ref.[4] could be used for calibration.

The vibrational dependence for $N_2^+(X,v)$ is very similar to the E_{cm}=12eV data of Kato et al[5], who reported relative measurements for (X,v≤3). The absolute value $\sigma(X,v=1)=19Å^2$ agrees well with the ≈20 $Å^2$ value deduced from a slight extrapolation of the data of Lindinger et al.[6]. The present results provide a previously unreported comparison between the effectiveness of vibrational and electronic excitation: for very similar internal energies, E_i=16.63eV for(X,v=4) and E_i =16.70eV for (A,v=0), the electronically excited state reacts <u>less</u> rapidly than the ground state.

N_2^+(X,v=1-3)and (A,v=0-3)ions are observed to react with cross-sections well beyond the Langevin limit(7.5$Å^2$ at E_{cm}=8eV), signifying that electron transfer occurs at distances larger than 1.1 Å. While charge transfer from N_2^+(X)can possibly be rationalised on the basis of a multiple-curve crossing model(cf.[7,8]),the interpretation of the A-state data should probably allow for collision-induced A→X transitions, implying a theoretical treatment involved at least three potential surfaces.

<u>Table 1</u> : N_2^+/Ar charge transfer cross-sections (in $Å^2$)for N_2^+ vibronic states as specified; E_{cm}= 8eV.

v =	0	1	2	3	4	5	6
N_2^+(X)	≤0.9	19±2	23±2	30±2	27±1	-	-
N_2^+(A)	12±1	17±2	17±3	15±3	3±3	≤5	≤10

REFERENCES

1. I.Nenner et al.,J.Chem.Phys.72(1980)6587
2. T.Baer et al.,J.Chem.Phys.70(1979)1585
3.W.B.Maier and R.F.Holland,J.Chem.Phys.70(1973)4501
4.P.Mahadevan and G.D.Magnuson,Phys.Rev.171(1968)103
5.T.Kato et al.,J.Chem.Phys.77(19 2)834
6.W.Lindinger et al.,Phys.Rev.A 23(1981)2319
7.E.Bauer et al.,J.Chem.Phys.51(1969)4173
8.M.Gérard-Aïn et al.,Int.J.Mass.Spectr.47(1983)167

a) LPCR,Orsay ; present address:LPPM, Bât.213 Univ.Paris-Sud, 91405 ORSAY.

b)LCAM, Bât.351, Univ. Paris-Sud, 91405 ORSAY

c)Dept.of Chemistry, U. North Carolina, Chapel Hill, N.C.27514, U.S.A.

DISSOCIATIVE CHARGE TRANSFER IN He^+-O_2 COLLISIONS: ENERGY SPECTRA AND ANGULAR DISTRIBUTIONS OF THE O^+ PRODUCT IONS

G. Bischof, P. Reinig and F. Linder

Fachbereich Physik, Universität Kaiserslautern
D-6750 Kaiserslautern, West Germany

In a high-resolution crossed-beam experiment the reaction $He^+ + O_2 \rightarrow He + O + O^+$ has been studied for collision energies $E(He^+)_{lab}$ = 1 - 100 eV. The O^+ product ions are detected in the plane of the crossed beams using a rotatable detector system with energy and mass analysis. O_2^+ product ions have not been found (< 1 % of O^+ yield) in accordance with more recent results from other investigations[1-4].

The energy spectra of the O^+ product ions show four distinct peaks (Fig. 1). The well-defined conditions of the present experiment allow a detailed kinematical analysis. As a result of this analysis, the long-distance energy-resonant charge transfer mechanism for this system is confirmed and the exit channels are identified as shown in table I.

The angular distribution of the O^+ product ions is strongly non-isotropic for all of the four exit channels. The intensity for channels I, II, III is peaked in the forward direction (direction of the H^+ beam), whereas channel V has a peak at an intermediate angle. These observations indicate a pronounced dependence of the charge transfer probability on the relative orientation of the O_2 molecular axis and the He^+ projectile trajectory.

The angle-integrated branching ratios for the four exit channels strongly depend on the initial He^+ collision energy (Fig. 2). At higher energies (100 eV), process II is the main channel, whereas at low collision energies (1 eV) process V is clearly dominant. The rather drastic variation of the branching ratios with the He^+ collision energy constitutes one of the main results of the present paper. An interpretation is proposed which is based on non-Franck-Condon behaviour of the charge transfer process at low energies.

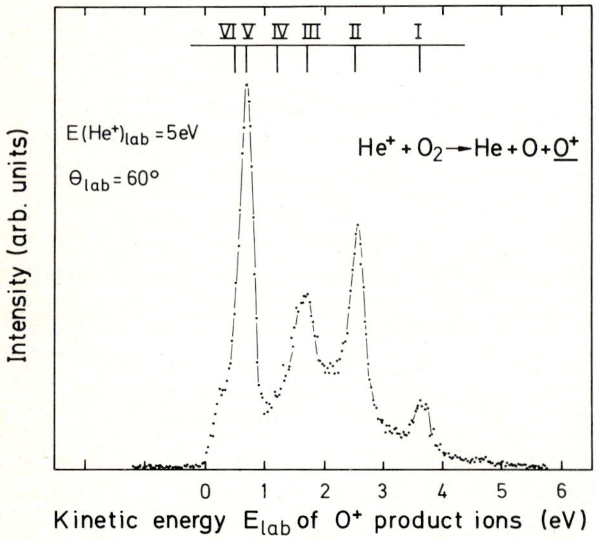

Fig. 1

Table I: Exothermic channels of the dissociative charge transfer reaction $He^+ + O_2 \rightarrow He + O + O^+$

	Product states	Energy E_i above O_2 ground state (eV)	$RE(He^+) - E_i$ (eV)	O^+ kinetic energy E_{cm}(eV)
I.*	$O(^3P) + O^+(^4S^o)$	18.73	5.86	2.93
II.*	$O(^1D) + O^+(^4S^o)$	20.70	3.89	1.95
III.*	$O(^3P) + O^+(^2D^o)$	22.05	2.54	1.27
IV.	$O(^1S^o) + O^+(^4S^o)$	22.92	1.67	0.84
V.*	$O(^3P) + O^+(^2P^o)$	23.75	0.84	0.42
VI.	$O(^1D) + O^+(^2D^o)$	24.02	0.57	0.29

*observed in the measurements

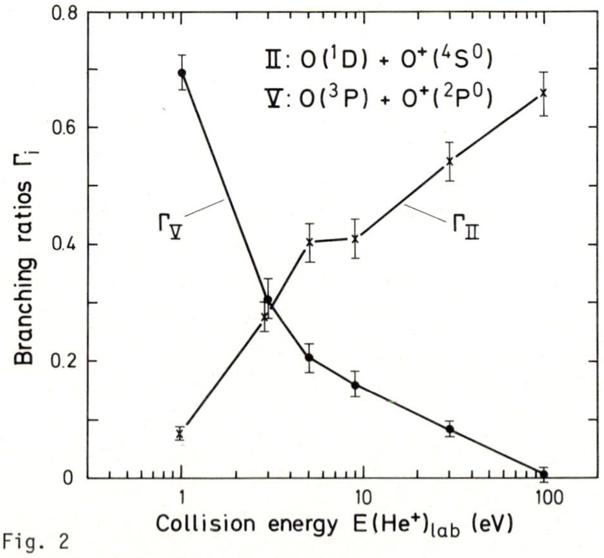

Fig. 2

References

1. R. Johnsen, J.A. Macdonald, M.A. Biondi, J. Chem. Phys. 66, 4718 (1977)
2. V.G.Anicich, J.B. Laudenschlager, W.T. Huntress, J.H. Futrell, J. Chem. Phys. 67, 4340 (1977)
3. W.R. Gentry in "Gas Phase Ion Chemistry", Academic Press 1979
4. G. Mauclaire, R. Derai, S. Fenistein, R. Marx, R. Johnsen, J. Chem. Phys. 70, 4023 (1979)

DISSOCIATIVE CHARGE EXCHANGE OF H_2^+, STUDIED WITH TRANSLATIONAL SPECTROSCOPY

D.P. de Bruijn, J. Neuteboom and J. Los
FOM-Institute for Atomic and Molecular Physics, Kruislaan 407, 1098 SJ Amsterdam, The Netherlands

V. Sidis
Université de Paris-Sud, Orsay

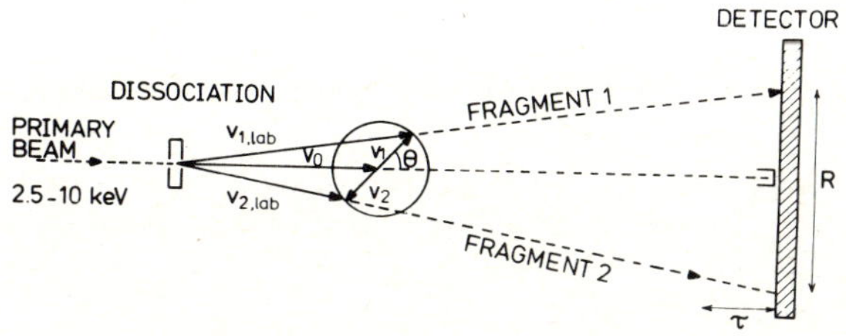

Fig. 1: Translational Spectroscopy with a time- and position-sensitive two-particle-detector.

In our experimental technique both fragments, arising from a dissociation in a fast beam, are detected in correlation by a channelplate detector. In this way the released kinetic energy in the c.o.m. fram (e_d) and the direction of the associated momentum, characterized by θ (fig. 1), can be measured simultaneously with high resolution [1].

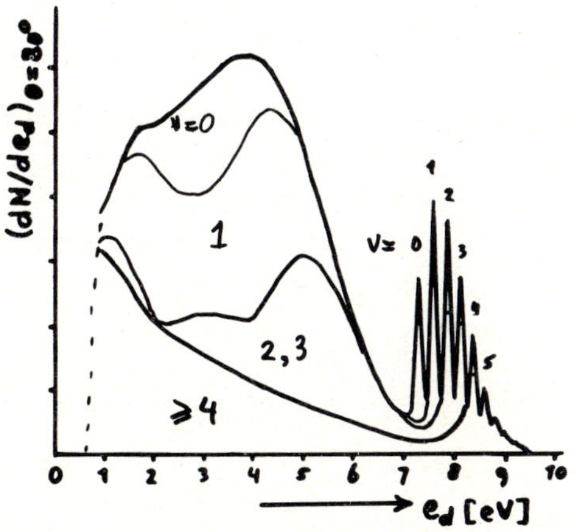

Fig. 2: Total and state-selected spectra for the released kinetic energy (e_d) in the dissociative charge exchange of H_2^+ and Mg at 5 keV.

This method has been used to study the dissociative charge exchange of H_2^+ with several targets (H_2, Ar, Mg, Na, Cs) in great detail. Depending on the ionisation energy of the target different excited states of H_2 will be favored. The following processes, leading to the formation of a H-H pair, have been identified experimentally:

1. Direct transition to the repulsive $b^3\Sigma_u^+$-state.
2. Radiative decay of the $a^3\Sigma_g^+$-state to the $b^3\Sigma_u^+$-state.
3. Predissociation of the $c^3\Pi_u$-state (by $b^3\Sigma_u^+$).
4. Predissociation of Rydberg-states such as the $d^3\Pi_u$ towards H(1s) + H(2p).

Process 1 (and 2) is illustrated in figure 2, where a Mg-target is used to give near resonant conditions in the charge exchange for low initial vibrations. To investigate the detailed behaviour of this process, experiments have been done with state-selected beams. An "asymptotic" theory, which describes this charge exchange very well, is presented at this conference by V. Sidis [2]. Processes 2 and 3 become dominant if targets like Cs and Na are used. Process 2 could positively be identified by measuring the c.o.m. deflection angle of the dissociating molecule. This interpretation is confirmed by the detection of the a→b radiation by Leventhal et al. [3].

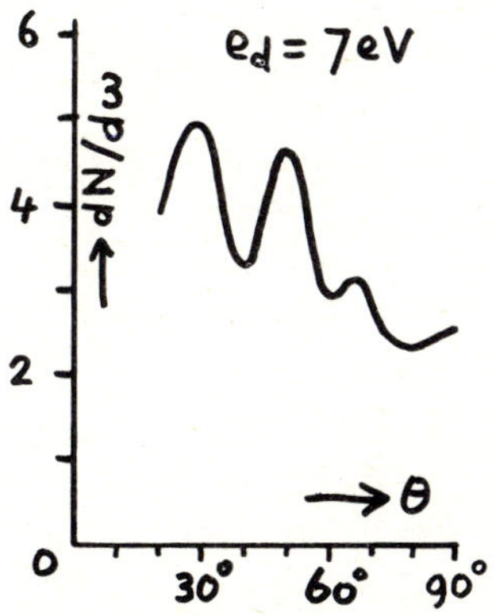

Fig. 3: Anisotropy of the dissociation process

$$H_2^+(v=\emptyset) + Cs \rightarrow H_2(c^3\Pi_u, v=\emptyset) + Cs^+ \rightarrow H(1s)+H(1s)+Cs^+$$

A very peculiar behaviour of the anisotropy of the process is observed mainly around e_d = 7 eV (fig. 3), where the predissociation of the lowest vibrational level of the $c^3\Pi_u$-state is located. The "oscillatory" pattern of the anisotropy can perhaps be explained by the different anisotropies of various rotational states. A separate contribution to this conference deals with the observed predissociating states (processes 3 and 4).

References

1. D.P. de Bruijn and J. Los, Rev.Sci.Instrum. 53 (7), 1020 (1982).
2. V. Sidis, contribution to this conference.
3. C.E. Burkhardt, J.J. Leventhal et al., J.Chem.Phys. 77 (3), 1354 (1982).

THEORY OF NEAR RESONANT-DISSOCIATIVE CHARGE EXCHANGE COLLISIONS AT KeV ENERGIES

V. Sidis
LCAM, Bât. 351, Université Paris-Sud, 91405 ORSAY Cedex, FRANCE
D. De Bruijn
FOM-Instituut Voor Atoom- En Molecuulfysica
1098 SJ AMSTERDAM, NETHERLANDS

A theoretical framework has been devised to study near resonant-dissociative charge exchange in H_2^+/Mg (or Alkali atom) collisions.

The vibration (n_i) and rotation ($\{J_i\} \equiv J_i, M_i, \Lambda_i$) of the diatom are assumed to be frozen during the charge exchange electronic transition and the collision proceeds along a straight line trajectory determined by the impact parameter $\vec{\rho}$. The charge exchange component of the system wavefunction thus writes

$$|\Psi_{ex}\rangle = (\sum_f \mathcal{B}_{if}(\vec{r},\vec{\rho})|f\rangle)\,|n_i,\{J_i\}\rangle$$

where the electronic transition amplitude is obtained from close coupling-impact parameter calculations for fixed internuclear distances (r) and space fixed orientations (Θ,Φ) of the diatom.

The relevant exchange interactions h_{if} (r,R,γ) between the initial (i) and final (f) channels as functions of the distance of approach (R) and relative orientation $\gamma=(\vec{r},\vec{R})$ are obtained from an extension of Smirnov's asymptotic theory[1]. As first suggested by Byklin *et al.*[2] LCAO forms of the active H_2 orbitals are used. Expressions are derived for the small-r ($r/2 \ll R$) case ($H_{if}^{<}$) when the molecule reacts as a whole, and for the large-r case ($H_{if}^{>}$) when each atom reacts individually; the two expressions are then joined using an overlap dependent switching function $\mathcal{S}$ (r) as :

$$h_{if} = \mathcal{S}H_{if}^{<} + (1-\mathcal{S})H_{if}^{>}$$

The γ dependence of h_{if} merely arises from a rotation of angle γ, in the triatom plane, of spherical harmonics quantized about the $\vec{r}$-axis

Calculations are performed on the processes

$$H_2^+ + Mg \rightarrow H_2^*(b^3\Sigma_u^+,\ c^3\Pi_g,\ a^3\Sigma_g^+) + Mg^+$$

with consideration of the a-b,c couplings arising from charge-dipole interactions and the R,γ-behaviour of the b and c levels due to charge-quadrupole interaction[3].

The continuous spectrum of dissociation products $Q_{cont}(\Theta)$ is obtained as the sum of the *direct* Q_b and *radiative* Q_a components defined as :

$$Q_b(\Theta) = \frac{1}{2}\int_0^{2\pi} d\Phi \int_0^{\infty} \rho d\rho |\langle\varepsilon_b|\mathcal{B}_{ib}|n_i\rangle|^2$$

$$Q_a(\Theta) = \sum_{na} \frac{1}{2}\left(\int_0^{2\pi} d\Phi \int_0^{\infty} \rho d\rho |\langle n_a|\mathcal{B}_{ia}|n_i\rangle|^2 \; F_{na}(\varepsilon_b)\right)$$

where $F_{na}(\varepsilon_b)$ describes the emission from a to b of a photon of frequency $\nu(\varepsilon_b)$ and the subsequent dissociation with energy ε_b

$$F_{na}(\varepsilon_b) = |M_{ab}|^2 \nu^3 \, d\nu/d\varepsilon_b \left(\int_{\nu_{min}}^{\nu_{max}} d\nu' |M_{ab}^2|\nu'^3\right)^{-1}$$

Similar expressions may be derived for the discrete part of the dissociation spectrum arising from predissociation of the c state.

Nice agreement is found between the present theory and experiment[4].

References

1. B.M. Smirnov, "Asymptotic methods in the theory of atomic collision" (Russian) (Atomizdat 1973)
2. V.I. Bylkin *et al.* Sov.Phys. JETP 92 540 (1971)
3. V. Sidis and C. Kubach J.Phys.B 11 2687 (1978) (see also Phys.Rev.A 23 110 & 119 (1981))
4. D. de Bruijn *et al.* in this Conference

Fig. 1 : Dissociation spectrum of H_2 triplet states formed by charge exchange in a H_2^+-Mg collision. (The vibrational population of the H_2^+ beam is assumed to be given by the Franck-Condon distribution for the $H_2(X,n=0) \rightarrow H_2^+(X,n_i)$ transition). The theoretical continuous spectrum represents the total cross section, differential in energy (ε) and solid angle (about Θ), of the dissociation products.

The theoretical discrete lines labelled c represent the total cross sections for populating vibrational levels n_c of the predissociative $c^3\Pi_u$ state per unit solid angle about Θ. The experimental data (...) from ref. 4 are normalized to the present theory.

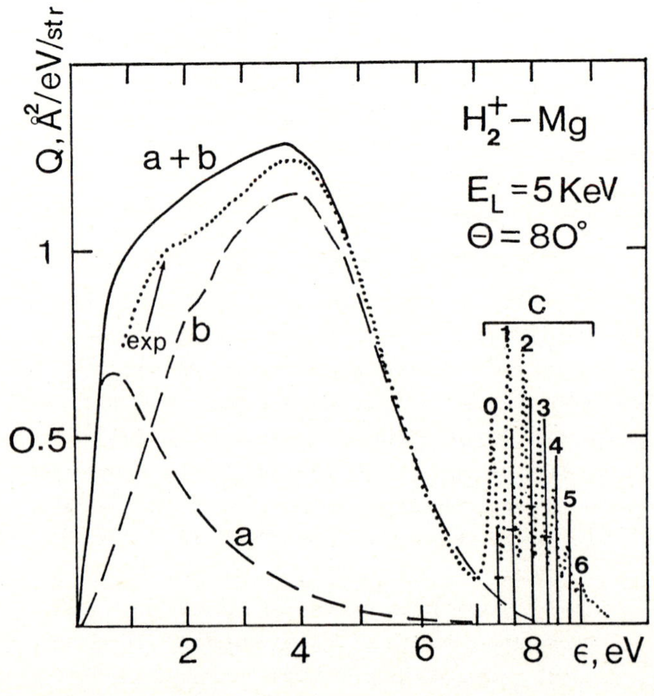

ELECTRON TRANSFER PROCESSES BETWEEN ELECTRONICALLY EXCITED Na AND SO_2

B.Auschwitz and K.Lacmann

Hahn-Meitner-Institut für Kernforschung Berlin
Bereich Strahlenchemie, Postbox, D-1000 Berlin 39, Fed. Republic of Germany

In a crossed beam experiment, we measured the angular distribution and translational energy loss of Na^+ ions formed in the reaction $Na(^2S, ^2P_{3/2}) + SO_2 \rightarrow Na^+ + SO_2^-$. The SO_2 beam came from a supersonic nozzle and the sodium beam had a kinetic energy between 20 and 60 eV.

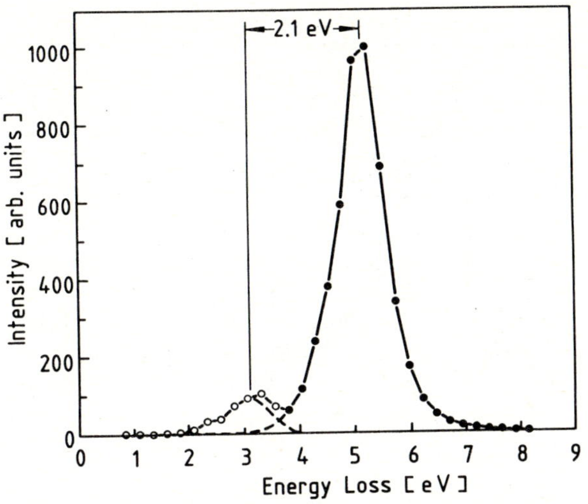

In Fig.1 is shown the Na^+ yield in the forward direction as a function of the energy loss ΔE at 28 eV(CM) with and without laser excitation. The shift of 2.1 eV equals the energy difference between the ^2S- and ^2P-states. Within experimental accuracy, the translational energy loss is reduced by the amount of excitation energy of Na. This suggests that there is no difference in the vibrational excitation of SO_2^- by collisions of ^2S and ^2P sodium atoms, although the respective median impact parameters are 2 and 4 Å . It can be shown that the cross section is increased 10-fold by the excitation.

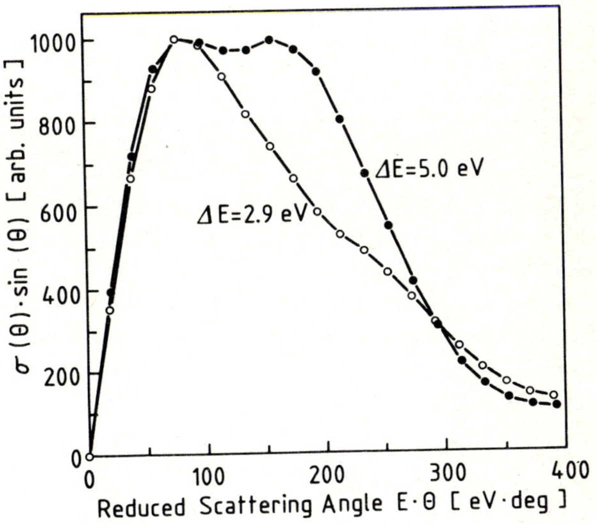

Fig.2 shows the polar differential cross-section as function of the reduced scattering angle, $E\theta$, measured at a fixed energy loss corresponding to the maxima of the spectra of Fig.1. The cross section for $Na(^2S)$ at 5 eV energy loss shows the typical double peak: the small-angle peak corresponds to covalent scattering (electron jumps as products separate) and the wide-angle peak to ionic scattering (electron jumps at the first encounter of crossing seam). With laser excited Na at $\Delta E = 2.9$ eV the ionic scattering peak is much smaller than the covalent peak. This effect is just the opposite to what has been observed with ground state atoms, where the ionic peak is always equal to or greater than the covalent peak[1]. This results from the vibrational motion of the molecular ion after the first passage of the crossing seam (bond stretching effect[1]). It will be shown that the observed "anti-bond stretching effect" results from a decrease in the crossing radius during the collision with excited atoms, which in turn leads to a decrease in the nonadiabatic transition probability at the second crossing and accordingly to a decrease in the ionic scattering.

References

1. a) K.Lacmann, Adv.Chem.Phys. 42, 513 (1980)
   b) A.W.Kleyn, J.Los and E.A.Gislason, Physics Reports 90, 1 (1982)

DIFFERENTIAL CROSS SECTIONS FOR ELECTRON TRANSFER WITH SELECTION OF THE IONIC EXIT CHANNEL FOR K, Cs + CH_3I COLLISIONS

A.J.F.Praxedes[+], M.J.P.Maneira and A.M.C.Moutinho[*]

Centro de Fisica Molecular das Universidades de Lisboa (INIC), Complexo I, Av.Rovisco Pais IST, 1000 Lisboa Portugal
+ Departamento de Fisica do Instituto Superior Tecnico (U.T.L.)
* Departamento de Fisica e Ciência dos Materiais da Faculdade de Ciências e Tecnologia (U.N.L.)

Differential cross sections for ion pair formation of M + CH_3I (M = K, Cs) were measured at laboratory energies ranging from 30 to 400 eV (see e.g. *Fig*.1). Time of flight as well as time correlation between positive and negative species were performed in order to discriminate the possible ionic exit channels $K^+ + CH_3I^-$ and $K^+ + I^-$. The experimental set up is an improved version of a similar arrangement mounted in a crossed molecular beam machine[1].

Electronic excitation of the K atom accompanied by substantial vibronic excitation of the target molecule had been reported[2]. Moreover total cross section measurements with mass analysis[3] showed that the CH_3I^- molecular ion is formed only at high velocities.

Measurements of time of flight and time differences between M^+ and the two possible negative ions showed indeed only the formation of I^- at low collision energies. For K collisions at E_{lab}= 268 eV (*Fig*.2) there is an enhancement at large τ in the region corresponding to CH_3I^-, although the main yield is I^-. This change is also related with the shift of the differential cross sections observed for large τ values at high collision energies. An increasing contribution of the excited state of CH_3I^- could account for these observations and it would predict large energy losses which had been measured in the case of collisions with sodium[4].

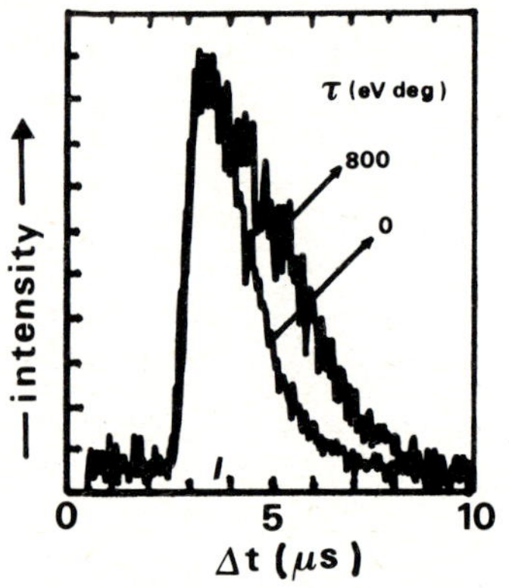

Fig.2 Spectrum of time of flight differences. The origin of time scale was arbitrarily chosen.

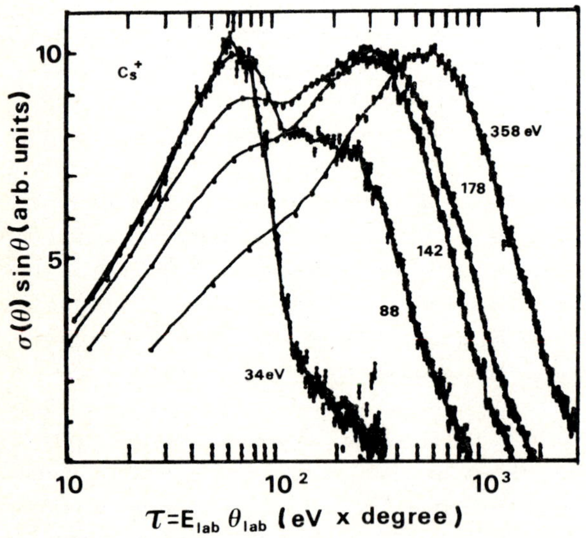

Fig.1 Polar differential cross section in the laboratory system for Cs^+. The alkali ion signal is corrected for the variation of the collision volume with θ_{lab} . All curves are scaled to the same maximum value of the peak. The experimental values are plotted with the corresponding error bar. The curve follows a moving average of five points.

A model for ion pair formation including neutral channels would improve the understanding of this behaviour and the angular dependence of the energy losses measured in the neutral exit channels[5].

This work has financial support of the Instituto Nacional de Investigação Cientifica and of the Junta Nacional de Investigação Cientifica e Tecnológica (under contract nº 138.79.74/I).

References

1. M.J.P.Maneira, A.J.F.Praxedes and A.M.C.Moutinho Proceedings of the 13th International Symposium on Rarefied Gas Dynamics (1982)
2. M.A.D.Fluendy, K.P.Lawley, C.Sholeen, D.Sutton Molecular Physics 42/1, 1 (1981)
3. A.W.Kleyn, M.M.Hubbers and J.Los (private communication)
4. M.R.Spalburg, Doctoral Scriptie Universiteit van Amsterdam (1981)
5. A.M.C.Moutinho, A.J.F.Praxedes, E.Cowen, M.A.D. Fluendy Contributed paper on this conference

ALIGNMENT IN MOLECULAR EXCITATION IN DISSOCIATIVE STATES OF H_2^+ AND H_3^+ †

O. Yenen, M. Natarajan and D.H. Jaecks

Behlen Laboratory of Physics, University of Nebraska, Lincoln, NE 68588

The electronic alignment in molecular excitation to specific dissociative states are determined from the polarization analysis of L_α intensities, in coincidence with the charged dissociation products observed at specific scattering angles.

I. 3.22 KeV H_2^+ - He Collisions

For the collision process $H_2^+ + He \rightarrow H_2^+ + He \rightarrow H^+ + H(2p) + He$, the polarization intensities $I(\beta)$ of L_α radiation are observed in coincidence with H^+ scattered at specific laboratory angles. The results are shown in Fig. 1. along with the least squares fit of the data to a general dipole intensity pattern. The measurements indicate that as the scattering angle is increased, the polarization distribution becomes increasingly "pinched" until at $\theta=\pm3.25°$. Beyond this angle the coincidence rates drop drastically. Also at this angle, the transformation of the velocities of dissociation products from the CM frame to the laboratory frame shows that the H_2^+ internuclear axis orientation is approximately perpendicular to the beam direction. Focusing our attention to this angle, we note that the polarization intensity of the L_α radiation shows the characteristics of $2p(m_\ell=\pm1)\rightarrow 1s$ transitions, using the H_2^+ internuclear axis as the quantization axis[1]. Thus the observed final state results from the excitation (at an H_2^+ internuclear separation of 1.5 a.u.) and decay of the $2p\pi_u$ state of H_2^+. As the excited H_2^+ recedes from the He target, the electronic wave function remains locked into the H_2^+ internuclear axis with a nodal plane perpendicular to the beam direction. Other orientations of H_2^+ axis lead to excitations of other states, since the space-fixed π-like wave function can be written as a coherent linear combination of body-fixed π and σ wave functions.

II. 4.83 KeV H_3^+ - He Collisions

The same technique is used for the collision process $H_3^+ + He \rightarrow H_3^{+*} + He \rightarrow H_2^+ + H(n=2) + He$, in which the polarized L_α photon is detected in coincidence with the H_2^+ scattered at a specific laboratory angle. At $\theta<1.25°$, using the line which joins the H-atom to the CM of H_2^+ as quantization axis, the polarization intensities $I(\beta)$ have the characteristics of the $2p(m_\ell=\pm1)$ state. Using the same axis, at $\theta>1.25°$, they show the characteristics of the $2p(m_\ell=0)$ state. This behaviour can be explained by considering the energy levels of the H_3^+ molecule within the context of (D3h) model (all equilateral triangle configuration)[2]. At $\theta=1.25°$, the transformation of velocities from the CM system to the laboratory frame gives a maximum kinetic energy of approximately 4.6 eV. Using the (D3h) energy curve, this corresponds to a distance of $.8\pm.03$ Å between protons of the H_3^+ molecule. At this internuclear separation, assuming a Franck-Condon type excitation from the ground state of H_3^+ to H_2^+ + He separated atom limit, one should not observe a large amount of coincidences at angles $\theta>1.91°$. In fact, this is verified within the experimental limits of the experiment. The very few observed coincidences result from collisions in which the H_3^+ CM is appreciably deflected and where the model probably breaks down.

References:

1. J. Macek and D.H. Jaecks, Phys. Rev. A4, 2281 (1971)
2. L.J. Schaad and W.Y. Hicks, J. Chem. Phys. 61, 1934 (1974)

†This work is supported by the National Science Foundation.

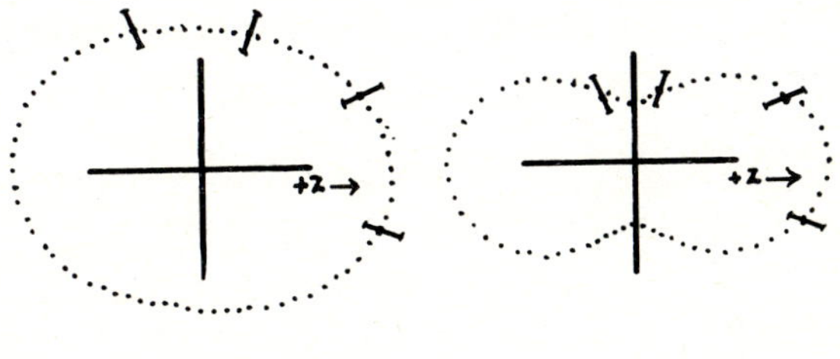

θ = 2.25°   θ = 3.25°

Fig. 1.

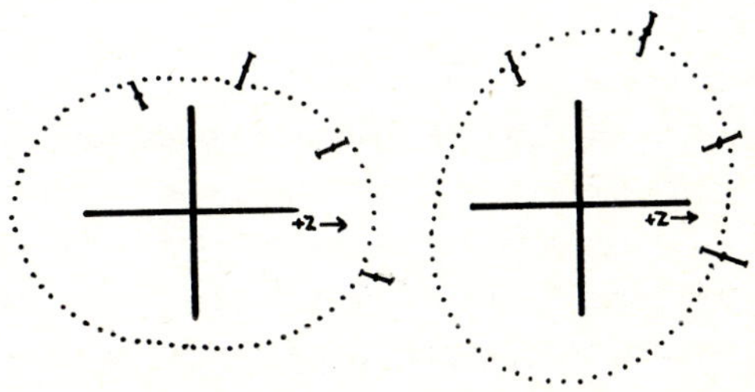

θ = 1.25°   θ = 1.5°

Fig. 2.

PREDISSOCIATION OF THE $c^3\Pi_u$- AND $d^3\Pi_u$-STATES OF H_2, STUDIED WITH TRANSLATIONAL SPECTROSCOPY

D.P. de Bruijn, J. Neuteboom and J. Los
FOM-Institute for Atomic and Molecular Physics, Kruislaan 407, 1098 SJ Amsterdam, The Netherlands.

T.R. Govers
Université de Paris-Sud, Orsay

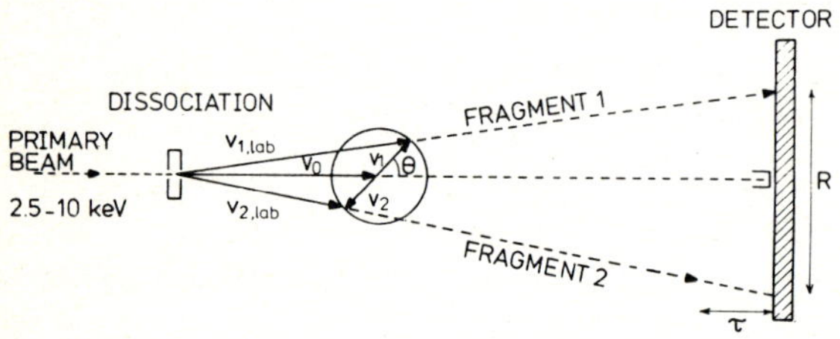

Fig. 1: Translational Spectroscopy with a time- and position-sensitive two-particle-detector.

In our experimental technique both fragments, arising from a dissociation in a fast beam, are detected in correlation by a channelplate detector. In this way the released kinetic energy in the c.o.m.-frame (e_d) and the direction of the associated momentum, characterized by θ (fig. 1), can be measured simultaneously with high resolution [1].

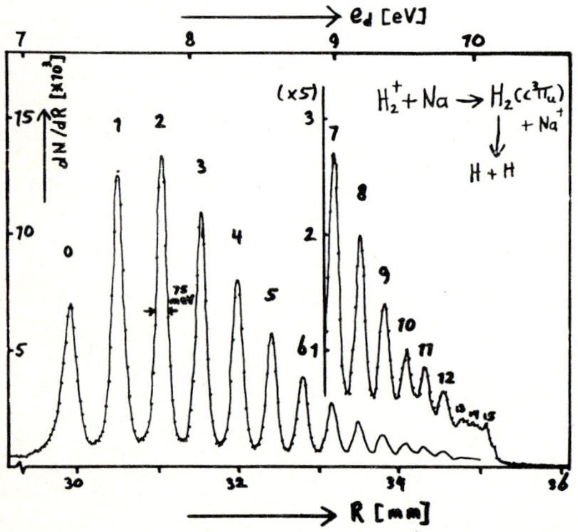

Fig. 2: Predissociation of the $c^3\Pi_u$-state of H_2, populated in a charge exchange of H_2^+ with Na at 6.9 keV.

Predissociating states, formed in a charge exchange collision, give rise to discrete e_d-spectra. This was first seen by Meierjohann and Vogler in 1977 [2] for the $c^3\Pi_u$-state of H_2. We have repeated this experiment with high resolution and were able to resolve most vibrational levels completely (fig. 2). A comparison with ab-initio calculations by Kolos [3] will be made. The width of the peaks is mainly caused by the finite length of the collision chamber (1 mm) and the rotational distribution. The observed rotational distribution depends strongly on the target chosen and is unexpectedly narrow (3-4 rotational levels) after "near-resonant" charge exchange (fig. 2). The predissociation lifetime of the lowest vibrational level has been estimated at about 2 nsec.

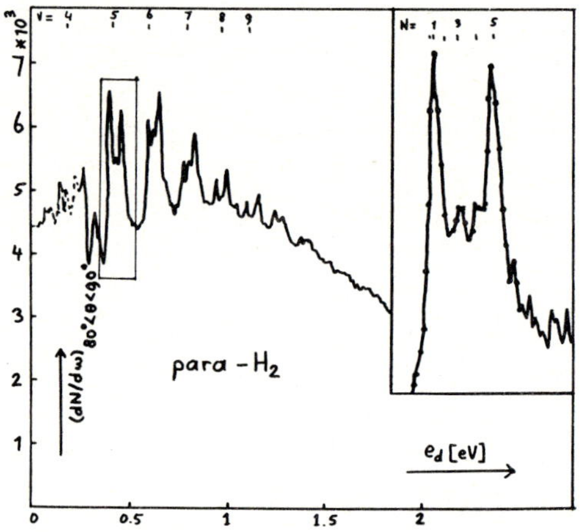

Fig. 3: Predissociation of the $d^3\Pi_u$-state of para-H_2 and other Rydberg states after charge exchange of para-H_2^+ with Cs at 5 keV.

The vibrational levels ≥ 4 of the $d^3\Pi_u$-state of H_2 are allowed to predissociate through the $e^3\Sigma_u$-state. The H(1s)-H(2p) pairs have a low kinetic energy ($e_d \leq 1.9$ eV) and thereby allow rotational resolution in our apparatus. This is illustrated in fig. 3 where para-H_2 is used to give only contributions of even rotational levels (quantum number N) to the predissociation. The sixth vibration (v = 5, shown expanded) shows an almost missing rotation (N = 3), while comparable phenomena are present for other vibrations.

References

1. D.P. de Bruijn and J. Los, Rev.Sci.Instrum. 53 (7), 1020 (1982).
2. M. Vogler and B. Meierjohann, Abstr.of papers, Xth ICPEAC Conf., Paris, 568 (1977).
3. W. Kolos and J. Rychlewski, J.of Molec.Spectr. 66, 428 (1977).

TRANSLATIONAL SPECTROSCOPY OF D_2^+ AND HD^+ FROM D_3^+ AND HD_2^+ DISSOCIATION

C. Cisneros, I. Alvarez, A. Morales, J. de Urquijo

Instituto de Física - UNAM Apdo. Post. 20364 01000 México, D.F.

A. Russek

Dept. of Physics, University of Connecticut-Storrs, Connecticut 06268

During the past two decades, intensive investigations have been undertaken in order to determine the mechanisms of collisional dissociation of molecular ions, but still little is known about the dissociation of HD_2^+, D_3^+ or H_3^+ ions. Recent translational spectroscopy measurements[1] have shown some interesting results about the electronic states of H_3^+.

By measuring the energy spectra of the HD^+ and D_2^+ ions arising from dissociative collisions of HD_2^+ and D_3^+ ions on H_2 and Ar, we have attempted to obtain more information about the dynamics of these collisions.

The experiment was performed in the same apparatus as described previously[2]. The primary-ion energy, E_o, was varied from 2000 up to 4000 eV in 500 eV steps. For each value of E_o, an energy profile of HD^+ and D_2^+ from HD_2^+ or D_2^+ from D_3^+ was obtained. This was done by means of the product-ion analyser, which accepted only ions scattered in the forward direction, or at different angles varying from 0° to 2°in the laboratory system. A typical product-ion energy profile is shown in Fig. 1.

From the kinematic relations between velocities in the laboratory system and the projectile centre of mass system, we can relate ΔE, the energy transferred from relative kinetic energy due to collision, E_L, the kinetic energy of an ejected product in the laboratory system, E_o, the initial kinetic energy of the projectile, W , the kinetic energy of the fragment relative to the centre of mass of the projectile, and θ, the scattering angle in the laboratory system, in such a way that ΔE_L, the energy spread of a product scattered at an angle θ is given by

$$\Delta E_L=(2\cos\theta/M^2)\{U^2m^2\cos^2\theta - UmM(U-2m(E_o-\Delta E))\}^{0.5} , \quad (1)$$

where $U=2M(E_o-\Delta E-W)$, m is the mass of the product and M is the mass of the incident ion.

Measurements of the energy spread of the products emitted in line with the beam enabled us to determine W.

The present results exhibit the following features:

(a) a broad energy distribution ΔE_L for all cases, this becoming larger as the primary ion energy was increased.

(b) a great difference in the energy distributions in dependence of the target. Distributions using H_2 as a target are wider than those obtained with Ar.

(c) a reasonable agreement of the change in ΔE_L for θ=0° was found for all combinations of projectile-product analysed.

(d) an average value for W of 13.6 eV, calculated from eq.(1) is being analysed.

(e) a departure from eq.(1) of the energy distributions obtained for θ=0.5° is under investigation.

References
1. S.C. Goh and J.B. Swan, Phys. Rev. A 24, 1624 (1981)
2. I. Alvarez, C. Cisneros and A. Russek, Phys. Rev. A 27, 77 (1982)

This work was partially supported by CONACYT, grant No. PCCBCNAL 100658

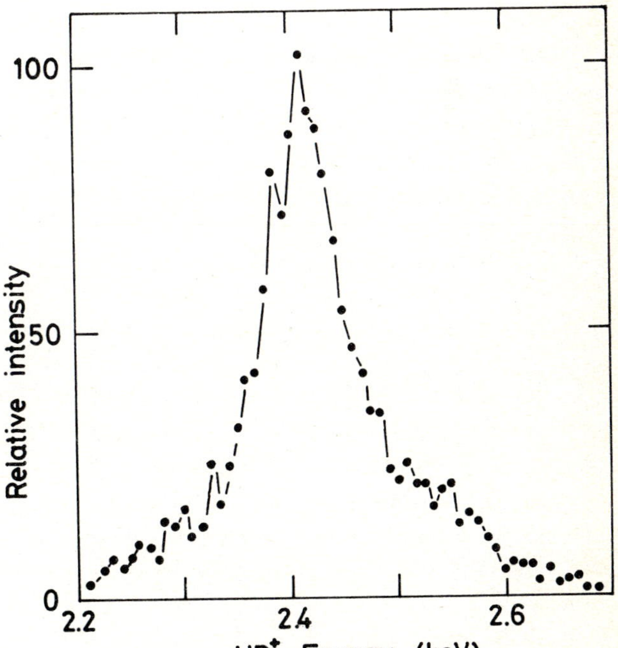

Fig. 1: The energy spectrum of HD^+ ions arising from the dissociation of 4 keV HD_2^+ ions on Ar.

DIFFERENTIAL POSITIVE-FRAGMENT CROSS SECTIONS OF HD_2^+ IN H_2 AT ENERGIES IN THE RANGE 1.5 TO 5 keV

A. Morales, J. de Urquijo, C. Cisneros, I. Alvarez

Instituto de Física - UNAM Apdo. Post. 20364 01000 México, D.F.

A. Russek

Dept. of Physics, University of Connecticut-Storrs, Connecticut 06268

Differential and total cross sections for the H^+, D^+, HD^+ and D_2^+ fragments due to collisions of HD_2^+ in H_2 have been measured and are reported here in Figs. 1 and 2.

A beam of HD_2^+ in the energy range 1.5 to 5 keV is passed through a H_2 target under the condition of single collision. The primary beam was extracted from a Colutron-type ion source fed with 75% of pure H_2-D_2 mixture and 25% Ar at a pressure of about 75 microns. After focusing, the beam is passed through a velocity filter and cylindrical bending plates. This well collimated beam current of HD_2^+ is finally incident on a differentially pumped gas cell containing pure H_2. Path lenghts and collimator apertures in the system ensure a root mean square angular resolution of better than 0.1°. The dissociation fragments H^+, D^+, HD^+ and D_2^+ were spatially and energy separated by means of a 45° parallel plate electrostatic analyser, and counted by a funnel type electron multiplier. All the results reported here were taken under the same source conditions.

Large scattering angles of the dissociation fragments were found as a result of different modes of dissociation of the HD_2^+ ion[1]. The observed spread in the spectra of both energy and laboratory angle seems to be consistent with translational spectroscopy measurements[2].

The angular distributions of the product ions indicate that a fraction of these comes from collision excitation to dissociating Rydberg states. These data suggest that the H-D isotopic mass difference affects the relative population of the collision fragments by influencing the internuclear geometry of the initial HD_2^+ projectile.

Plausible arguments will be presented and published elsewhere in order to explain the observed HD_2^+ dissociation.

References

1. I. Alvarez, C. Cisneros, J.A. Ray, C.F. Barnett and A. Russek, submitted to Phys. Rev. A
2. C. Cisneros, I. Alvarez, A. Morales, J. de Urquijo, and A. Russek, this Conference

This work was partially supported by CONACYT, grant No. PCCBCNAL 100658

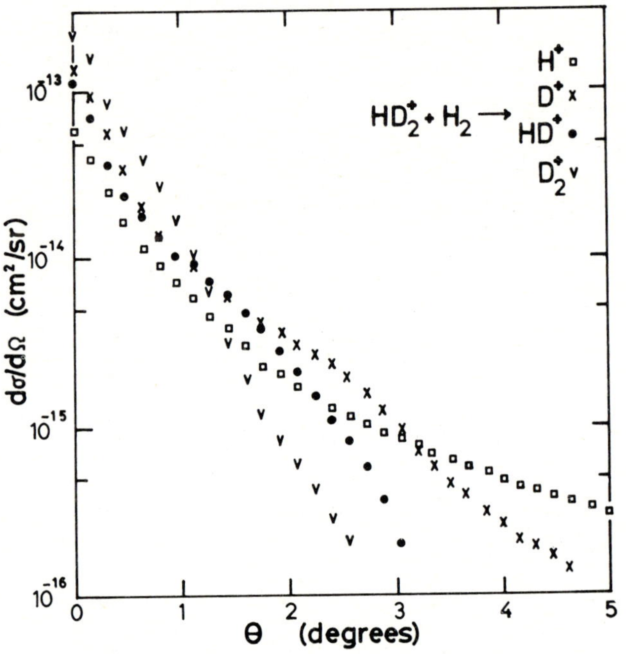

Fig. 1: Differential scattering cross sections as a function of the scattering angle for positive fragments at an energy of 4.15 keV, due to collisions of HD_2^+ in H_2 gas

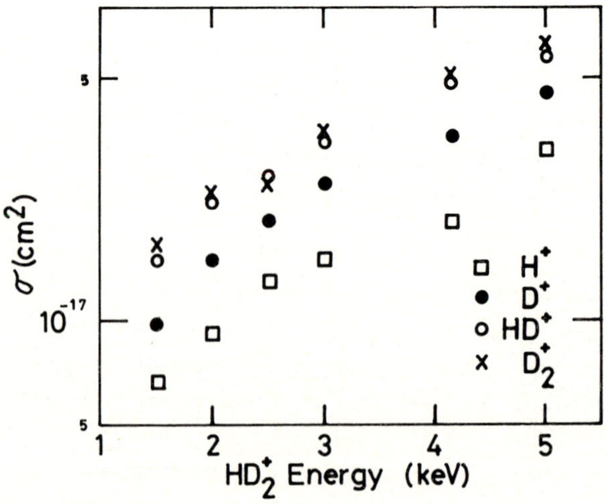

Fig. 2: Total formation cross sections of H^+, D^+, HD^+ and D_2^+ for HD_2^+ dissociative collisions, obtained from integration of the differential scattering cross sections.

COLLISIONAL ELECTRON-ION AND ION-ION RECOMBINATION IN A DENSE NEUTRAL AND ION GAS

M. R. Flannery

School of Physics, Georgia Institute of Technology, Atlanta, GA 30332

We present a basic microscopic theory of collisional electron ion recombination

$$e + A^{+} + M \rightarrow A + M$$

and of ion-ion recombination

$$X^{+} + Y^{-} + M \rightarrow [XY] + M$$

in a gas of atoms or molecules M. Deactivation occurs via binary collision of e, X^{+} and Y^{-} with M, and, for sufficiently high ion densities $N^{\pm}$, via ion-ion collisions. The proposed theory can be generalized to cover ion-molecule reactions (as ion-atom association) in a dense gas.

The full basic theory will be described and implemented to provide the first basic variation, apart from Monte Carlo computer simulations and various phenomenological methods [1], of the recombination rate α with gas density N. The effect of increase in ionization $N^{\pm}$ will also be discussed.

We also show for the first time that the recombination process may be regarded as proceeding by microscopic diffusion in both configuration R-space and in internal energy E-space of the recombining $(e - A^{+})$ or $(X^{+} - Y^{-})$ pair. The appropriate equations will be solved and compared with the full treatment.

This research is sponsored by the U.S. Air Force Office of Scientific Research under Grant No. AFOSR-80-0055.

References

1. M. R. Flannery, Phil. Trans. Roy. Soc. A 304, 447 (1982).

DISSOCIATIVE RECOMBINATION OF $e + H_3^+$. AN ANALYSIS OF REACTION PRODUCT CHANNELS

H. H. Michels and R. H. Hobbs

United Technologies Research Center, East Hartford, Connecticut 06108 USA

Accurate *ab initio* calculations of the ground and excited H_3 hypersurfaces have been carried out within a configuration-interaction (CI) framework. These surfaces have been examined for geometries appropriate for an analysis of the reaction products of dissociative-recombination (DR) of $e + H_3^+$. The primary purpose of this work was to examine the various product channels for an energy range of 1-5 eV, which is of interest in the analysis of volume dependent reactions in magnetron and other similar hydrogen ion source devices.

Accurate wavefunctions were constructed in C_{2v} and $D_{\infty h}$ symmetries using Slater-type orbital basis sets and large CI expansions. The results in C_{2v} symmetry are shown in Fig. 1 which indicate that direct recombination occurs through the $^2A_1$ resonance state which connects diabatically to $H_2^+ + H^-$. This resonance state lies $\sim$5.4 eV above H_3^+ in its ground vibrational state but couples effectively with H_3^+ with three quanta of vibrational energy. Under these conditions the predicted

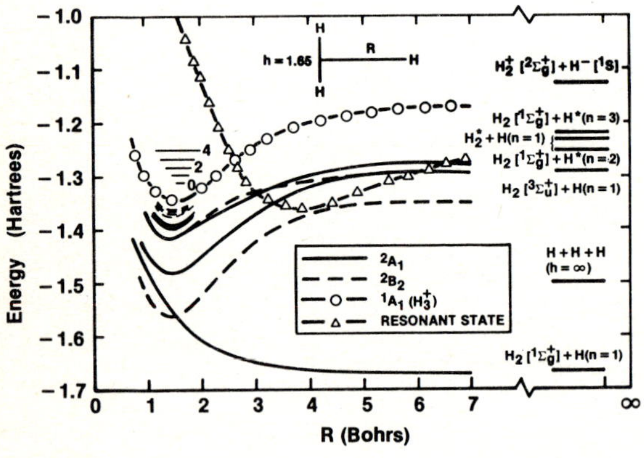

Fig. 1: Potential energy curves for H_3.

products are $H^*(n = 2) + H_2\ (X^1\Sigma_g^+)$, where any excess vibrational energy in the H_3^+ ions is transferred to the product H_2 molecule.

The formation of H_3^+ via the reaction, $H_2^+ + H_2 \rightarrow H_3^+ + H$, is exothermic for H_3^+ ($v \leq 4$). Thus, a likely recombination pathway is $e + H_3^+\ (v = 3,4) \rightarrow H_3\ (^2A_1$ resonance$) \rightarrow H_2^\ddagger\ (v = 0,1) + H^*(n = 2)$. The pertinent recombination energetics are indicated in Fig. 2. Direct dissociative recombination of H_3^+ ions

$e + H_3^+(\nu) \rightarrow H_3 \rightarrow H_2[X\ ^1\Sigma_g^+(\nu)] + H^*(n=2)$		ΔE (eV)
0	0	+0.91
1	0	+0.52
2	0	+0.13
3	0	−0.26
4	0	−0.64
4	1	−0.12

$H_2^+ + H_2 \rightarrow H_3^+(\nu=0) + H$ ΔE = −1.73 eV

$\rightarrow H_3^+(\nu=4) + H$ ΔE = −0.17 eV

Fig. 2: $e + H_3^+$ recombination energetics.

in the ground vibrational state is unlikely for low collisional energies ($\lesssim$ 6 eV) and indirect capture processes must be examined for this case. A similar result has been found for this system by Kulander and Guest.[1]

References

1. K. C. Kulander and M. F. Guest, J. Phys. B: At. Mol. Phys. 12, L501 (1979).

Supported in part by AFOSR under Contract F49620-81-C-0022.

VIBRATIONAL PREDISSOCIATION OF Ne...I_2(B) VAN DER WAALS MOLECULE: A QUASICLASSICAL TREATMENT

G. Delgado-Barrio, P. Mareca and P. Villarreal*

Instituto de Estructura de la Materia, C. S. I. C., Serrano 119, Madrid-6, Spain. *Permanent adress: Departamento de Química Física y Química Cuántica, Facultad de Ciencias, Universidad Autónoma de Madrid, Cantoblanco, Madrid-34, Spain.

A quasiclassical study of the vibrational predissociation

$$Ne...I_2(B^3\Pi_{O_u^+}, v) \xrightarrow{\Gamma_v} N_e\text{-}I_2(B^3\Pi_{O_u^+}, v'<v)$$

is reported here. The potential was approximated by the sum of three atom-atom Morse type functions with $D = 40\ cm^{-1}$, $\alpha = 1.5\ \AA^{-1}$, $\bar{R} = 4,36\ \AA$ for Rach Ne-I interaction and for I_2 we have used the same parameters previously employed (1).

The method has been already described elsewhere[1]. Essentially, it takes into account the following quantal conditions: total angular momentum $J = 0$; starting vibrorotational energy corresponding to (v_1, j=0) level, and ground triatomic energy as the starting one of the complex. When the parameters already mentioned are used, the approximate quantal treatment of Beswick et al.[2] leads to a value of $E_o = -64\ cm^{-1}$ for that magnitude, takeing as energetic origin the diatomic vibrational energy corresponding to the v_1 quantum number. This value agrees with the experimental one of Blazy et al[3]. The duration of trajectories with v_1 fixed was fitted to an exponential law[4], from which the Γ's are obtained.

In figure 1, the rates Γ are plotted versus v for the $N_e...I_2$ complex. At this moment, we have no notice of experimental Γ values for this complex. So that, we have also represented in that figure the quasiclassical rates for the $H_e...I_2$ molecule[1,5], wich are in good agreement with the experimental data of Sharfin et al.[6] The rates for both complexes follow a similar behaviour, being the Γ values for Ne...I_2 higleer than the corresponding ones for He...I_2. Thus, the complex of I_2 with He is more stable than with Ne. Anyway, the rates for the Ne case show a peaking as for the case, but now it is shigted at v = 52 while the maximum for the is allocated at v = 56. Also, a soft minimum for Ne appears at v ≃ 39, which has no equivalent for He. We think that these features correspond to the different dissociation energies in the van der Waals bond, there being an earlier closing of predissociation channels for the Ne case than for the He one. This fact can be stressed from table I, where vibrational distribution of the I_2 fragments, depending on the v value, are shown. They were determined by the use of a box-quantization procedure. For v = 36, the most effective channel is $\Delta v = -1$, but at v = 39 the channel $\Delta v = -2$ is already more than a factor of two larger than the previous one. As higher v values are considered, the distributions show a gradual closing of the channel $\Delta v = -2$ together with an increasing of the following ones. Incidentally, for v = 53, the $\Delta v = -3$ channel is already dominant and the $\Delta v = -4$ is comparable.

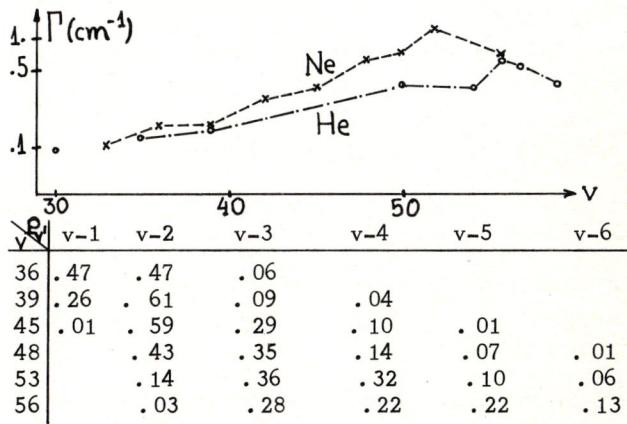

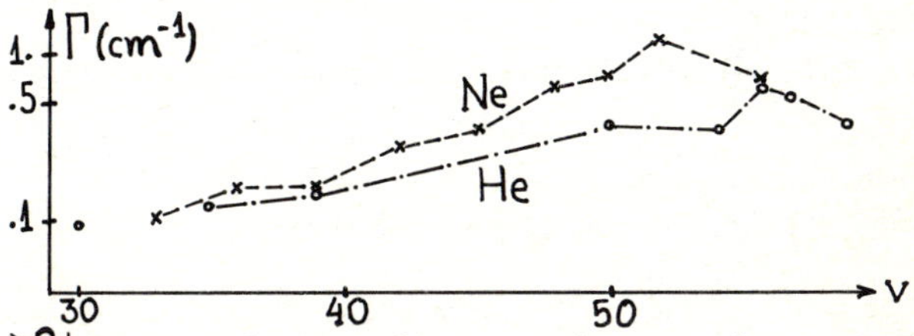

v \ $P_{v'}$	v-1	v-2	v-3	v-4	v-5	v-6
36	.47	.47	.06			
39	.26	.61	.09	.04		
45	.01	.59	.29	.10	.01	
48		.43	.35	.14	.07	.01
53		.14	.36	.32	.10	.06
56		.03	.28	.22	.22	.13

Vibrational distributions $P_{v'}$, of the I_2 fragment depending on the initial vibrational level v of the I_2 in the complex

References

1. G. Delgado-Barrio, P. Villarreal and G. Albelda, J. Chem. Phys., 78, 280 (1983).
2. J. A. Beswick and G. Delgado-Barrio, J. Chem. Phys., 73, 3653(1980).
3. J. A. Blazy, B. M. DeKoven, T. D. Russel and D. H. Levy, J. Chem. Phys., 72, 2439(1980).
4. S. B. Woodruff and D. L. Thompson, J. Chem. Phys., 71, 376(1979).
5. J. A. Beswick, G. Delgado-Barrio, P. Villarreal and P. Mareca, Faraday Disc. Chem. Soc, 73, 406(1982)
6. W. Sharfin, P. Kroger and S. C. Wallace, Chem. Phys. Lett., 85, 81(1982).

VIBRATIONAL PREDISSOCIATION OF THE T-SHAPE He...I_2(B) VAN DER WAALS MOLECULE: A CLOSE-COUPLING CALCULATION.

P. Villarreal*, G. Delgado Barrio and P. Mareca.

Instituto de Estructura de la Materia, C.S.I.C.
Serrano, 119. MADRID - 6 (Spain)

*Permanent adress: Departamento de Química Física y Química Cuántica.
Centro Coordinado CSIC-UAM. Facultad de Ciencias.
Universidad Autónoma de Madrid. CANTOBLANCO Madrid,34 (Spain)

Numerical calculations of close coupling equations to study vibrational predissociation (VP) of the T-shaped He...I_2(B) van der Waals molecule are reported here.This m olecule is one of the most carefully studied system in the context of triatomic van der Waals molecules[1].This VP process, that occurs on a single electronic potential energy, surface, can be described by

$$He...I_2(B^3\Pi 0_u^+,v) \xrightarrow{\Gamma} He + I_2(B^3\Pi 0_u^+,\ v' < v)$$

where the rate Γ, depending on the vibrational excitation v of the I_2 subunit, is related to the lifetime of the complex by the well-known expression $\tau = \hbar/2\Gamma$. Experimentally, a quasi-quadratic dependence[2] of the rates on v is found in the range $10\leq v\leq 45$. However, the rates deviate[3] from that law in the range $50\leq v\leq 64$, showing a maximum at $v\simeq 57$. From a theoretical viewpoint, the VP rate can be identified as the half-widths of the resulting resonances for the He-I_2(B) collision. The relevant close-coupling equations, in the T-shape configuration, have been described elsewhere[4]. In this work, we take a Morse function to describe the diatomic I_2 potential, while the van der Waals interaction is described as the sum of two atom-atom Morse type functions. The parameters are same already used[5]. In the framework of the "isolated narrow resonances"[6], the energy dependent S-matrix has a formal expression in the neighbourhood of a resonance

$$S(E) = S^{bg}(E)\left(\mathbb{1} - \frac{iA}{E-\bar{E}+i\Gamma/2}\right) \qquad (1)$$

where $\bar{E}$, Γ denote the position and width of the resonance, A is a almost energy independent matrix which can be expressed as $A = gg^+$, being g a complex vector with components verifying the requirement $\sum_1^N |g_i|^2 = \Gamma$, and N is the number of open channels. The "background" matrix S^{bg} is the corresponding S-matrix in absence of the resonance. Several methods have been proposed to estimate S^{bg}, many of them resting uppon the assumption of a functional energy dependence with adjustable parameters. As opposite we estimate directly S^{bg} by carrying out an artificial close-coupling calculation in which the closed channel supporting the resonance is neglected. After a little algebra, expression (1) leads to the following relation

$$\sum_{k=1}^{N} RE\,\{(S^{bg+}S)_{kk}\} = N - \frac{\Gamma^2/2}{(E-\bar{E})^2+\Gamma^2/4}$$

where RE {...} denotes "real part", from which a lorentzian fitting provides us the parameters $\bar{E}$ and Γ. We have represented the theoretical full-widths obtained, as a function of v, in Fig. 1. They are compared in the same figure with the experimental data of Sharfin et al.[3]. As can be seen, the general behaviour agrees fairly well with the experimental one up to v = 60, reaching a maximum at v = 58 as compared with that of Sharfin et al. (v = 57). This value is also in good agreement with previous calculations[5]. However, a pronounced minimum appears at v = 62. It is not accounted in the experimental data[3], but can be due to the difficulty of detecting so low values of Γ in this region of diatomic vibrational excitation. At any rate, from a theoretical viewpoint, this feature is conected with the large anharmonicity of the I_2 in this region. In fact, an interaction of configuration among the different resonances can modify their positions and widths, and makes feasible that the behaviour of the rates becomes oscillatory.

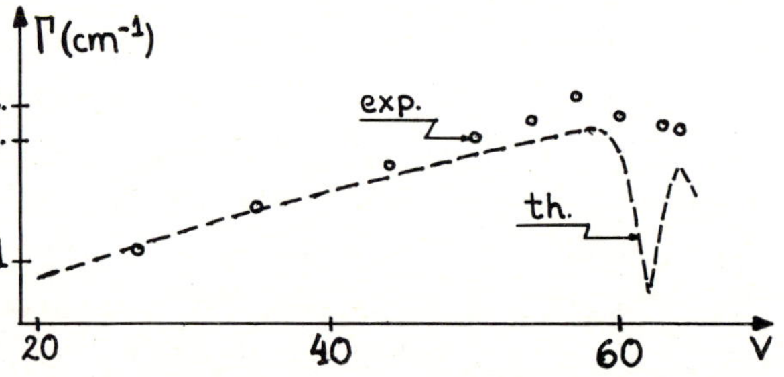

Figure 1.- Theoretical full-widths versus v in the range $20\leq v\leq 65$ together with the experimental data.

References.

1. a) D.H. Levy, Adv. Chem. Phys., 47(I), 323 (1981);
   b) J.A. Beswick and J. Jortner, Adv. Chem. Phys., 47 (I), 363 (1981)
2. R.E. Smalley, D.H. Levy, and L. Wharton, J. Chem.Phys 64, 3266 (1976).
3. W. Sharfin, P. Kroger and S.C. Wallace, Chem. Phys. Lett., 85, 81 (1982).
4. J.A. Beswick, G. Delgado Barrio and J. Jortner, J. Chem. Phys., 70, 3895 (1979).
5. G. Delgado Barrio, P. Villarreal, P. Mareca and G. Albelda, J. Chem. Phys., 78, 280 (1983).
6. C.J. Ashton, J.M. Hutson and M.S. Child (to be published)

COMPLEX-COORDINATE COUPLED-CHANNEL FORMALISM TO VIBRATIONAL AND ROTATIONAL PREDISSOCIATION OF VAN DER WAALS MOLECULES*

Shih-I Chu# and Krishna K. Datta

Department of Chemistry, University of Kansas, Lawrence, Kansas 66045 USA

There is currently much interest in the study of spectroscopy, structure and predissociation dynamics of weakly bound van der Waals (vdW) molecules[1,2,3]. Reliable anisotropic potential surfaces for several vdW complexes, particularly the rare gas-H_2 systems, are now becoming available. Recently we have developed two practical approaches for the accurate determination of vibrational and rotational predissociation lifetimes of vdW complexes. These are the complex-coordinate coupled-channel (CCCC) methods, one formulated in the space-fixed (SF), the other in the body-fixed (BF) frames of coordinates. The SFCCC theory[4,5] is more appropriate for the treatment of weak-coupling vdW complexes (such as Ar-H_2), whereas the BFCCCC theory[6] more natural for strong-coupling complexes (such as Ar-HCl). The utility and advantage of the CCCC methods may be summarized as follows: 1) It is an ab initio method (given a defined 'exact' Hamiltonian). 2) Only bound state structure calculations are required and no asymptotic conditions need to be enforced. 3) The resonance energies are obtained directly from eigenvalue analysis of appropriate non-Hermitian matrices, the imaginary parts of the complex eigenvalues being related directly to lifetimes of the vdW complexes. 4) It is applicable to many (open and closed) channel problems involving multiply coupled continua. These methods have been applied successfully to a number of vdW molecules[4-7].

In this conference, we shall present the CCCC theories and results for a recent study of the Ar-H_2 and Ar-HD vdW systems[8]. This study is prompted by the recent first successful experimental determination of the rotational predissociation (RP) line widths by McKellar[3] from the (improved) IR-high resolution spectra of Ar-HD and Kr-H_2 systems. The direct comparision between theoretical and experimantal results allows a critical assessment on the quality of the theoretical methods as well as the reliability of the potential surfaces used in the calculations.

The space-fixed complex-coordinate coupled-channel (SFCCCC) method is applied to the accurate determination of the level energies and widths (lifetimes) of predissociating states of Ar-HD(v=1, j=2) van der Waals molecule, using the Carley-Le Roy $BC_3(6,8)$ potential[9] for Ar-H_2. The calculated widths are in good harmony with the experimental data of McKellar (Fig. 1) and in excellent agreement with the close-coupling scattering results of Hutson and Le Roy[10].

*Work supported by DOE and ACS-PRF.

#Alfred P. Sloan Foundation Fellow.

References

1. For a recent review, see D. H. Levy, Advan. Chem. Phys. 47, 323 (1981).
2. J. A. Beswick and J. Jortner, Advan. Chem. Phys. 47, 363 (1981).
3. A. R. W. McKellar, Faraday Disc. Chem. Soc., 73, 000 (1982).
4. S. I. Chu, J. Chem. Phys. 72, 4772 (1980).
5. S. I. Chu and K. K. Datta, J. Chem. Phys. 76, 5307 (1982).
6. S. I. Chu, Chem. Phys. Lett. 88, 213 (1982).
7. K. K. Datta and S. I. Chu, Chem. Phys. Lett. 87, 357 (1982).
8. K. K. Datta and S. I. Chu, Chem. Phys. Lett. (in press, 1983).
9. R. J. Le Roy and J. C. Carley, Advan. Chem. Phys. 42, 353 (1980).
10. J. M. Hutson and R. J. Le Roy, Faraday Disc. Chem. Soc. 73, 000 (1982).

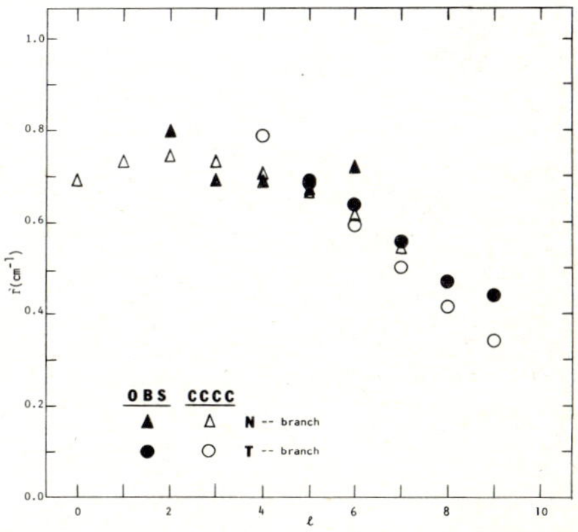

Figure 1. Comparison of calculated (SFCCCC) and experimental widths Γ for predissociating states of Ar - HD(v=1, j=2). ℓ is the relative orbital angular momentum.

STUDY OF THE REACTION OF NA*(3P) + HCL BY THE CROSSED MOLECULAR BEAM METHOD

H. Schmidt[+], M. F. Vernon, P. S. Weiss, M. H. Covinsky, and Y. T. Lee

Materials and Molecular Research Division, Lawrence Berkeley Laboratory
and Department of Chemistry, University of California, Berkeley, CA 94720

The reactions

$$Na^*(3P) + HCl \text{ -- } NaCl + H + E_{reac}(4.6 \text{ kcal/mole}) \quad (I)$$
$$+ E_{elec}(48.5 \text{ kcal/mole})$$
$$Na(3S) + HCl \text{ -- } NaCl + H + E_{reac}(4.6 \text{ kcal/mole}) \quad (II)$$

have been studied by the crossed beam method in the energy range of E_{cm}=3.8 kcal/mole - 17 kcal/mole.

The measurements of the formation of products were carried out with a crossed beam apparatus[1]. The Na atoms in the scattering region are excited by tuning a frequency stabilized single mode dye laser to the $3^2S_{1/2}$,F=2 -- $3^2P_{3/2}$,F=3 hyperfine transition. The optical pumping process maintains a steady state population of the atoms(30%) in the excited $3^2P_{3/2}$ state. Polarized laser light is used to produce aligned or oriented Na atoms[2] and to measure the change of the reactive scattering signal as a function of this alignment in space as well as the direction of the orientation. Angular distributions of reactively scattered NaCl and elastically or superelastically scattered Na are measured by a rotatable ultrahigh vacuum mass spectrometric detector.

Figure 1 shows the laboratory angular distribution of NaCl(m/e=58) for a collision energy of 5.6 kcal/mole. The squares are the data for reaction (I) and the triangles are the data for reaction (II). The measured intensities have been corrected for the different fluxes of Na*(3P) and Na(3S) atoms, resulting from the optical pumping process, in the

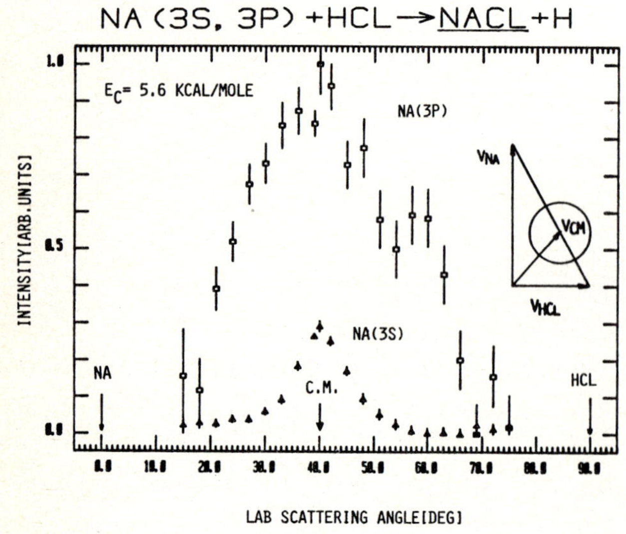

Fig. 1

scattering volume. In the right hand side of fig.1 a nominal Newton diagram is shown. The circle refers to the possible maximum velocity of NaCl, when all the available energy in reaction (I) is released in translational energy. It is remarkable that the signals from the excited state reaction(I) are more than three times larger than the ground state reaction(II). The angular distribution of reaction(II) only shows the laboratory angular distribution of the centroid of the system. Reaction(I) has a considerable intensity at a scattering angle of 60° near the kinematic angular limit. This shows that the electronic energy of Na*(3P) is converted very efficiently into translational energy of the products NaCl and H when the NaCl is scattered in the backward direction with respect to the incoming Na*(3P) atoms. Time of flight measurements of the NaCl velocity distribution are currently in progress to deconvolute complete center of mass product angular and energy distributions in order to allow a more detailed discussion of the reaction dynamics in relation to the potential energy surface.

The dependence of the reaction(I) on the alignment of the Na p-orbital with respect to the relative velocity was studied at a laboratory scattering angle of 60°. Within an error of 10% a dependence was not found. Measurements with oriented sodium atoms which are prepared by using left or right hand circularly polarized laser light, showed no circular asymmetry. For the Na-Cl-H system, the degree of symmetry is apparently too low for the reaction probability to depend on the initially prepared charge distribution of the sodium. The study of the systems Na*(3P) + Cl_2 and Na*(3P) + O_2 will show whether with a higher degree of symmetry the polarization of p-orbitals of Na atoms will influence the reaction cross sections.

+ Fachbereich Physik, Freie Universitaet Berlin, Berlin, Germany, F.R.

References

1 C.H. Becker, P. Casavecchia, P. W. Tiedemann, J. J. Valentini, and Y. T. Lee
J. Chem. Phys. 73, 2833 (1980)

2. A. Fischer, I. V. Hertel, Z. Phys. A304, 103 (1982)

M(IIa)-X_2 CHEMIIONISATION AS A FUNCTION OF COLLISION ENERGY

U. Ross, H.-J. Meyer, T. Schulze and D. Beck

Faculty of Physics, University of Bielefeld, Germany

Group IIa metal atom-halogen molecule reactions at thermal energy (~ 0.08 eV) are known to yield a wealth of products,

$$M + X^2 \rightarrow MX(X^2\Sigma) + X, \ (+ 2.1 \text{ eV}) \quad (1)$$

$$\rightarrow MX^* + X \quad , (MX^*: A^2\Pi, B^2\Sigma, C^2\Pi) \quad (2)$$

$$\rightarrow MX^+ + X^- \quad , (+ 0.7 \text{ eV}), \quad (3)$$

(data in parentheses relating to Ba-Cl_2) which reveal a fascinating diversity of collision dynamics in simple systems. This study on the dependence of M-X_2 chemiionisation on collision energy is concerned with channel (3) which diabatically correlates with the deep well stabilizing the MX_2 ground state.

An M atom beam is prepared by the seeding method to vary collision energy E. Its velocity v(M) is determined by time of flight. Mass spectrometric detection provides a relative measure of its number density n(M). It is crossed by the halogen beam which is operated at nearly effusive, constant oven conditions. A channeltron multiplier views the beam intersection volume collecting virtually all reaction products MX^+ of flux $\dot{N}_+$. This basically suffices to determine the energy dependence of the total integral chemiionisation cross section $\sigma_J(E)$ from

$$\sigma_J(E) = \text{const} \frac{\dot{N}_+}{n(M)|\vec{v}(M)+\vec{v}(X_2)|}, \quad E = \frac{\mu}{2}(v^2(M) + v^2(X_2)),$$

the constant velocity $v(X_2)$ being calculated from the X_2 oven temperature. High detection sensitivity permits *very* dilute conditions on both beams. Facing the ion detector a photomultiplier may be installed to monitor the read or blue chemiluminescence.

$\sigma_J(E)$ has been measured for the partners M = Ca, Sr, Ba and X = Cl_2, Br_2 as well as a number of polyatomic molecules. Markedly different behaviour is found for nearly all atom-diatom systems. Fig.1 shows a particularly intriguing result, the strongly peaking cross section of Ba-Cl_2. Somewhat attenuated and broader in energy a similar "resonance" is also present at E = 0.6 eV in (slightly) endoergic Sr-Cl_2 right above apparent threshold. In more endoergic Ca-Cl_2 it is missing. The origin of this effect of concerted action in two electron transfer will be discussed. Its presence explains why previous work on Ba-Cl_2 using a *thermal* beam at fixed nominal E = 0.08 eV has lead to suggest an activation threshold. But it would not appear to prove the suggestion correct. The slow increase of the ion yield at high energy in Fig. 1 is thought to be of statistical nature indicating a tightly coupled, possibly long lived intermediate. Similar behaviour is found with polyatomic molecules where

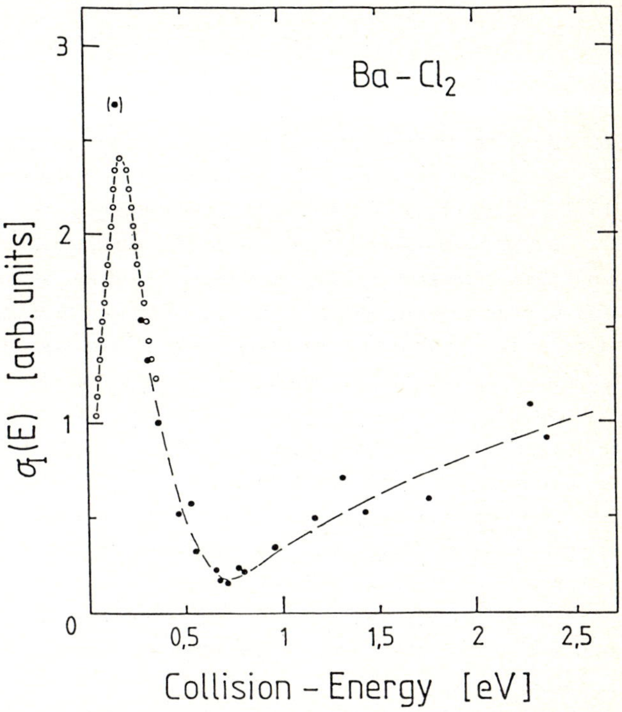

Fig.1. Ba-Cl_2 chemiionisation yield $\sigma_J(E)$ as a function of collision energy. Full circles have been measured by seeding. Open circles were obtained from a thermal Ba beam by TOF analysis of Ba and $BaCl^+$. The energy resolving power (~ 0.06 eV FWHM) for the latter data is *larger* than the spacing of points. The two sets of data are *crudely* connected according to equal slope in their region of overlap.

it clearly correlates with the number of degrees of freedom of the decomposing complex and permits explanation by RRKM arguments. Interestingly, first observations on the chemiluminescence yield in Sr-Cl_2 show that it tends to conform to the ion output in the corresponding slow rise at high energy, but is missing in the resonance region.

Apparent thresholds of endoergic chemiionisation are evaluated in Tab.I to yield preliminary lower limits of the energy of MX^+ formation $D_0(MX^+)$ and - using literature data on $D_0(MX)$ - upper limits of the ionisation energy J(MX).

Tab.I

	$D_0(MX^+)$[eV] ≳	J(MX)[eV] ≲
SrCl	4.21 ± 0.04	5.63 ± 0.06
SrBr	3.49 ± 0.05	5.6 ± 0.2
CaCl	4.14 ± 0.04	6.05 ± 0.08
CaBr	3.36 ± 0.1	6.1 ± 0.2

COLLISION ENERGY DEPENDENCE OF VIBRATIONAL/ROTATIONAL DISTRIBUTION OF BaBr PRODUCED IN THE CROSSED BEAM REACTION Ba + CH_3Br

Toshiaki Munakata, Yutaka Matsumi, and Takahiro Kasuya

Institute of Physical and Chemical Research, Hirosawa, Wako, Saitama 351, Japan

We present a crossed molecular beam experiment on the exothermic reaction:

$Ba + CH_3Br \rightarrow BaBr + CH_3$, $\Delta H_0^0 = -16.7$ kcal/mol.

The vibration/rotation populaitons of reaction product BaBr are probed by means of laser induced fluorescence (LIF) as a function of reagent collision energy.

The present molecular beam apparatus consists of two vacuum chambers: a CH_3Br beam source chamber, and a crossed beam reaction chamber. The supersonic CH_3Br beam is generated in a 10 Hz repetitive pulse of 1 ms duration through a conventional solenoid valve. After being collimated by a skimmer, the beam enters into the reaction chamber. The translational energy of CH_3Br beam is varied by changing the seeding ratio in H_2 carrier gas. A time-of-flight method (TOF) with a mechanical chopper and a mass spectrometer is employed in the measurement of the translational energy.

The effusive Ba beam is generated in an oven assembly situated in a corner of the reaction chamber. The Ba vapor is evaporated in a cylindrical crucible, which is heated to 1100 K by thermal radiation from a heating tube. The heating tube is enclosed in a couple of radiation shield tubing and a water cooled jacket. The Ba beam is collimated by a sequence of slits which are well aligned on the heater, the radiation shields and the cooling jacket.

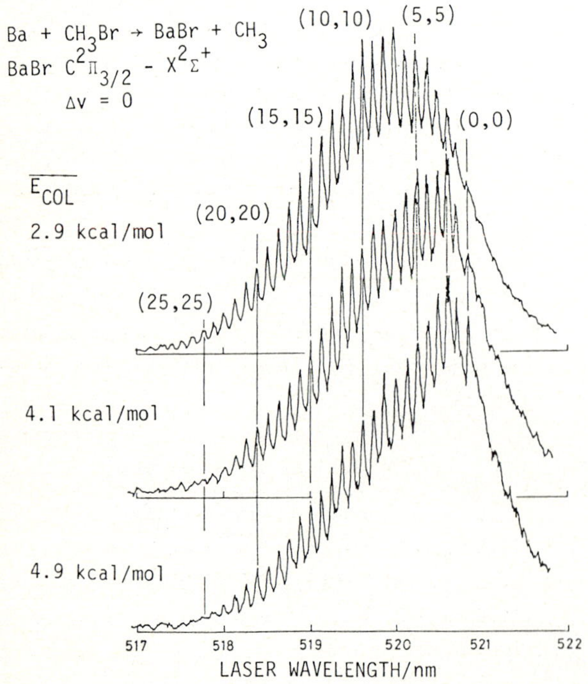

Fig. 1. LIF spectra of nascent BaBr produced at different mean collision energies, E_{COL}.

The internal states of the reaction product, BaBr, are probed by a pulsed, YAG laser-pumped tunable dye laser. The dye laser light at around 520 nm wavelength excites the BaBr fluorescence of the $\Delta v = 0$ band in the $C^2\Pi_{3/2} - X^2\Sigma^+$ system.

The LIF spectra observed at different collision energies are represented in Fig. 1. The most distinct feature is that the high vibrational side looses intensity with collision energy. A quantitative information on product internal state distribution is derived through a simulation of the observed LIF spectrum. In Fig. 2, the vibrational population, N_v, obtained from the simulation are plotted vs. vibrational quantum number, v. It is found from Fig. 2 that the vibrational excitation decreases remarkably with an increase of collision energy: the vibrational quantum on which N_v is peaked shifts from 12 down to 3 even for a small increment of collision energy from 2.9 to 4.9 kcal/mol.

Dynamical properties specific to the present reaction may be deduced when the experimental vibrational population is rearranged into 'surprisal' plots. The surprisal analysis suggests that a transition takes place between two reaction paths in accordance to an increase of collision energy: one is the reaction path through which BaBr is preferably produced in vibrationally excited states at around v = 12, and the other, the path through which internally cold CH_3 and statistically' excited BaBr are produced. Branching ratio into the former path decreases drastically with an increase of collision energy. The two reaction paths may possibly be related to a reaction mechanism based on two collision configurations: the former is the reaction through migratory encounter, and the latter, the reaction by direct attack of Ba on Br end.

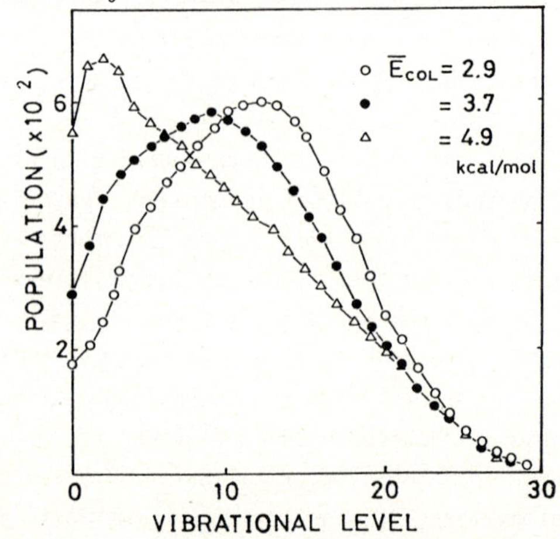

Fig. 2. Vibrational population of BaBr.

DYNAMICS OF VAN DER WAALS BOND EXCHANGE

D. R. Worsnop, S. J. Buelow, and D. R. Herschbach

Department of Chemistry, Harvard University, Cambridge, Massachusetts 02138

Angular and velocity distributions of products have been measured for the thermal energy $Xe + Ar_2$ reaction[1] and the analogous $Kr + Ar_2$ and $Kr + Xe_2$ reactions. As illustrated in Fig. 1, the velocity vector distribution of the scattered products shows both forward scattering peaks (at 50° relative to the attacking Xe atom) and a backward peak (at 180°). There also is a pronounced "hole" in the forward direction (at 0°). These features are in nearly quantitative agreement with predictions of a refined "sequential impulse model" (SIM) based on pairwise hard sphere interactions.[2] This SIM model has previously been applied to high energy (>20 eV) ion/molecule reactions,[3] but since the key parameter is the ratio of collision energy to well depth (E/D >4 here), it is also appropriate for the van der Waals domain.

The experiment employs two supersonic beams crossed at 90° and detects scattered products with a rotatable mass spectrometer equipped with a time-of-flight analyzer. The expansion conditions of the dimer beam were carefully controlled to suppress larger clusters. Because of the large E/D ratio, conservation of energy and momentum constrain the scattered diatom product to a narrow circular band centered on the center-of-mass velocity. The fit of the data to this kinematic "fingerprint" uniquely identifies the bimolecular exchange reaction.[1]

Figure 2 compares the angular distribution of scattered XeAr with predictions of SIM. The dashed curve shows the SIM prediction. The latter assumes that A + BC interact via sequential hard sphere elastic collisions (A off B, then B off C); reaction occurs if the final relative velocity of A and B or A and C corresponds to an energy less than the bond strength of AB or AC. As shown in Fig. 2, the model

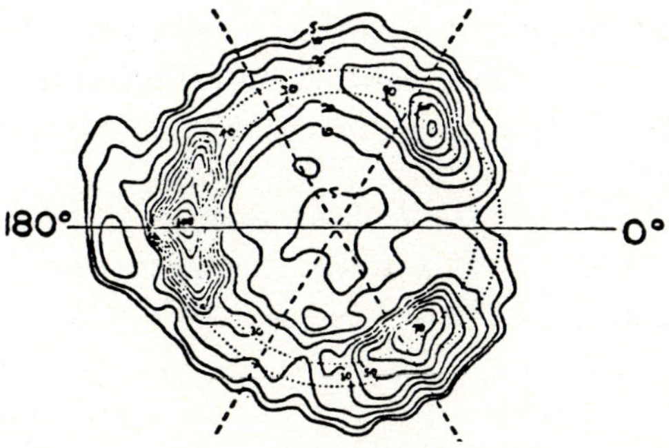

Fig. 1: Intensity contour map for XeAr reactive scattering of $Xe + Ar_2$ in center-of-mass system. The incident Xe atom travels in the 0° direction. The dotted circular band shows the kinematically allowed scattering region for nominal parent beam velocities.

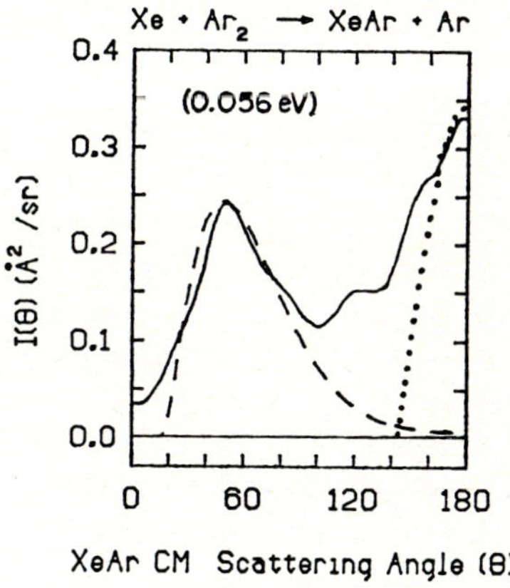

Fig. 2: Angular distribution of XeAr from $Xe + Ar_2$ in center-of-mass system. Solid curve is experimental; dashed predicted by SIM; dotted predicted for "knock-out" process.

predicts the location of the forward scattering peak. The "hole" at 0° occurs because at these collision energies, the forward scattered XeAr contains more internal energy than its bond strength and hence dissociates.

The backward peak is consistent with a simple knock-out process (not originally included in SIM), shown by the dotted curve. In this, A knocks B away, and the AC bond forms directly.

The $Kr + Ar_2$ and $Kr + Xe_2$ data are qualitatively similar to $Xe + Ar_2$, except that the backward channel dominates, as predicted by SIM. Furthermore, the E^{-2} dependence of the total cross section measured for the Xe and $Kr + Ar_2$ systems is in accord with SIM. The strong inverse energy dependence reflects the fact that collision induced dissociation, which is not measurable, is the major channel at these collision energies.[4] The He or $Ne + Ar_2$ and $Ar + Xe_2$ systems showed no discernable product signals; SIM indeed predicts the cross sections are at least an order of magnitude smaller than for $Xe + Ar_2$.

In the extended form considered here, SIM offers a useful asymptotic model for any $A + BC \rightarrow AB + C$ exchange reaction when the ratio E/D becomes large.

References

1. Preliminary results in D. R. Worsnop, S. J. Buelow and D. R. Herschbach, J. Phys. Chem. 85, 3024 (1981).
2. B. H. Mahan, W. E. W. Ruska and J. S. Winn, J. Chem. Phys. 65, 3888 (1976).
3. B. H. Mahan and W. E. W. Ruska, ibid. 65, 5044 (1976).
4. D. A. Dixon and D. R. Herschbach, Ber. Bunsenges. phys. Chem. 81, 145 (1977).

CHEMILUMINESCENCE IN REACTION OF POTASSIUM DIMERS WITH OXYGEN MOLECULES

J. Berlande, J. Cuvellier, P. de Pujo, and J.M. Mestdagh

Service de Physique des Atomes et des Surfaces, CEN/Saclay, 91191 Gif-sur-Yvette Cedex, France

At thermal energies, alkali atoms in their ground state do not react with oxygen molecules since the formation of alkali oxide would be an endoergic process ($\Delta E \sim 2eV$). On the other hand, as indicated by Figger et al[1], light emission is observed when alkali dimers interact with the same molecules since, for the case of potassium dimers, the reaction $K_2+O_2 \rightarrow KO_2 + K$ (1) is exoergic by some 1.3 - 1.8 eV and therefore one or both of the products might be in an excited state (there is some uncertainty regarding the binding energy of KO_2 and the ionic character of the K-O_2 bond [2]).

We have studied in a crossed beam apparatus, for the energy range 0.15 - 0.45 eV, the luminescent channel for process (1) where the potassium atoms are created in a radiative state. Potassium dimers are formed during the expansion of a high pressure ($\sim$ 150 Torr) potassium vapor through a 0.2 mm nozzle. Details of the apparatus and experimental techniques are given elsewhere[3]. It has been observed, that in the spectral range 2500-8500 Å, atomic light, originating from the first resonance lines of potassium (at 7665-7699 Å) represents the major part of the total luminescence. No other atomic line is observed. Surprisingly the ratio of the resonance line intensities is 3.2 ± 0.3 whatever the collision energy; it is larger than 2, the statistical weight. The cross section, for the process $K_2+O_2 \rightarrow K^* + KO_2$ (KO_2 being excited or not) is found nearly energy independent which confirms the exoergicity of the process and gives some indications regarding the binding energy of KO_2.

The absolute value of this cross section has been determinated by comparision with the process : $K(4P_{3/2})$ + He $\rightarrow K(4P_{1/2})$ + He which allows to calibrate the optical system and whose cross section is well known[4]. It is found to be 2.10^{-18} cm^2 ($^{+6}_{-1}$) : the major uncertainties come from the measurements of the density of potassium dimers and the density of excited $K(4P_{3/2})$ atoms needed for the calibration.

Process(1) proceeds most probably via a harpooning mechanism as it is the case when halogen molecules are substituted to O_2[5]. Since the crossing of the covalent and ionic curves for ground state reactants occurs around 4-5 Å, where the electron jump probability is high, most of the potassium atoms formed are in their ground state, in agreement with the small cross section measured in this work.

References

1. M. Figger, A. Kowalski, X.H. Zhu, Contributed paper- "50 years Dynamics of Chemical Reactions" Berlin oct. 1981- ed: Hahn-Meitner-Institut für Kernforschung-Berlin.
2. R.R. Herm and D.R. Herschbach, J. Chem.Phys. 52, 5780 (1970).
   L. Andrews and R.R. Smardzewski, J. Chem. Phys. 58, 2258 (1973).
3. J. Cuvellier, J.M. Mestdagh, J. Berlande, P. de Pujo, and A. Binet, Rev. Phys. Appl. 16, 679 (1981).
   J. Berlande, J. Cuvellier, P.de Pujo, J.M. Mestdagh, to be published.
4. J.M. Mestdagh, Thesis, Paris 1982 -unpublished.
5. W.S. Struve, J.R. Krenos, D.L. McFadden and D.R. Herschbach, J. Chem. Phys. 62, 404 (1975).

A NEW QUANTUM MECHANICAL THEORY OF CHEMICAL REACTIONS

E. Ficocelli Varracchio

Department of Chemistry, Centro Chimica Plasmi, C.N.R., University of Bari, 70100 Bari, Italy

Much progress, during the last few years, has been made in the quantum mechanical approach to the theory of chemical reactions, mostly through the introduction of "suitable" (natural, hyperspherical,...) coordinate systems (c.s.)[1]. These are characterized by a smooth change, when going from reactants to products of the process, and allow extracting, rather easily, the S-matrix of the rearrangement collision, from the total scattering wave function of the dynamical computation. Unfortunately, the enforcement of such c.s. requires a very detailed topological study of the electronic energy surface, prior to any dynamical computation. This aspect of the present quantal approaches is undesirable because it remarkably encumbers numerical applications and, besides, it establishes an apparent division between the subreactive and rearrangement regimes, which should, instead, be studied unitarily.

In order to circumvent the above problems, we are presently formulating a new quantum mechanical approach to the theory of chemical reactions[2]. The starting point of our formulation is based on a change of "representation" for the conventional system Hamiltonian (H), in such a way that, in the new representation, it may display "simultaneously" information on both the reactants and products of the process. Then the coupling between the entrance and exit channels will be already present in the structure of the new Hamiltonian. This will allow studying the dynamics by using conventional c.s. and, therefore, all needs for resorting to preliminary detailed topological studies of the potential energy surface will be eliminated.

We are developing the new theory for collisions involving an atom and a diatomic molecule, according to

$$A(\vec{k}_i) + BC(\alpha_i) \dashrightarrow AC(\beta_f) + B(\vec{k}_f) \qquad (1)$$

characterized by a system Hamiltonian that can be written, in conventional notation, as

$$H = H_\alpha^o + V_\alpha = H_\beta^o + V_\beta \qquad (2)$$

corresponding to a partitioning into a α (initial) and β (final) channels, respectively. The change of representation that we suggest for H, in (2), is effected by means of a "unitary transformation", according to

$$\hat{H} = U^{-1} H U \qquad (3)$$

with $\hat{H}$ the transformed Hamiltonian of the theory. In particular, the U operator has the physical effect of leading to "bound" molecules, starting from free nuclei, and it can be explicitly built by using models available in the literature[3].

We have evaluated the transformation (3) exactly and we have obtained a final form for $\hat{H}$, that can conveniently be expressed as the following sum of terms

$$\hat{H} = \hat{H}_o + \hat{V}_{dir} + \hat{V}_{diss} + \hat{V}_{reac} + \hat{V}_{free} \qquad (4)$$

In (4), $\hat{H}_o$ refers to the kinetic energies of the free atoms and molecules, respectively, while the $\hat{V}$ potentials explicitly refer to the channels (subreactive, dissociative, reactive, atom-atom scattering) available to the system. Explicit equations for all the terms in (4) have been obtained in full[2].

As an example, we give explicitly the $\hat{V}_{reac}$ term, which is responsible for the rearrangement (1), in a first order collision. This term can be written as

$$\hat{V}_{reac} = V_\alpha - \Delta_{BC}^b V_\alpha \qquad (5)$$

with V_α defined in (2) and Δ_{BC}^b a quantity characteristic of the transformation (3), essentially defined as a projection operator onto the "bound" subspace ($|\alpha_n\rangle$) of the BC molecule, i.e.

$$\Delta_{BC}^b = \Sigma_n^b |\alpha_n\rangle\langle\alpha_n| \qquad (6)$$

Details on the theory and its physical interpretation will be considered at the Conference.

References

1. "Atom-molecule collision theory", R.B. Bernstein ed. (Plenum Press, N.Y., 1979)
2. E. Ficocelli Varracchio, (submitted for publication)
3. M.D. Girardeau, J. Math. Phys. 16, 1901 (1975)

RESONANCES AND INTERFERENCES IN THE THEORY OF CHEMICAL REACTIONS

V.Aquilanti, S.Cavalli, G.Grossi, and A.Laganà

Dipartimento di Chimica dell'Università, 06100 Perugia, Italy

Recent progress in the theory of elementary chemical reactions has been favoured by the systematic introduction of the hyperspherical coordinate formalism,[1] both in analytical work and in its numerical implementations by several groups.

Hyperspherical coordinates are useful for the mapping of potential energy surfaces, and properties have recently been derived[2] for the harmonic expansions needed in the description of a reactive collision as a three-body quantum mechanical problem. However, most applications (including ours,[3-5] where references to other papers can be found) deal with the simplified situation that atoms are constrained to move on a line.

The work reported here concerns the application of adiabatic and diabatic models to understand interference effects and the resonance structure which has emerged in previous numerical work. In particular, it has been found that the introduction of semiclassical techniques within the hyperspherical coordinate approach leads to accurate approximations, often amenable to simple physical interpretation.

Following the scheme outlined previously,[3] adiabatic energy levels at fixed values of the radial hyperspherical coordinate have been calculated for several symmetrical collinear chemical reactions using a technique which exploits the eigenfunction expansion method[2] for potential energy surfaces. Typically, as the hyperradius varies from values lower than the saddle point to larger ones, one is faced with a problem of quantization in a single or in a double well. Uniform semiclassical quantization procedures have been tested against exact numerical calculations for a variety of systems, and have been found to be as accurate as required in most applications.

It had already been shown[4] that the simplest approach to the description of reaction probabilities, i.e. the assumption of full adiabatic decoupling, leads to extremely accurate results for a few cases (hydrogen exchange in I + HI and isotopic variants, Mu exchange in H + MuH): in these cases, neglecting coupling in the adiabatic representation allows the dynamics to be treated simply in terms of scattering from the potentials generated by the adiabatic levels as the hyperradius varies. Reaction probabilities, including interference effects and resonance position and widths, are then obtained using purely semiclassical techniques for phaseshifts and resonances.

The inadequacy of the strictly adiabatic model to cover the general case suggests that we look explicitly for the introduction of nonadiabatic coupling within the formalism. Our eigenfunction expansion method for potential energy surfaces is particularly well suited for the computation of the elements of the matrix which describes nonadiabatic coupling.

An extensive numerical study of the properties of this matrix for a variety of systems has been performed: it has been found, as anticipated,[4] that nonadiabatic coupling is localized at potential ridges, i.e. where the short range behaviour typical of the collision complex changes to that which pertains to reactants and products. In general, adiabatic levels have consequently a smooth dependence on the hyperradius, except near the ridge, where nonadiabatic coupling induces wells and barriers.

Exact calculations for the coupling matrix elements have been used to assess semiclassical estimates. The latter are based on arguments borrowed from spectroscopy of diatomics and atom-atom collision theory. The general qualitative agreement becomes quantitative when the coupling is strong: a Landau-Zener type of theories for curve crossing is then applicable. The analogy which emerges here is between the hyperradius and the usual Born-Oppenheimer interatomic distance of a diatomic system: therefore, for example, positions and widths of scattering resonances for an atom plus diatom reaction, are the analog of those necessary for calculating the predissociation pattern of a diatomic molecule.

The adaptation to present situations of semiclassical theories developed for the latter problem is therefore very promising: resonance positions not too accurately predicted by the adiabatic model, may be shifted considerably when nonadiabatic effects are included. In particular, for H + FH (and its isotopic variant D + FD), practically exact resonance positions and widths are obtained by a diabatic model, i.e. by assuming a fully nonadiabatic behaviour at ridge.

1) U.Fano, Phys.Rev. A24, 2402 (1981)
2) V.Aquilanti, G.Grossi, and A.Laganà, J.Chem.Phys. 76 1587 (1982)
3) V.Aquilanti, G.Grossi and A.Laganà, Chem.Phys.Lett. 93, 174 (1982)
4) V.Aquilanti, S.Cavalli and A.Laganà, Chem.Phys.Lett. 93, 179 (1982)
5) V.Aquilanti, S.Cavalli, G.Grossi and A.Laganà, in Theoretical Chemistry 2, edited by A.Gamba, North-Holland (1983), in press.

DEPENDENCE OF STATE-TO-STATE REACTIVE SCATTERING ON RELATIVE ANGULAR MOMENTUM

S.H. Suck, R.W. Emmons, and C. Klein

Department of Physics and Graduate Center for Cloud Physics Research
University of Missouri-Rolla, Rolla, Missouri 65401, USA

The method of distorted-wave Born approximation has been highly successful in treating a-few-nucleon transfer reactions in nuclear physics. Considering the relatively new introduction to molecular reactions, it is of great interest to examine its validity in various applications. In the present study we limit ourselves to the study of product rotational state distributions in the atom-diatomic molecule reactive scattering of $F + H_2 \to HF + H$. It is found that the predicted first peak positions in the product rotational state distributions show good correlation with observation. In addition, the trend of the peak position shift toward the lower rotational angular momentum state of product molecule with the enhancement of the product state vibrational manifold is predicted to be in precise agreement with observation. From the present study, we find that angular momentum matching plays a significant role in characterizing the product rotational state distributions. The peak position in the distribution is found to occur where the angular momentum match is most favorable.

Contribution of the direct process to reactive scattering is examined as a function of collision energy. Use of the angular momentum space distribution that appears in the formalism of the distorted-wave Born approximation is found to be helpful in explaining the cause of broadening in backward peaked state-to-all states reactive scattering angular distributions with the increase of collision energy. As the collision energy increases, the peak position of the product rotational state distribution in a given vibrational manifold is found to shift relatively slowly toward the higher rotational angular momentum states of the product molecule HF in the $F + H_2 \to HF + H$ reaction, thereby showing the preference of larger angular momentum transfer during the reaction.

Finally, the importance of transferred angular momentum in direct reactive scattering processes involving atom-diatomic-molecule systems is discussed. The proton transfer reaction of $H_e + H_2^+ \to H_eH^+ + H$ is found to generally occur via the largest angular momentum transfer. The presence of orbiting resonance in a reactive scattering system involving charge transfer is also presented. The resonance structure of state-to-state angular distribution is found to exist in a limited domain of scattering angles, showing the coexistence of the resonance and direct reaction process, and the interference between the two. It will be of great interest to examine in the future how the coupling of states[1] will reveal additional characteristics which are not predicted by the present DWBA method.[2]

References

1. S. H. Suck, Phys. Rev. A27, 187 (1983).
2. S. H. Suck, Phys. Rev. A15, 1893 (1977).

AN INVARIANT IMBEDDING APPROXIMATE SOLUTION - TYPE ALGORITHM FOR SOLVING COUPLED EQUATIONS FOR SCATTERING

F. Mrugała
Institute of Physics, Nicolas Copernicus University, Toruń, Poland

The problem of interest is a system of differential equations

$$\left[\frac{d^2}{dx^2} + B(x)\right]\Psi(x) = \Phi(x)$$

with an N×N symmetric coupling matrix B and with an inhomogenuity term, $\Phi(x)$, being an N-dim. vector function.

The sought information is contained in the four N×N matrices $L^{(i)}(x',x'')$ $(i=1,2,3,4)$ and in the two N-dim. vectors $Q(x',x'')$ and $T(x',x'')$. In terms of these quantities the following relations holds for any solution $\Psi(x)$ in an interval $[x',x'']$:

$$\begin{pmatrix} \dot\Psi(x') \\ \dot\Psi(x'') \end{pmatrix} = \begin{pmatrix} L^{(1)}(x',x'') & L^{(2)}(x',x'') \\ L^{(3)}(x',x'') & L^{(4)}(x',x'') \end{pmatrix} \begin{pmatrix} \Psi(x') \\ \Psi(x'') \end{pmatrix} + \begin{pmatrix} Q(x',x'') \\ T(x',x'') \end{pmatrix}$$

where $\dot\Psi$ stands for the derivative with respect to x.

The algorithm for finding the matrices $L^{(i)}(x',x'')$ and the vectors $Q(x',x'')$, $T(x',x'')$ is constructed as follows:

1. choose a sequence $\{x_k\}_{k=0}^{M}$ of equally spaced points in the interval $[x',x'']$; i.e, $x_k=x_0+kh$, $x_0=x'$, $x_M=x''$, $h=(x''-x')/M$ with M even,
2. initialize the propagation loop for the N×N matrices z, l and r and for the N-dim. vectors t and q:
$$z_0^{-1}=0 \qquad t_0 = 0 \qquad r_1 = -\frac{1}{h} + \frac{h}{3}B_0$$
$$l_1 = \frac{1}{h} \qquad q_1 = -\frac{h}{3}\Phi_0 \quad ,$$
3. perform the subsequent propagation steps:
$$z_k = -6 + \left[0.125 + \frac{h^2}{48}B_k\right]^{-1} - z_{k-1}^{-1} \qquad \text{for } k=1,3,5,\ldots M-1$$
$$z_k = 2 - \frac{2h^2}{3}B_k - z_{k-1}^{-1} \qquad \text{for } k=2,4,6,\ldots M$$
$$t_k = \left[0.125 + \frac{h^2}{48}B_k\right]^{-1}\frac{h^2}{6}\Phi_k + z_{k-1}^{-1}t_{k-1} \qquad \text{for } k=1,3,5,\ldots M-1$$
$$t_k = \frac{2h^2}{3}\Phi_k + z_{k-1}^{-1}t_{k-1} \qquad \text{for } k=2,4,6,\ldots M$$
$$\left.\begin{aligned} l_k &= l_{k-1}z_{k-1}^{-1} \\ q_k &= q_{k-1} - l_k t_{k-1} \\ r_k &= r_{k-1} + h\, l_{k-1} l_k^T \end{aligned}\right\} \quad \text{for } k=2,3,4,\ldots M$$
where $B_k = B(x_k)$, $\Phi_k = \Phi(x_k)$
4. determine the matrices $L^{(i)}$ and the vectors Q and T at the end of the propagation loop:
$$L^{(1)} = r_M \qquad L^{(2)} = l_M$$
$$L^{(3)} = -l_M^T \qquad L^{(4)} = \frac{1}{h}\left(z_M + \frac{h^2}{3}B_M - 1\right)$$
$$Q = q_M \qquad T = \frac{1}{h}\left(t_M - \frac{h^2}{3}\Phi_M\right) \quad .$$

The invariant imbedding nature of this algorithm ensures complete numerical stability even in the troublesome cases when the coupling matrix B has large negative eigenvalues.

The error of the calculated quantities is propertional to h^4.

The part concerning the determination of the matrix $L^{(4)}$ is already well known as the log-derivative method [1] which was found to be very usefull in the inelastic scattering calculations [2]. The present generalization of this method enables applications to other types of scattering problems such as reactive collisions or final state interaction problems (like photodissociation, photoionization etc.). Moreover, it is expected that it will be very advantageous to hybridize the new method with the existing approximate potential methods for solving these problems [3-4].

A generalization of the presented algorithm to systems with first derivative coupling is given in [5].

References

1. B.R. Johnson, J.Comput.Phys. 13, 445 (1973)
2. L.D.Thomas, M.H.Alexander, B.R.Johnson, W.A. Lester, Jr., J.C.Light, K.McLenithan, G.A.Parker, M.J.Redmon, T.G.Schmalz, D.Secrest, R.B. Walker, J.Comput.Phys. 41, 407 (1981)
3. J.C.Light, R.B.Walker, J.chem.Phys. 65, 4272(1976)
4. B.I.Schneider, H.S.Taylor, J.Chem.Phys. 77, 379 (1982)
5. F.Mrugała, D.Secrest, The generalized log-derivative method for inelastic and reactive collisions, submitted to J.Chem.Phys.

ATOM CAPTURE AND ATOM LOSS AT ASYMPTOTICALLY HIGH SPEEDS

M. Breinig, G. J. Dixon, P. Engar, S. B. Elston, and I. A. Sellin

University of Tennessee, Knoxville TN 37996, USA, and
Oak Ridge National Laboratory, Oak Ridge TN 37830 USA

Ion-atom collisions at asymptotically high speeds involving charge transfer or projectile or target ionization are the subject of many recent theoretical investigations. In the asymptotic velocity regime a perturbation expansion is supposed to provide an increasingly exact description of these processes. However, in current electron capture and loss experiments the asymptotic velocity regime can barely be reached. Furthermore, angular distribution measurements for charge transfer and projectile ionization processes become increasingly difficult at projectile speeds over 10 a.u. In experiments involving the capture of a whole atom from a molecule by a projectile ion or the loss of an atom from a projectile molecular ion, projectiles with kinetic energies on the order of 100 eV/u have asymptotically high speeds compared with characteristic vibrational speeds, while the laboratory scattering angles are measurably different from 0 deg. Doubly and triply differential cross section measurements become possible, thus revealing information about the scattering amplitudes.

For 100-300 eV/u Ar^+ and Kr^+ projectiles, we have measured the singly differential cross section $d\sigma/dv_H$ for capture of an H-atom from a CH_4 molecule into states lying in the vibrational continuum of the electronic ground state and into predissociative states of the projectile hydride molecule.[1] The proton yield into a cone of half angle Θ_o about the beam direction exhibits a peak when the outgoing speed of the captured H-atom v_H equals the outgoing speed of the projectile ion v. Preliminary angular distribution measurements show that this peak is present only when $\Theta_o > \Theta_c$, where Θ_c is the critical angle for capture via the Thomas double scattering mechanism. Structure at equal velocity intervals above and below v arises from capture into predissociative states (see Figure 1). For Θ_o = 1.6 deg, $d\sigma/dv_H$ scales with projectile energy as $E^{-2.7}$ for Ar^+ projectiles and as $E^{-3.5}$ for Kr^+ projectiles.

In atom loss experiments, the shapes and yields of proton loss peaks were measured for 100-1300 eV/u H_2^+, HeH^+, ArH^+ and KrH^+ molecular ions passing through various gas targets. Yields are only weakly dependent on the target gas and increase with projectile energy. Shapes and width are independent of projectile and target. For projectile energies above ∿ 175 eV/u, prominent features in the wings of the proton loss peaks indicate excitation of dissociative states of the projectile molecule. The positions of these features are independent of projectile velocity or target gas, but the production cross sections are strongly energy- and target-dependent.

This research was supported in part by the National Science Foundation, Physics Division, and by the Fundamental Interactions Branch, Division of Chemical Sciences, Office of Basic Energy Sciences, U. S. Department of Energy, under contract W-7405-eng-26.

Reference

1. M. Breinig, G. J. Dixon, P. Engar, S. B. Elston, and I. A. Sellin, submitted for publication to Phys. Rev. Lett.

Figure 1: Energy spectra of protons into a cone of half angle Θ_o = 1.6 deg (broken line) and Θ_o = 2.6 deg (solid line) centered about the beam direction after capture by 6.06 KeV Ar^+ projectiles on CH_4. A linear background has been subtracted.

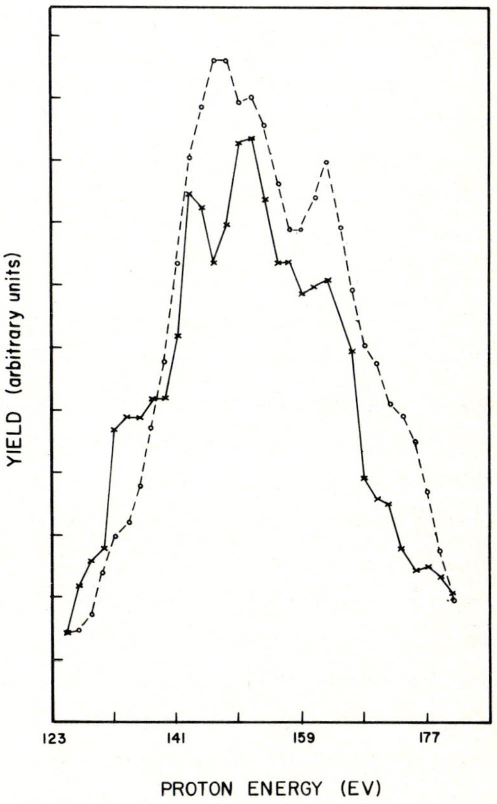

PROTON - OXYGEN POTENTIAL HYPERSURFACES RELEVANT TO VIBRATIONAL ENERGY TRANSFERS DURING COLLISIONS

F.A. Gianturco[*] and V. Staemmler[+]

[*]Dept. of Chemistry,University of Rome,Rome,Italy.

[+]Lehrstuhl fuer Theoretische Chemie,Ruhr-Universitaet Bochum, 4630 Bochum 1 , FRG .

The study of ion-molecule reactions under beam conditions has been recently enjoying an increased amount of interest because of the important role that such reactions appear to play in molecular astrophysics and as intermediates of more complex unimolecular protonation reactions in the gas phase.

In the present computational study we have looked in detail at a rather complicated diatomic target, the open-shell O_2 molecule, for which several potential energy surfaces are present and are likely to play a role in the low-energy regime that is being sampled by the recent collisional experiments by proton beams[1]. The reason for this study stems from the apparently anomalous behaviour of the latter diatomic system when compared with the results for other, similar systems (e.g. N_2,CO,NO) since the O_2-H^+ experiments show a much larger transfer of proton energy to the molecular rotovibrational degrees of freedom than any of the other diatomics, also showing a marked preference in this case for vibrational inelasticity with respect to the excitation of rotational modes[2].

The first ionisation potential of O_2 is smaller than that of the H atom and therefore the possibility exists that after the collision between O_2 and H^+ the exit channel might contain either of the possible ionic systems, each in one of its several lower-lying electronic states and distributed over the corresponding vibrational manifold. Our detailed SCF calculations of various 'cuts' of the most important potential energy surfaces indicate the existence of both avoided and allowed curve crossings, depending on the symmetry involved, and surmise this finding as being the major cause for the different O_2 behaviour[3]. In the present study the full set of hypersurfaces is examined for both the (O_2 - H^+) and the (O_2^+ - H) systems at various relative orientations (symmetries) and as function of the vibrational coordinates. The molecular target is seen to deform over a definite range of proton-molecule distances, a range which can be reasonably sampled under the conditions of the experiments and therefore the O_2 appears to undergo what might be called a 'space localized' charge transfer process while still remaining on a specific, adiabatic surface. As a consequence of the electronic distortion the target can then undergo strong deformation and therefore show the special behaviour seen in the experiments.

Several clarifying features of various regions of the relevant hypersurfaces will be presented at the Meeting and will provide some computational evidence for the mechanism suggested above, one that is likely to be of more general occurence than previously thought[4].

References

1.T.Ellenbroek,U.Gierz,M.Noll and J.P.Toennies,J.Phys. Chem.,86,1153(1982).

2.U.Gierz,M.P.I. fuer Stromungsforschung,Bericht 9/1980, Goettingen,FRG.

3.F.A.Gianturco,U.Gierz andJ.P.Toennies,J.PhysB,14,667(1981).

4. W.R.Gentry,H.Udseth and C.F.Giese,Chem.Phys.Lett.36,671 (1975).

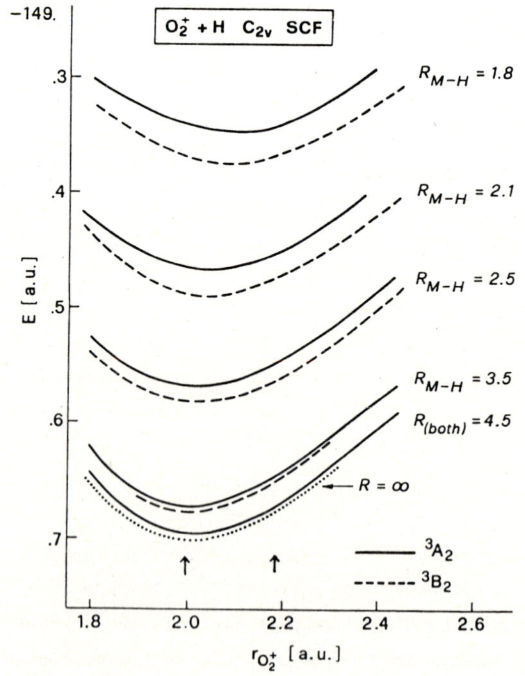

Fig.1- Perpendicular approach of neutral H to O_2^+ , on the $^3A_2$ and the $^3B_2$ surfaces, as function of the internal coordinate. The arrows indicate SCF equilibrium distances for O_2(right arrow) and for O_2^+ (left arrow).

A NEW METHOD FOR THE MEASUREMENT OF VIBRATIONAL STATE SELECTED ION-MOLECULE REACTIONS AT THERMAL ENERGIES

D. van Pijkeren, J. van Eck and A. Niehaus

Fysisch Laboratorium, Rijksuniversiteit Utrecht, the Netherlands

We have developed a new method that allows to study the dependence of ion-molecule reaction cross sections on the reactant vibrational state at thermal energies. Other existing methods[1,2] have until present been limited to collision energies above a few tenths of an eV, where the influence of vibrational energy is already partially masked by the influence of kinetic energy.

In our method the molecular ion is formed in a certain vibrational state by photoionization, and the photoelectron belonging to this event is measured in coincidence with the secondary ion created - with a certain probability - in a collision of the primary ion with some target atom or molecule present in the ionization-reaction volume. We call this method "PESICO" (photoelectron-secondary ion-coincidence)-method.

The experimental setup consists of mainly three parts. (i) For the formation of molecular ions by photoionization a dc-discharge lamp, operated with the rare gases, is used, which delivers the corresponding UV-resonance lines; (ii) For the energy analysis of the photoelectrons a cylindrical mirror analyzer of ∿1% resolution and an angle of acceptance of 2% of 4π is used; (iii) The different types of ions present in the ionization-reaction volume are mass analyzed in a time of flight spectrometer, the time zero being determined by an ion extraction pulse.

The method that allows to determine relative reaction probabilities W(v) for different vibrational states (v) of the molecular ion in thermal collisions with some molecule is as follows. An electron, detected at an energy corresponding to formation of an ion in vibrational state (v), triggers, after some delay-time during which a reaction of the ion may take place, the ion extraction pulse, and leads via the time of flight analysis to an identification of the ion. The probability that the ion is a certain reaction product is proportional to W(v) of that reaction, and the signal at the product ion mass may be written as

$$S(v) = G.F(v).W(v) \qquad (1)$$

with G some constant, and F(v) the probability to form the reactant ion in state (v). S(v) is the signal correlated to the detected electron. There is also a certain probability for an uncorrelated ion of the product ion mass to be extracted by the extraction pulse. This leads to an uncorrelated signal S'(v) which is superimposed on S(v), and which is a pure reflection of the "singles" electron spectrum:

$$S'(v) = C.F(v) \qquad (2)$$

C is another constant S'(v) is determined directly by coupling to each detected electron a second extraction pulse, which is delayed long enough to guarantee "statistical conditions" in the reaction volume. The ion signal at the product ion mass belonging to this second pulse is the uncorrelated signal S'(v). Calling $S_{tot}(v)$ the signal belonging to the first extraction pulse - $S_{tot}(v) = S(v) + S'(v)$ - the relative reaction probabilities W(v) can now be obtained as

$$(S_{tot}(v) - S'(v))/S'(v) = const.W(v) \qquad (3)$$

Fig. 1 shows the relative reaction probabilities W(v) obtained in this way for three reactions of $H_2^+(v)$,

$$H_2^+(v) + H_2(0) \rightarrow H_3^+ + H \qquad (4a)$$

$$H_2^+(v) + Ne \rightarrow NeH^+ + H \qquad (4b)$$

$$H_2^+(v) + He \rightarrow HeH^+ + H \qquad (4c)$$

whereby (v) varies from 0 to 8. Reactions (4b) and (4c) are endothermic for v < 3, and v < 4, respectively[3]. The observed increase of W(v) for v = 2 in case of (4b) is consistent with the small endothermicity of only 20 meV.

References

1. T. Kato, K. Tanaka and I. Koyano, J. Chem. Phys. 77, 337 (1982)
2. S.L. Anderson, T. Turner, B.H. Mahan and Y.T. Lee, J. Chem. Phys. 77, 1842 (1982)
3. P. Rosmus and E.A. Reinsch, Z. Naturforsch. 35A, 1066 (1980)

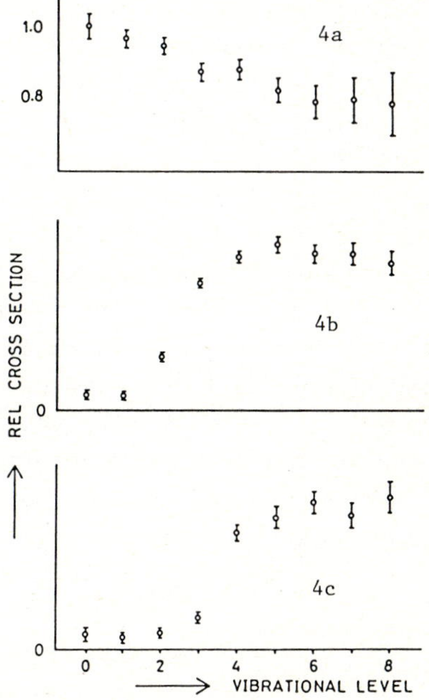

Fig. 1.

B $^2\Delta$ - STATE NH^+ (ND^+) EMISSION FROM REACTIONS OF METASTABLE N^+ IONS WITH $H_2(D_2)$: SPECTROSCOPY AND DYNAMICS.

I. Kusunoki
Research Institute for Scientific Measurements
Tohoku University, Sendai, Japan

and Ch. Ottinger
Max-Planck-Institut für Strömungsforschung
Göttingen, Germany

Chemiluminescent ion-molecule reactions of N^+ with H_2 and D_2 were studied spectroscopically in a beam/gas arrangement, at collision energies 1-9 eV_{CM}. Weak emission was observed between 3000 and 5000 Å, which was identified as the B $^2\Delta$ - X $^2\Pi$ transition of NH^+ (ND^+) ions formed in bimolecular collisions. From the observed dependence of the emission intensity on the collision energy, metastable N^+ ions, very likely in the ^1D state, can be identified as the reactants. Of the NH^+(B-X) transition only the (0,0) band at 4350 Å has previously been reported [1]. In this work 7 additional bands were observed and analyzed, and spectroscopic constants were derived. From these, RKR potential energy curves and Franck-Condon factors were calculated. Synthetic spectra were then generated by computer. Comparing these with the observed spectrum a strongly R-dependent transition moment function M(R) was derived. From M(R) and the RKR potential curves, relative band oscillator strengths were finally calculated for the 8 observed bands of NH^+(B-X) as well as, using the isotopic rules, for 10, less well resolved bands of ND^+(B-X). For the 3 eV_{CM} spectra, excited state vibrational level populations were then determined such that the calculated and the observed spectra matched. The apparatus spectral response and resolution were included in this step.

In Fig. 1 an experimental spectrum and the best-fit simulation are shown. The former includes NH(A-X) chemiluminescene previously reported by us [2] as well as the H(β) line from a secondary process. The NH^+(B) rotational level population was in the simulation derived from the widths of the bands. It is well described by a 7000 K Boltzmann distribution or a linear surprisal with Θ_R = 3.0. The vibrational distributions of NH^+(B) and ND^+(B), for 3 eV_{CM}, are shown in Fig. 2. They are compared with the distributions calculated for a linear surprisal having the parameters λ_v as given. Fig. 2 shows an unusual, unexplained dynamical isotope effect: The distribution of the higher vibrational levels is inverted for ND^+, but not for NH^+. - A rough estimate of the absolute luminescene reaction cross section gives 1 $Å^2$ (within a factor of 3), assuming a 1% metastable N^+ beam fraction.

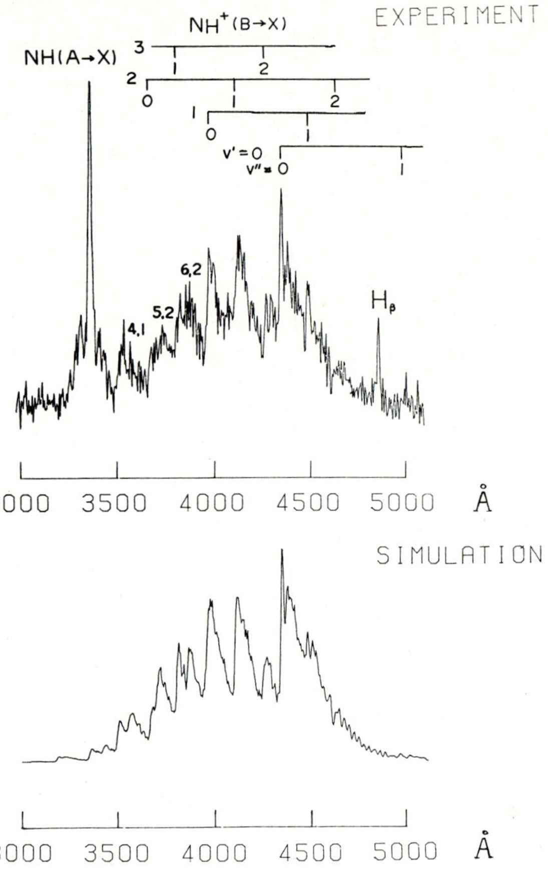

Fig.1: Spectra of chemiluminescene from ion-molecule reaction $N^+(^1D) + H_2 \longrightarrow NH^+(B\ ^2\Delta) + H$ at 3 eV_{CM}; resolution 17 Å FWHM.

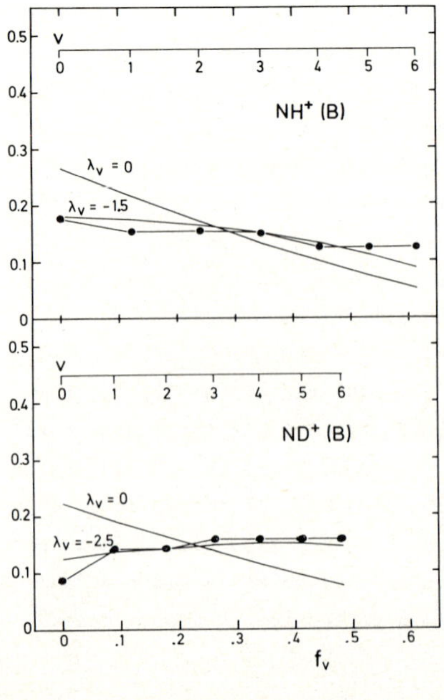

Fig.2: Vibrational product state distributions at 3eV_{CM}, as derived from chemiluminescene spectra. Error limits $\leq \pm$ 10% for v=0, increasing to $\leq \pm$ 30% for v > 3.

References

1. R. Colin and A.E. Douglas, Can.J.Phys. 46, 61 (1968)
2. I. Kusunoki and Ch. Ottinger, J.Chem.Phys. 70, 699 and 710 (1979)

COLLISIONAL RELAXATION AND REACTION OF VIBRATIONALLY EXCITED O_2^+

H. Böhringer*), M. Durup-Ferguson**), E.E. Ferguson
Aeronomy Laboratory, NOAA, Boulder, Colorado 80303

Vibrationally excited $O_2^+(X^2\pi g)$ ions created in an electron impact source and injected into a flow-tube survive more than 10^5 collisions in He or Ne buffer gas. The excited states with $v\geq1$ and $v\geq2$ are selectively detected by their enhanced charge transfer reaction rate with Xe and SO_2. Using this means to distinguish the different vibrational levels ("monitor ion technique") the rate constants, k_{q1} and k_{q2}, of quenching of $O_2^+(v=1)$ and $O_2^+(v=2)$ by various neutrals have been measured. The constants k_{q1} and k_{q2} and the corresponding numbers of collisions needed for quenching, Z_1 and Z_2, obtained at thermal energy are given in the table.

Table: Quenching rate constants, k_{q1} and k_{q2}, and numbers of collisions, Z_1 and Z_2, needed for quenching of $O_2^+(v=1)$ and $O_2^+(v=2)$, resp. α is the polarizability of the quenching neutral and $D(O_2^+-x)$ the bond-energy of the O_2^+-neutral complex.

quenching neutral	k_{q1} $\|cm^3 s^{-1}\|$	k_{q1} $\|cm^3 s^{-1}\|$	Z_1	Z_2	α $\|Å^3\|$	$D(O_2^+-x)$ $\|eV\|$
He	$\leq$2(-15)*	$\leq$6(-15)	$\geq$2.8(5)	$\geq$9.3(4)	0.205	∿0.026
Ne	$\leq$1.3(-14)	$\leq$2(-14)	$\geq$3.2(4)	$\geq$2.1(4)	0.395	∿0.1
D_2	6.5(-13)	2.6(-12)	1700	430	0.808	≲0.2
Ar	1(-12)	2.5(-12)	710	280	1.64	∿0.3
N_2	1.9(-12)	5.4(-12)	420	150	1.76	0.25
H_2	2.5(-12)	5(-12)	610	305	0.808	≲0.2
Kr	1.1(-11)	1.7(-11)	69	45	2.48	-
CO	4.4(-11)	6.5(-11)	21	14	1.95	-
CO_2	1(-10)	2(-10)	8.7	4.4	2.59	0.42
SF_6	1.1(-10)	2(-10)	8.8	4.9	4.48	-
O_2	3(-10)	4(-10)	2.5	1.9	1.60	0.42
SO_2	5.7(-10)	-	2.1	-	3.78	-
CH_4	6(-10)	-	1.9	-	2.56	-
H_2O	1.2(- 9)	-	1.2	-	1.45	>0.45

*) $a(-b)=a\cdot10^{-b}$

The results show that except for He and Ne quenching gas the rate constants for relaxation are very large compared to the corresponding values for neutral-neutral collisions (for quenching of $O_2(v=1)$ by N_2, $Z_1=1\cdot10^5$ and by CO_2, $Z_1=2.5\cdot10^4$). 2 $k_{q1}\approx k_{q2}$ for all cases but $O_2^{+*}-O_2$ collisions where $k_{q1}\approx k_{q2}$. There appears an obvious correlation of the quenching rate constant with the bond energy of the corresponding ion-quenching neutral complex, $D(O_2^+-x)$, and the polarizability, α, of the quenching neutral. For the quenching of $O_2^+(v=2)$ by CO_2, Kr and O_2 the rate constants have also been measured as a function of kinetic energy from 0.04 to 0.3 eV. For CO_2 and Kr quenching gas a decrease of the rate constant with increasing energy was observed while for quenching by O_2 the rate constant slightly increased with energy and was almost identical to the rate constant for symmetric charge-transfer of O_2^+ with O_2 in the investigated energy range.

The large values of k_{q1} and k_{q2} , the energy dependence, and the correlation with the ion-neutral complex bond energy indicates that the quenching process proceeds through an intermediate complex. The enhancement of the quenching rate constant through an intermediate complex is promoted by the strong ion-induced dipole interaction in ion-neutral collisions. The mechanisms of quenching in the cases studied here (except for He and Ne quenching gas) appears to be analogous to the well studied process of vibrational predissociation of van der Waals molecules. In both cases the relaxation process first involves the transfer of vibrational energy from the excited molecule or molecular ion into the van der Waals or ion-neutral-complex bond which is rapidly followed by the dissociation of the complexes. This interpretation is supported by the present results that show some features predicted by theoretical models for vibrational predissociation of van der Waals molecules.

An increase of the quenching rate constant with the number of vibrational quanta in $O_2^+(v)$ is observed as predicted by theory. Implications from earlier studies also show that the quenching rate constant increases with decreasing vibrational frequency of the excited molecular ion. A mechanism of deactivation other than V-T energy transfer could also be operative in the studied processes. There is a possibility of an almost resonant V-V transfer in the case of $O_2^{+*}-CO_2$ collisions and of V-R transfer in $O_2^{+*}-H_2$, D_2 collisions.

For the reaction $O_2^++CH_4$ which has been studied as a function of kinetic energy previously three products $CH_3O_2^+$, CH_3^+, and CH_4^+ have been observed. Formation of the latter two products is endothermic by ≃0.2 eV and 0.6 eV. At thermal energy the ground state and $O_2^+(v=1)$ ions react only very slowly to give $CH_3O_2^+$ even though the production of CH_3^+ is exothermic for ions with v=1. It was found, however, that the $O_2^+(v=1)$ ions are quenched by about every second collision with CH_4. The $O_2^+(v=2)$ and $O_2^+(v=3)$ ions react almost at the collision rate with CH_4 to give mainly CH_3^+ and CH_4^+ at thermal energy. Quenching of the vibrational levels v=2 and v=3 should be more efficient than quenching of v=1, but is not observed because the reactive processes dominate over the quenching.

*) Permanent address: Max-Planck-Institut für Kernphysik, Heidelberg, Germany

**) Permanent address: Laboratoire de Rêsonance Electronique et Ionique Universitê de Paris-Sud Orsay, France

Some production and destruction reactions of protonated formic acid, $HC(OH_2)_2^+$

H. Villinger, A. Saxer, E. Ferguson[a)], H. Bryant[b)], R. Richter and W. Lindinger
Institut für Experimentalphysik, Abt. Atom- und Molekülphysik, Universität Innsbruck, A 6020 Innsbruck
Austria

The ion $(CH_3O_2)^+$ has the cluster form $CH_3^+ . O_2$ when produced in the ternary association reaction[1,2] of CH_3^+ with O_2, which can be deduced from the occurance of switching reactions of this ion when colliding with CO or N_2. The dissociation energy $D(CH_3^+ - NH_3) \sim 4$ eV and the trends in the values of the rate coefficients of the ternary association reactions of CH_3^+ with various neutrals, including O_2 and NH_3, allow us to estimate the heat of formation ΔH_f $(CH_3^+ .O_2)$ to lie between 170 and 262 kcal/mole. These limits are considerably higher than the heat of formation for $(CH_3O_2)^+$ = 109 kcal/mole, obtained from appearance potential measurements by Munson and Franklin[3], who assumed this ion to be protonated formic acid. In order to investigate the possibilities of the existence of different isomeric forms of $(CH_3O_2)^+$ we performed various drift tube experiments with $(CH_3O_2)^+$ formed in two binary reactions, one being the fast proton transfer

$$COOH_2^+ + COOH_2 \rightarrow (CH_3O_2)^+ + COOH \qquad (1)$$

and the second being the extremely slow reaction

$$O_2^+ + CH_4 \rightarrow (CH_3O_2)^+ + H. \qquad (2)$$

For both cases (1) and (2) the following identical results were obtained.

Arrhenius plots of the rate coefficients obtained (in the energy regime from thermal to 1.5 eV KE_{cm}) for the breakup of $(CH_3O_2^+)$ in nonreactive collisions with Ar showed a dissociation barrier of (22 ± 6) kcal/mole and an energy independent ionic breakup pattern of $HCO^+ : H_3O^+$ = 50% : 50%. These results indicated the identity of $(CH_3O_2^+)$ produced in (1) and (2) for which additional support was obtained from isotopic exchange[4] reactions of the ions produced in reactions (1) and (2) with D_2O. In each case at near thermal energies H atoms were exchanged for D atoms in two consecutive steps, according to

$$(CH_3O_2)^+ + D_2O \rightarrow (CH_2DO_2)^+ + HDO \qquad (3a)$$

and

$$(CH_2DO_2)^+ + D_2 \rightarrow (CHD_2O_2)^+ + HDO, \qquad (4)$$

but the third H-atom was not exchanged even at high D_2O additions. This finding is consistent with the assumption of the $(CH_3O_2^+)$ ions having the structure $HC(OH)_2^+$, a structure, which is reached in reaction (2) only after considerable atomic rearrangement, explaining why the reaction proceeds so slowly.

The investigation of the energy dependence of reaction (3) shows a fast total rate coefficient k_3 = 7×10^{-10} $cm^3 sec^{-1}$ at thermal energy, decreasing to 3×10^{-10} cm^3 sec^{-1} at $KE_{cm} \simeq 0.7$ eV, followed by a sharp increase to about 7×10^{-10} cm^3 sec^{-1} at $KE_{cm} \sim 1$ eV. Besides the product channel indicated in (3a) the proton transfer channel

$$HC(OH)_2^+ + D_2O \rightarrow HD_2O^+ + HCOOH \qquad (3b)$$

is also observed at elevated energies, becoming dominant over the isotope exchange channel at energies where the total rate coefficient increases again after passing through a minimum. The rate coefficient k_5 for the proton transfer

$$HC(OH)_2^+ + H_2O \rightarrow H_3O^+ + HCOOH \qquad (5)$$

is immeasureably slow at room temperature, rising steeply to values of 2×10^{-10} cm^3 sec^{-1} at KE_{cm} = 0.7 eV and 7×10^{-10} cm^3 sec^{-1} at KE_{cm} = 1 eV, thus matching exactly the proton transfer channel in reaction (3). This increase is consistent with the differences in the proton affinities of water and formic acid. The present investigation proves the existence of a stable isomeric form $HC(OH)_2^+$ resulting from the binary reactions (1) and (2) and having a considerably lower heat of formation than $CH_3^+ . O_2$ produced in ternary association of CH_3^+ with O_2.

This work was supported by Project S-18/07.

References

1. D. Smith and N.G. Adams, Chem. Phys. Lett., 54 (1978) 535
2. D. Smith and N.G. Adams, Int. J. Mass Spectrom. Ion Phys., 23 (1977) 123
3. M.S.B. Munson and J.L. Franklin, J. Phys. Chem., 73 (1969) 4328
4. H. Villinger, R. Richter and W. Lindinger, Int. J. Mass Spectrom. Ion Phys. in press

Permanent address: a) Ae ronomy Lab., NOAA, Boulder Colo, 80303
b) Univ. New Mexico, Albuquerque NM 87131.

ION MOLECULES ASSOCIATION AT VERY LOW TEMPERATURES (20 K AND 70 K) : $N_2^+ + 2N_2 \rightarrow N_4^+ + N_2$ AND $O_2^+ + 2O_2 \rightarrow O_4^+ + O_2$

B.R. Rowe, J.B. Marquette, G. Dupeyrat

Laboratoire d'Aérothermique du C.N.R.S. 4ter, route des Gardes 92190 Meudon, France

The research in the field of ion-molecule reactions has been stimulated since it has been recognized that complex molecules can be formed by this way in dense interstellar clouds[1] ; but study of ion-molecule reactions at the extremely low temperature of interstellar clouds is a formidable challenge from the experimental point of view. A new technique is presented here which completely overcomes the problem of condensation of neutral reactants on the wall which is inherent in other experiments[2, 3].

Experimental : Our technic uses the uniform supersonic flows generated by de Laval nozzles in the SR3 rarefied wind tunnel of our laboratory. This wind tunnel has an exceptional size which allows to generate such flow at very low pressure. Actually ions are generated in the buffer flow by an electron beam : these ions react in the flow and are sampled by a mass spectrometer. The variation of the signal (reactant and product ions) versus distance from the beam yields the rate coefficient.

Two nozzles were used : one with a cryogenic cooling of the reservoir generates a flow with a Mach number of 3.9 and a 20 K temperature. The other with no cooling and a Mach number of 4 gives a 70 K temperature.

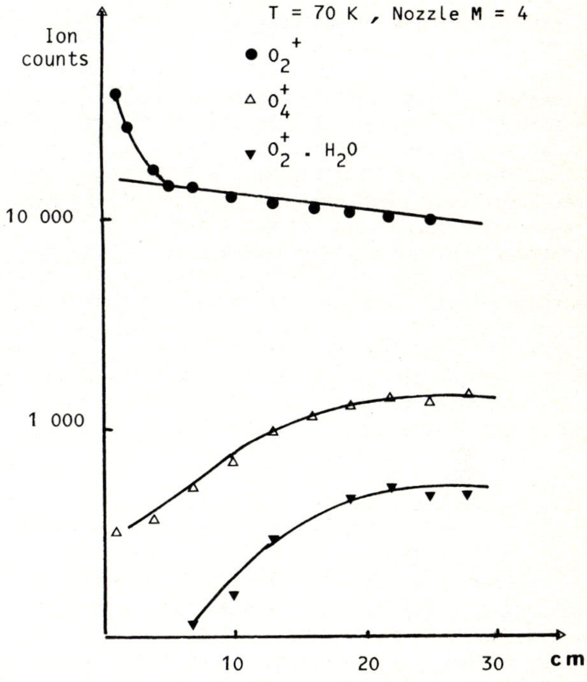

Fig.1 distance from electron beam

Results : The rate coefficient of radiative association can be linked to the rate of three-body association[4]. Two association reactions $N_2^+ + 2\,N_2 \rightarrow N_4^+ + N_2$ and $O_2^+ + 2\,O_2 \rightarrow O_4^+ + O_2$ were first studied in order to have a comparison for the 70 K results[3] ; fig. 1 shows a typical plot of ion densities versus distance to the ion source : the rate coefficient is given by the linear part of the O_2^+ decrease. On fig. 2, our results are compared with recent values by H. Böhringer and F. Arnold[3] and are in very good agreement at 70 K. The 20 K values completely confirm that the power law is no longer valid at very low temperature. These new results will certainly stimulate theoretical modeling of such reactions.

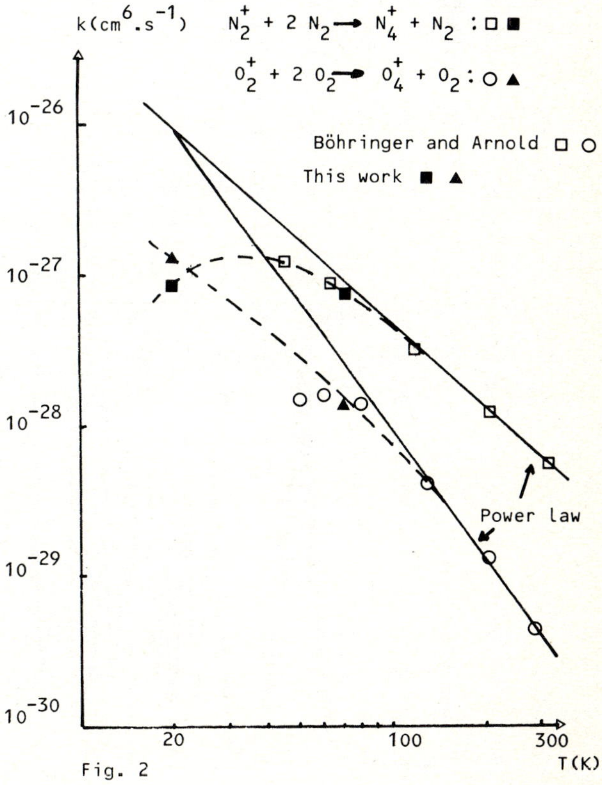

Fig. 2

Conclusions : The technics presented here has allowed the determination of reaction rate with nitrogen and oxygen at 20 K. But other condensable gases could be used and this technic has potentially the same chemical versatility than a flowing afterglow.

References :

1. E. Herbst and W. Klemperer, Ap. J., 185, 505 (1973)
2. J.A. Luine and G.H. Dunn, XII ICPEAC (1981)
3. H. Böhringer and F. Arnold, J. Ch. Ph. 77, 5534 (1982)
4. E. Herbst, Ap. J., 237, 462 (1980)

INITIALISATION OF ION REACTIONS INSIDE METHANE CLUSTERS BY ELECTRON IMPACT

A. Ding and J. Hesslich

Hahn-Meitner-Institut, Bereich Strahlenchemie, 1 Berlin 39, Germany

Reactions of atomic and molecular ions have been observed in the gas phase[1] and in the liquid state[2]. Due to recent advances of molecular beam techniques experiments are now possible which constitute a gradual transition between gas and liquid phase environments. By allowing ion molecule reactions within a molecular aggregate (cluster) of limited size it is possible to observe reactions which have not yet been observed with conventional techniques in the gas phase.

Methane clusters have been generated by supersonic expansion of a methane argon mixture from a room temperature nozzle of 30 μm diameter. Total pressures of 4 to 15 bar had to be employed in order to generate sufficiently large clusters. Room temperature expansion of pure methane did not yield any detectable amount of clusters. A time-of-flight method has been used to analyze the size distribution. A more detailed description of the experimental method has been given in a previous paper[3].

Fig. 1 shows a mass spectrum obtained in such an experiment. Besides the masses and fragments of the monomers (Ar^+ from Ar and CH_4^+, CH_3^+, CH_2^+, CH^+, C^+ from CH_4) several series of products formed by the aggregation of a molecular ion and a shell of methane molecules have been observed. This ion is either directly derived from an atomic or a molecular member of the cluster (Ar^+, CH_4^+), or it is a product from an ion molecule reaction (CH_5^+, $C_2H_4^+$, and possibly ArH^+). As the background pressure in the ionisation region is very low ($<10^{-6}$ torr), reactions of the clusters with the background gas can be ruled out. Therefore it has to be concluded that the products have been formed within the cluster, <u>after</u> the ionisation process:

$$(CH_4)_n^+ \longrightarrow CH_5^+(CH_4)_m + (n-m-1)CH_4 + CH_3 \;;m<n-1 \qquad (1)$$

$$(CH_4)_n^+ \longrightarrow C_2H_4^+(CH_4)_m + (n-m-2)CH_4 + 2H_2 \;;m<n-2 \qquad (2)$$

Process (1), whose products are known to be stable[4], corresponds to the ion molecule reaction

$$CH_4^+ + CH_4 \longrightarrow CH_5^+ + CH_3 \qquad (3)$$

while (2) is the equivalent of the reaction

$$CH_4^+ + CH_4 \longrightarrow C_2H_4^+ + 2H_2 \qquad (4)$$

Experiments have been performed to investigate whether pure methane clusters or mixed methane-argon clusters are the precursors of the reactive processes (1) and (2). When using krypton instead of argon as the carrier gas a sequence of clusters of the type $Kr_n(CH_4)_m$;$n<30, m<4$ is observed, but no reaction. We conclude from this that expansion of krypton-methane mixtures predominantly yield krypton clusters (with a few methane molecules attached to it) while argon-methane mixtures predominantly form methane clusters. This might be explained by the different strength of the van-der-Waals forces of these substances.

Gas phase reactions involving collisions of methane with CH^+ and CH_2^+, respectively, are the only ones which form $C_2H_4^+$ with a reasonable reaction rate[5]. The only observed product of the reaction of CH_4^+ with CH_4 is CH_5^+ in spite of the fact that both processes (3) and (4) are exothermic. This is probably caused by the complex dynamics of reaction (4) where 4 bonds are broken and 3 are newly formed in the process of the reaction. Statistical theories predict a strong dependence of the lifetime on the number of bonds in the molecular aggregate. Thus by increasing the cluster size the lifetime of the intermediate complex will increase by several orders of magnitude allowing process (2) to have a reasonable yield.

<u>References</u>

1) Ion-Molecule-Reactions, J.L.Franklin ed. Vol. 1,2, Butterworths, London (1972)
2) Ions and Ion Pairs in Organic Reactions, M. Szwarc ed., Vol.2, John Wiley, New York (1974)
3) A. Ding and J. Hesslich, Chem.Phys.Lett. 94,54(1983)
4) K. Hiraoka, P. Kebarle, J.Am.Chem.Soc. 97,4179(1975)
5) D. Smith and N.G. Adams, Int.J.Mass.Spectrom.Ion Phys. 23,123(1977)

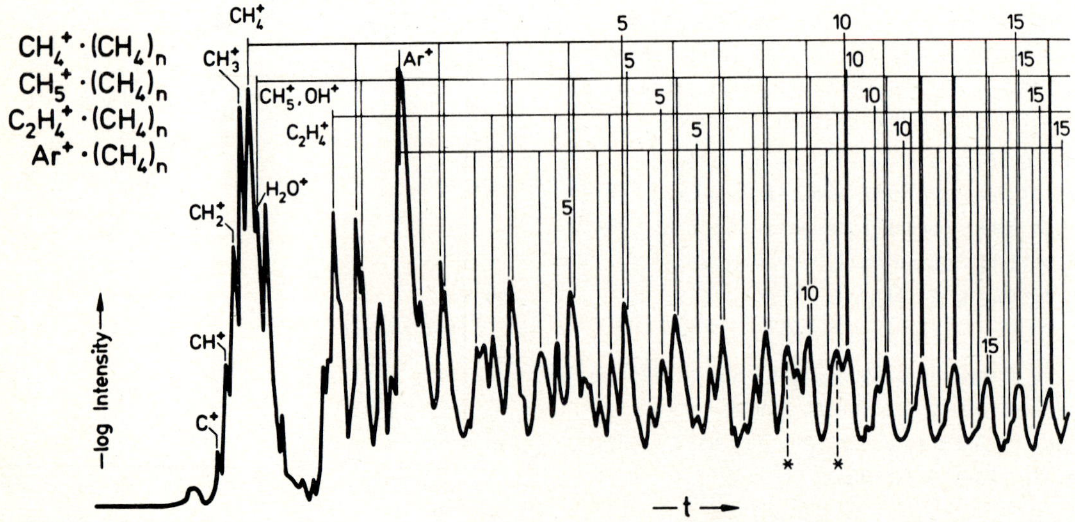

fig 1: time-of-flight mass spectrum of products formed after ionisation of methane clusters

PRODUCT STATE DISTRIBUTION OF HF IN THE REACTION F + H_2 AT ENHANCED COLLISION ENERGIES

J.Barnes, J.C.Polanyi, W.Reiland and D.Thomas

University of Toronto, Department of Chemistry, Toronto M5S 1A1, Canada

In a beam-gas experiment the effect of the reagent energy on the product vibrational and rotational state distribution of HF in the reaction F + H_2 has been studied by using the infrared chemiluminescence method [1].

The collision energy was controlled by changing the thermal energy of the supersonic H_2-beam in the range of 2 to 16 kcal/mole. The F atoms were produced in a pure CF_4 discharge. Both gas phase and surface relaxation of the HF (v',J') product molecules were eliminated by using low reagent flows and by cryopumping (45K) both the discharge and reaction products.

In addition HF molecules originating from surface reaction (remarkable population in v'=1 and 2 up to J'~10) was distinguished from that produced in gas phase collisions.

By increasing the collision energy a significant change in the relative vibrational state distribution has been observed, favouring v'=1 relative to v'=2 and 3. At the highest energies v'=4 becomes populated.

Although the mean energy going into the rotational degrees of freedom is slightly enhanced by going from low to higher energies, most of the translational energy is transferred into relative translational energy of the products.

The change in the shape of the rotational state distribution for v'=2 by going from 2 to 8 kcal/mole will be discussed along with other experimental [2] and theoretical [3] results.

References

1. D.S. Perry and J.C. Polanyi, Chem.Phys 12, (1976) 419.
2. R.K. Sparks, C.C. Hayden, K. Shobatake, D.M. Neumark and Y.T. Lee, in Horizons of Quantum Chemistry: Proc. III. Congr.of Quant.Chemistry, ed. K. Fukui and B. Pullmann (Reidel, Boston), 1980, pp.91-105.
3. M.J. Redmon and R.E. Wyatt, Chem.Phys. Lett.,63, (1979) 209.

CHARGE TRANSFER FROM RYDBERG ATOMS: M AND L DEPENDENCE OF FINAL STATE DISTRIBUTIONS

Alan D. MacKellar+ and Richard L. Becker*

+University of Kentucky, Lexington, KY 40506 USA and *Oak Ridge National Laboratory, Oak Ridge, TN 37830 USA

The results of recent experiments involving charge transfer collisions of Ar^+ and Na^+ ions with Rydberg states of Na atoms have been reported.[1] A basic assumption of the measurement technique, field ionization of the captured electron, is that the final state is in a low magnetic sublevel m'. If levels |m'| > 2 are present, the analysis becomes much more difficult because the field ionization process involves diabatic crossings.[2] From experimental[3] and theoretical[4] work there is a suggestion that, in a classical picture, the orbit of the electron contains the internuclear axis, implying that only low m' levels are present. However, a detailed analysis of final m' distributions has been lacking.

Using the classical trajectory Monte Carlo (CTMC) method we have recently studied, for several projectile velocities, total charge transfer cross sections and final n' distributions as a function of initial ℓ value.[5] For the Na(28d) target state good agreement with experiment[1] has been obtained in the velocity region $s = v/v_n \gtrsim 1$. We have extended the classical calculations to include final m' state distributions for any final n' and ℓ'. The results are given for initial n=28, ℓ=2, and s=1 by the solid curve in Fig. 1. It is seen that the distribution is rather wide, with a fwhm at |m'| = 5. We repeated the calculation with the second Euler angle describing the initial electron orientation set at $\theta=\pi/2$ and 0, which implies m=0 and |m| = ℓ, respectively. In Fig. 1 the dashed curve shows the m' distribution for m=0. It gives a fwhm at |m'| = 1.7. In the Table we give preliminary results in % of the total charge transfer cross section, as a function of |Δm| for two speeds s=1 and 1.658, and initial |m| = 0 and 2. For |Δm| = 0 only one value of m' contributes. It can be seen that a significant fraction of the total cross section comes from |Δm| > 8.

We have also calculated final charge transfer ℓ' distributions for s=1 and n=28, ℓ=2. Fig. 2 shows broad distributions peaked at ℓ'=12 (Δℓ=10) for the case in which $-2 \leq m \leq 2$ (solid curve) and at ℓ'=11 for m=0 (dashed curve).

*Sponsored by USDOE under contract W-7405-eng-76 with the Union Carbide Corp.

References

1. R.G. Rolfes and K.B. MacAdam, J. Phys. B 15, 4591 (1982). MacAdam and Rolfes, this conference.
2. T.H. Jeys et al., Phys. Rev. A 26, 335 (1982).
3. M. Breinig et al., Phys. Rev. A 25, 3015 (1982).
4. C. Bottcher, Phys. Rev. Lett. 48, 85 (1982).
5. A.D. MacKellar and R.L. Becker, Bull. Am. Phys. Soc. 27, 747 (1982).

$\sigma^{CT}(\lvert\Delta m\rvert)/\sigma^{CT}_{TOTAL}$ (%)				
|Δm|	s=1.0		s=1.658	
	m=0	|m|=2	m=0	|m|=2
0	14	3	16	5
1	20	5	18	7
2	12	4	16	7
3	9	8	10	8
4	8	9	8	6
5	6	6	5	7
6	5	4	6	6
7	4	5	4	6
8	3	4	3	4
≥ 9	19	52	14	44

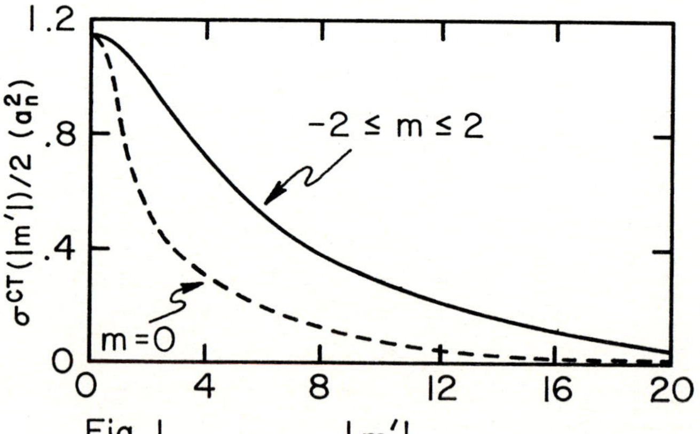

Fig. 1

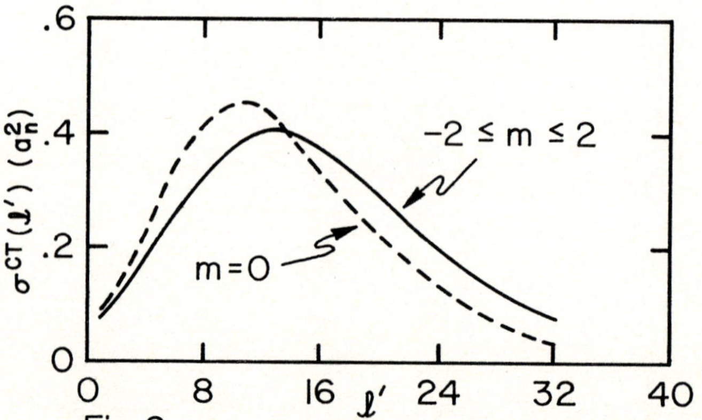

Fig. 2

CHARGE TRANSFER FROM RYDBERG ATOMS: VELOCITY DEPENDENCE OF FINAL-STATE DISTRIBUTIONS

Keith B. MacAdam and Richard G. Rolfes

University of Kentucky, Lexington, KY 40506, USA

Final-state distributions that result from charge-transfer collisions of Ar^+ and Na^+ ions with laser-excited Na(25s) atoms have been measured. Collision energies ranged from 580 to 2000 eV corresponding to projectile speeds v_I from 0.572 to 1.400 times the rms electronic orbital speed v_i in the initial state ($\tilde{v}=v_I/v_i$). Fast Rydberg atoms of the projectile species that were produced by charge transfer in the forward direction were field ionized in a dispersive stripper-analyzer,[1] which was calibrated by injection of fast Na atoms that were selectively laser-excited to Rydberg states. The energy dependence of the partial cross section for 25s→n'=34 charge transfer was measured, and all final-state distributions were normalized to the resulting values at n'=34.[2] Smoothed distributions for Ar^++Na(25s) charge transfer at several energies are presented in Fig. 1 as a function of Stark principal quantum number n_S.[3] Although the peaks are broad, Δn_S=11-19, they represent rather narrow concentrations of projectile excitation energy. The distribution at 800 eV has a fwhm of 17 meV, only 0.11% of the 15.7-eV excitation energy above the ArI ground state. Relative total charge-transfer cross sections were calculated from the areas under the curves of the final-state distribution. The results are plotted in Fig. 2 as a function of the reduced velocity $\tilde{v}$. The figure also contains the results of CTMC calculations[4-6] and experimental results for p+H(1s)→H(all n')+p.[7] The vertical scale units $a_n^2=n^4a_0^2$ allow simultaneous display of data for a variety of initial states. Our experimental results have been normalized to the Banks CTMC value at $\tilde{v}$=1. The Brinkman-Kramers approximation[8] for n→n' transfer gives a value 152 times greater than CTMC at $\tilde{v}$=1. It is shown with the same normalization as our data. CTMC calculations predict a cross section that is nearly flat for $\tilde{v}$<1, but our data show a continuing increase down to $\tilde{v}$=0.6. This work was supported in part by the NSF under Grant PHY-8205346.

References

1. K.B. MacAdam and R.G. Rolfes, Rev. Sci. Instr. 53, 592 (1982).
2. K.B. MacAdam and R.G. Rolfes (submitted to J. Phys.B)
3. R.G. Rolfes and K.B. MacAdam, J. Phys. B 15 4591 (1982).
4. D. Banks, Ph.D. Thesis, University of Stirling(1972).
5. D. Banks, K.S. Barnes, and J.M. Wilson, J. Phys. B 9, L141 (1976).
6. A.D. MacKellar and R.L. Becker, Bull. Am. Phys. Soc. 27, 747 (1982).
7. G.W. McClure, Phys. Rev. 148, 47 (1966).
8. L.J. Dubé and J.S. Briggs, J. Phys. B 14, 4595 (1981).

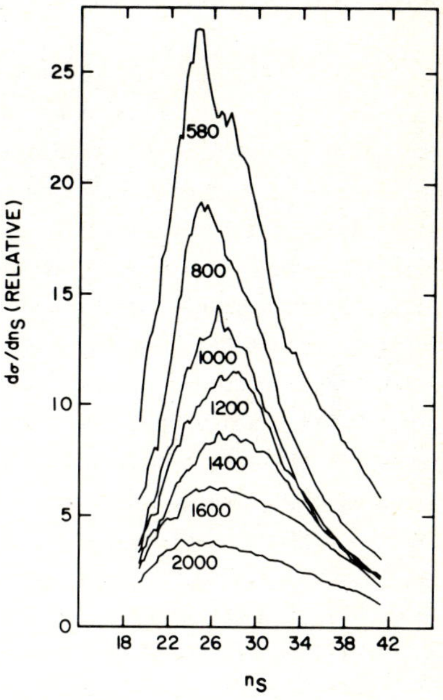

Fig. 1

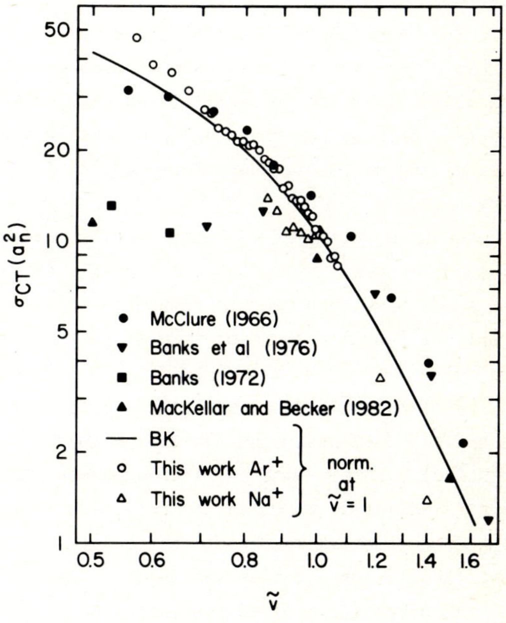

Fig. 2

PENNING IONIZATION IN COLLISIONS OF RYDBERG Rb ATOMS

M. Kimura and R. E. Olson

Physics Department, University of Missouri-Rolla, Rolla, MO 65401, U.S.A.

Recently, there has been experimental interest[1,2] in Penning Ionization (PI) collisions of two Rydberg atoms:

$$A^* + B^* \rightarrow A + B^+ + e$$
$$\rightarrow AB^+ + e$$

Theoretical interpretation and prediction of PI cross sections are meager. Hence, a primary goal of this work is to develop a theory which is applicable to PI collisions between two Rydberg atoms and interpret the results of experimental measurements.[1,2]

Using the classical version of Miller's PI theory,[3] we have proposed a simplified form for the energy width $\Gamma(R)$, which represents the decay from the discrete state into a continuum state.[4] The form of $\Gamma(R)$ is simply given by the product of the one-electron capture matrix element and the bound-free overlap matrix element. For the one-electron capture matrix element, we have adopted the semiempirical form presented by Olson, Smith and Bauer.[5] The bound-free overlap matrix element was evaluated using hydrogenic wavefunctions with quantum defect corrections for the bound state and a Coulomb wavefunction for the continuum state.

The magnitude of the bound-free overlap matrix increases as the principal quantum number n describing the bound state decreases at fixed ejected electron energy. This bound-free overlap matrix is the controlling factor to the determination of PI cross section.

We have performed PI cross section calculations on some Rb^{**}-Rb^{**} systems, since several of these processes have been measured experimentally.[2] Our results along with the experimental data are shown in Table 1.

Qualitatively, the results of the experiment can be explained by using the discussion above. Since the principal quantum number of the bound state decreases (i.e., ionization potential becomes larger), the bound-free overlap matrix element becomes larger so that the PI cross section increases.

We have also applied the theory to PI of the Na^{**}-Na^{**} system. The details of the results will be discussed.

Table 1. PI Cross Section for $Rb^{**}(5p) + Rb^{**}(n) \rightarrow Rb + Rb^+ + e^-$ (where n=6d, 7d, 8s)

	Cross Section ($x10^{-13}cm^2$)	
Initial State	Experimental[2]	Theory
6d	8.4 ± 1.8	6.05
7d	4.4 ± 1.8	5.01
8s	4.4 ± 1.4	5.12

Acknowledgement

Work supported by the Office of Naval Research.

References

1. V. S. Kushawaha and J. J. Leventhal, Phys. Rev. A 25, 349 (1982).
2. M. Cheret, L. Barbier, W. Lindinger and R. DeLoche, (to be published).
3. W. H. Miller, J. Chem. Phys. 52, 3560 (1970).
4. M. Kimura and R. E. Olson, (to be published).
5. R. E. Olson, F. T. Smith and E. Bauer, Appl. Opt. 10, 1848 (1971).

KINEMATICS OF COLLISIONAL IONIZATION IN Xe(nf)-SF_6 INTERACTIONS†

C. Higgs, M. P. Slusher, K. A. Smith, F. B. Dunning, and R. F. Stebbings

Department of Physics and Department of Space Physics and Astronomy
and the Rice Quantum Institute, Rice University, Houston, TX 77251

Negative ions formed in reactions of the type

$$X^{**} + SF_6 \rightarrow X^+ + SF_6^-$$

show long lifetimes against autodetachment.[1] This has lead to speculation that these negative ions may be stabilized by a post-attachment interaction with the Rydberg atom ion core.

To investigate this possibility we are studying in detail the kinematics of Xe(nf)-SF_6 collisions using the apparatus shown in fig. 1. Xe($^3P_0$) metastable atoms in a collimated, thermal energy beam are laser-excited in zero electric field to nf states in the presence of SF_6. After a known interaction time (~2-10 μsec), the collisionally-produced Xe^+ ions are directed out of the interaction region by a small pulsed electric field of ~2 μsec duration. The Xe^+ ions then drift through a field-free region prior to detection by a position sensitive detector (PSD) that measures both the time and position of their arrival. These measurements, and the known apparatus geometry and extraction pulse characteristics, provide information on the kinematics of the interaction.

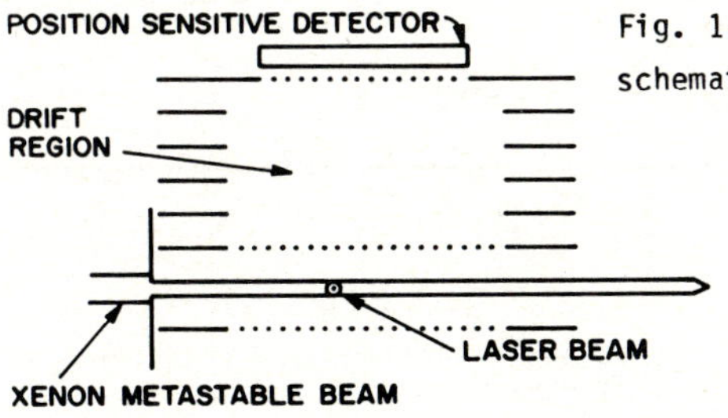

Fig. 1. Apparatus schematic.

Figure 2a shows Xe^+ arrival-time spectra obtained with Xe(17f) and Xe(60f). Figure 2b shows arrival-time spectra calculated, for particular assumed collisional behaviors, using a Monte-Carlo procedure. The widths of the measured arrival-time distributions suggest that electron transfer is not accompanied by significant energy transfer to the core ion. The Xe(17f) data indicate that collisions lead to isotropic elastic scattering of the core ion, whereas the narrower width of the Xe(60f) data points to predominantly forward scattering. Similar conclusions are reached by analysis of the spatial distribution of the Xe^+ ions at the PSD.

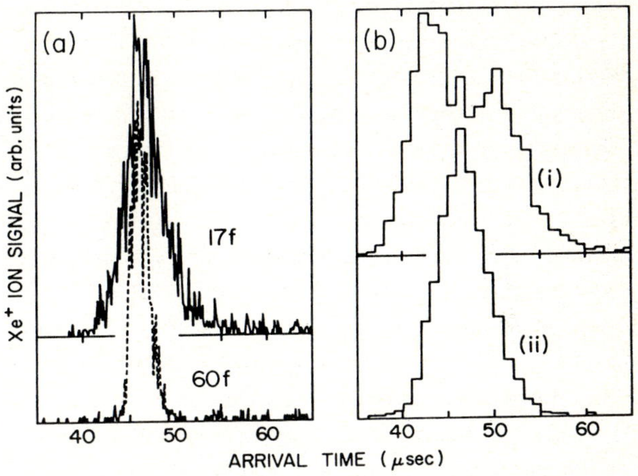

Fig. 2. a. Arrival-time spectra for Xe(17f) and Xe(60f). b. Arrival-time spectra calculated assuming isotropic Xe^+ scattering and (i) 50 meV energy transfer, (ii) no energy transfer.

The preliminary data indicate that few electron transfer reactions lead to significant energy transfer to the core ion, implying that the observed SF_6^- ion stability is a consequence of the very low initial translational energy of the Rydberg electron,[2] rather than a post-attachment interaction. The greater Xe^+ scattering at low n is expected since electron transfer then occurs at relatively small impact parameters where the Coulomb force between the product ions is larger.

References

1. C. E. Klots, J. Chem. Phys., 66, 5420 (1977); G. W. Foltz, et al., J. Chem. Phys., 67, 1352 (1977); J. P. Astruc et al., J. Phys. B, 12, L377 (1978); I. Dimicoli and R. Botter, J. Chem. Phys., 74, 2355 (1981).
2. L. G. Christophorou, Adv. in Electron. and Electron Phys., 46, 55 (1978).

† Research supported by the NSF under grant PHY 81-08452 and the Robert A. Welch Foundation.

EXCITATION AND IONIZATION OF HIGH RYDBERG ATOMS IN COLLISION WITH POLAR MOLECULES--CLASSICAL CALCULATIONS

S. C. Preston and N. F. Lane

Physics Department, Rice University, Houston, TX 77251

Full three-dimensional (3D) classical Monte-Carlo calculations have been carried out of cross sections for the Δn, $\Delta \ell$, ionization, and ΔJ processes

$$R(n\ ) + M(J) \rightarrow R(n'\ell') + M(J')$$
$$\rightarrow R^{+} + e + M(J')$$

where R, M refer to the atom and molecule, respectively and where the orientations of the atomic and molecular angular momenta are averaged prior to the collision. The 3D calculation is similar to the 2D approach described previously,[1] and many qualitative features of the results are the same. Hamilton's equations of motion are solved, subject to randomly chosen initial values of the coordinates and impact parameter. The molecule is represented as a rigid rotor, and the electron-molecule interaction is approximated by a cut-off dipole potential of the form $-[\vec{d}\cdot\vec{r}/r^3]C(r)$, where $\vec{r}$ locates the electron relative to the molecule, d is the molecular dipole moment, and C(r) smoothly forces the potential to zero for $r \ll r_c$. All our results are insensitive to choices of cut-off radius in the range $0.5 \lesssim r_c \lesssim 4.0\ a_0$. Moreover, for all the cases described here, we have found that the electron-molecule interaction can be ignored outside an interaction "bubble" of radius $r_b \sim 200\ a_0$. The trajectory of molecule, relative to the fixed atomic nucleus is taken to be a straight-line. We have shown that for all cases considered here, recoil of the molecule and of the atomic nucleus are unimportant. The atomic states are taken to be hydrogenic. A simple "binning" prescription is used in assigning final state n, ℓ, and J quantum numbers, according to the proximity of the final electron energy, angular momentum and molecular energy.

In order to illustrate the nature of our results, we show in Fig. 1a ΔJ cross sections for thermal-velocity collisions of Xe(25ℓ) and HF(J=2) for several values of ℓ from 0 to 24. The rates calculated from these cross sections are approximately a factor of 1/3 smaller than experiment.[2] At the present time we have no explanation for the disagreement. In Fig. 1b we compare $J \rightarrow J-1$ cross sections for collisions of Xe(25s) and HF(J) with and without the Coulomb filed of the atomic nucleus in order to illustrate its influence on energy transfer between the atom and molecule; the effect is seen to be small. By contrast, we find the ionization cross section for Xe^{**} collisions with HF to be grossly in error when the Coulomb field of the atomic nucleus is turned off. These and other results will be further described in the presentation.

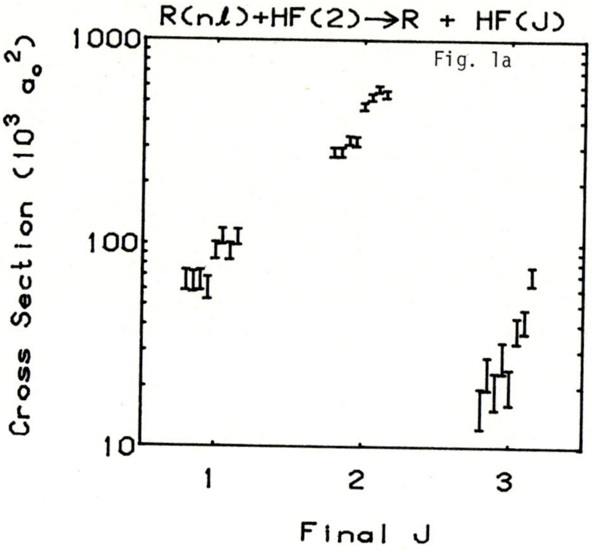

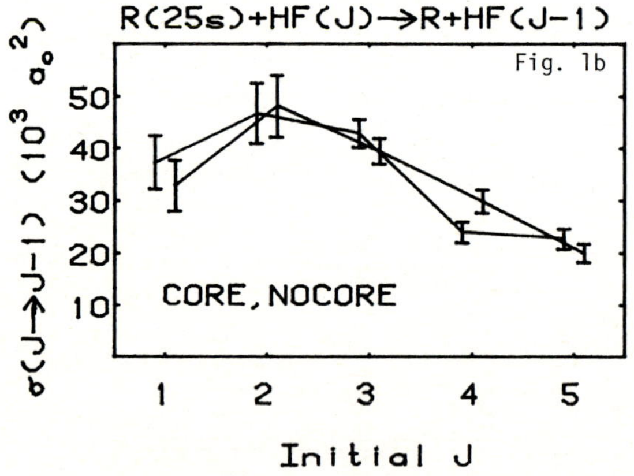

Fig. 1: J-changing cross sections for collisions of Xe(n=25) and HF
(a) J=2→J' transitions for n=25, ℓ=0,1,2,3,10,15,20,24 (data offset for clarity order left to right)
(b) J→J-1 transitions for n=25, =0 Coulomb "CORE" and "NO CORE" (order left to right).

Supported in part by the U.S. Department of Energy, Office of Basic Energy Sciences, and by the Robert A. Welch Foundation.

References

1. S. C. Preston, N. F. Lane, and G. J. Hatton, XII ICPEAC Abstracts, p. 1124.

2. see review by F. B. Dunning and R. F. Stebbings in Ann. Rev. Phys. Chem. 33, 173 (1982); C. Higgs, K. A. Smith, G. B. McMillian, F. B. Dunning and R. F. Stebbings, J. Phys. B 14, L285 (1981).

DETECTION OF HIGH-RYDBERG ATOMS AND MOLECULES BY CO_2- LASER RADIATION

A.A.Perov, A.Yu.Zayats, A.N.Stepanov, A.P.Simonov

Karpov Institute of Physical Chemistry, Ul.Obukha IO, IO7I2O Moscow, USSR

Among various methods used for investigation of the highly-excited atom and molecule properties[I] the ionization of such particles by IR radiation is comparatively novel one. In the present work, we report on the photoionization of the Rydberg atoms and molecules by CO_2 laser radiation with a photon energy of 0.II7 eV (IO.6 μm band) enough to ionize excited states with quantum numbers $n \geqslant II$.

The experiments have been carried out using a mass-spectrometer with a two-chamber ion source[2]. In the first chamber atoms and molecules were excited by electron impact. Between the chambers there was a parallel plate I5 x 20 mm^2 analyser. Electric field between the plates which could be varied from 0 up to 8 kV/cm was used to analyse the Rydberg states of different n, as well as to remove the charged particles. The second chamber was placed 35 mm from the first one. It was possible to irradiate the particles in the second chamber by a cw CO_2 laser beam transverse to the mass-spectrometer axis. Highly-excited atoms and molecules may also ionize near metallic surfaces of the collimating apertures or by collisions with the parent gas particles. Therefore when necessary to discriminate the signal due to the laser irradiation against total signal the laser beam was modulated by a chopper and the ion current was detected by a phase-lock amplifier.

The photoionization has been studied for Rydberg states of the following species: (i) rare gas atoms, (ii) N,C,O,H and F atoms formed via dissociative excitation of N_2,CO,NH_3 and NF_3 molecules, (iii) H_2,CO,N_2 and some other di- and triatomic molecules. It follows from the excitation functions (Fig.I) that the ion current increment due to laser irradiation decreases with increasing E_e. Some results are presented in Fig.I. For example, in case of Ar** $I_1/I=3.2$ at $E_e=25$ eV and $I_1/I=2.9$ at $E_e=50$ eV (here I_1 and I are the ion current with the laser on and off, respectively). This effect is much more pronounced for molecules (except for H_2): at E_e=I5eV, i.e. near the threshold in the region of the sharp maximum of the excitation

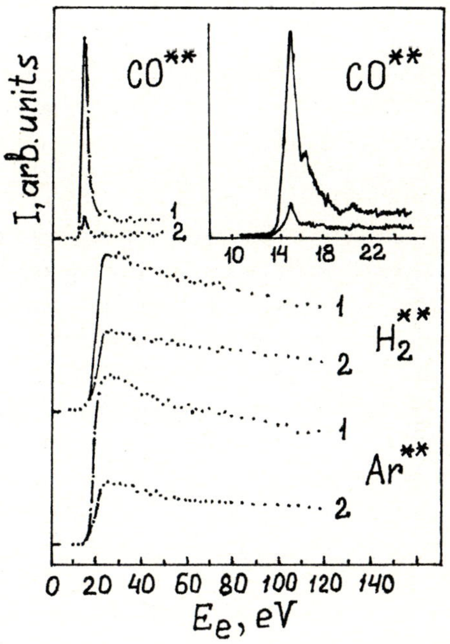

Fig.I. Excitation functions with (I) and without (2) CO_2 - laser irradiation.

function[3] $I_I/I=8$ - IO but at E_e=25 eV $I_1/I=3$.

Such behaviour may be due to increasing of the photoionization cross section with the angular momentum of excited electron. High-Rydberg states with large angular momenta are excited by electrons with the energy near the excitation threshold[4].

Using the Kramers formula[5] for the photoionization cross section we can found the n-distribution of Rydberg states. This distribution is the first derivative of the photoionization current dependence on the analysing electric field. Such distributions were established for a number of rare gas atoms and atomic fragments N,C,O,F from CO,N_2, and NF_3 molecules for $I6 \leqslant n \leqslant 4O$.

References

I.B.M.Smirnov,Uspekhi Fiz.Nauk.I3I,577 (I980).

2.A.Yu.Zayats,A.A.Perov,A.P.Simonov,Khim.Fiz. No.3,333 (I983).

3.S.E.Kupriyanov,A.A.Perov,A.Yu.Zayats,A.N. Stepanov,Pisma v Zh.Tekh.Fiz.7,No.I4,86I, (I98I).

4.U.Fano,J.Phys.B.7,L.4OI (I974).

5.H.A.Kramers,Philos.Mag.46,836 (I923).

MODEL POTENTIAL CALCULATIONS FOR THE GROUND AND EXCITED (RYDBERG) STATES OF THE Li_2^+, Na_2^+, K_2^+ IONS : CORE POLARIZATION EFFECTS

A. Henriet and F. Masnou-Seeuws

Laboratoire des Collisions atomiques et moléculaires, 91405 ORSAY CEDEX FRANCE

The model potential method of Valiron et al[1] is used to compute the molecular states of the alkali dimer cations from the ground state up to the states correlated to Li(5p)+Li^+, Na(5p)+Na^+, K(6s)+K^+. The one electron-two cores Schrödinger equation :

$$\left(T + V(r_A)+V(r_B)+V_3(\vec{R},\vec{r_A},\vec{r_B})+ \frac{1}{R} - \frac{\alpha_d}{R^4}\right) \Psi = E\Psi \qquad (1)$$

is solved using a basis of generalized Slater orbitals in prolate spheroïdal coordinates. In (1), r_A and r_B are the distances of the electron to the two cores A and B respectively, R is the internuclear distance, α_d the dipole polarizability of the cores. V(r) is a central model potential, attractive in the core region, and fitted by M. Klapisch[2] so as to reproduce the experimental spectrum of the alkali atom. V_3 is a tensorial cross polarization term[3] which is represented by the effective operator

$$\frac{\alpha_d}{R^3}\left(\frac{\vec{r_A}.\vec{R}}{r_A^3}\chi(\frac{r_A}{\rho}) + \frac{\vec{r_B}.\vec{R}}{r_B^3}\chi(\frac{r_B}{\rho}\right) \qquad (2)$$

The function χ is an arbitrary cut off function ; we chose :

$$\chi(u) = \left(1-\exp(-u)^6\right)^{1/2}$$

The molecular results are thus sensitive to the choice of the cut-off radius ρ, especially for the ground state. We have shown that ρ cannot easily be deduced from the polarization part of a potential V(r) fitted to the experimental atomic energies. In contrast, the photoionization cross-sections of the Na and K atoms strongly depend upon the choice of the cut-off radius in the core polarization term[4,5]. We present on the table results for the depth D_e and position R_e of the well of the ground state obtained by solving equation (1)

(i) by neglecting V_3 ($\rho=\infty$)

(ii) by choosing ρ values compatible with the photoionization cross sections of Weisheit and Dalgarno[5]

Li_2^+ $X^2\Sigma$	De(eV)	Re(Å)		De(eV)	Re(Å)
			Exp[6]	1.27 + 0.02	
this work (i)	1.30	3.08	(ii)ρ=1u.a.	1.28	3.09
Ref 7 (i)	1.32	3.06	(ii)ρ=1.10	1.30	3.10
Na_2^+ $X^2\Sigma$					
			Exp[7]	0.986	3.60+0.05
this work (i)	1.02	3.55	(ii)ρ=2.35	0.980	3.60
Ref 7 (i)	1.02	3.56	(ii)ρ=1.27	0.993	3.59
K_2^+ $X^2\Sigma$					
experiment[10]	0.85+0.1		Exp[9]	0.794	4.4
this work (ii) ρ= 4.22	0.813	4.44	(ii)ρ=3.4	0.800	
Ref 7 (i)	0.91	4.31	(ii)ρ=1.86	0.823	4.49

It is clear for Li_2^+ and Na_2^+ that the results obtained neglecting V_3 are outside the experimental error bar. In contrast, our calculations with the above mentionned choice for ρ agree very well with experiment. For K_2^+, the experimental determination is less certain, but could indicate that the cut-off radius 4.22 of Weisheit is too large. Our results are compared with recent pseudo potential calculations including core polarization effects[7].

The potential curves for excited states present interesting structures and avoided crossings. We have represented on the figure in the Na_2^+ case the variations of $n_{eff}=\sqrt{2}(E-\frac{1}{R}+\frac{\alpha_d}{R^4})^{-1/2}$ as a function of the internuclear distance.

References

1. Valiron et al. J.Phys.B 12 53 (1979)
2. Klapisch Comput. Phys. Commun 2 239 (1971)
3. Bottcher and Dalgarno Proc.Roy.Soc. A340 187 (1974)
4. Caves and Dalgarno J. Quant.Spectros.Radiat.Transfer 12 1539 (1972)
5. Weisheit Phys.Rev. A5 1621 (1972)
6. Mathur et al. Chem.Phys.Letters 56 336 (1978)
7. Fuentealba et al. Chem.Phys.Letters 89 418 (1982)
8. Martin S. et al. Chem.Phys.Letters 87 235 (1982)
9. Leutwyler et al. Chem.Phys. 48 253 (1980)
10. Foster, Leckenly, Robbins J.Phys.B 2 478 (1969)

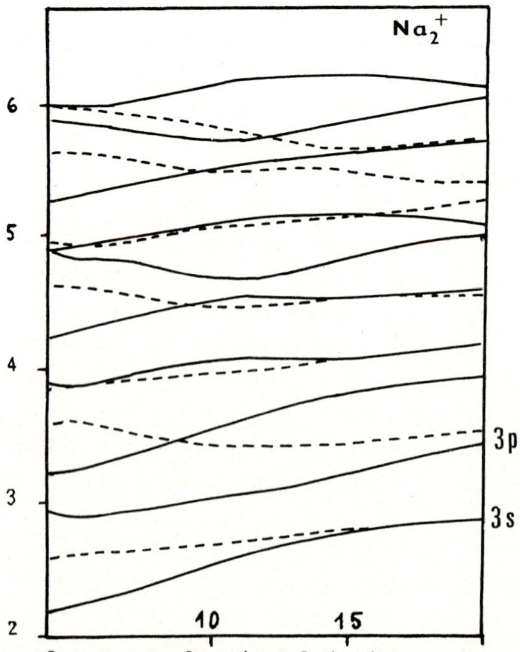

Variation of n_{eff} as a function of the internuclear distance R for the $^2\Sigma_g$ (full line) and $^2\Sigma_u$ (dash line) states of Na_2^+.

Electric Field-Ionization of Rydberg States of Fast Foil-Excited Sulphur Ions

E. P. Kanter, D. Schneider, and Z. Vager*
Physics Division, Argonne National Laboratory, Argonne, IL 60439

The absolute yield of beam-foil-excited Rydberg states is measured, using field-ionization techniques, for 125-MeV sulphur ions. The results are compared with yields inferred from recent measurements of delayed K x-rays.

The basic technique used here has been described in a previous publication.[1] The 125-MeV S^{14+} ions were obtained from the Argonne tandem-linac accelerator, the carbon target thicknesses were 5- and 10-μg/cm^2. A small positive bias voltage (typically ~+200 V) was applied to the target.

A typical electron energy spectrum, at 0° observation angle, is shown in Fig. 1.

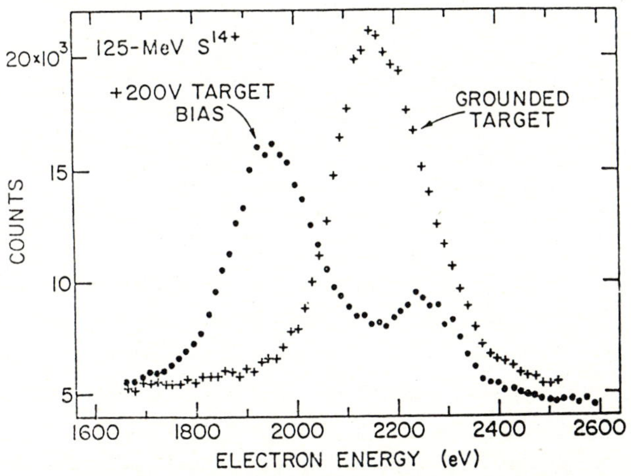

Fig. 1. Electron energy spectra at 0° observation angle for 125-MeV S^{14+} ions incident upon a 5-μg/cm^2 carbon foil. The spectrum with +200 volts bias on target reveals the presence of high-Rydberg atoms which are field-ionized in the spectrometer. The resulting electrons form the peak which does not shift with target bias.

The spectrum shows a peak due to convoy electrons.[2] At an apparent energy of slightly higher than that corresponding to beam-velocity electrons one observes a peak, which has been identified previously.[1] This peak results from field ionization of high Rydberg atoms in the field of the electron spectrometer. Assuming the classical ionization threshold:

$$F_c(\text{V/cm}) \approx \frac{10^9 Z^3}{4 n^4}$$

, then for a core charge of 15^+, we expect atoms with principal quantum numbers in the range $250 \lesssim n \lesssim 650$ to contribute to this peak. The absolute yield per incident ion of Rydberg atoms was determined. We find a yield of about $(1.04 \pm 0.02) \times 10^{-5}$ and $(1.17 \pm 0.02) \times 10^{-5}$ atoms/ion for 5- and 10-μg/cm^2 carbon foils, respectively. The errors quoted are statistical.

If one assumes an n^{-3} distribution of final states, then our data, for the 5-μg/cm^2 target, imply a distribution $P(n) = (1.52 \pm 0.03)n^{-3}$. To compare this yield to the recently published[3] delayed Lyman-α yield requires an estimate of the angular momenta of the Rydberg states formed. Cascade calculations give values of $P(n) = (30 \pm 15)\, n^{-3}$ for a similar beam energy and target thickness. It is this large calculated yield which is the basis of the argument which has been used against last-layer capture.[4] In order to explain these large yields, and the surprising target thickness dependence which they observe, Betz et al. have proposed a multi-step capture process.[4] This is a factor of 10-20 greater than the direct ionization results we find. This discrepancy is an indication of the inadequacy of such cascade calculations, particularly the n, ℓ dependence of the assumed initial population.

This work was supported by the U.S. Department of Energy, Office of Basic Energy Sciences, under Contract No. W-31-109-Eng-38.

---

*Weizmann Institute of Science, Rehovot, Israel.

References

1 Z. Vager, B. J. Zabransky, D. Schneider, E. P. Kanter, Gu Yuan Zhuang and D. S. Gemmell, Phys. Rev. Lett. 48 (1982) 592.

2 For a recent review, see M. Breinig et al., Phys. Rev. A 25 (1982) 3015.

3 J. Rothermel, H.-D. Betz, F. Bell, and V. Zacek, Nucl. Inst. and Meth. 194 (1982) 341.

4 H.-D. Betz, D. Röschenthaler, and J. Rothermel, Phys. Rev. Lett. 50 (1983) 34.

COLLISIONAL ANGULAR-MOMENTUM MIXING OF Na RYDBERG STATES BY A K^+ ION BEAM *

W. W. Smith+, P. Pillet++, R. Kachru, N. H. Tran and T. F. Gallagher

Molecular Physics Laboratory, SRI International, Menlo Park, California 94025

We have studied the n-dependence of the loss cross sections from Na nd (n=16-27) Rydberg states due to collisional angular-momentum mixing by 300 and 630 eV K^+ ions in a crossed-beam experiment. Final-state analysis was performed by selective field ionization (SFI) so that information about the n,l,m distribution resulting from the collisions could be inferred. The cross sections scale approximately as n^5 down to n=18, where this scaling appears to break down. Estimated absolute cross sections are consistent with previous work by MacAdam et al[1] and are three to four orders of magnitude larger than those previously observed in thermal-energy collisions with neutral Xe[2,3]. Our apparatus is similar to that of Ref. 2, to which we have added a simple thermionic alkali ion source and an ion-optical system. Yellow and blue pulsed dye lasers, linearly polarized, excite the Na atoms in two steps to the nd(J=3/2 and 5/2) states. The resulting Rydberg atoms are then detected by using a gated integrator with a variable time delay during which collisions with the K^+ ions can occur. The collisionally-depleted population of the nd states is monitored as a function of the product (It) of ion-beam current with time delay and the nd population is found to decay exponentially as a function of (It). The loss cross section is found from the decay constant.

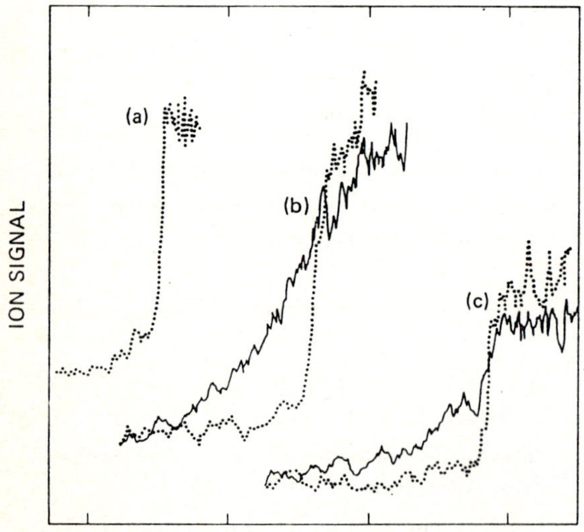

Figure 1. Plot of signal from field ionization of (a) Na 23d, (b) 22d and (c) 21d Rydberg states. Dotted curves: K^+ ion beam off; solid curves: ions on. Horizontal axis gives relative ionizing field.

Figure 1 shows a composite plot of the ion signal as a function of the field-ionization pulse amplitude. Models for final-state analysis by SFI discussed in Refs. 2,3 indicate that the sharp thresholds in the "ions-off" curves of Figure 1 correspond to ionization of l=2, m_l=0, with higher thresholds corresponding to l=2, m_l= 1 and 2. l-mixing collisions corresponding to "low-m" (m_l=0,1, l > 2) lower the SFI threshold below that for the m=0 d states, as shown in the solid curves (b) and (c). There is population from nd excitation extending all the way down to the (n+1) threshold; we interpret this as evidence for a fairly uniform distribution among all the allowed low m, l > 2 final states of the initial n, produced by the collision with the ion. The increase in the SFI Na^+ signal above the initial sharp m=0 thresholds implies probable collisional excitation of "high m" states, as seen previously[1].

The n^5 scaling may be interpreted in terms of the maximum electric field amplitude seen by the Rydberg atom during the collision with a slow incoming Coulomb projectile, for a straight line trajectory at large impact parameter b. If the maximum electric field exceeds that required to merge the nd state with the higher n,l states of the Stark manifold, then l-mixing to low-m states may be expected for a suitable range of collision times. Since Stark-crossing and merging fields scale as n^{-5} and are proportional to $1/b^2$, the l-mixing cross section ($\propto b^2$) should go inversely to this or like n^5.

We find that application of a small "DC" electric field to the interaction region during the collisions, sufficient to merge the nd, m=0 Stark states with the higher-l Stark states of the same n, substantially reduces the effects of collisional redistribution into low m, high l states.

*Research supported by ONR and NSF.

+On sabbatic leave, Fall 1982 from Physics Department, The University of Connecticut, Storrs, CT 06268.

++ Permanent address: Laboratoire Aime' Cotton, C.N.R.S. II, Batiment 505, 91045 Orsay Cedex, France.

References:

1. K. B. MacAdam, R. Rolfes and D.A. Crosby, Phys. Rev. A 24, 1286 (1981).
2. R. Kachru, T. F. Gallagher, F. Gounand, K. A. Safinya and W. Sandner, Phys. Rev. A 27, 795 (1983).
3. See also F. G. Kellert, C. Higgs, K. A. Smith, G. F. Hildebrandt, F. B. Dunning and R. F. Stebbings, J. Chem. Phys. 72, 6312 (1980);and F. G. Kellert et al, Phys. Rev. A 23, 1127 (1981).

COLLISIONAL TRANSFERS IN SODIUM RYDBERG STATES

A.L. ROCHE*, P. VALIRON** and F. MASNOU-SEEUWS***

* Laboratoire de Photophysique Moléculaire+, Bât. 213
Université de Paris-Sud, 91405 - ORSAY Cedex (France)

** Groupe d'Astrophysique, CERMO B.P. 68
38402 - SAINT-MARTIN-D'HERES Cedex (France)

*** Laboratoire des Collisions Atomiques et Moléculaires, Bât. 351
Université de Paris-Sud, 91405 - ORSAY Cedex (France).

Many experiments concerning thermal collisions between a Rydberg alkali A and a rare gas B have been performed, which consist in measuring the quenching cross section of a nd level of A[1]. It is assumed that the population transfer is due to the "ℓ-mixing" phenomenon :

$$A_{n\ell} + B \rightarrow A_{n\ell'} + B \qquad \ell',\ell \geqslant 2$$

and that the $A_{n\ell}$, $\ell \geqslant 2$ manifold can be considered as isolated during the collision.

Molecular calculations for the Rydberg states of the NaNe molecule have been performed by Dolan and Masnou-Seeuws [2] who extended the model potential method of Valiron *et al* [3]. They found significant departures from the Fermi model of a free electron scattered by the rare gas. In particular, they showed that the $\sigma(n+1)$p states are subject to avoided crossings with molecular levels correlated to $n\ell$, $\ell \geqslant 2$, so that it is possible to predict the collisional transfer

$$Na_{(n+1)p} + B \rightarrow Na_{n\ell'} + B \qquad \ell' \geqslant 2$$

An experiment has been performed [4] which measured the selective population transfer between Na6p and Na5d atomic levels due to collisions with He, Ne, Ar perturbers. The interpretation was performed in the framework of a two step collision model :

(i) at small internuclear distances, in a narrow region around the crossing distance $R = R_x = 6.5$ a.u., the population is transferred from the σ6p molecular level to a σ level correlated to the Na 5ℓ, $\ell \geqslant 2$ manifold, and hereafter called σ5x. A two state Landau-Zener model was used to treat the collision problem.

(ii) at large internuclear distances, the "ℓ-mixing" phenomenon is assumed to share the population between the levels of the Na 5ℓ, $\ell \geqslant 2$ manifold.

As the experimental cross-sections were only qualitatively reproduced in such a model, we present here a more elaborate theoretical treatment considering the R dependance of the exact radial coupling between the Rydberg states of the NaHe and NaNe molecules.

(i) at small internuclear distances, we could show that a two state model is justified, and that no important departure from the Landau-Zener model is obtained by considering a full quantal treatment of the collision with exact radial coupling. The experimental cross-section for Na-Ne collisions is then well reproduced.

(ii) however, in the NaHe case (see the figure), other avoided crossings occur at large internuclear distances between the molecular σ6p and σ5x states, with radial coupling leading to important collisional transfers. It can be seen from table I that such crossings increase very much the cross section.

	NaHe	NaNe	
(a)	26	9.7	experimental
(b)	6	7	two state model
(c)	5	8	two state model
(d)	17		two state model

Table I. Comparison between the experimental and calculated cross sections (Å^2) for collisional transfer from 6p to 5d, 5f, 5g group of levels, (b) Landau-Zener model, (c) quantal calculations with exact radial coupling at $R = R_x$, (d) introduction of crossings at large internuclear distances in the quantal treatment.

It is then no longer possible to consider a two state model at large internuclear distances. An analysis is in progress which will be presented at the conference.

References
1. T.F. Gallagher, Electronic and Atomic Collisions, p. 473 (North Holland, 1980)
2. P. Valiron, R. Gayet, R. Mc Carroll, F. Masnou-Seeuws and M. Philippe, J. Phys. B 12, 53 (1979)
3. M.E. Dolan and F. Masnou-Seeuws, J. Phys. B, 14 L. 583-9 (1981)
4. F. Masnou-Seeuws, J. Boulmer, T. Maurin, A.L. Roche, and P. Valiron, J. Phys. B 15, 2989 (1982).

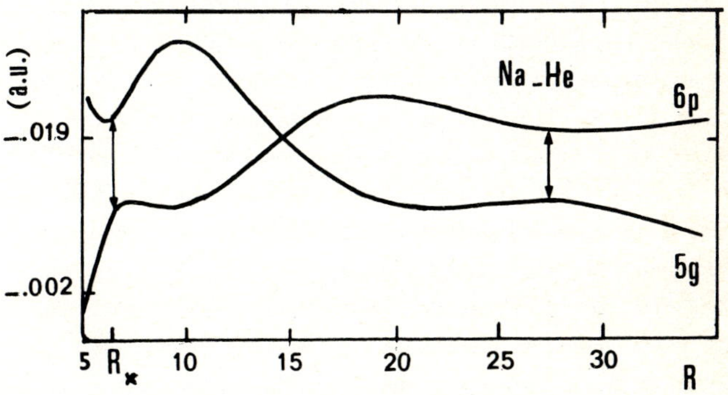

Ionisation energy curves for the σ6p and σ5x states of NaHe.

+ Laboratoire associé à l'Université Paris-Sud.

QUASI-ELASTIC STATE-CHANGING COLLISIONS OF HIGH-RYDBERG ATOMS WITH HEAVY RARE-GAS ATOMS

Yutaka Sato* and Michio Matsuzawa*+

Department of Engineering Physics* and Institute for Laser Science +
The University of Electro-Communications
Chofu-ga-oka, Chofu-shi, Tokyo 182, Japan

In previous reports[1,2,3], we evaluated the ℓ-changing cross sections of the high-Rydberg atoms with lighter rare-gas atoms based on the free electron model. For the lower n, say, $n \lesssim 25$, the formula adopted for the evaluation of the cross sections is not applicable to collisions with heavy rare-gas atoms because the higher order terms neglected in the modified effective range theory as input data to the free electron model still play appreciable roles. However, we note that for the high n, say, $n \gtrsim 25$, the same formula can be also applied to collisions with heavy rare-gas atoms because the above mentioned higher order terms become negligible.

Quite recently, experimental data on the title processes for the high n, say, $25 \lesssim n \lesssim 45$, have appeared[4,5,6]. Therefore to assess the validity of the free electron model quantitatively in this region of n, we have evaluated the cross sections $\sigma(n\ell \rightarrow n'\ell')$ for the quasi-elastic state-changing collisions of the high Rydberg atoms with the heavier rare-gas atoms such as

$$Na^{**}(nd)+B \rightarrow Na^{**}(n\ell'>2)+B \quad (1)$$

$$Xe^{**}(nf)+B \rightarrow Xe^{**}(n\ell'>3)+B \quad (2)$$

$$Rb^{**}(ns)+B \rightarrow Rb^{**}(n-3,\ell'>2)+B \quad (3)$$

$$Na^{**}(ns)+Ar \rightarrow Na^{**}(n-1,\ell'>1)+Ar \quad (4)$$

$$Na^{**}(np)+Ar \rightarrow Na^{**}(n-1,\ell'>1)+Ar \quad (5)$$

where B is Kr or Xe. In collisions with lighter rare-gas atoms, we have found[3] that the quasi-elastic state-changing collisions can be classified into two cases according to a parameter $u_{min}=n^2Q_{min}a_0$, ie, i) $u_{min}<1$, ii) $1<u_{min}\lesssim n$ where $Q_{min}(=\Delta E/\hbar V)$ is the minimum momentum transfer. Here ΔE is the energy defect of the state-changing collisions process and V is the relative velocity. In case ii), we have seen that the final states with lower ℓ' are suppressed in contrast with case i). This comes from the fact that a larger momentum transfer causes a larger angular momentum change inevitably. In the examples studied here except for (1), we have found the same trend as has been seen in collisions with the lighter rare-gas atoms because these processes are classified as case ii). However, at present there is no experimental data to be quantitatively compared with this theoretical prediction on the angular momentum distribution.

To compare with experimental data available at present, we have calculated the ℓ-mixing cross sections, ie, $\sigma_{\ell\ mix}(\Delta n)=\Sigma_{\ell'}\sigma(n\ell\rightarrow n'\ell')$. Our theoretical model shows that the cross sections for the title processes to other final channels than those given in Eq.(1)-(5) are negligibly small. Therefore we may compare our theoretical $\sigma_{\ell\ mix}(\Delta n)$ with the experimental data. Table 1 shows the typical comparison for process (2) and a reasonable agreement between theory and experiment. More detailed discussion will be made at the conference.

Table 1: Comparison of the thermally averaged theoretical ℓ-mixing cross sections with the experimental quenching ones for $Xe^{**}(nf)^a$+Xe system at 300 K.

n	$\sigma_q(10^{-12}\ cm^2)^b$	$\sigma_{\ell\ mix}(10^{-12}\ cm^2)$
24	4.4±1.8	3.93
32	3.9±1.7	———
35	———	3.35
37	3.5±1.4	———
45	———	2.34

a: The quantum defect δ_f of Xe**(nf) is taken to be 0.055.
b: Ref.4.

References

1. M. Matsuzawa, J. Phys. B 12 3743 (1979)
2. K. Sasano and M. Matsuzawa, J. Phys. B 14 L91 (1981)
3. K. Sasano, Y. Sato and M. Matsuzawa, Phys. Rev. A in press
4. C. Higgs, K. A. Smith, F.B. Dunning and R.F. Stebbings, J. Chem. Phys. 75 745 (1981)
5. M. Hugon, B. Sayer, P. B. Fournier and F. Gounand, J. Phys. B 15 2391 (1982)
6. M. Chapelet, J. Boulmer, J. C. Gauthier and J. F. Delpech, J. Phys. B 15 3455 (1982)

QUENCHING OF RUBIDIUM RYDBERG STATES : A TOOL FOR INVESTIGATION OF THE ELASTIC SCATTERING OF ULTRA-LOW ENERGY ELECTRONS (E$\sim$0.01 eV) BY RUBIDIUM ATOMS.

M. Hugon, F. Gounand, P.R. Fournier and J. Berlande

Service de Physique des Atomes et des Surfaces, CEN/Saclay, 91191 Gif-sur-Yvette Cedex, France

It has been shown that the collisional quenching of alkali Rydberg states by neutral perturbers proceeds mainly via the strong outer electron-perturber interaction[1]. Consequently the determination of the quenching cross sections can provide information about the elastic scattering of very low energy electrons by the perturbing atom[2], process which is generally difficult to investigate directly.

In the present work, we investigate the collisional quenching of very highly excited nS rubidium atoms by ground state rubidium, Rb(5S), atoms in the range of $34 \leq n \leq 43$[3]. The quenching cross sections are measured using a time-resolved selective field ionization technique[1]. The Rydberg nS states are populated by stepwise pulsed laser excitation in a cell. The experimental results obtained at T=400K are presented in figure 1.

On the other hand, the quenching cross sections Q_{Rb} (nS) are calculated using a theoretical approach, which only takes into account the valence electron-Rb(5S) interaction (first Born approximation with a Fermi pseudo-potential[4]). The calculated values of Q_{Rb}(nS) are proportional to the cross section for the elastic scattering of free electrons by Rb(5S) averaged over the velocity distribution of the Rydberg electron in the nS state[1], $\langle \sigma_{e^- -Rb} \rangle_{nS}$. For $n \sim 40$, the mean kinetic energy of the Rydberg electron is about 0.01 eV. In this range of energies, there is only one available calculation of $\sigma_{e^- -Rb}$[5]. The $\langle \sigma_{e^- -Rb} \rangle_{nS}$ values vary approximately from 2700 to $3100 \pi a_0^2$ for $32 \leq n \leq 41$. Figure 1 indicates that the results of the calculation using the Born+ Fermi potential approach are in good agreement with our measurements. However we get a similar agreement when taking for $\langle \sigma_{e^- -Rb} \rangle_{nS}$ a constant value of $2900 \pi a_0^2$. As the energy distribution of the Rydberg electron is broad and as the range of n investigated here ($34 \leq n \leq 43$) leads to a weak variation of the average energy $\sim 1/2n^2$ of the Rydberg electron, our method does not allow to check the variation of $\sigma_{e^- -Rb}$ as a function of the electron energy. But the agreement obtained clearly shows that the $\sigma_{e^- -Rb}$ value in the range $E \sim 0.01$ eV is in good agreement with the predicted value. The present study represents the first experimental estimation of the cross section for the elastic scattering of ultra-low energy electrons ($E \sim 0.01$ eV) by rubidium atoms.

References

1. M. Hugon, B. Sayer, P.R. Fournier and F. Gounand, J. Phys. B : At. Mol. Phys. 15, 2391 (1982).
2. K. Sasano and M. Matsuzawa, J. Phys. B: At. Mol. Phys. 14, L91 (1981).
3. M. Hugon, F. Gounand, P.R. Fournier and J. Berlande, submitted to J. Phys. B: At. Mol. Phys. (1983).
4. A.P. Hickman, Phys. Rev. A 19, 994 (1979).
5. L.C. Balling, Phys. Rev. 179, 78 (1969).

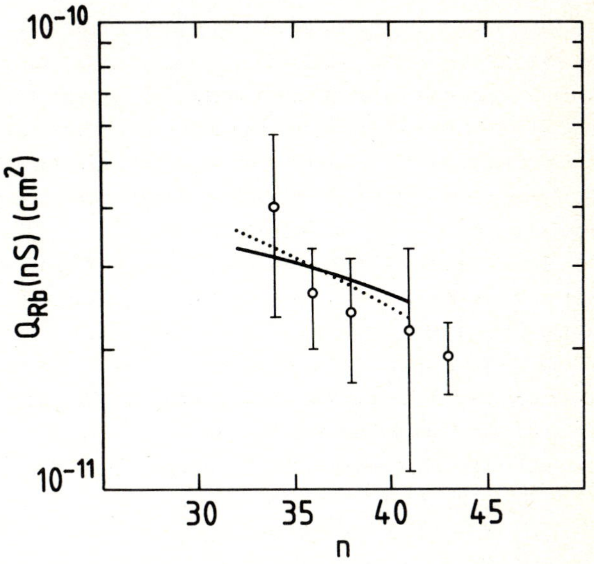

Figure 1 : Quenching cross sections for the rubidium nS states by Rb(5S). Experiment : ($\overline{\underline{\circ}}$). Calculation using the Born + Fermi potential approach by taking for $\langle \sigma_{e^- -Rb} \rangle_{nS}$: a) the calculated cross sections (——), b) a constant value of $2900 \pi a_0^2$ (.....) (see text).

APPLICATION OF THE FADDEEV WATSON EXPANSION TO THERMAL COLLISIONS OF RYDBERG ATOMS WITH NEUTRAL PARTICLES

E de Prunelé

Service de Physique des Atomes et des Surfaces, 91191 Gif-sur-Yvette Cedex, France

The Faddeev Watson expansion (FWE) for the T operator is applied to the study of thermal collisions between Rydberg atoms and neutral atoms. These collisions are considered as a three-body problem (the perturber, the Rydberg electron and its parent core). The core-perturber interaction is neglected for kinematics reasons. The evaluation of the FWE first and second order terms is made tractable by using an appropriate separable potential for the Rydberg electron-perturber interaction. In FWE second order term the t operator associated with the Rydberg atom (Rydberg electron + parent core) is approximated by the Coulomb potential.

This approach allow us to discuss the off shell-on shell approximation and to estimate the importance of taking into account explicitly the binding interaction of the Rydberg atom in the expression of the T operator. It also provides an indirect test of the impulse approximation which underlies previous work[1,2].

Detailed calculation for the process :

$$\mathrm{Rb}(n,\ \ell = 0) + \mathrm{He} \longrightarrow \mathrm{Rb}\ (n',\ \ell') + \mathrm{He} \qquad (1)$$

will be presented at the conference.

The contribution of the FWE second order term to the scattering amplitude is found especially significant when both the momentum transfers involved in the collision are large and the ratios ℓ/n , ℓ'/n' are small; the contribution decreases as n increases. Table I shows our results for the total depopulation cross sections where the cross sections for the process (1) at 520°K have been summed over final states ($n'\ell' \neq n\ell$). It can be seen that the approach suggested here is in good agreement with experimental results [3]. The improvement over the on shell Impulse Approximation is quite singificant for n=12. However other terms of the FWE could also contribute for low n values.

	n=12	n=18
On shell Impulse Approximation	265	336
F W E (first + second) order term	146	313
Experiment [3]	110 ± 20	325±65

Table I. Unit are 10^{-16} cm^2

References

1. M. Matsuzawa, J. Phys. B: Atom. Molec. Phys. 12, 3743 (1979).
2. A.P. Hickman, Phys. Rev. A 18, 1339 (1978).
3. M. Hugon, F. Gounand, P.R. Fournier and J. Berlande, J. Phys. B : Atom. Molec. Phys. 13, 1585 (1980).

SODIUM D STATE FINE-STRUCTURE IN AN ELECTRIC FIELD, AND FINE-STRUCTURE QUANTUM BEAT ECHOES

T. H. Jeys, M. C. Copel, G. B. McMillian, F. B. Dunning, and R. F. Stebbings

Department of Physics and Department of Space Physics and Astronomy
and the Rice Quantum Institute, Rice University, Houston, TX 77251

We have investigated the electric field dependence of certain sodium 34d fine structure intervals using a field ionization quantum beat technique.[1] In addition an "echo" in the quantum beat signal has been observed.

The experimental method is similar to that previously reported,[1] except that a small constant electric field ($<$ 0.22 V cm^{-1}) was applied in order to perturb the fine structure levels. Sodium atoms are photoexcited, in a small electric field, to a coherent superposition of 34d $^2D_{3/2}$ and 34d $^2D_{5/2}$ states. This superposition state is allowed to evolve for a variable time t' (0-9 μsec) after which the electric field is incremented by 0.6 V cm^{-1} in 10 nsec. This field transition stops the relative time development of the various $|m_\ell|$ components of the coherently excited state.[1] At a fixed time of 10 μsec after laser excitation the various $|m_\ell|$ components are identified by selective field ionization and, as a function of time of the field transition, exhibit pronounced quantum beats. The measured quantum beat frequencies are shown in fig. 1 and are in good agreement with those expected on the basis of energy level separations calculated using a numerical technique.[2]

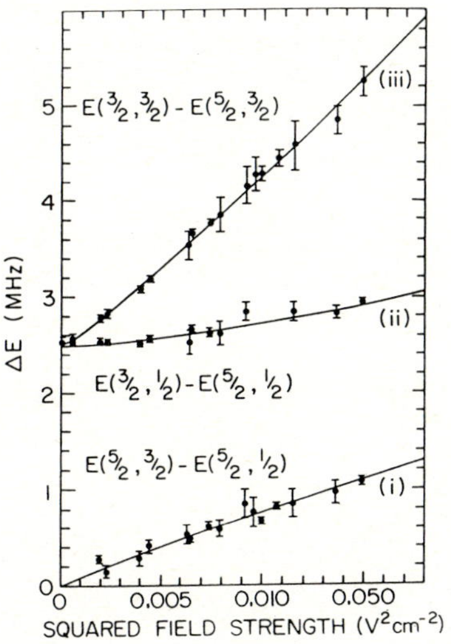

Fig. 1. Energy separations for the $|J,m_J\rangle$ states indicated (— theory, • experiment).

Note that the level separations do not show a simple quadratic dependence on the field strength.

Under ideal conditions the quantum beat signal will persist for the lifetime of the excited atoms.

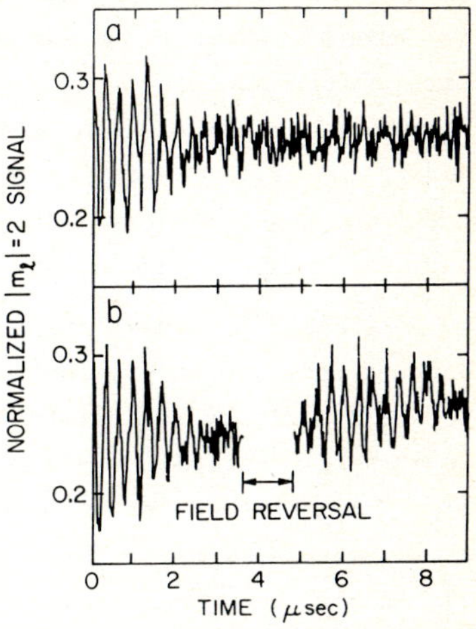

Fig. 2. a. Ionization signal vs. the time of transition to intermediate field.
b. Ionization signal with field reversal at ~4 μsec.

However, the presence of small inhomogeneities in the electric field results in different atoms experiencing slightly different electric field strengths. Since the temporal evolution of an atom depends on its local electric field, a small inhomogeneity will cause a gradual damping of the collective quantum beat signal (see Fig. 2a).

Upon reversal of the field (but not the inhomogeneity) those atoms that were initially exposed to the greater field are now exposed to the lesser field. Thus, field reversal leads to an eventual reestablishment of the initial phase conditions and the appearance of an echo in the quantum beat signal as illustrated in Fig. 2b. Observation of quantum beat echoes at late times suggests their use to study dephasing in collisions with target gases introduced into the excitation region.

Acknowledgments

This material is based upon work supported by the National Science Foundation under Grant No. PHY81-08452 and the Robert A. Welch Foundation.

1. Jeys, T. H., et al., Phys. Rev. A, 23, 3065 (1981).
2. Zimmerman, M. L., et al., Phys. Rev. A, 20, 2251 (1979).

COLLISIONS OF RYDBERG ATOMS IN AN ELECTRIC FIELD: CALCULATIONS USING HYDROGENIC WAVE FUNCTIONS IN PARABOLIC COORDINATES*

A. P. Hickman

Molecular Physics Laboratory, SRI International, Menlo Park, CA 94025, USA

The theory of collisions of Rydberg atoms with rare gas atoms is generalized to include the effects of an applied static electric field. Two effects occur. The first is that the energy differences between states change with the electric field strengths. The second effect is that the form of the wave function changes. In the field free case, the hydrogenic wave functions are normally obtained by separating variables in spherical coordinates, to obtain the well known functions $\Psi(r,\theta,\phi)$. In an electric field, it is appropriate to separate variables in parabolic coordinates. The resulting eigenfunctions have different spatial distributions than the spherical states.

Recent experiments[1,2] have probed systems where both effects may be expected to play a role, and it has been difficult to assess the importance of each effect separately. For the atoms and field strengths that have been investigated, it has been evident that the mixing cross sections would be diminished because of the greater energy difference between states induced by the field. It has thus been impossible to isolate the effect of the changing spatial distribution of the wave function. We have developed a theoretical model that allows the two effects to be analyzed independently, and will describe an investigation of the effects of the electron's spatial distribution.

In the field-free case, transitions between the nearly-degenerate states of Rydberg atoms induced by collisions with rare gas atoms have been treated by using the Born approximation for the nuclear motion, and a Fermi pseudopotential to model the interaction of the excited electron with the collision partner.[3,4] We treat the collision in the presence of an electric field by using the same two approximations, but with initial and final electronic states defined in terms of the parabolic wavefunctions that are appropriate in an electric field.

Cross sections are obtained for a variety of specific transitions of Rydberg atoms (n=10,15,20) caused by collisions. The calculations are strictly valid for hydrogen, but exhibit general features that are expected to apply to other Rydberg atoms as well. The results show a dramatic effect on the cross section for particular transitions depending on the degree of overlap of the initial and final charge distributions. In addition, it is observed that transitions are favored that involve small changes in the azimuthal quantum number m. This tendency is explained by analogy to optical selection rules.

*Work supported by AFOSR.

References

1. M.P. Slusher, C. Higgs, K.A. Smith, F. B. Dunning, and R. F. Stebbings, Phys. Rev. A 26, 1350 (1982).
2. R. Kachru, T. F. Gallagher, F. Gounand, K. A. Safinya, and W. Sandner, Phys. Rev. A 27, 795 (1983).
3. A. P. Hickman, Phys. Rev. A 18, 1339, (1978).
4. A. P. Hickman, Phys. Rev. A 23, 87 (1981).

ANALITICAL EXPRESSIONS FOR BORN ELECTRON IMPACT EXCITATION CROSS SECTIONS BETWEEN RYDBERG STATES $nl \to nl'$

I.Beigman*, L.Bureyeva**, M.Syrkin*

*P.N.Lebedev Physical Institute of the USSR Academy of Sciences, Leninsky Pr.,53, Moscow, USSR
**Scientific Council on Spectroscopy of the USSR Academy of Sciences, Pr.Sapunova,13-15, Moscow, USSR

The Born approximation is exact in the limit of high energies of incident electrons ($E \gg Ry$) and provides a reliable basis for more complicated theoretical approaches. Born excitation cross sections for transitions $nl \to nl'$ were considered elsewhere (for example, see[1-3]).

The aim of present work is the further development of analitical methods for Born cross sections. The Born exitation cross section can be written in the form:

$$\sigma(l \to l') = 8\pi a_0^2 \left(\frac{Ry}{E}\right) \cdot \int_{K-K'}^{K+K'} \sum_{\text{æ}} (2\text{æ}+1)(2l'+1) \begin{pmatrix} l & l' & \text{æ} \\ 0 & 0 & 0 \end{pmatrix}^2 \frac{dq}{q^3} R_{\text{æ}}^2(q) \qquad (1)$$

where K and K' are momenta of the incident and the scattered electrons, respectively; $R_{\text{æ}}(q)$ is the radial integral. Here we will consider the case $l' > l$. The main contribution to the sum (1) is due to the term with the minimal æ.

For $R_{\text{æ}}(q)$ the following expression is obtained:

$$R_{2m+\Delta}(q) = \prod_{i=0}^{2m+\Delta-1} \left[1 - (l'-i)^2/n^2\right]^{1/2} \cdot j_m(qn^2)\, J_{m+\Delta}(qn^2) \qquad (2)$$

where $j_m(x) = \sqrt{\frac{\pi}{2x}} J_{m+1/2}(x)$; $J_{m+\Delta}$ is the Bessel function; the value Δ is equal to zero or unity. The formula (2) gives the same result as that obtained in[2] for the case $l \ll n$. Besides, the equation (2) is exact at the limit $l \to 0$ and provides the values which practically agree with results of numerical calculations up to $qn^2 = 1 \div 3$.

For the excitation cross section the following expressions are obtained

for $\Delta l = 1$

$$\sigma(l \to l+1) = \frac{6\pi a_0^2 n^4 (l+1)}{2l+1} \left[1 - \left(\frac{l+1}{n}\right)^2\right] \frac{Ry}{E} \cdot \ln\left[\frac{\gamma \sqrt{E/Ry}}{n^2(\Delta E/Ry)}\right]; \qquad \gamma = 1{,}9$$

for $\Delta l > 1$

$$\sigma(l \to l+\text{æ}) = \frac{8(2\text{æ}+1)(2l+2\text{æ}+1)}{(E/Ry)} \cdot \begin{pmatrix} l & l+\text{æ} & \text{æ} \\ 0 & 0 & 0 \end{pmatrix}^2 \cdot \prod_{i=0}^{\text{æ}-1} \left[1 - \left(\frac{l+\text{æ}-i}{n}\right)^2\right] \frac{K_{\text{æ}}}{\text{æ}^4} \qquad (3)$$

where $K_{\text{æ}}$ has a weak dependence on n and l. For the case $l \ll n$ the function $K_{\text{æ}}$ is equal to 0,45; 0,42; 0,41; 0,40 for æ =2; 3; 4; 5, respectively.

References

1. I.Percival, D.Richards, Adv.Atom.Mol.Phys. 11,1 (1975)
2. M.Matsuzawa, Phys.Rev.A 9,241 (1974)
3. H.van Regemorter, Hoang Binh Dy, M.Prud'homme J.Phys.B 12,1053 (1979)

QUANTUM STATE DISTRIBUTIONS IN RYDBERG ATOMIC BEAMS CREATED BY ELECTRON IMPACT.

R. BARBE, J.-P. ASTRUC, A. LAGREZE, J.-P. SCHERMANN

Laboratoire de Physique des Lasers - Centre Scientifique et Polytechnique
Université Paris-Nord - 93430 VILLETANEUSE (France)

When a beam of atoms in Rydberg states is produced by electron bombardment [(1)(2)(3)], a wick distribution of n and ℓ quantum numbers is obtained. The n distribution can be experimentally determined by means of field ionization, but the ℓ distribution can only be inferred through a phenomenological model. The model of Ref.2 which only considered two classes of atoms (low ℓ and high ℓ) is here extended.

Measurements show that the n distribution approximately extends in 15-50 range (depending upon the time of flight between creation and detection) and, thus, ℓ from 0 up to 50. The corresponding $N > 1000$ states, coupled each other by collisions or radiative transitions, lead to a non-tractable N equation system. We have chosen a system, described by a 42×42 matrix, obtained by grouping the n values (from 10 to 54) in seven classes, and the ℓ values in six classes (P states, S and D states and four other ℓ-classes).

After their creation by electron impact, the low ℓ (S, P and D) Rydberg atoms suffer collisions with electrons, ground state atoms and are submitted to black-body radiation [(4)], inducing transfers between the different (n,ℓ) classes. Inside the electron gun, electron-Rydberg atoms collisions transfer from $\ell \leqslant 2$ values to high ℓ values with a proportionality factor equal to the degeneracy i.e. $2\ell + 1$, independantly from the $\Delta\ell$ transfer value.

From radiative transition rates calculated for each $n,\ell \to n',\ell\pm1$ couple, lifetimes and thermal radiation influence are determined. After integration over the velocity distribution of thermal [(1-3)] or supersonic beams [(5)], the model parameters (collision cross-sections) are obtained by fitting with the experimental results. An example of (n,ℓ) distribution under experimental conditions of Ref.6 and parameters of Ref.2 is given in the figure.

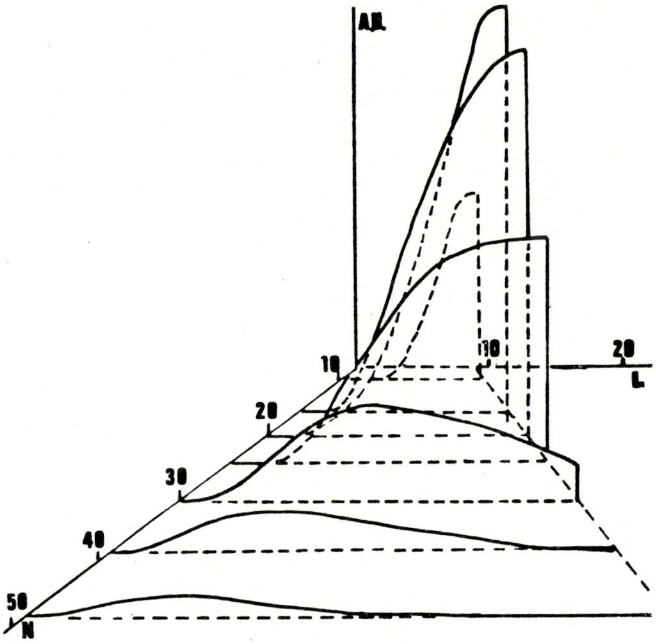

n,ℓ distribution of an effusive argon beam at 7 cm from the gun exit (gun temperature 700°K, beam chamber temperature 300°K).

REFERENCES.

1. H. HOTOP, A. NIEHAUS, J. Chem. Phys. 47 2505 (1967)
2. J.A. SCHIAVONE, D.E. DONOHUE, D.R. HERRICK, R.S. FREUND, Phys. Rev. A. 16,48 (1977).
3. C.A. KOCHER, C.L. SHEPARD, J. Chem. Phys. 74 379 (1981).
4. E.J. BEITING, G.F. HILDEBRANDT, F.G. KELLERT, G.W. FOLTE, K.A. SMITH, F.B. DUNNING, R.F. STEBBINGS, J. Chem. Phys. 70 3551 (1979).
5. I. DIMICOLI, Thèse de Doctorat (Université d'Orsay - 1980) unpublished.
6. J.-P. ASTRUC, R. BARBE, J.-P. SCHERMANN, J. Phys. B 12 L 377 (1979).

ELECTRON SCATTERING FROM A POTENTIAL IN A RADIATION FIELD

Robin Shakeshaft

Physics Department, University of Southern California, Los Angeles, CA 90089, USA.

I consider an electron scattering from a potential V in the presence of a monochromatic, spatically homogeneous, classical field of frequency ω, amplitude $\vec{a}$, and vector potential $\vec{A}(t) = \vec{a}\cos(\omega t)$. The electron has mass μ and is incident with an average velocity $\vec{v}$ and energy $E = \frac{1}{2}\mu v^2$. Following Rosenberg[1] it is convenient to introduce the three parameters: $\delta_1 = ea/(\mu vc)$, $\delta_2 = \hbar\omega/E$, $\delta_3 = v/c$. I assume that $\delta_3 \ll 1$ so that recoil and relativistic effects are negligible. The effective strength of the coupling of the electron with the field is δ_1/δ_2.

If $\delta_1/\delta_2 \gg 1$ the calculation of the cross section for stimulated absorption (or emission) presents, in general, a very difficult problem. This is because many photons can be emitted and absorbed so that a perturbation expansion (in powers of δ_1/δ_2) is not useful; nor is a decomposition of the electron motion into partial waves helpful since the electron can exchange many quanta of angular momenta with the radiation field so that many partial waves are coupled together. (The problem does simplify tremendously, however, in the soft photon regime $\delta_2 \ll 1$.) I have formulated a method which is applicable when δ_1 is not too large compared to unity; the coupling strength δ_1/δ_2 may be arbitrary. The method consists of expanding the state vector, $|\phi(t)\rangle$, of the electron as[2]

$$|\phi(t)\rangle = e^{ig(t)} \sum_n (-1)^n \, |f_n\rangle \, e^{-in\omega t} ,$$

$$g(t) = -Et/\hbar + (e/\hbar c\omega)\,(\vec{v}\cdot\vec{a})\sin(\omega t);$$

the vectors $|f_n\rangle$ satisfy an infinite set of coupled integral equations. Roughly $2\delta_1$ coupled equations must be solved to achieve convergence.

I applied the method to the case of a separable potential $V = \lambda|b\rangle\langle b|$ where, in coordinate space,

$$\langle\vec{r}|\vec{b}\rangle = a_0^{-3/2}\, e^{-(r/R)^2},$$

and where $R=a_0$ (the Bohr radius) and $\lambda = e^2/a_0$. I numerically solved the truncated set of $(2N+1)$ coupled integral equations for the $|f_n\rangle$ with $-N \leq n \leq N$. I calculated the integrated cross sections, σ_ν, for stimulated absorption of ν photons (ν is negative for emission).

N	σ_{-1}	σ_0	σ_1
1	1.75(-1)	3.81(-1)	3.97(-1)
2	9.99(-2)	3.10(-1)	2.78(-1)
3	9.66(-2)	3.07(-1)	2.72(-1)
4	9.76(-2)	3.07(-1)	2.72(-1)

Table 1. Cross sections σ_ν in units a_0^2.

To indicate the rate of convergence with N, I have tabulated some results in Table 1 for $\hbar\omega$ = 0.12eV, E = 1eV, and a photon flux of 10^{16} W/m^2. For these values, $\delta_1 \approx 4.5$, $\delta_2 = 0.12$, $\delta_3 \approx 0.002$, and $\delta_1/\delta_2 \approx 37.2$.

Note that any potential can be approximated by a truncated sum of separable potentials, and the calculation can be readily extended to this case.

References

1. L. Rosenberg, Phys. Rev. A 23, 2283 (1981)
2. N.M. Kroll and K.M. Watson, Phys. Rev. A8, 804 (1973).

ELECTRON-ATOMIC-HYDROGEN ELASTIC EXCHANGE COLLISIONS IN THE PRESENCE OF A LASER FIELD

F. Trombetta and C. J. Joachain
Physique Théorique, Université Libre de Bruxelles, Belgium

and

G. Ferrante
Istituto di Fisica dell'Università, Palermo, Italy

We have analyzed the modification of the elastic exchange amplitude in fast electron-atomic-hydrogen collisions, due to the presence of a laser field. This field is treated in the dipole approximation and is taken as single mode. It has been suggested recently[1], on the basis of a first order treatment of the electron-electron interaction, that the presence of a laser field reduces exchange effects in electron-atom collisions. In the present work we use an improved model in which i) more accurate laser-perturbed atomic wave functions are used and ii) the collision dynamics is described by using a second-order exchange amplitude[2] which, in the absence of the field, contains the leading contribution to elastic exchange scattering at all momentum transfers[3].

Values of the ratio

$$R = \frac{\left(\frac{d\sigma}{d\Omega}\right)_{LF}^{ex} \Big/ \left(\frac{d\sigma}{d\Omega}\right)_{LF}^{d}}{\left(\frac{d\sigma}{d\Omega}\right)_{FF}^{ex} \Big/ \left(\frac{d\sigma}{d\Omega}\right)_{FF}^{d}}$$

involving direct (d) and exchange (ex) differential cross sections assisted by the laser field (LF) and in the field-free (FF) case will be presented for various incident electron energies and scattering angles.

References

1. G. Ferrante, C. Leone and F. Trombetta, J. Phys. B 15, L475 (1982)
2. F. W. Byron, Jr., C. J. Joachain and R. M. Potvliege, J. Phys. B 14, L609 (1981)
3. C. J. Joachain, Comments At. Mol. Phys. 6, 69 (1977)

ELASTIC ELECTRON-HELIUM COLLISIONS IN ELECTRIC FIELD

C.Foglia

Sezione Teorica - Istituto di Fisica dell'Università and Gruppo Nazionale
Struttura della Materia del CNR - 43100 Parma - Italy

In this work we are concerned with electron-Helium elastic scattering at low and medium projectile energies, in a weak electric field $\vec{\mathcal{E}}$.

Differential cross sections for the elastic process e^--He(1s) $\rightarrow$ e^--He(1s) have been computed, using the integral expressions of scattering amplitude in the eikonal approximation previously derived by the author for scattering of electrons by an N-electrons neutral atom[1]

$$T_{fi} = \frac{(-1)^{N+1} m c e^2 \, 2^{3N-2} \, \pi^{N-2}}{\hbar^2 \, \Gamma(-i\eta)}$$

$$\times \sum_h c_h \int \frac{g(z^2) \exp(i\vec{q}\cdot\vec{R})}{R(R-z)^{i\eta N}} \, d\vec{R}$$

$$\times \sum_{j=1}^{N} \mathcal{D}_{hj} \mu_{hj} \left(\frac{\partial}{\partial \mu_{hj}^2}\right)^2 \left[\int_0^\infty d\lambda_j \, \lambda_j^{-i\eta-1} \right.$$

$$\left. \int_0^1 d\chi_j (1-\chi_j)^{-1} \exp\left(i\vec{q}_j^{\,\prime}\cdot\vec{R} - \Lambda_{hj} R\right)\left(1 - \frac{\lambda_j \chi_j}{\Lambda_{hj}}\right) \right]$$

$$\times \prod_{\substack{s=1 \\ s\neq j}}^{N} \mathcal{D}_{hs} \mu_{hs} \left(\frac{\partial}{\partial \mu_{hs}^2}\right)^2 \int_0^\infty d\lambda_s \, \lambda_s^{-i\eta}$$

$$\int_0^1 d\chi_s (1-\chi_s)^{-1} \chi_s \Lambda_{hs}^{-1} \exp\left(i\vec{q}_s^{\,\prime}\cdot\vec{R} - \Lambda_{hs} R\right) \Big|_{\vec{\gamma}_i = \vec{\gamma}_s = 0}$$

and applied with success to scattering by Hydrogen atoms[2].

The results are compared with other theoretical calculations and experimental data.

References

1. C.Foglia, Phys.Lett. 65A, 99 (1978)
2. C.Foglia, Z.Phys.A 301, 17 (1981)

EFFECT OF THE LASER POLARIZATION ON ELECTRON-ATOM COLLISIONS

P.Cavaliere and C.Leone

Istituto di Fisica-Facoltà di Ingegneria-Viale delle Scienze-Palermo-Italy

G.Ferrante

Istituto di Fisica dell'Università-Via Archirafi 36-Palermo-Italy

In this communication we present the effects of the laser polarization on electron-hydrogen collisions,considering the simplest cases of linearly and circularly polarized light.

The laser polarization lightly affects the projectile wavefunction,while play a more significant role to determine the dressed wavefunction of the atom.Using first order time dependent perturbation theory,the Stark effect occurs only for the degenerate states.

For linear polarization,the wavefunction of each substate is linear combination of the wavefunctions of the unperturbed multiplets and may be expanded in virtual states with energy E_n+shw,being E_n the unperturbed energy,hw the photon energy and s an integer number.

In circular polarization,the laser splits in energy the degenerate states,for example the n=2 state is splitted in 9 states,with energies and probabilities depending on the laser parameters.This result may be understood taking into account that,going over to a coordinate system rotating as the laser electric field,the non stationary problem is transformed into a stationary one,where the atom is acted upon by a constant electric and magnetic field.

The main differences induced on the differential cross sections(DCS)by the laser polarization are:

i)For linear polarization the DCS exhibits a factored structure like for free-free transition where the effect of the field is essentially buried inside the Bessel functions.

instead,for circular polarization this structure is not obtained as the laser parameters enter the argument of the Bessel function as well as the energy shifts and their probabilities of excitation.

spectrum exhibits peaks equally spaced corresponding to multiphoton emissions and absorptions. Also, for circular polarization, we have a discretization of the final energy of the scattered particles, but the each term substates generate a spectrum like one for linear polarization; thus the resulting spectrum is more complex than the previous case.

PARTICLE ATOM SCATTERING IN THE PRESENCE OF A NONRESONANT LASER FIELD

S. Bivona and D. Valenza
Istituto di Fisica della Facoltà di Ingegneria, Parco d'Orleans 90128 Palermo Italy
and
G. Ferrante
Istituto di Fisica dell'Università via Archirafi,36 90123 Palermo Italy

The theory of particle-atom collisions in the presence of a laser field compared with laser-assisted potential scattering is faced with the additional problem of taking into account the laser-induced restructuring of atomic levels. While the influence of the laser field on the incident particle motion is taken into account exactly, the similar problem in the case of the atom is instead quite involved, and has no general solution.

Here we report on an investigation concerned with the thory of particle-atom collisions, when the modifications due to the presence of the laser are taken into account accurately though approximately. In particular, we consider elastic and inelastic collisions of a particle p with hydrogen-like and alkaly atoms in the presence of a linearly polarized laser field.

First Born Approximation is used. The relative projectile-target motion is described exactly by nonrelativistic Volkov states, while the laser-modified target wavefunctions are generally determined within a 2nd-order time-dependent perturbation theory treatment, amounting to a 2nd AC Stark effect. This is done both for states with and without l-degeneracy. A second order treatment of the laser modification can be considered to provide a satisfactory approximation, as the laser modified wave functions contains almost all of the expected features: energy shifts, photon replicas, dynamic polarizability, and finally, terms accounting for atomic transitions as due to the direct laser-atom interaction during (but not dependent on) the collision.

In the case of hydrogenic atoms to have analytical laser modified wavefunctions for states with principal quantum number $n \geq 3$, we resort to a first order AC Stark effect treatment. In that case, a 2nd order treatment is possible only numerically. However, the results for n=1,2 give a fairly comprehensive insight into the basic modifications for this kind of collisions. The wavefunction for the states of the alkalies and for the ground state of the hydrogen atom are:

$$\Psi = \varphi_n(\underline{\xi}) \exp\left\{-i\frac{\tilde{E}_n t}{\hbar}\right\} \exp\left\{i\alpha_n \sin 2\omega t\right\} \exp\left\{-2\gamma_n \sin^2\omega t\right\}$$

with $\varphi_n(\underline{\xi})$ the unperturbed time independent wave function and α and γ two quantities strictly related to the atom dynamic polarizability. $\tilde{E}_n = E_n + 2\hbar\omega\alpha_n$; E_n the unperturbed energy of the state n and ω the angular frequency of the e.m. field taken as $\underset{\sim}{E} = \underset{\sim}{E}_0 \sin\omega t$. The excitated states of the hydrogen-like atoms are:

1) $\Psi_n = \sum_{\nu=1}^{\ell} C_{nk,\nu}\, \varphi_{n\nu}(\underline{\xi}) \exp\left\{-i\frac{E_n t}{\hbar} - i\rho_{n,k}\cos\omega t\right\}$

$\rho_{n,k}$ the first order Stark parameter;

2) $\Psi_\pm = \frac{1}{\sqrt{2}}\left\{\varphi_{2s} \pm \varphi_{2p}\right\} \exp\left\{\pm i\rho\cos\omega t\right\} \exp\left\{-i\frac{\tilde{E}t}{\hbar}\right\} \exp\left\{i\bar{\alpha}\sin 2\omega t\right\} \times \exp\left\{-2\bar{\gamma}\sin^2\omega t\right\}$

$\bar{\gamma} = \frac{1}{2}(\gamma_{2s} + \gamma_{2p})$; $\bar{\alpha} = \frac{1}{2}(\alpha_{2s} + \alpha_{2p})$; $\tilde{E} = E_2 + 2\hbar\omega\bar{\alpha}$ $(n = 2)$

The FBA differential cross sections are obtained as $\frac{d\sigma_{if}}{d\Omega} = \sum_s \left(\frac{d\sigma_{if}}{d\Omega}\right)_s$; $\left(\frac{d\sigma_{if}}{d\Omega}\right)_s = |C_s|^2 \left(\frac{d\sigma_{if}}{d\Omega}\right)_s^{FBA}$ where $\left(\frac{d\sigma_{if}}{d\Omega}\right)_s^{FBA}$ is the FBA field free cross section for the same process and $|C_s|^2$ are appropriate coefficients dependent on the particular scattering event considered. For some particular cases these are given below.

A) Hydrogen-like target 1) inelastic scattering from the ground state to a final state $|f\rangle$ $(|f\rangle \neq \Psi_\pm)$

$$(C_s)_1 = \sum_k J_k(a) \left(\frac{\alpha_i + \gamma_f}{a}\right)^k J_{2k+s}\left(\sqrt{\rho_f^2 + \lambda_{if}^2}\right) e^{2ik\varphi}$$

$$a = \sqrt{\alpha_i^2 - \gamma_f^2} \; ; \; \varphi = tg^{-1}(\rho_f/\lambda_{fi}) \; ; \; \lambda_\alpha = \frac{Q\, \underline{P}_\alpha \cdot \underline{E}_0}{\hbar M \omega^2} \; , \; (\alpha = i, f)$$

$$\lambda_{fi} = \lambda_f - \lambda_i \; ; \; \frac{P_f^2}{2M} = \frac{P_i^2}{2M} + \tilde{E}_i - \tilde{E}_f + s\hbar\omega$$

Q and M are the charge and the mass of the particle; J_k is the bessel function of order k (integer)

2) scattering from ground state to $\Psi_\pm$

$$(C_s)_2 = \sum_k J_k(a) \left(\frac{\alpha + \gamma}{a}\right)^k J_{2k+s}\left(\sqrt{\rho_f^2 + \lambda_{fi}^2}\right) e^{\mp 2ik\varphi} \; ; \; a = \sqrt{\alpha^2 - \gamma^2} \; ; \; \alpha = \alpha_i - \bar{\alpha} \; ; \; \gamma = \gamma_i + \bar{\gamma}$$

B) Alkali atoms target 3) inelastic scattering

$$(C_s)_3 = \sum_k J_{s-2k}(\lambda_{fi}) J_k(B) \left(\frac{\alpha_{fi} + \gamma_{fi}}{B}\right)^k$$

$$B = \sqrt{\alpha_{fi}^2 - \gamma_{fi}^2} \; ; \; \alpha_{fi} = \alpha_f - \alpha_i \; ; \; \gamma_{fi} = \gamma_f + \gamma_i$$

ELECTRON ATOM SCATTERING IN THE FIELD OF A CO_2 LASER

P.J. Curry, W.R. Newell and A.C.H. Smith

Department of Physics and Astronomy, University College London, Gower Street, London WC1E 6BT, England.

We wish to report measurements of free-free scattered signals of electrons which have absorbed or emitted photons from or to the radiation field of a high power CO_2 laser. The present results are compared with other work in the field[1,2].

A Ferranti 400 watt C.W. CO_2 laser is employed in conjunction with a high-resolution hemispherical electron spectrometer. The instrument is capable of measuring differential cross sections for scattering angles in the range 0° to 140° with operational resolutions of 60 meV or better depending on the current required. A parallel plate analyser has also been installed at a fixed angle as a reference signal monitor. The experiment is run on-line to an LSI-11 mini-computer system via a CAMAC interface.

Two spectra are shown in figures 1 and 2. In both cases the incident electron energy is 10eV, the scattering angle is 140°, the target atom is argon and the laser beam is focussed to a spot size of about 0.75mm diameter at the interaction region. The laser power was 280W and the data acquisition times were 210 and 93 minutes respectively.

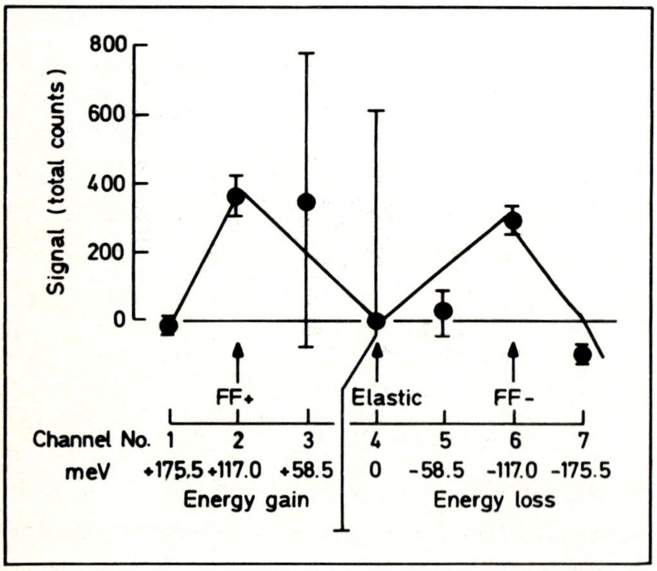

Figure 1

In figure 1 the energy change range spans in 7 channels the complete region were the FF+ (energy gain), elastic and FF- (energy loss) peaks should be observed. The signal is the difference between the scattered electron counts with the laser beam on and then off. The laser-beam-on counts are normalised to the laser-beam-off elastic channel counts before the subtraction (this is the reason why the elastic channel signal is identically zero). The large error bars in channels 3,4 and 5 are of course due to the large number of counts from elastic scattering. Signals are clearly observed at an energy gain of +117 meV and energy loss of -117meV corresponding to free-free transitions. Apart from a small barely- significant signal at channel 7, none of the other channels exhibit significant signals. An estimate of the value of the free-free factor Γ^2 in the Kruger-Schulz[3] formula together with an approximate knowledge of the beam interaction geometry enables us to predict a ratio of FF signal/elastic signal of 2×10^{-3}. This value compares favourably with the ratios $(1.99 \pm 0.29) \times 10^{-3}$ and $(1.61 \pm 0.21) \times 1^{-3}$ for the FF+ and FF- peaks respectively which we have measured.

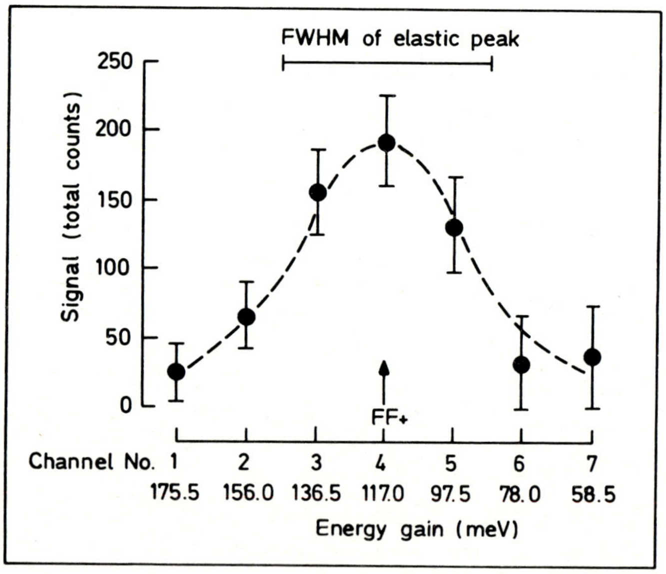

Figure 2

In the second spectrum only the energy gain range spanning the FF+ peak is covered. The FF+ feature is seen to be centred close to +117 meV and its width is the same as the width of the elastic peak (59 meV) as expected.

References

1. Andrick and Langhans, 1978, J. Phys. B11, p.2355.
2. Weingartshofer et al, 1979, Phys. Rev. A19, p.2372.
3. Kruger and Schulz, 1976, J. Phys. B9, p. 1899.

RADIATIVE ELECTRON-ATOM COLLISION CROSS SECTIONS

L. Dimou and F.H.M. Faisal

Fakultät für Physik, Universität Bielefeld, D-4800 Bielefeld 1
Federal Republic of Germany

Electron-atom scattering in a laser field is investigated with the help of an exactly soluble steady-state multichannel scattering model.[1] The radiative scattering amplitudes are shown to relate in an exact way to the scattering amplitude in the field free case with the replacement of the free-channel momenta $K_j = [2(E-\epsilon_j)]^{1/2}$ by the dressed channel momenta

$$s_{jn} = \sum_m K_{jm} \frac{1}{2} \int_o^\pi d\Theta \sin\Theta \, J^2_{m-N} (K_{jm} \alpha_o \sin\Theta)$$

where $\alpha_o = \frac{F}{\omega^2}$, F = peak field strength.

Fig. 1 shows the total radiative scattering signal at an off resonant frequency $(\frac{\omega_o - \omega}{\omega_o} = 0.0333)$ as a function of incident electron energy (in reduced units). A strong structure at the excitation threshold of the atom is seen which is due to the field modified threshold cusp. A repitition of the same is also noticeable at an energy $\approx \omega$ away from the threshold. In Fig. 2 the total cross section at a fixed incident electron energy (reduced unit 5) and as a function of laser frequency is displayed. The rich structure is essentially due to the exchange of photons by the electrons via the scattering resonance and a (negative energy) compound state. The resonance at $\omega \approx 6.11$ is associated with one photon emission resonance with a compound state at $E_c = -1.455$. The other structures are due to the presence of a scattering resonance and due to higher order resonant processes.

The model solution is compared with several current approximations for radiative collisions, such as the low-frequency and the high-frequency approximations and their region of validity is discussed. The steady state laser-atom ensemble distribution is derived in presence of decay processes such as photoionisation, spontaneous emission and transit time losses and compared with available limiting cases. The radiative scattering signal for elastic, inelastic and total cross sections are defined generally through

$$\frac{dS^{(N)}}{d\Omega} = Tr[\hat{\rho}\, \hat{f}]$$

for cross sections with N-photon absorption (emission) where $\hat{\rho}$ is the steady state density matrix and $\hat{f}$ is the fundamental scattering matrix.

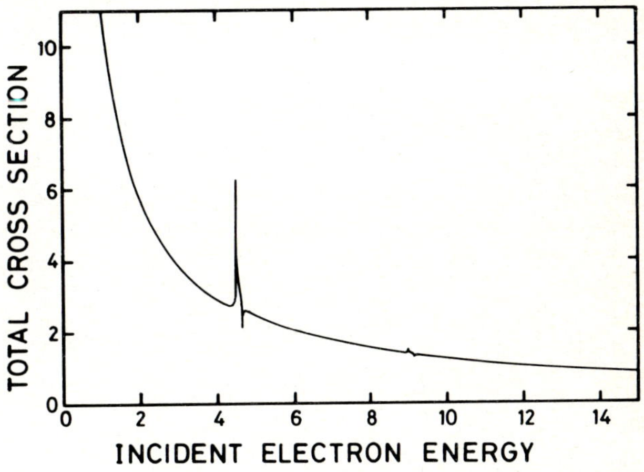

Fig. 1: Modification of threshold cusp in radiative collision.

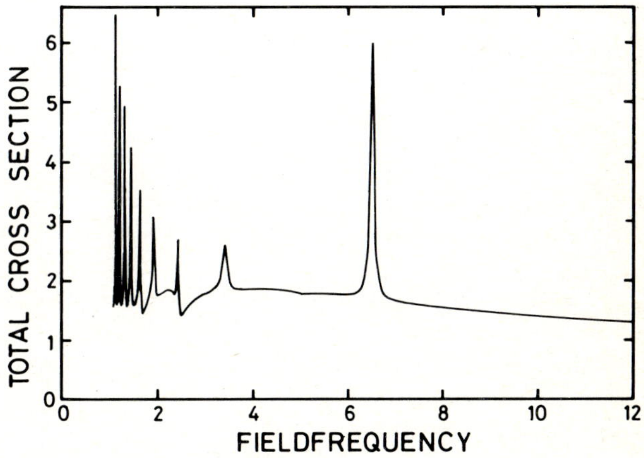

Fig. 2: Resonances in radiative (single and multi-photon) scattering cross sections.

References

1. F.H.M. Faisal (1982) Electron-Atom Scattering in Laser Field in "Photon-Assisted Collision and Related Topics" ed. N.K. Rahman and C. Guidotti, Harwood Academic Publishers N.Y.

SIMULTANEOUS ELECTRON-PHOTON EXCITATION OF ATOMS

S. Jetzke, F. H. M. Faisal, R. Hippler, H. O. Lutz

Fakultät für Physik, Universität Bielefeld, F. R. Germany

The inelastic scattering of electrons with atoms in the presence of an electro-magnetic field (EMF), for instance

$$e^- + h\omega + H(1s) \to e^- + H(2s)$$

is investigated within the frame of second order perturbation theory. The transition amplitude for the excitation of an atom from its groundstate $|i\rangle$ to an excited state $|f\rangle$ is given by ($e = \hbar = m = 1$)

$$A_{i\to f} = a^{12}_{i\to f} + a^{21}_{i\to f}$$

$$a^{kl}_{i\to f} = \left(\frac{1}{2\pi}\right)^3 \cdot \left(\frac{1}{i}\right)^2 \cdot \langle\psi_f(r_1)\exp(i\vec{k}_f\vec{r}_2)| \int_{-\infty}^{+\infty} dt_1 \int_{-\infty}^{t_1} dt_2 \, \tilde{V}_k(t_1)\tilde{V}_l(t_2)|\psi_i(r_1)\exp(i\vec{k}_i\vec{r}_2)\rangle \tag{1}$$

with

$$V_1 = -\frac{1}{|\vec{r}_1|} + \frac{1}{|\vec{r}_1-\vec{r}_2|} \tag{2.1}$$

$$V_2 = -\frac{1}{c}\vec{A}\cdot(\hat{p}_1+\hat{p}_2) + O(\vec{A}^2) \tag{2.2}$$

$$\vec{A} = 2a\vec{\varepsilon}\cdot\cos\omega t \tag{2.3}$$

The symbols have the usual significance.

From eq. 1 we get:

$$\frac{d\sigma}{d\Omega} = 8\pi\,\alpha\, I\cdot k_f/k_i\{(\vec{q}\cdot\vec{\varepsilon})/q^2\}^2\cdot f_A{}^2 \tag{3.1}$$

$$\frac{d\sigma}{d\theta_e} = \int\frac{d\sigma}{d\Omega}\,d\phi_e \tag{3.2}$$

$$\sigma = 8\pi\,\alpha\, I\cdot\frac{k_f}{k_i}\cdot f_A{}^2\cdot\pi/k_i{}^2 \left\{\cos^2\theta_\gamma\left(\frac{k_i{}^2-k_f{}^2}{k_ik_f}\ln\frac{k_i+k_f}{k_i-k_f}+2\right)+\frac{1}{2}\sin^2\theta_\gamma\left(\frac{k_i{}^2+k_f{}^2}{k_ik_f}\ln\frac{k_i+k_f}{k_i-k_f}-2\right)\right\} \tag{4}$$

with

$$f_A = \sum\!\!\!\!\!\!\int_j D_{fj}\{1/(\varepsilon_f-\omega-\varepsilon_j)+1/(\varepsilon_i+\omega-\varepsilon_j)\}D_{ji} \tag{5}$$

D_{km}: dipole transition matrix element between states $|k\rangle$ and $|m\rangle$.

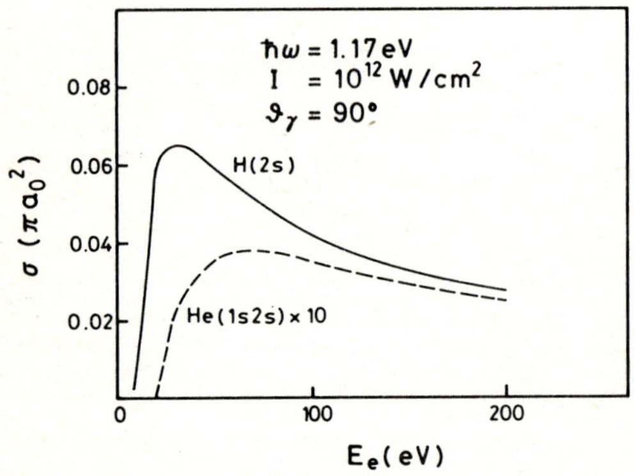

Fig. 1: Total cross section for excitation of H(2s) and H(1s2s) vs. energy of incident electrons.

For atomic hydrogen eq. 5 can be evaluated analytically[1]. In this work we have carried out the summation numerically. (The sum was truncated for $j = 50$ and continuum states were neglected at all.)

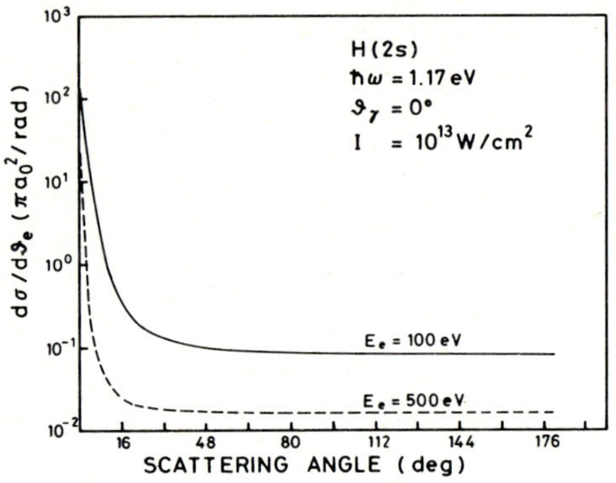

Fig. 2: Differential cross section for H(2s) excitation for two energies of incident electrons vs. electron scattering angle.

Figure 1 shows the total cross section for the excitation of H(2s) and He(1s2s). (The He-wavefunctions were taken from Chan[2].) The results for He are about one order of magnitude smaller than those for H.

Figure 2 displays the differential cross section, integrated over the azimuthal angle, for a H(2s) excitation.

References

1. N. R. Rahman, F. H. M. Faisal, J. Phys. B 11, 2003 (1978)
2. F. T. Chan, S. T. Chen, Phys. Rev. A 10, 4, 1151 (1974)

QUADRUPOLE EFFECT IN THE EXCITATION OF He BY ELECTRON IMPACT IN THE PRESENCE OF LASER

R.S. Pundir and K.C.Mathur

Department of Physics,University of Roorkee, Roorkee 247667, INDIA

The excitation of He (1^1S - 2^1S) by simultaneous electron and photon impact is studied in the dipole plus quadrupole approximation for wave number region 186103 - 186107 cm^{-1}. Significant changes in the joint excitation cross section are obtained due to multipole interference (between dipole and quadrupole) in the wave number region close to 3^1D intermediate state.

The scattering amplitude for the joint excitation of He atom, in which only one photon is absorbed from the laser beam, is given by[1] (a.u. are used)

$$F-G= -4\pi Q_o(z)d_1\Big[\sum_m (w_{mg}-w-i\Gamma_m)^{-1}\langle f|0|m\rangle \cdot \langle m|H'^{o}|g\rangle + \sum_n (w_{nf}+w+i\Gamma_n)^{-1}\langle f|H'^{o}|n\rangle\langle n|0|g\rangle\Big] \quad (1)$$

with $d_1=(1/q^2-1/2p^2)$, $\vec{q} = \vec{p}' - \vec{p}$

$Q_o(z) = J_o(z)$ exp (iz cos ϕ_q), $z=E_o q \sin\theta_q w^{-2}$,

$w_{mg}=w_m-w_g$ and $0 = \exp(i\vec{q}.\vec{r}_1)+\exp(i\vec{q}.\vec{r}_2)$

F and G are the direct and exchange amplitudes. w_m and w_n are the atomic frequencies of the intermediate states $|m\rangle$ and $|n\rangle$ respectively. $\vec{r}_1$ and $\vec{r}_2$ are the position vectors of atomic electrons. θ_q and ϕ_q are the polar angles of $\vec{q}$. Γ_n is the natural width of the nth level. E_o is the amplitude of the electric field of the laser and w is the laser frequency. The laser -atom interaction H'^{o} in the dipole plus quadrupole approximation, for circularly polarised light, is defined as[2]

$$H'^{o}= -i(2\pi\alpha wF)^{1/2}(R_1+R_2)$$

$$R_j = -(4\pi/3)^{1/2} r_j\ Y_{11}(\hat{r}_j) - (\pi/15)^{1/2} i k_{ph} r_j^2 \cdot Y_{21}(\hat{r}_j) \quad (2)$$

$\vec{k}_{ph}$ is the wave vector of photon. To perform the infinite summation in equation (1) we include $2^1P, 3^1P, 3^1D, 4^1P$ and 4^1D states exactly and account for the remaining states through the use of closure.

The total cross section (σ) for the excitation of He is given by

$$\sigma = \int (d\sigma/d\Omega)d\Omega \text{ where } d\sigma/d\Omega=(p'/p)\,|F-G|^2 \quad (3)$$

The wave number variation of σ at an incident electron energy of 200 eV and laser intensity of 10^{11} Watt/cm^2 is shown in Fig.1.

The curves DQ and D represent our results in the dipole plus quadrupole and pure dipole approximations respectively. The difference between DQ and D is small in the wave number region away from 3^1D quadrupole resonance showing that for this region the major contribution comes due to the dipole term. Close to 3^1D resonance we observe first a decrease in σ and then an increase showing that quadrupole term becomes important. The interference between dipole and quadrupole leads to significant change in σ. Detailed results will be presented at conference.

References

1. K.C.Mathur, IEEE J.QE-17,2233(1981)
2. Y.Flank,G.Laplanche, M.Jaouen and A. Rachman, J.Phys., B9, L409(1976)

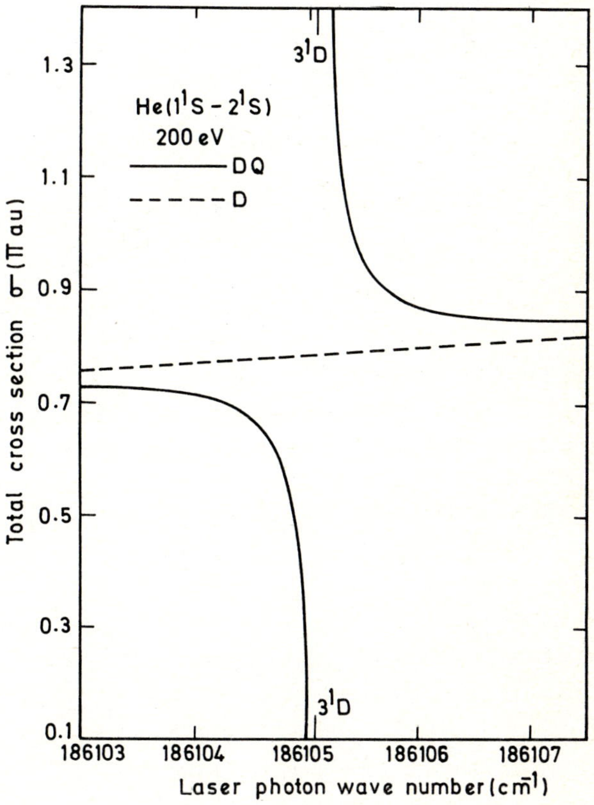

LASER ASSISTED ELECTRON IMPACT IONIZATION

M. Zarcone[+], D.L. Moores[++] and M.R.C. McDowell[+]

[+]Department of Mathematics, Royal Holloway College, Egham Hill, Egham, Surrey TW20 OEX, England

[++]Department of Physics and Astronomy, University College London, Gower Street, London WC1 E6BT, England

We will present results of a theoretical study of the (e,2e) reaction in Helium in the planar asymmetric (Ehrhardt) geometry both in the field free core and in the presence of a "Laser". Results at one energy have already been published[1].

The Laser field has been treated as single mode homogeneous and in the dipole approximation. Moreover the laser has been assumed off resonance with any atomic transition frequency and with an electric field intensity much smaller than the characteristic interatomic field.

The electron scattering is treated in first Born approximation, while the initial and final states of the target are calculated in

(a) a 3 state close coupling model 1s,2s,2p

(b) a 6 states close coupling model (a) + 3s, 3p, 3d.

The presence of the laser is found to affected significantly the shape of the angular distribution of the ejected electrons when photon exchanges occur. This change is shown to be determined by the behaviour of the Bessel functions occurring in a Kroll-Watson-type formula.

A secondary aim of the work is to obtain more accurate field free results at an impact energy of 500 eV to establish the difference between this model and the second Born approximation[2] model accurately.

Results will also be presented for laser assisted positron impact ionisation of helium.

References

1. M. Zarcone, D.L. Moores and M.R.C. McDowell, J.Phys.B, At.Mol.Phys. 16 L11 (1983).
2. H. Ehrhardt, M. Fischer, K. Yung, F.W. Byron,Jr. C.J. Joachain and B. Piraux, Phys. Rev. LeH. 48 1807 (1982).

LASER MODIFIED ELECTRON SCATTERING FROM A SLOWLY IONIZING ATOM*

Emilio Fiordilino and Marvin H. Mittleman

The City College of the City University of New York

Previous theories and experiments on electron-atom scattering in the field of a resonant laser were concerned only with lasers of very low intensity (<10 W/cm^2). In that case the only significant process introduced by the presence of the laser was a single photon emission or absorption via the resonant laser-atom interaction. Now that more intense tunable lasers are available, more interesting experiments are possible but the more intense laser can ionize the atom before or during the scattering process. We have allowed for this by incorporating the ionization channel into the solution for the wave function of the atom in the laser field. It is then possible to see absorptions and emissions of more than a single photon by the electron-atom combination.

The general expression for the cross section contains many terms. There are off-shell electron-atom scattering T matrices and interference between T matrices for different processes. As a simplification we have specialized to the problem of inelastic scattering from a Na target. The laser is tuned for the ground to $4P_{1/2}$ transition and the final state is chosen to be $4S_{1/2}$, with the electron absorbing two photons from the laser. We further simplify the calculation by having an electron energy which is high enough to approximate the T matrices by their Born approximations. In that case the distinction between on and off shell T matrices disappears but we still retain interference terms between different scattering processes.

Our numerical results show structure in the differential scattering cross section in a range of scattering angle between $0^\circ - 5^\circ$ for an initial energy of 40 eV. This should be readily observable with a laser intensity of the order of 10^9 W/cm^2. The details which will be presented depend upon the geometry of the laser polarization.

*This work was supported by O.N.R. Contract N00014-26-C0014.

THEORY OF LASER-INDUCED CHEMI-IONIZATION

H. P. Saha[*], J. S. Dahler and S. E. Nielsen[‡]

Dept. of Chemistry, University of Minnesota[*], Minneapolis, MN 55455
Chemistry Lab III, H. C. Ørsted Institute[‡]
University of Copenhagen, 2100 Copenhagen Ø, DENMARK

We recently constructed a classical path theory of laser-induced (LI) chemiionization processes[1], which closely paralleled earlier classical path theories of free-field (FF) associative ionization and Penning ionization. Here further connections are established between the theories of LI and FF collisional events and, in particular, it is demonstrated that the quantal formalism appropriate to LI processes differs in no essential way from that for comparable FF events. However, in place of the electronic energy operator which plays the central role in theories of FF chemi-ionization, it is the electric dipole moment operator (and the associated transition amplitude for photo-ionization) that dominates the stage in LI chemi-ionization.

The theory presented here can be extended to many other situations, including laser-induced non-ionizing collisional transitions. However, our present considerations will be limited to photoionization processes with the schematic representation

$$A + B \rightarrow (A\cdot\cdot B) \xrightarrow{\hbar\omega} \left\{ \begin{array}{l} A + B^+ + e^-, \text{ LIPI} \\ \\ AB^+ + e^-, \text{ LIAI} \end{array} \right\}$$

Events corresponding to the two different final states indicated in here will be called laser-induced Penning ionization (LIPI) and laser-induced associative ion-ization (LIAI), respectively. Weiner and his coworkers[2] have conducted crossed beam experiments with alkali atoms that involved laser-induced processes such as these. The laser frequency ω is chosen to lie below the ionization limits of both A and B and is not resonant with any electronic transitions of these two atoms. However, the photoionization limit of A••B may be exceeded during a binary encounter because the electronic states are collisionally distorted. Simple energetic considerations establish that the energy of the ejected electron and the post-collisional heavy-particle motions depend not only upon the experimentally controllable laser frequency and relative kinetic energy of the colliding atoms, but on the internuclear separation at which photoionization occurs.

Formulas are derived for several scattering cross sections descriptive of the energy spectra and angular distributions of reaction products. The principle limit-ations of the theory are its reliance upon the two-state approximation and its neglect of Born-Oppenheimer terms. The wave equation descriptive of the nuclear motions associated with the initial electronic state involves an operator which couples to the ionized final state. It is assumed that this coupling can be adequately represented by the local approximation. Furthermore, we replace the imaginary part of this complex valued local potential with its average over all internuclear orientations. Finally, the complexity of the cross section formulas and the number of matrix elements that must be computed are greatly reduced by assuming that the (magnitude of) relative angular momentum of the two nuclei is collisionally invariant.

The heavy-particle ($A+B^+$) energy-angle double-differential cross section and the associated heavy-particle (or photoelectron) energy distribution are found to depend in the manner $a + b\cos^2\theta_\alpha$ on the cosine of the angle, θ_α, between the laser polarization and the initial direction of relative motion. This polarization dependence has not been treated by earlier theories nor has it been observed experi-mentally.

The need for solving the heavy-particle scattering equations of the "exact" quantal theory can be eliminated by replacing the associated wavefunctions and phase shifts with suitable semiclassical approximations. Explicit formulas are derived for the corresponding semiclassical cross sections. Exploratory calculations, based upon these formulas, have been conducted for Na-Na collisions.

References

1. J. S. Dahler, R. E. Turner, S. E. Nielsen, J. Phys. Chem. 86, 1065 (1982)

2. P. Polak-Dingels, J. F. Delpach and J. Weiner, Phys. Rev. Lett. 44, 1663 (1980); J. Weiner and P. Polak-Dingels, J. Chem. Phys. 74, 508 (1981)

ELECTRON SPECTROSCOPY STUDY OF IONIZATION IN AN ALKALI VAPOR RESONANTLY EXCITED WITH A C.W. LASER

J.M. Bizau[+], B. Carré[++], P. Dhez[+], D. Ederer[+++], J.C. Keller[*], P. Koch[**], J.L. Le Gouët[*], J.L. Picqué[*], G. Spiess[++], F. Wuilleumier [+],

L.U.R.E., Université Paris-Sud, 91405 Orsay, Cedex, France

Ionization of an alkali vapor excited with a laser to the first excited atomic state, has received an increasing attention since :

1/ high ionization rate (∼ 90%) have been measured in dense vapors (10^{16} at/cm^3) of Na, Li, Ba, Sr,Ca, within short times.

2/ new processes as collisional ionization of excited states or laser-assisted collisional ionization (for high intensity excitation) have been observed in Na, Li Rb vapors.

3/ related phenomena of interest like the formation of highly excited states by energy pooling collisions of two resonant states appear to be involved in the ionization mechanisms.

In dense medium, electron-impact ionization developps to a great efficiency from initial seeding processes. At lower densities (∼10^{13} at/cm^3), electron-impact ionization remains negligible, and under weak excitation intensity, associative or Penning ionization of excited states are the dominant channels.

In the present experiment, ionization of Na vapor (10^{12} - 10^{13} cm^{-3}) excited to the 3p state by a c.w. dye laser, is observed via energy analysis of the emitted electrons by the irradiated volume. Beside Na_2^+ and Na^+ detection techniques used in previous experiments, the recording of electron energy spectra provides a rich and resolved information on : 1) the different ionization mechanisms, 2) further heating of free electrons in super elastic collisions with resonant states.

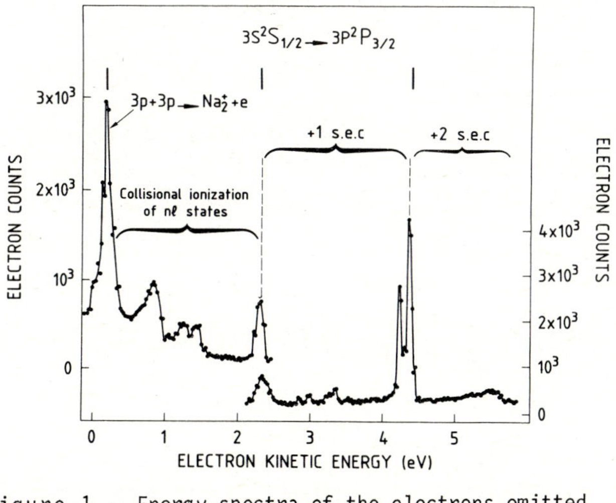

Figure 1 - Energy spectra of the electrons emitted from a C.W. laser excited Na vapor. From 0 to 2.1 eV an accelerating grid was used. The data are not corrected for electron spectrometer transmission. Electrons from associative ionization and penning ionization of nl states are also observed after 1 or 2 super elastic collisions with 3p state.

The experimental set-up consists in a weakly collimated effusive beam of Na, excited with a c.w. dye laser in the source region of a electron spectrometer (Cylindrical Mirror Analyser). Synchrotron radiation from A.C.O storage ring can be used for simultaneous measurement of atomic density and energy calibration, through inner-shell photoionization of Na. Energy distribution of the electrons is analyzed primarily between 0 and 2.1eV. The observed structure is attributed to electrons from purely collisional origin :

1/ two reproducible peaks distant of 0.13 eV are assigned to associative ionization of two 3p states. They correspond to the formation of final Na_2^+ in two different sets of vibrational levels (close respectively to v = 13 and v = 0).

2/ Penning ionization of high lying nl and 3p states, the nl states 3d, 4p, 5s, 4d, 4f, 5d, 6s being populated from energy pooling collisions between 3p states or further radiative decay.

Following spectra show the same patterns in the energy distribution between 2.1 eV and 4.2 eV, 4.2 and 6.3 eV, which correspond to sequential heating of the primary electrons (+ 2.1 eV) in super elastic collisions with the 3p state. Cross sections of the different collisional processes are estimated from the results. Similar studies have been performed on Ba vapor excited to the 6p state,and have also revealed a complex energy distribution of electrons from collisional origin .

+ Laboratoire de Spectroscopie Atomique et Ionique, Université Paris-Sud, Orsay.

++ Service de Physique des Atomes et des Surfaces, CEN. Saclay.

* Laboratoire Aimé Cotton, CNRS, Orsay

+++ N.B.S., Washington

** S.U.N.Y, Stony Brook.

References

1. T.B. Lucatorto and T.J. McIlrath, Phys.Rev. Lett.37 428 (1976).
2. M. Allegrini, G. Alzetta, A. Kopystynska and L. Moi, Opt. Comm. 19, 96 (1976).
3. B. Carré, F. Roussel, P. Breger and G. Spiess, J.Phys. B 14, 4271 (1981).
4. J.L. Le Gouët, J.L. Picqué, F. Wuilleumier, J.M.Bizau, P. Dhez, P. Koch and D. Ederer, Phys. Rev. Lett. 48 600 (1982).

ELECTRON SPECTROSCOPY STUDY OF IONIZATION IN AN ALKALI VAPOR EXCITED WITH A HIGH INTENSITY LASER

J.M. Bizau[+], B. Carré[++], P. Gérard[+], J.C. Keller[*], J.L. Le Gouët[*], J.L. Picqué[*], F. Roussel[++], G. Spiess[++], F. Wuilleumier[+]

L.U.R.E., Université Paris-Sud, 91405 Orsay Cedex, France

When an alkali vapor is excited with a high intensity laser to the first excited state, radiative processes of ionization i.e. photoionization or laser-induced collisional ionization developp in the medium beside the purely collisional ionization processes which are observed under weak intensity excitation.

The reported experiment extents the electron spectrometry study of ionization in a Na vapor excited to the 3p state to the case of an high intensity pulsed excitation.

Electron spectroscopy is proven to allow a clear resolution of the different ionization processes as well as of further heating of free electrons in one superelastic collision with the resonant states.

The experimental set-up and techniques have been described in joined abstract refered.

A pulsed dye laser ($\tau = 1\mu s$, $I_L \sim 10^5$ W/cm^2) is used to excite the low density vapor (10^{12}-10^{13} cm^{-3}) to the 3p $^2P_{1/2}$ or 3p $^2P_{3/2}$ levels, in the source region of an electron spectrometer (CMA). The energy distribution of the electrons emitted by the irradiated volume is recorded between 0 and 2.1 eV, with use of an accelerating grid. A rich structure is observed in the energy distribution which is compared to the one observed in the c.w. low intensity case. Peaks are attributed respectively to :

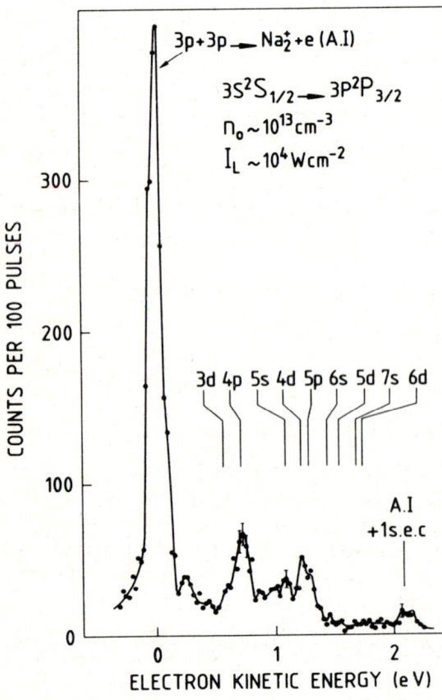

Figure 1 - Energy spectrum of the electrons emitted by the irradiated volume, between 0 and 2.1 eV. The peaks are attributed to associative Ionization 3p + 3p and photoionization of nl states.

1) associative ionization of two 3p states.
2) photoionization of 3d, 3p, 4d and to a smaller rate, of 5s and 6s states; the nl states being formed by energy-pooling collision of two resonant states. For the intensity used, photoionization dominates over Penning ionization nl + 3p which contributes to the signal for the same energies and which is observed in the c.w. low intensity case. Important heating of the free electrons in superelastic collisions with 3p state (+2.1eV) is proven to take place within the 1 μs pulse and for moderate atomic densities $n_o \sim 10^{13}$ at/cm^3.

Counting rates on the different peaks are compared for $3P_{1/2}$ and $3P_{3/2}$ resonances. Off-resonance spectra are also recorded since presenting marked structure of smaller amplitude . Laser-induced collisional effects are discussed from the results and comparison between high intensity and c.w low intensity cases.

A last experiment, in which c.w laser and pulsed laser beams are superposed, has given preliminary results. It has to check electrostatic trapping of slow electrons in the ionized medium.

+ Laboratoire de Spectroscopie Atomique et Ionique, Université Paris-Sud Orsay.
++ Service de Physique des Atomes et des Surfaces, CEN/SACLAY.
* Laboratoire Aimé Cotton, C.N.R.S., Orsay.

References

J.M. Bizau, B. Carré, P. Dhez, D. Ederer, J.C. Keller, P. Koch, J.L. Le Gouët, J.L. Picqué, G. Spiess, F. Wuilleumier, XIII ICPEAC 1983, preceding abstract.

He^{2+} + H COLLISIONS IN PRESENCE OF A LASER FIELD

L. Méndez, L.F. Errea and A. Riera

Departamento de Química Física y Química Cuántica.
Centro Coordinado CSIC-UAM.Universidad Autónoma de Madrid.
CANTOBLANCO (Madrid,34) (Spain).

We have calculated the charge transfer cross section for the process He^{2+} + H (1s) → He^{+}(n=2) + H^{+}, in presence of a laser field $\mathcal{E}$. We have employed the formalism of Copeland and Tang[1] which is based on the following assumptions:

1) A semiclassical description is used for the collision process. The wavefunction $\Psi(\underline{r},t)$ which represents the colliding system is solution of the equation

$$(H_{el} + H_{int})\Psi = i(\delta\Psi/\delta t) \qquad (1)$$

where H_{el} is the electronic hamiltonian.

2) The rotating wave and dipole approximations are used

$$H_{int} = - \underline{\mathcal{E}}\ \underline{r}\ e^{i\omega t} \qquad (2)$$

where $\underline{r}$ is the electron position vector.

3) The solution of eq. (1) is expressed in terms of a two-state molecular expansion

$$\Psi(\underline{r},t)=a_1(t)\chi_1\left[\underline{r},R(t)\right]\exp\left[-i\int_0^t E_1 dt'\right] + a_2(t)\chi_2\left[\underline{r},R(t)\right] \times \exp\left[-i\int_0^t E_2\, dt'\right] \qquad (3)$$

where χ_1, χ_2 are the molecular wavefunctions of energies E_1, E_2.

4) The dynamical coupling $\langle\chi_1|\frac{\delta}{\delta t}|\chi_2\rangle$ and the Stark broadening are neglected.

5) Perturbation theory is used. The averaged total cross section in this approximation, for a $\Sigma - \Sigma$ transition, is, if nuclear rectilinear trajectories are used with impact parameter b and relative nuclear velocity v

$$\sigma = (2\pi/3)\, \mathcal{E}^2 \int_0^\infty db\, b \left[|A|^2 + |B|^2\right] \qquad (4)$$

$$A = \int_{-\infty}^{\infty} \frac{b}{R}\, \mu \exp\left[-i\int_0^t (E_1-E_2-\omega)dt'\right] dt \; ,$$

$$B = \int_{-\infty}^{\infty} \frac{vt}{R}\, \mu \exp\left[-i\int_0^t (E_1-E_2-\omega)dt'\right] dt \qquad (5)$$

In a recent publication[2],the importance of the laser induced process to measure the density of α particles in tokamaks was pointed out. However the dynamical treatment of this calculations contained several important errors.

We have also considered the influence of $2p\sigma$-$2p\pi$ and $2p\sigma$-$2s\sigma$ transitions. Our calculated values of cross sections are an order of magnitude (or more) smaller than those $2p\sigma$-$3d\sigma$ transitions.

We have also checked further approximations:

I) The stationary phase approximation[3] (SPA)

II) Uniform approximation[4] (UA).

The validity of the approximations is illustrated in Fig.1

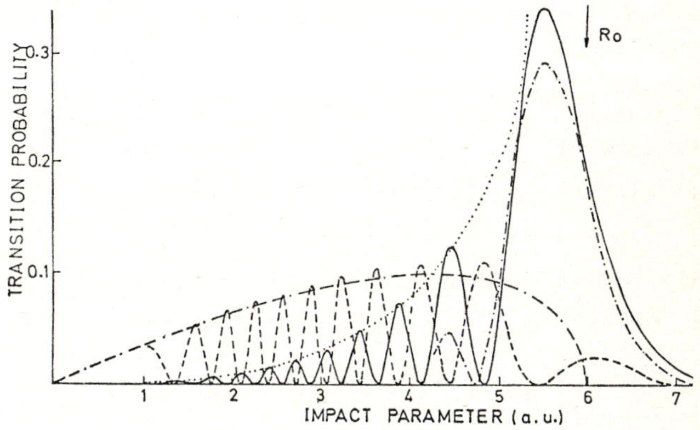

Fig. 1. $2p\sigma$-$3d\sigma$ laser induced transition probabilities for v = 1.55 x 10^7 cm/sec, λ= 3000 Å and laser intensity I = 0.1 TW/cm^2; —— $b\mathcal{E}^2|A|^2$;----- $\mathcal{E}^2|B|^2$ (eq.(5)); ····· and -·-·-·-- non oscillatory factors of $b\mathcal{E}^2|A|^2$ and $b\mathcal{E}^2|B|^2$ (SPA); -·-·- $b\mathcal{E}^2|A|^2$ (UA)

We also present the values of the total cross sections in figures 2 and 3 for a laser intensity I = 0.1 TW/cm^2. Our results differ significantly from those of ref. 2,in particular, the cross section peaks near the laser wavelength corresponding to the second harmonic of the Nd glass laser.

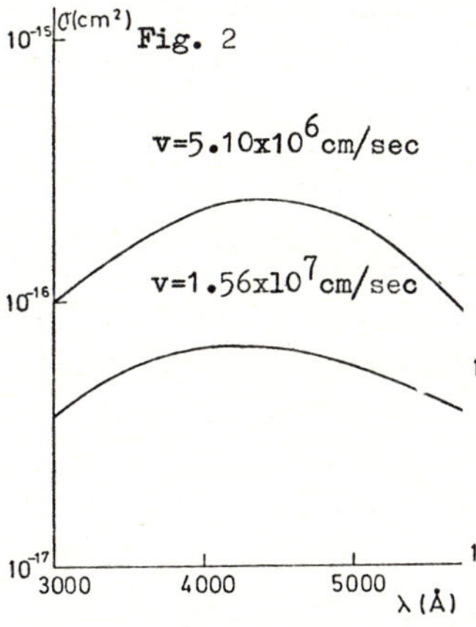

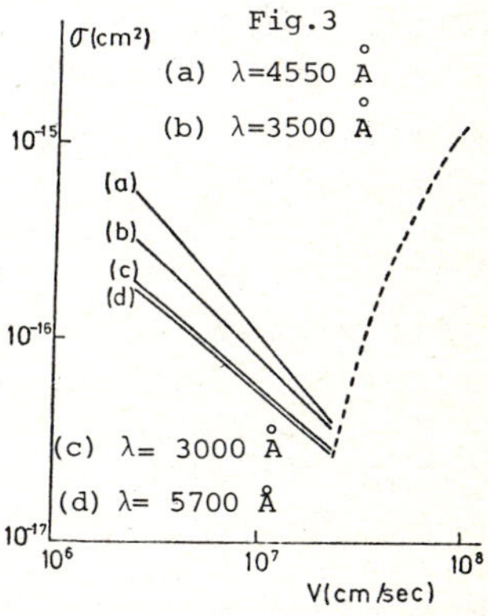

References

1. D.A. Copeland and C.L. Tang, J. Chem. Phys., 65,3161 (1976).
2. J.F. Seely, J. Chem. Phys., 75, 3321 (1981).
3. H. and B.S. Jeffreys, "Methods of Mathematical Physics". Cambridge University Press. Cambridge (1962).
4. M.S. Child, "Molecular Collision Theory". Academic Press. London (1974).

CHARGE TRANSFER IN THE PRESENCE OF A MAGNETIC FIELD

S. Bivona and B. Spagnolo

Istituto di Fisica della facoltà di Ingegneria, Parco d'Orleans 90128 Palermo Italy

and

G. Ferrante

Istituto di Fisica dell'Università, via Archirafi 36 90123 Palermo, Italy

Symmetric resonant charge transfer in atomic collisions in the presence of a homogeneous constant magnetic field is investigated within the impact parameter method and the atomic two-state expansion. The novelties brought about by the presence of the magnetic field are discussed in detail. The most peculiar new feature is the apparence of a magnetic field-dependent phase factor in the process wave function, which is responsible for severe reductions of the total cross section at moderatly strong fields.

The process wave function results to be:

$$\Psi(\underset{\sim}{r},\underset{\sim}{R},t)=\left\{c_i(t)\varphi_i(\underset{\sim}{r})+c_f(t)\varphi_f(\underset{\sim}{r}-\underset{\sim}{R})\exp\left[-\frac{i|e|}{2\hbar c}(\underset{\sim}{B}\wedge\underset{\sim}{R})\cdot\underset{\sim}{r}\right]\right\}\cdot e^{-iE_0t/\hbar}$$

φ the ground state wave function with energy E_0 of a one-electron atom embedded in the magnetic field $\underset{\sim}{B}$; $\underset{\sim}{R}$ the interionic separation, and the other simbols having the usual meaning.

Following standard procedures, the transition probability and the cross section for symmetric resonant charge transfer are obtained, respectively, as:

$$P(\underset{\sim}{v},\underset{\sim}{R_0},\underset{\sim}{B})=\sin^2\eta \quad ; \quad Q=2\pi\int_0^\infty P(\underset{\sim}{v},\underset{\sim}{R_0},\underset{\sim}{B})R_0\,dR_0$$

$$\eta=\frac{1}{\hbar v}\int_{-\infty}^{+\infty}\frac{W_{BA}-FW_{AA}}{1-F^2}\,dz \quad ; \quad W_{BA}=\int\varphi_B^*\exp[-ig]\,W\varphi_A\,d^3\underset{\sim}{r}$$

$$W_{AA}=\int\varphi_A^*W\varphi_A\,d^3\underset{\sim}{r} \quad ; \quad F=\int\varphi_A^*\exp[ig]\,\varphi_B\,d^3\underset{\sim}{r}$$

$g=\frac{e}{2\hbar c}(\underset{\sim}{B}\wedge\underset{\sim}{R_0})\cdot\underset{\sim}{r}$; R_0 is the impact parameter; $\underset{\sim}{v}$ is the incident ion velocity taken along the magnetic field; W stands for the ion-atom interaction in the initial channel; $\varphi_A\equiv\varphi_i(\underset{\sim}{r})$; $\varphi_B\equiv\varphi_f(\underset{\sim}{r}-\underset{\sim}{R})$.

Numerical evaluation of the transition probability and the total cross section is performed for the cases $H^+ + H \rightarrow H + H^+$ and $Rb^+ + Rb \rightarrow Rb + Rb^+$. For hydrogen, reductions of the cross sections up to 34% are found at relative velocity 4.8 10^7 cm/sec and magnetic field $B = 0.5\,B_0$ $(B_0 = 2.35\times10^5\,T)$, while for rubidium up to 43% at v=7.2 10^6 cm/sec and $B = 0.05\,B_0$.

The magnetic field enters the problem also through a "dressing" of the atomic states. With the values of the strength considered, which affect only moderately the atomic energies, the dressing is found to have only secondary importance in determining the absolute values of the total cross section, which are mainly controlled by large impact parameters. For rubidium instead the dressing results to be important in modifying the oscillatory structure of the total cross section, which is originated at small impact parameters.

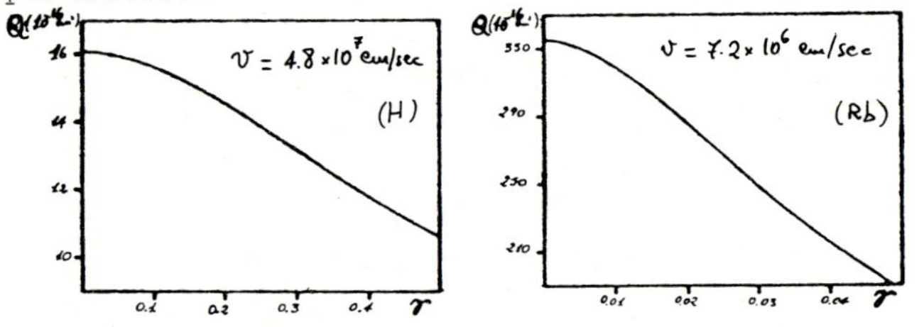

Fig.I (a,b):Cross sections vs. $\gamma = B/B_0$ at a given velocity

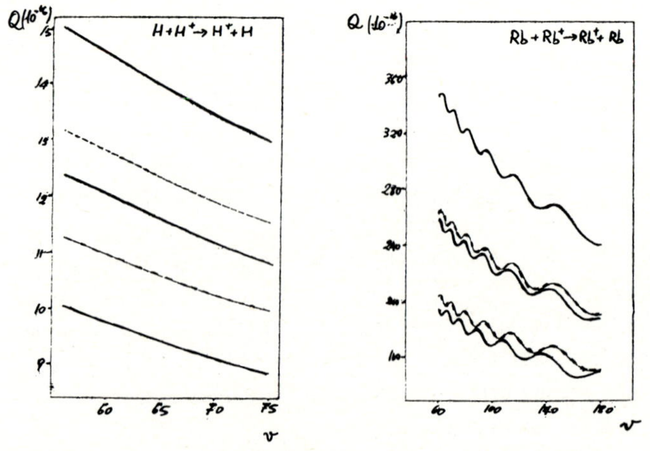

Fig.2 (a,b):Cross sections vs. velocity at three different values of γ . The broken curves are calculated using undressed atomic wave functions.

APPLICATION OF THE EIGENVALUE METHOD TO CARE AND RAIC

E.J. Robinson

Physics Department, New York University, New York, NY 10003 USA

Atom-atom collisions in laser fields have received considerable study.[1] These processes are often divided into two categories - collisionally-aided radiative excitation (CARE) and radiatively-assisted inelastic excitation (RAIC). CARE describes the reaction $A + B + \hbar\omega \rightarrow A^* + B$. Active atom A undergoes a transition from A to A^* as a result of the combined effect of the photon and the collision with B, which remains in its initial state. RAIC describes the reaction $A^* + B + \hbar\omega \rightarrow A + B^{**}$, where A and B may be dissimilar. The system is prepared with B in its ground state, A in an excited state. The combined collision-radiation interaction induces a transition to composite state B^{**} (excited), A (ground), with a photon nearly resonant with the energy difference between B^{**} and A^*. Since both partners change state, the process is impossible without the collision. (The charge-photon potential is a separate one-particle operator for each atom.) The excitation indicated by the CARE reaction could occur without the collision, but has negligible probability if the system detuning is large compared to noncollisional widths.

We follow the usual theories of these processes which treat the electronic motion in a two-level model and the nuclear motion as a classical straight line path. The internal time-dependent Schrödinger equation is a pair of coupled first-order differential equations for the state amplitudes. Setting $\hbar = 1$, and assuming $a_1 = 1$, $a_2 = 0$ as $t \rightarrow -\infty$, these are of the form

$$i\dot{a}_1 = A\, f(t)\, e^{i\phi(t)}\, e^{i\Delta t}\, a_2, \qquad (1a)$$

$$i\dot{a}_2 = A\, f(t)\, e^{-i\phi(t)}\, e^{-i\Delta t}\, a_1, \qquad (1b)$$

where $\int_{-\infty}^{\infty} f(t)\, dt = 1$, A measures the strength of the interaction, Δ is proportional to the detuning, and ϕ is real. It is always possible to choose ϕ so that f does not change sign.

The similarity of the structure of these equations to those that describe the interaction of non-colliding two-level systems with coherent pulses suggests that analogous methods for solution may be appropriate.

For given f, ϕ and Δ, $a_2(\infty)$ vanishes for certain A. The positions of the transition amplitude nodes may be calculated by solving an eigenvalue problem for A^2. One rewrites Eqs. (1) as uncoupled 2nd-order equations, defines $z = \int_{-\infty}^{t} f(t') - \frac{1}{2}$, chooses $a_2\, e^{\frac{-i(\Delta t+\phi)}{2}}$ and obtains

$$-b_2'' - \left[\frac{1}{4} \left(\frac{\Delta}{f}+\phi'\right)^2 + \frac{i}{2} \left(\frac{\Delta f'}{f^2} - \phi''\right) \right] b_2 = A^2 b_2, \qquad (2)$$

where prime indicates $\frac{d}{dz}$. The eigenvalues, A_n^2, are those A^2 for which $b_2 = 0$ as $z \rightarrow +\frac{1}{2}$, and are associated with eigenfunctions $b_n(z)$. The b_n form a complete orthogonal set, which may be normalized, so that $\int_{-\frac{1}{2}}^{\frac{1}{2}} b_n b_m\, dz = \delta_{mn}$. The eigenvalues are given by

$$A_n^2 = - \int_{-\frac{1}{2}}^{\frac{1}{2}} \left\{ \frac{d^2 b_n}{dz^2}\, b_n + b_n^2 \left[\frac{1}{4}\left(\frac{\Delta}{f} + \phi'\right)^2 + \frac{i}{2} \left(\frac{\Delta f'}{f^2} - \phi''\right)\right] \right\} dz. \qquad (3)$$

Eq. (3) may be used, with b_n replaced by a trial function, to calculate A_n^2 in a variational approximation when the problem is not exactly solvable.

In addition, one may express the transition amplitude via an expansion of the b_n, using the Green's function $G(\lambda,z,z')$, which satisfies

$$\left[-\frac{d^2}{dz^2} - \left\{\frac{1}{4}\left(\frac{\Delta}{f} + \phi'\right)^2 + \frac{i}{2} \left(\frac{\Delta f'}{f^2} - \phi''\right)\right\} - \lambda\right] G = \delta(z-z'). \qquad (4)$$

The transition amplitude is given by

$$b_2(\tfrac{1}{2}) = b_2^{(1)}(\tfrac{1}{2})\, R(0)/R(A^2),$$

where $R(\lambda) = \lim_{\substack{z \rightarrow -\frac{1}{2} \\ z \rightarrow +\frac{1}{2}}} \frac{\partial}{\partial z'} G(\lambda,z,z')$, and $b_2^{(1)}(\frac{1}{2})$ is the approximation to $b_2(\frac{1}{2})$ obtained by first-order perturbation theory. The Green's function is given by

$$G(\lambda,z,z') = \sum \frac{b_n(z)\, b_n(z')}{A_n^2 - \lambda},$$

and may itself be approximated variationally if the exact set of b_n is not known.

This work was supported by the U.S. Office of Naval Research.

1. A comprehensive list of references is included in P.R. Berman and E.J. Robinson, "Combined Radiation Field-Collisional Excitation of Atoms", appears in Photon-Assisted Collisions, N.K. Rahman and C. Guidotti, Eds. Harwood Academic Publishers, (Chur, London, New York, 1982).

2. The first reference to eigenvalues of A^2 is E.J. Robinson, Phys. Rev. A 24, 2239 (1981).

OPTICAL COLLISIONS IN STRONG LASER FIELDS

D.S.Bakaev, Yu.A. Vdovin, V.M. Yermachenko, S.I. Yakovlenko

Moscow Engineering Physics Institute, Moscow, USSR, 115409

The collision of a two-level (three-level) atom with buffer gas atom in a strong laser field is considered and the modification of the collision dynamics by the presence of a strong radiation field, predicted in paper[1], and experimentally confirmed in paper[2] is investigated. We study the effect of binary collisions on the resonance fluorescence spectrum and resonance Raman spectrum of an atom, the signal field absorption lineshape function[3]. Some aspects of the problem were considered in papers[4-6].

The collision dynamics becomes effected by the field in the regime $\Omega' \geqslant \tau^{-1}$, where $\Omega' = \sqrt{\Omega^2+\Delta^2}$, Ω is the Rabi frequency, Δ is the detuning of the incident laser frequency off resonance and τ is a duration of a collision.

The dressed-atom representation is used. The collision change of the density matrix can be written in the form of

$$\left(\frac{\partial}{\partial t}\langle i,n|\rho|j,n'\rangle\right)_{col} = -\Gamma_{ij}^{i'j'}\langle i',n|\rho|j',n'\rangle,$$

where i,j=1,2 are the dressed-atom states, n is a number of photons, $\Gamma_{ij}^{i'j'}$ is the relaxation matrix, connected with the collision S matrix

$$\Gamma_{ij}^{kl} = \langle(\delta_{ik}\delta_{jl} - S_{ik}S_{jl}^{*})\rangle_{av},$$

where the classical path dressed-state matrix element must be averaged over impact parameter, velocity and all directions relative to the laser field.

The S matrix was calculated for an interaction $u = Cr^{-n}$ in some limits and Γ matrix was obtained. The relaxation matrix element $\Gamma_{11}^{11} = \Gamma_{22}^{22}$ determines, for example, the collision width of the central component of the fluorescence spectrum and the term Γ_{12}^{12} determines the width of two other components. The same elements and the matrix elements Γ_{11}^{12} determine the signal field absorption lineshape function.

It was found that for the strong field $\Omega \gg \tau^{-1}$ all Γ_{ij}^{kl} except Γ_{12}^{12} tend to zero and

$$\Gamma_{12}^{12} = \left(\frac{\Delta}{\Omega'}\right)^{2/n-1}\left[\gamma + i\delta\,\mathrm{sign}(\Delta)\right],$$

where γ and δ are the collision broadening and shift of the spectral line inside the impact region. In the limit $\Omega \gg \Delta$ Γ_{12}^{12} also tends to zero and there is the narrowing of the spectral line to the collisionless value. In this limit there is the symmetrization of the relaxation matrix elements relative to the atomic states parametres.

$$\Gamma_{3j}^{3j} = \left|\frac{C_{tj}}{C_{ba}}\right|^{2/n-1}\left[\gamma - i\delta\,\mathrm{sign}(C_{tj}C_{ba})\right],$$

where $C_{tj} = C_t - C_j$; $C_{ba} = C_b - C_a$; C_a , C_b , C_t are the interaction constants for the resonant levels a , b and for the third nonresonant level t .
The matrix elements Γ_{3j}^{3j} determine the lineshape of three-level resonance Raman scattering.

References

1. V.S. Lisitsa, S.I. Yakovlenko, Zh.Eksp. Teor. Fiz. 68, 479 (1975)
2. J.L. Carlsten, A. Szoke, M.C. Raymer, Phys. Rev. A15, 1029 (1977)
3. D.S. Bakaev, Yu.A. Vdovin, V.M. Yermachenko, S.I. Yakovlenko, Zh. Eksp. Teor. Fiz. 83, 1297 (1982)
4. K. Burnett, J. Cooper, P.D. Kleiber, A. Ben-Reuven, Phys. Rev. A25, 1345 (1982)
5. G. Nienhuis, J. Phys. B15, 535 (1982)
6. C. Cohen-Tannoudji, S. Reynayd, J. Phys. B10, 345 (1977)

NONADIABATIC THEORY OF FINAL STATE DISTRIBUTIONS IN LASER ASSISTED ATOMIC COLLISIONS

P. S. Julienne and F. H. Mies

National Bureau of Standards, Washington, D.C. 20234, USA

Absorption of polarized light during an atomic collision produces a coherent distribution of final m_j states. Thus, light emitted by the products is polarized.[1,2] Subsequent collisions mix the final states and reduce the degree of polarization. We use the close coupled theory of scattering in a radiation field[3,4] to calculate both the radiative cross sections, $\sigma_\nu(m_j)$, which describe the initial production of final states, and the depolarization cross sections, $\sigma(m_j \rightarrow m_j')$, which describe the destruction of final state coherences. The close coupled formulation simultaneously accounts for both the radiative transition and the nonadiabatic mixing among the molecular states. The specific numerical calculations are for the simplest possible case, where two 'S atoms collide in the presence of laser radiation to produce 'S and 'P final state atoms ($j=1$, $m_j=0,\pm1$). The Sr + Ar system is well studied experimentally,[2,5,6] and our calculations based on SrAr ab initio potentials reproduce the features of the Sr + Ar profile, polarization ratio, and depolarization cross section.

A molecular formulation is used in which the initial and final state Hamiltonians are blocked according to total molecular angular momentum and parity, which are conserved for collisions in the absence of radiation. We assume a weak radiation field, which induces transitions according to the selection rules, $J \rightarrow J$, $J \pm 1$ and $p \rightarrow -p$. The cross sections $\sigma_\nu(m_j)$ depend on coherent sums of the radiative S matrix elements for $\Delta J = 0, \pm 1$ transitions. When the two atoms are close together, the electronic angular momentum projection rotates with the molecular axis, whereas when the atoms are far apart, the angular momentum is fixed in space. The switching between these two situations, respectively described by molecular Hund's coupling cases (a) and (e), occurs in the vicinity of some switching distance R_s where the Coriolis coupling between Born-Oppenheimer states in the body fixed frame is comparable to their electronic splitting. For the case of absorption in the far wings of the profile, where the laser frequency ν is detuned from the asymptotic transition frequency ν_∞, only collisions with small impact parameters contribute to $\sigma_\nu(m_j)$, and a perturbation theory in the body fixed frame can be developed for $\sigma_\nu(m_j)$. On the other hand, line core absorption, where ν is close to ν_∞, is more nearly described by absorption in the space fixed frame during collision with large impact parameters. For intermediate frequencies, interferences between absorption to different final states can introduce features in the polarization ratio which are quite sensitive to the molecular potentials.

We also make predictions for models of laser assisted collisional energy transfer (LICET)[1,3] with 'S + 'P final states. This case differs from the Sr + Ar case in that the dipole transition moment varies as R^{-3} asymptotically, instead of being constant. There are significant differences between the two cases with respect to polarization ratio and nonadiabatic effects on the total cross section.

References

1. P. R. Berman, Phys. A22, 1838, 1848 (1980).
2. P. Thomann, K. Burnett, and J. Cooper, Phys. Rev. Lett. 45, 1325 (1980).
3. F. H. Mies, in Theoretical Chemistry: Advances and Perspectives, edited by D. Henderson (Academic, New York, 1981), Vol. 6B, pp. 127-198.
4. P. S. Julienne, Phys. Rev. A26, 3299 (1982).
5. J. Alford, K. Burnett, and J. Cooper, 6th International Conference on Spectral Line Shapes, Boulder, Colorado, (1982).
6. H. Harima, Y. Kukuzo, K. Tachibana, and Y. Urano, J. Phys. B14, 3069 (1981).

PHASE CONTROL OF ATOMIC SCATTERING STATES IN TWO-PHOTON RADIATIVE COLLISIONS

Munir H. Nayfeh and G. B. Hillard

Department of Physics
University of Illinois at Urbana-Champaign
1110 West Green Street
Urbana, IL 61801

We consider the two-photon interaction of two colliding atoms having transition frequencies ω_1 and ω_3 with a pulsed laser field of frequency ω which is not resonating with any of the transitions of either atom. It has been established that when $\omega \simeq (\omega_1+\omega_3)/2$ the system of the two colliding atoms absorbs two photons and the atoms end up individually excited at ω_1 and ω_3 (two-photon radiative collision process). Our analysis shows that when one of the atoms receives both photons (one atom is radiatively active), the collisional interaction provides for the required coupling for the process, but at the same time causes some dephasing of the interaction (e.g. C_6/R^6) which limits the effectiveness of the coupling to a small range of internuclear separations. On the other hand when both atoms are radiatively active, we find that the collision provides an additional dephasing term which is proportional to the intensity of the laser (a.c. stark-collisional effect), thus making the overall C_6 of the interaction modulated by the intensity and detuning of the laser. This additional dephasing can be used to control and even eliminate the pure collisional dephasing (phase resonance), thus allowing the system to interact coherently over a wide range of internuclear separations (two-atom coherence). At phase resonance, the normal highly asymmetric lineshape for the process becomes symmetric. Moreover, accompanying this symmetry we find a large enhancement of the cross section. The enhancement is caused by the contribution from small impact parameters which, far away from phase resonance, would not have contributed.

DIPOLE MOMENTS AND INTERACTION POTENTIALS IN COLLISION-INDUCED ABSORPTION

A.Raczyński

Institute of Physics, Nicholas Copernicus University, 87-100 Toruń,Poland

In a collision of two or more atoms a transient dipole moment is induced which can interact with an incoming electromagnetic wave giving rise to absorption process. Thus the induced absorption spectra can be an important source of information about the interaction potentials and the induced dipole moments.

We have computed the absorption spectrum for some mixtures of inert gases in the binary collision approximation treating the dynamics strictly quantum-mechanically[1,2]. It is given by the formula

$$A(\omega)=A_2(\omega)\cdot\rho_A\,\rho_B$$

$$A_2 = 4\pi^2\omega/3\hbar c\ 1/V \sum_{i,j}(P_i-P_j)\ |\mu_{ji}|^2\,\delta(\omega-\omega_{ji}), \qquad (1)$$

(V - volume, i,j- two-particle translational states, P - Boltzmann factor, μ - induced dipole moment, ρ_A - density of gas A).

In order to calculate the sum in eq.(1) the partial decomposition has been performed, the one-dimensional Schrödinger equation has been solved numerically for different partial waves and different wave vectors and then the dipole moment matrix elements, the integral over wave vectors and finally the sum over partial waves have been calculated.

We have taken the usual form of the dipole moment

$$\mu(R) = \mu_0\ \exp(-R/\rho) + C_7/R^7 . \qquad (2)$$

For accurate interaction potentials it is generally possible to choose the parameters in eq.(2) so that a good agreement with experiment is achieved (Table 1).

The values of the dipole moment parameters obtained through the spectral moment analysis[3] are good enough to yield a good qualitative but not always quantitative agreement with experiment.

The dispersion-interaction R^{-7} term seems essential for obtaining good results.

Table 1

Absorption coefficient A_2 (in $10^{-6}cm^{-1}amagat^{-2}$) for Ne+Ar for the interaction potential of Candori et al.[4] and for the values of parameters μ_0=354 D, ρ=0.318 Å, C_7=-13.2 DÅ^7 (column I). The experimental values (column II) were taken from the graph in the paper of Bosomworth and Gush[5].

$\omega(cm^{-1})$	I	II	$\omega(cm^{-1})$	I	II
25	1.32	1.00	125	6.05	5.80
50	3.93	3.55	150	5.20	4.85
75	5.81	5.85	175	4.22	3.85
90	6.31	6.40	200	3.30	3.10
100	6.40	6.40	250	1.87	1.80
110	6.35	6.25	300	1.00	1.05

The results are sensitive to small changes of the dipole moment parameters.

The influence of the interaction potential is smaller but significant e.g. for the mixture He+Ar different Lennard-Jones potentials known in the literature yielded the absorption coefficients which differed even by 20-25% .

Our results and conclusions agree with those obtained independently by Birnbaum et al.[6].

References

1. A.Raczyński,Chem.Phys.72,321 (1982).
2. A.Raczyński,Chem.Phys.Lett.,to be published.
3. J.L.Hunt and J.D.Poll,Can.J.Phys.56,950 (1978).
4. R.Candori,E.Pirani and F.Vecchocattivi, Chem.Phys.Lett.30,202 (1982).
5. D.R.Bosomworth and H.P.Gush,Can.J.Phys.43, 729 (1965).
6. G.Birnbaum,M.S.Brown and L.Frommhold,Can.J. Phys.59,1544 (1981).

MULTIPHOTON-ASSISTED RADIATIVE COLLISIONS IN A STRONG FIELD REGIME

P. Pillet+, R. Kachru, N. H. Tran, W. W. Smith‡, and T. F. Gallagher

Molecular Physics Laboratory, SRI International, Menlo Park, CA 94025

There have been numerous studies of laser assisted collisions, and in spite of the fact that power densities well in excess of 1 MW/cm^2 have been used, these experiments all fall in the weak field regime amenable to perturbation theory treatments.[1]

Recently an experiment analogous to the laser induced collision experiments has been performed using Na Rydberg atoms and microwave radiation,[2] in which because the large dipole moment of Rydberg atoms, it was possible to observe radiatively induced collisions at microwave power levels of only a few $watt/cm^2$. As higher microwave powers can be easily obtained by using a cavity, it appears possible to reach the strong field regime for radiative collisions of Rydberg atoms interacting with microwaves.

We report here the observation of radiative collisions in a strong field regime. We have observed the multiphoton energy transfer processes $Na(ns) + Na(ns) \rightarrow Na(np) + Na\ ((n-1)p) + mh\nu$, where m is as high as 4 and which occur when the atoms are in an electric field of a strength such that the energy difference between the levels (n-1)p and ns is equal to that between the levels ns and np added to the energy $mh\nu$ of m microwave photons as shown in the Figure 1.

To interpret these results we have developed a theory of radiative collisions in a strong field regime. The approach is a generalization of that of Autler and Townes[3] to a multilevel system. We first treat the isolated atom "dressed" by the microwave field and then, as the duration of one collision (~ 1 ns) is long compared to the microwave period, the collisions of these dressed atoms as resonant dipole-dipole collisions.[4]

Along with the experimental cross section we also show in Figure 2 the calculated relative cross sections of resonant collisions, one-photon, two-photon, and three-photon assisted collisions. This work is supported by the Air Force Office of Scientific Research under contract F49620-79-C-0212.

+ Permanent Address: Laboratoire Aimé Cotton, C.N.R.S. II, Bâtiment 505, 91405 Orsay Cedex, France

‡ Permanent Address: Physics Department, The University of Connecticut, Storrs, Connecticut 06268, USA

References
1. S. E. Harris and J. C. White, IEEE QE-13, 972 (1977).
2. R. Kachru, N. H. Tran, and T. F. Gallagher, Phys. Rev. Lett. 49, 191 (1982).
3. S. H. Autler and Townes, Phys. Rev. 100, 703 (1955).
4. T. F. Gallagher, K. A. Safinya, F. Gounand, J. F. Delpech, W. Sandner, and R. Kachru, Phys. Rev. A 25, 1905 (1982).

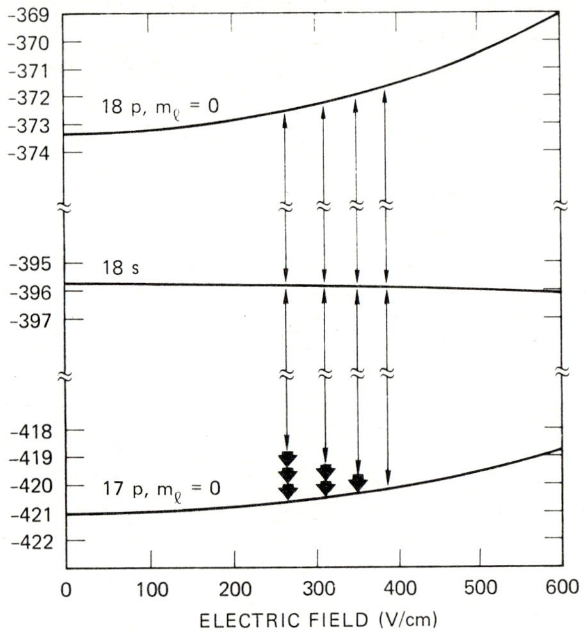

Fig. 1.: Energy level diagram relevant to the 0-3 photon assisted collisions, with the short arrows indicating microwave photons.

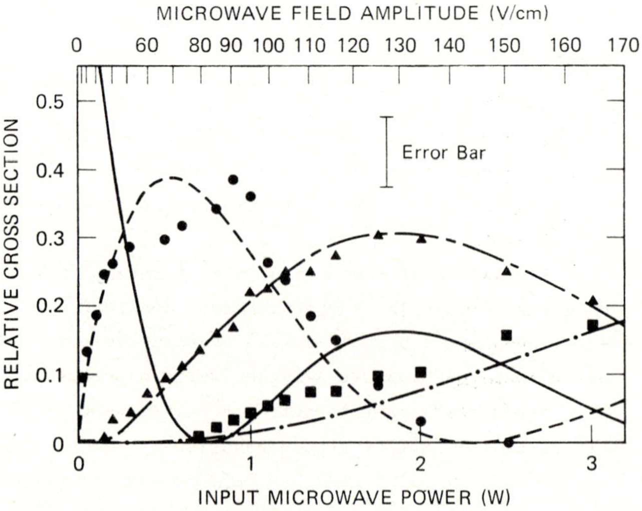

Fig. 2: Variations of relative cross-sections. 0-photon (——), 1-photon (●, - -), 2-photon (▲, - —), 3-photon (■, —·—), the points are experimental, the lines theoretical.

TEMPERATURE DEPENDENCE OF THE COLLISION INDUCED ABSORPTION OF CO_2 AT 0.091 CM^{-1}

G.Kouroupetroglou* and G.Boudouris

National Technical University,Physics Lab. A',Zografou Campus,Athens 624,Greece

In a previous publication[1] we reported results on collision indused microwave absorption in CO_2 obtained using a sensitive cavity technique at a frequency of 2.7 GHz.Square and cubic density dependence of the absorption was thus first observed for such a low frequency.Our results were,however,limited to a single temperature,about 293 K.We have now obtained measurements in CO_2 for a temperature range from 293 K to 358 K at the same frequency. The extension of the measurements to such a low frequency point is of increased theoretical interest for the interpretation of collision indused absorption (CIA) in non-dipolar gases but on the other hand,accurate measurement at lower microwave frequencies become progressively difficult[2]. The temperature dependence of CIA is of particular interest in the case of the CO_2 as this molecule has a relatively large quadrupole moment and a more difficult theoretical analysis should be applied.

A very sensitive cavity method has been used for measuring the weak CIA.The difference between the Q measured when the cavity is filled with the lossy gas and when it is filled with the lossles reference gas is a direct measure of ε''. The shift in the frequency of the resonance provides an accurate measure of the density of the gas in the cavity.The basic apparatus has been descibed earlier[1] .In the present series of measurements the 2.7 GHz cavity was placed in a temperature regulated pressure vessel. As a reference gas we used Ar.Both the Ar and the CO_2 were commercially available of 99.998% purity.

In Fig.1 are plotted several representative curves

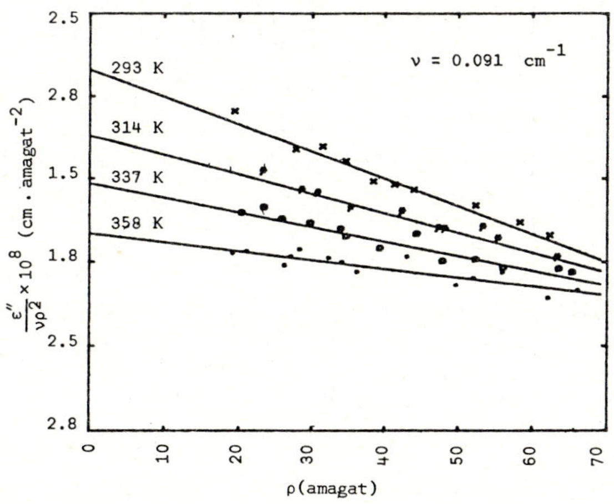

Fig.1: Plots $\varepsilon''/\nu\rho^2$ as a function of density ρ.

* Now with the University of Athens,Electronics Lab., 104 Solonos str.,Athens,Greece.

of $\varepsilon''/\nu\rho^2$ versus the density ρ, of CO_2 , for the temperature range 293 K to 358 K, showing that the data at each temperature may be represented by the expression:

$$\frac{\varepsilon''}{\nu\rho^2} = \varepsilon_2''/\nu + (\varepsilon_3''/\nu)\,\rho$$

where ε_2''/ν is the intercept and ε_3''/ν is the slope of the straight line fitted to the data at each temperature (ν is the frequency in cm^{-1}). In Table 1 are listed the values of these intercepts and slopes.The cubic

Table 1

T K	$\varepsilon_2''/\nu \times 10^8$ cm·amagat^{-2}	$-\varepsilon_3''/\nu \times 10^{11}$ cm·amagat^{-3}
293	2.16±0.05	16.6±1.0
314	1.76±0.04	11.8±0.9
337	1.47±0.04	8.7±0.9
358	1.17±0.04	5.3±0.9

coefficient is negative over the temperature range. The error limits given are estimated from the scatter in the points on the plots.

The values found for ε_2''/ν and ε_3''/ν were plotted on a log-log scale.For the square coefficients a straight line of slope -3.02 and for the cubic coefficients a straight line of slope -5.6 are the best fits.We can then represent the data in the form:

$$\frac{\varepsilon''}{\nu\rho^2} = 2.69(0.01)*10^{-8}\,(T/273)^{-3.02(0.1)} - 25.6(1)*10^{-11}(T/273)^{-5.6(0.5)}$$

This formula represents well the data of the collision induced absorption of CO_2 at 2.7 GHz over the whole region of temperatures and pressures studied.

Further analysis of the results will lead us to the calculation of the quadrupole moment of CO_2 ,determine two- and three-body relaxation times and test the available theoretical models given in the literature.

References: 1:G.Kouroupetroglou and G.Boudouris,J.Phys. Let. 42,L441(1981)

2:G.Birnbaum,B.Guillot and S.Bratos,Adv.in Chem.Physics 51,49(1982)

DYNAMICS OF POSITRONIUM IN A LASER FIELD

F.H.M. Faisal and P.S. Ray

Fakultät für Physik, Universität Bielefeld, D-4800 Bielefeld
Federal Republic of Germany

Dynamics of Ps-atoms in a coherent radiation field has recently become of considerable interest. Experimental results[1,2] on the break-up of Ps, which are also of interest in testing QED, requires careful theoretical analysis for its interpretation. A theory of laser-Ps interaction taking into account the photon induced recoil and related effects (which are of significance due to the very light mass of Ps) on to the c.m. motion of Ps is developed. The equation of motion of the system is

$$i\hbar \frac{\partial\Psi}{\partial t} = [-\frac{\hbar^2}{2M}\nabla_R^2 + (-\frac{\hbar^2}{2\mu}\nabla_r^2 - \frac{e^2}{r}) + H'(t)]\,\Psi \qquad (1)$$

$$H'(t) = -\,e(\vec{E}_0.\vec{r})\;2\cos(\omega t+\delta)\,\cos(\vec{k}.\vec{R}) \qquad (2)$$

where $\vec{R}$ and $\vec{r}$ denote the c.m. and relative coordinates respectively, $\Psi(\vec{R},\vec{r},t)$ is the total wave function, M and μ are the total and reduced masses of the system, $\vec{k}$ and ω are the wave vector and frequency of the laser photons and E_0 is the (peak) laser field strength. We make a generalisation of the usual Floquet[3,4] development of the total wave function

$$\Psi(\vec{R},\vec{r},t) = \sum_{nm}\psi_{nm}(\vec{r})\; e^{i(\vec{K}+n\vec{k}).\vec{R}+im(\omega t+\delta)-i\frac{E}{\hbar}t} \qquad (3)$$

$$\psi_{nm}(\vec{r}) = \sum_j a^j_{nm}\,\phi_j(\vec{r}) \qquad (4)$$

where $\vec{K}$ is the wave vector of the c.m.motion, E the total energy, and $\phi_j(\vec{r})$ are the internal Ps-wave functions with the energy eigenvalues ε_j. The probability amplitudes a^j_{nm} are determined by the stationary equations

$$[E - (\frac{\hbar^2(\vec{K}+n\vec{k})^2}{2\mu} + \varepsilon_j + m\hbar\omega)]\,a^j_{nm} = \sum_{j'} 2e\vec{E}_0\cdot(\vec{r})_{jj'}\cdot \cdot[\,a^{j'}_{n+1,m+1} + a^{j'}_{n+1,m-1} + a^{j'}_{n-1,m+1} + a^{j'}_{n+1,m+1}] \qquad (5)$$

The system of Floquet equations shown above are those obtained in the dipole approximation with respect to the relative motion of the Ps. There is no difficulty in writing down the eqations in the general multipolar case.

The first order as well as higher order matrix elements of the dipole operator for Ps can be obtained from that of H-atom by the following scaling rule :

$$[\,M_l^{(N)}\,(\tfrac{\omega}{2})\,]_{Ps} = 2^{(l+1)N-1}\,[\,M_l^{(N)}\,(\,\omega\,)\,]_H \qquad (6)$$

where N is the order of the multiphoton radial matrix element and l is the multipole order (l=1 dipole).

The free-free matrix elements of r play a significant role in the break-up problem, which are obtained from the formula of Gordon[5], where we deal with the singularity in the diagonal (continuum-continuum)matrix elements according to the principal value prescription.

Preliminary calculations for the excitation of $1^3S\rightarrow2^3P$ at a field intensity E_0=51.4 V/cm have been performed. At this lower field strength the excitation reaches maximum between 9-10 nsecs and the maximum population transferred is nearly $\frac{1}{3}$ of the ground state population. At the relatively higher field strength (E_0=514 V/cm) the system exhibits Rabi oscillation with increasing mean probability of excitation up to several nanoseconds. We have also noted significant modification of the MeV γ-quanta emission due to annihilation as a function of laser pulse interaction time. Theoretical results of multiphoton break-up reaction (ionisation) of Ps will also be presented and compared with the recently obtained data of Chu and Mills[1].

References

1. S.Chu and A.P.Mills, Phys.Rev.Lett.48,1333 (1982).
2. F.H.M.Faisal and P.S.Ray, J.Phys.B.14,L715 (1981).
3. J.H.Shirley, Phys.Rev. 138, B979 (1965).
4. F.H.M.Faisal, Nuovo Cim. 33B 775 (1976).
5. W.Gordon, Ann.d.Physik 2 1031 (1929).

SOFT-PHOTON INTERACTIONS GENERATING SYMMETRY-BROKEN PRE- AND POSTCOLLISION STATES OF MOLECULES

Peter Pfeifer
Fakultät für Chemie, Universität Bielefeld, D-4800 Bielefeld, Fed. Rep. Germany

In a collision experiment, formally described in terms of a free and an interacting Hamiltonian, H_0 and $H=H_0+V$, one measures only observables A that are constants of the motion of H_0, such as translation energy, angular momentum, or internal excitation of a projectile. (For simplicity, we exclude rearrangements.) The theoretical background of this limitation is[1] that only asymptotic constants, i.e. observables A for which the limits $\lim_{t\to\mp\infty} e^{iHt}Ae^{-iHt}$ exist, are open to experimental measurement in the remote past or future, and that every such A can be uniquely decomposed into a constant of the motion of H_0 plus an operator A' for which $\lim_{t\to\mp\infty} e^{iHt} \times A'e^{-iHt} = 0$. It follows that the stationary states of H_0 exhaust the possible pre- and postcollision states ρ_{in} and ρ_{out};[2] and that the scattering operator S, $\rho_{out} = S\rho_{in}S^{\dagger}$, indeed does not lead out of this class.[1]

This leads to the following symmetry dilemma: Suppose H_0 is the full nonrelativistic Hamiltonian for a molecule. Since it is invariant under inversion of all coordinates, no constant of the motion of H_0 can correspond to handedness of the molecule.[3] Thus, neither reactive nor elastic collisions should ever yield chiral molecular states. In reality the opposite is true: Optical isomers can never be brought into states of definite parity, as the inversion symmetry would suggest, but are always either left- or right-handed. So handedness is a "perfect quantum number";[3] it obeys a superselection rule; and it is even conserved in all elastic, and many reactive collisions.

A clue to the formal mechanism of such symmetry breaking is: If the interaction V in H is not suitably localized, the asymptotic time evolution need not be governed by H_0 and constants of the motion of H_0 may fail to be asymptotic constants of H.[4] New, interaction-created asymptotic constants may then dictate states ρ_{in} and ρ_{out} with symmetries other than those of H_0.[5]

It will be shown here that, under conditions stated below, the interaction between a molecule and the (quantized) radiation field is of such a persistent type. Indeed,[6] spontaneous photon emission from an excited molecular level undergoes a critical slowing down (with a line shape $\omega^{-1}(\ln \omega)^{-2}$ as the frequency ω goes to 0) when radiative corrections push the bare, excited level all the way to the ground level (this is nearly the case for the lowest inversion doublet of ammonia). Moreover,[6] when the upper level would be shifted below the ground level, the molecule acquires two inversion-noninvariant ground states which are separated by a superselection rule because they are dressed with infinitely many soft photons so as to carry opposite (transverse) magnetic fields of monopole-like long range. Analysis of the asymptotic constants (and their limits) of the underlying Weisskopf-Wigner type Hamiltonian yields, in the scattering framework, a corresponding spontaneous symmetry breakdown and a soft-photon cloud for the available pre- and postcollision states. A dressing transformation[7] allows the study of multiphoton processes for these mirror-image states, and the conclusion that no finite number of photons can effect transitions between them. Experimental consequences will be discussed.

References

1. P. Pfeifer and R.D. Levine, J. Chem. Phys. 78, in press (1983).
2. H. Primas, "Chemistry, Quantum Mechanics, and Reductionism", Lect. Notes Chem. 24 (Springer, Berlin, 1981), Chap. 5. Also Ref. 4, example 4.
3. P. Pfeifer, in: "Symmetries and Properties of Nonrigid Molecules", J. Maruani, Ed. (Elsevier, Amsterdam, 1983), in press.
4. P. Pfeifer, Adv. Chem. Phys. 53, 341 (1983).
5. J.V. Corbett, J. Math. Phys. 16, 271 (1975).
6. P. Pfeifer, Phys. Rev. A 26, 701 (1982).
7. L. Davidovich and H.M. Nussenzveig, in: "Foundations of Radiation Theory and Quantum Electrodynamics", A.O. Barut, Ed. (Plenum, New York, 1980), p. 83.

COLLISION INDUCED DIPOLE RADIATION OF NORMAL HYDROGEN GAS IN THE FREQUENCY RANGE OF THE COSMIC BACKGROUND

Joachim Schaefer and Wilfried Meyer

Max-Planck-Institut für Astrophysik, Garching, Fed. Rep. Germany,
Fachbereich Chemie, Universität Kaiserslautern, Fed. Rep. Germany.

Spontaneous emission intensities of collision induced dipole radiation of normal hydrogen gas have been calculated in first order time dependent perturbation approximation(1). In the numerical procedure we have used an ab initio vibrotor potential (M80) of H_2-H_2 for calculating bound state energies and eigenfunctions of the dimers and wave functions of the scattering system, and we have used the induced dipole moments of the collision complex determined in ab initio calculations with estimated errors of a few percent.

The continuum-bound emission contributes a rich spectral structure to the total emission spectrum(Fig.1), especially at low temperatures, while the continuum-continuum radiation provides a smooth continuum underground which is dominant at higher temperatures(Fig.2). The results have been derived by assuming a Maxwellian distribution of relative kinetic energies for the collision system.

The calculations gave converged results in the frequency range up to about 200 cm^{-1}, and for temperatures up to about 40 K. However, the energy resolution in the calculations was not sufficient to properly resolve the profiles of the lines below 20 cm^{-1}.

Although laboratory measurements are missing for collision induced hydrogen radiation at the desired temperature range, we claim that this type of radiation has been observed by H.P. Gush in a rocket-borne experiment intended to measure the Wien tail of the cosmic background radiation (CBR). The measured spectrum found there could be roughly explained by assuming that an excess over a thermal radiation curve of about 2.5 K is due to a hydrogen cloud of about 6 - 7 K, though the expected abundance of molecular hydrogen in that direction is orders of magnitude smaller. Fig. 3 shows the measured spectrum and the upper and lower bound of the expected thermal radiation. On the other hand, submillimeter radiation of Giant Molecular Clouds should have a significant contribution of dimer hydrogen emission(3).

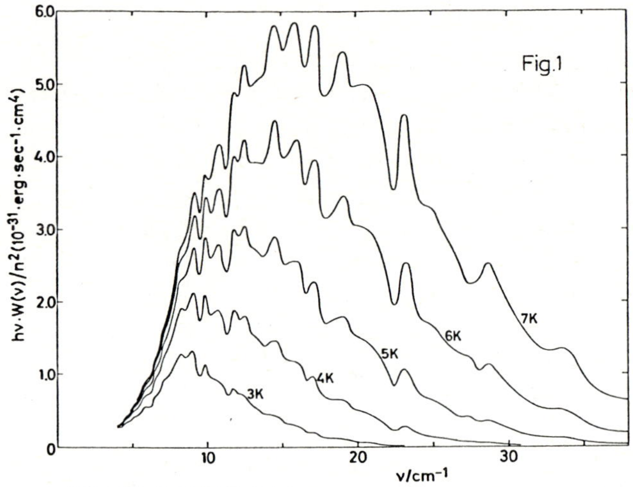

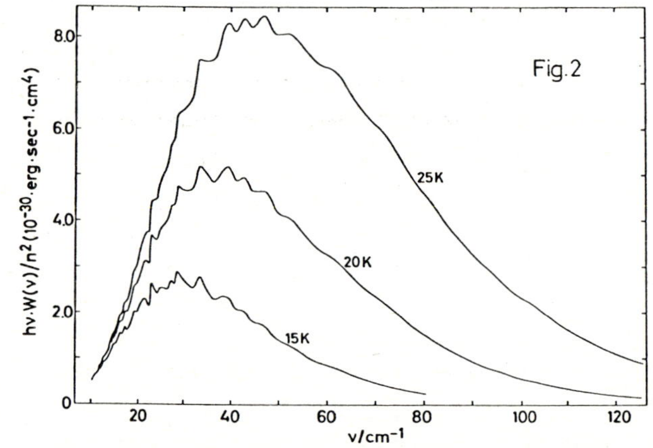

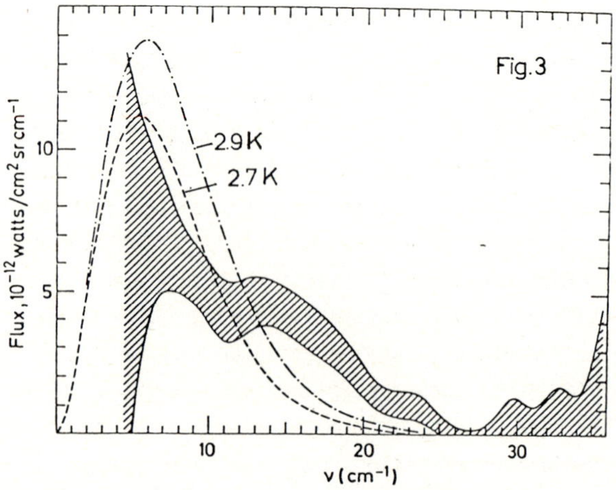

References:

(1) Simple systems have been handled in this formalism for instance by a) R.O. Doyle, J. Quant. Spectr. Radiat. Transt. 8, 1555 (1968) b) K.M. Sando and A. Dalgarno, Mol. Phys. 20, 103 (1971)

(2) H.P. Gush, Phys. Rev. Lett. 47, 745 (1981)

(3) G. Wiedemann, S. Drapatz, J. Schaefer, Frühjahrstagung der Deutschen Astronomischen Gesellschaft 1983, Konstanz.

A NEW ELECTRON-ELECTRON MULTI-COINCIDENCE APPARATUS

M. Völkel and W. Sandner

Fakultät für Physik, Hermann-Herder-Str.3, 7800 Freiburg, BRD

In this contribution, we present a brief description of a new electron-electron multi-coincidence apparatus. It was primarily designed to measure angular correlations between scattered electrons and Auger electrons, following inner shell ionization. First experimental results are presented at this conference [1].

The general setup is shown on fig.1. The incident electron beam of 0.2...50 keV energy intersects the target gas beam which effuses from a 0.2 mm diameter nozzle (not shown). Scattered electrons are detected by a 30° parallel plate analyzer, with the energy resolution typically set at 1.2%. The analyzer can be rotated around interaction volume in the range $15° < \theta_s < 145°$, thereby determining the energy and momentum transfer of scattered electrons.
The Auger electrons are detected in a cylindrical mirror analyzer with typically 0.25% resolution. It collects all electrons being emitted into a conical surface with 39.2° half opening angle. In polar coordinates, with $\theta = 0$ being the incident beam direction and $\phi = 0$ being the scattering plane we detect angles $50.8° < \theta < 129.2°$, $\phi = \arccos \{\cos(39.2°)/\sin\theta\}$. These directions are imaged onto a circular channel plate detector with 12 individual sector shaped anodes, allowing to detect a total of 12 mean emission directions $\{\theta_i, \phi_i\}$ simultaneously and indepentently.

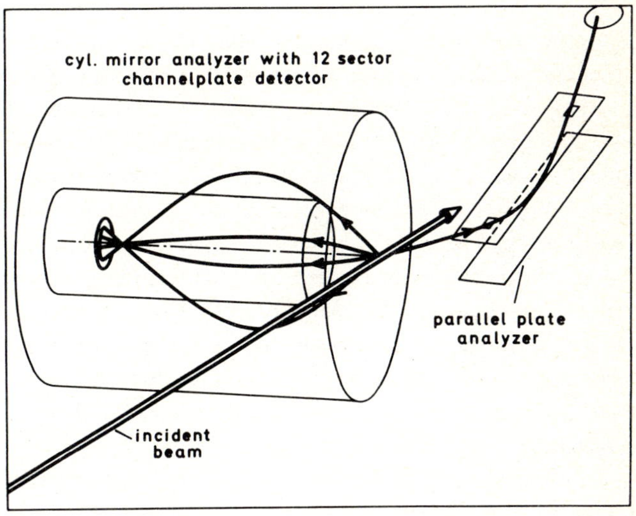

Fig.1: General setup

The angular correlation between scattered electrons and any of the 12 Auger detection lines is performed by the 12 : 1 multi-coincidence electronics (fig.2). It uses only one start-stop timing unit (50 ps resolution), with simultaneous digital position decoding of each accepted stop pulse. Time and position informations are utilized by a microcomputer to accumulate 12 separate coincidence spectra. The long time coincidence experiments are run and controlled by a separate master computer.

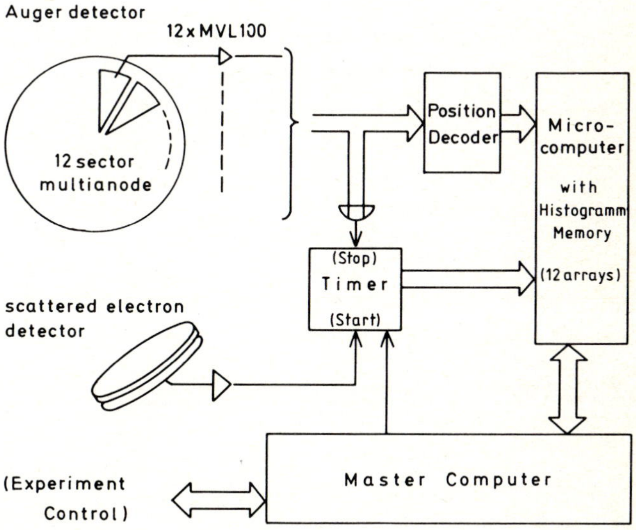

Fig.2: 12 : 1 - Coincidence Electronics

An essential feature of the new apparatus is the small coincidence width of about 1nsec. It has been achieved by consequent compensation of the electron flight time differences occuring inside the two energy analyzers. Details of the compensation methods are published separately [2].

References:

1. M. Völkel and W. Sandner, this book
2. M. Völkel and W. Sandner, J.Phys.E.Sci.Instr., in press

REDUCTION OF KINEMATIC LINE BROADENING IN THE AUGER ELECTRON SPECTRA OF FAST ION BEAMS

R. Bruch, A. Gutenkunst and K.H. Müller

Fakultät für Physik, Universität Freiburg, 7800 Freiburg, West Germany

Recently we proposed a method for the reduction of kinematic line broadening of collision induced projectile Auger decays using a refocusing technique[1]. Here we report the first measurement of projectile electron emission spectra where kinematic line broadening is drastically reduced by refocusing the image of the electron spectrometer. The experimental set-up is shown schematically in Fig. 1. Basically the apparatus consists of an electrostatic parallel-plate analyzer ($\alpha = 30^\circ$) and a refocusing unit. Some properties of this electron spectrometer facility should be noted. First, the analyzer can be rotated to measure angular distributions, second the target (gas, vapour, surface or foil) can be moved in and out of the focal region to measure delayed spectra and decay curves, third the exit slit of the analyzer can be translated to partially compensate kinematic peak broadening.

As a testing case of the proposed refocusing method we studied Auger-electron emission from 280 keV LiI(1s2ℓ2ℓ') states excited by $Li^+ + CH_4$ impact. For these core-excited states the transition energies are well known[2]. This facilitates the establishment of the electron energy scale in the source particle and laboratory frames.

The effect of refocusing is illustrated in Fig. 2 where the energy resolution of a specific Auger line is plotted versus the refocusing distance. From this figure it is clearly seen that the peak width decreases with increasing exit slit position and has a minimum close to the theoretically predicted refocusing distance[3].

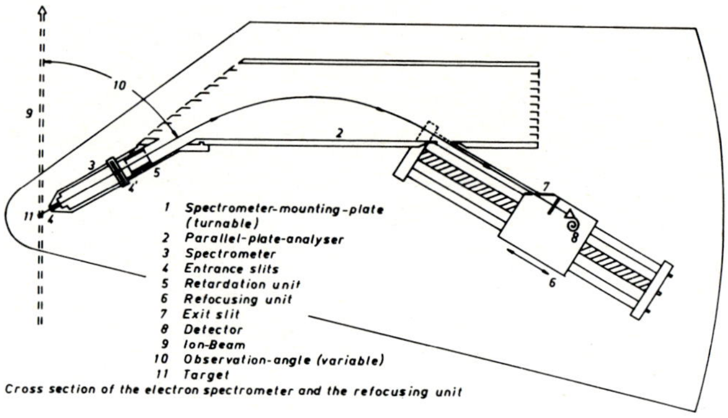

Fig. 1: Schematic diagram of the electrostatic parallel-plate analyzer.

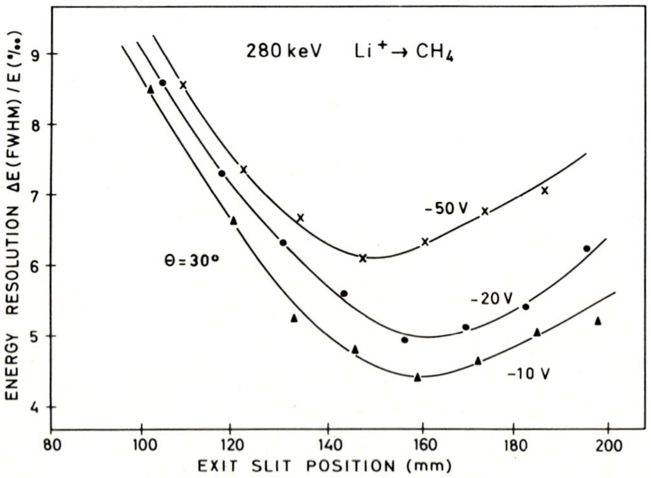

Fig. 2: Experimental energy resolution of the spectrometer as a function of the refocusing position.

References

1. P. Bachmann, A. Eberlein and R. Bruch, J. Phys. E 15, 207 (1982)
2. M. Rødbro, R. Bruch and P. Bisgaard, J. Phys. B 12, 2413 (1979)
3. A.Gutenkunst, Diplom-Thesis, Universität Freiburg (1982)

A COMPACT HIGH CURRENT PULSED ELECTRON GUN WITH SUB-NANOSECOND ELECTRON PULSE WIDTHS

M. A. Khakoo[+] and S. K. Srivastava

Jet Propulsion Laboratory, California Institute of Technology, Pasadena, California 91109

Pulsed electron guns with the capability of generating sharp electron pulses with nanosecond resolution are commonly used in delayed coincidence lifetime experiments[1] and in time-of-flight studies.[2] However, to generate sub-nanosecond wide electron pulses with adequate current intensities to enable such measurements to be made over reasonable experimental times is difficult especially when the observed collision process is of a low cross section value (10^{-18} - 10^{-20} cm^2). In such a case it is necessary to have not only adequately sharp electron pulses (sub-nanosecond fall-times) but also high electron intensities (peak pulsed currents of several tens of micro-amperes).

We have developed a pulsed electron gun capable of generating 700 pS wide (full width) electron pulses with peak instantaneous currents of 50 μA in height. Figure 1 shows a schematic diagram of the pulsed electron gun together with other components. The gun was magnetically collimated to increase the current intensity by a factor of 20 over conventional electrostatic guns and was double pulsed to obtain the required sub-nanosecond wide pulses.

The magnetic field, provided by two coils (Figure 1) of 1000 turns, inside diameter 3 cms and operated at 0.2-0.3 A, was approximately 150 Gauss. The electrons were thermionically generated by a tungsten hairpin filament F. The double pulsing method involving pulsing the negatively biased filament housing (-19 eV w.r.t. filament) H by a 20 V and 100 nS wide voltage pulse (risetime 10 nS) (Figure 2) and then sweeping the resulting burst of electrons across the aperture A_1 and A_2 (1 mm diameter) by the deflector P_1 by a delayed output from the same pulse generator. Both pulsed elements, i.e. F and P_1, were impedance matched to the pulsing electronics and pulsing coaxial cables (50Ω) by using capacitive coupling and 50Ω termination.

The size of the electron pulse widths was determined by observing the delayed fluorescence of the $1^1S \rightarrow 2^1P$ transitions in helium (584Å; lifetime 0.55 nS ± 4%)[4] in coincidence with the passage of the electron pulse through the collision region (Figures 1). The resulting time spectrum of 584Å photons is shown in Figure 2 and was deconvoluted,[5] assuming a Gaussian temporal profile for the electron pulse, to determine the electron pulse width to within 20%. The delayed coincidence calibration method was checked for resonance radiation trapping by repeating the calibration at lower values of gas beam pressures and for interference from $3^1P \rightarrow 1^1S$ photons by lowering the incident electron energy to below 22 eV.

[+]NASA-NRC Resident Research Associate.
Work is supported by NASA.

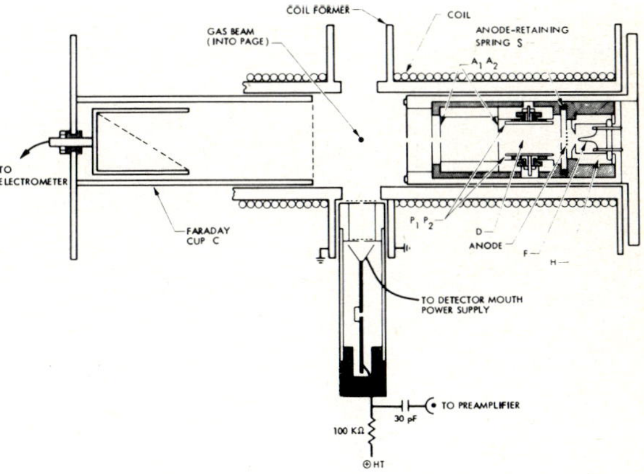

Fig. 1: A schematic diagram of the electron gun and the photon detection system used in the delayed coincidence calibration measurements. See text for description of labelling.

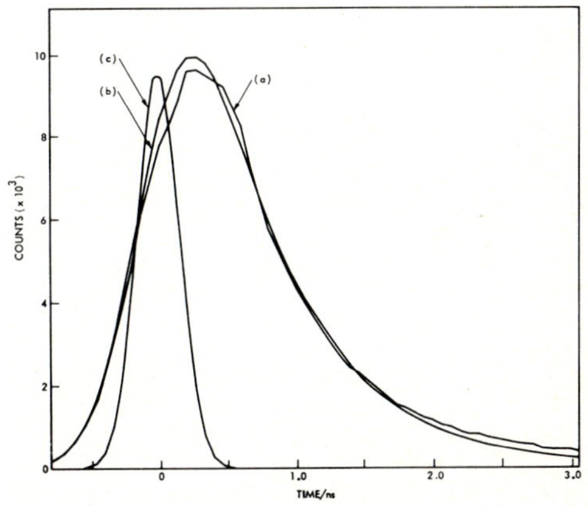

Fig. 2: (a) Typical delayed photon time spectrum observed in the calibration measurements photons. (b) Fitted function[5] using non-linear least squares (c) Deconvoluted temporal profile of the electron beam at the collision region.

References

1. A. Corney, Atomic and Laser Spectroscopy (Clarendon Press, Oxford, England) p. 173 (1977).
2. J. A. D. Stockdale, V. E. Anderson, A. E. Carter and L. Deleanu, J. Chem. Phys. 63 3886 (1975).
3. R. E. Kennerly, Rev. Sci. Inst. 46 1682 (1977).
4. M. T. Anderson and F. Weinhold, Phys. Rev. A 9 118 (1974).
5. R. E. Imhof and F. H. Read, Rep. Prog. Phys. 40 1 (1977).

ACO: A SOURCE OF MONOENERGETIC ELECTRONS FOR COLLISIONS WITH ATOMS AND MOLECULES

J.P. Ziesel[a], D. Field[b*], P.M. Guyon[c], T.R. Govers[d], M. Lavollée[a], O. Dutuit[c]

* For correspondence

Many processes in e-atom and e-molecule collisions are sharply resonant with electron kinetic energy. These include electronically, vibrationally and/or rotationally inelastic collisions and dissociative attachment. All these processes are important in a wide range of topics including atmospheric chemistry and physics, laser and plasma physics, the interstellar medium, fusion research and silicon device technology. In order accurately to characterize the resonances the electronic energy width should be smaller than the resonance width. Conventional electrostatic analyzers seem now, with a width of 10 meV, to be at their practical limit. There are good theoretical reasons for believing that the photoionization technique is the most promising for creating electron currents of useful magnitude for resolution below 10 meV. For the observation of rotational structure in e-molecule scattering a resolution of ∿1 meV (8 cm^{-1}) is required and this should in principle prove feasible using a photoionization source. The idea was first tested by W. Chupka who photoionized Ar with the continuum of a helium lamp; A. Gallagher and co-workers have photoionized metastable Ba atoms using a He-Cd laser. More recently we have developed a source to photoionize Xe at threshold using synchroton radiation. The initial energy resolution is determined by a combination of the synchrotron radiation bandwidth and the electric field across the photoionization region.

A transmission apparatus has been built to test this source at the ACO storage ring. The apparatus consists of a photoionization chamber, a lens system, and a collision chamber. In the photoionization chamber Xe is photoionized at the $^2P_{3/2}$ threshold at 1022.1 Å. The electrons thus formed are extracted by a 1.000 V cm^{-1} field into the lens system. The beam is then accelerated (or decelerated) to the requisite collision energy and focussed into the collision chamber, which contains the target gas. The transmitted current is measured with a channeltron detector. The attenuation due to scattering follows the Beer-Lambert law

$$I = I_o \exp - (NQ\ell)$$

where I, I_o are transmitted currents in the presence and absence of target gas, respectively, ℓ is the length of the chamber, N is the number density of target gas and Q the total scattering cross-section. The transmission method is well known to be a very sensitive technique for the detection of resonances (G.J. Schulz, Rev.Mod. Phys., 45, (1973), 378).

To test the apparatus we have looked at known resonances in N_2 at 2.3 eV and at 11.48 eV, in Ar at 11.08 eV, in He at 19.36 eV and in O_2 from 50 meV to 1 eV. All these resonances have been detected and their shapes are in good agreement with the work of others. The results of this pilot experiment are very encouraging since in a short time and without taking any special precautions we have been able to obtain an energy resolution of 30 meV, and have shown the potential of the apparatus, especially in the very low energy regime below 1 eV. Further experiments with improvements designed to upgrade the energy resolution are planned before the ICPEAC XIII meeting and we hope to present results on the scattering of low energy electrons (< 1 eV) by diatomic and polyatomic molecules such as HCl, CH_4 and CF_4.

a) L.C.A.M. Université Paris-Sud, Orsay
b) University of Bristol, U.K.
c) L.R.E.I. Université Paris-Sud, Orsay.

A NEW, INTERMEDIATE ENERGY, ELECTRON ENERGY LOSS SPECTROMETER.

G. Gerson B. de Souza, A. C. de A. e Souza, R. de B. Faria
Instituto de Química da Universidade Federal do Rio de Janeiro
Cidade Universitária - 21.910 - Rio de Janeiro - RJ - Brasil

A new, variable angle (0-60°), intermediate energy (0,5-3,0 keV), low resolution (1,0 eV), large energy-loss (0-500 eV) electron spectrometer has been designed and constructed.

The spectrometer, shown schematically in figure 1, is a crossed-beam apparatus, featuring a Möllenstedt-type electron velocity analyzer[1] and a triode gun as the electron source.

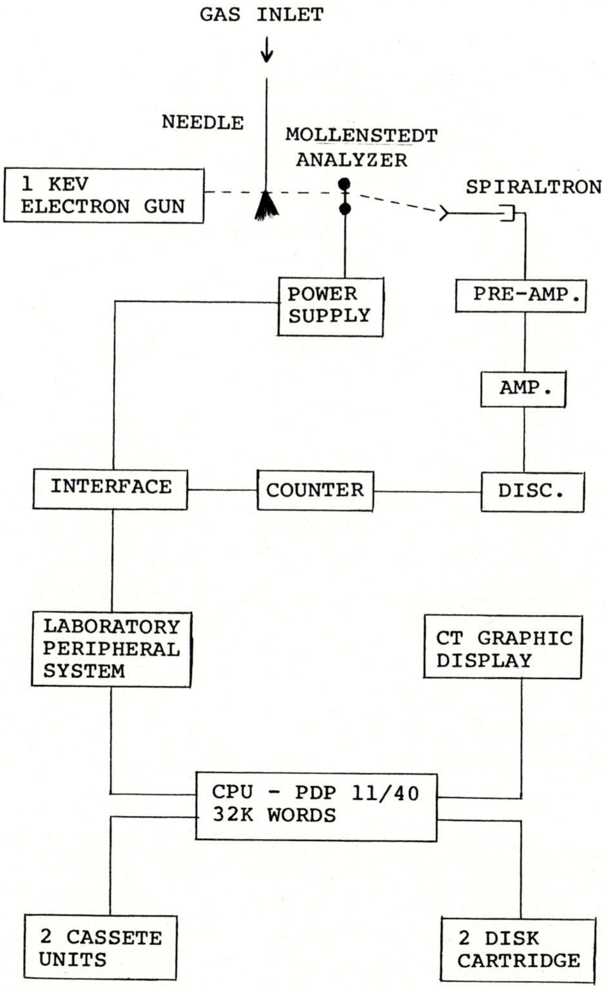

Figure 1: Schematic diagram of the Spectrometer

A versatile data acquisition system, based on a PDP 11-40 minicomputer has been designed and incorporated into the spectrometer.

Figure 2 shows the angular behavior of the spectra obtained for Nitrogen at 1 keV incident energy and θ= 1,2,3 and 4 degrees.

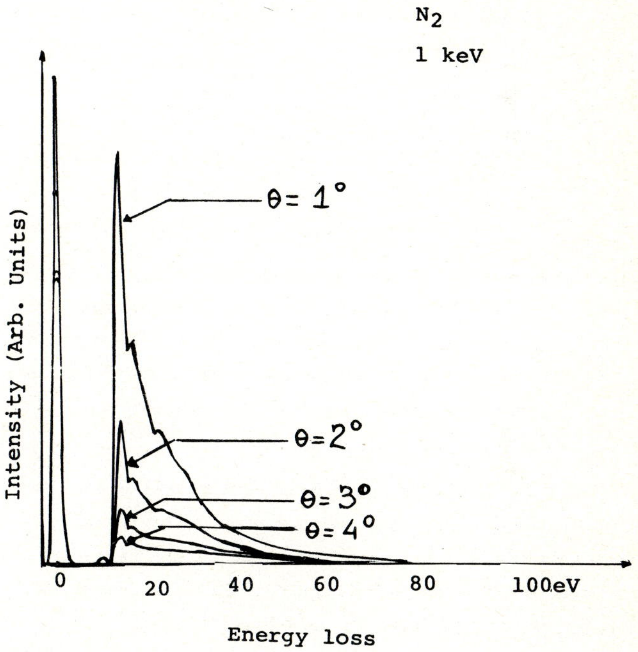

Figure 2 : Electron energy loss spectra for Nitrogen, 1 keV incident energy.

Financial support from the Conselho Nacional de Desenvolvimento Cientifico e Tecnologico (CNPq), Financiadora de Estudos e Projetos (FINEP) and Fundação Universitaria Jose Bonifacio (FUJB) is gratefully acknowledged.

Reference:

1. E.M.A. Peixoto, G.G.B. de Souza and M.C. A. Santos, Optik 53,405 (1979).

USE OF A SOFT X-RAY DIAGNOSTIC FOR HIGHLY CHARGED ION CURRENT MEASUREMENT

J.J. Bonnet, M. Bonnefoy and A. Fleury

CNAM Laboratoire de Physique des Collisions Atomiques 292 rue Saint-Martin 75141 Paris Cedex 03

and

S. Bliman, S. Dousson, D. Hitz and B. Jacquot

DRF-CENG 85X 38041 Grenoble Cedex - France

We describe an alternative technique for routine monitoring of highly charged ion currents. Our device uses soft X-ray emitted after charge exchange collision of the incident beam with a low density gas target.

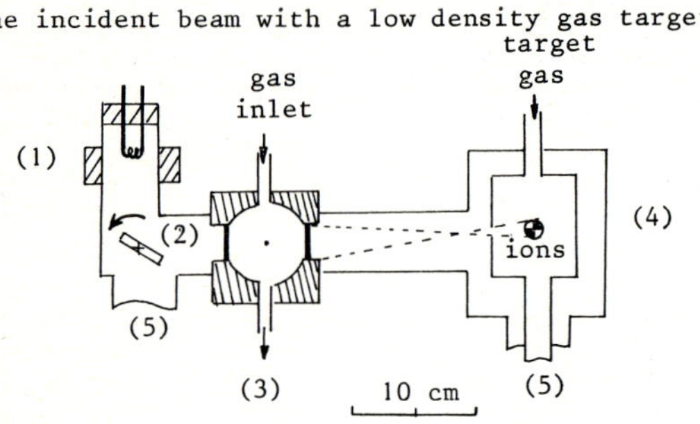

Fig. 1: Schematic of the experimental set-up : (1) miniaturised electron gun (2) rotating interchangeable anticathode (3) gas flow proportional counter supplied with two Makrofol side windows (2 μm thick) (4) differentialy pumped target gas all (5) pumping port.

Design considerations included the desire of a compact physical package and a good sensitivity down to at least 10^7 ions s^{-1}. The requirements are filled by the system illustrated on Figure 1. Precise energy calibration of the counter can be achieve with the X-ray tube which allows placement of a variety of anticathode to provide many characteristic lines in the 100-2 000 eV spectral range.

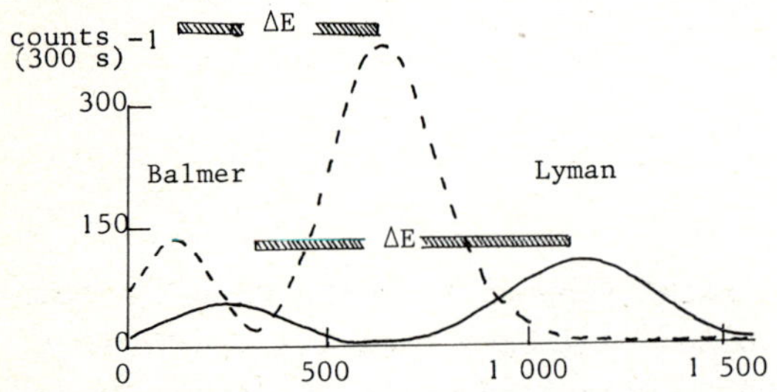

Fig. 2: Pulse height spectra : $O^{8+} + H_2$ collision (dashel line), $Ne^{10+} + H_2$ collision (full line).

Typical pulse height spectra corresponding to O^{8+}(10 KV) + H_2 and Ne^{10+}(10 KV) + H_2 collisions are illustrated on Figure 2. Both Balmer and Lyman series are in the viewing region of the X-ray detector. Measurement of the photon energy difference ΔE (see Figure 2) gives a reliable identification of the incident ionic species. Such monitoring can be extended to many others highly charged ions and is very useful for qualitative identification of very low intensity ion beams even after mass analysing.

Reference

1. R. Geller and B. Jacquot, Nucl. Instr. Meth. 184,293 (1981)

COMPACT PLANE-CRYSTAL X-RAY SPECTROMETER

W. Jitschin, B. Wisotzki, U. Werner, H. O. Lutz

Fakultät für Physik, Universität Bielefeld, F. R. Germany

In the field of energetic ion-atom collisions considerable interest has concentrated on few-electron (highly ionized) systems and multiple ionization. In such studies it is often necessary to employ high resolution detection devices in order to resolve the corresponding x-ray lines from neighboring lines. Gaseous or solid-state detectors generally have low resolution for energies below 10 keV, and the use of Doppler-tuned instruments is limited to special cases. Crystal spectrometer can yield high resolution and may be used over a broad x-ray energy range. Therefore, a rugged and portable instrument has been built for applications in accelerator-based experiments.

The instrument employs a plane diffraction crystal and simple mechanics (Fig. 1). X-rays enter through a collimating Soller slit; they hit a crystal where they are reflected according to Bragg's law; the deflected x-rays are detected in a small proportional counter. Tuning to different wavelengths is achieved by rotating crystal and proportional counter by a Θ, 2Θ drive which is operated by a computer controlled stepping motor.

A crucial point for good energy resolution is the necessary high x-ray collimation. Using a commercial Soller slit system an angular divergence of ca. 0.07° (FWHM) was achieved in the x-ray energy range investigated (1 to 7 keV). The reflectivity peak of customary crystals has a smaller width; also the accuracy of the mechanical drive is better than 0.01°. Thus the instrumental resolution is mainly determined by the 0.07° Soller slit collimation and amounts to e.g. 800 at $\Theta = 45^{\circ}$. This is better than the natural width of most x-ray lines (Fig. 2,3). The absolute efficiency of the instrument was measured to correspond to an effective solid angle of $3 \cdot 10^{-7}$ steradian at 4 keV x-ray energy.

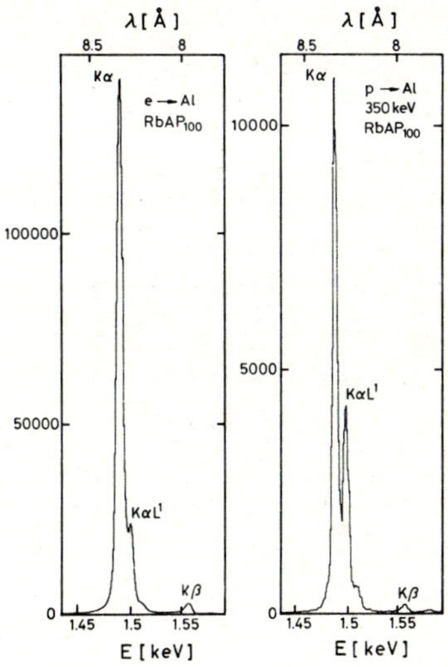

Fig. 2: Recorded Al K-spectra for 14 keV electron and 350 keV proton impact, RbAP crystal.

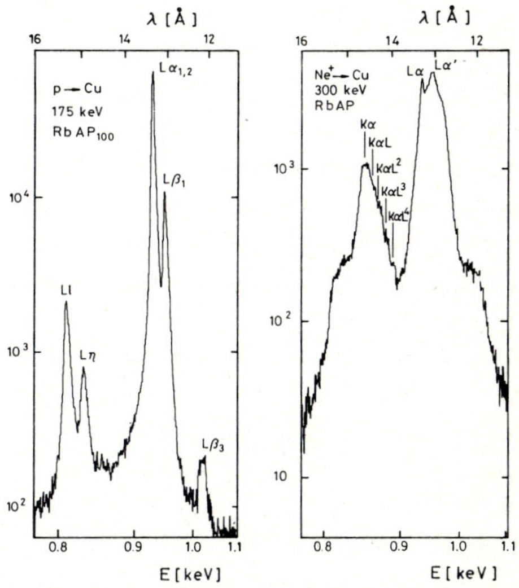

Fig. 3: Recorded Cu L-spectra for 175 keV proton and 300 keV Ne^{+} impact, RbAP crystal.

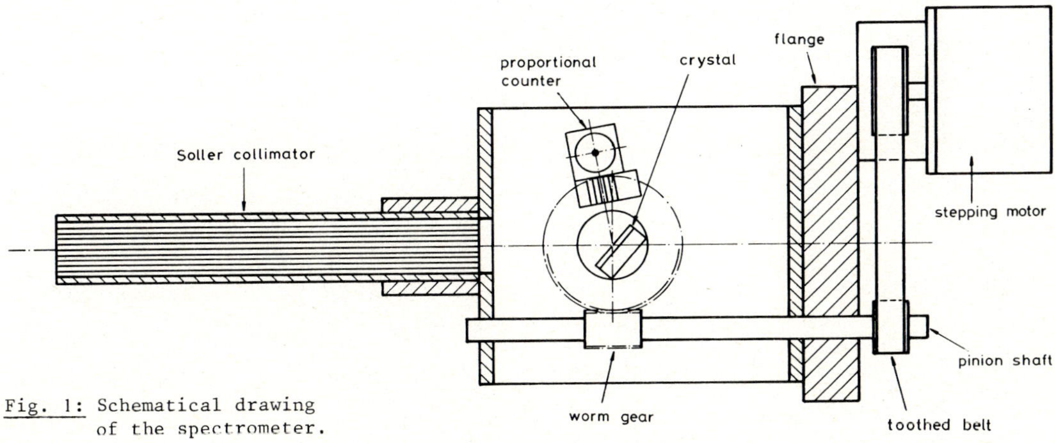

Fig. 1: Schematical drawing of the spectrometer.

X-RAY SPECTROSCOPY OF TOKAMAK PLASMAS: AN EXPERIMENTAL AND THEORETICAL STUDY*

E. Källne and J. Källne

Harvard Smithsonian Centre-Astrophysics, Cambridge, Massachusetts 02138 USA

A.K. Pradhan

Department of Physics, University of Windsor, Windsor, Ontario, Canada N9B 3P4

High resolution measurements made at the Alcator C tokamak (MIT) of the x-radiation from ions in the Helium isoelectronic sequence, Sulphur, Chlorine and Argon, are reported and their relevance to plasma diagnostics is pointed out.[1,2] Systematic variations in the line intensities, originating from transitions to the ground state $1s^2\ ^1S_0$ from the n = 2 states 2^1P, $2^3P_{1,2}$ and 2^3S_1, are studied as a function of plasma parameters as well as with nuclear charge Z. The observations are interpreted by means of, and compared with, theoretical calculations employing accurate excitation rates and other atomic data.

The averaged density and the temperature of the plasma, Ne and Te, is $1\text{-}2\times10^{14}$ cm^{-3} and 1-2 keV respectively. The Z-dependence of the line ratios is obtained for approximately constant plasma conditions that are deduced to be in nearly coronal equilibrium. The present measurements are the first of their kind and it is hoped that these will lead to the development of useful diagnostics for fusion plasmas. Fig. 1 shows a sample observed spectrum.

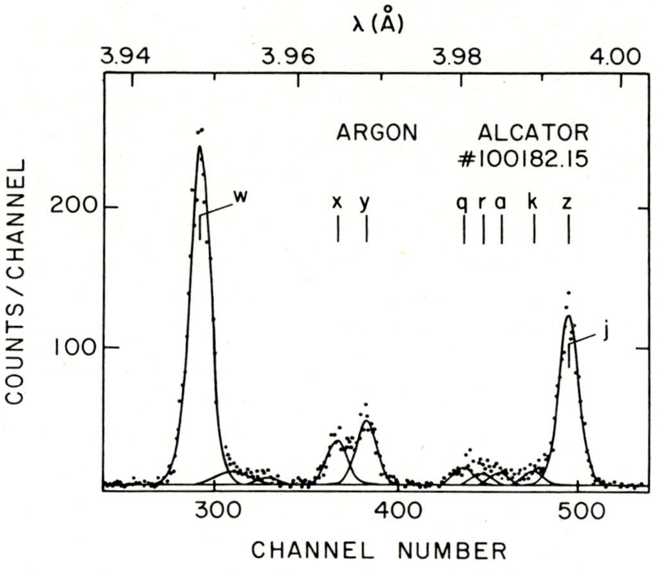

Fig. 1

* As part of the TOXRASD (Tokamak X-Ray Spectroscopy Diagnostics) program.

This work was supported by the U.S. Department of Energy and the Natural Sciences and Engineering Research Council of Canada.

References

1. E. Källne, J. Källne and A.K. Pradhan, Phys. Rev. A, in press (1983).
2. Ibid, Phys. Rev. A (Rapid Commun.), submitted, (1983).

A NEW TECHNIQUE FOR THE STUDY OF CHARGE CHANGING COLLISIONS USING RECOIL IONS

L. Liljeby, H. Cederquist, H. Danared, G. Astner, A. Bárány and A. Johnson
Research Institute of Physics, S-104 05 Stockholm, Sweden

The feasibility of using recoil ions produced by high-energy projectiles for collision studies was demonstrated by Cocke and coworkers [1,2]. In their set-up a pulsed beam from a tandem Van de Graaf is led through a gas target. Recoil ions with very low energy (eV) are produced, extracted by electrostatic fields and brought to a second gas target where the collisions under study occur. In a typical reaction both projectile and target atom change their charge. Often the net change of charge is not zero, i.e. electrons are lost. Hence, it is necessary to determine the charge of both the outgoing projectile and the (secondary) recoil which is also extracted by an electric field. The charges of the incoming projectile and the recoil are determined by the use of time-of-flight techniques; the time-of-flight for particles of equal mass after acceleration in an arbitrary electric field is proportional to $(\text{charge})^{-\frac{1}{2}}$. In order to determine also the charge of the projectiles after a collision, these were sent through an electrostatic analyser. If the energy loss (gain) in the collision is neglected, particles with equal initial charge/final charge ratio will pass the analyser at the same voltage. When the analyser is swept, one can collect data for all charges simultaneously in a list-mode experiment, registering the two flight times and the analyser setting.

We have copied this apparatus for an experiment [3] now in operation at the Stockholm 225-cm cyclotron, using 110 MeV $^{12}C^{6+}$ ions as primary projectiles. With the currents available (typically 30 pA after strong collimation and external pulsing to $\Delta t \simeq 1.5\ \mu s$) this leads to running times of many hours, in order to get reasonable statistics. It is clear that this experiment is severely hampered by what might be called the duty cycle of the electrostatic analyser. This is open to a certain charge only a fraction of time roughly equal to the relative energy resolution of the analyser, in our case $\simeq 1/30$. In order to improve the overall efficiency of the apparatus, we have modified it in the following way: The electrostatic analyser was replaced by an electrode system giving a field parallel to the secondary projectiles and accelerating them further after the collision. Hence, particles that have equal charge before the collision, but change their charge with unequal amounts are affected differently and become separated in time of flight.

Fig.1 shows a part of the time spectrum using argon in the two gas cells. Between the peaks representing events with no change in charge in the second cell, peaks occur corresponding to charge changing collisions, e.g. 5 → 4, 5 → 3 and 5 → 2. The measuring time for the data in the figure was 10 minutes. Coincidences with the secondary recoil ions have not yet been combined with this new technique.

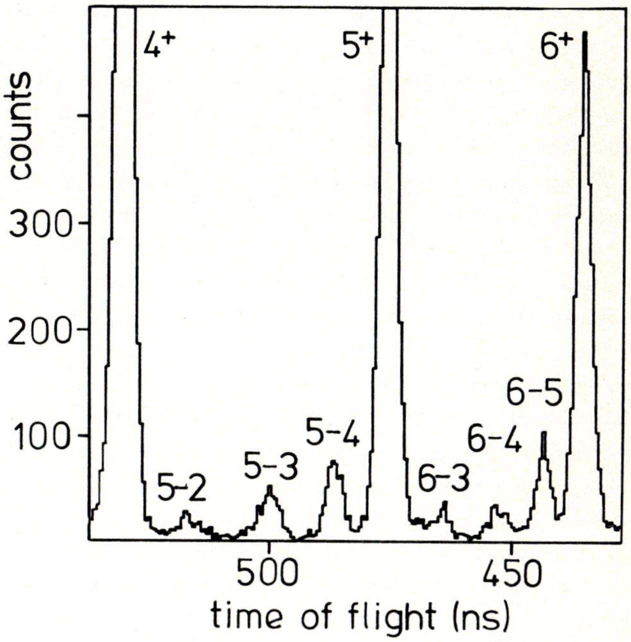

Fig.1: Part of a time-of-flight spectrum for Ar recoil ions produced in collisions with 110 MeV $^{12}C^{6+}$ ions. Between the peaks (4^+, 5^+, 6^+) with no change of charge in the second gas cell peaks occur corresponding to charge changing events. The time resolution is 4 ns.

References

1. C.L. Cocke, R. DuBois, T.J. Gray, E. Justiniano and C. Can, Phys. Rev. Letters 46, 1671 (1981)
2. E. Justiniano, C.L. Cocke, T.J. Gray, R.D. DuBois and C. Can, Phys. Rev. A24, 2953 (1981)
3. G. Astner, A. Bárány, H. Cederquist, H. Danared, P. Hvelplund, A. Johnson, H. Knudsen, L. Liljeby and L. Lundin, Physica Scripta, in press.

CONSTRUCTION PERFORMANCE OF A POSITION-SENSITIVE PARTICLE DETECTOR WITH DELAY LINE READOUT

B. Liu*, W. Enders, D. Liesen, R. Schulze

Gesellschaft für Schwerionenforschung, D-6100 Darmstadt

*Permanent Address: Institute of Modern Physics, Academia Sinica, Lanzhou, China

For several reasons the study of inner-shell excitation processes especially in collisions with very small impact parameters has been intensified in the last few years. Since in such collisions besides the scattered primary particles also recoiling target atoms can give a considerable contribution to the scattering intensity under the corresponding lab-angles (typically $\geq 15°$), an unambiguous determination of the kinematics is necessary for a proper determination of the impact parameter. Additionally, a high angular resolution of the particle detector is desirable in order to observe possible structures in the impact parameter dependence of the excitation processes.

Both conditions can be fulfilled by a position-sensitive parallel plate detector similar to that one developed by P. Fuchs et al.[1]. The detector consists of a 2 μ thick metalized Hostaphan foil and an especially designed delay line. The delay line is built up from 54 concentric circles of 1 mm width and 1 mm distance from each other and its upper compared to its lower part is shifted by half a period. Both the foil and the delay line are cut into two halves so that the two corresponding regions of the total azimuth are independently covered. The total delay for each half amounts to 180 nsec which corresponds to a specific propagation time of about 140 psec/cm. Without any change of the total delay time and thus with the same resolution also 4 quarters of the whole azimuth can be measured separately, namely just by turning the foil 90° with respect to the delay line. The distance between foil and delay line is 5 mm and the whole assembly is separated from the vacuum by an additional 2 μ Hostaphan foil (25 mm in front of the detector foil) which is supported by a Ni grid.

After first tests with 5.5 MeV α particles from an Am source, several measurements have been performed in the Stripperhalle. It turned out during these experiments that for x-ray-scattered particle coincidences an optimal performance could be reached using Isobuthylen gas at 12 Torr pressure and applying -800 V to the foil and 0 V to the delay line. Fig. 1 shows the energy signals of 1.4 MeV/N U projectiles taken at the detector foil and at the delay line. The pulses from the preamplifier were almost equal in height (~ 300 mV) with a rise time of about 8 nsec. A change of the sign of the high voltage at the foil resulted in a drastic decrease of the height of the delay line pulses. Fig. 2 gives the time spectrum (which corresponds to the angular dependence of the scattering intensity) which was obtained also with 1.4 MeV/N U projectiles. Both curves (the upper was taken without and the lower with coincidence condition) show the same characteristic features corresponding to the ring structure of the delay line and showing a rather good angular resolution. However, the orgin of this structure is not yet fully understood.

[1]P. Fuchs et al., GSI Scientific report 1977, p. 195

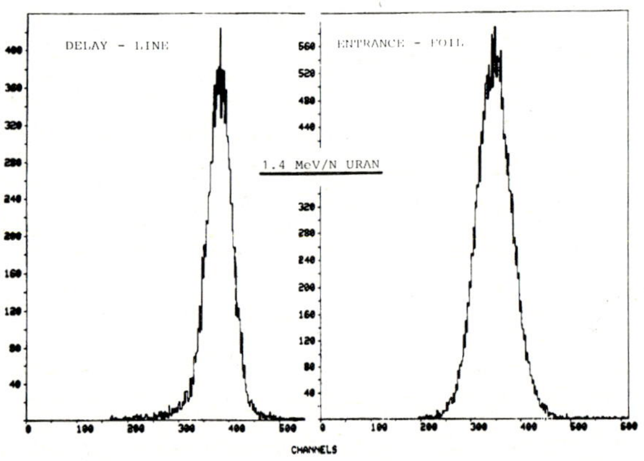

Fig. 1: Energy spectra of 1.4 MeV/N U projectiles.

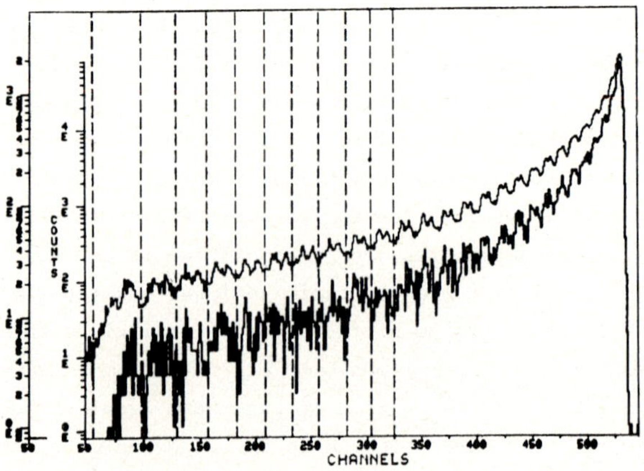

Fig. 2: Time spectra without (upper curve) and with (lower curve) coincidence condition.

AN ALTERNATIVE METHOD FOR THE MEASUREMENT OF THE EFFICIENCY OF AN ION DETECTOR

F. Brouillard, S. Chantrenne, W. Claeys, A. Cornet and P. Defrance

Institut de Physique Corpusculaire, Université Catholique de Louvain
B-1348 Louvain-la-Neuve - BELGIUM

The straightforward method for calibrating the efficiency (γ) of an ion detector is to observe the counting rate (N) produced by a beam of known intensity (i). Then $\gamma = Nq/i$.

But there is an upper limit to the counting rate of a detector (typically 10^5 counts/sec) and the corresponding current intensity of the order of 10^{-15}A is not easy to measure precisely. It is therefore valuable to have a means of attenuating the intensity of the ion beam in a reliable way.

In the method reported here, that attenuation is achieved by sweeping the ion beam, fast and repeatedly, across a narrow slit located in front of the detector.

By fast we mean that the sweep time T across the slit is smaller than the response time of the detector, so that only one pulse can be produced during one sweep. The probability that one pulse actually takes place is related to i, T and γ in the following way.

First, the probability P(n) that n ions pass through the slit is given by the Poisson formula

$$P(n) = \frac{1}{n!} \left(\frac{T}{\tau}\right)^n e^{-T/\tau}.$$

Where τ is the average time between two ions in the beam :

$$\tau = \frac{q}{i}.$$

Second, the probability p(n) that one pulse is delivered when n ions reach the detector (in a time smaller than the response) is clearly :

$$p(n) = 1 - (1-\gamma)^n.$$

The probability π of one pulse is therefore :

$$\pi = \sum_{n=o}^{\infty} (1-(1-\gamma)^n) \frac{1}{n!} \left(\frac{T}{\tau}\right)^n e^{-T/\tau} = 1 - e^{-\gamma \frac{T}{\tau}}.$$

If the sweep is repeated with frequency ν, the counting rate of the detector will therefore be :

$$N = \nu \left(1 - e^{-\gamma \frac{Ti}{q}}\right). \qquad (1)$$

The efficiency is thus obtained from the measurement of N/ν and i.

The experimental set-up is shown in figure 1. The slit is located at the bottom of a Faraday cup, that is used for the measurement of i. An electrostatic deflector is used for sweeping the ion beam. The time T can be calculated from the geometry and voltage features, but it can also be measured directly by replacing the detector by a second Faraday cup and measuring the current i' transmitted through the slit. This measurement can be performed with large intensities :

$$T = \frac{i'}{\nu i}.$$

The method has been used to measure the detection efficiency of channelplates for 3 keV protons. The measurement was repeated at different values of i. The behaviour of N/ν versus i predicted by (1) was verified as shown in figure 2.

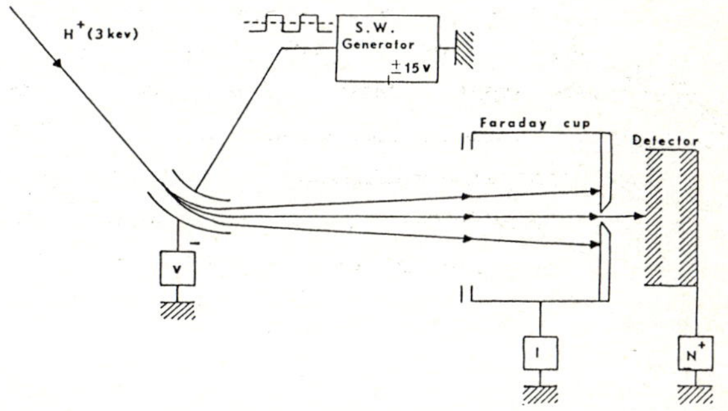

Fig. 1

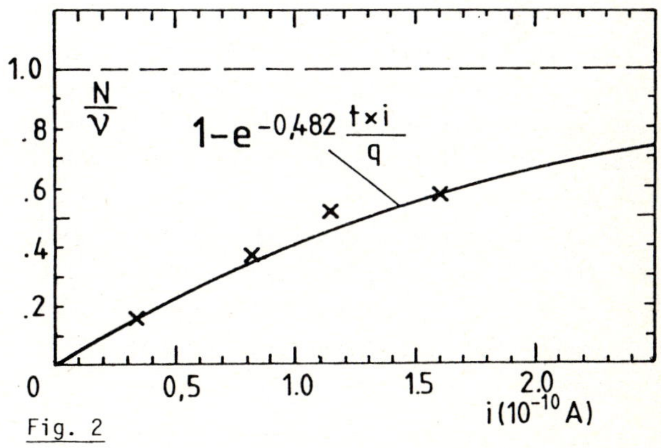

Fig. 2

MEASUREMENT OF DISSOCIATION FRACTION IN AN ELECTRON IMPACT HEATED ATOMIC HYDROGEN OVEN

G. A. Glass, L. H. Andersen, R. Holmes, J.-P. Rozet, D. A. Taylor,
R. S. Thoe*, S. B. Elston, M. Breinig, and I. A. Sellin

University of Tennessee, Knoxville TN 37996, U.S.A. and
Oak Ridge National Laboratory, Oak Ridge TN 37830 U.S.A.

An electron impact heated hydrogen dissociation oven for use in producing atomic hydrogen inside a gas cell for ion-atom collision experiments utilizes a 1.2 kW electron beam focused by a spherical lens system to heat the oven to temperatures in excess of 2400 K. High dissociation fractions are found with cell pressures $\gtrsim$ 5 mTorr. The compact design of the oven allows placement of an electron spectrometer close to the interaction region (length $\sim$ 1.5 cm) as well as direct measurement of the oven temperature by means of an optical pyrometer.

The magnetic fields produced by the electron beam ($\lesssim$ 40 milligauss) are orders of magnitude smaller than the fields associated with low-voltage, high-current heating methods. This fact offers distinct advantages when relatively low-energy electrons ejected in ion-atom collisions are to be observed.

One widely used means of determining the dissociation fraction in the oven involves comparison of the double electron capture signals with the oven cold and with the oven hot.[1] However, another method for determining this fraction has been devised. The dissociation fraction can be obtained by direct measurement of the ratio of yields of single capture to and single loss from bound states of a projectile ion along with ratios of other published cross sections.[2] The results and the limits of applicability of this method will be presented.

This research was supported in part by the Fundamental Interactions Branch, Division of Chemical Sciences, Office of Basic Energy Sciences, U. S. Department of Energy, under contracts DOE DE-AS05-79ER10512 and W-7405-eng-26.

References

1. J. E. Bayfield, Rev. Sci. Instr. 40, 869 (1969).
2. T. V. Goffe, M. B. Shah, and H. B. Gilbody, J. Phys. B 12, 3763 (1979).

* Present address: Lawrence Livermore Laboratory, Building 415, Livermore CA 94550, USA.

TOTAL ELECTRON-ATOM AND ELECTRON-MOLECULE SCATTERING CROSS-SECTIONS BY PHOTOELECTRON SPECTROSCOPY

Vijay Kumar, E. Krishnakumar and K.P. Subramanian
Physical Research Laboratory
Ahmedabad 380 009, India

It has previously been demonstrated by Kumar and Krishnakumar[1] that the powerful technique of photoelectron spectroscopy could be used to measure the electron-atom, electron-molecule scattering cross-sections at low electron energies. Since the experimental set up[1] was not specifically built for the electron scattering studies, it had the following limitations: 1) Only a few gas targets with larger cross-sections could be studied; 2) the scattering in the accelerating and analyzing regions in the photoelectron spectrometer could not be eliminated entirely; 3) only relative values of scattering cross-sections could be obtained. In view of this, a new experimental set up has been designed and fabricated which takes care of all these limitations.

The new photoelectron spectrometer is basically a cylindrical mirror analyzer and the ionizing source is a combination of VUV light source and an one-meter near normal incidence monochromator. Two separate high speed differential pumping arrangements have been used for fast evacuation of accelerating/analyzing region and electron detector (channeltron) region respectively. The pressure measurements in the ionizing region are made in an absolute fashion.

The experiment consists of measuring the intensities of the peaks in the photoelectron spectra of a source gas, e.g. some heavy monoatomic gas like xenon or krypton, at a constant pressure. The photoelectrons thus produced by the source gas get scattered by the target gas whose cross-section is to be determined. The electron scattering is measured as a function of pressure of the target gas. Each peak in the photoelectron spectrum of the source gas provides one point in the electron energy scale. Complete scanning of the electron energy is done by varying the energy of the ionizing radiation.

Results on total electron-helium scattering cross-sections would be presented in the conference.

References

1. Vijay Kumar and E. Krishnakumar, J. Electron Spectr. Rel. Phenom. 24, 1 (1981)

NEW CONCEPTS FOR TREATING RELATIVISTIC COLLISIONS

Felix T. Smith

1030 Palo Alto Avenue, Palo Alto, CA 94301, USA

It is surprising that ways have not been found to extend many of the productive calculational techniques of nonrelativistic dynamics, classical and quantal, into the relativistic regime by such means as an analytic expansion in powers of $\beta_i=|u_i|/c$.

A new fundamental examination[1] shows that the root of the difficulty has been the usual, but incorrect, formulation of the coordinate vector of space-time location as $\rho_i^o=(\vec r_i, ct_i)$--a form that is not covariant, but that changes from spacelike to timelike as t_i increases from zero. If time is measured as t_o from the cosmological big-bang singularity, $R=ct_o$ is a natural time-varying unit for scaling lengths, $\gamma_i=|r_i|/R$, and ρ_i can be written in the securely covariant form $\rho_i=R(\hat r_i\sinh\eta_i, \cosh\eta_i)$, $\tanh\eta_i=\gamma_i$, comparable to the 4-velocity $\upsilon_i=c(\hat u_i\sinh\epsilon_i, \cosh\epsilon_i)$, $\tanh\epsilon_i=\beta_i$.

These single-particle vectors are always timelike. Interparticle 4-vectors are always spacelike,

$$\rho_{ij}=\rho_i-\rho_j=R|2\sinh(\eta_{ij}/2)|(\hat r_{ij}\cosh h_{ij},\ \sinh h_{ij}), \quad (1)$$

$$\upsilon_{ij}=\upsilon_i-\upsilon_j=c|2\sinh(\epsilon_{ij}/2)|(\hat u_{ij}\cosh e_{ij},\ \sinh e_{ij}). \quad (2)$$

Self-consistent covariant definitions can be given for the vectors $\rho_{o(ij)}$, $\upsilon_{o(ij)}$ describing the motion of the center of mass. Their connection with their nonrelativistic analogs can be systematically established.

In the 4-vectors ρ_i and υ_i, R and c are scalar invariants, and so is $t_o=R/c$, the <u>progress time</u> for Hamiltonian dynamics and the Schrödinger equation. This scalar time is distinct from the frame-dependent observer times of different particle world-lines $t_i=(R/c)\cosh\eta_i$, in each case a single component of the 4-vector ρ_i.

Sommerfeld showed[2] that velocity vectors υ_i terminate on a sphere of imaginary radius ic in Minkowski space--i.e., a pseudosphere with constant negative curvature in noneuclidean geometry. The same geometry applies to the location vectors ρ_i. The interparticle vectors ρ_{ij} are chords through this pseudospherical surface; the corresponding arcs have a geodesic length $R\eta_{ij}$, an invariant generalization of the nonrelativistic length $|r_{ij}|$.

It is useful to consider the dynamics of particles entirely in the 3-space of the pseudosphere, describing the unit 4-vector ρ_i/R by the dimensionless 3-vector $\vec\eta_i=(\eta_i,\theta_i,\phi_i)$. The velocity 4-vector υ_i=/c, similarly, is described by the dimensionless 3-vector $\vec\epsilon_i=(\epsilon_i,\zeta_i,\omega_i)$. By comparing components of $\dot\rho_i$ and υ_i we can get formulas connecting $\vec\eta_i$ and $\vec\epsilon_i$,

$$t_o\dot\eta=(\hat r\cdot\hat u)\cosh\eta\,\sinh\epsilon-\sinh\eta\,\cosh\epsilon\,,$$

$$t_o\sinh\eta\,\dot\theta=\sinh\epsilon\,(\cos\theta\,\sin\zeta\,\cos(\omega-\phi)-\sin\theta\,\cos\zeta\,),$$

$$t_o\sinh\eta\,\sin\theta\dot\phi=\sin\epsilon\,\sin\zeta\,\sin(\omega-\phi).$$

With these variables, Hamiltonian dynamics and the Schrödinger equation can be rewritten for particles in the covariant language of special relativity, and the array of familiar nonrelativistic calculation techniques can be extended into the relativistic regime. The formulation of Hamilton's equations in this noneuclidean space will be illustrated, and an approach to relativistic corrections to the Schrödinger equation will be explored. Applications to collision problems will be suggested, and wider implications of the new point of view will be discussed.

<u>Footnotes</u>

[1] Felix T. Smith, to be published.

[2] A. Sommerfeld, Phys. Z. <u>10</u>, 826 (1909).

A FAST ASYMPTOTIC PACKAGE

H.E. Saraph and M.J. Seaton
Department of Physics and Astronomy, University College London, WC1E 6BT, UK

The coupled integro differential equations

$$(h + k^2 + \underset{\sim}{U})\,\underset{\sim}{\Psi} = 0,\quad h = \frac{d^2}{dr^2} - \frac{\ell(\ell+1)}{r^2} + \frac{2z}{r} \qquad (1)$$

that describe a collision process simplify, as the radial distance from the scattering centre increases, first to coupled differential equations

$$(h + k^2 + \underset{\sim}{V})\,\underset{\sim}{F} = 0,\quad \underset{\sim}{V} = \sum_{\lambda=1}^{\Lambda} \underset{\sim}{B}^{(\lambda)}\, r^{-\lambda-1} \qquad (2)$$

and then to the uncoupled equations

$$(h + k^2)\, f = 0,\quad r \to \infty \qquad (3)$$

that have well-known solutions. The scattering parameters are obtained by fitting the solutions of (1) to functions having the correct asymptotic forms, given by solutions of (3). The integration of the integro-differential equations (1) requires highly sophisticated techniques. Therefore, as soon as a distance r_a outside the ion core is reached, where equation (2) is valid, one switches to less demanding techniques. The problem now is to generate solutions of (2) that have correct asymptotic forms and good linear independence, and to obtain the collision parameters by fitting at r_a.

Asymptotic expansions for the solutions of (2) are usually not valid at r_a but are valid at some larger distance r_b making it necessary to integrate equations (2) numerically. When the distance $r_b - r_a$ is large this can be very time-consuming and cause numerical problems. For an up-to-date discussion of techniques and inherent problems see reference [1].

However, it is often found that $\underset{\sim}{V}$ is a small perturbation and hence that the functions $\underset{\sim}{F}$ do not differ by more than a few percent at r_a from the known functions f, solutions of (3). Using a Green's function expression for F

$$\begin{aligned}\underset{\sim}{F} &= f + f_1 \int_p^r f_2\, \underset{\sim}{V}\, \underset{\sim}{F}\, dx - f_2 \int_q^r f_1\, \underset{\sim}{V}\, \underset{\sim}{F}\, dx \\ &= f + f_1\, \underset{\sim}{\alpha} - f_2\, \underset{\sim}{\beta},\end{aligned} \qquad (4)$$

where f, f_1, f_2 are solutions of (3) and where $f_1 (d f_2/dr) - f_2 (d f_1/dr) = 1$. When $\underset{\sim}{V}$ is a small perturbation, $\underset{\sim}{\alpha}$ and $\underset{\sim}{\beta}$ are small and may be calculated using perturbation theory. For open channels

$$\underset{\sim}{\alpha}(r) = -\int_r^\infty f_2\, \underset{\sim}{V}\, f\, dx,\quad \underset{\sim}{\beta}(r) = +\int_r^\infty f_1\, \underset{\sim}{V}\, f\, dx. \qquad (5)$$

The case of closed channels will also be discussed. By transforming to integration contours in the complex x plane, one obtains integrands which are exponentially decaying and the integrals can be evaluated efficiently [2].

Results obtained with this approach are encouraging, cutting the time spent on the asymptotic solutions by an order of magnitude. A comprehensive account of the method will be presented.

References

[1] D.W. Norcross, 'Numerical methods for asymptotic solutions of the scattering equations' in 'Atoms in Astrophysics' (Eds: P.G. Burke, W. Eissner, D.G. Hummer and I.C. Percival). Plenum, New York (1983).

[2] N.C. Sil, M.A. Crees and M.J. Seaton: to be published.

ON THE CALCULATION OF QUANTAL MATRIX ELEMENTS OF A FUNCTION BETWEEN UNBOUND STATES

Staffan Yngve

Institute of Theoretical Physics, University of Uppsala, Thunbergsvägen 3, S-752 38 Uppsala, Sweden

Consider the one-dimensional Schrödinger equation

$$\frac{d^2u(x)}{dx^2} + [k^2 - U_c(x)]\, u(x) = 0 , \quad x_0 < x < +\infty \quad (1)$$

and let

$$Q^2_{k,c}(x) = k^2 - U_c(x) + \text{a possible k-independent modification} \quad (2)$$

c being a parameter in the potential (e.g. the angular momentum quantum number if $\hbar^2 U_c(x)/(2m)$ is the effective potential of a radial problem). It is assumed that $U_c(x) \to 0$ as $x \to +\infty$. Consider the solution $u_{k,c}(x)$ of (1), which satisfies the boundary condition $u_{k,c}(x) \to 0$ as $x \to x_0$, where x_0 may be $-\infty$ (for which case we must assume $U_c(-\infty) > k^2$). This boundary condition determines the solution $u_{k,c}(x)$ uniquely except for a constant factor, which we determine by the requirement that $u_{k,c}(x)$ shall have an amplitude equal to unity for large positive values of x and furthermore that $u_{k,c}(x)$ [for real values of c] shall be positive for sufficently large negative values of x.

We shall consider the matrix element

$$\langle k,c|f(x)|k',c'\rangle = \int_{x_0}^{+\infty} u_{k,c}(x)\, f(x)\, u_{k',c'}(x)\, dx , \quad (3)$$

where $f(x)$ is a so far unspecified function. In [1] an approximate phase-integral formula for the matrix element in (3) is obtained, which for the lowest order of approximation reads

$$\langle k,c|f(x)|k',c'\rangle = \tfrac{1}{2}(kk')^{\frac{1}{2}} \int_{x_t}^{\infty} Q^{-\frac{1}{2}}_{k,c}(x)\, Q^{-\frac{1}{2}}_{k',c'}(x)\, f(x)$$

$$\times\, e^{i[w^{(1)}_{k,c}(x) - w^{(1)}_{k',c'}(x)]}\, dx, \quad k > k' , \quad (4)$$

where

$$w^{(1)}_{k,c}(x) = \int_{x_t}^{x} Q_{k,c}(x)\, dx , \quad (5)$$

and where x_t satisfies

$$Q^2_{k,c}(x_t) = 0 . \quad (6)$$

The right-hand member of (4) and the corresponding formula in higher order approximations may be evaluated numerically by the aid of a computer program available at our institute or in some cases even analytically using the saddle-point method.

For the particular case

$$U_c(x) = a^2c^2e^{-ax} , \quad -\infty < x < +\infty , \quad (7)$$

$$f(x) = e^{-\gamma\, ax} , \quad \mathrm{Re}\,\gamma > 0 , \quad (8)$$

the author has shown that the matrix element (3) may be evaluated exactly as

$$\langle k,c|e^{-\gamma ax}|k',c'\rangle =$$

$$= (4\pi^2 a)^{-1} \left[\frac{2\pi k}{a} \sinh\left(\frac{2\pi k}{a}\right) \frac{2\pi k'}{a} \sinh\left(\frac{2\pi k'}{a}\right)\right]^{\frac{1}{2}}$$

$$\times\, e^{-\gamma \ln c^2}\, \Gamma(\gamma+i(k'+k)/a)\, \Gamma(\gamma+i(k'-k)/a)$$

$$\times\, \Gamma(\gamma-i(k'+k)/a)\, \Gamma(\gamma-i(k'-k)/a) . [\Gamma(2\gamma)]^{-1}$$

$$\times\, e^{i(k'/a)\ln(c'^2/c^2)}$$

$$\times\, {}_2F_1(\gamma+i(k'+k)/a), \gamma+i(k'-k)/a;\ 2\gamma;\ 1-c'^2/c^2) . \quad (9)$$

For the particular case $c' = c$ we have

$$\langle k,c|e^{-\gamma ax}|k',c\rangle =$$

$$= (4\pi^2 a)^{-1} \left[\frac{2\pi k}{a} \sinh\left(\frac{2\pi k}{a}\right) \frac{2\pi k'}{a} \sinh\left(\frac{2\pi k'}{a}\right)\right]^{\frac{1}{2}}$$

$$\times\, e^{-\gamma \ln c^2}\, \Gamma(\gamma+i(k'+k)/a)\, \Gamma(\gamma+i(k'-k)/a)$$

$$\times\, \Gamma(\gamma-i(k'+k)/a)\, \Gamma(\gamma-i(k'-k)/a)/\Gamma(2\gamma) . \quad (10)$$

Particular cases of (9) have earlier been obtained by Jackson and Mott [2], Mies [3], Eno and Balint-Kurti [4].

If c' and c are real and if $\gamma = \alpha + i\beta$, where $\alpha > 0$ and β is real, we obtain the matrix element $\langle k,c|e^{-\alpha ax}\cos(\beta ax)|k',c'\rangle$ by taking the real part of the right-hand member of (9) and $\langle k,c|e^{-\alpha ax}\sin(\beta ax)\, k',c'\rangle$ by taking the imaginary part of the same member.

In an investigation comparing exact and phase-integral matrix elements it is shown with the aid of (9) that the approximate phase-integral matrix elements of $e^{-\alpha ax}\cos(\beta ax)$ and $e^{-\alpha ax}\sin(\beta ax)$ in general become less accurate with increasing β and eventually break down .

References

1. N. Fröman, P.O. Fröman and F. Karlsson, Molec. Phys. 38, 749 (1979).
2. J.M. Jackson and N.F. Mott, Proc. Soc. A137, 703 (1932).
3. F.H. Mies, J. Chem. Phys. 40, 523 (1964).
4. L. Eno and G.G. Balint-Kurti, Chem. Phys. 23, 295 (1977).

THE NORMALIZED WAVEFUNCTION CLOSE TO THE CENTRE OF FORCE FOR BOUND AND UNBOUND STATES

Nanny Fröman, Per Olof Fröman and Staffan Yngve

Institute of Theoretical Physics, University of Uppsala, Thunbergsvägen 3, S-752 38 Uppsala, Sweden

Consider a non-relativistic particle moving in a spherically symmetric potential, which behaves as $-Z\hbar^2/(m\, a_0\, r)$ close to $r=0$, where in atomic physics a_0 is equal to the Bohr radius. The state considered corresponds to the angular momentum quantum number l and the energy E. The radial Schrödinger equation is written

$$\frac{d^2u}{dr^2} + Q^2(r,E)\, u = 0 \tag{1}$$

where, with obvious notations,

$$Q^2(r,E) = \frac{2m}{\hbar^2}\left(E - V(r) - \frac{\hbar^2 l(l+1)}{2mr^2}\right) =$$

$$- \frac{l(l+1)}{r^2} + \frac{2Z}{a_0 r} + a_0^{-2} \sum_{n=0}^{\infty} b_n (r/a_0)^n , \tag{2}$$

the last member representing $Q^2(r,E)$ close to $r=0$.

We normalize the real, unbound wavefunction according to

$$\int_0^\infty u(r,E)\, u(r,E')\, dr = \delta(E-E') , \tag{3}$$

and the bound state wavefunction according to

$$\int_0^\infty u^2(r,E_{n'})\, dr = 1 , \tag{4}$$

where $E_{n'}$ is an eigenvalue with the radial quantum number n'. Using results concerning the normalization integral obtained in [1], formula (7.28) in [2] and an appropriate connection formula (after generalization to phase-integral approximation of arbitrary order), we obtain the normalized radial wavefunction in the vicinity of $r=0$ as

$$u(r,E) = \left(\frac{2mc_l}{\hbar^2}\right)^{\frac{1}{2}} r^{l+1} \text{ (1 + higher powers of r) for unbound states,} \tag{5a}$$

and

$$u(r,E_{n'}) = \left(\frac{2mc_l}{\hbar^2} \frac{dE_{n'}}{dn'}\right)^{\frac{1}{2}} r^{l+1} \text{ (1 + higher powers of r) for bound states} \tag{5b}$$

where $dE_{n'}/dn'$ is to be obtained by means of spectroscopic data, and where

$$c_l = \underset{r=0}{\mathrm{Res}} \frac{1}{[u_l(r,E)]^2} , \tag{6}$$

$u_l(r,E)$ being chosen such that $u_l(r,E)/r^{l+1}$ tends to 1 as r tends to zero. The coefficients c_l are obtained with the aid of eqs. (1) and (2) as

$$c_l = \frac{(2Z/a_0)^{2l+1}}{[(2l+1)!]^2} \left(\prod_{s=0}^{l} \left\{ 1 + \left[\frac{b_0}{Z^2} - \frac{l(l+1)}{2} \frac{b_1}{Z^3}\right] s^2 + \right.\right.$$

$$\left.\left. + \frac{3}{2} \frac{b_1}{Z^3} s^4 \right\} + r_l \right) , \tag{6a}$$

where

$$r_0 = r_1 = 0 \tag{6b}$$

$$r_2 = 18\left[\frac{b_2}{Z^4} + \frac{b_3}{Z^5} + \left(\frac{b_1}{Z^3}\right)^2\right] \tag{6c}$$

$$r_3 = 45\left[6 \frac{b_2}{Z^4} + 22 \frac{b_3}{Z^5} + 13\left(\frac{b_1}{Z^3}\right)^2\right] + 675\left[4 \frac{b_4}{Z^6} + \frac{b_1}{Z^3} \frac{b_2}{Z^4} + \right.$$

$$\left. + 6 \frac{b_5}{Z^7}\right] + 45 \frac{b_0}{Z^4}\left[14 \frac{b_2}{Z^4} + 18 \frac{b_3}{Z^5} + 27\left(\frac{b_1}{Z^3}\right)^2\right] . \tag{6d}$$

For s-states we recognize the usual Fermi-Segrè formula [3].

As an application we have considered bound and unbound states for an exactly soluble Schrödinger equation (cf. Ref.4), for which r_l turns out to be zero. We have shown that our formula for the normalized wavefunction close to $r=0$ is in complete agreement with the exact formula for the bound-state wavefunction of that model, while for the unbound case our formula is not exact but very accurate for sufficiently small values of l. For bound states in the model considered, the figure below illustrates the relative error of formulas given in [5] and [6], respectively, and the exactness of the formula obtained in the present work. Corresponding formulas in [7] and [8] applied to the first order for the model considered give the correct behaviour of the normalization up to the order of magnitude b_1/Z^3, whereas going to the next order does not introduce a significant correction.

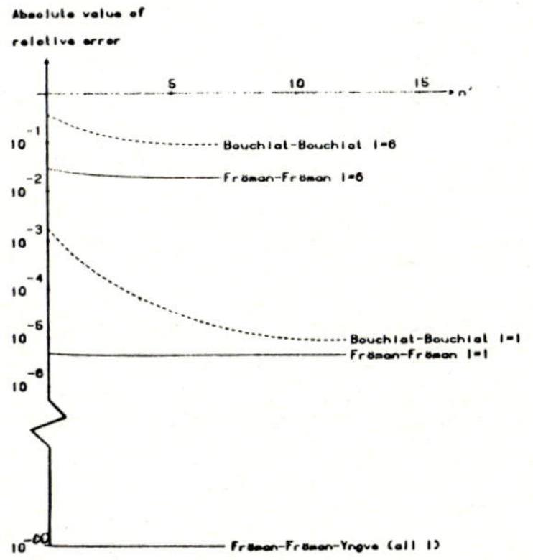

Details on the model and the parameter values used are found in [6]

References

1. S. Yngve, J. Math. Phys. 13, 324-331 (1972).
2. N. Fröman and P.O. Fröman, JWKB-Approximation. North-Holland Publishing Company, Amsterdam 1965.
3. N. Fröman and P.O. Fröman, Phys. Rev. A6 2064-2067 (1972).
4. U. Myhrman, J. Math. Phys. 21, 1732-1739 (1980).
5. M.A. Bouchiat and C. Bouchiat, J. Physique 35, 899-927 (1974).
6. N. Fröman and P.O. Fröman, J. Physique 42, 1491-1504 (1981).
7. J. McEnnan, L. Kissel and R.H. Pratt, Phys. Rev. A13 532-559 and 2325 (1976).
8. Z.R. Iwinski, Y.S. Kim and R.H. Pratt, Phys. Rev. C19. 1924-1937 (1979), A22, 1358-1360 (1980).

ASYMPTOTIC EXPANSIONS OF COULOMB WAVEFUNCTIONS OBTAINED BY MEANS OF THE SADDLE-POINT METHOD

Staffan Yngve

Institute of Theoretical Physics, University of Uppsala Thunbergsvägen 3, S.752 38 Uppsala, Sweden

The differential equation satisfied by the Coulomb wavefunction is $[\eta = \pm Z/(ka_0)]$

$$\frac{d^2u(r)}{dr^2} + \left(k^2 - \frac{2k\eta}{r} - \frac{l(l+1)}{r^2}\right) u(r) = 0 , \tag{1}$$

where $\eta = +|\eta|$ corresponds to a repulsive Coulomb potential and $\eta = -|\eta|$ to an attractive potential, a_0 being the Bohr radius.

The saddle-point method has been applied to the Coulomb wavefunctions e.g. in [1]. The treatment in our investigation differs from previous treatments such as [1] e.g. at the following three important points:

1) The quantities kr, η and l are no longer restricted to real values, but they may be complex.
2) The saddle-point method is in a consistent way applied to give a correct asymptotic behaviour of the approximation in the limit of very large values of kr.
3) Higher-order contributions are brought along. These contributions are of extreme numerical importance.

When l is small and r is situated on the positive axis, where $k^2 - 2k\eta/r - l(l+1)/r^2 > 0$ we obtain

$$F_l(\eta,kr) \mp i\, G_l(\eta,kr) = \mp\, iQ_\eta^{-\frac{1}{2}}(kr) \exp\{\pm i[w_1 - \tfrac{1}{2}\pi l + \sigma_l]\}$$

$$\times \{1 \pm iw_3 - \tfrac{1}{2}(Y_3 + w_3^2) \pm i(w_5 - \tfrac{1}{2}w_3 Y_3 - \tfrac{1}{6}w_3^3) - \tfrac{1}{2}(Y_5 - \tfrac{3}{4}Y_3^2 + 2\, w_3 w_5 - \tfrac{1}{2}w_3^2 Y_3 - \tfrac{1}{12} w_3^4) \pm i(w_7 - \tfrac{1}{2}w_3 Y_5 - \tfrac{1}{2}w_5 \times Y_3 - \tfrac{3}{8}w_3 Y_3^2 - \tfrac{1}{2}w_3^2 w_5 + \tfrac{1}{12} w_3^3 Y_3 + \tfrac{1}{120} w_3^5) - \tfrac{1}{2}(Y_7 - \tfrac{3}{2}Y_3 Y_5 + \tfrac{5}{8}Y_3^3 + 2\, w_3 w_7 + w_5^2 - \tfrac{1}{2}w_3^2 Y_5 - w_3 w_5 Y_3 + \tfrac{3}{8}w_3^2 Y_3^2 - \tfrac{1}{3}w_3^3 w_5 + \tfrac{1}{24}w_3^4 Y_3 + \tfrac{1}{360} w_3^6) + \ldots\} = \mp i\, Q_\eta^{-\frac{1}{2}}(1 + Y_3 + Y_5 + Y_7 + \ldots)^{-\frac{1}{2}} \exp\{\pm i(w_1 + w_3 + w_5 + w_7 + \ldots)\}, \tag{2}$$

where in the usual way the upper signs and the lower signs should be taken together, respectively, and where

$$Q_\eta(kr) = dw_1/d(kr) = [1 - 2\eta/(kr)]^{\frac{1}{2}} , \tag{3a}$$

$$w_1 = krQ_\eta - \eta \ln(kr - \eta + krQ_\eta) + \eta , \tag{3b}$$

$$w_3 = \frac{1}{12\eta} - \frac{1}{12\eta\, Q_\eta^3}\left(1 - \frac{3\eta}{kr} + \frac{9\eta^2}{2(kr)^2}\right) + \frac{l(l+1)}{2\eta}(1 - Q_\eta) , \tag{3c}$$

$$Y_3 = \frac{1}{Q_\eta}\frac{dw_3}{d(kr)} = \frac{1}{8\eta^2 Q_\eta^6}\left(\frac{4\eta^3}{(kr)^3} - \frac{3\eta^4}{(kr)^4}\right) - \frac{l(l+1)}{2(kr)^2 Q_\eta^2} , \tag{3d}$$

$$w_5 = \frac{1}{360\eta^3}\left(1 - \frac{1}{Q_\eta^9}\right) + \frac{1}{1920\eta^3 Q_\eta^9}\left(\frac{48\eta}{kr} - \frac{168\eta^2}{(kr)^2} + \frac{280\eta^3}{(kr)^3} + \frac{510\eta^4}{(kr)^4} - \frac{630\eta^5}{(kr)^5} + \frac{315\eta^6}{(kr)^6}\right) + \frac{l(l+1)}{16\eta^3 Q_\eta^5}\left(-\frac{4\eta^3}{(kr)^3} + \frac{3\eta^3}{(kr)^4}\right) + \frac{[l(l+1)]^2}{12\eta^3}\left(-1 + \frac{1}{Q_\eta} - \frac{\eta}{krQ_\eta} - \frac{\eta^2}{2(kr)^2 Q_\eta}\right) , \tag{3e}$$

$$Y_5 = \frac{1}{Q_\eta}\frac{dw_5}{d(kr)} = \frac{1}{16\eta^4 Q_\eta^{12}}\left(-\frac{24\eta^5}{(kr)^5} + \frac{22\eta^6}{(kr)^6} - \frac{21\eta^7}{(kr)^7} + \frac{63\eta^8}{8(kr)^8}\right) + \frac{l(l+1)}{16\eta^4 Q_\eta^8}\left(\frac{12\eta^4}{(kr)^4} - \frac{16\eta^5}{(kr)^5} + \frac{9\eta^6}{(kr)^6}\right) - \frac{[l(l+1)]^2}{8(kr)^4 Q_\eta^4} \tag{3f}$$

$$w_7 = \frac{1}{1260\eta^5} + \frac{1}{\eta^5 Q_\eta^{15}}\left(-\frac{1}{1260} + \frac{\eta}{84kr} - \frac{13\eta^2}{168(kr)^2} + \frac{143\eta^3}{504(kr)^3} - \frac{143\eta^4}{224(kr)^4} + \frac{143\eta^5}{160(kr)^5} - \frac{503\eta^6}{192(kr)^6} + \frac{83\eta^7}{64(kr)^7} - \frac{903\eta^8}{512(kr)^8} + \frac{633\eta^9}{512(kr)^9} - \frac{1899\eta^{10}}{5120(kr)^{10}}\right) + \frac{l(l+1)}{\eta^5 Q_\eta^{11}}\left(\frac{3\eta^5}{4(kr)^5} - \frac{11\eta^6}{16(kr)^6} + \frac{21\eta^7}{32(kr)^7} - \frac{63\eta^8}{256(kr)^8}\right) + \frac{[l(l+1)]^2}{60\eta^5}\left(-1 + \frac{1}{Q_\eta^7}\right) + \frac{[l(l+1)]^2}{60\eta^5\, Q_\eta^7}\left(-\frac{7\eta}{kr} + \frac{35\eta^2}{2(kr)^2} - \frac{35\eta^3}{2(kr)^3} + \frac{35\eta^4}{8(kr)^4} - \frac{161\eta^5}{8(kr)^5} + \frac{309\eta^6}{16(kr)^6}\right) + \frac{[l(l+1)]^3}{30\eta^5}\left(1 - \frac{1}{Q_\eta^3}\right) + \frac{[l(l+1)]^3}{10\eta^5\, Q_\eta^3}\left(\frac{\eta}{kr} - \frac{\eta^2}{2(kr)^2} - \frac{\eta^3}{6(kr)^3} - \frac{\eta^4}{8(kr)^4}\right) \tag{3g}$$

$$Y_7 = \frac{1}{Q_\eta}\frac{dw_7}{d(kr)} = \frac{1}{16\,\eta^6\, Q_\eta^{18}}\left(\frac{180\eta^7}{(kr)^7} - \frac{39\eta^8}{2(kr)^8} + \frac{205\eta^9}{(kr)^9} - \frac{825\eta^{10}}{4(kr)^{10}} + \frac{1899\eta^{11}}{16(kr)^{11}} - \frac{1899\eta^{12}}{64(kr)^{12}}\right) + \frac{l(l+1)}{16\,\eta^6\, Q_\eta^{14}}\left(-\frac{60\eta^6}{(kr)^6} + \frac{54\eta^7}{(kr)^7} - \frac{169\eta^8}{2(kr)^8} + \frac{63\eta^9}{(kr)^9} - \frac{315\eta^{10}}{16(kr)^{10}}\right) + \frac{[l(l+1)]^2}{16\,\eta^6\, Q_\eta^{12}}\left(\frac{28\eta^6}{(kr)^6} - \frac{47\eta^7}{(kr)^7} + \frac{103\eta^8}{4(kr)^8}\right) - \frac{1}{16(kr)^6}\frac{[l(l+1)]^3}{Q_\eta^6} . \tag{3h}$$

The expansions in the third member of (2) have the property of Wronskian conservation for any order of approximation. We recognize (2) with (3) as the modified higher-order phase-integral approximation of [2] corresponding to the omission of the centrifugal barrier in the first-order approximation. We similarly obtain saddle-point approximations not subject to the restriction of small l, which we recognize as the higher-order phase-integral approximations corresponding to the modification $l(l+1) \rightarrow (l+\frac{1}{2})^2$ in the first-order approximation. In the investigation we also obtain the phase-integral approximations of the Coulomb wavefunctions in different parts of the complex plane (apart from the vicinity of the pole and the zeros of the function $k^2 - 2k\eta/r - l(l+1)/r^2$).

References

1. I. Bloch, M.H. Hull jr., A.A. Broyles, W.G. Bouricius, B.E. Freeman and G. Breit, Phys. Rev. 80, 553 (1950).
2. N. and P.O. Fröman, Ann. Phys. (N.Y.) 83, 103 (1974).

ADIABATIC AND DIABATIC MOLECULAR STATES OF H_2 AT SMALL INTERNUCLEAR DISTANCES

F. Borondo, A. Macías and A. Riera.

Departamento de Química Física y Química Cuántica. Centro Coordinado.
CSIC-UAM. Universidad Autónoma de Madrid.
CANTOBLANCO (Madrid, 34) (Spain)

We have performed a full CI calculation of the energies of the first five adiabatic $^1\Sigma_g^+$ states of H_2, which are presented in Fig. 1. In this calculation we have used a basis set consisting of five 1s, three $2p_z$ and two of each $2p_x$ and $2p_y$. The radial couplings between these states have been calculated (exactly) using the method proposed by Macías and Riera[1], and are presented in Fig.2

In Fig. 1 one can observe the existence of a infinite series of pseudocrossings which was explained for the first time by Davidson[2] in terms of diabatic states with character $\sigma_g\sigma_g'$ and $\sigma_u\sigma_u'$. However a quantitative treatment of the diabatic states is still lacking.

We have calculated accurately the energies and (electrostatic and radial) couplings of these states using the block-diagonalization method[3]. We present in Fig. 1 the energy curves of the first four $\sigma_g\sigma_g'$ and the first two $\sigma_u\sigma_u'$ diabatic states superimposed to those of the adiabatic states.

We have also calculated the energies and couplings of the first five $^1\Sigma_u^+$ adiabatic states of H_2, using a basis set of five 1s, four $2p_z$ and two of each $2p_x$ and $2p_y$ GTO's centered on each nuclei. The corresponding total energies are displayed in Fig. 3, where one can appreciate the existence of another series of pseudocrossings in the energy curves. In this case we have found that these avoided crossings do not correspond quantitatively to molecular orbital diabatic structures, rather to structures with covalent or ionic character. We have calculated the corresponding $^1\Sigma_u^+$ diabatic energies by diagonalization of the radial couplings between adiabatic states.

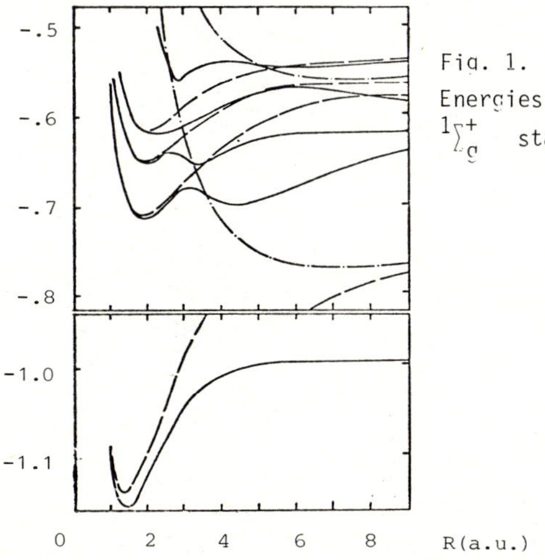

Fig. 1. Energies for $^1\Sigma_g^+$ states

—— adiabatic; ---- diabatic $\sigma_g\sigma_g'$; -·-·-·diabatic $\sigma_u\sigma_u'$

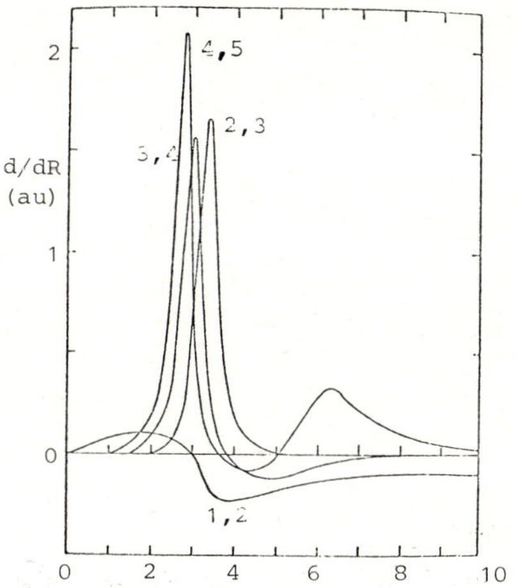

Fig. 2. Radial couplings $^1\Sigma_g^+$ states

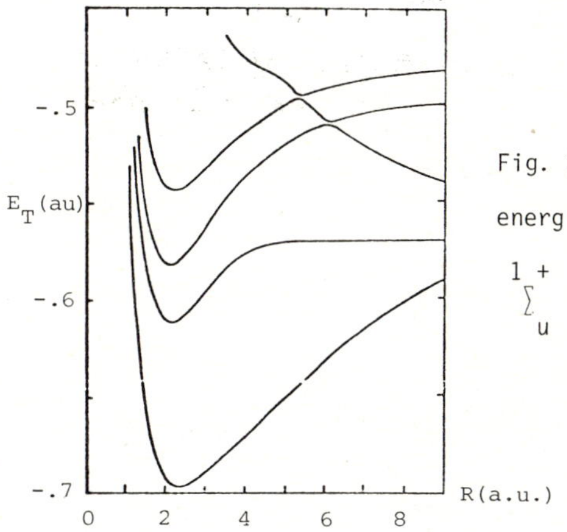

Fig. 3. Adiabatic energies for $^1\Sigma_u^+$ states

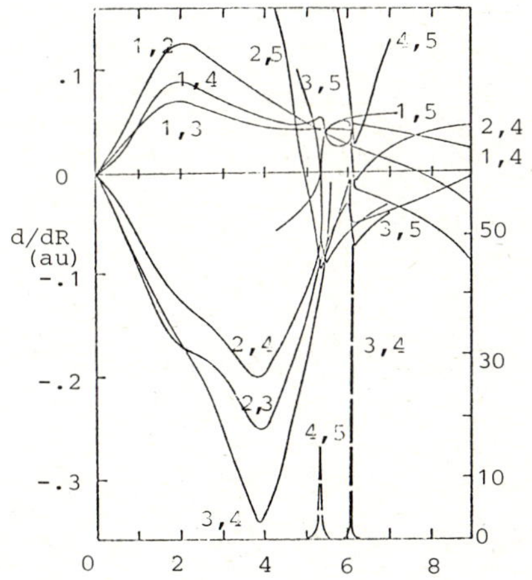

Fig. 4. Radial couplings $^1\Sigma_u^+$ states

--- right hand scale

—— left hand scale

References

1. A. Macías and A. Riera, J. Phys. B., 10,861 (1977); 11, 1077 (1978).
2. E.R. Davidson, J. Chem. Phys., 35, 1189 (1961).
3. V. López, A. Macías, R.D. Piacintini, A. Riera and M. Yáñez, J. Phys. B., 11, 2889 (1978).

LARGE ORDER PERTURBATION THEORY FOR ION H_2^+ AND FOR STARK EFFECT IN HYDROGEN

R.J. Damburg and R.Kh. Propin

Institute of Physics, Latvian Academy of Sciences 229021 Riga, Salaspils, USSR

The nature of Rayleigh-Schrödinger perturbation theory for electronic terms of H_2^+ and Stark effect in hydrogen will be discussed.

1. Eigenvalues of H_2^+ at large distances R between nuclei can be written

$$E_{g,u} = E_0 \mp \frac{1}{2}\Delta E + \delta E$$

At $n_3 = n_2 = m = 0$

$$\delta E = 4R^3 e^{-2R-2}\left(1 + \frac{4(\ln 2R - \psi(1)) - 5}{2R} + O(R^{-2})\right)$$

$$\psi(x) = \frac{d}{dx}\ln\Gamma(x)$$

When calculating $\frac{E_g + E_u}{2} = \sum_{N=0} \frac{E_N}{R^N} + \delta E,$

the sum should be terminated just before the term smallest in magnitude, but δE should be taken into account.

Table I

$n_3 = n_2 = m=0$

R	$\frac{1}{2}(E_g+E_u) - E_0$ [1]	N	δE
6	0.1028×10^{-2}	11	0.1115×10^{-2}
8	0.4286×10^{-4}	14	0.4756×10^{-4}
10	0.1535×10^{-5}	18	0.1634×10^{-5}
11	0.2644×10^{-6}	19	0.2894×10^{-6}

2. At small values of the field strenght F, the level width Γ for $n_1=n_2=m=0$ is

$$\Gamma = \frac{4}{F} e^{-\frac{2}{3F}} \sum_{N=0} a_N F^N$$

Asymptotic in N formula for a_N is obtained

$$a_N = -\frac{18}{\pi}\left(\frac{3}{2}\right)^N \Gamma(N+2)\Big\{1 + \frac{12[\psi(N+1) + \ln 6 - \psi(1)] - 241 - 27(-1)^N}{18(N+1)} + \frac{22511 + 729(-1)^N - 2352[\psi(N) + \ln 6 - \psi(1)]}{324(N+1)N} + \dots\Big\}$$

Table II

N	a_N num [2]	a_N asymp
19	-2.0379×10^{22}	-2.02918×10^{22}
20	-5.2622×10^{23}	-5.15078×10^{23}
29	-1.47869×10^{38}	-1.481747×10^{38}
30	-6.11042×10^{39}	-6.084146×10^{39}
39	-2.82103×10^{55}	-2.826363×10^{55}
40	-1.59489×10^{57}	-1.592570×10^{57}
49	-6.33156×10^{73}	-6.340607×10^{73}
50	-4.53661×10^{75}	-4.533740×10^{75}

3. A further comparison of the results obtained by asymptotic formulae with numerical data will be given.

1. J.M. Peek, J.Chem. Phys. 43, 3004 (1965)
2. H.J. Silverstone, E. Harrel and Ch. Grot Phys. Rev. 24 A, 1925 (1981)

THRESHOLD BREAKUP OF THREE COULOMB-INTERACTING PARTICLES: ANALYSIS WITH WANNIER THEORY AND JACOBI COORDINATES**

Jim Feagin

Behlen Lab., Univ. Nebraska-Lincoln, USA

In order to extend the Wannier threshold theory to arbitrary-mass 3-particle systems, and in particular to heavy-ion systems such as $2H^+ + e^-$ or $2H^+ + H^-$, I have reformulated the Rau-Peterkop[1,2] analysis of Wannier's threshold theory in terms of relative (Jacobi) coordinates. My results include arbitrary values of the system angular momentum L with respect to the system center of mass.

A useful connection with diabatic states of a "molecular" hamiltonian is easily established upon rotation to a body-fixed frame with z-axis along the line (of length R) joining particles 1 and 2 of like sign. In the Coulomb (asymptotic) zone, which determines the threshold energy dependence of the 3-particle wavefunction and, therefore, the cross section, the Coriolis interactions generated by the transformation to the body-fixed frame fall off as R^{-2}, whereas the Coulomb interaction, which maintains particle 3 of unlike sign along the line joining 1 and 2, falls off only as R^{-1}. Thus the wavefunction simply factors into a function describing the internal system motion and a rotation matrix determined by L describing the orientation of the body-fixed z axis. Consequently, I find by following Rau's prescription that the total cross section near threshold is independent of the system's angular momentum L, although angular distributions are not. The angular distributions may be easily analyzed, however, in terms of the rotation matrices.

Specializing to two-electron systems, my results are in agreement with the predictions of Greene and Rau[3] and Stauffer[4] in particular, with the nodal structure of the wavefunction along the Wannier ridge. For symmetric systems, where $Z_1 = Z_2$ and $M_1 = M_2$ but otherwise arbitrary, my results for S-wave cross sections are in agreement with Klar's results[5] derived with hyperspherical coordinates. Attempts to extend the hyperspherical analysis to arbitrary angular momentum L have been hampered by a cumbersome angular momentum algebra for arbitrary-mass systems. An additional advantage of the Jacobi coordinates is that they allow mass and charge dependences of the Wannier parameters to be easily traced.

For asymmetric systems, such as $e^+ + e^- + H^+$ in which the center of charge and center of mass of $e^+ - H^+$ do not coincide, the determination of the threshold law proves to be more straightforward using hyperspherical coordinates rather than Jacobi. By replacing the two Jacobi coordinates R and z (z locates the position of e^- relative to the center of mass of $e^+ - H^+$ along the body-fixed z axis) by the hyperspherical coordinates $\mu_{12}\mathcal{R}^2 = \mu_{12}R^2 + \mu_{12,3}z^2$† and $\alpha = \tan^{-1}[(\mu_{12,3}/\mu_{12})^{\frac{1}{2}} z/R]$, I find that near threshold $\sigma_L \propto E^{2.651}$ for infinite proton mass. A change in the exponent of order m_e/m_H occurs for finite proton masses. These results agree with Klar[6] but disagree with the numerical studies of Dimitrijevic and Grujic who found that the threshold exponent varies between 1.64 and 2.49 as the proton mass is varied from 1836 m_e to 10^{10} m_e.[7]

As a final application, I have analyzed into partial waves the scattering amplitude for breakup of an arbitrary 3-particle system. Although the Wannier theory, based on the asymptotic form of the wavefunction, cannot alone provide the coefficients in the partial-wave expansion, it does provide their energy dependence. Such an analysis is more than academic, since the production of H^- in the collisional breakup of H_3^+ has been observed.[8] Analyses of angular distributions and their comparison with experiment will prove to be the ultimate test of the Wannier theory against other threshold theories (cf. Temkin's analysis[9]).

**This work was begun with the generous support of the Alexander von Humboldt Foundation and continues with that of the US Department of Energy.

References

1. A.R.P. Rau, Phys. Rev. A4, 207 (1971).
2. R. Peterkop, J. Phys. B4, 513 (1971).
3. C.H. Greene and A.R.P. Rau, J. Phys. B16, 99 (1983).
4. A.D. Stauffer, Phys. Lett. 91A, 114 (1982).
5. H. Klar, Z. Phys. A307, 75 (1982).
6. H. Klar, J. Phys. B14, 4165 (1981).
7. M.S. Dimitrijević and P. Grujić, J. Phys. B16, 297 (1983).
8. D. Montgomery and D.A. Jaecks, to be published.
9. A. Temkin, Phys. Rev. Lett. 49, 365 (1982).

†Here μ_{12} is the reduced mass of m_1 and m_2, while $\mu_{12,3}$ is the reduced mass of m_3 and the center of mass of m_1 and m_2.

STIELTJES METHODS FOR PHOTOABSORPTION AND IONIZATION--A MERGING OF SPECTRAL AND COLLISION THEORIES[1]

M.R. Hermann and P.W. Langhoff

Department of Chemistry, Indiana University, Bloomington, IN 47405, USA

Hilbert space provides an appropriate setting for theoretical and computational studies of discrete and continuum Schrödinger states. In spite of this, rather different techniques are generally employed in the two cases. There are distinct advantages, however, in treating the discrete and continuum portions of Hamiltonian spectra from the common perspective of a unified computational approach. In the present report an account is given of Stieltjes methods devised for this purpose,[2] and some illustrative calculations are presented. Additional computational applications are reported separately in this Proceedings.[3]

Stieltjes approximations to the discrete and continuum states of self-adjoint operators are defined as eigenfunctions in n-term Cauchy-Lanczos basis sets. Choice of an appropriate test function, generally of physical significance, specifies the appropriate basis set, which is obtained by stable recursive procedures. The eigenvalues of Stieltjes functions are seen to be generalized Gaussian or Radau quadrature points of the spectral density formed by projection with the test function on the Schrödinger spectrum. Consequently, the Stieltjes functions are sure to appear in the dense portions of the spectrum, and their norms to provide the corresponding spectral quadrature weights. In finite orders, the spatial characteristics of Stieltjes functions correspond to spectral averages in the neighborhood of the quadrature points over Schrödinger states. The spectral content of an individual Stieltjes function is obtained in closed form in terms of orthogonal polynomials without explicit reference to or computations of the underlying Schrödinger states. Convergence ($n\to\infty$) is obtained in the discrete spectral region to Schrödinger eigenstates of finite norm, whereas in the essential portion of the spectrum the functions converge to scattering states of improper (infinite) norm. Renormalization in the conventional Dirac delta-function sense is accomplished in finite orders, in spite of the absence of an asymptotic spatial function. Connections with matrix partitioning and optical-potential theory indicate finite-order Stieltjes functions provide exact Schrödinger states over local portions of configuration space.

Figure 1 illustrates the convergence of Stieltjes functions in the regular p-wave continuum of atomic hydrogen. Similar convergence is obtained at other scattering energies and in other ℓ-wave channels from the Stieltjes development. The spectral composition of the Stieltjes functions of Fig. 1 are found to narrow with increasing order and to approach delta functions in the limit.

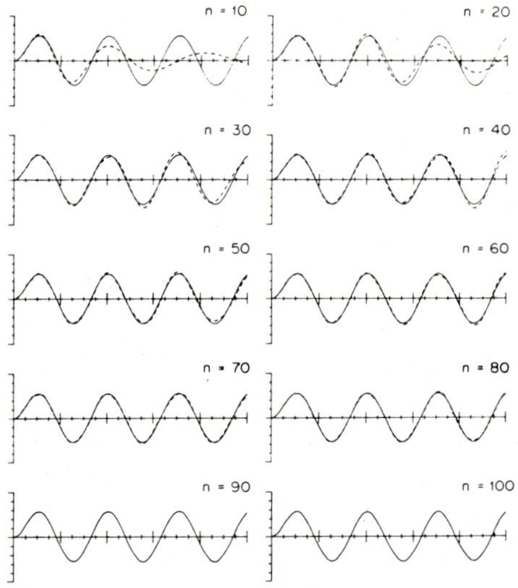

Fig. 1: Stieltjes function (- - -) representations of Coulomb p waves (——) at 2 a.u. energy.

The Stieltjes method is particularly advantageous when applied to molecules, since body-frame point group symmetry functions can be employed throughout. Consequently, graphical representations of molecular scattering continua can be constructed in this manner as useful diagnostics of partial-channel cross sections. Studies have been completed of all channels in H_2, N_2, CO, O_2, NO, H_2O, H_2S, CO_2, H_2CO, H_2CS, and CS_2 molecules, results that will be reported subsequently elsewhere.[3]

References

[1]Work supported in part by Grant #CHE-8002080 from the National Science Foundation.

[2]M.R. Hermann and P.W. Langhoff, Int. J. Quantum Chem. 23, 135 (1983).

[3]G.H.F. Diercksen and P.W. Langhoff, Abstract in this Proceedings.

CLASSICAL-TRAJECTORY MONTE CARLO CALCULATION OF NEGATIVE-MUON CAPTURE*

James S. Cohen

Theoretical Division, Los Alamos National Laboratory, Los Alamos, NM 87545 USA

The classical-trajectory Monte Carlo (CTMC) method[1-3] has been used to describe the slowing down and capture of negative muons by hydrogen atoms.[4] The reactions of concern are

$$\mu^- + H(1s) \rightarrow \begin{cases} \mu^- + H(n\ell) & \text{(inelastic)} \quad (1a) \\ \mu^- + e + H^+ & \text{(ionization)} \quad (1b) \\ e + H_\mu(n\ell) & \text{(capture)} \quad (1c) \end{cases}$$

Cross sections for these processes are required over a wide energy range and no single quantum-mechanical method is practical over the entire range. Since molecular ($p\mu^-e$) and nuclear-symmetry effects do not occur and since the muon is generally captured in a high-lying Rydberg orbital, classical mechanics is expected to have a wider range of validity for μ^-+H scattering than for p+H scattering where it has been quite successful.

The CTMC procedure has three stages: (1) Monte Carlo sampling of initial conditions assuming a microcanonical ensemble for the 1s hydrogen atom, (2) numerical integration of Hamilton's equations for the 3-body system, and (3) identification of the final state in the asymptotic region. From the results of a statistically large number of trajectories, the cross section for a reaction R is given by

$$\sigma_R = (N_R/N_{tot})\,\pi b_{max}^2 , \qquad (2)$$

where N_R is the number of times R occurred in N_{tot} trajectories chosen with impact parameters in $(0,b_{max})$.

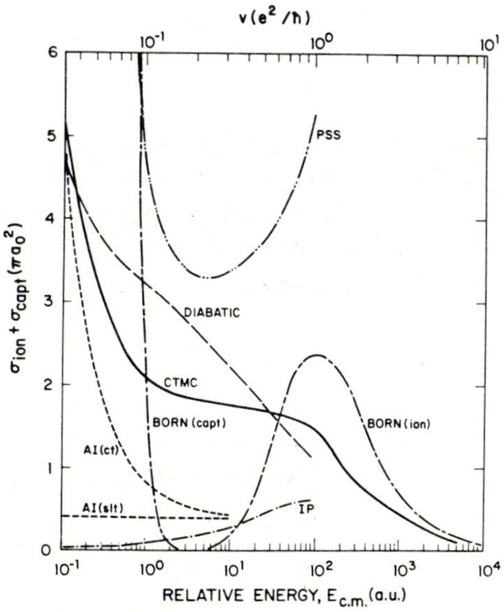

Fig. 1: Comparison of present (CTMC) total reactive (ionization + capture) cross section for μ^-+H with the Born approximation, adiabatic ionization (AI) with straight line and curved trajectories, impact parameter (IP) method, perturbed-stationary-state (PSS) method, and diabatic-state method.

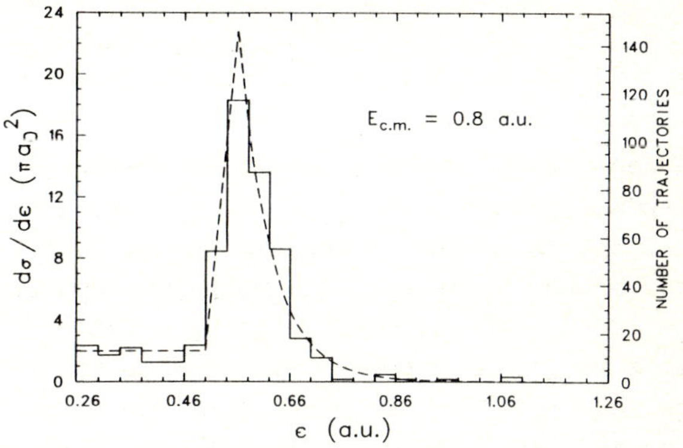

Fig. 2: Histogram of trajectory results (full line) for energy-loss cross section (broken line is a fit which is exponential at ε>0.56 a.u.).

In Fig. 1 the CTMC total reactive (ionization + capture) cross section is shown and compared with results of several other calculations for μ^-+H collisions. The statistical error of the CTMC cross sections is within 6% and the overall uncertainty is thought to be within 30%. The capture cross section (not shown separately) decreases rapidly at collision energies exceeding the ionization potential of the target hydrogen atom.

At high velocities ($v \gg 1$ a.u.) the ionization cross section is similar to that for p+H, as predicted by the Born approximation, but at low-to-intermediate velocities the behavior is quite different. In slow collisions the negative muon effectively shields the electron from the nucleus and consequently the electron is emitted with low kinetic-energy (E_e) only weakly dependent on the collision energy. The energy-loss ($\varepsilon = E_e + I_a$) distribution obtained for a collision at relative energy 0.8 a.u. is shown in Fig. 2. The low-energy behavior is akin to quantum-mechanical adiabatic ionization. While true that this shielding plays a large role, it must be recognized that the actual dynamics also plays an important role and the cross sections obtained are significantly larger than predicted by adiabatic ionization.

*Work performed under the auspices of the U.S. D.O.E.

1. R. Abrines and I. C. Percival, Proc. Phys. Soc. 88, 861 (1966); 88, 873 (1966).
2. R. E. Olson and A. Salop, Phys. Rev. A 16,(1977).
3. J. S. Cohen, Phys. Rev. A 26, 3008 (1982).
4. J. S. Cohen, Phys. Rev. A 27, 167 (1983).

POSITRONIUM FORMATION FROM MUONIUM

Lali Chatterjee*

Department of Physics, Jadavpur University, Calcutta-700032, India

S. Bhattacharyya

Gokhale College, Calcutta - 700020, India

Formation of positronium is generally studied through positron-atom or positron molecule scattering. We report an investigation of the formation of positronium by weak decay of the nucleus of the Muonium atom. The rate of the free muon decay is $R_f = .455 \times 10^{+6}$ sec.^{-1} & $R_{(e+e-)} = 0.006\ R_f$. Apart from its role as a novel method of positronium formation, this process is of fundamental interest as a weak transition between two similar electrodynamic systems.

We consider hence

$$(\mu^+ e^-) \longrightarrow (e^+ e^-) + \nu_e + \nu_\mu \qquad (1)$$

where the role of the electron is essentially that of a spectator to the weak interaction;- present, initially bound to the decaying lepton, and finally bound to the product positron.

The probability of all four final state leptons escaping free, exceeds that of positronium formation, due to the large energy release (100 MeV) involved in muon decay. However the process of positronium formation retains a finite probability and contributes mainly for positrons emitted at the lower end of the energy spectrum, and for those regions of phase space where the positron and electron have the same order of momentum and the angle between them is small. The initial and final states are treated field theoretically in the manner used previously by us[1], and for the weak decay we use the (V-A) four fermi Hamiltonian. Thus the matrix element can be written (after taking Vacuum Expectation values)

$$M_{fi} = \frac{G}{\sqrt{2}} \int \bar{U}_\mu(\underline{k})\, O_\alpha\, U_p(\underline{p})\, \bar{U}_{\nu_\mu}(\underline{m})\, O_\alpha\, U_{\nu_e}(\underline{l})\, g(\underline{k}_1,\underline{k}_2) \times g_f^*(\underline{k},\underline{p},\underline{l},\underline{m})\, \delta^3(\underline{k}-\underline{p}-\underline{m}-\underline{l})\, d^3\underline{k}_1\, d^3\underline{p}\, (2\pi) \qquad (2)$$

with $O_\alpha = \gamma_\alpha(1+\gamma_5) \qquad (3)$

$g_f(\underline{k},\underline{p},\underline{l},\underline{m})$ and $g_i(\underline{k}_1,\underline{k}_2)$ represent the final and initial system wave functions in momentum space. Specifically

$$g_f(\underline{k},\underline{p},\underline{l},\underline{m}) = (2\pi)^3 \phi_f\left(\frac{\underline{k}-\underline{p}}{2}\right) \delta^3(\underline{k}+\underline{p}+\underline{m}+\underline{l}-\underline{Q}_c)$$

$$g_i(\underline{k}_1,\underline{k}_2) = (2\pi)^3\, \phi_{1s}(\underline{k}_1)\, \delta^3(\underline{k}_1+\underline{k}_2-\underline{P}_c)$$

$\phi_f\left(\frac{\underline{k}-\underline{p}}{2}\right)$ and $\phi_{1s}(\underline{k}_1)$ refer to the positronium ground state wave-function, and to muonium wave function (Hydrogen type).

$\underline{Q}_c, \underline{P}_c$ = momentum of Centre of mass of final and initial systems.

The decay rate is obtained after integrating over bound state momenta, summing and averaging over spins, and integrating over phase space of final particles.

We obtain finally the dominant contribution to the rate of formation of positronium from muonium (terms neglected are of order at least $e^2 = (1/137)^2$ smaller) :

$$R = 0.0061255\ R_f$$

with $R_f = 0.455 \times 10^6\ \text{sec}^{-1}$

So, $R = 0.00278 \times 10^6\ \text{sec}^{-1}$

L. Chatterjee thanks C.S.I.R. for award of Research Associateship.

References

1. L. Chatterjee and S.Bhattacharyya Physics Letter 93A No.7 (1983) 360.
   S. Bhattacharyya, L. Chatterjee and T. Roy. Physica 106C (1981) 135.
   T. Roy. N. Clin. Letter 5,

* Mailing Address : Dr. Mrs. Lali Chatterjee
84/SB, Block 'E',
New Alipur,
Calcutta - 700053, India.

MESOMOLECULE FORMATION STUDIED QUANTUM ELECTRODYNAMICALLY

Lali Chatterjee*

Physics Department, Jadavpur University, Calcutta - 700032, India

Sujata Bhattacharyya

Gokhale College, Calcutta - 700020

Mesomolecule formation is of crucial importance in the investigation of sub-barrier fusion of light nuclei by muon catalysis[1]. Experiments determining the fusion yield involve the various meso-molecule formation rates in a highly mixed form. The very few theoretical calculations which were originally performed were repeated only by the Russian group[2]. The resonance effects found experimentally and theoretically for some of the heavier Hydrogen isotopes of exotic molecules provided the main motivation for investigating this phenomenon from a different angle.

We have calculated mesomolecule formation in the framework of QED using a field theoretic formalism[3], and working in coulomb gauge.

We consider the specific process of $(p\mu^- d)^+$ formation from atomic Hydrogen

$$(d\mu^-) + H \rightarrow (p\mu^- d)^+ + e^- \qquad (1)$$

The free electron carries away the energy released. The interaction of the proton of Hydrogen with the deuteron and muon of the incoming muonic atom are considered. The neutral $(d\mu^-)$ system is dimensionally 200 times smaller than the electronic orbit of Hydrogen. For formation of the $(p\mu^- d)^+$ system the relavent interactions occurs within the electron cloud. The electron's interaction is much less than that of the proton so it is neglected. Essentially the electron behaves as a spectator to the formation and participates only through the final state kinematics when it carries off some of the energy released.

The dominant term of the S matrix in the Coulomb gauge is

$$H_c = \int \frac{\sigma_p(x)\,\sigma_d(x')\,d^3x\,d^3x'}{4\pi|x-x'|} - \int \frac{\sigma_p(x)\,\sigma_{\mu^-}(x')\,d^3x\,d^3x'}{4\pi|x-x'|} \qquad (2)$$

where $\sigma_p(x)$, $\sigma_d(x')$, $\sigma_\mu(x)$ are the charge densities of the proton, deuteron and muon respectively.

$$M_{fi} = \frac{1}{2!}\langle \Psi_f | H_{int} | \Psi_i \rangle = M_{pd} + M_{p\mu} \qquad (3)$$

The decay rate is then obtained as

$$\lambda_{p\mu-d} = 2\pi \int \frac{|M_{fi}|^2 d^3Q\, d^3K}{(2\pi)^6} \qquad (4)$$

(M_{fi} is summed over spins)

where $\underline{Q}$, $\underline{K}$ are the external momenta of the outgoing molecular ion and electron.

The wavefunction of the molecular ion is taken in the Adiabatic Approximation, and space integrals over muon and nuclear coordinates are performed using error and exponential integrals.

Taking binding energy of the molecule as 214 eV and that of the atom as 13.6 eV, and performing the phase space integrations, we obtain the $(p\mu^- d)^+$ formation rate as $\lambda_{p\mu d} = 3.56 \times 10^6\ sec^{-1}$ for thermal energies of the incoming $(d\mu^-)$. This result compares well with other theoretical values and with experiment.

The same formulation may be applied to other heteronuclear mesomolecules and may be extended to homonuclear mesomolecules.

References

1. S.S. Gerstein and L.I. Penomerev : Muon Phys. Ac. Vol. 3, 1977.
2. L.I. Ponomarev and M.P. Faifman : Sov. Phys. JETP 44 (1976) 1886.
3. S. Bhattacharyya, L. Chatterjee and T. Roy : Physica 106C (1981).

* Mailing Address : Dr. Mrs. Lali Chatterjee
84/SB, Block 'E'
New Alipur
Calcutta-700053, India.

MESIC MOLECULES AND MUON CATALYZED FUSION

A.K. Bhatia and Richard J. Drachman

Laboratory for Astronomy and Solar Physics, NASA/Goddard Space Flight Center, Greenbelt, MD 20771

The process of muon-catalyzed fusion[1] involves very tightly bound molecules whose nuclei are isotopes of hydrogen held together by muons rather than electrons. The scale of such mesic molecules depends on the muon Bohr radius, $a_0/m_\mu \cong 250$ fm and the rate of spontaneous fusion reactions between the nuclei can become very large. Early enthusiasm for muon-catalyzed fusion as a practical energy source faded when detailed investigation revealed that it would be unlikely for a muon to catalyze more than about two successive reactions before being lost by capture or decay.

More recently, however, the process of resonant mesic molecule formation has been discovered for the $(dd\mu)$ and $(dt\mu)$ molecules.[2] This can very greatly accelerate the rate of mesic molecule formation, but it depends critically on the very loosely bound excited J=1, v=1 states of these molecules. The best theoretical values of these binding energies are by Melezhik, et al.[3] and are as follows: For $(dd\mu)$ $E = 1.91 \pm 0.05$ eV and for $(dt\mu)$ $E = 0.64 \pm 0.05$ eV.

We have recently applied standard variational methods to examine the spectrum of the $(pp\mu)$ and $(dd\mu)$ molecules, using the symmetric Euler angle technique.[4] The trial functions have the forms:

$$\Psi_{J=0} = (f + \tilde{f})\,\mathfrak{D}_0^{0+}$$

$$\Psi_{J=1} = -\cos(\theta_{12}/2)(f + \tilde{f})\,\mathfrak{D}_1^{1+} - \sin(\theta_{12}/2)(f - \tilde{f})\,\mathfrak{D}_1^{1-}$$

where $\tilde{f}(r_1, r_2, r_{12}) = f(r_2, r_1, r_{12})$ and

$$f(r_1, r_2, r_{12}) = e^{-\gamma(r_1+r_2)} \sum_{\ell \geq J, n, m=0} C^{\ell m n}\, r_1^{\ell}\, r_2^{m}\, r_{12}^{n}$$

The $\mathfrak{D}$ functions are defined in Ref. 4. In Table I we give binding energies for the J=0 states of these mesic molecules, calculated with up to 125 terms in the expansion of f. These improve upon the existing variational results, and are compared to the best non-variational results of Melezhik, et al.[3] The extension of these calculations to the J=1 manifold should not be difficult and results will be presented. The fact that these variational calculations bound the exact energies will be helpful in estimating the accuracy of the results. Eventually, these wave functions can be used to calculate the formation rates of mesic molecules.

We recommend the excellent recent review of the entire subject by Bracci and Fiorentini.[1]

Method	$pp\mu$(v=0)	$dd\mu$(v=0)	$dd\mu$(v=1)
Variational (present)	253.15	325.06	35.71
Perturbation (Ref. 3)	253.55	324.99	35.66
Truncation (Ref. 3)	252.95	325.04	35.80

Table 1: Binding energies in eV of the mesic molecules $(pp\mu)$ and $(dd\mu)$ for J=0. The v=0 entries are for the ground states and v=1 indicates the first excited state.

References

1. L. Bracci and G. Fiorentini, Physics Reports 86, 169 (1982) and references herein.
2. E.A. Vesman, Sov. Phys. JETP Letters 5, 113 (1967).
3. V.S. Melezhik, et al., Sov. Phys. JETP 52, 353 (1981).
4. A.K. Bhatia and A. Temkin, Rev. Mod. Phys. 36, 1050 (1964); Phys. Rev. 137, A1335 (1965).

THE ATOMIC AND MOLECULAR DATA BASE AT BELFAST AND DARESBURY

K.M. Aggarwal, K.A. Berrington, J.G. Hughes, F.J. Smith and M. Elder*.

School of Physics and Mathematical Sciences,
The Queen's University of Belfast, N. Ireland.

*SERC Daresbury Laboratory, Daresbury, Warrington.

Numerical data on Atomic and Molecular Physics is being collected and stored in a data base developed jointly by the School of Physics and Mathematical Sciences at the Queen's University of Belfast and by the U.K. S.E.R.C. Daresbury Laboratory. The work is supported in part by the UKAEA, Culham Laboratory, and by the S.E.R.C. There are two main parts to the data base. In the first part we store theoretical data produced form computer programs but which are too bulky for publication. These data are stored on demountable disks and tapes. The second part is concerned with the storage of recommended data such as cross sections and rate coefficients; it often uses average values of cross sections stored in the first part of the data base. This second part of the data base is available on-line. The two parts of the data base are linked by a user-friendly interactive program on the Daresbury AS/7000. We describe work on these two parts in more detail below.

Theoretical Data

There is now a large amount of accurate atomic data being produced by computer programs. Although such calculations are published in the scientific literature, usually journal space limits the amount of useful information that can be displayed and such data once located, must be converted to machine readable form before it can be used for display purposes or for further calculations. We have therefore built a data base system to store, manipulate and display such data[1].

In this first part of our data base we store electron-atom, electron-ion, photoionization and atomic structure data directly from the available program packages, such as the R-matrix and IMPACT packages. The data are stored at several levels, to cater for the different requirements of the users. For example, in electron ion or atom scattering, electron collision rates are given for a large range of temperatures. The basic cross section data are also available. At the most basic level, reactance matrices for the collision are stored from which detailed information about the collision can be obtained, e.g. angular distribution and resonance structure.

Recommended Data

The second part of the data base holds recommended cross sections and rate coefficients. Using the bibliographies produced by the IAEA in Vienna, published experimental and theoretical data are stored and assessed. The recommended values are chosen on the advice of both experimentalists and theoreticians, sometimes using scaling laws or extrapolations when no data is available. So far data collection has been completed on the ionization of light atoms and ions[2,3]. Work is advanced on charge transfer of the same ions with atomic hydrogen and on interatomic potentials; some work has started on electorn excitation[4].

A full list of ions for which data are currently stored, the type of data and the specific energy/temperature range etc. will be available from the authors during the conference. We also intend to display excitation rate coefficients for a large number of atoms and ions in a wide temperature range for the benefit of the user community. We welcome atomic collision data from atomic physicists and encourage all potential users to use the data base facilities.

References

1. K.A. Berrington, K.M. Aggarwal and M. Elder, January 1983, "A data management system for the storage, manipulation and display of atomic collision data", available from the authors.

2. J.G. Hughes, F.J. Smith, Physica Scripta 23, 197, 1981.

3. K.L. Bell, H.B. Gilbody, J.G. Hughes, A.E. Kingston and F.J. Smith, UKAEA report:"Atomic and molecular data for fusion, Part 1", CLM-R216.

4. K.M. Aggarwal, Mon. Not. R. Astr. Soc. 202, 15p - 20p, 1983.

ION AND ELECTRON COLLISION PROCESSES OF IMPORTANCE IN THE JOVIAN AND SATURNIAN MAGNETOSPHERIC PLASMAS

R. E. Johnson

Department of Nuclear Engineering and Engineering Physics
University of Virginia, Charlottesville, VA 22901 USA

One of the significant findings of the Voyager missions was the existence of plasmas in the Jovian and Saturnian magnetospheres which were rich in ions heavier than helium and hydrogen: (e.g., O^+, O^{++}, S^+, S^{++}, Na^+, K^+).[1] The source of these ions is presumed to be the surface of the ice covered satellites, the volcanoes on Io and, possibly, the atmospheres of Titan or the planets themselves. Even prior to the Voyager encounters R. A. Brown had discovered a cloud of neutral Na atoms and singly ionized sulfur in Jupiter's inner magnetosphere near the orbit of Io.[2] Ground-based measurements have also confirmed the existence of an extended neutral oxygen cloud in the vicinity of Io.[2]

Because of these findings, considerable interest has developed in those electron and ion impact cross sections which determine ionic composition and energy dissipation processes in the Jovian and Saturnian magnetospheric plasmas. Although the electron impact ionization cross sections for many of the neutrals are reasonably well known, the cross sections for charge exchange between neutrals and singly and doubly ionized species of Na, K, O and S, which are important in the vicinity of Io, have not been determined accurately.[3] In addition, accurate electron impact excitation cross sections for many of the ionized species are of particular interest for plasma diagnostics. Finally, as the neutrals observed most likely originated in form of molecule species (e.g., SO_2 from volcanic activity on Io) there is interest in ionization, dissociation, and charge exchange cross sections involving molecular species. This presentation will be a review of the relevant information on ionic composition and ion and electron energy spectra in the Jovian and Saturnian magnetospheric plasmas and the cross section data of interest for understanding the behavior of these plasmas.

References

1. See e.g., Science 204 (1979)
2. R. A. Brown and W. H. Ip, Science 213, 1493 (1981)
3. R. E. Johnson and D. F. Strobel, J. Geophys. Res. 87, 10,385 (1982)

RESONANT PHOTOEMISSION NEAR 3D THRESHOLDS IN LANTHANUM : THEORETICAL PREDICTIONS

F.Combet Farnoux

ERA "Spectroscopie Atomique et Ionique" Université Paris-Sud ; Bâtiment 350 - 91405 Orsay (France)

New measurements[1] of XAS (X absorption spectra) have been obtained recently with an improved resolution in the neighborhood of 3d edges for La metal, as shown in Fig.1. They show the existence of two main lines corresponding to transitions $3d_{5/2,3/2} \rightarrow 4f$, which have been identified as $3d^9 4f$ $^3D_1$ and $^1P_1$ states, the first one located below the M_5 threshold whereas the second one is above, in agreement with recent XPS measurements[2]. The ratio $^1P_1/^3D_1 = 1.55$ agrees well with the value 1.6 obtained from intensity ratios in absorption and emission[3], but disagrees totally with the statistical ratio 2/3. Since these lines are respectively located below the M_5 and M_4 thresholds, resonant photoemission measurements should be very useful to provide extra information about their decay channels and this anomalous behavior of the intensity ratio. In their absence, I present here theoretical investigations supported by calculations of the various partial cross sections σ_μ and the deexcitation widths Γ_μ in the various open channels μ for both $^3D_1$ and $^1P_1$ lines, using the general formalism that I have recently developed[4] and according to which:

$$\sigma_{\mu n} = |Q^{\circ}_{\mu}|^2 \left[1 + \alpha^2 \frac{q^2+1}{\varepsilon^2+1} + 2\alpha \frac{\varepsilon q - 1}{\varepsilon^2 + 1} \right]$$

$$\alpha_{\mu n} = \frac{1}{Q^{\circ}_{\mu}} \frac{H_{\mu n}}{V_{nt}} \langle nE | r | 0 \rangle ; n=1 \text{ for } {}^3D_1,\ n=2 \text{ for } {}^1P_1$$

$$|nE\rangle = \frac{1}{V_{nt}} \sum_{\mu} V_{n\mu} |\mu E\rangle \text{ and } \Gamma_{nt} = V^2_{nt} = \sum_{\mu} |H_{\mu n}|^2 = \sum_{\mu} \Gamma_{\mu n}$$

Q°_{μ} and $H_{\mu n}$ being respectively the dipole and the decay matrix element in each channel μ, whereas q and ε are the well known Fano absorption parameters.

These calculations are based on an Hamiltonian H which introduces the spin-orbit interaction: $H = H_o + H^{SO}$ H_o being a zero-order central field Hamiltonian (potential of the ground state configuration of xenon-like La^{3+} configuration).Since the predominant open channels μ involve 4s,4p and 4d ionisations, the continuum energy is too large to make really apparent the asymmetry of the corresponding lines, because of a rather small Q°_{μ} value. However, two types of decay channels are involved: direct recombination channels and Auger ones. I have shown that, whereas the Auger deexcitation channels are dominant for the $^3D_1$ excited state below both thresholds, both types of decay channels are important to a comparable extent for the $^1P_1$ state located between the M_5 and M_4 edges.

Considering both direct recombination and Auger channels, I have determined both $\sigma_{\mu n}$ and $\Gamma_{\mu n}$ for each line. Within this purpose, I have determined the spin-orbit matrix elements between $^1P_1$ and $^3D_1$ (either p^5Ed or p^5Es, or d^9Ep or d^9Ef) configurations with different continuum energies E and I have diagonalized the corresponding matrices. This diagonalisation results in a linear transformation of the p^5Ed or p^5Es, or d^9Ef or d^9Ep ($^1P_1$ and $^3D_1$) channels into two new final state channels having slightly different binding energies (corresponding to the fine structure splitting of the ionic core) . The continuum $|nE\rangle$ in interaction with the resonant line n considered is a linear combination of these new final states, according to the third formula in the opposite column.

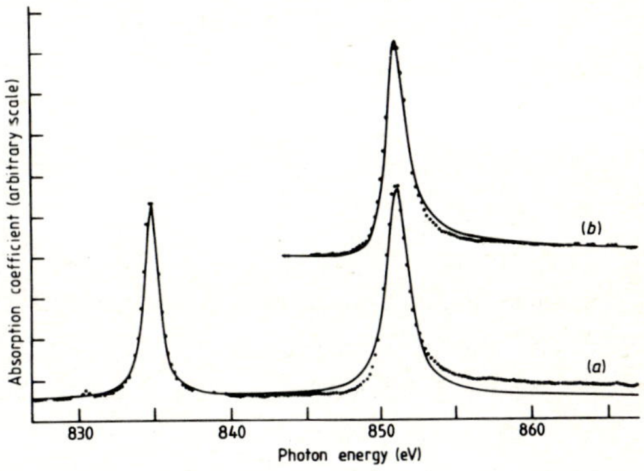

Fig.1 : 3d absorption spectrum of La metal, as presented in reference 1 (See caption there)

My calculations lead to the following conclusions:

1°) The $^3D_1/^1P_1$ intensity ratio in absorption is very different from the statistical M_5/M_4 ratio of 1.5 and this result has to be compared with the anomalous L_3/L_2 white-line ratio observed in the 3d transition[5] metals.

2°) The linewidths of these resonant states are much greater than those determined previously for the 3d hole states[6] ($3d^9$ $^2D_{5/2,3/2}$ configurations).

3°) Because of the importance of both exchange and spin-orbit effects, the discrepancy between the linewidths of the two lines is due not only to the extra Coster-Kronig transition $M_4M_5N_{6,7}$, but also to the various direct recombination channels into which the $^1P_1$ state decays more intensively than the $^3D_1$ state.

1. R.C.Karnatak,J.M.Esteva and J.P.Connerade J.Phys.B 14 , 4747 (1981)
2. J.M.Esteva,R.C.Karnatak,J.C.Fuggle and G.A.Sawatzky Pys.Rev.Lett. In press (1983)
3. J.P.Connerade and R.C.Karnatak J.Phys.F 11,1539(1981)
4. F.Combet Farnoux Phys.Rev. A 25 , 287 (1982)
5. R.D.Leapman and L.A.Grunes Phys.Rev.Lett.45,397(1980)
6. E.J.McGuire Phys.Rev. A 5 ,1043 (1972)

THRESHOLD CROSS SECTIONS OF EXCITED ATOMIC STATES

N.B.Avdonina*, A.Z.Devdariani**

*Leningrad Institute of Refrigerating Industry, Lomonosov st.9, Leningrad, SU. 169002

**Leningrad State University, Pervomay pr.100, Leningrad SU 168904

The photoionization of excited states of manyelectron alkali metal atoms at the narrow small energy region of the outgoing electron is considered. The threshold cross section as a function of the principal quantum number n for nd^2D states of Rb and Cs atoms have been calculated. We have used the central Hartree--Fock potential as well as the Quantum Defect Theory. In both cases it is shown that the threshold cross sections drastically differ from those obtained in the simple hydrogen model.[1]

The total photoionization cross section from the state with quantum numbers (n, l) is given by [2]

$$\sigma_{nl} = \frac{4\pi^2}{3}\alpha a_0^2 \frac{\varepsilon + J_{nl}}{2l+1}\left[l R^2_{n,l-1} + (l+1) R^2_{n,l+1}\right], \quad (1)$$

where α is the fine-structure constant, a_0 is the Bohr radius, a_o-the ionization potential. $R_{n,l\pm1}$ are the radial dipole matrix elements depended on the radial parts of the active electrons wave functions before and after the photoionization. At the threshold they can be obtained using the Burgess and Seaton formula [3]

$$R_{n,l\pm1} \sim \cos\pi\left(n_l^* + \Delta_l(\varepsilon) + X_l\right). \quad (2)$$

Δ_l is the quantum defect of nl level, n^* is the effective quantum number and X_l is an empirical parameter. With the principal quantum number n growing the values of cos function for $nd \rightarrow \varepsilon p$ photoionization transitions have been shown to increase too both in Rb and Cs. As to the $nd \rightarrow \varepsilon f$ transitions dipole matrix elements become smaller with the growth of n. In the case of Cs $R_{n,l+1}$ decrease so rapidly that the total threshold cross sections have pronounced minimum at $n \approx 8$.

At the more accurate Hartree-Fock approximation for $R_{n,l\pm1}$ the calculations were carried out in a similar manner as it was done previously for analysis of σ_{nl} dependence on ε. Fig.1 exhibits the results for the total photoionization cross section for Cs in terms of the principal quantum number n at the energy near the threshold. The existence of the strong short range core potential leads to the essential deviations of our results from the hydrogenlike atoms cross sections even for comparatively high values of n as the Quantum Defect Theory predicts. It can be explained assuming that the transitions of electrons from the orbit determined by the orbital quantum number l on the orbit $l_1 = l \pm 1$ occurs with the highest probability in the turning point nearest to the nuclei where the quasiclassical orbits are most close to each other and the deviation of the atomic potential from the Coulomb one is maximal.

Our results have also some deviations from those obtained using the formula (2). Specifically in the Hartree-Fock approximation wiselike for the method of quantum defect the threshold photoionization cross section for Cs has the minimum but in the first case it is essentially less pronounced.

The data obtained exhibit invalidity of description of the excited states photoionization of manyelectron atoms near the threshold using the hydrogen model.

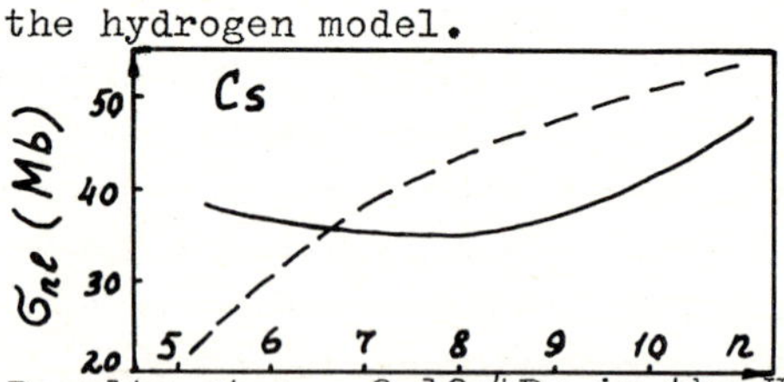

Fig.1:Results at $\varepsilon = 9\cdot10^{-4}$Ry in the Hartree--Fock(full line) and hydrogen(broken line) models.

References

1. L.Bureyeva, Astronomical Journal 45, 1215 (1968)
2. A.Z.Devdariani, A.H.Klucharev, Optics and Spectroscopy 42, 1204 (1977).
3. A.Burgess, M.J.Seaton, Rev.Mod.Phys. 30, 992 (1958)
4. N.B.Avdonina, M.Ya.Amusia, submitted to J.Phys.B.

OBSERVATION OF ℓ-STRUCTURE IN HYDROGENIC STATES OF HIGH-Z FOUR-ELECTRON IONS PRODUCED BY ION-SOLID COLLISIONS

A. E. Livingston

Department of Physics, University of Notre Dame, Notre Dame, Indiana, 46556, U.S.A.

Hydrogenic states in beryllium-like ions are sensitive to strong polarization effects due to the large polarizabilities of their lithium-like atomic cores. This results in resolvable ℓ-structure splittings through the spectroscopic observation of emission from $\Delta n=0$ transitions involving such states. The fast-ion-foil collisional excitation method efficiently excites hydrogenic states, or low-lying Rydberg states, in few-electron ions. We have used this excitation technique to initiate a study of the structures of $1s^22sn\ell$, $n>4$ states in beryllium-like ions with $Z\geq14$. We present here our observations of the $\Delta n=1$ transitions n=5-6 and n=6-7 in four-electron Si^{10+}(Z=14).

In Fig. 1 we show a portion of the emission spectrum from 45 MeV silicon ions excited by a 20μg/cm^2 carbon foil, obtained at the Notre Dame Injector/Tandem Accelerator Laboratory. The n=6-7 transition in beryllium-like Si XI at 1022Å shows a wide, partially-resolved structure that results from the highly polarizable lithium-like core. In Fig. 2 we show a high resolution scan through the structure of the n=5-6 transition in Si XI at 616Å.

A detailed interpretation of these observed structures is complicated by the expected influence of $1s^22pn'\ell'$ states, which have been found to strongly perturb both the ℓ-state and j-state structures of $1s^22sn\ell$ levels at lower Z[1]. Furthermore, the effects of core penetration[2] and dynamical correlation[3] may be important in these Rydberg state structures. The analysis of these data is underway in conjunction with the extension of these four-electron measurements to both higher-Z and lower-Z systems, in order to reveal the Z-dependent effects of the various contributions to the hydrogenic state structures.

1. K. Bockasten, et al. Ark. Fysik 37, 563 (1968).
2. A. Dalgarno and P. Shorer, Phys.Rev.A20, 1307(1979).
3. L.J. Curtis, Phys. Rev. A23, 362 (1981).

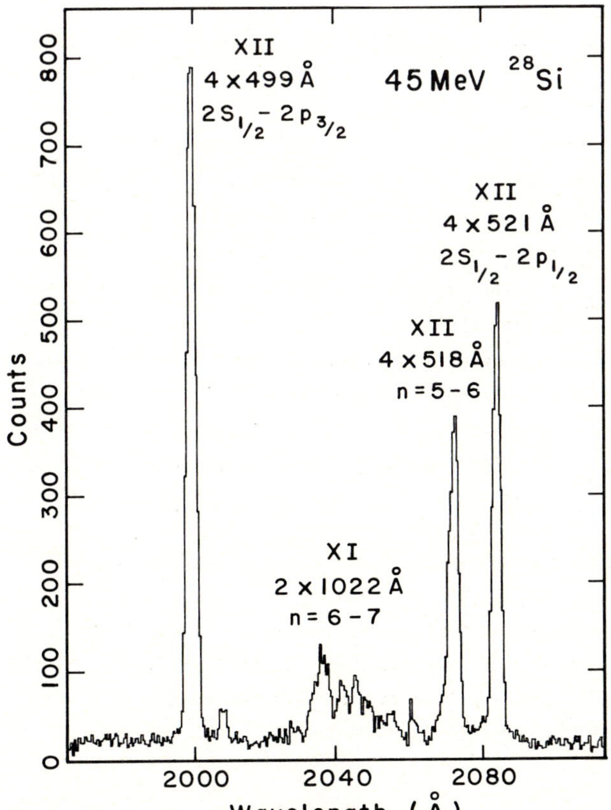

Fig. 1. The n=6-7 transition of Si XI.

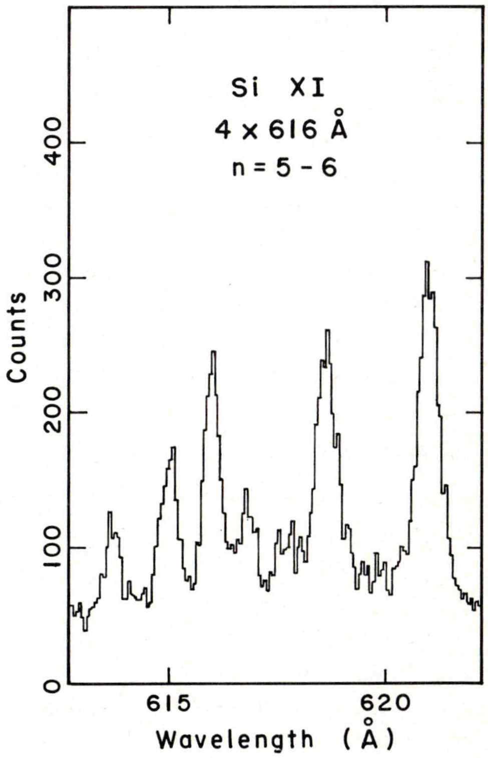

Fig. 2. Resolved ℓ-structure in n=5-6 transition of Si XI.

CHARGE TRANSFER AND FINE-STRUCTURE EXCITATION IN COLLISIONS OF O^{3+} WITH ATOMIC HYDROGEN AT THERMAL ENERGIES

S. Bienstock*, A. Dalgarno* and E. Roueff†

*Harvard-Smithsonian Center for Astrophysics, Cambridge, MA 02138
†Observatoire de Meudon, Paris, France

Charge transfer reactions are important processes both in astrophysical plasmas and in plasmas associated with magnetically confined fusion devices. In the former, charge transfer of multiply charged ions with atomic hydrogen and helium competes with ionization (recombination) processes as a mechanism for increasing (decreasing) the state of ionization of the plasmas. Applications to magnetic fusion include the formation of neutral hydrogen beams, the plasma heating by the beams, the transport of impurity ions, the plasma cooling by emission, and plasma diagnostics.

The current quantal theory of charge transfer[1,2,3] has been successful in predicting total cross sections as well as cross sections for charge transfer into specific atomic states of the product ions. This theory is extended here to allow for the occurrence of fine-structure excitation simultaneously with charge transfer. The method leads to specific predictions of the fine-structure population of the product ions, and of the relative intensities of the resulting emission lines. As an application, cross sections at 1 eV are reported for $O^{3+}(^2P^o_{\frac{1}{2},3/2})+H(^2S_{\frac{1}{2}})\rightarrow O^{2+}(^3P^o_{0,1,2},\,^3D_{1,2,3},\,^3S_1,\,^1P_1,\,^1P^o_1)+H^+$ and the rate coefficient for formation of $O^{2+}(2p3p\ ^3D_1)$ is compared with a recently obtained value[4] based on astrophysical considerations.

Adiabatic molecular potentials $\varepsilon_{ii}(R)$[5] and radial coupling matrix elements $A_{ij} = \langle\psi^a_i|\frac{d}{dR}|\psi^a_j\rangle$ for each molecular symmetry were transformed to a diabatic representation[1] in which charge transfer and excitation are induced by the off-diagonal elements of a symmetric potential energy matrix $\underset{\sim}{V}^{\Lambda S}(R)$. Spin-orbit couplings were included for states arising from the same atomic configuration.

The radial partial wavefunctions satisfy the Schrodinger equations[6]

$$\left(\frac{d^2}{dR^2}-\frac{\ell(\ell+1)}{R^2}+k^2_\gamma\right)F^J_{\gamma\ell}(R) = \frac{2\mu}{\hbar^2}\sum_{\gamma'\ell'} W^J_{\gamma\ell,\gamma'\ell'}(R)F^J_{\gamma'\ell'}(R) \quad (1)$$

The space-fixed diabatic interaction potential matrix elements are given by[6]

$$W^J_{\gamma\ell,\gamma'\ell'} = \sum_{\Lambda S} G^{\Lambda SJ}_{\gamma\ell,\gamma'\ell'}\, V^{\Lambda S}_{LS_1S_2,L'S'_1S'_2}(R) \quad (2)$$

where $V_{LS_1S_2,L'S'_1S'_2}(R)$ is a matrix element of the diabatic potential $\underset{\sim}{V}^{\Lambda S}(R)$, γ is the set of indices $(LS_1)j_1S_2j_{12}$, and $G^{\Lambda SJ}_{\gamma\ell,\gamma'\ell'}$ is a coefficient resulting from the transformations which relate[6] the different basis sets used.

The molecular states included separate to:

1. $O^{2+}(^3P^o) + H^+(^1S_0)$
2. $O^{2+}(^3P^o_1) + H^+(^1S_0)$
3. $O^{2+}(^3P^o_2) + H^+(^1S_0)$
4. $O^{2+}(^3D_1) + H^+(^1S_0)$
5. $O^{2+}(^3D_2) + H^+(^1S_0)$
6. $O^{2+}(^3D_3) + H^+(^1S_0)$
7. $O^{2+}(^3S_1) + H^+(^1S_0)$
8. $O^{3+}(^2P^o_{\frac{1}{2}}) + H(^2S_{\frac{1}{2}})$
9. $O^{3+}(^2P^o_{3/2}) + H(^2S_{\frac{1}{2}})$

The calculation involved 141 partial waves. A calculation is in progress which includes $O^{2+}(^1P_1)$ and $O^{2+}(^1P^o_1)$.

Cross-sections in 10(-16) cm^2 for E = 1.00 eV
(Energy measured relative to $^2P^o_{\frac{1}{2}}$ state)

States I → F	Cross-section	States I → F	Cross-section
8 1	0.0071	9 1	0.0038
8 2	0.0195	9 2	0.0134
8 3	0.0193	9 3	0.0289
8 4	9.14	9 4	7.76
8 5	14.9	9 5	13.8
8 6	17.8	9 6	21.2
8 7	8.42	9 7	11.6
		9 8	17.8

Detection of the radiation emitted by specific states of the product ions is a useful diagnostic probe of the mechanism by which charge transfer proceeds. The photon cascades which follow the electron capture into singlet and triplet states of O^{2+} have been worked out.[7] Some of the lines are a common feature of planetary nebulae. Using a detailed model of these objects, a rate coefficient of 2.1×10^{-9} cm^3 s^{-1} has been determined[4] for the process $O^{3+}(^2P^o_{\frac{1}{2},3/2}) + H(^2S) \rightarrow O^{2+}(2p3p\ ^3D_1) + H^+$ at 1 eV. This is in good agreement with the value of 2.5×10^{-9} cm^3 s^{-1} which follows from the present work.

References

1. T.G. Heil, S.E. Butler and A. Dalgarno, Phys. Rev. A23, 110 (1981).
2. S. Bienstock, T.G. Heil, C. Bottcher and A. Dalgarno, Phys. Rev. A25, 2850 (1982).
3. S. Bienstock, T.G. Heil and A. Dalgarno, Phys. Rev. A, in press (May 1983).
4. G.A. Shields, A. Dalgarno and A. Sternberg, Phys. Rev. A, in press (1983).
5. S.E. Butler, T.G. Heil and A. Dalgarno, Phys. Rev. A, in press (1983).
6. G. Chambaud, J.M. Launay, B. Levy, P. Millie, E. Roueff and F. Tran Minh, J. Phys. B13, 4205 (1980).
7. A. Dalgarno and A. Sternberg, Mon. Not. R. Astr. Soc. 200, 77p (1982).

RADIATIVE DECAY OF Li-LIKE IONS FOLLOWING CHARGE EXCHANGE OF 3.3 KeV/AMU IONS WITH H_2

S. Bliman+, M. Bonnefoy°, J.J. Bonnet°, S. Dousson+, A. FLeury°
D. Hitz+, E. Knystautas⁑, M.F. Politis° and F. Zadworny+

+ DRFC et DRF, CENG, 85 X 38041 Grenoble Cedex - France
° C.N.A.M. Laboratoire de Physique des Collisions Atomiques 292 rue Saint-Martin 75141 PARIS CEDEX 03
⁑ Département de Physique (CRAM), Université Laval, Quebec Canada G1K4P4

It has been checked experimentally that when He-like ions capture are electron from H_2, in a low velocity collision (here v is typically 0.35 v_o), the total single electron capture cross-section writes $\sigma_{q,q-1} = 8.1 \cdot 10^{-16}\ q\ (cm^2)$. This is seen for C, N, O and Ne ions [1].

From these measurements coupled with the observation of radiative decay of lithium like C and oxygen ions, it has been possible to assign where the capture goes : in the case of carbon, at 0.35 v_o most of the capture goes to n = 3 [2] and in the case of oxygen to n = 4 at the same velocity [3].

If we consider the case of nitrogen, at the same velocity (0.35 v_o), it is seen that in the capture collision process $N^{5+} + H_2 \rightarrow N^{4+}(n,l) + H_2^+$, the transfered electron is shared between n = 3 and n = 4 [Fig. 1]

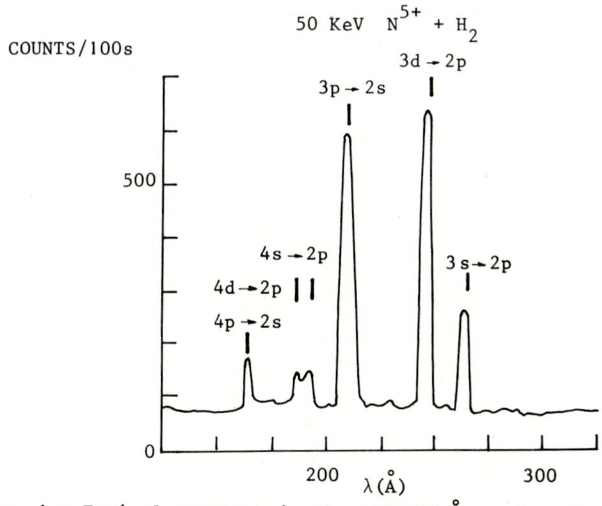

Fig. 1 : Typical spectrum in the 100-300 Å wavelength range.

This agrees with preliminary theoretical results by M. Gargaud et al [These proceedings] [4] .

To study the process : $Ne^{8+} + H_2 \rightarrow Ne^{7+}(n,l) + H_2^+$, we have used two spectrometers : a plane crystal Bragg diffractometer covering the range 10-100 Å and a V.U.V. grazing incidence spectrometer covering the range 80-550 Å. Typical spectra obtained at v = 0.35 v_o are given in Fig. 2a and 2b.

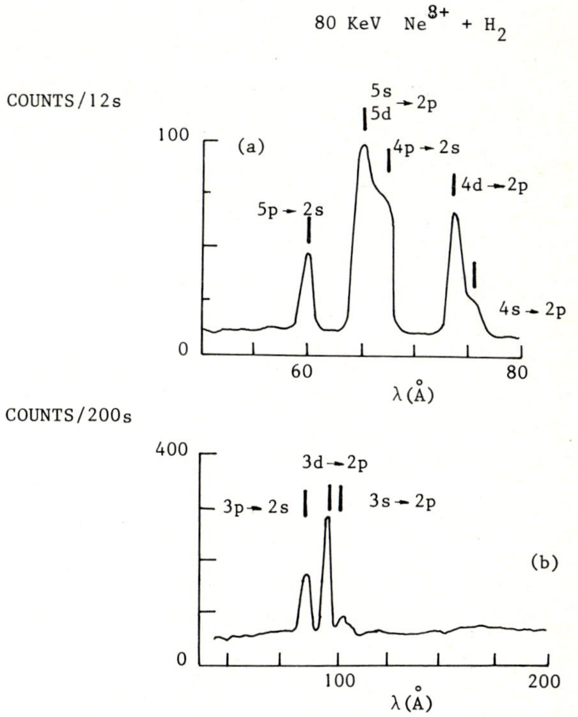

Fig. 2 : Typical uncorrected spectra :
a) 40-90 Å range, obtained with a plane crystal spectrometer.
b) 90-500 Å range, obtained with a grating mono chromator.

It is seen that in this capture the electrons are shared between n = 4 and n = 5.

In all of these collisional systems, considering relative branching ratios and transition probabilities from the count rates in the different lines, relative population are calculated.

The systematic study concerning the spectrometry of lithium-like ions after charge exchange indicates a strong varying behaviour from one collisional pair to another.

References

1. S. Bliman et al, Phys. Rev. A 23, 1703 (1981)
2. S. Bliman et al, J. Phys. B Lett. (1983)
3. R. Baptist et al, Phys. Lett. 93A, 185, (1983)
4. M. Gargaud et al, These proceedings (1983).

STUDY OF THE H_3 MOLECULES PRODUCED BY COLLISION OF FAST H_3^+ BEAMS WITH ARGON

M. J. Gaillard, A. G. de Pinho, J-C. Poizat, J. Remillieux and R. Saoudi
Institut de Physique Nucléaire (and IN2P3), Université Claude Bernard Lyon-1,
43, Bd du 11 Novembre 1918, 69622 Villeurbanne Cedex, France

The existence of long-lived H_3 molecules has been previously reported by several authors in experiments using H_3^+ beams [1]. More recently Herzberg discovered new spectral lines in hollow cathode discharge experiments with hydrogen, that were assigned to H_3 molecules [2].

We have studied the formation of the neutral H_3 molecule from fast H_3^+ ion beams (133 to 1000 keV/u) and the ionization of the produced neutral molecule. Two Argon gas targets, 170 cm apart, are used to study this $H_3^+ \rightarrow H_3 \rightarrow H_3^+$ collisional sequence. A permanent magnet, located after the first target, removes all the charged species from the beam entering the second gas target. The detection of an H_3^+ ion emerging from the second gas target is the signature of the formation of H_3 in the first target and of its survival until ionization in the second target. In contrast with the previous results using H_3^+ beams, our results are the first to present no ambiguity since the H_3 and HD molecules are separately identified by our detection system [3].

Our experimental method consists in measuring the variation of the yield of the neutralization-ionization sequence as a function of the Argon thickness in one of the gas cell, the pressure in the other cell being kept constant. The calculated yield involves the following cross sections : σ_c (electron capture by H_3^+), σ_ℓ (electron loss by H_3), σ_d^+ (dissociation of H_3^+) and σ_d^0 (dissociation of H_3). The dissociation cross section σ_d^+ was determined by measuring the molecular transmission yield of H_3^+ through the first gas target and we assumed that σ_d^0 is approximatively equal to σ_d^+. The cross sections σ_c and σ_ℓ were then calculated by fitting the experimental production yield of the $H_3^+ \rightarrow H_3 \rightarrow H_3^+$ sequence with its calculated expression.

The most remarkable result deals with the formation of H_3 : the electron capture cross section of H_3^+ is extremely small ($\sim 10^{-24}$ cm^2 at 400 keV/u), many orders of magnitude less than this of H^+ in the same gas. Furthermore σ_c decreases very rapidly (as $v^{-10.4}$) with the beam velocity v. These two observations are consistent with the conclusion that the long-lived states ($\tau \geq 3.10^{-7}$s) which are measured in this experiment are high Rydberg states of H_3.

A longitudinal electric field, which could reach 50 KV/cm, was applied on the H_3 beam, between the two gas cells, in order to measure the fraction of the H_3 molecules that is destroyed by Stark ionization. We observed that a large fraction of the H_3 molecules are ionized by fields as low as a few KV/cm, and that less than 30 % survive to a field of 50 KV/cm. With the same experimental configuration we then measured the field ionization of H_2 and HeH. As expected the flux of H_2 is not modified by the electric field, whereas a large fraction of the HeH molecules are field ionized. It is well known that the neutral HeH molecule is unstable in its ground state and has long-lived Rydberg states. The similar behaviour of H_3 and HeH in a strong electric field demonstrates the weakly bound character of the outer electron in the H_3 molecule produced by collisional neutralization of fast H_3^+ ions.

Since a permanent magnet was used to deflect the charged species between the two gas cells, all the projectiles experienced a motional electric field of a few KV/cm. In this field very high Rydberg states ($n \gtrsim 10$) of H_3 are destroyed by field ionization. Furthermore the long-lived Rydberg molecules which survived to predissociation and auto-ionization during their flight between the two targets must have a rather large angular momentum. In conclusion the various cross sections measured in these experiments show that we have observed H_3 Rydberg states in a narrow band of n and ℓ values.

References

1. M.F. Devienne, C.R. Acad. Sc. (Paris) 267 B, 1279 (1968); C.F. Barnett et al. Phys. Rev. A 5, 2120 (1972); T. Nagasaki et al. Phys. Lett. 38 A, 381 (1972).
2. G. Herzberg, J. Chem. Phys. 70, 4806 (1979).
3. N.V. de Castro Faria, M.J. Gaillard, J-C. Poizat, and J. Remillieux, Ann. Israel Phys. Soc. 4, 134 (1981).

A METHOD FOR ESTIMATING PROPERTIES OF SMALL CLUSTERS: THEORETICAL ASPECTS

Ray Hefferlin (1), G. V. Zhuvikin (2), Ken Caviness and Penny Duerksen (1)

(1) Southern College, Collegedale, TN USA
(2) Institute of Physics, Leningrad State University, Leningrad 198904 USSR

Elsewhere in this Conference[1] we report on results obtained in the estimation of properties of "clusters" of two atoms. Here we report on the extension of the method to larger clusters and on the theoretical bases of the method.

The method consists of the following: (1) Select a two-dimensional form of the periodic system of the atoms. (2) Consider it as a matrix. (3) Form the direct product of this matrix with itself N times, where N is the number of atoms initially presumed to be in a cluster, thus obtaining an architecture of matrix elements in a space of 2N dimensions.[2] (4) Select classes of clusters which are presumed to have similar properties. (5) Apply graphical, statistical and surface-fitting techniques to interpolate from known tabulated numerical values of any property, to those which are unknown, for clusters in the class.[1] (6) Move from one class to another, and/or from one N to another, as needed.

We now explain how to isolate three classes of clusters which have been used in step (4), above. (a) Isovalent clusters: all atoms which differ only in the principal quantum number, n, of the active electron are isovalent. Values of most ground-state properties of isovalent atoms are highly monotonic in n. Clusters in which any or all atoms are replaced by isovalent atoms are then suitable for use of step (5), above. (b) Multiplets: all atoms which differ only in how many s or p electrons there are in a given outermost shell (e.g. H and He or B through Ne), and all atoms in a transition-metal or rare-earth sequence (e.g. La through Yb) are here called a multiplet. Values of most ground-state properties vary along a multiplet of atoms in much the same way as along another multiplet of atoms which differs only in the value of n. Clusters of atoms, one or more of which is substituted by others in its multiplet, should then be compared with other clusters whose atoms come from multiplets differing only in the values n_1, n_2, etc. (c) Isolectronic clusters are appropriate subjects for the use of step (5), above.

In two-atom "clusters," classes (1) and (2) both have the result that each cluster (e.g. B_2, BAl, etc. or B_2, BC, etc.) is located in a square lattice on a plane. In three-atom clusters, they have the result that each cluster (B_3, B_2Al, etc. or B_3, B_2C, etc.) is represented by a point in a cubic lattice in three-space. Within this cubic lattice, all isoelectronic clusters are in a triangular lattice on a plane. In four-atom clusters, the result is that each cluster is a point in a cubic lattice in four-space. Within this cubic lattice, all isoelectronic clusters are in a tetragonal lattice in three-space (more specifically, as the Ga atoms in a GaAs crystal). The location of any chosen cluster of N atoms and n_e electrons in polar coordinates has the effect that the radial coordinate describes the extent to which the cluster differs from homonuclearity (e.g. B_3) or differs from isovalent forms of homonuclearity (e.g. BAlGa). The angle variables describe the extent to which there exist one or more pairs or triplets of homonuclear atoms (e.g. B_2C_2) or of the isovalent equivalents (e.g. $BAlC_2$). The coordinates have been completely worked out for clusters of up to three atoms and have been partly isolated for clusters of four atoms. The only obstacles to further progress are (i) the slowness of the mind to analyze multi-dimensional lattices and (ii) the different geometric shapes in which the actual cluster atoms may arrange themselves (e.g. isomers) and their relation to reduncancies in the lattices here described.

This entire method has been cast in group-theory notation for two-atom "clusters."[3] It is here stated that the extension to clusters with N atoms has been found possible.

Analysis of charged two-atom "clusters" has been reported previously.[2,4] Extension to N-atom clusters is trivial, methodologically.

References

1. "On Ground States, at Potential Minima, of Neutral Quasi-Molecules formed from Neutral Atoms or Ions."
2. Ray Hefferlin and Henry Kuhlman, J. Quant. Spectr. Rad. Transfer 24, 379 (1980)
3. G. V. Zhuvikin and Ray Hefferlin, Vestnik Leningradskovo Universiteta, scheduled for August, 1983
4. Ray Hefferlin and Wendy Innis, J. Quant. Spectr. Rad. Transfer 29, 97 (1983)

Current addresses

3. Ken Caviness
Physics Department
Lowell University
Lowell, MA 01854

4. Penny Duerksen
Department of Biochemistry
Emory School of Medicine
Atlanta, GA 30322

THE ROLE OF THE TILT ANGLE IN FLUORESCENT MEASUREMENTS OF THE GUEST-HOST SYSTEM IN ANISOTROPIC MOLECULAR LIQUIDS

J.FISZ

Institute of Physics,N.Copernicus University,Toruń,Poland

H.W.KUNERT

Institute of Physics,Technical University,60-965 Poznań,Poland

In several papers the role of the tilt angle in fluorescent measurements of the guest-host systems in anisotropic molecular liquids has not been considered yet.Usually,many authors assumed only two values of the tilt angle: $\varkappa=0$ i.e.homogeneous alignment and $\varkappa=\pi/2$ i.e. homeotropic alignment.

Here,using the Wigner rotation matrices of second rank $D^{(2)}_{mn}(\alpha,\beta,\gamma)$ and taking into consideration the general orientation of the direction with respect to the plates of the probe liquid crystal cells or biological membranes,see Fig1 we express the depolarization ratios $P_{\parallel}(\varepsilon,\varkappa,t) = I_{zz}(\varepsilon,\varkappa,t)/I_{zy}(\varepsilon,\varkappa,t)$ (eq.1) and $P_{\perp}(\varepsilon,\varkappa,t) = I_{xz}(\varepsilon,\varkappa,t)/I_{xy}(\varepsilon,\varkappa,t)$ (eq.2) by $\varepsilon,\varkappa$ and t. The left hand sides of eqs.1 and 2 are determined experimentally and the right hand sides are expressed by $\varepsilon,\varkappa$ and t.The ε is Euler angle betwen Y_L and Y_c as shown Fig.2.Starting from eq.1 we have (3)

$$A_oW_o(t) + A_1W_1(t) + A_2W_2(t) + A_3S_A + A_4S_E + 1 - P_{\parallel} = 0$$

for an instantaneous light pulse,and (4)

$$A_o\overline{W}_o(t) + A_1\overline{W}_1(t) + A_2\overline{W}_2(t) + A_3S_A + A_4S_E + 1 - \overline{P}_{\parallel} = 0$$

for an continuous illumination,where

$$A_o = P_2(\cos\varkappa)\{4 + P_{\parallel}(t)[B + 2P_2(\cos\varkappa)]\}$$

$$A_1 = 3\sin^2\varkappa - 2BP\cos^2\varkappa + \tfrac{3}{2}P_{\parallel}(t)\sin^2 2\varkappa$$

$$A_2 = \tfrac{3}{2}\sin^4\varkappa[2 - P_{\parallel}(t)] + BP_{\parallel}(t)\left[1 - \tfrac{1}{2}\sin^2\frac{\varkappa}{2}\right]$$

$$A_3 = 2P_2(\cos\varkappa)[1 - P_{\parallel}(t)],\quad B = 3\cos 2\varepsilon\,\sin^2\varkappa$$

$$A_4 = 2P_2(\cos\varkappa) + P_{\parallel}(t)[P_2(\cos\varkappa) + \tfrac{1}{2}B]$$

$W_L(t) = \langle D^{(2)*}_{Lo}(\Omega^{o}_{LA}) D^{(2)*}_{Lo}(\Omega^{t}_{LE})\rangle$ is the correlation function that describes the molecular dynamics, and

$$\overline{W}_L(t) = \int_0^\infty W_L(t)F(t)\,dt,\quad \overline{P}_{\parallel}(\varepsilon,\varkappa,t) = \frac{\overline{I}_{zz}(\varepsilon,\varkappa,t)}{\overline{I}_{zy}(\varepsilon,\varkappa,t)}$$

$$\overline{I}_{zj}(\varepsilon,\varkappa,t) = \int_0^\infty F(t)\,I_{zj}(\varepsilon,\varkappa,t)\,dt,\ (j= z,y)$$ and $F(t)$

is the probability that an initially excited molecule remains excited at time t.If a fluorescent molecule has a single exponential decay with a life time τ_f then $F(t) = \tau_f^{-1}\exp(-t/\tau_f)$

The Ω_{LA} and Ω_{LE} define the orientation of the absorption dipole transition at time zero and the emission transition dipole at time t with respect to the laboratory system.S_A and S_E describe the order parameters of the absorption and emission transition moments,respectivelly.

From eq.3 or 4 the five independent quantities W_o, W_1, W_2, S_A and S_E or $\overline{W}_o, \overline{W}_1, \overline{W}_2$,can be obtained in experiments if ε is varied five times.From this quantities it is possible to determine the order parameters and the relaxation times of the guest molecules.

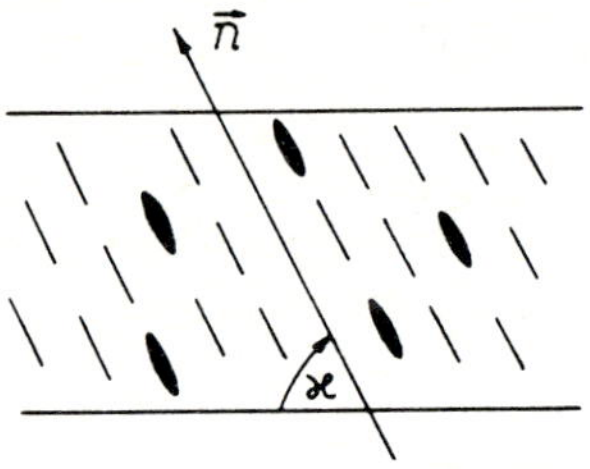

Fig1.The orientation of the direction with respect to the plates of the sample(| -represent the molecules of the anisotropic system and ⬮ -fluorescent guest).

Acknowledgements

We thank Dr hab.Z.Salamon for helpful discussions.

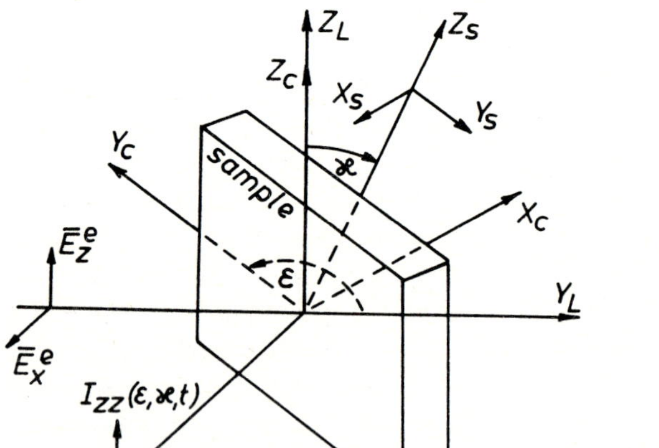

Fig.2.Experimental geometry: X_L,Y_L,Z_L is a laboratory coordinate system, X_c,Y_c,Z_c -of the sample and X_S,Y_S,Z_S-of the molecular direction. The exciting light $\vec{E}^e_j$ travels along Y_L,polarized parallel to the j-axis,j=x,z,the fluorescence intensity $I_{zk}(\varepsilon,\varkappa,t)$,polarized parallel to the k-axis,k=x,z is detected along Y_L.

CALCULATION OF COLLISION STRENGTHS FOR FINE-STRUCTURE TRANSITIONS IN THE SPECTRA OF RARE GAS ATOMS

F. Alan Bayes

Department of Physics and Astronomy,
University College London, Gower Street,
LONDON WC1 6BT

Calculations are in progress to determine collision strengths for the $2p^5\ ^2P^0$ ($J = 1/2 \rightarrow J = 3/2$) transition in the ground state of Ne^+ and the corresponding $3p^5\ ^2P^0$ ($J = 1/2 \rightarrow J = 3/2$) transition in Ar^+. These transitions give lines observed in the IR spectra of various astronomical objects.

The calculations involve solving coupled integro-differential equations using IMPACT, a computer program developed at UCL. The reactance matrices produced by IMPACT are transformed to intermediate coupling and used to solve the quantum defect expressions

$$R + \tan \pi\nu = \phi$$

to obtain bound states.

We can improve the accuracy of our results by comparing calculated bound states with the experimental levels tabulated in Moores (1948) and adjusting the fitting parameters to minimise the difference. This method was first used by Seaton (1958) and by Saraph and Seaton (1971). The method we use is an improved version of the older ones.

It is hoped that this method will prove to be useful in the calculations of other fine-structure in IR spectra.

References

1. Moores, C.E., 1948. Atomic Energy Levels, NSRDS - NDS 35, Vol. 1.
2. Seaton, M.J., 1958. Rev. Mod. Phys., 30, 992.
3. Saraph, H., Seaton, M.J., 1971, Phil. Trans., 271, 1.

ANOMALOUS EFFECTS IN VERY SMALL ANGLE ELECTRON POTASSIUM DIFFERENTIAL SCATTERING MEASUREMENTS

K. Rubin and I. Efremov

The City College of CUNY, New York, USA

For several years we have been investigating the differential scattering of low energy electrons (4ev) from potassium in a crossed beam configuration[1], with observations made on the recoiling atoms.[1,2] We present experimental data obtained at extremely small atomic scattering angles (2×10^{-5} rad) within the profile of the incident atom beam. These data show an oscillatory structure which cannot be accounted for in a plane wave description of the scattering. We suggest an explanation in which the colliding particles are described as wave packets and the observed structure in the data arises from interference between the incoming wave packet and the scattered wave. To illustrate this, consider the scattering of a beam of particles from a fixed scattering center with the incoming beam described in terms of a density matrix ρ. It can be shown that the ratio, $R(\hat{k})$, of the time integrated ensemble averages of the scattered to the unscattered currents in some direction $\hat{k}$, within the profile of the incoming beam, is given by

$$\Big\{-4\pi\,\mathrm{IM}\Big[\int\!\!\int f(\vec{k}\cdot\vec{k}')\langle\vec{k}|\rho|\vec{k}'\rangle\delta(E_k - E_{k'})k'dE_{k'}d\Omega_{k'}k^2dk\Big] + \Big[\int\!\!\int\!\!\int f^*(\vec{k}\cdot\vec{k}')f(\vec{k}\cdot\vec{k}'')\langle\vec{k}'|\rho|\vec{k}''\rangle \times \delta(E_k - E_{k'})\delta(E_{k''} - E_k)k'k''dE_{k'}dE_{k''} \times d\Omega_{k'}d\Omega_{k''}k^2dk\Big]\Big\} \Big/ \int\langle\vec{k}|\rho|\vec{k}\rangle k^2dk \qquad (1)$$

The first term is due to the interference between the incoming and scattered waves; the second term is the usual differential scattering. If the off diagonal elements of are not zero, interference will occur between the scattered and incoming packets at angles other than the forward direction. If $\langle\vec{k}|\rho|\vec{k}'\rangle = \langle\vec{k}|\rho|\vec{k}\rangle\delta(\vec{k}-\vec{k}')$ i.e., if the beam is equivalent to a set of incoherent plane waves, Eq. (1) will reduce to the usual plane wave result. Even in a crossed-beams experiment, where the situation is more complicated, since both the atomic and electron beams must be described in terms of wave packets and recoil must be taken into account, the interference effects described above will remain if the off diagonal elements of are not zero. When R is calculated from Eq. (1), with <u>all</u> the experimental parameters included, the solid curve shown in Fig. 1 is obtained, if the atom beam is described as a set of incoherent plane waves. The experimental values of R, also shown

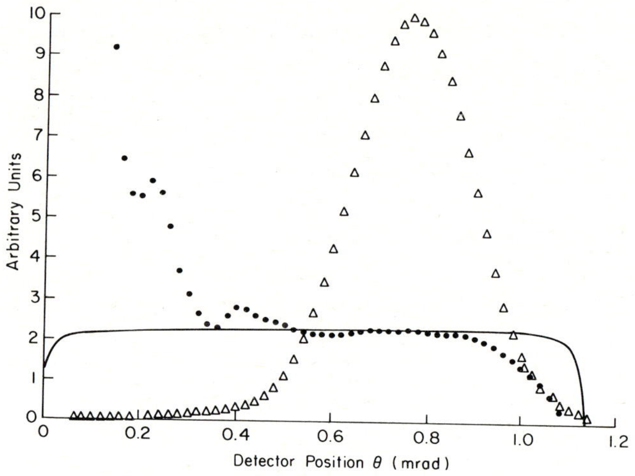

Fig. 1 Calculated value of R (solid line), experimental values of R (circles), and experimental atom beam profile (triangles).

in the figure (solid circles), have been normalized to the calculated value at the center of the beam profile.

It is clear that the data differ dramatically from the calculated curve, particularly by the presence of oscillations in the experimental results. It cannot be over emphasized that these oscillations are not due to oscillations in the differential cross section but rather to the coherence properties of the beams as described above.

The data indicate that there are fundamental questions concerning the proper quantum mechanical description of particle beams which can be answered by experiments of the kind we have performed. For example, it might be possible to use these interference effects to observe the relative phase of the scattering amplitude as a function of angle, since the interference terms involve products of the scattering amplitudes at different angles.

<u>References</u>

1. K. Rubin, VII International Conference on the Physics of Electronic and Atomic Collisions, Amsterdam, The Netherlands, edited by L.N. Branscomb et al. (North-Holland, Amsterdam (1971) pp 91-93.

2. K. Rubin and I. Efremov, Coherence and Correlation in Atomic Physics, (Plenum Press) (1980) pp 651-662.

EXCITATION OF THE Mg II RESONANCE DOUBLET BY MONOENERGETIC ELECTRONS

I.P.Zapesochnyi, A.I.Imre, V.I.Frontov*, A.N.Gomonaj*, A.I.Dashchenko*

Institute of Nuclear Researches, Ukrainian Academy of Sciences, USSR

*Uzhgorod State University, USSR

Excitation of ^2P levels of Mg^+ by monoenergetic electrons in energy interval from threshold to 15 eV was studied by spectroscopy method in crossed electron and ion beams.

The Mg^+ ions produced an a discharge source, formed into a beam by ion-optical system, separated from the Mg atoms by a 90^0 electrostatic selector, passed through collisional region, and were detected by Faraday cup.

The value of the ion beam current in the collisional region at 1 keV energy was $\sim 4\ \mu A$. The ion beam at $P \sim 10^{-8}$ Torr was crossed at a right angle by ribbon electron beam with 0,5 mm x 6 mm dimensions. The electron beam energy homogeneity was realised by means of 90^0 electrostatic selector and was equal $\Delta E_{1/2} \sim$ 0,3 eV. The value of electron current in (4÷20) eV energy interval ranged within (4÷30) μA. Spectral separation of the emission was performed by a high aperture ratio diffraction monochromator with a reverse linear dispersion of 2 nm/mm.

For the useful signal extraction from the background the modulation technique of crossed beams was used. The intensity of the useful signal was $\sim 6\ sec^{-1}$ at signal/background ratio of 1/5. The excitation function measurement was performed by measuring - control system with mini-computer [1]. The electrons' energy scanning in the given interval with 0,1 eV step, the gathering of experimental data, their processing and accumulation by recycle measurements were performed automatically according to an assigned managing programm.

Results of measurements are shown in the figure by dots. The statistical error of the measurements is shown by vertical bars equal in length to the 90% confidence interval. The solid line is the result of processing of experimental data by digital filtration [2]. The experimental results are normalised for the results [3] at (12÷15) eV electron energy.

The experimental curve reveals several features in the form of resonances with energetic width of (0,4÷0,5) eV. These features cannot be explained by cascade contribution from the higher ion levels. They can be attributed to resonance capture of bombarding electrons by Mg^+ ions, with the formation of short-lived autoinizing states of Mg I. Their decay into excited ion levels increases the population of these levels:

$$Mg^+(3S)+e \rightarrow Mg^{**}(n_1l_1,\ n_2l_2) \rightarrow Mg^{+*}(nl)+e,$$

where e is an Auger-electron.

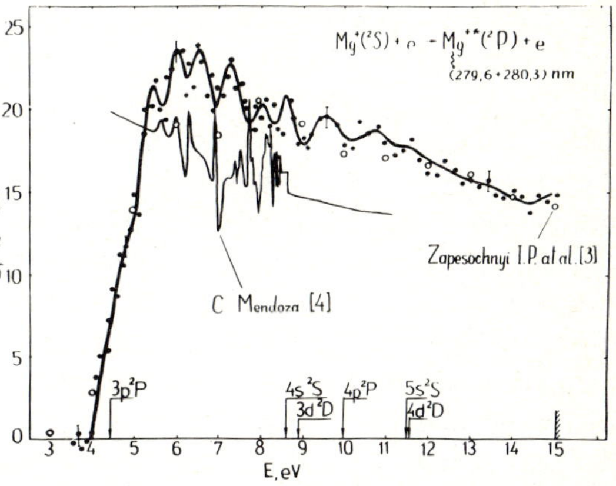

Finally, let's turn attention to the fact that a definite correlation is observed between the structure and theoretical calculation in a four-state close-coupling approximation [4], also given in the figure.

References

1. A.I.Dashchenko, V.I.Frontov, F.F.Papp, S.G.Solomchenko. Abstract of Papers, VIII AUCPEAC, 1981, Leningrad, p.292.
2. R.K.Otnes, L.Enochson. Applied time series analysis, vol.1. Basing techniques. 1978, New-York.
3. I.P.Zapesochnyi, V.A.Kel'man, A.I.Imre, A.I.Dashchenko, and F.F.Danch. Zh.Eksp.Teor. Fiz., 1975, 69, p.1948.
4. C.Mendoza. J.Phys.B: At.Mol.Phys., 1981, 14, p.2465.

NEAR THRESHOLD ELECTRON IMPACT EXCITATION OF POTASSIUM ATOM LOWER LEVELS

F.F.Papp, N.I.Romanjuk, O.B.Shpenik

Institute for Nuclear Researches (Uzhgorod), USSR

The optical excitation functions (EF) and polarisation of alkali metal resonance lines emission were studied previously [1].However, the obtained energy resolution of the electron beam gave no possibility to study the features on EF near threshold range. Still, some resonance and threshold features were revealed in electron-potassium atom scattering cross-sections [2].

Present paper deals with resonances on optical EF of potassium atoms by electron impact. A monoenergetic electron beam was produced by trochoidal electron monochromator (TEM). The elctron current was about 10^{-8}A, while energy distribution (electron energy spread, FWHM) did not exceed 0,04-0,05 eV.The potassium atoms were excited inside the vapour loaded cell, with the TEM attached. The emission of excited atoms was observed at a right angle to the electron beam direction and was projected onto the entrance slit of monochromator, which was placed along the direction of electron beam. The spectral line emission, extracted by photomultiplier photocathode, cooled by liquid nitrogen. A super sensitive system, operating in a single-electron counting mode, was applied for the weak photon flux registration. A multichannel analyser (MCA) with a complementary reversible counter and slow time - analysis device was used for the data storage. Energy scanning of incident electron beam was carried out in cyclic mode with 0,01 eV step and was synchronized with MCA memory channel switching. The exposition time for a single data point was provided, so that the certain statistical dispersion in the maximum of EF had not exceeded 3-5%. Energy scale calibration was made according to the spectroscopic threshold of potassium resonance line λ 7664 Å with accuracy of 0,02 eV. Optical EF for lowest members of principal, sharp and diffusion series of potassium atom were studied from excitation threshold to several eV above. For example,Fig.1 and 2 show the EF for the first two members of principal series. Most EF have a sudden increase in intensity at the threshold, and the first maximum does not exceed electron beam energy spread. Several features, such as the insignificant maxima, edges and irregularities were observed on the excitation functions of some lines. One can suggest, however, that the structure in the vicinity of the excitation thresholds of 3^2D and 5^2S levels, observed on the resonance line excitation function (Fig.1), is caused by the resonances [2], as well as threshold effects. This is confirmed, on one hand, by small width of this feature, (comparable with electron energy spread), it's close position by the excited 3^2D level and, on the other hand, by its shape, which is characteristic for the threshold feature (an almost symmetrical minimum), (see insertion in Fig.1).

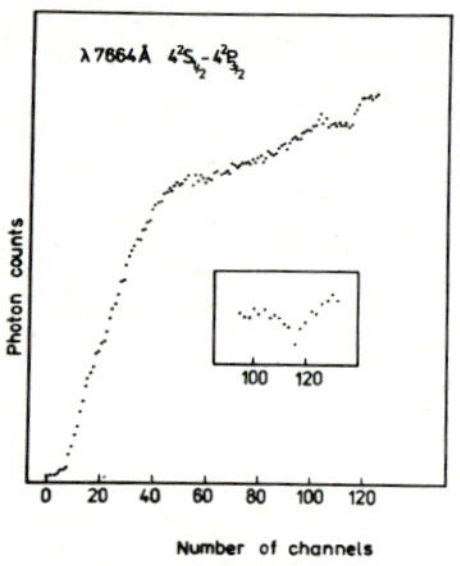

Fig.1

Comparison of the excitation functions for the first two members of potassium atom (Fig.1 and 2) displays sharp differences in their initial sections. In the case of resonance line an almost linear increase in cross-section near the threshold with insignificant edges at E=1,8; 1,9 eV is characteristic,whereas for λ 4044/47 Å lines the cross-section near threshold has a small rate of increase, which is interrupted by practically jump-like increase of cross-section at E=3,44 eV, and later on the curve remains almost parallel to energy axis. The upper members both sharp and diffusion series, display an almost jump-like increase in cross-section in the near-threshold region, what can be attributed to the threshold feature near the opening of inelastic channels, as well as to the resonances. At the conference the EF for a number of other spectral transitions will be presented, and more profound analysis of them will be given.

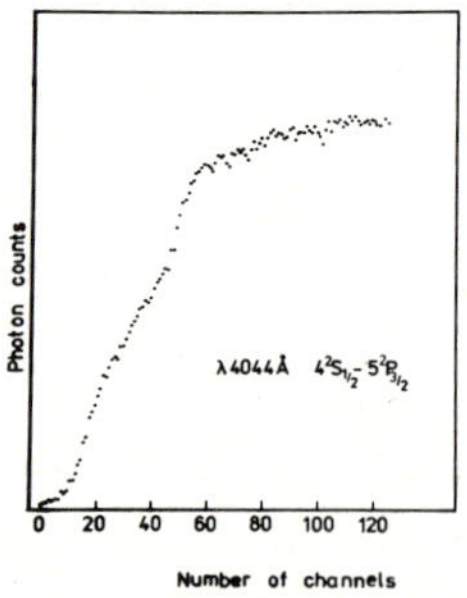

Fig.2

1. S.T.Chen, A.C.Gallagher. Phys.Rev.A. 1978, 17, No 2, 551-560.
2. M.Eyb, J.Phys.B. 1976, 9, No 1, 101-110.

ULTRASOFT X-RAY EMISSION OF BARIUM STUDY IN ELECTRON-ATOMIC COLLISIONS

V.S.Vukstich, Yu.V.Zhmenyak, I.P.Zapesochnyi

Institute for Nuclear Researches, Uzhgorod, USSR

The investigation of ultrasoft X-Ray emission of barium in energy range from 40 to 140 eV first was carried out. Experimental plant included a grazing-incidence vacuum monochromator and automatic system for control of experiment and processing the results of measurement [1].

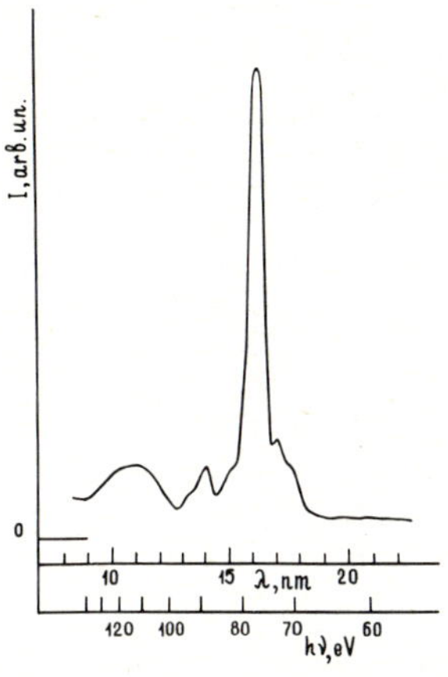

It was shown, that all observed emissions (see Fig.) caused via removing of one or two electrons from $4d^{10}$-subshell of barium. Most intensive emission at hν = 76 eV has excitation threshold equal to (97±2) eV and agrees well with energy of vacance in $4d^{10}$-subshell. This X-Ray emission correspondes to $4d^9 5p^6$–$4d^{10} 5p^5$ diagram transition, because the difference of the binding energies for $4d^{10}$- and $5p^6$-subshells is also equal to 76 eV [2].

The dependence of this emission on the energy of bombarding electrons in the range from the threshold of process to 1000 eV detaily was measured. It has peculiarite, attributed to additional making of 4d-vacance at the expence of nonradiative decay of autoionizing states of Ba I.

More weak emission lines with energy 88 and 69-73 eV adjacent to line with energy 76 eV from high and low energy sides, they are, accordingly, ionic and atomic satellites of diagram transition 4d 5p. Ionic satellite (hν = 88 eV) arise from making of double vacance in $4d^{10}$-subshell and it correspondes to $4d^8 5p^6$ – $4d^9 5p^5$ transition. Atomic satellites (hν = 69-73 eV) are related to the excitation of one from 4d-electrons in connected states ($4d^9 5p^6 nl$ – $4d^{10} 5p^5 nl$ transitions).

Broad emission, corresponding to the quant energy from 100 to 130 eV with maximum at 112 eV excite special interest. Last value of energy coincides with excitation threshold for this emission. We note, that the energy corresponding to this maximum (i.e. 112 eV) in continuous spectrum, is in good agreement with maximum of giant resonance in photoabsorption spectrum of atomic barium [3]. We may assume that broad emission discovered in our experiments is due to photones emission of barium atoms under influence of bombarding electrons because of virtual excitation of 4d-electrons in ef-states of continuous spectrum. Formation of this (so named "polarisation")bremsstrahlung is due to dinamical polarisation of 4d-subshell under of bombarding electrons.

R e f e r e n c e s

1. V.S.Vukstich, Yu.V.Zhmenyak, E.N.Postoy, I.P.Zapesochnyi. Ukrain.Phys.Journal, 1980, 25, 2008.
2. K.N.Huang, M.Aouagi, M.H.Chen, B.Crasemann, H.Mark. Atom Data and Nucl.Data Table, 1976, 18, 243.
3. P.Rabe, K.Radler, H.-W.Wolff. Proc.Intern. Conf. on Vacuum UV Radiat.Phys., Hamburg, 1974, 247.

ELECTRON IMPACT EXCITATION OF THULIUM AND GADOLINIUM ATOMS

L.L.Shimon

Uzhgorod State University, Uzhgorod, 294000, USSR

The crossed electron-atom beam method was used to determine effective excitation cross sections for 50 most intensive transitions in Tm and Gd atoms. The incident electron energy varied from the threshold up to 300 eV. The excitation pecularities for thulium atoms are generated via the instability factor an open 4f-subshell (the ground state configuration of thulium $4f^{13}6s^2\ ^2F_{7/2,5/2}$). In the case of gadolinium atoms the excitation pecularities were caused as result of existence an additional 5d-electron between the open 4f-subshell and $6s^2$ outer electrons. Basing upon the previous data for Yb, Eu and Sm [1-3] it is possible to present a complete picture of rare-atom excitation process depending both the outer subshell structures and making-up of the 4f-subshell.

1.Thulium. The magnitude of the effective cross-sections in maximum of excitation curve vary from $1,5\cdot10^{-16}$ cm^2 to $4\cdot10^{-18}$ cm^2.In contrast to the character of excitation for europium and ytterbium that have half-packed and fully-packed 4f-subshells, respectively, the efficiencies of spectral line excitations for Tm tend to the same value. This effect was caused by different mechanisms of excitation both the valence $6s^2$ electrons and electrons from the unpacked 4f-subshell. Besides the direct and exchange excitation cross-sections are also close in value. One point out that in the case of ytterbium [1] and europium [2] the corresponding cross-sections for direct and exchange excitation processes differ by one or two order in magnitude.

Exclusively for the spectral lines in Tm I corresponding to transitions from the most low-lying levels (see Fig.) one observe characteristic excitation functions caused by excitation of outer $6s^2$-electrons (with or without exchange) and inner 4f-electrons.The strong mixing effect for highly excited states ($E_{exc.} > 3$ eV) results that the excitation of 6s and 4f-electrons in some cases have the same behaviour (i.e. an uniform excitation function). And vice versa the components of the same multiplet may posess different types of excitation functions. Excitation of thulium and samarium atoms discover a close analogy [3].

2.Gadolinium. The most intensive spectral lines of gadolinium atoms are caused by excitation of 6s-6p transitions in analogy with other lanthanides. However, the cross sections of the above mentioned lines of gadolinium located in the small interval of values, namely from $2\cdot10^{-17}$ cm^2 to $2\cdot10^{-18}$ cm^2. They are smaller than that in other lanthanides. This is due to great density of terms and complicated spectra of gadolinium because of existence of the 5d-electron almost in all known configurations of Gd I.

The close location of the energy levels generate a great number of excitation mechanisms for spectral lines due to favourable conditions for mixing of excited states. Therefore a general form of excitation functions is characteristic for spectral lines of Gd I independently from a transition type (see Fig., curves 4 and 5).

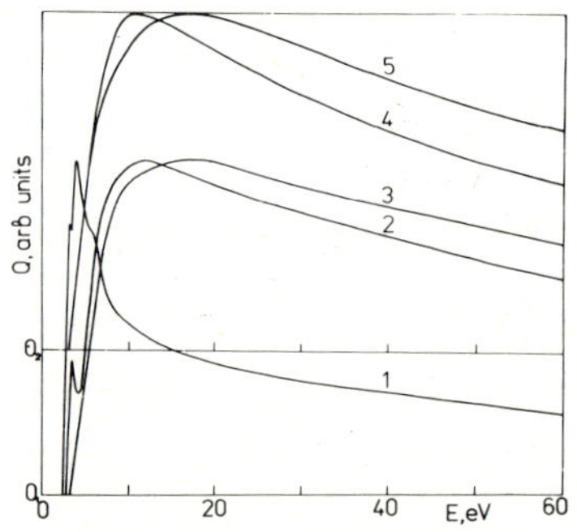

Spectral line excitation functions for Tm (1-3) and Gd (4-5)

1 - 506,1 nm, $4f^{13}(^2F_{7/2})6s6p(^3P_2)(3\frac{1}{2},2)_{7/2}$;

2 - 438,6 nm, $4f^{12}(^3H_5)5d_{3/2}6s^2(5,1\frac{1}{2})_{5/2}$;

3 - 374,4 nm, $4f^{13}(^2F_{7/2})6s6p(^1P_1)(3\frac{1}{2},1)_{7/2}$;

4 - 422,6 nm, $4f^7 5d6s6pJ\ ^9F_7$;

5 - 405,8 nm, $4f^7 5d6s6pJ\ ^9D_4$.

1. L.L.Shimon, N.V.Golovchak, I.I.Garga and I.V.Kurta. Opt.Spectr.,(1981), 50, 1037
2. N.V.Golovchak, I.I.Garga, L.L.Shimon. Opt.Spectr., (1978), 44, 23
3. L.L.Shimon, N.V.Golovchak, I.I.Garga. Opt.Spectr., (1981), 51, 36.

EXCITATION OF METASTABLE LEVELS OF NOBLE GAS ATOMS BY ELECTRON IMPACT IN INTERSECTING BEAMS

[+] O.B.Shpenik, [+] A.N.Zavilopulo, [†] A.V.Snegursky, [*] I.I.Fabricant

[+] Institute for Nuclear Researches (Uzhgorod), USSR
[†] Uzhgorod State University (Uzhgorod), USSR
[*] Institute of Physics (Riga), USSR

A new method is proposed in the present report for investigation of electron excitation of atomic and molecular long lived states. The method is based on spatial, energetic and time distributions determination of excited particles. The report also presents a survey of studies, carried out on excitation of metastable levels for all noble gas atoms.

Experimental setup includes (see Fig.1) a supersonic beam nozzle source with high spatial assignment (about 1°), 127° electron selector together with metastable atoms detector-channeltron, placed in the scattering chamber. Information about the process was extracted by means of measuring the yield of metastable atoms scattered in the different angles towards the primary direction of neutral atoms (drift angle Θ). Data collection and processing were completed by a heading measurement complex, including a multichannel pulse analyser and microcomputer. Information output was performed on a X-Y plotter.

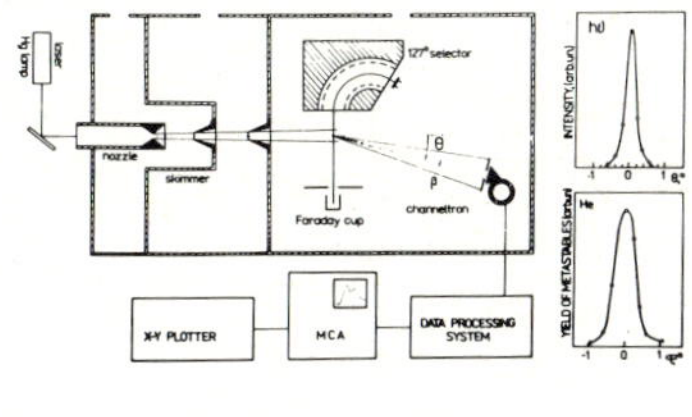

Fig.1

The differential and integral excitation cross sections were measured, corresponding to the He($2^{3,1}S$) and Ne, Ar, Kr, Xe(ns$[3/2]_2$,ns$'[1/2]_0$ atoms formation. The velocity distributions of metastables were also studied.

Differential cross sections were measured in incident energy region from threshold to 500 eV. One can see from Fig.2 that differential cross sections show features, displayed in sudden increase of cross section at definite drift angles Θ. As to the case of heavier atoms - Ar, Kr and Xe - the dynamics of features behavior is expressed weaker than for He and Ne atoms. This can be attributed to their larger mass and, consequently, to lower angular resolution of the detector β. The kinematic analysis of the process made possible the conclusion that the revealed features were connected with the presence of transmitted momentum from electrons to neutral atoms. The dynamics of features displaying confirms the fact, that the greater is the incident electron energy, the lower is the value of momentum transmitted. That is in good agreement with velocity distribution data for all metastable atoms studied.

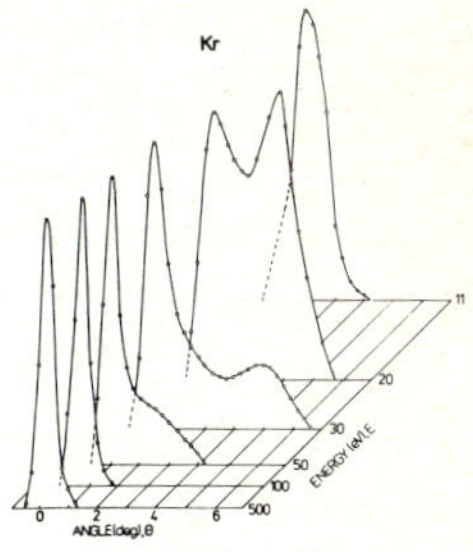

Fig.2

Energy dependencies of differential cross sections were measured for all considered atoms in the near threshold energy region. Two types of structure were revealed (see Fig.2). In the first case the position of features does not change at the energy scale, if the angle is changed. The comparison of such positions with these of features observed on total cross sections [1] made possible to refer the features of this type to the resonances connected with the negative-ion formation.

In the second case one observes the shift of structure with the drift angle variation. Features of this type were interpreted as kinematic ones, due to the existence of threshold scattering energies for the given drift angle Θ.

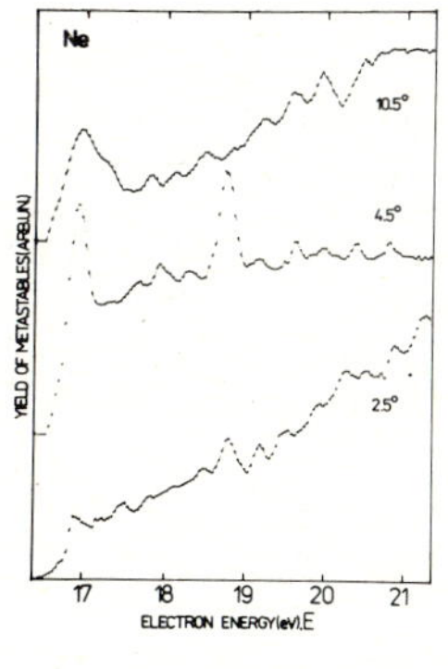

Fig.3

Total cross sections of the metastable atoms formation were defined by means of integrating the measured differential cross sections over all scattering angles. This procedure made possible to obtain the energy dependencies of total cross sections from threshold to 500 eV. Detailed analysis of the features observed should be presented at the conference, together with the results on total excitation cross sections.

1. Brunt J.N.H., King G.C., Read F.H. J.Phys.B. 1976, 9, 2195; 1977, 10, 433.

THE CALCULATION OF EXCITATION OF 2S AND 2P LEVELS OF He^+ ION BY ELECTRON IMPACT

M.I.Haysak, V.I.Lengyel, V.T.Navrotsky, E.P.Sabad
Uzhgorod State University /Uzhgorod/, USSR

The process of inelastic electron - He^+ ion scattering is considered. The basic aim is study of influence of autoionizing states of He atom, which converges to the threshold n=3 of He^+, on the excitation cross-sections σ(1S - 2S) and σ(1S - 2P), since it is known that the latest experimental data [1] show the resonances in these cross-sections. The calculations are carried out by means of modified Feshbach method[2-3], in which the coupling between closed and open channels is taken into account in a first-order approximation in interaction with taking into account 1S, 2S and 2P states of He^+. Closed channels are described by set of 24 basic functions of autoionizing states of He.

Both total and differential cross-sections are calculated. Total cross-section of excitation σ(1S - 2P) coincides well with the experimental data [1], but cross-section of (1S - 2S) - excitation differs from the experiment by factor 1,5 - 2 [3]. These conclusions agree with those, obtained earlier by different methods [4]. The differential cross-sections (see fig.1 and fig.2) are characterized by strong angular asymmetry with preferential scattering in the direction of large angles, which is due to influence of resonances. This circumstance is manifested most prominently in (1S - 2S) - excitation.

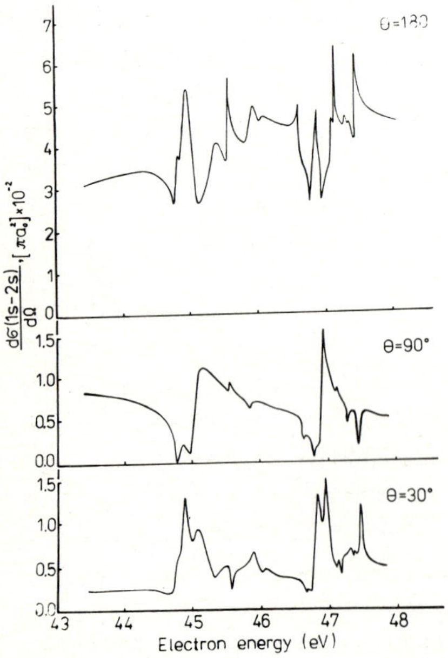

Fig.1 Differential cross-sections for the (1S - 2S) - excitation of He^+ ion by electron impact.

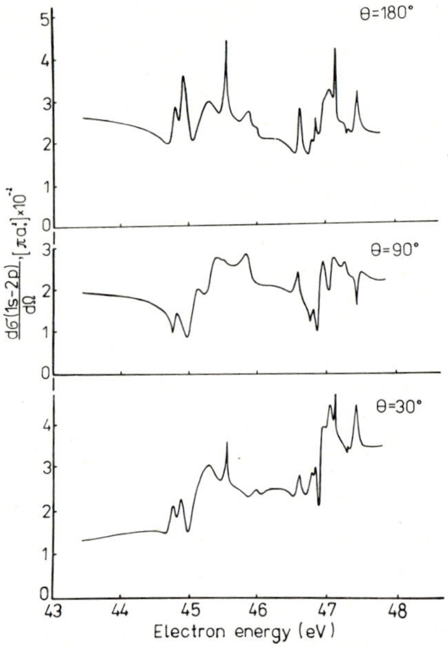

Fig.2 Differential cross-sections for the (1S - 2P) - excitation of He^+ ion by electron impact.

---

1. A.I.Imre, N.I.Semenyuk, A.I.Daschenko. Reports of XIII All-union conf. on electron -atomic collisions, Leningrad, 1981
2. V.V.Balashov, S.I.Grishanova, I.M.Kruglova, V.S.Senashenko. Opt. Spectrosc. (USSR), 88, 859 (1970)
3. M.I.Haysak, V.I.Lengyel, V.T.Navrotsky, E.P.Sabad. Ukr.Phys.J. (USSR), 25, 1329 (1980); 27, 1617 (1982)
4. M.J.Seaton. Adv.Atom.Molec.Phys., 11, 83 (1975)

CALCULATION OF DI-ELECTRON RECOMBINATION CROSS-SECTION ON He^+- ION

O.I.Zatsarinny, V.I.Lengyel, E.P.Sabad
Uzhgorod State University /Uzhgorod/, USSR

The process of di-electron recombination (DR) was actively investigated during past twenty years. It was shown that it becomes an important additional recombination mechanism in high-temperature [1]. Theoretical calculations of DR were almost entirely devoted to the determination of the velocity of this process and also to its influence on electron impact excitation of ions. The theoretical or experimental data on DR cross-section have not been published yet, though some experiments are being prepared at the moment [2]. It is quite clear, that DR cross-section contains much more information about physics of the process.

In the present work the calculation of DR cross-section of He^+-ion is carried out for the first time with taking into account 50 low-lying auto-ionizing states (AIS) with the following configurations: 2snl, 2pn'l'(n,n'≤5) The DR through these AIS is effective only for excited helium atom levels 1snl with n≤6. All these final states were taken into account in the calculation. The calculation of DR cross-section was carried out according to the expression, which was obtained in [3]:

$$\sigma(i\varepsilon \rightarrow s\gamma) =$$

$$\frac{1}{g_i g_e}\frac{2\pi}{q^2}\frac{4}{3}K^3\sum_{is}\left|\langle s\gamma|D|i\varepsilon\rangle+\sum_{d}\frac{\langle s\gamma|D|d\rangle\langle d|V|i\varepsilon\rangle}{\varepsilon+E_i-E_d+\frac{i}{2}\Gamma_d}\right|^2.$$

The symbol $|i\varepsilon\rangle$ denotes the state of ion + electron system with energy ε and momentum q in continuous spectrum, symbol $|s\gamma\rangle$ means system "bound state of atom plus photon" with wave-vector K, symbol $|d\rangle$ denoting AIS with energy E_d. Atomic unit system is used. Quantities g_i, g_e denote statistical weights of ion and electron, $\Gamma(d)=\Gamma_a(d)+\Gamma_r(d)$ are the widths of the AIS, D is dipol operator, V is the operator of interelectron interaction. Symbol $\sum_{is}$ denotes summation over all nonspecified quantum numbers of initial and final states.

Both AIS wave-functions and those of bound states of helium were calculated using the superposition configuration method. Basic functions were chosen in a form of antisymmetric products of a coulomb functions with $Z=2$. With the help of these functions all dipole matrix elements were calculated. The autoionizing widths were taken from [4].

The result of calculations of total cross-section DR on helium ion are presented on fig.1. It concicts of series of a narrow resonances, which are much larger than the cross-section of a direct recombination ($10^{-6}\pi a_0^2$). Dotted line represent the cross-section of DR, averaged on electron energy uncertainity on half-altitude equal to 1 eV, which is characteristic to the process of electron-helium ion scattering. For comparison we have shown on the same fig. the excitation cross-section of He^+(1s-2p) by electron impact.

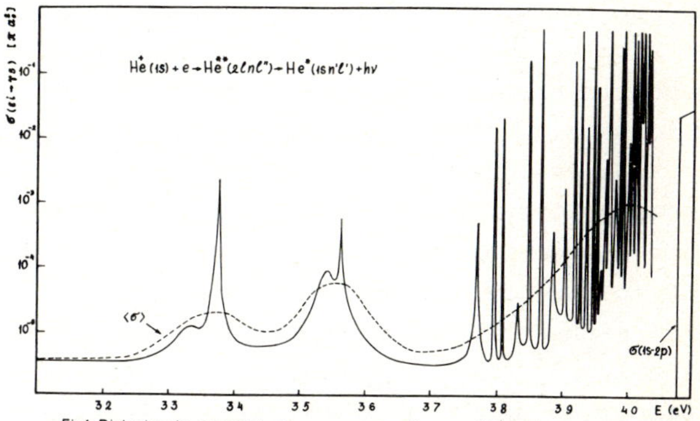

Fig.1. Dielectronic recombination cross section on He^+ under n=2 threshold

Since the emission wave-length at DR is close to the wave-length of transition He^+(1s-2p), this process is manifested in the experimental spectroscopic investigation of DR as an erosion of excitation threshold of He^+(1s-2p) and as one or several maximums below excitation threshold. However, from these results it follows also that the preliminary experimental estimates of DR cross-section [2] are over-estimated. We have to note that the calculated relative contribution of lower AIS to the total velocity of DR coinsides within 10% with the calculations of velocities by strong coupling method repoted in [1].

1. M.J.Seaton and P.J.Storey, in Atomic Processes and Applications, edites by P.G.Burke and B.L.Moiseiwtsch (North-Holland, Amsterdam, 1976), chap.6.
2. A.I.Imre, N.I.Semenyuk, A.I.Kuz'ma, I.P.Zapesochny. Repots of XIII All-union conf.on electron-atomic collisions, Leningrad,1981.
3. B.W.Shore, Astropys.J.,158, 1205(1969).
4. M.J.Conneely, L.Lipsky, J.Phys.B: Atom. Molec. Phys., 11, 4135 (1978).

THE INFLUENCE OF ELECTRON CAPTURE ON THE ELASTIC CROSS-SECTION OF ELECTRON - Mg^+- ION SCATTERING

V.I.Lengyel, V.T.Navrotsky, E.P.Sabad
Uzhgorod State University /Uzhgorod/, USSR
Institute for Nuclear Research, Academy of Sciences of Ukrainian SSR, USSR

During last few years much experimental information was obtained concerning the necessity of taking into account the autoionizing states for the description of electron-atom or electron-ion scattering. For instance, the study of photoabsorption spectrum of Mg in the far ultra-violet region [1] has shown that the photoionization cross-section of Mg is dominated by resonances due to formation and subsequent Auger-decay of autoionizing states of Mg of the type $1s^2 2s^2 2p^6 3pns\ ^1P^\circ$. Therefore it can be supposed that the elastic electron-Mg^+-ion scattering also will be dominated by electron capture which leads to a formation of autoionizing states and subsequent Auger-decay.

The calculation of differential elastic cross-section of electron - Mg^+-ion in energy region from threshold to 4,43 eV is carried out in the present work with the use of modified Feshbach method [2] proposed by Balashov et all. [3] for the investigation of photoionization of He. For the calculation of nonresonant part of scattering amplitude the programme "IMPACT" [4] was used. The wave-functions of autoionizing states were shosen in a form of multiconfiguration wave-functions, which were constructed in a form of combinations of products of one-electron functions of Mg^+. We have taken into account 45 $^{1,3}S^e$, $^{1,3}P^\circ$, $^{1,3}D^e$, $^{1,3}F^\circ$ autoionizing states.

On the figure energy dependence of the elastic differential cross-sections at different angles is shown. The positions of the calculated autoionizing states are marked.

As it is seen from the figure the main contribution to the elastic cross-section is made by $^3P^\circ$, $^3D^e$, $^{1,3}F^\circ$ resonances. These resonances define practically completely the magnitude of the elastic cross-section. It is especially clearly seen, that the structure of cross-section is influenced by wide triplet $3p3d\ ^3F^\circ$ resonance with the width $\Gamma = 0,264$ eV at the scattering angle equal to 180°. At the same angle $^3D^e$ resonances leads to appearance of narrow maxima, while $^3P^\circ$ resonances leads to a constructive interference picture in a form of narrow simmetric peaks, which become especially pronounced at $\theta = 150°$, when the contribution of $^{1,3}F^\circ$ resonances diminishes.

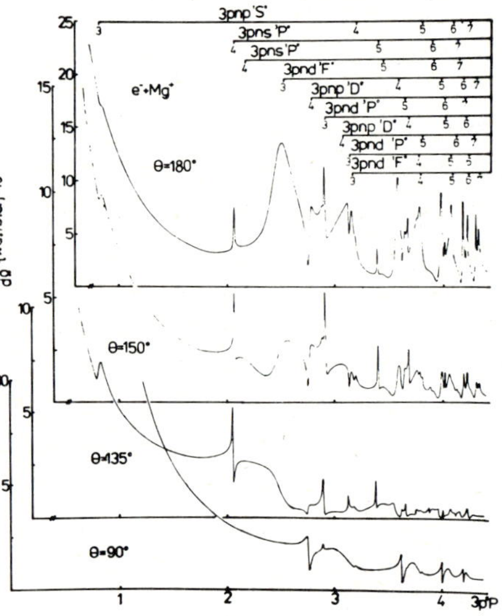

Figure 1. Elastic differential cross-section of electron - Mg^+- ion scattering at the different angles. On the X-axis energy E is shown in eV.

1. M.A.Baig, J.P.Connerade, Proc.Roy.Soc.Lond. A364, 353 (1978)
2. M.I.Haysak, V.I.Lengyel, V.T.Navrotsky, E.P.Sabad, Ukr.Phys.J. (USSR), 25, 1329 (1980); 27, 1617 (1982)
3. V.V.Balashov, S.I.Grishanova, I.M.Kruglova, V.S.Senashenko, Opt. Spectrosc. (USSR), 88, 859 (1970)
4. M.A.Crees, M.J.Seaton, P.M.H.Wilson, Computer.Phys.Commun., 15, 23 (1978)

THE AUTOIONIZING STATES OF A MAGNESIUM ATOM

O.I.Zatsarinny, V.I.Lengyel, V.T.Navrotsky, E.P.Sabad, M.Salak

Uzhgorod State University /Uzhgorod/, USSR

P.J.Safarik State University /Kosice/, Tshechoslovakia

The aim of this work was the calculation of energies of autoionizing states (AIS) of Mg in the region of first ionization threshold and their corresponding autoionizing widths. The latter are almost completely absent in literature. The calculation is carried out in a version of Feshbach method, proposed by Balashov with coll. [1] for the investigation of resonance photoionization process of He, and which was successfully developed in our earlier papers [2] for the investigation of resonance effects in electron-ion collisions. The energies and widths of $^{1,3}S^e$, $^{1,3}P^o$, $^{1,3}D^e$, $^{1,3}F^o$ AIS of Mg in the region of first ionization threshold are obtained. The results for some AIS are given in table.

The calculation revealed two interesting pecularities of AIS spectrum. Firstly, there is no unabiguous experimental identification of 3p3d ^1F AIS - at the moment [3-4]. This calculation gave a value for the energy of this AIS, which differs from data, given in both [3] and in [4]. But good coincidence with the experimental values and with other theoretical calculations for other $LS\pi$ - series of AIS allows to draw a conclusion that indentification of 3p3d $^1F^o$ AIS of Mg, which was given in [3-4] is, perhaps, not quite precise.

Secondly, it is established, that the lowest $^1D^e$ AIS Mg is not $3p^2$, but 3p4p AIS with the energy 10,71 eV. The state $3p^2\ ^1D^e$ is eroded after taking into account the interaction with the 3snd configurations and ceases to exist as a well-defined state. This confirms the results of experiment [3], and confirms the calculation, which was carried out in a close-coupling approximation [5-6].

The results, which were obtained, have shown, that this method is very effective for the calculation of characteristics of AIS of many-electron atoms.

1. V.V.Balashov, S.I.Grishanova, I.M.Kruglova, V.S.Senashenko, Opt. Spectrosc. (USSR), 88, 859 (1970)
2. M.I.Haysak, V.I.Lengyel, V.T.Navrotsky, E.P.Sabad, Ukr.Phys.J. (USSR), 25, 1329 (1980); 27, 1617 (1982)
3. D.Rassi, V.Pejcev, T.W.Ottley, K.J.Ross, J.Phys.B:At.Mol.Phys., 10, 2931 (1977)
4. R.Okasaka, K.Fukuda, J.Phys.B:At.Mol.Phys., 15, 347 (1982)
5. C.Mendoza, J.Phys.B:At.Mol.Phys., 14, 87(1981)
6. P.G.Burke, D.L.Moores, J.Phys.B:At.Mol.Phys., 1, 575 (1968)
7. G.Mehlman-Ballofet, J.M.Esteva, Astr.J., 157, 945 (1969)

Table. Characteristics of some AIS of Mg: energies E are given in eV, widths Γ - in eV, N_{eff} is effective quantum numbers.

State	THEORY						EXPERIMENT		
	Present calculation			[5]	[6]		[3]	[4]	[7]
	E	Γ	N_{eff}	N_{eff}	E	Γ	E	E	E
$3p^2\ ^1S^e$	8,46	0,051	1,896	1,971	8,67	0,017	8,45		
3p4p $^3S^e$	10,53	0,001	2,823	2,848					
3p4s $^1P^o$	9,80	0,580	2,362	2,396	9,75	0,430	9,81		9,86
3p4s $^3P^o$	9,71	0,007	2,318	2,341	9,71	0,003	9,54		
3p4p $^1D^e$	10,71	0,324	2,987	2,907			10,71		
3p4p $^3D^e$	10,41	0,023	2,726	2,716					
3p3d $^1F^o$	10,80	0,104	3,076				9,995	9,82	
3p3d $^3F^o$	10,12	0,264	2,532					10,03	

Author Index

Abouaf R. 295, 296
Achenbach C. 200
Achiba Y. 68
Adam M.-Y. 44
Afrosimov V.V. 546, 565
Aggarwal K.M. 196, 731
Ajello J.M. 267, 282
Akagi Y. 157
Allan R.J. 513
Allison A.C. 48
Alonso E.V. 365
Alston S. 97
Altman J. 162, 163
Alvarez I. 635, 636
Amamiya S. 569
Ambrose R. 162
Amme R.C. 335
Amundsen P.A. 508
Anan A. 145
Andersen L.H. 379, 393, 394, 416, 564, 714
Andersen N. 399, 458
Andersen T. 399, 458, 459
Andrä H.J. 515
Andresen P. 49
Andrews M.C. 428
Andriamonje S. 434, 437
Angel G.C. 495
Anholt R. 434, 437
Ankudinov V.A. 611
Anselmann R. 54
Antar A. 475
Antoni T. 241
Aquilanti V. 648
Arai S. 283
Archirel P. 624
Arcuni P. 383, 384
Arianer H. 208
Armour E.A.G. 304, 311
Ashurov A.R. 174
Astner G. 416, 711
Astruc J.-P. 676
Aubert-Frécon M. 317
Aumayr F. 484
Auschwitz B. 631
Avakov G.V. 174
Avaldi L. 430
Avdonina N.B. 734
Avrillier S. 324
Awaya Y. 387, 388, 465, 470, 471
Aynacioglu A.S. 359
Azria R. 295, 296

Bachau H. 385
Back C. 145
Back C.G. 135
Bae Y.K. 19, 24
Baer T. 627
Bähring A. 460
Bak J.F. 422
Bakaev D.S. 694
Baker O.K. 434, 437
Balashov V.V. 23, 139, 182
Banda H. 364, 439
Baragiola R.A. 365, 372, 391
Bárány A. 516, 572, 711
Barat M. 573
Barbe R. 676
Barbier L. 328
Barnes J. 659
Barrachina R.O. 375
Barrette J. 472
Bartschat K. 103, 149, 515
Barzen K. 339, 340
Barzick B. 78
Basalaev A.A. 546, 565
Basbas G. 428
Basu M. 183
Baum G. 144
Baumgärtel H. 73, 298
Baus J. 615
Bayes F.A. 741
Beck D. 643
Becker K. 259, 260
Becker R. 200
Becker R.L. 660
Becker U. 31
Becker U. 455
Bederson B. 83, 248
Beigman I. 675
Belic D.S. 210
Belkic Dz. 500
Bely-Dubau F. 208, 215
Benka O. 468
Benz A. 615
Berdermann E. 435
Berends R. 586
Berényi D. 390
Berger O. 146
Bergnes C. 353
Berinde A. 464
Berkner K.H. 525
Berkowitz J. 57
Berlande J. 603, 646, 671
Berman M. 230, 231
Bernstein E.M. 443, 529
Berrington K.A. 108, 192, 731
Berry S.D. 379, 381, 582
Beyer H.F. 417
Beyer H.-J. 141
Beyer W. 344
Beynon J.H. 288
Bhatia A.K. 730
Bhattacharyya S. 160, 300, 523, 728, 729
Bichsel H. 357
Bienstock S. 736
Bilau R. 448
Bischof G. 354, 628
Bisling P.G.F. 91
Bivona S. 681, 692
Bizau J.M. 27, 689, 690
Bjørnelund S.K. 564
Blankenship D.M. 350
Bliman S. 549, 708, 737
Blum K. 149, 515
Bocvarski V. 258
Boesten L. 245
Böhringer H. 655
Bonanno R. 327
Bonnefoy M. 549, 708, 737
Bonnet J.J. 549, 708, 737
Bordenave-Montesquieu D. 353
Borondo F. 522, 723
Borsella E. 74
Bosch F. 364, 435, 436, 439
Boskamp E. 461
Bossler J. 363
Bottcher C. 199, 211, 555
Botter R. 52
Bouby L. 294, 295
Boudouris G. 699
Boulmer J. 327
Boutonnet A. 353
Brandt D. 386, 427, 527
Bransden B.H. 115, 567
Brazuk A. 484, 551
Breinig M. 378, 379, 381, 582, 651, 714
Brenton G. 288
Breuckmann B. 168
Briand J.P. 208
Briggs J.S. 440, 496, 498, 528
Brion C.E. 39, 279, 280
Broad J.T. 60
Bronner W. 71
Brouillard F. 713
Brown C.J. 308
Bruch R. 383, 384, 469, 535, 536, 704
Bruijn D.P. de 629, 630, 634
Brutschy B. 73
Bryant H. 656
Buck U. 584, 585
Buckman S.J. 105, 106, 128
Budenholzer F.E. 593
Buelow S.J. 645
Bugrov G.E. 530
Bureyeva L. 675
Burgdörfer J. 378
Burke P.G. 1, 2, 103, 108, 149, 206, 234
Burkhard M. 381
Burkov S.M. 10
Burns D.J. 133
Burns W.S. 582
Bussert W. 28, 619
Butler K. 203
Byron Jr. F.W. 165

Cadez I. 255
Callaway J. 114
Can C. 563
Carnovale F. 39
Carré B. 27, 689, 690
Cartwright D.C. 121, 272
Castleman Jr. A.W. 58, 291
Cavaliere P. 680
Cavalli S. 648
Caviness K. 739
Cederbaum L.S. 230, 231
Cederquist H. 393, 416, 711
Chaguri E. 94
Champion R.L. 401
Chandra N. 251
Chang E.S. 236, 237, 256
Chantrenne S. 713
Charles P. 208
Chatterjee L. 160, 300, 523, 728, 729
Chaturvedi R.P. 428
Chebanier de Guerra A. 314
Chen Y.S. 480, 588
Cherepkov N.A. 12, 38
Cheret M. 328
Chetioui A. 441, 442, 537, 538
Chevallier P. 537
Chhaya V.M. 118
Chu Shih-I 61, 75, 641
Chung K.T. 469
Chupka W.A. 67
Church D.A. 462, 582
Chutjian A. 187, 188, 272
Ciortea C. 464
Cisneros C. 635, 636
Claeys W. 333, 334, 713
Clapis P. 475
Clark C.W. 107
Clark M. 386, 427, 443, 527
Clemente M. 435
Cocke C.L. 478
Coggiola M.J. 19, 24, 329, 620
Cohen J.S. 332, 727
Cole K. 627
Colin A. 238
Collins L.A. 216, 217, 257
Colson S.D. 67
Combet Farnoux F. 11, 733
Comer J. 261, 262, 264
Compton R.N. 65
Cook J.P.D. 179
Cooper C.D. 65
Copel M.C. 673
Cornet A. 333, 334, 713
Cosby P.C. 55, 56
Covinsky M.H. 642
Cowan E. 608
Cramon K.M. 532, 533
Crandall D.H. 204
Crothers D.S.F. 516, 526
Crowe A. 136, 137, 138
Csanak G. 121, 140
Cubric D. 268
Curry P.J. 244, 682
Cuvellier J. 603, 646
Cvejanovic D. 268, 269
Cvejanovic S. 258, 268, 269

Dababneh M.S. 303
Dagnac R. 353
Dahl P. 374, 598
Dahler J.S. 336, 361, 688
Dalgarno A. 48, 736
Damburg R.J. 724
Danared H. 416, 711
Danjo A. 130, 198
Daoud A. 153
Dashchenko A.I. 743
Dassen H.W. 260
Datta K.K. 75, 641
Datta S. 509, 524
Datz S. 211, 393
Davidson S. 330
Daviel S. 39, 280
De Alti G. 32
De Bruijn D. 630
Deb N.C. 194
Decleva P. 32
Defrance P. 334, 713
Dehmer J.L. 42, 43, 69, 70
Dehmer P.M. 69, 70
Delgado-Barrio G. 639, 640
Delos J.B. 406, 407
Delwiche J. 44, 51, 273
Demidov V.I. 331
Desai H.S. 118
Desai H.S. 95, 143
Deshmukh P.C. 16
Deuring A. 301
Deutsch H. 276
Devdariani A.Z. 331, 345, 734
Dhal S.S. 80
Dhez P. 27, 689
Di Martino V. 178, 277
Dickinson A.S. 513
Diedrich F. 62
Diercksen G.H.F. 46
Dijkkamp D. 550, 551, 552
Dill D. 34
Dilling W. 594
Dillon M.A. 117, 265
Dimicoli I. 52
Dimou L. 683
Ding A. 658
Diserens M.J. 205
Dishoeck E.F. van 48, 556
Dittner P.F. 211
Divoux S. 423, 424
Dixon G.J. 651
Dmitriev I.S. 530
Dohmann H.D. 445, 467
Domcke W. 230
Dousson S. 549, 573, 708, 737
Doverspike L.D. 401
Drachman R.J. 730
Dreizler R.M. 453, 456, 510, 553
Drentje A.G. 550, 551
Dubau J. 208, 214, 215
Dubé L.J. 534, 535, 536
DuBois R.D. 413, 414, 487
Duerksen P. 739
Duggan J.L. 428, 431
Duguet A. 152, 153
Duncan A.J. 172
Dunn G.H. 204, 210
Dunn K.F. 495
Dunning F.B. 663, 673
Dupeyrat G. 657
Düren R. 316, 343
Durup-Ferguson M. 655
Duthoit P. 574
Dutuit O. 53, 706
Dyall K.W. 168

Ebel F. 491
Eberly J.H. 76
Eck J. van 126, 127, 338, 653
Eckhardt M. 170
Ederer D. 27, 689
Edwards A.K. 57
Efremov I. 742
Ehrhardt H. 120, 240, 241
Eichler J.K.M. 499, 534, 554
Eisum N.H. 532
Elakhovsky D.V. 346
Elberfeld W. 590
Elder M. 731
Elford M.T. 489
Elliott D.S. 33, 62
Elston S.B. 378, 379, 582, 651, 714
Eminyan M. 135, 145
Emmons R.W. 649
Enders W. 712
Engar P. 651
Enulescu A. 464
Erkovitch O.S. 504
Ermolaev A.M. 514, 567
Errea L.F. 559, 691
Esaulov V.A. 409, 410, 411, 412
Eschen F. 91
Estep L. 162
Estrada H. 230

Fabricant I.I. 747
Fainberg Yu.A. 530, 531
Faisal F.H.M. 683, 684, 700

Fantoni R. 74, 178, 277
Fargher H.E. 113
Faria R. de B. 707
Fastrup B. 598
Faucher P. 214
Faust M. 485
Feagin J. 99, 440, 725
Felsmann M. 91
Ferch J. 90, 301
Ferguson E. 655, 656
Ferrante G. 678, 680, 681, 692
Ferray M. 603
Ficocelli Varracchio E. 243, 309, 647
Field D. 706
Filipovic D. 85
Finck K. 429
Fink M. 253, 254
Fiordilino E. 687
Fisz J. 740
Flaig H.-J. 170
Flaks I.P. 395
Flannery M.R. 637
Fleury A. 549, 708, 737
Floeder K. 301
Fluendy M.A.D. 608, 609
Fluerasu D. 464
Focke P. 376
Foglia C. 679
Folkmann F. 532, 533
Fon W.C. 107
Fontaine M.F. 238
Fournier P.G. 288
Fournier P.R. 671
Freitas L.C.G. 108, 222
Fricke B. 355, 356
Friedrich B. 623
Frischkorn H.J. 381
Fritsch W. 502
Frohlich H. 51, 53, 627
Fröman N. 719
Fröman P.O. 719
Fromme D. 301
Frontov V.I. 743
Frost M. 393, 394
Fujiwara K. 497
Fukuyama T. 285, 618
Fukuzawa F. 569, 581
Furst J.E. 128, 129
Furst M.L. 450
Fussen D. 333, 334
Futrell J.H. 291, 623

Gaillard M.J. 738
Gallagher A. 330, 599
Gallagher T.F. 29, 668, 698
Ganz J. 28, 619
Garcia J.D. 360
Gardner L.D. 59
Gargaud M. 574
Garibotti C.R. 375
Gasteren E.M. van 126
Gauyacq D. 67
Gauyacq J.P. 266, 297, 411
Gayet R. 560
Gealy M.W. 335
Geddes J. 483
Geiger J. 81
Geltman S. 82
Genz H. 167
Gérard P. 27, 690
Ghosh A.S. 183, 221
Gianturco F.A. 591, 652
Giardini-Guidoni A. 74, 178, 277
Gibner P.S. 480
Gien T.T. 190
Gierz U. 583
Gilbody H.B. 420, 483, 495, 547, 548
Gillen K.T. 329, 620
Gillespie G.H. 402
Girard P. 296
Girardeau M.D. 309
Gislason E.A. 226
Glass G.A. 714
Gochitashvili M.R. 493, 611
Godunov A.L. 380
Goeke J. 147
Goffe T.V. 413
Goldberg I.B. 13
Goldberger A.L. 481
Golden D.E. 134
Goldstein C. 208
Gomez R.D. 302
Gomonaj A.N. 743
Gonsior B. 423, 424
González Lepera E. 370, 376
Gordeev Yu.S. 451, 452, 550, 552
Gorriz M. 528
Gounand F. 671
Govers T.R. 53, 627, 634, 706
Graham W.G. 443, 525
Granitza B. 301
Gray T.J. 563
Greenland P.T. 474, 501
Greenwald K. 263
Gregory D.C. 204
Gressett J.D. 428
Griffin D.C. 199
Groeneveld K.-O. 379, 381
Groh W. 417, 576
Grosse E. 438
Grosser J. 337
Grossi G. 648
Grouard J.P. 409, 412
Grujic P. 173, 310
Grum-Grzhimailo A.N. 139, 182
Grün N. 322, 455, 557
Guberman S.L. 286
Guisti-Suzor A. 5
Gupta G.P. 358
Guschina N.A. 477
Gutenkunst A. 704
Güttner F. 364, 439
Guyon P.-M. 51, 53, 627, 706

Haberland H. 344
Hahn Y. 197, 202, 212
Hall J.M. 563
Hall R.I. 409, 412
Halpern J.B. 50
Hammond P. 105, 106, 185
Hanaki H. 577
Hanashiro H. 198
Hanne G.F. 147
Hansen J.E. 17
Hanssen J. 558, 560
Hara S. 227
Harel C. 558, 560
Harrison M.F.A. 205
Harth K. 28
Hartung H. 355, 356
Haruyama Y. 569, 581
Hasselbrink E. 316, 343
Hasted J.B. 540
Hatano Y. 283, 299, 606
Hausamann D. 344
Havener C.C. 479
Hayden H.C. 450
Hayes M.A. 189
Haysak M.I. 748
Hebel N. 240
Heckman V. 587
Heckmann P.H. 469, 536
Heer F.J. de 84, 177, 550, 551, 552
Hefferlin R. 323, 739
Hefter U. 41
Hege U. 397, 398
Heideman H.G.M. 126, 127, 338
Heil T.G. 555
Helm H. 55, 56, 275
Hemert M.C. van 48, 556
Heni M. 293
Henne A. 456
Henriet A. 317, 666
Henry R.J.W. 187, 193
Hermann M.R. 726
Hermann V. 54
Herschbach D.R. 645
Hertel I.V. 460, 599, 600, 601, 602
Hesslich J. 658
Hickman A.P. 674
Higgs C. 663
Hillard G.B. 696
Hino T. 198
Hippler R. 63, 172, 363, 429, 448, 463, 485, 684

Hirayama T. 201
Hirokawa R. 520
Hirooka T. 618
Hitachi A. 471
Hitchcock A.P. 280
Hitz D. 549, 573, 708, 737
Ho T.-S. 61, 554
Hobbs R.H. 638
Hock G. 367, 390
Hoffmann D.H.H. 446, 578, 579
Hoffmann R. 436, 472
Hofmann D. 379, 381
Hofmann M. 58
Holland D.M.P. 42
Holmes R. 714
Holøien E. 721
Horiuchi T. 577
Horsdal-Pedersen E. 478
Hoshiba K. 154, 274
Hotop H. 28, 120, 619
Howorka F. 612
Hsieh Y.-F. 302, 303
Huber B.A. 539
Hubin-Franskin M.-J. 44, 51, 273
Huddle J.R. 386
Huennekens J. 330
Huestis D.L. 55
Hughes J.G. 516, 731
Hugon M. 671
Humberston J.W. 308
Hummer C.R. 133
Humpert H.-J. 63
Hunton D.E. 58
Huq M.S. 401
Huster R. 168
Hvelplund P. 393, 394, 416, 564

Iga I. 94
Igoshin V.I. 589
Iida Y. 39
Iketaki Y. 155
Il'in R.N. 400, 404
Illenberger E. 293, 298
Imre A.I. 743
Imre K. 93
Imschweiler J. 604
Ingwersen H. 419, 575
Inokuti M. 151
Inouye H. 403, 494
Ishiguro E. 520
Ishihara T. 18, 184
Ito K. 51
Ito S. 154, 274
Itoh A. 371, 389, 581
Itoh Y. 397, 398
Iwai T. 541, 542
Izawa M. 494

Jackson W.M. 50
Jacob B. 453
Jacquot B. 549, 708
Jaduszliwer B. 83, 248
Jaecks D.H. 481, 633
Jain A. 222, 246, 247, 252, 307
Jakacky, Jr., J. 587, 614
Jakubassa-Amundsen D.H. 377, 508
James G.K. 169
Jamieson G. 600, 601, 602
Janda R.S. 402
Jansons M.L. 347
Jaraudias J. 52
Jelenkovic B. 612
Jepsen L. 399
Jetzke S. 684
Jeys T.H. 673
Jhanwar B.L. 225
Jitschin W. 150, 429, 463, 709
Joachain C.J. 165, 678
Johnson A. 711
Johnson B.M. 443, 472
Johnson D. 428
Johnson R.E. 732
Johnson W.R. 17
Jolly A. 441
Jones K.W. 443, 472
Jones P.R. 329
Jong M.A.M. de 127
Joshipura K.N. 95
Jost K. 91
Julienne P.S. 695
Jung K. 120, 240, 241
Jungen Ch. 67
Junker B.R. 102
Jureta J. 185, 258, 268, 269
Jureta J. 333, 334
Justiniano E. 436, 575

Kaase H. 170
Kabachnik N.M. 23, 466
Kabilan A.P. 88
Kachru R. 29, 668, 698
Kádár I. 390
Kahlert H.J. 539
Källne E. 710
Källne J. 710
Kambara T. 387, 388, 465, 470, 471
Kamber E.Y. 540
Kaminsky A.K. 368
Kamke B. 599
Kamke W. 599
Kanamori Y. 569, 581
Kaneko Y. 541, 542, 543, 544, 545, 622, 626
Kani K. 130
Kano S. 154, 274
Kanter E.P. 382, 667
Karashima S. 373, 580
Karemera M. 482
Kase M. 387, 388, 465, 470, 471
Kasuya T. 644
Katayama D.H. 605
Kato T. 625
Kauppila W.E. 302, 303
Kazansky A.K. 219, 220, 233, 287, 349
Kelbch S. 419
Keller F. 11
Keller J. 327
Keller J.C. 27, 689, 690
Kenefick R.A. 462, 582
Keski-Rahkonen O. 171
Kessel Q. 475
Kessler B. 146
Kessler J. 146, 147
Khakhaev A.D. 346
Khakoo M.A. 289, 705
Khare S.P. 92, 225
Kienle P. 435
Kihara S. 155
Kikiani B.I. 493, 611
Kimman J. 40, 177
Kimura K. 68
Kimura M. 518, 519, 568, 662
Kimura M. 541, 542
King G.C. 104, 105, 106, 185, 278
King R.F. 488
King S.J. 138
Kingston A.E. 2, 108, 191, 192, 206
Kita S. 494
Klar H. 172
Kleber M. 447, 590
Klein C. 649
Klein H. 200
Kleinpoppen H. 141, 150, 172, 485
Kleyn A.W. 408, 607
Kloepping W.H. 134
Knudsen H. 393, 394, 416, 564
Knystautas E. 737
Kobayashi N. 541, 542, 544, 545, 622, 626
Kocbach L. 433, 498
Koch P. 27, 66, 689
Kochem K.-H. 240
Kocur P.M. 431
Koenig W. 364, 435, 439
Kohl D.A. 253
Kohl J.L. 59, 195
Kohlhase A. 584
Kohno I. 471
Koike F. 405, 556, 571
Koike T. 130
Koizumi H. 606

Kolb D. 356, 722
Kolokolov N.B. 331
Komarov I.V. 349
Komatsu K. 369
Kondo Y. 198
Kondow T. 77, 285, 618, 621
Konomi I. 577
Konovalova Zh.M. 530
Konrad M. 351
Koot W. 408
Korsch H.J. 596
Koschar P. 381
Koshevoi I.D. 531
Kouchi N. 283
Kouroupetroglou G. 699
Koyano I. 625
Kozhuharov Ch. 364, 435, 439
Kral'kina E.A. 530
Krause J. 447
Krause L. 586
Krishnakumar E. 715
Kronast W. 22
Krueger W. 337
Krug J. 301
Kruit P. 64
Kubota H. 618
Kuchitsu K. 77, 285, 618, 621
Kuen I. 612
Kuhlman H. 323
Kulander K.C. 521
Kumagai H. 387, 388, 465, 470, 471
Kumar M. 156
Kumar V. 715
Kunert H.W. 740
Kunikeev Sh.D. 380
Kuprianov N.L. 589
Kuptsov S. 475
Kurawaki J. 284
Kurepa M. 258
Kuroki K. 281
Kusakabe T. 577
Kusunoki I. 654
Kutina R.E. 57
Kvale T.J. 350
Kviszhinadze R.V. 611
Kwan Ch.K. 302, 303
Kwiatkowski G. 293
Kyrölä E. 76

Lacmann K. 631
Lafyatis G.P. 195
Laganà A. 648
LaGattuta K.J. 197, 202, 212
Lagreze A. 676
Lahiri J. 8
Lahmam-Bennani A. 152, 153
Lahtinen J. 171
Lamanna U.T. 243, 591
Lamoureux M. 161
Land D.J. 425, 432
Landau M. 409, 412
Lane N.F. 332, 664
Langhoff P.W. 46, 263, 726
Lapicki G. 426, 428, 431
Larsen B.J. 532
Larsen M. 336
Larsen O.G. 561
Latimer C.J. 488
Latz R. 381
Laughlin C. 75
Launay J.M. 227, 228, 229
Laurent H. 208
Laurent M. 573
Lavender M. 304
Lavollée M. 40, 53, 627, 706
Lavrov V.M. 493, 610
Le Dourneuf M. 109, 110, 111, 227, 228, 229
Le Gouet J.L. 689, 690
Le Rouzo H. 35, 36
Le Sech C. 317
Leal E.P. 45
Leclerc B. 273
Lee C.C. 593
Lee C.J. 13
Lee M.T. 45, 94, 222
Lee Y.T. 207
Lee Y.T. 642
Leeuw P.E. van der 408
Lefebvre-Brion H. 35, 44
Legagneux-Piquemal P. 442, 537, 538
Legouet J.L. 27
Leisin O. 617
Leiter K. 276
Lemaire J. 52
Lengyel V.I. 748, 749, 750, 751
Lennon M. 547, 548
Leone C. 680
Leroi G.E. 43
Leuchs G. 33, 62
Leung K.T. 279
Levin V.G. 37, 174
Levy B. 624
Lewis Z.V. 596
Leyh B. 44
Liesen D. 444, 445, 467, 712
Liljeby L. 393, 394, 416, 564, 711
Lin C.D. 502, 511
Lindeman T.G. 58
Linden van den Heuvell H.B. van 126, 127
Linder F. 54, 78, 351, 354, 397, 398, 628
Lindinger W. 656
Lineberger W.C. 41
Linnartz K.P.J. 338
Lisini A. 32
Liu B. 444, 445, 712
Livingston A.E. 735
Locht R. 290
Lohmann B. 180
Lomsadze R.A. 493, 610
Lorent V. 333
Los J. 408, 607, 629, 634
Loulergue M. 208
Löw W. 167
Lozhkin K.O. 546, 565
Lubell M.S. 145
Lucchese R.R. 45
Lüdde H.J. 453, 510, 553
Luijks G. 72
Lunt S. 609
Lutz H.O. 63, 150, 363, 429, 448, 463, 485, 684, 709
Lutz N. 364, 439

MacAdam K.B. 661
Macek J. 99
MacGillivray W.R. 124
Macias A. 101, 421, 522, 723
MacKellar A.D. 660
Magunov A.I. 182
Maksimova O.B. 466
Malegat L. 238
Mandal C.R. 509, 524
Maneira M.J.P. 632
Manson S.T. 8, 9, 13, 14, 15, 16, 151, 414
Maor D. 444
Mareca P. 639, 640
Mariani D.R. 66
Marinkovic B. 85
Märk T.D. 275, 276, 291
Marquette J.B. 657
Martin B. 364, 439
Martin N.L.S. 175
Martin R.L. 332
Martinov S.V. 493
Masche C. 90, 301
Masnou-Seeuws F. 314, 317, 666, 669
Massaro P.A. 98
Masuda T. 569
Mathur D. 239
Mathur K.C. 305, 358, 685
Matsumi Y. 644
Matsumoto A. 198, 541, 542, 569
Matsuo T. 387, 388, 465, 470, 622
Matsuzawa M. 670
Matthias E. 33
Mawhorter R. 253, 254
Mazumder P.S. 183
McAbee T. 426, 427
McCann J.F. 526
McCarroll R. 513, 574

McCarthy I.E. 115, 277
McClelland J.J. 254
McConkey J.W. 259, 260
McCullough R.W. 547, 548
McCurdy C.W. 177, 232
McDaniel F.D. 428, 431
McDowell M.R.C. 512, 686
McEachran R.P. 89
McGregor I. 172
McGuire J.H. 166, 362, 499
McKoy V. 45
McLaughlin D.J. 202
McMillian G.B. 673
McPherson A. 125
Mead R.D. 41
Meckbach W. 370, 376
Mehlhorn W. 22, 168
Mehta R. 428, 431
Meier F. 391
Melchert F. 492
Méndez L. 559, 691
Mendizabal R. 421
Menzel W. 168
Meron M. 443, 472
Meskhi G.G. 477
Mestdagh J.M. 341, 603, 646
Meyer E. 460
Meyer F.E. 422
Meyer F.W. 562
Meyer H. 585
Meyer H.-J. 643
Meyer W. 702
Meyerhof W.E. 434, 437, 438
Michel Ch. 438
Michels H.H. 638
Mies F.H. 695
Mikoushkin V.M. 395
Milazzo M. 430
Mileev V.N. 380
Miller P.D. 211, 428
Mitchell I.V. 430
Mitsuke K. 621
Mittleman M.H. 687
Mizuno J. 18, 184
Mo O. 315
Moak C.D. 211
Moede M. 144
Moiseiwitsch B.L. 112
Mokler P.H. 417, 436, 444, 446, 578, 579
Molitoris J.D. 434, 437, 438
Möller R. 55, 56
Møller S.P. 422
Momigny J. 290
Montague R.G. 205
Montmagnon J.L. 409, 412
Moore C.F. 384
Moores D.L. 123, 203, 402, 686
Moorman L. 338
Morales A. 635, 636
Morenzoni E. 434, 437, 438
Morgan L.A. 218
Morgan T.J. 204, 210
Morgenstern R. 461
Morgner H. 615, 616, 617, 619
Morikawa Y. 201
Morin P. 44
Morita K. 157
Morita M. 283
Mosulishvili N.O. 493
Moutinho A.M.C. 608, 632
Mrugala F. 650
Msezane A.Z. 9, 21, 187, 193
Mueller D.W. 204, 210
Mukhamedzhanov A.M. 174
Mukherjee S.C. 457, 509, 524
Müller A. 200, 417, 492, 576
Muller H.G. 64
Müller K.H. 704
Müller M. 30
Müller O. 81
Müller R. 241
Müller-Fiedler R. 120
Munakata T. 644

Nagai N. 577
Nagata T. 490
Nagata T. 77
Nakamura H. 47, 520
Nakamura T. 544, 545
Namiki M. 470
Naslenas E. 79
Natarajan M. 481, 633
Navrotsky V.T. 748, 750, 751
Nayfeh M.H. 696
Neill P.A. 136
Neitzke H.-P. 458, 459
Nemirovsky I.B. 376
Nenner I. 44, 51
Nesbet R.K. 82, 218
Nescovic N. 211
Nessi M. 438
Neudatchin V.G. 37
Neumann H. 335
Neuteboom J. 629, 634
Nevostrueva I.A. 531
Newell W.R. 244, 682
Newman D.S. 104
Newman J.H. 480, 588
Neynaber R.H. 486
Nickel J.C. 93
Niehaus A. 186, 461, 653
Nielsen S.E. 336, 361, 688
Nikolaev V.S. 415, 530
Nikulin V.K. 477, 552, 566
Nishimura F. 281, 369
Nishimura H. 130
Nishizawa A. 569
Noble C.J. 218, 234
Nogueira J.C. 94
Noll M. 583
Noll T. 604
Nolte G. 158, 371, 379, 389, 418
Norcross D.W. 250

O C.-S. 582
Oda K. 201
Oda N. 281, 283, 369, 476
Oesterlin P. 71
Ogawa T. 284
Ogurtsov G.N. 395
Oh S.D. 14
Ohnesorge W. 62
Ohshima S. 285
Ohtani S. 198, 541, 542
Ojha P.C. 1
Okamoto T. 403
Okuno K. 541, 542, 543
Olivier J.L. 290
Olson R.E. 518, 519, 662
Onda K. 235
Ondrey G.S. 49
Ono T. 201
Onodera N. 245
Oppen G. von 359
Orel A.E. 521
Orient O.J. 292
Osimitsch S. 150
Østergaard K. 422
Ottinger Ch. 654
Ovchinnikov S.Yu. 451, 452
Oza D.H. 114

Padhy B. 100
Padial N.T. 250
Paikeday J.M. 96
Paixao F.J. da 45, 121
Pan Guang-Yan 177
Pande A.K. 306
Panov M.N. 546, 565
Papernov S.M. 347
Papp F.F. 744
Pareek P. 20
Park J.T. 350
Parker K. 323
Parr A.C. 42, 43
Parrish R. 323
Pascale J. 312, 313, 341, 342, 518
Pavlitchenkov A.V. 37
Peach G. 3, 512
Peacher J.L. 350
Pedrazzini G. 462
Pejcev V. 85
Pelamourgues L. 109, 110
Pelayo F. 421
Perov A.A. 665
Peschina J. 200
Pesnelle A. 325, 326

Petersen J.B.B. 422
Peterson J.R. 19, 24
Peterson R.S. 428
Petrella G. 591
Pfeifer P. 701
Pfeng E. 467
Phillips T. 584
Picart J. 324
Pichou F. 409, 412
Picqué J.L. 27, 689, 690
Pijkeren D. van 653
Pillet P. 29, 668, 698
Piltch N.D. 33
Pindzola M.S. 199
Pinho A.G. de 738
Piraux B. 165
Piticu I. 464
Platten H. 158, 371, 389
Poizat J.-C. 738
Polanyi J.C. 659
Poliakoff E.D. 43, 69
Politis M.F. 737
Pollack E. 587
Pommier J. 573
Ponce V.H. 372
Ponomarenko V.V. 219
Popov Yu.V. 181
Popova A.M. 504
Popova M.I. 368
Poppe D. 595, 596
Pradhan A.K. 213, 710
Pratt R.H. 7, 13, 14, 161
Pratt S.T. 69, 70
Praxedes A.J.F. 608, 632
Preston S.C. 664
Price J.L. 431
Priddy K. 323
Propin R.Kh. 724
Prunelé E. de 672
Pujo P. de 603, 646
Pulay P. 253
Pundir R.S. 305, 685
Pyle R.V. 525

Quarles C.A. 162, 163

Raczynski A. 697
Rademann K. 73
Radojevic V. 16
Rahal H. 319
Rai D.K. 100, 122, 592
Raith B. 423, 424, 469, 536
Raith W. 90, 144, 301
Rajgara F.A. 239
Ramayya A.V. 364, 439
Rao N.S. 143
Raoult M. 35
Raseev G. 35, 36, 44
Ratcliffe C.A. 413
Ray P.S. 700
Rayburn L.A. 428
Read F.H. 105, 106, 185, 278
Redd E. 350
Reddish T. 262
Reeves T. 440
Register D.F. 87, 93, 140
Reichert E. 148
Reihl H. 150
Reiland W. 600, 659
Reinig P. 354, 628
Remillieux J. 738
Rensfelt K.-G. 416
Reuss J. 72
Richard E.G. 288
Richard-Viard M. 53
Richter A. 167
Richter H. 473
Richter R. 656
Ricz S. 390
Riehle F. 159
Riera A. 101, 315, 421, 522, 559, 691, 723
Riley J.L. 128, 129
Rille E. 350
Rinn K. 492
Risley J.S. 125, 479
Rivarola R.D. 503, 507
Roberts M.J. 113
Robinson E.J. 693
Roche A.L. 669
Rolfes R.G. 661
Roller Z. 371, 389
Romanjuk N.I. 744
Ron A. 13, 14
Rosendorff S. 119
Roser R. 475
Ross K.J. 169, 175
Ross U. 643
Rothe E.W. 49
Roueff E. 736
Roussel F. 690
Rouze N. 125, 479
Rowe B.R. 657
Roy A. 239
Roy A.C. 164, 176
Roy D. 273
Roy K. 517
Rozet J.P. 441, 442, 537, 538, 714
Rubin K. 135, 145, 742
Rubino R. 475
Rubtsov V.I. 331
Rudd M.E. 413
Ruf M.-W. 28, 619
Rumble, Jr., J. 223, 224
Runge S. 319, 320, 325, 326
Russek A. 597, 614, 635, 636

Sabad E.P. 748, 749, 750, 751
Sabelli N.H. 226
Saeed K. 172
Saha H.P. 688
Sakharov V.I. 400, 404
Sakisaka M. 577
Salak M. 751
Salehkoutahi S. 162
Salin A. 507, 560
Salvini S. 234
Salzborn E. 200, 417, 491, 492, 576
Salzmann D. 7
Samoilov A.V. 552
Samoylov A.V. 566
Samson J.A.R. 20
Sandhya Devi K.R. 360
Sandner W. 142, 703
Saoudi R. 738
Saraph H.E. 717
Sasao M. 569
Sato H. 518, 519, 568
Sato K. 68
Sato K.N. 569
Sato Y. 403
Sato Y. 670
Saxena S. 358
Saxer A. 656
Schader J. 381
Schadt W. 419, 436
Schaefer J. 594, 702
Schartner K.-H. 170
Schauer M.M. 379
Scheid W. 322, 455, 557
Schellhorn M. 71
Schermann C. 409, 412
Schermann J.-P. 676
Schiwietz G. 158, 371, 379, 389
Schlachter A.S. 525, 576
Schlemmer P. 120
Schmidt E. 26, 31
Schmidt H. 642
Schmidt M. 30
Schmidt-Böcking H. 379, 418, 419, 436, 472, 575
Schmoranzer H. 339, 340, 604
Schneider B.I. 216, 217, 257
Schneider D. 371, 382, 383, 384, 389, 667
Schneider T. 371, 389
Schönfeldt W.A. 417, 436, 446, 578, 579
Schröder H. 26
Schröder W. 144
Schuch B. 417, 576
Schuch R. 419, 436, 463, 472, 575
Schulz C.P. 600, 601, 602
Schulz M. 575
Schulz P.A. 41
Schulze R. 712

Schulze T. 643
Schulze W. 722
Schwab A. 301
Schwier H. 63
Scott N.S. 103, 149
Scott P. 2
Seaman M. 323
Seaton M.J. 717
Seaver M. 67
Secrest D. 584
Seiberle H. 617
Sellin I.A. 378, 379, 381, 582, 651, 714
Sen Gupta K. 523
Senashenko V.S. 23, 380
Senger B. 366
Sengler W. 356
Sepp W.-D. 355, 356, 454
Serenkov I.T. 400, 404
Sewell E.C. 137
Shafroth S.M. 386, 426, 427, 443, 527
Shah M.B. 420
Shakeshaft R. 677
Shanker R. 448
Sharma A.K. 505, 506
Shaw D.A. 278
Shaw K. 323
Shchegolev V.A. 390
Shemansky D. 267
Shergin A.P. 477
Sheridan J.R. 613
Shibata H. 281, 369, 388, 465
Shimon L.L. 746
Shingal R. 517
Shinsaka K. 606
Short R.T. 582
Shpenik O.B. 744, 747
Shurigina Yu.A. 504
Shvegzhda Zh.L. 347
Sidhanta J. 457
Sidis V. 629, 630
Sidorovich V.A. 396, 415
Siegel A. 28, 619
Sil N.C. 164, 194
Simonov A.P. 665
Simons D.G. 425
Simony P.R. 499
Sin Fai Lam L.T. 87
Sinapius G. 301
Singh C.S. 122
Singh S.N. 86
Skalinski P. 586
Skapa H. 364, 439
Slevin J. 135, 145
Slusher M.P. 663
Smirnov Yu.F. 37
Smith A.C.H. 205, 244, 682
Smith F.J. 731
Smith F.T. 716
Smith K.A. 480, 588, 663
Smith S.J. 302, 303
Smith S.J. 33
Smith W.W. 450, 668, 698
Snegursky A.V. 747
Snyder R. 597
Soff G. 17
Sohn W. 240
Solarte E. 54
Solovyev E.A. 404
Soltani J. 364, 439
Sonntag B. 26, 31
Sørensen A.H. 422
Southworth S.H. 42
Souza A.C. de A. e 707
Souza G.G.B. de 707
Spagnolo B. 692
Spalburg M.R. 607
Specht H.J. 575
Spence D. 117, 265
Spiess G. 689, 690
Spruit M. 40
Srajer Dj. 269
Srivastava B.B. 80
Srivastava M.K. 86, 505, 506
Srivastava R. 100
Srivastava S.K. 271, 272, 289, 292, 705
Stachura Z. 436, 444, 446, 578
Staemmler V. 652
Stamatovic A. 275, 291
Standage M.C. 124
Starace A.F. 99
Stauffer A.D. 89
Stearns J.W. 525
Stebbings R.F. 480, 588, 663, 673
Steenman-Clark L. 214
Stefanov B. 249
Stegmaier J. 617
Stein T.S. 302, 303
Stelbovics A.T. 115
Stelson P.H. 211
Stepanov A.N. 665
Steph N.C. 134
Stephan C. 441, 442, 538
Stephan K. 275, 276, 291
Stephens J. 34
Stevens W.J. 223
Stich W. 510
Stirling W. 145
Stockdale J.A.D. 65
Stöckli M. 478
Stöffler W. 383, 384
Stoller Ch. 438
Stolte S. 72
Stolterfoht N. 158, 371, 379, 389, 418
Strakhova S.I. 10
Straten P. van der 461
Stubbs R.J. 261
Stuit D. 117
Subramanian K.P. 715
Suck S.H. 649
Suemitsu H. 20
Sugahara H. 130
Sulik B. 390
Sur S. 221
Süzer S. 298
Suzuki E. 299
Suzuki H. 154, 155, 157, 198, 201, 274, 392
Suzuki H. 570
Suzuki K. 77
Suzuki M. 77
Swenson J.K. 386, 427, 527
Syrkin M. 675
Szmytkowski C. 242
Szücs S. 482

Tabché-Fouhaile A. 51
Takagi S. 541, 542, 569
Takahashi A. 130
Takahashi J. 387, 388, 465, 470
Takahashi N. 283
Takayanagi T. 154, 155, 157, 274
Takenouchi S. 476
Takuma H. 154, 274
Tambe B.R. 13, 15
Tanaka H. 245
Tanaka K. 625
Tanaka Y. 606
Tang F.C. 145
Tang K.T. 321
Tang S.Y. 486
Tang Xiao 50
Tanis J.A. 426, 443, 525, 529
Tassen-Got L. 442
Taulbjerg K. 496, 561
Tawara H. 198, 541, 542
Tayal S.S. 156, 191, 192
Taylor D.A. 714
Taylor K.T. 6, 107
Teillet-Billy D. 294, 296, 297
Temkin A. 235
Teplova Ya.A. 504, 530, 531
Terao M. 482
Teubner E. 322
Teubner P.J.O. 128, 129
Thoe R.S. 714
Thomas B. 582
Thomas D. 659
Thompson D.G. 246, 307
Thylwe K.-E. 352
Tietz J.V. 61
Timmer C. 210
Tino A. 83, 248
Tip A. 64
Tiribelli R. 178
Tischer H. 343

Tittes H.-U. 600, 601, 602
Titze B. 49
Tiwary S.N. 4, 116, 206
Tobita T. 626
Toburen L.H. 413, 414, 487
Toennies J.P. 321, 583
Toepfer A. 453
Tokoro N. 476
Tonuma T. 471
Toriumi M. 299
Toshima N. 497, 570
Toten A. 428
Träbert E. 469, 536
Trajmar S. 87, 93, 140, 270, 271, 272
Tran Minh N. 324
Tran N.H. 29, 668, 698
Tripathi A.N. 86, 156
Tripathi D.N. 306, 592
Trombetta F. 678
Tronc M. 238, 295
Truhlar D.G. 223, 224
Tserruya I. 472
Tsertos H. 435
Tsurita Y. 569
Tsurubuchi S. 541, 542
Tubergen T. v. 84
Tunnell L. 563
Turner J.L. 232
Turner M. 145

Uddin M.N. 303
Uggerhøj E. 422
Ukai M. 606
Ullrich J. 419
Urakawa J. 387, 388, 465, 470
Urquijo J. de 635, 636

Vager Z. 667
Valance A. 318, 319
Valenza D. 681
Valiron P. 669
Vallée O. 324
Valluri S.R. 455
Van Zyl B. 335
Varga D. 390
Varghese S.L. 563
Vasilakis A. 145
Vdovin Yu.A. 694
Végh J. 390
Vernhet D. 442, 538
Vernon M.F. 642
Vervaat M.G.M. 607
Vidal R. 370, 376
Vijayshri 92
Villarreal P. 639, 640
Villinger H. 656
Vo Ky Lan 227, 228, 229

Völkel M. 142, 703
Vollmer W. 147
Vorobiev N.F. 530
Voss H. 26
Vu Ngoc Tuan 410, 411
Vukstich V.S. 745
Vuskovic L. 85, 87, 270

Wada A. 392
Wada K. 392
Wagenaar R.W. 84
Wagner W. 435
Wague A. 25
Wakiya K. 154, 155, 157, 198, 201, 274, 392
Walter O. 231
Walther H. 62
Walther L. 91
Wang D.-W. 462
Wang M.S. 7
Wang R.-G. 117, 265
Wang T.S. 406, 407
Warczak A. 445, 446, 578, 579
Warner C.D. 105
Watanabe S. 109, 110, 111
Watanabe T. 18, 184, 373, 497, 570, 580
Watel G. 325
Water W. van de 66
Watkin S. 135
Watson R.L. 462, 468
Watts M.F. 495
Webb C.J. 124
Weber G. 359
Weber W. 168
Weigold E. 179, 180
Weiner J. 327
Weiss P. 83, 248
Weiss P.S. 642
Wellenstein H.F. 152, 153
Welsh J.A. 605
Wendin G. 27
Werner H.-Ch. 418
Werner U. 709
Westbrook G. 162
Westerveld W.B. 125, 479
Wetzel H.-E. 26, 31
Weyhreter M. 78
Wheeler R.M. 428
Wiel M.J. van der 40, 64
Wiesemann K. 539
Wijngaarden W. van 259
Wilkie F.G. 548
Wille U. 448, 449
Williams I.D. 483
Williams J.F. 131, 132, 209
Williams O.M. 489
Willis S.L. 512
Wilson M. 169

Wilson S.R. 428
Winogradow A. 31
Winter H. 484, 551
Winter T.G. 511
Wisotzki B. 709
Wohrer K. 441
Wolcke A. 147
Wolf H. 148
Wolf R. 485
Wölfli W. 438
Wollenweber P. 339, 340
Woolsey J.M. 135
Worsnop D.R. 645
Woude R.L. van der 551
Wübker W. 146
Wuilleumier F. 27, 689, 690
Wutte U. 484

Yagisawa H. 348
Yagishita A. 387, 388
Yakovlenko S.I. 694
Yánez M. 421
Yelets I.S. 233, 287
Yencha A.J. 618
Yenen O. 481, 633
Yermachenko V.M. 694
Yin R.Y. 7
Yngve S. 718, 719, 720
York T. A. 261, 264
Yoshino M. 198

Zabransky B.J. 382
Zadworny F. 737
Zagidullin M.V. 589
Zagrebin A.L. 345
Zaitsev Yu.V. 346
Zaitseva O.I. 139
Zampieri G.E. 391
Zander A.R. 428
Zapesochnyi I.P. 743, 745
Zarcone M. 686
Zatsarinny O.I. 749, 751
Zavilopulo A.N. 747
Zayats A.Yu. 665
Zeippen C.J. 6
Zhmenyak Yu.V. 745
Zhuvikin G.V. 739
Ziesel J.P. 294, 295, 296, 706
Zimmermann G. 616
Zimmermann P. 30
Zinoviev A.N. 451, 452
Zitzewitz P.W. 301
Zoller P. 33
Zoran V. 464
Zubek M. 104
Zvirblis P. 79
Zwakhals C.J. 186